Mathematical Reasoning *for Elementary Teachers*

Fourth Edition

Calvin T. Long
Washington State University

Duane W. DeTemple
Washington State University

Boston San Francisco New York
London Toronto Sydney Tokyo Singapore Madrid
Mexico City Munich Paris Cape Town Hong Kong Montreal

Publisher: **Greg Tobin**
Acquisitions Editor: **Carter Fenton**
Associate Project Editor: **Joanne Ha**
Associate Editor: **RoseAnne Johnson**
Senior Marketing Manager: **Becky Anderson**
Marketing Assistant: **Maureen McLaughlin**
Senior Production Supervisor: **Peggy McMahon**
Production Services: **WestWords, Inc.**
Senior Manufacturing Buyer: **Evelyn Beaton**
Managing Editor: **Karen Wernholm**
Technical Art Rendering: **Techsetters, Inc.**
Senior Design Supervisor: **Barbara T. Atkinson**
Text and Cover Design: **Button & Sherman Design**
Cover Illustration: **Rob Colvin**
Software Development: **David Malone and Rebecca Williams**

Library of Congress Cataloging-in-Publication Data

Long, Calvin T.
Mathematical reasoning for elementary teachers / Calvin Long, Duane DeTemple—4th ed.
p. cm.
Includes index.
ISBN 0-321-28696-0
1. Mathematics—Study and teaching (Elementary) I. DeTemple, Duane W. II. Title.

QA135.6.L66 2006
510—dc22 2004050250

2 3 4 5 6 7 8 9 10 - QWT - 06

Exercise Sets:

The focus on mathematical reasoning continues in the problem sets in the sections labeled **Thinking Critically, Teaching Concepts, Thinking Cooperatively, Making Connections, Communicating, From State Student Assessments, Using a Calculator,** and **Using a Computer.**

(a) $\frac{}{20}$ (b) $\frac{7}{}$ (c) $\frac{}{17}$ (d) $\frac{30}{}$

21. Decide whether each statement is *true* or *false*. Explain your reasoning in a brief paragraph.

(a) There are infinitely many ways to replace two fractions with two equivalent fractions that have a common denominator.

(b) There is a unique least common denominator for a given pair of fractions.

(c) There is a least positive fraction.

(d) There are infinitely many fractions between 0 and 1.

22. How many different rational numbers are in this list?

$$\frac{27}{36}, 4, \frac{21}{28}, \frac{24}{6}, \frac{3}{4}, \frac{-8}{-2}$$

Teaching Concepts

23. Go on a "fraction safari" to the grocery store (or in a magazine or newspaper). Find at least five fractions. For each fraction, describe the unit, the numerator, and the denominator.

24. Like most fifth graders, Dana likes pizza. When given the choice of $\frac{1}{4}$ or $\frac{1}{6}$ of a pizza, Dana says, "Since 6 is bigger than 4 and I'm really hungry, I'd rather have $\frac{1}{6}$ of the pizza." Write a dialogue including useful diagrams to clear up Dana's misconception about fractions.

25. When asked to illustrate the concept of $\frac{2}{3}$ with a colored-region diagram, Shanti drew this figure. How would you respond to Shanti?

Thinking Critically

26. What fraction represents the part of the whole region that has been shaded? Draw additional lines to make your answer visually clear. For example, $\frac{2}{6}$ of the regular hexagon on the left is shaded, since the entire hexagon can be subdivided into six congruent regions as shown on the right.

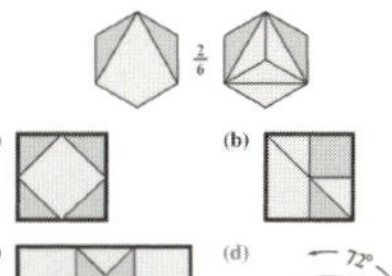

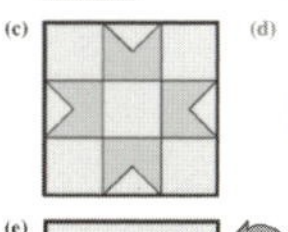

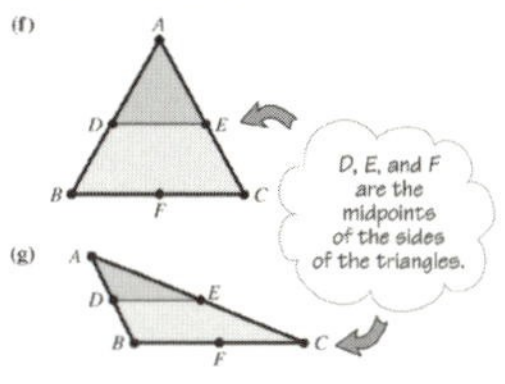

27. Start with a square piece of paper, and join each corner to the midpoint of the opposite side, as shown in the left-hand figure below. Shade the slanted square that is formed inside the original square. What fraction of the large square have you shaded? It will help to cut the triangular pieces as shown on the right and reattach them to form some new squares.

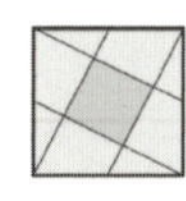

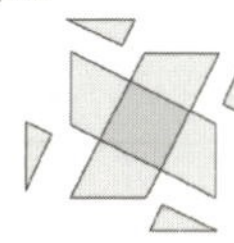

median is between the two starting fractions, since $\frac{1}{4} < \frac{1}{3} < \frac{1}{2}$.

(a) Find the mediant of $\frac{3}{4}$ and $\frac{4}{5}$, and show that it is between the two given fractions.

(b) Show that the mediant is always between the given two fractions; that is, show that if $\frac{a}{b} < \frac{c}{d}$, then

$$\frac{a}{b} < \frac{a+c}{b+d} < \frac{c}{d}.$$

(c) Carefully explain why the result of the property shown in part (b) guarantees that there are infinitely many fractions between any two given fractions.

32. The row of Pascal's triangle (see Section 3 of Chapter 1) shown below has been separated by a vertical bar drawn between the 15 and the 20, with 3 entries of the row to the left of the bar and 4 entries to the right of the bar.

1 6 15 | 20 15 6 1

Notice that $\frac{15}{20} = \frac{3}{4}$. That is, the two fractions formed by the entries adjacent to the dividing bar and by the number of entries to the left and the right of the bar are equivalent fractions. Was this an accident, or is this a general property of Pascal's triangle? Investigate the property further by placing the dividing bar in new locations and examining other rows of Pascal's triangle.

Thinking Cooperatively

The following two problems develop the concept of a fraction using pattern blocks, a popular and versatile manipulative that is used successfully in many elementary school classrooms. The following four shapes should be available to students working in groups of two or three:

Hexagon

Trapezoid

Triangle

Rhombus

33. Use your pattern blocks to answer these questions.

(a) Choose the hexagon as the unit. What fraction is each of the other three pattern blocks?

(b) Choose the trapezoid as the unit. What fraction is each of the other three pattern blocks?

(c) Choose the rhombus as the unit. What fraction is each of the other three pattern blocks?

34. (a) A hexagon and six triangles are used to form a six-pointed star. Choose the star as the unit. Next, form pattern-block shapes corresponding to each of these

Making Connections

36. Francisco's pickup truck has a 24-gallon gas tank and an accurate fuel gauge. Estimate the number of gallons in the tank at these readings.

(a)

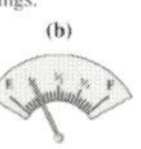
(b)

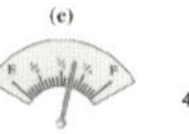
(c)

37. If 153 of the 307 graduating seniors go on to college, it is likely that a principal would claim that $\frac{1}{2}$ of the class is college bound. Give simpler convenient fractions that approximately express the data in these situations.

(a) Esteban is on page 310 of a 498-page novel. He has read _______ of the book.

(b) Myra has saved \$73 toward the purchase of a \$215 plane ticket. She has saved ________ of the amount she needs.

(c) Nine students in Ms. Evaldo's class of 35 students did perfect work on the quiz. ________ of the class scored 100% on the quiz.

(d) The Math Club has sold 1623 of the 2400 raffle tickets. It has sold ________ of the available tickets.

40. Fr...

rolling a single die and having more than one spot appear. Give fractions that express the probability of the following events:

(a) Getting a head in the flip of a fair coin.

(b) Drawing a face card from a deck of cards.

(c) Rolling an even number on a single die.

(d) Drawing a green marble from a bag that contains 20 red, 30 blue, and 25 green marbles.

Chapter Summary:

The chapter summary has been re-designed and revised for more ease of use for students as they prepare for tests and quizzes.

Chapter Summary

Key Concepts

Section 8.1 Algebraic Expressions and Equations

- Algebraic reasoning. Algebraic reasoning... terms by following these steps: introduce... equations, solve equations, and interpret... the original problem or pattern.
- Variables. Variables represent quantities... depend on varying choices of related qua... alized properties, express relationships, ...
- Algebraic expressions. An algebraic exp... variables, numbers, and operation symbo...
- Equation. Setting two algebraic expressi... one another creates an equation.
- Solution of equations. The values of the... the solutions of the equations and form ...

Section 8.2 Functions

- Function. A function is a rule that assign... *D*. The set *D* is the domain of the functi... element of the domain *D*, then the value...
- Range. The range of a function is the set...
- Representations of a function. Functions... formula; a table of values; an arrow diag... is output; a graph.
- Linear function. A function of the form $f($...

Section 8.3 Graphing Functions in the Cartesian Plane

- Cartesian coordinates. The Cartesian co... axes, with the horizontal axis typically c... called the *y*-axis. Any point $P(x, y)$ in t... coordinates.
- Distance formula. The distance betw... $PQ = \sqrt{(x_2 - x_1)^2 + (y_2 - y_1)^2}$.
- Line segment. $\overline{PQ}$ denotes the line segm...
- Slope. The slope of the line or line segm... is the ratio of the line's or line segment's... when $y_1 \neq y_2$. Slopes are not defined for... are parallel precisely when they have the...
- Equation of a line. The equation of a ... slope–intercept, and two-point forms. ... described by the general form of the equ...
- Nonlinear function. Nonlinear functions ... points are plotted and then connected by a smooth line to approximate the graph of the function. Graphing calculators and other technology can be effectively employed to make more accurate graphs.

Vocabulary and Notation

Section 8.1
Constant
Variable
Domain of a variable
Algebraic expression
Evaluation of an algebraic expression
Equation
Conditional equation
Solution set of an equation

Section 8.2
Function, f
Domain of a function
Value of a function at x, $f(x)$
Range of a function
Representations of a function
Formula, table of values, arrow diagram, machine, graph
Linear function, $f(x) = mx + b$

Section 8.3
Cartesian plane
Coordinate axes (x-axis and y-axis)
Origin
Coordinates
Quadrants
Distance formula
Line segment, $\overline{PQ}$
Line, $\overleftrightarrow{PQ}$
Slope of line segment, slope of line
Parallel lines
Equation of lines
Point–slope form, $y - y_1 = m(x - x_1)$
Slope–intercept form, $y = mx + b$
Two-point form, $y - y_1 = \frac{y_2 - y_1}{x_2 - x_1}(x - x_1)$
General, $Ax + By + C = 0$
Nonlinear function

Chapter Review Exercises

Section 8.1

1. Let a, b, and c denote the current ages of Alicia, Ben, and Cory, respectively. Write expressions in the variables a, b, and c that express the given quantity.
   (a) Alicia's age in 5 years
   (b) The fact that Ben is younger than Cory
   (c) The difference in age between Ben and Cory
   (d) The average age of Alicia, Ben, and Cory
2. Let a, b, and c denote the current ages of Alicia, Ben, and Cory as in problem 1. Write equations in the variables a, b, and c that express the given relationship.
   (a) Alicia will be 11 years old in 2 more years.
   (b) Three years ago, Ben was half of Cory's age last year.
   (c) Alicia's age is the average of Ben's and Cory's ages.
   (d) The average age of Alicia, Ben, and Cory is 10.
3. A "hex–square train" is made with toothpicks, alternating between hexagonal and square "cars" in the train. For example, three-car and four-car hex–square trains are shown in the next column.

(a) Evidently 14 toothpicks are required to form a three-car train. Give a formula for the number of toothpicks required to form an n-car train. (*Suggestion:* Take separate cases for trains with an odd or an even number of cars.)

(b) A certain hex–square train requires 102 toothpicks. How many hexagons and how many squares are in the train?

Section 8.2

4. For each table, decide whether it is the table of a function $y = f(x)$. If not, state why not. If so, give the domain and range.

(a)

x	3	6	7	8	3
y	4	3	5	1	0

To my wife and constant helpmate, Jean.

C.T.L.

To my wife, Janet, and my daughters, Jill and Rachel.

D.W.D.

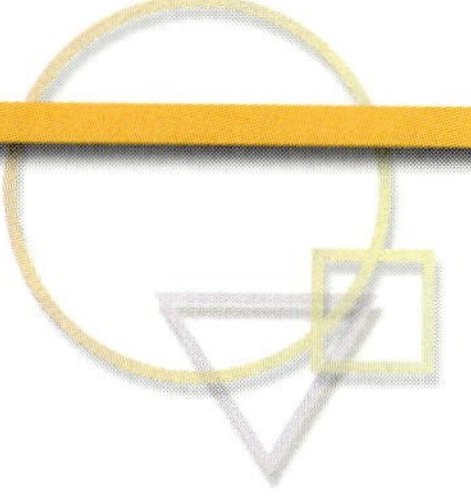

Contents

Preface

To The Student

You may be wondering what to expect from a college course in mathematics for prospective elementary or middle school teachers. Will this course simply repeat arithmetic and other material that you already know, or will the subject matter be new and interesting? We will try to answer that question here and at the same time provide a useful orientation to the text.

PROBLEM SOLVING AND MATHEMATICAL REASONING

The entire first chapter of this text is devoted to developing skills in problem solving and critical thinking, a theme that is continued throughout the book. At first, problem solving may seem daunting, but as you gain experience and begin to acquire an arsenal of strategies, you will become increasingly comfortable and will begin to find the challenge of solving a unique problem stimulating and even fun. Quite often, and much to their surprise, this has been the experience of students in our classes as they successfully match wits with challenging problems and gain insight that leads to even more success.

You should not expect to see instantly into the heart of a problem or to know immediately how it can be solved. The text contains many exercises that check your understanding of basic concepts and build basic skills, but you will continually encounter problems characterized by the following couplet:

Problems worthy of attack
prove their worth by hitting back

These problems are not unreasonably hard. (Indeed, many would be suitable with only a minor modification for use in classes you will subsequently teach.) However, they do require thought. Expect to try a variety of approaches, be willing to discuss possibilities with your classmates, and form a study group to engage in cooperative problem solving. Don't be afraid to try and perhaps fail, but then try again. This is the way mathematics is done, even by professionals, and as you gain experience, you will increasingly feel the real pleasure of success. Also, you will greatly improve your thinking and problem-solving skills if you take the time to write carefully worded solutions that explain your methods and reasoning. Similarly, it will help to engage in verbal mathematical discourse with your instructor and with other students. Finally, remember as little as possible, but be able to figure out as much as possible. Mechanical skills learned by rote without understanding are soon forgotten and guarantee failure, both for you now, and for your students later. Conversely, the ability to think creatively makes it more likely that the task can be successfully completed.

HOW TO READ THIS BOOK

Mathematics is not a spectator sport.
Learning is an inside job.

No mathematics textbook can be read passively. To understand the concepts and to benefit from the examples, you must be an active participant in a dialog with the text. Often, this means that you need to check a calculation, make a drawing, take a measurement, construct a model, or use a calculator or computer. If you first attempt to answer on your own questions raised in the examples, the solutions written in the text will be more meaningful and useful than they would be without your personal involvement.

Many of the problems are fully or partially answered in the back of the book. These include all of the chapter review problems and chapter test problems. These answers give you an additional source of worked examples, but again, you will benefit most fully by attempting to solve the problems on your own (or in a study group) before you check your reasoning by looking up the answer provided in the text. Other special features are described later in the Preface of the book.

To The Instructor

The principal goals of this text are to impart mathematical reasoning skills and a positive attitude to those who aspire to be elementary or middle school teachers. To help meet these goals, we have made a concerted effort to involve students in mathematical learning experiences that are intrinsically interesting, often surprising, and even aesthetically pleasing. With enhanced skill at mathematical reasoning and a positive attitude toward mathematics come confidence and an increased willingness to learn the mathematical content, skills, and effective teaching techniques necessary to become a competent teacher of mathematics.

Problem solving (read "mathematical reasoning") is stressed throughout the text. This emphasis begins in Chapter 1 on problem solving and continues not only in the discussion of the various topics, but, perhaps even more importantly, in the problem sets within the sections labeled "Thinking Critically," "Teaching Concepts," "Thinking Cooperatively," "Making Connections," "Communicating," "Using a Calculator," "Using a Computer," and "From State Student Assessments." In fact, the text is replete with activities, investigations, and a host of problems with results and answers that are attractive, surprising, and unexpected, yet are designed to engage the students in thoughtfully doing mathematics.

It is worth noting that in our own classes we have found it extremely profitable to spend considerable time (up to five weeks) on Chapter 1. This effort has gone a long way toward changing student attitudes and promoting their ability to reason mathematically. A course that begins and continues with an extensive study of the number systems and algorithms of arithmetic is not attractive or interesting to students who feel that they already know these things and have found them dull and boring. The material in Chapter 1, by contrast, and the many problems in the problem sets are new, stimulating, and not what students have previously experienced. Aside from enhancing interest, we have found that the extensive time spent on Chapter 1 develops attitudes and skills that make it possible to deal much more quickly with the usual material on number systems, algorithms, and all the subsequent ideas important to the teaching of mathematics in elementary schools. Note, however, that some instructors prefer to intersperse topics from Chapter 1 throughout their courses as they cover subsequent chapters.

GUIDING PHILOSOPHY AND APPROACH

The content and processes of mathematics are presented in an appealing and logically sound manner with these three principal goals in mind:

- to develop positive attitudes toward mathematics and mathematics teaching,
- to develop mathematical knowledge and skills, with particular emphasis on problem solving and mathematical reasoning, and
- to develop excellent teachers of mathematics.

In short, the goals of this text are to implement the recommendations of the National Council of Teachers in Mathematics (NCTM) *Principles and Standards for School Mathematics (*the *Principles and Standards)* published in the year 2000. The *Principles and Standards* are to "ensure quality, indicate goals, and promote positive changes in mathematics education in grades preK–12." Our approach to achieving the goals of this text reflects the recommendations of the *Principles and Standards 2000,* which we cite throughout the text.

Aside from mastering content and skills, teachers often pattern their own teaching after the ways they have been taught. This text models effective teaching by emphasizing

- activities,
- manipulatives,
- investigations,
- written projects,
- discussion questions,
- appropriate use of technology,

and, above all else,

- problem solving and mathematical reasoning.

It is our hope that this book will provide future elementary school and middle school teachers with the positive attitudes and mathematical skills they need to convey the beauty, usefulness, and power of mathematics to their own students.

PREREQUISITE MATHEMATICAL BACKGROUND

This text is for use in mathematics content courses for prospective elementary and middle school teachers. We assume that the students enrolled in these courses have completed two years of high school algebra and one year of high school geometry. We do not assume that the students will be highly proficient in algebra and geometry, but we presuppose that they have a basic knowledge of those subjects and reasonable arithmetic skills. Typically, students bring widely varying backgrounds to these courses, and this text is written to accommodate this diversity.

NEW IN THIS EDITION

- **Reasoning Mathematically** Chapter 1, Thinking Critically, has been significantly enhanced by adding a new Section 1.5, Reasoning Mathematically. The new section deals with the logic of doing mathematics—inductive reasoning, representational reasoning, mathematical statements, and deductive reasoning—to help students understand some of the more subtle aspects of doing mathematics.
- **Professional Development** Throughout the text, we have made a greater effort to stress professional development. In particular, we have increased the number of examples with solutions written so as to highlight Pólya's four-step approach to problem solving. We have also added a substantial number of problems in the Teaching Concepts category of the problem sets. These problems not only deal with subtle points that future teachers will

have to deal with in their own elementary school classrooms, but also cause students in the classes using this text to carefully clarify their own thinking on these issues. Finally, we have substantially revamped the Hands On features at the beginning of each chapter to make them even more effective as introductions to the material to be covered in the chapter. Some 50% of the Hands On features are completely new.

- **Statistics** Chapter 9 has been improved by updating data sets, introducing more problems where students generate their own data and by adding discussions of the standardized normal distribution, or z curve, as well as of z scores and percentiles.
- **Chapter Summaries** All the chapter summaries have been revised to make them more helpful to students. All the summaries now have the same format and more complete explanations of terms and symbols
- **Chapter Tests** All chapter tests have been reorganized in order to more accurately reflect what students might expect on an actual test in class. Changing the order so that the problems are not sequenced as they occur in the text creates a more realistic experience to better promote student learning

CONTENT FEATURES

- **Problem Solving** We begin the text with an extensive introduction to problem solving in Chapter 1. This theme is continued throughout the text in special problem-solving examples and is featured in the exercises in the problems grouped under the headings "Thinking Critically," "Teaching Concepts," "Thinking Cooperatively," "Making Connections," "Communicating," "Using a Calculator," "Using a Computer," and "From State Student Assessments." New in this fourth edition is Section 1.5, *Reasoning Mathematically*. This section deals with the logic of doing mathematics at a depth appropriate for elementary school teachers.

NEW!

- **Number Systems** Chapters 2, 3, 5, 6, and 7 focus on the various number systems and make use of discussion, pictorial and graphical representations, and manipulatives to promote an understanding of the systems, their properties, and the various modes of computation. Students are given plenty of opportunity for drill and practice, as well as for individual and cooperative problem solving, reasoning, and communication.
- **Number Theory** Chapter 4 contains much material that is new and interesting to students. Notions of divisibility, divisors, multiples, greatest common divisors, and least common multiples are developed first via informative diagrams and then through the use of manipulatives, sets, prime-factor representations, and the Euclidean algorithm. A final section contains interesting applications to zip codes and the numbering of credit cards in common use.
- **Algebraic Reasoning and Representation** Although algebraic notions are used earlier in the text, Chapter 8 gives a careful, but readable, discussion of algebraic ideas needed in elementary and middle school: variables, algebraic expressions and equations, functions, and graphing. All of these notions increasingly appear in texts for elementary school students and, even more so, in middle school texts. Schoolteachers must be conversant with algebraic ideas to be comfortable teaching from these new texts.
- **Statistics** Chapter 9, on statistics, is designed not only to give the students an appreciation of the basic measures and graphical representations of data, but also of the uses and misuses of statistics. This last issue is particularly important since we are confronted daily with a stream of "facts and figures" seductively intended to influence our thinking. The informed citizen needs to be aware of the legitimate predictive and descriptive power of statistics, as well as the way statistics can mislead. This section has been modified in this fourth edition to include updated data sets, additional problems where the students gener-

NEW!

ate their own data sets, and discussion of the standardized normal distribution, as well as z scores and percentiles.

- **Probability** In Chapter 10, we first study empirical probability—probability based on experience and repeated trials. This study prepares the way for the subsequent examination of the elements of theoretical probability of an event based on counting and other a priori considerations. The great surprise for the students is how closely the results agree, particularly when the number of trials is large. Of course, it is necessary to consider various methods of counting in order to compute theoretical probabilities. At the same time, counting is an important topic in its own right, and we have made it accessible through the use of tree diagrams, Venn diagrams, and careful explanation of the use of the words *or* and *and.*
- **Geometry** The creative and inductive nature of geometric discovery is emphasized in Chapters 11, 12, 13, and 14, and students who have acquired a distaste for the subject from a previous heavily axiomatic and deductive course will often see geometry in a positive new way. The text's approach to geometry is constructive and visual. Students are often asked to draw, cut, fold, paste, count, and so on, making geometry an experimental science. While the traditional construction and measurement tools continue to have a place, the visual and dynamic scope of geometry is enhanced with computer geometry software, such as *The Geometer's Sketchpad,* which can be purchased for use with this text. Problem solving and applications permeate the geometry chapters, and sections on tilings and symmetry provide an opportunity to highlight the aesthetic and artistic aspects of geometry. Examples are often taken from culturally diverse sources.
- **End-of-Text Material**
  - Appendix A, "Manipulatives in the Mathematics Classroom," offers a brief overview of the use of manipulatives. It provides succinct answers to such questions as What are manipulatives? Why are they used? and When should they be used?
  - Appendix B, "Spreadsheets," provides an introduction to spreadsheet software, highlighting the features that are especially useful in the mathematics classroom. In particular, it is shown how number sequences and statistical calculations can be treated with spreadsheet technology.
  - Appendix C, "Graphing Calculators," provides detailed examples to help illustrate some of the capabilities of TI graphing calculators that are useful in the middle school classroom. For example, it is shown how functions can be entered, graphed, and analyzed; how lists of data can be entered, examined statistically, and represented graphically; and how programs can be written and used to solve problems.
  - Appendix D, "A Brief Guide to *The Geometer's Sketchpad,*" provides an introduction to some of the basic features of this popular geometry software. Students are shown how the toolbox and menu commands are used to construct, measure, and transform geometric figures. The dynamic capability of the software, which allows figures to be manipulated, is used to discover and examine properties of classes of geometric shapes.
  - The useful Appendix E, "Resources," identifies sources useful to both instructors of this course and their students. Among the sources are addresses for the National Council of Teachers of Mathematics, including their meeting schedules and the *Principles and Standards for School Mathematics.* There is also a list of relevant journals and Internet sites on topics of interest, such as educational technology, Fibonacci numbers, fractions, geometry, history, integer sequences, pi (π), and prime numbers.
  - In "Answers to Selected Problems," exercise problems identified with a colored problem number are answered at the back of the student text. All answers are provided in the Instructor's Edition of the text.
  - The "Mathematical Lexicon." Many of the words, prefixes, and suffixes forming the vocabulary of mathematics are derived from words and word roots from Latin, Greek,

and other languages. The lexicon serves as an aid to learning and understanding the language of mathematics.

TOPICS OF SPECIAL INTEREST

The text includes several topics that students find especially interesting. These topics provide students with stimulating opportunities to hone such mathematical reasoning skills as problem solving, pattern recognition, algebraic representation, and calculator and computer usage. The following topics are threaded into several chapters and problem sets:

- **The Fibonacci and Lucas Numbers, and the Golden Ratio** The Fibonacci numbers (1, 1, 2, 3, 5, 8, . . .) have surprised and fascinated people over the ages and continue to serve as an unlimited source of mathematical and pedagogical examples. Information and problems dealing with the Fibonacci numbers, the closely related Lucas numbers, and the Golden Ratio are found on pages 20–21, 37, 62, 113, 151, 216–220, 224, 233, 244–245, 253–254, 266–267, 276, 293, 312–313, 359, 447, 519, 520, 563, 653, 671, 954, 955, and 956. It is not always obvious that there is a connection to the Fibonacci numbers, and much of the charm of such exercises consists in the surprise of discovery in unexpected places.
- **Pascal's Triangle** This well-known triangular pattern that has roots in ancient China has unexpected applications to counting the number of paths through a square lattice and combinatorics, and is replete with patterns awaiting discovery. In this text, references to Pascal's triangle are found on pages 28–34, 39, 70, 99, 357, 363, and 620.
- **Triangular Numbers** The numbers in the third column of Pascal's triangle (1, 3, 6, 10, 15, 21, . . .) appear in almost countless unexpected contexts. The pages with connections to these numbers are 25–27, 36–37, 38, 39–40, 55–56, 63, 113, 133, 311–312, 337, 481, 492, 520, and 955.
- **Magic Squares and Other Magic Patterns** See pages 8, 36, 38–39, 60–61, 112–113, 133–134, 311, 323, 377, and 392.

PROFESSIONAL DEVELOPMENT FEATURES

The teacher of mathematics should be aware of the historical development of mathematics, have some knowledge of the principal contributors to mathematics, and realize that mathematics continues to be a lively area of research. Moreover, the teacher must always be alert to ways to foster mathematical power in the elementary or middle school classroom. The text contains a number of features that prospective teachers will find to be valuable resources.

- "Highlights from History" illustrate the contributions individual men and women have made to mathematics and provide a cultural, historical, and personal perspective on the development of mathematical concepts and thought.
- "Into the Classroom" provides insights into teaching the topics in this text to elementary and middle school children. These tips often come from current elementary or middle school teachers.
- "School Book Pages" are taken from actual elementary or middle school textbooks and show how topics from the text are made meaningful to schoolchildren. They also show that the topics in the text are central to the elementary or middle school curriculum.
- "Did You Know" provides examples of recent advancements in mathematics, often demonstrating the important role the subject has in today's world and in our personal lives. This feature also highlights excerpts from mathematical literature that have relevance to the classroom teacher.

- "Just for Fun" shows the lighter side of mathematical problem solving, with engaging puzzles and teasers that have an element of surprise and humor.
- "Window on Technology" contains explanations of special applications of graphing calculators, spreadsheets, and geometry software.

Except for Chapter 1, each chapter is consistently and meaningfully structured according to the following pattern:

NEW and IMPROVED!

- **Chapter opener** A class activity called "Hands On" introduces the chapter topic. In this fourth edition, we have made a number of changes in these activities in order to even more effectively introduce the material covered in this chapter. Indeed, some 50% of the Hands On features in this edition are new. The Hands On activities are followed by "Connections" features that reveal the interconnections among the various parts of mathematics previously discussed and between mathematics and the real world.

NEW and IMPROVED!

- **Examples** are often presented in a *problem-solving mode,* asking the reader to independently obtain a solution that can be compared with the solution presented in the text. Solutions are frequently structured in the Pólya four-step format and the number of such solutions has been increased in this fourth edition.
- **Figures and tables** A large number of figures and tables reinforce the concepts, problems, and solutions.
- **From the NCTM *Principles and Standards*** Extensive excerpts from the *Principles and Standards* help students understand the timeliness and relevance of topics.
- **Think clouds** These notes serve as quick reminders and clarify key points in discussions.
- **Cooperative investigations** provide activities, open-ended problems, and opportunities for *cooperative learning.*
- **Problem sets** are organized according to the following categories:
  - "Understanding Concepts" exercises provide drill and reinforce basic concepts.

NEW and IMPROVED!

  - "Teaching Concepts" exercises pose questions that cause prospective teachers to carefully consider how they might go about clarifying subtle and often misunderstood points for students in classes they may soon teach. The act of answering these questions often forces the prospective teachers to think more deeply about them and come to a better understanding of the subtilities involved. The number of such examples has been increased in this edition.
  - "Thinking Critically" exercises offer problem-solving practice related to the section topic. Many of these problems can be used as classroom activities or with small groups.
  - "Thinking Cooperatively" exercises provide cooperative problem-solving experiences specifically for small groups.
  - "Making Connections" exercises apply the section concepts to solving real-life problems and to other parts of mathematics.
  - "Communicating" exercises offer students opportunities to write about mathematics and to investigate mathematics as a language.
  - "Using a Calculator" exercises provide problems that are best solved with a calculator. These problems are highlighted by a suitable icon, . Problems suitable for a graphing calculator have been highlighted by a suitable icon, .
  - "Using a Computer" exercises give students the opportunity to use various types of software. Again, a suitable icon, , indicates when the use of a computer would be helpful or desirable.
  - "From State Student Assessments" exercises provide examples of problems from tests now in use in many states to assess student progress. These problems help make future

teachers aware of the types of knowledge that their future students will be asked to master.
  - "For Review" exercises offer students continual reinforcement of concepts covered earlier in the text.
- **Epilogue** This is a brief concluding essay that discusses the importance of the materials just covered and provides a helpful summarizing overview.
- **End-of-Chapter Material** Each chapter closes with the following features:

NEW and IMPROVED!

  - Chapter Summary
    - Key Concepts This feature has been resturctured with a common format and more complete explanations in order to make it more helpful to students reviewing the chapter.
    - Vocabulary and Notation
  - Chapter Review Problems

NEW and IMPROVED!

  - Chapter Test The problems in these tests have been reordered to more accurately reflect what the stuudents are likely to experience in class.

COURSE FLEXIBILITY

Course Options

This text contains ample material for at least two semester-length courses. At Washington State University, elementary education majors are required to take two three-semester hour courses, with the option for an elective third course that is particularly suited to the needs of upper elementary and middle school teachers. Our text is used in all three courses. The suggestions following are for single semester-length courses, but instructors should have little difficulty selecting material that fits the coverage needed for courses in a quarter system:

- A first course, *Problem Solving and Numbers Systems,* covers Chapters 1 through 7. Our own first course devotes at least five weeks to Chapter 1. The problem-solving skills and enthusiasm developed in this chapter make it possible to move through most of the topics in Chapters 2 through 7 more quickly than usual. However, as noted earlier, some instructors prefer to intersperse topics from Chapter 1 among topics covered later in their courses. There is considerable latitude in which topics an instructor might choose to give lighter or heavier emphasis.
- A second course, *Algebra, Statistics, Probability, and Basic Geometry,* covers Chapters 8 through 12, with the optional inclusion of computer geometry software. (Appendix D gives a brief introduction to "The Geometer's Sketchpad.")
- An alternative second course, *Informal Geometry,* covers Chapters 11 through 14, with the optional inclusion of computer geometry software.

Once the basic notions and symbolism of geometry have been covered in Sections 11.1 and 11.2, the remaining chapters in geometry can be taken up in any order. Section 11.3, on figures in space, should receive some coverage before the instructor takes up surface area and volume in Section 12.4.

INSTRUCTOR SUPPLEMENTS

INSTRUCTOR'S EDITION

- Special version of the text
- Provides answers to both even- and odd-numbered exercises at the back of the text
- ISBN: 0–321–28739–8

INSTRUCTOR'S GUIDE

- Presents ideas, suggestions, and tips for planning for each chapter and their various sections. Includes a discussion of the "Hands On" activities and a collection of transparency masters for classroom use.
- ISBN: 0–321–28740–1

INSTRUCTOR'S SOLUTIONS MANUAL

- Provides complete solutions to all of the problems in the text
- ISBN: 0–321–28736-3

PRINTED TEST BANK

- Contains tests for each chapter of the text
- Answer keys are included
- ISBN: 0–321–28731–2

TESTGEN WITH QUIZMASTER

- Enables instructors to build, edit, print, and administer tests
- Features a computerized bank of questions developed to cover all topics
- Available on a dual-platform Windows/Macintosh CD-ROM
- ISBN: 0–321–28734–7

VIDEOTAPE SERIES

- Features an engaging team of lecturers
- Provides comprehensive coverage of each section and topic in the text
- ISBN: 0–321–28730–4

STUDENT SUPPLEMENTS

STUDENTS' SOLUTIONS MANUAL

- Provides detailed, worked-out solutions to all exercises that are answered in the back of the student text
- ISBN: 0–321–28738–X

DIGITAL VIDEO TUTOR

- Complete set of digitized videos for student use at home or on campus
- Ideal for distance learning or supplemental instruction
- ISBN: 0–321–28729–0

ADDISON-WESLEY MATH TUTOR CENTER

- Provides tutoring through a registration number packaged with a new textbook or purchased separately
- Staffed by college mathematics instructors
- Accessible via toll-free telephone, toll-free fax, e-mail, and the Internet (*www.aw-bc.com/tutorcenter*)

ACTIVITIES SUPPLEMENTS

ACTIVITIES MANUAL by Dolan, Williamson, Muri

- Provides hands-on manipulative-based activities keyed to the text that involve future elementary school teachers in discovering concepts, solving problems, and exploring mathematical ideas
- Activities can be adapted for use with elementary students at a later time
- ISBN: 0–321–28732–0

INSTRUCTOR'S RESOURCE GUIDE to accompany the ACTIVITIES MANUAL (*online supplement*)

- Contains answers for all activities, as well as additional teaching suggestions for some activities
- ISBN: 0–321–28728–2

MyMathLab is a series of text-specific, easily customized on-line courses for Addison-Wesley textbooks in mathematics and statistics. MyMathLab is powered by CourseCompass™—Pearson Education's on-line teaching and learning environment—and by MathXL®—our on-line homework, tutorial, and assessment system. MyMathLab gives you the tools you need to deliver all or a portion of your course on-line, whether your students are in a lab setting or working from home. MyMathLab provides a rich and flexible set of course materials, featuring free-response exercises that are algorithmically generated for unlimited practice and mastery. Students can also use on-line tools, such as video lectures, animations, and a multimedia textbook, to independently improve their understanding and performance. Instructors can use MyMathLab's homework and test managers to select and assign on-line exercises correlated directly to the textbook, and they can import TestGen tests into MyMathLab for added flexibility. MyMathLab's on-line gradebook—designed specifically for mathematics and statistics—automatically tracks students' homework and test results and gives the instructor control over how to calculate final grades. Instructors can also add offline (paper-and-pencil) grades to the gradebook. MyMathLab is available to qualified adopters. For more information, visit our website at www.mymathlab.com, or contact your Addison-Wesley sales representative.

MathXL® is a powerful on-line homework, tutorial, and assessment system that accompanies your Addison-Wesley textbook in mathematics or statistics. With MathXL, instructors can create, edit, and assign on-line homework and tests, using algorithmically generated exercises correlated to your textbook at the objective level. All student work is tracked in MathXL's on-line gradebook. Students can take chapter tests in MathXL and receive personalized study plans based on their test results. The study plan diagnoses weaknesses and links students directly to tutorial exercises for the objectives they need to study and retest. Students can also access supplemental animations and video clips directly from selected exercises. MathXL is available to qualified adopters. For more information, visit our website at www.mathxl.com, or contact your Addison-Wesley sales representative.

MathXL TUTORIALS ON CD® (ISBN: 0–321–28735–5)

This interactive tutorial CD-ROM provides algorithmically generated practice exercises that are correlated to the exercises in the textbook at the objective level. Every practice exercise is accompanied by an example and a guided solution designed to involve students in the solution process. Selected exercises may also include a video clip to help students visualize concepts. The software tracks student activity and scores and can generate printed summaries of students' progress.

www.interactmath.com
Get practice and tutorial help online! This interactive tutorial website provides algorithmically generated practice exercises that correlate directly to the exercises in the textbook. Students can retry an exercise as many times as they like with new values each time for unlimited practice and mastery. Every exercise is accompanied by an interactive guided solution that provides helpful feedback for incorrect answers, and students can also view a worked-out sample problem that steps them through an exercise similar to the one they're working on.

ACKNOWLEDGMENTS

We would like to thank the following individuals who reviewed either the current or previous editions of our text:

Richard Anderson-Sprecher
University of Wyoming

James E. Arnold
University of Wisconsin–Milwaukee

Bill Aslan
Texas A & M University–Commerce

James K. Bidwell
Central Michigan University

Martin V. Bonsangue
California State University–Fullerton

James R. Boone
Texas A & M University

Peter Braunfeld
University of Illinois–Urbana

Jane Buerger
Concordia College

Louis J. Chatterly
Brigham Young University

Phyllis Chinn
Humboldt State University

Lynn Cleary
University of Maryland

Max Coleman
Sam Houston State University

Dana S. Craig
University of Central Oklahoma

Lynn D. Darragh
San Juan College

Allen Davis
Eastern Illinois University

Gary A. Deatsman
West Chester University

Sheila Doran
Xavier University

Arlene Dowshen
Widener University

Stephen Drake
Northwestern Michigan College

Joseph C. Ferrar
Ohio State University

Marjorie A. Fitting
San Jose State University

Gina Foletta
Northern Kentucky University

Grace Peterson Foster
Beaufort County Community College

Ward Heilman
Bridgewater State College

Fay Jester
Pennsylvania State University

Wilburn C. Jones
Western Kentucky University

Carol Juncker
Delgado Community College

Jane Keiser
Miami University

Greg Klein
Texas A & M University

Mark Klespis
Sam Houston State University

Martha Ann Larkin
Southern Utah University

Charlotte K. Lewis
University of New Orleans

Jim Loats
Metropolitan State College of Denver

Catherine Louchart
Northern Arizona University

Jennifer Luebeck
Sheridan College

Eldon L. Miller
University of Mississippi

Carla Moldavan
Berry College

F. A. Norman
University of North Carolina–Charlotte

Jon Odell
Richland Community College

Anthony Piccolino
Montclair State College

Buddy Pierce
Southeastern Oklahoma University

Jane Pinnow
University of Wisconsin–Parkside

Robert Powers
University of Northern Colorado

Tamela D. Randolph
Southeast Missouri State University

Craig Roberts
Southeast Missouri State University

Jane M. Rood
Eastern Illinois University

Lisa M. Scheuerman
Eastern Illinois University

Julie Sliva
San Jose State University

Carol J. Steiner
Kent State University

Richard H. Stout
Gordon College

Kimberly Vincent
Washington State University

Many of the best aspects of the book are due to the creative suggestions of others. Also, we would especially like to thank Aron Cummings, who helped prepare the graphing calculator programs.

C.T.L.
D.W.D.

ABOUT THE AUTHORS

Both Calvin Long and Duane DeTemple have been extensively involved in mathematics education throughout their careers. They have taught the content for elementary teachers at Washington State University for many years. In addition, they have served as invited speakers at international, national, and regional mathematics education meetings; conducted extensive in-service programs for elementary, middle school, and high school teachers; taught in and directed numerous summer institutes designed for teachers; served as educational consultants to publishers, to the National Science Foundation, to the National Assessment of Educational Progress, to the Washington State Superintendent of Public Instruction, and to other organizations; and served on numerous committees of the National Council of Teachers of Mathematics (NCTM) and the Mathematical Association of America (MAA). Calvin Long also served on NCTM and MAA committees that drew up guidelines for the preparation for teachers of mathematics.

Chapter 1

Thinking Critically

1.1 An Introduction to Problem Solving
1.2 Pólya's Problem-Solving Principles
1.3 More Problem-Solving Strategies
1.4 Additional Problem-Solving Strategies
1.5 Reasoning Mathematically

Hands On The Gold Coin Game

Material Needed

15 yellow markers (preferably circular, and yellow if possible) for each pair of students.

Directions

This is a two-person game. Each pair of players is given 15 gold coins (markers) on the desktop. Taking alternate turns, each player removes one, two, or three coins from the desktop. The player who takes the last coin wins the game. Play several games, with each player alternately playing first. Try to devise a winning strategy, first individually as you play and then thinking jointly about how either the first or second player can play so as to force a win.

Questions to Consider

1. To discover a winning strategy, it might be helpful to begin with fewer coins. Start with just 7 coins and see if it is possible for one player or the other to play in such a way as to guarantee a win. Try this several times, and do not move on to question 2 until the answer is clear from this number of coins.
2. This time, start with 11 coins on the desktop. Is it now possible for one player or the other to force a win? Play several games until both you and your partner agree that there is a winning strategy, and see how the player using that strategy should play.
3. Extend the strategy you developed in step 2 to the original set of 15 coins.
4. Would the strategy work if you began with 51 markers? Explain carefully and clearly.

Variation

Devise a similar game in which the player taking the last coin *loses* the game, and explain how one player or the other can force a win for your new game.

Connections Mathematics Is Problem Solving

One of the most prominent features of current efforts to reform and revitalize mathematics instruction in American schools is the recommendation that such instruction should stress problem solving and quantitative reasoning. That this emphasis continues is borne out by the fact that it appears as the first of the process standards in NCTM's *Principles and Standards for School Mathematics,* published in 2000. The conviction is that children need to learn to *think* about quantitative situations in insightful and imaginative ways, and that mere rote memorization of seemingly arbitrary rules for computation is largely unproductive.

Of course, if children are to learn problem solving, their teachers must themselves be competent problem solvers and teachers of problem solving. Professor Robert Davis, a prominent mathematics educator, once said, "All too often we involve our students in a rhetoric of conclusions when, in fact, we ought to be involving them in a rhetoric of inquiry." This is true not only of elementary and secondary school students, but of college students as well. Thus, the purpose of this chapter, and indeed of this entire book, is to help you to think more critically and thoughtfully about mathematics, and to be more comfortable with mathematical reasoning and discourse.

from The NCTM Principles and Standards

Problem Solving Standard

Instructional programs from prekindergarten through grade 12 should enable all students to—

- *build new mathematical knowledge through problem solving;*
- *solve problems that arise in mathematics and in other contexts;*
- *apply and adapt a variety of appropriate strategies to solve problems;*
- *monitor and reflect on the process of mathematical problem solving.*

Problem solving is the cornerstone of school mathematics. Without the ability to solve problems, the usefulness and power of mathematical ideas, knowledge, and skills are severely limited. Students who can efficiently and accurately multiply but who cannot identify situations that call for multiplication are not well prepared. Students who can both develop *and* carry out a plan to solve a mathematical problem are exhibiting knowledge that is much deeper and more useful than simply carrying out a computation. Unless students can solve problems, the facts, concepts, and procedures they know are of little use. The goal of school mathematics should be for all students to become increasingly able and willing to engage with and solve problems.

Problem solving is also important because it can serve as a vehicle for learning new mathematical ideas and skills (Schroeder and Lester 1989). A problem-centered approach to teaching mathematics uses interesting and well-selected problems to launch mathematical lessons and engage students. In this way, new ideas, techniques, and mathematical relationships emerge and become the focus of discussion. Good problems can inspire the exploration of important mathematical ideas, nurture persistence, and reinforce the need to understand and use various strategies, mathematical properties, and relationships.

1.1 AN INTRODUCTION TO PROBLEM SOLVING

When the children arrived in Frank Capek's fifth grade class one day, this "special" problem was on the blackboard.

> *Old MacDonald had a total of 37 chickens and pigs on his farm. All together they had 98 feet. How many chickens were there and how many pigs?*

After organizing the children into problem-solving teams, Mr. Capek asked them to solve the problem. "Special" problems were always fun and the children got right to work. Let's listen in on the group with Mary, Joe, Carlos, and Sue.

> *"I'll bet there were 20 chickens and 17 pigs," said Mary.*
>
> *"Let's see," said Joe. "If you're right there are 2 × 20 or 40 chicken feet and 4 × 17 or 68 pig feet. This gives 108 feet. That's too many feet."*
>
> *"Let's try 30 chickens and 7 pigs," said Sue. "This should give us fewer feet."*
>
> *"Hey," said Carlos. "With Mary's guess we got 108 feet, and Sue's guess gives us 88 feet. Since 108 is 10 too much and 88 is 10 too few, I'll bet we should guess 25 chickens—just halfway between Mary's and Sue's guesses!"*

These children are using a **guess and check** strategy. If their guess gives an answer that is too large or too small, they adjust the guess to get a smaller or larger answer as needed. This can be a very effective strategy. By the way, is Carlos's guess right?

Let's look in on another group.

"Let's make a table," said Nandita. "We've had good luck that way before."

"Right, Nani," responded Ann. "Let's see. If we start with 20 chickens and 17 pigs, we have 2 × 20 or 40 chicken feet and 4 × 17 or 68 pig feet. If we have 21 chickens. . . ."

This is a powerful refinement of guess and check.

Chickens	Pigs	Chicken Feet	Pig Feet	Total
20	17	40	68	108
21	16	42	64	106
22	15	44	60	104
.	.	.	.	.
.	.	.	.	.
.	.	.	.	.

Making a table to look for a pattern is often an excellent strategy. Do you think the group with Nandita and Ann will soon find a solution? How many more rows of the table will they have to fill in? Can you think of a shortcut?

Mike said, "Let's draw a picture. We can draw 37 circles for heads and put two lines under each circle to represent feet. Then we can add two extra feet under enough circles to make 98. That should do it."

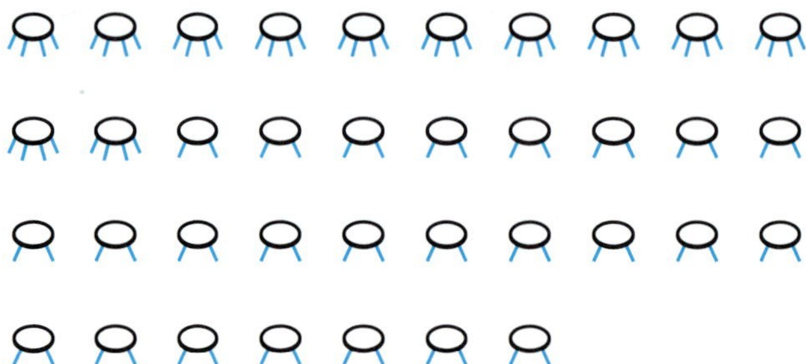

Drawing a picture is often a good strategy. Does it work in this case?

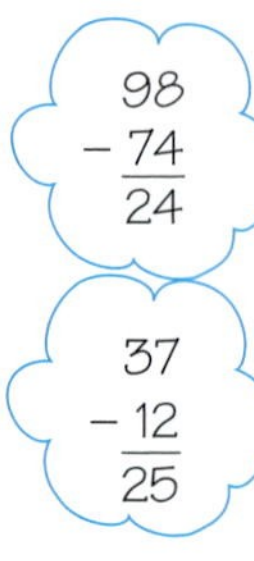

"Oh! The problem is easy," said Jennifer. "If we have all the pigs stand on their hind legs then there are 2 × 37 or 74 feet touching the ground. That means the pigs must be holding 24 front feet up in the air. This means there must be 12 pigs and 25 chickens!"

It helps if you can be ingenious like Jennifer, but it is not essential, and children *can* be taught strategies like the following:

Guess and check

Make a table

Look for a pattern

Draw a picture

These and other useful strategies will be discussed later, but for now let's try some problems on our own.*

EXAMPLE 1.1 GUESSING TONI'S NUMBER

Toni is thinking of a number. If you double the number and add 11, the result is 39. What number is Toni thinking of?

Solution 1

Guessing and checking

Guess 10.	$2 \cdot 10 + 11 = 20 + 11 = 31.$	This is too small.
Guess 20.	$2 \cdot 20 + 11 = 40 + 11 = 51.$	This is too large.
Guess 15.	$2 \cdot 15 + 11 = 30 + 11 = 41.$	This is a bit large.
Guess 14.	$2 \cdot 14 + 11 = 28 + 11 = 39.$	This checks!

Toni's number must be 14.

Solution 2

Making a table and looking for a pattern

Trial Number	Result Using Toni's Rule	
5	$2 \cdot 5 + 11 = 21$	
		2 larger
6	$2 \cdot 6 + 11 = 23$	
		2 larger
7	$2 \cdot 7 + 11 = 25$	
		2 larger
8	$2 \cdot 8 + 11 = 27$	
⋮	⋮	⋮

We need to get to 39 and we jump by 2 each time we take a step of 1. Therefore, we need to take

$$\frac{39 - 27}{2} = \frac{12}{2} = 6$$

more steps; we should guess $8 + 6 = 14$ as Toni's number as before.

EXAMPLE 1.2 GUESSING AND CHECKING

(a) Place the digits 1, 2, 3, 4, and 5 in these circles so that the sums across and vertically are the same. Is there more than one solution?

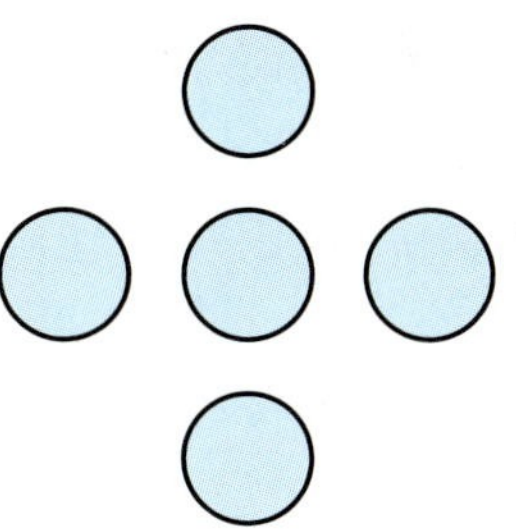

*Note that algebra students might solve this problem by solving the equation $2x + 4(37 - x) = 98$, where x denotes the number of chickens. But this approach is not available to fifth graders and it is certainly not as quick as Jennifer's solution!

(b) Can part (a) be accomplished if 2 is placed in the center? Why or why not?

Solution

(a) Using the guess and check strategy, suppose we put the 3 in the center circle. Since the sums across and down must be the same, we must pair the remaining numbers so that they have equal sums. But this is easy since $1 + 5 = 2 + 4$. Thus, one solution to the problem is as shown here.

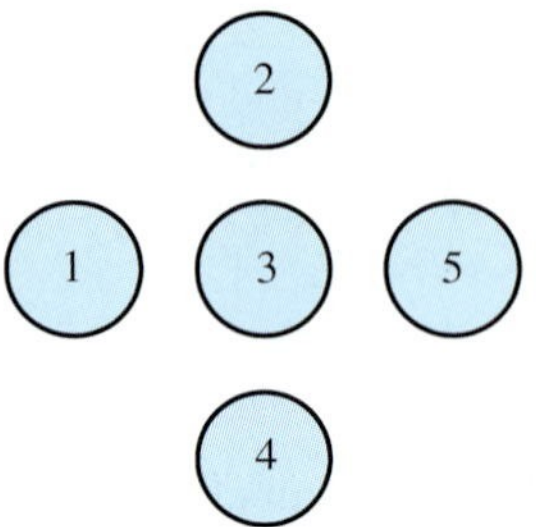

Checking further, we find other solutions like these.

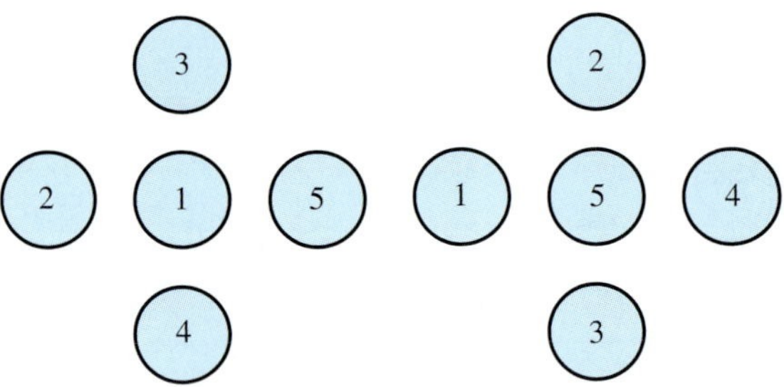

(b) What about putting 2 in the center? The remaining digits are 1, 3, 4, and 5, and these cannot be grouped into two pairs with equal sums since one sum is necessarily odd and the other even. Therefore, there is no solution with 2 in the center circle.

Problem Set 1.1

1. Levinson's Hardware has a number of bikes and trikes for sale. There are 27 seats and 60 wheels all told. Determine how many bikes there are and how many trikes.

(a) Use the guess and check strategy to find a solution.

(b) Complete this table to find a solution.

(c) Find a solution by completing this diagram.

Bikes	Trikes	Bike Wheels	Trike Wheels	Total
17	10	34	30	64
18	9	36	27	63
.	.	.	.	.
.	.	.	.	.
.	.	.	.	.

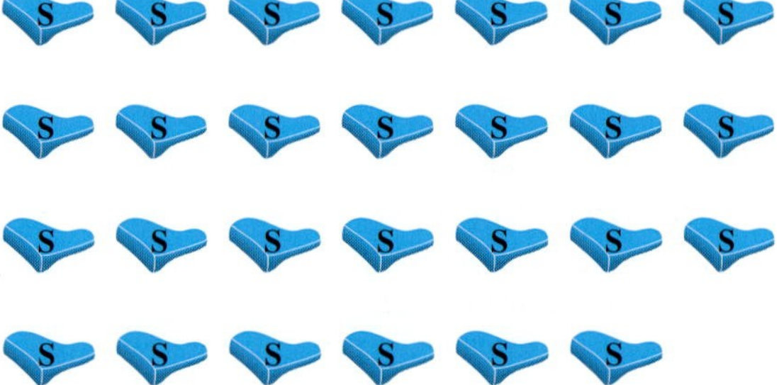

(d) Would Jennifer's method work for this problem? Explain briefly.

2. (a) Mr. Aiken has 32 18-cent and 27-cent stamps all told. The stamps are worth $7.65. How many of each kind of stamp does he have?

(b) Summarize your solution method in one or two *carefully* written sentences.

3. Make up a problem similar to problems 1 and 2.

4. (a) Place the digits 4, 6, 7, 8, and 9 in the circles to make the sums across and vertically equal 19.

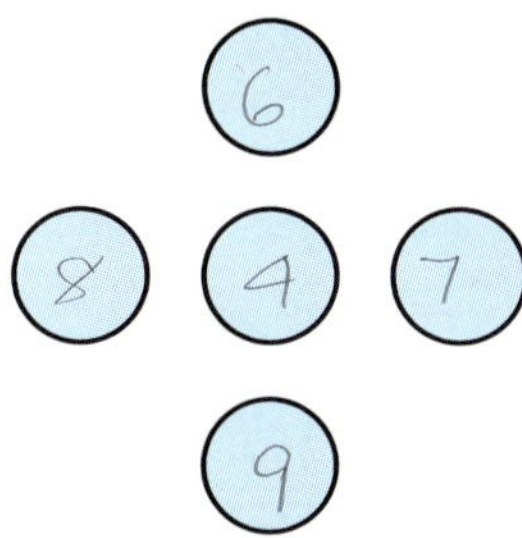

(b) Is there more than one answer to part (a)? Explain briefly.

5. Who am I? If you multiply me by 5 and subtract 8, the result is 52.

6. Who am I? If you multiply me by 15 and add 28, the result is 103.

7. Make up a problem like problems 5 and 6.

8. Melissa has nine coins with a total value of 48 cents. What coins does Melissa have?

9. (a) Using each of 1, 2, 3, 4, 5, and 6 once and only once, fill in the circles so that the sums of the numbers on each of the three sides of the triangle are equal.

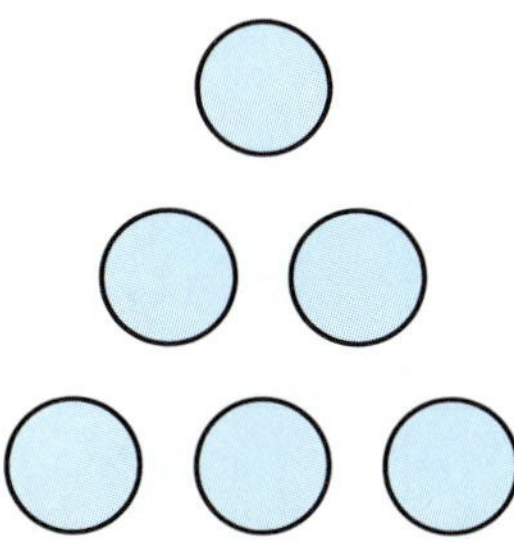

(b) Does part (a) have more than one solution?

(c) Write up a brief but careful description of the thought process you used in solving this problem.

10. In this diagram, the sum of any two horizontally adjacent numbers is the number immediately below and between them. Using the same rule of formation, complete these arrays.

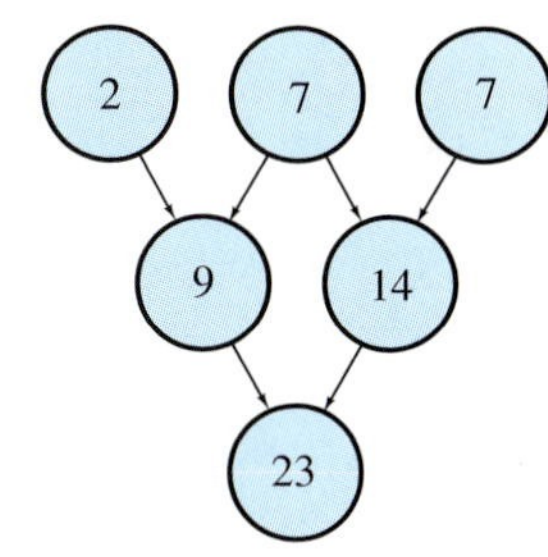

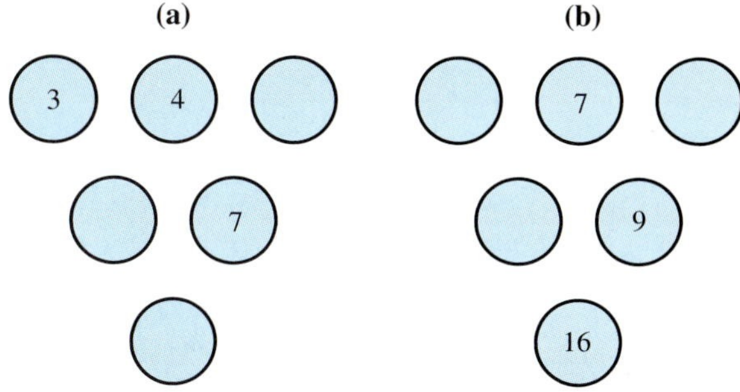

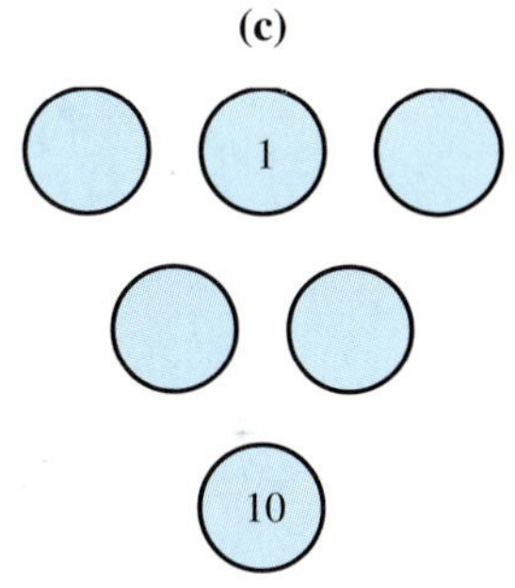

(d) Is there more than one solution to part (a), (b), or (c)?

11. Study the sample diagram. Note that

$$2 + 8 = 10,$$
$$5 + 3 = 8,$$
$$2 + 5 = 7, \text{ and}$$
$$3 + 8 = 11.$$

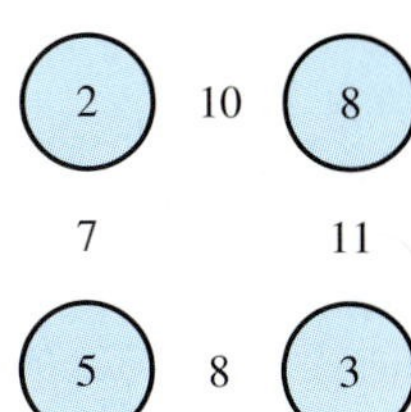

If possible, complete each of these diagrams so that the same pattern holds.

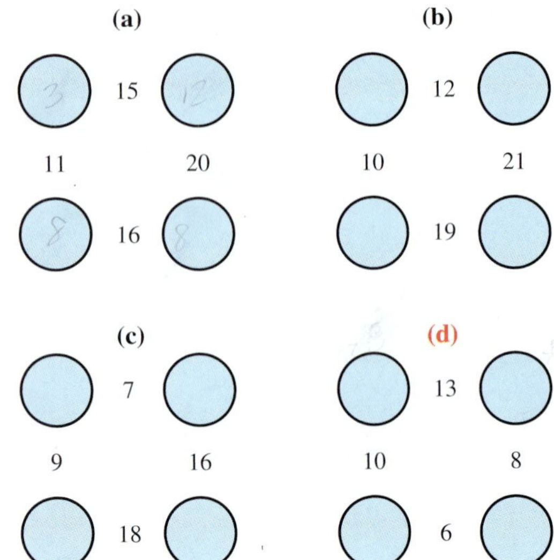

12. Study this sequence of numbers: 3, 4, 7, 11, 18, 29, 47, 76. Note that $3 + 4 = 7$, $4 + 7 = 11$, $7 + 11 = 18$, and so on. Use the same rule to complete these sequences.

(a) 1, 2, 3, ______, ______, ______, ______

(b) 2, ______, 8, ______, ______, ______, ______

(c) 3, ______, ______, 13, ______, ______, ______

(d) 2, ______, ______, ______, ______, 26

(e) 2, ______, ______, ______, ______, 11

13. **(a)** Use each of the numbers 2, 3, 4, 5, and 6 once and only once to fill in the circles so that the sum of the numbers in the three horizontal circles equals the sum of the numbers in the three vertical circles.

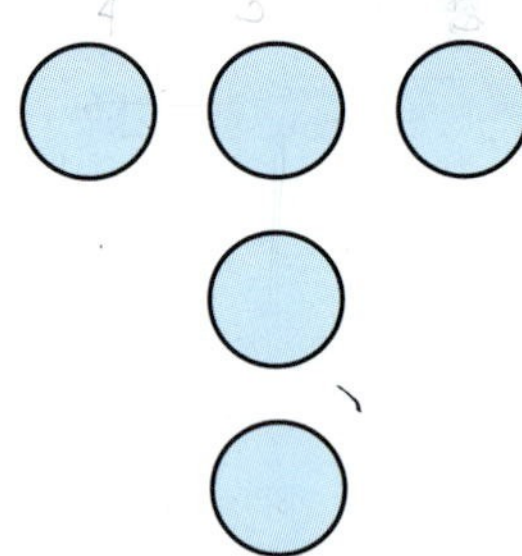

(b) Can you find more than one solution?

(c) Can you have a solution with 3 in the middle of the top row? Explain in two *carefully* written sentences.

14. **(a)** This is a magic square. Compute the sums of the numbers in each row, column, and diagonal of the square and write your answers in the appropriate circles.

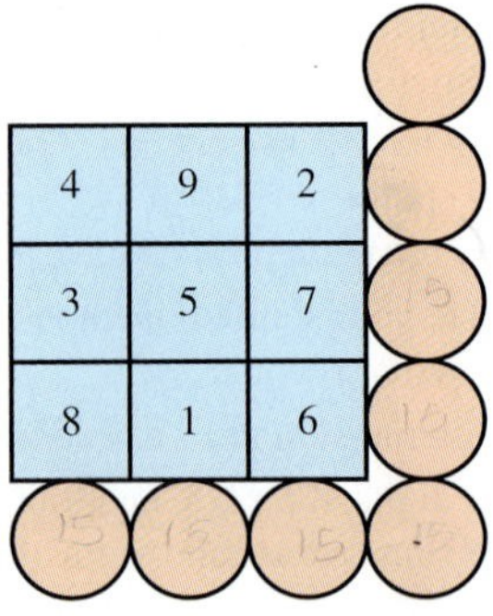

4	9	2
3	5	7
8	1	6

(b) Interchange the 2 and 8 and the 4 and 6 in the array in part (a) to create this magic *subtraction* square. For each row, column, and diagonal, add the two end entries and subtract the middle entry from this sum.

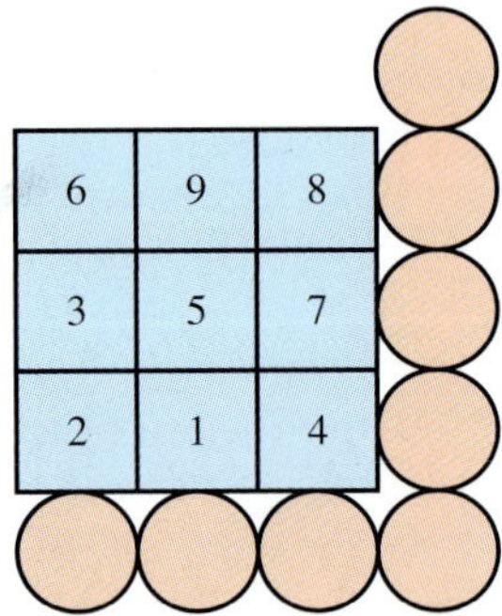

6	9	8
3	5	7
2	1	4

15. **(a)** Write the digits 0, 1, 2, 3, 4, 5, 6, 7, and 8 in the small squares to create another magic square. (*Hint:* Relate this to problem 14. Also, you may want to write these digits on nine small squares of paper that you can move around easily to check various possibilities.)

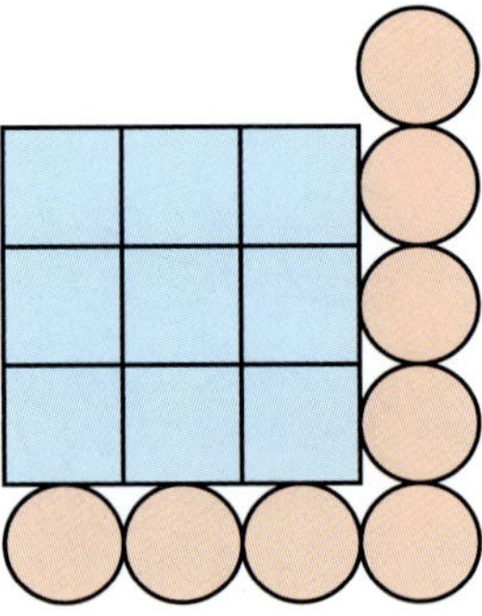

(b) Make a magic subtraction square using the numbers 0, 1, 2, 3, 4, 5, 6, 7, and 8.

16. Make up a guess and check problem of your own and solve it.

1.2 PÓLYA'S PROBLEM-SOLVING PRINCIPLES

Strategies

- Guess and check.
- Make an orderly list.
- Draw a diagram.

In *How to Solve It,** George Pólya identifies four principles that form the basis for any serious attempt at problem solving. He then proceeds to develop an extensive list of questions that teachers should ask students who need help in solving a problem, questions students can and should ask themselves as they seek solutions to problems.

Pólya's First Principle: Understand the problem

This principle seems too obvious to be mentioned. However, students are often stymied in their efforts to solve a problem because they don't understand it fully, or even in part. Teachers should ask students such questions as the following:

- Do you understand all the words used in stating the problem? If not, look them up in the index, in a dictionary, or wherever they can be found.
- What are you asked to find or show?
- Can you restate the problem in your own words?
- Is there yet another way to state the problem?
- What does *(key word)* really mean?
- Could you work out some numerical examples that would help make the problem clear?
- Could you think of a picture or diagram that might help you understand the problem?
- Is there enough information to enable you to find a solution?
- Is there extraneous information?
- What do you really need to know to find a solution?

Pólya's Second Principle: Devise a plan

Devising a plan for solving a problem once it is fully understood may still require substantial effort. But don't be afraid to make a start—you may be on the right track. There are often many reasonable ways to try to solve a problem, and the successful idea may emerge only gradually after several unsuccessful trials. A partial list of strategies includes the following:

- guess and check
- make an orderly list
- think of the problem as partially solved
- eliminate possibilities
- solve an equivalent problem
- use symmetry
- consider special cases
- use direct reasoning
- solve an equation
- look for a pattern
- draw a picture
- think of a similar problem already solved
- solve a simpler problem
- solve an analogous problem
- use a model
- work backward
- use a formula
- be ingenious!

Skill at choosing an appropriate strategy is best learned by solving many problems. As you gain experience, you will find choosing a strategy increasingly easy—and the satisfaction of making the right choice and having it work is considerable! Again, teachers can turn the above list of strategies into appropriate questions to ask students in helping them learn the art of problem solving.

---

*George Pólya, How to Solve It (Princeton, NJ: Princeton University Press, 1988).

HIGHLIGHT from HISTORY

George Pólya (1887–1985)

How does one most efficiently proceed to solve a problem? Can the art of problem solving be taught or is it a talent possessed by only a select few? Over the years, many have thought about these questions, but none so effectively and definitively as the late George Pólya, and he maintained that the skill of problem solving can be taught.

Pólya was born in Hungary in 1887 and received his Ph.D. in mathematics from the University of Budapest. He taught for many years at the Swiss Federal Institute of Technology in Zurich and would no doubt have continued to do so but for the advent of Nazism in Germany. Deeply concerned by this threat to civilization, Pólya moved to the United States in 1940 and taught briefly at Brown University and then, for the remainder of his life, at Stanford University. He was extraordinarily capable both as a mathematician and as a teacher. He also maintained a lifelong interest in studying the thought processes that are productive in both learning and doing mathematics. Indeed, among the numerous books that he wrote, he seemed most proud of *How to Solve It* (1945), which has sold over a million copies and has been translated into at least 21 languages. This book,

along with his two two-volume treatises, *Mathematics and Plausible Reasoning* (1954) and *Mathematical Discovery* (1962), form the definitive basis for much of the current thinking in mathematics education and are as timely and important today as when they were written.

Pólya's Third Principle: Carry out the plan

Carrying out the plan is usually easier than devising the plan. In general, all you need is care and patience, given that you have the necessary skills. If a plan does not work immediately, be persistent. If it still doesn't work, discard it and try a new strategy. Don't be discouraged; this is the way mathematics is done, even by professionals.

Pólya's Fourth Principle: Look back

Much can be gained by looking back at a completed solution to analyze your thinking and ascertain just what the key was to solving the problem. This is how we gain "mathematical power," the ability to come up with good ideas for solving problems never encountered before. The French mathematician and philosopher Henri Poincaré (1854–1912) put this rather strongly when he wrote,

> *Suppose I apply myself to a complicated calculation and with much difficulty arrive at a result. I shall have gained nothing by my trouble if it has not enabled me to foresee the results of other analogous calculations, and to direct them with certainty, avoiding the blind groping with which I had to be content the first time.*

Clearly, Poincaré felt that merely solving a problem was essentially meaningless if he did not also gain experience and insight that increased his "mathematical power." Often the connection between very dissimilar problems is tenuous at best. Yet, in working on a problem, something may be lurking in the back of your mind from a previous effort that says, "I'll bet if . . . ," and the plan does indeed work!

Questions to ask yourself in looking back after you have successfully solved a problem include the following:

- What was the key factor that allowed me to devise an effective plan for solving this problem?
- Can I think of a simpler strategy for solving this problem?

- Can I think of a more effective or powerful strategy for solving this problem?
- Can I think of *any* alternative strategy for solving this problem?
- Can I think of any other problem or class of problems for which this plan of attack would be effective?

Looking back is an often overlooked but extremely important step in developing problem-solving skills.

Let's now look at some examples.

Guess and Check

PROBLEM-SOLVING STRATEGY 1 Guess and Check

Make a guess and check to see if it satisfies the demands of the problem. If it doesn't, alter the guess appropriately and check again. When the guess finally checks, a solution has been found.

Students often feel that it is not "proper" to solve a problem by guessing. And they are right if the guess is not accompanied by a check. However, a process of guessing, checking, altering the guess if it does not check, guessing again in light of the preceding check, and so on is a legitimate and effective strategy. When a guess finally checks, there can be no doubt that a solution has been found. If we can be sure that there is only one solution, *the* solution has been found. Moreover, the process is often quite efficient and it may be the only approach available.

EXAMPLE 1.3 GUESS AND CHECK

In the first diagram the numbers in the big circles are found by adding the numbers in the two adjacent smaller circles as shown. Complete the second diagram so that the same pattern holds.

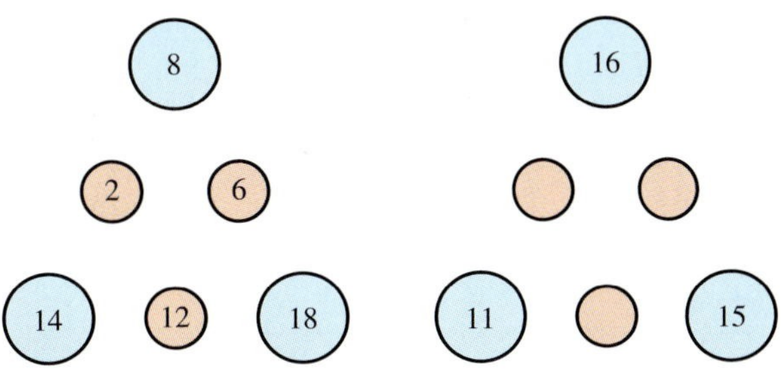

Solution *Understand the Problem*

Considering the example, it is pretty clear that we must find three numbers—a, b, and c—such that

$$a + b = 16, \quad a + c = 11, \quad \text{and} \quad b + c = 15.$$

How should we proceed?*

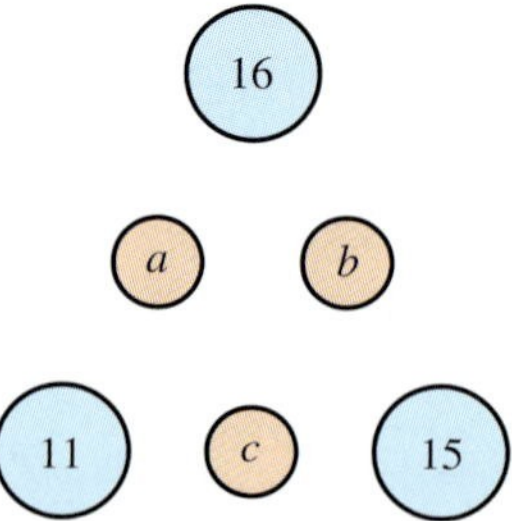

Devise a Plan

Let's try the *guess and check* strategy. It worked on several problems somewhat like this in the last problem set. Also, even if the strategy fails, it may at least suggest an approach that will work.

Carry Out the Plan

We start by guessing a value for a. Suppose we guess that a is 10. Then, since $a + b$ must be 16, b must be 6. Similarly, since $b + c$ must be 15, c must be 9. But then $a + c$ is 19 instead of 11 as it is supposed to be. This does not check.

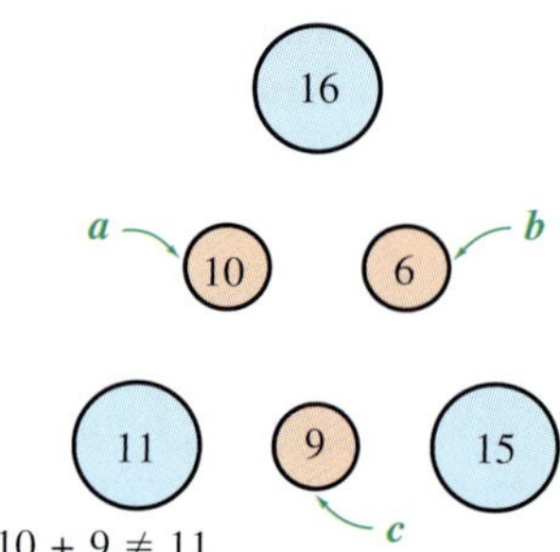

Since 19 is too large, we try again with a smaller guess for a. Guess that a is 5. Then, b is 11 and c is 4. But then $a + c$ is 9, which is too small, but by just a little bit. We should guess that a is just a bit larger than 5.

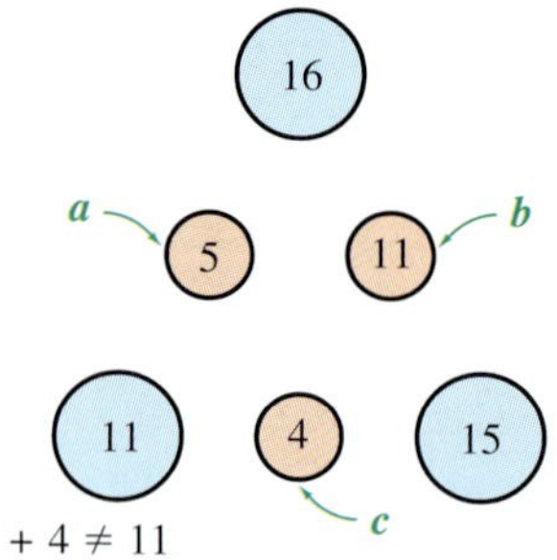

*Students who know algebra could solve this system of simultaneous equations, but elementary school students don't know algebra.

Guess that $a = 6$. This implies that b is 10 and c is 5. Now $a + c$ is 11 as desired, and we have the solution.

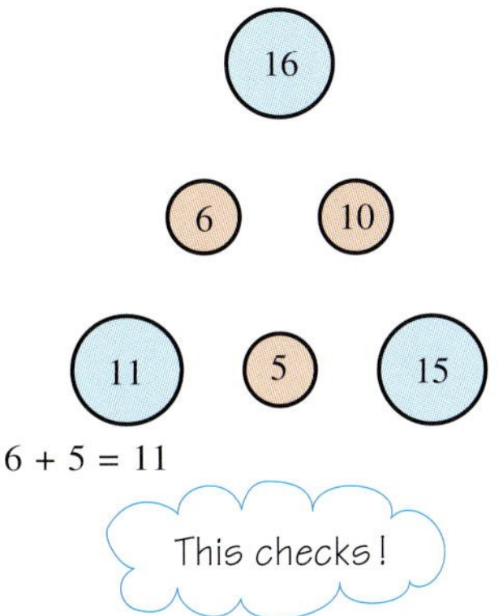

Look Back

Guess and check worked fine. Our first choice of 10 for a was too large, so we chose a smaller value. Our second choice of 5 was too small, but quite close. Choosing $a = 6$, which is between 10 and 5 but quite near 5, we obtained a solution that checked. Surely this approach would work equally well on other similar problems.

But wait. Have we fully understood this problem? Might there be an easier solution?

Look back at the initial example and at the completed solution to the problem. Do you see any special relationship between the numbers in the large circles and those in the small circles?

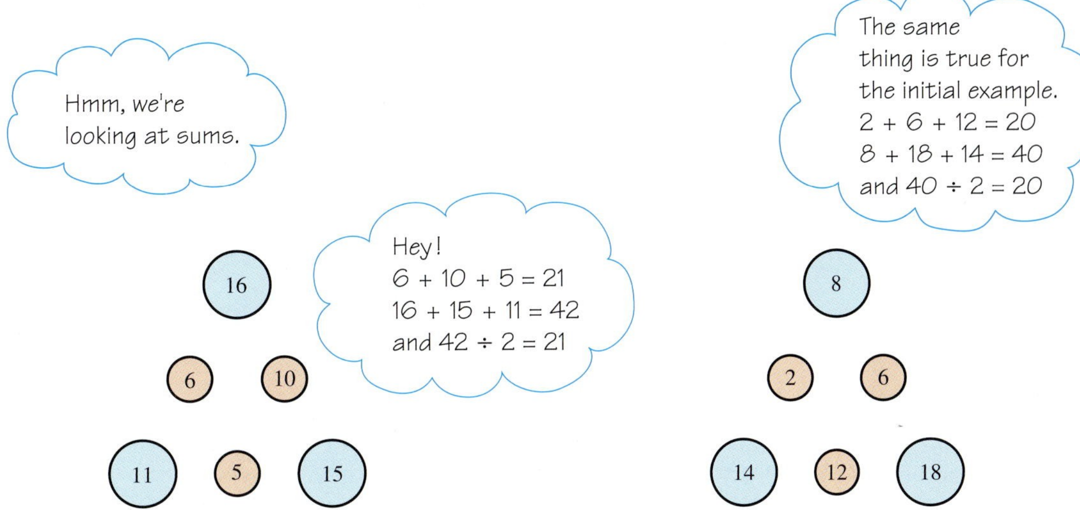

That's interesting; the sum of the numbers in the small circles in each case is just half the sum of the numbers in the large circles. Could we use this to find another solution method?

Sure! Since $16 + 15 + 11 = 42$ and $a + b + c$ is half as much, $a + b + c = 21$. But $a + b = 16$, so c must equal 5; that is,

$$c = 21 - 16 = 5,$$
$$b = 21 - 11 = 10, \quad \text{and}$$
$$a = 21 - 15 = 6.$$

This is much easier than our first solution and, for that matter, the algebraic solution. Quickly now, does it work on this diagram? Try it.

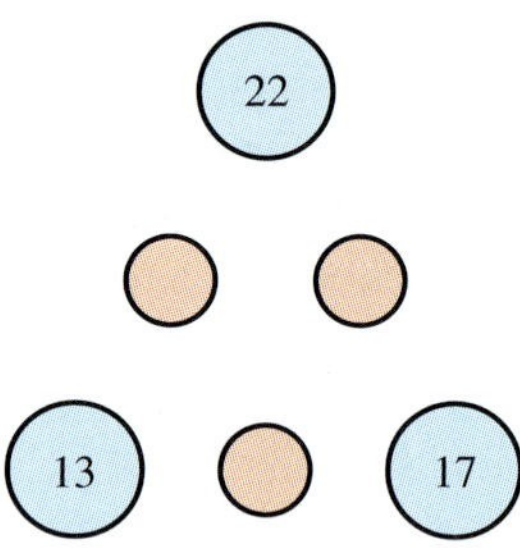

But there's one more thing. Do you understand *why* the sum of the numbers in the little circles equals half the sum of the numbers in the big circles? This diagram might help.

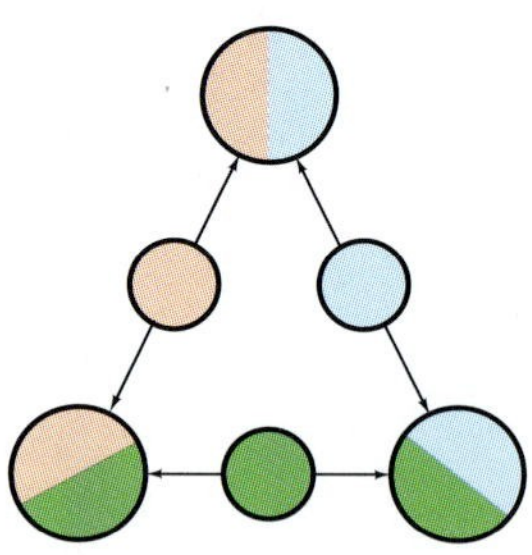

from **The NCTM Principles and Standards**

Apply and Adapt a Variety of Appropriate Strategies to Solve Problems

Of the many descriptions of problem-solving strategies, some of the best known can be found in the work of Pólya (1957). Frequently cited strategies include using diagrams, looking for patterns, listing all possibilities, trying special values or cases, working backward, guessing and checking, creating an equivalent problem, and creating a simpler problem. An obvious question is, How should these strategies be taught? Should they receive explicit attention, and how should they be integrated with the mathematics curriculum? As with any other component of the mathematical tool kit, strategies must receive instructional attention if students are expected to learn them. In the lower grades, teachers can help children express, categorize, and compare their strategies. Opportunities to use strategies must be embedded naturally in the curriculum across the content areas. By the time students reach the middle grades, they should be skilled at recognizing when various strategies are appropriate to use and should be capable of deciding when and how to use them.

SOURCE: Reprinted with permission from Principles and Standards for School Mathematics, *pages 53–54.* *Standards are listed with the permission of the National Council of Teachers of Mathematics (NCTM). NCTM does not endorse the content or validity of these alignments.*

Make an Orderly List

PROBLEM-SOLVING STRATEGY 2 Make an Orderly List

For problems that require consideration of many possibilities, make an orderly list or a table to make sure that no possibilities are missed.

Sometimes a problem is sufficiently involved that the task of sorting out all the possibilities seems quite forbidding. Often these problems can be solved by making a carefully structured list so that you can be sure that all of the data and all of the cases have been considered. Consider the next example.

EXAMPLE 1.4 MAKE AN ORDERLY LIST

How many different total scores could you make if you hit the dartboard shown with three darts?

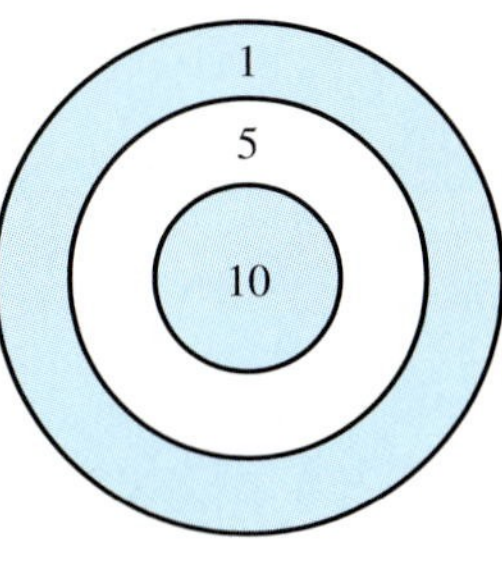

Solution

Understand the Problem

Three darts hit the dartboard and each scores a 1, 5, or 10. The total score is the sum of the scores for the three darts. There could be three 1s, two 1s and a 5, one 5 and two 10s, and so on. The fact that we are told to find the total score when throwing three darts at a dartboard is just a way of asking what sums can be made using three numbers, each of which is either 1, 5, or 10.

It's often helpful to restate the problem in a different way.

Devise a Plan

If we just write down sums hit or miss, we will almost surely overlook some of the possibilities. Using an orderly scheme instead, we can make sure that we obtain all possible scores. Let's make such a list. We first list the score if we have three 1s, then two 1s and one 5, then two 1s and no 5s, and so on. In this way, we can be sure that no score is missed.

Carry Out the Plan

Number of 1s	Number of 5s	Number of 10s	Total Score
3	0	0	3
2	1	0	7
2	0	1	12
1	2	0	11
1	1	1	16
1	0	2	21
0	3	0	15
0	2	1	20
0	1	2	25
0	0	3	30

The possible total scores are listed.

Look Back

Here the key to the solution was in being very systematic. We were careful first to obtain all possible scores with three 1s, then two 1s, then no 1s. With two 1s there could be either a 5 or a 10 as shown. For one 1 the only possibilities are two 5s and no 10s, one 5 and one 10, or no 5s and two 10s. Constructing the table in this orderly way makes it clear that we have not missed any possibilities.

Draw a Diagram

PROBLEM-SOLVING STRATEGY 3 Draw a Diagram

Draw a diagram or picture that represents the data of the problem as accurately as possible.

The aphorism "A picture is worth a thousand words" is certainly applicable to solving many problems. Language used to describe situations and state problems often can be clarified by drawing a suitable diagram, and unforeseen relationships and properties often become clear. As with the problem of the pigs and chickens on Old MacDonald's farm, even problems that do not appear to have pictorial relationships can sometimes be solved using this technique. Would you immediately draw a picture in attempting to solve the problem in the next example? Some would and some wouldn't, but it's surely the most efficient approach.

EXAMPLE 1.5 DRAW A DIAGRAM

In a stock car race the first five finishers in some order were a Ford, a Pontiac, a Chevrolet, a Buick, and a Dodge.

(a) The Ford finished seven seconds before the Chevrolet.
(b) The Pontiac finished six seconds after the Buick.
(c) The Dodge finished eight seconds after the Buick.
(d) The Chevrolet finished two seconds before the Pontiac.

In what order did the cars finish the race?

Solution

Understand the Problem

We are told how each of the cars finished the race relative to one other car. The question is, "Can we use just this information to determine the order in which the five cars finished the race?"

Devise a Plan

Imagine the cars in a line as they race toward the finish. If they do not pass one another, this is the order in which they will finish the race. We can draw a line to represent the track at the finish of the race and place the cars on it according to the conditions of the problem. Mark the line off in time intervals of one second. Then, using the first letter of each car's name to represent the car, see if we can line up *B*, *C*, *D*, *F*, and *P* according to the given information.

Carry Out the Plan

Here is a line with equally spaced points to represent one-second time intervals. Pick some point and label it *C* to represent the Chevrolet's finishing position.

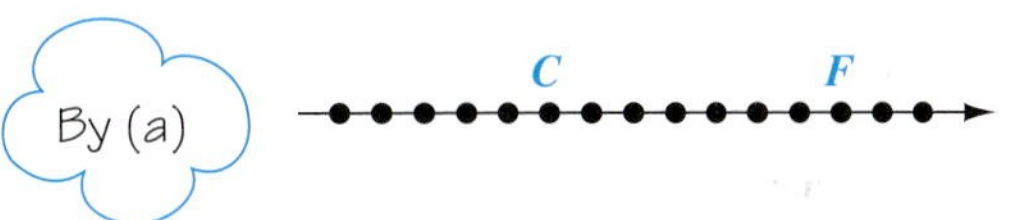

Then *F* is seven seconds ahead of *C* by condition (a) as shown above. Conditions (b) and (c) cannot yet be used since they do not relate to the positions of either *C* or *F*. However, (d) allows us to place *P* two seconds behind (to the left of) *C* as shown.

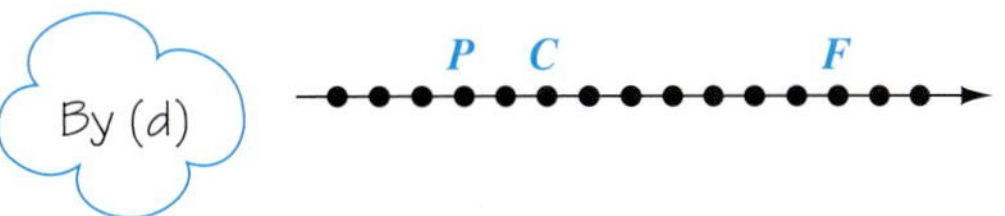

Since (b) relates the finishing position of *B* to *P*, we place *B* six seconds ahead of *P*.

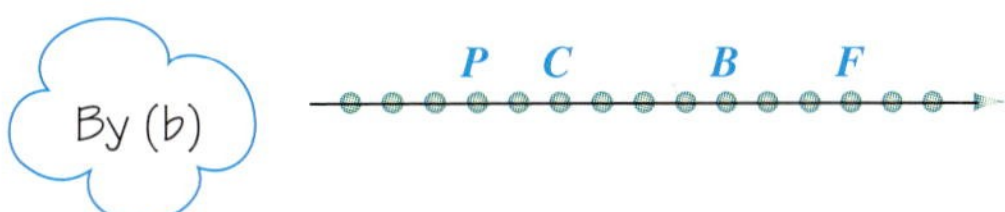

Similarly, (c) relates the finishing positions of *D* and *B* and allows us to place *D* eight seconds behind (to the left of) *B*. Since this accounts for all the cars, a glance at the last diagram reveals the order in which the cars finished the race. We repeatedly drew the line for pedagogical purposes in order to show the placement of the cars as each new condition was used. Ordinarily, all the work would be done on a single line since it is not necessary to show what happens at each stage as we did here.

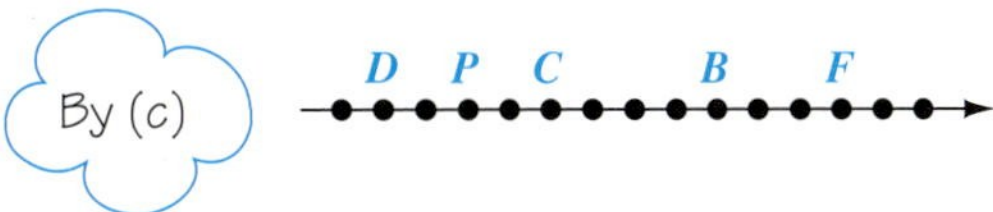

> *Look Back*
>
> Like the problem of the pigs and chickens on Old MacDonald's farm, this problem did not immediately suggest drawing a picture. However, having seen pictures used to solve these problems will help you to see how pictures can be used to solve other, even vaguely related, problems.

Problem Set 1.2

1. Diedre is thinking of a number. If you multiply it by 5 and add 13, you get 48. Could Diedre's number be 10? Why or why not?
2. John is thinking of a number. If you divide it by 2 and add 16, you get 28. What number is John thinking of?
3. Lisa is thinking of a number. If you multiply it by 7 and subtract 4, you get 17. What is the number?
4. Vicky is thinking of a number. Twice the number increased by 1 is 5 less than 3 times the number. What is the number? (*Hint:* For each guess, compute two numbers and compare.)
5. In Mrs. Garcia's class they sometimes play a game called **Guess My Rule.** The student who is It makes up a rule for changing one number into another. The other students then call out numbers and the person who is It tells what number the rule gives back. The first person in the class to guess the rule then becomes It and gets to make up a new rule.

   (a) For Juan's rule the results were

Numbers chosen	2	5	4	0	8
Numbers Juan gave back	7	22	17	−3	37

   Could Juan's rule have been, "Multiply the chosen number by 5 and subtract 3"? Could it have been, "Reduce the chosen number by 1, multiply the result by 5, and then add 2"? Are these rules really different? Discuss briefly.

   (b) For Mary's rule, the results were

Numbers chosen	3	7	1	0	9
Numbers Mary gave back	10	50	2	1	82

   What is Mary's rule?

   (c) For Peter's rule, the results were

Numbers chosen	0	1	2	3	4
Numbers Peter gave back	7	10	13	16	19

   Observe that the students began to choose the numbers in order starting with 0. Why is that a good idea? What is Peter's rule?

6. As in Example 1.3, the numbers in the big circles are the sums of the numbers in the two small adjacent circles. Place numbers in the empty circles in each of these arrays so that the same scheme holds.

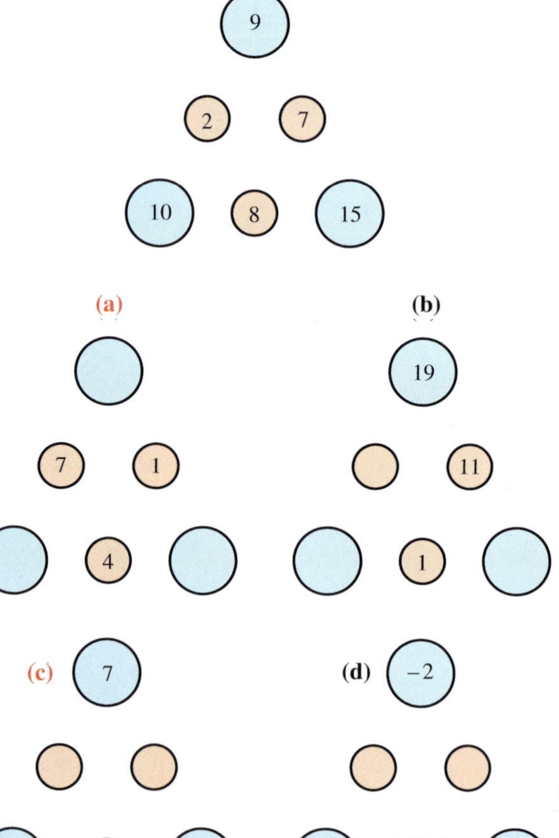

7. How many different amounts of money can you pay if you use four coins including only nickels, dimes, and quarters?
8. How many different ways can you make change for a 50-cent coin using quarters, nickels, dimes, and pennies?
9. List the three-digit numbers that can be written using each of the digits 2, 5, and 8 once and only once.
10. List the four-digit numbers that can be written using each of 1, 3, 5, and 7 once and only once.
11. When Anita made a purchase she gave the clerk a dollar and received 21 cents in change. Complete this table to show what Anita's change could have been.

Number of Dimes	Number of Nickels	Number of Pennies
2	0	1

12. Julie has 25 pearls. She put them in three velvet bags with an odd number of pearls in each bag. What are the possibilities?
13. A rectangle has an area of 120 cm^2. Its length and width are whole numbers.
    (a) What are the possibilities for the two numbers?
    (b) Which possibility gives the smallest perimeter?
14. The product of two whole numbers is 96 and their sum is less than 30. What are the possibilities for the two numbers?
15. Peter and Jill each worked a different number of days, but earned the same amount of money. Use these clues to determine how many days each worked:

    Peter earned $20 a day.

    Jill earned $30 a day.

    Peter worked five more days than Jill.
16. Bob can cut through a log in one minute. How long will it take Bob to cut a 20-foot log into 2-foot sections? (*Hint:* Draw a diagram.)
17. How many posts does it take to support a straight fence 200 feet long if a post is placed every 20 feet?
18. How many posts does it take to support a fence around a square field measuring 200 feet on a side if posts are placed every 20 feet?
19. Albright, Badgett, Chalmers, Dawkins, and Ertl all entered the primary to seek election to the city council. Albright received 2000 more votes than Badgett and 4000 fewer than Chalmers. Ertl received 2000 votes fewer than Dawkins and 5000 votes more than Badgett. In what order did each person finish in the balloting?
20. Nine square tiles are laid out on a table so that they make a solid pattern. Each tile must touch at least one other tile along an entire edge. The squares all have sides of length one.

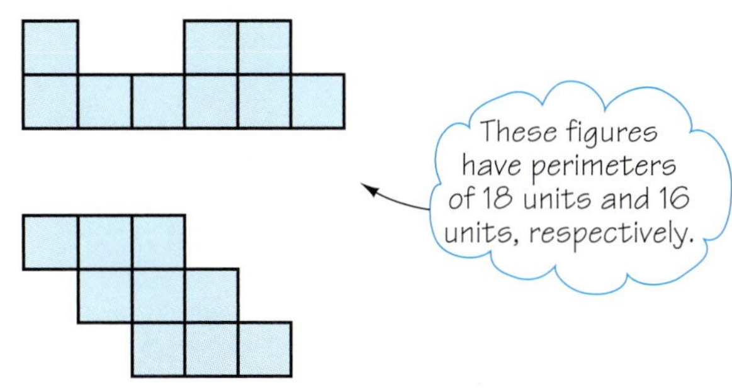

These figures have perimeters of 18 units and 16 units, respectively.

    (a) What are the possible perimeters of the figures that can be formed? (The perimeter is the distance around the figure.)
    (b) Which figure has the least perimeter?
21. A 9-meter by 12-meter rectangular lawn has a concrete walk 1 meter wide all around it outside the lawn. What is the area of the walk?

Teaching Concepts

22. Read one of the following from NCTM's *Principles and Standards for School Mathematics.*
    (a) Problem Solving Standard for Grades Pre-K–2, pages 116–121
    (b) Problem Solving Standard for Grades 3–5, pages 182–187
    (c) Problem Solving Standard for Grades 6–8, pages 256–261

    Write a critique of the standard you read, emphasizing your own reaction. How do the recommendations compare with your own school experience?

From State Student Assessments

23. (Florida, Grade 5)
    Pat is making a key ring by stringing beads on a leather strip. He may choose one of the following 2 colors for the leather strip.

    BLACK WHITE

    Then he may choose one of the following 3 colors for the beads.

    RED BLUE YELLOW

    How many different color combinations can Pat make?

    A. 8 B. 6 C. 4 D. 2

24. (Washington State, Grade 4)
Four students create their own Good Fitness Games. The students run their fastest and do as many sit-ups and pull-ups as they can. Their results follow. The students want to pick an overall winner. They decide that all the events are equally important. Tell who you think the overall winner is. Explain your thinking using words, numbers, or pictures.

	50-Meter Dash	Sit-Ups	600-Meter Run	Pull-Ups
Sarah	10 seconds	42	3 minutes, 15 seconds	4
Jan	7 seconds	37	3 minutes, 50 seconds	2
Angel	8 seconds	38	3 minutes, 20 seconds	6
Mike	9 seconds	27	3 minutes, 30 seconds	8

1.3 MORE PROBLEM-SOLVING STRATEGIES

Look for a Pattern

Strategies

- Look for a pattern.
- Make a table.
- Use a variable.
- Consider special cases.
- Solve an equivalent problem.
- Solve an easier similar problem.
- Argue from special cases.

PROBLEM-SOLVING STRATEGY 4 **Look for a Pattern**

Consider an ordered sequence of particular examples of the general situation described in the problem. Then carefully scrutinize these results, looking for a pattern that may be the key to the problem.

It is no overstatement to assert that this strategy is the most important of all problem-solving strategies. In fact, mathematics is often characterized as the study of patterns, and patterns occur in some form in almost all problem-solving situations. Think about the problems we have already considered and you will see patterns everywhere—numerical patterns, geometrical patterns, counting patterns, listing patterns, rhetorical patterns—patterns of all kinds.

Some problems, like those in the next examples, are plainly pattern problems, but looking for a pattern is almost never a bad way to start solving a problem.

EXAMPLE 1.6 LOOK FOR PATTERNS IN NUMERICAL SEQUENCES

Continue these numerical sequences. Fill in the next three blanks in each part.

(a) 1, 4, 7, 10, 13, _____, _____, _____
(b) 19, 20, 22, 25, 29, _____, _____, _____
(c) 1, 1, 2, 3, 5, _____, _____, _____
(d) 1, 4, 9, 16, 25, _____, _____, _____

Solution

Understand the Problem

In each case we are asked to discover a reasonable pattern suggested by the first five numbers and then to continue the pattern for three more terms.

Devise a Plan

Questions we might ask ourselves and answer in search of a pattern include the following: Are the numbers growing steadily larger? steadily smaller? How is each number related to its predecessor? Is it perhaps the case that a particular term depends on its two predecessors? on its three predecessors? Perhaps each term depends in a special way on the number of the term in the sequence; can we notice any such dependence? Are the numbers in the sequence somehow special numbers that we recognize? This is rather like playing *Guess My Rule*. Let's see how successful we can be.*

Carry Out the Plan

In each of (a), (b), (c), and (d), the numbers grow steadily larger. How is each term related to the preceding term or terms in each case? Are the terms related to their numbered place in the sequence? Do they have a special form we can recognize?

(a) For the sequence in part (a), each number listed is 3 greater than its predecessor. If this pattern continues, the next three numbers will be 16, 19, and 22.

(b) Here the numbers increase by 1, by 2, by 3, and by 4. If we continue this scheme, the next three numbers will be 5 more, 6 more, and 7 more than their predecessors. This would give 34, 40, and 47.

(c) If we use the idea of (a) and (b) for this sequence, we should check how much greater each entry is than its predecessor. The numbers that must be added are *0*, *1*, *1*, and *2*; that is, *0* + 1 = 1, *1* + 1 = 2, *1* + 2 = 3, and *2* + 3 = 5. This just amounts to adding any two consecutive terms of the sequence to obtain the next term. Continuing this scheme, we obtain 8, 13, and 21.

The sequence 1, 1, 2, 3, 5, 8, 13, 21, . . . , where we start with 1 and 1 and add any two consecutive terms to obtain the next, is called the **Fibonacci sequence** and the numbers are called the **Fibonacci numbers.** We denote the Fibonacci numbers by $F_1 = 1$, $F_2 = 1$, $F_3 = 2$, $F_4 = 3, \ldots, F_n =$ the nth Fibonacci number, and so on. In particular, observe that

$$F_3 = 2 = 1 + 1 = F_2 + F_1,$$
$$F_4 = 3 = 2 + 1 = F_3 + F_2,$$
$$F_5 = 5 = 3 + 2 = F_4 + F_3,$$

and so on. In general, any particular entry in the sequence is the sum of its two predecessors. The Fibonacci numbers first appeared in A.D. 1202 in the book *Liber Abaci* by Leonardo of Pisa (Fibonacci), the leading mathematician of the thirteenth century.

(d) Here 4 is 3 larger than 1, 9 is 5 larger than 4, 16 is 7 larger than 9, and 25 is 9 larger than 16. The terms seem to be increasing by the next largest **odd** number each time. Thus, the next three numbers should probably be 25 + 11 = 36, 36 + 13 = 49, and 49 + 15 = 64. Alternatively, in this case, we may recognize that the numbers 1, 4, 9, 16, and 25 are special numbers. Thus, $1 = 1^2$, $4 = 2^2$, $9 = 3^2$, $16 = 4^2$, and $25 = 5^2$. The sequence appears to be just the sequence of **square numbers.** The sixth, seventh, and eighth terms are just $6^2 = 36$, $7^2 = 49$, and $8^2 = 64$ as before.

Look Back

In all four sequences, we checked to see how much larger each number was than its predecessor. In each case we were able to discover a pattern that allowed us to write the

*Actually, there is a touchy point here. To be strictly accurate, any three numbers you choose in each case can be considered correct. There are actually infinitely many different rules that will give you any first five numbers followed by any three other numbers. What we seek here are relatively simple rules that apply to the given numbers and tell how to obtain the next three in each case.

next three terms of the sequence. In part (d), we also noted that the first term was 1^2, the second term was 2^2, the third term was 3^2, and so on. Thus, it was reasonable to guess that each term was the square of the number of its position in the sequence. This allowed us to write the next few terms with the same result as before.

We have already seen in earlier examples how making a table is often an excellent strategy, particularly when combined with the strategy of looking for a pattern. Like drawing a picture or making a diagram, making a table often reveals unexpected patterns and relationships that help to solve a problem.

Make a Table

PROBLEM-SOLVING STRATEGY 5 Make a Table

Make a table reflecting the data in the problem. If done in an orderly way, such a table will often reveal patterns and relationships that suggest how the problem can be solved.

EXAMPLE 1.7 MAKE A TABLE

(a) Draw the next two diagrams to continue this dot sequence.

• ⁚• ⁚•• ⁚••• ____, ____

(b) How many dots are in each figure?

____, ____, ____, ____, ____, ____

(c) How many dots would be in the one hundredth figure?

(d) How many dots would be in the one millionth figure?

Solution

Understand the Problem

What is given?

In part (a), we are given an ordered sequence of arrays of dots. We are asked to recognize how the arrays are being formed and to continue the pattern for two more diagrams. In part (b), we are asked to record the number of dots in each array in part (a). In parts (c) and (d) we are asked to determine specific numerical terms in the sequence of part (b).

Devise a Plan

In part (a), we are asked to continue the pattern of a sequence of arrays of dots. As with numerical sequences, our strategy will be to see how each array relates to its predecessor or predecessors, hoping to discern a pattern that we can extend two more times.

In part (b), we will simply count and record the numbers of dots in the successive arrays in part (a).

In parts (c) and (d), we will study the numerical sequence of part (b) just as we did in Example 1.6, hoping to discern a pattern and understand it sufficiently well that we can determine its one hundredth and one millionth terms.

Carry Out the Plan

In part (a), we observe that the arrays of dots are similar, but that each one has one more two-dot column than its predecessor. Thus, the next two arrays are

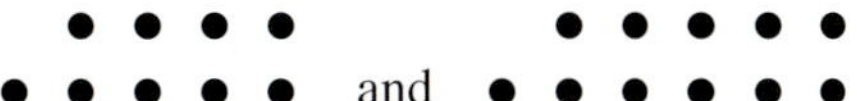

For part (b), we count the dots in each array of part (a) to obtain

$$1, 3, 5, 7, 9, 11, \ldots$$

These are just the odd numbers, and we could write out the first one million odd numbers and so answer parts (c) and (d). But surely there's an easier way.

Let's review how the successive terms were obtained. A table may help.

Reviewing this carefully, we finally experience an Aha!

Number of Entry	Entry
1	$1 = 1$
2	$3 = 1 + 2$
3	$5 = 1 + 2 + 2 = 1 + 2 \times 2$
4	$7 = 1 + 2 + 2 + 2 = 1 + 3 \times 2$
5	$9 = 1 + 2 + 2 + 2 + 2 = 1 + 4 \times 2$

The *second* term is $1 + 1 \times 2$. (2 − 1)

The *third* term is $1 + 2 \times 2$. (3 − 1)

The *fourth* term is $1 + 3 \times 2$. (4 − 1)

The number of 2s added is one less than the number of the term. Therefore, the one hundredth term is

$$1 + 99 \times 2 = 199$$

(100 − 1)

and the one millionth term is

$$1 + (1{,}000{,}000 - 1) \times 2 = 1 + 999{,}999 \times 2 = 1{,}999{,}999.$$

Look Back

The basic observation was that each diagram could be obtained by adding a column of two dots to its predecessor. Hence, the successive terms in the numerical sequence were obtained by adding 2 to each entry to get the next entry. Using this notion, we examined the successive terms and discovered that any entry could be found by subtracting 1 from the number of the entry, doubling the result, and adding 1. But this last sentence is rather cumbersome, and we have already seen that using *symbols* can make it easier to write mathematical statements. If we use n for the number of the term, the above sentence can be translated into this mathematical sentence:

$$e_n = (n - 1) \times 2 + 1 = 2n - 2 + 1 = 2n - 1.*$$

2n − 1 is the nth odd number.

*Symbols like this are often called *variables*, and the formula $e_n = 2n - 1$ is an example of a function. These notions are of considerable importance in mathematics and will be discussed in detail in Chapter 8.

School Book Page

Problem Solving in Grade Three

Chapter 6
Lesson
9

Problem Solving

Compare Strategies: Look for a Pattern and Draw a Picture

You Will Learn
how to solve the same problem using different strategies

Learn

Plan a class picnic for 29 people! Decide how many blankets to bring. Each blanket seats 4 people. How many blankets do you need?

Maura's Way

I made a table and looked for a pattern.

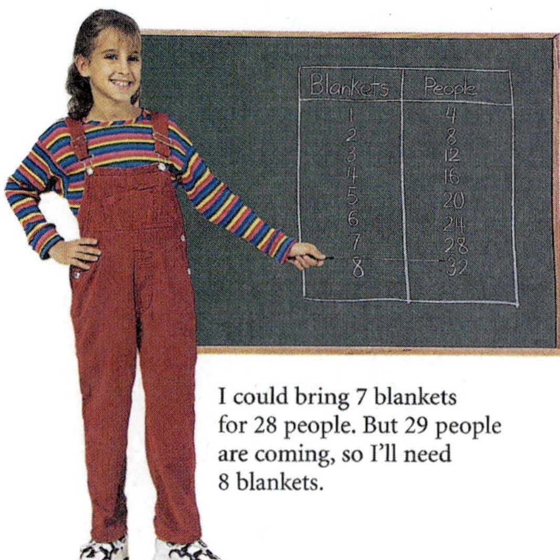

Blankets	People
1	4
2	8
3	12
4	16
5	20
6	24
7	28
8	32

I could bring 7 blankets for 28 people. But 29 people are coming, so I'll need 8 blankets.

Freddy's Way

I drew a picture.

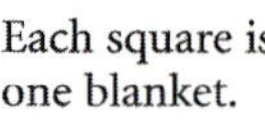

Each square is one blanket.

Each x is one person.

I drew 29 xs. Then I counted to see how many blankets were used.

There were 8 squares. So, I'll need to bring 8 blankets.

Talk About It

1. How did finding a pattern help Maura to solve the problem?
2. How did Freddy's picture help him find an answer?

262 Chapter 6 • More Multiplication Facts

SOURCE: From Scott Foresman–Addison Wesley Math, *Grade 3, p. 262, by Randall I. Charles et al.*

Questions for the Teacher

1. Pose three additional questions you might ask your students if you were teaching this class.
2. You are planning a party for 34 people. You must buy paper plates in packages of 8. How many packages of plates will you need?
   (a) Show the table you might make to answer this question.
   (b) Show a picture you might draw to answer this question.

Here e_n, read "e sub n," is the formula that gives the nth entry in the sequence of part (b). All we have to do is replace n by 1, 2, 100, and so on, to find the first entry, the second entry, the one hundredth entry, and so on. Thus,

$$e_1 = 2 \times 1 - 1 = 1,$$
$$e_2 = 2 \times 2 - 1 = 3,$$
$$e_{100} = 2 \times 100 - 1 = 199,$$

and so on.

Calculator Note

Many calculators have a "constant" function. Pressing the keys 1 [+] 2 [=] [=] [=] or, on some calculators, 2 [+] 1 [=] [=] [=], causes the calculator to add 2 to 1 three times. With such calculators, the successive terms of the sequence of part (b) in the preceding example can easily be obtained. Use this function to show that 49 is the twenty-fifth odd number.

Use a Variable

In the preceding example, and even earlier, we saw how using symbols or **variables** often makes it easier to express mathematical ideas and so to solve problems.

EXAMPLE 1.8 USE A VARIABLE—GAUSS'S TRICK

Carl Gauss (1777–1855) is generally acknowledged as one of the three greatest mathematicians of all time. When Gauss was just a young schoolboy, the teacher instructed the students in his class to add all the numbers from 1 to 100, expecting this to take a long time. To the teacher's surprise, young Gauss completed the task in about half a minute.

Solution

Gauss's strategy was to use a variable. If

$$s = 1 + 2 + 3 + \ldots + 100,$$

then

$$s = 100 + 99 + 98 + \ldots + 1.$$

1 + 100 = 101
2 + 99 = 101
3 + 98 = 101
⋮
100 + 1 = 101

Therefore, adding these two expressions for s, we obtain

$$2s = 101 + 101 + 101 + \ldots + 101,$$
$$2s = 100 \times 101,$$

a sum with 100 terms

and

$$s = \frac{100 \times 101}{2} = 5050.$$

DEFINITION *A Variable*

A **variable** is a letter that can represent any of the numbers of some set of numbers.

Sometimes we want to use a variable in representing the general term in a sequence. Thus, as we saw in Example 1.7, $2n - 1$ is the nth odd number. In the expression $2n - 1$, n is the variable and it can be replaced by any natural number.

Other times we want to find the numerical replacement for a variable that makes a statement true. For example, if we want to know which odd number 85 is, we need to determine n such that

$$2n - 1 = 85.$$

This implies that $2n = 86$, so $n = 43$. Hence, 85 is the forty-third odd number.*

> **PROBLEM-SOLVING STRATEGY 6 Use a Variable**
>
> Often a problem requires that a number be determined. Represent the number by a variable, and use the conditions of the problem to set up an equation that can be solved to ascertain the desired number.

EXAMPLE 1.9 USE A VARIABLE TO DETERMINE A GENERAL FORMULA

Look at these corresponding geometrical and numerical sequences.

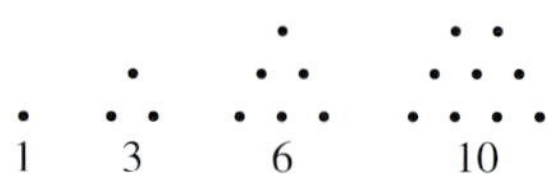

For fairly obvious reasons, the numbers 1, 3, 6, and 10 are called **triangular numbers** (the diagram with a single dot is considered a *degenerate triangle*). The numbers 1, 3, 6, and 10 are the first four triangular numbers. Find a formula for the nth triangular number.

Solution

Understand the Problem

Having just gone through a similar problem in Example 1.7, we understand that we are to find a formula for t_n, the nth triangular number.

Devise a Plan

The geometrical and numerical sequences in the statement of the problem suggest that we look for a pattern. How is each diagram related to its predecessor? How is each triangular number related to its predecessor? We'll try to guess patterns.

Carry Out the Plan

We add a diagonal of two dots to the first diagram to obtain the second, a diagonal of three dots to the second diagram to obtain the third, and so on. Thus, the next two diagrams should be as shown here.

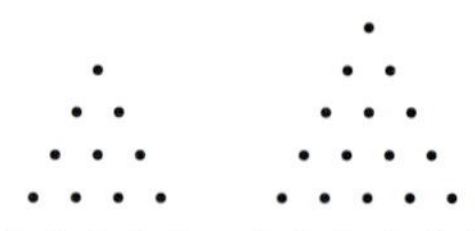

*These notions will be discussed in greater detail in Chapter 8.

Highlight from History

Carl Friedrich Gauss (1777–1855)

Sometimes called the "prince of mathematicians" and always rated as one of the three greatest mathematicians who ever lived, Carl Friedrich Gauss was born of poor parents in Braunschweig, Germany, in 1777. His father was a self-righteous lout; his mother, an intelligent, but barely literate, housewife. Gauss, on the other hand, was a genius who began making significant mathematical discoveries while still in his teens. Gauss contributed many original ideas across the spectrum of science and engineering, but he characterized mathematics as the "queen of the sciences."

Numerically, we add 2 to the first triangular number to obtain the second, 3 to the second triangular number to obtain the third, and so on. To make this even more clear, we can construct the following table.

Number of Entry	Entry
1	$t_1 = 1$
2	$t_2 = 1 + 2 = 3$
3	$t_3 = 1 + 2 + 3 = 6$
4	$t_4 = 1 + 2 + 3 + 4 = 10$
5	$t_5 = 1 + 2 + 3 + 4 + 5 = 15$

Indeed, it appears that

$$t_n = 1 + 2 + 3 + \cdots + n.$$

Then, using Gauss's trick, we obtain

$$t_n = n + (n - 1) + (n - 2) + \cdots + 1.$$

$1 + n = n + 1$
$2 + (n - 1) = n + 1$
$3 + (n - 2) = n + 1$
$\vdots$
$n + 1 = n + 1$

So,

$$\begin{aligned} 2t_n &= (n + 1) + (n + 1) + (n + 1) + \cdots + (n + 1) \\ &= n(n + 1) \end{aligned}$$

A sum with n terms

and

$$t_n = \frac{n(n + 1)}{2}$$

as required.

Look Back

Looking back, we note that the key to our solution lay in considering the sequence of special cases t_1, t_2, t_3, t_4, and t_5 and in looking for a pattern. This approach deserves special recognition as a problem-solving strategy. We call it **considering special cases.**

Consider Special Cases

> **PROBLEM-SOLVING STRATEGY 7 Consider Special Cases**
> In trying to solve a complex problem, consider a sequence of special cases. This will often show how to proceed naturally from case to case until one arrives at the case in question. Alternatively, the special cases may reveal a pattern that makes it possible to solve the problem.

Pascal's Triangle

One of the most interesting and useful patterns in all of mathematics is the numerical array called "Pascal's triangle."

Consider the problem of finding how many different paths there are from A to P on the grid shown in Figure 1.1 if you can only move *down* along edges in the grid. If we start to trace out paths willy-nilly, our chances of finding all possibilities are not good. A better approach is to notice that any path from A to P must contain four moves downward and to the left along the edges of small squares in the grid and six moves downward and to the right. Thus, the question becomes one of finding the number of different ways we can arrange four Ls (for left) and six Rs (for right) in order. This, by the way, illustrates another important problem-solving strategy, the strategy of **solving an equivalent problem.** The idea is to find a problem that is equivalent to the original problem that may be easier to solve. Here, finding how many ways you can put four Ls and six Rs in order using the strategy *make an orderly list* is easier than the original problem of finding the number of paths from A to P on the grid. Even this approach with four Ls and six Rs is sufficiently complicated that it is probably better to see if we can find yet another strategy.

Just for Fun

For Careful Readers

1. How much dirt is there in a hole 10 meters long, 5 meters wide, and 3 meters deep?
2. The outer track of a phonograph record is a circle of radius 6 inches. The unused center of the disc is a circle of radius 3 inches. If the record has 20 grooves per inch, how far, to the nearest inch, does the needle travel while the record plays from beginning to end?
3. Joni has two U.S. coins in her pocket. One of the coins is not a quarter. The total value of the coins is 26¢. What are the two coins?

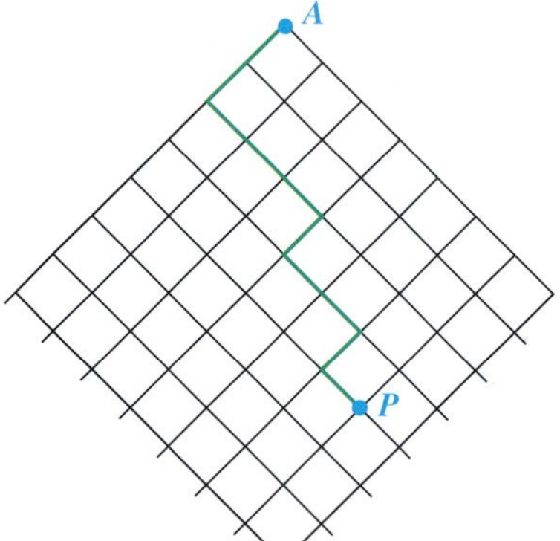

Figure 1.1
A path from A to P

A strategy that is often helpful is to **solve an easier similar problem.** It would certainly be easier if P were not so far down in the grid. Consider the easier similar problem of finding the number of paths from A to E in Figure 1.2.

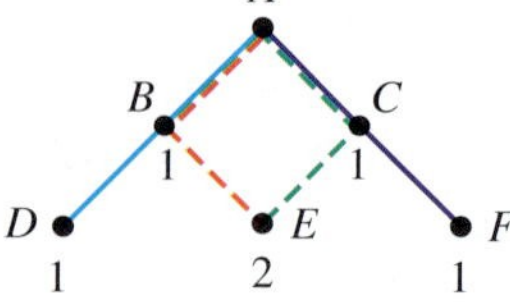

Figure 1.2
Solving an easier similar problem

Or, consider solving a similar *set* of easier problems all at once. How many different paths are there from A to each of B, C, D, E, and F? (Note that this is an example of considering a series of special cases.) Clearly, there is only one way to go from A to each of B and C, and we indicate this by the 1s under B and C in Figure 1.2. Also, the only route to D is through B, so there is only one path from A to D as indicated. For the same reason, there is one path from A to F. On the other hand, there are two ways to go from A to E—one route through B and one through C. We indicate this by placing a 2 under E on the diagram. This certainly doesn't solve the original problem, but it gives us a start and even suggests how we might proceed. Consider the diagram in Figure 1.3.

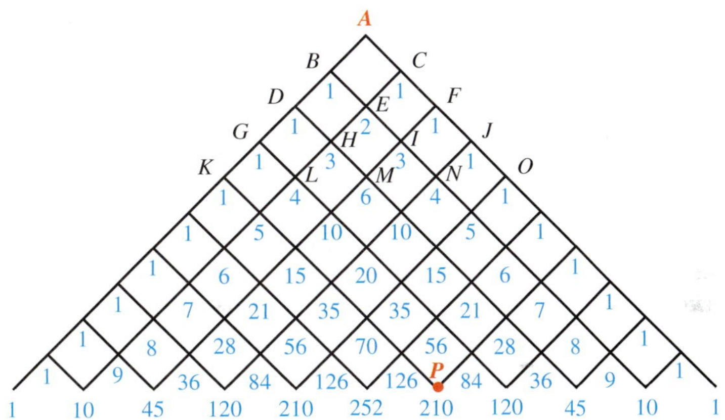

Figure 1.3
The number of paths from A *to* P

Having determined the number of paths from A to each of B, C, D, E, and F, could we perhaps determine the number of paths to G, H, I, and J and then continue on down the grid to eventually solve the original problem?

1. Always moving downward, the only way to get to G from A is via D. But there is only one path to D and only one path from D to G. Thus, there is only one path from A to G, and we enter a 1 under G on the diagram as shown.
2. The only way to get to H from A is via D or E. Since there is only one path from A to D and one from D to H, there is only one path from A to H via D. However, since there are two paths from A to E and one path from E to H, there are two paths from A to H via E. The number of paths from A to H is the number via D plus the number via E for $1 + 2 = 3$ paths, and we enter 3 under H on the diagram as shown.
3. The arguments for I and J are the same as for H and G, so we enter 3 and 1 under I and J, respectively, on the diagram.
4. But this reveals a very nice pattern that enables us to solve the original problem with ease. There can be only one path to any edge vertex on the grid since we have to go straight down the edge to get to such a vertex. For interior points on the grid, however, we can always reach the grid by paths through the points immediately above and to the left and right of such a point. Since there is only one path from each of these points to the point in question, the total number of paths to this point is the *sum* of the number of paths to these preceding two points. Thus, we easily generate the number of paths from A to any given point in the grid by simple addition. In particular, there are 210 different paths from A to P, as we initially set out to determine.

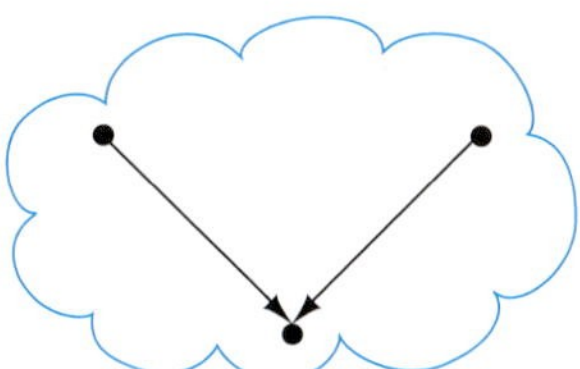

Without the grid and with an additional 1 at the top to complete a triangle, the number array in Figure 1.4 is called **Pascal's triangle.**

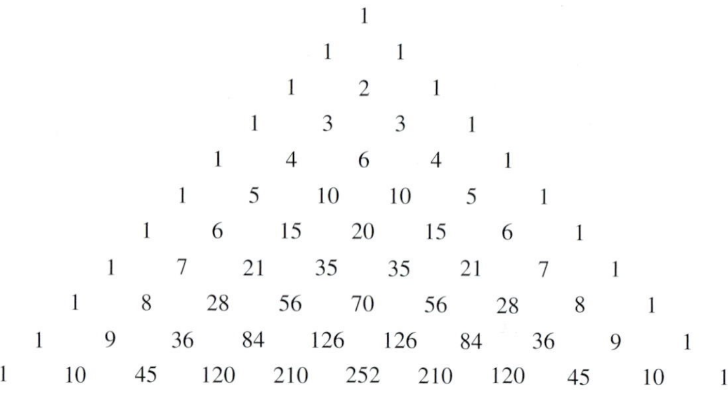

Figure 1.4
Pascal's triangle

The array is named after the French mathematician Blaise Pascal (1623–1662), who showed that these numbers play an important role in the theory of probability. However, the triangle was certainly known in China as early as the twelfth century. An interesting and clear depiction of the famous triangle from a fourteenth-century manuscript is shown in Figure 1.5.

Pascal's triangle is rich with remarkable patterns and is also extremely useful.

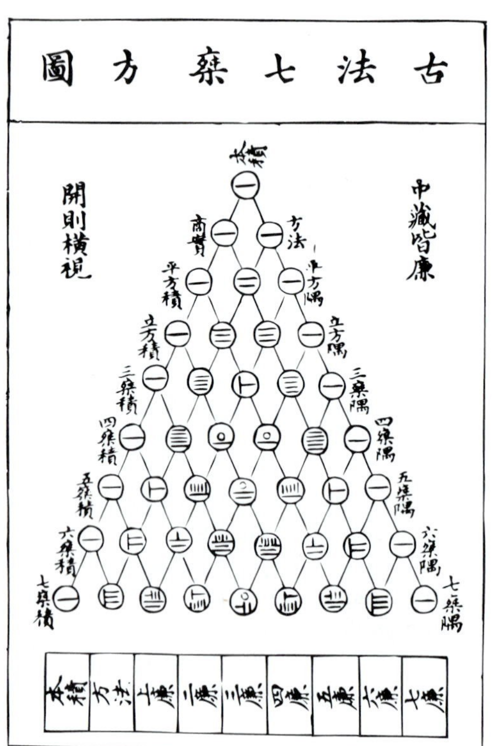

Figure 1.5
Pascal's triangle from Chu Shih-Chieh's Ssu Yuan Yii Chien, *A.D. 1303*

Before discussing the patterns, we observe that it is customary to call the single 1 at the top of the triangle the zeroth row (since, for example, in the path-counting problem discussed above, this 1 would represent a path of length 0). For consistency, we will also call the initial 1 in any row the zeroth element in the row, and the initial diagonal of 1s the zeroth diagonal. (See Figure 1.6.) Thus, 1 is the zeroth element in the fourth row, 4 is the first element, 6 the second element, and so on.

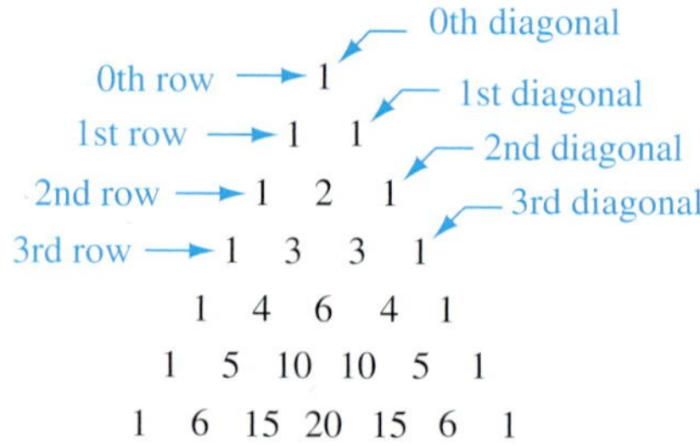

Figure 1.6
Numbered rows and diagonals in Pascal's triangle

from **The NCTM Principles and Standards**

Reasoning and Proof Standard

Instructional programs from prekindergarten through grade 12 should enable all students to—

- *recognize reasoning and proof as fundamental aspects of mathematics;*
- *make and investigate mathematical conjectures;*
- *develop and evaluate mathematical arguments and proofs;*
- *select and use various types of reasoning and methods of proof.*

During grades 3–5, students should be involved in an important transition in their mathematical reasoning. Many students begin this grade band believing that something is true because it has occurred before, because they have seen several examples of it, or because their experience to date seems to confirm it. During these grades, formulating conjectures and assessing them on the basis of evidence should become the norm. Students should learn that several examples are not sufficient to establish the truth of a conjecture and that counterexamples can be used to disprove a conjecture. They should learn that by considering a range of examples, they can reason about the general properties and relationships they find.

Mathematical reasoning develops in classrooms where students are encouraged to put forth their own ideas for examination. Teachers and students should be open to questions, reactions, and elaborations from others in the classroom. Students need to explain and justify their thinking and learn how to detect fallacies and critique others' thinking. They need to have ample opportunity to apply their reasoning skills and justify their thinking in mathematics discussions. They will need time, many varied and rich experiences, and guidance to develop the ability to construct valid arguments and to evaluate the arguments of others. There is clear evidence that in classrooms where reasoning is emphasized, students do engage in reasoning and, in the process, learn what constitutes acceptable mathematical explanation (Lampert 1990; Yackel and Cobb 1994, 1996).

SOURCE: Reprinted with permission from Principles and Standards for School Mathematics, *pages 56 and 188.* *Standards are listed with the permission of the National Council of Teachers of Mathematics (NCTM). NCTM does not endorse the content or validity of these alignments.*

EXAMPLE 1.10 FIND A PATTERN IN THE ROW SUMS OF PASCAL'S TRIANGLE

Consider special cases.

Look for a pattern.

(a) Compute the sum of the elements in each of rows zero through four of Pascal's triangle.
(b) Look for a pattern in the results of part (a) and guess a general rule.
(c) Use Figure 1.4 to check your guess for rows five through eight.
(d) Give a convincing argument that your guess in part (b) is correct.

Solution

Understand the Problem

For part (a), we must add the elements in the indicated rows. For part (b), we are asked to discover a pattern in the numbers generated in part (a). For part (c), we must compute the sums for four more rows and see if the results obtained continue the pattern guessed in part (b). In part (d), we are asked to argue convincingly that our guess in part (b) is correct.

Devise a Plan

Part (a) is certainly straightforward; we must compute the desired sums. To find the pattern requested in part (b), we should ask the question "Have we ever seen a similar problem before?" The answer, of course, is a resounding "yes"—all the problems in this section, but particularly Example 1.6, have involved looking for patterns. Surely, the techniques that succeeded earlier should be tried here. Appropriate questions to ask and answer include the following: "How are the successive numbers related to their predecessors?" "Are the numbers special numbers that we can easily recognize?" "Can we relate the successive numbers to their numbered location in the sequence of numbers being generated?" Answering these questions should help us make the desired guess. For part (c) we will compute the sums of the elements in rows five through eight to see if these numbers agree with our guess in part (b). If they *don't* agree, we will go back and modify our guess. If they *do* agree we will proceed to part (d) and try to make a convincing argument that our guess is correct. About all we have to go on is the fact that the initial and terminal elements in each row are 1s and that the sum of any two consecutive elements in a row is the element between these two elements but in the next row down.

Carry Out the Plan

(a) $1 = 1$
$1 + 1 = 2$
$1 + 2 + 1 = 4$
$1 + 3 + 3 + 1 = 8$
$1 + 4 + 6 + 4 + 1 = 16$

(b) It appears that each number in part (a) is just twice its predecessor. The numbers are just

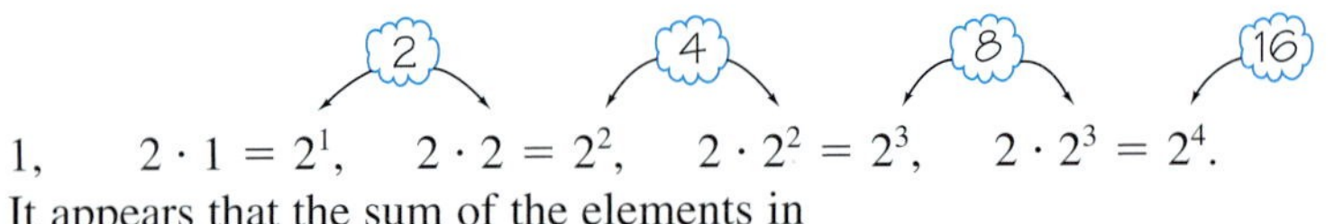

$1, \quad 2 \cdot 1 = 2^1, \quad 2 \cdot 2 = 2^2, \quad 2 \cdot 2^2 = 2^3, \quad 2 \cdot 2^3 = 2^4.$

It appears that the sum of the elements in

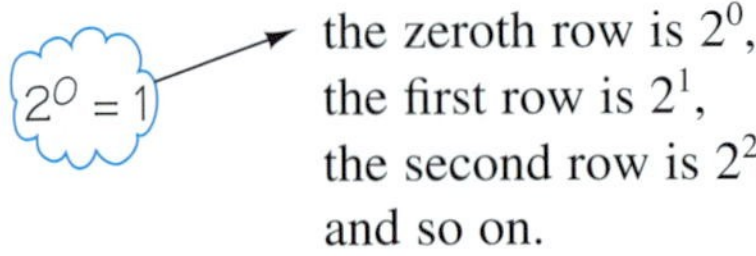

the zeroth row is 2^0,
the first row is 2^1,
the second row is 2^2,
and so on.

Our guess is that the sum of the elements in the nth row is 2^n.

(c) Computing these sums for the next four rows, we have

fifth row	$1 + 5 + 10 + 10 + 5 + 1 = 32 = 2^5$
sixth row	$1 + 6 + 15 + 20 + 15 + 6 + 1 = 64 = 2^6$
seventh row	$1 + 7 + 21 + 35 + 35 + 21 + 7 + 1 = 128 = 2^7$
eighth row	$1 + 8 + 28 + 56 + 70 + 56 + 28 + 8 + 1 = 256 = 2^8$

Since these results do not contradict our guess, we proceed to try to make a convincing argument that our guess is correct.

(d) What happens as we go from one row to the next? How is the next row obtained? Consider the third and fourth rows shown below. The arrows show how the fourth row is obtained from the third, and we see that each of the 1, 3, 3, and 1 in the third row appears *twice* in the sum of the elements in the fourth row. Since this argument would hold for any two consecutive rows, the sum of the numbers in any row is just twice the sum of the numbers in the preceding row. Hence, from above, the sum of the numbers in the ninth row must be $2 \times 2^8 = 2^9$, in the tenth row it must be $2 \cdot 2^9 = 2^{10}$, and so on. Thus, the result is true in general as claimed.

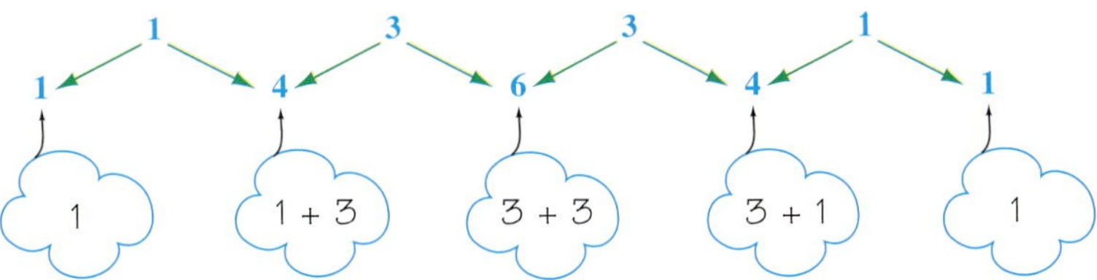

Look Back

Several aspects of our solution merit special comment. Beginning with the strategies of considering special cases and looking for a pattern, we were led to guess that the sum of the elements in the nth row of the triangle is 2^n. In an attempt to argue that this guess was correct, we considered the special case of obtaining the fourth row from the third. This showed that the sum of the elements in the fourth row was twice the sum of the elements in the third row. *Since the argument did not depend on the actual numbers appearing in the third and fourth rows, but only on the general rule of formation of the triangle,* it would hold for any two consecutive rows and so actually proves that our conjecture is correct. This type of argument is called **arguing from a special case** and is an important problem-solving strategy.

In considering Pascal's triangle we found it helpful to use three additional problem-solving strategies that are well worth highlighting.

PROBLEM-SOLVING STRATEGY 8 Solve an Equivalent Problem

Try to recast the problem in totally different terms that change it into a new, different, but completely equivalent, problem. Solving the equivalent problem also solves the original problem, and it may be easier.

PROBLEM-SOLVING STRATEGY 9 Solve an Easier Similar Problem

Instead of attempting immediately to solve a problem in general or for a reasonably large value like 10 or 20, try first to solve it for small values like 2 or 3, or even 1, 2, and 3. This may show how to solve the problem for larger values.

PROBLEM-SOLVING STRATEGY 10 Argue from Special Cases

A convincing argument for a general case can often be made by discussing a special case, but only using those features of the special case that are typical of the general case.

Exploring Number Patterns on a Spreadsheet

Spreadsheets were originally designed with business applications in mind, but they are also ideally suited to creating and investigating number patterns. Instructions to use a spreadsheet, together with several mathematical applications, are given in Appendix B and the accompanying exercises.

As a specific example of a spreadsheet application for a number pattern exploration, let's generate rows 0, 1, 2, . . . , 10 of Pascal's triangle.

Steps to generate Pascal's triangle on a spreadsheet:

Column A. Select cell A1, type 1 and click the accept button (or just press **Return** or **Enter**). The remaining 1s in column A can be entered by the familiar **Copy** and **Paste** commands: select cell A1 and **Copy,** and then select the cell range A2:A11 and **Paste.** (Alternatively, you may use the **Fill** command, or the "copy handle" at the lower right corner of the cell.) This fills in the first 11 cells of column A with 1s.

Column B. Enter a 0 in cell B1. Select cell B2, enter the formula "=B1 + A1", and then click the accept button (or press **Return** or **Enter**). Fill in the rest of column B with **Copy** and **Paste:** select cell B2 and **Copy,** and then select cells B3:B11 and **Paste.** The entries 0, 1, 2, . . . , 10 will now appear in column B.

Columns C through K. Select the cell range B1:B11 and **Copy.** Then select the rectangular cell range with cell C1 at the upper left and cell K11 at the lower right. Executing **Paste** will complete Pascal's triangle.

Once Pascal's triangle has been created on the spreadsheet, it becomes easy to explore some of its patterns. To be specific, let's investigate the sum of the entries of each row and the sum of the squares of the entries of each row. To accomplish this, enter the formula "=SUM(A1:K1)" in cell L1. Next, copy this formula and paste it into cells L2 through L11. The sums of the rows, 1, 2, 4, 8, . . . , are then displayed in column L. The sums of the squares of the entries of each row can be placed in column M in an analogous way. Enter the function "=SUMSQ(A1:K1)" in cell M1, and then copy it into cells M2:M11. Can you describe where to find the entries 1, 2, 6, 20, . . . within Pascal's triangle? Can you check your guess by adding additional rows to Pascal's triangle using spreadsheet procedures?

B2 =B1+A1

Pascal

	A	B	C	D	E	F	G	H	I	J	K	L	M
1	1	0	0	0	0	0	0	0	0	0	0	1	1
2	1	1	0	0	0	0	0	0	0	0	0	2	2
3	1	2	1	0	0	0	0	0	0	0	0	4	6
4	1	3	3	1	0	0	0	0	0	0	0	8	20
5	1	4	6	4	1	0	0	0	0	0	0	16	70
6	1	5	10	10	5	1	0	0	0	0	0	32	252
7	1	6	15	20	15	6	1	0	0	0	0	64	924
8	1	7	21	35	35	21	7	1	0	0	0	128	3432
9	1	8	28	56	70	56	28	8	1	0	0	256	12870
10	1	9	36	84	126	126	84	36	9	1	0	512	48620
11	1	10	45	120	210	252	210	120	45	10	1	1024	184756

Did You Know?

Aha!

Experimental psychologists like to tell a story about a professor who investigated the ability of chimpanzees to solve problems. A banana was suspended from the center of the ceiling, at a height that the chimp could not reach by jumping. The room was bare of all objects except several packing crates placed around the room at random. The test was to see whether a lady chimp would think of first stacking the crates in the center of the room, and then of climbing on top of the crates to get the banana.

The chimp sat quietly in a corner, watching the psychologist arrange the crates. She waited patiently until the professor crossed the middle of the room. When he was directly below the fruit, the chimp suddenly jumped on his shoulder, then leaped into the air and grabbed the banana.

The moral of this anecdote is: A problem that seems difficult may have a simple, unexpected solution. In this case the chimp may have been doing no more than following her instincts or past experience, but the point is that the chimp solved the problem in a direct way that the professor had failed to anticipate.

This quotation is from the introduction to Martin Gardner's delightful book *Aha! Insight* (Scientific American, Inc./W. H. Freeman and Company, 1978). It is full of interesting, thought-provoking problems as well as insightful solutions and analyses of the problem-solving process.

Problem Set 1.3

1. Look for a pattern and fill in the next three blanks with the most likely choices for each sequence.

(a) 2, 5, 8, 11, ____, ____, ____

(b) $-5, -3, -1, 1$, ____, ____, ____

(c) 1, 1, 3, 3, 6, 6, 10, ____, ____, ____

(d) 1, 3, 4, 7, 11, ____, ____, ____

(e) 2, 6, 18, 54, ____, ____, ____

2. (a) Draw three diagrams to continue this dot sequence.

```
•    • •    • • •
•,   • •,   • • •,
```

(b) What number sequence corresponds to the pattern of part (a)?

(c) What is the tenth term in the sequence of part (b)? the one hundredth term?

(d) Which even number is $2n$?

(e) What term in the sequence is 2402?

(f) Compute the sums $2 + 4 + 6 + \cdots + 2402$. (*Hint:* Use Gauss's trick.)

3. (a) Fill in the blanks to continue this dot sequence in the most likely way.

```
                •           • •           • • •
     •         • •         • • •         • • • •
•,   • •,     • • •,      • • • •,      ______, ______
```

(b) What number sequence corresponds to the sequence of dot patterns of part (a)?

(c) What is the tenth term in the sequence of part (b)? the one hundredth term?

(d) Which term in the sequence is 101? (*Hint:* How many 3s must be added to 2 to get 101?)

(e) Compute the sum $2 + 5 + 8 + \cdots + 101$. (*Hint:* Use Gauss's trick and the result of part (d).)

4. Sequences like 2, 5, 8, . . . , where each term is greater (or less) than its predecessor by a constant amount, are called **arithmetic** (a-rith-me′-tic) **progressions.** Find the number of terms in each of these arithmetic progressions.

(a) 5, 7, 9, . . . , 35

(b) -4, 1, 6, . . . , 46

(c) 3, 7, 11, . . . , 67

5. Compute the sum of each of these arithmetic progressions.

(a) $5 + 7 + 9 + \cdots + 35$

(b) $-4 + 1 + 6 + \cdots + 46$

(c) $3 + 7 + 11 + \cdots + 67$

(d) $1 + 7 + 13 + \cdots + 73$

6. Consider the arithmetic progression 2, 9, 16, 23, . . . , 86.

(a) How many 7s must be added to 2 to obtain 86?

(b) Compute the sum $2 + 9 + 16 + \cdots + 86$.

(c) What is the nth number in the progression?

(d) Compute the sum

$$2 + 9 + 16 + \cdots + (7n - 5).$$

7. **(a)** Fill in the blanks to continue this sequence of equations.

$$1 = 1$$
$$1 + 2 + 1 = 4$$
$$1 + 2 + 3 + 2 + 1 = 9$$
$$1 + 2 + 3 + 4 + 3 + 2 + 1 = 16$$
$$\_\_\_\_\_\_\_\_\_\_ = \_\_\_\_\_\_$$
$$\_\_\_\_\_\_\_\_\_\_ = \_\_\_\_\_\_$$

(b) Compute this sum.

$1 + 2 + 3 + \cdots + 99 + 100 + 99$
$+ \cdots + 3 + 2 + 1 = \_\_\_\_\_\_$

(c) Fill in the blank to complete this equation.

$1 + 2 + 3 + \cdots + (n - 1)$
$+ n + (n - 1) + \cdots + 3$
$+ 2 + 1 = \_\_\_\_\_\_$

8. **(a)** Fill in the blanks to continue this sequence of equations.

$$1 = 0 + 1$$
$$1 + 3 + 1 = 1 + 4$$
$$1 + 3 + 5 + 3 + 1 = 4 + 9$$
$$\_\_\_\_\_\_ = \_\_\_\_\_\_$$
$$\_\_\_\_\_\_ = \_\_\_\_\_\_$$

(b) What expression, suggested by part (a), should be placed in the blank to complete this equation?

$1 + 3 + 5 + \cdots + (2n - 3) + (2n - 1)$
$+ (2n - 3) + \cdots + 5 + 3 + 1 = \_\_\_\_\_\_$

(*Hint:* The number preceding n is $n - 1$.)

9. Writers of standardized tests often pose questions like "What is the next term in the sequence 2, 4, 8, . . . ?"

(a) How would you answer this question?

(b) Evaluate the expressions 2^n, $n^2 - n + 2$, and $n^3 - 5n^2 + 10n - 4$ in the following chart by replacing n successively by 1, 2, 3, and 4.

n	1	2	3	4
2^n				
$n^2 - n + 2$				
$n^3 - 5n^2 + 10n - 4$				

(c) In light of the results in (b), what criticism would you make of the test writer who would write a test question like that above? Compare the wording above with that in problem 1 of this problem set.

10. Here is the start of a 100 chart.

1	2	3	4	5	6	7	8	9	10
11	12	13	14	15	16	17	18	19	20
21	22	23	24	25	26	27	28	29	30

Shown below are parts of the chart. Without extending the chart, determine which numbers should go in the lavender squares.

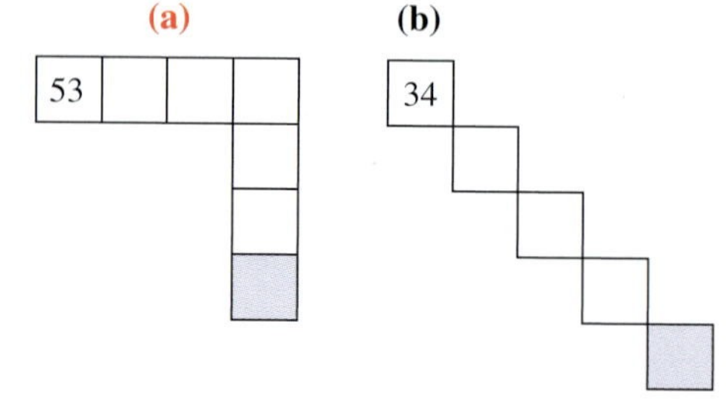

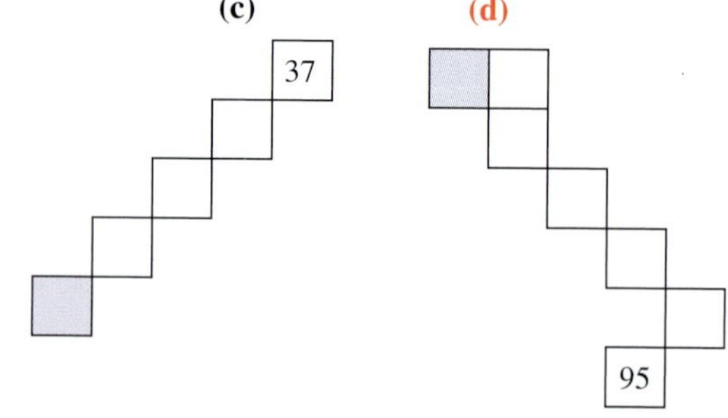

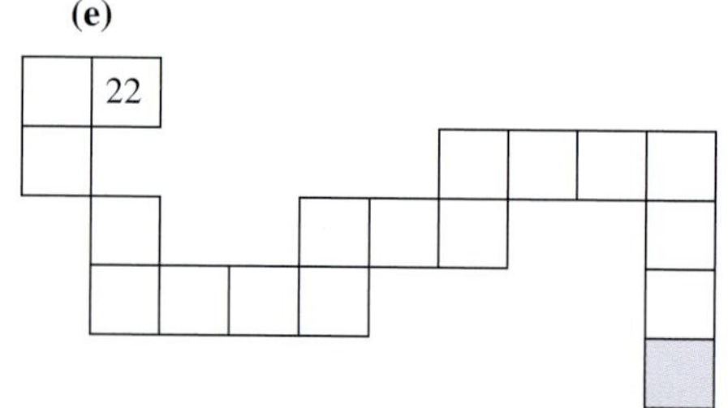

11. Five blue and five red discs are lined up in the B B B B B R R R R R arrangement shown.

Switching just two adjacent discs at a time, what is the least number of moves you can make to achieve the B R B R B R B R B R arrangement shown here?

(*Hint:* How many moves are required to rearrange B B R R to B R B R? B B B R R R to B R B R B R? and so on.)

12. **(a)** Complete the next two of this sequence of equations.

$$1 = 1$$
$$1 - 4 = -3$$
$$1 - 4 + 9 = 6$$
$$1 - 4 + 9 - 16 = -10$$
________________ = ________
________________ = ________

(b) Write the seventh and eighth equations in the sequence of equations of part (a). (*Hint:* Have you encountered the number sequence 1, 3, 6, 10, . . . before?)

(c) Write general equations suggested by parts (a) and (b) for even n and for odd n where n is the number of the equation.

13. **(a)** How many rectangles are there in each of these figures? (*Note:* Rectangles may measure 1 by 1, 1 by 2, 1 by 3, and so on.)

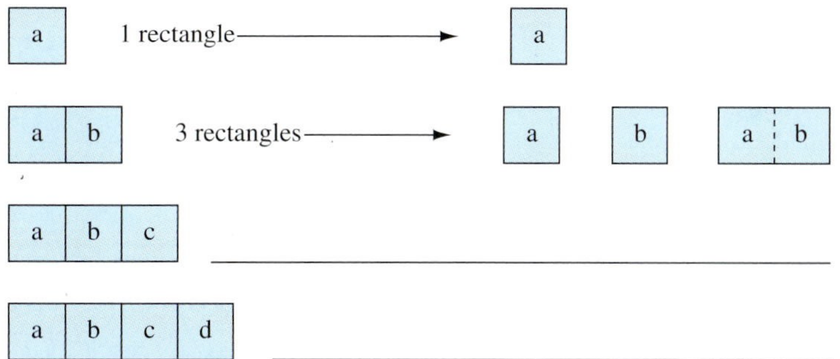

(b) How many rectangles are in this figure? ________

a	b	c	d	e	f	g	h	i	j	k	l	m	n	o	p

(c) How many rectangles are in a $1 \times n$ strip like that in part (b)?

(d) Argue that your guess in part (c) is correct, giving a lucid and careful write-up. (*Hint:* How many of each type of rectangle begin with each small square?)

14. **(a)** In how many ways can you exactly cover this diagram with "dominoes" that are the size of two small squares?

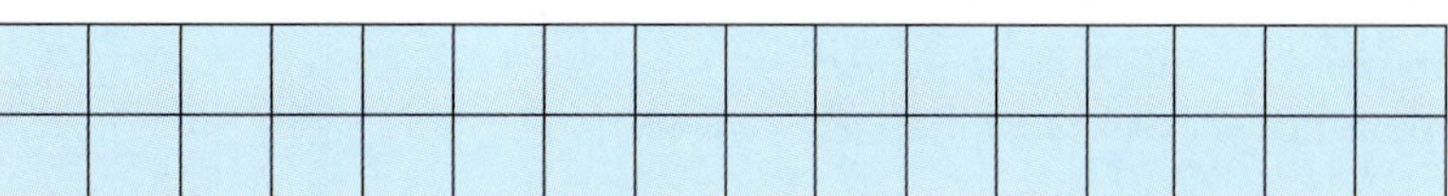

(*Hint:* This is too complicated as is. Consider this sequence of simpler arrays and look for a pattern.)

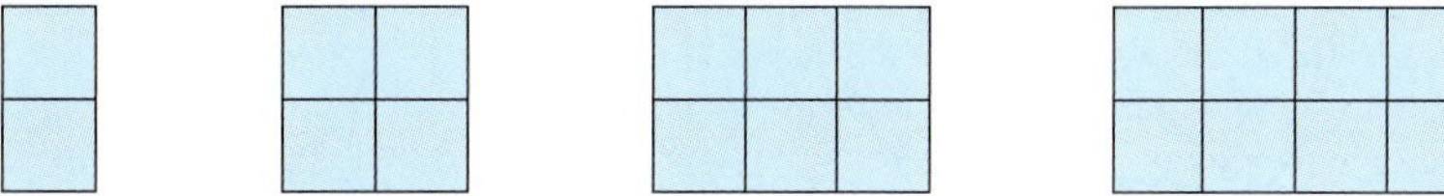

(b) Argue carefully that your solution in part (a) is correct.

(*Hint:* A covering must start with one vertical domino or two horizontal dominoes.)

15. If one must always move downward along the lines of the grid shown, how many different paths are there from point A to each of these points?

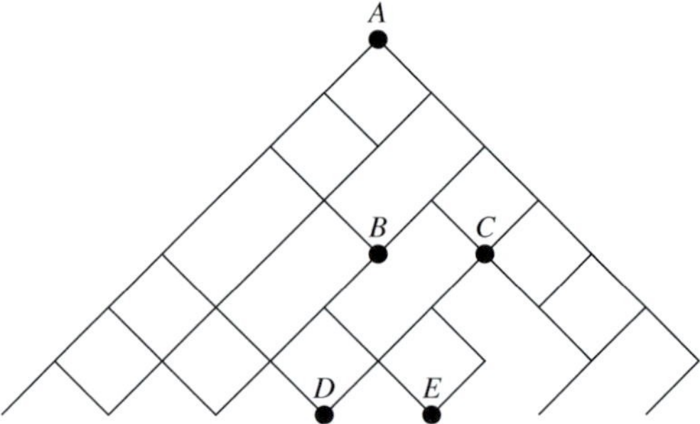

16. If one must always move upward or to the right on each of the grids shown, how many paths are there from A to B?

(a)

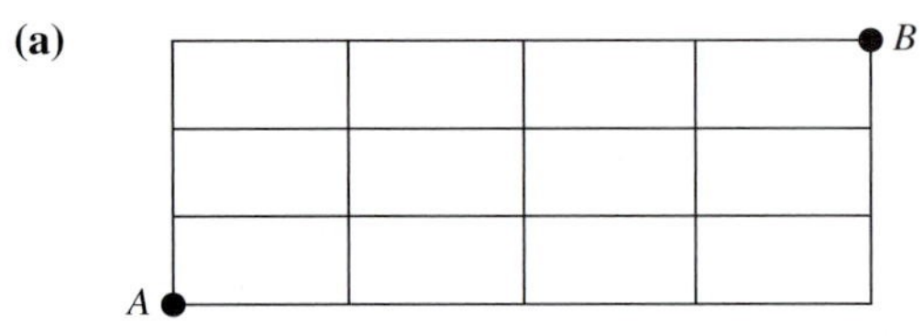

(b)

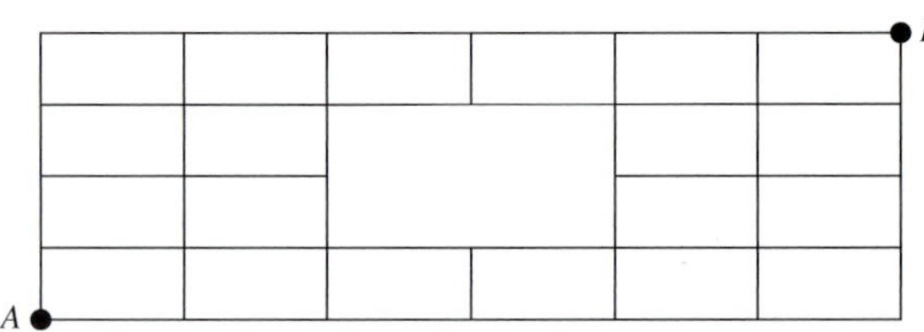

17. If one must follow along the paths of the following diagram in the direction of the arrows, how many paths are there from C to E?

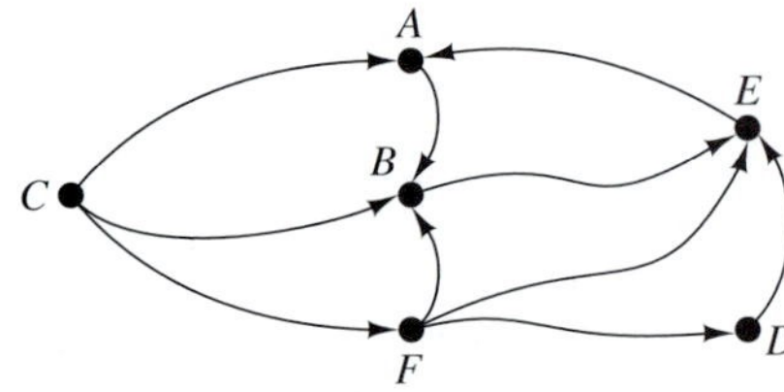

18. How many chords are determined by joining dots on a circle if there are

(a) 4 dots? **(b)** 10 dots?

(c) 100 dots? **(d)** n dots?

(e) Argue that your solution to part (d) is correct.

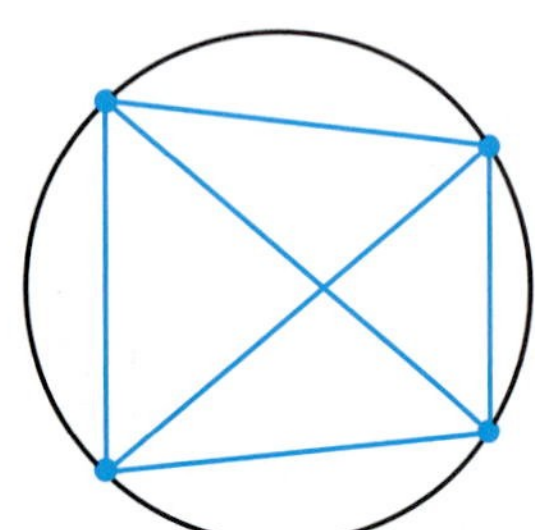

19. (a) How many games are played in a round-robin tournament with 10 teams if every team plays every other team once?

(b) How many games are played if there are 11 teams?

(c) Is this problem related to problem 18 of this problem set? If so, how?

20. Here is an addition table.

+	0	1	2	3	4	5	6	7	8	9
0	0	1	2	3	4	5	6	7	8	9
1	1	2	3	4	5	6	7	8	9	10
2	2	3	4	5	6	7	8	9	10	11
3	3	4	5	6	7	8	9	10	11	12
4	4	5	6	7	8	9	10	11	12	13
5	5	6	7	8	9	10	11	12	13	14
6	6	7	8	9	10	11	12	13	14	15
7	7	8	9	10	11	12	13	14	15	16
8	8	9	10	11	12	13	14	15	16	17
9	9	10	11	12	13	14	15	16	17	18

(a) Find the sum of the entries in these squares of entries from the addition table.

2	3
3	4

5	6
6	7

11	12
12	13

15	16
16	17

Look for a pattern and write a clear and simple rule for finding such sums almost at a glance.

(b) Find the sum of the entries in these squares of entries from the table.

4	5	6
5	6	7
6	7	8

10	11	12
11	12	13
12	13	14

14	15	16
15	16	17
16	17	18

(c) Write a clear and simple rule for computing these sums.

(d) Write a clear and simple rule for computing the sum of the entries in any square of entries from the addition table.

21. Consider the following sequence of equations.

$$1 = 1$$
$$3 + 5 = 8$$
$$7 + 9 + 11 = 27$$
$$13 + 15 + 17 + 19 = 64$$
$$\_\_\_ + \_\_\_ + \_\_\_ + \_\_\_ + \_\_\_ = \_\_\_$$

(a) Fill in the blanks to continue the sequence of equations.

(b) Guess a formula for the number on the right of the nth equation.

(c) Check that the expression $n^2 - n + 1$ generates the first number in the sum on the left of each equation and that $n^2 + n - 1$ generates the last number in these sums.

(d) Use the result of part (c) to prove that your guess to part (b) is correct. (*Hint:* How many terms are in the sum on the left of the nth equation?)

22. We have already considered the triangular numbers,

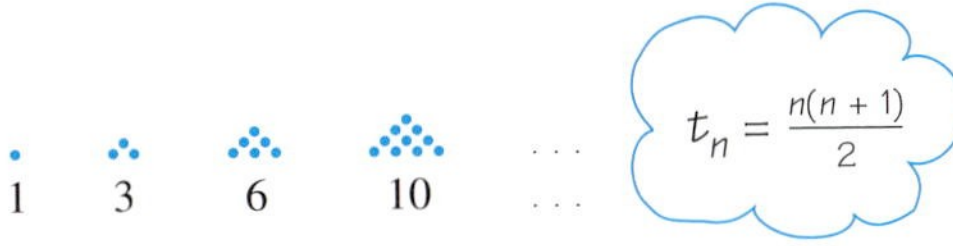

and the square numbers,

(a) Draw the next two figures to continue this sequence of dot patterns.

(b) List the sequence of numbers that corresponds to the sequence of part (a). These are called **pentagonal numbers.**

(c) Complete this list of equations suggested by parts (a) and (b).

$$1 = 1$$
$$1 + 4 = 5$$
$$1 + 4 + 7 = 12$$
$$1 + 4 + 7 + 10 = 22$$
$$\_\_\_\_\_\_ = \_\_\_\_$$
$$\_\_\_\_\_\_ = \_\_\_\_$$

Observe that each pentagonal number is the sum of an arithmetic progression.

(d) Compute the tenth term in the arithmetic progression 1, 4, 7, 10, ...

(e) Compute the tenth pentagonal number.

(f) Determine the nth term in the arithmetic progression 1, 4, 7, 10, ...

(g) Compute the nth pentagonal number, p_n.

23. **(a)** The **hexagonal numbers** are associated with this sequence of dot patterns. Complete the next two diagrams in the sequence.

(b) Write the first five hexagonal numbers.

(c) What is the tenth hexagonal number?

(d) Compute a formula for h_n, the nth hexagonal number.

24. Compute the sum of the numbers in the "handle" of each "hockey stick" in Pascal's triangle. Explain what you observe. Does the pattern always appear to hold?

25. Find and correct the error in the rod numerals appearing in the Chinese rendition of Pascal's triangle shown in Figure 1.5.

26. **(a)** Compute the square root of the product of the six elements surrounding an element in Pascal's triangle. In particular, do this for the six entries surrounding each of 4, 15, and 35.

(b) Does the limited amount of data from part (a) suggest a general conjecture? What appears to be true in general?

Using a Computer

The next two problems are not difficult to do by hand or with a calculator. At the same time, they are much more quickly done using a spreadsheet on a computer. See Appendix B for help with using a spreadsheet. Sample commands that may be suitable for your spreadsheet are given below in the problems.

27. In the following array the first row is just the infinite sequence of whole numbers. (On your spreadsheet enter 1 in A1. Enter "=1 + A1 " in B1 and use **Copy** and **Paste** or the **Fill** commands to extend the first row at least as far as Z1.) Each entry in subsequent rows is the sum of the three consecutive entries in the row above the desired entry, starting with the entry immediately above. Thus, $9 = 2 + 3 + 4$, $45 = 12 + 15 + 18$, and so on. (Enter "=A1 + B1 + C1" in A2 and use **Copy** and **Paste** or the **Fill** commands to extend row two to the right and also to extend the array downward and to the right.)

1	2	3	4	5	6	____	____	____	...
6	9	12	15	18	____	____	____	____	...
27	36	45	54	63	____	____	____	____	...
108	____	____	____	____	____	____	____	____	...

(a) Extend the array by filling in at least the blanks shown.

(b) Find a simpler rule than stated in the problem for producing the same array.

(c) Give a rule for determining the first entry in each row independent of the remaining columns in the array.

(d) Give a rule for determining the entries in row two independent of the rule of formation in the statement of the problem or the rule determined in part (b). Do the same for row three, row four, and the nth row.

28. In the array shown, the first row is just the infinite sequence of whole numbers and the first column is an infinite sequence of 1s. (On your spreadsheet, enter 1 in A1 and use **Copy** and **Paste** or the **Fill** commands to extend the first column downward. Extend the first row to the right by entering "=1 + A1" in B1 and using **Copy** and **Paste** or the **Fill** commands.) Each other entry in the array is the sum of the second entry in the preceding column and the entry immediately above the desired entry. Thus, $9 = 3 + 6$, $22 = 6 + 16$, and so on. (In your spreadsheet, enter "=A$2 + B1" in B2 and use **Copy** and **Paste** or the **Fill** commands to extend the array both down and to the right.)

1	2	3	4	5	6	____	____	____	...
1	3	6	10	15	____	____	____	____	...
1	4	9	16	25	____	____	____	____	...
1	5	12	22	____	____	____	____	____	...
1	____	____	____	____	____	____	____	____	...
____	____	____	____	____	____	____	____	____	...
...	...	...	...	...	...	...	...	...	...

(a) Extend the array by filling in the blanks shown.

(b) What kind of sequences are the columns in the array?

(c) Determine the tenth entry in the fourth column, and the nth entry.

(d) Note that the second row of the array is just the sequence of triangular numbers and the third row is the sequence of square numbers. Can you identify the sequences of numbers in the fourth row? the fifth row? Suggestion: Refer to problems 22 and 23 of this problem set.

From State Student Assessments

29. (Washington State, Grade 4)
Look at the following list of numbers. Describe two different patterns you see in these numbers.

9 18 27 36 45 54 63 72 81 90

30. (Michigan, Grade 4)
Justin created the following number pattern.

0, 1, 4, 13, 40, ______

What is the rule to find the next number in the pattern?

A. Add 1 to the last number.
B. Add 9 to the last number.
C. Triple the last number and add 1.
D. Double the last number and add 3.

31. (Illinois, Grade 5)

 + ◆ = 10
◆ − = 2
Each ◆ has the same value.
Each has the same value.
What is the value of the ◆?

1.4 ADDITIONAL PROBLEM-SOLVING STRATEGIES

Working Backward

Strategies

- Work backward.
- Eliminate possibilities.
- Use the pigeonhole principle.

> PROBLEM-SOLVING STRATEGY 11 **Work Backward**
>
> Start from the desired result and work backward step-by-step until the initial conditions of the problem are achieved.

Many problems require that a sequence of events occurs that results in a desired final outcome. These problems at first seem obscure and intractable, and you may be tempted to try a guess and check approach. However, it is often easier to work backward from the end result to see how the process would have to start to achieve the desired end. To make this more clear, consider the following example.

EXAMPLE 1.11 **WORKING BACKWARD—THE GOLD COIN GAME**

This is a two-person game. Place 15 golden coins (markers) on a desktop. The players play in turn and, on each play, can remove one, two, or three coins from the desktop. The player who takes the last coin wins the game. Can one player or the other devise a strategy that guarantees a win? *Note:* This is the problem of the Hands On activity at the beginning of the chapter, where you were expected to arrive at the solution by repeatedly trying simpler cases. We think that you will find the approach here much more effective and insightful.

Solution *Understand the Problem*

The assertion that the game is played with gold coins is just so much window dressing. What is important is that the players start with 15 objects; that they can remove one, two, or three objects on each play; and that the player who takes the last object wins the game. The question is, how can one play in such a way that he or she is sure of winning?

Devise a Plan

Since it is not clear how to begin to play or how to continue as the play proceeds, we turn the problem around to see how the game must end. We then work backward step-by-step to see how we can guarantee that the game ends as we desire.

Carry Out the Plan

We carry out the plan by presenting an imaginary dialogue that you could have with yourself to arrive finally at the solution to the problem.

Q. What must be the case just before the last person wins the game?
A. There must be 1, 2, or 3 markers on the desk.
Q. So how can I avoid leaving this arrangement for my opponent?
A. Clearly, I must leave at least 4 markers on the desk in my next-to-last move. Indeed, if I leave precisely 4 markers, my opponent must take 1, 2, or 3, leaving me with 3, 2, or 1. I can remove all of these on my last play to win the game.
Q. So how can I be sure to leave precisely 4 markers on my next-to-last play?
A. If I leave 5, 6, or 7 markers on my previous play, my opponent can leave *me* with 4 markers and he or she can then win. Thus, I must be sure to leave my opponent 8 markers on the previous play.
Q. All right. So how can I be sure to leave 8 markers on the previous play?
A. Well, I can't leave 9, 10, or 11 markers on the previous play or my opponent can take 1, 2, or 3 markers as necessary and so leave me with 8 markers. But then, as just seen, my opponent can be sure to win the game. Therefore, at this point, I must leave 12 markers on the desk.
Q. Can I be sure of doing this?
A. Only if I play first and remove 3 markers the first time. Otherwise, I have to be lucky and hope that my opponent will make a mistake and still allow me to leave 12, 8, or 4 markers at the end of one of my plays. The following outlines the play if I get to play first.

- I take 3 markers, leaving 12.
- My opponent takes 1, 2, or 3 markers, leaving 11, 10, or 9.
- I take 3, 2, or 1 marker as needed to make sure that I leave 8.
- My opponent takes 1, 2, or 3 markers, leaving 7, 6, or 5.
- I take 3, 2, or 1 marker as needed to assure that 4 markers are left on the desk.
- My opponent takes 1, 2, or 3 markers, leaving 3, 2, or 1.
- I take the remaining markers and win the game!

Look Back

In looking back, it is important to ask such questions as these.

- What was the key feature that led me to eventually solve this problem?
- Could I use this strategy to solve variations of this problem? For example, suppose the game started with 21 coins or 37 coins or, in general, with n coins.
- Suppose that each player could take up to 5 coins at a time. How would that affect the strategy?

- Could I use the successful strategy I just employed to solve other similar (or not so similar) problems?
- Could I devise other problems for which this strategy would lead to a solution?

Working backward is a must in many problem-solving situations, and particularly so with a problem like this. The desired strategy to win the game described is not at all clear. Here the strategy of working backward is somewhat similar to considering special cases and looking for a pattern. It is as if we started with just a few markers where the strategy was more apparent and gradually increased the number of markers until we reached the given number of 15. Working backward is a powerful strategy that ought to be in every problem solver's repertoire.

Just for Fun

How Many Pages in the Book?

If it takes 867 digits to number the pages of a book starting with page 1, how many pages are in the book?

Eliminate Possibilities

One way of determining what must happen in a given situation is to determine what the possibilities are and then to eliminate them one by one. If you can eliminate all but one possibility in this way, then that possibility must prevail. Suppose that either John, Jim, or Yuri is singing in the shower. Suppose also that you are able to recognize both John's voice and Yuri's voice, but that you do not recognize the voice of the person singing in the shower. Then the person in the shower must be Jim. This is another important problem-solving strategy that should not be overlooked.

> **PROBLEM-SOLVING STRATEGY 12 Eliminate Possibilities**
>
> Suppose you are guaranteed that a problem has a solution. Use the data of the problem to decide which outcomes are impossible. Then at least one of the possibilities not ruled out must prevail. If all but one possibility can be ruled out, then it must prevail.

Of course, if you use this strategy on a problem and **all** possibilities can be correctly ruled out, the problem has no solution. Don't be misled. It is certainly possible to have problems with no solution! Consider the problem of finding a number such that 3 more than twice the number is 15, and 6 more than 4 times the number is 34. This problem has no solution, since the first condition is only satisfied by 6 and the second is only satisfied by 7. Yet $6 \neq 7$.

However, if we know that a problem has a solution, it is sometimes easier to determine what can't be true than what must be true. In this approach to problem solving one eliminates possibilities until the only possibility left must yield the desired solution.

Consider the following problem.

EXAMPLE 1.12 ELIMINATE POSSIBILITIES

Beth, Jane and Mitzi play on the basketball team. Their positions are forward, center, and guard. Given the following information, determine who plays each position.

(a) Beth and the guard bought a milkshake for Mitzi.
(b) Beth is not a forward.

Solution *Understand the Problem*

Given the clues (a) and (b), we are to determine which girl plays each position.

Devise a Plan

The problem is confusing. Perhaps, if we make a table of possibilities, we can use the clues to eliminate some of them and so arrive at a conclusion.

Carry Out the Plan

The table showing all possibilities is as follows:

	Beth	Jane	Mitzi
forward			
center			
guard			

Using the clues in the order given, we see from (a) that neither Beth nor Mitzi is the guard, and we put Xs in the table to indicate this. But then it is clear that Jane is the guard, and we put an O in the appropriate cell to indicate this as well.

	Beth	Jane	Mitzi
forward			
center			
guard	X	O	X

But, if Jane is the guard, she is not the center or forward, and so we also X out these cells.

	Beth	Jane	Mitzi
forward		X	
center		X	
guard	X	O	X

This appears to be all the information we can get from condition (a), so we turn to condition (b) which tells us that Beth is not the forward. After we X out this cell, it becomes clear that Beth is the center. So we place an O in the Beth/center cell and an X in the Mitzi/center cell. This leaves the forward position as the only possibility for Mitzi, so we place an O in the Mitzi/forward cell. The completed table is as shown here,

	Beth	Jane	Mitzi
forward	X	X	O
center	O	X	X
guard	X	O	X

School Book Page

Problem Solving in Grade Five

Check

Problem Solving
Understand
Plan
Solve
Look Back

Use logical reasoning to solve the problem.

1. Anna, Gary, Mark, and Tina are from Alabama, Georgia, Mississippi, and Tennessee. None comes from a state that begins with the same letter as his or her name. Neither Anna nor Tina is from Georgia. Gary is from Tennessee.

   Which person comes from each state?

	AL	GA	MS	TN
Anna				
Gary				
Mark				
Tina				

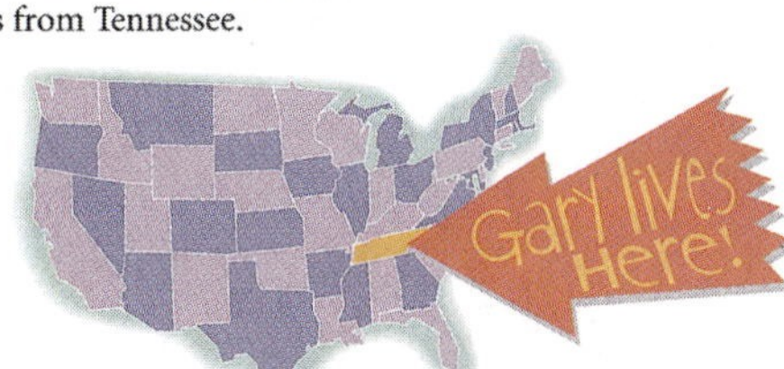

Problem Solving Practice

- Use Objects/Act It Out
- Draw a Picture
- Look for a Pattern
- Guess and Check
- Use Logical Reasoning
- Make an Organized List
- Make a Table
- Solve a Simpler Problem
- Work Backward

Choose a Tool

Use logical reasoning or any strategy to solve each problem.

2. **Logic** Ivan always answers in riddles. Daniella asked him what his address is on Chestnut Street. Ivan answered "It's a 3-digit number. The digit in the tens place is twice the digit in the ones place. The digit in the hundreds place is 3 times as great as the tens digit." What is Ivan's address?

3. **Time** Chantelle gets home from school each day at 4:40 P.M. On her way home she first walks to her friend's home. This takes 15 minutes. After chatting for 15 min, she walks to the library in 10 min. She stays there 45 min, doing her homework. The walk home from the library takes 15 min. What time does she leave school? What strategy did you use to solve this problem?

4. **Write Your Own Problem** Write a problem using the data shown in the table.

5. **Journal** When you can use either logical reasoning or draw a picture, which do you prefer? Give your reasons.

Pizza Toppings				
	Cheese	Mushroom	Sausage	Broccoli
Ali	yes	no	no	no
Ben	no	no	yes	no
Cass	no	no	no	yes
Dee	no	yes	no	no

PROBLEM SOLVING PRACTICE

Skills Practice Bank, page 573, Set 6 Lesson 9-9

SOURCE: From Scott Foresman–Addison Wesley Math, Grade 5, p. 427, by Randall I. Charles et al.

Questions for the Teacher

1. Solve problem 1 above. Students may wonder if it makes a difference in which order they use the clues. Does the order matter?
2. In order to help your students solve problem 2, would it help to ask if 1 could serve as the units digit of Ivan's house number (guess and check)? Could 2 be the units digit of the house number? Could any digit other than 1 be the units digit?
3. Formulate your own answer to problem 4.

and we conclude that Mitzi plays forward, Beth plays center, and Jane plays guard. Note that, in order to make the step-by-step process of eliminating possibilities more clear, we repeatedly redrew the table as the solution progressed. However, all the work could have been done in a single table and normally would have been.

Look Back

In this problem, we were confronted with data not easily analyzed. To bring order into this chaos, it seemed reasonable to make a table allowing for all possibilities and then to use the given statements to decide which possibilities could be ruled out. In this way we were able to delete possibilities systematically until the only remaining possibilities completed the solution.

Eliminating possibilities is often a successful approach to solving a problem.

The Pigeonhole Principle

If 101 guests are staying at a hotel with 100 rooms, can we make any conclusion about how many people there are in a room? It is probable that a number of the rooms are empty since some of the guests probably include married couples, families with children, and friends staying together to save money. But suppose most of the hotel's guests desire single rooms. How many such persons could the hotel possibly accommodate? If there were just one person per room, all 100 rooms would be occupied with one person left over. Thus, if *all* 101 guests are to be accommodated, there must be at least two persons in one of the rooms. To summarize,

> *If 101 guests are staying in a hotel with 100 guest rooms, then at least one of the rooms must be occupied by at least two guests.*

This reasoning is essentially trivial, but it is also surprisingly powerful. Indeed, it is so often useful that we name it **the pigeonhole principle,** which is stated here.

PROBLEM-SOLVING STRATEGY 13 The Pigeonhole Principle

If m pigeons are placed into n pigeonholes and $m > n$, then there must be at least two pigeons in one pigeonhole.

For example, if we place three pigeons into two pigeonholes, then there must be at least two pigeons in one pigeonhole. To make this quite clear, consider all possibilities as shown here:

Pigeonhole Number 1	Pigeonhole Number 2
3 pigeons	0 pigeons
2 pigeons	1 pigeon
1 pigeon	2 pigeons
0 pigeons	3 pigeons

In every case there are at least two pigeons in one of the pigeonholes.

A second useful way to understand this reasoning is to try to avoid the conclusion by spreading out the pigeons as much as possible. Suppose we start by placing one pigeon into each pigeonhole as indicated below. Then we have one more pigeon to put into a pigeonhole, and it must go in either hole number one or hole number two. In either case, one of the holes must contain a second pigeon and the conclusion follows.

Pigeonhole Number 1	Pigeonhole Number 2
1	1

EXAMPLE 1.13 USING THE PIGEONHOLE PRINCIPLE

An electrician working in a tight space in an attic can barely reach a box containing twelve 15-amp fuses and twelve 20-amp fuses. If her position is such that she cannot see into the box, how many fuses must she select to be sure that she has at least two fuses of the same strength?

Solution

Understand the Problem

The box contains twelve 15-amp fuses and twelve 20-amp fuses. The electrician is in a tight spot and can barely reach the box into which she cannot see. In one attempt, she wants to select enough fuses to be sure that she has at least two fuses of the same strength. We must determine how many fuses she must select to ensure the desired result.

Devise a Plan

Let's consider possibilities. To make sure we don't miss one, we make an orderly list.

Carry Out the Plan

If the electrician chooses two fuses, she must have

two 15-amp fuses	and	zero 20-amp fuses, or
one 15-amp fuse	and	one 20-amp fuse, or
zero 15-amp fuses	and	two 20-amp fuses.

Two fuses are **not** enough; she might get one of each kind. But if she selects a third fuse, it must either be a third 15-amp fuse, a second 15-amp fuse, a second 20-amp fuse, or a third 20-amp fuse. In any case, she has two fuses of the same strength and the condition of the problem is satisfied. Therefore, she needs to select only three fuses from the box.

Look Back

We certainly solved the problem considering possibilities. But might there be an easier solution? Choosing fuses of two kinds is much like putting pigeons into two pigeonholes. Thus, if we select three fuses, at least two must be the same kind by the pigeonhole principle, and we have the same result as before. Also, be sure to read the Into the Classroom feature on the next page.

Into the Classroom

Make use of Incorrect Student Responses

Observe that the number "twelve" in the statement of the preceding problem is misleading (only two fuses of each kind in the box are really needed), and this causes many students to respond that the answer is thirteen. Often, students make incorrect responses that serve as good springboards to useful classroom discussions. Rather than just saying that the response of thirteen is incorrect, a good teacher may say something like "Well, Pete, that's not quite right, but could you think of a question related to this problem for which thirteen *is* the correct answer?" This response not only corrects Pete, but also gives him an immediate opportunity to redeem himself in the eyes of the class. The ensuing discussion will not only inform Pete, but also enhance the understanding of the entire class. Other question that might be discussed are as follows: What questions might be asked for which fourteen is the correct answer? How many fuses must the electrician select in order to be sure that she has one 15-amp fuse? How many fuses must the electrician select to be sure that she has twelve 20-amp fuses? Also, one might repeat the problem with twelve 15-amp fuses, twelve 20-amp fuses, and twelve 30-amp fuses. The possibilities are almost limitless.

Highlight from History

Henri Poincaré (1854–1912)

The most remarkable mathematician at the beginning of the twentieth century was Henri Poincaré, born in Nancy, France, on August 29, 1854. Poincaré was the last person to take all of mathematics as his province. It is certainly true today that this would be impossible for any single individual and, even in 1880, it was generally believed that Gauss was the last mathematician of whom this could be said. But it was true of Poincaré, whose work was monumental—including over 500 landmark papers and some 30 important books.

Among Poincaré's many interests was the psychology of mathematical invention and discovery, or, as we have phrased it in this text, the psychology of problem solving. Poincaré thought and wrote extensively about this subject. One idea he expressed is akin to Pólya's look-back principle. A person learns to solve problems by solving problems and each time remembering the key to the solution so that it can be utilized again and again to solve other problems. As in any other endeavor, the key to success in problem solving is practice.

Problem Set 1.4

1. Play this game with a partner. The first player marks down 1, 2, 3, or 4 tallies on a sheet of paper. The second player then adds to this by marking down 1, 2, 3, or 4 more tallies. The first player to exceed a total of 30 loses the game. Can one player or the other devise a surefire winning strategy? Explain carefully.

2. Consider this mathematical machine.

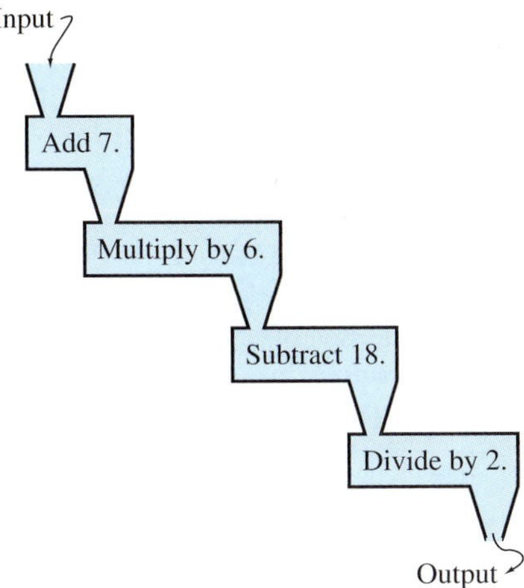

(a) What number would you have to use as input if you wanted 39 as the output?

(b) What would you have to input to obtain an output of 57?

(c) Describe a strategy for attacking this problem different from the one you used for parts (a) and (b).

3. Josh wanted to buy a bicycle but didn't have enough money. After telling his troubles to Sam Slick, Sam said, "I can fix that. See that fence? Each time you jump that fence, I'll double your money. There's one small thing though. You must give me \$32 each time for the privilege of jumping." Josh agreed, jumped the fence, received his payment from Sam Slick, and paid Sam \$32. Repeating the routine twice more, Josh was distressed to find that, on the last jump, after Sam had made his payment to Josh, Josh had only \$32 with which to pay Sam and so had nothing left. Sam, of course, went merrily on his way, leaving Josh wishing that he had known a little more about mathematics.

(a) How much money did Josh have before he made his deal with Sam?

(b) Suppose the problem is the same, but this time Josh jumps the fence five times before running out of money. How much money did Josh start with this time?

4. Lola is thinking of a number. If you multiply her number by 71, add 29, and divide by 2, you obtain 263. What is Lola's number? Solve this problem by working backward.

5. A stack of 10 cards numbered 0, 1, 2, . . . , 9 in some order lies face up on a desk. Form a new stack as follows: Place the top card face up in your hand, place the second card face up **under** the first card, place the third card face up **on top** of the new stack, place the fourth card face up **on the bottom** of the new stack, and so on. Arrange the cards in the original stack so that they appear in the new stack numbered in increasing order from the top down.

6. Moe, Joe, and Hiram are brothers. One day, in some haste, they left home with each one wearing the hat and coat of one of the others. Joe was wearing Moe's coat and Hiram's hat. Whose hat and coat was each one wearing?

7. Lisa Kosh-Granger likes to play number games with the students in her class since it improves their skill at both mental arithmetic and critical thinking. Solve each of these number riddles she gave to her class.

(a) I'm thinking of a number.

The number is odd.
It is more than 1 but less than 100.
It is greater than 20.
It is less than $5 \cdot 7$.
The sum of its digits is 7.
It is evenly divisible by 5.

What is the number? Was all the information needed?

(b) I'm thinking of a number.

The number is not even.
The sum of its digits is divisible by 2.
The number is a multiple of 11.
It is greater than $4 \cdot 5$.
It is a multiple of 3.
It is less than $7 \cdot 8 + 23$.

What is the number? Is more than one answer possible?

(c) I am thinking of a number.

The number is even.
It is not divisible by 3.
It is not divisible by 4.
It is not greater than 9^2.
It is not less than 8^2.

What is the number? Is more than one answer possible?

8. Saturday afternoon, Aaron, Boyd, Carol, and Donna stopped by the soda fountain for treats. Altogether, they ordered a chocolate malt, a strawberry milkshake, a banana split, and a double-dip walnut ice cream cone. Given the following information, who had which treat?

(a) Both boys dislike chocolate.

(b) Boyd is allergic to nuts.

(c) Carol bought a malt and a milkshake for Donna and herself.

(d) Donna shared her treat with Boyd.

9. Four married couples belong to a bridge club. The wives' names are Kitty, Sarah, Josie, and Anne. Their husbands' names (in some order) are David, Will, Gus, and Floyd.

Will is Josie's brother.

Josie and Floyd dated some, but then Floyd met his present wife.

Kitty is married to Gus.

Anne has two brothers.

Anne's husband is an only child.

Use this table to sort out who is married to whom.

	Kitty	Sarah	Josie	Anne
David				
Will				
Floyd				
Gus				

10. **(a)** Katrina chose one of the numbers 1, 2, 3, . . . , 1024 and challenged Sherrie to determine the number by asking no more than 10 questions to which Katrina would respond truthfully either "yes" or "no." Determine the number she chose if the questions and answers are as follows:

Questions	*Answers*
Is the number greater than 512?	no
Is the number greater than 256?	no
Is the number greater than 128?	yes
Is the number greater than 192?	yes
Is the number greater than 224?	no
Is the number greater than 208?	no
Is the number greater than 200?	yes
Is the number greater than 204?	no
Is the number greater than 202?	no
Is the number 202?	no

(b) In part (a), Sherrie was able to dispose of 1023 possibilities by asking just 10 questions. How many questions would Sherrie have to ask to determine Katrina's number if it is one of 1, 2, 3, . . . , 8192? if it is one of 1, 2, 3, . . . , 8000? Explain briefly but clearly. (*Hint:* Determine the differences between 512, 256, 128, 192, and so on.)

(c) How many possibilities might be disposed of with 20 questions?

11. **(a)** How many students must be in a room to be sure that at least two are of the same sex?

(b) How many students must be in a room to be sure that at least six are boys or at least six are girls?

12. **(a)** How many people must be in a room to be sure that at least two people in the room have the same birthday (not birth date)? Assume that there are 365 days in a year.

(b) How many people must be in a room to be sure that at least three have the same birthday?

13. In any collection of 11 natural numbers, show that there must be at least two whose difference is evenly divisible by 10. Helpful question: When is the difference of two natural numbers divisible by 10?

14. **(a)** In any collection of seven natural numbers, show that there must be two whose sum or difference is divisible by 10. (*Hint:* Try a number of particular cases. Try to choose numbers that show that the conclusion is false. What must be the case if the sum of two natural numbers is divisible by 10?)

(b) Find six numbers for which the conclusion of part (a) is false.

15. Show that, if five points are chosen in or on the boundary of a square with a diagonal of length $\sqrt{2}$ inches, at least two of them must be no more than $\frac{\sqrt{2}}{2}$ inches apart. (*Hint:* Consider this figure.)

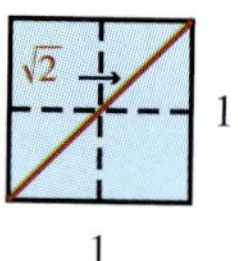

16. Show that if five points are chosen in or on the boundary of an equilateral triangle with sides 1 meter long, at least two of them must be no more than $\frac{1}{2}$ meter apart.

17. Think of 10 cups with one marble in the first cup, two marbles in the second cup, three marbles in the third cup, and so on. If the cups are arranged in a circle in any order whatsoever, show that some three adjacent cups in the circle must contain a total of at least 17 marbles.

18. A fruit grower packs apples in boxes. Each box contains at least 240 apples and at most 250 apples. How many boxes must be selected to be certain that at least three boxes contain the same number of apples?

19. Show that at a party of 20 people, there are at least two people with the same number of friends at the party. Presume that the friendship is mutual. (*Hint:* Consider the following three cases: (i) Everyone has at least one friend at the party; (ii) precisely one person has no friends at the party; (iii) at least two people have no friends at the party.)

20. Argue convincingly that at least two people in New York have precisely the same number of hairs on their heads. (*Hint:* You may need to determine a reasonable figure for the number of hairs on a human head.)

Teaching Concepts

21. Read one of the following from NCTM's *Principles and Standards for School Mathematics.*

(a) Reasoning and Proof Standard for Grades Pre-K–2, pages 122–126

(b) Reasoning and Proof Standard for Grades 3–5, pages 188–192

(c) Reasoning and Proof Standard for Grades 6–8, pages 262–267

Write a critique of the standard you read, emphasizing your own reaction. How do the recommendations compare with your own school experience?

From State Student Assessments

22. (Washington State, Grade 4)
Dan baked some cookies. Sam took half of the cookies. Then Sue took half of the remaining cookies. Later, Lisa took half of the cookies that were left. When Dan came home, he saw only three cookies. Tell how you could figure out how many cookies Dan baked altogether. Explain your thinking using words, numbers, or pictures.

23. (Washington State, Grade 4)
Emily, Mei, and Andrew go to the same camp. They each like different games. Use the information in the figure above to find out which game Mei likes best. Their favorites are tug-of-war, rope skipping, and relay race.

Emily's favorite game does *not* use a rope.

Andrew does *not* like tug-of-war.

Which is Mei's favorite game?

A. Tug-of-war

B. Relay race

C. Rope skipping

Figure for Problem 23

Tug-of-war

Rope skipping

Relay race

1.5 REASONING MATHEMATICALLY

In this concluding section on critical thinking, we will consolidate and extend our strategies of problem solving. In particular, we will discuss the following topics, each of which will be helpful in the chapters to come.

- inductive reasoning;
- representational reasoning;
- mathematical statements;
- deductive reasoning.

Inductive Reasoning

We use inductive reasoning to draw a general conclusion based on information obtained from specific examples. For example, think about the bears you've seen, maybe in zoos, in pictures in magazines, or perhaps even in the wild. Based on these experiences, you would likely draw the conclusion that bears are brown or black, or even white if you've seen a polar bear. Here's a mathematical example. Consider the square numbers 4, 9, 16, 25, and 36. Notice that they are either multiples of 4 or one larger than a multiple of 4. We can check that this property of squares is also true for other squares—say, 49, 64, 81, and 100. Even $1^2 = 1$ passes our check, since 1 is one larger than 4 times 0. Thus, inductive reasoning leads us to the generalization that the square of any whole number is either a multiple of 4 or one larger than a multiple of 4.

> **DESCRIPTION** *Inductive Reasoning*
> **Inductive reasoning** is drawing a conclusion based on evidence obtained from specific examples. The conclusion drawn is called a **generalization.**

Inductive reasoning is a powerful way to create and organize information. However, it only suggests what *seems* to be true, since we have not yet checked that the property holds for *all* examples. For example, there is a rare type of bear in southeastern Alaska called the blue bear. Though it is a genetic variant of the black bear, its fur is dark blue. Thus, the existence of the blue bear tells us that the statement "all bears are black, brown, or white" is false. Such an example, one that disproves a statement, is called a **counterexample.** It is interesting to notice that a proof of a generalization requires us to demonstrate that a certain property holds for every possible case, but a generalization can be proved to be false by finding just *one* counterexample.

EXAMPLE 1.14 USING INDUCTIVE REASONING IN MATHEMATICS

Examine the generalizations made below. Test the validity of each generalization with additional evidence. If you believe that the generalization is valid, try to offer additional reasons why this is so. If you believe that the generalization may be false, search for a counterexample.

(a) Consider three consecutive integers, such as 8, 9, and 10. Exactly one of these three numbers is a multiple of 3. Similarly, each of the consecutive triples 33, 34, and 35 and 121, 122, and 123 includes precisely one multiple of 3. Thus, in any string of three consecutive integers, it is probably true that exactly one is a multiple of 3.

(b) Place n points on a circle. Next, join each pair of these points with a line segment (that is, with a chord of the circle), where no more than two chords intersect at a single point. As shown in the diagrams below, the number of regions in the circle doubles with each additional point placed on the circle, giving the sequence 1, 2, 4, and 8.

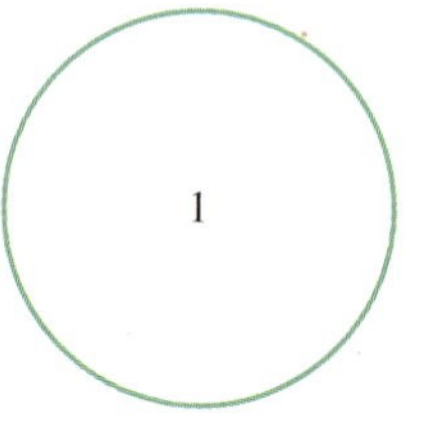

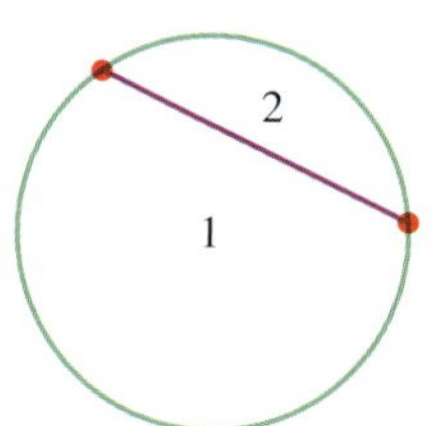

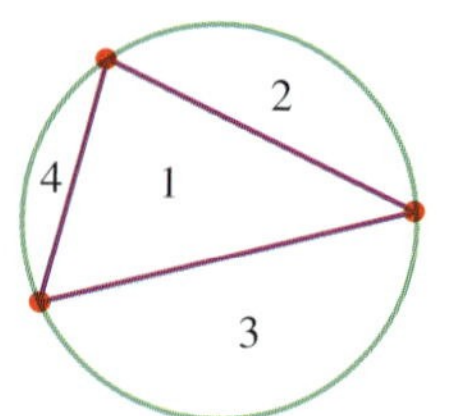

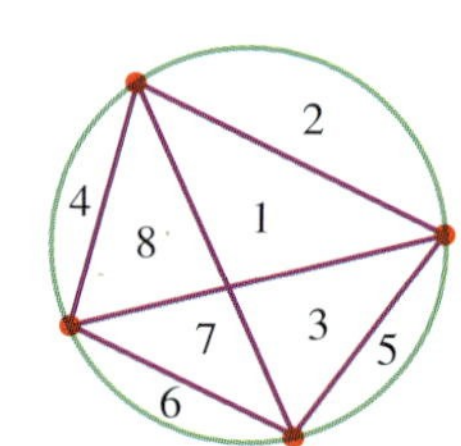

In conclusion, it is probably true that there are 2^{n-1} regions created by drawing chords between n points on a circle.

(c) $3 \times 439 = 1317$, and the sum of the digits of the product is $1 + 3 + 1 + 7 = 12$, a multiple of 3. Similarly, $3 \times 2687 = 8061$, and the sum of digits of this product is $8 + 0 + 6 + 1 = 15$, again a multiple of 3. Thus, it is probably true that the sum of the digits of any whole-number multiple of 3 is also a multiple of 3.

Solution

(a) Additional examples support the general conclusion. If the whole numbers are written in the form **0** 1 2 **3** 4 5 **6** 7 8 **9** 10 11 **12** 13 . . . , with the multiples of 3 shown in bold, we see that any three numbers in succession include exactly one of the bold numbers. Thus, the assertion *appears* to be true.

(b) Let's draw a sketch of the next two cases, with $n = 5$ and 6 points on the circle.

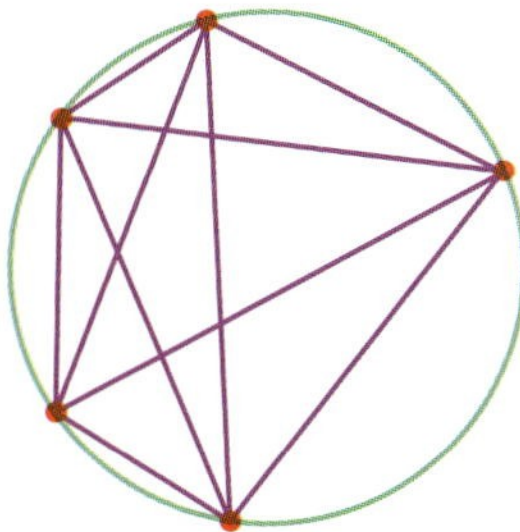
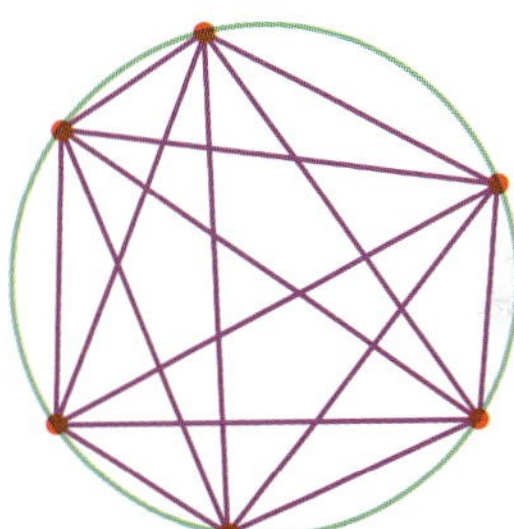

When $n = 5$ points are placed on the circle, we see there are $16 = 2^{5-1}$ regions, which supports the generalization. However, the doubling pattern breaks down with $n = 6$ points on the circle since we do not get 32 regions as we had expected. How many regions do the chords joining six points on a circle create?

(c) Additional examples all exhibit the same property. In Chapter 4, we will prove that the sum of the digits of any whole number divisible by 3 is also divisible by 3.

PROBLEM-SOLVING STRATEGY 14 Use Inductive Reasoning

- Observe a property that holds in several examples.
- Check that the property holds in other examples. In particular, attempt to find an example where the property does not hold (that is, try to find a counterexample).
- If the property holds in every example, state a generalization that the property is *probably* true in general.

A generalization that seems to be true, but has yet to be proved, is called a **conjecture.** Once a conjecture is given a proof, it is called a **theorem.**

Representational Reasoning

In mathematics, a *representation* is an object that captures the essential information needed for understanding and communicating mathematical properties and relationships. Often, the representation conveys information visually, as in a diagram, a graph, a map, or a table.

At other times, the representation is symbolic, such as a letter denoting a variable or an algebraic expression or equation. The representation can also be a physical object, such as a paper model of a cube or an arrangement of pebbles to represent a whole number. The following excerpt from the NCTM *Principles and Standards for School Mathematics* points out that a representation can even be simply a mental image.

from **The NCTM Principles and Standards**

Representation Standard for Grades 3–5

Instructional programs from prekindergarten through grade 12 should enable all students to—

- *create and use representations to organize, record, and communicate mathematical ideas;*
- *select, apply, and translate among mathematical representations to solve problems;*
- *use representations to model and interpret physical, social, and mathematical phenomena.*

In grades 3–5, students need to develop and use a variety of representations of mathematical ideas to model problem situations, to investigate mathematical relationships, and to justify or disprove conjectures. They should use informal representations, such as drawings, to highlight various features of problems; they should use physical models to represent and understand ideas such as multiplication and place value. They should also learn to use equations, charts, and graphs to model and solve problems. These representations serve as tools for thinking about and solving problems. They also help students communicate their thinking to others. Students in these grades will use both external models—ones that they can build, change, and inspect—as well as mental images.

Here is an example to illustrate how a representation can reveal and explain a property. Consider the "pyramidal sums" shown in this list:

$$1 = 1$$
$$1 + 2 + 1 = 4$$
$$1 + 2 + 3 + 2 + 1 = 9$$
$$1 + 2 + 3 + 4 + 3 + 2 + 1 = 16$$

Using inductive reason, we would likely draw the conclusion that the nth pyramidal sum is the square number n^2. That is,

$$1 + 2 + 3 + \cdots + (n - 1) + n + (n - 1) + \cdots + 3 + 2 + 1 = n^2.$$

To see why this is so, let's recall that any square number can be represented as a square pattern of dots. For example, 5^2 is represented by the 5×5 array shown in Figure 1.7. If we sum the numbers of dots along each diagonal of the square, as seen at the right of the figure, we see very clearly why $1 + 2 + 3 + 4 + 5 + 4 + 3 + 2 + 1$ is equal to the square number 5^2. In our "mind's eye," where we can create a mental image of the most general case, we see why any pyramidal sum is a square number.

In the next example, we make dot drawings with colored pencils on squared paper in order to make some more discoveries about number patterns through the use of representations.

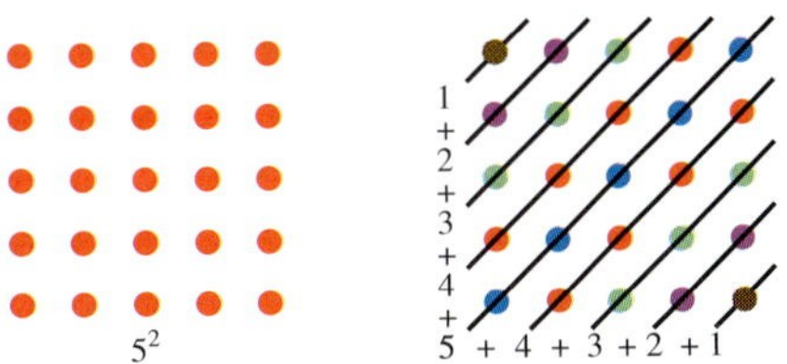

Figure 1.7 Any square number is a pyramidal sum

EXAMPLE 1.15 USING DOT REPRESENTATIONS TO DISCOVER NUMBER PATTERNS

Recall that the triangular and square numbers can be represented with dot patterns as shown here.

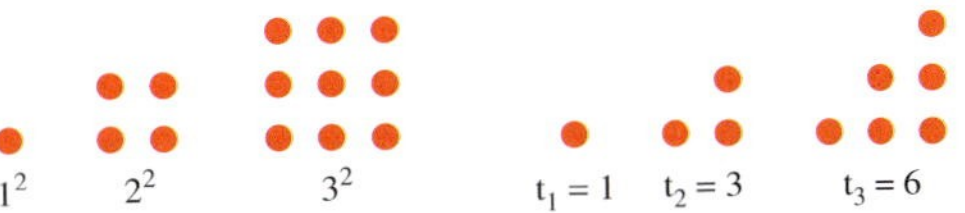

(a) Draw dot figures to show that $t_3 + t_3 = 3 \cdot 4$ and $t_4 + t_4 = 4 \cdot 5$. By inductive reasoning, why can you conclude that $t_n = \dfrac{n(n+1)}{2}$?

(b) Draw dot figures to show that $t_3 + t_4 = 4^2$ and $t_4 + t_5 = 5^2$. What generalization can you make?

(c) Draw dot figures to show that $7^2 = 1 + 8t_3$ and $9^2 = 1 + 8t_4$. [*Hint:* The 1 is the center dot in your square.] By inductive reasoning, what do you think is the value of $1 + 8t_{50}$? of $1 + 8t_n$?

Solution

(a) We generalize that two dot patterns, each representing t_n, can form an n by $n + 1$ rectangle. Therefore,

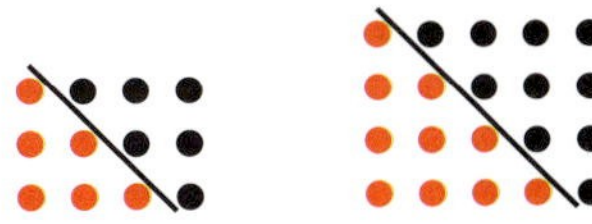

$$t_n + t_n = n(n+1), \quad \text{or} \quad t_n = \frac{n(n+1)}{2}.$$

(b) The patterns shown suggest that the patterns representing t_{n-1} and t_n can be arranged to form an n-by-n square. We conclude that $t_{n-1} + t_n = n^2$.

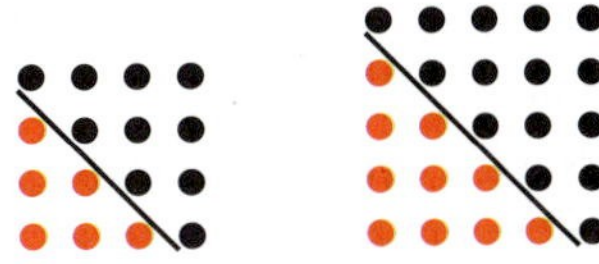

(c) Eight triangular dot patterns together with a single dot can be arranged to form a square with an odd number of dots in each row. Inductive reasoning suggests that $1 + 8t_{50} = 101^2 = 10{,}201$ and, in general, that $1 + 8t_n = (2n + 1)^2$.

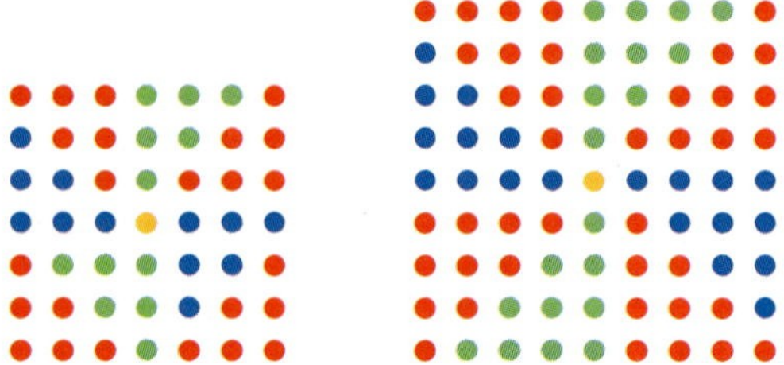

Mathematical Statements

In mathematics, a **statement** is a declarative sentence that is either true or false, but not both. Thus, "2 + 3 = 5" is a true statement, and "pigs can fly" is a false statement. "Shut the door" is not a statement since it is not a declarative sentence. The sentence "This sentence is a false statement" is declarative, but not a mathematical statement since it is neither ture nor false. (If it is true, then it is false, and if it is false, then it is true—so it is neither true nor false!)

Two statements, say p and q, are often combined to form another statement by using **and** and **or** operations.

- **And:** "p and q" is true when statement p and statement q are both true, and false when either or both of p and q are false.
- **Or:** "p or q" is true when either statement p or statement q (or even both) is true, and false only when both p and q are false.

This description can also be given a representation known as a **truth table,** as shown in Figure 1.8.

p	*q*	*p* and *q*	*p* or *q*
T	T	T	T
T	F	F	T
F	T	F	T
F	F	F	F

Figure 1.8
The truth table for "p and q" and "p or q"

Yet another way to combine two statements is to form a **conditional statement** or **"if . . . then" statement.**

- **If . . . then:** "If p then q" is true when the truth of statement p guarantees the truth of statement q.

For example, you have probably encountered the conditional statement "If a whole number has 0 as its last digit, then it is divisible by 10." Often, "if p then q" is read as "p implies q" and is written $p \rightarrow q$. The truth table for $p \rightarrow q$ is shown in Figure 1.9.

p	q	$p \rightarrow q$
T	T	T
T	F	F
F	T	T
F	F	T

Figure 1.9
The truth table for p → q, which shows that **p → q** *is false when, and only when, p is true and q is false*

The big surprise in Figure 1.9 is that $p \rightarrow q$ is true when p is false, independent of whether q is true or false. Thus, "if the moon is made of green cheese, then pigs can fly" is a true conditional statement! Fortunately, the moon is not made of green cheese, and so we don't need to be on the lookout for flying pigs.

Many theorems of mathematics have the "if p then q" form.

EXAMPLE 1.16 PROVING AN "IF . . . THEN" STATEMENT

Prove the following theorem:

If n is any whole number, then n^2 is either a multiple of 4 or one larger than a multiple of 4.

Solution

Understand the Problem

We must show that *any* whole number n, when squared, has one of two forms: it is either a multiple of 4, so that $n^2 = 4j$ for some whole number j, or else $n^2 = 4k + 1$ for some whole number k. We have earlier used inductive reasoning that supported the truth of the theorem.

Devise a Plan

The whole numbers are 0, 1, 2, 3, 4, 5, 6, . . . , alternating between even and odd. Their squares are 0, 1, 4, 9, 16, 25, 36, . . . , alternating between multiples of 4 and numbers that are one larger than a multiple of 4. This suggests that we consider two cases: when n is even and when n is odd. We also know that when n is even, it can be written as $n = 2r$ for some whole number r, and when n is odd it can be written as $n = 2s + 1$ for some whole number s. Thus, we need to consider the two cases $(2r)^2$ and $(2s + 1)^2$.

Carry Out the Plan

When $n = 2r$, we see that $n^2 = (2r)^2 = 4r^2$. That is, $n^2 = 4j$, where $j = r^2$. Similarly, when $n = 2s + 1$, we have $n^2 = (2s + 1)^2 = 4s^2 + 4s + 1 = 4(s^2 + s) + 1$. That is, $n^2 = 4k + 1$ for $k = s^2 + s$.

$(2s + 1)^2 = (2s + 1)(2s + 1)$
$= 4s^2 + 2s + 2s + 1$
$= 4s^2 + 4s + 1$

Look Back

Let's take a closer look at the squares of the odd numbers—that is, the numbers 1, 9, 25, 49, 81, ... These numbers are not only one larger than a multiple of 4, they're even one larger than a multiple of 8. This was acutally shown earlier in Example 1.15 (c): using dot pattern representations, we discovered that $(2n + 1)^2 = 8t_n + 1$, where t_n is the nth triangular number. We can also use dot pattern representations to visualize why the square of an even number is four times another square. The following figure shows that $10^2 = 4 \times 5^2$.

Deductive Reasoning

Suppose that we have a collection of true statements. If we can argue on the basis of these statements that another statement must also be true, then we are using **deductive reasoning.** Example 1.16 is an example of deductive reasoning, since we showed that if a number is a square whole number, then it is a multiple of 4 or one larger than a multiple of 4.

The list of true statements we begin with are known as the **premises,** or **hypotheses,** of the argument. The new true statement that we obtain is called the **conclusion** of the argument.

Here is another example of deductive reasoning.

Hypothesis:	***Statement 1.***	If you wish to become a successful elementary school teacher, then you must become proficient in mathematical reasoning.
	Statement 2.	You wish to become a successful elementary school teacher.
Conclusion:	***Statement 3.***	You must become proficient in mathematical reasoning.

In symbols, if we let p denote the statement "You wish to become a successful elementary school teacher" and q denote the statement "You need to become proficient in mathematical reasoning," then we have reached the conclusion using the **rule of direct reasoning.**

Rule of Direct Reasoning

Hypotheses: $\begin{cases} \text{If } p \text{ then } q \\ p \text{ is true} \end{cases}$

Conclusion: Therefore, q is true

The rule of direct reasoning may seem straightforward, but it is sometimes used incorrectly. Here is an example of invalid reasoning called the "fallacy of the converse":

If I am a good person, then nothing bad will happen to me.
Nothing bad has happened to me.
Therefore, I am a good person.
} INVALID REASONING

This argument is invalid since even if $p \rightarrow q$ is true and q is true, nothing can be said about whether p is true or not. This concept is clearly shown in the truth table of Figure 1.9.

On the other hand, the truth table does show that if $p \rightarrow q$ is true and q is *false,* then p is also necessarily false. Thus, we have the very useful **rule of indirect reasoning.**

Rule of Indirect Reasoning

Hypotheses:	If p then q q is false
Conclusion:	Therefore, p is false

The rule of indirect reasoning is often used to give a **proof by contradiction.** That is, if we take a statement p and derive from it a false statement q, then we know that p is false. The following checkerboard-tiling problem is a classic example of indirect reasoning and proof by contradiction.

EXAMPLE 1.17 TILING A CHECKERBOARD WITH DOMINOS

It is easy to cover the 64 squares of an 8-by-8 checkerboard with 32 dominos, where each domino covers two adjacent squares, one red and one black, of the checkerboard. But what if two diagonally opposite squares are removed? Can the 62 remaining squares be covered with 31 dominos?

Solution

If you believed that the modified board could be tiled, you would search for an arrangement of 31 dominos that covers all 62 squares. After a few unsuccessful trials, you might begin to have doubts. Let's use indirect reasoning, or reasoning by contradiction, and assume (though you haven't found it) that there is an arrangement of 31 dominos that covers the modified board. Notice that each domino, whether vertical or horizontal, covers one red and one black square, so that an equal number of red and black squares must be covered. However, the two squares we've removed have the same color (black in the foregoing diagram). This means that we have left behind 30 squares of one color and 32 squares of the other color. This situation contradicts our observation that any arrangement of the dominos covers an equal number of squares of the two colors. Thus, the assumption that a domino tiling exists leads to a falsehood, so our assumption must also be false. That is, there is no tiling of the modified checkerboard with dominos.

The rule of indirect reasoning is also misused with some frequency. Here is an example of what is called the "fallacy of the inverse":

If I am wealthy, then I am happy.
I am not wealthy.
Therefore, I am not happy.

} **INVALID REASONING**

An examination of the truth table in Figure 1.9 shows that, if $p \to q$ is true and p is false, q can be either true or false.

Deductive reasoning provides another important strategy for problem solving.

PROBLEM-SOLVING STRATEGY 15 Use Deductive Reasoning

- To show that a statement q is true, look for a statement p such that $p \to q$ is true. If p is true, you can deduce that q is also true by direct reasoning.
- To show that a statement p is false, show that $p \to q$, where q is false. That is, show that assuming p leads to a contradiction. Then p is false by indirect reasoning.

In this section, we have highlighted inductive, representational, and deductive reasoning. These modes of mathematical thought will be used throughout the remainder of this book and will be supplemented with some additional methods, such as proportional reasoning in Chapter 7, algebraic reasoning in Chapter 8, statistical reasoning in Chapter 9, and geometrical reasoning in Chapters 11–14.

Problem Set 1.5

1. **(a)** Compute the products 9×9, 79×9, 679×9, and 5679×9.

(b) Use inductive reasoning (don't calculate yet) to describe in words what you expect are the values of these products: 45679×9, 345679×9, 2345679×9, and 12345679×9.

(c) Use a calculator to see if your inductive reasoning was correct in part (b).

2. **(a)** Compute the value of these expressions: $1 \times 9 + 2$, $12 \times 9 + 3$, and $123 \times 9 + 4$.

(b) Use inductive reasoning (don't calculate yet) to describe in words what you expect are the values of these expressions: $1234 \times 9 + 5$, $12345 \times 9 + 6$, $123456 \times 9 + 7$, $1234567 \times 9 + 8$, $12345678 \times 9 + 9$, and $123456789 \times 9 + 10$.

(c) Use a calculator to check that your inductive reasoning was correct in part (b).

3. **(a)** Compute the value of these expressions: $1 \times 8 + 1$, $12 \times 8 + 2$, and $123 \times 8 + 3$.

(b) Use inductive reasoning (don't calculate yet) to describe in words what you expect are the values of these expressions: $1234 \times 8 + 4$, $12345 \times 8 + 5$, $123456 \times 8 + 6$, $1234567 \times 8 + 7$, $12345678 \times 8 + 8$, and $123456789 \times 8 + 9$.

(c) Use a calculator to check that your inductive reasoning was correct in part (b).

4. Consider a three-digit number *abc*, where the digit *a*, in the hundreds position, is larger than the units digit *c*. Now reverse the order of the digits to get the number *cba*. Subtract your two three-digit numbers, and let the difference be *def*. Finally, reverse the digits of *def* to form the number *fed*, and add this number to *def*. Altogether, you should do the following addition and subtraction steps with your beginning number *abc*:

$$\begin{array}{r} abc \\ -\ cba \\ \hline def \end{array} \qquad \begin{array}{r} def \\ +\ fed \\ \hline ???? \end{array}$$

Choose several examples of three digit numbers *abc*, and carry out the subtraction and addition steps. Use inductive reasoning to make a generalization about this process.

5. **(a)** Compute these products:

$1 \times 1089 =$ ________
$2 \times 1089 =$ ________
$3 \times 1089 =$ ________
$4 \times 1089 =$ ________
$5 \times 1089 =$ ________
$6 \times 1089 =$ ________
$7 \times 1089 =$ ________
$8 \times 1089 =$ ________
$9 \times 1089 =$ ________

(b) Did you have to compute all the products in part (a) to be pretty sure that you knew what all the answers would be? Explain briefly.

(c) Do you see any other interesting patterns in part (a)? Explain briefly.

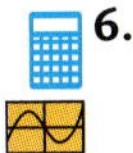

6. **(a)** **Palindromes.** A number palindrome is a number like 242 or 3113 that reads the same both forward and backward. A famous palindrome in words, attributed to Napoleon, is "Able was I ere I saw Elba." Complete the activity of the flow chart. This can be done with a calculator. If you use a calculator, be sure to record your results at each step as shown. Try this procedure several times, using different starting numbers.

(b) It takes four steps for the process to stop if we start with 372. Will it stop if we start with 98? If so, in how many steps?

(c) Do you think the process will always stop?

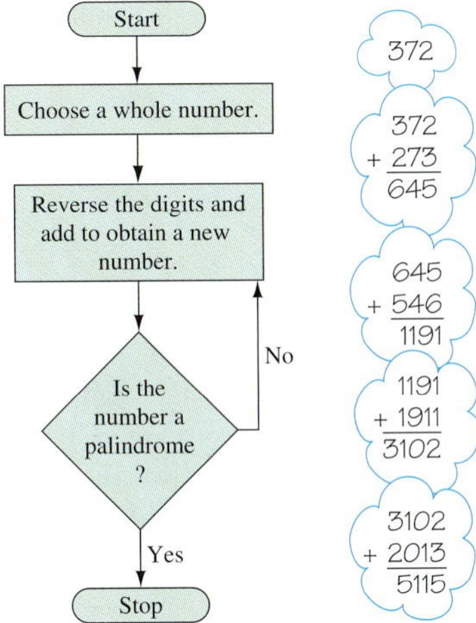

7. **(a)** **Kaprekar's Number.** Complete this activity. If you use your calculator, be sure to record the intermediate steps as shown below.

(b) Will the process stop if you start with 1999? If so, in how many steps?

(c) If you start with any four-digit number, do you think the process will always stop?

(d) What happens if you start with a three-digit number? five-digit number? Explain briefly.

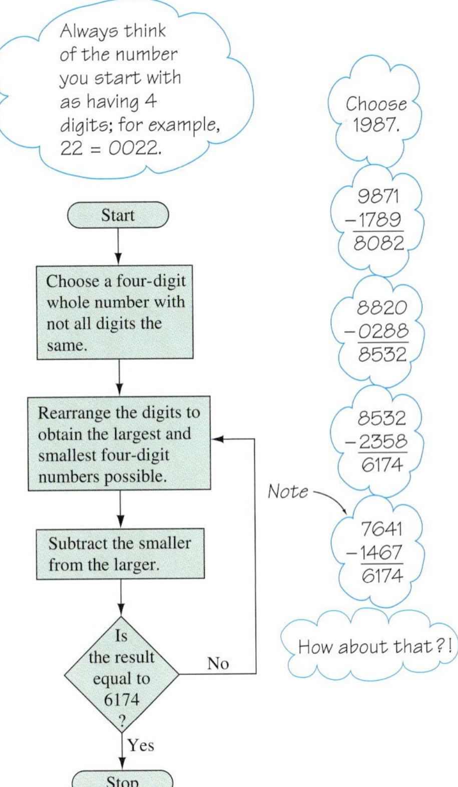

8. Use a ruler to draw two line segments and label three points on each line as shown.

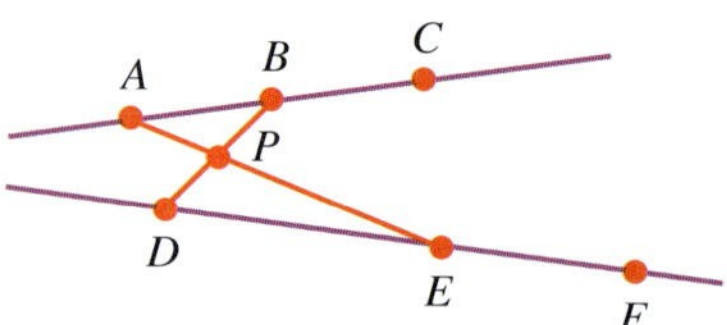

The segments $\overline{AE}$ and $\overline{BD}$ (shown in red) intersect to define point P. In the same way, draw the segments $\overline{AF}$ and $\overline{CD}$ to define the intersection point Q. Finally, draw the segments $\overline{BF}$ and $\overline{CE}$ to define the intersection point R. What seems to be special about the relative positions of the three points P, Q, and R? Use inductive reasoning to make a generalization, and test your conclusion by drawing a new pair of lines and points.

9. The given figure suggests that the number of pieces into which a pie is divided is doubled with each additional cut across the pie. Do you believe this is a valid generalization, or can you find a counterexample?

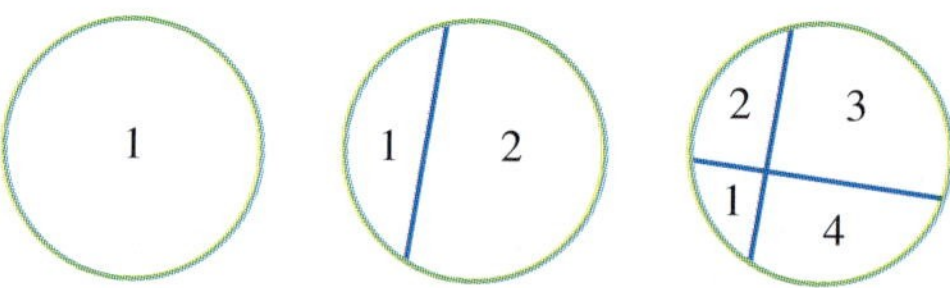

10. Suppose that logs are stacked on level ground. All of the logs in the bottom row must be side by side, and any log in an upper row must touch two lower logs. The ways to stack n circular logs are shown below for $n = 1, 2, 3,$ and 4 logs, showing that the respective numbers of stacking arrangements are 1, 1, 2, and 3.

(a) Do the Fibonacci numbers $F_1 = 1, F_2 = 1, F_3 = 2, F_4 = 3, F_5 = 5, F_6 = 8, \ldots$, first encountered in Example 1.6 (c), correctly count the number of ways to stack n logs? Investigate this generalization by drawing arrangements of 5 and 6 logs.

(b) Use inductive reasoning to investigate the number of ways to stack logs at most two layers high, where, as before, the bottom row of logs must be side by side and logs in the second row must rest on two logs of the bottom row.

11. Suppose a flagpole is erected on one of n blocks, with all of the blocks to the right (if any) used to attach guy wires and an equal number of blocks to the left of the pole also used as points of attachment of guy wires. Any block can be used to attach at most one guy wire. Here are the three permissible arrangements of a flagpole and guy wires on a row of four blocks.

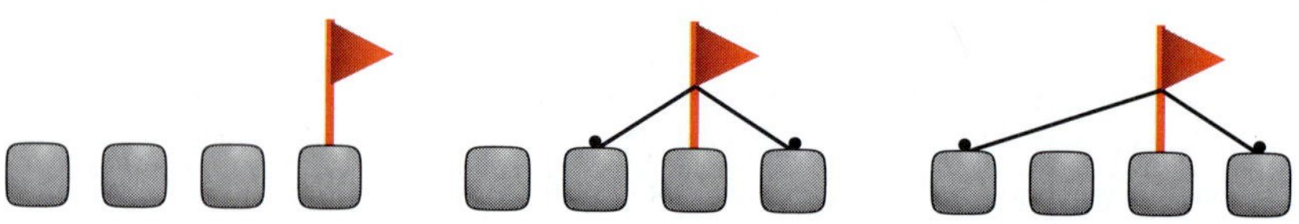

Use inductive reasoning to discover the number of arrangements of a flagpole and guy wires on a row of n blocks.

12. Consider the three triangles of pennies shown, each pointing upward. It is easy to see that the triangle of three pennies on the left can be inverted (that is, made to point downward) by moving one penny.

(a) What is the smallest number of pennies that must be moved in order to invert the 6-penny triangle?

(b) Show that the 10-penny triangle can be inverted by moving just three pennies.

(c) What do you think is the minimum number of pennies that must be moved to invert the 15-penny triangle? Experiment to see if you are justified in drawing this conclusion.

---

13. (a) The fifth pentagonal number P_5 is the sum of the arithmetic progression 1, 4, 7, 10, 13. (See problems 4 and 22(b), Problem Set 1.3.) Why does the dot pattern representation displayed below show that $P_5 = 1 + 4 + 7 + 10 + 13$? Give your answer in words and in a diagram drawn on triangular dot paper.

(b) Extend your dot representation of part (a) to show that P_6 is 51.

(c) Color a triangular dot pattern to show that $P_5 = 5 + 3t_4$, where $t_4 = 10$ is the fourth triangular number.

(d) Use triangular dot paper to show that $P_6 = 6 + 3t_5$.

(e) By inductive reasoning, and recalling that $t_{n-1} = \frac{1}{2}(n - 1)n$, obtain a formula for the nth pentagonal number $P_n = 1 + 4 + 7 + \cdots + (3n - 2)$.

14. Consider the trapezoid numbers $1 + 1 + 1$, $1 + 2 + 2 + 2 + 1$, $1 + 2 + 3 + 3 + 3 + 2 + 1, \ldots$, where the nth trapezoidal number is $1 + 2 + \cdots + n + n + n + \cdots + 2 + 1$.

(a) Create a dot pattern representation of the trapezoidal numbers.

(b) Notice that $1 + 1 + 1 = 1 \times 3$, $1 + 2 + 2 + 2 + 1 = 2 \times 4$, and $1 + 2 + 3 + 3 + 3 + 2 + 1 = 3 \times 5$. Show how the dots in a representation of a trapezoidal number can be rearranged into a rectangle with two more columns than rows.

(c) Using inductive reasoning, what general formula gives the nth trapezoidal number?

15. A log in the woods is just long enough to have space for seven frogs on top. Suppose there are three green frogs on the left and three red frogs on the right, with one empty space between. The frogs wish to change ends of the log, so that the three green frogs are on the right and the three red frogs are on the left. A frog can either move into an empty adjacent space or hop over one adjacent frog to an empty space on the opposite side of the frog being jumped over.

(a) Represent the frog problem with a physical model made from colored counters and a row of squares drawn on paper as shown here:

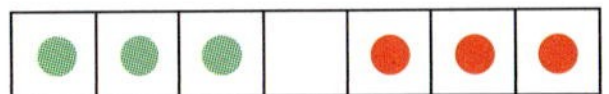

Use your model to show that the frogs can exchange ends of the log in 15 moves. This exercise may take several trials to find an efficient order in which to move the frogs. [*Suggestion:* Work in pairs, with one partner moving the counters and the other partner counting the number of moves.]

(b) On a three-space log that has space for one frog of each color and a single space between, the frogs can trade ends in three moves. What is the minimum number of moves to trade ends on a five-space log, where two green frogs start on the left and two red frogs start on the right?

(c) Examine your answers for the minimum number of moves to have $n = 1, 2,$ or 3 green frogs trade ends with the same number of red frogs on the other end of the log, and with just one extra space between them. What conjecture can you make for the case of n frogs of each color separated by one extra space? [*Hint:* There is a connection between the frog-jumping problem and the trapezoidal numbers of problem 14.]

16. Each of the numbers $2^5 - 2 = 32 - 2 = 30$, $3^5 - 3 = 243 - 3 = 240$, and $4^5 - 4 = 1024 - 4 = 1020$ is a multiple of 30.

(a) Use a calculator or spreadsheet (see the Window on Technology found in Section 1.3, or see Appendix C) to calculate additional values of $n^5 - n$. Do these examples continue to suggest that every number of the form $n^5 - n$ is a multiple of 30?

(b) It can be shown that $n^5 - n = n(n^4 - 1) = n(n^2 - 1)(n^2 + 1) = n(n - 1)(n + 1)(n^2 + 1)$. Explain why this factorization lets you deduce that $n^5 - n$ is divisible by 3.

(c) Suppose that d is the units (that is, the rightmost) digit of n. What is the units digit of n^5? Use a calculator to compute the fifth powers of the digits 0, 1, 2, 3, 4, 5, 6, 7, 8, and 9, and then explain how to deduce that $n^5 - n$ is divisible by 10.

17. Use deductive reasoning to show that if n is a multiple of 3, then n^2 is a multiple of 9. [*Note:* A multiple of k is a number of the form ks, where s is a whole number.]

18. Use indirect reasoning to show that if n^2 is odd, then n is also odd.

Into the Classroom

Teaching Problem Solving

Here are some bits of advice for teaching problem solving:

- Set out now to develop a store of interesting and challenging problems appropriate to the skill levels of the students in your class. This process should continue over your professional lifetime. A good start is to keep your present textbook and to use many of the problems you find here as is or modified to be more accessible to your students.
- Really listen to your students. They often have good ideas. They also frequently find it difficult to express their ideas clearly. Listening carefully and helping your students to communicate clearly are skills you should continually seek to improve.
- Don't be afraid of problems you don't already know how to solve. Say, "Well, I don't know. Let's work together and see what we can come up with." You needn't be the oracle who "knows all." Indeed, many students will be excited and motivated by the prospect of "working with my teacher" to solve a problem. They also learn not to be afraid to tackle the unknown, to make mistakes, and yet to persevere until a solution is finally achieved.
- Don't be too quick to help your students or to simply tell them how to solve a problem.
- Develop a long list of leading questions you can ask of your students, and that they can ask themselves, to help clarify their thinking and eventually arrive at a solution.
- Don't be afraid of having students in your class who are brighter and better at solving problems than you are. Just be happy to have such students; give them all the encouragement you can, and use them to help teach the others. Your ego should not be on the line—we have all taught students who are brighter than we are!

Epilogue Fascination with Mathematics—Past and Present

The fascination of individuals with mathematical problems, both real world and fanciful, goes back at least as far as recorded history. The reason for the interest in real-world problems, then and now, is obvious. The successful conduct of many human activities requires mathematical understanding. Thus, much of ancient mathematics was developed to meet particular needs and to enable individuals to accomplish desired tasks. The not-so-simple matter of making a reliable calendar, for example, requires a reasonably sophisticated understanding of mathematics, and the Babylonians had this ability by approximately 4700 B.C. Similarly, many practical notions from geometry, including formulas for areas and volumes of geometrical figures and the properties of right triangles, were known at least 1000 years before Pythagoras and Euclid.

But it is also true that much of mathematics was studied for its own sake, simply because it was interesting. It is even true that many of the problems still popular today are of extraordinarily ancient origin. One of the most interesting of ancient mathematical documents is the Rhind papyrus, purchased by the Scottish Egyptologist Henry Rhind in a small shop in Egypt in 1858. The scroll dates from approximately 1650 B.C. and is something of a mathematical handbook. However, it also contains a number of fanciful problems including this cryptic set of data.

Estate		
Houses	7	7^1
Cats	49	7^2
Mice	343	7^3
Heads of wheat	2401	7^4
Hekat measures	16,807	7^5

The problem is not explained, but one can guess that the challenge was to compute the sum

$$7 + 7^2 + 7^3 + 7^4 + 7^5. \leftarrow 19{,}607$$

This problem reappears in A.D. 1202 in *Liber Abaci* by the celebrated thirteenth-century mathematician Leonardo of Pisa, or Fibonacci. In translation, Fibonacci's rendition of the problem runs as follows:

> *There are seven old women on the road to Rome. Each woman has seven mules; each mule carries seven sacks; each sack contains seven loaves; with each loaf are seven knives and sheaths. How many are there in all on the road to Rome?*

The modern version of the problem is the following well-known nursery rhyme:

As I was going to St. Ives,
I met a man with seven wives,
Each wife had seven sacks,
Each sack had seven cats,
Each cat had seven kits.
Kits, cats, sacks, and wives,
How many were going to St. Ives?

It has been conjectured that the original problem in the Rhind papyrus might have been "An estate consisted of seven houses; each house had seven cats; each cat ate seven mice; each mouse ate seven heads of wheat; each head of wheat when planted would produce seven hekats of grain. How many of all these items were in the estate?" However that may be, this is a problem, still popular in children's literature, that was already ancient when Fibonacci copied it nearly 800 years ago. One even wonders if the O'Henry twist in the nursery rhyme version of the problem (where the answer is 1, and not $7 + 7^2 + 7^3 + 7^4 + 7^5 = 19{,}607$) might have existed in some form in the ancient Egyptian version.

In any case, mathematical problems and puzzles have piqued the curiosity and challenged the ingenuity of individuals for millennia, and this is no less true today. Introducing problem solving into the curriculum improves students' skills and their ability to think creatively and carefully; it also greatly enhances both student and teacher enjoyment of the entire educational process.

Finally, listing all possible strategies for solving problems is not possible. Those presented in this chapter are some of the most basic. But all have variants that we have not mentioned, all can be mixed and matched in many ways, and other methods, approaches, ideas, and schemes will occur both to you and to your students over the years.

Chapter Summary

Key Concepts

Section 1.1 An Introduction to Problem Solving

- General introduction: a classroom discussion.

Section 1.2 Pólya's Problem-Solving Principles

- Understand the problem. Make sure that the words, conditions, and desiderata of a problem are completely understood.
- Devise a plan. Carefully consider the problem and think of a possible approach to finding a solution.
- Carry out the plan. Attempt to carry out your plan. If it doesn't work out, try again. If it still doesn't solve the problem, modify the plan and try yet again.

- Look back. It is important to reconsider your thinking to see what led you to the solution. This is how you gain "mathematical power."
- Problem-solving strategy of "guess and check." Guessing is a good way to start to gain understanding of a problem, but guessing gives a solution or solutions *only* if each guess checks.
- Problem-solving strategy of "make an orderly list." This strategy helps avoid the omission of possibilities.
- Problem-solving strategy of "draw a diagram." A diagram often clarifies a problem.

Section 1.3 More Problem-Solving Strategies

- Look for a pattern. Patterns often suggest what the answer to a problem should be.
- Make a table. A table helps you search for patterns and eliminate possibilities that do not meet all of the criteria required of a solution.
- Use a variable. A symbol can often be used to represent and determine a number that is the answer to a problem.
- Consider special cases. What is true in the general case must also be true in a special case. What is true in a special case *suggests* what is true in the general case.
- Solve an equivalent problem. Often, a problem can be recast in an equivalent, but different, form that is easier to solve.
- Solve an easier, similar problem. Solving an easier, but similar, problem often suggests how to solve the problem at hand.
- Argue from special cases. An argument made from a special case, but depending on general principles (not properties applying only to the special case), often serves to prove the general case.

Section 1.4 Additional Problem-Solving Strategies

- Work backward. If you can't see how to start a solution, perhaps you can start from the desired conclusion and work backward to the beginning of the problem.
- Eliminate possibilities. If all possibilities but one can be ruled out, that possibility must be so.
- Use the pigeonhole principle. If you have items to consider that can be placed in different categories with more items than categories, then at least two items must be in some one category. This fact may lead to a solution to your problem.

Section 1.5 Reasoning Mathematically

- Inductive reasoning. This type of reasoning entails drawing a general conclusion on the basis of consideration of special cases. It does not necessarily yield truth, but certainly suggest what may be true.
- Representational reasoning. Various physical, pictorial, or even mental representations often make a problem more clear and make it possible to find a solution.
- Mathematical statements. These are declarative statements that are either true or false.
- If . . . then. A statement of the form "if p is true, then q is necessarily true."
- Deductive reasoning. This type of reasoning entails drawing necessary conclusions from given information to arrive at a proof or a solution to a problem.
- Rule of direct reasoning. If p implies q, then q can be shown to be true by showing that p is true.
- Rule of indirect reasoning. If p implies q and q is false, then p is false. This is the basis for proof by contradiction.

Vocabulary and Notation

Section 1.1

Guess and check
Make a table
Look for a pattern
Draw a picture

Section 1.2

Pólya's principles
Problem-solving strategies: guess and check, make an orderly list, draw a diagram
Guess my rule

Section 1.3

Pattern
Table
Variable
Special case
Equivalent problem
Fibonacci number, F_n
Fibonacci sequence
Triangular number, t_n
Pascal's triangle
Arithmetic sequence

Section 1.4

Work backward
Eliminate possibilities
Pigeonhole principle

Section 1.5

Counterexample
Conjecture
Theorem
Statement
If p then q
Premise, hypothesis
Conclusion

Chapter Review Exercises

Sections 1.1 and 1.2

1. Standard Lumber has 8-foot and 10-foot two-by-fours. If Mr. Zimmermann bought 90 two-by-fours with a total length of 844 feet, how many were 8 feet long? Give two solutions, (a) and (b), using different strategies.
2. (a) Using each of 1, 2, 3, 4, 5, 6, 7, 8, and 9 once and only once, fill in the circles in this diagram so that the sum of the three-digit numbers formed is 999.

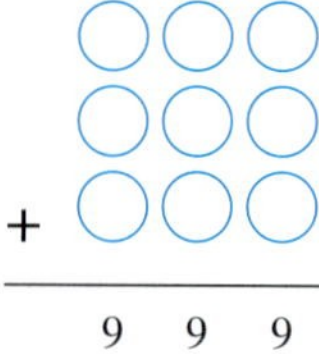

   (b) Is there more than one solution to this problem? Explain briefly.

   (c) Is there a solution to this problem with the digit 1 not in the hundreds column? Explain briefly.
3. Bill's purchases at the store cost $4.79. In how many ways can Bill receive change if he pays with a five-dollar bill?
4. How many three-letter code words can be made using the letters a, e, i, o, and u at most once each time?
5. A flower bed measuring 8 feet by 10 feet is bordered by a concrete walk 2 feet wide. What is the area of the concrete walk?
6. Karen is thinking of a number. If you double it and subtract 7, you obtain 11. What is Karen's number?
7. (a) Chanty is It in a game of *Guess My Rule.* If you give her a number, she uses her rule to determine another number. The numbers the other students gave Chanty and her responses are as shown. Can you guess her rule?

Student Input	Chanty's Responses
2	8
7	33
4	18
0	−2
3	13
⋮	⋮

   (b) Can you suggest a better strategy for the students to use in attempting to determine Chanty's rule? Explain briefly.

Section 1.3

8. Study this sequence: 2, 6, 18, 54, 162, ... Each number is obtained by multiplying the preceding number by 3. The sequences in (a) through (e) are formed in the same way, but with a different multiplier. Complete each sequence.
   (a) 3, 6, 12, ____, ____, ____
   (b) 4, ____, 16, ____, ____, ____
   (c) 1, ____, ____, 216, ____, ____
   (d) 2, ____, ____, ____, 1250, ____
   (e) 7, ____, ____, ____, ____, 7

9. Because of the high cost of living, Kimberly, Terry, and Otis each hold down two jobs, but no two have the same occupation. The occupations are doctor, engineer, teacher, lawyer, writer, and painter. Given the following information, determine the occupations of each individual.
   (a) The doctor had lunch with the teacher.
   (b) The teacher went fishing with Kimberly, who is not the writer.
   (c) The painter is related to the engineer.
   (d) The doctor hired the painter to do a job.
   (e) Terry lives next door to the writer.
   (f) Otis beat Terry and the painter at tennis.
   (g) Otis is not the doctor.

10. (a) Write down the next three rows to continue this sequence of equations.

$$2 = 1^3 + 1$$
$$4 + 6 = 2^3 + 2$$
$$8 + 10 + 12 = 3^3 + 3$$
$$\_\_\_\_\_\_ = \_\_\_$$
$$\_\_\_\_\_\_ = \_\_\_$$
$$\_\_\_\_\_\_ = \_\_\_$$

   (b) Write down the tenth row in the sequence in part (a).

11. (a) How many terms are in the arithmetic progression 7, 10, 13, 16, . . . , 79?
   (b) Compute the sum of the terms in part (a).

12. (a) A **geometric progression** is a sequence of numbers wherein each term is a constant multiple of the preceding term. Thus, 3, 6, 12, 24, . . . , 3072 is a geometric progression since each term is twice its predecessor.
   Analyzing these terms, we have

$$3 = 3,$$
$$6 = 2^1 \cdot 3,$$
$$12 = 2 \cdot 6 = 2(2^1 \cdot 3) = 2^2 \cdot 3,$$
$$24 = 2 \cdot 12 = 2(2^2 \cdot 3) = 2^3 \cdot 3,$$
$$\vdots$$

   Which term in the sequence is 3072?
   (b) Let $S = 3 + 6 + 12 + 24 + \cdots + 1536 + 3072$ denote the sum of the progression. Compute S.
   (c) Note that $2S = 6 + 12 + 24 + \cdots + 3072 + 6144$.
   (d) Note that $2S - S = S$, and use (b) and (c) to compute S a second time.

13. Compute the sum of this geometric progression:

$$5, 15, 45, \ldots, 295245$$

14. Consider a circle divided by n chords in such a way that every chord intersects every other chord interior to the circle and no three chords intersect in a common point. Complete the forthcoming table and answer these questions.
   (a) Into how many regions is the circle divided by the chords?
   (b) How many points of intersection are there?
   (c) Into how many segments do the chords divide one another?

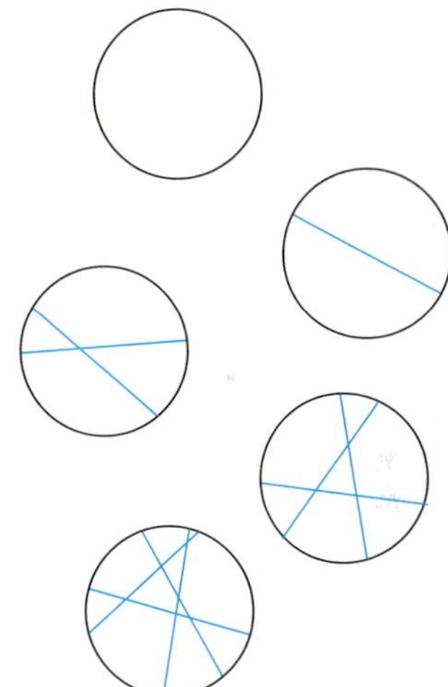

Number of Chords	Number of Regions	Number of Intersections	Number of Segments
0	1	0	0
1	2	0	1
2	4	1	4
3			
4			
5			
6			
⋮			
n			

15. Recall Pascal's triangle:

1
1 1
1 2 1
1 3 3 1
1 4 6 4 1
1 5 10 10 5 1
1 6 15 20 15 6 1
1 7 21 35 35 21 7 1
1 8 28 56 70 56 28 8 1
1 9 36 84 126 126 84 36 9 1
. . . . .

Consider the pattern of circled and squared entries in the following diagram. Compute the product of the circled entries in Pascal's triangle and the product of the squared entries for several placements of this pattern in the triangle. Does your work suggest a plausible conjecture? Explain briefly. Find other hexagonal patterns of entries with the observed property.

16. Compute the following sums associated with Pascal's triangle.

(a) $1 + 1 \cdot 2$

(b) $1 + 2 \cdot 2 + 1 \cdot 2^2$

(c) $1 + 3 \cdot 2 + 3 \cdot 2^2 + 1 \cdot 2^3$

(d) What do these sums suggest? Explain briefly.

(e) Compute the sums

$1 + 1 \cdot 3$,

$1 + 2 \cdot 3 + 1 \cdot 3^2$, and

$1 + 3 \cdot 3 + 3 \cdot 3^2 + 1 \cdot 3^3$.

(f) What do (a), (b), (c), (d), and (e) together suggest? What might you do to check your guess further? Explain briefly.

Section 1.4

17. How many cards must be drawn from a standard deck of 52 playing cards to be sure that

(a) at least two are of the same suit?

(b) at least three are of the same suit?

(c) at least two are aces?

18. How many books must you choose from among a collection of 7 mathematics books, 18 books of short stories, 12 chemistry books, and 11 physics books to be certain that you have at least 5 books of the same type?

Section 1.5

19. (a) Compute these products:

$67 \times 67 =$ ________

$667 \times 667 =$ ________

$6667 \times 6667 =$ ________

(b) Guess the result of multiplying 6,666,667 by itself. Are you sure your guess is correct? Explain in one *carefully* written sentence.

20. (a) Compute these products:

$1 \times 142{,}857 =$ ________

$2 \times 142{,}857 =$ ________

$3 \times 142{,}857 =$ ________

$4 \times 142{,}857 =$ ________

$5 \times 142{,}857 =$ ________

(b) Predict the product of 6 and 142,857. Now calculate the product and see if your prediction was correct.

(c) Predict the result of multiplying 7 times 142,857, and then compute this product.

(d) What does part (c) suggest about apparent patterns? Explain.

21. Draw three line segments l, m, and n from a common point O, and draw two triangles $\triangle ABC$ and $\triangle A'B'C'$ with corresponding vertices on l, m, and n, respectively, as shown. Let P, Q, and R be the points where the lines $\overline{AB}$ and $\overline{A'B'}$, $\overline{AC}$ and $\overline{A'C'}$, and $\overline{BC}$ and $\overline{B'C'}$, respectively, intersect. What seems to be true about P, Q, and R?

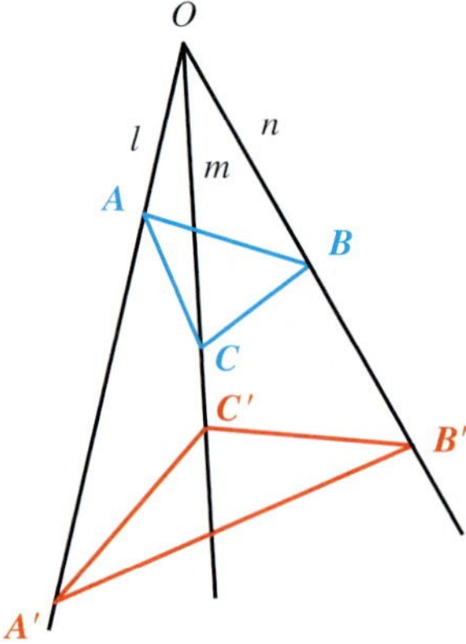

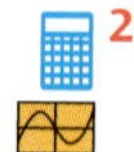

22. (a) **Collatz's Problem.** Complete this activity in the illustration at the top of the next page. This can be done easily on a calculator, but, if you use one, be sure to write down all the steps as shown here. Try this for a number of different starting values. Do you think the process will always stop?

(b) How many steps are required if you start with 9?

Figure for Problem 22

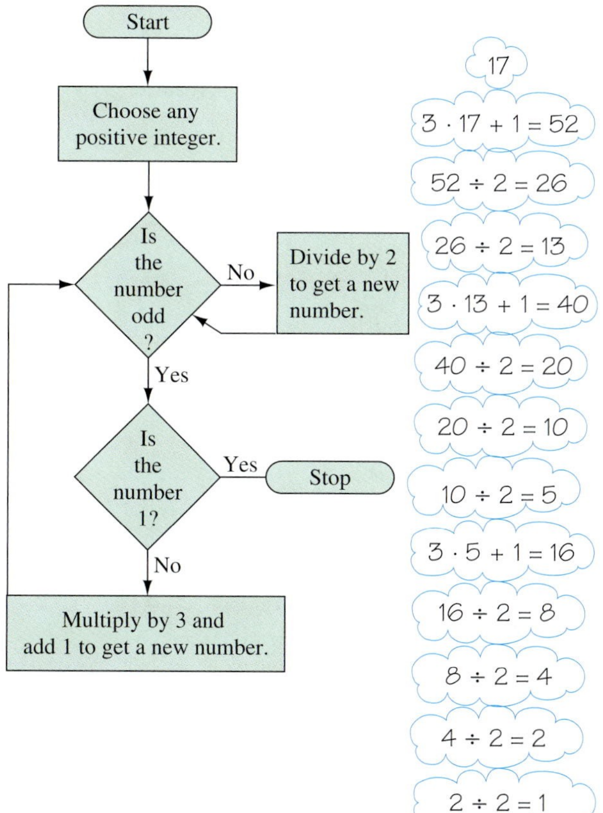

23. In Example 1.15, we used representations to show that the square of an odd number is one more than a multiple of eight. Use the fact from Example 1.15 that, if n is odd, then $n = 2s + 1$ for some whole number s and $n^2 = 4(s^2 + s) + 1$. Note that $s^2 + s = s(s + 1) = 2\frac{s(s + 1)}{2}$. Argue that $\frac{s(s + 1)}{2}$ is a whole number and hence that $n^2 = 8q + 1$ for some whole number q—i.e., that, again, n^2 is one larger than a multiple of eight.

24. In this section, we saw that the sum of the digits of a multiple of three is also a multiple of three. Is the same thing true of multiples of six? Why or why not?

Chapter Test

1. If five pigeons are placed into two pigeonholes, what is the least possible number of pigeons in the pigeonhole with the greatest number of pigeons?

2. Perform these multiplications as quickly as possible:

$2345679 \times 9 =$ _____

$1345679 \times 9 =$ _____

$1245679 \times 9 =$ _____

$1235679 \times 9 =$ _____

$1234679 \times 9 =$ _____

$1234579 \times 9 =$ _____

$1234569 \times 9 =$ _____

$1234568 \times 9 =$ _____

3. Consider the arithmetic progression whose first four terms are 1, 3, 5, and 7.

(a) Determine the one-hundredth term in this progression.

(b) Determine the sum of the first one hundred terms in this progression.

4. (a) Write the next two lines in this sequence of equations.

$$2 = 2 = 0^2 + 2 \cdot 1^2$$
$$2 + 5 + 2 = 9 = 1^2 + 2 \cdot 2^2$$
$$2 + 5 + 8 + 5 + 2 = 22 = 2^2 + 2 \cdot 3^2$$

__________________ = __________

__________________ = __________

(b) Write the tenth line in the sequence of part (a).

5. A frog is in a well 12 feet deep. Each day he climbs up 3 feet, and each night he slips back 2 feet. How many days will it take the frog to get out of the well?

6. Determine the sum of the elements in the first 10 rows in Pascal's triangle—i.e., rows zero through nine.

7. While three watchmen were guarding an orchard, a thief slipped in and stole some apples. On his way out, he met the three watchmen, one after another, and to each in turn he gave half the apples he had and two besides. In this way he managed to escape with one apple. How many had he stolen originally?

8. Consider the following equations:

$$1 = 0 + 1 = 1 - 0;$$

$$2 + 3 + 4 = 1 + 8 = 9 - 0;$$

$$5 + 6 + 7 + 8 + 9 = 8 + 27 = 36 - 1;$$

$$10 + 11 + 12 + 13 + 14 + 15 + 16 = 27 + 64 = 100 - 9.$$

(a) Continue this sequence for two more equations.

(b) What is the tenth row in the sequence?

(c) What is the nth row in the sequence?

9. (a) Make a magic square using each of the numbers 2, 7, 12, 27, 32, 37, 52, 57, and 62 once, and only once.

(b) Make a magic subtraction square using each of the numbers in part (a) once, and only once.

10. Note that

$$S_1 = \frac{1}{1 \cdot 2} = \frac{1}{2},$$

$$S_2 = \frac{1}{1 \cdot 2} + \frac{1}{2 \cdot 3} = \frac{2}{3},$$

$$S_3 = \frac{1}{1 \cdot 2} + \frac{1}{2 \cdot 3} + \frac{1}{3 \cdot 4} = \frac{3}{4},$$

$$\ldots.$$

Use inductive reasoning to show that

$$S_n = \frac{1}{1 \cdot 2} + \frac{1}{2 \cdot 3} + \cdots + \frac{1}{n(n+1)} = \frac{n}{n+1}.$$

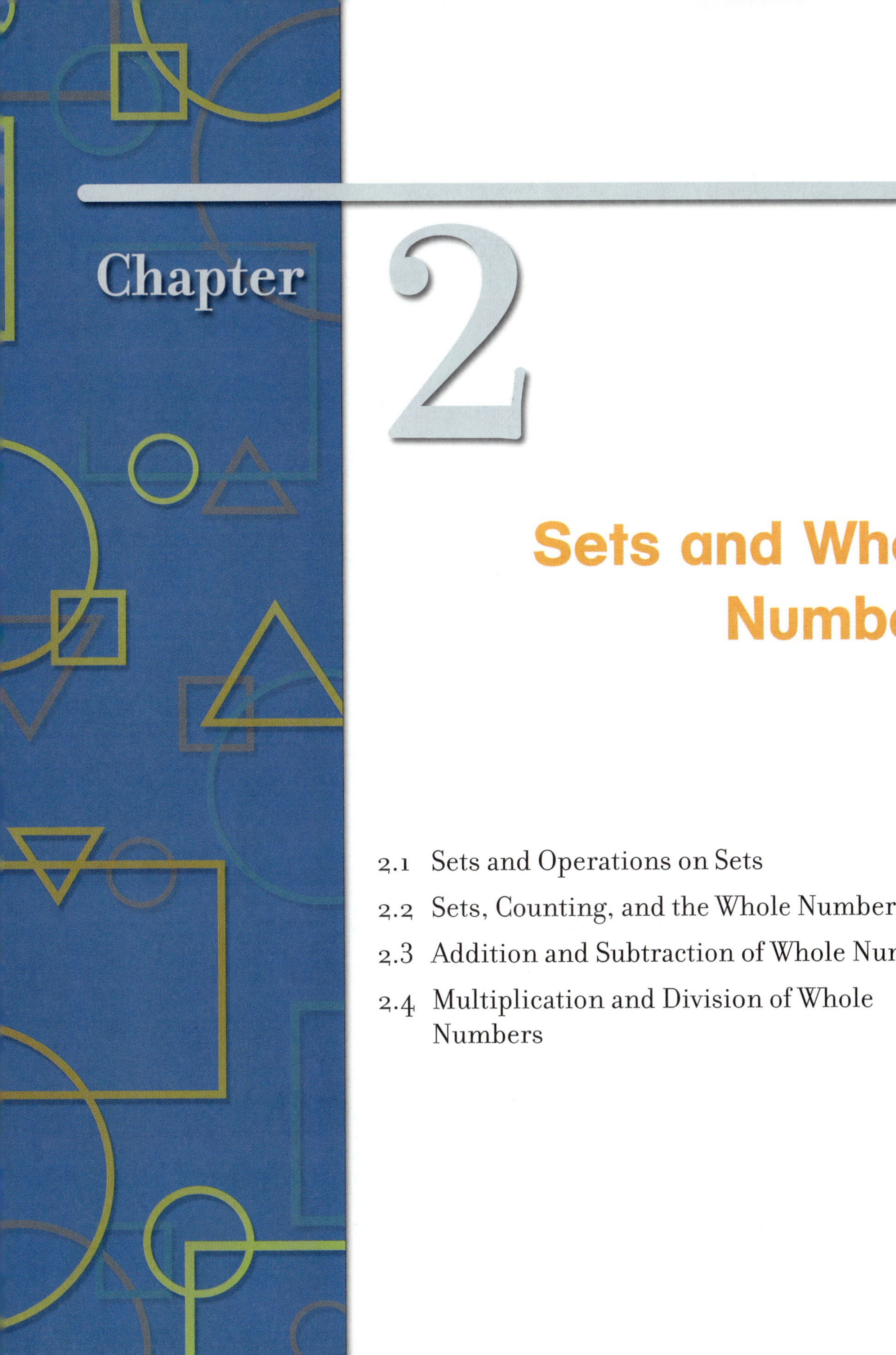

Chapter 2

Sets and Whole Numbers

Attribute Tic-Tac-Toe

Materials Needed

A set of 12 attribute pieces, defined by the properties of shape (hexagon, triangle, circle), color (blue, red), and size (large, small).

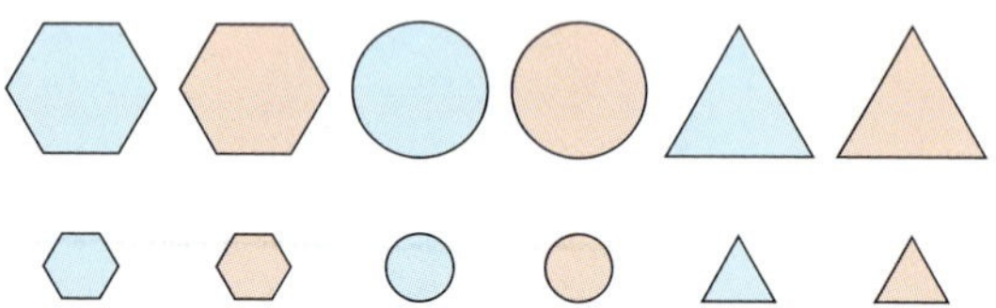

The game board is a 3-by-4 squared rectangle, with squares large enough to contain the attribute pieces.

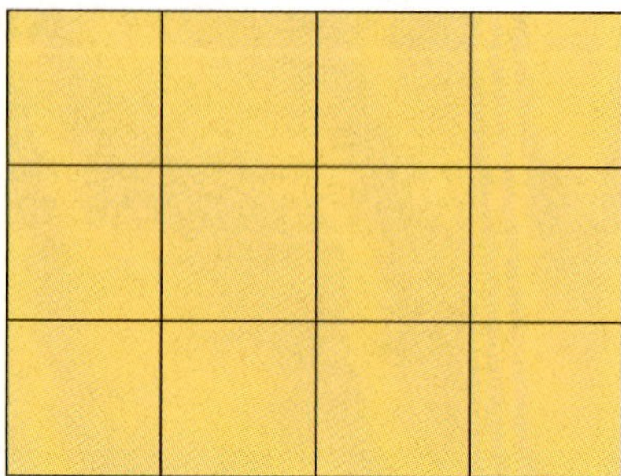

Directions

Attribute Tic-Tac-Toe is similar to tic-tac-toe, except that the two opposing players alternate choosing which attribute piece is played on any open square of the game board. The object of the game is to be the first player to complete a row, column, or diagonal with pieces sharing a common attribute. In the example shown here, the large red triangle can be played to create a diagonal of three pieces sharing the triangle shape attribute. There are four diagonals to consider on the game board. Play several games, and then try one of the following variations.

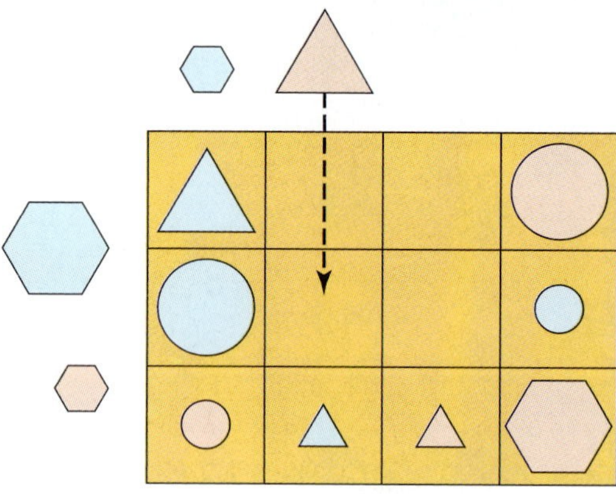

Variation 1. Play as before, but with this twist on the rules: your *opponent* chooses the attribute piece you are to place on the game board.

Variation 2 (Avoidance Attribute Tic-Tac-Toe). Play with the original rules, but try to *avoid* creating a common attribute of the pieces in a complete row, column, or diagonal. Your opponent wins by pointing out any common attribute you have made, but must do so before making another play. Is it possible for the game to end in a tie?

Connections From Sets to the Properties and Arithmetic of Whole Numbers

When you can measure what you are speaking about and express it in numbers, you know something about it; but when you cannot measure it, when you cannot express it in numbers, your knowledge is of a meager and unsatisfactory kind.

—SIR WILLIAM THOMSON (LORD KELVIN)*

In this chapter, we introduce the concept of **sets.** Sets were first introduced in the late nineteenth century and therefore are a quite recent addition to mathematical thought. Even so, sets play several important roles in mathematics and even in other sciences, as we will soon discover. We will see that sets provide a framework of organization, in which objects

**Popular Lectures and Addresses* (New York: Macmillan and Co., 1891, 1894).

HIGHLIGHT *from* HISTORY

The Early Origins of Mathematics

"The history of mathematics should really be the kernel of the history of culture."
—GEORGE SARTON

The transition from gathering food to producing food occurred some 10,000 years ago, marking the change from the Paleolithic to the Neolithic age. Agriculture required farmers to stay in one place for long periods, so people designed and constructed permanent dwellings. This change gave rise to problems dealing with quantity and form: laying out fields, tending flocks, measuring the amount of grain that should be stored over the winter, and knowing whether excess grain could be traded to neighboring villages. Thus, the two important branches of mathematics—number and geometry—have origins concurrent with the dawn of civilization.

Until about 600 B.C., mathematics was pursued primarily for its practical, decorative, and religious values. Problems in land apportionment, interest payments, and tax rates required solution for the development of commerce. Many problems of a mathematical nature were successfully solved during the construction of irrigation canals, temples, and pyramids. Interest in astronomy grew out of the need for calendars sufficiently accurate to forecast flood and growing seasons. Number systems and notations were developed, and even some empirically derived formulas from algebra and geometry were known. However, little use was made of symbolism, scant attention was given to abstraction and general methods, and nowhere was the notion of proof or even informal justification to be found.

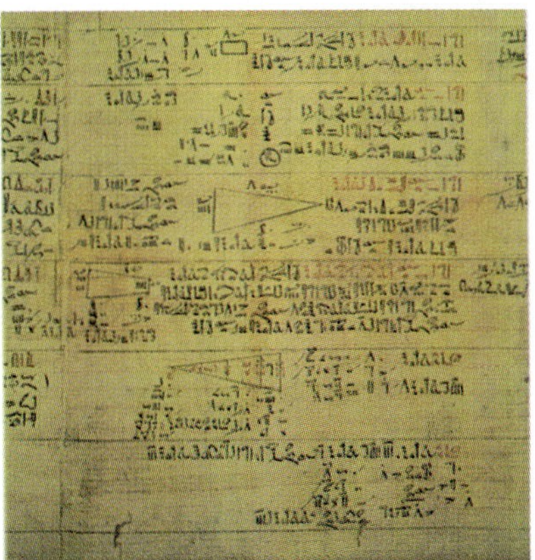

A portion of the Rhind papyrus that discusses the measurement of the area of a triangle and the slopes of pyramids. The papyrus, which contains 85 mathematical problems, was copied c. 1575 B.C. by the scribe Ahmes from a work written almost three centuries earlier.

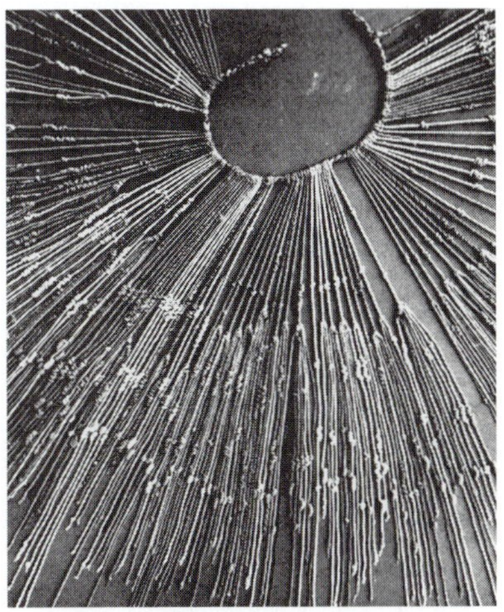

The Incas recorded and communicated quantitative data with quipus. A quipu is an assemblage of cords, with numerical values determined by the colors of the cords, the way the cords are spaced and interconnected, and the type and placement of knots tied in the cords. The quipu shown is in the Museo Nacional de Antropologia y Arqueologia, Lima, Peru.

can be grouped according to characteristic properties that either admit or exclude objects for membership in a given set. Of special interest is the number of objects within a set, giving rise to the **whole numbers,** 0, 1, 2, 3, ... The relations between and operations performed on sets give meaning to the ordering of the whole numbers and form the basis of the fundamental arithmetic operations applied to the whole numbers: addition, subtraction, multiplication, and division.

2.1 SETS AND OPERATIONS ON SETS

The notion of a set originated with the German mathematician Georg Cantor (1845–1918) in the last half of the nineteenth century. Sets have now become indispensable to nearly every branch of mathematics. Sets make it possible to organize, classify, describe, and communicate. For example, each number system—the whole numbers, the integers, the *from* The NCTM Principles and Standards

Standards for Grades 3–5

Most students enter grade 3 with enthusiasm for, and interest in, learning mathematics. In fact, nearly three-quarters of U.S. fourth graders report liking mathematics (Silver, Strutchens, and Zawojewski 1997). They find it practical and believe that what they are learning is important. If the mathematics studied in grades 3–5 is interesting and understandable, the increasingly sophisticated mathematical ideas at this level can maintain students' engagement and enthusiasm. But if their learning becomes a process of simply mimicking and memorizing, they can soon begin to lose interest. Instruction at this level must be active and intellectually stimulating and must help students make sense of mathematics.

SOURCE: Reprinted with permission from Principles and Standards for School Mathematics, *page 143.*

rational numbers, and the real numbers—is best viewed as a set together with a list of the operations and properties that the numbers in the system possess.

Intuitively, a set is a collection of objects. An object that belongs to the collection is called an **element** or **member** of the set. Words like *collection, family,* and *class* are frequently used interchangeably with "set." In fact, a set of dishes, a collection of stamps, and the class of 2008 at Harvard University are all examples of sets.

Cantor requires that a set be **well defined.** This means two things. First, there is a **universe** of objects that are allowed into consideration. Second, any object in the universe is either an element of the set or is not an element of the set. For example, "the first few presidents of the United States" is not a well-defined set, since "few" is a matter of varying opinion. On the other hand, "the first three presidents of the United States" does provide an adequate verbal description of a set, with the understanding that the universe is all people who have ever lived.

There are three ways to define a set:

Word Description: The set of the first three presidents of the United States
Listing in Braces: {George Washington, Thomas Jefferson, John Adams}
Set Builder Notation: $\{x \mid x$ is one of the first three presidents of the United States$\}$

The order in which elements in a set are listed is arbitrary, so listing Jefferson, the third president, before Adams is permissible. However, each element should be listed just once. In set builder notation, $\{x \mid x \ldots\}$ is read "the set of all x such that x is. . . ." The letter x used in set builder notation can be replaced with any convenient letter. The letter x is a variable representing any member of the universe, and the set contains exactly those members of the universe which meet the defining criteria that are listed to the right of the vertical bar. In some books, a colon replaces the vertical bar.

Capital letters $A, B, C, \ldots$ are generally used to denote sets. Membership is symbolized by $\in$, so that if P designates the above set of three U.S. presidents, then John Adams $\in P$, where $\in$ is read "is a member of." The symbol $\notin$ is read "is not a member of," so James Monroe $\notin P$. It is sometimes useful to choose letters that suggest the set being designated. For example, the set of **natural,** or **counting,** numbers will be written

$$N = \{1, 2, 3, \ldots\},$$

where the ellipsis ". . ." indicates "and so on."

HIGHLIGHT from HISTORY

Georg Cantor (1845–1918)

Georg Cantor was born in St. Petersburg, Russia, but at age 12 moved with his family to Germany. Cantor excelled in mathematics, and in 1867 completed his doctorate from the prestigious University of Berlin. His research soon led him to sets and the comparison of the size of various infinite sets. Today, this work is regarded as fundamental to mathematical thought, but at the time of its development it generated considerable controversy. In particular, Leopold Kronecker exhibited an unpleasant animosity toward Cantor and his work. This stifled Cantor's dream of a professorship at the University of Berlin, and may also have contributed to the onset of the mental breakdowns that plagued Cantor from age 40 until his death at age 73 in a mental hospital. Eventually, Cantor's work was widely recognized and appreciated. The mathematician David Hilbert proclaimed that "no one shall expel us from the paradise which Cantor has created for us."

EXAMPLE 2.1 DESCRIBING SETS

Each set below is taken from the universe N of the natural numbers, and has been described either in words, by listing in braces, or with set builder notation. Provide the two remaining types of description for each set.

(a) The set of natural numbers greater than 12 and less than 17.
(b) $\{x \mid x = 2n \text{ and } n = 1, 2, 3, 4, 5\}$
(c) $\{3, 6, 9, 12, \ldots\}$
(d) The set of the first 10 odd natural numbers.
(e) $\{1, 3, 5, 7, \ldots\}$
(f) $\{x \mid x = n^2 \text{ and } n \in N\}$

Solution

(a) $\{13, 14, 15, 16\}$
$\{n \mid n \in N \text{ and } 12 < n < 17\}$
(b) $\{2, 4, 6, 8, 10\}$
The set of the first five even natural numbers.
(c) The set of all natural numbers that are multiples of 3.
$\{x \mid x = 3n \text{ and } n \in N\}$
(d) $\{1, 3, 5, 7, 9, 11, 13, 15, 17, 19\}$
$\{x \mid x = 2n - 1 \text{ and } n = 1, 2, \ldots, 10\}$
(e) The set of the odd natural numbers.
$\{x \mid x = 2n - 1 \text{ and } n \in N\}$
(f) $\{1, 4, 9, 16, 25, \ldots\}$
The set of the squares of the natural numbers.

Venn Diagrams

Sets can be represented pictorially by **Venn diagrams,** named for the English logician John Venn (1834–1923). The universal set, which we denote by U, is represented by a rectangle. Any set within the universe is represented by a closed loop lying within the rectangle. The

A Venn diagram is an example of how a representation can aid the teaching and understanding of a concept. The Representation Standard is discussed in detail in the NCTM's *Principles and Standards for School Mathematics.*

region inside the loop is associated with the elements in the set. An example is given in Figure 2.1, which shows the Venn diagram for the set of vowels $V = \{a, e, i, o, u\}$ in the universe $U = \{a, b, c, \ldots, z\}$.

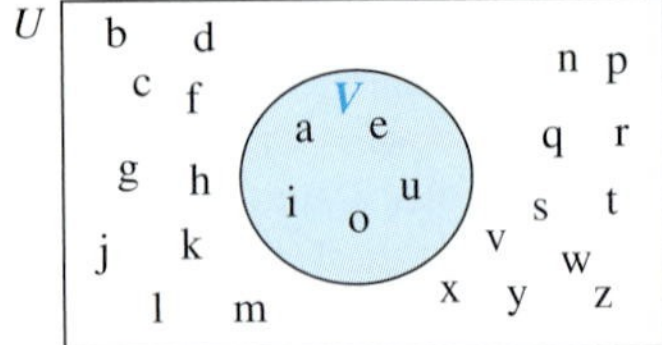

Figure 2.1
The Venn diagram showing the set of vowels in the universe of the 26-letter alphabet

Coloring, shading, or cross-hatching are also useful devices to distinguish the sets in a Venn diagram. For example, set A is blue in Figure 2.2, and the set of elements of the universe that do not belong to A is colored red.

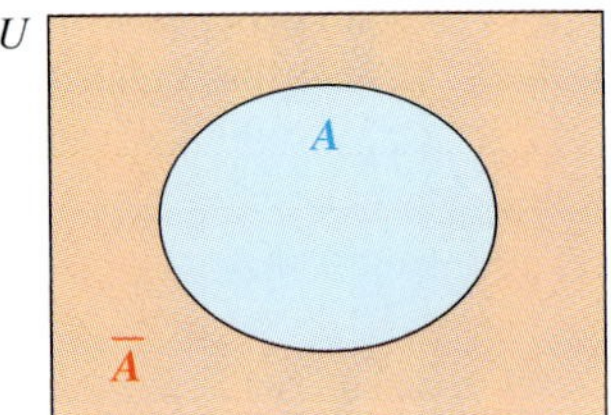

Figure 2.2
The Venn diagram of set A (shown in blue), its complement $\overline{A}$ (shown in red), and the universal set U (the region within the rectangle)

The elements in the universe that are not in set A form a set called the **complement** of A, which is written $\overline{A}$.

DEFINITION *The Complement of a Set A*
The **complement of set A,** written $\overline{A}$, is the set of elements in the universal set U that are not elements of A. That is,

$$\overline{A} = \{x \mid x \in U \text{ and } x \notin A\}.$$

EXAMPLE 2.2 FINDING SET COMPLEMENTS

Let $U = N = \{1, 2, 3, \ldots\}$ be the set of natural numbers. For the sets E and F below, find the complementary sets $\overline{E}$ and $\overline{F}$.

(a) $E = \{2, 4, 6, \ldots\}$
(b) $F = \{n \mid n > 10\}$

Solution

(a) $\overline{E} = \{1, 3, 5, \ldots\}$. That is, the complement of the set E of even natural numbers is the set $\overline{E}$ of odd natural numbers.
(b) $\overline{F} = \{1, 2, 3, 4, 5, 6, 7, 8, 9, 10\}$.

Relationships and Operations on Sets

Consider several sets, labeled $A, B, C, D, \ldots$, whose members all belong to the same universal set U. It is useful to understand how sets may be related to one another, and how two or more sets can be used to define new sets.

DEFINITION *Subset*
The set A is a **subset** of B, written $A \subseteq B$, if, and only if, every element of A is also an element of B.

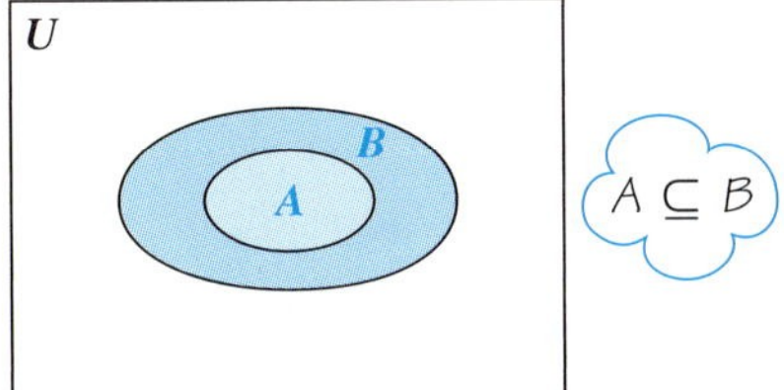

Figure 2.3
A Venn diagram when it is known that A is a subset of B

If A is a subset of B, every element of A also belongs to B. In this case it is useful in the Venn diagram to place the loop representing set A within the loop representing set B, as shown in Figure 2.3.

If two sets A and B have precisely the same elements, then they are **equal** and we write $A = B$. If $A \subseteq B$, but $A \neq B$, we say that A is a **proper subset** of B and write $A \subset B$. If $A \subset B$, there must be some element of B that is not also an element of A; that is, there is some x for which $x \in B$ and $x \notin A$.

DEFINITION *Intersection of Sets*
The **intersection** of two sets A and B, written $A \cap B$, is the set of elements common to both A and B. That is,

$$A \cap B = \{x \mid x \in A \text{ and } x \in B\}.$$

For example, $\{a, b, c, d\} \cap \{a, d, e, f\} = \{a, d\}$. The symbol $\cap$ is a special mathematical symbol called a **cap.**

Two sets with no element in common are said to be **disjoint.** The intersection of disjoint sets is then the set with no members, which is called the **empty set.** The empty set is given the special mathematical symbol $\varnothing$, which should not be mistaken for the Greek letter ϕ (phi). In symbols, two sets C and D are disjoint if, and only if, $C \cap D = \varnothing$. See Figure 2.4.

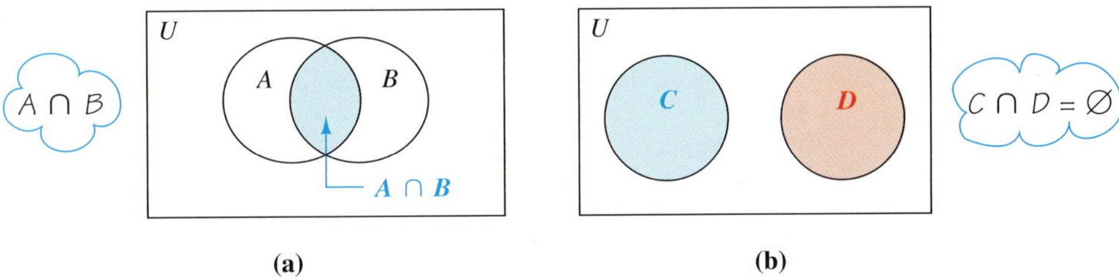

Figure 2.4
(a) *A Venn diagram whose shaded region shows the intersection A ∩ B*
(b) *A Venn diagram for sets C and D that are disjoint. That is, C ∩ D = ∅*

DEFINITION *Union of Sets*

The **union** of sets A and B, written $A \cup B$, is the set of all elements that are in A or B. That is,

$$A \cup B = \{x \mid x \in A \text{ or } x \in B\}.$$

For example, if $A = \{a, e, i, o, u\}$ and $B = \{a, b, c, d, e\}$, then $A \cup B = \{a, b, c, d, e, i, o, u\}$. Elements such as "a" and "e" that belong to both A and B are listed just once in $A \cup B$. The word *or* in the definition of union is used in the inclusive sense of "and/or." The symbol for union is the **cup,** $\cup$. The cup symbol must be carefully distinguished from the letter U used to denote the universal set. See Figure 2.5.

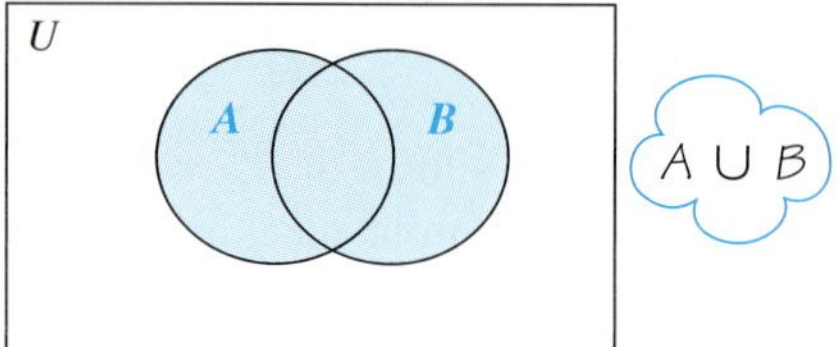

Figure 2.5
The shaded region corresponds to the union, A ∪ B, *of sets* A *and* B.

EXAMPLE 2.3 PERFORMING OPERATIONS ON SETS

Let $U = \{p, q, r, s, t, u, v, w, x, y\}$ be the universe, and let $A = \{p, q, r\}$, $B = \{q, r, s, t, u\}$, and $C = \{r, u, w, y\}$. Locate all 10 elements of U in a three-loop Venn diagram, and then find the following sets:

(a) $A \cup C$ **(b)** $A \cap C$ **(c)** $A \cup B$ **(d)** $A \cap B$
(e) $\overline{B}$ **(f)** $\overline{C}$ **(g)** $A \cup \overline{B}$ **(h)** $A \cap \overline{C}$

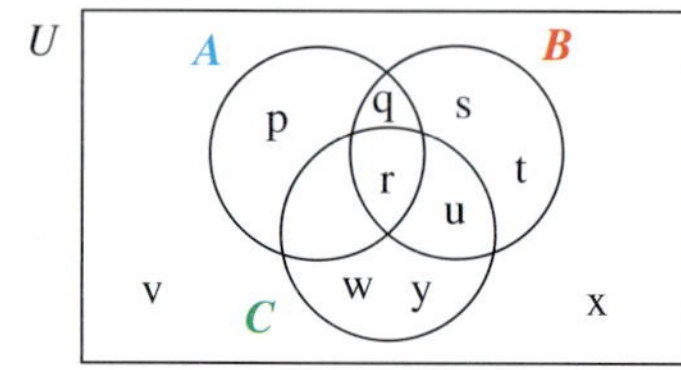

Solution

(a) $A \cup C = \{p, q, r, u, w, y\}$ **(b)** $A \cap C = \{r\}$
(c) $A \cup B = \{p, q, r, s, t, u\}$ **(d)** $A \cap B = \{q, r\}$
(e) $\overline{B} = \{p, v, w, x, y\}$ **(f)** $\overline{C} = \{p, q, s, t, v, x\}$
(g) $A \cup \overline{B} = \{p, q, r, v, w, x, y\}$ **(h)** $A \cap \overline{C} = \{p, q\}$

Using Sets for Problem Solving

The notions of sets and their operations are often used to understand a problem and communicate its solution. Moreover, Venn diagrams provide a visual representation useful for understanding and communication.

EXAMPLE 2.4 USING SETS TO SOLVE A PROBLEM IN COLOR GRAPHICS

The cathode-ray tube (CRT) on a color monitor uses three types of phosphors, each of which, when excited by an electron beam, produces one of three colors—red, blue, or green. By exciting different combinations of phosphors, a wider range of colors is possible. Use a Venn diagram to show what combinations are possible.

Solution

Introduce three loops in a Venn diagram—one for each of the component colors. As shown below, eight colors can be achieved. Most computers also allow the intensity of each color to be specified. In this way, many more colors can be obtained.

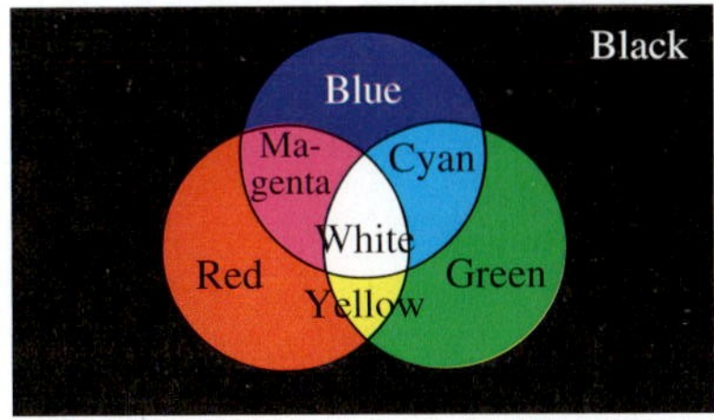

Notice that if red and green are excited, also including blue will produce white. Similarly, if red is excited, including both green and blue will again produce white. In symbols, this becomes

$$(R \cap G) \cap B = R \cap (G \cap B).$$

Since the result is independent of where the parentheses are placed, this illustrates what is called the **associative property** of the intersection operation. Since the order in which the intersections are performed has no effect on the outcome, we do not need parentheses and it is meaningful to write $R \cap G \cap B$. Similarly, union is associative, and it is meaningful to write $R \cup G \cup B$ without parentheses.

Into the Classroom

Sets Education for Elementary School Children

Sets provide a theoretical basis for defining the whole numbers. The operations and properties of the system of whole numbers can be defined and understood through their connection with corresponding operations and properties of sets. However, it is not appropriate to use *abstract* notions of set theory to teach young children. Instead, the teacher should use activities with physical collections of objects to give concrete, hands-on experiences with set ideas and associated number concepts. It is critically important to use an approach that has meaning in the child's world. Colored chips, attribute pieces, tiles, beans, and so on all become useful concrete embodiments of elements that may be organized and classified into sets—sets that can be seen, touched, and manipulated. The older child can later move successfully to pictorial representation. Later still, the student will become comfortable dealing with abstract models, represented entirely in words and symbols.

The associative property is just one example of a number of useful properties that hold for set operations and relations. The properties listed in the following theorem can be proved by reasoning directly from the definitions given earlier. They can also be justified by considering appropriately shaded Venn diagrams.

THEOREM *Properties of Set Operations and Relations*

1. Transitivity of inclusion

$$\text{If } A \subseteq B \text{ and } B \subseteq C, \text{ then } A \subseteq C.$$

2. Commutativity of union and intersection

$$A \cup B = B \cup A$$
$$A \cap B = B \cap A$$

3. Associativity of union and intersection

$$A \cup (B \cup C) = (A \cup B) \cup C$$
$$A \cap (B \cap C) = (A \cap B) \cap C$$

4. Properties of the empty set

$$A \cup \varnothing = \varnothing \cup A = A$$
$$A \cap \varnothing = \varnothing \cap A = \varnothing$$

5. Distributive properties of union and intersection

$$A \cap (B \cup C) = (A \cap B) \cup (A \cap C)$$
$$A \cup (B \cap C) = (A \cup B) \cap (A \cup C)$$

EXAMPLE 2.5 VERIFYING PROPERTIES WITH VENN DIAGRAMS

(a) Verify the distributive property $A \cap (B \cup C) = (A \cap B) \cup (A \cap C)$.
(b) Show that $A \cup B \cap C$ is not meaningful without parentheses.

Solution

(a) To shade the region corresponding to $A \cap (B \cup C)$, we intersect the A loop with the loop formed by the overlapping circles of $B \cup C$.

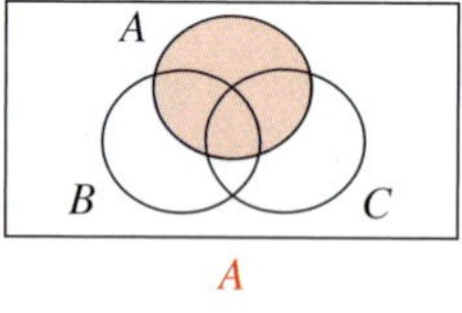

A $\cap$

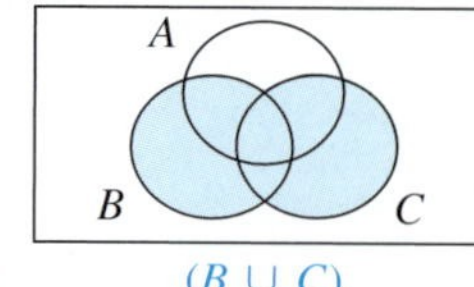

$(B \cup C)$ $=$

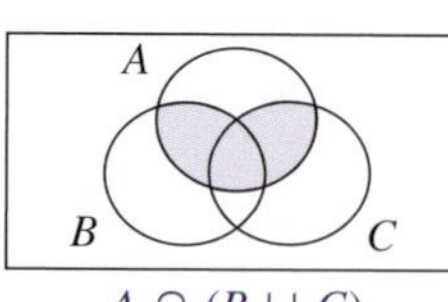

$A \cap (B \cup C)$

The shaded regions are the same.

Next, we combine the two almond-shaped regions of $A \cap B$ and $A \cap C$ to form $(A \cap B) \cup (A \cap C)$.

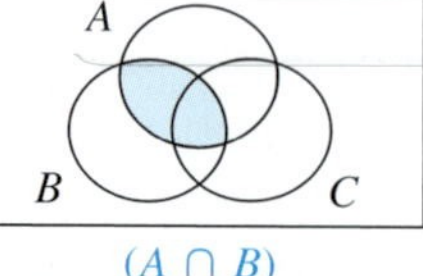

$(A \cap B)$ $\cup$

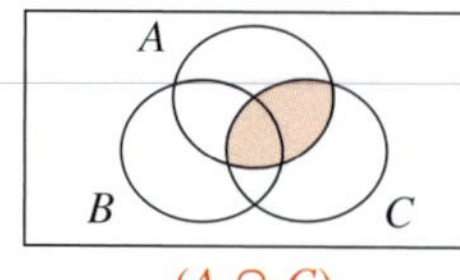

$(A \cap C)$ $=$

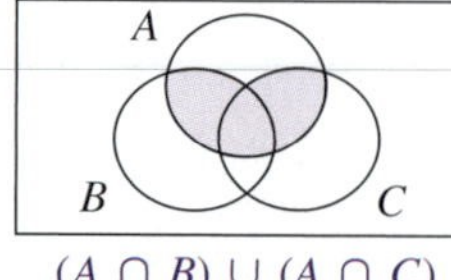

$(A \cap B) \cup (A \cap C)$

The shaded regions at the right agree, which verifies that intersection distributes over union; that is,

$$A \cap (B \cup C) = (A \cap B) \cup (A \cap C).$$

(b) If the union is taken first, we have the expression $(A \cup B) \cap C$, corresponding to the region shaded on the left in the following figure. If the intersection is taken first, we have the expression $A \cup (B \cap C)$, corresponding to the region shaded in the Venn diagram on the right. The shaded regions are different, so in general $(A \cup B) \cap C \neq A \cup (B \cap C)$.

$(A \cup B) \cap C$ $A \cup (B \cap C)$

We can also consider the explicit case $A = \{a\}$, $B = \{a, b\}$, $C = \{b, c\}$, for which $(A \cup B) \cap C = \{b\}$ and $A \cup (B \cap C) = \{a, b\}$. We see that $(A \cup B) \cap C \neq A \cup (B \cap C)$, since $\{b\} \neq \{a, b\}$. Such an example that shows that a statement is false is called a **counterexample.**

Problem Set 2.1

Understanding Concepts

Problem numbers in color indicate that the answer to this problem can be found at the back of the book.

1. Write the following sets by listing their elements.
   - **(a)** The set of states in the United States that border Nevada.
   - **(b)** The set of states in the United States whose names begin with the letter *M*.
   - **(c)** The set of states in the United States whose names contain the letter *Z*.
2. Write the following sets.
   - **(a)** The set of letters used in the sentence "List the elements in a set only once."
   - **(b)** The set of letters that are needed to spell these words: *team, meat, mate,* and *tame.*
3. Let $U = \{1, 2, 3, \ldots, 20\}$. Write these sets by listing the elements in braces.
   - **(a)** $\{x \in U \mid 7 < x \leq 12\}$
   - **(b)** $\{x \in U \mid x \text{ is odd and } 4 \leq x \leq 13\}$
   - **(c)** $\{x \in U \mid x \text{ is divisible by } 3\}$
   - **(d)** $\{x \in U \mid x = 3n \text{ for some } n \in N\}$
   - **(e)** $\{x \in U \mid x = n^2 \text{ for some } n \in N\}$
4. Write these sets in set builder notation, where $U = \{1, 2, \ldots, 20\}$.
   - **(a)** $\{11, 12, 13, 14\}$
   - **(b)** $\{6, 8, 10, 12, 16\}$
   - **(c)** $\{4, 8, 12, 16, 20\}$
   - **(d)** $\{2, 5, 10, 17\}$
5. Use set builder notation to write the following subsets of the natural numbers.
   - **(a)** The even natural numbers larger than 12.
   - **(b)** The squares of the odd numbers larger than or equal to 25.
   - **(c)** The natural numbers divisible by 3.

6. Discuss whether or not the following groups of elements describe well-defined sets.

(a) Moscow, Lima, Paris, Duluth

(b) Moscow, ID; Lima, MT; Paris, TX; Duluth, GA

(c) The smart students in my math class.

(d) The students in my class with a 3.5 GPA.

Trucker rolls to right city, but wrong state

DULUTH, Minn. (AP)

A truck driver made a longer trip than he bar–gained for when he ended up in the right town – but the wrong state.

SOURCE: "Trucker rolls to right city, but wrong state," The Daily Evergreen, *Vol. 97, No. 92, January 21, 1991. Reprinted by permission of* The Daily Evergreen.

7. Decide if the following set relationships are *true* or *false*.

(a) $\{s, c, r, a, m, b, l, e, d\} = \{a, b, c, d, e, l, m, r, s\}$

(b) $\{2\} \subset \{1, 2, 3\}$

(c) $\{1, 2, 3\} = \{3, 1, 2\}$

(d) $\{2\} \subseteq \{1, 2, 3\}$

(e) $\{1, 2, 3\} \subseteq \{3, 2, 1\}$

(f) $\{3\} \subset \{2\}$

8. Let $U = \{a, b, c, d, e, f, g, h\}$, $A = \{a, b, c, d, e\}$, $B = \{a, b, c\}$, and $C = \{a, b, h\}$. Locate all eight elements of U in a three-loop Venn diagram, and then list the elements in the following sets.

(a) $B \cup C$ **(b)** $A \cap B$ **(c)** $B \cap C$

(d) $A \cup B$ **(e)** $\overline{A}$ **(f)** $A \cap C$

(g) $A \cup (B \cap C)$

9. Let $L = \{6, 12, 18, 24, \ldots\}$ be the set of multiples of 6 and let $M = \{45, 90, 135, \ldots\}$ be the set of multiples of 45.

(a) Use your calculator to find four more elements of set M.

(b) Describe $L \cap M$.

(c) What is the smallest element in $L \cap M$?

10. Let $G = \{n \mid n \text{ divides } 90\}$ and $D = \{n \mid n \text{ divides } 144\}$. In listed form, $G = \{1, 2, 3, 5, 6, 9, 10, 15, 18, 30, 45, 90\}$.

(a) Find the listed form of the set D.

(b) Find $G \cap D$.

(c) Which element of $G \cap D$ is largest?

11. Draw and shade Venn diagrams that correspond to the following sets.

(a) $A \cap B \cap C$ **(b)** $A \cup (B \cap \overline{C})$

(c) $(A \cap B) \cup C$ **(d)** $\overline{A} \cup (B \cap C)$

(e) $A \cup B \cup C$ **(f)** $\overline{A} \cap B \cap C$

12. For each part, draw a Venn diagram whose loops for sets A, B, and C show that the conditions listed must hold.

(a) $A \subseteq C, B \subseteq C, A \cap B = \varnothing$

(b) $C \subseteq (A \cap B)$

(c) $(A \cap B) \subseteq C$

13. If $A \cup B = A \cup C$, is it necessarily true that $B = C$? Give a proof, or provide a counterexample.

14. Let U be the set of natural numbers 1 through 20, let A be the set of even numbers in U, and let B be the set of numbers in U that are divisible by 3.

(a) Find $\overline{A \cap B}$, $\overline{A} \cup \overline{B}$, $\overline{A \cup B}$, and $\overline{A} \cap \overline{B}$.

(b) What two pairs of the four sets of part (a) are equal?

15. The 12 shapes in the Venn diagram below are described by the following attributes.

Shape: circle, hexagon, or triangle

Size: small or large

Color: red or blue

Let C, H, and T denote the respective sets of circular, hexagonal, and triangular shapes. Similarly, let S, L, R, and B denote the sets of small, large, red, and blue shapes, respectively. The set A contains the shapes that are small and not a triangle, so $A = S \cap \overline{T}$.

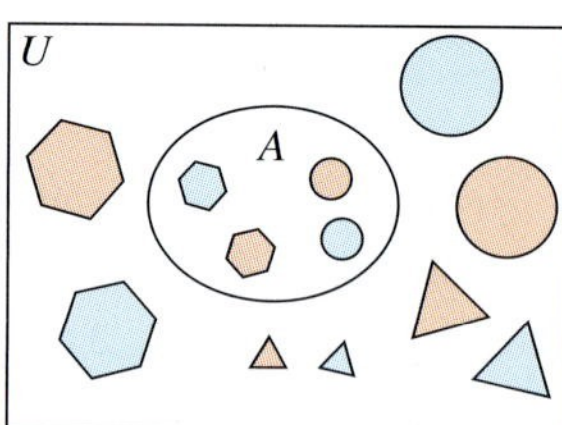

Describe which shapes are found in these sets.

(a) $R \cap C$ **(b)** $L \cap H$ **(c)** $T \cup H$

(d) $L \cap T$ **(e)** $B \cap \overline{C}$ **(f)** $H \cap S \cap R$

16. Let C, H, T, S, L, R, and B denote the sets of attribute shapes described in problem 15. The set of small hexagons can be written $S \cap H$ in symbolic form. Express these sets in symbolic form.

(a) The large triangles.

(b) The blue polygonal (that is, noncircular) shapes.

(c) The small or triangular shapes.

(d) The shapes that are either blue circles or are red.

Teaching Concepts

17. A large group activity can effectively teach set concepts to children. For example, let the classroom be the universe of the students. Use lengths of rope (about 30 feet, adjusting for the size of the class) to form Venn diagram loops on the floor. Place signs in the loops to identify the sets being considered. Once the signs are placed, have the children stand in the appropriate loop or overlapped regions of two or more loops, depending on the defining property of the set. For example, if loop A corresponds to the girls in the class, then the girls should stand inside the loop and the boys should stand outside the loop. Now create other interesting classroom Venn diagram activities.
   - **(a)** Define sets B and C for which B is a subset of C.
   - **(b)** Define sets D and E that are disjoint.
   - **(c)** Define sets F, G, and H where you believe there is a good likelihood that students will be found in all eight regions of the classroom Venn diagram.

18. A student is told that $A \cup B$ is the set of all elements that belong to A or to B. However, when asked to find the union of $A = \{a, b, c, d\}$ and $B = \{c, d, e, f, g, h\}$, the student says that the answer is $\{a, b, e, f, g, h\}$. How is the word "or" in the definition of set union being misunderstood? Give the student an everyday example where "or" is used in the mathematical sense.

Thinking Critically

19. Circular loops in a Venn diagram divide the universe U into distinct regions.

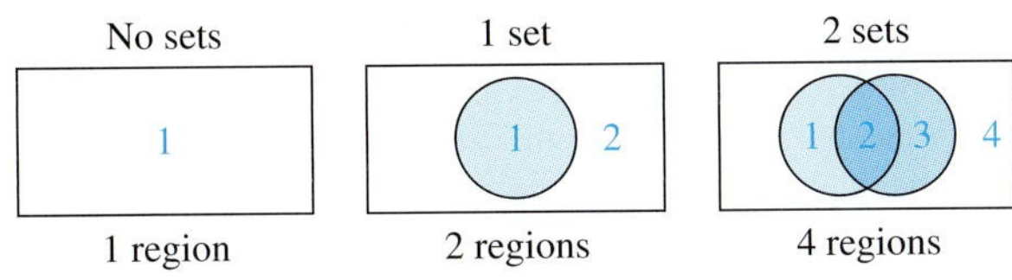

   - **(a)** Draw a diagram for three circles that gives the largest number of regions.
   - **(b)** How many regions do the four circles define in this figure?

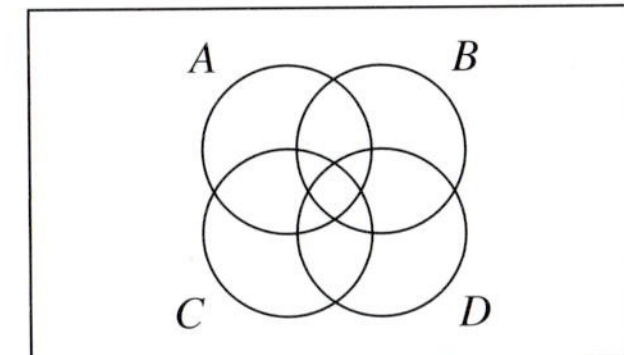

   - **(c)** Verify that the four-circle diagram is missing a region corresponding to $A \cap \overline{B} \cap \overline{C} \cap D$. What other set has no corresponding region?
   - **(d)** Will this Venn diagram allow for all possible combinations of four sets? Explain briefly, but carefully.

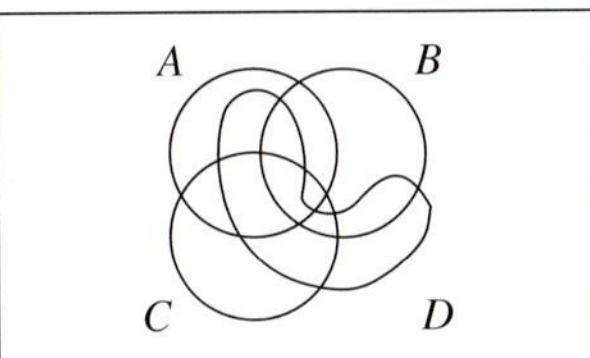

20. The English mathematician Augustus DeMorgan (1806–1871) showed that

$$\overline{A \cap B} = \overline{A} \cup \overline{B} \quad \text{and} \quad \overline{A \cup B} = \overline{A} \cap \overline{B}.$$

These identities are now called DeMorgan's Laws.
   - **(a)** Use a sequence of Venn diagrams to explain, in words and pictures, how to shade the regions corresponding to $\overline{A \cap B}$ and $\overline{A} \cup \overline{B}$. The final shaded regions obtained should be identical, justifying DeMorgan's First Law.
   - **(b)** Carefully explain how to justify DeMorgan's Second Law by shading Venn diagrams.

21. Each of the objects depicted in problem 15 is described by its attributes: shape (three choices), size (two choices), and color (two choices). The number of pieces in a full set of shapes can be varied by changing the number of attributes and the number of possible choices of an attribute. How many pieces are contained in these attribute sets?
   - **(a)** Shape: circle, hexagon, equilateral triangle, isosceles right triangle, rectangle, square
     Size: large, small
     Color: red, yellow, blue
   - **(b)** Use the attributes in part **(a)**, but also include this attribute:
     Thickness: thick, thin

Thinking Cooperatively

22. Obtain (or make) a 12-piece attribute shape set as described in problem 15, one set per small group. Let C, H, T, S, L, R, and B denote the sets of attribute shapes described in problem 15. Each loop in the Venn diagrams below corresponds to one of the above sets. Only some of the 12 attribute pieces are depicted in each figure. Determine how to label the loops. Is there more than one correct way to label each diagram?

(a)

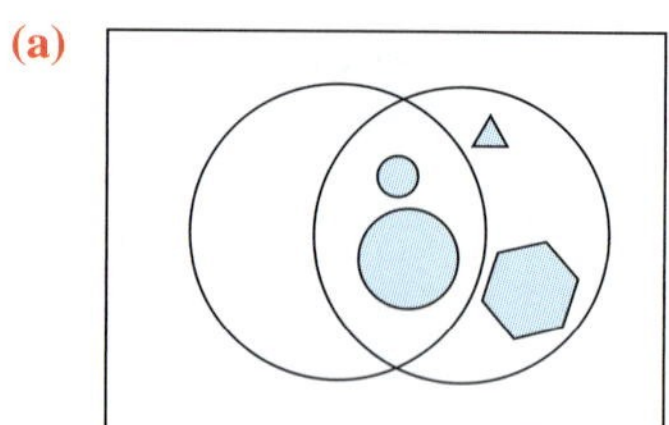

(b)

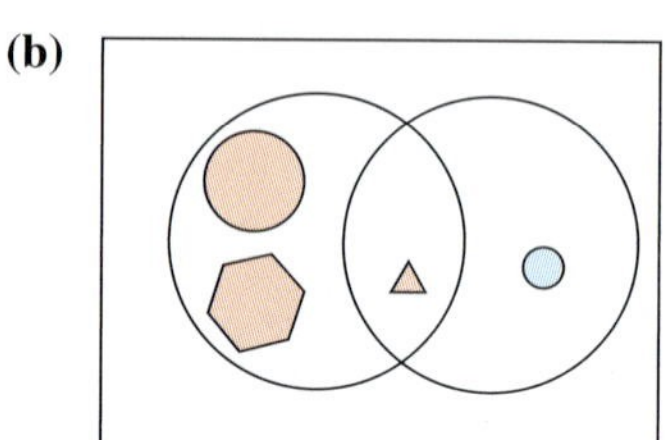

(c)

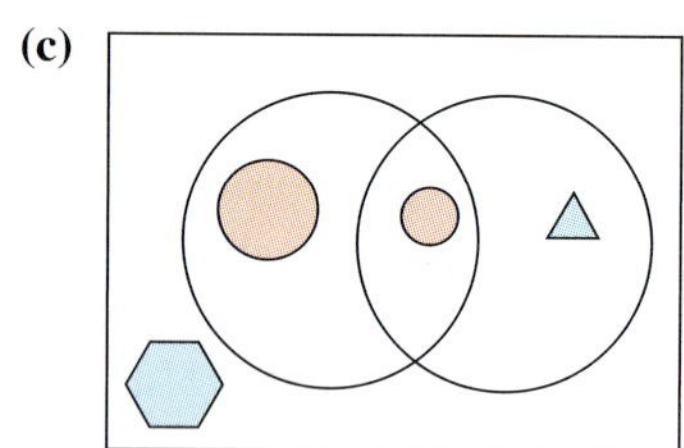

(d)

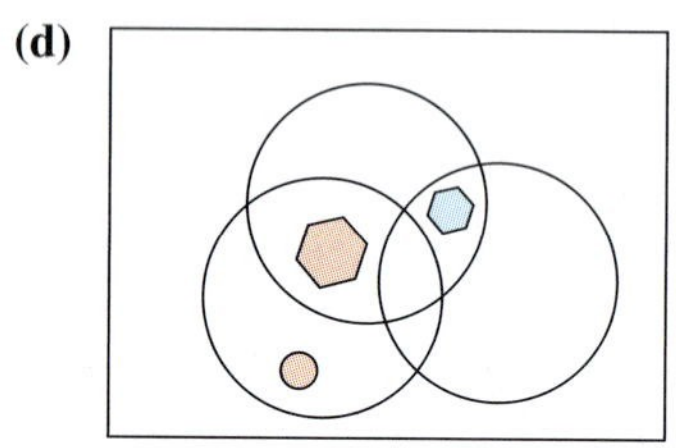

(e) Divide the class into groups. Have each group invent its own "missing label" puzzles similar to those shown. In turn, have each group challenge the other groups to label its puzzles.

23. Attribute cards can be made by drawing various shapes on cards, varying the figure shown, the color used, and so on. Here are some examples of a few attribute cards.

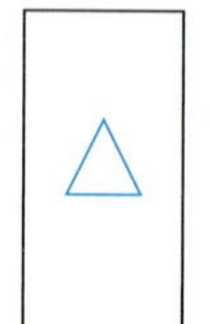 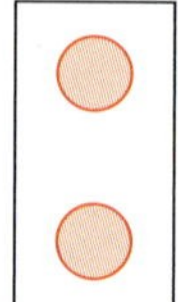 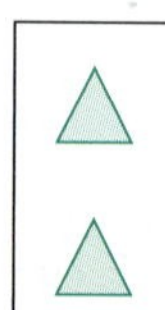 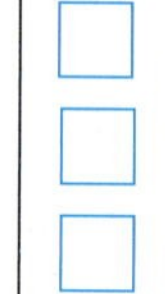 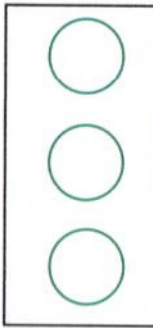

(a) Conjecture how many cards make a full deck. Explain how you get your answer.

(b) Form groups of three to five students. Each group is to design and make a set of attribute cards, drawing simple figures on small rectangles of cardstock with colored pens.

(c) Exchange decks of attribute cards among the cooperative groups. Shuffle each deck and turn over just a few of the cards. Can each group predict the number of cards in the complete deck? Carefully explain the reasoning used.

Making Connections

24. If a penny (P) and a nickel (N) are flipped, the possible outcomes can be listed by forming all of the subsets that could represent heads. There are four possible outcomes: $\varnothing$, $\{P\}$, $\{N\}$, and $\{P, N\}$. For example, $\varnothing$ is the outcome for which both coins land on tails.

(a) List the possible outcomes if a penny, nickel, and dime (D) are flipped. How many outcomes (that is, subsets of $\{P, N, D\}$) did you find?

(b) List the subsets of $\{P, N, D, Q\}$, where Q represents a quarter. How many subsets did you find?

(c) How many of the subsets $\{P, N, D, Q\}$ contain Q? How many do not contain Q?

(d) Use inductive reasoning to describe the number of subsets of a set with n elements.

25. **The ABO System of Blood Typing.** Until the beginning of the twentieth century, it was assumed that all human blood was identical. About 1900, however, the Austrian-American pathologist Karl Landsteiner discovered that blood could be classified into four groups according to the presence of proteins called antigens. This discovery made it possible to transfuse blood safely. A person with the antigens A, B, or both A and B has the respective blood type A, B, or AB. If neither antigen is present, the type is O. Draw and label a Venn diagram that illustrates the ABO system.

26. **The Rh System of Blood Typing.** In 1940, Karl Landsteiner (see problem 25) and the American pathologist Alexander Wiener discovered another protein that coats the red blood cells of some persons. Since the initial research was on rhesus monkeys, a person with the protein is classified as Rh positive (Rh+), and a person whose blood cells lack the protein is Rh negative (Rh−). Draw and label a Venn diagram illustrating the classification of blood in the eight major types A+, A−, B+, B−, AB+, AB−, O+, and O−.

Communicating

27. The word *set* has been given precise mathematical meaning. The same word is also used as part of our ordinary language. For example, you may own a set of golf clubs. Make a list of other examples where *set,* or *family, aggregate, class,* or *collection,* is used, and discuss to what extent their usage corresponds to the mathematical concept of *set.*

28. Many of the terms introduced in this section also have meaning in nonmathematical contexts. For example, *complement, union,* and *intersection* are used in ordinary speech. Other terms, such as *transitive, commutative, associative,* and *distributive,* are rare in everyday usage, but there are closely related words that are quite common, such as *transit, commute, associate,* and *distribute.* Discuss the differences and similarities in meanings of the words, and word roots, of the terms introduced in this section.

From State Student Assessments

29. (Washington State, Grade 4)
Ms. Yonan took a survey in her class to see how many students like hamburgers, pizza, or hot dogs. The results of the survey are shown in the chart.

NUMBER OF STUDENTS WHO LIKE:		
Only 1 of These Foods	Only 2 of These Foods	All 3 of These Foods
14	10	4

Look at the following three Venn diagrams. Choose the diagram that best represents the results of this survey.

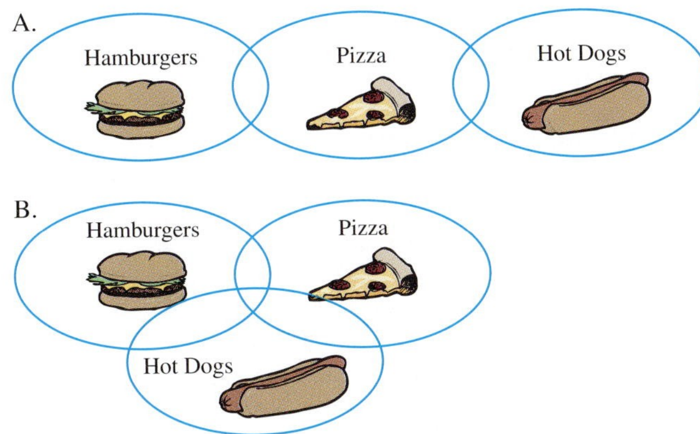

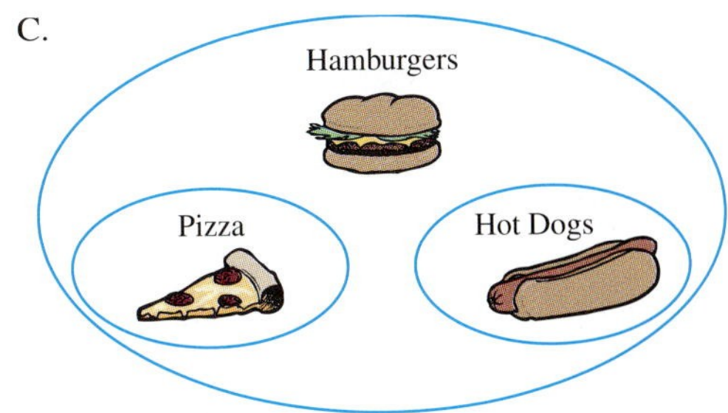

30. (Delaware, Grade 5)
Look at the Venn diagram below.

- Add the following numbers to your Venn diagram: 20, 24, 30, 36.
- Explain why you put each number where you did.

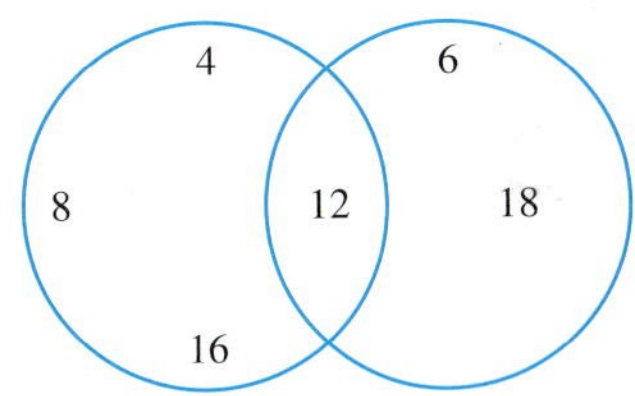

For Review

31. Show that the natural numbers 1, 2, . . . , 15 can be arranged in a list so that the sum of each adjacent pair is a square number. Write your solution to illustrate Pólya's four steps of problem solving.

32. A power company intends to number its power poles from 1 to 10,000. Each numeral is formed by gluing stamped-metal digits to the poles. How many metal 2s should the company order? Write your solution to illustrate Pólya's four steps of problem solving.

2.2 SETS, COUNTING, AND THE WHOLE NUMBERS

If you attend a student raffle, you might hear the following announcement when the entry forms are drawn:

> *"The student with ticket 50768-973 has just won second prize—four tickets to the big game this Saturday."*

There are three types of numbers.

This sentence contains three numbers, each of a different type, and each serving a different purpose.

First, a number can be an **identification,** or **nominal number,** such as the ticket number in the example. A nominal number is a sequence of digits used as a name or label. Telephone numbers, social security numbers, account numbers with stores and banks, serial numbers, and driver's license numbers are just a few examples of the use of numbers for identification and naming. The role such numbers play in contemporary society has expanded rapidly with the advent of computers.

The next type of number used by the raffle announcer is an **ordinal number.** The words *first, second, third, fourth,* and so on are used to describe the relative position of objects in an ordered sequence. Thus, ordinal numbers communicate location in an ordered collection. "First class," "second rate," "third base," "Fourth of July," "fifth page," "sixth volume," and "twenty-first century" are all familiar examples of ordinal numbers.

The final use of a number by the raffle announcer is to tell how many tickets had been won. That is, the prize is a set of tickets, and *four* tells us *how many* tickets are in the set. More generally, a **cardinal number** of a set is the number of objects in the set. Thus, cardinal numbers help communicate the basic notion of "how many."

It should be noticed that numbers, of whatever type, can be expressed verbally (in a language) or symbolically (in a numeration system). For example, the number of moons of Mars is "two" in English, "zwei" in German, and "dos" in Spanish. Symbolically, we could write 2 in the Indo-Arabic (or Hindu-Arabic*) system or II in Roman numerals. Numeration, as a system of symbolic representation of numbers, is closely related to algorithms for computation. Numeration systems, both historic and contemporary, are described in Chapter 3.

In the remainder of this section, we explore the notion of cardinal numbers more fully. Of special importance is the set of whole numbers, which can be viewed as the set of cardinal numbers of finite sets.

One-to-One Correspondence and Equivalent Sets

Suppose, when you come to your classroom, each student is seated at his or her desk. You would immediately know, *without counting,* that the number of children and the number of occupied desks are the same. This example illustrates the concept of a **one-to-one correspondence** between sets, in which each element of one set is paired with exactly one element of the second set, and each element of either set belongs to exactly one of the pairs.

DEFINITION *One-to-One Correspondence*

A **one-to-one correspondence** between sets A and B is a pairing of the elements of A with the elements of B in such a way that each element of A and B belongs to one, and only one, pair.

Figure 2.6 illustrates a one-to-one correspondence between the sets of board members and offices of the Math Club.

*"Indo" refers to India, the region of origination of the symbols. "Hindu" refers to the predominant religion of India.

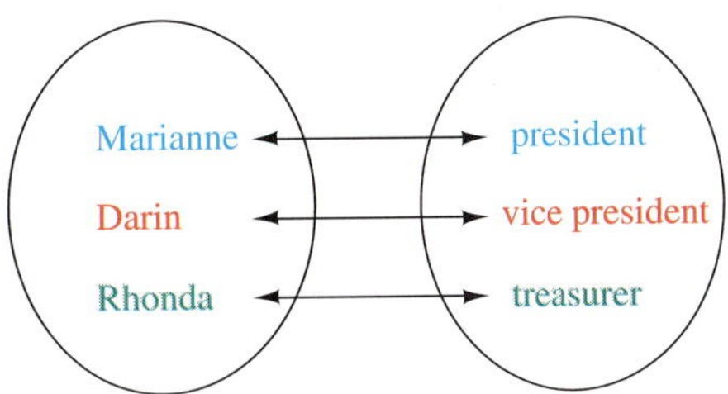

Figure 2.6
A one-to-one correspondence between the sets of board members {Marianne, Darin, Rhonda} and offices {president, vice president, treasurer} of the Math Club

Because there is a one-to-one correspondence between the sets {m, d, r} and {p, v, t}, we say that they are **equivalent sets** and write {m, d, r} $\sim$ {p, v, t}.

More generally we have the following definition.

DEFINITION *Equivalent Sets*
Sets A and B are **equivalent** if there is a one-to-one correspondence between A and B. When A and B are equivalent, we write $A \sim B$. We also say that equivalent sets **match.** If A and B are not equivalent we write $A \nsim B$.

It is easy to see that equal sets match. To see why, suppose that $A = B$. Since each x in A is also in B, the natural matching $x \leftrightarrow x$ is a one-to-one correspondence between A and B. Thus $A \sim B$. On the other hand, there is no reason to believe equivalent sets must be equal. For example, $\{\square, \bigstar\} \sim \{1, 2\}$, but $\{\square, \bigstar\} \neq \{1, 2\}$.

EXAMPLE 2.6 INVESTIGATING SETS FOR EQUIVALENCE

Let $A = \{x \mid x$ is a moon of Mars$\}$
$B = \{x \mid x$ is a former U.S. president whose last name is Adams$\}$
$C = \{x \mid x$ is one of the Brontë sisters of nineteenth-century literary fame$\}$
$D = \{x \mid x$ is a satellite of the fourth-closest planet to the sun$\}$

Which of these relationships, $=$, $\neq$, $\sim$, and $\nsim$, holds between distinct pairs of the four sets?

Solution

It is useful to write the sets in listed form:

$A = D =$ {Deimos, Phobos}, $B =$ {John Adams, John Quincy Adams}, and $C =$ {Anne, Charlotte, Emily}.

Therefore we see that

$$A \neq B, A \neq C, A = D, B \neq C, B \neq D, C \neq D,$$
$$A \sim B, A \nsim C, A \sim D, B \nsim C, B \sim D, \text{ and } C \nsim D.$$

The Whole Numbers

The sets {a, b, c}, {□, ○, △}, {Mercury, Venus, Earth}, and {Larry, Moe, Curly} are distinct, but they do share the property of "threeness." The English word *three* and the Indo-Arabic numeral 3 are used to identify this common property of all sets that are equivalent to $\{1, 2, 3\}$. In a similar way, we use the word *two,* and the symbol 2, to convey the idea that all of the sets that are equivalent to the set $\{1, 2\}$ have the same cardinality.

Some sets are quite large and it is difficult to know how large n must be for the set to be equivalent to the set $\{1, 2, 3, \ldots, n\}$. For example, the set of people alive in the world in the year 2000 would require n to be around 6×10^9 (that is, n is about six billion). Even so, this is an example of a finite set. In general, a set is said to be **finite** if it is either the empty set or is equivalent to a set $\{1, 2, 3, \ldots, n\}$ for some natural number n. Sets that are not finite are called **infinite.** For example, the set N of all of the natural numbers is an infinite set. It is usually easy to know whether a set is finite or infinite, but not always. In Chapter 4, we'll see Euclid's clever proof that the set of prime numbers $\{2, 3, 5, 7, 11, 13, 17, 19, 23, 29, 31, \ldots\}$ is an infinite set. Many of these prime numbers come in pairs of adjacent odd numbers known as **twin primes,** such as $(3, 5)$, $(5, 7)$, $(11, 13)$, $(17, 19)$, and $(29, 31)$. However, it is unknown if the set of twin primes is finite or infinite.

The whole numbers, as defined below, allow us to classify any finite set according to how many elements the set contains. It is useful to adopt the symbol $n(A)$ to represent the cardinality of the finite set A. If A is not the empty set, then $n(A)$ is a counting number. The cardinality of the empty set, however, requires a new name and symbol: we let **zero** designate the cardinality of the empty set and write $0 = n(\varnothing)$.

Use $n(A)$ to denote the number of elements in set A.

A common error is made by omitting the $n(\ )$ symbol: be sure not to write "$A = 4$" when your intention is to state that $n(A) = 4$.

DEFINITION *The Whole Numbers*

The **whole numbers** are the cardinal numbers of finite sets—that is, the numbers of elements in finite sets. If $A \sim \{1, 2, 3, \ldots, m\}$, then $n(A) = m$, and $n(\varnothing) = 0$, where $n(A)$ denotes the cardinality of set A. The set of whole numbers is written $W = \{0, 1, 2, 3, \ldots\}$.

EXAMPLE 2.7 DETERMINING WHOLE NUMBERS

For each set find the whole number that gives the number of elements in the set.

(a) $M = \{x \mid x \text{ is a month of the year}\}$
(b) $A = \{\text{a, b, c}, \ldots, \text{z}\}$
(c) $B = \{n \in N \mid n \text{ is a square number smaller than } 200\}$
(d) $Z = \{n \in N \mid n \text{ is a square number between 70 and 80}\}$
(e) $S = \{0\}$

Solution

(a) $n(M) = 12$, since $M \sim \{1, 2, 3, \ldots, 12\}$, and this amounts to counting the elements in M.
(b) $n(A) = 26$
(c) $n(B) = 14$, since $14^2 = 196 \in B$, but $15^2 = 225 \notin B$. Thus, $B = \{1, 4, 9, 16, 25, 36, 49, 64, 81, 100, 121, 144, 169, 196\}$ and B contains 14 elements.
(d) $n(Z) = 0$, since $8^2 = 64 < 70$, but $9^2 = 81 > 80$. Therefore, there are no square numbers between 70 and 80 and $Z = \varnothing$.
(e) $n(S) = 1$, since the set $\{0\}$ contains one element. This shows that zero is *not* the same as "nothing"!

HIGHLIGHT *from* HISTORY

The Origin of Zero

It required thousands of years of mathematical thought for the concept of zero to emerge and its importance to be recognized. Zero, as is true of any whole number, expresses a quantity. Thus, zero *is* something, and it should not be mistaken as synonymous with nothing. (Ask any student who gets a zero on a test if it means nothing to his or her grade!) The contemplation of the void was an important aspect of ancient Eastern philosophy, and it is possible to trace the idea, and even the word *zero,* to this tradition. The Hindu word *sunya,* meaning "void" or "empty," was translated into Arabic as *sifr.* In its Latinized form *sifr* became *zephirum,* from which the word *zero* ultimately evolved.

Just for Fun

Red and Green Jelly Beans

Caralee has two jars of jelly beans, labeled R and G. The jelly beans in one jar, R, are all red, and those in the other jar, G, are all green. Caralee removed 20 red jelly beans, mixed them up with the green jelly beans in jar G, and then without looking grabbed 20 jelly beans from the mixed jar and put them back in jar R. How does the number of green jelly beans in jar R compare to the number of red jelly beans in jar G?

Physical and Pictorial Representations for Whole Numbers

Here are just a few physical and pictorial representations useful for explaining whole- and natural-number concepts. It should be noticed that we have not yet introduced representations that require grouping, number base, or place value. In the next chapter, additional representations are introduced that help illustrate connections to numeration and algorithms for the operations of arithmetic.

Tiles Tiles are congruent squares, each about 2 centimeters ($\frac{3}{4}$ inch) on a side. They should be sufficiently thick to be easily picked up and moved about. Colored plastic tiles are available from suppliers, but tiles are easily handmade from vinyl tile or cardboard. Of course, beans, circular discs, and other objects can be used as well. However, square tiles can be arranged into rectangular patterns, and such patterns reveal many of the fundamental properties of the whole numbers. See Figure 2.7.

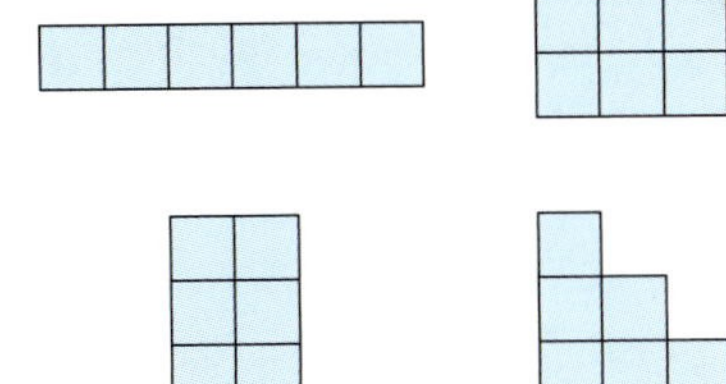

Figure 2.7
Some representations of "six" with square tiles

Cubes Cubes are much like tiles, but they can form both three-dimensional and two-dimensional patterns. Several attractive versions are commercially available. Unifix™ Cubes can be snapped together to form linear groupings. Multilink™ Cubes permit planar and spatial patterns, as we see in Figure 2.8.

Figure 2.8
Some representations of "12" with cubes

Number Strips and Rods Colored strips of cardboard or heavy paper, divided into squares, can be used to demonstrate and reinforce whole-number properties and operations. The squares should be ruled off, and colors can be used to visually identify the number of squares in a strip, as indicated in Figure 2.9. Number strips can be used nearly interchangeably with Cuisenaire® rods, which have been used effectively for many years. The colors shown in the figure correspond to those of Cuisenaire® rods. Whole numbers larger than 10 are illustrated by placing strips, or rods, end-to-end to form a "train."

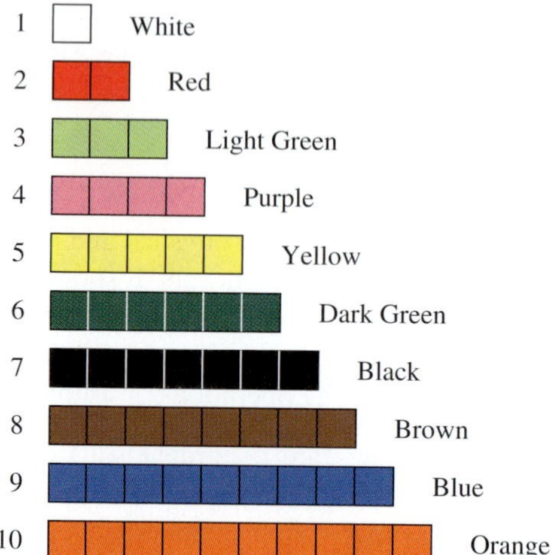

Figure 2.9
Number strips for the natural numbers 1 through 10

Number Line The number line is a pictorial model in which two distinct points on a line are labeled 0 and 1, and then the remaining whole numbers are laid off in succession with the same spacing (see Figure 2.10). Any whole number is interpreted as a distance from 0, which can be shown by an arrow. The number-line model is particularly important because it can also be used to visualize the number systems that are developed later in this book.

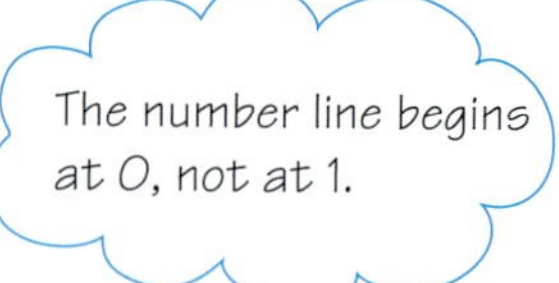

The number line begins at 0, not at 1.

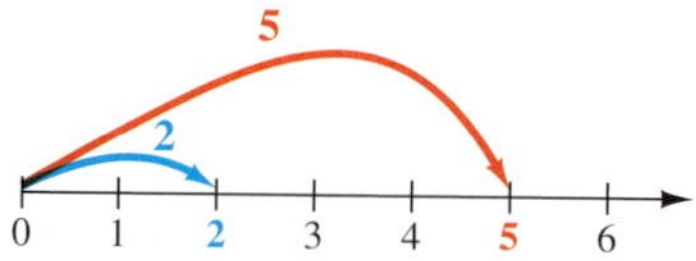

Figure 2.10
Illustrating "two" and "five" on the number-line model of the whole numbers

In this way, it can be seen how the whole numbers are extended to the integers, and how the integers are related to the rational and real number systems.

Ordering the Whole Numbers

We often wish to relate the number of elements in two given sets. For example, if each child in the class is given one cupcake, and there are some cupcakes left over, we would know that there are more cupcakes than children. Notice that children have been matched to a *proper* subset of the set of cupcakes.

HIGHLIGHT *from* HISTORY

The Equality and Inequality Symbols

Many of the mathematical symbols we commonly use today originated in Renaissance times or shortly thereafter. The sign $=$ was introduced in 1557 by Robert Recorde (1510–1558) of Cambridge. It appeared in *The Whetstone of Witte,* the first English book on algebra. According to Recorde, no two things are as alike as two parallel lines, and hence the symbol of equality. The symbols $>$ and $<$, for "greater than" and "less than," first appeared in 1631 in a posthumous publication of Thomas Harriot (1560–1621). The equality sign in Harriot's book was very long, looking like this: ══════.

The order of the whole numbers can be defined in the following way.

DEFINITION *Ordering the Whole Numbers*

Let $a = n(A)$ and $b = n(B)$ be whole numbers, where A and B are finite sets. If A matches a *proper* subset of B, we say that ***a* is less than *b*** and write $a < b$.

The expression $b > a$ is read "b is greater than a" and is equivalent to $a < b$. Also, $a \leq b$ means "a is less than or equal to b." The sets A and B used in checking the definition are arbitrary, and it is sometimes useful to choose $A \subset B$.

EXAMPLE 2.8 SHOWING THE ORDER OF WHOLE NUMBERS

Show that $4 < 7$ using **(a)** sets; **(b)** tiles; **(c)** rods; and **(d)** the number line.

Solution

(a) The following diagram shows that a set with 4 elements matches a proper subset of a set with 7 elements. Therefore, $4 < 7$.

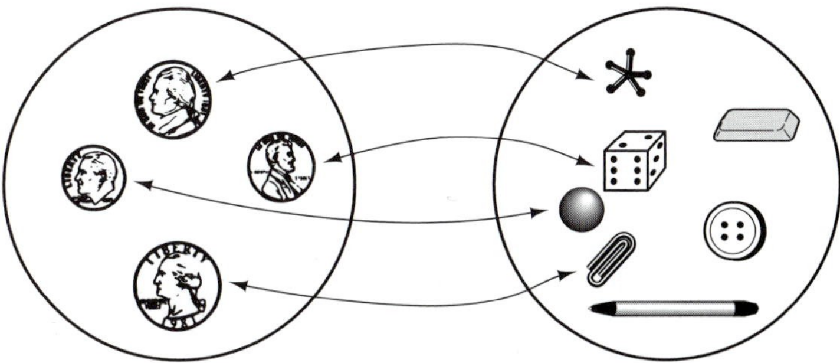

(b) By setting tiles side-by-side, it is seen that the 4 colored tiles match a proper subset of the 7 uncolored tiles.

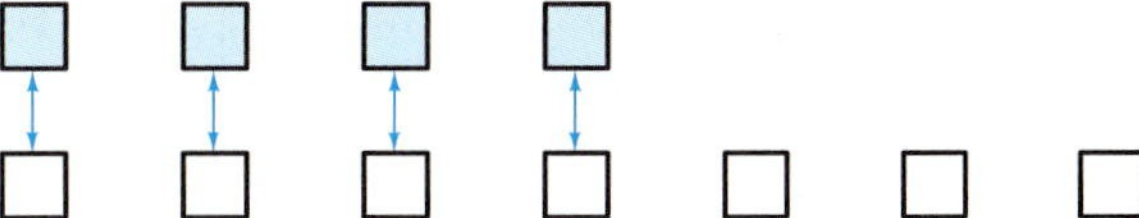

(c) With rods, the order of whole numbers is interpreted by comparing the lengths of the rods.

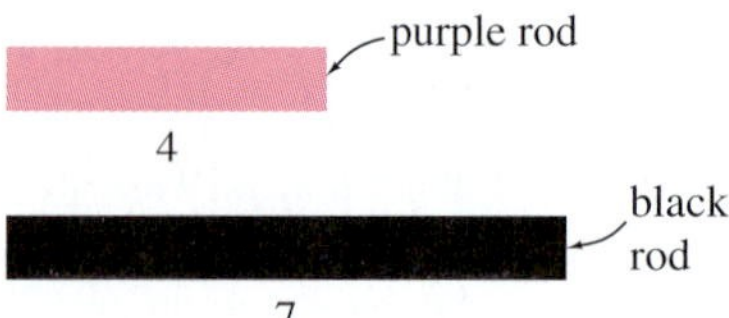

Numbers larger than 10 are compared by forming side-by-side trains of rods.

(d) On the number line, $4 < 7$ because 4 is to the left of 7.

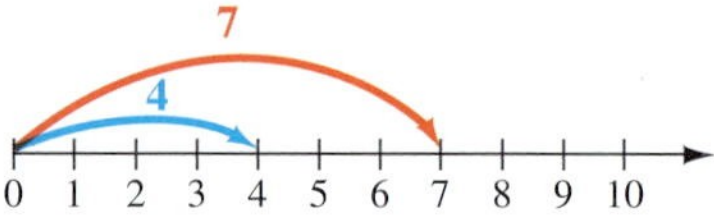

Problem Solving with Whole Numbers and Venn Diagrams

Venn diagrams and the associated concept of a whole number can often be used as a practical problem-solving tool, as we illustrate in the next two examples.

EXAMPLE 2.9 SOLVING A CLASSIFICATION PROBLEM

In a recent survey, the 60 students living in Harris Hall were asked about their enrollments in science, engineering, and humanities classes. The results were as follows:

24 are taking a science class.
22 are taking an engineering class.
17 are taking a humanities class.
5 are taking both science and engineering classes.
4 are taking both science and humanities classes.
3 are taking both engineering and humanities classes.
2 are taking classes in all three areas.

How many students are not taking classes in any of the three areas? How many students are taking a class in just one area? Using a Venn diagram, indicate the number of students in each region of the diagram.

Solution Let S, E, and H denote the set of students in science, engineering, and humanities classes, respectively. Since $n(S \cap E \cap H) = 2$, begin by placing the 2 in the region of the Venn diagram corresponding to the subset $S \cap E \cap H$. Then, by comparing $n(S \cap E \cap H) = 2$ and $n(E \cap H) = 3$, we conclude that there is one student who is taking both engineering and humanities, but not a science class. This allows us to fill in the 1 in the Venn diagram. Analogous reasoning leads to the entries within the loops of the following Venn diagram.

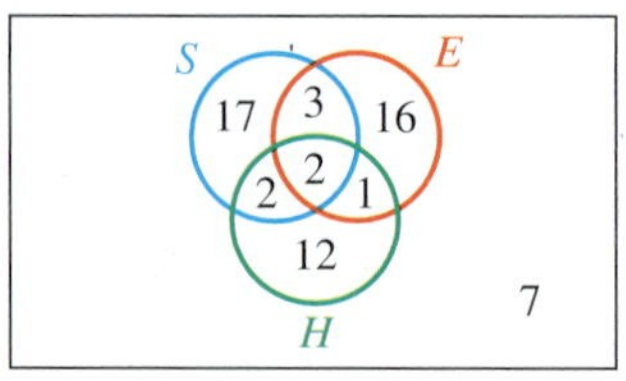

The values within the loops account for 53 of the 60 students, so it follows that 7 students are not taking classes in any of the three areas. Also, since 17 + 16 + 12 = 45, we conclude that 45 students are taking a class in just one area.

Richard Hamming

Another application of Venn diagrams is to describe a method of error detection and correction used commonly in our digital age. Information, including text, pictures, and even music, is now typically stored and transmitted either electronically or with radio waves. This is accomplished by breaking the information up into "words" that can be thought of as strings of pulses represented by a sequence of 0s and 1s. In the parlance of computing, coding words into 0s and 1s is called **binary coding,** and each 0 or 1 is called a **binary digit,** or a **bit.** Most realistic applications require longer strings, but even the strings of four 0s and 1s—0000, 0001, 0010, . . . , 1111—create a vocabulary of 16 "four-bit" words.

It is hoped that transmission errors will be infrequent, but even if a single 0 or 1 is misread, the intended word will be mistakenly replaced with another word. In 1948, Richard Hamming (1915–1998), a mathematician at Bell Laboratories, proposed an efficient method to detect errors in a binary code word. As an added bonus, the Hamming code can even correct the error if just one bit is in error. Hamming's idea for four-bit codes is described with the three-loop Venn diagrams shown in Figure 2.11. First, the 0s and 1s in the code word are placed in regions a, b, c, and d (in that order—this is important) in the three-loop Venn diagram. Then either a 0 or a 1 is placed in each of the regions e, f, and g so that *the number of 1s in each loop is even.*

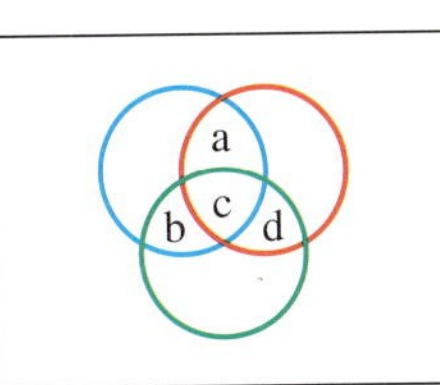

4-bit word abcd

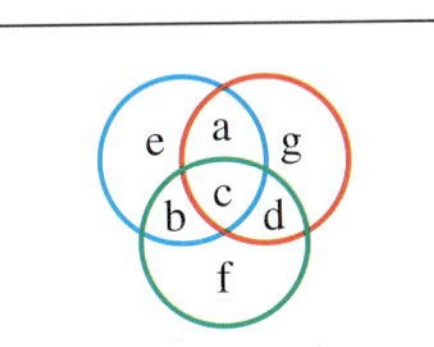

7-bit error-correcting word abcdefg

Figure 2.11
Hamming's scheme extends a four-bit code word abcd to a seven-bit error-correcting code word abcdefg by choosing e, f, and g so that each loop contains an even number of 1s.

As an example, consider the four-bit word 1001. To convert it to the corresponding seven-bit error-correcting word, we place the 0s and 1s into the regions abcd in that order in the three-loop Venn diagram. Then, a 1 is placed in region e so that the loop containing e now has two 1s. Similarly, a 1 is placed in region f to have two 1s in that loop. Finally, since the loop containing region g already has an even number of 1s, a 0 is placed in region g. Altogether, we see in Figure 2.12 that the four-bit "word" 1001 is extended to the seven-bit Hamming-coded word 1001110.

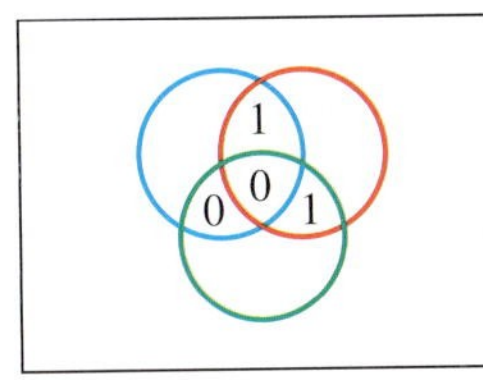

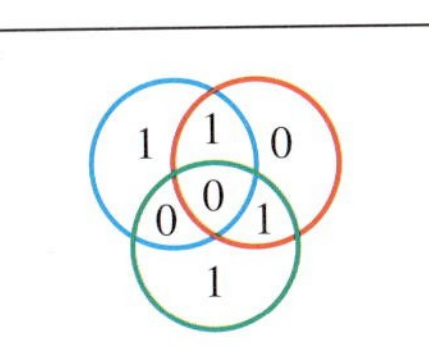

Figure 2.12
The four-bit word 1001 is extended to the Hamming seven-bit error-correcting word 1001110

The additional three digits allow us to check if the word has been transmitted correctly, and if not, the error can be corrected under the assumption that just one of the seven bits is incorrect. The following example shows how this works.

EXAMPLE 2.10 CORRECTING ERRORS WITH THE HAMMING CODE

The following seven-bit code words have *at most* an error in a single bit. Determine if an error is present and, if so, correct the error.

(a) 1101110
(b) 1111000
(c) 0110101
(d) 1000101

Solution The seven-digit words are each placed into three-loop diagrams, as shown below.

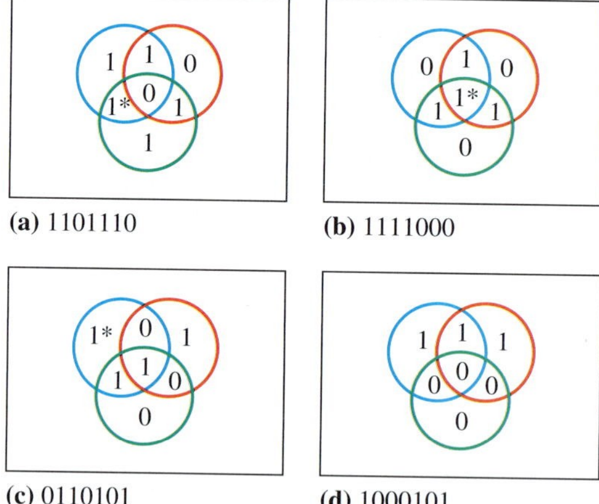

In the (a) diagram, both the left and lower loops have an odd number of 1s, but the rightmost loop has an even number of 1s. Therefore, the bit in the intersection of the left and lower loops, and in the complement of the right loop, is incorrect. We detect that the digit 1 marked with the asterisk in region b is incorrect, so the corrected seven-bit word is 1001110 (as shown in Figure 2.12).

In the (b) diagram, all three loops are "bad" in the sense that each contains an odd number of 1s. Since only one digit can be changed, we see that it must be the 1 (marked with the asterisk) in the c region, where all three loops intersect. The corrected seven-bit word is therefore 1101000.

In the (c) diagram, only the upper left loop has an odd number of 1s, so the 1 (again marked by the asterisk) must be changed to a 0, giving the corrected word 0110001.

All three loops in the (d) diagram are "good," since each contains an even number of 1s. Thus, no single error has occurred, and 1000101 is a correct seven-bit word.

Providing error correction came at the cost of requiring code words to be seven bits long rather than just four bits in length. However, it is often a reasonable price to pay to have assurance that information is correctly transmitted. In compact discs about a third of the bits are devoted to error correction. However, the system works so well that up to 14,000 consecutive errors can be corrected.

Problem Set 2.2

Understanding Concepts

1. Classify by type—cardinal, ordinal, or nominal—the numbers that appear in these sentences.
   (a) On June 13, Mary was promoted to first vice president.
   (b) On the eleventh hole, Joe's second shot with a 6-iron went 160 yards.
   (c) Erin's bowling partner left the 7 pin in the sixth frame, which added 9 to her score.
2. For each pair of sets, decide if the sets are equivalent to one another.
   (a) $\{1, 2, 3, 4, 5\}$ and $\{x \mid x$ is a letter in the phrase "PANAMA BANANA MAN"$\}$
   (b) {a, b, c} and {w, x, y, z}
   (c) {o, n, e} and {t, w, o}
   (d) $\{0\}$ and $\varnothing$
3. Let A, B, and C be finite sets, with $A \subset B \subseteq C$ and $n(B) = 5$.
   (a) What are the possible values of $n(A)$?
   (b) What are the possible values of $n(C)$?
4. Use counting to determine the whole number that corresponds to the cardinality of these sets.
   (a) $A = \{x \mid x \in N \text{ and } 20 \leq x < 35\}$
   (b) $B = \{x \mid x \in N \text{ and } x + 1 = x\}$
   (c) $C = \{x \mid x \in N \text{ and } (x - 3)(x - 8) = 0\}$
   (d) $D = \{x \mid x \in N,\ 1 \leq x \leq 100, \text{ and } x \text{ is divisible by both 4 and 6}\}$
5. Let $A = \{n \mid n \text{ is a cube of a natural number and } 1 \leq n \leq 100\}$

   $B = \{s \mid s \text{ is a state in the United States that borders Mexico}\}$

   Is $A \sim B$?
6. Let $N = \{1, 2, 3, 4, \ldots\}$ be the set of natural numbers and $S = \{1, 4, 9, 16, \ldots\}$ be the set of squares of the natural numbers. Then $N \sim S$, since we have the one-to-one correspondence $1 \leftrightarrow 1, 2 \leftrightarrow 4, 3 \leftrightarrow 9, 4 \leftrightarrow 16, \ldots n \leftrightarrow n^2$. (This example is interesting, since it shows that an infinite set can be equivalent to a proper subset of itself.) Show that each of the following pairs of sets are equivalent by carefully describing a one-to-one correspondence between the sets.
   (a) The whole numbers and natural numbers: $W = \{0, 1, 2, 3, \ldots\}$ and $N = \{1, 2, 3, 4, \ldots\}$.
   (b) The sets of odd and even natural numbers: $D = \{1, 3, 5, 7, \ldots\}$ and $E = \{2, 4, 6, 8, \ldots\}$.
   (c) The set of natural numbers and the set of powers of 10: $N = \{1, 2, 3, 4, \ldots\}$ and $\{10, 100, 1000, \ldots\}$
7. Decide which of the following sets are finite.
   (a) {grains of sand on all the world's beaches}
   (b) {whole numbers divisible by the number 46,182,970,138}
   (c) {points on a line segment that is one inch long}
8. The figure below shows two line segments, L_1 and L_2. The rays through point P give, geometrically, a one-to-one correspondence between the points on L_1 and the points on L_2—for example, $Q_1 \leftrightarrow Q_2$.

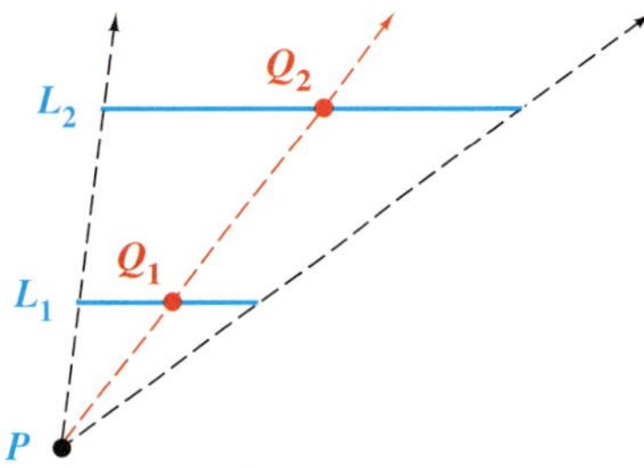

   Use similar geometric diagrams to show that the following figures are equivalent sets of points. Describe all possible locations of point P.
   (a) Two concentric circles

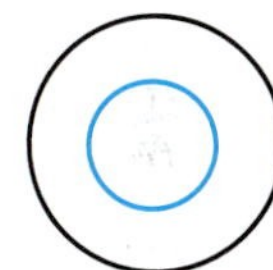

   (b) A circle and an inscribed square

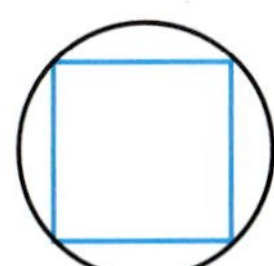

   (c) A triangle and its circumscribing circle

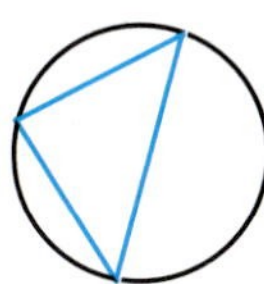

   (d) A semicircle and its diameter

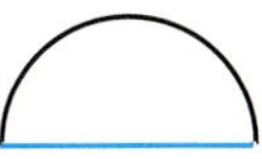

9. Decide whether the following statements are true or false. If false, give a counterexample; that is, give two sets that satisfy the hypothesis, but not the conclusion of the statement. A and B designate finite sets.
   (a) If $A \subseteq B$, then $n(A) \leq n(B)$.
   (b) If $n(A) < n(B)$, then $A \subset B$.
   (c) If $n(A \cup B) = n(A)$, then $B \subseteq A$.
   (d) If $n(A \cap B) = n(A)$, then $A \subseteq B$.
10. Let A and B be finite sets.
   (a) Explain why $n(A \cap B) \leq n(A)$.
   (b) Explain why $n(A) \leq n(A \cup B)$.
   (c) Suppose $n(A \cap B) = n(A \cup B)$. What more can be said about A and B?
11. A survey of 700 households revealed that 300 had only a TV, 100 had only a computer, and 100 had neither a TV nor a computer. How many households have both a TV and a computer? Use a two-loop Venn diagram to find your answer.
12. Let $U = \{1, 2, 3, \ldots, 1000\}$. Also, let F be the subset of numbers in U that are multiples of 5 and let S be the subset of numbers in U that are multiples of 6. Since $1000 \div 5 = 200$, it follows that $n(F) = n(\{5 \cdot 1, 5 \cdot 2, \ldots, 5 \cdot 200\}) = 200$.
   (a) Find $n(S)$ by using a method similar to the one above which showed that $n(F) = 200$.
   (b) Find $n(F \cap S)$.
   (c) Label the number of elements in each region of a two-loop Venn diagram with universe U and subsets F and S.
13. Finish labeling the number of elements in the regions in the Venn diagram shown, where the subsets A, B, and C of the universe U satisfy the conditions listed.

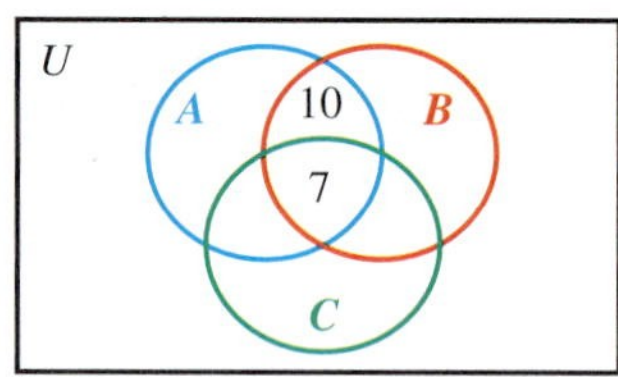

$n(U) = 100$
$n(A) = 40$
$n(B) = 50$
$n(C) = 30$
$n(A \cap B) = 17$
$n(B \cap C) = 12$
$n(A \cap C) = 15$
$n(A \cap B \cap C) = 7$

14. A poll of students showed that 55 percent liked basketball, 40 percent liked soccer, 55 percent liked football, 25 percent liked both basketball and soccer, 20 percent liked both soccer and football, 20 percent liked both basketball and football, and 10 percent liked all three sports. Use a Venn diagram to answer these questions: What percentage of students likes only one sport? What percentage does not like any of the three sports?
15. Number tiles can be arranged to form patterns that relate to properties of numbers.
   (a) Arrange number tiles to show why 1, 4, 9, 16, 25, . . . are called the "square" numbers.
   (b) Arrange number tiles to show why 1, 3, 6, 10, 15, . . . are called "triangular" numbers.
16. Enter 0s and 1s in three-loop Venn diagrams to extend the following four-bit code words to seven-bit Hamming code words.
   (a) 0101
   (b) 1100
   (c) 0100
17. At most one bit is in error in the following seven-bit Hamming code words. By entering 0s and 1s in three-loop Venn diagrams, determine whether the word is correct. If the word is incorrect, correct the error.
   (a) 1110101
   (b) 0011111
   (c) 1101000
   (d) 1111011

Teaching Concepts

18. Describe how you might teach a very young child the idea of "color" and the words such as *blue* or *red* that describe particular colors. Do you see any similarities in teaching the idea of color and the idea of "number," including such words as *two* or *five*?
19. Suggest a method to convey the notion of "zero" to a young child.
20. Imagine yourself teaching a third grader the transitive property of "less than," where $a < b$ and $b < c$ allow you to conclude $a < c$. (A proof is outlined in problem 24.) Write an imagined dialogue with the student, using number strips (or colored rods) as a manipulative.

Thinking Critically

21. Show that set equivalence satisfies the following properties. Carefully describe the one-to-one correspondences you choose.
   (a) The reflexive property: $A \sim A$
   (b) The symmetric property: If $A \sim B$, then $B \sim A$.
   (c) The transitive property: If $A \sim B$ and $B \sim C$, then $A \sim C$.

22. **(a)** How many ways can three balls of different colors—yellow, red, and green—be put in a ball can?

(b) How many different one-to-one correspondences are there between the two sets $\{1, 2, 3\}$ and $\{y, r, g\}$?

(c) Explain why a dozen dyed eggs, all of different colors, can be placed in a carton in $12 \times 11 \times 10 \times 9 \times 8 \times 7 \times 6 \times 5 \times 4 \times 3 \times 2 \times 1$ ways. [*Suggestion:* How many ways can you put the first egg into the carton? How many ways can you put the second egg into the carton? . . . How many ways are left to put the last egg into the carton?]

(d) How many different one-to-one correspondences are there between the two sets $\{1, 2, 3, \ldots, 12\}$ and $\{a, b, c, d, e, f, g, h, i, j, k, l\}$?

23. Consider a three-element set $C = \{a, b, c\}$. It has one subset with no elements—namely, $\varnothing$; three subsets with one element—$\{a\}$, $\{b\}$, $\{c\}$; three subsets with two elements—$\{a, b\}$, $\{b, c\}$, $\{a, c\}$; and one subset with three elements—$\{a, b, c\}$. This fills in row $n = 3$ of the following table, where the entry in row n and column r gives the number of ways to choose a subset of r elements from a set of n elements.

(a) Find all of the subsets of each of the four sets $\varnothing$, $A = \{a\}$, $B = \{a, b\}$, and $D = \{a, b, c, d\}$, and fill in the table through row $n = 4$.

(b) What pattern of numbers do you see in the table? Use the pattern to fill in the next two rows.

(c) A steering committee for the class play consists of Amy, Byron, Clea, Don, Edie, and Franco. How many ways can they choose a three-member subcommittee to arrange publicity?

24. Use the ideas of sets and the definition of the order relation for the whole numbers to justify the transitive property of "less than." That is, if k, l, and m are whole numbers satisfying $k < l$ and $l < m$, then $k < m$. (*Suggestion:* Consider sets satisfying $K \subset L \subset M$.)

25. Evelyn's Electronics Emporium hired Sloppy Survey Services (SSS) to poll 100 households at random. Evelyn's report from SSS contained the following data on ownership of a TV, VCR, or stereo:

TV only	8
TV and VCR (at least)	70
TV, VCR, and stereo	65
Stereo only	3
Stereo and TV (at least)	74
No TV, VCR, or stereo	4

Evelyn, who assumes anyone with a VCR also has a TV, is wondering if she should believe the figures are accurate. Should she?

26. At a school with 100 students, 35 students were taking Arabic, 32 Bulgarian, and 30 Chinese. Twenty students take only Arabic, 20 take only Bulgarian, and 14 take only Chinese. In addition, 7 students are taking both Arabic and Bulgarian, some of whom also take Chinese. How many students are taking all three languages? None of these three languages?

27. A political polling organization sent out a questionnaire that asked the following question: "Which taxes—income, sales, or excise—would you be willing to have raised?" Sixty voters' opinions were tallied by the office clerk:

Tax	Number Willing to Raise the Tax
Income	20
Sales	28
Excise	29
Income and sales	7
Income and excise	8
Sales and excise	10
Unwilling to raise any tax	5

The clerk neglected to count how many, if any, of the 60 voters are willing to raise all three of the taxes. Can you help the pollsters? Explain how.

28. There are 40 students in the Travel Club. They discovered that 17 members have visited Mexico, 28 have visited Canada, 10 have been to England, 12 have visited both Mexico and Canada, 3 have been only to England, and 4 have been only to Mexico. Some club members have not been to any of the three foreign countries and, curiously, an equal number have been to all three countries.

(a) How many students have been to all three countries?

(b) How many students have been only to Canada?

29. Letitia, Brianne, and Jake met at the mall on December 31. Letitia said that she intends to come to the mall every third day throughout the next year. Brianne said that she intends to be there every fourth day, and Jake said he would be there every fifth day.

	$r = 0$	$r = 1$	$r = 2$	$r = 3$	$r = 4$	$r = 5$	$r = 6$
$n = 0$							
$n = 1$							
$n = 2$							
$n = 3$	1	3	3	1			
$n = 4$							
$n = 5$							
$n = 6$							

Table for Problem 33

Letitia said that she knew she would be at the mall a total of 121 days, since 365 ÷ 3 is 121 with a remainder of 2. Brianne said that she'd be at the mall 91 days, since 365 ÷ 4 is 91 with a remainder of 1. Moreover, Brianne said that of those 91 days, she would expect to see Letitia 30 times, since they will both be coming every 12 days and 365 ÷ 12 is 30 with a remainder of 5.

(a) How many days will all three friends meet at the mall in the next year?

(b) Use reasoning similar to that of Letitia and Brianne to construct a three-loop Venn diagram that shows the number of times the three friends go to the mall in all of the possible combinations.

(c) How many days in the year will Jake be at the mall by himself?

(d) How many days in the year will none of the three be at the mall?

30. (a) The two one-bit words, 0 and 1, could be coded as the two-bit words 00 and 11, respectively. Could an error in one of the two bits be detected? Could it be corrected?

(b) Suppose the one-bit words 0 and 1 are coded as three-bit words 000 and 111, respectively. Could an error in at most one of the three bits be detected? Could the error be corrected?

31. A four-loop Venn diagram is shown below, where the "loops" for sets *A* and *B* are the triangular regions above the diagonals of the rectangle representing the universe, (that is, *A* is the region above the diagonal that runs from the bottom left corner of the rectangle to the top right corner, and *B* is the region above the diagonal that runs from the bottom right corner to the top left corner) and the loops for sets *C* and *D* are ovals. The 11 regions labeled a, b, . . . , k belong to two or more of the four sets, and the regions w, x, y, and z belong inside the respective loop *A*, *D*, *C*, or *B* and no other. Each of the $2^{11} = 2048$ 11-bit code words can be extended to an error-correcting 15-bit word by entering the first 11 bits in regions a, b, . . . , k and then entering a 0 or 1 into regions w, x, y, and z so that each of the four loops contains an even number of 1s.

(a) Verify that the 11-bit word 10011011100 has the 15-bit Hamming code 100110111001010.

(b) Correct the one incorrect bit in the 15-bit Hamming-coded word 001100111100100.

(c) There is at most one incorrect bit in the word 110011101101101. Determine whether the word is correct, and if not, correct the error.

A, B, C, *and* D

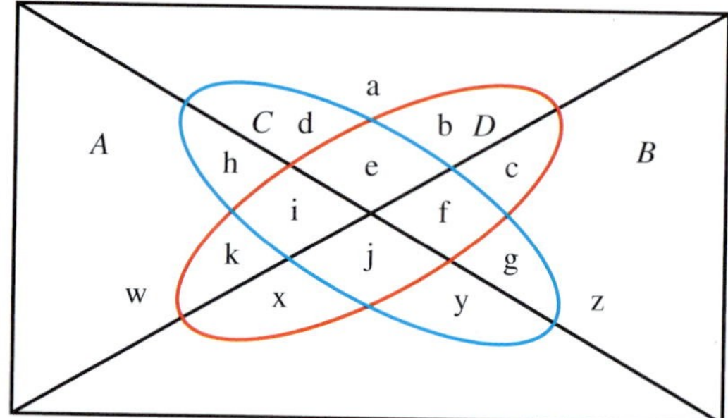

Making Connections

32. In the late twentieth century, modern society entered what some observers have called "the information age." People are now accustomed to being identified by number as often as by name. Make a list of your own identification numbers—social security, credit card, telephone, and so on. It may be helpful to look through your wallet!

33. Blood tests of 100 people showed that 45 had the A antigen and 14 had the B antigen (see problem 25 of Section 2.1). Another 45 had neither antigen and so are of type O. How many people are of type AB, having both the A and B antigens? Draw and label a Venn diagram that shows the number of people with blood type A, B, AB, and O.

Communicating

34. The English words for the whole numbers are *zero, one, two, three,* and so on. Make similar lists in Spanish, French, and other languages. For example, your list in German would begin *null, eins, zwei, drei,* . . . . Do you notice any common roots of the words?

From State Student Assessments

35. (Washington State, Grade 4)
What numbers do W, X, and Y probably represent on the number line?

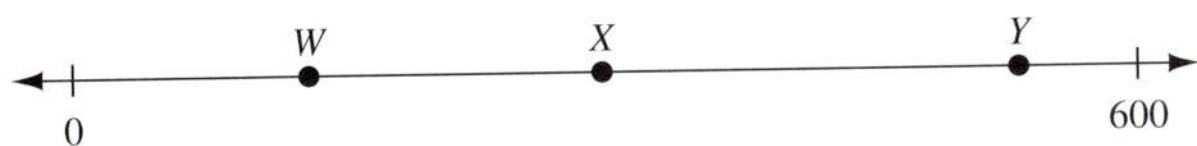

A. $W = 100, X = 200, Y = 500$
B. $W = 150, X = 300, Y = 400$
C. $W = 150, X = 300, Y = 525$

For Review

36. Let $U = \{0, 1, 2, \ldots, 25\}$, $A = \{n \mid n \text{ is even}\}$, $B = \{m \mid m = n^2 \text{ for some } n \in U\}$, and $C = \{p \mid p \in U \text{ and } p \text{ divides into 48 with no remainder}\}$. Make a Venn diagram that shows the location of all the elements of U.

37. The elements of a universal set $U = \{\text{a, b, c, d, e, f, g}\}$ with two subsets A and B are shown in the following Venn diagram.

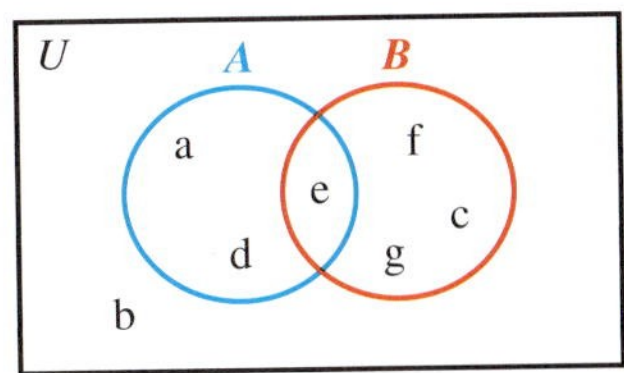

List the elements in the following sets.

(a) $\overline{A}$ (b) $A \cap B$
(c) $A \cup \overline{B}$ (d) $A \cap \overline{B}$
(e) $\overline{A} \cap \overline{B}$ (f) $\overline{A \cup B}$
(g) $\overline{A \cap B}$ (h) $(A \cup B) \cap \overline{A \cap B})$

38. Draw a three-loop Venn diagram for each set given, shading the region corresponding to the set.

(a) $(A \cap \overline{B}) \cup C$
(b) $(A \cup B \cup C)$
(c) $(A \cup B \cup C) \cap (\overline{A \cap B \cap C})$

2.3 ADDITION AND SUBTRACTION OF WHOLE NUMBERS

In this section we introduce the operations of addition and subtraction on the set of whole numbers $W = \{0, 1, 2, 3, \ldots\}$. In each operation, two whole numbers are combined to form another whole number. Because *two* whole numbers are added to form the sum, addition is called a **binary operation.** Similarly, subtraction is defined on a pair of numbers, so subtraction is also a binary operation.

The definitions of addition and subtraction are accompanied by a variety of conceptual models that give the operations both intuitive and practical meaning. A corresponding variety of physical and pictorial representations are described that reinforce these conceptual notions at the concrete and visual levels. It is vitally important that children be able to interpret and express the operations and their properties through manipulatives and visualization. Activities with these representations prepare them to understand the algorithms of computation and build confidence in their ability to select the appropriate operations for problem solving.

The Set Model of Whole-Number Addition

The whole numbers answer the basic question "How many?" For example, if Alok collects baseball cards and A is the set of cards in his collection, then $a = n(A)$ is the number of cards he owns. Suppose his friend Barbara also has a collection of baseball cards, forming a set B with $b = n(B)$ cards. If Alok and Barbara decide to combine their collections, the new collection would be the set $A \cup B$ and would contain $n(A \cup B)$ cards. The **addition,** or **sum,** of a and b can be defined as the number of cards in the combined collection. That is, the sum of two whole numbers a and b is given by $a + b = n(A \cup B)$, where A and B are disjoint sets, $a = n(A)$, and $b = n(B)$.

Addition answers the question "How many elements are in the union of two disjoint sets?" Figure 2.13 illustrates how the set model is used to show that $3 + 5 = 8$.

First, disjoint sets A and B are found, with $n(A) = 3$ and $n(B) = 5$. Since $n(A \cup B) = 8$, we have shown that $3 + 5 = 8$. By using physical objects such as beans to form the sets, there is no question about disjointness.

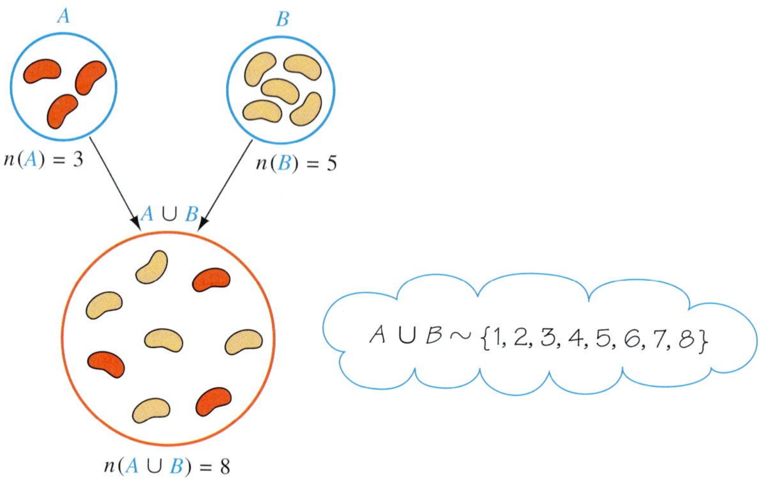

Figure 2.13
Showing $3 + 5 = 8$ with the set model of addition

Here is the general definition.

DEFINITION *The Addition of Whole Numbers*
Let a and b be any two whole numbers. If A and B are any two disjoint sets for which $a = n(A)$ and $b = n(B)$, then the **sum of *a* and *b*,** written $a + b$, is given by

$$a + b = n(A \cup B).$$

The expression $a + b$ is read "a plus b" and a and b are called the **addends** or **summands**.

$$\mathbf{a} + \mathbf{b}$$
↑ ↑ addends

EXAMPLE 2.11 USING THE SET MODEL OF ADDITION

The University Math Club membership includes 14 women and 11 men. The club members major in math, in physics, or in both of these areas.

(a) How many students belong to the Math Club?
(b) If 21 club members major at least in math, and 6 at least in physics, how many have double majors in both?

Solution

(a) If C denotes the set of members of the club, then $C = M \cup W$, where M and W denote the sets of men and women members, respectively. Since M and W are disjoint sets, the number of club members is

$$n(C) = n(M \cup W) = n(M) + n(W) = 11 + 14 = 25.$$

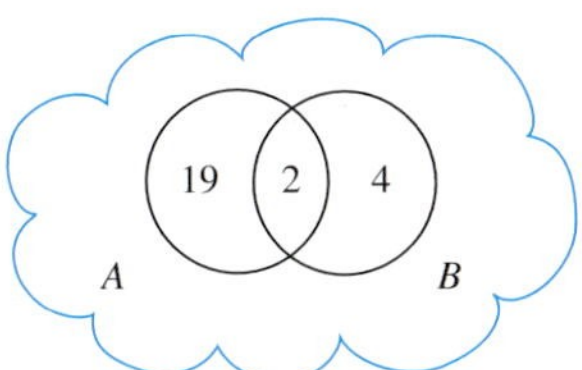

(b) Let A denote the set of club members with a math major, and B the set of club members with a physics major. Since $n(A) + n(B) = 21 + 6 = 27$ is 2 more than the 25 members of the club, 2 club members belong to both sets A and B. That is, $n(A \cap B) = 2$ club members have a double major. Notice the equation $n(A \cup B) = n(A) + n(B) - n(A \cap B)$ (see problem 24 of the problem set for Section 2.3).

The set model of addition can be illustrated with manipulatives such as those described in the last section. Many addition facts and patterns become evident when they are discovered and visualized by means of concrete representations. An example is the triangular numbers $t_1 = 1, t_2 = 3, t_3 = 6, \ldots$ introduced in Chapter 1. Recall that t_n denotes the nth triangular number, where the name refers to the triangular pattern that can be formed with t_n objects. By using number-tile patterns, as shown in Figure 2.14, we see that the sum of any two successive triangular numbers forms a square number. Indeed, we have the general formula

$$t_{n-1} + t_n = n^2, n = 1, 2, \ldots.$$

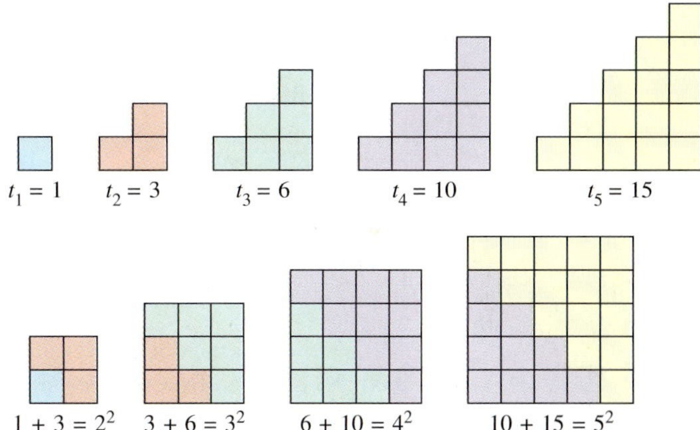

Figure 2.14
The sum of two successive triangular numbers is a square number

The Measurement Model of Addition

On the number line, whole numbers are geometrically interpreted as distances. Addition can be visualized as combining two distances to get a total distance. Right-pointing arrows are used to indicate distances. Figure 2.15 illustrates $3 + 5$ on the number line. It is important to notice that the two distances are not overlapping, and the tail of the arrow representing 5 is placed at the head of the arrow representing 3. The result of the addition is 8, which is indicated on the number line by circling the 8.

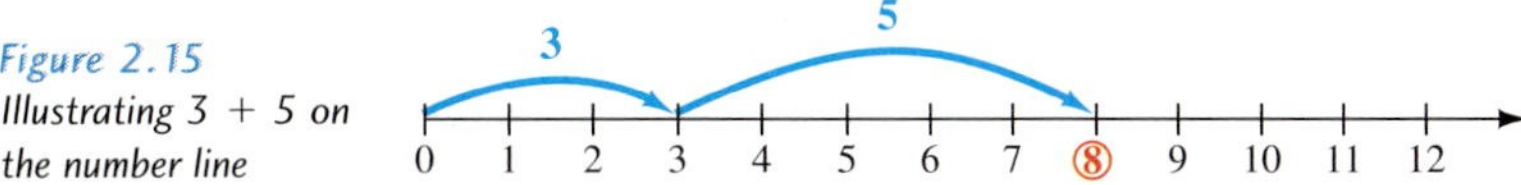

Figure 2.15
Illustrating 3 + 5 on the number line

Properties of Whole-Number Addition

The sum of any two whole numbers is also a whole number, so we say that the set of whole numbers has the **closure property** under addition. This property is so obvious for addition that it may seem unnecessary to mention it. However, the set of whole numbers is not closed under subtraction or division. Even under addition, many subsets of whole numbers do not have the closure property. For example, if $D = \{1, 3, 5, \ldots\}$ denotes the subset of the odd whole numbers, then D does *not* have the closure property under addition. (For instance, the sum of 1 and 3 is not in D.) On the other hand, the set of even whole numbers, $E = \{0, 2, 4, 6, \ldots\}$, is closed under addition; this is so because the sum of any two even whole numbers is also an even whole number.

Some other important properties of whole-number addition correspond to properties of operations on finite sets. For example, the commutative property of union, $A \cup B = B \cup A$, proves that $a + b = b + a$ for all whole numbers a and b. Similarly, the associative property $A \cup (B \cup C) = (A \cup B) \cup C$ tells us that $a + (b + c) = (a + b) + c$. Since $A \cup \varnothing = \varnothing \cup A = A$, we obtain the additive-identity property of zero, $a + 0 = 0 + a = a$. Zero is said to be the **additive identity** because of this property.

The properties in the following theorem can all be proved using the definition of whole-number addition and what we already know about operations on sets.

THEOREM *Properties of Whole-Number Addition*

Closure Property	If a and b are any two whole numbers, then $a + b$ is a unique whole number.
Commutative Property	If a and b are any two whole numbers, then $a + b = b + a$.
Associative Property	If a, b, and c are any three whole numbers, then $a + (b + c) = (a + b) + c$.
Additive-Identity Property of Zero	If a is any whole number, then $a + 0 = 0 + a = a$.

The properties of whole-number addition can also be illustrated with physical and pictorial models of the whole numbers. Figure 2.16 shows how the associative property can be demonstrated with number strips.

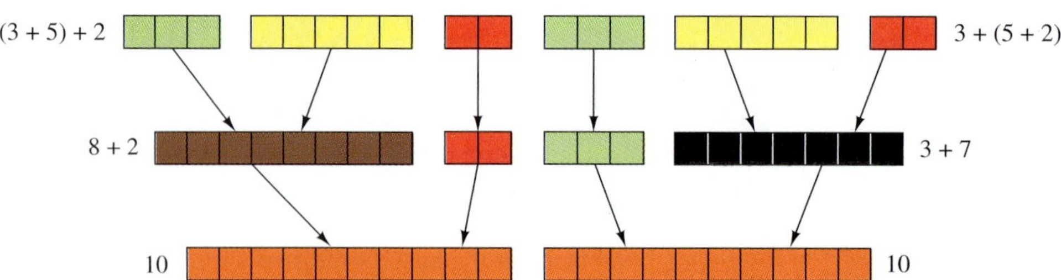

Figure 2.16
Illustrating the associative property $(3 + 5) + 2 = 3 + (5 + 2)$ with number strips or Cuisenaire® rods

The addition properties are very useful when adding several whole numbers, since we are permitted to rearrange the order of the addends and the order in which pairs of addends are summed.

Cooperative Investigation

Counting Cars and Trains

Materials Needed

A set of Cuisenaire® rods (or colored number strips) for each cooperative group of three or four students. The rods, which have lengths from 1 to 10 centimeters, are color coded as follows.

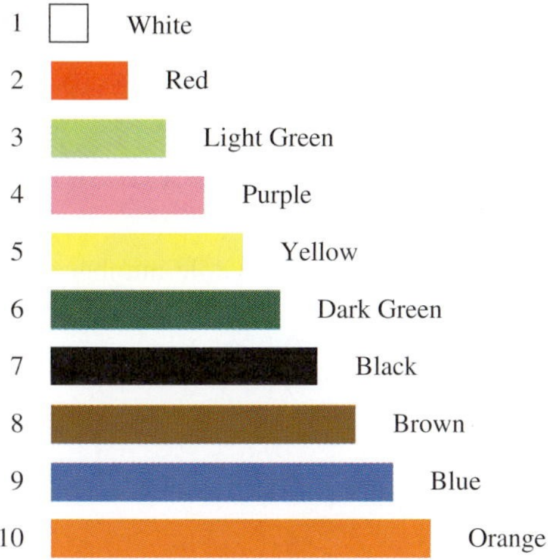

Directions

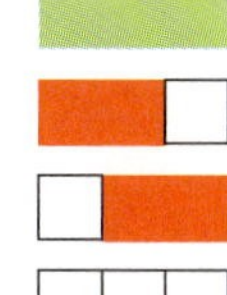

Form a train by placing one or more rods end-to-end: each rod in a train is a car. For example, in the figure to the right there are four ways to form trains that have the same overall length as the light green (LG) rod.

The order in which the cars appear is taken into account, so we consider the red–white and white–red trains as two different trains (imagine that the engine is the rightmost car, and the caboose is the leftmost car).

1. How many trains can you form that have the same length as the purple rod?
2. How many trains can you form that have the same length as the yellow rod?
3. What pattern do you observe in the number of trains of length 3, 4, and 5? On the basis of this pattern, predict the number of trains of length 8.
4. How many trains can be formed of length n, where n is any whole number and you use cars of lengths 1 up to n? Can you justify your conjecture? Can you give more than one justification?
5. A train made up of only red and white cars is called an RW-train. Answer questions 1 through 4 for RW-trains.
6. Call a train with no white cars a $\overline{\text{W}}$-train. Answer questions 1 through 4 for $\overline{\text{W}}$-trains.

EXAMPLE 2.12 | USING THE PROPERTIES OF WHOLE-NUMBER ADDITION

(a) Which property justifies each of the following statements?

(i) $8 + 3 = 3 + 8$

(ii) $(7 + 5) + 8 = 7 + (5 + 8)$

(iii) A million plus a quintillion is not infinite.

(b) Justify each equality below:

$$\begin{aligned}(20 + 2) + (30 + 8) &= 20 + [2 + (30 + 8)] && \text{(i)}\\ &= 20 + [(30 + 8) + 2] && \text{(ii)}\\ &= 20 + [30 + (8 + 2)] && \text{(iii)}\\ &= (20 + 30) + (8 + 2) && \text{(iv)}\end{aligned}$$

Solution

(a) (i) Commutative property, (ii) associative property, (iii) the sum is a whole number by the closure property, and is therefore a finite value.
(b) (i) associative property, (ii) commutative property, (iii) associative property, (iv) associative property.

EXAMPLE 2.13 ILLUSTRATING PROPERTIES ON THE NUMBER LINE

What properties of whole-number addition are shown below?

(a)

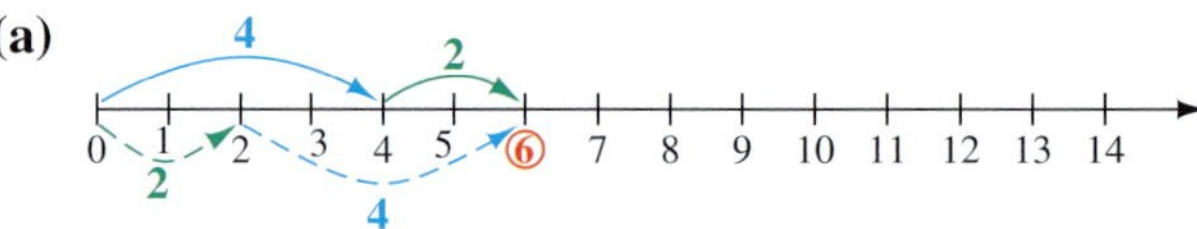

(b)

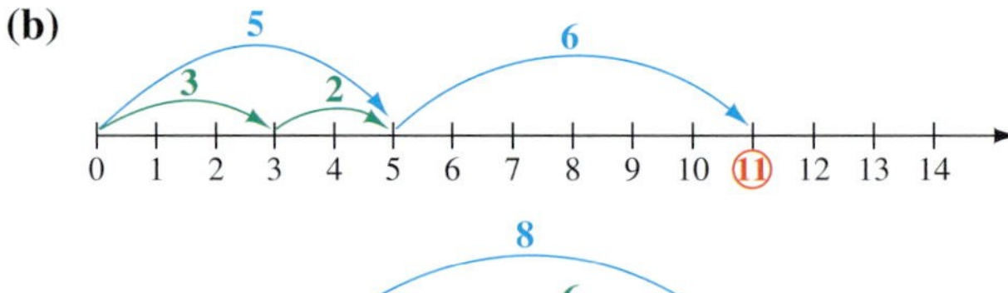

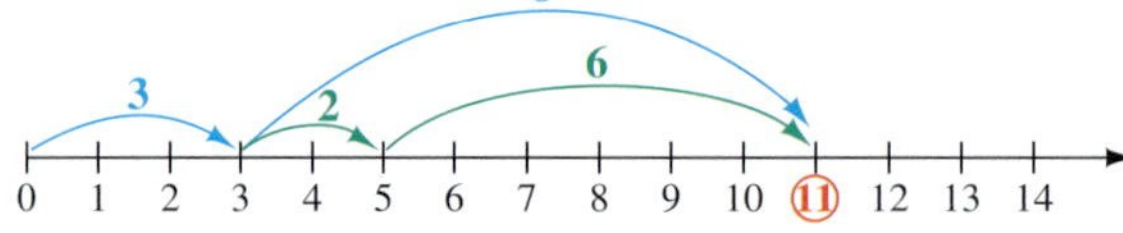

(c)

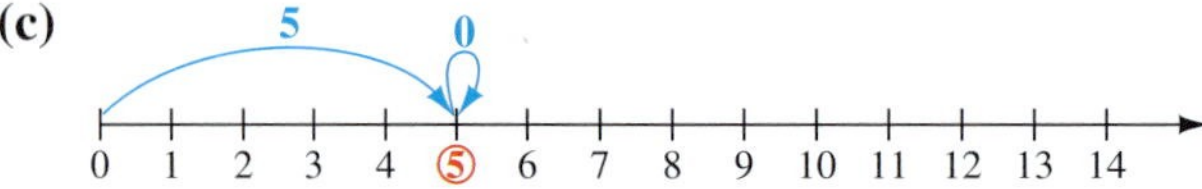

Solution

(a) The commutative property: $4 + 2 = 2 + 4$.
(b) The associative property: $(3 + 2) + 6 = 3 + (2 + 6)$.
(c) The additive-identity property: $5 + 0 = 5$.

Subtraction of Whole Numbers

Subtraction can be defined in a brief statement.

DEFINITION *Subtraction of Whole Numbers*
Let a and b be whole numbers. The **difference,** written $a - b$, is the unique whole number c such that $a = b + c$. That is, $a - b = c$ if, and only if, there is a whole number c such that $a = b + c$.

The expression $a - b$ is read "a minus b." The a is the **minuend** and b is the **subtrahend.**

Into the Classroom

Using Addition Properties to Learn Addition Facts

As children learn addition, the properties of whole-number addition should become a habit of thought from the very beginning. Indeed, the properties are useful even for learning and recalling the sums of one-digit numbers, as found in the addition table shown. The commutative property $a + b = b + a$ means the entries above the diagonal repeat the entries below the diagonal. Also, the first column of addition with zero is easy by the additive-identity property. The next four columns, giving addition with 1, 2, 3, and 4, can be learned by "counting on." For example, $8 + 3$ is viewed as "8 plus 1 makes 9, plus 1 more makes 10, plus 1 more makes 11." The diagonal entries $1 + 1 = 2, 2 + 2 = 4, \ldots$ are the "doubles," which are readily learned by knowing how to count by twos: $2, 4, \ldots, 18$. The remaining entries in the table can be obtained by combining "doubles" and "counting on." For example, $6 + 8$ is viewed as $6 + (6 + 2)$, which is $(6 + 6) + 2$; knowing that 6 doubled is 12 and counting on 2 gives the answer of 14.

Other effective strategies include the following:

Making tens: For example, $8 + 6 = (8 + 2) + 4 = 10 + 4 = 14$.
Counting back: For example, 9 is 1 less than 10, so $6 + 9$ is 1 less than $6 + 10 = 16$, giving 15. In symbols, this amounts to

$$6 + 9 = (6 + 10) - 1 = 16 - 1 = 15.$$

+	0	1	2	3	4	5	6	7	8	9
0	0	1	2	3	4	5	6	7	8	9
1	1	2	3	4	5	6	7	8	9	10
2	2	3	4	5	6	7	8	9	10	11
3	3	4	5	6	7	8	9	10	11	12
4	4	5	6	7	8	9	10	11	12	13
5	5	6	7	8	9	10	11	12	13	14
6	6	7	8	9	10	11	12	13	14	15
7	7	8	9	10	11	12	13	14	15	16
8	8	9	10	11	12	13	14	15	16	17
9	9	10	11	12	13	14	15	16	17	18

Look for patterns. They also help you learn the facts!

$a - b = c$

c is the difference of a and b
b is the subtrahend
a is the minuend

This definition relates subtraction to addition, and is the definition most easily extended to the integers, rational numbers, and real numbers.

Since $8 = 5 + 3$, the definition tells us that $8 - 5 = 3$. However, the practical value of subtraction is not revealed in the definition above. To understand the nature and value of subtraction, we will introduce four conceptual models: **take-away, missing**

addend, comparison, and **number-line** (or **measurement**). The following four problems illustrate each of the four conceptual models.

Take-away:
Eroll has \$8 and spends \$5 for a ticket to the movies. How much money does Eroll have left?

Missing addend:
Alice has read 5 chapters of her book. If there are 8 chapters in all, how many more chapters must she read to finish the book?

Comparison:
Georgia has 8 mice and Tonya has 5 mice. How many more mice does Georgia have than Tonya?

Number-line:
Mike hiked up the mountain trail 8 miles. Five of these miles were hiked after lunch. How many miles did Mike hike before lunch?

In all four problems, the answer is 3 because we know the addition fact $8 = 5 + 3$. For example, Eroll's \$8, when written as \$5 + \$3, shows that a \$5 movie ticket and the \$3 still in his pocket account for all of the \$8 he had originally. The cashier at the box office "took away" 5 of Eroll's 8 dollars, leaving him with \$3. Thus, the problem is an example of the **take-away model** of subtraction. Alice, having read 5 chapters, wants to know how many more chapters she must read; the emphasis now is on what number is to be added to 5 to get 8, so the second problem illustrates the **missing-addend model** of subtraction. Similarly, the remaining problems illustrate the **comparison model** and the **number-line** (or **measurement**) **model** of subtraction.

The four basic conceptual models of the subtraction $8 - 5$ are visualized as follows.

Take-Away Model

1. Start with 8 objects.

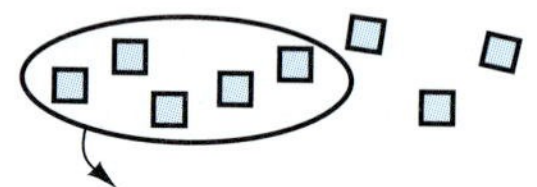

2. Take away 5 objects.
3. How many objects are left?

Missing-Addend Model

1. Start with 5 objects.
2. How many more objects are needed to give a total of 8 objects?

Comparison Model

1. Start with two collections, with 8 objects in one collection and 5 in the other.

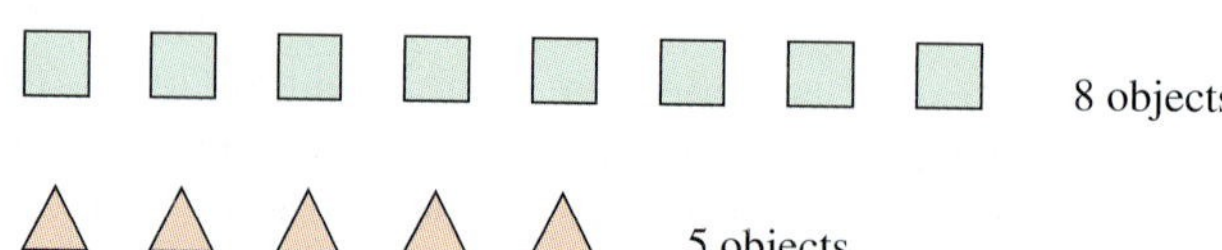

2. How many more objects are in the larger collection?

Number-Line Model

1. Move forward (to the right) 8 units.

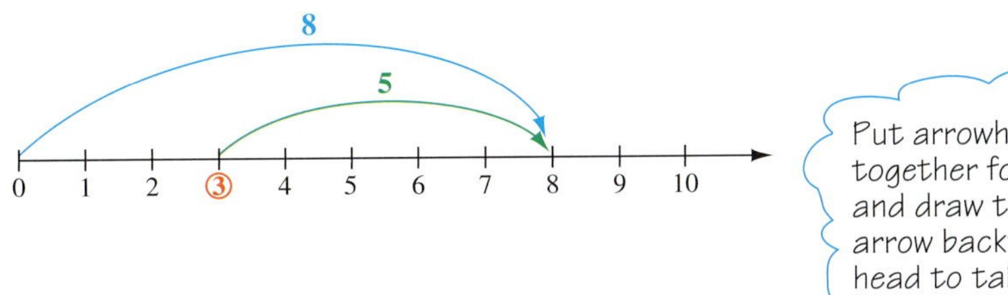

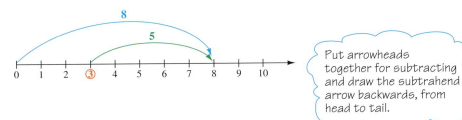

Put arrowheads together for subtracting and draw the subtrahend arrow backwards, from head to tail.

2. Remove a jump to the right of 5 units.
3. What is the distance from 0?

Notice that the head of the arrow representing 5 is positioned at the head of the arrow representing 8 and is drawn backwards, since it is subtracted. The result of the subtraction, 3, is shown by circling the 3.

It is important to notice that the whole numbers 8 and 5 are each represented by a right-pointing arrow. Since the outcome of the subtraction $8 - 5$ is 3, the arrow from 0 to 3 on the number line is also right-pointing.

More generally, any whole number is represented on a number line as a right-pointing arrow. In Chapter 5, the whole numbers are enlarged to include the positive and negative integers. There, right-pointing arrows represent the positive integers (natural numbers), and left-pointing arrows represent the negative integers.

EXAMPLE 2.14 IDENTIFYING CONCEPTUAL MODELS OF SUBTRACTION

Identify the conceptual model of subtraction that best fits these problems.

(a) Mary got 43 pieces of candy trick-or-treating on Halloween. Karen got 36 pieces. How many more pieces of candy does Mary have than Karen?

(b) Mary gave 20 pieces of her 43 pieces of candy to her sick brother, Jon. How many pieces of candy does Mary have left?

(c) Karen's older brother, Ken, collected 53 pieces of candy. How many more pieces of candy would Karen need to have as many as Ken?

(d) Ken left home and walked 10 blocks east along Grand Avenue trick-or-treating. The last 4 blocks were after crossing Main Street. How far is Main Street from Ken's house?

Solution

(a) Comparison model

(b) Take-away model

(c) Missing-addend model

(d) Number-line model

The set of whole numbers is not closed under subtraction. For example, $2 - 5$ is undefined, since there is no whole number n that satisfies $2 = 5 + n$. Similar reasoning shows that subtraction is not commutative, so the order in which a and b are taken is important. Neither is subtraction associative, which means that parentheses must be placed with care in expressions involving subtractions. For example,

$$5 - (3 - 1) = 5 - 2 = 3, \text{ but } (5 - 3) - 1 = 2 - 1 = 1.$$

Highlight from History

The Origins of the + and − Symbols

The + and − symbols first appeared in an arithmetic book by Johannes Widman, published in Leipzig in 1489. The signs were not used as symbols of operation, but simply to indicate excess or deficiency. Widman states, "What is −, that is minus, what is + , that is more." The + symbol is thought to be a contraction of the Latin word *et,* which often indicated addition. The minus symbol may be a contraction of the abbreviation $\overline{\text{m}}$ for minus. The + and − symbols were used to designate algebraic operations by the Dutch mathematician Vander Hoecke in 1514.

Cooperative Investigation

Diffy The process in this activity is sometimes called "Diffy." The name comes from the process of taking successive differences of whole numbers, and the activity provides an interesting setting for practicing skills in subtraction. The activity can be done on a spreadsheet (see Example B.2 of Appendix B) or with a graphing calculator program (see Appendix C). Work in pairs to check each other's calculations.

Directions

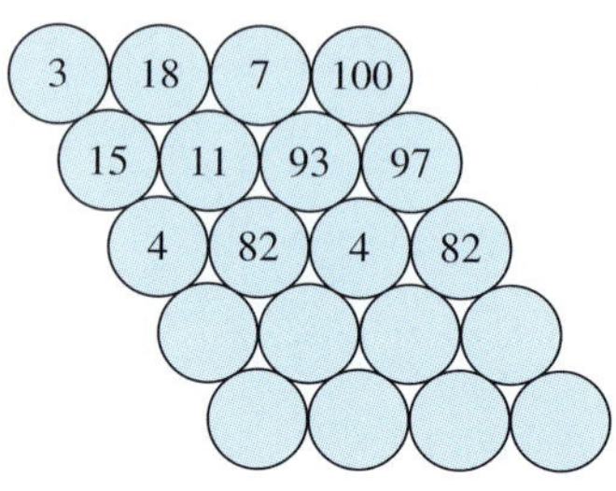

Step 1. Make an array of circles as shown and choose four whole numbers to place in the top four circles.

Step 2. In the first three circles of the second row write the differences of the numbers above and to the right and left of the circle in question, always being careful to subtract the smaller of these two numbers from the larger. In the fourth circle of the second row place the difference of the numbers in the first and fourth circles in the preceding row, again subtracting the smaller number from the larger.

Step 3. Repeat Step 2 to fill in successive rows of circles in the diagram. You may stop if you obtain a row of zeros.

Step 4. Replace the four numbers in the top row, and then repeat Steps 1, 2, and 3 several times, each time replacing the four numbers in the top row with different numbers.

Questions

1. Do you think the process will always stop?
2. Can you find four numbers such that the process terminates at the first step? the second step? the third step? Try several sets of starting numbers.
3. On the basis of your work so far, what do you guess is the largest number of steps needed for the process to stop?
4. Can you find four starting numbers such that the process requires eight steps to reach termination?
5. Try Diffy with the starting numbers 17, 32, 58, and 107.

Problem Set 2.3

Understanding Concepts

1. Let $A = \{\text{apple, berry, peach}\}$, $B = \{\text{lemon, lime}\}$, and $C = \{\text{lemon, berry, prune}\}$.
   (a) Find (i) $n(A \cup B)$, (ii) $n(A \cup C)$, and (iii) $n(B \cup C)$.
   (b) In which case is the number of elements in the union *not* the sum of the number of elements in the individual sets?
2. Let $n(A) = 5$, $n(B) = 8$, and $n(A \cup B) = 10$. What can you say about $n(A \cap B)$?
3. Let $n(A) = 4$ and $n(A \cup B) = 8$.
   (a) What are the possible values of $n(B)$?
   (b) If $A \cap B = \varnothing$, what is the only possible value of $n(B)$?
4. Draw number strips (or rods) to illustrate
   (a) $4 + 6 = 10$;
   (b) $2 + 8 = 8 + 2$; and
   (c) $3 + (2 + 5) = (3 + 2) + 5$.
5. Illustrate each of these additions with a number-line diagram. Circle the value on the number line that corresponds to the sum.
   (a) $3 + 5$
   (b) $5 + 3$
   (c) $4 + 2$
   (d) $0 + 6$
   (e) $3 + (5 + 7)$
   (f) $(3 + 5) + 7$
6. Make up a word problem that uses the set model of addition to illustrate $30 + 28$.
7. Make up a word problem that uses the measurement model of addition to illustrate $18 + 25$.
8. Which of the following sets of whole numbers is closed under addition? If the set is not closed, give an example of two elements from the set whose sum is not in the set.
   (a) $\{10, 15, 20, 25, 30, 35, 40, \ldots\}$
   (b) $\{1, 2, 3, \ldots, 1000\}$
   (c) $\{0\}$
   (d) $\{1, 5, 6, 11, 17, 28, \ldots\}$
   (e) $\{n \in N \mid n \geq 19\}$
   (f) $\{0, 3, 6, 9, 12, 15, 18, \ldots\}$
9. What properties of addition are used in these equalities?
   (a) $14 + 18 = 18 + 14$
   (b) 12,345,678 + 97,865,342 is a whole number
   (c) $18 + 0 = 18$
   (d) $(17 + 14) + 13 = 30 + 14$
   (e) $(12 + 15) + (5 + 38) = 50 + 20$
10. An easy way to add $1 + 2 + 3 + 4 + 5 + 6 + 7 + 8 + 9 + 10$ is to write the sum as $(1 + 10) + (2 + 9) + (3 + 8) + (4 + 7) + (5 + 6) = 11 + 11 + 11 + 11 + 11 = 55$.
    (a) Compute $1 + 2 + 3 + \cdots + 20$. Describe your procedure.
    (b) What properties of addition are you using to justify why your procedure works?
11. Illustrate each of these subtractions with a number-line diagram. Remember to put the heads of the minuend and subtrahend arrows together.
    (a) $7 - 3$
    (b) $7 - 4$
    (c) $7 - 7$
    (d) $7 - 0$
12. If it is known that $5 + 9 = 14$, then it follows that $9 + 5 = 14$, $14 - 9 = 5$, and $14 - 5 = 9$. Many elementary texts call such a group of four basic facts a "fact family."
    (a) What is the fact family that contains $5 + 7 = 12$?
    (b) What is the fact family that contains $12 - 4 = 8$?
13. (a) Draw four number-line diagrams to illustrate each of the number facts in the fact family (see problem 12) $5 + 9 = 14$, $9 + 5 = 14$, $14 - 9 = 5$, and $14 - 5 = 9$.
    (b) Repeat part (a), but for the fact family that contains $11 - 4 = 7$.
14. Which subtraction model—take-away, missing addend, comparison, or measurement—corresponds best to each problem below?
    (a) Ivan solved 18 problems and Andreas solved 13 problems. How many more problems has Ivan solved than Andreas?
    (b) On Malea's 12-mile hike to Mirror Lake, she came to a sign that said she had 5 miles left to reach the lake. How far has Malea walked so far?
    (c) Jake has saved \$45 toward the \$60 CD player he wants to buy. How much money does he have to add to his savings to purchase the player?
    (d) Joshua has a book of 10 pizza coupons. If he used 3 coupons to buy pizza for Friday's party, how many coupons does he have left?
15. Make up a word problem that corresponds well to each of the four conceptual models for subtraction.
    (a) Take-away

(b) Missing addend
(c) Comparison
(d) Measurement (number-line)

16. Jeff must read the last chapter of his book. It begins on the top of page 241 and ends at the bottom of page 257. How many pages must he read?

17. Parentheses must be inserted to make the expression $6 - 3 + 2 - 1$ unambiguous. For example, $(6 - 3) + (2 - 1) = 3 + 1 = 4$ or $(6 - (3 + 2)) - 1 = (6 - 5) - 1 = 1 - 1 = 0$. For each expression below, insert parentheses to see how many different values can be obtained.

(a) $9 - 5 - 3 - 1$
(b) $9 + 7 - 2 + 1$
(c) $5 + 7 + 3 + 2$

18. Notice that $(6 + (8 - 5)) - 2 = 7$, but a different placement of parentheses on the left side would give the statement $(6 + 8) - (5 - 2) = 11$. Place parentheses to turn these into *true* statements.

(a) $8 - 5 - 2 - 1 = 2$
(b) $8 - 5 - 2 - 1 = 4$
(c) $8 - 5 - 2 - 1 = 0$
(d) $8 + 5 - 2 + 1 = 12$
(e) $8 + 5 - 2 + 1 = 10$

19. Each circled number is the sum of the adjacent row, column, or diagonal of the numbers in the square array.

6	4	(10)
2	5	(7)
(8)	(9)	(11)

Fill in the missing entries in these patterns.

(a)

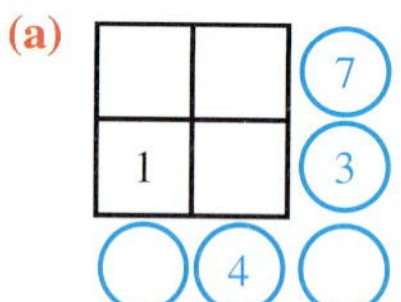

(b)

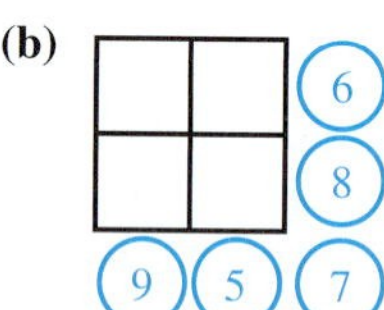

20. The first figure below shows that the numbers 1, 2, 3, 4, 5, 6 can be placed around a triangle in such a way that the three numbers along any side sum to 9, which is shown circled. Arrange the numbers 1, 2, 3, 4, 5, 6 to give the sums circled in the next three figures.

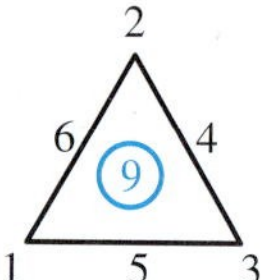

(a)

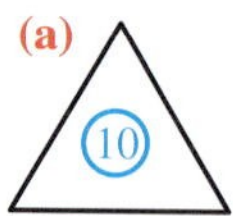

(b)

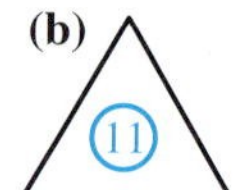

(c)

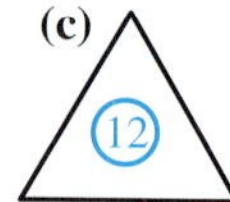

Teaching Concepts

21. Suppose that Andrea told you that "Blake and I started out with the same number of marbles, but I gave him one of mine. Now Blake has one more marble than me." Why is Andrea's reasoning incorrect? How can you address her misunderstanding?

22. The addition operation can be represented with beans and loops of string, as illustrated in Figure 2.13. In words and pictures, describe how you would use this representation to illustrate the properties of whole-number addition to a youngster. For example, your figure illustrating the associative property will be something like the number-strip diagram in Figure 2.16.

23. Invent ways to use a double-six set of dominoes to teach children basic addition facts and addition properties. Notice that a blank half of the domino can represent zero.

Thinking Critically

24. If $A = \{a, b, c, d\}$ and $B = \{c, d, e, f, g\}$, then $n(A \cup B) = n(\{a, b, c, d, e, f, g\}) = 7$, $n(A) = 4$, $n(B) = 5$, and $n(A \cap B) = n(\{c, d\}) = 2$. Since $7 = 4 + 5 - 2$, this suggests that

$$n(A \cup B) = n(A) + n(B) - n(A \cap B).$$

Use Venn diagrams to justify this formula for arbitrary finite sets A and B.

25. Since $12 \times 16 = 192$ and $5 \times 40 = 200$, it follows that among the first 200 natural numbers $\{1, 2, \ldots, 200\}$ there are 16 that are multiples of 12, and 40 that are multiples of 5. Just 3 are multiples of *both* 5 and 12, namely 60, 120, and 180. Use the formula of problem 24 to find how many natural numbers in the set $\{1, 2, \ldots, 200\}$ are divisible by *either* 12 or 5 or both.

26. **A Super Magic Square from India.** Magic squares and other magic configurations offer a way to give children considerable practice in addition and subtraction, with the bonus of having fun while doing so. Not counting reflection and rotation symmetries, there are 880 magic squares of size 4 by 4. The one shown on the left below, which was discovered in India around the eleventh or twelfth century, is one of the most magic of all. Its rows, columns, and diagonals all sum to the magic constant 34, but you should be able to find a great many other patterns of four small squares whose entries also sum to the magic constant. For example, the four-corner squares do so, as can be indicated by the shading pattern on the right. Use similar shading patterns to describe other patterns of four squares that sum to the magic constant.

7	12	1	14
2	13	8	11
16	3	10	5
9	6	15	4

27. **Adams's Magic Hexagon.** In 1957 Clifford W. Adams discovered a magic hexagon, in which the sum of the numbers in any "row" is 38. Fill in the empty cells of the partially completed hexagon shown, using the whole numbers 6, 7, . . . , 15, to re-create Adams's discovery. When completed, each cell will contain one of the numbers 1, 2, . . . , 19.

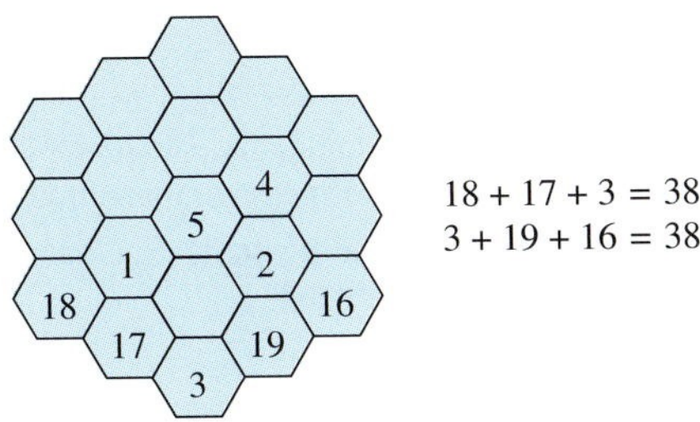

28. **A Magic Hexagram.** The numbers 1, 2, . . . , 12 can be entered into the 12 regions of the diagram below so that each of the six rows of five triangles (shown by the arrows) contains numbers that sum to the magic constant 33. The first six numbers have been put into place, so you are challenged to enter the last six numbers, 7 through 12, to finish this magic hexagram discovered in 1991 by the mathematicians Brian Bolt, Roger Eggleton, and Joe Gilks.

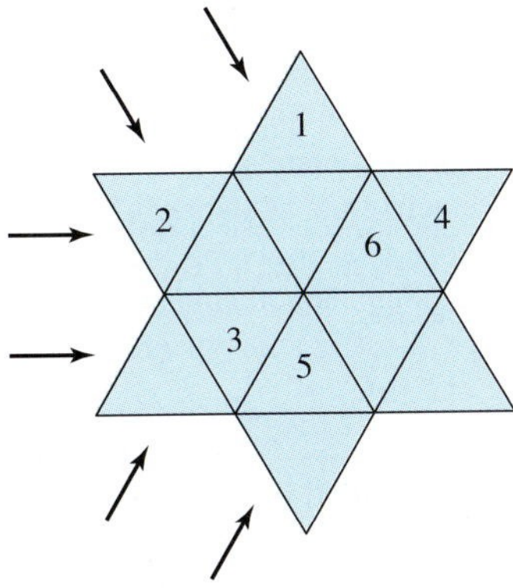

29. There is a nonempty subset of the whole numbers that is closed under subtraction. Find this subset.

30. The set C contains 2 and 3 and is closed under addition.

(a) What whole numbers must be in C?

(b) What whole numbers may not be in C?

(c) Are there any whole numbers definitely not in C?

(d) How would your answers to (a), (b), and (c) change if 2 and 4 instead of 2 and 3 were contained in C?

31. The triangular numbers $t_1 = 1$, $t_2 = 3$, $t_3 = 6$, . . . are shown in Figure 2.14.

(a) Complete the following table of the first 15 triangular numbers:

n	1	2	3	4	5	6	7	8	9	10	11	12	13	14	15
t_n	1	3	6	10											

(b) The first 10 natural numbers can be expressed as sums of triangular numbers. For example, $1 = 1$, $2 = 1 + 1$, $3 = 3$, $4 = 1 + 3$, $5 = 1 + 1 + 3$, $6 = 6$, $7 = 1 + 6$, $8 = 1 + 1 + 6$, $9 = 3 + 6$, and $10 = 10$. Show that the natural numbers 11 through 25 can be written as a sum of triangular numbers. Use as few triangular numbers as possible each time.

(c) Choose five more numbers at random (don't look at your table!) between 26 and 120, and write each of them as a sum of as few triangular numbers as possible. What is the largest number of triangular numbers needed?

(d) On July 10, 1796, the 19-year-old Carl Friedrich Gauss wrote in his notebook, "EUREKA! NUM = △ + △ + △." What theorem do you think Gauss had proved?

32. **Fibonacci Sums.** Recall the Fibonacci numbers 0, 1, 1, 2, 3, 5, 8, 13, 21, 34, 55, 89, 144, ... Sameer claims that any natural number can be written as a sum of two or more distinct Fibonacci numbers. For example, $29 = 21 + 5 + 3$ and $55 = 34 + 21$. Do you think Sameer might be right? Investigate this possibility yourself and offer your opinion to support or refute Sameer's contention.

Thinking Cooperatively

33. The following table represents a partially complete scrambled addition table. The rows and columns have both been mixed up. See if your cooperative team can complete the table.

+	5					2				3
3										
					18					
				12						
			5			6				
							0			
			8						14	
5										
						3				
8								16		

34. Have your cooperative team make up its own incomplete scrambled addition-table puzzle, similar to the one in problem 33. It may be helpful to use two colors, to separate the entries that are given from the entries that are determined by what is given. Trade puzzles among groups and see which presents the most challenge.

35. Three numbers (not necessarily all different) are determined by rolling three dice. By using either two or three of these numbers (not just one), together with parentheses and + and − symbols, the objective of the game is to form expressions equalling the seven numbers 0, 1, 2, 3, 4, 5, and 6 on a hex board as shown. The team filling in all seven positions on the board first is the winning team. If no team fills in all seven positions, the dice are rolled again. For example, a roll of 3, 3, and 4 would allow you to fill in 0 (as $3 - 3$), 1 (as $4 - 3$), 2 (as $(3 + 3) - 4$), 4 (as $(3 - 3) + 4$ or $(3 + 4) - 3$, say), and 6 (as $3 + 3$). Your team would hope to fill in the 3 and 5 on the next roll of the dice. Play the game in class, and practice with these questions:

(a) Show that you should be able to win the game with the roll 1, 4, 5.

(b) Can you win the game with a single roll of 2, 3, 5?

(c) Can you win the game with a single roll of 2, 4, 6?

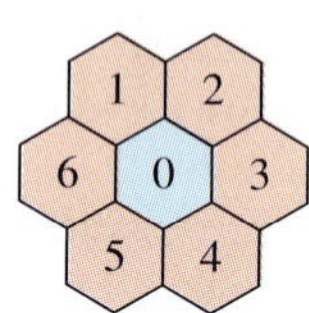

Making Connections

36. Addition and subtraction problems arise frequently in everyday life. Consider such activities as scheduling time, making budgets, sewing clothes, making home repairs, planning finances, modifying recipes, and making purchases. Make a list of five addition and five subtraction problems you have encountered at home or in your work. Which conceptual model of addition or subtraction corresponds best to each problem?

From State Student Assessments

37. (Massachusetts, Grade 4)
Corey worked these problems.

$2 + 8 = 10$	$6 + 6 = 12$	$10 + 4 = 14$
$12 + 2 = 14$	$8 + 4 = 12$	$2 + 6 = 8$

When Corey finished, he said, "I think that the sum of ANY two numbers is always an even number." Maya looked at Corey's work. She thought a bit and said, "I think that the sum of any two ODD numbers is always an odd number."

(a) Is Corey correct? Explain the reasons for your answer using pictures, numbers, or words.

(b) Is Maya correct? Explain the reasons for your answer using pictures, numbers, or words.

38. (Massachusetts, Grade 4)
Which of the following problems CANNOT be solved using the number sentence below?

$$15 - 8 = \square$$

A. Siu Ping had 15 trading cards. She gave 8 to Jim. How many does she have now?

B. Siu Ping needs 15 more trading cards than Jim has. Jim has 8. How many does Siu Ping need?

C. Siu Ping has 15 trading cards. Jim has 8. How many more does Siu Ping have than Jim?

D. Siu Ping needs 15 trading cards. She has 8. How many more does she need?

For Review

39. The Venn diagram shows the number of elements in each region. Find the numbers of elements in the sets indicated.

(a) $n(A \cap B)$ **(b)** $n(B \cup C)$ **(c)** $n(\overline{A} \cup C)$
(d) $n(B \cap C)$ **(e)** $n(A \cap \overline{C})$ **(f)** $n(\overline{A \cup B \cup C})$

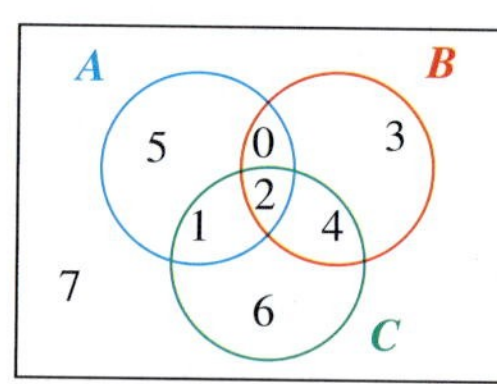

40. A school of 300 students has 100 students in each of algebra, geometry, and statistics, with 30 students enrolled in each pair of classes and 10 students enrolled in all three classes. How many students are taking only geometry? How many students are not taking any of these three classes?

2.4 MULTIPLICATION AND DIVISION OF WHOLE NUMBERS

Multiplication as Repeated Addition

Misha has an after-school job at a local bike factory. Each day she has a 3-mile round-trip walk to the factory. At her job she assembles 4 hubs and wheels. How many hubs and wheels does she assemble in 5 afternoons? How many miles does she walk to and from her job each week?

These two problems can be answered by repeated addition. Misha assembles

$$4 + 4 + 4 + 4 + 4 = 20$$

Sum of 5 fours, written $5 \cdot 4$

hubs and wheels. Repeated addition is illustrated by the set diagram shown in Figure 2.17.

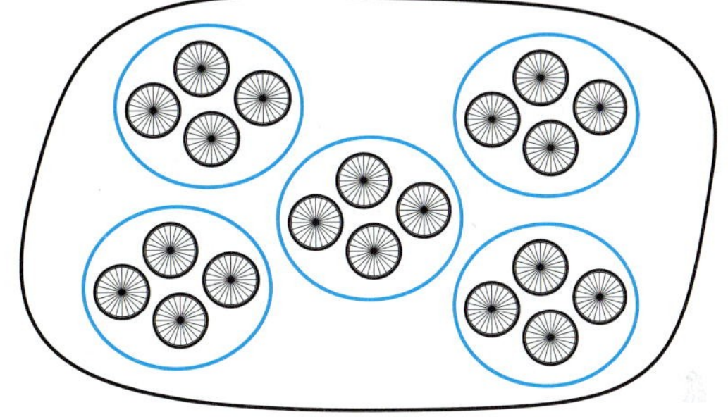

Figure 2.17
A set model shows that 5 times 4 is 20 since $4 + 4 + 4 + 4 + 4 = 20$ by repeated addition

Misha walks

$$3 + 3 + 3 + 3 + 3 = 15$$

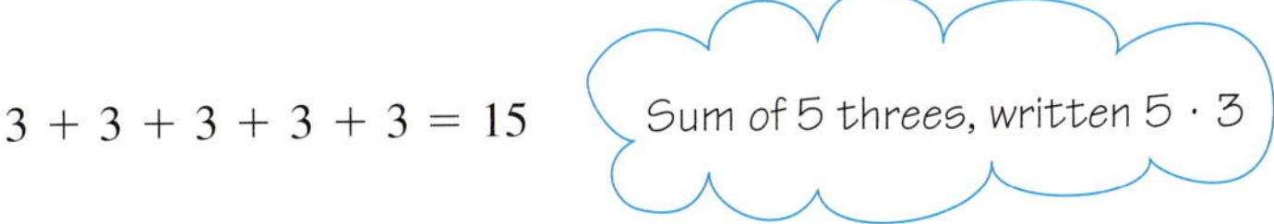

miles each week. This is illustrated by the number-line model in Figure 2.18.

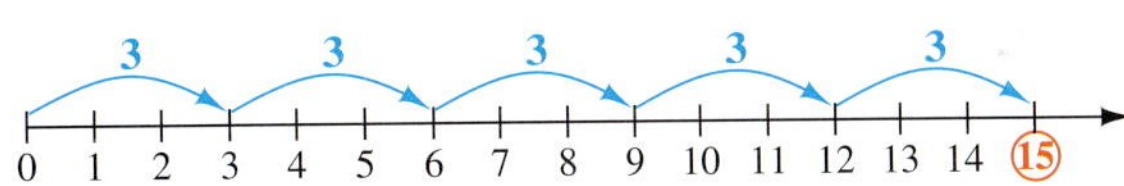

Figure 2.18
Number-line model to show 5 times 3 is 15

If Misha only worked one day in the week, she would assemble $1 \cdot 4 = 4$ wheels. If she were sick all week and missed work entirely, she would not assemble any wheels; therefore $0 \cdot 4 = 0$.

Since multiplication is defined for all pairs of whole numbers, and the outcome is also a whole number, we see that multiplication is a binary operation on the set of whole numbers.

DEFINITION *Multiplication of Whole Numbers as Repeated Addition*
Let a and b be any two whole numbers. Then the **product** of a and b, written $a \cdot b$, is defined by

$$a \cdot b = \underbrace{b + b + \cdots + b}_{a \text{ addends}} \text{ when } a \neq 0$$

and by

$$0 \cdot b = 0.$$

The dot symbol for multiplication is often replaced by a cross, $\times$ (not to be mistaken for the letter x), or by a star, * (asterisk), the symbol computers use most often. Sometimes no symbol at all is used, or parentheses are placed around the factors. Thus, the expressions

$$a \cdot b, \quad a \times b, \quad a * b, \quad ab, \quad \text{and} \quad (a)(b)$$

all denote the multiplication of a and b. Each whole number, a and b, is a **factor** of the product $a \cdot b$, and often $a \cdot b$ is read "a times b."

In addition to the set model and number-line model, multiplication can also be represented by the *array model,* the *rectangular area model,* the *multiplication tree model,* and the *Cartesian product model.* Each of these models provides a useful conceptual and visual representation of the multiplication operation. For a given context in a problem, it is frequently obvious which model provides the most natural representation, but often there are several models that are each a reasonable choice.

The Array Model for Multiplication

Suppose Lida, as part of her biology research, planted 5 rows of bean seeds and each row contains 8 seeds. How many seeds did she plant in her rectangular plot?

The 40 seeds Lida planted form a 5-by-8 rectangular array, as shown in Figure 2.19. In the array model, the numbers of rows and columns in the rectangular array easily identify the factors in the multiplication. Also, the result of the product is found by counting the number of discrete objects in the array.

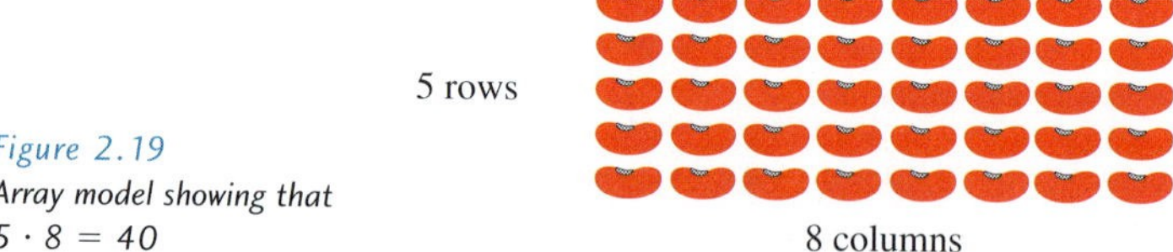

Figure 2.19
Array model showing that $5 \cdot 8 = 40$

The Rectangular Area Model for Multiplication

Janet wants to order square ceramic tiles to cover the floor of her 4-foot by 6-foot hallway. If the tiles are each 1 square foot, how many will she need to order? The rectangular area model, as shown in Figure 2.20, shows that 24 tiles are required.

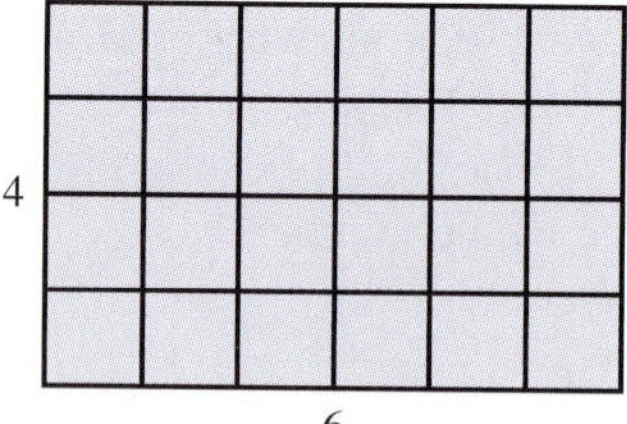

Figure 2.20
Rectangular area model showing that $4 \times 6 = 24$

In this model, the dimensions of the rectangle correspond to the factors, and the area of the rectangle corresponds to the value of the product. We will show in later chapters that the rectangular area model for multiplication is especially important because it extends from whole-number multiplication to the multiplication of rational numbers, integers, and even real numbers. For example, if Janet's floor measured $4\frac{1}{2}$ by 6 feet, the array would show that Janet needs $4\frac{1}{2} \times 6 = 27$ tiles, where some of the tiles need to be cut.

The Multiplication Tree Model

Melissa has a box of 4 flags, colored red, yellow, green, and blue. How many ways can she display two of the flags on a flagpole? It is useful to think of the two decisions Melissa must make. First, she must select the uppermost flag, and next, she must choose the lower flag from those remaining in the box. There are 4 choices for the uppermost flag, and for *each* of these choices there are 3 choices for the lower flag. The multiplication tree shown in Figure 2.21 represents the product 4×3.

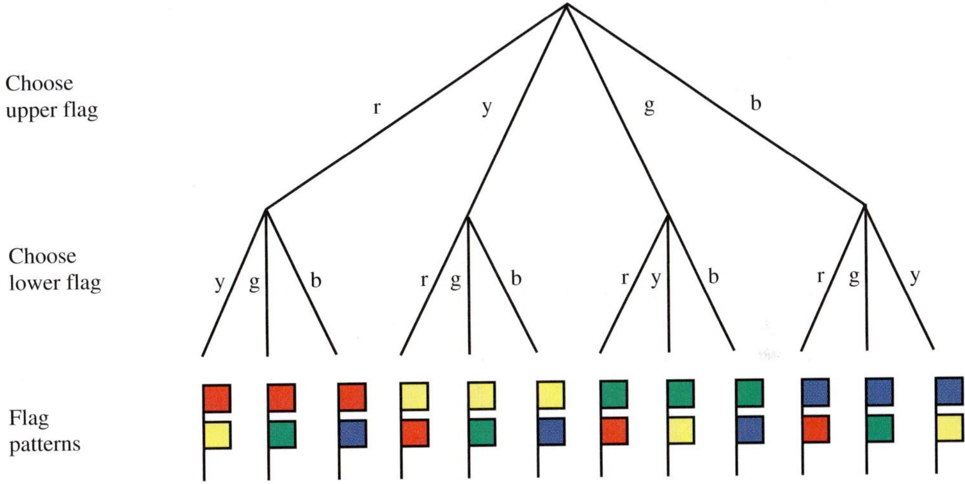

Figure 2.21
A multiplication tree showing that $4 \times 3 = 12$

The multiplication tree in Figure 2.21 looks more like a tree if you turn it upside down. Often, multiplication trees grow to the right.

The Cartesian Product Model of Multiplication

At Sonya's Ice Cream Shop, a customer can order either a sugar or a waffle cone and one of four flavors of ice cream—vanilla, chocolate, mint, or raspberry. Any ice cream cone order can be written as an **ordered pair (a, b),** where the first component a of the ordered pair indicates the type of cone and the second component b of the ordered pair indicates the flavor. For example, if $C = \{s, w\}$ and $F = \{v, c, m, r\}$ are the respective sets of cone types and flavors, then an order for raspberry ice cream in a waffle cone corresponds to the ordered pair (w, r). The total set of ice cream orders can be pictured in a rectangular array much like the figures used to model multiplication as a rectangular array.

		Flavor			
		v	c	m	r
Type of cone	s	(s, v)	(s, c)	(s, m)	(s, r)
	w	(w, v)	(w, c)	(w, m)	(w, r)

The set of ordered pairs shown in the table is called the **Cartesian product** of C and F, written $C \times F$, where the cross symbol $\times$ should not be confused with the letter x. The name "Cartesian" honors the French mathematician and philosopher René Descartes (1596–1650). Here is the general definition of the Cartesian product of two sets.

DEFINITION *Cartesian Product of Sets*
The **Cartesian product** of sets A and B, written $A \times B$, is the set of all ordered pairs whose first component is an element of set A and whose second component is an element of set B. That is,

$$A \times B = \{(a, b) \mid a \in A \text{ and } b \in B\}.$$

At Sonya's, there are $n(C) = 2$ ways to choose the type of cone and $n(F) = 4$ ways to choose the flavor of ice cream. This gives $n(C \times F) = 2 \cdot 4 = 8$ ways to order an ice cream cone. More generally, the Cartesian product of sets gives us an alternative way to define the multiplication of whole numbers.

ALTERNATIVE DEFINITION *Multiplication of Whole Numbers via the Cartesian Product*
Let a and b be whole numbers, and let A and B be any sets for which $a = n(A)$ and $b = n(B)$. Then $a \cdot b = n(A \times B)$.

Since $\varnothing \times B = \varnothing$ (there are no ordered pairs in $\varnothing \times B$, since no first component can be chosen from $\varnothing$) and $0 = n(\varnothing)$, this definition of multiplication is consistent with the earlier definition of multiplication by 0; that is, $0 \cdot b = b \cdot 0 = 0$.

EXAMPLE 2.15 USING THE CARTESIAN PRODUCT MODEL OF MULTIPLICATION

To get to work, Juan either walks, rides the bus, or takes a cab from his house to downtown. From downtown, he either continues on the bus the rest of the way to work or catches the train to his place of business. How many ways can Juan get to work?

Solution

If $A = \{\text{w, b, c}\}$ is the set of possibilities for the first leg of his trip, and $B = \{\text{b, t}\}$ is the next set of choices, then Juan has altogether

$$3 \cdot 2 = n(A \times B) = n(\{(\text{w, b}), (\text{w, t}), (\text{b, b}), (\text{b, t}), (\text{c, b}), (\text{c, t})\}) = 6$$

ways to commute to work. We notice that it doesn't matter whether or not A and B have an element in common.

Properties of Whole-Number Multiplication

It follows from the definition that the set of whole numbers is **closed under multiplication:** the product of any two whole numbers is a unique whole number. We also observed earlier that $0 \cdot b = 0$ and $1 \cdot b = b$ for all whole numbers b.

By rotating a rectangular array through 90°, we interchange the number of rows and the number of columns in the array, but we do not change the total number of objects in the array. Thus $a \cdot b = b \cdot a$, which demonstrates the **commutative property of multiplication.** An example is shown in Figure 2.22.

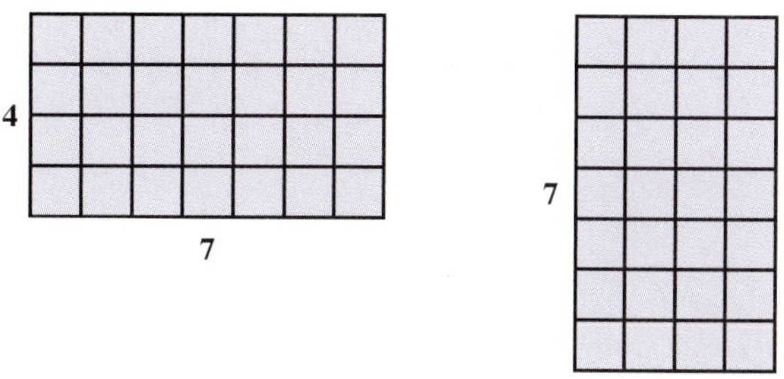

Figure 2.22
Multiplication is commutative:
$4 \cdot 7 = 7 \cdot 4$

Since we already know that $1 \cdot b = b$ for any whole number b, the commutative property tells us that $1 \cdot b = b \cdot 1 = b$. This makes 1 a **multiplicative identity.**

A three-dimensional model, as shown in Figure 2.23, shows that multiplication is **associative,** so $a \cdot (b \cdot c) = (a \cdot b) \cdot c$. This means that an expression such as $4 \cdot 3 \cdot 5$ is meaningful without parentheses: the product is the same for both ways parentheses can be placed.

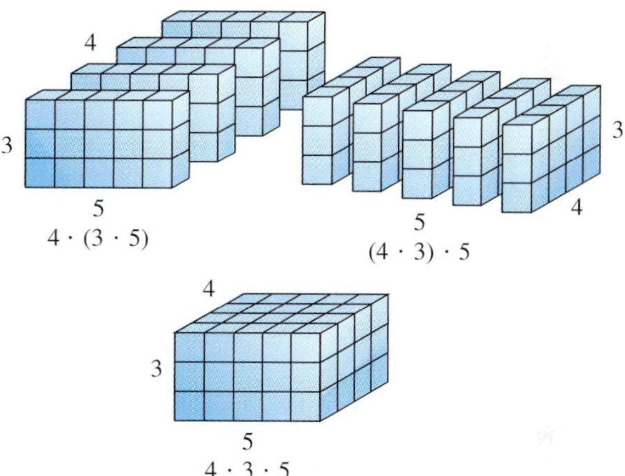

Figure 2.23
Multiplication is associative:
$4 \cdot (3 \cdot 5) = (4 \cdot 3) \cdot 5$

There is one more important property, the **distributive property,** that relates multiplication and addition. This property is the basis for the multiplication algorithm discussed in the next chapter. The distributive property can be nicely visualized by the rectangular area model. Figure 2.24 illustrates that $4 \cdot (6 + 3) = (4 \cdot 6) + (4 \cdot 3)$; that is, the factor 4 *distributes* itself over each term in the sum $6 + 3$.

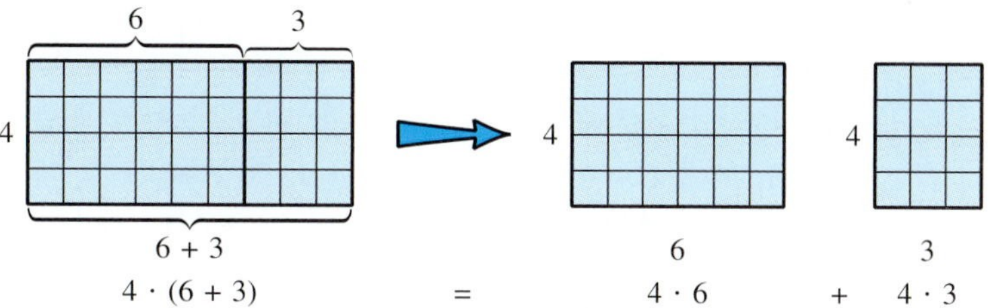

Figure 2.24
Multiplication is distributive over addition: $4 \cdot (6 + 3) = (4 \cdot 6) + (4 \cdot 3)$

Here is a summary of the properties of multiplication on the whole numbers. Each property can be proved from the definition of multiplication of whole numbers.

THEOREM *Properties of Whole-Number Multiplication*

Closure Property	If a and b are any two whole numbers, then $a \cdot b$ is a unique whole number.
Commutative Property	If a and b are any two whole numbers, then $a \cdot b = b \cdot a$.
Associative Property	If a, b, and c are any three whole numbers, then $a \cdot (b \cdot c) = (a \cdot b) \cdot c$.
Multiplicative Identity Property of One	The number 1 is the unique whole number for which $b \cdot 1 = 1 \cdot b = b$ holds for all whole numbers b.
Multiplication-by-Zero Property	For all whole numbers b, $0 \cdot b = b \cdot 0 = 0$.
Distributive Property of Multiplication over Addition	If a, b, and c are any three whole numbers, then $a \cdot (b + c) = (a \cdot b) + (a \cdot c)$ and $(a + b) \cdot c = (a \cdot c) + (b \cdot c)$.

EXAMPLE 2.16 MULTIPLYING TWO BINOMIAL EXPRESSIONS

(a) Use the properties of multiplication to justify the formula $(a + b)(c + d) = ac + ad + bc + bd$. The factors $(a + b)$ and $(c + d)$ are called **binomials,** since each factor has two terms.

(b) Visualize $(2 + 4)(5 + 3)$ with an area model drawn on squared paper.

(c) Illustrate the expansion $(a + b)(c + d) = ac + ad + bc + bd$ with the area model of multiplication.

(d) Visualize $16 \cdot 28 = (10 + 6) \cdot (20 + 8) = 200 + 80 + 120 + 48$ with an area diagram.

(e) Show that $(x + 7) \cdot (2x + 5) = 2x^2 + 5x + 14x + 35$ with an area diagram.

Solution

(a)

$(a + b)(c + d) = (a + b)c + (a + b)d$	Distributive property
$= ac + bc + ad + bd$	Distributive property
$= ac + ad + bc + bd$	Commutative property of addition

(b)

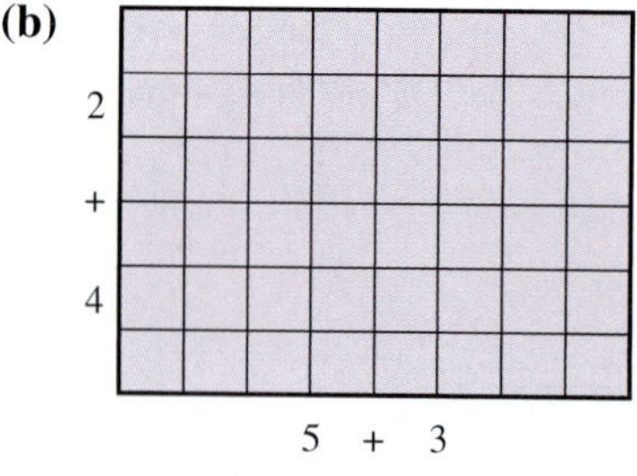

=

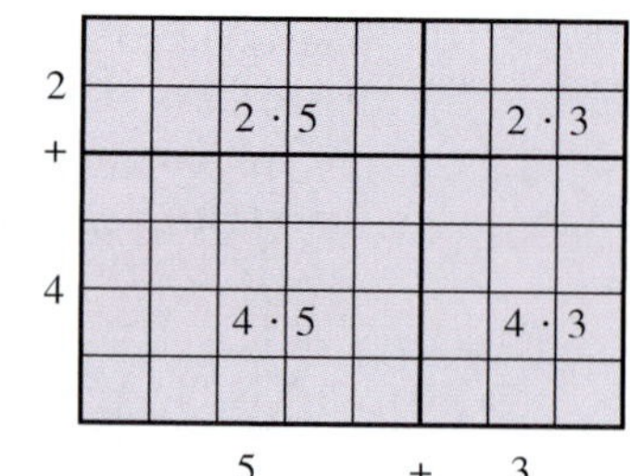

(c)

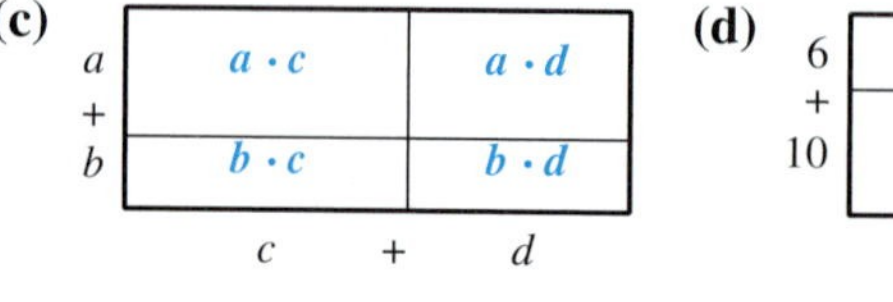

(d)

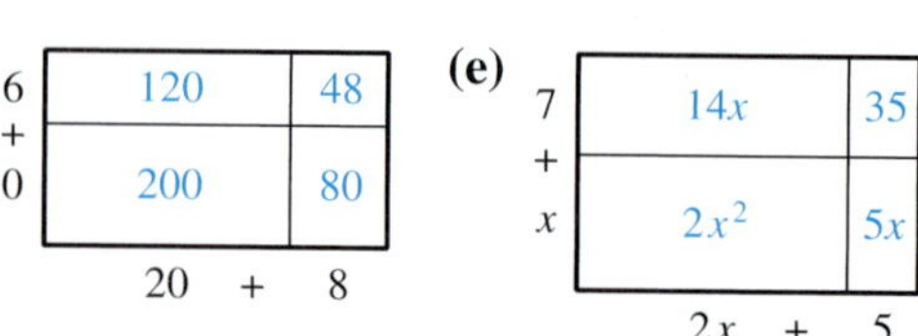

Division of Whole Numbers

There are three conceptual models for the division $a \div b$ of a whole number a by a nonzero whole number b: the **repeated-subtraction** model, the **partition** model, and the **missing-factor** model.

The Repeated-Subtraction Model of Division

Ms. Rislov has 28 students in her class whom she wishes to divide into cooperative learning groups of 4 students per group. If each group requires a set of Cuisenaire® rods, how many sets of rods must Ms. Rislov have available? The answer, 7, is pictured in Figure 2.25, and is obtained by counting how many times groups of 4 can be formed, starting with 28. Thus, $28 \div 4 = 7$. The repeated-subtraction model can be realized easily with physical objects; the process is called **division by grouping.** Since groups of 4 are being "measured out" of the class of 28, repeated subtraction is also called **measurement division.**

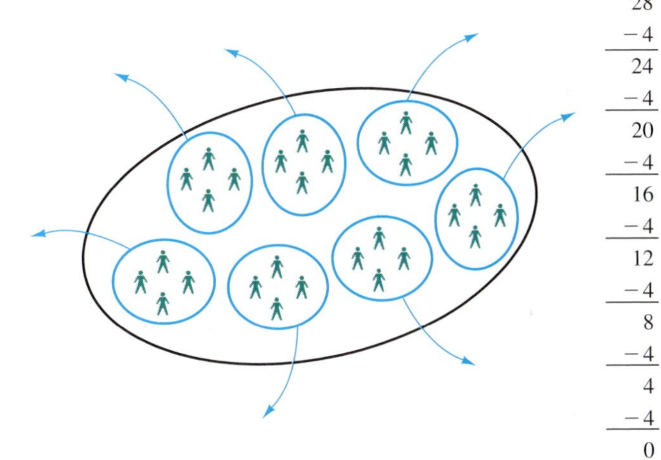

Figure 2.25
Division as repeated subtraction: $28 \div 4 = 7$ because seven 4s can be subtracted from 28

The Partition Model of Division

When Ms. Rislov checked her supply cupboard, she discovered she had only 4 sets of Cuisenaire® rods to use with the 28 students in her class. How many students must she assign to each set of rods? The answer, 7 students in each group, is depicted in Figure 2.26. The partition model is also realized easily with physical objects, in which case the process is called **division by sharing** or **partitive division.**

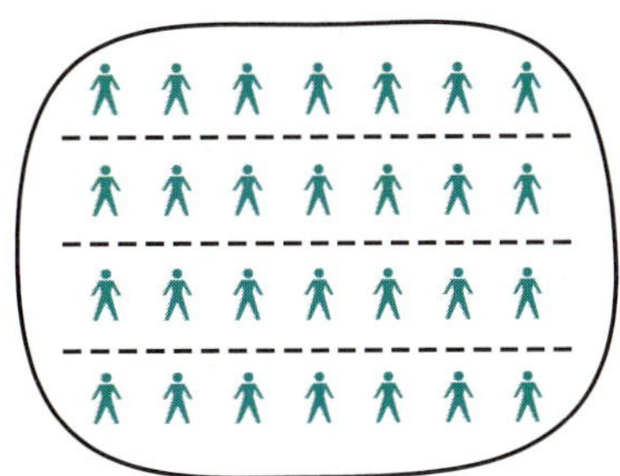

Figure 2.26
Division as a partition: $28 \div 4 = 7$ because when 28 objects are partitioned into 4 equal-sized sections, there are 7 objects in each partition

The Missing-Factor Model of Division

In the repeated-subtraction model, $28 \div 4 = 7$ because, when grouped by fours, seven groups will be formed. That is, $28 = 4 + 4 + 4 + 4 + 4 + 4 + 4 = 7 \cdot 4$. However, in the partition model, $28 \div 4 = 7$ because $28 = 7 + 7 + 7 + 7 = 4 \cdot 7$. In both cases, the division $28 \div 4$ can be viewed as finding the factor c for which $28 = 4 \cdot c$ or $28 = c \cdot 4$. The missing-factor model is usually the concept adopted to define division formally.

DEFINITION *Division in Whole Numbers*
Let a and b be whole numbers with $b \neq 0$. Then $a \div b = c$ if, and only if, $a = b \cdot c$ for a unique whole number c.

The symbol $a \div b$ is read "a divided by b," where a is the **dividend** and b is the **divisor.** If $a \div b = c$, then we say that ***b* divides *a*** or is a **divisor** of a, and c is called the **quotient.**

$$a \div b = c$$

dividend → a, divisor → b, quotient → c

Division is also symbolized by a/b or $\frac{a}{b}$; the slash notation is used by most computers.

Any multiplication fact with nonzero factors, say $3 \cdot 5 = 15$, is related to four equivalent facts that form a **fact family,** such as

$$3 \cdot 5 = 15, \qquad 15 \div 3 = 5,$$
$$5 \cdot 3 = 15, \quad \text{and} \quad 15 \div 5 = 3.$$

A fact family can be represented with the rectangular array model, as shown in Figure 2.27 with the arrays of colored discs.

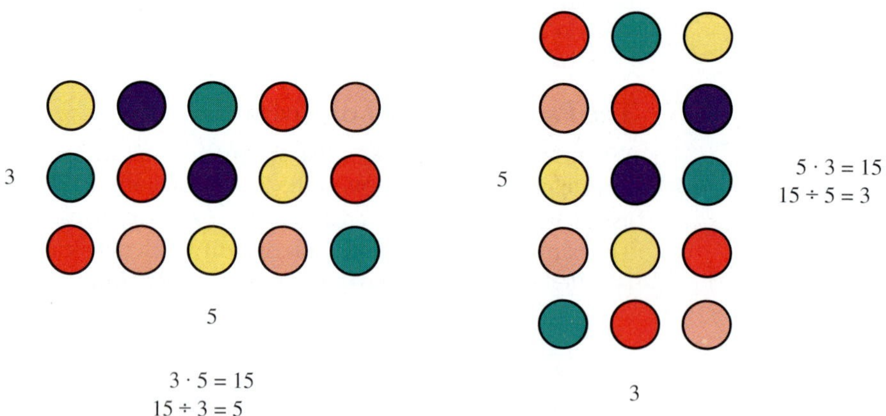

Figure 2.27
Rectangular arrays, formed with colored discs, illustrate the fact family $3 \cdot 5 = 15$, $15 \div 3 = 5$, $5 \cdot 3 = 15$, and $15 \div 5 = 3$

The three division models are nicely illustrated in the School Book Page in this section. Studying these examples and the accompanying Questions for the Teacher will help you understand the meaning of division.

School Book Page

Modeling Division in Grade Four

Chapter 4
Lesson
7

Reviewing the Meaning of Division

You Will Learn
three ways to think about division

Learn

Could you ride a bike with no brakes, no handlebars, and only one wheel? The Andover One-Wheelers do it all the time! It's a club for unicycle riders.

The Andover One-Wheelers are from Andover, New Hampshire.

Example 1

You can think of division as sharing.

Suppose 24 riders form 3 circles. How many riders are in each circle?

Find $24 \div 3$.

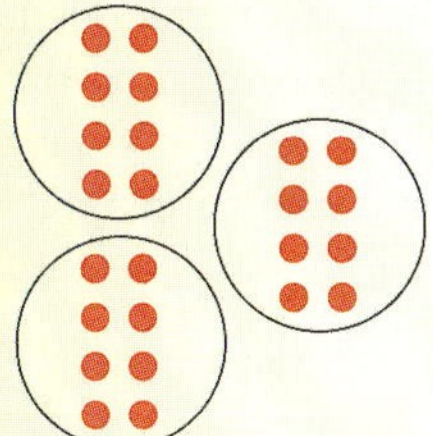

8 riders are in each circle.

Example 2

You can think of division as repeated subtraction.

If 24 riders form circles with 6 riders in each circle, how many circles are there?

Find $24 \div 6$.

Subtract 6 from 24 until you have 0. Count how many times you subtracted.

$24 - \mathbf{6} = 18$
$18 - \mathbf{6} = 12$
$12 - \mathbf{6} = 6$
$6 - \mathbf{6} = 0$

$24 \div 6 = 4$
There are 4 circles.

Example 3

You can think of division as the opposite of multiplication.

Suppose 24 riders form 2 rows with an equal number of riders in each row. How many riders are in each row?

$24 \div 2 = ?$
$2 \times ? = 24$
$2 \times 12 = 24$
$24 \div 2 = 12$

So, there are 12 riders in each row.

Talk About It

What multiplication fact can help you find $36 \div 4$?

166 Chapter 4 • Multiplication and Division Concepts and Facts

SOURCE: From Scott Foresman–Addison Wesley, Math Grade 4, *p. 166, by Randall I. Charles et al. Copyright © 2002 Pearson Education, Inc. Reprinted with permission.*

Questions for the Teacher

1. The red arrays of dots represent unicycle riders. What other arrangement of dots in each loop might a fourth grader form to make the pictorial model correspond more closely to the problem posed in Example 1?
2. Describe how to use a set of red counters to illustrate the repeated-subtraction model in Example 2 to fourth graders.
3. How would you use 24 red counters to illustrate the missing-factor model in Example 3?
4. What help could you offer to fourth graders who wish to respond to the "Talk About It" question at the bottom of the School Book Page?

EXAMPLE 2.17 **COMPUTING QUOTIENTS WITH MANIPULATIVES**

Suppose you have 78 number tiles. Describe how to illustrate $78 \div 13$ with the tiles, using each of the three basic conceptual models for division.

Solution

(a) **Repeated subtraction.** Remove groups of 13 tiles each. Since 6 groups are formed, $78 \div 13 = 6$.

(b) **Partition.** Partition the tiles into 13 equal-sized parts. Since each part contains exactly 6 tiles, $78 \div 13 = 6$.

(c) **Missing factor.** Use the 78 tiles to form a rectangle with 13 rows. Since it turns out there are 6 columns in the rectangle, $78 \div 13 = 6$.

Division by Zero Is Undefined

The definition of division tells us that $12 \div 3 = 4$, since 4 is the unique whole number for which $12 = 3 \times 4$. Division *into zero* is also defined. For example, $c = 0$ is the unique whole number for which $0 = 5 \times c$, so $0 \div 5 = 0$. However, division *by zero* (or "$a \div 0$") is not defined for any whole number a. The reason for this can best be explained by taking the cases $a \neq 0$ and $a = 0$ separately.

Case 1: $a \neq 0$. In this case, "$a \div 0$" would be equivalent to finding the missing factor c that makes $a = c \cdot 0$. But $c \cdot 0 = 0$ for all whole numbers c, so there is no solution when $a \neq 0$. Thus, $a \div 0$ is undefined for $a \neq 0$.

Case 2: $a = 0$. In this case "$0 \div 0$" is equivalent to finding a *unique* whole number c for which $0 = 0 \cdot c$. This equation is satisfied by every choice for the factor c. Since no *unique* factor c exists, the division of 0 by 0 is also undefined.

In the division $a \div b$, there is no restriction on the dividend a. For all $b \neq 0$, we have $0 \div b = 0$, since $0 = b \cdot 0$.

The multiplicative identity 1 has two simple, but useful, relationships to division:

$$\frac{b}{b} = 1 \quad \text{for all } b, \quad \text{where } b \neq 0;$$

$$\frac{a}{1} = a \quad \text{for all } a.$$

Division with Remainders

Consider the division problem $27 \div 6$. There is no whole number c that satisfies $27 = c \cdot 6$, so $27 \div 6$ is not defined in the whole numbers. That is, the set of whole numbers is *not closed* under division.

By allowing the possibility of a **remainder,** we can extend the division operation. Consider $27 \div 6$, where division is viewed as repeated subtraction. Four groups of 6 can be removed from 27. This leaves 3, which are too few to form another group of 6. This can be written

$$27 = 4 \cdot 6 + 3.$$

Here 4 is called the **quotient** and 3 is the **remainder.** This information is also written

$$27 \div 6 = 4 \text{ R } 3,$$

where R separates the quotient from the remainder. A representation of division with a remainder is shown in Figure 2.28.

Cooperative Investigation

The Krypto Game

The Krypto game, which is available commercially or can be handmade, consists of 25 cards numbered 1 through 25. Each player, or cooperative group, is dealt a hand of 5 cards plus a sixth card that designates the target number. Here is an example:

Hand: | 21 | 5 | 17 | 8 | 3 | Target: | 12 |

The object of the game is to combine the cards in the hand, using any of the operations $+$, $-$, $\times$, and $\div$ and using each card once and only once, to obtain the target number. Here is one solution:

$$21 - 5 = 16$$
$$16 \div 8 = 2$$
$$2 + 3 = 5$$
$$17 - 5 = 12 \leftarrow \text{Target}$$

That is, $17 - (((21 - 5) \div 8) + 3) = 12$. Other solutions are $((17 + 8) \div 5) + (21 \div 3) = 12$ and $(5 + 17 + 8 + 3) - 21 = 12$.

In this game, guess and check and working backward are useful problem-solving strategies. Parentheses are needed to show the order in which operations are performed.

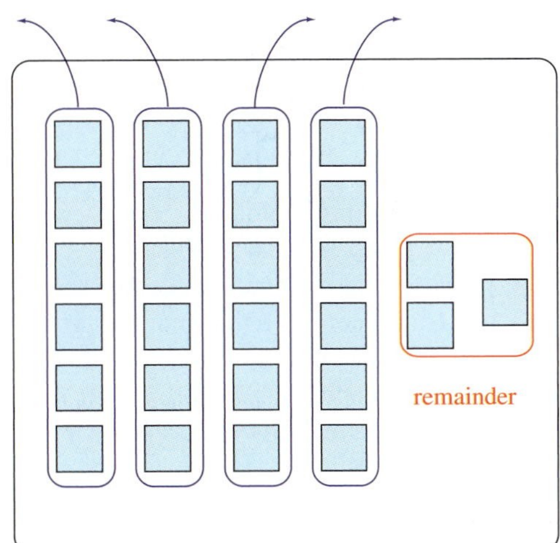

Figure 2.28
Representing division with a remainder: $27 \div 6 = 4$ R 3

In general, we have the following result.

THEOREM *The Division Algorithm*
Let a and b be whole numbers with $b \neq 0$. Then there is a unique whole number q called the **quotient** and a unique whole number r called the **remainder** such that

$$a = q \cdot b + r, 0 \leq r < b.$$

It is common to write

$$a \div b = q \text{ R } r$$

if $a = q \cdot b + r$, $0 \leq r < b$. The quotient q is the largest whole number of groups of b objects that can be formed from a objects, and the remainder r is the number of objects that are left over. The remainder is 0 if, and only if, b divides a according to the definition of division in whole numbers. This important case is explored fully in Chapter 4.

EXAMPLE 2.18 USING THE DIVISION ALGORITHM TO SOLVE THE MARCHING BAND PROBLEM

Mr. Garza was happy to see that so many students in the school band had turned out for the parade. He had them form into rows of 6, but it turned out that just 1 tuba player was in the back row. To his dismay, when he re-formed the band into rows of 5, there was still a lone tuba player in the back row. In desperation, Mr. Garza had the band reassemble into rows of 7. To his relief, every row was filled! How many students are marching in the parade band?

Solution

If we let n denote the number of students marching in the band, then we know that when n is divided by 6 the remainder is 1. Thus, $n = 6q + 1$ for some whole number $q = 0, 1, 2, \ldots$. This means n is somewhere in the list of numbers that are 1 larger than a multiple of 6:

$$1, 7, 13, 19, 25, 31, 37, 43, 49, 55, 61, 67, 73, 79, 85, 91, 97, 103, \ldots.$$

Similarly, $n \div 5$ has a remainder of 1, so n is also a number in this list:

$$1, 6, 11, 16, 21, 26, 31, 36, 41, 46, 51, 56, 61, 66, 71, 76, 81, 86, 91, 96, 101, \ldots.$$

Finally, 7 is a divisor of n, so n is one of these numbers:

$$7, 14, 21, 28, 35, 42, 49, 56, 63, 70, 77, 84, 91, 98, 105, \ldots.$$

Comparing the three lists, we find that 91 is the smallest number common to all three lists, so it is possible there are 91 band members. By extending the lists (it helps to notice that the numbers common to the first two lists, 1, 31, 61, 91, . . . , jump ahead by 30 each step), we find that the next smallest number in all three lists is 301, since $301 = 50 \cdot 6 + 1 = 60 \cdot 5 + 1 = 43 \cdot 7$. It seems pretty unlikely the band is this large, so we conclude that there are 91 members of the band. Mr. Garza has arranged them in 13 rows, since $91 \div 7 = 13$.

from The NCTM Principles and Standards

Understanding Meanings of Operations and How They Relate to One Another

In grades 3–5, students should focus on the meanings of, and relationship between, multiplication and division. It is important that students understand what each number in a multiplication or division expression represents. For example, in multiplication, unlike addition, the factors in the problem can refer to different units. If students are solving the problem 29×4 to find out how many legs there are on 29 cats, 29 is the number of cats (or number of groups), 4 is the number of legs on each cat (or number of items in each group), and 116 is the total number of legs on all the cats. Modeling multiplication problems with pictures, diagrams, or concrete materials helps students learn what the factors and their product represent in various contexts.

Students should consider and discuss different types of problems that can be solved using multiplication and division. For example, if there are 112 people traveling by bus and each bus can hold 28 people, how many buses are needed? In this case, 112 ÷ 28 indicates the number of groups (buses), where the total number of people (112) and the size of each group (28 people in each bus) are known. In a different problem, students might know the number of groups and need to find how many items are in each group. If 112 people divide themselves evenly among four buses, how many people are on each bus? In this case, 112 ÷ 4 indicates the number of people on each bus, where the total number of people and the number of groups (buses) are known. Students need to recognize both types of problems as division situations, should be able to model and solve each type of problem, and should know the units of the result: is it 28 buses or 28 people per bus? Students in these grades will also encounter situations where the result of division includes a remainder. They should learn the meaning of a remainder by modeling division problems and exploring the size of remainders given a particular divisor. For example, when dividing groups of counters into sets of 4, what remainders could there be for groups of different sizes?

SOURCE: Reprinted with permission from Principles and Standards for School Mathematics, *page 151.* *Standards are listed with the permission of the National Council of Teachers of Mathematics (NCTM). NCTM does not endorse the content or validity of these alignments.*

Exponents and the Power Operation

Instead of writing $3 \cdot 3 \cdot 3 \cdot 3 \cdot 3$ we can follow a notation introduced by René Descartes and write 3^5. This operation is called "taking 3 to the fifth power." The general definition is described as follows.

DEFINITION *The Power Operation for Whole Numbers*
Let a and m be whole numbers, where $m \neq 0$. Then ***a* to the *m*th power,** written a^m, is defined by

$$a^1 = a, \quad \text{if } m = 1,$$

and

$$a^m = \overbrace{a \cdot a \cdot \cdots \cdot a}^{m \text{ factors}}, \quad \text{if } m > 1.$$

The number a is called the **base,** m is called the **exponent** or **power,** and a^m is called an **exponential expression.** Special cases include squares and cubes. For example, 7^2 is read "7 squared," and 10^3 is read "10 cubed." On most computers, 7^2 and 10^3 would be typed in as 7 ^ 2 and 10 ^ 3, where the circumflex ^ separates the base from the exponent. Calculators often have a [^] or [y^x] key to compute powers. For example, 3 [y^x] 2 [=] will give the answer 9.

EXAMPLE 2.19 | WORKING WITH EXPONENTS

Compute the following products and powers, expressing your answers in the form of a single exponential expression a^m.

(a) $7^4 \cdot 7^2$ **(b)** $6^3 \cdot 6^5$ **(c)** $2^3 \cdot 5^3$
(d) $3^2 \cdot 5^2 \cdot 4^2$ **(e)** $(3^2)^5$ **(f)** $(4^2)^3$

Solution

(a) $7^4 \cdot 7^2 = (7 \cdot 7 \cdot 7 \cdot 7) \cdot (7 \cdot 7) = 7 \cdot 7 \cdot 7 \cdot 7 \cdot 7 \cdot 7 = 7^6$

(b) $6^3 \cdot 6^5 = (6 \cdot 6 \cdot 6) \cdot (6 \cdot 6 \cdot 6 \cdot 6 \cdot 6)$
$= 6 \cdot 6 \cdot 6 \cdot 6 \cdot 6 \cdot 6 \cdot 6 \cdot 6 = 6^8$

(c) $2^3 \cdot 5^3 = (2 \cdot 2 \cdot 2) \cdot (5 \cdot 5 \cdot 5) = (2 \cdot 5) \cdot (2 \cdot 5) \cdot (2 \cdot 5)$
$= (2 \cdot 5)^3 = 10^3$

(d) $3^2 \cdot 5^2 \cdot 4^2 = (3 \cdot 3) \cdot (5 \cdot 5) \cdot (4 \cdot 4) = (3 \cdot 5 \cdot 4) \cdot (3 \cdot 5 \cdot 4)$
$= (3 \cdot 5 \cdot 4)^2 = 60^2$

(e) $(3^2)^5 = (3)^2 \cdot (3)^2 \cdot (3)^2 \cdot (3)^2 \cdot (3)^2$
$= (3 \cdot 3) \cdot (3 \cdot 3) \cdot (3 \cdot 3) \cdot (3 \cdot 3) \cdot (3 \cdot 3)$
$= 3 \cdot 3 \cdot 3 \cdot 3 \cdot 3 \cdot 3 \cdot 3 \cdot 3 \cdot 3 \cdot 3 = 3^{10}$

(f) $(4^2)^3 = (4^2) \cdot (4^2) \cdot (4^2) = (4 \cdot 4) \cdot (4 \cdot 4) \cdot (4 \cdot 4)$
$= 4 \cdot 4 \cdot 4 \cdot 4 \cdot 4 \cdot 4 = 4^6$

Example 2.19 reveals that multiplication of exponentials follows useful patterns that can be used to shorten calculations. For example, $7^4 \cdot 7^2 = 7^{4+2}$ and $6^3 \cdot 6^5 = 6^{3+5}$ are two examples of the general rule $a^m \cdot a^n = a^{m+n}$ for multiplying exponentials with the same base. Similarly, $2^3 \cdot 5^3 = (2 \cdot 5)^3$ is a case of $a^n \cdot b^n = (a \cdot b)^n$, and $(3^2)^5 = 3^{2 \cdot 5}$ is an example of $(a^m)^n = a^{m \cdot n}$.

THEOREM *Multiplication Rules of Exponentials*

Let a, b, m, and n be whole numbers, where $m \neq 0$ and $n \neq 0$. Then

(i) $a^m \cdot a^n = a^{m+n}$;
(ii) $a^m \cdot b^m = (a \cdot b)^m$;
(iii) $(a^m)^n = a^{m \cdot n}$.

Proof of (i): $a^m \cdot a^n = a^{m+n}$

$$a^m \cdot a^n = \underbrace{a \cdot a \cdot \cdots \cdot a}_{m \text{ factors}} \cdot \underbrace{a \cdot a \cdot \cdots \cdot a}_{n \text{ factors}}$$
$$= \underbrace{a \cdot a \cdot \cdots \cdot a}_{m + n \text{ factors}}$$
$$= a^{m+n}.$$

The proofs of (ii) and (iii) are similar.

If the formula $a^m \cdot a^n = a^{m+n}$ were extended to allow $m = 0$, it would state that $a^0 \cdot a^n = a^{0+n} = a^n$. This suggests that it is reasonable to define $a^0 = 1$ when $a \neq 0$.

DEFINITION *Zero as an Exponent*

Let a be any whole number, $a \neq 0$. Define $a^0 = 1$.

Highlight from History

Symbols for Multiplication, Division, and Power

William Oughtred introduced the symbol × for multiplication in 1631. Gottfried Leibniz objected to the ×, writing to Bernoulli in 1698 that it "... is easily confounded with *x*." Leibniz preferred the dot symbol ·, which Thomas Harriot had introduced in the same year, 1631. The symbol ÷ was introduced in 1659 by the Swiss mathematician J. H. Rohn. It was adopted into the English-speaking countries by John Wallis and others, but on the European continent the colon symbol a : b of Leibniz was the symbol of regular choice. The slanted line /, and the horizontal line, gradually became more common. Exponential notation became common with its systematic use by Descartes in *La Geometrie.* With computers, where it is best to keep expressions on one line, the circumflex ^ is used: thus, 4 ^ 7 denotes 4^7.

To see why 0^0 is not defined, notice that there are two conflicting patterns:

$$3^0 = 1, 2^0 = 1, 1^0 = 1, 0^0 = ?$$
$$0^3 = 0, 0^2 = 0, 0^1 = 0, 0^0 = ?$$

The multiplication formula $a^m \cdot a^n = a^{m+n}$ can also be converted to a corresponding division fact. For example,

$$a^{5-3} \cdot a^3 = a^{(5-3)+3} = a^5$$

so

$$a^5/a^3 = a^{5-3}.$$

In general, we have the following theorem.

THEOREM *Rules for Division of Exponentials*
Let a, b, m, and n be whole numbers, where $m \geq n > 0$, $b \neq 0$, and $a \div b$ is defined. Then

$$\textbf{(i)}\ \ b^m/b^n = b^{m-n}$$

and

$$\textbf{(ii)}\ \ (a^m/b^m) = (a/b)^m.$$

EXAMPLE 2.20 **WORKING WITH EXPONENTS**

Rewrite these expressions in exponential form a^m.

(a) $5^{12} \cdot 5^8$ **(b)** $7^{14}/7^5$ **(c)** $3^2 \cdot 3^5 \cdot 3^8$
(d) $8^7/4^7$ **(e)** 2^{5-5} **(f)** $(3^5)^2/3^4$

Solution

(a) $5^{12+8} = 5^{20}$ **(b)** $7^{14-5} = 7^9$ **(c)** $3^{2+5+8} = 3^{15}$
(d) $(8/4)^7 = 2^7$ **(e)** $2^0 = 1$ **(f)** $3^{5 \cdot 2-4} = 3^6$

Problem Set 2.4

Understanding Concepts

1. What multiplication fact is illustrated in each of these diagrams? Name the multiplication model that is illustrated.

(a)

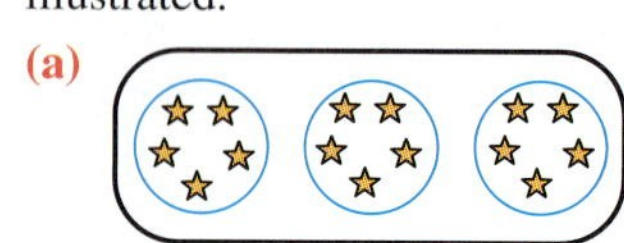

(b)

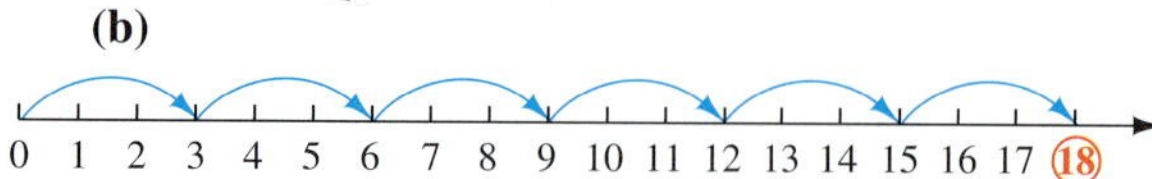

(c)

(d)

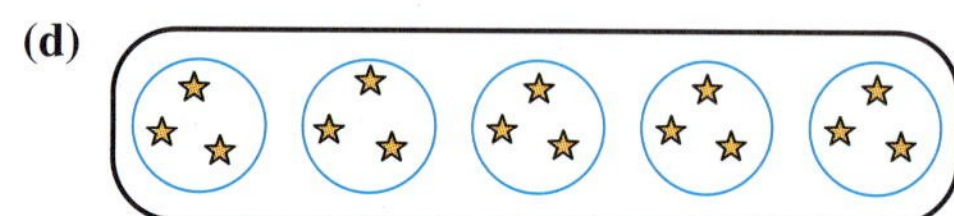

(e)

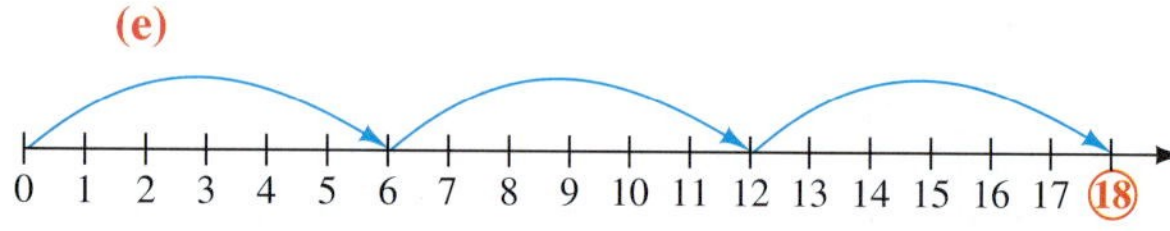

(f) (g)

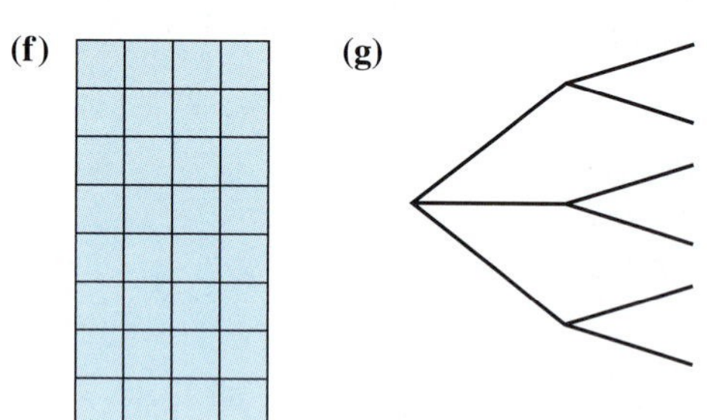

2. Discuss which model of multiplication—set model (repeated addition), number-line (measurement), array, rectangular area, multiplication tree, or Cartesian product—best fits the following problems:

   (a) A set of dominoes came in a box containing 11 stacks of 5 dominoes each. How many dominoes are in the set?

   (b) Marja has 3 skirts that she can "mix or match" with 6 blouses. How many outfits does she have to wear?

   (c) Harold hiked 10 miles each day until he crossed the mountains after 5 days. How many miles was his hike?

   (d) Ace Widgit Company makes 35 widgits a day. How many widgits are made in a 5-day workweek?

   (e) Janet's sunroom is 9 by 18 feet. How many 1-square-foot tiles does she need to cover the floor?

   (f) Domingo rolls a die and flips a coin. How many outcomes are there?

3. Multiplication as repeated addition can be illustrated on most calculators. For example, $4 \cdot 7$ is computed by 7 [+] 7 [+] 7 [+] 7 [=]. Each press of the [+] key completes any pending addition and sets up the next one, so the intermediate products $2 \cdot 7 = 14$ and $3 \cdot 7 = 21$ are displayed along the way. Many calculators have a "constant" feature, which enables the user to avoid having to reenter the same addend over and over. For example, [+] 734 [=] [=] [=], or 734 [+] [+] [+], may compute $3 \cdot 734$; it all depends on how your particular calculator operates.

   (a) Explain carefully how repeated addition is best accomplished on your calculator.

   (b) Use repeated addition on your calculator to compute these products. Check your result by using the [×] key.

   (i) $4 \cdot 9$ **(ii)** 7×536

   (iii) $6 \times 47{,}819$ **(iv)** $56{,}108 \times 6$ (What property may help?)

4. The Cartesian product of finite and nonempty sets can be illustrated by the intersections of a crossing-line pattern, as shown.

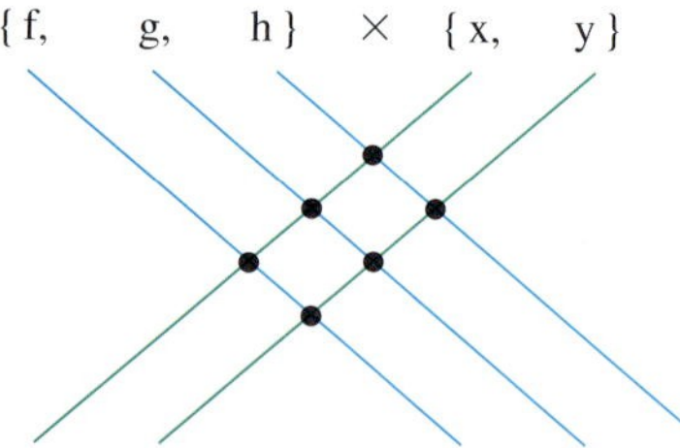

   (a) Explain why the number of intersection points in the crossing-line pattern for $A \times B$ is $a \cdot b$, where $a = n(A)$ and $b = n(B)$.

   (b) Draw the crossing-line pattern for {□, △} × {♥, ♦, ♣, ♠}.

5. Which of the following sets of whole numbers are closed under multiplication? Explain your reasoning.

(a) $\{1, 2\}$ **(b)** $\{0, 1\}$ **(c)** $\{0, 2, 4\}$

(d) $\{0, 2, 4, \ldots\}$ (the even whole numbers)

(e) $\{1, 3, 5, \ldots\}$ (the odd whole numbers)

(f) $\{1, 2, 2^2, 2^3\}$ **(g)** $\{1, 2, 2^2, 2^3, \ldots\}$

(h) $\{1, 7, 7^2, 7^3, \ldots\}$

6. Which of the following subsets of the whole numbers $W = \{0, 1, 2, \ldots\}$ are closed under multiplication? Explain carefully.

(a) $\{0, 1, 2, 3, 4, 6, 7, \ldots\}$ (that is, the whole numbers except for 5)

(b) $\{0, 1, 2, 3, 4, 5, 7, 8, \ldots\}$

(c) $\{0, 1, 4, 5, 6, \ldots\}$

7. What properties of whole number multiplication justify these equalities?

(a) $4 \cdot 9 = 9 \cdot 4$

(b) $4 \cdot (6 + 2) = 4 \cdot 6 + 4 \cdot 2$

(c) $0 \cdot 439 = 0$ **(d)** $7 \cdot 3 + 7 \cdot 8 = 7 \cdot (3 + 8)$

(e) $5 \cdot (9 \cdot 11) = (5 \cdot 9) \cdot 11$ **(f)** $1 \cdot 12 = 12$

8. What property of multiplication is illustrated in the diagrams at the bottom of the page?

9. Use the rectangular area model to illustrate each of the following statements. Make drawings similar to Figure 2.24.

(a) $(2 + 5) \cdot 3 = 2 \cdot 3 + 5 \cdot 3$

(b) $3 \cdot (2 + 5 + 1) = 3 \cdot 2 + 3 \cdot 5 + 3 \cdot 1$

(c) $(3 + 2) \cdot (4 + 3) = 3 \cdot 4 + 3 \cdot 3 + 2 \cdot 4 + 2 \cdot 3$

10. The FOIL method is a useful way to recall how to expand the product of two binomials $(a + b) \times (c + d)$: multiply the First terms, the Outer terms, the Inner terms, and the Last terms of the binomials and sum the four products to obtain the formula $(a + b) \times (c + d) = ac + ad + bc + bd$. Carefully explain how the rectangular area diagram below justifies the FOIL method.

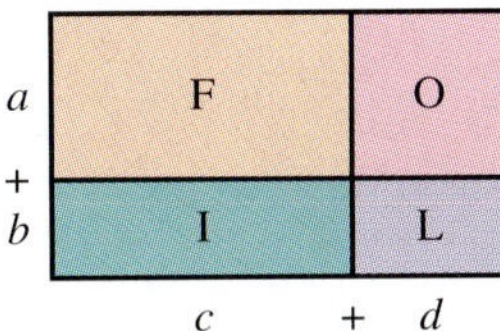

11. Use the figure below to show that the product of trinomials, $(a + b + c) \cdot (d + e + f)$, can be written as a sum of nine products.

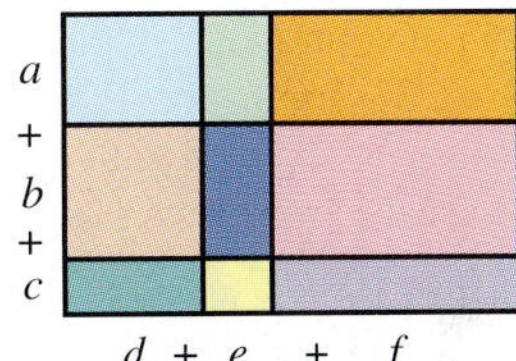

12. Modify Figure 2.23 to show how the associative property $3 \cdot (2 \cdot 4)$ can be illustrated with arrays of cubes in three-dimensional space.

13. Shannon bought 18 nuts and 18 bolts. The bolts were 86¢ each, and the nuts were 14¢ each. The store clerk computed the bill as shown below. How did Shannon already know the answer by simple mental math?

```
   4           3
  86          14
× 18        × 18          1 1
 688         112        15.48
 86           14       + 2.52
1548         252        18.00
```

Diagrams for Problem 8

(a)

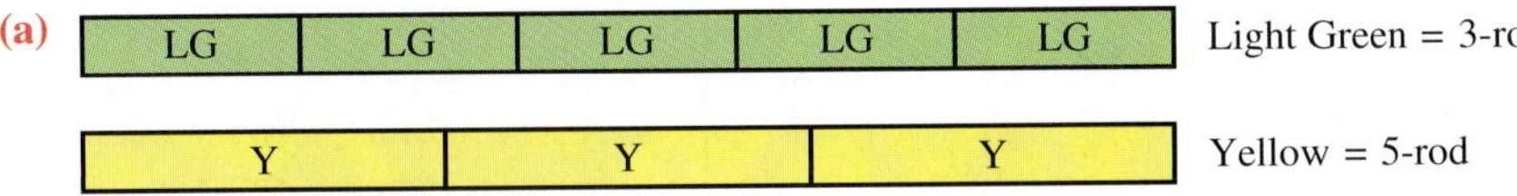

(b)

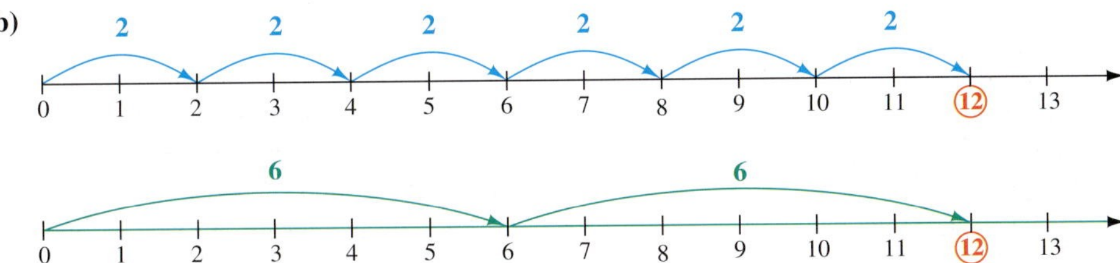

14. What properties of multiplication make it easy to compute these values mentally?

(a) $7 \cdot 19 + 3 \cdot 19$

(b) $24 \cdot 17 + 24 \cdot 3$

(c) $36 \cdot 15 - 12 \cdot 45$

15. What division facts are illustrated below?

(a)

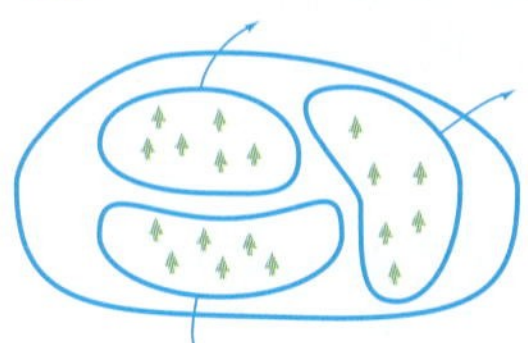

(b)

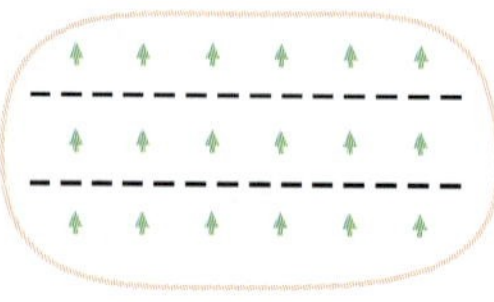

16. A 2-by-3 rectangular array is associated with the fact family $2 \cdot 3 = 6$, $3 \cdot 2 = 6$, $6 \div 2 = 3$, and $6 \div 3 = 2$. What fact family is associated with each of these rectangular arrays?

(a) 4 by 8 (b) 6 by 5

17. Discuss which of three conceptual models of division—repeated subtraction, partition, missing-factor—best corresponds to the following problems. More than one model may fit.

(a) Preston owes \$3200 on his car. If his payments are \$200 a month, how many months will Preston make car payments?

(b) An estate of \$76,000 is to be split among 4 heirs. How much can each heir expect to inherit?

(c) Anita was given a grant of \$375 to cover expenses on her trip. She expects that it will cost her \$75 a day. How many days can she plan to be gone?

18. Use repeated subtraction on your calculator to compute the following division problems, where remainders are possible. Be sure to take advantage of the "constant" feature of your calculator.

(a) $78 \div 13$ (b) $832 \div 52$

(c) $96 \div 14$ (d) $548{,}245 \div 45{,}687$

19. Solve for the unknown whole number in the following expressions.

(a) $y \div 5 = 5$ R 4

(b) $20 \div x = 3$ R 2

20. Rewrite the following in the form of a single exponential.

(a) $3^{20} \cdot 3^{15}$ (b) $4^8 \cdot 7^8$ (c) $(3^2)^5$

(d) $x^7 \cdot x^9$ (e) $y^3 \cdot z^3$ (f) $(t^3)^4$

21. Write the following as 2^m for some whole number m.

(a) 8 (b) $4 \cdot 8$ (c) 1024 (d) 8^4

22. Find the exponents that make the following equations true.

(a) $3^m = 81$ (b) $3^n = 531{,}441$

(c) $4^p = 1{,}048{,}576$ (d) $2^q = 1{,}048{,}576$

Teaching Concepts

23. Large-group kinesthetic activities are often an effective way to deepen understanding of basic concepts. For example, have the children hold hands in groups of three and count the number of groups to illustrate the concept of division by three with the repeated-subtraction (grouping) model of division. Carefully describe analogous large-group activities that illustrate

(a) Multiplication as repeated addition

(b) Multiplication as an array

(c) Partitive division

(d) Missing-factor division

24. When the class was asked to pose a meaningful problem that corresponded to the division $14 \div 3$, Peter's problem was answered by 4, Tina's by 5, and Andrea's by 4 with a remainder of 2. Carefully explain how all three children may be correct, by posing three problems of your own that give their answers.

25. Discuss how division can be modeled with number strips. Write a brief essay that includes several examples, each illustrated with carefully drawn figures.

Thinking Critically

26. The binary operation ★ is defined for the set of shapes {○, □, △} according to the following table. For example, △ ★ □ = ○.

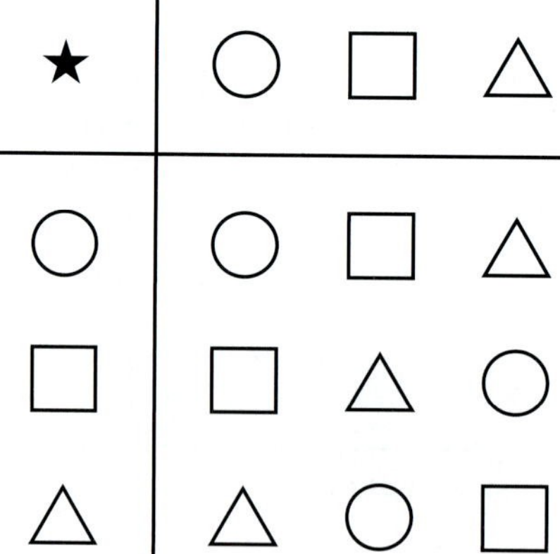

★	○	□	△
○	○	□	△
□	□	△	○
△	△	○	□

Is the operation closed? commutative? associative? Is there an identity shape?

27. (a) Verify that the three numbers in every row, column, and diagonal in the square below have the same product, making it a *magic multiplication square.*

8	256	2
4	16	64
128	1	32

(b) Write each number in the magic multiplication square as a power of two, and then explain how the multiplication square can be obtained from a related magic addition square.

(c) Create a magic multiplication square with the numbers 1, 3, 9, 27, 81, 243, 729, 2187, and 6561.

28. (a) The Math Club wants to gross \$500 on its raffle. Raffle tickets are \$2 each and 185 tickets have been sold so far. The answer is 65. What is the question?

(b) A total of 67 eggs was collected from the hen house and placed in standard egg cartons. If the answer is 5, what is the question? If the answer is 7, what is the question?

29. There are 318 folding chairs being set up in rows in an auditorium, with 14 chairs being put in each row starting at the front of the auditorium.

(a) If the answer is 22, what is the question?

(b) If the answer is 10, what is the question?

30. (a) Verify that $4 \cdot (5 - 2) = 4 \cdot 5 - 4 \cdot 2$.

(b) Prove that multiplication distributes over subtraction. That is, for all whole numbers a, b, and c, with $b \geq c$, show that $a \cdot (b - c) = a \cdot b - a \cdot c$.

31. A clock chimes on the hour, once at one o'clock, twice at two o'clock, and so on. How many times does it chime every 24-hour period?

32. How many total spots are on the 28 dominoes of a "double-six" domino set? (*Suggestion:* Imagine you have *two* sets of dominoes, and each domino of one set has been matched with its "complementary" domino from the other set. The figure below shows two pairs of complementary dominoes.)

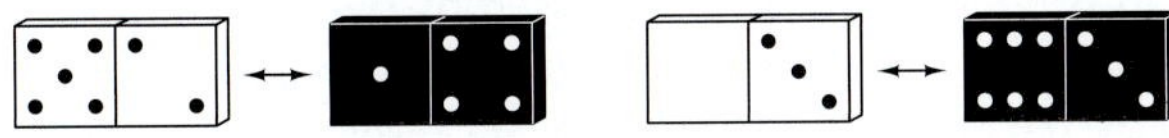

33. A certain whole number less than 100 leaves the remainders 1, 2, 3, and 4 when divided, respectively, by 2, 3, 4, and 5. What is the whole number?

Thinking Cooperatively

34. Each team is given a game sheet as shown. The teacher rolls 3 dice (or a die 3 times) to determine three numbers x, y, and z. Each team uses whole-number operations, including power, involving all three numbers exactly once, to obtain expressions equal to as many of the numbers 0, 1, 2, . . . , 18 as possible. The team filling all regions of the game sheet first is the winner. If no team has won at the end of the first roll, the numbers x, y, and z are replaced by rolling the dice again. Each team now fills in remaining empty positions on the game sheet using the new set of three numbers. The dice are rolled again until there is a winning team. Example: Suppose the dice turn up the numbers 2, 3, 3. Show that at least 14 of the regions on the board can be filled.

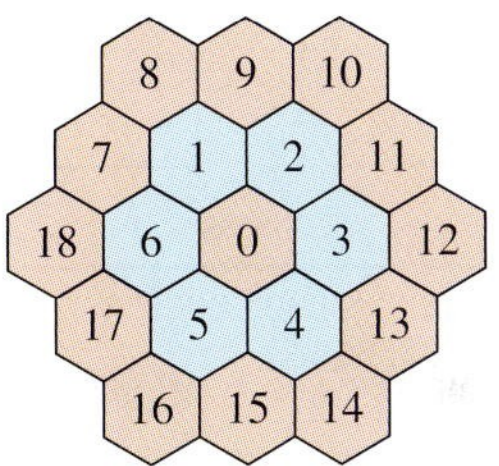

Using a Calculator

35. Let T_n denote the sum of the first n triangular numbers, so that

$$T_n = 1 + 3 + 6 + 10 + \cdots + t_n,$$

where

$$t_n = 1 + 2 + 3 + \cdots + n.$$

Now observe that

$$T_1 = 1 = 1 \cdot 2 \cdot \tfrac{3}{6};$$
$$T_2 = 1 + 3 = 4 = 2 \cdot 3 \cdot \tfrac{4}{6};$$
$$T_3 = 1 + 3 + 6 = 10 = 3 \cdot 4 \cdot \tfrac{5}{6}.$$

(a) See if the pattern continues to hold for T_4, $T_5, \ldots, T_{12}$.

(b) What is T_{100}?

36. **A Magic Sum-of-Products Square.** Consider the square shown. This is a magic addition square, since the sum of the three numbers in any row, column, or diagonal is the same magic constant 30. Show that the square also has this amazing property: the sum of the row products (that is, $11 \times 2 \times 17 + 16 \times 10 \times 4 + 3 \times 18 \times 9$), the sum of the column products, and the sum of the diagonal products are all the same.

11	2	17
16	10	4
3	18	9

37. The eighth triangular number is $t_8 = (8 \cdot 9)/2 = 36$, which is also a square number, since $6^2 = 36$. Verify that the triangular numbers t_{49}, t_{288}, and t_{1681} are also square numbers. Recall that $t_n = n(n+1)/2$.

From State Student Assessments

38. (Texas, Grade 5)

Ricky is trying to decide which outfit to wear to a party. His choices are shown in the diagram.

Pants

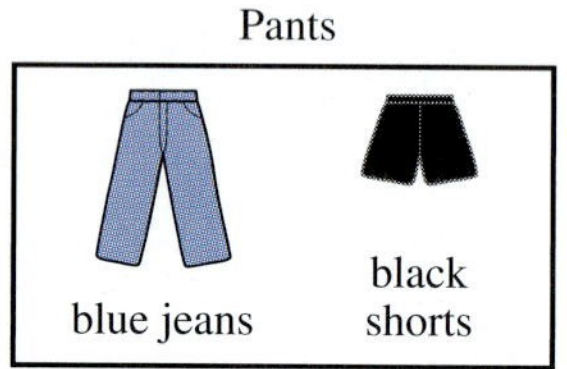

Shirts

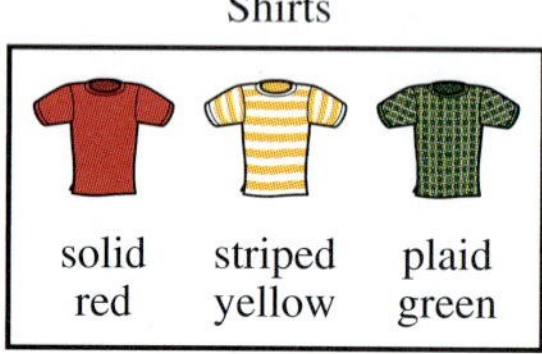

Shoes

How many possible different outfits can Ricky create if he chooses 1 pair of pants, 1 shirt, and 1 pair of shoes?

A. 7 B. 8 C. 10 D. 12

39. (Illinois, Grade 5)

Which expression fits the diagram below?

A. $4 \times (3 \times 3)$
B. $3 \times (4 \times 4)$
C. $3 + (3 \times 4)$
D. $3 + (3 + 4)$
E. $3 \times (3 + 4)$

40. (Michigan, Grade 4)

Robert is cutting oranges for his soccer teammates. He will cut each orange into 4 pieces. Each player will get 2 pieces. There are 20 players on the team. How many oranges does Robert need to cut?

A. 5 B. 8 C. 10 D. 12

For Review

41. What addition–subtraction fact family contains $2 + 7 = 9$?

42. Let $n(A) = 2$, $n(B) = 7$, and $n(A \cup B) = 9$. What is $n(A \cap B)$?

43. What basic facts are illustrated by these number-strip trains?

Epilogue From Counting to Numbers

The simplest mathematical idea of all is counting, a skill possessed by our ancient human ancestors even before the rise of civilization. The value of counting as a survival tool is also apparent in terms of evolutionary development. Indeed, recent research* has identified areas of the brain associated with basic number processing and even a vestigial number line. Remarkably, mathematical brain structures have also been found in pigeons, rats, chimpanzees, and other animals. Humans, being in possession of complex language

*Two very readable accounts of this remarkable research are recommended: Stanislas Dehaene's *The Number Sense: How the Mind Creates Mathematics* (New York: Oxford University Press, 1997) and Keith Devlin's *The Math Gene: How Mathematical Thinking Evolved and Why Numbers Are Like Gossip* (New York: Basic Books, 2000).

and symbolic skills, have the means to refine rudimentary and approximate number ideas into a sophisticated and precise system of numbers and calculations. This chapter has examined the first steps of this development. The idea of counting was taken to a new level of abstraction, namely the whole numbers. The arithmetic operations—addition, subtraction, multiplication, division—were then developed, together with the properties of these operations.

Chapter Summary

Key Concepts

Section 2.1 Sets and Operations on Sets

- Set. A set is a collection of objects from a stated universe. A set can be described verbally, by a list, or with set builder notation.
- Venn diagram. Sets can be visualized with Venn diagrams, in which the universe is a rectangle and closed loops correspond to sets. Elements of a set are associated with points within the loop corresponding to that set.
- Set operations and relations. Sets are combined and related by the following means: set complement $(\overline{A})$, subset $(A \subseteq B)$, proper subset $(A \subset B)$, intersection $(A \cap B)$, and union $(A \cup B)$.

Section 2.2 Sets, Counting, and the Whole Numbers

- Types of number. Nominal numbers name objects (e. g., an ID number), ordinal numbers indicate position (e.g., 5th place), and cardinal numbers indicate the number of elements in a set (e.g., There are 9 justices of the U.S. Supreme Court).
- Equivalence of sets. Two sets are equivalent if there is a one-to-one correspondence (or matching) of the elements of the two sets.
- Whole numbers. The whole numbers are the cardinal numbers of finite sets, with zero being the cardinal number of the empty set. The whole numbers can be represented and visualized by a variety of manipulatives and diagrams, including tiles, cubes, number strips, rods, and the number line.
- Order of the whole numbers. The whole numbers are ordered, so that m is less than n if a set with m elements is a proper subset of a set with n elements.

Section 2.3 Addition and Subtraction of Whole Numbers

- Addition of whole numbers. Addition of whole numbers is defined in the set model by $a + b = n(A \cup B$, where $a = n(A)$, $b = n(B)$, and A and B are disjoint finite sets. Addition can also be visualized on the number line with the measurement model.
- Closure property. The sum of two whole numbers is a whole number.
- Commutative property. For all whole numbers a and b, $a + b = b + a$.
- Associative property. For all whole numbers a, b, and c, $a + (b + c) = (a + b) + c$.
- Zero property of addition. Zero is an additive identity, so $a + 0 = 0 + a = a$ for all whole numbers a.
- Subtraction of whole numbers. $a - b = c$, where c is the unique whole number for which $a = b + c$.
- Conceptual models of subtraction. These models are the take-away model, the missing-addend model, the comparison model, and the number-line model.

Section 2.4 Multiplication and Division of Whole Numbers

- Multiplication. Multiplication is defined as a repeated addition, so that $a \cdot b = b + b + \cdots + b$, where there are a addends.
- Models of multiplication. These models are the array model, the rectangular area model, the multiplication tree model, and the Cartesian product $[a \cdot b = n(A \times B)]$. Here, $A \times B$ denotes the Cartesian product of sets A and B.
- Properties of multiplication. For all whole numbers a, b, and c, $a + b$ is a whole number (the closure property), $a \cdot b = b \cdot a$ (the commutative property), $a \cdot (b \cdot c) = (a \cdot b) \cdot c$ (the associative property), $a \cdot (b + c) = a \cdot b + a \cdot c$ (the distributive property), $1 \cdot a = a \cdot 1 = a$ (1 is a multiplicative identity), and $0 \cdot a = a \cdot 0 = 0$ (multiplication-by-zero property).
- Division. $a \div b, b \neq 0$, is defined if, and only if, there is a unique whole number c such that $a = b \cdot c$.
- Conceptual models of division. These models are the repeated-subtraction (grouping) model, the partition (sharing) model, and the missing-factor models where the missing factor c in the multiplication equation $a = b \cdot c$ is calculated.
- Division with remainder. Given any whole numbers a and b, with $b \neq 0$, there is a quotient q and a remainder r so that $a = b \cdot q + r$, where $0 \leq r < b$.
- Powers and exponents. For any positive whole number a (the base) and any whole number m (the exponent or power), the exponential expression a^m is defined by $a^m = a \cdot a \cdot \ldots \cdot a$, where there are m factors of a. Also, $a^0 = 1$.

Vocabulary and Notation

Section 2.1

Universe $\boldsymbol{U}$
Element of $\in$
List in braces $\{\mathbf{a}, \mathbf{b}, \ldots\}$
Set builder notation
  $\{\boldsymbol{x} \mid \boldsymbol{x}$ is $(\textit{condition})\}$
Venn diagram
Complement $\overline{\boldsymbol{A}}$
Subset $\boldsymbol{A} \subseteq \boldsymbol{B}$
Equal sets $\boldsymbol{A} = \boldsymbol{B}$
Proper subset $\boldsymbol{A} \subset \boldsymbol{B}$
Intersection $\boldsymbol{A} \cap \boldsymbol{B}$
Disjoint sets $\boldsymbol{A} \cap \boldsymbol{B} = \varnothing$
Empty set $\varnothing$
Union $\boldsymbol{A} \cup \boldsymbol{B}$
Transitive property of set inclusion
Commutative property of union and intersection
Associative property of union and intersection
Distributive properties of union and intersection

Section 2.2

Nominal (naming, identification) number
Ordinal number
Cardinal number
One-to-one correspondence
Equivalent sets $\boldsymbol{A} \sim \boldsymbol{B}$
Finite set
Infinite set
Whole number $\boldsymbol{n}(\boldsymbol{A})$
Zero $\mathbf{0} = \boldsymbol{n}(\varnothing)$
Number tiles, cubes, strips, rods
Number line
Less than: $\boldsymbol{a} < \boldsymbol{b}$; greater than: $\boldsymbol{a} > \boldsymbol{b}$

Section 2.3

Binary operation
Addition (sum) $\boldsymbol{a} + \boldsymbol{b}$
  Addend
  Summand
  Set model: $\boldsymbol{a} + \boldsymbol{b} = \boldsymbol{n}(\boldsymbol{A} \cup \boldsymbol{B})$ for $\boldsymbol{A} \cap \boldsymbol{B} = \varnothing$
  Measurement (number line) model

Properties of addition
- Closure, commutative, associative, and additive-identity property of zero

Subtraction (difference) $a - b$
- Minuend
- Subtrahend
- Take-away model
- Missing-addend model
- Comparison model
- Measurement (number-line) model

Addition–subtraction fact family

Section 2.4

Multiplication (product) $a \cdot b$
- Factor
- Repeated-addition model
- Array model
- Rectangular area model
- Multiplication tree model

Cartesian product of sets $A \times B$

Cartesian product model of multiplication
- $a \cdot b = n(A \times B)$, where $a = n(A)$, $b = n(B)$

Properties of multiplication
- Closure, commutative, and associative property
- Multiplicative identity property of one
- Multiplication-by-zero property
- Distributivity over addition

Division $a \div b$ (or a/b)
- Repeated-subtraction (grouping or measurement) model
- Partition (sharing) model
- Missing-factor model
- Dividend
- Divisor

Multiplication–division fact family

Division with remainders
- Quotient
- Remainder

Division algorithm $a \div b = q \text{ R } r$ or
- $a = bq + r, 0 \leq r < b$

Power operation

Base and exponent a^m

Exponential expression

Chapter Review Exercises

Section 2.1

1. Let $U = \{n \mid n$ is a whole number and $2 \leq n \leq 25\}$, $S = \{n \mid n \in U$ and n is a square number$\}$, $P = \{n \mid n \in U$ and n is a prime number$\}$, ($n > 1$ is *prime* when it is divisible by just two different numbers: itself and 1), and $T = \{n \mid n \in U$ and n is a power of 2$\}$
   (a) Write S, P, and T in listed form.
   (b) Find the following sets: $\overline{P}$, $S \cap T$, $S \cup T$, and $S \cap \overline{T}$.
2. Draw a Venn diagram of the sets S, P, and T in problem 1.
3. Replace each box $\square$ with one of the symbols $\cap$, $\cup$, $\subset$, $\subseteq$, or $=$ to give a correct statement for general sets A, B, and C.
   (a) $A \square A \cup B$
   (b) If $A \subseteq B$ and A is not equal to B, then $A \square B$
   (c) $A \square (B \cup C) = (A \cap B) \cup (A \cap C)$
   (d) $A \square \varnothing = A$

Section 2.2

4. Let $S = \{\text{s, e, t}\}$ and $T = \{\text{t, h, e, o, r, y}\}$. Find $n(S)$, $n(T)$, $n(S \cup T)$, $n(S \cap T)$, $n(S \cap \overline{T})$, and $n(T \cap \overline{S})$.
5. Show that the set of square natural numbers less than 101 is in one-to-one correspondence with the set $\{\text{a, b, c, d, e, f, g, h, i, j}\}$.
6. Show that the set of cubes $\{1, 8, 27, \ldots\}$ is equivalent to the set of natural numbers.
7. Label the number of elements in each region of a two-loop Venn diagram for which $n(U) = 20$, $n(A) = 7$, $n(B) = 9$, and $n(A \cup B) = 11$.

Section 2.3

8. Explain how to illustrate $5 + 2$ with
   (a) the set model of addition;
   (b) the number-line (measurement) model of addition.
9. What properties of whole-number addition are illustrated on the number lines below?
   (a)

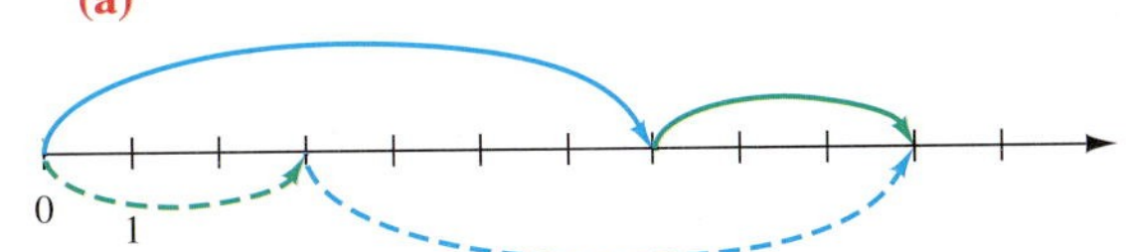

(b)

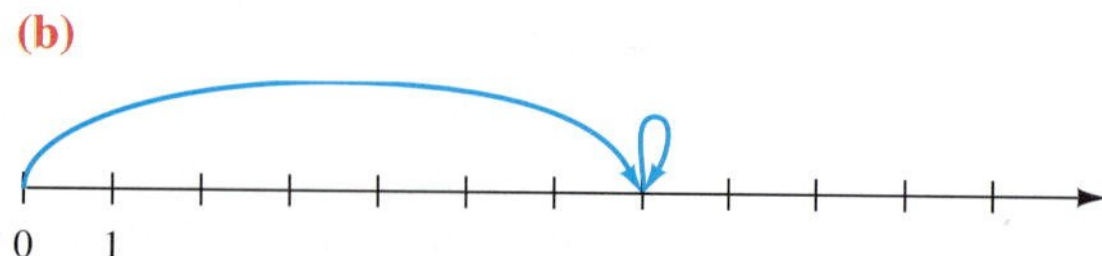

10. Draw figures that illustrate $6 - 4$

(a) using sets

(b) using the number line

Section 2.4

11. Draw representations of the product 4×2 using the given representation.

(a) A set diagram

(b) An array of discrete objects

(c) A rectangular area

(d) A multiplication tree

(e) A number-line diagram

12. Let $A = \{p, q, r, s\}$ and $B = \{x, y\}$.

(a) Find $A \times B$.

(b) What product is modeled with $A \times B$?

13. Whiffle balls 2 inches in diameter are packed individually in cubical boxes and then in cartons of 3 dozen balls. What are the dimensions of a suitable rectangular carton?

14. A drill sergeant lines up 92 soldiers in rows of 12, except for a partial row in the back. How many rows are formed? How many soldiers are in the back row?

15. Draw figures that illustrate the division problem $15 \div 3$, using these models:

(a) repeated subtraction (grouping objects in a set)

(b) partition (sharing objects)

(c) missing factor (rectangular array)

Chapter Test

1. What operation on whole numbers is being illustrated in the following diagrams?

(a)

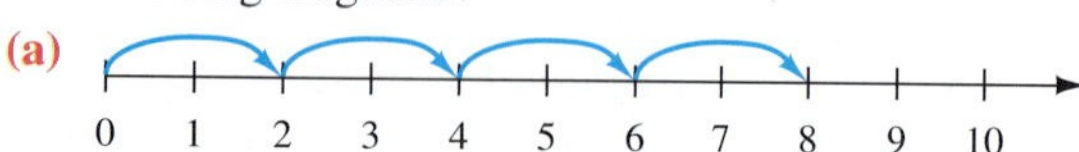

(b)

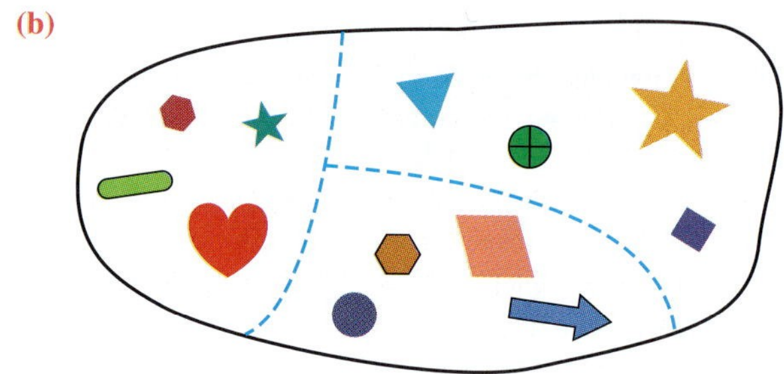

(c)

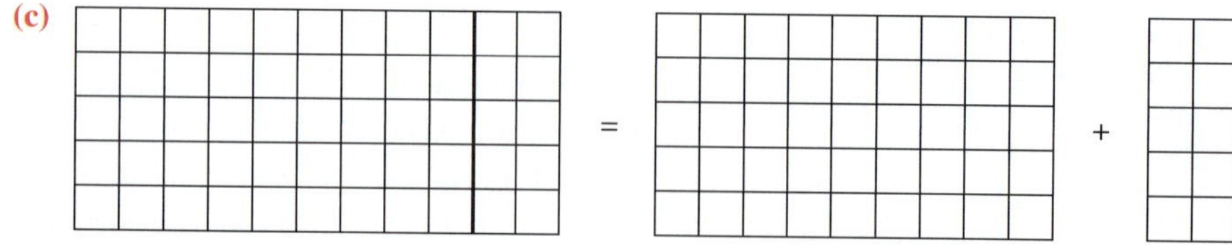

(d)

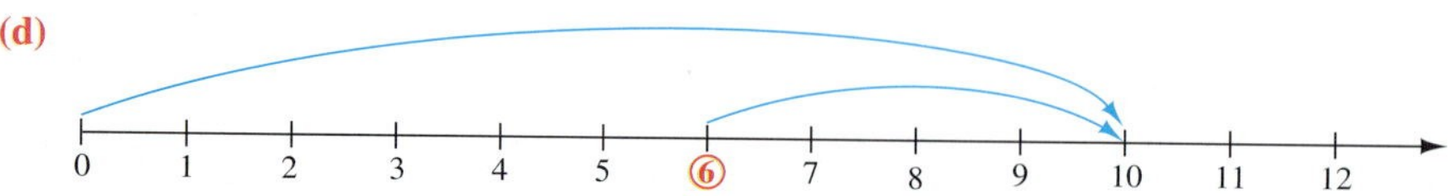

2. The following sentence contains three types of numbers: "On the *15th* of April, Joe Taxpayer sent in form *1040* and a money order for *\$253*."

   Name and describe the three kinds of numbers.

3. Let $S = \{1, 2, 4, 8, 16, \ldots\}$ denote the set of powers of 2.
   (a) Is S closed under multiplication?
   (b) Is S closed under addition?

   Explain why you answered as you did.

4. Explain why $5 < 8$, using the definition of whole-number inequality given in terms of sets.

5. Let "&" denote the binary operation on the set W of whole numbers defined by

$$a \,\&\, b = a + b + ab.$$

   (a) Is W closed under &?
   (b) Is & commutative?
   (c) Is & associative?
   (d) Is there an identity element for & in W?

6. Discuss which conceptual model of subtraction you feel best corresponds to each of the following problems.
   (a) On Monday, Roberto hiked 11 miles to the lake. On Friday at noon he had hiked 6 miles back down the trail. How much farther does Roberto have to hike to get back to the trailhead?
   (b) Kerri has 56 customers on her paper route, but her manager left her only 48 papers. How many extra papers should she have brought over before starting her delivery?
   (c) All but 7 of the 3 dozen picnic plates were used. How many people came to the picnic?

7. What property of whole numbers justifies the following equalities?
   (a) $4 + (6 + 2) = (4 + 6) + 2$
   (b) $8 \cdot (4 + x) = 8 \cdot 4 + 8x$
   (c) $3 + 0 = 0 + 3 = 3$
   (d) $2 \cdot (8 \cdot 5) = (2 \cdot 8) \cdot 5$

8. Shade regions in Venn diagrams that correspond to the following sets.
   (a) $A \cap (B \cup C)$
   (b) $A \cup B$
   (c) $(A \cap \overline{B}) \cup (B \cap \overline{A})$

9. Show how to illustrate the following properties of whole-number multiplication with the rectangular array model.
   (a) $2 \cdot (4 + 3) = 2 \cdot 4 + 2 \cdot 3$
   (b) $2 \cdot 5 = 5 \cdot 2$

10. Let $n(A \times B) = 21$. What are all of the possible values of $n(A)$?

11. Let $A = \{\text{w, h, o, l, e}\}$, $B = \{\text{n, u, m, b, e, r}\}$, $C = \{\text{z, e, r, o}\}$. Find:
    (a) $n(A \cup B)$
    (b) $n(B \cap \overline{C})$
    (c) $n(A \cap C)$
    (d) $n(A \times C)$

12. Let A and B be two sets in the universe $U = \{\text{a, b, c}, \ldots, \text{z}\}$. If $n(A) = 12$, $n(B) = 14$, and $n(A \cup B) = 21$, find $n(A \cap B)$ and $n(\overline{A \cap B})$.

13. Suppose $A \cap B = A$. What can you say about $A \cap \overline{B}$?

14. Althea made 5 gallons of root beer and wishes to bottle it in 10-ounce bottles.
    (a) How many bottles does she need? Remember that a quart contains 32 ounces.
    (b) Discuss which model of division—grouping or sharing—corresponds best to your answer for part **(a)**.

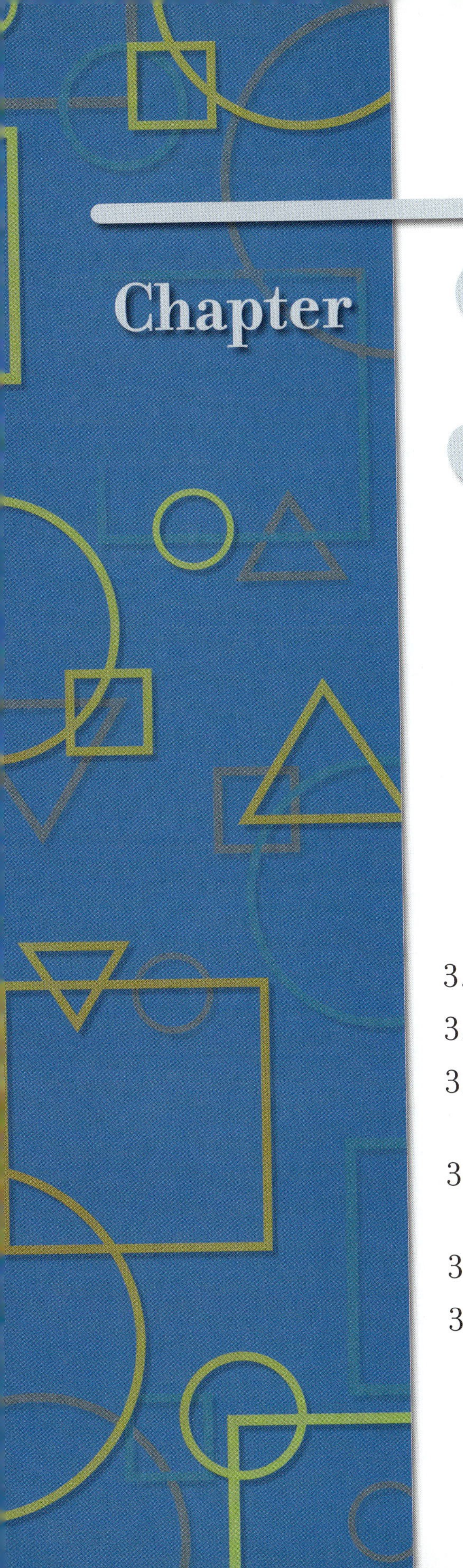

Chapter 3

Numeration and Computation

Numbers from Rectangles

Materials Needed

1. One rectangle of each of these shapes for each student.

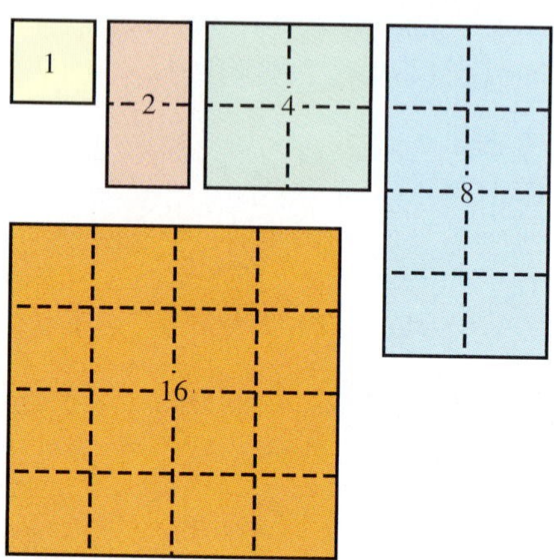

2. One record sheet like this for each student.

	16	8	4	2	1
0	0	0	0	0	0
1	0	0	0	0	1
18	1	0	0	1	0
19	1	0	0	1	1

	16	8	4	2	1
20	1	0	1	0	0
21					
38					
39					

Directions

Step 1. Use the rectangles to determine representations of each of the numbers 0, 1, 2, . . . , 39 as a sum of the numbers 1, 2, 4, 8, or 16, with each of these numbers used at most once.

Step 2. For each representation determined in Step 1, record the numbers (rectangles) used by placing a 0 or a 1 in the appropriate columns of the record sheet. The rows for 0, 1, 18, 19, and 20 have been done for you.

(a) Do all the numbers from 0 through 39 have such a representation?

(b) What additional numbers could be represented if you had a 32 rectangle?

(c) Describe any interesting patterns you see on your record sheet.

Connections — Counting and Calculating

In Chapter 2, we introduced whole numbers as the cardinal numbers of finite sets. For example, if A is the set shown, then $n(A) = 5$; that is, the number five is an abstract idea that represents the count of the elements in A and in each set equivalent to A. Historically, the notion of number developed over many years from people's need to count objects—sheep, goats, arrows, warriors, beads, and so on. Indeed, the need to count collections of concrete objects to develop an understanding of numbers was not just true historically; it remains equally true for individuals today. Research has shown that it is essentially impossible for children to understand the abstract notion of "five" without counting many sets of five objects—five beans, five fingers, five pennies, or five concrete objects of any kind. Thus, the need to use a variety of manipulative devices in the elementary school classroom to help children develop an understanding of numerousness—or number—cannot be ignored.

In this chapter we consider how we write numbers and how this is reflected in the methods used to perform calculations both by hand and with calculators and computers. We begin with a brief look at some of the earlier numeration systems and then consider the Indo-Arabic system we use today.

It is essential for children's understanding that these notions be introduced with extensive use of hands-on manipulative devices and copious pictorial representations. To do otherwise is to condemn students to rote memorization and meaningless manipulation of symbols. Put positively, students who are allowed to work with appropriate devices—to handle, to manipulate, and to experience—are much more likely to develop real understanding of concepts and skills and to develop enthusiasm for the study of mathematics.

3.1 NUMERATION SYSTEMS PAST AND PRESENT

To appreciate the power of the Indo-Arabic (or Hindu-Arabic) numeration system, or decimal system, as we call it, it is important to know something about numeration systems of the past. Just as the idea of number historically arose from the need to determine "how many," the demands of commerce in an increasingly sophisticated society stimulated the development of convenient symbolism for writing numbers and methods for calculating. The symbols for writing numbers are called **numerals,** and the methods for calculating are called **algorithms.** Taken together, any particular system of numerals and algorithms is called a **numeration system.**

The earliest means of recording numbers consisted of creating a set of tallies—marks on stone, stones in a bag, notches in a stick—one for one for each item being counted. Indeed, the original meaning of the word *tally* was a stick with notches cut into it to record debts owed or paid. Often such a stick was split in half, with one half going to the debtor and the other half to the creditor. It is still common practice today to keep count by making tallies or marks with the minor, but useful, refinement of marking off the tallies in groups of five. Thus,

is much easier to read as 22 than

But such systems for recording numbers were much too simplistic for large numbers and for calculating.

Of the various systems used in the past, we consider only four here—the Egyptian, the Roman, the Babylonian, and the Mayan systems.

The Egyptian System

As early as 3400 B.C., the Egyptians developed a system for recording numbers on stone tablets using hieroglyphics (Table 3.1). This system was based on 10, as is our modern system, and probably for the same reason. That is, we humans come with a built-in "digital" calculator, as it were, with 10 convenient keys.

The Egyptians had symbols for the first few powers of 10, and then combined symbols additively to represent other numbers. (See Table 3.1.)

TABLE 3.1 Egyptian Symbols for Powers of 10

Power of 10	$10^0 = 1$	$10^1 = 10$	$10^2 = 100$	$10^3 = 1000$	$10^4 = 10{,}000$	$10^5 = 100{,}000$	$10^6 = 1{,}000{,}000$
Egyptian Symbol	I	∩	?				
Description	a staff	a yoke	a scroll	a lotus flower	a pointing finger	a fish	an amazed person

The system was cumbersome not only because it required numerous symbols to represent even relatively small numbers, but also because computation was awkward. Thus, 336 would be written as

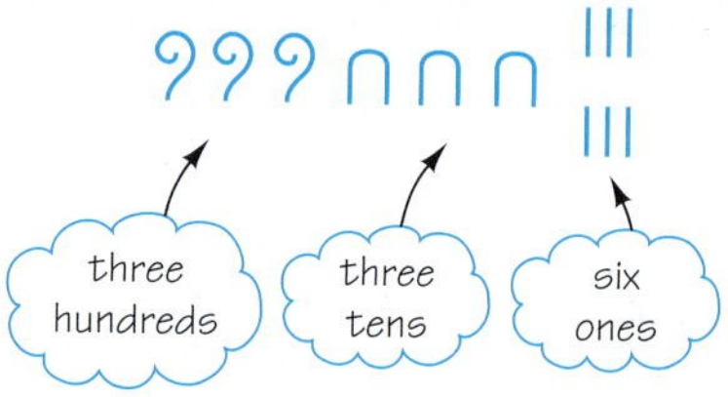

and 125 was written as

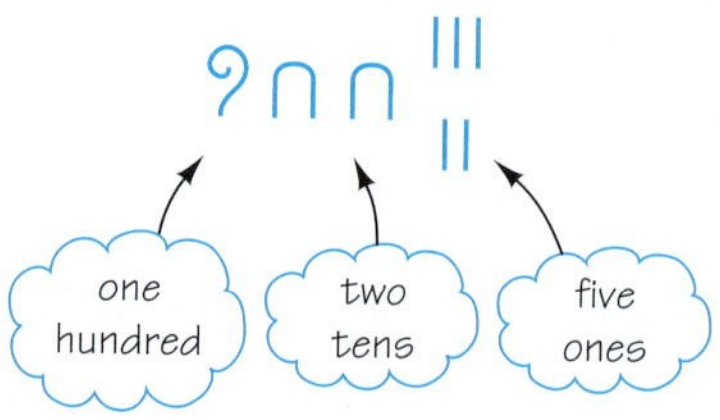

Then 336 + 125 was found by combining all the symbols to obtain

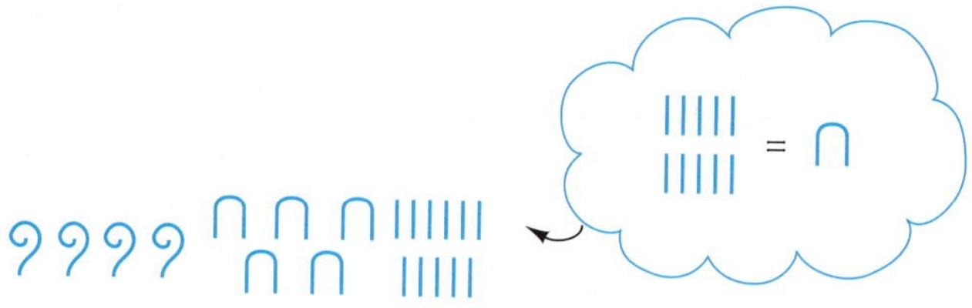

and then replacing 10 of the symbols for 1 by one symbol for 10 to finally obtain

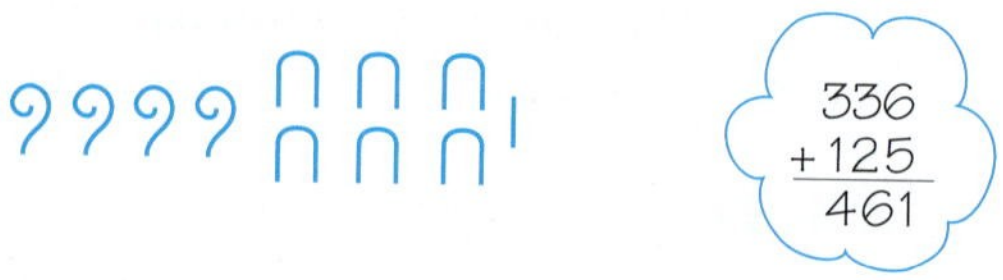

representing 461. You can easily imagine how cumbersome multiplication (as repeated addition) and division (as repeated subtraction) would have been in this system.

The Roman System

The Roman system of numeration is already somewhat familiar from its current usage on the faces of analog watches and clocks, on cornerstones, and on the façades of buildings to record when they were built. Originally, the system was completely additive, like the Egyptian system, with the familiar symbols shown in Table 3.2.

TABLE 3.2 Roman Numerals and Their Modern Equivalents

Roman Symbol	I	V	X	L	C	D	M
Modern Equivalent	1	5	10	50	100	500	1000

Using these symbols, 1959 originally was written as

M D CCCC L V IIII.

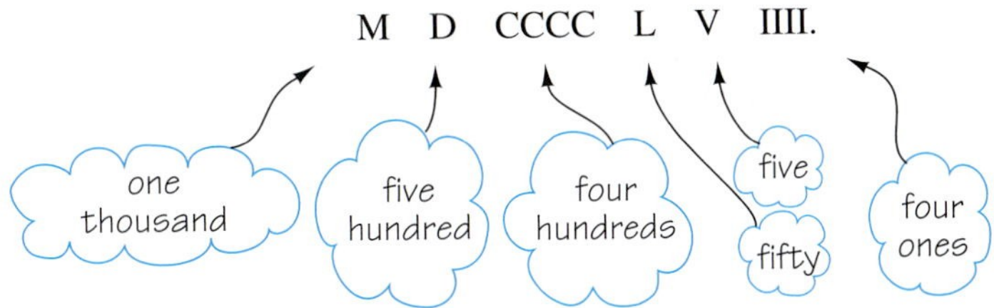

Later, a subtractive principle was introduced to shorten the notation. If a single symbol for a lesser number was written to the left of a symbol for a greater number, then the lesser number was to be *subtracted* from the greater number. In particular, the common representations were as shown here:

IIII was written as IV.

VIIII was written as IX.

XXXX was written as XL.

CCCC was written as CD.

DCCCC was written as CM.

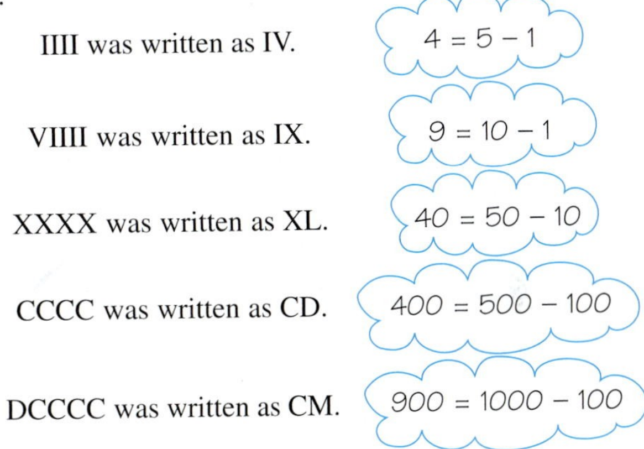

Using this principle, 1959 could then be written more succinctly as follows:

M CM L IX

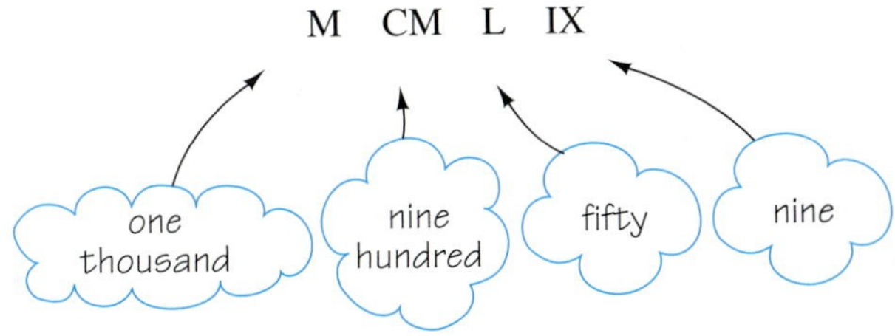

As with the Egyptian system, arithmetic* calculations using Roman numerals were quite cumbersome. For this reason, much of the commercial calculation of the time was performed on devices such as abacuses, counting boards, sand trays, and the like.

*When used as a noun, "arithmetic" is pronounced ə-rith′mə-tik. When used as an adjective, it is pronounced ar′ith-met′ik.

The Babylonian System

Developed about the same time as the Egyptian system, and much earlier than the Roman system, the Babylonian system was more sophisticated than either in that it introduced the notion of **place value,** in which the position of a symbol in a numeral determined its value. In particular, this made it possible to write numerals for even very large numbers by using very few symbols. Indeed, the system utilized only two symbols, ▼ for 1 and ❮ for 10, and combined these additively to form the "digits" 1 through 59. Thus,

❮❮▼ and , ❮❮❮▼▼▼▼

respectively, represented 21 and 34. Beyond 59, the system was positional to base sixty (a sexagesimal system), where the positions from right to left represented multiples of successive powers of 60 and the multipliers (digits) were the composite symbols for 1 through 59. Thus,

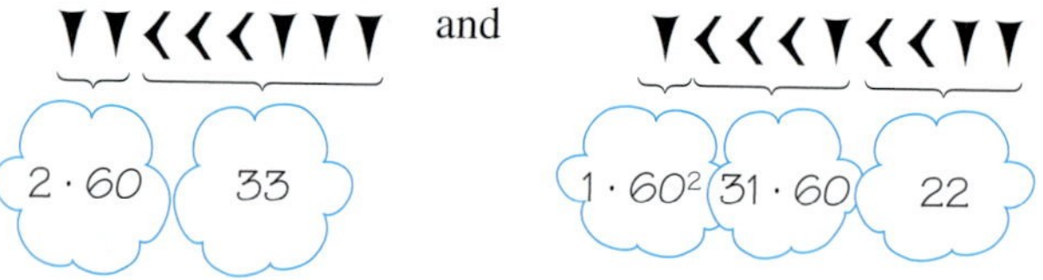

represented $2 \cdot 60^1 + 33 = 153$ and $1 \cdot 60^2 + 31 \cdot 60^1 + 22 = 5482$.

A difficulty with the Babylonian system was the lack of a symbol for zero. In writing numerals, scribes simply left a space if a certain position value was not to be used, and, since spacing was not always uniform, this often made it necessary to infer values from context. For example, the Babylonian notation for 83 and 3623 originally were, respectively,

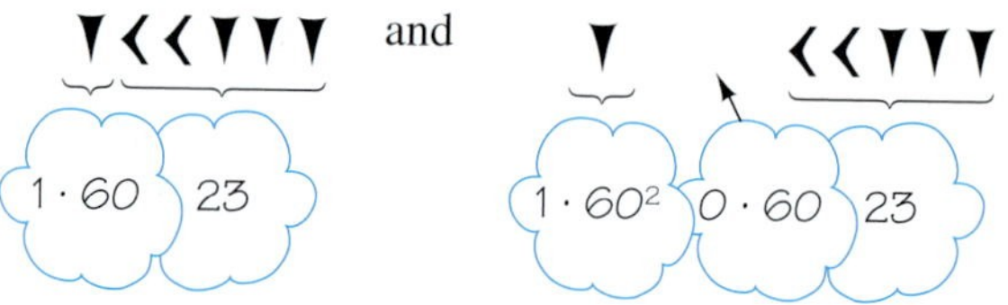

and these could easily be confused if the spacing were not clear. Indeed, there was no way even to indicate missing position values on the extreme right of a numeral, so ▼ could represent 1 or $1 \cdot 60^1 = 60$ or $1 \cdot 60^2 = 3600$, and so on. Eventually the Babylonians employed the symbol ▲ as a place holder to indicate missing position values, though they never developed the notion of zero as a number. Using this symbol, ▼❮❮▼▼▼ was clearly understood as $1 \cdot 60 + 23 = 83$ and ▼▲❮❮▼▼▼ was unmistakably

$$1 \cdot 60^2 + 23 = 3623.$$

The Mayan System

One of the most interesting of the ancient systems of numeration was developed by the Mayans in the region now known as the Yucatan Peninsula, in southeastern Mexico. As early as A.D. 200, these resourceful people had developed a remarkably advanced society. They were the first Native Americans to develop a system of writing and to manufacture paper and books. Their learned scholars knew more about astronomy than was known at that time anywhere else in the world. Their calendar was very accurate, with a 365-day year and a leap year every fourth year. In short, many scholars believe that the Mayans very early developed the most sophisticated society ever attained by early residents of the Western Hemisphere.

Like their other achievements, the Mayan system of numeration was unusually advanced. Actually there were *two* systems. Both systems were positional systems like the Babylonian system and our present base ten, or decimal, system, and both contained a symbol for zero. One system was a base twenty, or vigesimal, system based on powers of 20 and utilizing rather involved hieroglyphics for the numerals 1 through 19. Since this system has not survived in written form, we describe the second system, which was a modification of the first, devised to facilitate computations related to the calendar. It has the advantage that $18 \cdot 20 = 360$ is much closer to the length of a year than $20 \cdot 20 = 400$.* Thus, the positions from the third on were for $18 \cdot 20$, $18 \cdot 20^2$, $18 \cdot 20^3$, and so on, rather than $20^2, 20^3, 20^4, \ldots$ as with the first system. Also, the second system used a simple set of symbols for the numerals 0 through 19 as shown in Table 3.3.

TABLE 3.3 Mayan Numerals for 0 through 19

Mayan Symbol	Modern Equivalent	Mayan Symbol	Modern Equivalent
	0		10
	1		11
	2		12
	3		13
	4		14
	5		15
	6		16
	7		17
	8		18
	9		19

The Mayans wrote their numerals in a vertical style with the unit's position on the bottom. For example, the number 43,487 would appear as

$$6 \cdot 18 \cdot 20^2 = 43{,}200$$
$$0 \cdot 18 \cdot 20 = 0$$
$$14 \cdot 20 = 280$$
$$7 \cdot 1 = \underline{\quad 7}$$
$$43{,}487$$

*The Mayan calendar was composed of 18 months of 20 days each, with five additional days not associated with any month and with provision for a leap year.

All that is needed to convert a Mayan numeral to modern notation is a knowledge of the digits and the value of the position each digit occupies as shown in Table 3.4.

TABLE 3.4 Position Values in the Mayan System

Position Level	Position Value
.	.
.	.
.	.
fifth	$18 \cdot 20^3 = 144{,}000$
fourth	$18 \cdot 20^2 = 7200$
third	$18 \cdot 20 = 360$
second	20
first	1

EXAMPLE 3.1 USING MAYAN NOTATION TO WRITE A NUMBER

Write 27,408 in Mayan notation.

Solution

We will not need the fifth, or any higher, position since $18 \cdot 20^3 = 144{,}000$ is already greater than 27,408. How many 7200s, 360s, 20s, and 1s are needed? This can be answered by repeated subtraction or, more simply, by division. From the arithmetic shown,

```
         3              16          2       8
7200)27,408     360)5808      20)48    1)8
     21,600          360         40      8
      5 808         2208          8      0
                    2160
                      48
```

it follows that we need three 7200s, sixteen 360s, two 20s, and eight 1s. Thus, in Mayan notation, 27,408 appears as

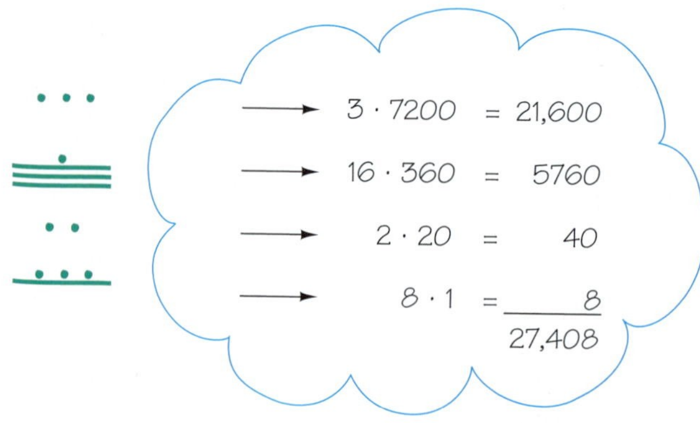

The ingenuity of this system, particularly since it involves the 18s, is easy to overlook. It turns out that such a positional system *will not work* unless the value of each position is a number that evenly divides the value of the next higher position. This fact is not at all obvious!

The Indo-Arabic System

Today, the most universally used system of numeration is the **Indo-Arabic,** or **decimal, system.** The system was named jointly for the East Indian scholars who invented it at least as early as 800 B.C. and for the Arabs who transmitted it to the Western world. Like the Mayan system, it is a positional system. Since its **base** is ten, it requires special symbols for the numbers zero through nine. Over the years, various notational choices have been made, as shown in Table 3.5.

TABLE 3.5 Symbols for Zero through Nine, Ancient and Modern

Tenth-century East Indian	०	१	२	३	४	५	६	७	८	९
Current Arabic	٠	١	٢	٣	٤	٥	٦	٧	٨	٩
Fifteenth-century European	0	1	2	3	4	5	6	7	8	9
Modern cursive	0	1	2	3	4	5	6	7	8	9
Modern print	0	1	2	3	4	5	6	7	8	9
Lighted scoreboard	0	1	2	3	4	5	6	7	8	9
Calculator display	0	1	2	3	4	5	6	7	8	9
Machine readable	0	1	2	3	4	5	6	7	8	9

The common symbols 0, 1, 2, 3, 4, 5, 6, 7, 8, and 9 are called **digits,** as are our fingers and toes, and it is easy to imagine the historical significance of this terminology. With these 10 symbols and the idea of positional notation, all that is needed to write the numeral for any whole number is the value of each digit and the value of the position the digit occupies in the numeral. In the decimal system, the positional values, as shown in Table 3.6, are well known.

TABLE 3.6 Positional Values in Base Ten

Position Names	...	Hundred Thousands	Ten Thousands	Thousands	Hundreds	Tens	Units
Decimal Form	...	100,000	10,000	1000	100	10	1
Powers of 10	...	10^5	10^4	10^3	10^2	10^1	10^0

As our number words suggest, the symbol 2572 means 2 thousands plus 5 hundreds plus 7 tens plus 2 ones. In so-called **expanded notation** we write

2000 + 500 + 70 + 2

$$\begin{aligned}2572 &= 2 \cdot 1000 + 5 \cdot 100 + 7 \cdot 10 + 2 \cdot 1\\ &= 2 \cdot 10^3 + 5 \cdot 10^2 + 7 \cdot 10^1 + 2 \cdot 10^0.\end{aligned}$$

The amount each digit contributes to the number is the value of the digit times the value of the position the digit occupies in the representation.

Physical Models for Positional Systems

The Classroom Abacus

The Indo-Arabic scheme for representing numbers derived historically from the use of counting boards and abacuses of various types to facilitate computations in commercial transactions. Such devices are still useful today in helping children to understand the basic concepts. One such device, commercially available for classroom use, is shown in Figure 3.1.

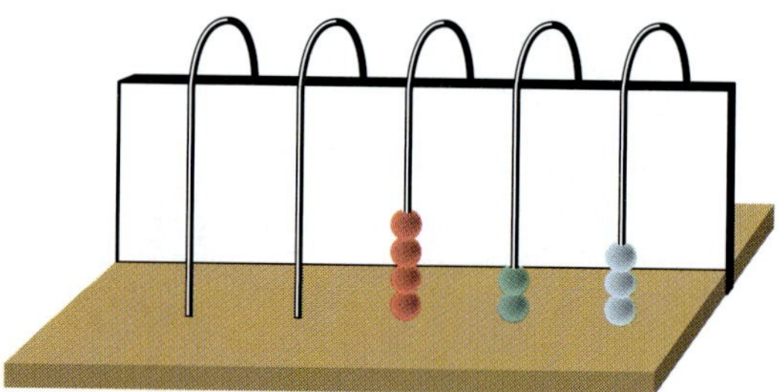

Figure 3.1
A classroom demonstration abacus

The device consists of a series of wire loops fixed into a wooden base and with a vertical shield affixed to the center of the base under the wire loops so that the back is hidden from the students. On each wire loop are beads that can be moved from behind the shield to the front and vice versa.

To demonstrate counting and positional notation, one begins by moving beads from behind the shield to the front on the wire to the students' right (the instructor's left), counting 1, 2, 3, and so on, as the beads are moved to the front of the shield. Once 10 beads on the first wire are counted, all 10 are moved back behind the shield, and the fact that 10 beads on this wire have all been counted once is recorded by moving a single bead to the front of the shield on the second wire. The count 11, 12, 13, and so on, then continues with a single bead on the first wire again moved forward with each count. When the count reaches 20, 10 beads on the first wire will have been counted a second time, and this is recorded by moving a second bead forward on the second wire and moving the beads on the first wire to the back again. Then the count continues. When the count reaches 34, for example, the children will see the arrangement of beads shown schematically in Figure 3.2 with three beads showing on the second wire and four beads showing on the first wire. A natural way to record the result is to write 34; that is, 3 tens and 4 ones. Note that we never leave 10 beads up on any wire.

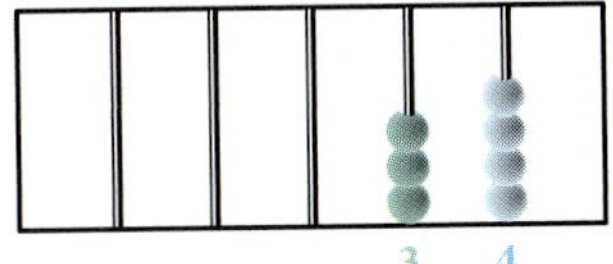

Figure 3.2
Thirty-four on the classroom abacus

What happens if we count to 100—that is, if we count 10 beads on the first wire 10 times? Since we move one bead on the second wire forward each time we count 10 beads on the first wire, we will have 10 beads showing on the second wire. But we record this on the abacus by moving one bead forward on the third wire and moving all beads on the second wire back behind the shield. Thus, each bead showing on the first wire counts for 1, each bead showing on the second wire counts for 10, each bead showing on the third wire counts for 100, and so on. If we count to 423, for example, the arrangement of beads is as illustrated earlier in Figure 3.1, and the count is naturally recorded as 423.

Approached in this way, the notion of place value becomes much more concrete. One can actually experience the fact that each bead on the second wire counts for 10, each bead on the third wire counts for 100, and so on—particularly if allowed to handle the

HIGHLIGHT *from* HISTORY

Fibonacci

The most talented mathematician of the Middle Ages was Leonardo of Pisa (ca. 1170–1250), the son of a Pisan merchant named Bonaccio. The Latin *Leonardo Filius Bonaccio* (Leonardo son of Bonaccio) was soon contracted to Leonardo Fibonacci. This was further shortened simply to Fibonacci, which is still popularly used today. The young Fibonacci was brought up in Bougie—still an active port in modern Algeria—where his father served for many years as customs manager. It was here and on numerous trips throughout the Mediterranean region with his father that Fibonacci became acquainted with the Indo-Arabic numerals and the algorithms for computing with them that we still use today. Fibonacci did outstanding original work in geometry and number theory and is best known today for the remarkable sequence

$$1, 1, 2, 3, 5, 8, 13, 21, 34, 55, 89, \ldots,$$

which bears his name. However, his most important contribution to Western civilization remains his popularization of the Indo-Arabic numeration system in his book *Liber abaci,* written in 1202. This book so effectively illustrated the vast superiority of this system over the other systems then in use that it soon was widely adopted not only for use in commerce, but also in serious mathematical studies. Its use so simplified computational procedures that its effect on the rapid growth of mathematics during the Renaissance and beyond can only be characterized as profound.

Into the Classroom

Mary Cavanaugh Discusses the Use of Manipulatives

Teaching for meaning in mathematics is a major focus of math educators. The work of Piaget and others suggests that students begin their exploration of mathematical concepts through hands-on experiences with manipulatives. Manipulatives appeal to several senses and are used to physically involve students in a learning situation. Students manipulate the objects with actions such as forming, ordering, comparing, tracing, joining, or separating groups. By these actions they gain an understanding of the meanings of number and various operations in mathematics.

Research points to multi-digit numeration as the pivotal concept that students must learn with the help of manipulatives before they can succeed in learning computational algorithms. A comfortable progression through representing numbers, trading, and computing with manipulatives is developmentally appropriate for providing a solid foundation for understanding arithmetic concepts. Premature introduction of paper-and-pencil procedures pushes students into memorizing a complex sequence of mathematical acts before the acts have meaning.

Manipulating objects allows students to develop internal images of number and the relative magnitude of number that can be recalled when needed. Pictures of concrete materials that follow manipulative experiences form a connection between the internal images, the pictures, and abstract symbols. The guidance of teachers is needed to help bridge the gap between what is learned from manipulatives and what is written using mathematical symbols.

A research recommendation, easily realized with manipulatives, is that teachers can and should expect students to enjoy the learning of mathematics. Concrete materials appeal to a variety of senses and motivate students. The frequent use of manipulatives in the classroom helps make the expectation of enjoyment evident.

SOURCE: From ScottForesman Exploring Mathematics, Grades 1–7, *by L. Carey Bolster et al. Copyright © 1994 Pearson Education, Inc. Reprinted with permission. Mary Cavanaugh is the coordinator of "Math, Science, and Beyond" for Solana Beach School District, Solana Beach, California.*

device and move the beads while counting. One can also see that the counting process on the abacus can proceed as far as desired, though more and more wire loops would have to be added to the device. It follows that any whole number can be represented in one, and only one, way in our modern system, using only the digits 0, 1, 2, 3, 4, 5, 6, 7, 8, and 9.

Other physical devices can also be used to illustrate positional notation, and many are even more concrete than the historical abacus mimicked above.

Sticks in Bundles

One simple idea for introducing positional notation is to use bundles of small sticks, which can be purchased inexpensively at almost any craft store. Single sticks are 1s. Ten sticks can be bound together in a bundle to represent 10. Ten bundles can be banded together to represent 100, and so on. Thus, 34 would be represented as shown in Figure 3.3.

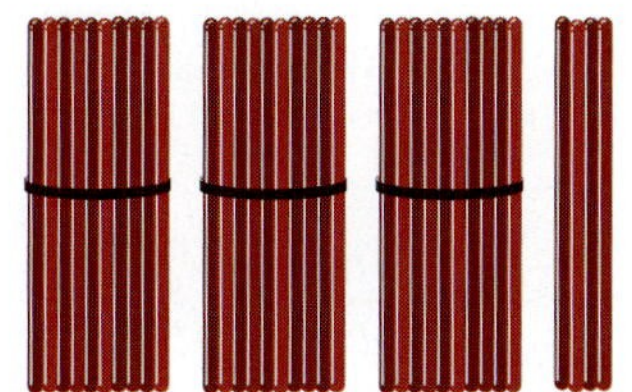

Figure 3.3
Representing 34 with sticks and bundles of sticks

Unifix™ Cubes

Here, single cubes represent 1s. Ten cubes snapped together to form a stick represent 10. Ten sticks bound together represent 100, and so on. Again, 34 would be represented as shown in Figure 3.4.

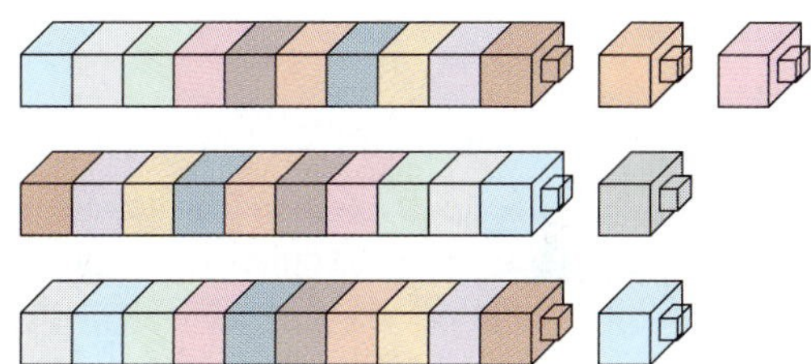

Figure 3.4
Representing 34 with Unifix™ cubes

Units, Strips, and Mats

Pieces for this concrete realization of the decimal system are easily cut from graph paper with reasonably large squares. A unit is a single square, a strip is a strip of 10 squares, and a mat is a square 10 units on a side, as illustrated in Figure 3.5. To make the units, strips, and mats more substantial and hence easier to manipulate, the graph paper can be pasted to a substantial tag board or copied on reasonably heavy stock. Using units, strips, and mats, 254 would be represented as shown in Figure 3.6.

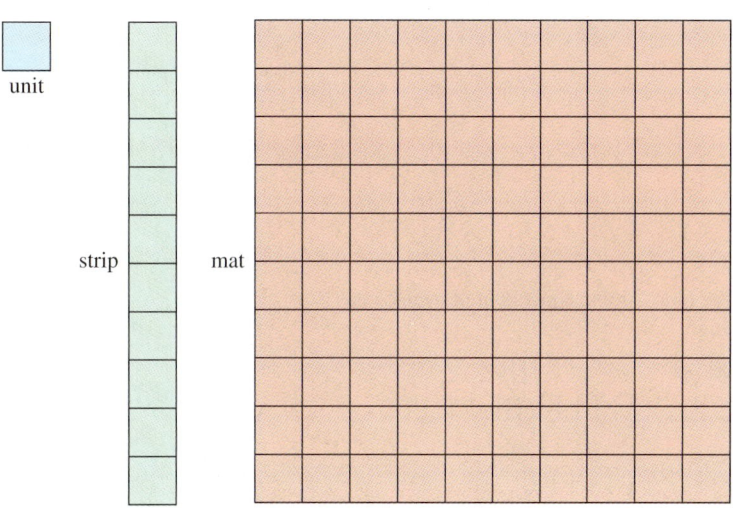

Figure 3.5
A unit, a strip, and a mat

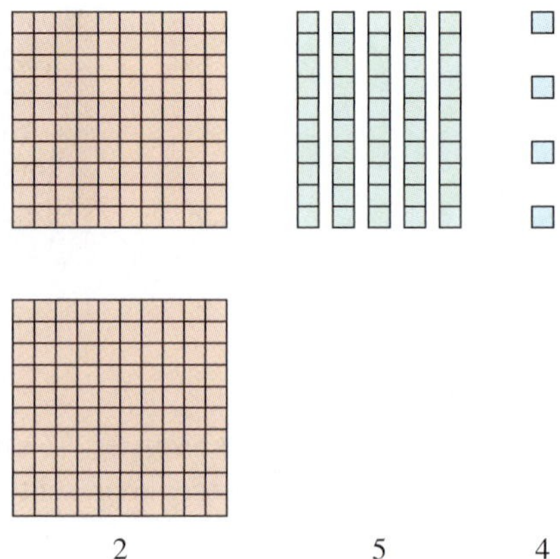

Figure 3.6
Representing 254 with units, strips, and mats

Base Ten Blocks

These commercially prepared materials include single cubes that represent 1s, sticks called "longs" made up of 10 cubes that represent 10, "flats" made up of 10 longs that represent 100, and "blocks" made up of 10 flats that represent 1000, as shown in Figure 3.7.

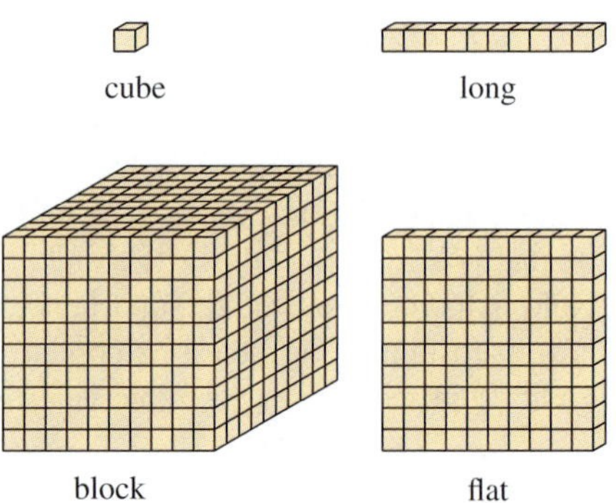

Figure 3.7
Base ten blocks for 1, 10, 100, and 1000

Problem Set 3.1

Understanding Concepts

1. Write the Indo-Arabic equivalent of each of the following.

(a) 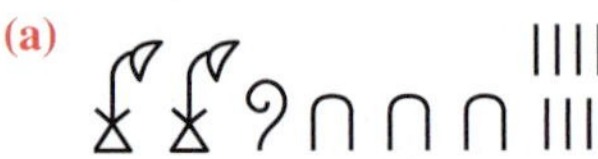

(b) 

(c) 

(d) MDCCXXIX

(e) DCXCVII

(f) CMLXXXIV

(g) ▼▲▲ **(h)** ▼▲▲▲▲

(i) ▼▼ ⟨⟨⟨ ▼▼

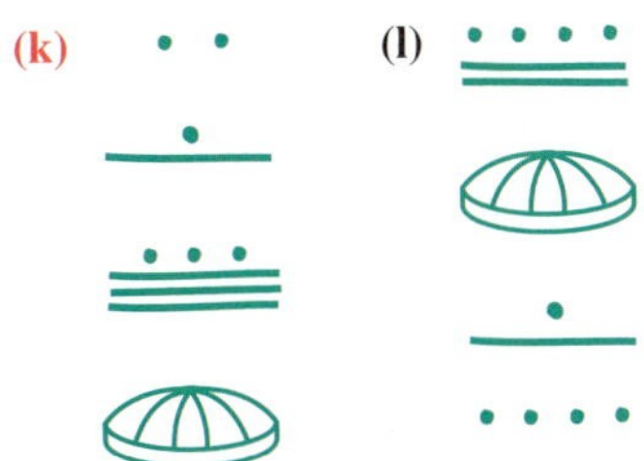

2. Write the following in Egyptian notation.
   (a) 11 (b) 597 (c) 1949
3. Write the following in Roman notation using the subtraction principle as appropriate.
   (a) 9 (b) 486 (c) 1945
4. Write these in Babylonian notation.
   (a) 251 (b) 3022 (c) 18,741
5. Write these in Mayan notation.
   (a) 12 (b) 584 (c) 12473
6. Add 𓍢𓍢𓍢 𓎆𓎆𓎆𓎆𓎆𓎆𓎆𓎆 ||| and 𓍢𓍢 𓎆𓎆𓎆𓎆𓎆𓎆𓎆𓎆 ||| using only Egyptian notation. Write your answer as simply as possible.
7. Add MDCCCXVI and MCCCLXIV using only Roman notation. Write your answer as simply as possible.
8. Add 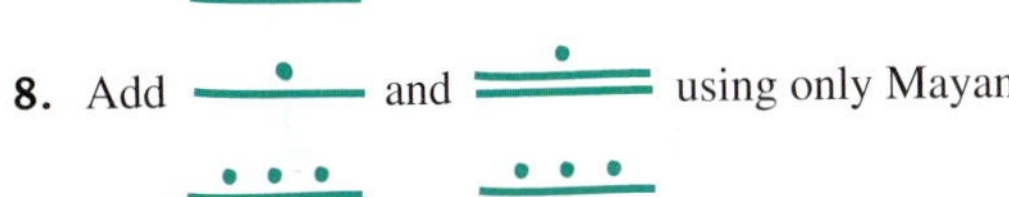 and using only Mayan notation. Write your answer as simply as possible.
9. (a) Write 2002, 2003, and 2004 in Roman numerals.
   (b) To date, what year has had the Roman numeral with the most symbols? Give your answer in both Roman and Indo-Arabic numerals.
10. The ancient Chinese "rod system" of numeration was a base ten positional system like our modern system with two possible arrangements for the digits 1 through 9:

1	2	3	4	5	6	7	8	9
𝍠	𝍡	𝍢	𝍣	𝍤	𝍥	𝍦	𝍧	𝍨
𝍩	𝍪	𝍫	𝍬	𝍭	𝍮	𝍯	𝍰	𝍱

For easy reading, the vertical arrangement was used in the columns corresponding to the even powers of 10 and the horizontal arrangement for the columns corresponding to the odd powers of 10. A blank space was used for the digit zero. Thus, 12,462 would appear as 𝍠𝍪𝍣𝍮𝍡. Write the modern equivalent of the following.
   (a) 𝍫𝍦𝍰𝍤 (b) 𝍪𝍥𝍰𝍣
   (c) 𝍮 𝍬𝍧 (d) 𝍢𝍯 𝍱𝍥
11. Write each of the following using the Chinese rod system of notation.
   (a) 2763 (b) 38,407 (c) 804 (d) 7561
12. Perform each of the following computations using the Chinese rod system of notation. Check using modern notation.
   (a) 𝍢𝍯𝍣 + 𝍡𝍰𝍠 (b) 𝍭𝍧𝍯𝍥 + 𝍣𝍮𝍡
   (c) 𝍮𝍡𝍪𝍣 - 𝍦𝍫𝍡 (d) 𝍧𝍯𝍤 - 𝍡𝍰𝍣
13. Draw a sketch to illustrate 452 using units, strips, and mats. (See Figures 3.5 and 3.6.) Use dots to represent units, vertical line segments to represent strips, and squares to represent mats.
14. Draw a sketch to illustrate 234 using base ten blocks.
15. Draw a sketch of the exposed side of a classroom abacus to illustrate 2475.
16. Write 24,872 and 3071 in expanded notation.
17. Suppose you have 3 mats, 24 strips, and 13 units for a total count of 553. Briefly describe the exchanges you would make to keep the same total count but have the smallest possible number of manipulative pieces.
18. Suppose you have 2 mats, 7 strips, and 6 units in one hand and 4 mats, 5 strips, and 9 units in the other.
   (a) Put all these pieces together and describe the exchanges required to keep the same total count but with the smallest possible number of manipulative pieces.
   (b) Explain briefly but clearly what mathematics the manipulation in part (a) represents.
19. Suppose you have 3 mats and 6 units on your desk and want to remove a count represented by 3 strips and 8 units.
   (a) Describe briefly, but clearly, the exchange that must take place to accomplish the desired task.
   (b) After removing the 3 strips and 8 units, what manipulative pieces are left on your desk?
   (c) What mathematics does the manipulation in part (a) represent?

Teaching Concepts

20. Recall that units, strips, and mats are typically made from graph paper (see Figure 3.5). How would you augment this simple set of manipulatives to make one representing 1000 for a lesson you might teach to a class of third graders?

21. Triana has 12 dollars and 33 dimes in her piggy bank, while Leona has 9 dollars and 11 dimes.

(a) If the girls pool their money, how many dollars and dimes will they have? Note that this question can be answered without adding; one can simply count on instead.

(b) If the girls buy their mother a bottle of perfume for \$25, how much money will they have left over?

(c) Describe any exchanges you made in determining the answer to part (b).

22. Jon works at odd jobs for Mr. Taylor. On four successive Saturdays, he worked 6 hours and 35 minutes, 4 hours and 15 minutes, 7 hours and 30 minutes, and 5 hours 20 minutes, respectively. All told, how many 8-hour days, how many hours, and how many minutes was this amount of time equivalent to? (Make sure that the number of hours in your answer is less than 8 and that the number of minutes is less than 60.)

23. Make up a problem along the lines of problem 21 but using hours, days, and weeks.

24. **(a)** How does a problem like problem 21 help students to understand positional notation?

(b) Would understanding the notion of exchanging help students studying calculation with fractions deal with a problem like this?

$$\begin{array}{r} 2\frac{3}{7} \\ +\ 2\frac{5}{7} \\ \hline \end{array}$$

Explain.

Thinking Critically

25. Suppose the discussion of the classroom abacus on pages 150 to 151 were repeated with "five" playing the role of "ten." Then 42 would be represented on the abacus as shown here. Moreover, it would be recorded as 132 and read as "one three two" and *not* as "one hundred thirty-two."

What numbers would be represented if the beads on the abacus described are as shown in these diagrams?

(a)

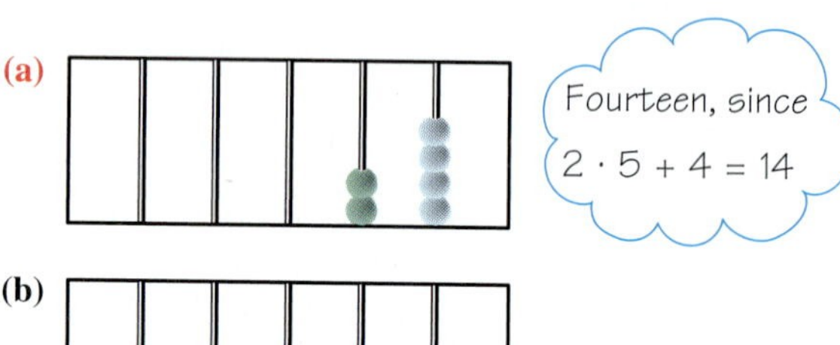

(b)

(c) (d) (e)

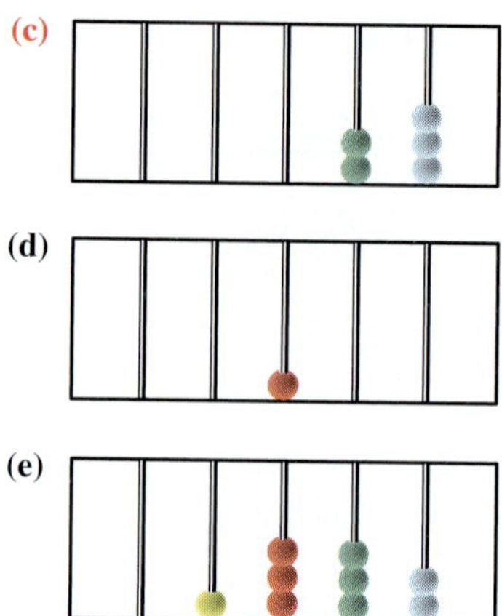

26. The number represented in the diagram of problem 25(b), would be recorded as 10 ("one zero"). How would the numbers for parts (a), (c), (d), and (e) naturally be recorded?

27. Suppose the abacus of problem 25 is configured as shown here.

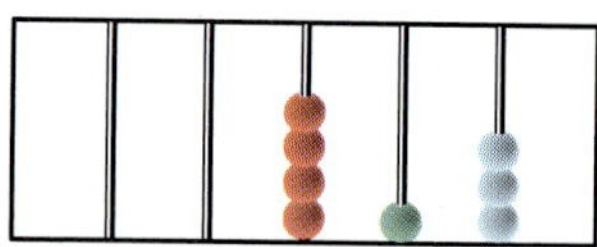

(a) What number does this arrangement represent?

(b) How would this number naturally be recorded?

(c) Suppose the number shown in part (a) is increased by 7. Draw a diagram to show how the abacus will then be configured.

28. Suppose the abacus of problem 25 is as shown here.

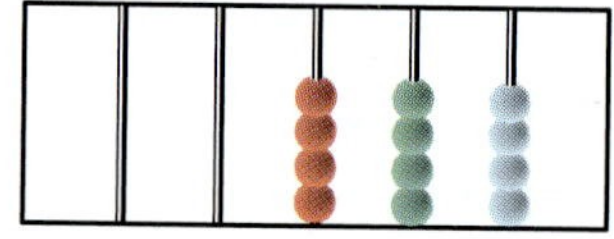

How will it appear if the number represented is increased by 1? Draw a suitable diagram.

29. Draw diagrams of the abacus of problem 25 for each of these numbers.

(a) five **(b)** seven **(c)** ten **(d)** twenty-five

(e) thirty-two **(f)** one hundred four

(g) one hundred twenty-five

(h) one hundred forty-seven

30. The rightmost wire on the abacus in problem 25 is called the units wire. What would be appropriate names for

(a) the second wire from the right?

(b) the third wire from the right?

31. If the number represented on the abacus in problem 25 is recorded as 132, what is the count?

From State Student Assessments

32. (Illinois, Grade 3)
Find the missing number that makes the sentence true.

500 + _____ + 9 = 569

○ 50

○ 60

○ 69

○ 90

33. (Connecticut, Grade 4)
What number is shown by the blocks in this picture?

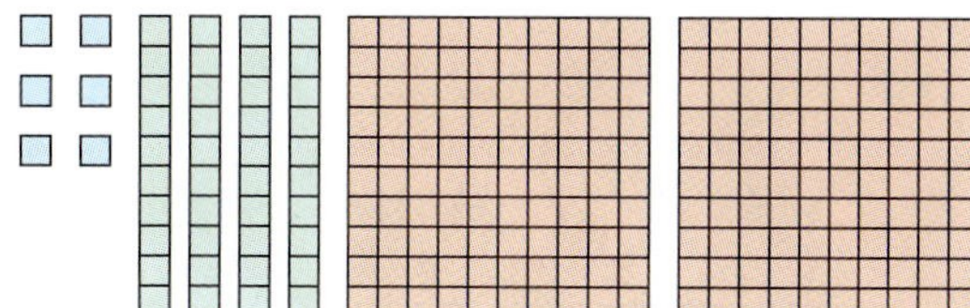

642

426

264

246

34. (Connecticut, Grade 4)
Which means the same as 385?

300 + 80 + 5

30 + 800 + 5

300 + 80 + 50

3 + 8 + 5

35. (Connecticut, Grade 4)
Which means the same as 2 tens and 18 ones?

28

38

318

2018

For Review

36. Given sets $U = \{1, 2, 3, 4, 5, 6, 7, 8, 9\}$, $A = \{1, 3, 5, 7, 9\}$, $B = \{1, 2, 3, 4\}$, and $C = \{2, 4, 6, 8\}$, complete the following.

(a) $A \cup B =$ _______

(b) $A \cap B =$ _______

(c) $\overline{A} =$ _______

(d) $A \cap C =$ _______

(e) $A \cap (B \cup C) =$ _______

(f) $(A \cap B) \cup (A \cap C) =$ _______

(g) $n(A) =$ _______, $n(C) =$ _______,
$n(A \cup C) =$ _______

(h) $n(A) =$ _______, $n(B) =$ _______,
$n(A \cup B) =$ _______,
$n(A) + n(B) - n(A \cap B) =$ _______

(i) Explain in two carefully written sentences the reason for the difference between the results of parts (g) and (h).

37. Name the property that justifies each of the following.

(a) $3x + 5x = (3 + 5)x$

(b) $4\pi + 7\pi = 7\pi + 4\pi$

(c) $3x + (4y + z) = 3x + (z + 4y)$

(d) $21 + 7 = 20 + (1 + 7)$

38. Write two subtraction equations corresponding to each of these equations.

(a) $3 + 7 = 10$

(b) $11 + 5 = 16$

(c) $21 + 19 = 40$

39. Write one addition equation and a second subtraction equation that correspond to each of these equations.

(a) $17 - 8 = 9$

(b) $23 - 15 = 8$

(c) $23 - 5 = 18$

3.2 NONDECIMAL POSITIONAL SYSTEMS

In the preceding section we discussed several numeration systems, including the decimal system in common use today. As already mentioned, the decimal system is a positional system based on ten. This is probably so because we have ten fingers and historically, as now, people often counted on their fingers. Suppose that there are little green aliens on Mars who have only one hand with five fingers. If Martians were to go through essentially the same process of developing a number system as occurred with the decimal system in ancient India, how would their system work?

Base Five Notation

For our purposes, perhaps the quickest route to understanding **base five notation** is to reconsider the abacus of Figure 3.1. This time, however, we allow only five beads to be moved forward on each wire. As before, we start with the wire to the students' right (our left) and move one bead forward each time as we count one, two, three, and so on. When we reach five, we have counted the first wire once, and we record this on the abacus by moving one bead forward on the second wire while moving all five beads on the first wire to the back. We continue to count, and when the count reaches ten, we will have counted all the beads on the first wire a second time. We move a second bead forward on the second wire and again move all the beads on the first wire to the back. Thus, each bead on the second wire counts for five; that is, the second wire is the "fives" wire. When the count reaches 19, the abacus will appear as in Figure 3.8, and

$$19 = 3 \cdot 5 + 4.$$

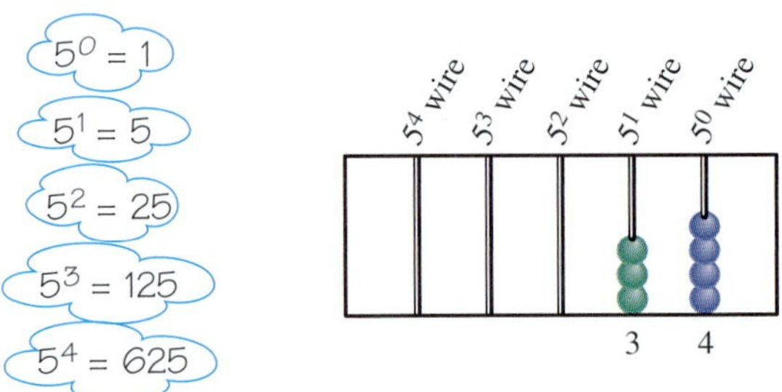

Figure 3.8
A count of 19 on a five-bead abacus

If we continue the count to 25, all the beads on the first wire will have been counted five times. This means that we have moved all five beads to the front on the *second* wire. This is recorded on the abacus by returning these beads to the back of the abacus and moving one bead forward on the *third* wire. Thus, the third wire becomes the $25 = 5^2$ wire. The count can be continued in this way, and it is apparent that any whole number will eventually be counted and can be recorded on the abacus with only 0, 1, 2, 3, or 4 beads per wire showing at the front; that is, we need only the digits 0, 1, 2, 3, and 4 in base five. For example, if we continue the counting process up to 113, the abacus will have three beads on the units wire, two on the fives wire, and four on the twenty-fives wire—that is,

$$113 = 4 \cdot 5^2 + 2 \cdot 5 + 3.$$

Just as we shorten $4 \cdot 10^2 + 2 \cdot 10 + 3$ to 423, the little green Martian might also shorten $4 \cdot 5^2 + 2 \cdot 5 + 3$ to 423. Thus, the three-digit sequence 423 can represent many different numbers depending on which base is chosen. To the average American, who doesn't even think about it, it means "four hundred twenty-three." To our Martian friend it means *flug globs zeit-tab*—that is, Martian for $4 \cdot 5^2 + 2 \cdot 5 + 3$. Note that our Martian friend certainly would *not* say "one hundred thirteen" since that is decimal, or base ten, language. To avoid confusion in talking about base five numeration, we agree that we will *not* say "four hundred twenty-three" when we read 423 as a base five numeral. Instead, we will say "four two three base five" and will write 423_{five}, where the subscript "five" indicates the base.

In base six, we understand that

$$423_{\text{six}} = 4 \cdot 6^2 + 2 \cdot 6 + 3,$$

HIGHLIGHT from HISTORY

Benjamin Banneker (1731–1806)

Benjamin Banneker's grandfather (known as Banna Ka and subsequently as Bannaky) was a slave who was bought and then freed by and married to Banneker's grandmother, who initially came from England to America as an indentured servant. The Bannakys had several children, including a daughter, Mary. When Mary Bannaky grew up, she purchased a slave whom she also married. Benjamin Bannaky (later Banneker) was born to this union in 1731.

Benjamin Banneker owed much of his education to his Quaker neighbors. As a teenager, he developed an interest in mathematics, science, astronomy, and mechanics. His diverse accomplishments include a mathematical study of the 17-year locust, the accurate prediction of solar eclipses, and an extraordinary wooden clock that is accepted as the first clock made entirely in America. Like another American inventor and scientist, Benjamin Franklin, Banneker published a popular almanac. It included antislavery essays, information on crops and tides, and astronomical data. He was one of three members of a commission that planned and surveyed the site for the federal capital at Washington, DC. When Banneker died in 1806, Thomas Jefferson occupied the White House in the city Banneker had helped to design.

and we read the numeral as "four two three base six." Doing that arithmetic, it becomes clear that $423_{\text{six}} = 159_{\text{ten}}$. As an additional example, 423_{twelve} should be read "four two three base twelve," and in expanded form, we have

$$423_{\text{twelve}} = 4 \cdot 12^2 + 2 \cdot 12 + 3 = 603_{\text{ten}}.$$

Unless expressly stated to the contrary, a numeral written *without* a subscript should be read as a base ten numeral.

Observe that, in base five, we need the digits 0, 1, 2, 3, and 4, and we need to know the values of the positions in base five. In base six, the digits are 0, 1, 2, 3, 4, and 5, and we need to know the positional values in base six. In base twelve, we use the digits 0, 1, 2, 3, 4, 5, 6, 7, 8, 9, T, and E, where T and E are, respectively, the digits for 10 and 11. For these bases, the positional values are given in Table 3.7.

TABLE 3.7 Digits and Positional Values in Bases Five, Six, and Twelve

Base b	Digits used	DECIMAL VALUES OF POSITIONS					
		...	b^4	b^3	b^2	b^1	units ($b^0 = 1$)
Five	0, 1, 2, 3, 4	...	$5^4 = 625$	$5^3 = 125$	$5^2 = 25$	$5^1 = 5$	$5^0 = 1$
Six	0, 1, 2, 3, 4, 5	...	$6^4 = 1296$	$6^3 = 216$	$6^2 = 36$	$6^1 = 6$	$6^0 = 1$
Twelve	0, 1, 2, 3, 4, 5, 6, 7, 8, 9, T, E	...	$12^4 = 20{,}736$	$12^3 = 1728$	$12^2 = 144$	$12^1 = 12$	$12^0 = 1$

EXAMPLE 3.2 | CONVERTING FROM BASE FIVE TO BASE TEN NOTATION

Write the base ten representation of 3214_{five}.

Solution We use the place values of Table 3.7 along with the digit values. Thus,

$$\begin{aligned} 3214_{\text{five}} &= 3 \cdot 5^3 + 2 \cdot 5^2 + 1 \cdot 5^1 + 4 \cdot 5^0 \\ &= 3 \cdot 125 + 2 \cdot 25 + 1 \cdot 5 + 4 \cdot 1 \\ &= 375 + 50 + 5 + 4 \\ &= 434. \end{aligned}$$

Therefore, "three two one four base five" is four hundred thirty-four.

EXAMPLE 3.3 CONVERTING FROM BASE TEN TO BASE FIVE NOTATION

Write the base five representation of 97.

Solution Recall that 97 without a subscript has its usual meaning as a base ten numeral.

We begin with the notion of grouping. Suppose that we have 97 beans and want to put them into groups of single beans (units), groups of five beans (fives), groups of five groups of five beans each (twenty-fives), and so on. The problem is to complete the grouping using the least possible number of groups. This means that we must use as many of the larger groups as possible. Diagrammatically, we have

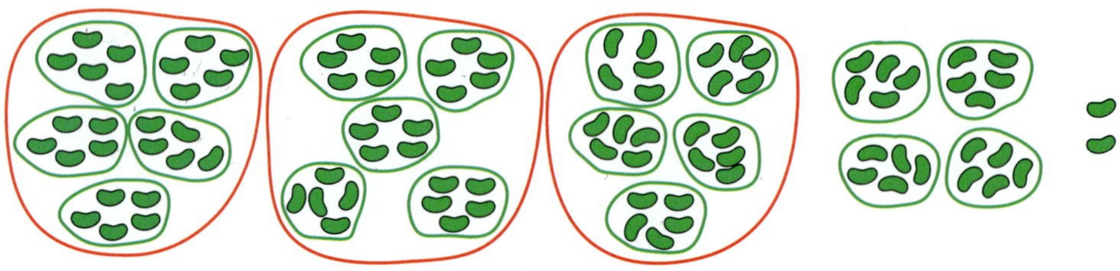

so $97_{\text{ten}} = 342_{\text{five}}$. Arithmetically, this corresponds to determining how many of each position value in base five are required to represent 97. This can be determined by successive divisions. Referring to Table 3.7 for position values, we see that 125 is too big. Thus, we begin with 25 and divide successive remainders by successively lower position values:

$$\begin{array}{r} 3 \\ 25\overline{)97} \\ 75 \\ \hline 22 \end{array} \qquad \begin{array}{r} 4 \\ 5\overline{)22} \\ 20 \\ \hline 2 \end{array} \qquad \begin{array}{r} 2 \\ 1\overline{)2} \\ 2 \\ \hline 0 \end{array}$$

These divisions reveal that we need three 25s, four 5s, and two 1s, or, in tabular form,

25s	5s	1s
3	4	2

Thus,

$$97_{\text{ten}} = 342_{\text{five}}.$$

Did You Know?

The Utility of Other Bases

At first thought, it might appear that positional numeration systems in bases other than base ten are merely interesting diversions. Quite to the contrary, they are extremely important in today's society. This is particularly true of the base two, or binary, system; the base eight, or octal, system; and the base sixteen, or hexadecimal, system. The degree to which calculators and computers have affected modern life is simply enormous, and the basic notion that allows these devices to operate is base two arithmetic. A switch is either on or off; a spot on a magnetic grid is either magnetized or not magnetized; either a spot on a compact disc is activated or it is not. All of these devices are capable of recording two states—either 0 or 1—and so can be programmed to do base two arithmetic and to record other data in code "words" consisting of strings of 0s and 1s. A drawback of base two notation is that numerals for relatively small numbers become quite long. Thus,

$$60_{\text{ten}} = 111100_{\text{two}},$$

and this makes programming a computer somewhat cumbersome. The notation is greatly simplified by using octal or hexadecimal notation, since these are intimately related to binary notation. Thus, triples of binary digits become single octal digits and vice versa, and quadruples of digits in binary notation correspond to single digits in hexadecimal. For example, since

$$111_{\text{two}} = 7 \text{ and } 100_{\text{two}} = 4,$$

it follows that

$$60_{\text{ten}} = 111100_{\text{two}} = 74_{\text{eight}}.$$

For this reason, computer programmers often work in octal or hexadecimal notation.

As a check, we note that

$$\begin{aligned} 342_{\text{five}} &= 3 \cdot 25 + 4 \cdot 5 + 2 \\ &= 75 + 20 + 2 \\ &= 97_{\text{ten}}. \end{aligned}$$

Problem Set 3.2

Understanding Concepts

1. In a long column, write the base five numerals for the numbers from zero through 25.

2. Briefly describe the pattern or patterns you observe in the list of numerals in problem 1.

3. Here are the base six representations of the numbers from zero through 35 arranged in a rectangular array. Briefly describe any patterns you observe in this array.

0	1	2	3	4	5
10	11	12	13	14	15
20	21	22	23	24	25
30	31	32	33	34	35
40	41	42	43	44	45
50	51	52	53	54	55

4. What would be the entries in the next two rows of the table in problem 3?

5. Write the base ten representations of each of the following.

(a) 413_{five} (b) 2004_{five} (c) 10_{five}
(d) 100_{five} (e) 1000_{five} (f) 2134_{five}

6. Write the base ten representations of each of the following.

(a) 413_{six} (b) 2004_{six} (c) 10_{six}
(d) 100_{six} (e) 1000_{six} (f) 2134_{six}

7. Write the base ten representations of each of these numbers. Remember that in base twelve, the symbols for the digits 10 and 11 are T and E, respectively.

(a) 413_{twelve} (b) 2004_{twelve} (c) 10_{twelve}
(d) 100_{twelve} (e) 1000_{twelve} (f) $2\text{TE}4_{\text{twelve}}$

8. Determine the base five representation for each of the following. Remember that a numeral with no subscript is understood to be in base ten.

(a) 362 (b) 27 (c) 5 (d) 25

9. Determine the base six representation for each of the following.

(a) 342 (b) 21 (c) 6 (d) 216

10. Determine the base twelve representation for each of the following.

(a) 2743 (b) 563 (c) 144 (d) 1584

11. Base two is a very useful base. Since it requires only two digits, 0 and 1, it is the system on which all calculators and computers are based.

(a) Make a table of position values for base two up as far as $2^{10} = 1024$.

(b) Write each of these in base ten notation.

(i) 1101_{two} (ii) 111_{two} (iii) 1000_{two} (iv) 10101_{two}

(c) Write each of these in base two notation.

(i) 24 (ii) 18 (iii) 2 (iv) 8

(d) Write the numbers from zero to 31 in base two notation in a vertical column, and discuss any pattern you observe in a short paragraph.

(e) In three or four sentences, compare part (d) of this problem with the "Hands On" activity at the beginning of the chapter.

12. (a) How can you use the five-bead classroom abacus to convince your students that every whole number can be represented *uniquely* (i.e., in one and only one way) in base five notation?

(b) How would you modify the drawing in Figure 3.8 to illustrate the base five representation of 3241_{five}?

Teaching Concepts

13. Explain how you would respond to one of your students who claims that the base five representation of 188_{ten} is 723_{five}. Note that $7 \cdot 5^2 + 2 \cdot 5 + 3 = 188_{ten}$.

14. (a) In trying to help your students to better understand positional notation, you might ask them what sort of number system a tribe of natives who counted on both their fingers and toes might have invented.

(b) If your students had trouble answering the question in part (a), what might you do to help?

Thinking Critically

15. Here's an interesting trick. Consider these pictures of cards you might make for use in your class.

1	3	5	7	9	11
13	15	17	19	21
23	25	27	29	31

2	3	6	7	10	11
14	15	18	19	22	
23	26	27	30	31	

4	5	6	7	12	13
14	15	20	21	22	
23	28	29	30	31	

8	9	10	11	12	13
14	15	24	25	26	
27	28	29	30	31	

16	17	18	19	20	21
22	23	24	25	26	
27	28	29	30	31	

(a) Record the first number of each card that has the day of the month on which you were born.

(b) Add the numbers in part (a).

(c) Surprised? Our experience is that elementary school students are, too, and that they immediately want to know how the trick works. See if you can discover the secret by carefully comparing the cards with your answer to problem 11, part (d).

16. (a) Add $11{,}111{,}111_{two}$ and 1_{two} in base two.

(b) What are the base two numerals for 2^n and $2^n - 1$? Explain briefly, describing a pattern.

17. Recall that the rows of Pascal's triangle (see Section 1.4) are numbered 0, 1, 2, 3, ... Thus, row five is 1, 5, 10, 10, 5, 1, which has four odd entries. Also, the base two representation of 5 is 101_{two}, with two 1s and $2^2 = 4$. Remarkably, if f is the number of 1s in the base two representation of n, then 2^f gives the number of odd entries in the nth row of Pascal's triangle. Verify that this is true for the following rows of Pascal's triangle.

(a) row $n = 7$, whose entries are 1, 7, 21, 35, 35, 21, 7, 1

(b) row $n = 8$, whose entries are 1, 8, 28, 56, 70, 56, 28, 8, 1

(c) row $n = 11$, whose entries are 1, 11, 55, 165, 330, 462, 462, 330, 165, 55, 11, 1

18. The two one-digit sequences of 0s and 1s are 0 and 1. The four two-digit sequences of 0s and 1s are 00, 10, 01, and 11.

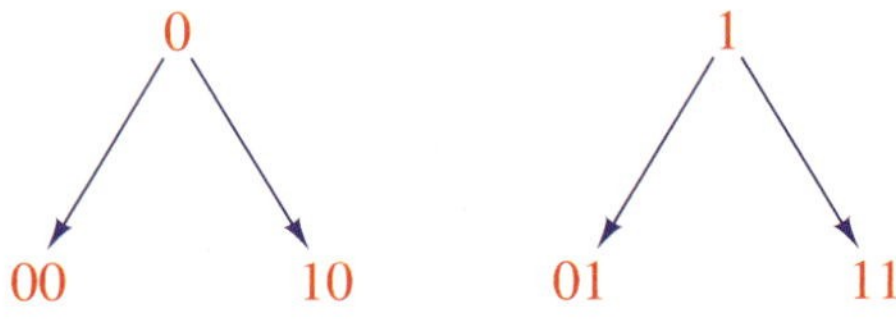

(a) Write down all of the three-digit sequences of 0s and 1s. Can you think of an easy way to do this? Explain.

(b) Write down all of the four-digit sequences of 0s and 1s. Can you think of an easy way to do this? Explain.

(c) How many five-digit sequences of 0s and 1s are there?

(d) How many n-digit sequences of 0s and 1s are there? Explain why this is so.

19. (a) Think of each three-digit sequence of 0s and 1s in problem 18(a) as a base two numeral. What are the base ten equivalents of these numerals?

$$011_{two} = 0 \cdot 2^2 + 1 \cdot 2 + 1 = 3_{ten}$$

(b) Think of each four-digit sequence of 0s and 1s in problem 18(b) as a base two numeral. What are the base ten equivalents of these numerals?

(c) Think of each n-digit sequence of 0s and 1s as a base two numeral. What do you think are the base ten equivalents of these numerals? Explain.

20. There are 1000 base ten numerals using three or fewer digits, namely, 0, 1, 2, . . . , 999. Another way to count these numerals is to note that there are 10 choices for the units digit, 10 choices for the tens digit, and 10 choices for the hundreds digit, giving $10 \cdot 10 \cdot 10 = 1000$ three-digit numerals as just noted. This counts such three-digit numerals as 006 and 083, which would be written simply as 6 and 83.

(a) How many base two numerals are there with three or fewer digits?

(b) Consider the three-element universal set $U = \{a, b, c\}$. The subset $A = \{a, c\}$ can be associated with the three-digit base two numeral 101 since A contains a, does not contain b, and does contain c. Similarly, the empty subset is associated with 000, and the subset {b} is associated with 010. In view of your answer to part (a), how many subsets are there of {a, b, c}?

(c) How many subsets are there of {a, b, c, d}?

(d) How many subsets are there of an n-element set?

21. The diagram below shows a Cuisenaire rod train of length four (see the "Cooperative Investigation" in Section 2.3), consisting of a red two-rod caboose at the left, a one-rod car, and a one-rod engine. This train can be associated with the three-digit base two numeral 011, where the digit 0 indicates a single car of length two, and each 1 indicates a break from one car to the next. Similarly, two consecutive 0s indicate a single car of length three, and three consecutive 0s indicate a single car of length four. Thus, two red rods form a train of length four associated with the base two numeral 010.

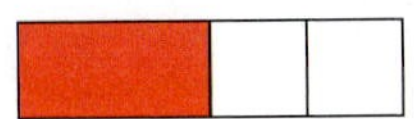

(a) Draw all trains of length four, and write their corresponding base two numerals. Explain how you know that there are $2^3 = 8$ of these trains.

(b) Using the same reasoning as in part (a), determine the number of different trains of length five.

(c) How many trains of length n can be formed if cars come in all lengths—1, 2, 3, . . . , n?

Using a Calculator

22. (a) Lyudmila discovered a method to convert a numeral in any base to its equivalent in base ten using her calculator. For example, to convert 423_{five}, she used the entry string

4 [×] 5 [+] 2 [=] [×] 5 [+] 3 [=]

to obtain the correct result of 113. Similarly, to convert 4075_{eight}, she used the string

4 [×] 8 [+] 0 [=] [×] 8 [+] 7 [=]
[×] 8 [+] 5 [=]

to obtain the correct result of 2109. Lyudmila isn't quite sure why her procedure works. Write an imaginary dialogue with her to help her understand why the method is valid.

(b) Would this method succeed on an ordinary four-function calculator?

For Review

23. Write two division equations corresponding to each of these multiplication equations.

(a) $3 \cdot 17 = 51$ (b) $11 \cdot 91 = 1001$

(c) $9 \cdot 121 = 1089$

24. Write a multiplication equation and a second division equation corresponding to each of these division equations.

(a) $341 \div 11 = 31$ (b) $455 \div 65 = 7$

(c) $124{,}857 \div 13{,}873 = 9$

25. Lida Lee has three sweaters, four blouses, and two pairs of slacks that mix and match beautifully. How many different outfits can she wear using these nine items of clothing?

26. (a) If n represents Shane Stagg's age six years ago, how old is Shane now?

(b) If m represents Shane's age now, how old was Shane six years ago?

Cooperative Investigation

A Remarkable Base Three Trick

Practice performing this trick with a partner until you both can do it with ease.

Mathematical magic certainly has a place in the classroom. Students find it interesting, fun, and highly motivational. Consider the following problem: You are given 12 coins that appear identical, but one of the coins is false and is either too heavy or too light—you don't know which. You have a balance scale and are to find the false coin and to determine whether it is heavy or light in just three weighings. This is a difficult problem that is usually solved by considering a whole series of cases. However, if you know base three arithmetic, you can determine the false coin so quickly right in front of your class that you appear to be a real wizard! Here's how it can be done.

1. Number the coins from 1 through 12.
2. With your back turned, ask your class (partner) to agree on the number of the coin it (he or she) wants to be false and to decide whether it is to be heavy or light.
3. Indicate that you are going to make three weighings, and ask the class (your partner) what the movement of the *left-hand* pan on each weighing will be. If the left-hand pan goes up, record a 2. If it balances, record a 1. If it goes down, record a 0. Now use the results of these weighings to form a three-digit base three numeral. The first weighing determines the *nines* digit, the second weighing determines the *threes* digit, and the third weighing determines the *units* digit. If the base ten number determined in this way is less than 13, it names the false coin. If the number is more than 13, then 26 minus the number generated names the false coin. You can tell, by knowing which is the false coin, whether it is heavy or light by noting how the left-hand pan moved on a weighing involving the false coin.

For example, suppose the class (your partner) chooses coin 7 as the false coin and decides that it should be heavy. Then on the three weighings, the left-hand pan goes down, goes up, and balances, and we generate $021_{\text{three}} = 7$, the number of the false coin. Moreover, since coin 7 was in the left-hand pan, which went down on the first weighing, coin 7 must be heavy. On the other hand, suppose that coin 7 is light. Then on the three weighings, the left-hand pan goes up, goes down, and balances, and we generate $201_{\text{three}} = 19$. But then $26 - 19 = 7$, so coin 7 is the false coin and must be light since the left-hand pan went up on the first weighing.

Note the addition using base three notation,

$$\begin{array}{r} 201 \\ +021 \\ \hline 222 \end{array}$$

and $222_{\text{three}} = 26_{\text{ten}}$

Our experience is that children like the trick very much and are strongly motivated to learn base three notation in order to pull the trick on their friends and parents. Note also that, to do the trick well, students must be able to perform mental calculations quickly—itself a worthy goal.

3.3 ALGORITHMS FOR ADDING AND SUBTRACTING WHOLE NUMBERS

Just for Fun

What's the Difference?

Perform these subtractions.

91	95
−19	−59
42	62
−24	−26
74	61
−47	−16
82	81
−28	−18
32	54
−23	−45

(a) What do you notice about the answer in each case?

(b) Can you predict the answers to all problems like this just by looking at the two digits involved?

(c) What seems to be true about the sums of the digits in the answers to all such two-digit problems?

Today's predominant opinion is that we spend much more time than necessary trying to teach youngsters the intricacies of pencil-and-paper calculations. The time would be much better spent helping students learn to think critically about quantitative situations, to develop thoughtful mathematical behavior as opposed to rote memorization of rules, and to learn effective techniques for solving thought-provoking problems while handling the arithmetic details by appropriate use of a calculator.

The calculation skills and essential knowledge for school mathematics include the following:

- knowledge of the one-digit arithmetic facts,
- the meaning of the arithmetic operations,
- the properties of numbers,
- the meaning of positional notation,
- the ability to perform mental arithmetic,
- the ability to use pencil-and-paper algorithms with small numbers,
- an understanding of why the algorithms give correct results,
- the ability to calculate correctly and efficiently with a calculator, and
- the ability to estimate the results of calculations.

We address these essentials here.

The Addition Algorithm

Consider the following addition.

$$\begin{array}{r} \scriptstyle 1 \\ 28 \\ +\ 45 \\ \hline 73 \end{array}$$

This process depends entirely on our positional system of notation. But why does it work this way? Why "carry" the "1"? Why add by columns? To many people, these procedures remain a great mystery—you add this way because you were told to—and it's simply done by rote with no understanding.

As noted earlier, children learn abstract notions by first experiencing them concretely with devices they can actually see, touch, and manipulate. Thus, one should introduce the addition algorithm with manipulatives like base ten blocks; sticks and bundles of sticks; units, strips, and mats; and abacuses, or, better yet, by using a number of these devices. For illustrative purposes here, we use units, strips, and mats.

After cutting out their strips and mats, students will be aware that there are 10 units on a strip and 10 strips, or 100 units, on a mat, and they will be aware that, in manipulating these materials, they can make these exchanges back and forth as needed.

EXAMPLE 3.4 MAKING EXCHANGES WITH UNITS, STRIPS, AND MATS

Suppose a number is represented by 15 units, 11 strips, and 2 mats. What exchanges must be made in order to represent the same number but with the smallest number of manipulative pieces?

Solution

Understand the Problem

We are given the mats, strips, and units indicated in the statement of the problem and are asked to make exchanges that reduce the number of loose pieces of apparatus while keeping the same number of units in all.

Devise a Plan

Since we know that 10 units form a strip and 10 strips form a mat, we can reduce the number of loose pieces by making these exchanges.

Carry Out the Plan

If we actually had in hand the pieces described, we would physically make the desired exchanges. Here in the text, we illustrate the exchanges pictorially.

We reduce the number of pieces by replacing 10 units by 1 strip and 10 strips by 1 mat. This gives 3 mats, 2 strips, and 5 units for 325 units as shown.

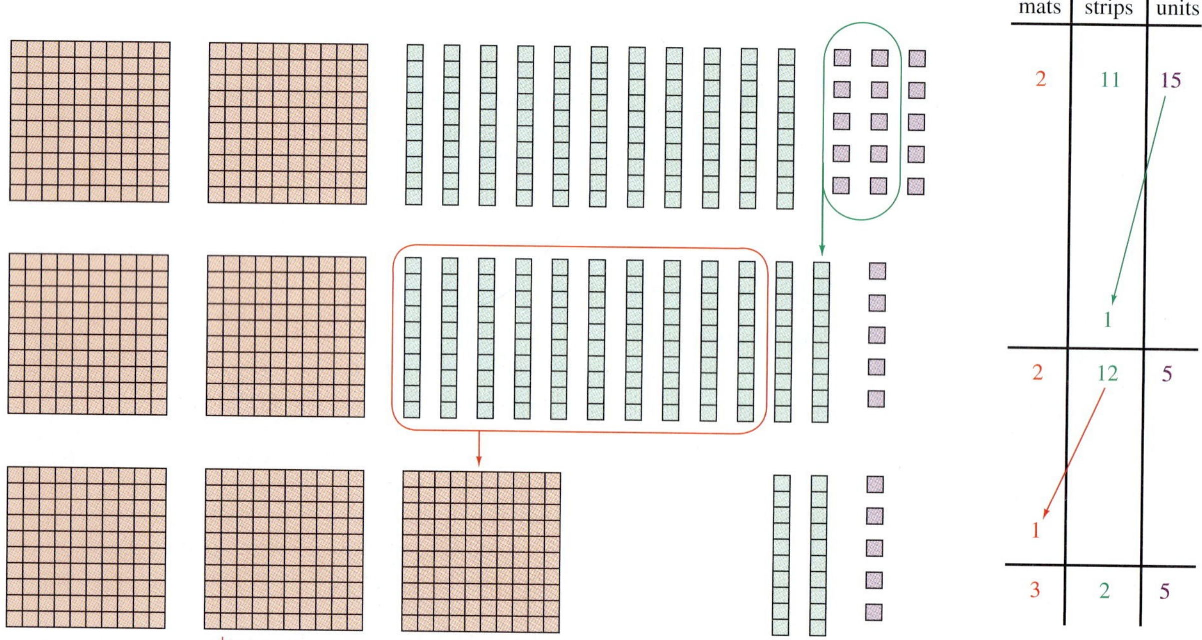

Look Back

No more combining can take place because it takes 10 units to make a strip and 10 strips to make a mat. Thus, we finish with 3 mats, 2 strips, and 5 units for a total of 325 units. Moreover, we have not changed the total number of units, since

$$2 \cdot 100 + 11 \cdot 10 + 15 = 200 + 110 + 15 = 325.$$

from **The NCTM Principles and Standards**

Place-value concepts can be developed and reinforced using calculators. For example, students can observe values displayed on a calculator and focus on which digits are changing. If students add 1 repeatedly on a calculator, they can observe that the units digit changes every time, but the tens digit changes less frequently. Through classroom conversations about such activities and patterns, teachers can help focus students' attention on important place-value ideas.

. . . . .

Students also develop understanding of place value through the strategies they invent to compute (Fuson et al. 1997). Thus, it is not necessary to wait for students to fully develop place-value understandings before giving them opportunities to solve problems with two- and three-digit numbers. When such problems arise in interesting contexts, students can often invent ways to solve them that incorporate and deepen their understanding of place value, especially when students have opportunities to discuss and explain their invented strategies and approaches.

. . . . .

As students encounter problem situations in which computations are more cumbersome or tedious, they should be encouraged to use calculators to aid in problem solving. In this way, even students who are slow to gain fluency with computation skills will not be deprived of worthwhile opportunities to solve complex mathematics problems and to develop and deepen their understanding of other aspects of number.

SOURCE: Reprinted with permission from Principles and Standards for School Mathematics, *pages 81, 82, 86, and 87. Copyright © 2000 by the National Council of Teachers of Mathematics, Reston, VA. All rights reserved. Standards are listed with the permission of the National Council of Teachers of Mathematics (NCTM). NCTM does not endorse the content or validity of these alignments.*

EXAMPLE 3.5 DEVELOPING THE ADDITION ALGORITHM

Find the sum of 135 and 243.

Solution *With Units, Strips, and Mats*

One hundred thirty-five is represented by 1 mat, 3 strips, and 5 units, and 243 is represented by 2 mats, 4 strips, and 3 units, as shown. All told, this gives a total of 3 mats, 7 strips, and 8 units. Therefore, since no exchanges are possible, the sum is 378. Note how this illustrates the column-by-column addition algorithm typically used in pencil-and-paper calculation.

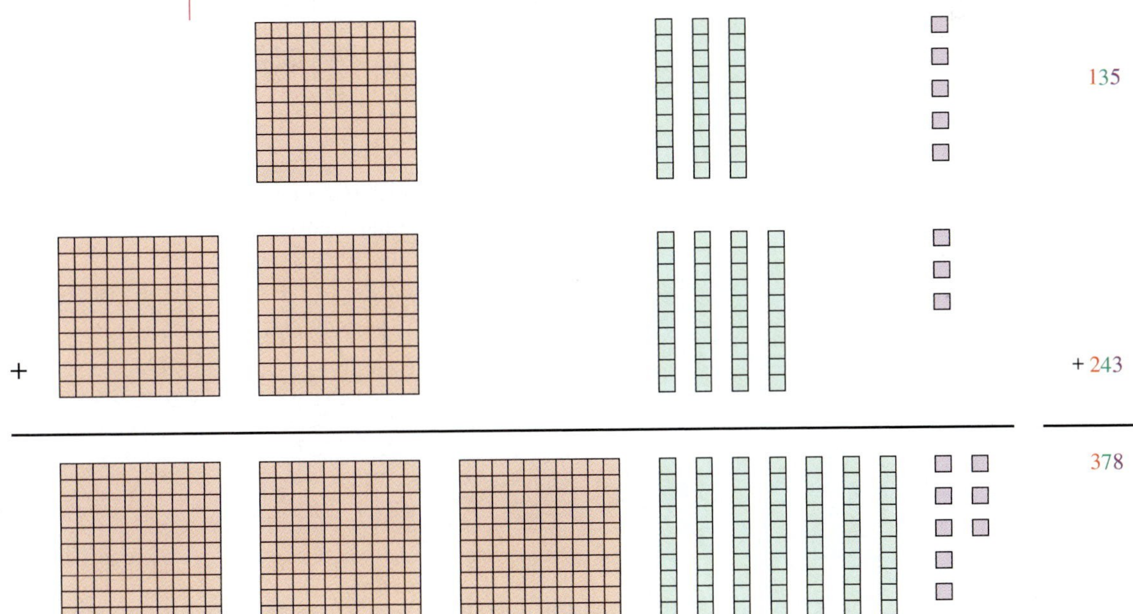

Highlight from History

The Origin of the Word Algorithm

The word *algorithm* derives from al-Khowarizmi (al-ko-war-iz-mi), the name of the eighth-century Arab scribe who wrote two books on arithmetic and algebra giving a very careful account of the Indo-Arabic system of numeration and methods of calculation. Though al-Khowarizmi made no claim to inventing the system, careless readers of Latin translations of his books began to attribute the system to him and call the new methods of calculation al-Khowarizmi or, carelessly, *algorismi.* Over time, the word became *algorithm* and came to mean any orderly, repetitive scheme.

With Place-Value Cards

A somewhat more abstract approach to this problem is to use **place-value cards**—that is, cards marked off in squares labeled 1s, 10s, and 100s from right to left, and with the appropriate number of markers placed in each square to represent the desired number. A marker on the second square is worth 10 markers on the first square; a marker on the third square is worth 10 markers on the second square; and so on. The addition of this example is illustrated as follows.

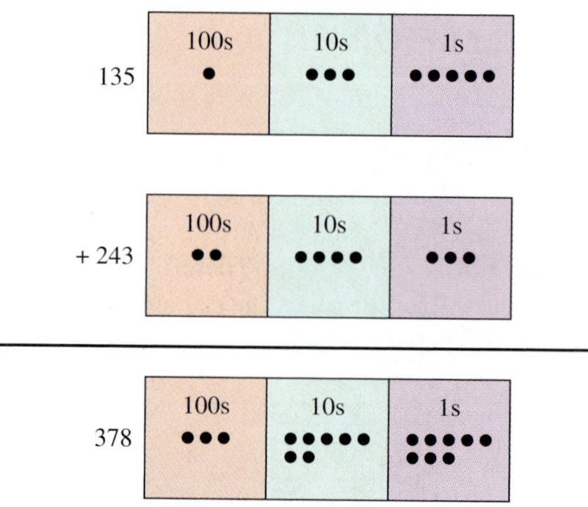

With Place-Value Diagrams and Instructional Algorithms

An even more abstract approach leading finally to the usual algorithm is provided by the following place-value diagrams and instructional algorithms.

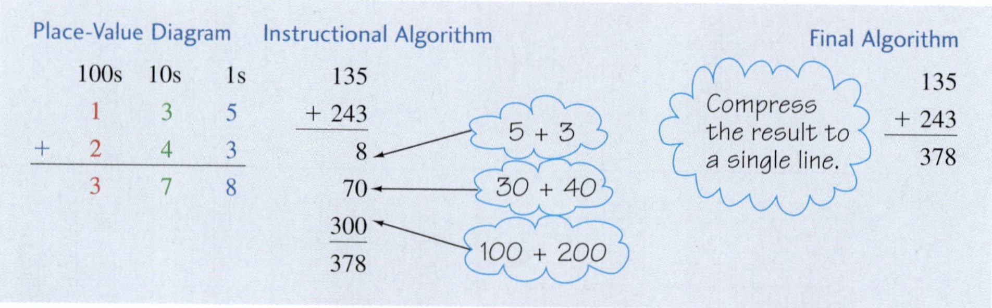

Notice how the degree of abstraction steadily increases as we move through the various solutions. If the elementary school teacher goes directly to the final algorithm, many students will be lost along the way. The approach from the concrete gradually moving toward the abstract is more likely to impart the desired understanding.

EXAMPLE 3.6 ADDING WITH EXCHANGING

Find the sum of 357 and 274.

Solution *With Units, Strips, and Mats*

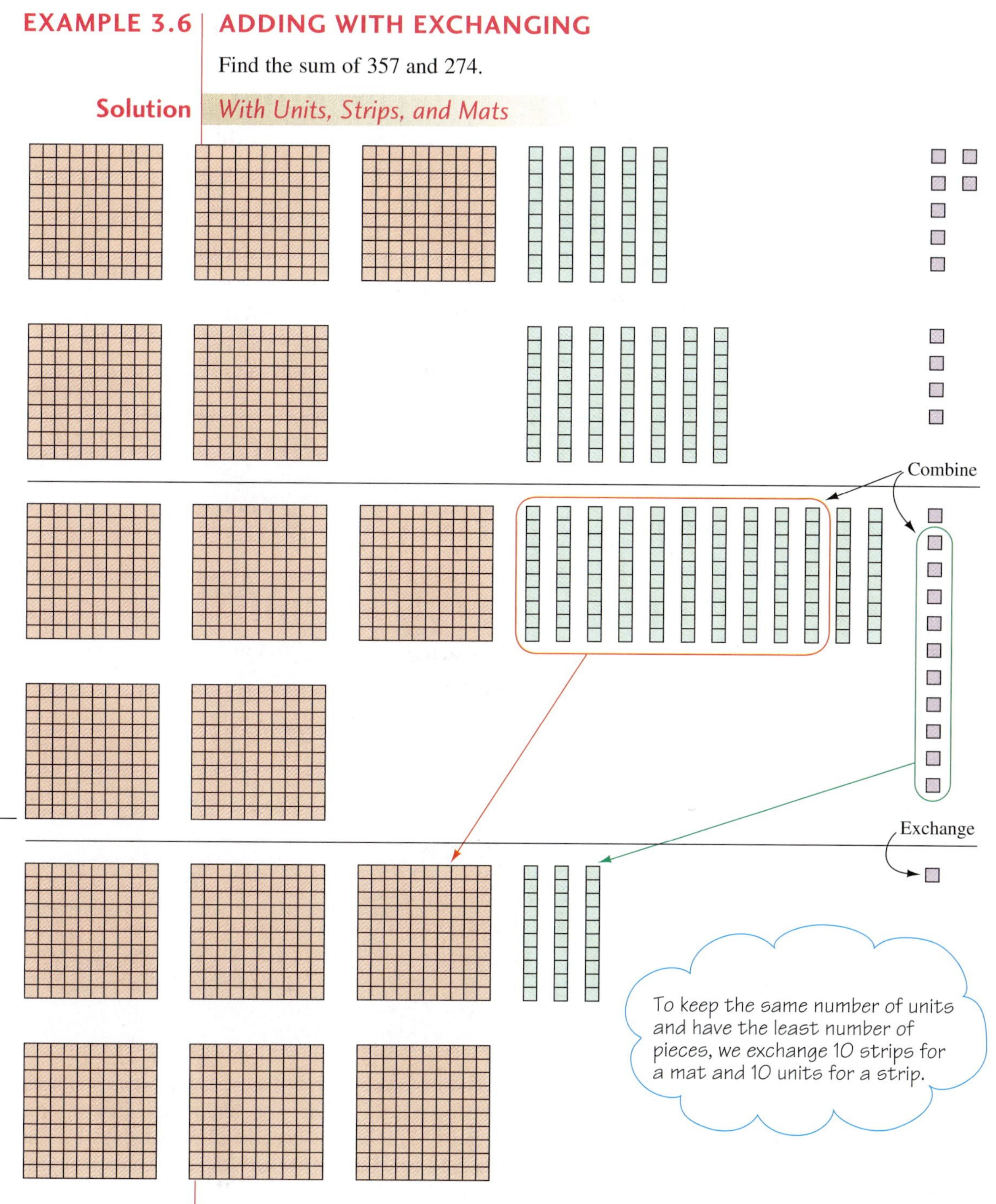

Combine
Exchange
To keep the same number of units and have the least number of pieces, we exchange 10 strips for a mat and 10 units for a strip.

Note how the above illustrates not only the usual column-by-column addition of pencil-and-paper arithmetic, but also "carrying," or, more appropriately stated, "exchanging."

The 1 "carried" from the first column to the second indicates the exchange of 10 units for one 10 (represented by 1 strip), and the 1 carried from the second column to the third indicates the exchange of 10 tens (i.e., 10 strips) for 100 (represented by one mat). Many teachers try to avoid the traditional word *carry,* since it encourages students to perform the algorithms by rote, with no understanding of what they are actually doing. Words like *exchange, trade,* and *regroup* are much more descriptive and actually describe what is being done.

With Place-Value Cards

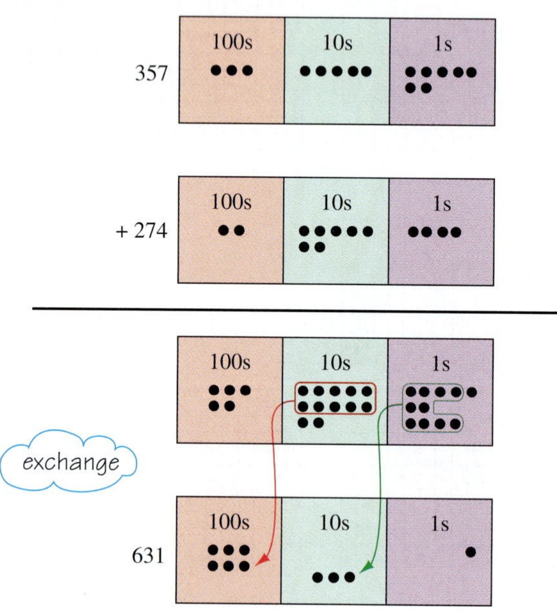

With Place-Value Diagrams and Instructional Algorithms

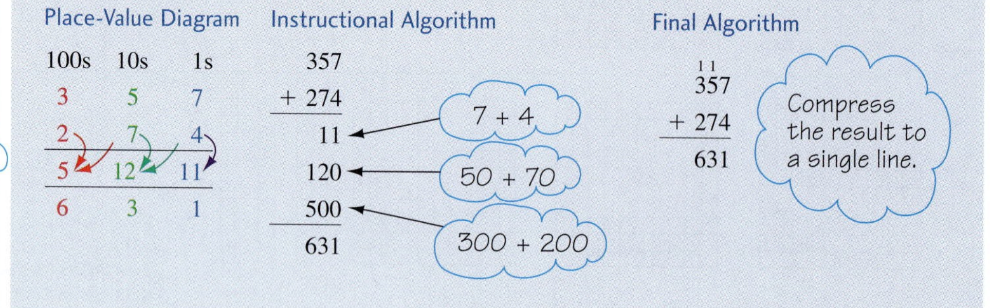

Notice again how the level of abstraction steadily increases as we move through these solutions. This example and the various solutions should make it clear to students that at no point are they "carrying a 1." It is always the case of exchanging 10 1s from the ones column for one 10 in the tens column, 10 10s from the tens column for one 100 in the hundreds column, and so on.

The Subtraction Algorithm

We can illustrate the subtraction algorithm in much the same way as the addition algorithm. For primary school children, the idea of subtraction is often understood in terms of "take away." Thus, if you have 9 apples and I take away 5, you have 4 left as shown in Figure 3.9.

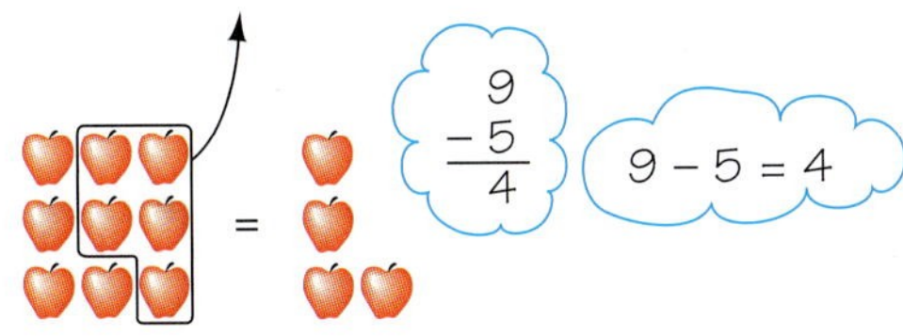

Figure 3.9
Subtraction as "take away"

EXAMPLE 3.7 SUBTRACTING WITHOUT EXCHANGING

Subtract 243 from 375.

Solution *With Units, Strips, and Mats*

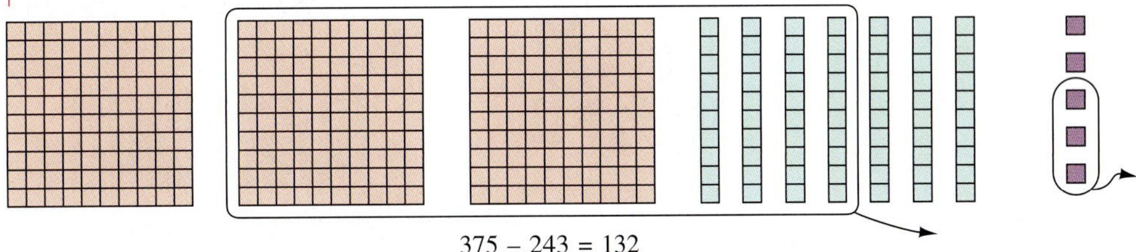

375 − 243 = 132

With Place-Value Cards

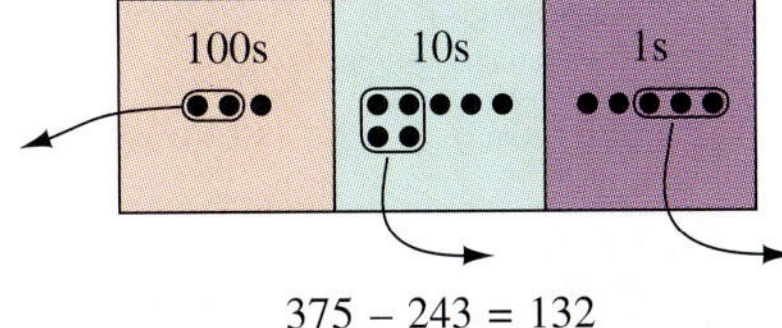

375 − 243 = 132

With Place-Value Diagrams and Instructional Algorithms

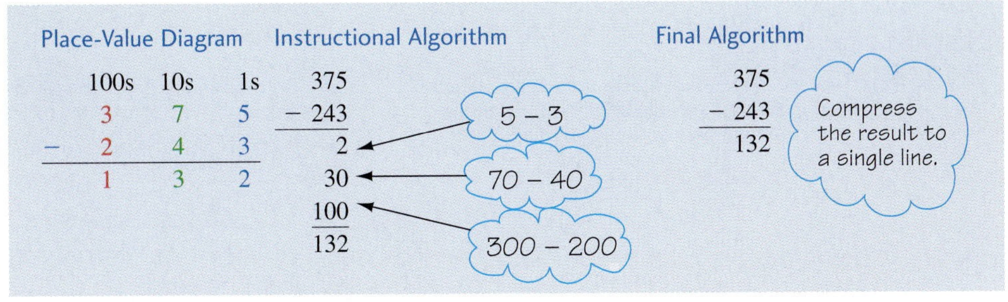

School Book Page

Subtracting with Manipulatives

Chapter 4 Lesson 8

Exploring Subtracting 3-Digit Numbers

Problem Solving Connection
- Use Objects/ Act It Out
- Look for a Pattern

Materials
place-value blocks

Explore

You can use place-value blocks to subtract 3-digit numbers.

Work Together

1. Find 163 – 71.
   a. Use place-value blocks to show 163.

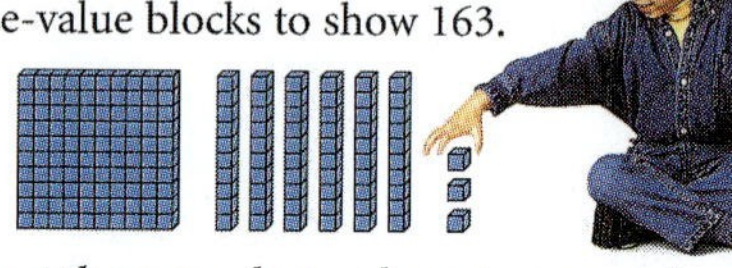

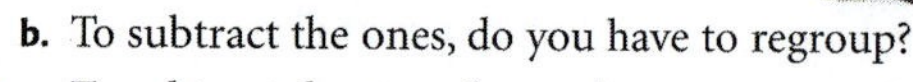

   b. To subtract the ones, do you have to regroup?
   c. To subtract the tens, do you have to regroup?

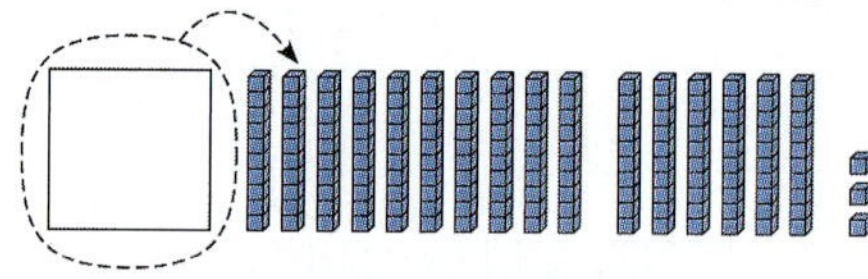

   d. What is 163 – 71?

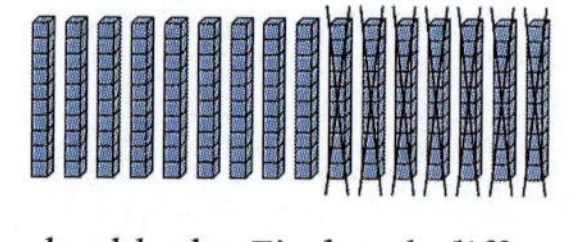

2. Use place-value blocks. Find each difference.
   a. 153 – 16 b. 166 – 139 c. 245 – 62

Remember
You can use addition to check your subtraction.

Talk About It

3. Would you have to regroup to find 151 – 81? Explain.
4. Frank says, "To find 432 – 191, you need to regroup a hundred." Do you agree? Explain.

166 Chapter 4 • Subtracting Whole Numbers and Money

Questions for the Teacher

1. In the lesson above, place-value blocks are used to illustrate regrouping in subtracting three-digit numbers. Carefully explain how you could use money—dollars, dimes, and pennies—to accomplish the same purpose.
2. On the sample page, the word "regroup" was used instead of "borrow." Why is it better to use words like *regroup, exchange,* and *trade* rather than *borrow* and phrases like "borrow 1 from the tens column"?

EXAMPLE 3.8 SUBTRACTING WITH EXCHANGING

Subtract 185 from 362.

Solution

With Units, Strips, and Mats

We start with 3 mats, 6 strips, and 2 units.

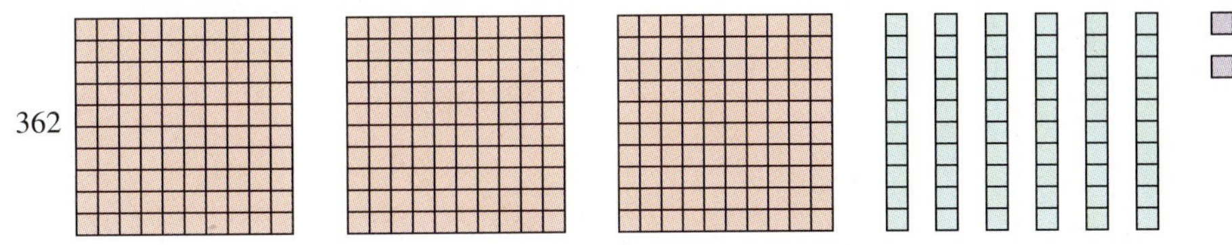

We want to take away 1 mat, 8 strips, and 5 units. Since we cannot pick up 5 units from our present arrangement, we exchange a strip for 10 units to obtain 3 mats, 5 strips, and 12 units.

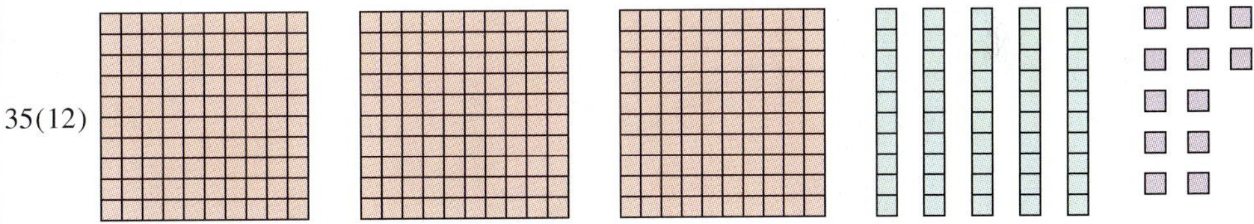

We can now take away 5 units, but we still cannot pick up 8 strips. Therefore, we exchange a mat for 10 strips to obtain 2 mats, 15 strips, and 12 units. Finally, we are able to take away 1 mat, 8 strips, and 5 units (that is, 185 units) as shown.

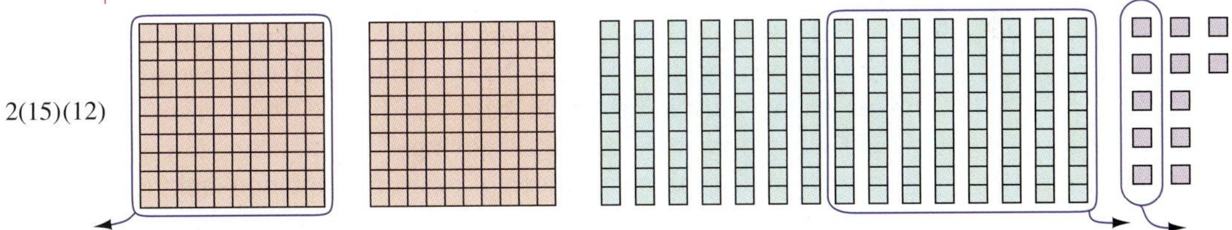

This leaves 1 mat, 7 strips, and 7 units. So

$$362 - 185 = 177.$$

With Place-Value Cards

We must take away 1 marker from the 100s square, 8 markers from the 10s square, and 5 markers from the 1s square. To make this possible, we trade 1 marker from the 10s square for 10 markers on the 1s square and 1 marker on the 100s square for 10 markers on the 10s square. Now, taking away the desired markers, we have 1 marker left on the 100s square, 7 markers left on the 10s square, and 7 markers left on the 1s square for 177 as shown on the next page.

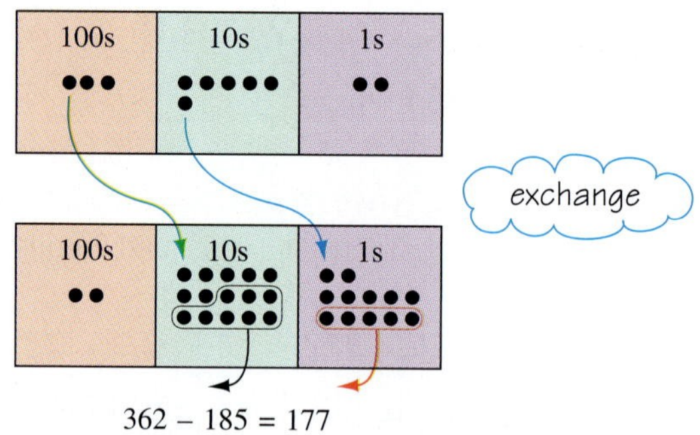

With Place-Value Diagrams and Instructional Algorithms

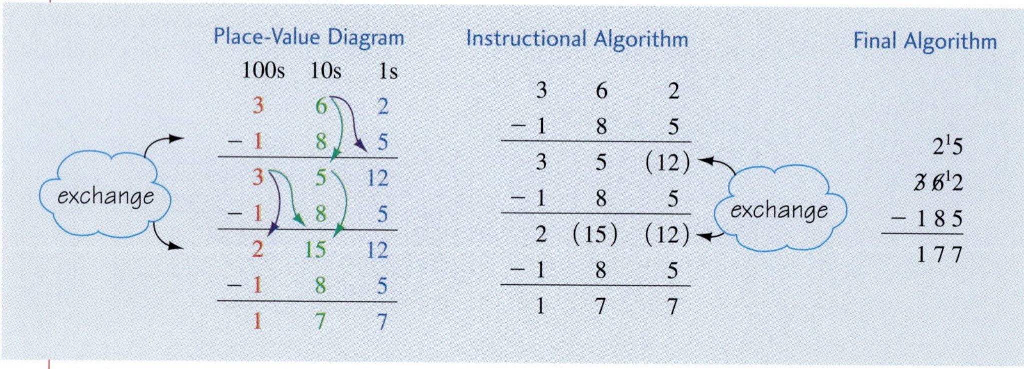

Algorithms in Other Bases

It is interesting to note that the algorithms for addition and subtraction are just as valid in base five, or any other base, as they are in base ten and that the notations can be developed with manipulatives just as for base ten. Indeed, it is reasonable to consider other bases right along with base ten. By studying bases together this way, students gain a greater understanding of the whole idea of positional notation and the related algorithms. In base five, for example, the place-value cards would have a units or 1s square, a 5s square, a 5^2 or 25s square, and so on. With sticks, we could use loose sticks, bundles of 5 sticks, bundles of 5 bundles of sticks, and so on. With units, strips, and mats, we would have units, strips with 5 units each, and mats with 5 strips per mat.

Adding in Base Five

The next example shows how the addition algorithm works in base five. Let's see how the process works using place-value cards.

EXAMPLE 3.9 ADDING IN BASE FIVE

Compute the sum of 143_{five} and 234_{five} in base five notation.

Solution *With Place-Value Cards*

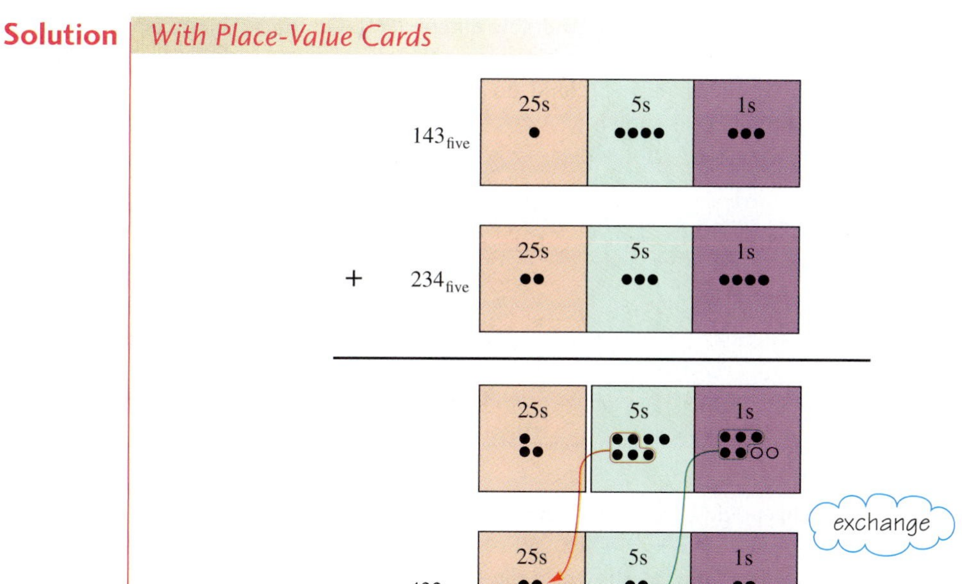

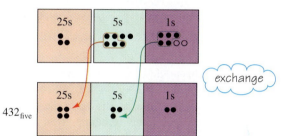

With Place-Value Diagrams and Instructional Algorithms

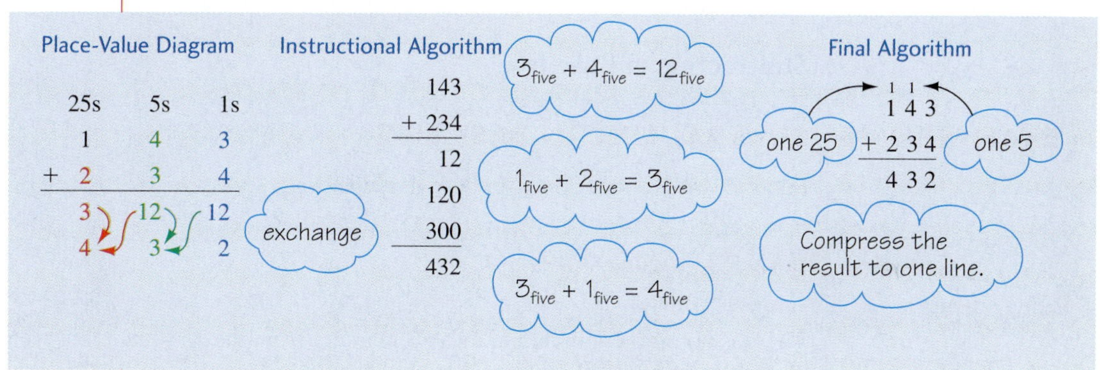

So the sum is 432_{five}.

To check the addition, we convert everything to base ten, in which we feel comfortable:

$$143_{five} = 1 \cdot 25 + 4 \cdot 5 + 3 \cdot 1 = 48_{ten}$$
$$234_{five} = 2 \cdot 25 + 3 \cdot 5 + 4 \cdot 1 = 69_{ten}$$
$$432_{five} = 4 \cdot 25 + 3 \cdot 5 + 2 \cdot 1 = 117_{ten}$$

The result is confirmed since $48_{ten} + 69_{ten} = 117_{ten}$.

The difficulty in doing base five arithmetic is unfamiliarity with the meaning of the symbols—we do not recognize at a glance, for example, that $8_{ten} = 13_{five}$. On the other hand, thinking of units, strips with 5 units per strip, and so on, it is not hard to think of 1 strip and 3 units as 8. In fact, doing arithmetic in another base gives good practice in mental arithmetic, which is a desirable end in itself. Of course, elementary school children memorize the addition and multiplication tables in base ten so that the needed symbols are readily recalled, and we could do the same thing here. The addition table in base five, for example, is as shown.

+	0	1	2	3	4
0	0	1	2	3	4
1	1	2	3	4	10
2	2	3	4	10	11
3	3	4	10	11	12
4	4	10	11	12	13

These numerals are in base five.

This could be used to make the above addition easier and more immediate. From the table, we see that $3 + 4 = 12_{\text{five}}$. So we write down the 2 and exchange the five 1s for a 5 in the fives column, as shown in the final algorithm in Example 3.9. Next, $4 + 3 = 12_{\text{five}}$, and the 1 from the exchange gives 13_{five}. Thus, we write down the 3 and exchange the five 5s for one 25 in the next column. Finally, $1 + 2 = 3$, and 1 from the exchange makes 4, so we obtain 432_{five} as before. However, there is no merit in memorizing the addition table in base five; it is far better just to think carefully about what is going on.

Subtracting in Base Five

EXAMPLE 3.10 SUBTRACTING IN BASE FIVE

Subtract 143_{five} from 234_{five} in base five notation.

Solution *With Place-Value Cards*

There is no problem in taking away 3 markers from the 1s square, but we cannot remove 4 markers from the 5s square without exchanging 1 marker on the 25s square for 5 markers on the 5s square. Taking away the desired markers, we are left with zero 25s, four 5s, and one 1 for 41_{five}.

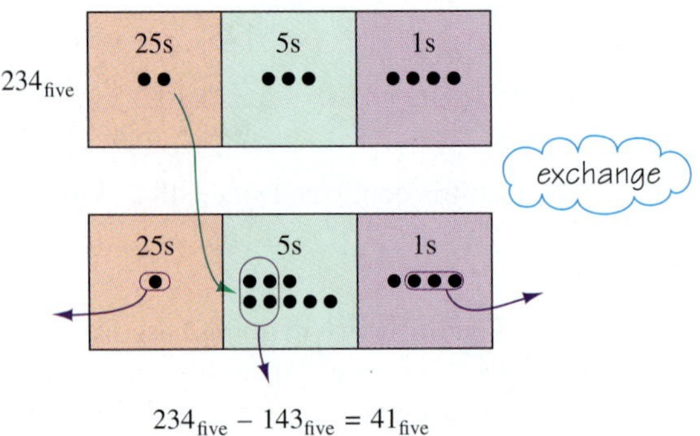

$234_{\text{five}} - 143_{\text{five}} = 41_{\text{five}}$

With Place-Value Diagrams And Instructional Algorithms

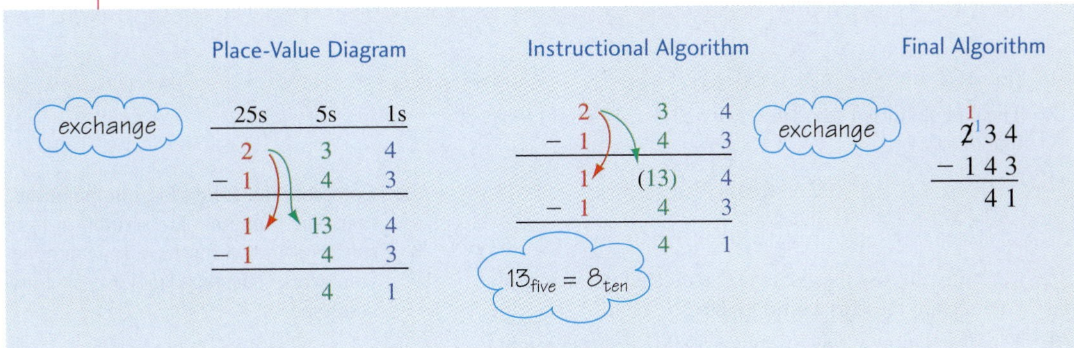

So the answer is 41_{five}.

As a check, we already know that $243_{five} = 69_{ten}$ and $143_{five} = 48_{ten}$. Since $69 - 48 = 21$ and $41_{five} = 4 \cdot 5 + 1 \cdot 1 = 21$, the calculation checks.

Problem Set 3.3

Understanding Concepts

1. Sketch the solution to 36 + 75 using
   (a) mats, strips, and units. Draw a square for a mat, a vertical line segment for a strip, and a dot for a unit.
   (b) place-value cards marked 1s, 10s, and 100s from right to left.

2. Use the instructional algorithm for addition to perform the following additions:
   (a) 23 + 44
   (b) 57 + 84
   (c) 324 + 78

3. The hand calculation of the sum of 279 and 84 involves two exchanges and might appear as follows.

```
 1 1
 279
  84
 ---
 363
```

   Carefully describe each of the exchanges. How would you explain these to a third grader?

4. While Sylvia was trying to balance her checkbook, her calculator battery went dead. When she added up the outstanding checks by hand, her work looked like this:

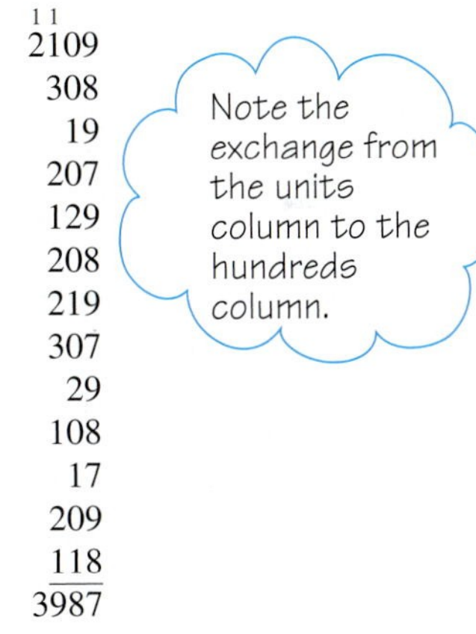

1 1
2109
308
19
207
129
208
219
307
29
108
17
209
118
3987
Note the exchange from the units column to the hundreds column.

   (a) Is the addition correct?
   (b) Discuss each of the exchanges shown.
   (c) Would you have proceeded like Sylvia did?
   (d) How else might the addition been carried out?
   (e) What would you say to a student of yours who did the computation as Sylvia did?

5. Sketch the solution to 275 − 136 using
   (a) mats, strips, and units.
   (b) place-value cards marked 1s, 10s, and 100s from right to left.

6. Use the instructional algorithm for subtraction to perform these subtractions:
   (a) 78 − 35
   (b) 75 − 38
   (c) 414 − 175

7. The calculation of the difference 523 − 247 might look like this:

$$\begin{array}{r} {}^{4\ 11} \\ \not{5}\,\not{2}\,{}^{1}3 \\ -\ 2\,4\,7 \\ \hline 2\,7\,6 \end{array}$$

Carefully discuss the exchanges indicated. How would you explain these to a third grader?

8. The calculation of the difference 30,007 − 1098 might look like this:

$$\begin{array}{r} {}^{2\ \ 9\ \ 9\ \ 9} \\ \not{3}\,{}^{1}\not{0}\,{}^{1}\not{0}\,{}^{1}\not{0}\,{}^{1}7 \\ -\ 1\,0\,9\,8 \\ \hline 2\,8,9\,0\,9 \end{array}$$

Carefully discuss the exchanges indicated.

9. Perform these additions and subtractions, being careful not to leave more than 59 seconds or 59 minutes in your answers.

   (a) 3 hours, 24 minutes, 54 seconds
   +2 hours, 47 minutes, 38 seconds

   (b) 7 hours, 56 minutes, 29 seconds
   +3 hours, 27 minutes, 52 seconds

   (c) 5 hours, 24 minutes, 54 seconds
   −2 hours, 47 minutes, 38 seconds

   (d) 7 hours, 46 minutes, 29 seconds
   −3 hours, 27 minutes, 52 seconds

10. The column-by-column addition of numbers can be justified as follows. State the property of the whole numbers that justifies each of these steps. We begin with expanded notation.

$36 + 52 = (3 \cdot 10 + 6) + (5 \cdot 10 + 2)$ expanded notation
$= 3 \cdot 10 + [6 + (5 \cdot 10 + 2)]$ (a) ________
$= 3 \cdot 10 + [(6 + 5 \cdot 10) + 2]$ (b) ________
$= 3 \cdot 10 + [(5 \cdot 10 + 6) + 2]$ (c) ________
$= 3 \cdot 10 + [5 \cdot 10 + (6 + 2)]$ (d) ________
$= (3 \cdot 10 + 5 \cdot 10) + (6 + 2)$ (e) ________
$= (3 + 5) \cdot 10 + (6 + 2)$ (f) ________
$= 8 \cdot 10 + 8$ addition facts
$= 88$ expanded notation

11. (a) In a single column, write the base four representations of the numbers from 0 to 15 inclusive.
    (b) Briefly discuss any pattern you noticed in part (a).

12. Complete the addition table for base four arithmetic shown by placing the sum of $a + b$ at the intersection of the ath row and the bth column. The subscript "four" may be omitted here.

+	0	1	2	3
0				
1				
2				
3				

13. Complete the following computations in base four notation. The numerals are written in base four, and the subscript indicating base four may be omitted from your work. Check your work by converting to base ten notation.

   (a) $\begin{array}{r} 231 \\ +\ 121 \\ \hline \end{array}$ (b) $\begin{array}{r} 303 \\ +\ \ 33 \\ \hline \end{array}$ (c) $\begin{array}{r} 1223 \\ +\ \ 231 \\ \hline \end{array}$

   (d) $\begin{array}{r} 333 \\ +\ 101 \\ \hline \end{array}$ (e) $\begin{array}{r} 32 \\ +\ 13 \\ \hline \end{array}$ (f) $\begin{array}{r} 302 \\ -\ 103 \\ \hline \end{array}$

   (g) $\begin{array}{r} 212 \\ -\ \ 33 \\ \hline \end{array}$ (h) $\begin{array}{r} 3102 \\ -\ 1033 \\ \hline \end{array}$

Teaching Concepts

14. How would you respond to one of your students who subtracted 229 from 2003 as

$$\begin{array}{r} {}^{1\ 9\ \ 9} \\ 2\,\not{0}\,\not{0}\,{}^{1}3 \\ -\ \ 2\,2\,9 \\ \hline 1\,7\,7\,4 \end{array}$$

saying that he or she had exchanged one of the 200 tens for 10 units?

15. In a long-distance relay race of 50 miles, the four runners on one team had the following times: 1 hour 2 minutes 23 seconds; 51 minutes 31 seconds; 1 hour 47 seconds; and 48 minutes 27 seconds. What was the total time for the team? Comment on the exchanges involved in solving this problem. (*Note:* The final answer should be stated with the least possible number of minutes and seconds.)

16. Read the Number and Operations Standard for Grades Pre-K–2, pages 78–88 in NCTM's *Principles and Standards for School Mathematics,* and write a critique of the standard, emphasizing your own reaction to the recommendations. How do the recommendations compare with your own school experience?

Thinking Critically

17. On his way to school, Peter dropped his arithmetic paper in a puddle of water, blotting out some of his work. What digits should go under the blots on these problems? (The base is ten.)

   (a) (b) (c)

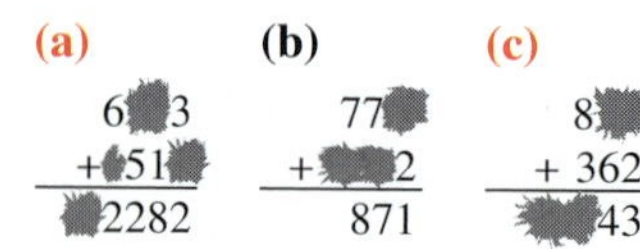

(d)

2 4 8■
− 1■2 2
■1■9

(e)

4■2
− 1843
■15■

(f)

34■5
−■748
■2■

18. Find the missing digits in each of these base ten addition problems.

(a)
_437
2_1
+ 347_
6_94

(b)
_721
901_
+ 71_3
_0_26

(c)
38_1
24_3
+ 512_
__5_9

(d)
5_4
612_
+ 8_1
76_6

19. Fill in the missing digits in each of these base ten subtraction problems.

(a)
_3_
− 2_1
594

(b)
3__4
− _346
175_

(c)
7_4_
− _5_4
808

(d)
63__4
− 2_12_
_6209

20. These additions and subtractions are written in different bases. Determine the base used in each case. (*Hint:* There may be more than one correct answer.)

(a)
231
+ 414
1200

(b)
231
+ 414
1045

(c)
231
+ 414
645

(d)
344
+ 143
1042

(e)
523
− 254
236

(f)
523
− 254
247

(g)
523
− 254
28E

(h)
1020
− 203
312

21. There is a rather interesting addition algorithm called the **scratch** method that proceeds as follows. Consider this sum.

2 834
576
4 835
2 743
10,988

Begin by adding from the top down in the units column. When you add a digit that makes your sum 10 or more, scratch out the digit as shown and make a mental note of the units digit of your present sum. Start with the digit noted and continue adding and scratching until you have completed the units column, writing down the units digit of the last sum as the units digit of the answer as shown. Now, count the number of scratches in the units column and, starting with this number, add on down the tens column repeating the scratch process as you go. Continue the entire process until all the columns have been added. This gives the desired answer. Explain why the algorithm works.

Thinking Cooperatively

For each of the next three problems, work with a partner. For each problem, make sure that the answers you give are agreed upon by your partner.

22. Give each pair of students a set of the ten digits, each printed on a square of heavy paper.

(a) Have the students choose and place the digits on the desktop to form the greatest four-digit number that is a multiple of six.

(b) Can any odd number be a multiple of six? Why or why not?

(c) Have the students choose and place the digits on the desktop to form the least six-digit number for which the sum of the digits is at least 20.

(d) Have the students choose and place the digits on the desktop to form the greatest six-digit number for which the sum of the digits is less than 20.

23. Form two four-digit numbers using each of 1, 2, 3, 4, 5, 6, 7, and 8 once, and only once, so that

(a) the sum of the two numbers is as large as possible.

(b) the sum of the two numbers is as small as possible.

(c) You can do better than guess and check on parts (a) and (b). Explain your solution strategy briefly.

(d) Is there only one answer to each of parts (a) and (b)? Explain in two sentences.

24. (a) Let s denote the sum of the numbers in problem 23, parts (a) and (b). Use your calculator to compute these products.

$1 \cdot s =$ ________
$2 \cdot s =$ ________
$3 \cdot s =$ ________
$4 \cdot s =$ ________
$5 \cdot s =$ ________
$6 \cdot s =$ ________
$7 \cdot s =$ ________
$8 \cdot s =$ ________
$9 \cdot s =$ ________
$10 \cdot s =$ ________

(b) Briefly discuss any patterns you notice in part (a).

(c) See if you can predict the product $41 \cdot s$, and then check the result with a calculator.

(d) Compute $14 \cdot s$, $23 \cdot s$, and $32 \cdot s$.

(e) Compute $6 \cdot s$, $17 \cdot s$, $28 \cdot s$, and $39 \cdot s$.

(f) What kind of sequences are 14, 23, 32, 41 and 6, 17, 28, 39?

(g) Briefly discuss the patterns you notice in part (d). Do these patterns continue for $50 \cdot s$, $59 \cdot s$, $68 \cdot s$, and so on?

(h) Briefly discuss the patterns you discern in part (e).

25. Form two four-digit numbers using each of the digits 1, 2, 3, 4, 5, 6, 7, and 8 precisely once so that:

(a) the difference of the two numbers is a natural number that is as small as possible.

(b) the difference of the two numbers is as large as possible.

(c) Briefly explain the strategy you used in solving this problem.

(d) Is there more than one solution to parts (a) and (b)? Why or why not?

Making Connections

26. Julien spent one hour and 45 minutes mowing the lawn and two hours and 35 minutes trimming the hedge and some shrubs. How long did he work all together?

27. (a) Mr. Arronson has four pieces of oak flooring left over from a job he just completed. If they are 3′ 8″, 4′ 2″, 6′ 10″, and 5′ 11″ long, what total length of flooring does he have left over? Make sure the number of inches in your answer is less than 12.

(b) If Mr. Arronson uses 9′10″ of the flooring left over in part (a) to make a picture frame, how much flooring does he have left over then?

28. After her dad gave her her allowance of 10 dollars, Ellie had 25 dollars and 25 cents. After buying a stuffed kangaroo for 14 dollars and 53 cents, including tax, how much money did Ellie have left?

29. It was 25 minutes after 5 in the morning when Ari began his paper route. If it took him an hour and three-quarters to deliver the papers, at what time did he finish?

From State Student Assessments

30. (Georgia, Grade 4)
If you change the digit 6 to a 9 in the number 56,907, what will be the difference?

A. three hundred

B. nine hundred

C. one thousand

D. three thousand

31. (Georgia, Grade 4)
Fay picked apples in the orchard. She picked 45 apples on Monday, 57 on Tuesday, and 39 on Wednesday. How many apples did she pick in all?

A. 84

B. 92

C. 102

D. 141

32. (Texas, Grade 4)
A hiker is climbing a mountain that is 6238 feet high. She stops to rest at 4887 feet. How many more feet must she climb to reach the top?

F. 2351 feet

G. 1451 feet

H. 1361 feet

I. 1351 feet

J. Not here

33. (California, Grade 4)
Solve: $619{,}581 - 23{,}183 = ?$
Solve: $6747 + 321{,}105 = ?$

For Review

34. Write two division equations and a second multiplication equation corresponding to each of these multiplication equations.

(a) $11 \cdot 91 = 1001$

(b) $7 \cdot 84 = 588$

(c) $13 \cdot 77 = 1001$

35. Write a multiplication equation and a second division equation corresponding to each of these division equations.

(a) $1001 \div 7 = 143$

(b) $323 \div 19 = 17$

(c) $899 \div 31 = 29$

36. If Jennifer Hurwitz had four sweaters, three blouses, and two pairs of slacks that all mixed and matched, how many different outfits consisting of a sweater, blouse, and pair of slacks could she wear?

37. Let $A = \{1, 2, 3, 4, 5\}$, $B = \{2, 4, 6, 8\}$, and $C = \{3, 4, 5, 6, 7\}$.

(a) Determine $A \cup B \cup C$, $A \cap B$, $A \cap C$, $B \cap C$, and $A \cap B \cap C$.

(b) Determine $n(A \cup B \cup C)$.

(c) Compute $n(A) + n(B) + n(C) - n(A \cap B) - n(A \cap C) - n(B \cap C) + n(A \cap B \cap C)$.

38. (a) Repeat problem 37 with three different sets A, B, and C of your own choosing.

(b) What general result do problems 37 and 38(a) suggest?

(c) Write a paragraph arguing that the result guessed in part (b) is true. (*Hint:* Consider an element x not in any of A, B, or C; y in A, but not B or C; and so on.)

3.4 ALGORITHMS FOR MULTIPLICATION AND DIVISION OF WHOLE NUMBERS

Multiplication and division in most ancient numeration systems were quite complicated. The decimal system makes these processes much easier, but many people still find them confusing. At least part of the difficulty is that the ideas are often presented as a collection of rules to be learned by rote, with little or no effort made to impart understanding. In this section, we endeavor to strip away some of the mystery.

Multiplication Algorithms

Multiplication is repeated addition. Thus, $2 \cdot 9$ means $9 + 9$, $3 \cdot 9$ means $9 + 9 + 9$, and so on. But repeated addition is slow and tedious, and easier algorithms exist. As with addition and subtraction, these should be introduced starting with concrete approaches and gradually becoming more and more abstract. The development should proceed through units, strips, and mats; place-value cards; classroom abacuses; and so on. Let's consider the product $9 \cdot 3$.

EXAMPLE 3.11 DEVELOPING THE MULTIPLICATION ALGORITHM

Compute the product of 9 and 3.

Solution

Using Units, Strips, and Mats

Since $9 \cdot 3 = 3 + 3 + 3 + 3 + 3 + 3 + 3 + 3 + 3$, we can illustrate this as shown with 9 rows of 3 units each. Simplifying the original array by appropriately exchanging units for strips, we eventually have 2 strips and 7 units, which is recorded as 27. Of course, elementary school children should actually handle the materials, making the necessary exchanges of units for strips.

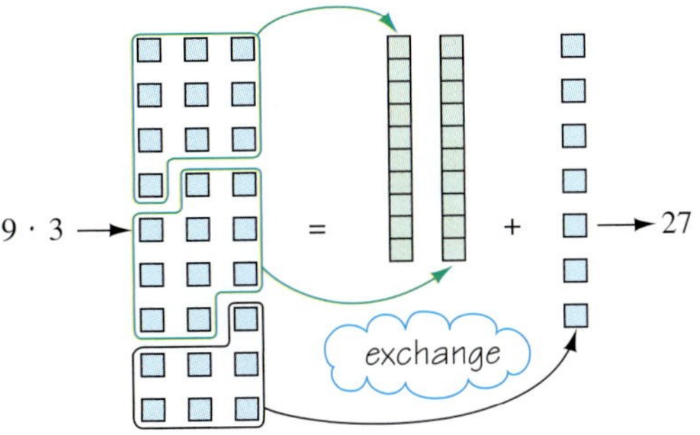

Using Place-Value Cards

With place-value cards, students should start with 9 rows of 3 markers each on the 1s square of their place-value cards as shown here. They should then exchange 10 markers on the 1s square for 1 marker on the 10s square as many times as possible, and record the fact that this gives 2 tens plus 7 units, or 27.

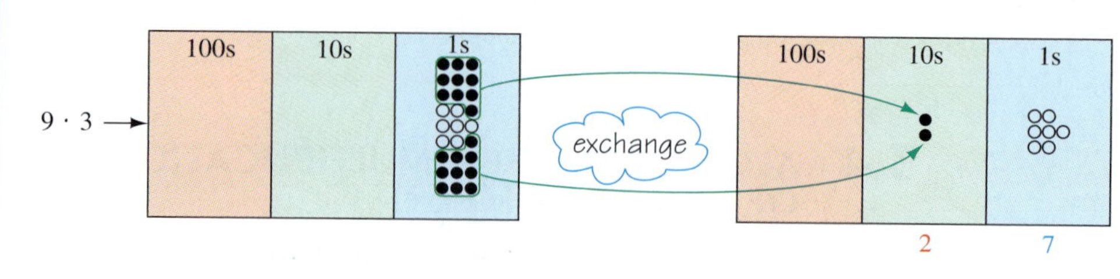

Using manipulatives as illustrated, children can experience, learn, and actually *understand* all the one-digit multiplication facts. It should not be the case that $9 \cdot 3 = 27$ is a string of meaningless symbols memorized by rote. This basic fact must be understood.

Once the one-digit facts are thoroughly understood *and memorized,* as they must be, even for intelligent use of a calculator, one can move on to more complicated problems. Consider, for example, $3 \cdot 213$. This could be illustrated with units, strips, and mats, but, for brevity, we will go directly to place-value cards, expanded notation, and then to an algorithm, as illustrated in Example 3.12. Again, we make use of the understanding that $3 \cdot 213 = 213 + 213 + 213$.

EXAMPLE 3.12 COMPUTING A PRODUCT WITH A MULTIDIGIT NUMBER

Compute the product of 3 and 213.

Solution *Using Place-Value Cards, Expanded Notation, and Algorithms*

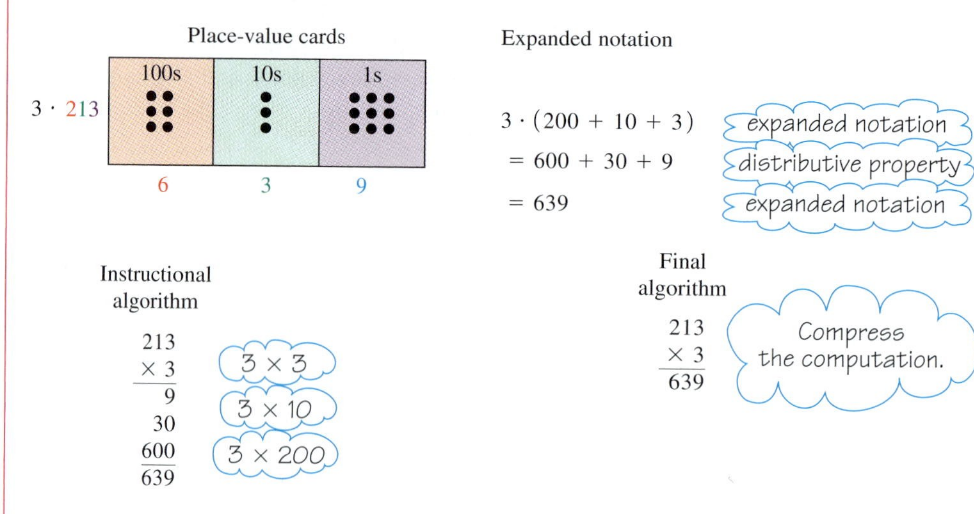

The product $3 \cdot 213$ did not require exchanging. Consider the product $4 \cdot 243$. Presentation of this product proceeds from the concrete representation to the final algorithm.

EXAMPLE 3.13 MULTIPLYING WITH EXCHANGING

Compute the product of 4 and 243.

Solution *Using Place-Value Cards, Expanded Notation, and Algorithms*

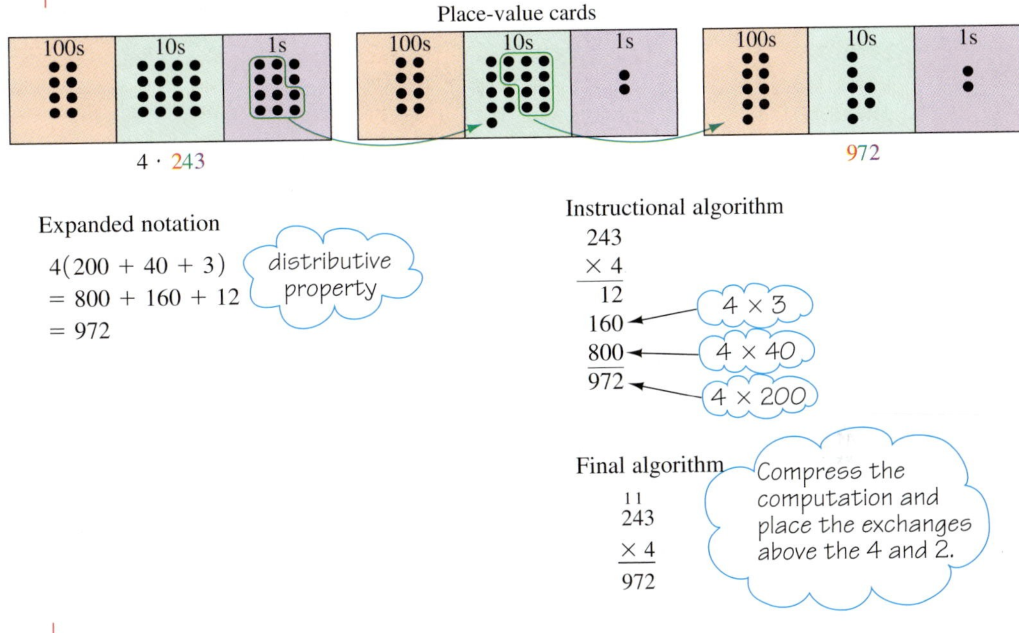

The preceding development represents many lessons. However, the various demonstrations should be clearly tied together, and enough time should be spent to ensure that each level of the chain of reasoning is understood before proceeding to the next.

Finally, consider the product 15 · 324. Unquestionably, the most efficient algorithm is that provided by the calculator. The pencil-and-paper algorithm, however, is not unusually difficult, as the following series of calculations show.

EXAMPLE 3.14 MULTIPLYING MULTIDIGIT NUMBERS

Compute the product of 15 and 324.

Solution *Using Expanded Notation, an Instructional Algorithm, and the Final Algorithm*

Expanded notation

$$
\begin{aligned}
15 \cdot 324 &= (10 + 5) \cdot 324 && \text{distributive property} \\
&= 10 \cdot 324 + 5 \cdot 324 \\
&= 10(300 + 20 + 4) + 5(300 + 20 + 4) && \text{expanded notation} \\
&= 3000 + 200 + 40 + 1500 + 100 + 20 && \text{distributive property} \\
&= 4860
\end{aligned}
$$

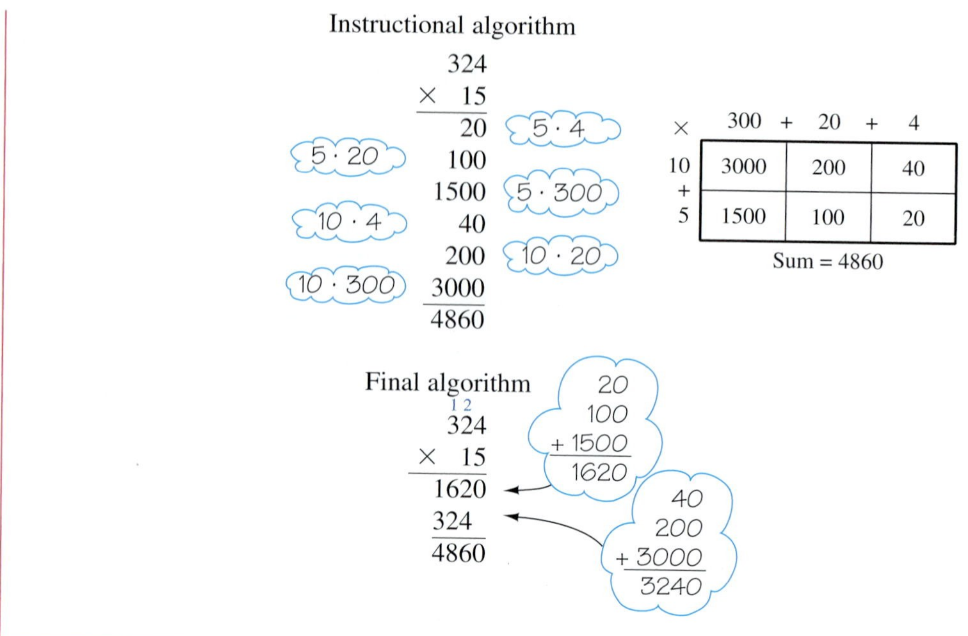

Multiplication in Other Bases

As with the algorithms for addition and subtraction, the multiplication algorithm depends on the idea of positional notation, but is independent of the base. The arithmetic is more awkward for bases other than ten, since we do not think in other bases as we do in base ten, but the ideas are the same. However, working in other bases not only enhances understanding of base ten, but also provides an interesting activity that helps to improve mental arithmetic skills. Let's consider an example.

from **The NCTM Principles and Standards**

Research suggests that by solving problems that require calculation, students develop methods for computing and also learn more about operations and properties (McClain, Cobb, and Bowers 1998; Schifter 1999). As students develop methods to solve multidigit computation problems, they should be encouraged to record and share their methods. As they do so, they can learn from one another, analyze the efficiency and generalizability of various approaches, and try one another's methods. In the past, common school practice has been to present a single algorithm for each operation. However, more than one efficient and accurate computational algorithm exists for each arithmetic operation. In addition, if given the opportunity, students naturally invent methods to compute that make sense to them (Fuson forthcoming; Madell 1985). The following episode, drawn from unpublished classroom observation notes, illustrates how one teacher helped students analyze and compare their computational procedures for division:

> Students in Ms. Sparks' fifth-grade class were sharing their solutions to a homework problem, 728 ÷ 34. Ms. Sparks asked several students to put their work on the board to be discussed. She deliberately chose students who had approached the problem in several different ways. As the students put their work on the board, Ms. Sparks circulated among the other students, checking their homework.
>
> Henry had written his solution:

$34 \times 10 = 340$
$34 \times 20 = 680$

$$\begin{array}{r} 680 \\ +\ 34 \\ \hline 714 \end{array} \qquad \begin{array}{r} 728 \\ -714 \\ \hline 14 \end{array}$$

Henry explained to the class, "Twenty 34s plus one more is 21. I knew I was pretty close. I didn't think I could add any more 34s, so I subtracted 714 from 728 and got 14. Then I had 21 remainder 14."

EXAMPLE 3.15 MULTIPLICATION IN BASE SIX

Compute the product $324_{six} \cdot 15_{six}$ in base six.

Solution

We use the base six positional values from Table 3.7 and the instructional algorithm of Example 3.14. To simplify notation, we will write all *numerals* in base six notation so that the subscripts will be omitted. Finally, the numeral *words* will have their usual base ten meaning. The work proceeds as shown, with descriptive comments in the think cloud.

$$\begin{array}{r} 324 \\ \times\ 15 \\ \hline 32 \\ 140 \\ 2300 \\ 40 \\ 200 \\ 3000 \\ \hline 10152 \end{array}$$

Think

$5 \times 4 = \text{twenty} = 3 \text{ sixes} + 2 \text{ units} = 3 \cdot 6^1 + 2 \cdot 6^0 = 32_{six}$

$5 \times 20 = 5 \times 2 \text{ sixes} = 10 \text{ sixes} = (1 \cdot 6 + 4) \cdot 6^1 = 1 \cdot 6^2 + 4 \cdot 6^1 + 0 \cdot 6^0 = 140_{six}$

$5 \times 300 = 5 \times 3 \text{ thirty-sixes} = 15 \text{ thirty-sixes} = (2 \cdot 6 + 3) \cdot 6^2 = 2 \cdot 6^3 + 3 \cdot 6^2 + 0 \cdot 6^1 + 0 \cdot 6^0 = 2300_{six}$

$10 \times 4 = 1 \text{ six} \times 4 = 4 \cdot 6^1 + 0 \cdot 6^0 = 40_{six}$

$10 \times 20 = 1 \text{ six} \times 2 \text{ sixes} = 2 \cdot 6^2 + 0 \cdot 6^1 + 0 \cdot 6^0 = 200_{six}$

$10 \times 300 = 1 \text{ six} \times 3 \text{ thirty-sixes} = 3 \cdot 6^3 + 0 \cdot 6^2 + 0 \cdot 6^1 + 0 \cdot 6^0 = 3000_{six}$

To check, we convert all numerals to base ten:

$$\begin{aligned} 324_{six} &= 3 \cdot 6^2 + 2 \cdot 6^1 + 4 \cdot 6^0 \\ &= 3 \cdot 36 + 2 \cdot 6 + 4 \cdot 1 \\ &= 108 + 12 + 4 = 124_{ten} \\ 15_{six} &= 1 \cdot 6^1 + 5 \cdot 6^0 \\ &= 1 \cdot 6 + 5 \cdot 1 \\ &= 6 + 5 = 11_{ten} \\ 10{,}152_{six} &= 1 \cdot 6^4 + 0 \cdot 6^3 + 1 \cdot 6^2 + 5 \cdot 6^1 + 2 \cdot 6^0 \\ &= 1 \cdot 1296 + 0 \cdot 216 + 1 \cdot 36 + 5 \cdot 6 + 2 \cdot 1 \\ &= 1296 + 36 + 30 + 2 = 1364_{ten} \end{aligned}$$

Since $11_{ten} \cdot 124_{ten} = 1364_{ten,}$ the check is complete.

Division Algorithms

An approach to division discussed in Chapter 2 was repeated subtraction. This ultimately led to the so-called **division algorithm,** which we restate here for easy reference.

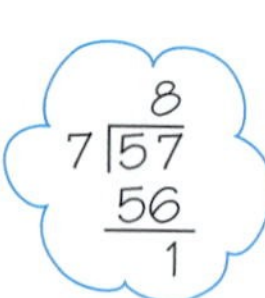

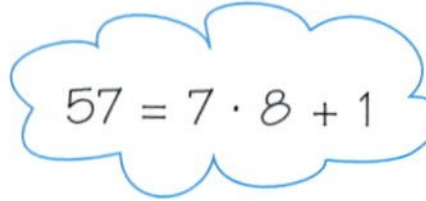

> **THEOREM** *The Division Algorithm*
> If a and b are whole numbers with b not zero, there exists precisely one pair of whole numbers q and r with $0 \leq r < b$ such that $a = bq + r$.

The Long-Division Algorithm

Suppose we want to divide 941 by 7. Doing this by repeated subtraction will take a long time, even with a calculator. But suppose we subtract several 7s at a time and keep track of the number we subtract each time. Indeed, since it is so easy to multiply a number by 10, 100, 1000, and so on, let's subtract hundreds of 7s, tens of 7s, and so on. The work might be organized like this:

7)941			
700	Subtract	100	7s
241			
70	Subtract	10	7s
171			
70	Subtract	10	7s
101			
70	Subtract	10	7s
31			
28	Subtract	4	7s
3		134	number of 7s subtracted

Since $3 < 7$, the process stops and we see that 941 divided by 7 gives a quotient of 134 and a remainder of 3. As a check, we note that $941 = 7 \cdot 134 + 3$. The above work could have been shortened if we had subtracted the three tens of 7s all at once and then the four 7s all at once like this:

7)941		
700	100	7s
241		
210	30	7s
31		
28	4	7s
3	134	

A slightly different form of this algorithm, sometimes called the **scaffold** algorithm, is obtained by writing the 100, 30, and 4 above the division symbol like this:

134	
4	
30	
100	
7)941	
700	7 · 100
241	
210	7 · 30
31	
28	7 · 4
3	

Note that 1, 3, and 4 are the quotients when 9, 24, and 31, respectively, are divided by 7.

Just for Fun

What's the Sum?

Perform these additions.

91 + 19	95 + 59	42 + 24
62 + 26	74 + 47	61 + 16
82 + 28	81 + 18	32 + 23
54 + 45		

(a) Investigate the results of dividing the answers to the above additions by 11.

(b) Can you predict the answers to all problems just by looking at the two digits involved?

Many teachers prefer the scaffold algorithm as the final algorithm for division, since it fully displays the mathematics being done. Others still cling to the following, which is easily obtained from the scaffold algorithm, even though it seriously masks the mathematics and forces many students to rely on rote memorization rather than understanding:

$$\begin{array}{r}
1\,3\,4 \\
7\overline{)9^{2}4^{3}1} \\
\underline{7} \\
24 \\
\underline{21} \\
31 \\
\underline{28} \\
3
\end{array}$$

A further example with a larger divisor may be helpful.

EXAMPLE 3.16 USING THE LONG-DIVISION ALGORITHM

Divide 28,762 by 307.

Solution Scaffold algorithm Standard algorithm

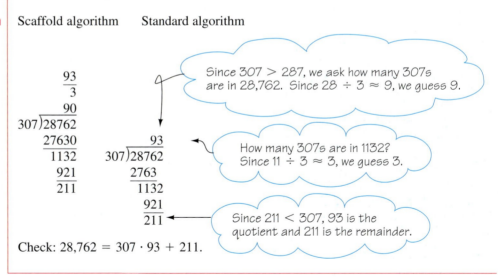

Scaffold algorithm:

$$\begin{array}{r}
93 \\
\underline{3} \\
90 \\
307\overline{)28762} \\
\underline{27630} \\
1132 \\
\underline{921} \\
211
\end{array}$$

Standard algorithm:

$$\begin{array}{r}
93 \\
307\overline{)28762} \\
\underline{2763} \\
1132 \\
\underline{921} \\
211
\end{array}$$

Since 307 > 287, we ask how many 307s are in 28,762. Since 28 ÷ 3 ≈ 9, we guess 9.

How many 307s are in 1132? Since 11 ÷ 3 ≈ 3, we guess 3.

Since 211 < 307, 93 is the quotient and 211 is the remainder.

Check: $28{,}762 = 307 \cdot 93 + 211$.

The Short-Division Algorithm

A division algorithm that is quite useful, even in this calculator age, is the **short-division** algorithm. This is a much simplified version of the long-division algorithm and is quite useful and quick when the divisor is a single digit. It can be developed using the scaffold method as above. Also, it follows directly from the long-division algorithm if that is already known.

EXAMPLE 3.17 USING THE SHORT-DIVISION ALGORITHM

Divide 2834 by 3 and check your answer.

Solution Consider the following divisions, which show how the short-division algorithm derives from the long-division algorithm:

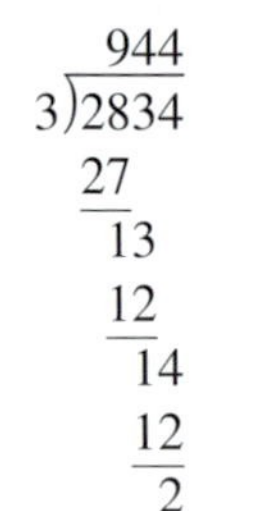

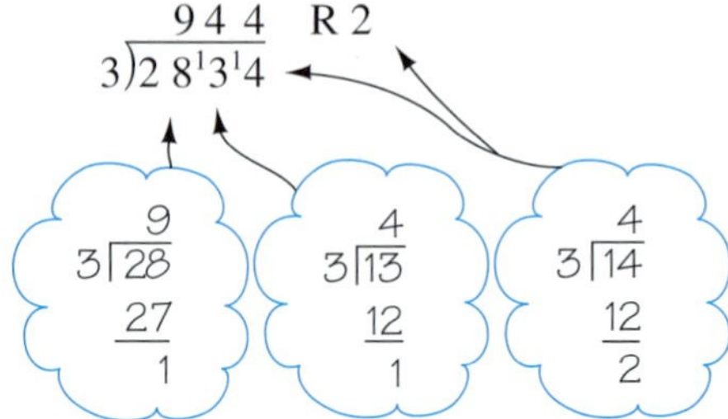

Problem Set 3.4

Understanding Concepts

(Note: Problems 1 through 11 are all base ten problems.)

1. (a) Make a suitable drawing of units and strips to illustrate the product $4 \cdot 8 = 32$.

   (b) Make a suitable sketch of place-value cards to illustrate the product $4 \cdot 8 = 32$.

2. Make a suitable sketch of place-value cards to illustrate the product $3 \cdot 254 = 762$.

3. (a) In the product shown, what does the red 2 actually represent?

```
   2
  274
×  34
 1 096
 8 22
 9,316
```

   (b) In the product shown in part (a), when multiplying $4 \cdot 7$, one "exchanges" a 2. What is actually being exchanged?

4. The diagram shown illustrates the product $27 \cdot 32$. Discuss how this is related to finding the product by the instructional algorithm for multiplication.

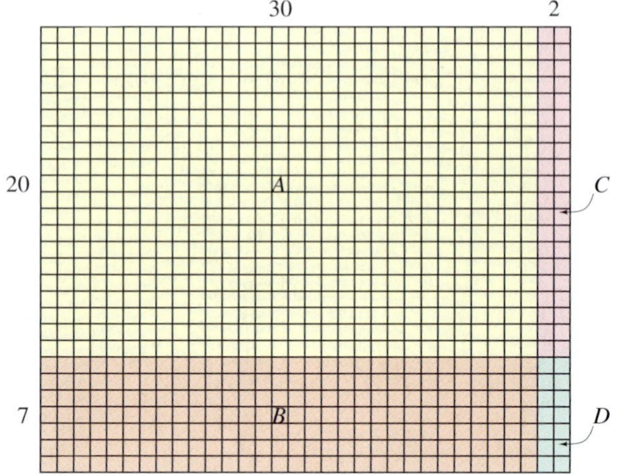

5. What property of the whole numbers justifies each step in this calculation?

$17 \cdot 4 = (10 + 7) \cdot 4$	expanded notation
$= 10 \cdot 4 + 7 \cdot 4$	(a) ________
$= 10 \cdot 4 + 28$	one-digit multiplication fact
$= 10 \cdot 4 + (2 \cdot 10 + 8)$	expanded notation
$= 4 \cdot 10 + (2 \cdot 10 + 8)$	(b) ________
$= (4 \cdot 10 + 2 \cdot 10) + 8$	(c) ________
$= (4 + 2) \cdot 10 + 8$	(d) ________
$= 6 \cdot 10 + 8$	one-digit addition fact
$= 68$	expanded notation

6. Draw a sequence of sketches of units, strips, and mats to illustrate dividing 429 by 3.

7. What calculation does this sequence of sketches illustrate? Explain briefly.

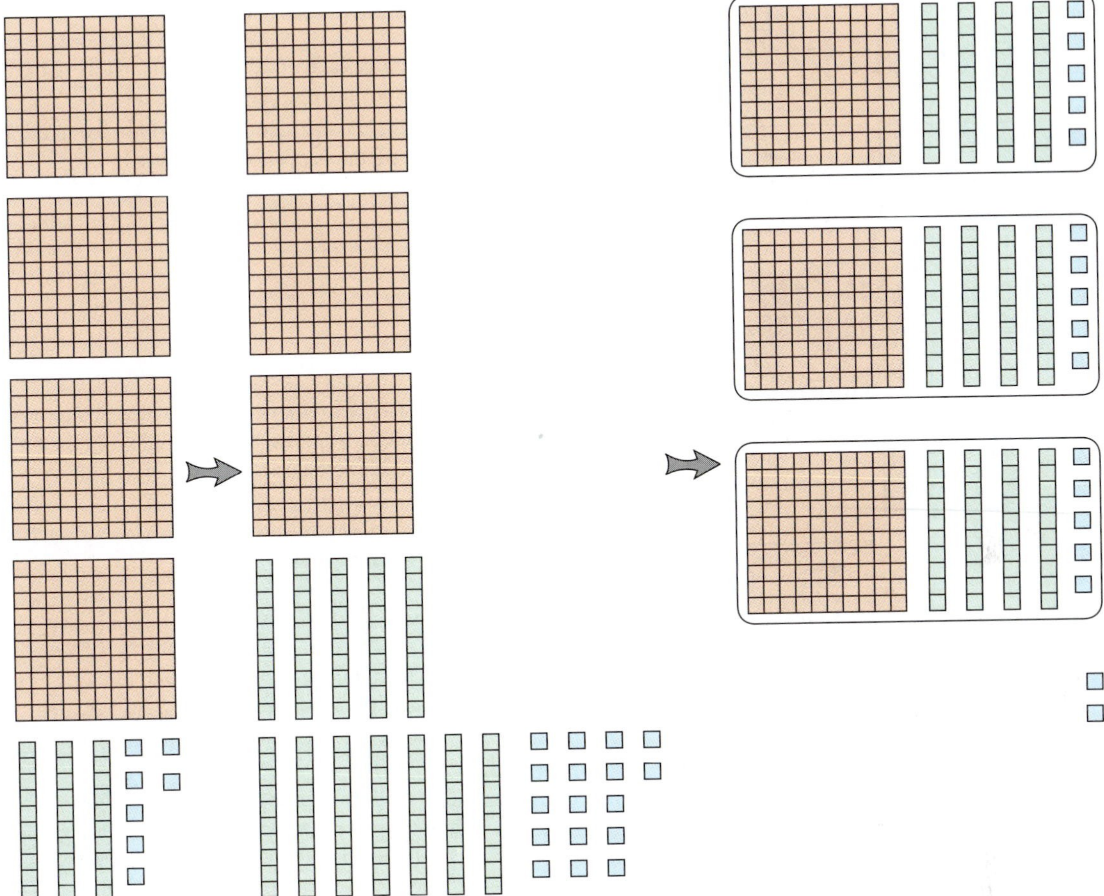

8. Multiply 352 by 27, using the instructional algorithm for multiplication.

9. Find the quotient q and a remainder r when a is divided by b, and write the result in the form $a = bq + r$ of the division algorithm for each of these choices of a and b.

(a) $a = 27, b = 4$

(b) $a = 354, b = 29$

(c) $a = 871, b = 17$

10. Perform each of the following divisions by the scaffold method. In each case, check your results by using the equation of the division algorithm.

(a) $351\overline{)7425}$

(b) $23\overline{)6814}$

(c) $213\overline{)3175}$

11. Use short division to find the quotient and remainder for each of these. Check each result.

(a) $5\overline{)873}$

(b) $7\overline{)2432}$

(c) $8\overline{)10{,}095}$

(Note: Problems 12 through 15 are in bases other than ten.)

12. Construct base five addition and multiplication tables.

13. What is being illustrated by the following sequence of sketches of place-value cards? Explain briefly.

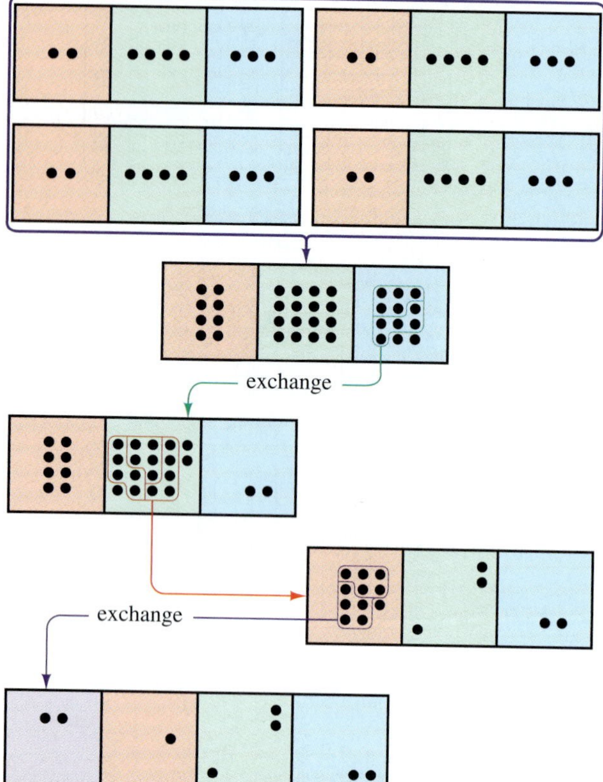

14. Carry out these multiplications, using base five notation. All the numerals are already written in base five, so no subscript is needed.

(a) $\begin{array}{r} 23 \\ \times\ 3 \\ \hline \end{array}$ (b) $\begin{array}{r} 432 \\ \times\ 41 \\ \hline \end{array}$ (c) $\begin{array}{r} 2013 \\ \times\ 23 \\ \hline \end{array}$

(d) Convert the numerals in parts (a), (b), and (c) to base ten, and check the results of your base five computations.

15. Carry out these divisions using base five notation. All numerals are already in base five, so the subscripts are omitted.

(a) $4\overline{)231}$ (b) $32\overline{)2342}$ (c) $213\overline{)34{,}122}$

(d) Convert the numerals in parts (a), (b), and (c) to base ten, and check the results of your base five computations.

Teaching Concepts

16. Problem 4 of this problem set shows a sketch to represent the product $27 \cdot 32$.

(a) Draw a similar diagram to represent the product $17 \cdot 23$.

(b) In expanded notation, the product appears as follows:

$$\begin{aligned}(10 + 7) \cdot (20 + 3) &= (10 + 7) \cdot 20 + (10 + 7) \cdot 3 \\ &= 10 \cdot 20 + 7 \cdot 20 + 10 \cdot 3 + 7 \cdot 3 \\ &= 200 + 140 + 30 + 21\end{aligned}$$

Identify each of 200, 140, 30, and 21 with the regions in your diagram for part (a).

(c) Using the instructional algorithm for multiplication, the product appears as follows:

$$\begin{array}{r} 23 \\ 17 \\ \hline 21 \\ 140 \\ 30 \\ 200 \\ \hline 391 \end{array}$$

This is clearly an alternative representation of the product, using expanded notation. In the final algorithm, the product appears as follows:

$$\begin{array}{r} {}^{2} \\ 23 \\ 17 \\ \hline 161 \\ 23 \\ \hline 391 \end{array}$$

Explain the 2 that is exchanged in terms of the instructional algorithm, and show how the numbers 161 and 23 derive from the instructional algorithm.

17. Another multiplication algorithm is the **lattice algorithm.** Suppose, for example, you want to multiply 324 by 73. Form a two-by-three rectangular array of boxes with the 3, 2, and 4 across the top and the 7 and 3 down the right side as shown in the accompanying figure. Now compute the products $3 \cdot 7 = 21$, $2 \cdot 7 = 14$, $4 \cdot 7 = 28$, $3 \cdot 3 = 9$, $2 \cdot 3 = 6$, and $4 \cdot 3 = 12$. Place the products in the appropriate boxes ($3 \cdot 7$ is in the 3 column and the 7 row, for example), with the units digit of the product below the diagonal in each box and the tens digit (if there is one) above the diagonal. Now add down the diagonals and add any "exchanges" to the sum above the next diagonal. The result of 23,652 is the desired product.

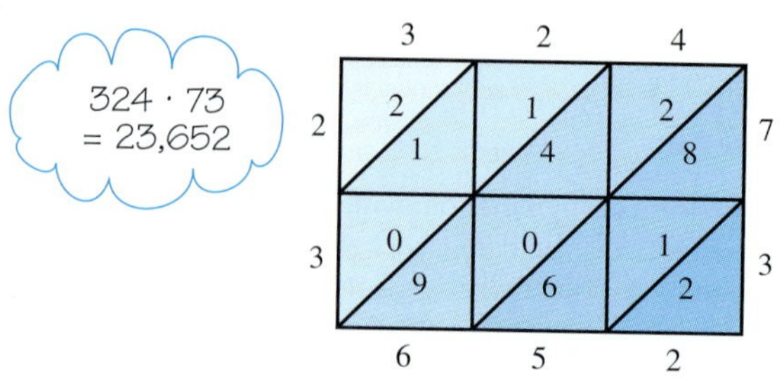

(a) Multiply 374 by 215 using the lattice algorithm.

(b) Write a short paragraph comparing the lattice algorithm with the standard pencil-and-paper algorithm.

(c) Would presenting this algorithm to students in a class you might teach help them gain a better understanding of positional notation? Explain.

18. Consider the following computation:

$$\begin{array}{r} 374 \\ \times\ 23 \\ \hline 748 \\ 1122 \\ \hline 8602 \end{array}$$

(a) Is the algorithm correct? Explain briefly.

(b) Multiply 285 by 362 using this method.

(c) What would you say to one of your students who multiplied multidigit numbers in the manner shown above? Would you insist that there is really just one correct way to perform the pencil-and-paper calculation? Explain.

Thinking Critically

19. The **Egyptian algorithm** for multiplication was one of the interesting subjects explained in the Rhind papyrus mentioned in Chapters 1 and 2. We will explain the algorithm by giving the following example. Suppose we want to compute 19 times 35. Successively doubling 35, we obtain this list:

→ $1 \cdot 35 = 35$
→ $2 \cdot 35 = 70$
$4 \cdot 35 = 140$
$8 \cdot 35 = 280$
→ $16 \cdot 35 = 560$

19 = 1 + 2 + 16

Adding the results in the indicated rows gives us 665 as the desired product.

(a) After carefully considering the above computation, write a short paragraph explaining how and why the process always works.

(b) This scheme is also known as the **duplation algorithm.** Use duplation to find the product of 24 and 71.

20. The **Russian peasant algorithm** for multiplication is similar to the duplation algorithm described in problem 19. To find the product of 34 and 54, for example, successively divide the 34 by 2 (ignoring remainders if they occur) and successively multiply 54 by 2. This gives the following lists:

~~34~~	~~54~~
17	108
~~8~~	~~216~~
~~4~~	~~432~~
~~2~~	~~864~~
1	1728
	1836

Now cross out the even numbers in the left-hand column and the companion numbers in the right-hand column. Add the remaining numbers in the right-hand column to obtain the desired product. To see why the process works, consider the products $34 \cdot 54 = 1836$ and $17 \cdot 108 = 1836$. Also, consider $8 \cdot 216$, $4 \cdot 432$, and $2 \cdot 864$.

(a) Why are $34 \cdot 54$ and $17 \cdot 108$ the same?

(b) Why are $17 \cdot 108$ and $8 \cdot 216$ different? How much do they differ?

(c) Why are $8 \cdot 216$, $4 \cdot 432$, $2 \cdot 864$, and $1 \cdot 1728$ all the same?

(d) Write a short paragraph explaining why the Russian peasant algorithm works.

(e) Use the Russian peasant algorithm to compute $29 \cdot 81$ and $11 \cdot 243$.

21. Earlier in this chapter, we showed how to convert a base ten numeral for a number into a numeral in another base. The process was described as it was in order to impart proper understanding. However, the conversion can be completed much more easily. The trouble with showing this new scheme (at least in the beginning) is that it can be carried out with absolutely no understanding. Using the earlier method, we find that $583_{\text{ten}} = 4313_{\text{five}}$. Now let's see how the new scheme works. Divide 583 by 5 to obtain a quotient and a remainder, then divide the quotient by 5 and record the next remainder, and so on like this:

Remarkably, $4313_{\text{five}} = 583_{\text{ten}}$, as we have just seen. But why does the method work? If we write each of the preceding divisions in the form of the division algorithm, we have

$$583 = 5 \cdot 116 + 3$$
$$116 = 5 \cdot 23 + 1$$

and

$$23 = 5 \cdot 4 + 3.$$

Combining these, we obtain

$$\begin{aligned} 583 &= 5 \cdot 116 + 3 \\ &= 5(5 \cdot 23 + 1) + 3 \\ &= 5^2 \cdot 23 + 5 \cdot 1 + 3 \\ &= 5^2(5 \cdot 4 + 3) + 5 \cdot 1 + 3 \\ &= 5^3 \cdot 4 + 5^2 \cdot 3 + 5 \cdot 1 + 3, \end{aligned}$$

which is the expanded form of 4313_{five}. Thus, $583_{\text{ten}} = 4313_{\text{five}}$ as noted. Write 482_{ten} in each of these bases and check by reconverting the numeral obtained to base ten.

(a) base seven (b) base four
(c) base two (d) base twelve

Thinking Cooperatively

Divide into small groups to work each of the next four problems. In each case, discuss possible strategies among your group and develop an answer agreed upon by the entire group.

22. Use each of 1, 3, 5, 7, and 9 once, and only once, in the boxes to obtain the largest possible product in each case.

(a) (b)

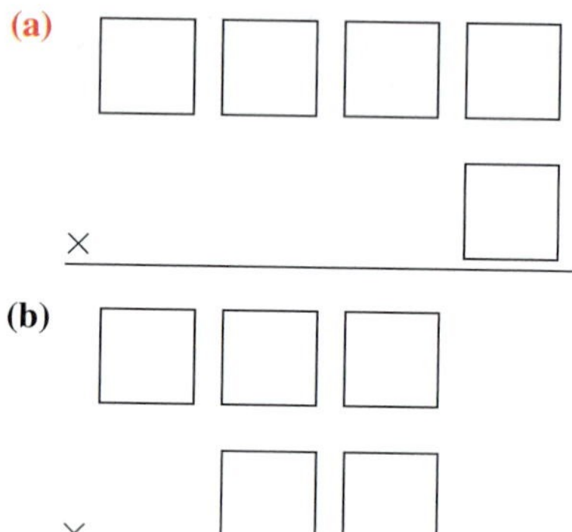

23. Use each of 1, 3, 5, 7, and 9 once, and only once, in the boxes to obtain the smallest possible product in each case.

(a)

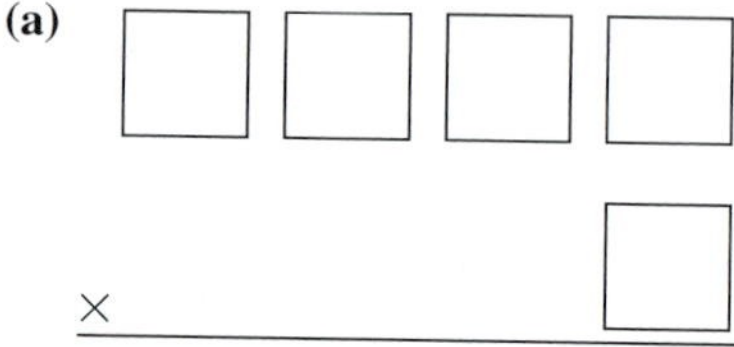

(b)

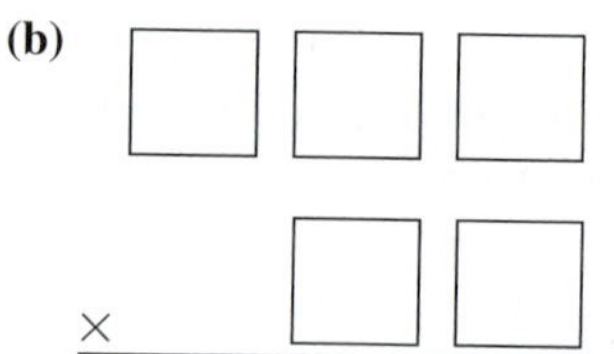

24. A druggist has a balance scale to weigh objects. She also has two 1-gram weights, two 3-gram weights, two 9-gram weights, and two 27-gram weights. She places an object to be weighed in one pan and her weights in the other.

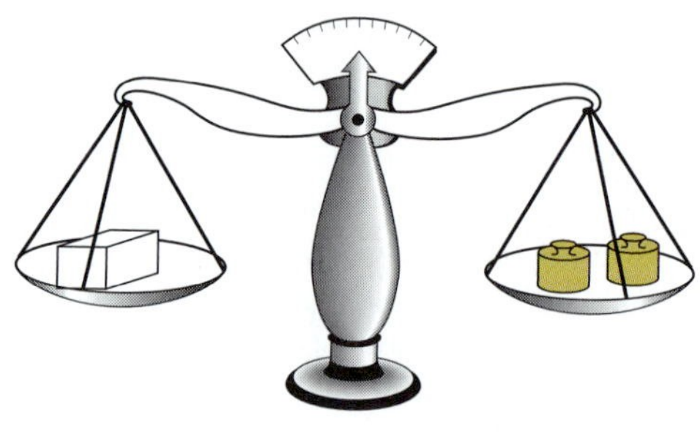

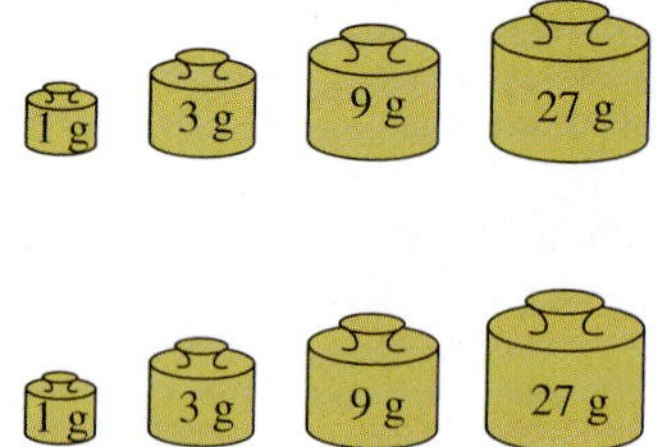

(a) What weights can she weigh in this way?

(b) What mathematics would this activity be illustrating if you did it in one of your classes?

25. Suppose you had only one of each type of weight and that you could put weights in both scales, some along with the object being weighed.

(a) What weights of objects could you possibly weigh in this way?

(b) Could you "weigh" a balloon filled with helium that was tied to one of the pans of the scales and exerted an upward pull of 11 grams on the pan? What balloons could you "weigh" in this way?

Making Connections

26. Alicia, Arturo, and Adam took part in a walk for a charity. Each had sponsors who agreed to contribute five dollars for each mile each student walked. Alicia walked 4 miles and had 6 sponsors. Arturo walked 7 miles and had 11 sponsors. Adam walked 8 miles and had 7 sponsors. In all, how much money was contributed to the charity on the students' behalf?

27. On Monday, Melody practiced for her piano lesson for 48 minutes. On each of Tuesday and Wednesday, she practiced for a full hour. On Thursday she practiced for only 35 minutes.

(a) All told, how many minutes had she practiced before her piano lesson on Friday?

(b) Express the time Melody practiced in hours and minutes using as few minutes as possible.

28. The students in the third-grade class at Franklin Grade School wanted to earn \$165 to buy a wheelchair for one of their classmates who had been seriously injured in an accident. The students planned to sell boxes of cookies at \$2.25 per box. How many boxes would they need to sell? (*Suggestion:* Express the amounts of money in pennies.)

Using a Calculator

29. **(a)** Compute 375×2433.

(b) Examine the following calculation, where certain zeros are suppressed.

×		24	33
		3	75
	1	1 24	75
	18	00	
		99	
	72		
	91	23	75

75 · 33
75 · 2400
300 · 33
300 · 2400

Note that this agrees with your answer to part (a). Use the same idea to compute the following product that would ordinarily exceed the capacity of your calculator:

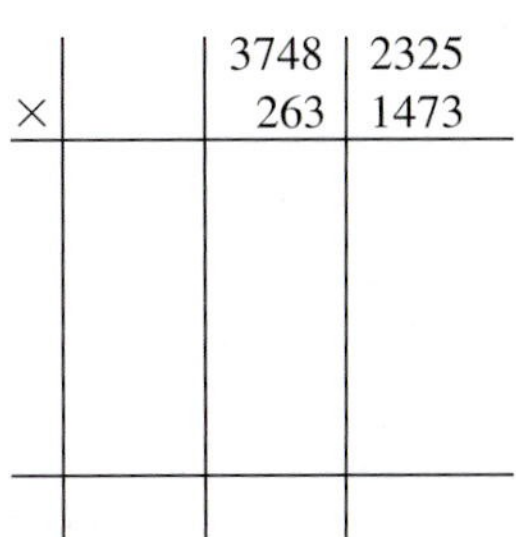

30. **Computing quotients on a calculator with Integer Divide.** Some calculators have an integer divide key, [INT ÷], that gives the quotient and remainder when one integer is divided by another. Use the integer divide key to find the quotient and remainder when

(a) 723 is divided by 37.

(b) 34,723 is divided by 508.

31. **Computing quotients and remainders on a standard calculator.** On most calculators, the remainder of a division of integers is expressed as the decimal part of the display. Thus, dividing 34,678 by 44 gives the answer 788.13636 on a calculator that displays eight digits. The quotient is the integer part of the displayed answer, and the decimal part of the display is $R \div 44$, where R is the remainder. To compute the remainder, subtract the quotient, 788, leaving 0.1363636 in the display (since the calculator actually works with greater accuracy than it shows). Since 0.1363636 represents $R \div 44$, just multiply by 44 to obtain the remainder. Actually, because of round-off error, your calculator may give 5.9999997, which you should interpret as 6 since R is a whole number. To check, note that $34{,}678 = 788 \cdot 44 + 6$. Note that other calculators may well give something different from 5.9999997 depending on their built-in accuracy, but it will be a number quite close to 6 and should be so interpreted. Without using the integer divide key, use your calculator to compute the quotient and remainder when

(a) 276,523 is divided by 511.

(b) 347,285 is divided by 87.

(c) 374,821 is divided by 357.

32. The integer divide capability of some calculators makes it particularly easy to change from base ten notation to notation in another base, using the method described in problem 21 above. For example, 583_{ten} can be converted to base five by entering the following string in the calculator and noting the remainder after each division:

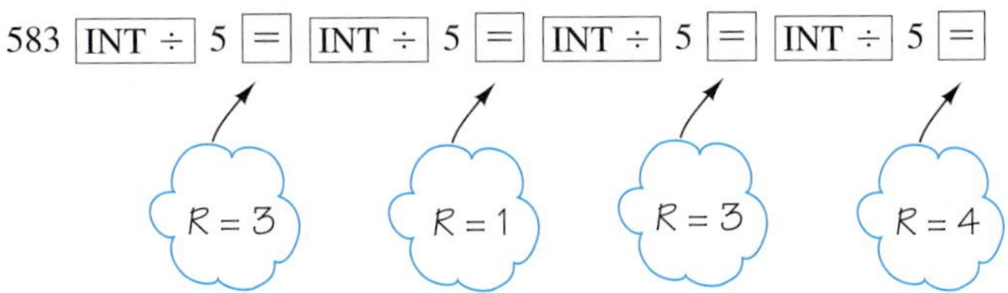

583 INT ÷ 5 = INT ÷ 5 = INT ÷ 5 = INT ÷ 5 =
R = 3
R = 1
R = 3
R = 4

Writing the remainders in reverse order, we find that the base five representation of 583_{ten} is 4313_{five}. Use a calculator (and the method of problem 31 if necessary) to determine the representation of 5781_{ten} in each of the following bases.

(a) base five **(b)** base seven **(c)** base twelve **(d)** base six

From State Student Assessments

33. (Washington State, Grade 4)

When Lori tries multiplying with her calculator, she gets the following results.

$$8 \times 3 = 34 \qquad 4 \times 2 = 18$$
$$9 \times 5 = 55 \qquad 8 \times 4 = 42$$

Lori knows her multiplication facts and knows that something is wrong with her calculator. She wants to multiply 9×6. Explain what is wrong with Lori's calculator and tell what she needs to do to get the correct answer. Use words, numbers, or pictures.

34. (Connecticut, Grade 4)

Which number fact goes with this picture?

☺	☺	☺
☺	☺	☺
☺	☺	☺
☺	☺	☺
☺	☺	☺

$5 + 3 =$

$5 \div 3 =$

$5 \times 3 =$

$5 - 3 =$

35. (Texas, Grade 4)

Which number sentence should **NOT** have a 6 in the box?

A. $18 \div \square = 3$

B. $28 \div 4 = \square$

C. $30 \div \square = 5$

D. $36 \div 6 = \square$

For Review

36. An alternative to using expanded form in explaining addition and subtraction algorithms is to write out in words what numerals mean. For example, we could write

232 = 2 hundreds + 3 tens + 2 ones

which we might call **word expanded form.** Then, we could write

$$\begin{array}{rl} 232 = & 2 \text{ hundreds} + 3 \text{ tens} + 2 \text{ ones} \\ +\ 465 = & 4 \text{ hundreds} + 6 \text{ tens} + 5 \text{ ones} \\ \hline & 6 \text{ hundreds} + 9 \text{ tens} + 7 \text{ ones} \\ & = 697. \end{array}$$

Use word expanded form to perform these additions and subtractions.

(a) $\begin{array}{r} 634 \\ +\ 163 \\ \hline \end{array}$ (b) $\begin{array}{r} 247 \\ +\ 332 \\ \hline \end{array}$ (c) $\begin{array}{r} 363 \\ +\ 532 \\ \hline \end{array}$

(d) $\begin{array}{r} 674 \\ -\ 122 \\ \hline \end{array}$ (e) $\begin{array}{r} 725 \\ -\ 413 \\ \hline \end{array}$ (f) $\begin{array}{r} 544 \\ -\ 432 \\ \hline \end{array}$

37. Note that

2 hundreds + 14 tens + 5 ones

= 2 hundreds + 10 tens + 4 tens + 5 ones

= 2 hundreds + 1 hundred + 4 tens + 5 ones

= 3 hundreds + 4 tens + 5 ones.

Now perform these additions and subtractions, using word expanded form.

(a) $\begin{array}{r} 374 \\ +\ 483 \\ \hline \end{array}$ (b) $\begin{array}{r} 264 \\ +\ 327 \\ \hline \end{array}$ (c) $\begin{array}{r} 724 \\ +\ 532 \\ \hline \end{array}$

(d) $\begin{array}{r} 418 \\ -\ 237 \\ \hline \end{array}$ (e) $\begin{array}{r} 367 \\ -\ 249 \\ \hline \end{array}$ (f) $\begin{array}{r} 642 \\ -\ 246 \\ \hline \end{array}$

38. In base five, the word expanded form of 231 is 2 twenty-fives + 3 fives + 1 one. Use the base five word expanded form to perform each of these additions and subtractions. The numerals are already written in base five.

(a) $\begin{array}{r} 213 \\ +\ 131 \\ \hline \end{array}$ (b) $\begin{array}{r} 332 \\ +\ 12 \\ \hline \end{array}$ (c) $\begin{array}{r} 142 \\ +\ 123 \\ \hline \end{array}$

(d) $\begin{array}{r} 231 \\ -\ 130 \\ \hline \end{array}$ (e) $\begin{array}{r} 344 \\ -\ 232 \\ \hline \end{array}$ (f) $\begin{array}{r} 342 \\ -\ 104 \\ \hline \end{array}$

39. Write 495_{ten} and 7821_{ten} in base six

(a) using the positional-value method and Table 3.7, in Section 3.2.

(b) using the short-division-with-remainder method of problem 21 above.

40. Perform these computations entirely in base five. The numerals are already written in base five, so the subscripts are omitted.

(a) $\begin{array}{r} 34 \\ +\ 23 \\ \hline \end{array}$ (b) $\begin{array}{r} 243 \\ +\ 22 \\ \hline \end{array}$ (c) $\begin{array}{r} 312 \\ -\ 21 \\ \hline \end{array}$

(d) $\begin{array}{r} 423 \\ -\ 234 \\ \hline \end{array}$ (e) $\begin{array}{r} 32 \\ \times\ 4 \\ \hline \end{array}$ (f) $\begin{array}{r} 241 \\ \times\ 22 \\ \hline \end{array}$

(g) $23\overline{)344}$ (h) $32\overline{)2341}$

Calvin and Hobbes **by Bill Watterson**

Cooperative Investigation

Magic in Base Three

Alternate performing this trick with a partner until you both can do it with ease.

Materials Needed

A deck of cards.

Discussion

This is an extremely effective card trick that surprises and excites students. They immediately want to know how the trick is done, and this provides great motivation for learning. Doing the trick well in front of a group requires only basic knowledge of base three numeration, a quick eye, and the ability to do just a small amount of arithmetic while keeping up a steady stream of chatter about mind reading and other psychic nonsense. For students, the activity is a great skill and understanding builder with many times the benefit of dull drill and practice. To pull the trick off, proceed as follows.

1	2	3
4	5	6
7	8	9
10	11	12
13	14	15
16	17	18
19	20	21
22	23	24
25	26	27

Step 1. Shuffle a deck of playing cards several times and then count off 27 cards from the top of the deck, setting the remaining 25 cards aside.

Step 2. Ask someone to volunteer as your assistant. With your back turned, ask the assistant to select one of the 27 cards and show it to the audience so that all can see. Then move off to the side where you cannot possibly see individual cards and ask the assistant to lay out the cards face up on the desk in the order shown. Note that the numbers in the drawing indicate only the order in which the cards are to be distributed. Thus, 1 might be the queen of spades, 2 the three of diamonds, and so on.

Step 3. Ask the assistant to tell you which column contains the chosen card and then to scoop up the cards by columns and place them face up in his or her hand. At this point you must be especially aware of the way in which the cards are picked up. Carefully watch where the column containing the selected card is placed. There are three positions in the assistant's hand. The column in question can be placed in

the 0 position—next to the palm
the 1 position—in the middle
the 2 position—away from the palm

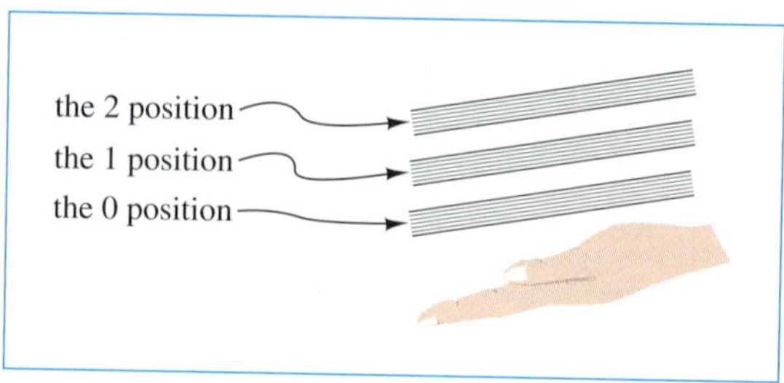

as indicated in the diagram. With this numbering, the first time the cards are picked up determines the units digit in the base three representation of a number n, and you make a mental note of 0, 1, or 2 as appropriate.

Step 4. After the cards are picked up, ask the assistant to turn the assembled pack of all 27 cards face down in his or her hand and then to redistribute them as in Step 2.

Step 5. Again ask the assistant to tell you in which column the selected card lies and then to pick them up as in Step 3. Carefully watch to see where the critical column is placed. This determines the 3s digit of n. Note 0, 3, or 6 as appropriate and add it to the number you are remembering from Step 3.

Step 6. Ask the assistant to repeat Steps 4 and 5 once more, thus determining the 9s digit of n. Note 0, 9, or 18 as appropriate and add it to the sum determined in Step 5. This determines n.

Step 7. Finally, ask the assistant to place the assembled cards face down in his or her hand and count down to the card numbered $n + 1$. Everyone will be surprised to see that this is the chosen card!

But there is more. What if you had 9 cards? 81 cards? 243 cards? Could the trick be modified so that it would still work? Could a similar trick be worked out for other bases? A little experimenting would suggest answers to these questions.

3.5 MENTAL ARITHMETIC AND ESTIMATION

The ability to make accurate estimates and do mental arithmetic is increasingly important in today's society. When buying several items in a store, it is helpful to know before you go to the cashier that you have enough money to pay for your purchase. It is also worthwhile to keep a mental check on the cashier to make sure that you are charged correctly. More importantly, as we increase our use of calculators and computers, it has become essential that we be able to tell if an answer is "about right." Because of their quickness and the neatness with which they display answers, it is tempting to accept as true whatever answer your calculator or computer gives. It is quite easy to enter an incorrect number or press an incorrect operation key and so obtain an incorrect answer. Care must be exercised, and this requires estimation skills.

The One-Digit Facts

It is essential that the basic addition and multiplication facts be memorized since all other numerical calculations and estimations depend on this foundation. At the same time, this should not be rote memorization of symbols. Using a variety of concrete objects, students should actually *experience* the fact that $8 + 7 = 15$, that $9 \cdot 7 = 63$, and so on. Moreover, rather than having children simply memorize the addition and multiplication tables, these should be learned by the frequent and long-term use of manipulatives, games, puzzles, oral activities, and appropriate problem-solving activities. In the same way, children learn the basic properties of the whole numbers which, in turn, can be used to recall some momentarily forgotten arithmetic fact. For example, $7 + 8$ can be recalled as $7 + 7 + 1$, $6 \cdot 9$ can be recalled as $5 \cdot 9 + 9$, and so on. In the same way, the properties of whole numbers along with the one-digit facts form the basis for mental calculation. Here are several strategies for mental calculation.

Easy Combinations

Always look for **easy combinations** in doing mental calculations. The next example shows how this works.

EXAMPLE 3.18 USING EASY COMBINATIONS

Use mental processes to perform these calculations.

(a) $35 + 7 + 15$ **(b)** $8 + 3 + 4 + 6 + 7 + 12 + 4 + 3 + 6 + 3$
(c) $25 \cdot 8$ **(d)** $4 \cdot 99$ **(e)** $57 - 25$ **(f)** $47 \cdot 5$

Solution

(a) Using the commutative and associative properties, we have

$$35 + 7 + 15 = 35 + 5 + 10 + 7 = 40 + 10 + 7 = 50 + 7 = 57$$

Think
35, 40, 50, 57
The answer is 57.

(b) Note numbers that add to 10 or multiples of 10.

$$8 + 3 + 4 + 6 + 7 + 12 + 4 + 3 + 6 + 3 = 56$$

Think
20, 30, 40, 50, 53, 56
The answer is 56.

(c) $25 \cdot 8 = 25 \cdot 4 \cdot 2 = 100 \cdot 2 = 200$

Think
25, 100, 200
The answer is 200.

(d) $4 \cdot 99 = 4(100 - 1) = 400 - 4 = 396$

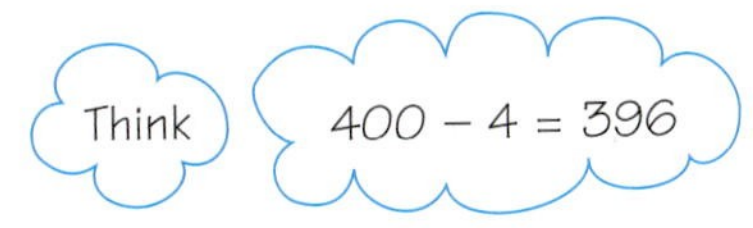
Think
400 – 4 = 396

(e) $57 - 25 = 50 - 25 + 7 = 25 + 7 = 32.$

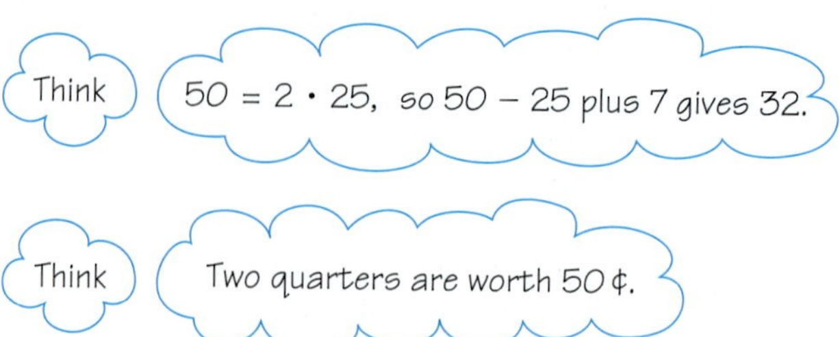
Think
50 = 2 • 25, so 50 − 25 plus 7 gives 32.
Think
Two quarters are worth 50¢.

(f) $47 \cdot 5 = 47 \cdot 10 \div 2 = 470 \div 2 = 235.$

Think
47, 470, 235

Adjustment

In parts (d) and (e) of Example 3.18, we made use of the fact that 99 and 57 are close to 100 and 50, respectively. This is an example of adjustment. **Adjustment** simply means that we modify numbers in a calculation to minimize the mental effort required.

EXAMPLE 3.19 USING ADJUSTMENT IN MENTAL CALCULATION

Use mental processes to perform these calculations.

(a) $57 + 84$ **(b)** $83 - 48$ **(c)** $286 + 347$
(d) $493 \cdot 7$ **(e)** $2646 \div 9$ **(f)** $639 \div 7$

Solution

(a)
$$\begin{aligned} 57 + 84 &= (57 + 3) + (84 - 3) \\ &= 60 + 81 = 60 + 80 + 1 \\ &= 140 + 1 = 141 \end{aligned}$$

Think
57 + 84, 60 + 81, 140, 141

(b) $83 - 48 = (83 + 2) - (48 + 2) = 85 - 50 = 35$

Think
83 − 48, 85 − 50, 35

(c) $286 + 347 = (286 + 14) + (347 - 14)$
$= 300 + 300 + 47 - 14$
$= 600 + 33 = 633$

Think 300, 647 − 14, 633

(d) $493 \cdot 7 = (500 - 7) \cdot 7 = 3500 - 49 = 3451$

Think (500 − 7) · 7, 3500 − 49, 3451

(e) $2646 \div 9 = (2700 - 54) \div 9 = 300 - 6 = 294$

Think ÷ 9, 2700 − 54, 300 − 6, 294

(f) $639 \div 7 = (630 + 7 + 2) \div 7 = 90 + 1 \text{ R } 2 = 91 \text{ R } 2$

Think ÷ 7, 630 + 7 + 2, 90 + 1 R 2, 91 R 2

Working from Left to Right

Because it tends to reduce the amount one has to remember, many expert mental calculators **work from left to right** rather than the other way around as in most of our standard algorithms.

EXAMPLE 3.20 WORKING FROM LEFT TO RIGHT

Use mental processes to perform these calculations.

(a) $352 + 647$ **(b)** $739 - 224$ **(c)** $4 \cdot 235$

Solution

(a) $352 + 647 = (300 + 50 + 2) + (600 + 40 + 7)$
$= (300 + 600) + (50 + 40) + (2 + 7)$
$= 900 + 90 + 9 = 999$

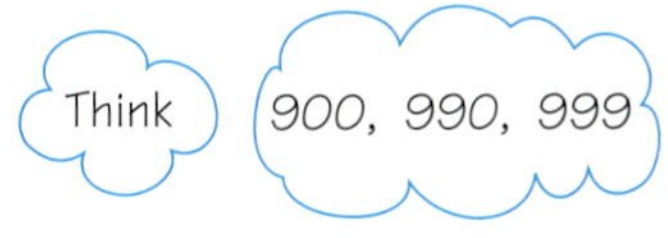

(b) $739 - 224 = (700 + 30 + 9) - (200 + 20 + 4)$
$= (700 - 200) + (30 - 20) + (9 - 4)$
$= 500 + 10 + 5 = 515$

Think 500, 510, 515

(c) $4 \cdot 235 = 4(200 + 30 + 5)$
$= 800 + 120 + 20$
$= 920 + 20$
$= 940$

Left-to-right methods often combine nicely with an understanding of positional notation to simplify mental calculation. Since $4200 = 42 \cdot 100$, for example, we might compute the sum

$$\begin{array}{r} 3700 \\ 900 \\ 2800 \\ +\ 5600 \\ \hline \end{array}$$

by thinking of

$$\begin{array}{r} 37 \\ 9 \\ 28 \\ +\ 56 \\ \hline \end{array}$$

Then, working from left to right, we think

Rounding

Often we are *not* interested in exact values. This is certainly true when *estimating* the results of numerical calculations, and it is often the case that exact values are actually unobtainable. What does it mean, for example, to say that the population of California in 2000 was 28,874,293? Even if this is supposed to be the actual count on a given day, it is almost surely in error because of the sheer difficulty in conducting a census. How many illegal immigrants were not counted? How many homeless people? How many transients? In gross terms, it is probably accurate to say that the population of California was approximately 29,000,000, or 29 million people. To obtain this figure, we **round** to the nearest million. This is accomplished by considering the digit in the hundred thousands position. If this digit is 5 or more, we increase the digit in the millions position by 1 and replace all the digits to the right of this position by 0s. If the hundred thousands digit is 4 or less, we leave the millions digit unchanged and replace all the digits to its right by 0s.

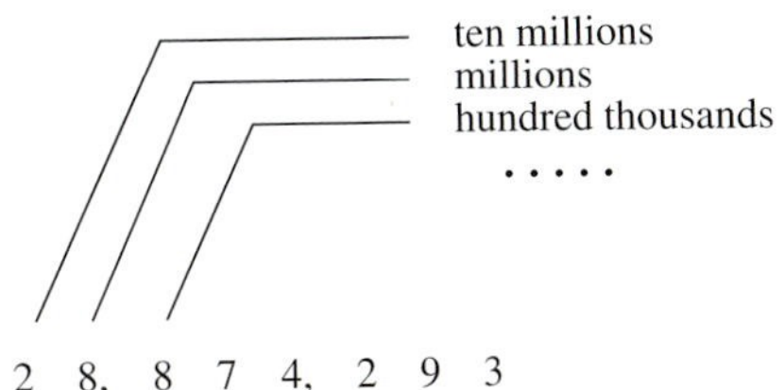

Rounding Using the 5-Up Rule

(i) Determine to which position you are rounding.
(ii) If the digit to the right of this position is 5 or more, add 1 to the digit in the position to which you are rounding. Otherwise, leave the digit unchanged.
(iii) Replace with 0s all digits to the right of the position to which you are rounding.

EXAMPLE 3.21 USING THE 5-UP RULE TO ROUND WHOLE NUMBERS

Round 27,250 to the position indicated.

(a) the nearest ten thousand
(b) the nearest thousand
(c) the nearest hundred
(d) the nearest ten

Solution

(a) The digit in the ten thousands position is 2. Since the digit to its right is 7 and $7 > 5$, we add 1 to 2 and replace all digits to the right of the rounded digit with 0s. This gives 30,000.
(b) This time 7 is the critical digit, and 2 is the digit to its right. Thus, to the nearest thousand, 27,250 is rounded to 27,000.
(c) Here, 2 is the digit in the hundreds position, and 5 is to its right. Thus, 27,250 is rounded to 27,300 to the nearest hundred.
(d) This time, 5 is the critical digit, and 0 is the digit on its right. Thus, no change need be made, because 27,250 is already rounded to the nearest ten.

from The NCTM Principles and Standards

Estimation serves as an important companion to computation. It provides a tool for judging the reasonableness of calculator, mental, and paper-and-pencil computations. However, being able to compute exact answers does not automatically lead to an ability to estimate or judge the reasonableness of answers, as Reys and Yang (1998) found in their work with sixth and eighth graders. Students in grades 3–5 will need to be encouraged to routinely reflect on the size of an anticipated solution. Will 7×18 be smaller or larger than 100? If 3/8 of a cup of sugar is needed for a recipe and the recipe is doubled, will more or less than one cup of sugar be needed? Instructional attention and frequent modeling by the teacher can help students develop a range of computational estimation strategies including flexible rounding, the use of benchmarks, and front-end strategies. Students should be encouraged to frequently explain their thinking as they estimate. As with exact computation, sharing estimation strategies allows students to access others' thinking and provides many opportunities for rich class discussions.

SOURCE: Reprinted with permission from Principles and Standards for School Mathematics, *pages 155–156.* *Standards are listed with the permission of the National Council of Teachers of Mathematics (NCTM). NCTM does not endorse the content or validity of these alignments.*

HIGHLIGHT *from* HISTORY

Emmy Noether (1882–1935)

Emmy Noether was born in Erlangen, Germany, to a family noted for mathematical talent. Much of her life was spent at the University of Göttingen, exploring, teaching, and writing about algebra. This university—where Carl Gauss had taught a century earlier—was the first in Germany to grant a doctoral degree to a woman. Yet Noether met with frustrating discrimination there. For many years she was denied appointment to the faculty; finally she was given an impressive title as "extraordinary professor"—with no salary. But her abilities overcame the obstacles that daunted many other women in mathematics. Her work in the 1920s brought invitations to lecture throughout Europe and in Moscow. In 1933, as the Nazi party came to power in Germany, Noether met with persecution not only as a woman but as an intellectual, a Jew, a pacifist, and a political liberal. She fled to the United States, where she taught and lectured at Bryn Mawr and Princeton until her death in 1935.

"How can it be allowed that a woman become . . . a professor . . . ? What will our soldiers think when they return to the University and find that they are expected to learn at the feet of a woman?"
—Faculty member at Göttingen, in 1918

". . . for two of the most significant sides of the theory of relativity, she gave at that time [1919] the genuine and universal mathematical formulation."
—Hermann Weyl, colleague at Göttingen

"In the judgement of the most competent living mathematicians, Fräulein Noether was the most significant creative mathematical genius thus far produced since the higher education of women began. In the realm of algebra . . . she discovered methods which have proved of enormous importance . . . ."
—Albert Einstein, 1935

"She was the most creative abstract algebraist in the world."
—Eric Temple Bell in *Men of Mathematics*

SOURCE: Biographical information is from Lynn Osen, Women in Mathematics *(MIT Press, 1974). Quotations are cited in that source, original references including anonymous faculty member, Constance Reid,* Hilbert *(Springer-Verlag, 1970, p. 143); Weyl,* Scripta Mathematica, *Vol. 3, 1935; Einstein,* New York Times, *May 4, 1935;* Men of Mathematics *(Simon and Schuster, 1965, p. 261). From Mathematics in Modules,* Intermediate Algebra, *A5, Teachers Edition. Reprinted by permission.*

Estimation

The ever increasing use of calculators and computers makes it essential that students develop skill at estimation. How large an answer should I expect? Is this about the right answer? These are questions students should ask and be able to answer. And they can be answered reasonably effectively on the basis of a good understanding of one-digit arithmetic facts and positional notation. To be effective, the estimator must also be adept at mental arithmetic.

Approximating by Rounding

Rounding is often used in finding estimates. The advantage of **approximating by rounding** is that it gives a single estimate that is reasonably close to the desired answer. The idea is to round the numbers involved in a calculation to the position of the leftmost one or two digits and to use these rounded numbers in making the estimate. For example, consider $467 + 221$. Rounding to the nearest hundred, we have

$$467 \approx 500 \quad \text{and} \quad 221 \approx 200,$$

where we use the symbol $\approx$ to mean "is approximated by." Thus, we obtain the approximation

$$467 + 221 \approx 500 + 200 = 700.$$

The actual answer is 688, so the approximation is reasonably good. Rounding to the nearest ten usually gives an even closer approximation if it is needed. Thus,

$$467 \approx 470, \qquad 221 \approx 220,$$

and $470 + 220$ gives the very close approximation 690.

EXAMPLE 3.22 APPROXIMATING BY ROUNDING

Round to the leftmost digit to find approximate answers to each of these calculations. Also, compute the exact answer in each case.

(a) $681 + 241$ **(b)** $681 - 241$
(c) $681 \cdot 241$ **(d)** $57{,}801 \div 336$

Solution To the nearest hundred, $681 \approx 700$ and $241 \approx 200$. Also, $57{,}801 \approx 60{,}000$ and $336 \approx 300$. Using these, we obtain the approximations shown.

	Approximation	Exact Answer
(a)	$681 + 241 \approx 700 + 200 = 900$	$681 + 241 = 922$
(b)	$681 - 241 \approx 700 - 200 = 500$	$681 - 241 = 440$
(c)	$681 \cdot 241 \approx 700 \cdot 200 = 140{,}000$	$681 \cdot 241 = 164{,}121$
(d)	$57{,}801 \div 336 \approx 60{,}000 \div 300 = 200$	$57{,}801 \div 336 = 172 \text{ R } 9$

Here the quotient is approximately 200.

School Book Page

Using Estimates

Chapter 2 Lesson 14

Estimating Sums and Differences

You Will Learn
how to estimate sums and differences of whole numbers and decimals

Vocabulary
front-end estimation
a way to estimate a sum by adding the first digit of each addend and adjusting the result based on the remaining digits

Learn

John collected 1,004,024 pennies. They were worth $10,040.24 and weighed almost 7,000 lb.

This table shows the weights and values of some pennies.

Weights and Values of Pennies		
Number	(lb)	($)
144	1	$1.44
288	2	$2.88
576	4	$5.76
1,152	8	$11.52

John Tregembo, from Plymouth, Michigan, collected his pennies from 1982 to 1995.

If you have 6 lb of pennies, about how many pennies do you have? You can find 6 pounds of pennies by subtracting 2 from 8 or by adding 4 and 2. You can estimate by rounding or **front-end estimation**.

Math Tip
You can use ≈ to show about or "approximately equal."

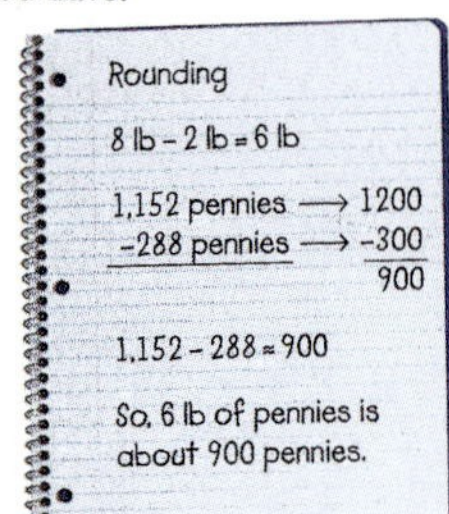

Rounding
8 lb − 2 lb = 6 lb
1,152 pennies ⟶ 1200
−288 pennies ⟶ −300
900
1,152 − 288 ≈ 900
So, 6 lb of pennies is about 900 pennies.

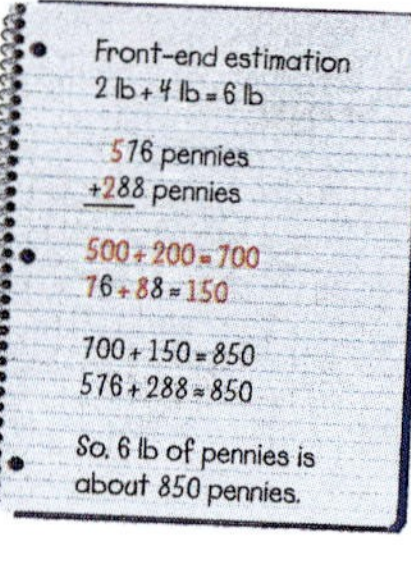

Front-end estimation
2 lb + 4 lb = 6 lb
576 pennies
+288 pennies
500 + 200 = 700
76 + 88 ≈ 150
700 + 150 = 850
576 + 288 ≈ 850
So, 6 lb of pennies is about 850 pennies.

Talk About It

Why was one estimate higher than the other?

Check

Estimate each sum or difference.

1. 520 + 375
2. 884 − 406
3. $7.07 − 3.59
4. 621 + 432 + 561
5. $9.82 + 3.12

6. **Reasoning** If you round $9.75 and $25.82 to the nearest dollar, will your estimated sum be more or less than the actual sum?

82 Chapter 2 • Whole Numbers and Decimals

SOURCE: From Scott Foresman–Addison Wesley Math, *Grade 5, p. 82, by Randall I. Charles et al. Copyright © 2002 Pearson Education, Inc. Reprinted with permission.*

Questions for the Teacher

1. The above rounding estimate for $1152 - 288$ was obtained by rounding to the nearest hundred as opposed to rounding to the leftmost digit. Which method seems preferable here? Explain.

2. Using "front-end estimation" twice in the addition problem in the notebook on the right above gives an even closer estimate than rounding. Explain why this is likely to be the case.

Problem Set 3.5

Understanding Concepts

1. Calculate the given expression mentally, using easy combinations. Write a sequence of numbers indicating intermediate steps in your thought process. The first one is done for you.

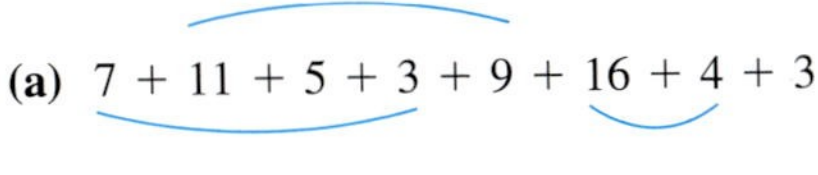

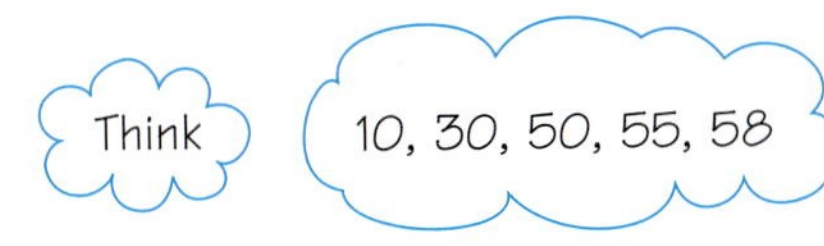

   (b) 6 + 9 + 17 + 5 + 8 + 12 + 3 + 6
   (c) 27 + 42 + 23 (d) 47 − 23
   (e) 48 · 5 (f) 21,600 ÷ 50

2. Calculate mentally using adjustment. Write down a sequence of numbers indicating intermediate steps in your thought process. The first one is done for you.
   (a) 78 + 64

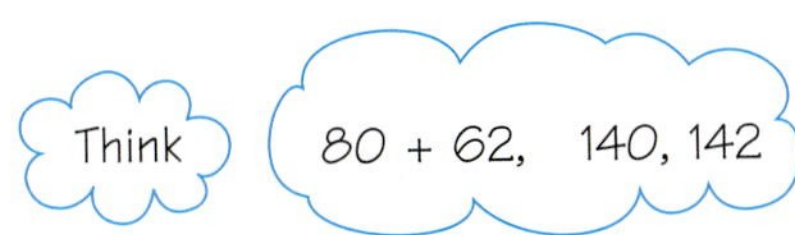

   (b) 294 + 177 (c) 306 − 168 (d) 294 − 102
   (e) 479 + 97 (f) 3493 ÷ 7 (g) 412 · 7

3. Perform these calculations mentally from left to right. Write down a sequence of numbers indicating intermediate steps in your thought process.
   (a) 425 + 362 (b) 363 + 274 (c) 572 − 251
   (d) 764 − 282 (e) 3 · 342
   (f) 47 + 32 + 71 + 9 + 26 + 32

4. Round 235,476 to the
   (a) nearest ten thousand.
   (b) nearest thousand.
   (c) nearest hundred.

5. Round each of these to the position indicated.
   (a) 947 to the nearest hundred.
   (b) 850 to the nearest hundred.
   (c) 27,462,312 to the nearest million.
   (d) 2461 to the nearest thousand.

6. Rounding to the leftmost digit, calculate approximate values for each of these:
   (a) 478 + 631 (b) 782 + 346 (c) 678 − 431
   (d) 257 · 364 (e) 7403 · 28 (f) 28,329 ÷ 43
   (g) 71,908 ÷ 824

7. Compute exact answers to parts (a) through (g) of problem 6.

8. Rounding to the nearest thousand and using mental arithmetic, estimate each of these sums and differences.

   (a)
   17,281
   6564
   12,147
   2481
   + 13,671

   (b)
   2734
   3541
   2284
   3478
   + 7124

   (c)
   28,341
   942
   2431
   4716
   + 12,824

   (d)
   4270
   − 1324

   (e)
   21,243
   − 7824

   (f)
   37,481
   − 16,249

   (g) Use your calculator to compute the exact value of the answers to parts (a) through (f).

9. Using rounding to the leftmost digit, estimate these products.
   (a) 2748 · 31 (b) 4781 · 342 (c) 23,247 · 357
   (d) Use your calculator to determine the exact values of the products in parts (a) through (c).

10. Use rounding to the leftmost digit to estimate the quotient in each of the following.
    (a) 29,342 ÷ 42 (b) 7431 ÷ 37
    (c) 79,287 ÷ 429
    (d) Use your calculator to determine the exact values of the quotients in parts (a) through (c).

Teaching Concepts

11. (a) Would rounding to the leftmost digit give a very good estimate of this sum? Why or why not?

    1478
    2395
    1492
    + 5481

    (b) What would you suggest that your students do to obtain a more accurate estimate?
    (c) Compute the accurate answer to the addition in part (a).

12. How would you respond to one of your students who rounded 27,445 to the nearest thousandth as follows: 27,445; 27,450; 27,500; 28,000?

13. Read the Number and Operations Standard for Grades 3–5, pages 148–156, from NCTM's *Principles and Standards for School Mathematics,* and write a short critique of the standard, emphasizing your own reaction. How do these recommendations compare with your own school experience?

Thinking Critically

14. Theresa and Fontaine each used their calculators to compute 357 + 492. Fontaine's answer was 749 and Theresa's was 849. Who was most likely correct? In two brief sentences, tell how estimation can help you decide whose answer was probably correct.

15. Use rounding to estimate the results of each of the following.

    (a) $\dfrac{452 + 371}{281}$ (b) $\dfrac{3 \cdot 271 + 465}{74 + 9}$

    (c) $\dfrac{845 \cdot 215}{416}$

    (d) Use your calculator to determine the exact answer to each problem in parts (a) through (c).

16. Sometimes the last digits of numbers can help you decide if calculator computations are correct.

    (a) Given that one of 27,453; 27,587; or 27,451 is the correct result of multiplying 283 by 97, which answer is correct?

    (b) In two brief sentences, tell how consideration of last digits helped you answer part (a).

17. Since 25,781; 24,323; 26,012; and 25,243 are all about the same size, about how large is their sum? Explain briefly.

Thinking Cooperatively

Solve each of the next three problems as a group, discussing the ideas and strategies that eventually led to correct solutions. Arrive at a solution agreed upon by all the members of your group.

18. Note that $(2)(4678) = 9356$. Place parentheses in each of the following strings of digits to make the equality true.

    (a) 2 4 6 7 8 = 16,272

    (b) 2 4 6 7 8 = 19,188

    (c) 2 4 6 7 8 = 19,736

    (d) 2 4 6 7 8 = 12,864

19. Place parentheses and plus signs in each string of digits to make these equalities true. (*Hint:* $(88) + (88) + (88) + (88) = 352$.)

    (a) 8 8 8 8 8 8 8 8 = 136

    (b) 8 8 8 8 8 8 8 8 = 17,776

    (c) 8 8 8 8 8 8 8 8 = 928

    (d) 8 8 8 8 8 8 8 8 = 9064

    (e) 8 8 8 8 8 8 8 8 = 8920

20. Place parentheses and division signs in each of these strings of digits so that the equalities are true. Remember that it is *not* generally the case that $(a \div b) \div c = a \div (b \div c)$. (*Hint:* $(844 \div (4 \div 2)) \div (2 \div 1) = 211$.)

    (a) 8 4 4 4 2 2 1 = 844,422

    (b) 8 4 4 4 2 2 1 = 42,221

    (c) 8 4 4 4 2 2 1 = 42

    (d) 8 4 4 4 2 2 1 = 38 R 46

Making Connections

21. Utah, Colorado, New Mexico, and Arizona are the only four states in the United States that meet at a common corner. The areas of the four states are, respectively, 82,168 square miles, 103,730 square miles, 121,365 square miles, and 113,642 square miles.

    (a) Mentally estimate the combined area of the four states, writing down a sequence of numerals to indicate your thought process.

    (b) Compute the actual sum of the areas in part (a).

22. At the time it was published, an atlas gave the population of Colorado as 3,849,400.

    (a) Approximately how many people per square mile lived in Colorado at that time? Write down a sequence of numerals to indicate your thought process.

    (b) Determine the answer to part (a) correct to the nearest unit.

23. The same atlas that provided the data in problem 22 gave the area and population of Delaware as 1955 square miles and 731,900 people. About how many times as many people per square mile lived in Delaware as in Colorado at the time the atlas was published? Write down a sequence of numerals to indicate your thought process.

24. While grocery shopping with $40, you buy the following items at the price listed:

2 gallons of milk	$2.29 a gallon
1 dozen eggs	$1.63 per dozen
2 rolls of paper towels	$1.21 per roll
1 5-pound pork roast	$1.98 per pound
2 boxes of breakfast cereal	$3.19 each
1 azalea	$9.95 each

(a) About how much will all of this cost?

(b) If you don't buy the azalea, about how much change should you receive?

From State Student Assessments

25. (Connecticut, Grade 4)
Joe needs to subtract 319 from 799. Which of the following would be BEST for Joe to use to ESTIMATE the difference?

700 − 300

700 − 400

800 − 300

800 − 400

26. (Washington State, Grade 4)
Solve 52 × 40.

A. 2800 B. 2080 C. 280 D. 208

27. (Washington State, Grade 4)
Estimate the answer. Show how you found your estimate.

$$9\overline{)820}$$

28. (Washington State, Grade 4)
Solve the problem.

```
 254
 +67
----
```

A. 211 B. 221 C. 311 D. 321

For Review

29. In a college mathematics class, all the students are also taking anthropology, history, or psychology, and some of the students are taking two, or even all three, of these courses. If (i) 40 students are taking anthropology, (ii) 11 students are taking history, (iii) 12 students are taking psychology, (iv) 3 students are taking all three courses, (v) 6 students are taking anthropology and history, and (vi) 6 students are taking psychology and anthropology,

(a) how many students are taking only anthropology?

(b) how many students are taking anthropology or history?

(c) how many students are taking history and anthropology, but not psychology?

30. Fill in the missing digits in each of these addition problems.

```
(a)      _742    (b)     2341    (c)    _21_
          41_            4_30            _0_
         69_3            1___            41_
     +   2_18       +    3_18         + 771_
     --------       ---------         ------
       _2,818          _3,100           9666
```

31. Fill in the missing digits in each of these subtraction problems.

```
(a)    27_4    (b)    7_01    (c)    _22_
     - _64_         - 192_         - 2333
     ------         ------         ------
         91           _8_9           1__9
```

32. Fill in the missing digits in each of these multiplication and division problems.

```
(a)      34_    (b)      3___7    (c)          1_
     ×    __        ×       1_           _61)3_24
     -------        ----------                6_
         6_4             ___296              ----
      ___8             _7___                 121_
     -------        ----------              ----
      __,57_          _6_,6__                 __0
```

Cooperative Investigation

Multiplication Tic-Tac-Toe

Materials Needed

1. A calculator for each student.
2. A multiplication tic-tac-toe handout for each pair of students.

Procedure

This is like regular tic-tac-toe except that on each play the player chooses two numbers from the given list and places his or her mark (X or O) on the square containing the product of the numbers chosen. Thus, in part (a), if the first player's symbol is X and 11 and 23 are chosen on the first play, an X is placed on 253 on the diagram, and the notation 11 · 23 is written in the square. As usual, the first player to get three Xs or Os in a row wins the game. If a player chooses a product not in a square, (s)he loses that turn.

(a) 11, 12, 15, 19, 23

345	132	285
228	209	437
253 $11 \cdot 23$ (crossed out)	276	180

(b) 13, 14, 17, 19, 21

221	266	273
247	238	182
323	399	357

(c) 7, 23, 341, 2706, 4123

94,829	2387	11,156,838
1,405,943	7843	18,942
28,861	62,238	922,746

Parts (d) and (e) are the same as parts (a) through (c) except that the numbers in the squares are the approximations of the true products, obtained by rounding each number to the leftmost position.

(d) 23, 27, 36, 47, 55

600 $\approx 23 \cdot 27$ (crossed out)	800	1800
2000	1500	1200
1000	2400	3000

(e) 143, 254, 361, 2391, 2511

40,000	800,000	300,000
600,000	200,000	1,200,000
120,000	6,000,000	900,000

(f) For parts (a) through (c), discuss how looking at the last digits in the numerals can help you play the game.

(g) For part (c), discuss how estimation can help you play the game.

3.6 GETTING THE MOST OUT OF YOUR CALCULATOR

Strategies

- Use algorithmic thinking.

It's a safe bet that almost everyone who reads these words is the owner of at least one calculator. It's also a safe bet that almost every reader knows how to add, subtract, multiply, and divide with his or her calculator. However, most calculators have useful features that are not well understood and are not employed by some users. In this section, we discuss some of these special features.

First, we note that there are three common types of logic that determine how your calculator operates—arithmetic (a´-rith-me´-tic), algebraic, and reverse Polish. **Reverse Polish notation** is a powerful system used on sophisticated scientific calculators and is not suitable for elementary school students. Machines utilizing **arithmetic logic** are too simplistic and again are not recommended for classroom use. Machines using **algebraic logic** (such machines are easily identified by the presence of left and right parenthesis keys) are the most appropriate for student use and have built-in features that make calculations easier and more natural. Of course, your instruction manual is the basic source of information about the operation of your particular calculator. Currently, a number of calculators are designed specifically for the elementary and middle school classroom. Among other things, these machines use algebraic logic, do arithmetic with both fractions and decimals, and do integer division with remainders. Most of the discussion here applies equally well to other algebraic machines, but your instruction manual is the final arbiter. We will also refer briefly to the graphing calculator (see Appendix C), which is becoming increasingly useful to both mathematics and science teachers in the upper elementary grades.

from **The NCTM Principles and Standards**

Electronic technologies—calculators and computers—are essential tools for teaching, learning, and doing mathematics. They furnish visual images of mathematical ideas, they facilitate organizing and analyzing data, and they compute efficiently and accurately. They can support investigation by students in every area of mathematics, including geometry, statistics, algebra, measurement, and number. When technological tools are available, students can focus on decision making, reflection, reasoning, and problem solving.

Students can learn more mathematics more deeply with the appropriate use of technology (Dunham and Dick 1994; Sheets 1993: Boersvan Oosterum 1990; Rojano 1996; Groves 1994). Technology should not be used as a replacement for basic understandings and intuitions; rather, it can and should be used to foster those understandings and intuitions. In mathematics-instruction programs, technology should be used widely and responsibly, with the goal of enriching students' learning of mathematics.

The existence, versatility, and power of technology make it possible and necessary to reexamine what mathematics students should learn as well as how they can best learn it. In the mathematics classrooms envisioned in *Principles and Standards,* every student has access to technology to facilitate his or her mathematics learning under the guidance of a skillful teacher.

SOURCE: Reprinted with permission from Principles and Standards for School Mathematics, *pages 24 and 25.* *Standards are listed with the permission of the National Council of Teachers of Mathematics (NCTM). NCTM does not endorse the content or validity of these alignments.*

Priority of Operations

Suppose you enter

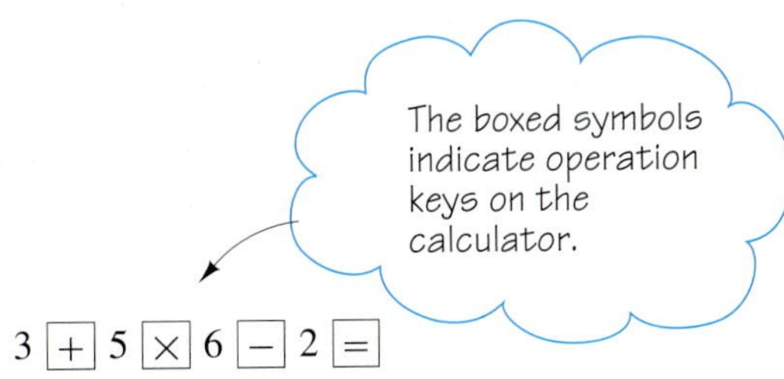

The boxed symbols indicate operation keys on the calculator.

3 [+] 5 [×] 6 [−] 2 [=]

into your calculator. A calculator with arithmetic logic would perform each operation in exactly the order entered and would give the answer 46; that is,

$$3 + 5 = 8$$
$$8 \times 6 = 48$$
$$48 - 2 = 46.$$

In contrast, a calculator with algebraic logic multiplies and divides *before* adding and subtracting. Also, if there are pending operations of equal priority, these calculators execute them from left to right. Since these priorities are those generally accepted in mathematics, algebraic machines correspond well to the standard rules and properties of arithmetic. Thus, the above sequence of entries in an algebraic calculator gives the answer 31, obtained as follows:

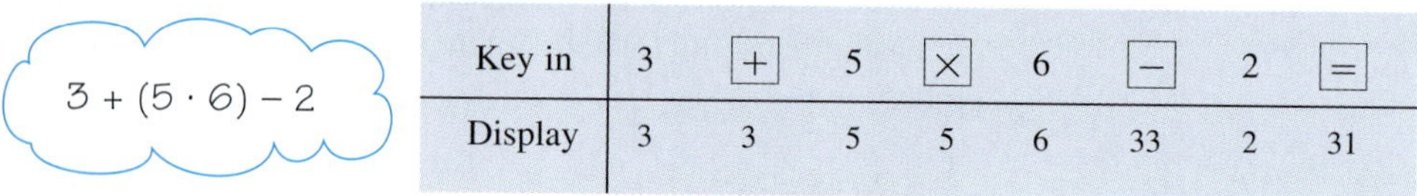

Key in	3	+	5	×	6	−	2	=
Display	3	3	5	5	6	33	2	31

When − was pressed, the calculator *first* multiplied 5 × 6 to get 30 and *then* completed the pending addition to 3, giving 33. When 2 and = were pressed, it then subtracted the 2 to give 31.

While we have not yet discussed the use of all of these keys, the priority of operations for machines with algebraic logic is shown in Table 3.8.

TABLE 3.8 Priority of Operations on Calculators with Algebraic Logic

Priority	Keys	Explanation
1	( )	Operations in parentheses are performed before other operations.
2	x^2 √	Operations performed on a single number.
3	^	Exponentiation.
4	× ÷	Multiplications and divisions are completed before additions and subtractions.
5	+ −	Additions and subtractions are completed last.
6	=	Terminates a calculation.

EXAMPLE 3.23 UNDERSTANDING THE ORDER OF CALCULATOR OPERATIONS

(a) Indicate what should be entered into a calculator with algebraic notation to compute

$$27 \div 3 + 24 \cdot 4.$$

(b) Actually key the sequence in part (a) into your calculator and complete the computation.

Solution

(a) Because of the priority of operations, it is necessary to key in only the following:

27 [÷] 3 [+] 24 [×] 4 [=].

(b) On a machine with algebraic logic, the calculator makes the following sequence of calculations:

$$27 \div 3 = 9, 24 \cdot 4 = 96, 9 + 96 = 105.$$

Using the Parentheses Keys [(] and [)]

It is possible to override the priority of operations built into the calculator by use of the parentheses keys. Suppose you want to compute

$$(789 + 364) \cdot (863 + 939).$$

This is accomplished by keying the sequence

[(] 789 [+] 364 [)] [×] [(] 863 [+] 939 [)] [=]

into the calculator. The desired answer is 2,077,706. If we keyed in

789 [+] 364 [×] 863 [+] 939 [=]

(the same entries, but omitting the parentheses), the calculator would show 315,860, determined by computing the product of 364 and 863 and then adding 789 and 939 in that order.

EXAMPLE 3.24 USING THE [=] AND PARENTHESES KEYS

Perform this computation on your calculator:

$$216 \div (3 + 24) \cdot 4.$$

Solution

Key in 216 [÷] [(] 3 [+] 24 [)] [×] 4 [=] to obtain 32. Remember that divide and multiply are operations of the same priority. Thus, the calculator first computes the sum in parentheses, divides 216 by that sum, and multiplies the result by 4. Using clearer notation, what the calculator is computing is

$$(216 \div (3 + 24)) \cdot 4$$

but the built-in priority of operations makes it unnecessary to key the expression into the calculator this way.

Before proceeding, several observations should be made.

1. It is always a good idea to begin each calculation by pressing the [ON] key. This key turns the calculator on. Equally importantly, it clears all preceding data from all parts of the calculator so that present work will not be rendered incorrect by the presence of unexpected and unwanted information from a preceding calculation.
2. Pressing the [=] key causes the calculator to complete all entered calculations up to that point. This can be used on occasion to simplify calculation. For example, to compute $(29 + 37) \div 11$, one could use parentheses or, alternatively, enter

   29 [+] 37 [=] [÷] 11 [=]

to obtain the correct answer of 6. As a check, key in

[(] 29 [+] 37 [)] [÷] 11 [=]

to see that you obtain the same answer.

3. Parentheses must always be entered in pairs; that is, for each left parenthesis entered, a right parenthesis must be entered later. Otherwise, when [=] is entered, an error message will appear in the display to inform you of an error in entering parentheses.

If, in the midst of a calculation, you obtain an error message of any kind, you must press the [ON] key to clear the machine. Then repeat the calculation, being careful to restructure your procedure to avoid the previous error.

It may be useful to consider one more example concerning priority of operations.

EXAMPLE 3.25 PRIORITIZING OPERATIONS

Compute $\frac{323 - 4 \cdot 38}{19}$.

Solution If you enter

323 [−] 4 [×] 38 [÷] 19 [=]

into the calculator, you obtain the incorrect answer 315. Because of the priority of operations, this sequence of commands computes $4 \cdot 38 \div 19$ and subtracts this from 323. However, the problem requires that the entire quantity $323 - 4 \cdot 38$ be divided by 19. This can be accomplished in several ways, but perhaps the following are easiest. Using parentheses, enter

[(] 323 [−] 4 [×] 38 [)] [÷] 19 [=]

Using [=] twice, enter

323 [−] 4 [×] 38 [=] [÷] 19 [=]

Key in each of these sequences and note that each gives the correct answer of 9.

Using the [x^2] and [√] Keys

Each of these keys causes the calculator to perform an operation on a single number. To compute

$11^2 = 121$

11^2, enter 11 [x^2].

Do each of these for practice.

To compute

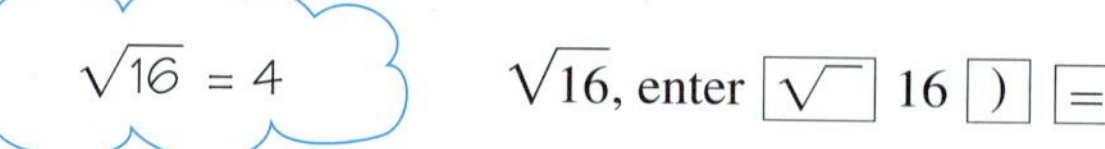

$\sqrt{16}$, enter [√] 16 [)] [=]

Note the order in each case.

since, when the [√‾] button is pressed, the display actually shows √‾ (, and this requires depressing the right parenthesis key and the equals key to complete the calculation. On some calculators, to compute $\sqrt{16}$, you must enter [√‾] [(] 16 [)] [=], and, on others, you simply enter 16 [√‾]. In what follows, we will use the notation first discussed above, but you should be sure to consult the instruction booklet for your calculator to see which you should use. Alternatively, just trying each possibility on your calculator should make your choice clear.

In each case, the operation is carried out immediately; it is not necessary to use the [=] key to complete the computation.

Using the [^] Key

To compute 5^3 we use the [^] key. Here, we have to tell the calculator what number we want to raise to a power and what power to raise it to. For 5^3, we enter 5 [^] 3 [=] to obtain 125. Check to see that the entry strings

2 [^] 5 [=]

and

2 [×] 2 [×] 2 [×] 2 [×] 2 [=]

both give 32 as the value of 2^5.

Note that, on some calculators, exponentiation is accomplished by use of a [y^x] key. On these calculators, 2^5 would be calculated by entering 2 [y^x] 5.

Using the Memory Keys—[M], [M+], [M−], [MR], and [R]

Most calculators with algebraic logic have a memory accessed with keys like [M], [M+], and [M−]. [M] places the number in the display in the memory while still keeping it in the display. The [M+] and [M−] keys respectively add the number in the calculator display to or subtract the number in the calculator display from the number in the memory while maintaining the number being added or subtracted in the display. If nothing has been placed in the memory, it will be assumed to contain a 0. Also, when a number has been placed in the memory, the display will show a small "m" to remind you that the memory is not empty. Keys like [MR], [RM], and [R], recall what is in the memory and show it in the display. These capabilities are useful in elementary ways but, with a little imagination, serve more surprising and powerful purposes. In the text that follows, we will use the notations [M+], [M−], and [MR], even though your calculator may use other designations and may not have all the possible capabilities of the [M+] and [M−] keys.

EXAMPLE 3.26 USING THE [M+] AND [MR] KEYS

Compute the quotient

$$\frac{\sqrt{158{,}404} - 200}{6019 - 5986}$$

without using parentheses.

Solution The idea is to compute 6019 − 5986 and store it in the memory. Then compute $\sqrt{158{,}404} - 200$ and divide the result by the number stored in the memory using the division, ÷, and recall memory, MR, keys. We enter the following to obtain the correct answer of 6:

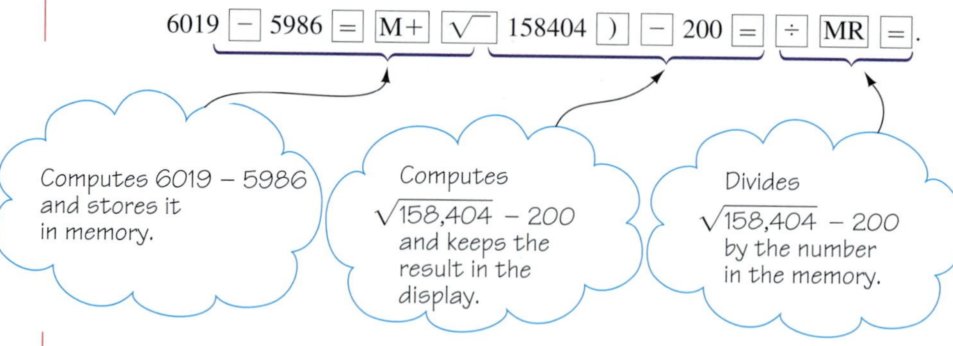

6019 − 5986 = M+ √ 158404) − 200 = ÷ MR =.
Computes 6019 − 5986 and stores it in memory.
Computes √158,404 − 200 and keeps the result in the display.
Divides √158,404 − 200 by the number in the memory.

EXAMPLE 3.27

USING THE M+ AND M− KEYS

Compute

$$329 \cdot 742 - 3 \cdot 8914 + 2 \cdot 2175.$$

Solution

The idea is to compute the products and add them to or subtract them from the contents of the memory. We then obtain the answer by using the MR key. The desired string entered into the calculator is

329 × 742 = M+ 3 × 8914 = M− 2 × 2175 = M+ MR.

This gives the correct answer of 221,726. However, in this simple case, it is easier to enter the string

329 × 742 − 3 × 8914 + 2 × 2175 =

to obtain 221,726 as before. Be sure to check both entry sequences to see that they give this result.

Using the INT÷ Key to Compute Integer Division with Remainders

For calculators designed for use in elementary school classrooms, it was thought appropriate to provide the capability of dividing one natural number by another and displaying the quotient as well as the remainder. Because of limited space in the display, there is a limitation on the size of the numbers that can be used. If either the quotient or the remainder is too large, an error message will appear in the calculator display, and the problem will have to be completed by some other method.

EXAMPLE 3.28

USING THE INT÷ KEY

(a) Using the INT ÷ key, compute the quotient and the remainder when 89,765 is divided by 78.

(b) Using the INT ÷ key, compute the quotient and the remainder when 897,654 is divided by 81.

Solution

(a) Entering 89765 [INT ÷] 78 [=] into your calculator, you see the quotient, 1150, and the remainder, 65, in the display. Use your calculator to check this result by showing that

$$1150 \cdot 78 + 65 = 89765.$$

(b) When you enter 897654 [INT ÷] 81 [=] into your calculator, the display may give an error message since the quotient is a five-digit number.

Using the Built-in Constant Function

Consider the arithmetic progression

$$3, 7, 11, 15, 19, \ldots, 67.$$

It is easy to use your calculator to compute all the terms in this progression. Simply enter 3 [+] 4 and then repeatedly press the key [=] to repeatedly add 4. This utilizes the built-in constant function of most calculators with algebraic logic. Note that, on some calculators, you may need to enter 4 [+] 3 and then repeatedly press [=].

EXAMPLE 3.29 FINDING THE SUM OF AN ARITHMETIC PROGRESSION

Compute the sum $3 + 7 + 11 + \cdots + 67$ of the first 17 terms in the arithmetic progression above.

Solution

1. If we remember young Gauss's trick on page 000, then we have

$$S = 3 + 7 + 11 + \cdots + 67,$$
$$S = 67 + 63 + 59 + \cdots + 3, \text{ and}$$
$$2S = 70 + 70 + 70 + \cdots + 70 = 17 \cdot 70.$$

So

$$S = \frac{17 \cdot 70}{2} = 595.$$

2. An alternative approach to this problem is to use the [M+] and the built-in constant function of your calculator. Thus, we enter

3 [M+] [+] 4 [=] [M+] [=] [M+] [=] . . . ,

repeating the [M+] [=] sequence until the display shows 67. We then press [M+] once more and [MR] to again obtain 595 as the answer.

EXAMPLE 3.30 FINDING THE SUM OF A GEOMETRIC PROGRESSION

Find the sum of the first 15 terms of the geometric progression whose first four terms are 4, 12, 36, 108.

Solution

The consecutive terms of the progression can be found on your calculator by entering 4 [×] 3 [=] and then pressing [=] repeatedly.

As with the arithmetic progression, this problem can also be solved on algebraic-notation machines with a built-in constant function. Thus, if we enter

4 [M+] [×] 3 [=] [M+] [=] [M+] [=] · · · [=] [M+] [MR],

where we use the [=] key 14 times (to add 15 terms), we obtain the desired answer of 28,697,812. Check this on your calculator.

EXAMPLE 3.31 REPEATEDLY ADDING THE SAME NUMBER TO DIFFERENT NUMBERS

Compute these sums: $3 + 17$, $7 + 17$, $9 + 17$, and $24 + 17$.

Solution

The built-in constant function capability of many calculators is useful in ways other than those shown in Examples 3.29 and 3.30. In the present case, we can proceed as follows: Entering 3 [+] 17 [=] yields the desired sum of 20. Now, instead of entering 7 [+] 17 [=], just enter 7 [=] to obtain 24, 9 [=] to obtain 26, and 24 [=] to obtain 41. In any case, if you have to add the same number to a large number of other numbers, this approach greatly simplifies the chore. Note that the same approach also works for subtraction.

EXAMPLE 3.32 REPEATEDLY MULTIPLYING SEVERAL DIFFERENT NUMBERS BY THE SAME NUMBER.

Compute these products: 5×7, 8×7, 20×7, and 30×7.

Solution

As in the preceding example, we simply enter 5 [×] 7 [=], 8 [=], 20 [=], and 30 [=] to obtain the desired products—35, 56, 140, and 210, respectively. Also, the same idea works for division.

Algorithmic Thinking

An approach to doing mathematics that is particularly important when working with calculators and computers is **algorithmic thinking**—the doing of mathematical tasks by means of a sequential, and often repetitive, set of steps. This methodology was illustrated modestly in the preceding examples explaining the repeated use of the [=] and [M+] keys. But much more can be done with your calculator to illustrate this approach to problem solving. To make the idea more clear, consider the following example.

EXAMPLE 3.33 GENERATING THE FIBONACCI SEQUENCE ALGORITHMICALLY

Develop an efficient algorithm for generating successive terms of the Fibonacci sequence (1, 1, 2, 3, 5, . . .), using the special capabilities of your calculator.

Solution

The Fibonacci numbers can be computed with the straightforward use of the [+] and [=] keys on your calculator. But this requires repeatedly entering the proper numbers. More efficient algorithmic approaches can be devised that require only entering one or

Automating Algorithms on a Graphing Calculator

Graphing calculators can display graphs of functions and plots of statistical data. (See Appendix C.) Graphing calculators are also **programmable,** which means that the steps of an algorithm can be entered into the calculator as a **program.** Several useful programs can be found in Appendix C. Programs for the calculator can be quite sophisticated, using a wide variety of input/output and control instructions; in fact, the calculator is really a small computer. Many upper elementary and middle school teachers are finding that learning to program a calculator can pay large dividends for classroom use. Many mathematical "What if?" explorations can be pursued using a program to automate the laborious calculations, allowing students to focus on the higher level aspects of the investigation.

Consider the following problem:

> *Given any two numbers* A *and* B, *generate the Fibonacci-type sequence in which each successive number of the sequence is the sum of the two preceding numbers. What seems to be happening to the ratio of successive numbers in the sequence?*

For example, when $A = B = 1$, we generate the Fibonacci sequence 1, 1, 2, 3, 5, 8, ... and compute the successive ratios $1/1, 2/1, 3/2, 5/3, 8/5, \ldots$. When $A = 1$ and $B = 3$, we generate the Lucas sequence, 1, 3, 4, 7, 11, 18, ... (named for Edouard Lucas, 1842–1891), and then examine the ratios $3/1, 4/3, 7/4, 11/7, \ldots$. This investigation would be tedious using an ordinary calculator, especially if we wanted to compute the sequence and ratios for lots of choices of A and B. The following program, however, written for the TI–73 (and other TI graphing calculators), allows the user to compute Fibonacci-type sequences easily. It is essential to try many choices of A and B to see what pattern emerges.

PROGRAM:FIB	[Name the program]
Input "INPUT A",A	[The user inputs the
Input "INPUT B",B	initial values of A and B]
$2 \rightarrow K$	[K counts the number of terms in the sequence so far]
Lbl 1	[Label a line in the program to return to later]
ClrScreen [ClrHome on TI–82]	[Clear the home screen]
Disp"NO. TERMS SO FAR",K	[Display the number of terms computed so far]
Disp "LAST 2 TERMS",A,B	[Display the last two values computed in the sequence]
Disp "RATIO",B/A	[Display the ratio of the last two terms]
Pause	[Pause to look at the displayed text and numbers]
$A + B \rightarrow C$	[Add the last two numbers of the sequence]
$B \rightarrow A$	[Move down the sequence to the next two values to be
$C \rightarrow B$	considered]
$K + 1 \rightarrow K$	[Update the number of terms computed so far]
Goto 1	[Return to the Label 1 line in the program]

The screen below shows the result of executing the program for $A = 1$, $B = 3$, and $N = 20$ which generates the first 20 Lucas numbers. The nineteenth and twentieth Lucas numbers are 9349 and 15,127, respectively, and their ratio, 15,127/9349, is 1.618034014.

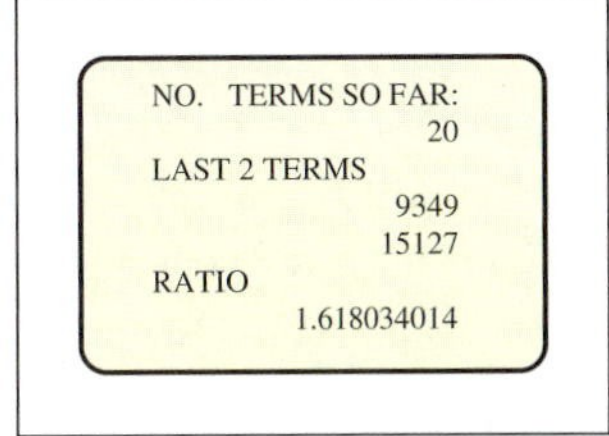

The ratio should begin to look familiar after you try other choices of the starting values A and B.

two numbers initially and then repetitively, using the special keys on your calculator to complete the task. (*Note:* Since not all calculators with algebraic logic operate exactly the same, great care must be exercised in devising an algorithm suitable for your machine.)

The following algorithm generates the Fibonacci numbers by making particularly effective use of the [MR] and [M+] keys found on most calculators. The table below gives the entry (a number or keystroke), the value x seen in the display, and the value M contained in the memory.

Entry	1	M+	+	M+	MR	+	M+	MR	+	M+	MR	+	...
x	1	1	1	1	2	3	3	5	8	8	13	21	...
M	0	1	1	2	2	2	5	5	5	13	13	13	...

The Fibonacci numbers generated by the algorithm are shown in red. Each repetition of the three-keystroke pattern [M+] [MR] [+] first repeats the last number displayed and then generates the next two numbers of the Fibonacci sequence. Since two of each three successive keystrokes generate a new Fibonacci number, the algorithm is highly efficient.

PROBLEM-SOLVING STRATEGY Use Algorithmic Thinking

Sometimes a problem or set of problems can be solved by devising a set of operations that can be carried out repetitively on a calculator or computer. Doing the problem by hand may be prohibitively time consuming.

Problem Set 3.6

Understanding Concepts

1. Use your calculator to compute each of the following.
   (a) $284 + 357$ **(b)** $357 - 284$ **(c)** $284 \cdot 357$
   (d) $284 \div 71$ **(e)** $781 - 35 + 24$
   (f) $781 - (35 + 24)$ **(g)** $781 - (35 - 24)$
   (h) $861 - 423 - 201$ **(i)** $861 - (423 + 201)$
2. Compute the following using your calculator.
   (a) $271 \cdot 365$ **(b)** $1183 \div 91$
   (c) $1024 \div 16 \div 2$ **(d)** $1024 \div (16 \div 2)$
3. Use your calculator to calculate each of these division problems.
   (a) $\dfrac{420 + 315}{15}$ **(b)** $423 + 315 \div 15$
   (c) $\dfrac{4441 + 2332}{220 + 301}$ **(d)** $\dfrac{16{,}157 + 17 \cdot 13}{722 - 291}$
4. Evaluate the following using your calculator.
   (a) 29^2 **(b)** $\sqrt{1849}$
   (c) $\sqrt{2569 - 1480}$ **(d)** $\sqrt{1444} - \sqrt{784}$
5. Write out an entry string to compute
   $$\frac{\sqrt{784} - 91 \div 13}{8 \cdot 49 - 11 \cdot 35}$$
   (a) using parentheses.
   (b) without using parentheses.
6. Use the [M+], [M−], and [MR] keys to compute
   $$\frac{4041 + 1237}{91} + \frac{3381 + 2331}{84} - \frac{2113 + 2993}{46}.$$
   (*Note:* You should be able to do this entirely with your calculator. Nothing need be (should be) written down but the answer.)
7. **(a)** Write out the expression you are evaluating if you enter 1831 [−] 17 [×] 28 [+] 34 [=] into a calculator with algebraic logic.
   (b) Write the numerical answer to part (a).
8. **(a)** Write out the expression you are evaluating if you enter
   42 [×] 34 [−] 14 [×] 6 [=] [÷] 28 [=]
   into your calculator.

(b) What expression does the following string evaluate?

42 [×] 34 [−] 14 [×] 6 [x^2] [=] [÷] 28 [=]

9. (a) Make your calculator count by 2s by entering

0 [+] 2 [=] [=] [=] . . . .

(b) Make your calculator generate the odd numbers by entering

1 [+] 2 [=] [=] [=] . . . .

10. (a) Make your calculator count by 17s by entering

0 [+] 17 [=] [=] [=] . . . .

(b) If you count by 17s, do you ever get to 323? If so, when?

11. Use the built-in constant function to generate the first 10 terms of the arithmetic progression whose first four terms are 2, 6, 10, and 14; that is, use the [=] key repeatedly.

12. Use the built-in constant function to generate the first 10 terms of the geometric progression whose first four terms are 3, 15, 75, and 375.

Teaching Concepts

13. (a) Use the [=] and [M+] keys to compute the sum of the arithmetic progression

$$5 + 8 + 11 + \cdots + 47.$$

(b) Compute the sum in part (a) by using Gauss's trick.

(c) Gauss's trick is useful and should be taught to students, but it is at least somewhat confusing. What is the most confusing part of the determination of the sum by Gauss's trick?

(d) How would you use the calculator calculation of part (a) to help your students to better understand Gauss's trick?

14. (a) Use the [=] and [M+] keys to compute the sum of the geometric progression

$$3 + 15 + 75 + \cdots + 234{,}375.$$

(b) Let S denote the sum required in part (a). Using this sum, write out the sum for $5S$. (Note that the common ratio for the geometric progression is 5.) Then $4S = 5S - S$ and $S = 4S \div 4$. Compute S by hand in this way.

(c) The hand calculation in part (b) is rather neat and should be taught to students. How would you use the calculator calculation of part (a) to help your students to better understand the hand calculation of part (b)?

15. (a) Compute 2^{20} by using the built-in constant function and mentally keeping track of the number of times you multiply by 2. (*Hint:* Enter 2 [×] 2 [=] [=] . . . , repeating the [=] for as long as necessary.)

(b) Use the [^] key on your calculator to compute 2^{20}.

(c) Can you calculate 2^{20} by hand? Note that 2 times 2 equals 4, times 2 equals 8, times 2 equals 16, times 2 equals 32. So $2^5 = 32$. Therefore, $2^{20} = 32 \cdot 32 \cdot 32 \cdot 32$. Complete the hand calculation of 2^{20}.

(d) Explain how you could use parts (a), (b), and (c) in a lesson you might teach to a fourth- or fifth-grade class.

Thinking Critically

16. (a) The Lucas numbers are 1, 3, 4, 7, 11, 18, . . . , where we start with 1 and 3 and, as with the Fibonacci numbers, add any two consecutive terms to obtain the next. Notationally, we will set $L_1 = 1, L_2 = 3, L_3 = 4, \ldots, L_n =$ the nth Lucas number, and so on, In Example 3.33, we developed an algorithm for successively generating the Fibonacci numbers. Modify the algorithm to obtain an algorithm that generates the Lucas numbers. (*Suggestion:* Try inserting a 3 after the first [M+] in the algorithm of Example 3.33.)

(b) Use the algorithm of part (a) to compute the first 10 Lucas numbers.

17. (a) Write the expression your calculator will evaluate if you enter the string

1 [+] [√] 5 [)] [=] [÷] 2 [=] [M+],

and tell where in your calculator the computed number (called the golden ratio) can be found.

(b) Write in general terms the series of expressions your calculator will evaluate if you enter the string

1 [+] [√] 5 [)] [=] [÷] 2 [=] [M+]
[×] [MR] [=] [=] [=] . . . ,

repeatedly pressing the [=] key. Enter in the following table the expressions and the whole numbers *nearest* (either smaller than or larger than) the computed numbers. The first line has been completed for you.

n	Expression	Nearest Whole Number
2	$\left(\frac{1+\sqrt{5}}{2}\right)^2$	3
3		
4		
5		
6		
7		
8		

(c) Make a conjecture based on the results in part (b).

(d) Compute $(1 + \sqrt{5})/2$ and, using the $\boxed{\wedge}$ key, check your conjecture in part (c) for $n = 35$. Note that $L_{35} = 20{,}633{,}239$ is the thirty-fifth Lucas number. See problem 16.

18. **(a)** Use the built-in constant function of your calculator to compute the values of the expressions in the following table, filling in the "Nearest Whole Number" column as you repeatedly press the $\boxed{=}$ key. The first line has been completed for you.

n	Expression	Nearest Whole Number
1	$\left(\frac{1+\sqrt{5}}{2}\right)/\sqrt{5}$	1
2	$\left(\frac{1+\sqrt{5}}{2}\right)^2/\sqrt{5}$	
3	$\left(\frac{1+\sqrt{5}}{2}\right)^3/\sqrt{5}$	
4	$\left(\frac{1+\sqrt{5}}{2}\right)^4/\sqrt{5}$	
5	$\left(\frac{1+\sqrt{5}}{2}\right)^5/\sqrt{5}$	
6	$\left(\frac{1+\sqrt{5}}{2}\right)^6/\sqrt{5}$	
7	$\left(\frac{1+\sqrt{5}}{2}\right)^7/\sqrt{5}$	
8	$\left(\frac{1+\sqrt{5}}{2}\right)^8/\sqrt{5}$	

(b) Make a conjecture based on the results obtained in part (a).

(c) Compute $(1 + \sqrt{5})/2$ and use your $\boxed{\wedge}$ key to check your conjecture for $n = 36$. Note that $F_{36} = 14{,}930{,}352$.

Thinking Cooperatively

For problems in this section, divide your class into groups of about four students each. Require that each group decide on a common answer to each part of each question.

19. **(a)** Use the definition of the Fibonacci and Lucas numbers (see problem 16) to complete this table.

n	1	2	3	4	5	6	7	8	9
F_n	1	1							
L_n	1	3							

(b) Compute the sums $F_1 + F_3, F_2 + F_4, F_3 + F_5, \ldots$. What result do these sums suggest?

(c) Compute the sums $L_1 + L_3, L_2 + L_4, L_3 + L_5, \ldots$. What result do these sums suggest?

(d) Compute the sums $F_1 + L_1, F_2 + L_2, F_3 + L_3, \ldots$. What result do these sums suggest?

20. **(a)** Use the function $y = \sqrt{5x^2 - 4}$ to compute the values of y in the following table. If the computed value of y is not a whole number, do not enter it into the table.

x	1	2	3	4	5	6	7	8	...
y									

(b) What values of x in part (a) yield whole-number values for y? Identify these values of x and y. (They should be familiar.)

(c) Guess what value of x will next yield a whole-number value for y. Also guess the y-value it should yield, and check your guess with your calculator.

21. **(a)** Use the function $y = \sqrt{5x^2 + 4}$ to compute the values of y in the following table. If the computed value of y is not a whole number, do not enter it into the table.

x	1	2	3	4	5	6	7	8	...
y									

(b) What values of x in part (a) yield whole-number values for y? Identify these values of x and y. (They should be familiar.)

(c) Guess what value of x will next yield a whole-number value for y. Also guess what y-value it should yield, and check your guess with your calculator.

From State Student Assessments

22. (Ohio, Grade 4)
Trisha bought 6 packages of ice cream bars, each costing \$2.29. Which buttons should she press on her calculator to find out the total cost of the ice cream bars?

Press 2 → Press · → Press 2 → Press 9 → Press x → Press 6 → Press =

Press 2 → Press · → Press 2 → Press 9 → Press ÷ → Press 6 → Press =

Press 6 → Press ÷ → Press 2 → Press · → Press 2 → Press 9 → Press =

For Review

23. Write a second addition equation and two subtraction equations equivalent to $18 + 17 = 35$.

24. Write two addition equations and a second subtraction equation equivalent to $27 - 9 = 18$.

25. Write a second multiplication equation and two division equations equivalent to $27 \cdot 11 = 297$.

26. Write two multiplication equations and a second division equation equivalent to $96 \div 12 = 8$.

27. Draw a rectangular array to illustrate the product $5 \cdot 7$.

28. Draw an appropriate diagram to illustrate the equation $5(3 + 4) = 5 \cdot 3 + 5 \cdot 4$—that is, to illustrate the distributive property for whole numbers.

Epilogue Calculating Today

In this chapter we have considered the art of writing numbers and performing calculations by using methods extending from ancient to modern times. Our algorithms range from pencil-and-paper procedures to the use of calculators and computers. History shows that the development of the art of calculation has been a tortuous task extending over several thousands of years, but that it has finally reached a stage of extraordinary speed and accuracy. Calculators perform ordinary arithmetic at the touch of a few buttons, and the speed of the latest high-speed computing machines is measured in nanoseconds.*

But speed is not the purview of machines alone. History is replete with the names of calculating prodigies who could perform the most astounding feats of mental arithmetic quickly and accurately. Be that as it may, with command of the ideas discussed in this chapter, every child can become a calculating prodigy in his or her own right.

The important notions covered in this chapter include the following:

- the basic number facts,
- positional notation in base ten,
- the basic algorithms,
- estimation and approximation, and
- the use of a calculator with algebraic notation.

These concepts put great arithmetic power within easy reach of everyone and make it possible for students to spend a great deal more time thinking about and doing more significant and meaningful mathematics. We consider some of these more important ideas starting in the next chapter when we examine notions from number theory.

Chapter Summary

Key Concepts

Section 3.1 Numeration Systems Past and Present

- Egyptian system. An additive system using hieroglyphic symbols.
- Roman system. An additive system using letters to represent numbers.
- Babylonian system. A base sixty positional system with just two symbols combined additively to represent the digits 1 through 59. Though a symbol to represent zero was

*A nanosecond is one billionth of a second.

not present originally, the system eventually acquired such a symbol as a placeholder, but not as a number.

- Mayan system. A positional system to base twenty for the first two positions while, from the third position on, the positions are $18 \cdot 20$, $18 \cdot 20^2, \ldots$. The digits are a rather simple set of well-chosen symbols.
- Indo-Arabic system. Also known as the Hindu-Arabic system, this is our present positional base ten system.
- Digits and expanded notation. For example, in base ten, the ten digits are 0, 1, 2, 3, . . . , 9, and $4073 = 4 \cdot 10^3 + 0 \cdot 10^2 + 7 \cdot 10^1 + 3 \cdot 10^0$ in expanded form.
- Physical models for positional systems. Abacus; sticks in a bundle; Unifix™ cubes; units, strips, mats; and base ten blocks.

Section 3.2 Nondecimal Positional Systems

- Positional systems. Notational systems like our present Indo-Arabic system but to bases other than just ten.
- Converting from base ten to another base. The method for converting from base ten notation to notation in another base.
- Converting from other bases to base ten. The method for converting from notation in some other base to base ten notation.

Section 3.3 Algorithms for Adding and Subtracting Whole Numbers

- Use of manipulative devices. Units, strips, and mats; place-value cards; place-value diagrams.
- Instructional algorithms. Greatly expanded algorithms that enhance student understanding of standard algorithms.
- Exchanging. Ten units can be exchanged for one ten, 10 tens can be exchanged for one hundred, and so on.
- Algorithms for additional and subtraction in other bases. Similar to base ten algorithms.

Section 3.4 Algorithms for Multiplication and Division of Whole Numbers

- Multiplying. Various algorithms for multiplying in base ten notation.
- Multiplication in nondecimal bases. Similar to base ten algorithms.
- Division. Various algorithms for dividing in base ten notation.
- Short-division algorithm. A neat algorithm for dividing by a single digit.

Section 3.5 Mental Arithmetic and Estimation

- Easy combinations. In doing a calculation, combine numbers that add to ten or a multiple of ten.
- Adjustment. Change numbers to ten or multiples of ten, with subsequent adjustment to allow for those changes.
- Work from left to right. This approach is often easier than working from right to left.
- Estimate. If an exact answer is not required, do mental arithmetic with rounded numbers.
- Rounding. The 5-up rule; rounding to the leftmost digit.

Section 3.6 Getting the Most out of Your Calculator

- Priority of operations. The built-in order in which a calculator performs operations.
- Parenthesis keys. Using these keys allows one to override a calculator's built-in order of performing operations, giving you full control of the calculation.
- Memory keys. These keys allow you to store a value for later use in a calculation.

- Function keys. These keys allow you to perform the usual arithmetic operations.
- Thinking algorithmically. Devising repetitive schemes for performing involved operations with fewer keystrokes, particular for recurring problems in which only the data change.

Vocabulary and Notation

Section 3.1

Egyptian hieroglyphics
Roman numerals
Babylonian numerals
Mayan numerals
Place-value manipulatives: sticks and bundles of sticks: Unifix™ cubes; base ten blocks; units, strips, and mats.

Section 3.2

Place value
Digits
Positional values

Section 3.3

Algorithms for addition and subtraction
Exchange
Place-value cards
Instructional algorithm
Addition and subtraction in nondecimal bases

Section 3.4

Algorithms for multiplication and division
Multiplication in nondecimal bases

Section 3.5

One-digit facts
Easy combinations
Adjustment
Working from left to right
Rounding
Estimation
Obtaining approximate answers by rounding

Section 3.6

Algebraic logic
Special keys: [( ], [ )], [(−)], [x^2], [^] (or [y^x]), [M], [MR], [M+], [M−], [INT ÷], [√‾]
Built-in constant function

Chapter Review Exercises

Section 3.1

1. Write the Indo-Arabic equivalent of each of these.
   (a)
   (b) 
   (c) MCMXCVIII
2. Write 234,572 in Mayan notation.
3. Suppose you have 5 mats, 27 strips, and 32 units, for a total count of 802. Briefly describe the exchanges that must be made to represent this number with the smallest possible number of manipulative pieces. How many mats, strips, and units result?

Section 3.2

4. Find the base ten equivalent of each of the following.
   (a) 101101_{two} (b) 346_{seven} (c) $2T9_{\text{twelve}}$
5. Write 287_{ten} as a numeral in each base indicated.
   (a) base five (b) base two (c) base seven

Section 3.3

6. Sketch the solution to 47 + 25 using mats, strips, and units. Draw a square for each mat, a vertical line segment for each strip, and a dot for each unit.
7. Use the instructional algorithm for addition to perform the following additions.

(a) 42 + 54 (b) 47 + 35 (c) 59 + 63

8. Use sketches of place-value cards to illustrate each of these subtractions.

(a) 487 − 275 (b) 547 − 152

9. Perform the following calculations in base five notation. Assume that the numerals are already written in base five.

(a) $\begin{array}{r} 2433 \\ +\ 141 \\ \hline \end{array}$ (b) $\begin{array}{r} 2433 \\ -\ 141 \\ \hline \end{array}$ (c) $\begin{array}{r} 243 \\ \times\ 42 \\ \hline \end{array}$

Section 3.4

10. Perform these multiplications using the instructional algorithm for multiplication.

(a) 4 × 357 (b) 27 × 642

11. Use the scaffold method to perform each of these divisions.

(a) $7\overline{)895}$ (b) $347\overline{)27{,}483}$

12. Use the short-division algorithm to perform each of these divisions.

(a) $5\overline{)27{,}436}$ (b) $8\overline{)39{,}584}$

13. Carry out each of these multiplications in base five. Assume that the numerals are already written in base five.

(a) 23 · 42 (b) 2413 · 332

14. Use the Russian peasant method to compute the product 42 · 35.

Section 3.5

15. Round 274,535

(a) to the nearest one hundred thousand.

(b) to the nearest ten thousand.

(c) to the nearest thousand.

16. Rounding to the leftmost digit, compute approximations to the answers to each of these expressions.

(a) 657 + 439 (b) 657 − 439

(c) 657 · 439 (d) 1657 ÷ 23

Section 3.6

17. Compute the following using your calculator:

$$\frac{\sqrt{1444} - 152 \div 19}{2874 - 2859}$$

18. (a) Compute the sum of this arithmetic progression.

$$4 + 11 + 18 + \cdots + 333$$

(b) Compute the sum of this geometric progression.

$$3 + 6 + 12 + 24 + \cdots + 24{,}576$$

19. (a) Compute each of the products in the following table and, for each product, place the integer closest to the result you obtain in the space provided in the table. (*Suggestion:* Compute $(1 + \sqrt{5})/2$ and store it in memory for repeated use. Then use the constant function feature of your calculator.)

Remember: The Lucas numbers are 1, 3, 4, 7, 11, 18, 29, ...

	Nearest integer
$\left(\frac{1+\sqrt{5}}{2}\right) \cdot 7$	
$\left(\frac{1+\sqrt{5}}{2}\right) \cdot 11$	
$\left(\frac{1+\sqrt{5}}{2}\right) \cdot 18$	
$\left(\frac{1+\sqrt{5}}{2}\right) \cdot 29$	

(b) Make a conjecture on the basis of part (a).

(c) Does your conjecture hold for all Lucas numbers? How about L_1, L_2, and L_3?

(d) How might you modify your conjecture in view of part (c)? Note that L_{34} = 12,752,043 and L_{35} = 20,633,239.

20. Devise a problem like problem 19 for the Fibonacci numbers.

Chapter Test

1. Make a schematic drawing using mats, strips, and units to illustrate the addition of 74 and 48. Draw squares for mats, straight line segments for strips, and dots for units.

2. Round 3,376,500 to the

(a) nearest million.

(b) nearest one hundred thousand.

(c) nearest ten thousand.

(d) nearest one thousand.

3. Place the digits 1, 3, 5, 7, and 9 in the proper boxes to achieve the maximum product.

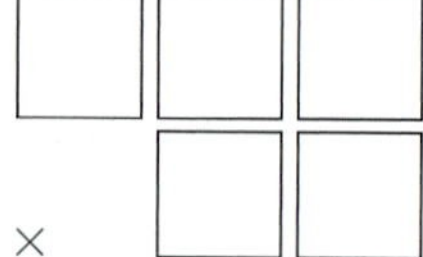

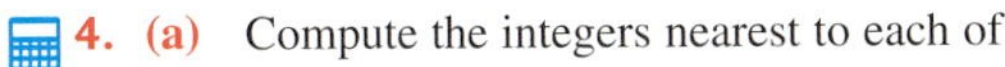

4. **(a)** Compute the integers nearest to each of

$$f_1 = \frac{1+\sqrt{2}}{\sqrt{8}},\ f_2 = \frac{(1+\sqrt{2})^2}{\sqrt{8}},$$
$$f_3 = \frac{(1+\sqrt{2})^3}{\sqrt{8}},\text{ and } f_4 = \frac{(1+\sqrt{2})^4}{\sqrt{8}}.$$

(b) Guess the most likely choices of nearest intergers for f_5 and f_6.

(c) Guess what rule (other than that of part (a)) might be used to obtain the successive values of f_n.

5. Write the base ten equivalents of each of the following.

(a) 21022_{three}

(b) 317_{eight}

(c) 4213_{five}

6. Fill in the missing digits in this addition problem.

```
  2_37
+ _22_
 _00_1
```

7. Fill in the missing digits in this subtraction problem.

```
  _23_
- 35_2
  4_94
```

8. Round each number to the leftmost digit to find an estimate for the sum

$$378 + 64 + 291 + 39 + 3871.$$

9. **(a)** Compute the pairs of quantities

1^3	and	1^2;
$1^3 + 2^3$	and	$(1+2)^2$;
$1^3 + 2^3 + 3^3$	and	$(1+2+3)^2$;
$1^3 + 2^3 + 3^3 + 4^3$	and	$(1+2+3+4)^2$.

(b) Compute the square roots of the answers to the computations of part (a).

(c) Guess a formula for $1^3 + 2^3 + \cdots + n^3$ and for $(1+2+3+\cdots+n)^2$.

10. Write 39,485 in Mayan notation.

11. Compute the sum of this arithmetic progression.

$$3 + 8 + 13 + \cdots + 123$$

12. Place the digits 0, 2, 4, 6, and 8 in the proper boxes to obtain the least product, given that 0 cannot be placed in either of the left-hand boxes.

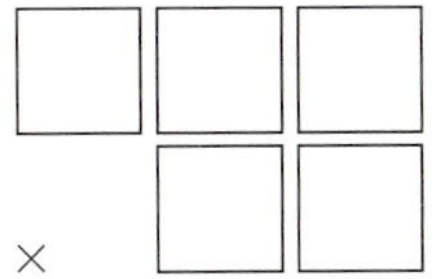

13. Write an algorithm to generate the sequence

$$2, 5, 7, 12, 19, \ldots,$$

where we start with 2 and 5 and add each two consecutive terms to obtain the next term.

14. Perform each of the following calculations entirely in base five. The numerals are already written in base five.

(a)
```
  242
+  43
```

(b)
```
  242
-  43
```

(c)
```
  242
×  43
```

15. Compute the sum of this geometric progression.

$$3 + 15 + 75 + \cdots + 1{,}171{,}875$$

16. Write 281_{ten} as a numeral in each of these bases.

(a) base five

(b) base two

(c) base twelve

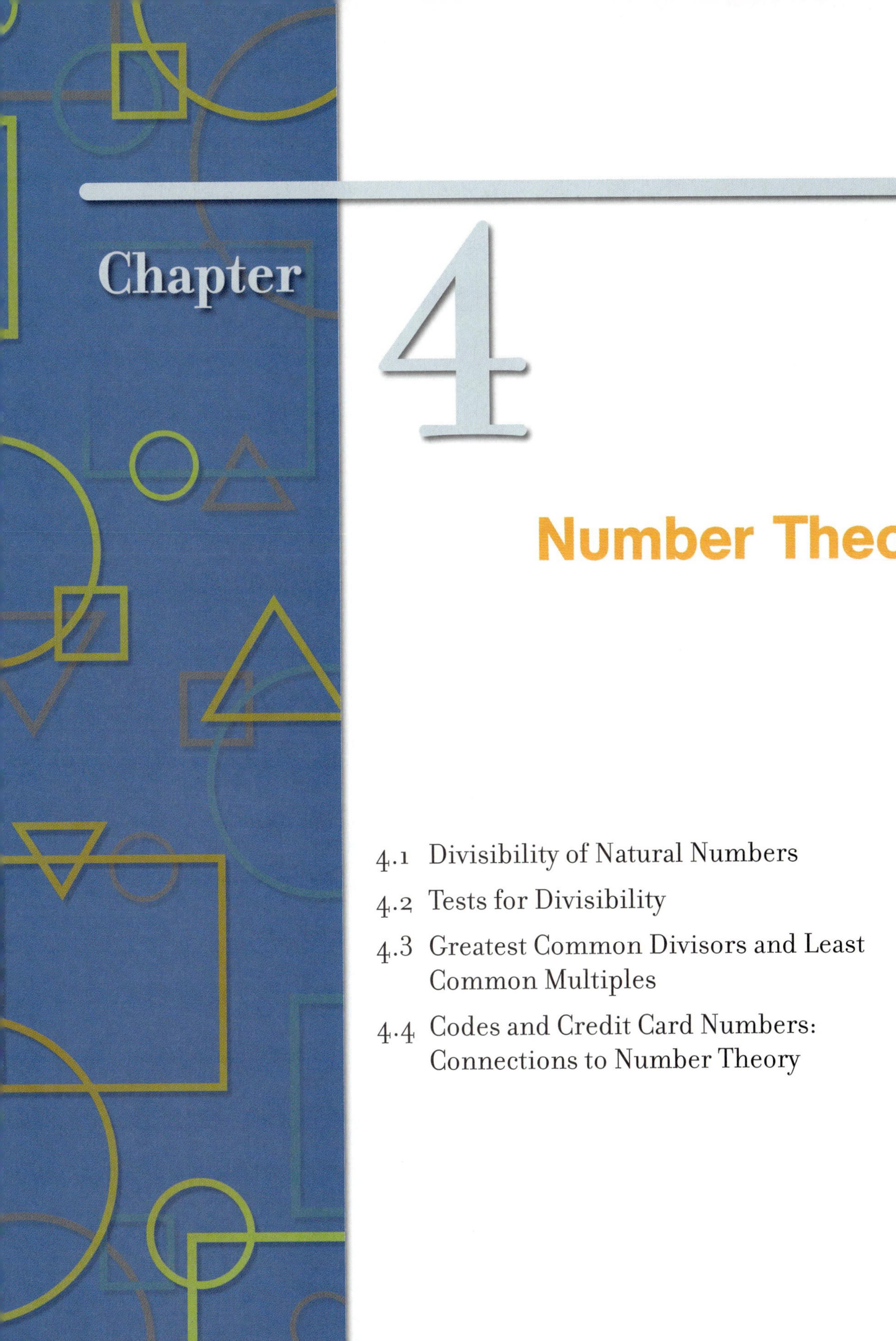

Chapter 4

Number Theory

Primes and Composites via Rectangular Arrays

Materials Needed

1. Twenty-five small cubes or number tiles for each student or small group of students.
2. One record sheet like this for each student.

Values of n	Dimensions of Rectangles	Number of Rectangles	
1			
2			
3			
*4	$1 \times 4,\ 2 \times 2,\ 4 \times 1$	3	1, 2, 4
5			
6			
7			
8			
9			
10			
11			
12			
19			
20			
21			
22			
23			
24			
25			

Directions

1. For each value of n, make up all possible rectangular arrays of n tiles. Then record on your record sheet the dimensions of each rectangle and the number of rectangles. For $n = 4$, we have the rectangles shown here, and we fill in the fourth row of the record sheet as shown above.

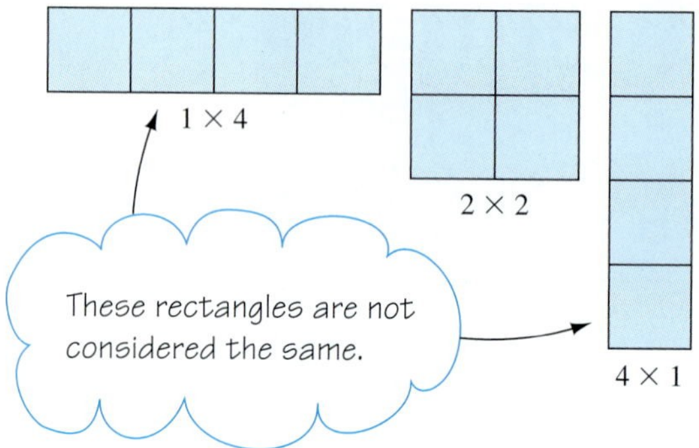

2. In Chapter 2, we used diagrams like this to illustrate the product $3 \cdot 5 = 15$.

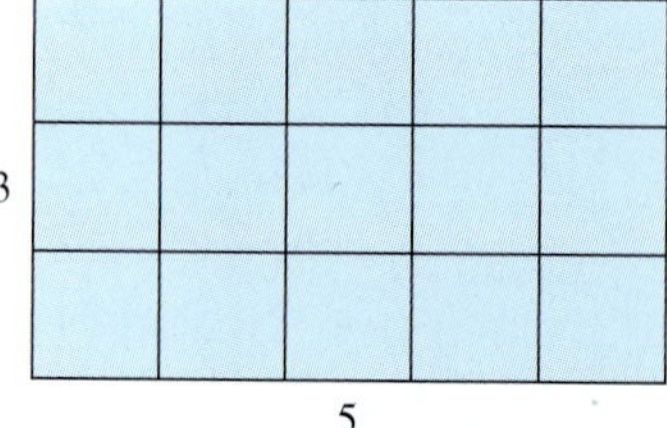

We call 3 and 5 **factors** of 15. Label the last column of your record sheet "Factors of n" and list the factors in increasing order for each value of n. The row for $n = 4$ is done for you.

3. Natural numbers that have exactly two factors are called **prime numbers.** Place a P along the left side of your record sheet next to each prime number.
4. Numbers with more than two factors are called **composite numbers.** Place a C along the left-hand side of your record sheet next to each composite number.
5. Is 1 a prime or composite number or neither? Why?
6. Put an asterisk just to the left of each n that has an odd number of factors. Do these numbers seem to share some other property? State a guess (conjecture) about natural numbers with an odd number of factors.
7. Carefully considering all the data on your record sheet, see if you can guess a number with just 7 factors. Check to see that your guess is correct.

from **The NCTM Principles and Standards**

Understanding of number develops in prekindergarten through grade 2 as children count and learn to recognize "how many" in sets of objects. A key idea is that a number can be decomposed and thought about in many ways. For instance, 24 is 2 tens and 4 ones and also 2 sets of twelve. Making a transition from viewing "ten" as simply the accumulation of 10 ones to seeing both as 10 ones *and* as 1 ten is an important first step for students toward understanding the structure of the base-ten number system (Cobb and Wheatley 1988). Throughout the elementary grades, students can learn about classes of numbers and their characteristics, such as which numbers are odd, even, prime, composite, or square.

Throughout their study of numbers, students in grades 3–5 should identify classes of numbers and examine their properties. For example, integers that are divisible by 2 are called *even numbers* and numbers that are produced by multiplying a number by itself are called *square numbers*. Students should recognize that different types of numbers have particular characteristics; for example, square numbers have an odd number of factors and prime numbers have only two factors.

Students can also work with whole numbers in their study of number theory. Tasks, such as the following, involving factors, multiples, prime numbers, and divisibility, can afford opportunities for problem solving and reasoning.

1. Explain why the sum of the digits of any multiple of 3 is itself divisible by 3.
2. A number of the form *abcabc* always has several prime-number factors. Which prime numbers are always factors of a number of this form? Why?

SOURCE: Reprinted with permission from Principles and Standards for School Mathematics, *pages 33, 151, and 217.*

Connections

The Fascination with Numbers

In Chapter 2, we considered the whole numbers, operations with whole numbers, and some of their properties. In Chapter 3, we studied various systems for writing whole numbers and a variety of algorithms for calculating. Initially, these ideas arose in response to people's needs—the need to count, the need to record counts, and the need of merchants to calculate in the course of doing business. However, very early on, people began to be fascinated with numbers themselves and their many interesting properties. For example, the Greeks called 6 a **perfect number** because

$$6 = 1 + 2 + 3$$

$1 \cdot 6 = 6$
$2 \cdot 3 = 6$

and 1, 2, and 3 are all the natural numbers that divide 6 evenly except for 6 itself.

It turns out that 28 also has this property, since

$$28 = 1 + 2 + 4 + 7 + 14$$

$1 \cdot 28 = 28$
$2 \cdot 14 = 28$
$4 \cdot 7 = 28$

and 1, 2, 4, 7, and 14 are all the natural numbers that divide 28 evenly except for 28 itself. The next two perfect numbers are 496 and 8128. As of this writing, 40 perfect numbers are known. It is conjectured that there are infinitely many perfect numbers, but there is no proof that this is so.

In this chapter, we consider divisibility properties of the natural numbers. These ideas are necessary and useful (in working with fractions, for example) and are a rich source of interesting problems and puzzles that can be used to heighten student interest in learning mathematics.

4.1 DIVISIBILITY OF NATURAL NUMBERS

Divides, Divisors, Factors, Multiples

In Chapter 3 we considered the division algorithm. If a and b are whole numbers with b not zero, when we divide a by b we obtain a unique quotient q and remainder r such that $a = bq + r$ and $0 \leq r < b$. Thus, the division

$$3\overline{)14} \quad 4 \text{ R } 2$$

is equivalent to the equation

$$14 = 3 \cdot 4 + 2.$$

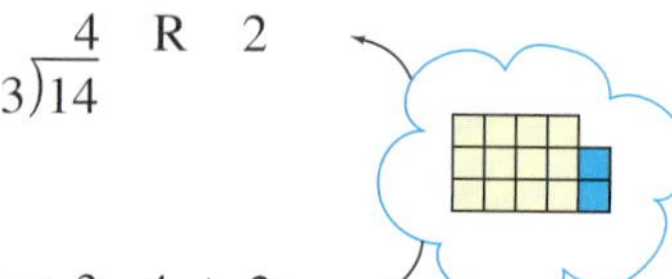

Of special interest in this chapter, because it is important in dealing with fractions, is the case when the remainder r is zero. Then $a = bq$ and we say that **b divides a evenly** or, more simply, **b divides a.** This is expressed in other terminology as indicated here.

> **DEFINITION** *Divides, Factor, Divisor, Multiple*
> If a and b are whole numbers with $b \neq 0$ and there is a whole number q such that $a = bq$, we say that b **divides** a. We also say that b is a **factor** of a or a **divisor** of a and that a is a **multiple** of b. If b divides a and b is less than a, it is called a **proper divisor** of a.

A useful model for the ideas "b divides a" and "a is a multiple of b" has already been provided by the array models for multiplication of natural numbers in Chapter 2. Thus, the five-by-seven rectangular array in Figure 4.1 illustrates the fact that 5 and 7 are both divisors of 35 and that 35 is a multiple of 5 and also of 7.

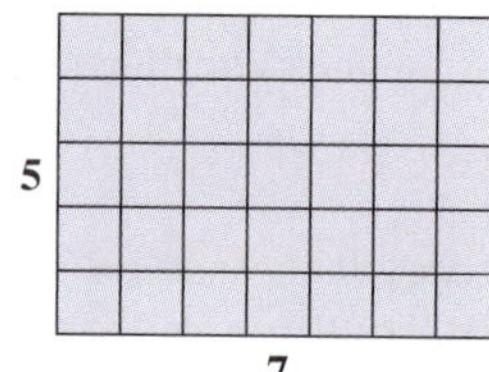

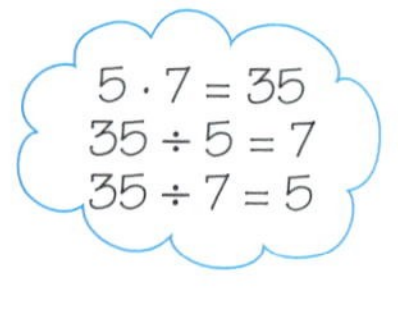

Figure 4.1
Array model showing that 5 and 7 are factors of 35 and that 35 is a multiple of both 5 and 7

EXAMPLE 4.1 THE MULTIPLES OF 2

Write the multiples of 2.

Solution

According to the above definition, the multiples of 2 are the numbers of the form $2q$, where q is a whole number. Taking $q = 0, 1, 2, 3, \ldots$, we obtain the multiples

$$0 = 2 \cdot 0, \quad 2 = 2 \cdot 1, \quad 4 = 2 \cdot 2, \quad 6 = 2 \cdot 3, \ldots,$$

which are just the **even** whole numbers. These multiples could also be obtained by starting with 0 and counting by 2s. Also, except for 0, the multiples of 2 can be illustrated by this series of rectangles:

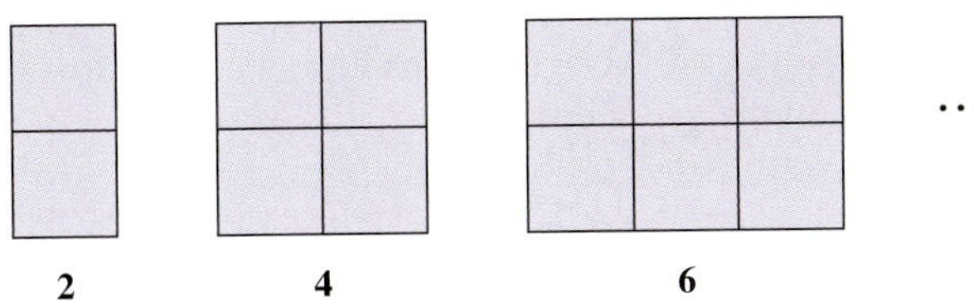

As in the preceding example, the multiples of 3 are

$$0 = 3 \cdot 0, \quad 3 = 3 \cdot 1, \quad 6 = 3 \cdot 2, \quad 9 = 3 \cdot 3, \ldots,$$

which can be obtained by starting with 0 and counting by 3s. In general, the multiples of m are

$$0 = m \cdot 0, m = m \cdot 1, 2m = m \cdot 2, 3m = m \cdot 3, \ldots,$$

obtained by starting with 0 and counting by m's.

EXAMPLE 4.2 THE DIVISORS, OR FACTORS, OF 6

List all the factors of 6.

Solution

Understand the Problem

We must find all the whole numbers b for which there is another whole number q for which $6 = bq$.

Devise a Plan

We saw in Chapter 2 and also in the Hands On in this chapter that rectangular arrays can be used to represent products. Thus, we consider the equivalent problem of finding all rectangular arrays with area 6.

Carry Out the Plan

By trial and error, we find that there are only four such rectangles as shown below, and accordingly the desired factors of 6 are 1, 2, 3, and 6. Finally, it is correct to say that 1, 2, 3, and 6 are factors, or divisors, of 6 and that 6 is a multiple of each of 1, 2, 3, and 6. It is also correct to say that 1 divides 6, 2 divides 6, 3 divides 6, and 6 divides 6.

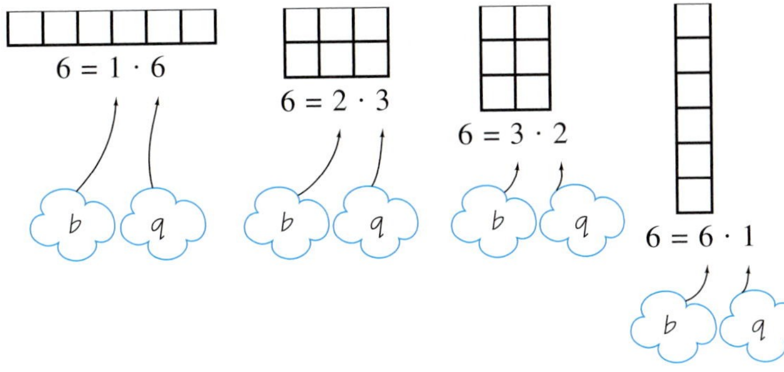

Look Back

The key to this solution was observing that products can be represented by rectangular arrays. Thus, we were able to consider the easier, equivalent problem of finding all rectangular arrays of area 6. We note, by the way, that if $6 = bq$, then both b and q are factors of 6. Also, as b runs through the factors of 6—1, 2, 3, 6—so does the quotient q, but in reverse order—6, 3, 2, 1.

Prime and Composite Numbers

Since $1 \cdot a = a$, 1 and a are *always* factors of a for every natural number a. For this reason, 1 and a are often called trivial factors of a, and $1 \cdot a$ and $a \cdot 1$ are trivial factorings. Some numbers, like 2, 3, 5, and 7, have only trivial factorings. Other numbers, like 6, have nontrivial factorings. The number 1 stands alone since it has only one factor, 1 itself. All this is summarized in the following definition.

DEFINITION *Primes, Composite Numbers, and Units*
A natural number that possesses exactly two different factors, itself and 1, is called a **prime number.** A natural number that possesses more than two different factors is called a **composite number.** The number 1 is called a **unit;** it is neither prime nor composite.

The primes are sometimes called the building blocks of the natural numbers, since every natural number other than 1 is either a prime or a product of primes. For example, consider a number like 180. This is composite since, for example, we can write $180 = 10 \cdot 18$. Moreover, 10 and 18 are both composite since $10 = 2 \cdot 5$ and $18 = 2 \cdot 9$. Now 2 and 5 are both primes and cannot be factored further. But $9 = 3 \cdot 3$, and 3 is a prime. We simply continue to factor a composite number into smaller and smaller factors and stop when we can proceed no further—that is, when the factors are all primes. In the case of 180, we see that

Did You Know?

Fascinating Fibonaccis

"For many years I have been captivated and intrigued by Fibonacci numbers. It has been enormously satisfying to share what I know about them with my young . . . students. Their responses have run the gamut from profound disbelief to patronizing good humor, sprinkled with scientific inquisitiveness. Invariably they ask for more information that they might take home to share with family and friends, demonstrating a confidence in others' interest in the subject and thus revealing their own. Unfortunately the information they seek has been deeply buried in scientific journals or aging mathematical literature, occasionally surfacing as one perfunctory page in a textbook or as a scholarly article in a popular periodical.

"It seemed to me that the time had come to collect and sort out what is currently known about this fascinating subject, to make it understandable and to excite the curiosity of believers and skeptics alike. What is so special about these numbers? Where did they come from? Why do they keep popping up in unlikely, unrelated places? Where might they be that no one has yet thought to look? What answers might they hold for the world, if not the universe?"

This is the preface to the book by middle school teacher Trudi Hammel Garland, *Fascinating Fibonaccis: Mystery and Magic in Numbers.* The book brims with great ideas that will not only stimulate your imagination, but also will do the same for your students.

SOURCE: From Fascinating Fibonaccis: Mystery and Magic in Numbers, *by Trudi Hammel Garland. Copyright © 1987 by Dale Seymour Publications. Reprinted with permission of Pearson.*

$$\begin{aligned}180 &= 10 \cdot 18 \\ &= 2 \cdot 5 \cdot 2 \cdot 9 \\ &= 2 \cdot 5 \cdot 2 \cdot 3 \cdot 3,\end{aligned}$$

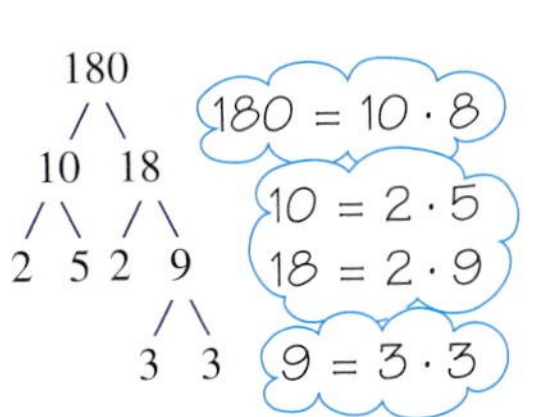

$180 = 2 \cdot 5 \cdot 2 \cdot 3 \cdot 3$

Figure 4.2
A factor tree for 180

and this is a product of primes.

A convenient way of organizing this work is to develop a **factor tree,** as shown in Figure 4.2, to keep track of each step in the process. But there are other ways to factor 180 as these factor trees show:

```
      180                 180
      / \                 / \
     2   90              3   60
         / \                 / \
       10    9              2   30
       / \  / \                 / \
      2   5 3  3               6   5
                              / \
                             2   3
```

$180 = 2 \cdot 2 \cdot 5 \cdot 3 \cdot 3$ $\qquad$ $180 = 3 \cdot 2 \cdot 2 \cdot 3 \cdot 5$

Another, perhaps even more convenient, scheme for finding the prime factors of a number should be mentioned as well. Repeatedly use prime divisors and short division until arriving at a quotient that is a prime. For 180, we might have these divisions:

```
     5           3           5
  3)15        3)9        2)10
  3)45       2)18        2)20
  2)90       2)36        3)60
 2)180      5)180       3)180
```

$180 = 2 \cdot 2 \cdot 3 \cdot 3 \cdot 5$ $\qquad$ $180 = 5 \cdot 2 \cdot 2 \cdot 3 \cdot 3$ $\qquad$ $180 = 3 \cdot 3 \cdot 2 \cdot 2 \cdot 5$

In making a factor tree, most of the arithmetic is done mentally. The short-division scheme makes for a bit more writing while still leaving a written record of the results. Thus, some find the short-division approach more helpful.

The most important thing about all the factorings shown for 180 is that, no matter what scheme is used and no matter how it is carried out, the *same prime factors always result.* That this is always the case is stated here without proof.

> **THEOREM** *Simple-Product Form of the Fundamental Theorem of Arithmetic*
> Every natural number greater than 1 is a prime or can be expressed as a product of primes in one, and only one, way apart from order.

The preceding theorem is the reason that we do not think of 1 as either a prime or a composite number. If 1 were considered a prime, the theorem would not be true. For example, we could multiply 180 by any number of factors of 1 and the prime factorization of 180 would not be unique.

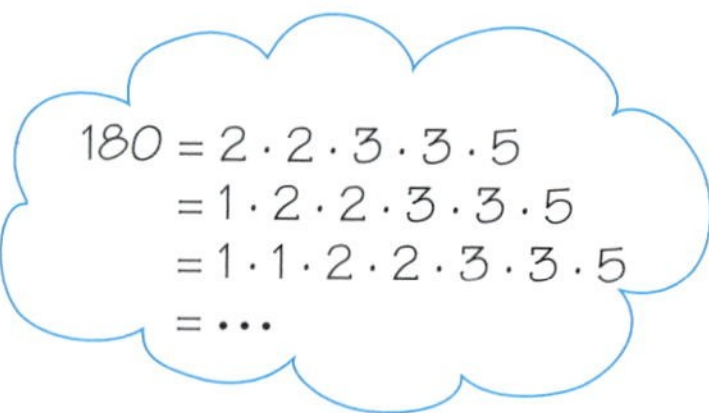

EXAMPLE 4.3 THE PRIME FACTORS OF 600

Represent 600 as a product of prime factors.

Solution

Using a factor tree:

```
         600
         / \
       30   20
      / \   / \
    15  2  2   10
    / \        / \
   3   5      2   5
```

Using short division:

```
    5
 5)25
 2)50
 2)100
 3)300
 2)600
```

$$600 = 2 \cdot 2 \cdot 2 \cdot 3 \cdot 5 \cdot 5$$

In Example 4.3, we can also write 600 as a product of primes by collecting like primes together and writing their products as powers; i.e., we write $600 = 2^3 \cdot 3^1 \cdot 5^2$. Since this could be done for any natural number, it is useful to give an alternative version of the **fundamental theorem of arithmetic.**

> **THEOREM** *Prime-Power Form of the Fundamental Theorem of Arithmetic*
> Every natural number n greater than 1 is a power of a prime or can be expressed as a product of powers of different primes in one, and only one, way apart from order. This representation is called the **prime-power representation of n.**

EXAMPLE 4.4 THE PRIME-POWER REPRESENTATION OF 675

Determine the prime-power representation of 675.

Solution

$$\begin{array}{r} 3 \\ 3\overline{)9} \\ 3\overline{)27} \\ 5\overline{)135} \\ 5\overline{)675} \end{array}$$

Using the short-division method, we see that $675 = 3^3 \cdot 5^2$.

The Divisors of a Natural Number

An important feature of the prime-power representations of numbers is that they make it possible to tell at a glance if one number divides another. In Example 4.4, we saw that $675 = 3^3 \cdot 5^2$. Suppose that r divides 675; then there is an integer s such that

$$rs = 675 = 3^3 \cdot 5^2.$$

Then those primes that divide r must be among the primes that divide 675, and they must appear to no higher power. For example, $r = 15 = 3^1 \cdot 5^1$ divides 675. Moreover, the quotient s must be $3^2 \cdot 5^1 = 45$ so that the product of r and s contains three 3s and two 5s. In fact, all the divisors of 675 can be written down by making a systematic list as shown here:

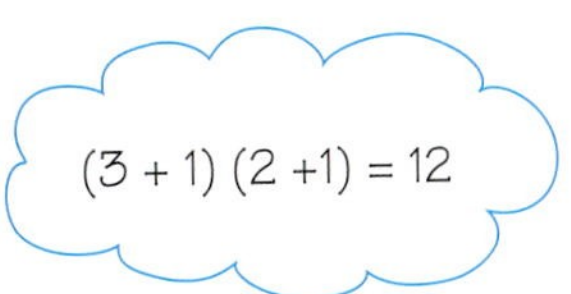

$$\begin{array}{rrr} 1 = 3^0 \cdot 5^0 & 5 = 3^0 \cdot 5^1 & 25 = 3^0 \cdot 5^2 \\ 3 = 3^1 \cdot 5^0 & 15 = 3^1 \cdot 5^1 & 75 = 3^1 \cdot 5^2 \\ 9 = 3^2 \cdot 5^0 & 45 = 3^2 \cdot 5^1 & 225 = 3^2 \cdot 5^2 \\ 27 = 3^3 \cdot 5^0 & 135 = 3^3 \cdot 5^1 & 675 = 3^3 \cdot 5^2 \end{array}$$

In determining r above, there are four choices for the exponent on 3—0, 1, 2, or 3—and three choices for the exponent on 5—0, 1, or 2. Thus, there are $3 \cdot 4 = 12$ possible factors of 675, as the above list shows. This argument can be repeated in general and the result stated as a theorem that tells which numbers divide a given number.

THEOREM *The Divisors of a Natural Number*
Let $n = p_1^{a_1} p_2^{a_2} \cdots p_r^{a_r}$ be the prime-power representation of n. Then m divides n if, and only if, $m = p_1^{b_1} p_2^{b_2} \cdots p_r^{b_r}$, where $0 \le b_1 \le a_1, 0 \le b_2 \le a_2, \ldots, 0 \le b_r \le a_r$. Moreover, the number of factors, or divisors, of n is given by $N = (a_1 + 1)(a_2 + 1) \ldots (a_r + 1)$.

EXAMPLE 4.5 THE DIVISORS OF 600

List all the divisors of 600.

Solution

The divisors must be all the numbers of the form $2^r \cdot 3^s \cdot 5^t$ with $0 \le r \le 3$, $0 \le s \le 1$, and $0 \le t \le 2$. We make a systematic list of the divisors as follows:

$1 = 2^0 \cdot 3^0 \cdot 5^0$	$5 = 2^0 \cdot 3^0 \cdot 5^1$	$25 = 2^0 \cdot 3^0 \cdot 5^2$
$2 = 2^1 \cdot 3^0 \cdot 5^0$	$10 = 2^1 \cdot 3^0 \cdot 5^1$	$50 = 2^1 \cdot 3^0 \cdot 5^2$
$4 = 2^2 \cdot 3^0 \cdot 5^0$	$20 = 2^2 \cdot 3^0 \cdot 5^1$	$100 = 2^2 \cdot 3^0 \cdot 5^2$
$8 = 2^3 \cdot 3^0 \cdot 5^0$	$40 = 2^3 \cdot 3^0 \cdot 5^1$	$200 = 2^3 \cdot 3^0 \cdot 5^2$
$3 = 2^0 \cdot 3^1 \cdot 5^0$	$15 = 2^0 \cdot 3^1 \cdot 5^1$	$75 = 2^0 \cdot 3^1 \cdot 5^2$
$6 = 2^1 \cdot 3^1 \cdot 5^0$	$30 = 2^1 \cdot 3^1 \cdot 5^1$	$150 = 2^1 \cdot 3^1 \cdot 5^2$
$12 = 2^2 \cdot 3^1 \cdot 5^0$	$60 = 2^2 \cdot 3^1 \cdot 5^1$	$300 = 2^2 \cdot 3^1 \cdot 5^2$
$24 = 2^3 \cdot 3^1 \cdot 5^0$	$120 = 2^3 \cdot 3^1 \cdot 5^1$	$600 = 2^3 \cdot 3^1 \cdot 5^2$

Note that the four choices for r, the two choices for s, and the three choices for t show that there are $4 \cdot 2 \cdot 3 = 24$ divisors of 600.

Two Questions about Primes

Since we have just seen how important the primes are as "building blocks" for the natural numbers, it is reasonable to ask the following two questions:

- How many primes are there?
- How does one determine whether a given number is a prime?

We answer these questions in the order asked.

There Are Infinitely Many Primes

The answer to the first question just posed is that there are infinitely many primes. To see this, we describe a step-by-step process that determines at least one new prime at each step. Since the process can be continued without end, it follows that the set of primes is infinite. The argument proceeds as follows.

Since 1 is the only natural number less than 2, the only factors of 2 are 1 and 2 itself. Thus, 2 is a prime. Consider $3 = 2 + 1$. Since 2 does not divide 3, the only divisors of 3 are 1 and 3. So 3 is also a prime. Consider $7 = 2 \cdot 3 + 1$. Recalling the division algorithm, we find that if you divide 7 by either 2 or 3, there is a remainder of 1. Thus, 2 does not divide 7, and 3 does not divide 7. But, by the fundamental theorem of arithmetic, 7 must have a prime divisor. Therefore, 7 has a prime factor different from both 2 and 3. In fact, 7 is itself a prime. Similarly, $43 = 2 \cdot 3 \cdot 7 + 1$ leaves a remainder of 1 when divided by 2, 3, or 7. Thus, 43 is divisible by a prime different from 2, 3, and 7. In fact, 43 is also a prime. We next consider $1807 = 2 \cdot 3 \cdot 7 \cdot 43 + 1$. As before, 1807 is not divisible by any of 2, 3, 7, or 43 (since all leave a remainder of 1), so there must exist a prime other than these primes. In this case, $1807 = 13 \cdot 139$, and both 13 and 139 are primes. In any event, if we multiply all the known primes together and add 1, we obtain a number that must be divisible by at least one new prime. Since this process can be continued *ad infinitum,* it follows that there is no end to the list of primes; that is, there are infinitely many primes.

HIGHLIGHT from HISTORY

Euclid of Alexandria

One of the great mathematicians of the ancient world was Euclid of Alexandria. Little is known of Euclid's life. The dates of his birth and death are not even known (though he lived about 300 B.C.), nor is his birthplace or any of the other little details one might find of interest. What is known is that he was a first-rate teacher and scholar at the great university at Alexandria under Ptolemy II and that he authored some dozen books on mathematics. The most important of these is his *Elements (Stoichia),* which set a new standard of rigor for mathematical thought that persists to this day. It is almost certain that no book, save the Bible, has been more used, studied, or edited than the *Elements,* and none has had a more profound effect on scientific thought.

This monumental work is divided into 13 books and is more popularly known for its remarkable treatment of geometry—an approach that set the pattern for what is taught in schools even today. However, books VII, VIII, and IX deal with number theory and contain many interesting results. Euclid was the first to prove that there are infinitely many primes, and his proof is essentially the one given in this text. He also invented the Euclidean algorithm for determining the greatest common divisor of two natural numbers, which we consider shortly. Another fascinating result is Euclid's formula for even perfect numbers. Recall that a number is considered perfect if it is the sum of its proper divisors. Thus, as observed earlier,

$$6 = 1 + 2 + 3$$

and

$$28 = 1 + 2 + 4 + 7 + 14$$

are the first two perfect numbers. The next three perfect numbers are 496, 8128, and 33,550,336. That Euclid was able to devise a formula for all even perfect numbers is truly remarkable.

So impressed was she by Euclid's work that the American poet Edna St. Vincent Millay was moved to write, "Euclid alone has looked at beauty bare!"

THEOREM *The Number of Primes*
There are infinitely many primes.

Determining Whether a Given Integer Is a Prime

We now answer the second question posed above. The argument we give uses indirect reasoning; i.e., it is a proof by contradiction. Observe that if n is composite, it must have at least one nontrivial factor (and hence at least one prime factor) not greater than its own square root. For, suppose $n = bc$ with $b > \sqrt{n}$ and $c > \sqrt{n}$. Then

$$n = bc > \sqrt{n} \cdot \sqrt{n} = n.$$

But this says that $n > n$, which is nonsense. Thus, either $b \leq \sqrt{n}$ or $c \leq \sqrt{n}$, and we have this theorem.

THEOREM *Prime Divisors of n*
If n is composite, then there is a prime p such that p divides n and $p \leq \sqrt{n}$ (that is, $p^2 \leq n$).

To use this theorem, we still need to know all the primes less than or equal to $\sqrt{n}$. Remarkably, a systematic method for determining all the primes up to a given limit was

School Book Page

Prime and Composite Numbers

Chapter 4 Lesson 14

Exploring Prime and Composite Numbers

Problem Solving Connection
- Look for a Pattern
- Use Logical Reasoning

Materials
hundred chart

Vocabulary
prime number whole number with exactly two different factors
composite number whole number greater than 1 that is not prime

Explore

Eratosthenes (*ehr uh TAHS thuh neez*) was a mathematician from Cyrene (now known as Libya) who lived about 2,200 years ago. He studied **prime numbers**. The number 3 is an example of a prime number. It has only two factors, 1 and 3.

Composite numbers have more than two factors. An example of a composite number is 8. Its factors are 1, 2, 4, and 8.

Cyrene was the capital of ancient Cyrenaica.

1 × 8 ● ● ● ● ● ● ● ●   2 × 4 ● ● ● ● / ● ● ● ●

You tell whether a number is prime by testing possible factors. There are no other factors for 7 besides 7 and 1. So, 7 is a prime number.

1 × 7 ● ● ● ● ● ● ●

Eratosthenes' Sieve is a process that "strains" out composite numbers and leaves prime numbers behind. Use it to find the prime numbers between 1 and 100.

Did You Know?
A computer has been used to find a prime number that, if printed in a newspaper, would fill 12 pages.

Work Together

1. Use a hundred chart. Follow the directions to cross out composite numbers and circle prime numbers.
   a. Cross out 1. It has only 1 factor.
   b. Circle 2, the least prime number. Cross out all numbers divisible by 2.
   c. Repeat this step with 3, the next prime number.
2. Continue this process until you reach 100.
3. List all of the circled numbers. There should be 25.

What pattern can you see in the list of prime numbers?

Questions for the Teacher

1. Is the discussion of the sieve of Eratosthenes above fully adequate? How are students to tell if 2, 3, 5, . . . are primes? Must they draw a rectangular array each time? Is there any other way to tell? How would you help them at this point?
2. Students will wonder why no more numbers are sieved out of the hundred chart after the multiples of 7. How would you help them to understand why this is so and why all the numbers not crossed out at this point are primes?

devised by the Greek mathematician Eratosthenes (276–195 B.C.). Aptly called the **sieve of Eratosthenes,** the method depends only on counting. Suppose we want to determine all the primes up to 100. Since 1 is neither prime nor composite, we write down all the whole numbers from 2 to 100. Note that 2 is a prime since its only possible factors are itself and 1. However, every second number after 2 is a multiple of 2 and therefore is composite. Thus, we delete, or "sieve out," these numbers from our list. The next number not deleted is 3. It must be a prime since it is not a multiple of 2, the only smaller prime. We now "sieve out" all multiples of 3 after 3 itself; that is, we strike from the list every third number after 3 whether it has been struck out before or not. The next number not already deleted must also be a prime since it is not a multiple of 2 or 3, the only smaller primes. Thus, 5 is a prime, and every fifth number after 5 must be deleted as a multiple of 5. In the same way, we determine that 7 is a prime and delete every seventh number after 7. Similarly, 11 is the next prime, but, when we delete every eleventh number after 11, no new numbers are deleted. In fact, all the remaining numbers at this point are primes, and the sieve, as shown in Figure 4.3, is complete.

Just for Fun

The Chinese Remainder Problem

Find the least positive integer having remainders of 2, 3, and 4 when divided by 5, 7, and 9, respectively. Problems like these were studied by Chinese mathematicians like Sun-Tse as early as the first century A.D.

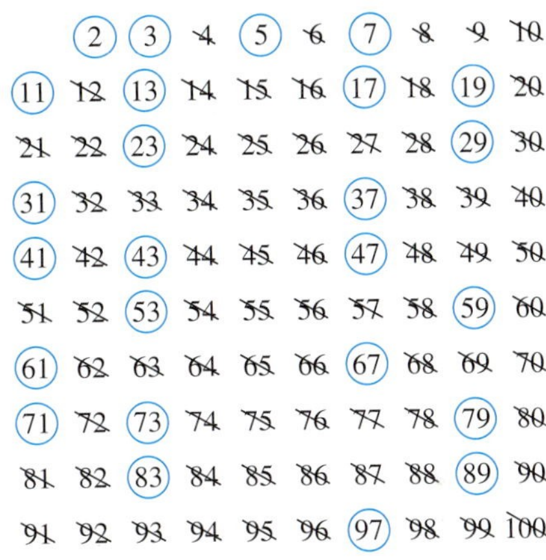

Figure 4.3 *The sieve of Eratosthenes for* $n = 100$

The primes have been circled to make them stand out. To see why the sieve is complete at this point, note that if $n \leq 100$, then $\sqrt{n} \leq \sqrt{100} = 10$. Hence, if n is composite and is in the list, it must have a prime factor less than or equal to 10 by the preceding theorem. Thus, the sieve is complete by the time we have deleted all multiples of 2, 3, 5, and 7.

EXAMPLE 4.6 DETERMINING THE PRIMALITY OF 439

Show that 439 is a prime.

Solution

Understand the Problem

We must show that 439 has no factors other than itself and 1.

Devise a Plan

By the preceding theorem, if 439 is composite (i.e., not a prime), then it must have a prime divisor p such that $p \leq \sqrt{439}$. We must show that this is impossible. Since we wish to show that something is impossible, perhaps we can use indirect reasoning.

Carry Out the Plan

Assume that 439 is composite. Then, by the preceding theorem, it must have a prime divisor p such that $p \leq \sqrt{439} \doteq 21$. From the sieve of Eratosthenes, we see that the only possibilities for p are 2, 3, 5, 7, 11, 13, 17, and 19. Putting 439 in the memory of our calculator and using MR repeatedly, we easily complete the desired divisions and determine that none of these primes divides 439. Since this contradicts the assumption that 439 is composite, that assumption is false. Thus, 439 is a prime.

Look Back

The key to this solution was simply noting that, if a whole number is composite, then it must have a prime divisor less than or equal to its own square root. Since $\sqrt{439} \doteq 21$, we easily ruled out all possibilities by use of a calculator.

Did You Know?

The largest prime known at this writing is $2^{24,036,583} - 1$, a huge number with 7,235,733 digits in its decimal representation. Normally, determining the primality of such a large number would be beyond the power of even the largest and fastest computers. However, the special form of this number, $2^n - 1$, makes it especially susceptible to attack by a relatively fast algorithm. Numbers of the form $2^n - 1$ are called Mersenne numbers, after the French monk Father Marin Mersenne, who discovered the first few primes of this form early in the seventeenth century. Checking such numbers for primality has now become a test of the speed of new computers and a pastime for computer buffs who are willing to let their personal computers work at the task for almost countless hours. If you want to keep up with the latest on these large primes, you can check the World Wide Web at *www.mersenne.org/13466917.htm.*

EXAMPLE 4.7 A PRIME DIFFERENT FROM 2, 3, 7, AND 13

Determine a prime different from 2, 3, 7, and 13 that divides $2 \cdot 3 \cdot 7 \cdot 13 + 1 = 547$.

Solution

Understand the Problem

The problem is clear enough. We must find a prime different from 2, 3, 7, and 13 that divides 547.

Devise a Plan

By now, we are beginning to gain more mathematical power as we remember past solutions. Indeed, the proof that there are infinitely many primes comes to mind. There, we had to find a prime different from 2; a prime different from 2 and 3; a prime different from 2, 3, and 7; and so on. Perhaps we can use a similar argument here.

Carry Out the Plan

To find a prime different from 2, 3, and 7, we considered the number $2 \cdot 3 \cdot 7 + 1 = 43$. If we adopt the same strategy here, we will consider $2 \cdot 3 \cdot 7 \cdot 13 + 1 = 547$. As before, from the division algorithm, it is clear that dividing 547 by 2, 3, 7, and 13 always leaves a remainder of 1. Therefore, 547 is either a prime or it has a prime factor $p \leq \sqrt{547} \doteq 23$.

Since 2, 3, 7, and 13 have already been checked, we see from the sieve of Eratosthenes that it only remains to check 11, 17, 19, and 23. Again, we place 547 in the memory of our calculator and, using MR, perform divisions of 547 by 11, 17, 19, and 23. Since none of these divides 547, it follows as in Example 4.6 that 547 is itself a prime.

Look Back

Having seen a similar problem before, we found it possible to use the same general approach to successfully solve this problem.

Prime-Power Factorizations with a Graphing Calculator

The algorithm to compute the prime factorization of a given whole number N is easy to describe in a general way:

Step 1. Set $P = 2$, the smallest possible prime that may be a factor of N.

Step 2. See if P divides N and, if so, continue to see if higher powers $P^2, P^3, \ldots$ also divide N. If the largest exponent is E (that is, P^E divides N, but P^{E+1} does not), then P is a prime factor of N with corresponding power E.

Step 3. Let the quotient N/P^E become the new value N to be factored, let $P + 1$ become the new lowest possible factor P of N to be checked out, and return to Step 2 (unless $P > \sqrt{N}$, in which case you are done).

Step 4. Continue to repeat Steps 2 and 3 until $P > \sqrt{N}$, at which time the program is done.

The program FACTOR for the T1-73 graphing calculator (or T1-8X or other machines) can be found in Appendix C. This program automates the algorithm outlined. It is made efficient by checking only for odd factors $P = 3, 5, 7, \ldots$ once the even prime factor $P = 2$ has been examined. The user's guide accompanying your graphing calculator will explain how to enter and edit the FACTOR program. Once you have the program available, it can be a helpful and fun program for you to use.

The result of factoring $N = 134{,}483{,}440$ is shown on the leftmost screen depicted below. Being careful to match each prime factor with its corresponding power, we see in just a few moments of calculator time that $N = 2^4 \times 5^1 \times 7^3 \times 13^2 \times 29^1$. The primes and corresponding powers can also be viewed in the List Editor: press ON 1 to quit the program and then press LIST. The side-by-side lists of prime factors and powers are then easily viewed, as seen in the rightmost screen below.

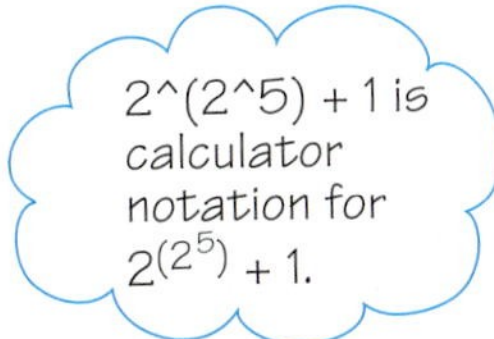

N=?134483440
PRIME FACTORS
{2 5 7 13 29}
POWERS
{4 1 3 2 1}

L1	L2	L3 2
2	4	------
5	1	
7	3	
13	2	
29	1	
------	------	

L2(1)=4

You might like to try factoring the quite large number $N = 2\text{^}(2\text{^}5) + 1$ (that is, $N = 4{,}294{,}967{,}297$). The great seventeenth-century mathematician Pierre Fermat thought this number was prime, since replacing the exponent 5 with any of 0, 1, 2, 3, or 4 gives the respective numbers 3, 5, 17, 257, and 65537, which are all prime. In about a minute and a half, your calculator will tell you that Fermat was mistaken. What is the factorization of N?

Problem Set 4.1

Understanding Concepts

1. Draw array diagrams to show that
   (a) 4 is a factor of 36.
   (b) 6 is a factor of 36.
2. Draw array diagrams to illustrate all the factorings of 35 taking order into account; that is, think of $1 \cdot 35$ as different from $35 \cdot 1$.
3. (a) List the first 10 multiples of 8, starting with $0 \cdot 8 = 0$.
   (b) List the first 10 multiples of 6, starting with $0 \cdot 6 = 0$.
   (c) Use parts (a) and (b) to determine the least natural number that is a multiple of both 8 and 6.

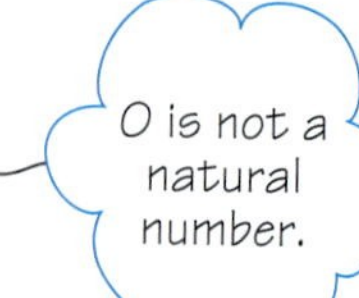

4. Complete this table of all factors of 18 and their corresponding quotients.

Factors of 18	1	2						
Corresponding quotients	18	9						

5. Construct factor trees for each of these numbers.
   (a) 72 (b) 126 (c) 264 (d) 550
6. Use short division to find all the prime factors of each of these numbers.
   (a) 700 (b) 198 (c) 450 (d) 528
7. (a) List all the divisors (factors) of 48.
   (b) List all the divisors (factors) of 54.
   (c) Use parts (a) and (b) to find the largest common divisor of 48 and 54.
8. (a) Determine the prime-power representations of both 136 and 102.
   (b) Determine the set of all divisors (factors) of 136.
   (c) Determine the set of all divisors of 102.
   (d) Determine the greatest divisor of both 136 and 102.
9. Determine the prime-power representation of each of these numbers.
   (a) 48 (b) 108 (c) 2250 (d) 24,750
10. Let $a = 2^3 \cdot 3^1 \cdot 7^2$.
   (a) Is $2^2 \cdot 7^1 = 28$ a factor of a? Why or why not?
   (b) Is $2^1 \cdot 3^2 \cdot 7^1 = 126$ a factor of a? Why or why not?
   (c) One factor of a is $b = 2^2 \cdot 3^1$. What is the quotient when a is divided by b?
   (d) How many different factors does a possess?
   (e) Make an orderly list of all of the factors of a.
11. To determine whether 599 is a prime, which primes must you check as possible divisors?
12. Use your calculator and the information from the sieve of Eratosthenes in Figure 4.3 to determine whether 1139 is prime. If it is not prime, give its prime factors.

Teaching Concepts

13. (a) If n is composite, is it true that all prime factors of n must not exceed $\sqrt{n}$?
   (b) How would you convince your class that it is sometimes the case that one or more prime factors of a composite number n are greater than $\sqrt{n}$ even though it is always the case that at least one prime factor of n must be less than $\sqrt{n}$?
14. (a) If n, b, and c are natural numbers and n divides bc, is it necessarily the case that n divides b or n divides c? Justify your answer.
   (b) How would you explain to your class that your answer to part (a) is correct? Would it be useful to discuss the role of counterexamples here?
15. Which of the following are true and which are false? Justify your answer in each case.
   (a) n divides 0 for every natural number n.
   (b) 0 divides n for every natural number n.
   (c) 1 divides n for every natural number n.
   (d) n divides n for every natural number n.
   (e) 0 divides 0.
   (f) How would you explain to your class that your answers to parts (a) through (e) are correct?

Thinking Critically

16. (a) Which of the primes less than 100 shown in the sieve of Eratosthenes (Figure 4.3) are adjacent to (i.e., one more or one less than) a multiple of 6?
   (b) What is the first prime that is not adjacent to a multiple of 6?
   (c) Show that every prime larger than 3 is necessarily adjacent to a multiple of 6. (*Hint:* By the division algorithm, every natural number n must leave a remainder r when divided by 6 with $0 \le r < 6$ (i.e., every natural number can be written in the form $n = 6q + r$, where $0 \le r < 6$). What are the possibilities when n is a prime?)

(d) Note that part (c) does *not* say that every multiple of 6 is adjacent to a prime. Find the first multiple of 6 that is not adjacent to a prime. (*Suggestion:* Extend the sieve of Eratosthenes up to 130 and recall that the multiples of 6 can be generated by using the built-in constant function on your calculator.)

17. (a) List all the distinct factors of $9 = 3^2$.

(b) Find three natural numbers (other than 9) that have precisely three distinct factors.

(c) Find three natural numbers that have precisely four distinct factors.

18. Earlier, we observed that 6 and 28 are perfect numbers—i.e., numbers equal to the sum of their proper divisors.

(a) Determine the prime-power representation of 496.

(b) Show that 496 is a perfect number.

(c) Determine the prime-power representation of 8128.

(d) Show that 8128 is a perfect number.

(e) Write 31 and 127 in the form $2^n - 1$.

(f) Recalling that 6 and 28 are perfect numbers and considering the results of parts (a), (b), (c), and (d), make a conjecture about the prime-power representation of an even perfect number.

(g) Given that 8191 is a prime, determine the prime factorization of 33,550,336 and show that this number is perfect. Does this result strengthen your confidence in your conjecture in part (f)?

19. A number is said to be **deficient** if the sum of its proper divisors is less than the number. Similarly, a number is said to be **abundant** if the sum of its proper divisors is greater than the number. Thus, 14 is deficient, since $1 + 2 + 7 < 14$, and 12 is abundant, since $1 + 2 + 3 + 4 + 6 > 12$. Classify each of the following as deficient or abundant.

(a) 10 (b) 18 (c) 20 (d) 16 (e) 468

(f) 2^n, where n is a natural number

(g) p, where p is a prime

20. A pair of natural numbers m and n are called **amicable** if the sum of the proper divisors of m equals n and the sum of the proper divisors of n equals m. Show that 220 and 284 are an amicable pair.

21. A prime p such that $2p + 1$ is also a prime is called a **Germain prime,** after the eminent nineteenth-century German mathematician Sophie Germain (see the Highlight from History on page 799). Which of 11, 13, 97, 241, and 359 are Germain primes?

22. Assume that a divides b and b divides c, where a, b, and c are natural numbers. Give a carefully written three-sentence argument showing that a divides c.

23. If p is a prime, b and c are natural numbers, and p divides bc, is it necessarily the case that p divides b or p divides c? Justify your answer.

24. Assume that p and q are different primes and that n is a natural number. If p divides n, and q divides n, argue briefly that pq divides n.

25. Let N_n be the natural number whose decimal representation consists of n consecutive 1s. For example, $N_2 = 11$, $N_7 = 1{,}111{,}111$, and $N_9 = 111{,}111{,}111$.

(a) Show that N_2 divides N_4, N_2 divides N_6, and N_2 divides N_8.

(b) Guess the quotient when N_{18} is divided by N_2, and check your guess by multiplication by hand.

26. Let N_n be as in problem 25.

(a) Show that N_3 divides N_6, N_3 divides N_9, and N_3 divides N_{12}.

(b) Guess the quotient when N_{18} is divided by N_3, and check your guess by multiplication by hand.

27. Let N_n be as in problems 25 and 26.

(a) Does N_3 divide N_5?

(b) Does N_3 divide N_7?

(c) Does N_3 divide N_{15}?

(d) Guess conditions on m and n which guarantee that N_m divides N_n.

28. Draw a square measuring 10 centimeters on a side. Draw vertical and horizontal line segments dividing the square into rectangles of areas 12, 18, 28, and 42 square centimeters, respectively. Where should A, B, C, and D be located?

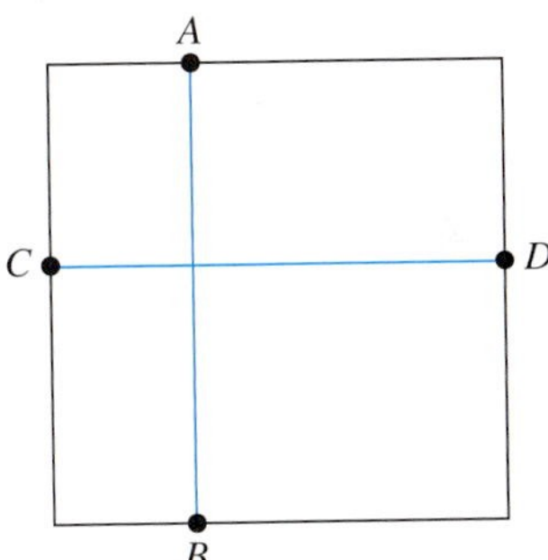

29. When the students arrived in Mr. Gowing's classroom one day, there were 28 three-by-five cards neatly propped up in the chalk tray. They were numbered from 1 through 28 and arranged in order with the numbers facing the chalkboard. After the students were seated, Mr. Gowing asked the first student in the first row of seats to go to the board and turn each card so that they all faced the classroom. He then asked the second student to go to the board and turn every second card back facing the board. The third student was to turn every third card, whether it had been turned before or not. Thus, for example, the sixth card would have been turned toward the class by the first student, toward the board by the second student, and toward the class again by the third student. The process continued until the twenty-eighth student turned the twenty-eighth card for the last time.

(a) When the process was complete, which numbers were left facing the students?

(b) Show why the answer to part (a) is as it is. (*Hint:* You may also want to make a table to exhibit what is happening.)

Thinking Cooperatively

30. Work with members of your cooperative group to complete this activity. The Fibonacci numbers 1, 1, 2, 3, 5, 8, 13, . . . were defined in Section 1.4, Example 1.6.

(a) Extend the following table so that it shows the first 30 Fibonacci numbers.

n	1	2	3	4	5	6	7	8	9	10	. . .
F_n	1	1	2	3	5	8	13	21	34	55	. . .

(b) Which Fibonacci numbers F_m are divisible by $F_3 = 2$? Carefully describe the pattern you discover.

(c) Which Fibonacci numbers F_m are divisible by $F_4 = 3$? by $F_5 = 5$? Describe the patterns you discover.

(d) If r is a natural number, conjecture which Fibonacci numbers are divisible by F_r on the basis of your observations in parts (b) and (c).

Using a Calculator

31. Use FACTOR to determine the prime-power representation of each of these numbers.

(a) 548 **(b)** 936 **(c)** 274 **(d)** 45,864

(e) Use the results of parts (a), (b), (c), and (d) to determine which, if any, of these numbers divide(s) another of these numbers.

32. Use FACTOR to determine the prime-power representation of each of these integers.

(a) 894,348 **(b)** 245,025 **(c)** 1,265,625

(d) Which of the numbers in parts (a), (b), and (c) are squares?

(e) What can you say about the prime-power representation of a square? Explain briefly.

(f) Guess how you might know from a glance at its prime-power representation that 93,576,664 is the cube of a natural number. Explain.

From State Student Assessments

33. (Michigan, Grade 4)
Mrs. Ramirez said: "Count the letters in your name. If the sum is a multiple of 4, you can line up first." Which of these students could line up first?

Dirk	Terrence	Juanita	Michael	Kayla	Shelby	Arnie

A. Dirk and Terrence B. Michael and Juanita
C. Arnie and Shelby D. Kayla and Dirk

34. (Oregon, Grade 5)
Which of the following is a prime number?
A. 37 B. 39 C. 72 D. 93

35. (California, Grade 5)
Determine the prime factors of all numbers through 50 and write the numbers as the product of their prime factors by using exponents to show multiples of a factor (e.g., $24 = 2 \times 2 \times 2 \times 3 = 2^3 \times 3$).

For Review

36. Let $S = \{1, 3, 5, \ldots\}$ be the set of all odd natural numbers. Which of the following are *true* and which are *false?* Justify your answer in each case.

(a) S is closed with respect to addition.

(b) S is closed with respect to multiplication.

(c) The commutative property for addition holds for elements of S.

(d) The commutative property for multiplication holds for elements of S.

(e) The associative property for addition holds for elements of S.

(f) The associative property for multiplication holds for elements of S.

(g) The distributive property for multiplication over addition holds for elements of S.

(h) S possesses an additive identity.

(i) S possesses a multiplicative identity.

37. **(a)** Compute the one hundredth term in the arithmetic sequence 2, 5, 8, 11, ...

(b) Compute the sum of the first one hundred terms in the sequence of part (a).

38. Write the most likely choices for the next three terms in each of these sequences.

(a) 3, 7, 11, 15, ____, ____, ____

(b) 1, 3, 7, 17, 41, ____, ____, ____

(c) 1, 3, 6, 10, ____, ____, ____

(d) 1, 0, 1, 0, 2, 0, 3, 0, 5, ____, ____, ____

(e) 3, 6, 12, 24, ____, ____, ____

(f) 2, 9, 13, 31, 57, 119, ____, ____, ____

39. Compute the sum of the terms in each of these arithmetic progressions.

(a) 2, 5, 8, . . . , 155

(b) 7, 12, 17, . . . , 152

(c) 3, 10, 17, . . . , 689

A Neat Fibonacci Trick

Materials Needed

1. A calculator for each student.
2. A Fibonacci sum record sheet for each student as below.

Directions

1. Working with two or three other students, start the activity by placing any two natural numbers into the first two rows of column 1 of the record sheet, and then complete the column by adding any two consecutive entries to obtain the next entry in the Fibonacci manner. Finally, add the 10 entries obtained and divide the sum by 11.
2. Repeat the process to complete all but the last column of the record sheet. Then look for a pattern and make a conjecture. Lastly, prove that your conjecture is correct by placing a and b into the first and second positions in the last column and then repeating the process as before. Your conjecture and proof should be agreed upon by all members of your group.

FIBONACCI SUM RECORD SHEET

	1	2	3	4	5	General Case
1						a
2						b
3						
4						
5						
6						
7						
8						
9						
10						
Sum						
Sum ÷ 11						

4.2 TESTS FOR DIVISIBILITY

Divisibility by 2, 3, 5, 9, and 10

It is convenient for both teachers and students to have simple tests that show when one number is divisible by another. Relatively easy tests exist for the primes 2, 3, 5, 7, 11, and 13; for 9, 10, and powers of 2; and for products of numbers in the preceding list. Indeed, tests for divisibility by every prime exist, but they become quite cumbersome and are not useful. In general, for larger primes, we are reduced to using technology and/or trial and error. Also, all the tests depend on using base-ten notation.

Because it is so useful in answering divisibility questions, we first develop a result about the divisibility of sums and differences of numbers. Since

$$84 = 66 + 18 = 3 \cdot 22 + 3 \cdot 6 = 3(22 + 6) = 3 \cdot 28$$

and

$$48 = 66 - 18 = 3 \cdot 22 - 3 \cdot 6 = 3(22 - 6) = 3 \cdot 16,$$

it is clear that 3 divides both 84 and 48, because it divides both 66 and 18. Moreover, since the same argument could be repeated in general, we have the following theorem.

THEOREM *Divisibility of Sums and Differences*
If n, a, and b are natural numbers and n divides both a and b, then n divides both $a + b$ and $a - b$.

Now consider an expression like

$$9276 = 927 \cdot 10 + 6 \qquad \text{or, equivalently,} \qquad 9276 - 927 \cdot 10 = 6.$$

Since 2 divides 10, it follows from the theorem on divisibility of sums and differences that 2 divides 9276 if, and only if, 2 divides 6. Therefore, 2 does divide 9276.

In general, when we divide any natural number n by 10, we obtain a quotient q and a remainder r (the units digit in the decimal representation of n) such that

$$n = 10q + r \qquad \text{or, equivalently,} \qquad n - 10q = r.$$

6 is the units digit in the decimal representation of 9276

Hence, as above, 2 divides n if, and only if, 2 divides r. Moreover, since 5 divides 10, these same equations ensure that 5 divides n if, and only if, 5 divides r. Thus, since 0, 2, 4, 6, and 8 are the only digits divisible by 2, and 0 and 5 are the only digits divisible by 5, we have the following theorem.

THEOREM *Divisibility by 2 and 5*
Let n be a natural number. Then n is divisible by 2 if, and only if, its units digit is 0, 2, 4, 6, or 8. Similarly, n is divisible by 5 if, and only if, its units digit is 0 or 5.

The test for divisibility by $10 = 2 \cdot 5$ follows immediately from the tests for divisibility by 2 and 5.

THEOREM *Divisibility by 10*
Let n be a natural number. Then n is divisible by 10 if, and only if, 2 and 5 divide n—that is, if, and only if, the units digit of n is 0.

Proof By definition, 10 divides n if, and only if, $10q = n$ for some q. But, since $2 \cdot 5 = 10$, this is so if, and only if, $2(5q) = n$ and $5(2q) = n$—that is, if, and only if, 2 divides n and 5 divides n. Thus, the divisibility tests for both 2 and 5 must be satisfied, and this is so if, and only if, the units digit of n is 0.

It is important to note that the preceding theorem is a special case of a much more general result.

THEOREM *Divisibility by Products*
Let a and b be natural numbers with no common factor other than 1. Then, if a divides c and b divides c, it follows that ab divides c.

Proof From Section 4.1, all the prime factors of a must appear in c and to at least as high a power as they appear in a. Similarly, all prime factors of b must appear in c and to at least as high a power as they appear in b. But, since a and b have no factors in common, they can have no prime factors in common, and this implies that all primes appearing in either a or b must appear in c and to at least as high a power as they appear in ab. Therefore, ab divides c, as claimed.

EXAMPLE 4.8 DIVISIBILITY BY PRODUCTS

Using the fundamental theorem of arithmetic, show that $18 \cdot 25$ divides 22,050.

Solution

Since $18 = 2^1 \cdot 3^2$, $25 = 5^2$, and $22{,}050 = 2^1 \cdot 3^2 \cdot 5^2 \cdot 7^2$, it is clear that 18 divides 22,050 and that 25 divides 22,050. Moreover, since 18 and 25 have no common factor other than 1, it follows from the preceding theorem that $18 \cdot 25$ divides 22,050. In fact, $22{,}050 \div (18 \cdot 25) = 72 = 49$.

THEOREM *Tests for Divisibility by 3 and 9*
A natural number is divisible by 3 if, and only if, the sum of its digits is divisible by 3. Similarly, a natural number is divisible by 9 if, and only if, the sum of its digits is divisible by 9.

Instead of a formal proof, we use an example to reveal why these tests hold. Consider $n = 27{,}435$, and let $s = 2 + 7 + 4 + 3 + 5$ denote the sum of the digits of n. Then

$$\begin{aligned} n - s &= 27{,}435 - (2 + 7 + 4 + 3 + 5) \\ &= (20{,}000 - 2) + (7000 - 7) + (400 - 4) + (30 - 3) \\ &= 2(10{,}000 - 1) + 7(1000 - 1) + 4(100 - 1) + 3(10 - 1) \\ &= 2 \cdot 9999 + 7 \cdot 999 + 4 \cdot 99 + 3 \cdot 9 \\ &= 9(2 \cdot 1111 + 7 \cdot 111 + 4 \cdot 11 + 3 \cdot 1) \\ &= 9q, \end{aligned}$$

$q = 2 \cdot 1111 + 7 \cdot 111 + 4 \cdot 11 + 3 \cdot 1$

and

$$n = s + 9q.$$

Therefore, by the theorem on divisibility of sums and differences, 9 divides n if, and only if, 9 divides s. Since $9 = 3 \cdot 3$, the same argument holds for divisibility by 3. Moreover, since in this case $s = 21$, n is divisible by 3 and not by 9.

EXAMPLE 4.9 DIVISIBILITY BY 2, 3, 5, 9, AND 10

State whether each of the following is *true* or *false* without actually dividing. Give a reason for your answer.

(a) 2 divides 43,826 **(b)** 3 divides 111,111 **(c)** 10 divides 26,785
(d) 9 divides 10,020,006 **(e)** 6 divides 111,111 **(f)** 5 divides $(287 + 78)$

Solution

(a) True, since the units digit of 43,826 is 6 (that is, even).
(b) True, since $1 + 1 + 1 + 1 + 1 + 1 = 6$ and 3 divides 6.
(c) False, since 10 divides a number if, and only if, its units digit is 0. Here it is 5.
(d) True, since $1 + 0 + 0 + 2 + 0 + 0 + 0 + 6 = 9$ and 9 divides 9.
(e) False. If 6 divides 111,111, then 2 divides 111,111 since 2 divides 6. But 2 does not divide 111,111 since the units digit is not even.
(f) True. Five does not divide 287 and 5 does not divide 78. Yet $287 + 78 = 365$, and 5 divides 365 since the units digit is 5.

Cooperative Investigation

Two Interesting Division Patterns

Materials Needed

1. Pencil and paper for each student.
2. A calculator for each student would be useful, but is not necessary.

Directions

1. Complete the following divisions.
2. After each set of quotients is determined, look for a pattern and state a conjecture.

Division Pattern Number 1

$99 \div 11 =$ ________

$9999 \div 11 =$ ________

$999{,}999 \div 11 =$ ________

$99{,}999{,}999 \div 11 =$ ________

Conjecture: ________

____

____

Test your conjecture by multiplying a suitable 15-digit number by 11.

Division Pattern Number 2

$11 \div 11 =$ ________

$1001 \div 11 =$ ________

$100{,}001 \div 11 =$ ________

$10{,}000{,}001 \div 11 =$ ________

Conjecture: ________

____

____

Test your conjecture by multiplying a suitable 14-digit number by 11.

Divisibility by 11

Proceeding by example, we now develop a test for divisibility by 11. Consider the numbers 8,571,937 and $t = 8 - 5 + 7 - 1 + 9 - 3 + 7 = (8 + 7 + 9 + 7) - (5 + 1 + 3)$, where 8, 7, 9, and 7 are the seventh, fifth, third, and first digits, respectively, in 8,571,937 and 5, 1, and 3 are the sixth, fourth, and second digits, respectively. Since

$$\begin{aligned}
8{,}571{,}937 - t &= 8{,}000{,}000 + 500{,}000 + 70{,}000 + 1000 + 900 + 30 + 7 - 8 + 5 - 7 + 1 \\
&\quad - 9 + 3 - 7 \\
&= 8(1{,}000{,}000 - 1) + 5(100{,}000 + 1) + 7(10{,}000 - 1) + 1(1000 + 1) \\
&\quad + 9(100 - 1) + 3(10 + 1) \\
&= 8 \cdot 999{,}999 + 5 \cdot 100{,}001 + 7 \cdot 9999 + 1 \cdot 1001 \\
&\quad + 9 \cdot 99 + 3 \cdot 11 \\
&= 8 \cdot 90{,}909 \cdot 11 + 5 \cdot 9091 \cdot 11 + 7 \cdot 909 \cdot 11 \\
&\quad + 1 \cdot 91 \cdot 11 + 9 \cdot 9 \cdot 11 + 3 \cdot 11 \\
&= 11 \cdot (8 \cdot 90{,}909 + 5 \cdot 9091 + 7 \cdot 909 + 1 \cdot 91 + 9 \cdot 9 + 3) \\
&= 11q,
\end{aligned}$$

$8{,}000{,}000 - 8 = 8(1{,}000{,}000 - 1)$

$500{,}000 + 5 = 5(100{,}000 + 1)$

etc.

By the preceding Cooperative Investigation

where q is a natural number, $8{,}571{,}937 - t = 11q$ is divisible by 11. Since $8{,}571{,}937 = t + 11q$, it follows from the theorem of the divisibility of sums and differences that 8,571,937 is divisible by 11 if, and only if, t is divisible by 11. In the present case, $t = 22$, so 8,571,937 is divisible by 11, as one can easily check.

Since the preceding argument is essentially general, we have the following theorem.

THEOREM *Divisibility Test for 11*

A natural number is divisible by 11 if, and only if, the difference of the sums of the digits in the even and odd positions in the number is divisible by 11.

EXAMPLE 4.10 | TESTING FOR DIVISIBILITY BY 11

Show that $n = 8{,}193{,}246{,}781{,}053{,}476{,}109$ is divisible by 11 and that $m = 76{,}124{,}738{,}465{,}372{,}103$ is not divisible by 11.

Solution

First consider n. Using mental arithmetic, we find that

$$8 + 9 + 2 + 6 + 8 + 0 + 3 + 7 + 1 + 9 = 53$$

and

$$1 + 3 + 4 + 7 + 1 + 5 + 4 + 6 + 0 = 31.$$

Since $53 - 31 = 22$ and 11 divides 22, it follows that 11 divides n. In fact,

$$8{,}193{,}246{,}781{,}053{,}476{,}109 \div 11 = 744{,}840{,}616{,}459{,}406{,}919.$$

Now consider m. Using mental arithmetic, we find that

$$7 + 1 + 4 + 3 + 4 + 5 + 7 + 1 + 3 = 35$$

and

$$6 + 2 + 7 + 8 + 6 + 3 + 2 + 0 = 34.$$

Since $35 - 34 = 1$ and 11 does not divide 1, it follows that 11 does not divide m. Here,

$$76{,}124{,}738{,}465{,}372{,}103 \div 11 = 6{,}920{,}430{,}769{,}579{,}282 \text{ R } 1.$$

HIGHLIGHT *from* HISTORY

Eratosthenes of Cyrene (276–195 B.C.)

Much has been written about the remarkable scholarly accomplishments of the ancient Greeks. Such accounts are sprinkled with names like Pythagoras, Thales, Euclid, Archimedes, Aristarchus, Eudoxus, Hypatia, Theano, and Apollonius. Another extraordinarily able contributor to Greek thought was Eratosthenes, who was remarkably gifted not only in mathematics, but also in poetry, astronomy, geography, history, philosophy, and even athletics. Eratosthenes was born in Cyrene, on the south shore of the Mediterranean, lived most of his early life in Athens, and, later, was invited to Alexandria by Ptolemy III of Egypt to tutor his son and to serve as librarian of the great Alexandrian University. We have already mentioned his remarkable sieve for determining all the prime numbers up to any given limit, but he is perhaps best known for his unusually accurate determination that the circumference of the earth is very nearly 25,000 miles—a figure much nearer the actual value than that of any of the ancients either before or after. In old age, Eratosthenes became almost totally blind, and it is said that he died of voluntary starvation in about 195 B.C.

A Combined Test for Divisibility by 7, 11, and 13

We now describe a test for divisibility by 7, 11, and 13 all at the same time. Consider a number like

$$n = 92{,}252{,}191{,}213.$$

Break n up into a series of three-digit numbers

$$092,\ 252,\ 191,\ 213$$

determined by the three-digit groups starting from the right in n. Now, as in the test for divisibility by 11, compute the sum of the numbers in the odd-numbered positions in the above list and the sum of the numbers in the even-numbered positions. This gives

$$\begin{array}{r} 092 \\ +\ 191 \\ \hline 283 \end{array} \qquad \text{and} \qquad \begin{array}{r} 252 \\ +\ 213 \\ \hline 465 \end{array}$$

as shown. The difference between these sums is

$$465 - 283 = 182,$$

and n will be divisible

by 7 if, and only if, 182 is divisible by 7,

by 11 if, and only if, 182 is divisible by 11, and

by 13 if, and only if, 182 is divisible by 13.

Since $182 = 7 \cdot 26$ and $182 = 13 \cdot 14$, it follows that 7 divides n and 13 divides n. On the other hand, 11 does not divide 182, so 11 does not divide n. Find the quotients when n is divided by 7 and by 13, and find the quotient and remainder when n is divided by 11.

EXAMPLE 4.11 DIVISIBILITY BY 7, 11, AND 13

Test to see if $n = 8{,}346{,}261{,}059{,}482{,}647$ is divisible by 7, 11, or 13.

Solution Breaking n up into three-digit numbers and computing the respective sums of the numbers in even positions and odd positions, we have

$$\begin{array}{r} 008 \\ 261 \\ 482 \\ \hline 751 \end{array} \qquad \text{and} \qquad \begin{array}{r} 346 \\ 059 \\ 647 \\ \hline 1052 \end{array}$$

The test number is $1052 - 751 = 301$. Since $7 \cdot 43 = 301$, 7 divides 301 and hence 7 divides n. However,

$$301 = 11 \cdot 27 + 4 \qquad \text{and} \qquad 301 = 13 \cdot 23 + 2.$$

Thus, 11 does not divide 301, and 13 does not divide 301.

A Test for Divisibility by Powers of 2

THEOREM *A Test for Divisibility by 4, 8, and Other Powers of 2*
Let n be a natural number. Then 4 divides n if, and only if, 4 divides the number named by the last two digits of n. Similarly, 8 divides n if, and only if, 8 divides the number named by the last three digits of n. In general, 2^r divides n if, and only if, 2^r divides the number named by the last r digits of n.

This theorem can be understood by looking at an example. Consider a number like 28,476,324. This can be written in the form

$$\begin{aligned} 28{,}476{,}324 &= 284{,}763 \cdot 100 + 24 \\ &= 4 \cdot 25 \cdot 284{,}763 + 24. \end{aligned}$$

Thus, by the theorem of divisibility of sums and differences, 4 divides 28,476,324 if, and only if, 4 divides 24. Therefore, 4 does indeed divide 28,476,324.

In general, when we divide any natural number n by 100, we obtain a quotient q and a remainder r such that

$$n = 100q + r = 4(25q) + r,$$

and by the argument above, 4 divides n if, and only if, 4 divides r.

A similar argument holds for divisibility by 8, since $n = 1000q + s$, where s is the number represented by the last three digits of n, and 8 divides 1000. For higher powers of 2, the test depends on the fact that 16 divides 10,000, 32 divides 100,000, and so on.

EXAMPLE 4.12 TESTING FOR DIVISIBILITY BY 4 AND 8

Test the following for divisibility by 4 and 8.

(a) 2452 **(b)** 3849 **(c)** 7672 **(d)** 39,000

Solution

Since 2 divides 4, 4 divides 8, 8 divides 16, and so on, divisibility of a number by any power of 2 guarantees divisibility by any lower power. Thus, it follows that if 2 does not divide n, then 2^k does not divide n for any k.

(a) Since $452 = 8 \cdot 56 + 4$, 8 does not divide 452. Thus, 8 does not divide 2452. However, 4 divides 52, so 4 divides 2452 and 2 divides 2452.

(b) Since 2 does not divide 9, it follows that 2 does not divide 3849, 4 does not divide 3849, and 8 does not divide 3849.

(c) Since $672 = 8 \cdot 84$, it follows that 8 divides 7672, 4 divides 7672, and 2 divides 7672.

(d) Since 8 divides 1000, it follows that 8 divides 39,000, 4 divides 39,000, and 2 divides 39,000.

Problem Set 4.2

Understanding Concepts

1. Test each number for divisibility by each of 2, 3, and 5. Do the work mentally.

   (a) 1554 **(b)** 1999

   (c) 805 **(d)** 2450

2. Use the results of problem 1 to decide which, if any, of the numbers in problem 1 are divisible by

   (a) 6. **(b)** 10.

   (c) 15. **(d)** 30.

3. Test each of these for divisibility by 7, 11, and 13.

   (a) 253,799 **(b)** 834,197 **(c)** 1,960,511

4. Use the results of problem 3 to decide which of the numbers in problem 3 are divisible by

   (a) 77. **(b)** 91.

   (c) 143. **(d)** 1001.

5. Is 1,927,643,001,548 divisible by 11? Explain briefly.

6. **(a)** At a glance, determine the digit d so that 87,543,24d is divisible by 4. Is there more than one answer? Explain.

   (b) Can you choose the digit d so that 87,543,24d is divisible by 8 and not 16? Explain.

7. Determine the digit d so that 6,34d,217 is divisible by 11.

8. **(a)** Fill in the missing digit so that 897,650,243,28_ is divisible by 6. Can this be done in more than one way?

(b) Fill in the missing digit so that the number in part (a) is divisible by 11.

9. A palindrome is a number that reads the same forward and backward, like 2,743,472.

(a) Give a clear, but brief, argument to show that every palindrome with an even number of digits is divisible by 11.

(b) Is it possible for a palindrome with an odd number of digits to be divisible by 11? Explain.

Teaching Concepts

10. A student claims that 157,163 is divisible by 3 since the last digit in the number is 3. Explain how you would correct the student's thinking.

11. A student claims that to multiply a two-digit number by 11, all you have to do is write a three-digit number whose first and last digits are the first and last digits of the two-digit number and whose middle digit is the sum of the digits of the two-digit number (e.g., $32 \cdot 11 = 352$).

(a) Is this ever true?

(b) Is this always true?

(c) If a three-digit number is such that the middle digit is the sum of its first and last digits, is the number divisible by 11?

(d) How would you discuss (a), (b), and (c) with a class of fourth graders?

12. Tia claims that 24 divides 420 since 6 and 4 both divide 420. Explain how you would correct Tia's thinking.

13. Max claims that 15 does not divide $177 + 48$ since 15 does not divide 177 and 15 does not divide 48.

(a) Is Max correct?

(b) How would you respond to Max?

Thinking Critically

14. Show that every number whose decimal representation has the form *abc,abc* is divisible by 7, 11, and 13.

15. **(a)** Reverse the digits in a two-digit number to obtain a second number, and then subtract the smaller of these two numbers from the larger. What possible numbers can result? Can you tell at a glance what number will result? Explain carefully.

(b) Repeat part (a) for three-digit numbers.

16. **(a)** Consider a six-digit number that is divisible by 27, like 142,857. Show by actually dividing that each of 428,571; 285,714; 857,142; 571,428; and 714,285 is also divisible by 27.

(b) 769,230 and 153,846 are two other numbers divisible by 27. See if the pattern of part (a) holds for these numbers.

(c) What general property is suggested by parts (a) and (b) of this problem? State it carefully.

(d) Find another six-digit number that is divisible by 27. Does it have the property you guessed in part (c)?

(e) Check to see that all the numbers in parts (a) and (b) are also divisible by 11. Is this true of all six-digit numbers divisible by 27?

Thinking Cooperatively

17. The first 30 Fibonacci numbers are displayed in the following table.

n	1	2	3	4	5	6	7	8	9	10
F_n	1	1	2	3	5	8	13	21	34	55

n	11	12	13	14	15	16	17	18	19	20
F_n	89	144	233	377	610	987	1597	2584	4181	6765

n	21	22	23	24	25	26	27	28	29	30
F_n	10,946	17,711	28,657	46,368	75,025	121,393	196,418	317,811	514,229	832,040

Do the following in cooperation with three or four other students. In each case, come to a consensus on the conjectures you make.

(a) In problem 30, Problem Set 4.1, you were asked to conjecture which Fibonacci numbers were divisible by 2, by 3, and by 5. If you did not do that problem, do it now.

(b) List the first few Fibonacci numbers that are divisible by 4.

(c) On the basis of the limited data of part (b), guess what must be true of n if 4 divides F_n.

(d) Use the results of part (a) to guess what must be true of n if 6 divides F_n.

(e) Note that 7 divides F_8, 7 divides F_{16}, and 7 divides F_{24}. Conjecture what must be true about n if 7 divides F_n.

(f) What do the results of parts (a) through (e) suggest about the divisibility of Fibonacci numbers by natural numbers?

Making Connections

18. A common error in banking is to make an interchange or transposition of some of the digits in a number involved in a transaction. For example, a teller may pay out \$43.34 on a check actually written for \$34.43 and hence be short by \$8.91 at the end of the day. Show that such a mistake always causes the teller's balance sheet to show an error, in terms of pennies, here 891, that is divisible by 9. (*Hint:* Recall that, in the proof of the divisibility test for 9, we showed that every number differs from the sum of its digits by a multiple of 9.)

19. If a teller's record of the day's work is out of balance by an amount, in pennies, that is a multiple of 9, is it necessarily the case that he or she has made a transposition error in the course of the day's work? Explain.

For Review

20. Use short division to determine the prime-power representation of these natural numbers.

(a) 8064 **(b)** 2700 **(c)** 19,602

21. Construct a factor tree for each of these natural numbers, and write their prime-power representations.

(a) 7000 **(b)** 6174 **(c)** 6237

22. How many factors does each number in problem 21 possess? Be sure to show your work on this problem.

23. What is the quotient when a is divided by b if $a = 2^3 \cdot 3^4 \cdot 5^1 \cdot 7^2$ and $b = 2^1 \cdot 3^2 \cdot 7^1$?

24. **(a)** Is $2^4 \cdot 3^2 \cdot 5^1 \cdot 7^2$ divisible by $2^3 \cdot 3^1 \cdot 7^2$? Why or why not?

(b) Is $2^4 \cdot 3^2 \cdot 5^1 \cdot 7^2$ divisible by $2^2 \cdot 5^2 \cdot 7^2$? Why or why not?

Cooperative Investigation

Representing Integers as Sums

Materials Needed

One copy of the sheet outlined below for each student.

Directions

Divide the class into groups of three or four students each, and have them carry out the directions on the handout.

Investigation 1. Representing Natural Numbers as a Sum of Consecutive Integers

Some numbers can be written as a sum of two or more consecutive positive integers. For example, $3 = 1 + 2$, and $15 = 4 + 5 + 6$ (alternatively, $15 = 1 + 2 + 3 + 4 + 5$). On the other hand, 2 cannot be written as the sum of two or more consecutive integers. Fill in the rest of this table, and then conjecture what pattern is revealed in your table.

n		*n*		*n*		*n*	
1	X	11		21		31	
2	X	12		22		32	
3	1 + 2	13		23		33	
4		14		24		34	
5		15		25		35	
6		16		26		36	
7		17		27		37	
8		18		28		38	
9		19		29		39	
10		20		30		40	

Investigation 2. Representing Numbers with Sums of Consecutive Evens or Odds

Some numbers can be written as a sum of two or more consecutive even or consecutive odd positive integers. For example, $6 = 2 + 4$, and $45 = 5 + 7 + 9 + 11 + 13$ (alternatively, $45 = 13 + 15 + 17$). On the other hand, 2 cannot be written as the sum of two or more consecutive integers that are all even or all odd. Fill in this table, and then conjecture what pattern is revealed in your table.

n		*n*		*n*		*n*	
1	X	11		21		31	
2	X	12		22		32	
3		13		23		33	
4		14		24		34	
5		15		25		35	
6	2 + 4	16		26		36	
7		17		27		37	
8		18		28		38	
9		19		29		39	
10		20		30		40	

4.3 GREATEST COMMON DIVISORS AND LEAST COMMON MULTIPLES

An architect is designing an elegant display room for an art museum. One wall is to be covered with large square marble tiles. To obtain the desired visual effect, the architect wants to use the largest tiles possible. If the wall is to be 12 feet high and 42 feet long, how large can the tiles be?

If the tiles measure 4 feet on a side, the possible height of the wall must be a multiple of 4 (i.e., 4 must be a divisor of 12). Indeed, the length of the side of a tile must be a divisor of both the height and length of the wall (i.e., a common divisor of both 12 and 42). Since the sets of divisors of 12 and 42 are $D_{12} = \{1, 2, 3, 4, 6, 12\}$ and $D_{42} = \{1, 2, 3, 6, 7, 14, 21, 42\}$, respectively, the tile size must be chosen from the set $D_{12} \cap D_{42} = \{1, 2, 3, 6\}$, the set of common divisors of both 12 and 42. Thus, if the tiles are to be as large as possible, they must measure 6 feet on a side since 6 is the largest of the common divisors of 12 and 42.

Considerations like these lead to the notion of the greatest common divisor of two natural numbers, defined formally as follows.

DEFINITION *Greatest Common Divisor*
Let m and n be natural numbers. The greatest natural number d that divides both m and n is called their **greatest common divisor** and we write $d = \text{GCD}(m, n)$.

We note that divisors are often called **factors,** that the greatest common divisor is often called the **greatest common factor,** and that $\text{GCD}(m, n)$ is often written as $\mathbf{GCF}(m, n)$.

Before proceeding further, we note in particular that greatest common divisors and least common multiples, play an important role in elementary school in helping students to understand the arithmetic of fractions.

Greatest Common Divisors by Intersection of Sets

This method, suggested by the above discussion, works well when the numbers involved are small and all the divisors of both numbers are easily written down. By way of explanation, it is probably best to do an example.

EXAMPLE 4.13 FINDING THE GREATEST COMMON DIVISOR BY INTERSECTION OF SETS

Find the greatest common divisor of 18 and 45.

Solution

Let D_{18} and D_{45} denote the sets of divisors of 18 and 45, respectively. Since

$$D_{18} = \{1, 2, 3, 6, 9, 18\} \qquad \text{and} \qquad D_{45} = \{1, 3, 5, 9, 15, 45\},$$

$D_{18} \cap D_{45} = \{1, 3, 9\}$ is the set of common divisors of 18 and 45. Thus, $\text{GCD}(18, 45) = 9$.

Greatest Common Divisors from Prime-Power Representations

The greatest common divisor of two numbers can also be found by using their prime-power representations. Consider

$$54 = 2^1 \cdot 3^3 \qquad \text{and} \qquad 45 = 3^2 \cdot 5^1.$$

Since $2^0 = 5^0 = 1$, we can write these representations so that they *appear* to be products of powers of the same primes. Thus, we have

$$54 = 2^1 \cdot 3^3 \cdot 5^0 \qquad \text{and} \qquad 45 = 2^0 \cdot 3^2 \cdot 5^1.$$

From the theorem of the divisors of a natural number (just preceding Example 4.5), the divisors of 54 are numbers of the form

$$d = 2^a \cdot 3^b \cdot 5^c \qquad \begin{cases} 0 \le a \le 1 \\ 0 \le b \le 3 \\ 0 \le c \le 0 \end{cases}$$

and the divisors of 45 are numbers of the form

$$d = 2^a \cdot 3^b \cdot 5^c. \qquad \begin{cases} 0 \le a \le 0 \\ 0 \le b \le 2 \\ 0 \le c \le 1 \end{cases}$$

Thus, we obtain the largest common divisor by choosing a, b, and c as large as possible while still satisfying both the above sets of inequalities. Hence, a must be the smaller of 0 and 1, b must be the smaller of 3 and 2, and c must be the smaller of 0 and 1. Therefore,

$$\text{GCD}(54, 45) = 2^0 \cdot 3^2 \cdot 5^0 = 9.$$

THEOREM *GCDs from Prime-Power Representations*

Let

$$a = p^{a_1} p^{a_2} \cdots p_r^{a_r} \qquad \text{and} \qquad b = p_1^{b_1} p_2^{b_2} \cdots p_r^{b_r}$$

be the prime power representations of a and b, respectively. Then

$$\text{GCD}(a, b) = p_1^{c_1} p_2^{c_2} \cdots p_r^{c_r},$$

where c_1 = the lesser of (a_1, b_1), c_2 = the lesser of (a_2, b_2), . . . , and c_r = the lesser of (a_r, b_r).

EXAMPLE 4.14 FINDING THE GREATEST COMMON DIVISOR BY USING PRIME-POWER REPRESENTATIONS

Compute the greatest common divisor of 504 and 3675.

Solution We first find the prime-power representation of each number:

$$\begin{array}{r} 3 \\ 3\overline{)9} \\ 7\overline{)63} \\ 2\overline{)126} \\ 2\overline{)252} \\ 2\overline{)504} \end{array} \qquad \begin{array}{r} 3 \\ 7\overline{)21} \\ 7\overline{)147} \\ 5\overline{)735} \\ 5\overline{)3675} \end{array}$$

$$504 = 2^3 \cdot 3^2 \cdot 5^0 \cdot 7^1 \qquad 3675 = 2^0 \cdot 3^1 \cdot 5^2 \cdot 7^2$$

Choosing the smaller of the exponents on each prime, we see that

$$\text{GCD}(504, 3675) = 2^0 \cdot 3^1 \cdot 5^0 \cdot 7^1 = 21.$$

As a check, we note that $504 \div 21 = 24$, that $3675 \div 21 = 175$, and that 24 and 175 have no common factor other than 1.

School Book Page

Fractions and Greatest Common Denominators

Chapter 7 Lesson 5

Simplest Form

You Will Learn
how to find the simplest form of a fraction

Vocabulary
simplest form
when the greatest common factor (GCF) of the numerator and denominator is 1

Learn

Did you ever wonder where the Dead Sea got its name? It's so salty that no fish can live in it. The Dead Sea is $\frac{24}{100}$, or $\frac{6}{25}$, salt. $\frac{6}{25}$ is in simplest form.

A fraction is in its **simplest form** if the greatest common factor (GCF) of the numerator and the denominator is 1.

You can find the simplest form of $\frac{18}{24}$ in more than one way.

One Way
Divide numerators and denominators by common factors until the GCF is 1.
$\frac{18 \div 2}{24 \div 2} = \frac{9}{12}$
$\frac{9 \div 3}{12 \div 3} = \frac{3}{4}$
$\frac{3}{4}$ is the simplest form for $\frac{18}{24}$.

Another Way
Divide both numerator and denominator by their GCF.
Factors of 18: 1, 2, 3, (6), 9, 18
Factors of 24: 1, 2, 3, 4, (6), 8, 12, 24
6 is the greatest common factor.
$\frac{18 \div 6}{24 \div 6} = \frac{3}{4}$

Did You Know?
The shore around the Dead Sea is 1,310 feet below sea level.

Talk About It

Which method is a short cut? Explain.

Check

Find the simplest form for each fraction.

1. $\frac{12}{16}$ 2. $\frac{6}{8}$ 3. $\frac{4}{10}$ 4. $\frac{10}{15}$ 5. $\frac{8}{12}$

6. **Reasoning** Explain why all fractions with a numerator of 1 are in simplest form.

310 Chapter 7 • Fractions and Mixed Numbers

Questions for the Teacher

1. Two different methods are shown above for writing a fraction in simplest form. Students may wonder why they both work. How would you explain to a student that they are equivalent?

2. To use the first of the above methods for simplifying a fraction, it is necessary to recognize common factors. How would you answer a student who asked you how to use the first method to simplify $\frac{143}{253}$? Could you use divisibility tests?

The Euclidean Algorithm

This method for finding greatest common divisors is found in Book IV of Euclid's *Elements,* written in about 300 B.C. It has the distinct advantage of working unfailingly no matter how large the two numbers are or how complicated the arithmetic is. It all depends on the division algorithm.

Dividing a by b, we obtain a quotient q and a remainder r such that

$$a = bq + r, \qquad 0 \leq r < b.$$

But then

$$r = a - bq.$$

Therefore, by the theorem of divisors of sums and differences, d divides a and d divides b if, and only if, d divides b and d divides r. Thus, a and b must have the same set of divisors as b and r and, hence, the same greatest common divisor.

THEOREM *The GCD and the Division Algorithm*
Let a and b be any two natural numbers and let q and r be determined by the division algorithm. Thus,

$$a = bq + r, 0 \leq r < b.$$

Then, $\text{GCD}(a, b) = \text{GCD}(b, r)$.

This theorem is the basis of the **Euclidean algorithm,** which we illustrate by determining $\text{GCD}(1539, 3144)$. We begin by dividing 3144 by 1539 and then continue by dividing successive divisors by successive remainders. It turns out that the last nonzero remainder is the desired greatest common divisor. Of course, these divisions can be performed by hand or with a calculator. (They are particularly easy using the INT ÷ key, as discussed in Section 3.6.) We obtain these divisions.

$$\begin{array}{r} 2\text{ R }66 \\ 1539\overline{)3144} \end{array} \quad \begin{array}{r} 23\text{ R }21 \\ 66\overline{)1539} \end{array} \quad \begin{array}{r} 3\text{ R }3 \\ 21\overline{)66} \end{array} \quad \begin{array}{r} 7\text{ R }0 \\ 3\overline{)21} \end{array}$$

In the form of the division algorithm, these results give

$$\begin{aligned} 3144 &= 2 \cdot 1539 + 66 \\ 1539 &= 23 \cdot 66 + 21 \\ 66 &= 3 \cdot 21 + 3 \\ 21 &= 3 \cdot 7 \end{aligned}$$

Hence, from the preceding theorem.

$$\text{GCD}(1539, 3144) = \text{GCD}(66, 1539) = \text{GCD}(21, 66) = \text{GCD}(3, 21) = 3$$

since 3 divides 21. Thus, the last nonzero remainder is the greatest common divisor of 1539 and 3144, as asserted.

HIGHLIGHT from HISTORY

Julia Robinson

The long struggle for women's rights is certainly reflected in mathematics. From the time of Hypatia (370?–A.D. 415) to the twentieth century, few women worked in mathematics—long considered a male domain. Though strides still need to be made, many women now ply the trade, and there is even an Association of Women in Mathematics (AWM). One of the most successful female mathematicians of our time was Professor Julia Bowman Robinson, solver of the tenth in David Hilbert's famous list of problems. Professor Robinson was born in St. Louis in 1919. Her graduate work in logic was done at the University of California where she received her Ph.D. in 1948. It was also at Berkeley that she met her husband-to-be, number theorist Raphael Robinson. After solving Hilbert's tenth problem in 1970, she was made a member of the prestigious National Academy of Sciences as well as the American Academy of Arts and Sciences and was promptly awarded the rank of Professor at UC–Berkeley. Professor Robinson served as President of the American Mathematical Society and was also active in the Association of Women in Mathematics. She died on July 30, 1985, hoping that she would not be remembered "as the first woman this or that"; instead, she stated, "I would prefer to be remembered, as a mathematician should, simply for the theorems I have proved and the problems I have solved." There is little doubt but that her hope will be realized.

THEOREM *The Euclidean Algorithm*

Let a and b be any two natural numbers. Using the division algorithm, determine natural numbers $q_1, q_2, \ldots, q_s$ and $r_1, r_2, \ldots, r_{s-1}$ such that

$$\begin{aligned} a &= bq_1 + r_1, & 0 &\le r_1 < b \\ b &= r_1q_2 + r_2, & 0 &\le r_2 < r_1 \\ r_1 &= r_2q_3 + r_3, & 0 &\le r_3 < r_2 \\ &\vdots \\ r_{s-3} &= r_{s-2}q_{s-1} + r_{s-1}, & 0 &\le r_{s-1} < r_{s-2} \\ r_{s-2} &= r_{s-1}q_s. \end{aligned}$$

Then, $\text{GCD}(a, b) = r_{s-1}$.

EXAMPLE 4.15 USING THE EUCLIDEAN ALGORITHM

Compute GCD(18,411, 1649) using the Euclidean algorithm.

Solution

Using the Euclidean algorithm, we have

$$1649\overline{)18{,}411}\ \ 11 \text{ R } 272 \qquad 272\overline{)1649}\ \ 6 \text{ R } 17 \qquad 17\overline{)272}\ \ 16 \text{ R } 0$$

and it follows that GCD(18,411, 1649) = 17.

The Least Common Multiple

In fourteenth-century France, the style of choral motet writing called *Ars Nova* was such that, while the tenor line was written in a repeated pattern of a fixed number of measures (an isorhythmic pattern), the other three parts were written in an isorhythmic pattern with a different number of measures. Moreover, the motets were designed so that the end of the motet, the end of a tenor pattern, and the end of a pattern for the other three parts all coincided. How short might such a motet be? Or, more generally, how long might such a motet need to be?

If the length of the tenor pattern is eight measures, the length of the motet must be a multiple of eight—8, 16, 24, 32, 40, 48, 56, 64, . . . measures. Also, if the length of the pattern of the other three parts is six measures, the possible lengths of the motet must be a multiple of six—6, 12, 18, 24, 30, 36, 42, 48, 54, 60, . . . measures. Indeed, the length of the motet must be a multiple of both six and eight and thus must appear in both lists. Such a number is called a **common multiple** of 6 and 8. Thus, in the present case, the possible lengths of the motet are the common multiples of 6 and 8—24, 48, 72, 96, . . . measures; and the shortest such motet is 24 measures long. Quite reasonably, the smallest number in a list of common multiples of two numbers is called their **least common multiple.**

In general, we have the following definition.

DEFINITION *Least Common Multiple*
Let a and b be natural numbers. The least natural number m that is a multiple of both a and b is called their **least common multiple,** and we write $m = \text{LCM}(a, b)$.

Did You Know?

Eureka!

"At last, shout of 'Eureka!' in age-old math mystery," read the headline in the *New York Times* on June 24, 1993. The article then proceeded to detail the announcement of the proof of Fermat's last theorem, a problem that had plagued mathematicians for over 350 years. It has been known since the time of the Babylonians (ca. 1600 B.C.) that there are infinitely many triples of natural numbers x, y, and z for which $x^2 + y^2 = z^2$. However, Pierre de Fermat, the greatest French mathematician of the seventeenth century, in 1637 noted in the margin of his copy of Diophantus's *Arithmetica* "that the equation $x^n + y^n = z^n$ has **no** solution in natural numbers if n is greater than two." He went on to write, ". . . and I have assuredly found an admirable proof of this, but the margin is too narrow to contain it." This produced a frenzy on the part of others to find the proof, and much first-rate mathematics was produced in the process. However, the proof resisted all efforts for over three centuries until Dr. Andrew Wiles, a British mathematician at Princeton University, using the most sophisticated of the accumulated methods, finally produced a proof. In view of the great effort required to construct it, it is seriously doubted that Fermat actually had a valid proof. Nevertheless, he had the insight to guess the result, and the efforts to provide a proof greatly enriched mathematics.

Least Common Multiples by Intersection of Sets

As with the greatest common divisor, the preceding discussion suggests a method for finding the least common multiple of two natural numbers. The following method is simple and works particularly well if the two numbers are not large. Let a and b be any two natural numbers. The sets of natural-number multiples of a and b are

$$M_a = \{a, 2a, 3a, \ldots\} \quad \text{and} \quad M_b = \{b, 2b, 3b, \ldots\}.$$

Therefore, $M_a \cap M_b$ is the set of all natural-number common multiples of a and b, and the least number in this set is the least common multiple of a and b. Incidentally, it should be noted that ab is clearly a common multiple of both a and b, so, in using this method to find $\text{LCM}(a, b)$, one need not extend the sets beyond this product.

EXAMPLE 4.16 FINDING A LEAST COMMON MULTIPLE BY SET INTERSECTION

Find the least common multiple of 9 and 15.

Solution

$$\begin{aligned} M_9 &= \{9, 18, 27, 36, 45, 54, 63, 72, 81, 90, \ldots\}; \\ M_{15} &= \{15, 30, 45, 60, 75, 90, 105, \ldots\}; \\ M_9 \cap M_{15} &= \{45, 90, \ldots\}. \end{aligned}$$

Since $M_9 \cap M_{15}$ is the set of all natural-number common multiples of 9 and 15, the least element of this set is $\text{LCM}(9, 15)$. Therefore, $\text{LCM}(9, 15) = 45$. Notice that $\text{GCD}(9, 15) = 3$, so $\text{GCD}(9, 15) \cdot \text{LCM}(9, 15) = 3 \cdot 45 = 135 = 9 \cdot 15$.

Least Common Multiples from Prime-Power Representations

Suppose we want to find the least common multiple of 54 and 45. Computing the prime-power representations of these numbers, we obtain

$$54 = 2^1 \cdot 3^3 \cdot 5^0 \quad \text{and} \quad 45 = 2^0 \cdot 3^2 \cdot 5^1,$$

where we use the zero exponents to make it *appear* that 54 and 45 are products of powers of the same primes. By the theorem just preceding Example 4.5, the multiples of 54 are numbers of the form

$$m = 2^a \cdot 3^b \cdot 5^c \qquad \begin{cases} a \geq 1 \\ b \geq 3 \\ c \geq 0, \end{cases}$$

and the multiples of 45 are numbers of the form

$$m = 2^a \cdot 3^b \cdot 5^c \qquad \begin{cases} a \geq 0 \\ b \geq 2 \\ c \geq 1 \end{cases}$$

We obtain the least common multiple of 54 and 45 by choosing a, b, and c as small as possible while still satisfying both the above sets of inequalities. It follows that we must choose a to be the larger of 1 and 0, b the larger of 3 and 2, and c the larger of 0 and 1. Hence,

$$m = 2^1 \cdot 3^3 \cdot 5^1 = 270 = \text{LCM}(54, 45).$$

Since this argument could be repeated in general, we have the following theorem.

THEOREM *LCMs from Prime-Power Representations*
Let

$$a = p_1^{a_1} p_2^{a_2} \cdots p_s^{a_s} \qquad \text{and} \qquad b = p_1^{b_1} p_2^{b_2} \cdots p_s^{b_s}$$

be the respective prime-power representations of a and b. Then

$$\text{LCM}(a, b) = p_1^{d_1} p_2^{d_2} \cdots p_s^{d_s},$$

where d_1 = the larger of (a_1, b_1), d_2 = the larger of (a_2, b_2), . . . , and d_s = the larger of (a_s, b_s).

EXAMPLE 4.17 FINDING LCMS AND GCDS BY USING PRIME-POWER REPRESENTATIONS

Compute the least common multiple and greatest common divisor of $r = 2^2 \cdot 3^4 \cdot 7^1$ and $s = 3^2 \cdot 5^2 \cdot 7^3$.

Solution

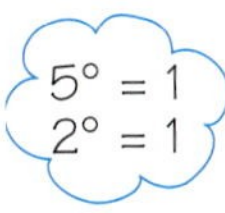

Both the least common multiple and the greatest common divisor will be of the form $2^a \cdot 3^b \cdot 5^c \cdot 7^d$. For the greatest common divisor, we choose the *smaller* of the two exponents with which each prime appears in r and s, and for the least common multiple, we choose the *larger* of each pair of exponents. Thus, since

$$r = 2^2 \cdot 3^4 \cdot 5^0 \cdot 7^1 = 2268 \text{ and } s = 2^0 \cdot 3^2 \cdot 5^2 \cdot 7^3 = 77{,}175,$$

we have that

$$\text{GCD}(r, s) = 2^0 \cdot 3^2 \cdot 5^0 \cdot 7^1 = 63$$

and

$$\text{LCM}(r, s) = 2^2 \cdot 3^4 \cdot 5^2 \cdot 7^3 = 2{,}778{,}300.$$

Since both exponents on each prime were used in finding $\text{GCD}(r, s)$ and $\text{LCM}(r, s)$, it follows that

$$rs = \text{GCD}(r, s) \cdot \text{LCM}(r, s).$$

Thus,

$$rs = 2268 \cdot 77{,}175 = 175{,}032{,}900$$

and

$$\text{GCD}(r, s) \cdot \text{LCM}(r, s) = 63 \cdot 2{,}778{,}300 = 175{,}032{,}900.$$

Since the last part of the solution in Example 4.17, showing that $rs = \text{GCD}(r, s) \cdot \text{LCM}(r, s)$, could be repeated in general, we have the following important theorem.

THEOREM $ab = GCD(a, b) \cdot LCM(a, b)$
If a and b are any two natural numbers, then $ab = \text{GCD}(a, b) \cdot \text{LCM}(a, b)$.

Least Common Multiples by Using the Euclidean Algorithm

As noted above, if a and b are natural numbers, then $\text{GCD}(a, b) \cdot \text{LCM}(a, b) = ab$. Thus,

$$\text{LCM}(a, b) = \frac{ab}{\text{GCD}(a, b)},$$

and we have already learned that $\text{GCD}(a, b)$ can always be found using the Euclidean algorithm.

EXAMPLE 4.18 FINDING A LEAST COMMON MULTIPLE BY USE OF THE EUCLIDEAN ALGORITHM

Find the least common multiple of 2268 and 77,175 by using the Euclidean algorithm.

Solution These are the same two numbers treated in Example 4.17. However, here our procedure is totally different. Using the Euclidean algorithm, we have these divisions:

$$2268\overline{)77{,}175} \quad 34 \text{ R } 63 \qquad\qquad 63\overline{)2268} \quad 36 \text{ R } 0$$

Since the last nonzero remainder is 63, $\text{GCD}(2268, 77{,}175) = 63$ and

$$\begin{aligned}\text{LCM}(2268, 77{,}175) &= \frac{2268 \cdot 77{,}175}{63}\\ &= 2{,}778{,}300,\end{aligned}$$

as before.

Using Technology to Find GCDs and LCMs

Graphing calculators often have built-in GCD and LCM programs. For example, on the TI-73, the functions **lcm** and **gcd** are found by pressing [MATH]. These work quickly at the press of just a few buttons and are very handy. Also, the program EUCLID in Appendix C finds both the GCD and the LCM of two numbers in a single series of steps. The GCD and LCM functions are also available on spreadsheets like Excel.

Problem Set 4.3

Understanding Concepts

1. Find the greatest common divisor of each of these pairs of numbers by the method of intersection of sets of divisors.
   (a) 24 and 27
   (b) 14 and 22
   (c) 48 and 72
2. Find the least common multiple of each of these pairs of numbers by the method of intersection of sets of multiples.
   (a) 24 and 27
   (b) 14 and 22
   (c) 48 and 72
3. Use the results of problems 1 and 2 to show that
   (a) $24 \cdot 27 = \text{GCD}(24, 27) \cdot \text{LCM}(24, 27)$.
   (b) $14 \cdot 22 = \text{GCD}(14, 22) \cdot \text{LCM}(14, 22)$.
   (c) $48 \cdot 72 = \text{GCD}(48, 72) \cdot \text{LCM}(48, 72)$.
4. Use the method based on prime-power representations to find the greatest common divisor and least common multiple of each of these pairs of numbers.

(a) $r = 2^2 \cdot 3^1 \cdot 5^3$ and $s = 2^1 \cdot 3^3 \cdot 5^2$

(b) $u = 5^1 \cdot 7^2 \cdot 11^1$ and $v = 2^2 \cdot 5^3 \cdot 7^1$

(c) $w = 2^2 \cdot 3^3 \cdot 5^2$ and $x = 2^1 \cdot 5^3 \cdot 7^2$

5. Use the Euclidean algorithm to find each of the following.

(a) GCD(3500, 550) and LCM(3500, 550)

(b) GCD(3915, 825) and LCM(3915, 825)

(c) GCD(624, 1044) and LCM(624, 1044)

Teaching Concepts

6. How would you respond to a student who claims that GCD$(0, n) = n$ for every whole number n?

7. Since n divides 0 for every natural number n, how would you respond to a student who claims that LCM$(0, n) = 0$ for every natural number n?

8. How would you respond to Tawana, who claims that mn is the least common multiple of m and n in every case?

9. Paulita says, "Why do we have to learn about greatest common divisors, anyway? It seems like a waste of time to me." How would you respond to Paulita?

10. Gerald says, "I can see that greatest common divisors can be useful, but how about least common multiples? They seem like just a lot of busywork to me." How would you respond to Gerald?

Thinking Critically

11. The following problem is from Claudia Zaslavsky's *The Multicultural Classroom* (Portsmouth, NH: Heineman, 1996, p. 113). "Challenge upper-grade students to calculate the interval in days between the two successive New Year's Days that coincide in both Mayan calendars—the 365-day everyday calendar and the 260-day ritual calendar. What is the greatest common divisor (GCD) of both numbers? What is the least common multiple?" Solve Claudia's problem.

12. The notions of the greatest common divisor and the least common multiple extend naturally to more than two numbers. Moreover, the prime-power method extends naturally to finding GCD(a, b, c) and LCM(a, b, c).

(a) If $a = 2^2 \cdot 3^1 \cdot 5^2$, $b = 2^1 \cdot 3^3 \cdot 5^1$, and $c = 3^2 \cdot 5^3 \cdot 7^1$, compute GCD$(a, b, c)$ and LCM(a, b, c).

(b) Is it necessarily true that GCD$(a, b, c) \cdot$ LCM$(a, b, c) = abc$?

(c) Find numbers r, s, and t such that GCD$(r, s, t) \cdot$ LCM$(r, s, t) = rst$.

13. Use the method of intersection of sets to compute the following.

(a) GCD(18, 24, 12) and LCM(18, 24, 12)

(b) GCD(8, 20, 14) and LCM(8, 20, 14)

(c) Is it true that GCD(8, 20, 14) · LCM(8, 20, 14) = 8 · 20 · 14?

14. (a) Compute GCD(6, 35, 143) and LCM(6, 35, 143).

(b) Is it true that GCD(6, 35, 143) · LCM(6, 35, 143) = 6 · 35 · 143?

(c) Guess under what conditions GCD$(a, b, c) \cdot$ LCM$(a, b, c) = abc$.

15. (a) Compute GCD(GCD(24, 18), 12) and LCM(LCM(24, 18), 12) and compare with the results of problem 13(a).

(b) Compute GCD(GCD(8, 20), 14) and LCM(LCM(8, 20), 14) and compare the results with the results of problem 13(b).

(c) In one or two written sentences, state a conjecture based on parts (a) and (b).

---

16. Cuisenaire® rods are colored rods 1 centimeter square and of lengths 1 cm, 2 cm, 3 cm, . . . , 10 cm as shown here.

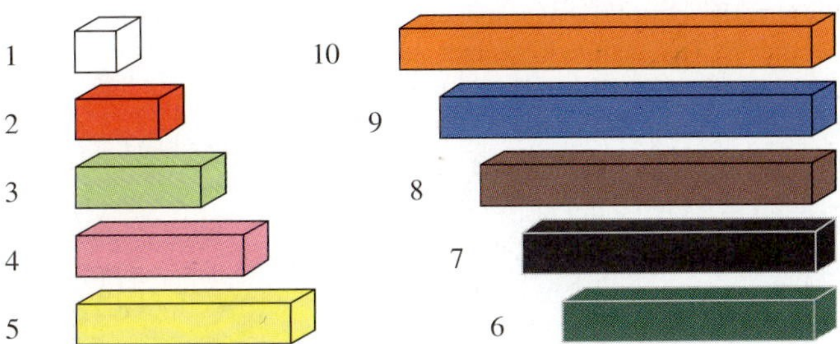

We say that a smaller rod *measures* a longer rod if a number of the smaller rods placed end to end are the same length as the given rod.

(a) Which rods will measure the 9-rod?

(b) Which rods will measure the 6-rod?

(c) What is the greatest common divisor of 9 and 6?

(d) Place the 10-rod and the 8-rod end to end to form an "18-train" as in the Cooperative Investigation box in Section 2.3, page 33. Which rods or trains will measure the 18-train?

(e) Place the 7-rod and the 5-rod end to end to form a 12-train. Which rods or trains will measure the 12-train?

(f) What is the greatest common divisor of 12 and 18?

(g) Briefly describe how you might use Cuisenaire® rods to demonstrate the notion of greatest common divisor to children.

---

17. Cuisenaire® rods are described in the preceding problem.

(a) What is the shortest train or length that can be measured by both 4-rods and 6-rods?

(b) What is the least common multiple of 4 and 6?

(c) What is the shortest train or length that can be measured by both 6-rods and-9 rods?

(d) What is the least common multiple of 6 and 9?

(e) Briefly describe how you could use Cuisenaire® rods to demonstrate the notion of least common multiple to children.

18. (a) Indicate how you could use a number line to illustrate the notion of greatest common divisor to children.

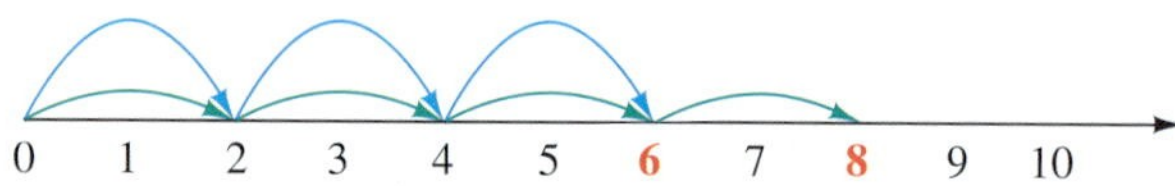

(b) Indicate how you could use a number line to illustrate the notion of least common multiple to children.

19. The greatest-common-divisor machine of Andres Zavrotsky (U.S. patent number 2 978 816, April 11, 1961) is described in Martin Gardner's *The Sixth Book of Mathematical Games from Scientific American* (San Francisco: W. H. Freeman, 1971). For example, build a 6-by-8 "billiard table" out of mirrors and shine a light at 45° from a corner *P*. The beam will bounce around the "table" and will eventually be absorbed at a corner. Let *Q* be the point on the long side of the "table" closest to *P* where the light beam "bounces." Then

$$\text{GCD}(6, 8) = (1/2) \cdot PQ = (1/2) \cdot 4 = 2.$$

Using graph paper, draw a similar diagram to determine GCD(9, 15).

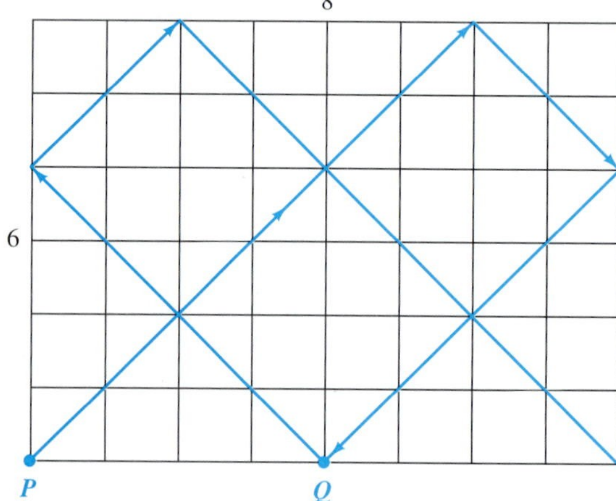

20. A simple graphical scheme for finding the greatest common divisor of two numbers is illustrated below. Suppose again that we want to compute GCD(6, 8). Draw a 6-by-8 rectangle and draw a diagonal from corner to corner.

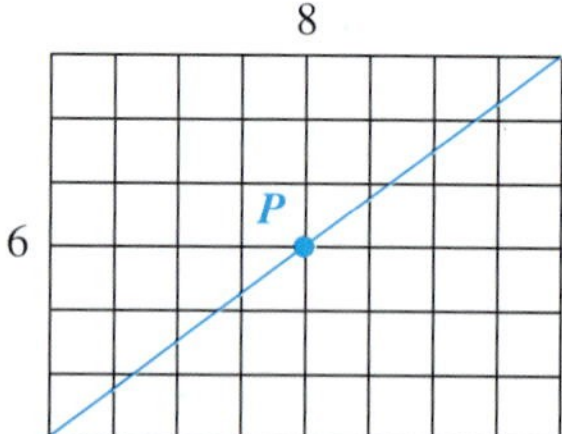

In this case, the diagonal passes through just one point *P* that is a corner of squares on the graph paper. *P* divides the diagonal into two parts, and 2 = GCD(6, 8). Use graph paper and draw a similar diagram to determine GCD(15, 25).

Thinking Cooperatively

21. Work through this problem as a group project. You will particularly want to consult with the members of your group as you respond to parts (e), (f), (g), and (h). Recall that the Fibonacci sequence was defined in Example 1.6, in Section 1.4. For easy reference, the first 30 numbers in the sequence are listed here. Also recall that we use the notation F_n to designate the nth Fibonacci number. Thus, $F_1 = F_2 = 1$, $F_3 = 2$, and so on.

n	1	2	3	4	5	6	7	8	9	10
F_n	1	1	2	3	5	8	13	21	34	55

n	11	12	13	14	15	16	17	18	19	20
F_n	89	144	233	377	610	987	1597	2584	4181	6765

n	21	22	23	24	25	26	27	28	29	30
F_n	10,946	17,711	28,657	46,368	75,025	121,393	196,418	317,811	514,229	832,040

(a) In problem 30 of Problem Set 4.1, you should have guessed that for $m > 2$, F_m divides F_n if, and only if, m divides n. Compute each of these quotients: $F_{12} \div F_6$, $F_{18} \div F_9$, and $F_{30} \div F_{15}$.

(b) If n is composite, must F_n be composite? Explain briefly.

(c) If n is a prime, must F_n be a prime? Explain briefly.

(d) Complete the following list.

$\text{GCD}(F_6, F_9) = \text{GCD}(8, 34) = 2 = F_3$ $\quad$ $\text{GCD}(6, 9) = 3$

$\text{GCD}(F_{14}, F_{21}) = \text{GCD}(377, 10{,}946) = 13 = F_7$ $\quad$ $\text{GCD}(14, 21) = 7$

$\text{GCD}(F_{10}, F_{15}) =$ $\quad$ $\text{GCD}(10, 15) =$

$\text{GCD}(F_{20}, F_{30}) =$ $\quad$ $\text{GCD}(20, 30) =$

$\text{GCD}(F_{16}, F_{24}) =$ $\quad$ $\text{GCD}(16, 24) =$

$\text{GCD}(F_{12}, F_{18}) =$ $\quad$ $\text{GCD}(12, 18) =$

(e) On the basis of the calculations in part (d), make a conjecture about $\text{GCD}(F_m, F_n)$.

(f) Test your conjecture by computing $\text{GCD}(F_{24}, F_{28})$.

(g) Does the result of part (f) prove that your conjecture in part (e) is correct?

(h) If $\text{GCD}(F_{16}, F_{20})$ were to equal 4, what could you conclude about your conjecture in part (e)?

(i) Actually compute $\text{GCD}(F_{16}, F_{20})$.

Making Connections

22. Fractions will not be discussed in this text until Chapter 6. However, in elementary school you should have learned that a fraction a/b can be written in simplest terms (i.e., as c/d, where c and d have no common factor other than 1) by dividing both a and b by $\text{GCD}(a, b)$. For example, $16/20 = 4/5$, since $\text{GCD}(16, 20) = 4$, $16 \div 4 = 4$, and $20 \div 4 = 5$. Compute the following:

(a) $\text{GCD}(6, 8)$ **(b)** $\text{GCD}(18, 24)$

(c) $\text{GCD}(132, 209)$ **(d)** $\text{GCD}(315, 375)$

Determine a fraction in simplest terms equivalent to each of these.

(e) $6/8$ **(f)** $18/24$ **(g)** $132/209$ **(h)** $315/375$

23. The front wheel of a tricycle has a circumference of 54 inches, and the back wheels have a circumference of 36 inches.

If points P and Q are both touching the sidewalk when Marja starts to ride, how far will she have ridden when P and Q first touch the sidewalk at the same time again?

24. Sarah Speed and Hi Velocity are racing cars around a track. If Sarah can make a complete circuit in 72 seconds and Hi completes a circuit in 68 seconds,

(a) how many seconds will it take for Hi to first pass Sarah at the starting line?

(b) how many laps will Sarah have made when Hi has completed one more lap than she has?

25. In a machine, a gear with 45 teeth is engaged with a gear with 96 teeth, with teeth A and B on the small and large gears, respectively, in contact as shown. How many more revolutions will the small gear have to make before A and B are again in the same position?

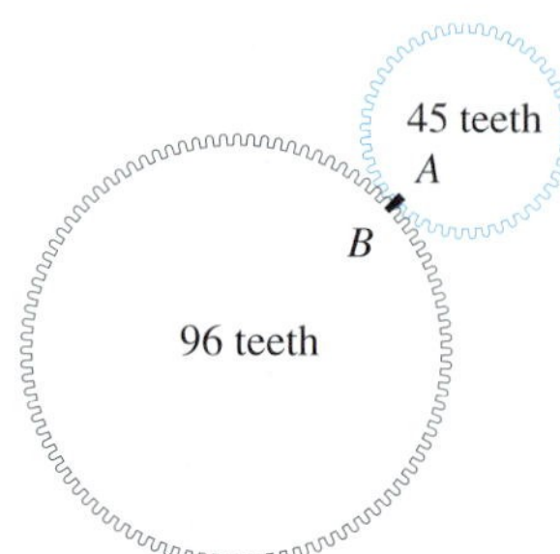

26. The musical notes A and E just below and above middle C have fundamental frequencies of 220 and 330 cycles per second (cps), respectively. Whenever these notes are sounded, overtones whose frequencies are multiples of the fundamental frequencies are also generated. A and E harmonize because many of their overtones have the same frequencies (660, 1320, 1980, . . .). The closer the least common multiple of the fundamental frequencies is to the fundamental frequencies, the better the harmony is.

(a) A, C, and C-sharp have frequencies of 220, 264, and 275 cps, respectively. Compute LCM(220, 264), LCM(220, 275), and LCM(264, 275).

(b) On the basis of part (a), which pair of these three notes will produce the most pleasing harmony?

(c) Compute GCD(220, 264), GCD(220, 275), and GCD(264, 275).

(d) How can the results of part (c) tell you if two notes harmonize nicely?

From State Student Assessments

27. (Texas, Grade 4)

Which is the least common multiple of 12 and 15?

A. 45 B. 48 C. 60 D. 72

Using a Graphing Calculator

For problems 28, 29, 30, and 32, consult Appendix C.

28. Compute the following, using DIVISORS.

(a) GCD(45, 48) (b) GCD(48, 54)

29. Use MULTPLS to compute each of these numbers.

(a) LCM(45, 48) (b) LCM(48, 54)

30. Use FACTOR to compute each of the following.

(a) The prime-power representation of 205800, 31460, and 25840.

(b) GCD(31460, 205800)

(c) LCM(31460, 205800)

(d) GCD(205800, 31460, 25840)

(e) LCM(205800, 31460, 25840)

31. Compute the following, using EUCLID.

(a) GCD(36461, 33269)

(b) GCD(16583, 16377)

(c) LCM(36461, 33269)

(d) LCM(16583, 16377)

For Review

32. (a) Check to see if $n^2 - 81n + 1681$ is prime for $n = 1, 2, 3, 4,$ and 5.

(b) On the basis of part (a), make a conjecture about the values of $n^2 - 81n + 1681$.

(c) Does the fact that your conjecture is true for $n = 1$, 2, 3, 4, and 5 mean that the conjecture is always true? Explain briefly.

(d) Check your conjecture for $n = 6$, 7, and 8. What do you conclude? Explain briefly.

(e) Check your conjecture for $n = 80$. How does this affect your belief in your conjecture? Explain.

(f) Check your conjecture for $n = 81$. What do you conclude? Explain briefly.

33. Let $n = 2 \cdot 5 \cdot 7 + 1$.

(a) Could 2, 5, or 7 divide n evenly? Why or why not?

(b) Is n prime or composite?

34. Let $n = 2 \cdot 3 \cdot 5 + 7 \cdot 11 \cdot 13$.

(a) Do any of 2, 3, 5, 7, 11, or 13 divide n evenly? Explain briefly.

(b) Must n have a prime divisor different from 2, 3, 5, 7, 11, or 13? Explain briefly.

(c) Is n prime or composite? (*Hint:* Use the program FACTOR on a graphing calculator.)

35. Let $n = 2 \cdot 3 \cdot 5 + 7 \cdot 11 \cdot 13 + 17 \cdot 19 \cdot 23$.

(a) Do any of 2, 3, 5, 7, 11, 13, 17, 19, or 23 divide n evenly? Explain briefly.

(b) Can you argue as in problem 34(b) that n must have a prime divisor different from 2, 3, 5, 7, 11, 13, 17, 19, and 23?

(c) Find the prime-power representation of n, using the program FACTOR on a graphing calculator.

4.4 CODES AND CREDIT CARD NUMBERS: CONNECTIONS TO NUMBER THEORY

Secret and not-so-secret codes have been used for centuries for both serious and recreational purposes. Aside from their considerable utility, the aura of mystery often associated with codes makes them attractive enrichment topics that constantly surprise and fascinate students. We discuss zip codes and credit card numbers here not only because they are attractive to students, but also because they provide an interesting connection to number theory. International Standard Book Number (ISBN) codes will be considered in the problem set at the end of this section.

Bar Codes on Envelopes and Postcards

We begin with a question: What is the purpose of the strings of long and short vertical lines or bars that appear today on virtually every piece of mail? Surely the markings are placed on the cards and envelopes for a purpose. They must mean something; they must convey some sort of information.

That means the string of marks must be a code of some sort. But what is secret about business reply cards and other pieces of mail? Why should there be coded information on each such card? Since it is unlikely that the information is intended to be kept secret, it is no doubt coded for some other purpose. Do you suppose that it is encoded in this way so that it is machine readable? Actually, given the ever increasing use of automation, coupled with the steady increase in the amount of mail being processed, it is quite likely that this has something to do with the automated sorting of mail. Consider the bar codes shown on the cards in Figure 4.4.

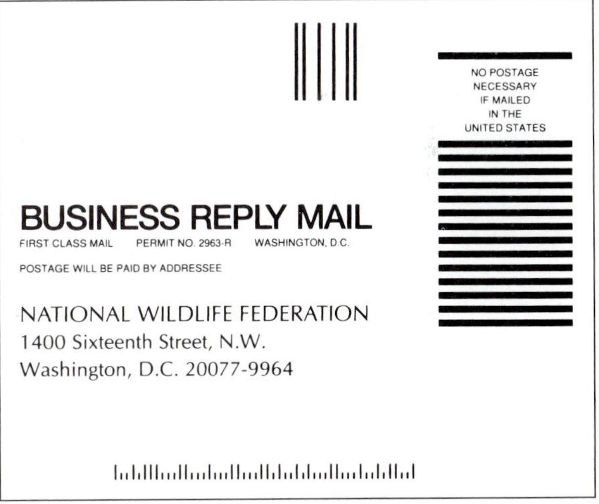

Figure 4.4
Bar codes on postcards

The first bit of information needed to deliver a piece of mail is the post office to which the mail must be sent, and this is determined from the five-digit zip code. Actually, we all have a nine-digit zip code, as shown on the postcards, with the first two of the last four digits indicating a portion of a delivery area (for example, a specific delivery route) and the last two digits indicating a specific business or organization, a specific building, or a specific portion (say, a part of a street) of a delivery route. Do you suppose the bar code in question gives the zip code in machine-readable form? Let's analyze the bar codes and see if we can associate them with the zip codes.

Did You Know?

Public-Key Encryption

Number theory is usually considered a part of pure mathematics—that is, mathematics studied for its own sake with no thought that it might be, or even could be, applied to real-world problems. Who would have guessed that the notions of primality and factoring would turn out to provide the basis for a simple and remarkably secure method of sending secret messages in code? The idea is to determine two 100-digit primes, p and q, and to publish the product pq for all to see. Using this "key," anyone can send you a message, but no one can read the message unless they know p and q; that is, unless they are able to factor the 200-digit product pq. While it is a three- or four-minute job to determine two 100-digit primes on one of today's fastest computers, it would take one of these same machines on the order of 10^9 years (that is, one billion years) to factor the product pq. Thus, unless someone has discovered a way to break the code without determining p and q—a very unlikely circumstance—it is effectively unbreakable. The importance of the existence of such a code in our modern society, with its need to keep masses of data (including financial records, industrial secrets, and computer databases of all sorts) secret, cannot be overstated.

To simplify the work, we repeat here the two zip codes and what we guess are the associated bar codes.

07713-0001

20077-9964

Looking for patterns and trying to associate the numbers with the bar code, we note that each string begins and ends with a long bar. In fact, this is always true, as you can easily check by finding other examples in your mailbox. These two bars probably tell the machine when the code starts and stops. Doing a count, we find that there are 52 bars in each code. Thus, deleting the start and stop bars, there are 50 bars to determine the zip code. Since there are nine numbers and a dash in the zip code, this may mean that all code groups are of length five. Thus, we break the bar code up into five-bar groups as shown below and try to associate the successive numbers in the zip code with the successive groups. As a check, we can use the fact that several digits are repeated both within and between the two codes. For the first code, we obtain the pairings shown. But this doesn't make sense because then 0 and the dash correspond to the same code group.

0 7 7 1 3 – 0 0 0 1

Indeed, given the four 0s in this code, it appears that no code group corresponds to the dash and that ||ııı corresponds to 0. This means that the last two code groups both correspond to 1s, so there must be additional information in the code. Let's look again at both codes, this time leaving out the dash. We obtain these pairings.

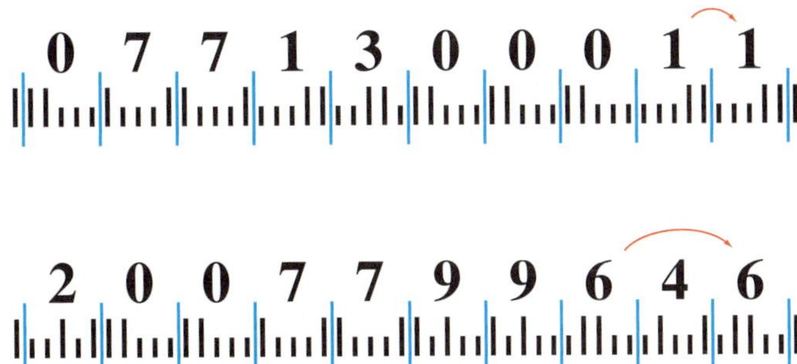

Note that each of 0, 1, and 7 appears more than once in these codes and that each time they appear, they have the same representation in the bar code! This is very strong evidence that we are correct, that the bar code represents the zip code, and that the code representations are as shown here. Of course, the last 1 and 6 in the above diagrams were determined by the earlier appearance of the same five-bar patterns in each zip code. Finally, it is not clear from the illustration what code group should correspond to each of 5 and 8, though it might be guessed with careful study of the following patterns. Otherwise, these code groups can be determined by looking at other pieces of mail whose zip codes contain these digits.

We first guessed that the bar code must indeed be a code and that it likely gives the zip code in machine-readable form. We then noticed that both codes have 52 bars, beginning and ending with a long bar. Ignoring the first and last bars, we had a bar code with 50 bars. Our guess was that each five consecutive bars represented a digit that, in turn, could be determined by looking at the printed zip code.

This gave us 10 digits and a remaining question regarding the tenth digit. Trying a variety of possibilities, we find that

$$0 + 7 + 7 + 1 + 3 + 0 + 0 + 0 + 1 + 1 = 20$$

and

$$2 + 0 + 0 + 7 + 7 + 9 + 9 + 6 + 4 + 6 = 50.$$

It appears that the tenth digit in each case, here 1 and 6, respectively, has been chosen so that the sum of all the digits is a multiple of 10. This suggests that the tenth digit is a *check digit* that is included so that incorrect codes can be detected by machine. In fact, this is the case, and, if an error is detected in this way, the piece of mail is automatically shunted aside for human inspection and correction of the zip code.

EXAMPLE 4.19 DETERMINING BAR CODES FROM ZIP CODES

Determine the bar code corresponding to each of these zip codes.

(a) 86004–7548
(b) 99164–1263

Solution

Careful consideration of the above pattern of bar code groups for digits (or by checking other pieces of mail) reveals that the code groups for 5 and 8, respectively, are ı|ı|ı and |ıı|ı. Thus, being sure to put in the long start and stop bars and remembering that the last code group must correspond to a digit that makes the sum of all the digits a multiple of 10, we obtain the following.

(a)

(b)

EXAMPLE 4.20 DETERMINING ZIP CODES FROM BAR CODES

Determine the zip code corresponding to each of these bar codes. In each case, compute the sum of the digits corresponding to the bar code, and be sure that the check digit makes all of the digits add to a multiple of 10.

(a)

(b)

Solution

(a) Discarding the start and stop bars, we partition the remaining bars into five-bar groups. Checking the above code group representations, we find that the zip code here is 87403–5671. Since the sum of these digits is 41, the check digit is 9, and this corresponds to |ı|ıı as shown.

(b) Proceeding as in part (a), the zip code is 02116–7574. Since the sum of these digits is 33, the check digit is 7, and this corresponds to |ııı| as shown.

Credit Card Numbers

As is certainly well known, almost everyone today has a credit card. Credit cards are used for purchases of all sorts—to buy items in local stores, to make purchases over the Internet, to make reservations at hotels and motels, and even to make charitable contributions. Except for paying for the smallest purchases, they almost seem to have made money obsolete.

With the heavy use of credit cards, it is imperative that their authenticity be easily and quickly verifiable. This is so in stores where the card is used in person, but it is even more critical when the transaction is made over the Internet or by telephone. For this reason, credit card numbers are very carefully determined so that identification is swift and sure.

One step taken to help in the verification of a card has been to associate the first one or several digits of the card number with the issuing company. Thus, numbers on VISA cards always begin with a 4. MasterCard numbers always start with 51, 52, 53, 54, or 55. Discover cards start with 6011, and American Express cards start with 34 or 37. If a person states that he or she has a certain type of card, but the beginning digits on the card are not correct, then it is either a phony card or the number has been misstated.

But there is more. One of the most common errors in typing a number when, for example, ordering something over the Internet is that of transposing two adjacent digits. The card-numbering scheme has been devised to immediately catch such an error so that the buyer can be quickly informed and can resubmit a corrected number. The following shows how the check for transposition errors is made.

Consider the following possibility for a credit card number. (Note that most credit card numbers have 16 digits, but even if they don't, the checking procedure is still the same.)

$$5490 \quad 1234 \quad 5678 \quad 9123$$

Step 1. Counting from the right, underline the digits in the even-numbered positions as shown.

$$\underline{5}\,4\,\underline{9}\,0 \quad \underline{1}\,2\,\underline{3}\,4 \quad \underline{5}\,6\,\underline{7}\,8 \quad \underline{9}\,1\,\underline{2}\,3$$

Step 2. Double the underlined digits in Step 1.

$$10 \quad 18 \quad 2 \quad 6 \quad 10 \quad 14 \quad 18 \quad 4$$

Step 3. Determine the sum of the *digits* (not the numbers themselves) of the numbers in Step 2.

$$1 + 0 + 1 + 8 + 2 + 6 + 1 + 0 + 1 + 4 + 1 + 8 + 4 = 37$$

Step 4. Determine the sum of the digits not underlined (i.e., those in the odd-numbered positions) in Step 1.

$$3 + 1 + 8 + 6 + 4 + 2 + 0 + 4 = 28$$

Step 5. Add the two numbers obtained in Steps 3 and 4.

$$37 + 28 = 65$$

The number 65 is a check sum that can be computed almost instantaneously by computer. It gives a quick check (see below) that a transposition has or has not been made. Immediately informing the card user that an error has been made enables the user to resubmit the correct number and continue the transaction. A card number with a valid check sum is then processed further to ensure that it corresponds to an active account.

But we still have not explained how the check sum reveals that a transposition has been made. To discover this, note that the following are all possibilities for correct credit card numbers.

(a) 4 5 3 7 6 0 0 1 2 4 8 3 3 2 7 9
(b) 3 4 4 1 8 4 2 4 5 5 3 0 0 1 6 6
(c) 5 1 2 0 0 0 5 3 6 4 0 0 2 1 1 6

Computing the check sum for these three numbers, we obtain 70, 60, and 30, respectively. If you have a credit card, compute the check sum for your card and compare it with the three check sums just computed. A glance reveals that all the check sums are multiples of 10, a requirement that must be satisfied if the number is a correct possibility for a valid credit card number.

EXAMPLE 4.21 COMPUTING A MISSING CREDIT CARD NUMBER DIGIT

What digit must replace d to make the following a possibility for a valid credit card number?

$$6\,0\,1\,1 \quad 3\,0\,0\,2 \quad 3\,2\,5\,1 \quad 6\,3\,0\,d$$

Solution Counting from the right, we find that the digits in the even-numbered positions are 0, 6, 5, 3, 0, 3, 1, and 6, and their respective doubles are 0, 12, 10, 6, 0, 6, 2, and 12. The sum of the digits in these numbers is 21, and the sum of the digits in the odd-numbered positions is $d + 3 + 1 + 2 + 2 + 0 + 1 + 0 = d + 9$. Thus, the check sum is $d + 30$, so d must be 0 if this sum is to be a multiple of 10.

Problem Set 4.4

Understanding Concepts

1. Determine the check digit for each of these zip codes.
   (a) 98243-3712
   (b) 02116-4780
   (c) 27109-4423

2. Write the bar code for each of these zip codes.
   (a) 99164-3113
   (b) 18374-2147
   (c) 38423-1747

3. Write the zip code given by each of these bar codes, unless the code is necessarily incorrect. If it is necessarily incorrect, tell why.
   (a)
   (b)
   (c)

4. The following are correct credit card numbers that could be assigned to customers. If they were used to activate accounts, what company would be the issuer in each case?
   (a) 5400 4211 1012 0012 (b) 3421 7002 1101 2307
   (c) 6011 2210 3200 4257 (d) 4020 3410 0012 2307

5. Which of the following could be valid credit card numbers?
   (a) 5123 0011 1201 5406 (b) 3400 2013 2416 0104
   (c) 2140 3761 4002 0109 (d) 4010 1011 2103 4578

6. In each of these, determine d so that the number is a valid credit card number possibility.
   (a) 4d01 1032 4761 2003 (b) 530d 1021 4630 1023

Teaching Concepts

7. (a) Eli says that he can see if the number

   5210 0123 1010 2304

   is a correct possibility for a credit card number by computing this sum:

   $4 + 0 + 3 + 4 + 0 + 2 + 0 + 2 + 3 + 4 + 1 + 0 + 0 + 2 + 2 + 1 + 0.$

   How would you respond to Eli? Is he correct?

   (b) Is the number a correct possibility for a credit card number?

8. Trudi says that she can check the number in problem 7 for correctness by computing this sum:

$$1 + 0 + 2 + 2 + 0 + 0 + 1 + 4 + 3 + 2 + 0 + 2 + 0 + 4 + 3 + 0 + 4.$$

How would you respond to Trudi? Is she correct?

9. Rebecca says that the zip code for her New York address is 92413-7612. Is Rebecca mistaken? How would you respond to her? How are zip codes distributed across the United States?

Thinking Critically

10. A blot covers the first few bars of this correct zip code. Determine the first five-bar code group.

11. Could the following possibly be a correct bar code for a zip code? Explain.

12. **(a)** How many ways can the two long bars and the three short bars in a zip code group be arranged in order? Explain how you determined your answer.

(b) Does your answer to part (a) suggest that the bar code equivalents of the digits forming a zip code are particularly appropriate? Explain.

Thinking Cooperatively

For the following problems, work with four or five other students. Your answers should be consensus answers for the entire group.

13. In transmitting a credit card number over the Internet, 96, with the 9 in an even position counting from the right, was transposed to 69; and 58, with the 5 also in an even position counting from the right, was transposed to 85. Would the check sum described above catch such an error?

14. In transmitting a credit card number over the Internet, if 09 was transposed to 90 or vice versa, would the error be caught by computing the check sum? Explain.

15. **(a)** How many pairs of digits can possibly cause a transposition error in transmitting a credit card number? Explain.

(b) Divide the possible error-causing transposition pairs between members of the group in order to share the work of checking all pairs to see which pairs cause errors that would not be caught by the check sum.

Making Connections

16. Since about 1972, all books published anywhere in the world have been given an identifying number called an International Standard Book Number (ISBN). These numbers greatly facilitate buying and selling books, inventory control, and so on. A typical ISBN is

0-13-257502-7

and, somewhat like a zip code, the last digit is a check digit. It works like this: There are 10 digits in the code, and the check digit is chosen so that 10 times the first digit plus 9 times the second digit, plus 8 times the third digit, . . . , plus 1 times the check digit is divisible by 11. Thus, for the above, $(10 \times 0) +$

$$(9 \times 1) + (8 \times 3) + (7 \times 2) +$$
$$(6 \times 5) + (5 \times 7) + (4 \times 5) +$$
$$(3 \times 0) + (2 \times 2) + (1 \times 7) = 143,$$

and 11 divides 143. The check digit (or any other digit) may be an X, denoting a 10.

(a) Which of these ISBNs is correct?

(i) 0-70-808228-7

(ii) 0-201-30722-7

(b) Supply the check digit to complete each of these correct ISBNs.

(i) 5-648-00738-

(ii) 3-540-11200-

Using a Computer

17. The following commands set up a spreadsheet to check an ISBN code.

(i) Put "10" in A1.

(ii) Put "= A1 − 1" in B1 and **COPY** and **PASTE,** or **FILL,** right to J1.

(iii) Put the digits of the ISBN code in A2 through J2. If X appears in the code, replace it by 10.

(iv) Put "= A1*A2" in A3 and **COPY** and **PASTE,** or **FILL,** right to J3.

(v) Put "= SUM(A3:J3)" in K3.

(vi) Put "= MOD(K3, 11)" in L3. This places in L3 the remainder when the number in K3 is divided by 11. If L3 contains 0, the code is correct.

Set up such a spreadsheet and use it to determine whether the following codes are correct. Note that once a sheet is set up, all you need do to immediately see if a given code is correct is enter the digits of the code in A2 through J2.

(a) 0-673-46483-0

(b) 0-321-01330-2

18. Use the computer program of problem 17 to determine the check digit that should be added to complete these nine-digit ISBN codes.

(a) 0-13-257-502

(b) 0-88133-836

For Review

19. (a) Write each of the following as fractions in simplest form:

$$1,\quad 1+\frac{1}{1},\quad 1+\frac{1}{1+\frac{1}{1}},\quad 1+\frac{1}{1+\frac{1}{1+\frac{1}{1}}},\ \text{and}$$

$$1+\frac{1}{1+\frac{1}{1+\frac{1}{1+\frac{1}{1}}}}.$$

(b) Make a conjecture based on the results of part (a). In particular, what would be the thirteenth fraction if this list were to be continued?

20. (a) Use your calculator to compute $(1+\sqrt{5})/2$ correct to the nearest ten thousandth.

(b) Use your calculator to compute the decimal equivalent of the fraction of your guess in problem 19(b) correct the nearest ten thousandth.

(c) To the nearest ten thousandth, what number do you obtain if you enter the string

1 [1/x] [+] 1 [=] [1/x] [+] 1 [=] · · ·

into your calculator, repeating the sequence [1/x] [+] 1 [=] thirteen times? *Note:* If you do not have a [1/x] key on your calculator, you will have to replace [1/x] in the input sequence by the entry string [^] 1 [(−)].

(d) Compare the result of part (c) with that of part (a).

(e) Make a conjecture on the basis of problem 19 and parts (a)–(d) of this problem.

21. (a) Use your calculator to compute $(1+\sqrt{5})/2$, correct to the nearest ten thousandth.

(b) Use your calculator to compute each of the following, correct to the nearest ten thousandth:

$$\sqrt{1},\quad \sqrt{1+\sqrt{1}},\quad \sqrt{1+\sqrt{1+\sqrt{1}}},$$

$$\sqrt{1+\sqrt{1+\sqrt{1+\sqrt{1}}}},$$

$$\text{and}\ \sqrt{1+\sqrt{1+\sqrt{1+\sqrt{1+\sqrt{1}}}}}$$

(c) To the nearest ten thousandth, what number do you obtain if you enter the string

1 [+] 1 [=] [√‾] [+] 1 [=] [√‾] · · ·

into your calculator, repeating the string [+] 1 [=] [√‾] nine times? Note that this algorithm successively evaluates the expressions in part (b). Also note that your calculator may require you to replace [√‾] by [√‾] [)] or by [√‾] [(] [)] in the entry string.

(d) Make a conjecture based on the results of parts (a)–(c).

Epilogue Number Theory, the Queen of Mathematics

Almost 200 years ago, Carl Gauss wrote, "Mathematics is the queen of the sciences, but number theory is the queen of mathematics." What Gauss was really saying is that our modern technology, and work in science in general, depends so heavily on mathematics that progress in these areas would be essentially impossible without mathematical skill. This assertion is increasingly true in the social sciences as well. Beyond that, Gauss was saying that mathematics is also intrinsically interesting and intellectually satisfying and that this is particularly true of number theory.

In this chapter, we have introduced the basic number-theoretic notions of divisibility, factoring, factors and multiples, primes and composite numbers, least common multiples, and greatest common divisors. These ideas not only are useful in other parts of mathematics and in disciplines like computer science, but they also provide interesting and stimulating motivational material for the elementary mathematics classroom. In particular, number theory is replete with interesting and challenging problems that provide additional opportunities to further develop problem-solving skills. It is increasingly the case that number-theoretic notions appear in elementary school texts and that these ideas must be understood by teachers.

Chapter Summary

Key Concepts

In what follows, unless expressly stated to the contrary, all symbols represent whole numbers.

Section 4.1 Divisibility of Natural Numbers

- Divisibility. If $b \neq 0$ and there exists a number q such that $a = bq$, then we say that b divides a.
- Factor, or divisor. If b divides a, b is called a factor, or divisor, of a.
- Proper divisor. A proper divisor is a divisor of a number that is less than the number.
- Multiple. If b divides a, a is called a multiple of b.
- Prime. A prime is a natural number with only two factors.
- Composite number. A composite number is a natural number with more than two factors. Alternatively, c is composite if $c = ab$ where a and b are natural numbers and neither a nor b equals 1.
- Unit. The number 1, which has only one divisor.
- Fundamental theorem of arithmetic. Every natural number greater than 1 can be represented as a product of primes in just one way, apart from order. Alternatively, every natural number greater than 1 can be represented in just one way, apart from order, as a product of powers of primes.
- All divisors of a number. All divisors of a number may be determined from the number's prime-power representation.
- Numerousness of the set of primes. There are infinitely many primes.
- Determining primality. Every natural number greater than 1 is either a prime or has a prime factor less than or equal to its own square root.
- The sieve of Eratosthenes. A method for determining all primes up to any given limit.

Section 4.2 Tests for Divisibility

- Divisibility of sums and differences. If n, a, and b are natural numbers and n divides both a and b, then n divides $a + b$ and $a - b$.
- Divisibility by products. If a divides c, b divides c, and a and b have no common factor other than 1, then ab divides c.
- Divisibility by 2 and 5. n is divisible by 2 provided that its units digit is 0, 2, 4, 6, or 8. n is divisible by 5 provided that its units digit is 0 or 5.
- Divisibility by 10. n is divisible by 10 provided that its units digit is 0.
- Divisibility by 3 or 9. n is divisible by 3 provided that the sum of the digits in its

decimal representation is divisible by 3. n is divisible by 9 provided that the sum of its digits is divisible by 9.

- Divisibility by 11. n is divisible by 11 provided that the difference of the sum of the digits in the even positions in the decimal representation of n and the sum of the digits in the odd positions is divisible by 11.
- Divisibility by powers of 2. n is divisible by 2^r provided that the number named by the last r digits of n is divisible by 2^r.

Section 4.3 Greatest Common Divisors and Least Common Multiples

- Greatest common divisor. If d is the greatest of all common divisors of a and b, it is called their greatest common divisor.
- Greatest common divisor of a and b by intersection of sets. The greatest number in the intersection of the sets of divisors of a and b.
- Greatest common divisor of a and b from prime-power representations. The product of prime powers p^s, where s is the lesser of the exponents (including 0) on p in the prime-power representations of a and b.
- The Euclidean algorithm. A surefire method for finding the greatest common divisor of a and b, using the division algorithm.
- Least common multiple. If m is the least of all common multiples of a and b, it is called their least common multiple.
- Least common multiple of a and b by the intersection of sets. The least number in the intersection of the sets of multiples of a and b.
- Greatest common multiple of a and b from prime-power representations. The product of prime powers p^t, where t is the greater of the exponents on p in the prime-power representations of a and b.
- Product of the greatest common divisor and least common multiple of a and b. This product equals ab.

Section 4.4 Codes and Credit Card Numbers: Connections to Number Theory

- Bar codes on mail. Zip codes in machine-readable form.
- Verification of valid credit card numbers. Schemes to check correctness of credit card numbers.

Vocabulary and Notation

Section 4.1

Factor, divisor
Proper divisor
Perfect number
Multiple
Prime
Composite number
Unit
Factor tree

Section 4.2

Divisibility by 2, 5, and 10
Divisibility by 11
Simultaneous test for divisibility by 7, 11, and 13
Divisibility by powers of 2

Section 4.3

$\text{GCD}(a, b)$, $\text{GCF}(a, b)$.
$\text{LCM}(a, b)$.

Section 4.4

Check sums

Chapter Review Exercises

Section 4.1

1. Draw rectangular diagrams to illustrate the factorings of 15, taking order into account; that is, think of $1 \cdot 15$ as different from $15 \cdot 1$.
2. Construct a factor tree for 96.
3. (a) Determine the set D_{60} of all divisors of 60.
   (b) Determine the set D_{72} of all divisors of 72.
   (c) Use $D_{60} \cap D_{72}$ to determine GCD(60, 72).
4. (a) Determine the prime-power representation of 1200.
   (b) Determine the prime-power representation of 2940.
   (c) Use parts (a) and (b) to determine GCD(1200, 2940) and LCM(1200, 2940).
5. Use information from the sieve of Eratosthenes in Figure 4.3 to determine whether 847 is prime or composite.
6. (a) Determine a composite natural number n with a prime factor greater than $\sqrt{n}$.
   (b) Does the n in part (a) have a prime divisor less than $\sqrt{n}$? If so, what is it?
7. Determine natural numbers r, s, and m such that r divides m and s divides m, but rs does not divide m.
8. Use the number $n = 3 \cdot 5 \cdot 7 + 11 \cdot 13 \cdot 17$ to determine a prime different from 3, 5, 7, 11, 13, or 17.

Section 4.2

9. Using mental methods, test each number for divisibility by 2, 3, 5, and 11.
   (a) 9310 (b) 2079
   (c) 5635 (d) 5665
10. Test each number for divisibility by 7, 11, and 13.
   (a) 10,197 (b) 9373
   (c) 36,751
11. Use the results of problem 9 to decide which of these are *true*.
   (a) 15 divides 9310 (b) 33 divides 2079
   (c) 55 divides 5635 (d) 55 divides 5665
12. Let $m = 3^4 \cdot 7^2$.
   (a) How many divisors does m have?
   (b) List all the divisors of m.
13. Determine d so that 2,765,301,2d3 is divisible by 11.
14. (a) Determine the least natural number divisible by both q and m if $q = 2^3 \cdot 3^5 \cdot 7^2 \cdot 11^1$ and $m = 2^1 \cdot 7^3 \cdot 11^3 \cdot 13^1$.
   (b) Determine the largest number less than the q of part (a) that divides q.

Section 4.3

15. (a) Find the greatest common divisor of 63 and 91 by the method of intersection of sets of divisors.
   (b) Find the least common multiple of 63 and 91 by the method of intersection of sets of multiples.
   (c) Demonstrate that GCD(63, 91) · LCM(63, 91) = 63 · 91.
16. If $r = 2^1 \cdot 3^2 \cdot 5^1 \cdot 11^3$, $s = 2^2 \cdot 5^2 \cdot 11^2$, and $t = 2^3 \cdot 3^1 \cdot 7^1 \cdot 11^3$, determine each of the following.
   (a) GCD(r, s, t)
   (b) LCM(r, s, t)
17. Determine each of the following using the Euclidean algorithm.
   (a) GCD(119790, 12100)
   (b) LCM(119790, 12100)
18. Seventeen-year locusts and 13-year locusts both emerged in 1971. When will these insects' descendants next emerge in the same year?

Section 4.4

19. Determine the check digit for each of these zip codes.
   (a) 99163-2781
   (b) 96002-4857
20. Write out the bar code for each of these zip codes.
   (a) 83004-2765
   (b) 12011-4302

---

21. Write out the zip code named by each of these bar codes. Which one, if either, is incorrect?
   (a)
   (b)

---

22. Which of the following could be valid credit card numbers?
   (a) 4210 3010 0411 2003
   (b) 5200 1002 3051 2126

23. In each of these, determine d and e so that the number is a valid credit card number possibility. Is there only one possibility in each case?
   (a) 6011 2010 de01 1412
   (b) d230 0110 2e41 1002

24. Using your own credit card, transpose two adjacent digits and see if the check sum detects the transposition.

Chapter Test

1. Using mental methods, test each of these for divisibility by 2, 3, 9, 11, and 13.
   (a) 62,418
   (b) 222,789

2. Let $m = 2^3 \cdot 5^2 \cdot 7^1 \cdot 11^4$ and $n = 2^2 \cdot 7^2 \cdot 11^3$.
   (a) Does r divide m if $r = 2^2 \cdot 5^1 \cdot 7^2 \cdot 11^3$? Why or why not?
   (b) How many divisors does m have?
   (c) Determine GCD(m, n).
   (d) Determine LCM(m, n).

3. Indicate whether each of these is *always true* (T) or *not always true* (F).
   (a) If a divides c and b divides c, then ab divides c.
   (b) If r divides s and s divides t, then r divides t.
   (c) If a divides b and a divides c, then a divides $(b + c)$.
   (d) If a does not divide b and a does not divide c, then a does not divide $(b + c)$.

4. Determine the check digit for the following zip code: 67001-2034.

5. Determine whether 281 is a prime.

6. Given that the check digit for each of these zip codes is 0, classify each as correct or incorrect and tell why.
   (a) 06992-7548
   (b) 84232-7612

7. Use the Euclidean algorithm to determine GCD(154, 553) and LCM(154, 553).

8. Use the Euclidean algorithm to determine each of the following.
   (a) GCD(13,534, 997,476)
   (b) LCM(13,534, 997,476)

9. Which of the following could be valid credit card numbers?
   (a) 4021 3340 0123 4301
   (b) 5302 0159 2002 1051

10. (a) Make a factor tree for 8532.
   (b) Write the prime-power representation of 8532.
   (c) Name the largest natural number smaller than 8532 that divides 8532.
   (d) Name the smallest number larger than 8532 that is divisible by 8532.

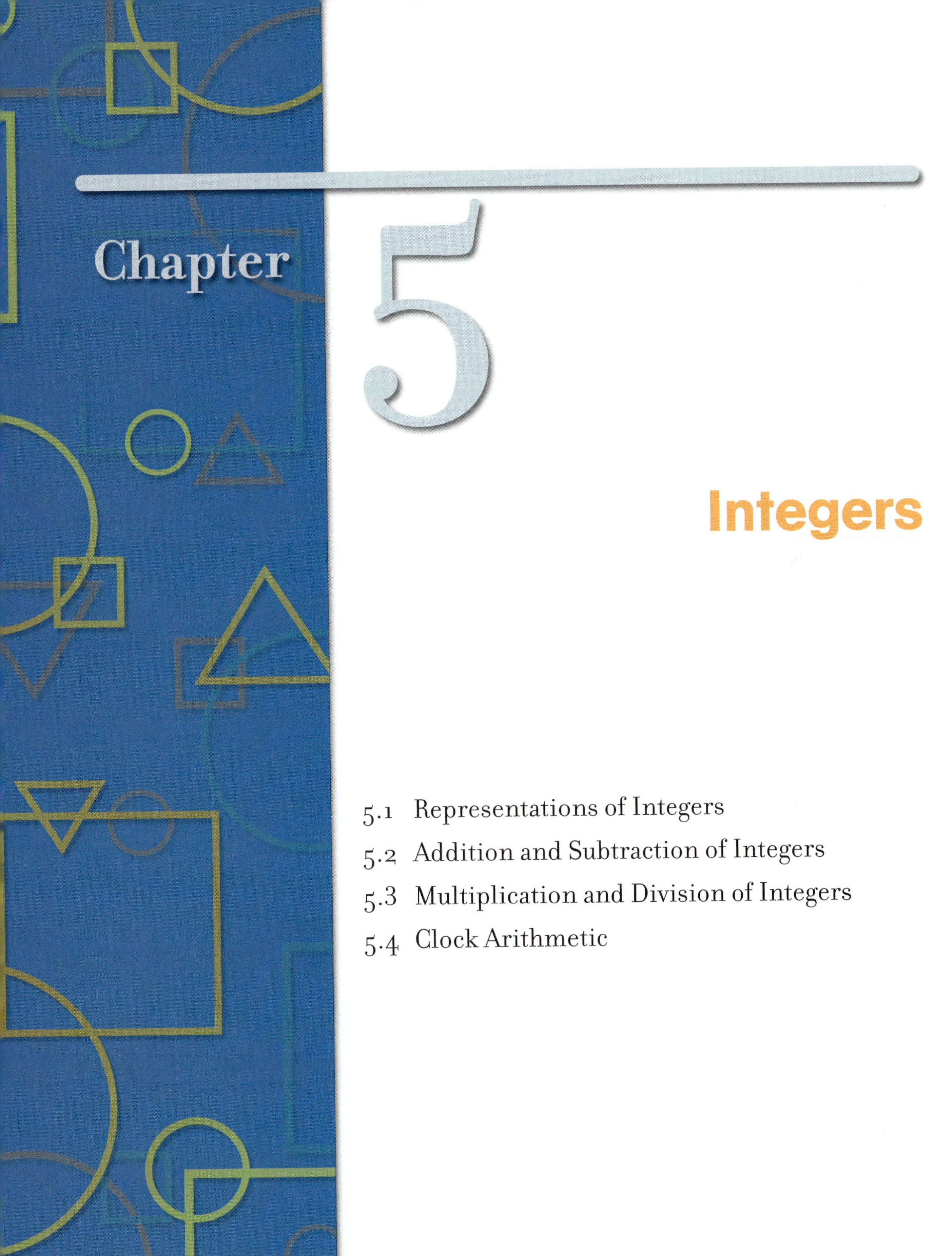

Chapter 5

Integers

Hands On Black–Red Nim

Materials needed

1. Eight black counters (markers) and eight red counters for each pair of students.
2. Sheets of paper for each pair of students to keep score on.

Directions

1. The students in each pair should arrange their markers in three piles, one with eight black markers, one with five red markers, and one with three red markers.
2. On each play, a player can take one, two, or three markers, but only from a single pile.
3. On each play, the player records his or her cumulative score, noting that each red marker cancels a black marker and vice versa. Thus if a player has a score of 4B and draws two red markers, the cumulative score is 2B. If (s)he has a score of 2B and draws two red markers, the cumulative score is 0. If the present cumulative score is 2B and the player draws three red markers, the cumulative score becomes 1R.
4. The players play alternately, and the one taking the last marker adds 6B to his or her cumulative score no matter what color the last marker taken was.
5. The player with the highest black score (or the lowest red score if neither has a black score) wins the game. The privilege of going first rotates between players from game to game.

Question

How would your strategy change if you could take any number of markers on any given play?

Connections A Further Extension of the Number System

In Chapter 2, we discussed the whole numbers and operations with whole numbers. In particular, we observed that the whole numbers and their operations were developed as a direct result of people's need to count. But a modern society has many quantitative needs aside from counting, and these needs often require numbers other than whole numbers.

In this chapter, we consider the set I of **integers,** which consists of

- the **natural numbers,** or **positive integers,** denoted by 1, 2, 3, . . . ;
- the number **zero,** 0; and
- the **negative integers,** denoted by $-1, -2, -3, \ldots$.

We note immediately that

$$N \subset W \subset I.$$

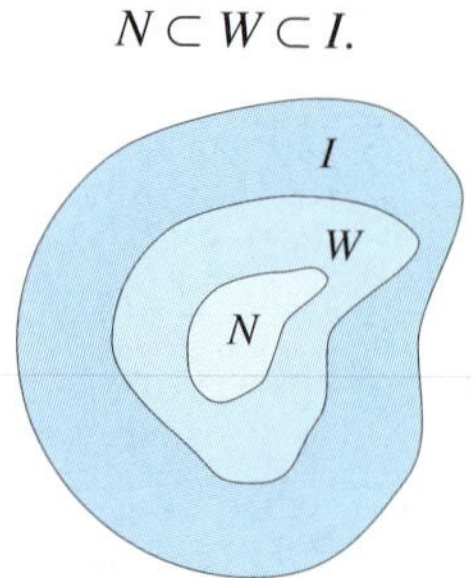

In today's society, these numbers are used to record debits and credits, profits and losses, changes in prices in the stock market, degrees above zero and degrees below zero in temperatures, yards lost and yards gained in football, points won or points "in the hole" in many card and board games, and so on.

As usual, we introduce the set of integers by means of manipulatives, pictures, and diagrams, with steadily increasing levels of abstraction to suggest, in turn, how these ideas might be presented to elementary school children.

5.1 REPRESENTATIONS OF INTEGERS

Businesspeople of all kinds regularly use the phrases "in the black" and "in the red" to indicate whether a given business has experienced a profit or a loss or to indicate whether a bank account still has money available or if it has been overdrawn. Since accountants often note these states of affairs with black and red ink, we adopt that same convention here.

Representing Integers by "Drops" of Colored Counters

Mark Garza gives each student in his class a collection of about 25 counters colored black on one side and red on the other. These are easily made by duplicating the desired shapes on red construction paper and then having the students cut them out and color one side of each counter black.

Mr. Garza has each student drop several counters on his or her desktop, match the black and red counters to each other, and record as the score for the drop the number of unmatched black or red counters. This could be viewed as recording the results of playing a game against two opponents and winning points (black counters) from one opponent and losing points (red counters) to the other opponent. The score is then the net gain or loss on a given play, and this can be used to represent positive or negative integers. For example, Figure 5.1 shows a drop of six black and four red counters for a score of 2B. We interpret this as representing *positive two* and write 2. Similarly, we interpret a score of 2R as *negative two* and write −2.

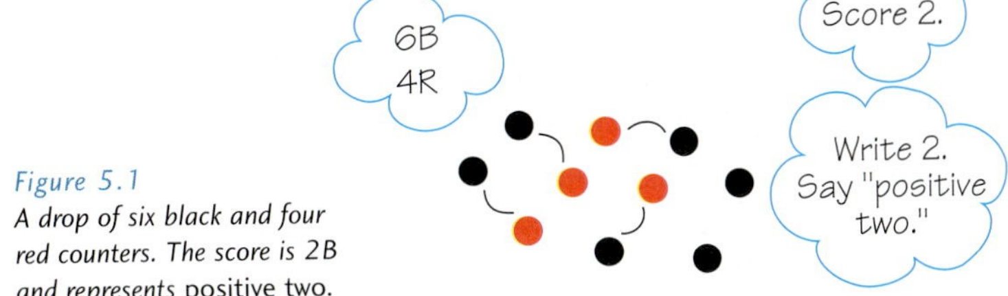

Figure 5.1
A drop of six black and four red counters. The score is 2B and represents positive two.

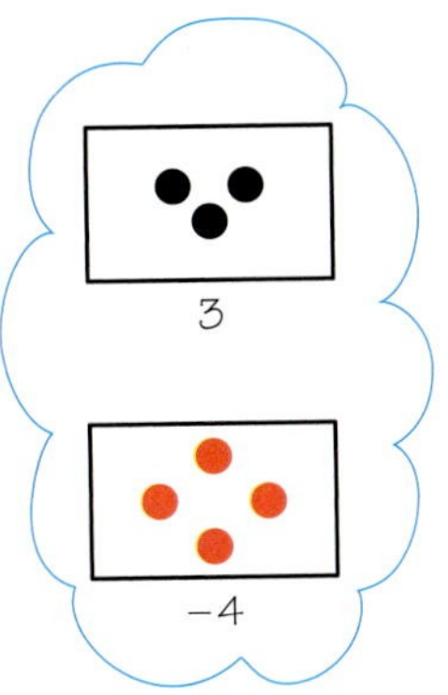

It is important to observe that different drops can result in the same score. For example, the two diagrams in Figure 5.2 both represent −2. Moreover, −2 could also be represented by a diagram with only 2 red counters. A drop with counters of only one color is the simplest of all such representations, and we will have occasion to use such drops in what follows. Mr. Garza discusses this at length with his students and asks particularly how to increase or decrease the number of counters in a drop without changing the score. A moment's reflection makes it clear that adding or taking away the same number each of black and red counters does not change the score of a drop. In summary, two drops of colored counters are *equivalent* and represent the same integer if one can be obtained from the other by adding (or deleting) the same number each of black and red counters. The simplest representation of the positive integer n is a drop of n black counters. The simplest representation of the negative integer $-n$ is a drop of n red counters.

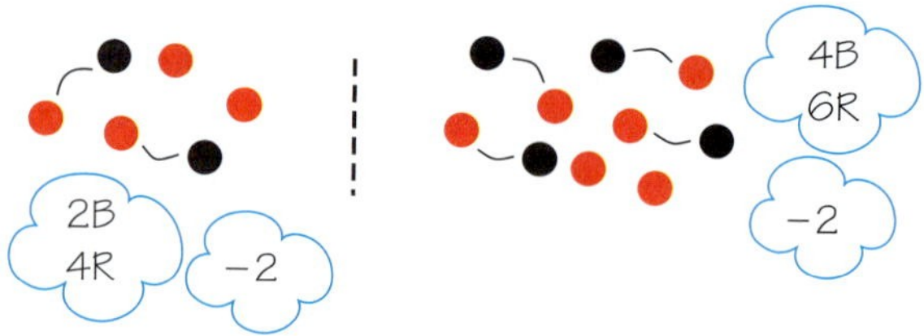

Figure 5.2
Two drops representing −2

A drop consisting of an equal number of black and red counters is of special interest. Since no counters remain unmatched, such a drop represents 0, which is therefore *neither positive nor negative.* Moreover, as we have just seen, adding such a drop to an existing drop does not change its score, and this corresponds numerically to the fact that

$$0 + n = n + 0 = n$$

for *any* integer n—positive, negative, or zero. Finally, any such drop can be viewed as a combination of two drops—one of s black counters and one of s red counters for some natural number s. Separately these drops would score sB and sR and represent s and $-s$, respectively. Since together they score 0, we have the following definition.

DEFINITION *The Integers*
The **positive integers** are the natural numbers.

The **negative integers** are the numbers $-1, -2, -3, \ldots$, where $-s$ is defined by the equality

$$s + (-s) = (-s) + s = 0.$$

The integer 0 is neither positive nor negative and has the property

$$0 + n = n + 0 = n$$

for every integer n. The **integers** consist of the positive integers, the negative integers, and zero.

A Comment on Notation

Observe that it is standard to use the same sign in writing $8 - 6$ and -5, for example. The expression $8 - 6$ is often read "8 minus 6," and the sign, −, is often called the *minus* sign. However, there is a double meaning in the usage that must be clearly understood. In $8 - 6$, the sign is used to indicate the *operation of subtraction,* and in writing -5, it is used to indicate the *negative,* or *additive inverse,* of 5. This distinction is made on your calculator with two separate keys, one for subtraction and one for negation. To compute $8 - 6$ on the calculator, one keys in the sequence

8 [−] 6 [=],

and to enter -5 in the display, one keys in the sequence

Indeed, keying in the sequence

sequentially shows 5, -5, 5, -5, and 5 in the calculator display, since the [(−)] key changes the sign of the number in the display each time it is pressed. This suggests that the negative of negative 5 is 5 (that $-(-5) = 5$), and we will see later that this is so. Indeed, as we will show, $-(-n) = n$ for every integer n.

Note that if you key [(−)] 2 [ENTER] into a TI-73 or similar calculator, the display shows $^{-}2$ rather than -2. And if 5 [−] [(−)] 2 [ENTER] is keyed in, the display shows $5 - {}^{-}2$ and the answer 7, indicating that subtracting the negative of 2 from 5 gives 7, as we will show later.

We observe that, like the TI-73, many texts for elementary school try to avoid the difficulty noted above by using a raised minus sign to indicate the negative of a number. Thus, they write $^{-}5$ in place of -5. And some even write $^{+}2$ in place of 2. Ultimately, however, these texts invariably change to the standard notation of -5 and 2 used here. We feel that it causes less confusion to do this at the outset, stressing that the context makes it clear when "subtract" is meant, as opposed to "the negative of."

Note that when a symbol like n is written to represent a number, it is *not* implied that n is necessarily positive. For example, n might represent negative 3 (that is, -3), and then $-n$ is $-(-3)$ or 3. *Thus, the negative of* n *is negative or positive according as* n *is positive or negative.* If we mean n to be positive, we must say so by calling it a natural number, a positive integer, a positive rational number, or simply a positive number. Otherwise, we must allow for the possibility that n represents either a positive or negative number or zero.

from **The NCTM Principles and Standards**

Middle-grades students should also work with integers. In lower grades, students may have connected negative integers in appropriate ways to informal knowledge derived from everyday experiences, such as below-zero winter temperatures or lost yards of football plays. In the middle grades, students should extend these initial understandings of integers. Positive and negative integers should be seen as useful for noting relative changes or values. Students can also appreciate the utility of negative integers when they work with equations whose solution requires them, such as $2x + 7 = 1$.

SOURCE: Reprinted with permission from Principles and Standards for School Mathematics, *pages 217–218.*

Finally, note that the negative of n is frequently called the *additive inverse* of n. The reason for this terminology will become apparent when we consider adding and, particularly, subtracting integers. It is also useful in a broader context, as, for example, when we consider the elementary school enrichment topic of clock arithmetic—a topic that is interesting to schoolchildren and gives them a better understanding of ordinary arithmetic.

DEFINITION *Additive Inverse*

For every integer n, the negative of n is also called **the additive inverse of n.** In somewhat more general terms, the additive inverse of n is an integer m such that

$$n + m = m + n = 0.$$

EXAMPLE 5.1 SCORING DROPS OF COLORED COUNTERS

Determine the score and the integer represented by each of these drops.

(a) 

(b) 

(c) 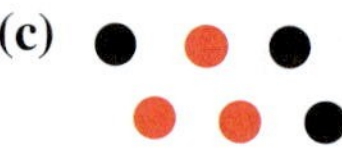

(d)

Solution

(a) Matching the two red counters with two of the black counters, we see that the score is 3B, representing 3.

(b) Matching the red counter with a black counter, we see that the score is 2B, representing 2.

(c) Matching the three red with the three black counters, we see that the score is 0; we have no red or black counters unmatched.

(d) Since there are no black counters, the score is 3R and the integer represented is -3.

EXAMPLE 5.2 DROPS FOR GIVEN INTEGERS

Illustrate two different drops for each of these integers.

(a) 2 **(b)** -3 **(c)** 0 **(d)** 5

Solution

(a) Here, the drops must have two more black counters than red counters. Two possibilities are

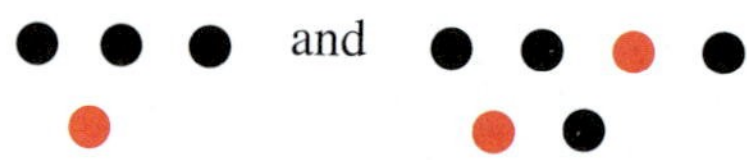

(b) Here, the drop must have three fewer black counters than red counters. Possibilities include

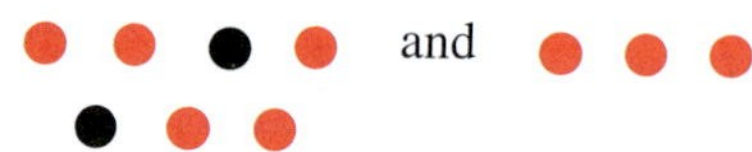

(c) Here, the drop must contain the same number of red and black counters or no counters at all. Possibilities include

and

(d) Here, the drop must have five more black than red counters. Possibilities include

and

Mail-Time Representations of Integers

Integers can also be represented in other real-life situations. The following examples illustrate how the mail delivery of a check or a bill affects your overall net worth, or the value of your assets at any given time.

At mail time, suppose that you are delivered a check for \$20. What happens to your net worth? Answer: It goes up by \$20.

At mail time, you are delivered a bill for \$35. What happens to your net worth? Answer: It goes down by \$35.

At mail time, you received a check for \$10 and a bill for \$10. What happens to your net worth? Answer: It stays the same.

EXAMPLE 5.3 **INTERPRETING MAIL-TIME SITUATIONS**

(a) At mail time, you are delivered a check for \$27. What happens to your net worth?

(b) At mail time, you are delivered a bill for \$36. Are you richer or poorer? By how much?

Solution

(a) Your net worth goes up \$27.

(b) Poorer. Your net worth goes down by \$36.

EXAMPLE 5.4 **DESCRIBING MAIL-TIME SITUATIONS FOR GIVEN INTEGERS**

Describe a mail-time situation corresponding to each of these integers.

(a) −42 **(b)** 75 **(c)** 0

Solution

(a) At mail time, the letter carrier brought you a bill for \$42. Are you richer or poorer, and by how much?

(b) At mail time, you were delivered a check for \$75. What happens to your net worth?

(c) Quite to your surprise, at mail time, the mail carrier skipped your house, so you received no checks and no bills. Are you richer or poorer, and by how much?

Number-Line Representations of Integers

We have already used a number line to illustrate whole numbers, and it can be used equally effectively to represent integers. Choose an arbitrary point on the number line for 0. Then successively measure out unit distances on each side of 0, and label successive points on the right of 0 with successive positive integers and points on the left with successive negative integers as shown in Figure 5.3.

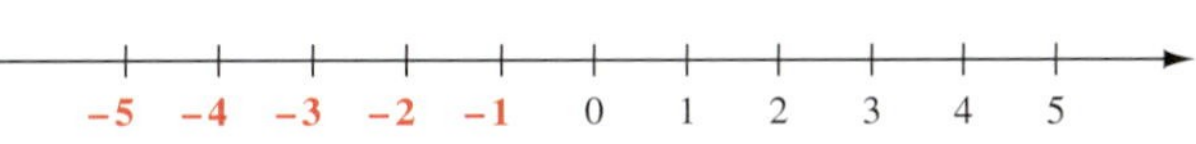

Figure 5.3
Representing integers on a number line

HIGHLIGHT from HISTORY

Edward A. Bouchet (1852–1918)

Edward Bouchet's list of firsts is impressive: the first black man to attend and graduate from Yale University; the first elected to Phi Beta Kappa, the oldest collegiate honor society; the first, in 1876, to earn a doctoral degree. He earned his Ph.D. for an application of mathematics to physics, "Measuring Refractive Indices." Instead of resting on his scholarly laurels, Dr. Bouchet moved to Philadelphia to spend 26 years teaching physics, chemistry, and mathematics at the Institute for Colored Youth. This quaker school, now known as Cheyney State College, was founded by another eminent black mathematician, Charles Reason. Bouchet's example has been followed by many black scholars, such as Dr. Elbert Cox, the first to earn a doctorate in pure mathematics (1925), who devoted his efforts to encouraging graduate work by students at Howard University. As a young student in his hometown, New Haven, Connecticut; as high school valedictorian; and as teacher, high school principal, and scholar, Bouchet's accomplishments and warm personality were a source of inspiration to countless young men and women, black and white.

SOURCE: From Mathematics in Modules, Algebra, A4, *Teacher's Edition. Reprinted by permission.*

This corresponds nicely to marking thermometers with degrees above zero and degrees below zero, and with the practice in most parts of the world (with the notable exceptions of North America and Russia) of numbering floors above ground and below ground in a skyscraper. It also corresponds to the countdown of the seconds to liftoff and beyond in a space shuttle launch. The count 9, 8, 7, 6, 5, 4, 3, 2, 1, liftoff!, 1, 2, 3, . . . is not really counting backward and then forward, but forward all the time. The count is actually

9 seconds before liftoff	-9
8 seconds before liftoff	-8
.	.
.	.
.	.
1 second before liftoff	-1
Liftoff!	0
1 second after liftoff	1
2 seconds after liftoff	2

and so on. This example is very real to space-age children and helps to make positive and negative numbers real and understandable.

It is also often helpful to represent positive and negative integers by curved arrows. For example, *an arrow from any point to a point 5 units to the right represents 5,* and *an arrow from any point to a point 5 units to the left represents -5,* as illustrated in Figure 5.4.

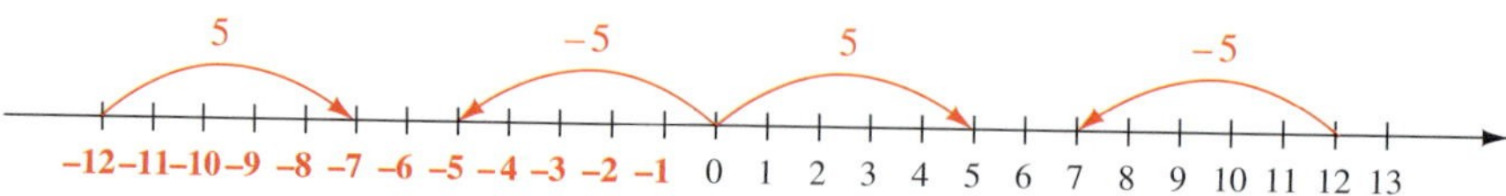

Figure 5.4
Using arrows to represent integers

Thus, the arrow for -5 has the same length as the arrow for 5, but is directed in the opposite direction, and vice versa. This is indicated in some elementary school texts by writing *opp 5* to indicate -5. Since opp means that you simply change the direction of the arrow, it follows that opp (opp n) $= n$ for any integer n. In our notation, this would be written $-(-n) = n$ for any integer n.

THEOREM *The Negative of the Negative of an Integer*
For every integer n, $-(-n) = n$.

EXAMPLE 5.5 DETERMINING NEGATIVES

Determine the negative of each of these integers.

(a) 4 **(b)** -2 **(c)** 0 **(d)** -320

Solution

(a) The negative of 4 is -4.
(b) The negative of -2 is $-(-2) = 2$.
(c) The negative of 0 is $-0 = 0$ since 0 is neither positive nor negative.
(d) The negative of -320 is $-(-320) = 320$.

Note that the negative of a negative integer is positive.

Absolute Value of an Integer

While discussing number-line representations, it is convenient to introduce the notion of absolute value of an integer. We have just observed that the integer 5 can be illustrated by the point numbered 5 on the number line. Similarly, -5 is represented by the point numbered -5. On the other hand, both 5 and -5 are five units from 0 on the number line, as shown in Figure 5.5. The **absolute value** of an integer n is defined to be the distance of the corresponding point on the number line from 0. We indicate the absolute value of n by writing $|n|$.

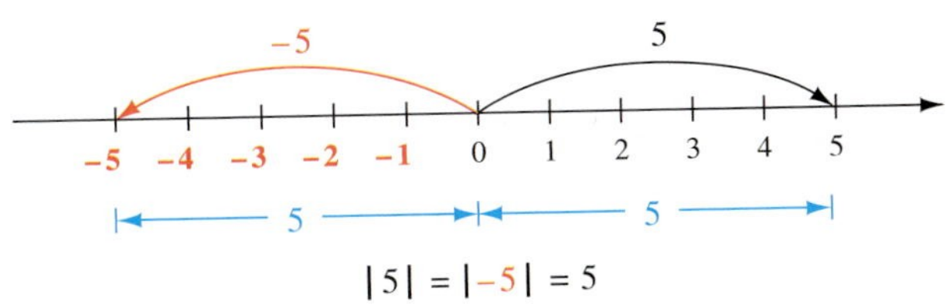

Figure 5.5
The absolute value of both 5 and -5 is 5.

EXAMPLE 5.6 DETERMINING ABSOLUTE VALUES

Determine the absolute values of these integers.

(a) -11 **(b)** 13 **(c)** 0 **(d)** -9

Solution

We plot the numbers on the number line and determine the distance of the points from 0.

School Book Page

Introducing Integers

9-1 Using Integers to Represent Quantities

You'll Learn ...

- to use integers to represent real-world quantities
- to find the opposite of an integer
- to find the absolute value of an integer

... How It's Used

Sailors need to know ocean and harbor depths to prevent their ships from running aground.

Vocabulary

negative numbers
origin
opposite numbers
integers
absolute value

▶ Lesson Link Most of the numbers you've studied so far have been greater than zero. Now you'll explore numbers that are less than zero. ◀

Explore Numbers Less than Zero

Hills and Valleys

In Plaquemines Parish, Louisiana, a borehole was drilled to 22,570 ft below sea level. Sea level is the average height of the earth's oceans.

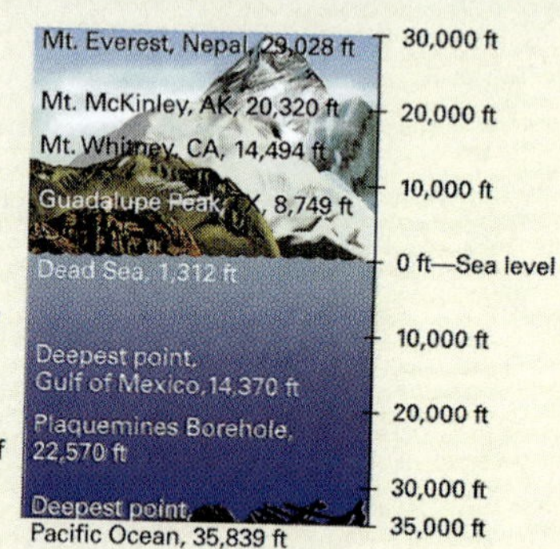

1. Which landmarks are below sea level? Above sea level?
2. Which is closer to sea level, the deepest point in the Gulf of Mexico or Mt. Whitney? How much closer?
3. Which is farther from sea level, Mt. McKinley or the Plaquemines borehole? How much farther?
4. Describe a way to show the difference between numbers of feet above sea level and below sea level.

Learn Using Integers to Represent Quantities

A vertical number line can be used to compare heights and depths.

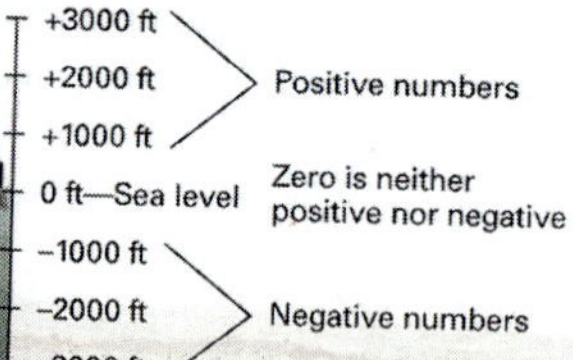

432 Chapter 9 • Integers

SOURCE: Scott Foresman–Addison Wesley Middle School Math, *Course 2, p. 432, by Randall I. Charles et al. Copyright © 2002 Pearson Education, Inc. Reprinted with permission.*

Questions for the Teacher

1. This lesson uses elevations above and below sea level to introduce negative numbers. What other situations illustrate positive and negative numbers?
2. Number lines help students understand negative and positive numbers. This lesson also uses *opposite numbers* and the number line, as shown in the following excerpt:

Opposite numbers are the same distance from zero. −2 and 2 are opposites because they are both 2 units from zero.

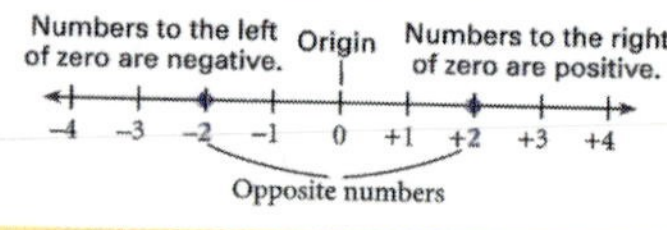

Could you use the operation of addition to help students understand opposites? Explain.

(a) $|-11| = 11 = -(-11)$ since -11 is 11 units from 0.
(b) $|13| = 13$ since 13 is 13 units from 0.
(c) $|0| = 0$ since 0 is 0 units from 0.
(d) $|-9| = 9 = -(-9)$ since -9 is 9 units from 0.

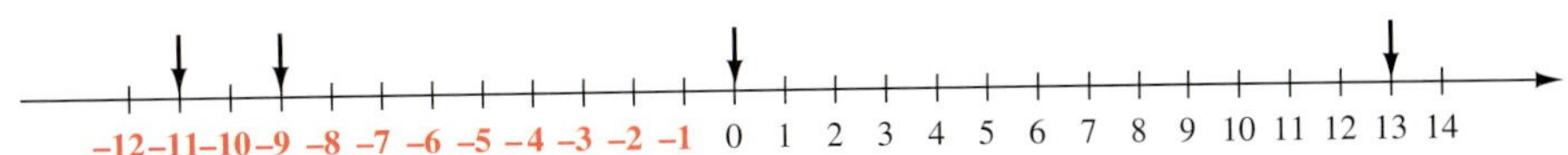

Careful examination of the results of Example 5.6 suggests an alternative definition that makes it possible to avoid drawing diagrams to find absolute values.

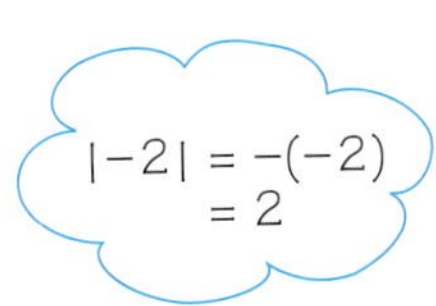

DEFINITION *Absolute Value of an Integer*
If a is an integer, then

$$|a| = \begin{cases} a \text{ if } a \text{ is positive or zero} \\ -a \text{ if } a \text{ is negative.} \end{cases}$$

EXAMPLE 5.7 DETERMINING ABSOLUTE VALUES

Determine the absolute values of these integers.

(a) -71 **(b)** 29 **(c)** 0 **(d)** -852

Solution

(a) Since -71 is negative, $|-71| = -(-71) = 71$.
(b) Since 29 is positive, $|29| = 29$.
(c) Since 0 is 0, $|0| = 0$.
(d) Since -852 is negative, $|-852| = -(-852) = 852$.

Problem Set 5.1

Understanding Concepts

1. Draw two colored-counter diagrams to represent each of these scores or integers.
   (a) 5B (b) 2R
   (c) 0 (d) -3 (e) 3
2. Draw a colored-counter diagram with the least number of counters to represent each of the following.
   (a) 3 (b) -4
   (c) 0 (d) 2
3. (a) Draw a colored-counter diagram to represent 5.
   (b) Draw what you would see if you turned over all the colored counters in your diagram for part (a). What integer would this new diagram represent?
   (c) We could quite reasonably call the diagram of part (b) the opposite of the diagram of part (a). Thus, we might reasonably call -5 the opposite of 5 and write opp 5. As noted earlier, some elementary texts actually use this terminology. What would these texts write in place of -17?
   (d) Using the opp idea with colored counters, write an argument that $-(-5) = 5$.
   (e) Use the colored-counter model to argue convincingly that $-(-n) = n$ for any integer n.
4. (a) Describe a mail-time situation that illustrates 14.
   (b) Describe a mail-time situation that illustrates -27.
5. At mail time, you are delivered a check for \$48 and a bill for \$31.

(a) Are you richer or poorer, and by how much?

(b) What integer does this situation illustrate?

6. **(a)** At mail time, you are delivered a check for \$27 and a bill for \$42. What integer does this situation illustrate?

(b) Describe a different mail-time situation that illustrates the same integer as in part (a).

7. Draw a number line and plot the points representing these integers.

(a) 0 **(b)** 4 **(c)** -4 **(d)** 8

(e) $(4 + 8)/2$

(f) Where is $(4 + 8)/2$ relative to 4 and 8?

8. What integers are represented by the curved arrow on each of these number-line diagrams?

(a)

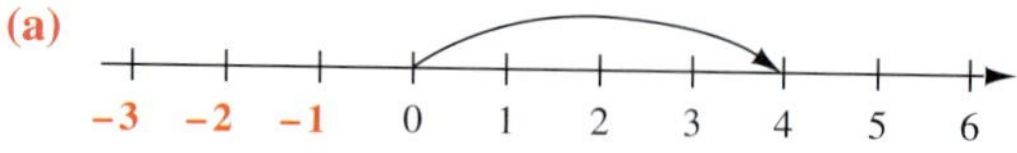

(b)

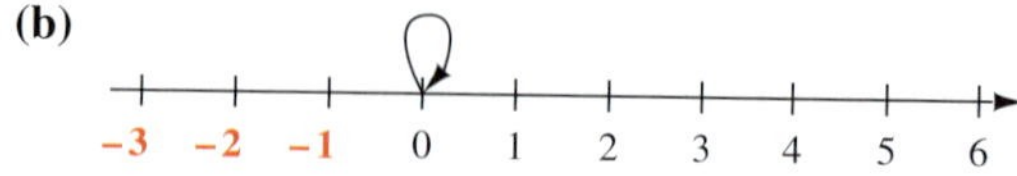

(c)

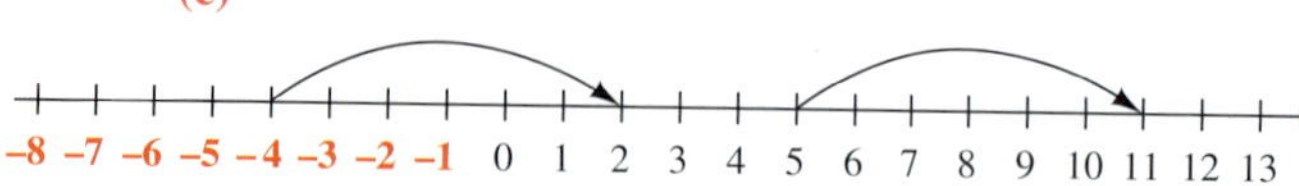

(d)

9. Draw number-line diagrams to represent each of these integers.

(a) 7 **(b)** 0 **(c)** -9 **(d)** 9

10. Find the absolute values of these quantities.

(a) 34 **(b)** $4 - 4$

(c) -76 **(d)** $17 - 5$

(e) How far is it between 5 and 17 on the number line?

11. For what values of x are these equations true?

(a) $|x| = 13$ **(b)** $|x| + 1 = 2$ **(c)** $|x| + 5 = 0$

12. Determine all pairs (x, y) of integer values of x and y for which $|x| + |y| = 2$.

Teaching Concepts

13. **(a)** How would you use drops of colored counters to help students understand that $-(-4) = 4$?

(b) How would you use drops of colored counters to help students understand that $-(-n) = n$ for every integer n (positive, negative, or zero)?

14. Althea claims that $-0 = 0$. How would you respond to Althea? Is she correct?

15. The definition of absolute value is often confusing to students. On the one hand, they understand that the absolute value of a number is always positive. On the other hand, the definition states that $|n| = -n$ sometimes. How would you explain this seeming contradiction?

16. Manuelita doesn't understand how $-0 = 0$ can possibly be true. How would you use a number line to help clear up her understanding?

17. Note that the distance between 5 and 24 on the number line is 19. Also, $|24 - 5| = |19|$ and $|5 - 24| = |-19| = 19$. What does this example suggest about the absolute value of the difference of two natural numbers in general?

18. Plot 7, 21, and $(7 + 21)/2$ on a number line. What does this example suggest about the relative locations of numbers a, b, and $(a + b)/2$ on the number line?

Thinking Critically

19. **(a)** What colored counters would have to be added to this array in order to represent -3?

(b) Could the question in part (a) be answered in more than one way?

(c) How many different representations of -3 can be made with 20 or fewer counters?

20. If all the counters are used each time, list all the integers that can be represented using

(a) 12 counters.

(b) 11 counters.

21. **(a)** If all the counters are used each time, describe the set of integers that can be represented using n counters.

(b) How many different integers are representable using n counters as in part (a)?

22. If some or all of the counters are used each time, describe the set of integers that can be represented

(a) using 12 counters.

(b) using 11 counters.

(c) using n counters.

Thinking Cooperatively

Do the next three problems with a group of two or three other students. At each step, discuss your solution with the other members of your group and determine a consensus answer for your group.

23. **(a)** What integers are represented by these arrays of counters?

(b) Considering the pattern of answers to part (a), what integer would be represented by a similar array with n rows and n columns?

24. **(a)** How many differently appearing rows of counters can you make using all of 20 counters each time? Remember that each counter has a black side and a red side. Two possibilities with only four counters are shown here.

(b) How many differently appearing rows of counters can you make with at least 1 and at most 20 counters?

(c) How many differently appearing rows of counters can you make with n counters (all used each time) if precisely two of the counters show red and the others show black? (*Suggestion:* Consider the special cases $n = 2, 3, 4$, and 5.)

25. **(a)** How many black and how many red counters are there in a triangular array like this but with 20 rows?

You should not actually need to make a diagram with 20 rows in order to answer this question.

(b) Write a brief explanation of your solution to part (a).

(c) Repeat part (a) but with a triangular array with 21 rows.

(d) What integers are represented by the triangular arrays in parts (a) and (c)?

(e) Make a table of integers represented by triangular arrays like those in parts (a) and (c), but with n rows for $n = 1, 2, 3, 4, 5, 6, 7$, and 8.

(f) Carefully considering the table of part (e), conjecture what integer is represented in a triangular array like those in parts (a) and (c), but with n rows, where n is any natural number. (*Suggestion:* Consider n odd and n even separately.)

Making Connections

26. In Europe, the floor of a building at ground level is called the ground floor. What in America is called the second floor is called the first floor in Europe, and so on.

(a) If an elevator in a tall building in Paris, France, starts on the fifth basement level below ground, B5, and goes up 27 floors, on which numbered floor does it stop?

(b) What would the answer to part (a) be if the building were located in New York?

27. If an elevator starts on basement level B3 and goes down to B6, how far down has it gone?

28. The Wildcats made a first down on the 33-yard line. On the next three plays, they lost 9 yards on a fumble, lost 6 yards when their quarterback was sacked, and completed a 29-yard pass play. Where was the line of scrimmage on the next play?

From State Student Assessments

29. (Massachusetts, Grade 8. Note that students were *not* allowed to use calculators to solve this problem.)

Use the number line below to answer the question.

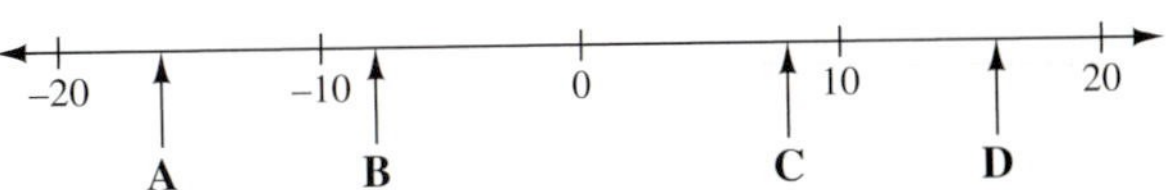

Which point represents the number $(-2)^4$?

A. Point A

B. Point B

C. Point C

D. Point D

For Review

30. How many different factors does each of these numbers have?

(a) $2310 = 2^1 \cdot 3^1 \cdot 5^1 \cdot 7^1 \cdot 11^1$

(b) $5{,}336{,}100 = 2^2 \cdot 3^2 \cdot 5^2 \cdot 7^2 \cdot 11^2$

31. If $a = 2^1 \cdot 3^3 \cdot 5^2$ and $b = 2^3 \cdot 3^2 \cdot 5^1 \cdot 7^1$, compute each of the following.

(a) $\text{GCD}(a, b)$ **(b)** $\text{LCM}(a, b)$

(c) $\text{GCD}(a, b) \cdot \text{LCM}(a, b)$ **(d)** $a \cdot b$

32. **(a)** Does c divide a if a is as in problem 31 and $c = 2^1 \cdot 3^4 \cdot 5^1$? Why or why not?

(b) Is d a multiple of b if b is as in problem 31 and $d = 2^4 \cdot 3^2 \cdot 5^2 \cdot 7^1$? Why or why not?

33. Determine the prime-power representation of each of these numbers.

(a) 1400 **(b)** 5445 **(c)** 4554

34. Determine each of the following, using the Euclidean algorithm.

(a) $\text{GCD}(4554, 5445)$ **(b)** $\text{LCM}(4554, 5445)$

5.2 ADDITION AND SUBTRACTION OF INTEGERS

Addition of Integers

In the preceding section, we introduced the integers by using devices like colored counters, mail-time stories, and number lines. In this section, we consider the addition and subtraction of integers, and it is helpful to students to use these same devices to illustrate how these operations should be performed in this enlarged number system.

Addition of Integers by Using Sets of Colored Counters

We discussed integers in the preceding section in terms of sets of colored counters. Also, in Chapter 2, addition of whole numbers was defined in terms of sets. If $a = n(A)$, $b = n(B)$, and $A \cap B = \emptyset$, then $a + b$ was defined as $n(A \cup B)$.

This idea works equally well for integers by using **sets of colored counters.** Notice that when working with actual sets of counters, the counters are necessarily different so that any two distinct sets A and B clearly satisfy the condition $A \cap B = \emptyset$. In what follows, we presume that the diagrams show actual physical sets of counters so that the condition $A \cap B = \emptyset$ is automatically satisfied. Moreover, we avoid using set notation by drawing a loop around two sets we wish to combine into a single set. Suppose, for example, that we wish to illustrate the addition of 8 and -3. We draw the diagram shown in Figure 5.6 and interpret this as a drop consisting of all the counters in the combined set.

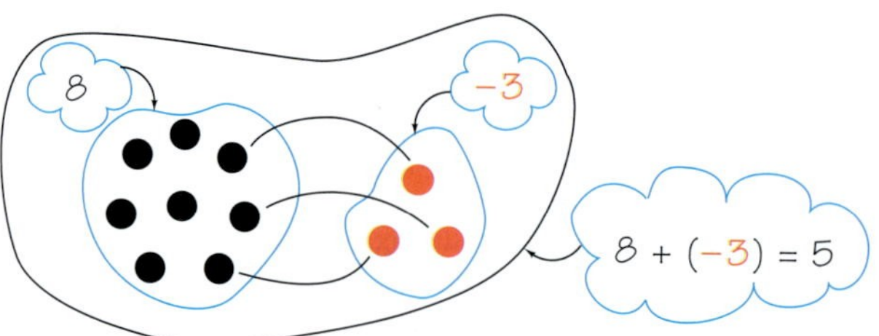

Figure 5.6
Diagram of colored counters illustrating $8 + (-3) = 5$

The score of this drop, 5B, illustrates that the desired sum is 5, as shown. Of course, in Figure 5.6, we used the simplest possible representation of 8 and -3. However, as shown in Figure 5.7, the result is the same if we use other, equivalent representations for 8 and -3. In both Figures 5.6 and 5.7, the total number of black counters exceeds the total number of red counters by 5. Thus, in each case, the score of the combined set is 5B, representing 5.

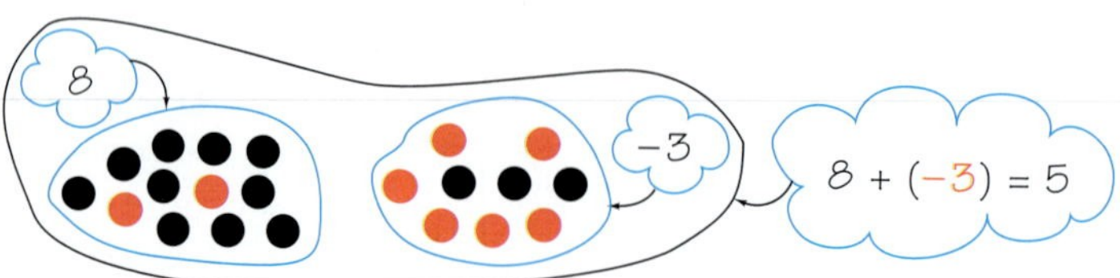

Figure 5.7
Another representation of $8 + (-3) = 5$ using sets of colored counters

EXAMPLE 5.8 REPRESENTING SUMS OF INTEGERS BY USING COLORED-COUNTER DIAGRAMS

Draw appropriate diagrams of colored counters to illustrate each of these sums.

(a) $(-3) + 5$ **(b)** $(-2) + (-4)$ **(c)** $5 + (-7)$ **(d)** $4 + (-4)$

Solution

(a) Using the simplest representations of -3 and 5, we draw this diagram.

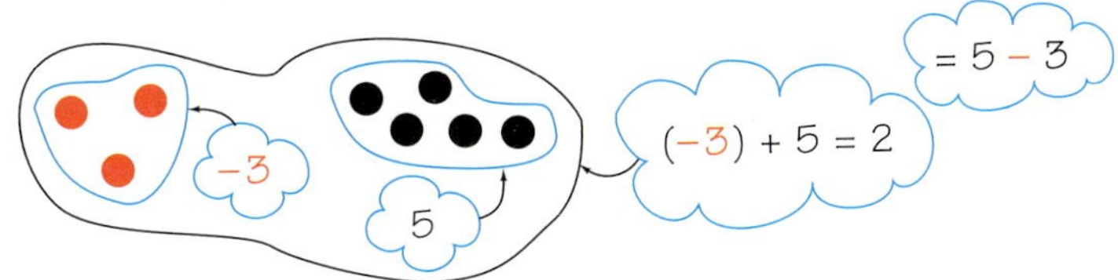

Since the combined set has a score of 2B, this represents 2. Thus, $(-3) + 5 = 2$, and we note that $2 = 5 - 3$ as well. Hence, $(-3) + 5 = 5 - 3 = 2$.

(b) This sum can be represented as shown.

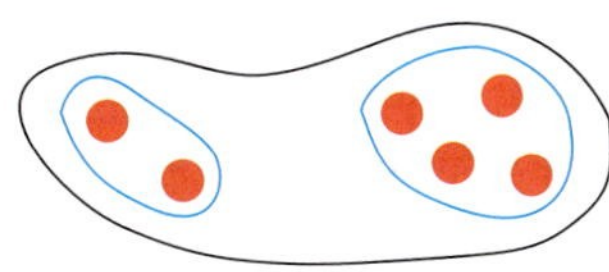

Since the combined set has a score of 6R, this illustrates the sum

$$(-2) + (-4) = -6,$$

and we note that $-6 = -(2 + 4)$. Thus,

$$(-2) + (-4) = -(2 + 4) = -6.$$

(c) $5 + (-7) = -2 = -(7 - 5)$.

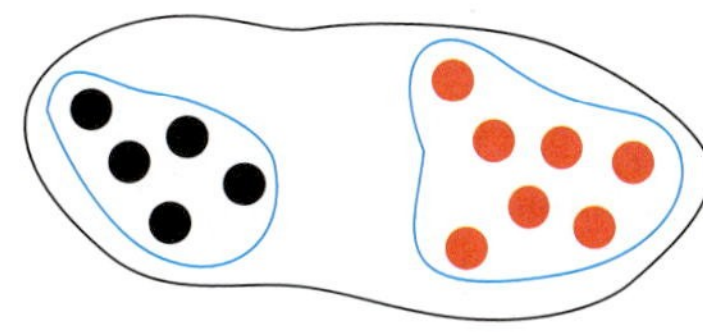

(d) $4 + (-4) = 0 = 4 - 4$.

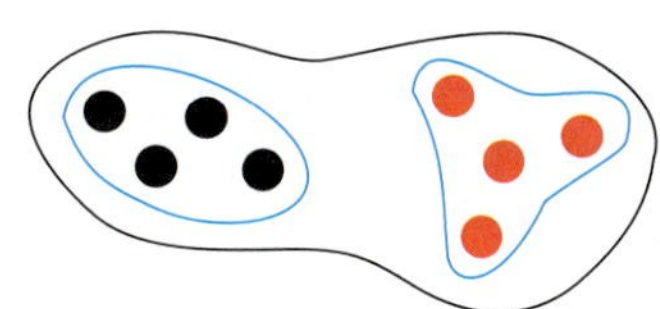

The results of Example 5.8 are entirely typical, and we state them here as a theorem.

THEOREM *Adding Integers*
Let m and n be positive integers so that $-m$ and $-n$ are negative. Then the following are true:

- $(-m) + (-n) = -(m + n)$.
- If $m > n$, then $m + (-n) = m - n$.
- If $m < n$, then $m + (-n) = -(n - m)$.
- $n + (-n) = (-n) + n = 0$.

EXAMPLE 5.9 ADDING INTEGERS

Compute these sums.

(a) $7 + 11$ **(b)** $(-6) + (-5)$ **(c)** $7 + (-3)$
(d) $4 + (-9)$ **(e)** $6 + (-6)$ **(f)** $(-8) + 3$

Solution

(a) $7 + 11 = 18$
(b) $(-6) + (-5) = -(6 + 5) = -11$
(c) Since $7 > 3$, $7 + (-3) = 7 - 3 = 4$.
(d) Since $4 < 9$, $4 + (-9) = -(9 - 4) = -5$.
(e) $6 + (-6) = 0$
(f) $(-8) + 3 = 3 + (-8)$, and since $8 > 3$, $3 + (-8) = -(8 - 3) = -5$. Therefore, $(-8) + 3 = -5$.

Like the properties of whole-number addition, except for the existence of additive inverses, the properties of integer addition derive from the corresponding properties of sets.

THEOREM *Properties of the Addition of Integers*
Let m, n, and r be integers. Then the following hold:

Closure Property	$m + n$ is an integer
Commutative Property	$m + n = n + m$
Associative Property	$m + (n + r) = (m + n) + r$
Additive-Identity Property of Zero	$0 + m = m + 0 = m$
Existence of Negative	$(-m) + m = m + (-m) = 0$

Proof Since we have defined integers in terms of unions of sets of colored counters, the first four properties in the theorem follow from the fact that, for any sets M, N, and R,

$$M \cup N \text{ is a set,}$$
$$M \cup N = N \cup M,$$
$$M \cup (N \cup R) = (M \cup N) \cup R, \text{ and}$$
$$\varnothing \cup M = M \cup \varnothing = M.$$

The existence of the negative, or additive inverse, of m for every integer m follows as a generalization of part (d) of Example 5.8.

Bringing something *to* you is adding.

Addition of Integers by Using Mail-Time Stories

A second useful approach to addition of integers is by means of **mail-time stories.**

At mail time, suppose you receive a check for \$13 and another check for \$6. Are you richer or poorer, and by how much? Answer: Richer by \$19. This illustrates that $13 + 6 = 19$.

EXAMPLE 5.10 ADDING INTEGERS BY USING MAIL-TIME STORIES

Write the addition equation illustrated by each of these stories.

(a) At mail time, you receive a bill for \$3 and a check for \$5. Are you richer or poorer, and by how much?
(b) At mail time, you receive a bill for \$2 and another bill for \$4. Are you richer or poorer, and by how much?
(c) At mail time, you receive a check for \$5 and a bill for \$7. Are you richer or poorer, and by how much?
(d) At mail time, you receive a check for \$4 and a bill for \$4. Are you richer or poorer, and by how much?

Solution

(a) Receiving a bill for \$3 and a check for \$5 makes you \$2 richer. This illustrates $-3 + 5 = 2$.
(b) Receiving a bill for \$2 and another bill for \$4 makes you \$6 poorer. This illustrates $(-2) + (-4) = -6$.
(c) Receiving a check for \$5 makes you richer by \$5, but receiving a bill for \$7 makes you \$7 poorer. The net effect is that you are \$2 poorer. This illustrates $5 + (-7) = -2$.
(d) Receiving a \$4 check and a \$4 bill exactly balances out, and you are neither richer nor poorer. This illustrates $4 + (-4) = 0$.

Note that these results are exactly the results of Example 5.8 above. Moreover, the arguments hold in general, and we are again led to the theorem immediately preceding Example 5.9.

Addition of Integers by Using a Number Line

Suppose we want to illustrate $5 + 4$ on a number line. The addition can be thought of as starting at 0 and counting five units to the right (in the positive direction on the number line) and then "counting on" four more units to the right. Figure 5.8 shows that this is the same as counting nine units to the right from 0 straight away. Thus, $5 + 4 = 9$.

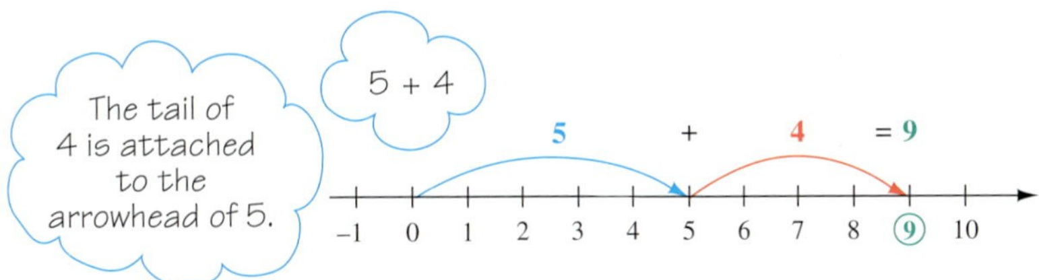

Figure 5.8
Illustrating $5 + 4 = 9$ on a number line

If 4 is depicted by counting four units to the right, then -4 should be depicted by counting four units to the left. Thus, the addition $5 + (-4)$ is depicted on the number line as in Figure 5.9.

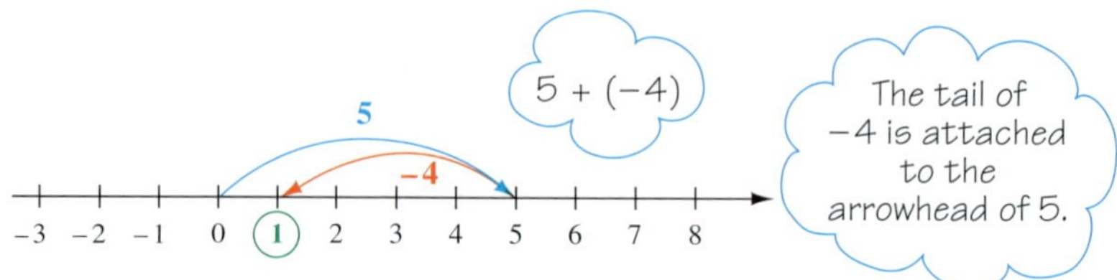

Figure 5.9
Illustrating $5 + (-4) = 1$ on a number line

We count five units to the right from 0 and then "count on" four units to the left in the direction of the -4 arrow. As seen on the diagram, this justifies $5 + (-4) = 1$.

EXAMPLE 5.11 ADDING INTEGERS ON A NUMBER LINE

What addition fact is illustrated by each of these diagrams?

(a)

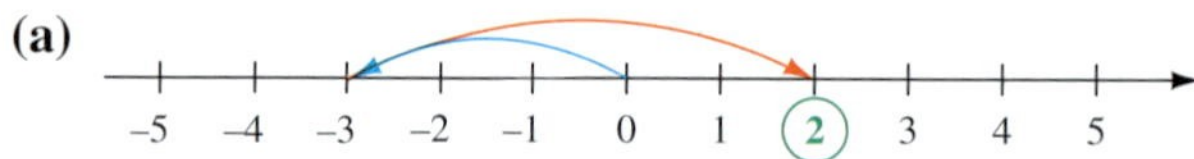

(b)

−7 −6 −5 −4 −3 −2 −1 0 1 2 3 4

(c)

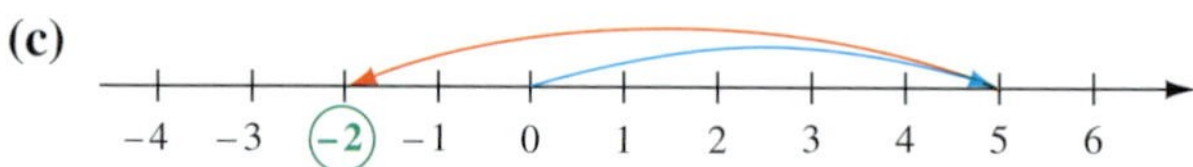

(d)

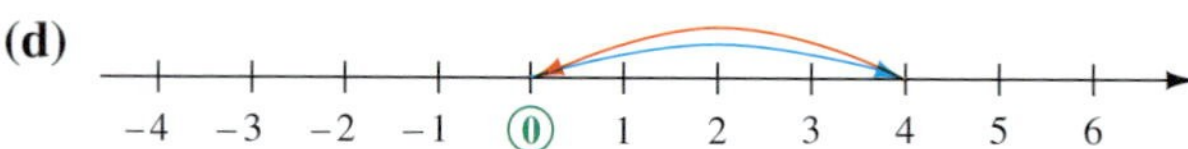

Solution Since counting in the direction indicated by an arrow means adding, these diagrams represent the following sums.

(a) $-3 + 5 = 2$ **(b)** $(-2) + (-4) = -6$
(c) $5 + (-7) = -2$ **(d)** $4 + (-4) = 0$

Note that these are again precisely the same results as in Example 5.8.

EXAMPLE 5.12 DRAWING A NUMBER-LINE DIAGRAM FOR A GIVEN SUM

Draw a number-line diagram to illustrate $(-7) + 4$.

Solution Since -7 is indicated by counting seven units to the left from 0 (in the direction indicated by the minus sign on the -7), and adding 4 is indicated by counting on to the right (in the positive direction since 4 is a positive integer), we have this diagram.

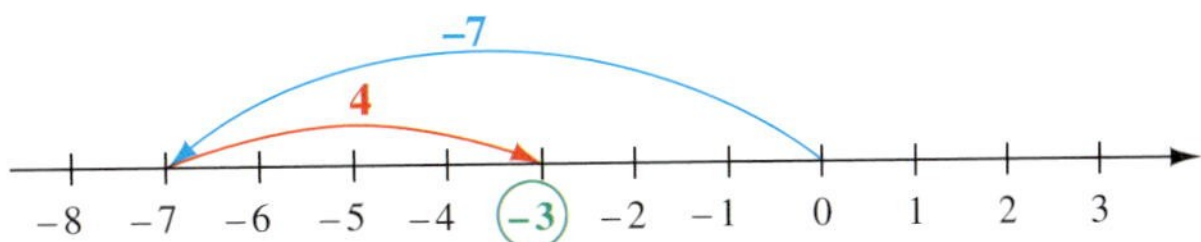

Thus, $(-7) + 4 = -3 = -(7 - 4)$.

Ordering the Set of Integers

Since the set with three black counters shown in Figure 5.10 contains fewer counters than the set with seven black counters, we say that *3 is less than 7* and write $3 < 7$. We also observe that $3 + 4 = 7$ and say that 7 is 4 more than 3.

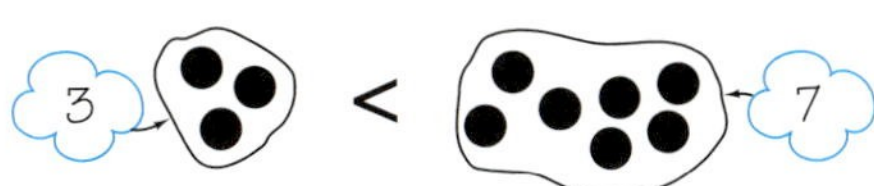

Figure 5.10
Comparing 3 and 7; $3 < 7$

This idea is also easily illustrated on a number line, as shown in Figure 5.11. In particular, we note again that $3 + 4 = 7$, and this implies that *3 is to the left of* 7 on a number line. With this in mind, we extend the notion of "less than" to the set of all integers. In

Into the Classroom

Nancy Rolsen on Comparing and Ordering Integers

Students seem to gain a better understanding of positive and negative numbers through active participation. When comparing and ordering integers, I give each student a large index card with an integer written on it and have them form a human number line by standing in the order of their integers. I put masking tape on the floor to make a number line and we mark off the integers in equal intervals. At the beginning of the unit I use the number line for comparing and ordering integers. Later I have students show addition of integers on this number line.

To introduce graphing ordered pairs of integers, I separate the room into four quadrants with masking tape and use students' desks as coordinate points. Then I call on students by giving the coordinates of their desks. I also give directions such as, "Will all students with y-coordinate of 2 stand." I note that a horizontal row of students is standing.

SOURCE: Scott Foresman–Addison Wesley Middle School Math, *Course 2, p. 428, by Randall I. Charles et al. Copyright © 2002 Pearson Education, Inc. Reprinted with permission.*

particular, if a is to the left of b on a number line, then there is a positive integer c such that $a + c = b$, and so $a < b$.

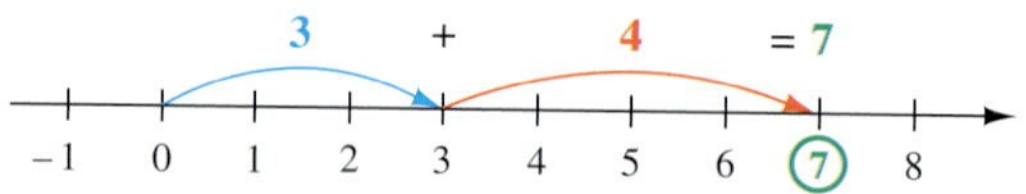

Figure 5.11
Comparing 3 and 7 on a number line; 3 < 7

DEFINITION *Less Than and Greater Than for the Set of Integers*
Let a and b be integers. We say that ***a* is less than *b*,** and write $a < b$, if, and only if, there is a positive integer c such that $a + c = b$. We say that ***b* is greater than *a*,** and write $b > a$, if, and only if, $a < b$.

Notions similar to "less than" and "greater than" are **"less than or equal to"** and **"greater than or equal to,"** and these mean just what they say. That is, we say that a is less than or equal to b and write $a \leq b$ if, and only if, $a < b$ or $a = b$. Similarly, we say that a is greater than or equal to b and write $a \geq b$ if, and only if, $a > b$ or $a = b$. Moreover, it is important to note that if a and b are any two points on a number line, either a is to the left of b, $a = b$, or a is to the right of b. Thus, the integers satisfy the so-called law of trichotomy.

THEOREM *The Law of Trichotomy*
If a and b are any two integers, then precisely one of these three possibilities must hold:

$$a < b \quad \text{or} \quad a = b \quad \text{or} \quad a > b.$$

Since the law of trichotomy holds for the integers, they are said to be *ordered*—that is, they can be lined up on a number line in order of increasing size.

EXAMPLE 5.13 ORDERING PAIRS OF INTEGERS

Place a less-than or greater-than sign in the circle as appropriate.

(a) 17 ○ 121 **(b)** 2 ○ −7 **(c)** −7 ○ −27 **(d)** 0 ○ −6

Solution

(a) $17 < 121$ since $17 + 104 = 121$.
(b) $2 > -7$ since $-7 + 9 = 2$.
(c) $-7 > -27$ since $-27 + 20 = -7$.
(d) $0 > -6$ since $-6 + 6 = 0$.

104, 9, 20, and 6 are all positive.

EXAMPLE 5.14 ORDERING A SET OF INTEGERS

Plot each of these integers on a number line and then list them in increasing order: −5, −9, 7, 0, 12, and −8.

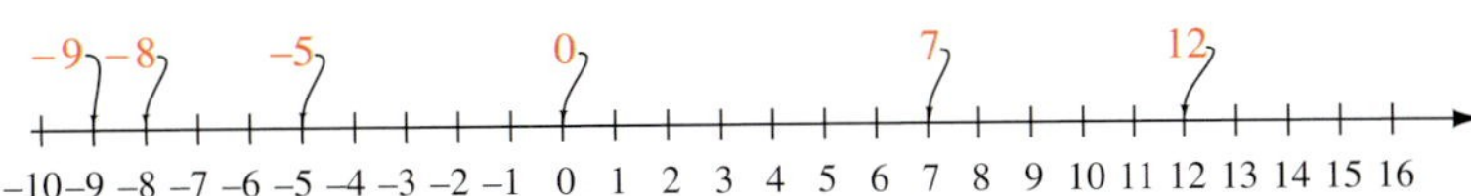

Solution

Reading from left to right, we see that

$$-9 < -8 < -5 < 0 < 7 < 12.$$

EXAMPLE 5.15 OPERATING WITH INEQUALITIES

If a, b, and c are integers and $a < b$, prove that $a + c < b + c$.

Solution

Understand the Problem

We are told that a, b, and c are integers with $a < b$. We're asked to show that $a + c < b + c$.

Devise a Plan

Imagine the following internal dialogue that reveals how a plan emerges:

Perhaps I should ask and answer some questions of myself to see if I can come up with a plan.

Say it in another way.

Q. What am I given?
A. That $a < b$, where a and b are integers.
Q. What must I show?
A. That $a + c < b + c$, where c is also an integer.
Q. What does it mean to say that $a + c < b + c$? Can I say this in another way?
A. I guess so. According to the definition, $a + c < b + c$ if, and only if, some natural number added to $a + c$ gives $b + c$.
Q. Okay. So how could I find such a natural number? Do I have anything to go on?
A. Well, I'm given that $a < b$. This means that there is some natural number r such that $a + r = b$. Perhaps I can use this.

Carry Out the Plan

Since $a < b$, there is some natural number r such that $a + r = b$. Therefore, add c to both sides:

$$a + r + c = b + c.$$

Then, by the commutative property for addition,

$$a + c + r = b + c.$$

This shows that adding the natural number r to $a + c$ gives $b + c$, and that means that $a + c < b + c$, as was to be shown.

Look Back

Basically, we obtained the solution just by making sure we understood the problem. Key questions answered were the following: What is given? What must be shown? Can we say all this in a different way (that is, what does $a < b$ mean, and so on)? With answers to these questions clearly in mind, a little arithmetic finished the job.

Subtraction of Integers by Using Sets of Colored Counters

As with the subtraction of whole numbers, one approach to the subtraction of integers is the notion of "take away." Consider the subtraction

$$7 - (-3).$$

Modeling 7 with counters, we must "take away" a representation of -3. Since any representation of -3 must have at least three red counters, we must use a representation of 7 with at least three red counters. The simplest representation of this subtraction is shown in Figure 5.12.

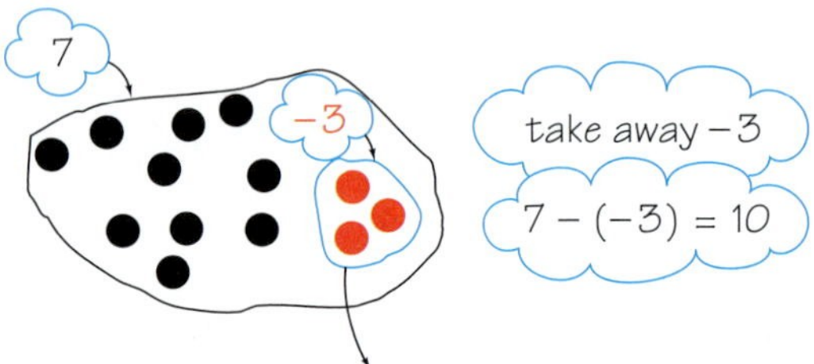

Figure 5.12
Colored-counter representation of $7 - (-3) = 10$

Thus, $7 - (-3) = 10$. Also, the result does not change if we use another representation of 7 with at least three red counters, since equivalent representations are obtained by adding (or deleting) the same number of counters of each color from a given representation. A second representation of $7 - (-3) = 10$ is shown in Figure 5.13.

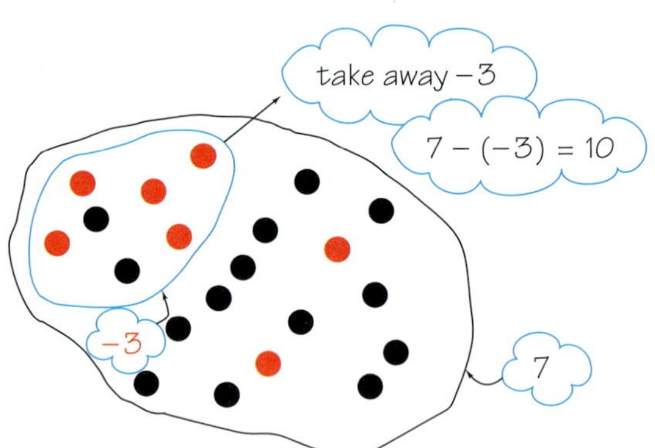

Figure 5.13
A second colored-counter representation of $7 - (-3) = 10$

EXAMPLE 5.16 | SUBTRACTING USING COLORED COUNTERS

Write out the subtraction equation illustrated by each of these diagrams.

(a)

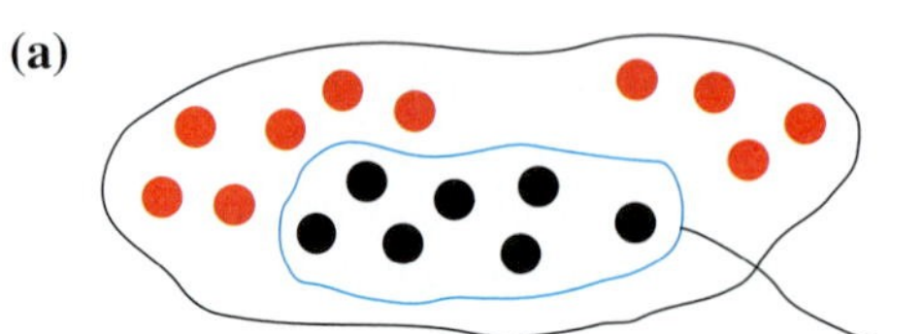

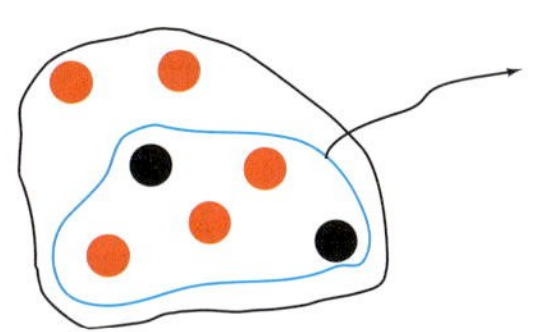

(c)

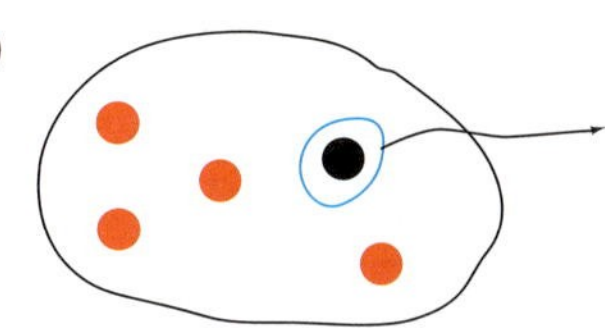

(d)

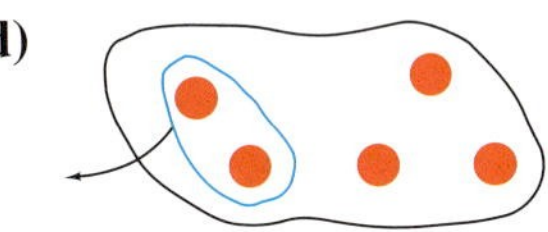

Solution

(a) Since the large loop contains 10 red counters and seven black counters, it represents −3. The small loop contains seven black counters and thus represents 7. If we remove the small loop of counters as indicated, we are left with 10 red counters, representing −10. Thus, this diagram illustrates the subtraction $(-3) - 7 = -10$.

(b) This diagram represents the subtraction $(-3) - (-1) = -2$.

(c) This diagram represents the subtraction $(-3) - 1 = -4$.

(d) This diagram represents the subtraction $(-5) - (-2) = -3$.

EXAMPLE 5.17 DRAWING DIAGRAMS FOR GIVEN SUBTRACTION PROBLEMS

Draw a diagram of colored counters to illustrate each of these subtractions, and determine the result in each case.

(a) $7 - 3$ **(b)** $(-7) - (-3)$ **(c)** $7 - (-3)$ **(d)** $(-7) - 3$

Solution

(a) Many different diagrams could be drawn, but the simplest is shown here.

(b) As in part (a), we can use a diagram with counters of only one color.

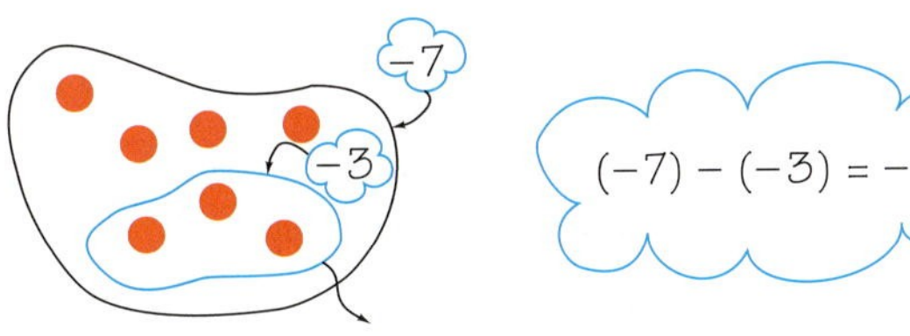

(c) This is illustrated in Figures 5.12 and 5.13.

(d) Here, in order to remove three black counters (that is, subtract 3), the representation for -7 must have at least three black counters. The simplest diagram is as shown.

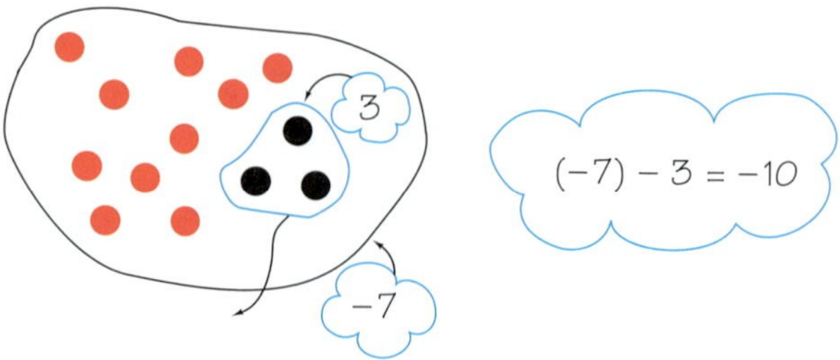

Notice that we can interpret any subtraction diagram as an addition diagram and vice versa. For example, if we interpret the diagram of Figure 5.14 below as taking away the four red counters, the diagram represents the subtraction

$$5 - (-4) = 9.$$

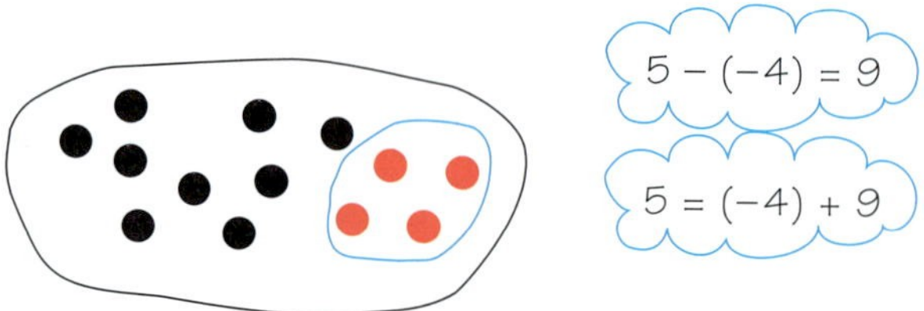

Figure 5.14
Diagram illustrating both $5 - (-4) = 9$ *and* $5 = (-4) + 9$

On the other hand, if we view it as a diagram showing the combining of the set of black counters with the set of red counters, it illustrates the addition

$$5 = (-4) + 9.$$

Since all the subtraction diagrams can be interpreted in this way, we have the following definition.

DEFINITION *Subtraction of Integers*
If a, b, and c are integers, then

$$a - b = c$$

if, and only if, $a = b + c$.

This was also the definition of subtraction of whole numbers. Thus, as before, we have a family of equivalent facts; that is,

$$a - b = c, \qquad a = b + c, \qquad a = c + b, \qquad \text{and} \qquad a - c = b$$

all express essentially the same relationship among the integers a, b, and c. If we know that any one of these equations is true, then all are true.

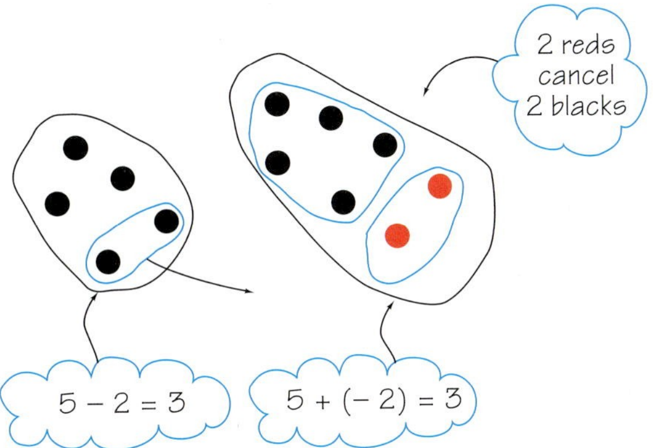

Figure 5.15
Subtracting by adding the negative

Another important fact about subtraction of integers is illustrated in Figure 5.15. Note that the diagram on the left illustrates the subtraction

$$5 - 2 = 3,$$

while the diagram on the right illustrates the addition

$$5 + (-2) = 3.$$

Since the effect of adding the negative of 2 to 5 is the same as subtracting 2 from 5, we see that

$$5 - 2 = 5 + (-2).$$

Moreover, since the opposite of an integer represented by a set of counters is always represented by the set with all the counters turned over, this proposition is true in general and can be expressed as a theorem.

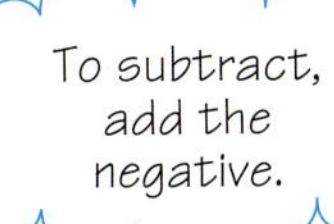

THEOREM *Subtracting by Adding the Negative*
Let a and b be any integers. Then

$$a - b = a + (-b).$$

Finally, since we have already seen that the set of integers is closed under addition, an immediate consequence of this theorem is that, unlike the set of whole numbers, *the set of integers is closed under subtraction.*

THEOREM *Closure Property for the Subtraction of Integers*
The set of integers is closed under subtraction.

EXAMPLE 5.18 | SUBTRACTING BY ADDING

Perform each of these subtractions as additions.

(a) $7 - 3$ **(b)** $(-7) - (-3)$ **(c)** $7 - (-3)$ **(d)** $(-7) - 3$

Solution We make use of the theorem stating that $a - b = a + (-b)$ for any integers a and b.

(a) $7 - 3 = 7 + (-3) = 4$
(b) $(-7) - (-3) = (-7) + 3 = -4$
(c) $7 - (-3) = 7 + 3 = 10$
(d) $(-7) - 3 = -7 + (-3) = -10$

Subtraction of Integers by Using Mail-Time Stories

For this model of subtraction to work, we must imagine a situation where checks and bills are immediately credited or debited to your account as soon as they are delivered, whether they are really intended for you or not. If an error has been made by the mail carrier, he or she must return and reclaim delivered mail and take it to the intended recipient. Thus,

bringing a check adds a positive number,
bringing a bill adds a negative number,
taking away a check subtracts a positive number, and
taking away a bill subtracts a negative number.

EXAMPLE 5.19 SUBTRACTION FACTS FROM MAIL-TIME STORIES

Indicate the subtraction facts that are illustrated by each of the following mail-time stories.

(a) The mail carrier brings you a check for $7 and takes away a check for $3. Are you richer or poorer, and by how much?
(b) The mail carrier brings you a bill for $7 and takes away a bill for $3. Are you richer or poorer, and by how much?
(c) The mail carrier brings you a check for $7 and takes away a bill for $3. Are you richer or poorer, and by how much?
(d) The mail carrier brings you a bill for $7 and takes away a check for $3. Are you richer or poorer, and by how much?

Solution

(a) You are $4 richer. This illustrates $7 - 3 = 4$.
(b) You are $4 poorer. This illustrates $(-7) - (-3) = -4$.
(c) You are $10 richer. This illustrates $7 - (-3) = 10$.
(d) You are $10 poorer. This illustrates $(-7) - 3 = -10$.

Note that these are precisely the same subtractions that were illustrated in Example 5.17 with diagrams of colored counters and in Example 5.18 by adding negatives.

Subtraction of Integers by Using the Number Line

The addition $5 + 3 = 8$ is illustrated on a number line in Figure 5.16.

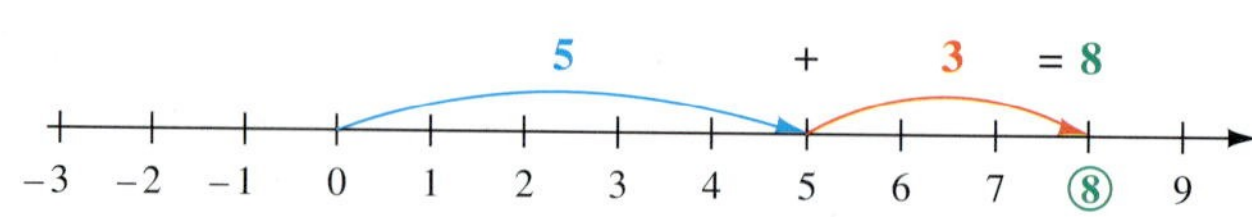

Figure 5.16
The sum $5 + 3 = 8$ on the number line

Now consider the subtraction

$$5 - 3.$$

In this instance, we start at 0 and count five units to the right as before, and then we count *backward* three units in the direction *opposite from that indicated by the 3 arrow.* Figure 5.17 below shows that the result is then 2, as we already know from the subtraction of whole numbers. In particular, this diagram accurately models the missing-addend approach to subtraction. Here, the dashed red arrow (for 2) shows what must be added to 3 to obtain 5. (Recall that arrows representing positive integers are directed from left to right, and those representing negative integers are directed from right to left.) Also, it might be helpful, when modeling subtraction by using arrows on a number line, to draw the arrow representing the subtrahend by starting with the head of this arrow at the head of the arrow representing the minuend and drawing from the arrowhead of the subtrahend backward toward the tail the necessary number of units.

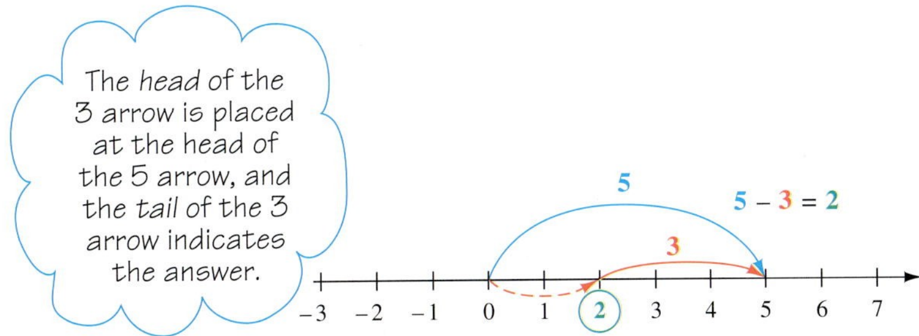

Figure 5.17
The subtraction 5 − 3 = 2 on the number line

EXAMPLE 5.20 DRAWING DIAGRAMS TO ILLUSTRATE SUBTRACTION ON THE NUMBER LINE

Illustrate each of these subtractions on a number line and give the result in each case.

(a) $7 - 3$ **(b)** $(-7) - (-3)$ **(c)** $7 - (-3)$ **(d)** $(-7) - 3$

Solution

(a)

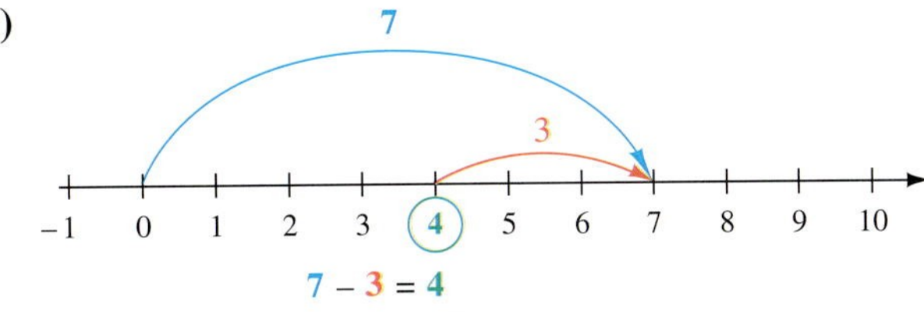

(b)

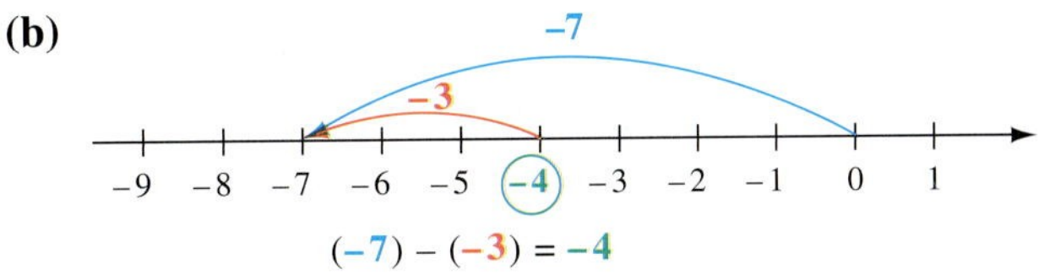

(c)

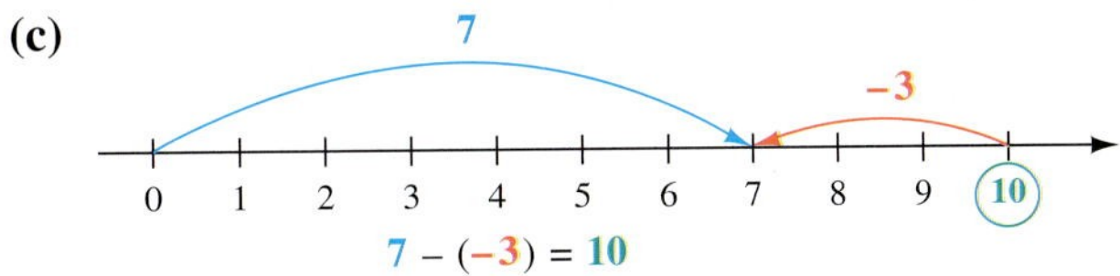

(d)

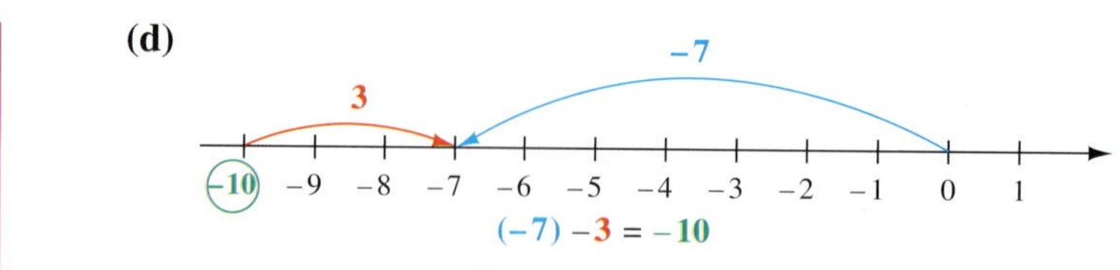

The number line is particularly useful in demonstrating the result

$$a - b = a + (-b)$$

noted earlier. The diagram illustrating that $(-7) - 3 = -10$ in part (d) of Example 5.20 is essentially the same as that in Figure 5.18 which illustrates that $(-7) + (-3) = -10$. Thus, as in the theorem.

$$(-7) - 3 = (-7) + (-3).$$

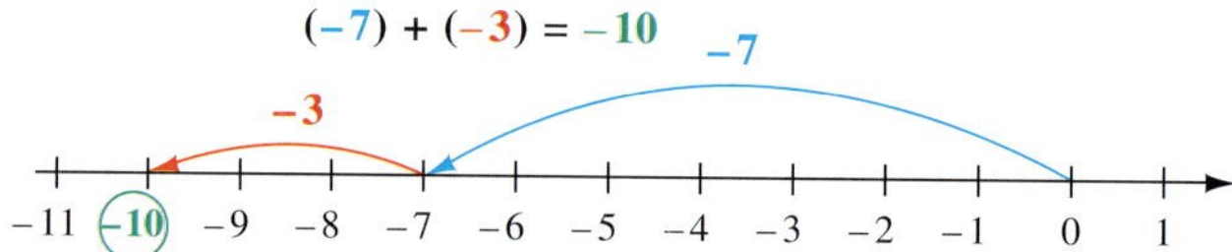

Figure 5.18
The sum $(-7) + (-3) = -10$ on the number line

Adding and Subtracting Integers with a Calculator

Calculator addition and subtraction of natural numbers has already been discussed. Integers can also be added and subtracted on a calculator, but care must be taken, since integers are both positive and negative.

Using the Negation Key, [(−)]

We observed earlier that to enter -5 into the display of your calculator, you enter

[(−)] 5.

Thus, $27 + (-83)$ can be found by keying into your calculator the string

27 [+] [(−)] 83 [=],

and the result is -56. Similarly, entering the string

[(−)] 71 [−] [(−)] 93 [=]

performs the subtraction

$$(-71) - (-93) = 22.$$

EXAMPLE 5.21 ADDING AND SUBTRACTING INTEGERS BY USING THE [(−)] KEY

Perform these computations by using the [(−)] key.

(a) $(-27) + (-95)$ **(b)** $(-27) - (-95)$ **(c)** $3250 + (-4729)$

School Book Page

Subtracting Integers

Subtracting Integers

9-5

▶ Lesson Link In the last lesson, you added integers. Now you'll explore integer subtraction. ◀

You'll Learn ...

- to subtract integers

... How It's Used

Business owners need to add and subtract integers to see whether their business is making or losing money.

Explore Subtracting Integers

What Difference Does It Make?

Materials: Algebra tiles

Subtracting Integers

- To model 2 − 3, use algebra tiles to model the first number in the difference.
- The second number in the difference tells you how many tiles to "take away." If you do not have enough to take away, add yellow-red pairs until you have the right number.
- Take away the number of tiles equal to the second number in the difference. Use the remaining tile(s) to write the difference.

2 − 3

Now there are 3 positive tiles to take away.

Take away 3 tiles.

2 − 3 = −1

1. Use algebra tiles to find each difference.

a. 5 − 3 **b.** −7 − (−2) **c.** −6 − (−4)

d. 2 − 3 **e.** 2 − (−3) **f.** −3 − (−5)

2. Why can you add yellow-red pairs to the original number of tiles?

3. When you subtract a negative integer from a number, how does the result compare to the original number? Explain why this happens.

Learn Subtracting Integers

You can model subtraction with algebra tiles. To subtract a number, take away that number of tiles. To subtract 5 − 3:

Start with 5 positive tiles.

Take away 3 tiles.

There are 2 tiles left.

5 − 3 = 2

9-5 • Subtracting Integers **455**

SOURCE: Scott Foresman–Addison Wesley Middle School Math, *Course 2, p. 455, by Randall I. Charles et al. Copyright © 2002 Pearson Education, Inc. Reprinted with permission.*

Questions for the Teacher

1. The above lesson continues on the next page with this poser.

> The Drop Zone™ Stunt Tower at Paramount's Great America® has a free-fall altitude change of −129 ft. It replaces the Edge™, which had a −60 ft change. How much farther do you free-fall in the Drop Zone? **−69 feet**

How would you help a student to understand that this translates into the equation $(-129) - (-60) = -69$, as desired by the teacher's edition of the text?

2. What would you say to a student who models the preceding problem with the equation $129 - 60 = 69$ and says that you free-fall 69 feet farther on the Drop Zone? Is the student wrong? Explain.

Solution In each case, we indicate the key string entered into the calculator and the resulting answer.

(a) [(–)] 27 [+] [(–)] 95 [=], −122
(b) [(–)] 27 [−] [(–)] 95 [=], 68
(c) 3250 [+] [(–)] 4729 [=], −1479

Problem Set 5.2

Understanding Concepts

1. Draw diagrams of colored counters to illustrate these computations, and state the answer in each case.
   (a) $8 + (-3)$ (b) $(-8) + 3$ (c) $(-8) - (-3)$
   (d) $8 - (-3)$ (e) $9 + 4$ (f) $9 + (-4)$
   (g) $(-9) + 4$ (h) $(-9) - (-4)$
2. Describe mail-time situations that illustrate each computation, and state the answer in each case.
   (a) $(-27) + (-13)$ (b) $(-27) - 13$
   (c) $27 + 13$ (d) $27 - 13$
   (e) $(-41) + 13$ (f) $(-41) - 13$
   (g) $(-13) + 41$ (h) $13 - 41$
3. Draw number-line diagrams that illustrate each computation, and state the answer in each case.
   (a) $8 + (-3)$ (b) $8 - (-3)$ (c) $(-8) + 3$
   (d) $(-8) - (-3)$ (e) $4 + (-7)$ (f) $4 - (-7)$
   (g) $(-4) + 7$ (h) $(-4) - (-7)$
4. Write each of these subtractions as an addition.
   (a) $13 - 7$ (b) $13 - (-7)$
   (c) $(-13) - 7$ (d) $(-13) - (-7)$
   (e) $3 - 8$ (f) $8 - (-3)$
   (g) $(-8) - 13$ (h) $(-8) - (-13)$
5. Perform each of these computations.
   (a) $27 - (-13)$ (b) $12 + (-24)$
   (c) $(-13) - 14$ (d) $-81 + 54$
   (e) $(-81) - 54$ (f) $(-81) - (-54)$
   (g) $(-81) + (-54)$ (h) $27 + (-13)$
   (i) $(-27) - 13$
6. By 2 P.M., the temperature in Cutbank, Montana, had risen 31° from a nighttime low of 41° below zero.
   (a) What was the temperature at 2 P.M.?
   (b) What computation does part (a) illustrate?
7. (a) If the high temperature on a given day was 2° above zero and the morning's low was 27° below zero, how much did the temperature rise during the day?
   (b) What computation does part (a) illustrate?
8. (a) If the high temperature for a certain day was 8° above zero and that night's low temperature was 27° below zero, how much did the temperature fall?
   (b) What computation does part (a) illustrate?
9. During the day, Sam's Soda Shop took in \$314. That same day, Sam paid a total of \$208 in bills.
   (a) Was Sam's net worth more or less at the end of the day? By how much?
   (b) What computation does part (a) illustrate?
10. During the day, Sam's Soda Shop took in \$284. Also, Sam received a check in the mail for \$191 as a refund for several bills that he had inadvertently paid twice.
   (a) Was Sam's net worth more or less at the end of the day? By how much?
   (b) What computation does part (a) illustrate?
   (c) If you think of the \$191 check as removing or taking away the bills previously paid, what computation does this represent? Explain.
11. Place a less-than or greater-than sign in each circle to make a *true* statement.
   (a) −117 ○ −24 (b) 0 ○ −4 (c) 18 ○ 12
   (d) 18 ○ −12 (e) −5 ○ 1 (f) −5 ○ −9
12. List these numbers in increasing order from least to greatest: −5, 27, 5, −2, 0, 3, −17.

Teaching Concepts

13. Keyshawn drew the following diagram to illustrate the subtraction of −3 from 7, arguing that you have to count from 7 on the number line 3 steps in the opposite direction from that indicated by −3.

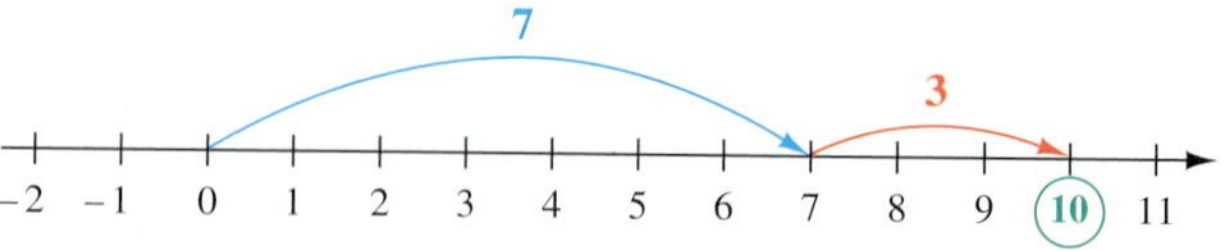

How would you respond to Keyshawn? Compare Keyshawn's response with the solutions to part (c) of both Example 5.19 and Example 5.20.

14. Melanie says that adding the negative, $-n$, of a number n to $m + n$ just exactly nullifies the effect of adding n to m. Is Melanie right? How would you respond to her?

15. Julliene answered "No" to the question "Is the inequality $10 \geq 7$ a correct assertion?" She maintained that the inequality should be $10 > 7$. Is Julliene correct? How would you respond to her?

16. Armand said that to represent subtracting -3 from 7 by using colored counters, you should add three black counters to seven black counters.

(a) Is Armand correct?

(b) Determine the sum $7 - (-3)$.

(c) How would you respond to Armand?

Thinking Critically

17. Which of the following are *true?*

(a) $3 < 12$ (b) $-3 < -12$

(c) $-3 < 12$ (d) $3 < -12$

18. If $a < b$, is $a \leq b$ true? Explain.

19. If $a \geq b$, is $a > b$ true? Explain.

20. Is $2 \geq 2$ true? Explain.

21. For which integers x is it true that $|x| < 7$?

22. For which integers x is it true that $|x| > 99$?

23. (a) Compute each of these absolute values.

(i) $|5 - 11|$ (ii) $|(-4) - (-10)|$

(iii) $|8 - (-7)|$ (iv) $|(-9) - 2|$

(b) Draw a number line and determine the distance between the points on the number line for each of these pairs of integers.

(i) 5 and 11 (ii) -4 and -10

(iii) 8 and -7 (iv) -9 and 2

(c) Since parts (a) and (b) are completely representative of the corresponding general cases, state a general theorem summarizing these results.

24. (a) Compute each of these pairs of expressions.

(i) $|7 + 2|$ and $|7| + |2|$

(ii) $|(-8) + 5|$ and $|-8| + |5|$

(iii) $|7 + (-6)|$ and $|7| + |-6|$

(iv) $|(-9) + (-5)|$ and $|-9| + |-5|$

(v) $|6 + 0|$ and $|6| + |0|$

(vi) $|0 + (-7)|$ and $|0| + |-7|$

(b) Since the results of part (a) are completely typical, place one of the signs $>, <, \geq$, or $\leq$ in the circle to make the following a *true* statement. For any integers a and b,

$$|a + b| \bigcirc |a| + |b|.$$

25. Let a and b be positive integers, with $a < b$. If c is a negative integer, prove that $ac > bc$.

(*Suggestion:* Try using specific numbers first.)

26. (a) Make a magic square using the numbers $-4, -3, -2, -1, 0, 1, 2, 3$, and 4.

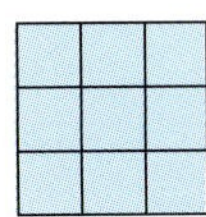

(b) Make a magic subtraction square using the numbers $-4, -3, -2, -1, 0, 1, 2, 3$, and 4.

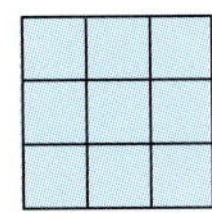

27. Place the numbers $-2, -1, 0, 1$, and 2 in the circles in the diagram so that the sum of the numbers in each direction is the same.

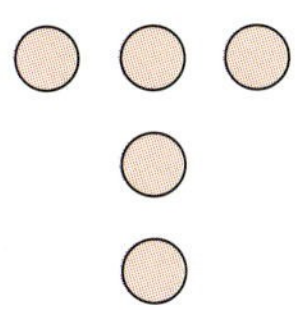

(a) Can this be done with 0 in the middle of the top row? If so, show how. If not, why not?

(b) Can this be done with 2 in the middle of the top row? If so, show how. If not, why not?

(c) Can this be done with -2 in the middle of the top row? If so, show how. If not, why not?

(d) Can this be done with 1 or -1 in the middle of the top row? If so, show how. If not, why not?

28. Perform these pairs of computations.

(a) $7 - (-3)$ and $(-3) - 7$

(b) $(-2) - (-5)$ and $(-5) - (-2)$

(c) Does the commutative law for subtraction hold for the set of integers? Explain briefly.

29. Does the associative law for subtraction hold for the set of integers (that is, is $a - (b - c) = (a - b) - c$ true for all integers a, b, and c)? If so, explain why. If not, give a counterexample.

30. The equation $a(b - c) = ab - ac$ expresses the distributive law for multiplication over subtraction. Is it true for all integers a, b, and c? If so, explain why. If not, give a counterexample.

Thinking Cooperatively

Do the next five problems in a group with two or three other students. At each step, discuss your solution with the other members of your group and determine a consensus answer for your group.

31. Recall that the triangular numbers are the numbers 1, 3, 6, 10, 15, . . . generated by the formula

$$t_n = \frac{n(n + 1)}{2}.$$

(a) Fill in the blanks to complete these equations and to extend the pattern.

$$1 = \underline{\qquad\qquad}$$
$$1 - 3 = \underline{\qquad\qquad}$$
$$1 - 3 + 6 = \underline{\qquad\qquad}$$
$$1 - 3 + 6 - 10 = \underline{\qquad\qquad}$$
$$\underline{\qquad\qquad} = \underline{\qquad\qquad}$$
$$\underline{\qquad\qquad} = \underline{\qquad\qquad}$$

(b) What would the nth equation in this pattern be? (*Hint:* You may have to consider two cases—n odd and n even.)

32. Recall that the Fibonacci numbers are the numbers $F_1 = 1$, $F_2 = 1$, $F_3 = 2$, $F_4 = 3$, $F_5 = 5$, . . . where we start with 1 and 1 and then add any two successive terms to obtain the next term in the sequence.

(a) Fill in the blanks to continue this pattern.

$F_1 = 1$

$F_1 + F_3 = 1 + 2 = 3$

$F_1 + F_3 + F_5 = 1 + 2 + 5 = 8$

$\underline{\qquad\qquad} = \underline{\qquad\qquad}$

$\underline{\qquad\qquad} = \underline{\qquad\qquad}$

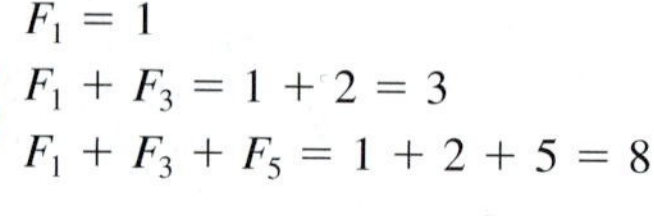

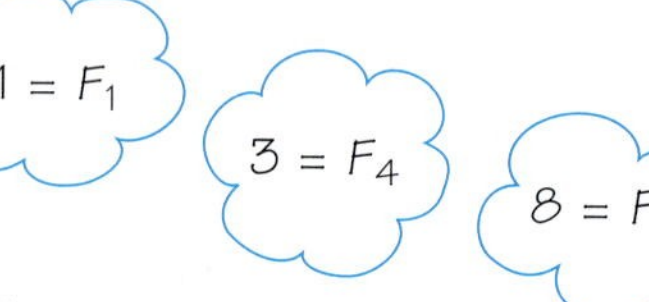

(b) Guess a formula for the general result suggested by part (a).

33. Continue each of these patterns two more steps and guess a general result in each case.

(a) $F_2 = 1$

$F_2 + F_4 = 1 + 3 = 4$

$F_2 + F_4 + F_6 = 1 + 3 + 8 = 12$

$\underline{\qquad\qquad} = \underline{\qquad\qquad}$

$\underline{\qquad\qquad} = \underline{\qquad\qquad}$

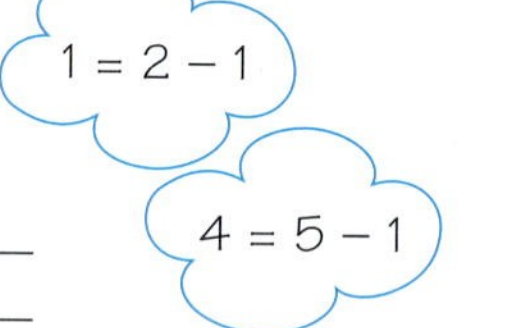

General result: ____________________

(b) $F_1 = 1$

$F_1 + F_2 = 1 + 1 = 2$

$F_1 + F_2 + F_3 = 1 + 1 + 2 = 4$

$\underline{\qquad\qquad} = \underline{\qquad\qquad}$

$\underline{\qquad\qquad} = \underline{\qquad\qquad}$

General result: ____________________

34. Note that the successive powers $(-1)^1 = -1$, $(-1)^2 = 1$, $(-1)^3 = -1$, and $(-1)^4 = 1$ alternate between 1 and -1. Thus, $(-1)^n = 1$ if n is even and -1 if n is odd. Using this idea, continue each of these patterns two more steps and guess a general result in each case.

(a) $F_1 = 1$

$F_1 - F_3 = 1 - 2 = -1$

$F_1 - F_3 + F_5 = 1 - 2 + 5 = 4$

$\underline{\qquad\qquad} = \underline{\qquad\qquad}$

$\underline{\qquad\qquad} = \underline{\qquad\qquad}$

General result: ____________________

(b) $F_2 = 1$

$F_2 - F_4 = 1 - 3 = -2$

$F_2 - F_4 + F_6 = 1 - 3 + 8 = 6$

$\underline{\qquad\qquad} = \underline{\qquad\qquad}$

$\underline{\qquad\qquad} = \underline{\qquad\qquad}$

General result: ____________________

(c) $1 = 1$

$1 - 1 = 0$

$1 - 1 + 2 = 2$

$1 - 1 + 2 - 3 = -1$

$1 - 1 + 2 - 3 + 5 = 4$

$1 - 1 + 2 - 3 + 5 - 8 = -4$

$1 - 1 + 2 - 3 + 5 - 8 + 13 = 9$

_______________ = _______________

_______________ = _______________

General result: ______________________________

(*Suggestion:* To guess the general result, think very carefully about the last three equations in part (c).)

---

35. **(a)** The general result guessed in problem 34, part (c), does not hold for $n = 1$ unless we define F_0 to be 0. Would this be consistent with the definition of the Fibonacci numbers as given in problem 32? Explain.

(b) Determine F_{-1}, F_{-2}, F_{-3}, and F_{-4} so that the pattern established by the definition of problem 32 still holds.

(c) Determine F_{-n} so that the definition of problem 32 holds for all integer values of n. (*Hint:* Use $(-1)^n$ as in problem 34.)

(d) Recall that the Lucas numbers are the numbers $L_1 = 1, L_2 = 3, L_3 = 4, \ldots$ where we start with 1 and 3 and then add any two successive terms to obtain the next term in the sequence. As in part (c), determine L_0 and L_{-n} so that this definition holds for all integer values of n.

Making Connections

36. One day Anne had the flu. At 8 A.M. her temperature was 101°. By noon her temperature had increased by 3°, and then it fell 5° by six in the evening.

(a) Write a single addition equation to determine Anne's temperature at noon.

(b) Write a single equation using both addition and subtraction to determine Anne's temperature at 6 P.M.

37. Vicky was 12 years old on her birthday today.

(a) How old was Vicky on her birthday seven years ago?

(b) How old will Vicky be on her birthday seven years from now?

(c) Write addition equations that answer both parts (a) and (b) of this problem.

38. Greg's bank balance was \$4500. During the month, he wrote checks for \$510, \$87, \$212, and \$725. He also made deposits of \$600 and \$350. What was his balance at the end of the month?

39. A ball is thrown upward from the top of a building 144 feet high. Let h denote the height of the ball above the top of the building t seconds after it was thrown. It can be shown that $h = -16t^2 + 96t$ feet.

(a) Complete this table of values of h.

t	h
0	0
1	80
2	
3	
4	
5	
6	
7	

(b) Give a carefully worded plausibility argument (not a proof) that the greatest value of h in the table is the greatest height the ball reaches.

(c) Carefully interpret (explain) the meaning of the value of h when $t = 7$.

40. The velocity of the ball in problem 39 in feet per second is given by the equation $v = -32t + 96$.

(a) Complete the table of values of v shown below.

(b) Carefully interpret (explain) the value of v when $t = 0$.

(c) Interpret the value of v when $t = 3$.

t	r
0	96
1	64
2	
3	
4	
5	
6	
7	

(d) Carefully interpret the meaning of the values of v for $t = 4, 5, 6$, and 7.

(e) Compare the values of v for $t = 0$ and 6, 1 and 5, and 2 and 4. What do these values tell you about the motion of the ball?

Using a Calculator

41. If we write $a - b + c - d$, it is understood to mean $(((a - b) + c) - d)$. The operations are performed from left to right as on a calculator with algebraic logic.

(a) Compute $1 - 2 + 3 - \cdots + 99$.

(b) Could you think of an easy way to complete part (a) without a calculator?

(c) Compute $1 - 2 + 3 - \cdots + 99 - 100$.

42. Use your calculator to perform these calculations.

(a) $3742 + (-2167)$ **(b)** $(-2751) + (-3157)$

(c) $(-2167) - 3742$ **(d)** $(-3157) - (-2751)$

(e) $-(3571 - 5624)$ **(f)** $-[49{,}002 + (-37{,}621)]$

From State Student Assessments

43. (California, Grade 4)
Identify on a number line the relative position of positive fractions, positive mixed numbers, and positive decimals to two decimal places.

True or false?

1. $-9 > -10$

2. $-31 < -29$

44. (Massachusetts, Grade 8)
Calculators are not allowed here because one can obtain the correct answer without understanding.)
Compute: $(-2)(-5)(-1) =$

For Review

45. Test the following numbers for divisibility by each of 2, 3, 5, and 11.

(a) 214,221 **(b)** 106,090 **(c)** 1,092,315

46. Test each of the following for divisibility by 7, 11, and 13.

(a) 965,419 **(b)** 1,140,997 **(c)** 816,893

47. Test each of the following for divisibility by 4, 6, and 8.

(a) 62,418 **(b)** 83,224 **(c)** 244,824

48. Show that each of 271,134; 427,113; 342,711; 134,271; 113,427; and 711,342 is divisible by 27.

49. Let r, s, t, and u be natural numbers. If $r = s + t$, u divides s, and u does not divide t, use the method of proof by contradiction to prove that u does not divide r.

5.3 MULTIPLICATION AND DIVISION OF INTEGERS

Multiplication of Integers

Just as the set of integers is an extension of the set of natural numbers, multiplication and division in the set of integers are direct extensions of these operations for natural numbers. Recall that multiplication of natural numbers was originally defined in Chapter 2 as repeated addition. Thus, by definition,

$$4 \cdot 3 = 3 + 3 + 3 + 3 = 12.$$

In like manner, we have that

$$4 \cdot (-3) = (-3) + (-3) + (-3) + (-3) = -12.$$

In general, if m and n are positive integers, so that $-m$ and $-n$ are negative, we have that

$$m \cdot (-n) = -mn.$$

This is all well and good, but this approach does not make clear what the products $(-m) \cdot (-n)$ and $(-m) \cdot n$ should be. To understand, we consider patterns of products.

Multiplication of Integers by Using Patterns of Products

Consider the following pattern of equalities. Note that the second factors in the successive products

$$4 \cdot 3 = 12$$
$$4 \cdot 2 = 8$$
$$4 \cdot 1 = 4$$
$$4 \cdot 0 = 0$$

decrease by 1 each time and that the successive results decrease by 4. Continuing the pattern, we obtain this sequence of equalities.

$$\begin{aligned} 4 \cdot 3 &= 12 \\ 4 \cdot 2 &= 8 \\ 4 \cdot 1 &= 4 \\ 4 \cdot 0 &= 0 \\ 4 \cdot (-1) &= -4 \\ 4 \cdot (-2) &= -8 \\ 4 \cdot (-3) &= -12 \end{aligned}$$

Since these results agree with what would be obtained by repeated addition, the pattern of products is an appropriate guide. In general, it suggests that

$$m \cdot (-n) = -mn.$$

positive times negative gives negative

Using this result, we can now construct the following two additional patterns that suggest what to make of the products $(-m) \cdot (-n)$ and $(-m) \cdot n$.

$$\begin{aligned} 3 \cdot (-3) &= -9 & (-3) \cdot (-3) &= 9 \\ 2 \cdot (-3) &= -6 & (-3) \cdot (-2) &= 6 \\ 1 \cdot (-3) &= -3 & (-3) \cdot (-1) &= 3 \\ 0 \cdot (-3) &= 0 & (-3) \cdot 0 &= 0 \\ (-1) \cdot (-3) &= 3 & (-3) \cdot 1 &= -3 \\ (-2) \cdot (-3) &= 6 & (-3) \cdot 1 &= -6 \\ (-3) \cdot (-3) &= 9 & (-3) \cdot 1 &= -9 \end{aligned}$$

Thus, the patterns suggest that

$$(-m) \cdot (-n) = mn \qquad \text{and} \qquad (-m) \cdot (n) = -mn.$$

negative times negative gives positive

negative times positive gives negative

Collecting these results together, we have the following theorem.

THEOREM *The Rule of Signs*
Let m and n be positive integers so that $-m$ and $-n$ are negative integers. Then the following are true:

- $m \cdot (-n) = -mn$;
- $(-m) \cdot n = -mn$;
- $(-m) \cdot (-n) = mn$.

HIGHLIGHT from HISTORY

Charlotte Angas Scott (1858–1931)

Charlotte Scott is an example of a woman who has overcome great obstacles in the field of mathematics. Born in England, she attended Girton College of Cambridge University at a time when women were barred from receiving degrees and from even attempting the examinations for honors. She took the exams "informally," however, and besides two first places won eighth place in math, a field "too difficult for women." As the honors lists were read, shouts for "*Scott of Girton!*" cheered her achievement. Scott received her doctoral degree from the University of London and was called to the United States to become the only woman on the founding faculty of Bryn Mawr College in Pennsylvania. A long record of scholarship, writing, and contributions to the field of analytic geometry followed. Scott's students found her classes exciting, her mathematical style "elegant," and her own gift for clear explanation combined nicely with her quick understanding of their "stupidity." In later years, her record at Cambridge opened new doors for women; and many of her female students joined her in the lists of distinguished mathematicians and teachers.

SOURCE: From Mathematics in Modules, Algebra, A3, Teacher's Edition. *Reprinted by permission.*

EXAMPLE 5.22 MULTIPLYING INTEGERS

Compute these products.

(a) $(-7)\cdot(-8)$ **(b)** $(-8)\cdot(-7)$

(c) $3\cdot(-7)$ **(d)** $(-7)\cdot 3$

(e) $[8\cdot(-5)]\cdot 6$ **(f)** $8\cdot[(-5)\cdot 6]$

(g) $(-4)\cdot[(-5)+7]$ **(h)** $(-4)\cdot(-5)+(-4)\cdot 7$

Solution

(a) $(-7)\cdot(-8)=7\cdot 8=56$

(b) $(-8)\cdot(-7)=8\cdot 7=56$

(c) $3\cdot(-7)=-(3\cdot 7)=-21$

(d) $(-7)\cdot 3=-(7\cdot 3)=-21$

(e) $[8\cdot(-5)]\cdot 6=[-(8\cdot 5)]\cdot 6=(-40)\cdot 6=-(40\cdot 6)=-240$

(f) $8\cdot[(-5)\cdot 6]=8\cdot[-(5\cdot 6)]=8\cdot(-30)=-(8\cdot 30)=-240$

(g) $(-4)\cdot[(-5)+7]=-(4\cdot 2)=-8$

(h) $(-4)\cdot(-5)+(-4)\cdot 7=4\cdot 5+[-(4\cdot 7)]=20+(-28)=-8$

The results of the preceding examples illustrate that the closure, commutative, associative, and distributive properties of multiplication hold for the set of integers. They also suggest that $1\cdot r=r\cdot 1=r$ and that $0\cdot r=r\cdot 0=0$ for every integer r. Since these results are typical, we have the following theorem.

Just for Fun

Three on a Bike?

Three teenagers arrived at a rental agency, each desiring to rent a bicycle for the day. There was only one bicycle left, but the teenagers finally agreed to share equally the \$30 charge for the one day and also to share equally in the bicycle's use. Later the manager of the agency decided that his charge was excessive and sent his helper to return \$5 to the teenagers. Being dishonest, the helper pocketed \$2 and only returned \$3 to the teenagers. Thus, each paid \$9 for the use of the bicycle. But \$27 plus the \$2 the helper took makes only \$29. What happened to the other dollar?

THEOREM *Multiplication Properties of Integers*

Let r, s, and t be any integers.

Closure Property rs is an integer.

Commutative Property $rs = sr$.

Associative Property $r(st) = (rs)t$.

Distributive Property $r(s + t) = rs + rt$.

Multiplicative-Identity Property of One $1 \cdot r = r \cdot 1 = r$.

Multiplicative Property of Zero $0 \cdot r = r \cdot 0 = 0$

Multiplication of Integers by Using Mail-Time Stories

Recall that in mail-time stories, the mail carrier bringing checks and bills corresponds to the addition of positive and negative numbers, respectively. Similarly, taking away checks and bills corresponds to subtracting positive and negative numbers, respectively.

Suppose the letter carrier brings five bills for \$11 each. Are you richer or poorer, and by how much? Answer: Poorer by \$55. Since this is repeated addition, this illustrates the product

$$5 \cdot (-11) = -55.$$

Suppose the mail carrier takes away four bills for \$13 each. Are you richer or poorer, and by how much? Answer: Richer by \$52. This illustrates the product

$$(-4) \cdot (-13) = 52$$

since 4 bills for the same amount are *taken away*.

EXAMPLE 5.23 **WRITING MAIL-TIME STORIES FOR MULTIPLICATION OF INTEGERS**

Write a mail-time story to illustrate each of these products.

(a) $(-4) \cdot 16$ **(b)** $(-4) \cdot (-16)$ **(c)** $4 \cdot (-16)$ **(d)** $4 \cdot 16$

Solution

(a) The letter carrier takes away four checks for \$16 each. Are you richer or poorer, and by how much? Answer: \$64 poorer; $(-4) \cdot 16 = -64$.

(b) The letter carrier takes away four bills for \$16 each. Are you richer or poorer, and by how much? Answer: \$64 richer; $(-4) \cdot (-16) = 64$.

(c) The letter carrier brings four bills for \$16 each. Are you richer or poorer, and by how much? Answer: Poorer by \$64; $4 \cdot (-16) = -64$.

(d) The letter carrier brings you four checks for \$16 each. Are you richer or poorer, and by how much? Answer: Richer by \$64; $4 \cdot 16 = 64$.

Multiplication of Integers by Using a Number Line

Think of

$$3 \cdot 4 \text{ as } 4 + 4 + 4,$$
$$3 \cdot (-4) \text{ as } (-4) + (-4) + (-4),$$
$$(-3) \cdot (4) \text{ as } -4 - 4 - 4, \text{ and}$$
$$(-3) \cdot (-4) \text{ as} -(-4) - (-4) - (-4).$$

These can be illustrated on the number line as follows.

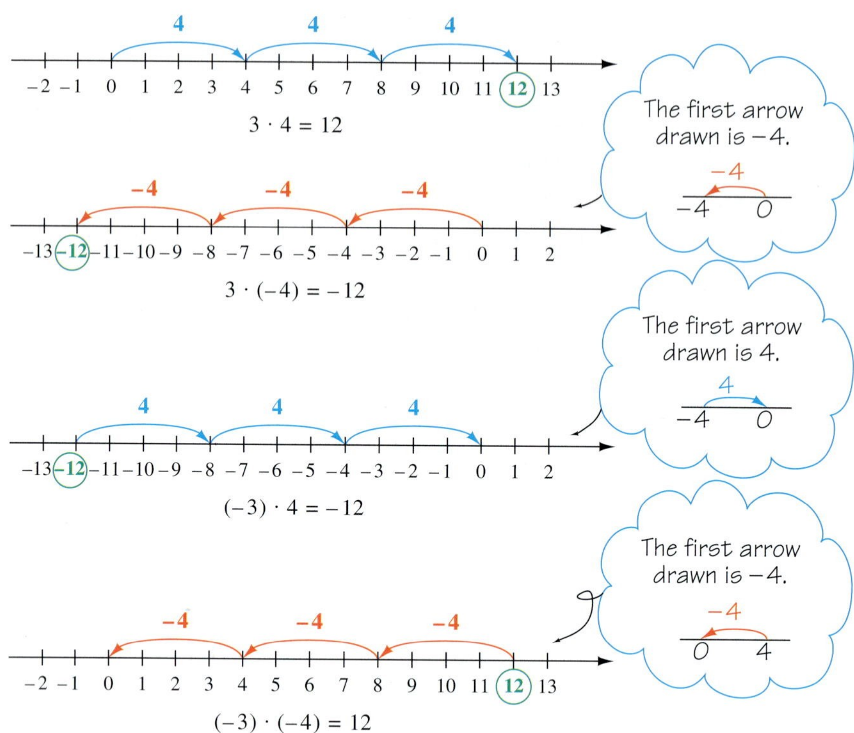

EXAMPLE 5.24 MULTIPLYING INTEGERS BY USING A NUMBER LINE

What products do each of these diagrams illustrate?

(a)

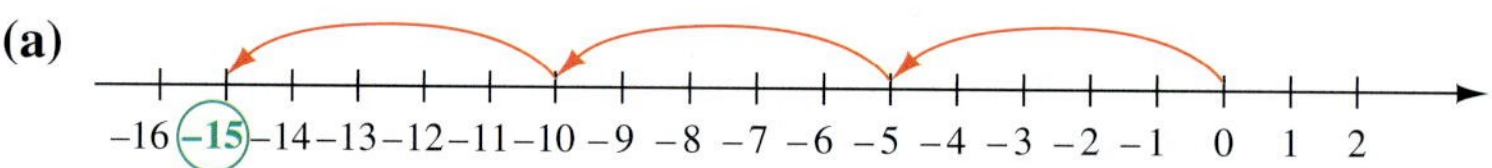

(b)

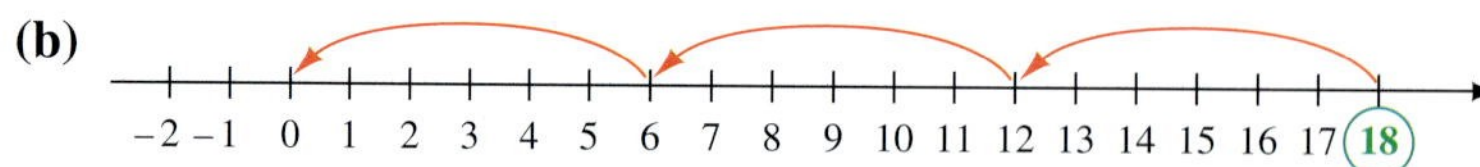

Solution

(a) Starting with the tail of the rightmost arrow at 0, we move in the direction of the arrows 5 units to the left, then 5 more units to the left, and finally 5 more units to the left. Thus, we are *adding* -5 to -5 to -5, and the result is $3 \cdot (-5) = -15$.

(b) Here, we start with the head of the leftmost arrow at 0 and move in the direction *opposite* to the direction of the arrows, 6 units to the right, 6 more units to the right, and finally 6 more units to the right. Since the arrows are directed from right to left, we are *subtracting* -6 and -6 and -6. The result is $(-3) \cdot (-6) = 18$.

Division of Integers

When discussing division of natural numbers, we considered families of facts. Thus, the equations

$$12 = 3 \cdot 4 \qquad 12 \div 3 = 4$$
$$12 = 4 \cdot 3 \qquad 12 \div 4 = 3$$

all express the same relationship between the numbers 12, 3, and 4. Indeed, one definition of division of whole numbers was the following:

If a, b, and c are whole numbers with $b \neq 0$, then $a \div b = c$ if, and only if, $a = bc$.

Thus, to find the quotient $143 \div 11$, we determine c such that $143 = 11 \cdot c$. Since $11 \cdot 13 = 143$, $143 \div 11 = 13$.

Applying the same ideas to integers, we have the following fact families.

(a) $12 = (-3) \cdot (-4) \qquad 12 \div (-3) = -4$
$12 = (-4) \cdot (-3) \qquad 12 \div (-4) = -3$

(b) $-12 = 3 \cdot (-4) \qquad (-12) \div 3 = -4$
$-12 = (-4) \cdot 3 \qquad (-12) \div (-4) = 3$

These equations are typical and, as in the case of the natural numbers, lead to the following definition of division in the set of integers.

DEFINITION *Division of Integers*
If a, b, and c are integers with $b \neq 0$, then $a \div b = c$ if, and only if, $a = b \cdot c$.

EXAMPLE 5.25 DIVIDING INTEGERS

Perform the following divisions.

(a) $28 \div 4$ **(b)** $28 \div (-4)$ **(c)** $(-28) \div 4$ **(d)** $(-28) \div (-4)$

Solution The solutions depend entirely on the preceding definition.

(a) $28 \div 4 = 7$ since $28 = 4 \cdot 7$.
(b) $28 \div (-4) = -7$ since $28 = (-4) \cdot (-7)$.
(c) $(-28) \div 4 = -7$ since $-28 = 4 \cdot (-7)$.
(d) $(-28) \div (-4) = 7$ since $-28 = (-4) \cdot 7$.

Since these results are entirely typical, we state here the rule of signs for division.

THEOREM *Rule of Signs for Division of Integers*
Let m and n be positive integers so that $-m$ and $-n$ are negative integers, and suppose that n divides m. Then the following are true:

- $m \div (-n) = -(m \div n)$;
- $(-m) \div n = -(m \div n)$;
- $(-m) \div (-n) = m \div n$.

Thus, *given that n divides m,* we see that

- a positive integer divided by a negative integer is a negative integer,
- a negative integer divided by a positive integer is a negative integer,
- a negative integer divided by a negative integer is a positive integer, and
- a positive integer divided by a positive integer is a positive integer.

EXAMPLE 5.26 PERFORMING DIVISION OF INTEGERS

If possible, compute each of these quotients.

(a) $(-24) \div (-8)$ **(b)** $24 \div (-8)$ **(c)** $48 \div 12$
(d) $(-48) \div 12$ **(e)** $(-57) \div 19$ **(f)** $(-12) \div 0$
(g) $(-51) \div (-17)$ **(h)** $28 \div (9 - 5)$ **(i)** $(27 + 9) \div (-4)$

Solution

We use the preceding theorem and the definition of division of integers.

(a) $(-24) \div (-8) = 3$. Check: $-24 = (-8) \cdot 3$.
(b) $24 \div (-8) = -3$. Check: $24 = (-8) \cdot (-3)$.
(c) $48 \div 12 = 4$. Check: $48 = 12 \cdot 4$.
(d) $(-48) \div 12 = -4$. Check: $-48 = 12 \cdot (-4)$.
(e) $(-57) \div 19 = -3$. Check: $-57 = 19 \cdot (-3)$.
(f) $(-12) \div 0$ is not defined, since there is no number c such that $-12 = 0 \cdot c = 0$.
(g) $(-51) \div (-17) = 3$. Check: $-51 = (-17) \cdot 3$.
(h) $28 \div (9 - 5) = 28 \div 4 = 7$. Check: $28 = 4 \cdot 7$.
(i) $(27 + 9) \div (-4) = 36 \div (-4) = -9$. Check: $36 = (-4) \cdot (-9)$.

Problem Set 5.3

Understanding Concepts

1. Perform these multiplications.
   (a) $7 \cdot 11$ (b) $7 \cdot (-11)$
   (c) $(-7) \cdot 11$ (d) $(-7) \cdot (-11)$
   (e) $12 \cdot 9$ (f) $12 \cdot (-9)$
   (g) $(-12) \cdot 9$ (h) $(-12) \cdot (-9)$
   (i) $(-12) \cdot 0$
2. Perform these divisions.
   (a) $36 \div 9$ (b) $(-36) \div 9$
   (c) $36 \div (-9)$ (d) $(-36) \div (-9)$
   (e) $(-143) \div 11$ (f) $165 \div (-11)$
   (g) $(-144) \div (-9)$ (h) $275 \div 11$
   (i) $72 \div (21 - 19)$
3. Write another multiplication equation and two division equations that are equivalent to $(-11) \cdot (-25{,}753) = 283{,}283$.
4. Write two multiplication equations and another division equation that are equivalent to $(-1001) \div 11 = -91$.
5. What computation does each of these mail-time stories illustrate?
   (a) The mail carrier brings you six checks for $13 each. Are you richer or poorer? By how much?
   (b) The mail carrier brings you four bills for $23 each. Are you richer or poorer? By how much?
   (c) The mail carrier takes away three bills for $17 each. Are you richer or poorer? By how much?
   (d) The mail carrier takes away five checks for $20 each. Are you richer or poorer? By how much?

6. What computation does each of these number-line diagrams represent?

(a)

(b)

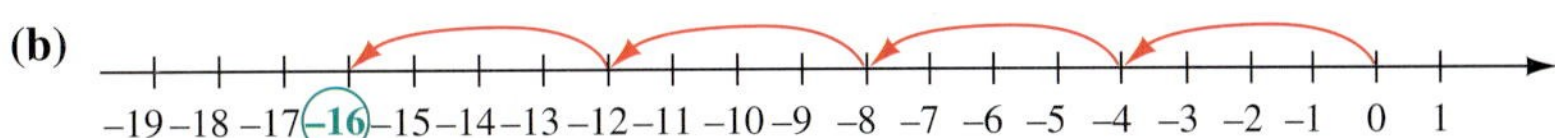

(c)

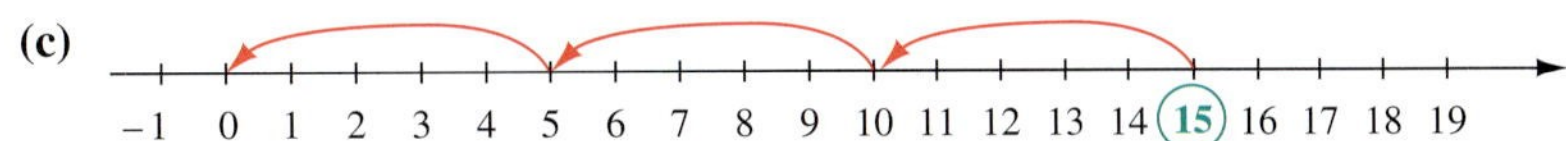

7. Draw a number-line diagram to illustrate $(-4) \cdot 3 = -12$.

Teaching Concepts

8. Marni claims that $\sqrt{4} = -2$ since $(-2) \cdot (-2) = 4$. Daniel claims that Marni is wrong, since everyone knows that $\sqrt{4} = 2$. How would you settle the disagreement between Marni and Daniel?

9. If we assume that the integers obey the distributive property as illustrated earlier in this section, then a proof of the rule of signs is available. Fill in the blanks to justify the steps of these proofs.

(a) $3 \cdot 0 = 0$ __________

$3 \cdot [5 + (-5)] = 0$ Definition of negative

$3 \cdot 5 + 3 \cdot (-5) = 0$ __________

$3 \cdot (-5) = -(3 \cdot 5)$ __________

(b) If a and b are any positive integers, then

$a \cdot 0 = 0$ __________

$a[b + (-b)] = 0$ __________

$ab + a(-b) = 0$ __________

$a \cdot (-b) = -(ab)$ __________

(c) $0 \cdot 5 = 0$ __________

$[3 + (-3)] \cdot 5 = 0$ Definition of negative

$3 \cdot 5 + (-3) \cdot 5 = 0$ __________

$(-3) \cdot 5 = -(3 \cdot 5)$ __________

(d) If a and b are any positive integers, then

$0 \cdot b = 0$ __________

$[a + (-a)] \cdot b = 0$ __________

$ab + (-a)b = 0$ __________

$(-a)b = -(ab)$ __________

(e) $(-3) \cdot 0 = 0$ __________

$(-3) \cdot [5 + (-5)] = 0$ __________

$(-3) \cdot 5 + (-3) \cdot (-5) = 0$ __________

$-(3 \cdot 5) + (-3) \cdot (-5) = 0$ By part (c)

$(-3) \cdot (-5) = 3 \cdot 5$ __________

(f) $(-a) \cdot 0 = 0$ __________

$(-a) \cdot [b + (-b)] = 0$ __________

$(-a)b + (-a)(-b) = 0$ __________

$-(ab) + (-a)(-b) = 0$ ______________________

$(-a)(-b) = ab$ ______________________

(g) Would parts (b), (d), and (f) be appropriate for an elementary school classroom? Why or why not?

(h) Might parts (a), (c), and (e) be appropriate for an upper elementary or middle school classroom? Discuss.

Thinking Critically

10. In the diagram below, the sum of the numbers in any two small circles is the number in the large circle between them.

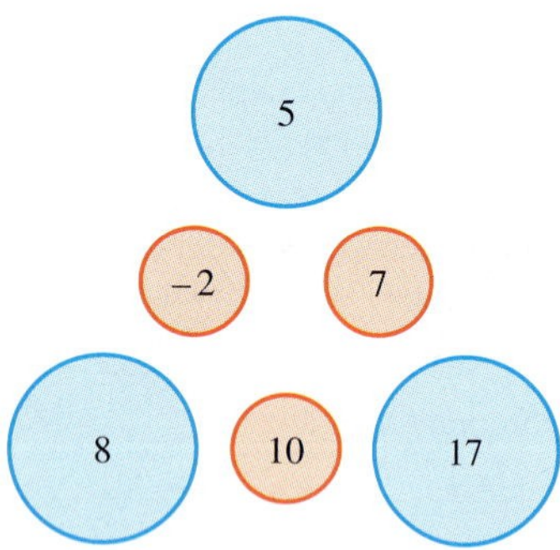

Complete the following so that the same pattern holds in each case.

(a)

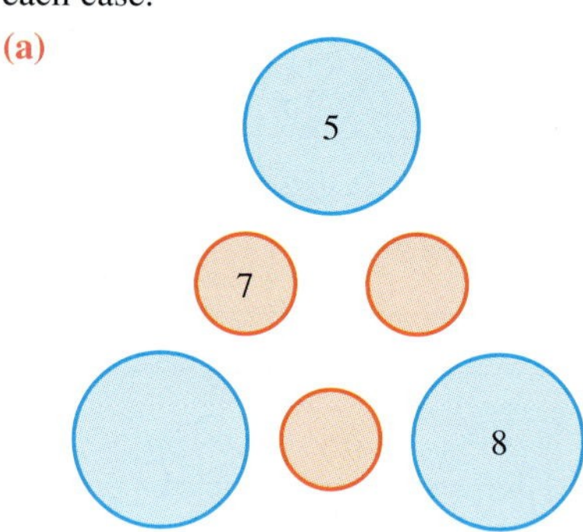

(b)

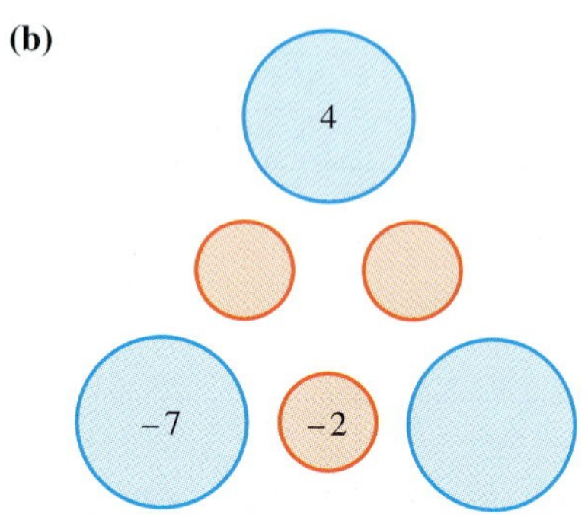

(c)

(d)

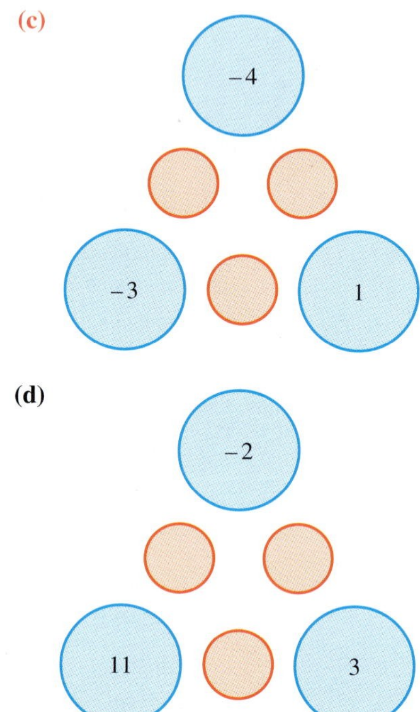

11. This problem is like the preceding one except that there are five large and five small circles. Again, the sum of the numbers in two small circles is the number in the large circle between them.

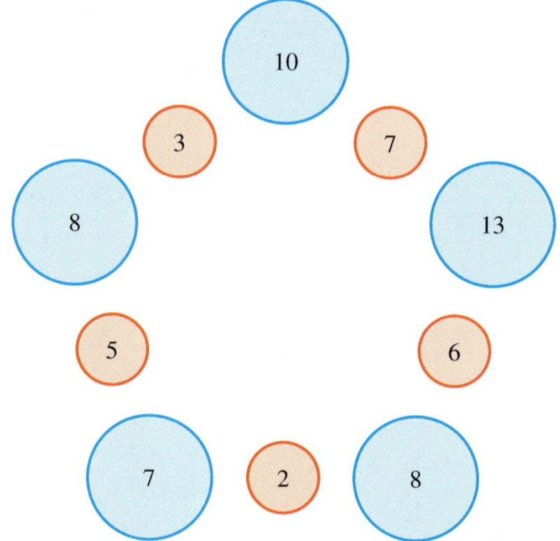

Complete each of the following so that the same pattern holds in each case. Indicate if the solution to each of these is unique.

(a)

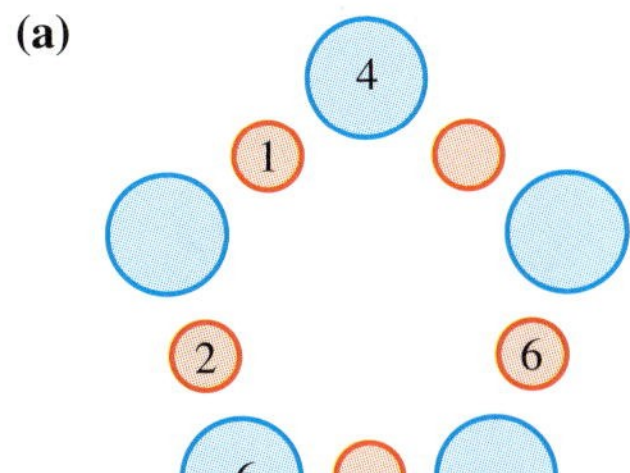

(b)

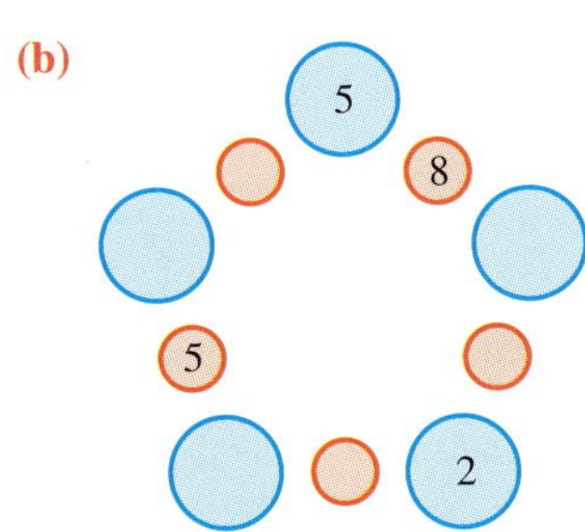

(c)

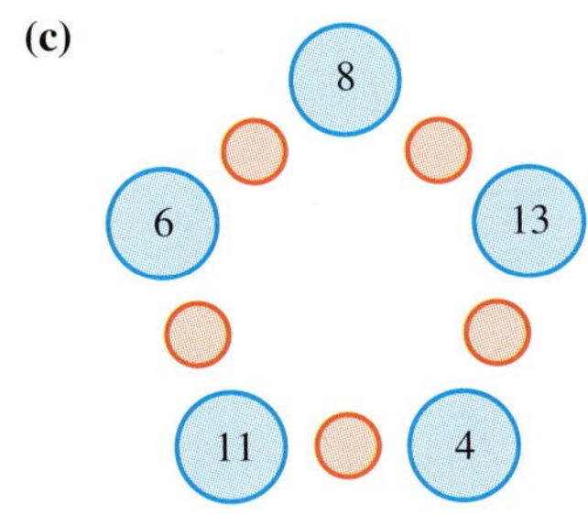

(d) Is a puzzle like these always solvable? Explain. (*Hint:* Let u, v, w, x, and y be the numbers in the small circles.)

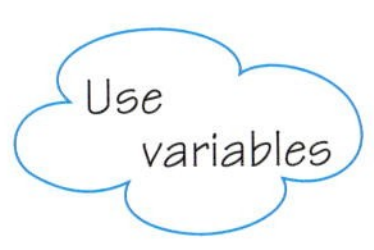

12. Consider a four-large-circle version of problem 10 as shown below.

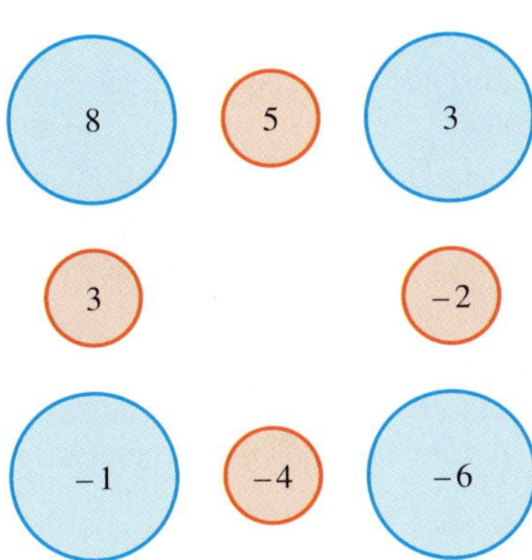

(a) Complete the following diagram so that the same relationships hold.

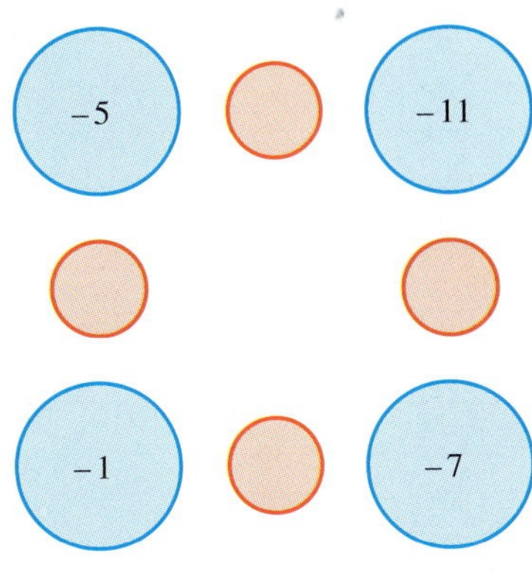

(b) Is a puzzle like this always solvable? Explain. (*Hint:* Let x, y, z, and w be the numbers in the small circles.)

(c) What must be the case to ensure that a puzzle with six large circles is solvable? Explain carefully.

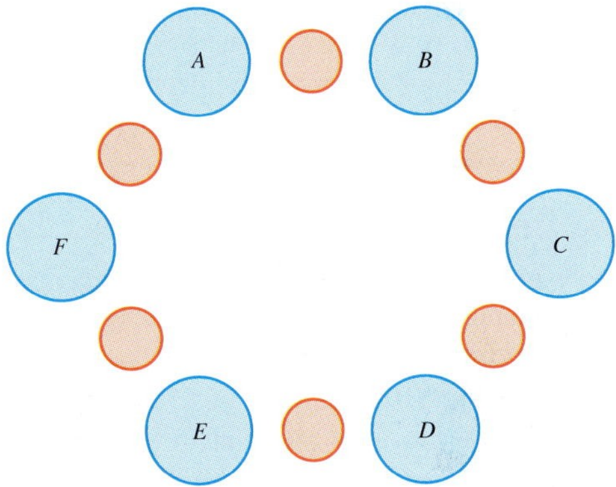

13. Use the numbers $-8, -6, -4, -2, 0, 2, 4, 6$, and 8 to make a magic square. What should the sum in each row, column, and diagonal be? What should the middle number be?

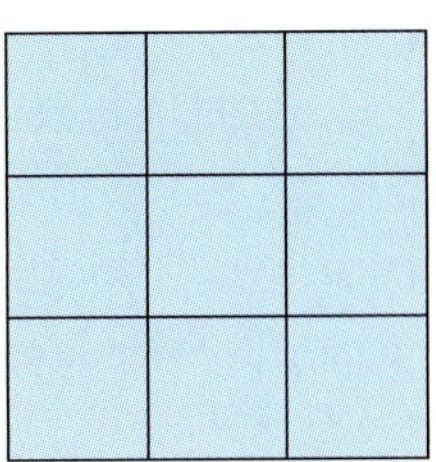

14. Use the numbers $-7, -6, -5, -1, 0, 1, 5, 6$, and 7 to make a magic square.

15. **(a)** Using each of $-3, -1, 1$, and 3 at most once, what sums can be found? For example, three such sums are

$3 + (-1) = 2$, $(-3) + (-1) = -4$, and $1 = 1$.

(b) What strategy did you use to solve part (a)? Explain.

r	s	$r + s$	$x = 2rs$	$y = r^2 - s^2$	$z = r^2 + s^2$	$x^2 + y^2$	z^2
2	1	3					
3	1	4					
4	1	5					
3	2	5					
5	1	6					
4	2	6					
6	1	7					
5	2	7					
4	3	7					
7	1	8					
6	2	8					
5	3	8					

Thinking Cooperatively

Work with two or three other students to complete this problem.

16. **(a)** Complete the table above for values of r and s such that $r + s = 3, 4, 5, 6, 7$, and 8, with $r > s > 0$, by dividing the work among members of the group.

(b) Discuss the results of part (a) and agree with your group on a conjecture the results suggest.

Making Connections

17. The members of the Pep Club tried to raise money by raffling off a pig, agreeing that if they actually lost money on the enterprise, they would share the loss equally.

(a) If they lost \$105 and there are 15 members in the club, how much did each club member have to pay?

(b) What arithmetic might be used to illustrate this story? Explain.

18. It cost \$39 each to buy sweatshirts for members of the Pep Club.

(a) If there are 15 members in the Pep Club, what was the total cost?

(b) What arithmetic might be used to illustrate this story? Explain.

Using a Calculator

19. Calculate each of these without using the [M+], [M−], or [MR] key on your calculator.

(a) $31 - 47 + 88 + 16 - 5$

(b) $57 + 165 \div (-11) + 17$

(c) $(47 + 81 - 56 + 9) \div (67 - 31 - 9)$

20. The Diffy process described in the Cooperative Investigation in Section 2.3 employed whole numbers. Explain why the process is essentially unchanged if we start with all negative or a mixture of positive and negative integers.

For Review

21. Use the built-in constant function on a calculator to compute the twentieth term in each of these sequences.

(a) 5, 8, 11, 14, 17, . . .

(b) 5, 10, 20, 40, 80, . . .

22. Let $a_1 = 1$, $a_2 = 2$, and $a_{n+2} = 2a_{n+1} + a_n$ for $n \geq 1$.

(a) Compute a_3, a_4, a_5, a_6, a_7, and a_8.

(b) Compute the quotients $\frac{a_2}{a_1}, \frac{a_3}{a_2}, \frac{a_4}{a_3}, \frac{a_5}{a_4}, \frac{a_6}{a_5}, \frac{a_7}{a_6}$, and $\frac{a_8}{a_7}$.

(c) Let $b_1 = 2$, $b_2 = 6$, and $b_{n+2} = 2b_{n+1} + b_n$ for all $n > 0$. Compute b_3, b_4, b_5, b_6, b_7, and b_8.

(d) Compute the quotients $\frac{b_2}{b_1}, \frac{b_3}{b_2}, \frac{b_4}{b_3}, \frac{b_5}{b_4}, \frac{b_6}{b_5}, \frac{b_7}{b_6}$, and $\frac{b_8}{b_7}$.

(e) Compute the decimal expansion of $1 + \sqrt{2}$.

(f) Consider the results of parts (b), (d), and (e) and briefly discuss what seems to be the case.

(g) Compute $a_1 + a_3$, $a_2 + a_4$, $a_3 + a_5$, $a_4 + a_6$, $a_5 + a_7$, and $a_6 + a_8$ and compare with the results of part (c). Make a conjecture based on your observations.

(h) Compute $(b_1 + b_3) \div 8$, $(b_2 + b_4) \div 8$, $(b_3 + b_5) \div 8$, $(b_4 + b_6) \div 8$, $(b_5 + b_7) \div 8$, and $(b_6 + b_8) \div 8$. Make a conjecture based on these computations.

(i) Compute $b_1 + 2a_1$, $b_2 + 2a_2$, $b_3 + 2a_3$, $b_4 + 2a_4$, $b_5 + 2a_5$, $b_6 + 2a_6$, $b_7 + 2a_7$, and $b_8 + 2a_8$. Make a conjecture based on these computations.

5.4 CLOCK ARITHMETIC

Mathematics is all around us. Making use of such naturally occurring mathematical ideas is an important and useful teaching strategy. What, for example, is the mathematics of an ordinary clock, as depicted in Figure 5.19?

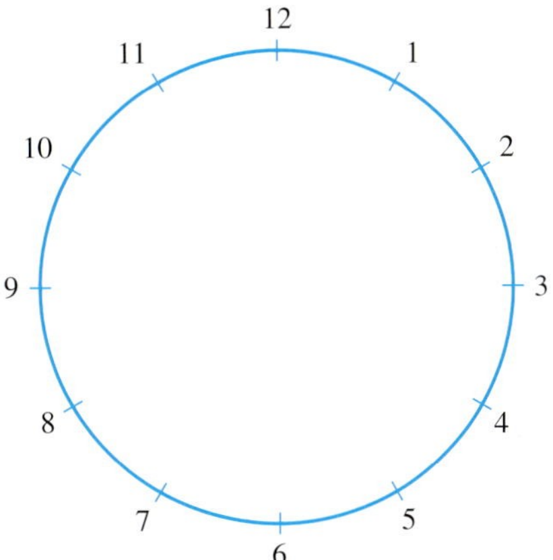

Figure 5.19
A 12-hour clock

Clock Addition and Multiplication

What is the point on the number line 9 units to the right of 7?

This is an addition problem that can be solved on the number line by "counting on." Starting at 0, we first count out to 7 and then count on 9 more to 16. Thus, $7 + 9 = 16$. Now, thinking of the clock, we might ask, "What time is it 9 hours after 7 o'clock?" Again, this can be answered by counting, but this time we count clockwise around the face of a clock. Starting at 12 on the clock diagrammed in Figure 5.19, we count to 7 and then count on 9 more to arrive at 4. Thus, in **12-hour clock arithmetic,** it is reasonable to say that 7 plus 9 is 4 and to write

$$7 +_{12} 9 = 4,$$

where we indicate 12-hour clock addition by the symbol $+_{12}$.

Proceeding in the same way, we find that $9 +_{12} 12 = 12 +_{12} 9 = 9$. Indeed, since counting on 12 steps takes us all the way around the clock, it is clear that $n +_{12} 12 = 12 +_{12} n = n$ for every n on the clock. Thus, 12 plays the role of zero in 12-hour clock arithmetic, and it will suit our purposes to replace 12 by 0 and to number the clock as shown in Figure 5.20. Thus, the numbers in the arithmetic are those in the set $T = \{0, 1, 2, 3, 4, 5, 6, 7, 8, 9, 10, 11\}$, and the arithmetic for the clock is determined as for whole numbers except that we count around the clock instead of on the number line.

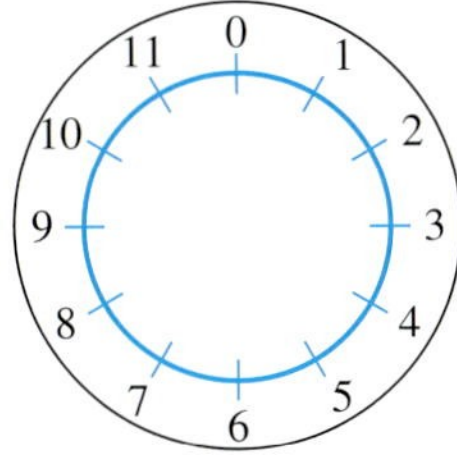

Figure 5.20
The 12-hour clock with 12 replaced by 0

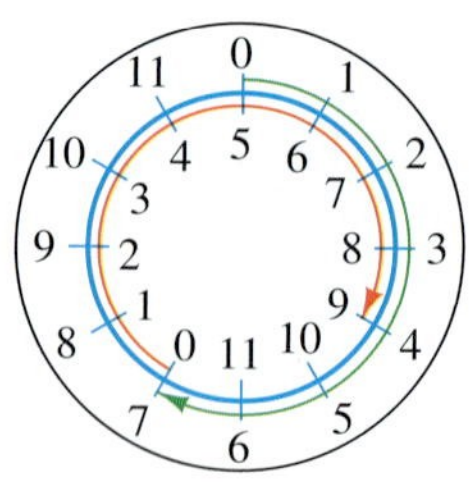

Figure 5.21
Illustrating $7 +_{12} 9 = 4$ with two rotatable discs

A simple, but effective, manipulative for visualizing addition (and subtraction) in clock arithmetic is easily made from two circular discs marked as shown in Figure 5.21 and joined at the center so that they can rotate against one another. (Paper plates, where one is trimmed down a bit, are reasonably durable and work well.) To add 9 to 7 in 12-hour clock arithmetic, for example, set the 0 on the inner circle under the 7 on the outer circle. Then count around to 9 on the inner circle. Since 9 on the inner circle lies under 4 on the outer circle, $7 +_{12} 9 = 4$ as noted above.

Now consider the sum $7 +_{12} 9 +_{12} 11 +_{12} 5 +_{12} 10 +_{12} 8 +_{12} 6$. Starting at 0 and counting to 7, then counting on 9 more, then 11 more, then 5 more, then 10 more, then 8 more, and then 6 more, gets us to 8. Thus, we have

$$7 +_{12} 9 +_{12} 11 +_{12} 5 +_{12} 10 +_{12} 8 +_{12} 6 = 8.$$

In whole number arithmetic this sum would be 56. Since $56 = 4 \cdot 12 + 8$, we would have counted four times around the clock and then on 8 more, obtaining the result shown. In a similar way, we determine that

$$7 \times_{12} 9 = 9 +_{12} 9 +_{12} 9 +_{12} 9 +_{12} 9 +_{12} 9 +_{12} 9 = 3$$

and, in ordinary whole number arithmetic, $7 \cdot 9 = 63 = 5 \cdot 12 + 3$. These considerations lead to the following definition.

DEFINITION *Computing 12-Hour Clock Sums and Products*
To compute sums and products in 12-hour clock arithmetic, perform the same computation in whole number arithmetic, divide by 12, and take the remainder, r, as the answer.

EXAMPLE 5.27 COMPUTING 12-HOUR CLOCK SUMS AND PRODUCTS

Perform the following computations in 12-hour clock arithmetic.

(a) $7 +_{12} 8$ **(b)** $3 +_{12} 9$ **(c)** $11 +_{12} 7 +_{12} 6 +_{12} 10$
(d) $7 \times_{12} 8$ **(e)** $4 \times_{12} 9$ **(f)** $5 \times_{12} 5 \times_{12} 5$

Solution

Using the division algorithm, we have the following.

(a) $7 + 8 = 15 = 1 \cdot 12 + 3$, so $7 +_{12} 8 = 3$.
(b) $3 + 9 = 12 = 1 \cdot 12 + 0$, so $3 +_{12} 9 = 0$.
(c) $11 + 7 + 6 + 10 = 34 = 2 \cdot 12 + 10$, so $11 +_{12} 7 +_{12} 6 +_{12} 10 = 10$.
(d) $7 \cdot 8 = 56 = 4 \cdot 12 + 8$, so $7 \times_{12} 8 = 8$.
(e) $4 \cdot 9 = 36 = 3 \cdot 12 + 0$, so $4 \times_{12} 9 = 0$.
(f) $5 \cdot 5 \cdot 5 = 125 = 10 \cdot 12 + 5$, so $5 \times_{12} 5 \times_{12} 5 = 5$.

It follows from the above definition that the properties for addition and multiplication of whole numbers also hold in 12-hour clock arithmetic. For example, in whole-number arithmetic, $(a + b) + c = a + (b + c)$. Since we obtain 12-hour clock sums by taking ordinary sums, dividing by 12, and taking the remainder as the answer, it follows that $(a +_{12} b) +_{12} c = a +_{12} (b +_{12} c)$. Since the other properties are treated in the same way, it follows that, in 12-hour clock arithmetic,

- the closure properties for both addition and multiplication hold,
- addition and multiplication are both commutative,
- addition and multiplication are both associative, and
- 0 is the additive identity and 1 is the multiplicative identity.

Clock Subtraction

Like whole-number subtraction, subtraction in clock arithmetic is defined in terms of addition. In whole-number subtraction, $a - b = c$ if, and only if, $a = b + c$. In like manner, we have this definition.

> **DEFINITION** *Clock Subtraction*
>
> Let $T = \{0, 1, 2, 3, \ldots, 11\}$. For all $a \in T$ and $b \in T$, $a -_{12} b = c$ if, and only if, $a = b +_{12} c$.

EXAMPLE 5.28 **12-HOUR CLOCK SUBTRACTION**

Compute the following clock "differences."

(a) $11 -_{12} 4$ **(b)** $4 -_{12} 7$ **(c)** $8 -_{12} 5$ **(d)** $5 -_{12} 8$

Solution

(a) Since $4 +_{12} 7 = 11$, it follows that $11 -_{12} 4 = 7$.
(b) Since $7 +_{12} 9 = 4$, it follows that $4 -_{12} 7 = 9$.
(c) Here, $8 = 5 +_{12} 3$, and so $8 -_{12} 5 = 3$.
(d) Since $8 +_{12} 9 = 5$, it follows that $5 -_{12} 8 = 9$.

Since any point on the clock can be reached from any other point by counting in either the clockwise direction or the counterclockwise direction, clock subtraction can always be performed as clock addition. For example, $5 -_{12} 9 = 8$, since if we start at 0 and count 5 in the clockwise direction and 9 in the counterclockwise direction, we arrive at 8. However, 8 can also be reached by starting at 0 and counting 5 in the clockwise direction and then 3 more in the clockwise direction. Thus,

$$5 -_{12} 9 = 5 +_{12} 3 = 8$$

as illustrated in Figure 5.22. Indeed, since the counts of 9 and 3 have to reach all the way around the circle, it is clear that $9 + 3 = 12$. This is always the case, and we have the following theorem.

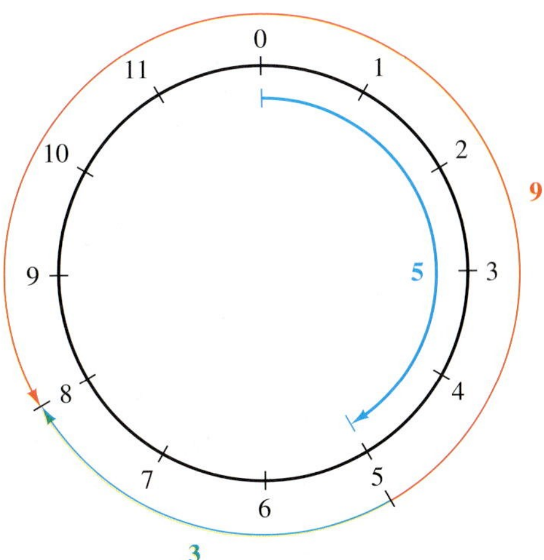

Figure 5.22
Illustrating clock subtraction as clock addition

THEOREM *Clock Subtraction as Clock Addition*
Let $T = \{0, 1, 2, 3, \ldots, 11\}$. For all a and b in T, $a -_{12} b = a +_{12} (12 - b)$.

EXAMPLE 5.29 SUBTRACTING BY ADDING IN CLOCK ARITHMETIC

Perform the following subtractions as additions.

(a) $3 -_{12} 8$ **(b)** $11 -_{12} 7$ **(c)** $9 -_{12} 11$ **(d)** $2 -_{12} 6$

Solution

(a) Since $12 - 8 = 4$, $3 -_{12} 8 = 3 +_{12} 4 = 7$. Check: $8 +_{12} 7 = 3$.
(b) Since $12 - 7 = 5$, $11 -_{12} 7 = 11 +_{12} 5 = 4$. Check: $7 +_{12} 4 = 11$.
(c) Since $12 - 11 = 1$, $9 -_{12} 11 = 9 +_{12} 1 = 10$. Check: $11 +_{12} 10 = 9$.
(d) Since $12 - 6 = 6$, $2 -_{12} 6 = 2 +_{12} 6 = 8$. Check: $6 +_{12} 8 = 2$.

In the preceding theorem and example, we saw that we can subtract b in 12-hour clock arithmetic by *adding* $12 - b$. We call $12 - b$ the **additive inverse** of b and note that we can subtract b from a by *adding the additive inverse of b to a* just as we can by adding the negative, or additive inverse, of b to a in integer arithmetic.

DEFINITION *The Additive Inverse in Clock Arithmetic*
Let $T = \{0, 1, 2, \ldots, 11\}$. If $a \in T$ and $a +_{12} b = b +_{12} a = 0$, then b is called the **additive inverse** of a. Also, a is the **additive inverse** of b.

Using this terminology, the preceding theorem can now be written as follows.

THEOREM *Clock Subtraction as Clock Addition*
To subtract b from a in clock arithmetic, add the additive inverse of b to a.

Since clock subtraction can always be performed as clock addition and clock arithmetic is closed under clock addition, it follows that clock arithmetic is also closed under clock subtraction. However, parts (c) and (d) of Example 5.28 make it clear that subtraction is not commutative in clock arithmetic, and the fact that $8 -_{12} (7 -_{12} 5) = 6$ and $(8 -_{12} 7) -_{12} 5 = 8$ makes it clear that clock subtraction is not associative. In summary,

- clock arithmetic *is* closed under addition,
- clock arithmetic *is* closed under subtraction,
- subtraction *is not* commutative in clock arithmetic, and
- subtraction *is not* associative in clock arithmetic.

At this point, the reader should briefly review the two theorems and the discussion of the negative just preceding Example 5.5 and Example 5.18. Comparing the earlier discussion of the negative with the present discussion makes it clear that, in clock arithmetic, the additive inverse plays the role of "the negative of," but here the terminology "the negative of" is clearly inappropriate. For example, the additive inverse of 3 in 12-hour clock arithmetic is 9. There is no -3 in this arithmetic. At the same time, studying clock arithmetic is both interesting and informative to middle school students. The system is similar to, yet different from, ordinary integer arithmetic, and studying these similarities and differences helps students to more fully understand ordinary arithmetic.

Division in Clock Arithmetic

Division without remainder is defined in clock arithmetic just as it is for whole numbers.

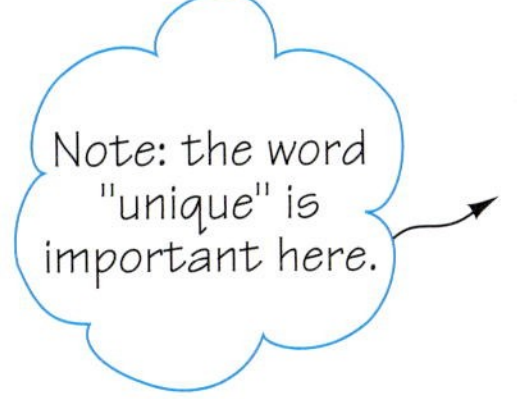
Note: the word
"unique" is
important here.

> **DEFINITION** *Clock Division*
>
> Let $T = \{0, 1, 2, 3, \ldots, 11\}$. For $a \in T$ and $b \in T$, we say that ***b* divides *a*** in clock arithmetic, and write $a \div_{12} b = c$, if, and only if, there exists a *unique* $c \in T$ such that $a = b \times_{12} c$.

EXAMPLE 5.30 12-HOUR CLOCK DIVISION

Perform these divisions if possible.

(a) $8 \div_{12} 5$ **(b)** $7 \div_{12} 8$ **(c)** $4 \div_{12} 10$

Solution

Division in clock arithmetic is greatly facilitated if one has a complete multiplication table as shown below.

12-Hour Clock Multiplication Table

$\times_{12}$	0	1	2	3	4	5	6	7	8	9	10	11
0	0	0	0	0	0	0	0	0	0	0	0	0
1	0	1	2	3	4	5	6	7	8	9	10	11
2	0	2	4	6	8	10	0	2	4	6	8	10
3	0	3	6	9	0	3	6	9	0	3	6	9
4	0	4	8	0	4	8	0	4	8	0	4	8
5	0	5	10	3	8	1	6	11	4	9	2	7
6	0	6	0	6	0	6	0	6	0	6	0	6
7	0	7	2	9	4	11	6	1	8	3	10	5
8	0	8	4	0	8	4	0	8	4	0	8	4
9	0	9	6	3	0	9	6	3	0	9	6	3
10	0	10	8	6	4	2	0	10	8	6	4	2
11	0	11	10	9	8	7	6	5	4	3	2	1

(a) From the table, 4 is the only element a in T such that $8 = 5 \times_{12} a$. Therefore, $8 \div_{12} 5 = 4$.

(b) From the table, there is no $c \in T$ such that $7 = 8 \times_{12} c$. Therefore, this division is not defined.

(c) From the table, $10 \times_{12} 4 = 10 \times_{12} 10 = 4$. Since there is *more than one number* $c \in T$ such that $10 \times_{12} c = 4$, this division is *not* defined.

Into the Classroom

Clock Arithmetic

Clock arithmetic should be considered a worthwhile enrichment topic for the elementary school classroom. Our experience is that children like mathematical ideas drawn from their immediate surroundings and that they are intrigued and interested by this arithmetic generated by an ordinary clock, an arithmetic that is both similar to and different from ordinary arithmetic. As usual, the arithmetic should be introduced via manipulatives (in this case, an ordinary clock or a device like that illustrated in Figure 5.21 and, later, diagrams of n-hour clocks for other values of n), and then students should be led to develop the resulting arithmetic pretty much on their own. The teacher should ask only occasional pertinent questions and make occasional suggestions as the development proceeds. Of special importance is the difference between the arithmetics generated by n-hour clocks when n is prime and when n is composite. In particular, n-hour clock arithmetic where n is a prime number is closed under division except for division by zero, and, as we have seen, this is not so when n is composite. Other interesting questions and properties of clock arithmetic that can be turned into classroom activities will be seen in Problem Set 5.4.

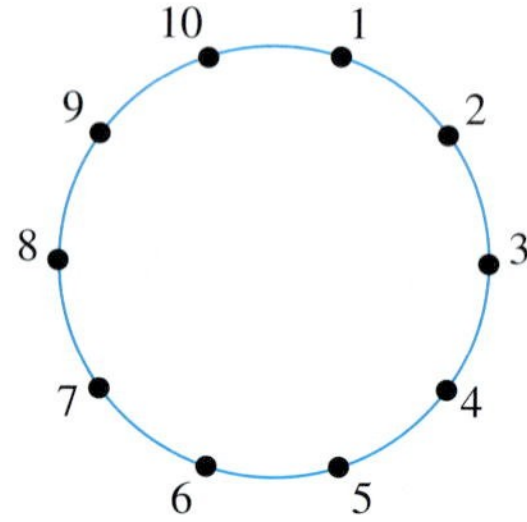

Figure 5.23
A 10-hour clock

Suppose we had a "10-hour" clock as shown in Figure 5.23 instead of a 12-hour clock. The resulting arithmetic is like 12-hour clock arithmetic, except that to find the 10-hour sum or product of two elements of

$$T = \{0, 1, 2, 3, 4, 5, 6, 7, 8, 9\},$$

we find the sum or product as in ordinary arithmetic, divide the answer by 10, and use the resulting remainder r as the answer in 10-hour clock arithmetic.

Subtraction and division are then defined in terms of addition and of multiplication as above.

EXAMPLE 5.31 CALCULATING IN 10-HOUR CLOCK ARITHMETIC

Perform each of these 10-hour clock computations.

(a) $6 +_{10} 9$ **(b)** $6 -_{10} 9$ **(c)** $6 \times_{10} 9$ **(d)** $6 \div_{10} 9$

Solution

Understand the Problem

We are asked to do various calculations in 10-hour clock arithmetic where the elements are $T = \{0, 1, 2, \ldots, 9\}$.

Devise a Plan

As in 12-hour clock arithmetic, we find the results for addition and multiplication in 10-hour clock arithmetic by finding the results in ordinary integer arithmetic, dividing by 10, and taking the remainder r as the desired answer. In 10-hour clock arithmetic, we can subtract b from a by adding the additive inverse of b, $10 - b$, to a. To find $a \div_{10} b$, we must see if there is one, and only one, number c in T such that $b \times_{10} c = a$.

Carry Out the Plan

(a) Since $6 + 9 = 15 = 10 \cdot 1 + 5$, $6 +_{10} 9 = 5$.

(b) $6 -_{10} 9 = 6 +_{10} (10 - 9) = 6 +_{10} 1 = 7$.

(c) Since $6 \cdot 9 = 54 = 10 \cdot 5 + 4$, $6 \times_{10} 9 = 4$.

(d) We must find a number c in T such that $9 \times_{10} c = 6$. If there is precisely one such number c, then $6 \div_{10} 9 = c$. Running through the values of c, we see that 4 is the only number in T whose product with 9 leaves a remainder of 6—i.e., $9 \times_{10} 4 = 6$—and so $6 \div_{10} 9 = 6$ as required.

Look Back

Basically, all we had to do was to make sure that we understood the problem. Beyond that, it was a matter of simple arithmetic.

Problem Set 5.4

Understanding Concepts

1. Compute these 12-hour clock sums.
   (a) $5 +_{12} 9$ (b) $8 +_{12} 7$ (c) $8 +_{12} 4$
   (d) $4 +_{12} 7$ (e) $10 +_{12} 10$ (f) $7 +_{12} 8$
2. Compute these n-hour clock sums, where $+_n$ denotes n-hour clock addition; for example, $5 +_7 3$ indicates the 7-hour clock addition of 5 and 3.
   (a) $3 +_5 4$ (b) $17 +_{26} 13$ (c) $2 +_{10} 7$
   (d) $7 +_9 4$ (e) $12 +_{16} 10$ (f) $2 +_7 6$
3. In 12-hour clock arithmetic, $a -_{12} b$ can be computed by starting at a and counting b steps *counterclockwise* around the clock. Use this method to compute each difference.
   (a) $9 -_{12} 7$ (b) $8 -_{12} 11$
   (c) $5 -_{12} 9$ (d) $8 -_{12} 10$
   (e) $2 -_{12} 11$ (f) $8 -_{12} 8$
4. Recall that 0 is the additive identity in 12-hour clock arithmetic and that b is the additive inverse of a if, and only if, $a +_{12} b = b +_{12} a = 0$. Compute the additive inverses of each of these numbers in 12-hour clock arithmetic.
   (a) 7 (b) 11 (c) 9 (d) 8
5. Compute each of these differences by adding. Be sure to show what you are adding each time.
   (a) $9 -_{12} 7$ (b) $8 -_{12} 11$ (c) $5 -_{12} 9$
   (d) $8 -_{12} 10$ (e) $2 -_{12} 11$ (f) $8 -_{12} 8$
6. Compute these products in 12-hour clock arithmetic.
   (a) $5 \times_{12} 7$ (b) $9 \times_{12} 11$ (c) $8 \times_{12} 9$
   (d) $8 \times_{12} 6$ (e) $4 \times_{12} 6$ (f) $4 \times_{12} 9$
7. Perform these divisions if they are defined. (*Suggestion:* See the table in Example 5.30.)
   (a) $5 \div_{12} 7$ (b) $7 \div_{12} 10$ (c) $8 \div_{12} 4$
   (d) $8 \div_{12} 5$ (e) $9 \div_{12} 5$ (f) $6 \div_{12} 11$
8. Two numbers are said to be **relatively prime** if their greatest common divisor is 1.
   (a) List the numbers in $T = \{0, 1, 2, \ldots, 11\}$ that are relatively prime to 12.
   (b) List the numbers in $T = \{0, 1, 2, \ldots, 11\}$ that are not relatively prime to 12.
   (c) Compare the results of parts (a) and (b) with the results of Example 5.30. What conjecture does this comparison suggest?
9. Construct complete addition and multiplication tables for 5-hour clock arithmetic.
10. Perform these computations in 5-hour clock arithmetic.
   (a) $3 +_5 4$ (b) $2 +_5 3$ (c) $4 +_5 4$
   (d) $3 \times_5 4$ (e) $2 \times_5 3$ (f) $4 \times_5 4$
   (g) $3 -_5 4$ (h) $2 -_5 3$ (i) $4 -_5 4$
   (j) $3 \div_5 4$ (k) $2 \div_5 3$ (l) $4 \div_5 4$
11. (a) What is the additive identity in 5-hour clock arithmetic? Why?
   (b) What is the multiplicative identity in 5-hour clock arithmetic? Why?
12. If $a +_5 b = b +_5 a = 0$, then ***b* is the additive inverse of *a*** and ***a* is the additive inverse of *b*** in 5-hour clock arithmetic. Why?
   (a) Compute the additive inverse of each of 1, 2, 3, 4, and 0 in 5-hour clock arithmetic.

As in 12-hour clock arithmetic, we can subtract b from a in 5-hour clock arithmetic by *adding* the additive inverse of b to a. Perform each of these subtractions in two ways—(i) by counting backward on a 5-hour clock and (ii) by adding the additive inverse of the number being subtracted.

(b) $2 -_5 4$ **(c)** $3 -_5 2$ **(d)** $1 -_5 3$

13. If $a \times_5 b = b \times_5 a = 1$, then ***a* is called the multiplicative inverse of *b*** and ***b* is called the multiplicative inverse of *a*** in 5-hour clock arithmetic, and we write $a = b^{-1}$ and $b = a^{-1}$.

(a) Which numbers in 5-hour clock arithmetic possess multiplicative inverses? (*Hint:* Check the results of problem 9 above.)

(b) Which numbers in 12-hour clock arithmetic possess multiplicative inverses? (*Hint:* Check the table in Example 5.30.)

14. Just as we can subtract in clock arithmetic by adding the additive inverse, we can divide by multiplying by the multiplicative inverse. Use the definition of clock division to compute each of the following.

(a) $4 \div_{12} 7$ **(b)** $3 \div_{12} 11$ **(c)** $3 \div_5 2$
(d) $2 \div_5 2$ **(e)** $2 \div_5 4$ **(f)** $4 \div_5 3$

Compute each of the quantities in parts (a) through (f) by multiplying by the multiplicative inverse; that is, compute each of these products.

(g) $4 \times_{12} 7^{-1}$ **(h)** $3 \times_{12} 11^{-1}$ **(i)** $3 \times_5 2^{-1}$
(j) $2 \times_5 2^{-1}$ **(k)** $2 \times_5 4^{-1}$ **(l)** $4 \times_5 3^{-1}$

15. Using the table in Example 5.30, solve the following equations in 12-hour clock arithmetic. The first one is done for you.

(a) $(3 \times_{12} y) +_{12} 7 = 4$
(b) $(7 \times_{12} y) -_{12} 4 = 8$
(c) $(y +_{12} 2) \div_{12} 11 = 3$
(d) $(2 \div_{12} y) -_{12} 4 = 3$

Solution:
$3 \times_{12} y = 4 +_{12} 5$
$3 \times_{12} y = 9$
$y = 3, 7, \text{ or } 11$

Teaching Concepts

16. In the military, the hours of the day are numbered from 1 through 24, starting at 1 A.M.

(a) What set of numbers would you be inclined to consider in the arithmetic associated with military time?

(b) Determine each of the following in military clock arithmetic.

(i) $17 +_{24} 8$ **(ii)** $8 +_{24} 16$ **(iii)** $2 -_{24} 14$
(iv) $23 -_{24} 5$ **(v)** $5 \times_{24} 7$ **(vi)** $23 \times_{24} 23$
(vii) $3 \div_{24} 11$ **(viii)** $11 \div_{24} 3$

17. What other notions from ordinary life might you use to illustrate other clock arithmetics?

18. (a) What ordinary arithmetic operations must be performed to compute $7 \times_{12} 11$?

(b) Compare the likely attractiveness to school students of computing $7 \times_{12} 11$ with that of solving the ordinary corresponding drill problem. Compute 7×11. Compute the quotient and remainder when 77 is divided by 12.

19. Falla claims that $8 \div_{12} 2 = 8$ since $8 = 4 \times_{12} 2$. How would you respond to Falla?

Thinking Critically

20. Powers in *n*-Hour Clock Arithmetic. Since $a^s = a \cdot a \cdots a$, with s factors of a, we can compute a^s in n-hour clock arithmetic in the usual way. That is, $a^s = r$ in n-hour clock arithmetic, where r is the remainder when a^s is divided by n. Moreover, it follows that the usual rules

$$a^s a^t = a^{s+t} \quad \text{and} \quad (a^s)^t = a^{st}$$

hold in clock arithmetic just as they do in ordinary arithmetic. Compute the given powers and products in 12-hour clock arithmetic. Do parts (e), (f), (g), and (h) in order, using the result of each part to compute the result required in the next part. (*Note:* To compute 3^5, for example, using your calculator, enter the string 3 [^] 5 [=]. Then reenter the answer and use [INT ÷] to determine the result in 12-hour clock arithmetic.)

(a) 3^3 **(b)** 3^4 **(c)** $3^3 \cdot 3^4$ **(d)** 3^{3+4}
(e) 4^2 **(f)** $(4^2)^5$ **(g)** $4^{2 \cdot 5}$ **(h)** 4^{10}

21. (a) Complete the following table. The first two rows have been completed for you. *Note:* In using your calculator to compute $2^n - 2$ in n-hour clock arithmetic, you may have to enter the string 2 [^] n [−] 2 [=], record the result in the second column of the table, and then *reenter* the result and use [INT ÷] to determine the correct result for column three.

n	$2^n - 2$	$2^n - 2$ in n-hour clock arithmetic
2	2	0
3	6	0
4		
5		
6		
7		
8		
9		
10		
11		
12		
13		

(b) What seems to be the case when n is a prime in the table of part (a)? Make a conjecture.

(c) Compute $2^{23} - 2$ in 23-hour arithmetic. Note that you cannot use the [INT ÷] key on some calculators to determine the remainder when dividing by 23 since the quotient is too large. Instead, enter the sequence 2 [^] 23 [−] 2 [=] [÷] 23 [=] [−] 364722 [=] [×] 23 [=]. (Did you actually have to perform the last six steps in this sequence to know what the remainder was in this case?) Does this result strengthen your belief in the conjecture you made in part (b)? Does it prove that it is correct?

22. **(a)** Repeat problem 21(a) with 2 replaced by 3.
(b) Make a conjecture on the basis of part (a).

23. You have little to go on, but make a conjecture generalizing the conjectures for problems 21 and 22. If you need more data, repeat problem 21 (a) and (b) with 2 replaced by 4, by 5, by 6, etc.

Thinking Cooperatively

In the next three problems, divide the work among the members of your group, discuss the results obtained, and determine conjectures agreed upon by the group.

24. Make a conjecture concerning the value of $1 \cdot 2 \cdot 3 \cdots (p - 1) + 1$ in p-hour clock arithmetic, where p is a prime. (*Suggestion:* Consider a number of examples using the first few primes.)

25. Make a conjecture concerning the value of $a^{p-1} - 1$ in p-hour clock arithmetic if p is a prime and p does not divide a.

26. Fill in the following table for a number of choices of a and b in $T = \{0, 1, 2, 3, \ldots, 11\}$, and endeavor to discover when $a \times_{12} y = b$ is solvable and how many solutions this equation has. (*Hint:* Use the table in Example 5.30.)

a	b	GCD $(a, 12)$	b/GCD $(a, 12)$	12/GCD $(a, 12)$	No. of solutions of $a \times_{12} y = b$

For Review

27. Draw rectangular diagrams to illustrate all the factorings of 12, taking order into account; for example, think of $1 \cdot 12$ as different from $12 \cdot 1$.

28. Use the method of intersection of sets to determine the following.
(a) GCD(60, 150)
(b) LCM(60, 150)

29. (a) Use a factor tree to determine all the prime divisors of 540.
(b) How can you tell at a glance if 540 is or is not divisible by 9?

30. (a) Write the prime-power representations of 540 and 600.
(b) Use part (a) to determine GCD(540, 600).
(c) Use part (a) to determine LCM(540, 600).

31. Does $2^3 \cdot 3^5 \cdot 7^2 \cdot 11^6$ evenly divide $2^4 \cdot 3^7 \cdot 5^2 \cdot 7^1 \cdot 11^8$? Explain briefly.

32. To determine whether 427 is a prime, which primes must you check as possible divisors?

Epilogue Developing the Number System

In the struggle to develop numbers, history shows that it took a very long time for people to develop the natural numbers, and even longer to develop the notion of zero. Even then, zero was first introduced only as a placeholder in positional systems like the Mayan, and it was not until much later that it was considered to be a number.

Somewhat surprisingly, negative numbers made their appearance on the mathematical scene before zero did. No evidence of the recognition of negative numbers, as distinct from subtrahends, appears in ancient Egyptian, Babylonian, Hindu, Chinese, or Greek mathematics. Still, the rules of signs, considered at length in this chapter, were established early on by considering such products as $(8 - 4) \cdot (7 - 5)$. The Chinese made use of such subtractions at least as early as 200 B.C., but the rules of signs in Chinese mathematics were not stated explicitly until A.D. 1299. The first mention of negative numbers in Western mathematics occurred in *Arithmetica* by the Greek mathematician Diophantus, in about A.D. 275, though spoken of there in disparaging terms. Diophantus called the equation $4x + 20 = 4$ *absurd,* since it would require that $x = -4$. The first substantial use of negative numbers occurred in the work of the Hindu mathematician Brahmagupta, in about A.D. 628, and after that time they appear in all Indian works on the subject.

In this chapter, we have used various devices to introduce the notions of negative numbers and the rules for operating with them. The fact that these numbers were considered absurd or nonnumbers by early mathematicians notwithstanding, negative numbers, like all the other numbers, are simply ideas in our minds. They are in constant use to solve problems that occur daily in modern technological society.

We have also introduced clock arithmetic in this chapter, not only because it is intrinsically interesting to school students, but also because it is similar to, and yet different from, integer arithmetic. Thus, studying clock arithmetic not only helps students more fully understand integer arithmetic, but also provides interesting (and disguised) drill in that arithmetic. Moreover, the concept of the multiplicative inverse in clock arithmetic (see problems 13 and 14 in Problem Set 5.4) is a precursor of the same notion that will reappear in the study of fractions and rational numbers in Chapter 6.

Chapter Summary

Key Concepts

In what follows, all symbols represent integers.

Section 5.1 Representations of Integers

- $-n$. The negative of n, defined by $n + (-n) = (-n) + n = 0$.
- Integers. $\ldots, -3, -2, -1, 0, 1, 2, 3, \ldots$.
- Additive inverse, negative. An integer m such that $n + m = m + n = 0$; i.e., $m = -n$.
- Drops of colored counters. Representations of integers with sets of black and red markers.
- Mail-time stories. Bringing a check for n dollars represents n. Bringing a bill for n dollars represents $-n$.
- Number-line representations. If n is positive, an arrow of length n oriented from left to right and placed anywhere on the number line represents n. If the arrow is oriented from right to left, it represents $-n$.
- The negative of the negative of an integer. $-(-n) = n$.
- Absolute value. The absolute value of a number r is the distance of r from 0 on the number line. Also, $|r| = r$ if $r \geq 0$, and $|r| = -r$ if $r < 0$.

Section 5.2 Addition and Subtraction of Integers

- Addition of integers. If r and s are integers, then $(-r) + (-s) = -(r + s)$. If $r > s > 0$, then $r + (-s) = r - s$. If $0 < r < s$, then $r + (-s) = -(s - r)$.
- Law of trichotomy. If r and s are integers, then either $r < s$, $r = s$, or $r > s$.
- Properties of addition. For integers r, s, and t, $r + s$ is an integer, $r + s = s + r$, $(r + s) + t = r + (s + t)$, and 0 is the additive identity, i.e., $0 + r = r + 0 = r$ for all integers r.
- Less than. r is less than s provided there is a *positive* integer t such that $r + t = s$.
- Greater than. r is greater than s provided s is less than r.
- Subtraction of integers. For integers r, s, and t, $r - s = t$ if, and only if, $r = s + t$. Also, $r - s = r + (-s)$.
- Closure property for subtraction. $r - s$ is always an integer.

Section 5.3 Multiplication and Division of Integers

- Multiplication of integers. If r and s are integers, then $r \cdot (-s) = -(r \cdot s)$, $(-r) \cdot s = -(r \cdot s)$, and $(-r) \cdot (-s) = r \cdot s$.

- Properties of multiplication. For integers r, s, and t, $r \cdot s$ is an integer, $r \cdot s = s \cdot r$, $(r \cdot s) \cdot t = r \cdot (s \cdot t)$, $r \cdot (s + t) = r \cdot s + r \cdot t$, $1 \cdot r = r \cdot 1 = r$, and $0 \cdot r = r \cdot 0 = 0$.
- Division of integers. For integers r and s, $r \div s = t$ if, and only if, $r = s \cdot t$ for some integer t. $r \div (-s) = -(r \div s)$, $(-r) \div s = -(r \div s)$, and $(-r) \div (-s) = r \div s$.

Section 5.4 Clock Arithmetic

- Clock arithmetic. The arithmetic inherent on a 12-hour clock.
- Clock addition and multiplication. Do the corresponding ordinary arithmetic, divide by 12, and take the remainder as the result.
- Clock subtraction. $r -_{12} s = t$ if, and only if, $r = s +_{12} t$. Also, $r -_{12} s = r +_{12} (12 - s)$. $12 - s$ is the additive inverse of s for every s in the arithmetic.
- Clock division. $r \div_{12} s = t$ if, and only if, s is the *only* element in the arithmetic for which $r = s \times_{12} t$. No such t may exist.
- n-hour clock arithmetic. The same as 12-hour clock arithmetic, but with 12 replaced by n.

Vocabulary and Notation

Section 5.1

Integer
Negative
Additive inverse
$|n|$

Section 5.2

Law of trichotomy

Section 5.3

Rule of signs

Section 5.4

$+_n$, $-_n$, $\times_n$, $\div_n$
Additive identity
Additive inverse

Chapter Review Exercises

Section 5.1

1. You have 15 counters colored black on one side and red on the other.

(a) If you drop them on your desktop and seven come up black and eight come up red, what integer is represented?

(b) If you drop them on your desktop and twice as many come up black as red, what number is being represented?

(c) What numbers are represented by all possible drops of the 15 counters?

2. (a) If the mail carrier brings you a check for \$12, are you richer or poorer, and by how much? What integer does this situation illustrate?

(b) If the mail carrier brings you a bill for \$37, are you richer or poorer, and by how much? What integer does this situation illustrate?

3. (a) 12° above zero illustrates what integer?

(b) 24° below zero illustrates what integer?

4. (a) List five different drops of colored counters that represent the integer -5.

(b) List five different drops of colored counters that represent the integer 6.

5. (a) Give a mail-time story that illustrates -85.

(b) Give a mail-time story that illustrates 47.

6. (a) What number must you add to 44 to obtain 0?

(b) What number must you add to -61 to obtain 0?

Section 5.2

7. What addition is represented by this diagram?

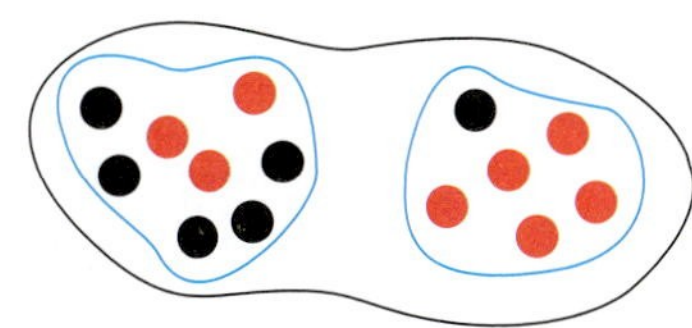

8. What subtraction is represented by this diagram?

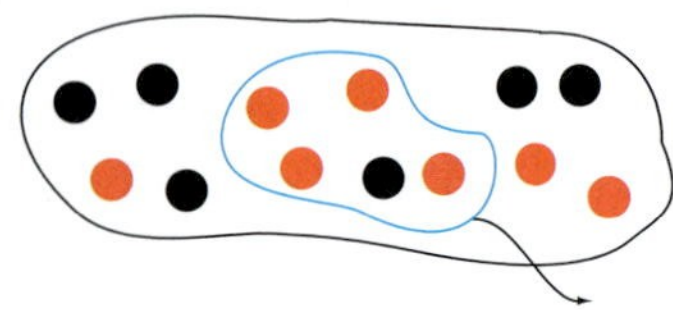

9. What additions and subtractions are represented by these mail-time stories?
   (a) At mail time, the letter carrier brings you a check for \$45 and a bill for \$68. Are you richer or poorer, and by how much?
   (b) At mail time, the letter carrier brings you a check for \$45 and takes away a bill for \$68 left previously. Are you richer or poorer, and by how much?

10. What additions and/or subtractions do these number-line diagrams represent?

(a)

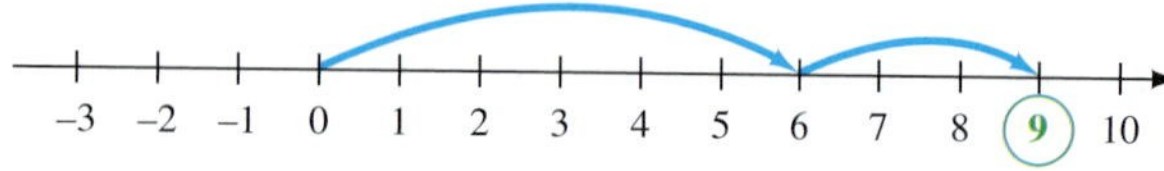

(b)

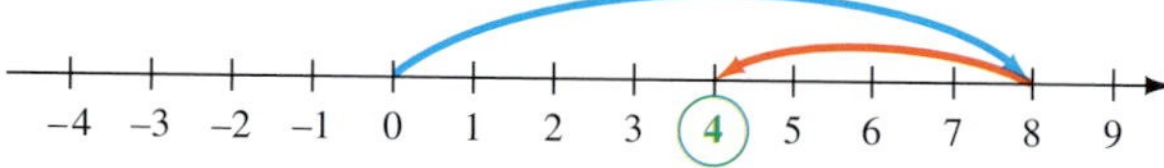

(c)

(d)

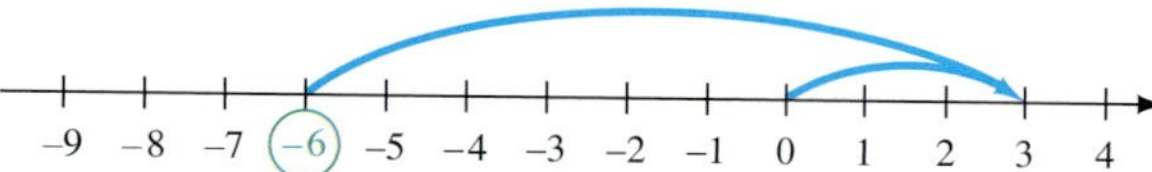

(e)

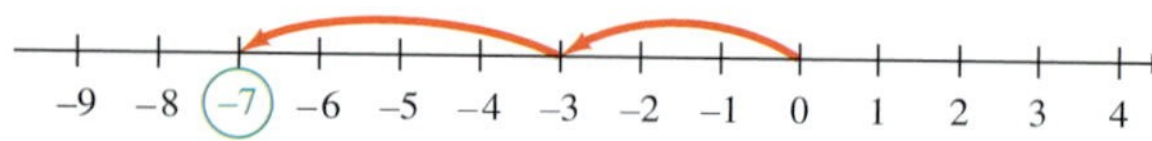

(f)

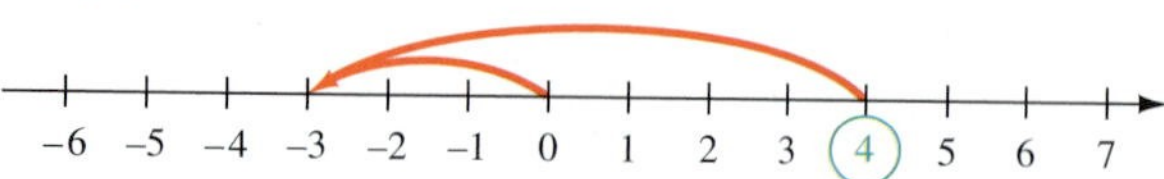

(g)

11. Perform these additions and subtractions.
   (a) $5 + (-7)$ (b) $(-27) - (-5)$
   (c) $(-27) + (-5)$ (d) $5 - (-7)$
   (e) $8 - (-12)$ (f) $8 - 12$

12. (a) If it is 15° below zero and the temperature falls 12°, what temperature is it?
   (b) What arithmetic does this situation illustrate?

13. (a) Dina's bank account was overdrawn by \$12. What was her balance after she deposited \$37 she earned working at a local pizza parlor?
   (b) What arithmetic does this situation illustrate?

14. (a) Plot these numbers on a number line: $-2, 7, 0, -5, -9$, and 2.
   (b) List the numbers in part (a) in increasing order.
   (c) Determine what integer must be added to each number in your list from part (b) to obtain the next.

Section 5.3

15. What products do these number-line diagrams represent?

(a)

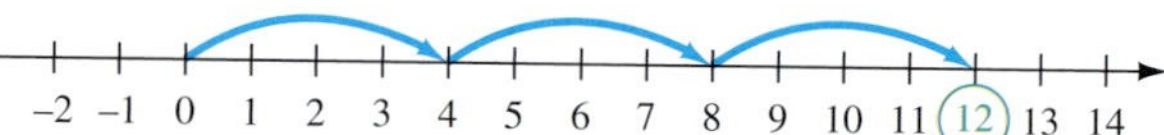

(b)

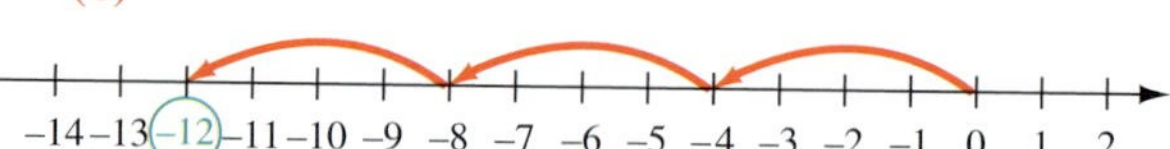

(c)

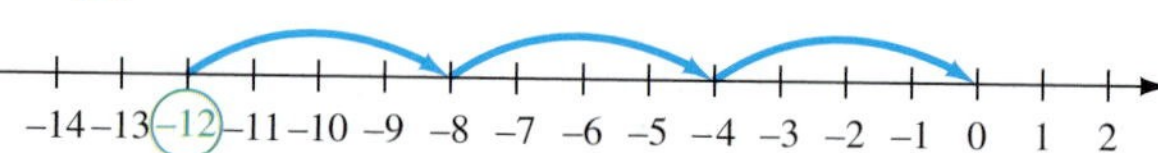

(d)

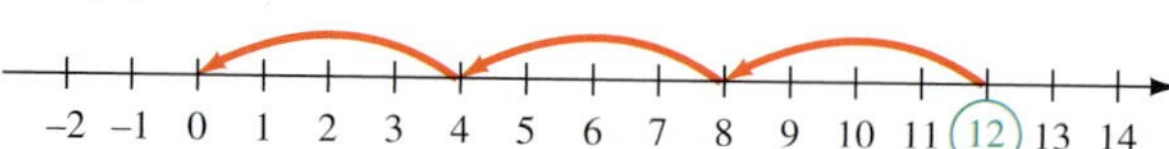

16. (a) Show that $3 \cdot 5 + 3 \cdot (-5) = 0$. (*Suggestion:* Begin with the fact that $3 \cdot 0 = 0$.)
   (b) If $3 \cdot 5 + 3 \cdot (-5) = 0$, what can you conclude about the product $3 \cdot (-5)$?
   (c) Show that $-(3 \cdot 5) + (-3) \cdot (-5) = 0$. (*Suggestion:* Begin with the fact that $0 \cdot (-5) = 0$, and use the result of part (a).)
   (d) If $-(3 \cdot 5) + (-3) \cdot (-5) = 0$, what can you conclude about the product $(-3) \cdot (-5)$?

17. Perform each of these computations.
   (a) $(-8) \cdot (-7)$ (b) $8 \cdot (-7)$
   (c) $(-8) \cdot 7$ (d) $84 \div (-12)$
   (e) $(-84) \div 7$ (f) $(-84) \div (-7)$

18. Write a mail-time story to illustrate each of these products.
(a) $7 \cdot (12)$ (b) $(-7) \cdot (13)$
(c) $(-7) \cdot (-13)$

19. If d divides n, prove that d divides $-n$, $-d$ divides n, and $-d$ divides $-n$.

20. If a and b are integers, the greatest common divisor, (GCD), of a and b is the largest positive integer dividing both a and b. Compute each of the following.
(a) GCD(255, −39) (b) GCD(−1001, 2651)

21. If n is an integer not divisible by 2 or 3, show that $n^2 - 1$ is divisible by 24. (*Hint:* By the division algorithm, n must be of one of these forms: $6q$, $6q + 1$, $6q + 2$, $6q + 3$, $6q + 4$, or $6q + 5$.)

Section 5.4

22. Perform the indicated clock calculations if they are defined.
(a) $4 +_{12} 9$ (b) $9 -_{12} 4$ (c) $4 \times_{12} 9$
(d) $4 \div_{12} 9$ (e) $9 +_{12} 8$ (f) $9 \times_{12} 0$
(g) $4 \div_{12} 0$ (h) $9 \div_{12} 7$ (i) $9 \times_{12} 7$

23. Perform these clock calculations.
(a) $5 +_7 6$ (b) $6 -_7 5$
(c) $6 \times_7 5$ (d) $6 \div_7 5$

24. List the numbers in 10-hour clock arithmetic for which 10-hour clock division is *not* defined.

Chapter Test

1. Mary Lou's checkbook balance was $129. What was it after she deposited $341 and then wrote checks for $13, $47, and $29? What arithmetic does this illustrate?

2. At mail time, if the mail carrier took away five bills for $27 each, are you richer or poorer, and by how much? What calculation does this illustrate?

3. (a) Use your calculator to compute these sums.

$$1 =$$
$$1 - 4 =$$
$$1 - 4 + 9 =$$
$$1 - 4 + 9 - 16 =$$
$$1 - 4 + 9 - 16 + 25 =$$

(b) Make a conjecture suggested by the pattern of part (a).

4. (a) What sums can be obtained using only the numbers −10, −5, −2, −1, 1, 2, 5, and 10, each at most once and without using any number with its double (i.e., you can't use 5 with 10, 1 with 2, −1 with −2, or −5 with −10)?
(b) Do the representations in part (a) appear to be unique?

5. Perform each of these computations.
(a) $(-7) + (-19)$ (b) $(-7) - (-19)$
(c) $7 - (-19)$ (d) $7 + (-19)$
(e) $(-6859) \div 19$ (f) $(-24) \cdot 17$
(g) $36 \cdot (-24)$ (h) $(-1155) \div (-11)$
(i) $0 \div (-27)$

6. Write a mail-time story to illustrate the subtraction $7 - (-4) = 11$.

7. The Fibonacci sequence is formed by adding any two consecutive numbers in the sequence to obtain the next number. If the same rule is followed in each of these sequences, correctly fill in the blanks.
(a) −5, −3, ____, ____, ____, ____
(b) 7, ____, 2, ____, ____, ____
(c) 6, ____, ____, ____, −12, ____

8. Tammie, Jody, and Nora formed a small club. After a pizza party celebrating the first anniversary of the club's existence, they owed the local pizzeria $27. The bill was paid and shared equally by the three girls. Was each one richer or poorer, and by how much? What arithmetic does this illustrate?

9. Perform the indicated clock calculations.
(a) $7 +_8 5$ (b) $7 +_{12} 5$ (c) $5 -_7 7$
(d) $7 \times_8 5$ (e) $7 \div_8 5$
(f) 7^5 (in 8-hour clock arithmetic)

10. Draw a number-line diagram to illustrate each of these calculations.
(a) $(-7) + 10$ (b) $10 - (-7)$ (c) $7 \cdot (-5)$

11. The least common multiple, (LCM), of integers a and b is the least positive integer divisible by both a and b. Compute LCM(−240, 54).

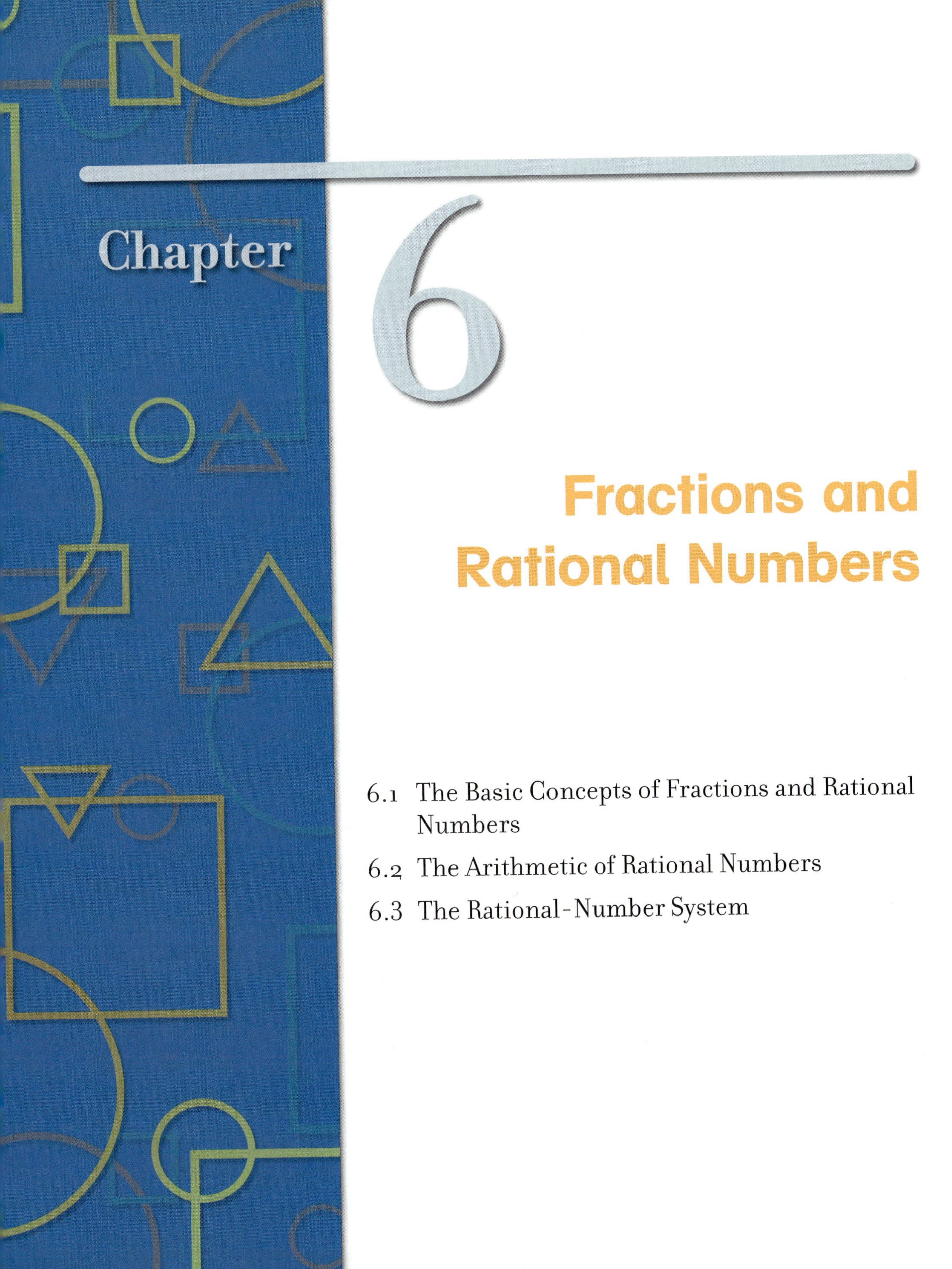

Chapter 6

Fractions and Rational Numbers

Hands On Folded Fractions

Materials Needed

Each student (or cooperating pair of students) should have about six paper squares (4- to 5-inch side length) and several colored pencils. Patty Paper (waxed paper that serves as meat-patty separators) works very well.

Example: Folding Quarters

A unit of area is defined as the area of a square.

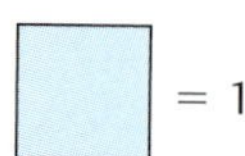

Fold the unit square in half twice vertically and then unfold. Since the four rectangles are congruent (that is, have the same shape and size), each rectangle represents the fraction $\frac{1}{4}$.

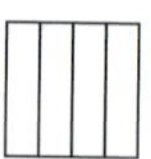

Coloring individual rectangles gives representations of the fractions $\frac{0}{4}, \frac{1}{4}, \frac{2}{4}, \frac{3}{4}$, and $\frac{4}{4}$. (Notice that we count regions, not folds or fold lines.)

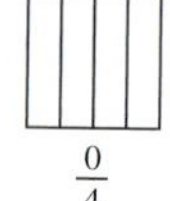
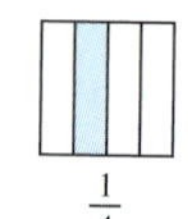
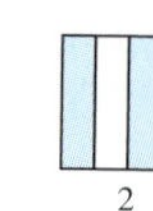
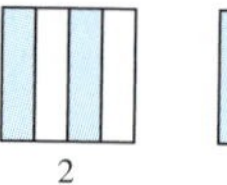
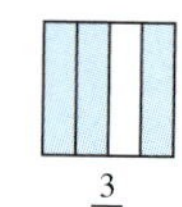
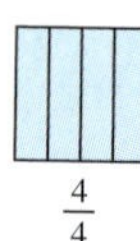

$\frac{0}{4}$ $\frac{1}{4}$ $\frac{2}{4}$ $\frac{3}{4}$ $\frac{4}{4}$

Quarters can also be obtained by other folding procedures. For example, we can make a vertical half fold followed by a horizontal half fold to create four small squares within the unit square. This technique illustrates why $\frac{1}{2} \times \frac{1}{2} = \frac{1}{4}$.

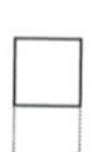
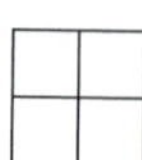

If one colored fraction pattern can be rearranged and regrouped to cover the same region of the unit as another fraction pattern, we say that the two fractions are equivalent. For example, the following rearrangement and regrouping shows that 2/4 is equivalent to 1/2. Therefore, we write $\frac{2}{4} = \frac{1}{2}$.

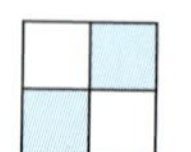
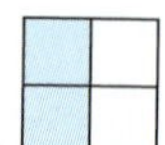
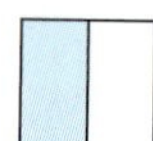

Fraction Folding Activities and Problems

1. **Eighths.** Fold a square into quarters as in the example, and also fold along the two diagonals.
   (a) Identify these fractions.

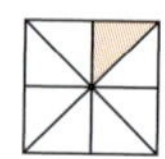
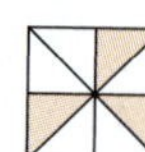
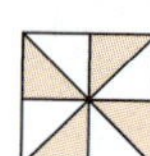

   (b) How many *different* ways can two of the eight regions be colored to give a representation of $\frac{2}{8}$? Two colorings are considered identical if one pattern can be rotated to become identical to the second pattern.

2. **Sixths.** "Roll" the paper square into thirds and flatten, and then fold in half in the opposite direction to divide the square into six congruent rectangles. This procedure illustrates that $\frac{1}{2} \times \frac{1}{3} = \frac{1}{6}$.

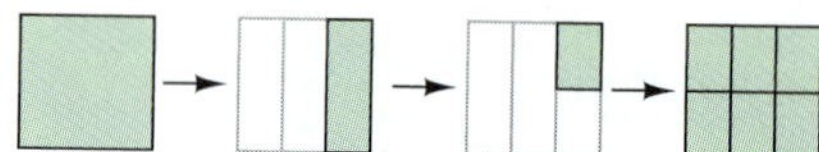

   How many *different* ways can you represent the fraction $\frac{3}{6}$? Remember that two patterns are different only if one cannot be rotated to become identical to another.

3. **Identifying fractions.** The leftmost shaded region shown in the figure below is bounded by segments joining successive midpoints of the sides of the square. What is the area of the shaded region? The region can be divided into smaller regions that are easy to rearrange into a pattern which shows that $\frac{1}{2}$ of the square is shaded.

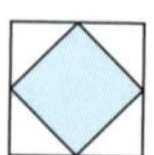
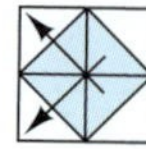
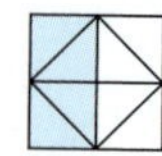

   Use similar multiple representations of the same amount of shaded region to identify the fractions represented by the following colored regions:
   (a) The shaded region obtained by joining the midpoints of the vertical edges of the square with the $\frac{1}{3}$ and $\frac{2}{3}$ points along the horizontal edges.

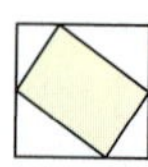

(b) The shaded region obtained by joining the $\frac{1}{3}$ and $\frac{2}{3}$ points along the vertical and horizontal edges.

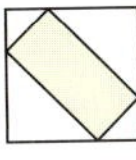

(c) The shaded square whose corners are the intersections of the segments joining the corners of the unit square to the midpoint of an opposite side of the unit square. (*Suggestion:* Rearrange the unshaded regions to create squares, each congruent to the shaded square.)

Connections

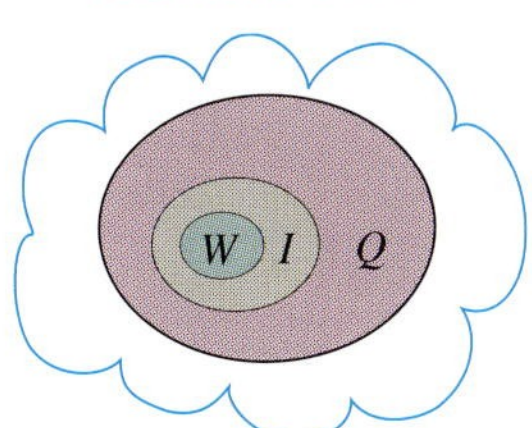

Numbers for Parts and Pieces

In Chapter 5 the set of whole numbers $W = \{0, 1, 2, \ldots\}$ was extended to the set of integers $I = \{\ldots, -2, -1, 0, 1, 2, 3, \ldots\}$. In the set of integers it becomes possible to solve problems that are more difficult or even impossible to solve in the whole-number system. In particular, the integers are closed under subtraction. Therefore, the equation $x + a = b$ has the unique solution $x = b - a$ for any integers a and b. However, equations like $ax = b, a \neq 0$, may not have a solution even in the integers. Thus, it is necessary to extend the integers to an even larger set of numbers, the rational-number system Q.

A rational number is represented by an ordered pair of integers a and b, $b \neq 0$, which is written as $\frac{a}{b}$ and called a fraction. Section 6.1 introduces the basic concepts of fractions and rational numbers. Pictorial and physical models are used to illustrate equivalent fractions, fractions in simplest form, common denominators, and inequalities. In Section 6.2, the arithmetic operations—addition, subtraction, multiplication, and division—are defined for rational numbers by introducing corresponding operations for fractions. The properties of the rational-number system are explored in the concluding Section 6.3. This section also discusses estimations and mental arithmetic, the use of calculators, and practical applications of the rational numbers.

The rational numbers have two new important properties that do not hold in the integers. First, *the rational numbers are closed under division.* For example, the equation $3x = 2$ has no solution in the integers. However, the division $2 \div 3$ is defined in the rational numbers, so the equation has the solution $x = \frac{2}{3}$. This property is why the letter Q is chosen to denote the rational numbers: a rational number is represented by a *quotient* of integers. The second new property is called the *density property of the rational numbers:* between any two rational numbers there are infinitely many more rational numbers. For example, between 0 and 1 there are no integers, but the rational numbers $\frac{1}{2}, \frac{2}{3}, \frac{1}{5}$, and $\frac{4}{17}$, and infinitely many others, lie between 0 and 1. These new properties bring powerful new tools to solving problems in both the mathematical and the real world.

6.1 THE BASIC CONCEPTS OF FRACTIONS AND RATIONAL NUMBERS

The whole numbers, as suggested by the word "whole," arise most often in *counting* problems, where the units or objects being counted cannot be subdivided into smaller parts. For example, Earth has one moon and Mars has two moons, but no planet can have some number of moons between one and two. If Earth's moon were to split into two parts, we would have two moons! In many situations, however, the objects or quantities of interest can be meaningfully subdivided. For example, when just one cookie remains, two young children may well display a good understanding of what "one half" of a cookie is all about.

Fractions were first introduced in *measurement* problems, to express a quantity that is less than a whole unit. Indeed, the word *fraction* comes from the Latin word *fractio,* meaning "the act of breaking into pieces." Figure 6.1 shows a measuring cup and a ruler, two common items on which fractions appear.

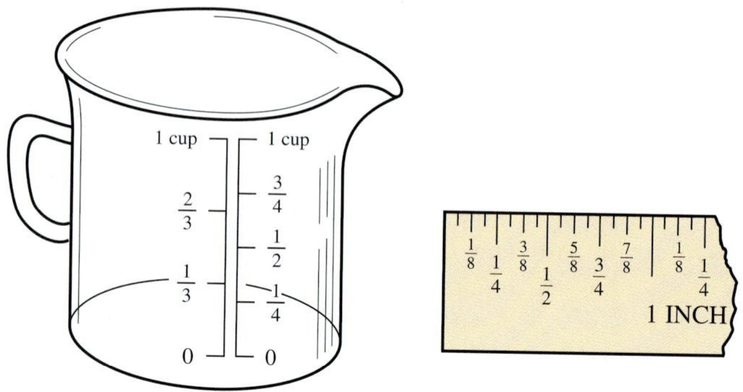

Figure 6.1
Fractions are used on measuring cups and rulers to indicate subdivisions of the basic unit

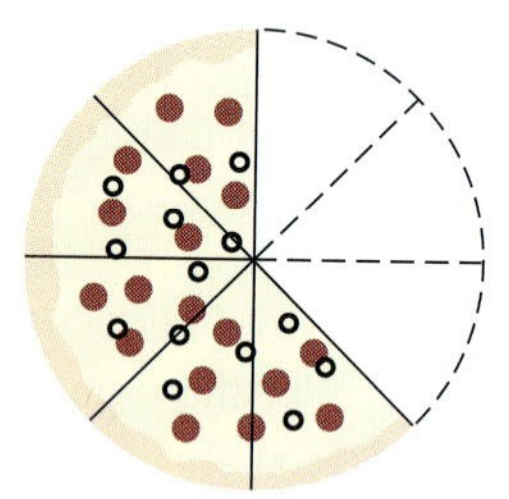

Figure 6.2
Choosing a whole pizza as the unit and subdividing it into 8 equal parts shows that $\frac{3}{8}$ of the pizza has been eaten and $\frac{5}{8}$ remains

Fractions may be used to indicate capacity, length, weight, area, or indeed any quantity for which it is sensible to subdivide the fundamental unit of measure into some number of equal-sized parts. To interpret the meaning of any fraction $\frac{a}{b}$, we must

- agree on the **unit** (for example, the unit is a cup, an inch, the area of the hexagon in a set of pattern blocks, a whole pizza, and so on);
- understand that the unit is **subdivided into *b* parts of equal size;** and
- understand that we are considering ***a* of the parts of the unit.**

The number *b* is called the **denominator** of the fraction, a word derived from the Latin *denominare,* meaning "namer." The number *a* is called the **numerator,** derived from the Latin *numeros,* meaning "number."

For example, suppose the unit is one pizza, as shown in Figure 6.2. The pizza has been divided into 8 parts, and 3 parts have been consumed. Using fractions, we can say that $\frac{3}{8}$ of the pizza was eaten, and $\frac{5}{8}$ of the pizza remains.

Hagar The Horrible Dik Browne

The notion of a subdivided unit of measure motivates the following formal definition of a fraction.

DEFINITION *Fractions*

A **fraction** is an ordered pair of integers a and b, $b \neq 0$, written $\frac{a}{b}$ or a/b. The integer a is called the **numerator** of the fraction, and the integer b is called the **denominator** of the fraction.

This definition, it should be noticed, permits the numerator or denominator to be a negative integer. For example, $\frac{-3}{-8}, \frac{-4}{12}, \frac{31}{-4}$, and $\frac{0}{1}$ are all fractions according to the definition just given. Both negative and positive integers are allowed to ensure that additive inverses exist in the rational-number system Q.

Any integer m is viewed as the fraction $\frac{m}{1}$. Usually the denominator 1 is not written explicitly. For example, we write 3 instead of $\frac{3}{1}$.

Models for Fractions

Several physical and pictorial representations are useful in the elementary school classroom to illustrate fraction concepts. These include colored regions, the set model, fraction strips, and the fraction number line. These models foster the development of deep understanding, as distinct from formulas and rules of calculation, which are devoid of intrinsic meaning. To be successful, a representation of a fraction must clearly answer these three questions:

- What is the unit?
- Into how many equal parts (the denominator) has the unit been subdivided?
- How many of these parts (the numerator) are under consideration?

Errors and misconceptions about fractions indicate that at least one of these three questions has not been properly answered or clearly considered. Perhaps the most frequently occurring error is the failure to identify the unit to which the fraction refers, since the notation a/b makes the numerator and denominator explicit, but the unit is left implicit.

Colored Regions

A shape is chosen to represent the unit and is then subdivided into subregions of equal size. A fraction is visualized by coloring some of the subregions. Three examples are shown in Figure 6.3. Colored-region models are sometimes called **area** models.

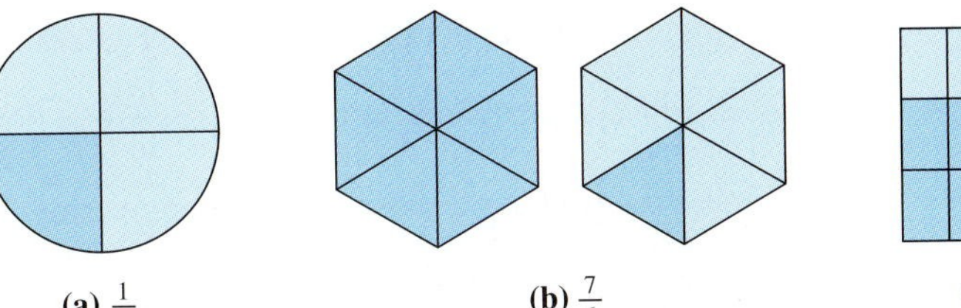

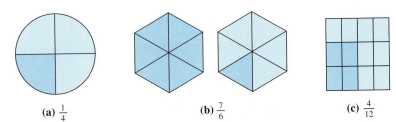

Figure 6.3
Some colored-region models for fractions

Cooperative Investigation

Exploring Fraction Concepts with Fraction Tiles

The basic concepts of fractions are best taught with a variety of manipulative materials. These can be purchased from education supply houses or, alternatively, can be homemade from patterns printed and cut from cardstock. The commercially made manipulatives are usually colored plastic, with the unit a square, a circle, a rectangle, or some other shape that can be subdivided in various ways to give denominators such as 2, 3, 4, 6, 8, and 12. Usually the shapes are also available in translucent plastic for demonstrations on an overhead projector. In the activity below, you will make and then work with a set of *fraction tiles.*

Materials

Each group of three or four students needs a set of fraction tiles, either a plastic set or a set cut from cardstock using a pattern such as the one shown (at reduced scale) to the right. The pieces at the right are lettered A through G, but can also be described by their color.

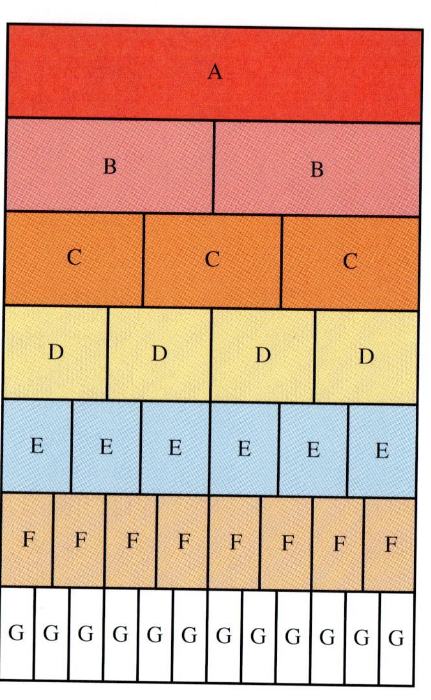

Directions

If a B tile is chosen as the unit, then the fraction 1/2 is represented by a D tile, which can be verified by comparing two D tiles and one B tile. Similarly, an A tile represents 2, and 2/3 is represented by two E tiles or by a C tile. Now use your fraction tile set to answer and discuss the questions below. Compare and discuss your answers within your group.

Questions

1. If an A tile is the unit, find tiles (one or more) to represent $\frac{1}{3}, \frac{2}{3}, \frac{3}{3}, \frac{4}{3}, \frac{2}{6}$, and $\frac{5}{12}$.
2. If a C tile is $\frac{1}{6}$, what fractions are represented by each of the other tiles?
3. If a B tile represents $\frac{3}{2}$, represent these fractions with tiles: $\frac{1}{2}, \frac{1}{1}, \frac{1}{4}, \frac{3}{1}$, and $\frac{3}{4}$.

The Set Model

The unit is a finite set of objects, U, divided naturally into $n(U)$ parts by its elements. Each subset A of U corresponds to the fraction $\dfrac{n(A)}{n(U)}$. For example, the set of 10 apples shown in Figure 6.4 contains a subset of 3 that are wormy. Therefore, we would say that $\frac{3}{10}$ of the apples are wormy. In Chapter 10 we will see that the set model of fractions is particularly useful in probability. An apple drawn at random from the 10 apples has a $\frac{3}{10}$ probability of being wormy. Sometimes, the set model is called the *discrete model.*

Figure 6.4
The set model depicts that $\frac{3}{10}$ of the apples are wormy

Fraction Strips

Here the unit is defined by a rectangular strip of cardstock. A fraction, such as $\frac{3}{6}$, is modeled by shading 3 of 6 equally sized subrectangles of the card. Sample fraction strips are shown in Figure 6.5. A set of fraction strips typically contains strips for the denominators 1, 2, 3, 4, 6, 8, and 12.

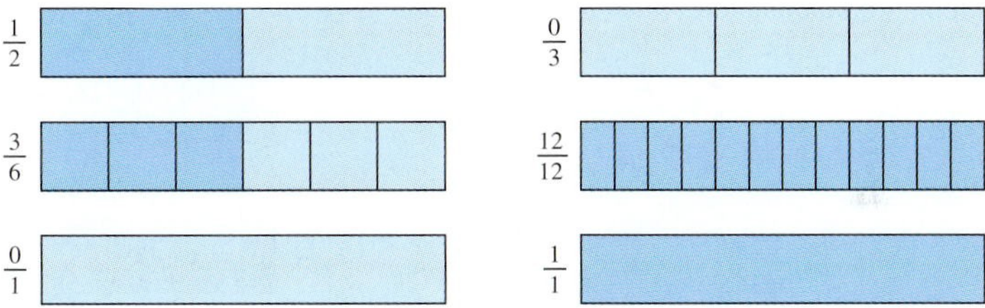

Figure 6.5
Examples of fractions modeled by fraction strips

The Number-Line Model

Once the points corresponding to 0 and 1 are assigned on a number line, all of the points corresponding to the integers are determined. The unit is the length of the line segment from 0 to 1 or, equivalently, the distance between successive integer points. A fraction such as $\frac{5}{4}$ is assigned to a point along the number line by subdividing the unit interval into 4 equal parts and then counting off 5 of these lengths to the right of 0. Typical fractions are shown in Figure 6.6. It should be noticed that the same point on the fraction number line can be named by different fractions. For example, $\frac{1}{2}$ and $\frac{2}{4}$ both correspond to the same distance. The number-line model has the advantage of allowing negative fractions, such as $-\frac{3}{4}$, to be represented.

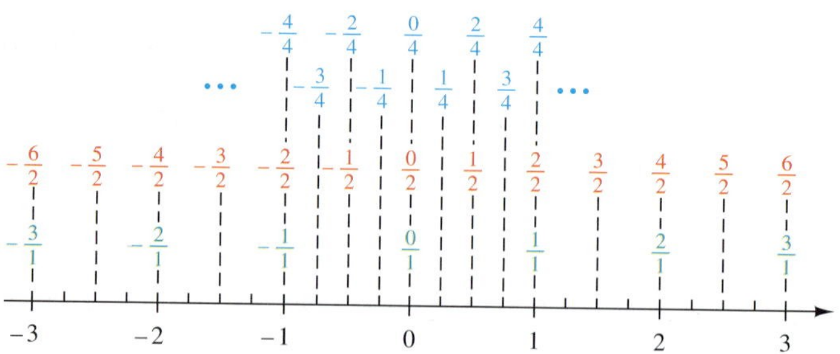

Figure 6.6
Fractions on the number-line model

Equivalent Fractions

In Figure 6.7, the fraction strip representing $\frac{2}{3}$ is further subdivided by the vertical dashed lines to show that $\frac{4}{6}, \frac{6}{9}$, and $\frac{8}{12}$ are other fractions that express the *same* shaded portion of a whole strip.

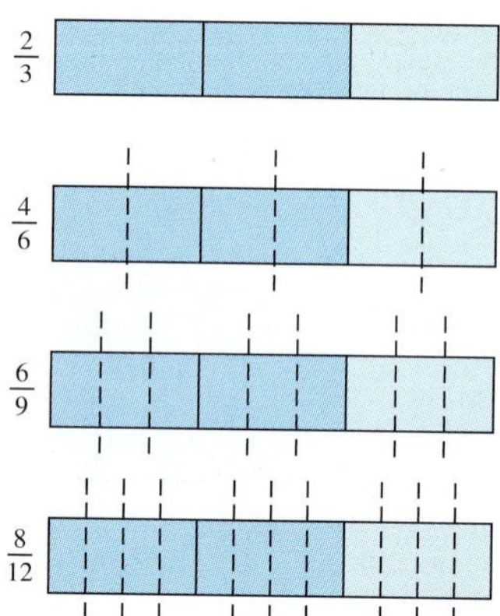

Figure 6.7
The fraction-strip model showing that $\frac{2}{3}, \frac{4}{6}, \frac{6}{9}$, and $\frac{8}{12}$ are equivalent fractions

Fractions that express the same quantity are called **equivalent fractions.** The equality symbol,=, is used to signify that fractions are equivalent, so we write

$$\frac{2}{3} = \frac{4}{6} = \frac{6}{9} = \frac{8}{12}.$$

The number of additional dashed lines between each vertical pair of solid lines shown in Figure 6.7 can be increased arbitrarily; therefore,

$$\frac{2}{3} = \frac{2 \cdot n}{3 \cdot n}$$

for $n = 1, 2, \ldots$. In this way, beginning with one fraction $\frac{a}{b}$, an infinite list, $\frac{2a}{2b}, \frac{3a}{3b}, \frac{4a}{4b}, \ldots$, of equivalent fractions is obtained.

In summary, we have the following important property.

PROPERTY *The Fundamental Law of Fractions*

Let $\frac{a}{b}$ be a fraction. Then

$$\frac{a}{b} = \frac{an}{bn}, \text{ for any integer } n \neq 0.$$

Equivalent fractions can also be obtained by dividing both the numerator and denominator by a common factor. For example, the numerator and denominator of $\frac{35}{21}$ are each divisible by 7, so by the fundamental law

$$\frac{35}{21} = \frac{5 \cdot 7}{3 \cdot 7} = \frac{5}{3}.$$

Now suppose we are given two fractions, say $\frac{3}{12}$ and $\frac{2}{8}$, and we wish to know if they are equivalent. From the fundamental law of fractions we know that

$$\frac{3}{12} = \frac{3 \cdot 8}{12 \cdot 8} = \frac{24}{96} \quad \text{and} \quad \frac{2}{8} = \frac{2 \cdot 12}{8 \cdot 12} = \frac{24}{96}.$$

Thus, $\frac{3}{12}$ and $\frac{2}{8}$ are equivalent fractions. More generally, two given fractions $\frac{a}{b}$ and $\frac{c}{d}$ are respectively equivalent to $\frac{ad}{bd}$ and $\frac{bc}{bd}$. These are equivalent if, and only if, the numerators are equal, so we get the following theorem.

THEOREM *The Cross-Product Property of Equivalent Fractions*

The fractions $\frac{a}{b}$ and $\frac{c}{d}$ are **equivalent** if, and only if, $ad = bc$. That is,

$$\frac{a}{b} = \frac{c}{d} \quad \text{if, and only if,} \quad ad = bc.$$

EXAMPLE 6.1 SHOWING THE EQUIVALENCE OF FRACTIONS

(a) Show that $\frac{6}{14} = \frac{9}{21}$.

(b) Show that $\frac{-23}{47} = \frac{2231}{-4559}$.

(c) Find m if $\frac{m}{6} = \frac{10}{15}$.

(d) Find *all* fractions that are equivalent to $\frac{2}{3}$.

Solution

(a) Since $6 \cdot 21 = 126 = 9 \cdot 14$, $\frac{6}{14} = \frac{9}{21}$ by the cross-product property.

(b) Using a calculator (or paper and pencil), we find that $(-23) \cdot (-4559) = 104{,}857$ and $2231 \cdot 47 = 104{,}857$.

(c) $\frac{m}{6} = \frac{10}{15}$ if, and only if, $m \cdot 15 = 10 \cdot 6$. Thus, $m = 60 \div 15 = 4$.

(d) Suppose $\frac{2}{3} = \frac{a}{b}$. Then $2b = 3a$. Since 3 is a factor of $3a$, 3 must also be a factor of $2b$. Since 3 is not a factor of 2, it must be a factor of b, so $b = 3 \cdot n$ for some nonzero integer n. This gives us $3a = 2(3 \cdot n)$, so $a = 2 \cdot n$. This shows that any fraction $\frac{a}{b}$ equivalent to $\frac{2}{3}$ can be written as $\frac{2 \cdot n}{3 \cdot n}$ for some integer n.

Altogether, the set of fractions equivalent to $\frac{2}{3}$ is

$$\left\{\ldots, \frac{-6}{-9}, \frac{-4}{-6}, \frac{-2}{-3}, \frac{2}{3}, \frac{4}{6}, \frac{6}{9}, \ldots\right\}.$$

Fractions in Simplest Form

Often it is preferable to use the simplest equivalent form of a fraction instead of a more complicated fraction. For example, it may be best to use $\frac{2}{3}$ instead of $\frac{400}{600}$, and to use $\frac{-3}{4}$ instead of $\frac{75}{-100}$.

Here is the definition of a fraction in **simplest** or **reduced form,** or in **lowest terms.** The term *reduced* is best avoided, since this mistakenly suggests that a reduced fraction is smaller than the unreduced one.

DEFINITION *Fractions in Simplest Form*

A fraction $\frac{a}{b}$ is in **simplest form** if a and b have no common divisor larger than 1 and b is positive.

There are several ways to determine the simplest form of a fraction $\frac{a}{b}$.

Method 1. Divide successively by common factors. Suppose we want to write $\frac{560}{960}$ in simplest form. Using the fundamental law of fractions, we successively divide both numerator and denominator by common factors. Since 560 and 960 are both divisible by 10, it follows that

$$\frac{560}{960} = \frac{56 \cdot 10}{96 \cdot 10} = \frac{56}{96}.$$

But both 56 and 96 are even (that is, divisible by 2), so

$$\frac{56}{96} = \frac{28 \cdot 2}{48 \cdot 2} = \frac{28}{48}.$$

Again, 28 and 48 are easily seen to be divisible by 4 so that

$$\frac{28}{48} = \frac{7 \cdot 4}{12 \cdot 4} = \frac{7}{12}.$$

Finally, since 7 and 12 have no common factor other than 1, $\frac{7}{12}$ is the simplest form of $\frac{560}{960}$. Indeed, one might quickly and efficiently carry out this simplification as shown here to obtain the desired result.

$$\frac{\overset{\overset{7}{\cancel{28}}}{\cancel{560}}}{\underset{\underset{12}{\cancel{48}}}{\cancel{960}}} \quad \text{or} \quad \frac{560}{960} = \frac{56}{96} = \frac{28}{48} = \frac{7}{12}$$

Method 2. Divide *a* and *b* by GCD(a, b). Using the ideas of Chapter 4, we determine that GCD$(560, 960) = 80$. Therefore, $\frac{560}{960} = \frac{7 \cdot 80}{12 \cdot 80} = \frac{7}{12}$.

Method 3. Divide by the common factors in the prime power representation of *a* and *b*. Using this method, we have

$$\frac{560}{960} = \frac{2^4 \cdot 5 \cdot 7}{2^6 \cdot 3 \cdot 5} = \frac{7}{2^2 \cdot 3} = \frac{7}{12}.$$

Method 4. Use a fraction calculator. Many classroom calculators are fraction capable. For example, on the TI-15 *Explorer,* the fraction $\frac{560}{960}$ can be entered by pressing 560 [*n*] 960 [d̄] [Enter]. To simplify, pressing [Simp] [Enter] gives 280/480 in the display. Pressing [Fac] displays the common factor 2 that was divided into both 560 and 960. Pressing [Fac] again returns 280/480 to the display, and [Simp] [Enter] gives 140/240, where a second factor of 2 has been divided out. Continuing the sequence [Simp] [Enter] three more steps gives the final result of 7/12. By making a list of the common factors, 2, 2, 2, 2, 5, the *Explorer* makes it easy to compute GCD$(560, 960) = 2 \cdot 2 \cdot 2 \cdot 2 \cdot 5 = 80$.

EXAMPLE 6.2 | SIMPLIFYING FRACTIONS

Find the simplest form of each fraction.

(a) $\frac{240}{72}$ **(b)** $\frac{-450}{1500}$ **(c)** $\frac{294}{-84}$ **(d)** $\frac{399}{483}$

Solution

(a) By Method 1, $\frac{240}{72} = \frac{120}{36} = \frac{60}{18} = \frac{10}{3}$, where the successive common factors 2, 2, and 6 were divided into the numerator and denominator.

(b) Since $1500 = 3 \cdot 450 + 150$ and $450 = 3 \cdot 150 + 0$, the Euclidean algorithm shows that GCD(450, 1500) = 150. Thus, by Method 2,

$$\frac{-450}{1500} = \frac{-3 \cdot 150}{10 \cdot 150} = \frac{-3}{10}.$$

(c) Using Method 3 this time, we find

$$\frac{294}{-84} = \frac{2 \cdot 3 \cdot 7^2}{(-2) \cdot 2 \cdot 3 \cdot 7} = \frac{7}{-2} = \frac{-7}{2}.$$

(d) If 399 [n] 483 [d̄] [Enter] is entered on the TI-15 *Explorer,* then pressing [Simp] [Enter] repeatedly gives first 133/161 and next 19/23. Therefore $\frac{399}{483} = \frac{19}{23}$. The common factors are 3 and 7, so GCD(399, 483) = $3 \cdot 7 = 21$.

Common Denominators

When working with two fractions, it is important to know how they can be replaced with equivalent fractions with the *same* denominator. For example, $\frac{5}{8}$ and $\frac{7}{10}$ can each be replaced by equivalent fractions with the **common denominator** $8 \cdot 10 = 80$ so that

$$\frac{5}{8} = \frac{5 \cdot 10}{8 \cdot 10} = \frac{50}{80} \quad \text{and} \quad \frac{7}{10} = \frac{7 \cdot 8}{10 \cdot 8} = \frac{56}{80}.$$

In the same way, any two fractions $\frac{a}{b}$ and $\frac{c}{d}$ can be rewritten with the common denominator $b \cdot d$, since $\frac{a}{b} = \frac{a \cdot d}{b \cdot d}$ and $\frac{c}{d} = \frac{c \cdot b}{d \cdot b}$.

It is sometimes worthwhile to find the common positive denominator that is as small as possible. Assuming $\frac{a}{b}$ and $\frac{c}{d}$ are in simplest form, we require a common denominator that is a multiple of both b and d. The least such common multiple is called the **least common denominator** and is the least common multiple of b and d. For the example $\frac{5}{8}$ and $\frac{7}{10}$, we would calculate LCM(8, 10) = 40, so 40 is the least common denominator. Therefore,

$$\frac{5}{8} = \frac{5 \cdot 5}{8 \cdot 5} = \frac{25}{40} \quad \text{and} \quad \frac{7}{10} = \frac{7 \cdot 4}{10 \cdot 4} = \frac{28}{40}.$$

The least common denominator can often be determined by mental arithmetic. Consider, for example, $\frac{5}{6}$ and $\frac{3}{8}$. With a little practice, it won't take long to notice that $4 \cdot 6 = 24$ and $3 \cdot 8 = 24$, and that 24 is the least common denominator. Thus, $\frac{5}{6} = \frac{20}{24}$ and $\frac{3}{8} = \frac{9}{24}$ when written with the least common denominator.

EXAMPLE 6.3 FINDING COMMON DENOMINATORS

Find equivalent fractions with a common denominator.

(a) $\frac{5}{6}$ and $\frac{1}{4}$

(b) $\frac{9}{8}$ and $\frac{-12}{7}$

(c) $\frac{14}{-16}$ and $\frac{11}{12}$

(d) $\frac{3}{4}$, $\frac{5}{8}$, and $\frac{2}{3}$

Solution

(a) By mental arithmetic, the least common denominator of $\frac{5}{6}$ and $\frac{1}{4}$ is 12. Therefore, $\frac{5}{6} = \frac{10}{12}$ and $\frac{1}{4} = \frac{3}{12}$. This can be shown with fraction strips.

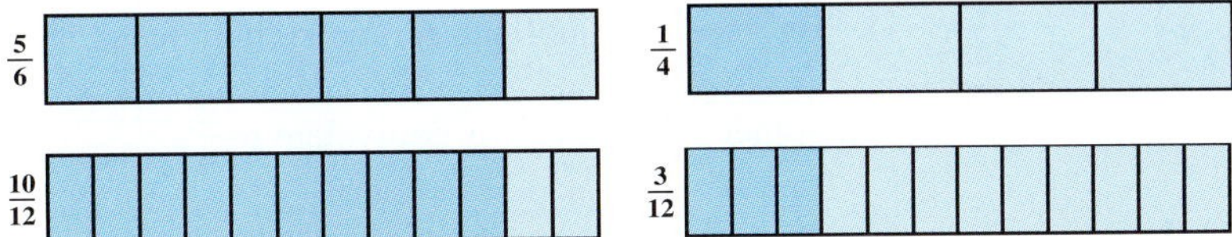

(b) Since 8 and 7 have no common factors other than 1, $8 \cdot 7 = 56$ is the least common denominator:

$$\frac{9}{8} = \frac{9 \cdot 7}{8 \cdot 7} = \frac{63}{56} \quad \text{and} \quad \frac{-12}{7} = \frac{-12 \cdot 8}{7 \cdot 8} = \frac{-96}{56}.$$

(c) First we simplify $\frac{14}{-16}$ to $\frac{-7}{8}$. Since 8 and 12 have 4 as their greatest common divisor, we see that $\text{LCM}(8, 12) = 8 \cdot 12/\text{GCD}(8, 12) = 96/4 = 24$. Thus,

$$\frac{14}{-16} = \frac{-7}{8} = \frac{-7 \cdot 3}{8 \cdot 3} = \frac{-21}{24} \quad \text{and} \quad \frac{11}{12} = \frac{11 \cdot 2}{12 \cdot 2} = \frac{22}{24}.$$

(d) With three fractions, it is still possible to use the product of all of the denominators as a common denominator. Since $4 \cdot 8 \cdot 3 = 96$, we have

$$\frac{3}{4} = \frac{3 \cdot 8 \cdot 3}{4 \cdot 8 \cdot 3} = \frac{72}{96}, \quad \frac{5}{8} = \frac{5 \cdot 4 \cdot 3}{8 \cdot 4 \cdot 3} = \frac{60}{96}, \quad \text{and} \quad \frac{2}{3} = \frac{2 \cdot 4 \cdot 8}{3 \cdot 4 \cdot 8} = \frac{64}{96}.$$

Alternatively, $\text{LCM}(4, 8, 3) = 24$, so we have

$$\frac{3}{4} = \frac{3 \cdot 6}{4 \cdot 6} = \frac{18}{24}, \quad \frac{5}{8} = \frac{5 \cdot 3}{8 \cdot 3} = \frac{15}{24}, \quad \text{and} \quad \frac{2}{3} = \frac{2 \cdot 8}{3 \cdot 8} = \frac{16}{24}.$$

This expresses the three fractions as equivalent fractions with the least common denominator.

Rational Numbers

We have seen that different fractions can nevertheless express the same amount or correspond to the same point on a number line. For example, the fraction strips in Figure 6.7 each show that two-thirds of a whole strip has been shaded, but the fraction that represents this amount can be any one of the choices $\frac{2}{3}, \frac{4}{6}, \frac{6}{9}, \ldots$. Similarly, in Figure 6.6 the single point on the number line at a distance of one-half of a unit to the left of the origin can be expressed by any of the equivalent fractions $\frac{-1}{2}, \frac{1}{-2}, \frac{-2}{4}, \frac{2}{-4}, \ldots$.

Numbers such as "two-thirds" and "negative one-half" that can be represented by fractions are examples of **rational numbers.** The following definition is suitable for use in the elementary school classroom.*

DEFINITION *Rational Numbers*
A **rational number** is a number that can be represented by a fraction a/b, where a and b are integers, $b \neq 0$. Two rational numbers are **equal** if, and only if, they can be represented by equivalent fractions.

The set of rational numbers is denoted by Q, where the letter Q reminds us that a fraction is represented as a quotient. A rational number such as $\frac{3}{4}$ can also be represented by $\frac{6}{8}$, $\frac{30}{40}$, or any other fraction that is equivalent to $\frac{3}{4}$.

EXAMPLE 6.4 REPRESENTING RATIONAL NUMBERS

How many different rational numbers are given in this list?

$$\frac{2}{5}, \quad 3, \quad \frac{-4}{-10}, \quad \frac{39}{13}, \quad \text{and} \quad \frac{7}{4}.$$

Solution Since $\frac{2}{5} = \frac{-4}{-10}$ and $3 = \frac{39}{13}$, there are three different rational numbers: $\frac{2}{5}$, 3, and $\frac{7}{4}$.

Ordering Fractions and Rational Numbers

By placing the fraction strips for $\frac{3}{4}$ and $\frac{5}{6}$ side by side, as shown in Figure 6.8(a), it becomes geometrically apparent that $\frac{3}{4}$ represents a smaller shaded portion of a whole strip than $\frac{5}{6}$. The comparison can be made even clearer by replacing the strips $\frac{3}{4}$ and $\frac{5}{6}$ by the equivalent strips $\frac{9}{12}$ and $\frac{10}{12}$, as shown in Figure 6.8(b). Since $9 < 10$, we see that $\frac{9}{12}$ is less than $\frac{10}{12}$, and therefore why $\frac{3}{4}$ is less than $\frac{5}{6}$.

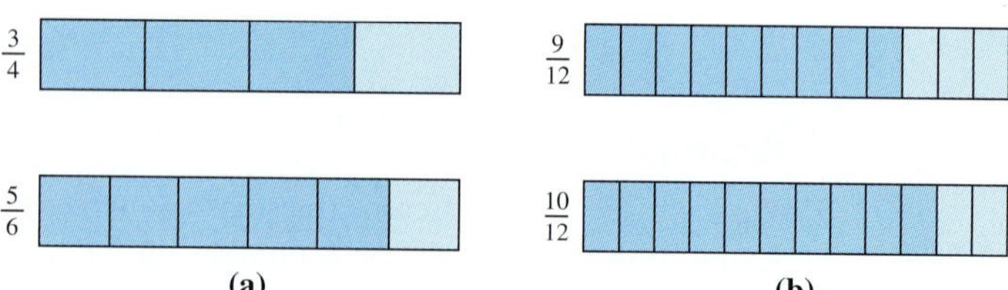

Figure 6.8
Showing $\frac{3}{4} < \frac{5}{6}$ *and* $\frac{9}{12} < \frac{10}{12}$ *with fraction strips*

*In advanced mathematics a **rational number** is defined to be a set containing all of the fractions that are equivalent to some given fraction. The informal definition given above, however, conveys the proper intuitive meaning of rational number.

More generally, let two rational numbers be represented by the fractions $\frac{a}{b}$ and $\frac{c}{d}$ with positive denominators b and d. These numbers can be compared by first rewriting them in equivalent form $\frac{ad}{bd}$ and $\frac{bc}{bd}$. Then $\frac{ad}{bd} < \frac{bc}{bd}$ if, and only if, $ad < bc$. This leads us to the following definition.

DEFINITION *Order Relation on the Rational Numbers*

Let two rational numbers be represented by the fractions $\frac{a}{b}$ and $\frac{c}{d}$, with b and d positive. Then $\frac{a}{b}$ is **less than** $\frac{c}{d}$, written $\frac{a}{b} < \frac{c}{d}$, if, and only if, $ad < bc$.

The corresponding relations less than or equal to, $\leq$; greater than, $>$; and greater than or equal to, $\geq$ are defined similarly.

EXAMPLE 6.5 COMPARING RATIONAL NUMBERS

Replace the box by the proper relation $<$, $=$, or $>$ for each pair of rational numbers.

(a) $\frac{3}{4} \square \frac{2}{5}$ **(b)** $\frac{15}{29} \square \frac{6}{11}$

(c) $\frac{2106}{7047} \square \frac{234}{783}$ **(d)** $\frac{-10}{13} \square \frac{22}{-29}$

Solution

(a) Since $3 \cdot 5 = 15 > 8 = 2 \cdot 4$, we have $\frac{3}{4} > \frac{2}{5}$.

(b) Since $15 \cdot 11 = 165 < 174 = 6 \cdot 29$, we have $\frac{15}{29} < \frac{6}{11}$.

(c) Using a calculator, we find that $2106 \cdot 783 = 1{,}648{,}998 = 7047 \cdot 234$. Thus, the two fractions are equivalent: $\frac{2106}{7047} = \frac{234}{783}$.

(d) First we write $\frac{22}{-29}$ as $\frac{-22}{29}$ so that its denominator is positive. Then, since $-10 \cdot 29 = -290 < -286 = -22 \cdot 13$, we conclude that $\frac{-10}{13} < \frac{22}{-29}$.

from **The NCTM Principles and Standards**

During grades 3–5, students should build their understanding of fractions as parts of a whole and as division. They will need to see and explore a variety of models of fractions, focusing primarily on familiar fractions such as halves, thirds, fourths, fifths, sixths, eighths, and tenths. By using an area model in which part of a region is shaded, students can see how fractions are related to a unit whole, compare fractional parts of a whole, and find equivalent fractions. They should develop strategies for ordering and comparing fractions, often using benchmarks such as 1/2 and 1. For example, fifth graders can compare fractions such as 2/5 and 5/8 by comparing each with 1/2—one is a little less than 1/2, and the other is a little more. By using parallel number lines, each showing a unit fraction and its multiples, students can see fractions as numbers, note their relationship to 1, and see relationships among fractions, including equivalence.

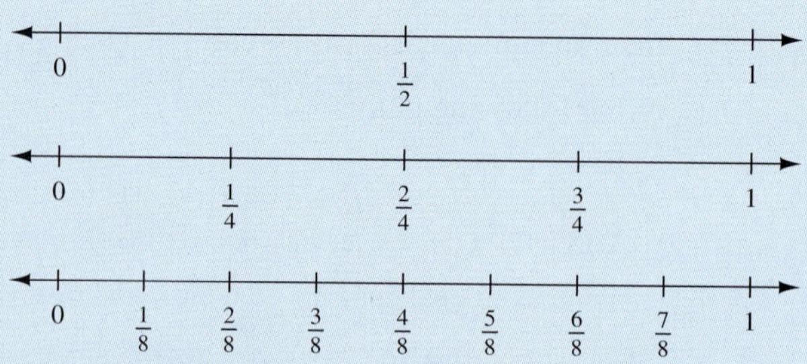

SOURCE: Reprinted with permission from Principles and Standards for School Mathematics, *page 150.*

Standards are listed with the permission of the National Council of Teachers of Mathematics (NCTM). NCTM does not endorse the content or validity of these alignments.

Problem Set 6.1

Understanding Concepts

1. What fraction is represented by the darker shaded portion of the following figures?

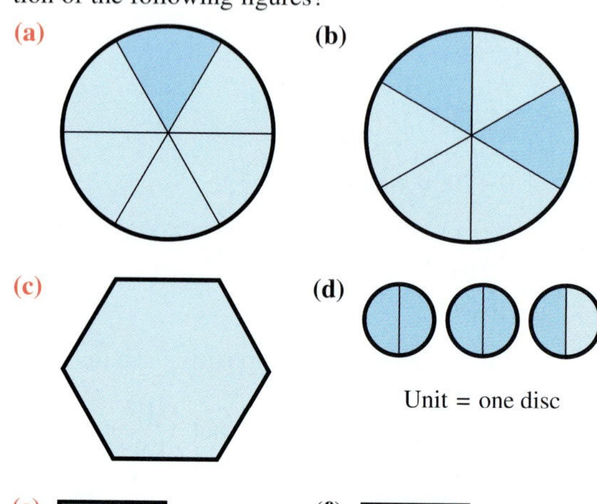

2. Subdivide and shade the unit octagons shown to represent the given fraction.

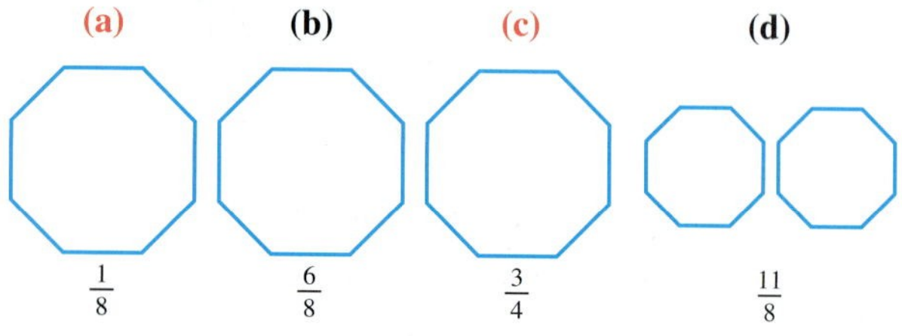

3. Consider the set shown below.

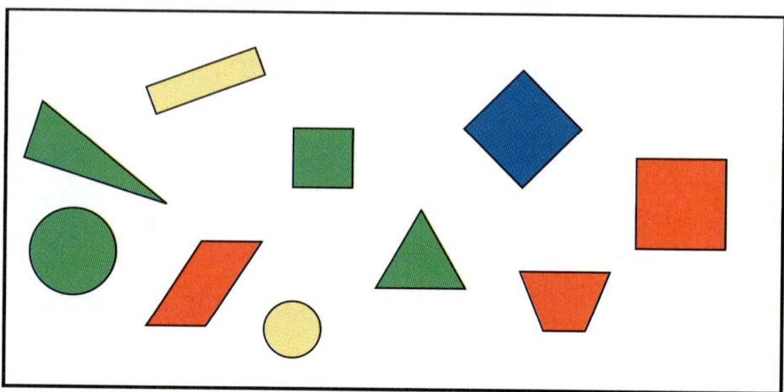

What fraction of the objects in the set

(a) are green?
(b) are squares?
(c) are rectangles?
(d) are quadrilaterals (four-sided polygons)?
(e) have no line of symmetry?

4. For each lettered point on the number lines below, express its position by a corresponding fraction. Remember, it is the number of subintervals in a unit interval that determines the denominator, not the number of tick marks.

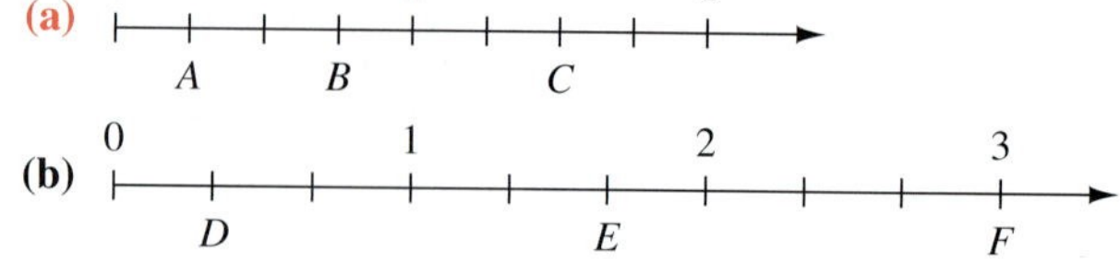

(c) 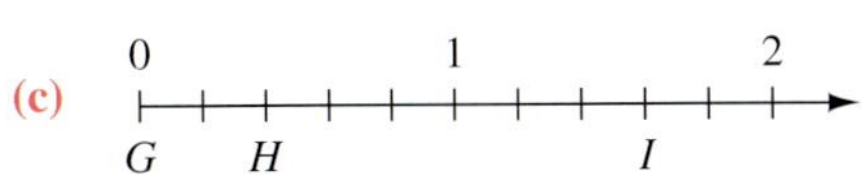

(d) −1 0 1

L K J

(e) −2 −1 0

P N M

5. Depict the fraction $\frac{4}{6}$ with the following models.

(a) Colored-region model (b) Set model

(c) Fraction-strip model (d) Number-line model

6. Express the following quantities by a fraction placed in the blank space.

(a) 20 minutes is __________ of an hour.

(b) 30 seconds is __________ of a minute.

(c) 5 days is __________ of a week.

(d) 25 years is __________ of a century.

(e) A quarter is __________ of a dollar.

(f) 3 eggs is __________ of a dozen.

(g) 2 feet is __________ of a yard.

(h) 3 cups is __________ of a quart.

7. In Figure 6.7, fraction strips show that $\frac{2}{3}, \frac{4}{6}, \frac{6}{9}$, and $\frac{8}{12}$ are equivalent fractions. Use a similar drawing of fraction strips to show that $\frac{3}{4}, \frac{6}{8}$, and $\frac{9}{12}$ are equivalent fractions.

8. What equivalence of fractions is shown in these pairs of colored-region models?

(a) (b) (c)

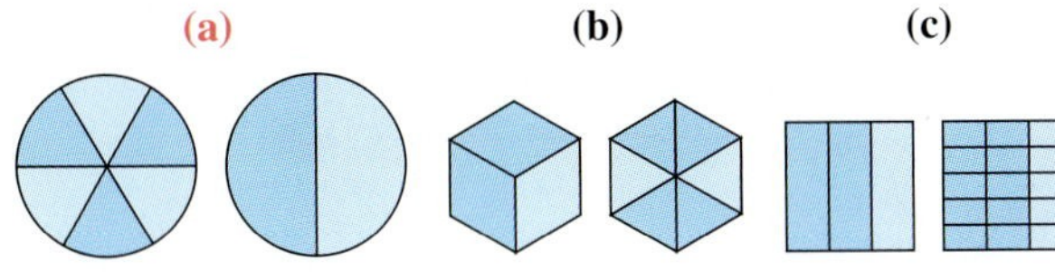

9. Find four different fractions equivalent to $\frac{4}{9}$.

10. Subdivide and mark the unit square on the right to illustrate that the given fractions are equivalent.

(a) (b) (c)

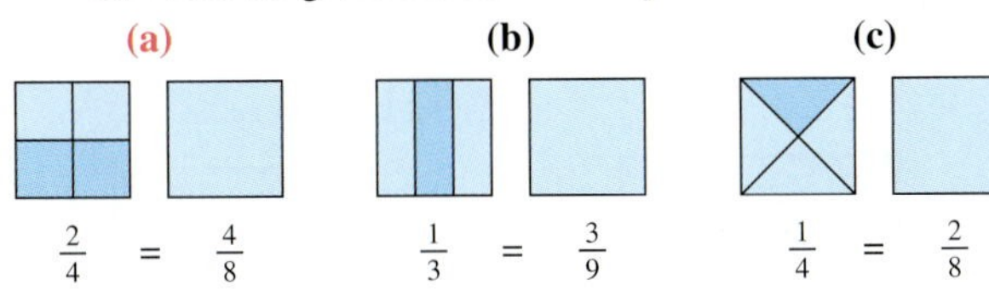

$\frac{2}{4} = \frac{4}{8}$ $\frac{1}{3} = \frac{3}{9}$ $\frac{1}{4} = \frac{2}{8}$

11. Fill in the missing integer to make the fractions equivalent.

(a) $\frac{4}{5} = \frac{\quad}{30}$ (b) $\frac{6}{9} = \frac{2}{\quad}$

(c) $\frac{-7}{25} = \frac{\quad}{500}$ (d) $\frac{18}{3} = \frac{-6}{\quad}$

12. Determine whether each set of two fractions is equivalent by calculating equivalent fractions with a common denominator.

(a) $\frac{18}{42}$ and $\frac{3}{7}$ (b) $\frac{18}{49}$ and $\frac{5}{14}$

(c) $\frac{9}{25}$ and $\frac{140}{500}$ (d) $\frac{24}{144}$ and $\frac{32}{96}$

13. Determine which of these pairs of fractions are equivalent.

(a) $\frac{78}{24}$ and $\frac{546}{168}$ (b) $\frac{243}{317}$ and $\frac{2673}{3487}$

(c) $\frac{412}{-864}$ and $\frac{-308}{616}$

14. (a) Is it true that $\frac{4 \cdot 3}{9 \cdot 3} = \frac{12}{27}$?

(b) Is it true that $\frac{4 \cdot 3}{9 \cdot 3} = \frac{4}{9}$?

(c) Is it true that $\frac{4 + 3}{9 + 3} = \frac{7}{12}$?

(d) Is it true that $\frac{4 + 3}{9 + 3} = \frac{4}{9}$?

15. Rewrite the following fractions in simplest form.

(a) $\frac{84}{144}$ (b) $\frac{208}{272}$ (c) $\frac{-930}{1290}$ (d) $\frac{325}{231}$

16. Find the prime factorizations of the numerators and denominators of these fractions and use them to express the fractions in simplest form.

(a) $\frac{96}{288}$ (b) $\frac{247}{-75}$ (c) $\frac{2520}{378}$

17. For each of these sets of fractions, determine equivalent fractions with a common denominator.

(a) $\frac{3}{11}$ and $\frac{2}{5}$ (b) $\frac{5}{12}$ and $\frac{2}{3}$

(c) $\frac{4}{3}, \frac{5}{8}$, and $\frac{1}{6}$ (d) $\frac{1}{125}$ and $\frac{-3}{500}$

18. For each of these sets of fractions, determine equivalent fractions with the least common denominator.

(a) $\frac{3}{8}$ and $\frac{5}{6}$ (b) $\frac{1}{7}, \frac{4}{5}$, and $\frac{2}{3}$

(c) $\frac{17}{12}$ and $\frac{7}{32}$ (d) $\frac{17}{51}$ and $\frac{56}{42}$

19. Order the rational numbers from least to greatest in each part.

(a) $\frac{2}{3}, \frac{7}{12}$ (b) $\frac{2}{3}, \frac{5}{6}$ (c) $\frac{5}{6}, \frac{29}{36}$

(d) $\frac{-5}{6}, \frac{-8}{9}$ (e) $\frac{2}{3}, \frac{5}{6}, \frac{29}{36}, \frac{8}{9}$

20. Find the numerator or denominator of these fractions to create a fraction that is close to $\frac{1}{2}$, but is slightly larger.

(a) $\frac{\quad}{20}$ (b) $\frac{7}{\quad}$ (c) $\frac{\quad}{17}$ (d) $\frac{30}{\quad}$

21. Decide whether each statement is *true* or *false*. Explain your reasoning in a brief paragraph.

(a) There are infinitely many ways to replace two fractions with two equivalent fractions that have a common denominator.

(b) There is a unique least common denominator for a given pair of fractions.

(c) There is a least positive fraction.

(d) There are infinitely many fractions between 0 and 1.

22. How many different rational numbers are in this list?

$$\frac{27}{36}, 4, \frac{21}{28}, \frac{24}{6}, \frac{3}{4}, \frac{-8}{-2}$$

Teaching Concepts

23. Go on a "fraction safari" to the grocery store (or in a magazine or newspaper). Find at least five fractions. For each fraction, describe the unit, the numerator, and the denominator.

24. Like most fifth graders, Dana likes pizza. When given the choice of $\frac{1}{4}$ or $\frac{1}{6}$ of a pizza, Dana says, "Since 6 is bigger than 4 and I'm really hungry, I'd rather have $\frac{1}{6}$ of the pizza." Write a dialogue including useful diagrams to clear up Dana's misconception about fractions.

25. When asked to illustrate the concept of $\frac{2}{3}$ with a colored-region diagram, Shanti drew this figure. How would you respond to Shanti?

Thinking Critically

26. What fraction represents the part of the whole region that has been shaded? Draw additional lines to make your answer visually clear. For example, $\frac{2}{6}$ of the regular hexagon on the left is shaded, since the entire hexagon can be subdivided into six congruent regions as shown on the right.

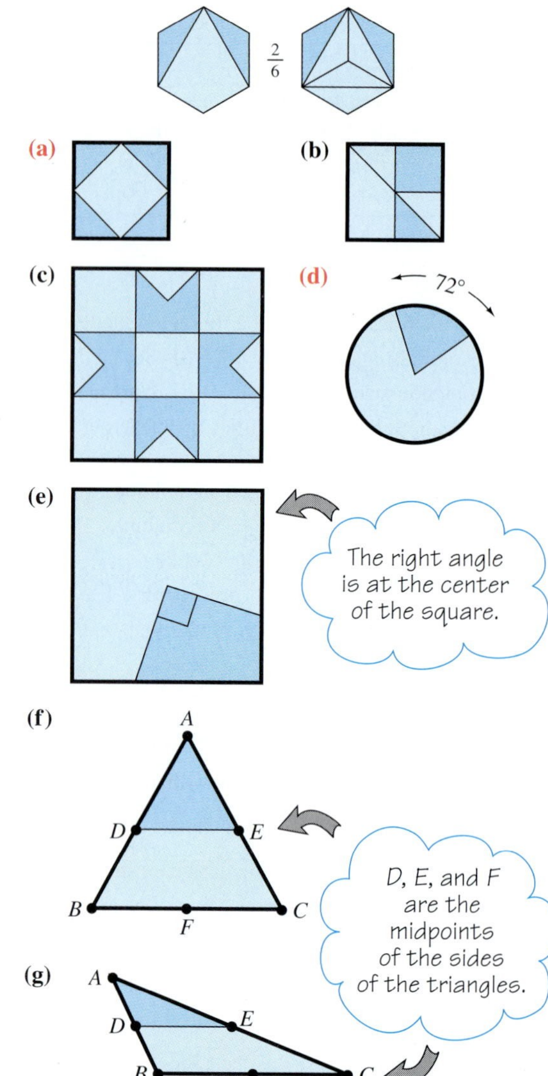

27. Start with a square piece of paper, and join each corner to the midpoint of the opposite side, as shown in the left-hand figure below. Shade the slanted square that is formed inside the original square. What fraction of the large square have you shaded? It will help to cut the triangular pieces as shown on the right and reattach them to form some new squares.

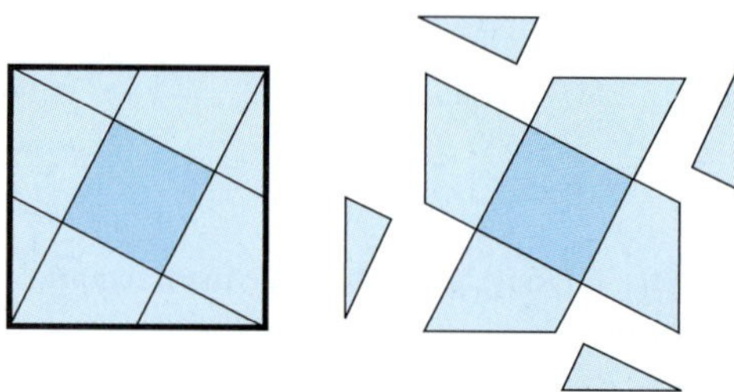

28. Start with an equilateral triangle cut from paper, and join each corner to the one-third point of the opposite side as shown in the left-hand figure below. Shade the small equilateral triangle that is formed inside the large triangle. What fraction of the large triangle have you shaded? It will help to cut the six triangular pieces as shown on the right and reattach them to form some new equilateral triangles.

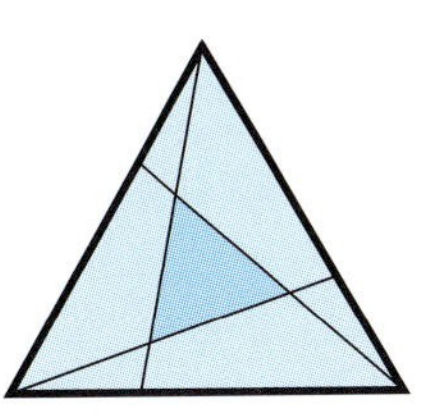

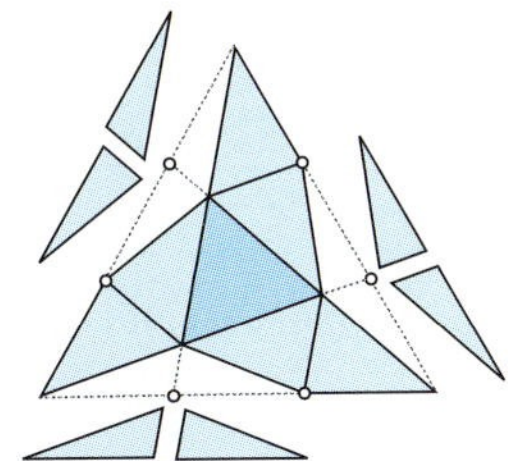

29. Solve this fraction problem from ancient India. "In what time will four fountains, being let loose together, fill a cistern, which they would severally fill in a day, in half a day, in a quarter and in a fifth part of a day?" (Reference: *History of Hindu Mathematics,* by B. Datta and A. N. Singh. Bombay, Calcutta, New Delhi, Madras, London, New York: Asia Publishing House, 1962, p. 234.)

30. **(a)** Andrei bought a length of rope at the hardware store. He used half of it to make a bow painter for his canoe and then used a third of the remaining piece to tie up a roll of carpet. He now has 20 feet of rope left. What was the length of rope Andrei purchased? Obtain your answer pictorially, by adding additional marks and labels to the drawing below.

Used for painter

(b) Answer part (a) with a new diagram, but with this revised data: Andrei used one-third of the rope to make a leash for his dog and one-fourth of the remaining rope to tie up a bundle of stakes. This left 15 feet of the rope unused. What was the original length of the rope?

31. The **mediant** of two fractions $\frac{a}{b}$ and $\frac{c}{d}$ is $\frac{a+c}{b+d}$, where b and d are positive denominators. For example, the mediant of $\frac{1}{4}$ and $\frac{1}{2}$ is $\frac{1+1}{4+2} = \frac{2}{6} = \frac{1}{3}$. Notice that the mediant is between the two starting fractions, since $\frac{1}{4} < \frac{1}{3} < \frac{1}{2}$.

(a) Find the mediant of $\frac{3}{4}$ and $\frac{4}{5}$, and show that it is between the two given fractions.

(b) Show that the mediant is always between the given two fractions; that is, show that if $\frac{a}{b} < \frac{c}{d}$, then

$$\frac{a}{b} < \frac{a+c}{b+d} < \frac{c}{d}.$$

(c) Carefully explain why the result of the property shown in part (b) guarantees that there are infinitely many fractions between any two given fractions.

32. The row of Pascal's triangle (see Section 3 of Chapter 1) shown below has been separated by a vertical bar drawn between the 15 and the 20, with 3 entries of the row to the left of the bar and 4 entries to the right of the bar.

1 6 15 | 20 15 6 1

Notice that $\frac{15}{20} = \frac{3}{4}$. That is, the two fractions formed by the entries adjacent to the dividing bar and by the number of entries to the left and the right of the bar are equivalent fractions. Was this an accident, or is this a general property of Pascal's triangle? Investigate the property further by placing the dividing bar in new locations and examining other rows of Pascal's triangle.

Thinking Cooperatively

The following two problems develop the concept of a fraction using pattern blocks, a popular and versatile manipulative that is used successfully in many elementary school classrooms. The following four shapes should be available to students working in groups of two or three:

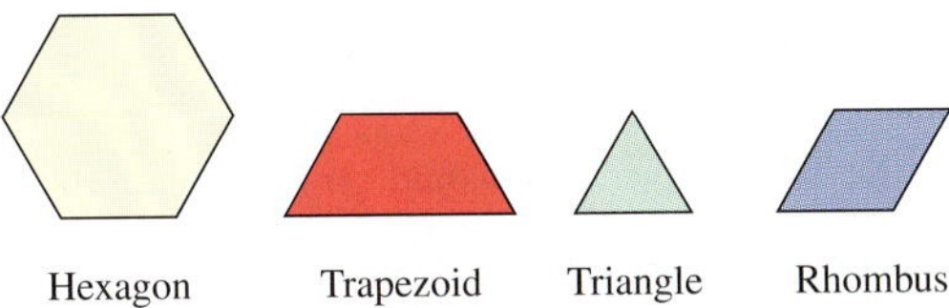

33. Use your pattern blocks to answer these questions.

(a) Choose the hexagon as the unit. What fraction is each of the other three pattern blocks?

(b) Choose the trapezoid as the unit. What fraction is each of the other three pattern blocks?

(c) Choose the rhombus as the unit. What fraction is each of the other three pattern blocks?

34. **(a)** A hexagon and six triangles are used to form a six-pointed star. Choose the star as the unit. Next, form pattern-block shapes corresponding to each of these fractions: $\frac{1}{6}, \frac{1}{4}, \frac{1}{3}, \frac{2}{3}$, and $\frac{3}{4}$. Use as few pieces as you can for each shape.

(b) Create a pattern-block shape that will be defined as the unit and for which you may form other pattern block shapes corresponding to the fractions $\frac{1}{8}$ and $\frac{5}{8}$.

35. Working in pairs, shuffle an ordinary 52-card deck and deal 26 cards each to you and your partner. Let the number of red and black cards in your hand be denoted by r and b, respectively. Similarly, let R and B respectively denote the number of red and black cards in your partner's hand. Now form the fractions r/b and B/R. Do you see any surprises? Can you explain what is happening? Work together with your partner to write a clear explanation.

Making Connections

36. Francisco's pickup truck has a 24-gallon gas tank and an accurate fuel gauge. Estimate the number of gallons in the tank at these readings.

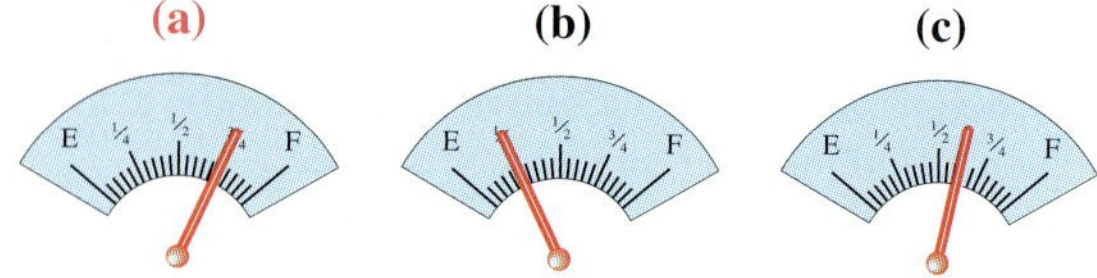

37. If 153 of the 307 graduating seniors go on to college, it is likely that a principal would claim that $\frac{1}{2}$ of the class is college bound. Give simpler convenient fractions that approximately express the data in these situations.

(a) Esteban is on page 310 of a 498-page novel. He has read __________ of the book.

(b) Myra has saved \$73 toward the purchase of a \$215 plane ticket. She has saved __________ of the amount she needs.

(c) Nine students in Ms. Evaldo's class of 35 students did perfect work on the quiz. __________ of the class scored 100% on the quiz.

(d) The Math Club has sold 1623 of the 2400 raffle tickets. It has sold __________ of the available tickets.

38. Lakeside High won 19 of its season's 25 baseball games. Rival Shorecrest High School lost just 5 of the 21 games it played. Can Shorecrest claim to have had the better season? Explain.

39. **Basketball Math.** If a basketball player makes m free throws in n attempts, the rational number $\frac{m}{n}$ gives a measure of the player's skill at the foul line. Suppose, going into the play-offs, Carol and Joleen have the following records over the first and second halves of the regular season. For example, Carol made 38 free throws in 50 attempts during the first half of the season.

$\frac{m}{n}$	Carol	Joleen
Nov.–Jan.	$\frac{38}{50}$	$\frac{30}{40}$
Jan.–Mar.	$\frac{42}{70}$	$\frac{14}{24}$
Entire Season		

(a) Who was the more successful free-throw shooter during the first half of the regular season? during the second half?

(b) Combine the data for each half season. For example, Carol hit 80 free throws in 120 attempts. Which player had the higher success rate over the entire season?

(c) Suppose the data in the table represent the success rate of two drugs: drug C versus drug J. Which drug is best in each half of the five-month trial period? Which drug looks most promising overall?

40. **Fractions in Probability.** If a card is picked at random from an ordinary deck of 52 playing cards, there are 4 ways it can be an ace, since it could be the ace of hearts, diamonds, clubs, or spades. To measure the chances of drawing an ace, it is common to give the probability as the rational number $\frac{4}{52}$. In general, if n equally likely outcomes are possible and m of these outcomes are successful for an event to occur, then the probability of the event is $\frac{m}{n}$. As another example, $\frac{5}{6}$ is the probability of rolling a single die and having more than one spot appear. Give fractions that express the probability of the following events:

(a) Getting a head in the flip of a fair coin.

(b) Drawing a face card from a deck of cards.

(c) Rolling an even number on a single die.

(d) Drawing a green marble from a bag that contains 20 red, 30 blue, and 25 green marbles.

(e) Drawing either a red or a blue marble from the bag of marbles described in part (d).

41. What fraction represents the probability that the spinner shown comes up (a) yellow? (b) red? (c) blue? (d) not blue?

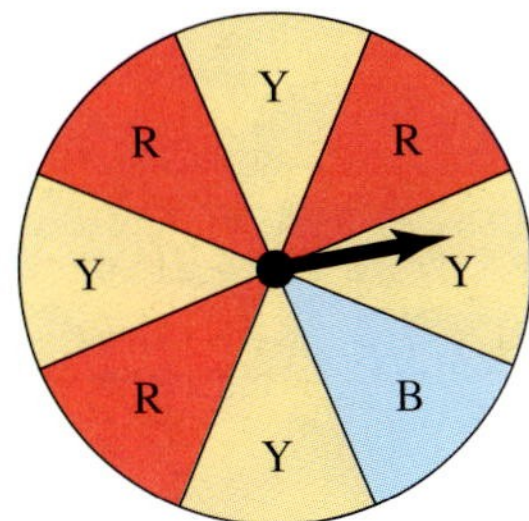

From State Student Assessments

42. (Minnesota, Grade 5)
Four children drew fraction pictures. Which drawing is $\frac{2}{3}$ shaded?

A. Aaron's **B.** Betsy's
C. Carla's **D.** Daisy's

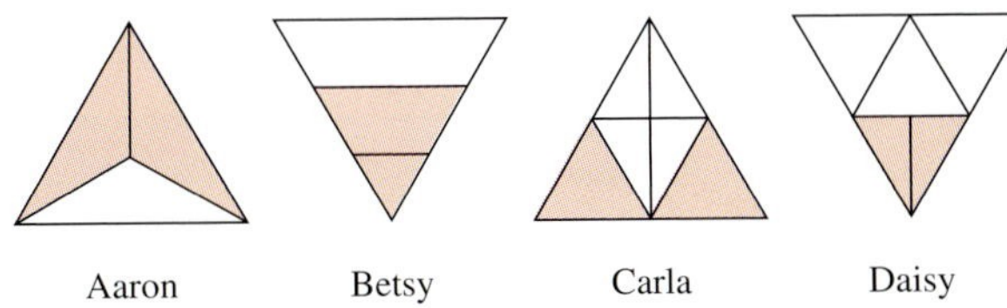

43. (Oregon, Grade 5)
Jules's eyes are covered. If he were to throw a dart at this board, how likely is it that a dart that hits the board would land on a shaded box?

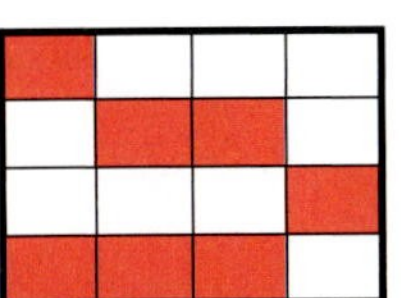

A. More than three-fourths the time
B. Less than half the time
C. About half the time
D. More than half the time

For Review

44. The Fibonacci sequence 1, 1, 2, 3, 5, . . . is formed by adding two successive terms to obtain the next one, beginning with 1 and 1. Suppose subtraction is used in place of addition, to get

$$1, 1, 0, 1, -1, 2, -3, \ldots$$

(a) Obtain the next ten terms of this sequence.

(b) If you know that $F_{24} = 46{,}368$ and $F_{25} = 75{,}025$ are the twenty-fourth and twenty-fifth Fibonacci numbers, what do you think the twenty-seventh and twenty-eighth numbers in the subtractive sequence are? Explain your reasoning.

45. (a) Find integers m and n that solve the equation $8m - 11n = 1$. It may be helpful to compare a list of multiples of 8 to a list of multiples of 11.

(b) Explain why it is not possible to find integers x and y that satisfy the equation $8x - 12y = 1$. (*Hint:* Is there an integer that divides the left side of the equation, but cannot divide the right side?)

46. Find the digit d so that $49{,}d84$ is divisible by 24. (*Hint:* Use divisibility tests for 3 and 8.)

6.2 THE ARITHMETIC OF RATIONAL NUMBERS

In this section we consider addition, subtraction, multiplication, and division in the set of rational numbers. The geometric and physical models of fractions serve to motivate the definition of each operation. The primary goal here is to learn how manipulatives and visualizations are used to convey the meaning of each arithmetic operation to children.

Addition of Rational Numbers

The sum of $\frac{3}{8}$ and $\frac{2}{8}$ is illustrated in two ways in Figure 6.9. The colored-region and number-line models both show that $\frac{3}{8} + \frac{2}{8} = \frac{5}{8}$. The colored-region visualization corresponds to the set model of addition, and the number-line visualization corresponds to the measurement model of addition.

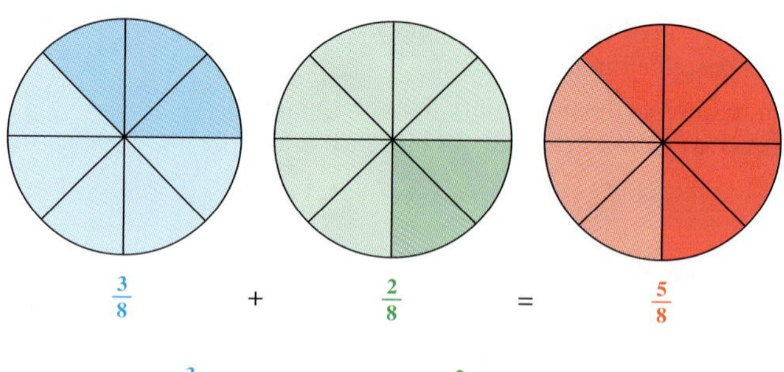

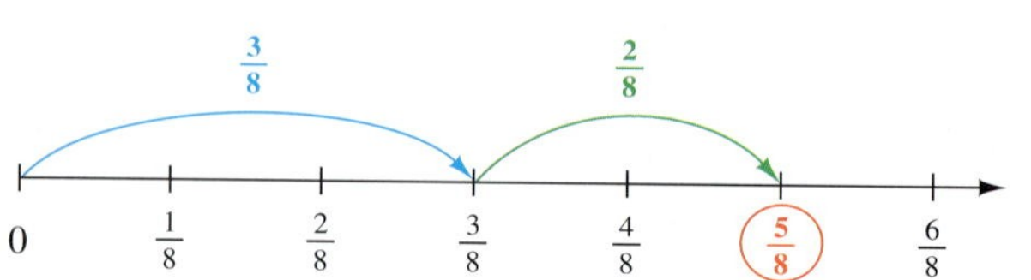

Figure 6.9
Showing $\frac{3}{8} + \frac{2}{8} = \frac{5}{8}$ with the colored-region and number-line models

The models suggest that the sum of two rational numbers represented by fractions with a common denominator should be found by adding the two numerators. This motivates the following definition.

DEFINITION *Addition of Rational Numbers*

Let two rational numbers be represented by fractions $\frac{a}{b}$ and $\frac{c}{b}$ with a common denominator. Then their **sum** is the rational number given by

$$\frac{a}{b} + \frac{c}{b} = \frac{a + c}{b}.$$

To add rational numbers represented by fractions with unlike denominators, we first rewrite the fractions with a common denominator. For example, to add $\frac{1}{4}$ and $\frac{2}{3}$, we rewrite them with 12 as a common denominator:

$$\frac{1}{4} = \frac{1 \cdot 3}{4 \cdot 3} = \frac{3}{12} \quad \text{and} \quad \frac{2}{3} = \frac{2 \cdot 4}{3 \cdot 4} = \frac{8}{12}.$$

According to the definition above, we then have

$$\frac{1}{4} + \frac{2}{3} = \frac{1 \cdot 3}{4 \cdot 3} + \frac{2 \cdot 4}{3 \cdot 4} = \frac{1 \cdot 3 + 2 \cdot 4}{4 \cdot 3} = \frac{3 + 8}{12} = \frac{11}{12}.$$

The procedure just followed can be modeled with fraction strips as shown in Figure 6.10. It is important to see how the function strips are aligned.

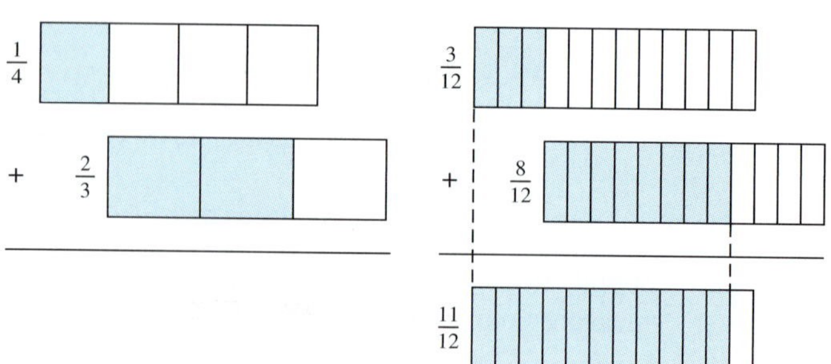

Figure 6.10
The fraction-strip model showing $\frac{1}{4} + \frac{2}{3} = \frac{3}{12} + \frac{8}{12} = \frac{11}{12}$

The same procedure can be followed to add any two rational numbers $\frac{a}{b}$ and $\frac{c}{d}$. We have

$$\frac{a}{b} + \frac{c}{d} = \frac{a \cdot d}{b \cdot d} + \frac{c \cdot b}{d \cdot b} = \frac{ad + bc}{bd}.$$

Rewrite the fractions with a common denominator.

Add the numerators, and retain the common denominator.

EXAMPLE 6.6 ADDING RATIONAL NUMBERS

Show the steps followed to compute these sums, as if you were showing a fifth grader.

(a) $\frac{3}{10} + \frac{4}{7}$ **(b)** $\frac{3}{8} + \frac{-7}{24}$

(c) $\left(\frac{3}{4} + \frac{5}{6}\right) + \frac{2}{3}$ **(d)** $\frac{3}{4} + \left(\frac{5}{6} + \frac{2}{3}\right)$

Solution

(a) $\frac{3}{10} + \frac{4}{7} = \frac{3 \cdot 7}{10 \cdot 7} + \frac{4 \cdot 10}{7 \cdot 10} = \frac{21}{70} + \frac{40}{70} = \frac{61}{70}.$

(b) $\frac{3}{8} + \frac{-7}{24} = \frac{3 \cdot 24}{8 \cdot 24} + \frac{8 \cdot (-7)}{8 \cdot 24} = \frac{72}{192} + \frac{-56}{192} = \frac{72 - 56}{192} = \frac{16}{192}.$

A better method is to use the least common denominator. Thus, $\frac{3}{8} + \frac{-7}{24} = \frac{3 \cdot 3}{8 \cdot 3} + \frac{-7}{24} = \frac{9}{24} + \frac{-7}{24} = \frac{2}{24}$. To see that the two answers are equivalent, we rewrite each fraction in simplest form:

$$\frac{16}{192} = \frac{8}{96} = \frac{4}{48} = \frac{1}{12} \quad \text{and} \quad \frac{2}{24} = \frac{1}{12}.$$

(c) The parentheses tell us to compute first

$$\frac{3}{4} + \frac{5}{6} = \frac{9}{12} + \frac{10}{12} = \frac{19}{12}.$$

Then we compute

$$\left(\frac{3}{4} + \frac{5}{6}\right) + \frac{2}{3} = \frac{19}{12} + \frac{2}{3} = \frac{19}{12} + \frac{8}{12} = \frac{19 + 8}{12} = \frac{27}{12},$$

which simplifies to $\frac{9}{4}$.

(d) The sum in parentheses is

$$\frac{5}{6} + \frac{2}{3} = \frac{5}{6} + \frac{4}{6} = \frac{9}{6} = \frac{3}{2}.$$

Then we have

$$\frac{3}{4} + \left(\frac{5}{6} + \frac{2}{3}\right) = \frac{3}{4} + \frac{3}{2} = \frac{3}{4} + \frac{6}{4} = \frac{9}{4}.$$

Parts (c) and (d) of Example 6.6 show that

$$\left(\frac{3}{4}+\frac{5}{6}\right)+\frac{2}{3}=\frac{3}{4}+\left(\frac{5}{6}+\frac{2}{3}\right),$$

since each side represents the rational number $\frac{9}{4}$. This result is a particular example of the associative property for the addition of rational numbers. The properties of the arithmetic operations on the rational numbers will be explored in Section 6.3.

Proper Fractions and Mixed Numbers

The sum of a natural number and a fraction is most often written as a **mixed number.** For example, $2+\frac{3}{4}$ is written $2\frac{3}{4}$ and is read "two and three-quarters." It is important to realize that it is the addition symbol, +, that is suppressed, since the common notation xy for multiplication might suggest, incorrectly, that $2\frac{3}{4}$ is $2\cdot\frac{3}{4}$. Thus, $2\frac{3}{4}=2+\frac{3}{4}$, not $\frac{6}{4}$.

A mixed number can always be rewritten in the standard form $\frac{a}{b}$ of a fraction. For example,

$$2\frac{3}{4}=2+\frac{3}{4}=\frac{8}{4}+\frac{3}{4}=\frac{11}{14}.$$

Mixed numbers and their equivalent forms as a fraction can be visualized this way:

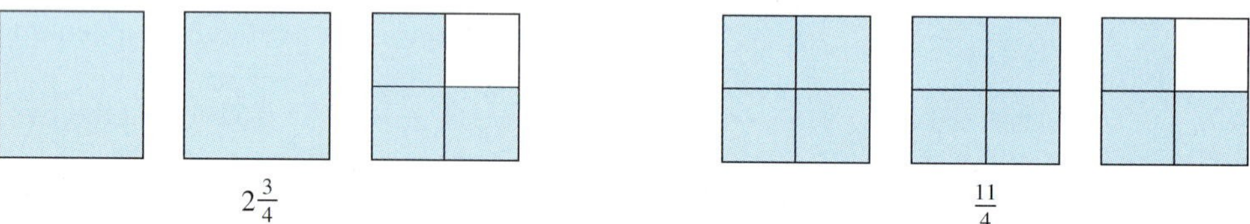
$2\frac{3}{4}$
$\frac{11}{4}$

Since $A\frac{b}{c}=\frac{Ac}{c}+\frac{b}{c}=\frac{Ac+b}{c}$, we see that it is an easy calculation to write a mixed number as a fraction: $A\frac{b}{c}=\frac{Ac+b}{c}$.

A fraction $\frac{a}{b}$ for which $0\le |a| < b$ is called a **proper fraction.** For example, $\frac{2}{3}$ is a proper fraction, but $\frac{3}{2}$, $\frac{-8}{5}$, and $\frac{6}{6}$ are not proper fractions. It is common, though not necessary, to rewrite fractions that are not proper as mixed numbers. For example, to express $\frac{439}{19}$ as a mixed number, we first use division with a remainder to find that $439 = 23\cdot 19 + 2$. Then we have

$$\frac{439}{19}=\frac{23\cdot 19+2}{19}=\frac{23}{1}+\frac{2}{19}=23+\frac{2}{19}=23\frac{2}{19}.$$

In mixed-number form, it is obvious that $23\frac{2}{19}$ is just slightly larger than 23; this fact was not evident in the original fraction form $\frac{439}{19}$. Nevertheless, it is perfectly acceptable

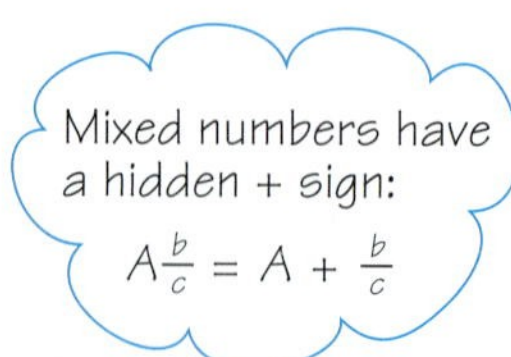
Mixed numbers have a hidden + sign:
$A\frac{b}{c} = A + \frac{b}{c}$

Just for Fun

The Sultan's Estate

When the sultan died, his will divided his possessions among his three sons. The sultan's eldest son was to receive one-half of his horses, his second son was to receive one-third of the horses, and his youngest son was to receive one-ninth of them. The sultan left 17 horses, and after struggling for some time trying to decide how to carry out the instructions of the will, the perplexed sons decided to turn to their wise uncle who had just ridden up on his own horse. The uncle, after a moment's reflection, said, "Let me contribute my own horse, making 18 altogether." The sons saw now that the eldest received $\frac{1}{2}$ of 18, or 9 horses. The middle son inherited 6 horses, since that is $\frac{1}{3}$ of 18, and the youngest son was given 2 horses, since $\frac{1}{9}$ of 18 is 2. Once the 9 + 6 + 2 = 17 horses were presented to the sons, the uncle returned home mounted on his own horse!

Cooperative Investigation

Fraction Magic in Pascal's Triangle

Pascal's triangle has several amazing fraction properties, one of which you will investigate in this activity. Work in small groups to check your work and discuss the patterns you discover.

Materials

Each group should have several photocopies of Pascal's triangle of numbers. As shown in the example below, the triangle should be arranged with vertical columns and extend at least through a dozen rows.

Directions

1. Choose any two entries in the second column of Pascal's triangle as the upper- and lower-left-hand corners of a rectangle. Let the upper-right corner of the rectangle be the value 1.
2. Find these three fractions, expressing each in simplest form:

   $\frac{c}{r}$, where c is the number of columns and r is the number of rows in the rectangle;

   $\frac{y}{z}$, where y and z are the last two numbers in the bottom row of the rectangle;

   $\frac{c}{s} + \frac{d}{t} + \cdots + \frac{h}{y} + \frac{1}{z}$, where $c, d, \ldots, h$, 1 are the numbers in the top row and $s, t, \ldots, y, z$ are the numbers in the bottom row of the rectangle.
3. Compare the three fractions, and describe any pattern you observe. Formulate a conjecture that describes your pattern, and give additional examples to support your conjecture.

Example

1											
1	1										
1	2	1									
1	3	3	1								
1	**4**	**6**	**4**	**1**							
1	5	10	10	5	1						
1	6	15	20	15	6	1					
1	7	21	35	35	21	7	1				
1	8	28	56	70	56	28	8	1			
1	**9**	**36**	**84**	**126**	126	84	36	9	1		
1	10	45	120	210	252	210	120	45	10	1	
1	11	55	165	330	462	462	330	165	55	11	1

The values $c = 4$ and $s = 9$ have been chosen from the second column, giving a rectangle with $c = 4$ columns and $r = 6$ rows. The last two entries in the bottom row are $y = 84$ and $z = 126$. The three fractions to be considered are

$$\frac{c}{r} = \frac{4}{6}$$

$$\frac{y}{z} = \frac{84}{126}$$

$$\frac{c}{s} + \frac{d}{t} + \frac{h}{y} + \frac{1}{z} = \frac{4}{9} + \frac{6}{36} + \frac{4}{84} + \frac{1}{126}$$

The completion of the example is left to you. You must still evaluate the sum, simplify all of the fractions, describe your observations, and formulate a conjecture.

to express rational numbers as "improper" fractions. In general, the fractional form $\frac{a}{b}$ is the more convenient form for arithmetic and algebra, and the mixed-number form is easiest to understand for practical applications. For example, it would be more common to buy $2\frac{1}{4}$ yards of material than to request $\frac{9}{4}$ yards.

EXAMPLE 6.7 WORKING WITH MIXED NUMBERS

(a) Give an improper fraction for $3\frac{17}{120}$.

(b) Give a mixed number for $\frac{355}{113}$.

(c) Give a mixed number for $\frac{-15}{4}$.

(d) Compute $2\frac{3}{4} + 4\frac{2}{5}$.

Solution

(a) $3\frac{17}{120} = \frac{3}{1} + \frac{17}{120} = \frac{3 \cdot 120 + 1 \cdot 17}{120} = \frac{360 + 17}{120} = \frac{377}{120}$.

This rational number was given by Claudius Ptolemy around A.D. 150 to approximate π, the ratio of the circumference of a circle to its diameter. It has better accuracy than $3\frac{1}{7}$, the value proposed by Archimedes in about 240 B.C.

(b) Using the division algorithm, we calculate that $355 = 3 \cdot 113 + 16$. Therefore,

$$\frac{355}{113} = \frac{3 \cdot 113 + 16}{113} = \frac{3}{1} + \frac{16}{113} = 3\frac{16}{113},$$

which corresponds to a point somewhat to the right of 3 on the number line. The value $\frac{355}{113}$ was used around A.D. 480 in China to approximate π; as a decimal, it is correct to six places!

(c) $\frac{-15}{4} = \frac{-(3 \cdot 4 + 3)}{4} = -\left(3 + \frac{3}{4}\right) = -3\frac{3}{4}$.

(d) $2\frac{3}{4} + 4\frac{2}{5} = 2 + 4 + \frac{3}{4} + \frac{2}{5} = 6 + \frac{15}{20} + \frac{8}{20} = 6 + \frac{23}{20} = 7\frac{3}{20}$.

Subtraction of Rational Numbers

Figure 6.11 shows how the take-away, measurement, and missing-addend conceptual models of the subtraction operation can be illustrated with colored regions, the number line, and fraction strips. In each case we see that $\frac{7}{6} - \frac{3}{6} = \frac{4}{6}$.

Subtraction of whole numbers and integers was defined on the basis of the missing-addend approach, which emphasizes that subtraction is the inverse operation to addition. Subtraction of rational numbers is defined in the same way.

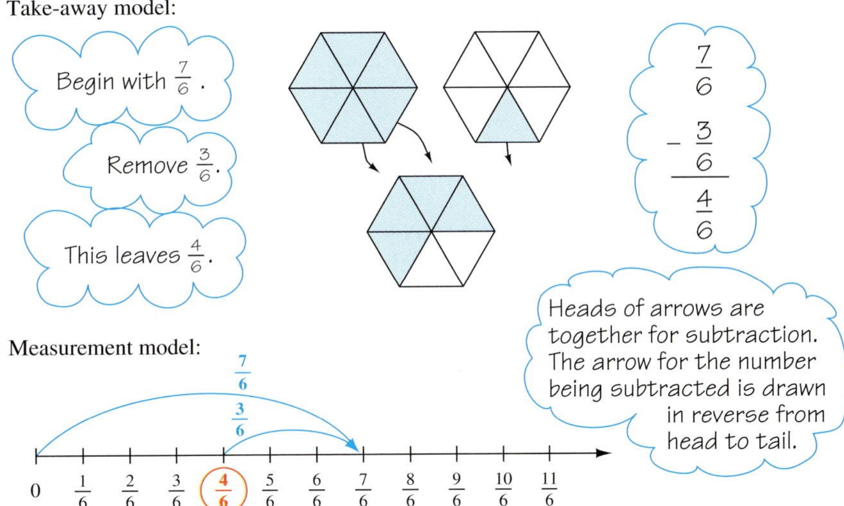

Begin with $\frac{7}{6}$.
Remove $\frac{3}{6}$.
This leaves $\frac{4}{6}$.

Heads of arrows are together for subtraction. The arrow for the number being subtracted is drawn in reverse from head to tail.

Missing-addend model:

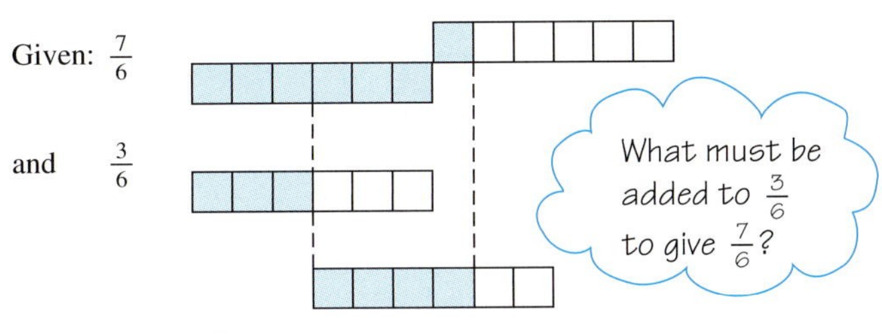

Given: $\frac{7}{6}$
and $\frac{3}{6}$
What must be added to $\frac{3}{6}$ to give $\frac{7}{6}$?

Since $\frac{3}{6} + \frac{4}{6} = \frac{7}{6}$, then $\frac{7}{6} - \frac{3}{6} = \frac{4}{6}$.

Figure 6.11
Models which show that $\frac{7}{6} - \frac{3}{6} = \frac{4}{6}$

DEFINITION *Subtraction of Rational Numbers*

Let $\frac{a}{b}$ and $\frac{c}{d}$ be rational numbers. Then $\frac{a}{b} - \frac{c}{d} = \frac{e}{f}$ if, and only if, $\frac{a}{b} = \frac{c}{d} + \frac{e}{f}$.

For rational numbers expressed by fractions $\frac{a}{b}$ and $\frac{c}{b}$ with the same denominator, the formula $\frac{a}{b} - \frac{c}{b} = \frac{a-c}{b}$ follows easily from the definition. For fractions with unlike denominators, it is necessary first to find a common denominator.

EXAMPLE 6.8 SUBTRACTING RATIONAL NUMBERS

Show, as you might to a student, the steps to compute these differences.

(a) $\frac{4}{5} - \frac{2}{3}$ **(b)** $\frac{103}{24} - \frac{-35}{16}$ **(c)** $4\frac{1}{4} - 2\frac{2}{3}$

Solution

(a) $\frac{4}{5} - \frac{2}{3} = \frac{4 \cdot 3}{5 \cdot 3} - \frac{5 \cdot 2}{5 \cdot 3} = \frac{12}{15} - \frac{10}{15} = \frac{12 - 10}{15} = \frac{2}{15}$.

(b) Since LCM(24, 16) = 48, the least common denominator, 48, can be used to give

$$\frac{103}{24} - \frac{-35}{16} = \frac{206}{48} - \frac{-105}{48} = \frac{206 - (-105)}{48} = \frac{311}{48}.$$

(c) $4\frac{1}{4} - 2\frac{2}{3} = \frac{17}{4} - \frac{8}{3} = \frac{17 \cdot 3}{4 \cdot 3} - \frac{4 \cdot 8}{4 \cdot 3} = \frac{51}{12} - \frac{32}{12} = \frac{51 - 32}{12} = \frac{19}{12} = 1\frac{7}{12}.$

Alternatively, subtraction of mixed numbers can follow the familiar regrouping algorithm:

$$\begin{array}{r} 4\frac{1}{4} \\ -2\frac{2}{3} \\ \hline \end{array} \;\rightarrow\; \begin{array}{r} 4\frac{3}{12} \\ -2\frac{8}{12} \\ \hline \end{array} \;\rightarrow\; \begin{array}{r} 3\frac{15}{12} \\ -2\frac{8}{12} \\ \hline 1\frac{7}{12} \end{array}$$

$4\frac{3}{12} = 3 + \frac{12}{12} + \frac{3}{12} = 3\frac{15}{12}$

This can also be visualized with the area model of fractions.

Highlight from History

Egyptian Fractions

The hieroglyphic numerals of ancient Egypt were described in Chapter 3. Fractions were likewise expressed with hieroglyphs. For example, the symbol ⬭ originally indicated $\frac{1}{320}$ of a bushel, but later it came to denote a fraction in general. Here are some hieroglyphic fractions and their modern equivalents:

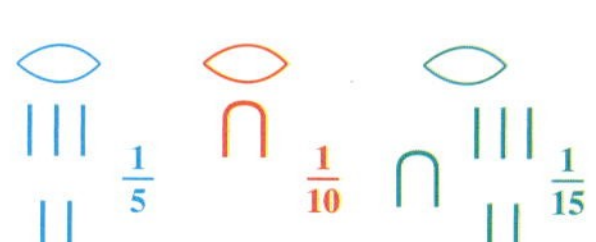

Some common fractions were denoted by special symbols:

With $\frac{2}{3}$ as an exception, Egyptian fractions were written as sums of fractions with numerator 1 and distinct denominators, though no summation sign appeared. For example, $\frac{2}{5}$ would be written

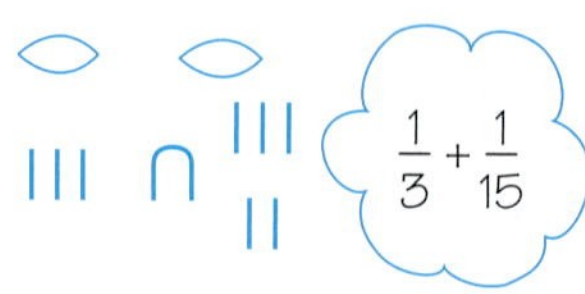

With more effort the Egyptians expressed $\frac{7}{29}$ as

$$\frac{1}{6} + \frac{1}{24} + \frac{1}{58} + \frac{1}{87} + \frac{1}{232}.$$

The Egyptians assumed every fraction could be written as a sum of distinct unit-numerator fractions, but no record has been found that justifies this assumption. In A.D. 1202, Leonardo of Pisa—otherwise known as Fibonacci—gave the first proof of this theorem: *Every fraction between 0 and 1 can be written as a sum of distinct fractions with numerator 1.*

Into the Classroom

Charlotte Jenkins Discusses Addition and Subtraction of Fractions

I use paper plate activities to introduce adding and subtracting fractions with different denominators. To find $\frac{5}{8} + \frac{1}{4}$, I give each student two paper plates. I ask students to draw lines to divide one plate into fourths and the other into eighths. Then I have them cut the plates to join $\frac{1}{4}$ of one to $\frac{5}{8}$ of the other, and ask them to describe the result. To find $\frac{5}{8} - \frac{1}{4}$, I give each student one paper plate. I ask students to draw lines to divide the plateinto eighths and shade five of the eighths. Then I have them cut $\frac{1}{4}$ of area and describe the shaded amount that is left. I have students repeat the activities to find other sums and differences.

SOURCE: Scott Foresman–Addison Wesley Middle School Math Course 1 Teacher's Edition, *p. 320, © 2002 Pearson Education, Inc. Reprinted with permission.*

Multiplication of Rational Numbers

In earlier chapters, the rectangular area model provided a useful visualization of multiplication for both the whole numbers and the integers. The area model also serves well to motivate the definition of multiplication of rational numbers.

In Figure 6.12(a), we see that $2 \cdot 3 = 6$, since shading a 2-by-3 rectangle covers exactly 6 units (that is, 1-by-1) squares. In (b), the vertical dashed lines divide the unit

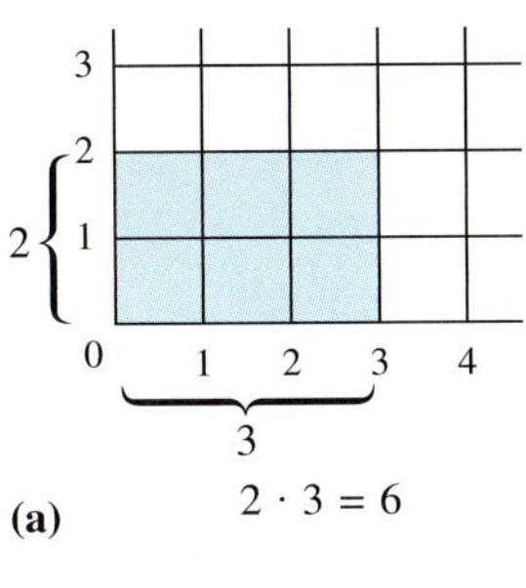

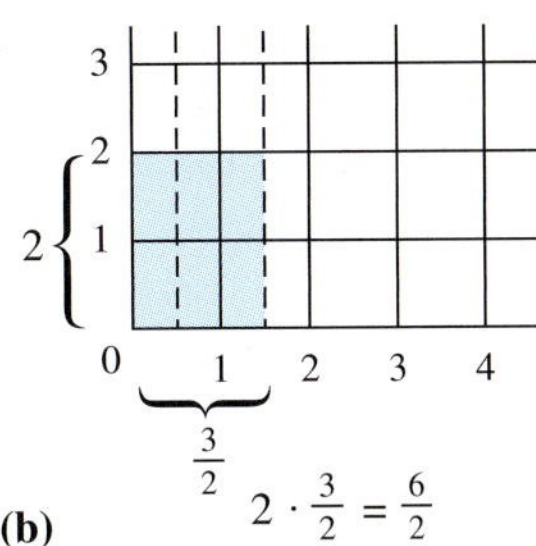

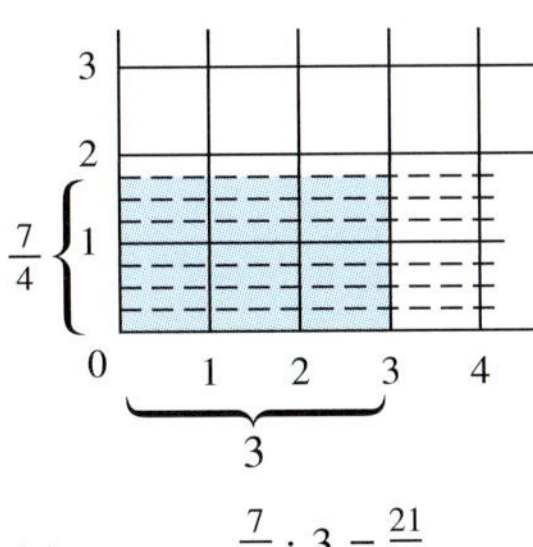

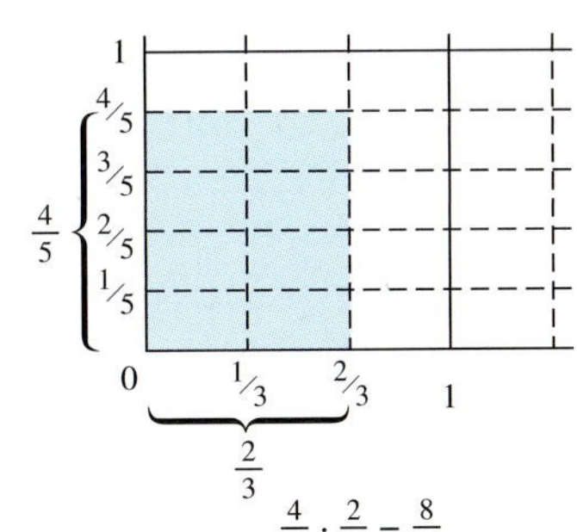

Figure 6.12
Extending the area model of multiplication to fractions

squares into congruent halves, and we see that $2 \cdot \frac{3}{2} = \frac{6}{2}$, since 6 half-units form the shaded rectangle. We also see that $2 \cdot \frac{3}{2} = \frac{3}{2} + \frac{3}{2} = \frac{6}{2}$, and so the concept of multiplication as repeated addition is retained when a whole number multiplies a rational number. Parts (c) and (d) of Figure 6.12 show that we can just as well multiply a rational number by either a whole number or another fraction. In general, if $\frac{a}{b}$ and $\frac{c}{d}$ are any two fractions, each unit square is divided into bd congruent rectangles and ac of the small rectangles are shaded. This shows that $\frac{ac}{bd}$ units are shaded. Thus, extending the area model of multiplication to fractions leads to the following definition.

DEFINITION *Multiplication of Rational Numbers*

Let $\frac{a}{b}$ and $\frac{c}{d}$ be rational numbers. Then their **product** is given by

$$\frac{a}{b} \cdot \frac{c}{d} = \frac{ac}{bd}.$$

Often a product, say $\frac{4}{5} \cdot \frac{2}{3}$, is read as four-fifths "of" two-thirds. The association between "of" and "times" is natural for multiplication by whole numbers and extends naturally to multiplication by rational numbers. For example, "I'll buy three *of* the half-gallon size bottles" is equivalent to buying $3 \cdot \frac{1}{2} = \frac{3}{2} = 1\frac{1}{2}$ gallons.

EXAMPLE 6.9 CALCULATING PRODUCTS OF RATIONAL NUMBERS

Practice your fraction multiplication skills with these examples. Look for ways to shorten your work, and express each answer in simplest form.

(a) $\frac{5}{8} \cdot \frac{2}{3}$ **(b)** $\frac{56}{88} \cdot \frac{-4}{7}$ **(c)** $3\frac{1}{7} \cdot 5\frac{1}{4}$

Solution It is useful to look for common factors in the numerator and denominator before doing any multiplications, applying the Fundamental Law of Fractions.

(a) $\frac{5}{8} \cdot \frac{2}{3} = \frac{5 \cdot 2}{8 \cdot 3} = \frac{5 \cdot 2}{2 \cdot 4 \cdot 3} = \frac{5}{4 \cdot 3} \cdot \frac{2}{2} = \frac{5}{12}.$

(b) $\frac{56}{88} \cdot \frac{-4}{7} = \frac{56 \cdot (-4)}{88 \cdot 7} = \frac{7 \cdot 8 \cdot (-4)}{8 \cdot 11 \cdot 7} = \frac{-4}{11} \cdot \frac{7 \cdot 8}{8 \cdot 7} = \frac{-4}{11}.$

(c) $3\frac{1}{7} \cdot 5\frac{1}{4} = \frac{22}{7} \cdot \frac{21}{4} = \frac{22 \cdot 21}{7 \cdot 4} = \frac{2 \cdot 11 \cdot 3 \cdot 7}{7 \cdot 2 \cdot 2} = \frac{11 \cdot 3}{2} \cdot \frac{2 \cdot 7}{7 \cdot 2} = \frac{33}{2} = 16\frac{1}{2}.$

EXAMPLE 6.10 COMPUTING THE AREA AND COST OF A CARPET

The hallway in the Bateks' house is a rectangle 4 feet wide and 20 feet long; that is, in yards it measures $\frac{4}{3}$ yards by $\frac{20}{3}$ yards. What is the area of the carpet in square yards?

School Book Page

Visualizing Fraction Multiplication

Multiplying by a Fraction

7-3

▶ **Lesson Link** In the last lesson, you learned to multiply fractions and mixed numbers by whole numbers. Now you'll multiply fractions and mixed numbers by fractions. ◀

You'll Learn ...

- to multiply a fraction by another fraction

... How It's Used

Consumers multiply fractions by fractions when calculating the price of an item with multiple discounts.

Explore Multiplying by a Fraction

Inner Sections

Materials: Colored pencils

Multiplying a Fraction by a Fraction

$\frac{2}{5} \times \frac{3}{4} = \frac{6}{20}$

- Draw a rectangle. Divide it vertically into equal sections. There should be as many sections as the denominator of the first number.
- Divide the rectangle horizontally into equal sections. There should be as many sections as the denominator of the second number.
- Color in a number of vertical strips equal to the numerator in the first number.
- Use a different color to shade a number of horizontal strips equal to the numerator in the second number.
- Describe the area where both colors overlap.

1. Model these problems.

a. $\frac{1}{2} \times \frac{1}{3}$ **b.** $\frac{1}{4} \times \frac{2}{5}$ **c.** $\frac{5}{6} \times \frac{2}{3}$

d. $\frac{2}{7} \times \frac{2}{7}$ **e.** $\frac{1}{5} \times \frac{3}{5}$ **f.** $\frac{1}{2} \times \frac{5}{8}$

2. Describe the pattern between the numerators in the problem and the numerator in the answer.

3. Describe the pattern between the denominators in the problem and the denominator in the answer.

4. Is your answer bigger or smaller than both of the fractions you started with? Why?

7-3 • Multiplying by a Fraction **375**

SOURCE: From Scott Foresman–Addison Wesley Middle School Math Course 1*, p. 375, by Randall I. Charles et al.*

Questions for the Teacher

1. The instructions for coloring the rectangle say, "Color in a number of vertical strips equal to the numerator in the first number." A student wonders what would happen if, when coloring the rectangle illustrating $\frac{2}{5} \times \frac{3}{4}$, the second and fourth strips were shaded yellow instead of the first and second strips. How would you respond?
2. What advantage is there to coloring the leftmost vertical strips and the lowermost horizontal strips?
3. On page 373 of the school book, the students discovered that $3 \times 1\frac{7}{12} = 4\frac{3}{4}$. Since the answer is bigger than both the fractions being multiplied, a student is perplexed about how to answer Question 4 of the School Book Page. How would you help the student out?

Mrs. Batek wants to know how much she will pay to buy carpet priced at $18 per square yard.

Solution Since $\frac{4}{3} \cdot \frac{20}{3} = \frac{80}{9} = 8\frac{8}{9}$, the area is $8\frac{8}{9}$ square yards, or nearly 9 square yards. This can be seen in the diagram below, which shows the hallway divided into six full square yards, six $\frac{1}{3}$-square-yard rectangular regions, and eight square regions that are each $\frac{1}{9}$ of a square yard. This gives the total area of $6 + \frac{6}{3} + \frac{8}{9} = 8\frac{8}{9}$ square yards. The cost of the carpet will be $8\frac{8}{9} \cdot 18 = \frac{80}{9} \cdot 18 = 160$ dollars.

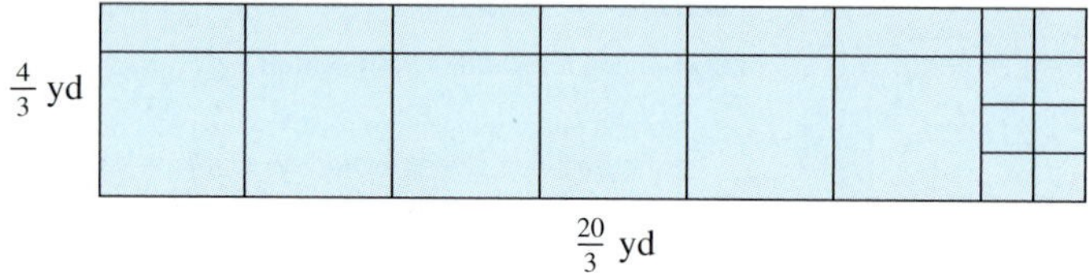

Division of Rational Numbers

As with division in whole numbers and integers, the missing-factor model is the basis for the definition of division of rational numbers.

> **DEFINITION** *Division of Rational Numbers*
>
> Let $\frac{a}{b}$ and $\frac{c}{d}$ be rational numbers, where $\frac{c}{d}$ is not zero. Then $\frac{a}{b} \div \frac{c}{d} = \frac{e}{f}$ if, and only if, $\frac{a}{b} = \frac{c}{d} \cdot \frac{e}{f}$.

This definition stresses that division is the inverse operation of multiplication. However, other conceptual models of division—as a measurement (that is, as a repeated subtraction or grouping) and as a partition (or sharing)—continue to be important. Indeed, it is these models that are best represented with manipulatives and diagrams, enabling the teacher to convey a deep understanding of what division of fractions really means and how it is used to solve problems. In the two examples that follow, the solution to meaningful problems is obtained by reasoning directly from the concept of division either as a grouping or as a sharing. Notice that no formula or computational rule is required to obtain the answer.

EXAMPLE 6.11 SEEDING A LAWN: ILLUSTRATING DIVISION OF RATIONAL NUMBERS WITH AN AREA MEASUREMENT MODEL

The new city park will have a $2\frac{1}{2}$-acre grass playfield. Grass seed can be purchased in large bags, each sufficient to seed $\frac{3}{4}$ of an acre. How many bags are needed? Will there be some grass seed left over to keep on hand for reseeding worn spots in the field?

Solution The answer will be given by determining the number of $\frac{3}{4}$-acre regions in the $2\frac{1}{2}$-acre field. That is, we must answer the division problem $2\frac{1}{2} \div \frac{3}{4}$. In the diagram below, the field is partitioned into four regions. Three of these regions each cover $\frac{3}{4}$ of an acre and therefore require a whole bag of seed. There is also a $\frac{1}{4}$-acre region that will use $\frac{1}{3}$ of a bag. Thus, a total of $2\frac{1}{2} \div \frac{3}{4} = 3\frac{1}{3}$ bags of seed are needed for the playfield. We conclude that four bags of seed should be ordered, which is enough for the initial seeding and leaves $\frac{2}{3}$ of a bag on hand for reseeding.

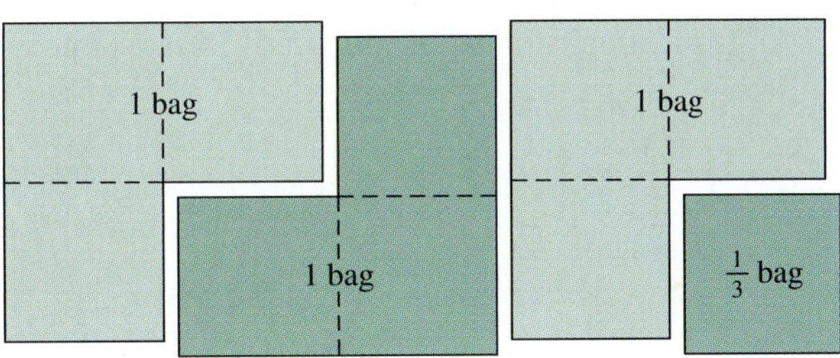

The next example uses the sharing, or partitive, model of division. This model works well when the divisor is a positive integer.

EXAMPLE 6.12 SHARING THE LOAD: ILLUSTRATING DIVISION OF RATIONAL NUMBERS WITH THE SHARING MODEL

Kayla, LeAnn, and Sarah are planning a weekend backpack trip and have a total of $7\frac{1}{2}$ pounds of freeze-dried food to carry. They want to share the load equally. How many pounds of food should each of them carry?

Solution First, represent the $7\frac{1}{2}$ pounds of food with the following diagram.

Next, partition the food into three equal parts.

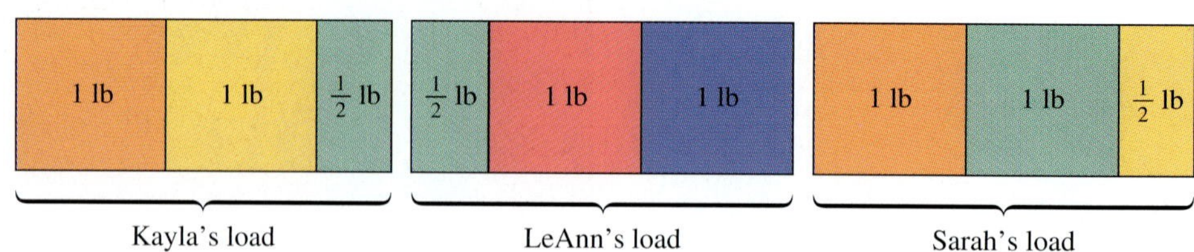

We see that each hiker should carry $2\frac{1}{2}$ pounds. The sharing model, as represented in the diagrams on the previous page, has shown us that $7\frac{1}{2} \div 3 = 2\frac{1}{2}$.

In the two examples just above, it was shown that $2\frac{1}{2} \div \frac{3}{4} = 3\frac{1}{3}$ and $7\frac{1}{2} \div 3 = 2\frac{1}{2}$. That is, $\frac{5}{2} \div \frac{3}{4} = \frac{10}{3}$ and $\frac{15}{2} \div \frac{3}{1} = \frac{5}{2}$. These examples suggest that a simple way to compute divided fractions is to "invert" the divisor by interchanging its numerator and denominator and then multiply. For example, to compute $\frac{5}{2} \div \frac{3}{4}$ we evaluate the multiplication $\frac{5}{2} \cdot \frac{4}{3} = \frac{20}{6} = \frac{10}{3}$. Similarly, $\frac{15}{2} \div \frac{3}{1}$ is found by the computation $\frac{15}{2} \div \frac{3}{1} = \frac{15}{2} \cdot \frac{1}{3} = \frac{15}{6} = \frac{5}{2}$. The **invert-and-multiply rule** for dividing fractions is true in general, since we have the following theorem.

THEOREM *The Invert-and-Multiply Algorithm for Division of Fractions and Rational Numbers*

$$\frac{a}{b} \div \frac{c}{d} = \frac{a}{b} \cdot \frac{d}{c}, \text{ where } \frac{c}{d} \neq 0.$$

Proof By the definition of division, we must verify that the product of $\frac{c}{d}$ and $\frac{a}{b} \cdot \frac{d}{c}$ is the dividend $\frac{a}{b}$. This is true, since $\frac{c}{d} \cdot \left(\frac{a}{b} \cdot \frac{d}{c}\right) = \frac{cad}{dbc} = \frac{acd}{bcd} = \frac{a}{b}$.

"Inverting" a nonzero divisor $\frac{c}{d}$ to obtain $\frac{d}{c}$ is also called **taking the reciprocal,** and $\frac{d}{c}$ is called the **reciprocal** of $\frac{c}{d}$.

DEFINITION *Reciprocal of a Rational Number*

The **reciprocal** of a nonzero rational number $\frac{c}{d}$ is $\frac{d}{c}$.

Since

$$\frac{c}{d} \cdot \frac{d}{c} = \frac{c \cdot d}{d \cdot c} = 1,$$

$\frac{d}{c}$ is also known as the **multiplicative inverse** of $\frac{c}{d}$. The importance of the multiplicative inverse is investigated in Section 6.3.

Sometimes the invert-and-multiply rule is mistaken for a definition of division of fractions. This is unfortunate; the algorithm does not relate directly to any conceptual model of division, and division becomes a meaningless and unpleasant process of following abstract and seemingly arbitrary rules.

The algorithm for division tells us that the rational numbers are closed under division, since every quotient $\frac{a}{b} \div \frac{c}{d}$ with $c \neq 0$ defines the unique fraction $\frac{ad}{bc}$. The whole numbers are not closed under division. For example, 3 is not divisible by 8. On the other hand, if we view 3 and 8 as rational numbers, then we see that

$$3 \div 8 = \frac{3}{1} \div \frac{8}{1} = \frac{3 \cdot 1}{1 \cdot 8} = \frac{3}{8}.$$

In general, if a and b are any integers, with $b \neq 0$, then interpreting a and b as *rational numbers* we find that

$$a \div b = \frac{a}{b}.$$

This shows that extending the set of integers to the rational numbers gives us a set of numbers for which division is closed.

EXAMPLE 6.13 DIVIDING RATIONAL NUMBERS

Show, as if you are working with a fifth grader, all of the steps needed to compute these division problems.

(a) $\frac{3}{4} \div \frac{1}{8}$ **(b)** $\frac{-7}{4} \div \frac{2}{3}$ **(c)** $3 \div \frac{4}{3}$

(d) $39 \div 13$ **(e)** $13 \div 39$ **(f)** $4\frac{1}{6} \div 2\frac{1}{3}$

Solution

(a) $\frac{3}{4} \div \frac{1}{8} = \frac{3}{4} \cdot \frac{8}{1} = \frac{24}{4} = 6.$ **(b)** $\frac{-7}{4} \div \frac{2}{3} = \frac{-7}{4} \cdot \frac{3}{2} = \frac{-21}{8}.$

(c) $3 \div \frac{4}{3} = \frac{3}{1} \cdot \frac{3}{4} = \frac{9}{4}.$ **(d)** $39 \div 13 = \frac{39}{1} \cdot \frac{1}{13} = \frac{39}{13} = 3.$

(e) $13 \div 39 = \frac{13}{1} \cdot \frac{1}{39} = \frac{13}{39} = \frac{1}{3}.$

(f) $4\frac{1}{6} \div 2\frac{1}{3} = \frac{25}{6} \div \frac{7}{3} = \frac{25}{6} \cdot \frac{3}{7} = \frac{25 \cdot 3}{6 \cdot 7} = \frac{25}{2 \cdot 7} = \frac{25}{14}.$

In the next example, the missing-factor model of division is used to solve a real-life problem. The reciprocal, being a multiplicative inverse, allows us to solve for the missing factor.

EXAMPLE 6.14 BOTTLING ROOT BEER: ILLUSTRATING DIVISION OF RATIONAL NUMBERS WITH THE MISSING-FACTOR MODEL

Ari is making homemade root beer. The recipe he followed nearly fills a 5-gallon glass jug, and he estimates it contains $4\frac{3}{4}$ gallons of root beer. He is now ready to bottle his root beer. How many $\frac{1}{2}$-gallon bottles can he fill?

Solution

Let x denote the number of half-gallon bottles required, where x will be allowed to be a fraction, since we expect that some bottle may be only partially filled. We must then solve the equation

$$x \cdot \frac{1}{2} = 4\frac{3}{4},$$

since this is the missing-factor problem that is equivalent to the division problem $4\frac{3}{4} \div \frac{1}{2}$.

Since

$$4\frac{3}{4} = \frac{19}{4} = \frac{19}{2} \cdot \frac{1}{2},$$

we see that the missing factor is $x = \frac{19}{2} = 9\frac{1}{2}$.

This is shown in the figure below. Ari will need 9 half-gallon bottles, and he will probably see if he can also find a quart bottle to use.

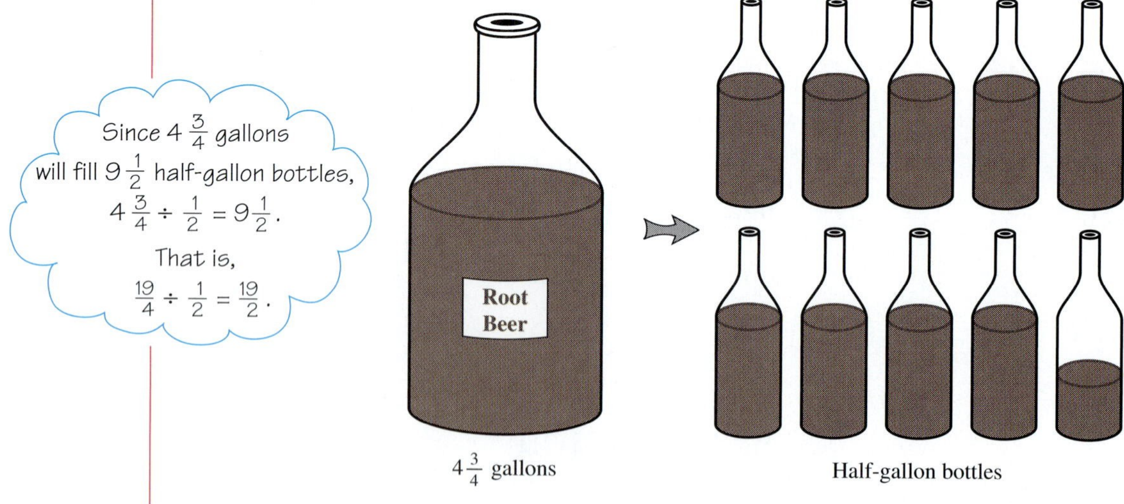

Since $4\frac{3}{4}$ gallons will fill $9\frac{1}{2}$ half-gallon bottles, $4\frac{3}{4} \div \frac{1}{2} = 9\frac{1}{2}$.
That is, $\frac{19}{4} \div \frac{1}{2} = \frac{19}{2}$.
Root Beer
$4\frac{3}{4}$ gallons
Half-gallon bottles

Using a Fraction Calculator

Some calculators, such as the Texas Instruments TI-15 *Explorer*™ calculator, are capable of performing "rational arithmetic." That is, fractions can be added, subtracted, multiplied, and divided, with the result displayed as a fraction. For example, suppose we wish to evaluate the expression

$$\left(2\frac{3}{4}+\frac{2}{3}\right) \div \left(\frac{4}{5}-\frac{1}{2}\right)$$

on the TI-15. The entry string
(2 Unit 3 n 4 d + 2 n 3 d) ÷ (4 n 5 d − 1 n 2 d) =
yields the fractional answer 410/36. Pressing Simp Enter expresses the answer in simplest form, 205/18. Pressing $U\frac{n}{d} \leftrightarrow \frac{n}{d}$ then converts the fraction to mixed-number form, $11\frac{7}{18}$. These instructions may need to be modified somewhat to work on a different model of calculator.

Did You Know?

Division of Fractions: Contrasting Teachers from the United States and China

The difference between the mathematical knowledge of the U.S. teachers and that of the Chinese teachers became more striking with the topic of division by fractions. The first contrast was presented in calculation. The interview question of this chapter asked the teachers to calculate $1\frac{3}{4} \div \frac{1}{2}$. The process of calculation revealed features of teachers' procedural knowledge and of their understanding of mathematics, as well as of their attitude toward the discipline.

In the two previous chapters* all teachers presented a sound procedural knowledge. This time, only 43% of the U.S. teachers succeeded in calculation and none of them showed an understanding of the rationale of the algorithm. Most of these teachers struggled. Many tended to confound the division-by-fractions algorithm with those for addition and subtraction or for multiplication. These teachers' procedural knowledge was not only weak in division with fractions, but also in other operations with fractions. Reporting that they were uncomfortable doing calculation with mixed numbers or improper fractions, these teachers also had very limited knowledge about the basic features of fractions.

All of the Chinese teachers succeeded in their calculations and many of them showed enthusiasm in doing the problem. These teachers were not satisfied by just calculating and getting an answer. They enjoyed presenting various ways of doing it—using decimals, using whole numbers, applying the three basic laws, etc. They went back and forth across subsets of numbers and across different operations, added and took off parentheses, and changed the order of operations. They did this with remarkable confidence and amazingly flexible skills. In addition, many teachers made comments on various calculation methods and evaluated them. Their way of "doing mathematics" showed significant conceptual understanding.

*on subtraction with regrouping and multidigit multiplication

SOURCE: Liping Ma, Knowing and Teaching Elementary Mathematics: Teachers' Understanding of Fundamental Mathematics in China and the United States *(Mahwah, NJ: Lawrence Erlbaum Associates, Publishers, 1999), pp. 80–81.*

Problem Set 6.2

Understanding Concepts

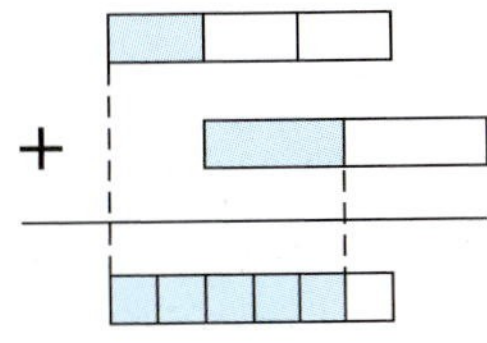

1. **(a)** What addition fact is illustrated by the following fraction-strip model?

   (b) Illustrate $\frac{1}{6}+\frac{1}{4}$ with the fraction-strip model.

(c) Illustrate $\frac{2}{3} + \frac{3}{4}$ with the fraction-strip model. (The sum will require two strips.)

2. Use the colored-region model to illustrate these sums, with the unit given by a circular disc.

(a) $\frac{2}{5} + \frac{6}{5}$ (b) $\frac{1}{4} + \frac{1}{2}$ (c) $\frac{2}{3} + \frac{1}{4}$

3. The points A, B, C, . . . , G, and H are equally spaced along the rational-number line.

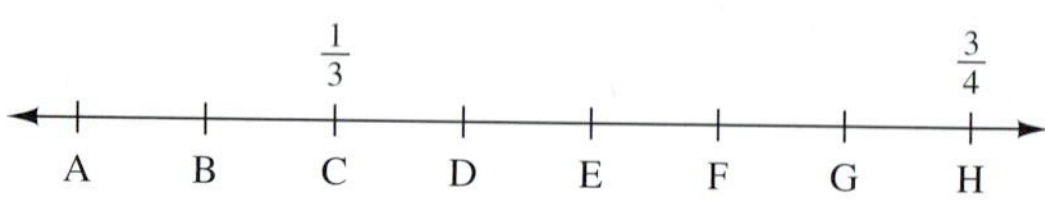

(a) What rational number corresponds to point G?

(b) What point corresponds to $\frac{1}{2}$?

(c) Are there lettered points that correspond to $\frac{1}{4}$? to $\frac{1}{6}$?

(d) Which lettered points are nearest to either side of $\frac{4}{7}$?

4. Represent each of these sums with a number-line diagram.

(a) $\frac{1}{8} + \frac{3}{8}$ (b) $\frac{1}{4} + \frac{5}{4}$ (c) $\frac{2}{3} + \frac{1}{2}$

5. Use the number-line model to illustrate these sums. Recall that negative fractions are represented by arrows that point to the left.

(a) $\frac{3}{4} + \frac{-2}{4}$ (b) $\frac{-3}{4} + \frac{2}{4}$ (c) $\frac{-3}{4} + \frac{-1}{4}$

6. Perform these additions. Express each answer in simplest form.

(a) $\frac{2}{7} + \frac{3}{7}$ (b) $\frac{6}{5} + \frac{4}{5}$ (c) $\frac{3}{8} + \frac{11}{24}$

(d) $\frac{6}{13} + \frac{2}{5}$ (e) $\frac{5}{12} + \frac{17}{20}$ (f) $\frac{6}{8} + \frac{-25}{100}$

(g) $\frac{-57}{100} + \frac{13}{10}$ (h) $\frac{213}{450} + \frac{12}{50}$

7. Express these fractions as mixed numbers.

(a) $\frac{9}{4}$ (b) $\frac{17}{3}$

(c) $\frac{111}{23}$ (d) $\frac{3571}{-100}$

8. Express these mixed numbers as fractions.

(a) $2\frac{3}{8}$ (b) $15\frac{2}{3}$

(c) $111\frac{2}{5}$ (d) $-10\frac{7}{9}$

9. (a) What subtraction fact is illustrated by this fraction-strip model?

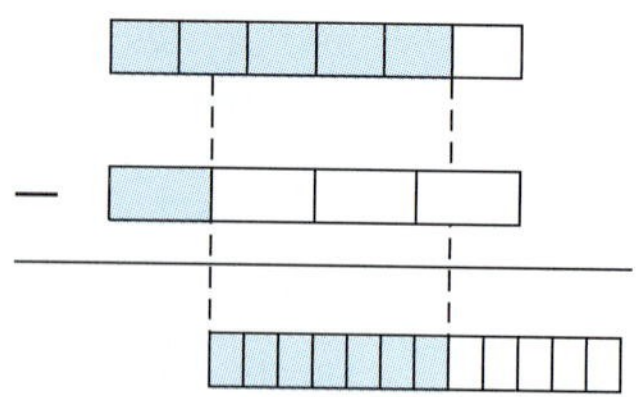

(b) Use the fraction-strip model to illustrate $\frac{2}{3} - \frac{1}{4}$.

10. (a) What subtraction fact is illustrated by this colored-region model?

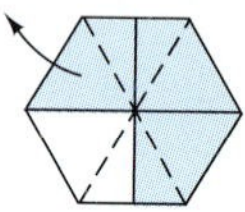
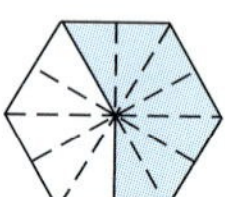

(b) Use a colored-region model to illustrate $\frac{2}{3} - \frac{1}{4}$.

11. (a) What subtraction fact is illustrated by this number-line model?

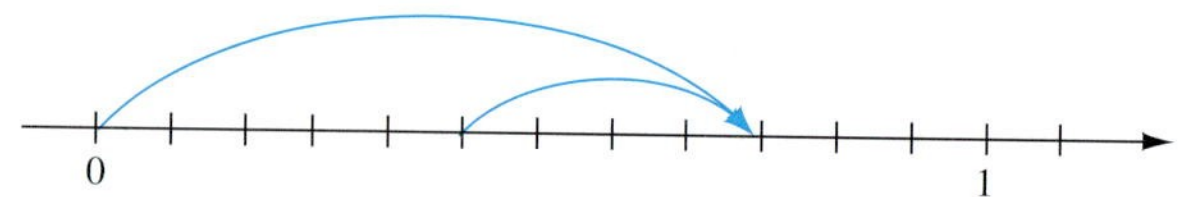

(b) Illustrate $\frac{2}{3} - \frac{1}{4}$ with the number-line model.

12. Compute these differences, expressing each answer in simplest form.

(a) $\frac{5}{8} - \frac{2}{8}$ (b) $\frac{3}{5} - \frac{2}{4}$ (c) $2\frac{2}{3} - 1\frac{1}{3}$

(d) $4\frac{1}{4} - 3\frac{1}{3}$ (e) $\frac{6}{8} - \frac{5}{12}$ (f) $\frac{1}{4} - \frac{14}{56}$

(g) $\frac{137}{214} - \frac{-1}{3}$ (h) $\frac{-23}{100} - \frac{198}{1000}$

13. What multiplication facts are illustrated by these rectangular area models?

(a)

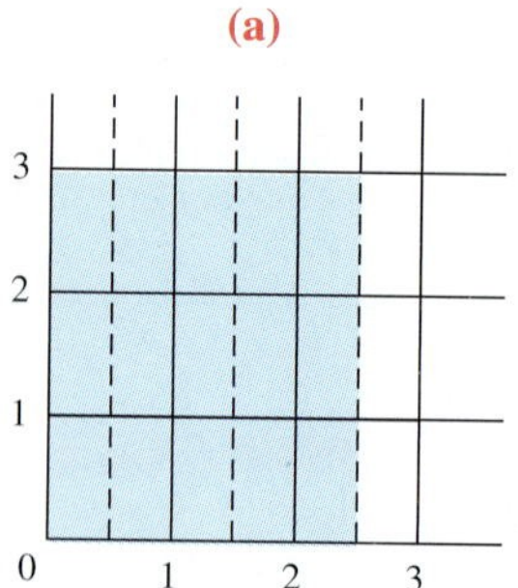

(b)

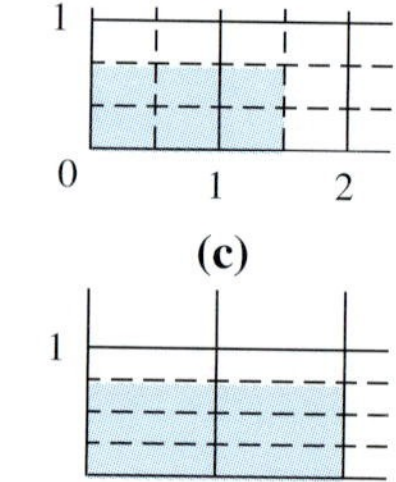

14. Illustrate these multiplications with the rectangular area model used in problem 13.

(a) $2 \times \frac{3}{5}$ (b) $\frac{3}{2} \times \frac{3}{4}$ (c) $1\frac{2}{3} \times 2\frac{1}{4}$

15. A rectangular plot of land is $2\frac{1}{4}$ miles wide and $3\frac{1}{2}$ miles long. What is the area of the plot, in square miles? Draw a sketch that verifies your answer.

16. Find the reciprocals of the following rational numbers.

(a) $\frac{3}{8}$ (b) $\frac{4}{3}$ (c) $2\frac{1}{4}$

(d) $\frac{1}{8}$ (e) 5 (f) 1

17. Compute these divisions, expressing each answer in simplest form.

(a) $\frac{2}{5} \div \frac{3}{4}$ (b) $\frac{6}{11} \div \frac{4}{3}$ (c) $\frac{100}{33} \div \frac{10}{3}$

(d) $2\frac{3}{8} \div 5$ (e) $3 \div 5\frac{1}{4}$ (f) $\frac{21}{25} \div \frac{7}{25}$

18. Compute the fraction with simplest form that is equivalent to the given expression.

(a) $\frac{2}{3} \cdot \left(\frac{3}{4} + \frac{9}{12}\right)$ (b) $\left(\frac{3}{5} - \frac{3}{10}\right) \div \frac{6}{5}$

(c) $\left(\frac{2}{5} \div \frac{4}{15}\right) \cdot \frac{2}{3}$

19. Set up and evaluate expressions to solve these map problems.

(a) Each inch on a map represents an actual distance of $2\frac{1}{2}$ miles. If the map shows Helmer as being $3\frac{3}{4}$ inches due east of Deary, how far apart are the two towns?

(b) A map shows that Spokane is 60 miles north of Colfax. A ruler shows that the towns are $3\frac{3}{4}$ inches apart on the map. How many actual miles are represented by each inch on the map?

20. An alternative, but equivalent, definition of rational-number inequality is the following:

$$\frac{a}{b} < \frac{c}{d} \text{ if, and only if, } \frac{c}{d} - \frac{a}{b} > 0.$$

Use the alternative definition to verify these inequalities.

(a) $\frac{2}{3} < \frac{3}{4}$ (b) $\frac{4}{5} < \frac{14}{17}$ (c) $\frac{19}{10} < \frac{99}{50}$

Teaching Concepts

The division of fractions is one of the most difficult concepts in the elementary school curriculum to teach. In each problem below, provide pictures and word descriptions that you believe would effectively convey the central ideas to a youngster. Some of the examples in the section can be used as models.

21. Sean has a job mowing grass for the city, using a riding mower. He can mow $\frac{5}{6}$ of an acre per hour. How long will he need to mow the 3-acre city park?

22. Angie, Bree, Corrine, Dot, and Elaine together picked $3\frac{3}{4}$ crates of strawberries. How many crates should each be allotted in order to distribute the berries evenly?

23. Gerry is making a pathway out of concrete stepping-stones. The path is 25 feet long, and each stone extends $1\frac{2}{3}$ of a foot along the path. By letting x denote the number of stones, Gerry knows that he needs to solve the equation $x \cdot 1\frac{2}{3} = 25$, but he isn't sure how to solve for x. Provide a careful explanation.

Thinking Critically

24. A bookworm is on page 1 of Volume 1 of a set of encyclopedias neatly arranged on a shelf. He (or she—how do you tell?) eats straight through to the last page of Volume 2. If the covers of each volume are $\frac{1}{8}$" thick and the pages are a total of $\frac{3}{4}$" thick in each of the volumes, how far does the bookworm travel?

25. Find the missing fractions in the Magic Fraction Squares below so that the entries in every row, column, and diagonal add to 1.

(a)

$\frac{1}{2}$	$\frac{1}{12}$	
$\frac{1}{4}$		

(b)

	$\frac{1}{3}$	$\frac{3}{5}$
		$\frac{2}{15}$

26. The positive rational numbers 3 and $1\frac{1}{2}$ are an interesting pair because their sum is equal to their product: $3 + \frac{3}{2} = \frac{6}{2} + \frac{3}{2} = \frac{9}{2}$ and $3 \cdot \frac{3}{2} = \frac{9}{2}$.

(a) Show that $3\frac{1}{2}$ and $1\frac{2}{5}$ have the same sum and product.

(b) Show that $2\frac{3}{5}$ and $1\frac{5}{8}$ have the same sum and product.

(c) If two positive rational numbers $\frac{a}{b}$ and $\frac{c}{d}$ have the same sum and product, what must be true of the sum of their reciprocals?

27. Three children had just cut their rectangular cake into three equal parts as shown to share, when a fourth friend joined them. Describe how to make one additional straight cut through the cake so all four can share the cake equally.

28. Start with any triangle, and let its area be the unit. Next, trisect each of its sides and join these points to the opposite vertices. This creates a dissection of the unit triangle, as shown in the figure below. Amazingly, the areas of these regions are precisely the fractions shown in the figure. For example, the inner lavender hexagon has area $\frac{1}{10}$, and the area of each green triangle is $\frac{1}{21}$.

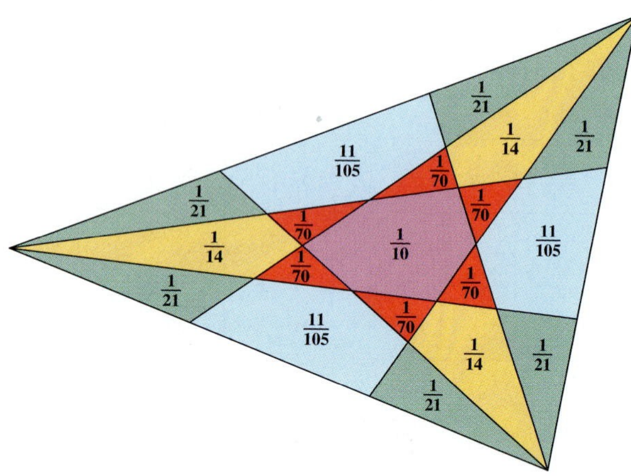

(a) Verify that the large triangle is a unit triangle, by summing the areas of the regions within the triangle.

(b) What is the area of the three-pointed star composed of the hexagon, the six red triangles, and the four yellow quadrilaterals?

(c) What is the area of the triangle composed of the hexagon and the three red triangles on every other side of the hexagon? (*Note:* Your answer can also be proved with the method suggested in problem 28 of Problem Set 6.1.)

29. Other Algorithms for Division. The invert-and-multiply algorithm, $\frac{a}{b} \div \frac{c}{d} = \frac{a}{b} \cdot \frac{d}{c}$, transforms a division of rational numbers into a multiplication. Here are two other useful algorithms:

Common denominators: $\frac{a}{b} \div \frac{c}{b} = \frac{a}{c}$.

Divide numerators and denominators:

$$\frac{a}{b} \div \frac{c}{d} = \frac{a \div c}{c \div d}.$$

Thus,

$$\frac{3}{19} \div \frac{15}{19} = \frac{3}{15} = \frac{1}{5} \text{ and } \frac{24}{35} \div \frac{6}{5} = \frac{24 \div 6}{35 \div 5} = \frac{4}{7}.$$

Choose one of the two new algorithms to perform these calculations.

(a) $\frac{7}{12} \div \frac{11}{12}$ **(b)** $\frac{4}{15} \div \frac{2}{3}$ **(c)** $\frac{19}{210} \div \frac{19}{70}$

30. With the exception of $\frac{2}{3}$, which was given the hieroglyph [hieroglyph], the ancient Egyptians attempted to express all rational numbers as a sum of different fractions, each with 1 as the numerator.

(a) Verify that $\frac{23}{25} = \frac{1}{2} + \frac{1}{3} + \frac{1}{15} + \frac{1}{50}$.

(b) Verify that $\frac{7}{29} = \frac{1}{6} + \frac{1}{24} + \frac{1}{58} + \frac{1}{87} + \frac{1}{232}$.

(c) Verify that $\frac{7}{29} = \frac{1}{5} + \frac{1}{29} + \frac{1}{145}$.

(d) If [symbol] represents $1\frac{1}{2}$, and [symbol] is interpreted as "one over," does the symbol [symbol] seem reasonable for $\frac{2}{3}$?

31. The Rhind (or Ahmes) papyrus of about 1650 B.C. opens with the words "Directions for Obtaining the Knowledge of All Dark Things." The papyrus solves 85 problems and includes a table expressing the fractions with numerator 2 and an odd denominator 5 through 101 as a sum of unit (numerator of 1) fractions. For example, the first two entries in the table are

$$\frac{2}{5} = \frac{1}{3} + \frac{1}{15} \text{ and } \frac{2}{7} = \frac{1}{4} + \frac{1}{28}.$$

(a) Verify the next entry of the table: $\frac{2}{9} = \frac{1}{5} + \frac{1}{45}$.

(b) Verify the general formula

$$\frac{2}{2n-1} = \frac{1}{n} + \frac{1}{n(2n-1)}.$$

(c) Let $n = 51$ in the formula of part (b) to find the last entry in the table.

32. Solve this problem found in the Rhind papyrus: "A quantity and its $\frac{1}{7}$th added together become 19. What is the quantity?"

Using a Calculator

33. DIFFY with Fractions. We previously played the DIFFY game, described in Section 2.3, with whole numbers. Here are the beginning lines of DIFFY where the entries are fractions. A new line is formed by subtracting the smaller fraction from the larger. The DIFFY graphing-calculator program (see Appendix C) can also be used for fraction entries.

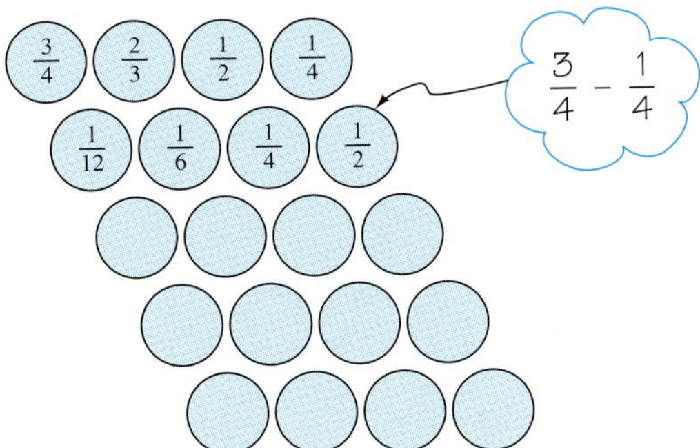

(a) Fill in additional lines of the DIFFY array started above. Does it terminate?

(b) Try fraction DIFFY with these fractions in your first row: $\frac{2}{7}, \frac{4}{5}, \frac{3}{2}$, and $\frac{5}{6}$.

(c) Suppose you know that DIFFY with whole-number entries always terminates with 0, 0, 0, and 0. Does it necessarily follow that DIFFY with fractions must terminate? Explain your reasoning carefully.

34. DIVVY. The process called DIVVY is like DIFFY (see problem 33) except that the larger fraction is *divided* by the smaller. The first few rows of a sample DIVVY are shown.

(a) Continue to fill in additional rows, using a calculator if you like.

(b) Try DIVVY with $\frac{2}{7}, \frac{4}{5}, \frac{3}{2}$, and $\frac{5}{6}$ in the first row.

Don't let complicated fractions put you off! Things should get better if you persist.

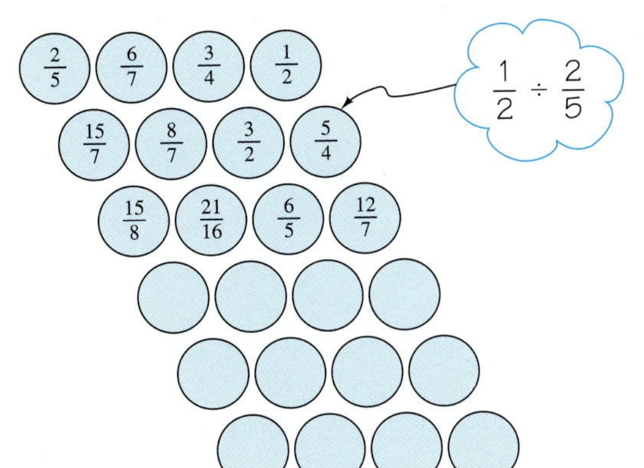

Making Connections

35. At a certain university, a student's senior thesis is acceptable if at least $\frac{3}{4}$ of the student's committee votes in its favor. What is the smallest number of favorable votes needed to accept a thesis if the committee has 3 members? 4 members? 5 members? 6 members? 7 members? 8 members?

36. A sign on a roll-end of canvas says that the canvas contains 42 square yards. The width of the canvas, which is easily measured without unrolling, is 14 feet (or $4\frac{2}{3}$ yards). What is the length of the piece of canvas, in yards?

37. Tongue-and-groove decking boards are each $2\frac{1}{4}$ inches wide. How many boards must be placed side by side to build a deck 14 feet in width?

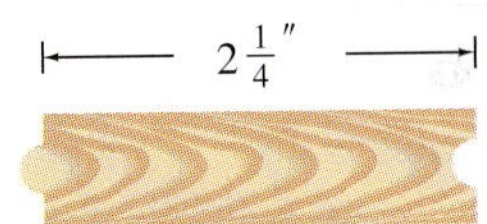

38. Six bows can be made from $1\frac{1}{2}$ yards of ribbon. How many bows can be made from $5\frac{3}{4}$ yards of ribbon?

39. Andre has 35 yards of material available to make aprons. Each apron requires $\frac{3}{4}$ yard. How many aprons can Andre make?

40. Gisela paid \$28 for a skirt that was "$\frac{1}{3}$ off." What was the original price of the skirt?

41. A soup recipe calls for $2\frac{3}{4}$ cups of chicken broth and will make enough to serve 8 people. How much broth is required if the recipe is modified to serve 6 people?

Communicating

42. For each fraction operation below, make up a realistic word problem whose solution requires the computation shown. Try to create an interesting and original situation.

(a) $\frac{19}{32} + \frac{1}{4}$ **(b)** $3\frac{1}{4} - 1\frac{1}{16}$

(c) $\frac{4}{5} \times \frac{7}{8}$ **(d)** $\frac{9}{10} \div \frac{3}{5}$

43. Respond to a student who asks, "Should fractions always be written in simplest form, or are there situations where it would be better not to simplify?"

From State Student Assessments

44. (Kentucky, Grade 4)
Mowing the Yard. The green shaded area of the picture below shows what part of the yard Jessie mowed in 30 minutes.

She wonders about how long it takes her to mow the whole yard. Write a note to Jessie

A. telling her about how long it takes to mow her whole yard and

B. explaining to her how you found your answer. Be sure to include a drawing of the yard in your explanation.

45. (Delaware, Grade 5)
Four children want to share seven brownies. They can cut the brownies to make the shares even. How many brownies should each child get? Please explain with words and/or pictures how you got your answer.

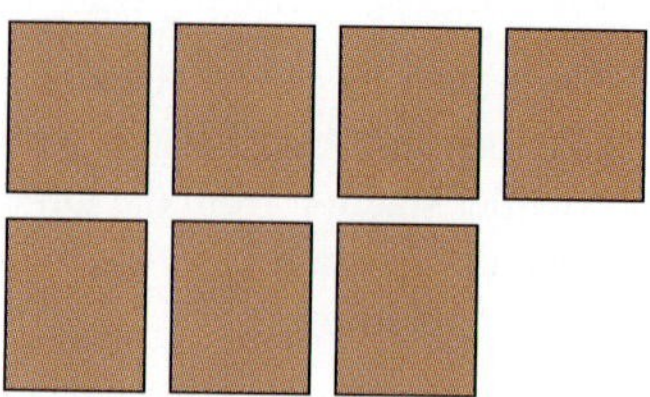

For Review

46. Represent these rational numbers by fractions in simplest form.

(a) $\frac{168}{48}$ (b) $\frac{945}{3780}$

47. Arrange these rational numbers in increasing order.

$$\frac{31}{90}, \frac{1}{3}, \frac{19}{60}, \text{ and } \frac{4}{13}.$$

48. Find the least common denominator of $\frac{4}{9}, \frac{5}{12}$, and $\frac{4}{15}$, and then find the sum of the three numbers.

6.3 THE RATIONAL-NUMBER SYSTEM

This section explores the properties of the rational numbers. Many properties will be familiar, since the integers have the same properties. However, we will also discover some important new properties of rational numbers that have no counterpart in the integers. This section also gives techniques for estimation and computation and presents additional examples of the application of rational numbers to the solution of practical problems.

Properties of Addition and Subtraction

To add two rational numbers, the definition introduced in the preceding section tells us to represent the rational numbers by fractions with a common denominator and then add the two numerators. For example, to add $\frac{5}{6}$ and $\frac{3}{10}$ we could use the common denominator 30; thus,

$$\frac{5}{6} + \frac{3}{10} = \frac{25}{30} + \frac{9}{30} = \frac{34}{30}.$$

We could also have used the common denominator 60 to find

$$\frac{5}{6} + \frac{3}{10} = \frac{50}{60} + \frac{18}{60} = \frac{68}{60}.$$

The two answers, namely, $\frac{34}{30}$ and $\frac{68}{60}$, are different fractions. However, they are equivalent fractions, and both represent the *same* rational number, $\frac{17}{15}$, when expressed by a fraction in simplest form. More generally, *any* two rational numbers have a unique rational number that is their sum. That is, the rational numbers are closed under addition.

It is also straightforward to check that addition in the rational numbers is commutative and associative. For example, $\frac{5}{6} + \frac{3}{10} = \frac{3}{10} + \frac{5}{6}$ and $\frac{3}{4} + \left(\frac{-1}{3} + \frac{2}{5}\right) = \left(\frac{3}{4} + \frac{-1}{3}\right) + \frac{2}{5}$. Similarly, 0 is the additive identity. For example, $\frac{7}{9} + 0 = \frac{7}{9}$, since $0 = \frac{0}{9}$.

The rational numbers share one more property with the integers, namely, the existence of **negatives,** or **additive inverses.** We have the following definition.

DEFINITION *Negative or Additive Inverse*

Let $\frac{a}{b}$ be a rational number. Its **negative,** or **additive inverse,** written $-\frac{a}{b}$, is the rational number $\frac{-a}{b}$.

For example, $-\frac{4}{7} = \frac{-4}{7}$. This is the additive inverse of $\frac{4}{7}$, since we see that

$$\frac{4}{7} + \left(-\frac{4}{7}\right) = \frac{4}{7} + \frac{-4}{7} = \frac{4 + (-4)}{7} = \frac{0}{7} = 0.$$

As another example, $-\left(-\frac{3}{4}\right) = -\left(\frac{-3}{4}\right) = \frac{-(-3)}{4} = \frac{3}{4}$, which illustrates the general property $-\left(-\frac{a}{b}\right) = \frac{a}{b}$.

A negative is also called an **opposite.** This term describes how a rational number and its negative are positioned on the number line: $-\frac{a}{b}$ is on the opposite side of 0 from $\frac{a}{b}$.

The properties of addition on the rational numbers are listed in the following theorem.

THEOREM *Properties of Addition of Rational Numbers*

Let $\frac{a}{b}$, $\frac{c}{d}$, and $\frac{e}{f}$ be rational numbers. The following properties hold.

Closure Property	$\frac{a}{b} + \frac{c}{d}$ is a rational number.
Commutative Property	$\frac{a}{b} + \frac{c}{d} = \frac{c}{d} + \frac{a}{b}$
Associative Property	$\left(\frac{a}{b} + \frac{c}{d}\right) + \frac{e}{f} = \frac{a}{b} + \left(\frac{c}{d} + \frac{e}{f}\right)$
Zero is an Additive Identity	$\frac{a}{b} + 0 = \frac{a}{b}$
Existence of Additive Inverses	$\frac{a}{b} + \left(-\frac{a}{b}\right) = 0$, where $-\frac{a}{b} = \frac{-a}{b}$

In the integers we discovered that subtraction was equivalent to the addition of the negative. The same result holds for the rational numbers. Since the rational numbers are closed under addition and every rational number has a negative, this means that the rational numbers are closed under subtraction.

THEOREM *Formulas for Subtraction of Rational Numbers*

Let $\frac{a}{b}$ and $\frac{c}{d}$ be rational numbers. Then $\frac{a}{b} - \frac{c}{d} = \frac{a}{b} + \left(-\frac{c}{d}\right) = \frac{ad - bc}{bd}$.

Subtraction is neither commutative nor associative, as examples such as $\frac{1}{2} - \frac{1}{4} \neq \frac{1}{4} - \frac{1}{2}$ and $1 - \left(\frac{1}{2} - \frac{1}{4}\right) \neq \left(1 - \frac{1}{2}\right) - \frac{1}{4}$ show. This means that subtraction requires that careful attention be given to the order of the terms and the placement of parentheses.

EXAMPLE 6.15 SUBTRACTING RATIONAL NUMBERS

Compute the following differences.

(a) $\frac{3}{4} - \frac{7}{6}$ **(b)** $\frac{2}{3} - \frac{-9}{8}$ **(c)** $\left(-2\frac{1}{4}\right) - \left(4\frac{2}{3}\right)$

Solution

(a) $\frac{3}{4} - \frac{7}{6} = \frac{3 \cdot 6}{4 \cdot 6} - \frac{4 \cdot 7}{4 \cdot 6} = \frac{18 - 28}{24} = \frac{-10}{24} = \frac{-5}{12} = -\frac{5}{12}$.

(b) $\frac{2}{3} - \frac{-9}{8} = \frac{2}{3} + \frac{9}{8} = \frac{2 \cdot 8}{3 \cdot 8} + \frac{3 \cdot 9}{3 \cdot 8} = \frac{16}{24} + \frac{27}{24} = \frac{43}{24}$.

(c) $\left(-2\frac{1}{4}\right) - \left(4\frac{2}{3}\right) = (-2 - 4) - \left(\frac{1}{4} + \frac{2}{3}\right) = -6 - \left(\frac{3}{12} + \frac{8}{12}\right) = -6\frac{11}{12}$.

Properties of Multiplication and Division

Multiplication of rational numbers includes all of the properties of multiplication for the integers. For example, let's investigate the distributive property of multiplication over addition by considering a specific case:

$$\frac{2}{5} \cdot \left(\frac{3}{4} + \frac{7}{8}\right) = \frac{2}{5} \cdot \left(\frac{6}{8} + \frac{7}{8}\right) = \frac{2}{5} \cdot \frac{13}{8} = \frac{26}{40}.$$

Add first, then multiply.

$$\frac{2}{5} \cdot \frac{3}{4} + \frac{2}{5} \cdot \frac{7}{8} = \frac{6}{20} + \frac{14}{40} = \frac{12}{40} + \frac{14}{40} = \frac{26}{40}.$$

Multiply first, then add.

These computations show that multiplication by $\frac{2}{5}$ distributes over the sum $\frac{3}{4} + \frac{7}{8}$. A similar calculation proves the general distributive property $\frac{a}{b} \cdot \left(\frac{c}{d} + \frac{e}{f}\right) = \frac{a}{b} \cdot \frac{c}{d} + \frac{a}{b} \cdot \frac{e}{f}$.

There is one important new property of multiplication of rational numbers that is *not* true for the integers: the existence of multiplicative inverses. For example, the nonzero rational number $\frac{5}{8}$ has the multiplicative inverse $\frac{8}{5}$, since

$$\frac{5}{8} \cdot \frac{8}{5} = 1.$$

This property does not hold in the integers. For example, since there is no integer m for which $2 \cdot m = 1$, the integer 2 does not have a multiplicative inverse in the set of integers.

THEOREM *Properties of Multiplication of Rational Numbers*

Let $\frac{a}{b}$, $\frac{c}{d}$, and $\frac{e}{f}$ be rational numbers. The following properties hold.

Closure Property $\frac{a}{b} \cdot \frac{c}{d}$ is a rational number.

Commutative Property $\frac{a}{b} \cdot \frac{c}{d} = \frac{c}{d} \cdot \frac{a}{b}$

Associative Property $\left(\frac{a}{b} \cdot \frac{c}{d}\right) \cdot \frac{e}{f} = \frac{a}{b} \cdot \left(\frac{c}{d} \cdot \frac{e}{f}\right)$

Distributive Property of Multiplication over Addition and Subtraction

$$\frac{a}{b} \cdot \left(\frac{c}{d} + \frac{e}{f}\right) = \frac{a}{b} \cdot \frac{c}{d} + \frac{a}{b} \cdot \frac{e}{f} \text{ and } \frac{a}{b} \cdot \left(\frac{c}{d} - \frac{e}{f}\right) = \frac{a}{b} \cdot \frac{c}{d} - \frac{a}{b} \cdot \frac{e}{f}$$

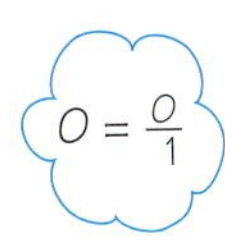

Multiplication by Zero $0 \cdot \frac{a}{b} = 0$

One is a Multiplicative Identity $1 \cdot \frac{a}{b} = \frac{a}{b}$

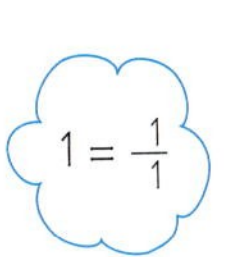

Existence of Multiplicative Inverse If $\frac{a}{b} \neq 0$, then there is a unique rational number, namely, $\frac{b}{a}$, for which $\frac{a}{b} \cdot \frac{b}{a} = 1$.

EXAMPLE 6.16 SOLVING AN EQUATION WITH THE MULTIPLICATIVE INVERSE

Consuela paid \$36 for a pair of shoes at the "one-fourth off" sale. What was the price of the shoes before the sale?

Solution

Let x be the original price of the shoes, in dollars. Consuela paid $\frac{3}{4}$ of this price, so $\frac{3}{4} \cdot x = 36$. To solve this equation for the unknown x, multiply both sides by $\frac{4}{3}$ to get

$$\frac{4}{3} \cdot \frac{3}{4} \cdot x = \frac{4}{3} \cdot 36.$$

But $\frac{4}{3} \cdot \frac{3}{4} = 1$ by the multiplicative inverse property, so

$$1 \cdot x = \frac{4 \cdot 36}{3} = 48.$$

Since 1 is a multiplicative identity we find that $1 \cdot x = x = 48$. Therefore, the original price of the pair of shoes was \$48.

In the last section, we discovered that the rational numbers are closed under division, since any division $\frac{a}{b} \div \frac{c}{d}$ (where $c \neq 0$) is given by the rational number $\frac{a \cdot d}{b \cdot c}$. Nevertheless, division in the rational numbers is neither commutative nor associative, which means it is important to place terms in their proper order and to use parentheses to avoid ambiguities. For example, $\frac{1}{3} \div \frac{1}{2} = \frac{2}{3}$, but $\frac{1}{2} \div \frac{1}{3} = \frac{3}{2}$, which shows that division is not commutative. Since division is not associative, an expression such as $2/3/4$ requires us to place parentheses to know if we wish to compute $(2/3)/4 = 2/12 = 1/6$ or $2/(3/4) = 8/3$.

Properties of the Order Relation

The order relation has many useful properties that are not difficult to prove.

THEOREM *Properties of the Order Relation on the Rational Numbers*

Let $\frac{a}{b}$, $\frac{c}{d}$, and $\frac{e}{f}$ be rational numbers.

Transitive Property

$$\text{If } \frac{a}{b} < \frac{c}{d} \text{ and } \frac{c}{d} < \frac{e}{f}, \text{ then } \frac{a}{b} < \frac{e}{f}.$$

Addition Property

$$\text{If } \frac{a}{b} < \frac{c}{d}, \text{ then } \frac{a}{b} + \frac{e}{f} < \frac{c}{d} + \frac{e}{f}.$$

Multiplication Property

$$\text{If } \frac{a}{b} < \frac{c}{d} \text{ and } \frac{e}{f} > 0, \text{ then } \frac{a}{b} \cdot \frac{e}{f} < \frac{c}{d} \cdot \frac{e}{f}.$$

$$\text{If } \frac{a}{b} < \frac{c}{d} \text{ and } \frac{e}{f} < 0, \text{ then } \frac{a}{b} \cdot \frac{e}{f} > \frac{c}{d} \cdot \frac{e}{f}.$$

Trichotomy Property Exactly one of the following holds:

$$\frac{a}{b} < \frac{c}{d}, \frac{a}{b} = \frac{c}{d}, \text{ or } \frac{a}{b} > \frac{c}{d}.$$

The Density Property of Rational Numbers

By the definition of inequality, we know that

$$\frac{1}{2} < \frac{2}{3}.$$

Alternatively, using 6 as a common denominator, we have

$$\frac{3}{6} < \frac{4}{6},$$

and with 12 as a common denominator, we have

$$\frac{6}{12} < \frac{8}{12}.$$

In the last form, we see that $\frac{7}{12}$ is a rational number that is between $\frac{6}{12}$ and $\frac{8}{12}$. That is,

$$\frac{1}{2} < \frac{7}{12} < \frac{2}{3},$$

as shown in Figure 6.13.

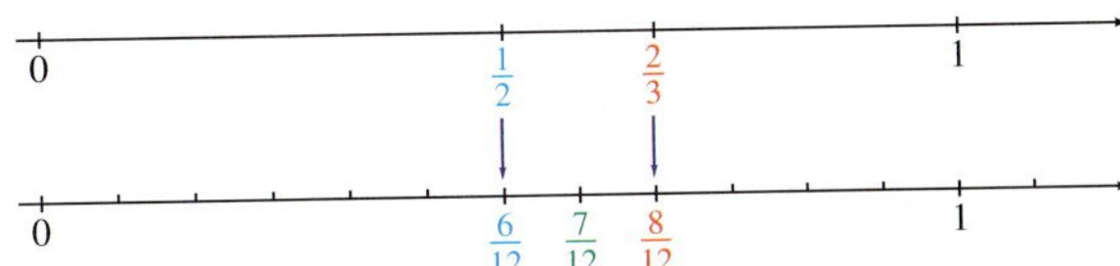

Figure 6.13

The rational number $\frac{7}{12}$ is between $\frac{1}{2}$ and $\frac{2}{3}$

The idea used to find a rational number that is between $\frac{1}{2}$ and $\frac{2}{3}$ can be extended to show that between *any* two rational numbers there is some other rational number. This interesting fact is called the **density property** of the rational numbers. The analogous property does not hold in the integers. For example, there is no integer between 1 and 2.

THEOREM *The Density Property of Rational Numbers*

Let $\frac{a}{b}$ and $\frac{c}{d}$ be any two rational numbers, with $\frac{a}{b} < \frac{c}{d}$. Then there is a rational number $\frac{e}{f}$ between $\frac{a}{b}$ and $\frac{c}{d}$; that is, $\frac{a}{b} < \frac{e}{f} < \frac{c}{d}$.

EXAMPLE 6.17 FINDING RATIONAL NUMBERS BETWEEN TWO RATIONAL NUMBERS

Find a rational number between the two given fractions.

(a) $\frac{2}{3}$ and $\frac{3}{4}$ **(b)** $\frac{5}{12}$ and $\frac{3}{8}$

Solution

(a) Using $2 \cdot 3 \cdot 4 = 24$ as a common denominator, we have $\frac{2}{3} = \frac{16}{24}$ and $\frac{3}{4} = \frac{18}{24}$. Since $\frac{16}{24} < \frac{17}{24} < \frac{18}{24}$, this shows that $\frac{17}{24}$ is one answer.

(b) If we use 48 as a common denominator, we have $\frac{5}{12} = \frac{20}{48}$ and $\frac{3}{8} = \frac{18}{48}$. Thus $\frac{19}{48}$ is one answer. Alternatively, we could use 480 as a common denominator, writing $\frac{5}{12} = \frac{200}{480}$ and $\frac{3}{8} = \frac{180}{480}$. This makes it clear that $\frac{181}{480}, \frac{182}{480}, \ldots, \frac{199}{480}$ are all rational numbers between $\frac{3}{8}$ and $\frac{5}{12}$. Using a larger common denominator allows us to identify even more rational numbers between the two given rationals.

Computations with Rational Numbers

To work confidently with rational numbers, it is important to develop skills in estimation, rounding, mental arithmetic, efficient paper-and-pencil computation, and the use of the calculator.

Estimations

In many applications, the exact fractional value can be rounded off to the nearest integer value; if more precision is required, values can be rounded to the nearest half, third, or quarter.

EXAMPLE 6.18 USING FRACTION ROUNDING TO CONVERT A BROWNIE RECIPE

Krishna's recipe, shown in the box, makes two dozen brownies. He'll need five dozen for the Math Day picnic, so he wants to adjust the quantities of his recipe. How should this be done?

Brownie Recipe (2 dozen)

- 4 squares chocolate
- $\frac{3}{4}$ cup butter
- 2 cups sugar
- 3 eggs
- 1 teaspoon vanilla
- 1 cup flour
- $1\frac{1}{4}$ cups chopped walnuts

Solution

Since Krishna needs $2\frac{1}{2}$ times the number of brownies given by his recipe, he will multiply the quantities by $\frac{5}{2}$. For example, $\frac{5}{2} \times 4 = 10$, so he will use 10 squares of chocolate. Similarly, he will use $\frac{5}{2} \times 2 = 5$ cups of sugar, $2\frac{1}{2}$ teaspoons of vanilla, and $2\frac{1}{2}$ cups of flour. On the other hand, $\frac{5}{2} \times \frac{3}{4} = \frac{15}{8} = 1\frac{7}{8}$, so Krishna will use "just short" of 2 cups of butter. Similarly, $\frac{5}{2} \times \frac{5}{4} = \frac{25}{8} = 3\frac{1}{8}$, so Krishna will use about 3 cups of chopped walnuts. Finally, $\frac{5}{2} \times 3 = \frac{15}{2} = 7\frac{1}{2}$, so he will use either 7 or 8 eggs.

Mental Arithmetic

By taking advantage of the properties, formulas, and algorithms associated with the operations, it is often possible to simplify the computational process. Some useful strategies are demonstrated in the following example.

EXAMPLE 6.19 COMPUTATIONAL STRATEGIES FOR RATIONAL-NUMBER ARITHMETIC

Perform these computations mentally.

(a) $53 - 29\frac{3}{5}$ **(b)** $\left(2\frac{1}{8} - 4\frac{2}{3}\right) + 7\frac{7}{8}$

(c) $\frac{7}{15} \times 90$ **(d)** $\frac{3}{8} \times 14 + \frac{3}{4} \times 25$

(e) $4\frac{1}{6} \times 18$ **(f)** $\frac{5}{8} \times \left(\frac{7}{10} \times \frac{24}{49}\right)$

Solution

(a) Adding $\frac{2}{5}$ to each term gives $53 - 29\frac{3}{5} = 53\frac{2}{5} - 30 = 23\frac{2}{5}$.

(b) $\left(2\frac{1}{8} - 4\frac{2}{3}\right) + 7\frac{7}{8} = \left(2 + 7 + \frac{1}{8} + \frac{7}{8}\right) - 4\frac{2}{3}$
$= 10 - 4\frac{2}{3} = 10\frac{1}{3} - 5 = 5\frac{1}{3}$.

(c) $\frac{7}{15} \times 90 = 7 \times \frac{90}{15} = 7 \times 6 = 42$.

(d) $\frac{3}{8} \times 14 + \frac{3}{4} \times 25 = \frac{3}{8} \times 14 + \frac{3}{8} \times 50$
$= \frac{3}{8} \times (14 + 50) = \frac{3}{8} \times 64$
$= 3 \times \frac{64}{8} = 3 \times 8 = 24$.

(e) $4\frac{1}{6} \times 18 = \left(4 + \frac{1}{6}\right) \times 18 = 4 \times 18 + \frac{1}{6} \times 18 = 72 + 3 = 75$.

(f) $\frac{5}{8} \times \left(\frac{7}{10} \times \frac{24}{49}\right) = \frac{5}{10} \times \frac{7}{49} \times \frac{24}{8} = \frac{1}{2} \times \frac{1}{7} \times 3 = \frac{3}{14}$.

Rational Numbers on a Calculator

Some calculators can perform arithmetic on fractional values and display the result as a fraction. Most calculators, however, perform calculations only in decimal form, and it takes special care to express the final answer as a fraction.

EXAMPLE 6.20 COMPUTING WITH RATIONAL NUMBERS ON A CALCULATOR

Use a calculator to find $\frac{3}{8} \times \left(\frac{-7}{12} + \frac{9}{5}\right)$.

Solution

On a fraction calculator. On the TI-73 *Explorer* the steps are

3 [b/c] 8 [➡] [×] [(] [(–)] 7 [b/c] 12 [➡] [+] 9 [b/c] 5 [➡] [)] [Enter]

which gives 219/480 in the display. Pressing [Simp] [Enter] gives the simplified form 73/160.

On a decimal calculator. We can first calculate a decimal answer by pressing

3 [÷] 8 [×] [(] [(–)] 7 [÷] 12 [+] 9 [÷] 5 [)] [=],

obtaining 0.45625. To express this as a fraction, we first observe that 8, 12, and 5 are the denominators of the terms that appear in the problem. Thus, the answer can be written as a fraction with $8 \cdot 12 \cdot 5 = 480$ as a denominator. Multiplying by 480 we get $0.45625 \times 480 = 219$, so the decimal answer 0.45625 is equivalent to the fraction 219/480; in simplest form this is 73/160.

We conclude with a real-life application of the use of rational numbers and their arithmetic.

EXAMPLE 6.21

DESIGNING WOODEN STAIRS

A deck is 4′ 2″ above the surface of a patio. How many steps and risers should there be in a stairway that connects the deck to the patio? The decking is $1\frac{1}{2}''$ thick, the stair treads are $1\frac{1}{8}''$ thick, and the steps should each rise the same distance, from one to the next. Calculate the vertical dimension of each riser.

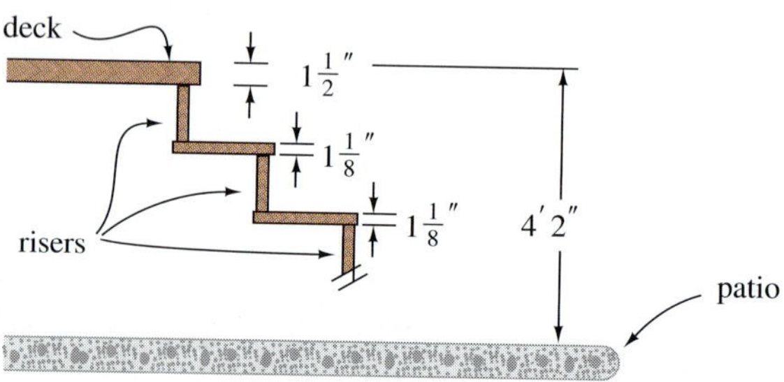

Solution

Understand the Problem

We have been given some important dimensions, including the thickness of the treads and the deck. We have *not* been told what dimension to cut the risers. This, we see from the figure, depends on the number of steps we choose, with a lower rise corresponding to a greater number of steps. We must choose a number of steps that feels natural to walk on. We must also be sure each step rises the same amount.

Devise a Plan

If we know what vertical rise from step to step is customary, we can first get a reasonable estimate of the number of steps to use. Once it is agreed what number of steps to incorporate in the design, we can calculate the height of each riser. The riser meeting the deck must be adjusted to account for the decking being thicker than the stair tread.

Carry Out the Plan

A brief survey of existing stairways shows that most steps rise 5″ to 7″ from one to the next. Since 4′ 2″ = 50″, $\frac{50}{5} = 10$, and $\frac{50}{7} = 7\frac{1}{7}$, 8, or perhaps 9, steps will work well, including the step onto the deck itself. Let's choose 8. Then $\frac{50''}{8} = 6\frac{2''}{8} = 6\frac{1''}{4}$; that is, each combination of riser plus tread is to be $6\frac{1''}{4}$. Since the treads are $1\frac{1''}{8}$ thick, the first seven risers require the boards to be cut $6\frac{1''}{4} - 1\frac{1''}{8} = 5\frac{1''}{8}$ wide. Since the decking is $1\frac{1''}{2}$ thick, the uppermost riser is $6\frac{1''}{4} - 1\frac{1''}{2} = 4\frac{3''}{4}$ wide.

Look Back

If we are concerned that the steps will take up too much room on the patio, we might use a steeper, 7-step design. Each riser plus tread would rise $7\frac{1}{7}''$. Experienced carpenters would think of this as a "hair" more than $7\frac{1}{8}''$ and cut the lower six risers from a board 8″ wide for a total rise of $7\frac{1}{8}''$. The last riser is cut to make any final small adjustment.

Problem Set 6.3

Understanding Concepts

1. Explain what properties of addition of rational numbers can be used to make this sum very easy to compute.

$$\left(3\frac{1}{5} + 2\frac{2}{5}\right) + 8\frac{1}{5}.$$

2. What properties can you use to make these computations easy?

(a) $\frac{2}{5} + \left(\frac{3}{5} + \frac{2}{3}\right)$ (b) $\frac{1}{4} + \left(\frac{2}{5} + \frac{3}{4}\right)$

(c) $\frac{2}{3} \cdot \frac{1}{8} + \frac{2}{3} \cdot \frac{7}{8}$ (d) $\frac{3}{4} \cdot \left(\frac{5}{9} \cdot \frac{4}{3}\right)$

3. Find the negatives (that is, the additive inverses) of the following rational numbers. Show each number and its negative on the number line.

(a) $\frac{4}{5}$ (b) $\frac{-3}{2}$ (c) $\frac{8}{-3}$ (d) $\frac{4}{2}$

4. Compute these sums of rational numbers. Explain what properties of rational numbers you find useful. Express your answers in simplest form.

(a) $\frac{1}{6} + \frac{2}{-3}$ (b) $\frac{-4}{5} + \frac{3}{2}$

(c) $\frac{9}{4} + \frac{-7}{8}$ (d) $\frac{3}{4} + \frac{-5}{8} + \frac{7}{-12}$

5. Compute these differences of rational numbers. Explain what properties you find useful. Express your answers in simplest form.

(a) $\frac{2}{5} - \frac{3}{4}$ (b) $\frac{-6}{7} - \frac{4}{7}$ (c) $\frac{3}{8} - \frac{1}{12}$

(d) $3\frac{2}{5} - \frac{7}{10}$ (e) $2\frac{1}{3} - 5\frac{3}{4}$ (f) $-4\frac{2}{3} - \frac{-19}{6}$

6. Calculate these products of rational numbers. Explain what properties of rational numbers you find useful. Express your answers in simplest form.

(a) $\frac{3}{5} \cdot \frac{7}{8} \cdot \left(\frac{5}{3}\right)$ (b) $\frac{-2}{7} \cdot \frac{3}{4}$

(c) $\frac{-4}{3} \cdot \frac{6}{-16}$ (d) $3\frac{1}{8} \cdot 2\frac{1}{5} \cdot (40)$

(e) $\frac{14}{15} \cdot \frac{60}{7}$ (f) $\left(\frac{4}{11} \cdot \frac{22}{7}\right) \cdot \left(\frac{-3}{8}\right)$

7. Find the reciprocals (that is, the multiplicative inverses) of these rational numbers. Show each number and its reciprocal on the number line.

(a) $\frac{3}{2}$ (b) $\frac{4}{-9}$

(c) $\frac{-4}{-11}$ (d) 5

(e) −2 (f) $2\frac{1}{2}$

8. Use the properties of the operations of rational-number arithmetic to perform these calculations. Express your answers in simplest form.

(a) $\frac{2}{3} \cdot \frac{4}{7} + \frac{2}{3} \cdot \frac{3}{7}$ (b) $\frac{4}{5} \cdot \frac{2}{3} - \frac{3}{10} \cdot \frac{2}{3}$

(c) $\frac{4}{7} \cdot \frac{3}{2} - \frac{4}{7} \cdot \frac{6}{4}$ (d) $\left(\frac{4}{7} \cdot \frac{2}{5}\right) \div \frac{2}{7}$

9. Justify each step in the following proof of the distributive property of multiplication over addition. In this case, the two addends have a common denominator, but this operation is always possible for any rational addends by using a common denominator for the two fractions.

$$\frac{a}{b} \cdot \left(\frac{c}{d} + \frac{e}{d}\right) = \frac{a}{b} \cdot \frac{c+e}{d} \qquad \text{(a) Why?}$$

$$= \frac{a \cdot (c+e)}{b \cdot d} \qquad \text{(b) Why?}$$

$$= \frac{a \cdot c + a \cdot e}{b \cdot d} \qquad \text{(c) Why?}$$

$$= \frac{a \cdot c}{b \cdot d} + \frac{a \cdot e}{b \cdot d} \qquad \text{(d) Why?}$$

$$= \frac{a}{b} \cdot \frac{c}{d} + \frac{a}{b} \cdot \frac{e}{d} \qquad \text{(e) Why?}$$

10. If $\frac{a}{b} \cdot \frac{4}{7} = \frac{2}{3}$, what is $\frac{a}{b}$? Carefully explain how you obtain your answer. What properties do you use?

11. Solve each equation for the rational number x. Show your steps, and explain what property justifies each step.

(a) $4x + 3 = 0$ **(b)** $x + \frac{3}{4} = \frac{7}{8}$

(c) $\frac{2}{3}x + \frac{4}{5} = 0$ **(d)** $3\left(x + \frac{1}{8}\right) = -\frac{2}{3}$

12. The Fahrenheit and Celsius temperature scales are related by the formula $F = \frac{9}{5}C + 32$. For example, a temperature of 20° Celsius corresponds to 68° Fahrenheit, since

$$\frac{9}{5} \cdot 20 + 32 = 36 + 32 = 68.$$

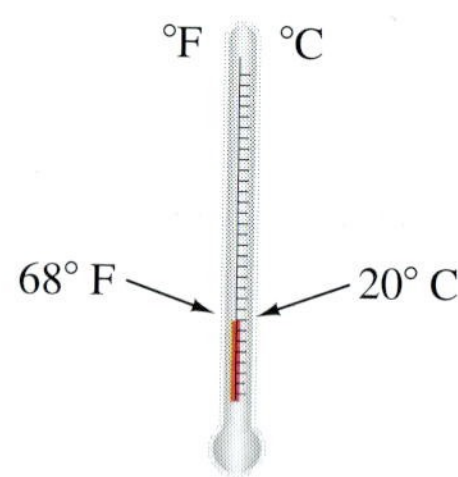

(a) Explain what properties of rational-number arithmetic permit you to deduce the equivalent formula $C = \frac{5}{9}(F - 32)$ relating the two temperature scales.

(b) Fill in the missing entries in this table.

°C	−40°		0°	10°	20°		
°F		−13°			68°	104°	212°

(c) Electronic signboards frequently give the temperature in both Fahrenheit and Celsius. What is the temperature if both readings are the same (that is, F = C)? the negative of one another (that is, F = −C)? Answer the latter question with an exact rational number and the approximate integer that would be seen on the signboard.

13. Arrange each group of rational numbers in increasing order. Show, at least approximately, the numbers on the number line.

(a) $\frac{4}{5}, -\frac{1}{5}, \frac{2}{5}$ **(b)** $\frac{-3}{7}, \frac{4}{7}, \frac{-5}{7}$

(c) $\frac{3}{8}, \frac{1}{2}, \frac{3}{4}$ **(d)** $\frac{-7}{12}, \frac{-2}{3}, \frac{3}{-4}$

14. Verify these inequalities.

(a) $\frac{-4}{5} < \frac{-3}{4}$ **(b)** $\frac{1}{10} > -\frac{1}{4}$ **(c)** $\frac{-19}{60} > \frac{-1}{3}$

15. The properties of the order relation can be used to solve inequalities. For example, if $-\frac{2}{3}x + \frac{1}{4} < -\frac{1}{2}$, then

$-8x + 3 < -6$	(multiply by 12)
$-8x < -9$	(subtract 3 from both sides)
$8x > 9$	(multiply by -1, which reverses the direction of inequality)
$x > \frac{9}{8}$	(divide both sides by 8)

That is, all rational numbers x greater than $\frac{9}{8}$, and only these, satisfy the given inequality. Solve these inequalities. Show all of your steps.

(a) $x + \frac{2}{3} > -\frac{1}{3}$ **(b)** $x - \left(-\frac{3}{4}\right) < \frac{1}{4}$

(c) $\frac{3}{4}x < -\frac{1}{2}$ **(d)** $-\frac{2}{5}x + \frac{1}{5} > -1$

16. Find a rational number that is between the two given rational numbers.

(a) $\frac{4}{9}$ and $\frac{6}{11}$ **(b)** $\frac{1}{9}$ and $\frac{1}{10}$

(c) $\frac{14}{23}$ and $\frac{7}{12}$ **(d)** $\frac{141}{568}$ and $\frac{183}{737}$

17. Find three rational numbers between $\frac{1}{4}$ and $\frac{2}{5}$.

18. In the "Hagar the Horrible" cartoon shown early in Section 6.1, Lucky Eddie is counting by eighths to 10 to begin the charge. Propose a method for Lucky Eddie to count with fractions that forever delays the charge. What property of rational numbers did you use?

19. For each given rational number, choose the best estimate from the list provided.

(a) $\frac{104}{391}$ is approximately $\frac{1}{3}, \frac{1}{4}$, or $\frac{1}{2}$.

(b) $\frac{217}{340}$ is approximately $\frac{1}{3}, \frac{1}{2}$, or $\frac{2}{3}$.

(c) $\frac{-193}{211}$ is approximately $-\frac{1}{2}, -1, 1$, or $\frac{1}{2}$.

(d) $\frac{453}{307}$ is approximately $\frac{3}{4}, 1, 1\frac{1}{3}$, or $1\frac{1}{2}$.

20. Use estimations to choose the best approximation of the following expressions.

(a) $3\frac{19}{40} + 5\frac{11}{19}$ is approximately 8, $8\frac{1}{2}$, 9, or $9\frac{1}{2}$.

(b) $2\frac{6}{19} + 5\frac{1}{3} - 4\frac{7}{20}$ is approximately 3, $3\frac{1}{3}$, $3\frac{1}{4}$, or 4.

(c) $17\frac{8}{9} \div 5\frac{10}{11}$ is approximately 2, 3, $3\frac{1}{2}$, or 4.

21. First do the following calculations mentally. Then use words and equations to explain your strategy.

(a) $\frac{1}{2} + \frac{1}{4} + \frac{3}{4}$ **(b)** $\frac{5}{2} \cdot \left(\frac{2}{5} - \frac{2}{10}\right)$

(c) $\frac{3}{4} \cdot \frac{12}{15}$ **(d)** $\frac{2}{9} \div \frac{1}{3}$

(e) $2\frac{2}{3} \times 15$ **(f)** $3\frac{1}{5} - 1\frac{1}{4} + 7\frac{4}{5}$

(g) $6\frac{1}{8} - 8\frac{1}{4}$ **(h)** $\frac{2}{3} \cdot \frac{7}{4} - \frac{2}{3} \cdot \frac{1}{4}$

22. **(a)** Hal owns $3\frac{1}{2}$ acres and just purchased an adjacent plot of $\frac{3}{4}$ acres. The answer is $4\frac{1}{4}$. What is the question?

(b) Janet lives $1\frac{3}{4}$ miles from school. On the way to school, she stopped to walk the rest of the way with Brian, who lives $\frac{1}{2}$ of a mile from school. The answer is $1\frac{1}{4}$. What is the question?

(c) A family room is $5\frac{1}{2}$ yards wide and 6 yards long. The answer is 33 square yards. What is the question?

(d) Clea made $3\frac{1}{2}$ gallons of ginger ale, which she intends to bottle in "fifths" (that is, bottles that contain a fifth of a gallon). The answer is 17 full bottles and $\frac{1}{2}$ of another bottle. What is the question?

23. Invent interesting word problems that lead you to the given expression.

(a) $\frac{3}{4} + \frac{1}{2}$ **(b)** $4\frac{1}{2} \div \frac{3}{4}$ **(c)** $7\frac{2}{3} \times \frac{1}{4}$

Teaching Concepts

24. When asked to evaluate the sum $\frac{1}{3} + \frac{3}{5}$, a student claimed that the answer, when simplified, is $\frac{1}{2}$. How do you suspect the student arrived at this answer? Describe what you might do to help this student's understanding.

25. When asked to evaluate the difference $\frac{9}{11} - \frac{3}{22}$, a student gave the answer $\frac{165}{242}$. What would you suggest to the student?

Thinking Critically

26. Problem 31 of the Rhind papyrus, if translated literally, reads as follows: "A quantity, its $\frac{2}{3}$, its $\frac{1}{2}$, its $\frac{1}{7}$, its whole, amount to 33." That is, in modern notation, $\frac{2}{3}x + \frac{1}{2}x + \frac{1}{7}x + x = 33$. What is the quantity?

27. Solve this problem from the Rhind papyrus: "Divide 100 loaves among five men in such a way that the share received shall be in arithmetic progression and that one seventh of the sum of the largest three shares shall be equal to the sum of the smallest two." (*Suggestion:* Denote the shares as s, $s + d$, $s + 2d$, $s + 3d$, and $s + 4d$.)

28. The ancient Egyptians measured steepness of a slope by the fraction $\frac{x}{y}$, where x is the number of hands of horizontal "run" and y is the number of cubits of vertical "rise." Seven hands form a cubit. Problem 56 of the Ahmes papyrus asks for the steepness of the face of a pyramid 250 cubits high and having a square base 360 cubits on a side. The papyrus gives the answer $5\frac{1}{25}$. Show why this is correct.

29. **The Law of the Lever.** One of Archimedes' (ca. 225 B.C.) great achievements was the law of the lever. The law can be described in terms of a condition under which weights hung from a beam pivoting at 0 will be in balance. For example, the two beams shown below are in balance.

$W_1, W_2,$ W_n

−3 0 5 | −2 0 3 6

5 3 | 9 4 1

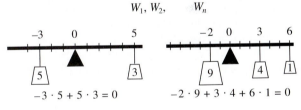

$-3 \cdot 5 + 5 \cdot 3 = 0$ $-2 \cdot 9 + 3 \cdot 4 + 6 \cdot 1 = 0$

In general, if weights . . . , are hung at positions $x_1, x_2, \ldots, x_n$, then the beam is balanced if, and only if, $x_1 \cdot W_1 + x_2 \cdot W_2 + \cdots + x_n \cdot W_n = 0$.

For each diagram (not drawn to scale) below, find the missing weight W or unknown position x that will balance the beam. You can expect rational-number answers.

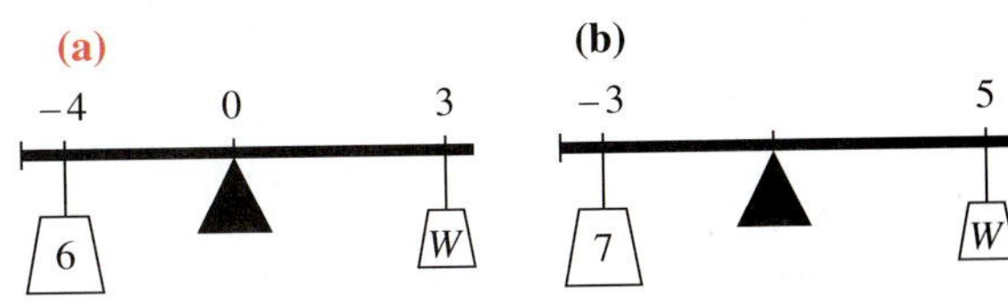

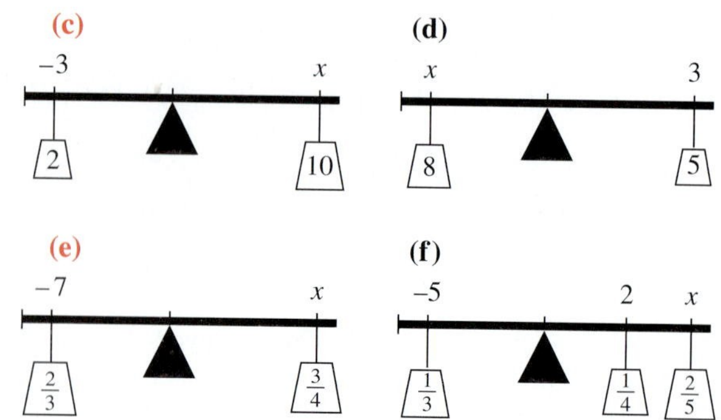

30. **(a)** Let x, y, and z be arbitrary rational numbers. Show that the sums of the three entries in every row, column, and diagonal in the Magic Square are the same.

$x - z$	$x - y + z$	$x + y$
$x + y + z$	x	$x - y - z$
$x - y$	$x + y - z$	$x + z$

(b) Let $x = \frac{1}{2}$, $y = \frac{1}{3}$, and $z = \frac{1}{4}$. Find the corresponding Magic Square.

(c) Here is a partial Magic Square that corresponds to the form shown in part (a). Find x, y, and z, and then fill in the remaining entries of the square.

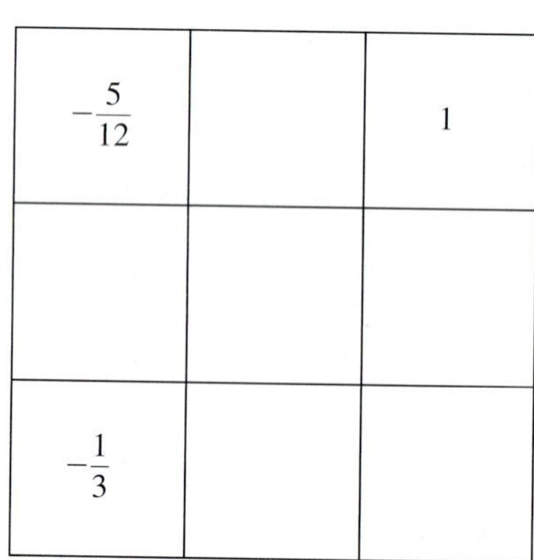

$-\frac{5}{12}$		1
$-\frac{1}{3}$		

(*Hint:* $2x = (x + y) + (x - y)$.)

31. Three-quarters of the pigeons occupy two-thirds of the pigeonholes in a pigeon house (one pigeon per hole), and the other one-quarter of the pigeons are flying around. If all the pigeons were in the pigeon holes, just five holes would be empty. How many pigeons and how many holes are there? Give an answer based on the following diagram, adding labels to the diagram and carefully describing your reasoning in words.

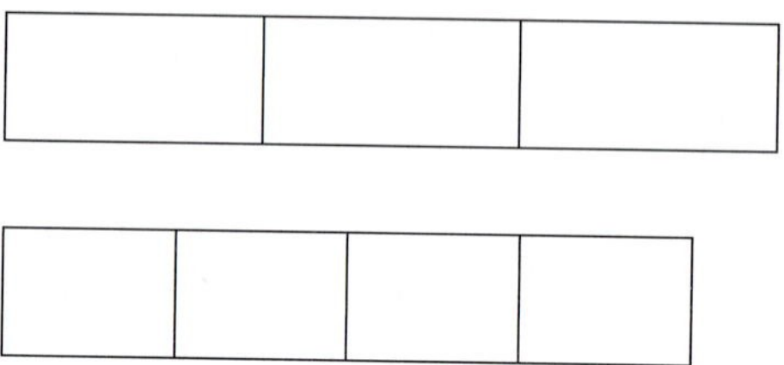

32. A bag contains red and blue marbles. One-sixth of the marbles are blue, and there are 20 more red marbles than blue marbles. How many red marbles and how many blue marbles are there?

(a) Give an answer based on the set diagram, adding labels and describing how to use the diagram to obtain an answer.

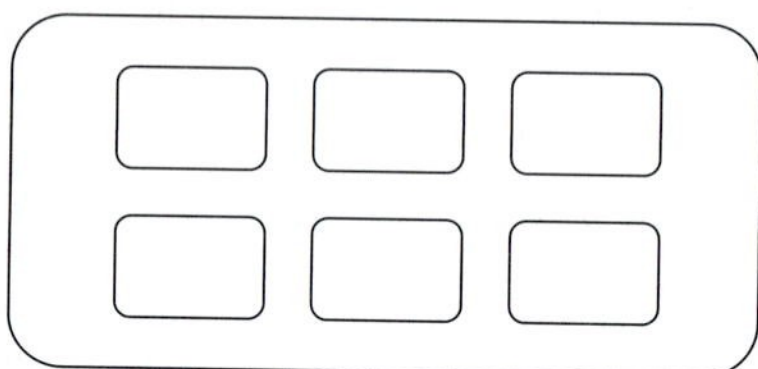

(b) Give an answer based on finding and solving equations.

Making Connections

33. The earth revolves around the sun once in 365 days, 5 hours, 48 minutes, and 46 seconds. Since this *solar* year is more than 365 days long, it is important to have 366-day-long *leap* years to keep the seasons in the same months of the calendar year.

(a) The solar year is very nearly 365 days and 6 hours, or $365\frac{6}{24} = 365\frac{1}{4}$ days, long. Explain why having the years divisible by 4 as leap years is a good rule to decide when leap years should occur.

(b) Show that a solar year is $365\frac{20{,}926}{24 \cdot 60 \cdot 60}$ days long.

(c) Estimate the value given in part (b) with $365\frac{20{,}952}{24 \cdot 60 \cdot 60}$, and then show that this can also be written as $365\frac{1}{4} - \frac{1}{100} + \frac{1}{400}$.

(d) What rule for choosing leap years is suggested by the approximation $365\frac{1}{4} - \frac{1}{100} + \frac{1}{400}$ to the solar year?
(*Hint:* 1900 was not a leap year, but 2000 was a leap year.)

34. Fractions have a prominent place in music, where the time value of a note is given as a fraction of a whole note. The note values and their corresponding fractions are shown in this table:

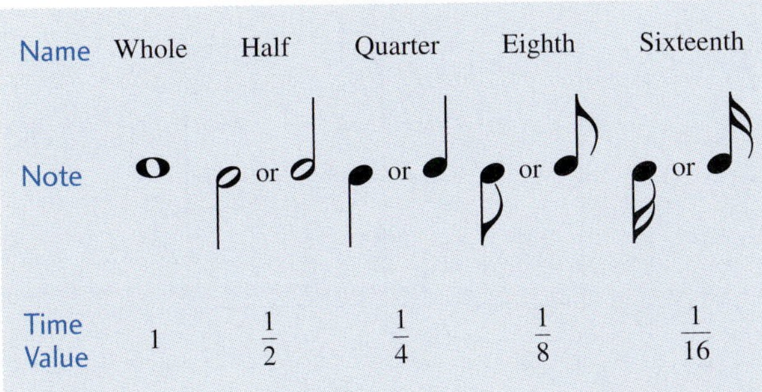

Name	Whole	Half	Quarter	Eighth	Sixteenth
Note		or	or	or	or
Time Value	1	$\frac{1}{2}$	$\frac{1}{4}$	$\frac{1}{8}$	$\frac{1}{16}$

The time signature of the music can be interpreted as a fraction that gives the time duration of the measure. Here are two examples:

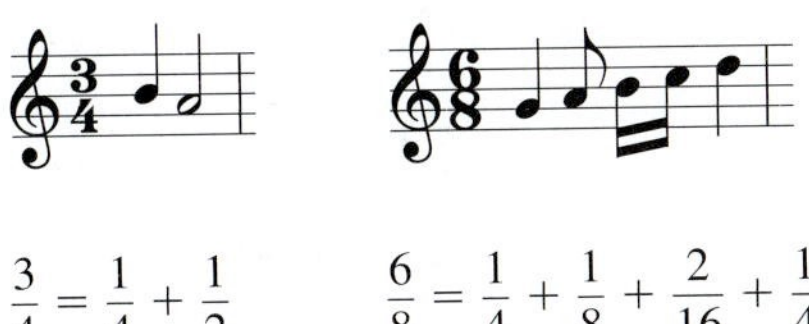

$\frac{3}{4} = \frac{1}{4} + \frac{1}{2}$ $\quad$ $\frac{6}{8} = \frac{1}{4} + \frac{1}{8} + \frac{2}{16} + \frac{1}{4}$

In each measure shown, fill in the upper number of the time signature by adding the note values shown in the measure.

(a)

(b)

(c)

35. Design a stairway that connects the patio to the deck (see Example 6.21).

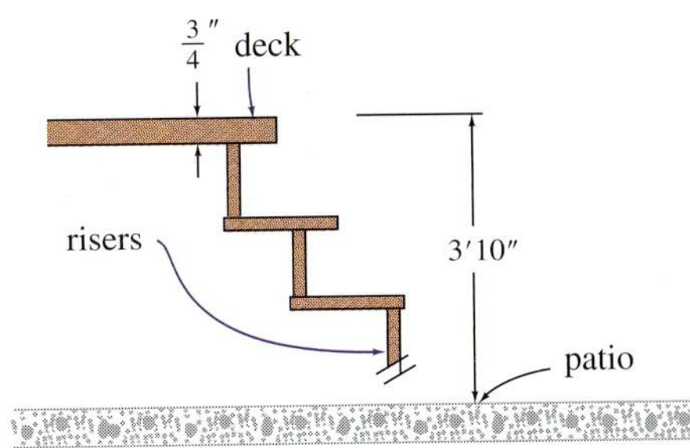

Assume the tread of each stair is $\frac{1}{2}''$ thick.

36. This recipe makes six dozen cookies. Adjust the quantities to have a recipe for four dozen cookies.

1 cup shortening	$1\frac{1}{2}$ cup sugar
1 tsp baking soda	3 eggs
3 cups unsifted flour	$\frac{1}{2}$ tsp salt
9 oz mincemeat	

37. Anja has a box of photographic print paper. Each sheet measures 8-by–10 inches. She could easily cut a sheet into four 4-by–5-inch rectangles. However, to save money, she wants to get six prints, all the same size, from each 8-by–10-inch sheet. Describe how she can cut the paper.

38. Krystoff has an 8′ piece of picture frame molding, shown in cross-section below.

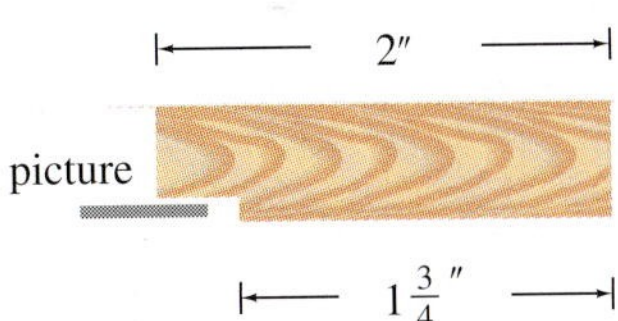

Is this 8′ piece sufficient to frame a 16″-by-20″ picture? Allow for saw cuts and some extra space to ensure that the picture fits easily into the frame.

Using a Calculator

39. Use a calculator to evaluate the following expressions. Express your answers in simplest form.

(a) $\frac{4}{5} \times \left(\frac{18}{25} - \frac{3}{4}\right)$

(b) $\left(\frac{-7}{3} + \frac{4}{9}\right) \div \frac{1}{5}$

40. Mathematicians from ancient times hoped to find a rational number $\frac{a}{b}$ equal to the square root of 2. A rather inaccurate choice is $\frac{3}{2}$, which is too large, since

$\left(\frac{3}{2}\right)^2 = \frac{9}{4} = 2 + \frac{1}{4}$. An improved estimate is obtained by replacing the fraction $\frac{a}{b}$ with the fraction $\frac{a + 2b}{a + b}$. For example, $\frac{3}{2}$ is replaced with $\frac{3 + 2 \cdot 2}{3 + 2} = \frac{7}{5}$. Since $\left(\frac{7}{5}\right)^2 = \frac{49}{25} = 2 - \frac{1}{25}$, we see that $\frac{7}{5}$ is a better approximation to $\sqrt{2}$ than $\frac{3}{2}$.

(a) Start with $\frac{7}{5}$, and use the replacement rule $\frac{a}{b} \Rightarrow \frac{a + 2b}{a + b}$ to obtain another rational number. Is it a better approximation for $\sqrt{2}$ than $\frac{7}{5}$?

(b) Use the replacement rule once more, starting with the rational number obtained in part (a). Do you obtain an even better approximation of $\sqrt{2}$?

(c) Set up a graphing calculator or spreadsheet program to automate the replacement rule $\frac{a}{b} \Rightarrow \frac{a + 2b}{a + b}$. Begin with any fraction you wish and carry out the rule several times. Describe any pattern or properties of the sequence of rational numbers you generate.

From State Student Assessments

41. (Minnesota, Grade 5)

Which figure shows $\frac{3}{5}$ shaded?

A.

B.

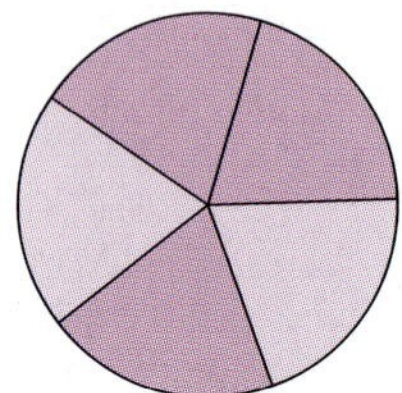

C.

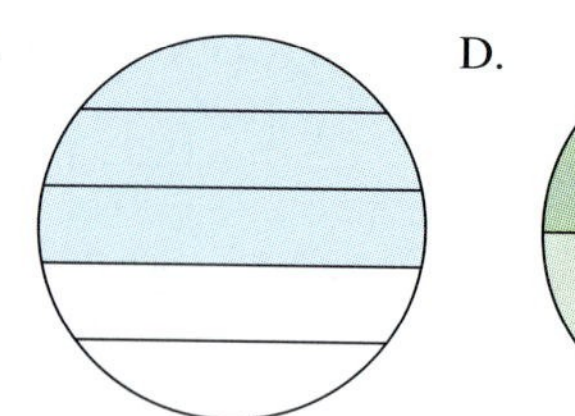

D.

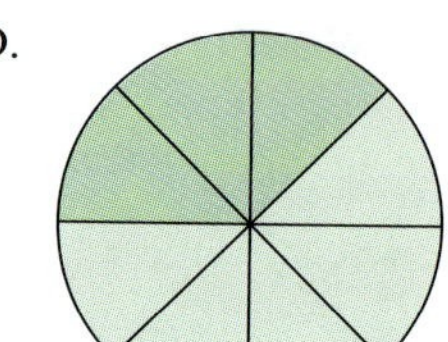

42. (Delaware, Grade 5)
John split a cake into four pieces. The pieces are *not* all the same shape. Do you believe the four pieces are the same size? Please explain.

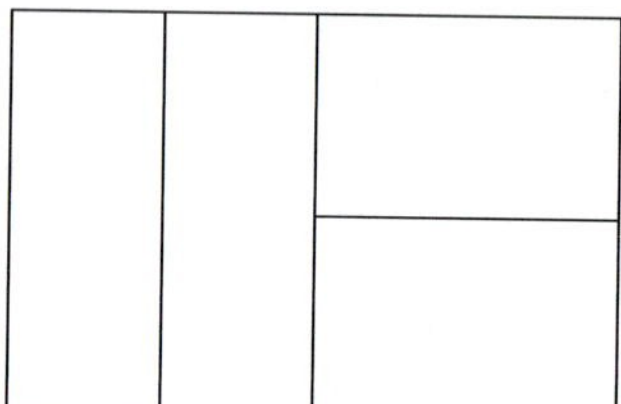

For Review

43. **(a)** The Ahmes papyrus is about 6 yards long and $\frac{1}{3}$ of a yard wide. What is its area in square yards?

(b) The Moscow papyrus, another source of mathematics of ancient Egypt, is about the same length as the Ahmes papyrus, but has $\frac{1}{4}$ the area. What is the width of the Moscow papyrus?

44. A TV, regularly priced at \$345, is offered on sale at "up to one-third off." What range of prices would you expect to pay?

45. The triangular shape shown to the right is $\frac{1}{4}$ of a whole figure. What does the whole figure look like? Give several answers.

Epilogue Fractions and Technology

The equation $x + m = n$ cannot always be solved in the whole numbers. However, by extending the whole numbers to the integers, this equation always has a unique solution. Computationally, going from whole-number to integer arithmetic comes at a modest cost, since the algorithms used to perform calculations on integers are essentially the same as those for whole numbers; the only added step is determining whether the answer is positive or negative.

In the rational-number system, the greatest gain is that the equation $rx + p = q$ can always be solved uniquely for the unknown x, whatever values are given for the rational numbers r, p, and q (although r must be nonzero).

However, going to rational-number arithmetic comes with a high cost in the complexity of the computations. For example, adding two rationals $\frac{a}{b} + \frac{c}{d}$ requires three integer multiplications and an integer addition to produce the answer $\frac{ad + bc}{bd}$. Even then, the answer may not be in simplest form. It is scarcely surprising that schoolchildren (and adults) often find fractions and the arithmetic of fractions difficult.

When calculators and computers first became widely available, there was some talk of diminishing, or even removing, fractions from the curriculum. After all, these machines were based on decimal fractions, not on common fractions. There are many good reasons this did not and should not occur, including the following:

- fractions, as expressions of ratios and rates, convey a dynamic sense of number that is absent in decimal fractions;
- the algebraic concepts used in rational-number arithmetic provide important readiness skills for learning algebra.

At the present time, technology is moving in a direction that supports teaching, learning, and using fractions. For example, inexpensive calculators with fraction arithmetic capabilities are now becoming common in the classroom. Also, computer algebra systems for both calculators and computers can now perform rational-number arithmetic in the blink of an eye to unlimited precision. With these devices, problem solving and critical thinking in the realm of rational numbers can actually be a source of pleasure.

In this chapter, we have discussed the properties and arithmetic of fractions and rational numbers, both from a concrete and pictorial point of view and from a more theoretical standpoint. We have also seen how the operations with fractions often depend on notions from number theory. Many of the examples illustrated the practical side of fractions, in which the solution of real-life problems led naturally to operations with rational numbers.

Chapter Summary

Key Concepts

Section 6.1 The Basic Concepts of Fractions and Rational Numbers

- Fraction. A fraction is an ordered pair of integers a and b, $b \neq 0$, written as $\frac{a}{b}$ or a/b.
- Representations. Fractions are represented physically and pictorially with colored regions, sets, fraction strips, and the number line.
- Equivalent fractions. Two fractions a/b and c/d that represent the same quantity (that is, the same rational number) are called equivalent, and we write $a/b = c/d$. In particular, $\frac{a}{b} = \frac{a \cdot n}{b \cdot n}$ for all nonzero integers n (the fundamental law of fractions), and $\frac{a}{b} = \frac{c}{d}$ if, and only if, $ad = bc$ (the cross-product property).
- Simplest form. Any given fraction can be written in simplest form by finding the equivalent fraction with the smallest positive denominator.
- Common denominator. Two or more fractions can be replaced with equivalent fractions with the same denominator.
- Rational number. A rational number is a number represented by a fraction a/b. The same rational number can also be represented by any fraction equivalent to a/b.

- Order of rational numbers. If two rational numbers are represented as a/b and c/d, where b and d are positive denominators, then $a/b < c/d$ if, and only if, $ad < bc$.

Section 6.2 The Arithmetic of Rational Numbers

- Addition. The sum of two rational numbers with a common denominator, represented by fractions a/b and c/b, is defined by $\frac{a}{b} + \frac{c}{b} = \frac{a+c}{b}$. Since two rational numbers that do not have a common denominator, represented by fractions a/b and c/d, can be written with the common denominator bd as ad/bd and bc/bd, respectively, the sum of fractions with unlike denominators is given by $\frac{a}{b} + \frac{c}{d} = \frac{ad+bc}{bd}$.
- Subtraction. Subtraction is defined using the missing-addend model, so that $\frac{a}{b} - \frac{c}{d}$ is defined to be the fraction $\frac{e}{f}$ for which $\frac{a}{b} = \frac{c}{d} + \frac{e}{f}$. The subtraction formula $\frac{a}{b} - \frac{c}{d} = \frac{ad-bc}{bd}$ follows from the definition of subtraction.
- Multiplication. Extending the area model of multiplication motivates the definition of multiplication of fractions: $\frac{a}{b} \cdot \frac{c}{d} = \frac{ac}{bd}$.
- Division. Division is defined using the missing-factor model of division, so that $\frac{a}{b} \div \frac{c}{d} = \frac{e}{f}$ if, and only if, $\frac{a}{b} = \frac{c}{d} \cdot \frac{e}{f}$.
- Invert-and-multiply algorithm. It follows from the definition of fraction division that $\frac{a}{b} \div \frac{c}{d} = \frac{a}{b} \cdot \frac{d}{c}$. However, this formula is not the definition of fraction division.
- Multiplicative inverse. A nonzero rational number a/b has a unique multiplicative inverse given by its reciprocal b/a, since the product of a fraction and its reciprocal is the multiplicative identity 1. That is, $\frac{a}{b} \cdot \frac{b}{a} = 1$.

Section 6.3 Rational-Number System

- Properties of the rational numbers. The rational numbers are closed under addition, addition is commutative and associative, multiplcation is commutative and associative, multiplication is distributive over addition, each nonzero rational number has a multiplicative inverse given by its reciprocal, and zero is the additive identity.
- Inverses. Every rational number a/b has an additive identity $-a/b$ given by $(-a)/b$, since $\frac{a}{b} + \frac{-a}{b} = 0$. Therefore, subtraction is equivalent to adding the inverse: $\frac{a}{b} - \frac{c}{d} = \frac{a}{b} + \left(-\frac{c}{d}\right)$.
- Density property. There is a rational number (indeed, infinitely many rational numbers) between any two distinct rational numbers.
- Computations with rational numbers. Estimation, rounding, mental arithmetic, paper-and-pencil calculations, and calculator and computer calculations are equally important for work with the fractions and rational numbers as for any other number system. In particular, the arithmetic of fractions provides important readiness skills for algebra.

Vocabulary and Notation

Section 6.1

Fraction
Numerator
Denominator
Models for fractions:
- Colored regions
- Sets
- Fraction strips
- Number line

Equivalent fractions
Cross-product property
Fundamental law of fractions
Simplest (or reduced) form
Common denominator
Least common denominator
Rational number
Set of rational numbers, Q
Order relation on the rational numbers

Section 6.2

Operations on fractions:
- Addition $\frac{a}{b} + \frac{c}{d}$
- Subtraction $\frac{a}{b} - \frac{c}{d}$
- Multiplication $\frac{a}{b} \cdot \frac{c}{d}$ or $\frac{a}{b} \times \frac{c}{d}$
- Division $\frac{a}{b} \div \frac{c}{d}$

Mixed number
Proper fraction
Invert-and-multiply algorithm
Reciprocal, or multiplicative inverse

Section 6.3

Negative (or opposite, or additive inverse)
Transitive property of order
Density property

Chapter Review Exercises

Section 6.1

1. What fraction is represented by the darker blue shading in each of the colored-region models shown? In (a) and (d), the unit is the region inside one circle. In (b) and (c), the units are the regions inside the hexagon and square, respectively.

(a)
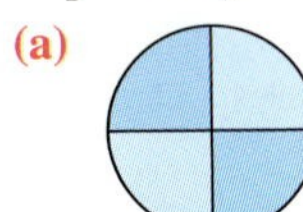

(b)
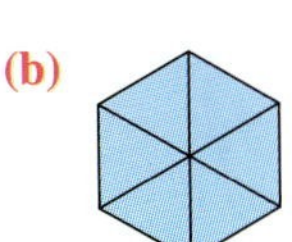

(c)
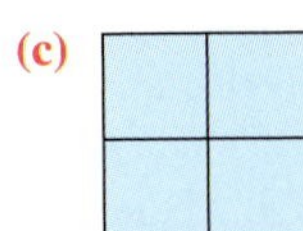

(d)
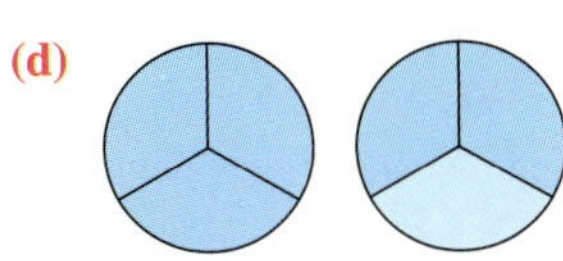

2. Label the points on the number line that correspond to these rational numbers:

(a) $\frac{3}{4}$ (b) $\frac{12}{8}$

(c) 1 (d) $2\frac{3}{8}$

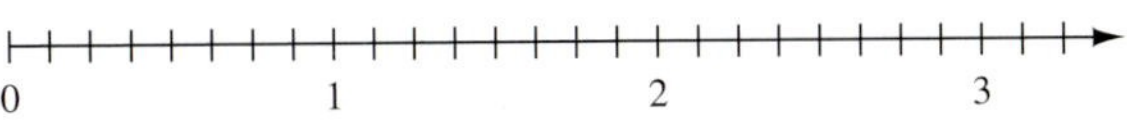

3. Express each rational number by a fraction in simplest form.

(a) $\frac{27}{81}$ (b) $\frac{100}{825}$

(c) $\frac{378}{72}$ (d) $\frac{3^5 \cdot 7^2 \cdot 11^3}{3^2 \cdot 7^3 \cdot 11^2}$

4. Order these fractions from smallest to largest.

$$\frac{1}{2}, \frac{13}{27}, \frac{25}{49}, \frac{13}{30}, \frac{26}{49}$$

5. Find the least common denominator of each set of fractions.

(a) $\frac{4}{9}, \frac{5}{12}$ (b) $\frac{7}{18}, \frac{5}{6}, \frac{1}{3}$

Section 6.2

6. Illustrate $\frac{3}{4} + \frac{7}{8}$ on the number line.

7. Illustrate $\frac{3}{4} - \frac{1}{3}$ with fraction strips.

8. Compute these sums and differences.

(a) $\frac{3}{8} + \frac{1}{4}$ (b) $\frac{2}{9} + \frac{-5}{12}$

(c) $\frac{4}{5} - \frac{2}{3}$ (d) $5\frac{1}{4} - 1\frac{5}{6}$

9. Illustrate these products by labeling and coloring appropriate rectangular regions.

(a) $3 \times \frac{1}{3}$ (b) $\frac{2}{3} \times 4$ (c) $\frac{5}{6} \times \frac{3}{2}$

10. Gina, Hank, and Igor want to share $2\frac{1}{4}$ pizzas equally. Draw an appropriate diagram to show how much pizza each should be given. What conceptual model of division should you use?

11. Each quart of soup calls for $\frac{2}{3}$ of a cup of pinto beans. How many quarts of soup can be made with 3 cups of beans? Draw an appropriate diagram to find your answer. What conceptual model of division are you using?

12. On a map, it is $7\frac{1''}{8}$ from Arlington to Banks. If the scale of the map is $1\frac{1''}{4}$ per mile, how far is it between the two towns?

Section 6.3

13. Perform these calculations, expressing your answers in simplest form.

(a) $\frac{-3}{4} + \frac{5}{8}$ (b) $\frac{4}{5} - \frac{-7}{10}$

(c) $\left(\frac{3}{8} \cdot \frac{-4}{27}\right) \div \frac{1}{9}$ (d) $\frac{2}{5} \cdot \left(\frac{3}{4} - \frac{5}{2}\right)$

14. Solve each equation. Give the rational number x as a fraction in simplest form.

(a) $3x + 5 = 11$ (b) $x + \frac{2}{3} = \frac{1}{2}$

(c) $\frac{3}{5}x + \frac{1}{2} = \frac{2}{3}$ (d) $-\frac{4}{3}x + 1 = \frac{1}{4}$

15. Find two rational numbers between $\frac{5}{6}$ and $\frac{10}{11}$.

16. Do the following calculations mentally. Explain your method.

(a) $1\frac{1}{3} + 2\frac{5}{12} + \frac{1}{4}$ (b) $\frac{6}{7} \cdot \frac{28}{3} \cdot \frac{5}{8}$ (c) $\frac{36}{5} \div \frac{9}{25}$

Chapter Test

1. Carefully explain why $\frac{a+b}{b} = \frac{c+d}{d}$ if, and only if, $\frac{a}{b} = \frac{c}{d}$.

2. Perform these calculations:

(a) $\frac{1}{3} + \frac{5}{8} - \frac{5}{6}$ (b) $\left(\frac{2}{3} - \frac{5}{4}\right) \div \frac{3}{4}$

(c) $\frac{4}{7} \cdot \left(\frac{35}{4} + \frac{-42}{12}\right)$ (d) $\frac{123}{369} \div \frac{1}{3}$

3. Describe how the following calculations can be performed efficiently with "mental math."

(a) $\frac{19}{111} \cdot \left(\frac{2}{3} + \frac{-4}{6}\right)$ (b) $\frac{5}{6} \cdot \frac{36}{15}$

(c) $\frac{5}{8} \cdot \left(\frac{9}{5} - \frac{1}{5}\right)$ (d) $\frac{2}{3} \cdot \frac{3}{4} \cdot \frac{4}{5} \cdot \frac{5}{6}$

4. An acre is $\frac{1}{640}$ of a square mile.

(a) How many acres are in a rectangular plot of ground $\frac{1}{8}$ mile wide by $\frac{1}{2}$ mile long?

(b) A rectangular piece of property contains 80 acres and is $\frac{1}{2}$ mile long. What is the width of the property?

5. (a) What is the density property of the rational numbers?

(b) Find a rational number between $\frac{3}{5}$ and $\frac{2}{3}$.

6. Illustrate $\frac{2}{3}$ (a) on the number line, (b) with a fraction strip, (c) with a colored-region model, and (d) with the set model.

7. What is the best approximate answer listed for each of these problems?

(a) $2\frac{1}{48} + 3\frac{1}{99} + 6\frac{13}{25}$ is approximately 11, $11\frac{1}{2}$, 12, or $12\frac{1}{4}$.

(b) $8 \cdot \left(2\frac{1}{2} + 3\frac{7}{15}\right)$ is approximately 40, 44, 48, or 56.

(c) $11\frac{9}{10} \div \frac{21}{40}$ is approximately 20, 23, 26, or 30.

8. (a) Define *multiplicative inverse.*

(b) Find the multiplicative inverses of these rational numbers: $\frac{3}{2}$, $\frac{-4}{5}$, and -5.

9. Give three different fractions, each equivalent to $-\frac{3}{4}$.

10. Solve the following equations and inequalities for all possible rational numbers x. Show all of your steps.

(a) $2x + 3 > 0$ (b) $\frac{3}{4}x + \frac{1}{2} = \frac{1}{3}$

(c) $\frac{5}{4}x > -\frac{1}{3}$ (d) $\frac{1}{2} < 4x + \frac{5}{6}$

11. (a) Define division in the rational-number system.
(b) Justify the invert-and-multiply algorithm; that is, prove that $\frac{a}{b} \div \frac{c}{d} = \frac{a}{c} \cdot \frac{d}{c}$, where $c \neq 0$.

12. (a) Invent a realistic problem whose solution requires the calculation $\frac{4}{5} \cdot \frac{2}{3}$.
(b) Make up a realistic problem that leads to $\frac{3}{8} \div \frac{3}{10}$.

13. (a) Define *additive inverse.*
(b) Find the additive inverses of the following rational numbers: $\frac{3}{4}$, $\frac{-7}{4}$, and $\frac{8}{-2}$.

14. Order these rational numbers from least to greatest.

$$\frac{16}{5}, \frac{2}{3}, -1\frac{1}{2}, 0, \frac{5}{8}, 3, -3$$

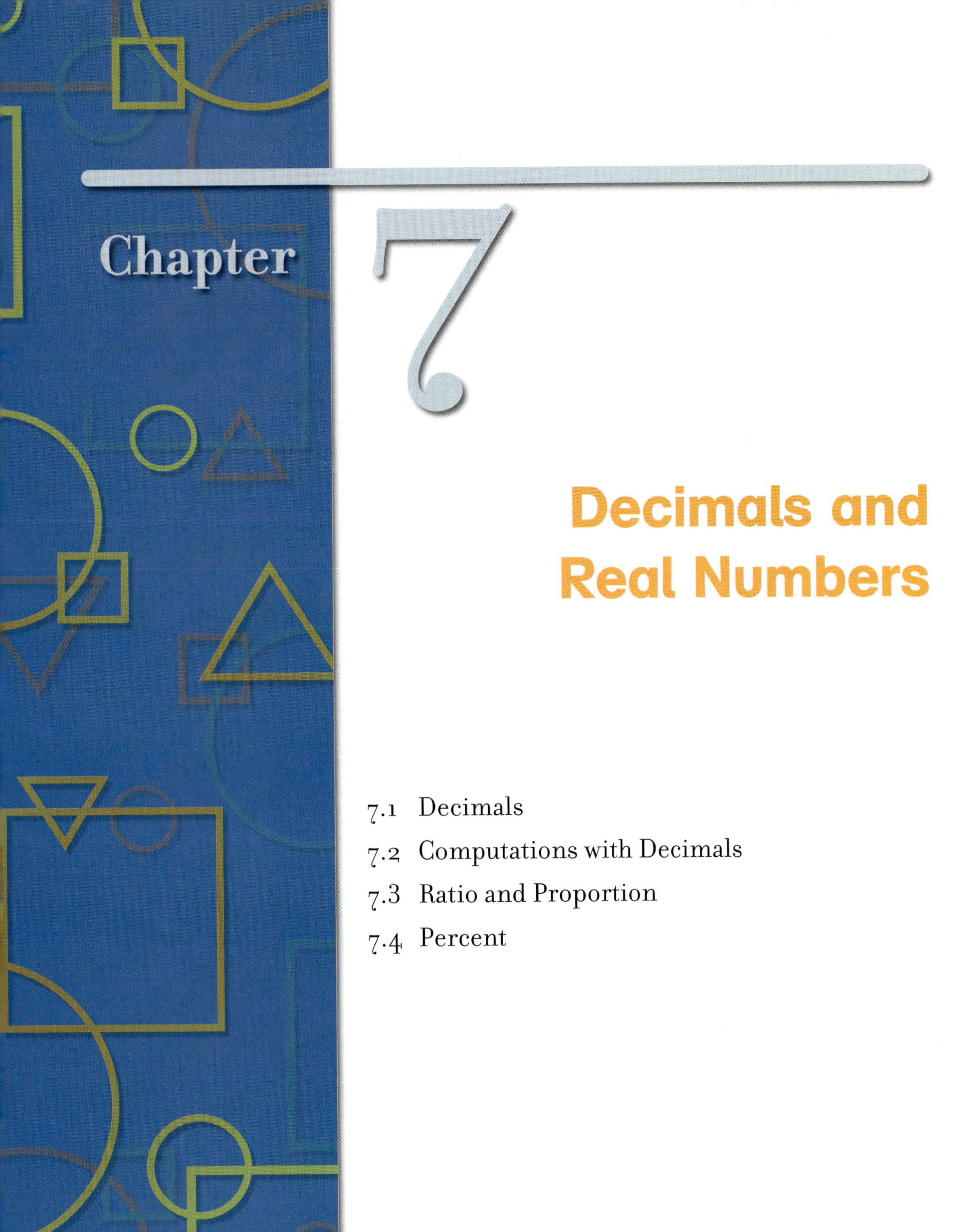

Chapter 7

Decimals and Real Numbers

Triangles and Squares

Materials Needed

One 3″ × 5″ card, scissors, a sharp pencil, and one sheet of plain white paper.

Directions

Step 1. Cut a triangle with unequal sides off the corner of the card and denote the lengths of the three sides by a, b, and c as shown.

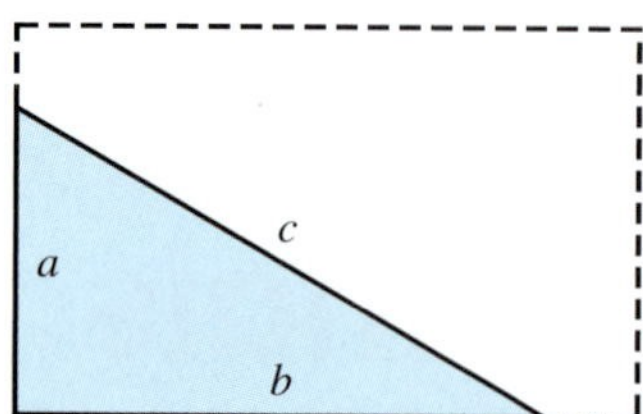

Step 2. With the lettering on the triangle always facing up, carefully trace around the triangle four times to form a big square with another square in the middle as shown here.

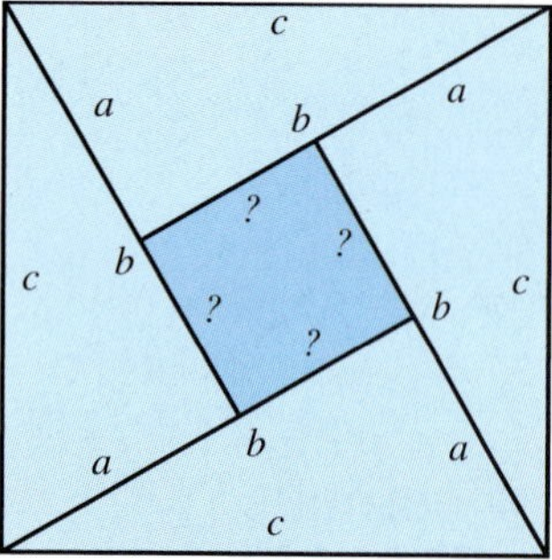

Step 3. Determine the dimensions of the small square in the center, and then find its area. Remember that, using the distributive property of multiplication over subtraction and the commutative property, we have

$$\begin{aligned}(x - y)(x - y) &= (x - y)x - (x - y)y \\ &= x^2 - yx - xy + y^2 \\ &= x^2 - 2xy + y^2.\end{aligned}$$

Step 4. The area of the large square is clearly four times the area of the triangle plus the area of the small square; it is also equal to c^2. Express these two ways of finding the area of the large square by an equation, and simplify it as much as possible.

Step 5. Relate the equation in Step 4 to the diagram shown here. What does the equation reveal about triangles with one square corner (that is, one right angle)? Explain briefly.

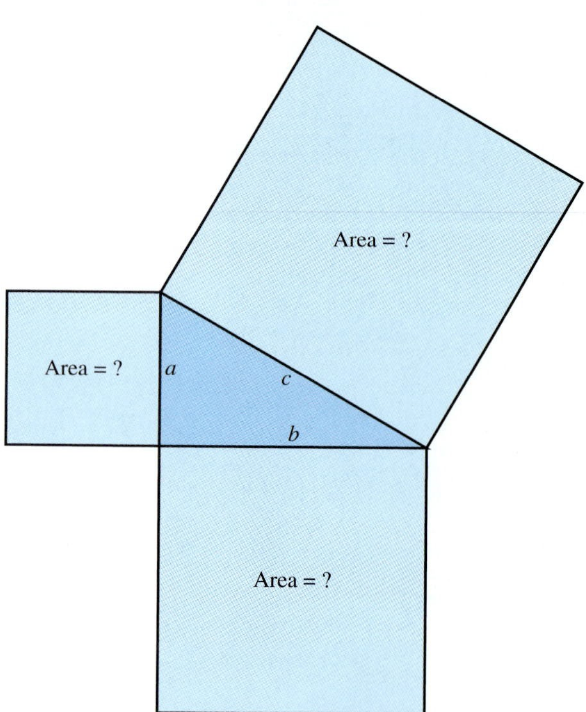

Step 6. If c is the length of the diagonal of a square one unit on a side, then the formula of Step 4 gives $c^2 = 2$. Remarkably, as we will see later, this number c is not a rational number.

Connections

Further Enlarging the Number System

In the preceding chapters, we have considered

- the set of natural numbers, N,
- the set of whole numbers, W,
- the set of integers, I, and
- the set of rational numbers, Q.

The fact that each succeeding set is an extension of the immediately preceding set can be expressed in a Venn diagram as shown in Figure 7.1.

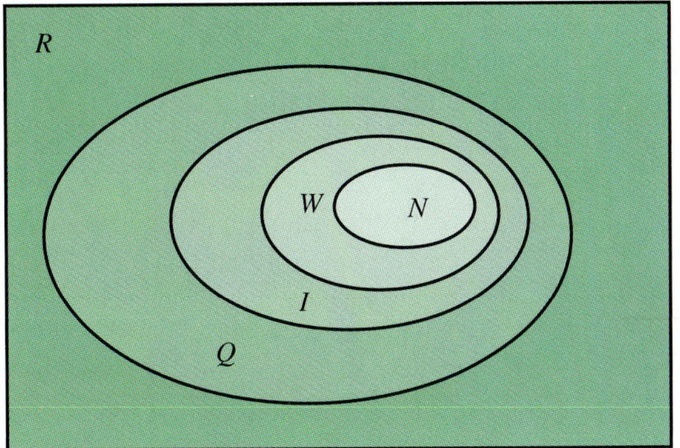

Figure 7.1
Venn diagram for the natural numbers, whole numbers, integers, and rational numbers

The following statements illustrate the relationships between these sets:

- The number 2 is a natural number, a whole number, an integer, and a rational number.
- The number 0 is a whole number, an integer, and a rational number; it is *not* a natural number.
- The number -5 is an integer and a rational number; it is not a whole number or a natural number.
- The number $3/4$ is a rational number; it is not an integer, a whole number, or a natural number.

From a practical point of view, each number system was created to meet a specific need:

- The natural numbers came into being as a result of people's need to count.
- The number 0, which, with the natural numbers comprises the set of whole numbers, was first introduced simply as a placeholder in positional number systems, but it also allows us to count the number of elements in the empty set and has other useful arithmetic properties.
- The integers allow us to keep track of debits and credits, gains and losses, degrees above zero and degrees below zero, and so on.
- The rational numbers are needed to make accurate measurements of lengths, areas, volumes, and other quantities.

From a more mathematical point of view, the natural numbers allow us to solve equations like

$$x - 3 = 0 \quad \text{and} \quad 7x - 21 = 0.$$

The whole numbers allow us to solve equations like

$$3x = 0 \quad \text{and} \quad (x - 5)(x - 7) = 0.$$

Extending the number system to include all integers allows us to solve equations like

$$x + 4 = 0 \quad \text{and} \quad 9x + 36 = 0.$$

Extending the number system to include the rational numbers allows us to solve equations like

$$3x - 4 = 0 \quad \text{and} \quad (2x + 1)(5x - 4) = 0.$$

In summary, the rational-number system permits us to deal with a wide range of practical and theoretical problems. However, remarkably, there remain simple questions that ought to have simple answers but do not if we do not extend the number system beyond the rationals. For example, if we stopped with the rationals, we would be unable to answer such a simple question as "How long is the diagonal of a square measuring 1 unit on a side?" (See Figure 7.2.) Thus, it is necessary to extend the number system at least once more. In this chapter, we consider the system of numbers called the real numbers.* We do so by means of decimals.

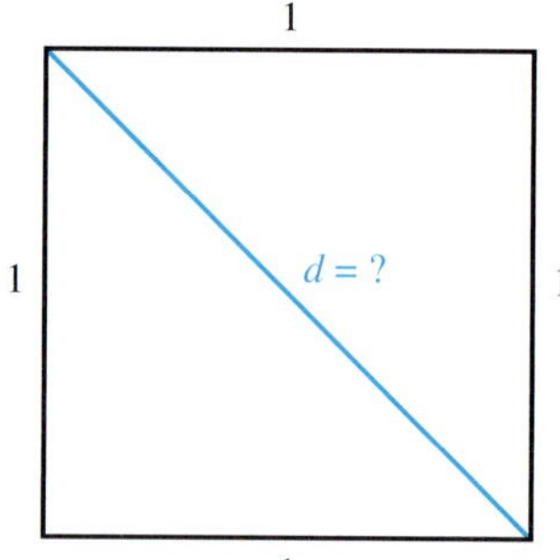

Figure 7.2
The length of the diagonal of a square 1 unit on a side is not a rational number.

7.1 DECIMALS

Since it is a positional system based on ten, we have used the term "decimal system" (from the Latin *decimus,* meaning "tenth") to refer generally to the Indo-Arabic system of numeration in common use today. However, more colloquially, people often refer to expressions like 0.235 or 2.7142 as **decimals,** as opposed to 24, 98, 0, or 2478, which they more often speak of as whole numbers or integers. In fact, both are part and parcel of the same system. Just as the **expanded form** of 2478 is

$$\begin{aligned} 2478 &= 2 \cdot 10^3 + 4 \cdot 10^2 + 7 \cdot 10^1 + 8 \cdot 10^0 \\ &= 2000 + 400 + 70 + 8, \end{aligned}$$

the expanded form of 0.235 is

$$\begin{aligned} 0.235 &= 2 \cdot \frac{1}{10^1} + 3 \cdot \frac{1}{10^2} + 5 \cdot \frac{1}{10^3} \\ &= \frac{2}{10} + \frac{3}{10} + \frac{5}{1000}, \end{aligned}$$

*The word *real,* when applied to numbers, is used to distinguish these numbers from those called *imaginary* in a final extension of the number system to the system of *complex numbers*. We do not consider the complex-number system in this book.

and the expanded form of 23.47 is

$$23.47 = 2 \cdot 10^1 + 3 \cdot 10^0 + 4 \cdot \frac{1}{10^1} + 7 \cdot \frac{1}{10^2}$$
$$= 20 + 3 + \frac{4}{10} + \frac{7}{100}.$$

Representations of Decimals

For beginning grade school students, it is helpful to introduce the study of decimals by considering concrete physical manipulative devices and pictorial representations. We discuss three such representations here.

Using Mats, Strips, and Units As in the introduction of fractions in Chapter 6, if a mat represents one unit, or 1, then a strip represents one-tenth, or $\frac{1}{10}$, and a unit represents one one-hundredth, or $\frac{1}{100}$, as shown in Figure 7.3. Thus, a display of 3 mats, 2 strips, and 5 units, as shown in Figure 7.4, represents $3 \cdot 1 + 2 \cdot \frac{1}{10} + 5 \cdot \frac{1}{100} = 3 + \frac{2}{10} + \frac{5}{100} = \frac{325}{100}$, which we write using the shorthand notation 3.25. Conversely, the expanded notation for 3.25 is $3 + \frac{2}{10} + \frac{5}{100}$. The decimal point is used to separate the integer part of the numeral from the fractional part.

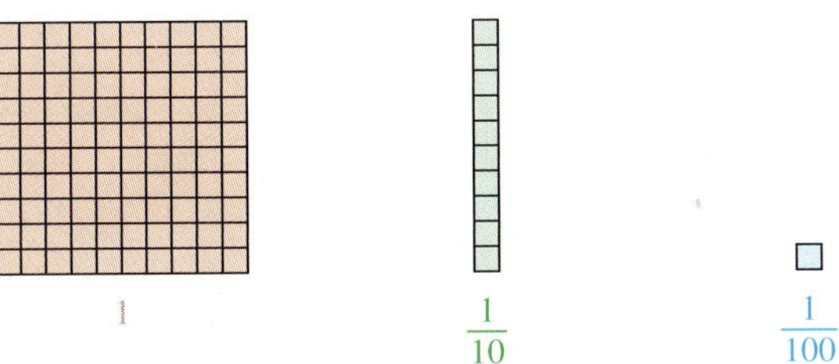

Figure 7.3
Representation of 1, $\frac{1}{10}$, and $\frac{1}{100}$ using a mat, a strip, and a unit

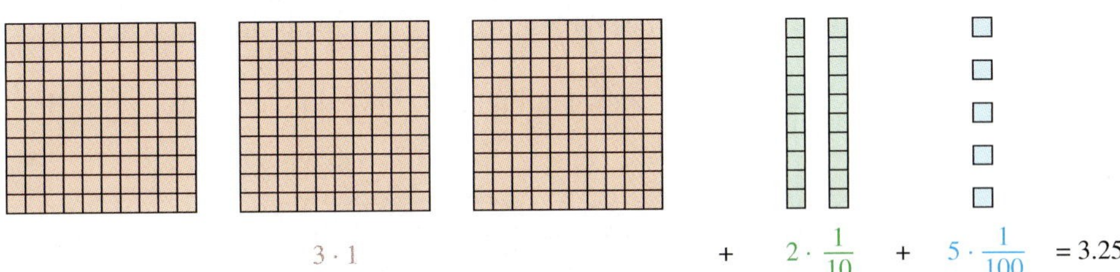

Figure 7.4
Representation of 3.25 using mats, strips, and units

Using base-ten blocks Consider the base-ten blocks depicted in Figure 7.5. Suppose the block is chosen as the unit. Then, since 10 flats are equivalent to a block, a flat represents $\frac{1}{10}$. Similarly, 10 longs are equivalent to a flat, and $10 \cdot 10 = 100$ longs are equivalent to

HIGHLIGHT *from* HISTORY

The Decimal Point

The development of mathematics over time has been paralleled by the development of good mathematical notation that facilitates doing mathematics. One of these developments was that of the Indo-Arabic, or decimal, system of notation, as already observed. Related to this is the use of a period (the decimal point) to separate the integer part and the fractional part of a numeral. Thus,

$$25.423 = 25\frac{423}{1000}.$$

As we have seen, use of the decimal point makes possible the easy extension of the algorithms for integer computation to decimal computation. Though it seems natural to us now, the decimal point was developed rather late in the history of mathematics, and even today its use is not completely standardized. To illustrate the development, we list here various notations of the past and present.

3.4813	modern American
3·4813	modern English
3,4813	modern continental Europe
3 | 4813	Rudolph, 1530
3 ⓪ 4 ① 8 ② 1 ③ 3 ④	S. Stevin, 1585
3 . 4 8 1 3	J. Beyer, 1603
3 4813	J. Beyer, 1603
$3^{(0)}4^{(1)}8^{(2)}1^{(3)}3^{(4)}$	R. Norton, 1608
34813 ④	W. Kalcheim, 1629

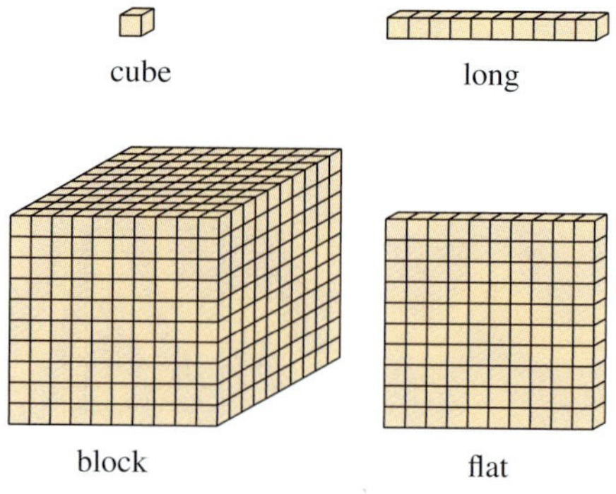

Figure 7.5
Base-ten blocks

a block. Thus, a long represents $\frac{1}{100}$. Finally, since there are $10 \cdot 10 \cdot 10 = 1000$ cubes in a block, a cube represents $\frac{1}{1000}$. With this in mind, a display of 2 blocks, 1 flat, 3 longs, and 2 cubes, as shown in Figure 7.6, would represent $2 \cdot 1 + 1 \cdot \frac{1}{10} + 3 \cdot \frac{1}{100} + 2 \cdot \frac{1}{1000} =$

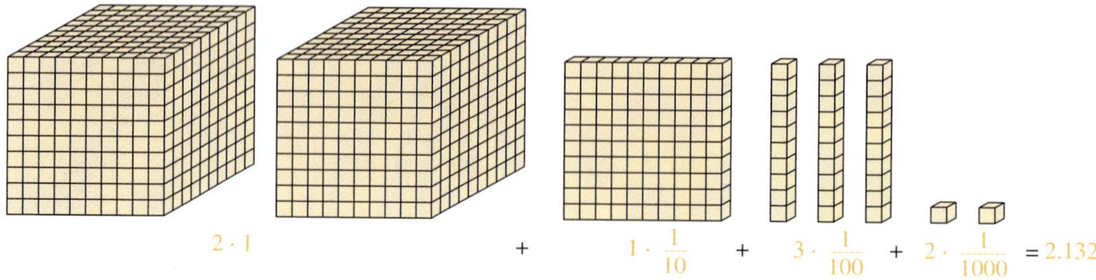

Figure 7.6
Representing 2.132 using base-ten blocks

$2 + \frac{1}{10} + \frac{3}{100} + \frac{2}{1000} = \frac{2132}{1000}$, which we write using the shorthand notation 2.132. Conversely, the expanded notation for 2.132 is $2 + \frac{1}{10} + \frac{3}{100} + \frac{2}{1000}$.

On the other hand, if the flat is chosen as the unit, then the above reasoning dictates that the block represents 10, the flat represents 1, the long represents $\frac{1}{10}$, and the cube represents $\frac{1}{100}$. In this case, the display in Figure 7.6 would represent $2 \cdot 10 + 1 \cdot 1 + 3 \cdot \frac{1}{10} + 2 \cdot \frac{1}{100} = 20 + 1 + \frac{3}{10} + \frac{2}{100}$, which we would write as 21.32.

Using dollars, dimes, and pennies A particularly apt manipulative, already familiar to children, is provided by money. Since 10 dimes are worth one dollar and one hundred pennies are also worth one dollar, a dollar represents 1, a dime represents $\frac{1}{10}$, and a penny represents $\frac{1}{100}$. Thus, the 3.25 of Figure 7.4 would be represented as in Figure 7.7. This

Figure 7.7
Representing 3.25 using dollars, dimes, and pennies

representation is particularly useful because students are already familiar with the fact that this set of coins is worth three dollars and 25 cents, as well as the fact that this is written \$3.25, thus making the understanding of the above expanded notation for 3.25 quite natural.

EXAMPLE 7.1 USING MONEY TO REPRESENT DECIMALS

How would you use money in attempting to explain the decimal 23.75 to elementary school students?

Solution

Use 2 10-dollar bills, 3 dollars, 7 dimes, and 5 pennies. The students will easily see that this collection of bills and coins is worth \$23.75, and this makes it relatively easy to explain the expanded notation

$$23.75 = 2 \cdot 10 + 3 \cdot 1 + 7 \cdot \frac{1}{10} + 5 \cdot \frac{1}{100}$$

since a dime is one-tenth of a dollar and a penny is one one-hundredth of a dollar.

School Book Page

Modeling Decimals

3-1 Place Value: Comparing and Ordering Decimals

You'll Learn ...

■ to compare and order decimals

... How It's Used

Environmental scientists need to compare the decimal values of pollutants that they find in the air or in water.

▶ **Lesson Link** You have worked with whole numbers. Now you'll begin to study decimals by deciding which of two decimals is greater. ◀

Explore Comparing and Ordering Decimals

Model Behavior

Materials: Graph paper, Colored pencils

Modeling Decimals

The graph paper shows a model of the number 1.47. To model a decimal:

- Color a complete 10-by-10 grid for each whole in the decimal.
- Draw another 10-by-10 grid next to the last complete grid.
- In this grid, color one 10-by-1 strip for each tenth in the decimal. Color a small square for each hundredth.

1.47

1. Draw a grid model for each decimal.

   a. 1.3 b. 1.29 c. 0.8 d. 1.30 e. 0.51 f. 0.99

2. What do you notice about the models for 1.3 and 1.30? What does this tell you about these numbers? Explain.
3. Rank the decimals from smallest to largest. Explain your reasoning.
4. The number 51 is greater than 8. Why is 0.8 greater than 0.51?
5. Could you use this method to model 1.354? Explain.

Learn Comparing and Ordering Decimals

The place value of each digit of a whole number is one-tenth of the value of the place to its left. Moving to the right of a decimal point, you can create the place values *tenths, hundredths, thousandths,* and so on.

106 *Chapter 3 • Number Sense: Decimals and Fractions*

SOURCE: Scott Foresman–Addison Wesley Middle School Math, *Course 2, p. 106, by Randall I. Charles et al.*

Questions for the Teacher

1. The above lesson models decimals. Each 10-by-10 square represents one unit. Each column of 10 small squares represents one-tenth. Each small square represents one one-hundredth. Since this is the language of fractions, how would you explain the notation 1.47 to a student?
2. How would you assist a student having trouble with problem 4 above?
3. How would you help a student understand how to answer problem 5 above? Would it help to have a set of base-ten blocks that includes a 1000 cube? Explain.

from **The NCTM Principles and Standards**

When students leave grade 5, they should be able to solve problems involving whole-number computation and should recognize that each operation will help them solve many different types of problems. They should be able to solve many problems mentally, to estimate a reasonable result for a problem, to efficiently recall or derive the basic number combinations for each operation, and to compute fluently with multi-digit whole numbers. They should understand the equivalence of fractions, decimals, and percents and the information each type of representation conveys. With these understandings and skills, they should be able to develop strategies for computing with familiar fractions and decimals.

By using the representations discussed, students can be led to understand that the shorthand notation for $\frac{1}{10}$ is 0.1, for $\frac{1}{100}$ is 0.01, for $\frac{1}{1000}$ is 0.001, for $\frac{2}{10}$ is 0.2, for $\frac{3}{100}$ is 0.03, etc., and to understand that each digit of a decimal numeral contributes an amount to the number being represented that depends both on the digit and on its position in the number as shown in Table 7.1.

TABLE 7.1 Positional Values in the Decimal System

	POSITION NAMES								
Form	. . .	Hundreds	Tens	Units	Tenths	Hundredths	Thousandths	Ten-Thousandths	. . .
Decimal Form	. . .	100	10	1	0.1	0.01	0.001	0.0001	. . .
Fractional Form	. . .	100	10	1	$\frac{1}{10}$	$\frac{1}{100}$	$\frac{1}{1000}$	$\frac{1}{10{,}000}$	. . .
Power of 10	. . .	10^2	10^1	10^0	10^{-1}	10^{-2}	10^{-3}	10^{-4}	. . .*

The pattern in expanded notation becomes even more clear if we use negative exponents.

Negative Exponents and Expanded Exponential Form

We have already observed that, for a natural number n,

$$a^n \text{ is shorthand for the product } \overbrace{a \cdot a \cdots a}^{n \text{ factors}}$$

and therefore

*See the upcoming discussion of negative exponents.

$$a^n \cdot a^m = \overbrace{a \cdot a \cdots a}^{n \text{ factors}} \cdot \overbrace{a \cdots a}^{m \text{ factors}} = a^{m+n}.$$

Moreover, if this rule is to hold for $m = 0$, then

$$a^n \cdot a^0 = a^{n+0} = a^n,$$

and we must define $a^0 = 1$ as we did in Chapter 2. Now suppose that n is a positive integer so that $-n$ is negative. What shall we mean by a^{-n}? If we require that the above rule of exponents holds for **negative exponents** as well as for 0, then

$$a^n \cdot a^{-n} = a^{n+(-n)} = a^0 = 1.$$

This implies that we should define a^{-n} to be $1/a^n$.

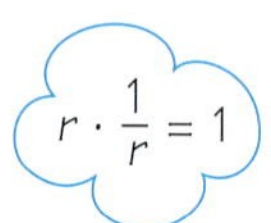

DEFINITION *Negative Numbers and Zero as Exponents*

If n is a positive integer and $a \neq 0$, then $a^0 = 1$ and $a^{-n} = \dfrac{1}{a^n}$.

Using this definition, we complete the last row of Table 7.1 as shown. We also write the decimals discussed at the beginning of this section in **expanded exponential form** as follows:

$$\begin{aligned} 0.235 &= 0 + 2 \cdot \frac{1}{10^1} + 3 \cdot \frac{1}{10^2} + 5 \cdot \frac{1}{10^3} \\ &= 2 \cdot 10^{-1} + 3 \cdot 10^{-2} + 5 \cdot 10^{-3}; \\ 23.47 &= 2 \cdot 10^1 + 3 \cdot 10^0 + 4 \cdot \frac{1}{10^1} + 7 \cdot \frac{1}{10^2} \\ &= 2 \cdot 10^1 + 3 \cdot 10^0 + 4 \cdot 10^{-1} + 7 \cdot 10^{-2}. \end{aligned}$$

In this form, the pattern of decreasing exponents is both clear and neat.

EXAMPLE 7.2 WRITING DECIMALS IN EXPANDED EXPONENTIAL FORM

Write each of these in expanded exponential form.

(a) 234.72 **(b)** 30.0012

Solution

(a) $234.72 = 2 \cdot 10^2 + 3 \cdot 10^1 + 4 \cdot 10^0 + 7 \cdot 10^{-1} + 2 \cdot 10^{-2}$

(b) $30.0012 = 3 \cdot 10^1 + 0 \cdot 10^0 + 0 \cdot 10^{-1} + 0 \cdot 10^{-2} + 1 \cdot 10^{-3} + 2 \cdot 10^{-4}$

Multiplying and Dividing Decimals by Powers of 10

There is an emphasis in elementary mathematics textbooks on what happens notationally (to the decimal point) when a number is multiplied by a power of 10. Consider the decimal

$$25.723 = 2 \cdot 10^1 + 5 \cdot 10^0 + 7 \cdot 10^{-1} + 2 \cdot 10^{-2} + 3 \cdot 10^{-3}.$$

If we multiply by 10^2 then, using the distributive property and the above rule for multiplying powers, we obtain

$$\begin{aligned}(10^2)(25.723) &= (10^2)(2 \cdot 10^1 + 5 \cdot 10^0 + 7 \cdot 10^{-1} + 2 \cdot 10^{-2} + 3 \cdot 10^{-3}) \\ &= 2 \cdot 10^{2+1} + 5 \cdot 10^{2+0} + 7 \cdot 10^{2+(-1)} + 2 \cdot 10^{2+(-2)} + 3 \cdot 10^{2+(-3)} \\ &= 2 \cdot 10^3 + 5 \cdot 10^2 + 7 \cdot 10^1 + 2 \cdot 10^0 + 3 \cdot 10^{-1} \\ &= 2572.3,\end{aligned}$$

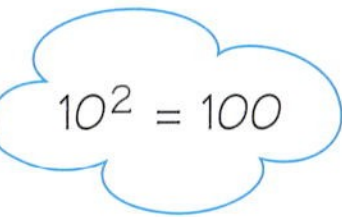

and the notational effect is to move the decimal point two places to the right. Note that 2 is both the exponent of the power of 10 we are multiplying by and the number of zeros in 100.

The result is analogous if we divide 25.723 by 10^2, except that the notational effect is to move the decimal point two places *to the left*. To see this, recall that we can divide by multiplying by the multiplicative inverse. Thus,

$10^{-2} = \frac{1}{10^2}$

$$\begin{aligned}(25.723) \div 10^2 &= (25.723) \cdot (1/10^2) \\ &= (25.723) \cdot 10^{-2} \\ &= (2 \cdot 10^1 + 5 \cdot 10^0 + 7 \cdot 10^{-1} + 2 \cdot 10^{-2} + 3 \cdot 10^{-3}) \cdot (10^{-2}) \\ &= 2 \cdot 10^{1+(-2)} + 5 \cdot 10^{0+(-2)} + 7 \cdot 10^{(-1)+(-2)} + 2 \cdot 10^{(-2)+(-2)} \\ &\quad + 3 \cdot 10^{(-3)+(-2)} \\ &= 2 \cdot 10^{-1} + 5 \cdot 10^{-2} + 7 \cdot 10^{-3} + 2 \cdot 10^{-4} + 3 \cdot 10^{-5} \\ &= 0.25723.\end{aligned}$$

These results are typical of the general case, which we state here as a theorem.

> **THEOREM** *Multiplying and Dividing Decimals by Powers of 10*
> If r is a positive integer, the notational effect of multiplying a decimal by 10^r is to move the decimal point r places to the right. The notational effect of dividing a decimal by 10^r (that is, multiplying by 10^{-r}) is to move the decimal point r places to the left.

EXAMPLE 7.3 | MULTIPLYING AND DIVIDING DECIMALS BY POWERS OF 10

Compute each of the following.

(a) $(10^3)(253.26)$ **(b)** $(253.26) \div 10^3$
(c) $(100)(34.764)$ **(d)** $(34.764) \div 10{,}000$

Solution

The preceding theorem gives the desired results.

(a) $(10^3)(253.26) = 253{,}260$
(b) $(253.26) \div 10^3 = 0.25326$
(c) $(100)(34.764) = (10^2)(34.764) = 3476.4$
(d) $(34.764) \div 10{,}000 = 34.764 \div (10^4) = 0.0034764$

Terminating Decimals as Fractions

From grade school, we know that

$$\frac{1}{3} = 0.333\ldots$$

where the ellipsis dots indicate that the string of 3s continues without end. Such a decimal is called a **nonterminating decimal.** A decimal like 24.357, which has a finite number of digits, is called a **terminating decimal.** The preceding discussion concerning expanded notation shows that every terminating decimal represents a rational number; that is, any terminating decimal can be represented by a fraction with integers in the numerator and denominator. For example, using expanded notation, we write

$$\begin{aligned} 24.357 &= 20 + 4 + \frac{3}{10} + \frac{5}{10} + \frac{7}{1000} \\ &= \frac{20{,}000}{1000} + \frac{4000}{1000} + \frac{300}{1000} + \frac{50}{1000} + \frac{7}{1000} \\ &= \frac{24{,}357}{1000}. \end{aligned}$$

Alternatively,

$$\begin{aligned} 24.357 &= 24.357 \cdot \frac{1000}{1000} \\ &= \frac{24{,}357}{1000} \end{aligned}$$

$a \cdot \frac{b}{c} = \frac{ab}{c}$

as before. In each case, the denominator is determined by the position of the rightmost digit. (In this case, the 7 is in the thousandths position.)

The preceding discussion shows how to convert a terminating decimal into a ratio of two integers, that is, into a fraction. Not all rational numbers have finite decimal expansions, but those that do can be converted to decimal form as follows. Consider

$$\frac{17}{40} = \frac{17}{2^3 \cdot 5^1}.$$

Since the prime-factor representation of the denominator contains only 2s and 5s, the fraction can be written so that the denominator is a power of 10. Thus,

Multiply numerator and denominator by 5^2.

$$\frac{17}{40} = \frac{17}{2^3 \cdot 5^1} = \frac{17 \cdot 5^2}{2^3 \cdot 5^3} = \frac{17 \cdot 25}{10^3} = \frac{425}{1000} = 0.425.$$

Since the numbers in the preceding discussion are typical of the general case, the results can be summarized as a theorem.

THEOREM *Terminating Decimals and Rational Numbers*
If a and b are integers with $b \neq 0$, if a/b is in simplest form, and if the prime-factor representation of b contains only 2s and 5s, then a/b can be represented as a terminating decimal and vice versa.

EXAMPLE 7.4 WRITING A TERMINATING DECIMAL AS A RATIO OF TWO INTEGERS

Express each of these in the form a/b, where the fraction is in simplest form.

(a) 31.75 **(b)** 4.112 **(c)** −0.035

Solution

(a) $31.75 = 31.75 \cdot \frac{100}{100} = \frac{3175}{100} = \frac{127}{4}$. Note that GCD(3175, 100) = 25.

(b) $4.112 = 4.112 \cdot \frac{1000}{1000} = \frac{4112}{1000} = \frac{514}{125}$. Note that GCD(4112, 1000) = 8.

(c) $-0.035 = -0.035 \cdot \frac{1000}{1000} = \frac{-35}{1000} = \frac{-7}{200}$. Note that GCD(35, 1000) = 5.

We just saw how to write 17/40 as a finite decimal by multiplying both numerator and denominator by powers of 2 or 5. But the task can also be accomplished by division. Thus, by hand or using a calculator, we have that

$$\frac{17}{40} = 0.425.$$

$\frac{17}{40} = 17 \div 40$

EXAMPLE 7.5 CONVERTING CERTAIN FRACTIONS TO DECIMALS

Convert each of these fractions to decimals by writing each as an equivalent fraction whose denominator is a power of 10. Check by dividing using a calculator.

(a) $\frac{37}{40}$ **(b)** $-\frac{29}{200}$

Solution

(a)
$$\begin{aligned}\frac{37}{40} &= \frac{37}{2^3 \cdot 5^1} = \frac{37 \cdot 5^2}{2^3 \cdot 5^3}\\ &= \frac{37 \cdot 25}{10^3}\\ &= \frac{925}{1000}\\ &= 0.925.\end{aligned}$$

$5^1 \cdot 5^2 = 5^3$

$5^2 = 25,\ 37 \cdot 25 = 925$

Also, by calculator, 37 ÷ 40 = 0.925.

(b)
$$\begin{aligned}-\frac{29}{200} &= -\frac{29}{2^3 \cdot 5^2} = -\frac{29 \cdot 5}{2^3 \cdot 5^3}\\ &= -\frac{145}{10^3}\\ &= -\frac{145}{1000}\\ &= -0.145.\end{aligned}$$

Also, by calculator, −(29 ÷ 200) = −0.145.

Nonterminating Decimals and Rational Numbers

Somewhat surprisingly, not all rational numbers have decimal expansions that terminate. For example, as noted earlier,

$$\frac{1}{3} = 0.333\ldots = 0.\overline{3}$$

Simon Stevin (1548–1620)

The most influential mathematician from the Low Countries in the sixteenth century was the Flemish mathematician Simon Stevin. Stevin served in the Dutch army as quartermaster and was something of an expert on military engineering and the building of fortifications. His most significant contribution to mathematics was the extension of the use of the positional Indo-Arabic system of numeration to the rational and real numbers, thus greatly facilitating numerical calculation. To the local Dutch, he was most famous for his invention of a carriage powered by sails and capable of carrying more than a score of people at speeds up to 20 miles per hour.

---

where the three dots indicate that the decimal continues *ad infinitum,* and the bar over the 3 indicates the digit or group of digits that repeats. To see why this is so, set

$$x = 0.333\ldots.^{*}$$

Then, using the preceding theorem, we obtain

$$10x = 3.333\ldots,$$

so that

$$10x - x = 9x = 3.$$

$$\begin{array}{r} 3.333\ldots \\ -\ 0.333\ldots \\ \hline 3.000\ldots \end{array}$$

But this implies that

$$x = \frac{3}{9} = \frac{1}{3}.$$

Notice that the decimal expansion of $1/3$ is a nonterminating, but *repeating,* decimal; that is, the stream of 3s repeats without end. In general, we have the following definition.

> **DEFINITION** *A Repeating Decimal*
> A nonterminating decimal that has the property that a digit or group of digits repeats *ad infinitum* from some point on is called a **periodic** or **repeating decimal.** The number of digits in the repeating group is called the **length of the period.**

Just as $0.333\ldots = .\overline{3}$ represents the rational number $1/3$, every repeating decimal represents a rational number.

EXAMPLE 7.6 REPEATING DECIMALS AS RATIONAL NUMBERS

Write each of these repeating decimals in the form a/b where a and b are integers and the fraction is in simplest form. Check by dividing a by b with your calculator.

---

*Actually, there is a touchy point here that we gloss over. The question is whether $0.333\ldots = 3/10 + 3/100 + 3/1000 + \ldots$ means anything at all since it is the sum of an *infinite* number of numbers. That the answer is "yes" really depends on ideas from calculus!

(a) $0.242424\ldots = 0.\overline{24}$ **(b)** $3.14555\ldots = 3.14\overline{5}$

Solution

(a) Let $x = 0.242424\ldots.$ (a decimal of period 2)
Then

Multiply by $10^2 = 100$ to move the decimal point two places to the right.

$$100x = 24.242424\ldots$$

and

$$100x - x = 99x = 24.$$

$24.242424\ldots$
$-\ 0.242424\ldots$
$24.000000\ldots$

But then

$$x = \frac{24}{99} = \frac{8}{33}.$$

Also, by calculator, $8 \div 33 \doteq 0.2424242$.

Here, the calculator answer is only approximate, since only finitely many digits can appear in the calculator display.

(b) Let $x = 3.14555\ldots.$

Then

Multiply by 10^2 to move the decimal point two places to the right where the period starts.

$$100x = 314.555\ldots$$

and

a decimal of period 1

$$1000x = 3145.555\ldots.$$

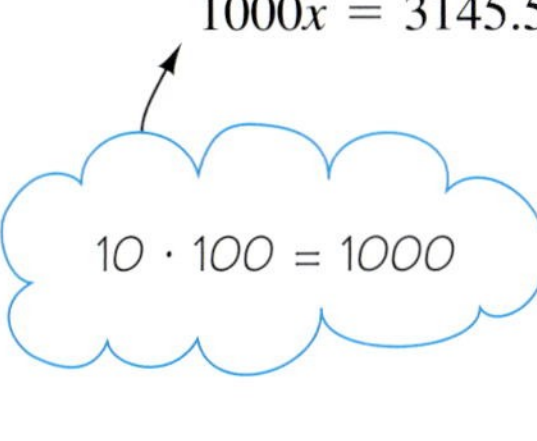

Multiply by 10 again to move the decimal point one more place (the length of the period) to the right.

Then

$$1000x - 100x = 900x = 2831$$

and

$$x = \frac{2831}{900}.$$

$3145.555\ldots$
$-\ 314.555\ldots$
$2831.000\ldots$

Also, by calculator, $2831 \div 900 \doteq 3.1455556$. Explain why the decimal is finite and ends with a 6.

Example 7.6 is typical of the general case, and so we have the following theorem.

THEOREM *Rational Numbers and Periodic Decimals*
Every repeating decimal represents a rational number a/b. If a/b is in simplest form, b must contain a prime factor other than 2 or 5. Conversely, if a/b is such a rational number, its decimal representation must be repeating.

"I THOUGHT IT WAS A BREAKTHROUGH, BUT IT WAS ONLY A MISPLACED DECIMAL."

Ordering Decimals

Elementary mathematics textbooks also lay stress on properly ordering decimals. Ordering decimals is much like ordering integers. For example, to determine the larger of 247,761 and 2,326,447, write both numerals as if they had the same number of digits; that is, write

$$0,247,761 \quad \text{and} \quad 2,326,447.$$

Then determine the first place from the *left* where the digits differ. It follows from the idea of positional notation that the larger integer is the integer with the larger of these two different digits. In the present case, the first digits differ, and so

$$0{,}247{,}761 < 2{,}326{,}447.$$

0 < 2

In a similar example,

$$34{,}716 < 34{,}723$$

1 < 2

since the first corresponding digits that differ are the 1 and the 2, and $1 < 2$.

Now consider the decimals

$$0.2346612359 \quad \text{and} \quad 0.2348999.$$

Suppose we multiply both numbers by 10^{10} so that both become integers. We obtain

$$2{,}346{,}612{,}359 \quad \text{and} \quad 2{,}348{,}999{,}000.$$

We order these as integers in the manner just discussed to obtain

$$2{,}346{,}612{,}359 < 2{,}348{,}999{,}000.$$

6 < 8

Thus,

$$(10^{10})(0.2346612359) < (10^{10})(0.2348999000).$$

But this implies that

$$0.2346612359 < 0.2348999,$$

Divide both sides by 10 10.

and we are done.

These arguments could be repeated in general with the following result.

If ab < ac and a > 0,
then b < c.

THEOREM *Ordering Decimals*

To order two positive decimals, adjoin zeros on the left if necessary so that there are the same number of digits to the left of the decimal point in both numbers, and then determine the first digits from the left that differ. The decimal with the lesser of these two digits is the lesser decimal.

EXAMPLE 7.7 ORDERING DECIMALS

In each case, decide which of the decimals represents the lesser number.

(a) 2.35714 and 2.35709 **(b)** 23.45 and $23.4\overline{5}$

Solution

(a) Here, the first digits from the left that differ are 1 and 0. Since $0 < 1$, it follows that $2.35709 < 2.35714$.

(b) Since $23.4\overline{5} = 23.4555\ldots$ and $23.45 = 23.45\overline{0}$, the first digits from the left that differ are 0 and 5. Since $0 < 5$, it follows that $23.45 < 23.4\overline{5}$.

EXAMPLE 7.8 ORDERING DECIMALS AND FRACTIONS

Arrange these numbers in order from least to greatest.

$$\frac{11}{24}, \quad \frac{3}{8}, \quad 0.37, \quad 0.4584, \quad 0.37666\ldots, \quad 0.4583$$

Solution

The easiest approach is to write all these numbers as decimals. Since

$$\frac{11}{24} = 0.458333\ldots \quad \text{and} \quad \frac{3}{8} = 0.375,$$

it follows that

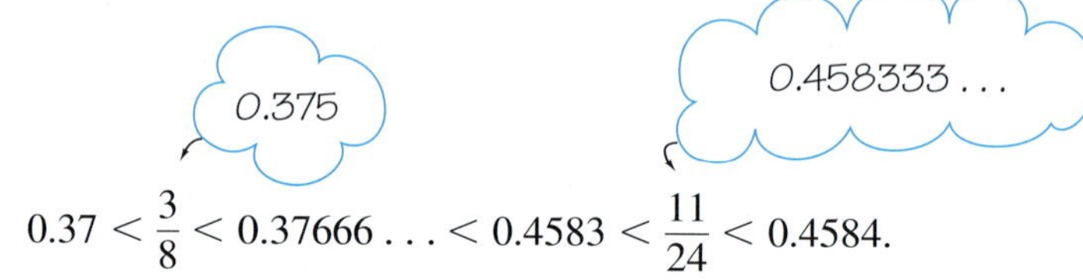

$$0.37 < \frac{3}{8} < 0.37666\ldots < 0.4583 < \frac{11}{24} < 0.4584.$$

The Set of Real Numbers

Earlier, we showed that

> *the decimal expansion of a rational number either terminates or is nonterminating and repeating.*

This raises an interesting question. What kind of numbers are represented by nonterminating, nonrepeating decimals like 0.101001000100001 . . . ? Such numbers cannot be represented as ratios of integers and so are called **irrational numbers.** This distinguishes them from the rational numbers considered in Chapter 6. The set consisting of all rational and irrational numbers is called the set of **real numbers.**

DEFINITION *Irrational and Real Numbers*
Numbers represented by nonterminating, nonperiodic decimals are called **irrational numbers.** The set R consisting of all rational numbers and all irrational numbers is called the set of **real numbers.**

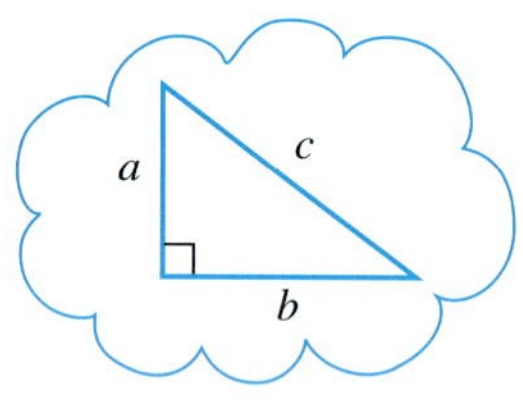
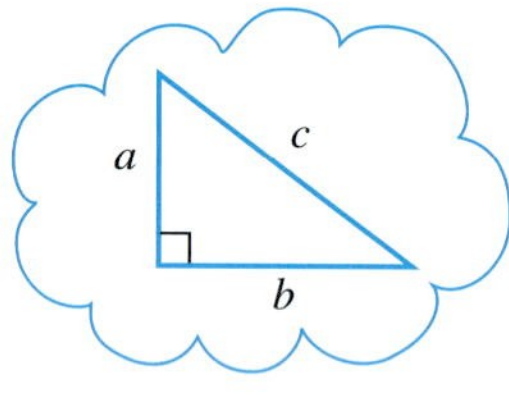

Thus, the set of real numbers is another extension of the number system. This number system contains all the earlier systems as illustrated by the Venn diagram in Figure 7.1.

Irrationality of $\sqrt{2}$

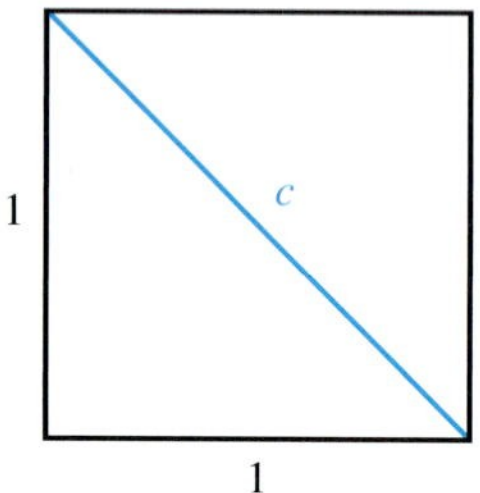

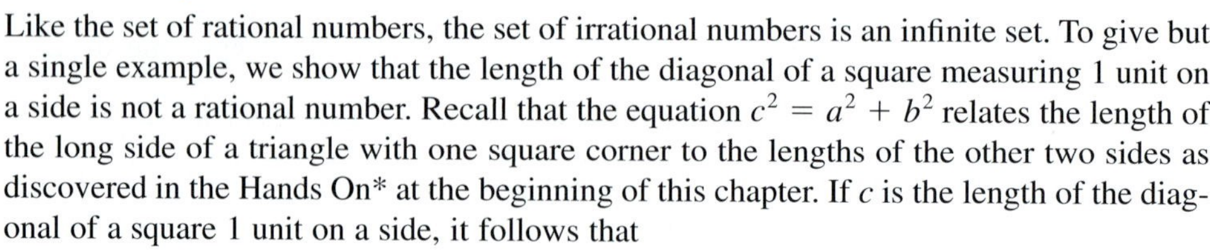

Like the set of rational numbers, the set of irrational numbers is an infinite set. To give but a single example, we show that the length of the diagonal of a square measuring 1 unit on a side is not a rational number. Recall that the equation $c^2 = a^2 + b^2$ relates the length of the long side of a triangle with one square corner to the lengths of the other two sides as discovered in the Hands On* at the beginning of this chapter. If c is the length of the diagonal of a square 1 unit on a side, it follows that

$$c^2 = 1^2 + 1^2 = 2.$$

*A triangle with one square corner (that is, a corner forming a 90° angle) is called a **right** triangle. The fact that $c^2 = a^2 + b^2$ for such a triangle is called the **Pythagorean theorem.**

HIGHLIGHT from HISTORY

Pythagoras and Irrational Numbers

The great Greek mathematician Pythagoras (ca. 585–497 B.C.) is best known because of the Pythagorean theorem considered in the Hands On at the beginning of this chapter. Actually, it is doubtful that Pythagoras discovered this theorem or even proved the result true. At best, the theorem may be due to one of the members of the religious, philosophical, and scientific society founded by Pythagoras, called the Pythagorean School. The Pythagoreans said that "all is number" (and by numbers they meant natural numbers) and made numbers the basis for their entire system of thought. It is true that the Pythagoreans considered ratios and were able to deal with situations like measurement that required rational numbers. However, the discovery by Hippasus of Metapotum that the diagonal and side of a square are incommensurable (that is, have no common unit of measure), which amounts to showing that $\sqrt{2}$ is not rational and so cannot be expressed as a ratio of integers, struck a serious blow to the Pythagorean system. The story goes that Hippasus made his discovery while on a sea voyage with his fellow Pythagoreans who were so distressed that Hippasus was promptly thrown overboard and drowned.

Recall that if $c^2 = 2$, then c is called the square root of 2, and we write $c = \sqrt{2}$. We now show that $\sqrt{2}$ is not a rational number.

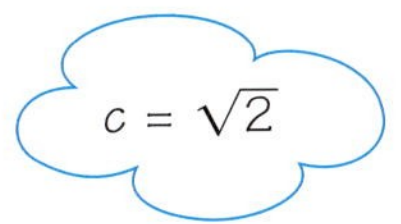

THEOREM *Irrationality of* $\sqrt{2}$

If $c^2 = 2$, then c is not a rational number.

Proof (By Contradiction) Suppose, to the contrary, that there is a rational number c for which $c^2 = 2$. Since any rational number can be expressed as a fraction in simplest form, let

$$c = \frac{u}{v}$$

where u and v have no common factor other than 1. But $c^2 = 2$, so

$$2 = \frac{u^2}{v^2}.$$

Thus,

$$2v^2 = u^2.$$

This says that 2 is a prime factor of u^2. Hence, by the fundamental theorem of arithmetic, 2 is a prime factor of u. But then $u = 2k$ for some integer k, and

$$2v^2 = u^2 = (2k)^2 = 4k^2.$$

This implies that

$$v^2 = 2k^2,$$

and so 2 is also a factor of v^2 and hence of v. But then u and v have a factor of 2 in common, and u/v is not in simplest form. This contradicts the fact that *u/v is in simplest form.*

In view of this contradiction, the assumption that c is rational that started this chain of reasoning must be false. Therefore, c is irrational as was to be proved.

EXAMPLE 7.9 PROVING A NUMBER IRRATIONAL

Show that $3 + \sqrt{2}$ is irrational.

Solution

Understand the Problem

We have just seen that $\sqrt{2}$ is irrational. We must show that $3 + \sqrt{2}$ is irrational.

Devise a Plan

Does the assertion even make sense? Is it possible that $3 + \sqrt{2}$ is rational? If so, then $3 + \sqrt{2} = s$, where s is some rational number. Perhaps we can use this equation, and the fact that we already know that $\sqrt{2}$ is irrational, to arrive at the desired conclusion.

Carry Out the Plan

Since $3 + \sqrt{2} = s$, it follows that $\sqrt{2} = s - 3$. But 3 and s are rational, and we know that the rational numbers are closed under subtraction. This implies that $\sqrt{2}$ is rational, and we know that that is not so. Therefore, the assumption that $3 + \sqrt{2} = s$, where s is rational, must be false. Hence, $3 + \sqrt{2}$ is irrational as was to be shown.

Look Back

Initially, it was not clear that the assertion of the problem had to be true, so we assumed briefly that it was not true. But this led directly to the contradiction that $\sqrt{2}$ was rational, and so the assumption that $3 + \sqrt{2}$ was rational had to be false; that is, $3 + \sqrt{2}$ had to be irrational as we were to prove.

The preceding argument could be repeated exactly with 3 replaced by any rational number r. Thus, since there are infinitely many rational numbers, the preceding example shows that there are infinitely many irrational numbers. Moreover, in a very precise sense, which we will not discuss here, there are many more irrational numbers than rational numbers. One such number that is well known, but not usually known to be irrational, is the number

$$\pi = 3.14159265\ldots$$

occurring in the formula $A = \pi r^2$ for the area of a circle of radius r and in the formula $C = 2\pi r$ for the circumference of a circle. Moreover, if the natural number m is not a perfect square, it can be shown that $\sqrt{m}$ is also irrational. Indeed, if m is not a perfect nth power, $\sqrt[n]{m}$ is irrational, where by $\sqrt[n]{m}$ we mean a number c such that $c^n = m$.

Real Numbers and the Number Line

When we discussed the length of the diagonal of the square measuring 1 unit on a side and proved that $\sqrt{2}$ is irrational, we made a tacit assumption that is easily overlooked. Simply put, the assumption was that to every possible length there corresponds exactly one real number that names the length in question. Figure 7.8 illustrates how we can accurately plot the length $\sqrt{2}$ on the number line.

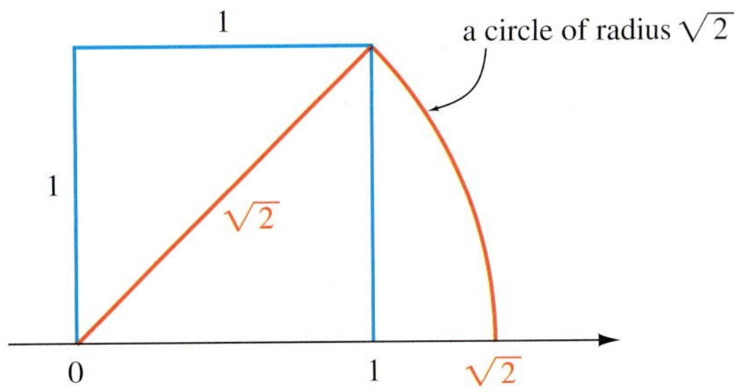

1

a circle of radius $\sqrt{2}$

1

$\sqrt{2}$

0 1 $\sqrt{2}$

Figure 7.8
$\sqrt{2}$ on the number line

Also, the decimal expansion of $\sqrt{2}$ can be found step by step as follows. Using a calculator and trial and error, we find that

$$1^2 = 1 < 2 < 4 = 2^2,$$
$$1.4^2 = 1.96 < 2 < 2.25 = 1.5^2,$$
$$1.41^2 = 1.9881 < 2 < 2.0164 = 1.42^2,$$
$$1.414^2 = 1.999396 < 2 < 2.002225 = 1.415^2,$$

and so on. Thus, the decimal expansion of $\sqrt{2}$ is

$$\sqrt{2} = 1.414\ldots$$

Note that this is determined by the lower estimates above.

where successive digits are found by catching $\sqrt{2}$ between successive units, between successive tenths, between successive hundredths, and so on. To show that this can be done for any real number is somewhat tricky, but the following theorem can be proved.

THEOREM *Real Numbers and the Number Line*
There is a one-to-one correspondence between the set of real numbers and the set of points on a number line. The absolute value of the number associated with any given point gives the distance of the point to the right or left of the 0.

Problem Set 7.1

Understanding Concepts

1. Write the following decimals in expanded form and in expanded exponential form.
   (a) 273.412 (b) 0.000723 (c) 0.20305
2. Write these decimals as fractions in lowest terms and determine the prime factorization of the denominator in each case.
   (a) 0.324 (b) 0.028 (c) 4.25
3. Write these fractions as terminating decimals.
   (a) $\frac{7}{20}$ (b) $\frac{7}{16}$ (c) $\frac{3}{75}$ (d) $\frac{18}{2^2 \cdot 5^4}$
4. Determine the rational fraction in lowest terms represented by each of these periodic decimals.
   (a) $0.321321\ldots = 0.\overline{321}$
   (b) $0.12414141\ldots = 0.12\overline{41}$
   (c) $3.262626\ldots = 3.\overline{26}$
   (d) $0.666\ldots = 0.\overline{6}$
   (e) $0.142857142857142857\ldots = 0.\overline{142857}$
   (f) $0.153846153846153846\ldots = 0.\overline{153846}$
5. (a) Give an example of a fraction whose decimal expansion is repeating and has a period of length 3.

(b) Use your calculator to check your answer to part (a). Is this check really complete? Explain.

(c) Check your answer to part (a) by some other means.

6. For each of these, write the numbers in order of increasing size (from least to greatest).

(a) 0.017, 0.007, $0.01\overline{7}$, 0.027

(b) 25.412, 25.312, 24.999, $25.41\overline{2}$

(c) $\frac{9}{25}$, 0.35, 0.36, $\frac{10}{25}$, $0.\overline{35}$

(d) $\frac{1}{4}, \frac{5}{24}, \frac{1}{3}, \frac{1}{6}$

7. Show that $\sqrt{3}$ is irrational. (*Hint:* Have you ever seen a similar result proved?)

8. Show that $3 - \sqrt{2}$ is irrational.

9. Show that $2\sqrt{2}$ is irrational.

Teaching Concepts

10. In discussing ordering decimals in this section, we used the fact that "if $ab < ac$ and $a > 0$, then $b < c$." The condition $a > 0$ is often not understood by students. Give an example to show that the condition is necessary.

11. Ian claims that since $6 < 8$, $6 \div 2 < 8 \div 2$. How would you respond to Ian? Would it be useful to include the entire class in your discussion with Ian? Explain.

12. Evita claims that if $a > b$, it is not always true that $a^2 > b^2$. Would it be useful to let Evita explain her reasoning to the entire class? How could she do so?

13. Marita claims that, since $1/3 = 0.333\ldots$, it must be the case that $1 = 0.999\ldots$.

(a) How would you respond to Marita and involve the entire class in the discussion?

(b) Write the repeating decimal 0.24999 . . . as a fraction. Do you think your students might be surprised at the result?

(c) Could you write the finite decimal 3.23 as a repeating decimal?

14. You may be surprised to discover that some rational numbers have two *different* decimal representations.

(a) What rational number is represented by the periodic decimal $0.0999\ldots = 0.0\overline{9}$?

(b) Write two other decimals that are different but represent the same rational number.

(c) What rational numbers possess two different decimal expansions? Explain briefly.

15. Compute the decimal expansions of each of the following.

(a) $\frac{1}{11}$ (b) $\frac{1}{111}$ (c) $\frac{1}{1111}$

(d) Guess the decimal expansion of $1/11111$.

(e) Check your guess in part (d) by converting the decimal of your guess back into a ratio of two integers.

(f) Carefully describe the decimal expansion of $1/N_n$ where $N_n = 111\ldots1$, with n 1s in its representation.

Thinking Critically

16. What rational number corresponds to each of these periodic decimals? (Refer to Example 7.6 if you need help.)

(a) $0.747474\ldots = 0.\overline{74}$

(b) $0.777\ldots = 0.\overline{7}$

(c) $0.235235235\ldots = 0.\overline{235}$

(d) If a, b, and c are digits, what rational numbers are represented by each of these periodic decimals? You should be able to guess these results on the basis of patterns observed in parts (a), (b), and (c).

(i) $0.aaa\ldots = 0.\overline{a}$

(ii) $0.ababab\ldots = 0.\overline{ab}$

(iii) $0.abcabcabc\ldots = 0.\overline{abc}$

17. Write the decimal representing each of these rational numbers without doing any calculation, or at most doing only mental calculation.

(a) $\frac{5}{9}$ (b) $\frac{22}{99}$ (c) $\frac{317}{999}$

(d) $\frac{17}{33}$ (e) $\frac{14}{11}$

(f) Check the answers in parts (a) through (e) using your calculator. Is this check foolproof? Explain.

18. (a) Give an example which shows that the sum of two irrational numbers is sometimes rational.

(b) Give an example which shows that the sum of two irrational numbers is sometimes irrational.

19. Give an example which shows that the product of two irrational numbers is sometimes rational.

20. Is $2/\sqrt{2}$ rational or irrational? Explain.

21. Give an example which shows that the quotient of two irrational numbers is sometimes rational.

22. What real numbers are represented by points A, B, and C in this diagram? Explain your answer in a brief paragraph.

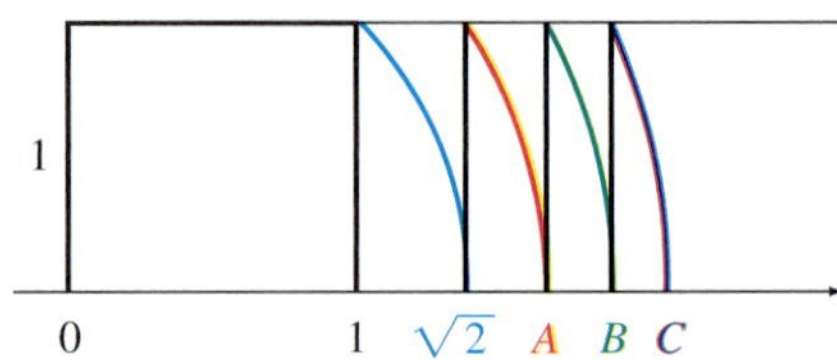

23. Find a rational number between the irrational numbers π and $2 + \sqrt{2}$.

24. **(a)** If a is an integer, what are the possibilities for the last digit of the decimal representation of a^2?

Last digit of a	0	1	2	3	4	5	6	7	8	9
Last digit of a^2										

(b) If b is an integer, what are the possibilities for the last digit of the decimal representation of $2b^2$?

Last digit of b	0	1	2	3	4	5	6	7	8	9
Last digit of $2b^2$										

(c) Using the results of parts (a) and (b), make a careful argument that $\sqrt{2} = a/b$, with a and b integers, is impossible.

Making Connections

25. **(a)** If it takes Suzan 1 minute to walk to her friend's house, how long, in minutes, will it take her to walk halfway to her friend's house?

(b) If Suzan is halfway to her friend's house, how long, in minutes, will it take her to walk half the remaining distance to her friend's house?

(c) If Suzan walks halfway to her friend's house and then half the remaining distance, how long, in minutes, will it take her to walk half the then remaining distance?

(d) In order to walk to her friend's house, Suzan must first walk halfway there, then walk half the remaining distance, then half the then remaining distance, then half the still remaining distance, and so on. Will Suzan ever be able to reach her friend's house?

(e) What must the sum of the times in parts (a), (b), (c), and (d) be?

(f) What does the answer to part (e) suggest about the reasonableness of the equality $1/3 = 0.333\ldots$?

26. **(a)** If $S = \frac{1}{2} + \frac{1}{2^2} + \frac{1}{2^3} + \cdots$, write $2S$ as a similar sum.

(b) Use part (a) to compute $S = 2S - S$.

(c) How does the answer to part (b) relate to part (e) of problem 25?

Thinking Cooperatively

27. Dividing the work between members of your group, compute the decimal representations of these rational numbers, and compute the prime-power representations of their denominators.

(a) $\frac{23}{6}$ **(b)** $\frac{7}{24}$ **(c)** $\frac{11}{9}$ **(d)** $\frac{23}{18}$ **(e)** $\frac{7}{22}$

(f) $\frac{23}{22}$ **(g)** $\frac{13}{110}$ **(h)** $\frac{311}{88}$ **(i)** $\frac{37}{220}$ **(j)** $\frac{29}{5500}$

(k) Discuss the results of (a) through (j) and arrive at a consensus prediction about how many digits there are between the decimal point and the repeating part of the decimal expansion of a rational number r/s.

Using a Computer

28. **(a)** Choose any two positive integers a and b and compute the first 20 terms in the sequence $f_1 = a$, $f_2 = b$, $f_3 = 2a + 2b$, $f_4 = 4a + 6b, \ldots$ where every term after the second is twice the sum of its two predecessors. This is most easily done on a spreadsheet.

(b) Compute the quotients $f_2/f_1, f_3/f_2, f_4/f_3, \ldots, f_{20}/f_{19}$.

(c) Compute $1 + \sqrt{3}$.

(d) Consider the results of parts (b) and (c), and then write up a brief analysis of what seems to be the case.

From State Student Assessments

29. (Connecticut, Grade 6)

(a) The shaded portion of this picture represents the number

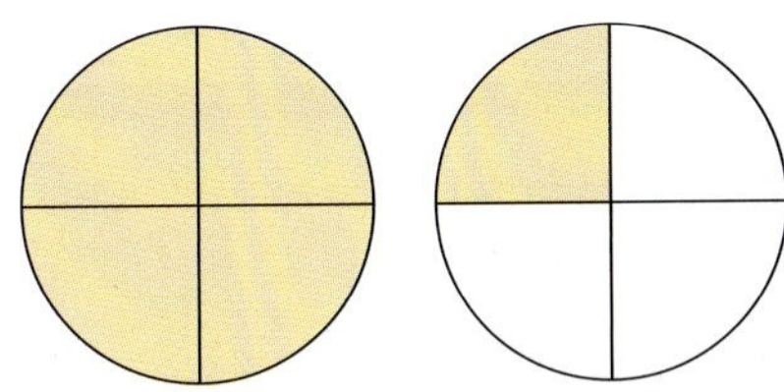

A. 1 B. 2 C. 1.5 D. 1.25

(b) Which of the following shows 1.63?

A.

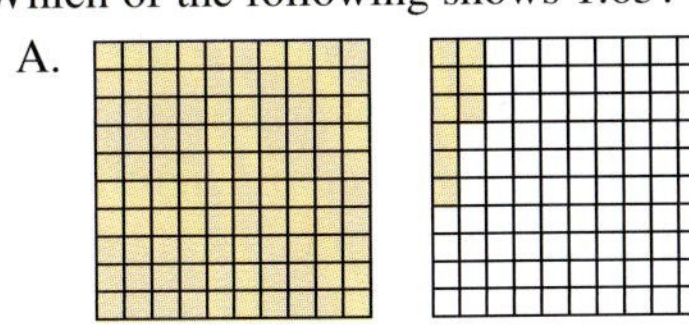

B.

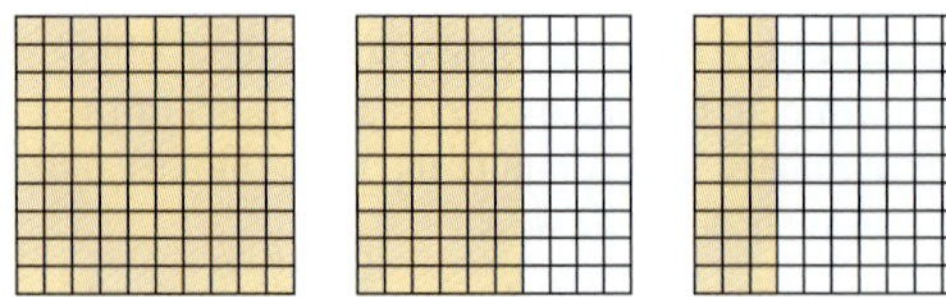

C.

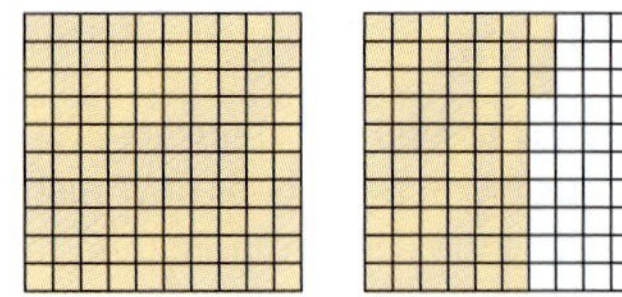

D.

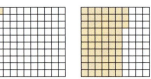

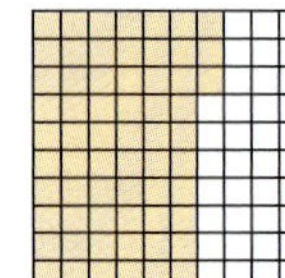

30. (Minnesota, Grade 5)

Use the picture below to answer the question. Radio station **WMTH** can be found at 93.7 on your radio dial. Which letter shown above represents where you would find **WMTH?**

A. letter A B. letter B

C. letter C D. letter D

For Review

31. Arrange these fractions in order from least to greatest: $3/5, 3/4, -2/3, 27/29, 7/8, 1/2$

32. The *mediant* of a/b and c/d is the fraction $(a+c)/(b+d)$.
Consider the following sequences of fractions, where each new sequence is obtained by adding to the previous sequence the mediants of that sequence.

$$\frac{0}{1}, \frac{1}{1}$$

$$\frac{0}{1}, \frac{1}{2}, \frac{1}{1}$$

$$\frac{0}{1}, \frac{1}{3}, \frac{1}{2}, \frac{2}{3}, \frac{1}{1}$$

(a) List the next two sequences found in this way.

(b) For any two adjacent fractions $\frac{a}{b}$ and $\frac{c}{d}$ in any of these sequences, compute $ad - bc$.

(c) Make a conjecture based on the results of part (b).

33. Perform the following computations leaving all answers as fractions in simplest form.

(a) $\frac{1}{2} + \frac{2}{3}$ **(b)** $\frac{1}{2} - \frac{2}{3}$ **(c)** $\frac{1}{2} \cdot \frac{2}{3}$

(d) $\frac{1}{2} \div \frac{2}{3}$ **(e)** $\frac{3}{4} \cdot \left(1 + \frac{3}{5}\right)$ **(f)** $\frac{3}{4} \div \left(1 - \frac{3}{5}\right)$

34. Find two rational numbers between $1/2$ and $1/3$.

35. In a certain population, $2/3$ of the men are married and $1/2$ of the women are married. What fraction of the adult population is unmarried?

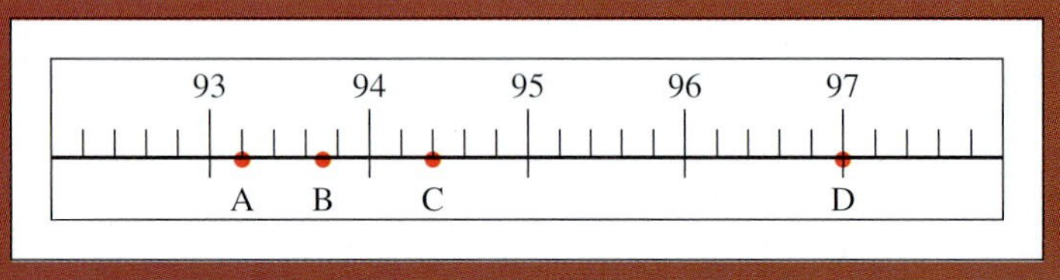

Picture for Problem 30

7.2 COMPUTATIONS WITH DECIMALS

It is essential that elementary school students develop a sense of *numeracy*—i.e., an idea of what numbers *mean* and what makes *sense* when dealing with numbers. For example, is it practically useful to consider a decimal like 0.7342061? The answer, in general, is *no*. Even in scientific work where considerable accuracy is required, a decimal correct to seven significant digits is rarely meaningful. It is much more likely that an approximation like 0.73, or 0.734, or even simply 0.7 is all that can be used effectively. This naturally leads to the question of approximation and rounding of decimals.

Rounding Decimals

Suppose a *couturière* wants to buy material to make into a dress for her upcoming fashion show. The material comes in 40-inch widths and she needs a piece 3.75 yards long. If the material costs \$15.37 per yard, how much will she have to pay? Using a calculator, we find that the price would be

$$(3.75)(\$15.37) = \$57.6375,$$

and the *couturière* would be charged \$57.64, the cost rounded to the nearest cent. The process of rounding here is precisely the same as it was for integers. Thus, 2.3254071 rounded

- to the nearest integer is 2,
- to the nearest tenth is 2.3,
- to the nearest hundredth is 2.33,

and so on. As before, we are using the 5-up rule.

RULE *The 5-Up Rule for Rounding Decimals*
To round a decimal to a given place, consider the digit in the next place to the right. If this digit is less than 5, leave the digit in the place under consideration unchanged and replace all digits to its right by 0s. If the digit to the right of the place in question is 5 or more, increase the digit in the place under consideration by 1 and replace all digits to the right by 0s.

EXAMPLE 7.10 ROUNDING DECIMALS

Round each of these to the indicated position.

(a) 23.2047 to the nearest integer
(b) 3.6147 to the nearest tenth
(c) 0.015 to the nearest hundredth

Solution

23.000 . . .
= 23

(a) Since we are asked to round 23.2047 to the nearest integer, we consider the digit to the right of 3. Since this digit is 2 and $2 < 5$, we leave the 3 unchanged and replace the digits to its right by 0s. Thus, 23.2047 rounded to the nearest integer is 23.

3.6000 . . .
= 3.6

(b) In rounding 3.6147 to the nearest tenth, we note that the digit to the right of the 6 is 1. Since $1 < 5$, we leave the 6 unchanged and replace the digits to its right by 0s. Thus, 3.6147 rounded to the nearest tenth is 3.6.

0.02000 . . .
= 0.02

(c) Here, the digit in question is 1 and the digit to its right is 5. Thus, we increase the 1 by 1 and replace the digits to its right by 0s. Hence, 0.015 rounded to the nearest hundredth is 0.02.

Adding and Subtracting Decimals

To do some repair work at the Louvre in Paris, plumber Jean Ferre estimated that he needed 17.5 meters of copper tubing. After a plan change, he decided that he needed an additional 15.75 meters of tubing. When he finished the job, Ferre discovered that he had 2.34 meters of tubing left over. How much tubing did Ferre purchase and how much did he use to complete the job? Answers to questions such as these require that we consider the addition and subtraction of decimals.

Suppose we wish to add 2.71 and 37.762. Converting these decimals to fractions, we have

$$2.71 = \frac{271}{100} = \frac{2710}{1000} \quad \text{and} \quad 37.762 = \frac{37{,}762}{1000}.$$

Thus,

$$\begin{aligned} 2.71 + 37.762 &= \frac{2710}{1000} + \frac{37{,}762}{1000} \\ &= \frac{40{,}472}{1000} \\ &= 40.472. \end{aligned}$$

To add decimals by hand, write the numbers in vertical style lining up the decimal points, and then add essentially just as we add integers. With a calculator, we enter the string

2.71 [+] 37.762 [=]

and automatically obtain the desired sum 40.472. The important thing with a calculator is to recognize that we are adding approximately 3 to approximately 38, so the sum should be approximately 41. The estimation and mental arithmetic should proceed right along with the calculator manipulation to be sure that we recognize if we have inadvertently made a calculator error.

Suppose now that we want to subtract 2.71 from 37.762. The calculation is much the same as above, and we have

$$\begin{aligned} 37.762 - 2.71 &= \frac{37{,}762}{1000} - \frac{2710}{1000} \\ &= \frac{35{,}052}{1000} \\ &= 35.052. \end{aligned}$$

As before, in hand calculation, we write the problem in vertical style lining up the decimal points, and then subtract essentially as we subtract integers.

If we do this operation with a calculator, we should estimate and perform mental calculation as well. Thus, we think,

and thus avoid gross calculator errors.

EXAMPLE 7.11 | ADDING AND SUBTRACTING DECIMALS

Compute each of these by estimating, by calculator, and by hand.

(a) 23.47 + 7.81 **(b)** 351.42 − 417.815

Solution

(a) By estimating: Approximately 23 plus approximately 8 gives approximately 31.

By calculator: 23.47 [+] 7.81 [=] 31.28

By hand:
$$\begin{array}{r} \overset{1\,1}{23.47} \\ +\ \ 7.81 \\ \hline 31.28 \end{array}$$

(b) By estimating: Approximately 350 minus approximately 400 gives approximately −50. By calculator: 351.42 [−] 417.815 [=] −66.395

By hand:
$$\begin{array}{r} \overset{3}{\not4}{}^{1}17.\overset{7}{\not8}{}^{1}15 \\ -\ 351.42 \\ \hline 66.395 \end{array}$$

Therefore, since 417.815 > 351.42, the desired answer is −66.395.

Multiplying Decimals

Tom Swift wanted to try out his new Ferrari on a straight stretch of highway. If he drove at 91.7 miles per hour for 15 minutes, how far did he go? Since 15 minutes equals 0.25 hours and distance traveled equals rate times elapsed time, Tom traveled $(91.7) \cdot (0.25)$ miles. For a more explicit answer, we need to be able to multiply decimals. Converting these decimals to fractions, we have

$$91.7 = \frac{917}{10} \quad \text{and} \quad 0.25 = \frac{25}{100}.$$

Thus,

$$\begin{aligned} (91.7) \cdot (0.25) &= \frac{917}{10} \cdot \frac{25}{100} \\ &= \frac{917 \cdot 25}{10 \cdot 100} \\ &= \frac{22{,}925}{1000} \\ &= 22.925, \end{aligned}$$

917
× 25
4585
1834
22,925
Three digits to the right of the decimal point since we are dividing by 1000 = 10³.

and the multiplication is performed just as the multiplication of integers, except for the placement of the decimal point. Indeed, by hand, it is common to write the following:

$$\begin{array}{r} 91.7 \\ \times\ 0.25 \\ \hline 4585 \\ 1834 \\ \hline 22.925 \end{array}$$

One digit to the right of the decimal point.
Two digits to the right of the decimal point.
Three digits to the right of the decimal point since 1 + 2 = 3.

Since this is typical of the general case, we summarize it as a theorem.

THEOREM *Multiplying Decimals*
To multiply two decimals, do the following:

1. Multiply as with integers.
2. Count the number of digits to the right of the decimal point in each number in the product, add these numbers, and call their sum t.
3. Finally, place the decimal point in the product obtained so that there are t digits to the right of the decimal point.

Finally, we note that since Tom's odometer gives mileage only to the nearest tenth, it actually shows that Tom traveled approximately 22.9 miles. Thus, the accuracy to three decimal places in the preceding multiplication is hardly useful to Tom. Also, Tom's estimated speed of 91.7 miles per hour as well as his measurement of the 15-minute time interval were almost surely not entirely accurate. Hence, the accuracy to three decimal places is surely not warranted; it is certainly much more reasonable in this case to round the answer to 22.9, or even to 23, miles.

EXAMPLE 7.12 MULTIPLYING DECIMALS

Compute these products by estimating, by calculator, and by hand.

(a) $(471.2) \cdot (2.3)$ **(b)** $(36.34) \cdot (1.02)$

Solution

(a) By estimating: Approximately 500 times approximately 2 gives approximately 1000.

By calculator: 471.2 [×] 2.3 [=] 1083.76

By hand:

$$\begin{array}{r} 471.2 \\ \times \quad 2.3 \\ \hline 14136 \\ 9424 \\ \hline 1083.76 \end{array}$$

two digits to the right of the decimal point

1 + 1 = 2

(b) By estimating: Approximately 36 times approximately 1 gives approximately 36.

By calculator: 36.34 [×] 1.02 [=] 37.0668

By hand:

$$\begin{array}{r} 36.34 \\ \times \quad 1.02 \\ \hline 7268 \\ 36340 \\ \hline 37.0668 \end{array}$$

Dividing Decimals

Tom Swift also has his own airplane. If Tom traveled 537.6 miles in 2.56 hours in his airplane, how fast did he travel? Again, since distance traveled equals rate times elapsed time, rate equals distance traveled divided by time. Therefore, we need to compute the quotient $537.6 \div 2.56$. Converting these decimals to fractions, we have

$$\begin{aligned} 537.6 \div 2.56 &= \frac{5376}{10} \div \frac{256}{100} \\ &= \frac{5376}{10} \cdot \frac{100}{256} \\ &= \frac{537{,}600}{2560} \\ &= \frac{53{,}760}{256} \\ &= 210. \end{aligned}$$

Thus, the problem is reduced to that of dividing 53,760 by 256, that is, to dividing integers. Recall that when confronted by a division like

$$2.56\overline{)537.6},$$

students are often told to move the decimal point in both the divisor and the dividend two places to the right so that the divisor becomes an integer. The preceding calculation with fractions justifies this rule, and, by hand, we have

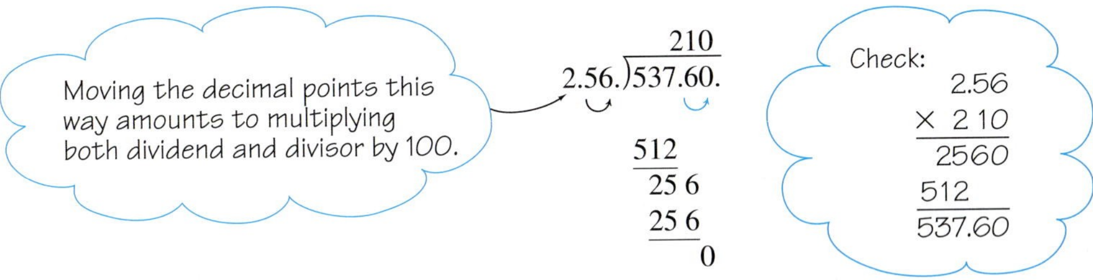

Moving the decimal points this way amounts to multiplying both dividend and divisor by 100.
210
2.56.)537.60.
512
25 6
25 6
0
Check:
2.56
× 2 10
2560
512
537.60

More on Periodic Decimals

Earlier, we claimed that the decimal expansion of a fraction a/b in simplest form, where b has prime factors other than 2 and 5, is necessarily nonterminating but repeating. We are now in a position to give a justification, and we do so by example.

EXAMPLE 7.13 **THE DECIMAL EXPANSION OF 3/7**

Use long division to find the decimal expansion of 3/7.

Solution Since $3/7 = 3 \div 7$, we obtain the desired decimal expansion by dividing 3 by 7. We have

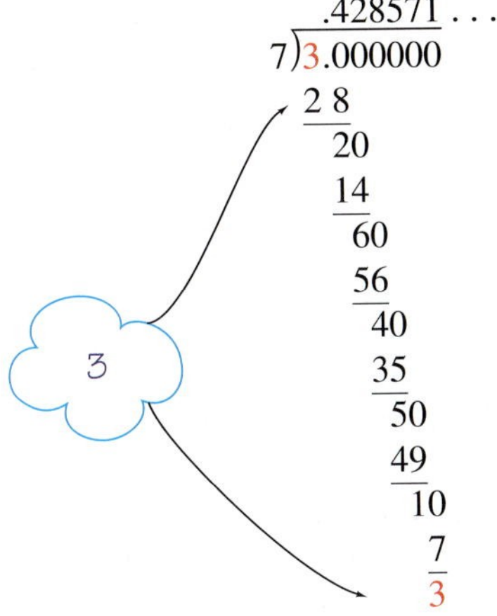

and, since the remainder at this stage is 3, the number with which we began, the division will continue *ad infinitum* with this repeating pattern. Thus,

$$\frac{3}{7} = 0.428571428571428571\ldots = 0.\overline{428571}.$$

Moreover, since we are dividing by 7, the only possible remainders at each step are 0, 1, 2, 3, 4, 5, and 6. Thus, if the division does not terminate, it must repeat after at most six steps. This is true in general and thereby essentially proves the theorem mentioned.

As a check of the preceding result, we compute the rational number determined by the decimal

$$x = 0.428571428571428571\ldots = 0.\overline{428571}.$$

It follows that

$$1{,}000{,}000x - x = 999{,}999x = 428{,}571.$$

So

$$x = \frac{428{,}571}{999{,}999} = \frac{3}{7}$$

Divide both numerator and denominator by 142,857.

as claimed.

Scientific Notation

Some products and quotients cannot be calculated directly on a nonscientific calculator because the results are either too large or too small. For example, attempting to calculate either

$$876{,}592 \cdot 7654 \quad \text{or} \quad 0.0011 \div 65{,}536$$

results in an error message on many calculators. On many scientific calculators, the display for these two calculations shows something like

6.7094 09

$6.7094 \times 10^9 = 6{,}709{,}400{,}000$

and

1.6785 −08,

$1.6785 \times 10^{-8} = 0.000000016785$

respectively, and these notations are understood to mean

$$6.7094 \times 10^9 \quad \text{and} \quad 1.6785 \times 10^{-8}.$$

What these calculators are doing is giving an **approximate** answer in each case, using a form of what is called **scientific notation.*** That is, they give the correct first few digits of the answer in each case as a number between 1 and 10 and then tell what power of 10 to multiply this by in order to correctly place the decimal point.

*The actual answers, respectively, are 6,709,435,168 and 0.000000016784667968 75 . . . .

In computing answers like

$$6{,}709{,}435{,}168 \quad \text{and} \quad 0.000000016784667968 75\ldots,$$

it is highly unlikely that we are interested in, or, because of inaccuracy of measurements, for example, that we are very sure of the last several digits of each result. Thus, the first answer is much more informative if we understand it as essentially 6.7 billion, or as 6.7×10^9, and the second answer is more readily appreciated as 1.7×10^{-8}. We usually *round off* the number either to the number of digits we are interested in or to the number of digits we are sure of and then use scientific notation. The digits we write before writing the power of 10 are called **significant digits.** Thus, we might write the first answer above as

$$6.7 \times 10^9 \text{ to two significant digits}$$

or

$$6.7094 \times 10^9 \text{ to five significant digits.}$$

Similarly, the second answer would be written as

$$1.7 \times 10^{-8} \text{ to two significant digits}$$

or

$$1.6785 \times 10^{-8} \text{ to five significant digits.}$$

DEFINITION *Scientific Notation*
To write a number in **scientific notation,** write it as the product of a number greater than or equal to 1 and less than 10, times the appropriate power of 10 to correctly place the decimal point. The digits in the number multiplied by the power of 10 are called **significant digits.**

EXAMPLE 7.14 WRITING NUMBERS IN SCIENTIFIC NOTATION

Write each of these in scientific notation, using the number of significant digits indicated.

(a) 93,000,000 using two significant digits
(b) 93,000,000 using three significant digits
(c) 0.000027841 using two significant digits
(d) 0.000027841 using three significant digits

Solution

(a) First write $93{,}000{,}000 = 9.3000000 \times 10^7$. Then round this to 9.3×10^7 to obtain the answer to two significant digits.
(b) This is the same as part (a), and we proceed as before except that we round the answer to 9.30×10^7 to obtain three significant digits.
(c) Here, $0.000027841 = 2.7841 \times 10^{-5}$, and this is rounded to 2.8×10^{-5} to obtain the answer correct to two significant digits.
(d) This is the same as part (c), but 2.7841×10^{-5} is rounded to 2.78×10^{-5} to obtain the answer correct to three significant digits.

Some calculators have a key marked [EXP], [E], or [EE], and one can calculate in exponential notation by using this key. For example, to calculate the product

$$(8.77 \times 10^7) \cdot (7.65 \times 10^3)$$

and write the answer to three significant figures, enter the string

8.77 [EXP] 7 [×] 7.65 [EXP] 3 [=].

In the display, we read something like 6.709 11, or perhaps 6.70905 E 11, and write this to three significant figures as 6.71×10^{11}.

EXAMPLE 7.15 CALCULATING USING SCIENTIFIC NOTATION

Compute each of these to three significant figures using scientific notation.

(a) $(2.47 \times 10^{-5}) \cdot (8.15 \times 10^{-9})$
(b) $(2.47 \times 10^{-5}) \div (8.15 \times 10^{9})$

Solution

(a) Enter 2.47 [EXP] [(−)] 5 [×] 8.15 [EXP] [(−)] 9 [=].
Read 2.013 − 13 in the display and record 2.01×10^{-13}.

(b) Enter 2.47 [EXP] [(−)] 5 [÷] 8.15 [EXP] 9 [=].
Read 3030.6748 in the display and record 3.03×10^{-15}.

Note, too, how the use of scientific notation can assist in mental approximation of calculator answers. For example, consider Example 7.15, part (a). Using only left-digit approximations and rules for multiplication of exponents, think

2 times 8 = 16 | times $10^{-5} \cdot 10^{-9} = 10^{-14}$ | gives 1.6×10^{-13}

so the above answer is probably correct. An even closer approximation can be obtained by using a two-digit approximation on the first number and one-digit approximation on the second. Think

2.5 times 8 = 20 | times 10^{-14} | gives 2.0×10^{-13}

and the approximation is quite close. Similarly, if we use the same approximation on part (b), we think

2.5 divided by 8 ≐ 0.3 = $3 \cdot 10^{-1}$ | times $10^{-5} \cdot 10^{9} = 10^{4}$ | gives 3×10^{3}

and the approximation is again quite good.

Problem Set 7.2

Understanding Concepts

1. Perform these additions and subtractions by hand.
   (a) $32.174 + 371.5$ (b) $371.5 - 32.174$
   (c) $0.057 + 1.08$ (d) $0.057 - 1.08$
2. Perform these multiplications and divisions by hand.
   (a) $(37.1) \cdot (4.7)$ (b) $(3.71) \cdot (0.47)$
   (c) $138.33 \div 5.3$ (d) $1.3833 \div 0.53$
3. Estimate the result of each of these computations mentally and then perform the calculations using a calculator.
   (a) $4.112 + 31.3$ (b) $31.3 - 4.112$
   (c) $(4.112) \cdot (31.3)$ (d) $31.3 \div 4.112$
4. By long division, determine the decimal expansion of each of these fractions.
   (a) $5/6$ (b) $3/11$ (c) $2/27$
5. Write each of these in scientific notation with the indicated number of significant digits.
   (a) 276,543,421 to three significant digits
   (b) 0.000005341 to two significant digits
   (c) 376,712.543248 to two significant digits

6. Calculate each of these with a suitable calculator and write the answer to three significant digits.
   (a) $0.0000127 \times 0.000008235$
   (b) $98{,}613{,}428 \times 5{,}746{,}312$
   (c) $0.0000127 \div 98{,}613{,}428$
   (d) $98{,}613{,}428 \div 0.000008234$

7. For each of these, estimate the answer and then calculate the result to three significant digits on a suitable calculator.
   (a) $(7.123 \times 10^5) \cdot (2.142 \times 10^4)$
   (b) $(7.123 \times 10^5) \div (2.142 \times 10^4)$
   (c) $(7.123 \times 10^5) \cdot (2.142 \times 10^{-9})$
   (d) $(7.123 \times 10^{-2}) \div (2.142 \times 10^8)$

Teaching Concepts

8. One can use the rectangular area model of multiplication to help students understand multiplication of decimals just as was done to promote understanding of multiplication of natural numbers and rational numbers. The following diagram illustrates the product 2.3×3.2.
   (a) Identify the colored regions in the diagram with the numbers shown in the hand calculation of the product.
   (b) How would you use the diagram to justify the exchange shown in the hand calculation?

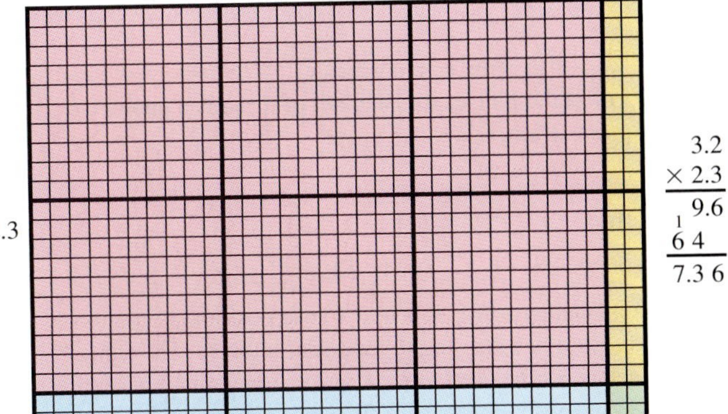

   (c) Discuss how the diagram helps to illustrate the rule for the proper placement of the decimal point in the final answer.
9. When asked to round 7.2447 to the nearest tenth, Toni proceeded as follows: 7.245, 7.25, 7.3. Toni seems to need help. How would you help her?
10. When multiplying 23.4 times 3.26, Andre is confused about where to place the decimal point. How would you help him?

Thinking Critically

11. Use these numbers to make a magic square. See problem 14 in Problem Set 1.1.
    0.123, 0.246, 0.369, 0.492, 0.615, 0.738, 0.861, 0.984, 1.107
12. Use the numbers in problem 11 to form a magic subtraction square. See problem 14 in Problem Set 1.1.
13. Use these numbers to make a magic square.
    7.02, 16.38, 11.70, 18.72, 2.34, 9.36, 4.68, 21.06, 14.04
14. Use the numbers in problem 13 to make a magic subtraction square.
15. The sum of the numbers in any two adjacent blanks is the number immediately below and between these two numbers. Complete each of these so that the same pattern holds. The first one has been completed for you.

    (a)
    2.107 1.3 4.26
    3.407 5.56
    8.967

    (b)
    21.06 3.21 ___
    ___ 5.00
    ___

    (c)
    ___ 0.041 ___
    2.415 ___
    7.723

    (d)
    ___ 1.414 ___
    ___ ___
    3.142

    (e) Can any of these be completed in more than one way? Explain.

16. Fill in the blanks so that each of these is an arithmetic progression.

(a) 3.4, 4.3, 5.2, ____, ____, ____

(b) −31.56, ____, −21.10, ____, ____, ____

(c) 0.0114, ____, ____, 0.3204, ____, ____

(d) 1.07, ____, ____, ____, −9.21, ____

17. Fill in the blanks so that each of these is a geometric progression.

(a) 2.11, 2.321, ____, ____, ____

(b) 35.1, ____, 1.404, −0.2808, ____

(c) 6.01, ____, ____, 0.75125, ____

18. In Chapter 3, we saw how to use positional notation to bases other than ten to represent integers. The same ideas can be used to represent rational and real numbers. Write a two- or three-page paper explaining how this would work. Be sure to include examples.

19. Recall that the Fibonacci numbers are the numbers 1, 1, 2, 3, 5, 8, 13, 21, 34, 55, 89, ...

(a) Compute the decimal expansion of 1/89 correct to ten decimal places. Note that 89 is the 11th Fibonacci number.

(b) Are you surprised at the decimal expansion of 1/89, particularly in the 7th, 8th, 9th, and 10th places?

(c) By hand, compute this sum.

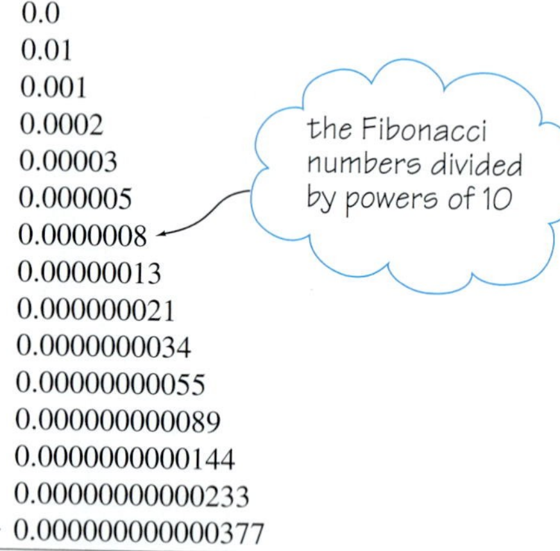

0.0
0.01
0.001
0.0002
0.00003
0.000005
0.0000008
0.00000013
0.000000021
0.0000000034
0.00000000055
0.000000000089
0.0000000000144
0.00000000000233
+ 0.000000000000377
the Fibonacci numbers divided by powers of 10

(d) Make a conjecture on the basis of the result in part (c). Recall that many decimals never end and consequently are actually sums of infinitely many terms.

(e) Does it make sense to set $F_0 = 0$?

Thinking Cooperatively

Do the next three problems with two or three other students. At each step, discuss your solution with the other members of the group and determine a consensus answer for the group.

20. The numbers 3.447, 2.821, 5.764, 3.2, 2.351, 3.001, and 4.2444 are placed in a circle in some order. Show that the sum of some three consecutive numbers must exceed 10.6. (*Hint:* Have you ever seen a similar problem before?)

21. Place numbers in the circles in these diagrams so that the numbers in the large circles are the sums of the numbers in the two adjacent smaller circles.

(a)

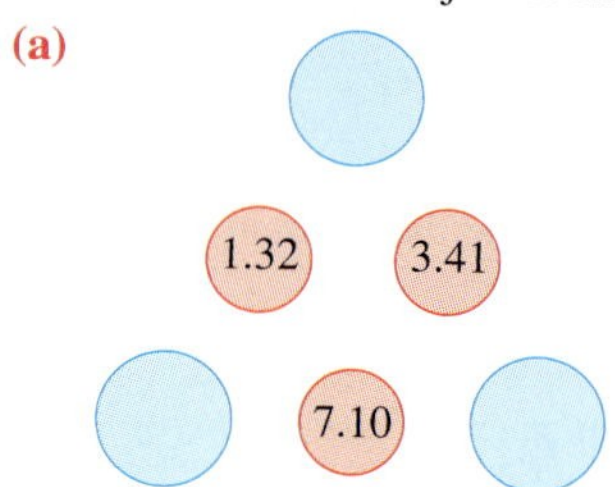

(b)

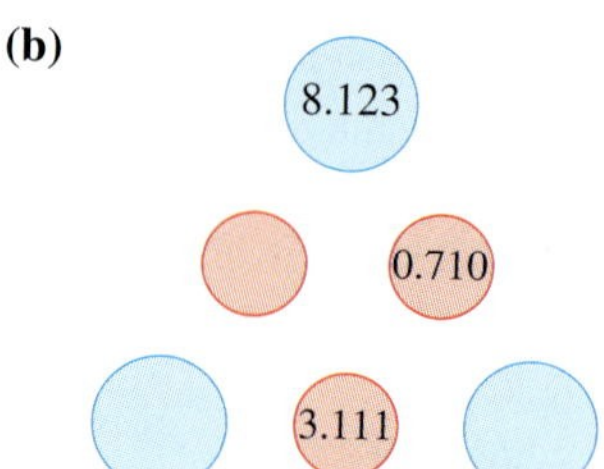

(c)

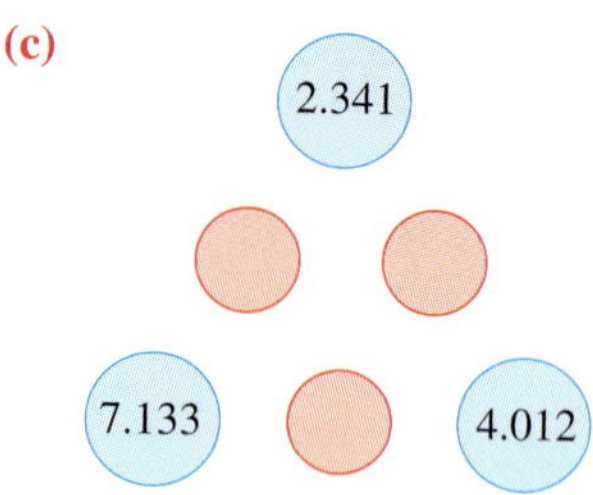

(d)

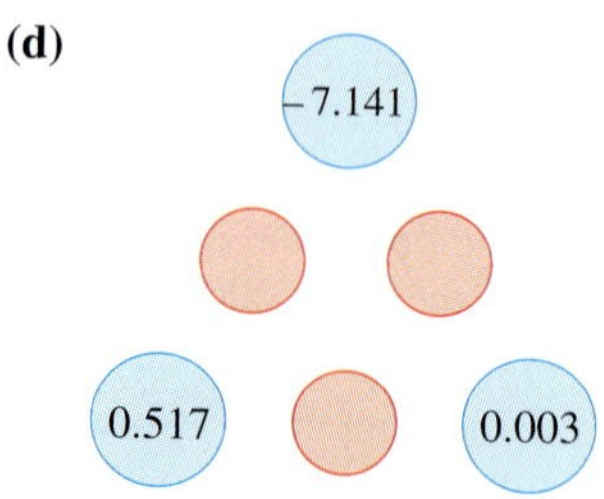

22. If possible, place numbers in the circles in these diagrams so that the numbers in the large circles exactly equal the sum of the numbers in the adjacent small circles.

(a)

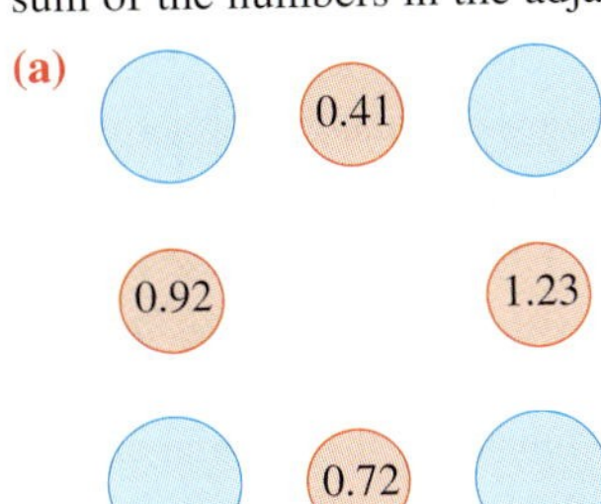

(b)

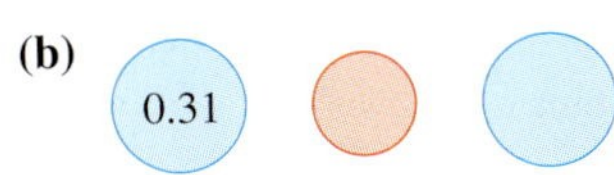

(c)

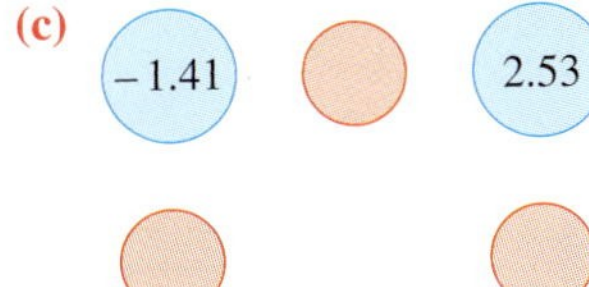

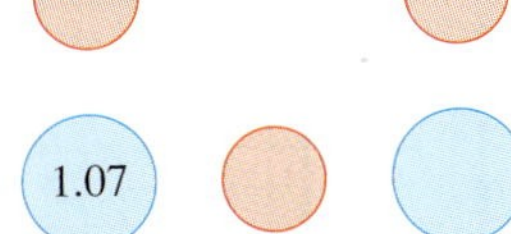

(d)

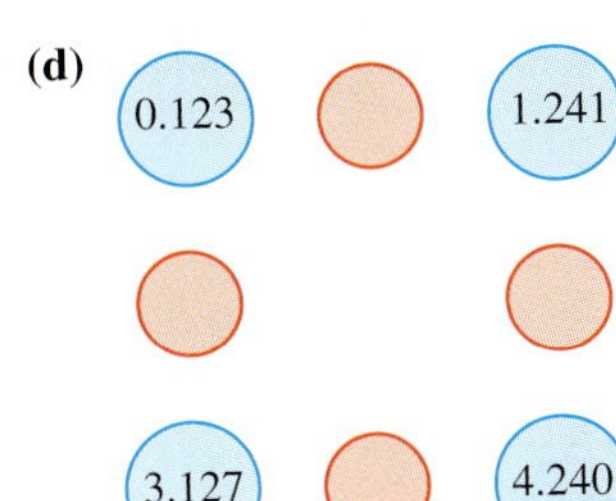

(e) What must be the case for these problems to be solvable? Explain.

(f) Is there more than one solution to all or any of these?

Making Connections

23. Kristina bought pairs of gloves as Christmas presents for three of her best friends. If the gloves cost \$9.72 a pair, how much did she spend for these presents?

24. Yolanda also bought identical pairs of gloves for each of her four best friends. If her total bill was \$44.92, how much did each pair of gloves cost?

25. Dante cashed a check for \$74.29 and then bought a Walkman for \$42.91 and a special pair of earphones for \$17.02. After paying for his purchases, how much did he have left?

26. A picture frame 2.25 inches wide surrounds a picture 17.5 inches wide by 24.75 inches high.

(a) What is the area of the picture frame?

(b) What is the area of the picture?

Using a Calculator

27. Among the cryptic notes of the Indian mathematician Srinivasa Ramanujan is the equation

$$\pi^4 = 97.409091\ldots.$$

This equation suggests that $\pi^4 \doteq 97.4\overline{09}$.

(a) Show that $97.4\overline{09} = 97\frac{1}{2} - \frac{1}{11}$.

(b) Use your calculator to calculate $\pi^4 - \left(97\frac{1}{2} - \frac{1}{11}\right)$ and determine just how good Ramanujan's approximation is.

28. (a) In Chapter 2, you were introduced to the process called DIFFY. Use DIFFY to complete this array. Remember that the first, second, and third circles in any row contain the differences (greatest minus least) of the numbers in the preceding row and that the fourth circle contains the difference of the elements in the first and fourth circles of the preceding row.

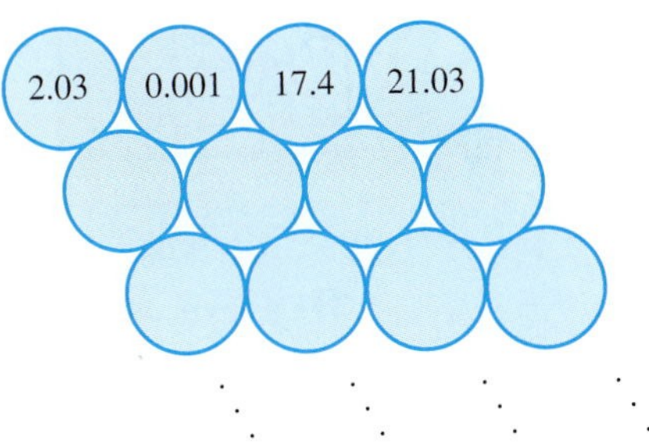

(b) Choose four more numbers and complete a second DIFFY array.

(c) Can you find four numbers that cause the process to continue for at least eight steps?

(d) Use 17.34, 31.62, 58.14, and 107.1 to complete a DIFFY process. For how many steps does the process continue?

(e) Do you think the process will always terminate?

29. We have found that the DIFFY process always terminates for natural numbers, integers, and rational numbers.

(a) Execute the graphing-calculator program DIFFY with the irrational numbers $\sqrt{2}$, $\sqrt{3}$, π, and $\sqrt{37}$. Does DIFFY terminate for these initial entries?

(b) Execute DIFFY with the initial entries 1, 1.839286755, 3.382975768, and 6.222262523. Does DIFFY terminate for these initial entries?

(c) How many steps did it take for DIFFY to terminate in part (b)? Considering this, do you think it might be possible to find four initial entries for which DIFFY would *never* terminate?

30. (a) For this problem, use DIVVY to complete this array.

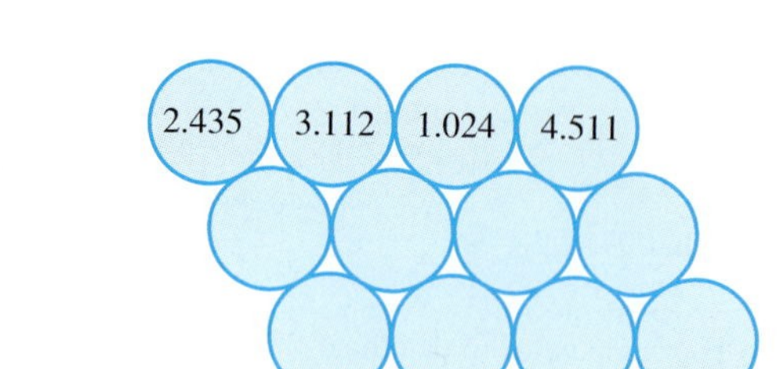

(b) Choose four positive real numbers and use DIVVY to complete this array.

(c) Do you think the process will always terminate? Explain.

31. (a) Execute the graphing-calculator program DIVVY with the initial irrational numbers $\sqrt{2}$, $\sqrt{3}$, π, and $\sqrt{37}$. Does DIVVY terminate for these numbers?

(b) Execute DIVVY with the initial entries 2, 3.578330778, 10.43223066, and 74.65994408. Does DIVVY terminate for these initial entries?

(c) How many steps did it take for DIVVY to terminate in part (b)? Would it be reasonable to guess that there might be four initial entries for which DIVVY would *never* terminate?

From State Student Assessments

32. (Oregon, Grade 5)
Non-Calculator:
Which number is between 3.7 and 3.8?

A. 3.81 B. 3.79 C. 3.68 D. 3.5

33. (Texas, Grade 4)
2.04 − 0.96 =

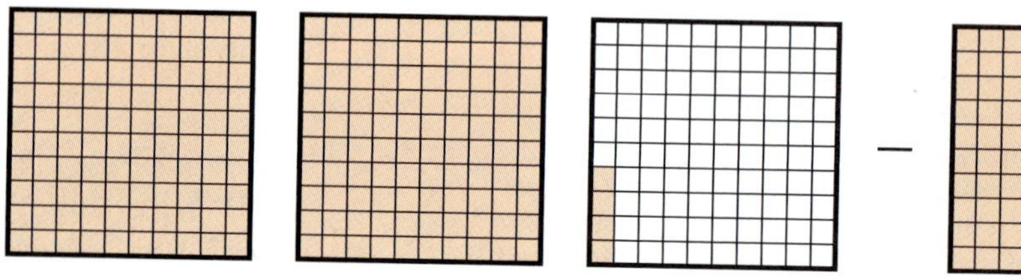

A. 1.08 B. 1.18 C. 1.92 D. 3.00

34. (Texas, Grade 4)
1.38 + 0.62 =

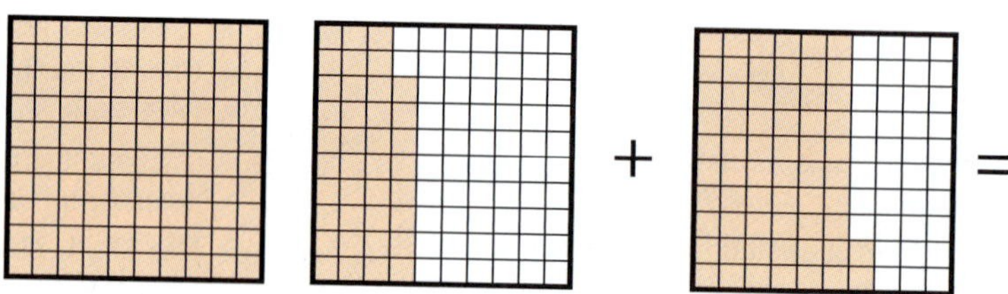

F. 0.76 G. 1 H. 1.9 J. 2

35. (Connecticut, Grade 6)
Jenny bought three cassette tapes for $5.99 each. What was the total cost of the three tapes before tax?

A. $6.02 B. $8.99 C. $17.77 D. $17.97

For Review

36. Write each of these in Mayan notation.

(a) 231 (b) 15,278 (c) 7142

37. Write each of these (now written in Mayan notation) in base-ten notation.

(a) (b)  (c)

38. Write each of these using Egyptian numerals.

(a) 547 (b) 2486 (c) 854

39. Convert each of these Egyptian numerals to their base-ten equivalents.

(a)

(b)

(c)

7.3 RATIO AND PROPORTION

Ratio

At basketball practice, Caralee missed 18 free throws out of 45 attempts. Since she made 27 free throws, we say that the **ratio** of the number missed to the number made was 18 to 27. This can be expressed by the fraction 18/27 or, somewhat archaically, by the notation 18 : 27. The notation 18 : 27 is read "18 to 27," as in the statement "The ratio of the number of free throws Caralee missed to the number she made was 18 : 27." We will always use the fraction notation in what follows.

Other ratios from Caralee's basketball practice are

- the ratio of the number of shots made to the number attempted—27/45,
- the ratio of the number of shots missed to the number attempted—18/45, and
- the ratio of the number of shots made to the number missed—27/18.

DEFINITION *Ratio*

If a and b are real numbers with $b \neq 0$, the **ratio of a to b** is the quotient a/b.

Ratios occur with great frequency in everyday life. If you use 10.4 gallons of gasoline in driving 400.4 miles, the efficiency of your car is measured in miles per gallon given by the ratio 400.4/10.4, or 38.5 miles per gallon. If Lincoln Grade School has 405 students and 15 teachers, the student–teacher ratio is the quotient 405/15. If Jose Varga got 56 hits in 181 times at bat, his batting average is the ratio 56/181. The number of examples that could be cited is almost endless.

EXAMPLE 7.16 DETERMINING RATIOS

Determine these ratios.

(a) The ratio of the number of boys to the number of girls in Martin Luther King High School if there are 285 boys and 228 girls

(b) The ratio of the number of boys to the number of students in part (a)

Solution

(a) The desired ratio is 285/228.
(b) Since the total number of students is 285 + 228 (that is, 513), the desired ratio is 285/513.

The ratio of the number of boys to the number of students in Martin Luther King High School was shown to be 285/513, or 285 to 513. This is certainly correct, but it is not nearly as informative as it would be if the ratio were written in simplest form. Thus,

$$\frac{285}{513} = \frac{5}{9},$$

285 ÷ 57 = 5
513 ÷ 57 = 9

and this says that 5/9 (or a little more than 1/2) of the students in Martin Luther King High School are boys. Expressing a ratio by a fraction in simplest form is often useful and informative.

EXAMPLE 7.17 EXPRESSING RATIOS IN SIMPLEST FORM

Express these ratios in simplest form.

(a) The ratio of 385 to 440
(b) The ratio 432/504

Solution

(a) The ratio of 385 to 440 is the quotient 385/440. Expressing this in simplest form, we have

$$\frac{385}{440} = \frac{7}{8}.$$

Thus, in simplest form, the ratio is 7 to 8.

(b) Expressing the quotient in simplest form, we have

$$\frac{432}{504} = \frac{6}{7}.$$

Thus, in simplest form, the ratio is 6/7, or 6 to 7.

EXAMPLE 7.18 DETERMINING A LESS OBVIOUS RATIO

If one-seventh of the students at Garfield High are nonswimmers, what is the ratio of nonswimmers to swimmers?

Solution

Understand the Problem

The desired ratio is the number of nonswimmers attending Garfield High School divided by the number of swimmers.

Devise a Plan

The needed numbers are not given in the problem, and we don't even know how many students attended the school. If, for example, there were 700 students attending the school, then $\frac{1}{7} \cdot 700 = 100$ would be swimmers and the remaining $600 = \frac{6}{7} \cdot 700$ would be swimmers. But suppose there are n students attending Garfield. Then $\frac{1}{7} \cdot n = \frac{n}{7}$ are nonswimmers and $\frac{6}{7}n = \frac{6n}{7}$ are swimmers. Perhaps we can use these expressions to determine the desired ratio.

Carry Out the Plan

Letting n denote the number of students attending Garfield High, we just saw that $\frac{n}{7}$ students are nonswimmers and $\frac{6n}{7}$ are swimmers. Therefore, the desired ratio is

$$\frac{\frac{n}{7}}{\frac{6n}{7}} = \frac{n}{7} \cdot \frac{7}{6n} = \frac{1}{6}.$$

Look Back

It turns out that we didn't need to know the actual number of students attending Garfield High, since we could use a variable and get general expressions for the numbers of swimmers and nonswimmers. Taking the quotient of these expressions, we found the desired ratio. Using a variable in this way is often a useful strategy. On the other hand, we didn't even need to use a variable. Assuming that there were 700 students attending the school, there would be 100 nonswimmers and 600 swimmers, and the desired ratio is again $\frac{100}{600} = \frac{1}{6}$.

Proportion

Ratios allow us to make clear comparisons when actual numbers sometimes make them more obscure. For example, at basketball practice, Caralee made 27 of 45 free throws attempted and Sonja made 24 of 40 attempts. Which player appears to be the better foul-shot shooter? For Caralee, saying that the ratio of shots made to shots tried is 27/45 amounts to saying that she made 3/5 of her shots. That is,

$$\frac{27}{45} = \frac{3}{5}.$$

$\frac{27}{45}$ and $\frac{3}{5}$ are equivalent fractions.

Similarly, for Sonja, the ratio of shots made to shots attempted is

$$\frac{24}{40} = \frac{3}{5},$$

$\frac{27}{45}$, $\frac{24}{40}$, and $\frac{3}{5}$ are all equivalent fractions.

and this suggests that the two girls are equally capable at shooting foul shots. Because of its importance in such comparisons, the equality of two ratios is called a **proportion.**

DEFINITION *Proportion*
If a/b and c/d are two ratios and

$$\frac{a}{b} = \frac{c}{d},$$

this equality is called a **proportion.**

From Chapter 6, we know that

$$\frac{a}{b} = \frac{c}{d}$$

for integers a, b, c, and d if, and only if, $ad = bc$. But essentially the same argument holds if a, b, c, and d are real numbers. This leads to the next theorem.

THEOREM *Conditions for a Proportion*
The equality

$$\frac{a}{b} = \frac{c}{d}$$

is a proportion if, and only if, $ad = bc$.

EXAMPLE 7.19 DETERMINING PROPORTIONS

In each of these, determine x so that the equality is a proportion.

(a) $\frac{28}{49} = \frac{x}{21}$ **(b)** $\frac{2.11}{3.49} = \frac{1.7}{x}$

Solution

We use the preceding theorem, which amounts to multiplying both sides of the equality by the product of the denominators, or "cross multiplying," as we often say.

(a)

$$\begin{aligned} \frac{28}{49} &= \frac{x}{21} \\ 28 \cdot 21 &= 49x \\ \frac{28 \cdot 21}{49} &= x \\ 12 &= x \end{aligned}$$

(b)

$$\begin{aligned} \frac{2.11}{3.49} &= \frac{1.7}{x} \\ 2.11x &= (1.7)(3.49) \\ x &= \frac{(1.7)(3.49)}{2.11} \\ x &\doteq 2.81 \end{aligned}$$

EXAMPLE 7.20 PROVING A PROPERTY OF PROPORTIONS

If

$$\frac{a}{b} = \frac{c}{d},$$

prove that

$$\frac{a+b}{b} = \frac{c+d}{d}.$$

Solution

Understand the Problem

We are given that $\frac{a}{b} = \frac{c}{d}$ is a proportion and are asked to show that $\frac{a+b}{b} = \frac{c+d}{d}$ is also a proportion.

Devise a Plan

Since it is not immediately clear what to do, we ask what it means to say that $\frac{a}{b} = \frac{c}{d}$ and $\frac{a+b}{b} = \frac{c+d}{d}$ are proportions. Perhaps this will put the problem in a form that is easier to understand and to work on. By the preceding theorem,

$$\frac{a}{b} = \frac{c}{d} \text{ if, and only if, } ad = bc,$$

and

$$\frac{a+b}{b} = \frac{c+d}{d}$$

if, and only if, $(a + b)d = b(c + d)$. Perhaps we can use the first of these equations to prove the second.

Carry Out the Plan

We want to show that $(a + b)d = b(c + d)$; that is, using the distributive property,

$$ad + bd = bc + bd.$$

Say it in a different way.

But we know that

$$ad = bc,$$

and adding bd to both sides of this equation gives

$$ad + bd = bc + bd.$$

Hence

$$\frac{a+b}{b} = \frac{c+d}{d}$$

as was to be shown.

Look Back

Here, our principal strategy was simply to ask, "What does it mean to say that $\frac{a}{b} = \frac{c}{d}$ and $\frac{a+b}{b} = \frac{c+d}{d}$ are proportions?" Answering this question allowed us to "say it in a different way"—that is, to state an equivalent problem that proved to be quite easy to solve. The strategy, **say it in a different way,** is often very useful.

Applications of Proportions

Suppose that a car is traveling at a constant rate of 55 miles per hour. Table 7.2 gives the distances the car will travel in different time periods.

TABLE 7.2 Distance Traveled in t Hours at 55 Miles per Hour

Time t	1	2	3	4	5	6	7	8
Distance d	55	110	165	220	275	330	385	440

The ratios d/t are all equal for the various time periods shown. That is,

$$\frac{55}{1} = \frac{110}{2} = \frac{165}{3} = \frac{220}{4} = \frac{275}{5} = \frac{330}{6}$$

and so on. Thus, each pair of ratios from the list forms a proportion. Indeed, $d/t = 55$ for every pair d and t. This is also expressed by saying that the distance traveled at a constant rate is proportional to the elapsed time. In the above instance,

$$d = 55t$$

for every pair d and t. The number 55 is called the **constant of proportionality.**

DEFINITION *y proportional to x*
If the variables x and y are related by the equation

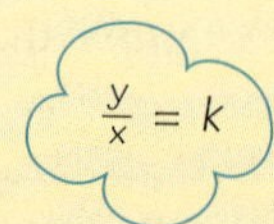

$$y = kx,$$

then **y is said to be proportional to x,** and k is called the **constant of proportionality.**

This situation is extremely common in everyday life. Gasoline consumed by your car is proportional to the miles traveled; the cost of pencils purchased is proportional to the number of pencils purchased; income from the school raffle is proportional to the number of tickets sold; and so on.

EXAMPLE 7.21 INCOME FROM THE SCHOOL RAFFLE

The sixth-grade class at Jefferson Middle School is raffling off a turkey as a moneymaking project. If the turkey cost \$22 and raffle tickets are sold for 75¢ each, how many tickets will have to be sold for the class to break even? How many tickets will have to be sold if the class is to make a profit of \$20?

Solution

If I represents income in dollars and N represents the number of tickets sold, then $I = 0.75N$; that is, I is proportional to N. To break even, the class must sell enough tickets so that

75¢ equals \$.75

$$22 = 0.75N.$$

Thus,

$$N = 22 \div 0.75 = 29.\overline{3}.$$

School Book Page

Ratio and Proportion

Chapter 12 Lesson 2

Patterns in Ratio Tables

You Will Learn
how to use ratio tables to write proportions

Vocabulary
equal ratios
ratios that give the same comparison

proportion
a statement that two ratios are equal

Learn

Softball, anyone? Courtney's and Taylor's team won a softball championship. The team often practices throwing in pairs. That's a ratio of:

$\frac{\text{1 softball}}{\text{2 players}}$

Courtney and Taylor play softball in Bethlehem, Pennsylvania.

Math Tip
What you know about equivalent fractions will help you work with equal ratios.

Example 1

How many softballs will be needed by 10 players? A table of equal ratios can help.

		(1 × 2)	(1 × 3)	(1 × 4)	(1 × 5)
Softballs	1	2	3	4	5
Players	2	4	6	8	10
		(2 × 2)	(2 × 3)	(2 × 4)	(2 × 5)

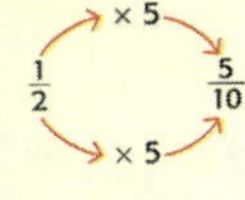

The ratio is 5:10. So, there are 5 softballs for 10 players.

You can also divide to find **equal ratios**.

Example 2

There are 8 teams with 72 players. Use equal ratios to find the number on 1 team.

$\frac{\text{teams}}{\text{players}} = \frac{8}{72} = \frac{1}{■}$

		(8 ÷ 2)	(8 ÷ 4)	(8 ÷ 8)
Teams	8	4	2	1
Players	72	36	18	9
		(72 ÷ 2)	(72 ÷ 4)	(72 ÷ 8)

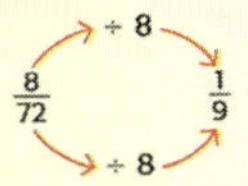

The ratio is 1:9. So, each team has 9 players.

A statement that two ratios are equal is called a **proportion**.

$\frac{1}{2} = \frac{5}{10}$ is a proportion. $\frac{8}{72} = \frac{1}{9}$ is a proportion.

Talk About It

How can ratio tables be used to find equal ratios?

530 Chapter 12 • Ratio, Percent, and Probability

SOURCE: Scott Foresman–Addison Wesley Math, *Grade 5*, p. 530, by Randall I. Charles et al., Copyright © 2002 Pearson Education, Inc. Reprinted with permission.

Questions for the Teacher

1. How would you assist a student who found it difficult to understand that the ratios 3 : 6 and 4 : 8 are equal?
2. How would you help a student who does not understand how to use division to determine whether two ratios are equal as suggested above?
3. An extension of Example 2 above might be to ask how many players would be on 12 teams. How would you help a student who had trouble answering this question? By the way, is it true that most baseball teams have only 9 players? How would you respond to a student who raised this last point?

To break even, the class will have to sell at least 30 tickets. In order to make a profit of \$20, enough tickets must be sold so that

$$42 = 0.75N;$$

that is,

$$N = 42 \div 0.75 = 56.$$

To make a profit of \$20, 56 tickets must be sold.

Recall for the next example that in geometry, two figures are said to be similar if they are the same shape but not necessarily the same size—i.e., one is a magnification of the other.

EXAMPLE 7.22 COMPUTING THE HEIGHT OF A TREE

Ms. Gulley-Pavey's fifth-grade class had been studying the concepts of ratio and proportion. One afternoon, she took her students outside and challenged them to find the height of a tree in the school yard. After a lively discussion, the students decided to measure the length of the shadow cast by a yardstick and that cast by the tree, arguing that these should be proportional. To help convince the class that this was so, Omari drew the picture shown. If the lengths of the shadows are 4′7″ and 18′9″, respectively, complete the calculation to determine the height of the tree.

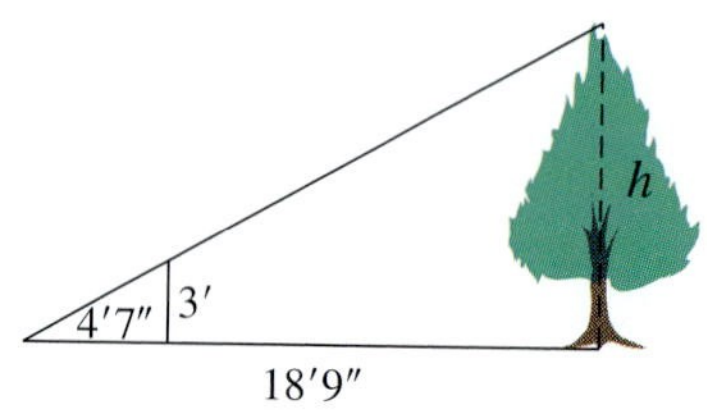

Solution

Since the two triangles shown in the diagram are similar, the lengths of the sides are proportional. Also, 7″ = 7/12 feet and 9″ = 9/12 feet. Therefore, we have the ratios

$$\frac{h}{3} = \frac{18\frac{9}{12}}{4\frac{7}{12}} = \frac{\frac{225}{12}}{\frac{55}{12}} = \frac{225}{12} \cdot \frac{12}{55},$$

The units of both ratios must be the same.

and it follows that

$$h = \frac{3 \cdot 225 \cdot 12}{12 \cdot 55} \doteq 12.27' \doteq 12'3''.$$

$0.27 \times 12 \doteq 3$

Thus, the tree was approximately 12′3″ tall.

Problem Set 7.3

Understanding Concepts

1. There are 10 girls and 14 boys in Mr. Tilden's fifth-grade class. What is the ratio of
   (a) boys to girls? (b) girls to students?
   (c) boys to students? (d) girls to boys?
   (e) students to girls? (f) students to boys?
2. Determine which of these are proportions.
   (a) $\frac{2}{3} = \frac{8}{12}$ (b) $\frac{21}{28} = \frac{27}{36}$
   (c) $\frac{7}{28} = \frac{8}{31}$ (d) $\frac{51}{85} = \frac{57}{95}$
   (e) $\frac{14}{49} = \frac{18}{60}$ (f) $\frac{20}{35} = \frac{28}{48}$
   (g) $\frac{1.5}{2.1} = \frac{11.5}{16.1}$ (h) $\frac{17.1}{6.2} = \frac{31.2}{9.7}$
   (i) $\frac{0.84}{0.96} = \frac{91.7}{104.8}$
3. Determine values of r and s so that each of these is a proportion.
   (a) $\frac{6}{14} = \frac{r}{21}$ (b) $\frac{8}{12} = \frac{10}{r}$ (c) $\frac{47}{3.2} = \frac{s}{7.8}$
4. Express each of these ratios as fractions in simplest form.
   (a) A ratio of 24 to 16
   (b) A ratio of 296 to 111
   (c) A ratio of 248 to 372
   (d) A ratio of 209 to 341
   (e) A ratio of 3.6 to 4.8
   (f) A ratio of 2.09 to 3.41
   (g) A ratio of 6.264 to 9.396
5. Collene had an after-school job at Taco Time at $5.50 per hour.
   (a) How much did she earn on Monday if she worked $3\frac{1}{2}$ hours?
   (b) On Tuesday she earned $27.50. How long did she work?
   (c) Show that the ratio of the time worked to the amount earned on Monday is equal to the ratio of the time worked to the amount earned on Tuesday; that is, show that these two ratios form a proportion.
   (d) Show that the ratio of the time worked on Monday to the time worked on Tuesday equals the ratio of the amount earned on Monday to the amount earned on Tuesday; that is, show that these two ratios also form a proportion.
6. David Horwitz bought four sweatshirts for $119.92. How much would it cost him to buy nine sweatshirts at the same price per sweatshirt?
7. If s is proportional to t and $s = 62.5$ when $t = 7$, what is s when $t = 10$?
8. The flagpole at Sunnyside Elementary School casts a shadow 9′8″ long at the same time that Mr. Schaal's shadow is 3′2″ long. If Mr. Schaal is 6′3″ tall, how tall is the flagpole to the nearest foot?
9. A kilometer is a bit more than six-tenths of a mile. If the speed limit along a stretch of highway in Canada is 90 kilometers per hour, about how fast can you travel in miles per hour and still not break the speed limit?

Teaching Concepts

10. How could you use measurement and a diagram like the following to help your students understand proportions and the phrase "is proportional to"?

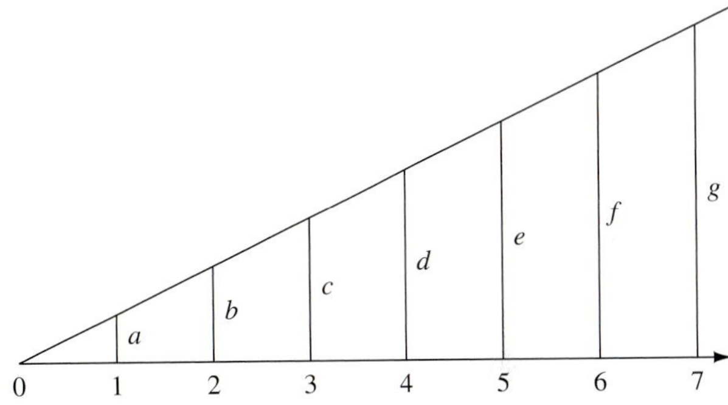

11. Some teachers use diagrams like the following, depicting $a : b = c : d$, or $a/b = c/d$, to help their students visualize proportions.

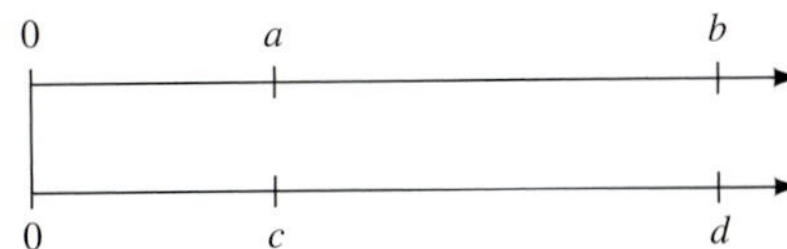

   (a) If $a = 6$ when $b = 24$, determine c when $d = 72$.
   (b) If $b = 2400$ when $a = 128$, determine d when $c = 512$.
   (c) If $b = 24.7$ when $a = 1.4$, determine d when $c = 4.5$.
12. Shiro claims that the area of a square is proportional to the length of a side. Is Shiro right? How would you respond to him?

13. Write a two-page summary of "On Being the Right Size," by J. B. S. Haldane, on pages 952–957 of *The World of Mathematics*, James R. Newman, ed. (New York: Simon and Schuster, 1956). In particular, make clear how ratio and proportion play a role in this study.

Thinking Critically

14. If a is to b as c is to d, that is, if

$$\frac{a}{b} = \frac{c}{d},$$

(a) show that b is to a as d is to c.

(b) show that $a - b$ is to b as $c - d$ is to d.

(c) show that a is to $a + b$ as c is to $c + d$.

(d) show that $a + b$ is to $c + d$ as b is to d.

15. (a) If y is proportional to x^2 and $y = 27$ when $x = 6$, determine y when $x = 12$.

(b) Determine the ratio of the y-values in part (a).

(c) If y and x are related as in part (a), what happens to the value of y if the value of x is doubled? Explain.

16. (a) If y is proportional to x^3 and $y = 32$ when $x = 12$, determine the value of y when $x = 6$.

(b) Determine the ratio of the y-values in part (a).

(c) If y and x are related as in part (a), what happens to the value of y if the value of x is doubled? tripled? quadrupled? Explain.

17. If y is proportional to $1/x$ and $y = 3.5$ when $x = 84$, determine y when $x = 14$. (*Hint:* $y = k(1/x)$.)

Thinking Cooperatively

18. Working with two or three other students and using a rectangular mirror and a measuring tape, take the needed measurements and discuss the various aspects of this problem as you seek a solution agreed upon by your entire group.

(a) You stand in front of a mirror on a wall and can just barely see your entire reflection. If your height is h and H is the vertical dimension of the mirror, determine the ratio h to H. (*Hint:* Make a suitable drawing.)

(b) Where should the mirror of minimum height H in part (a) be located on the wall?

(c) Does the distance you stand from the wall make a difference in your answers to parts (a) and (b)? Explain.

19. Working with your group, hang a screen-door spring about 20 inches long from a support like a portable coat hanger, and hang from the end of the spring a tin can or other container capable of holding up to a 2-pound weight. Using a meterstick, measure as carefully as possible the *amount of stretch* of the spring if a half-pound weight is hung from the spring, a 1-pound weight is hung, and a 2-pound weight is hung. What can you conclude about the ratios of the weights to the amount of stretch? Decide on a common answer for your group.

20. Working with your group, make a pendulum from a long piece of string (say, 150 centimeters long) by tying three or four heavy washers onto the end of the string. Mark the string at 10-centimeter intervals starting from the center of the washers. Hold the pendulum from a suitable support and measure the time t in seconds for the pendulum to make 10 complete swings for various lengths l of string from the support to the center of the washers.

(a) Complete the following table, computing l to the nearest hundredth, and arrive at a consensus conclusion for the results observed.

l	t	$\sqrt{l}$
10		
20		
30		
40		
50		

(b) Predict how long it will take for 10 swings if the pendulum is 100 centimeters long. Then check your guess by actually timing 10 swings. Is the time obtained about what you expect?

Making Connections

21. Celeste won a 100-meter race with a time of 11.6 seconds, and Michelle came in second with a time of 11.8 seconds. Given that each girl ran at a constant rate throughout the race, determine the ratio of Celeste's speed to Michelle's in simplest terms.

22. On a trip of 320 miles, Sunao's truck averaged 9.2 miles per gallon. At the same rate, how much gasoline would his truck use on a trip of 440 miles?

23. Which is the best buy in each case?
   (a) 32 ounces of cheese for 90¢ or 40 ounces of cheese for \$1.20
   (b) A gallon of milk for \$2.21 or two half gallons at \$1.11 per half gallon
   (c) A 16-ounce box of bran flakes at \$3.85 per box or a 12-ounce box at \$2.94 per box

24. The ratio of Dexter's salary to Claudine's is 4 to 5. If Claudine earns \$3200 per month, how much of a raise will Dexter have to receive to make the ratio of his salary to Claudine's 5 to 6?

25. The ratio of boys to girls in Ms. Zombo's class is 3 to 2. In Mr. Stolarski's class it is 4 to 3. If there are 30 students in Ms. Zombo's class and 28 students in Mr. Stolarski's class, what is the ratio of boys to girls in the combined classes?

Using a Calculator

26. If you have not already completed the Window on Technology box in Section 3.6, complete the following.
   (a) The golden ratio, $(1 + \sqrt{5})/2$, dates from the Pythagorean school of Greek mathematics of about 600 B.C. Compute the decimal representation of the golden ratio.
   (b) Compute the ratios F_{n+1}/F_n for $n = 1, 2, 3, \ldots, 10$, where F_n denotes the nth Fibonacci number, as defined in Example 1.6, part (c).
   (c) Compute the ratios L_{n+1}/L_n for $n = 1, 2, 3, \ldots, 10$ where L_n denotes the nth Lucas number, as defined in problem 16, Problem Set 3.6.
   (d) Carefully considering the results of parts (b) and (c), what seems to be the case about the computed ratios?

27. (a) In the diagram at the top of the next column, use a metric ruler to *very carefully* measure the lengths AC, BC, CD, BE, and DE (even trying to estimate to the nearest tenth of a millimeter, i.e., to the nearest hundredth in the decimal representation of each length). Then compute the ratios

$$\frac{AC}{BC}, \frac{BC}{CD}, \frac{CD}{DE}, \frac{AD}{DC}, \text{ and } \frac{BE}{ED}.$$

   Given the difficulty in obtaining really accurate measurements, what seems to be true of all of these ratios?
   (b) Using a photocopier with an enlarging capability, obtain a larger copy of the diagram in part (a) and repeat part (a) for the new diagram. The larger diagram should make for more accurate measurements and hence more accurate values for the ratios. Do the results now seem more consistent?

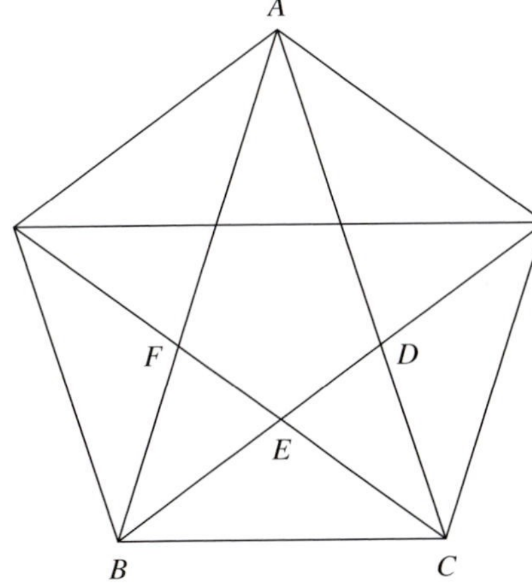

28. If a 12″ pepperoni pizza from Ricco's costs \$9.56, what should a 14″ pizza from Ricco's cost?

29. Suppose your car uses 8.7 gallons of gas traveling 192 miles. Determine approximately how many gallons it would use traveling a distance of 305 miles.

30. If it takes $1\frac{1}{3}$ cups of sugar to make a batch of cookies, how much sugar would be required to make four batches?

31. If three equally priced shirts cost a total of \$59.97, how much would seven shirts cost at the same price per shirt?

32. If 12 erasers cost \$8.04 and Mrs. Orton bought \$14.07 worth of erasers, how many erasers did she buy?

33. One day Kenji Okubo took his class of sixth graders outside and challenged them to find the distance between two rocks that could easily be seen, one above the other, on the vertical face of a bluff near the school. After some discussion, the children decided to hold a rod in a vertical position at a point 100′ from the base of the cliff.

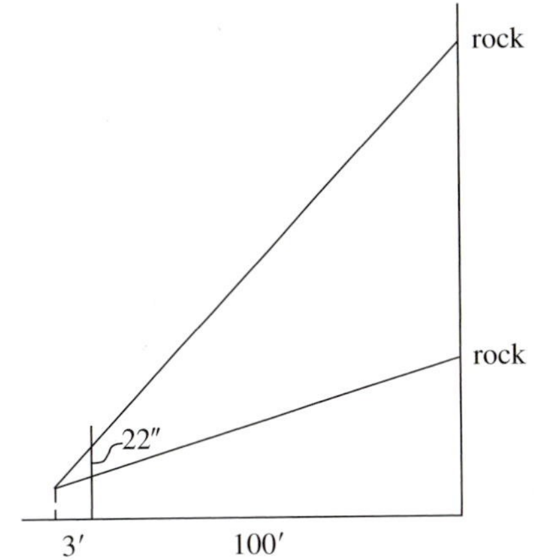

They also decided to mark the points on the rod where the lines of sight of a student standing 3′ farther from the cliff and looking at the rocks cut the rod. If the marks on the rod were 22″ apart, what was the distance between the two rocks? (*Caution:* Convert all measurements to feet.)

From State Student Assessments

34. (Colorado, Grade 4)

Marcus has two different size boxes of raisins. He counts 62 raisins in the 4 ounce box.

A. Estimate how many raisins are in the 16 ounce box.

B. Explain your estimate and show your work so Marcus can see how you decided.

35. (Minnesota, Grade 5)
Use the drawing below to answer the question.

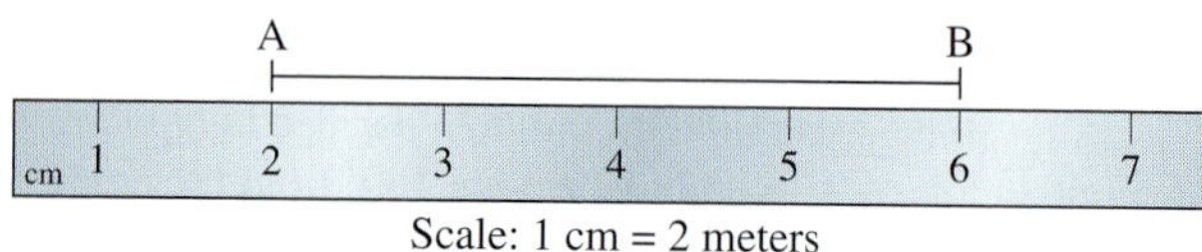

The segment AB is taken from a scale drawing. What is the actual length that this segment represents?

A. 4 meters

B. 6 meters

C. 8 meters

D. 12 meters

36. (Connecticut, Grade 6)
20 out of every 100 high school dropouts are gifted and talented students. Which of the following also describes this situation?

A. 4/25 of the dropouts are gifted and talented.

B. 1/4 of the dropouts are gifted and talented.

C. 1/5 of the dropouts are gifted and talented.

D. 5/1 of the dropouts are gifted and talented.

For Review

37. Perform these computations.

(a) $4 + (-7)$ **(b)** $8 - (-5)$

(c) $(-5) + (-7)$ **(d)** $(-8) + (-5)$

(e) $8 - (-4)$ **(f)** $(-3)(-5)$

(g) $12 \div (-3)$ **(h)** $(-28) \div 4$

(i) $(-28) \div (-4)$ **(j)** $(-495) \div 11$

(k) $57 \div (-19)$ **(l)** $[4 + (-15)] \div 11$

38. Find the prime-power representation of 521,752.

39. If $a = 3^6 \cdot 5^2 \cdot 11^3$, how many divisors does a have?

40. Does $b = 3^2 \cdot 5 \cdot 11^3$ divide a of problem 39? Explain.

41. Find the smallest integer greater than b of problem 40 that is divisible by b.

42. Find the largest integer less than b of problem 40 that divides b.

7.4 PERCENT

Percent

It is essentially impossible to live in today's society and not be conversant with the notion of **percent.**

- What percent interest do you pay on the outstanding balance on your credit card?
- What percent interest do you pay on the amount borrowed to buy a new car? Does this add up to very much money?
- If you are required to make a down payment of 11% of the purchase price of \$175,000 when buying a home, how much do you have to put down?
- How rapidly will your retirement grow if you make periodic payments into an account paying 5% interest compounded quarterly?

Percent (from the Latin *per centum*, meaning "per hundred") is one of the most important ratios in school mathematics. Thus, 50% is the ratio 50/100, and this is quickly reduced to the fraction 1/2 or written as the decimal 0.50. Thus, if I have \$98 and give you 50% of what I have, I give you

$$\frac{1}{2} \cdot \$98 = \$49, \quad \text{or} \quad 0.5 \times \$98 = \$49.$$

The "of" in the preceding sentence translates into "times." Thus,

$$50\% \text{ of} \quad \text{means} \quad 50\% \times,$$
$$\frac{1}{2} \text{ of} \quad \text{means} \quad \frac{1}{2} \times,$$

and

$$0.5 \text{ of} \quad \text{means} \quad 0.5 \times.$$

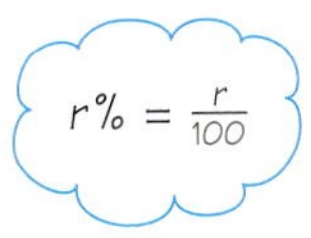

DEFINITION *Percent*

If r is any nonnegative real number, then r percent, written $r\%$, is the ratio

$$\frac{r}{100}.$$

Since $r\%$ is defined as the ratio $r/100$ and the notational effect of dividing a decimal by 100 is to move the decimal point two places to the left, it is easy to write a given percent as a decimal. For example, $12\% = 0.12$, $25\% = 0.25$, $130\% = 1.3$, and so on. Conversely, $0.125 = 12.5\%$, or $12\frac{1}{2}\%$; $0.10 = 10\%$; $1.50 = 150\%$; and so on.

EXAMPLE 7.23 EXPRESSING DECIMALS AS PERCENTS

Express these decimals as percents.

(a) 0.25 **(b)** 0.333 . . . **(c)** 1.255

Solution

(a) $0.25 = 25\%$
(b) $0.33\ldots = 33.333\ldots\%$
(c) $1.255 = 125.5\%$

$= 33\frac{1}{3}\%$

EXAMPLE 7.24 EXPRESSING PERCENTS AS DECIMALS

Express these percents as decimals.

(a) 40% **(b)** 12% **(c)** 127%

Solution

(a) $40\% = 0.40$
(b) $12\% = 0.12$
(c) $127\% = 1.27$

$40\% = \frac{40}{100} = 0.40$

EXAMPLE 7.25 EXPRESSING PERCENTS AS FRACTIONS

Express these percents as fractions in lowest terms.

(a) 60% **(b)** $66\frac{2}{3}\%$ **(c)** 125%

Solution

(a) By definition, 60% means 60/100. Therefore,

$$60\% = \frac{60}{100} = \frac{3}{5}.$$

(b) $$66\frac{2}{3}\% = \frac{66\frac{2}{3}}{100} = \frac{\frac{200}{3}}{100} = \frac{2}{3}.$$

(c) $$125\% = \frac{125}{100} = \frac{5}{4}, \text{ or } 1\frac{1}{4}.$$

EXAMPLE 7.26 EXPRESSING FRACTIONS AS PERCENTS

Express these fractions as percents.

(a) $\frac{1}{8}$ **(b)** $\frac{1}{3}$ **(c)** $\frac{16}{5}$

Solution 1 *Using Proportions*

Since percents are ratios, we can use variables to determine the desired percents.

(a) Suppose $1/8 = r\% = r/100$. Then

$$r = 100 \cdot \frac{1}{8} = 12.5,$$

and

$$r\% = 12.5\%.$$

(b) If $1/3 = s\% = s/100$, then

$$s = \frac{100}{3} = 33\frac{1}{3},$$

$= 33\frac{1}{3}\%$

and

$$s\% = 33\frac{1}{3}\%.$$

(c) If $16/5 = u\% = u/100$, then

$$u = \frac{16}{5} \cdot 100 = 320,$$

and

$$u\% = 320\%.$$

Solution 2 *Using Decimals*

Here, we write the fractions as decimals and then as percents.

(a) By division,

$$\frac{1}{8} = 0.125 = 12.5\%.$$

(b) Here,

$$\frac{1}{3} = 0.333\ldots = 33.\bar{3}\% = 33\frac{1}{3}\%.$$

(c) $\frac{16}{5} = 3.2 = 320\%.$

Applications of Percent

Use of percents is commonplace. Three of the most common types of usages are illustrated in the next three examples.

EXAMPLE 7.27 CALCULATING A PERCENTAGE OF A NUMBER

The Smetanas bought a house for \$175,000. If a 15% down payment was required, how much was the down payment?

Solution 1 *Using an Equation*

The down payment is 15% of the cost of the house. Thus, if d is the down payment,

$$\begin{aligned} d &= 15\% \times \$175{,}000 \\ &= 0.15 \times \$175{,}000 \\ &= \$26{,}250. \end{aligned}$$

Solution 2 *Using Ratio and Proportion*

The size of the down payment is proportional to the cost of the house; that is, $d/175{,}000$ is a ratio that is equivalent to the ratio 15%. The following diagram makes the desired proportion visually more clear.

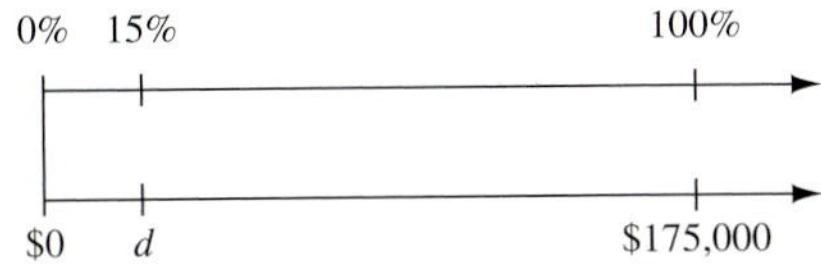

Thus,

$$\frac{d}{\$175{,}000} = \frac{15}{100} = 0.15,$$

and

$$\begin{aligned} d &= 0.15 \times \$175{,}000 \\ &= \$26{,}250. \end{aligned}$$

School Book Page

Percent in Grade Five

Chapter 7 Lesson 12

Understanding Percent

You Will Learn
how to use percents to name amounts less than or equal to 1

Vocabulary
percent
a way to compare a number with one hundred

Learn

In a recent survey, 100 students were asked which household chore they liked the least. The list shows which jobs the students mentioned.

How can you name the part of those surveyed who liked washing dishes the least?

Chores we Like the Least

Washing dishes	31
Laundry	20
Yard work	19
Cleaning room	17
Babysitting	13

Did You Know?
Percent means per 100. It comes from the Latin *per centum*.

Mia wrote a fraction to describe the part. $\frac{31}{100}$

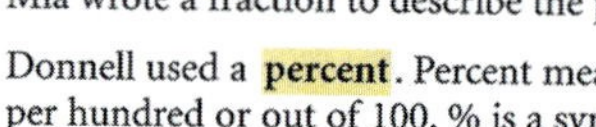

Donnell used a **percent**. Percent means per hundred or out of 100. % is a symbol for percent. 31%

José thought of the number of students as 31 out of 100. 0.31

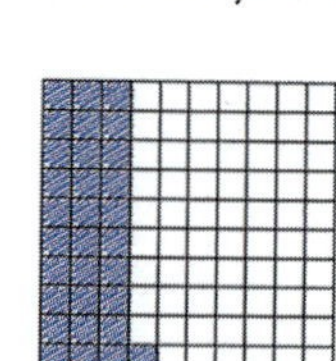

Talk About It

Look at the grid shown above. What fraction and percent name the unshaded part?

Check

Write the hundredths fraction and the percent shaded in each picture.

1.  2.  3.

4. **Reasoning** Doing laundry was the chore least liked by 20 of the 100 students. Use a grid to show the percent who named laundry.

SOURCE: Scott Foresman–Addison Wesley Math, *Grade 5*, p. 328, by Randall I. Charles et al. Copyright © 2002 Pearson Education, Inc. Reprinted with permission.

Questions for the Teacher

1. This introductory lesson on percent focuses on a survey of 100 people and a rectangular array of 100. Does this seem to you a good way to begin? Explain.
2. This lesson continues with these questions. How would you help your students to answer them?

Estimation Estimate the percent of each figure that is shaded.

19. 20. 21.

3. How would you respond to a child who says that 25% is the same as 1/4? Would you welcome such an observation? Why or why not?

EXAMPLE 7.28 CALCULATING A NUMBER OF WHICH A GIVEN NUMBER IS A GIVEN PERCENTAGE

Soo Ling scored 92% on her last test. If she got 23 questions right, how many problems were on the test?

Solution 1 *Using a Variable*

Let n denote the number of questions on the test. Since Soo Ling got 23 questions correct for a score of 92%, we know that 23 is 92% of n; that is,

$$23 = 92\% \times n = 0.92n.$$

Hence,

$$n = \frac{23}{0.92} = 25.$$

Solution 2 *Using Ratio and Proportion*

As seen in the diagram, the ratio of 23 to the number of questions on the test must be the same as the ratio 92%.

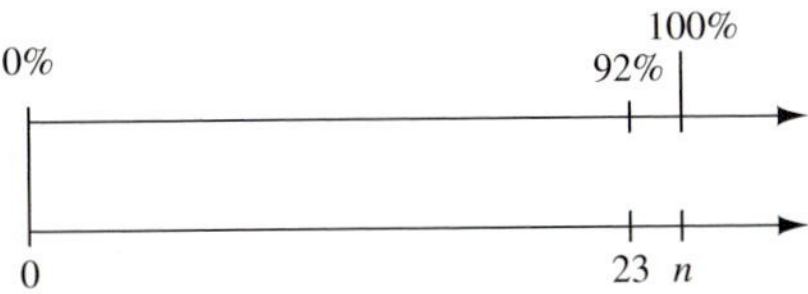

Thus,

$$\frac{23}{n} = \frac{92}{100} = 0.92.$$

Therefore,

$$23 = 0.92 \times n,$$

and

$$n = 23 \div 0.92 = 25$$

as before.

EXAMPLE 7.29 CALCULATING WHAT PERCENTAGE ONE NUMBER IS OF ANOTHER

Tara got 28 out of 35 possible points on her last math test. What percentage score did the teacher record in her grade book for Tara?

Solution 1 *Using the Definition*

Tara got 28 thirty-fifths of the test right. Since

$$\frac{28}{35} = 0.80 = 80\%,$$

the teacher recorded 80% in her grade book.

Solution 2 *Using Ratio and Proportion*

Let x be the desired percentage. Then, as above, the diagram helps one visualize the proportion

$$\frac{x}{100} = \frac{28}{35}.$$

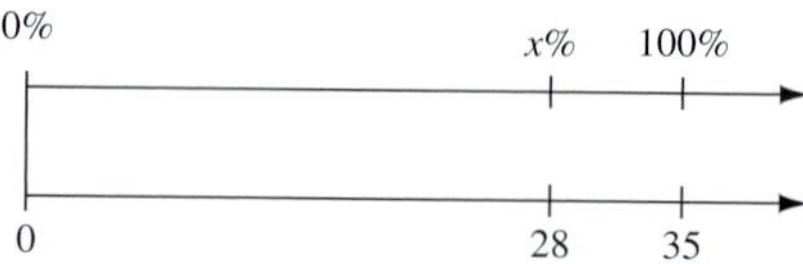

Thus,

$$x = \frac{28 \cdot 100}{35} = 80\%.$$

Compound Interest

If you keep money in a savings account at a bank, the bank pays you interest at a fixed rate (percentage) for the privilege of using your money. For example, suppose you invest \$5000 for a year at 7% interest. How much is your investment worth at the end of the year? Since the interest earned is 7% of \$5000, the interest earned is

$$7\% \times \$5000 = 0.07 \times \$5000 = \$350,$$

and the value of your investment at the end of the year is

$$\begin{aligned} \$5000 + \$350 &= \$5000 + 0.07 \times \$5000 \\ &= \$5000 \cdot (1.07) \\ &= \$5350. \end{aligned}$$

If you leave the total investment in the bank, its value at the end of the second year is

$$\begin{aligned} \$5350 + 0.07 \times \$5350 &= \$5350 \cdot (1.07) \\ &= \$5000 \cdot (1.07)(1.07) \\ &= \$5000 \cdot (1.07)^2 \\ &= \$5724.50. \end{aligned}$$

from above

Similarly, at the end of the third year, your investment would be worth

$$\$5000 \cdot (1.07)^3 = \$6125.22$$

to the nearest penny. In general, it would be worth

$$\$5000 \cdot (1.07)^n$$

at the end of n years. This is an example of **compound interest,** where the term "compound" implies that each year you earn interest on all the interest earned in preceding years as well as on the original amount invested (the **principal**).

Usually, interest is compounded more than once a year. Suppose the \$5000 investment just discussed was made in a bank that pays interest at the rate of 7% compounded semiannually, that is, twice a year. Since the rate for a year is 7%, the rate for half a year

Into the Classroom

Anne Lawrence Discusses Percent

To illustrate percent in a fun way, I purchase packages of M & M's® or Smarties®. I buy a package for each group of students, or, if the packages are small, one package for each student.

Initially each group (or each student) determines the total number of candies in the package. Then, working with the ratio $\frac{\text{part}}{\text{whole}}$, they determine what fraction of the total each color represents. Students compute $\frac{\text{part}}{\text{whole}}$ to obtain a decimal, and then they

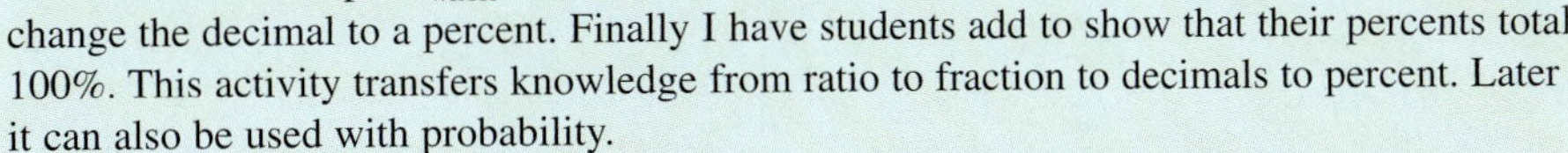

change the decimal to a percent. Finally I have students add to show that their percents total 100%. This activity transfers knowledge from ratio to fraction to decimals to percent. Later it can also be used with probability.

SOURCE: *Scott Foresman–Addison Wesley Middle School Math*, Course 2, p. 382D, by Randall I. Charles et al.

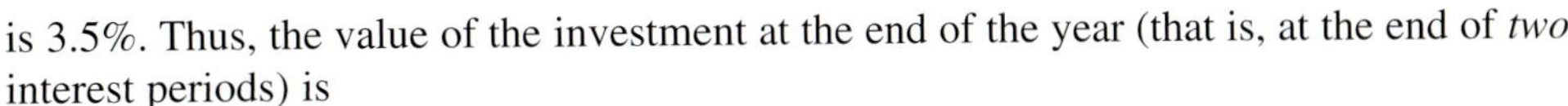

is 3.5%. Thus, the value of the investment at the end of the year (that is, at the end of *two* interest periods) is

$$\$5000(1.035)^2 = \$5356.13,$$

and the values at the end of two years and three years, respectively, are

$$\$5000(1.035)^4 = \$5737.62$$

and

$$\$5000(1.035)^6 = \$6146.28.$$

These can be easily calculated using the [^] key or the built-in constant function of your calculator.

Compounding more and more frequently is to your advantage, and some banks are now compounding monthly or even daily.

The above calculations are typical and are summarized in this theorem.

$\frac{r}{t}$ is the rate per interest period and nt is the number of interest periods.

THEOREM *Calculating Compound Interest*

The value of an investment of P dollars at the end of n years, if interest is paid at the annual rate of r% compounded t times a year, is

$$P\left(1 + \frac{r/t}{100}\right)^{nt}.$$

EXAMPLE 7.30 COMPUTING THE COST OF DEBT

Many credit card companies charge 18% interest compounded monthly on unpaid balances. Suppose your card was "maxed out" at your credit limit of $10,000 and that you were unable to make any payments for two years. Aside from penalties, how much debt would you owe, based on compound interest alone?

Solution Since the interest is computed at 18% compounded monthly, the effective rate per month is 18%/12 = 1.5%, and the number of interest periods in two years is $12 \cdot 2 = 24$. Thus, your debt to the nearest penny would be

$$\$10{,}000(1.015)^{24} = \$14{,}295.03.$$

If the debt went unpaid for six years, you would owe

$$\$10{,}000(1.015)^{72} = \$29{,}211.58.$$

This is almost triple what you originally owed! The above calculations can be made by repeatedly multiplying \$10,000 by 1.015, using the built-in constant feature of your calculator. Even more easily, you can compute $\$10{,}000(1.015)^{24}$ directly by entering the following string into your calculator:

10000 [×] 1.015 [^] 24 [=].

The Mathematics of Growth

Population growth occurs in exactly the same way that an investment grows if it is earning compound interest. Suppose, for example, that the population in the Puget Sound region in northwest Washington is approximately 2.2 million and that it is growing at the rate of 5% per year. In one year the population will be approximately

$$2.2(1.05) = 2.3$$

million. If the growth continues unabated, in 14 years it will be approximately

$$2.2(1.05)^{14} = 4.4$$

million, twice what it is today. Given the fact that the area has already experienced several years of water shortages, these figures are cause for concern among officials in the area.

Like population growth, prices of commodities also rise with inflation as an investment grows by drawing compound interest.

Did You Know?

It's Only \$1000

The accompanying article appeared in the August 9, 1993, issue of *Newsweek*. This is a dramatic example of the power of compound interest—particularly at a rate as high as 24%. By the way, is the Wilsons' calculation correct? If the bond was purchased on January 1, 1865, compute its value on January 1, 1993.

If Nevada officials are welshing, it's understandable. In April, Allen and Kathy Wilson of Montello, Nev., sued the state to cash in a \$1,000 state bond issued in 1865 with a yearly interest rate of 24 percent. Why wouldn't the state pay up? With interest compounded, according to the couple's calculations, the bond is now worth \$657 trillion. *The Wilsons, who collect old documents like stock certificates and investigate if they're redeemable, were willing to settle for a lesser amount: say, a paltry \$54 million. Last week a state judge ruled the bond had to have been cashed by 1872. "I'm glad the people who brought this absurd lawsuit won't get a plugged nickel," said State Treasurer Bob Seale.*

EXAMPLE 7.31 PRICING A CAR

If the economy were to experience a steady inflation rate of 2.5% per year, what would be the price of a new car in five years if the same-quality car sells today for \$18,400?

Solution Using the same formula as in computing compound interest, we find that the price of the car five years from now would be approximately

$$\$18{,}400(1.025)^5 \doteq \$20{,}818.$$

Problem Set 7.4

Understanding Concepts

1. Write each of these ratios as percents.
   (a) $\frac{3}{16}$ (b) $\frac{7}{25}$ (c) $\frac{37}{40}$
   (d) $\frac{5}{6}$ (e) $\frac{3.24}{8.91}$ (f) $\frac{7.801}{23.015}$
   (g) $\frac{1.6}{7}$ (h) $\frac{\sqrt{2}}{\sqrt{6}}$
2. Write each of these as percents.
   (a) 0.19 (b) 0.015 (c) 2.15 (d) 3
3. Write each of these as fractions in simplest form.
   (a) 10% (b) 25% (c) 62.5% (d) 137.5%
4. Compute each of the following.
   (a) 70% of 280 (b) 120% of 84
   (c) 38% of 751 (d) $7\frac{1}{2}$% of \$20,000
   (e) .02% of 27,481 (f) 1.05% of 845
5. Compute each of these mentally.
   (a) 50% of 840 (b) 10% of 2480
   (c) 12.5% of 48 (d) 125% of 24
   (e) 200% of 56 (f) 110% of 180
6. What percentage of the markers in each of these arrays are Xs?

(a)	(b)
OXOXOXOXOX	XOXOXOXOXO
XOXOXOXOXO	OOOXXXXXXX
OXOXOXOXOX	XXXOOOXXXX
XOXOXOXOXO	XXXXXXXOXX
OXOXOXOXOX	OXOXOXXXXX
XOXOXOXOXO	XOXXOXXXOX

7. Without counting, estimate the percentage of markers in each array that are Xs.

(a)	(b)
XXXXXXXXXX	XXXXXXXXXX
OXXXXXXXXX	XXXXXXXXXX
OOXXXXXXXX	OOOOOOOOOO
OOOXXXXXXX	XXXXXXXXXX
OOOOXXXXXX	OOOOOOOOOO
OOOOOXXXXX	XXXXXXXXXX

8. If the 20-mm by 20-mm square shown represents 100%, draw rectangles 20 mm wide that represent each of these percentages.

   (a) 75% (b) 125% (c) 200% (d) 20%
9. What percentage of each of these figures is red?

(a) (b)

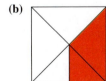

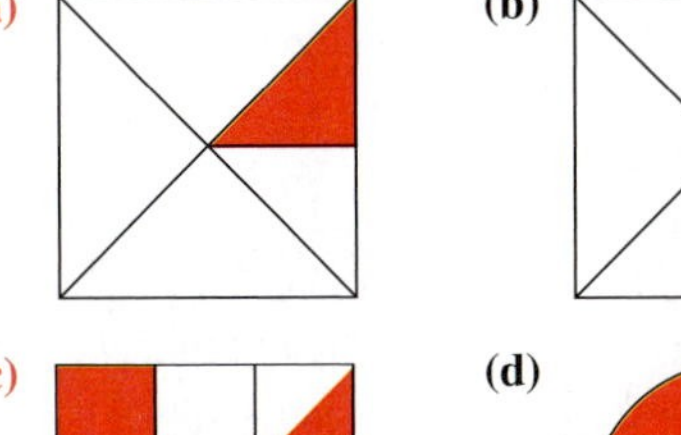

(c)

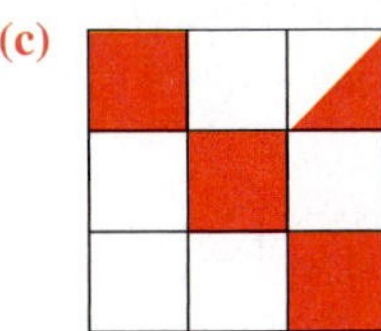

(d)

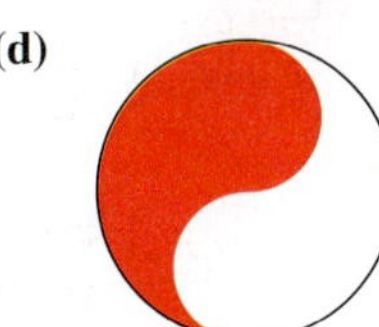

10. How many small squares must be shaded to represent the given percentages of the large square shown?

(a) 100% **(b)** 0% **(c)** 25% **(d)** 87%

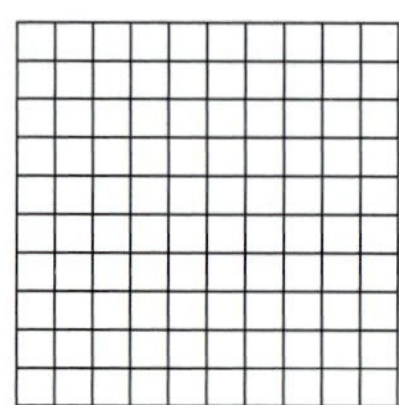

11. How many of the small rectangles must be shaded to represent the given percentages of the large rectangle shown?

(a) 50% **(b)** 25% **(c)** 20% **(d)** 37.5%

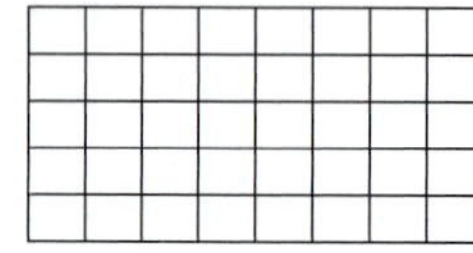

12. Mentally convert each of these to a percent.

(a) $\frac{7}{28}$ **(b)** $\frac{11}{33}$

(c) $\frac{72}{144}$ **(d)** $\frac{44}{66}$

13. Mentally estimate the number that should go in the blank to make each of these *true*.

(a) 27% of ____ equals 16.

(b) 4 is ____% of 7.5.

(c) 41% of 120 = ____.

Teaching Concepts

14. When asked about his performance in an upset victory in a football game, the quarterback said that he gave 110% of effort. Briefly discuss the reasonableness of this assertion.

15. In a stockholders' meeting, the chief executive officer said that the company had earned 110% of the previous year's profits. Briefly discuss the reasonableness of this assertion.

16. Hari says that his dad has told him about the "rule of 72," which states that an investment at r% interest compounded annually will double in value in $72 \div r$ years. Does the rule seem to hold for an investment at 7% compounded annually? How would you respond to Hari?

Thinking Critically

17. In a given population of men and women, 40% of the men are married and 30% of the women are married. What percentage of the adult population is married?

18. A garage advertised car repairs at 10% off the usual price—5% on parts and 5% on labor. The Consumer Protection Agency chided the garage for false advertising. Was the CPA correct? Explain.

19. After declining to do a "shady" job for Mayor Pigg for 25% of the take, Chester Slocum finally agreed to the deal when the Mayor offered him 25% of 25% of the take. What does this say about Chester's understanding of arithmetic? Explain briefly.

20. During the first half of a basketball game, the basketball team at Red Cloud High School made 60% of their 40 field goal attempts. During the second half, they scored on only 25% of 44 attempts from the field. To the nearest 1%, what was their field goal shooting percentage for the entire game?

21. During the first half of a basketball game, Skeeter Thoreson missed all five of her field goal attempts. During the second half, she hit 75% of her 16 attempts from the field. What was her field goal shooting percentage for the game?

Making Connections

22. Tabata's Furniture calculates the retail price of furniture they sell by marking up its cost at wholesale a full 100%.

(a) If they had a sale with all items marked down 20% from the retail price, what percentage profit did they actually make on each item sold during the sale?

(b) If they had a sale with all items marked down 50% from the retail price, what percentage profit did they make on each item sold during the sale?

23. A merchant obtains the retail price of an item by adding 20% to the wholesale price. Later he has a sale and marks every item down 20% from the retail price. Is the sale price the same as the wholesale price? Explain briefly. (*Hint:* Consider a specific item whose wholesale price is $100.)

24. Acme Electric in Boise, Idaho, purchases hot water heaters at wholesale for $185 each and sells them after marking up the wholesale price 45%. Since Idaho has a 5% sales tax, how much would you have to pay Acme for one of its water heaters?

25. Connie's Craft Corner sells its card-stamping materials after marking them up 60% over their wholesale price. At a recent sale, it marked its merchandise down 20%.

(a) What is the sale price of items that originally cost Connie's $46?

(b) Including sales tax of 7%, how much must a customer pay Connie's for the items in part (a)?

(c) What percentage profit does Connie's ultimately make on items sold in the sale?

26. There is \$100 in the cash register at the start of a cashier's shift at a supermarket. At the end of the shift, the register contains \$800. If the state has a 5% sales tax, how much of the \$800 is for tax?

27. The mortgage company requires an 11% down payment on houses it finances. If the Sumis bought a house for \$158,000, how much down payment did they have to make?

28. Mr. Swierkos invested \$25,000 in a mutual fund. If his broker deducted a 6% commission before turning the rest of the money over to the mutual fund, and the value of each share increased by 18% during the year, what percentage return on his investment did Mr. Swierkos realize at the end of the first year?

29. Kneblemann's Fine Clothes was having a sale with all merchandise marked down 20%.

(a) How much would Bethany have to pay for a dress originally priced at \$78?

(b) If Natasha paid \$84 for a suit at Kneblemann's sale, what was the original price of the suit?

(c) Explain a quick and easy way to calculate the sale price of items at Kneblemann's sale by using a calculator.

30. Show that the sale price of items marked down 15% is the same as 85% of the original retail price.

31. Find the value of each of these investments at the end of the time period specified.

(a) \$2500 invested at $5\frac{1}{4}\%$ interest compounded annually for seven years

(b) \$8000 invested at 7% interest compounded semiannually for 10 years

Using a Calculator

32. How much would you have to invest at 6% interest compounded annually in order to have \$16,000 at the end of five years? (*Note:* If P is the amount invested, then $16{,}000 = P \cdot (1.06)^5$.)

33. Suppose you place \$1000 in a bank account on January 1 of each year for nine years. If the bank pays interest on such accounts at the rate of 5.3% interest compounded annually, how much will your investment be worth on January 1 of the tenth year?

34. Successive powers of $1 + \frac{r}{100}$ are easily computed by using the built-in constant feature of your calculator. Since the value of an investment at r% interest compounded annually for n years is $P(1 + \frac{r}{100})^n$, determine the least number of years for the value of an investment to double at each of these rates.

(a) 5% (b) 7% (c) 14% (d) 20%

(e) The "rule of 72" states that you divide 72 by the interest rate to obtain the doubling time for an initial investment. This rule is used by many bankers to give a crude approximation of the time for the value of an investment to double at a given rate. Does it seem like a reasonable rule? Explain briefly.

35. (a) If the population of Oregon in 1999 was 3,300,000 and it is increasing at the rate of 4% per year, what will the population be in the year 2010? Give your answer correct to the nearest one hundred thousand.

(b) Using the same assumptions as in part (a), what was the population of Oregon in 1990? (*Hint:* Use the built-in constant feature of your calculator to divide repeatedly by 1.04.)

Using a Computer

Do this problem on a spreadsheet.

36. (a) At age 25, Sally Saver began to contribute \$1000 per year to a retirement fund that pays 8% interest compounded annually. What will be the accumulation in Sally's account when she retires at age 65? (*Suggestion:* Enter "Year" in A1, "Sally" in B1, "\$1000" in B2, and "= 1000 + (1.08)*B2" in B3 and **COPY** and **PASTE** or **FILL** down to B41.)

(b) Larry Livitup began his retirement savings at age 50, paying \$6000 per year into an account also paying 8% interest compounded annually. What is Larry's accumulation when he retires at age 65? (*Suggestion:* Enter "Larry" in C1 of the spreadsheet of part (a), "\$6000" in C2, and "= 6000 + (1.08)*C2" in C3 and **COPY** and **PASTE** or **FILL** down to C16.)

(c) Sally deposited only $40 \cdot \$1000 = \$40{,}000$ into her retirement account, and Larry deposited $15 \cdot \$6000 = \$90{,}000$ into his. Who apparently has the better plan?

From State Student Assessments

37. (Oregon, Grade 5)

Richard answered 22 of the 25 questions on this week's Language Arts test correctly. About what percent of the total number of questions did he answer correctly?

A. 22 percent
B. 40 percent
C. 80 percent
D. 90 percent

38. (Massachusetts, Grade 8)

Use the advertisement below to answer the question.

ROXBURY BIKE SHOP
ONE DAY SALE – SATURDAY ONLY
All bikes must go!!

9:00 a.m. – 10% off originally-marked price
10:00 a.m. – 10% off 9:00 price
11:00 a.m. – 10% off 10:00 price

AND SO ON UNTIL
ALL BIKES ARE SOLD

Mr. Howard bought a bike with an originally-marked price of \$400. What was the price of the bike at 12:15 P.M.?

A. \$262.44

B. \$240.00

C. \$291.60

D. \$280.00

For Review

39. Write each of these fractions in simplest form.

(a) $\frac{51}{69}$ **(b)** $\frac{143}{1001}$ **(c)** $-\frac{38}{57}$

40. Perform each of these additions and subtractions. Be sure to write your answer in simplest form.

(a) $\frac{1}{12}+\frac{1}{4}$ **(b)** $\frac{1}{5}+\frac{1}{12}$

(c) $\frac{8}{15}-\frac{17}{45}$ **(d)** $\frac{3}{143}-\frac{1}{91}$

(e) $\frac{11}{84}+\frac{5}{96}$ **(f)** $\frac{21}{60}-\frac{7}{75}$

41. Perform each of these multiplications and divisions leaving your answers as fractions in simplest form.

(a) $\frac{27}{44}\cdot\frac{22}{81}$ **(b)** $\frac{176}{247}\cdot\frac{95}{99}$

(c) $\frac{33}{35}\div\frac{22}{63}$ **(d)** $\frac{159}{286}\div\frac{102}{253}$

(e) $\frac{25}{44}\cdot\frac{68}{75}\cdot\frac{11}{16}$ **(f)** $\frac{169}{289}\div\frac{104}{221}$

42. Determine r, s, and t to make each of these equalities *true*.

(a) $\frac{25}{65}=\frac{r}{26}$ **(b)** $\frac{86}{99}=\frac{129}{s}$ **(c)** $\frac{4}{3}\div t=\frac{7}{6}$

Epilogue The Number Systems of Arithmetic

With this chapter, we have completed our study of the number systems of arithmetic. There is one more extension—to the complex numbers—that is needed for the study of algebra and other more advanced courses, but it is inappropriate for teaching on the elementary school level.

Table 7.3, on the next page, provides a summary of the number systems we have considered and the properties that each system satisfies.

It is important to note the following:

- Neither the commutative property for subtraction nor that for division holds in any of these systems.
- Only the rational numbers, the real numbers, and the prime-hour clock arithmetic system have multiplicative inverses for all nonzero (nonadditive identity) elements. Hence, only these systems are closed under division except for division by zero.

This table is convenient, but just knowing the properties is *not* enough. It is at least as important that teachers know conceptual and physical models appropriate to illustrate the properties and operations for each number system. They must also know how to use the systems to model and solve problems, including real-world problems. Understanding and critical thinking must be the goal of mathematical instruction—not memorizing properties and procedures by rote!

TABLE 7.3 Number Systems and Their Properties

Properties	Natural numbers	Whole numbers	Integers	Rational numbers	Real numbers	*n*-hour clock arithmetic, *n* composite	*n*-hour clock arithmetic, *n* prime
Closure property for addition	•	•	•	•	•	•	•
Closure property for multiplication	•	•	•	•	•	•	•
Closure property for subtraction			•	•	•	•	•
Closure property for division except for division by zero				•	•		•
Commutative property for addition	•	•	•	•	•	•	•
Commutative property for multiplication	•	•	•	•	•	•	•
Commutative property for subtraction							
Commutative property for division							
Associative property for addition	•	•	•	•	•	•	•
Associative property for multiplication	•	•	•	•	•	•	•
Associative property for subtraction							
Associative property for division							
Distributive property for multiplication over addition	•	•	•	•	•	•	•
Distributive property for multiplication over subtraction	•	•	•	•	•	•	•
Contains the additive identity, 0		•	•	•	•	•	•
Contains the multiplicative identity, 1	•	•	•	•	•	•	•
The multiplication property for 0 holds		•	•	•	•	•	•

Properties	Natural numbers	Whole numbers	Integers	Rational numbers	Real numbers	n-hour clock arithmetic, n composite	n-hour clock arithmetic, n prime
Each element possesses an additive inverse			•	•	•	•	•
Each element except 0 possesses a multiplicative inverse				•	•		•

Chapter Summary

Key Concepts

Section 7.1 Decimals

- Decimal. In general, the Indo-Arabic system of positional notation for numerals. Colloquially, nonintegers represented in positional notation to base ten.
- Expanded form. As with integers—e.g., $2.437 = 2 \cdot 10^0 + 4 \cdot 10^{-1} + 3 \cdot 10^{-2} + 7 \cdot 10^{-3}$.
- Manipulatives to represent decimals. One can use mats, strips, and units; base-ten blocks; dollars, dimes, and pennies; place-value cards; and so on.
- Zero and negative numbers as exponents. If $a \neq 0$, then $a^0 = 1$ and $a^{-r} = \frac{1}{a^r}$.
- Multiplying decimals by powers of 10. If r is a natural number, multiplying a number by 10^r notationally moves the decimal point r places to the right. Multiplying by 10^{-r} moves the decimal point r places to the left.
- Terminating decimals. Decimal representations with finitely many terms. These decimals represent rational numbers r/s where the prime factors of s are only 2s or 5s.
- Repeating decimals. Decimal representations with infinitely many terms, where one term or a series of terms repeats *ad infinitum.*
- Nonterminating, nonrepeating decimals. Decimals with infinitely many terms but with no pattern of digits that repeats *ad infinitum.* These decimals represent irrational numbers.
- Real numbers. The set of all numbers that can be represented as decimals. Alternatively, the set of all rational and irrational numbers.
- Ordering decimals. Write the decimals so that they have the same number of digits to the left of the decimal point (by adjoining zeros to one or the other as needed). Determine the first digits from the left where the two decimals differ. The decimal having the lesser of these two digits represents the lesser number.
- $\sqrt{2}$ is irrational. $\sqrt{2}$ and infinitely many other numbers are irrational.

Section 7.2 Computations with Decimals

- Rounding decimals. Use the 5-up rule.
- Adding and subtracting decimals. Align decimal points and then add or subtract as integers.
- Multiplying decimals. Multiply as integers. Let r denote the total number of digits to the right of the decimal points in the two decimals. Place the decimal point r places from the right in the answer.
- Dividing decimals. Multiply both dividend and divisor by the appropriate power of ten to make the divisor an integer. Using the long-division algorithm, divide as integers, placing the decimal point above the decimal point in the dividend.
- Scientific notation. A decimal written as the product of a number r, with $1 \leq r < 10$, and a power of ten to correctly place the decimal point.
- Significant digits. The digits in a decimal r when r is written in scientific notation. r may be rounded to the nearest tenth, hundredth, thousandth, etc.

Section 7.3 Ratio and Proportion

- Ratio. A fraction, usually with positive numerator and denominator.
- Proportion. The equality of two ratios. $\frac{a}{b} = \frac{c}{d}$ if, and only if, $ad = bc$.
- y is proportional to x. If $y = kx$ for some constant k, then y is said to be proportional to x. The number k is called the constant of proportionality.

Section 7.4 Percent

- Percent. For any real number, $r\% = \frac{r}{100}$.
- Interest. The percentage rate paid to borrow money or earned on an investment.
- Compound interest. Interest calculated so that the interest paid in any given year is paid on both the original amount and on the interest earned in prior years.
- Calculating compound interest. The amount at the end of n years of A dollars earning $r\%$ interest compounded t times a year is $A \cdot \left(1 + \frac{r}{100t}\right)^{nt}$.
- The rule of 72. A "rule of thumb" which says that an investment earning $r\%$ interest will double in value in approximately $\frac{72}{r}$ years.

Vocabulary and Notation

Section 7.1

Decimal
Expanded form
Terminating decimal
Repeating decimal
Nonterminating, nonrepeating decimal
Irrational number
Real number

Section 7.2

Scientific notation
Significant digits

Section 7.3

Ratio
Proportion
y is proportional to x

Section 7.4

Percent
Interest
The rule of 72

Chapter Review Exercises

Section 7.1

1. Write these decimals in expanded exponential form.
   (a) 273.425 (b) 0.000354
2. Write these fractions in decimal form.
   (a) $\frac{7}{125}$ (b) $\frac{6}{75}$ (c) $\frac{11}{80}$
3. Write these decimals as fractions in simplest form.
   (a) 0.315 (b) 1.206 (c) 0.2001
4. Arrange these numbers in order from least to greatest:
   $\frac{4}{12}$, 0.33, 0.3334, $\frac{5}{13}$, $\frac{2}{66}$.
5. Write these numbers as fractions in simplest form.
   (a) $10.\overline{363}$ (b) $2.1\overline{42}$
6. Suppose $a = 0.2020020002000020000002\ldots$, continuing in this way with one more zero between each successive pair of 2s. Is this number rational or irrational? Explain briefly.
7. Using only mental arithmetic, determine the numbers represented by these base-ten numerals as fractions in simplest form.
   (a) $0.222\ldots = 0.\overline{2}$ (b) $0.363636\ldots = 0.\overline{36}$

Section 7.2

8. Perform these computations by hand.
   (a) $21.734 + 3.2145 + 71.24$
   (b) $23.471 - 2.89$
   (c) 35.4×2.37
   (d) $24.15 \div 3.45$
9. Compute the following using a calculator.
   (a) $31.47 + 3.471 + 0.0027$
   (b) $31.47 - 3.471$
   (c) 31.47×3.471
   (d) $138.87 \div 23.145$
10. Write estimates of the results of these calculations, and then do the computing accurately with a calculator.
    (a) $47.25 + 13.134$
    (b) $52.914 - 13.101$
    (c) 47.25×13.134
    (d) $47.25 \div 13.134$
11. Write each of these in scientific notation using four significant digits.
    (a) 24,732,654 (b) 0.000012473
12. Using an appropriate calculator and scientific notation, perform each of these calculations and write the results using scientific notation with three significant digits.
    (a) $(2.74 \times 10^5) \cdot (3.11 \times 10^4)$
    (b) $(2.74 \times 10^5) \div (3.11 \times 10^{-4})$
13. Show that $3 - \sqrt{2}$ is irrational.
14. Show that the sum of two irrational numbers can be rational. (*Hint:* Consider problem 13.)
15. What can you say about the decimal expansion of an irrational number?
16. (a) A wall measures 8.25 feet by 112.5 feet. What is the area of the wall?
    (b) If it takes 1 quart of paint to cover 110 square feet, how many quarts of paint must be purchased to paint the wall in part (a)?
17. Give an example of a fraction whose decimal is repeating and has a period of length 4.
18. Use your calculator to determine the periodic decimal expansions of the following numbers. Remember that the calculator will necessarily round off decimals, so don't be misled by the last digit in the display if it seems to break a pattern.
    $$\frac{5}{18}, \frac{41}{333}, \frac{11}{36}, \frac{7}{45}, \frac{13}{80}$$
    (a) Determine where the period starts in each case.
    (b) See if you can guess a rule for determining when the period of the decimal form of a fraction a/b in simplest form begins. (*Hint:* Consider the prime factorization of b.)

Section 7.3

19. Maria made 11 out of 20 free throw attempts during a basketball game. What was the ratio of her successes to failures on free throws during the game?
20. Determine which of these are proportions.
    (a) $\frac{775}{125} = \frac{155}{25}$
    (b) $\frac{31}{64} = \frac{15}{32}$
    (c) $\frac{9}{24} = \frac{12}{32}$
21. If Che bought 2 pounds of candy for \$3.15, how much would it cost him to buy 5 pounds of candy at the same price per pound?
22. It took Donnell 7.5 gallons of gas to drive 173 miles. Assuming that he gets the same mileage per gallon, how much gasoline will he need to travel 300 miles?
23. If y is proportional to x and $y = 7$ when $x = 3$, determine y when $x = 5$.

24. If a flagpole casts a shadow 12′ long when a yardstick casts a shadow 10″ long, how tall is the flagpole?

Section 7.4

25. Convert each of these to percents.

(a) $\frac{5}{8}$ (b) 2.115 (c) 0.015

26. Convert each of these percents to decimals.

(a) 28% (b) 1.05% (c) $33\frac{1}{3}\%$

27. If the sales tax is calculated at 7.2%, how much tax is due on a $49 purchase?

28. If a tax of $6.75 is charged on an $84.37 purchase, what is the sales tax rate?

29. Referring to problem 19 above, what percent of free throws attempted did Maria make during the game?

30. If you invest $3000 at 8% interest compounded every three months (quarterly), how much is your investment worth at the end of two years?

Chapter Test

1. When did you invest $5000 in a bank paying $5\frac{3}{4}\%$ interest compounded annually if it is worth $6612.60 now?

2. Give an example of a fraction whose decimal expansion is repeating with a period of 3.

3. If you invest $2000 now in a bank paying 4.2% interest compounded semiannually, what is the least whole number of years you must leave your investment in the bank in order to withdraw at least $4000?

4. Write these fractions in decimal form.

(a) $\frac{84}{175}$ (b) $\frac{24}{99}$ (c) $\frac{7}{11}$

5. Suppose that you borrow $1000 now at 9% compounded monthly. If you make no payments in the meantime, how much will you owe at the end of two years?

6. Write each of these decimals as a fraction in simplest form.

(a) $0.454545\ldots = 0.\overline{45}$

(b) $31.5555\ldots = 31.\overline{5}$

(c) $0.34999\ldots = 0.34\overline{9}$

7. Without doing the hand or calculator calculation, determine how many digits should appear to the right of the decimal point in the product 21.432×3.41.

8. The Pirates won 17 of their 32 hockey games.

(a) What was the ratio of their wins to losses?

(b) What percentage of their games did they win?

9. Compute the product $(2.34 \times 10^6) \cdot (3.12 \times 10^{-19})$ using an appropriate calculator. Write your answer correct to three significant digits.

10. Mr. Spence put $1425 down on a car selling for $9500. What percent of the purchase price did the dealer require as down payment?

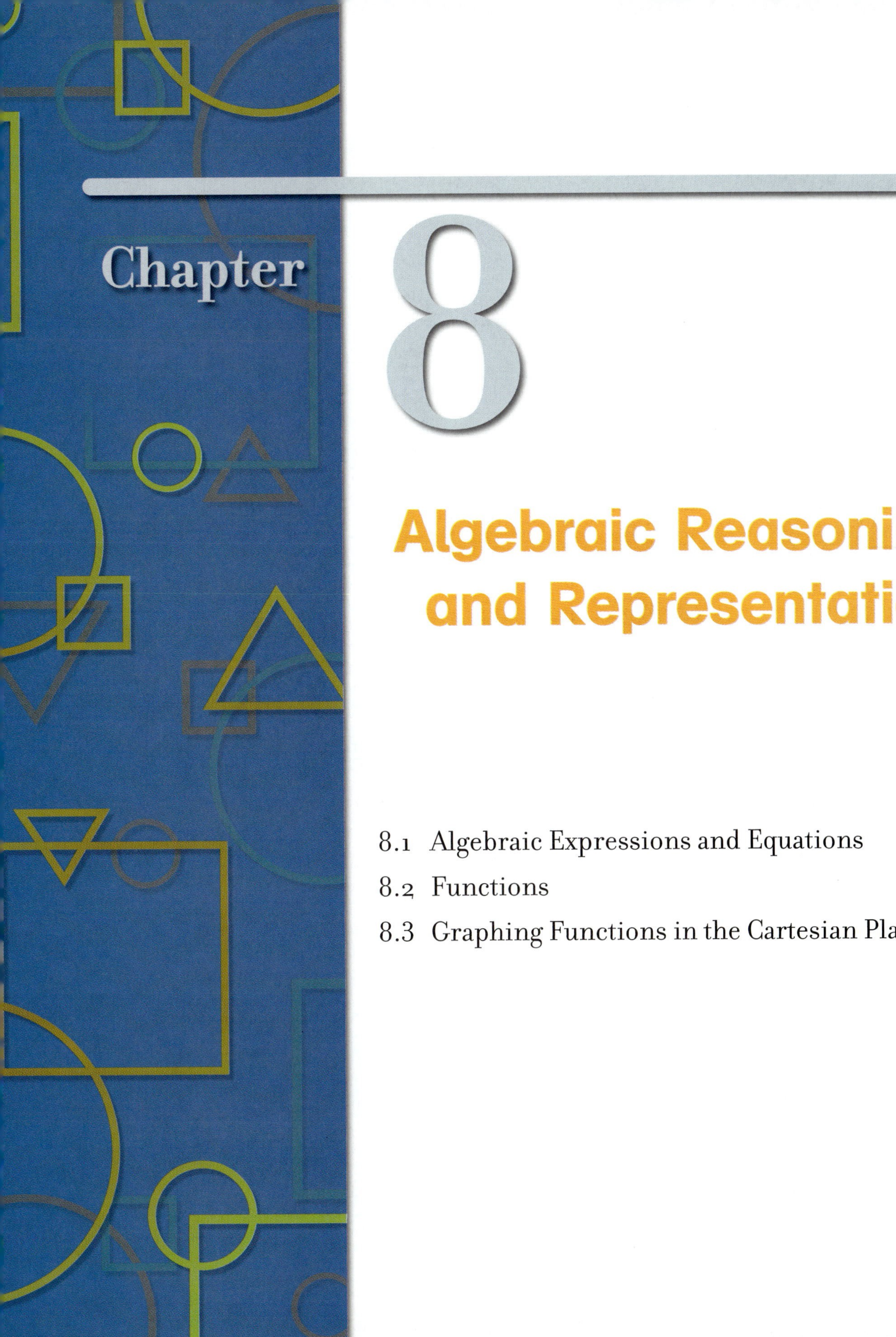

Chapter 8

Algebraic Reasoning and Representation

The Equation Balance Scale

Materials Needed

Each student needs an enlarged copy of the materials sheet shown below. Cut out the "unknowns" marked "x," the unit squares, and the rectangle depicting the Equation Balance Scale.

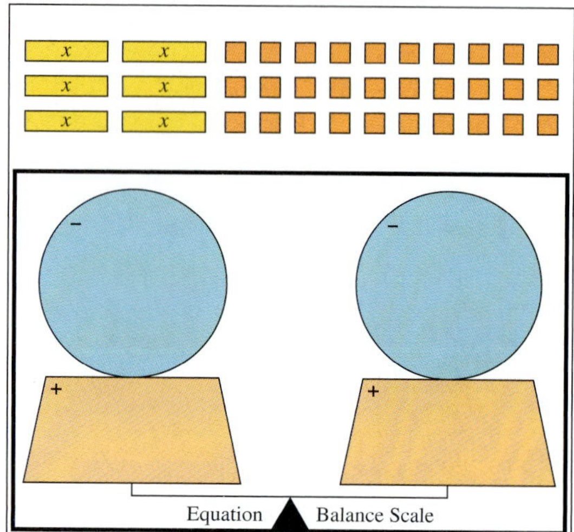

Using the Equation Balance Scale

An algebraic equation such as $3x - 5 = x + 7$ is represented on the Equation Balance Scale by placing three "x" shapes in the + weight area and five unit squares in the negative "balloon" on the left side of the scale. On the right side, an "x" and seven unit squares are placed in the + weight area. Thinking of the scale as balanced gives the following initial representation of the equation.

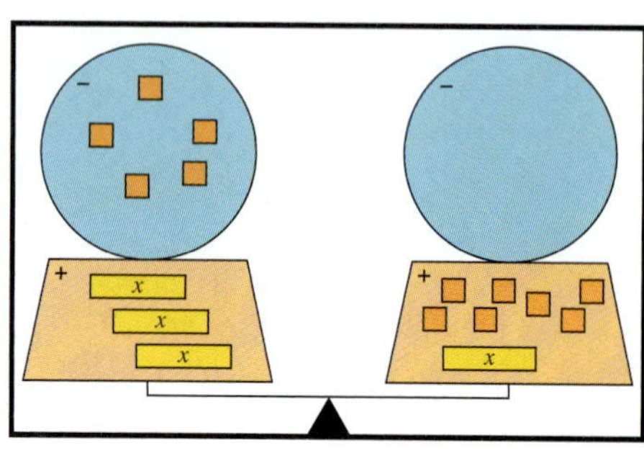

$3x - 5 = x + 7$

To solve the equation, we manipulate the pieces to maintain the balance, and modify the equation to reflect these manipulations. We can begin by removing an "x" from each side of the balance.

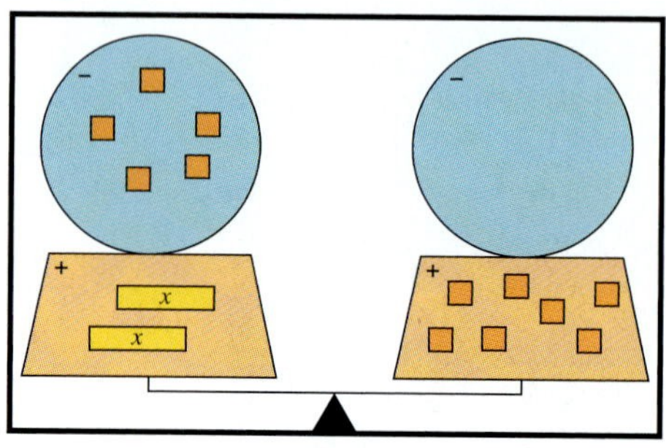

$2x - 5 = 7$

Next, add five unit squares to both sides.

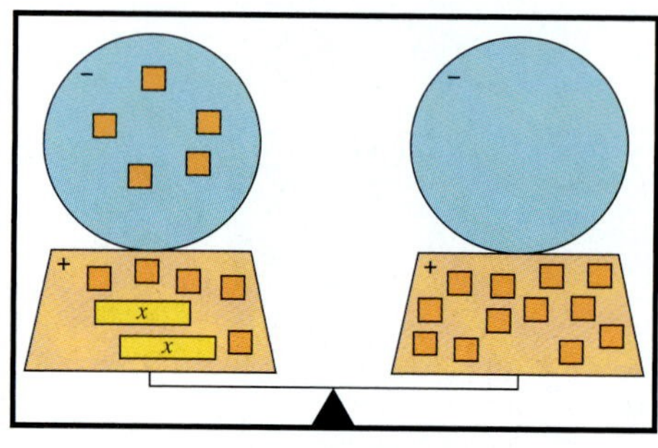

$(2x - 5) + 5 = 7 + 5$

On the left side of the balance, the five + weight units are matched by the five − lifting units in the balloon, so we get the following diagram and equation.

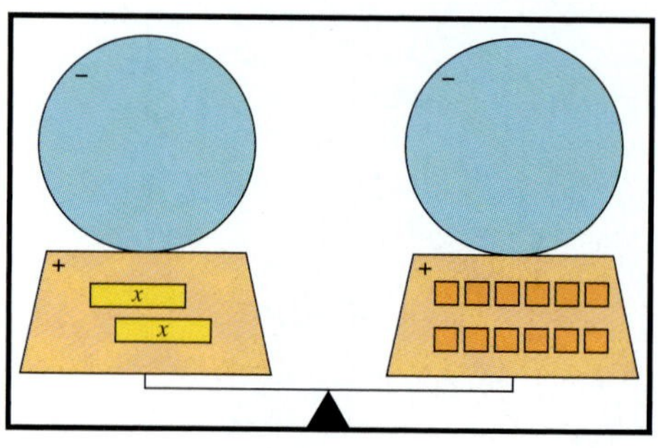

$2x = 12$

An equivalent manipulation would be to move the five negative units in the balloon on the left to the positive weight on the right. That is, $2x - 5 = 7$ becomes $2x = 7 + 5$, so we have $2x = 12$ as before.

Finally, we see that there are two rows of identical arrangements of an "x" on the left and two rows of six unit squares on the right. Therefore, we divide the number of pieces on each side of the balance by 2, and we have the following solution.

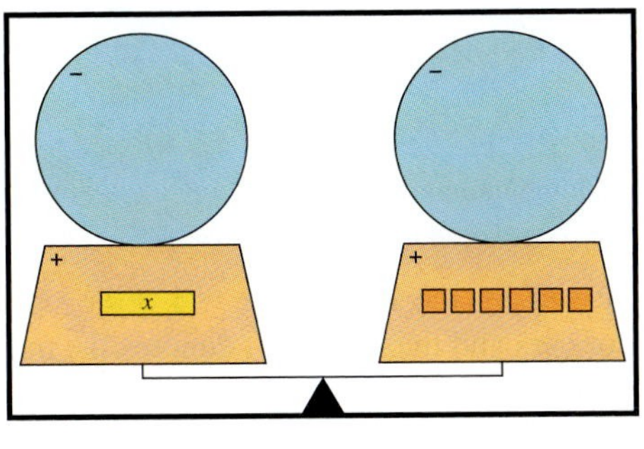

$x = 6$

Equations to Solve with the Equation Balance Scale

Work in pairs or small groups to solve the equations given below, using your Equation Balance Scale. Carefully describe each step of your solutions, and show the new equation that expresses the step just completed.

1. $x + 4 = 9$
2. $x + 6 = 1$
3. $3x + 5 = 14$
4. $4x - 3 = 2x + 5$
5. $2x - 1 = 5 - x$
6. $2 - x = 10 - 3x$

Connections

What Is Algebra?

The question "What is algebra?" has several answers. For example, "Algebra is generalized arithmetic" and "Algebra is a language" are frequently offered. Others describe algebra as "the mathematics of patterns." Each of these descriptions alludes to the key idea that algebra is an extension of arithmetical thinking. In particular, algebra plays these three roles in mathematical reasoning:

- **Algebra describes generality.**
  For example, $5 + 8 = 8 + 5$ is a fact about the commutativity of addition for a particular pair of whole numbers, but the fundamental idea is that 5 and 8 can be replaced with *any* two whole numbers. This can be said in words, but the idea is conveyed most effectively with the algebraic statement "$m + n = n + m$ for any two whole numbers m and n." Here m represents *any* whole number, and the same is true of n. We indicate this by saying that m and n are *generalized* whole numbers.

- **Algebra solves problems.**
  Consider this problem: *In two more years, Angie will be twice the age of her three-year-old brother. How old is Angie?* To solve this problem algebraically, let the symbol x represent Angie's unknown age, a placeholder for a number we wish to determine. The problem stated in words is then equivalent to the symbolic statement $x + 2 = 2(3 + 2)$. That is, $x + 2 = 10$ and so $x = 10 - 2 = 8$. Thus, the equation has been solved to reveal that Angie is currently eight years old. Here the letter x played the role of an *unknown*.

- **Algebra reveals and explains patterns.**
  The figure below shows the first three "trapezoid trains" constructed from toothpicks.

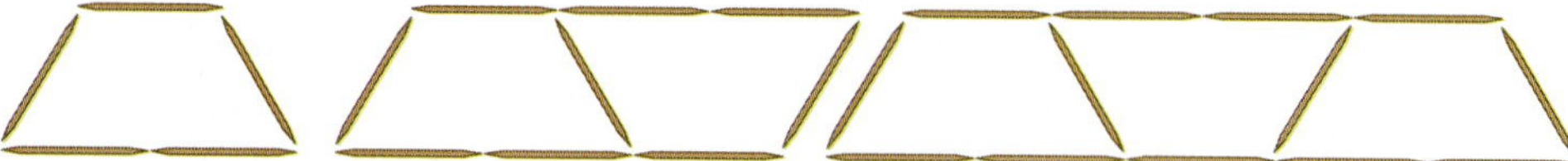

Consider the following question.

> *What is the relationship between the number of cars (that is, trapezoids) in the train and the number of toothpicks required to form the train?*

This relationship can be expressed by the equation $t = 1 + 4c$, where the variable c denotes the number of cars and the variable t denotes the number of toothpicks.

In Section 8.1, the language of algebra is investigated in more depth. The central ideas of variables, algebraic expression, equations, and solutions are explored, with applications to problem solving and the description of patterns. In Section 8.2, the dynamic aspect of algebra—its ability to describe change and make predictions—is enlarged by taking up the fundamental concept of a function. Functions are represented in several ways, but of special importance is the graph of a function. The concluding Section 8.3 examines how the properties of a function are reflected in the shape of its graph.

from The NCTM Principles and Standards

Algebra Standard for Grades 3–5

Instructional programs from prekindergarten through grade 12 should enable all students to—	In grades 3–5 all students should—
Understand patterns, relations, and functions	• describe, extend, and make generalizations about geometric and numeric patterns; • represent and analyze patterns and functions, using words, tables, and graphs.
Represent and analyze mathematical situations and structures using algebraic symbols	• identify such properties as commutativity, associativity, and distributivity and use them to compute with whole numbers; • represent the idea of a variable as an unknown quantity using a letter or a symbol; • express mathematical relationships using equations.
Use mathematical models to represent and understand quantitative relationships	• model problem situations with objects and use representations such as graphs, tables, and equations to draw conclusions.
Analyze change in various contexts	• investigate how a change in one variable relates to a change in a second variable;

- identify and describe situations with constant or varying rates of change and compare them.

SOURCE: Reprinted with permission from Principles and Standards for School Mathematics, page 158. Standards are listed with the permission of the National Council of Teachers of Mathematics (NCTM). NCTM does not endorse the content or validity of these alignments.

8.1 ALGEBRAIC EXPRESSIONS AND EQUATIONS

In this section we discuss

- the meaning and uses of variables;
- how to introduce variables to describe a problem or pattern;
- how to form algebraic expressions involving variables;
- how to create equations by setting one expression equal to another;
- how to solve equations, either by evaluating the unknowns or by obtaining additional relationships between the variables that explain the structure of a pattern;
- how to interpret the solution of equations as a solution of a problem or a description of a pattern.

It is helpful to view algebraic thinking as a circle of connected ideas, beginning and ending with the pattern or problem that is being investigated. The essential steps are shown in Figure 8.1 and are developed in more detail throughout this section.

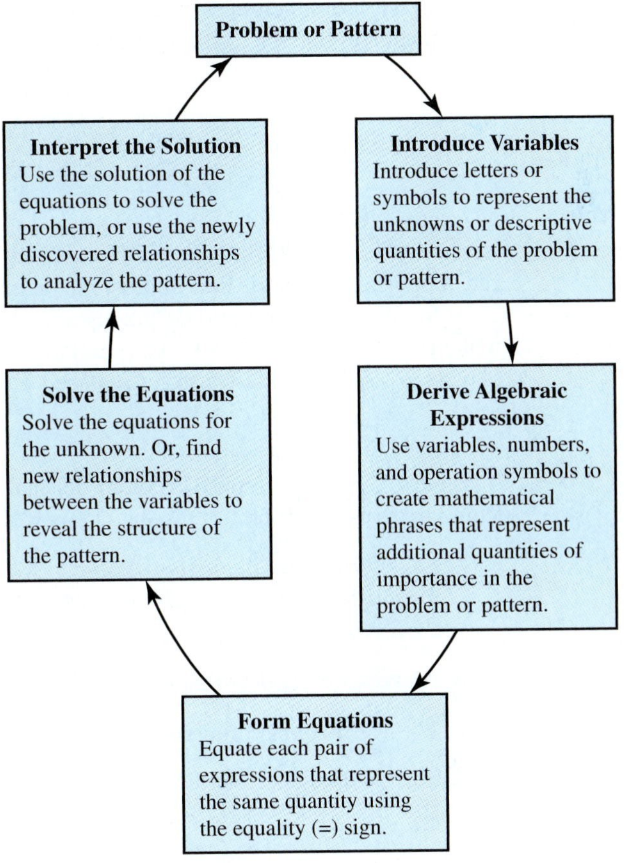

Figure 8.1
The steps in algebraic reasoning

Variables

Mathematical expressions in elementary arithmetic involve fixed values called **constants.** For example, there are 12 in a dozen, there are four sides of a rectangle, and $2 + 5 = 7$. Constants describe static situations in which change, generality, and variability are not under consideration. Mathematical expressions in algebra still include constants, but also include quantities that are unknown, that can change, or whose value depends on how other, related values are determined or selected. Changeable quantities are referred to as **variables** and are denoted by a symbol. Usually the symbols are letters, such as the x in the equation $x - 3 = 5$, but sometimes other symbols are used. For example, consider asking what number must be placed in the placeholder indicated by the box in the equation $\square - 3 = 5$.

Introducing variable quantities imparts a dynamic aspect to algebraic reasoning that goes far beyond the static reasoning of simple computation. Variables are used in at least four different ways in algebra:

Variables Describe Generalized Properties For example, the distributive property of multiplication over addition in the real numbers is succinctly stated as

$$a(b + c) = ab + ac, \text{ for all real numbers } a, b, \text{ and } c.$$

Here, the symbols a, b, and c are *generalized variables,* used to describe a general property or pattern. A generalized variable represents an arbitrary member of the set of elements for which the property or pattern holds.

Variables Express Relationships Jolie was born on her three-year-old sister Kendra's birthday. How are their ages related? Letting J and K denote their respective ages, we have several choices to express the age relationship:

$J = K - 3$	("Jolie is three years younger than Kendra.")
$K = J + 3$	("Kendra is three years older than Jolie.")
$K - J = 3$	("The difference in age between Kendra and her younger sister Jolie is three years.")

Variables Serve as Unknowns in Equations For example, in the early grades students can be asked to find the number to place in the box that makes the sentence $\square + 5 = 9$ true. In the middle grades a student can be asked to find all values of x for which $(2x - 6)(x - 1) = 0$. The second example shows why it is appropriate to think of x as a variable, since it can be replaced either with 3 or with 1 to make the equation true.

Variables Express Formulas Problem 50 of the 85 mathematical problems found in the Rhind papyrus (ca. 1650 B.C.) is stated this way by the scribe Ahmes:

> *Example of a round field of diameter 9. What is the area? Take away 1/9 of the diameter; the remainder is 8. Multiply 8 times 8; it makes 64. Therefore, the area is 64.*

Thus, Ahmes is proposing that the area A of a circle of diameter d is given by the formula $A = \left(d - \frac{1}{9}d\right)^2$. Other problems in the papyrus implicitly provide other useful formulas—some exact and others (like the circle area formula) only approximate. Many of these formulas continue to be of value in modern times, and of course many new ones have been derived as well. Several useful formulas are collected in Table 8.1. Unlike Ahmes's interpretation, their statements are made clear by the use of variables.

The following example gives you an opportunity to check your understanding of the meaning and uses of variables.

TABLE 8.1 Some Commonly Used Formulas

Topic	Variables	Formula
Distance	d = distance traveled r = rate of travel (speed) t = time of travel	$d = rt$
Interest (compounded annually)	A = current amount P = initial principal i = annual interest rate (decimal) n = years elapsed	$A = P(1 + i)^n$
Triangular numbers	$t_n = 1 + 2 + \cdots + n$ (sum of the first n integers) n = number of terms in the sum	$t_n = \frac{1}{2}n(n + 1)$
Perimeter of a rectangle	P = perimeter L = length W = width	$P = 2L + 2W$
Area of a rectangle	A = area L = length W = width	$A = LW$
Circumference of a circle	C = circumference r = radius	$C = 2\pi r$
Area of a circle	A = area r = radius	$A = \pi r^2$

EXAMPLE 8.1 IDENTIFYING THE ROLE OF VARIABLES

In each part below, identify the variable and the role it plays.

(a) The length of a rectangle is twice its width w, and the area of the rectangle is 18 square feet. What is the width w of the rectangle?

(b) The associative law of addition in the set of integers states that

$$a + (b + c) = (a + b) + c, \text{ for all integers } a, b, \text{ and } c.$$

(c) A single taxpayer with an adjusted gross income of X dollars in 2003, where X is between \$28,400 and \$68,800, owes the Internal Revenue Service \$3910 + (0.25)(X − \$28,400).

(d) Milos is twice as tall as his little sister, Anke, so $m = 2a$, where m and a denote the respective heights of the siblings.

Solution

(a) w has the role of an unknown that satisfies the equation $w(2w) = 18$, or $w^2 = 9$. Since w is positive, the value of the unknown, namely, $w = 3$ feet, can be determined from the equation.

HIGHLIGHT *from* HISTORY

From al-jabr *to Algebra*

Beginning in about the twelfth century, mathematics in Latin Europe was greatly stimulated when translations of Arabic mathematical texts became available. Two ninth-century works of Muhammad ibn Musa al-Khowarizmi (or Mohammed the son of Moses from Khorezm) were especially important. One text, translated into Latin as *Algoritmus de numero Indorum,* introduced the Indo-Arabic decimal positional numeration system into Europe and gave us the word *algorithm* (see the Highlight from History box in Section 3.3). The other text, *Hisāb al-jabr w'al-muqā-bala,* was translated as *Liber algebrae et almucabola.* Although it was scarcely original, it became a widely used source of information on linear and quadratic equations and explains how the word *algebra* originated from the Arabic *al-jabr.* The literal meaning of the title of al-Khowarizmi's book is "The calculation of reduction and confrontation," which alludes to moving a negative quantity on one side of an equation to the opposite side and to combining like terms on the same side of an equation. al-Khowarizmi used the terms *root* and *square* instead of what we would write as x and x^2. The familiar letter symbols of algebra were not introduced until the very end of the sixteenth century, and only fully entered the mathematical language with the work of Descartes, Newton, and Leibniz. Today letters are used so extensively that algebra is sometimes jokingly referred to as "the intensive study of the last three letters of the alphabet."

(b) a, b, and c play the role of generalized variables in the set of integers.
(c) X has the role of a variable in the formula used to compute the federal income tax owed.
(d) m and a are variables in the equation that expresses a relationship between the heights of the children.

Algebraic Expressions

A **numerical expression** is any representation of a number that involves numbers and operation symbols. For example, $3 + 8$ and $22 \div 2$ are both numerical expressions for 11. An **algebraic expression** is a representation that involves variables, numbers, and operation symbols. For example, $4x + 8y$ is an algebraic expression. It might represent the cost of taking x children and y adults to a play, where children's tickets are \$4 each and adults' tickets are \$8 each.

DEFINITION *Algebraic Expression*
An **algebraic expression** is a mathematical expression involving variables, numbers, and operation symbols.

In the examples that follow, you are challenged to create algebraic expressions.

EXAMPLE 8.2 | FORMING ALGEBRAIC EXPRESSIONS

For each situation, form algebraic expressions that represent the requested values.

(a) The cost of every item in the store is increased by 15 cents. What is the cost of an item that used to cost c dollars? What is the old cost of an item that now costs d dollars?

(b) There was 3% inflation each of the last two years. What is the current price of an item that cost p dollars last year? If an item is q dollars today, what was its price two years ago? Assume the costs exactly followed the inflation rate.

(c) There are 900 seats in the school auditorium. If s seats are needed for school staff, how many tickets for seats can be given to each of g graduating seniors? (*Hint:* Use the "round down" or "floor function" defined by $\lfloor x \rfloor$ = largest integer less than or equal to x. The floor function rounds downward to the nearest integer no larger than x. For example, $\lfloor 3.1416 \rfloor = 3$, $\lfloor 5 \rfloor = 5$, and $\lfloor 7.75 \rfloor = 7$.)

(d) The electric power company charges $5 a month plus 7¢ per kilowatt-hour of electricity used. What is the monthly cost to use K kilowatt-hours?

(e) A student earned a credit hours of A work, b credit hours of B work, and so on. What is the student's GPA? Let each A contribute four points, each B three points, and so on.

Solution

(a) $c + 0.15$, $d - 0.15$

(b) $1.03p$, $q/(1.03)^2$

(c) There are $900 - s$ tickets to be evenly distributed to the g graduates, so each student can be given the largest whole number of tickets that is no larger than $\frac{900 - s}{g}$. That is, $\left\lfloor \frac{900 - s}{g} \right\rfloor$.

(d) $5 + .07K$

(e) The number of grade points earned is $4a + 3b + 2c + d$. The number of credit hours is $a + b + c + d + f$. Therefore, the GPA is given by the quotient

$$\frac{4a + 3b + 2c + d}{a + b + c + d + f}.$$

The **domain** of a variable is the set of values for which the expression is meaningful. For example, if n denotes the number of students in a room, the domain is the set of whole numbers. If x is the width of a rectangle, the domain is the set of positive real numbers. An algebraic expression is **evaluated** by replacing each of its variables with particular values from the domain of the variable. For example, consider again the expression $4x + 8y$ that gives the cost of x children and y adults attending the theater. Here the domain of each of the variables x and y is the set of whole numbers. When evaluated at $x = 5$ and $y = 3$, the expression has the value $4 \cdot 5 + 8 \cdot 3 = 44$.

In the next example, you are asked to derive algebraic expressions, then to check them by choosing appropriate values for the variables, and finally to evaluate the expressions when given values of the variables.

EXAMPLE 8.3 FORMING, CHECKING, AND EVALUATING ALGEBRAIC EXPRESSIONS

Toothpicks can be used to form "cross" patterns, such as the three shown below. Additional crosses are formed by following the same pattern of construction. What is the number of squares in the nth cross? How many toothpicks are required to construct the nth cross?

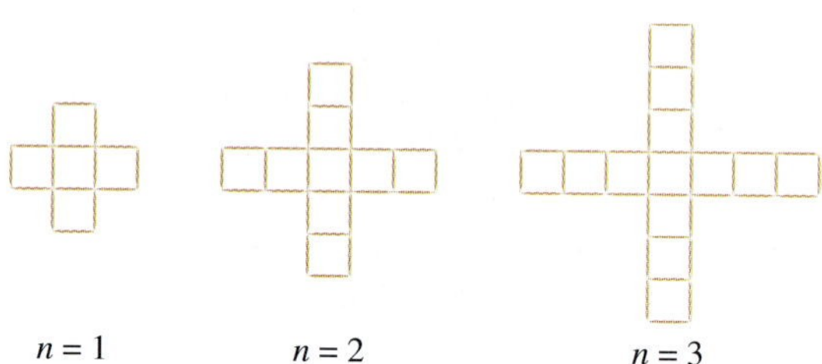

Solution

Understand the Problem

We seek two expressions that each involve the variable n. One is to give the number of squares in the nth cross, and the other expression must give the number of toothpicks required to form the pattern. We see that n also gives the number of squares in each of the four "arms" of the cross.

Devise a Plan

Each new pattern is created by adding an additional square to each of the four arms of the preceding pattern. That is, the number of squares is increased by four in each new pattern. This suggests that we look for an expression of the form $a + 4n$, where we will need to determine the value of the constant a. Similarly, adding one square to an arm of the cross requires three additional toothpicks, so adding a square to each of the four arms requires 12 additional toothpicks. This suggests we look for an expression of the form $b + 12n$ to give the number of toothpicks in the nth cross, where the constant b will need to be determined.

Carry Out the Plan

In the first cross pattern, $n = 1$. Evaluating $a + 4n$ at $n = 1$ gives $a + 4$, and since there are 5 squares in the pattern we get the equation $a + 4 = 5$. That is, $a = 1$ and the number of squares in the nth cross is given by $1 + 4n$. The constant b in the expression $b + 12n$ can be determined similarly: there are 16 toothpicks in the $n = 1$ pattern, so we must have $b + 12 = 16$. That is, $b = 4$ and the number of toothpicks in the nth pattern is given by the expression $4 + 12n$.

Look Back

The plan that has been followed will apply to any sequence where each new term is a fixed number larger than the preceding term. That is, we have an arithmetic sequence. For example, to find an expression that gives the nth term of the sequence 3, 9, 15, 21, . . . , we would look for an expression of the form $c + 6n$, since 6 is the common difference of the sequence. Choosing $n = 1$, the constant c must satisfy the condition $c + 6 = 3$. That is, $c = 3 - 6 = -3$. Thus, the nth term is given by the expression $6n - 3$.

Equations and Their Solution

Two algebraic expressions with the same value form an equation, symbolized with the equal sign, =, placed between the expressions.

> **DEFINITION** *Equation*
> An **equation** is a mathematical expression stating that two algebraic expressions have the same value. The equality sign, =, indicates that the expression on the left side has the same value as the expression on the right side of the symbol.

Every equation is one of two types, either an identity or a conditional equation. An **identity** is an equation that is true for all evaluations of the variables from their domains. For example, $(x + y)^2 = x^2 + 2xy + y^2$ is an identity in the real numbers, since the expressions on both sides of the equality sign have the same value for all real numbers x and y. On the other hand, if only certain values of the variables give equality, then the equation is **conditional.** In this case, x is referred to as the **unknown** and one hopes that all of the values of x can be determined from the equation. For example, $x^2 = 9$ is a conditional equation in the unknown x with the two solutions 3 and -3. If the domain of x is more restricted, say to the positive numbers, there would be just one solution, $x = 3$. Some equations, such as $x^2 + 5 = 0$ in the domain of all real numbers, may not have a solution. Determining the set of values that make an equation true is an important step in algebraic reasoning.

> **DEFINITION** *Solution Set of an Equation*
> The **solution set** of an equation is the set of *all* values in the domain of the variables that satisfy the given equation.

EXAMPLE 8.4 SETTING UP AND SOLVING AN EQUATION

Larry has exam scores of 59, 77, 48, and 67. What score does he need on the next exam to bring his average for all five of the exams to 70?

Solution

Let the variable s denote Larry's minimum needed score on the fifth exam. The domain of s is the set of numbers between 0 and 100. Using the variable s, we find that the expression giving Larry's average over all five tests is

$$\frac{59 + 77 + 48 + 67 + s}{5}.$$

Since Larry is looking for the lowest score s that will give him a 70 average, we form the equation

$$\frac{59 + 77 + 48 + 67 + s}{5} = 70.$$

To solve this equation, multiply each side of the equation by 5 and add the sum of the first four test scores to get

$$251 + s = 350.$$

Finally, subtract 251 from both sides of the equation to solve for s.

$$s = 350 - 251 = 99.$$

Larry must hope for a 99 or 100 on the last test.

The solution just presented shows a typical procedure for solving an equation. Namely, we are allowed to perform the same operation on both sides of an equation if the new equation has the same solution set as the initial equation. Such equations with identical solution sets are called **equivalent equations.** Naturally, the hope is to obtain an equivalent equation such as $s = 99$ for which the solution set is obvious. It should be clear that adding to or subtracting from both sides of an equation gives an equivalent equation. Multiplication and division are also allowed, but *only* by nonzero values. More complex operations are allowed if care is taken that the equations are indeed equivalent. For example, consider the equation $\sqrt{2x^2 - 1} = x$. If both sides of the equation are squared, we get the new equation $2x^2 - 1 = x^2$, which is equivalent to $x^2 = 1$. The new equation therefore has two solutions, $x = 1$ and $x = -1$. Only the positive solution $x = 1$ is a solution of $\sqrt{2x^2 - 1} = x$, however. This is because it is always assumed any square root is nonnegative, and so the original equation requires the x also to be nonnegative. It is good practice when solving equations to evaluate the expressions in the original equations with the values in the proposed solution set to check that equality really does hold.

This section concludes with one more example that illustrates the steps in algebraic reasoning shown in Figure 8.1.

EXAMPLE 8.5 SOLVING A RATE PROBLEM: TOM AND HUCK WHITEWASH A FENCE

Two years ago it took Tom 8 hours to whitewash the fence. Last year, Huck took just 6 hours to whitewash the fence. This year, Tom and Huck have decided to work together, so they'll have time left in the afternoon to angle for catfish. How long will the job take the two boys?

Solution

Let T denote the time Tom and Huck together need to whitewash the fence, in hours. That is, T is the unknown. Since Tom can whitewash the fence in 8 hours, he can whitewash $\frac{1}{8}$ of the fence per hour. In T hours, Tom will have whitewashed $\frac{T}{8}$ of the fence.

Similarly, Huck whitewashes at $\frac{1}{6}$ of the fence per hour, and Huck therefore can whitewash $\frac{T}{6}$ of the fence in T hours. Working together, the entire fence will be whitewashed when

$$\frac{T}{8} + \frac{T}{6} = 1.$$

Multiplying both sides by 48 gives the equivalent equation

$$6T + 8T = 48.$$

That is, $14T = 48$, so $T = \frac{48}{14} = 3\frac{3}{7}$. Working together, Tom and Huck can white-wash the fence in just less than three and a half hours and will enjoy an afternoon of fishing.

Problem Set 8.1

Understanding Concepts

1. Classify the quantities in each part as a constant or a variable.
   (a) The number of feet in a mile
   (b) The number of hours of daylight in a day
   (c) The price of a gallon of gasoline
   (d) The speed of light in empty space
   (e) The distance from the earth's center to the moon

2. In each of the situations below, classify the role of the variables as one of the following types: generalized; expressing a relationship; expressing a formula; an unknown.
   (a) For all real numbers x, y, and z, $x(y + z) = xy + xz$.
   (b) Elena is 4 inches shorter than her husband, Joe, so $E = J - 4$.
   (c) To construct a circular flowerbed covering 400 square feet of ground, the radius r of the bed must satisfy $\pi r^2 = 400$.
   (d) Tickets to the play are \$5 for adults and \$3 for children. If A adults and C children attend the play, the total proceeds are $5A + 3C$.

3. Alicia, Boris, Carlos, Dan, and Xavier all collect Pokémon™ cards. If x denotes the number of cards in Xavier's collection, form expressions for the number of cards in the other children's collections, using the information provided. For example, Alicia has 6 more cards than Xavier, so Alicia has $x + 6$ cards.
   (a) Boris is 3 cards short of having twice the number of cards as Xavier.
   (b) Carlos has 2 more than half the number of cards owned by Xavier.
   (c) Dan has the same number of cards as Alicia and Xavier combined.

4. Penny is p years old. Form algebraic expressions with the variable p that represent the ages requested.
   (a) Penny's age in 5 years
   (b) Penny's age 8 years ago
   (c) The *current* age of Penny's little brother, who in 2 more years will be half of Penny's age
   (d) The *current* age of Penny's mother, who, 3 years ago, was 4 times Penny's age

5. Five children, say, A, B, C, D, and E, are in a line. A writes a number on a slip of paper and hands it to B. B squares the number received and writes that value on a new slip of paper that is handed to C. In a similar way, C adds 5 to the number obtained from B and passes that number on to D. D multiplies the number by 7 and writes it on a slip of paper that is handed to E.
   (a) If A writes a 3 on a slip of paper, what number will be handed to E?
   (b) If child A puts x on the first slip of paper, what algebraic expression is handed to child E?
   (c) If E is given the number 35, what number did A write on the initial slip of paper?
   (d) Suppose the children line up in the order A, D, B, C, and E. If A writes y on a slip of paper, what expression is eventually handed to E?

6. As with problem 5, children are in a line and each is assigned one operation to perform. If the first child writes x on a slip of paper and the last child is handed a slip on which the expression $(3 - 5x)^3 + 4$ appears, describe how many children are in the line and what single operation each performs.

7. The first three trains in a sequence of trapezoid trains are shown below. Verify that the number t of toothpicks required to form a train with c trapezoidal cars is given by the formula $t = 1 + 4c$.

8. A rectangular table seats 6 people, 1 person on each end and 2 on each of the longer sides. Thus, two tables placed end to end seat 10 people.
   (a) How many people can be seated if n tables are placed in a line end to end?
   (b) How many tables, set end to end, are required to seat 24 people?

9. The first three cases in a sequence of "hexstars" are shown below. Imagine that the patterns are formed with toothpicks.

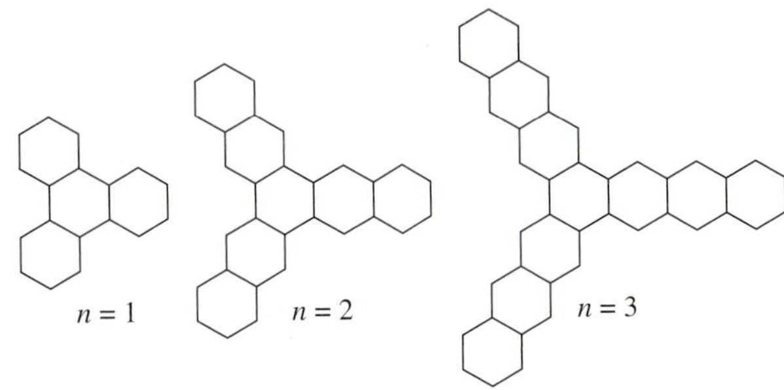

(a) How many additional hexagons need to be added to the pattern for $n = 3$ to create the pattern for $n = 4$?

(b) Derive an expression in the variable n that gives the number of hexagons in the nth hexstar.

(c) How many additional toothpicks are needed to enlarge the hexstar for $n = 3$ to obtain the hexstar for $n = 4$?

(d) Derive an expression in the variable n that gives the number of toothpicks in the nth hexstar.

10. Carefully show all of the steps needed to solve each equation below.

(a) $5x - 3 = 17$

(b) $3x + 8 = 7x - 4$

11. Bernie weighs 90 pounds plus half his own weight. What does Bernie weigh?

12. The figure below shows an addition pyramid, in which each number is the sum of the two numbers immediately below.

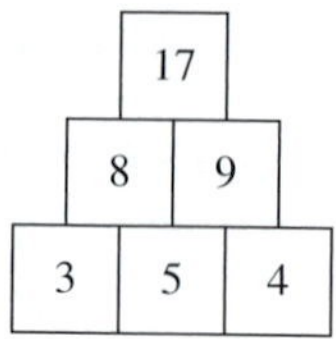

For each of the incomplete addition pyramids below, show that the values in each square can be determined by forming and solving an equation in the unknown variable shown.

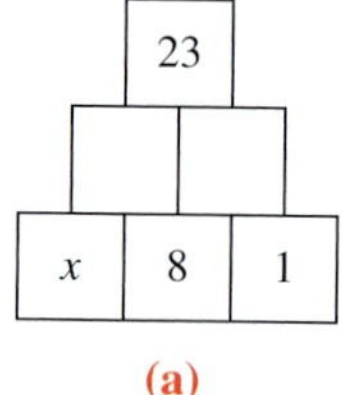

(a)

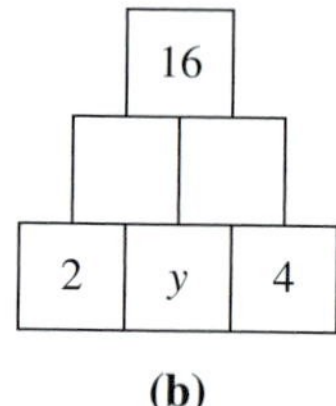

(b)

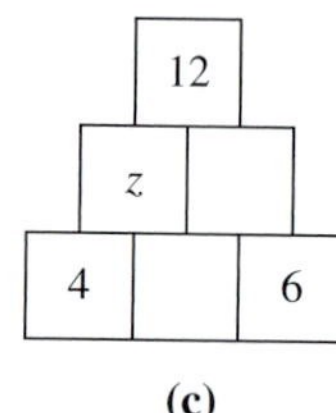

(c)

13. In the circle patterns below, each number in a large circle is the sum of the numbers in the two adjacent small circles. The pattern on the left is complete, but the pattern on the right needs to be completed. Do so by letting x denote the value in one of the small circles. Now obtain expressions and equations that let you solve for x and determine the values in all three small circles.

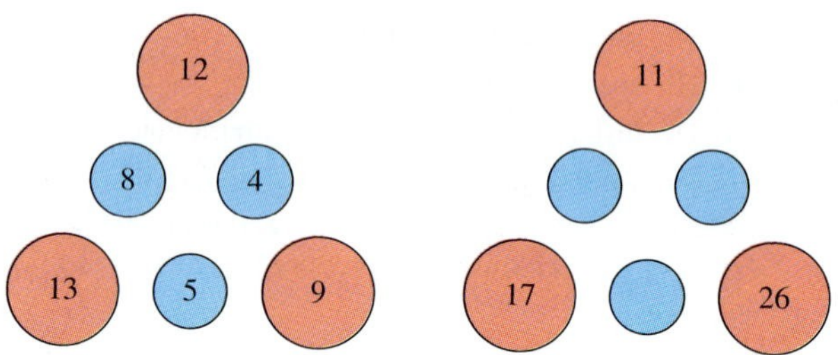

14. Find the numbers to place in each of the five small circles shown to the right so that each number given in a large circle is the sum of the numbers in the two adjacent small circles. Do so by letting one of the values in a small circle be denoted by an unknown and then obtaining and solving an equation for the unknown.

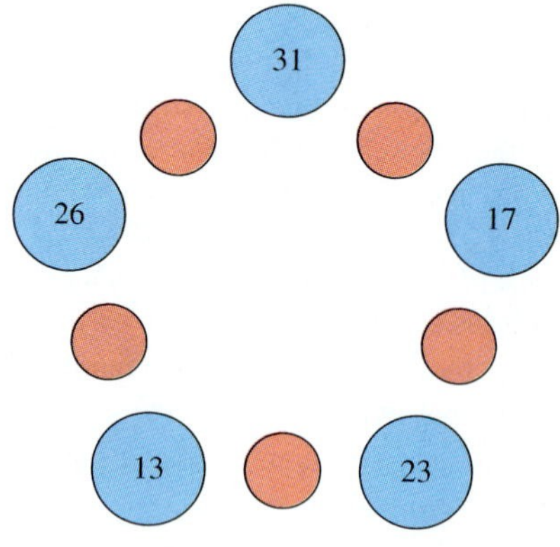

---

15. Here is a "proof" showing that $1 = 2$. Identify what error has been made.

Step 1. Let $x = 1$ and $y = 1$.

Step 2. Multiply both equations by y to get $xy = y$ and $y^2 = y$.

Step 3. Since both xy and y^2 equal y, then $xy = y^2$.

Step 4. Subtract x^2 from both sides to get $xy - x^2 = y^2 - x^2$.

Step 5. Factor both sides to get $x(y - x) = (y + x)(y - x)$.

Step 6. Divide both sides by $y - x$ to get $x = y + x$.

Step 7. By substitution, $1 = 1 + 1$, so $1 = 2$.

$(y + x)(y - x)$
$= y(y - x) + x(y - x)$
$= y^2 - yx + xy - x^2$
$= y^2 - x^2$

16. Consider the pair of equations $y - 3x = 2$, $y + 4 = 6x$.
   (a) Solve each equation for y as an expression in x.
   (b) Equate the expressions found in part (b) to obtain an equation in the variable x.
   (c) Solve the equation in part (b) for the unknown x.
   (d) Solve for y.

17. Little Red Riding Hood rode her bike 15 miles to Grandma's house, arriving in 3 hours. When she discovered the Big Bad Wolf, she turned around and scampered home at 15 miles per hour. The *distance* = *rate* × *time* formula will be helpful to answer these questions.
   (a) What was Little Red Riding Hood's average speed to Grandma's house?
   (b) How long did it take to get home?
   (c) How long was the round-trip?
   (d) How many hours did the round-trip take?
   (e) What is the average speed for the round-trip?
   (f) Is the average speed for the round-trip equal to the average of the speeds to and from Grandma's house? Why is this?

18. Old MacDonald has 100 chickens and goats in the barnyard. Altogether, there are 286 feet. How many chickens and how many goats are in the barnyard? Set up and solve equations to find your answer.

Teaching Concepts

19. A student thinks that the expressions $(a + b)^2$ and $a^2 + b^2$ are equal to one another. How would you explain to the student why this is not so?

20. A student has noticed that every squared natural number is either a multiple of 4 or one larger than a multiple of 4, but isn't sure if this is always true. Use algebraic reasoning to explain to the student why this is always true. (*Suggestion:* Every even natural number has the form $2k$ for some natural number k, and every odd natural number has the form $2k + 1$ for some natural number k.)

21. During a lesson on factoring natural numbers, Marcia said, "I think when you factor an odd number, you always get odd factors."
   (a) Is Marcia correct?
   (b) How could you convince Marcia and the rest of the class that Marcia is or is not correct?

Thinking Critically

22. Let n be a two-digit whole number with this special property: if the sum of the digits is added to the product of the digits, the result is n. For example, $n = 29$ has the special property, since $2 + 9 + 2 \times 9 = 11 + 18 = 29$. Find all two-digit numbers with this property.

23. Huong spent 90 cents at the store, buying some pencils at 15 cents each and some erasers at 6 cents each. He purchased several more erasers than pencils. How many pencils and erasers did he buy? Use algebraic reasoning to find your answer.

24. Imagine that the square $n \times n$ grid below is constructed by using toothpicks to form the edges of the small squares in the pattern.

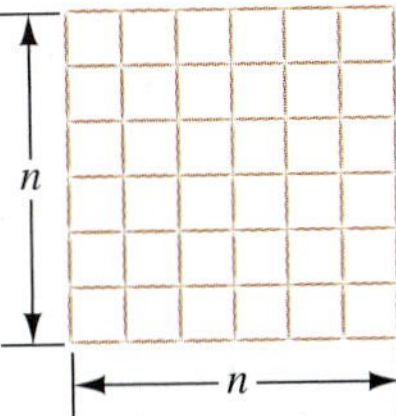

   (a) How many toothpicks are required to make the pattern? (*Suggestion:* How many vertical toothpicks are used?)
   (b) There are clearly n^2 small squares in the $n \times n$ grid. Explain why $4n^2 + 4n$ is twice the number of toothpicks needed to form the pattern.

25. Imagine that the triangular pattern shown below is constructed from toothpicks.
   (a) Find the number of small triangles in the cases $m = 1, 2, 3,$ and 4, and then give an expression for the number of small triangles in the triangular grid of size m.
   (b) How many toothpicks are needed to form the mth triangular grid pattern? (*Suggestion:* First determine the number of horizontal toothpicks in the pattern.)

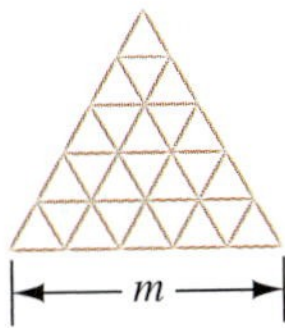

26. A commuter travels to work at an average speed of u and returns home at an average speed v. Show that the average speed for the round-trip is $\frac{2uv}{u + v}$.

27. Suppose that n^3 sugar cubes are stacked to form an $n \times n \times n$ cube, as shown. The large cube is now painted yellow. Depending on where a sugar cube is positioned in the large cube, it may have one or several of its six faces painted yellow. For example, if $n = 1$,

all six faces are painted. For larger n, there are some sugar cubes with unpainted faces.

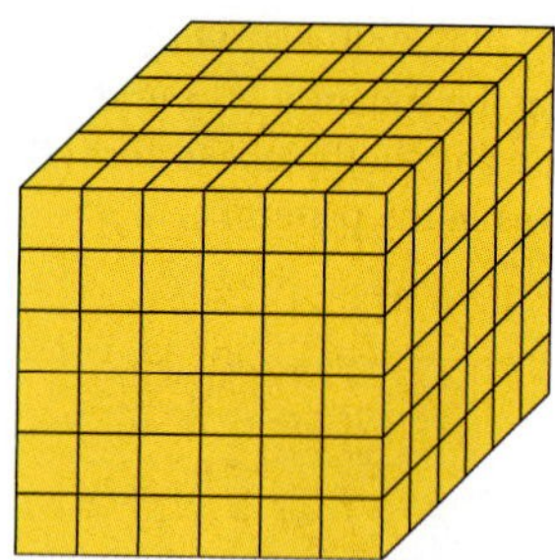

(a) Describe how the faces of the sugar cubes in a $2 \times 2 \times 2$ cube are painted.

(b) Count the number of sugar cubes in a $3 \times 3 \times 3$ case with 0, 1, 2, or 3 painted faces. Do you account for all of the sugar cubes that form the large cube?

(c) Give expressions in the variable n, $n \geq 3$, for the number of sugar cubes in the $n \times n \times n$ cube with 0, 1, 2, or 3 painted faces, respectively.

(d) The four expressions obtained in part (c), when added together, should give n^3 for all n larger than or equal to 3. Why? Use algebraic simplification to check your identity in the variable n.

28. Dana received grades of 78, 86, and 94 on the first three exams. For an A, the average on four exams must be at least 90. What can Dana do to get an A?

29. Find the solution set in the real numbers of each of these equations.

(a) $2(3x - 6x + 1) = 3(1 - 2x) - 1$

(b) $x(x - 2) = 2(x^2 - x + 1)$

30. A rectangle is 3 times as long as it is wide and has the same perimeter as a square whose area is 4 square feet larger than that of the rectangle. What are the dimensions of both the rectangle and the square?

Using a Calculator

31. The formula for the area of a circle given implicitly in the Rhind papyrus (ca. 1650 B.C.) is $A = \left(\frac{16}{9}\right)^2 r^2$, where r denotes the radius of the circle.

(a) How much error, in square feet, is made if this formula is used to measure the area of a circle of radius 100 feet instead of the exact formula $A = \pi r^2$?

(b) Archimedes (ca. 300 B.C.) suggested that π can be approximated by $22/7$. How much error is made computing the area of a circle of radius 100 feet if the formula $A = \left(\frac{22}{7}\right)r^2$ is used instead of the exact formula?

32. An initial amount of \$100 is deposited into a bank paying 4% annual interest, so that the account balance after n years is given by $100(1 + .04)^n$.

(a) How much money is in the account after 10 years? after 20 years?

(b) In approximately how many years will the initial amount of \$100 have doubled to \$200? Evaluate the expression above for several reasonable choices of n to find your answer.

(c) The "rule of 72" says that, at least roughly, the time to double your money at i percent annual interest is given by the expression $72/i$. Does this agree with your answer to part (b)?

Using a Computer

33. The square grid of size n requires $2n(n + 1)$ toothpicks to form it, and the triangular grid requires $\frac{3}{2}m(m + 1)$ (see problems 24 and 25). Use a spreadsheet (or a calculator) to find the smallest number of toothpicks that can make a square grid and then be reassembled to make a triangular grid with no toothpicks left over. (It can be shown that there are infinitely many ways to form both a square and triangular pattern with the same number of toothpicks in each; the next two smallest cases use 16,380 and 3,177,720 toothpicks, respectively.)

Making Connections

34. Rug remnants can be purchased for \$3.60 per square yard, and the edges can be finished at a cost of 12¢ per foot. Let L and W respectively denote the length and width of a rectangular remnant, given in feet. Give an expression for the cost, in dollars, to buy and finish an L-by-W remnant.

35. Zal is paid \$18 an hour for a 40-hour workweek, but is paid time and a half for overtime work. Give expressions for Zal's pay if he works t hours. You'll need to distinguish between the cases where t is less than and greater than 40.

36. An ISP (Internet service provider) charges \$10 per month plus 20¢ per hour of time on-line, where fractions of hours are rounded up to the nearest full hour. What is the cost to be on-line t hours during a month? (*Suggestion:* Use the "round up" or "ceiling function" $\lceil x \rceil$, which gives the smallest integer greater than or equal to x.)

37. The "girth" of a package is its distance around. A package can be mailed through the U.S. Postal Service only if its girth plus length does not exceed 108 inches.

(a) Form an expression for the girth plus length of a rectangular box whose shorter sides measure x and y inches and whose length is z inches.

(b) A rectangular box is 4 feet long. If the box has dimensions in the ratio $x:2x:3x$, what is the largest it can be and still be mailed through the USPS?

From State Student Assessments

38. (Oregon, Grade 5)
Tyler and Rory both collect baseball cards. Rory has 8 more than twice as many as Tyler. If T is the number of cards that Tyler has, which choice shows the number of cards that Rory has?

A. $16 - T$

B. $2T - 8$

C. $2T + 8$

D. $8 - 2$

39. (Massachusetts, Grade 4)
If ○○ = □, which of the following is true?

A. ○○○ = □

B. ○○ = □□

C. ○○○○○ = □□

D. ○○○○○○ = □□□

For Review

40. Express the repeating decimal 3.14141414 . . . as a fraction.

41. Megan hiked up a $5\frac{1}{4}$-mile trail in $3\frac{1}{2}$ hours. What was her average speed?

42. A gear wheel with 24 teeth is engaged with another gear wheel with 36 teeth, which in turn is engaged with a third gear wheel with 18 teeth. If the first wheel revolves at 300 rpm (revolutions per minute), how fast is the third wheel turning?

8.2 FUNCTIONS

In this section we discuss how the value of a variable can depend on the value of another variable or variables. The rule that connects these variables is called a **function.** The concept of a function adds an important dynamic aspect to algebraic reasoning, providing a way to describe change and make predictions.

Defining Functions

Imagine yourself at the service station, filling your tank with gasoline priced at \$1.799 per gallon. If you purchase 10 gallons of gas, your bill will be \$17.99, and if you put 13.28 gallons in your tank, your bill (rounded to the penny) will be \$23.89. The general principle is that your bill that day at the station is a *function* of the amount of gasoline you purchase. If we let the variable x represent the number of gallons of gasoline purchased and let the variable y represent the final bill in dollars, then y is given by the simple rule $y = 1.799x$. Nicely enough, the pump carries out this calculation right before our eyes and, as the values of x whirl by on one display of the pump, we can simultaneously watch the corresponding values of y on another display.

It is easy to make a list of functions used in our lives. Here are a few examples:

- The amount of postage on a first-class letter is a function of the weight of the envelope rounded upward to the nearest ounce.
- The recommended amount of lawn fertilizer to be applied is a function of the number of square feet of lawn area.
- The amount of federal income tax you will owe is a function of your taxable income the previous year.

These examples should help you understand the general definition of a function.

DEFINITION *Function*
A **function** on a set D is a rule that associates to each element $x \in D$ precisely one value y. The set D is called the **domain** of the function.

The definition of function requires that precisely one value y be assigned to each x value in the domain. This should seem reasonable, since it would seem strange to buy 10 gallons of gas and be given two bills, say one for \$17.99 and another bill for \$18.75.

A function is often denoted by a letter such as f, and we write $y = f(x)$ to indicate that the value of the variable y is determined by the value of the variable x by the function f. Often a function is denoted by a word or abbreviation to help remember its definition. For example, the function that computes the square root of a positive number is frequently denoted by Sqrt, so that Sqrt(49) = 7.

A function can also be viewed as a set of ordered pairs $\{(x, y) | x \in D$ and $y = f(x)\}$. Since just one y value is associated to each x, a set of ordered pairs represents a function when, and only when, there are no two ordered pairs with the same x value, but different y values.

If a function assigns the value y to an element x in the domain, then y is called the **image,** or **value,** of f at x. For example, if f denotes the function giving our bill at the gasoline station and we have purchased $x = 10$ gallons of gas, then $17.99 = f(10)$ is the value of the function at $x = 10$. We can also write that $f(x) = 1.299 \cdot x$ and $D = \{x | x \geq 0\}$, which describes the gasoline-buying function by an algebraic equation and the domain of permissible values of x.

The set of all image values is called the **range** of the function.

DEFINITION *Range of a Function*

The **range** of a function f on a set D is the set of images of f. That is,

$$\text{range} f = \{y | y = f(x) \text{ for some } x \in D\}.$$

For example, a first-class letter mailed in the United States costs \$0.37 for the first ounce and \$0.23 for each additional ounce. Therefore, the range of the function that gives the cost of mailing a first-class letter is {\$0.37, \$0.60, \$0.83, . . .}.

HIGHLIGHT from HISTORY

Two Women from Early Mathematics: Theano (ca. sixth century B.C.) and Hypatia (A.D. 370–415)

The early history of mathematics mentions few women's names. Indirect evidence, however, suggests that at least some women of ancient times had access to mathematical knowledge, and likely made contributions to it. For example, the misnamed "brotherhood" of Pythagoreans, at the insistence of Pythagoras himself, included women in the Order both as teachers and as scholars. Indeed, Theano, the wife and former student of Pythagoras, assumed leadership of the School at the death of her husband. Theano wrote several treatises on mathematics, physics, medicine, and child psychology.

Hypatia was the first woman to attain lasting prominence in mathematics history. Her father, Theon, was a professor of mathematics at the Alexandrian Museum. Theon took extraordinary interest in his daughter and saw to it that she was thoroughly educated in arts, literature, science, philosophy, and, of course, mathematics. Hypatia's fame as a mathematician was secured in Athens, where she studied with Plutarch the Younger and his daughter Asclepigenia. Later she returned to the university at Alexandria, where she lectured on Diophantus's *Arithmetica* and Apollonius's *Conic Sections* and wrote several treatises of her own. The account of Hypatia's life of accomplishments in mathematics, astronomy, and teaching ends on a tragic note, for in March of 415 she was seized by a mob of religious fanatics and brutally murdered.

EXAMPLE 8.6 FINDING THE RANGE OF A FUNCTION

Let f be the function defined by the formula $f(x) = x(10 - x)$ on the domain $D = \{1, 2, 3, 4, 5, 6, 7, 8, 9, 10\}$. Find the range of f.

Solution

The image of f at $x = 1$ is $f(1) = 1 \cdot (10 - 1) = 9$. Similarly, $f(2) = 2 \cdot 8 = 16$, $f(3) = 3 \cdot 7 = 21$, $f(4) = 4 \cdot 6 = 24$, $f(5) = 5 \cdot 5 = 25$, $f(6) = 6 \cdot 4 = 24$, $f(7) = 7 \cdot 3 = 21$, $f(8) = 8 \cdot 2 = 16$, $f(9) = 9 \cdot 1 = 9$, and $f(10) = 10 \cdot 0 = 0$. Thus, the range of f is $\{0, 9, 16, 21, 24, 25\}$. Some image values, such as 9, 16, and 21, correspond to more than one x in the domain, but any x yields a unique y-value in the range.

One of the most popular ways to introduce function concepts to elementary school children is to play the game "Guess My Rule."

EXAMPLE 8.7 GUESSING ERICA'S RULE

Erica is "it" in a game of Guess My Rule. As the children pick an input number, Erica tells what number her rule gives back as shown in this table. Can you guess Erica's rule?

Children's Choice, x	Result of Erica's Rule, y
2	−1
5	8
6	11
0	−7
1	−4

Solution

Understand the Problem

The rule Erica has chosen is a function—given a value of the input variable x, she uses her function to determine the value of the output variable y that she reveals to the class. We must guess her function. We wish to express the formula for y in terms of a variable x or n as in the preceding examples.

Devise a Plan

The children's choices are somewhat random. Perhaps a pattern will become more apparent if we arrange their values in order of increasing size. We anticipate that Erica's function is given by a formula.

Carry Out the Plan

Rearranging the children's choices in order of increasing size, we have the following table.

Children's Choice, x	Result of Erica's Rule, y
0	−7
1	−4
2	−1
5	8
6	11

When $x = 0$ is the input, the formula reads $y = -7$. This suggests that -7 is a separate term in the formula; when $x = 0$ all the other terms are zero. Also, we observe that when x increases by 1 from 0 to 1, from 1 to 2, and from 5 to 6, the output number increases by 3. This suggests that the formula also contains the term $3x$, since this quantity increases by 3 each time x increases by 1. Combining these observations, we guess that Erica's function (rule), E, is given by the formula

$$E(x) = 3x - 7.$$

Checking, we see that $E(0) = 3 \cdot 0 - 7 = -7$, $E(1) = 3 \cdot 1 - 7 = -4$, $E(2) = -1$, $E(5) = 8$, and $E(6) = 11$ as in Erica's table. When challenged, Erica reveals that we have guessed correctly.

Look Back

Erica's rule is really a function—given a value of x, her rule returns a single value for y. We guessed the rule by arranging the data in a more orderly way, noting that her rule associated -7 with 0, and that the output number increased by 3 each time the input number increased by 1. Thus, we correctly guessed Erica's function to be $E(x) = 3x - 7$.

Describing and Visualizing Functions

There are several useful ways to describe and visualize functions.

- **Functions as formulas.** Consider, for example, a circle of radius r, where the variable r is any positive number. The formula $area(r) = \pi r^2$ defines the function *area* that expresses the area of the circle as a function of the radius r. Similarly, the formula $circum(r) = 2\pi r$ defines the function *circum* that gives the circumference of a circle of radius r. Several interesting formulas were encountered in Chapter 1, each of which can be viewed as a function defined on the domain N, the set of natural numbers. For example, the nth odd number is given by the formula $h(n) = 2n - 1$, so that the fiftieth odd number is $h(50) = 2(50) - 1 = 99$. Another example is the sum $1 + 2 + 3 + \cdots + n$ of the first n natural numbers (that is, the nth triangular number), which is given by the formula $t(n) = n(n + 1)/2$. Thus, the sum of the first 100 natural numbers is $1 + 2 + 3 + \cdots + 99 + 100 = t(100) = 100(100 + 1)/2 = 100 \cdot 101/2 = 5050$. Functions that are defined on the domain N of the natural numbers are called **sequences,** and it is common to use the notation f_n instead of $f(n)$. For example, the nth triangular number is given by $t_n = n(n + 1)/2$, and in particular $t_{100} = 5050$.
- **Functions as tables.** The following table gives the grades of three students on an essay question.

Student	Grade
Raygene	8
Sergei	7
Leticia	10

No algebraic formula connects the student to the grade. Even so, as long as the table assigns a unique grade to each student, then a function is completely described. The domain is $D = \{\text{Raygene, Sergei, Leticia}\}$.

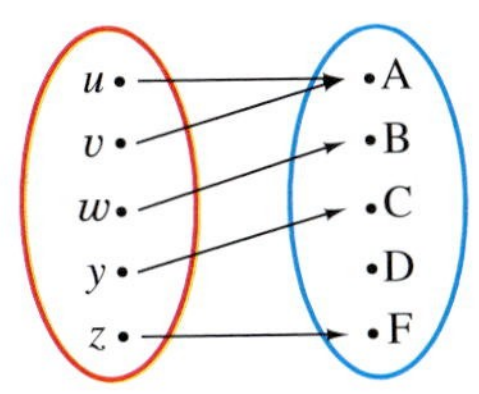

Figure 8.2
A function represented with an arrow diagram

- **Functions as arrow diagrams.** In an arrow diagram, two loops are formed, with one loop representing the domain and the second loop representing a set that contains the values of the function and possibly other points as well. Every point in the domain loop must have exactly one arrow which extends from that point to one of the points in the second loop. Figure 8.2 shows an arrow diagram in which the students Ursula, Vincent, Whitney, Yolanda, and Zach have been assigned grades on their class project. We see that Ursula and Vincent both received As and that no student received a D grade.
- **Functions as machines.** Viewing a function as a machine gives students an attractive dynamic visual model. The machine has an input hopper that accepts any domain element x, and an output chute that gives the image $y = f(x)$. A few function machines are shown in Figure 8.3.

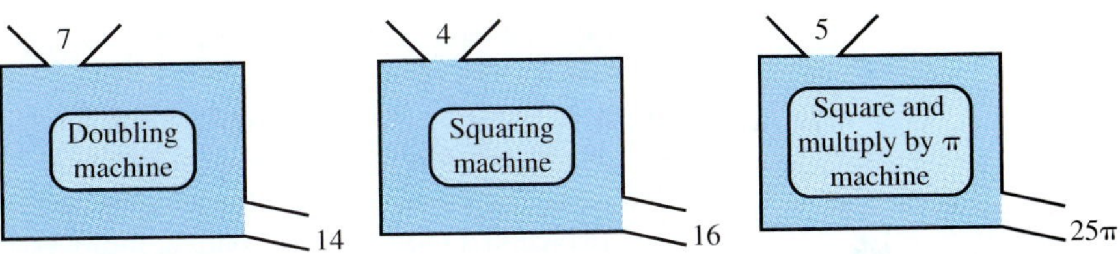

Figure 8.3
Three function machines

- **Functions as graphs.** A function whose domain and range are sets of numbers can be graphed on a set of x- and y-axes: if $f(x) = y$, plot the points (x, y) for all x in the domain. The Fibonacci sequence $F_1 = 1, F_2 = 1, F_3 = 2, F_4 = 3, \ldots$ and the doubling function $y = 2x$ are plotted in Figure 8.4.

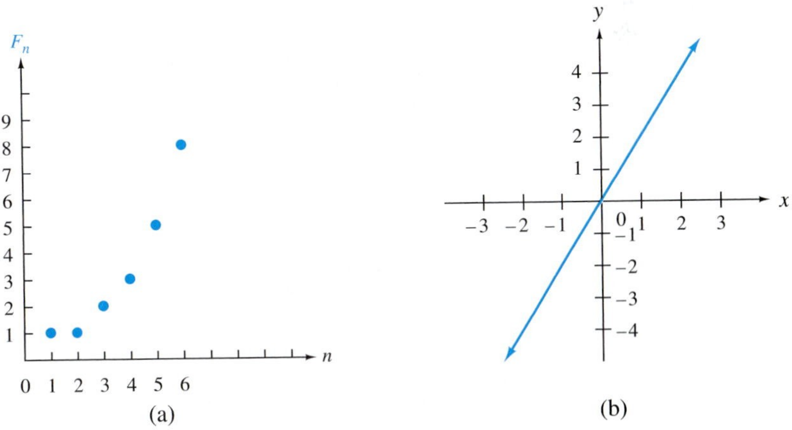

Figure 8.4
Graphs of (a) the Fibonacci sequence and (b) the doubling function

Computers and graphing calculators are very useful tools to create graphs and investigate the properties of the function represented by a graph. However, it is still instructive to make graphs by hand, using graph paper, rulers, and colored pencils. Begin by making a table of selected values of x from the domain and calculating the corresponding image values y. The table of values will help you to choose an appropriate range of x- and y-values to include on the axes. Next plot the points (x, y) from your table onto your graph. Finally, fill in the rest of the graph by smoothly connecting the points.

EXAMPLE 8.8 CALCULATING THE COST OF OWNING AND DRIVING A CAR

A survey of car owners shows that the monthly cost (in dollars) to own and drive an automobile is given by the function $f(x) = 0.31x + 175$. Here, x represents the number of miles driven throughout the month and 175 represents the monthly cost of ownership that is independent of the miles driven (insurance, vehicle license fees, and so on).

(a) Make a table that shows the cost of having a car that is driven 0, 100, 200, . . . , 1000 miles per month.

(b) Use the table of values in part (a) to draw a graph that shows the cost of driving a car for up to 1000 miles in a month.

(c) Use your graph to estimate the corresponding limit on the number of miles driven in a month if your budget limits your car expenditures to $350.

Solution

(a)

Miles, x	0	100	200	300	400	500	600	700	800	900	1000
Cost, y	175	206	237	268	299	330	361	392	423	454	485

(b)

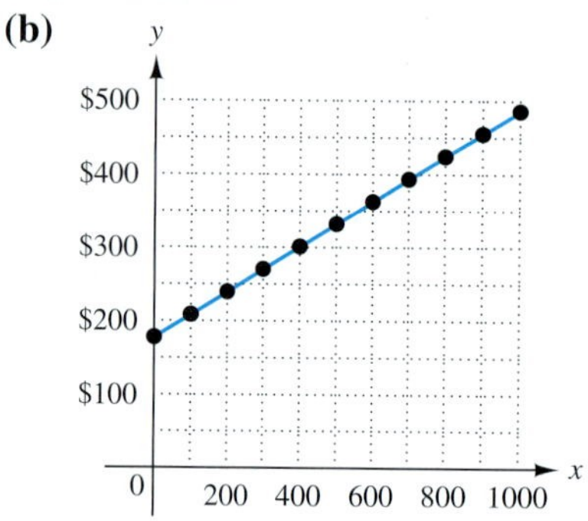

(c) The graph indicates that about 570 miles can be driven. A more exact value can be obtained from the equation $350 = 0.31x + 175$. Subtracting 175 from both sides and then dividing by 0.31 shows that $x = \frac{350 - 175}{0.31} \doteq 565$.

The points of the graph of the function $f(x) = 0.31x + 175$ in Example 8.8 lie along a line. More generally, the graph of any function of the form $f(x) = mx + b$ is a line, where m and b are constants. For this reason, such functions are called **linear functions.** In the next section, additional attention is given to lines and their equations. The graphs of some nonlinear functions are also investigated briefly; graphing calculators and computers can be quite helpful for such functions.

Cooperative Investigation

Vases and Graphs

Materials

Students will work in small groups of three or four. Each group needs a ruler, measuring cups, a container of water*, pencils, and several sheets of graph paper. Each group will work with a vase, so at least one vase is required per group. The vases should all have circular cross sections, but a large variety of shapes is important to provide a variety of graph shapes. Be sure to include a cylindrical vase with vertical sides. The rulers are used to measure the level of water, so translucent glass vases are best, or vases with sufficiently large openings so the ruler can be read easily.

Directions

Each group works with one vase at a time. Water is added 1 cup at a time, with the height of water in the vase measured after each cup is added. For small vases, add just a half or even a quarter cup of water at a time. Enter the data in a table of values at the top of a sheet of graph paper. Finally, plot the data points in a graph on the sheet, with the quantity of water in the vase on the horizontal axis and the height of the water on the vertical axis. Connect the points with a smooth curve. Also include a sketch of the vase on the sheet. When the graph is completed, groups trade vases and repeat the above steps on a new sheet of graph paper. This time, however, the group should first sketch the predicted shape of the graph they expect to obtain. Next, the actual graph is made from the experimental data and compared with the predicted graph.

Discussion Question

What type of vase has a straight-line graph?

*If water is inconvenient, it is also possible to use beans, puffed rice, or other available dry materials.

Problem Set 8.2

Understanding Concepts

1. Which of the following describe(s) functions on the set A? If your answer is "not a function," explain why not.
   (a) A is the set of students in a university. Each student is associated with the classes in which he or she is enrolled.
   (b) A is the set of classes offered by a university. Each class is associated with the number of students enrolled in the class.
   (c) A is the set of students in a classroom. Each student is associated with the grade earned on the first midterm.

(d) *A* is the set of letter grades A through F earned by the 25 students in a math class. Each grade is associated with a student who earned that grade on the first midterm.

2. Is a one-to-one correspondence from set *A* to set *B* a function defined on *A*? If so, what is the range of the function?

3. Decide which of these arrow diagrams represent(s) a function with domain *A*. If the diagram does not define a function, explain why. If a function is defined, give the range of the function.

(a)

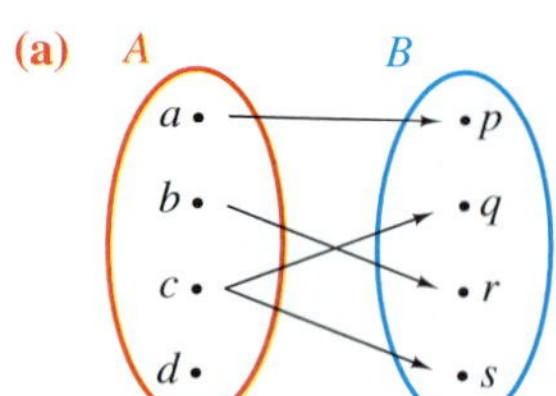

(b)

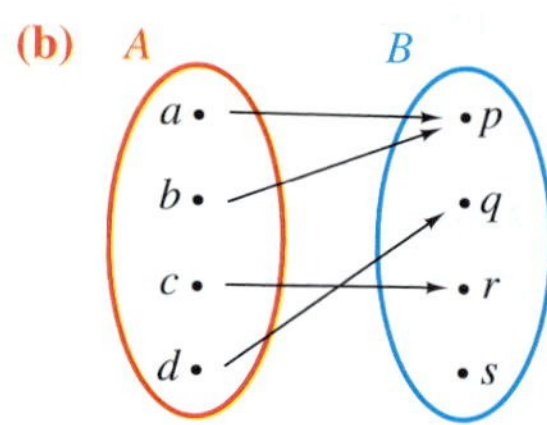

(c)

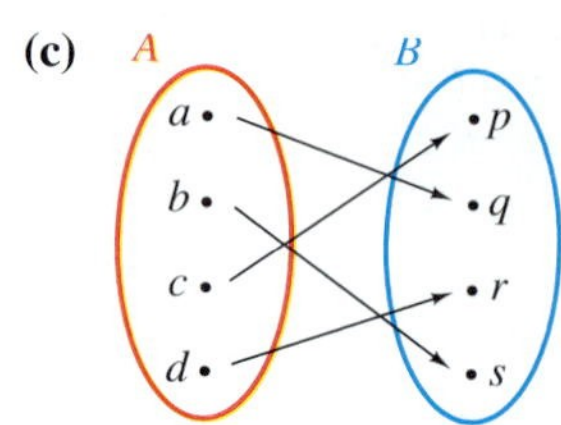

4. Draw an arrow diagram for each of the given sets of ordered pairs, where in each diagram the left loop represents the set $A = \{1, 2, 3, 4, 5\}$ and the right loop represents the set $B = \{1, 2, 3, 4\}$. For each set, explain why the set of ordered pairs does or does not represent a function with domain *A*.

(a) $\{(1, 3), (2, 3), (1, 2), (3, 2), (4, 3), (5, 4)\}$

(b) $\{(1, 2), (2, 1), (3, 3), (4, 2), (5, 2)\}$

(c) $\{(1, 4), (2, 3), (4, 1), (5, 2)\}$

5. Each of the axes systems shows a plot of points. Identify those plots that represent the graph of a function $y = f(x)$. If the plot cannot be a graph of a function, explain why not. If a function graph is depicted, give both the domain and the range of the function.

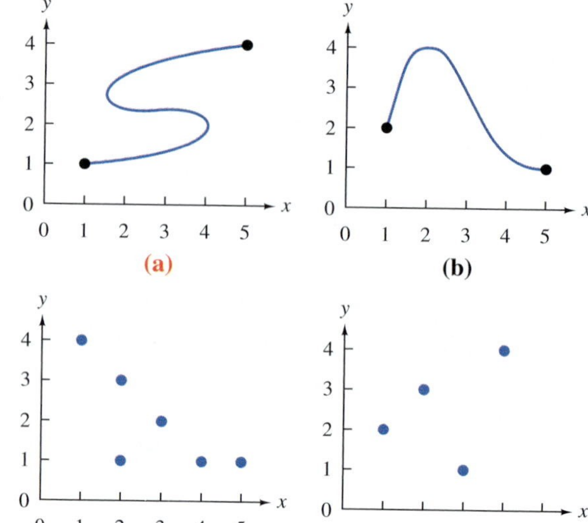

6. Let *f* be the function given by the formula $f(x) = 8 - 2x$ on the domain $D = \{x | x = 2, 3, 4, 5\}$.

(a) Make a table of the *x*- and *y*-values.

(b) What is the range of *f*?

(c) Sketch the graph of *f*.

7. Let *g* be the function given by the formula $g(x) = 2x - 2$, defined for all the real (decimal) numbers *x* in the interval $-2 \leq x \leq 3$.

(a) Make a table of values of $g(x)$ for $x = -2, -1, 0, 1, 2$, and 3.

(b) Plot the points in your table of part (a), and then sketch the entire graph.

(c) What is the range of *g*?

8. Let *g* be the function from $S = \{0, 1, 2, 3, 4\}$ to the whole numbers *W* given by the formula $g(x) = 5 - 2x + x^2$.

(a) Find $g(0)$, $g(1)$, $g(2)$, $g(3)$, $g(4)$.

(b) What is the range of *g*?

9. Let *h* be the function defined by $h(x) = x^2 - 1$, where the domain is the set of real numbers.

(a) Find $h(2)$.

(b) Find $h(-2)$.

(c) If $h(t) = 15$, what are the possible values of *t*?

(d) Find $h(7.32)$.

10. Karalee made a trip to the store one afternoon. Her trip is shown on the next page, where her distance (in miles) from home is graphed as a function of the hours past noon. Refer to the graph to answer these questions about her trip:

(a) What time did she leave home?

(b) When did she realize she forgot her checkbook?

Graph for Problem 10

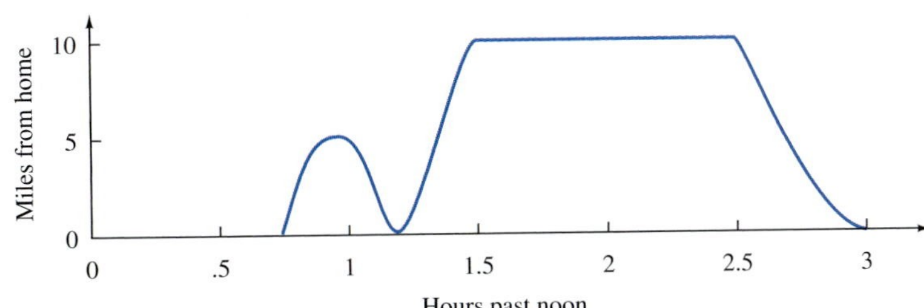

(c) What did she do about the forgotten checkbook?

(d) When did Karalee park the car at the store?

(e) How long did she shop?

(f) How far away was the store?

(g) Did Karalee encounter slower traffic going to or coming from the store? Explain how you can determine this from the graph.

11. Four graphs are shown, labeled G1, G2, G3, and G4. Which of these graphs matches the function described?

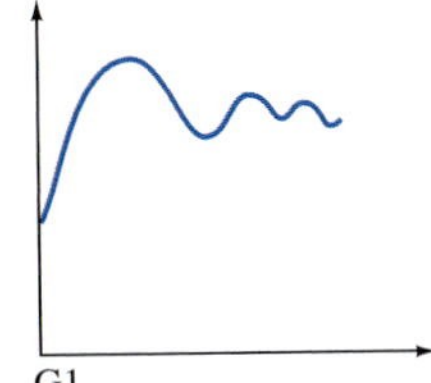
G1

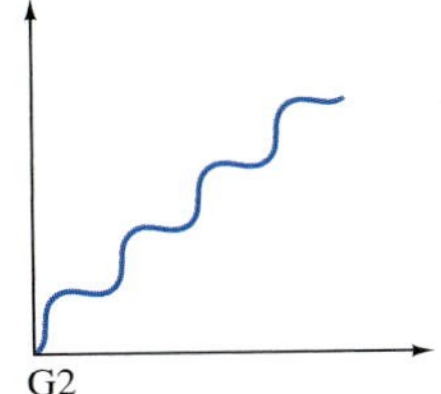
G2

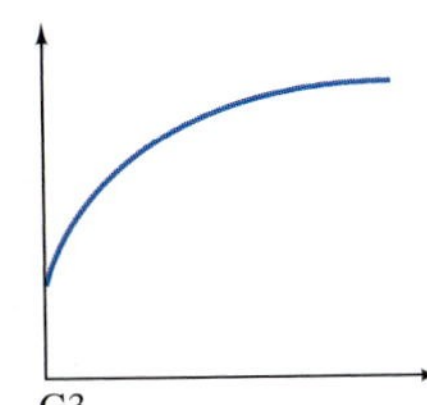
G3

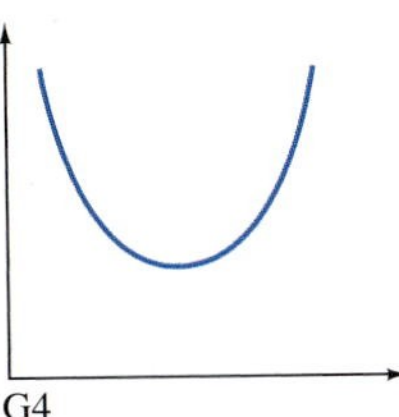
G4

(a) Temperature of a pan of water placed on a hot burner

(b) Height of a flag being run up the flagpole

(c) Height of a weight hung from a Slinky®

(d) Length of the shadow of a flagpole throughout a sunny day in Dallas

12. Make simple approximate sketches of graphs (such as those in problem 11) that correspond to these functions of time.

(a) Temperature of a forgotten cup of hot tea

(b) Perceived pitch of a train whistle as the train passes

(c) Height of the water in a bathtub, during the time someone takes a bath

(d) Hours of daylight in Chicago during a calendar year; that is, domain = $\{1, 2, \ldots, 365\}$

13. Consider the "double-square" rectangle of height x and width $2x$, as shown.

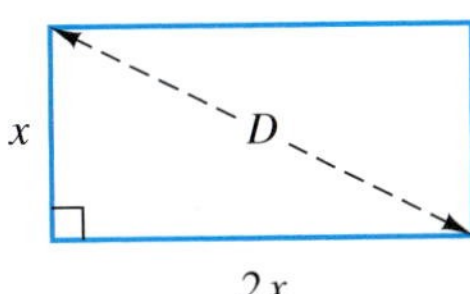

(a) Find the function $A(x)$ that gives the area of a double-square of width x.

(b) Find the function $P(x)$ that gives the perimeter of a double-square of width x.

(c) Use the Pythagorean theorem to obtain the function $D(x)$ for the length of the diagonal of a double-square of width x.

14. Solve each of these Guess My Rule games.

(a)

Guess	4	7	2	0	10	1
Response	9	12	7	5	15	6

(b)

Guess	4	2	0	7	3	9
Response	10	6	2	16	8	20

(c)

Guess	4	5	7	3	0	10
Response	17	26	50	10	1	101

15. Che is "it" in a game of Guess My Rule. The students' inputs and Che's outputs are shown in the table below.

Input given to Che	5	2	4	0	−2	−5
Output reported by Che	24	3	15	−1	3	24

(a) Guess Che's rule (function), $y = C(x)$.

(b) Antonio says that Che's rule is $C(x) = x^2 - 1$, but Claudette claims that it is $C(x) = (x + 1)(x - 1)$. Who is correct, Antonio or Claudette? Explain.

16. Suppose two function machines are hooked up in sequence, so the output chute of machine g empties into the input hopper of machine f. Such a coupling of machines, which is defined if the range of g is a subset of the domain of f, is called the **composition** of f and g and can be written $F(x) = f(g(x))$.

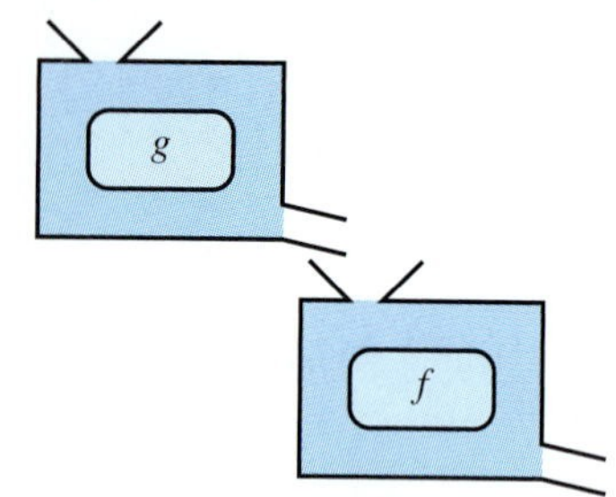

(a) Suppose f is the doubling function $f(x) = 2x$ and g is the "add 3" function $g(t) = t + 3$. Then $F(4) = f(g(4)) = f(4 + 3) = f(7) = 2 \times 7 = 14$. Evaluate $f(g(x))$ for $x = 0, 1, 2$, and 3.

(b) Suppose the doubling and "add 3" function machines are coupled in reverse order to define the composition of g and f given by $G(x) = g(f(x))$. Then $G(4) = g(f(4)) = g(2 \times 4) = g(8) = 8 + 3 = 11$. Evaluate $g(f(x))$ for $x = 0$, 1, 2, and 3.

17. The statement "There are 6 times as many students as professors at this university" can be written symbolically as $S = 6P$, where S denotes the number of students and P denotes the number of professors. Some years ago, incoming freshmen at a certain university were challenged to convert the given statement to an equation, and a large percentage mistakenly answered with "$6S = P$." Convert each of these statements into simple formulas, using the letters suggested.

(a) There are 10 times as many female nurses as male nurses at the hospital. (F and M)

(b) Sean has 4 more dollars than Gunther. (S and G)

(c) With 10 more points, LeAnn's score will be twice that of Itzhak. (L and I)

Teaching Concepts

18. Function concepts can be taught to young children by incorporating appropriate manipulatives and activities. For example, suppose a student plays the role of a function machine whose inputs are lengths of Unifix™ cubes. When handed a length of five cubes, the student returns a length of eight cubes. When given a length of two cubes, the student returns a length of five cubes.

(a) What function is the student apparently evaluating? How could it be tested? How many cubes should be put into the machine so that the output is 10 cubes snapped together?

(b) Describe a function machine, again employing Unifix™ cubes, that illustrates the addition function $f(a, b) = a + b$ on the domain of pairs of natural numbers.

19. Create an activity, using commonly available manipulatives, that will illustrate a function concept. Use the Unifix™ cube activity described in problem 18 as an example, but be original.

Thinking Critically

20. Shaquita noticed the following pattern when summing the squares of the successive natural numbers.

$$S_1 = 1^2 = 1 = 2 \cdot 3 \cdot 4/24$$
$$S_2 = 1^2 + 2^2 = 5 = 4 \cdot 5 \cdot 6/24$$
$$S_3 = 1^2 + 2^2 + 3^2 = 14 = 6 \cdot 7 \cdot 8/24$$

(a) Does Shaquita's pattern hold for S_5 and S_6?

(b) Find a formula that gives Shaquita's function S_n.

21. The 3×3 square array shown below contains nine unit 1×1 squares, four 2×2 squares, and one 3×3 square, making $1 + 4 + 9 = 14$ squares in all.

(a) Fill in the partially completed table below counting the squares of various sizes in a given square array.

(b) Make a conjecture about the number of squares, S_6, that can be found in a 6×6 array.

(c) Give a formula for S_n, the number of squares in an $n \times n$ array.

22. Investigate the sum-of-the-cubes function $C_n = 1^3 + 2^3 + 3^3 + \cdots + n^3$. Make a table of values of C_n and see if you can discover a connection between C_n and the triangular number function t_n given by $t_n = n(n + 1)/2$. (*Suggestion:* Enter values of t_n in a new row of your table for easy comparison.)

Diagram for Problem 21

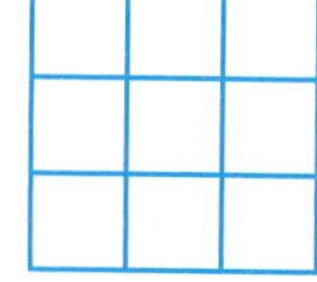

Table for Problem 21a

Size of the Array	Number of Squares of Size 1 × 1	2 × 2	3 × 3	4 × 4	5 × 5	Total Number of Squares
1 × 1	1					1
2 × 2	4	1				5
3 × 3	9	4	1			14
4 × 4						
5 × 5						

23. The bank robbers are 9 miles north of Dodge City and are continuing to head north at 8 miles per hour on their horses. If the posse begins the chase on fresher horses at 10 miles per hour, how many hours will it take for the posse to catch the thieves? Show your answer with a graph that plots both the "robber" function $9 + 8t$ and the "posse" function $10t$, which give the distance north of Dodge after t hours of the chase.

Making Connections

Linear functions. *A function f is called* **linear** *if there are two numbers m and b such that $f(x) = mx + b$ for all x in the domain of the function. Some useful linear functions are investigated in problems 24 through 28.*

24. **Hooke's Law.** If a (small) weight w is suspended from a spring, the length L of the stretched spring is a linear function of w. Find m and b in the formula $L = mw + b$ if the unstretched spring has length 10″ and a weight of 2 pounds stretches it to 14″.

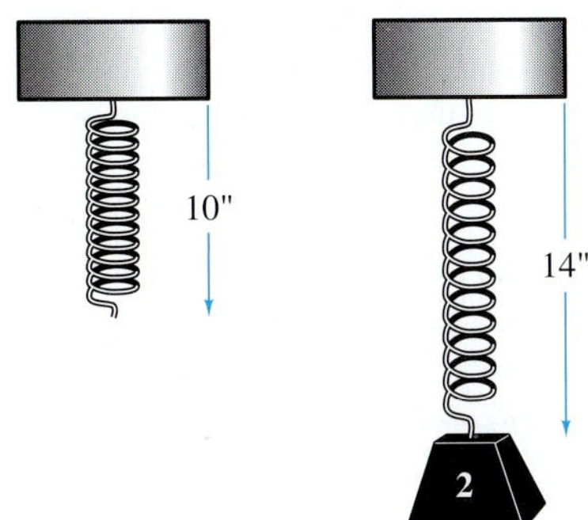

25. **Temperature Conversion.** The temperature at which water freezes is 32° Fahrenheit and 0° Celsius, and the temperature at which water boils is 212° Fahrenheit and 100° Celcius. Find the constants m and b in the formula $F = mC + b$ to express the Fahrenheit temperature F as a function of the Celsius temperature C.

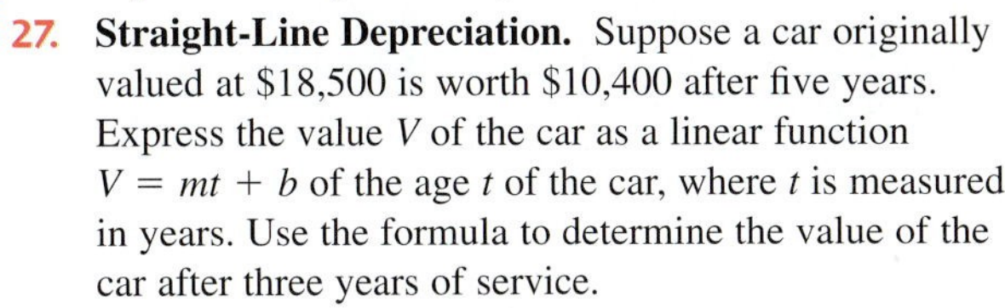

26. **Speed of a Dropped Object.** If we neglect the resistance of air, the (downward) speed of an object t seconds after being dropped is given by $v = 32t$ feet per second. How many seconds does it take an object to reach a downward speed of 100 miles per hour? Recall that there are 5280 feet in one mile, since you'll need to express 100 mph in feet per second.

27. **Straight-Line Depreciation.** Suppose a car originally valued at \$18,500 is worth \$10,400 after five years. Express the value V of the car as a linear function $V = mt + b$ of the age t of the car, where t is measured in years. Use the formula to determine the value of the car after three years of service.

28. **The Lightning Distance Function.** The speed of sound is about 760 miles per hour. Assuming a lightning flash takes no appreciable time to be seen, and the corresponding peal of thunder is heard after t seconds, show that $d = t/5$ gives the approximate distance d (in miles) to the lightning strike.

29. **Taxi Fare.** The Evergreen Taxi Company charges \$2.15 for the first mile and \$1.25 for each additional mile (or fraction thereof). What is the charge for these trips?

    (a) 4.3 miles (b) 7.8 miles

30. **First-Class Postage.** First-class postage in the United States is 37 cents for the first ounce and 23 cents for each additional ounce or fraction of an ounce. What does it cost to mail a package weighing

    (a) $1\frac{1}{2}$ ounces? (b) half a pound?

    (c) 2 ounces?

31. The plot below and to the left shows the number of hours of daylight throughout the year from January to December, where each graph depends on the latitude. For example, the blue curve is for 30° north latitude and applies well to Houston, Texas, and Jacksonville, Florida.

    (a) Which curve gives the number of hours of daylight of Quito, Ecuador?

    (b) Which curve gives the number of hours of daylight of Anchorage, Alaska?

    (c) Why do you know that the red curve corresponds to a latitude north of the Arctic Circle?

    (d) What are the times of the year when every place on earth has 12 hours of daylight?

    (e) Approximately when in the year is the number of daylight hours the largest? the smallest? What are these events called?

Graph for Problem 31

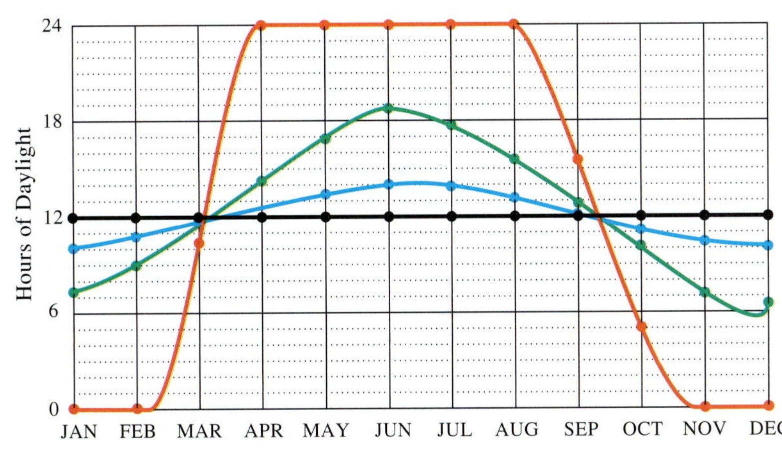

Table for Problem 32

Schedule X—Use if your filing status is single.

If TAXABLE INCOME		THEN The TAX Is		
Is Over	But Not Over	This Amount	Plus This %	Of the Excess Over
\$0	\$7,000	\$0.00	10%	\$0.00
\$7,000	\$28,400	\$700.00	15%	\$7,000
\$28,400	\$68,800	\$3,910.00	25%	\$28,400
\$68,800	\$143,500	\$14,010.00	28%	\$68,800
\$143,500	\$311,950	\$34,926.00	33%	\$143,500
\$311,950	–	\$90,514.50	35%	\$311,950

Using a Calculator

32. Tax Rate Schedules. Federal income tax rates for a single person are given in Schedule X. Use it to find the taxes owed for Demitrius and Cyndi, both of whom are single.

(a) Demitrius had a taxable income of \$41,162.

(b) Cyndi had a taxable income of \$134,520.

33. Windchill Temperature. In windy, cold weather, the increased rate of heat loss makes the temperature feel colder than the actual air temperature. Weather reports often take account of the effect of the wind by reporting the windchill temperature (WCT), which is determined as a function of both the air temperature T (in degrees Fahrenheit) and the wind speed V (in miles per hour). In November 2001, a new formula was adopted, replacing a less meaningful formula derived in the 1940s. The new formula, used for temperatures up to 50°F and wind speeds above 3 mph, is

$$\text{WCT} = 35.74 + 0.6215T - 35.74 \cdot V^{0.16} + 0.4275 \cdot T \cdot V^{0.16}$$

(a) Verify that a temperature of 10°F in a wind of 20 miles per hour produces a windchill temperature of −9°F.

(b) A weather forecaster claims that a wind of 36 miles per hour gives a WCT of −36°F. What is the air temperature, to the nearest degree?

34. Day-of-Week Function. The following function gives the number Y_1 of the day of the week ($1 = $ Sunday, $2 = $ Monday, etc.) as a function of the date (D, M, Y) in the Gregorian calendar introduced in 1582. The three variables in the function are

$D = $ number of the day of the month,

$M = $ number of the month (with $13 = $ January, $14 = $ February, $3 = $ March, . . . , $12 = $ December), and

$Y = $ year.

The function, in graphing calculator form, is

$$Y_1 = 7 * \text{fPart}((D + 2M + \text{iPart}(.6M + .6) + Y + \text{iPart}(Y/4) - \text{iPart}(Y/100) + \text{iPart}(Y/400) + 2)/7)$$

The functions "fPart" (fractional part) and "iPart" (integer part) are built-in functions on your graphing calculator. As an example, to find the day of the week of July 4, 1776, enter the function Y_1 and then return to the home screen and enter on one line

$$7 \rightarrow \text{M:}4 \rightarrow \text{D:}1776 \rightarrow \text{Y:}Y_1.$$

Pressing ENTER returns a 5, showing us that the day of the week was Thursday. Using 2nd ENTRY, you can edit the entry lines to investigate new dates.

(a) Verify that President Lincoln signed the Emancipation Proclamation on Friday, January 1, 1863. (*Reminder:* For January use $M = 13$.)

(b) What day of the week was July 20, 1969, the day astronaut Neil Armstrong became the first person to set foot on the moon?

(c) What day of the week was January 1, 2001, the first day of the new millennium?

35. Enter the value $G = (1 + \sqrt{5})/2$ (the golden ratio) into the memory of your calculator. Now enter these two functions into your calculator:

$$Y_1 = \text{iPart}(X*G) \text{ and } Y_2 = \text{iPart}(X*G^2)$$

Here "iPart" stands for "integer part" and rounds any positive decimal number downward to the nearest whole number.

(a) Evaluate the two functions to check the entries given in the table below, and then fill in the rest of the table.

X	1	2	3	4	5	6	7	8	9	10	11	12
Y_1	1	3	4	6	8							
Y_2	2	5	7	10	13							

(b) On the basis of your table, what connection do you see between the functions $Y_1(X)$ and $Y_2(X)$?

(c) Suppose that R_1 and R_2 denote the respective ranges of Y_1 and Y_2. On the basis of the pattern in your table of values, what does $R_1 \cap R_2$ seem to be? How about $R_1 \cup R_2$?

From State Student Assessments

36. (New Jersey, Grade 4)
Mrs. Thompson's class recently completed a lesson on number patterns. Each student had to write a rule to describe a pattern of numbers and list some numbers in the pattern. Two examples are shown on the right.

(a) Tony's pattern is shown at right. Write a rule to describe his pattern.

1, 4, 5, 8, 9, . . .

(b) Write your own rule for a number pattern. Also, write the first five numbers in your pattern.

Rule: Start with 1. Add 1 to the first number, add 2 to the second number, add 3 to the third number, and so on.
Pattern: 1, 2, 4, 7, 11, . . .

Rule: Start with 2. Multiply each number by 2 to get the next number in the pattern.
Pattern: 2, 4, 8, 16, 32, . . .

For Review

37. A rectangular pen is three times as long as it is wide. It requires 80 feet of fencing to enclose. What are the dimensions of the rectangle?

38. Create an expression in the variable n that will give the nth term of the sequence whose first five terms are 5, 11, 17, 23, and 29.

39. Find values for the constants a, b, and c so that the expression $an^2 + bn + c$ has the values 1, 0, 1, and 4 when n has the respective values 0, 1, 2, and 3. What is the value of the expression when n is 4? when n is 5?

8.3 GRAPHING FUNCTIONS IN THE CARTESIAN PLANE

In this section we explore the graphs of functions from a geometric point of view. In particular, we investigate how the shape of a graph reveals the properties of the corresponding function. Of special interest are functions of the form $f(x) = mx + b$, where m and b are constants. We will see that the graph of a function with this form is a line in the plane, explaining why functions of this type are called linear functions. The process of going from a function to its graph can also be reversed. For example, given a line in the plane, we will find a function whose graph is the given line. The section concludes with a brief look at some nonlinear functions. We begin by introducing the coordinate plane, which is also called the Cartesian plane in honor of René Descartes (see the next Highlight from History box).

The Cartesian Coordinate System

Consider the two perpendicular number lines illustrated in Figure 8.5. Any point in the plane can be uniquely located by giving its distance to the right or left of the vertical number line and its distance above or below the horizontal number line. In Figure 8.5, the point P is 5 units to the right of the vertical number line and 3 units above the horizontal number

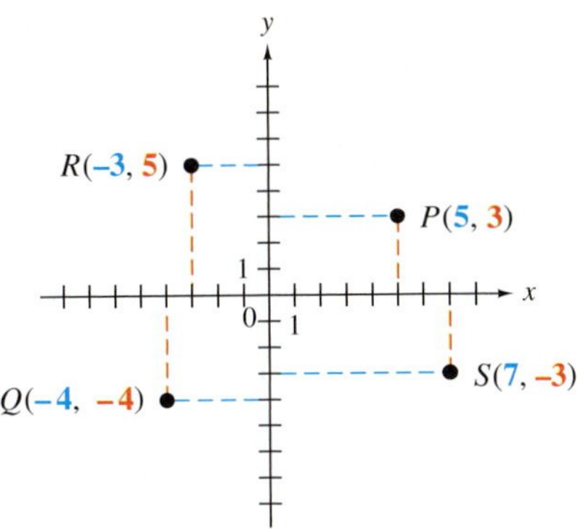

Figure 8.5
The Cartesian coordinate system

line, and there is only one such point. Thus, P is identified by the ordered number pair $(5, 3)$ and we sometimes write $P(5, 3)$ as shown. Other times we will just write $(5, 3)$.

The two number lines are called **coordinate axes** or just **axes.** Typically the horizontal number line is called the ***x*-axis** and the vertical number line is called the ***y*-axis.** The two numbers in the number pair (a, b) that locate a point are called the **coordinates** of the point. The first number in the pair is called the ***x*-coordinate** and gives the distance of the point to the right or left of the vertical axis—that is, in the direction of the x-axis. The second number in the ordered pair is called the ***y*-coordinate** and gives the distance of the point above or below the horizontal axis—that is, in the direction of the y-axis. The axes divide the plane into four regions or **quadrants** numbered I, II, III, and IV counterclockwise from the upper-right-hand quadrant. A point

- lies in quadrant I if both its coordinates are positive,
- lies in quadrant II if the first coordinate is negative and the second coordinate is positive,
- lies in quadrant III if both coordinates are negative, and
- lies in quadrant IV if the first coordinate is positive and the second coordinate is negative.

The point $(0, 0)$ where the axes intersect is called the **origin** of the coordinate system. These notions are summarized in Figure 8.6.

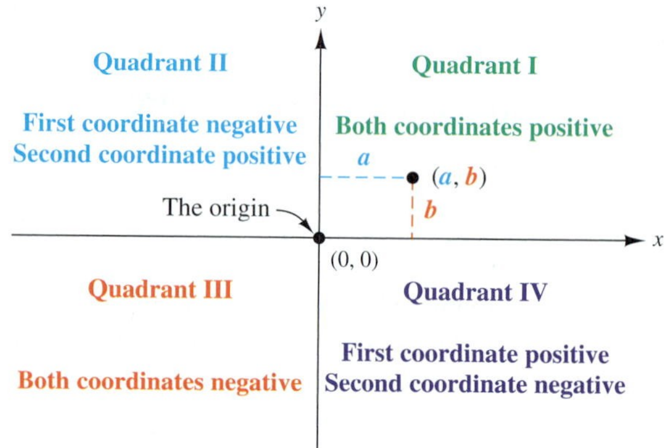

Figure 8.6
Salient features of the Cartesian coordinate system

EXAMPLE 8.9 PLOTTING POINTS

The diagram below shows the partial outline of a house.

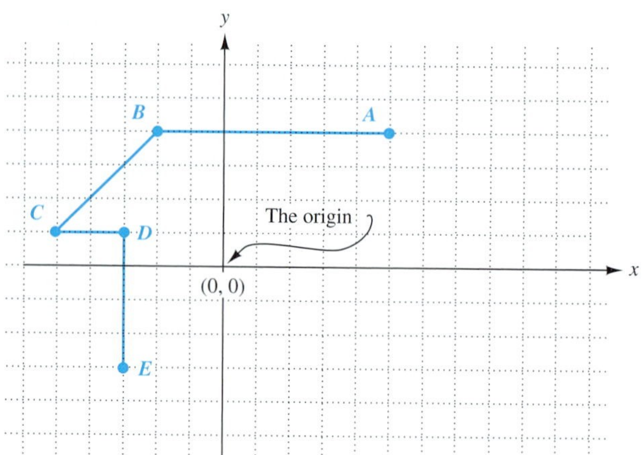

HIGHLIGHT from HISTORY

René Descartes and Coordinate Geometry

It is often difficult to ascribe the development of any particular body of mathematics to its originator. The fact is that mathematics is the cumulative result of the efforts of numerous individuals working over hundreds or even thousands of years. The invention of coordinate geometry is customarily ascribed to René Descartes (1596–1650), one of the leading seventeenth-century mathematicians, who did indeed make great strides in commingling the ideas of geometry and algebra as explained in a book titled *La géométrie*. However, Descartes never thought of an ordered pair (a, b) as the coordinates of a point in the plane. Thus, the terminology "Cartesian product" and "Cartesian coordinate system," ascribing these ideas to Descartes, is largely misplaced. The idea of coordinates goes back at least as far as the Greek Apollonius of Perga, in the third century B.C., and was utilized by Nicole Oresme (1323?–1382), the French Bishop of Lisieux, and by others. The idea was also known to the amateur, but inspired, Pierre de Fermat, a contemporary of Descartes, and was certainly popularized by the Dutch mathematician Frans van Schooten (1615–1700) in his *Geometria a Renato Des Cartes* (*Geometry by René Descartes*) in 1649. It is probably not unreasonable to say that our modern ideas of coordinate geometry were inspired by Descartes, but organized and popularized by Schooten.

(a) Give the coordinates (that is, the ordered pair) naming each of A, B, C, D, and E.

(b) Plot the points $F(8, -3)$, $G(8, 1)$, $H(10, 1)$, $I(7, 4)$, $J(6, 4)$, $K(6, 5)$, and $L(5, 5)$.

(c) Draw the segments $\overline{EF}$, $\overline{FG}$, $\overline{GH}$, $\overline{HI}$, $\overline{IJ}$, $\overline{JK}$, $\overline{KL}$, and $\overline{LA}$.

Solution

(a) A is 5 units to the right of $(0, 0)$ (that is, in the x-direction from the origin) and 4 units above $(0, 0)$ (that is, in the y-direction from the origin). Thus, A is the point $(5, 4)$. The x-coordinate of A is 5 and the y-coordinate of A is 4. Similarly, we determine that the coordinates of the other points are $B(-2, 4)$, $C(-5, 1)$, $D(-3, 1)$, and $E(-3, -3)$.

(b) The point $F(8, -3)$ is 8 units to the right of $(0, 0)$ and 3 units *below* $(0, 0)$. Similarly, the other points are located as shown.

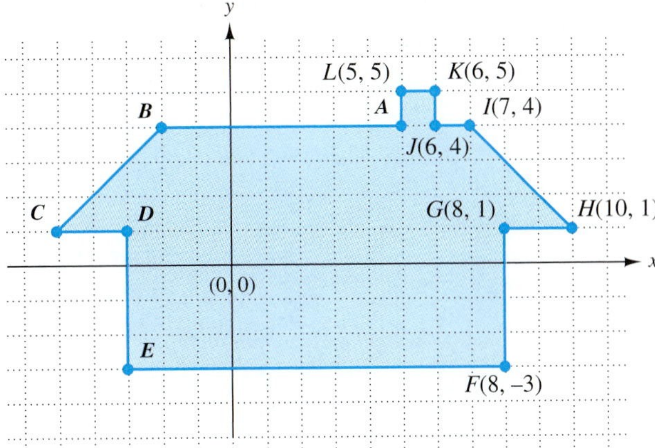

(c) After we have drawn the designated line segments, the completed figure forms the outline of a house.

The Distance Formula

Consider the points $P(2, 5)$ and $Q(7, 8)$ shown in Figure 8.7.

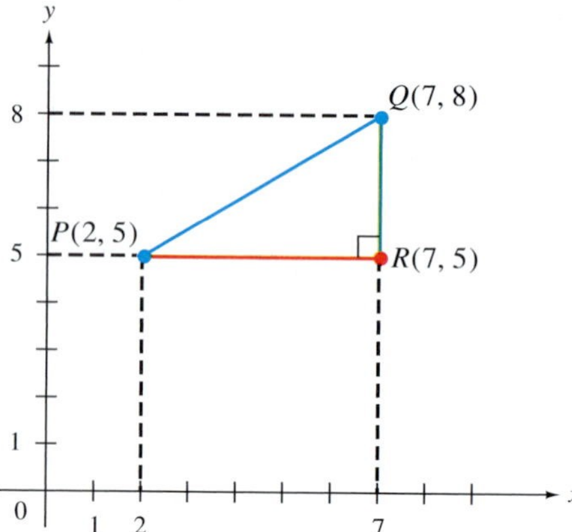

Figure 8.7
The distance between points P and Q is found by using the Pythagorean theorem for the right triangle PQR

The vertical line through the upper point and the horizontal line through the lower point intersect at the point $R(7, 5)$. The two points $P(2, 5)$ and $R(7, 5)$ lie on the horizontal line of points whose y-coordinates are $y = 5$. The distance between P and R, which is denoted by PR, is then just the absolute value of the difference in the x-coordinates of the points. That is, $PR = |7 - 2| = 5$. Similarly, the points $Q(7, 8)$ and $R(7, 5)$ both are on the vertical line for which $x = 7$, so the distance between Q and R is $QR = |8 - 5| = 3$.

To find the distance between P and Q, we use the fact that PQR is a right triangle with hypotenuse PQ and legs PR and RQ. The Pythagorean theorem then tells us that $PQ^2 = PR^2 + RQ^2$. That is,

$$\begin{aligned} PQ &= \sqrt{PR^2 + RQ^2} \\ &= \sqrt{(7 - 2)^2 + (8 - 5)^2} \\ &= \sqrt{25 + 9} = \sqrt{34} \doteq 5.8. \end{aligned}$$

The procedure above can be followed in exactly the same way beginning with any two points $P(x_1, y_1)$ and $Q(x_2, y_2)$. This gives us the following theorem.

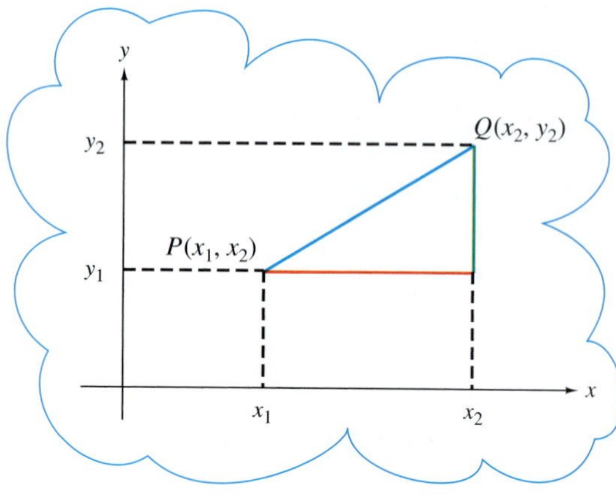

THEOREM *The Distance Formula*
Let P and Q be the points (x_1, y_1) and (x_2, y_2). Then the distance between P and Q is

$$PQ = \sqrt{(x_2 - x_1)^2 + (y_2 - y_1)^2}.$$

In the distance formula, it does not make any difference which point is chosen for P and which is chosen for Q. For example, the distance between $(4, 7)$ and $(1, 3)$ is given by both

$$\begin{aligned}\sqrt{(4-1)^2+(7-3)^2} &= \sqrt{3^2+4^2}\\ &= \sqrt{9+16} = \sqrt{25} = 5\end{aligned}$$

and

$$\begin{aligned}\sqrt{(1-4)^2+(3-7)^2} &= \sqrt{(-3)^2+(-4)^2}\\ &= \sqrt{9+16} = \sqrt{25} = 5.\end{aligned}$$

EXAMPLE 8.10 PROVING THAT A TRIANGLE IS ISOSCELES

Prove that the triangle with vertices $R(1, 4)$, $S(5, 0)$, and $T(7, 6)$ is isosceles (that is, that two sides have the same length).

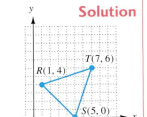

Solution

We compute the length of the three sides.

$$\begin{aligned} RS &= \sqrt{(1-5)^2+(4-0)^2} = \sqrt{16+16} = \sqrt{32};\\ RT &= \sqrt{(1-7)^2+(4-6)^2} = \sqrt{36+4} = \sqrt{40};\\ ST &= \sqrt{(5-7)^2+(0-6)^2} = \sqrt{4+36} = \sqrt{40}.\end{aligned}$$

Since $RT = ST$, it follows that $\triangle RST$ is isosceles.

$\overleftrightarrow{PQ}$ is a line, $\overline{PQ}$ is a segment, and PQ is the length of $\overline{PQ}$.

The unique line determined by two points P and Q in the plane is denoted by $\overleftrightarrow{PQ}$. The points on the line that are between P and Q form the **line segment** with endpoints P and Q. The line segment is denoted by $\overline{PQ}$, and the length of the line segment PQ is the distance between its endpoints. It is important to notice that the notation makes a distinction between the geometric object $\overline{PQ}$ (a set of points in the plane) and its length PQ (a nonnegative real number).

Slope

Consider the lines l, m, n, and p in Figure 8.8. The properties that distinguish between two lines are their location on the coordinate system and their direction or steepness. The direction or steepness of a line is the same as that of any segment of the line and this leads to the notion of **slope.**

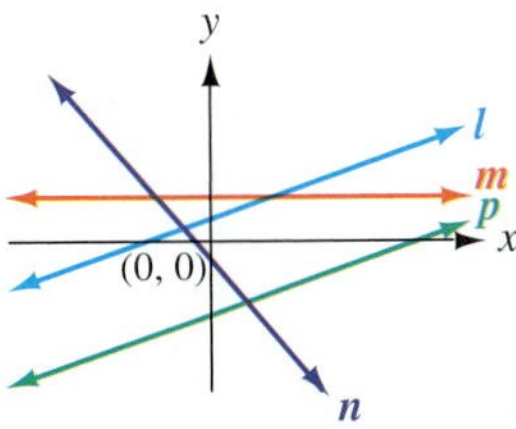

Figure 8.8
Lines in the plane

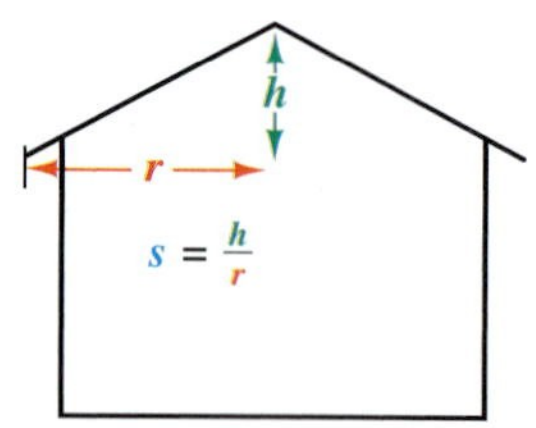

Figure 8.9
Slope on a roof

Carpenters use the ratio

$$s = \frac{h}{r},$$

where h is the vertical distance a roof rises and r is the horizontal distance over which the rise takes place, to compute the slope of a roof as shown in Figure 8.9. This is sometimes expressed by saying that the slope of a roof is "the rise over the run." Surveyors use the same idea when calculating the slope of a road. If a road rises 5 feet while moving forward horizontally 100 feet, the road has a slope of 0.05. In surveying, slopes are usually expressed as percents. Thus, a grade with a slope of 0.05 is said to be a 5% grade.

We also use the idea *rise over run* to determine the slope of a line segment. Consider the points $P(3, 5)$ and $Q(9, 7)$ shown in Figure 8.10. In moving from P to Q one moves up 2 units while moving to the right 6 units. The rise over the run gives a slope of $\frac{1}{3}$, indicated by the letter m. In this case,

$$m = \frac{7-5}{9-3} = \frac{2}{6} = \frac{1}{3}.$$

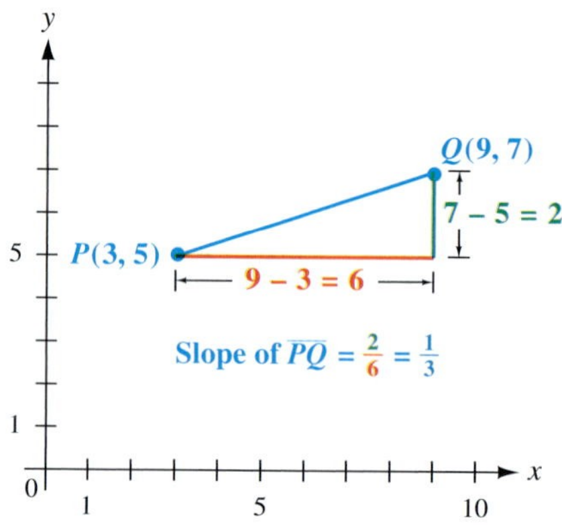

Figure 8.10
Slope of a segment

In general, the reasoning is similar and leads to this definition of slope.

DEFINITION *Slope of a Line Segment*
Let $P(x_1, y_1)$ and $Q(x_2, y_2)$, with $x_1 \neq x_2$, be two points. Then the **slope of the line segment** $\overline{PQ}$ is given by

$$m = \frac{y_2 - y_1}{x_2 - x_1}.$$

If $x_1 = x_2$ in the preceding definition, then $x_2 - x_1 = 0$ and $\overline{PQ}$ is vertical. Since division by zero is undefined, we must say that *a vertical segment has no slope* or, equivalently, that *the slope of a vertical segment is undefined.*

If $y_1 = y_2$ in the preceding definition, then $\overline{PQ}$ is horizontal and $m = 0$. Thus, saying that a line segment is horizontal is the same as saying that it has zero slope.

In computing the slope of a segment, it makes no difference which point is chosen as P and which is chosen as Q. However, once the choice is made, one must stick with it and always subtract *in the same direction* in both numerator and denominator. For example, in computing the slope of the segment in Figure 8.10, we identified P and Q as $(3, 5)$ and $(9, 7)$, respectively. But this could have been reversed to obtain

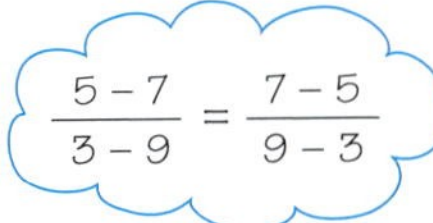

$$m = \frac{5 - 7}{3 - 9} = \frac{-2}{-6} = \frac{1}{3}$$

as before.

Finally, in Figure 8.10 the slope was positive and the segment sloped upward to the right. This is always the case for segments with positive slopes. If $y_1 = y_2$ in the definition of slope, then the slope is 0 and the segment $\overline{PQ}$ is necessarily horizontal. If the slope of a segment is negative, the segment slopes *downward to the right.*

EXAMPLE 8.11 FINDING THE SLOPES OF LINE SEGMENTS

The points $P(-4, 3)$, $Q(5, 6)$, $R(5, -1)$, and $S(-1, -3)$ are the vertices of the four-sided polygon $PQRS$ shown in the figure below. Find the slope of each side of the polygon.

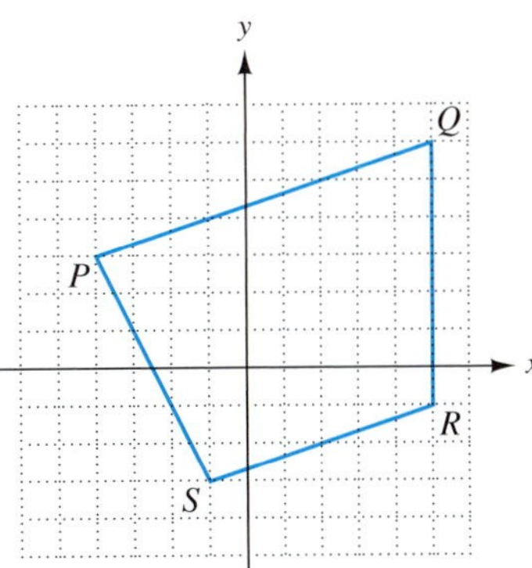

Solution

Using the slope formula $m = \dfrac{y_2 - y_1}{x_2 - x_1}$, we find

$$m(\overline{PQ}) = \frac{6 - 3}{5 - (-4)} = \frac{3}{9} = \frac{1}{3},$$

$$m(\overline{RS}) = \frac{(-3) - (-1)}{(-1) - (5)} = \frac{-2}{-6} = \frac{1}{3}, \text{ and}$$

$$m(\overline{PS}) = \frac{(-3) - (3)}{(-1) - (-4)} = \frac{-6}{3} = -2.$$

The side $\overline{QR}$ is vertical, since both R and Q have the same x-coordinate, 5. Therefore, $\overline{QR}$ has undefined slope.

The **slope of a line** is the slope of any segment on the line, as in the following definition.

DEFINITION *Slope of a Line*
The **slope of a line** is the slope of any segment on the line.

As noted earlier, the slope of a line is a measure of the line's steepness on a coordinate system. Thus, if lines are equally steep, they are parallel, and, if they are parallel, then they are equally steep. In terms of slopes, we have the following theorem.

THEOREM *Condition for Parallelism*
Two segments (lines) are parallel if, and only if, they have same slope.

The segments PQ and RS in Example 8.11 have the same slope, so they are an example of parallel line segments. Since the four-sided polygon $PQRS$ has a pair of parallel sides, we have shown that it is a trapezoid.

Equations of Lines

Using the tools of coordinate geometry it is now possible to write equations whose graphs are lines.

We begin by considering a particular example.

EXAMPLE 8.12 DETERMINING THE EQUATION OF A LINE THROUGH (2, 3) WITH SLOPE 4/3

Derive an equation of the line through point $P(2, 3)$ and having slope $4/3$.

Solution *Understand the Problem*

There is one, and only one, line through the point $(2, 3)$ and having slope $4/3$. One can draw the line by plotting the point $P(2, 3)$ and then plotting the point $Q(5, 7)$ 3 units to the right and 4 units above P. The line segment through these points must have slope $4/3$, so the line through these points must be the desired line. Suppose $R(x, y)$ is *any* point on the line. We must find an equation involving the variables x and y that is satisfied by those, and only those, points that lie on the line.

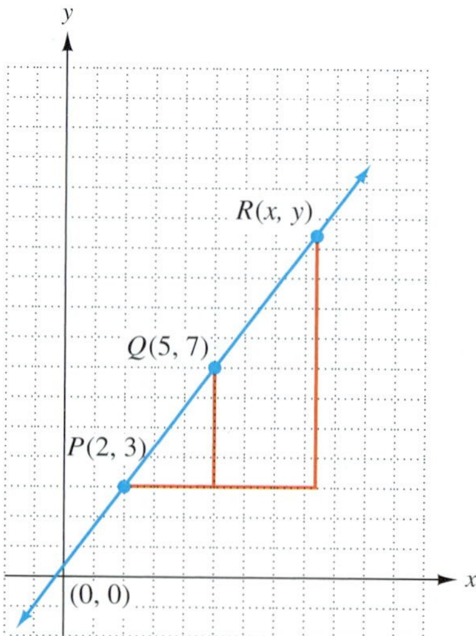

HIGHLIGHT *from* HISTORY

Maria Agnesi (1718–1799)

Maria Gaetana Agnesi was a child prodigy. Her father, Pietro Gaetana, who was a professor of mathematics at the University of Bologna in Italy, recognized his daughter's talent and encouraged her studies. Before she was 13, Agnesi spoke many languages, including Greek, Hebrew, and Latin. As a young woman, she gave talks on mathematics and philosophy to adult friends of her parents during parties at their home.

Agnesi's famous two-volume book, *Analytical Institutions,* took her 10 years to write. It includes discussions of algebra, geometry, and calculus. In it, she describes the *Witch of Agnesi,* a curve first proposed by the French mathematician Fermat and studied extensively by Agnesi herself. Because *Analytical Institutions* was so clearly written, it was translated into French and English and used as a textbook. It is the first work of such stature written by a woman that has survived.

According to some accounts, Pope Benedict XIV appointed Agnesi to teach at the University of Bologna around 1750; however, it is possible that she never actually taught there. It is certain that Agnesi retired from mathematics just as her intellectual powers were at their peak. When in her forties, she decided to devote the remaining years of her life to helping the sick and poor. Maria Gaetana Agnesi died at the age of 81, leaving the world a rich scholastic and humanitarian legacy.

SOURCE: From Portraits for Classroom Bulletin Boards: Women Mathematicians, *text by Virginia Slachman. Copyright © 1990 by Dale Seymour Publications, an imprint of Pearson Learning Group, a division of Pearson.*

Devise a Plan

Since the slope of a line can be determined by *any* two points on the line, $R(x, y)$ is on the line if, and only if,

$$\text{slope } \overline{PR} = \frac{4}{3}.$$

Perhaps we can use this fact to derive the desired equation.

Carry Out the Plan

Since

$$\text{slope } \overline{PR} = \frac{y-3}{x-2},$$

it follows that R is on the line in question if, and only if,

$$\frac{y-3}{x-2} = \frac{4}{3}$$

or, alternatively,

$$y - 3 = \frac{4}{3}(x-2).$$

Hence, this must be the desired equation.

Look Back
Since there is one, and only one, line through a given point and having a given slope, $R(x, y)$ is on the desired line if, and only if, slope $\overline{PR} = 4/3$. By expressing slope $\overline{PR}$ in terms of x and y, we obtained the desired equation.

The preceding example is typical, and the result can be stated as a theorem that can be proved by exactly the same argument as that of Example 8.12.

THEOREM *Point–Slope Form of the Equation of a Line*
The equation of the line through $P(x_1, y_1)$ and having slope m is

$$y - y_1 = m(x - x_1).$$

This is called the **point–slope form** of the equation of a line.

As this theorem suggests, there are several forms of the equation of a line. Another particularly useful form is stated in the next theorem. First we note that if a line crosses the y-axis at the point $(0, b)$, b is called the ***y*-intercept** of the line.

THEOREM *Slope–Intercept Form of the Equation of a Line*
The **slope–intercept form** of the equation of a line is

$$y = mx + b,$$

where m is the slope and b is the y-intercept.

Proof Since b is the y-intercept, the line passes through the point $(0, b)$. Also, it has slope m. Therefore, by the point–slope form of the equation of a line, the desired equation is

$$y - b = m(x - 0)$$

or, equivalently,

$$y = mx + b$$

as claimed.

The theorem just proved started with a line in the plane, defined by its slope m and its y-intercept, and deduced that its y-values are given by the function $y = f(x) = mx + b$. In the opposite direction, we see that the graph of any function of this form is a line of slope m that intersects the y-axis at $y = b$. For example, the graph of the function $f(x) = -\frac{x}{2} + 3$ has slope $-\frac{1}{2}$ and intersects the y-axis at $y = 3$. It has now become clear why functions of the form $f(x) = mx + b$ are called linear functions: their graphs are straight lines.

Linear functions have this important characterizing property: Since $y_2 - y_1 = m(x_2 - x_1)$for a constant m, the change $y_2 - y_1$ in the output values of the function is proportional to the corresponding change $x_2 - x_1$ of input values.

EXAMPLE 8.13 USING THE SLOPE–INTERCEPT FORM OF THE EQUATION OF A LINE

Write the equations of the following lines with the slope and y-intercept as indicated. Also, draw each line on a coordinate system.

(a) $m = -3, b = 5$
(b) $m = 0, b = -4$

Solution

(a) Using the above theorem, we obtain the equation $y = -3x + 5$. To draw the line, we plot the point $(0, 5)$ and the point $(1, 2)$, which is 3 units *below* and 1 unit to the right of $(0, 5)$. Then we draw the line through these two points.

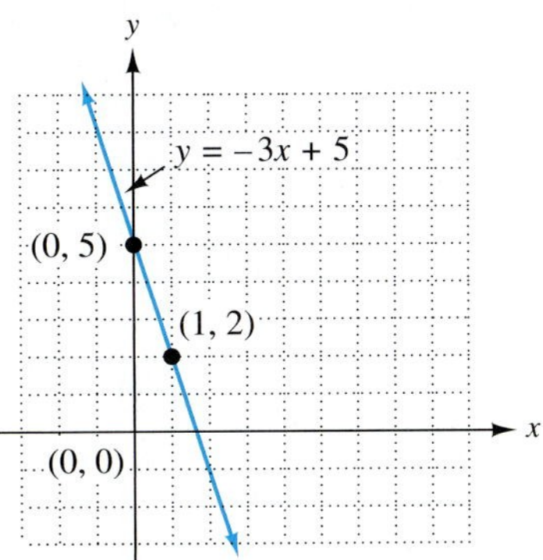

(b) This line goes through the point $(0, -4)$ and has slope 0. Therefore, using the slope–intercept form, we obtain the equation

$$y = 0x + (-4),$$

or just

$$y = -4.$$

This says that the line is horizontal and that a point is on this line if, and only if, its y-coordinate is -4. The x-coordinates of these points are unrestricted. The line is as shown.

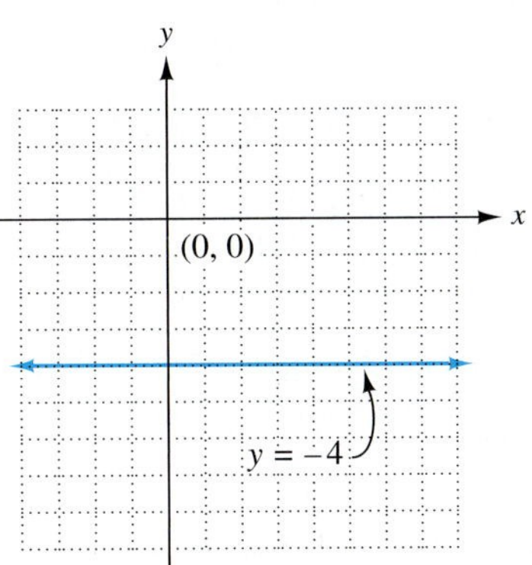

The preceding example suggests that *every* horizontal line has an equation of the form $y = b$; the slope is 0 and the x-values of points on the lines are left unrestricted. By analogy, *every* vertical line has an equation of the form $x = a$; there is *no* slope and the y-values of points on the lines are left unrestricted.

EXAMPLE 8.14 DETERMINING THE EQUATION OF A LINE THROUGH TWO POINTS

Determine the equation of the line through $P(-2, 3)$ and $Q(6, 7)$.

Solution

The slope of the desired line is the slope of $\overline{PQ}$:

$$\text{slope } \overline{PQ} = \frac{7 - 3}{6 - (-2)} = \frac{4}{8} = \frac{1}{2}.$$

We now use the point–slope form of the equation of the line, using either P or Q. Using $P(-2, 3)$, we have

$$y - 3 = \frac{1}{2}(x - (-2)), \text{ or, equivalently, } y - 3 = \frac{1}{2}(x + 2).$$

This equation can be rewritten in the form $y = \frac{1}{2}x + 1 + 3 = \frac{1}{2}x + 4$, which from the slope–intercept form tells us that the line intersects the y-axis at $y = 4$. Using the point $Q(6, 7)$, we have

$$y - 7 = \frac{1}{2}(x - 6).$$

This equation can also be rewritten in slope–intercept form, becoming $y = \frac{1}{2}x - 3 + 7 = \frac{1}{2}x + 4$, just as before. It makes no difference which point, P or Q, is used.

The equation of the line through any two points $P(x_1, y_1)$ and $Q(x_2, y_2)$ can be found in the same way as shown in Example 8.14. Assuming $\overline{PQ}$ is not a vertical segment, it has the slope given by $m = \frac{y_2 - y_1}{x_2 - x_1}$. The point–slope equation of the line, using point $P(x_1, y_1)$, is $y - y_1 = m(x - x_1)$, so we have the following theorem.

THEOREM *Two-Point Form of the Equation of a Line*
The equation of the line through $P(x_1, y_1)$ and $Q(x_2, y_2)$, where $x_1 \neq x_2$, is

$$y - y_1 = \left(\frac{y_2 - y_1}{x_2 - x_1}\right)(x - x_1).$$

This is called the **two-point form** of the equation of a line.

School Book Page

Graphing in Grade Five

Chapter 12 Lesson 3

Exploring Equal Ratios

Problem Solving Connection
- Make a Table
- Look for a Pattern

Materials
grid paper

Vocabulary
coordinates
an ordered number pair used in graphing

Explore

Liz keeps herself busy. She enjoys making porcelain dolls and their clothes. She also swims each week.

Liz is from Marcus, Iowa.

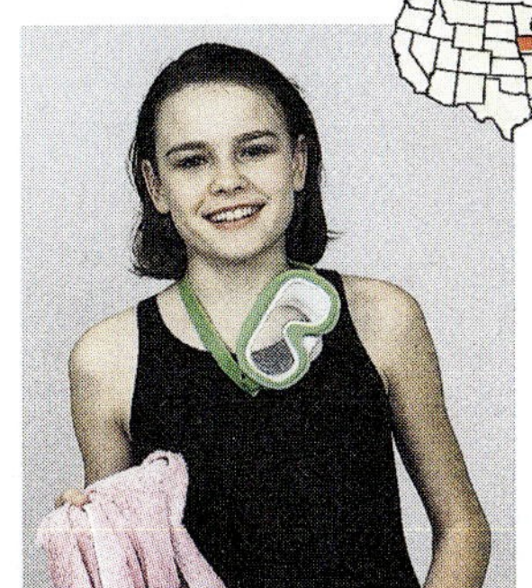

Work Together

Liz swims one day a week. She works on her dolls three days a week.

1. What is the ratio of doll-making days to swimming days in one week?
2. Copy and complete the ratio table that shows this ratio up to 6 weeks. Show each ratio as an ordered pair.

Ratio of Doll-Making Days to Swimming Days						
Week	1	2	3	4	5	6
Doll-Making	3	6		12		
Swimming	1					
Ordered Pairs	(3,1)					

3. Create a graph and plot the **coordinates** in the ratio table. What will you label each axis on this graph?

Did You Know?
Fred Newton swam the longest distance on record in 1930. He swam 1,826 mi down the Mississippi River.

Talk About It

4. What happens when ordered pairs for equal ratios are graphed? What kind of pattern do you see?
5. Explain why the ratios $\frac{3}{2}, \frac{6}{4}, \frac{9}{6}, \frac{12}{8}$, and $\frac{15}{10}$ would lie on the same line if they were graphed.

532 Chapter 12 • Ratio, Percent, and Probability

SOURCE: Scott Foresman–Addison Wesley Math, *Grade 5*, p. 532, by Randall I. Charles et al.

Questions for the Teacher

1. This lesson emphasizes an important relationship between equal ratios and graphing. Describe the graph the students will obtain in response to item 3 above. What would be the case if all these points were connected by line segments? If (x, y) were a number pair associated in the table, determine an equation relating x and y.
2. This lesson can be connected to the game Guess My Rule (see problem 5, Problem Set 1.3 and Example 8.7 of this chapter). Max is "it" and, in the following set of ordered pairs, the first entries are the numbers the students chose and the second entries are the corresponding numbers Max gave back: $(1, 5)$, $(4, 14)$, $(0, 2)$, $(3, 11)$, and $(5, 17)$. Graph the ordered pairs and give Max's rule. Explain the connection between Max's rule, the graph, and the above lesson.

The condition $x_1 \neq x_2$ in the theorem ensures that the given points P and Q are not on a vertical line, and therefore the line through P and Q has a defined slope. To include the possibility of a vertical line in the plane, we consider the **general equation** of the line,

$$Ax + By + C = 0,$$

where A, B, and C are constants and A and B aren't both 0. When $B = 0$, this equation becomes $Ax + C = 0$, so $x = -\frac{C}{A}$. That is, the solution set of the general equation is the set of points on the vertical line that intersects the x-axis at the point $\left(-\frac{C}{A}, 0\right)$. If $B \neq 0$, the general equation can be rewritten in the equivalent form

$$y = mx + b,$$

where $m = -\frac{A}{B}$ and $b = -\frac{C}{B}$. In this case, the solution set is the graph of the linear function $f(x) = mx + b$. For example, the general equation $3x - 12y - 5 = 0$ can be rewritten as $y = \frac{3}{12}x - \frac{5}{12} = \frac{1}{4}x - \frac{5}{12}$. We now see that the line has slope $\frac{1}{4}$ and intersects the y-axis at $-\frac{5}{12}$.

Graphing Nonlinear Functions

We have seen that it is unexpectedly easy to graph a linear function, since we have discovered that its graph is a straight line. For example, the linear function $f(x) = -2x + 7$ can be graphed by observing that the two points $P(0, 7)$ and $Q(1, 5)$ are on the graph of the function. The rest of the graph is obtained by drawing the line through P and Q.

A function *not* of the form $f(x) = mx + b$ is called a **nonlinear function.** Some examples of nonlinear functions are $g(x) = x^2 - 3x + 5$, $h(x) = \frac{x}{x^2 + 2}$, and $k(x) = 2^x$. The graphs of these functions, each created with a graphing calculator, are shown in Figure 8.11.

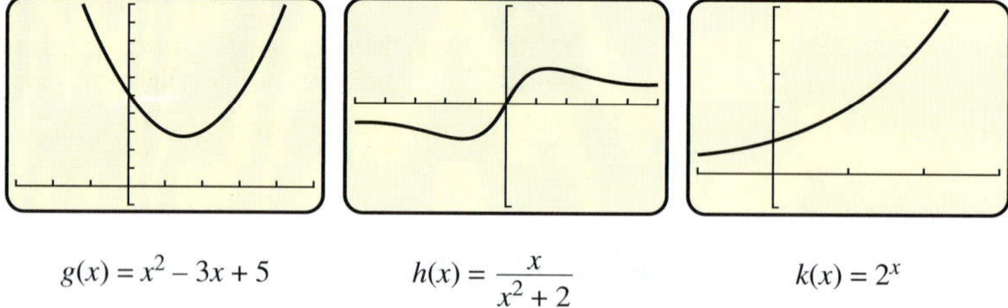

$g(x) = x^2 - 3x + 5$ $h(x) = \frac{x}{x^2 + 2}$ $k(x) = 2^x$

Figure 8.11
The graphs of three nonlinear functions

As expected, none of the graphs is a straight line.

To graph nonlinear functions by hand, it is necessary to make a table of *many* points—not just two!—that are on the graph. Plotting these points will then suggest the shape of the graph that can be approximated by smoothly connecting the points with a curve. This procedure is used in Example 8.15.

EXAMPLE 8.15 | MAKING THE BIGGEST ANIMAL PENS

The third-grade class wants to make pens for its pet rabbits and guinea pigs. A parent has donated 24 feet of chain-link fencing material, which will be used to make two side-by-side rectangular pens along a wall, as shown in the figure. What dimensions of x and w will give the pens their largest total area?

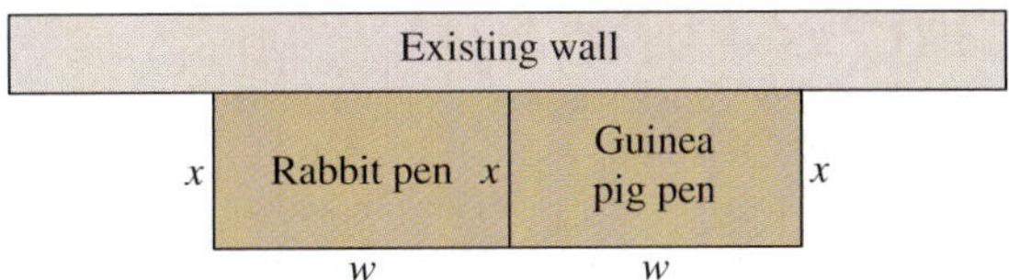

Solution As seen in the figure, the pens form a rectangle x feet wide and $2w$ feet long. Therefore, the total area of the pens is given by $A = 2wx$. The pens will require $3x + 2w$ feet of fencing. Since 24 feet of fencing is available, this gives the equation $3x + 2w = 24$, which can be rewritten as $2w = 24 - 3x$. Thus, the area A of the pens is given by the following equation of x.

$$A = 2wx = (24 - 3x)x = 24x - 3x^2.$$

We can now see how the total area A of the pens is given as the value of the function $f(x) = 24x - 3x^2$. Since the variable x is a length of a side of the pen, we must have $x \geq 0$. Also, the three sides perpendicular to the wall, each of length x, must use less than the 24 feet of available fence, so $3x \leq 24$, or $x \leq 8$. Therefore, the domain of the function is $D = \{x | 0 \leq x \leq 8\}$.

The following table of values of the function indicates that the largest area occurs when $x = 4$. The total area of the two pens is 48 square feet and $2w = 24 - 3 \cdot 4 = 12$. Therefore, $w = 6$, so each pen is 4 feet by 6 feet.

x	0	1	2	3	4	5	6	7	8
Area A	0	21	36	45	48	45	36	21	0

The following graph of the function $y = 24x - 3x^2$ also shows that the maximum area occurs when $x = 4$.

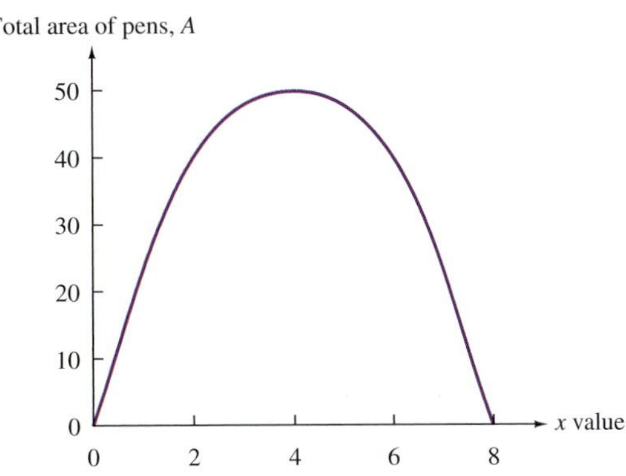

The table of values of a function and its graph can be created easily with a graphing calculator. The function in Example 8.15 is also discussed in the Window on Technology box in this section.

The Open-Topped Box Problem

Materials

Each group of four to seven students needs eight sheets of centimeter-squared grid paper, scissors, tape, and two sheets of graph paper.

Directions to Make an Open-Topped Box from a Rectangle

Each group will make open-topped boxes by cutting x-by-x-sized squares from the corners of a 16-by-21-centimeter rectangle cut from the centimeter-squared grid paper. Fold the rectangular sides of the box upward and tape the corners to create the box. The group should make a set of boxes corresponding to $x = 1, 2, 3, 4, 5, 6$, and 7 centimeters.

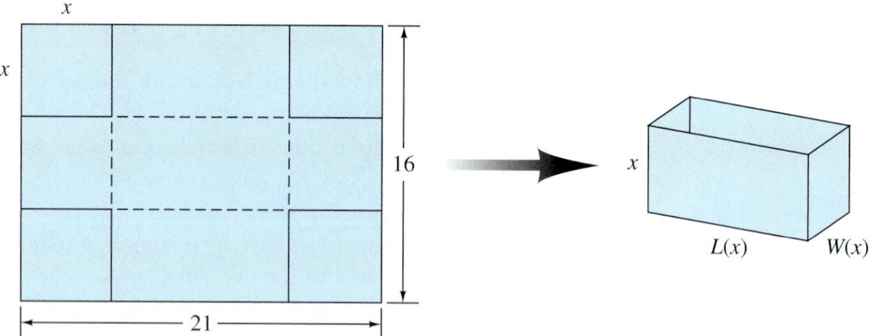

Activities and Questions

1. Guess which box has the largest volume. Which box apparently has the smallest volume?
2. Fill in the table below. In the last two rows, compute the area $B(x)$ of the bottom of your box and the total area $S(x)$ of the sides of your box.

x	1	2	3	4	5	6	7
L							
W							
V							
B							
S							

3. Divide your group into two subgroups. One subgroup graphs the volume of the box as a function of x for $1 \leq x \leq 7$ and, on the basis of its graph, decides which value of x results in the open-topped box of largest volume. The other subgroup graphs the two functions $B(x)$ and $S(x)$ (on the same set of axes) and reports on the value of x for which $B(x) = S(x)$. As a group, report a connection on the two reported values of x from the subgroups.
4. (If graphing calculators are available) Express each of $V(x)$, $B(x)$, and $S(x)$ as functions of x. Then enter these functions on a graphing calculator to graph the functions $V(x)$, $B(x)$, and $S(x)$.

Investigating Functions with a Graphing Calculator

A calculator of any type is helpful to compute the values of a function $y = f(x)$ for various choices of the domain variable x. For dealing with more complex functions, however, the graphing calculator provides a truly powerful tool for problem solving and classroom demonstrations. An introduction to the graphing calculator is found in Appendix C.

Functions are entered into the graphing calculator using the [Y =] key to open the function editor. Once a function is entered, set the x- and y-ranges of values by using the [WINDOW] key. Finally, display the graph by pressing the [GRAPH] key. Further explorations of the function are aided with the [TRACE] and [ZOOM] features of the calculator.

As an example, consider the function $f(x) = x(24 - 3x)$. This function arose in Example 8.15, where the values of the function represent the total area of the two animal pens being constructed. Since the area can't be negative, the function will be defined on the domain $0 \le x \le 8$. The three basic steps to graphing the function are shown in these three screens.

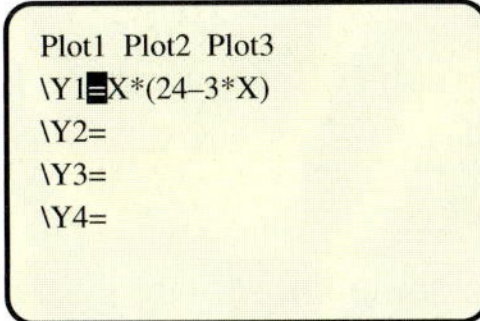

Press [Y =] *and enter the formula of the function.*

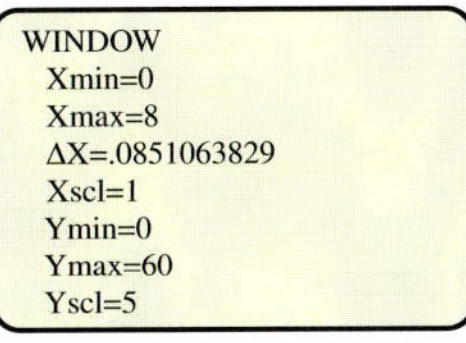

Press [WINDOW] *and enter the range of x and y values to be shown on the graph.*

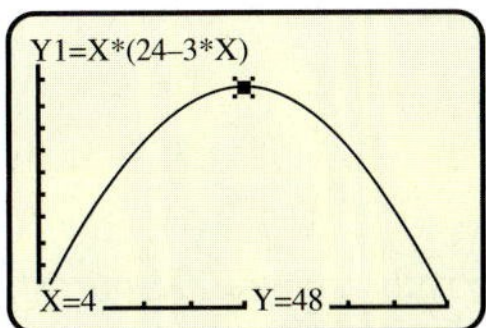

Press [GRAPH] *to display the graph.* [TRACE] *is used to find the maximum area.*

It is also easy to make a table of values of the function. Press [2nd] [TBLSET] to set up the table parameters, and then press [2nd] [TABLE] to see the table of values of the function.

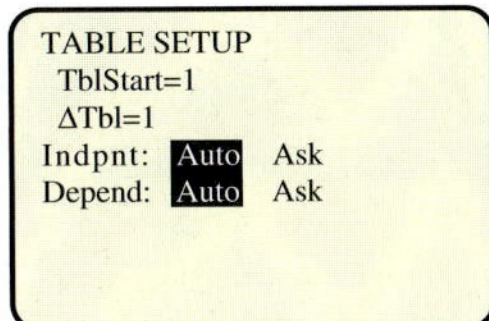

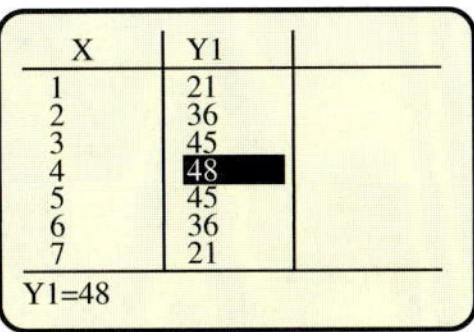

The graphing calculator makes it a simple process to see that the maximum total area of the two pens is 48 square feet and occurs when x is 4 feet.

Problem Set 8.3

Understanding Concepts

1. Plot and label the following points on a Cartesian coordinate system drawn on a sheet of graph paper.
   (a) $P(5, 7)$ (b) $Q(5, -7)$ (c) $R(-5, 7)$
   (d) $S(-5, -7)$ (e) $T(0, 5)$ (f) $U(7, 1)$
   (g) $V(0, -5.2)$ (h) $W(-7, 1)$ (i) $X(0, 0)$
2. Plot these points and connect them in order with line segments: $(1, 1)$, $(1, 11)$, $(4, 13)$, $(5, 15)$, $(6, 13)$, $(7, 12)$, $(10, 11)$, $(9, 10)$, $(6, 9)$, $(4, 7)$, and $(1, 1)$.
3. (a) Plot the points $(5, 0)$, $(4, 3)$, $(3, 4)$, $(0, 5)$, $(-3, 4)$, $(-4, 3)$, $(-5, 0)$, $(-4, -3)$, $(-3, -4)$, $(0, -5)$, $(3, -4)$, and $(4, -3)$.
   (b) What do you observe about the points in part (a)?
4. Compute the distance between these pairs of points.
   (a) $(-2, 5)$ and $(4, 13)$ (b) $(3, -4)$ and $(8, 8)$
   (c) $(0, 7)$ and $(8, -8)$ (d) $(3, 5)$ and $(2, -4.3)$
5. (a) Prove that $R(1, 2)$, $S(7, 10)$, and $T(5, -1)$ are the vertices of a right triangle. (*Hint:* Show that the square of one side is the sum of the squares of the other two sides.)
   (b) Draw the triangle RST of part (a) on graph paper.
6. Compute the slopes of the line segments determined by these pairs of points. In each case tell whether the segment is vertical, horizontal, slopes upward to the right, or slopes downward to the right.
   (a) $P(1, 4)$, $Q(3, 8)$
   (b) $R(-2, 5)$, $S(-2, -6)$
   (c) $U(-2, -3)$, $V(-4, -7)$
   (d) $C(3, 5)$, $D(-3, 5)$
   (e) $E(1, -2)$, $F(-2, -5)$
   (f) $G(-2, -2)$, $H(4, -5)$
7. Determine b so that the slope of $\overline{PQ}$ is 2, where P and Q are the points $(b, 3)$ and $(4, 7)$, respectively.
8. Determine d so that the slope of $\overline{CD}$ is undefined if C and D are the points $(d, 3)$ and $(-5, 5)$, respectively.
9. Determine a if the point $(a, 3)$ is on the line $2x + 3y = 18$.
10. (a) Determine two different points on the line $3x + 5y + 15 = 0$.
    (b) Draw the graph of the line in part (a) on a coordinate system.
11. Graph each of these lines on a single coordinate system and label each line.
    (a) $3x + 5y = 12$ (b) $6x = -10y + 12$
    (c) $5y - 3x = 15$ (d) $6x + 10y = 24$
    (e) What do you conclude about the lines of parts (a) and (d)?
12. (a) Determine two points on the line $4x + 2y = 6$.
    (b) Use the points determined in part (a) to compute the slope of the line.
    (c) Solve the equation in part (a) for y in terms of x and thereby again determine the slope of the line, as well as the y-intercept. (*Hint:* Solving for y in terms of x gives the slope–intercept form of the equation of a line.)
13. In each case determine k so that the line is parallel to the line $3x - 5y + 45 = 0$.
    (a) $7x + ky = 21$ (b) $kx - 8y - 24 = 0$
    (c) $y = kx + 5$ (d) $x = ky + 5$
14. Draw the graphs of each of these linear functions.
    (a) $y = 2x - 3$ (b) $y = 0.5x + 2$
    (c) $y = -3x$
15. (a) On the same coordinate system draw the graphs of these three linear functions: $y = 4x$, $y = 4x + 5$, and $y = 4x - 3$.
    (b) Briefly discuss the graphs in part (a).
16. (a) Draw the graph of $3x + 2y + 6 = 0$.
    (b) Does the equation of part (a) define y as a function of x? If so, identify the function.
    (c) Draw the graph of the equation $5x - 3y - 15 = 0$.
    (d) Does the equation of part (c) define y as a function of x? If so, identify the function.
    (e) Does the equation of part (c) define x as a function of y? If so, define the function.
17. (a) On a single set of axes, draw the graphs of $y = 2x + 3$, $y = 2(x - 3) + 3$, and $y = 2(x + 4) + 3$.
    (b) Compare the graphs of part (a).
18. Graph each of these functions.
    (a) $y = x^2$ (b) $y = (x - 2)^2$ (c) $y = (x + 3)^2$
    (d) Discuss the relationship between the graphs of parts (a), (b), and (c).
19. Graph each of these functions.
    (a) $y = x^2 - 4x + 4$
    (b) $y = x^2 + 6x + 9$
    (c) $y = x^2 + 4x + 4$
    (d) Could you write each of these functions in a different, more concise form? (*Hint:* Consider problem 18.)

20. **(a)** On a single set of axes, graph the equations $y = x^2$, $y = x^2 + 4$, and $y = x^2 - 3$.

(b) Compare the graphs of part (a).

21. Use graphing techniques to find

(a) the minimum value of the quantity $x^2 + 10x$ and the value of x at which it occurs.

(b) the maximum value of $8x - 2x^2$ and the value of x at which it occurs.

(c) the minimum value of $2x^2 - 4x + 10$ and the value of x at which it occurs.

Teaching Concepts

22. Kristi has a large detailed map showing the winding 6-mile-long trail up to the top of Pyramid Peak. She knows she started up the trail at 8 A.M., hiked at a steady rate, and reached the summit at noon. Kristi would like to make a graph of her distance up the trail during the hike, so she can see on the map where she was at each hour and half hour along the trail. There are mileage markers on the map, but she is confused about how the graph will look because the trail was very crooked. Carefully describe how you would work with Kristi to help her accomplish her goals.

23. Children often think all graphs are linear. Devise an activity where measurements are made and entered into a table, and graphs are made which clearly show that some data vary linearly and other related data vary nonlinearly.

Thinking Critically

24. Find an equation for each of the lines **(a)**, **(b)**, **(c)**, and **(d)** shown on the coordinate system below.

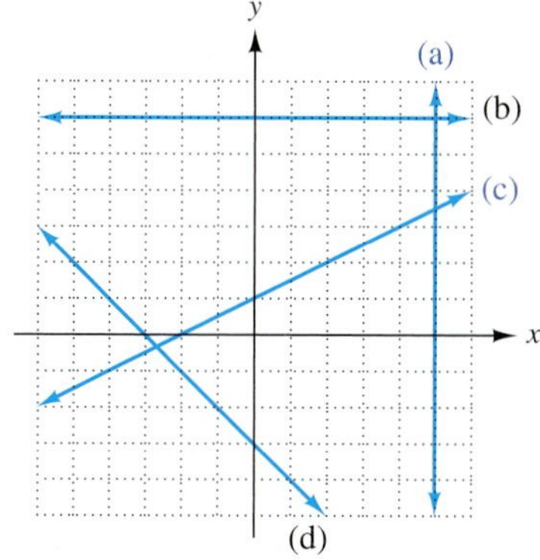

25. If $A(0, 0)$, $B(3, 5)$, $C(r, s)$, and $D(7, 0)$ are the vertices of a parallelogram, determine r and s.

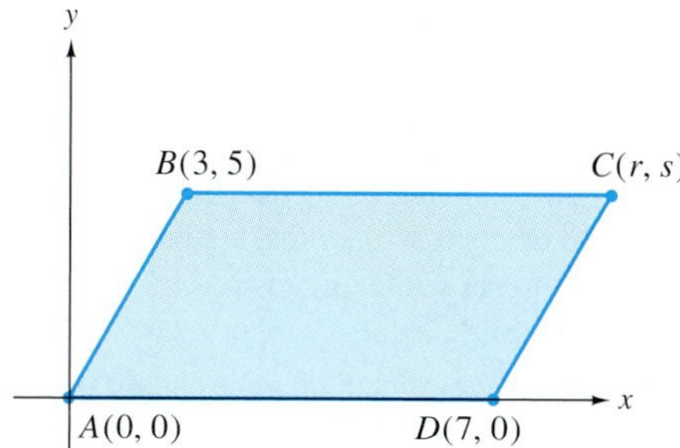

26. **Slopes of Perpendicular Lines.** Let the line l_1 shown below be neither vertical nor horizontal. The triangle along l_1 with a horizontal leg of length a and a vertical leg of length b shows that the line l_1 has slope $m_1 = \frac{b}{a}$. Line l_2 is perpendicular to l_1 and can be viewed as the line obtained by rotating line l_1 about the intersection point P through 90°.

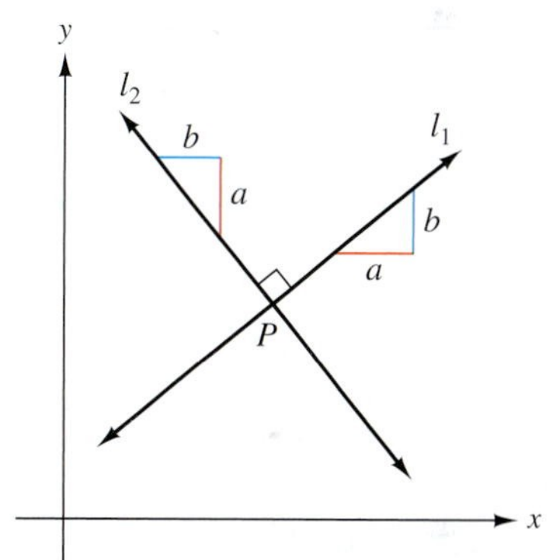

(a) Find the slope m_2 of line l_2 and show that $m_1 = -\frac{1}{m_2}$ or, equivalently, $m_1 m_2 = -1$.

(b) If l_1 is vertical or horizontal, what can be said about the slope of any line perpendicular to l_1?

27. In each part below, you are given an equation of a line and a point. Find the equation of the line through the given point that is perpendicular to the given line. (The slope of the perpendicular line is the negative reciprocal of the slope of the given line if the given line is neither vertical nor horizontal, as described in problem 26.)

(a) $y = 2x$, $P(0, 0)$

(b) $y = 3x + 5$, $Q(1, 2)$

(c) $y = -\frac{2}{3}x + 7$, $R(4, -1)$

(d) $2y + 6x - 5 = 0$, $S(0, 3)$

28. A square $ABCD$ in the coordinate plane has vertices at $A(2, 3)$, $B(6, 2)$, $C(r, s)$, and $D(u, v)$. Find all possible choices of r, s, u, and v.

29. Suppose the third graders in Example 8.15 want to make a third pen for baby chicks, still using the 24 feet of fence. That is, there are now to be three side-by-side pens along the wall, each measuring x feet by w feet. Draw a figure showing the pens along the wall, with appropriate labels, and then make a table of values and plot a graph to show the dimensions that give the pens the largest possible area.

30. A square in the coordinate plane is shown below, with its vertices at $A(1, 0)$, $B(1, 1)$, $C(0, 1)$, and $D(0, 0)$. Since the squared distance between points $P(x, y)$ and $Q(a, b)$ is given by the formula $PQ^2 = (x - a)^2 + (y - b)^2$, we see that $PA^2 = (x - 1)^2 + (y - 0)^2 = x^2 - 2x + 1 + y^2$ is the squared distance between P and vertex A of the square.

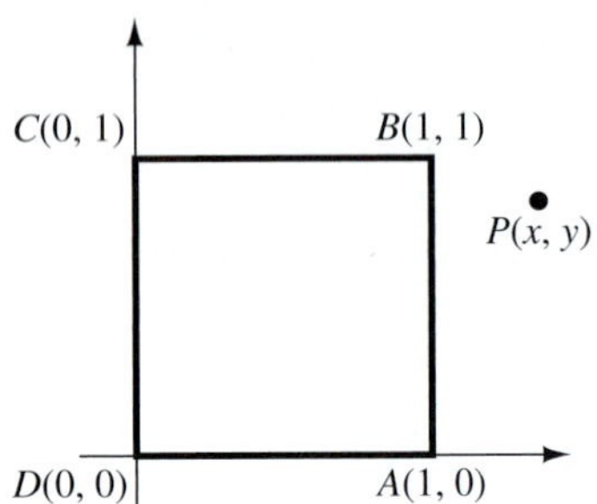

(a) Find PB^2, PC^2, and PD^2.

(b) Describe the set of points P in the coordinate plane for which $PA^2 + PC^2 = PB^2 + PD^2$. (Yes, the answer is a surprise!)

31. **The Witch of Agnesi.** Maria Agnesi (see the latter Highlight from History box in this section) introduced an interesting curve in the Cartesian plane. The curious name of the curve came about when an Italian word was incorrectly translated as "witch." The curve can be drawn on a piece of graph paper this way: Use a compass to draw a circle of radius 2 centered at the point $(0, 2)$. Then draw a horizontal line through the point $(0, 4)$ where the circle crosses the y-axis. The point P on the curve is constructed by drawing a vertical line segment between the two horizontal lines and joining the upper endpoint U of the vertical segment to the origin $(0, 0)$. Finally, draw the horizontal line through the point V where the segment intersects the circle. The intersection of the vertical line through U and the horizontal line through V gives point P on Agnesi's curve. More points on the curve, such as Q and R, are constructed in the same way as P.

(a) Use graph paper, a ruler, and a compass to construct points along the curve above the x-values $-7, -6, -5, \ldots, 0, \ldots, 6, 7$. Connect your

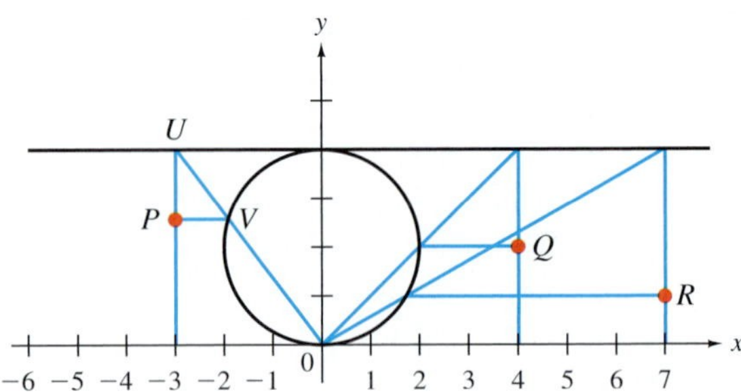

points with a smooth curve that gives you an accurate "Witch of Agnesi."

(b) Make a table of x- and y-values of the function $f(x) = \dfrac{64}{16 + x^2}$ for x-values $-7, -6, -5, \ldots, 0, \ldots, 6, 7$. Plot these points on a new sheet of graph paper, connecting the points to create a smooth curve representing the graph of the function.

(c) Compare the curves made in parts (a) and (b). What would you guess is the name of your curve in part (b)?

32. The water tanks I, II, III, and IV are shown as follows:

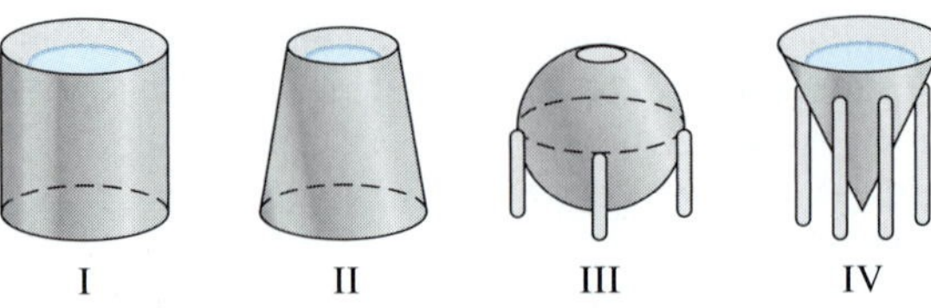

Match each tank to its corresponding graph that shows how the height h of water (plotted on the vertical axis) depends on the volume v of water (plotted on the horizontal axis) contained in the tank.

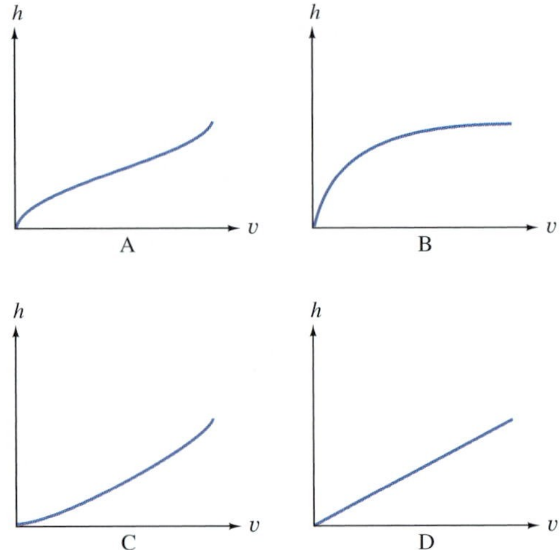

Thinking Cooperatively

33. Follow the directions given in this section's Cooperative Investigation activity, The Open-Topped Box Problem, but this time start with a 12-centimeter by 12-centimeter square cut from centimeter-squared grid paper. Work in small groups to cut small x-by-x squares from the corners of the starting square to create an open-topped box x centimeters high.

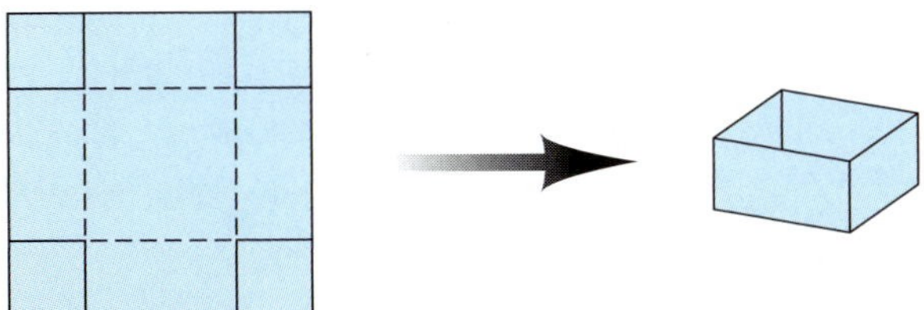

(a) Within your group, make boxes corresponding to $x = 1, 2, 3, 4,$ and 5 centimeters. Guess which box has the largest volume.

(b) Derive a formula for the volume $V(x)$ of the box as a function of x, and make a graph of the volume function. On the basis of your graph, what value of x will give the box of largest volume?

(c) Derive a formula for the area $B(x)$ of the bottom of the box and the total area $S(x)$ of the sides of the box. Plot the graphs of $B(x)$ and $S(x)$ on the same axes, over the domain $0 \leq x \leq 6$. At what value of x is $B(x) = S(x)$?

(d) Do you see a connection between your answers to parts (b) and (c)? Discuss within your group, and form a careful description of your observations.

Making Connections/Using a Calculator

34. (a) ADA-Approved Ramps. The Americans with Disabilities Act states, "The maximum slope of a ramp in new construction shall be 1:12. The maximum rise for any run shall be 30 in." What is the minimum amount of run for a rise of 30 inches?

(b) Highway Design. Highway 195 into Lewiston, Idaho, undergoes an elevation difference of 1800 feet in 7 miles. What is the average percent slope of the Lewiston grade?

35. World Population. The table below gives the world population (in billions) every 20 years since 1900, and an estimate of the population for 2020.

Year	1900	1920	1940	1960	1980	2000	2020
Pop.	1.6	1.9	2.3	3.0	3.7	6.0	7.6

(a) Make a graph of the world population as given in the table.

(b) Letting x denote the number of decades (10-year periods) since 1900, plot the graph of the function $y = (1.45)1.14^x$ on the graph drawn in part (a). Does the formula mimic the population data well in your opinion?

(c) If a graphing calculator is available, use it to plot the table and the function given in part (b) on the same screen to compare the two graphs.

36. Jennifer has a new car and is curious what gas mileage she can expect. She has kept records for the last five fill-ups, showing the miles traveled and the gallons of gas purchased.

Miles	165	263	232	113	204
Gallons	8.5	13.2	11.8	6.4	9.6

(a) Make a plot of the five data points, with gallons on the horizontal axis and miles on the vertical axis.

(b) Use a ruler to draw the line that seems to best fit the data. This line is called the **linear regression line.**

(c) Estimate the slope of the regression line in part (b) to determine Jennifer's gas mileage.

(d) Linear Regression with a Graphing Calculator. Enter the gallons and miles data into two lists, L1 and L2, on a graphing calculator (consult the User's Guide for your calculator as needed). Plot the five data points using the STATS PLOT menu. Then use commands under the STAT menu to have the calculator compute the regression line (i.e., the line of best fit). Graph the regression line and see how closely it fits the data. Were the graph of your line and its slope estimated by hand in part (c) close to the graph and slope determined by your calculator?

From State Student Assessments

37. (Colorado, Grade 4)
This is a game board.

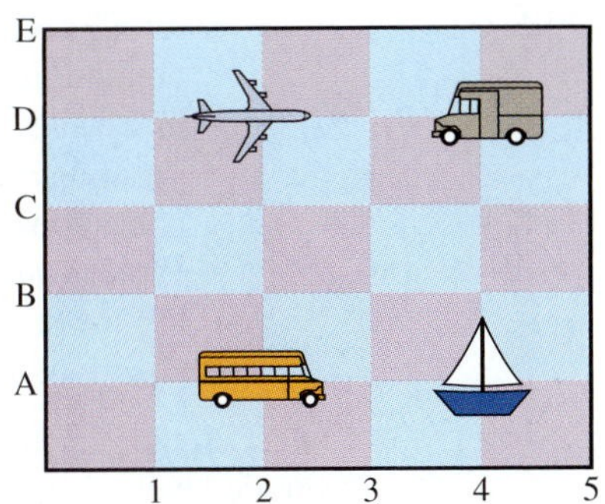

Which object is located at (2, D)?

A. The plane

B. The truck

C. The bus

D. The boat

38. (Colorado, Grade 4)
Look at the grid below. Point X is identified by the ordered pair (7, 6).

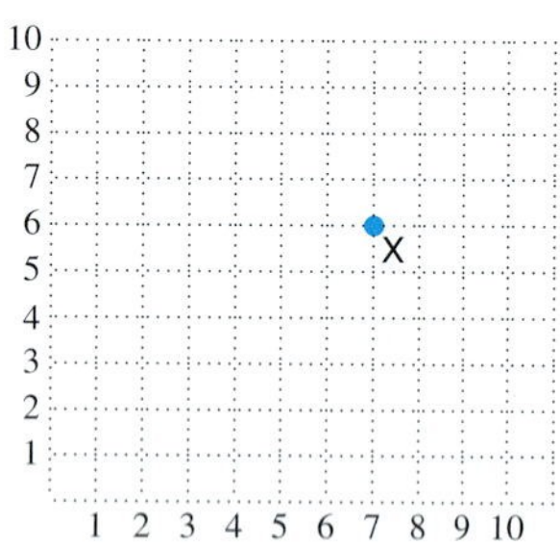

Which set of ordered pairs identifies three points that will form a straight line when connected?

A. (2, 1), (5, 5), and (8, 7)
B. (2, 7), (4, 5), and (7, 4)
C. (3, 2), (6, 5), and (9, 8)

For Review

39. Let $f(x) = 1/x$ be defined for all real numbers x in the interval $0.2 \le x \le 5$.

(a) Find $f(0.2)$, $f(1)$, $f(5/3)$, $f(4)$, and $f(5)$.

(b) If $f(r) = 4$, $f(s) = 0.5$, $f(t) = 2$, $f(u) = \frac{4}{7}$, and $f(v) = \frac{1}{\pi}$, find r, s, t, u, and v.

(c) What is the range of f?

40. Propose simple functions for the following Guess My Rule games.

(a)

Guess	5	2	−3	4	−2
Response	−11	−5	5	−9	3

(b)

Guess	5	2	−3	4	−2
Response	40	10	0	28	−2

Epilogue Developing Algebraic Reasoning

Algebraic reasoning and representation are fundamental to *all* levels of mathematics. The concepts and skills that have been developed in this chapter, together with the associated vocabulary and notations, are algebraic ideas that should be fostered in students in the elementary mathematics classroom, beginning with the earliest grades. The NCTM's *Principles and Standards for School Mathematics* states,

> In grades 3–5, algebraic ideas should emerge and be investigated as students—
>
> - identify or build numerical and geometric patterns;
> - describe patterns verbally and represent them with tables or symbols;
> - look for and apply relationships between varying quantities to make predictions;
> - make and explain generalizations that seem to always work in particular situations;
> - use graphs to describe patterns and make predictions;
> - explore number properties;
> - use invented notation, standard symbols, and variables to express a pattern, generalization, or situation.

Very similar algebraic ideas are expressed in the standards appropriate for the K–2 and 6–8 grade levels.

Chapter Summary

Key Concepts

Section 8.1 Algebraic Expressions and Equations

- Algebraic reasoning. Algebraic reasoning is used to solve problems and understand patterns by following these steps: introduce variables, derive algebraic expressions, form equations, solve equations, and interpret the solution of the equations in the context of the original problem or pattern.
- Variables. Variables represent quantities that are unknown, that can change, or that depend on varying choices of related quantities. In particular, variables describe generalized properties, express relationships, express formulas, and serve as unknowns.
- Algebraic expressions. An algebraic expression is a mathematical expression involving variables, numbers, and operation symbols.
- Equation. Setting two algebraic expressions that represent the same quantity equal to one another creates an equation.
- Solution of equations. The values of the variables for which the equations are true are the solutions of the equations and form a set called the solution set of the equations.

Section 8.2 Functions

- Function. A function is a rule that assigns exactly one value to each element x in a set D. The set D is the domain of the function. If the function is denoted by f, and x is an element of the domain D, then the value assigned to x is denoted by $f(x)$.
- Range. The range of a function is the set of all values assigned by the function.
- Representations of a function. Functions can be represented by the following means: a formula; a table of values; an arrow diagram; a machine into which x is input and $f(x)$ is output; a graph.
- Linear function. A function of the form $f(x) = mx + b$, where m and b denote constants.

Section 8.3 Graphing Functions in the Cartesian Plane

- Cartesian coordinates. The Cartesian coordinate system consists of two perpendicular axes, with the horizontal axis typically called the x-axis and the vertical axis typically called the y-axis. Any point $P(x, y)$ in the plane is uniquely described by its x and y coordinates.
- Distance formula. The distance between $P(x_1, y_1)$ and $Q(x_2, y_2)$ is given by $PQ = \sqrt{(x_2 - x_1)^2 + (y_2 - y_1)^2}$.
- Line segment. $\overline{PQ}$ denotes the line segment with endpoints P and Q.
- Slope. The slope of the line or line segment through the points $P(x_1, x_2)$ and $Q(y_1, y_2)$ is the ratio of the line's or line segment's "rise over run"; that is, slope $(\overline{PQ}) = \dfrac{x_2 - x_1}{y_2 - y_1}$ when $y_1 \neq y_2$. Slopes are not defined for vertical segments or lines. Lines or segments are parallel precisely when they have the same slope.
- Equation of a line. The equation of a nonvertical line can be given in point–slope, slope–intercept, and two-point forms. All lines, including vertical lines, can be described by the general form of the equation of a line, namely, $Ax + By + C = 0$.
- Nonlinear function. Nonlinear functions are graphed by making a table of values. These points are plotted and then connected by a smooth line to approximate the graph of the function. Graphing calculators and other technology can be effectively employed to make more accurate graphs.

Vocabulary and Notation

Section 8.1

Constant
Variable
  Domain of a variable
Algebraic expression
  Evaluation of an algebraic expression
Equation
  Conditional equation
  Solution set of an equation

Section 8.2

Function, f
  Domain of a function
  Value of a function at x, $f(x)$
  Range of a function
Representations of a function
  Formula, table of values, arrow diagram, machine, graph
Linear function, $f(x) = mx + b$

Section 8.3

Cartesian plane
  Coordinate axes (x-axis and y-axis)
  Origin
  Coordinates
  Quadrants
Distance formula
Line segment, $\overline{PQ}$
Line, $\overleftrightarrow{PQ}$
Slope of line segment, slope of line
Parallel lines
Equation of lines
  Point–slope form, $y - y_1 = m(x - x_1)$
  Slope–intercept form, $y = mx + b$
  Two-point form, $y - y_1 = \frac{y_2 - y_1}{x_2 - x_1}(x - x_1)$
  General, $Ax + By + C = 0$
Nonlinear function

Chapter Review Exercises

Section 8.1

1. Let a, b, and c denote the current ages of Alicia, Ben, and Cory, respectively. Write expressions in the variables a, b, and c that express the given quantity.

(a) Alicia's age in 5 years

(b) The fact that Ben is younger than Cory

(c) The difference in age between Ben and Cory

(d) The average age of Alicia, Ben, and Cory

2. Let a, b, and c denote the current ages of Alicia, Ben, and Cory as in problem 1. Write equations in the variables a, b, and c that express the given relationship.

(a) Alicia will be 11 years old in 2 more years.

(b) Three years ago, Ben was half of Cory's age last year.

(c) Alicia's age is the average of Ben's and Cory's ages.

(d) The average age of Alicia, Ben, and Cory is 10.

3. A "hex–square train" is made with toothpicks, alternating between hexagonal and square "cars" in the train. For example, three-car and four-car hex–square trains are shown in the next column.

(a) Evidently 14 toothpicks are required to form a three-car train. Give a formula for the number of toothpicks required to form an n-car train. (*Suggestion:* Take separate cases for trains with an odd or an even number of cars.)

(b) A certain hex–square train requires 102 toothpicks. How many hexagons and how many squares are in the train?

Section 8.2

4. For each table, decide whether it is the table of a function $y = f(x)$. If not, state why not. If so, give the domain and range.

(a)

x	3	6	7	8	3
y	4	3	5	1	0

(b)

x	3	6	7	8	4
y	4	3	5	1	2

(c)

x	3	6	7	8	4
y	4	3	3	4	3

5. Solve these two Guess My Rule games. Carefully describe a function that agrees with the table of values.

(a)

x	4	2	0	5	1
y	14	8	2	17	5

(b)

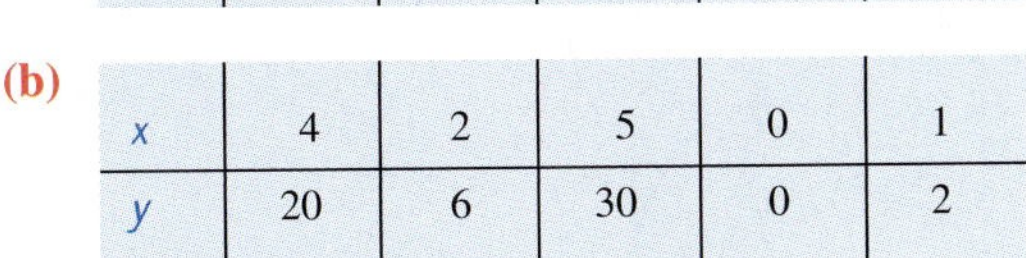

x	4	2	5	0	1
y	20	6	30	0	2

6. Let f be a function defined by the formula $f(x) = 2x(x - 3)$.

(a) Find $f(3)$, $f(0.5)$, and $f(-2)$.

(b) If $f(x) = 0$, what are the possible values of x?

7. When light passes into two face-to-face layers of glass, it is either reflected or transmitted when it strikes the glass surfaces. There is 1 path with no internal reflections, 2 paths with 1 internal reflection, and 3 paths with 2 internal reflections.

(a) Show that there are 5 paths with 3 internal reflections.

(b) Conjecture a formula that gives the number of paths that have n internal reflections.

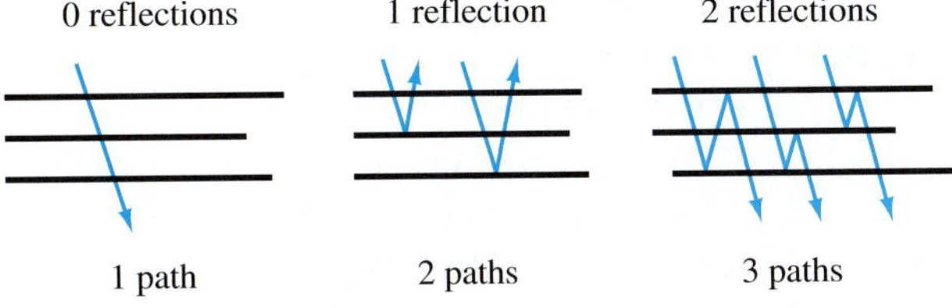

Section 8.3

8. (a) Plot the seven-sided polygon $ABCDEFG$ on graph paper, where the coordinates of the vertices (corners) of the polygon are $A(4, 2)$, $B(3, 3)$, $C(0, 3)$, $D(-2, 2)$, $E(-3, -2)$, $F(0, -2)$, and $G(4, 0)$.

(b) Point A is in the first quadrant, and point G is on the positive x-axis. Give similar descriptions for the other vertices of the polygon.

(c) The slope of side $\overline{AB}$ is -1, since the "rise over run" ratio is $\frac{-1}{1}$. Find the slopes of the other six sides of the polygon, when defined. Are there any parallel sides?

9. Plot the hexagon $ABCDEF$ on graph paper, where the vertices of the hexagon are $A(0, 0)$, $B(7, 0)$, $C(16, 12)$, $D(7, 24)$, $E(0, 24)$, and $F(-9, 12)$.

(a) Find the lengths AB, AC, AD, AE, AF, and CF.

(b) Draw all the diagonals of the hexagon, using the symmetry of the figure, and label the lengths of all the sides and diagonals. Why do you think this figure is called an "integer hexagon"?

10. Find an equation of each of the lines described below:

(a) The line through $R(3, 4)$ with slope 2

(b) The line through $T(6, -1)$ and $U(-2, 5)$

(c) The line of slope 3 that intersects the y-axis at $y = -4$

11. Make a table of values, plot the points in your table, and then connect the points to graph the function over the indicated domain.

(a) $f(x) = 8 + 2x - x^2$, for $-3 \le x \le 5$.

(b) $g(x) = \dfrac{2x^2 - 8}{x^2 + 2}$, for $-4 \le x \le 4$.

12. The Fibonacci number sequence is given by $F(1) = 1$, $F(2) = 1$, $F(3) = 2, \ldots$, where in general $F(n) = F(n - 1) + F(n - 2)$ for $n \ge 3$. That is, the Fibonacci numbers are defined on the domain of the positive integers.

(a) Discuss how the domain for the Fibonacci numbers can be enlarged to include 0 and the negative integers. For example, why is $F(0) = 0$ a good way to define the function at 0?

(b) Answer (a) again, but for the Lucas numbers $L(1) = 1$, $L(2) = 3$, $L(3) = 4, \ldots$, where $L(n) = L(n - 1) + L(n - 2)$.

Chapter Test

1. Solve each equation, clearly showing each step.
   (a) $5x - 7 = x + 9$
   (b) $\dfrac{5 + 7x}{5x - 2} = 2$
2. Consider the function $f(x) = 2x^2 + 4x$.
   (a) Graph the function.
   (b) Determine the smallest value of the function, and the corresponding value of x at which this minimum value occurs.
3. Classify each equation as linear or nonlinear. If linear, rewrite the equation in the form $f(x) = mx + b$ of a linear function and identify the slope of the line and the point at which the line intersects the y-axis.
   (a) $6x = 2y + 8$
   (b) $y = 3x^2 + 4$
   (c) $y - 2x = -x + 5$
4. Create algebraic expressions for the quantities described in each part. Be sure to first introduce and define appropriate variables.
   (a) The cost of monthly phone service from a company that charges \$5 per month plus 12¢ per minute
   (b) The monthly cost of using a checking account from a bank that charges \$8 per month and 15¢ per check
   (c) The cost of taking a group of adults and children to the fair, where the admission fee is \$4 for adults and \$2.50 for children
5. Write the equation of each line **(a), (b),** and **(c)** shown below.

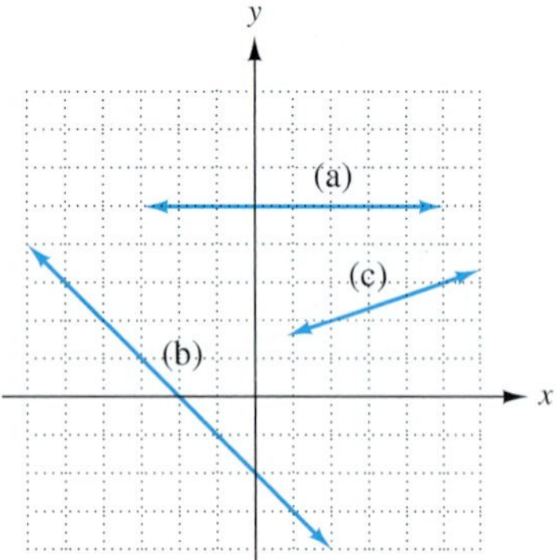

6. Match these graphs to the hike described. Each graph shows the elevation as a function of time into the hike.
   (a) A hike up Mount Shasta and return to base camp.
   (b) A round-trip hike into the Grand Canyon.
   (c) A hike along the wilderness beach in Olympic National Park.

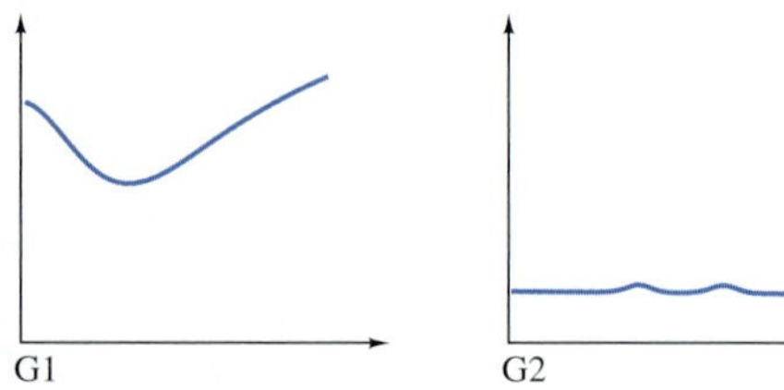

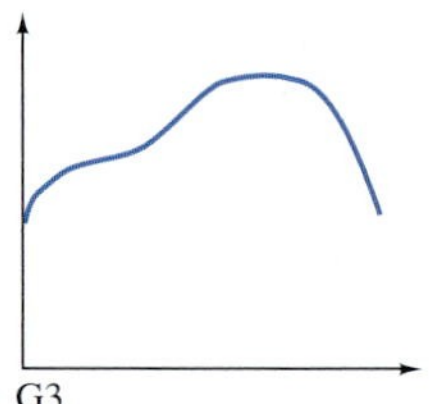

7. A bee starts in cell 1 of the honeycomb shown below and moves by crawling over cell walls. Assume that the bee always moves to the right, from one cell to one of the two adjacent cells.

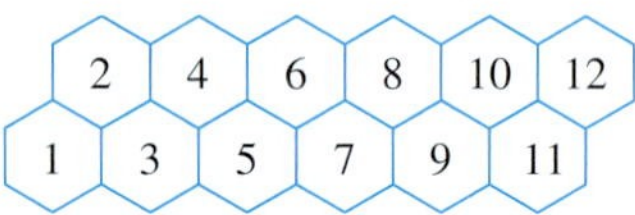

There is one path to cell 2, two paths to cell 3, and three paths to cell 4.
   (a) How many paths are there to cell 5?
   (b) Let F be the function for which $F(n)$ is the number of paths from cell 1 to cell n. Explain why $F(n + 2) = F(n + 1) + F(n)$ for $n = 1, 2, 3, \ldots$.
   (c) How many paths can the bee follow to cell 12?
8. Let $P(4, 2)$, $Q(7, 6)$, $R(3, 9)$, and $S(a, b)$ be the vertices of a square.
   (a) What are the coordinates a and b?
   (b) What is the length of each side of the square?
9. Let f be the function given by $f(x) = x^2 - 4x + 5$ on the domain $\{1, 2, 3, 4, 5\}$. Give the range of the function.
10. The following dot patterns show that n^2, the square of any positive integer n, can be written as the sum of two successive triangular numbers. For example, $1 = 1 + 0$, $4 = 3 + 1$, $9 = 6 + 3$, and in general $n^2 = t_n + t_{n-1}$, where $t_n = 1 + 2 + \cdots + n = \frac{1}{2}n(n + 1)$ denotes the nth triangular number.

Let d_n denote the number of dots in the nth diamond pattern, as shown to the right.

(a) Draw the patterns for $n = 5$ and 6 on graph paper, and then make a table of values of d_n for $n = 1, 2, 3, 4, 5$, and 6.

(b) Shade triangles in your figure to show that $d_5 = 1 + 4t_4$ and $d_6 = 1 + 4t_5$.

(c) Part (b) generalizes to show that $d_n = 1 + 4t_{n-1}$ for all n. Now show that $d_n = 1 + 2(n-1)n$.

(d) Shade triangles on your figure to show that $d_5 = t_5 + t_4 + t_4 + t_3$ and $d_6 = t_6 + t_5 + t_5 + t_4$.

(e) Part (c) generalizes to show that $d_n = t_n + t_{n-1} + t_{n-1} + t_{n-2}$. Now show that $d_n = n^2 + (n-1)^2$.

(f) Show that the two formulas obtained in parts (c) and (e) are equivalent.

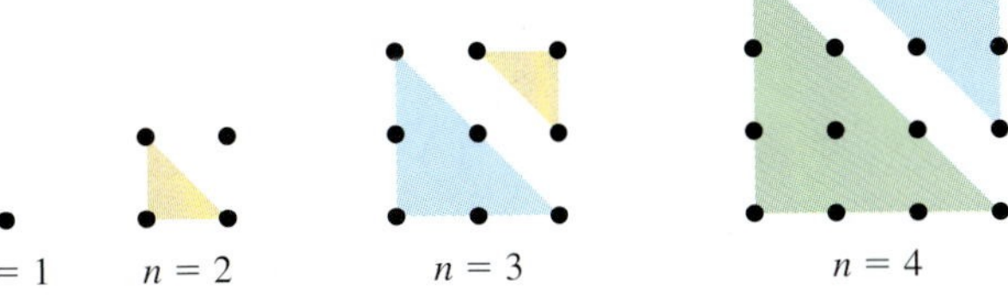

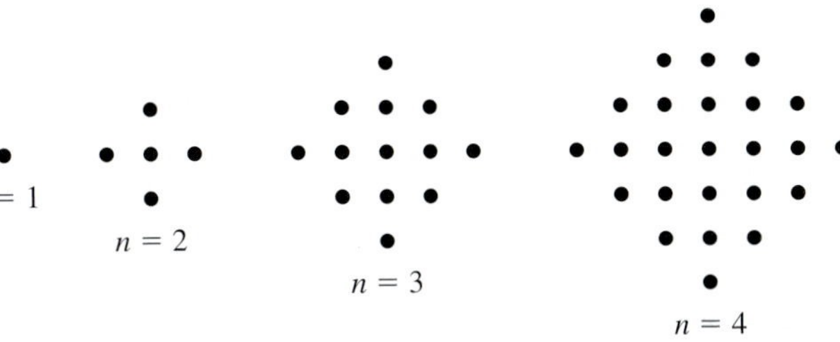

Chapter 9

Statistics: The Interpretation of Data

How Many Beans in the Bag?

Materials Needed

1. Three or four opaque bags containing about 200 beans each (this number being known only to the instructor), with the same number of beans in each bag.
2. A bright colored marker to go along with each bag.

Directions

1. Divide the class into groups of about ten students each, and give each group a bag of beans.
2. The students in each group mark 25 beans with their colored marker and return the marked beans to their bag.
3. The beans in the bag are then thoroughly mixed.
4. Each student, without peeking, removes five beans from the bag (without replacement) and notes the number of marked beans obtained.
5. After all the students in a group have made their selection, the collection of all beans selected makes up the sample for the group.
6. Each group determines the fraction of marked beans in its sample (the number of marked beans in the sample divided by the total number of beans in the sample) and uses this figure to estimate the original number of beans in the bag. (Let n denote the number of beans in the bag; then $25/n$ is the fraction of marked beans in the bag.)
7. The students repeat steps 2 through 6, but this time, in step 2, they mark 50 beans and return them to the bag. The students then compare the estimate obtained this time with that obtained the first time, and discuss which estimate is likely to be the better one.
8. Finally, the instructor should reveal the actual number of beans in each of the bags and discuss the outcomes with the class.

Connections

The Onslaught of Information

Every day, Americans are confronted by a deluge of "facts" and figures. The public media are replete with assertions like the following:

"Fully 18% of Americans currently live below the poverty line."
"Studies show that 62% of teenagers in America are sexually active."
"According to the most recent Gallup poll, only 39% of Americans approve of the way the president is handling his job."
"Eight out of 10 dentists surveyed prefer Whito Toothpaste."

How are such figures obtained? Are they accurate? Are they reliable? Are they misleading? Do polling organizations actually check with every American before issuing such statements? These are serious questions since the figures quoted often form the basis not only for individual decisions, but also for decisions made by the government—decisions that affect us all. The fact is that many such statements are reliable, while others are questionable at best.

In this chapter, we consider how assertions like these are generated and with what degree of confidence they can be believed. Without such understanding, it is necessarily the case that much of what goes on in daily life must remain a mystery to be accepted or rejected on the basis of whim, impression, or faith—a situation that is clearly less than satisfactory.

As a foundation for understanding, we must first know how data are collected, interpreted, and presented to the public.

9.1 THE GRAPHICAL REPRESENTATION OF DATA

Dot Plots

In a class for prospective elementary school teachers, the final examination scores for the students were as shown in Table 9.1. This table shows the **data** simply recorded in a list.

TABLE 9.1 Final Examination Scores in Mathematics for Elementary School Teachers, Section 1

79	78	79	65	95	77	49
91	63	58	78	96	74	68
71	86	91	94	79	69	86
62	78	77	88	67	78	84
69	53	79	75	64	89	77

Just scanning the data gives some idea of how the class did, but it is more revealing to organize the data by representing each score by a dot placed above a number line as in Figure 9.1. Data depicted in this way are called a **dot plot,** or sometimes a **line plot.** The dot plot makes it possible to see at a glance that the scores range from 49 through 96; that most scores are between 60 and 80, with a large group between 75 and 80; and that the "typical" score is probably about 77 or 78. It also reveals that scores like 49 and 53 are quite atypical. Data organized and displayed on a dot plot are much easier to interpret than raw data.

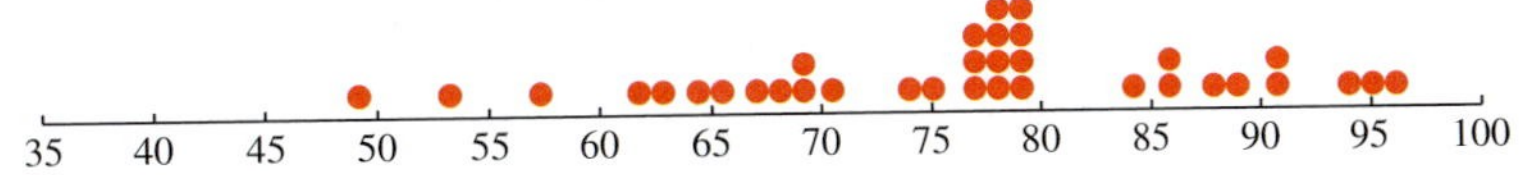

Figure 9.1
Dot plot of the final examination scores in Mathematics for Elementary School Teachers, Section 1

from **The NCTM Principles and Standards**

Prior to the middle grades, students should have had experiences collecting, organizing, and representing sets of data. They should be facile both with representational tools (such as tables, line plots, bar graphs, and line graphs) and with measures of center and spread (such as median, mode, and range). They should have had experience using some methods of analyzing information and answering questions, typically about a single population.

In grades 6–8, teachers should build on this base of experience to help students answer more-complex questions, such as those concerning relationships among populations or samples and those about relationships between two variables within one population or sample. Toward this end, new representations should be added to the students' repertoire. Box plots, for example, allow students to compare two or more samples, such as the heights of students in two different classes. Scatterplots allow students to study related pairs of characteristics in one sample, such as height versus arm span among students in one class. In addition, students can use and further develop their emerging understanding of proportionality in various aspects of their study of data and statistics.

Stem and Leaf Plots

Stem and leaf plots for displaying data are very similar to dot plots and are particularly useful for comparison between two sets of data.

To draw a stem and leaf plot for the data in Table 9.1, we let the tens digits of the scores be the stems and let the units digits be the leaves. Thus, the scores 79, 78, and 79 are represented by

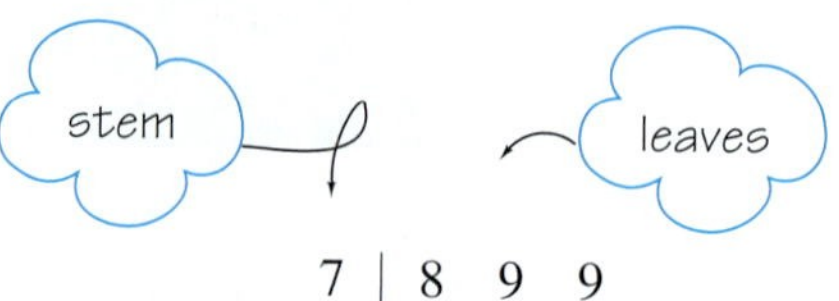

The completed plot appears in Figure 9.2.

Stem	Leaves
4	9
5	3 8
6	2 3 4 5 7 8 9 9
7	1 4 5 7 7 7 8 8 8 8 9 9 9 9
8	4 6 6 8 9
9	1 1 4 5 6

Figure 9.2
Stem and leaf plot of the final examination scores in Mathematics for Elementary School Teachers, *Section 1*

The stem and leaf plot gives much the same visual impression as the dot plot and allows a similar interpretation.

To compare two sets of similar data, it is useful to construct stem and leaf plots on the same stem. Figure 9.3 shows such a plot for final examination scores in Sections 1 and 2 of Mathematics for Elementary School Teachers.

Section 2		Section 1
3	4	9
9 8 7 5	5	3 8
8 8 5 5 5 5 3 1	6	2 3 4 5 7 8 9 9
5 5 4 4 3 0	7	1 4 5 7 7 7 8 8 8 8 9 9 9 9
9 7 6 4 4 2 0 0 0	8	4 6 6 8 9
6 5 5 0	9	1 1 4 5 6
0 0 0	10	

Figure 9.3
Stem and leaf plots of the final examination scores in Mathematics for Elementary School Teachers, *Sections 1 and 2*

In this figure, it is easy to see that, while the two classes are quite comparable, Section 2 had a wider range of scores with one lower and several higher than those in Section 1.

Histograms

Another common tool for organizing and summarizing data is a **histogram.** A histogram for the data in Table 9.1 is shown in Figure 9.4. In a histogram, scores are grouped in intervals, and the number of scores in each interval is indicated by the height of the rectangle constructed above the interval. The number of times any particular data value occurs is called its **frequency.** Similarly, the number of data values in any interval is the **frequency of the interval.** Thus, the vertical axis of a histogram indicates frequency, and the horizontal axis indicates data values or ranges of data values.

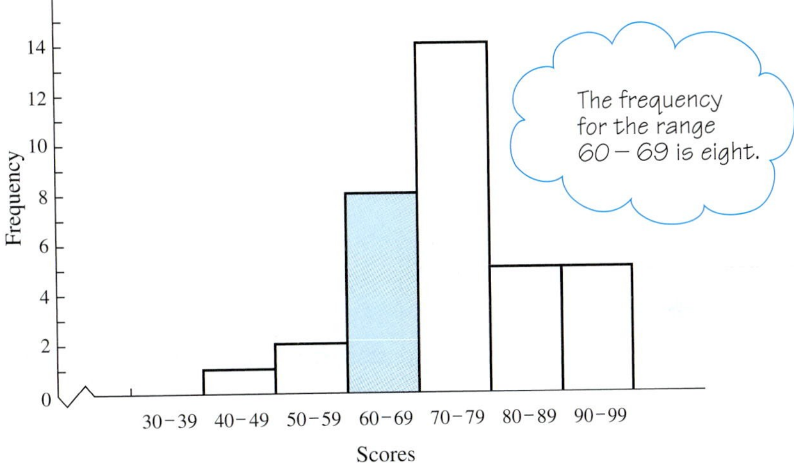

Figure 9.4
Histogram of the final examination scores in Mathematics for Elementary School Teachers, Section 1

We lose some detail with the histogram. For example, the histogram in Figure 9.4 does not show how many students scored exactly 71. However, it has the advantage of giving a compact and accurate summary that is particularly useful with large collections of data that could not be conveniently represented using dot or stem and leaf plots.

Histograms are useful for giving visual summaries of data that are either discrete or that vary continuously. For example, people's heights vary continuously. You may say that your height is 5′4″ (or 64″), but that is not exactly true. Instead, it is true only *to the nearest inch.* Consider the data in Table 9.2, which gives the heights of the 80 boys at Eisenhower High School. The heights were measured to the nearest inch, so the numbers

TABLE 9.2 **Heights to the Nearest Inch of Boys at Eisenhower High School Arranged in Increasing Order**

64	67	68	69	69	70	71	72
65	67	68	69	69	70	71	72
66	68	68	69	69	70	71	72
66	68	68	69	69	70	71	72
66	68	68	69	69	70	71	72
67	68	68	69	69	70	71	72
67	68	68	69	70	70	71	73
67	68	69	69	70	70	71	73
67	68	69	69	70	70	71	74
67	68	69	69	70	70	71	74

given already represent *grouped data;* that is, a measurement of 66 was recorded if a boy's height was judged to be between 65.5 and 66.5 inches. Even a height of almost exactly 66.5 inches was grouped into the 66- or 67-inch class as deemed most appropriate by the person doing the measuring.

A histogram representing this data is shown in Figure 9.5. Note that in both Figure 9.4 and Figure 9.5, the sum of the heights of the rectangles gives the number of data values.

Grouping data into classes and displaying the data in a histogram is a useful visualization of the characteristics of the data set. In drawing a histogram, the scales should be chosen so that all the data can be represented. Also, the number of classes into which the data is grouped should not be so few that it hides too much information and not so numerous that one loses the visual advantage of constructing the diagram in the first place.

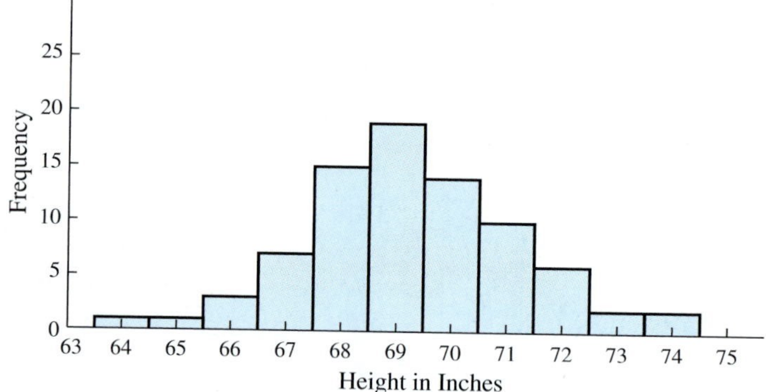

Figure 9.5
Histogram of the heights of boys at Eisenhower High School

Line Graphs

A **line graph** for the data in Figure 9.5 is constructed by joining the midpoints of the tops of the rectangles in the figure by line segments. (See Figure 9.6.) Without the rectangles, which would not ordinarily be drawn, the line graph appears as in Figure 9.7. Since the vertical axis represents the frequency with which measurements occur, the line graph of a set of data like this is often called a **frequency polygon.**

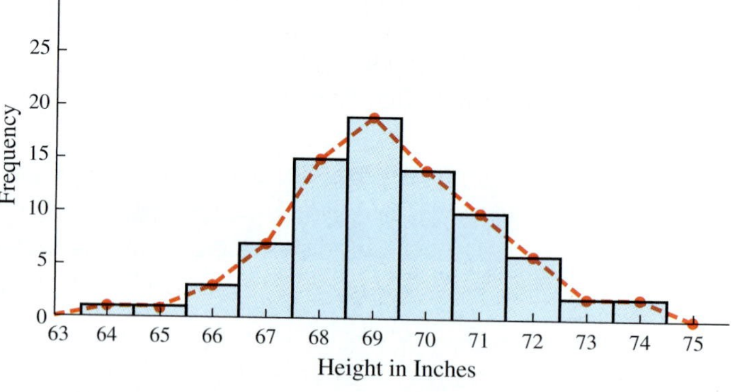

Figure 9.6
Histogram and line graph of the heights of boys at Eisenhower High School

A line graph, or frequency polygon, is particularly appropriate for representing data that vary continuously since this variation is strongly suggested by the sloping line segments. The rectangles in the histogram of Figure 9.5 tend to obscure the fact that the heights of boys represented by the rectangle centered at 67 range from 66.5″ to 67.5″. On the other hand, if the data really are discrete, they are probably better represented by a histogram than a frequency polygon. Since, in constructing the frequency polygon from the histogram, we added and subtracted small triangles of equal area, the area under the frequency polygon,

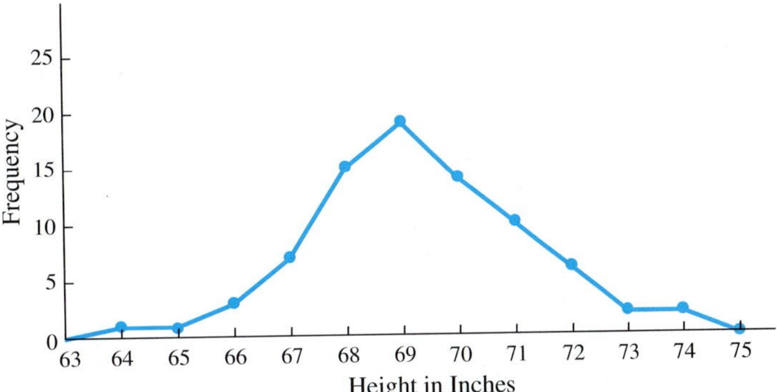

Figure 9.7
Line graph (frequency polygon) of the heights of boys at Eisenhower High School

or line graph, still gives the number of data, values and the area under the graph and above a given interval gives the number of boys whose heights fall in that interval.

Line graphs not necessarily related to histograms are particularly effective when they are used to indicate trends over periods of years—trends in the stock market, trends in the consumption of electrical energy, and so on. For example, consider the data in Table 9.3 which gives the yearly consumer expenditure for food in the United States at five-year intervals from 1950 through 2005. These data are represented visually by the line graph in Figure 9.8. In this case, the points on the graph are determined by the year and the food expenditure for that year. The points are then connected by straight line segments.

TABLE 9.3 Consumer Expenditures for Food in the United States in Billions of Dollars

Year	1950	1955	1960	1965	1970	1975	1980	1985	1990	1995	2000	2005
Expenditures	44.0	53.1	66.9	81.1	110.6	167.0	264.4	345.4	440.6	499.1	559.4	622.5

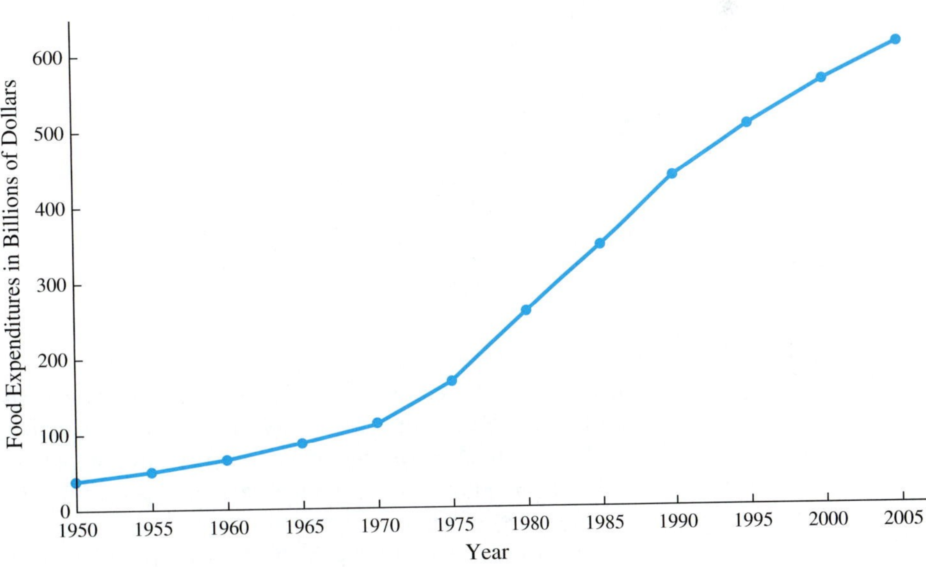

Figure 9.8
U.S. consumer expenditures for food in billions of dollars

One advantage of a line graph is that it makes it possible to estimate data values not explicitly given otherwise.

EXAMPLE 9.1 ESTIMATING DATA VALUES FROM A LINE GRAPH

Using the graph in Figure 9.8, estimate the U.S. consumer expenditure for food in 1972.

Solution

Understand the Problem

The graph gives the expenditure at five-year intervals. We are asked to estimate the expenditure for 1972.

Devise a Plan

Having Figure 9.8 already simplifies our task. The graph suggests that the total expenditure grows steadily each year, and, while the growth is almost surely not "straight-line growth" between data points as indicated by the diagram, the straight line joining the data points for 1970 and 1975 surely approximates the actual growth. If we draw a vertical line from the point representing 1972 on the horizontal axis, the height of the line segment should give us the approximate expenditure for 1972.

Carry Out the Plan

The point on the horizontal axis representing 1972 is two-fifths of the way from 1970 to 1975. Draw a vertical line from this point to the line graph. Then draw a horizontal line from the point where this line cuts the line graph to the vertical axis. This determines the point on the vertical axis that gives approximately $140 billion as the value of U.S. consumer expenditure for food in 1972.

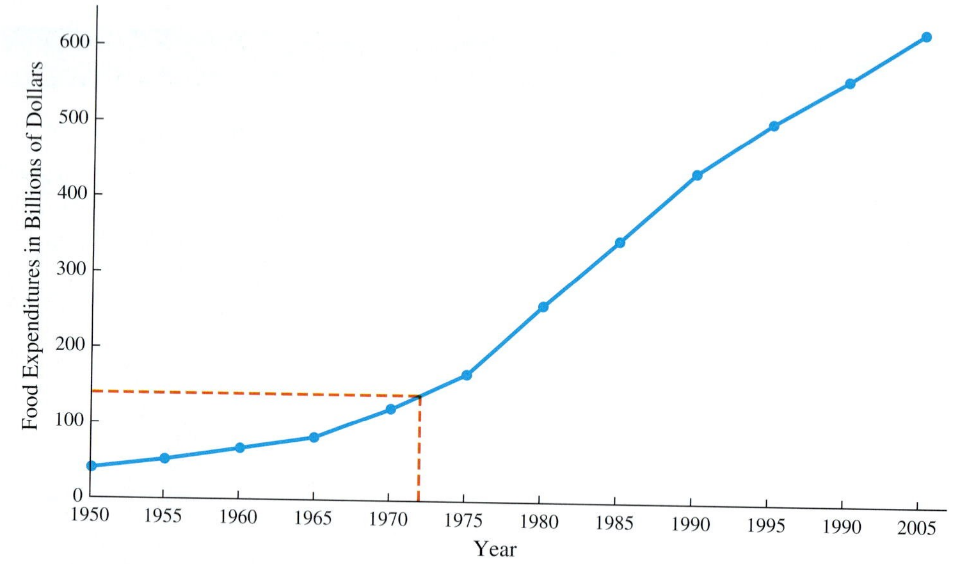

Look Back

The solution was achieved by noting that the line graph, which gives the appropriate values of expenditures every five years, suggests that the expenditures steadily increase and that the actual values for intervening years no doubt lie relatively close to the straight line segments joining the given data points. Indeed, it appears that a curved line through the data points might give an even better approximation somewhat less than $140 billion. However, as an estimate, $140 billion is reasonably accurate.

Score	Grade
90–100	A
80–89	B
70–79	C
60–69	D
0–59	F

Bar Graphs

Bar graphs, similar to histograms, are often useful in conveying information about so-called categorical data where the horizontal scale represents some nonnumerical attribute. For example, consider the final examination scores for Mathematics for Elementary School Teachers, Section 1, as listed in Table 9.1. Suppose that the instructor determines grades as indicated in the accompanying table. Then members of the class were awarded 5 As, 5 Bs, 14 Cs, 8 Ds, and 3 Fs. If we indicate grades on the horizontal scale and frequency on the vertical scale, we can construct the bar graph shown in Figure 9.9. In general, the rectangles in a bar graph do not abut, and the horizontal scale may be designated by any attribute—grade in class, year, country, city, and so on. As usual, however, the vertical scale will denote frequency—the number of items in the given class.

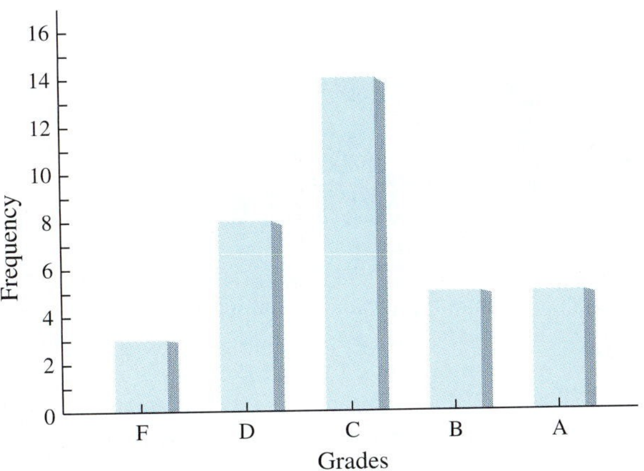

Figure 9.9
Bar graph of the final examination grades in Mathematics for Elementary School Teachers, *Section 1*

Bar graphs are useful in displaying data concerning nonnumerical items. As the following example shows, they are also useful in comparing data concerning two or more similar groups of items.

EXAMPLE 9.2 COMPARING GRADES IN TWO MATHEMATICS CLASSES BY MEANS OF A BAR GRAPH

Draw a suitable bar graph to make a comparison of the grades in Mathematics for Elementary School Teachers, Sections 1 and 2.

Solution

The desired bar graph can be obtained by drawing two adjacent bars (rectangles) for each letter grade, with a suitable indication of which bars to associate with each section. If we use blue bars for Section 1 and pink bars for Section 2, a suitable graph might look like this. (See Figure 9.3 on the next page for the scores in Section 2. These scores merit 7 As, 9 Bs, 6 Cs, 8 Ds, and 5 Fs, using the same scale as for Section 1 above.)

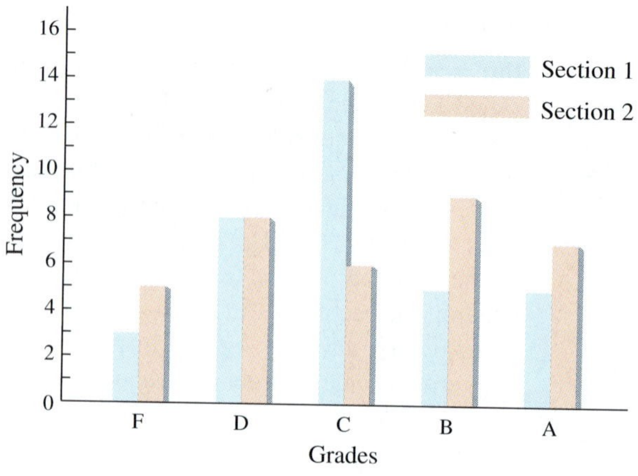

Into the Classroom

Clem Boyer Comments on Statistics in Elementary School

In learning statistics, graphing, and probability, students are learning about the real world. When they work with tables, charts, and graphs and use the language and notation of graphing mathematics, they are developing important real-life skills in reading, interpreting, and communicating information. Working with statistics prepares students to deal with the endless number of statistics in today's world. In using probability to predict outcomes, they are learning how to cope mathematically with the uncertainties in the real world. Students should plan and carry out the collection and organization of data to satisfy their curiosity about everyday living. They need to construct, read, and interpret simple maps, tables, charts, and graphs. In doing these things, they find out how to present information about the numerical data.

In order for students to manage statistics in this age of technology, it is important for them to learn to find measures of central tendency (mean, median, and mode) and measures of dispersion (range and deviation). Further, students need to recognize the basic uses and misuses of statistical representation and inference in order for them to be wise consumers.

All of these skills in working with data improve students' ability to interpret the data they read and hear about every day. Being able to use the terminology when displaying data will help students communicate their findings.

Learning probability has applications in the real world too. Students find out how to identify situations in which immediate past experience does not affect the likelihood of future events. Their lives are enriched when they can see how mathematics is used to make predictions regarding election results, business forecasts, and sporting events.

SOURCE: From ScottForesman Exploring Mathematics, *Grades 1–7, by L. Carey Bolster et al. Copyright © 1994 Pearson Education, Inc. Reprinted with permission. Clem Boyer is the former Coordinator of Mathematics, K–12, for the District School Board of Seminole County in Sanford, Florida.*

Pie Charts

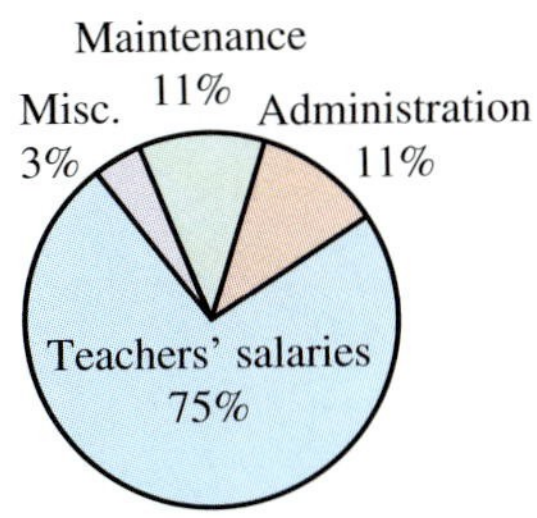

Figure 9.10
Percent of each tax dollar expended by Mile High School District by category

Another pictorial method for conveying information is a **pie chart,** also sometimes called a **circle plot.** For example, the pie chart in Figure 9.10 shows the parts of the budget Mile High School District used for various purposes. The number of degrees in the angular measure of each part of the chart is the appropriate fraction or percentage of 360°. Thus, the angular sector for the portion representing administration measures

$$0.11 \times 360^\circ = 39.6^\circ,$$

and so on. As shown here, pie charts are most often used to show how a whole (total budget, total revenues, total sources of oil, and so on) is divided up.

EXAMPLE 9.3 A LOTTERY PIE CHART

The 2003 volume of the *Statistical Abstract of the United States* shows that several states took in the following gross amounts of money from sales of lottery tickets during the years 1980–2002. Amounts are listed in billions of dollars for the various lottery games: Instant—18.5; Three-digit—5.3; Four-digit—2.9; Lotto—9.6; Other—5.6. Draw a pie chart showing the gross income the states received from each game.

Solution

Taking the sum of the gross proceeds from the various games, we find that the total gross income from all games was 41.9 billion dollars. Draw a circle and divide it into sectors whose central angles have measures equal to the appropriate fraction of 360°. To the nearest degree, the central angles for the various games are as follows:

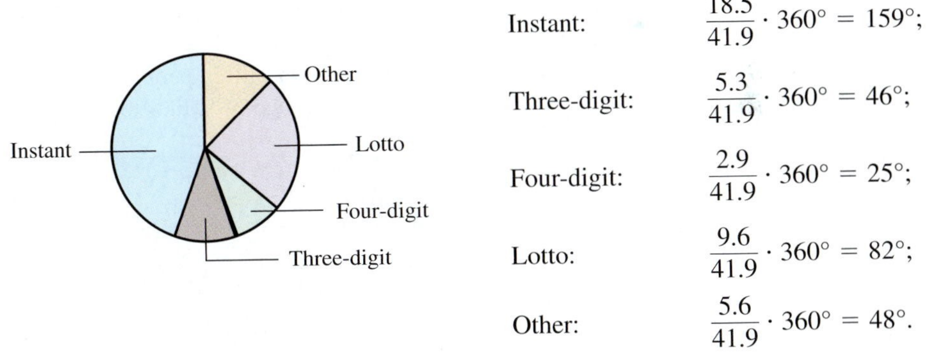

Instant: $\dfrac{18.5}{41.9} \cdot 360^\circ = 159^\circ;$

Three-digit: $\dfrac{5.3}{41.9} \cdot 360^\circ = 46^\circ;$

Four-digit: $\dfrac{2.9}{41.9} \cdot 360^\circ = 25^\circ;$

Lotto: $\dfrac{9.6}{41.9} \cdot 360^\circ = 82^\circ;$

Other: $\dfrac{5.6}{41.9} \cdot 360^\circ = 48^\circ.$

Thus, the required pie chart is as shown.

Increasingly, pie charts and other pictorial representations of data are drawn by computer and colored to give a more pleasing effect to the eye. If the pie chart is drawn in perspective, as if seen from an angle as in Figure 9.11, the central angles are no longer completely accurate. However, the pie chart still gives a good visual understanding of the apportionment of the whole being discussed. Also, in Figure 9.11, the pieces of the pie are separated slightly to produce a more pleasing visual effect.

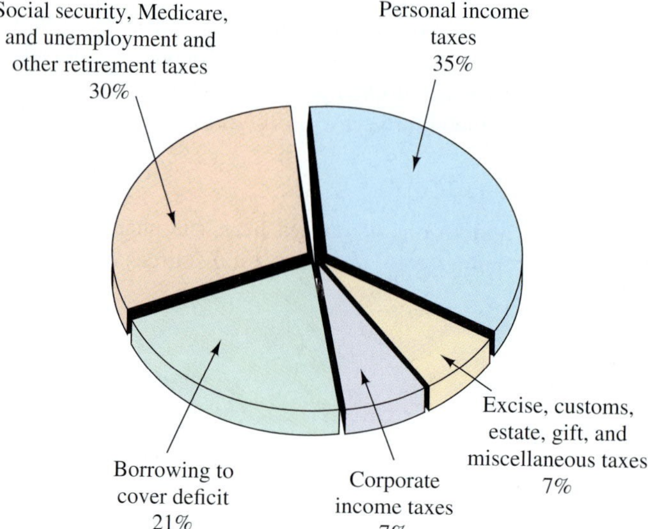

Figure 9.11
Pie chart showing U.S. government sources of revenue for fiscal year 1991.
SOURCE: 1992 IRS Form 1040 instruction booklet

Pictographs

A **pictograph** is a picture of a set of small figures or icons used to represent data, and often to represent trends. Usually, the icons are suggestively related to the data being represented. Consider the pictograph presentation of past and projected growth in world population in Figure 9.12. The pictograph accurately indicates that the world population more than doubled over the 40-year period from 1950 to 1990. It also suggests that, while the rate of increase is expected to diminish, the population will almost double again in the next 60 years.

To make it possible to correctly interpret a pictograph, it is necessary to include a key that indicates the value or amount each small icon represents. For example, in Figure 9.12 each icon represents one billion people. In making a pictograph, the determination of the key depends on the range of values to be represented. The key chosen must

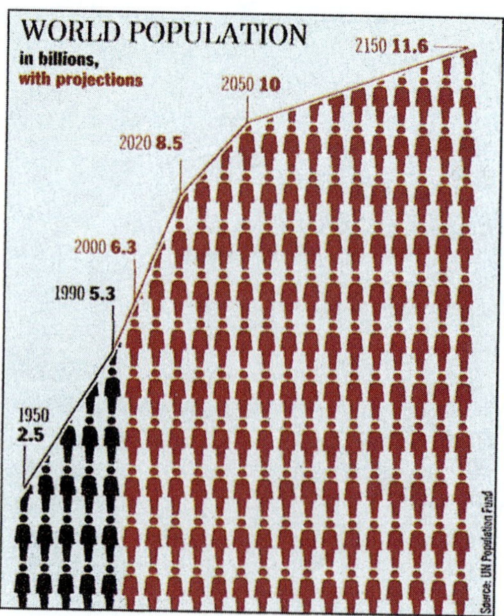

Figure 9.12
Pictograph of world population growth: ♟ = 1 billion
SOURCE: Graph, "World Population," from Time, June 1, 1992, p. 54. Copyright © 1992 Time, Inc. Reprinted by permission.

School Book Page

Analyzing Data in Grade Three

Chapter 1 Lesson 1

Reading Pictographs

You Will Learn
how to read a pictograph

Vocabulary

pictograph
a graph that uses pictures, or symbols, to show data

data
information used to make calculations

key
part of a pictograph that tells what each symbol shows

symbol
a picture in a pictograph that shows a given number of objects

Learn

What's the best thing to put on a pizza? Third graders at Shoreline Elementary School in Whitehall, Michigan, had lots of ideas. How many students chose pepperoni?

A **pictograph** can help you compare **data**.

Favorite Pizza Toppings	
Sausage	⊛ ◐
Vegetables	◐
Extra cheese	⊛ ⊛ ⊛ ◐
Pepperoni	⊛ ⊛ ⊛ ⊛ ⊛ ◐

Title — Symbol — Key: ⊛ = 2 votes

The **key** tells you that each **symbol** shows 2 votes. Since ⊛ shows 2 votes, ◐ shows 1 vote. So, 11 students chose pepperoni.

How can you tell what each symbol shows in a pictograph?

Check

Use the pictograph to answer each question.

1. Which pizza topping had the fewest votes?
2. Which pizza topping had 7 votes?
3. **Reasoning** If 10 students voted for onions, how many symbols would there be for onions? Explain.

Chapter 1 • Data, Graphs, and Facts Review

Questions for the Teacher

1. What problem might students have in interpreting the above pictograph? How might you help them to achieve understanding?
2. What difficulties might students have in drawing a pictograph? What help might you need to give?
3. What is the key bit of information needed to understand a pictograph?

be sufficiently small that the resulting pictograph is large enough to show the desired detail, but not so small that the pictograph becomes unwieldy.

Choosing Good Visualizations

Each of the graphical representations discussed is appropriate to summarize and present data so that the reader can visualize frequencies and determine trends. The various representations are more appropriate in some instances than others, and most are subject to serious distortion if the intent is to mislead the reader. For example, the pictograph in Figure 9.13 represents the oil consumption in the United States for the years 1994 and 2004.

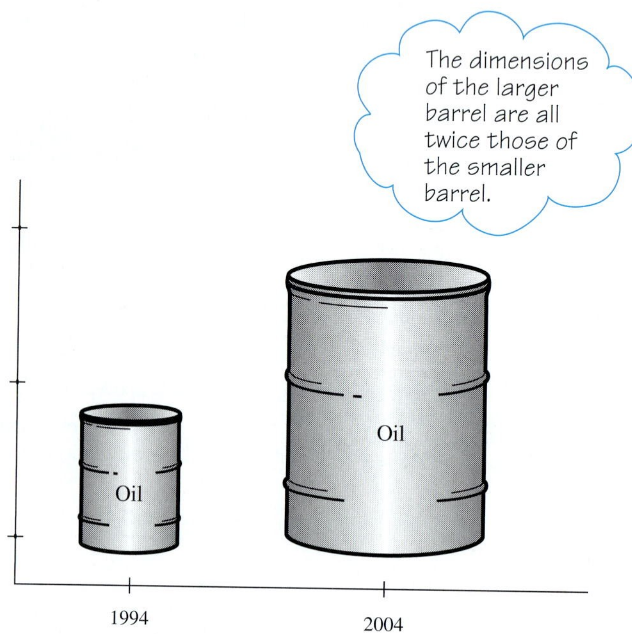
The dimensions of the larger barrel are all twice those of the smaller barrel.
Oil
Oil
1994
2004

Figure 9.13
Oil consumption in the United States in 1994 and 2004

While the vertical scale honestly indicates that approximately twice as much oil was used in 2004 as in 1994, the pictograph is misleading because the volume of the larger barrel is *eight times* the volume of the smaller barrel. The casual reader is quite likely to get a badly distorted idea of the relative amount of oil used in the two years. Of course, that may be exactly what the person who constructed the pictograph intended. Have you ever seen such distortions on television? in the newspaper? in advertisements? Be observant the next time you see such a diagram.

Particularly suitable uses for the various visual representations of data described in this section are summarized as follows.

- Dot plots: summarizing relatively small sets of data—grades in a class, heights of students in a class, birth months of students in a class, and so on.
- Stem and leaf plots: for essentially the same purposes as line plots; especially useful in comparing small data sets.
- Histograms: summarizing information in large sets of data that can be naturally grouped into intervals.
- Line graphs: summarizing trends over time.
- Pie charts: representing relative amounts of a whole.
- Pictographs: summarizing relative amounts, trends, and data sets; useful in comparing quantities.

Statistical Plotting with a Graphing Calculator

It is still worthwhile for children to draw simple statistical plots by hand, using graph paper and assorted drawing tools such as rulers, protractors, and colored pencils. The "hands-on" approach can go far to convey a solid understanding of what it means to visualize data and what steps are required to obtain a useful visual representation. Once the basic plotting concepts are understood, however, it is instructive to see how technology can be employed. Plots drawn with a graphing calculator, spreadsheet, or other computer software offer a number of advantages:

- It becomes possible to handle large realistic data sets, such as the data gathered by a survey or the result of a scientific experiment.
- It is easy to edit and sort the data and to modify the style of plot to find the best visualization of the data set.

Plotting and charting software is now widely available and is contained in every spreadsheet and many word processing programs. The graphing calculator also provides a statistical plotting tool, one that is especially convenient for use in the classroom. Appendix C provides a brief description of the procedures followed to create a statistical plot.

Here is a sample data set and some of the statistical plots that may be obtained with the TI-73 graphing calculator.

Insect Species

It has been estimated that there are about 751,000 insect species, including 112,000 species of moths and butterflies, 103,000 species of bees and wasps, and 290,000 species of beetles, with the remaining 246,000 species falling into smaller classes. The two TI-73 screens just below show how the four insect categories and their corresponding numbers of species (in thousands) have been entered into two lists INSCT and SPECI. (*Note:* The TI-73 allows user-named lists of up to five letter or number symbols, starting with any letter.) The right-hand screen is a view of the Stat Plot editor, showing that a pie chart has been selected, with the percentages within each category to be displayed in the chart.

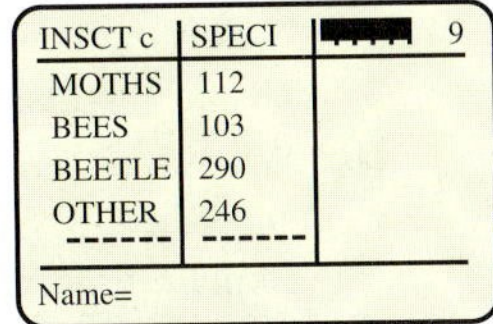

List Editor Screen

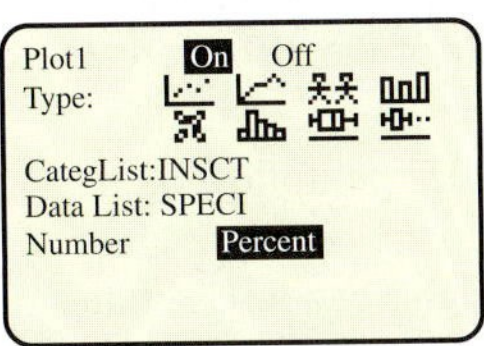

Stat Plot Edit Screen

The resulting pie chart of the data is shown below at the left. We could also have chosen to represent the data with a bar graph or pictograph, as shown in the middle and at the right, respectively.

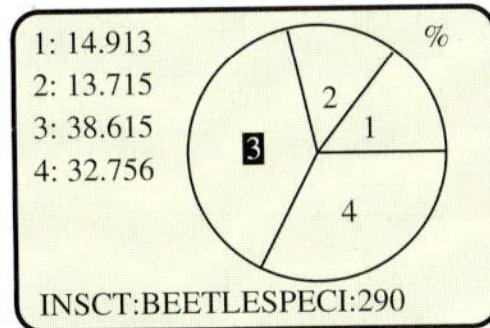

Pie Chart

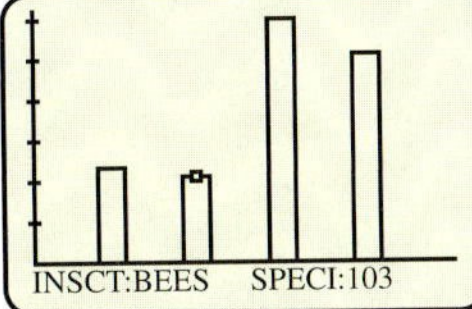

Bar Graph

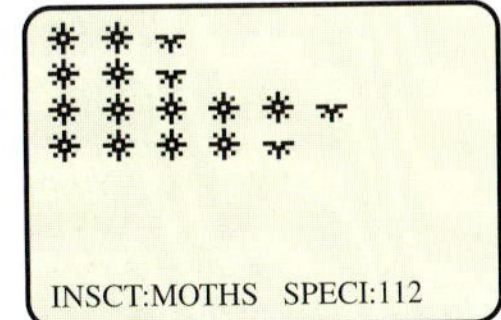

Pictograph

From the pie chart, we see that nearly 40% of the 751,000 different species of insects are beetles! The eminent geneticist J. B. S. Haldane once commented that "God must have loved beetles; He made so many of them."

Problem Set 9.1

Understanding Concepts

1. The scores below were obtained on the final examination in an introductory mathematics class of 40 students.

98	80	98	76	79	94	71	45	89	71
62	61	95	77	83	49	65	58	56	89
66	87	74	64	75	58	72	75	48	88
75	51	84	76	95	69	61	69	33	86

(a) After scanning the data, what do you think the "typical" or "average" score is?

(b) Make a dot plot to organize these data.

(c) After looking at the dot plot, what seems to be the "typical" score?

(d) Do you identify any scores that seem to be particularly atypical of this data set? Explain.

(e) Write a two- or three-sentence description of the results of the final examination.

2. Make a stem and leaf plot of the data in problem 1.

3. Make a histogram for the data in problem 1, using the ranges 20–29, 30–39, . . . , 90–99 on the horizontal scale.

4. At the same time that heights of the boys at Eisenhower High School were studied, heights of the girls were also studied.

(a) Draw a histogram to summarize the following data set, which gives the heights to the nearest inch of the 75 girls at Eisenhower High School. Use intervals one unit wide centered at the whole-number values 56, 57, . . . , 74.

57	62	63	64	66
58	62	63	64	66
60	62	63	64	66
60	62	63	65	66
61	62	63	65	66
61	62	63	65	66
61	62	63	65	66
61	63	64	65	66
61	63	64	65	66
61	63	64	65	66
62	63	64	65	67
62	63	64	65	67
62	63	64	65	70
62	63	64	65	70
62	63	64	66	73

(b) Write two or three sentences describing the distribution of the heights of the girls.

5. (a) Draw a frequency polygon for the data in problem 4 by joining the midpoints of the tops of the rectangles of the histogram by straight line segments.

(b) What does the area under the frequency polygon and between the scores 60.5 and 64.5 represent? Explain briefly.

6. The scores on the first, second, and third tests given in a class in educational statistics as the term progressed are shown.

First test:	92,	80,	73,	74,	93,	75,	76,	68,	61,	76,
	83,	94,	63,	74,	76,	86,	82,	70,	65,	74,
	83,	87,	98,	77,	67,	64,	87,	96,	62,	64
Second test:	52,	65,	84,	91,	86,	76,	73,	52,	68,	79,
	88,	94,	98,	84,	53,	59,	63,	66,	77,	81,
	94,	81,	64,	56,	96,	58,	64,	57,	83,	87
Third test:	97,	91,	61,	67,	72,	81,	63,	56,	53,	59,
	43,	56,	64,	78,	93,	99,	84,	84,	61,	56,
	73,	77,	57,	46,	93,	87,	93,	78,	46,	87

(a) Draw three separate, but parallel, dot plots for these three sets of scores.

(b) Write a three- or four-sentence analysis of these dot plots suggesting what happened during the term to account for the changing distribution of scores.

7. Together, the Smiths earn $64,000 per year, which they spend as shown.

Taxes	$21,000
Rent	$10,800
Food	$5,000
Clothes	$2,000
Car payments	$4,800
Insurance	$5,200
Charity	$7,000
Savings	$6,000
Misc.	$2,200

Draw a pie chart to show how the Smiths spend their yearly income.

8. This pie chart indicates how the city of Metropolis allocates its revenues each year.

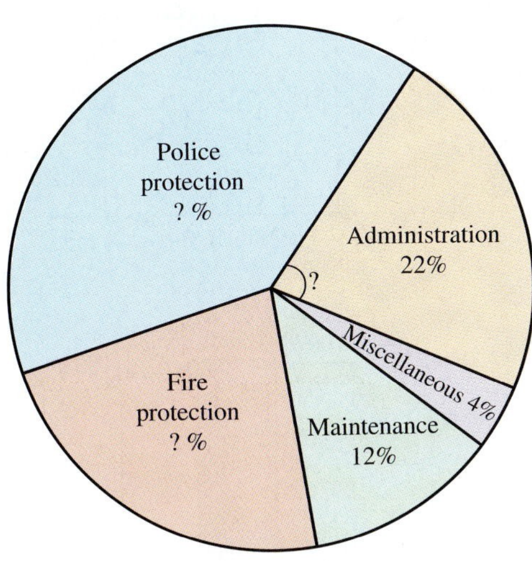

(a) What is the measurement of the central angle of the sector representing administrative expense?

(b) Using a protractor to measure the angle, determine what percent of the city budget goes for police protection.

(c) How does the city's expenditure for maintenance compare with its expenditure for police protection?

(d) How do the expenditures for administration and fire protection compare?

9. This bar graph shows the distribution of grades on the final examination in a class in English literature.

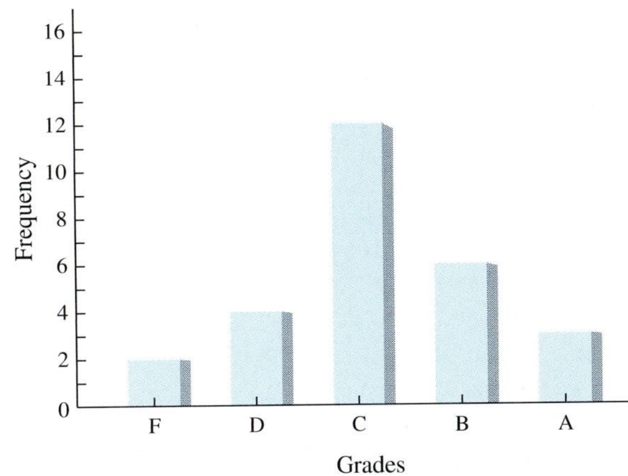

(a) From the bar graph, determine how many students in the class got Cs.

(b) How many more students got Bs than Ds?

(c) What percent of the students earned As?

10. (a) Go to a busy campus parking lot and record the number of cars that are predominately white, black, red, gray, green, and "other."

(b) Make a bar graph to display and summarize your data.

(c) Based on part (a), if you were to stand on a busy street corner and watch 200 cars go by, how many would you expect to be predominately white?

11. (a) Make a bar graph for the data in the table shown.

KICK-STARTING THE SUMMER

Being first matters. A quick look at the last five movies that launched the season.

Film	Year	Revenue* (millions)
'The Mummy'	1999	**$413.5**
'Gladiator'	2000	**457.2**
'The Mummy Returns'	2001	**430.0**
'Spider-Man'	2002	**821.7**
'X2: X-Men United'	2003	**404.5**

*Worldwide figures to date

Source: From Newsweek, *May 10, 2004.*

(b) The table seems to be somewhat deficient. Would it have been more meaningful to have included data for the film that appeared second on the market in each of the years shown? Why or why not?

Teaching Concepts

12. Collecting their own data actively engages students in the study of statistical notions. This not only heightens student interest, but also imparts a sense of meaning and reality to the study of statistics. Name four activities you deem particularly suitable for a class of elementary students that would involve the collection and representation of data.

13. Buy a small package of M&M's with mixed colors. Open the package and pour out the M&M's.

(a) How many M&M's of each color are in the package?

(b) Make a bar graph of the data from part (a).

(c) Make a pictograph to display the data from part (a).

(d) Would it be reasonable to guess that most packages of M&M's contain about twice as many yellow as green candies?

14. (a) Roll two dice fifty times and record the number of times (the frequency) you obtained each score.

(b) Draw a bar graph for the data of part (a) showing frequency on the vertical axis and score of the horizontal axis.

15. The following is from page 11 of the teacher's edition of *Scott Foresman–Addison Wesley Math,* Grade 5, by Randall I. Charles et al., copyright © 2002 Pearson Education, Inc. Reprinted with permission.

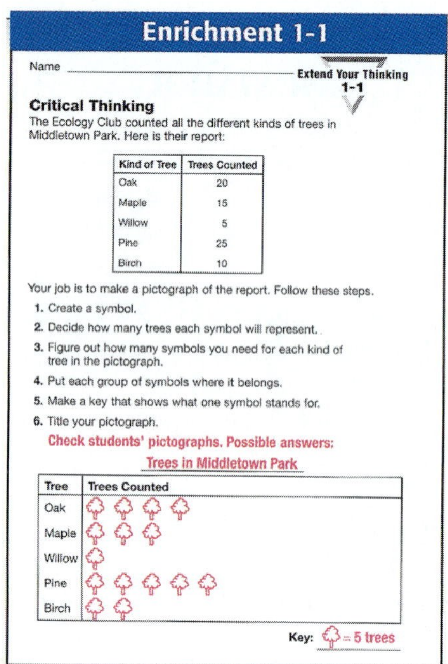

Enrichment 1-1

Name ______________ Extend Your Thinking 1-1

Critical Thinking

The Ecology Club counted all the different kinds of trees in Middletown Park. Here is their report:

Kind of Tree	Trees Counted
Oak	20
Maple	15
Willow	5
Pine	25
Birch	10

Your job is to make a pictograph of the report. Follow these steps.

1. Create a symbol.
2. Decide how many trees each symbol will represent.
3. Figure out how many symbols you need for each kind of tree in the pictograph.
4. Put each group of symbols where it belongs.
5. Make a key that shows what one symbol stands for.
6. Title your pictograph.

Check students' pictographs. Possible answers:

Trees in Middletown Park

Tree	Trees Counted
Oak	
Maple	
Willow	
Pine	
Birch	

Key: = 5 trees

(a) The teacher's edition shows a possible correct answer. How many small icons should the plot show for the pine category if the key is such that each icon (small tree symbol) represents 10 trees?

(b) Make your own pictograph of these data, with each icon representing two trees.

Thinking Critically

Data are often presented in a way that confuses, or even purposely misleads, the viewer.

16. (a) Discuss briefly why the television evening news might show histogram (A) below rather than (B) in reporting stock market activity for the last seven days. Is one of these histograms misleading? Why or why not?

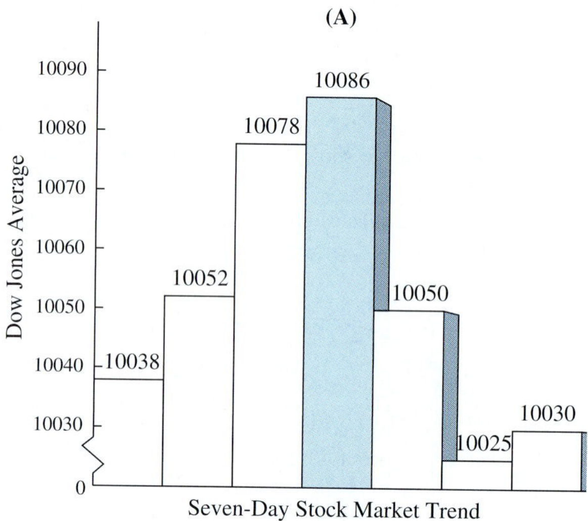

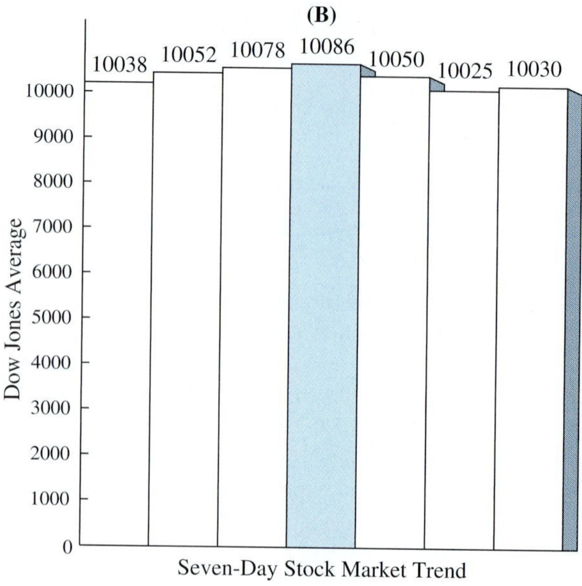

(b) What was the percentage drop in the Dow Jones Average from the fourth to the fifth day as shown in the preceding histograms? Should an investor worry very much about this 36-point drop in the market?

(c) Was the Dow Jones Average on day five approximately half of what it was on day four, as suggested by histogram (A)?

17. Longlife Insurance Company printed a brochure with the following pictographs showing the growth in company assets over the 10-year period 1990–1999.

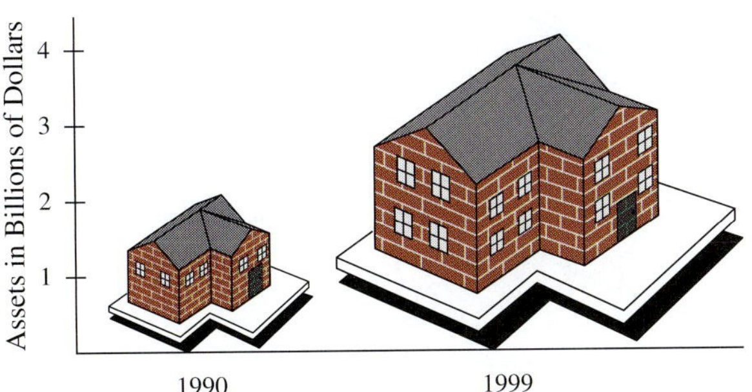

(a) Do the pictographs accurately indicate that the assets were \$2 billion in 1990 and \$4 billion in 1999, or might one assume from the pictographs that the assets were actually much greater in 1999 then in 1990? Explain briefly.

(b) The larger building shown is just twice the height of the smaller, and the two buildings are similar as geometrical drawings. Do these drawings accurately convey the impression that the assets of Longlife Insurance Company just doubled during the 10-year period? Explain your reasoning. What is the ratio of the volume of the large building to the volume of the small building? (*Suggestion:* Suppose both buildings were rectangular boxes, with the dimensions of the second just twice those of the first.)

(c) Would it have been more helpful (or honest) to print the actual asset value for each year on the front of each building?

18. Some merchandisers take advantage of optical illusions just as some pollsters, advertisers, and others do.

(a) Which of the cans depicted here seems to have the greater volume? Note that the diagrams are drawn to scale.

(b) Actually compute the volumes of the cans.

(c) Of these two cans, which one are you more likely to see in the grocery store? Why do you suppose this is so?

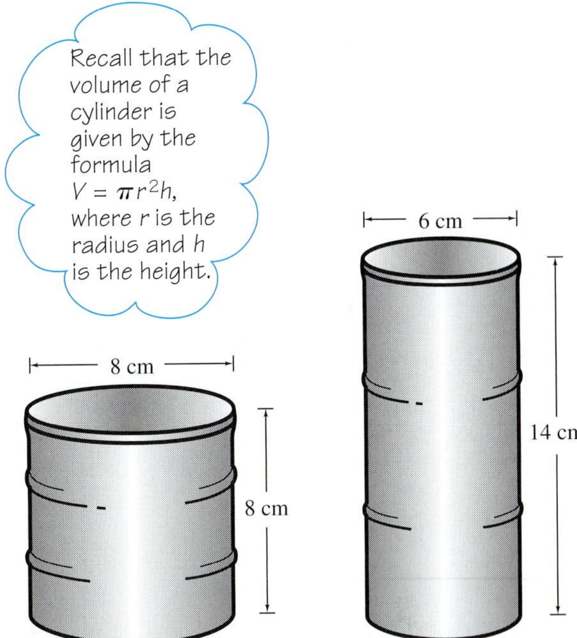

Thinking Cooperatively

This exercise is best done as a class activity.

19. Our measure of a "yard" was originally the length from the tip of the nose to the fingertip of the outstretched arm of an English king. Working in small groups, use a tape measure to measure this length to the nearest inch for each student in class. Record the data on the chalkboard (and for later use in Problem Set 9.2) in two sets—one set for men and one set for women. Divide the class into several small groups.

(a) Members of one or two small groups each make a dot plot for the data in each of the two sets.

(b) Members of one or two small groups each make a double stem and leaf plot for the data in each of the two sets.

(c) Members of one or two small groups each make a histogram for the data in each of the two sets.

(d) Members of one or two small groups each make a frequency polygon for the data in each of the two sets.

(e) Members of one or two small groups each make a pictograph for the data in each of the two sets.

(f) As a class, discuss the various representations of the data. Which ones seem most informative? What conclusions are suggested regarding the length of a "yard?" Discuss briefly.

Making Connections

20. This line graph from the 2003 edition of *The Digest of Educational Statistics* shows that nearly all public schools had Internet access by 2001.

(a) Determine from the graph the year that approximately 50% of all public schools had Internet access.

(b) In what year did approximately 90% of the schools have Internet access?

(c) In 2001, only approximately 90% of all classrooms had Internet access. What type of classrooms likely would not have had access? That is, what subjects would likely have been taught in the 10% of classrooms that did not have Internet access?

Figure 31. – Percent of all public schools and instructional rooms having Internet access: Fall 1994 to Fall 2001

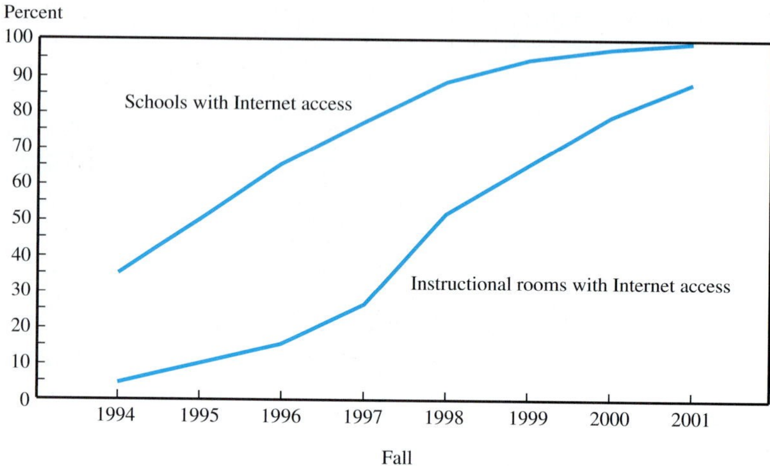

SOURCE: U.S. Department of Education, National Center for Education Statistics, Fast Response Survey System, Internet Access in Public Schools and Classrooms: 1994–2001.

21. This pie chart from the *Digest of Education Statistics, 2002,* shows the percentage of persons 25 years and older in 2001 that attained various levels of education.

(a) Assuming that these figures remain relatively constant from year to year, approximately what percentage of high school graduates who go to college fail to earn a bachelor's degree? (*Hint:* Assume that people who earn master's, professional, and doctor's degrees must first earn a bachelor's degree.)

(b) What percentage of high school graduates earn a bachelor's degree?

(c) What percentage of the people who earn bachelor's degrees eventually earn a doctor's degree?

Figure 5. – Highest level of education attained by persons 25 and older. March 2001

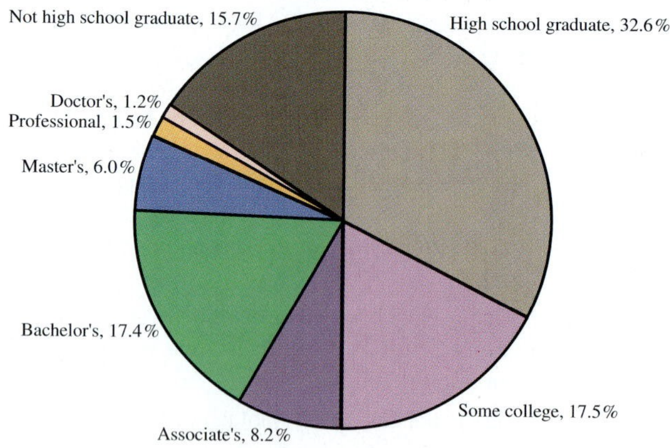

Total Persons age 25 and over = 177.0 million

NOTE: Detail may not sum to totals due to rounding.

SOURCE: U.S. Department of Commerce, Bureau of the Census, *Current Population Survey, unpublished data.*

22. The graph at the top of the next page indicates Google's net income or loss for the years 1999 through 2003 and first three months of 2004.

(a) Supposing that Google's income continued to come in at the present rate, what would its net income for 2004 be?

(b) Draw a line graph (including your estimate for 2004) for Google's income for 1999 through 2004.

(c) About what percentage increase in net income did Google experience in 2003 over 2002?

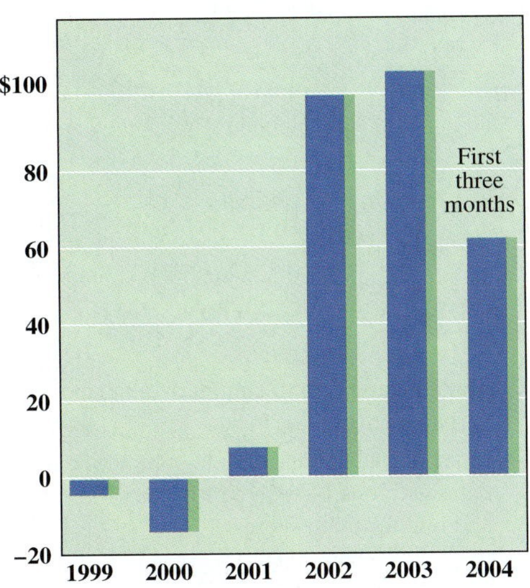

Source: From Newsweek, May 10, 2004.

(d) If the assumption in part (a) proved to be correct, what would Google's percentage increase in net income in 2004 over 2002 be?

23. Mathematics has been traditionally viewed as a "male" subject. That this is less and less the case is made clear by the graphs below and at the top of the next column.

(a) From the graphs, determine how many times as many women earned Ph.D.s in mathematics during the 1990s as during the 1890s.

(b) Determine how many men earned Ph.D.s during the 1890s and during the 1990s.

(c) Determine the ratio of the number of men to the number of women earning Ph.D.s in mathematics during the 1890s and during the 1990s.

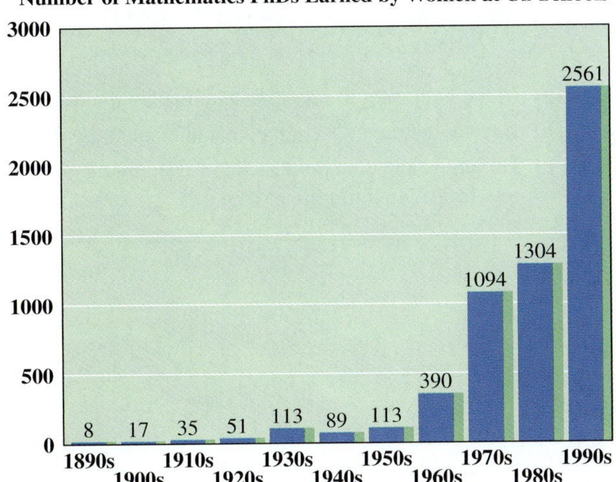

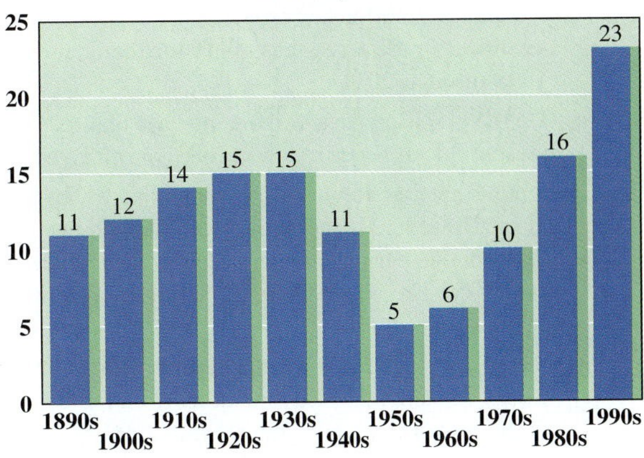

24. The graphs shown display projected payout and income for the Social Security fund for the years 1998 through 2032.

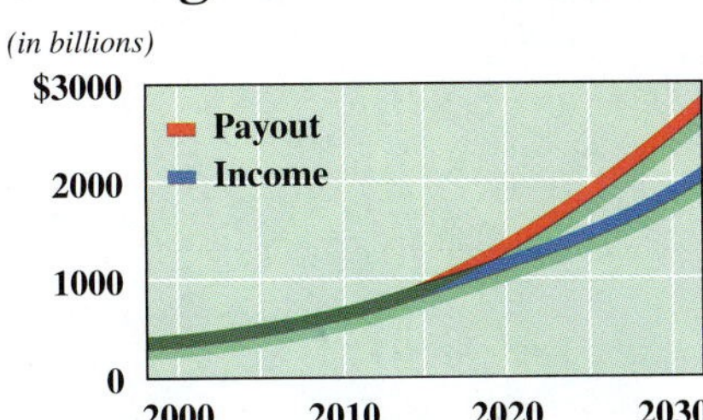

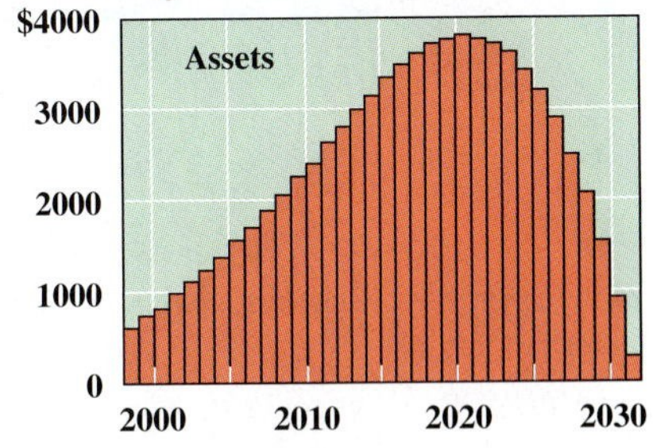

Estimated Operations of the Oasi and DI trust funds, excluding Interest earned.
Infographics: © Jared Schneidman, Design

SOURCE: From Reader's Digest, December 1998, p. 75.

(a) The payout and income line graphs show that payout and income are expected to be essentially the same for the years 1998 though 2015 and that payout should exceed income for the period 2015 through 2032. That being the case, what must be the explanation for the expected growth in assets as shown in the assets bar graph for the interval 1998 through 2020?

(b) Do the graphs make a convincing case that the present Social Security laws need to be changed?

25. Line graphs are certainly useful in indicating trends over time. However, extrapolating (i.e., projecting values beyond the period covered by the graph) can be very risky. The observed trend simply may not continue.

(a) Suppose you had guessed the number of PCs that would be shipped worldwide in 2001 on the basis of the part of the graph shown up to 2000. What might you reasonably have guessed the figure would be for 2001?

(b) By how much would you have overestimated the number of PCs shipped in 2001, given the graph as shown?

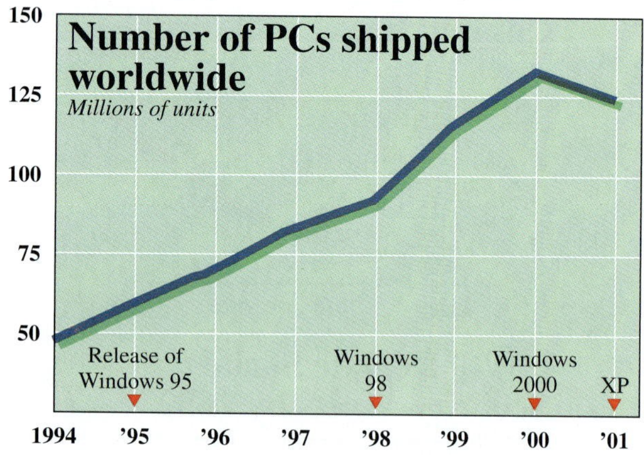

SOURCE: *Gartner Dataquest.*

Using a Computer

For the following problems, use a statistical package from your computer laboratory.

26. Consider the SAT scores recorded here.

Student	Verbal Score	Math Score
Dina	502	444
Carlos	590	520
Rosette	585	621
Broz	487	493
Coleen	585	602
Deiter	481	572
Darin	605	599
Luana	547	499

(a) Make a bar graph to summarize the above data for verbal scores.

(b) Make a double bar graph to summarize the above data, with one of each pair of bars for verbal scores and one for math scores.

27. For fiscal year 2002, federal expenditures were divided as follows:

Social programs—14%

Physical, human, and community development—14%

Net interest on debt—14%

Defense, veterans, and foreign affairs—24%

Social security, medicare, and other retirement—32%

Law enforcement and general government—2%

(a) Make a pie chart to reflect these data.

(b) Make a bar graph that reflects these data.

28. The numbers of successful field goals kicked by the various NFL kickers during the first 12 games of the 2004 season are as shown.

19	21	8	8	16	18
24	22	14	14	27	18
21	24	15	14	18	20
14	18	17	13	13	15
14	19	23	19	22	15

(a) Make a dot plot to organize and display the data.

(b) Draw a histogram to organize and display the data.

(c) Do any of the data points seem to be largely atypical? Explain.

29. Make a line graph to graphically display these data.

Population of Washington in Millions	2.10	2.32	2.63	2.83	3.12	3.45	3.80	4.18	4.81	5.76
Year	1960	1965	1970	1975	1980	1985	1990	1995	2000	2005

30. Use the computer to draw a pie chart for the data from problem 7, and measure the angles to see how it compares with your drawing for problem 7.

From State Student Assessments

31. (Illinois, Grade 3)

This graph shows the results of a classroom vote on favorite pets.

How many more students voted for dogs than for cats?

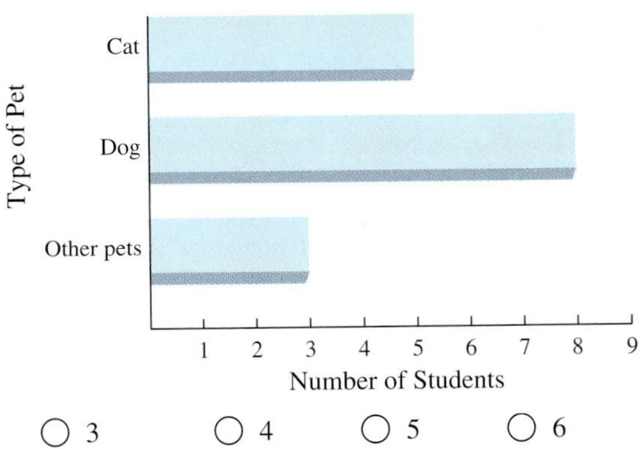

○ 3 ○ 4 ○ 5 ○ 6

32. (Michigan, Grade 4)

Miguel has a birdfeeder in his backyard. He made the picture graph below to show how many of each kind of bird he has seen. Miguel has seen 10 purple finches.

Number of Birds Seen

Black-Capped Chickadees	
Cardinals	
Goldfinches	
Purple Finches	

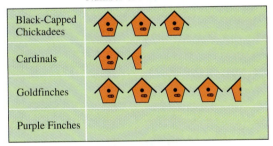

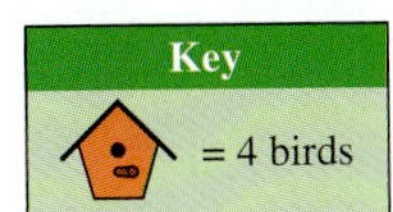

What should the picture graph show for purple finches?

A.  B. 

C.  D. 

33. (Minnesota, Grade 5)

Use the following pie graph to answer the question.

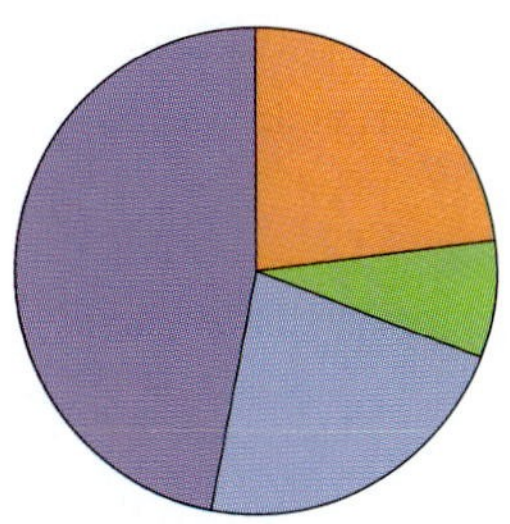

Which set of numbers would best fit this pie graph?

○ A. 54, 8, 30, 8

○ B. 47, 23, 8, 22

○ C. 51, 17, 15, 17

○ D. 27, 26, 24, 23

34. (Michigan, Grade 4)

Directions: Solve the following problem. There may be more than one way to answer correctly. Show as much of your work as possible.

Tyler read 9 books.

Lauren read 6 books.

Kyle read 5 books.

Emily read 12 books.

Use the data and make a graph. Write three questions that could be answered by using the data on this graph.

35. (Massachusetts, Grade 4)

Use the information in the line graph below to answer the question.

The graph shows the weight gain of a puppy during its first week of life. Which is NOT true about the weight of the puppy?

A. The puppy gained 2 ounces between Day 3 and Day 4.

B. The puppy weighed $1\frac{1}{2}$ pounds on Day 5.

C. The puppy's weight doubled during the first week.

D. The puppy's weight tripled during the first week.

Weight of Puppy During First Week

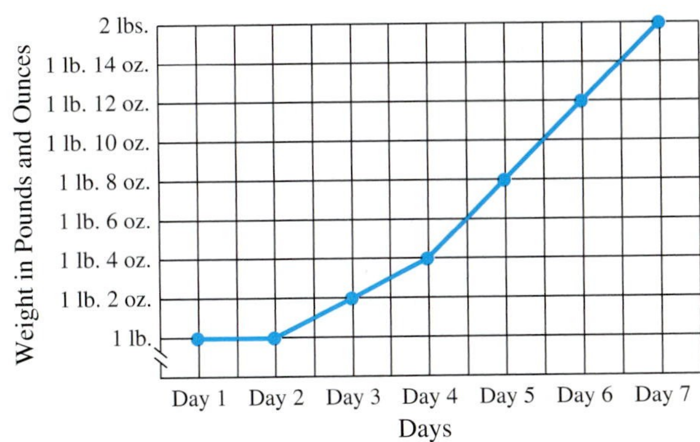

For Review

36. Write these fractions in decimal form.

(a) $\frac{1}{8}$ (b) $\frac{7}{40}$ (c) $\frac{17}{250}$ (d) $\frac{7}{20}$

37. Write these fractions in decimal form.

(a) $\frac{1}{9}$ (b) $\frac{3}{7}$ (c) $\frac{3}{14}$ (d) $\frac{7}{15}$

38. Write these decimals as fractions in reduced form.

(a) 0.375 (b) 0.3125 (c) 0.444 (d) 0.33

39. Write these decimals as fractions in reduced form.

(a) $3.7444\ldots = 3.7\overline{4}$ (b) $.\overline{02}$

(c) $.0\overline{2}$ (d) $31.\overline{72}$

(e) $4.7\overline{314}$ (f) $.431\overline{23}$

9.2 MEASURES OF CENTRAL TENDENCY AND VARIABILITY

Measures of Central Tendency

Consider the data of Table 9.1 and the corresponding dot plot of Figure 9.1 on page 525. The dot plot is a considerable improvement over the disorganized raw data for assessing the performance of the class. We can see at a glance that most of the grades lie between 62 and 96, with a large cluster between 75 and 80. We also see that the lowest grade is 49 and the highest grade is 96. But even more definitive information might be desired. For example,

- What is the "typical" grade for the class?
- How did most of the students do?
- Did many of the students perform markedly differently from the bulk of the class?

There are several different possibilities for answering these questions. First consider the following data sets, representing grades on tests, and their corresponding line plots.

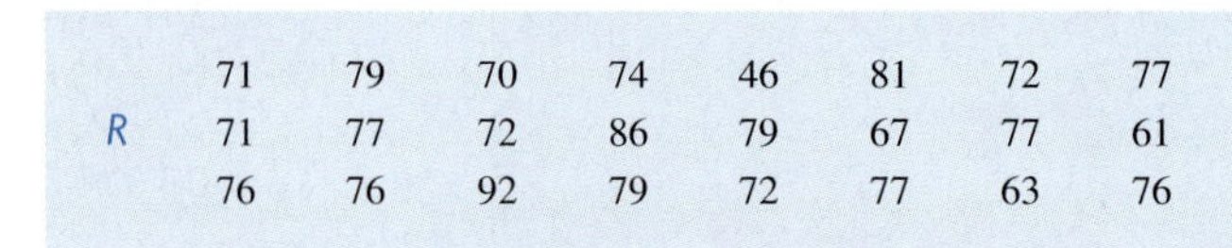

	71	79	70	74	46	81	72	77
R	71	77	72	86	79	67	77	61
	76	76	92	79	72	77	63	76

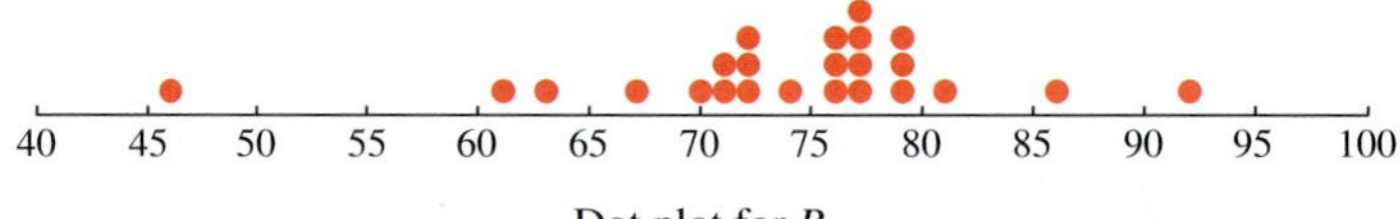

Dot plot for R

	72	77	75	75	67	76	69	76	71
S	68	77	79	82	73	69	76	68	
	69	71	78	72	79	74	73	73	

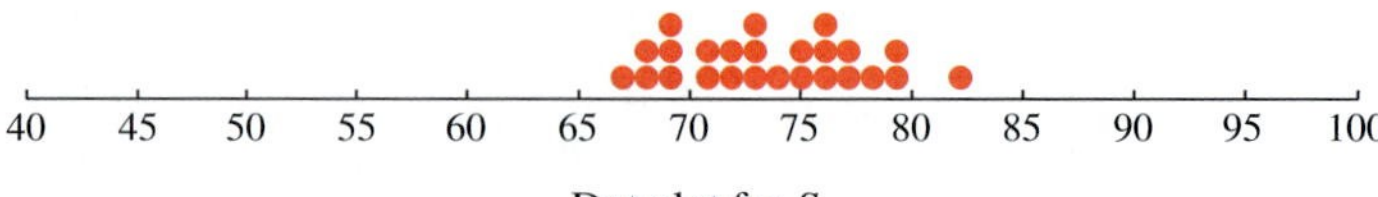

Dot plot for S

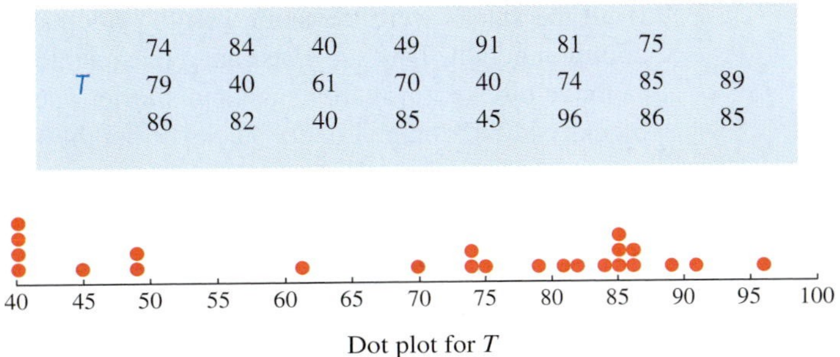

	74	84	40	49	91	81	75	
T	79	40	61	70	40	74	85	89
	86	82	40	85	45	96	86	85

Dot plot for T

The dot plots help to identify the following characteristics of the data sets.

1. The scores in data set R seem to cluster about 76, even though they range all the way from 46 to 92 and are generally rather widely spread. Apparently, some of the students did very well, while others did quite poorly. If we had to choose a single grade as typical of the entire class, it would probably be about 76.
2. The scores in data set S are much less spread out than those of R, ranging only from 67 to 82. Thus, all the students did reasonably well, with none outstandingly good and none outstandingly poor. The grades seem to cluster about 74, and we would probably select this grade as reasonably typical of the entire class.
3. The scores in data set T are very widely spread, ranging all the way from 40 to 96. Clearly, a number of students did very poorly, while a substantial number did quite well. Here, it is more difficult to select a single grade as typical. If we ignore the very poorest grades, we may want to choose 85 as typical. But there are many scores that differ widely from this.

Apparently, two features naturally arise when it comes to analyzing a data set:

- the typical, or central, value of the data, and
- the dispersion, or spread, of the data about the central value.

We now consider various standard approaches to identifying and quantifying these features.

The Mean

The dot plots just considered allow one to develop an intuitive, but not very precise, notion of the typical, or central, value of a set of data. One very useful and precisely defined central value is the **mean,** frequently called the **arithmetic mean** or **average.** An effective manipulative device for introducing this notion to students that is quite independent of, and different from, the dot plots just considered is provided by a simple set of blocks. For example, consider the data 7, 5, 7, 3, 8, and 6. Arrange a number of blocks in six stacks containing 7, 5, 7, 3, 8, and 6 blocks, respectively, as follows, and ask what the typical or average height of all the stacks is.

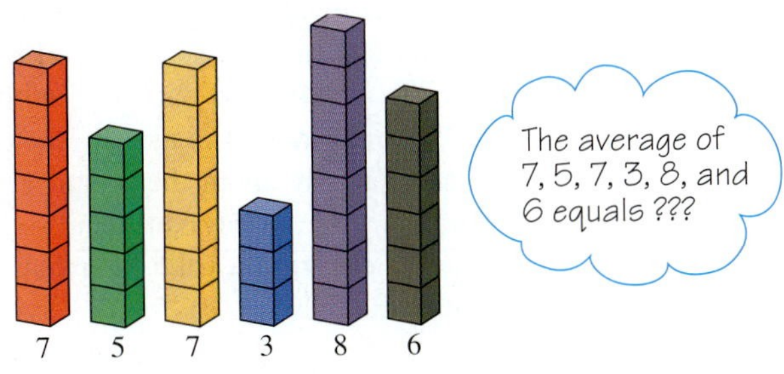

The average of
7, 5, 7, 3, 8, and
6 equals ???
7 5 7 3 8 6

If all the stacks were the same height, the answer would be obvious—it would be their common height. This suggests that a reasonable approach to answer the question might be to move blocks from taller stacks to shorter ones in an effort to even them up. Indeed, the blocks can be arranged in six stacks of height 6 as shown, and this suggests that the average height is 6.

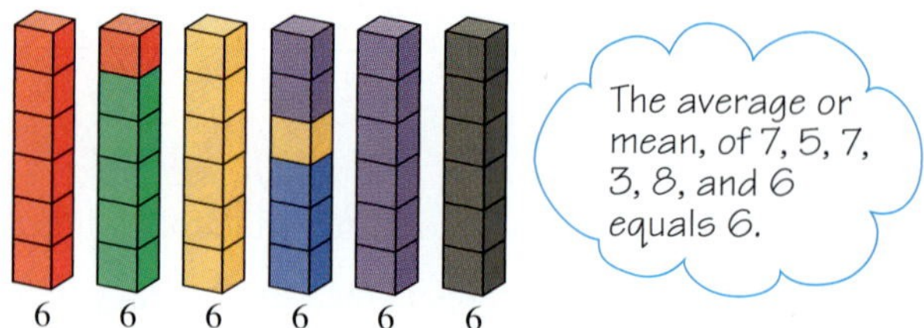

6 6 6 6 6 6
The average or mean, of 7, 5, 7, 3, 8, and 6 equals 6.

Arithmetically, the number of blocks in the two arrangements has not changed, so

$$6 \cdot 6 = 7 + 5 + 7 + 3 + 8 + 6, \text{ and } 6 = \frac{7 + 5 + 7 + 3 + 8 + 6}{6}.$$

Thus, the average or typical height of the original stacks is found by finding the sum of their heights and dividing by the number of stacks. This naturally leads to the following definition.

DEFINITION *The Mean of a Set of Data*

The **mean,** or **average,** of a collection of values is $\bar{x} = S/n$, where S is the sum of the values and n is the number of values.

For the data set R, we compute the mean by adding all scores and dividing by 24, the number of scores. Thus,

$$\begin{aligned}\bar{x} &= (71 + 79 + 70 + 74 + 46 + 81 + 72 + 77 + 71 + 77 + 72 + 86 \\ &\quad + 79 + 67 + 77 + 61 + 76 + 76 + 92 + 79 + 72 + 77 + 63 + 76)/24 \\ &\doteq 73.8.\end{aligned}$$

This is reasonably close to our informal feeling that 76 is reasonably representative of the scores in the data set. Actually, 73.8 is somewhat smaller than expected, and this shows that the mean of a data set is sensitive to atypical data values like 46. In this case, the mean of the data with 46 omitted is 75—very close to our informal determination.

For the data sets S and T, we find that the means are respectively 73.6 and 71.2. For S, the mean gives a very good estimate of what we intuitively felt was the typical or central value. For T, the mean of 71.2 seems unduly low, and this is again a reflection of the fact that the mean can be strongly affected by the presence of extremely atypical values like 40, 40, 40, 40, 45, 49, and even 61. Without these values, the mean is a much more expectable 82.6. At any rate, it is usually the case that, in most data sets, most data values cluster reasonably closely about the mean. This is particularly true if we have criteria for deciding when data values are too atypical so that we may delete them from consideration. We develop such a criterion a little later but, for now, we consider other frequently used measures of central values.

The Median

The **median** of a collection of values is the middle value in the collection when the values are arranged in order of increasing size, or the average of the two middle values in case the number of values is even.

> **DEFINITION** *The Median of a Set of Data*
> Let a collection of n data values be written in order of increasing size. If n is odd, the **median,** denoted by $\hat{x}$, is the middle value in the list. If n is even, $\hat{x}$ is the average of the two middle values.

EXAMPLE 9.4 DETERMINING A MEDIAN

Determine the median of the data in data set R on page 546.

Solution

The scores in R are arranged in order in the line plot of R. Since there are 24 scores, the median is the average of the twelfth and thirteenth scores. Thus, we see that $\hat{x} = 76$.

Note that the median in the preceding example not only closely approximates the mean, but also agrees reasonably well with our intuitive idea of the typical value of the collection of scores.

Also, it follows from the definition that the median is a data value if the number of data values is odd and is *not* necessarily a data value if the number of values is even. Thus, the median of the nine scores

24, 25, 25, 27, 29, 31, 32, 34, 37

is 29, the fifth score, while the median of the 10 scores

42, 42, 43, 44, 44, 46, 47, 47, 47, 49

is 45, the average of the two middle scores.

The Mode

Another value often taken as "typical" of a set of data is the value occurring most frequently. This value is called the **mode.** From the definition, it is clear that there may be more than one mode. For example, data set S has three modes, 69, 73, and 76. Even so, it may be the case that a mode gives a better indication of the typical value of the data than either the median or the mean. Moreover, unlike the mean, neither the median nor the mode is affected by the existence of extremely atypical values.

> **DEFINITION** *A Mode of a Set of Data*
> A **mode** of a collection of values is a value that occurs the most frequently. If two or more values occur equally often and more frequently than all other values, there are two or more modes.

EXAMPLE 9.5 DETERMINING MEANS, MEDIANS, AND MODES

Determine the mean, median, and mode for each of the data sets *R*, *S*, and *T* on pages 546 and 547, and discuss which measures are most representative of the respective data sets.

Solution Let $\bar{x}_R$, $\hat{x}_R$, $\bar{x}_S$, $\hat{x}_S$, $\bar{x}_T$, and $\hat{x}_T$ denote the means and medians of *R*, *S*, and *T*, respectively. We have already seen that $\bar{x}_R = 73.8$ and that $\hat{x}_R = 76$. Since the mode is the most frequently occurring score (or the several such scores if they occur equally often and more frequently than all other scores), the mode of *R* is 77. In this case, all three measures are reasonably representative of the scores making up the data set.

Since *S* has 25 scores,

$$\begin{aligned}\bar{x}_S &= (72 + 77 + 75 + 75 + 67 + 76 + 69 + 76 + 71 + 68 + 77 + 79 + 82 \\ &\quad + 73 + 69 + 76 + 68 + 69 + 71 + 78 + 72 + 79 + 74 + 73 + 73)/25 \\ &\doteq 73.6.\end{aligned}$$

Also, $\hat{x}_S$ is the middle, or thirteenth, score. Thus, counting on the line plot, we find that $\hat{x}_S = 73$. Finally, *S* has three modes, 69, 73, and 76, since each occurs three times and more often than any other score. We observe that the mean, the median, and the middle mode all seem to reasonably represent the set of scores in S.

Finally, since *T* contains 23 scores,

$$\begin{aligned}\bar{x}_T &= (74 + 84 + 40 + 49 + 91 + 81 + 75 + 79 + 40 + 61 + 70 + 40 \\ &\quad + 74 + 85 + 89 + 86 + 82 + 40 + 85 + 45 + 96 + 86 + 85)/24 \\ &\doteq 71.2, \\ \hat{x}_T &= 79,\end{aligned}$$

and the mode of *T* is 40. As noted earlier, *T* is difficult to characterize. The mean, $\bar{x}_T$, is strongly affected by the several very low scores and so does not seem to fairly represent the data set. Similarly, the mode of 40 is clearly not representative. In this case, the median seems to be the most representative value.

EXAMPLE 9.6 DETERMINING AN AVERAGE

All 12 players on the Uni Hi basketball team played in their 78-to-65 win over Lincoln. Jon Highpockets, Uni Hi's best player, scored 23 points in the game. How many points did each of the other players average?

Solution

Understand the Problem

The problem is to determine averages when we are not explicitly given the data values. What we do know is that Uni Hi scored 78 points, that Jon Highpockets scored 23 of the points, and that all 12 players on the team played in the game.

Devise a Plan

Since the average score for each of the 11 players other than Jon is the sum of their scores divided by 11, we must determine the sum of their scores.

Carry Out the Plan

Since Uni Hi scored a total of 78 points and Jon scored 23, the total for the rest of the team must have been $78 - 23 = 55$ points. Therefore, the average number of points for these players is $55 \div 11 = 5$.

Look Back

The solution depended on knowing the definition of average. The real question was how many points were scored by all the players on Uni Hi's team other than Jon and how many such players there were. But those figures, and hence the solution to the problem, were easily obtained by subtraction.

EXAMPLE 9.7 DETERMINING A TYPICAL VALUE FOR A SET OF DATA

The owner/manager of a factory earned \$850,000 last year. The assistant manager earned \$48,000. Three secretaries earned \$18,000 each, and the other 16 employees each earned \$27,000.

(a) Prepare a dot plot of the salaries of the people deriving their income from the factory.

(b) Compute the mean, median, and mode of the salaries of the people deriving their income from the factory.

(c) Which is most typical of the salaries of those associated with the factory—the mean, median, or mode?

Solution

(a) The dot plot is shown here.

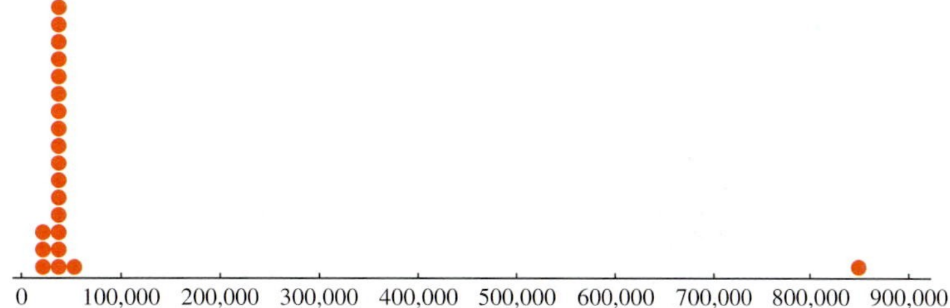

(b) The mean is

$$\bar{x} = \frac{18{,}000 + 18{,}000 + 18{,}000 + 27{,}000 + \cdots + 27{,}000 + 48{,}000 + 850{,}000}{21}$$

$$\doteq \$65{,}905.$$

The median $\hat{x}$ is the eleventh in the ordered list of salaries. Thus,

$$\hat{x} = 27{,}000 \text{ dollars.}$$

The mode is the most frequently occurring salary. Thus,

$$\text{mode} = 27{,}000 \text{ dollars.}$$

(c) The mean is clearly not typical of the salary most workers at the factory earn. The value of $\bar{x}$ is unduly affected by the huge salary earned by the owner/manager. Here, the median and mode are the same and are more typical of salaries of those deriving their income from the factory since \$27,000 is the salary of 16 of the 21 people. Note that this last sentence is really an argument that, in this case, the most typical value is the mode. That the mode and median here are equal is incidental.

While the mean is the most commonly used indicator of the typical value of a data set, the preceding example makes it clear that this choice can be quite misleading. As will be seen in the problem set for this section, it is easy to construct examples where the median is the most typical value, and other examples, like the preceding, where the mode is most typical.

Measures of Variability

The most useful analysis of data would reveal both the center (typical value) and the *spread,* or *variability,* of the data. We now consider how the spread of data is determined. The simplest measure is the **range,** the difference between the smallest and largest data values. This certainly tells something about how the data occurs, but it is often misleading, particularly if the data set contains a few extremely low or high values that are quite atypical of most of the other values. A better understanding is obtained by determining **quartiles.**

Somewhat imprecisely, the lower quartile Q_L of a set of data arranged in order of increasing size and having median $\hat{x}$ is the median value of the data values *less than* $\hat{x}$. Similarly, the upper quartile Q_U is the median of the data values *greater than* $\hat{x}$. The difficulty with this definition is that it is easily misunderstood. For example, for the data set

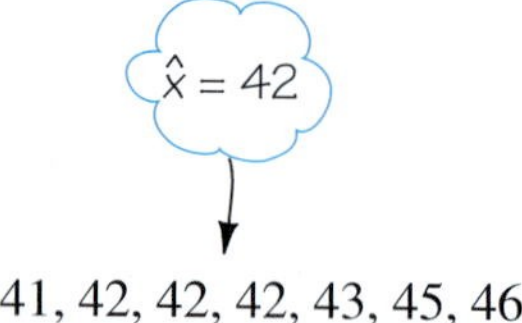

$$40, 41, 42, 42, 42, 43, 45, 46,$$

$\hat{x} = (42 + 42)/2 = 42$ and lies between the two 42s as shown. Thus, $Q_L = 41.5$ is the median of 40, 41, 42, and 42. It is *not* the median of the scores 40 and 41. Similarly, $Q_U = 44$ is the median of 42, 43, 45, and 46 and *not* the median of the scores 43, 45, and 46. More precisely, we have the following definition.

DEFINITION *Upper and Lower Quartiles*
Consider a set of data arranged in order of increasing size. Let the number of data values, n, be written $n = 2r$ for n even, or $n = 2r + 1$ for n odd, for some integer r. In either case, the **lower quartile,** denoted by Q_L, is the median of the first r data values. Also, the **upper quartile,** denoted by Q_U, is the median of the last r data values.*

EXAMPLE 9.8 DETERMINING QUARTILES

Determine $\hat{x}$, Q_L, and Q_U for these data sets. Note that these values sometimes are and sometimes are not data values.

(a) $A = \{12, 7, 14, 15, 9, 11, 10, 11, 0, 8, 17, 5\}$
(b) $B = \{27, 14, 13, 12, 26, 22, 24, 22, 23, 19, 10, 19, 22\}$
(c) $C = \{16, 22, 20, 15, 12, 14, 16, 14, 21\}$
(d) $D = \{32, 26, 29, 25, 26, 27, 29, 28, 29, 25\}$

*The precise definition of quartiles is not entirely standardized. For example, the lower and upper quartiles for a set of $2r + 1$ data values are often taken as the medians of the first and last $r + 1$ (rather than r) values, respectively, in the ordered list.

Solution

(a) First, we order the values in A to obtain the following list:

$$\begin{array}{cccccccccccc} 0 & 5 & 7 & 8 & 9 & 10 & 11 & 11 & 12 & 14 & 15 & 17 \end{array}$$

with arrows marking Q_L (between 7 and 8), $\hat{x}$ (between 10 and 11), and Q_U (between 12 and 14).

Since there are 12 data values, the median is the average of the sixth and seventh values, i.e., $\hat{x} = \dfrac{10 + 11}{2} = 10.5$, as indicated above. Since $2r = 12$, we have $r = 6$. Thus, $Q_L = \dfrac{7 + 8}{2} = 7.5$ is the median of the six data values preceding $\hat{x}$, and $Q_U = \dfrac{12 + 14}{2} = 13$ is the median of the six data values following $\hat{x}$.

(b) The ordered list for B is as follows:

$$\begin{array}{ccccccccccccc} 10 & 12 & 13 & 14 & 19 & 19 & 22 & 22 & 22 & 23 & 24 & 26 & 27 \end{array}$$

with arrows marking Q_L (between 13 and 14), $\hat{x}$ (at 22), and Q_U (between 23 and 24).

Thus, it follows that $\hat{x} = 22$ is the middle data value. Since $2r + 1 = 13$, we have $r = 6$. Thus, $Q_L = \dfrac{13 + 14}{2} = 13.5$ is the median of the six data values preceding $\hat{x}$, and $Q_U = \dfrac{23 + 24}{2} = 23.5$ is the median of the six data values greater than $\hat{x}$.

(c) The ordered set for C is as shown here:

$$\begin{array}{ccccccccc} 12 & 14 & 14 & 15 & 16 & 16 & 20 & 21 & 22 \end{array}$$

with arrows marking Q_L (between 14 and 14), $\hat{x}$ (at 16), and Q_U (between 20 and 21).

Since there are nine data values, $\hat{x} = 16$, the middle data value. Also, $2r + 1 = 9$, so $r = 4$, $Q_L = \dfrac{14 + 14}{2} = 14$ is the median of the four data values less than $\hat{x}$, and $Q_u = 20.5$ is the median of the four data values greater than $\hat{x}$.

(d) The ordered data set for D is as follows:

$$\begin{array}{cccccccccc} 25 & 25 & 26 & 26 & 27 & 28 & 29 & 29 & 29 & 32 \end{array}$$

with arrows marking Q_L (at 26), $\hat{x}$ (between 27 and 28), and Q_U (at 29).

Since there are 10 data points, $\hat{x}$ is the average of the fifth and sixth values. Also, since $2r = 10$, we have $r = 5$. Thus, $Q_L = 26$ is the third of the five data points less than $\hat{x}$, and $Q_U = 29$ is the third of the five data points greater than $\hat{x}$.

It follows from the definition that approximately 25% of the data values are less than or equal to Q_L, approximately 25% lie between Q_L and $\hat{x}$, approximately 25% lie between $\hat{x}$ and Q_U, and approximately 25% are greater than or equal to Q_U. Thus, approximately

50% of the values in a data set lie between Q_L and Q_U, and the length of this interval, called the **interquartile range** and denoted by **IQR,** provides a good measure of the spread of the data. In fact, if a data value falls below Q_L by more than $1.5 \cdot \text{IQR}$ or above Q_U by more than $1.5 \cdot \text{IQR}$, it is called an **outlier** and is often ignored when analyzing the data set. Thus, "outlier" is the term applied to those values referred to earlier that are so atypical of the values in a data set that they are often ignored.*

DEFINITION *Outlier*
An **outlier** in a set of data is a data value that is *less than* $Q_L - (1.5 \cdot \text{IQR})$ or *greater than* $Q_U + (1.5 \cdot \text{IQR})$.

EXAMPLE 9.9 THE MEDIAN, THE QUARTILES, THE INTERQUARTILE RANGE, AND THE OUTLIERS

Determine the median, the quartiles, the interquartile range, and the outliers for data set *R* on page 546.

Solution

The values in *R* are ordered in the line plot on page 546. Since *R* contains 24 data values, the median is the average of the twelfth and thirteenth data values, i.e., $\hat{x} = (76 + 76)/2 = 76$. Also, $2r = 24$, so $r = 12$ and Q_L and Q_U are the medians of the first and last 12 data values, respectively. Thus, $Q_L = (71 + 71)/2 = 71$ and $Q_U = (77 + 79)/2 = 78$. Therefore, the interquartile range is $\text{IQR} = 78 - 71 = 7$. Finally, since

$$Q_L - 1.5 \cdot \text{IQR} = 71 - (1.5 \cdot 7) = 71 - 10.5 = 60.5$$

and

$$Q_U + 1.5 \cdot \text{IQR} = 78 + (1.5 \cdot 7) = 78 + 10.5 = 88.5,$$

it follows that 46 and 92 are outliers.

Symbolically, if the 24 points shown represent the ordered data values in *R*, then Q_L, $\hat{x}$, Q_U, the interquartile range, and the outliers are as shown.

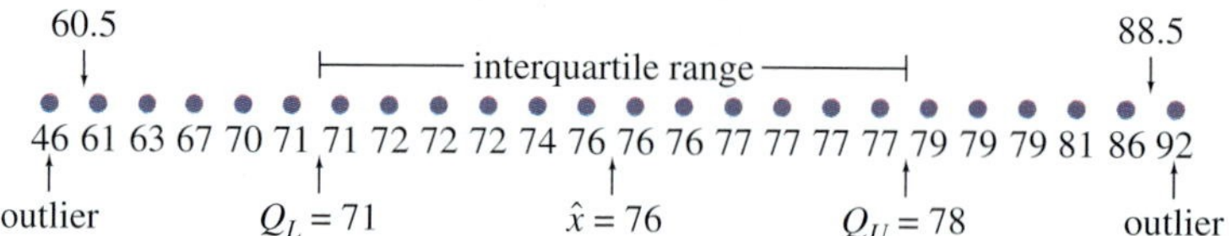

Box Plots

The least and greatest scores, or the **extremes,** along with the lower and upper quartiles and the median, give a concise numerical summary, called the **5-number summary,** of a set of data. Since the median of the data in Example 9.9 is 76 and the extremes are 46 and 92, the 5-number summary is 46 − 71 − 76 − 78 − 92. Moreover, a **box plot,** often called a **box and whisker plot,** gives a vivid graphical representation of the 5-number summary.

*As with the median, the definitions of interquartile range and outlier are not entirely standardized. The choices made here are among the most common.

DEFINITION *Box Plot*
A **box plot** consists of a central box extending from the lower to the upper quartile, with a line marking the median and line segments, or whiskers, extending outward from the box to the extremes.

For example, the box plot for Example 9.9 is shown in Figure 9.14.

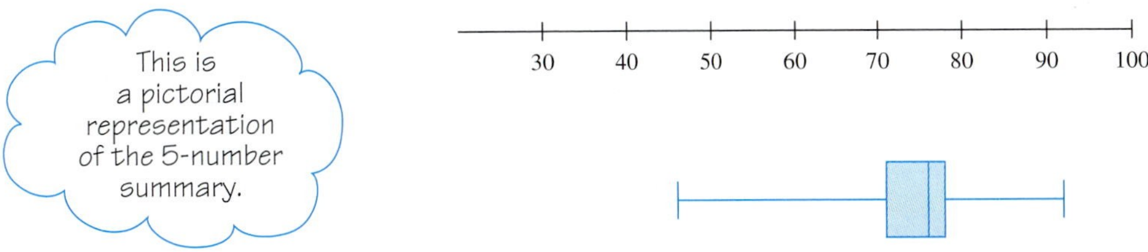

Figure 9.14
Box plot for data set R

An additional advantage of box plots is that they make it possible to make useful comparisons between data sets containing widely differing numbers of values. This is made clear in the next example.

EXAMPLE 9.10 MAKING BOX PLOTS FOR COMPARISONS

The data below are the final scores of male and female students in Calculus I. Draw box plots to compare the distribution of women's scores with the distribution of men's scores.

Women's scores: 95, 79, 53, 78, 71, 88, 77, 80, 79, 79

Men's scores: 84, 85, 53, 77, 66, 81, 79, 59, 65, 61, 81, 68,
68, 80, 76, 87, 85, 74, 92, 76, 70, 85, 55, 79,
74, 80, 73, 48, 66, 83, 48, 60, 87, 58, 64, 78,
82, 69, 76, 83, 94, 86, 73, 85, 75, 69, 49, 52,
59, 68, 65, 75, 31, 69, 73, 56, 95

Solution

To make the plots, we need the 5-number summaries. First arrange the scores in order of increasing size.

Women's scores: 53, 71, 77, 78, 79, 79, 79, 80, 88, 95

Men's scores: 31, 48, 48, 49, 52, 53, 55, 56, 58, 59, 59, 60,
61, 64, 65, 65, 66, 66, 68, 68, 68, 69, 69, 69,
70, 73, 73, 73, 74, 74, 75, 75, 76, 76, 76, 77,
78, 79, 79, 80, 80, 81, 81, 82, 83, 83, 84, 85,
85, 85, 85, 86, 87, 87, 92, 94, 95

For the women, the extreme scores are 53 and 95 and the median is 79, the average of the fifth and sixth scores. The lower quartile is 77, the median of the first five

School Book Page

Statistics in Grade Five

Chapter 1 Lesson 4

Range, Mode, and Median

You Will Learn
how to find the range, mode, and median for a set of data

Vocabulary

range
difference between greatest number and least number

mode
number that occurs most often

median
middle number when data are in order

Learn

Suppose you wanted to buy a mountain bike. The line plot below shows the prices of different mountain bikes.

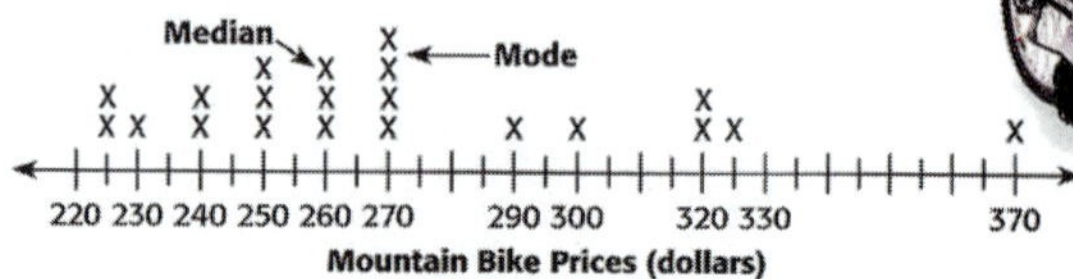

range greatest number ($370) – least number ($225): $145

mode number that occurs most often in a set of data: $270

median middle number when data are ordered: $260

There can be more than one mode. If the number of data items is even, the median is halfway between the two middle numbers.

You can find the mode and median of the test scores in this line plot.

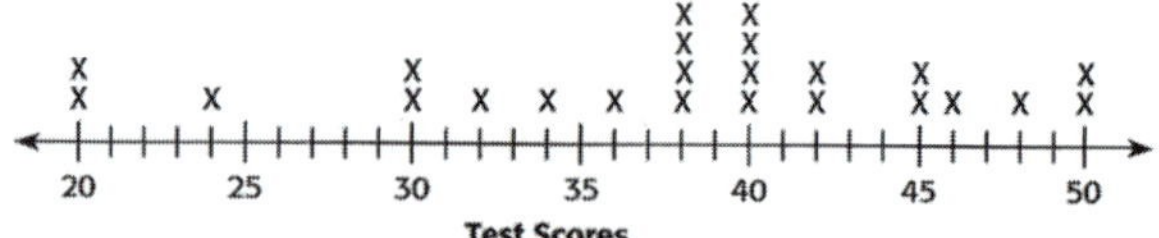

There are two modes: 38 and 40. There are 24 scores, so the median is halfway between the 12th and 13th scores. The median score is 39.

Problem Solving Hint
To find the median, put data items in order from least to greatest.

Talk About It

Will a set of data always have a range, a mode, and a median? Explain.

Check

Using Data Use the Data File on page 6 to answer **1–4.**

1. Compare the ranges for girls' and boys' arm spans.
2. Which plot has only two modes?
3. Is the median of the girls' or boys' arm spans greater?
4. **Reasoning** Do boys or girls tend to have longer arm spans?

16 Chapter 1 • Data, Graphs, and Facts Review

SOURCE: From Scott Foresman–Addison Wesley Math, Grade 5, p. 16, by Randall I. Charles et al.

Questions for the Teacher

1. If you were teaching this lesson, what points would you try to bring out? Does the range accurately reflect the spread of the data for either of the two data sets shown in the line plots? Are any of the data values outliers?
2. Choose a single number for each data set that you intuitively feel is most typical of the data set. Explain the reasons for your choice.
3. Discuss how well you feel the medians and modes typify the values in the two data sets.
4. The problems at the bottom of the above page refer to an earlier activity where the students measured each other's arm spans and then analyzed the data. How do you think students would react to such an activity?

scores. The upper quartile is 80, the median of the last five women's scores. Thus, the 5-number summary of the women's scores is

$$53 - 77 - 79 - 80 - 95.$$

Similarly, the 5-number summary of the men's scores is

$$31 - 64 - 74 - 81 - 95.$$

These summaries give the box plots shown.

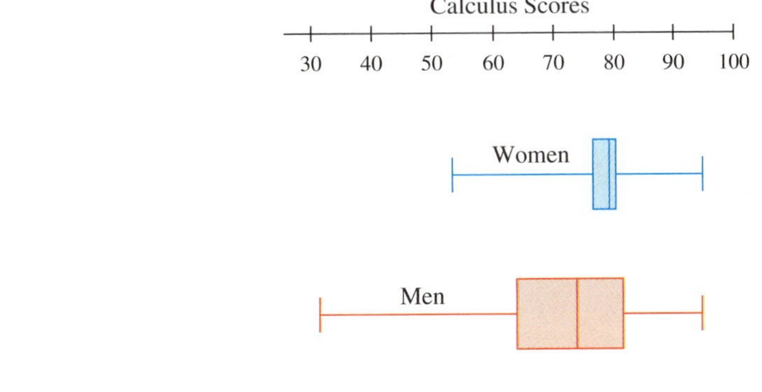

The box plots of Example 9.10 give a precise visual comparison of the performances of female and male students in calculus, even though the numbers of students are quite different. One might reasonably speculate as to why the distributions of grades differ as they do. (For example, it is almost invariably the case that larger data sets have both smaller and larger extreme values, as suggested here.) Nevertheless, one rather clear indication is that the female students in the class were no less able than the male students.

The Standard Deviation

We have already observed that the range is one measure of the spread of a data set. It is not a very precise measure, however, since it depends only on the extreme data values, which may differ markedly from the bulk of the data. This deficiency is largely remedied by the 5-number summary and its visualization by a box plot. However, an even better measure of variability is the **standard deviation.** (See problems 7 and 8 in the problem set at the end of this section. These problems are answered in the answer section in the text. There, we discuss other possibilities for measuring variability, as well as a reason for choosing the standard deviation as defined here.)

DEFINITION *The Standard Deviation of a Set of Data*
Let $x_1, x_2, x_3, \ldots, x_n$ be the values in a set of data and let $\bar{x}$ denote their mean. Then

$$s = \sqrt{\frac{(\bar{x} - x_1)^2 + (\bar{x} - x_2)^2 + \cdots + (\bar{x} - x_n)^2}{n}}$$

is the **standard deviation.***

*For important technical reasons, professional statisticians replace the n in the denominator of the fraction under the square root sign by $n - 1$. The difference is small, however, so to avoid confusion, we use the definition given here.

EXAMPLE 9.11 COMPUTING A STANDARD DEVIATION

Compute the mean and standard deviation for this set of data.

35	42	61	29	39

Solution

$$\begin{aligned}\bar{x} &= (35 + 42 + 61 + 29 + 39)/5 \\ &\doteq 41.2. \\ s^2 &= [(41.2 - 35)^2 + (41.2 - 42)^2 + (41.2 - 61)^2 + (41.2 - 29)^2 \\ &\quad + (41.2 - 39)^2]/5. \\ &\doteq 116.96.\end{aligned}$$

Thus,

$$s \doteq \sqrt{116.96} \doteq 10.8.$$

It turns out that there is an easier formula for calculating standard deviations that, on a calculator, requires only the $\boxed{\sqrt{}}$ and $\boxed{x^2}$ keys in addition to the usual keys for arithmetic. It is only a matter of messy manipulation (which we do not reproduce here) to show that

$$s = \sqrt{\frac{x_1^2 + x_2^2 + \cdots + x_n^2}{n} - \bar{x}^2},$$

where $x_1, x_2, \ldots, x_n$ are the data values and $\bar{x}$ is their mean.

EXAMPLE 9.12 ALTERNATIVE CALCULATION OF THE STANDARD DEVIATION

Calculate the mean and standard deviation of the data set in Example 9.11 using the alternative formula just given.

Solution

The mean is calculated as in Example 9.11. Then

$$s = \sqrt{\frac{35^2 + 42^2 + 61^2 + 29^2 + 39^2}{5} - 41.2^2} \doteq 10.8.$$

Scientific and business calculators frequently have built-in statistics routines that make the calculation of means and standard deviations even simpler. Alternatively, these computations can be performed automatically on a computer using a spreadsheet or other appropriate software. It is important to note that calculator and computer programs for computing standard deviations offer two alternatives that differentiate between the standard deviation of a population and the standard deviation of a sample. The standard deviation we have discussed is the standard deviation of a population. If it is not clear which one you are obtaining when you compute a standard deviation by machine, compute both and use the smaller of the two values obtained. For example, for the data values 1, 2, and 3, the options obtained are 1 and 0.82. For our present purpose, you should use 0.82.

Just as the mean is an indication of the typical value of a set of data, the standard deviation is a measure of the typical deviation of the values from the mean.* If the standard deviation is large, the data are more spread out; if it is small, the data are more concentrated near the mean. This is immediately apparent from the dot plots of data sets R, S, and T discussed earlier. In Figure 9.15, these dot plots are reproduced again, with the

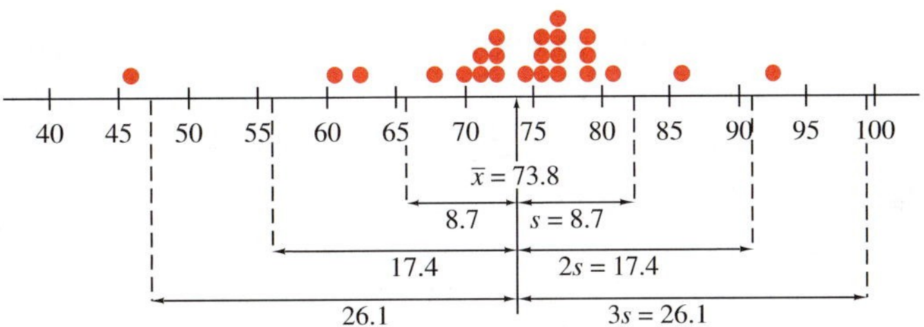

Dot plot for R: All but five data points lie within one standard deviation of the mean. Only one data point lies more than three standard deviations from the mean.

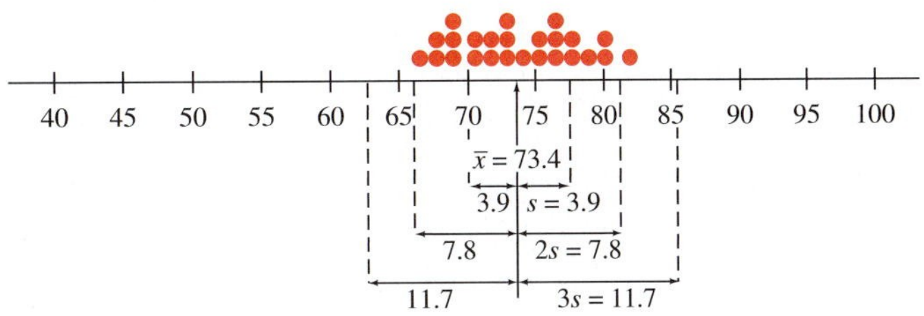

Dot plot for S: All but 10 data points lie within one standard deviation of the mean. Only one data point lies more than two standard deviations from the mean.

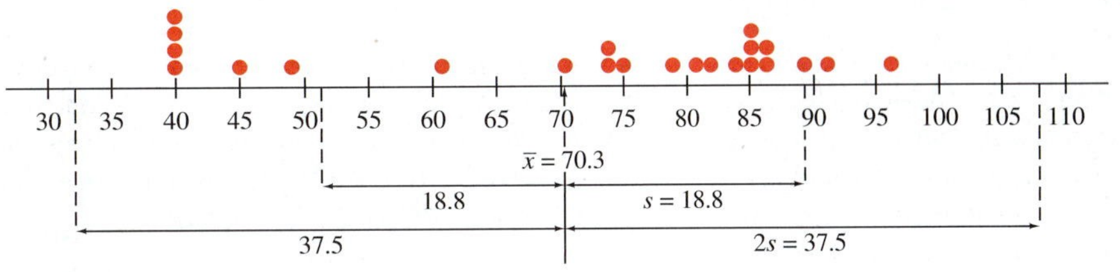

Dot plot for T: All but nine data points lie within one standard deviation of the mean. None lies beyond two standard deviations of the mean.

Figure 9.15
Dot plots for R, S, and T showing the location of the mean and the spread of the data relative to the standard deviation

*The **variance** $v = s^2$ of a set of data is also a good measure of the variability of the data. However, the more commonly used measure is the standard deviation.

addition of the location of the mean as well as the spread of each data set relative to its standard deviation. A most important fact is that, for most data sets, most data values fall within one standard deviation of the mean, and almost none lie as far as three standard deviations from the mean.

EXAMPLE 9.13 DETERMINING THE FRACTION OF DATA VALUES NEAR THE MEAN

Compute the fraction (expressed as a percent) of the data values in Example 9.11 that falls

(a) within one standard deviation of the mean.
(b) within two standard deviations of the mean.

Solution

41.2 − 10.8 = 30.4

41.2 + 10.8 = 52.0

(a) In Example 9.11, $\bar{x} = 41.2$ and $s \doteq 10.8$. Thus, those entries within one standard deviation of the mean lie between 30.4 and 52.0. A count reveals that 3 of the entries fall in this range, and we have

$$\frac{3}{5} = .60\ldots \doteq 60\%.$$

41.2 − 2(10.8) = 19.6
41.2 + 2(10.8) = 62.8

(b) Those data values within two standard deviations of the mean must lie between 19.6 and 62.8. This includes all of the data values, and we have

$$\frac{5}{5} = 1\ldots \doteq 100\%.$$

Did You Know?

A How-To Manual for Lying

"The secret language of statistics, so appealing in a fact minded culture, is employed to sensationalize, inflate, confuse, and oversimplify. Statistical methods and statistical terms are necessary in reporting the mass of data of social and economic trends, business conditions, "opinion" polls, the census. But without writers who use the words with honesty and understanding and readers who know what they mean, the result can only be semantic nonsense.

This book is a sort of primer in how to use statistics to deceive. It may seem altogether too much like a manual for swindlers. Perhaps I can justify it in the manner of the retired burglar whose published reminiscences amounted to a graduate course in how to pick a lock and muffle a footfall: The crooks already know these tricks; honest men must learn them in self defense."

Darrell Huff's book *How to Lie With Statistics* is at once a delightful and informative treatise on the art of lying with statistics.

SOURCE: Darrell Huff, How to Lie With Statistics *(New York: W. W. Norton, Inc., 1954), pp. 8 and 9.*

Problem Set 9.2

Understanding Concepts

1. Determine the mean, median, and mode for this set of data.

18	27	17	19	21	24	18	15
23	18	17	14	22	19	27	30

2. (a) Compute the mean, median, and mode for this set of data.

69	81	77	69	64	85	81	73	79
74	70	78	86	80	71	79	77	70
67	70	79	80	71	67	69	79	81

(b) Draw a dot plot for the data in part (a).

(c) Does either the mean, median, or mode seem typical of the data in part (a)?

(d) Might it be reasonable to suspect that the data in part (a) actually come from two essentially different populations (say, daily incomes from two entirely different companies)? Explain your reasoning.

3. (a) Compute the quartiles for the data in problem 2.

(b) Give the 5-number summary for the data in problem 2.

(c) Draw a box plot for the data in problem 2.

(d) Determine the interquartile range, IQR, for the data in problem 2.

(e) Identify any outliers in the data of problem 2.

4. (a) Draw side-by-side box plots to compare students' performances in class A and class B if the final grades are as shown here.

Class A:	91,	63,	65,	73,	65,	86,
	96,	75,	75,	79,	84,	72,
	80					
Class B:	87,	72,	95,	89,	69,	79,
	56,	64,	66,	67,	89,	47

(b) Briefly compare the performances in the two classes on the basis of the box plots in part (a).

(c) Determine the interquartile range, IQR, for each of the classes in part (a).

(d) Identify any outliers in the classes of part (a).

5. Use the data in problem 1 to answer the following.

(a) Compute the mean.

(b) Compute the standard deviation.

(c) What percent of the data is within one standard deviation of the mean?

(d) What percent of the data is within two standard deviations of the mean?

(e) What percent of the data is within three standard deviations of the mean?

6. (a) Choose an appropriate scale and draw a dot plot for this set of measurements of the heights in centimeters of two-year-old ponderosa pine trees.

22.2	23.5	22.5	22.6	23.0	22.8
22.4	22.2	23.0	23.3	23.9	22.7

(b) Compute the mean and standard deviation for these data.

(c) What percent of the data is within one standard deviation of the mean?

(d) What percent of the data is within two standard deviations of the mean?

(e) What percent of the data is within three standard deviations of the mean?

Teaching Concepts

7. Shawn asks why the sum of the differences of the data values from the mean,

$$(\bar{x} - x_1) + (\bar{x} - x_2) + \cdots + (\bar{x} - x_n),$$

isn't used as a measure of variability.

(a) Compute this sum for the data set in problem 6(a).

(b) How would you respond to Shawn's question?

8. During the classroom discussion resulting from Shawn's question in problem 7, Leona suggests that the sum of the absolute values of the differences of the data values from the mean be used as a measure of variability since the terms are all positive and thus cannot cancel each other out.

(a) Compute the sum of the absolute values of the differences in problem 7, part (a).

(b) Compute the mean of the absolute values of differences in problem 7, part (a); i.e., divide the sum in part (a) of this problem by 12.

(c) How would you respond to Leona?

Thinking Critically

9. On June 1, 1998, the average age of the 33 employees at Acme Cement was 47 years. On June 1, 1999, three of the staff aged 65, 58, and 62 retired and were replaced by four employees aged 24, 31, 26, and 28. What was the average age of the employees at Acme Cement on June 1, 1999?

10. **(a)** Compute the mean and standard deviation for these data.

28	34	41	19	17	23

(b) Add 5 to each of the values in part (a) to obtain 33, 39, 46, 24, 22, and 28. Compute the mean and the standard deviation for this new set of values.

(c) What properties of the mean and standard deviation are suggested by parts (a) and (b)?

11. Compute the mean and standard deviation for the data represented by each of these two histograms.

(a)

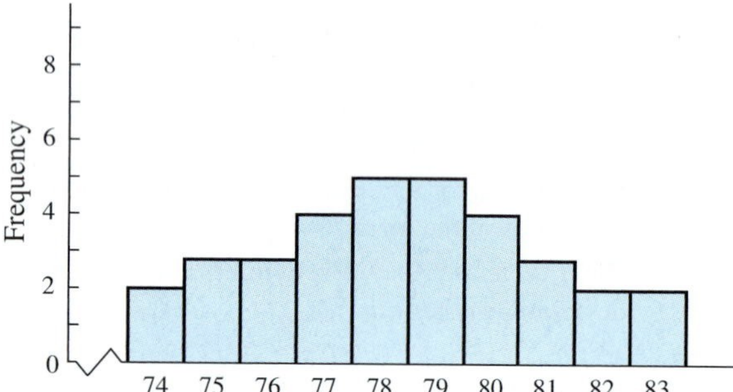

(b)

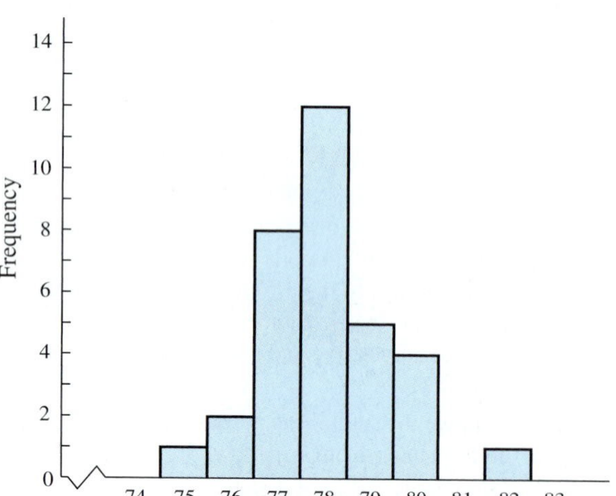

(c) Briefly explain why the standard deviation for the data of part (b) is less than that for part (a).

12. Does the mean, median, or mode seem to be the most typical value for this set of data? Explain briefly. (*Suggestion:* Draw a line plot.)

42	47	38	16	45	41	16	48	44

13. **(a)** Determine the mean, median, and mode of the data in the dot plot shown.

(b) Does the mean, median, or mode seem to be the most typical of these data? Explain briefly.

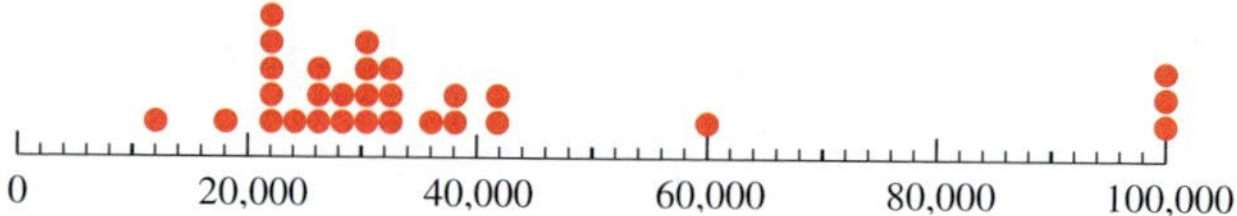

14. Produce sets of data that satisfy these conditions.

(a) mean = median < mode

(b) mean = mode < median

(c) median = mode < mean

15. **(a)** What can you conclude if the standard deviation of a set of data is zero? Explain.

(b) What can be said about the standard deviation of a set of data if the values all lie very near the mean? Explain.

16. Let Q_L, $\hat{x}$, and Q_U denote the lower quartile, median, and upper quartile, respectively, of a set of data.

(a) Create a set of data with the property that exactly 25% of the data lie in each of these ranges.

$$x < Q_L, \quad Q_L < x < \hat{x}, \quad \hat{x} < x < Q_U, \quad Q_U < x$$

(b) Create a set of data for which it is not true that 25% of the data lies in the ranges specified in part (a).

17. A collection of data contains 10 values consisting of a mix of ones, twos, and threes.

(a) If $\bar{x} = 3$ what is the data set?

(b) If $\bar{x} = 2$, what are the possibilities for the data set?

(c) If $\bar{x} = 1$ what is the data set?

(d) Could $\bar{x} = 1$ and $s \neq 0$ for this data set? Explain.

18. Compute the mean of each of these collections of data.

(a) $A = \{27, 38, 25, 29, 41\}$

(b) $B = \{27, 38, 25, 29, 41, 32\}$

(c) $C = \{27, 38, 25, 29, 41, 32, 32\}$

(d) $D = \{27, 38, 25, 29, 41, 32, 32, 32, 32, 32, 32\}$

(e) What conclusion is suggested by the calculations in parts (a) through (d)?

(f) Guess the mean of the following set of data, and then do the calculation to see if your guess is correct:

$$E = \{27, 38, 25, 29, 41, 60, 4, 60, 4\}$$

(g) What general result does the calculation in part (f) suggest?

19. **(a)** The mean of each of these collections of data is 45.

$$R = \{45, 35, 55, 25, 65, 20, 70\}$$
$$S = \{45, 35, 55, 25, 65, 20, 70, 45, 45\}$$
$$T = \{45, 35, 55, 25, 65, 20, 70, 80, 10\}$$

Which of R and S has the smaller standard deviation? No computation is needed; justify your response with a single sentence.

(b) Like the means of R and S in part (a), the mean of T is 45. Is the standard deviation for this set the same as that for S? Note that both of these sets have the same number of entries. Explain your conclusion.

20. According to Garrison Keillor, all the children in Lake Wobegon are above average. Is this assertion just a joke, or is there a sense in which it could be true?

21. If the mean of $A = \{a_1, a_2, \ldots, a_{30}\}$ is 45 and the mean of $B = \{b_1, b_2, \ldots, b_{40}\}$ is 65, compute the mean of the combined data set. (*Hint:* The answer is not 55.)

Thinking Cooperatively

Divide the class into groups of three or four students each. Make sure that all members of each small group agree on the answers required of their group.

22. As we have seen before, it often helps to understand a concept if it can be visualized using an appropriate manipulative. Work with about three other students to carry out the following activity.

(a) Suppose you want to demonstrate the idea of the mean of a set of data to fourth graders. Using a set of blocks, form stacks of heights 5, 1, 4, 7, 6, and 7. Now move blocks from higher stacks to lower stacks in an effort to form six stacks all of the same height. Can this be done? If so, how many blocks are in each stack?

(b) Determine the mean of 5, 1, 4, 7, 6, and 7.

(c) Comparing the results of parts (a) and (b), what do you conclude?

(d) How would you elaborate on the above to make the idea of the mean clear to your students? Explain carefully.

(e) How would the above idea work with these data: 5, 1, 4, 7, 11, and 7? Discuss briefly.

23. Consider the data collected in problem 19, Problem Set 9.1.

(a) Members of one or two small groups each determine the mode or modes, the mean, and the standard deviation of the data for men.

(b) Members of one or two small groups each determine the mode or modes, the mean, and the standard deviation of the data for women.

(c) Members of one or two small groups each determine the 5-number summaries and draw side-by-side box plots for the two sets of data.

(d) As a class, discuss the results obtained in (a), (b), and (c) and decide on a consensus opinion of the most appropriate length of a "yard" and on a reasonable range in which the length might lie, on the basis of the data considered. Do the results differ markedly for men and women?

24. In ancient times, the cubit was taken as the length of the human arm from the tip of the elbow to the tip of the middle finger (generally understood to vary from about 17 to about 21 inches). Working in groups, repeat problem 23, but for the cubit rather than the "yard."

25. Working in a group with about three other students, toss seven pennies 30 times and record the number of heads each time.

(a) Determine the mean and standard deviation of the data obtained.

(b) Determine what percent of the data lies within one standard deviation of the mean.

(c) What percent of the data differs from the mean by more than two standard deviations?

26. **(a)** Using a metric ruler as a straightedge and for measuring, *carefully* draw a rectangle with what you deem the most aesthetically pleasing proportions (length and width). Before making your final drawing, you may want to make several freehand sketches to help you decide which shape you prefer most. After making your final drawing, measure the length and width as accurately as possible and determine the ratio of the length of the long side to the length of the short side, accurate to one decimal place. Finally, write your ratio on the chalkboard.

(b) After all the ratios are noted on the chalkboard, determine their mean and standard deviation.

(c) As a class, decide on a single number that is most representative of the data of part (a).

(d) The ratio $(1 + \sqrt{5})/2 = 1.618\ldots$ is called the Golden Ratio. (See problem 17(a), Problem Set 3.6.) A Golden Rectangle is a rectangle for which the ratio of the length of the long side to that of the short side is the Golden Ratio. The ancient Greeks thought that the Golden Rectangle was the most aesthetically pleasing of all. Does your class seem to agree?

(e) Finally, determine what percent of the ratios lie within one standard deviation of the mean determined in part (b).

27. **(a)** Working with a group of three or four other students, use a plastic 30-centimeter ruler to measure individual "drop distances." Working in pairs in the group, one student holds the ruler vertically by the tip and lets it drop between the thumb and forefinger of the other student. The object is to see how far the ruler drops before the second student can catch it. After a few free trials, repeat the process five times, measuring as accurately as possible the distance the ruler falls each time. Record these five measurements.

(b) Compute the mean "drop distance" for yourself and for your group, and record the latter mean on the chalkboard.

(c) Determine the mean of the means on the chalkboard, as well as their standard deviation.

(d) Discuss the results of part (c) with the entire class and determine a consensus opinion of the distance that seems most typical for the members of your class. Also determine a range that encompasses the "drop distances" of most of the class.

Making Connections

28. **(a)** From the data in the bar graph shown is it possible to determine the average median income in 2000 for men 25 years old and older with education not exceeding a Master's degree? Explain briefly.

(b) What legitimate conclusions can you make on the basis of the bar graph shown?

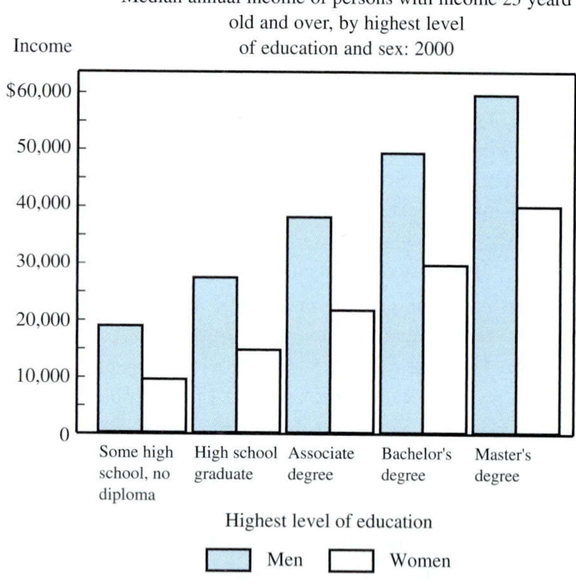

SOURCE: U.S. Department of Commerce, Bureau of the Census, Current Population Reports, *Series P-60, "Money Income in the United States: 2000."*

29. **(a)** Use the data in the pie chart shown to determine the average proceeds from the various types of state-run lottery games in 1964–1995.

(b) Determine the percentage of the profits from lotteries that is used to finance education.

(c) Why do you suppose the majority of the proceeds from lottery games is used to finance education?

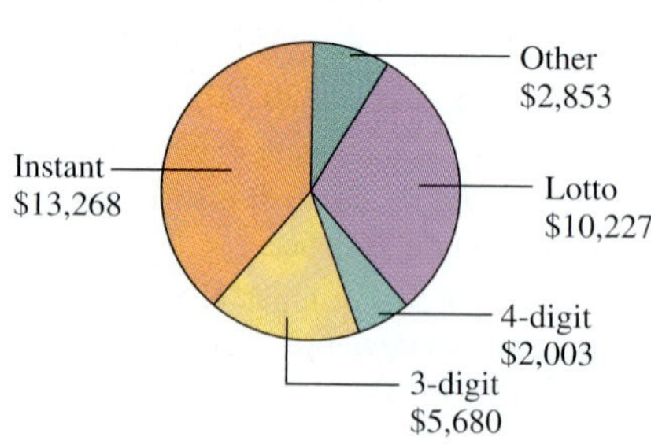

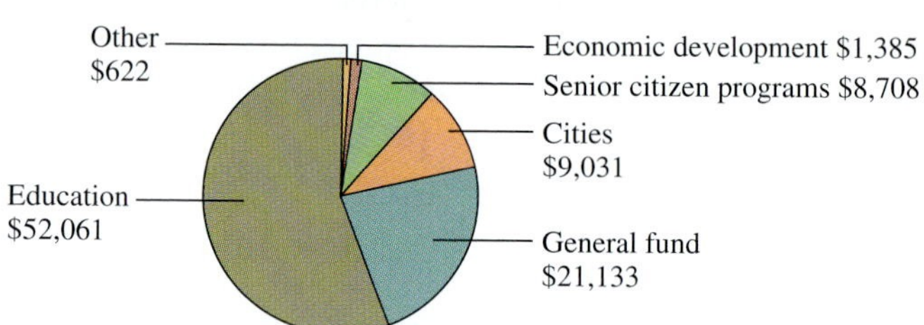

SOURCE: Chart prepared by the U.S. Bureau of the Census. For data, see table 499. Statistical Abstract of the United States, *1997, 11th ed, Washington, D.C.: 1997, Figure 9.2.*

Using a Calculator or Computer

30. Use a suitable calculator or a spreadsheet or other software on a computer to determine the following for the data in Table 9.1.

(a) $\bar{x}$ **(b)** $\hat{x}$ **(c)** the mode **(d)** s

(e) Q_L **(f)** Q_U **(g)** the box plot

31. Repeat problem 30 for the data in Problem Set 9.1, problem 28, on the success of NFL field-goal kickers in 2004.

From State Student Assessments

32. (Illinois, Grade 4)

The science class held a frog jumping contest. What is the median distance the frog jumped?

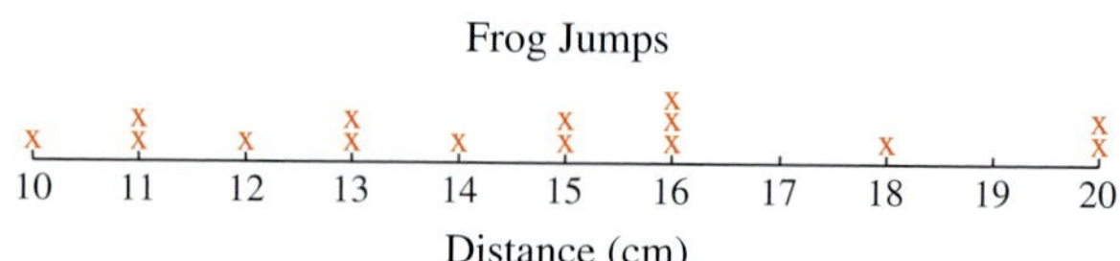

A. 11 B. 15

C. 16 D. 20

33. (Minnesota, Grade 5)

The members of the Spanish club sold calendars as a fund raiser. Below is a list of each person's sales.

Amy	6
Chris	9
Raul	15
Ali	9
Sonya	13
Ker	7
Maya	12
Allesandro	18
Nicky	10

What was the mean calendar sale for a member of the Spanish club?

○ A. 9

○ B. 10

○ C. 11

○ D. 12

34. (Texas, Grade 5)
Terry picked 156 tomatoes from the 12 tomato plants in his garden. On the average, how many tomatoes did each plant produce?

A. 11

B. 12

C. 13

D. 14

E. Not Here

35. (Pennsylvania, Grade 5)
A total of 106,789 people attended the Rose Bowl football game. If the average ticket price was $21.52, **about** how much money did the Rose Bowl take in?

A. $20,000

B. $200,000

C. $2,000,000

D. $20,000,000

For Review

36. Find two rational numbers between 5/8 and 6/8.

37. Write the following real numbers in order from the least to the greatest. Be sure to identify the two numbers that are equal.

$$\frac{3}{4}, 0.74\overline{9}, 0.7\overline{49}, 0.74\overline{09}, 0.749, 0.74949$$

38. Given that $\sqrt{3}$ is irrational and r is rational, prove that $\sqrt{3} + r$ is irrational.

39. Given that $\sqrt{3}$ is irrational and r is rational, prove that $\sqrt{3} \div r$ is irrational.

9.3 STATISTICAL INFERENCE

Populations and Samples

In statistics, a **population** is a particular set of things or properties about which one desires information. If the desire is to determine the average yearly income of all adults in the United States, the population is the set of *all* adults in the United States. If one is only concerned about the average yearly income of adults in Nevada, the population is the set of *all* adults in Nevada. Other examples of populations are

- all boys in Eisenhower High School in Yakima, Washington,
- all lightbulbs manufactured on a given day by Acme Electric Company,
- all employees of AT&T,

and so on. One might want to determine

- the average height of boys in Eisenhower High School,
- the average life of lightbulbs produced by Acme Electric Company, or
- the average cost of medical care for employees of AT&T.

Since it is often impractical or impossible to check each member of a population, the idea of statistics is to study a **sample** or subset of the population and to make inferences about the entire population on the basis of the study of the sample.

If the goal of accurate estimation of population characteristics is to be achieved, the population must be carefully identified and the sample appropriately chosen. Indeed, lack of care in identifying populations and choosing samples was precisely the source of some of the most spectacular errors of the past in the use of statistics. When a sample does not accurately reflect the composition of the entire population, it is not surprising that attributes of the sample do not accurately indicate attributes of the population.

For example, suppose a study of the heights of boys at Eisenhower High School is desired. Instead of measuring each boy in the school, it is decided to study a sample of just 20 of the boys. If the sample were to be selected by choosing every fourth name out of an alphabetical listing of all 80 boys, would it likely be representative of the entire population? Probably, since there is likely little or no connection between last names and heights. However, a sample consisting of the members of the basketball team is clearly not representative since basketball players tend to be unusually tall. It turns out that the best approach to sampling is to use a **random sample.**

DEFINITION *A Random Sample*
A **random sample** of size r is a subset of r individuals from the population chosen in such a way that every such subset has an equal chance of being chosen.

For example, suppose an urn contains a mixture of red and white beans, and you want to estimate what fraction of the beans are red by selecting a sample of 20 beans. You proceed by mixing the beans thoroughly and then, with your eyes closed, selecting 20 beans. Since each subset of 20 beans has an equal chance of being selected, the sample is indeed random.

Other schemes also work well.

Suppose AT&T wishes to study the employees at one of its plants by selecting a random sample of 20 employees and asking them to respond to a questionnaire. One way to obtain a random sample would be to put the names of all the employees on tags, place the tags in a large container, mix the tags thoroughly, and then have someone close his or her eyes and select a sample of 20 tags. The employees whose tags are chosen constitute the random sample.

BIZARRE SEQUENCE OF COMPUTER-GENERATED RANDOM NUMBERS

Another way to obtain the sample is to use a sequence of digits chosen in such a way that each digit is equally likely to be any one of the 10 possibilities, and the choice of each digit is independent of the choice of every other digit. Such a sequence would be a **random sequence of digits.** One way to select such a random sequence is to construct a simple spinner* with 10 36° sectors numbered 0, 1, 2, 3, 4, 5, 6, 7, 8, and 9 as shown in Figure 9.16. Spinning the spinner repeatedly produces a string of digits. Since the result of each spin is independent of the result of every other spin, each digit is equally likely to be selected and the digit sequence is random.

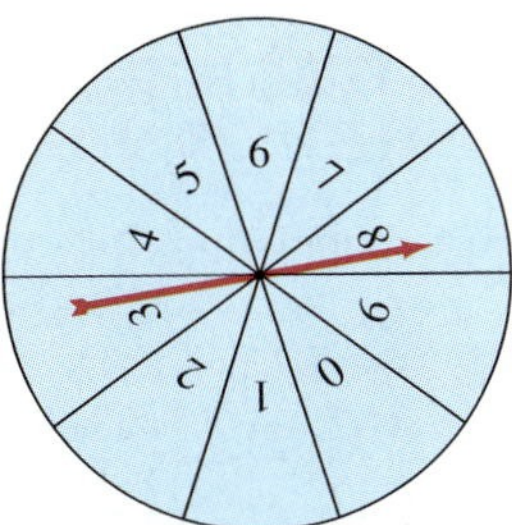

Figure 9.16
A simple random digit generator

The desired sample of AT&T employees can now be obtained as follows (suppose the plant in question has 9762 employees):

1. Assign each employee a four-digit number from among 0001, 0002, . . . , 9762.
2. Generate a random sequence of four-digit numbers by repeatedly spinning the spinner four times. Since each digit is equally likely to appear on any spin, each four-digit number is equally likely to appear on any four spins. If the spinning process generates 0000, a four-digit number greater than 9762, or any number already obtained, simply ignore it and continue to generate more four-digit numbers. When 20 appropriate numbers have been generated, they can be used to identify the 20 employees to be included in the sample.

Finally, as you might expect, many calculators have built-in statistics routines that will generate random numbers, and computer software exists for the same purpose. In Appendix C, the program for this purpose is RANDOM. RANDOM asks you to indicate the smallest and largest numbers you will allow and how many random numbers in that interval it should generate. It then proceeds to give you the desired number of random numbers, each containing 10 digits, in the interval specified—one each time you press ENTER. Perhaps the easiest way to obtain a string of single random digits is to specify 0 as the minimum number allowed and 10,000,000,000 as the maximum number. Then specify 10 or 20 as the number of desired numbers and proceed to read off the digits in order in each number as it appears. This will give a sequence of random digits.

Population Means and Standard Deviations

Professional statisticians find it helpful to use different symbols for means and standard deviations of populations and of samples. For populations, the mean and standard

*For a description of a superior, and more accurate, spinner, see Figure 10.2.

deviation are denoted by the Greek letters μ and σ (mu and sigma), respectively. Thus, for a population of size N, the population mean and population standard deviation are given by

$$\mu = \frac{x_1 + x_2 + \cdots + x_N}{N} \quad \text{and} \quad \sigma = \sqrt{\frac{(x_1 - \mu)^2 + (x_2 - \mu)^2 + \cdots + (x_N - \mu)^2}{N}}$$

where $x_1, x_2, \ldots, x_N$ are *all* the numbers in the population.

EXAMPLE 9.14 COMPUTING A POPULATION MEAN AND STANDARD DEVIATION

Consider a population that consists of the numbers shown.

64	65	68	67	59	66	63	66	64	62	65	66	63	66	66	63
63	64	62	67	63	61	60	64	64	63	63	65	64	65	66	63
65	62	63	65	61	64	63	64	62	69	65	65	64	64	63	64
66	65	64	64	63	64	66	67	69	63	65	63	64	64	64	65
67	68	64	62	66	62	64	61	65	62	65	62	65	62	66	63

Use a spreadsheet or other suitable software on your computer or the built-in statistics routine on a suitable calculator to compute μ and σ for this population.

Solution Since there are 80 numbers in the population,

$$\mu = \frac{64 + 65 + \cdots + 63}{80} \doteq 64.2$$

and

$$\sigma = \sqrt{\frac{(64 - 64.2)^2 + (65 - 64.2)^2 + \cdots + (63 - 64.2)^2}{80}} \doteq 1.9.$$

Estimating Population Means and Standard Deviations

Suppose we wish to know the mean and standard deviation of some large or inaccessible population. Since these statistics, μ and σ, may be difficult or impossible to compute, we estimate them with the sample mean

$$\bar{x} = \frac{x_1 + x_2 + \cdots + x_n}{n}$$

and the sample standard deviation

$$s = \sqrt{\frac{(x_1 - \bar{x})^2 + (x_2 - \bar{x})^2 + \cdots + (x_n - \bar{x})^2}{n}}$$

of a suitably chosen sample $x_1, x_2, \ldots, x_n$ of size n.

EXAMPLE 9.15 ESTIMATING A POPULATION MEAN AND STANDARD DEVIATION

(a) Estimate the mean and standard deviation of the population in Example 9.14 by using a spinner, as illustrated in Figure 9.16, or the program RANDOM to select a random sample of size 10.

(b) Compare the results of part (a) with the results obtained in Example 9.14.

Solution

(a) Suppose your spinner generates the digit sequence 5, 5, 2, 9, 1, 0, 4, 5, 3, 1, 2, 4, 1, 9, 4, 6, 6, 9, 1, 7. Using these two at a time, we obtain the two-digit numbers

55	29	10	45	31
24	19	46	69	17

Since all are different, these determine the random sample shown here.

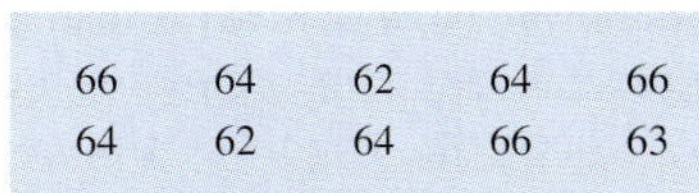

66	64	62	64	66
64	62	64	66	63

The fifty-fifth number in the data set is 66, and so on.

Thus, the mean of the numbers in the sample is

$$\bar{x} = (66 + 64 + 62 + 64 + 66 + 64 + 62 + 64 + 66 + 63)/10 = 64.1.$$

The variance is

$$\begin{aligned} v &= s^2 \\ &= [(64.1 - 66)^2 + (64.1 - 64)^2 + (64.1 - 62)^2 \\ &\quad + (64.1 - 64)^2 + (64.1 - 66)^2 + (64.1 - 64)^2 \\ &\quad + (64.1 - 62)^2 + (64.1 - 64)^2 + (64.1 - 66)^2 \\ &\quad + (64.1 - 63)^2]/10 \doteq 2.09, \end{aligned}$$

and the sample standard deviation is

$$s = \sqrt{v} \doteq 1.45.$$

(b) We observe that $\bar{x}$ and s for the sample are reasonable approximations for μ and σ, respectively, for the population as determined in Example 9.14.

Suppose we were to repeat the preceding example, but with a random sample of size 15. For the resulting sample, $\bar{x}$ and s should be slightly better approximations to μ and σ,

Highlight from History

R. A. Fisher (1890–1962)

Statistics, a relative newcomer to the mathematical scene, was developed largely during the last one hundred years. One of the most influential personalities in this development was Sir Ronald A. Fisher, a British geneticist and statistician. To remove the effect of differences in soil fertility of different plots of ground used in testing the effects of various fertilizers on plant growth, Fisher introduced the idea of selecting plots by a random process and showed how to correctly compare results from such randomized experiments. Fisher also contributed other new ideas to statistics, and his books and professional papers did much to shape statistics into an organized science.

respectively, from Example 9.14 than the values obtained above. In fact, it is generally true that larger samples tend to yield better approximations to population characteristics.

Distributions

We return now to the data of Table 9.2, in which the population consisted of all boys in Eisenhower High School. A histogram of the boys' heights to the nearest inch appears in Figure 9.5, and a line graph for the same data is shown in Figure 9.7.

Since the heights of the columns in the histogram represent the number, or frequency, of the measurements in each range (63.5–64.5, 64.5–65.5, and so on) and the width of each column is one, the total area of all the columns in the histogram is 80, the total number of boys in the population.

The **relative frequency** of the measurements in each range in Figure 9.5 is the fraction (expressed as a decimal) of the total number of boys represented in that range. If the heights of the column in the histogram are determined by relative frequency, the diagram remains the same except for the designation on the vertical scale as shown in Figure 9.17. Also, since the width of each column is one, the *area* of each column gives the fraction of the population whose heights fall in that range. Moreover, the area of the first three columns gives the fraction of the population with heights ranging from 63.5 to 66.5 inches, and the total area of the histogram is one.

As noted earlier, histograms often are representations of grouped data which make it appear that all the data values in a given range are the same. As in the present case, this

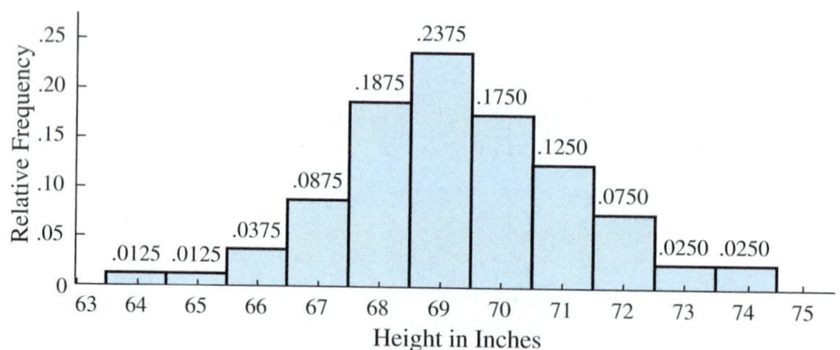

Figure 9.17
Histogram of Figure 9.5, but with the vertical scale denoting relative frequency

is frequently not so. The boys' heights are listed to the nearest inch, whereas people's heights actually vary continuously. A truer representation of such data is provided by a line graph, or frequency polygon, as in Figure 9.7. If the vertical scale represented relative frequency rather than frequency, the graph would appear unchanged as shown in Figure 9.18. Also, since such a diagram can be obtained from a histogram by deleting and adding small triangles of equal area, the area under the **relative-frequency polygon** is still one, and the area of that portion of the polygon from, say, 63.5 to 66.5, equals the fraction of the population of boys whose heights lie in this range.

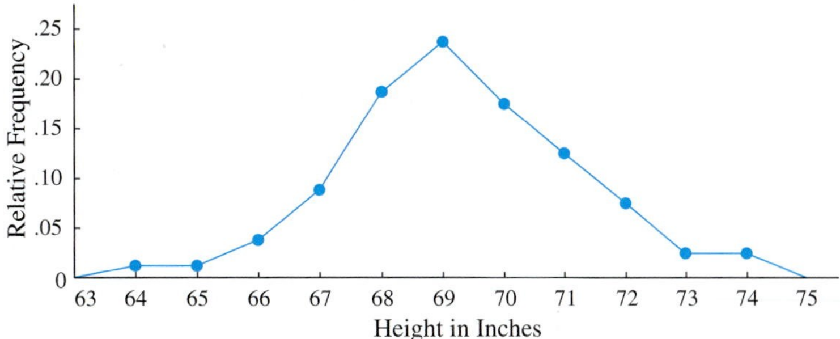

Figure 9.18
Relative-frequency polygon of the heights of boys in Eisenhower High School

In addition, had the measurements been taken more and more closely and the ranges in the histogram made narrower and narrower, the tops of the columns in the histogram, as well as the corresponding relative-frequency polygon, would have more and more closely approximated a smooth bell-shaped curve, called a **normal distribution,** as shown in Figure 9.19. Here also the area under the curve and above the interval between a and b indicates the relative frequency, or fraction, of the boys measured having heights between a and b. This fraction indicates the *likelihood* or *probability* that a boy chosen at random from the population would have a height in the given range.

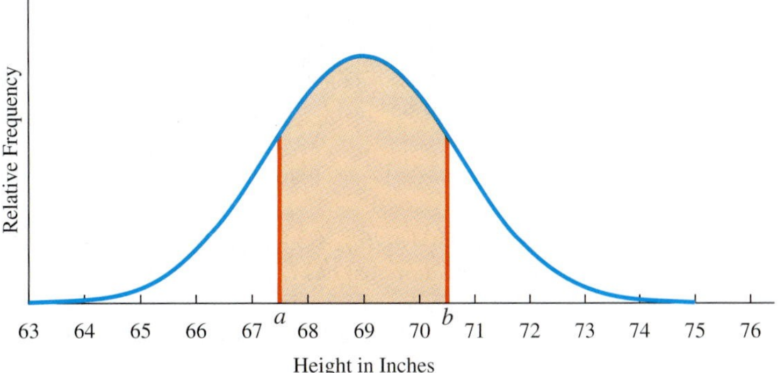

Figure 9.19
Normal distribution of the heights of boys in Eisenhower High School

For many populations, the distribution of the measurements of the property being considered will be a continuous (and often normal) curve as in Figure 9.19. However, in other cases, the observations are not continuous, are not normal, or are discrete. If the observations are discrete, the distribution "curve" is just a histogram.

DEFINITION *A Distribution Curve*
A curve or histogram that shows the relative frequency of the measurements of a characteristic of a population that lies in any given range is a **distribution curve.** The area under such a curve or histogram is always 1.

Knowing the distribution of a population frequently allows one to say with some precision what the average value is and what percentage of the population lies within different ranges. In particular, the normal distribution has been studied in great detail, and it can be shown that very nearly 68% of the population lies within one standard deviation of the mean, very nearly 95% of the population lies within two standard deviations of the mean, and very nearly 99.7% (or virtually *all*) of the population lies within three standard deviations of the mean, as illustrated in Figure 9.20. Using the language of probability, we would say that the probability that a given data value lies within one standard deviation of the mean is 0.68, the probability that a given data value lies within two standard deviations of the mean is 0.95, and the probability that any given data value lies within three standard deviations of the mean is 0.997 (virtually 100%). Considerations like these are what make it possible for very carefully designed polls and other studies to claim that their results are accurate to within a given tolerance, say 3%.

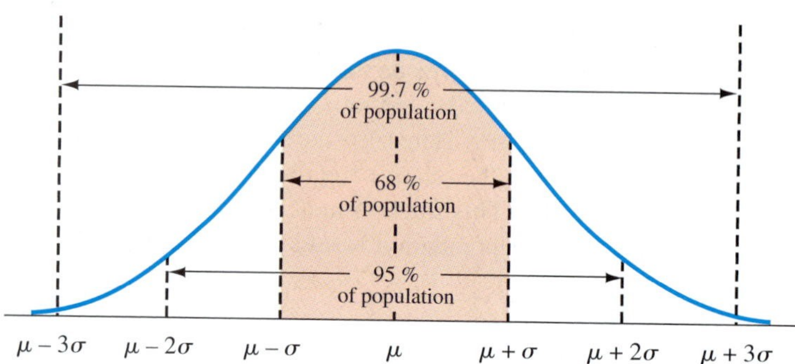

Figure 9.20
Percent of data within one, two, and three standard deviations of the mean of a normal distribution

THEOREM *The 68–95–99.7 Rule for Normal Distributions*
For a population that has a normal distribution, about 68% falls within one standard deviation of the mean, about 95% falls within two standard deviations of the mean, and about 99.7% falls within three standard deviations of the mean.

It turns out that many populations are normally distributed, or approximately so. Thus, the 68–95–99.7 rule is approximately true for these populations and for samples from these populations. For samples, the approximation is increasingly accurate for increasingly large sample sizes.

z -Scores and Percentiles

As noted previously, the normal distribution has been studied deeply and with great care. Some additional facts are as follows:

- The graph of the distribution is symmetric about a vertical line drawn through the mean; i.e., if the curve were folded along this line, the two halves of the curve would exactly match each other. Thus, the area under the curve to the left of the mean equals the area under the curve to the right of the mean, and it follows that the mean of the distribution is also its median.
- The maximum height of the curve occurs at the mean, so the mean also equals the mode.
- Since the scale of the vertical axis is relative frequency, the area under the entire curve is one. Also, the area under the curve over various intervals has been carefully tabulated, thus making such probability statements as the 68–95–99.7 rule possible.

One stratagem that makes it possible to effectively compare two different normal distributions has been the creation of the so-called **standardized** form of the distribution (also called the **z curve**). This comparison is accomplished by altering the scale on the horizontal axis by the transformation

$$z = \frac{x - \mu}{\sigma}$$

while maintaining the frequency scale on the vertical axis. This transformation does not materially alter the shape of the distribution, as can be seen in Figure 9.21, which is the standard form of the distribution in Figure 9.20.

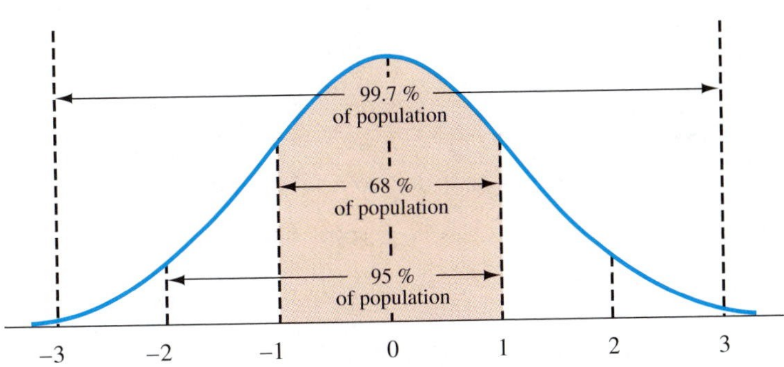

Figure 9.21
Standardized form of a normal distribution illustrating the 68–95–99.7 rule

Note that *any* normal distribution in standard form will have mean 0 and standard deviation 1. To see this, set $x = \mu$; then $z = \frac{\mu - \mu}{\sigma} = 0$. And if we set $x = \mu + \sigma$ (i.e., x is one standard deviation above the mean), then $z = \frac{(\mu + \sigma) - \mu}{\sigma} = \frac{\sigma}{\sigma} = 1$. Thus, in standard form, 68% of the population lies between -1 and 1, 95% of the population lies between -2 and 2, and 99.7% of the population lies between -3 and 3, as shown in Figure 9.21.

Normal distributions in standard form may appear tall and skinny or short and fat, depending on their standard deviations (see Figure 9.22), but the foregoing statements remain valid in every case.

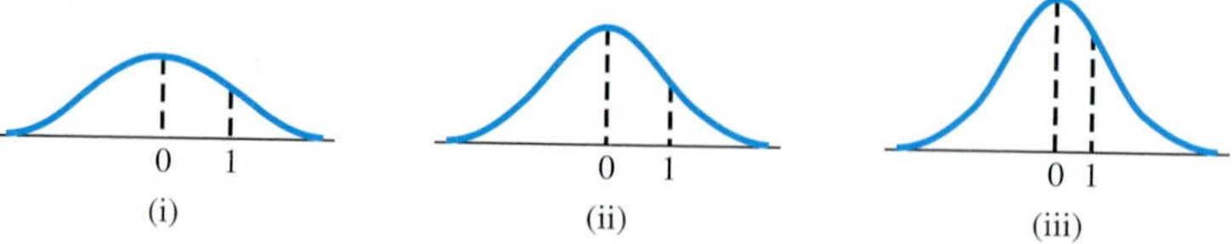

Figure 9.22
Three standard normal distributions with standard deviations $\sigma_1 > \sigma_2 > \sigma_3$ for (i), (ii), and (iii)

The preceding considerations have led to the notion of a **z score,** or **standard score.** If one wants to determine a characteristic of a population that is probably normally distributed (weights of full-term newborn babies, heights of senior boys attending Eisenhower High School, heights of senior girls attending Eisenhower High School, miles per gallon of 2004 Jeep Grand Cherokees, etc.) by a sampling procedure, it is often the case that each observation in the sample is converted to a z score as follows:

DEFINITION *z Score*

If x is an observation in a set of data, the z score corresponding to x is given by $z = \dfrac{x - \overline{x}}{s}$.

By using z, it is possible to tell if an observation is only fair, quite good, or rather poor. For example, a z score of 2 on a national test would be considered quite good, since it is 2 standard deviations above the mean. Indeed, we will see that the z score of 2 lies in the upper 2.5% of the population.

EXAMPLE 9.16 CALCULATING z SCORES

Convert these data to a set of z scores:

66 64 62 64 66 64 62 64 66 63

Solution

$$\overline{x} \doteq \frac{66 + 64 + 62 + 64 + 66 + 64 + 62 + 64 + 66 + 63}{10} = 64.1.$$

Using the alternative formula for calculating the standard deviation given on page 558, we obtain

$$s = \sqrt{\frac{66^2 + 64^2 + 62^2 + 64^2 + 66^2 + 64^2 + 62^2 + 64^2 + 66^2 + 63^2}{10} - 64.1^2}$$
$$\doteq 1.45.$$

Then the different z scores are

$$\frac{66 - 64.1}{1.45} \doteq 1.31, \qquad \frac{64 - 64.1}{1.45} \doteq -0.07,$$
$$\frac{62 - 64.1}{1.45} \doteq -1.45, \quad \text{and} \quad \frac{63 - 64.1}{1.45} \doteq -0.76.$$

Consider Example 9.16 further. From Table 9.4, pages 585 and 586, by indicating area (and hence probability) under the portion of a standard normal distribution to the left of a given point, it is possible to determine that 90.32% of the population lies to the left of (below) 1.31 and that only 7.35% of the population lies below −1.45. Considerations like these lead to the notion of a **percentile.**

DEFINITION *Percentile*
For *any* frequency distribution, the rth percentile is the number r, with $0 \leq r \leq 100$, such that $r\%$ of the data in a data set of $r\%$ of a population is less than or equal to r.

EXAMPLE 9.17 DETERMINING A PERCENTILE OF A POPULATION

Use Table 9.4, pages 585, and 586, to determine the percentile corresponding to the z score of −0.76 in a standard normal population.

Solution Searching down the left column of Table 9.4 on page 585, we come to the row marked −0.7. Moving across this row to the column headed .06, we find . 2236, the entry for −0.76. This tells us that 22.36% of the population lies to the left of −0.76.

EXAMPLE 9.18 DETERMINING A PERCENTILE OF A SAMPLE

Determine the percentile corresponding to 63 in the data set of Example 9.16.

Solution Since three of the ten scores in the data set are less than or equal to 63, and $3/10 = 0.30 = 30\%$, 63 is at the 30th percentile.

Note that the score of 63 in Example 9.16 gave a z score of −0.76. But the percentiles in Examples 9.17 and 9.18 are different. Why should this be so? The answer is that one percentile is for a population and the other for a sample. The one for the sample should approximate the one for the population, but that does not mean they should be equal. The population mean and standard deviation were 64.2 and 1.9, respectively, while for the sample, $\bar{x}$ and s were 64.1 and 1.45. Thus, the population shows more spread, and it is not surprising that the z score −0.76 is further to the left relative to the population as a whole and that there is, therefore, relatively less of the population to its left then to the left of 63 in the sample.

EXAMPLE 9.19 DETERMINING THE PERCENTAGE OF THE POPULATION IN AN INTERVAL

Use Table 9.4 on pages 585 and 586 to show that 34% of a normally distributed population lies between the z scores −0.44 and 0.44.

Solution Proceeding as in Example 9.17, we find that 33% of the population has a z score that is less than or equal to −0.44. Similarly, we find that 67% of the population lies to the left of 0.44. Thus, by subtraction, $67\% - 33\% = 34\%$ of the population lies between −0.44 and 0.44, as we were to show.

Cooperative Investigation

Hand Spans

Materials Needed

1. A metric ruler for each cooperative group.
2. A calculator with a built-in statistics package for each group.

Directions

1. Divide the students in the class into groups of 8 to 10 students.
2. In each group, the students measure the hand span (thumbtip to little fingertip) of each student in the group to the nearest millimeter and record the data, both on their own papers and on the chalkboard.
3. In each group, the students compute the mean of the hand-span measurements for their group.
4. Several students now independently compute the mean μ and standard deviation σ of the data on the chalkboard (the population). If their computations do not agree, someone has made an error.
5. Briefly conduct a class discussion on how well the means of the various groups (samples) approximate the mean of the entire class (the population).
6. Each individual student should now compute the percentage of the population measurements that lies within one, two, and three standard deviations of the population mean.
7. Discuss how well the results of Step 6 compare with the 68–95–99.7 rule for the normal distribution. Does it seem reasonable to expect that measurements of hand spans are normally distributed?
8. Repeat the entire investigation, but include only the measurements for hand spans of female students. Discuss how the removal of the data for male students affects the results of the investigation.

Problem Set 9.3

Understanding Concepts

1. Describe the population that should be sampled to determine each of the following.
   - **(a)** The percentage of freshmen in U.S. colleges and universities in 2004 who earn baccalaureate degrees within 10 years.
   - **(b)** The percentage of U.S. college and university football players in 2004 who earn baccalaureate degrees within 10 years.
   - **(c)** The fraction of people in the United States who feel that they have adequate police protection.
   - **(d)** The fraction of people in Los Angeles who feel that they have adequate police protection.
   - **(e)** Would you include children in the population you describe in parts (c) and (d)? people in mental institutions? known criminals?
2. Polls are often conducted by telephone. Might such a technique bias the results of the poll? Explain briefly.
3. Suppose a poll is conducted by face-to-face interviews, but the names of the interviewees are selected at random from names listed in the telephone book. Would such a poll yield valid results? Discuss briefly.
4. The registrar at State University wants to determine the percentages of students (a) who live at home, (b) who live in apartments, and (c) who live in dormitories. There are 25,000 students in the university, and the registrar proposes to select a sample of 100 students by choosing every 250th name from the list of all students arranged in alphabetical order.
   - **(a)** What is the population?
   - **(b)** Is the sample random? Explain.

5. In performing a study of college and university faculty attitudes in the United States, investigators first divided the population of all colleges and universities into groups according to size—25,000 or more students, 10,000 to 24,999 students, 3000 to 9999 students, and fewer than 3000 students. Using their judgment, they then chose two schools from each group and asked each school to identify a random sample of 100 of their faculty.
   (a) Was this a good way to obtain a statistically reliable (that is, random) sample of faculty? Why or why not?
   (b) What is the population? Are there four distinct populations? Discuss briefly.
6. The Honorable J. J. Wacaser, United States Representative, recently sent a questionnaire to his constituents to determine their opinion on several bills being considered by the House of Representatives. Discuss how representative of the voters in his district the responses to his poll are likely to be.
7. Discuss how Representative Wacaser (see problem 6) might actually choose a random sample of voters in his district.
8. To determine the average life of lightbulbs it manufactures, a company chooses a sample of the bulbs produced on a given day by selecting and testing to failure every 100th bulb.
   (a) What is the population?
   (b) Is this a good way to select a sample? Why or why not?
   (c) Is the sample random? Why or why not?
9. Choose a representative sample of 20 students in your college or university, and ask how many hours each person in your sample watches television each week.
   (a) Describe how you chose your sample to ensure that it was representative of your entire student body.
   (b) On the basis of your sample, estimate how many hours of television most students on your campus watch each week.
   (c) Combine your data with those of all the other students in your class, and determine a revised estimate of the number of hours per week each student in your college or university watches television.
10. Describe and perform a study to determine how many of the students at your college or university have seen the movie *Gone With the Wind.*
11. For a population with a normal distribution with mean 24.5 and standard deviation 2.7,
   (a) about 68% of the population lies between what limits?
   (b) about 95% of the population lies between what limits?
   (c) about 99.7% of the population lies between what limits?
12. Convert these data sets into sets of z scores.
   (a) 17 22 21 19 23 19
   (b) 2 7 3 6 5 8
13. (a) Compute the sum of the z scores in problem 12(a).
   (b) Compute the sum of the z scores in problem 12(b).
14. (a) What percentile is the data value 22 in the data set of problem set 12(a)?
   (b) What percentile is the data value 6 in the data set of problem 12(b)?

Use Table 9.4 on pages 585 and 586 to determine the answers to problems 15 and 16.

15. A population is normally distributed.
   (a) What percentage of the population lies to the left of the population z score -1.75?
   (b) What percentage of the population lies to the left of the population z score 0.26?
16. (a) What percentage of the population lies between the population z scores -1.75 and 1.75?
   (b) What percentage of the population lies between the population z scores -0.67 and 0.67?

Teaching Concepts

17. Marita claims that tossing a single die will produce a random sequence of the digits 1, 2, 3, 4, 5, and 6. How would you respond to Marita?
18. After the class discussion of Marita's claim (problem 17), Mark asserts that you could generate a random sequence of the numbers 2, 3, . . . , 12 by repeatedly tossing a pair of dice. How would you respond to Mark?
19. Prompted by the discussions engendered by Marita's and Mark's claims (problems 17 and 18), Rebecca claims that you could generate a random sequence of two-digit numbers by repeatedly tossing a red die and a green die and recording the result on the red die as the first digit and the result on the green die as the second digit to form a two-digit number. How would you respond to Rebecca?

Thinking Critically

20. A large university was charged with sexual bias in admitting students to graduate school. Admissions were by departments, and the figures are as shown.
   (a) Compute the percentages of men and women applicants admitted by the school as a whole.
   (b) Do the figures in part (a) suggest that sexual bias affected admission of students?

(c) Compute the percentages of men and women applicants admitted by each department.

(d) Do the figures in part (c) suggest that sexual bias affected admission to the various departments?

	MEN		WOMEN	
Department	Number of Applicants	Number Admitted	Number of Applicants	Number Admitted
1	373	22	341	24
2	560	353	25	17
3	325	120	593	202
4	191	53	393	94
5	417	138	375	131
6	825	512	108	89
Totals	2691	1198	1835	557

(e) Carefully, but briefly, explain this apparent contradiction. (*Hint:* See "Just for Fun, A Probability Paradox," in Section 10.3.)

21. Suppose you generate a sequence of 0s and 1s by repeatedly rolling a die and recording a 0 each time an even number comes up and a 1 each time an odd number comes up. Is this a random sequence of 0s and 1s? Explain.

22. A TV ad proclaims that a study shows that 8 out of 10 dentists surveyed prefer Whito Toothpaste. How could it possibly make such a claim if, in fact, only 1 dentist out of 10 actually prefers Whito?

23. Two sociologists mailed out questionnaires to 20,000 high school biology teachers. On the basis of the 200 responses they received, they claimed that fully 72% of high school biology teachers in the United States believe the biblical account of creation. Is their claim justified by this survey? Explain.

24. Consider the numbers 1, 2, 4, 8, and 16.

(a) Compute their mean, sometimes called the arithmetic mean.

(b) Compute their **geometric mean;** that is, compute $\sqrt[5]{1 \cdot 2 \cdot 4 \cdot 8 \cdot 16}$.

(c) Does the arithmetic mean or the geometric mean seem more typical of this sequence of numbers?

25. Katja can paddle a canoe 3 miles per hour in still water. The water in a stream flows at the rate of 1 mile per hour. Thus, when she is going upstream in the canoe, Katja's effective speed is 2 miles per hour, and when she is going downstream, it is 4 miles per hour.

(a) What is Katja's average speed (total distance ÷ total time) if she paddles 4 miles upstream and back?

(b) Compute the harmonic mean of 2 and 4—Katja's effective speeds upstream and downstream. The **harmonic mean** of several numbers is the reciprocal of the mean of their reciprocals.

(c) Compute the mean, i.e., the arithmetic mean, of 2 and 4.

(d) Discuss whether the arithmetic or harmonic mean seems the more appropriate average in this problem.

26. In the shoe business, which average of foot sizes is most important—the mean, median, or mode?

27. Consider the data set {6, 11, 10, 8, 12, 8}, where all the data are just 11 less than those in problem 12(a). It's as if the data were drawn from the same distribution moved 11 units to the left.

(a) What would you expect the z scores for this new set of data to be?

(b) Actually compute the z scores for this new set of data.

28. Show that the sum of the z scores for any set of data is equal to zero. *Suggestion:* Argue from a special case, say, using the data set {1, 2, 3}, but do not actually compute the z scores; i.e., use the strategy "Argue from a special case."

29. Consider the data set {34, 44, 42, 38, 46, 38}, where all the data are just twice what they were in problem 12(a).

(a) What would you expect the z scores of this data set to be?

(b) Compute the z scores for this data set.

(c) How do you account for the results in part (b)?

Thinking Cooperatively

30. Divide the class into groups of three or four students each and give each group eight pennies. Have each student in each group thoroughly shake and toss the pennies five times and record the number of heads obtained each time. Then have a member of the group record on the chalkboard the total number of times zero heads, one head, . . . , eight heads were obtained by the group. Let the population to be studied be the combined set of data obtained by the entire class.

(a) Have each group construct a relative-frequency polygon (see Figures 9.17 and 9.18) for the population. Since the polygon shows the distribution of the population, have each group determine a consensus opinion as to whether the distribution approximates a normal distribution.

(b) Have each group compute the population mean and standard deviation.

(c) Have each group determine whether the 68–95–99.7 rule holds for this population.

31. Mai Ling claims that the spinner of Figure 9.16 likely would fail to generate a truly random sequence of digits because, in order to spin it, a person would likely hold it still and start each spin with the pointer in the same position (likely horizontal) each time. But always starting with the pointer in the same position would almost surely cause the spinner to favor some digits over others. How would you respond to Mai Ling?

Making Connections

32. A fish biologist is studying the effect of recent management practices on the size of the cutthroat trout population in Idaho's Lochsa River. The biologist and her helpers first catch, tag, and release 200 such trout on a given day. Two weeks later, the team catches 150 cutthroat trout and determines that only three of these fish were tagged two weeks earlier. Estimate the size of the cutthroat population in the Lochsa River. (*Hint:* Determine the fraction of tagged cutthroat in the river in two different ways if there are x cutthroat in the river.)

From State Student Assessments

33. (Kentucky, Grade 5)

Which Bag Is It? Mr. Carew put the number of marbles shown below into three bags. He brought the bags to class so his students could experiment with them. Christopher and Stephanie each chose one of the bags and pulled a marble from that bag 100 times, putting the marble back into the bag each time.

A. These are Christopher's results:

Red marble: 11 times Green marble: 89 times

Which bag is most likely to be the one that Christopher chose? Explain why.

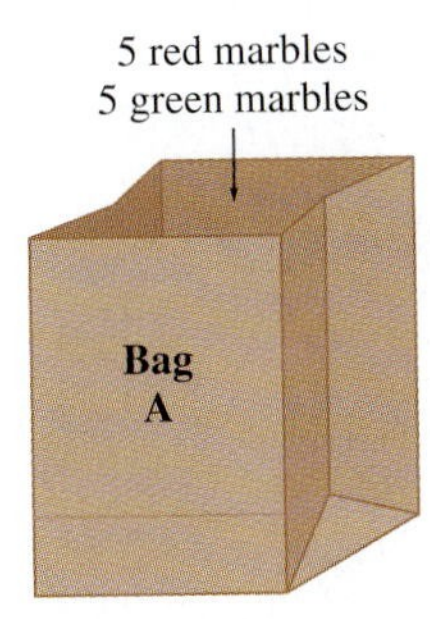

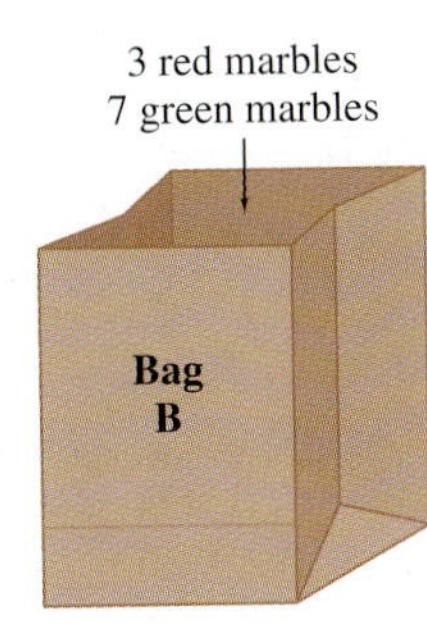

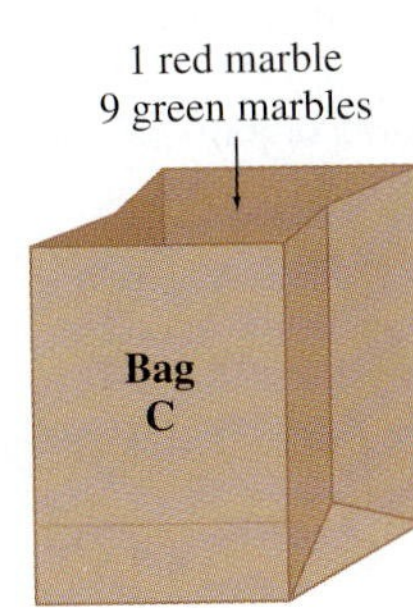

B. These are Stephanie's results:

Red marble: 32 times Green marble: 68 times

Which bag is most likely to be the one that Stephanie chose? Explain why.

For Review

34. The following data are yields in pounds of hops.

3.4	4.4	4.8	4.5	5.1
4.8	5.5	4.7	3.5	3.6

(a) Compute the mean for these data.

(b) Compute the standard deviation for these data.

35. The 12 players on Uni Hi's basketball team averaged 5 points each during the first half of a game against Roosevelt. They averaged 7 points each for the entire game.

(a) How many points did Uni Hi score during the game?

(b) What was the average score for each player during the second half?

36. (a) What is the median of the data in problem 34?

(b) Draw a box plot for the data in problem 34.

37. Determine the mode for the data in problem 34.

38. (a) Compute the mean of 21, 25, 27, 20, and 22.

(b) What is the mean of 21, 25, 27, 20, 22, and 23?

(c) Were you sure of the answer to part (b) before doing the calculation? Explain.

(d) What is the mean of 21, 25, 27, 20, 22, 20, and 26?

(e) Were you sure of the answer to part (d) before doing the calculation? Explain.

Cooperative Investigation

Using Samples to Approximate Characteristics of Populations

The chart shown contains 100 integers (the population) displayed in such a way that they can be represented by a number pair (a, b). For example, entry $(2, 7)$ is the integer 24, and entry $(7, 3)$ is 26.

	0	1	2	3	4	5	6	7	8	9
0	21	22	20	24	22	29	25	21	27	17
1	25	12	28	21	22	17	28	18	18	26
2	19	17	23	29	19	16	24	24	25	19
3	22	17	26	11	31	19	14	20	23	17
4	26	13	30	26	18	23	37	24	27	28
5	14	15	25	20	24	18	20	30	35	21
6	18	30	22	20	20	23	27	26	33	13
7	24	21	23	26	28	19	28	29	31	23
8	21	27	22	25	21	16	23	27	16	25
9	23	22	24	22	16	15	19	24	25	20

For this investigation, parts (a) through (g) should be executed by each cooperative group of two or three students. Part (h) should involve the entire class.

(a) Use a spinner as in Figure 9.16 or the graphing calculator program RANDOM to generate five number pairs (a, b) to determine a sample of five numbers from the above table. Compute the mean and standard deviation of your 5-number sample.

(b) Repeat part (a), but with a sample of size 10.

(c) On the basis of parts (a) and (b), give two estimates for each of the population mean and standard deviation.

(d) Record your means for parts (a) and (b) on the chalkboard.

(e) Determine $\bar{x}_5$, the mean of the means of the samples of size 5, and $\bar{x}_{10}$, the mean of the means of samples of size 10, from the chalkboard. Also compute s_5 and s_{10}, the standard deviations of the means of the samples of size 5 and size 10, respectively.

(f) Using the result of part (e) again to estimate the population mean and standard deviation.

(g) Using a spreadsheet or other suitable technology, calculate the actual population mean and standard deviation for the entire population. Alternatively, these figures can be given to the class by the instructor.

(h) Briefly discuss the results of parts (a) through (g).

Epilogue The Information Age

In today's world, we are often confronted with masses of data that must be organized and summarized to be understood. We are also constantly bombarded with numerical information prepared by other people—pollsters, politicians, special interest groups, federal and state governments, foundations, testing agencies, and others—that we must be able to evaluate if we are to be informed citizens. As H. G. Wells once wrote, "Statistical thinking will one day be as necessary for efficient citizenship as the ability to read and write." Statistical thinking requires that we be able to read graphs and charts of all kinds, understand diagrams, and properly understand that, while sample results vary from sample to sample, a random sample of the appropriate size can yield very accurate information about a population. At the same time, it is equally important that we be able to discern how graphs, charts, and statistics can misinform, either intentionally or unintentionally.

In this chapter, we have discussed many aspects of statistics—the organization, interpretation, and display of data; the meaning of the mean, median, and mode as indications of a "typical" value of a set of data; notions like the range, box plots, and the standard deviation as indications of the spread of a set of data; and the notion of a population and how samples can be used to provide information about populations. We have also given consideration to distributions in general, to normal distributions in particular, and to z scores and percentiles. Statistics is a powerful and useful tool, but it must be used with understanding and with care.

Chapter Summary

Key Concepts

Section 9.1 The Graphical Representation of Data

- Data. Measurements of the characteristic of a sample that is chosen in order to study the same characteristic of a population.
- Dot plot. A representation of data that uses dots above a number line to represent data values.
- Stem and leaf plot. A representation of data in which the first one or more digits of each data value constitute the stem, and the remaining digits constitute the leaves.
- Histogram. A representation of data in which the height of the bar at a point on the number line represents the number of data points in an interval about the point or the relative frequency of the data in the interval.
- Frequency. The number of times a data value appears in a data set.
- Frequency of an interval. The number of times a data value appears in a given interval along a number line.
- Line graph. A graph formed by joining data points.
- Frequency polygon. A line graph formed by joining the midpoints of the tops of bars of a histogram.
- Bar graph. A histogramlike representation in which the bars are drawn above points on the horizontal axis corresponding to nonnumerical categories.
- Pie chart. A circle divided into sectors whose sizes correspond to the various percentages of a whole devoted to various categories.
- Pictograph. A representation of data in which the frequency is indicated by icons chosen to bear some relationship to the data being represented.
- Misleading representations. Ways in which representations of data can be made to confuse.

Section 9.2 Measures of Central Tendency and Variability

- Mean. The sum of the data values in a sample divided by the number of data values.
- Median. The middle data value for a sample with an odd number of data values, and the average of the two middle values for a sample with an even number of data values.
- Mode. The most frequently occurring data value.

- Range. The difference between the greatest and least data values in a sample.
- Quartile. Roughly speaking, the points Q_L and Q_U such that 25% of the data does not exceed Q_L and 25% of the data is not less than Q_U.
- Interquartile range. $\text{IQR} = Q_U - Q_L$.
- Outlier. A data value less than $\overline{x} - 1.5 \cdot \text{IQR}$ or greater than $\overline{x} + 1.5 \cdot \text{IQR}$.
- Box plot. A representation showing the least data value, Q_L. $\overline{x}$, Q_U, and the greatest data value.
- Standard deviation. A measure of the spread of the data in a sample. The formula for s is $s = \sqrt{\dfrac{(x_1 - \overline{x})^2 + (x_2 - \overline{x})^2 + \cdots + (x_s - \overline{x})^2}{n}}$, where $x_1, x_2, \ldots, x_n$ are the data values in the sample.
- Alternative formula for the standard deviation. $s = \sqrt{\dfrac{x_1^2 + x_2^2 + \cdots + x_n^2}{n} - \overline{x}^2}$.

Section 9.3 Statistical Inference

- Population. The set about which information is desired.
- Sample. A subset of the population chosen to give an accurate estimate of the characteristic of a population about which information is desired.
- Random sample. A subset chosen from a population in such a way that every such subset has a equal chance of being chosen.
- Random sequence of digits. A sequence of digits chosen in such a way that, at each step, each digit has an equal chance of being chosen.
- Random numbers. Numbers formed by successively choosing the appropriate number of digits from a random sequence of digits.
- Relative frequency. The frequency of the occurrence of a data value in a sample, expressed as a percentage.
- Relative-frequency polygon. A histogram in which the heights of the bars indicate relative frequency.
- Distribution. A curve or histogram showing the relative frequency of the measurements of a characteristic of a population.
- Normal distribution. A special bell-shaped distribution valid for many populations.
- The 68–95–99.7 rule. In a normal distribution, 68% of the population lies within one standard deviation of the mean, 95% of the population lies within two standard deviations of the mean, and 99.7% of the population lies within three standard deviations of the mean.
- z score. If x is a data value in a sample, the corresponding z score is $z = \dfrac{x - \overline{x}}{s}$.
- Percentile. A number such that a given percentage of a sample or distribution is less than or equal to that number.

Vocabulary and Notation

Section 9.1

Data
Dot plot
Stem and leaf plot
Histogram
Frequency of an interval
Line graph
Frequency polygon
Bar graph
Pie chart
Pictograph

Section 9.2

Mean, $\overline{x}$
Median, $\hat{x}$
Mode
Range
Quartiles, Q_L and Q_U
Interquartile range, IQR
Outlier
Box plot
Standard deviation

Section 9.3

Random sample
Random sequence of digits
Random numbers
Relative frequency
Relative-frequency polygon
The 68–95–99.7 rule
z score
Percentile

Chapter Review Exercises

Section 9.1

1. The following are the numbers of hours of television watched during a given week by the students in Mrs. Karnes' fourth-grade class.

17	8	17	13	16	13	8	9	17	7
8	7	14	14	11	13	11	13	11	17
12	15	11	10	12	13	9	21	19	12

   (a) Make a dot plot to organize and display these data.
   (b) From the dot plot, estimate the average number of hours per week the students in Mrs. Karnes' class watch television.

2. Make a stem and leaf plot to organize and display the data in problem 1.

3. Choosing suitable scales, draw a histogram to summarize and display the data in problem 1.

4. The following are the numbers of hours of television watched during the same week as in problem 1, but by the students in Ms. Stevens' accelerated fourth-grade class.

13	8	9	11	11	12	8	9
11	11	6	8	9	11	11	6
8	9	11	11	6	8	9	11

Prepare a double stem and leaf plot to display and compare the number of hours of television watched by Mrs. Karnes' and Ms. Stevens' classes during the given week.

5. (a) Draw a line graph to show the trend in the retail price index of farm products as shown in this table.

1965	1970	1975	1980	1985	1990	1995	2000	2005
35	42	64	88	104	134	168	184	190

   (b) Using part (a), estimate the retail price index for farm products in 1972.
   (c) Using part (a), estimate what the retail price index for farm products will be in 2010.

6. Find five numbers such that four of the numbers are less than the mean of all five.

7. Draw a pie chart to accurately illustrate how the State Department of Highways spends its budget if the figures are as shown: Administration—12%; New Construction—36%; Repairs—48%; Miscellaneous—4%.

8. (a) Criticize this pictograph, designed to suggest that the administrative expenses for Cold Steel Metal appear to be less than double in 2002 than in 2001 even though, in fact, the administrative expenses did double.

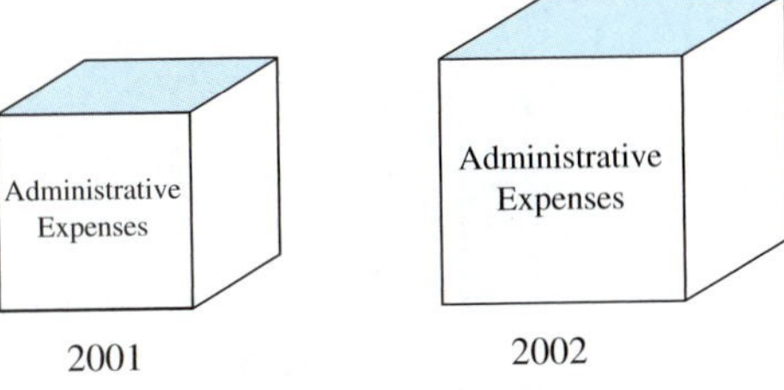

   (b) If the administrators are challenged by the stockholders, can they honestly defend the pictograph? (*Hint:* Measure the cubes very carefully with a metric ruler and compute their volumes.)

9. What would you need to know in order to take this statement seriously? "A survey shows that the average medical doctor in the United States earns $185,000 annually."

Section 9.2

10. Compute the mean, median, mode, and standard deviation for the data in problem 1.

11. (a) Compute the quartiles for the data in problem 1.
    (b) Give the 5-number summary for the data in problem 1.
    (c) Identify any outliers in the data of problem 1.
    (d) Compute the quartiles for the data in problem 4.
    (e) Give the 5-number summary for the data in problem 4.
    (f) Identify any outliers in the data of problem 4.
    (g) Using the same scales, draw side-by-side box plots to compare the data in problems 1 and 4.

12. Compute the mean and standard deviation for the data represented in these two histograms.

(a)

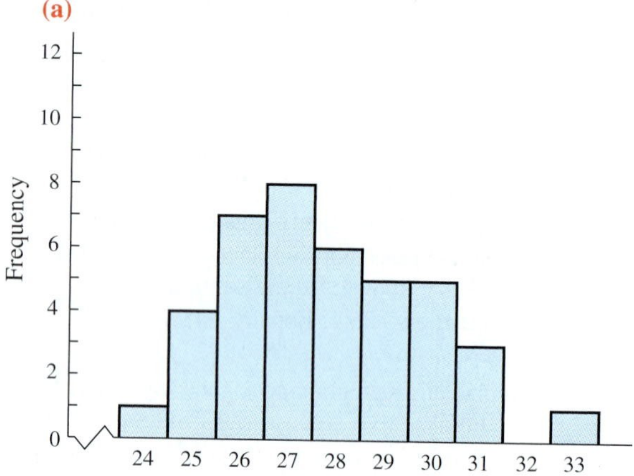

(b)

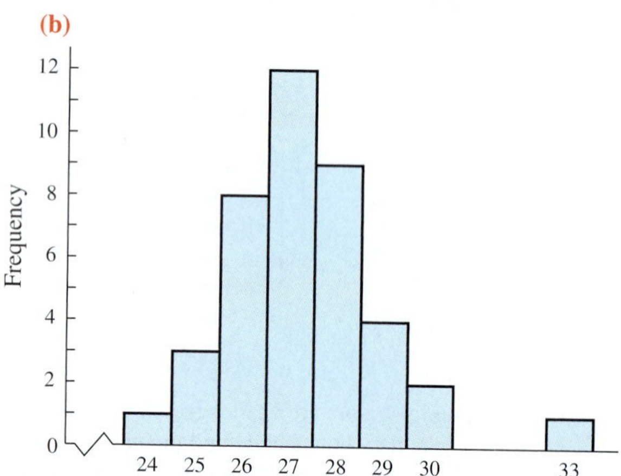

(c) Briefly explain the results of your computations in parts (a) and (b).

13. Three students were absent when the remaining 21 students in the class took a test on which their average score was 77. When the three students took the test later on, their scores were 69, 62, and 91. Taking these grades into account, what was the new average of all the test scores?

14. Mr. Renfro's second-period Algebra I class of 27 students averaged 75 on a test, and his fourth-period class of 30 students averaged 78 on the same test. What was the average of all the second- and fourth-period test scores?

Section 9.3

15. In a study of drug use by college students in the United States, the investigators chose a sample of 200 students from State University. Was this an appropriate choice for the study? Explain.

16. Suppose you want to ascertain by a sampling procedure what percentage of the people in the United States are unemployed. How might you describe the population that should be sampled? Should every person residing in the United States be included in the population? Discuss briefly.

17. Discuss briefly the biases that are inherent in telephone polls.

18. Discuss briefly the biases that are inherent in samples obtained by voluntary responses to questionnaires like those sent out by members of Congress to their constituents.

19. Describe two different ways in which a random sample of 100 of the 10,000 students at State University can be obtained.

20. Suppose that only one dentist out of 10 actually prefers Whito Toothpaste over all other brands. By taking many random samples of size 10, might it be possible eventually to obtain a sample in which eight out of 10 dentists in the sample preferred Whito? Explain.

21. Determine the z scores for the data set $\{7, 9, 6, 12, 15, 7, 9\}$.

22. What percentile is 12 in problem 21?

23. Use Table 9.4 on pages 585 and 586 to determine what percentile corresponds to the population z score 1.65.

24. Use Table 9.4 on pages 585 and 586 to determine what percentage of a standardized normal population lies between the population z scores -0.9 and 0.9.

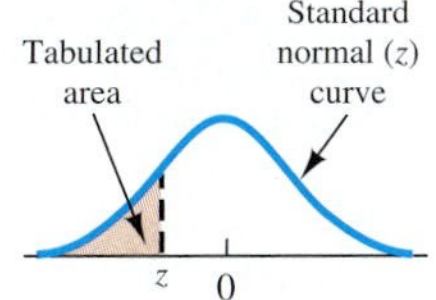

TABLE 9.4 Areas to the Left of z Scores for a Standardized Normal Distribution

z*	.00	.01	.02	.03	.04	.05	.06	.07	.08	.09
−3.8	.0001	.0001	.0001	.0001	.0001	.0001	.0001	.0001	.0001	.0000
−3.7	.0001	.0001	.0001	.0001	.0001	.0001	.0001	.0001	.0001	.0001
−3.6	.0002	.0002	.0001	.0001	.0001	.0001	.0001	.0001	.0001	.0001
−3.5	.0002	.0002	.0002	.0002	.0002	.0002	.0002	.0002	.0002	.0002
−3.4	.0003	.0003	.0003	.0003	.0003	.0003	.0003	.0003	.0003	.0002
−3.3	.0005	.0005	.0005	.0004	.0004	.0004	.0004	.0004	.0004	.0003
−3.2	.0007	.0007	.0006	.0006	.0006	.0006	.0006	.0005	.0005	.0005
−3.1	.0010	.0009	.0009	.0009	.0008	.0008	.0008	.0008	.0007	.0007
−3.0	.0013	.0013	.0013	.0012	.0012	.0011	.0011	.0011	.0010	.0010
−2.9	.0019	.0018	.0018	.0017	.0016	.0016	.0015	.0015	.0014	.0014
−2.8	.0026	.0025	.0024	.0023	.0023	.0022	.0021	.0021	.0020	.0019
−2.7	.0035	.0034	.0033	.0032	.0031	.0030	.0029	.0028	.0027	.0026
−2.6	.0047	.0045	.0044	.0043	.0041	.0040	.0039	.0038	.0037	.0036
−2.5	.0062	.0060	.0059	.0057	.0055	.0054	.0052	.0051	.0049	.0048
−2.4	.0082	.0080	.0078	.0075	.0073	.0071	.0069	.0068	.0066	.0064
−2.3	.0107	.0104	.0102	.0099	.0096	.0094	.0091	.0089	.0087	.0084
−2.2	.0139	.0136	.0132	.0129	.0125	.0122	.0119	.0116	.0113	.0110
−2.1	.0179	.0174	.0170	.0166	.0162	.0158	.0154	.0150	.0146	.0143
−2.0	.0228	.0222	.0217	.0212	.0207	.0202	.0197	.0192	.0188	.0183
−1.9	.0287	.0281	.0274	.0268	.0262	.0256	.0250	.0244	.0239	.0233
−1.8	.0359	.0351	.0344	.0336	.0329	.0322	.0314	.0307	.0301	.0294
−1.7	.0446	.0436	.0427	.0418	.0409	.0401	.0392	.0384	.0375	.0367
−1.6	.0548	.0537	.0526	.0516	.0505	.0495	.0485	.0475	.0465	.0455
−1.5	.0668	.0655	.0643	.0630	.0618	.0606	.0594	.0582	.0571	.0559
−1.4	.0808	.0793	.0778	.0764	.0749	.0735	.0721	.0708	.0694	.0681
−1.3	.0968	.0951	.0934	.0918	.0901	.0885	.0869	.0853	.0838	.0823
−1.2	.1151	.1131	.1112	.1093	.1075	.1056	.1038	.1020	.1003	.0985
−1.1	.1357	.1355	.1314	.1292	.1271	.1251	.1230	.1210	.1190	.1170
−1.0	.1587	.1562	.1539	.1515	.1492	.1469	.1446	.1423	.1401	.1379
−0.9	.1841	.1814	.1788	.1762	.1736	.1711	.1685	.1660	.1635	.1611
−0.8	.2119	.2090	.2061	.2033	.2005	.1977	.1949	.1922	.1894	.1867
−0.7	.2420	.2389	.2358	.2327	.2296	.2266	.2236	.2206	.2177	.2148
−0.6	.2743	.2709	.2676	.2643	.2611	.2578	.2546	.2514	.2483	.2451
−0.5	.3085	.3050	.3015	.2981	.2946	.2912	.2877	.2843	.2810	.2776
−0.4	.3446	.3409	.3372	.3336	.3300	.3264	.3228	.3192	.3156	.3121
−0.3	.3821	.3783	.3745	.3707	.3669	.3632	.3594	.3557	.3520	.3483
−0.2	.4207	.4168	.4129	.4090	.4052	.4013	.3974	.3936	.3897	.3859
−0.1	.4602	.4562	.4522	.4483	.4443	.4404	.4364	.4325	.4286	.4247
−0.0	.5000	.4960	.4920	.4880	.4840	.4801	.4761	.4721	.4681	.4641

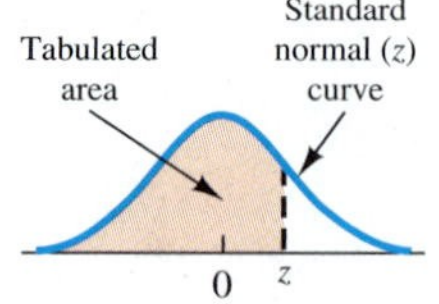

TABLE 9.4 continued

z*	.00	.01	.02	.03	.04	.05	.06	.07	.08	.09
0.0	.5000	.5040	.5080	.5120	.5160	.5199	.5239	.5279	.5319	.5359
0.1	.5398	.5438	.5478	.5517	.5557	.5596	.5636	.5675	.5714	.5753
0.2	.5793	.5832	.5871	.5910	.5948	.5987	.6026	.6064	.6103	.6141
0.3	.6179	.6217	.6255	.6293	.6331	.6368	.6406	.6443	.6480	.6517
0.4	.6554	.6591	.6628	.6664	.6700	.6736	.6772	.6808	.6844	.6879
0.5	.6915	.6950	.6985	.7019	.7054	.7088	.7123	.7157	.7190	.7224
0.6	.7257	.7291	.7324	.7357	.7389	.7422	.7454	.7486	.7517	.7549
0.7	.7580	.7611	.7641	.7673	.7704	.7734	.7764	.7794	.7823	.7852
0.8	.7881	.7910	.7939	.7967	.7995	.8023	.8051	.8078	.8106	.8133
0.9	.8159	.8186	.8212	.8238	.8264	.8289	.8315	.8340	.8365	.8389
1.0	.8413	.8438	.8461	.8485	.8508	.8531	.8554	.8577	.8599	.8621
1.1	.8643	.8665	.8686	.8708	.8729	.8749	.8770	.8790	.8810	.8830
1.2	.8849	.8869	.8888	.8907	.8925	.8944	.8962	.8980	.8997	.9015
1.3	.9032	.9049	.9066	.9082	.9099	.9155	.9131	.9147	.9162	.9177
1.4	.9192	.9207	.9222	.9236	.9251	.9265	.9279	.9292	.9306	.9319
1.5	.9332	.9345	.9357	.9370	.9382	.9394	.9406	.9418	.9429	.9441
1.6	.9452	.9463	.9474	.9484	.9495	.9505	.9515	.9525	.9535	.9545
1.7	.9554	.9564	.9573	.9582	.9591	.9599	.9608	.9616	.9625	.9633
1.8	.9641	.9649	.9656	.9664	.9671	.9678	.9686	.9693	.9699	.9706
1.9	.9713	.9719	.9726	.9732	.9738	.9744	.9750	.9756	.9761	.9767
2.0	.9772	.9778	.9783	.9788	.9793	.9798	.9803	.9808	.9812	.9817
2.1	.9821	.9826	.9830	.9834	.9838	.9842	.9846	.9850	.9854	.9857
2.2	.9861	.9864	.9868	.9871	.9875	.9878	.9881	.9884	.9887	.9890
2.3	.9893	.9896	.9898	.9901	.9904.	.9906	.9909	.9911	.9913	.9916
2.4	.9918	.9920	.9922	.9925	.9927	.9929	.9931	.9932	.9934	.9936
2.5	.9938	.9940	.9941	.9943	.9945	.9946	.9948	.9949	.9951	.9952
2.6	.9953	.9955	.9956	.9957	.9959	.9960	.9961	.9962	.9963	.9964
2.7	.9965	.9966	.9967	.9968	.9969	.9970	.9971	.9972	.9973	.9974
2.8	.9974	.9975	.9976	.9977	.9977	.9978	.9979	.9979	.9980	.9981
2.9	.9981	.9982	.9982	.9983	.9984	.9984	.9985	.9985	.9986	.9986
3.0	.9987	.9987	.9987	.9988	.9988	.9989	.9989	.9989	.9990	.9990
3.1	.9990	.9991	.9991	.9991	.9992	.9992	.9992	.9992	.9993	.9993
3.2	.9993	.9993	.9994	.9994	.9994	.9994	.9994	.9995	.9995	.9995
3.3	.9995	.9995	.9995	.9996	.9996	.9996	.9996	.9996	.9996	.9997
3.4	.9997	.9997	.9997	.9997	.9997	.9997	.9997	.9997	.9997	.9998
3.5	.9998	.9998	.9998	.9998	.9998	.9998	.9998	.9998	.9998	.9998
3.6	.9998	.9998	.9999	.9999	.9999	.9999	.9999	.9999	.9999	.9999
3.7	.9999	.9999	.9999	.9999	.9999	.9999	.9999	.9999	.9999	.9999
3.8	.9999	.9999	.9999	.9999	.9999	.9999	.9999	.9999	.9999	1.0000

Chapter Test

1. Suppose the average American spends 40% of his or her income paying taxes. If this information were to be shown with a pie chart, what should the central angle be for this portion of the chart?
2. Compute z scores for 42, 86, and 80 from problem 3 as part of the entire data set.
3. Prepare a dot plot to summarize and visually display these data.

42	86	80	90	74	84	86	80	63	92
93	81	95	78	70	41	66	76	87	88
75	88	87	78	89	85	77	87	81	57

4. Compute the following for the data in problem 3:
   (a) the mean
   (b) the median
   (c) the mode
   (d) the standard deviation
5. Prepare a stem and leaf plot to summarize the data in problem 3.
6. (a) Sketch a box plot for the data in problem 3.
   (b) Would you say that these data contain any outliers? Explain.
7. In making inferences based on samples, why is it important to choose random samples?
8. What does it mean to say that a sample is a random sample? Be brief, but lucid.
9. What percentile is 84 in the data set of problem 3?
10. Suppose that Nanda obtains scores of 77%, 79%, and 72% on her first three tests in French. What total score must she earn on her last two tests in order to average at least 80% in the course?

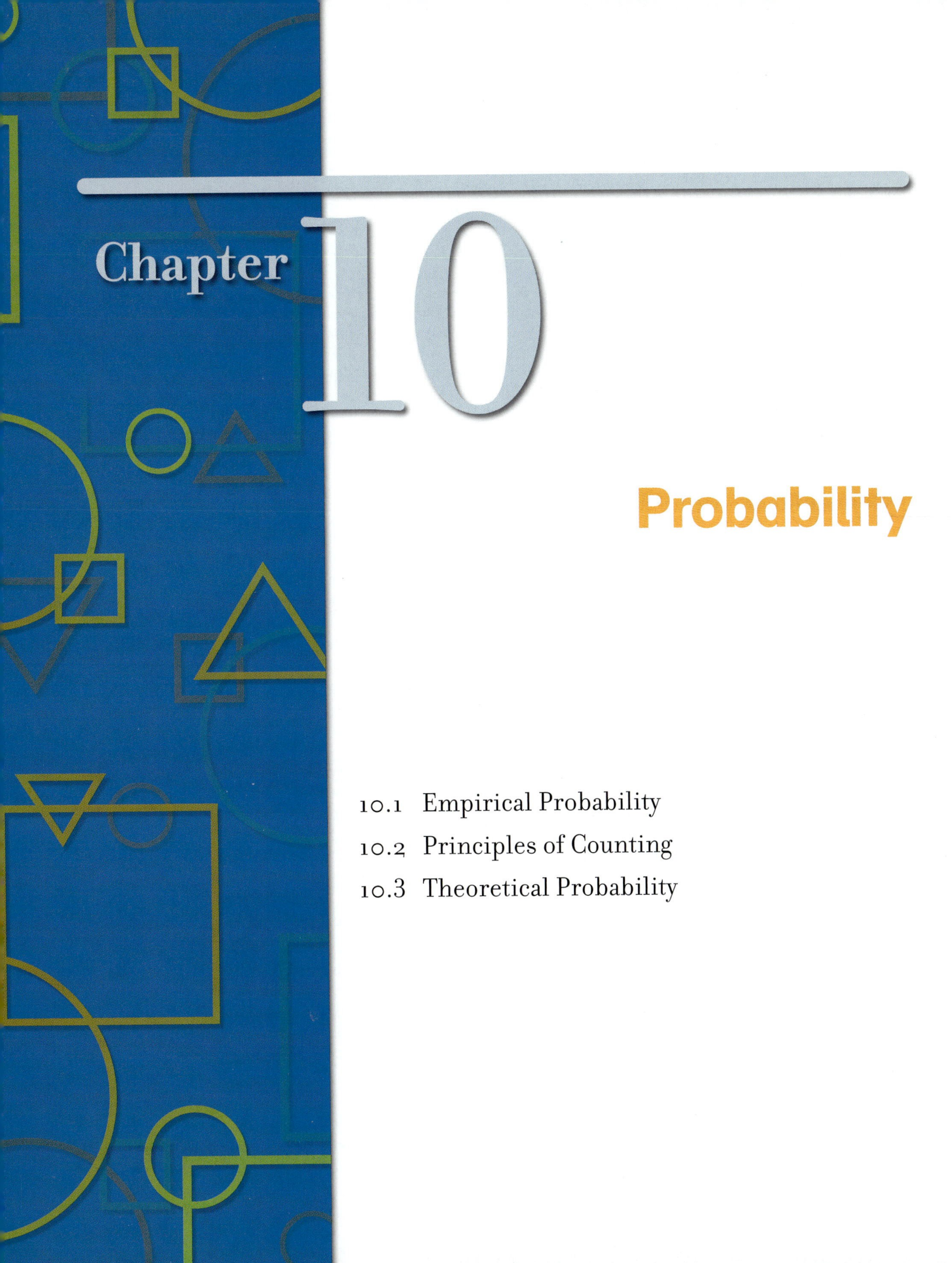

Chapter 10

Probability

Hands On: Are Pennies Fair?

Materials Needed

1. One penny, preferably new, for each student.
2. Tables or level-topped desks for the students to work upon.

Directions

Each student stands his or her penny on edge on the table and raps the tabletop sharply to cause the penny to fall. This is repeated five times, noting each time whether the penny falls heads up or tails up. When all the students have completed this task, record the data on the chalkboard and compute the fraction of heads obtained on all trials.

Is the result surprising? What does it suggest about the likelihood that a penny set on edge on a table will fall heads up if the table is rapped sufficiently hard that the penny falls?

Connections

From Polls to Probability

In the preceding chapter, we considered the problem of estimating a characteristic of a population (the residents of the United States, the students at a certain university, the lightbulbs produced by a specific company, etc.) by studying the same characteristic of a sample or subset of individuals chosen from the population. For example, suppose that a poll of 1000 residents of the United States reveals that 610 favor a certain bill before Congress. On the basis of this result, the pollster will likely claim that this is true of 61 percent of *all* residents of the United States, with a possible error of, say, $\pm 3\%$. That is, between 58 and 64 percent of the residents favor the bill. Can such a claim be justified? Is it believable? Sometimes the answer to this question is certainly a resounding "no," but, for a study based on a properly chosen random sample and other proven results from the theory of probability, the answer is a reassuring "yes."

In this chapter, we study the notion of probability. As usual, we use concrete devices and experiences to gain an intuitive feel for the notion and to set the stage for a more mathematical approach considered later. Indeed, our approach is more than just pedagogical since there are two different, but valid, views of probability that correspond to the approaches taken here.

Empirical Probability

The first view of probability is that it is a measure of what happens in the long run. In the Hands On activity described above, unless something very unusual occurred, you discovered that pennies stood on edge on a tabletop do not fall heads up and tails up equally often. Rather, the penny falls heads up approximately nine-tenths of the time. In terms of empirical probability, we say that pennies stood on edge on a tabletop fall heads up with probability 9/10. This figure, based on past experience, allows us to predict that about 90 heads will appear in the next 100 repetitions of the coin experiment.

Theoretical Probability

Another view of probability is that, in many cases, the probability of an event can be defined by carefully analyzing the experiment about to be performed. For example, a quality die (*dice* is the plural of die) is made in the shape of a perfect cube with its weight evenly distributed, so that any one of its six symmetrical faces is equally likely to come up. In this case, it seems reasonable to assign the probability of 1/6 that, say, a five will be rolled. Indeed, each of the six outcomes—1, 2, 3, 4, 5, or 6—is equally likely to occur, and therefore each outcome has probability 1/6. This notion of probability is called **theoretical,** or **mathematical, probability,** or simply **probability.** Since theoretical probability is assigned before any experiments are performed, it is also call *a priori* probability, in contrast with the name *a posteriori* probability sometimes used to indicate empirical probability.

Empirical probability is examined in Section 10.1. Probabilities of events are determined on the basis of conducting experiments or examining data. Section 10.2 focuses on methods of counting, preparing the way to make the calculations needed in the concluding Section 10.3 on theoretical probability.

10.1 EMPIRICAL PROBABILITY

Suppose a penny is stood on its edge on a table, and the table is given a sharp rap to topple the coin. The experiment has two outcomes, since the coin will land either heads or tails up. If, in 100 experiments, the outcome heads occurs 91 times, we say that the empirical probability of obtaining a head is 91/100. In symbols, this is written $P_e(H) = 0.91$, where H denotes the event that the coin lands heads upward. More generally, we have the following definition.

DEFINITION *Empirical Probability*

Suppose an experiment with a number of possible outcomes is performed repeatedly—say, n times—and that a specific outcome E occurs r times. The **empirical** (or experimental or experiential) **probability,** denoted by $P_e(E)$, that E will occur on any given trial of the experiment is given by

$$P_e(E) = \frac{r}{n}.$$

If only a few experiments have been conducted, the empirical probability can vary widely and will not be a good indicator of how to expect future outcomes of the experiment to occur. However, as the number of experiments increases, the variation decreases. This important fact, which we state without proof, is called the law of large numbers. It is also known as Bernoulli's theorem, honoring the contribution of Jakob Bernoulli (1654–1705) to the theorem.

THEOREM *The Law of Large Numbers*

If an experiment is performed repeatedly, the empirical probability of a particular outcome more and more closely approximates a fixed number as the number of trials increases.

from **The NCTM Principles and Standards**

Students in grades 3–5 should begin to learn about probability as a measurement of the likelihood of events. In previous grades, they will have begun to describe events as certain, likely, or impossible, but now they can begin to learn how to quantify likelihood. For instance, what is the likelihood of seeing a commercial when you turn on the television? To estimate probability, students could collect data about the number of minutes of commercials in an hour.

Students should also explore probability through experiments that have only a few outcomes, such as using game spinners with certain portions shaded and considering how likely it is that the spinner will land on a particular color. They should come to understand and use 0 to represent the probability of an impossible event and 1 to represent the probability of a certain event, and they should use common fractions to represent the probability of events that are neither certain nor impossible. Through these experiences, students encounter the idea that although they cannot determine an individual outcome, such as which color the spinner will land on next, they can predict the frequency of various outcomes.

SOURCE: Reprinted with permission from Principles and Standards for School Mathematics, *page 181. Copyright © 2000 by the National Council of Teachers of Mathematics, Reston, VA. All rights reserved. Standards are listed with the permission of the National Council of Teachers of Mathematics (NCTM). NCTM does not endorse the content or validity of these alignments.*

A striking example of the law of large numbers was provided by John Kerrich, an English mathematician imprisoned by the Germans during the Second World War. To while away the time, Kerrich decided to toss a coin 10,000 times and to recompute $P_e(H)$, the empirical probability of getting a head, after each toss. His first 10 tosses yielded 4 heads, so $P_e(H)$ equalled 0.4 at that point. After 30 tosses $P_e(H)$ was 0.57, after 100 tosses it was 0.44, after 1000 tosses it was 0.49, and after that it varied up and down slightly, but stayed very close to 0.5. After 10,000 tosses, the number of heads obtained was 5067, for an empirical probability of $P_e(H) \doteq 0.51$. During the experiment, relatively long sequences of consecutive heads and of consecutive tails occurred. Nevertheless, in the long run, the law of large numbers prevailed, and the intuitive guess of 1/2 was borne out. The variation in $P_e(H)$ is effectively illustrated in Figure 10.1.

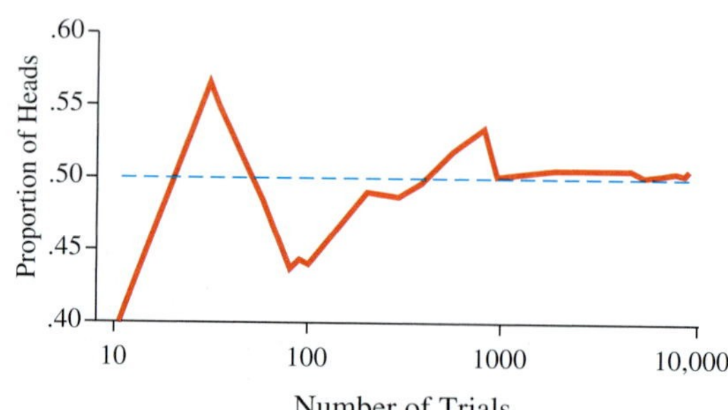

Figure 10.1
Ratio of the number of heads to the number of tosses in Kerrich's coin-tossing experiment
SOURCE: Adaptation of Figure 2 from Statistics, *second edition, by David Freedman, Robert Pisani, Roger Purves, and Ani Adhikari, p. 250. Copyright © 1991 by W. W. Norton & Company, Inc; copyright © 1978 by David Freedman, Robert Pisani, Roger Purves, and Ani Adhikari. Reprinted by permission of W. W. Norton & Company, Inc.*

EXAMPLE 10.1 DETERMINING AN EMPIRICAL PROBABILITY

Open a book at random and note the number of the right-hand page. Do this 20 times and determine the empirical probability that the number of the right-hand page is divisible by 3.

Solution

Since every third right-hand page is numbered with a number that is divisible by 3, we would expect that the empirical probability should be about 1/3. Here are results obtained on checking 20 pages chosen at random.

349	69	267	407	133
395	269	123	331	373
155	235	187	273	401
297	83	852	263	303

HIGHLIGHT from HISTORY

The Mathematical Bernoullis

Like talent of any kind, mathematical talent of the highest order appears in individuals with relative infrequency. This makes it most remarkable that such talent appeared in three generations of the Bernoulli family of Basel, Switzerland. Nicolaus Bernoulli (1623–1708) was a merchant and not a mathematician. However, two of his sons, four of his grandsons, and two of his great-grandsons were mathematicians of the highest order. The Bernoullis were fiercely competitive and often quarreled among themselves about priority in mathematical discoveries. Nevertheless, they contributed enormously to the development of mathematics, including contributions to the theory of probability by Jakob and Johann, sons of the merchant Nicolaus, and by Daniel, one of Johann's three mathematically talented sons. In particular, Jakob was the first individual to clearly conceptualize and prove the law of large numbers, the basis for the notion of empirical probability.

Jakob Bernoulli

Since a number is divisible by 3 if, and only if, the sum of its digits is divisible by 3, we determine that seven of these page numbers are divisible by 3. Hence,

$$P_e(\text{a right-hand page number is divisible by } 3) = \frac{7}{20} = 0.35,$$

a very close approximation to $1/3$.

Connections with Statistics

Most of what was said about statistics in Chapter 9 can be rephrased in terms of empirical probability. For example, saying that 28% of the items in a sample possess a certain property is the same as saying that the empirical probability that an item in the sample possesses a property A is $P_e(A) = 0.28$. Moreover, since population properties are estimated by properties of samples, one would go on to say that the empirical probability that an individual in the population possesses the property is also 0.28. Results of surveys and polls, summaries of data, and averages of all kinds also can be interpreted as empirical probabilities. Batting averages are empirical probabilities. Percentages of shots made in basketball are empirical probabilities. The life insurance industry is based on empirical probabilities derived from mortality tables.

EXAMPLE 10.2 COMPUTING EMPIRICAL PROBABILITY FROM A HISTOGRAM

Suppose you wanted to study the heights of high school boys in the United States and decided to use Eisenhower High School as typical. Use Figure 9.17 on page 570 to determine the empirical probability that the height in inches of a high school boy is in the range $67.5 < h < 70.5$.

Solution

The percentages of boys with heights in the ranges 67.5–68.5, 68.5–69.5, and 69.5–70.5 are, respectively, 18.75%, 23.75%, and 17.5%. Thus, the total percentage of boys with heights in the range 67.5–70.5 is $18.75 + 23.75 + 17.5$, or 60%. Therefore,

60% $= 0.60 = r/n$, where r is the number of boys with heights in the desired range and n is the number of boys in the high school. Since P_e also equals r/n, it follows that $P_e = 0.60$.

Additional Examples of Computing Empirical Probabilities

Empirical probabilities can be determined either from existing data or from data gathered from an experiment.

EXAMPLE 10.3

COMPUTING EMPIRICAL PROBABILITY FROM DATA

The final examination scores of students in a precalculus class are as shown. Compute the empirical probability that a student chosen at random from the class had a score in the 70s.

67	56	76	84	36	50	84
47	59	54	79	100	48	80
60	100	100	68	79	95	81
98	76	33	73	83	77	67

Solution

Since six of the 28 students scored in the 70s, the empirical probability that a student chosen at random had a score in the 70s is $6/28 \doteq 0.21$.

EXAMPLE 10.4

DETERMINING EMPIRICAL PROBABILITIES FROM AN EXPERIMENT

Christine has five pennies. She is curious how often she should expect to see at most one head when all five coins are flipped onto the floor. To find an answer, she repeatedly flips the five pennies and counts the number of heads that turn up. After repeating the experiment 50 times, she obtains the following frequency table.

Number of heads	0	1	2	3	4	5
Frequency	1	7	13	16	11	2

On the basis of Christine's data, what is the empirical probability that a flip of five coins results in at most one head?

Solution

The data show that exactly 1 head appeared on 7 of the trials, and no heads (that is, all tails) appeared once. This means the outcome of at most 1 head occurred $1 + 7 = 8$ times in the 50 trials, giving an empirical probability of $P_e(\text{at most 1 head}) = 8/50 = 0.16$.

Discovering General Properties of Probability

Often, an event of interest is described by coupling simpler outcomes with the word "and" or the word "or." This is called a **compound event.** The next two examples illustrate how the probability of a compound event is related to the individual probabilities of the events that have been combined to form the compound event.

EXAMPLE 10.5 COMPUTING EMPIRICAL PROBABILITY AND THE WORD *OR*

Roll a pair of dice 50 times and compute these empirical probabilities.

(a) $P_e(7)$ **(b)** $P_e(11)$ **(c)** $P_e(7 \text{ or } 11)$
(d) Show that $P_e(7 \text{ or } 11) = P_e(7) + P_e(11)$.

Solution

Actually performing the experiment, we obtained these results. From the data, the desired empirical probabilities are as shown.

2	3	4	5	6	7	8	9	10	11	12
|		|||	||||	~~||||~~ ~~||||~~	~~||||~~ ~~||||~~ ||	~~||||~~ |||	||||	||||	||	||

(a) $P_e(7) = 12/50 = 0.24$
(b) $P_e(11) = 2/50 = 0.04$
(c) $P_e(7 \text{ or } 11) = 14/50 = 0.28$
(d) $P_e(7 \text{ or } 11) = 0.28 = 0.24 + 0.04 = P_e(7) + P_e(11)$.

Example 10.5 suggests that the empirical probability of a compound event A or B is given by the formula $P_e(A \text{ or } B) = P_e(A) + P_e(B)$. However, it is important to check that no trial can result in a tally mark recorded for *both* events A and B at the same time. Two events that cannot both happen together on a single trial are called **mutually exclusive events.** This term and some others useful in probability are contained in the following list of definitions.

DEFINITIONS *The Terminology of Probability*

Outcome: a result of one trial of an experiment

Sample Space: the set of all outcomes of an experiment

Event A: a set A of some of the outcomes of an experiment (that is, a subset A of the sample space)

Mutually Exclusive Events A and B: two events A and B such that the occurrence of A precludes the occurrence of B, and vice versa (that is, A and B are disjoint: $A \cap B = \varnothing$).

Using these terms, the following property was illustrated in Example 10.5.

PROPERTY *Empirical Probability of Mutually Exclusive Events*
If A and B are mutually exclusive events, then $P_e(A \text{ or } B) = P_e(A) + P_e(B)$.

The next example shows how to modify the formula for non–mutually exclusive events.

Into the Classroom

Vera Holliday Comments on Teaching Probability in Her Classroom

Westlane Middle School
Indianapolis, Indiana

Approximately one week prior to beginning our unit on probability, my students are introduced to a simple game called "Something for Nothing." I prepare a bag of 90 yellow marbles and 10 blue marbles. For three days, I have each student pick a marble from the bag. The results are recorded on a tally form provided for them, and, after each student has drawn a marble, the marbles are collected and not returned to the bag. If they pick a blue marble from the bag, they receive a specified number of extra credit points. For the next three days, the same process is followed but this time students immediately return the marble to the bag before the next student picks a marble. I continue this game through the study of probability and use it as the basis for discussing the meaning of probability.

SOURCE: *Middle School Math,* Course 1, Volume 2 (Teacher's Edition), p. 622 © 1999 Pearson Education, Inc. Reprinted with permission.

EXAMPLE 10.6 COMPUTING THE EMPIRICAL PROBABILITY OF NON–MUTUALLY EXCLUSIVE EVENTS

A penny and a dime are flipped. Determine the empirical probability that the dime shows a head or both coins land with the same side up. Use the data collected in the table, which shows the outcomes of 50 flips of the pair of coins.

		Dime	
		Head	Tail
Penny	Head	卌 卌 ||||	卌 ||||
	Tail	卌 卌	卌 卌 卌 ||

Solution

Let A denote the event "the dime shows a head," and let B denote the event "both coins land with the same side up." All but the nine trials at the upper right of the table are occurrences of the event A or B, so $P_e(A \text{ or } B) = 41/50 = 0.82$. On the other hand, the empirical probability of having the dime show a head is $P_e(A) = 24/50 = 0.48$, since the 24 tallies in the left column of the table are occurrences of event A. Similarly, both coins showed heads 14 times and both showed tails 17 times, telling us that the empirical probability that the two coins show the same side is $P_e(B) = 31/50 = 0.62$. Then $P_e(A) + P_e(B) = 0.48 + 0.62 = 1.10$, which is 0.28 larger than $P_e(A \text{ or } B)$.

To identify the meaning of 0.28, we compute the empirical probability that both a head appeared on the dime *and* the two coins showed the same side. This compound event, A and B, corresponds to the 14 tallies in the upper left of the table, so the empirical probability is $P_e(A \text{ and } B) = 14/50 = 0.28$. This is exactly the amount by which $P_e(A) + P_e(B)$ exceeded $P_e(A \text{ or } B) = 0.82$, so we have the formula

$$P_e(A \text{ or } B) = P_e(A) + P_e(B) - P_e(A \text{ and } B).$$

The same reasoning can be applied to any two non–mutually exclusive events, giving the following general property of empirical probability.

> **PROPERTY** *Empirical Probability of A or B for Non–Mutually Exclusive Events*
> If A and B are any two events, then $P_e(A \text{ or } B) = P_e(A) + P_e(B) - P_e(A \text{ and } B)$.

This formula is still correct even when A and B are mutually exclusive events, since in that case $P_e(A \text{ and } B) = 0$ and the formula simplifies to

$$P_e(A \text{ or } B) = P_e(A) + P_e(B).$$

EXAMPLE 10.7 COMPUTING EMPIRICAL PROBABILITY AND THE WORD *AND*

Determine the empirical probability of obtaining two heads on a single toss of two coins, that is, the probability of obtaining a head on the first coin **and** a head on the second coin. Does the probability turn out to be about what you would expect? So that you can tell the coins apart, use a penny and a dime.

Solution

It is instructive to generate your own data, filling in a frequency table similar to the one shown here, which reproduces the data in Example 10.6. As before, let A denote the event that a head appears on the dime. This time, let B denote the event that a head appears on the penny. Using the data in the table, we find that $P_e(A) = 24/50 = 0.48$, $P_e(B) = 23/50 = 0.46$, and $P_e(A \text{ and } B) = 14/50 = 0.28$. Since $(0.48) \cdot (0.46) \doteq 0.22 \approx 0.28$, we have the approximate equation

		Dime	
		Head	Tail
Penny	Head	14	9
	Tail	10	17

$$P_e(A \text{ and } B) \approx P_e(A) \cdot P_e(B).$$

In Example 10.7, it was important to understand that the outcome of one coin had no influence on the other. Events with this property are called **independent events** according to the following definition.

> **DEFINITION** *Independent Events*
> Events A and B are **independent events** if the occurrence or nonoccurrence of event A does not affect the occurrence or nonoccurrence of event B and vice versa.

For an example of *dependent* events, you might imagine flipping a dime and penny that are glued tail to tail. If A and B are the respective events that the dime and penny land with head facing upward, it is clear that the occurrence of A affects the occurrence of B.

The property of independent events that was illustrated in Example 10.7 can be stated this way.

> **PROPERTY** *Empirical Probability of Independent Events A and B*
> If A and B are independent events, then $P_e(A \text{ and } B) \approx P_e(A) \cdot P_e(B)$.

Cooperative Investigation

Strings and Loops

Materials

Six pieces of string per student, all pieces the same length (about 7 inches)

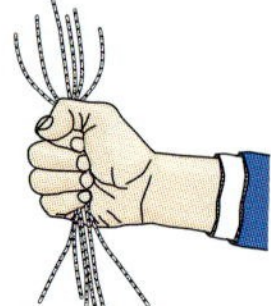

Procedure

Students work in pairs.

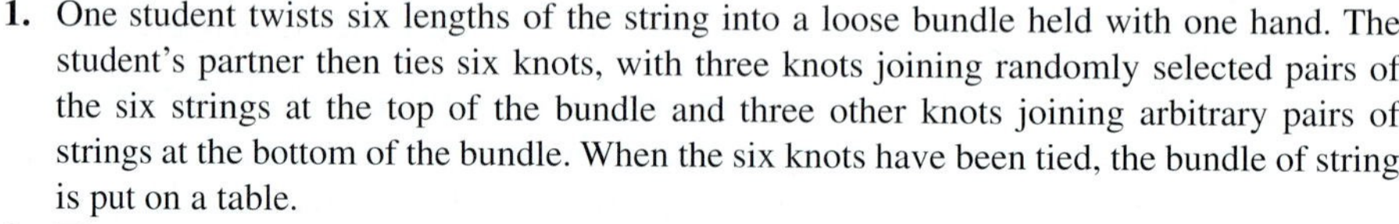

1. One student twists six lengths of the string into a loose bundle held with one hand. The student's partner then ties six knots, with three knots joining randomly selected pairs of the six strings at the top of the bundle and three other knots joining arbitrary pairs of strings at the bottom of the bundle. When the six knots have been tied, the bundle of string is put on a table.

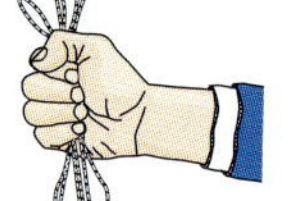

2. The partners reverse roles and tie six knots in another six-string bundle.
3. The bundles are taken apart to identify what pattern of loops has been created by the six knots. There are three possible loop patterns:

   T: three small two-string loops;
   M: one medium four-string loop and one small two-string loop;
   L: one large six-string loop.

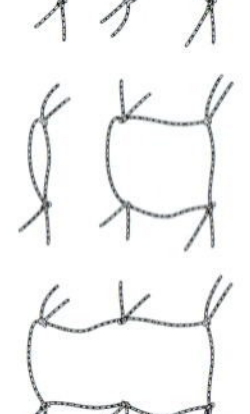

Class Project

Collect the data from all pairs of students and have each pair estimate the empirical probabilities of $P_e(T)$, $P_e(M)$, and $P_e(L)$. Which pattern seems most likely to occur? Which seems least likely? Are you surprised?

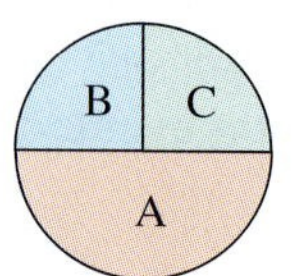

Empirical Probability and Geometry

An excellent spinner like the one shown in Figure 10.2 can be made by gluing an appropriately marked paper disc on top of a metal-rimmed price tag readily obtainable at any stationery store. A round toothpick is forced through the center to act as an axle. Good results can be obtained as follows. Hold the spinner upside down and not too tightly in the left hand and up at eye level. Spin the spinner by twisting briskly between the thumb and forefinger of the right hand. Stop the spinner while it is still spinning rapidly by pinching it between the thumb and forefinger of the right hand with the forefinger on top. Before spinning the spinner, mark a vertical line on the tip of your right thumb with a pen. Then, when the spinner is pinched, let go with the left hand, turn the spinner over with the right hand, and note and record in which region the mark on your thumb lies. One can also use a standard spinner as illustrated in Figure 9.16. The "arrow" can be just a partially unbent paper clip, spun about the point of a pencil held vertically at the center of the disc.

Figure 10.2
Making a spinner

EXAMPLE 10.8 DETERMINING EMPIRICAL PROBABILITY GEOMETRICALLY

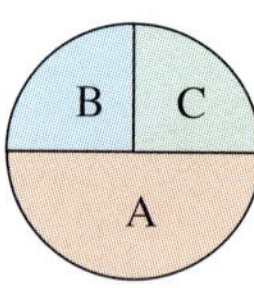

The spinner shown is spun and stopped as just described.

(a) What do you intuitively feel the probability is that your thumb mark will fall in region A?

(b) Spin the spinner 20 times and record the number of times your thumb mark falls in each region. Then compute $P_e(\mathrm{A})$, $P_e(\mathrm{B})$, and $P_e(\mathrm{C})$, the empirical probabilities that the mark falls in regions A, B, and C, respectively.

Solution

(a) Since the arc length associated with regions B and C is one-quarter of the circumference of the spinner and that of region A is one-half the circumference, it is reasonable to guess that $P_e(\mathrm{A}) \doteq 1/2$ and $P_e(\mathrm{B}) \doteq P_e(\mathrm{C}) \doteq 1/4$.

(b) Actually spinning the spinner, denote the number of times the mark falls in A, B, and C, respectively, by $n(\mathrm{A})$, $n(\mathrm{B})$, and $n(\mathrm{C})$. Then

$$P_e(\mathrm{A}) = \frac{n(A)}{20}, \quad P_e(\mathrm{B}) = \frac{n(B)}{20}, \quad \text{and} \quad P_e(\mathrm{C}) = \frac{n(C)}{20}.$$

As guessed in part (a), these ratios should approximate 1/2, 1/4, and 1/4, or 0.5, 0.25, and 0.25 respectively.

Simulation

Simulation is a method for determining answers to real problems by conducting experiments whose outcomes are analogous to the outcomes of the real problems. Consider, for example, a couple interested in understanding how many boys or girls they might anticipate if they decide to have children. In this case, we assume

- that the birth of either a boy or a girl is equally likely, and
- that the sex of one child is completely independent of the sex of any other child.

These assumptions suggest tossing a coin since the occurrence of a head or a tail is equally likely and what happens on one toss of the coin is completely independent of what happens on any other toss.

EXAMPLE 10.9 USING SIMULATION TO DETERMINE EMPIRICAL PROBABILITY

Use simulation to determine the empirical probability that a family with three children contains at least one boy and at least one girl.

Solution

Using the assumptions above, we can simulate the real problem by repeatedly tossing three coins. Here are the results of such an experiment.

TTH	TTH	HHT	HHT	TTH
TTT	HTT	HTT	HHT	HHT
HHH	HHT	HHT	HHT	HHH
TTT	HHH	HTT	HHT	HHT
HHT	TTH	TTH	HHT	TTH

The empirical probability based on these results is

$$P_e = 20/25 = 0.80.$$

Random Numbers and Simulation on a Graphing Calculator

Performing an experiment with physical objects and conducting a probabilistic study of a real-world problem may be tedious and time consuming, or even impossible or prohibitively expensive. For example, imagine the difficulties faced by a fire department that wishes to study its response times to a fire alarm. Fortunately, there is a simple and inexpensive alternative, namely, using a computer or calculator to mimic the experiment. A computer model can incorporate random locations of the fire, random fire intensities, random times of day, and random traffic conditions. The computer study can help determine where a new fire station would be best located and what equipment should be available at the station. Such a computer analysis is called a **computer model** or **simulation.** Simulations can also be carried out with a graphing calculator, providing an especially convenient tool for classroom investigations and demonstrations.

The basic random-number commands on a graphing calculator are **rand** (*random*) and **randInt** (*random integer*), found under the PRB (*probability*) submenu of the MATH menu. The **rand** command returns a random decimal number between 0 and 1, and **randInt** (*lower, upper*) returns a random integer n such that $lower \leq n \leq upper$. A new random number is obtained each time ENTER is pressed.

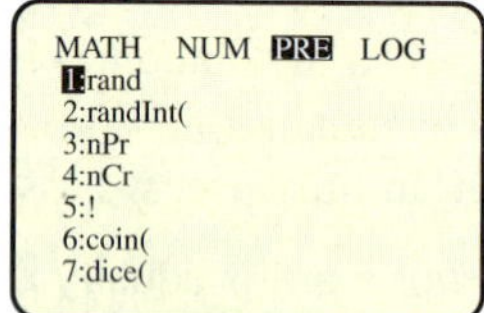

The Probability Menu of the TI-73 Graphing Calculator

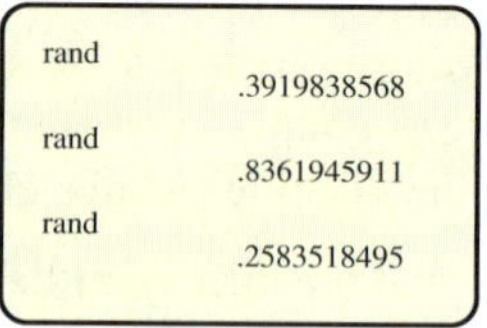

Random Numbers

For many applications, it is necessary to modify **rand** or to use **randInt** to create a random number of the required type. For example, tossing a fair coin can be simulated with **iPart**(**2*rand**) or, alternatively, with **randInt**(**0,1**). Either command returns a 0 or a 1, which can be interpreted as a tail or a head, respectively.

To sum the total spots on a roll of a pair of dice, we could use the command **randInt**(**1,6**) + **randInt**(**1,6**).

The TI-73 has some built-in functions that make it especially easy to flip any number of coins and roll any number of dice. The command **coin**(n) will simulate flipping a coin n times by returning a random list of n 0s and 1s. The command **dice**(r,d) simulates r rolls of d dice, returning a list of the r sums of spots on the d dice.

```
randInt(0,1)     0
randInt(0,1)     1
randInt(0,1)     0
randInt(0,1)     1
randInt(0,1)     1
randInt(0,1)     0
```

Simulated Tosses of One Coin

```
randInt(1,6) + ran
dInt(1,6)
                  9
randInt(1,6) + ran
dInt(1,6)
                  7
```

Rolling a Pair of Dice

```
dice(5,3)
  {16 14 9 10 11}
dice(5,3)
     {9 5 17 6 13}
dice(5,3)
     {10 8 8 11 12}
```

Simulating Rolls of Three Dice Five Times on the TI-73

Problem Set 10.1

Understanding Concepts

1. Refer to Table 9.1 on page 525 and determine the empirical probability that a student in the class obtains a grade of 79.
2. **(a)** Refer to Table 9.2 on page 526 and determine the empirical probability that a boy at Eisenhower High School is between 69.5 and 70.5 inches tall.

   (b) What is the empirical probability that one of the boys at Eisenhower High School is between 70.5 and 73.5 inches tall?
3. Prepare a 3″ × 5″ card by writing the numbers 1, 2, 3, and 4 on it as shown. Show the card to 20 college or university students, and ask them to choose a number and tell you their choice. Record the results on the back of the card and then compute $P_e(3)$, the empirical probability that a person chooses 3. Are you surprised at the result? Explain briefly.

   1 2 3 4
4. **(a)** On a 3″ × 5″ card, write the digits 1, 2, 3, 4, and 5 as shown. Show the card to 20 different students, and ask them to select a digit on the card and tell you which digit they selected. Record the results on the back of the card and compute the probabilities $P_e(1)$, $P_e(2)$, $P_e(3)$, $P_e(4)$, and $P_e(5)$.

   1 2 3 4 5

   (b) Compute $P_e(1) + P_e(2) + P_e(3) + P_e(4) + P_e(5)$.

   (c) Did you need to perform the actual calculations in part (b) to be sure what the answer would be? Explain briefly.

   (d) If you were to repeat part (a) with 100 different students, how many do you think would select the digit 4?
5. **(a)** Toss three coins 20 times and determine the empirical probability of obtaining three heads.

   (b) Using the data from part (a), determine the empirical probability of *not* obtaining three heads.

   (c) Using the data from part (a), determine the empirical probability of obtaining two heads and a tail.

   (d) Could part (b) of this question be easily determined from your answer to part (a)? Explain.
6. **(a)** Make an orderly list of all possible outcomes resulting from tossing three coins. (*Hint:* Think of tossing a penny, a nickel, and a dime.)

   (b) Does your listing in part (a) give you reason to believe that the result of problem 5(a) is about as expected? Explain briefly.
7. A computer is programmed to simulate experiments and to compute empirical probabilities. Match at least one of the computed probabilities with each of the descriptive sentences listed.

   (a) $P_e(A) = 0$ **(b)** $P_e(B) = 0.5$

   (c) $P_e(C) = -0.5$ **(d)** $P_e(D) = 1$

   (e) $P_e(E) = 1.7$ **(f)** $P_e(F) = 0.9$

   (i) This event occurred every time.

   (ii) There was a bug in the program.

   (iii) This event occurred often, but not every time.

   (iv) This event never occurred.

   (v) This event occurred half the time.
8. **(a)** Drop five thumbtacks on your desktop and determine the empirical probability that a tack dropped on a desktop will land point up.

   (b) Repeat part (a) but with 20 thumbtacks.

   (c) Give your best estimate of the number of thumbtacks that would land point up if 100 tacks were dropped on the desktop.
9. **(a)** Roll two dice 20 times and compute the empirical probability that you obtain a score of 8 on a single roll of two dice.

   (b) List the ways you can obtain a score of 8 on a roll of two dice. (*Hint:* Think of rolling a red die and a green die.)

   (c) Does the result of part (b) suggest that the result of part (a) is about right? (*Hint:* How many ways can a red die and a green die come up on a single roll?)
10. Construct a spinner with three regions A, B, and C as in Figure 10.2, but such that you would expect $P_e(A) \approx 1/2$, $P_e(B) \approx 1/3$, and $P_e(C) \approx 1/6$ on, say, 18 spins. What size angles determine the regions A, B, and C?

11. Construct a spinner as in Figure 10.2, but with the circle marked as shown here.

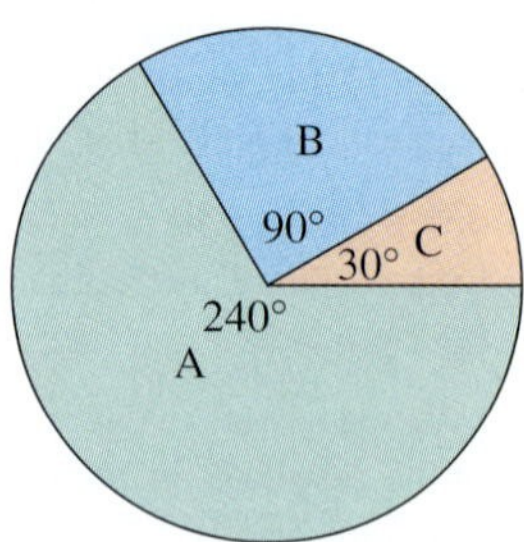

(a) Spin the spinner 20 times and compute $P_e(A)$, $P_e(B)$, and $P_e(C)$.

(b) Are the results of part (a) about as you expected? Explain briefly.

12. A die is rolled repeatedly until a 6 is obtained. Perform the experiment 10 times and determine the experimental probability that 6 first appears on the second roll.

13. (a) A die is rolled repeatedly until a 6 is obtained. Repeat this experiment 10 times and record the results. Estimate how many rolls it should take to obtain a 6.

(b) Might it take 10 rolls to obtain a 6? Explain briefly.

(c) Might it take 100 rolls to obtain a 6? Why or why not?

14. From the data in problem 13, part (a), compute the empirical probability that a 6 first appears on the fourth roll of the die.

15. Consider a spinner similar to the one shown. The head of the spinner arrow falls in the outer ring, and the tail of the arrow falls in the inner ring. Let A be the event that the head of the arrow lands in a red sector, and let B denote the event that the tail of the arrow lands in a yellow sector.

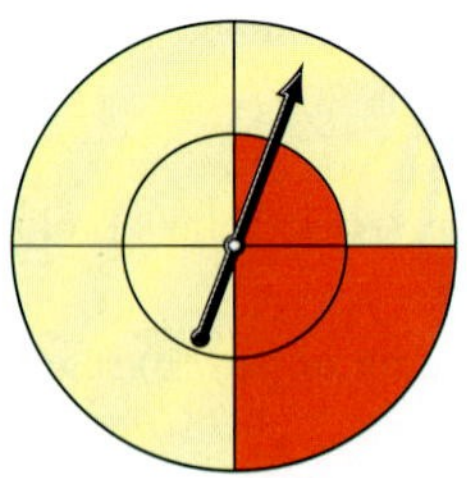

(a) Assuming that you were to conduct a large number of spins, give estimates for each of the empirical probabilities $P_e(A)$, $P_e(B)$, $P_e(A \text{ or } B)$, and $P_e(A \text{ and } B)$.

(b) What formula can be used to check your answers to part (a)?

(c) Are the events A and B mutually exclusive? Explain why or why not.

(d) Are the events A and B independent? Explain why or why not.

Teaching Concepts

16. Myrna has just completed an experiment in math class and claims that the experimental probability that an event occurs is 1.2. How would you respond to Myrna?

17. Kenji is doing an experiment in mathematics class. He has flipped a coin 10 times and has obtained a tail on each trial. He says that his luck is sure to change and that he will surely obtain a head on the next trial.

(a) How would you respond to Kenji?

(b) Have you ever heard individuals in gaming situations make claims similar to Kenji's? Are they justified? Explain.

Thinking Critically

18. Suppose that an experiment is conducted 100 times.

(a) If event A never occurs, what is $P_e(A)$? Explain briefly.

(b) If event A occurs every time, what is $P_e(A)$? Explain briefly.

(c) What range of values is possible for $P_e(A)$? Explain briefly.

19. Roll a pair of dice 20 times and count the number of times you get a 7 and the number of times you get an 8.

(a) Compute $P_e(7)$.

(b) Compute $P_e(8)$.

(c) Compute $P_e(7 \text{ or } 8)$.

(d) Compute $P_e(7) + P_e(8)$.

(e) Explain why the results of parts (c) and (d) are as they are.

20. Consider the experiment of shuffling a deck of playing cards and selecting a card at random. Repeat this experiment 20 times and note the result each time. Let $P_e(R)$ denote the empirical probability that a card is red, let $P_e(F)$ denote the empirical probability that a card is a face card, let $P_e(R \text{ or } F)$ denote the empirical probability that a card is red or is a face card, and let $P_e(R \text{ and } F)$ denote the empirical probability that a card is red and is a face card. Compute these probabilities.

(a) $P_e(R)$ (b) $P_e(F)$

(c) $P_e(R \text{ or } F)$ (d) $P_e(R \text{ and } F)$

(e) $P_e(R) + P_e(F) - P_e(R \text{ and } F)$

(f) Compare the results of parts (c) and (e). Do these results suggest a general property? Explain.

21. Consider an experiment of simultaneously tossing a single coin and rolling a single die. Repeat the experiment 20 times and compute these empirical probabilities.

(a) $P_e(H)$, the empirical probability that a head occurs

(b) $P_e(5)$, the empirical probability that a 5 occurs

(c) $P_e(H \text{ and } 5)$, the empirical probability that a head and a 5 occur simultaneously

(d) $P_e(H) \cdot P_e(5)$

(e) Should the results of parts (c) and (d) be about the same? Explain briefly.

22. Make up a three-card deck consisting of two aces and a queen. An experiment consists of shuffling the cards and inspecting the card on top and the card on the bottom. Let A denote the event that the top card is an ace, and let B denote the event that the bottom card is an ace. Repeat the experiment 30 times, keeping notes on the occurrence of the events A and B.

(a) Use your data to compute the empirical probabilities $P_e(A)$, $P_e(B)$, $P_e(A) \cdot P_e(B)$, and $P_e(A \text{ and } B)$.

(b) Do you expect that $P_e(A \text{ and } B)$ should be approximately equal to $P_e(A) \cdot P_e(B)$? Explain why or why not.

23. You are shown three cards: one black on both sides, one white on both sides, and one black on one side and white on the other. One card is selected and shown to be black on one side.

(a) Considering the possibilities, what do you think is the likelihood or probability that the selected card is black on the other side as well?

(b) Conduct the experiment just described by selecting at random one of the three cards and looking at *one side only* of the card selected. If it is white, ignore it. If it is black, record it, and record the color of the other side of the card. Repeat the experiment 10 times, carefully shuffling the cards between experiments. Compute the empirical probability that the second side of the card is black given that the first side is black. Does your experiment tend to confirm or refute your guess in part (a)?

24. Two dice are rolled 20 times. Compute $P_e(13)$, the empirical probability of obtaining a score of 13. Explain the result very briefly.

25. Two dice are rolled 20 times. Compute

$$P_e(2) + P_e(3) + P_e(4) + \cdots + P_e(12).$$

Explain the result very briefly.

Thinking Cooperatively

26. Work in a room whose floor is tiled with squares. Each student is given a thin wooden skewer (used to barbecue), cut to the length of the side of a tile. An experiment consists of throwing the skewer onto the floor and seeing whether or not the skewer crosses one of the parallel lines separating rows of floor tile. The lines separating columns of floor tiles will not be considered. Each student repeats the experiment 20 times and computes the empirical probability $P_e(C)$ that a throw crosses one of the lines. According to a calculation of Comte de Buffon (1707–1788), $P_e(C)$ is approximately $2/\pi$, or about 0.64.

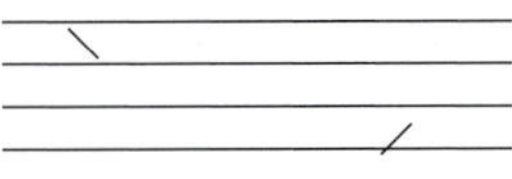

(a) Do you expect any one student to have an empirical probability close to $2/\pi$?

(b) Combine the data of individual students and compute a class value of $P_e(C)$. Do you expect to get better agreement with $2/\pi$?

27. Cover a large corkboard or bulletin board with a rectangular pattern of tangent circles as shown. An experiment consists of throwing a dart and recording whether the dart lands within a circle or not. Use circles about 1 inch in diameter, so while it can be certain that the dart lands in the board, it is uncertain whether the dart lands inside any circle. If darts are inconvenient, throw small pins onto the horizontal board and determine whether the point of the pin lies in the interior of a circle or not.

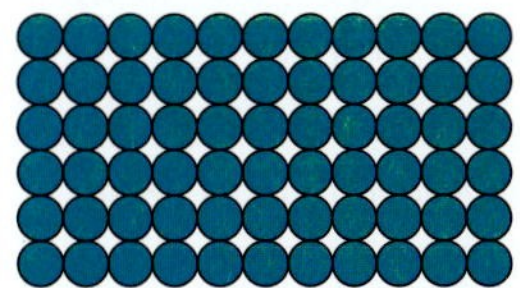

(a) Repeat the experiment numerous times and compute the empirical probability of hitting within one of the circles on the board.

(b) Does your empirical probability seem to be a good approximation of the ratio of the area of a circle to the area of the square that just contains the circle?

Making Connections

28. Barry Bonds of the San Francisco Giants hit his seventy-third home run on October 7, 2001, the last day of the 162-game regular season. This eclipsed the 1998 record of Mark McGwire of the St. Louis Cardinals. Bonds was followed by Sammy Sosa, who hit 64 home runs during the season.

(a) What was the empirical probability that Bonds would hit at least one home run during any given game of the regular season during 2001?

(b) On the basis of your answer to part (a), how many home runs would you expect Bonds to hit during any stretch of nine games during the regular season?

(c) Answer parts (a) and (b) for Sammy Sosa.

29. The four blood types are O, A, B, and AB. These occur with the empirical probabilities $P_e(O) = 0.45$,

$P_e(A) = 0.41$, $P_e(B) = 0.10$, and $P_e(AB) = 0.04$. What is the probability that, for a married couple,

(a) both spouses are type O?

(b) both spouses are type AB?

(c) one spouse is type O and the other type B? (Be careful; the answer is *not* 0.045.)

30. An abbreviated mortality table is shown below. It shows, for example, that of every 10,000 people living at birth, 9806 live to see their tenth birthday.

Age	Living at Beginning of Year	Age	Living at Beginning of Year
0	10,000	50	8762
10	9806	60	7699
20	9666	70	5592
30	9480	80	2626
40	9241	90	468

Use the table to compute the empirical probability that

(a) a newborn lives to be 10 years old.

(b) a newborn lives to be 90 years old.

(c) a person now 20 lives to be 50.

(d) a person now 30 will die before reaching age 70.

Using a Calculator

The following problems can be investigated with the COINTOSS *for the graphing calculator program found in Appendix C. Alternatively, other statistical programs may be available for you to use.*

31. Flip four coins to simulate the birth of four children in a family. Use 100 repetitions to determine the empirical probability there that are two boys and two girls.

32. Consider the experiment of flipping 10 coins n times. Draw line graphs showing the number of heads that appear for these values of n.

(a) $n = 10$

(b) $n = 50$

(c) $n = 100$

(d) If 100 coins were flipped one million times, sketch the general shape you would anticipate the histogram would take. (Don't attempt to simulate this on your calculator, since the computation time would be unreasonable.)

From State Student Assessments

33. (Texas, Grade 5)
Susan, Dorothy, and Allison play on the same softball team. They keep a record of how many times each girl goes to bat and how many hits she gets. The chart shows how each girl did last week.

SOFTBALL HITS RECORD

Name	Number of Times at Bat	Number of Hits
Dorothy	8	2
Susan	10	4
Allison	14	9

What is the probability that Susan will get a hit the next time she goes to bat?

A. $\frac{1}{8}$ B. $\frac{4}{11}$ C. $\frac{2}{5}$ D. $\frac{2}{3}$

34. (Illinois, Grade 5)
A table shows the result of 100 spins. The spinner has 6 equal sections. Which drawing probably shows the way the spinner was drawn?

Color	Number of Times This Color Was Spun
Orange	32
White	16
Gray	34
Tan	18

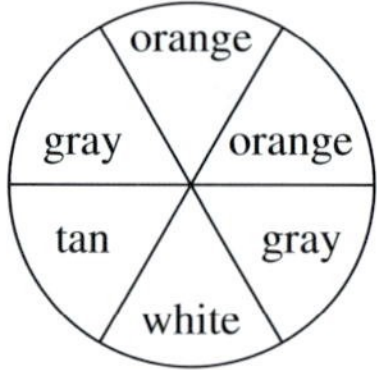

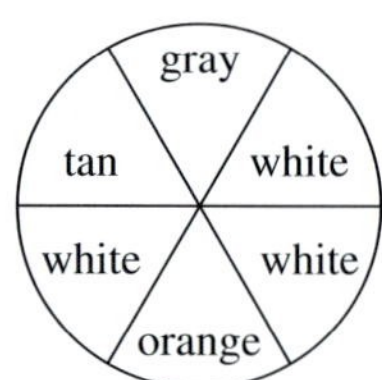

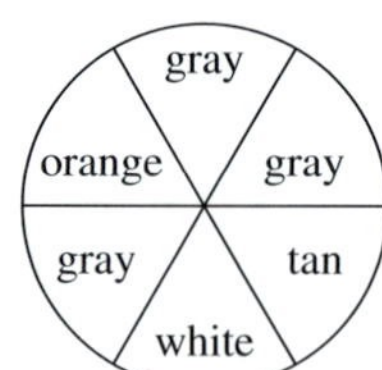

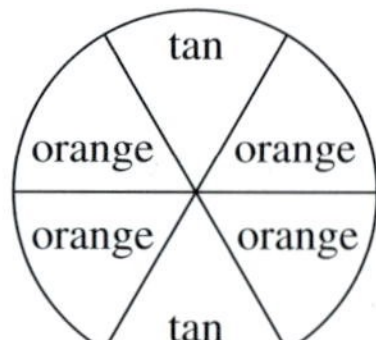

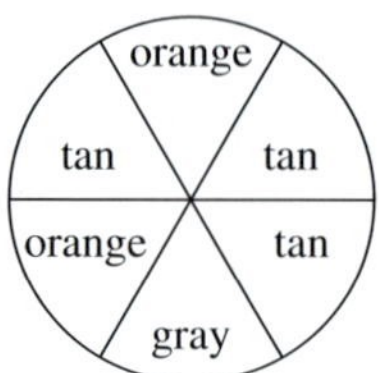

For Review

35. Compute the mean, median, and mode for these values.

30 28 34 33 29 28 27 31

36. Compute the standard deviation for the data in problem 35.

37. What percent of the data in problem 35 lies within one standard deviation of the mean? two standard deviations? three standard deviations?

38. If a population with a normal distribution has mean 28 and standard deviation 8.4, what is the probability that an individual from the population falls in the range $11.2 \le x \le 44.8$?

10.2 PRINCIPLES OF COUNTING

Kerrich's experiment, discussed in the last section, showed that the empirical probability of obtaining a head on a single toss of a coin was very nearly $1/2$. This is to be expected, since there are only two ways a tossed coin can land, and either heads or tails seems equally likely to come up. Similarly, if you repeatedly rolled a single die and computed the empirical probability of obtaining a 5, it would closely approximate $1/6 \doteq 0.17$. Again, this is expected since there are six faces on a die and these are very nearly equally likely to come up. In Example 10.7, the empirical probability of obtaining two heads on a single throw of two coins roughly approximated $1/4$, and HH is one of four equally likely possibilities: HH, HT, TH, TT.

These and several of the examples and problems in the preceding section were designed to show that the empirical probability of an event could be predicted on the basis of a straightforward analysis of the possible outcomes of an experiment. Thus, if we desire the probability that two heads and a tail appear in any order when three coins are tossed, we can list the possible outcomes—

HHH	HHT	HTH	THH
HTT	THT	TTH	TTT

—and observe that in three of the eight equally likely outcomes, two heads and a tail appear in some order. Hence, we would expect the empirical probability to be about $3/8 = 0.375$.

Considerations like these show that there is a close correlation between empirical and theoretical probability. Indeed, it is simply an extension of the law of large numbers to assert that the empirical probability of an event more and more closely approximates the theoretical probability, provided the theoretical probability is correctly determined. In particular, it is important, as demonstrated in Example 10.7, that the equally likely outcomes of an experiment be properly identified. If this is not done successfully, the theoretical probabilities will almost surely be incorrect.

In Section 10.3, we will consider theoretical probability in some detail. First, however, it is important to develop some principles of counting. We will not want to have to list all of the equally likely outcomes each time we solve a problem. Also, since sample spaces are *sets* and events are *subsets,* much of the discussion will be stated in terms of sets and subsets. In particular, if A is any set, then $n(A)$ will denote the number of elements of A as in Chapter 2.

The words *or* and *and* play key roles in counting, and a proper understanding of these roles makes solving counting problems much easier. We begin by considering the role of *or*.

Counting and the Word *Or*

Consider the question "How many diamonds or face cards are in an ordinary deck of playing cards?" Of course, we can simply count the 13 diamonds and then proceed to count the other nine face cards (kings, queens, and jacks) that are not diamonds. Thus, the answer of 22 is easily found. But there is another way to determine this sum. It seems more involved, but reveals an important pattern that greatly facilitates solving more complex problems. Let D denote the set of diamonds in the deck of cards, and let F denote the set of face cards. The two sets are shown in a Venn diagram in Figure 10.3, where we see there are $n(D) = 13$ diamonds, $n(F) = 12$ face cards, and $n(D \cap F) = 3$ cards that are both diamonds and face cards. The 22 cards that are diamonds or face cards are not given by $n(D) + n(F) = 13 + 12 = 25$, since the three cards in the intersection of the sets will be counted twice. To compensate, the number of cards in the intersection can be subtracted, so $n(D) + n(F) - n(D \cap F) = 13 + 12 - 3 = 22$ gives the correct count, $n(D \cup F) = 22$. Thus, we have the formula $n(D \text{ or } F) = n(D) + n(F) - n(D \cap F)$, where the word *or* is used to indicate the union of the sets.

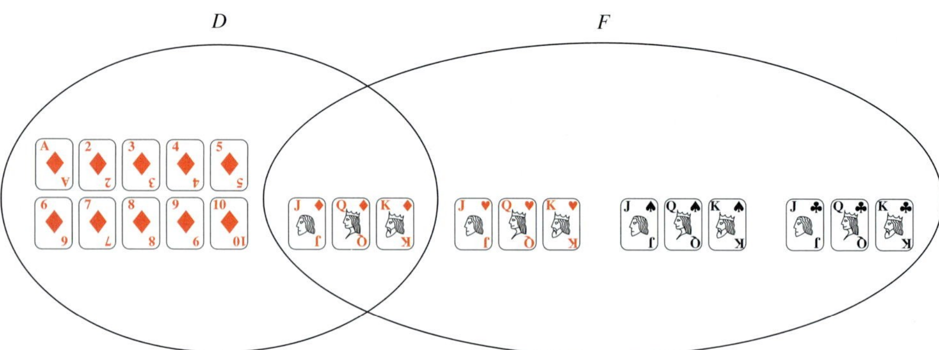

Figure 10.3
$n(D \cup F) = n(D) + n(F) - n(D \cap F)$

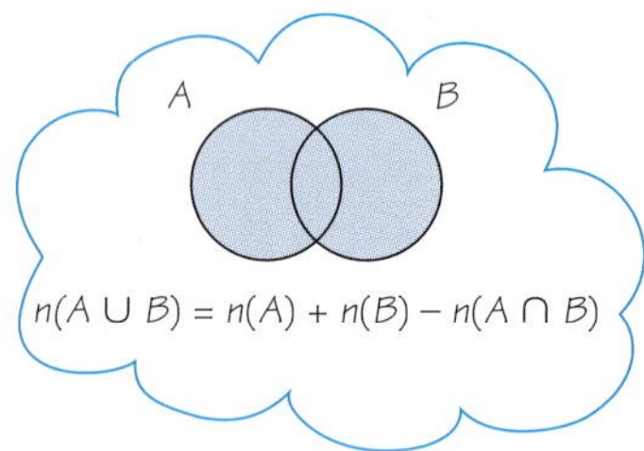

The same argument applies more generally to any two sets A and B and proves the following theorem.

THEOREM *The Addition Principle of Counting*
If A and B are events, then $n(A \text{ or } B) = n(A) + n(B) - n(A \cap B)$.

EXAMPLE 10.10 COUNTING AND *OR*

In how many ways can you select a red card or an ace from an ordinary deck of playing cards?

Solution 1

As above, there are 26 red cards in a deck (including two aces) as well as two black aces. Thus, the desired answer is 28.

Solution 2

Let R denote the set of red cards and let A denote the set of aces. Then $n(R) = 26$, $n(A) = 4$, $n(R \cap A) = 2$, and

$$n(R \text{ or } A) = n(R) + n(A) - n(R \cap A)$$

$$= 26 + 4 - 2$$
$$= 28$$

as before.

A very important special case of the addition principle of counting is illustrated by Example 10.11. Recall from Section 10.1 that two events A and B are **mutually exclusive** if no outcome belongs (or "is favorable") to both A and B. That is, $A \cap B = \varnothing$.

EXAMPLE 10.11 COUNTING AND *OR* WHEN EVENTS ARE MUTUALLY EXCLUSIVE

Determine the number of ways of obtaining a score of 7 or 11 on a single roll of two dice.

Solution 1

One way to solve this problem is simply to list all possible outcomes when rolling two dice and then to count those that are favorable. Thinking of rolling a red die and a green die makes it clear that there are 36 possible outcomes as shown below and that eight (see circled pairs) are favorable—six outcomes yield a score of 7 and two yield a score of 11. Thus,

$$n(7 \text{ or } 11) = 8 = 6 + 2 = n(7) + n(11).$$

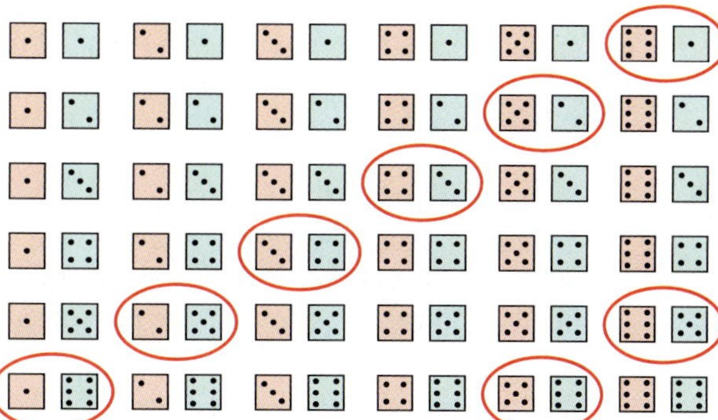

Solution 2

Let F be the set of favorable outcomes when rolling the dice, let D be the set of outcomes yielding 7, and let E be the set of outcomes yielding 11.

Since $F = D \cup E$, we may use the addition principle of counting to obtain

$$\begin{aligned} n(F) &= n(D \cup E) \\ &= n(D \text{ or } E) \\ &= n(D) + n(E) - n(D \cap E). \end{aligned}$$

But $n(D \cap E) = 0$, since $D \cap E = \varnothing$. Thus,

$$n(F) = n(D) + n(E) = 6 + 2 = 8$$

as before.

The preceding solution illustrates that

$$n(D \text{ or } F) = n(D \cup F) = n(D) + n(F)$$

when D and F are mutually exclusive events, that is, when $D \cap F = \varnothing$ and therefore $n(D \cap F) = 0$. This result is general and can be formalized as follows.

THEOREM *The Addition Principle of Counting for Mutually Exclusive Events*
If A and B are mutually exclusive events, then $n(A \text{ or } B) = n(A) + n(B)$.

EXAMPLE 10.12 | CHOOSING A CHOCOLATE

A box of 40 chocolates contains 14 cremes, 16 caramels, and 10 chocolate-covered nuts. In how many ways can you select a creme or a caramel from the box?

Solution

Let C denote the set of cremes and let C^* denote the set of caramels. Then $C \cap C^* = \varnothing$, and so, by the addition principle for the mutually exclusive events C and C^*,

$$n(C \text{ or } C^*) = 14 + 16 = 30.$$

Thus, the number of ways of choosing a creme or a caramel is 30.

EXAMPLE 10.13 | DETERMINING HOW MANY ARE ON AN AIRPLANE

On an airplane from Frankfurt to Paris, all the people speak only French or German. If 71 speak French, 85 speak German, and 29 speak both French and German, how many people are on the plane?

Solution 1

Let F denote the set of French speakers on the plane and let G denote the set of German speakers. Of course, a person who speaks both French and German belongs to *both* sets F and G. Since all people on the plane speak only French or German, it follows that the number of persons on the plane is $n(F \cup G)$. But, by the addition principle of counting,

$n(F \cup G) = n(F \text{ or } G)$

$$\begin{aligned} n(F \cup G) &= n(F) + n(G) - n(F \cap G) \\ &= 71 + 85 - 29 = 127. \end{aligned}$$

Thus, 127 people are on the plane.

Solution 2

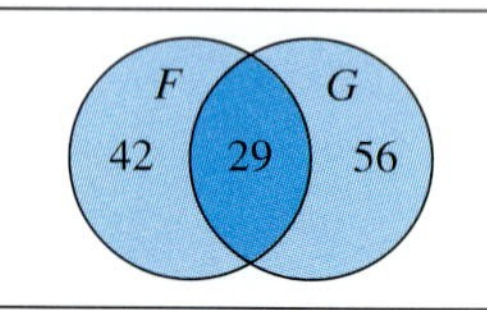

Let F and G be as above and consider the Venn diagram shown. Starting with the innermost region of the diagram, one fills in the appropriate numbers. Since 29 people fall in the set $F \cap G$, the common region of the two circles, and 71 are in the F circle, then $71 - 29 = 42$ people must fall inside the F circle, but outside the common region, as shown. Similarly, $85 - 29 = 56$ people must fall inside the G circle, but outside the common region. Thus, finally,

$$n(F \cup G) = 42 + 29 + 56 = 127$$

as before.

Counting and the Word *And*

Figure 10.4
A marble bag holds three red marbles and one green marble

A bag of marbles contains three red marbles and one green marble, as shown in Figure 10.4. Consider the following **two-stage experiment:**

Stage 1 Draw a marble from the bag and place it in your left hand.
Stage 2 Keeping the first marble in your left hand, draw a second marble from the three remaining in the bag and place it in your right hand.

If we denote the three red marbles as r_1, r_2, and r_3, and the green marble as g, one possible outcome of the two-stage experiment is r_1r_3, meaning that the red marble r_1 is placed in the left hand and the red marble r_3 is placed in the right hand. Of course, several other outcomes are also possible, and so we ask the following question: In how many ways can this two-stage experiment be analyzed?

One strategy is to **make an orderly list.**

r_1r_2	r_1r_3	r_2r_3	r_2r_1	r_3r_1	r_3r_2
r_1g	r_2g	r_3g	gr_1	gr_2	gr_3

The list shows that there are 12 outcomes. Note that outcome r_1r_2 is not the same as outcome r_2r_1; i.e., order makes a difference.

A second strategy is to make a **possibility tree,** as shown in Figure 10.5. (Curiously, trees are most often drawn sideways or upside down.) Following each "limb" from its "root" to its "leaf" corresponds to one of the 12 possible outcomes of the two-stage experiment.

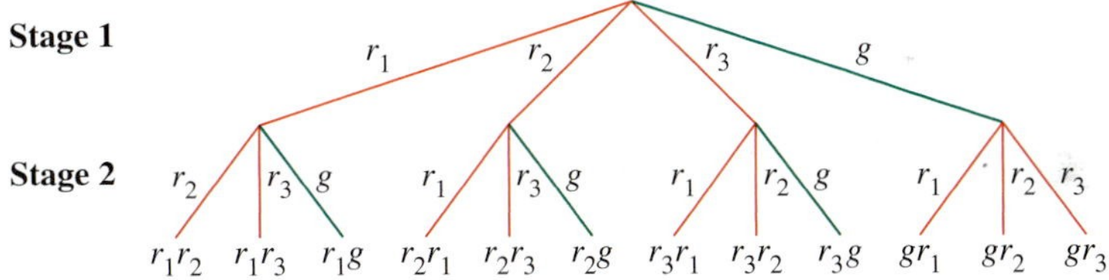

Figure 10.5
A possibility tree for drawing two marbles without replacement

Let's next ask a somewhat different question: In how many ways will both the left hand *and* the right hand hold a red marble? In symbols, if A denotes the set of first-stage outcomes in which a red marble is put into the left hand, and B denotes the set of second-stage outcomes in which a red marble is put into the right hand, then we want to count $n(A \text{ and } B)$. From the list of possibilities, we see that the first row of outcomes corresponds to a red marble in each hand, so $n(A \text{ and } B) = 6$. There is, however, a second way to obtain the answer by carefully examining the possibility tree in Figure 10.5. First, there are three ways to perform Stage 1 in which a red marble is placed in the left hand. That is, $n(A) = 3$. Next, *given* that one of the red marbles has been placed in your left hand, there are two ways to put a second red marble in your right hand. This number is symbolized by $n(B|A) = 2$ and is read as "the number of ways B can occur given that A has already occurred is 2." Observing that $n(A \text{ and } B) = 6 = 3 \cdot 2 = n(A)n(B|A)$, we have illustrated the following general counting principle.

THEOREM *The Multiplication Principle of Counting*
Let A be a set of outcomes of Stage 1, and B a set of outcomes of Stage 2. Then the number of ways, $n(A \text{ and } B)$, that A and B can occur in a two-stage experiment is given by

$$n(A \text{ and } B) = n(A)n(B|A),$$

where $n(B|A)$ denotes the number of ways B can occur given that A has already occurred.

It is important to notice that, in the preceding discussion, $n(B) = 9$. That is, of the 12 outcomes of the two-stage experiment, there are 9 cases where the right-hand marble is red. In three of these cases, the left-hand marble drawn first was green. Since $n(B) \neq n(B|A)$, we see that A and B are dependent events.

In other two-stage experiments, the number of outcomes of the event B does not depend on whether or not event A has occurred. That is, A and B are *independent events* by the definition given in Section 10.1. In symbols, $n(B) = n(B|A)$ for independent events. This formula gives us a special case of the multiplication principle.

THEOREM *The Multiplication Principle of Counting for Independent Events*
Let A be a set of outcomes of Stage 1, and B a set of outcomes of Stage 2. If A and B are independent events, then the number of ways, $n(A \text{ and } B)$, that A and B can occur in a two-stage experiment is given by

$$n(A \text{ and } B) = n(A)n(B).$$

The following two examples give practice using the multiplication principles of counting.

EXAMPLE 10.14 COUNTING THE NUMBER OF WAYS TO DRAW TWO ACES

How many ways, from an ordinary deck of 52 cards, can two aces be drawn in succession if

(a) the first card drawn is replaced in the deck, the cards are reshuffled, and then the second card is drawn?

(b) the first card is drawn, but not replaced in the deck, and then the second card is drawn?

Solution

Understand the Problem

In both parts of the problem, we are asked in how many ways we can draw two aces from an ordinary deck of 52 playing cards. There is a difference, however. In part (a), after the first card is drawn, it is replaced and the deck is reshuffled before the second card is drawn. In the second case, the first card is *not* replaced before the second card is drawn.

Devise a Strategy

Let A be the event of choosing an ace on the first draw, and let B be the event of choosing an ace on the second draw. Because of the replacement and reshuffling in part (a), it is clear that the choice of the first card cannot possibly affect the choice of the second card.

Thus, the two events are clearly independent. However, this is not the case in part (b). If, for example, the ace of clubs is chosen on the first draw, there remain only three possible favorable choices for the second draw, instead of four. Thus, in part (b), A and B are *dependent* events. For part (a), we can use the formula $n(A \text{ and } B) = n(A) \cdot n(B)$, and for part (b), we must use the formula $n(A \text{ and } B) = n(A) \cdot n(B|A)$.

Carry Out the Plan

For part (a), $n(A \text{ and } B) = n(A) \cdot n(B) = 4 \cdot 4 = 16$, since there are four ways to choose an ace on each draw. For part (b), however, $n(A \text{ and } B) = n(A) \cdot n(B|A) = 4 \cdot 3 = 12$, since, if an ace is chosen on the first draw, there remain only three favorable choices for the second draw.

Look Back

Considering the physical aspects of both parts of the problem, it became clear that events A and B were independent in part (a) and dependent in part (b). Thus, we were able to use the appropriate formula in each case to determine the correct answers.

EXAMPLE 10.15 DETERMINING THE NUMBER OF CODE "WORDS"

How many five-letter code "words" can be formed

(a) if repetition of letters is allowed?
(b) if repetition of letters is not allowed?

Solution

Understand the Problem

Since we are talking about words in code, a word need not look like a word. Thus, *arefg* is a perfectly acceptable code word. Moreover, to determine a five-letter code word, we must choose a first letter *and* a second letter *and* a third letter *and* a fourth letter *and* a fifth letter. The question then is "In how many ways can we do all these things?"

Devise a Plan

Since we have to do a first thing *and* a second thing *and* a third thing, . . . , *and* a fifth thing, the word "and" suggests that we use one of the multiplication principles of counting, extended to handle a five-stage experiment.

Carry Out the Plan

(a) We must choose letters to place in these blanks.

____ ____ ____ ____ ____

(b) If repetition of letters is allowed, then the choice of any letter does not affect the choice of any other letter. Thus, the choices are independent and there are 26 ways to fill each blank, so

$$
\begin{aligned}
&n(\text{five-letter code words with repetition of letters allowed}) \\
&\quad = 26 \cdot 26 \cdot 26 \cdot 26 \cdot 26 \\
&\quad = 26^5 \\
&\quad = 11{,}881{,}376.
\end{aligned}
$$

(b) If repetition is not allowed, there are still 26 choices for the first letter, but only 25 for the second, 24 for the third, 23 for the fourth, and 22 for the fifth. Thus, the number of code words in this case is

$$\begin{aligned} &n(\text{five-letter code words without repetition of letters}) \\ &\quad = 26 \cdot 25 \cdot 24 \cdot 23 \cdot 22 \\ &\quad = 7{,}893{,}600. \end{aligned}$$

Look Back

The key to working this problem was the use of the word "and" in describing what actually must be done to accomplish the desired task—that is, choose a first letter *and* choose a second letter, *and* . . . , *and* choose a fifth letter. Thus, using the multiplication principle repeatedly, we calculated

$$\begin{aligned} &n(\text{ways to choose a five-letter code word}) = \\ &\quad n(\text{ways to choose first letter}) \\ &\cdot n(\text{ways to choose second letter} \mid \text{first letter}) \\ &\cdot n(\text{ways to choose third letter} \mid \text{first two letters}) \\ &\cdot n(\text{ways to choose fourth letter} \mid \text{first three letters}) \\ &\cdot n(\text{ways to choose fifth letter} \mid \text{first four letters}). \end{aligned}$$

Finally, it was necessary to decide whether the stages were independent (as in (a)) or dependent (as in (b)).

Combinations and Permutations

In Example 10.15, we were required to find the number of five-letter code words under certain conditions. Thus, *abcde* and *acdbe* are both acceptable code words and are different since the order of the letters is different, even though the sets $\{a, b, c, d, e\}$ and $\{a, c, d, b, e\}$ are the same. These distinctions are formalized in the following definition.

HIGHLIGHT *from* HISTORY

Leonhard Euler (1707–1783)

Another of the titans of mathematics was the Swiss mathematician Leonhard Euler (pronounced "Oiler"), born in Basel in 1707 and a contemporary of the second generation of the remarkable Bernoulli family. Euler is the most prolific mathematician who ever lived, publishing 886 papers and books during his lifetime and posthumously for an average of approximately 800 pages of new mathematics per year. He was dubbed "Analysis Incarnate" by the French academician François Arago, who declared that Euler could produce mathematics without apparent effort "just as men breathe and eagles sustain themselves in the air." Euler went blind in 1766, to the considerable distress of his many friends and colleagues. However, aware that the condition was coming on, Euler taught himself to write his complicated formulas on a large slate with his eyes closed. Then with a scribe to write down the explanations of his formulas and with all the facts and formulas of the then known mathematics safely tucked away in his memory, his work continued unabated until his death.

On September 13, 1783, having earlier outlined on his slate the calculation of the orbit of the newly discovered planet Uranus, he called for his grandson to be brought in. While playing with the child he suffered a stroke and, with the words "I die," he quietly passed away.

DEFINITION *Combinations and Permutations*
Let U be a set of objects. A subset of U with r objects is called a **combination of r objects of U.** An ordered sequence of r objects from U is called a **permutation of r objects of U.**

EXAMPLE 10.16 CHOOSING A SOCIAL COMMITTEE AND OFFICERS FOR THE MATH CLUB

There are five members of the Math Club.

(a) In how many ways can the slate of officers, a president and a treasurer, be chosen?

(b) In how many ways can the two-person Social Committee be chosen?

Solution For convenience, suppose the five members of the Math Club form the set $U = \{a, b, c, d, e\}$.

(a) Choosing the officers is the same as choosing a sequence of two members of U, first the president and then the treasurer. That is, we wish to know the number of permutations of five objects taken two at a time. By the multiplication principle of counting, there are five choices for the president, and, once the president is chosen, there are four remaining members from which to choose the treasurer. Thus, the number of ways to choose the two officers is 20. As a check, we can make a list of the 20 possible slates of officers, where the first person in each pair is president and the second person the treasurer:

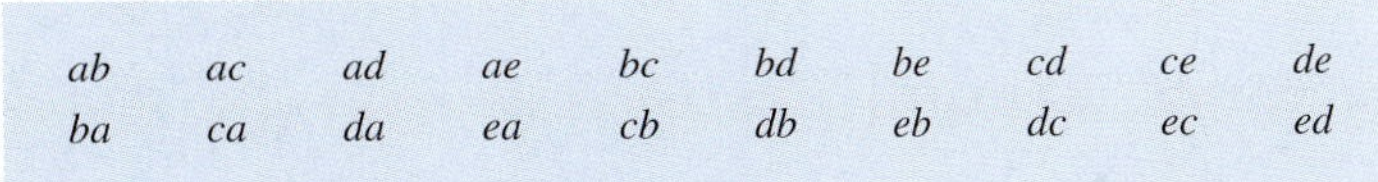

ab	*ac*	*ad*	*ae*	*bc*	*bd*	*be*	*cd*	*ce*	*de*
ba	*ca*	*da*	*ea*	*cb*	*db*	*eb*	*dc*	*ec*	*ed*

(b) Unlike the officer selection, the members of the Social Committee form an unordered subset of two elements. For example, a committee consisting of a and b is the same committee if b had been chosen first and then a. Thus, the number of two-person committees is the number of combinations of five objects taken two at a time. Thus, there are 10 committees, as shown in the following list:

$\{a, b\}$ $\{a, c\}$ $\{a, d\}$ $\{a, e\}$ $\{b, c\}$ $\{b, d\}$ $\{b, e\}$ $\{c, d\}$ $\{c, e\}$ $\{d, e\}$

In counting problems, the actual permutations or combinations of a set of elements taken r at a time are not important, but their number is. In this context, the important questions are as follows:

- How many permutations of n things taken r at a time are there?
- How many combinations of n things taken r at a time are there?

Before answering these questions, it will be helpful to introduce just a bit of new notation. As in Example 10.15, part (b), products like $26 \cdot 25 \cdot 24 \cdot 23 \cdot 22$ o r

$7 \cdot 6 \cdot 5 \cdot 4 \cdot 3 \cdot 2 \cdot 1$ frequently appear. Since these are somewhat tedious to write out, we use the shorthand notation

$$7! = 7 \cdot 6 \cdot 5 \cdot 4 \cdot 3 \cdot 2 \cdot 1,$$

where 7! is read **"7 factorial."** Thus, for example,

$$5! = 5 \cdot 4 \cdot 3 \cdot 2 \cdot 1.$$

Moreover,

$$26 \cdot 25 \cdot 24 \cdot 23 \cdot 22 = \frac{26 \cdot 25 \cdot 24 \cdot 23 \cdot 22 \cdot (21!)}{(21!)} = \frac{26!}{21!},$$

where $21! = 21 \cdot 20 \cdot 19 \cdots 3 \cdot 2 \cdot 1$.

In general, we define $n!$ for every integer $n \geq 0$.

DEFINITION *The Factorial, n!*

Let n be a whole number. Then ***n* factorial,** or ***n*!,** is defined by

$$n! = n \cdot (n-1) \cdot (n-2) \cdots 1 \text{ for } n \geq 1$$

and

$$0! = 1.$$

That 0! is defined to be 1 may seem strange, but we justify this part of the definition shortly.

EXAMPLE 10.17 MANIPULATING FACTORIALS

Compute each of these expressions.

(a) $1!, 2!, 3!, 4!$ **(b)** $4 \cdot 3!$ **(c)** $(4 \cdot 3)!$

(d) $4! + 3!$ **(e)** $4! - 3!$ **(f)** $\frac{8!}{5!}$

(g) $\frac{8!}{7!}$ **(h)** $\frac{8!}{8!}$ **(i)** $\frac{8!}{0!}$

Solution

(a) $1! = 1, 2! = 2 \cdot 1 = 2, 3! = 3 \cdot 2 \cdot 1 = 6, 4! = 4 \cdot 3 \cdot 2 \cdot 1 = 24$

(b) $4 \cdot 3! = 4 \cdot (3 \cdot 2 \cdot 1) = 4! = 24$

(c) $(4 \cdot 3)! = 12! = 479{,}001{,}600$, using a calculator

(d) $4! + 3! = 4 \cdot 3! + 3! = 5 \cdot 3! = 30$

(e) $4! - 3! = 4 \cdot 3! - 3! = 3 \cdot 3! = 18$

(f) $\frac{8!}{5!} = \frac{8 \cdot 7 \cdot 6 \cdot 5 \cdot 4 \cdot 3 \cdot 2 \cdot 1}{5 \cdot 4 \cdot 3 \cdot 2 \cdot 1} = 8 \cdot 7 \cdot 6 = 336$

(g) $\frac{8!}{7!} = \frac{8 \cdot 7 \cdot 6 \cdot 5 \cdot 4 \cdot 3 \cdot 2 \cdot 1}{7 \cdot 6 \cdot 5 \cdot 4 \cdot 3 \cdot 2 \cdot 1} = 8$

(h) $\frac{8!}{8!} = 1$

(i) $\frac{8!}{0!} = \frac{8!}{1} = 8 \cdot 7 \cdot 6 \cdot 5 \cdot 4 \cdot 3 \cdot 2 \cdot 1 = 40{,}320$

With the factorial notion in hand, we now return to the problem of determining the number of permutations of n things taken r at a time and the number of combinations of n things taken r at a time. The following notation is helpful.

NOTATION *$P(n, r)$ and $C(n, r)$*
$P(n, r)$ denotes the **number of permutations of n things taken r at a time.**
$C(n, r)$ denotes the **number of combinations of n things taken r at a time.**

Some textbooks use the notations nPr and nCr to denote permutations and combinations, respectively. This notation is also commonly used to designate built-in functions on a graphing calculator to compute the numbers of permutations and combinations.

To derive formulas for $P(n, r)$, $P(n, n)$, and $C(n, r)$, let's reconsider the "code words" problem of Example 10.15. In part (b), we are to determine the number of different five-letter code words with repetition not allowed. Since the order in which letters appear in a code word certainly matters, this is precisely the problem of determining the number, $P(26, 5)$. Since there are 26 choices for the first letter, 25 choices for the second letter, and so on, the answer is

$$P(26, 5) = 26 \cdot 25 \cdot 24 \cdot 23 \cdot 22.$$

Five factors since we must choose five letters

Repeating the argument in general, we have

$$P(n, r) = n(n-1)(n-2)\cdots(n-r+1).$$

Here, there are r factors since we are choosing r letters.

Note also that

$$\begin{aligned} P(n, r) &= \frac{n(n-1)\cdots(n-r+1)(n-r)\cdots 1}{(n-r)(n-r-1)\cdots 1} \\ &= \frac{n!}{(n-r)!}. \end{aligned}$$

Setting $r = n$ in the above two formulas, we obtain

$$P(n, n) = n! \quad \text{and} \quad P(n, n) = \frac{n!}{(n-n)!} = \frac{n!}{0!}.$$

Since these must be the same, it follows that we should define 0! to be 1 as on page 614.

Suppose we wanted 26-letter code words without repetition; that is, suppose we wanted to compute $P(26, 26)$. Repeating the above argument, we have that

- the first letter can be chosen in 26 ways,
- the second letter can be chosen in 25 ways,
- the third letter can be chosen in 24 ways,

$\vdots$

- the last letter can be chosen in 1 way.

Since the product of these numbers gives the desired result, the number of permutations of 26 things taken all at a time is

$$P(26, 26) = 26 \cdot 25 \cdot 24 \cdots 1 = 26!.$$

Repeating the argument for a general n, we have

$$P(n, n) = n!.$$

Lastly, consider again the problem of determining the number of permutations of 26 things taken five at a time. One way to determine such a permutation is to view it as a two-stage process:

Stage 1 choose, without regard to order, the five letters to appear in the permutation, *and*

Stage 2 determine the order in which the letters are to appear.

We can choose the five letters in $C(26, 5)$ ways, and the five objects can be put in order in $5!$ ways. Thus, by the multiplication principle of counting,

$$P(26, 5) = C(26, 5) \cdot 5!.$$

Dividing both sides of the equation by $5!$, we obtain

$$C(26, 5) = \frac{P(26, 5)}{5!} = \frac{26 \cdot 25 \cdot 24 \cdot 23 \cdot 22}{5!}.$$

Five factors in both numerator and denominator, since $5! = 5 \cdot 4 \cdot 3 \cdot 2 \cdot 1$

This argument could be repeated in general to give

$$C(n, r) = \frac{n(n-1)\cdots(n-r+1)}{r!}.$$

r factors in both numerator and denominator

The formulas we have discovered can be collected to give the following important theorem.

THEOREM *Formulas for P(n, r) and C(n, r)*
Let n and r be natural numbers with $0 < r \leq n$. Then

$$P(n, r) = n(n-1)(n-2)\cdots(n-r+1),$$
$$P(n, n) = n!,$$

n

and

$$C(n, r) = \frac{n(n-1)(n-2)\cdots(n-r+1)}{r!}.$$

EXAMPLE 10.18 COMPUTING P(n, r) AND C(n, r)

Compute each of the following.

(a) $P(7, 2)$ **(b)** $P(8, 8)$ **(c)** $P(25, 2)$
(d) $C(7, 2)$ **(e)** $C(8, 8)$ **(f)** $C(25, 2)$

Solution

(a) $P(7, 2) = 7 \cdot 6 = 42$
(b) $P(8, 8) = 8! = 40{,}320$
(c) $P(25, 2) = 25 \cdot 24 = 600$
(d) $C(7, 2) = \dfrac{7 \cdot 6}{2!} = 21$
(e) $C(8, 8) = \dfrac{8!}{8!} = 1$

(f) $C(25, 2) = \dfrac{25 \cdot 24}{2 \cdot 1} = 300$

EXAMPLE 10.19 PERMUTATIONS OF FOUR RED FLAGS, THREE BLUE FLAGS, AND TWO GREEN FLAGS

How many ways can you run four red flags, three blue flags, and two green flags up a pole if the flags are indistinguishable except for color?

Solution

Understand the Problem

We must put four red flags, three blue flags, and two green flags in order. Since the flags are indistinguishable except for color, interchanging flags of the same color will not make any difference; that is, the arrangement

R R B G B G R B R

will not change if we interchange red flags among themselves, or blue flags among themselves, or green flags among themselves. Apparently, the only way to obtain a different arrangement is to choose different locations in which to place the red, blue, and green flags.

Devise a Plan

How can we determine in how many ways we can choose the four places for red flags, the three places for blue flags, and the two places for green flags? This is just the number of ways we can choose four of the nine places to receive red flags *and* three of the remaining five places to receive blue flags *and* two of the remaining two places to receive green flags. The words "and" in the preceding sentence are the key. They suggest that we use the multiplication principle of counting.

Carry Out the Plan

1. The number of ways we can choose four of the nine places to receive red flags is $C(9, 4)$, the number of combinations of nine things (spaces) taken four at a time.
2. Having chosen the four places to receive red flags, we must now choose three of the remaining five places to receive blue flags, and this can be done in $C(5, 3)$ ways.
3. This leaves two places from which we must choose the places to receive the two green flags, and this can be done in $C(2, 2)$ ways.

But we have to do Step 1 *and* Step 2 *and* Step 3, and so, by the multiplication principle of counting, the desired answer is just the product of the number of ways we can complete each step. This is

$$C(9, 4) \cdot C(5, 3) \cdot C(2, 2) = \frac{9 \cdot 8 \cdot 7 \cdot 6}{4!} \cdot \frac{5 \cdot 4 \cdot 3}{3!} \cdot \frac{2 \cdot 1}{2!}$$

$$= \frac{9!}{4!\,3!\,2!}.$$

Some arithmetic shows that this value is 1260.

Look Back

The solution just given for the flags problem would work just as well if we had different numbers of red, blue, and green flags, or even if we also had some yellow flags. For example, suppose there are 25 flags (and a tall flagpole!), including seven red, nine blue, five green, and four yellow flags. Assuming the flags of any one color are indistinguishable, we could run the flags up the pole to form

$$\frac{25!}{7!\,9!\,5!\,4!}$$

different color patterns. This expression can be evaluated on a scientific or graphing calculator, showing that there are nearly three trillion color patterns possible with the 25 flags!

EXAMPLE 10.20 DETERMINING THE NUMBER OF ZIP CODE GROUPS

As seen in Section 4 of Chapter 4, the bar code commonly seen under the address on pieces of mail is the zip code in machine-readable form. When broken down, it turns out that each code group consists of two long bars and three short bars—for example, ıılıl. How many different code groups can be formed in this way?

Solution This is just the number of ways one can choose the positions of the two long bars from the five positions available. By the preceding theorem, the result is

$$C(5, 2) = \frac{5 \cdot 4}{2 \cdot 1} = 10.$$

The result is opportune, since the 10 code groups are just adequate to represent the 10 digits—0, 1, 2, . . . , 9—needed to express a zip code.

Problem Set 10.2

Understanding Concepts

1. Two dice are thrown. Determine the number of ways a score of 4 or 6 can be obtained. For example, $4 = 3 + 1 = 1 + 3 = 2 + 2$, and so on.
2. Two dice are thrown. Determine the number of ways to obtain a score of at least 4. (*Hint:* In how many ways can you fail to obtain a score of at least 4? How many outcomes are possible all told?)
3. A coin is tossed and a die is rolled.
   (a) In how many ways can the outcome consist of a head and an even number?
   (b) In how many ways can the outcome consist of a head or an even number?
4. There are 26 students in Mrs. Pietz's fifth-grade class at the International School of Tokyo. All of the students speak either English or Japanese, and some speak both languages. If 18 of the students can speak English and 14 can speak Japanese,
   (a) how many speak both English and Japanese?
   (b) how many speak English, but not Japanese? (*Hint:* Draw a Venn diagram.)
5. All of the 24 students in Mr. Walcott's fourth-grade class at the International School of Tokyo speak either English or Japanese. If 11 of the students speak only English and nine of the students speak only Japanese, how many speak both languages?
6. In how many ways can you draw a club or a face card from an ordinary deck of playing cards?
7. In how many ways can you select a red face card or a black ace from an ordinary deck of playing cards?
8. (a) How many four-digit natural numbers can be named using each of the digits 1, 2, 3, 4, 5, and 6 at most once?

(b) How many of the numbers in part (a) begin with an odd digit? (*Hint:* Choose the first digit first.)

(c) How many of the numbers in part (b) end with an odd digit? (*Hint:* Choose the first digit first and the last digit second.)

9. (a) If repetition of digits is not allowed, how many three-digit numbers can be formed using the digits 1, 2, 3, 4, and 5?

(b) How many of the numbers in part (a) begin with either of the digits 2 or 3?

(c) How many of the numbers in part (a) are even?

10. Construct a possibility tree to determine all three-letter code words using only the letters a, b, and c without repetition.

11. If four coins are tossed, construct a possibility tree to determine in how many ways one can obtain two heads and two tails.

12. A bag contains three red marbles r_1, r_2, and r_3 and two green marbles g_1 and g_2. Two marbles are drawn in succession without replacing the marble drawn first.

(a) Construct a possibility tree for this two-stage experiment.

(b) Let A be the first-stage event that the marble drawn first is green, and let B be the second-stage event that the second marble drawn is red. List the outcomes of the two-stage experiment that are in the compound event A and B.

(c) Verify that $n(A \text{ and } B) = n(A)n(B|A)$.

13. Evaluate each of these expressions.

(a) $7!$ (b) $9! - 7!$ (c) $9! \div 7!$

(d) $9! + 7!$ (e) $7 \cdot 7!$ (f) $0!$

14. Evaluate each of the following.

(a) $P(13, 8)$ (b) $P(15, 15)$ (c) $P(15, 2)$

(d) $C(13, 8)$ (e) $C(15, 15)$ (f) $C(15, 2)$

15. How many four-letter code words can be formed using a standard 26-letter alphabet

(a) if repetition is allowed?

(b) if repetition is not allowed?

16. How many five-digit numbers can be formed with the first three digits odd and the last two digits even

(a) if repetition of digits is allowed?

(b) if repetition of digits is not allowed?

17. The Chess Club has six members. In how many ways

(a) can all six members line up for a picture?

(b) can the club choose a president and secretary?

(c) can the club choose three members to attend the regional tournament, with no regard to order?

18. How many different signals can be sent up on a flagpole if each signal requires three blue and three yellow flags and the flags are identical except for color?

19. (a) How many different arrangements are there of the letters in TOOT?

(b) How many different arrangements are there of the letters in TESTERS?

Teaching Concepts

20. Melanee understands that $2! = 2 \cdot 1$, $4! = 4 \cdot 3 \cdot 2 \cdot 1$ and, in general, that $n! = n(n-1)\cdots 1$. That is, she understands that to compute $n!$, you take the product of all the numbers starting with n, $n-1$, $n-2$, and so on until finally reaching 1. But this leaves her considerably confused when we define $0! = 1$. How, she asks, can you start with 0 and reduce the number by one repeatedly until 1 is reached? It just doesn't make sense. What would you say to Melanee, and to the rest of the class, for that matter, since many of them no doubt share Melanee's view?

21. Tyrone understands the formulas

$$P(n, r) = n(n-1)\cdots(n-r+1)$$

for the number of permutations of n things taken r at a time and

$$C(n, r) = \frac{n(n-1)\cdots(n-r+1)}{r(r-1)\cdots 1}$$

for the number of combinations of n things taken r at a time and can apply them correctly. But he is still confused by problems like problem 19, where one is asked to find the number of orderings (permutations) of several items that are not all distinct. What would you say or do to help Tyrone?

22. Write out a careful argument showing that

$$C(n, r) = C(n, n-r)$$

for whole numbers n and r with $0 \le r \le n$.

23. Write out a careful argument showing that $C(n, r) = \dfrac{n!}{r!(n-r)!}$.

24. Bai was able to write out the careful and correct argument required in problem 23. But he notices that then $C(n, n) = \frac{n!}{n!0!}$, while from the theorem on page 616, $C(n, n) = \frac{n!}{n!} = 1$. For this to make sense, he observes that we are again led to define $0! = 1$. How do you respond to Bai?

Thinking Critically

25. Two dice are thrown.

(a) In how many ways can the score obtained be even?

(b) In how many ways can the score be a multiple of 5?

(c) In how many ways can the score be a multiple of 3? (*Hint:* All possibilities when two dice are rolled are shown in Example 10.11.)

26. In how many ways can you arrange the nine letters *a*, *a*, *a*, *a*, *b*, *b*, *b*, *c*, and *c* in a row?

27. How many of the arrangements in problem 26 start with *a* and end with *b*?

28. (a) In how many of the 120 arrangements of *a*, *b*, *c*, *d*, and *e* does *b* immediately follow *a*? (*Hint:* Think of *ab* as a single symbol.)

(b) In how many of the 120 possible arrangements of *a*, *b*, *c*, *d*, and *e* are *a* and *b* adjacent?

(c) In how many arrangements of *a*, *b*, *c*, *d*, and *e* does *a* precede *e*?

29. Sets of six cards are selected without replacement from an ordinary deck of playing cards.

(a) In how many ways can you choose a set of six hearts?

(b) In how many ways can you select a set of three hearts and three spades?

(c) In how many ways can you select a set of six hearts or six spades?

30. How many different sequences of 10 flips of a coin result in five heads and five tails?

31. The Debate Team has six girls and five boys. In how many ways can a four-person team be selected if there are to be two boys and two girls put on the team?

32. Compute the values of $C(n, r)$ in this table and extend the table two more rows. Do you need the formula to extend the table? Explain.

$$C(0, 0)$$
$$C(1, 0) \quad C(1, 1)$$
$$C(2, 0) \quad C(2, 1) \quad C(2, 2)$$
$$C(3, 0) \quad C(3, 1) \quad C(3, 2) \quad C(3, 3)$$
$$C(4, 0) \quad C(4, 1) \quad C(4, 2) \quad C(4, 3) \quad C(4, 4)$$

33. A bag contains eight marbles of assorted colors, of which just one is red. Use the symbol $C(n, r)$ to answer these questions.

(a) In how many ways can a subset of any five marbles be chosen?

(b) In how many ways can a subset of five marbles be chosen, none red?

(c) In how many ways can a subset of five marbles be chosen, including the red marble?

(d) What formula expresses the fact that your answer to part (a) is the sum of your answers to parts (b) and (c)?

(e) Create a "marble story" to derive the formula $C(10, 4) = C(9, 4) + C(9, 3)$.

Thinking Cooperatively

34. **Making Rod Trains.** In small cooperative groups, investigate making trains with Cuisenaire rods, as described in the Cooperative Investigation activity in Section 2.3. In particular, use the rods to form all possible trains of length five. For example, here are four ways to form a train of length five. Also notice these trains contain varying numbers of cars (that is, rods), from one car to four cars.

(a) In how many total ways can trains of length five be formed?

(b) How many length-five trains contain one car? two cars? three cars? four cars? five cars?

(c) In the following figure, each dashed segment can either be left as is or made solid. Use this figure to explain why there are $2 \times 2 \times 2 \times 2$, or 2^4, trains of length 5.

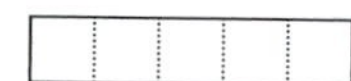

(d) To make a three-car train of length five, two of the four dashed lines must be made solid to show a separation between adjacent cars. Discuss why this means there are $C(4, 2)$, or six, trains of length five that each have three cars.

(e) Carefully explain how to find a formula giving the number of trains of length *n*.

(f) Carefully explain how to find a formula, in terms of $C(m, t)$ for appropriate choices of *m* and *t*, that gives the number of trains of length *n* that have *c* cars.

Making Connections

35. An electrician must connect a red, a white, and a black wire to a yellow, a blue, and a green wire in some order. How many different connections are possible?

36. **(a)** Mrs. Ruiz has 13 boys and 11 girls in her class. In how many ways can she select a committee to organize a class party if the committee must contain three boys and three girls?

(b) Lourdes, a girl, and Andy, a boy, always fight. How many ways can Mrs. Ruiz select the committee of part (a) if she does not want both Lourdes and Andy on the committee? Note that Lourdes can be on the committee and Andy not on the committee or vice versa.

37. In the state of Washington, each automobile license plate shows three letters followed by three digits, or three digits followed by three letters. How many different license plates can be made

(a) if repetition of digits and letters is allowed?

(b) if repetition of digits and letters is not allowed?

38. A four-bit code word is any sequence of four digits, where each digit is either a 0 or a 1. For example, 0100 and 1011 are four-bit words.

(a) How many different four-bit code words are there?

(b) How many different six-bit code words, such as 001011, are there?

(c) If a vocabulary of 1000 code words is required, how long must the words be?

Using a Calculator

Scientific and graphing calculators have built-in functions to compute factorials $n!$, permutations $P(n, r)$, and combinations $C(n, r)$. For example, there may be a key labeled ! or $n!$ on a scientific calculator. A graphing calculator will have functions !, nPr, and nCr under the Math and Probability menus. For example, 5! ENTER will give 120, 5 nPr 3 ENTER will yield 60, and 5 nCr 3 ENTER will yield 10.

39. Use a calculator to evaluate the following expressions.

(a) 9! **(b)** 11! **(c)** $P(7, 5)$

(d) $P(8, 6)$ **(e)** $C(7, 4)$ **(f)** $C(9, 5)$

40. Use a calculator to evaluate the number of ways

(a) 10 people can get in a single line for a photograph.

(b) to place six books on a shelf, choosing from a set of 10 books.

(c) to choose a set of eight hearts from a deck of cards.

(d) to choose your six "lucky numbers" in a lottery, choosing from the numbers 1 through 44.

From State Student Assessments

41. (Texas, Grade 5)

Ricky is trying to decide which outfit to wear to a party. His choices are shown in the diagram.

Pants

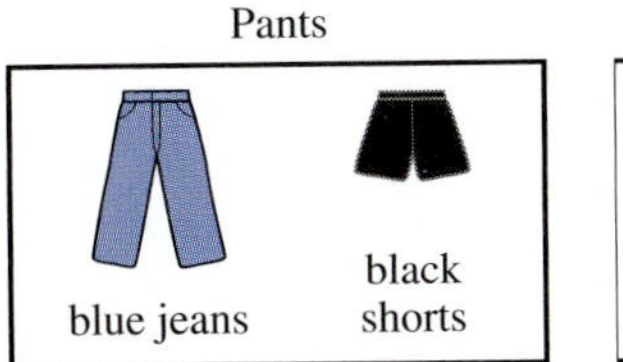

Shirts

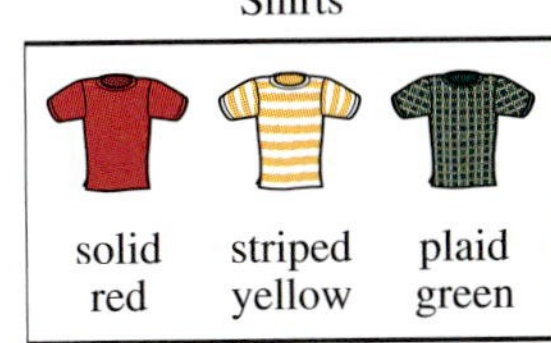

Shoes

How many possible different outfits can Ricky create if he chooses 1 pair of pants, 1 shirt, and 1 pair of shoes?

F. 7 G. 8 H. 10 I. 12

For Review

42. Toss two dice 20 times and determine the empirical probability of obtaining a score of 4.

43. Francine has lost 20 times in a row while playing roulette. She feels that her luck is bound to change soon and so begins to bet heavily. Briefly discuss Francine's reasoning.

44. Rashonda is 12 years old and gets a \$10 allowance each week. She typically spends \$5.50 for entertainment, \$3.50 for snacks, and \$1 for miscellaneous purchases. Draw a pie chart to show how Rashonda spends her allowance.

45. The final scores in Professor Kane's educational psychology class were 53, 77, 82, 82, 86, 67, 77, 64, 72, 68, 60, 74, 56, 57, 82, 81, 88, and 90.

(a) Display these data using a dot plot.

(b) Determine the mean, median, and mode for these data.

(c) Compute the 5-number summary for these data.

(d) Display the result of part (c) using a box plot.

10.3 THEORETICAL PROBABILITY

We now turn our attention to the study of **theoretical probability,** which we refer to simply as **probability.** The terminology of Section 10.1 for empirical probability—outcome, sample space, event, mutually exclusive events, dependent events and independent events—will continue to be important to describe the key concepts. In this section, we will make one important new assumption:

Each outcome of an experiment is as likely to occur as any other outcome.

When this can be assumed, the outcomes are said to be **equally likely.** For example, flipping a fair coin has two equally likely outcomes, a head or a tail. Drawing a card at random from a standard deck has 52 equally likely outcomes. As an example of outcomes that are not equally likely, consider flipping two fair coins. It is *not true* that two heads, two tails, and a head and a tail are the three equally likely outcomes. If the sides showing, say, on a penny and a quarter, are listed as HH, HT, TH, and TT, it is seen that there are four equally likely outcomes of the two-coin experiment.

When the outcomes in the sample space S of an experiment are equally likely, the probability of an event $E \subset S$ is given by the following definition.

DEFINITION *Theoretical Probability of an Event*
Let S, the **sample space,** denote the set of **equally likely outcomes** of an experiment, and let E, an **event,** denote a subset of outcomes of the experiment. Let $n(S)$ and $n(E)$ denote the number of outcomes in S and E, respectively. The **probability** of E, denoted by $P(E)$, is given by

$$P(E) = \frac{n(E)}{n(S)}.$$

Be aware that the phrase *equally likely* is of critical importance in the above definition. If the outcomes in the sample space are not equally likely, the definition of probability simply doesn't make sense. For example, suppose the experiment consists of rolling two dice. It would not do to think of the sample space as the 11 outcomes $\{2, 3, 4, \ldots, 12\}$ since these outcomes are not equally likely. The difficulty is that there is only one way to obtain a 2, two ways to obtain a 3, three ways to obtain a 4, . . . , six ways to obtain a 7, . . . , and only one way to obtain a 12. Thus, the event of obtaining a 3 is twice as likely as the event of obtaining a 2, and so on. However, if we mistakenly thought of the sample space as the above set with each of 2, 3, 4, . . . , 11, and 12 equally likely, we would obtain

$$P(2) = P(3) = \cdots = P(12) = \frac{1}{11},$$

a manifest absurdity. In fact, as we have seen before, there are 36 equally likely ways two dice can come up, so that

$$P(2) = \frac{1}{36}, P(3) = \frac{2}{36}, \cdots, P(12) = \frac{1}{36}.$$

Moreover, these probabilities are closely approximated by the corresponding empirical probabilities.

The definition of probability assumes that we can count outcomes. This implies that the sample space is finite. Thus, our discussion of theoretical probability will be largely restricted to the finite case. It is possible to treat the infinite case as well, as with geometric probability, but the general theory is beyond the scope of this text. The needed techniques of counting are perhaps best understood by considering several examples.

EXAMPLE 10.21 DETERMINING THE PROBABILITY OF ROLLING A SCORE OF 11 WITH TWO DICE

Compute the probability of obtaining a score of 11 on a single roll of two dice.

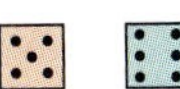

Solution Here the sample space S is the set of all 36 equally likely outcomes illustrated in Example 10.11. Let E denote the event of rolling a score of 11. Since only two of the outcomes in the sample space result in 11,

$$P(11) = \frac{n(E)}{n(S)} = \frac{2}{36} = \frac{1}{18}.$$

EXAMPLE 10.22 COMPUTING THE PROBABILITY OF ROLLING A SUM OF 5 OR 8 WITH TWO DICE

Determine the probability of rolling a sum of a 5 *or* an 8 on a single roll of two dice.

Solution 1 As we have just seen, the sample space S is the set of all 36 ways two dice can come up. The favorable outcomes are as shown here, where F is the event of rolling a 5 and E is the event of rolling an 8.

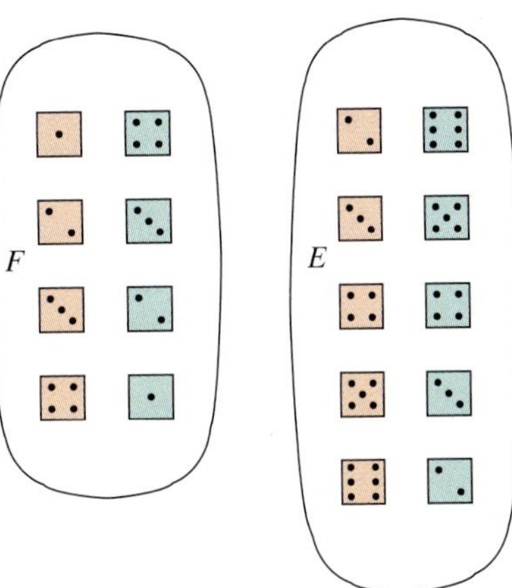

It follows that

$$P(5 \text{ or } 8) = \frac{4 + 5}{36} = \frac{1}{4}.$$

Solution 2 Note that if 5 is rolled, then 8 is not, and vice versa. Thus, rolling 5 and rolling 8 are mutually exclusive events, and the *or* in the statement of the problem reminds us that we can use the addition principle of counting mutually exclusive events. Thus,

$$P(5 \text{ or } 8) = \frac{n(5 \text{ or } 8)}{n(S)} = \frac{n(5) + n(8)}{n(S)} = \frac{4 + 5}{36} = \frac{1}{4}$$

as before. Notice that

$$P(5 \text{ or } 8) = \frac{4}{36} + \frac{5}{36} = P(5) + P(8).$$

EXAMPLE 10.23 COMPUTING THE PROBABILITY OF OBTAINING A FACE CARD OR A DIAMOND

Determine the probability of obtaining a face card *or* a diamond if a card is drawn at random from an ordinary deck of playing cards.

Solution 1 The sample space S is the set of all 52 cards in the deck. Let D denote the event of selecting a diamond, and let F denote the event of selecting a face card. Since there are 13 diamonds (including face cards) and 9 nondiamond face cards,

$$P(D \text{ or } F) = \frac{13 + 9}{52} = \frac{22}{52} = \frac{11}{26}.$$

Solution 2 Using the addition principle of counting, we know that

$$P(D \text{ or } F) = \frac{n(D \text{ or } F)}{n(S)} = \frac{n(D) + n(F) - n(D \cap F)}{n(S)} = \frac{13 + 12 - 3}{52} = \frac{22}{52} = \frac{11}{26}$$

Three face cards are diamonds.

as before.

EXAMPLE 10.24 DETERMINING PROBABILITIES WITH RESTRICTIVE CONDITIONS

All 24 students in Mr. Henry's preschool are either three or four years old, as shown in the table at the top of the next page. A student is selected at random.

	Age Three	Age Four
Boys	8	3
Girls	6	7

(a) What is the probability that the student is three years old?
(b) What is the probability that the student is three years old, given that a boy was selected?

Solution Let B denote the set of boys, T the set of three-year-olds, and S the set of all students in the class.

(a) $P(T) = \dfrac{n(T)}{n(S)} = \dfrac{14}{24} = \dfrac{7}{12}.$

(b) If we know that a boy was selected, then the sample space is not the set of all students in the class, but *all boys in the class.* Similarly, the set of favorable outcomes is the set of all boys in the class who are three years old. Thus, the desired probability is

$$P(T|B) = \frac{8}{11}.$$

This is an example of so-called **conditional probability,** and the standard notation $P(T|B)$ is read as "the probability of T given that B has occurred." In this case, $P(T|B)$ is the probability of selecting a three-year-old given that a boy has been selected.

EXAMPLE 10.25 USING THE MULTIPLICATION PRINCIPLE OF COUNTING FOR INDEPENDENT EVENTS

A red die and a green die are rolled. What is the probability of obtaining an even number on the red die *and* a multiple of 3 on the green die?

Solution The sets of favorable outcomes on the red and green dice, respectively, are $R = \{2, 4, 6\}$ and $G = \{3, 6\}$, and the sample space is the set of all 36 ways the two dice can come up. The word "and" is a broad hint to use one of the multiplication principles of counting. Since the dice are independent, the number of favorable outcomes is

$$n(R \text{ and } G) = n(R) \cdot n(G) = 3 \cdot 2 = 6.$$

R and G are clearly independent.

Hence, the desired probability is $P = 6/36 = 1/6$. Notice that

$$P(R \text{ and } G) = \frac{6}{36} = \frac{3}{6} \cdot \frac{2}{6} = P(R) \cdot P(G).$$

EXAMPLE 10.26 SELECTING BALLS WITHOUT REPLACEMENT

An urn contains three identical red and two identical white balls. Two balls are drawn one after the other without replacement.

(a) What is the probability that the first ball is red?
(b) What is the probability that the second ball is red given that the first ball is red?
(c) What is the probability that both balls are red?

Solution 1

(a) Since there are initially five balls and three are red, $P(\text{1st ball is red}) = 3/5$.
(b) Since a red ball has already been selected on the first draw, the count changes for the selection of the second ball. For the second selection, there remain four balls of which two are red. Thus,

$$P(\text{2nd ball is red} \mid \text{1st ball is red}) = \frac{2}{4} = \frac{1}{2}.$$

read as "given that"

(c) In order to select two red balls, it is necessary to select a red ball on the first draw *and* a red ball on the second draw. By the multiplication principle of counting for dependent events, the number of possible selections of two balls is $5 \cdot 4 = 20$, and the number of possible selections of two red balls is $3 \cdot 2 = 6$. Hence,

$$\begin{aligned} P(\text{2 red balls}) &= \frac{6}{20} \\ &= \frac{3 \cdot 2}{5 \cdot 4} \\ &= \frac{3}{5} \cdot \frac{2}{4} \\ &= P(\text{1st ball is red}) \cdot P(\text{2nd ball is red} \mid \text{1st ball is red}). \end{aligned}$$

Solution 2

A second way to solve this problem is to label the balls r_1, r_2, r_3, w_1, and w_2 so that we can identify the different red and white balls. With the balls identified in this way, we make an orderly list of all possible selections of two balls. In this listing, for example, the entry (r_1, r_2) indicates that red ball number 1 was chosen first and red ball number 2 was chosen second.

(r_1, r_2)	(r_1, r_3)	(r_1, w_1)	(r_1, w_2)	← all selections with r_1 first
(r_2, r_1)	(r_2, r_3)	(r_2, w_1)	(r_2, w_2)	← all selections with r_2 first
(r_3, r_1)	(r_3, r_2)	(r_3, w_1)	(r_3, w_2)	← and so on
(w_1, r_1)	(w_1, r_2)	(w_1, r_3)	(w_1, w_2)	
(w_2, r_1)	(w_2, r_2)	(w_2, r_3)	(w_2, w_1)	

Counting possibilities, we have

(a) $P(\text{1st ball is red}) = \dfrac{12}{20} = \dfrac{3}{5}$,

HIGHLIGHT from HISTORY

Srinivasa Ramanujan (1887–1920)

Perhaps the most exotic and mysterious of all mathematicians was Srinivasa Ramanujan, born to a high-caste family of modest means in Kumbakonam in southern India in 1887. Largely self-taught, Ramanujan wrote a letter in 1913 to the eminent English mathematician G. H. Hardy at Cambridge University in which he included a list of formulas he had discovered. Of the formulas, Hardy wrote, "(such formulas) defeated me completely . . . a single look at them is enough to show that they were written by a mathematician of the highest class. They must be true because, if they were not true, no one would have had the imagination to invent them. Finally, (you must remember that I knew nothing about Ramanujan, and had to think of every possibility), the writer must be completely honest, because great mathematicians are commoner than thieves and humbugs of such incredible skill."

In any event, Hardy arranged for Ramanujan to come to England in 1914, and the two collaborated intensively for the next three years, with Ramanujan using his unorthodox methods and fantastic intuition to come up with deep and totally unexpected results that Hardy, with his considerable intellectual power and formal training, then proved. Ramanujan fell ill in 1917 and returned to India, where he died in 1920.

The English mathematician Littlewood once remarked that every positive integer was one of Ramanujan's friends. Once, during Ramanujan's illness, Hardy visited him in the hospital in Putney. Trying to find a way to begin the conversation, Hardy remarked that he had come to the hospital in cab number 1729 and that he could not imagine a more uninteresting number, to which Ramanujan replied, "No, it is a very interesting number; it is the smallest number that can be written as the sum of two cubes in two different ways!"*

*$1729 = 9^3 + 10^3 = 1^3 + 12^3$

(b) $P(\text{2nd ball is red} \mid \text{1st ball is red}) = \frac{6}{12} = \frac{1}{2}$, and

(c) $P(\text{2 red balls}) = \frac{6}{20}$

$= P(\text{1st ball red}) \cdot P(\text{2nd ball red} \mid \text{1st ball red})$

as before.

EXAMPLE 10.27 **COMBINATIONS AND PROBABILITY**

There are 10 boys and 13 girls in Mr. Fleck's fourth-grade class and 12 boys and 11 girls in Mrs. Patero's fourth-grade class. A picnic committee of six people is selected at random from the total group of students in both classes.

(a) What is the probability that all the committee members are girls?
(b) What is the probability that all the committee members are girls given that all come from Mr. Fleck's class?
(c) What is the probability that the committee has three girls and three boys?
(d) What is the probability that the committee has three girls and three boys given that Mary Akers and Ann-Marie Harborth are on the committee?

Solution

(a) Since there are six committee members and 46 students in all, the total number of possible committees is $C(46, 6) = 46 \cdot 45 \cdot 44 \cdot 43 \cdot 42 \cdot 41/6!$. Since

Just for Fun

A Probability Paradox

Consider the hats containing colored balls on the tables shown. The balls in the hats on tables *A* and *B* are combined and placed in the hats on table *C*. Let $P(R|G)$ denote the probability of randomly drawing a red ball from a gray hat and let $P(R|B)$ denote the probability of randomly drawing a red ball from a brown hat. Compute and compare $P(R|G)$ and $P(R|B)$ for each of tables *A*, *B*, and *C*. Is the result surprising?

Table *A*

Table *B*

continued

there are 24 girls, the number of committees with six girls is $C(24, 6) = 24 \cdot 23 \cdot 22 \cdot 21 \cdot 20 \cdot 19/6!$. Therefore,

$$P(\text{committee has six girls}) = \frac{24 \cdot 23 \cdot 22 \cdot 21 \cdot 20 \cdot 19/6!}{46 \cdot 45 \cdot 44 \cdot 43 \cdot 42 \cdot 41/6!}$$

$$= \frac{24 \cdot 23 \cdot 22 \cdot 21 \cdot 20 \cdot 19}{46 \cdot 45 \cdot 44 \cdot 43 \cdot 42 \cdot 41}$$

$$= \frac{76}{5289}$$

$$\doteq 0.014.$$

Look for common factors in the numerator and denominator to simplify the fraction.

(b) Here, the population is the set of students in Mr. Fleck's class. Thus, P(six girls on committee | all committee members are chosen from Mr. Fleck's class)

$$= \frac{C(13, 6)}{C(23, 6)}$$

$$= \frac{13 \cdot 12 \cdot 11 \cdot 10 \cdot 9 \cdot 8/6!}{23 \cdot 22 \cdot 21 \cdot 20 \cdot 19 \cdot 18/6!}$$

$$= \frac{13 \cdot 12 \cdot 11 \cdot 10 \cdot 9 \cdot 8}{23 \cdot 22 \cdot 21 \cdot 20 \cdot 19 \cdot 18}$$

$$= \frac{52}{3059}$$

$$\doteq 0.017.$$

(c) The committee is chosen from all the students, so the sample space contains $C(46, 6)$ possible committees. Since the three boys must be chosen from among the 22 boys in both classes *and* the three girls must be chosen from among the 24 girls in both classes, we have, by the multiplication principle of counting for independent events, that the number of committees with three girls *and* three boys is $C(22, 3) \cdot C(24, 3)$.

Thus,

P(three girls and three boys on the committee)

$$= \frac{C(22, 3)C(24, 3)}{C(46, 6)}$$

$$= \frac{22 \cdot 21 \cdot 20}{3 \cdot 2 \cdot 1} \cdot \frac{24 \cdot 23 \cdot 22}{3 \cdot 2 \cdot 1} \cdot \frac{6 \cdot 5 \cdot 4 \cdot 3 \cdot 2 \cdot 1}{46 \cdot 45 \cdot 44 \cdot 43 \cdot 42 \cdot 41}$$

$$= \frac{1760}{5289} \doteq 0.333.$$

$$\frac{1}{C(46, 6)} = \frac{1 \cdot 2 \cdot 3 \cdot 4 \cdot 5 \cdot 6}{46 \cdot 45 \cdot 44 \cdot 43 \cdot 42 \cdot 41}$$

(d) Since Ann-Marie and Mary are on the committee, choosing the committee requires choosing only four more students. Thus, the sample space consists of the $C(44, 4)$ students chosen from among the 44 children other than Ann-Marie and Mary. Also, the number of favorable cases is found by selecting one more girl from among the 22 other girls in $C(22, 1)$ ways *and* selecting the three boys in $C(22, 3)$ ways. Thus,

Just for Fun

Continued

Table C

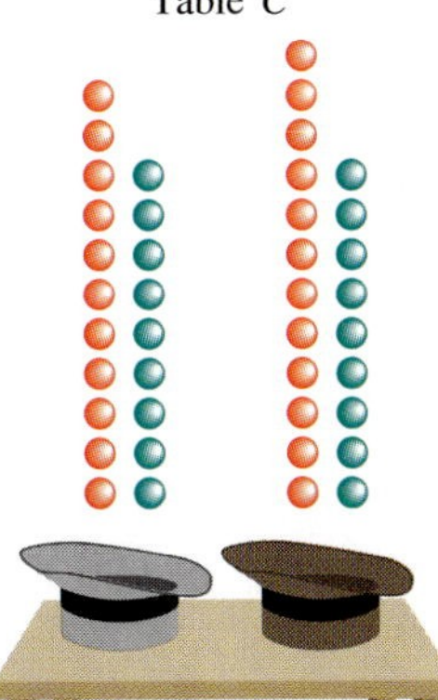

SOURCE: Adapted illustration by Ilil Arbel from "Mathematical Games" by Martin Gardner from *Scientific American*, March 1976, page 120.

P(committee has three girls and three boys | Mary and Ann-Marie are on the committee)

$$= \frac{C(22,1)C(22,3)}{C(44,4)}$$

$$= \frac{22}{1} \cdot \frac{22 \cdot 21 \cdot 20}{3 \cdot 2 \cdot 1} \cdot \frac{4 \cdot 3 \cdot 2 \cdot 1}{44 \cdot 43 \cdot 42 \cdot 41}$$

$$= \frac{440}{1763} \doteq 0.250.$$

$$\frac{1}{C(44,4)} = \frac{4 \cdot 3 \cdot 2 \cdot 1}{44 \cdot 43 \cdot 42 \cdot 41}$$

Complementary Events

Suppose a card is drawn from a deck, and A is the event "draw a face card." The set of outcomes *not* in A is called the complementary event, denoted by $\overline{A}$. Since there are 12 face cards in the 52-card deck, we see that

$$n(A) = 12, \qquad n(\overline{A}) = 40, \qquad P(A) = \frac{12}{52}, \quad \text{and} \quad P(\overline{A}) = \frac{40}{52}.$$

More generally, if A and $\overline{A}$ are events such that $A \cup \overline{A} = S$ and $A \cap \overline{A} = \varnothing$, then A and $\overline{A}$ are called **complementary events.** Moreover, $n(A) + n(\overline{A}) = n(S)$, and this implies that

$$P(A) + P(\overline{A}) = 1.$$

$$\frac{n(A)}{n(S)} + \frac{n(\overline{A})}{n(S)} = \frac{n(S)}{n(S)}$$

$$P(A) + P(\overline{A}) = 1$$

Alternatively, this implies that

$$P(A) = 1 - P(\overline{A}).$$

In ordinary language, suppose obtaining A is considered success. Then obtaining $\overline{A}$ is failure, and we have that

$$P(\text{success}) = 1 - P(\text{failure}).$$

THEOREM *Probability of Complementary Events*

Let A and $\overline{A}$ be complementary events; that is, $A \cup \overline{A} = S$ and $A \cap \overline{A} = \varnothing$. Then

$$P(A) = 1 - P(\overline{A}).$$

Equivalently, the probability of success in an experiment is 1 minus the probability of failure.

The preceding theorem is often useful as this example shows.

EXAMPLE 10.28 USING COMPLEMENTARY PROBABILITY

Compute the probability of obtaining a score of at least 4 on a single roll of 2 dice.

Solution

Here, success is obtaining a 4 or a 5 or a 6 or . . . or a 12. The probability of doing this is the sum of all the individual probabilities, and determining these requires considerable computation. However, failure occurs if we obtain either a sum of 2 or 3, and the probability of doing this is much easier to compute. Since S contains 36 equally likely outcomes and we fail by rolling two 1s, a 1 and a 2, or a 2 and a 1 in order, the probability of failure is $3/36 = 1/12$. Thus, the desired probability is $1 - 1/12 = 11/12$.

Properties of Probability

The preceding examples revealed a number of important properties of probability, which can be collected in the following theorem.

THEOREM *Properties of Probability**

1. $P(A) = 0$ if, and only if, A cannot occur.
2. $P(A) = 1$ if, and only if, A always occurs.
3. For any event A, $0 \leq P(A) \leq 1$.
4. For any events A and B,
$$P(A \text{ or } B) = P(A) + P(B) - P(A \text{ and } B).$$
5. If A and B are mutually exclusive events, then
$$P(A \text{ or } B) = P(A) + P(B).$$
6. For any events A and B,
$$P(A \text{ and } B) = P(A)P(B|A).$$
7. If A and B are independent events, then
$$P(A \text{ and } B) = P(A)P(B).$$
8. If E and $\overline{E}$ are complementary events, then $P(E) + P(\overline{E}) = 1$.

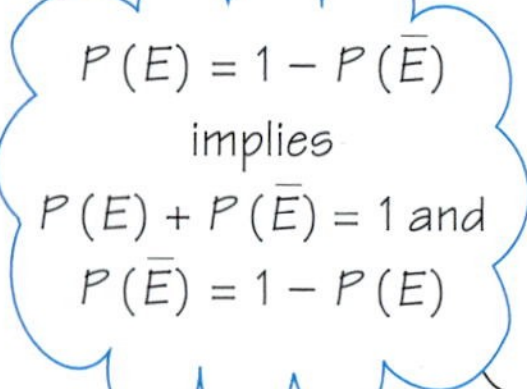
$P(E) = 1 - P(\overline{E})$
implies
$P(E) + P(\overline{E}) = 1$ and
$P(\overline{E}) = 1 - P(E)$

Proof

Let S be the sample space.

Divide through by $n(S)$.

1. Event A cannot occur if, and only if, $n(A) = 0$. Therefore, $P(A) = n(A)/n(S) = 0$ if, and only if, A cannot occur.
2. A always occurs if, and only if, $n(A) = n(S)$. Therefore, $P(A) = n(A)/n(S) = n(S)/n(S) = 1$ if, and only if, A always occurs.
3. An event A never occurs, sometimes occurs, or always occurs. Therefore, $0 \leq n(A) \leq n(S)$, and, dividing by $n(S)$, we obtain
$$0 \leq P(A) \leq 1.$$

4–7. These are general results from earlier representative examples.

8. Since $E \cup \overline{E} = S$ and $E \cap \overline{E} = \varnothing$, it follows that $n(S) = n(E) + n(\overline{E})$ and hence that $1 + P(E) + P(\overline{E})$.

*It is worth remembering that, in discussing probability in this chapter, we are limiting our discussions to situations where the sample space is finite. The infinite case can be successfully treated, and all the properties remain true except for properties 1 and 2. In this infinite case, it is possible that $P(A) = 0$ and yet event A can still occur. Similarly, it is possible that $P(A) = 1$ and yet event A never occurs. These anomalies, however, need not concern us here.

School Book Page

A Probability Investigation in Grade Four

Chapter 12
Lesson
7

Exploring Fairness

Problem Solving Connection
- Use Objects/ Act It Out
- Use Logical Reasoning

Materials
number cubes, labeled 1–6

Vocabulary
outcome
the result of an action or event
fair
a game is fair if each player has an equal chance of winning
equally likely
just as likely to happen as not

Explore

"What do I like to do for entertainment? Play video games!" answers Reginald. Sometimes Reginald plays video games with a friend.

In a fair game, players start out with the same chance of winning.

Reginald lives in Little Rock, Arkansas.

Work Together

1. Play a game of Match Me! with a partner. One person tosses a number cube. The partner then tosses another number cube, trying to match the number from the first toss. Repeat 10 times. Tally your scores. Change roles and play again.

Scoring
The first person gets 1 point when the cubes do not match. The partner gets 1 point when the cubes do match.

2. Play a game of Odds and Evens with a partner. Decide who will be "odds" and who will be "evens." Take turns tossing a number cube 10 times. Record each toss.

Scoring
"Odds" gets 1 point for each outcome of 1, 3, or 5. "Evens" gets 1 point for each outcome of 2, 4, or 6.

3. Which of these two games is fair to both partners? Which game is not fair? Explain your answer.

Talk About It

4. What is a "fair" game? Try to use the words *likely* or *unlikely* in your answer.
5. How could you make the unfair game more fair?

542 Chapter 12 • Dividing by 2-Digit Divisors and Probability

SOURCE: From Scott Foresman–Addison Wesley Math, *Grade 4, p. 542, by* Randall I. Charles, et al. *Copyright © 2002 Pearson Education, Inc. Reprinted with permission.*

Questions for the Teacher

1. Kendra observes that, in the first activity, just one number makes a match and five numbers make a mismatch. Therefore, reasons Kendra, the Match Me! game can be made fair by giving the second person five attempts to roll a matching number. How would you suggest Kendra check the fairness of this version of Match Me!?
2. Jolie and Kent have invented a new game—Checker Match Me! Six red and six black checkers are put in a bag. Each player in turn draws a checker from the bag (without replacement). The first player gets a point if the checkers have different colors. The second player gets a point if the checkers have the same color. Jolie and Kent think the game is fair, since there are equally many checkers of each color. How would you suggest that Jolie and Kent investigate the fairness of their game?

The following example uses several of the properties of probability to reexamine the Strings and Loops Cooperative Investigation found in Section 10.1. This time, our point of view is theoretical, rather than empirical, probability.

EXAMPLE 10.29 DETERMINING THE THEORETICAL PROBABILITIES OF THE STRINGS AND LOOPS ACTIVITY

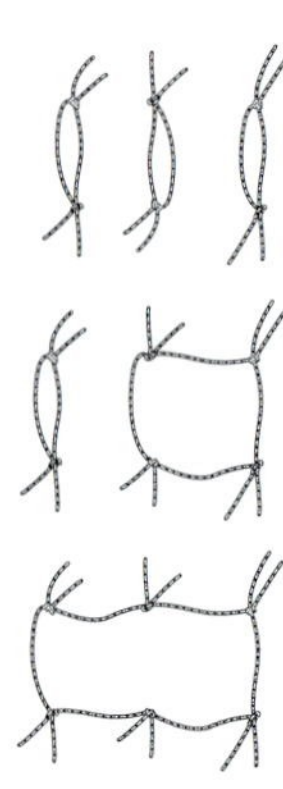

Six pieces of string of equal length are held in a bundle in one person's hand. A second person ties three knots at each end of the bundle of strings, where each knot joins a randomly selected pair of strings. After the six knots are tied, the bundle is examined to see what pattern of loops has been formed. There are three possibilities:

T: three small two-string loops
M: one medium four-string loop and one small two-string loop
L: one large six-string loop

What are the probabilities **(a)** $P(T)$, **(b)** $P(L)$, and **(c)** $P(M)$ of these events?

Solution

(a) It doesn't matter how the knots at the top of the bundle are tied, since the knots always yield the same pattern shown at the left in the figure below. Event T can be viewed as a three-stage experiment: $T = A$ and B and C, where A, B, and C are the successive events that a loop is tied by the first, second, and third knot. There are six strings to choose from when tying the first knot on the bottom, so the pair to be tied can be chosen in $C(6, 2) = \frac{6 \cdot 5}{2} = 15$ ways. Clearly, three pairs result in forming a loop, so $P(A) = \frac{3}{15} = \frac{1}{5}$. After A has occurred, there are four strings left from which to choose a pair. Therefore, there are $C(4, 2) = \frac{4 \cdot 3}{2} = 6$ pairs that can be chosen to be tied, and two of these pairs from a loop. That is, $P(B \mid A) = \frac{2}{C(4, 2)} = \frac{2}{6} = \frac{1}{3}$. Since the last knot tied is certain to form a loop, we have $P(C \mid A \text{ and } B) = P(C) = 1$. It then follows from the formula on conditional probability that

$$P(T) = P(A \text{ and } B \text{ and } C) = P(A) \cdot P(B \mid A) \cdot P(C) = \frac{1}{5} \cdot \frac{1}{3} \cdot 1 = \frac{1}{15}.$$

(b) For event L to occur, only the last of the three knots tied on the bottom of the bundle can form a loop. That is, $P(L) = P(\overline{A} \text{ and } \overline{B} \text{ and } C)$. By the formula for complementary probabilities, $P(\overline{A}) = 1 - \frac{1}{5} = \frac{4}{5}$. When $\overline{A}$ has occurred, there are four strings remaining from which to choose the next pair to tie. Two of these pairs form a loop, so

$$P(\overline{B} \mid \overline{A}) = 1 - P(B \mid \overline{A}) = 1 - \frac{1}{3} = \frac{2}{3}.$$

As in part (a), C is independent of the outcomes of A and B, and $P(C) = 1$. Thus,

$$P(L) = P(\overline{A} \text{ and } \overline{B} \text{ and } C) = P(\overline{A}) \cdot P(\overline{B} \mid \overline{A}) \cdot P(C) = \frac{4}{5} \cdot \frac{2}{3} \cdot 1 = \frac{8}{15}.$$

(c) The mutually exclusive events T, M, and L include all possible outcomes of the string experiment. Thus, $S = T \cup M \cup L$, where S is the sample space. Using the results of parts (a) and (b), we obtain

$$1 = P(S) = P(T \text{ or } M \text{ or } L) = P(T) + P(M) + P(L) = \frac{1}{15} + P(M) + \frac{8}{15},$$

so

$$P(M) = 1 - \left(\frac{1}{15} + \frac{8}{15}\right) = \frac{6}{15} = \frac{2}{5}.$$

Most people are surprised to learn that obtaining three small loops is quite rare, and a bit more than half the time a single large loop is formed.

Odds

When someone speaks of the **odds** in favor of an event E, they are comparing the likelihood that the event will happen with the likelihood that it will not happen. Consider an urn containing four blue balls and one yellow ball. If a ball is chosen at random, what are the odds that the ball is blue? Since a blue ball is four times as likely to be selected as a yellow ball, it is typical to say that the odds are 4 to 1. The odds are actually the ratio $4/1$, but, when quoting odds, one usually writes $4:1$, which is read "four to one." This is the basis for the following definition.

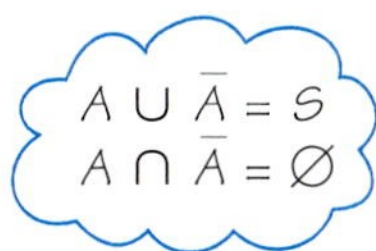

DEFINITION *Odds*

Let A be an event and let $\overline{A}$ be the complementary event. Then the **odds in favor** of A are $n(A)$ to $n(\overline{A})$, and the **odds against** A are $n(\overline{A})$ to $n(A)$.

EXAMPLE 10.30 DETERMINING THE ODDS IN FAVOR OF ROLLING A 7 OR 11

In the game of craps, one wins on the first roll of the pair of dice if a 7 or 11 is thrown. What are the odds of winning on the first roll?

Solution

Let W be the set of outcomes that result in 7 or 11. Since $W = \{(1, 6), (2, 5), (3, 4), (4, 3), (5, 2), (6, 1), (5, 6), (6, 5)\}$ and there are 36 ways two dice can come

$\frac{8}{28} = \frac{2}{7}$

up, $n(W) = 8$, and $n(\overline{W}) = 36 - 8 = 28$. Thus, the odds in favor of W are 8 to 28 or, more simply, 2 to 7.

EXAMPLE 10.31 DETERMINING PROBABILITIES FROM ODDS

If the odds in favor of event E are 5 to 4, compute $P(E)$ and $P(\overline{E})$.

Solution

$\frac{n(E)}{n(\overline{E})}$ in lowest terms is 5/4.

Since E and $\overline{E}$ are complementary, $n(S) = n(E) + n(\overline{E})$. Since the odds in favor of E are 5 to 4, $n(E) = 5k$ and $n(\overline{E}) = 4k$ for some integer k. Therefore,

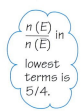

$$P(E) = \frac{n(E)}{n(S)} = \frac{n(E)}{n(E) + n(\overline{E})} = \frac{5k}{5k + 4k} = \frac{5}{9}$$

and

$$P(\overline{E}) = \frac{n(\overline{E})}{n(S)} = \frac{n(\overline{E})}{n(E) + n(\overline{E})} = \frac{4k}{5k + 4k} = \frac{4}{9}.$$

EXAMPLE 10.32 ODDS FROM PROBABILITIES

Given $P(E)$, determine the odds in favor of E and the odds against E.

Solution

The odds in favor of E are

$$\frac{n(E)}{n(\overline{E})} = \frac{n(E)/n(S)}{n(\overline{E})/n(S)}$$

$$= \frac{P(E)}{P(\overline{E})}$$

$P(\overline{E}) = 1 - P(E)$

$$= \frac{P(E)}{1 - P(E)}.$$

This last ratio would be expressed as a ratio of integers a/b in lowest terms, and the odds quoted would be stated as a to b.

The odds against E are given by the reciprocal of the ratio giving the odds in favor of E. Thus, the odds against E are

$$\frac{1 - P(E)}{P(E)} = \frac{b}{a}$$

$= \frac{P(\overline{E})}{P(E)}$

and are quoted as b to a.

Expected Value

At a carnival, you are offered the chance to play a game that consists of rolling a single die just once. If you play, you win the amount in dollars shown on the die. If you play the game several times, how much would you expect to win? Of course, you may be lucky and win \$6 on each of a series of rolls. However, because of the law of large numbers, you would *expect* to roll a 6 only about 1/6 of the time. Since this is true for each of the numbers on the die, you should expect to win on average approximately

$$\begin{aligned}\frac{1}{6}\cdot 1 + \frac{1}{6}\cdot 2 + \frac{1}{6}\cdot 3 + \frac{1}{6}\cdot 4 + \frac{1}{6}\cdot 5 + \frac{1}{6}\cdot 6 \\ = \frac{1}{6}\cdot(1+2+3+4+5+6) \\ = \frac{1}{6}\cdot 21 = \$3.50\end{aligned}$$

per roll. If it costs you \$4 to play the game, the carnival confidently expects players to *lose* 50¢ per game on the average. Thus, the carnival stands to make a handsome profit if a large number of patrons play the game each night.

The preceding discussion introduces the notion of **expected value.**

DEFINITION *Expected Value of an Experiment*
Let the outcomes of an experiment be a sequence of real numbers (values), $v_1, v_2, \ldots, v_n$, and suppose the outcomes occur with respective probabilities $p_1, p_2, \ldots, p_n$. Then the **expected value** of the experiment is

$$e = v_1p_1 + v_2p_2 + \cdots + v_np_n.$$

EXAMPLE 10.33 WINNING AT ROULETTE

An American roulette wheel has 38 compartments around its rim. Two of these are colored green and are numbered 0 and 00. The remaining compartments are numbered from 1 to 36 and are alternately colored black and red. When the wheel is spun in one direction, a small ivory ball is rolled in the opposite direction around the rim. When the wheel and the ball slow down, the ball eventually falls in any one of the compartments with equal likelihood if the wheel is fair. One way to play is to bet on whether the ball will fall in a red slot or a black slot. If you bet on red, for example, you win the amount of the bet if the ball lands in a red slot; otherwise, you lose. What is the expected win if you consistently bet \$5 on red?

Solution Since the probability of winning on any given try is 18/38 and the probability of losing is 20/38, your expected win is

$$\begin{aligned}\frac{18}{38}\cdot 5 + \frac{20}{38}\cdot(-5) &= \frac{90-100}{38} \\ &\doteq -0.26.\end{aligned}$$

On average, you should expect to lose 26¢ per play. Is it any wonder that casinos consistently make a handsome profit?

In the preceding example, it was pretty clear that you should expect to lose slightly more often than win. This next example is less clear.

EXAMPLE 10.34 DETERMINING THE EXPECTED VALUE OF AN UNUSUAL GAME

Suppose you are offered the opportunity to play a game that consists of a single toss of three coins. It costs you \$21 to play the game, and you win \$100 if you toss three heads, \$20 if you toss two heads and a tail, and nothing if you toss more than one tail. Would you play the game?

Solution

Many people would play the game hoping to "get lucky" and roll HHH frequently. But is this reasonable? What is your expected return? The expected value of the game is

$$\frac{1}{8}\cdot 100+\frac{3}{8}\cdot 20+\frac{3}{8}\cdot 0+\frac{1}{8}\cdot 0=\frac{160}{8}=\$20.$$

P (3 heads)
P (2 heads and 1 tail)
P (1 head and 2 tails)
P (3 tails)

Thus, on average you should expect to win \$1 less than it costs you to play the game each time. Unless the excitement is worth at least \$1, you should not play the game.

Geometric Probability

In Section 10.1, we considered the empirical probability of a spinner marked like that in Figure 10.6 stopping at any particular place. Here, $P_e(A)$, $P_e(B)$, and $P_e(C)$ all turn out to be approximately equal to $1/3 = 0.\overline{3}$. This is not surprising since the three arcs bordering regions A, B, and C are equally long. More generally, we would define the theoretical **geometric probability** of a region on the spinner to be the ratio of the region's corresponding arc length to the circumference of the circle. Equivalently, the probability of stopping a spinner on a sector is the ratio of the measure of the central angle of the sector to 360°.

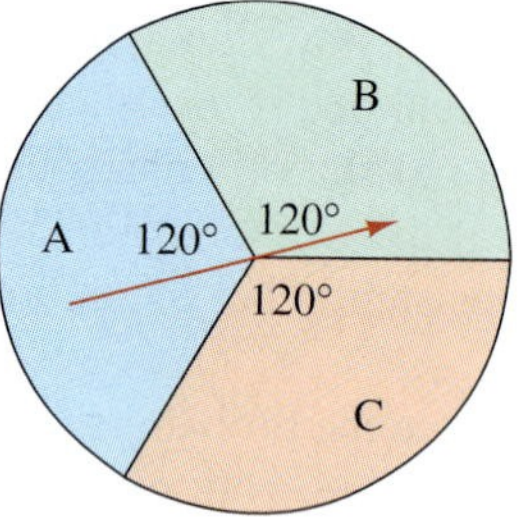

Figure 10.6
A spinner with equally likely outcomes

Similarly, consider the diagram of Figure 10.7. The probability that a point chosen at random in the square (say, by using a sequence of random numbers) will belong to region A should be $1/4$, since one-fourth of the *area* of the region lies in A.

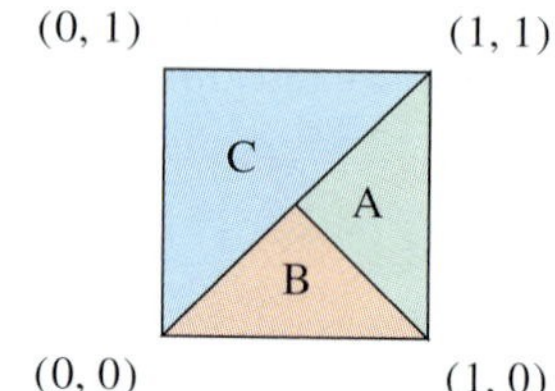

Figure 10.7
Geometric probability as a ratio of areas

Did You Know?

State Lotteries

B.C. BY JOHNNY HART

State lotteries are becoming increasingly numerous and popular. Over \$20 billion worth of tickets were sold in 1990, and the figure continues to grow. The big draw is the chance to become an instant multimillionaire, but the chance of doing so is extremely small. Your expected return for each one-dollar bet is only about 50 cents. In fact, your chances of winning are much better in Las Vegas or Atlantic City, where the expected return on a one-dollar bet is 85 to 95 cents. Indeed, the probabilities are constant in the big casinos, and knowledgeable gamblers rely on the law of large numbers to predict with considerable accuracy their average winnings over the long term. But the probabilities are constantly changing in a lottery so that this calculation is quite complex. Indeed, it can be shown that, as the lottery jackpot increases and thousands of people rush to buy tickets, the probability of any ticket winning drops so much that the expected value of a dollar bet actually falls below 50 cents. State lotteries constantly produce substantial revenues, and the only return for almost all players is the pleasure of imagining themselves becoming instantly wealthy.

EXAMPLE 10.35 DETERMINING THE GEOMETRIC PROBABILITY OF A CARNIVAL GAME

At the carnival, a "double your money" game is played by tossing a quarter onto a large table that has been ruled into a grid of squares of the same size. If your quarter lands entirely within any square, you win back two quarters, but if the coin touches a grid line, you lose the quarter. If a quarter is 2.5 centimeters in diameter and the squares have sides 6 centimeters long, should you play the game?

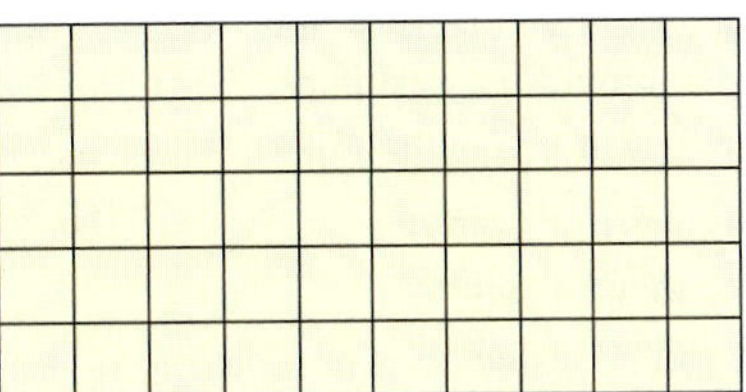

Solution The sample space S can be considered as the points in a 6-by-6-centimeter square. To win, the center of the quarter must land at least 1.25 centimeters from each side of the large square. That is, the winning region W is a 3.5-centimeter square centered in the larger square, as shown in the diagram. The probability of a win is

$$P(W) = \frac{\text{area of small square}}{\text{area of large square}} = \frac{3.5^2}{6^2} = \frac{12.25}{36} = \frac{49}{144} \doteq 0.34.$$

We can now compute the expected value of the game, remembering that a win gives us a net gain of \$0.25 (we must subtract the cost of playing from the \$0.50 won) and a loss is $-\$0.25$:

$$\text{Expected value} = \frac{49}{144} \cdot \$0.25 + \frac{95}{144} \cdot (-\$0.25) \doteq -\$0.08.$$

We expect to lose, and the operator to win, about 8¢ per play.

Problem Set 10.3

Understanding Concepts

1. (a) Explicitly list all the outcomes for the experiment of tossing a penny, a nickel, a dime, and a quarter.
   (b) Determine the probability $P(\text{HHTT})$ of obtaining a head on each of the penny and nickel and a tail on each of the dime and quarter in the experiment of part (a).
   (c) Determine the probability of obtaining two heads and two tails in the experiment of part (a).
   (d) Determine the probability of obtaining at least one head in the experiment of part (a). (*Hint:* Note that the complementary event consists of obtaining four tails.)
2. Determine the probability of obtaining a total score of 3 or 4 on a single throw of two dice.
3. A family has two children, including at least one boy. What is the probability that the other child is a girl? *Note:* The answer is *not* one-half.
4. Acme Auto Rental has three red Fords, four white Fords, and two black Fords. Acme also has six red Hondas, two white Hondas, and five black Hondas. If a car is selected at random for rental to a customer,
   (a) what is the probability that it is a white Ford?
   (b) what is the probability that it is a Ford?
   (c) what is the probability that it is white?
   (d) what is the probability that it is white given that the customer demands a Ford?
5. Five black balls numbered 1, 2, 3, 4, and 5 and seven white balls numbered 1, 2, 3, 4, 5, 6, and 7 are placed in an urn. If one is chosen at random,
   (a) what is the probability it is numbered 1 or 2?
   (b) what is the probability that it is numbered 5 or that it is white?
   (c) what is the probability that it is numbered 5 given that it is white?
6. Mrs. Ricco has seven brown-eyed and two blue-eyed brunettes in her fifth-grade class. She also has eight blue-eyed and three brown-eyed blondes. A child is selected at random.
   (a) What is the probability that the child is a brown-eyed brunette?
   (b) What is the probability that the child has brown eyes or is a brunette?
   (c) What is the probability that the child has brown eyes given that he or she is a brunette?
7. Suppose that you randomly select a two-digit number (that is, one of 00, 01, 02, . . . , 99) from a sequence of random numbers obtained by repeatedly spinning a spinner. What is the probability that the number selected
   (a) is greater than 80?
   (b) is less than 10?
   (c) is a multiple of 3?
   (d) is even or is less than 50?
   (e) is even and is less than 50?
   (f) is even given that it is less than 50?
8. In a certain card game, you are dealt two cards face up. You then bet on whether a third card dealt is between the other two cards. (For example, a 10 is between a 9 and a queen, and so on.) What is the probability of winning your bet if you are dealt
   (a) a 5 and a 7?
   (b) a jack and a queen?
   (c) a pair of 9s?
   (d) a 5 and a queen?
9. In playing draw poker, a flush is a hand with five cards all in one suit. You are dealt five cards and can throw away any of these and be dealt more cards to replace them. If you are dealt four hearts and a spade, what is the probability that you can discard the spade and be dealt a heart to fill out your flush?
10. You are dealt three cards at random from an ordinary deck of playing cards. What is the probability that all three are hearts? (*Hint:* How many ways can you select three cards at random from the deck? How many ways

can you select three cards at random from among the hearts?)

11. Consider a spinner made as in Figure 10.2, but marked and shaded as indicated here. The spinner is spun and grasped between your thumb and forefinger while it is still spinning rapidly.

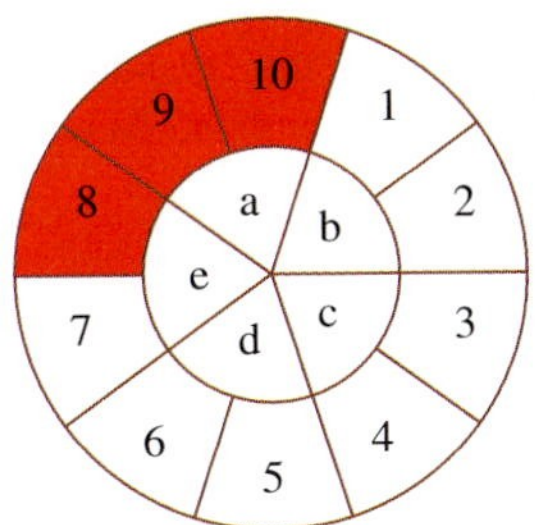

(a) What is the probability that the mark on your thumb is on the shaded area?

(b) What is the probability that your mark falls in regions 1 or 2?

(c) What is the probability that your mark falls in regions 10 or 6?

(d) What is the probability that your mark lands between the two radii that determine region e?

(e) What is the probability that the mark falls in region 8 given that it falls in the shaded area?

(f) What is the probability that it falls in a region marked by a vowel given that it falls in an odd-numbered region?

(g) What is the probability that it falls in a region marked by a vowel or an odd number?

12. A dartboard is marked as shown.

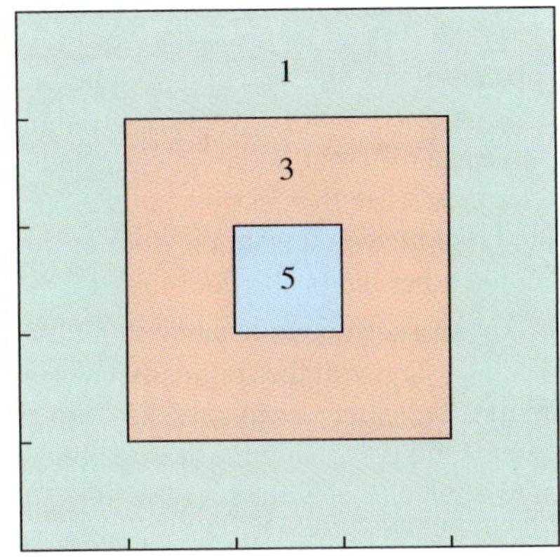

Josie is good enough that she always hits the dartboard with her darts, but, beyond that, the darts hit in random locations. If a single dart is thrown, compute these probabilities.

(a) $P(1)$

(b) $P(3)$

(c) $P(5)$

(d) If Josie wins the number of dollars indicated by the number of the region in which her dart falls, how much is her expected win (the expected value)?

(e) Suppose it costs Josie \$2 each time she throws a dart. Should she play darts as in part (d)? Explain.

13. A dartboard is marked like the spinner in problem 11. The radius of the inner circle is 1, and that of the outer circle is 2. What is the probability that a dart hitting the board at random

(a) hits in region b?

(b) misses region b, given that it hits the inner circle?

(c) hits in region b given that it hits in a, b, or c?

(d) hits in region 1 given that it hits in a, 1, or 2?

14. Two dice are thrown.

(a) What are the odds in favor of getting a score of 6?

(b) What are the odds against getting a 6?

(c) What is the probability of getting a 6?

15. If $P(A) = 2/5$, compute the odds in favor of A resulting from a single trial of an experiment.

16. If $P(A) = 1/2$, $P(B) = 1/3$, $P(C) = 1/6$, and A, B, and C are mutually exclusive, compute the odds in favor of A or C resulting from a single trial of an experiment.

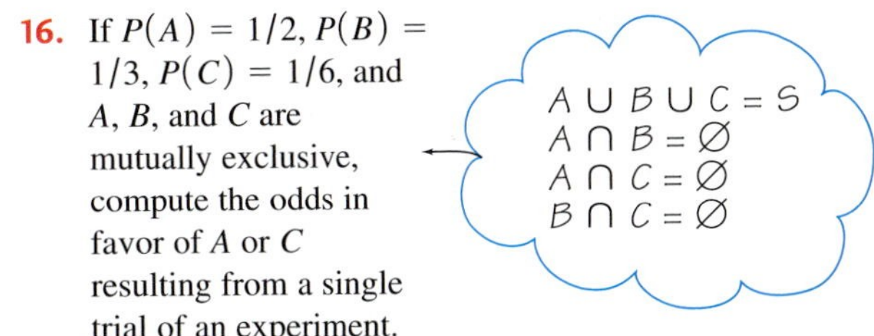

17. Compute the expected value of the score when rolling two dice.

18. A game consists of rolling a pair of dice. You win the amounts shown for rolling the score shown in the table below.
Compute the expected value of the game.

Roll	2	3	4	5	6	7	8	9	10	11	12
\$ Won	4	6	8	10	20	40	20	10	8	6	4

Table for Problem 18

Teaching Concepts

19. Burgerville has only vanilla, strawberry, and chocolate ice cream. Jeraldo claims that the probability is $1/3$ that a person buying an ice cream cone at Burgerville will buy a chocolate cone. How would you respond to Jeraldo, and what related points would you want to discuss with the entire class?

20. A fifth grader tells you she is certain it will rain sometime over the weekend. It seems she heard on the weather forecast that there is a 50% chance of rain both Saturday and Sunday, and, of course, $50\% + 50\% = 100\%$, a certainty. Write an imagined dialogue with the student to convince her that it may not rain after all.

21. The situation in problem 20 can be simulated by repeatedly tossing a coin, since the probability of obtaining a head on each toss is $\frac{1}{2} = 50\%$.

(a) Is it certain that a person will obtain at least one head on two tosses of a single coin?

(b) What is the probability of obtaining at least one head on three successive tosses of a coin?

(c) What is the probability of obtaining at least one head on ten successive tosses of a coin?

(d) What is the probability of *ultimately* obtaining at least one head?

Thinking Critically

22. Two balls are drawn at random from an urn containing six white and eight red balls.

(a) Compute the probability that both balls are white, using combinations. Recall that

$$C(n, r) = \frac{n(n-1)(n-2)\cdots(n-r+1)}{r!}.$$

(b) Compute the probability that both balls are red.

23. An urn contains eight red, five white, and six green balls. Four balls are drawn at random.

(a) Compute P (all four are red).

(b) Compute P (exactly two are red and exactly two are green).

(c) Compute P (exactly two are red or exactly two are green).

24. Consider the set of all five-letter code words without repetition of letters. Recall that

$$P(n, r) = n(n-1)(n-2)\cdots(n-r+1).$$

(a) What is the probability that a code word begins with the letter a?

(b) What is the probability that, in a code word, c is immediately followed by d?

(c) What is the probability that a code word starts with a vowel and ends with a consonant?

(d) In how many of the original set of five-letter code words are c and d adjacent?

25. Six dice are rolled. What is the probability that all six numbers—1, 2, 3, 4, 5, and 6—are obtained? (*Hint:* This can happen in more than one way. For example, 1, 2, 3, 4, 5, 6, and 6, 5, 1, 2, 3, 4 are just two of the possibilities.)

26. What is the probability that the six volumes of Churchill's *Second World War* appear in the correct order if they are randomly placed on a shelf?

27. What is the probability that a randomly dealt five-card hand from a deck of playing cards will contain

(a) exactly two aces?

(b) at least two aces?

28. If seven dice are tossed, what is the probability that every number will appear? (*Hint:* In how many ways can the number that appears twice be chosen?)

Thinking Cooperatively

29. Modify the Strings and Loops experiment described in Example 10.29 by using eight lengths of string instead of six.

(a) How many patterns of loops can be formed by tying eight knots, four at each end, of the eight-string bundle?

(b) Perform the experiment and combine the data from the class. Use the data to estimate the empirical probabilities that four small loops are formed and one large loop is formed.

(c) Determine the theoretical probability of obtaining four small loops.

(d) Determine the theoretical probability of obtaining one large loop.

30. On a popular TV game show, the contestant is asked to select one of three doors. Behind one door is a very valuable prize. Behind another door is a so-so, not very valuable prize. Behind the third door is a joke prize. After the contestant selects a door, but before the prize behind the selected door is revealed, the show's host always opens a different door, which reveals one of the two lesser prizes. He then asks the contestant if he or she would like to switch from the door originally chosen and pick the other unopened door instead.

(a) Poll the class to see how many think it is better to switch or stay with the originally chosen door. Now simulate the game, first using the "don't switch" strategy and then using the "switch doors" strategy. An easy way to play the game is to use three paper cups, respectively containing a quarter, a nickel, and a penny to represent the three types of prizes. Which strategy seems to work best?

(b) In small groups, discuss the mathematical probabilities of the two strategies. For example, what is the probability that a contestant following the "don't switch" strategy will get the most valuable prize?

What is the probability for a contestant following the "switch doors" strategy?

Using a Calculator

31. There are $7 \times 7 = 7^2$ ways in which two people's birthdays can fall on the days of the week. They will have birthdays on different days in $7 \cdot 6$ ways, so the probability that two people have birthdays on different days of the week is $P(W_2) = \frac{7 \cdot 6}{7 \cdot 7} = \frac{6}{7}$. By complementary probabilities, the probability that two people were born on the same day of the week is (not surprisingly) $P(\overline{W}_2) = 1 - P(W_2) = 1 - \frac{6}{7} = \frac{1}{7}$.

(a) Find the probability $P(W_3)$ that three people's birthdays are on different days. Then find the probability $P(\overline{W}_3)$ that, in a group of three people, at least two share a birthday on the same day of the week.

(b) Find the probabilities requested in part (a), but for four people.

32. Overlooking February 29, there are 365 days on which a person's birthday can occur. In a group of five people, their birthdays can occur in 365^5 ways, and their birthdays can be on all different dates of the year in $P(365, 5) = 365 \cdot 364 \cdot 363 \cdot 362 \cdot 361$ ways. Thus, the probability that the five birthdays are distinct is

$$P(Y_5) = \frac{365 \cdot 364 \cdot 363 \cdot 362 \cdot 361}{365 \cdot 365 \cdot 365 \cdot 365 \cdot 365} = \frac{P(365, 5)}{365^5} \doteq 0.97.$$

By complementary probabilities, the probability that at least two of the five have birthdays on the same day of the year is

$$P(\overline{Y}_5) = 1 - \frac{365 \cdot 364 \cdot 363 \cdot 362 \cdot 361}{365 \cdot 365 \cdot 365 \cdot 365 \cdot 365} = 1 - \frac{P(365, 5)}{365^5} \doteq 0.03.$$

(a) Calculate $P(Y_{23})$, the probability that there are no common birthdays in a group of 23 people. Then calculate $P(\overline{Y}_{23})$, the probability that at least two people in a group of 23 share a birthday.

(b) Make a table of values $P(\overline{Y}_n)$ for $n = 20, 21, \ldots, 30$ that gives the probabilities, in a group of n people, that at least two people have birthdays on the same day of the year. (*Suggestion:* Define Y1 = 1 − 365 nPr X/365^X. Under [2nd] [TBLSET] (Table Set), enter TblStart = 20 and ΔTbl = 1. View the table by pressing [2nd] [TABLE].)

Making Connections

33. In one unfortunate shipment, 10% of the portable tape players manufactured by Imperfect Electronics had defective switches, 5% had defective batteries, and 2% had both defects. If you purchased a tape player from this shipment, what is the probability that your player

(a) has a defective switch or a defective battery?

(b) has a good switch, but a defective battery?

(c) has both a good switch and a good battery?

34. California originally operated a 6/49 lottery, meaning that the grand prize went to a player (or players) who picked the same six numbers that were later drawn at random from the set of numbers 1 through 49. In 1990, California went to a 6/53 lottery, but later the state went to a 6/51 lottery.

(a) Find the probability of winning a 6/49 lottery.

(b) Find the probability of winning a 6/53 lottery.

(c) Find the probability of winning a 6/51 lottery.

(d) Why do you think California changed to a 6/53 lottery?

(e) Why do you think the state went to a 6/51 lottery?

35. In the casino game Keno, the player purchases a ticket and marks eight of the "spots" numbered 1 through 80. Every 20 minutes or so, the casino randomly draws 20 balls from a drum of 80 numbered balls. If sufficiently many of the player's spots are among the 20 numbers, the player wins. Usually the player must have five winning spots to receive a prize, and a larger prize is awarded if six, seven, or all eight winning spots are marked.

(a) What is the probability of marking exactly five winning spots?

(b) What is the probability of marking exactly six winning spots?

(c) What is the probability of marking exactly seven winning spots?

(d) What is the probability of marking exactly eight winning spots?

From State Student Assessments

36. (Washington State, Grade 4)
Special cakes are baked for May Day in France. A small toy is dropped into the batter for each cake before baking. Whoever gets the piece of cake with the toy in it is "king" or "queen" for the day.
Which cake below would give you the best chance of finding the toy in your piece?

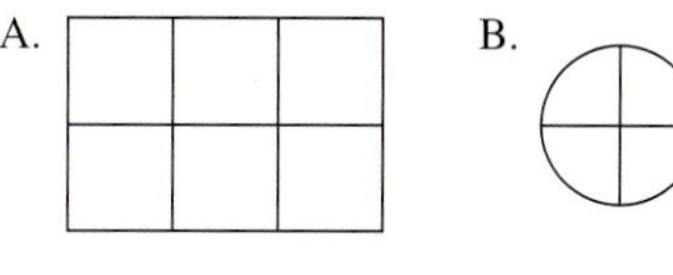

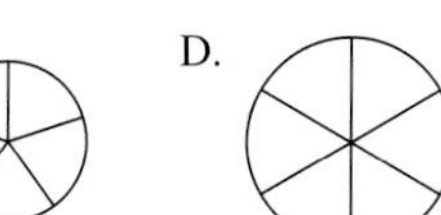

37. (Illinois, Grade 5)
A supply box has only three markers left: blue, purple, and green. If, without looking, two markers are pulled together, how likely is it that they will be blue and green?
○ 2/5 ○ 2/2 ○ 3/5 ○ 1/2 ○ 1/3

For Review

38. In how many ways can you choose four marbles of the same color from an urn containing seven yellow and eight blue marbles?

39. In how many ways can you choose two yellow and three blue marbles from the urn of problem 38?

40. How many five-letter code words can be made without repetition of letters if vowels and consonants must alternate?

41. How many different permutations are there of the letters a, a, a, a, b, b, c, c, c, c, and c?

Epilogue Two Views of Probability

A seventeenth-century Frenchman, the Chevalier de Méré, thought that the event of obtaining at least one 1 on four rolls of a die and the event of getting at least two 1s on 24 rolls of a pair of dice were equally likely. He reasoned as follows:

First event:

- On one roll of a die, there is a $1/6$ chance of getting a 1.
- Therefore, on four rolls of a die, the chance of getting at least one 1 is $4 \cdot (1/6) = 2/3$.

Second event:

- On one roll of a pair of dice, the chance of a 2 is $1/36$.
- Therefore, on 24 rolls of a pair of dice, the chance of at least one 2 is $24 \cdot (1/36) = 2/3$.

In each case, the chance was two-thirds. However, long experience with many trials showed that the first event was slightly more likely than the second, and this difficulty became known as the Paradox of the Chevalier de Méré. De Méré asked his friend, the mathematician and philosopher Blaise Pascal, about the problem, and Pascal in turn sought the help of the jurist and amateur mathematician Pierre de Fermat. Together, Pascal and Fermat were able to solve the problem, showing that the first event occurs with probability 0.518 and that the second occurs with probability 0.491.

In fact, the study of probability substantially began with the French and in direct response to questions about games of chance, even though it had been considered briefly in the sixteenth century by such mathematicians as the Italian Cardano. Today, probability and statistics play a critical role in society—in public-opinion polls, in evaluating experimental data of all sorts, in quality control in manufacturing, in studying the behavior of atoms and of subatomic particles, and in countless other ways.

In this chapter, we have considered the basic facts about probability. As revealed in the little vignette about the Chevalier de Méré, we have seen that there are two different, but related, notions of probability—empirical probability and theoretical probability. The first view of probability is that chance can be measured on the basis of long-run experience. It was the experience of the Chevalier that event number one was slightly more likely than event number two, even though his analysis suggested otherwise. The assertion based on experience was an expression of empirical probability—namely,

$$P_e(\text{event number one}) > P_e(\text{event number two}).$$

The paradox was cleared up by Pascal and Fermat using strict mathematical principles that formed the genesis of theoretical probability. Much of the thrust of this chapter has been to show that these two notions are closely intertwined. Recall that the law of large numbers asserts that the empirical probability of an event more and more closely approaches the theoretical probability of the event as the number of trials is made larger and larger.

We have given attention to developing the ideas of probability in both senses and have seen in Chapter 9 how the interpretation of data, based on probability principles, can lead to surprisingly accurate conclusions. Without this kind of careful analysis, however, action based on collected data can lead to disastrous results.

Chapter Summary

Key Concepts

Section 10.1 Empirical Probability

- Empirical probability. The fraction of the number of times an event occurs over a large number of trials.
- The law of large numbers. The empirical probability of an event more and more closely approximates a fixed number as the number of trials increases.
- Outcome. The result of a single trial of an experiment.
- Sample space. The set of all possible outcomes of an experiment.
- Event. A subset of the sample space.
- Compound event. An event consisting of a combination of two or more events.
- Mutually exclusive events. Events A and B such that $A \cap B = \varnothing$.
- Empirical probability of compound events. For any events A and B, $P_e(A \text{ or } B) = P_e(A) + P_e(B) - P_e(A \text{ and } B)$. If $A \cap B = \varnothing$, then $P_e(A \text{ or } B) = P_e(A) + P_e(B)$.
- Independent events. Events A and B such that the occurrence or nonoccurrence of A does not affect that of B, and vice versa.
- Empirical probability of independent events. If A and B are independent, then $P_e(A \text{ and } B) = P(A) \cdot P_e(B)$.
- Simulation. A method of determining answers to real problems by conducting experiments with outcomes analogous to those of the real problems.

Section 10.2 Principles of Counting

- Addition principle of counting. For any events A and B, $n(A \text{ or } B) = n(A) + n(B) - n(A \text{ and } B)$. If $A \cap B = \varnothing$, then $n(A \text{ or } B) = n(A) + n(B)$.
- Multiplication principle of counting. For any events A and B, $n(A \text{ and } B) = n(A) \cdot n(B \mid A)$. For independent events A and B, $n(A \text{ and } B) = n(A) \cdot n(B)$.
- Combination. A subset of r things taken from a set of n things.
- Permutation. An ordered subset of r things taken from a set of n things.
- n factorial. The product $n! = n(n-1)(n-2)\cdots 1$; $0! = 1$.

Section 10.3 Theoretical Probability

- Equally likely events. Each outcome of an experiment is as likely to occur as any other.
- Theoretical probability. The number of ways an event can occur divided by the number of possible outcomes of an experiment.
- Complementary events. Events A and $\overline{A}$ such that $A \cap \overline{A} = \varnothing$ and $A \cup \overline{A} = S$.
- Event of probability 0. $P(A) = 0$ if, and only if, A cannot occur.
- Event of probability 1. $P(A) = 1$ if, and only if, A always occurs.

- Range of probabilities. For any A, $0 \le P(A) \le 1$.
- Probability of A or B. For any events A and B, $P(A \text{ or } B) = P(A) + P(B) - P(A \text{ and } B)$. If A and B are mutually exclusive, $P(A \text{ or } B) = P(A) + P(B)$.
- Probability of A and B. For any events A and B, $P(A \text{ and } B) = P(A) \cdot P(B|A)$. For independent events, $P(A \text{ and } B) = P(A) \cdot P(B)$.
- Probability of complementary events. For complementary events A and $\overline{A}$, $P(A) + P(\overline{A}) = 1$.
- Odds. If A and $\overline{A}$ are complementary, the odds in favor of A are $n(A)$ to $n(\overline{A})$.
- Expected value. If the numerical outcomes of an experiment are $u_1, u_2, \cdots, u_s$ and they occur with respective probabilities $p_1, p_2, \cdots, p_s$, then the expected value of the experiment is $e = p_1u_1 + p_2u_2 + \cdots + p_su_s$.
- Geometric probability. Probability depending on the geometry of an experiment.

Vocabulary and Notation

Section 10.1

Outcome
Sample space
Event
Compound event
Independent event
$P_e(A)$
Simulation

Section 10.2

"Or"
"And"
Possibility tree
Dependent events
$P(B|A)$
Independent events
Combination
Permutation
$n! = n(n-1)\cdots 1; 0! = 1$
$P(n, r) = n(n-1)\cdots(n-r+1)$
$C(n, r) = n(n-1)\cdots(n-r+1)/r!$

Section 10.3

Sample space, S
Equally likely events
Complementary events
Odds
Expected value

Chapter Review Exercises

Section 10.1

1. Toss four coins 20 times and determine the empirical probability of obtaining three heads and one tail.

2. (a) Roll three dice 20 times and determine the empirical probability of obtaining a total score of 3 or 4.

   (b) From the data in part (a), determine the empirical probability of obtaining a score of at least 5.

3. From the data of problem 2, determine

$$P_e(5 \text{ or } 6 \text{ or } 7 \mid 5 \text{ or } 6 \text{ or } 7 \text{ or } 8 \text{ or } 9).$$

4. Conduct a survey of 20 randomly chosen college students at your college or university and determine the empirical probability that chocolate is the favorite flavor of ice cream.

5. (a) Drop five thumbtacks on a tabletop 20 times and determine the empirical probability that precisely three of the tacks land point up.

   (b) From the data of part (a), determine the empirical probability that two or three of the five tacks in part (a) land point up.

6. (a) A die is rolled repeatedly until a 5 or a 6 appears. Perform this experiment 10 times and estimate the number of rolls required.

   (b) From the data in part (a), compute the empirical probability that it takes precisely five rolls to obtain a 5 or 6 for the first time.

7. Shuffle a deck of cards and select a card at random. Return the card to the deck, shuffle, and draw again for a total of 20 trials.

(a) Compute P_e(ace or heart).

(b) Compute P_e(ace and heart).

(c) Compute P_e(ace | heart).

8. Number a set of 3″ by 5″ note cards from 1 to 10. Shuffle the deck thoroughly and deal the cards face up on a table while at the same time counting the number of cards dealt. If the number of the card is the same as the number of cards dealt, say you have a match. For example, if the sixth card dealt is the card with a 6 on it, you have a match. Perform the experiment 25 times and compute the empirical probability that a match occurs.

Section 10.2

9. (a) Three coins are tossed. Make an orderly list of all possible outcomes.

   (b) In how many ways can you obtain two heads and one tail?

10. (a) How many ways can you select nine players for a baseball team from among 15 players if any player can play any position?

    (b) How many ways can you select the team in part (a) if only two players can pitch and only three others can catch? Note that these five players can also play all other positions.

11. (a) All of the 90 students in Ferry Hall speak at least one of French, English, or German. If 38 speak English, 24 speak French and English, 27 speak German and English, and 17 speak German, French, and English, how many speak German or French?

    (b) How many of the students in part (a) speak French and English, but not German?

12. (a) How many ways can you select two clubs from an ordinary deck of playing cards?

    (b) How many ways can you select two face cards from an ordinary deck of playing cards?

    (c) How many ways can you select two clubs or two face cards from an ordinary deck of playing cards?

13. (a) How many five-letter code words can be made if repetition of letters is not allowed?

    (b) How many of the five-letter code words in part (a) start and end with vowels?

    (c) How many of the code words of part (a) contain the three-letter sequence *aef*?

14. (a) In how many ways can the letters in STREETS be placed in recognizably different orders?

    (b) In how many of the orderings of part (a) are the two *E*s adjacent?

    (c) How many of the orderings in part (a) begin with a *T*?

Section 10.3

15. (a) Explicitly list all elements in the sample space if two coins and a die are tossed.

    (b) Compute $P(\text{T, T, 5})$, the probability of getting two tails on the coins and 5 on the die in part (a).

16. Compute $P(5 \mid \text{T, T})$, the probability of obtaining 5 on the die given that both coins came up tails in problem 15.

17. Compute the probability of obtaining a sum of at most 11 on a single roll of two dice.

18. An urn contains five white, six red, and four black balls. Two balls are chosen at random.

    (a) What is the probability that they are the same color?

    (b) What is the probability that both are white?

    (c) What is the probability that both are white given that they are the same color?

19. Consider a spinner made as shown in Figure 10.2, but marked as indicated here. Compute these probabilities.

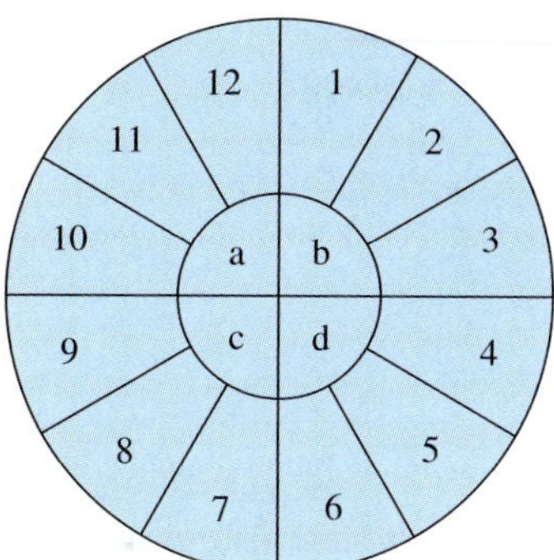

(a) $P(b \text{ and } 8)$ (b) $P(b \text{ or } 8)$ (c) $P(b \mid 8)$

(d) $P(b \text{ and } 2)$ (e) $P(b \text{ or } 2)$ (f) $P(2 \mid b)$

20. Three coins are tossed.

    (a) What are the odds in favor of getting two heads and one tail?

    (b) What are the odds in favor of getting three heads?

21. (a) If $P(A) = 0.85$, what are the odds in favor of A occurring on any given trial?

    (b) If the odds in favor of A are 17 to 8, determine $P(A)$.

22. You play a game where you win the amount shown with the probability shown.

    $P(\$5) = 0.50 \quad P(\$10) = 0.25 \quad P(\$20) = 0.10$

    (a) What is the expected value of the game?

    (b) If it costs you \$10 to play the game of part (a), is it wise to play? Explain.

23. A census taker was told by a neighbor that a family of five lived in the next house—two parents and three children. When the census taker visited the house, he was greeted by a girl. What is the probability that the other two children were both boys? Explain briefly.

Chapter Test

1. Calculate each of the following.
   (a) $7!$ (b) $\frac{9!}{6!}$ (c) $\frac{8!}{(8-8)!}$ (d) $7 \cdot 6!$
   (e) $P(8, 5)$ (f) $P(8, 8)$ (g) $C(9, 3)$
   (h) $C(9, 9)$
2. An urn contains five yellow, four blue, and eight green marbles.
   (a) In how many ways can one select five green marbles?
   (b) In how many ways can one select five yellow and five green marbles?
   (c) In how many ways can one select five yellow or five green marbles?
3. If you select five marbles from the urn in problem 2, what is the probability that two are yellow given that three are green?
4. We claim that the probability is about 0.6 that a person shown a card with the numbers 1, 2, 3, and 4 printed on it will, when asked, choose the number 3. How can such a probability be calculated? Explain briefly.

1 2 3 4

5. If $P(E) = 0.35$, what are the odds of obtaining E on a single trial of an experiment?
6. How many four-letter code words can be made using the letters a, b, c, d, e, f, and g
   (a) with repetition allowed?
   (b) with repetition not allowed?
7. How many of the code words in problem 6, part (b),
   (a) begin with a vowel?
   (b) have c and d adjacent?
8. What kind of probability is used when an assertion like "The probability that penicillin will cure a case of strep throat is 0.9" is made? Explain briefly.
9. In Mrs. Spangler's calculus class, all of the students are also studying one or more foreign languages. If
   27 students study French,
   29 students study German,
   17 students study Chinese,
   12 students study German and French,
   3 students study German and Chinese,
   2 students study French and Chinese, and
   1 student studies French, German, and Chinese,
   (a) how many students are in Mrs. Spangler's class?
   (b) how many students in the class study Chinese only?
   (c) how many students study French and German, but not Chinese?
10. What are the odds in favor of selecting a yellow marble if a single marble is drawn from the urn described in problem 2?

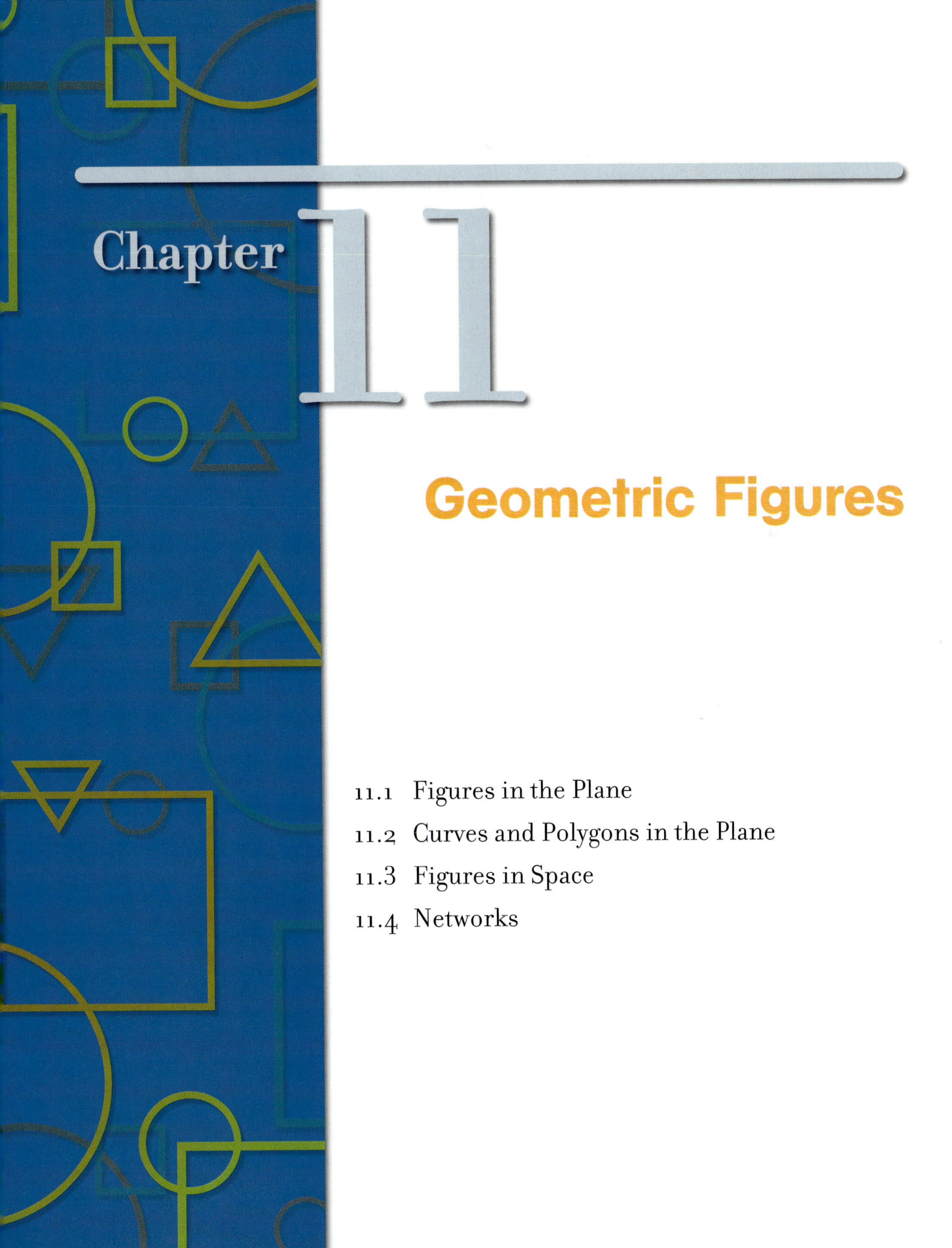

Chapter 11

Geometric Figures

11.1 Figures in the Plane
11.2 Curves and Polygons in the Plane
11.3 Figures in Space
11.4 Networks

Investigating Triangles via Paper Folding

Materials Needed

Paper (thin and colorful paper similar to origami paper works well), scissors, rulers, protractors, pencils, and tape.

Directions

Many important geometric concepts, figures, and relationships can be investigated with paper folding. In this activity, you will first practice some basic constructions using paper folding—creating perpendicular lines, angle bisectors, and perpendicular bisectors. You will then use these basic constructions to discover some properties of triangles and quadrilaterals. (A quadrilateral is a polygon with four sides.)

The Basic Folds

Use your measuring tools to check that these constructions work as claimed.

1. **Perpendicular line.** Draw a line ℓ and place a point P on a sheet of paper. Crease the paper so the line is folded onto itself and the crease passes through the point P. Unfold your paper and then use your protractor to check that the crease makes a $90°$ angle with the line ℓ.

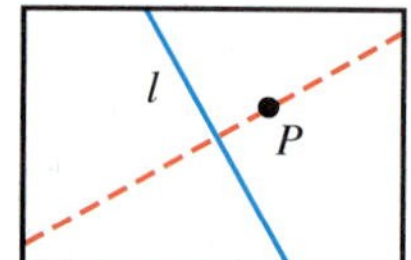

Perpendicular line

2. **Angle bisector.** Draw two rays from a point A, and make a crease through A so that one ray is folded onto the other ray. Unfold your paper and use your protractor to show that the crease makes angles of equal size with the two rays.

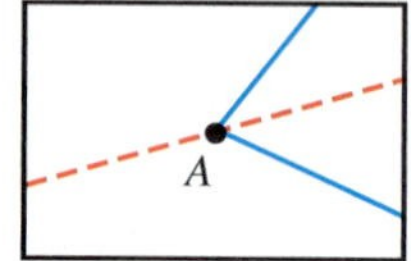

Angle bisector

3. **Perpendicular bisector.** Draw a line segment joining two points C and D. Crease your paper so that point C is folded on top of point D. Unfold your paper and label the point where the crease crosses the segment $\overline{CD}$ as point M. Use your measuring tools to check that M is the same distance from both C and D and that the crease is perpendicular to the segment.

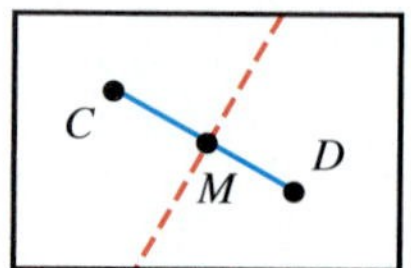

Perpendicular bisector

Exploring Properties of Triangles

The basic folds make it simple to construct some interesting lines and line segments associated with a triangle, as shown in the following diagram:

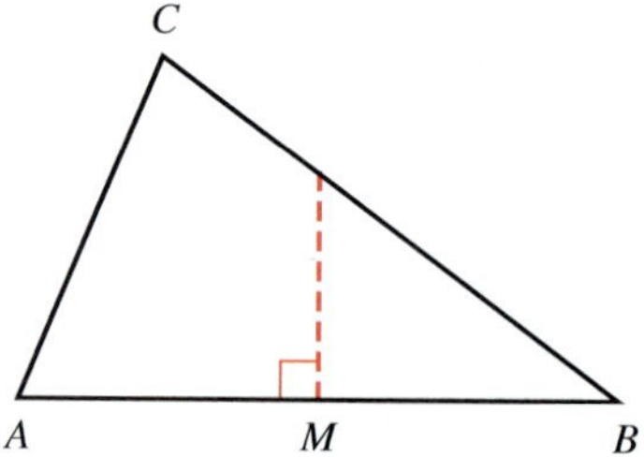

Midpoint M and perpendicular bisector of side $\overline{AB}$

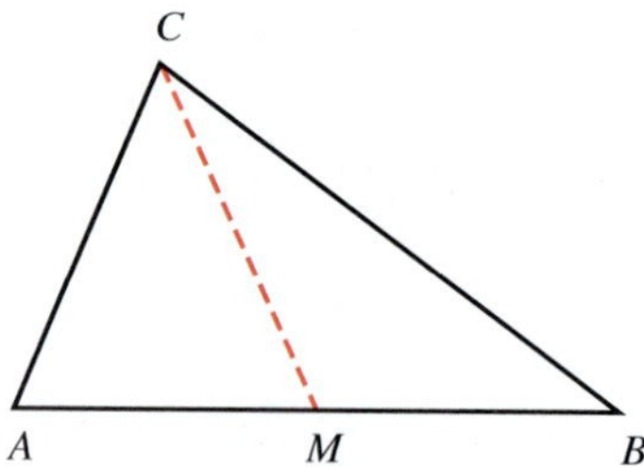

Median through vertex C to midpoint of side $\overline{AB}$

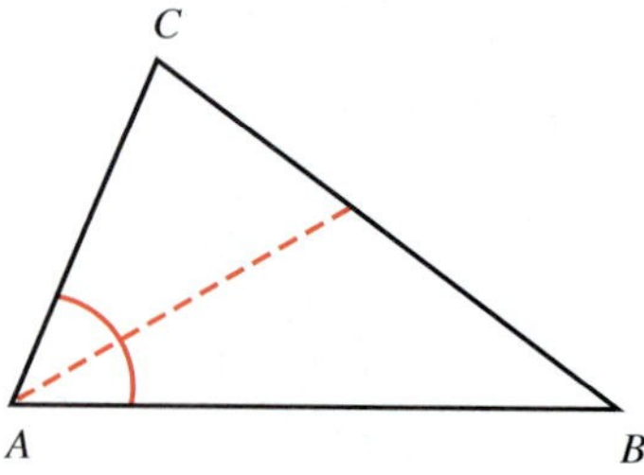

Angle bisector at vertex A

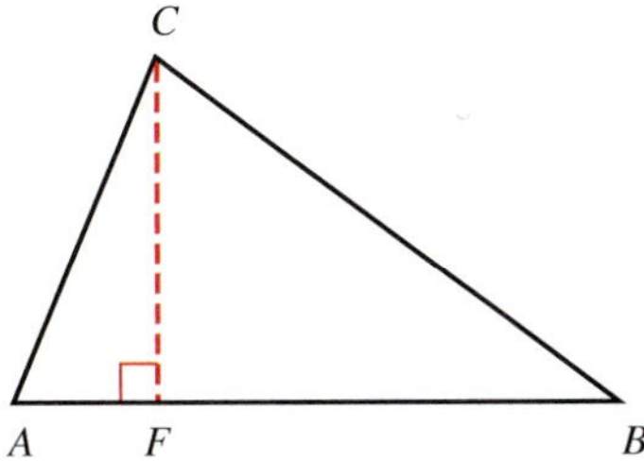

Altitude from vertex C perpendicular to side $\overline{AB}$

1. **The Midpoint of a Hypotenuse of a Right Triangle.** Cut off the corner of a rectangular sheet of paper to create a right triangle ABC, where the 90° angle is at vertex C. Use folding to construct the midpoint M of the hypotenuse $\overline{AB}$. Now find the perpendicular bisectors of legs $\overline{AC}$ and $\overline{BC}$. What do you find interesting? How do the respective distances from M to A, B, and C compare with one another?
2. **The Intersection of the Perpendicular Bisectors of the Sides of a Triangle.** Cut out a triangle from a sheet of paper. It may have any shape at all, but begin with a triangle with no angle larger than 90°. Next, construct the perpendicular bisectors of each of the sides of the triangle. What is interesting about how your lines intersect? Repeat your investigation with a triangle with one angle larger than 90°. Tape your unfolded triangle onto a large sheet of paper and use a ruler to extend the crease lines. Describe what property you observe.
3. **The Intersection of the Medians of a Triangle.** Use short creases to find the midpoints of all three sides of a paper triangle, and then make creases to construct all three medians of your triangle. What special property do you discover?
4. **The Intersection of the Angle Bisectors of a Triangle.** Use paper folding to construct the angle bisectors of all three angles of a paper triangle. What special property do you observe?
5. **The Intersection of the Altitudes of a Triangle.** Use paper folding to construct the altitudes through each of the vertices of a triangle. What seems special about your three lines? If your triangle has an angle larger than 90°, you will want to tape your unfolded triangle to a large sheet of paper and use a ruler to extend your crease lines.
6. **The Angle Sum and Area of a Triangle.** Use folding to construct the altitude $\overline{CF}$ of a paper triangle ABC. Measure the height $h = CF$ and the length of the base $b = AB$ of your triangle. Next, fold all three vertices of your triangle to F, as shown below.

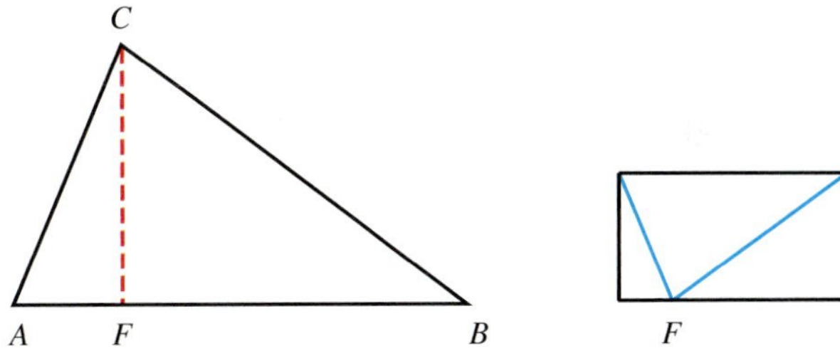

Why is the new figure a rectangle? (A rectangle is a four-sided polygon with a 90° angle at each vertex.) How do the lengths of the sides of your rectangle compare with the height and base of your triangle? How is the area of your rectangle related to the area of your triangle? What has your folding revealed about the sum of the measures of the angles of your triangle?

Connections

The Nature of Informal Geometry

This chapter is the first of four dealing with topics in geometry that have importance to the teacher of elementary school mathematics. To understand the approach in these chapters it is necessary to know what we mean by geometry—and what we do not mean. Geometry for us means the informal study of shape. The three words in this description—*informal, shape,* and *study*—warrant some discussion.

Informal. Until about 600 B.C. geometry was pursued in response to practical, artistic, and religious needs. The pyramids of ancient Egypt (ca. 2500 B.C.) and the Stonehenge observatory (ca. 2800 B.C.) provide evidence that complex problems of form and measurement were solved even in Neolithic times. Over the centuries people built and interacted with a variety of patterns, objects, and structures. The shapes that recurred most often were named and some of their properties were discovered. Considerable knowledge of geometry was accumulated, but mathematics was not yet an organized and independent discipline, and the ideas of proof and deduction were still absent.

In the period 600–300 B.C., Thales, Pythagoras, Zeno, Eudoxus, Euclid, and others organized this accumulated knowledge and experience, and transformed geometry into a theoretical science. Utilitarian considerations gave way to abstraction and general methods. With Euclid's *Elements,* geometry became a formal system in which geometric theorems were deduced logically from a list of statements called axioms that were accepted without proof. To many people, geometry is restricted to a Euclidean formalism in which exacting standards of proof and logical development must be met.

In this text, however, we return to learning by trusting our intuition and experience. Geometric facts are discovered by explorations with pictorial representations and physical models, with little attention given to the overall logical structure. This models the levels of learning geometry that were identified by the van Hieles' research (see Into the Classroom in Section 11.1). However, there will be many opportunities to verify patterns and conjectures by examining the consequences of properties and facts that have already been accepted.

Shape. Marjorie Senechal* points out that "shape" is an undefinable term, partly because we must leave room for new shapes as they are discovered. For example, fractals and CAT scan images are shapes of current interest and importance, made possible by a combination of mathematics and computers. For us, the shapes of interest will most often be figures such as polygons, curves, cubes, and spheres, which are familiar in classical geometry.

Study. Like the objectives of geometers of ancient times, our goals are to recognize differences and similarities among shapes; to analyze the properties of a shape or class of shapes; and to model, construct, and draw shapes in a variety of ways. These goals are inseparably intertwined, but it will be seen that the discussion follows three threads of development: *classification, analysis,* and *representation.*

Materials for Explorations

Many examples will be presented in the form of an *exploration.* First, you will represent a shape, perhaps with a drawing or physical model, that satisfies the stated conditions. You are next asked to discover, analyze, and describe the properties of the shape. Often you will not want to read further until you have followed the directions and made some discoveries for yourself; only then should you read on to see if the patterns and relationships you have uncovered agree with those discussed in the text.

The following tools and materials will be useful to draw, construct, or create the shapes you will explore:

- colored pencils
- ruler (best if marked in both inches and millimeters)
- compass (be sure it is of good quality)
- tape
- glue
- protractor
- drafting triangles (30°–60°–90° and 45°–45°–90°)
- scissors
- unlined paper
- graph paper
- dot paper in both square and triangular patterns
- patty paper (waxed meat-patty separating sheets)

*See "Shape," a chapter of *On the Shoulders of Giants: New Approaches to Numeracy,* L. A. Steen, ed. (Washington, DC: National Academy Press, 1990).

A variety of manipulatives are available from commercial suppliers and are of great value in the study of geometry. Hopefully, you will have access to such items as the following:

- geoboards
- tangrams
- pattern blocks
- geometric solids (wood or plastic)
- pentominoes
- reflective drawing tools such as a Mira®

Geometric figures can also be created and investigated with geometry software. One such program, *The Geometer's Sketchpad®*, is described in the Window on Technology box in Section 11.1. Additional information can be found in Appendix D.

from **The NCTM Principles and Standards**

Geometry in Grades Pre-K–2

Children begin forming concepts of shape long before formal schooling. The primary grades are an ideal time to help them refine and extend their understandings. Students first learn to recognize a shape by its appearance as a whole or through qualities such as "pointiness." They may believe that a given figure is a rectangle because "it looks like a door."

Pre-K–2 geometry begins with describing and naming shapes. Young students begin by using their own vocabulary to describe objects, talking about how they are alike and how they are different. Teachers must help students gradually incorporate conventional terminology into their descriptions of two- and three-dimensional shapes. However, terminology itself should not be the focus of the pre-K–2 geometry program. The goal is that early experiences with geometry lay the foundation for more-formal geometry in later grades. Using terminology to focus attention and to clarify ideas during discussions can help students build that foundation.

Teachers must provide materials and structure the environment appropriately to encourage students to explore shapes and their attributes. For example, young students can compare and sort building blocks as they put them away on shelves, identifying their similarities and differences. They can use commonly available materials such as cereal boxes to explore attributes or shapes of folded paper to investigate symmetry and congruence. Students can create shapes on geoboards or dot paper and represent them in drawings, block constructions, and dramatizations.

Students need to see many examples of shapes that correspond to the same geometrical concept as well as a variety of shapes that are nonexamples of the concept. Through class discussions of such examples and nonexamples, geometric concepts are developed and refined.

Students also learn about geometric properties by combining or cutting apart shapes to form new shapes. Interactive computer programs provide a rich environment for activities in which students put together or take apart (compose and decompose) shapes. Technology can help all students understand mathematics, and interactive computer programs may give students with special instructional needs access to mathematics they might not otherwise experience.

11.1 FIGURES IN THE PLANE

The shapes in the picture gallery of plane figures (page 653) and the picture gallery of space figures (page 696) are each highly complex when viewed as a whole, but underlying this complexity is an orderly arrangement of simpler parts. In this section we consider the most basic shapes of geometry: points, lines, segments, rays, and angles. We will also be introduced to a large number of notations and terms that are essential for the communication of geometric concepts and relationships.

Points and Lines

A point on paper is represented by a dot. A point on a television monitor is represented by a small rectangle of phosphors, called a pixel, that glows when excited by a beam of electrons. Neither a dot nor a pixel is an exact representation of a geometric point. In the mind's eye, dots and pixels are decreased in size until they become ideal **points**—that is, just locations in space. On paper we still draw dots to represent points, and we label the points with uppercase letters as shown.

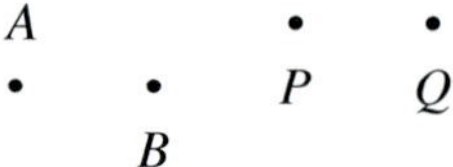

A **line,** like a point, is undefined, but its meaning is suggested by representations such as a tightly stretched thread, a laser beam, or the edge of a ruler. We assume that any two points determine one and only one line that contains the two points. Lines will often be denoted with lowercase letters such as l and m. If A and B are two points, then the line through A and B is denoted by $\overleftrightarrow{AB}$. On paper, lines can be drawn with either a ruler or a **straightedge.** A straightedge is like a ruler, but without any marks on it.

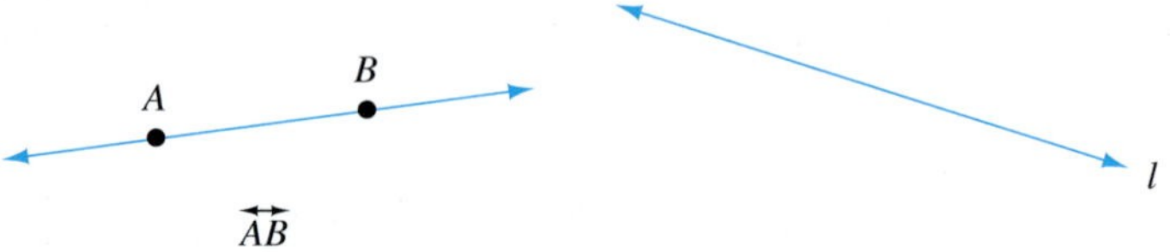

The arrows in the drawings and in the notation $\overleftrightarrow{AB}$ indicate that lines extend infinitely far in two directions.

Three or more points usually determine several lines, but if they lie on just one line, then we say the points are **collinear,** as shown in Figure 11.1.

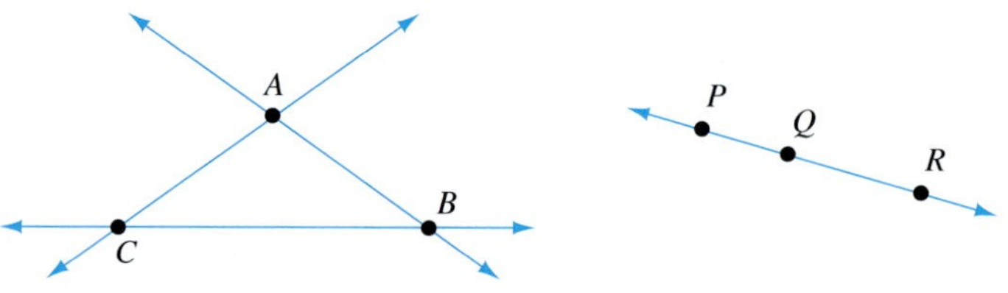

Figure 11.1
Three points determine either three lines or one line

Three noncollinear points determine a **plane,** which is yet another undefined term used to describe a set of points that idealize a flat space such as a tabletop. In this section and the next, we consider only sets of points that belong to a single plane. Subsets of a plane are called **plane figures** or **plane shapes.** In Section 11.3 we explore solid shapes, in which not all of the points belong to a single plane.

A Picture Gallery of Plane Figures

Fissures in a gelatinous preparation of tin oil

Butterfly wings

M. C. Escher's sketch of a wall mosaic in the Alhambra

A snow crystal

A fractal, an example of a complex, beautiful image created with a computer

The pattern of seeds in the head of a sunflower

Dynamic-Geometry Software

Several newly developed computer programs are now available to create geometric figures and investigate their properties. One of these programs, *The Geometer's Sketchpad,** is described in Appendix D.

Geometry programs allow the user to construct and label nearly any geometric shape. This includes points, lines, segments, and rays, and more sophisticated shapes such as circular sectors and polygon interiors. Existing elements of the figure can also be used to construct new elements of the figure. For example, given a line segment, the software can quickly construct the midpoint of the segment. Or, given a line and a point, the software will easily construct the line through the point that is parallel to the selected line. Similarly, points of intersection, perpendiculars, angle bisectors, and so on are quickly and easily constructed—certainly a welcome contrast to classical compass and ruler constructions. Moreover, once a figure has been created, its lengths, angles, area, perimeter, and other properties can be measured to high precision by the software itself.

The most exciting aspect of the new software is its dynamic nature. By clicking and dragging on a point or line of the figure, the figure can be manipulated into new shapes. For example, a line segment can be stretched or moved to new positions. If a midpoint of the segment has been constructed, the manipulation will continuously move the midpoint so that it is *always* a midpoint of the moving segment. Likewise, if a line perpendicular or parallel to the manipulated line has been constructed, these lines will automatically move to continue to be *always* parallel or perpendicular to the manipulated line.

Let's look at an example that demonstrates the power of dynamic-geometry software. First, a quadrilateral *ABCD* (that is, a four-sided polygon) and its interior (shown in light blue) have been constructed, as shown in the accompanying figure. The midpoints *E*, *F*, *G*, and *H* of the four sides have also been constructed with the software and joined with line segments to create the quadrilateral *EFGH* and its interior (shown in darker blue). It would appear that *EFGH* has some special properties, so we have used the measurement functions of the software to measure several angles, lengths, and areas. The measurements we have made indicate that the opposite sides of *EFGH* have the same length and the opposite angles have the same size. Moreover, we have used the calculation capability of the software to show that adjacent angles of *EFGH* have angle measures that sum to 180° and that the area of *ABCD* is twice that of *EFGH*. To check that these properties were not accidental, we have used the software to manipulate the shape of *ABCD* to assume the new shape shown to the right. Since the length, angle, and area properties continue to hold for the manipulated figure, we become increasingly confident that the geometry software has revealed general principles.

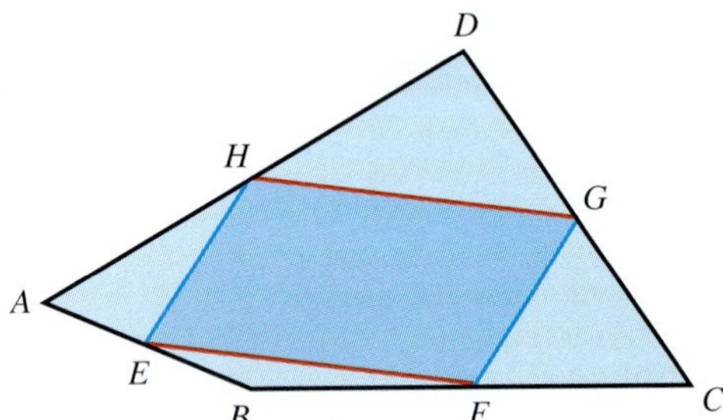

m $\overline{GH} = 3.85$ cm	m $\angle FEH = 58.06°$
m $\overline{EF} = 3.85$ cm	m $\angle HGF = 58.06°$
m $\overline{HE} = 2.71$ cm	m $\angle EHG = 121.94°$
m $\overline{FG} = 2.71$ cm	m $\angle GFE = 121.94°$
Area $HGFE = 8.85$ cm^2	m $\angle FEH$ + m $\angle EHG = 180.00°$
Area $ABCD = 17.70$ cm^2	$\dfrac{(\text{Area } ABCD)}{(\text{Area } HGFE)} = 2.00$

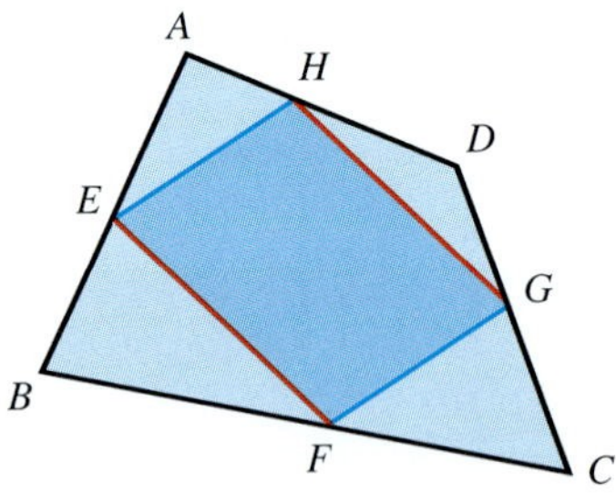

m $\overline{GH} = 3.05$ cm	m $\angle FEH = 68.82°$
m $\overline{EF} = 3.05$ cm	m $\angle HGF = 68.82°$
m $\overline{HE} = 2.66$ cm	m $\angle EHG = 111.18°$
m $\overline{FG} = 2.66$ cm	m $\angle GFE = 111.18°$
Area $HGFE = 7.56$ cm^2	m $\angle FEH$ + m $\angle EHG = 180.00°$
Area $ABCD = 15.11$ cm^2	$\dfrac{(\text{Area } ABCD)}{(\text{Area } HGFE)} = 2.00$

*A free demonstration version can be downloaded from Key Curriculum Press's Web site, www.keypress.com.

Two lines in the plane that do not have a point in common are called **parallel.** We write $l \parallel m$ if l and m are parallel lines. Two distinct lines p and q in a plane that are not parallel must have a single point in common, called their **point of intersection,** and we write $p \nparallel q$. Three lines can intersect at zero, one, two, or three points. In the case of one point of intersection, the three lines are said to be **concurrent.** A line that intersects two other lines is called a **transversal.** The different ways in which two or three lines can be arranged in the plane are shown in Figure 11.2.

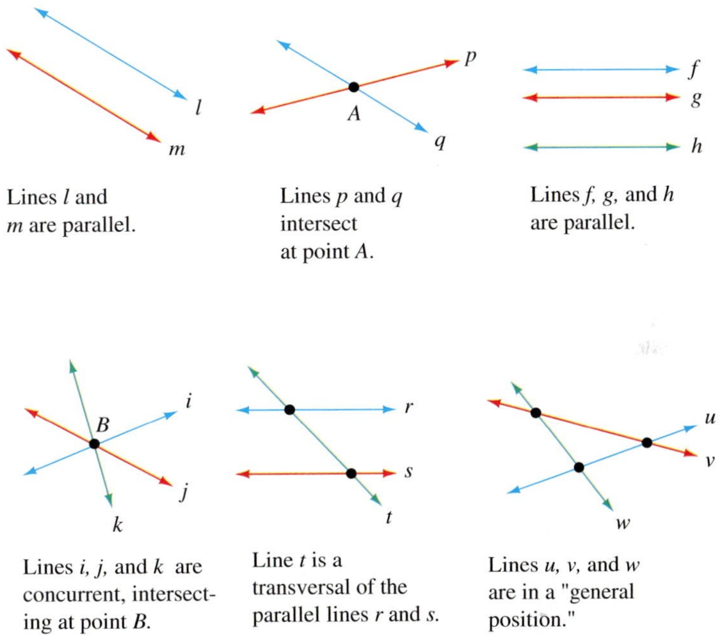

l
m
p
A
q
f
g
h
Lines *l* and *m* are parallel.
Lines *p* and *q* intersect at point *A*.
Lines *f, g,* and *h* are parallel.
i
B
j
k
r
s
t
u
v
w
Lines *i, j,* and *k* are concurrent, intersecting at point *B*.
Line *t* is a transversal of the parallel lines *r* and *s*.
Lines *u, v,* and *w* are in a "general position."

Figure 11.2
The possible arrangements of two and three lines in a plane

EXAMPLE 11.1 | EXPLORING COLLINEARITY AND CONCURRENCY

(a) Place three circular objects (for example, coins and cups) of different size on a sheet of paper and trace them to create three circles, labeled C_1, C_2, and C_3. Next, place a ruler tightly against one side of the objects you used to trace C_1 and C_2 and draw the line l that just touches both circles. Move the ruler to the other side of your objects to draw the line m, as shown in the accompanying figure. The two lines you've drawn are called the external tangents to the circles C_1 and C_2. Label their point of intersection as P.

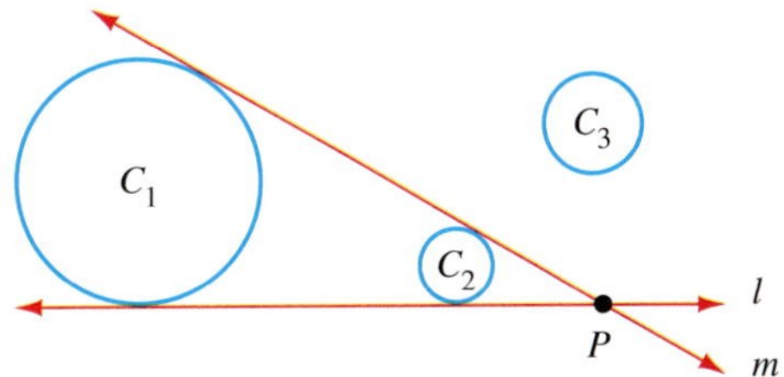

In the same way, use your ruler to draw the two lines externally tangent to C_2 and C_3, and let Q be their point of intersection. Finally, draw the external tangents of C_1 and C_3, and let R be their point of intersection. What conjecture do you have concerning P, Q, and R?

(b) Trace around a cup bottom (or use a compass) to draw an accurate circle. Then use a ruler to draw any three lines that are tangent to the circle, at points labeled X, Y, and Z, and that intersect in pairs at the points labeled A, B, and C. Finally, draw the lines $\overleftrightarrow{AX}$, $\overleftrightarrow{BY}$, and $\overleftrightarrow{CZ}$ (in the accompanying figure). What conjecture can you make about $\overleftrightarrow{AX}$, $\overleftrightarrow{BY}$, and $\overleftrightarrow{CZ}$?

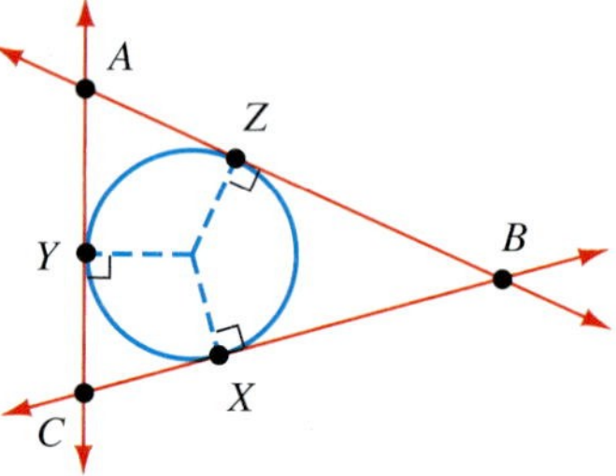

Solution

(a) P, Q, and R are collinear.

(b) $\overleftrightarrow{AX}$, $\overleftrightarrow{BY}$, and $\overleftrightarrow{CZ}$ are concurrent.

Line Segments and the Distance Between Points

Let A and B be any two points. The line $\overleftrightarrow{AB}$ can be viewed as a copy of the number line. That is, every point on $\overleftrightarrow{AB}$ corresponds to a unique real number and every real number corresponds to a unique point on $\overleftrightarrow{AB}$. If A and B correspond to the real numbers x and y, respectively, then the absolute value, $|x - y|$, gives the **distance** between A and B. We denote this distance by AB.

AB with no overbar denotes the length of segment $\overline{AB}$.

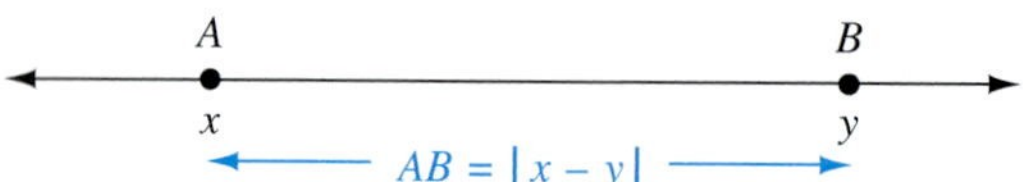

The points on the line $\overleftrightarrow{AB}$ that are between A and B, together with A and B themselves, form the **line segment** $\overline{AB}$. Points A and B are called the **endpoints** of $\overline{AB}$, and the distance AB is the **length** of $\overline{AB}$. It is important to see that the overbar used in the notation distinguishes the real number AB from the line segment $\overline{AB}$.

Just for Fun

Arranging Points at Integer Distances

The following arrangement of six points is remarkable because all 15 segments between pairs of points have integer length.

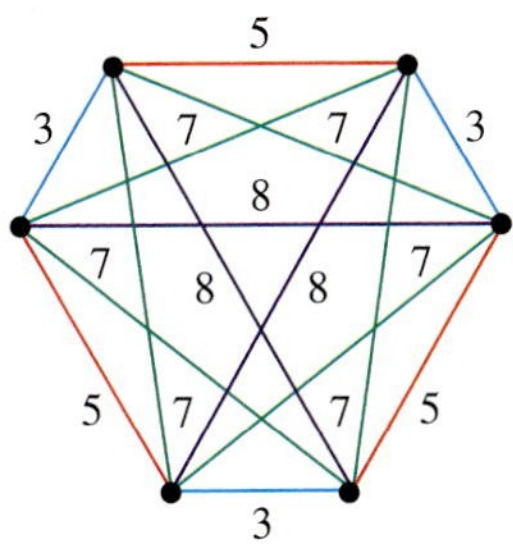

Find an arrangement of four points whose six distances between pairs of points are 2″, 2″, 3″, 4″, 4″, 4″. It will help your search to cut six straws to the lengths given.

Two segments $\overline{AB}$ and $\overline{CD}$ are said to be **congruent** if they have the same length. This is symbolized by writing $\overline{AB} \cong \overline{CD}$. Thus, $\overline{AB} \cong \overline{CD}$ if, and only if, $AB = CD$. The point M in $\overline{AB}$ that is the same distance from A and B is called the **midpoint** of $\overline{AB}$. This information is summarized in Figure 11.3.

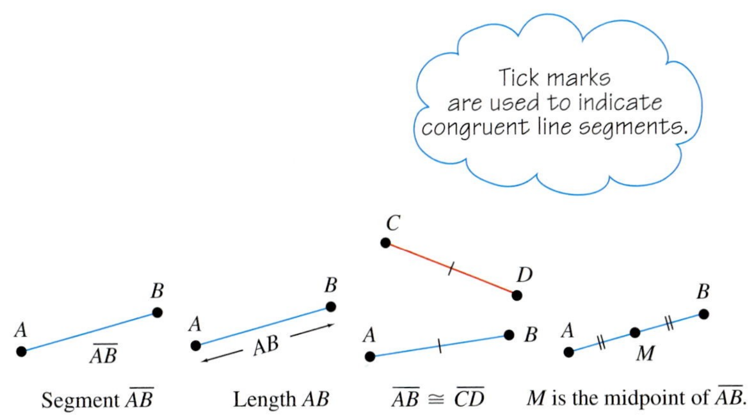

Figure 11.3
A segment $\overline{AB}$*, its length* AB*, congruent segments, and the midpoint* M *of* $\overline{AB}$

Rays, Angles, and Angle Measure

A **ray** is a subset of a line that contains a point P, called the **endpoint** of the ray, and all points on the line lying to one side of P. If Q is any point on the ray other than P, then $\overrightarrow{PQ}$ denotes the ray. The union of two rays with a common endpoint is an **angle.** If the rays are $\overrightarrow{AB}$ and $\overrightarrow{AC}$, then the angle is denoted by $\angle BAC$. The common endpoint of the two rays is called the **vertex** of the angle and is the middle letter in the symbol for the angle (e.g., A in $\angle BAC$). The points B and C not at the vertex can be written in either order, so that $\angle CAB$ denotes the same angle as $\angle BAC$. The rays $\overrightarrow{AB}$ and $\overrightarrow{AC}$ are called the **sides** of the angle. See Figure 11.4.

An angle whose sides are not on the same line partitions the remaining points of the plane into two parts, the **interior** and the **exterior** of the angle. The points along a line segment that joins an endpoint on side $\overrightarrow{AB}$ to an endpoint on $\overrightarrow{AC}$ are all interior points of $\angle BAC$.

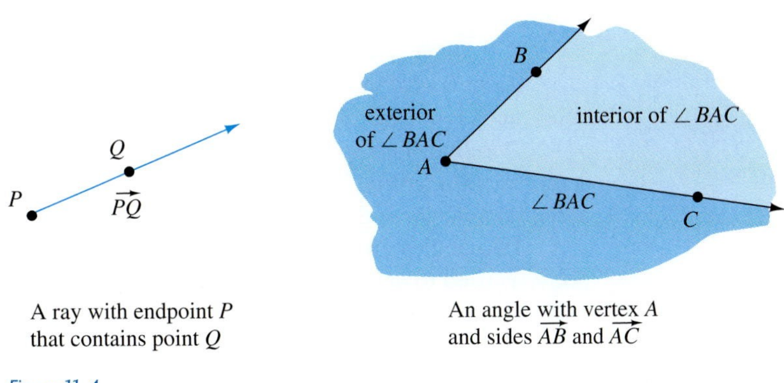

Figure 11.4
A ray $\overleftrightarrow{PQ}$ *and an angle* $\angle$BAC

m(∠A) denotes the measure of the angle ∠A with vertex at point A.

If $\angle BAC$ is the only angle with its vertex at A, it is common to write $\angle A$ in place of $\angle BAC$. When more than one angle has a vertex at A, it is essential to use the full three-letter symbol. Sometimes it is useful to number the angles that appear in a drawing and refer to $\angle 1$, $\angle 2$, $\angle 3$, and so on.

The size of an angle is measured by the amount of rotation required to turn one side of the angle to the other by pivoting about the vertex. The **measure of an angle** is generally given in **degrees,** where there are 360° in a full revolution. The measure of $\angle A$ is denoted by $m(\angle A)$. If the rotation is imagined to pass through the interior of the angle, the measure is a number between 0° and 180°. Unless stated otherwise, $m(\angle A)$ is the measure of $\angle A$ not larger than 180°.

An angle of measure 180° is a **straight angle,** an angle of measure 90° is a **right angle,** and an angle of measure 0° is a **zero angle.** Angles measuring between 0° and 90° are **acute,** and angles measuring between 90° and 180° are **obtuse.**

In some applications the measure of interest corresponds to the rotation through the exterior of the angle and is therefore a number between 180° and 360°. An angle with measure greater than 180°, but less than 360°, is called a **reflex angle.**

The classification of angles according to their measure is summarized in Figure 11.5.

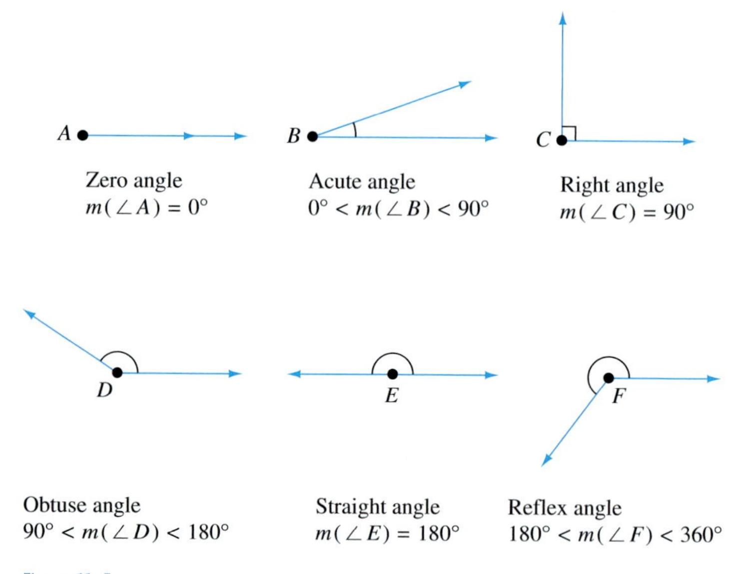

Figure 11.5
The classification of angles by their measure

Right angles in drawings are indicated by a small square placed at the vertex. A circular arc is required to indicate reflex angles.

Two lines l and m that intersect at right angles are called **perpendicular** lines. This is indicated in writing by $l \perp m$. Similarly, two rays, two segments, or a segment and a ray are perpendicular if they are contained in perpendicular lines.

Two angles are **congruent** if, and only if, they have the same measure. The symbol $\cong$ is used to denote the congruence of angles. Thus,

$$\angle P \cong \angle Q \text{ if, and only if, } m(\angle P) = m(\angle Q).$$

The **protractor** is used both to measure angles and to draw angles having a given measure. The protractor and other traditional tools useful for drawing and measuring are

Did You Know?

A First-Person Account of Life on a Plane

I call our world Flatland, not because we call it so, but to make its nature clearer to you, my happy readers, who are privileged to live in Space.

Imagine a vast sheet of paper on which straight Lines, Triangles, Squares, Pentagons, Hexagons, and other figures, instead of remaining fixed in their places, move freely about, on or in the surface, but without the power of rising above or sinking below it, very much like shadows—only hard and with luminous edges—and you will then have a pretty correct notion of my country and countrymen. Alas, a few years ago, I should have said "my universe": but now my mind has been opened to a higher view of things.

More than a century ago, scholar and theologian Edwin Abbott anonymously published this delightful, satiric tale of the inhabitants of a two-dimensional world. Isaac Asimov called it "to this day, the best introduction one can find into the manner of perceiving dimensions."

SOURCE: Edwin Abbott, Flatland: A Romance of Many Dimensions *(New York: Barnes & Noble, Inc., 1983), p. 3.*

shown in Figure 11.6. Increasingly, these tools are being supplemented and replaced by the available geometry software for the computer.

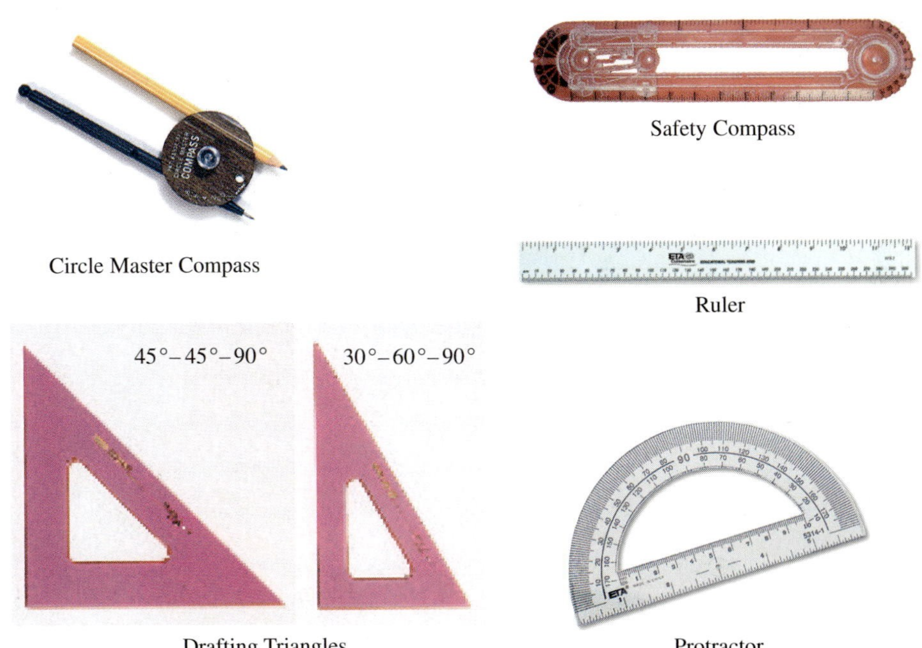

Figure 11.6
Some useful tools for measuring and drawing geometric figures

EXAMPLE 11.2 | EXPLORING ANGLES AND DISTANCES IN A CIRCLE

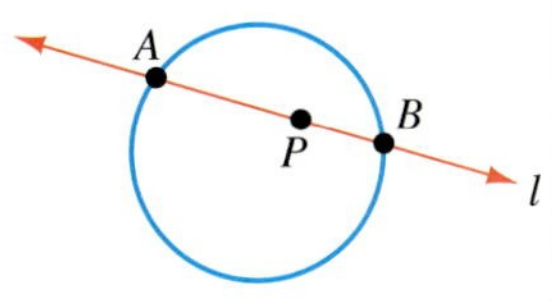

(a) Draw a large circle and choose any point P, other than the center, inside the circle. Any line l through P intersects the circle in two points, say A and B. Measure the distances AP and PB (to the nearest millimeter) and compute the product $AP \cdot PB$. Draw several other lines through P and measure the distances

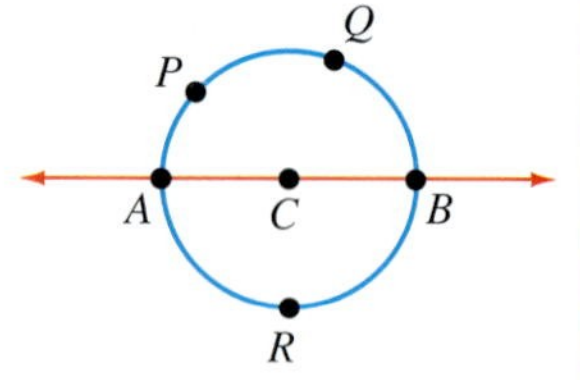

of the two segments. Which line through P makes the product of distances, $AP \cdot PB$, as large as possible?

(b) Draw a circle with center C. Draw a line through C, and let A and B denote its intersections with the circle. Choose any three points P, Q, and R on the circle other than A or B. Use a protractor to measure $\angle APB$, $\angle AQB$, and $\angle ARB$. What general result does this suggest?

Solution

(a) For every choice of l, the product $AP \cdot PB$ is the same. Therefore, no line through P gives a larger product than any other line.

(b) Each angle is a right angle. This is one of geometry's earliest theorems, attributed to Thales of Miletus (ca. 600 B.C.).

Pairs of Angles and the Corresponding-Angles Theorem

As shown in Figure 11.7, two angles are **complementary** if the sum of their measures is 90°. Similarly, two angles are **supplementary** if their measures sum to 180°.

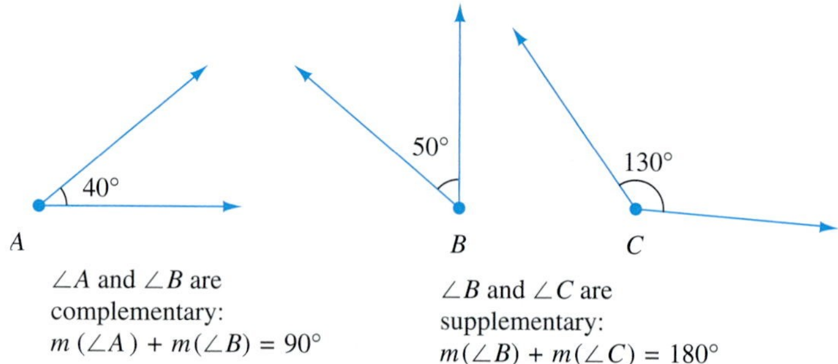

Figure 11.7
Examples of complementary and supplementary angles

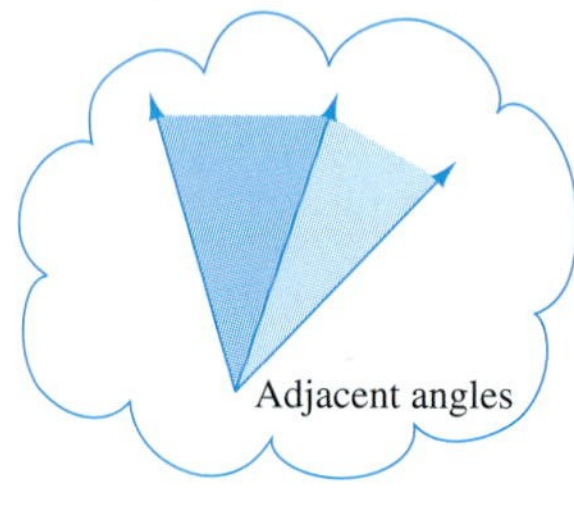

Two angles that have a common side and nonoverlapping interiors are called **adjacent** angles. Supplementary and complementary angles frequently occur as adjacent angles, as shown in Figure 11.8.

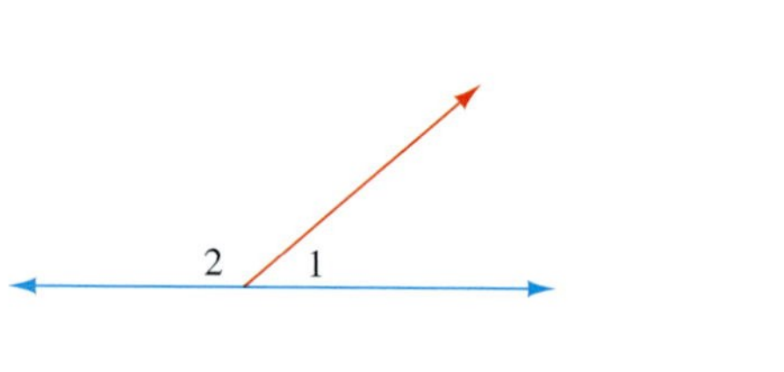

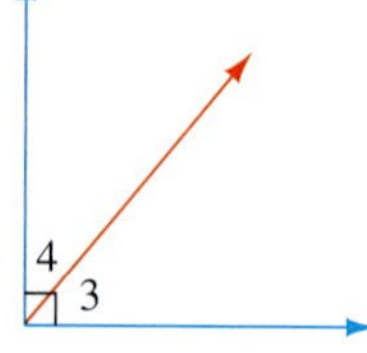

$\angle 1$ and $\angle 2$ are adjacent supplementary angles.

$\angle 3$ and $\angle 4$ are adjacent complementary angles.

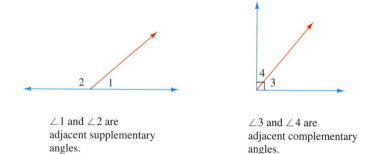

Figure 11.8
Adjacent supplementary and complementary angles

A pair of nonadjacent angles formed by two intersecting lines is called **vertical angles,** as shown in Figure 11.9. Since $\angle 1$ and $\angle 2$ are supplementary, we know that $m(\angle 1) + m(\angle 2) = 180°$. Likewise, $\angle 2$ and $\angle 3$ are supplementary, so we also have

HIGHLIGHT from HISTORY

Thales of Miletus (ca. 640–546 B.C.)

Classical Greek mathematics developed in a succession of schools where a collection of scholars worked with a great leader. The first of these was the Ionian school, founded by Thales of Miletus. Thales traveled extensively and resided for some time in Egypt, where he learned Egyptian mathematics. He compared the length of the shadow cast by a pyramid to that cast by a vertically held rod and was able to calculate the height of the pyramid. Thales is credited with several propositions in geometry and was instrumental in introducing the notions of abstraction and proof into mathematics.

Thales was known as a statesman, engineer, businessman, philosopher, astronomer, and mathematician. There are a number of anecdotes told about Thales. For example, there is the story of the recalcitrant mule that, when carrying a pack of salt, would lie down in the river to dissolve, and lighten, his load. Thales cleverly broke the mule of this habit by loading him with sponges. In Aristophanes' play *The Clouds,* the wisdom of one of the characters is made clear when he is referred to as "a veritable Thales."

$m(\angle 2) + m(\angle 3) = 180°$. Comparing these two equations shows that $m(\angle 1) = m(\angle 3)$. This proves another theorem of Thales.

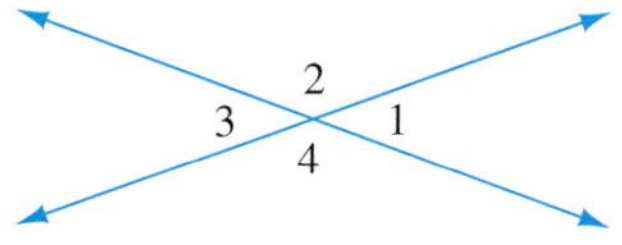

$\angle 1$ and $\angle 3$ are vertical angles.
$\angle 2$ and $\angle 4$ are vertical angles.

Figure 11.9
Intersecting lines form two pairs of vertical angles

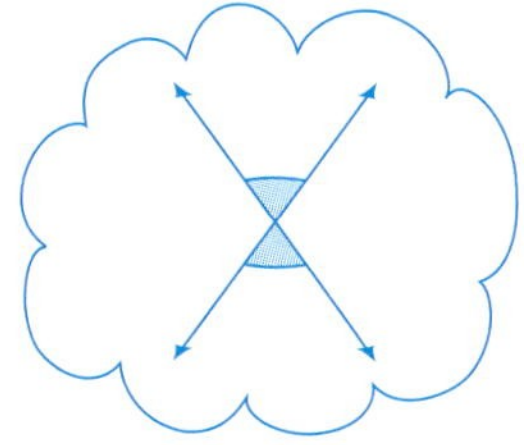

THEOREM *Vertical-Angles Theorem*
Vertical angles have the same measure.

Now consider the angles formed when two lines l and m are intersected at two points by a transversal t. There are eight angles formed, in four pairs of **corresponding angles.**

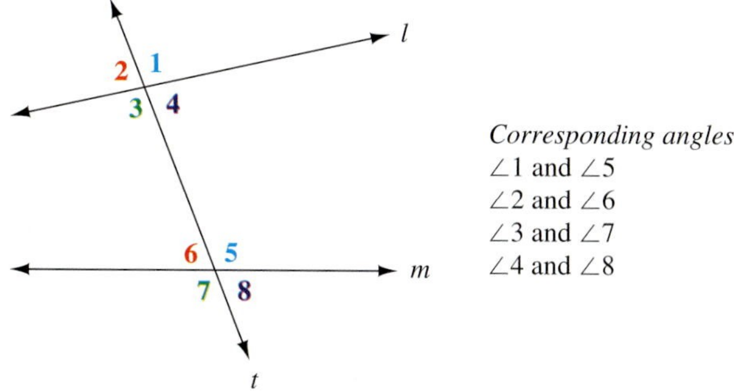

Corresponding angles
$\angle 1$ and $\angle 5$
$\angle 2$ and $\angle 6$
$\angle 3$ and $\angle 7$
$\angle 4$ and $\angle 8$

A case of special importance occurs when l and m are parallel lines, as shown in Figure 11.10. It would appear that each pair of corresponding angles is a pair of congruent angles. Conversely, if any one pair of corresponding angles is a congruent pair of angles, then the lines l and m appear to be parallel. We will accept the truth of these observations, giving us the **corresponding-angles property.** Many formal treatments of Euclidean geometry introduce the corresponding-angles property as an axiom.

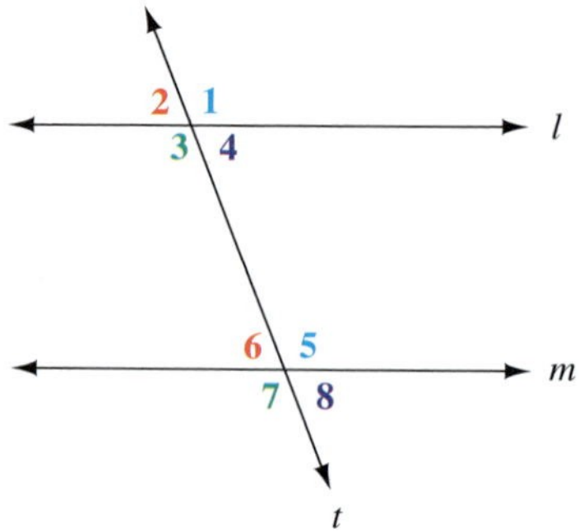

Figure 11.10
Lines l and m are parallel if, and only if, the angles in some corresponding pair have the same measure

PROPERTY *Corresponding-Angles Property*

- If two parallel lines are cut by a transversal, then the corresponding angles have the same measure.
- If two lines in the plane are cut by a transversal and some pair of corresponding angles has the same measure, then the lines are parallel.

EXAMPLE 11.3 | USING THE CORRESPONDING-ANGLES PROPERTY

(a) Lines l and m are parallel and $m(\angle 6) = 35°$. Find the measures of the remaining seven angles.

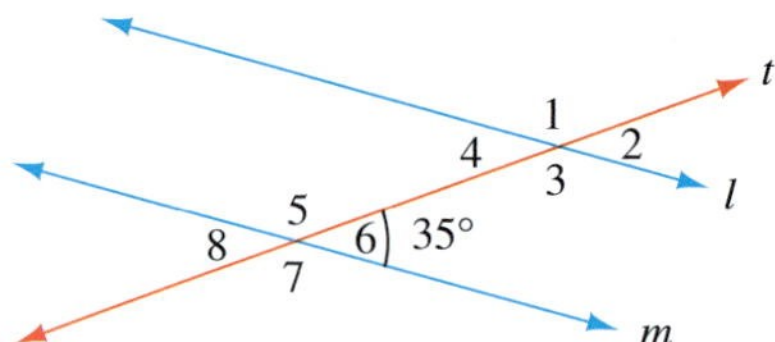

(b) Lines t and j intersect at P and form an angle measuring 122°. Describe how to use a protractor and straightedge to draw a line k through Q that is parallel to line j.

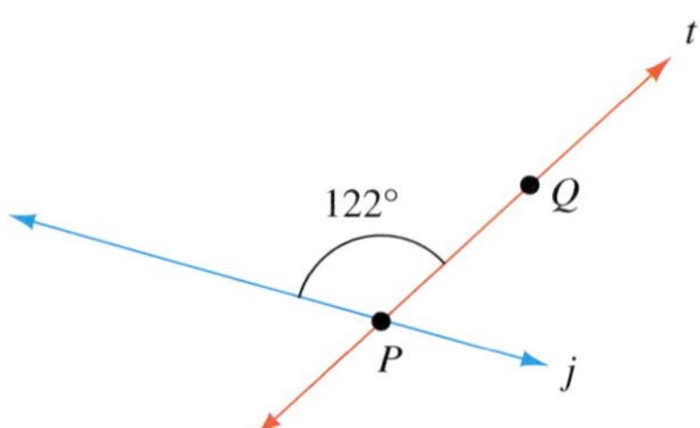

Solution

(a) Since $\angle 8$ and $\angle 6$ are vertical angles, $m(\angle 8) = 35°$. Also $\angle 5$ and $\angle 7$ are supplements of $\angle 6$, and so $m(\angle 5) = m(\angle 7) = 180° - 35° = 145°$. By the corresponding-angles property, $m(\angle 1) = m(\angle 5) = 145°$, $m(\angle 2) = m(\angle 6) = 35°$, $m(\angle 3) = m(\angle 7) = 145°$, and $m(\angle 4) = m(\angle 8) = 35°$.

(b) Use the protractor to form the corresponding angle measuring 122° at point Q.

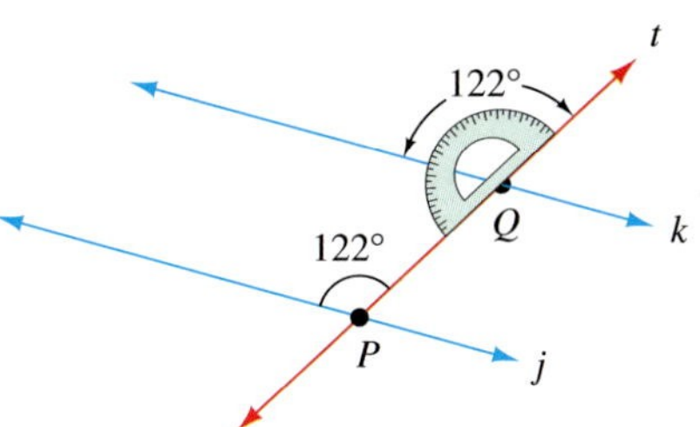

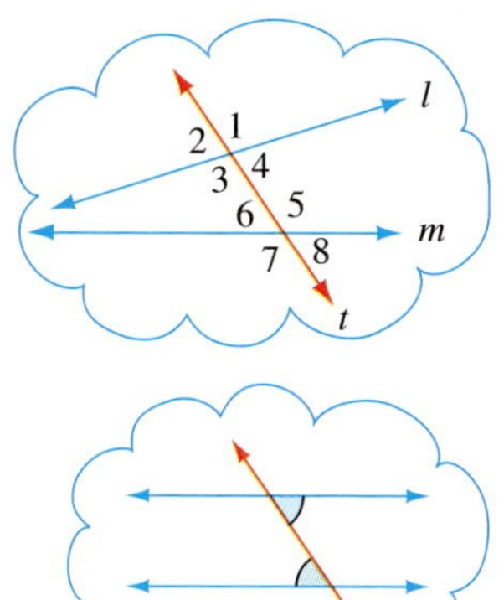

The pair of angles $\angle 4$ and $\angle 6$ between l and m, but on opposite sides of the transversal t, is called **alternate interior angles.** Since $\angle 2$ and $\angle 4$ are vertical angles, they are congruent by the vertical-angles theorem. Thus, the corresponding angles $\angle 2$ and $\angle 6$ are congruent if, and only if, the alternate interior angles $\angle 4$ and $\angle 6$ are congruent. This gives the following consequence of the corresponding-angles property.

THEOREM *Alternate-Interior-Angles Theorem*
Two lines cut by a transversal are parallel if, and only if, a pair of alternate interior angles is congruent.

The Measure of Angles in Triangles

If a triangle ABC is cut from paper, and its three corners are torn off, it is soon discovered that the three pieces will form a straight angle along a line l (see Figure 11.11). Thus, $m(\angle 1) + m(\angle 2) + m(\angle 3) = 180°$ and we have a physical demonstration that the sum of the measures of the angles of a triangle is 180°.

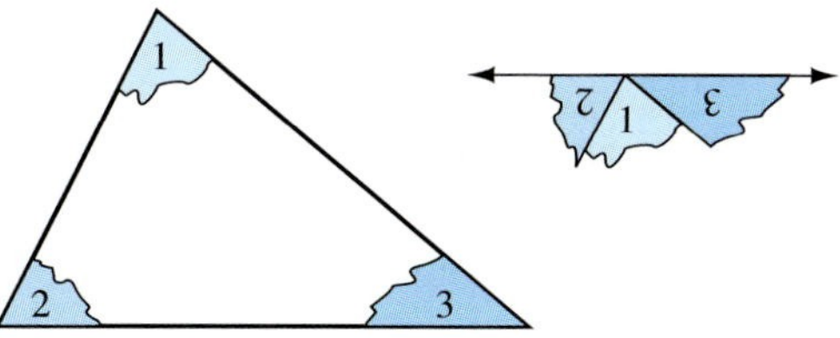

Figure 11.11
The torn corners of a triangle cut from paper can be placed along a line to show that $m(\angle 1) + m(\angle 2) + m(\angle 3) = 180°$

The alternate-interior-angles theorem can be used to prove the result.

Highlight from History

The Babylonian Origins of Degree Measure

There is no doubt that degree measure, with 360 degrees composing a full turn, had its origin in ancient Babylonia. At one time it was suggested that the Babylonians thought a year was 360 days long, which would mean that the earth would advance one degree per day as it revolved in its circular orbit about the sun. This explanation must be dismissed, however, since the early Babylonians were fully aware that a year exceeded 360 days.

A more plausible explanation has been suggested by Otto Neugebauer, an authority on Babylonian mathematics and science. In the early Sumerian period, time was often measured by how long it took to travel a Babylonian mile (a Babylonian mile was about 7 modern miles). In particular, a day turned out to be the time required to travel 12 Babylonian miles. Since the Babylonian mile, being quite long, had been subdivided into 30 equal parts for convenience, there were $(12)(30) = 360$ parts in a day's journey. The complete turn of the earth each day was therefore divided into 360 parts, giving rise to degree measure of an angle.

THEOREM *Sum of Angle Measures in a Triangle*
The sum of the measures of the angles in a triangle is 180°.

Proof: Consider the line l through point A that is parallel to the line $m = \overleftrightarrow{BC}$.

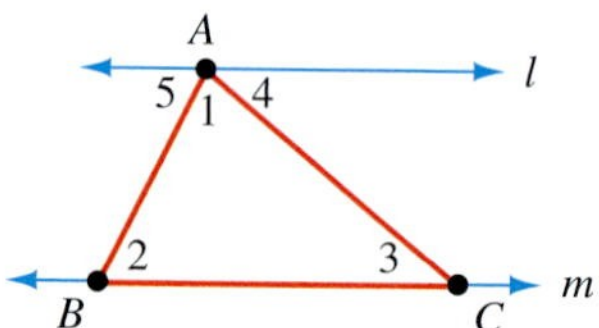

Line $\overleftrightarrow{AB}$ is a transversal to l and m for which $\angle 5$ and $\angle 2$ are alternate interior angles. Thus, $m(\angle 5) = m(\angle 2)$ by the alternate-interior-angles theorem. Similarly, $\angle 4$ and $\angle 3$ are alternate interior angles for the transversal $\overleftrightarrow{AC}$, so $m(\angle 4) = m(\angle 3)$. Since $\angle 5$, $\angle 1$, and $\angle 4$ form a straight angle at vertex A, we know that $m(\angle 5) + m(\angle 1) + m(\angle 4) = 180°$. Thus, by substitution, $m(\angle 2) + m(\angle 1) + m(\angle 3) = 180°$.

EXAMPLE 11.4 MEASURING AN OPPOSITE EXTERIOR ANGLE OF A TRIANGLE

In the accompanying figure, $\angle 4$ is called an **exterior angle** of triangle PQR, and $\angle 1$ and $\angle 2$ are its **opposite interior angles.** Show that the measure of the exterior angle is equal to the sum of the measures of the opposite interior angles; that is, show that $m(\angle 4) = m(\angle 1) + m(\angle 2)$.

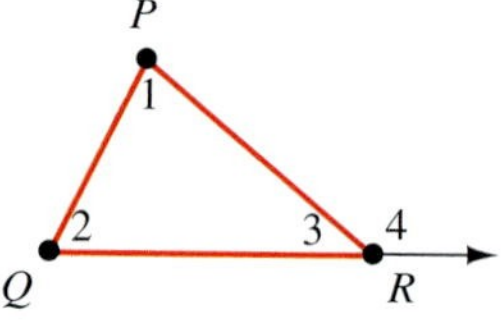

Solution By the preceding theorem, we have $m(\angle 1) + m(\angle 2) + m(\angle 3) = 180°$. Also, $\angle 3$ and $\angle 4$ are supplementary, so that $m(\angle 3) + m(\angle 4) = 180°$. Therefore, $m(\angle 1) + m(\angle 2) + m(\angle 3) = m(\angle 3) + m(\angle 4)$. Subtracting $m(\angle 3)$from both sides of this equation gives $m(\angle 1) + m(\angle 2) = m(\angle 4)$.

Directed Angles

Until now, we have measured angles without regard to the *direction*—clockwise or counterclockwise—one side rotates until it coincides with the second side. Often, it is useful to specify one side as the **initial side** and the other side as the **terminal side.** Angles are then measured by specifying the number of degrees to rotate the initial side to the terminal side. Mathematicians usually assign a positive number to counterclockwise turns, and a negative number to clockwise turns. Angles that specify an initial and final side and a direction of turn are called **directed angles.** Some examples are shown in Figure 11.12, where the arrows on the circular arcs indicate the direction of turn. Notice that an angle measure of $-90°$ could also be assigned the measure $+270°$.

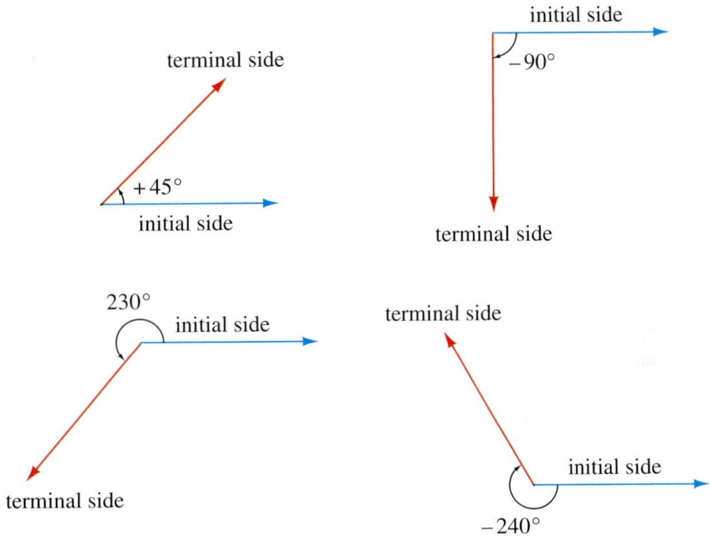

Figure 11.12
Directed angles, measured positively for counterclockwise turns and negatively for clockwise turns

EXAMPLE 11.5 MEASURING DIRECTED ANGLES

Patty Pathfinder's trip through the woods to Grandmother's house started and ended in an easterly direction, but zigzagged through Wolf Woods to avoid trouble. The first two angles Patty turned through are 45° and $-60°$, as shown. Use a protractor to measure the three remaining turns. What is the sum of all five directed angles? Explain your surprise or lack of surprise.

Into the Classroom

Activity-Based Learning and the van Hiele Levels

From kindergarten onward, geometry is learned best through hands-on activities. A successful teacher will take advantage of the enjoyment children experience when working with colored paper, straws, string, crayons, toothpicks, and other tangible materials. Children learn geometry by doing geometry as they construct two- and three-dimensional shapes, combine their shapes to create attractive patterns, and build interesting space figures using plane shapes. By its nature, informal geometry provides unlimited opportunities to construct shapes, designs, and structures that all work to capture a child's interest.

According to pioneering research of the van Hieles in the late 1950s, the knowledge children construct for themselves through hands-on activities is essential to learning geometry. Dr. Pierre van Hiele and his late wife, Dr. Dina van Hiele-Geldof, both former mathematics teachers in the Netherlands, theorized that learning geometry progresses through five levels, which can be described briefly as follows.

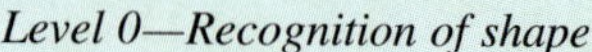

Level 0—Recognition of shape

Children recognize shapes holistically. Only the overall appearance of a figure is observed, with no attention given to the component parts of the figure. For example, a figure with three curved sides would likely be identified as a triangle by a child at Level 0. Similarly, a square tilted point downward may not be recognized as a square.

Level 1—Analysis of single shapes

Children at Level 1 are cognizant of the component parts of certain figures. For example, a rectangle has four straight sides that meet at "square" corners. However, at Level 1 the interrelationships of figures and properties are not understood.

Level 2—Relationships among shapes

At Level 2 children understand how common properties create abstract relationships among figures. For example, a square is both a rhombus and a rectangle. Also, simple deductions can be made about figures, using the analytic abilities acquired at Level 1.

Level 3—Deductive reasoning

The student at Level 3 views geometry as a formal mathematical system and can write deductive proofs.

Level 4—Geometry as an axiomatic system

This is the abstract level, reached only in high-level university courses. The focus is on the axiomatic foundations of a geometry, and no dependence is placed on concrete or pictorial models.

Ongoing research supports the thesis that students learn geometry by progressing through the van Hiele levels. This text—by means of hands-on activities, and examples and problems that require constructions and drawings—promotes the spirit of the van Hiele approach. However, it is the elementary school classroom teacher who must bring geometry to life for his or her students, by creating interesting activities that support each child's progression through the first three van Hiele levels.

Solution $m(\angle 1) = 120°$, $m(\angle 2) = -155°$, and $m(\angle 3) = 50°$. The sum of all five directed angles is $45° - 60° + 120° - 155° + 50° = 0°$. This is not surprising, since Patty's path started and stopped in the same direction and her path didn't make any loops.

Problem Set 11.1

Understanding Concepts

1. Use symbols to name each of the figures shown. If more than one symbol is possible, give all possible names.

(a)

(b)

(c)

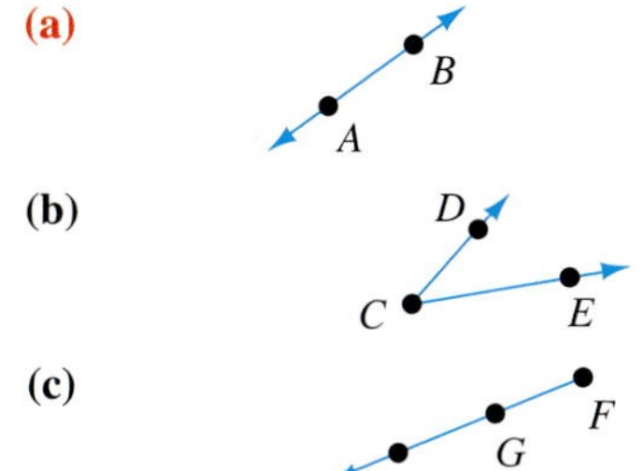

(d)

2. The points E, U, C, L, I, and D are shown below.

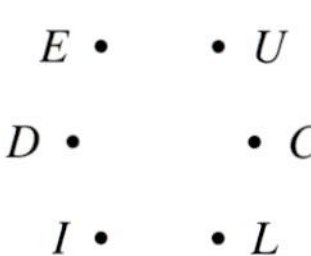

Draw the following figures:

(a) $\overleftrightarrow{EU}$ (b) $\overrightarrow{CL}$ (c) $\overline{ID}$

3. Trace the 5-by-5 square lattice shown, and draw the line segment $\overline{AB}$.

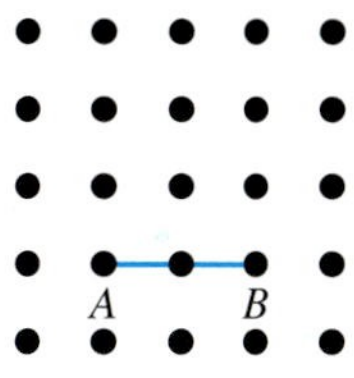

Use colored pencils to circle all of the points C of the lattice that make $\angle BAC$

(a) a right angle, (b) an acute angle,
(c) an obtuse angle, (d) a straight angle, and
(e) a zero angle.

4. (a) How many nonzero angles are shown in the following figure? Give the three-letter symbol, such as $\angle APB$, for each angle.

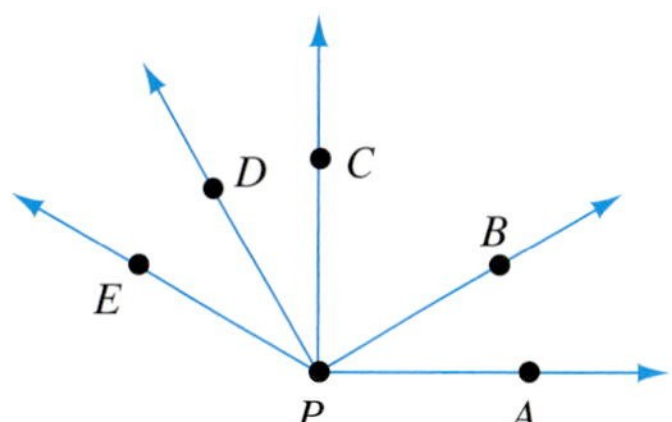

(b) Measure each angle you identified in part (a), and classify it as acute, right, or obtuse.

5. In the figure shown, $\angle BXD$ is a right angle and $\angle AXE$ is a straight angle.

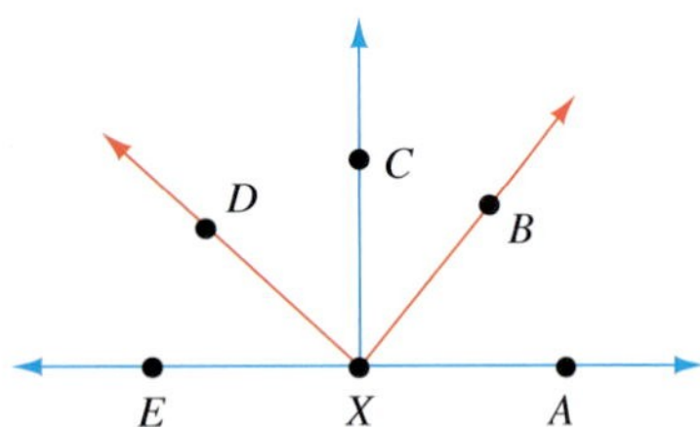

If $m(\angle BXC) = 35°$ and $m(\angle BXE) = 132°$, explain how you can determine the measures of $\angle AXB$, $\angle CXD$, and $\angle DXE$ without using a protractor.

6. The point P shown below is the intersection of the two ***external* tangent** lines to a pair of circles. (See Example 11.1.) The point Q is the intersection of the two ***internal* tangent** lines to two circles.

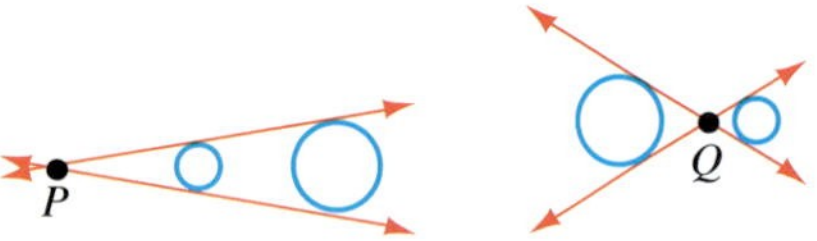

Draw three circles C_1, C_2, and C_3, all of different radii and with no circle containing or intersecting either of the other circles. Let P be the intersection of the external tangent lines of C_1 and C_2, and let Q and R be the respective intersections of the internal tangent lines of the pairs of circles C_2, C_3 and C_1, C_3. What conclusion is suggested by your drawing?

7. Two intersecting circles determine a line, as shown below on the left. Draw three circles, where each circle intersects the other two circles as in the example shown on the right. Next draw the three lines determined by each pair of intersecting circles. What conclusion is suggested by your drawing?

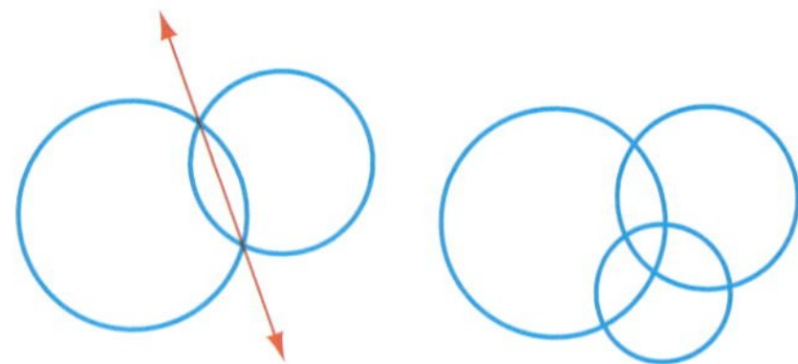

8. Use a compass (or carefully trace around a cup bottom or other circular object) to draw three circles of the same size through point A. Let B, K, and L be the other points of intersection of pairs of circles. Now draw a fourth circle of the same size as the other three that passes through B and creates points of intersection M and N as shown.

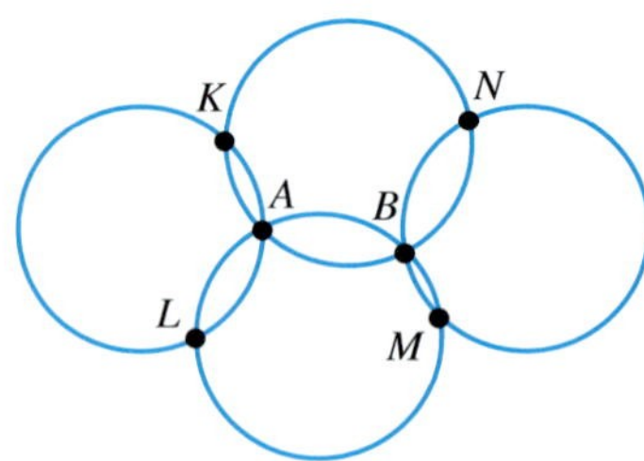

(a) Draw the lines $\overleftrightarrow{KL}$ and $\overleftrightarrow{MN}$. What can you say about these two lines?

(b) Draw the lines $\overleftrightarrow{LM}$ and $\overleftrightarrow{KN}$. What can you say about these two lines?

(c) Draw the line $\overleftrightarrow{AB}$. What connection does it seem to have to any of the lines drawn earlier?

9. Draw a circle with center at point A and a second circle of the same radius with center at point B, where the radius is large enough to cause the circles to intersect at two points C and D. Let M be the point of intersection of $\overline{CD}$ and $\overline{AB}$.

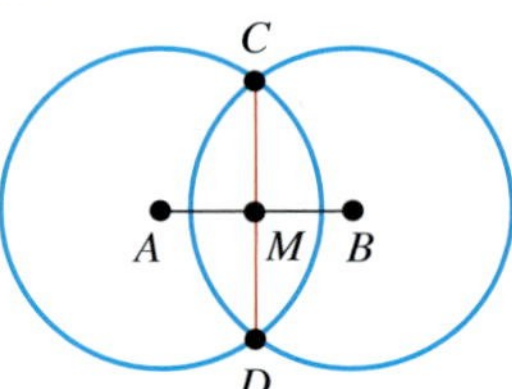

(a) Use a protractor to measure the angles at M. What can you say about how $\overline{CD}$ and $\overline{AB}$ intersect?

(b) Use a ruler to measure $\overline{MA}$ and $\overline{MB}$. What can you say about point M?

10. Draw two circles, and locate four points on each circle. Draw the segments as shown.

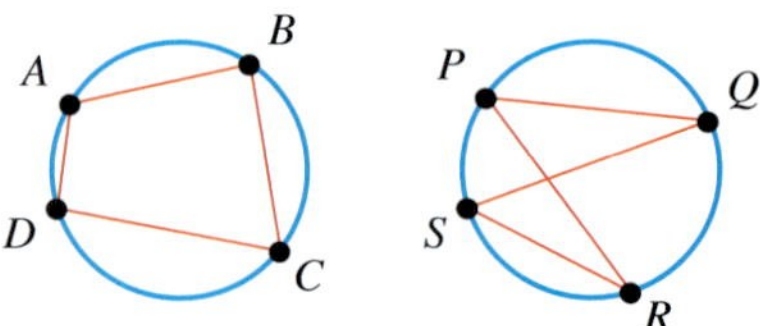

(a) Carefully measure the angles with vertices at A, B, C, and D with a protractor. What relationships do you see on the basis of your measurements?

(b) Measure the angles with vertices at P, Q, R, and S, and discuss what relationships hold among the angles in that figure.

11. Draw a circle, labeling its center as O. Draw an angle whose vertex is at O and whose sides intersect the circle at points A and B.

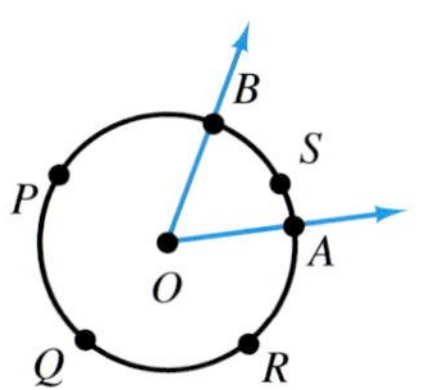

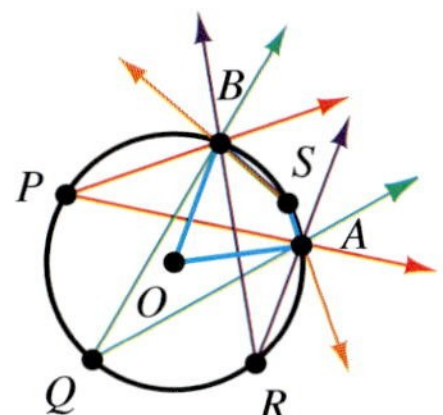

Let P, Q, and R be points on the circle that are in the exterior of $\angle AOB$. Let S be a point on the circle that is in the interior of $\angle AOB$.

(a) Use a protractor to find $m(\angle AOB)$, $m(\angle APB)$, $m(\angle AQB)$, and $m(\angle ARB)$. Describe how the angles compare with one another.

(b) Use a protractor to measure $\angle ASB$, where S is on the smaller arc between A and B. How is $\angle ASB$ related to the angles you considered in part (a)?

12. The hour and minute hands of a clock form a zero angle at noon and midnight. Between noon and midnight, how many times do the hands again form a zero angle?

13. How many degrees does the minute hand of a clock turn through

(a) in sixty minutes? **(b)** in five minutes?

(c) in one minute?

How many degrees does the hour hand of a clock turn through

(d) in sixty minutes?

(e) in five minutes?

14. Find the angle formed by the minute and hour hands of a clock at these times.

(a) three o'clock **(b)** six o'clock

(c) 4:30 **(d)** 10:20

15. The lines l and m are parallel. Find the measures of the numbered angles shown.

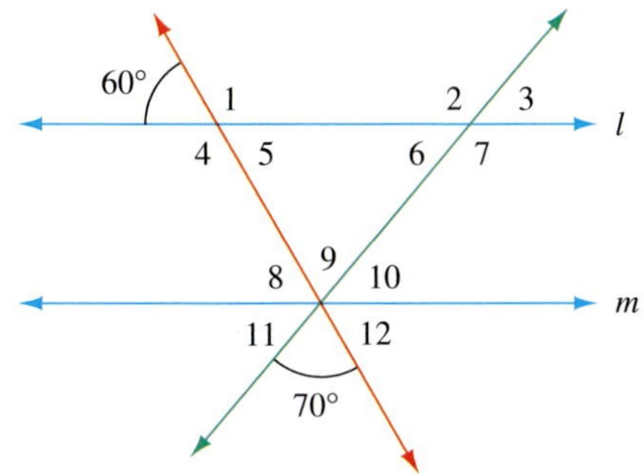

16. Determine the measure of $\angle P$ if $\overrightarrow{AB}$ and $\overrightarrow{CD}$ are parallel.

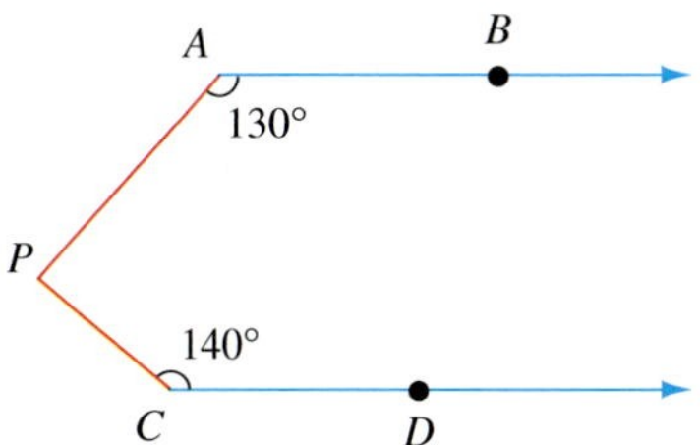

17. Find the measures of the numbered angles in the triangles shown.

(a)

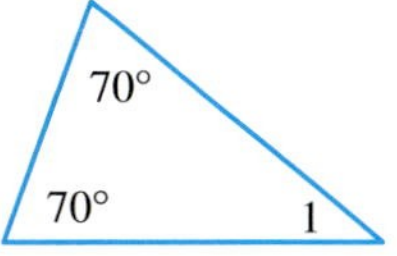

(b)

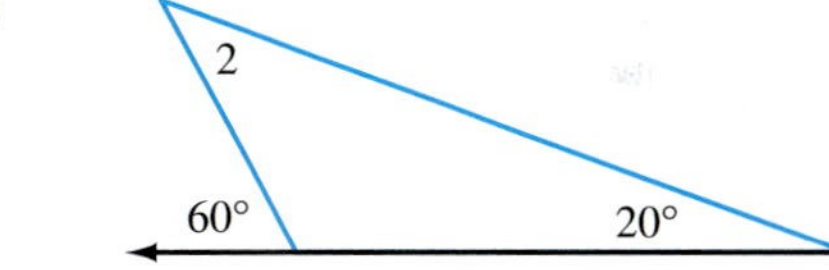

(c)

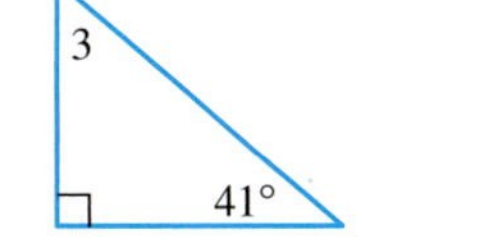

(d)

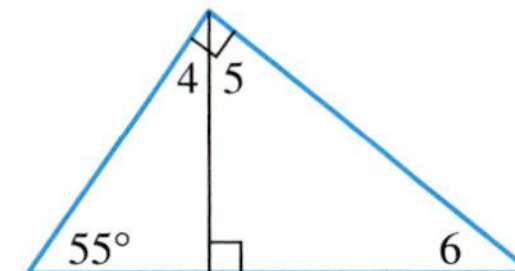

18. Find the measures of the interior angles of the following triangles.

(a)

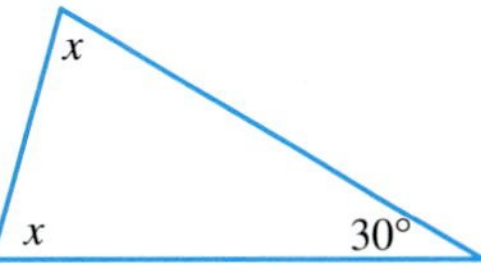

(b)

(c)

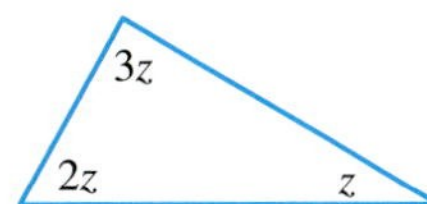

19. **(a)** Can a triangle have two obtuse angles? Why?

(b) Can a triangle have two right angles? Why?

(c) Suppose no angle of a triangle measures more than 60°. What do you know about the triangle?

20. A hiker started heading due north, then turned to the right 38°, then turned to the left 57°, and next turned to the right 9°. To resume heading due north, what turn must the hiker make?

Teaching Concepts

21. Youngsters (and college students, too!) often learn geometric concepts best when working with common three-dimensional objects, say a cardboard shoebox. The corners of a box model points, and each edge of the box represents a line segment. Additional lines can be drawn on the box with a ruler. Create a lesson that investigates several of the concepts of this section, using a shoebox as a manipulative. Do you see some congruent segments? Some right angles? and so on.

22. **Treasure Hunt.** Create a treasure hunt game in which groups of children are given starting points and directions in the classroom. Each group uses protractors and rulers to follow directions that, when accurately carried out, will lead them to "treasures" hidden throughout the room. For example, the directions might start, "Begin at point A facing the front of the room. Turn 40° and go 8 feet to arrive at point B. Turn −120° and go 20 feet to. . . ." In particular, show a diagram of a classroom, a starting point, and a route leading to the location T of the hidden treasure. Give the corresponding directions for the route.

Thinking Critically

23. Five lines are drawn in the plane.

(a) What is the smallest number of points of intersection of the five lines?

(b) What is the largest number of points of intersection?

(c) If m is an integer between the largest and smallest number of intersection points, can you arrange the lines to have m points of intersection?

24. Three noncollinear points determine three lines, as was shown in Figure 11.1.

(a) How many lines are determined by four points, no three of which are collinear?

(b) How many lines are determined by five points, no three of which are collinear?

(c) How many lines are determined by n points? Assume that no three points are collinear.

25. The arrangement of five points shown has two lines that each pass through three of the points.

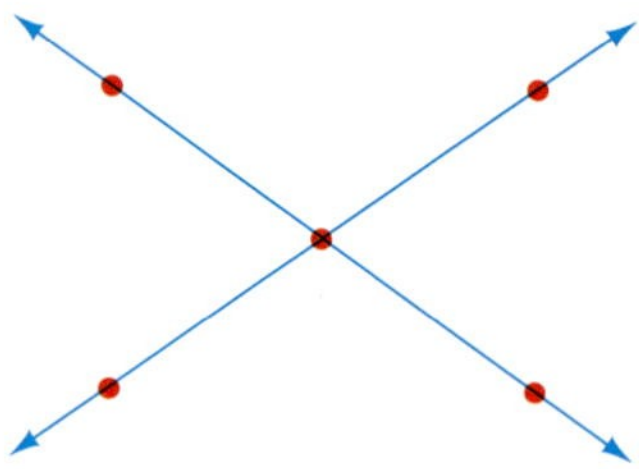

(a) Find an arrangement of six points that has four lines that each pass through three of your six points.

(b) Find an arrangement of seven points that has six lines that each pass through three of your seven points.

26. Suppose that a large triangle is drawn and a pencil is placed along an edge. What property of triangles is illustrated by the sequence A through H of slides and turns shown below? Explain in a carefully worded paragraph.

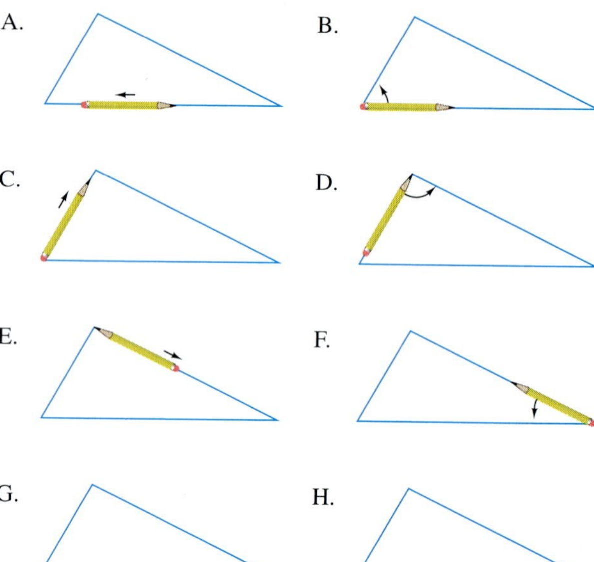

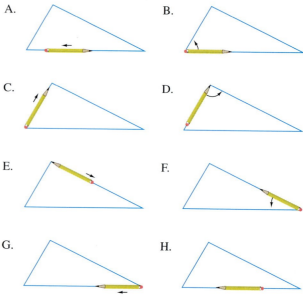

27. Use a ruler to draw a large four-sided polygon (a *quadrilateral*) such as the one shown here. Starting with a pencil laid along one side of the quadrilateral, slide the pencil to a corner, rotate it to the next side, slide the pencil to the next corner, and so on. (See problem 26 above.)

(a) What direction will the pencil point to when it returns to the initial side? What does this say about the sum of the measures of the angles at the four corners of a quadrilateral?

(b) Draw a diagonal across the quadrilateral to form two triangles. What does the sum of angle measures in the triangles tell you about the sum of angle measures in the quadrilateral?

(c) Cut the quadrilateral out with scissors, and rip off the four corners. What angle can be covered with the four pieces?

28. In "taxicab geometry," the points are the corners of a square grid of "city blocks" in a plane. In the figure shown, the shortest trip from A to B must cover five blocks, and so the **taxi distance** from A to B is five. A "taxi segment" is the set of points on a path of shortest taxi distance from one point to another, and so $\{A, W, X, Y, Z, B\}$ is a taxi segment from A to B.

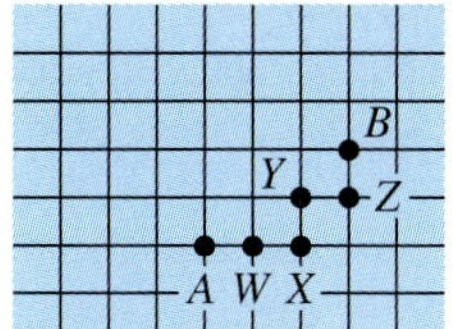

(a) How many taxi segments join A and B?

(b) Find all points that are at a taxi distance of five from A. Does your "taxi circle" look like a circle drawn with a compass?

(c) Use pencils of different colors to draw the concentric taxi circles of taxi radius 1, 2, 3, 4, 5, and 6. Describe the pattern you see.

29. The incident (incoming) and reflected (outgoing) rays of a light beam make congruent angles with a flat mirror.

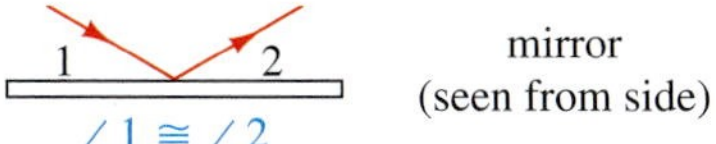

Suppose two mirrors are perpendicular to one another. Show that after the second reflection, the outgoing ray is parallel to the incoming ray. (*Hint:* Show that $\angle 5$ and $\angle 6$ are supplementary. Why does this imply that the rays are parallel?)

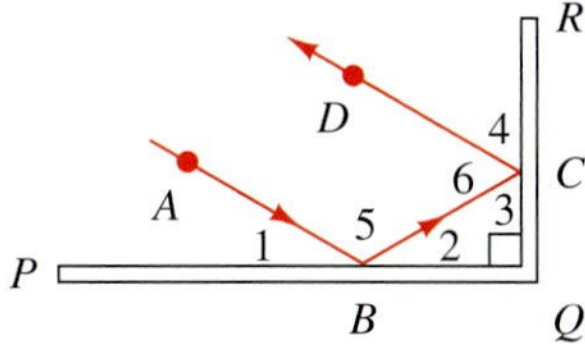

Making Connections

30. Spirals, in two directions, create the seed pattern in the head of the sunflower shown in the picture gallery of plane figures in this section. Carefully count the number of spirals in each direction. Have you encountered these two numbers before? They belong to a famous sequence of numbers.

31. Why do the hands on a clock turn in the direction we call "clockwise"? It will help to think of the type of clock first used and where it originated.

32. An explorer made the following trip from base camp.

First leg: north 2 miles

Second leg: southeast 5 miles

Third leg: west 6 miles

Fourth leg: south 1 mile

(a) Make a scale drawing of the trip.

(b) Show the angle the explorer turned through to go from one leg of the journey to the next.

(c) Estimate the compass heading and approximate distance the explorer needs to follow to return most directly to base camp.

33. A plumb bob (i.e., a small weight on a string) suspended from the center of a protractor can be used to measure the angle of elevation of a treetop. If the string crosses the protractor's scale at the angle marked P, what is the measure of the angle of elevation?

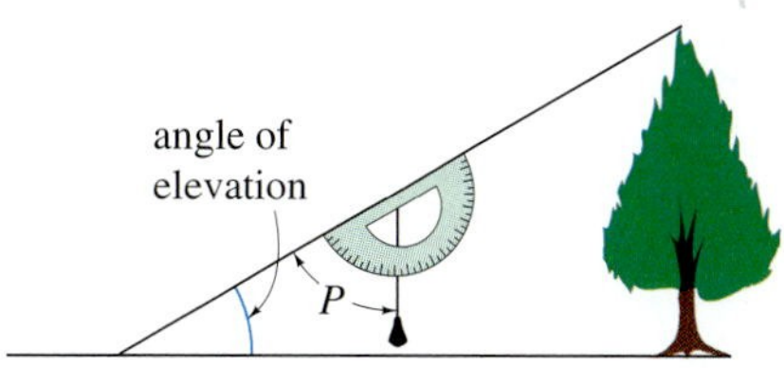

34. (a) How many degrees does the earth turn in one hour?

(b) How many degrees does the earth turn in one minute?

(c) On a clear night with a full moon, it can be observed that the earth's rotation makes it appear that the moon moves a distance of its own diameter in two minutes of time. What angle does a diameter of the moon make as seen from the earth?

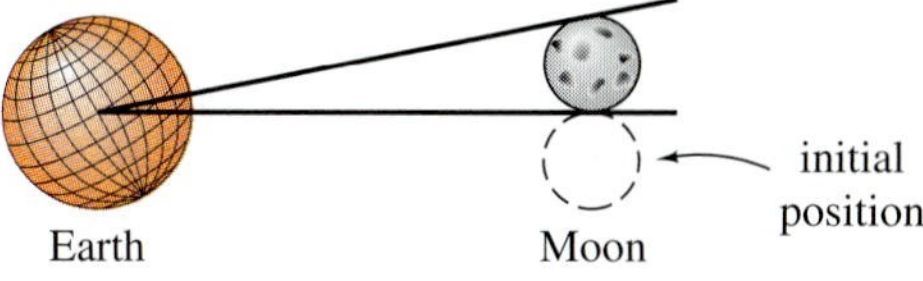

35. Suppose Polaris (the "pole star") is at an angle of elevation of 37° above the horizon. Explain how this information can be used to estimate your latitude, which is $m(\angle EOP)$ on the accompanying diagram. N is the north pole, O is the earth's center, E is the point on the equator directly south of your position P, H is a point on the horizon to the north, and $\overrightarrow{NS_1}$ and $\overrightarrow{PS_2}$ are parallel rays to the distant star Polaris.

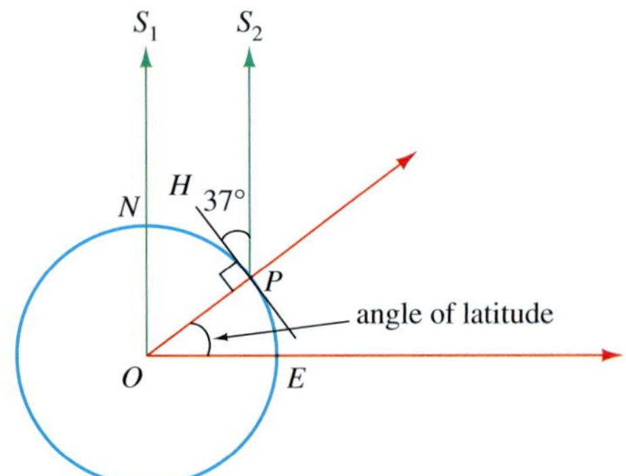

36. At the winter solstice in the northern hemisphere (on or near December 21 each year), the earth's equatorial plane is tilted 23.5° away from its plane of revolution (the *ecliptic plane*) about the sun. At noon in the northern hemisphere, the angle from the horizon to the sun (the *solar altitude*) can be used to estimate the latitude of the point P of observation.

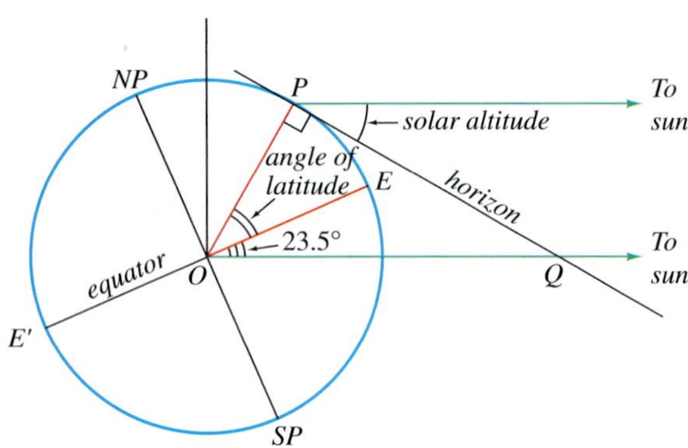

(a) Suppose the solar altitude is 42° at noon on December 21. What is the latitude?

(b) Juneau, Alaska, is approximately at latitude 58°. What is the solar altitude at noon on the winter solstice?

(c) What is the latitude of the Arctic Circle, above which the sun is below the horizon on the winter solstice?

Using a Calculator

37. Machinists, engineers, astronomers, and others often require angle measurements that are accurate to a fraction of a degree. Sometimes decimal fractions are used, such as 38.24°. It is also common to follow the historic idea of first subdividing a degree into 60 equal parts called **minutes** (from the medieval Latin *pars minuta prima,* meaning first minute [mī-nōōt′] part) and next subdividing a minute into 60 equal parts called **seconds** (*partes minutae secundae,* meaning second minute part). For example 24°13′46″ is read "24 degrees, 13 minutes, and 46 seconds." The following computation shows how to convert to decimal degrees, using the facts that $1' = \frac{1°}{60}$ and $1'' = \frac{1°}{3600}$:

$$24°13'46'' = 24° + \left(\frac{13}{60}\right)^{\circ} + \left(\frac{46}{3600}\right)^{\circ}$$
$$\doteq (24 + 0.217 + 0.013)° = 24.230°.$$

Here is how to convert to degrees–minutes–seconds from a decimal measure:

$$38.24° = 38° + (0.24)(60)' = 38° + 14.4'$$
$$= 38° + 14' + (0.4)(60)''$$
$$= 38°\,14'\,24''.$$

Use your calculator to convert the following angle measures from decimal to degrees–minutes–seconds, or the reverse.

(a) 58° 36′ 45″ **(b)** 141° 50′ 03″

(c) 71.32° **(d)** 0.913°

Using a Computer

38. Use geometry software to draw any triangle ABC.

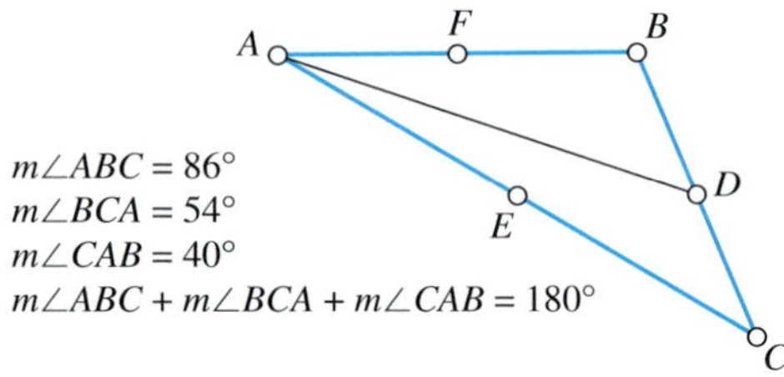

(a) Measure the three angles of the triangle. (Use the Measure Menu on *Geometer's Sketchpad.*) Sum the three angle measures, using the Measure Menu/Calculate . . . command. Click and drag a vertex or side of the triangle to manipulate the triangle to assume new shapes. What happens to the sum of the angles?

(b) Select the three sides of the triangle. (Click on the sides.) Then use Construct Menu/Point at Midpoint command to construct the midpoints of the three sides. The figure above shows the segment $\overline{AD}$, called a **median** of the triangle. Construct two medians, and construct the point G at which your two medians intersect. Use the Measure Menu/Distance command to measure the distances GA and GD. Then use the Measure/Calculate . . . command to compute the ratio GA/GD. Similarly, measure GB and GE and compute the ratio GB/GE. What conjecture can you make? Manipulate the triangle to lend support to your conjecture.

(c) Construct the third median of the triangle, and propose a theorem about the medians of any triangle. Manipulate the triangle to lend support to your hypothesis.

39. Draw a circle. From a point P outside of your circle, draw three lines that each intersect the circle at two points. Let A, B, C, A', B', and C' be the intersection points as shown. The segments $\overline{AB'}$ and $\overline{A'B}$ intersect to determine a point Q. Similarly, let $\overline{BC'}$ and $\overline{B'C}$, and $\overline{AC'}$ and $\overline{A'C}$, determine the respective points R and S.

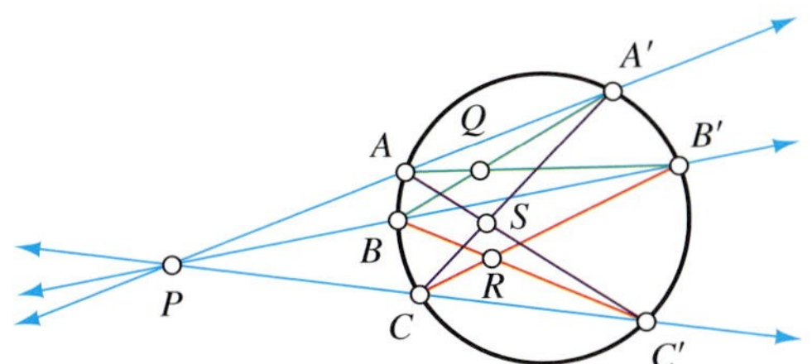

(a) What conjecture can you make concerning Q, R, and S? Drag P and the lines to investigate your conjecture.

(b) Discuss how the line $\overleftrightarrow{QR}$ can be used to construct the rays from P that are tangent to the circle.

40. Draw two squares that share a common vertex at A. Label the vertices of the squares $ABCD$ and $AB'C'D'$ in counterclockwise order.

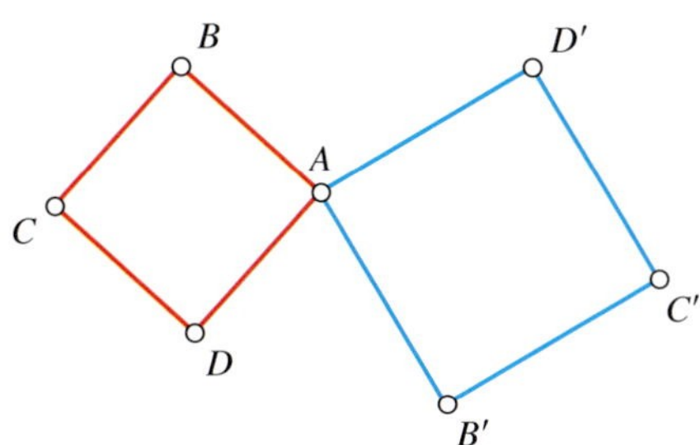

(a) Draw the lines $\overleftrightarrow{BB'}$ and $\overleftrightarrow{DD'}$, and let P denote their intersection point. Conjecture how the lines cross.

(b) Draw the line $\overleftrightarrow{CC'}$. Discuss how this line crosses the two lines drawn in part (a).

(c) Draw the line $\overleftrightarrow{AP'}$. At what angles does it intersect the lines drawn before?

Communicating

41. The word *acute* has a nonmathematical meaning, as in "acute appendicitis." Similarly, *obtuse* might be used to say that someone's argument is stupid or unclear. Look these words up in a dictionary and comment on why their mathematical and nonmathematical meanings are related.

From State Student Assessments

42. (Washington State, Grade 4)
Raul is going to a friend's house. Raul remembers that his friend's house is on a street parallel to Southport. On which street does Raul's friend live?

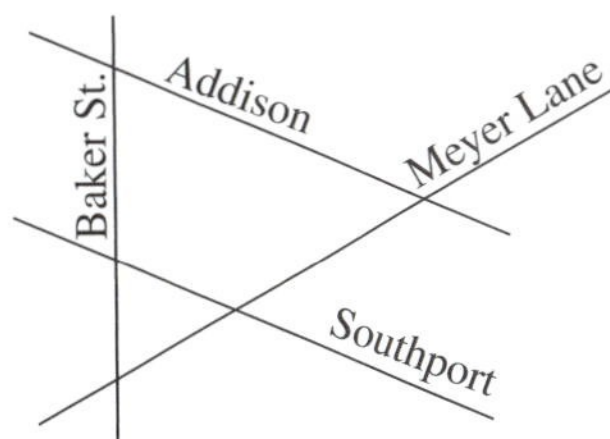

A. Addison B. Baker Street
C. Meyer Lane

43. (Massachusetts, Grade 4)
To answer parts (a) through (d), connect dots on the dot patterns shown below. Always connect the dots to draw closed shapes with STRAIGHT SIDES.

(a) On the dot pattern labeled Part a, draw a shape that has EXACTLY ONE right angle. DRAW A RING around the RIGHT ANGLE.

(b) On the dot pattern labeled Part b, draw a shape that has NO right angles.

(c) On the dot pattern labeled Part c, draw a shape that has at least ONE acute angle. DRAW A RING around the ACUTE ANGLE.

(d) On the dot pattern labeled Part d, draw a shape that has EXACTLY TWO right angles. DRAW A RING around each RIGHT ANGLE.

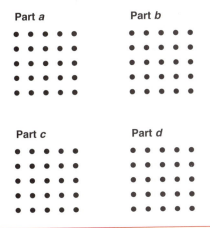

Part *a* **Part *b***

Part *c* **Part *d***

11.2 CURVES AND POLYGONS IN THE PLANE

Curves and Regions

A **curve** in the plane can be described informally as a set of points that a pencil can trace without lifting until all points in the set are covered. A more precise definition is required for advanced mathematics, but this intuitive idea of curve meets our present needs. If the pencil never touches a point more than once, then the curve is **simple.** If the pencil is lifted at the same point at which it started tracing the curve, then the curve is **closed.** If the common initial and final point of a closed curve is the only point touched more than once in tracing the curve, then the curve is a **simple closed curve.** We require that a curve have both an initial and a final point, and so lines, rays, and angles are *not* curves for us.

Several examples of curves are shown in Figure 11.13.

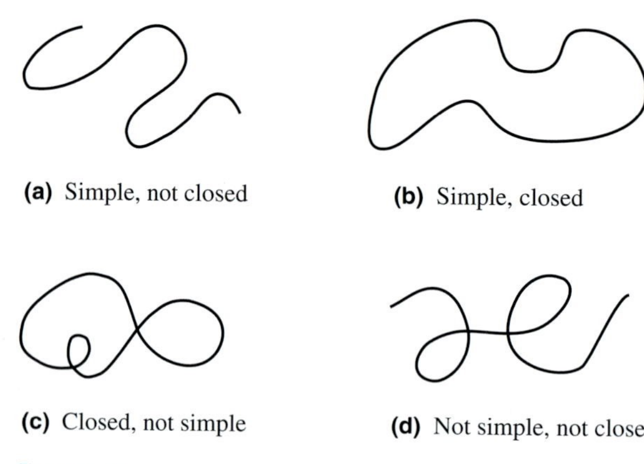

Figure 11.13
The classification of curves

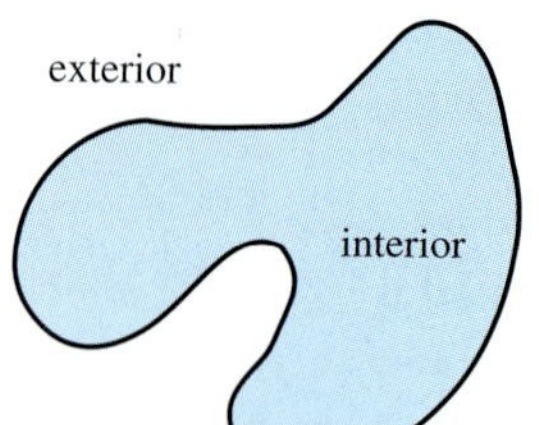

Figure 11.14
A simple closed curve and its interior and exterior

Any simple closed curve partitions the points of the plane into three disjoint pieces: the curve itself, the interior, and the exterior, as shown in Figure 11.14. This property of a simple closed curve may seem very obvious, but in fact it is an important theorem of mathematics.

THEOREM *Jordan Curve Theorem*
A simple closed plane curve partitions the plane into three disjoint subsets: the curve itself, the interior of the curve, and the exterior of the curve.

The French mathematician Camille Jordan (1838–1922) was the first to recognize that such an "obvious" result needed proof. To see why the theorem is difficult to prove (even Jordan's own proof was incorrect!), try to determine whether the points G and H are inside or outside the very crinkly, but still simple, closed curve shown in Figure 11.15.

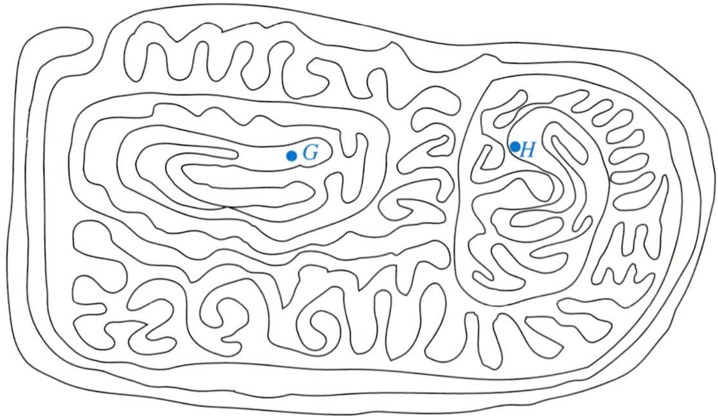

Figure 11.15
Is G in the interior or exterior of this simple closed curve? What about H?

EXAMPLE 11.6 DETERMINING THE INTERIOR POINTS OF A SIMPLE CLOSED CURVE IN THE PLANE

Devise a method to determine whether a given point is in the interior or the exterior of a given simple closed curve.

Solution

Think of the curve as a fence. If we jump over the fence, we either go from the interior to the exterior of the curve or vice versa. Now draw any ray from the given point.

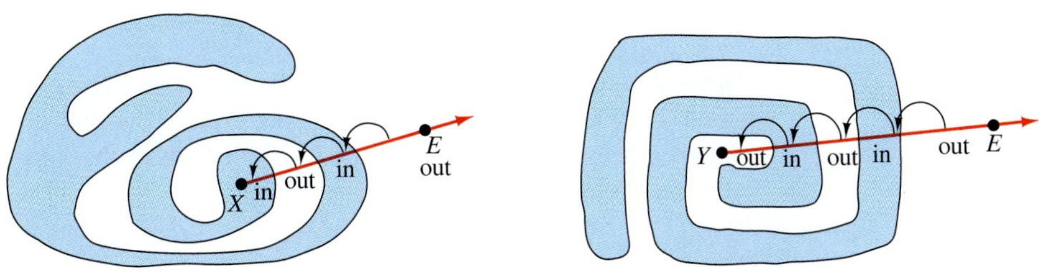

Start at an exterior point E on the ray and see how many times the fence is crossed as you move to the endpoint of the ray. If the curve is crossed an odd number of times to reach the point X from the exterior point E, then the given point X is an interior point. An even number of crossings from E to reach a point Y means that Y is an exterior point. It should now be simple to verify that G is exterior and H is interior to the curve in Figure 11.15.

The interior and exterior of a simple closed curve are also called **regions.** More generally, the complement of some system of lines, rays, and curves will be composed of one or more regions. For example, a line partitions the plane into two regions called half planes. An angle, if not zero or straight, partitions the plane into two regions called the interior and the exterior of the angle.

EXAMPLE 11.7 COUNTING REGIONS IN THE PLANE

Count the number of regions into which the plane is partitioned by the following shapes.

(a) a figure 8 **(b)** a segment **(c)** two nonintersecting circles
(d) a square and its two diagonals **(e)** a pentagram
(f) any simple nonclosed curve

Solution

(a) 3 **(b)** 1 **(c)** 3 **(d)** 5 **(e)** 7 **(f)** 1

Convex Curves and Figures

The interior of an angle has the property that the segment between any two interior points does not leave the interior. This means that the interior of an angle is a convex figure according to the following definition.

> **DEFINITION** *Convex and Concave Figures*
> A figure is **convex** if, and only if, it contains the segment $\overline{PQ}$ for each pair of points P and Q contained in the figure. A figure that is not convex is called **concave.**

Several convex and concave shapes are shown in Figure 11.16. To show that a figure is concave, it is enough to find two points P and Q within the figure whose corresponding line segment $\overline{PQ}$ contains at least one point not in the figure. A concave figure is sometimes called a **nonconvex** figure.

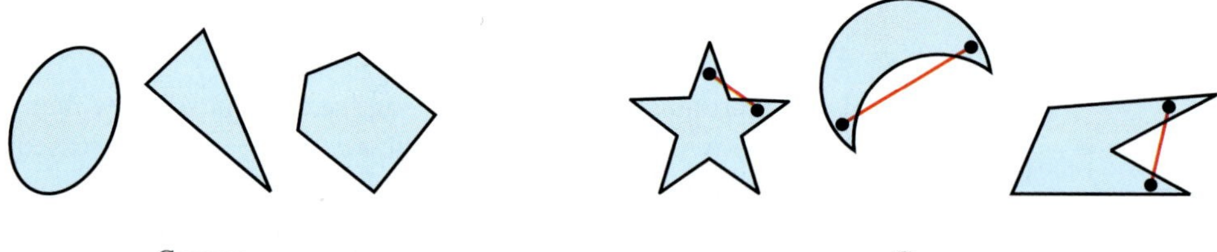

Figure 11.16
Convex and concave plane figures

Polygonal Curves and Polygons

A curve that consists of a union of finitely many line segments is called a **polygonal curve.** The endpoints of the segments are called **vertices,** and the segments are the **sides** of the polygonal curve. A **polygon** is a simple closed polygonal curve. The interior of a polygon is called a **polygonal region.** A **convex polygon** is a polygon whose interior is convex. Figure 11.17 illustrates examples of polygonal curves.

Just for Fun

Triangle Pick-Up-Sticks

The 16 toothpicks shown below form eight triangular regions, where each toothpick borders either one or two of the regions. Remove a certain number of toothpicks and attempt to leave behind two triangular regions in each case. There can be no toothpicks remaining that do not border at least one of the two triangles. Which case gives you the most trouble?

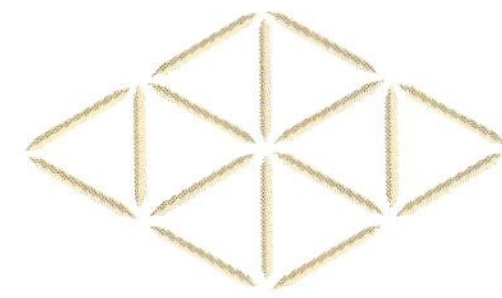

a. Remove 6 toothpicks
b. Remove 7 toothpicks
c. Remove 8 toothpicks
d. Remove 9 toothpicks
e. Remove 10 toothpicks
f. Remove 11 toothpicks

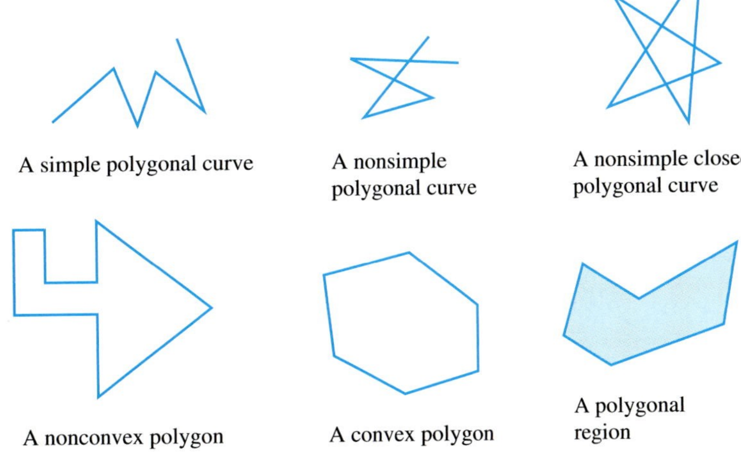

Figure 11.17
The classification of polygonal curves

Polygons are named according to the number of sides or vertices they have. For example, a polygon of 17 sides is sometimes called a *heptadecagon* (*hepta* = seven, *deca* = 10). With more directness it can also be called a 17-gon. The common names of polygons are shown in Table 11.1.

TABLE 11.1 Names of Polygons

Polygon	Number of Sides	Example
Triangle	3	
Quadrilateral	4	
Pentagon	5	
Hexagon	6	
Heptagon	7	

continued

TABLE 11.1 Names of Polygons (continued)

Polygon	Number of Sides	Example
Octagon	8	
Nonagon (or enneagon)	9	
Decagon	10	
n-gon	*n*	

The rays along two sides with a common vertex determine an **angle of the polygon.** For a convex polygon the interiors of these angles include the interior of the polygon. The angles are also called **interior angles,** as shown in Figure 11.18. An angle formed by replacing one of these rays with its opposite ray is an **exterior** angle of the polygon. The two exterior angles at a vertex are congruent by the vertical-angles theorem. The interior angle and either of its adjacent exterior angles are supplementary.

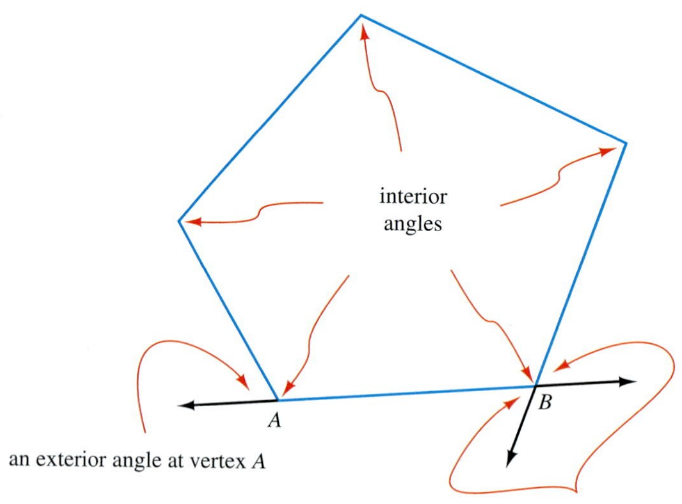

Figure 11.18
Angles in a convex polygon

In the following theorem we consider the interior angle and *one* of its supplementary exterior angles at each vertex of a convex polygon.

THEOREM *Sums of the Angle Measures in a Convex Polygon*

(a) The sum of the measures of the exterior angles of a convex polygon is 360°.

(b) The sum of the measures of the interior angles of a convex n-gon is $(n - 2)180°$.

Proof:

(a) Imagine a walk completely around a polygon. At each vertex, we must turn through an exterior angle. At the conclusion of the walk we are heading in the same direction as we began, so our total turn is through 360°. If $\angle 1', \angle 2', \ldots, \angle n'$ denote the exterior angles, this shows that $m(\angle 1') + m(\angle 2') + \cdots + m(\angle n') = 360°$.

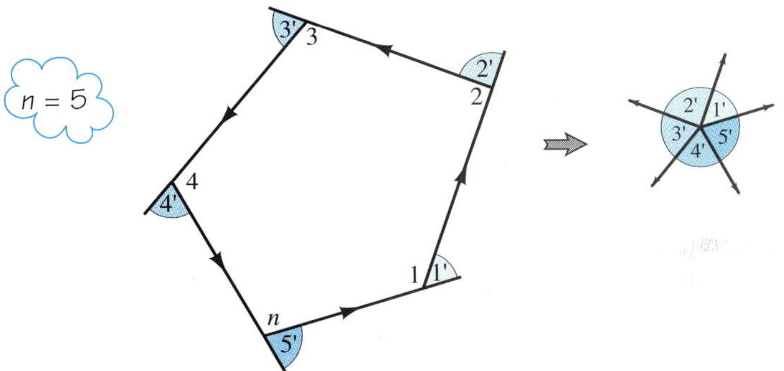

(b) Since an interior and exterior angle at a vertex are supplementary, we have the equations $m(\angle 1) = 180° - m(\angle 1')$, $m(\angle 2) = 180° - m(\angle 2'), \ldots,$ $m(\angle n) = 180° - m(\angle n')$. Adding these n equations gives us, by part (a),

$$\begin{aligned} m(\angle 1) + m(\angle 2) + \cdots + m(\angle n) & \\ &= n \cdot 180° - [m(\angle 1') + m(\angle 2') + \cdots + m(\angle n')] \\ &= n \cdot 180° - 360° = (n - 2) \cdot 180°, \end{aligned}$$

where we need the result of part (a) in the second equality.

EXAMPLE 11.8 | FINDING THE ANGLES IN A PENTAGONAL ARCH

Find the measures $3x$, $8x$, y, and z of the interior and exterior angles of the pentagon *PENTA*.

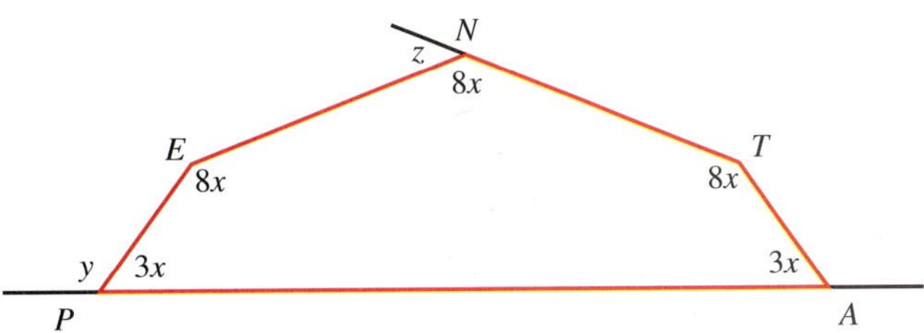

Solution By the theorem just proved, we know that the sum of the measures of the five interior angles is

$$3x + 8x + 8x + 8x + 3x = (5 - 2)180°.$$

That is, $30x = 3 \cdot 180°$, and so $x = (3 \cdot 180°/30) = 18°$. Thus, the interior angles at P and A measure $3 \cdot 18° = 54°$ and at E, N, and T measure $8 \cdot 18° = 144°$. The measures of the exterior angles are $y = 180° - 54° = 126°$ and $z = 180° - 144° = 36°$.

In a nonconvex polygon, some of the interior angles are reflex angles, with measures larger than 180°. Nevertheless, it can be proven (see problems 26 and 27 in Problem Set 11.2) that the sum of the measures of the n interior angles is still given by $(n - 2)180°$.

THEOREM *Sum of Interior Angle Measures of a General Polygon*
The sum of the measures of the interior angles of any n-gon is $(n - 2)180°$.

An example of a nonconvex heptagon is shown in Figure 11.19.

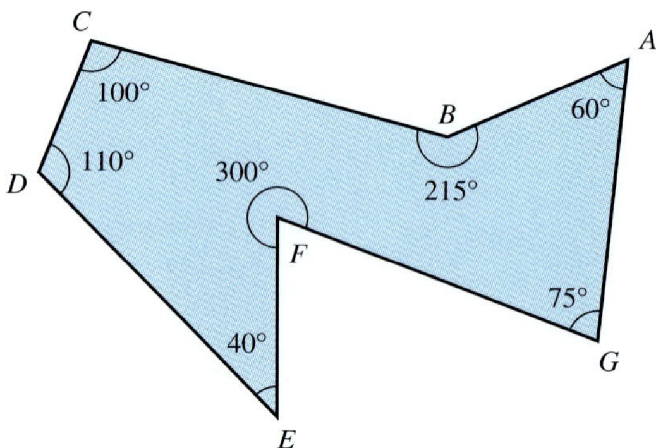

Figure 11.19
The sum of the interior angle measures of the 7-gon is 60° + 215° + 100° + 110° + 40° + 300° + 75° = 900° = (7 − 2) · 180°

EXAMPLE 11.9 MEASURING THE ANGLES IN A FIVE-POINTED STAR

The reflex angles at each of the five "inward" points of the star shown have three times the measure of the angles of the "outward" points. What is the measure of the angle at each point of the star?

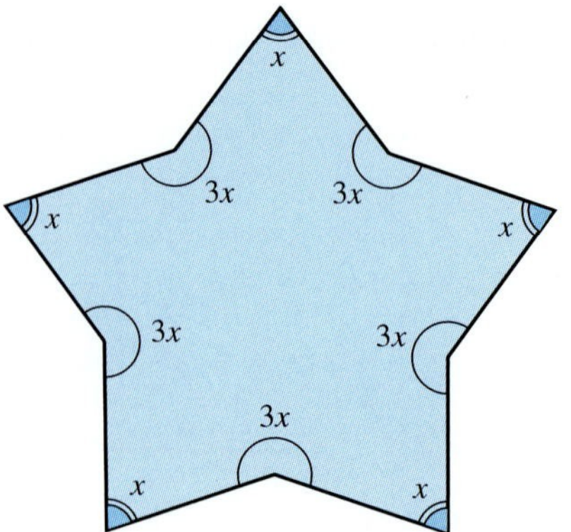

Solution The sum of the measures of all 10 interior angles of the star is $5x + 5 \cdot (3x)$, or $20x$. Since the star is a decagon (or 10-gon), the sum must equal $(10 - 2) \cdot 180° = 1440°$. This gives us the equation $20x = 1440°$. Solving for x shows that $x = 1440° \div 20 = 72°$. Thus, each acute interior angle measures 72°, and each of the reflex angles at the inward points measures $3x = 216°$.

A walk around any closed curve, simple or nonsimple, that returns to the starting point and to the same orientation as the walk began must have turned through some integer multiple of 360°. It is customary to measure turns to the left (counterclockwise) as positive and turns to the right (clockwise) as negative.

> **THEOREM** *The Total-Turn Theorem*
> The total turn around any closed curve is an integral multiple of 360°.

To help determine the total turn, draw a point S at an arbitrary point along the curve and lay a pencil at that point, with the point of the pencil oriented in the direction of travel. Now trace the curve with the pencil, and count the net number of rotations the pencil has made when it returns to its initial position at point S. For a polygonal curve, the pencil turns only when it reaches a vertex of the curve.

EXAMPLE 11.10 FINDING TOTAL TURNS

Find the total turn for each of the following closed curves.

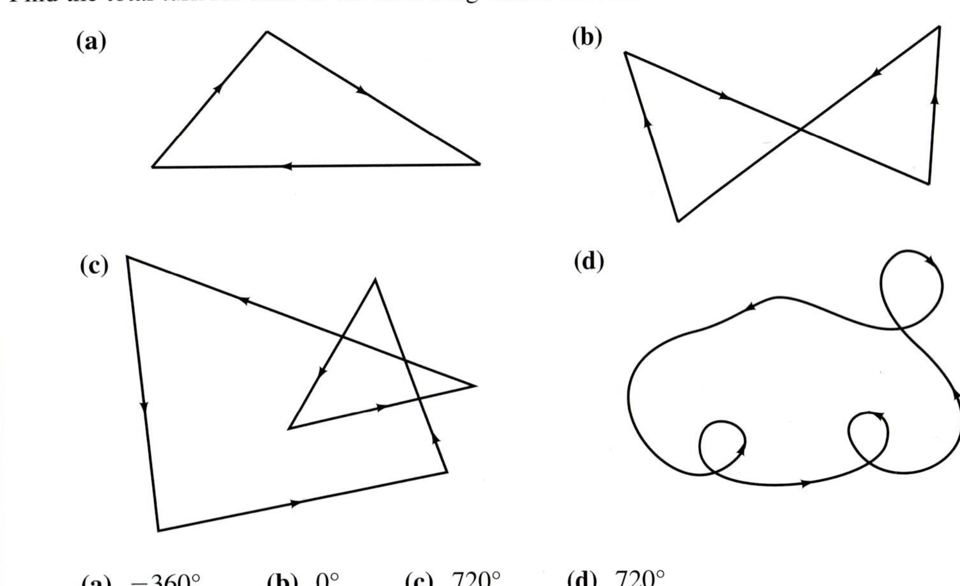

Solution **(a)** −360° **(b)** 0° **(c)** 720° **(d)** 720°

Triangles

Triangles are classified by the measures of their angles or sides, as shown in Table 11.2.

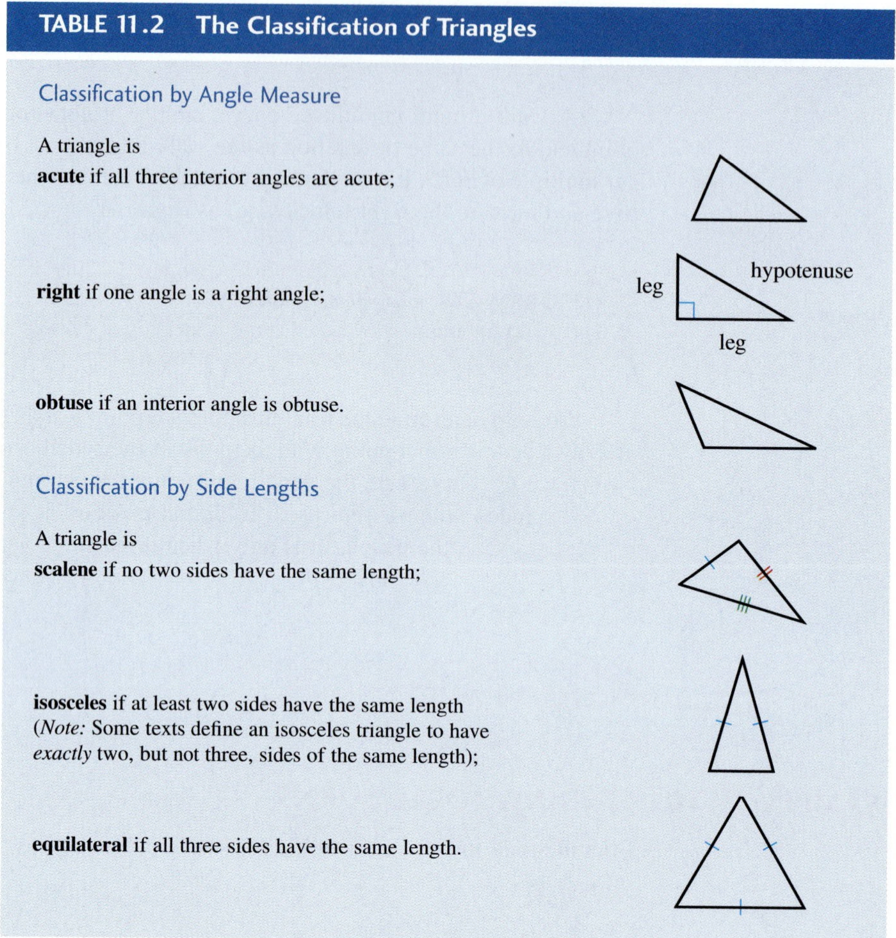

TABLE 11.2 The Classification of Triangles

Classification by Angle Measure

A triangle is

acute if all three interior angles are acute;

right if one angle is a right angle;

obtuse if an interior angle is obtuse.

Classification by Side Lengths

A triangle is

scalene if no two sides have the same length;

isosceles if at least two sides have the same length (*Note:* Some texts define an isosceles triangle to have *exactly* two, but not three, sides of the same length);

equilateral if all three sides have the same length.

EXAMPLE 11.11 CLASSIFYING TRIANGLES

There are a number of triangles in the figure shown with vertices at A, B, C, D, E, and F. Classify the triangles according to Table 11.2. Use the corner of an index card to check for right angles, and use a ruler or mark on the edge of an index card to check for congruent sides.

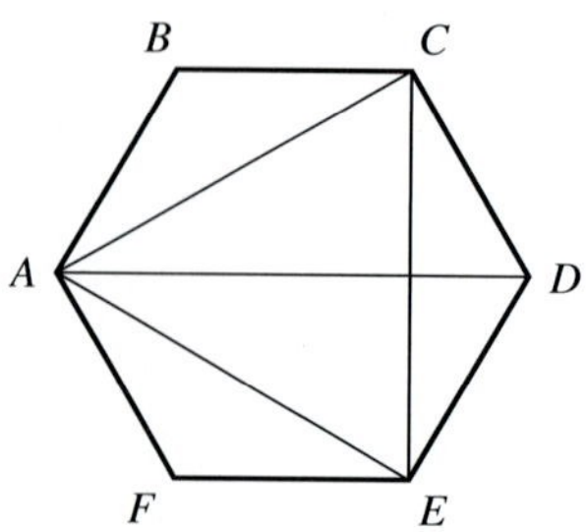

Solution Acute: $\triangle ACE$; Right: $\triangle ACD$ and $\triangle AED$; Obtuse: $\triangle ABC$, $\triangle CDE$, and $\triangle AFE$; Scalene: $\triangle ACD$ and $\triangle AED$; Isosceles: $\triangle ABC$, $\triangle CDE$, $\triangle AFE$, and $\triangle ACE$; Equilateral: $\triangle ACE$.

Quadrilaterals

The four-sided polygons are classified as shown in Table 11.3. This classification allows a parallelogram to be described as a trapezoid, and similarly a square is a rectangle and a rectangle is a parallelogram.

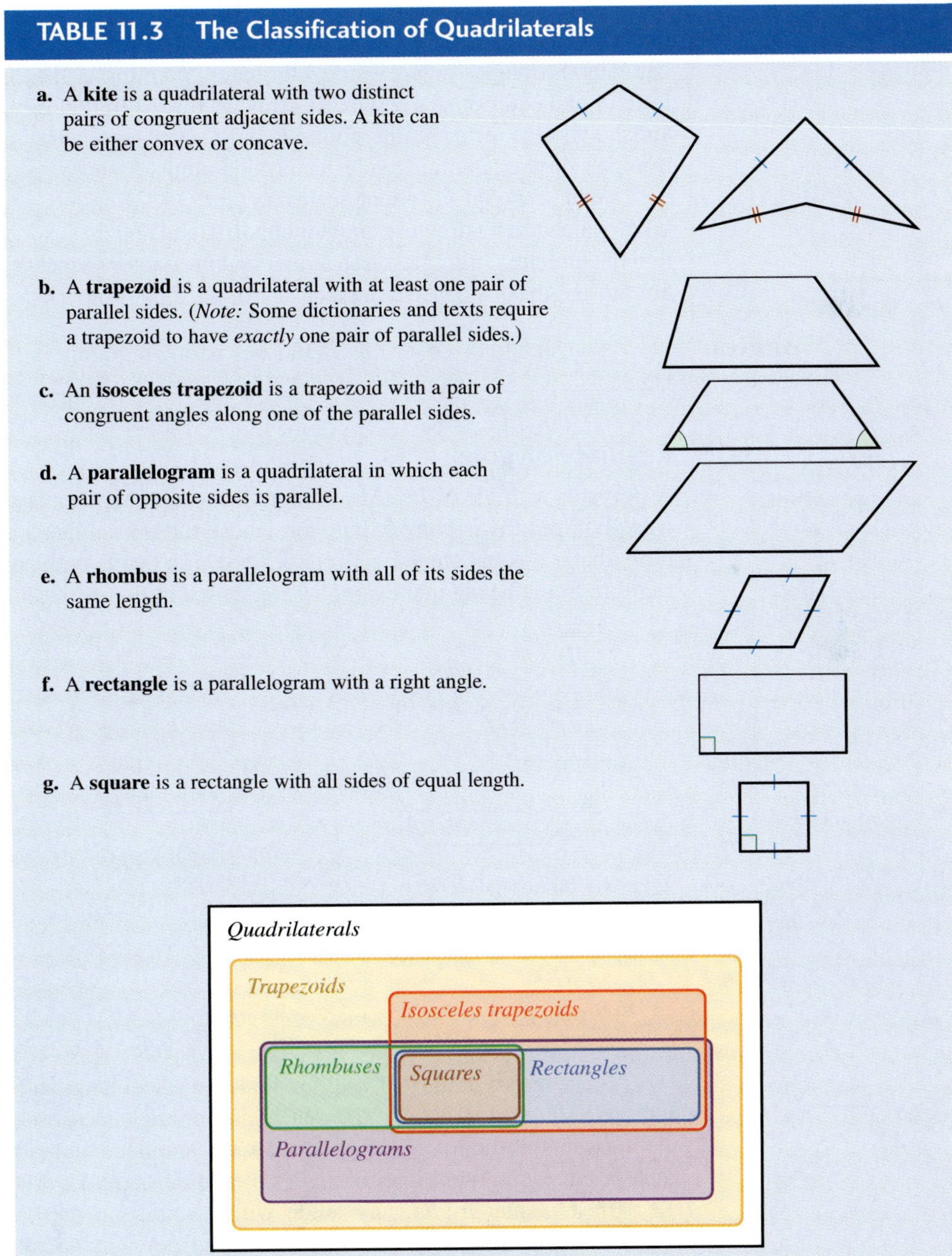

TABLE 11.3 The Classification of Quadrilaterals

a. A **kite** is a quadrilateral with two distinct pairs of congruent adjacent sides. A kite can be either convex or concave.

b. A **trapezoid** is a quadrilateral with at least one pair of parallel sides. (*Note:* Some dictionaries and texts require a trapezoid to have *exactly* one pair of parallel sides.)

c. An **isosceles trapezoid** is a trapezoid with a pair of congruent angles along one of the parallel sides.

d. A **parallelogram** is a quadrilateral in which each pair of opposite sides is parallel.

e. A **rhombus** is a parallelogram with all of its sides the same length.

f. A **rectangle** is a parallelogram with a right angle.

g. A **square** is a rectangle with all sides of equal length.

The classification hierarchy is summarized by a Venn diagram. Notice that the squares are the intersection of the rhombus and rectangle loops. Arranging figures in classes that are subsets of one another can be very useful. For example, suppose we wish to show that the points A, B, C, and D are the vertices of a square. Step 1 may be to show that one pair of sides is parallel, telling us that $ABCD$ is a trapezoid. Step 2 may show that the remaining pair of opposite sides is parallel, and now we know that $ABCD$ is a parallelogram. If Step 3 shows that $ABCD$ is a kite and Step 4 shows that $\angle A$ is a right angle, then we correctly deduce that $ABCD$ is a square.

EXAMPLE 11.12 EXPLORING QUADRILATERALS

Draw a convex quadrilateral $ABCD$. Use a compass to erect equilateral triangles on each side, alternately pointing to the interior and exterior of the quadrilateral. Two such triangles are shown here, determining points P and Q. The equilateral triangles on the remaining sides determine points R and S.

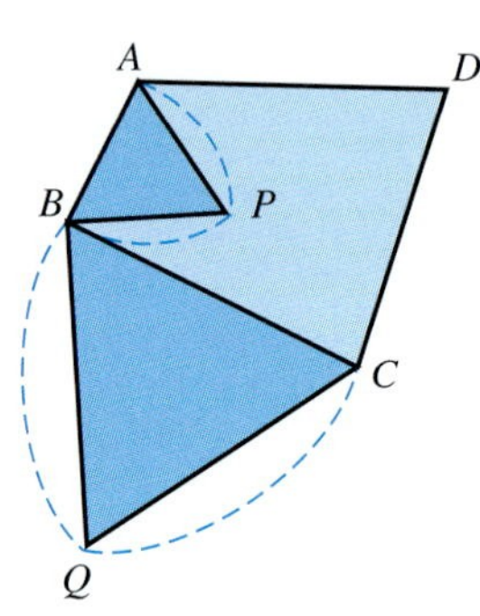

What can you say about quadrilateral $PQRS$? Support your conjecture by drawing another quadrilateral and its system of equilateral triangles. Use a ruler and protractor to measure the lengths of sides and the measure of the angles of $PQRS$.

Solution

You should discover that $PQRS$ is a parallelogram.

To construct an equilateral triangle with side $\overline{AB}$, draw a circular arc centered at A through B and another arc centered at B through A. The arcs intersect at the point P needed to complete the equilateral triangle ABP.

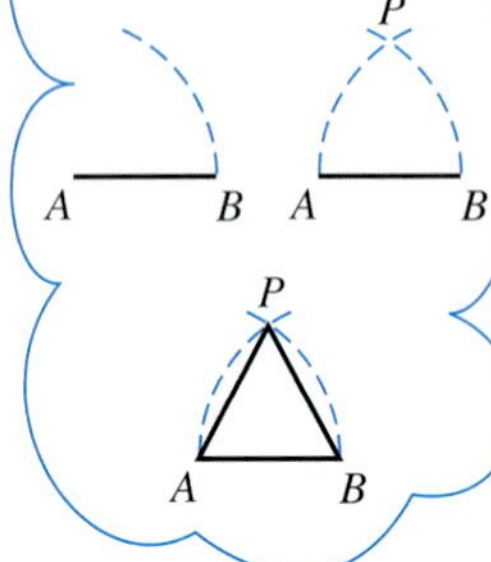

Regular Polygons

A polygon with all of its sides congruent to one another is **equilateral** (that is, "equal sided"). Similarly, a convex polygon whose interior angles are all congruent is **equiangular** (that is, "equal angled"). A convex polygon that is both equilateral and equiangular is **regular.** Some hexagonal examples are shown in Figure 11.20.

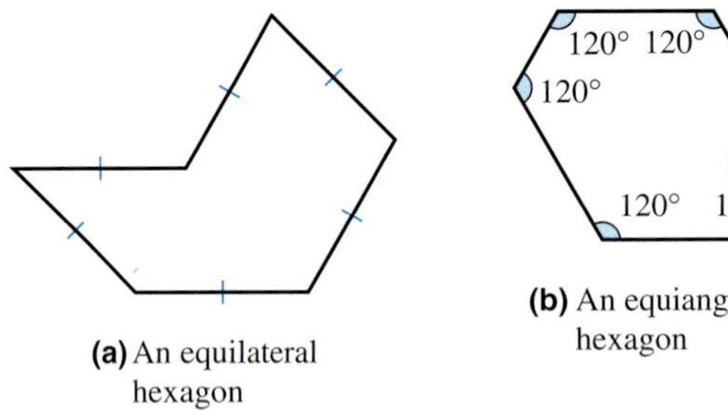

(a) An equilateral hexagon

(b) An equiangular hexagon

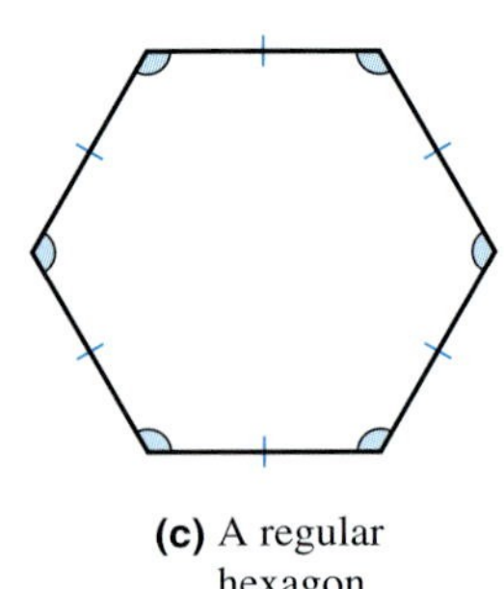

(c) A regular hexagon

Figure 11.20
Hexagons that are equilateral, equiangular, and regular

Since the six congruent interior angles in an equiangular hexagon have measures that add up to $(6 - 2) \cdot 180° = 720°$, each interior angle measures $720°/6 = 120°$. Similarly, the measures of the interior angles of an equiangular n-gon add up to $(n - 2) \cdot 180°$, so each of the n congruent interior angles measures $(n - 2) \cdot 180°/n$.

In a regular n-gon, any angle with a vertex at the center of a regular polygon and sides containing adjacent vertices of the polygon is called a **central angle** of the polygon. The following formulas give the measures of the exterior, interior, and central angles of a regular polygon.

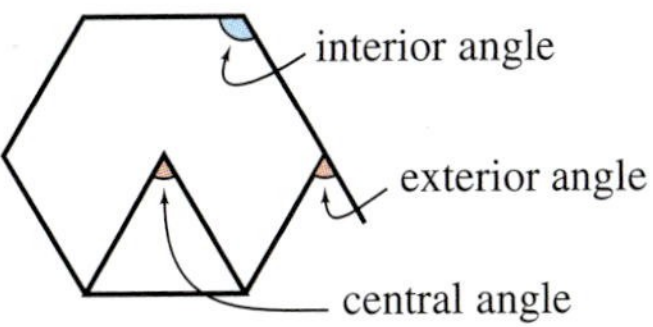

THEOREM *Angle Measure in a Regular n-gon*

In a regular n-gon,

- each interior angle has measure $(n - 2) \cdot 180°/n$;
- each exterior angle has measure $360°/n$;
- each central angle has measure $360°/n$;

It is useful to observe that an interior and an exterior angle are supplementary. Thus, the measure of an interior angle is also given by $180° - 360°/n$.

EXAMPLE 11.13 WORKING WITH ANGLES IN REGULAR *N*-GONS

(a) The Baha'i House of Worship in Wilmette, Illinois, has the unusual floor plan shown below. What are the measures of $\angle ABC$ and $\angle AOB$?

(b) Suppose an archeologist found a broken piece of pottery as shown on the right below. If the angle measures 160° and it is assumed the plate had the form of a regular polygon, how many sides would the complete plate have had?

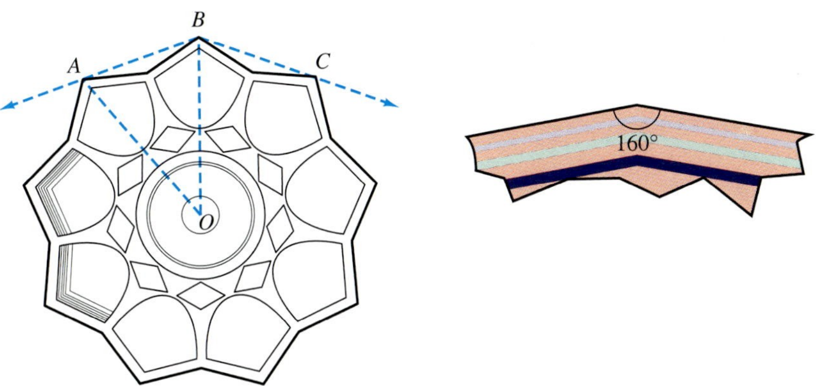

Solution

(a) $\angle ABC$ is the interior angle of a regular 9-gon and it has measure $(9 - 2) \cdot 180°/9 = 140°$. $\angle AOB$ is a central angle of a regular 9-gon, so its measure is $360°/9 = 40°$.

(b) The corresponding exterior angle measures 20°. Since $20° = 360°/n$, it is seen that $n = 360°/20° = 18$. Under the assumptions stated, the plate would have had 18 sides.

Circles

By definition, a **circle** is the set of all points in the plane that are at a fixed distance—the **radius**—from a given point—the **center.** The **chord, diameter, tangent line, arc, sector,** and **segment** of a circle are shown in Figure 11.21. The word *radius* is used in two ways: it is both a segment from the center to a point on the circle and the length of such a segment. Likewise, *diameter* is both a segment and a length. The interior of a circle is a **disc.**

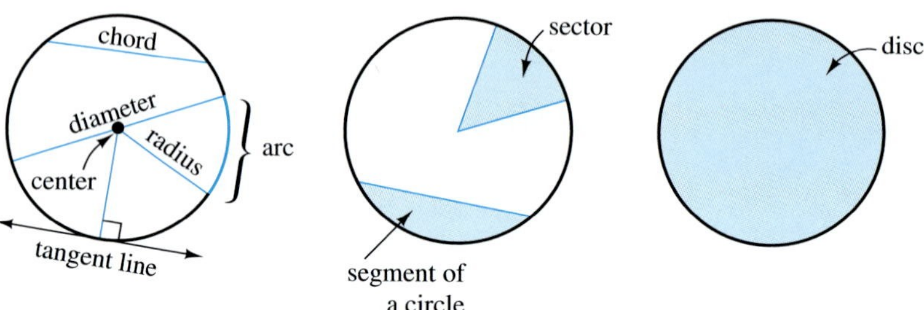

Figure 11.21
The parts of a circle

EXAMPLE 11.14 EXPLORING PERPENDICULAR CHORDS IN A PAIR OF CIRCLES

Use a compass (or trace around the bottom of a cup) to draw two congruent intersecting circles. Let A, B, C, and D denote the respective centers and intersection points of the circles. Draw any line m through C, and denote its intersection points with the two circles as R and O. Similarly, draw the line n through D that is perpendicular to m. Let H and M denote the points at which line n intersects the circles.

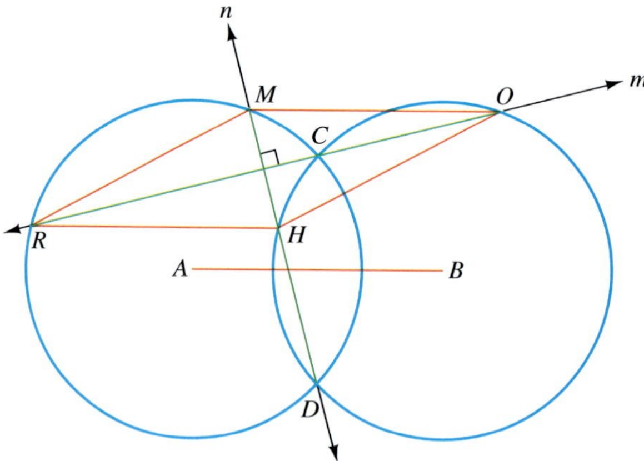

(a) Use measurement tools to describe the shape of the quadrilateral $RHOM$. What connection to the distance AB between the circles' centers do you discover?

(b) Use a protractor to measure the angles $\angle RAM$, $\angle MAC$, and $\angle CBO$. What relationship do you discover?

Solution

(a) $RHOM$ is a rhombus whose sides have the same length as the distance AB between the centers of the circles.

(b) Your measurements should show that $m(\angle RAM) = m(\angle MAC) + m(\angle CBO)$.

From Paper Discs to Polygons

Materials Needed

1. Two paper discs per student, each 7 to 8 inches (or 18 to 20 centimeters) in diameter. All discs used in a group should be the same size.
2. Drawing and measuring tools (pencils, protractors, rulers).

Directions

Work in groups of four. There are two sets of explorations, each using one paper disc per student.

Explorations with the First Paper Disc

1. Make a light pencil mark on the disc that you think estimates the center of the disc. To check how close you are, lightly (do not make a heavy crease) fold the disc in half. Undo your fold and lay the disc out flat. Is your mark along the diameter you have folded? Now fold the disc in half once more in a new direction and unfold. Is your mark at the intersection of the two diameters? Clearly mark the true center of your disc, and label it O.
2. Fold across a chord of the disc so that the folded arc of the circle passes through the center O, as shown. Without unfolding the first fold, make a second fold across a chord that has the same endpoint as the first chord and again with the circular arc passing through point O. Finally, fold the remaining arc of the circle. Does it also pass through point O? What kind of a triangle seems to have been created? Check out your guess by measuring the three sides and the three angles of the triangle. Compare with others in your group.

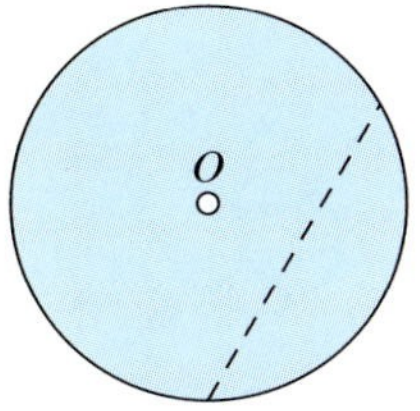

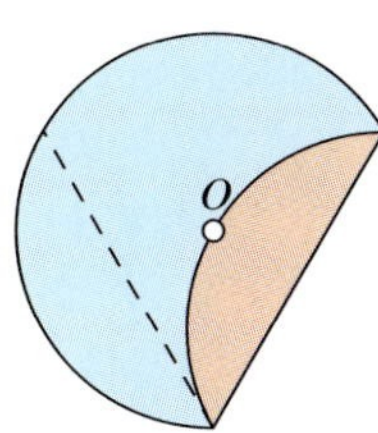

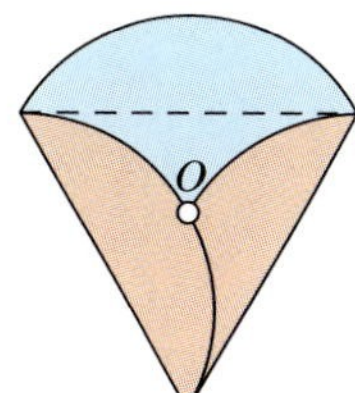

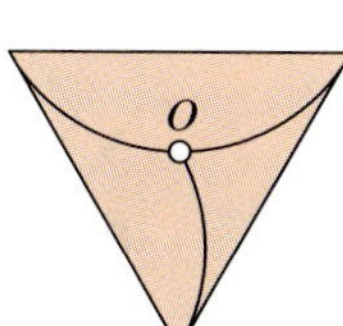

3. Find the midpoints of the sides of your triangle (how can this be done with folding?) and mark them with a pencil. Fold a vertex of your triangle to the midpoint on the opposite side. What polygon have you created? Without unfolding, fold another vertex of a triangle to the marked midpoint on the opposite side. What is the polygon now? Fold the third vertex to the midpoint marked on the opposite side. What is the name of this polygon?

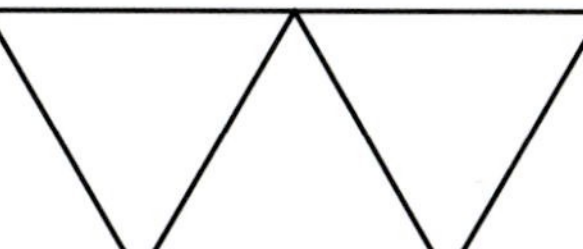
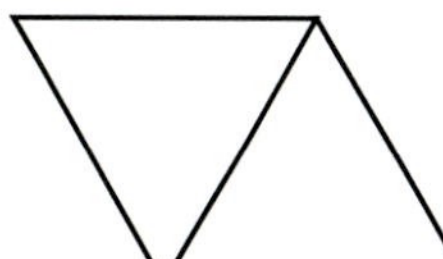
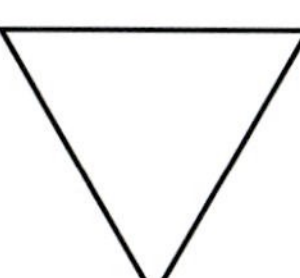

4. Unfold your disc to return to the large equilateral triangle made in Exploration Step 2. Now fold each vertex to the center point O. Describe the polygon you have now created.

Explorations with the Second Paper Disc

5. Mark a point well away from the center of the paper disc, and label it H. Fold two arcs, sharing a common endpoint, so that both arcs pass through point H. Next, fold the remaining arc of the disc to form a triangle. Did the third arc you folded also pass through point H? Compare with others in your group.

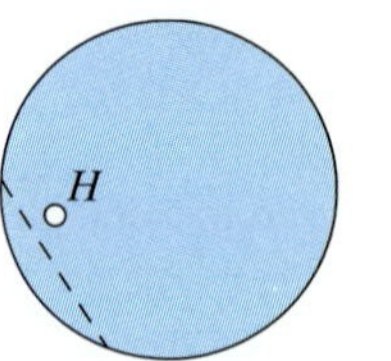

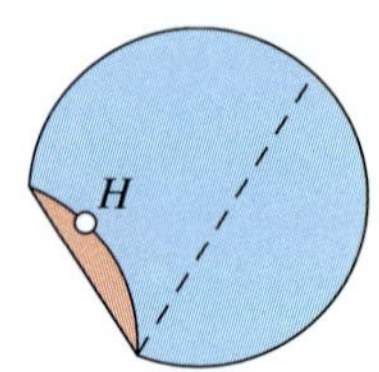

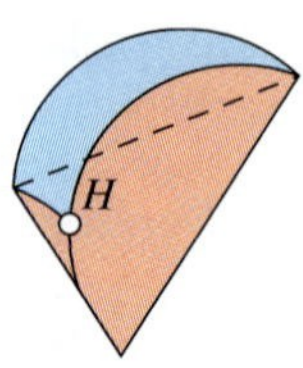

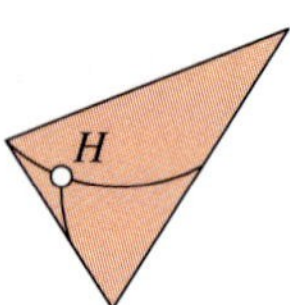

6. Unfold your triangle, and use a ruler to draw the chord that begins at a vertex of the triangle and passes through point H. Similarly, draw the chords through H from the other two vertices of the triangle. At what angle does each chord intersect the opposite side of the triangle? Use a protractor to measure the angle, and compare your answer with others in your group.

Problem Set 11.2

Understanding Concepts

1. If the figure shown has the property listed, place a check in the table below.

(a)
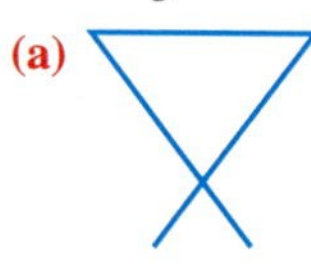

(b)

(c)

(d)
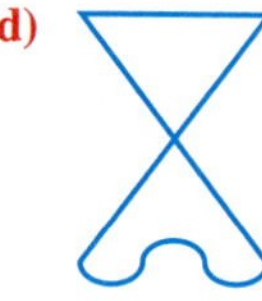

(e)

(f)

(g)

(h)
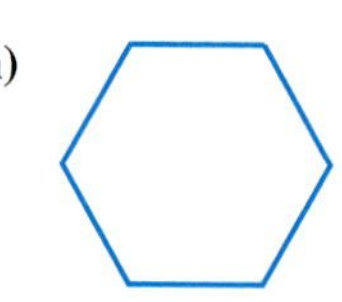

(i)

(j)

(k)
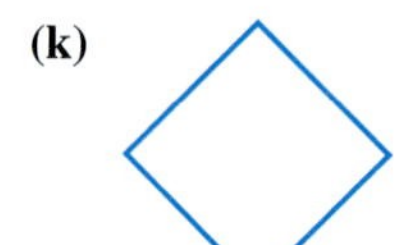

(l)
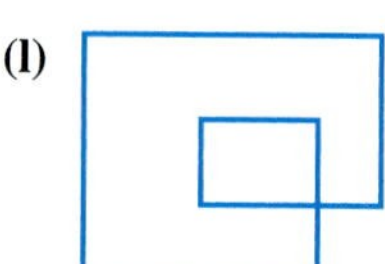

	(a)	(b)	(c)	(d)	(e)	(f)	(g)	(h)	(i)	(j)	(k)	(l)
Simple curve												
Closed curve												
Polygonal curve												
Polygon												

2. Draw figures that satisfy the given description.
   (a) a nonsimple closed four-sided polygonal curve
   (b) a concave pentagon
   (c) an equiangular quadrilateral
   (d) a convex octagon
3. Can a line cross a simple closed plane curve 99 times? Explain why or why not.
4. Classify each region as convex or concave.

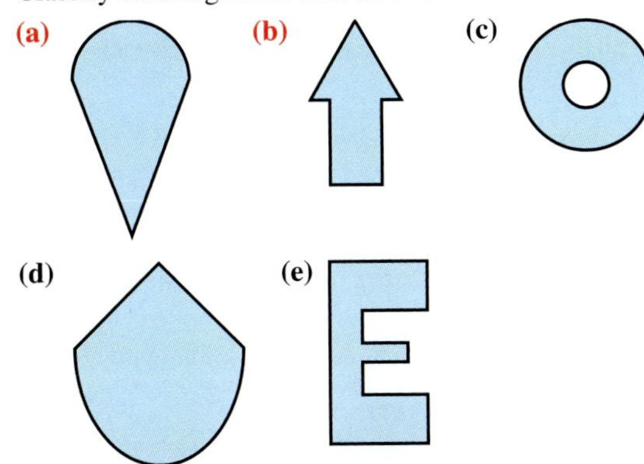

5. Imagine stretching a rubber band tightly around each figure shown. Shade the region within the band with a colored pencil. Is the shaded region always convex?

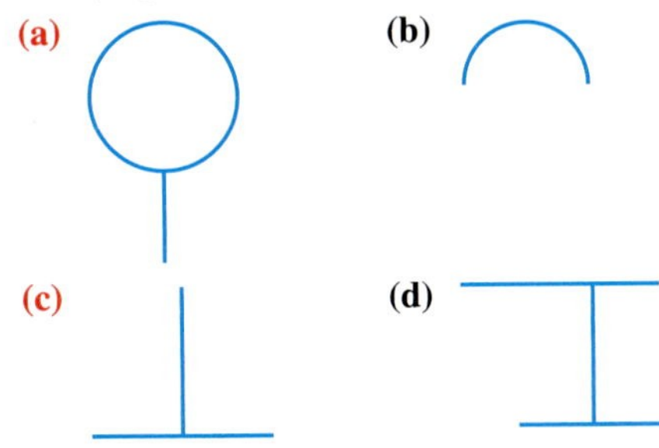

6. How many different regions in the plane are determined by these figures?

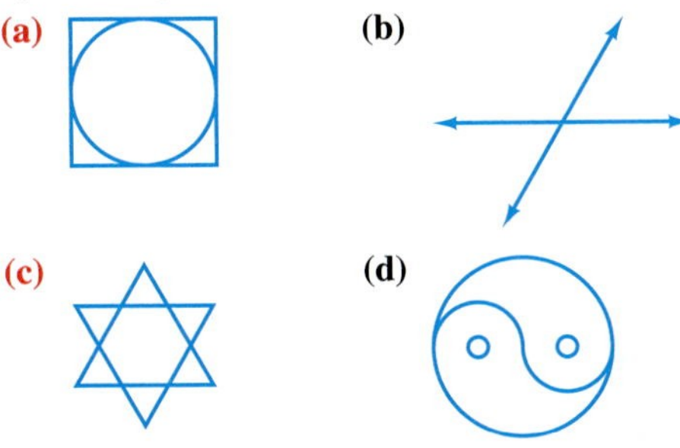

7. Determine the measures of the interior angles of this polygon.

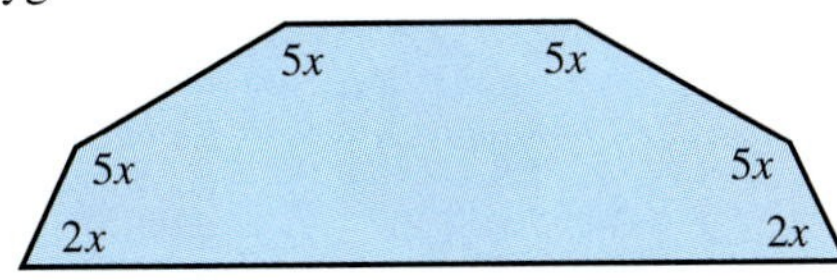

8. Calculate the measures of the angles in this concave polygon.

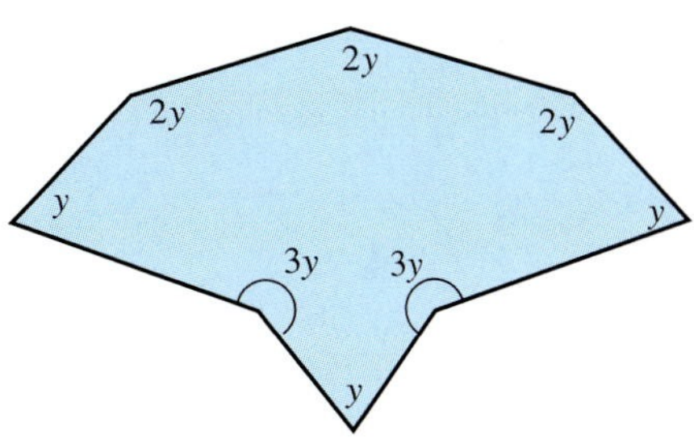

9. A **lattice polygon** is formed by a rubber band stretched over the nails of a geoboard. Find the sum of the measures of the interior angles of the following lattice polygons.

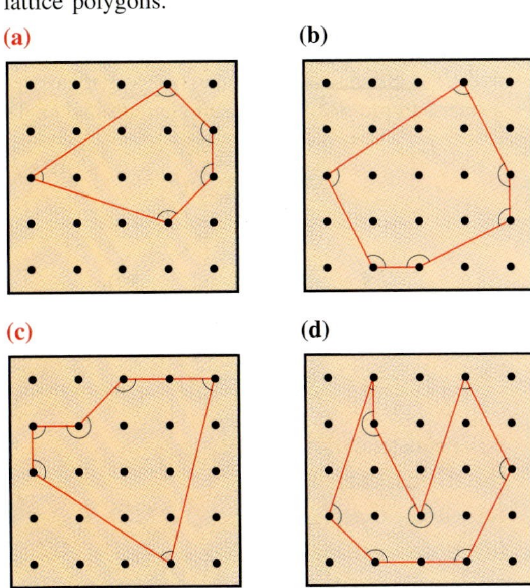

10. Draw lattice polygons (see problem 9) on squared dot paper whose interior angles have the given sum of their measures. Place arcs that indicate the interior angles, as in problem 9.
    (a) 180° (b) 1080°
    (c) 1440° (d) 1800°
11. The interior angles of an n-gon have an average measure of 175°.
    (a) What is n?
    (b) Suppose the polygon has flexible joints at the vertices. As the polygon is flexed to take on new shapes, what happens to the average measure of the interior angles? Explain your reasoning.
12. What is the amount of total turn for the following closed curves? Assign a positive measure to counterclockwise turning.

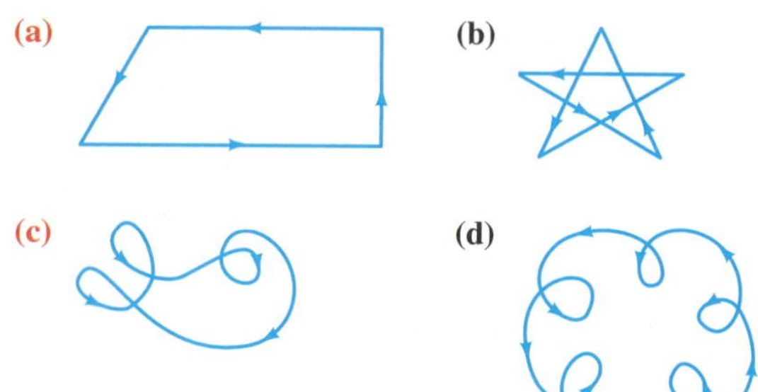

13. Suppose you walk north 10 paces, turn left through 24°, walk 10 paces, turn left through 24°, walk 10 paces, and so on.
    (a) Will you return to your starting point?
    (b) What is the shape of the path?

14. Fill in the missing vertices to give the type of triangle required, choosing vertices from A, B, C, D, E, F, G, and H. There may be more than one way to answer. Use a ruler and protractor to measure lengths and angles.

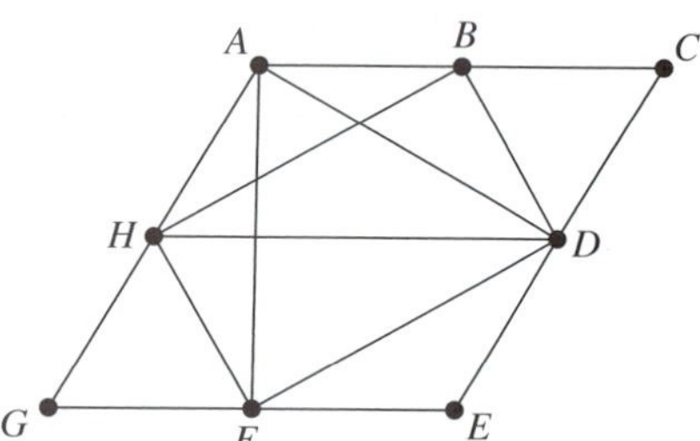

    (a) Equilateral triangle D __ __
    (b) Right triangle F __ __
    (c) Obtuse triangle F __ __
    (d) Isosceles triangle E __ __
    (e) Acute triangle H __ __

15. Refer to the figure shown in problem 14 to give the type of quadrilateral required, filling in vertices from A, B, C, D, E, F, G, and H.
    (a) Rhombus A __ __ __
    (b) Rectangle B __ __ __
    (c) Isosceles trapezoid A __ __ __
    (d) Nonisosceles trapezoid G __ __ __
    (e) Kite E __ __ __

16. For each regular n-gon shown, give the measures of the interior, exterior, and central angles.

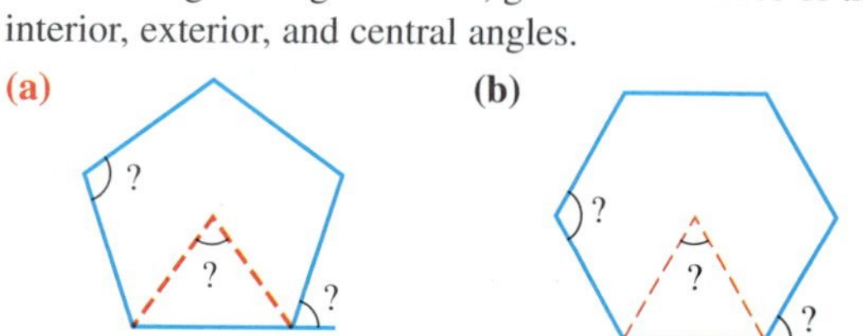

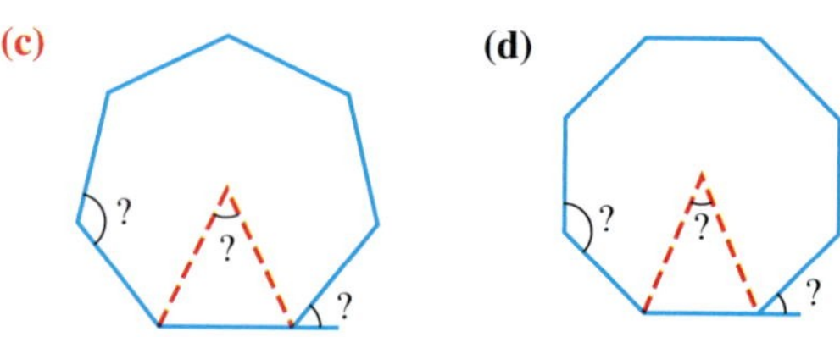

17. (a) A regular n-gon has exterior angles of measure 15°. What is n?
    (b) A regular n-gon has interior angles each measuring $172\frac{1}{2}°$. What is n?

Teaching Concepts

18. **Pattern-Block Shapes.** Pattern blocks are an especially effective manipulative to explore polygons and angles. The six shapes of pattern blocks are shown below.

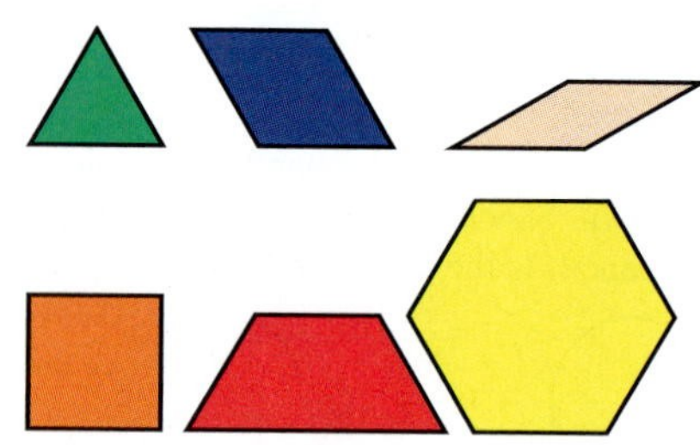

Which shapes
    (a) are quadrilaterals?
    (b) are parallelograms?
    (c) are rhombuses?
    (d) contain an obtuse angle?
    (e) are regular polygons?

19. **Pattern-Block Angles.** The interior angles of pattern blocks (see problem 18) can be investigated by creating designs such as those shown below.

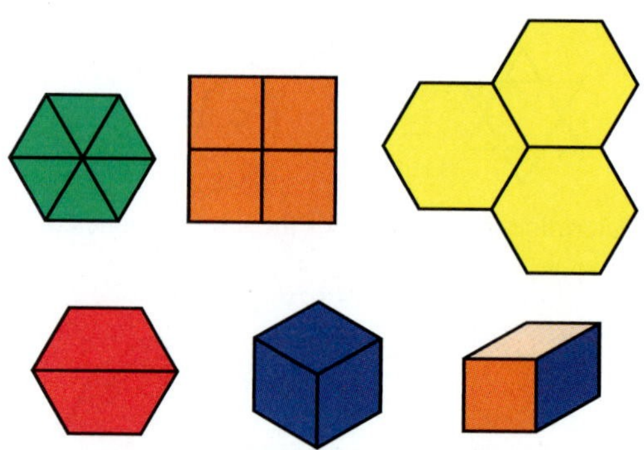

Explain how to use these designs to determine the measures of the interior angles

(a) of the equilateral triangle, square, and regular hexagon.

(b) of the red trapezoid.

(c) of the blue rhombus.

(d) of the small rhombus.

Thinking Critically

20. A brave, but not mathematically inclined, knight at point K wants to rescue the princess at point P. The evil king has not only confined the princess to the castle tower, but has also erected an enormously long impenetrable stone wall that winds and twists all over the countryside. There is also a dangerous dragon at point D. Looking out the one small window in the castle tower, the princess has created the map shown below. Only portions of the stone wall are visible from her window, but the mathematically knowledgeable princess knows the entire wall forms a simple closed curve.

(a) Is the princess worried that the dragon will find the knight?

(b) Is she hopeful the knight will come to her rescue?

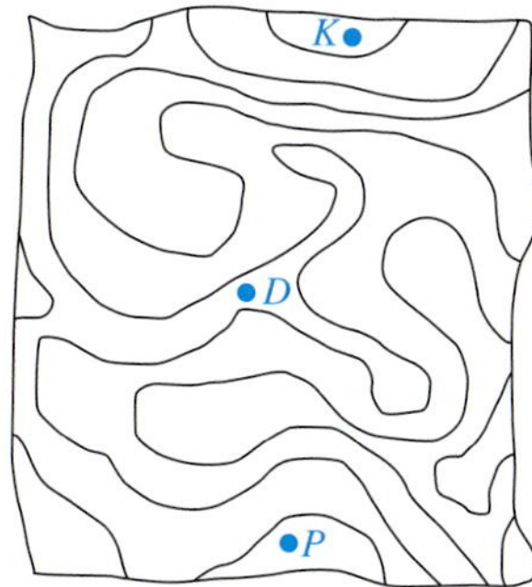

21. The closed curves shown below are not simple, since they have *double* points at $A, B, C, \ldots$. If the upper curve is followed completely around, the double points are encountered in the order $ABCCBA$, which we can rewrite in zigzag form as ${}^{A}{}_{B}{}^{C}{}_{C}{}^{B}{}_{A}$.

(a) Write the sequence of double points encountered when following the curve with double points D, $E, \ldots, K$, and then rewrite the points in zigzag form.

(b) Draw several more examples of closed curves with double points. Label the points and then give the sequences (also in zigzag form) in which the double points are encountered by following the curve. What property do you observe about the zigzag patterns?

(c) Do you think you can find a closed curve with four double points A, B, C, and D that are encountered in the order $ABDCCADB$? Explain, using your discovery in part (b).

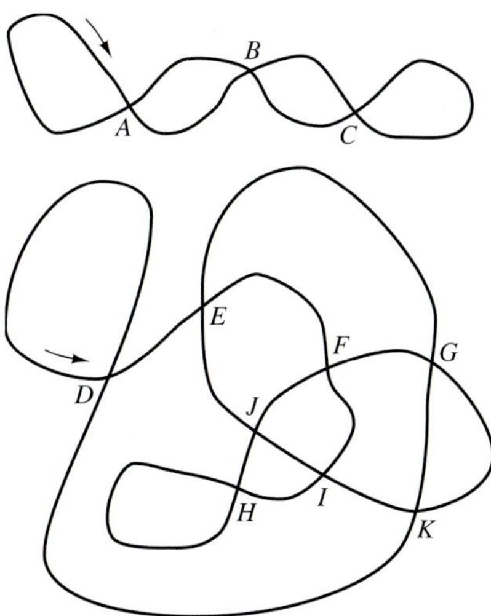

22. A goat is tethered at the corner of an 80-foot by 30-foot rectangular barn. If the rope is 50 feet long, describe the region the goat can reach.

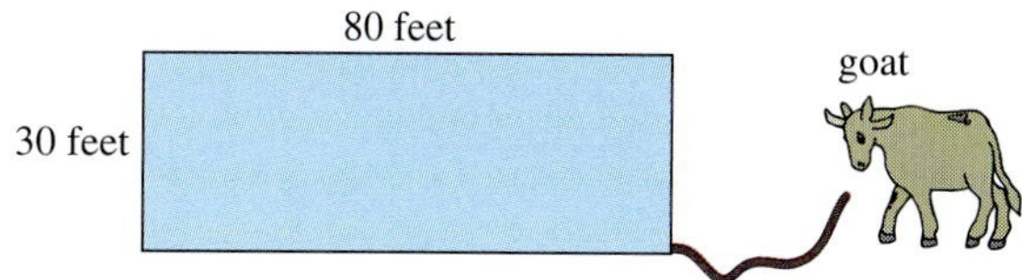

23. (a) A boat B is anchored at point A. Describe, in words and a sketch, the region where the boat can drift due to wind and currents.

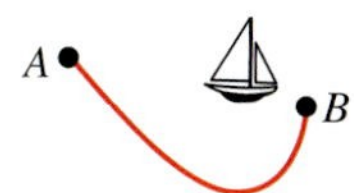

(b) Suppose a second anchor at point C has been set, as shown in the next diagram. Describe, in words and a sketch, the region to which the boat is now confined.

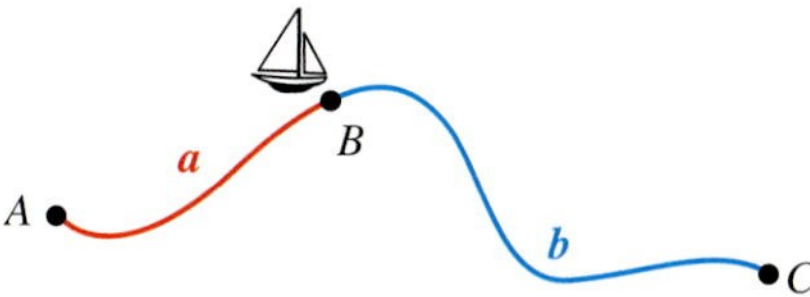

24. The segment $\overline{AB}$ is to be completed to become a side of a triangle ABC.

Describe, in words and sketches, the set of points C so that

(a) $\triangle ABC$ is a right triangle and $\overline{AB}$ is a leg;

(b) $\triangle ABC$ is a right triangle and $\overline{AB}$ is a hypotenuse

(*Hint:* For (b), proceed experimentally using the corner of a sheet of paper as a right angle);

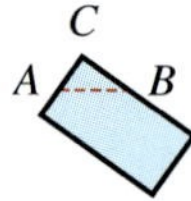

(c) $\triangle ABC$ is an acute triangle;

(d) $\triangle ABC$ is an obtuse triangle.

25. If an interior angle of a polygon has measure m, then $360° - m$ is called the measure of the **conjugate angle** at that vertex. Find a formula that gives the sum of the measures of the conjugate angles of an n-gon, and give a justification for your formula. As an example, the measures of the conjugate angles in this pentagon add up to 1260°.

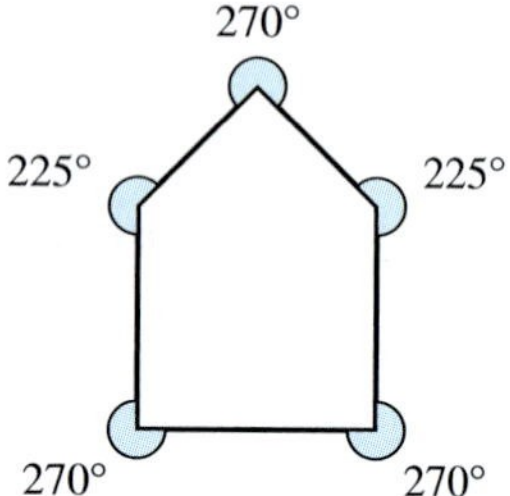

26. The heptagonal region shown on the left has been broken into five triangular regions by drawing four nonintersecting diagonals across the interior of the polygon, as shown on the right. In this way we say that the polygon is triangulated by diagonals.

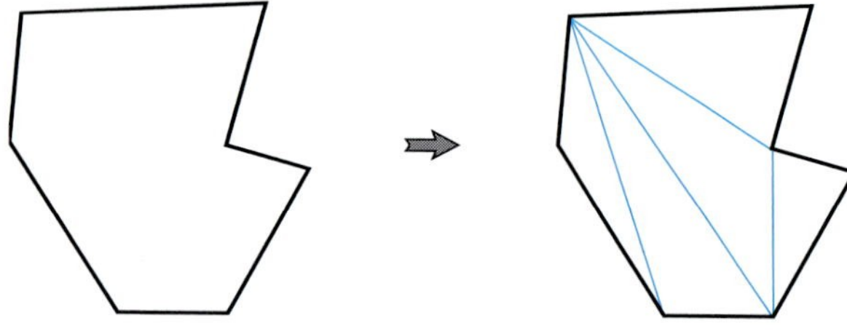

(a) Investigate how many diagonals are required to triangulate any n-gon.

(b) How many triangles are in any triangulation of an n-gon by diagonals?

(c) Explain how a triangulation by diagonals can give a new derivation of the formula $(n - 2) \cdot 180°$ for the sum of the measures of the interior angles of any n-gon.

27. The polygon shown contains a point S in its interior that can be joined to any vertex by a segment that remains inside the polygon. Drawing all such segments produces a triangulation of the interior of the polygon. (Compare with problem 26.)

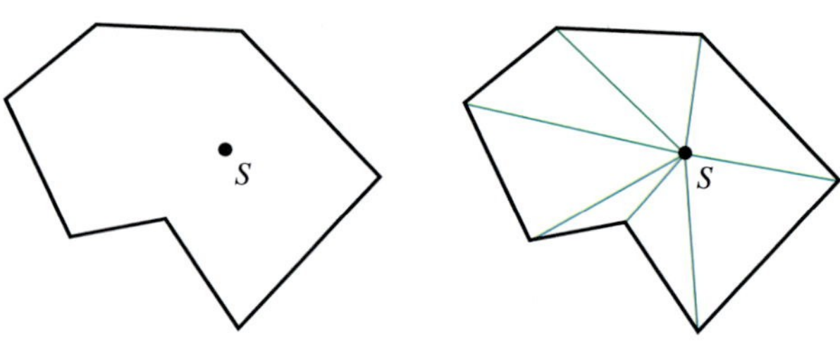

If an n-gon contains such a point S, explain how the corresponding triangulation can be used to derive the $(n - 2) \cdot 180°$ formula for the sum of interior angle measures.

28. **(a)** Find the sum of the angle measures in the five-pointed star shown on the left below. Explain how the total-turn theorem can be used to obtain your answer.

(b) What is the measure of the angle in each point of the pentagram shown on the right?

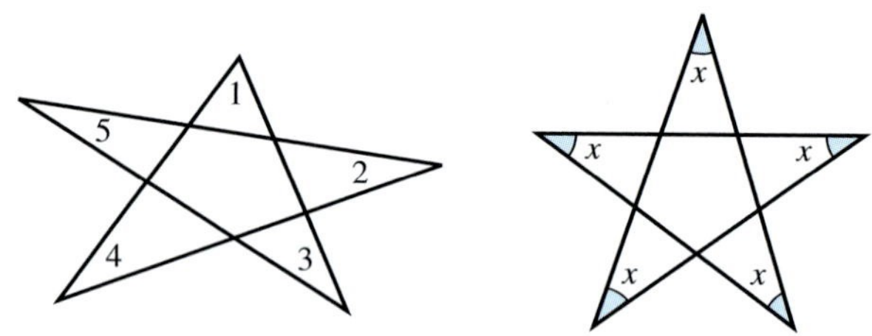

29. The wall mosaic from the Alhambra shown in the picture gallery of plane figures in Section 11.1 contains two nested stars, each with 16 outward points, as shown below.

(a) What is the measure of each angle at the points of the tan outer star?

(b) What is the measure of each angle at three points of the blue inner star? (*Hint:* The inner star is really two 8-pointed stars.)

30. What is the largest number of regions you can form with a system of 10 circles?

31. Regions can be formed in a circle by drawing chords, no three of which are concurrent. If C chords are drawn and they intersect in I points, determine a formula for the number of pieces P (that is, the regions) that are formed inside the circle.

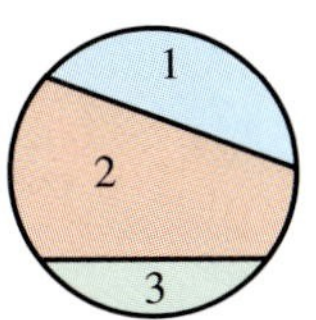

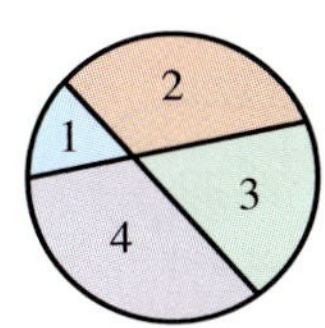

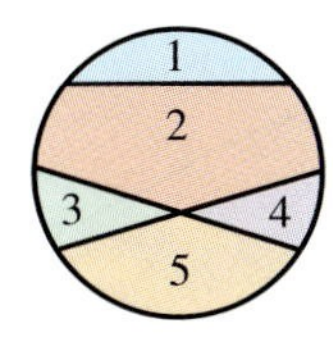

P	C	I
3	2	0
4	2	1
5	3	1

32. Let $ABCD$ be a parallelogram.

(a) Prove that $\angle A$ and $\angle B$ are supplementary.

(b) Prove that $\angle A \cong \angle C$ and $\angle B \cong \angle D$.

33. Let $PQRS$ be a convex quadrilateral for which $\angle P \cong \angle R$ and $\angle Q \cong \angle S$.

(a) Prove that $\angle P$ and $\angle Q$ are supplementary.

(b) Prove that $PQRS$ is a parallelogram.

34. Let F and G be convex figures.

(a) Prove that $F \cap G$ is also a convex figure.

(b) Is $F \cup G$ necessarily a convex figure? Explain.

Thinking Cooperatively

35. Work in pairs, with one partner using a black pencil and the other partner a red pencil. Each partner draws a closed curve on his or her own clean sheet of paper. The papers are exchanged, and each partner draws a second closed curve. The newly drawn curve should cross the previously drawn curve several times. However, it cannot pass through an intersection of the curve first drawn, nor can it pass through a point of intersection of the previously drawn curve and itself a second time. Circle the points where the red and black curves cross one another.

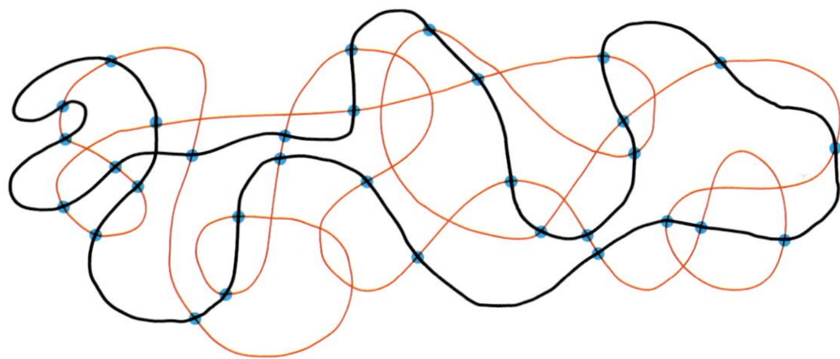

The example shows 32 points at which the red and black curves cross one another.

(a) Count the number of crossing points of the red and black curves that are circled. Compare your number with the number of crossing points counted by other partners, and redo the drawing and counting to gather more evidence. What kind of numbers seems to occur? What kinds of numbers apparently never occur?

Using fresh sheets of paper, draw new closed curves. The curves will partition the plane into regions. Using the same pencil, show that each region can either be shaded or left blank to create a pattern so that regions that share a border have opposite shading. Regions that share only a point are allowed to have the same shading.

(b) Trade papers, and draw a second closed curve of the opposite color, as done for part (a). Is your result discovered from part (a) now more obvious? Discuss and explain.

36. Work in pairs, with one partner using a black pencil and the other partner a red pencil. Each partner draws a *simple* closed curve on his or her own clean sheet of paper. The papers are exchanged, and a second simple closed curve is drawn. The newly drawn curve should cross the previously drawn curve several times. At each intersection point, the newly drawn curve must cross from the inside to the outside of the previously drawn curve or vice versa (it cannot just touch and turn away). Each partner then classifies and marks each region of the plane by its type:

✓ : Interior to both curves
× : Exterior to both curves
□ : Interior to black and exterior to red curve
■ : Interior to red and exterior to black curve

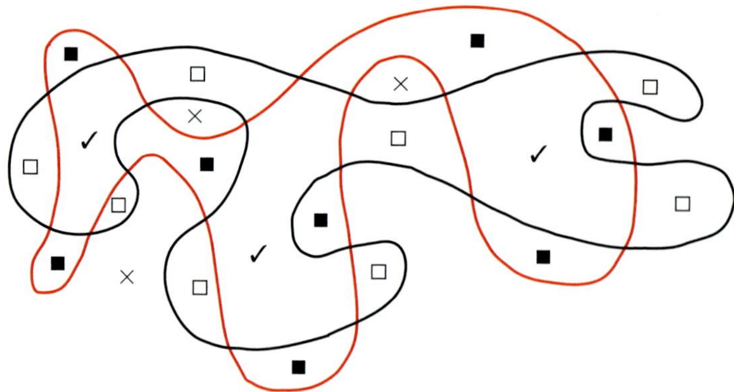

Count the number of regions of each of the four types. Compare your counts with the values obtained by other partners, and draw several additional examples to provide additional data. What connections do you see among the numbers of regions of each type in any crossing pattern created by two simple closed curves? Make a conjecture, and create additional examples to investigate your conjecture.

Making Connections

37. Access to underground utility cables, water pipes, and storm drains is usually provided by circular holes covered by heavy metal circular covers. What unsafe condition would be present if a square shape were used instead of the circular one?

38. The valve stems on fire hydrants are usually triangular or pentagonal. Fire trucks carry a special wrench with a triangular or pentagonal hole that fits over the valve stem.

(a) Why are squares and regular hexagons not used? (*Hint:* What is the shape of the jaws of ordinary adjustable wrenches?)

(b) Why are squares and regular hexagons the standard shape found in bolt heads and nuts?

Using a Computer

Use geometry software for these problems. For example, see Appendix D.

39. Draw two squares $ABCD$ and $AB'C'D'$ that share a common vertex at A. The labeling of the vertices is in the same direction (say counterclockwise) about the centers Q and E. Construct the respective midpoints R and S of $\overline{BD'}$ and $\overline{B'D}$.

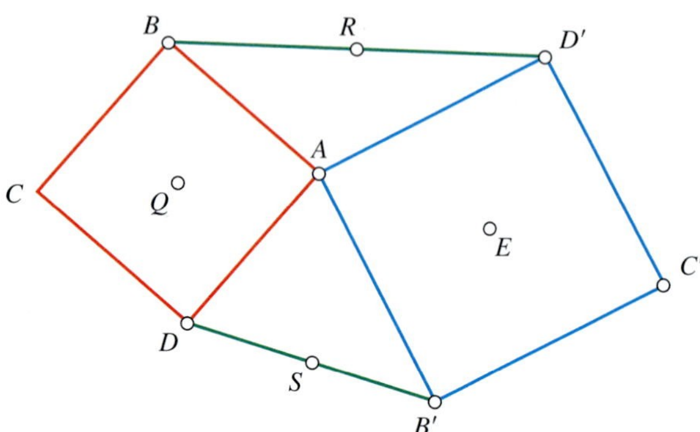

(a) What conclusion can you make about the quadrilateral $SQRE$?

(b) Does it make any difference if the squares overlap?

40. Construct a general quadrilateral $ABCD$ and the midpoints K, L, M, and N of its sides. The quadrilateral $KLMN$ is called the **medial quadrilateral** of $ABCD$.

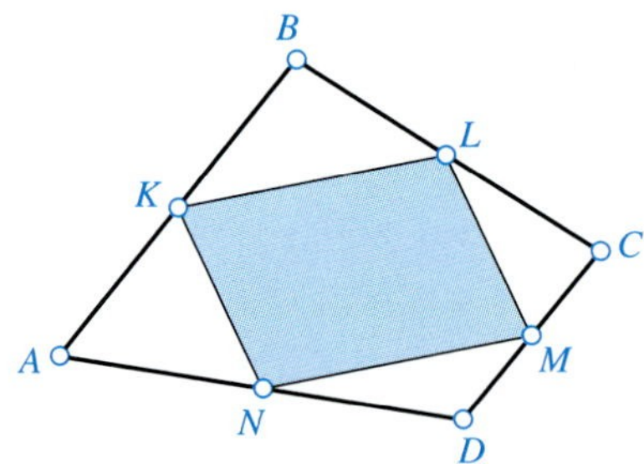

(a) What type of quadrilateral does $KLMN$ appear to be? Explore with your geometry software.

(b) Construct two lines, perpendicular to one another. Next, construct a quadrilateral $ABCD$ with vertices A and C on one of your lines, and vertices B and D on the other line. Finally, construct the medial quadrilateral $KLMN$ of $ABCD$. What type of quadrilateral does the medial quadrilateral now appear to be? Explore with your software.

(c) Drag one of the vertices of the quadrilateral $ABCD$ drawn in part (b) until the medial quadrilateral $KLMN$ appears to be a square. What type of quadrilateral does $ABCD$ appear to be? Explore with your software.

From State Student Assessments

41. (Massachusetts, Grade 4)
What rule did Ray use to sort the shapes below into two groups?

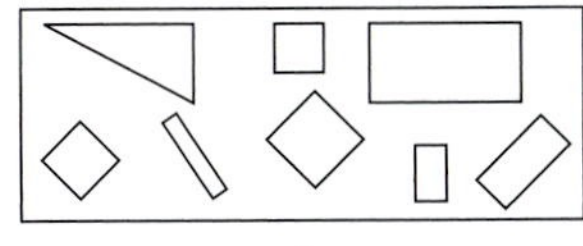

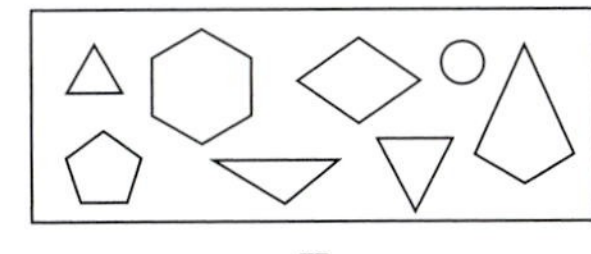

A. I = three sides, II = four sides
B. I = big shapes, II = small shapes
C. I = shapes with right angles II = shapes without right angles
D. I = shapes with four or more angles, II = shapes with less than four angles

42. (Massachusetts, Grade 4)
These shapes are quadrilaterals.

These shapes are not quadrilaterals.

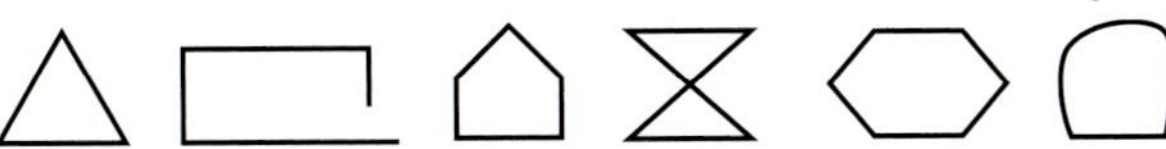

(a) Write a complete definition for a quadrilateral. Be sure to tell all the important ideas about what makes a quadrilateral.

(b) This is a rhombus:

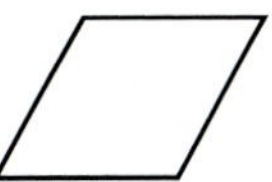

A rhombus is a special kind of quadrilateral. Explain what makes it different from other quadrilaterals.

For Review

43. Find two triangles, one acute and one obtuse, whose interior angle measures are each an integer multiple of 36°. Give drawings of each type of triangle.

44. In the figure below, $\angle APC$ and $\angle BPD$ are right angles. Show that $\angle 1 \cong \angle 3$.

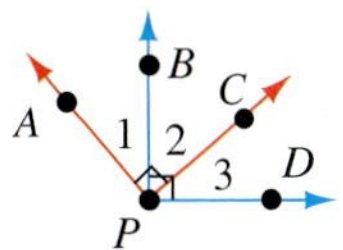

45. Prove that the two acute angles in a right triangle are complementary.

46. Let $\overline{PQ} \parallel \overline{AB}$ and $\overline{RQ} \parallel \overline{AC}$. Find the measures of $\angle 1$, $\angle 2, \ldots, \angle 8$. Explain how you found your answers.

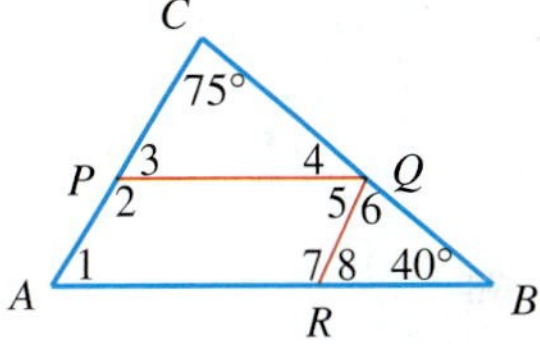

11.3 FIGURES IN SPACE

The picture gallery of space figures on page 000 shows several interesting examples of shapes whose points do not belong to a single plane. Intuitively, we think of space as three dimensional. For example, the shape of a shoebox requires us to know not just width and length, but height as well. In this section we classify, analyze, and represent some of the basic figures in space.

Planes and Lines in Space

There are infinitely many planes in space. Each plane partitions the points of space into three disjoint sets: the plane itself and two regions called **half-spaces.** Two planes are either **parallel** or intersect in a line, as shown in Figure 11.22.

The angle between two intersecting planes is called the **dihedral angle** (*di* = two, *hedral* = face of a geometrical form). A dihedral angle is measured by measuring an angle whose sides lie in the planes and are perpendicular to the line of intersection of the planes. Some examples of dihedral angles and their measures are shown in Figure 11.23.

A Picture Gallery of Space Figures

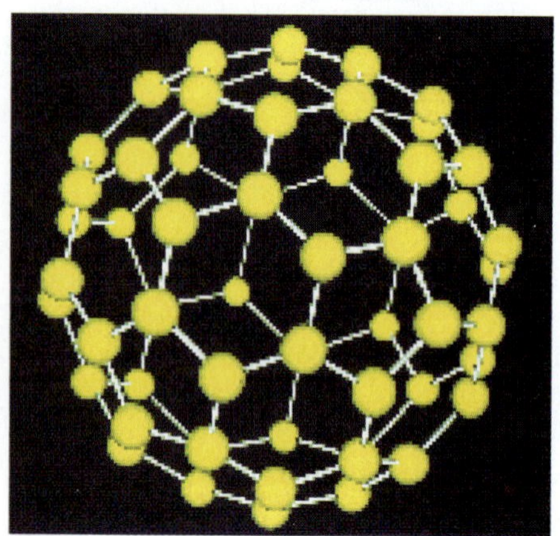

A buckyball (named for Buckminster Fuller), the third form of pure carbon

A seashell (left) and a computer-drawn ideal representation

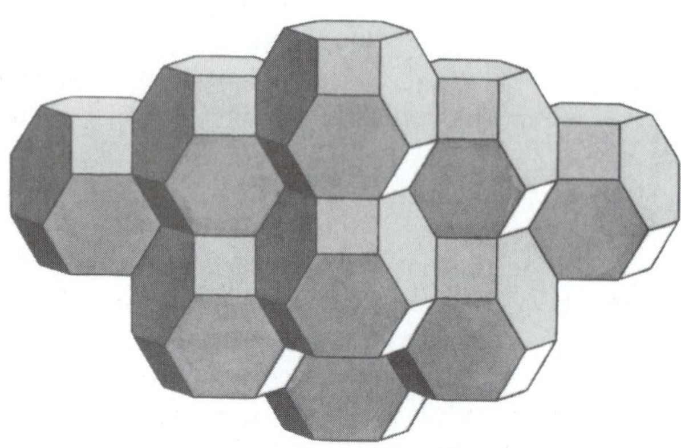

A partial filling of space by truncated octahedra

DVODECEDRON PLANVS
VACVVS.
PACIOLI. De Divina Proportione, 1509 (FACSIMILE NYPL RARE BOOK ROOM)

Leonardo da Vinci's drawings of an icosahedron and a dodecahedron for Fra Luca Pacioli's Di Divina Proportione

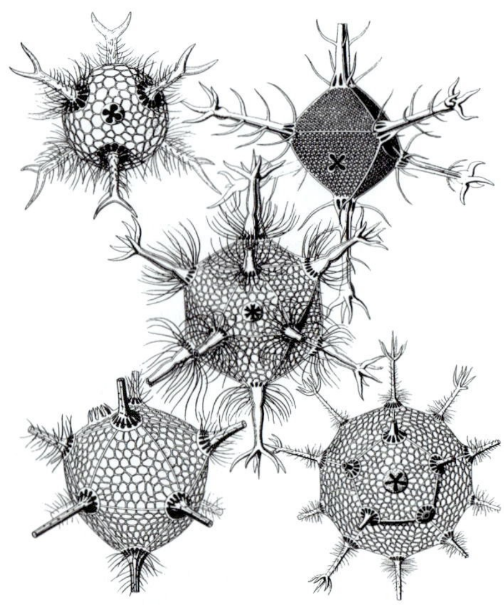

Skeletons of microscopic radiolaria

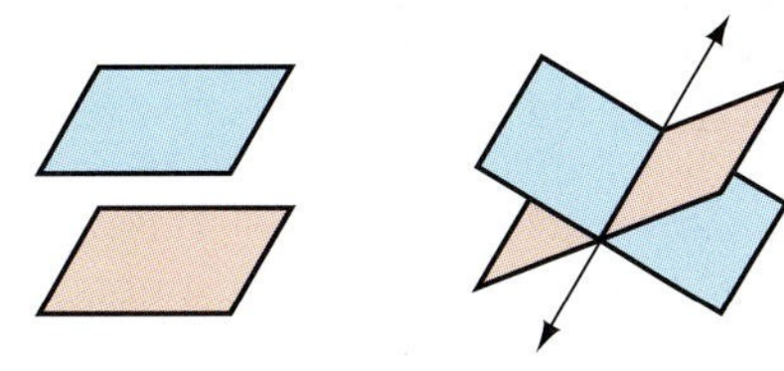

Figure 11.22
Parallel and intersecting planes

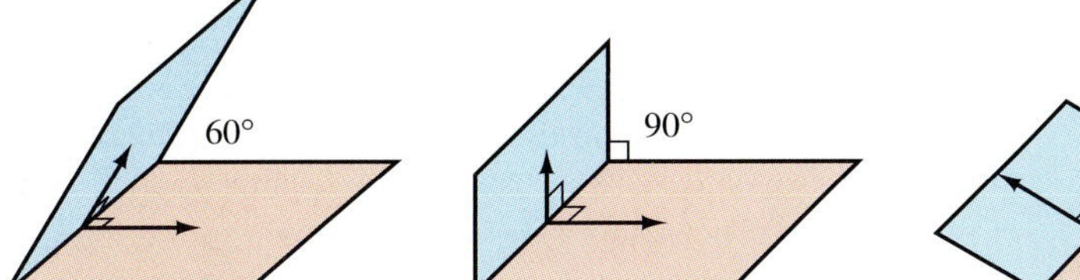

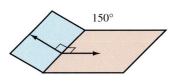

Figure 11.23
Dihedral angles and their measures

Two nonintersecting lines in space are **parallel lines** if they belong to a common plane. Two nonintersecting lines that do not belong to a common plane are called **skew lines.** A line l that does not intersect a plane P is said to be **parallel to the plane.** A line m is **perpendicular to a plane** Q at point A if every line in the plane through A intersects m at a right angle. Diagrams illustrating these terms are shown in Figure 11.24.

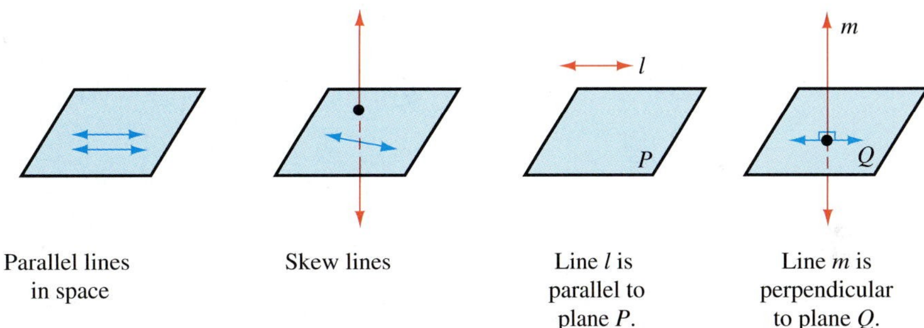

Figure 11.24
Lines and planes in space

Curves, Surfaces, and Solids

The intuitive concept of a curve can be extended from the plane to space by imagining figures drawn with a "magic" pencil whose point leaves a visible trace in the air. Two examples are shown in Figure 11.25, a helix (corkscrew) and a space octagon whose sides are 8 of the 12 edges of a cube.

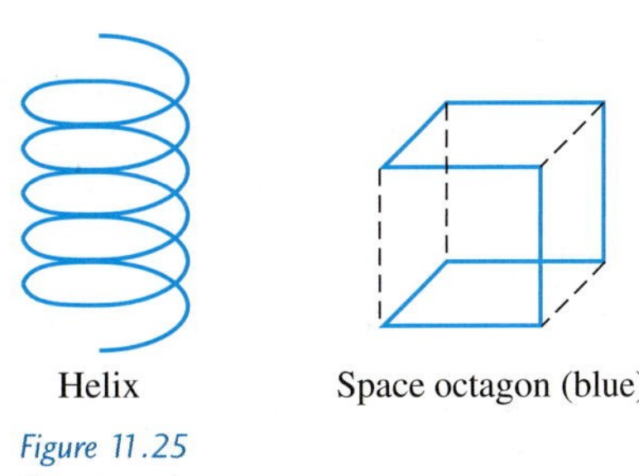

Figure 11.25
Two curves in space

A **sphere** is the set of points at a constant distance from a single point called the **center.** A sphere partitions the remaining points of space into two disjoint regions, namely the points inside the sphere and the points outside the sphere. Any surface without holes and that encloses a hollow region—its interior—is called a **simple closed surface.** An easy check to see if a figure is a simple closed surface is to imagine what shape it would take if it were made of stretchy rubber: if it can be "blown up" into a sphere, then it is a simple closed surface.

The union of all points on a simple closed surface and all points in its interior forms a space figure called a **solid.** For example, the shell of a hardboiled egg can be viewed, overlooking its thickness, as a simple closed surface; the shell, together with the white and yolk of the egg, forms a solid.

A simple closed surface is **convex** if the line segment that joins any two of its points contains no point that is in the region exterior to the surface; that is, the solid bounded by the surface is a convex set in space. The sphere, soup can, and box shown in Figure 11.26 are all convex. The potato skin surface shown in the figure is not convex, however, since it is possible to find two points on this surface for which the line segment connecting them contains points in the exterior region.

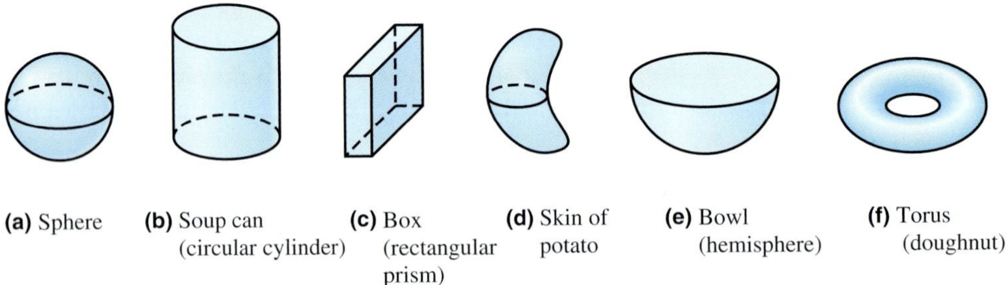

(a) Sphere **(b)** Soup can (circular cylinder) **(c)** Box (rectangular prism) **(d)** Skin of potato **(e)** Bowl (hemisphere) **(f)** Torus (doughnut)

Figure 11.26
(a), (b), (c), and (d) are simple closed surfaces; (e) is a nonclosed surface; and (f) is closed, but not simple.

Polyhedra

Joining plane polygonal regions from edge to edge forms a simple closed surface called a **polyhedron.**

> **DEFINITION** *Polyhedron*
> A **polyhedron** is a simple closed surface formed from planar polygonal regions. Each polygonal region is called a **face** of the polyhedron. The vertices and edges of the polygonal regions are called the **vertices** and **edges,** respectively, of the polyhedron.

Polyhedra (*polyhedra* is the plural of polyhedron) are named according to the number of faces. For example, a **tetrahedron** has four faces, a **pentahedron** has five faces, a **hexahedron** has six faces, and so on. Several polyhedra are pictured in Figure 11.27. The word *truncated* in the name of a polyhedron means that the polyhedron is formed by removing the corners of another polyhedron. For example, removing the eight corners of a cube creates a truncated cube with six hexagonal faces and eight triangular faces. The word *stellated* in the name of a polyhedron means that pyramids have been erected on the faces of another polyhedron.

The most spectacular polyhedral shapes on earth are the Egyptian and Mayan pyramids. Egyptian pyramids have a square base and four congruent triangular faces that meet

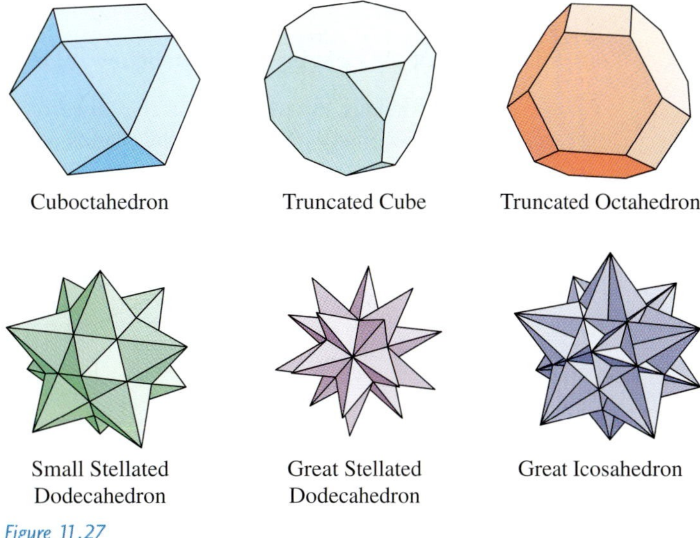

Figure 11.27
Examples of polyhedra

at a common vertex. The Mayan pyramids have a stepped form. In geometry, a **pyramid** can have any polygonal region as a base. Triangular faces rise from the base edges to meet at a common vertex, called the **apex** of the pyramid, a point that is not in the plane of the base. Examples of pyramids and their names are shown in Figure 11.28.

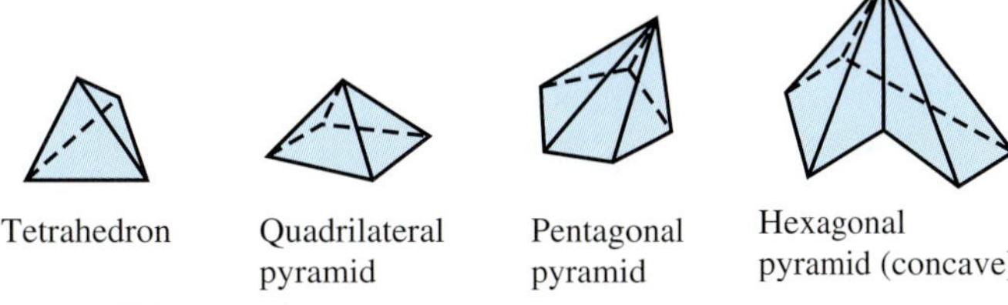

Figure 11.28
Pyramids and their names

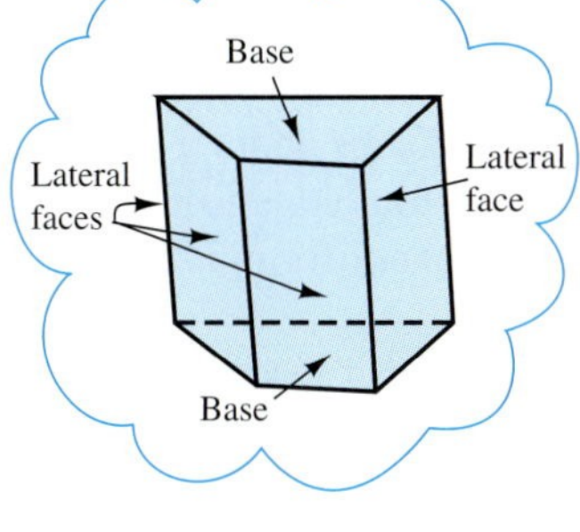

Another commonly occurring polyhedral shape is a **prism.** A prism has two **bases** that are congruent polygonal regions lying in parallel planes; the lateral faces joining the bases are all parallelograms. If the lateral faces of a prism are all rectangles, it is a **right prism;** otherwise, it is an **oblique prism** and the edges between the bases are not perpendicular to the plane of the base. Examples of prisms and their names are shown in Figure 11.29.

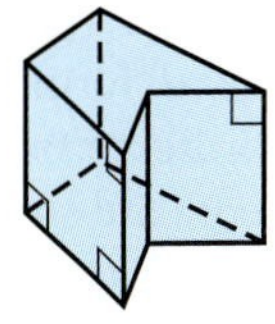
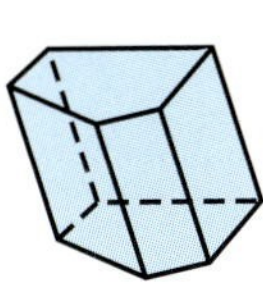

A right triangular prism

An oblique triangular prism

A right pentagonal prism (concave)

An oblique pentagonal prism

Figure 11.29
Right and oblique prisms

EXAMPLE 11.15 DETERMINING ANGLES AND PLANES IN A HEXAGONAL PRISM

The bases of the right prism shown below are regular hexagonal regions.

(a) What are the measures of the dihedral angles at which the faces intersect?
(b) How many pairs of parallel planes contain the faces of this prism?

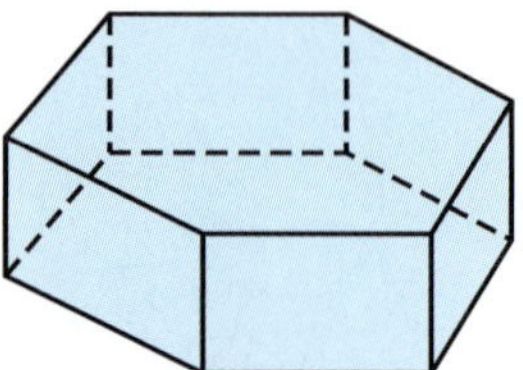

Solution

(a) Since the prism is a right prism, the dihedral angles made by either base to any lateral face measure 90°. The lateral faces meet at 120° angles, since this is the measure of the interior angles of a regular hexagon.
(b) The opposite sides of a regular hexagon are parallel, so there are three pairs of parallel planes containing the opposite lateral faces of the prism. A fourth pair of parallel planes contains the hexagonal bases of the prism.

Regular Polyhedra

DEFINITION *Regular Polyhedron*
A **regular polyhedron** is a polyhedron with these properties:

- the surface is convex;
- the faces are congruent regular polygonal regions;
- the same number of faces meet at each vertex of the polyhedron.

The most commonly seen regular polyhedron is the cube: the six faces are congruent squares, and three squares meet at each of the eight vertices. The cube is the only regular polyhedron with square faces, since if we were to attempt to put four squares about a single vertex, their interior angle measures would add up to 360°. That is, four edge-to-edge squares with a common vertex lie in a common plane and therefore cannot form a "corner" figure of a regular polyhedron.

Similar reasoning with equilateral triangles shows that corner figures in space can be formed with either three, four, or five congruent copies of the triangle. However, a convex corner cannot be formed with six or more equilateral triangles. Likewise, there is just one way to form a corner with congruent regular pentagons, and it is impossible to form a corner figure from regular n-gons for any $n \geq 6$. The possible corner figures of regular polyhedra are shown in Figure 11.30.

Each of the five corner figures depicted in Figure 11.30 can be completed to form a regular polyhedron. These are shown in Table 11.4, which also shows patterns called **nets.** Models of the polyhedra can be made by cutting the net from heavy paper, folding, and gluing. It helps to include flaps on every other outside edge of the net; these are then coated with glue and tucked under the adjoining face, forming a sturdy model.

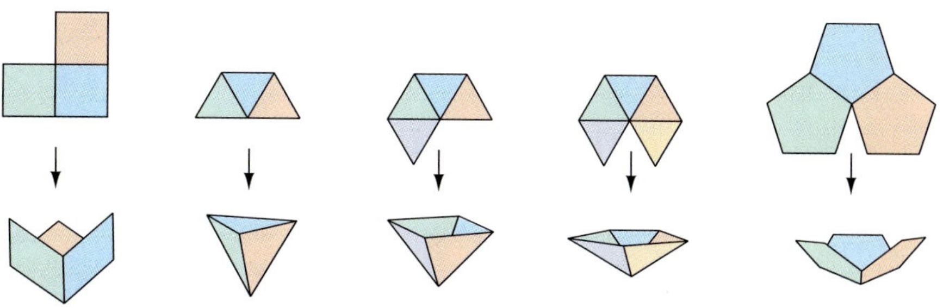

Figure 11.30
The five ways to form corner figures with congruent regular polygons

TABLE 11.4 The Five Regular Polyhedra

Polyhedron Name	Face Polygons	Net	Model
Cube	6 squares		
Tetrahedron	4 equilateral triangles		
Octahedron	8 equilateral triangles		
Icosahedron	20 equilateral triangles		
Dodecahedron	12 regular pentagons		

The five regular polyhedra were known to the ancient Greeks. They are described in Plato's book *The Republic,* and so the shapes are often referred to as the **Platonic solids.** Theaetetas (ca. 415–369 B.C.), a member of the Platonic school, is credited with the first proof that there are no regular polyhedra other than the five known to Plato.

Euler's Formula for Polyhedra

The name given to a polyhedron usually indicates its number of faces. For example, an octahedron has eight faces. A more complete description of a polyhedron includes its number of vertices and edges. The following notation will be useful:

$F =$ the number of faces of a polyhedron;

$V =$ the number of vertices of a polyhedron;

$E =$ the number of edges of a polyhedron.

For a regular octahedron, we have $F = 8$, $V = 6$, and $E = 12$.

Just for Fun

The Woven Cube

Cut three rectangular strips of paper, each strip a different color. Fold each along the dotted lines shown. Now weave the three strips into a rigid cube. The opposite faces should be the same color, and the end squares of the strips should be neatly tucked inside.

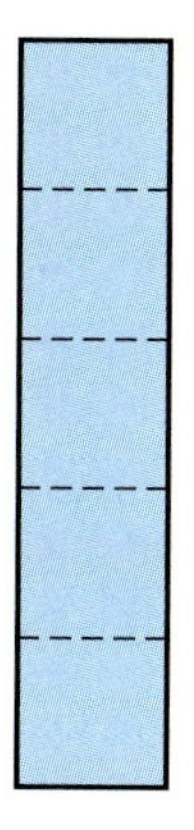

(The regular polyhedra can all be woven with suitably cut and folded paper strips. Patterns and directions may be found in *Build Your Own Polyhedra,* by Peter HIlton and Jean Pedersen, Addison-Wesley Publishing Company, 1988.)

The Envelope Tetrahedron Model

Diagrams and photos of polyhedra are certainly useful, but physical models that can be seen and touched are much better. The construction of models of polyhedra is a worthwhile classroom activity; useful geometric principles are learned as students create beautiful and interesting shapes. Skeletal models are formed easily from drinking straws joined by thin string run through the straws and tied at the vertices. Paper models, in which a carefully drawn net of the polyhedron is cut, folded, and glued, can be colored in interesting ways.

Here is a quick way to construct a regular tetrahedron from an ordinary envelope.

1. Glue the flap of the envelope down.
2. Fold the envelope in half lengthwise, forming a crease $\overline{AB}$ along the centerline.
3. Fold a corner point C upward from corner D, so that C determines point E on the centerline. Once E is found, flatten out the fold.
4. Fold the envelope straight across at E, and then cut the envelope off at the height of E.
5. Make sharp folds along $\overline{DE}$ and $\overline{CE}$.
6. Open up the envelope by pulling the two sides of the envelope at E apart; a regular tetrahedron should appear!

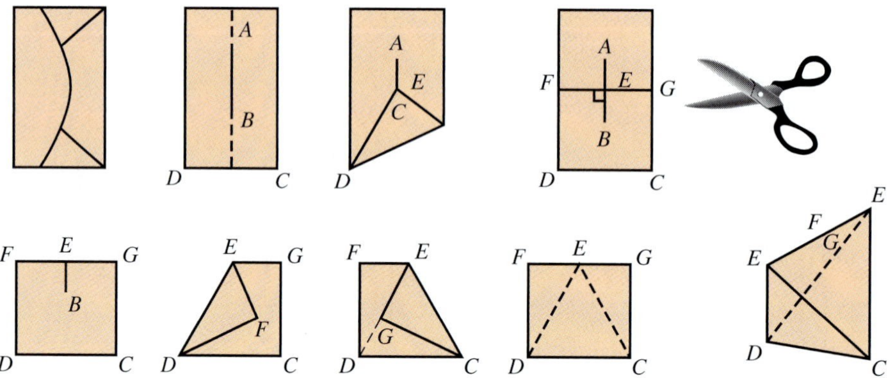

After completing your model, justify the construction procedure.

In 1752 the great Swiss mathematician Leonhard Euler discovered that the number of faces F, the number of vertices V, and the number of edges E are related to one another. Euler was unaware that he had rediscovered a formula found about 1635 by the French mathematician–philosopher René Descartes.

EXAMPLE 11.16 | DISCOVERING EULER'S FORMULA

Let V, F, and E denote the respective number of vertices, faces, and edges of a polyhedron. What relationship holds between V, F, and E?

Did You Know?

A Flexible Polyhedron

When a paper model of a polyhedron is constructed by cutting, folding, and gluing a pattern net as shown in Table 11.4, the partially completed model is quite flexible. However, when the last face is glued in place no flexibility remains. The French mathematician Augustin-Louis Cauchy (1789–1857) conjectured that all polyhedra are rigid, and in 1813 proved that all *convex* polyhedra are indeed rigid. For over 150 years no one could show that concave polyhedra must also be rigid. In 1978 the issue was resolved in an unexpected direction: Robert Connelly of Cornell University constructed a polyhedron from 18 triangular faces that is noticeably flexible! Several other flexible polyhedra are now known. The simplest one, with just 14 faces, was found by Klaus Steffen. It is pictured below, together with its pattern. All of the flexible polyhedra discovered so far share a remarkable property: as the surface flexes, the volume of the enclosed region remains the same!

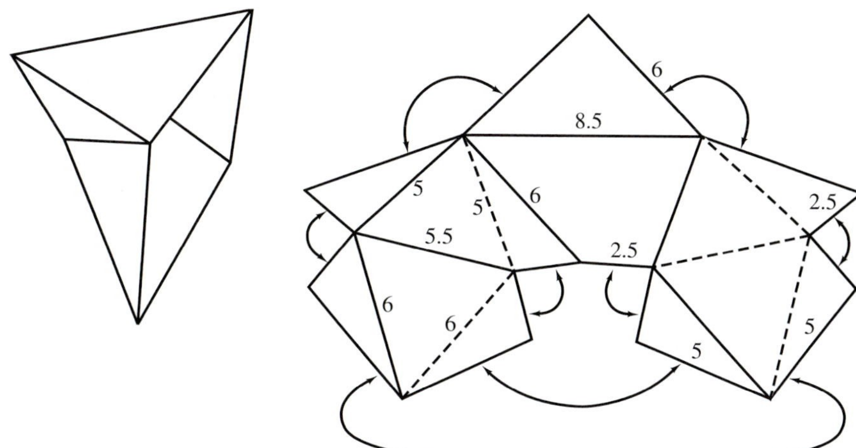

SOURCE: Figure 21 and Figure 22 from Mathematics Magazine, *Volume 52, Number 5, November 1979, p. 281. Copyright © 1979 by the Mathematical Association of America. Reprinted by permission.*

Solution

Understand the Problem

The numbers V, F, and E are not independent of one another. The goal is to uncover a formula that relates the three numbers corresponding to *any* polyhedron.

Devise a Plan

Formulas are often revealed by seeing a pattern in specific cases. By making a table of values of V, F, and E, we have a better chance to see what this pattern may be. To be confident that the pattern holds for all polyhedra, we need to examine polyhedra of varied kinds.

Carry Out the Plan

A pentagonal pyramid, a hexagonal prism, a "house," and a truncated icosahedron are shown below. The truncated icosahedron, formed by slicing off the corners of an icosahedron to form pentagons, may look familiar; it is a common pattern on soccer balls. It is also the pattern of a buckyball, as shown in the picture gallery of space figures at the beginning of this section.

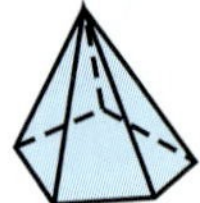
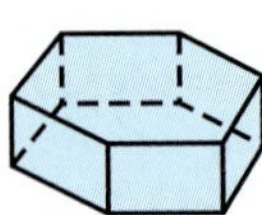
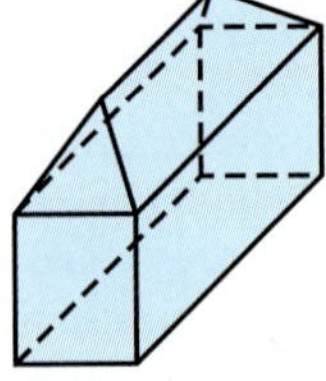
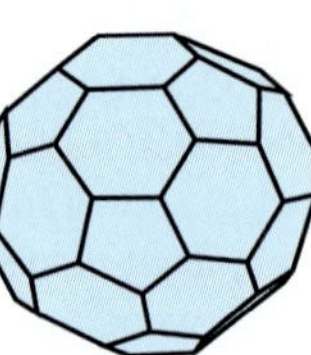

The following table lists the number of vertices, faces, and edges for these polyhedra, as well as for some of the regular polyhedra depicted in Table 11.4.

Polyhedron	V	F	E
Pentagonal pyramid	6	6	10
Hexagonal prism	12	8	18
"House"	10	9	17
Cube	8	6	12
Tetrahedron	4	4	6
Octahedron	6	8	12
Truncated icosahedron	60	32	90

The table reveals that the sum of the number of vertices and faces is 2 larger than the number of edges. That is, $V + F = E + 2$.

Look Back

If any two values of V, F, and E are known, the remaining value can be found using Euler's formula $V + F = E + 2$. For example, the dodecahedron has $F = 12$ pentagonal faces. The product $5 \cdot 12$ counts *twice* the number of edges, since each edge borders two of the pentagonal faces. Thus, $E = 5 \cdot 12/2 = 30$ for the dodecahedron. Euler's formula can now be used to compute the number of vertices. Solving for V, we get $V = E + 2 - F = 30 + 2 - 12 = 20$, so a dodecahedron has 20 vertices.

The evidence gathered in Example 11.16 supports the following theorem. A proof of the theorem is given in Section 11.4.

THEOREM *Euler's Formula for Polyhedra*

Let V, F, and E denote the respective number of vertices, faces, and edges of a polyhedron. Then

$$V + F = E + 2.$$

School Book Page

Exploring Solids in Grade Two

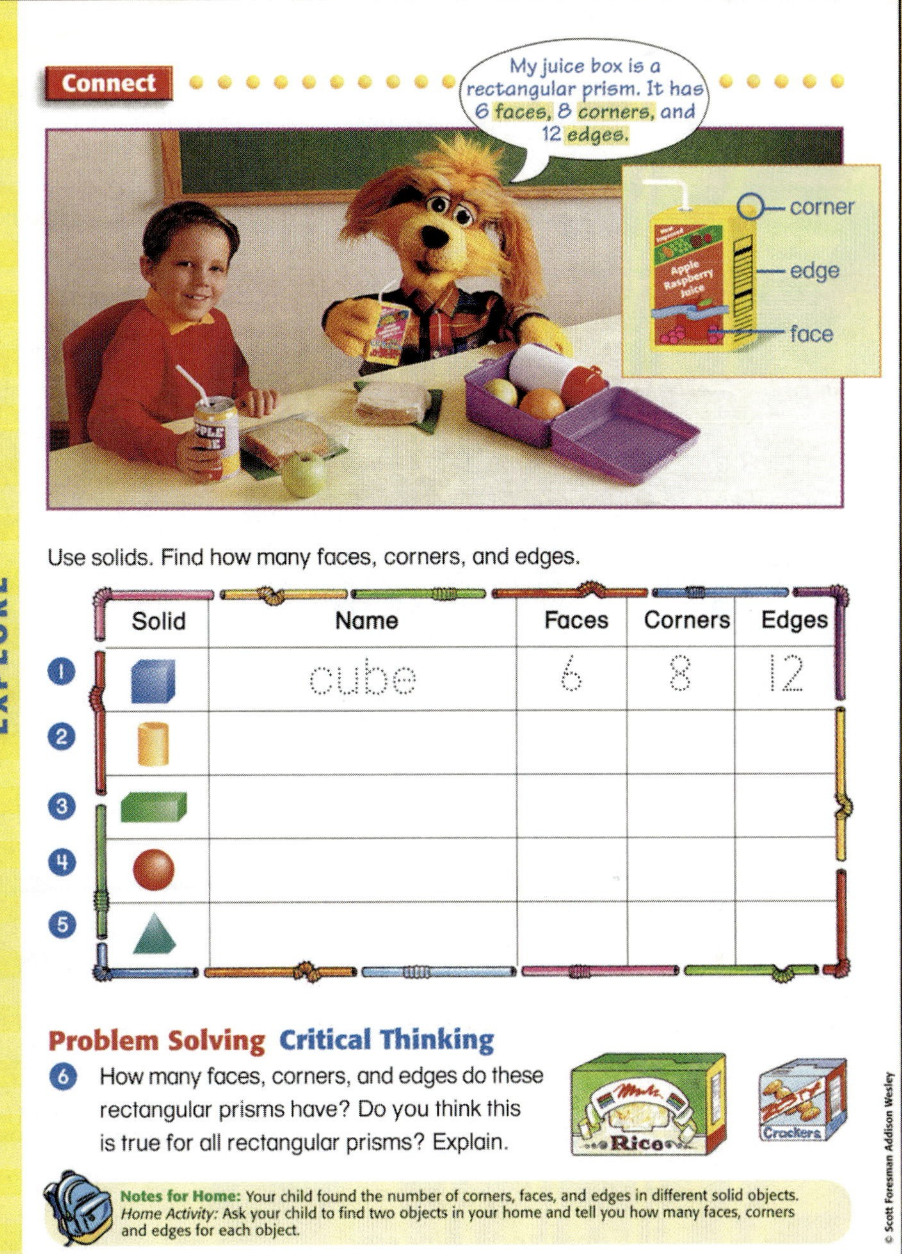

EXPLORE

Connect

My juice box is a rectangular prism. It has 6 faces, 8 corners, and 12 edges.

Use solids. Find how many faces, corners, and edges.

	Solid	Name	Faces	Corners	Edges
1		cube	6	8	12
2					
3					
4					
5					

Problem Solving Critical Thinking

6 How many faces, corners, and edges do these rectangular prisms have? Do you think this is true for all rectangular prisms? Explain.

Notes for Home: Your child found the number of corners, faces, and edges in different solid objects. *Home Activity:* Ask your child to find two objects in your home and tell you how many faces, corners and edges for each object.

446 four hundred forty-six

SOURCE: From Scott Foresman–Addison Wesley Math, *Grade 2, p. 446, by Randall I. Charles et al. Copyright © 2002 Pearson Education, Inc. Reprinted with permission.*

Questions for the Teacher

1. All the students have identified the second solid as a cylinder, but some have filled in the respective numbers of faces, corners, and edges with 2, 0, 0 while others think it should be 3, 0, 2. Which answer is correct, and how would you explain the mistakes being made by some of the students?
2. In *Notes for Home* at the bottom of the School Book Page, students will have little difficulty finding examples of rectangular prisms in their homes. What are other objects they may look for that have some of the other shapes shown on the School Book Page?

EXAMPLE 11.17 SEARCHING FOR POLYHEDRA WITH HEXAGONAL FACES

The Epcot Center in Florida is the site of one of the world's largest geodesic domes. The surface of the 165-foot-diameter structure is covered with both hexagons and pentagons. Similarly, the microscopic frame of the radiolarian (a unicellular planktonic organism) is covered by pentagons and hexagons. These shapes suggest the following question: *Can all the faces of a polyhedron be hexagonal?* Show that this is not possible, even if the hexagons need not all be congruent and are permitted to be irregular.

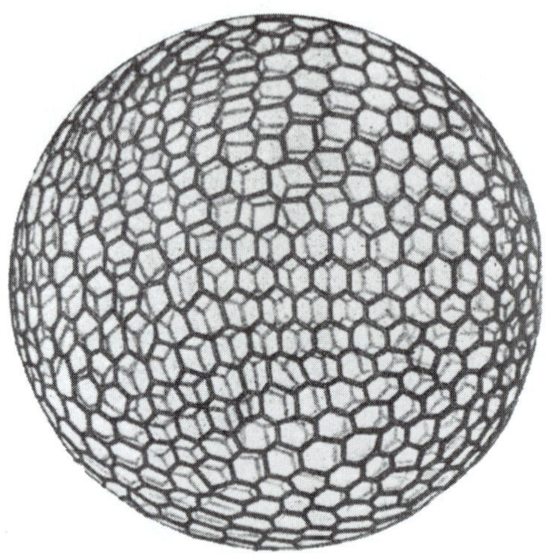

Solution Suppose, to the contrary, that hexagonal faces can form a polyhedron. As usual, let F, E, and V denote the number of faces, edges, and vertices. Since each face is bordered by 6 edges and each edge touches 2 faces, we obtain the formula $6F = 2E$. Thus, we have $F = \frac{2E}{6} = \frac{E}{3}$. Moreover, each of the E edges has 2 ends, so there are $2E$ ends of edges altogether. Since at least 3 ends of edges meet at each of the V vertices, we see that $3V \leq 2E$, or, equivalently, $V \leq \frac{2E}{3}$. Adding $V \leq \frac{2E}{3}$ to $F = \frac{E}{3}$, we get the inequality $V + F \leq \frac{2E}{3} + \frac{E}{3} = E$. But this contradicts Euler's formula, which tells us that $V + F = E + 2 > E$. We conclude that there is no polyhedron whose faces are all hexagons.

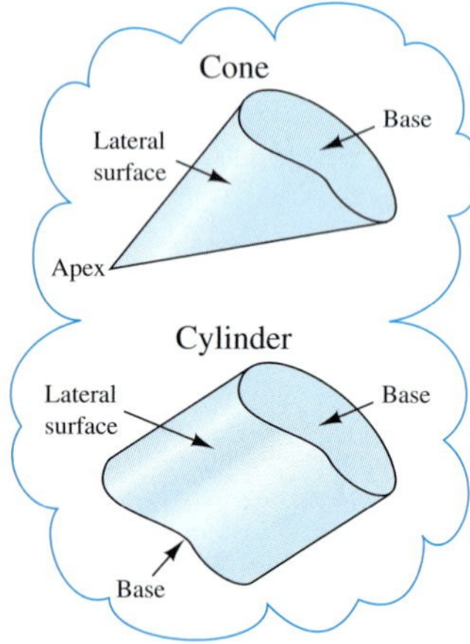

Cones and Cylinders

Cones and cylinders are simple closed surfaces that generalize pyramids and prisms, respectively. A **cone** has a **base** that is a region bounded by a simple closed curve in a plane. The curved **lateral surface** is generated by the line segments that join one point—the **apex** (or **vertex**)—not in the plane of the base to the points of the curve bounding the base. A **right circular cone,** an **oblique circular cone,** and a **general cone** are shown in Figure 11.31. The line segment $\overline{AB}$ through the apex A of a cone that intersects the plane of the base perpendicularly at B is called the **altitude** of the cone.

A **cylinder** is the surface generated by translating the points of a simple closed region in one plane to a parallel plane. Examples are shown in Figure 11.32. The points joining corresponding points on the curves bounding the bases form the **lateral surface.** If the line segments joining corresponding points in the two bases are perpendicular to the planes of the bases, it is a **right cylinder.** Cylinders that are not right cylinders are **oblique cylinders.**

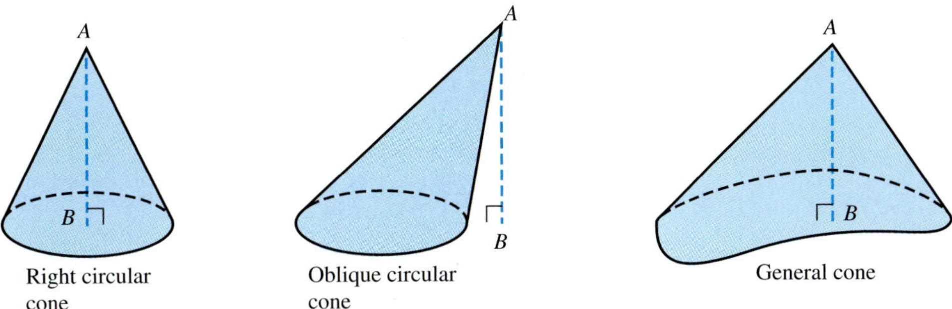

Figure 11.31
A right circular cone, an oblique circular cone, and a general cone

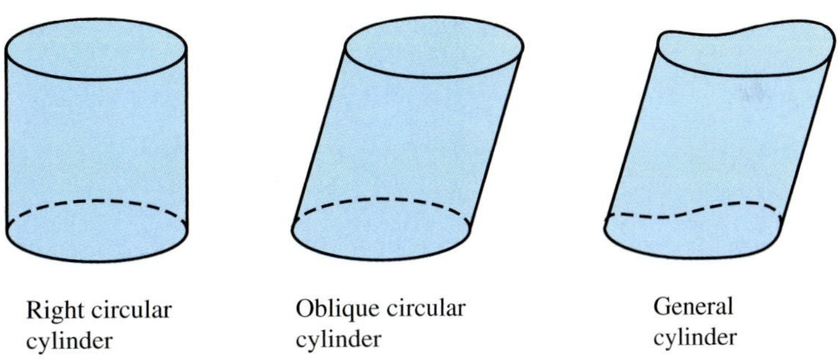

Figure 11.32
A right circular cylinder, an oblique circular cylinder, and a general cylinder

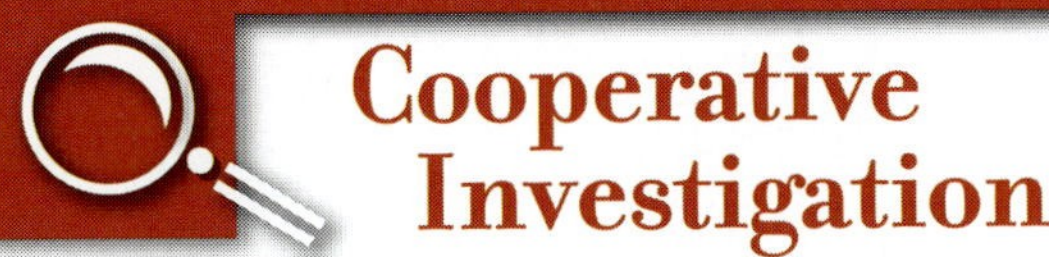

Cooperative Investigation

From Paper Discs to Polyhedra

Materials Needed

Five congruent (about 4 to 5 inches in diameter) paper discs per person and glue sticks.

Directions

In the Cooperative Investigation: From Paper Discs to Polygons (see Section 11.2), it was discovered that a paper disc can be folded into an equilateral triangle by folding three arcs to the center O of the disc.

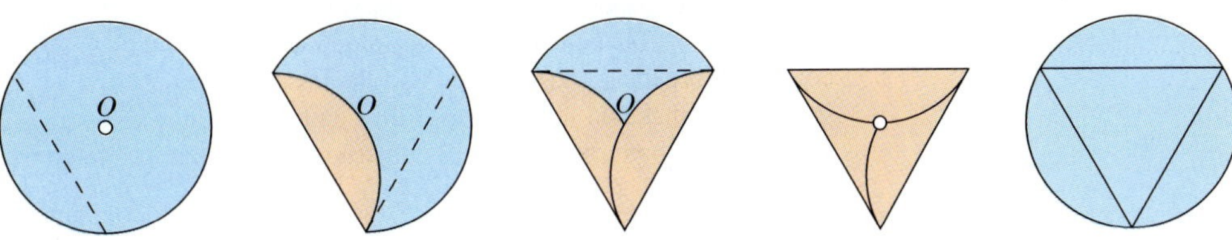

In this investigation, you will work in groups of four to fold and glue paper discs to construct polyhedra.

Activities

Each group (not individuals) will construct the polyhedra described as follows:

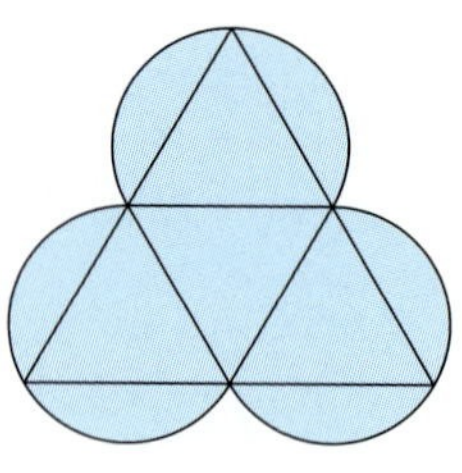

1. Glue four folded discs together to form the pattern shown here. The circular segments between triangles are folded upward and glued together along their common edge. Now fold the remaining six circular segments upright, apply glue, and make a regular tetrahedron.

2. Use a glue stick to construct *two* copies of the four-triangle pattern shown here. Join the two patterns to make a regular octahedron.

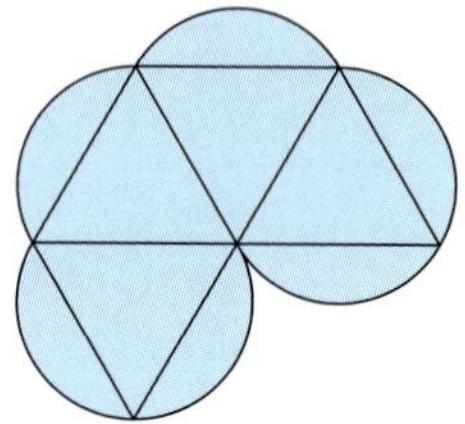

3. Work in cooperation with another four-person group to construct an icosahedron. Each group can make a five-disc pattern as shown here. What other pattern can each group make to complete the icosahedron?

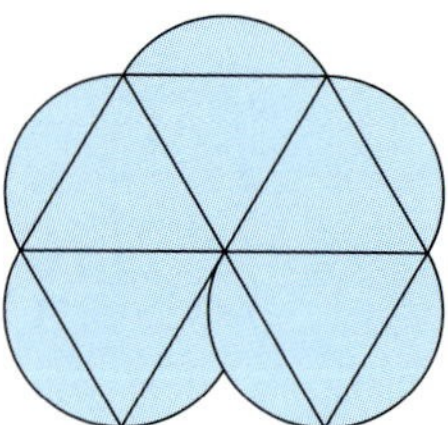

4. Construct a double pyramid with six folded discs.
5. Convex polyhedra with congruent equilateral triangular faces are called **deltahedra.** Four deltahedra have been constructed in Activities 1 through 4. There are eight deltahedra in all—see if you can find the remaining four types by constructing new deltahedra with folded discs. Compare your discoveries with other groups.
6. Fold a stack of four paper discs to form four congruent *scalene* triangles. Glue the four discs to form the pattern shown. Does this pattern form an irregular tetrahedron?

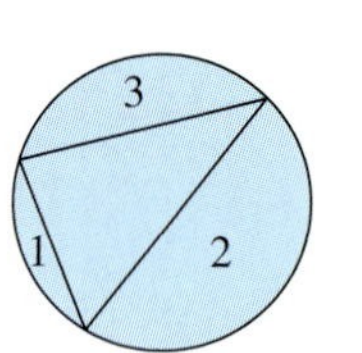

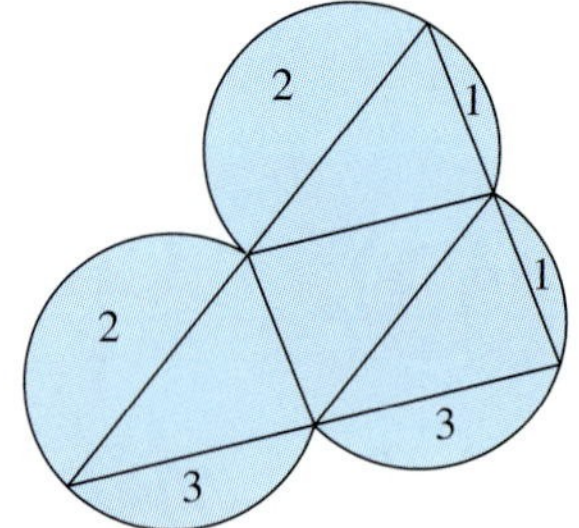

7. Fold eight congruent scalene triangles. Turn four of the discs upside down (it's helpful to make the eight triangles from paper discs of two colors, turning over the discs of one of the colors). Use two folded discs of each color to make the pattern shown, and then use the remaining discs to make an identical four-disc pattern. Can the two four-disc patterns be used to construct an irregular octahedron?

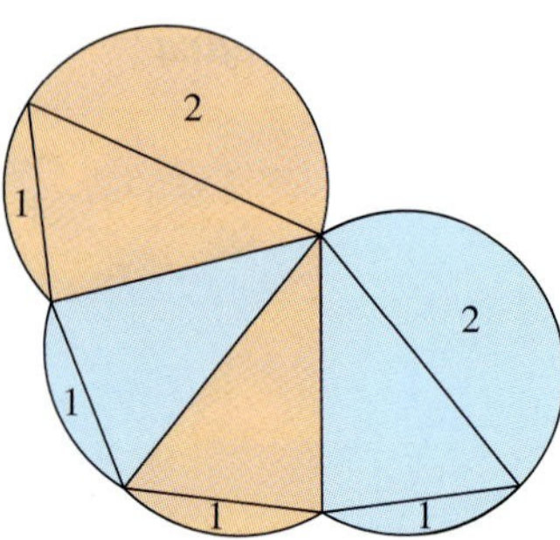

Problem Set 11.3

Understanding Concepts

1. Which of the following figures are polyhedra?

(a) (b) (c)

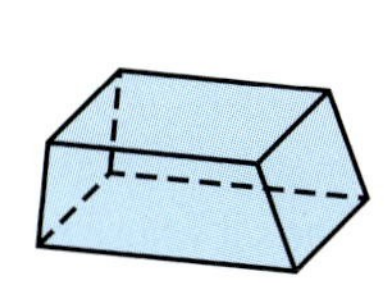
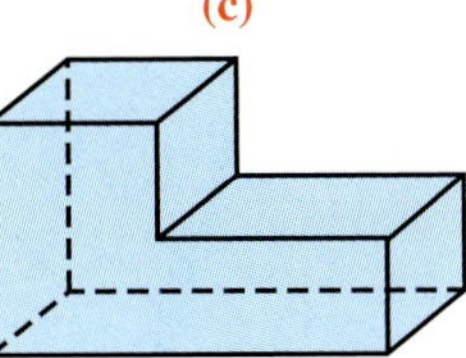

(d) (e) (f)

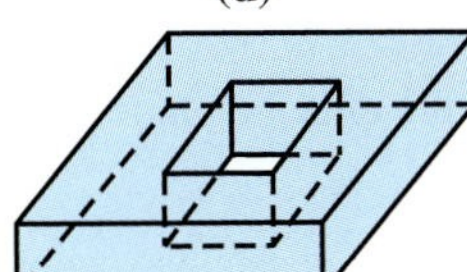
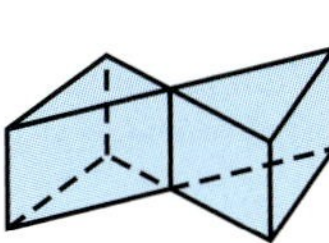
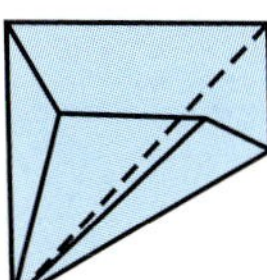

2. Name each of these surfaces.

(a) (b) (c) (d)

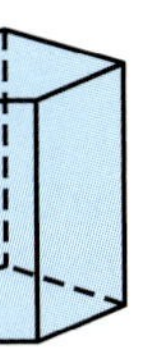
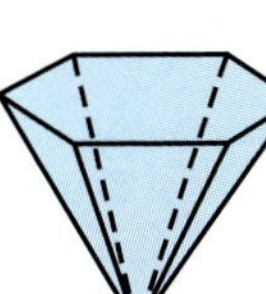
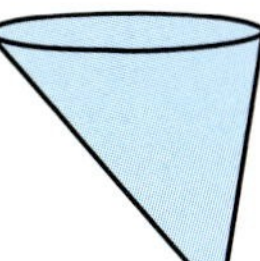
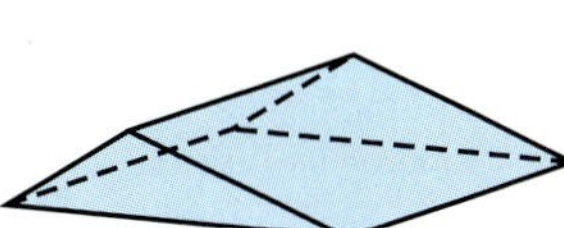

(e) (f) (g)

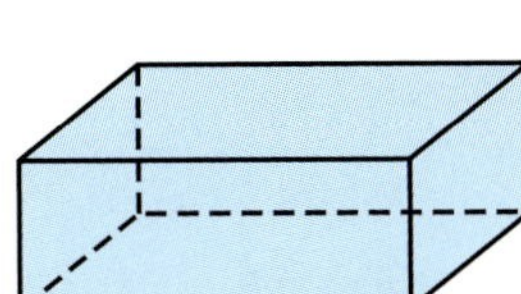
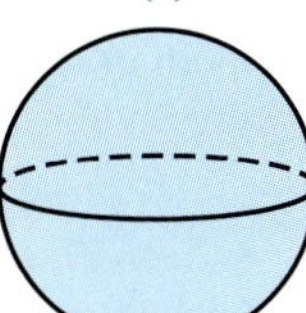
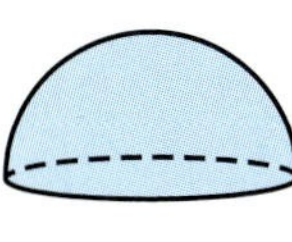

3. A tetrahedron is shown below.
   (a) How many planes contain its faces?
   (b) Name all of the edges.
   (c) Name all of the vertices.
   (d) Name all of the faces.

4. Draw freehand pictures of the following figures, using dashed lines to indicate hidden edges. Don't just copy—transfer the image in your mind to paper.
   (a) cube (b) tetrahedron (c) square pyramid
   (d) pentagonal right prism
   (e) oblique hexagonal prism
   (f) octahedron
   (g) right circular cone
5. It's easy to make two triangles from five toothpicks, touching only at the ends (see figure on next page). Show how to make four triangles from six toothpicks.

Figure for Problem 5

6. A cube with vertices A, B, C, D, E, F, G, and H is shown in the accompanying figure. Vertices D, E, G, and H are the vertices of a tetrahedron inscribed in the cube.

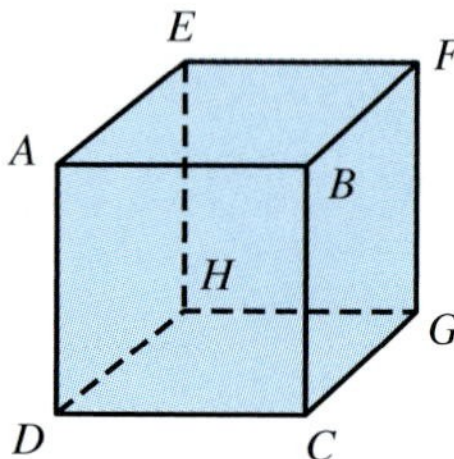

(a) Trace the cube in one color and then draw the tetrahedron $DEGH$ in a different color.

(b) Three of the faces of the tetrahedron $DEGH$ are subsets of the faces of the cube. Find a tetrahedron inscribed in the cube that has none of its faces in the planes of the faces of the cube. Sketch your tetrahedron and the surrounding cube.

7. The pattern shown on the left folds up to form the cube on the right.

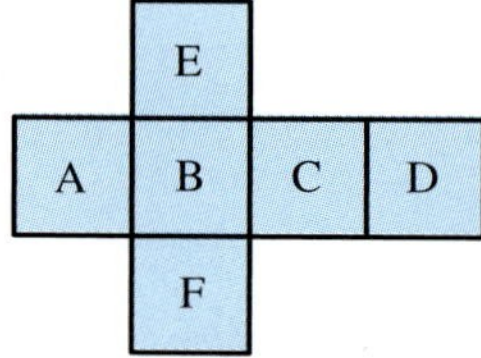

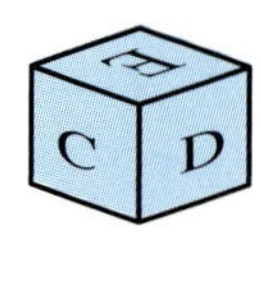

Sketch the letter, *in its correct orientation,* that should appear on each blank face shown below, where the same pattern is used.

(a)

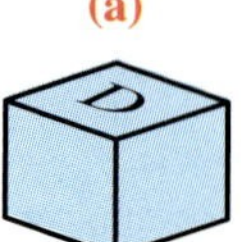

(b)

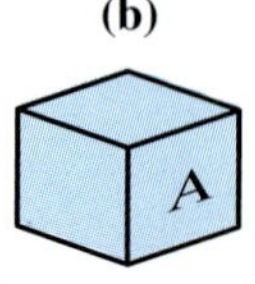

(c)

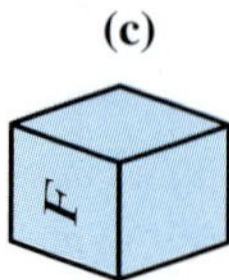

8. The apex of the pyramid shown below is at the center of the cube shown by the dashed lines. What is the dihedral angle that each lateral face of the pyramid makes

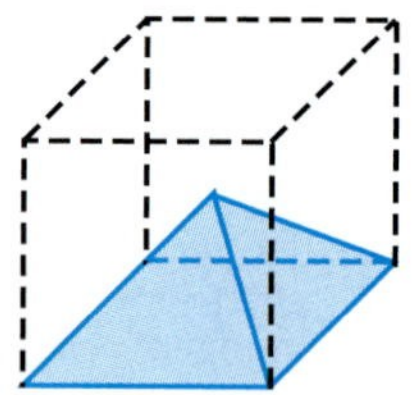

(a) to the square base?

(b) to an adjacent lateral face? (*Hint for (b):* It will help to imagine that the cube is filled with six congruent pyramids whose apexes coincide at the cube's center.)

9. The figure shown here is a right prism whose bases are regular pentagons.

(a) What is the measure of the dihedral angle between each lateral face and a base?

(b) What is the measure of the dihedral angle between adjacent lateral faces?

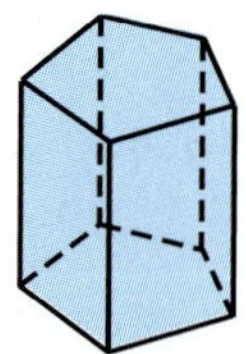

10. Sketch a cube. Color the eight vertices either red or blue (or mark R or B at the vertices) so that no plane through any four vertices of the cube has the same color at all four vertices.

11. The numbers in the 2-by-3 grid of squares correspond to the pattern of stacked cubes shown in the **isometric drawing** on the right.

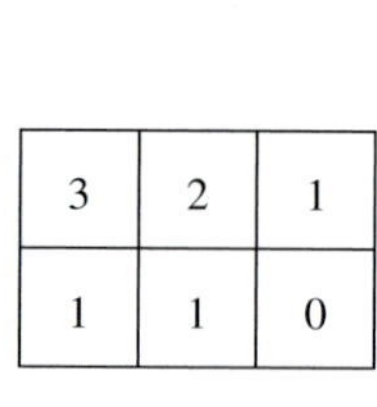

3	2	1
1	1	0

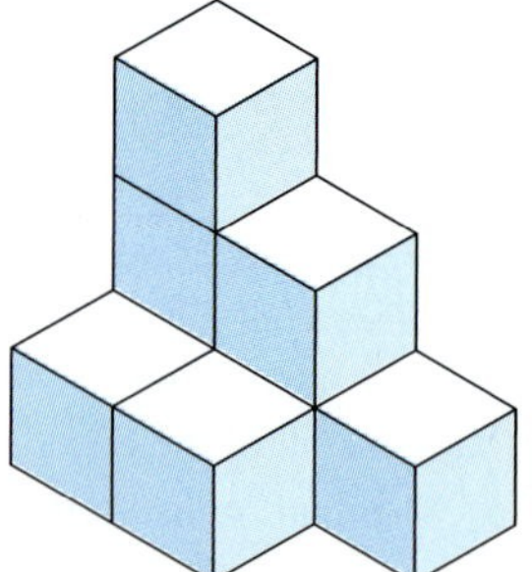

Make an isometric drawing of these patterns.

(a)

2	3	2
2	2	1

(b)

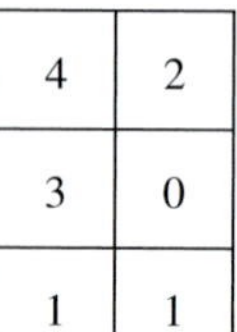

4	2
3	0
1	1

12. Are any of the regular polyhedra

(a) prisms?

(b) pyramids?

13. Verify Euler's formula for

(a) a pyramid with a hexagonal base.

(b) a prism with octagonal bases.

(c) the regular icosahedron. (*Suggestion:* Modify the counting method used for the regular dodecahedron in the Look Back step of Example 11.16.)

14. Complete the table below using Euler's formula. These four polyhedra are representative of a class of 13 polyhedra discovered by Archimedes.

Polyhedron	NUMBER OF VERTICES, FACES, AND EDGES V	F	E
Truncated tetrahedron	12	8	—
Truncated dodecahedron	—	32	90
Snub cube	24	—	60
Great rhombicosidodecahedron	120	62	—

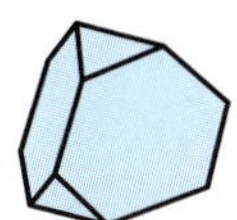

Truncated tetrahedron

Truncated dodecahedron

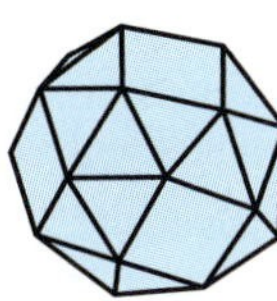

Snub cube

Great rhombicosidodecahedron

15. A **double pyramid** (or **dipyramid**) is a polyhedron with triangular faces arranged about a plane polygon and extending both upward to an upper apex and downward to a lower apex.

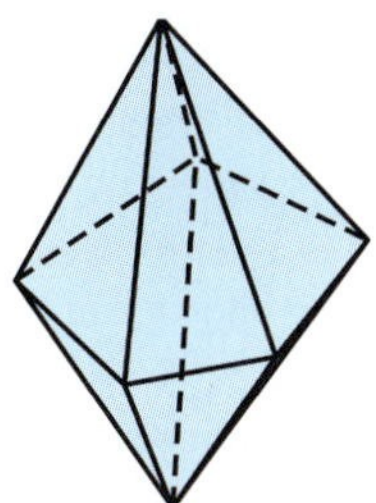

(a) Find the number of vertices, edges, and faces for the pentagonal double pyramid shown. Then verify that Euler's formula is satisfied.

(b) Repeat part (a), but for the double pyramid built from a polygon with 20 sides.

16. An **antiprism** has two base polygons with the same number of sides, together with triangular lateral faces.

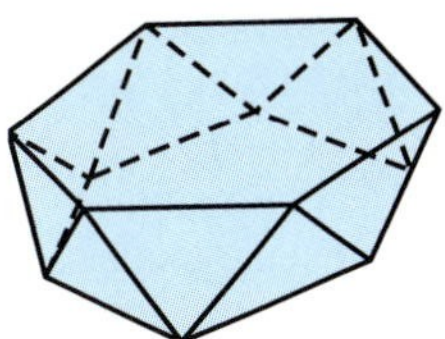

(a) Find the number of vertices, edges, and faces for the hexagonal antiprism shown. Then verify that Euler's formula is satisfied.

(b) Repeat part (a), but for an antiprism whose bases are 23-gons.

17. Draw nets corresponding to the following polyhedra.

(a) A square pyramid with equilateral triangular faces

(b) A truncated tetrahedron (See the figure in problem 14.)

18. Check whether Euler's formula holds for the figures below. If not, explain what assumption required for Euler's formula is not met.

(a)

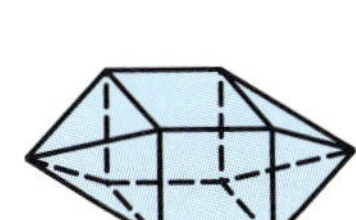

(b)

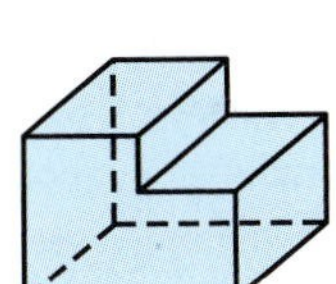

(c)

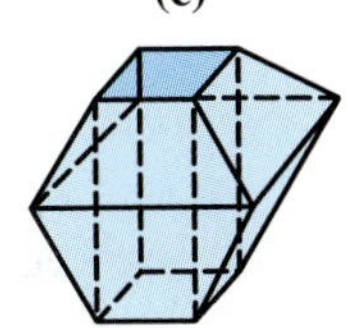

(An octahedron with a square hole)

Teaching Concepts

19. **Shape Search.** Go on a geometry walk; search for interesting three-dimensional shapes you find around campus, at the grocery store, in local sculptures, in nature, . . . , wherever! Photograph or make accurate drawings of three or four shapes that you find especially interesting, and then write careful descriptions that discuss the properties of your shapes. For example, if your shape is a polyhedron, count its vertices, faces, and edges and verify Euler's formula. If possible, include some shapes that can be contributed to a class collection. For example, many products at the store are packaged in ingenious boxes.

20. **Make a Shape.** Children (and college students, too) can learn important principles of geometry by constructing their own three-dimensional shapes. Often, just ordinary materials such as toothpicks and mini-marshmallows suffice, say, to make one of the regular polyhedra or even a figure as complex as a Buckyball (the truncated icosahedron). Instructions can be found readily from several interesting sites on the Internet. A well-equipped math lab might also include some commercially manufactured kits that can be used to create fun shapes, such as Polydrons™ or the Zome System. Your challenge: make an interesting shape, and write a brief report about its properties and why you find it interesting.

Thinking Critically

21. Let V, E, and F denote the number of vertices, edges, and faces, respectively, of a polyhedron.

(a) Explain why $2E \geq 3F$. (*Hint:* Every face has at least three sides, and every edge borders two faces.)

(b) Explain why $2E \geq 3V$. (*Hint:* Every vertex is the endpoint of at least three edges.)

(c) Show that every polyhedron has at least six edges. (*Hint:* Add the inequalities of parts (a) and (b), and use Euler's formula.)

(d) Use (a) and (b) to prove that no polyhedron can have seven edges. (*Hint:* Use Euler's formula.)

(e) Show that there are polyhedra with 6, 8, 9, 10, . . . edges.

22. A convex polyhedron with five faces is called a **pentahedron.** Find and sketch two different types of pentahedra. (*Hint:* One is "easy as pie.")

23. A polyhedron with six faces is a **hexahedron.** For example, a cube is a hexahedron.

(a) Draw a pyramid that is a hexahedron.

(b) Draw a double pyramid (see problem 15) that is a hexahedron.

24. Suppose a skeletal model of a convex polyhedron is made, where just the edges of the polyhedron are outlined. If the model is viewed in perspective from a position just outside the center of a face, the edges of that face form a bounding polygon inside which the remaining edges are seen. The resulting pattern of edges is called a **Schlegel diagram,** named for the German mathematician Viktor Schlegel, who invented the diagram in 1883. Schlegel diagrams for the cube and dodecahedron are shown below.

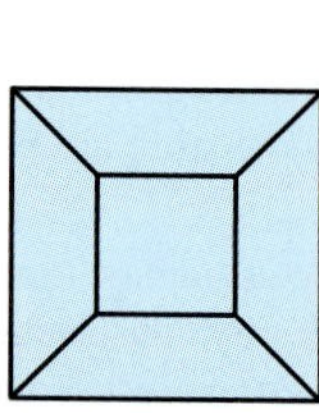

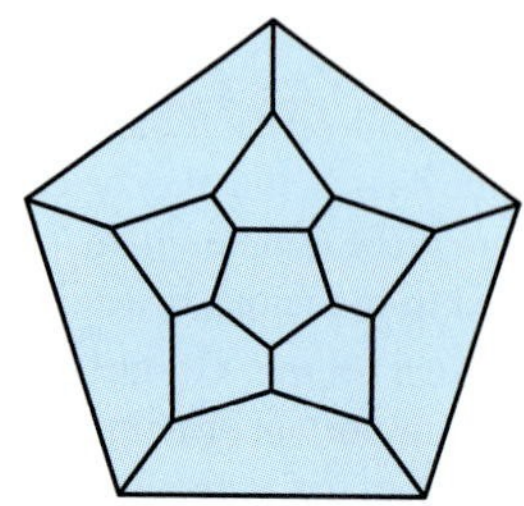

Draw Schlegel diagrams for

(a) the tetrahedron

(b) the octahedron

(c) the icosahedron (*Hint:* Start by drawing a large equilateral triangle, and keep in mind that each vertex must touch five edges.)

25. The dihedral angles of the regular polyhedra are given in the following table.

Regular Polyhedron	Measure of Dihedral Angle (in degrees and minutes)
Cube	90°
Tetrahedron	70° 32′
Octahedron	109° 28′
Dodecahedron	116° 34′
Icosahedron	138° 11′

(a) How many tetrahedra can be placed around a common edge without overlap? What is the size of the gap that remains?

(b) Why is the cube the only regular polyhedron that will completely fill space?

Thinking Cooperatively

26. Five congruent squares can be joined along their edges to form 12 distinct shapes known as **pentominoes.** Pentominoes were invented by mathematician and electrical engineer Solomon Golomb in 1953 in a talk to the Harvard Mathematics Club. They are named by the letters they somewhat resemble.

(a) Which of the 12 pentominoes on the next page can be folded to form a cube-shaped box with an open top?

(b) Copy the T pentomino onto paper, cut the pattern with scissors, and construct an open-topped cubical box by taping edges. Now choose a target pentomino (not the T since it's too easy!), and sketch it on the bottom of the box. Then, use scissors to cut the box apart to form the targeted pentomino shape. (*Note:* This activity was created by Marion Walter of the University of Oregon, who makes open cubical boxes from the bottoms of clean milk cartons, discarding the tops.)

27. A **hexomino** is formed by joining six congruent squares along their edges. There are 35 different hexominoes in all, including the five shown on the next page.

(a) Hexomino (i) is a net for a cube. Which of the other four hexominoes shown can be folded to form a cube?

(b) There are 11 hexominoes that can be folded to form a cube. Try to find all 11 shapes. Be careful

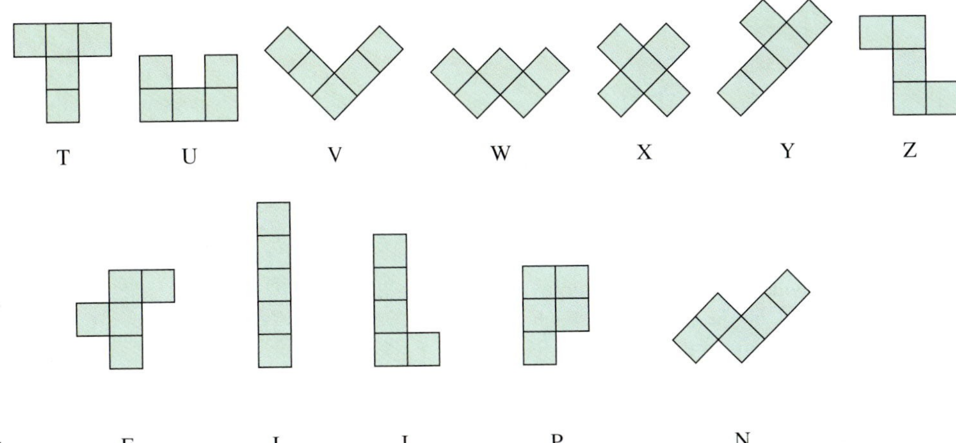

Figure for Problem 26

SOURCE: Figure 10, "The pentominoes" from Polyominoes, by Solomon W. Golomb, page 23. (Charles Scribner's Sons, 1965, Revised Edition, Princeton University Press, 1994.)

not to repeat a shape; two congruent shapes may at first appear to be different when one is rotated or flipped over.

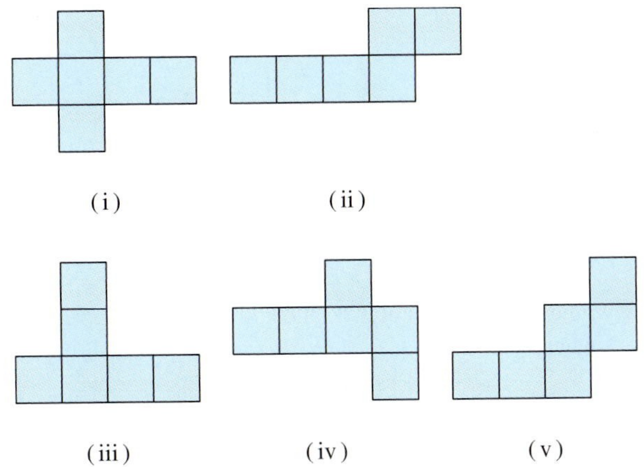

28. A net for a square pyramid is shown on the left. A net for a more general quadrilateral pyramid is shown on the right. The dot at P in each net locates the point in the plane of the base directly beneath the apex of the pyramid.

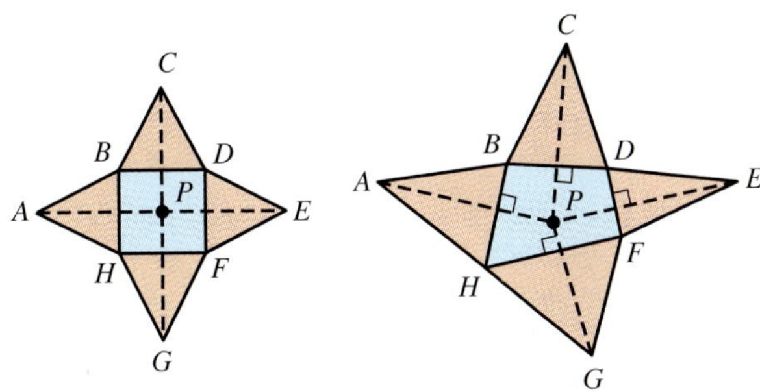

(a) Explain why $AB = BC$, $CD = DE, \ldots, GH = HA$ in the nets and why the dashed lines from P are perpendicular to the sides of the base polygon.

(b) Draw a convex polygon, a point P in its interior, and the rays from P that are perpendicular to the sides of the polygon. Choose a point A on one of the rays, and use a compass to draw the arc at vertex B that constructs point C as the intersection of the arc with the next ray. (See the figure.) Continue to draw circular arcs, completing your pattern. Finally, cut, fold, and tape your pattern to construct your paper pyramid.

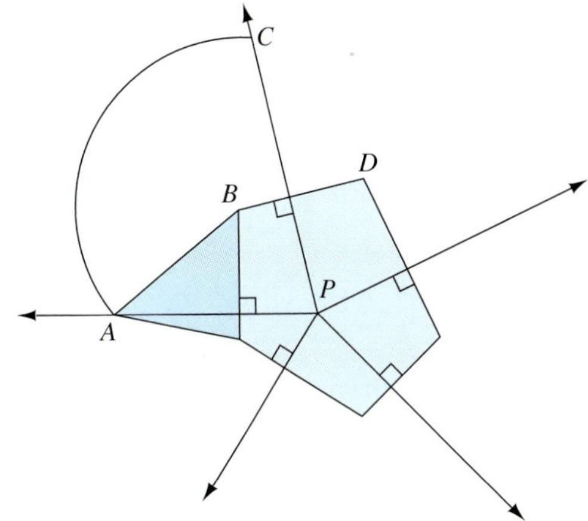

Making Connections

29. The round door shown on the next page has a problem. What facts of space geometry create a difficulty?

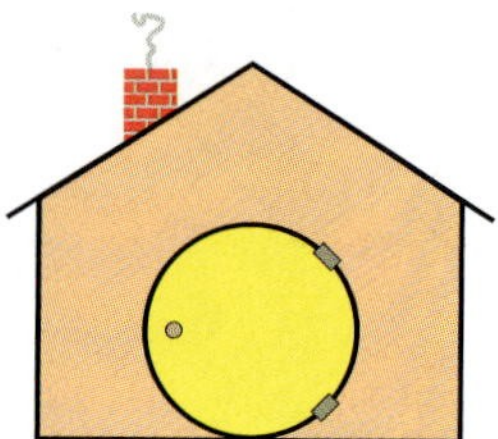

What is important about the placement of door hinges?

30. The ancient Greeks divided physical space into five parts: the universe, earth, air, fire, and water. Each of these was associated with one of the five regular polyhedra. Investigate what correspondence was made.

31. Biologists and physical scientists frequently become involved with the analysis of shape and form. For example, many viruses have an icosahedral structure, and the crystalline structures of minerals are often polyhedral forms of considerable beauty. An example of a pyrite crystal is shown here.

Browse through your school library and see what three-dimensional shapes are receiving interest and attention. Report on your findings.

From State Student Assessments

32. (Washington State, Grade 4)
Look at the cube below.

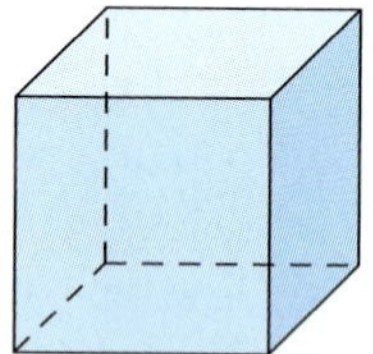

Now look at figures A, B, and C.

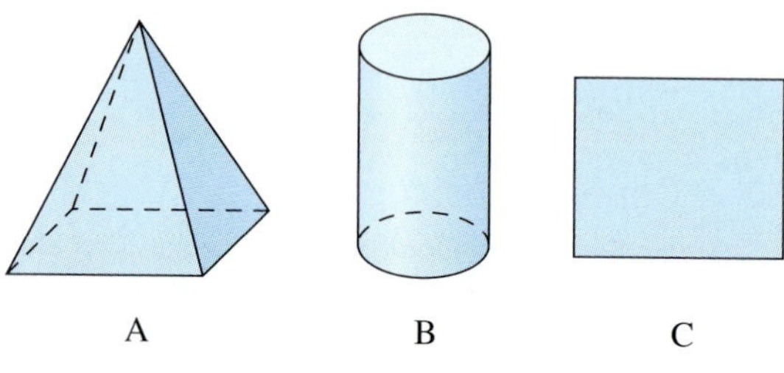

Choose one figure. Tell what that figure has in common with the cube. Explain your answer using words, numbers, or pictures. Now choose a different figure. Tell something it has in common with the cube. Explain your answer using words, numbers, or pictures.

33. (Illinois, Grade 5)
Angela wants to construct a cylinder from paper shapes. What shapes will she need?

A. 2 circles and 1 rectangle
B. 4 triangles and 1 square
C. 2 triangles and 1 rectangle
D. 2 hexagons and 2 rectangles
E. 3 circles and 3 triangles

34. (Massachusetts, Grade 4)
Which shape CAN be folded to form an OPEN box?

A. 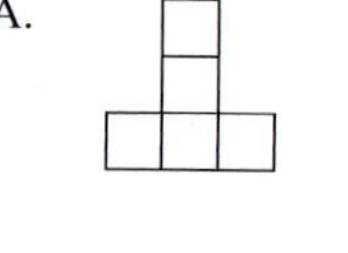

B.

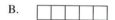

C.

D.

For Review

35. Find the measures of x, y, and z in the following figure, where $l \parallel m$.

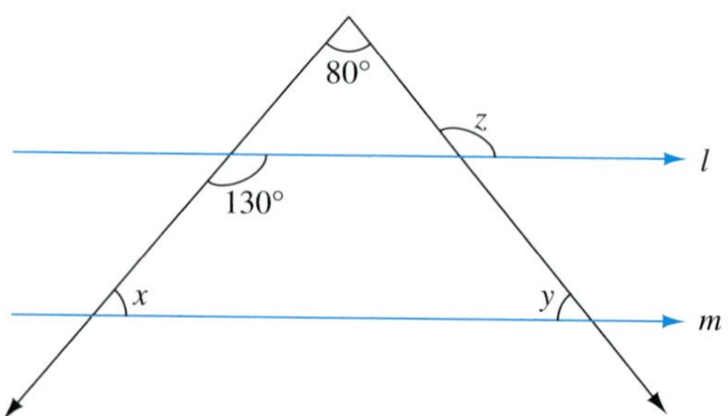

36. A triangle has no diagonals, and a convex quadrilateral has two diagonals.

(a) Fill in the entries in the table on the next page.
(b) Describe a pattern that you see in the table.
(c) How many diagonals are in a convex dodecagon?
(d) How many diagonals are in a convex 100-gon?

Table for Problem 36(a)

	n					
	3	4	5	6	7	8
Number of diagonals in convex n-gon	0	2				

37. It is easy to join the 12 points of a 3-by-4 square array by a polygon. Here is one way.

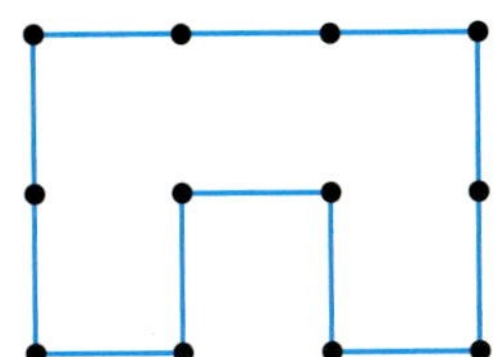

Find polygons that join all of the points of these square arrays.

(a) 4-by-6

(b) 5-by-7

11.4 NETWORKS

The Königsberg Bridge Problem

Leonhard Euler (1707–1783) (see biography in Section 10.2) lived for a short time in the East Prussian city of Königsberg, now called Kaliningrad in the Russian Federation. The Pregel River flows through the city, forming two islands. At the time of Euler there were seven bridges connecting the islands to one another and to the two sides of the river. Euler's own drawing is reproduced in Figure 11.33. The islands are labeled A and D, the shores of the river are labeled B and C, and the bridges are labeled a, b, c, d, e, f, and g.

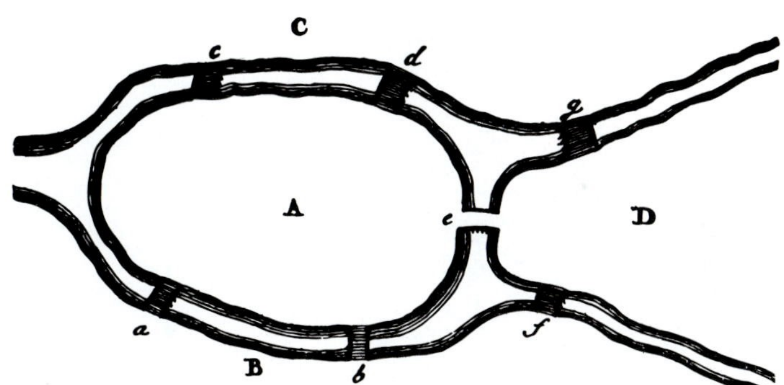

Figure 11.33
The seven bridges of Königsberg in the early 1700s

It was common for Königsbergers to take Sunday walks, and people wondered if it was possible to walk over all seven bridges without crossing any bridge more than one time. Euler solved the now famous Königsberg Bridge Problem to illustrate the ideas of what he called "the geometry of position," and what today is called topology. His most important step was to associate each of the four land masses with a point—A, B, C, or D. For each bridge joining one land mass to another, he drew a curve from one point to the other corresponding point. Euler's representation of the problem, shown in Figure 11.34, is a system of points and curves known as a **network.**

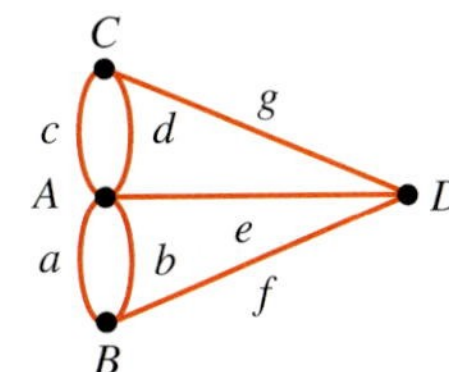

Figure 11.34
The network corresponding to the Königsberg Bridge Problem

The distances between points and the precise shape of the curves joining points are of no importance; what is important is that there are two bridges between A and B, that there is no bridge connecting B and C, and so on. Because the network contains all of the problem's relevant information, Euler was able to phrase the Königsberg Bridge Problem in a new way:

> *Without lifting your pencil, can you trace over* ***all*** *edges of the network exactly once?*

Euler realized that a deeper understanding of the problem would be gained if the question were asked for general networks, not just the one corresponding to the Königsberg bridges.

DEFINITION *Network*
A **network** consists of two finite sets:

- a set of **vertices,** represented by a set of points in the plane,

and

- a set of **edges** that join some of the pairs of vertices, represented by joining the corresponding points in the plane by a curve.

Some additional examples of networks are shown in Figure 11.35. Any edge of a network must always have its endpoints at vertices of the network, and there must be no other vertex along the edge. In particular, a point at which two edges cross one another is not a vertex of the network. For example, network (2) of Figure 11.35 has just the six vertices shown by the large dots.

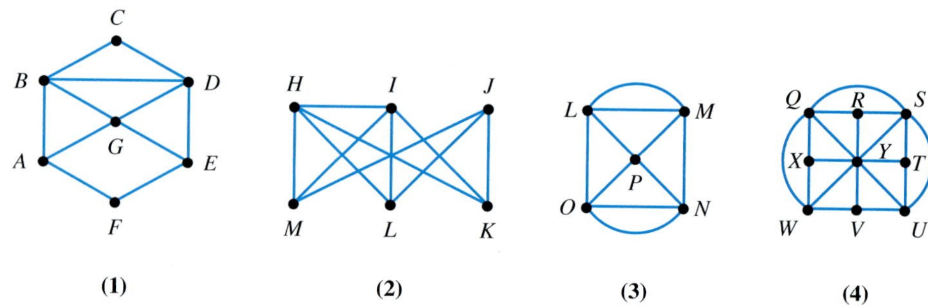

Figure 11.35
Four examples of networks

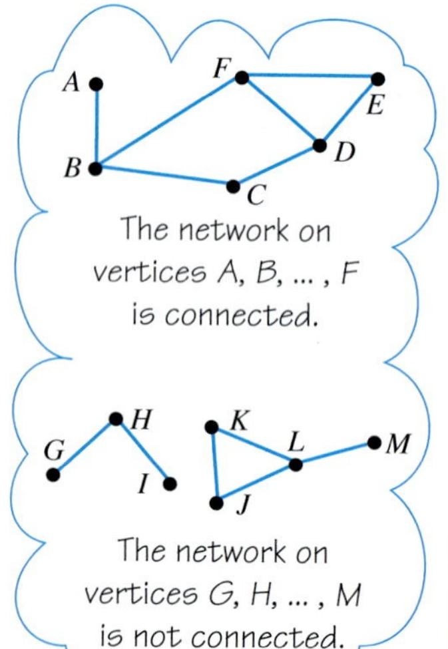

A **path** in a network is a curve traced by following a sequence of edges in the network, where no edge is retraced, but vertices can be revisited. If every pair of vertices of a network can be joined by some path, then we say that the network is **connected.** For example, each of the four networks in Figure 11.35 is connected. The Königsberg Bridge Problem is then equivalent to asking if there is a path that covers each edge of a network once and only once.

DEFINITION *Euler Paths and Traversable Networks*
An **Euler path** in a network is a path that traverses each edge once and only once. A network is **traversable** if, and only if, it has an Euler path.

In the next example, you will discover that classifying the vertices of a network as even or odd is most helpful in a search for an Euler path.

DEFINITION *Degree, Even, and Odd Vertices*

- The **degree** of a vertex is the number of edges emanating from the vertex.
- A vertex is **odd** if it has odd degree.
- A vertex is **even** if it has even degree.

For example, vertices A and E of network (1) of Figure 11.35 are odd, since they have degree 3. Vertices B and C are even, with respective degrees 4 and 2.

EXAMPLE 11.18 WHEN IS A NETWORK TRAVERSABLE?

(a) Which of the networks in Figure 11.35 are traversable? Experiment by tracing a path on a sheet of paper laid over the network.
(b) What is the number of odd vertices in each of the networks?
(c) Do you see any connection between the traversability of a network and the number of odd vertices?

Solution

(a) Network (1) is traversable; one path is *ABCDEFAGBDGE*.
Network (2) is not traversable.
Network (3) is traversable; one path is *LMNOLMPNOPL*.
Network (4) is not traversable.
(b) Network (1) has 2 odd vertices: A and E.
Network (2) has 4 odd vertices: J, K, L, and M.
Network (3) has 0 odd vertices.
Network (4) has 6 odd vertices: Q, R, S, T, V, and X.
(c) The networks with 0 or 2 odd vertices are traversable.
The networks with 4 or 6 odd vertices are not traversable.

Euler observed that each time a path passes through a vertex, it uses two edges: one to enter the vertex and another to exit. Except for the beginning and ending vertices, all of the vertices of a traversable network must therefore be even vertices, and it is impossible to traverse a network with more than two odd vertices. If there are two odd vertices, these are necessarily the endpoints of the Euler path. If there are no odd vertices, the Euler path must terminate at the same vertex as it started, and the path forms a closed curve passing over each edge exactly one time.

This reasoning shows that if a network is traversable, then it has zero or two odd vertices. With more effort, the converse can also be shown: any connected network with zero or two odd vertices is traversable. This is a celebrated result of Euler.

THEOREM *Euler's Traversability Theorem*

A connected network is traversable if, and only if, it has either no odd vertices or two odd vertices. If it has no odd vertices, any Euler path is a closed curve that ends at the same vertex it started from. If the network has two odd vertices, these vertices are the endpoints of any Euler path.

Since the Königsberg Bridge network has four odd vertices, there is no Euler path. Königsbergers must recross some bridge on their walk.

The Königsberg Bridge Problem and the traversability of networks may appear to have no practical importance, but in fact there are many useful applications to real problems. Here are two examples:

- What route can a telephone company inspection crew follow to check all of its lines without having to go over any section twice?
- What route can a street sweeper follow to clean all of the city streets and not have to travel over any blocks that have already been cleaned?

Counting Vertices, Edges, and Regions in Planar Networks

DEFINITION *Planar Network*

A network is **planar** if it can be drawn in the plane without any intersection points of its edges other than endpoints.

For example, the network with four vertices and edges between each pair of distinct vertices is planar, as shown in Figure 11.36. There are no intersections of the edges. The network on five vertices with edges between each pair of distinct vertices is not planar, since it is impossible to arrange all of the edges so that no two of them intersect. By removing just one edge, you can arrange the remaining nine edges between five vertices to form a planar network. Try it!

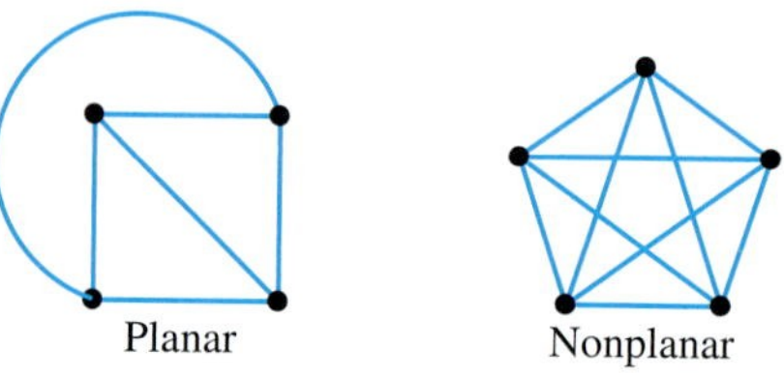

Figure 11.36
A planar and a nonplanar network

Any connected planar network partitions the plane into disjoint regions. It is interesting to count the number of vertices V, the number of edges E, and the number of regions R that correspond to a connected network.

EXAMPLE 11.19 COUNTING VERTICES, EDGES, AND REGIONS

The connected network below has $V = 7$ vertices, $E = 9$ edges, and partitions the plane into $R = 4$ regions (the unbounded region is counted).

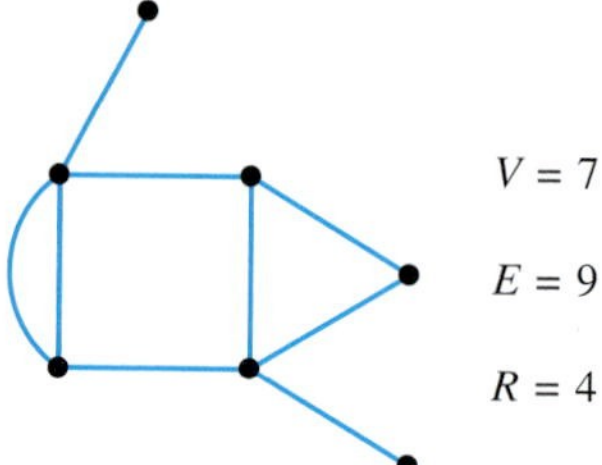

The values 7, 4, and 9 of V, R, and E, respectively, have been entered into the table. Next, count V, E, and R for each of the following networks, and enter their values into the table.

	V	R	E
	7	4	9
(1)			
(2)			
(3)			

(1) (2) (3)

Do you see a pattern? Draw more examples to check your conjecture.

Solution You should discover that the sum of the number of vertices and the number of regions is 2 more than the number of edges.

The exploration in Example 11.19 leads to the conjecture that the formula $V + R = E + 2$ holds for every connected planar network. An example is shown in Figure 11.37.

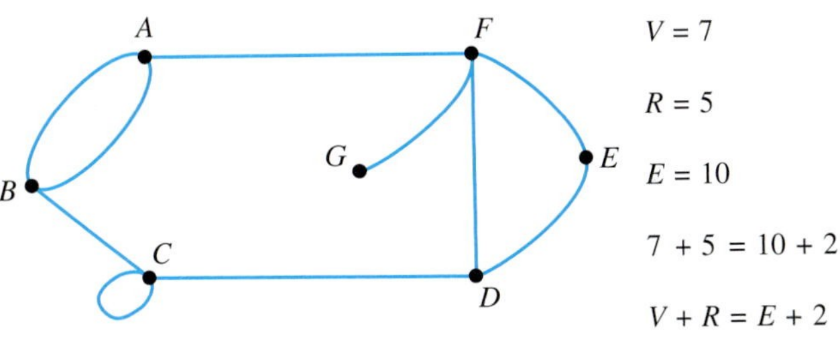

Figure 11.37
Euler's formula $V + R = E + 2$ *holds for any connected planar network with* V *vertices,* R *regions, and* E *edges.*

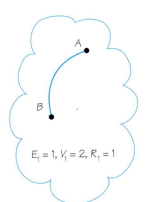

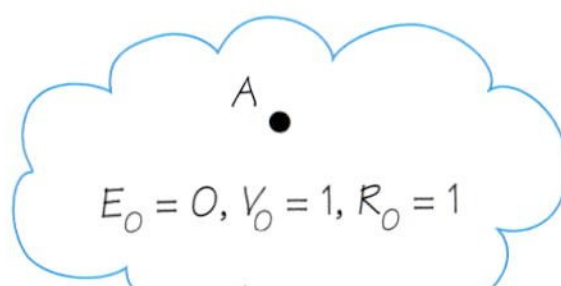

THEOREM *Euler's Formula for Connected Planar Networks*

Let V be the number of vertices, R the number of regions, and E the number of edges of a connected planar network. Then V, R, and E satisfy **Euler's formula:**

$$V + R = E + 2.$$

A

B

$E_1 = 1, V_1 = 2, R_1 = 1$

Proof:* An informal justification for Euler's formula rests on the idea of beginning with a single point and then appending one edge at a time until any given network is completely drawn. At each step, the new edge must maintain the connectivity of the network.

*Optional

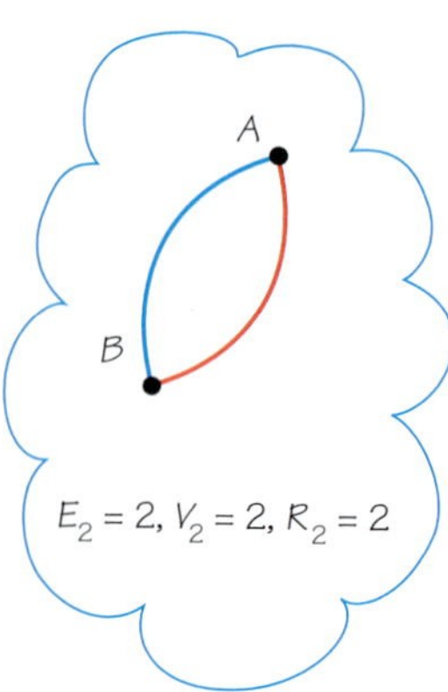

To keep the discussion concrete, consider drawing the network shown in Figure 11.37. Draw vertex A to start. This minimal network has 1 vertex ($V_0 = 1$), 1 region ($R_0 = 1$), and no edges ($E_0 = 0$). Thus, Euler's formula $V_0 + R_0 = E_0 + 2$ is satisfied, since $1 + 1 = 0 + 2$. Now add the vertex B and one edge joining A with B. The new network increases the number of vertices by $1(V_1 = V_0 + 1)$, increases the number of edges by $1(E_1 = E_0 + 1)$, and leaves the number of regions unchanged ($R_1 = R_0$). Adding 1 to each side of the equation $V_0 + R_0 = E_0 + 2$ gives us $(V_0 + 1) + R_0 = (E_0 + 1) + 2$. This can be rewritten as $V_1 + R_1 = E_1 + 2$, so Euler's formula continues to hold. Next, draw the second edge from A to B. This increases both the number of edges and number of regions by 1. Since adding 1 to both sides of $V_1 + R_1 = E_1 + 2$ gives $V_1 + (R_1 + 1) = (E_1 + 1) + 2$, we see that Euler's formula $V_2 + R_2 = E_2 + 2$ holds for the network consisting of the vertices A and B and the two edges joining these vertices. Continuing in this way, each new edge appended to the existing network preserves the validity of Euler's formula. In particular, the formula holds when the network is completed.

EXAMPLE 11.20 SOLVING THE PIZZA PROBLEM

Suppose C cuts are made across a circular pizza, and there are I points of intersection of pairs of cuts. Assume no two cuts intersect on the bounding circle, and no three cuts intersect at the same point inside the pizza. How many pieces P of pizza are there?

Solution

Understand the Problem

It helps to examine a particular case, such as the one shown here. The cuts are drawn to satisfy the conditions of the problem. In the drawing we can see that 4 cuts and 3 intersections of cuts result in 8 pieces of pizza. Our goal is to see if we can find a formula that gives us the number P in terms of the variables C and I.

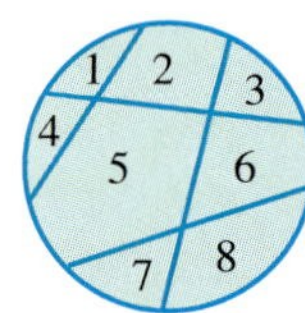

C = 4 cuts
I = 3 intersections of cuts
P = 8 pieces of pizza

Devise a Plan

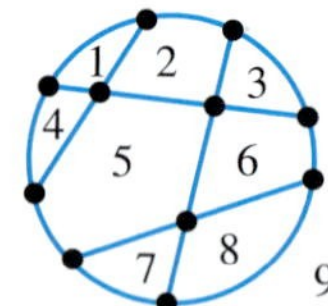

The cut-up pizza can be viewed as a connected planar network as shown at the left. If we can determine the numbers V and E of vertices and edges, Euler's formula will allow us to solve for the number R of regions of the network. Thus, our plan is to relate V and E to the numbers C and I. In Euler's formula, R includes the region outside the pizza, which is not a piece of pizza. Therefore, the number of pieces of pizza is given by $P = R - 1$.

Carry Out the Plan

Each cut forms 2 vertices on the circle bounding the pizza, and each intersection of cuts gives one vertex of the network inside the circle. Altogether, this gives $V = 2C + I$ vertices in the network. To count the number of edges in the network, let's first suppose that there are no intersecting cuts. Each cut then forms 2 edges on the circle and is itself an

The Game of Sprouts

The English mathematicians John Conway and Michael Paterson invented a game called Sprouts in 1967, in which two players take turns drawing new edges to an evolving network on n given initial vertices. It is permissible to draw an edge that returns to the starting vertex; such an edge is called a loop. There are just four simple rules:

1. Each new edge forms either a loop or joins two different vertices.
2. A new vertex must be created along the new edge.
3. The new edge cannot cross itself or any previously drawn edge, or pass through a vertex.
4. No vertex can have a degree larger than 3.

The first five moves of a Sprouts game on $n = 2$ initial vertices are shown below. The player who made the fifth move wins, since the other player cannot make a legal move.

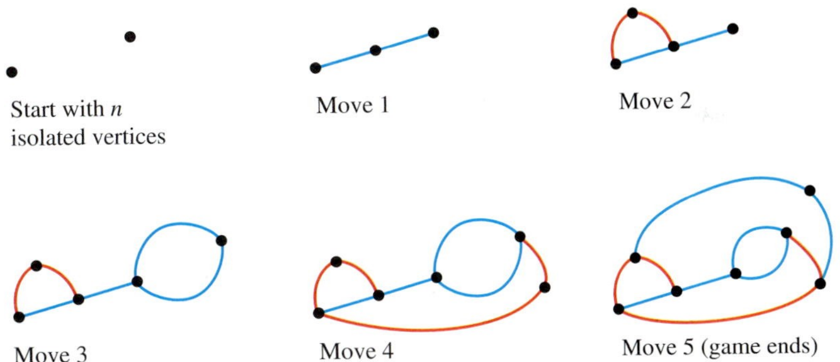

(a) Play several games on $n = 2$ initial vertices. At most, how many moves do the games take?
(b) Play several games on $n = 3$ initial vertices. What is the largest number of moves possible before the games end?
(c) Look for a pattern of the largest number of moves possible in a Sprouts game on n initial vertices.
(d) What is the smallest number of moves to end a game on n initial vertices?

SOURCE: Sprouts is described in Martin Gardner's book Mathematical Carnival *(New York: Knopf, 1975).*

edge, giving $3C$ edges of the network. If we next suppose that some of the cuts intersect, it is seen that each intersection creates 2 additional edges of the network not yet counted. Altogether then, there are $E = 3C + 2I$ edges in the network. Solving for R in Euler's formula $V + R = E + 2$, we get

$$R = E - V + 2 = (3C + 2I) - (2C + I) + 2 = C + I + 2.$$

Since $P = R - 1$, we arrive at the final formula:

$$P = C + I + 1.$$

Look Back

In the example drawn on the previous page, there are $C = 4$ cuts intersecting in $I = 3$ points. Since $C + I + 1 = 4 + 3 + 1 = 8$, we now see why we counted $P = 8$ pieces of pizza, and obtain a check that our formula is correct.

Connecting Euler's Formulas for Planar Networks and for Polyhedra

In Section 11.3, it was discovered without proof that the number of vertices, faces, and edges of a polyhedron are related by Euler's formula $V + F = E + 2$. It is natural to wonder if this formula for polyhedra is related to the very similar formula $V + R = E + 2$ for connected planar networks. To understand the connection, imagine that each edge of a polyhedron is replaced with a segment of a rubber band. The rubber-band skeleton of the polyhedron can then be stretched and flattened to form a connected planar network with V vertices and E edges. This is shown in the case of a cube in Figure 11.38.

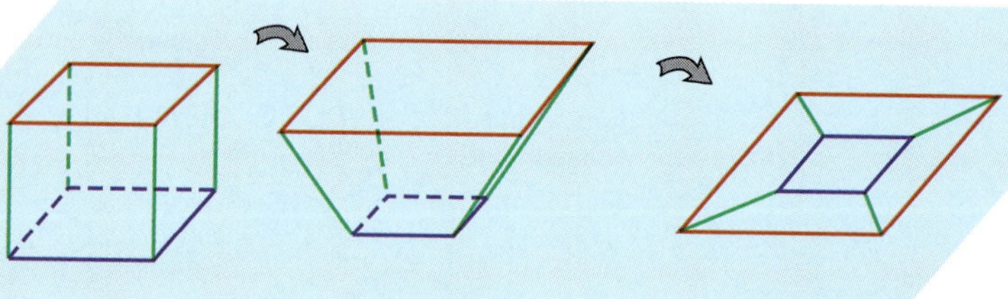

Figure 11.38
The skeleton of edges of any polyhedron, such as the cube shown, can be stretched and flattened to form a planar network

There is a one-to-one matching between the regions of the network (including the unbounded region) and the faces of the polyhedron. Thus, $R = F$, and we conclude that the two Euler formulas are equivalent to one another.

Problem Set 11.4

Understanding Concepts

1. Andre has met everyone but Emma, Coralee has met everyone but Bianca, and Emma has met only Coralee and Diego. Which of these networks represents this information? Which do not? Explain how you make your decision.

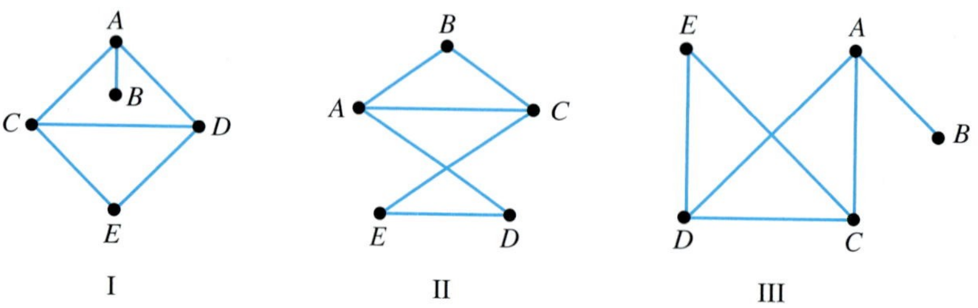

2. Decide which of the following networks are traversable. Give an Euler path for each traversable network.

(a) A B C H D G F E

(b) 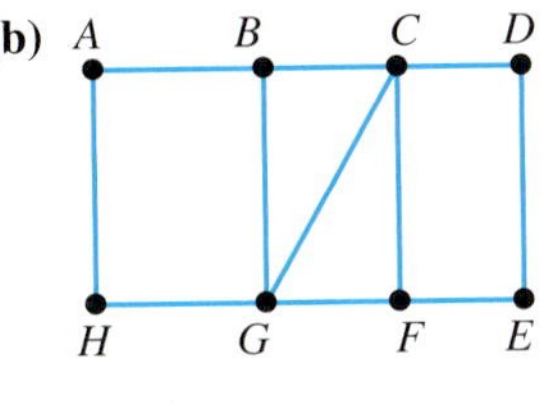

(c) A B F C G E D

(d) 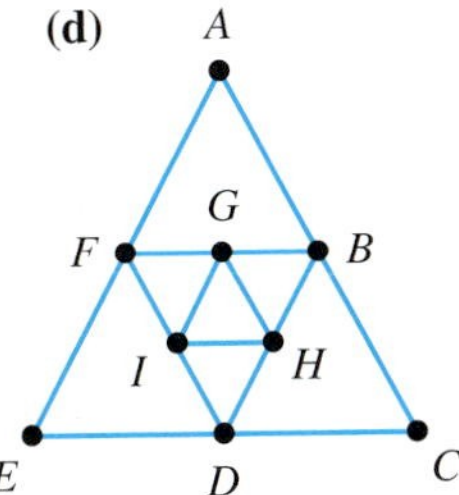

(e) 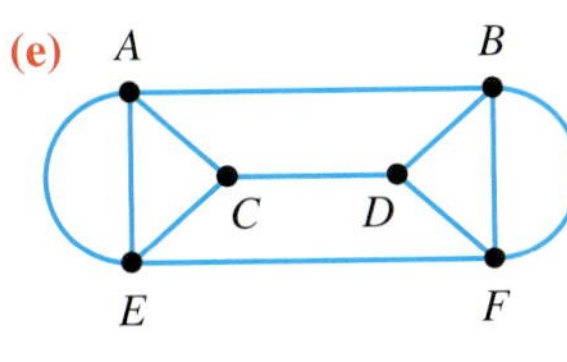

(f) 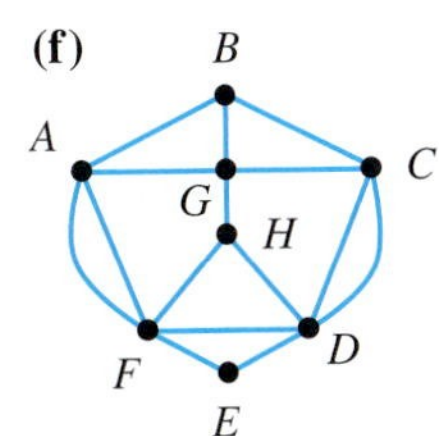

3. Find an Euler path for each of these networks.

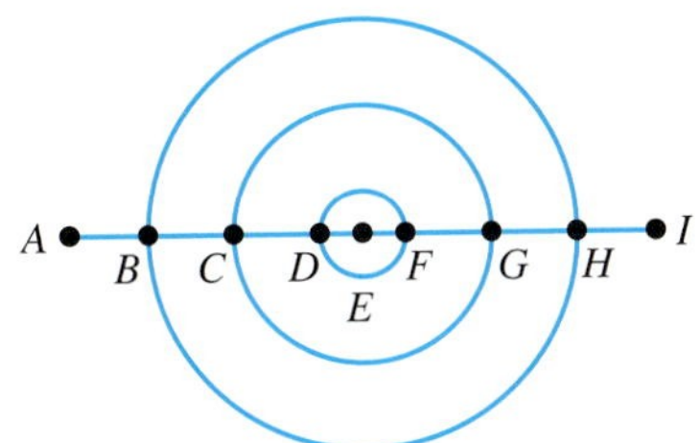

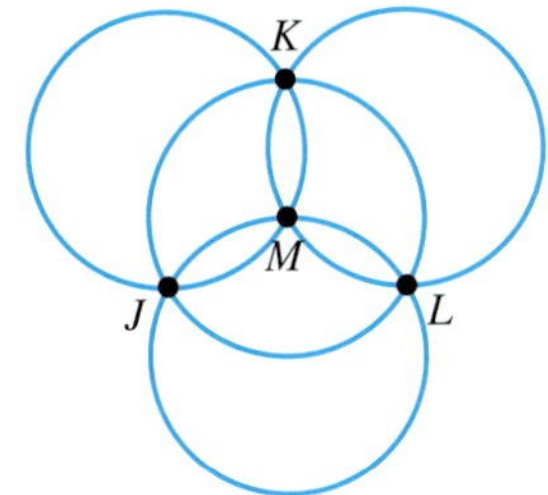

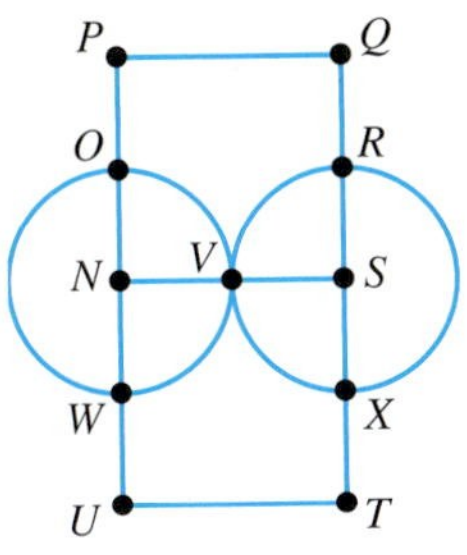

4. (a) The **total degree** D of a network is the sum of the degrees of all of the vertices. For example, the network in problem 2(a) has total degree $D = 2 + 4 + 2 + 4 + 2 + 4 + 2 + 4 = 24$. This network also has $E = 12$ edges. Find D and E for the remaining networks shown in problem 2.

(b) Guess how the total degree D is related to the number of edges, E, in any network. Test your conjecture on several networks of your own choosing.

5. A connected network has the following degrees at its vertices: 2, 2, 4, 8, 3, 6, 6, and 1.

(a) Is the network traversable?

(b) How many edges does the network have? (*Hint:* See problem 4(b).)

6. Here are two more examples from Euler's paper on the Königsberg Bridge Problem.

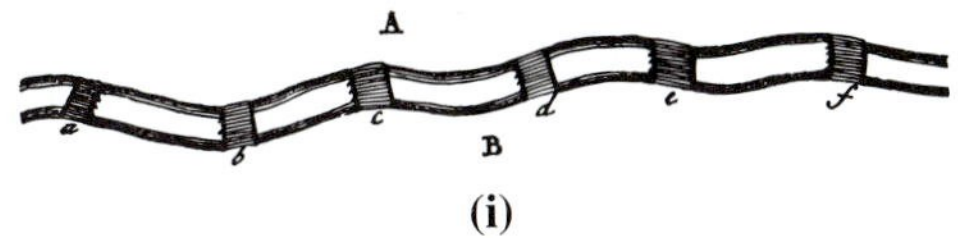

(i)

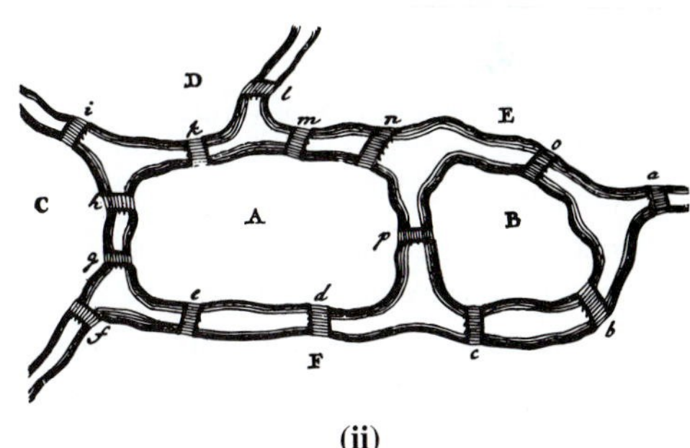

(ii)

(a) Draw the networks corresponding to (i) and (ii).

(b) Is network **(i)** traversable? Why?

(c) Is network **(ii)** traversable? Explain your reasoning.

7. (a) Draw the network that corresponds to the following system of bridges and land masses.

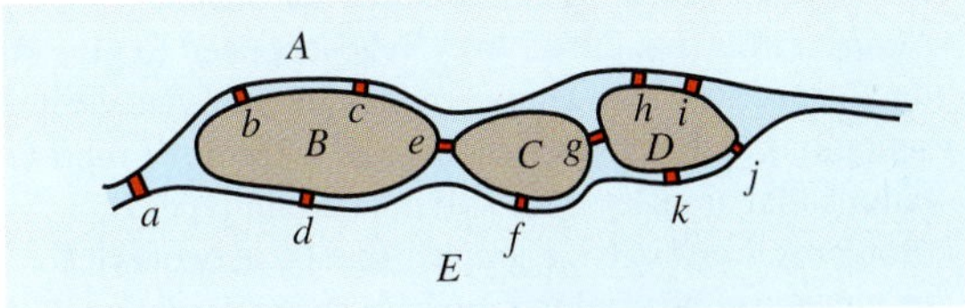

(b) Explain why the network is not traversable.

(c) What is the smallest number of new bridges required to form a traversable network? Where should the bridge(s) be placed?

8. A floor plan of a house is shown on the following page. There are five rooms A, B, C, D, E and the outside O, connected by the doorways a, b, c, d, e, f, and g.

Figure for Problem 8

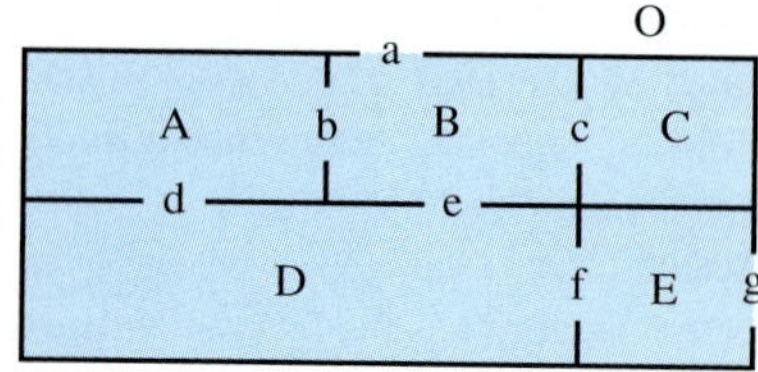

(a) Draw a network whose vertices are labeled A, B, C, D, E, and O and whose edges correspond to the doorways.

(b) Is it possible to walk through each doorway exactly once? If so, where must you begin and end your walk?

9. A connected planar network is shown below.

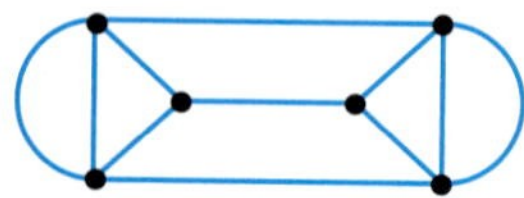

(a) What are the numbers V of vertices, R of regions, and E of edges?

(b) Does Euler's formula hold for this network?

10. (a) Draw a connected planar network whose 10 edges separate the plane into 6 regions.

(b) Can you draw a connected planar network with 10 edges that separates the plane into 12 regions? Explain.

Teaching Concepts

11. When examining a network for traversability, Karinna has decided that she can overlook all vertices of degree 2. Is she justified to make this simplification? Give a careful explanation of why or why not.

12. **Networks as Representations.** According to the NCTM's *Principles and Standards for School Mathematics* (pp. 295–296), "Teachers need to give students experiences in using a wide range of visual representations and introduce them to new forms of representations that are useful for solving certain types of problems. Vertex–edge graphs [that is, networks], for instance, can be used to represent abstract relationships among people or objects in many different kinds of situations." As an example, in the network shown, each vertex represents a student group at a school, and an edge between two vertices indicates that at least one student belongs to both groups.

(a) Use the network representation to decide how many different times are needed so that five groups can meet and no student has more than one meeting at a time.

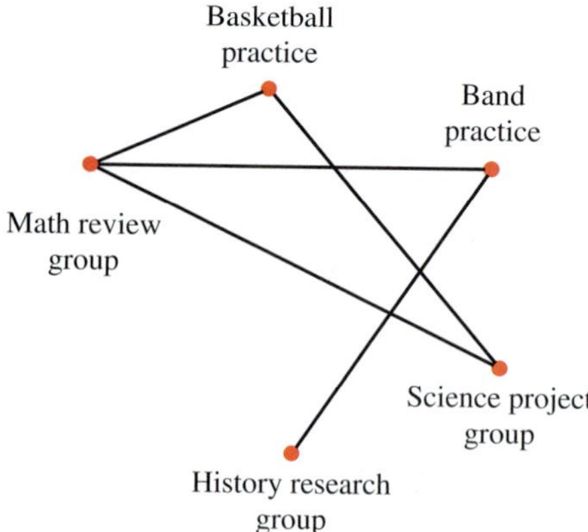

(b) Find another example of representation of a problem situation that uses networks. Carefully describe how the vertices and edges are related to the problem, and include a specific example that would be of interest to schoolchildren.

Thinking Critically

13. A connected network that does not contain a closed path of distinct edges is called a **tree.**

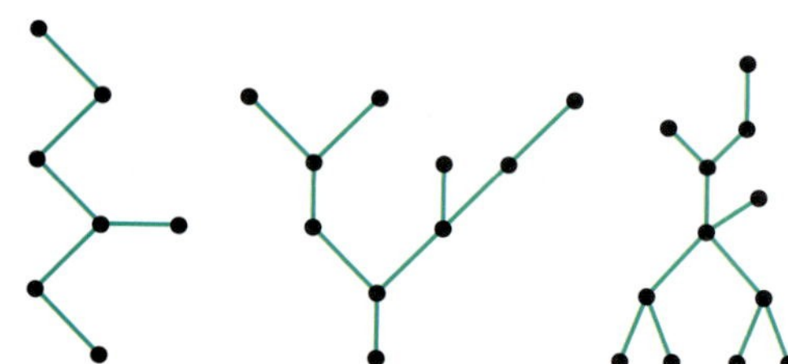

Trees

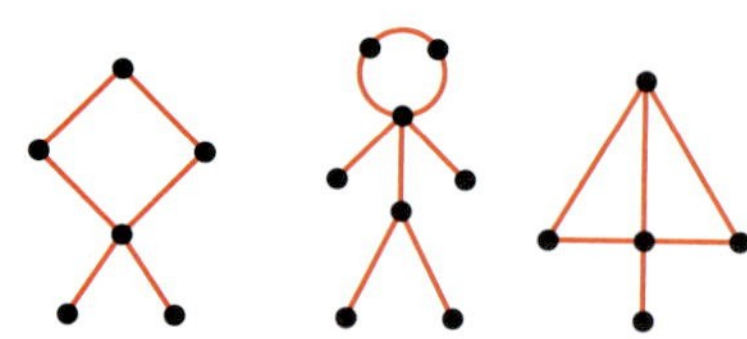

Not trees

(a) Suppose a network contains two vertices that can be joined by two different paths with no edges common to both paths. Why is it impossible for this network to be a tree?

(b) The **diameter** of a tree is the largest number of edges required to join any two vertices by a path. For example, the leftmost tree above has diameter 5. Find the diameters of the two other trees drawn above.

(c) Let V and E denote the respective number of vertices and edges of a tree. What formula relates V and E? Prove your result.

14. What trees are traversable?

15. Let D represent the sum of the degrees of all of the vertices of a network, and let E denote the number of edges in the network. (D is the total degree of the network; see problem 4.) Imagine that the network is a highway map, with the vertices representing towns and the edges representing highways connecting towns. Suppose each town puts up a city-limits sign along each highway that leaves town.

(a) Show that the total number of signs is given both by D and by $2E$, so that you obtain the equation $D = 2E$.

(b) Is it possible to construct a network whose total degree D is 17?

16. In problem 4, you discovered that the total degree D (the sum of degrees of all the vertices of the network) is given by $D = 2E$, where E is the number of edges. Suppose the degrees at the even vertices are $e_1, e_2, \ldots, e_m$, and the degrees at the odd vertices are $d_1, d_2, \ldots, d_n$. Thus, $2E = D = e_1 + e_2 + \cdots + e_m + d_1 + d_2 + \cdots + d_n$.

(a) Explain why $d_1 + d_2 + \cdots + d_n$ is an even integer.

(b) Since $d_1 + d_2 + \cdots + d_n$ is even, explain why n is even. Since n is the number of odd vertices in the network, you've proved the following result:

The number of vertices of odd degree in a network is always even.

17. Prove that the number of people at a party who have shaken hands an odd number of times is an even number. (*Hint:* See problem 16.)

18. The connected network below has six odd vertices, so it cannot be traced without lifting the pencil.

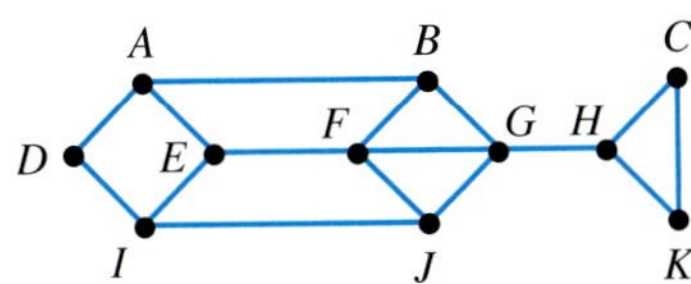

(a) Show that the network can be traced in three strokes; that is, the pencil is lifted twice and then placed at a different vertex.

(b) If a connected network has $2m$ odd vertices, with $m > 0$, explain why its edges can be traced in m strokes. (*Suggestion:* Add, temporarily, $m - 1$ new edges to the network.)

19. Suppose that you are asked to trace a connected network, without lifting your pencil, so that each edge is traced exactly twice. Is this always possible regardless of the number of odd vertices? Explain why or why not.

20. The planar network shown below on the vertices A, B, C, D, F, G, H, I, J, and K, is not connected; indeed, it is made up of $P = 3$ connected pieces.

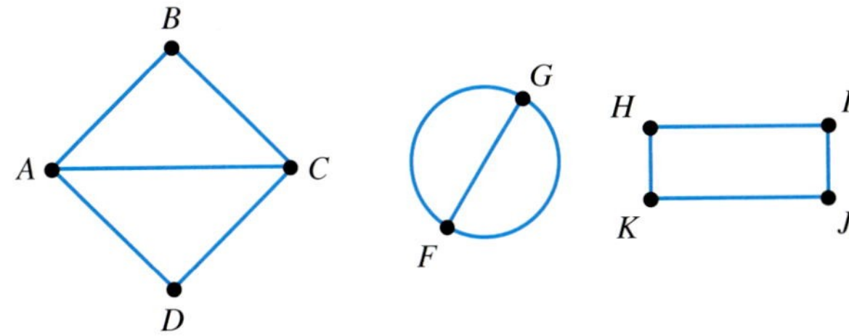

For this network, $V = 10$, $E = 12$, and $R = 6$, and we see that $V + R = 16$, but $E + 2 = 14$.

(a) Draw several more examples of disconnected planar networks. For each network count V, R, E, and P, where P is the number of connected pieces in your network. Can you guess an Euler formula that relates V, R, E, and P?

(b) Prove your conjecture stated in part (a). (*Suggestion:* Add new edges to connect the network.)

21. One circle separates the plane into 2 regions; 2 circles that cross one another in 2 points separate the plane into 4 regions; and 3 circles form 8 regions when each pair of circles crosses at 2 points and no more than 2 circles intersect at any point.

(a) What is the number of regions determined by 4 circles? Assume that each pair of circles intersects at 2 points, and no 3 circles intersect at any 1 point.

(b) How many regions are formed by n intersecting circles? (*Hint:* There are $C(n, 2) = \frac{1}{2}n(n - 1)$ pairs of circles. Now use Euler's formula and mimic the solution of Example 11.20.)

Thinking Cooperatively

22. These three trees (see problem 13 for the definition of a tree) are *isomorphic* (*iso* = same, *morph* = form) to one another, since the positions of the vertices can be rearranged by bending (but not breaking) to make all three networks look exactly alike.

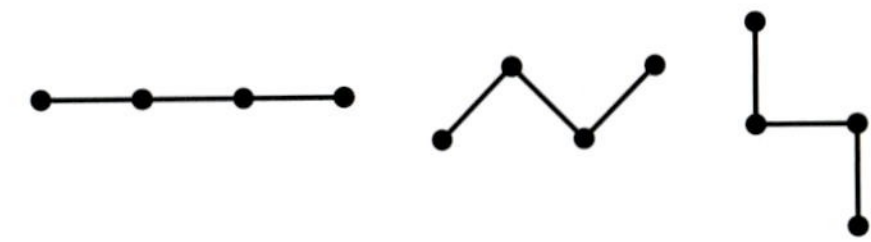

Similarly, these two trees are isomorphic:

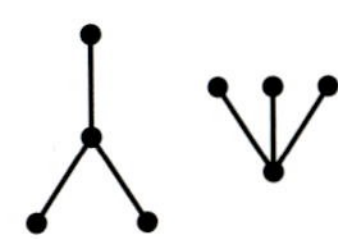

On the other hand, each tree in the first group is *non-isomorphic* to each tree in the second group. One way to see this is to note that each tree in the first group has vertices of degrees 1, 2, 2, and 1, but the lower trees each have vertices of degrees 1, 1, 1, and 3. It is easy to check that any tree with 4 vertices must be isomorphic to one of the two types shown above.

Work in small groups, and compare results between groups, to carry out these investigations.

(a) Find the 3 nonisomorphic trees on 5 vertices.

(b) Find the 6 nonisomorphic trees on 6 vertices.

(c) Find the 11 nonisomorphic trees on 7 vertices.

23. In 1958, the American mathematician David Gale invented the two-person game **Bridg-It.** The game is played on isometric dot paper, with rows alternating between two colors of dots (say, black and red, as shown). The players take turns drawing horizontal or vertical edges between adjacent vertices of their color. The object of the game is to create a path from one side of the board to the other.

(a) If the red player started the game, who should be the winner of the game shown? What if the black player played first?

(b) Play a few games of Bridg-It.

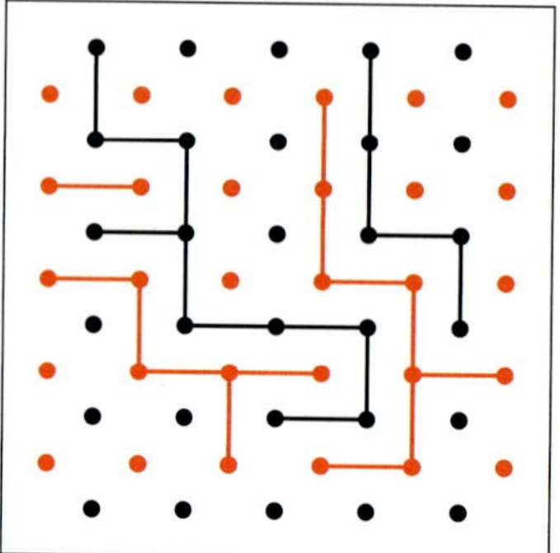

Making Connections

24. An 8-pin connecting terminal is wired as shown on the left. An electrical engineer claims that 5 of the 12 connecting wires can be eliminated. On the right-hand diagram draw 7 of the 12 original wires that provide for the same current flows as the original circuit.

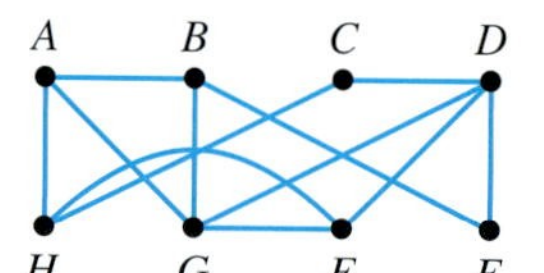

A B C D

H G F E

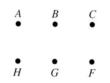

SOURCE: From Donald Bushaw, Max Bell, Henry O. Pollak, Maynard Thompson, and Zalman Usiskin, A Sourcebook of Applications of School Mathematics. © 1980 NCTM.

25. Chemists frequently use a type of network called a *structural formula* to show the bonds linking the atoms of a molecule. For example, three-dimensional models of methane, CH_4, and ethane, C_2H_6, and their corresponding structural formulas are shown here.

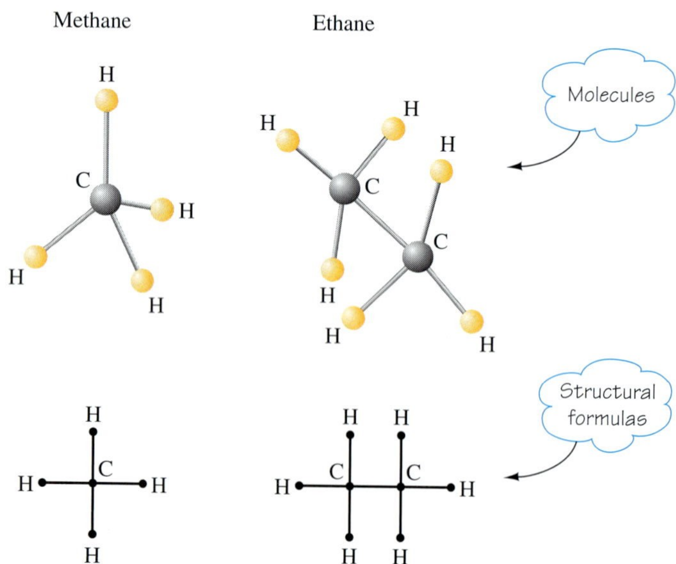

(a) What is the degree (valence) of each carbon atom? of each hydrogen atom?

(b) Sketch the structural formula of propane, C_3H_8.

(c) There are two forms of butane, C_4H_{10}. Their "skeletons," showing the four carbon atoms and the bonds between them, are as follows:

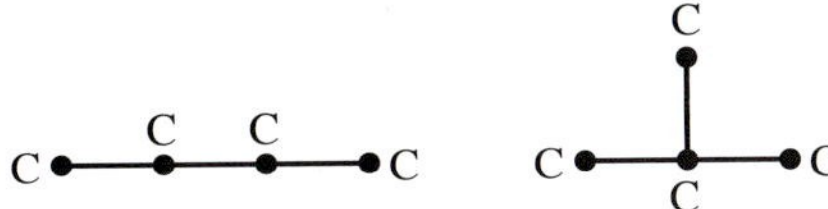

Add the hydrogen atoms and bonds to these skeletons to obtain the structural formulas for the two forms of butane.

(d) The next molecule in the series of alkanes is pentane, which has five carbon atoms. Sketch the three forms of the skeleton of pentane, and add hydrogen atoms and bonds to complete the structural formulas. Does each form have the same number of hydrogen atoms?

(e) Make reasonable guesses for the chemical formulas of hexane, heptane, and octane, the next three hydrocarbons in the alkane series.

From State Student Assessments

26. (New Jersey, Grade 8)
A computer network is to be set up so that

- the supervisor can communicate with every terminal, and
- each worker can communicate with the supervisor and exactly two coworkers. Which network meets these requirements?

A. 

B. 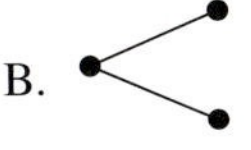

C.

D.

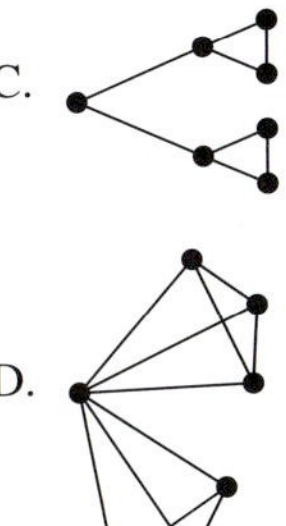

For Review

27. Draw a parallelogram, and then draw outward-facing squares along each edge of your parallelogram. What type of quadrilateral is formed by joining the successive center points of the squares?

28. The following equilateral nonagon (9-gon) can be used to tile the plane in many ways, including a spiral pattern similar to the one in Figure 13.27. Find the measures of all of the interior angles.

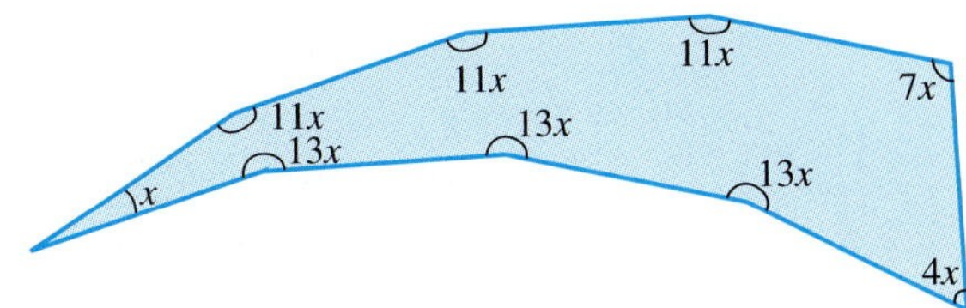

29. In the figure shown below, lines l and m are parallel. Find the measures x, y, and z of the indicated angles.

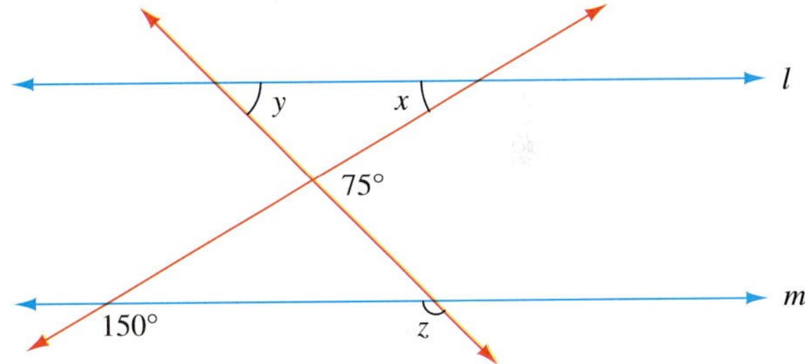

Epilogue Visualization

After struggling with a difficult problem or abstract concept, a student may suddenly smile and announce, "I see it now," or perhaps exclaim, "I've got the picture." Information conveyed in visual form can easily be superior to the same information described by a thousand, or even ten thousand, well-chosen words. Our ability to visualize is only partly dependent on the acuity of our eyesight: even more important is the mind's eye, which sharpens our perception by providing us with skills and abilities to identify, analyze, and classify shape.

In recent years the computer has enabled us to see in directions unimaginable just a short time ago. This development is a continuation of technological advances that redefine what is observable. Nearly four hundred years ago, Galileo's telescope detected the moons of Jupiter. Today, optical and radio telescopes reveal quasars at the edge of the universe. A hundred years after Galileo, the universe of the very small became observable with Leeuwenhoek's invention of the microscope. Today, tunnelling microscopes produce images in which single atoms are distinguishable.

Learning to understand and interpret new images is an exciting challenge. Fortunately, many of the concepts and experiences first encountered in elementary geometry prepare the way to meet this challenge. In this chapter we have introduced many of the basic notions of geometry: point, line, plane, curve, surface, angle, distance between points, measure of an angle, region, and space. In the chapters that follow, these basic notions are developed in more depth as we encounter the ideas of measurement, transformations, tilings, symmetries, congruence, constructions, and similarity.

Chapter Summary

Key Concepts

Section 11.1 Figures in the Plane

- Line. Two distinct points A and B uniquely determine a line $\overleftrightarrow{AB}$.
- Line segment. A line segment $\overline{AB}$ is the set of points on the line $\overleftrightarrow{AB}$ that are between the endpoints A and B.
- Length of a segment. The length of the line segment $\overleftrightarrow{AB}$ is the distance AB between the points A and B.
- Ray. A ray $\overrightarrow{AB}$ is the set of points including the endpoint A, the segment $\overline{AB}$, and all of the points P of the line $\overleftrightarrow{AB}$ for which B is between A and P.
- Angle. An angle $\angle ABC$ with vertex B is the union of two rays $\overrightarrow{BA}$ and $\overrightarrow{BC}$ that form the sides of the angle.
- Angle measure. The measure of an angle, $m(\angle BAC)$, is the number of degrees of turn to rotate side $\overrightarrow{BA}$ onto side $\overrightarrow{BC}$, turning about the vertex B.
- Supplementary and complementary angles. Two angles are complementary or supplementary if the sum of their measures is respectively 90° or 180°.
- Classification of angles. An angle is acute, right, or obtuse when its measure is respectively less than, equal to, or greater than 90°. A zero angle, straight angle, or reflex angle is an angle whose measure is respective 0°, 180°, or greater than 180°.
- Corresponding-angle criterion for parallel lines. Two lines crossed by a transversal line are parallel (that is, do not intersect) if, and only if, a pair of corresponding angles is congruent angles (that is, the angles have the same measure).
- Angle sum in a triangle. The sum of the measures of the three interior angles of a triangle is 180°.

Section 11.2 Curves and Polygons in the Plane

- Classification of curves. Curves are simple (non-self-intersecting), closed (no endpoints), or simple closed (simple and closed).
- Polygonal curves. A polygonal curve is a curve formed by sequentially joining points called vertices with line segments called sides of the polygonal curve.
- Polygon. A polygon is a simple closed polygonal curve. A polygon with 3, 4, 5, 6, 7, 8,

9, 10, . . . , n sides is called a triangle, quadrilateral, pentagon, hexagon, heptagon, septagon, octagon, nonagon (or enneagon), decagon, . . . , n-gon, respectively.

- Classification of triangles by side lengths. Triangles are scalene (all sides of different lengths), isosceles (at least two sides of the same length), or equilateral (all sides of the same length).
- Classification of triangles by angle measures. Triangles are acute (all angles acute), right (one angle a right angle), or obtuse (one angle obtuse).
- Classification of quadrilaterals. A quadrilateral is a trapezoid (two sides parallel), a parallelogram (two pairs of opposite parallel sides), a rhombus (parallelogram with all sides of the same length), a rectangle (parallelogram with all right angles), a square (all sides of the same length and all right angles), or a kite (two pairs of adjacent congruent sides).
- Jordan curve theorem. A simple closed curve partitions the points of the plane not on the curve into two disjoint regions: the interior of the curve and the exterior of the curve.
- Total-turn theorem. The total turn made when traversing any closed curve is an integer multiple of 360°.
- Sum of angle measures in a polygon. The sum of measures of the interior angles of an n-gon is $(n - 2)180°$.
- Regular polygon. A polygon is regular if its sides are congruent and its interior angles are congruent. In a regular n-gon, the exterior and central angles each measure $360°/n$, and each interior angle measures $(n - 2)180°/n$.

Section 11.3 Figures in Space

- Simple closed surface. A simple closed surface is a surface without holes or boundary edges that encloses a region called its interior.
- Polyhedron. A polyhedron (*pl.* polyhedra) is a simple closed surface formed by planar polygonal regions. The polygons are called faces, and the sides and vertices of the faces are called the edges and vertices, respectively, of the polyhedron. Polyhedra include prisms, pyramids, and the five regular polyhedra (tetrahedron, cube, octahedron, dodecahedron, and icosahedron).
- Euler's formula. The numbers V of vertices, F of faces, and E of edges of a polyhedron are related by the formula $V + F = E + 2$.
- Curved surfaces. Spheres, hemispheres, cones, prisms, and so on, are curved surfaces.

Section 11.4 Networks

- Network. A network is a set of vertices and a set of edges that join some of the pairs of vertices.
- Degree of a vertex. The degree of a vertex is the number of edges that have the vertex as an endpoint. A vertex is even or odd if its degree is respectively an even or odd number.
- Connected network. A network is connected if every two vertices can be joined by a path (a sequence of vertices with edges joining adjacent vertices in the sequence.)
- Traversable network. A network is traversable if there is an Euler path covering each edge of the network precisely one time. Euler showed that a connected network is traversable when the number of vertices of odd degree is either two or zero.
- Planar network. A network is planar if it can be drawn so that no two edges cross one another, so that the network partitions the plane into regions.
- Euler's formula for networks. The numbers V of vertices, R of regions, and E of edges of a connected planar network are related by the formula $V + R = E + 2$.

Vocabulary and Notation

Section 11.1

Point, $A, B, \ldots$
Line l, $\overleftrightarrow{AB}$
Collinear points
Plane
Plane figure
Parallel lines, $l \parallel m$
Point of intersection
Concurrent lines
Transversal
Distance between points, AB
Line segment, $\overline{AB}$
Endpoints
Congruent line segments, $\overline{AB} \cong \overline{CD}$
Midpoint
Ray, $\overrightarrow{AB}$
Angle, $\angle ABC$
Vertex of an angle
Measure of an angle, $m(\angle ABC)$
Zero, acute, right, obtuse, straight, and reflex angles
Perpendicular lines, $l \perp m$
Congruent angles, $\angle A \cong \angle B$
Complementary angles
Supplementary angles
Adjacent angles
Vertical angles
Corresponding angles
Alternate interior angles
Interior and exterior angles of a triangle
Directed angle
  Initial and terminal side

Section 11.2

Curve
  Simple, closed, and simple closed
Jordan curve theorem
  Interior and exterior region of a simple closed curve
Convex figure
Concave figure (nonconvex figure)
Polygonal curve
Vertices
Sides
Polygon
  Pentagon, hexagon, heptagon, octagon, nonagon (or enneagon), decagon, . . . , n-gon
Polygonal region
Convex polygon
Interior and exterior angles of a polygon
Triangle: acute, right, obtuse, scalene, isosceles, and equilateral
Quadrilateral: kite, trapezoid, isosceles trapezoid, parallelogram, rhombus, rectangle, and square
Regular polygon
Equilateral
Equiangular
Central angle
Circle
  Diameter, radius, center, chord, tangent line, arc, sector, segment, and disc

Section 11.3

Plane
  Parallel and intersecting planes, dihedral angle, half-space, skew lines, and perpendicular planes
Sphere
Center
Simple closed surface
Solid
Polyhedron
  Vertex, face, and edge
  Pyramid
  Prism, right prism, and oblique prism
  Regular polyhedron
Cone (right circular and oblique circular)
Cylinder (right and oblique)
Tetrahedron
Pentahedron
Hexahedron
Apex (vertex)
Base
Net of a polyhedron
Platonic solids
Lateral surface
Altitude

Section 11.4

Network
  Vertices, edges, path, degree of a vertex, odd or even vertex, and connected network
Traversable network
  Euler path
Planar network

Chapter Review Exercises

Section 11.1

1. Let $ABCD$ be the quadrilateral shown below.

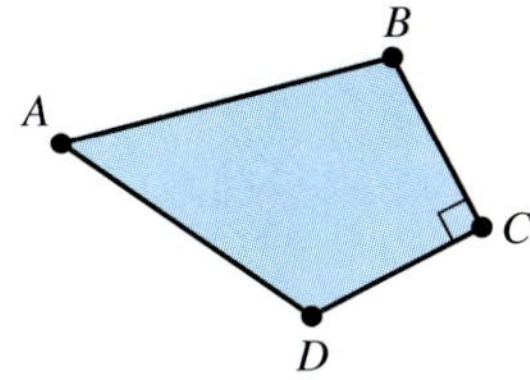

Give symbols for the following:

(a) The line containing the diagonal through C

(b) The diagonal containing B

(c) The length of the side containing A and D

(d) The angle *not* containing D

(e) The measure of the interior angle at C

(f) The ray that has vertex at D and is perpendicular to a side of the quadrilateral

2. For the quadrilateral shown in problem 1, which angle(s) appear to be

(a) acute?

(b) right?

(c) obtuse?

3. An angle measures 37°. What is the measure of

(a) its supplementary angle?

(b) its complementary angle?

4. Lines l and m below are parallel. Find the measures p, q, r, and s of the angles shown.

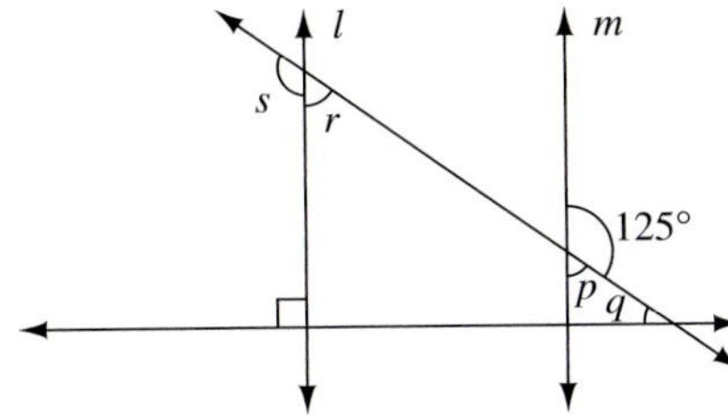

5. Find the measures x, y, and z of the angles in the following figure.

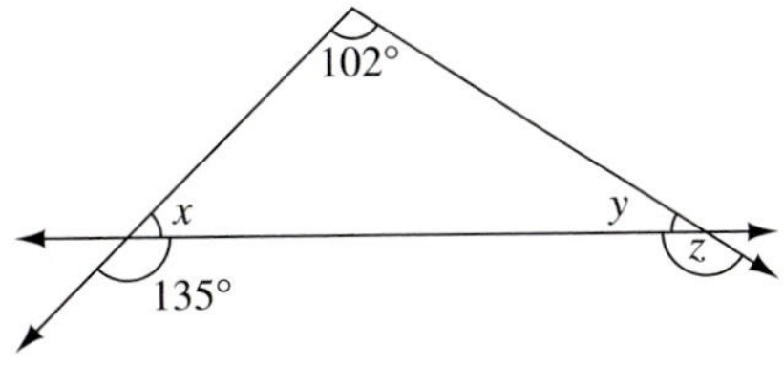

Section 11.2

6. Match each curve to one of the descriptions below.

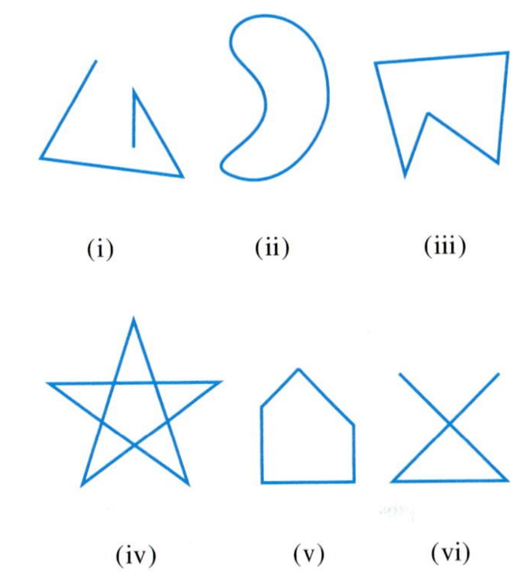

(a) nonconvex nonsimple polygonal curve

(b) nonclosed simple curve

(c) nonsimple nonclosed polygonal curve

(d) convex polygon

(e) simple closed nonconvex nonpolygonal curve

(f) nonconvex polygon

7. For each part, carefully explain your reasoning.

(a) Can a triangle have two obtuse angles?

(b) Can a convex quadrilateral have three obtuse interior angles?

(c) Is there an "acute" quadrilateral (that is, a quadrilateral whose angles are all acute)?

8. Find the measures of the interior angles of this polygon.

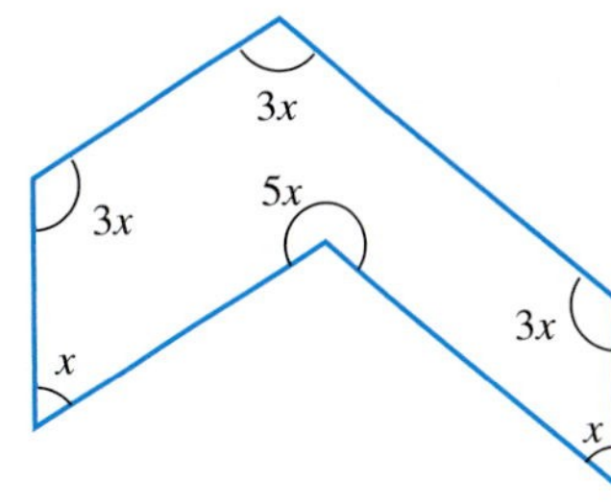

9. A turtle walks along the path $ABCDEFGA$ in the direction of the arrows (see figure on following page), returning to the starting point and initial direction. What total angle does the turtle turn through?

Figure for Exercise 9

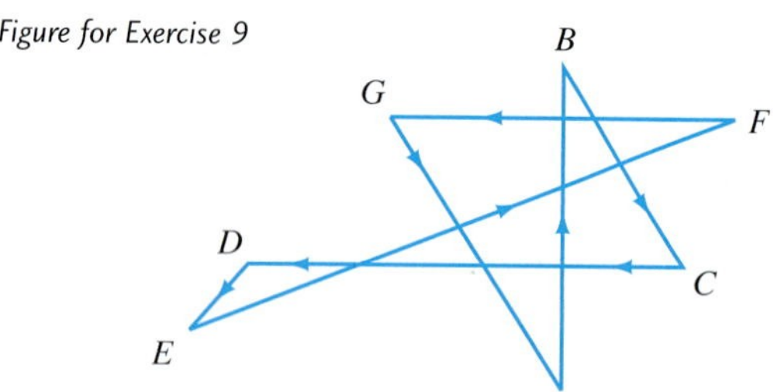

Section 11.3

10. Let $ABCDEFGH$ be the vertices of a cube, as shown below.

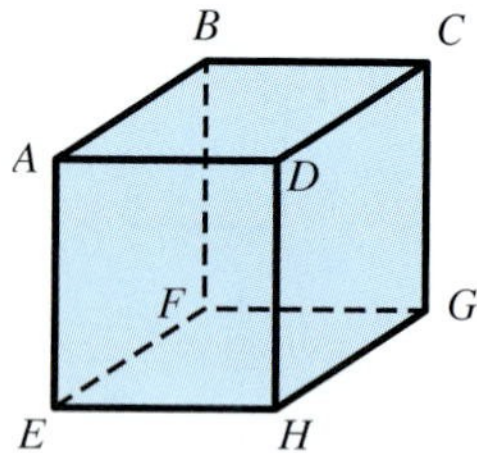

(a) How many planes are determined by the faces of the cube?

(b) Which edges of the cube are parallel to edge $\overline{AB}$?

(c) Which edges of the cube are contained in lines that are skew to the line $\overleftrightarrow{AB}$?

(d) What is the measure of the dihedral angle between the plane containing $ABCD$ and the plane containing $ABGH$?

11. Name the following surfaces in space.

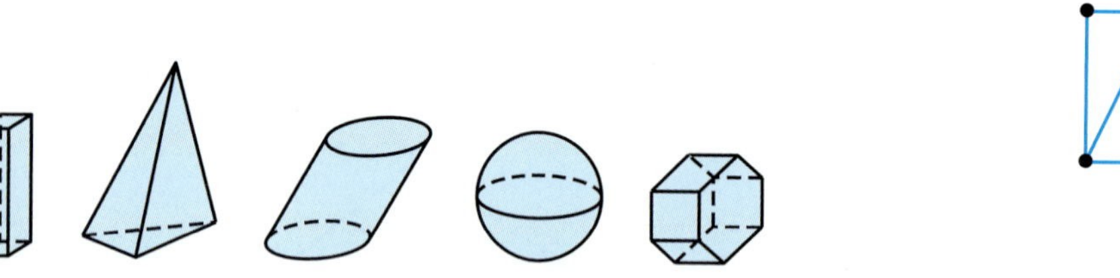

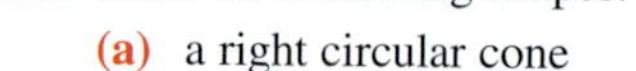

12. Draw the following shapes.

(a) a right circular cone

(b) a pentagonal prism

(c) a nonconvex quadrilateral pyramid

13. (a) Draw a regular octahedron.

(b) Using your drawing in part (a), count the number of vertices, faces, and edges of the octahedron, and then verify that Euler's formula holds for the octahedron.

14. A polyhedron has 14 faces and 24 edges. How many vertices does it have?

Section 11.4

15. (a) Explain why the network shown is not traversable.

(b) Name two vertices that, if connected by a new edge, would make the resulting network traversable. Then list the vertices of an Euler path.

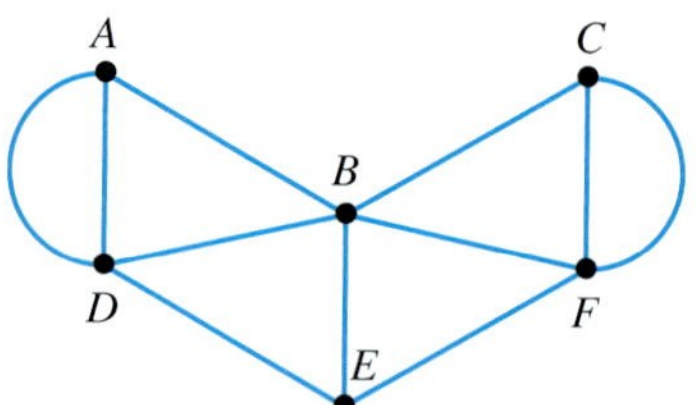

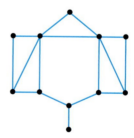

16. Is there a walk that crosses each of the bridges shown below exactly once? Explain your reasoning, using an appropriate network.

17. Count the number of vertices, regions, and edges in the following network, and then verify that Euler's formula holds.

Figure for Exercise 16

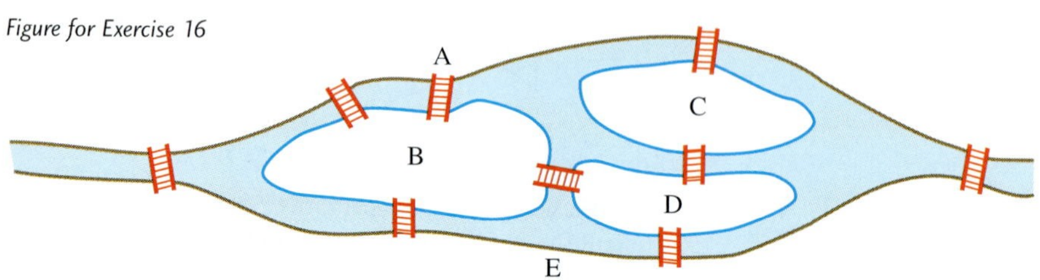

Chapter Test

1. The average interior angle measure of a convex polygon is 174°. What is the number of sides of the polygon? Explain how you found your answer.

2. Sketch an example of each of the following types of curves.
   (a) a simple closed curve
   (b) a convex heptagon
   (c) a nonclosed simple polygonal curve
   (d) a closed nonsimple polygonal curve

3. A cube octahedron has eight triangular faces and six square faces.

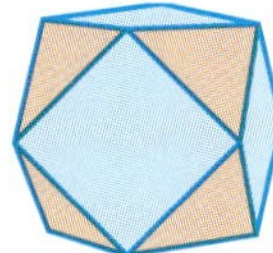

   (a) What is the number of edges of a cube octahedron? Explain how you do your counting.
   (b) What is the number of vertices of a cube octahedron? Explain how you found your answer.

4. Let P, Q, R, S, and T be the points shown. Draw and label the following figures.

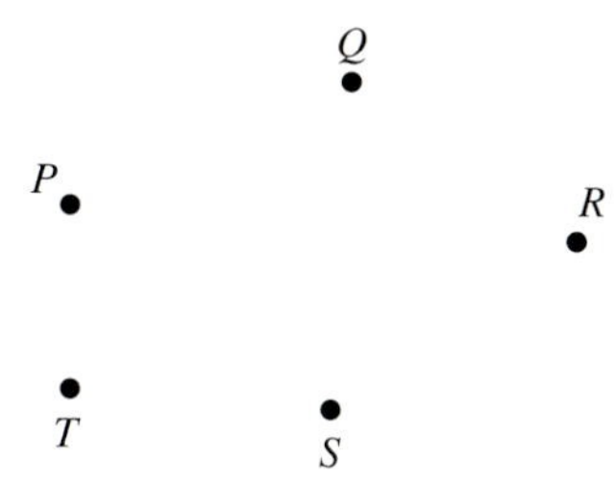

   (a) $\overrightarrow{PQ}$ (b) $\overleftrightarrow{PR}$ (c) $\overline{SQ}$ (d) $\angle PST$

5. (a) A connected planar network with 11 edges partitions the plane into 7 regions. How many vertices does this network have?
   (b) Draw a connected planar network with 11 edges and 7 regions.

6. At which vertices of the polygon does the interior angle appear to be

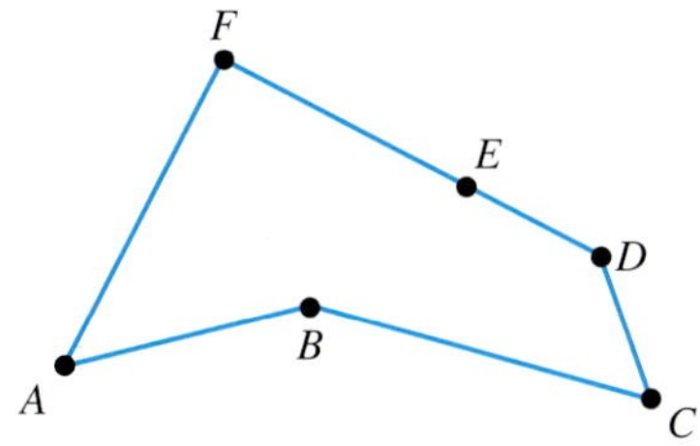

   (a) acute? (b) right? (c) obtuse?
   (d) straight? (e) reflex?

7. Decide whether each statement below is *true* or *false*.
   (a) Every square is a rhombus.
   (b) Some right triangles are obtuse.
   (c) All equilateral triangles are isosceles.
   (d) All squares are kites.

8. What is the measure of the angles in the points of this symmetric eight-point star? Explain how you found your answer.

9. For each part, name the regular polygon with the stated property.
   (a) The central angle has measure 36°.
   (b) The exterior angles each measure 45°.
   (c) The interior angles each measure 140°.

10. Pyramids are erected outward to both bases of a heptagonal prism.
   (a) Sketch the surface that is described.
   (b) Directly count the number of vertices, faces, and edges of the resulting polyhedron.
   (c) Verify that Euler's formula is satisfied.

11. A right prism has bases bounded by regular pentagons. What is the dihedral angle at which two adjacent lateral faces meet?

12. Find the angle measures r, s, and t in the figure shown, where lines l and m are parallel.

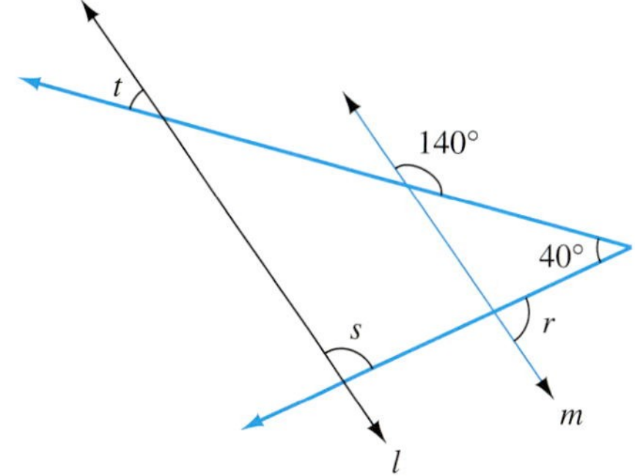

13. Consider the following networks.

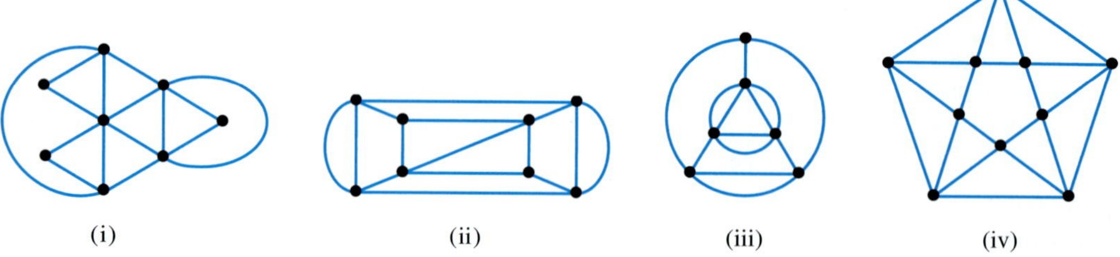

(a) Which networks are not traversable?

(b) Which networks are traversable starting at an arbitrary vertex?

(c) Which networks are traversable starting at some, but not every, vertex?

(d) Which network becomes traversable when only one edge is added to the network? Describe the new edge.

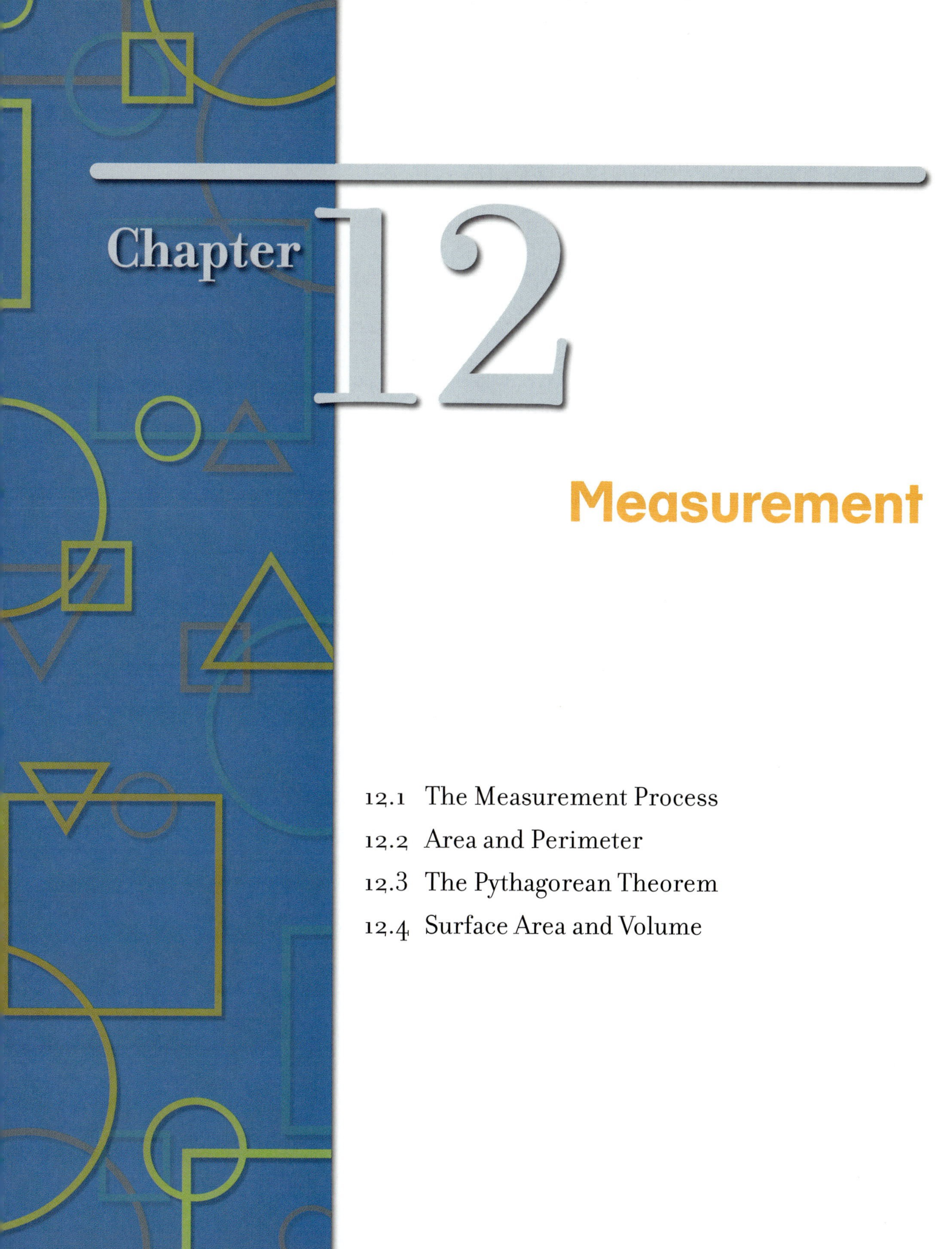

Chapter 12

Measurement

Hands On: The Metric Measurement and Estimation Tournament

Directions

Set up the six event stations with the materials described in the forthcoming table. For each event, put a copy of the directions for that event at the event's station. Divide the class into six teams, with each team assigned to one of the event stations and give the team score sheet found below. Each team then competes in its assigned event, records its scores on the score sheet, and rotates to their next event station. When all teams have completed all of the events, the true distances between cities are made known for the Distance Flier event, and the true areas of the foot patterns are revealed for the Big Foot event. After the teams have completed filling out their team score sheet, the teams are ranked in each event according to the lowest average error made by the team. For each event, the most accurate team is awarded 10 points, with the first and second runner-up teams given 7 and 3 points, respectively. The team with the most total points for all six events wins the tournament.

Materials and Directions for the Events

Event	Materials	Directions
String Cut	Ball of string, scissors, die, centimeter ruler	In turn, each team member rolls the die and cuts a piece of string whose length in centimeters is estimated to be ten times the number on the die. Measure the actual length of the string and record the error for each team member. Then compute and record the average team error.
Shot Put	Foam ball (or wad of newspaper), 10-meter-long tape measure (if a metric tape measure is unavailable, mark meters on a tape measure that shows units in feet)	Each team member throws the ball like a shot put, and estimates the distance of the throw in meters. The true distance is measured, and the error made in the estimates is recorded for each team member. Compute and record the average team error.
Weight Lift	Scale (reading in grams), collection of objects of various weights (rocks, canned goods, books, etc.)	Each team member chooses an object and estimates its weight in grams. The objects are then weighed, the errors recorded, and the average team error computed.
Pour It On	Bag of rice (or small dry beans), die, bowl, measuring cup marked in milliliters	In turn, each team member rolls the die to get a value $d = 1, 2, 3, 4, 5,$ or 6. The team member then pours rice into the bowl, attempting to fill it with $d \times 100$ milliliters of rice. Finally, pour the rice in the bowl into the measuring cup to measure the actual volume of rice. Determine the individual error made, and the team's average error.
Distance Flier	Map of the United States. (or a state), paper bag containing cards showing the names of about a dozen cities on the map (whose true distances apart are on an overhead transparency shown at the end of the tournament to determine the teams' errors and rankings in the event)	Each student draws two cards from the bag and estimates the distances between the cities in kilometers. The individual errors and average team error are determined when the instructor reveals the true distances between the cities.

Big Foot Area 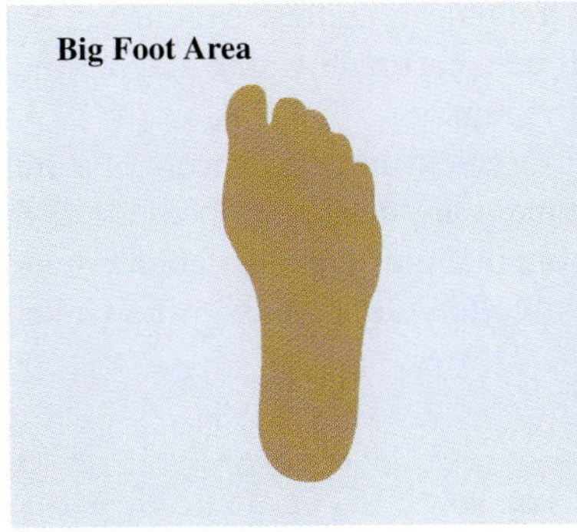	Centimeter-squared paper, die, collection of six foot shapes whose areas are known only to the class instructor (who reveals the areas at the end of the activity to determine the teams' errors and rankings in the event.	A die is rolled to select the team's foot pattern. The pattern is traced onto centimeter-squared paper, and the entire team estimates its area in square centimeters. The team's error is determined when the instructor reveals the true areas of the foot shapes.

Team Score Sheet

Event		Team Members					Total of errors	Average error	Tournament points
String Cut	Estimate								
	Actual								
	Error								
Shot Put	Estimate								
	Actual								
	Error								
Weight Lift	Estimate								
	Actual								
	Error								
Pour It On	Estimate								
	Actual								
	Error								
Distance Flier	City A								
	City B								
	Est. dist.								
	True dist.								
Big Foot	# of foot measured is _____	Estimated area of foot = cm^2					Error of measurement =		
		Actual area of foot = cm^2							

Connections The Principles and Processes of Measurement

Measurement played a limited, but important, role in Chapter 11. Only two elements were measured: line segments, measured by the distance between their endpoints; and angles, measured by the degrees of rotation needed to turn one side to the other. In this chapter, we introduce more general notions of measurement of geometric figures. All curves, not just segments, will be given a length. Plane regions will be measured by area and perimeter. Space figures will be measured by surface area and volume.

We begin by discussing the general process of measurement and the concept of a unit of measurement. The two principal systems of measurement are then described: the U.S. Customary (English) System, used in the United States, but in almost no other country; and the International System (metric), used by all countries worldwide, including the United States.

12.1 THE MEASUREMENT PROCESS

The geometry of the Babylonians and ancient Egyptians always had a practical purpose, and often this purpose was dependent on having knowledge of size and capacity. It was important to know the areas of fields, the volumes of granaries, and so on. Many engineering projects gave rise to geometric problems concerned with magnitudes. To be specific, suppose a canal has a given trapezoidal cross-section and a known length. If we know how much volume of earth one worker can dig in one day, how many workers are needed to excavate the canal in a given period of time?

Determining size requires that a comparison be made with a **unit.** For example, the volume of a canal could be expressed in "worker-days," where a worker-day is the volume one person can excavate in one day's labor. The worker-day is thus a unit of volume. It is analogous to the original definition of acre, which was the area of land that could be plowed in one day with one team of oxen.

In early times, units of measurement were defined more for convenience than accuracy. For example, many units of length correspond to parts of the human body, some of which are shown in Figure 12.1. The hand, span, foot, and cubit all appear in early records of Babylonia and Egypt.

Many of these units later became standardized and persist today. For example, horses are still measured in hands, where a hand is now 4 inches. Originally, an inch was the length of 3 barleycorns placed end to end.

The **measurement process** can be viewed as a sequence of steps.

The Measurement Process

(i) Choose the property (such as length, area, volume, capacity, temperature, time, or weight) of an object or event that is to be measured.
(ii) Select an appropriate unit of measurement.
(iii) Use a measurement device to "cover," "fill," "time," or otherwise provide a comparison of the object with the unit.
(iv) Express the measurement as the number of units used.

In Example 12.1, a **tangram** piece is chosen to be the unit of area measurement. Tangrams originated in ancient China and continue to be a versatile manipulative in the classroom. The instructions below show how to make a set of tangram pieces by folding

from **The NCTM Principles and Standards**

Measurable Attributes and the Processes of Measurement in Grades Pre-K–2

Children should begin to develop an understanding of attributes by looking at, touching, or directly comparing objects. They can determine who has more by looking at the size of piles of objects or identifying which of two objects is heavier by picking them up. They can compare shoes, placing them side by side, to check which is longer. Adults should help young children recognize attributes through their conversations. "That is a *deep* hole." "Let's put the toys in the *large* box." "That is a *long* piece of rope." In school, students continue to learn about attributes as they describe objects, compare them, and order them by different attributes. Seeing order relationships, such as that the soccer ball is bigger than the baseball but smaller than the beach ball, is important in developing measurement concepts.

Teachers should guide students' experiences by making the resources for measuring available, planning opportunities to measure, and encouraging students to explain the results of their actions. Discourse builds students' conceptual and procedural knowledge of measurement and gives teachers valuable information for reporting progress and planning next steps. The same conversations and questions that help students build vocabulary help teachers learn about students' understandings and misconceptions. For example, when students measure the length of a desk with rods, the teacher might ask what would happen if they used rods that were half as long. Would they need more rods or fewer rods? If students are investigating the height of a table, the teacher might ask what measuring tools would be appropriate and why.

Although a conceptual foundation for measuring many different attributes should be developed during the early years, linear measurements are the main emphasis. Measurement experiences should include direct comparisons as well as the use of nonstandard and standard units. For example, teachers might ask young students to find objects in the room that are about as long as their foot or to measure the length of a table with connecting cubes. Later, they can supply standard measurement tools, such as rulers to measure classroom plants, and use those measurements to chart the plants' growth.

SOURCE: Reprinted with permission from Principles and Standards for School Mathematics, *page 103.* *Standards are listed with the permission of the National Council of Teachers of Mathematics (NCTM). NCTM does not endorse the content or validity of these alignments.*

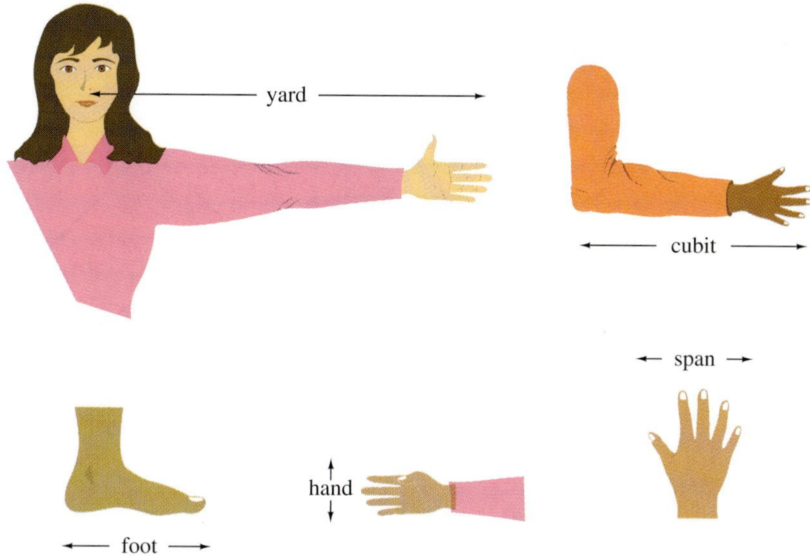

Figure 12.1
Examples of traditional units of length based on the human body

and cutting a square sheet of paper. A more sturdy set of tangrams can be cut from cardboard or vinyl tile, using the paper shapes as templates.

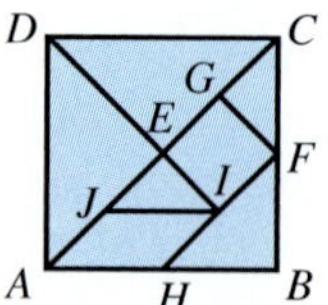

The seven tangram pieces.

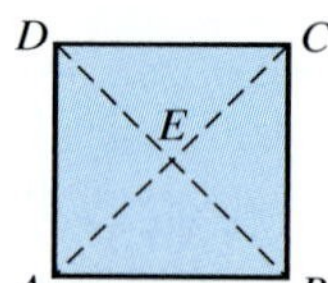

Step 1: Fold the two diagonals of the square *ABCD*. Let *E* denote its center.

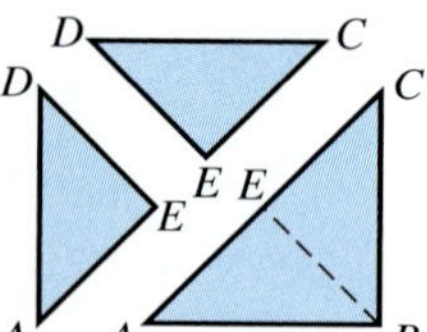

Step 2: Cut out the tangram pieces *CDE* and *ADE*, leaving triangle *ABC*.

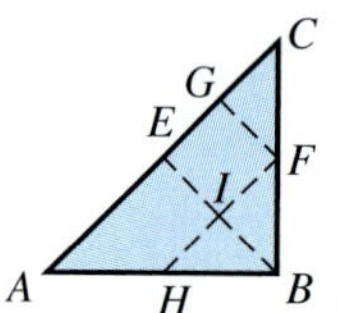

Step 3: Fold *B* and *C* to center point *E* to create folds *FG* and *FH*.

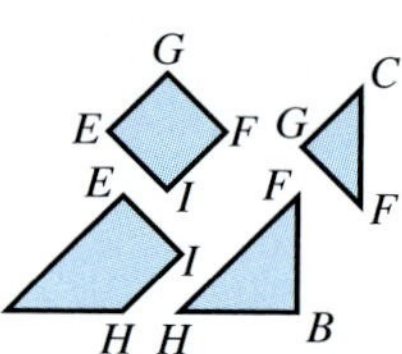

Step 4: Cut out tangram pieces *BFH*, *CFG*, and *FGEI*, leaving trapezoid *AHIE*.

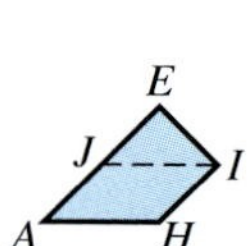

Step 5: Fold *E* to *H* to create fold *IJ*.

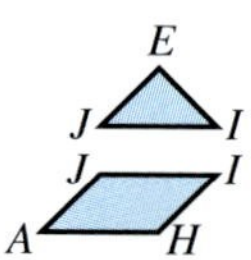

Step 6: Cut out tangram pieces *AHIJ* and *EIJ*.

EXAMPLE 12.1 INVESTIGATING TANGRAM MEASUREMENTS

Label the tangram pieces I, II, . . . , VII as shown. Use shape I, the small isosceles right triangle, as the unit of "one tangram area" (abbreviated 1 tga) to measure the following:

(a) the area of each of the tangram pieces,
(b) the area of the "fish," and
(c) the area of the circle that circumscribes the square. Are your measurements exact or only approximate?

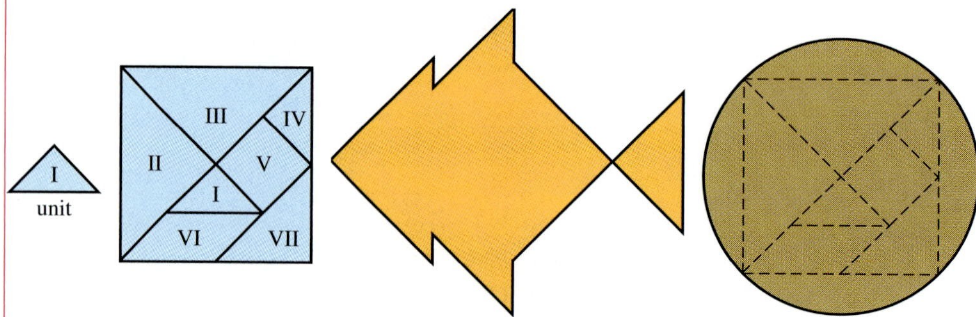

Solution

(a) Each tangram piece can be covered by copies of the unit shape I, giving the exact measurements in the table.

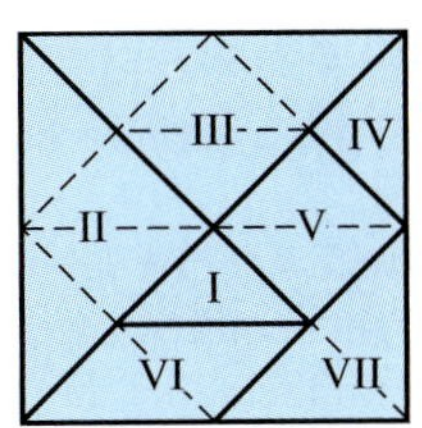

Piece	Area
I, IV	1 tga
II, III	4 tga
V, VI, VII	2 tga

(b) The fish is covered by the seven tangram pieces, so its area is exactly $(4 + 2 + 1 + 2 + 1 + 4 + 2)$ tga = 16 tga.

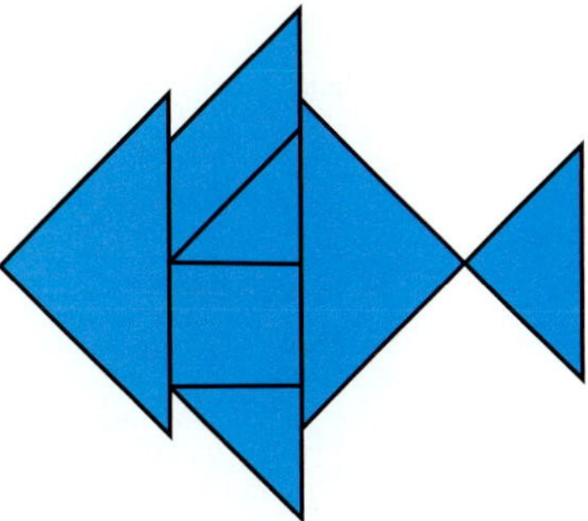

(c) The circle can be covered by the seven tangram pieces, together with 12 additional copies of the unit shape. This shows that the circle's area is between 16 tga and 28 tga. Thus, we might estimate the area at about 25 tga.

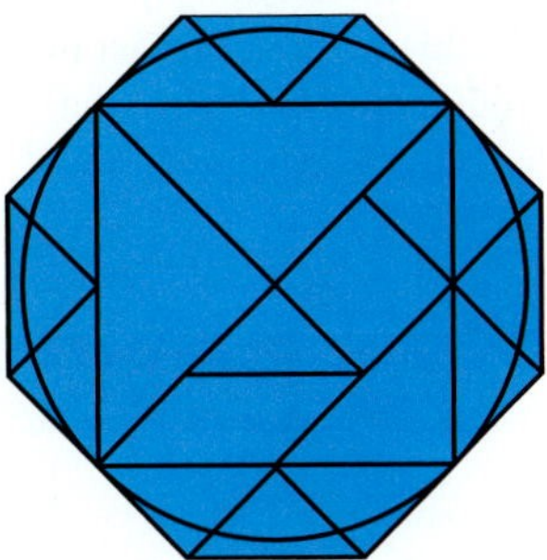

An important practical purpose of measurement is communication. By agreeing on common units of measurement, people are able to express and interpret information about size, quantity, capacity, and so on. Historically, as commerce developed and goods were traded over increasingly large distances, the need for a standard system of units became more and more apparent. In the seventeenth and eighteenth centuries, the rise of science and the beginning of the industrial revolution gave further impetus to the development of universal systems of measurement.

The U.S. Customary, or "English," System of Measures

The English system arose from a hodgepodge of traditional informal units of measurement. Table 12.1 lists some of the units of **length** with this system. The ratios comparing one unit of length with another are clearly the result of accident, not planning. Learning the customary system requires extensive memorization, and using the system involves computations with cumbersome numerical factors.

TABLE 12.1 Units of Length in the Customary System

Unit	Abbreviation	Equivalent Measurement in Feet
Inch	in	$\frac{1}{12}$ ft
Foot	ft	1 ft
Yard	yd	3 ft
Rod	rd	$16\frac{1}{2}$ ft
Furlong*	fur	660 ft
Mile	mi	5280 ft

*The *furlong* is a shortening of "furrow long," revealing its origin in agriculture.

Area is a measure of the region bounded by a plane curve. Any shape could be chosen as a unit, but the square is the most common. The size of the square is arbitrary, but it is natural to choose the length of a side to correspond to a unit measure of length. Areas are therefore usually measured in square inches, square feet, and so on. A moderate-sized house may have 1800 square feet of floor space, a living room carpet may cover 38 square yards, and a national forest may cover 642 square miles. An exception to this pattern is the acre; 640 acres have a total area of 1 square mile. Some common units of area are listed in Table 12.2. The superscript-2 notation shows a square unit; for example, ft^2 indicates square feet.

TABLE 12.2 Units of Area in the Customary System

Unit	Abbreviation	Equivalent Measure in Other Units
Square inch	in^2	$\frac{1}{144}$ ft^2
Square foot	ft^2	144 in^2, or $\frac{1}{9}$ yd^2
Square yard	yd^2	9 ft^2
Acre	acre	$\frac{1}{640}$ mi^2, or 43,560 ft^2
Square mile	mi^2	640 acres, or 27,878,400 ft^2

The ratios comparing one unit of area with another can be visualized, as shown in Figure 12.2. We see that the area of a 3-ft by 3-ft square is obtained by the multiplication

$3 \text{ ft} \times 3\text{ft} = 3 \times 3 \times \text{ft} \times \text{ft} = 9 \text{ ft}^2$. *When computing with dimensioned quantities, it is essential to retain the units in all equations and expressions.* For example, it is correct to write 12 in = 1 ft; without the dimensions, this equation would be incorrect, since $12 \neq 1$. Omitting the units in expressions is a common source of errors.

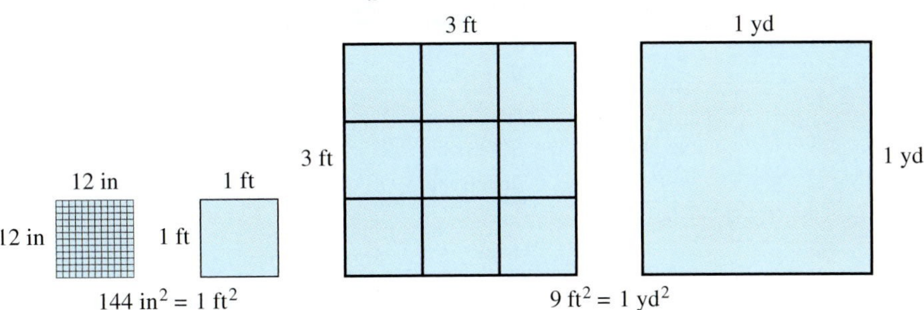

Figure 12.2
Comparing units of area measure

Volume is the measure of space taken up by a solid in three-dimensional space. The unit, as shown in Table 12.3, is the volume of a cube whose side length is one of the standard units of length. For example, the displacement of the pistons in a car engine may be 327 cubic inches, which is abbreviated 327 in^3.

TABLE 12.3 Units of Volume in the Customary System

Unit	Abbreviation	Equivalent Measure in Other Units
Cubic inch	in^3	$\frac{1}{1728} \text{ ft}^3$
Cubic foot	ft^3	1728 in^3, or $\frac{1}{27} \text{ yd}^3$
Cubic yard	yd^3	27 ft^3

The ratios comparing units of volume are illustrated in Figure 12.3.

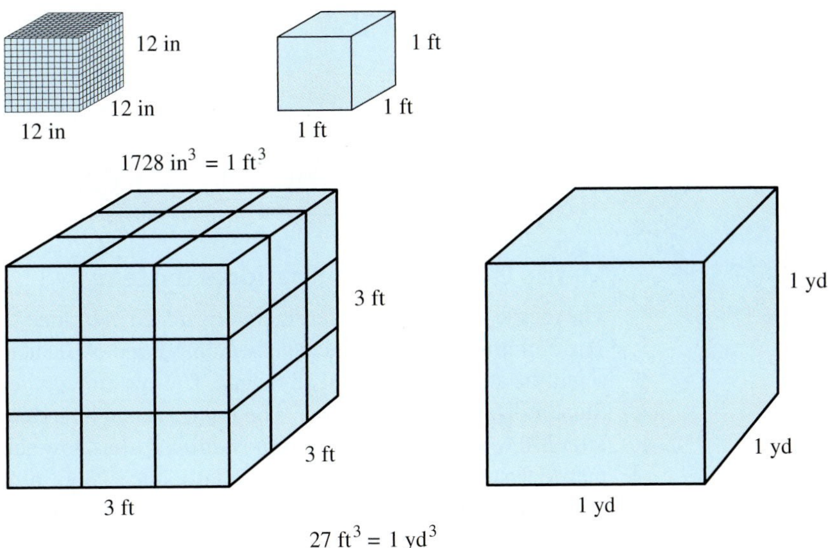

Figure 12.3
Comparing units of volume measure

Capacity is the volume that can be held in a container such as a bottle, pan, basket, or tank. Capacity is often expressed in in^3, ft^3, or yd^3, but other units are also in common use. Some common examples are shown in Table 12.4.

TABLE 12.4 Units of Capacity in the U.S. Customary System

Unit	Abbreviation	Equivalent Measure in Other Units
Teaspoon	tsp	$\frac{1}{3}$ tablespoon
Tablespoon	T or tbl	2 fl. oz.
Fluid Ounce	fl. oz.	$\frac{1}{8}$ cup
Cup	C	8 fl. oz., or $\frac{1}{4}$ quart
Quart	qt	4 cups, or $\frac{1}{4}$ gallon
Gallon	gal	4 quarts, or 231 cubic inches
Bushel	bu	2150.42 cubic inches

The bushel is a dry measure, unlike the liquid measures of the other units in Table 12.4. It has no direct relation to the gallon. In the thirteenth century, a bushel was any large amount, so a lover might shower his sweetheart with a "bushel of kisses."

Metric Units: The International System

The metric system of measurement originated in France shortly after the revolution of 1789. The definitions of the units have been modified over succeeding years, taking advantage of scientific and technological advances. The system was codified in 1981 by the International Standardization Organization. The International System of Units, also called the **SI system** after its French name, *Système Internationale,* has now achieved worldwide acceptance. The metric system has been a legal standard since 1866 in the United States; indeed, the customary units were *defined* in terms of metric units in 1893. In the 1970s, a movement to replace customary units with metric units was unsuccessful. About the same time, most other English-speaking countries, including Great Britain, Canada, Australia, and New Zealand, did change to metric units. Even day-to-day measurements in those countries—speed limits, distances between cities, and amounts in recipes—were replaced with metric units.

The principal advantage of the metric system—other than its universality—is the ease of comparison of units. The ratio of one unit to another is always a power of 10, which ties the metric system conveniently to the base-ten numeration system. This makes it quite simple to convert a measurement in one metric unit to the equivalent measurement in another metric unit.

Each power of 10 is given a prefix that modifies the fundamental unit. For example, the factor 1000 (that is, 10^3) is expressed by the prefix *kilo*. Thus, a kilometer is 1000 meters. Similarly, the factor $\frac{1}{100}$ (that is, 10^{-2}) is expressed by the prefix *centi*. Therefore, a centimeter is $\frac{1}{100}$ of a meter. The more commonly used prefixes and their symbols are listed in Table 12.5.

TABLE 12.5 The SI Decimal Prefixes

Prefix	Factor	Symbol
kilo	$1000 = 10^3$	k
hecto	$100 = 10^2$	h
deka (or deca)	$10 = 10^1$	da
(none for basic unit)	$1 = 10^0$	(none)
deci	$0.1 = 10^{-1}$	d
centi	$0.01 = 10^{-2}$	c
milli	$0.001 = 10^{-3}$	m
micro	$0.000001 = 10^{-6}$	μ (Greek *mu*)

Length

The fundamental unit of length in the SI system is the **meter,** abbreviated by the symbol m. The unit symbol is always written last in SI, so there can be no confusion with the prefix *milli,* which is also given the symbol m. For example, one thousandth of a meter is a millimeter, written as 1 mm. There is no space between the first and second m, and there are no periods between or after the symbols. When typed, the symbols are always in roman font, not italic.

The most commonly used metric units of length are listed in Table 12.6.

TABLE 12.6 Metric Units of Length

Unit	Abbreviation	Multiple or Fraction of 1 Meter
1 kilometer	1 km	1000 m
1 hectometer	1 hm	100 m
1 dekameter	1 dam	10 m
1 meter	1 m	1 m
1 decimeter	1 dm	0.1 m
1 centimeter	1 cm	0.01 m
1 millimeter	1 mm	0.001 m
1 micrometer (or micron)	1 μm	0.000001 m

The prefix (for example, the "c" in the notation cm) in a metric measurement can be replaced with its corresponding numerical factor. For example,

$$251 \text{ cm} = 251 \times 10^{-2} \text{ m} = 2.51 \text{ m}.$$

Similarly, a power of 10 can be replaced with the corresponding prefix, as in

$$0.179 \text{ m} = 179 \times 10^{-3} \text{ m} = 179 \text{ mm}.$$

Some metric measurements are shown in Figure 12.4. Since the items shown in Figure 12.4 differ so dramatically in size, the scale for each image (except for the nickel) is greatly reduced or enlarged. The length shown as representing one micrometer is intended to make it clear that waves of red and violet light, the colon bacillus, and the smallpox virus are so tiny as to be determinable only by using sophisticated scientific equipment.

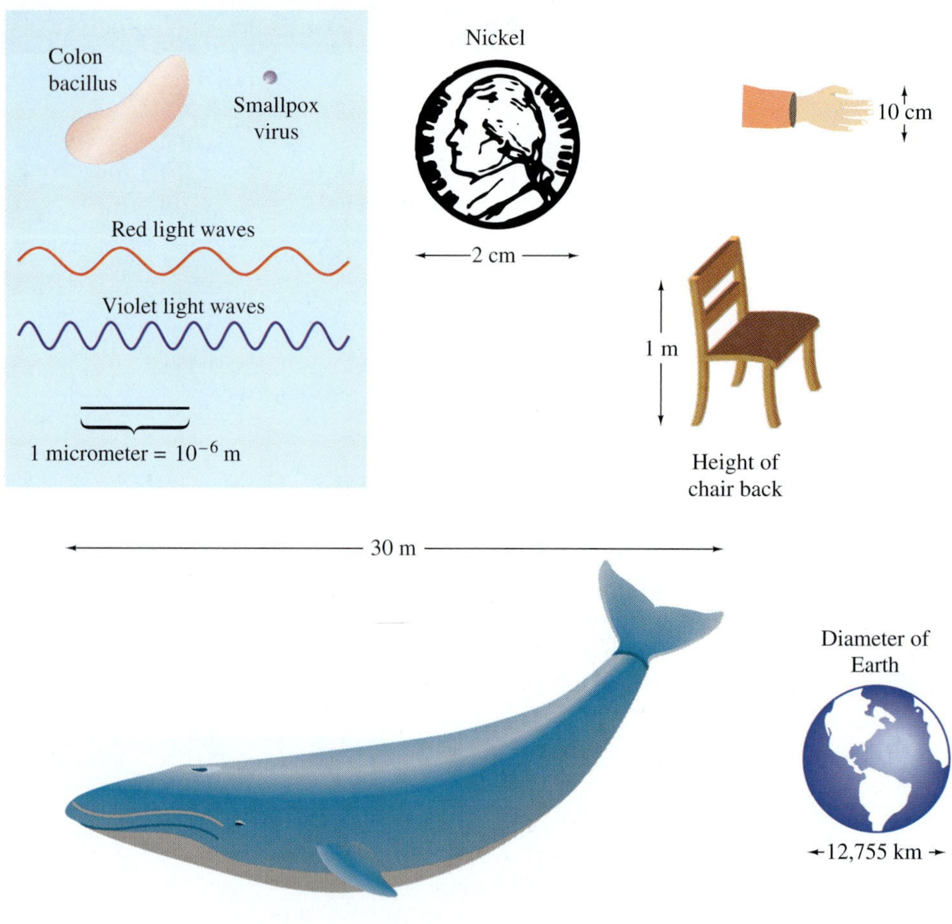

Figure 12.4
Examples of metric measurements of length

EXAMPLE 12.2 | DETERMINING TINY LENGTHS IN FIGURE 12.4

Using a metric ruler, measure the following elements from Figure 12.4 as carefully as possible and give the measurements in micrometers and meters.

(a) The length of the colon bacillus
(b) The diameter of the smallpox virus
(c) The length of one wave of red light (i.e., the distance from one peak to the next)
(d) The length of one wave of violet light

Solution

(a) The length of the image of the colon bacillus is approximately 19 mm. Since the length of the unit representing one micrometer is approximately 14 mm, the length of the colon bacillus is approximately

$$\frac{19}{14}\,\mu\text{m} \doteq 1.4\,\mu\text{m} = 1.4 \times 10^{-6}\text{ m}.$$

(b) Estimating as closely as possible, it appears that the diameter of the image of the smallpox virus is 1.5 mm. Thus, the actual length of the diameter is approximately

$$\frac{1.5}{14}\,\mu\text{m} \doteq 0.1\,\mu\text{m} = 10^{-7}\text{ m}.$$

(c) The length of the image of a single wave of red light is approximately 11 mm. Thus, the actual length is approximately

$$\frac{11}{14}\,\mu\text{m} \doteq 0.8\,\mu\text{m} = 0.8 \times 10^{-6}\text{ m}.$$

(d) The length of the image of a single wave of violet light is approximately 5.5 mm. Thus, the the actual length is approximately

$$\frac{5.5}{14}\,\mu\text{m} \doteq 0.4\mu\text{m} = 0.4 \times 10^{-6}\text{ m}.$$

HIGHLIGHT from HISTORY

Redefining the Meter

The meter was originally defined in the 1790s as one ten-millionth of the distance from the North Pole to the equator. Making the measurement, however, presented impossible difficulties. Not only was the earth quite unlike a perfect sphere, but political turmoil in revolutionary France led to the arrest of government surveyors as royalist spies, who narrowly escaped the guillotine. In 1889, the meter was fixed as the distance between two marks on a platinum-iridium bar, but scientists were unsatisfied since measurements could be no more accurate than one part in a million. In 1960 the meter was again redefined, becoming 1,650,763.73 wavelengths of the reddish-orange light emitted by krypton 86, a rare atmospheric gas. The new meter was accurate to 4 parts per billion, but still was an irritant to scientists measuring quantities like continental drift and the distance to the moon. In 1983, the meter was redefined yet again, in a way that connects length to time. The new definition specifies the meter as the distance traveled by light in space in 1/299,792,458 of a second. One second of time, which can be precisely measured with atomic clocks, is defined as the duration of 9,193,631,770 vibration cycles of the Cesium 133 atom. The definition invokes a sacred tenet of physics, that the speed of light in space is a universal constant, namely, 299,792,458 meters per second.

EXAMPLE 12.3 CHANGING UNITS IN THE METRIC SYSTEM

Convert these measurements to the unit shown.

(a) 1495 mm = ____ m
(b) 29.4 cm = ____ mm
(c) 38,741 m = ____ km

Solution

(a) $1495 \text{ mm} = 1495 \times 10^{-3} \text{ m} = 1.495 \text{ m}$
(b) $29.4 \text{ cm} = (294 \times 10^{-1}) \times 10^{-2} \text{ m} = 294 \times 10^{-3} \text{ m} = 294 \text{ mm}$
(c) $38{,}741 \text{ m} = 38.741 \times 10^{3} \text{ m} = 38.741 \text{ km}$

Area

Area is usually expressed in square meters (m^2) or square kilometers (km^2). Another common unit is the hectare. A **hectare** (ha) is the area of a 100-m square; that is, $1 \text{ ha} = 10{,}000 \text{ m}^2$. See Table 12.7.

The floor space of a classroom might typically be about one **are** (pronounced "air"). A hectare is about 2.5 acres, so the area of farmland is measured in hectares in metric countries.

TABLE 12.7 Metric Units of Area

Unit	Abbreviation	Multiple or Fraction of 1 Square Meter
1 square centimeter	1 cm^2	0.0001 m^2
1 square meter	1 m^2	1 m^2
1 are (1 square dekameter)	1 a	100 m^2
1 hectare (1 square hectometer)	1 ha	$10{,}000 \text{ m}^2$
1 square kilometer	1 km^2	$1{,}000{,}000 \text{ m}^2$

Volume and Capacity

Small volumes are typically measured in cubic centimeters (abbreviated cm^3). Large volumes are often measured in cubic meters (m^3). A convenient unit of capacity is the **liter,** defined as a cubic decimeter. Thus, the liter (abbreviated either as l or L, with L preferred in the United States to avoid confusion with the numeral 1) is the volume of a cube whose sides each measure $1 \text{ dm} = 10 \text{ cm}$; see Figure 12.5. Therefore, since $10 \text{ cm} \times 10 \text{ cm} \times 10 \text{ cm} = 1000 \text{ cm}^3$, a liter is also 1000 cubic centimeters. Recalling that *milli* is the prefix for $\frac{1}{1000}$, a milliliter (mL or ml) is the same as one cubic centimeter:

$$1 \text{ L} = 1 \text{ liter} = 1000 \text{ cm}^3;$$
$$1 \text{ mL} = 1 \text{ milliliter} = 1 \text{ cm}^3.$$

Large plastic bottles of soda usually contain 2 liters, while a typical soft drink can contains about 354 milliliters. A child's dose of cough medicine may be 3 mL. A recipe may call for 0.5 liters of water. In metric countries, gasoline is priced by the liter, and to fill a car's gas tank takes about 40 to 60 liters.

Weight and Mass

The **weight** of an object is the force exerted on the object by gravity. For example, a brick on the surface of the earth may weigh 6 pounds, but on the surface of the moon it would

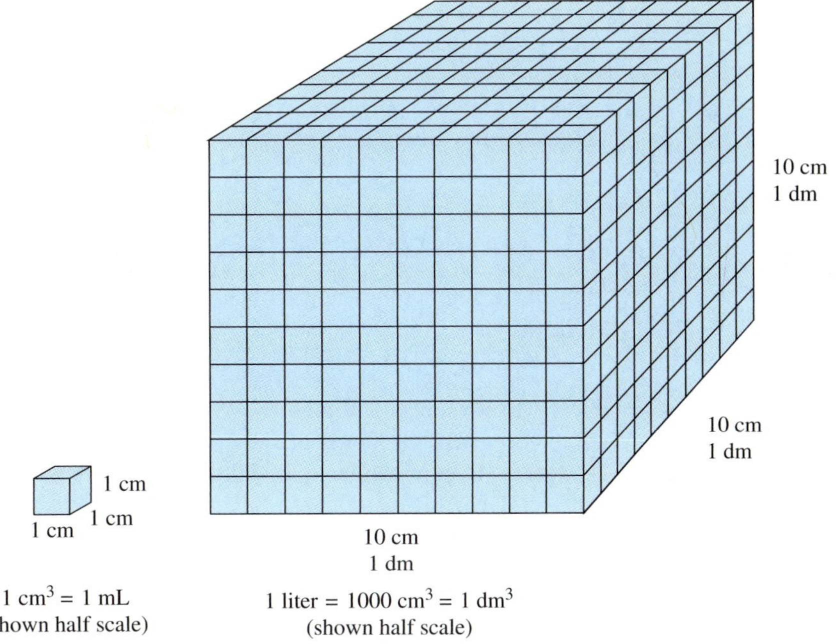

Figure 12.5
A liter is a cubic decimeter, or, equivalently, 1000 cubic centimeters

only weigh about 1 pound. During the journey from earth to moon, the brick would weigh nearly nothing at all. Nevertheless, an astronaut would not care to be hit by a weightless brick, since the brick never loses its **mass.**

In science, the distinction between mass and weight is very important: mass is the amount of matter of an object, and weight is the force of gravity on the object. But on the surface of the earth and in everyday-life situations, the weight of an object is proportional to its mass. That is, the mass of an object is accurately estimated by weighing it.

The U.S. customary unit of weight is the familiar pound. Lighter weights are often given in ounces, and very heavy weights are given in tons:

$$16 \text{ ounces (oz)} = 1 \text{ pound (lb)};$$
$$2000 \text{ pounds} = 1 \text{ ton}.$$

Table 12.8 lists some metric units of weight. The base unit of weight in the metric system is the **kilogram,** which is the weight of one liter of water. Since there are 1000 cubic centimeters in one cubic decimeter (one liter) and a liter of water weighs one kilogram, one cubic centimeter of water weighs 0.001 of a kilogram or, equivalently, one gram.

TABLE 12.8 Metric Units of Weight

Unit	Abbreviation	Multiples of Other Metric Units
1 milligram	1 mg	0.001 g
1 gram	1 g	0.001 kg
1 kilogram	1 kg	1000 g
1 metric ton	1 t	1000 kg

One milligram is approximately the weight of a grain of salt. It is a common measure of vitamins and medicines. A gram is approximately the weight of half a cube of sugar. Canned goods and dry packaged items at the grocery store will usually be weighed in grams. In metric countries, larger food items, such as meats, fruits, and vegetables, are priced by the kilogram. A kilogram is about 2.2 pounds.

EXAMPLE 12.4 ESTIMATING WEIGHTS IN THE METRIC SYSTEM

Match each item to the approximate weight of the item taken from the list that follows.

(a) Nickel
(b) Compact automobile
(c) Two-liter bottle of soda
(d) Recommended daily allowance of vitamin B–6
(e) Size D battery
(f) Large watermelon

List of weights: 2 mg, 2 kg, 100 g, 1200 kg, 9 kg, 5 g

Solution

(a) 5 g **(b)** 1200 kg **(c)** 2 kg **(d)** 2 mg **(e)** 100 g
(f) 9 kg

Temperature

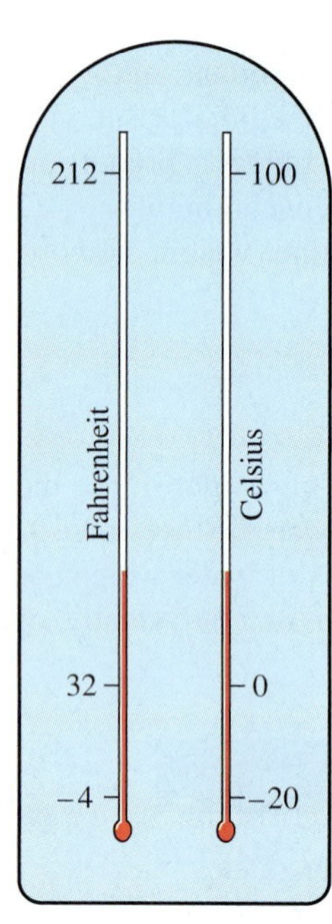

Figure 12.6 *Thermometers showing Fahrenheit and Celsius scales*

There are two commonly used scales to measure temperature. According to the **Fahrenheit scale,** 32°F represents the freezing point of water and 212°F the boiling point of water. Thus, the Fahrenheit scale introduces 180 degrees of division between the freezing and boiling temperatures. The **Celsius scale** divides this temperature range into 100 degrees: the freezing point is 0° Celsius, and the boiling point is 100° Celsius.

The thermometers shown side by side in Figure 12.6 can be used to derive a formula that relates F degrees Fahrenheit to the equivalent temperature in C degrees Celsius. First, notice that $F - 32$ changes from 0 to 180 between freezing $(F = 32)$ and boiling $(F = 212)$, so $(F - 32)\frac{100}{180}$ changes from 0 to 100 between freezing and boiling. This is the same as the Celsius temperature scale, so $C = (F - 32)\frac{100}{180}$. The same reasoning shows that $C\left(\frac{180}{100}\right) + 32$ changes from 32 to 212 as the Celsius temperature C changes from freezing $(C = 0)$ to boiling $(C = 100)$, so this expression gives the temperature F in degrees Fahrenheit. That is, the temperature scales are related by the formulas

$$C = (F - 32)\frac{100}{180} \quad \text{and} \quad F = \frac{180}{100}C + 32.$$

Unit Analysis

It is often of interest to express a measurement given in one unit by the equivalent measurement in a new unit. A procedure known as **unit analysis** (or **dimensional analysis**) can help arrange the calculation to make it clear if the factors comparing units are used as

School Book Page

Temperature Measurement in Grade Four

Chapter 11
Lesson 15

Temperature

You Will Learn
how to read Celsius and Fahrenheit thermometers

Vocabulary
degrees Celsius metric units of temperature
degrees Fahrenheit customary units of temperature

Learn

During the opening ceremonies at the 1994 Olympics, the thermometer read 14° Fahrenheit, or −10° Celsius. It was the coldest Winter Olympics ever!

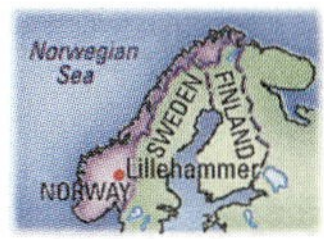

The 1994 Winter Olympics were held in Lillehammer, Norway.

Degrees Celsius are metric units of temperature and are written °C.

Degrees Fahrenheit are customary units of temperature and are written °F.

The thermometer's scale is 2°.

Freezing point is 0°C, or 32°F. We read temperatures that go below 0° as minus. This thermometer shows minus 10° C, the temperature at Lillehammer.

Did You Know?
Normal human body temperature is 98.6°F, or 37°C.

Talk About It

What Celsius temperature is equal to 50°F? How would you describe this temperature?

514 Chapter 11 • Decimals and Metric Measurement

SOURCE: From Scott Foresman–Addison Wesley Math, *Grade 4, p. 514, by Randall I. Charles et al. Copyright © 2002 Pearson Education, Inc. Reprinted with permission.*

Questions for the Teacher

1. Since water freezes at 32° Fahrenheit and 0° Celsius, a child thinks a degree Fahrenheit is bigger than a degree Celsius. How would you respond?
2. The cold temperature in Norway at the 1994 Winter Olympics provides a context to understand 14° Fahrenheit and its metric equivalent, −10° Celsius. Give Celsius temperatures for these contexts:
   (a) normal body temperature; (b) room temperature; (c) a cup of hot tea; (d) bath water; (e) Death Valley on a hot summer day; (f) a cold glass of water.
3. A student plans to visit her uncle in Toronto, Canada, and hopes to go sledding, since the temperature is 20 degrees outside. What discussion would you suggest?

multipliers or divisors. The idea of unit analysis can be explained by an example. Suppose a distance has been given as 3.75 miles, and you would like the distance in yards. You recall that 1 mi = 5280 ft and 3 ft = 1 yd. These equations can also be written as $1 = \frac{5280 \text{ ft}}{1 \text{ mi}}$ and $1 = \frac{1 \text{ yd}}{3 \text{ ft}}$. Therefore,

$$3.75 \text{ mis} = 3.75 \cancel{\text{mis}} \times \frac{5280 \cancel{\text{ft}}}{1 \cancel{\text{mi}}} \times \frac{1 \text{ yd}}{3 \cancel{\text{ft}}} = \frac{3.75 \times 5280}{3} \text{ yds.}$$

Since $\frac{3.75 \times 5280}{3} = 6600$, we see that 3.75 miles = 6600 yards.

Unit conversion, even using unit analysis, should always be accompanied by careful reasoning. Always ask, "Is this answer reasonable? Does this answer agree with an approximate mental estimation?" Mistakes with unit conversions have led to some unfortunate disasters. For example, in 1999, engineers mistook a measurement of force given in pounds to be given in the metric unit of newtons of force. As a result, over four times the correct amount of thrust was applied to the *Mars Climate Orbiter,* sending the $125 million space probe to an early demise in the atmosphere of the Red Planet.

EXAMPLE 12.5 COMPUTING SPEED AND CAPACITY WITH UNIT ANALYSIS

(a) A cheetah can run 60 miles per hour. What is the speed in feet per second?

(b) A fish tank at the aquarium has the shape of a rectangular prism 2 m deep by 3 m wide by 3 m high. What is its capacity in liters?

Solution

(a) $60 \frac{\text{mi}}{\text{hr}} = 60 \frac{\cancel{\text{mi}}}{\cancel{\text{hr}}} \times \frac{5280 \text{ ft}}{1 \cancel{\text{mi}}} \times \frac{1 \cancel{\text{hr}}}{60 \cancel{\text{min}}} \times \frac{1 \cancel{\text{min}}}{60 \text{ sec}} = \frac{60 \times 5280 \text{ ft}}{60 \times 600 \text{ sec}} = 88 \frac{\text{ft}}{\text{sec}}.$

(b) Recall that a liter is a cubic decimeter, and *deci* is the prefix for one tenth. Therefore, the volume of the tank is

$$(2 \text{ m}) \times (3 \text{ m}) \times (3 \text{ m}) = 18 \text{ m}^3 = 18 \text{ m}^3 \times \left(\frac{10 \text{ dm}}{1 \text{ m}}\right)^3 = 18 \times \cancel{\text{m}^3} \times 10^3 \times \frac{\text{dm}^3}{\cancel{\text{m}^3}}$$

$$= 18{,}000 \cancel{\text{dm}^3} \times \frac{1 \text{ liter}}{\cancel{\text{dm}^3}} = 18{,}000 \text{ liters.}$$

Problem Set 12.1

Understanding Concepts

1. For each object listed below, make a list of measurable properties.
   (a) A bulletin board (b) An extension cord
   (c) A file box (d) A table
2. Suppose you are designing a house. Give examples of measurements you believe are important to consider. For example, the height of the house may be needed to satisfy a zoning regulation. Discuss examples of measurements of (a) length, (b) area, and (c) volume and capacity. What units are appropriate?
3. Find the area of each tangram figure shown, where a unit is the area of the small isosceles right triangle.

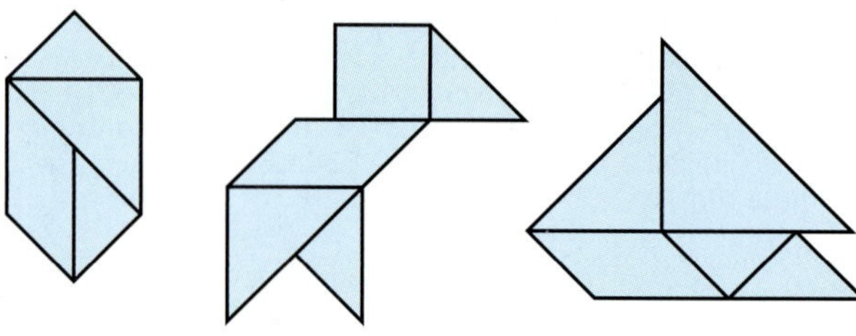

4. Let a *pen* be the area of a penny.
   (a) Estimate the area of a 4″ × 6″ card in pens.
   (b) Discuss why pens are a difficult unit of area to use.

5. Arrange these solids in a list according to volume, from smallest to largest. Are there any ties?

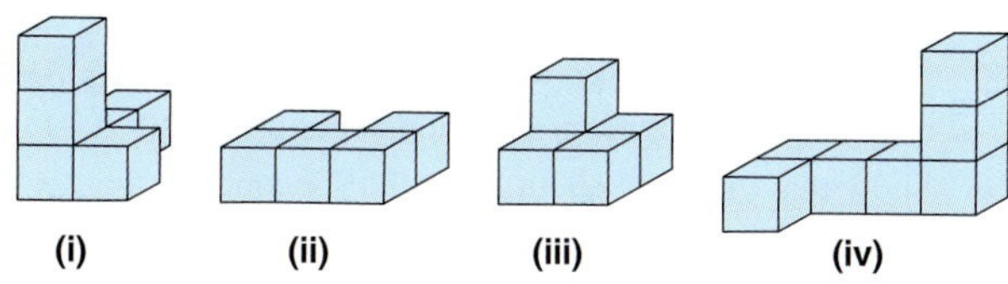

6. (a) Verify that an acre contains 43,560 square feet. Show your computation.
   (b) A square lot contains 1 acre. What is the length of each side, to the nearest foot?

7. (a) A football field is 120 yards long (including the end zones) and 160 feet wide. What is the area of a football field in acres?
   (b) A soccer field measures 110 meters by 70 meters. What is its area in ares? in hectares?

8. A measurement of 12.6 cm is assumed to be precise to the nearest last digit. That is, the true length is between 12.55 cm and 12.65 cm. Find the minimum and maximum true values for the following measurements.
   (a) A distance of 166 kilometers from Portland to Eugene
   (b) A piece of notebook paper 27.9 centimeters long
   (c) A pencil lead 0.50 millimeters in diameter
   (d) A diving board 3.5 meters over the water

9. A small bottle of Perrier sparkling mineral water contains 33 cL.
   (a) What is the volume in milliliters?
   (b) Will three small bottles fill a 1-liter bottle?

10. Fill in the blanks.
   (a) 58,728 g = ____ kg
   (b) 632 mg = ____ g
   (c) 0.23 kg = ____ g
   (d) A cubic meter of water weighs kg.

11. Give the most reasonable answer listed in each part.
   (a) A newborn baby weighs about: 8.3 kg, 3.5 kg, 750 g, 1625 mg.
   (b) A compact car weighs about: 5000 kg, 2000 g, 1200 kg, 50 kg.
   (c) The recommended daily allowance of vitamin C is: 250 g, 60 mg, 0.3 kg, 0.002 mg.

12. In each of the following, select the most reasonable metric measurement.
   (a) The height of a typical center in the National Basketball Association is: 6.11 m, 3 m, 95 cm, 212 cm.
   (b) The diameter of a coffee cup is about: 50 m, 50 mm, 500 mm, 5 km.
   (c) A coffee cup has a capacity of about: 8 L, 8 mL, 240 mL, 500 mL.

13. Use a metric ruler to measure these items.
   (a) The size of a sheet of standard notebook paper
   (b) The length and width of the cover of this textbook
   (c) The diameter of a nickel
   (d) The perimeter of (distance around) your wrist

14. The dimensions of Noah's Ark are given in the Bible as 300 cubits long, 50 cubits wide, and 30 cubits high. Give the dimensions in (a) meters and (b) feet. Use a meter stick and ruler to measure your own cubit, as shown in Figure 12.1.

15. The unit of distance in China is the *li*, traditionally about a third of a mile, but redefined more recently to be 500 meters. Other measures in China are the *yin*, *zhang*, *bu*, *chi*, and *cun*, where

    1 li = 15 yin, 1 yin = 10 zhang, 1 zhang = 2 bu,
    1 bu = 5 chi, and 1 chi = 10 cun.

   (a) How many chis are in a li?
   (b) How many cuns are in a yin?
   (c) What is 3 chi when expressed in meters?
   (d) A paper clip is a little over 3 cm long. How would its length most likely be expressed in Chinese units?

Teaching Concepts

16. Children often enjoy word games, and this exercise can provide a fun way to learn some of the metric prefixes. For example, the prefix for 10^{-12} in the metric system is pico, so what are 10^{-12} boos? One picoboo (peekaboo), of course. Now try these.
   (a) 10 millipedes
   (b) 10^{-6} phones
   (c) 10^{6} phones
   (d) 10 cards
   (e) 1000 pains
   (f) 2000 mockingbirds

17. Most of us can visualize the length of an inch, a foot, a yard, and even a football field because we have repeatedly experienced these measurements. To develop the same feeling for metric measurements of length, use a metric ruler and meterstick or metric tape measure to determine each of the following in metric units.
   (a) The length of a new pencil
   (b) The length of your shoe
   (c) Your own height

(d) The height and width of a door

(e) The length of a yardstick

(f) The length of a football field

18. Write a one-paragraph report that explains why the words *ounce* and *inch* are related.

19. Write a one-paragraph report that explains why the abbreviation for pound is *lb.*

Thinking Critically

20. Pints, quarts, and gallons are part of a larger "doubling" system of capacity measure.

1 jigger = 2 mouthfuls	1 pint = 2 cups
1 jack = 2 jiggers	1 quart = 2 pints
1 jill = 2 jacks	1 bottle = 2 quarts
1 cup = 2 jills	1 gallon = 2 bottles
	1 pail = 2 gallons

(a) How many mouthfuls are in a jill? a cup? a pint?

(b) Suppose one mouthful, one jigger, one jack, . . . , and one gallon are poured into an empty pail. Does the pail overflow, is it exactly filled, or is there room for more? (*Hint:* Draw an empty pail. Put one gallon in, then one bottle, and so on.)

21. The weight of diamonds and other precious gemstones is given in *carats,* where 1 carat = 200 mg. The largest diamond discovered thus far is the Cullinan, found in 1906 at the Premier mine in South Africa. It weighed 3106 carats. Using the conversion 1 kg = 2.2 lb, estimate the weight of the Cullinan diamond in pounds.

Thinking Cooperatively

22. Nearly 4700 years ago, the Great Pyramid of Khufu was built to astonishing accuracy using the measuring unit of the cubit (see Figure 12.1). For further accuracy, the cubit was divided into seven *palms,* and each palm was further subdivided into four *digits* (finger widths). Longer distances were measured by the *hayt,* equal to 100 cubits.

(a) Use a meterstick to measure the cubit (elbow to fingertip distance, to the nearest centimeter) and palm (distance across four fingers, to the nearest millimeter) of 10 classmates, and make a histogram of your data. Compute the average and standard deviation of the cubit and palm measurements. Does it seem accurate that seven palms are in a cubit?

(b) Give reasonable ancient Egyptian measurements for the height of the ceiling in your classroom, the length of a piece of notebook paper, and the length of a football field.

Using a Calculator

23. Verify that a hectare is about $2\frac{1}{2}$ acres. Use the approximate conversion 1.6 km ≐ 1 mile, and show all of your steps.

24. (a) Use the conversion 1 in ≐ 2.54 cm to calculate the number of cubic inches in a liter.

(b) Which volume is larger, 6.2 liters or 327 cubic inches?

25. A fortnight is 2 weeks. Convert a speed of 25 inches per minute to its equivalent in furlongs per fortnight.

26. Show that there are about 30 million seconds in a year.

Making Connections

27. Here is a metric recipe for spaghetti sauce, except that the prefixes (if any) of some of the measures have been obliterated by some previous spills. Fill in the correct prefix, or indicate that there is no prefix needed, for these ingredients in the recipe.

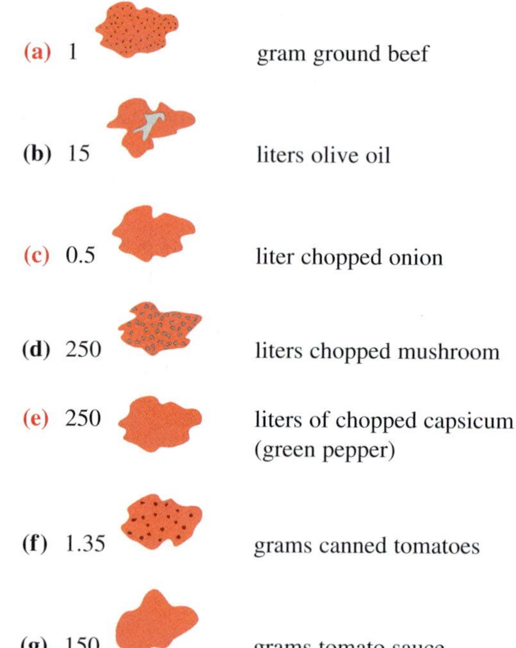

(a) 1 ___ gram ground beef

(b) 15 ___ liters olive oil

(c) 0.5 ___ liter chopped onion

(d) 250 ___ liters chopped mushroom

(e) 250 ___ liters of chopped capsicum (green pepper)

(f) 1.35 ___ grams canned tomatoes

(g) 150 ___ grams tomato sauce

28. Metric countries rate the fuel efficiency of a car by the number of liters of gasoline required to drive 100 kilometers. If a car takes 9 liters per 100 kilometers, what is its efficiency in miles per gallon? Use the conversions 1 gal ≐ 3.7854 L and 1 mile ≐ 1.6 km.

29. A light-year is the distance light travels in empty space in one year.

(a) Light travels at a speed of 186,000 miles per second. The star nearest the sun is Proxima Centauri, in the constellation Centaurus, whose distance from the earth is four light-years. What is the distance to Proxima Centauri in miles?

(b) In metric measurements, the speed of light is 3.00×10^8 meters per second. Verify that a light-year is about 10^{16} meters.

30. An herbicide is bottled in concentrated form. A working solution is mixed by adding 1 part concentrate to 80 parts water.

(a) How many liquid ounces of concentrate should be added to 5 gallons of water?

(b) How many liters of water should be added to 65 milliliters of concentrate?

31. Lumber is measured in board feet, where a board foot is the volume of a piece of lumber one foot square and one inch thick.

(a) How many board feet are in a two-by-four (2″ by 4″) that is 10 feet long? The volume of a rectangular solid is length times width times height.

(b) Lumber is priced in dollars per thousand board feet. Suppose two-by-fours 10 feet long are $690 per thousand board feet. What is the cost of 144 two-by-fours, each 10 feet long?

32. How far, in miles, did Captain Nemo's submarine *Nautilus* travel under the sea, according to the title of Jules Verne's well-known novel? (*Hint:* Find the definition of *league.*)

33. On July 23, 1983, Air Canada Flight 143 was on the ground in Montréal. The pilot had 7682 liters of fuel on board, but knew the Boeing 767 would need 22,300 kilograms to reach its destination, Edmonton, Alberta. Since airliners measure fuel by weight, not volume, the pilot asked for the weight of a liter of fuel and was told it was 1.77 kilograms.

(a) Calculate how many liters of fuel were added to the plane's tanks.

(b) Actually, 1.77 is the number of pounds, not kilograms, of fuel per liter. There are really just 0.803 kilograms per liter. How much fuel should have been added to the plane's tanks?

[Yes, the plane ran out of fuel, but the pilots managed to glide 22 miles to a safe landing in Winnepeg.]

From State Student Assessments

34. (Texas, Grade 3)
Which is the best estimate of the capacity of a baby bottle?

F. 1 milliliter
G. 1 cup
H. 1 liter
J. 1 gallon

35. (Connecticut, Grade 4)

ABOUT how many footstools would be the same height as the chair? 2 3 4 5

36. (Connecticut, Grade 4)

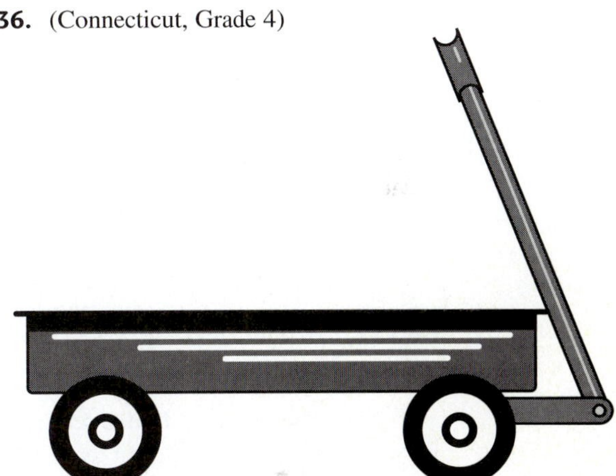

Use your ruler to measure the length of the wagon in this picture to the NEAREST centimeter.

5 centimeters 8 centimeters
12 centimeters 15 centimeters

37. (Washington State, Grade 4)
Your class project is to build a bird feeder.

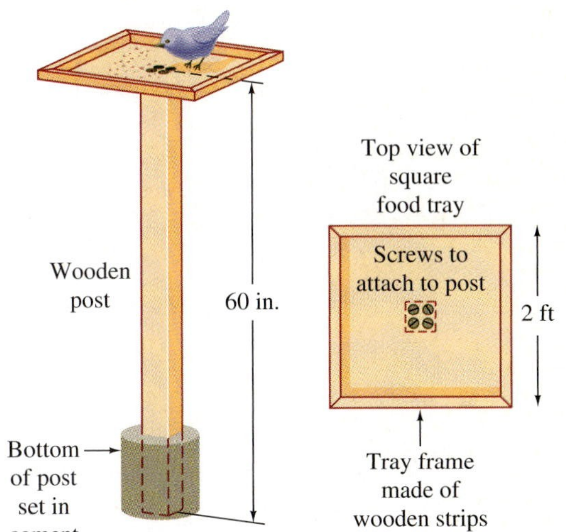

Item	Cost per Unit
wooden post	\$2.50 per foot
Wooden strips for tray frame	\$1.00 per foot
tray bottom	\$6.00
tools, screws, nails, wood glue, and cement mix	Loaned or donated by parents

Explain how you could use the information given to find the total cost of materials. Use words, numbers, or pictures.

For Review

38. What are the measures of the interior angles of this triangle?

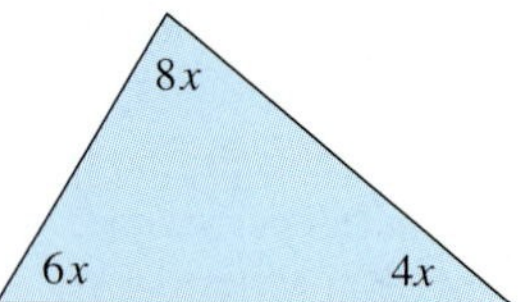

39. Prove that the measure of an exterior angle of a triangle is greater than the measure of either of the opposite interior angles: $m(\angle 3) > m(\angle 1)$ and $m(\angle 3) > m(\angle 2)$.

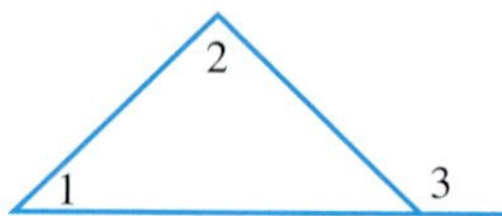

40. **(a)** Make a freehand (pencil only—no ruler or other drawing tools) drawing of a regular octahedron.

(b) Beside your drawing of the octahedron, draw a truncated octahedron. Do this by removing the corners of the octahedron to form squares, which reduces the triangular faces of the octahedron to regular hexagonal faces.

12.2 AREA AND PERIMETER

Measurements in Nonstandard Units

The number of units required to cover a region in the plane is the **area** of the region. Usually squares are chosen to define a **unit of area,** but any shape that tiles the plane (that is, covers the plane without gaps or overlaps) can serve equally well. Working with a nonstandard unit allows students to discover important general principles of the measurement process.

EXAMPLE 12.6 MAKING MEASUREMENTS IN NONSTANDARD UNITS

Find the area of each figure A, B, C, and D in terms of the unit of area shown at the right.

(a)

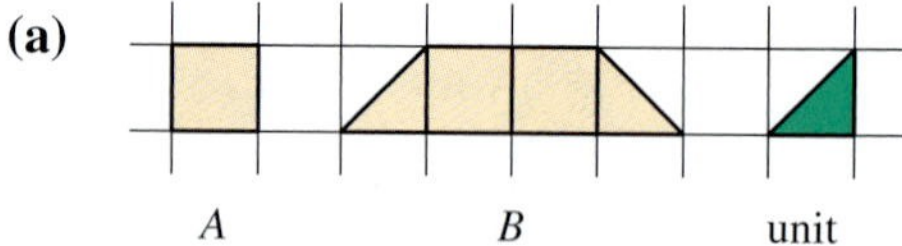

(b)

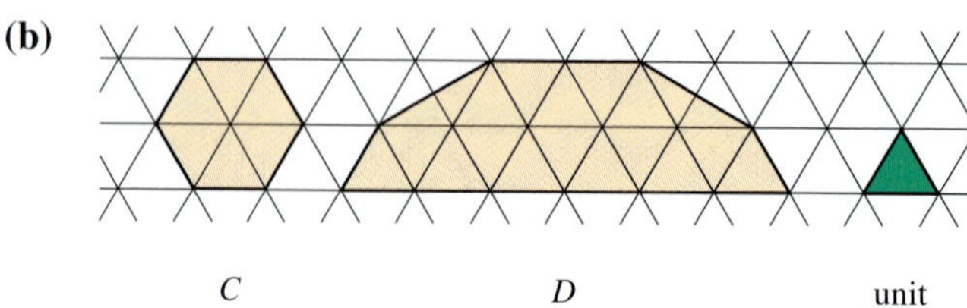

Solution

(a) The full square A can be covered by 2 of the unit shapes, so area$(A) = 2$ units. Region B is covered by 6 units, so area$(B) = 6$ units.

(b) The hexagon C is covered by 6 of the triangular units, so area$(C) = 6$ units. Region D cannot be covered directly by the triangular units, although it is evident that the area of D is between 16 and 20 units. To find the exact area, remove and then rejoin a triangular piece as shown below to form a new shape D' of the same area as D. That is, area$(D') =$ area(D). Since D' can be covered by 18 triangular units, it follows that area$(D) = 18$ units.

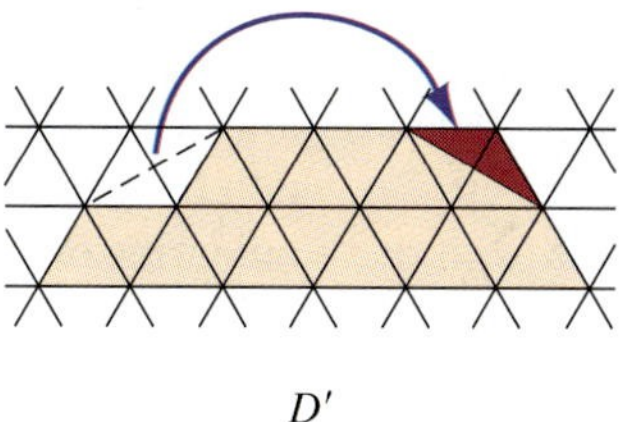

A hexagonal region.

A hexagonal region dissected into three subregions.

The Congruence and Addition Properties of Area

The solution just given in the example uses two important principles for the calculation of areas. First, if two figures are **congruent**—that is, have the same size and shape—then the two figures also have the same area. Second, if a figure is **dissected**—that is, if the figure is partitioned into nonoverlapping subregions that cover the larger figure—then the area of the larger figure is the sum of the areas of the subregions. The following **congruence and addition properties** of area will be used repeatedly throughout the rest of this chapter.

PROPERTIES *The Congruence and Addition Properties of Area*

Congruence property

If region R is congruent to region S, then the two regions have the same area:

$$\text{area}(R) = \text{area}(S).$$

Addition property

If a region R is dissected into nonoverlapping subregions $A, B, \ldots, F$, then the area of R is the sum of the areas of the subregions:

$$\text{area}(R) = \text{area}(A) + \text{area}(B) + \cdots + \text{area}(F).$$

These properties are illustrated in Figure 12.7.

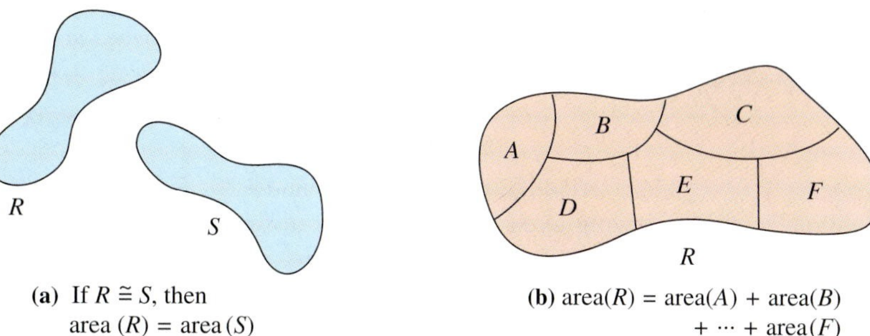

(a) If $R \cong S$, then area $(R) =$ area (S)

(b) area$(R) =$ area$(A) +$ area$(B) + \cdots +$ area(F)

Figure 12.7

The congruence property and the addition property of area

The congruence and addition properties show that rearranging the pieces of a figure forms a new figure with the same area as the original figure.

EXAMPLE 12.7 SOLVING LEONARDO'S PROBLEMS

Leonardo da Vinci (1452–1519) once became absorbed in showing how the areas of certain curvilinear (curved-sided) regions could be determined and compared among themselves and with rectangular regions. The pendulum and the ax are two of the examples he included in notes for his book *De Ludo Geometrico* (roughly meaning "Fun with Geometry"), which he never completed. The dots show the centers of the circular arcs that form the boundaries of the regions.

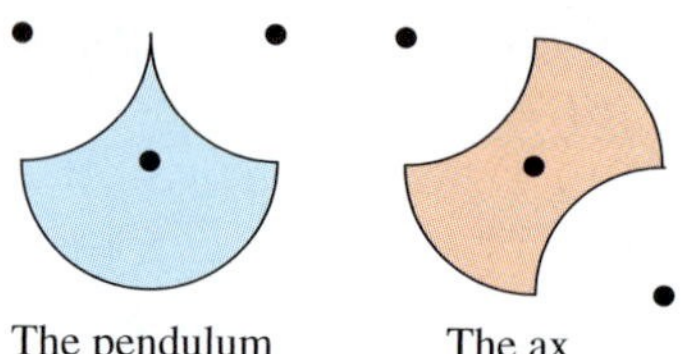

The pendulum The ax

If the arcs forming the pendulum and the ax have radius 1, show that the areas of both figures are equal to that of a 1-by-2 rectangle.

Solution After inscribing the figures in a square, we then use the congruence and addition properties of area to rearrange the subregions to form a 1-by-2 rectangle.

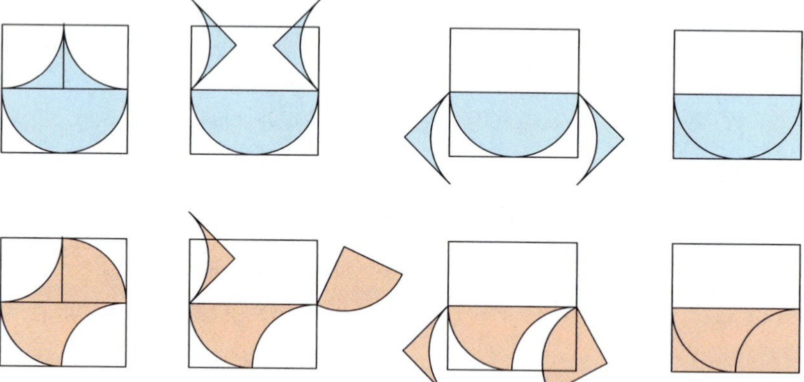

It is surprising to learn that the pendulum and ax both have an area of 2 square units.

SOURCE: These examples and others are described in a booklet by Herbert Wills III, Leonardo's Dessert, *National Council of Teachers of Mathematics, Reston, VA, 1985.*

Unlike Examples 12.6 and 12.7, most area measurement problems are answered by giving a reasonable *estimate* of the area. Units of square shape are easy to subdivide into smaller squares to give a more precise estimation.

EXAMPLE 12.8 INVESTIGATING THE AREA OF A CYCLOIDAL ARCH

Imagine rolling a wheel along a straight line, with a reflector at point P on the rim. Point P traces an arch-shaped curve. In the early seventeenth century Galileo investigated this curve and named it the **cycloid.** Discover for yourself the conjecture Galileo made about how the area of the cycloidal arch compares to the area of the circle used to generate the arch.

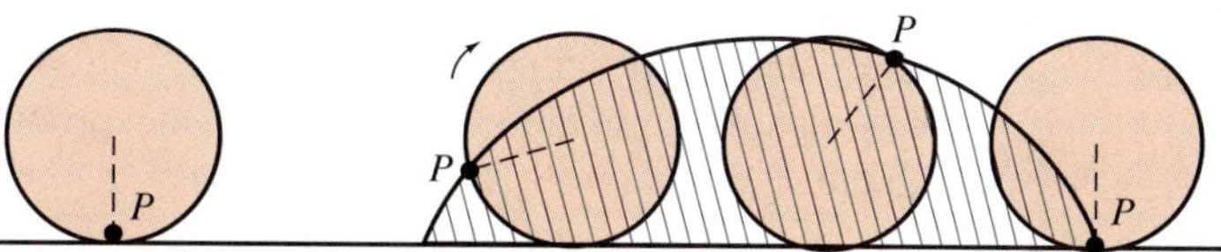

Solution

Understand the Problem

It is visually clear that the cycloidal arch has an area that is much larger than that of the circle. Our goal is to guess the ratio area(cycloid)/area(circle) that gives the comparison between the areas.

Devise a Plan

The areas of the arch and circle must both be measured in some unit of area. For example, we can use squares of size U, where the diameter of the circle is equal to the sum of four side lengths of U. For better accuracy, we can also use small square units of size u, where the side length of u is half that of U.

Carry Out the Plan

The circle and arch are overlaid by a square grid, with squares of unit area U.

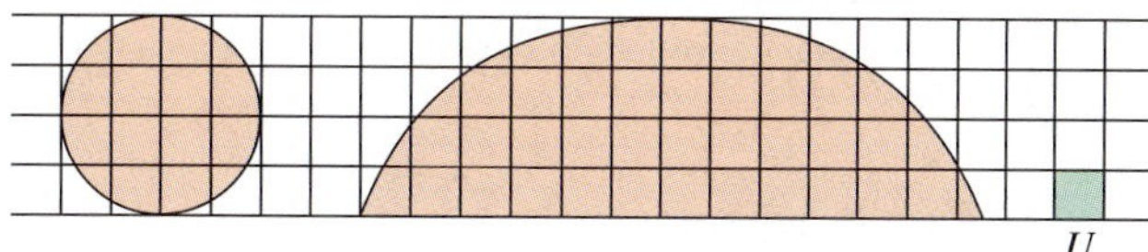

The circle is entirely within 16 squares of size U, but does not entirely cover about one unit of area in each of the four corners. Thus, we estimate that area(circle) $\doteq$ 12 U. Similarly, we see that area(cycloid) $\doteq$ 37 U is a reasonable estimate.

Better accuracy is given by the grid of squares of unit area u. The diagram below leads us to the estimated area(circle) $\doteq$ 50 u and area(cycloid) $\doteq$ 149 u.

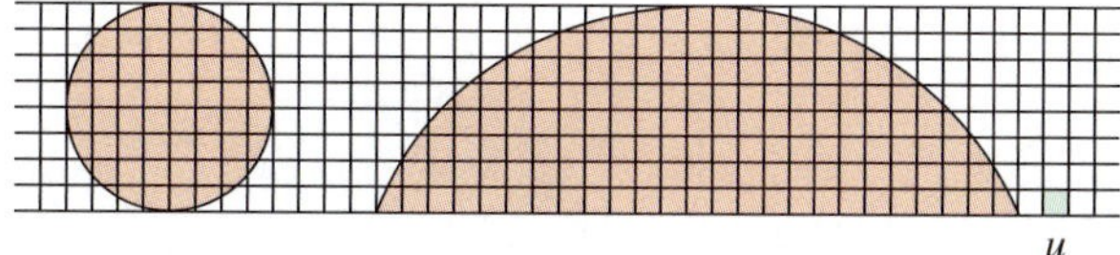

Both $\frac{37\ U}{12\ U}$ and $\frac{149\ u}{50\ u}$ are nearly 3, which in fact was Galileo's conjecture. The correctness of Galileo's conjecture was proved in 1634 by Gilles Persone de Roberval.

Look Back

The finer grid of squares gave us additional precision in our measurements, but this required considerably more time and effort to obtain. The measurement process nearly always requires us to make a judgment about how to balance the conflicting needs of precision versus cost.

Just for Fun

How to Cover a Long Hole with a Short Board

Can you cut the board shown into just two pieces to exactly cover the 60-cm by 12-cm hole?

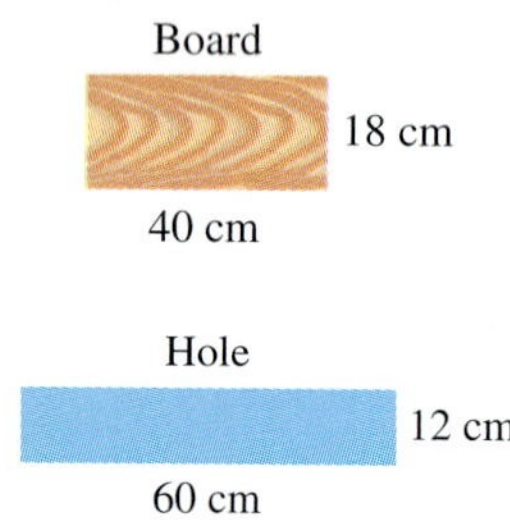

Areas of Polygons

Rectangles

A 3-cm by 5-cm rectangle can be covered by 15 unit squares when the unit square is 1 cm^2, as shown in Figure 12.8. Similarly a 2.5-cm by 3.5-cm rectangle can be covered by six whole units, five half-unit squares, and one quarter-unit square, giving a total area of 8.75 square centimeters. This is also the product of the width and the length, since $2.5 \text{ cm} \times 3.5 \text{ cm} = 8.75 \text{ cm}^2$.

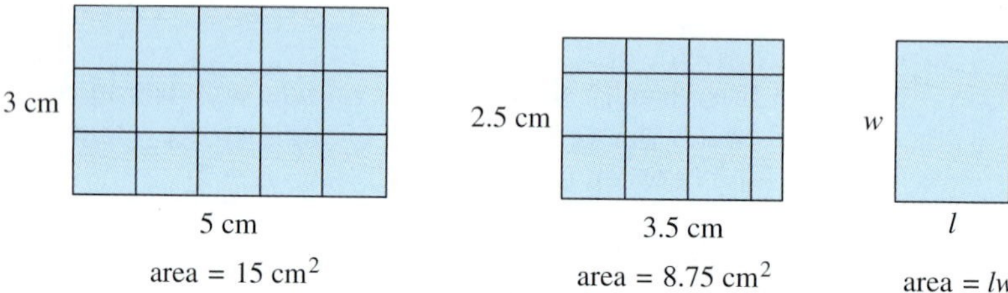

Figure 12.8
The area of a rectangle is the product of its length and width

For any rectangle, the formula for the area A is as follows.

FORMULA *Area of a Rectangle*
A rectangle of length l and width w has area A given by the formula $A = lw$.

Parallelograms

Suppose a parallelogram has a pair of opposite sides b units long, and these sides are h units apart; an example is shown in Figure 12.9. We say that b is the **base** of the parallelogram and h is the **altitude,** or **height.** (Unless the parallelogram is a rectangle, the altitude is *not* the same as the length of the other two sides of the parallelogram.) Removing and replacing a right triangle T forms a rectangle of the same area as the parallelogram. The rectangle has length b and width h, so its area is bh. Therefore, the area of the parallelogram in Figure 12.9 is also bh.

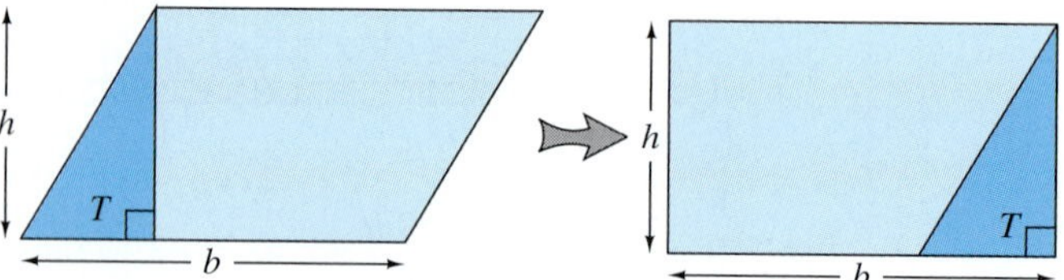

Figure 12.9
A parallelogram of base b and altitude h has the same area as a b by h rectangle. Therefore, the area of the parallelogram is bh.

Any parallelogram with base b and altitude h can be dissected and rearranged to form a rectangle of length b and width h in a similar way to that shown in Figure 12.9. (See problem 35 in Problem Set 12.2 for a more general case.) This yields the following formula.

FORMULA *Area of a Parallelogram*

A parallelogram of base b and altitude h has area A given by $A = bh$.

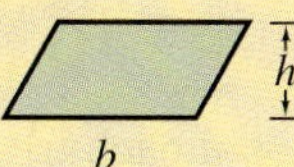

EXAMPLE 12.9 USING THE PARALLELOGRAM AREA FORMULA

Find the area of each parallelogram, and then compute the lengths x and y.

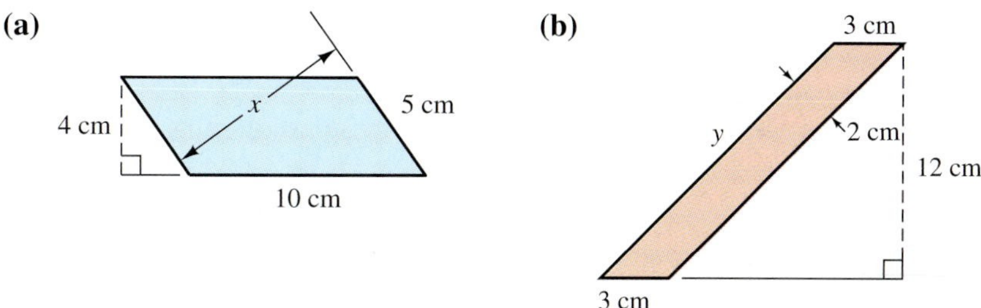

Solution

(a) The parallelogram has base 10 cm and height 4 cm, so its area is $A = (10 \text{ cm})(4 \text{ cm}) = 40 \text{ cm}^2$. If the side of length 5 cm is considered the base, then x is the corresponding height and $A = (5 \text{ cm})x$. Since $A = 40 \text{ cm}^2$, we find $x = 40 \text{ cm}^2/5 \text{ cm} = 8 \text{ cm}$.

(b) The procedure for (a) is followed. The area is $A = (3 \text{ cm})(12 \text{ cm}) = 36 \text{ cm}^2$. Viewing the side of length y as the base with corresponding altitude 2 cm, we have $36 \text{ cm}^2 = y(2 \text{ cm})$. Therefore, $y = 36 \text{ cm}^2/2 \text{ cm} = 18 \text{ cm}$.

Triangles

Figure 12.10 shows that a triangle of base b and altitude h can be dissected and rearranged to form a parallelogram of base $\frac{b}{2}$ and altitude h. The formula $\frac{1}{2}bh$ for the area of the triangle then follows from the area formula already derived for the parallelogram.

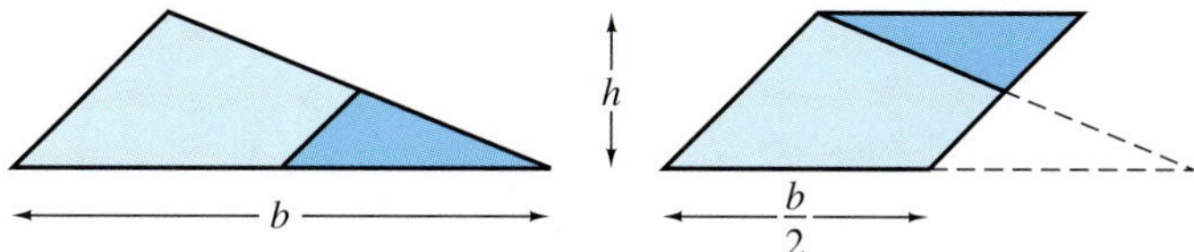

Figure 12.10
A triangle of base b *and altitude* h *can be dissected and rearranged to form a parallelogram of base* $\frac{b}{2}$ *and altitude* h.

FORMULA *Area of a Triangle*

A triangle of base b and altitude h has area $A = \frac{1}{2}bh$.

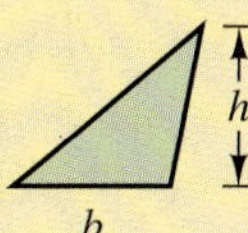

Any side of a triangle can be considered as the base, so there are three pairs of bases and altitudes.

EXAMPLE 12.10 USING THE TRIANGLE AREA FORMULA

Find the area of each triangle and the distances v and w.

(a)

(b)

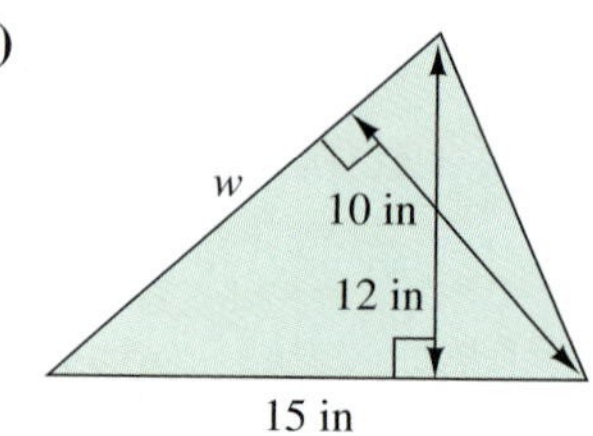

Solution

(a) The formula $A = \frac{1}{2}bh$ shows that the area of the triangle is $A = \frac{1}{2}(10\text{ cm}) \cdot (7\text{ cm}) = 35\text{ cm}^2$. If the side of length 14 cm is considered the base, then the corresponding altitude is v. Since $A = \frac{1}{2}(14\text{ cm}) \cdot (v)$, $v = A/(7\text{ cm}) = (35\text{ cm}^2)/(7\text{ cm}) = 5\text{ cm}$.

(b) The area of the triangle is $A = \frac{1}{2}(15\text{ in}) \cdot (12\text{ in}) = 90\text{ in}^2$. Considering the side of length w as the base, we find that the corresponding altitude is 10 in and $A = \frac{1}{2}w(10\text{ in})$. Therefore, $w = (90\text{ in}^2)/(5\text{ in}) = 18\text{ in}$.

Trapezoids

There are several ways to derive the formula for the area of a trapezoid with altitude h and bases a and b. For example, a diagonal drawn through the trapezoid dissects it into two triangles. The altitudes of both triangles are h, and their bases are a and b, so the areas of the triangles are $\frac{1}{2}ah$ and $\frac{1}{2}bh$. Adding the areas gives the following formula, as also shown in Figure 12.11.

FORMULA *Area of a Trapezoid*

A trapezoid with bases of length a and b and altitude h has area $A = \frac{1}{2}(a + b)h$.

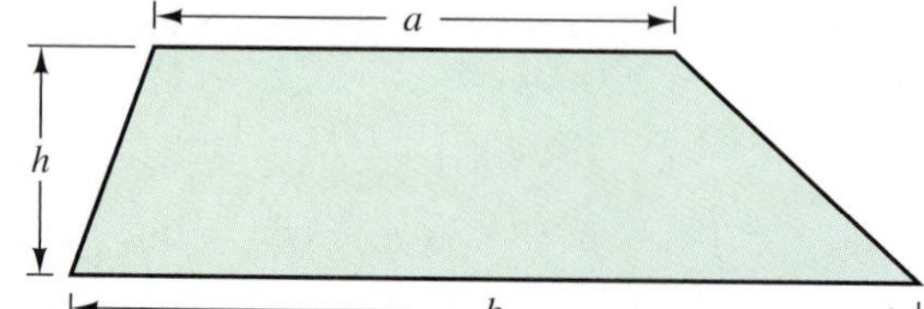

Figure 12.11
The area of a trapezoid is $\frac{1}{2}(a + b)h$

EXAMPLE 12.11 FINDING THE AREAS OF LATTICE POLYGONS

A polygon formed by joining the points of a square array is called a **lattice polygon.** Lattice polygons are easy to draw on dot paper, or they can be formed with rubber bands on a geoboard. Find the area of the lattice polygons shown, where the unit of area is the area of a small square of the array.

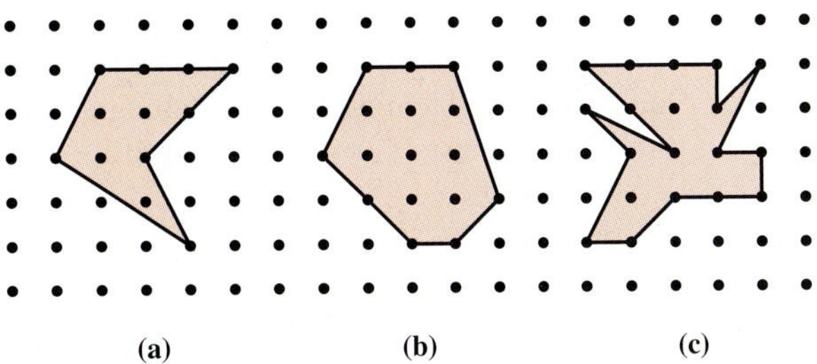

Solution

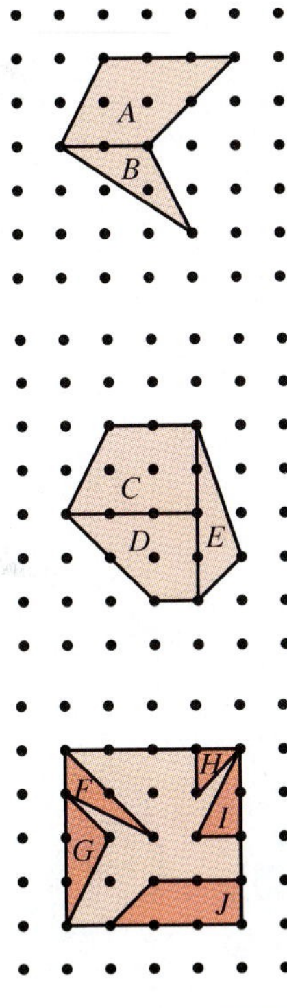

(a) A horizontal line dissects the polygon into a trapezoid A of area $\frac{1}{2}(3 + 2) \cdot (2) = 5$ and a triangle B of area $\frac{1}{2}(2)(2) = 2$. The area of the polygon is therefore 7.

(b) The lattice hexagon can be dissected into trapezoids C and D and triangle E. The total area of the hexagon is therefore

$$\frac{1}{2}(2 + 3) \cdot (2) + \frac{1}{2}(3 + 1) \cdot (2) + \frac{1}{2}(4) \cdot (1) = 11.$$

Other dissections of the hexagon can be used, but the total area will always be the same.

(c) We could solve the problem in the same way as parts (a) and (b), but there is another useful technique: construct a square about the polygon, and then subtract the areas of the regions F, G, H, I, and J. Therefore the area of the polygon is

$$16 - \left(1 + 1\frac{1}{2} + \frac{1}{2} + 1 + 2\frac{1}{2}\right) = 9\frac{1}{2}.$$

Length of a Curve

The length of a polygonal curve is obtained by summing the lengths of its sides. The length of a nonpolygonal curve is measured, or at least estimated, by calculating the length of an approximating polygonal curve with vertices on the given curve. The accuracy of the estimation is improved by using an approximating polygonal curve with more vertices, as shown in Figure 12.12.

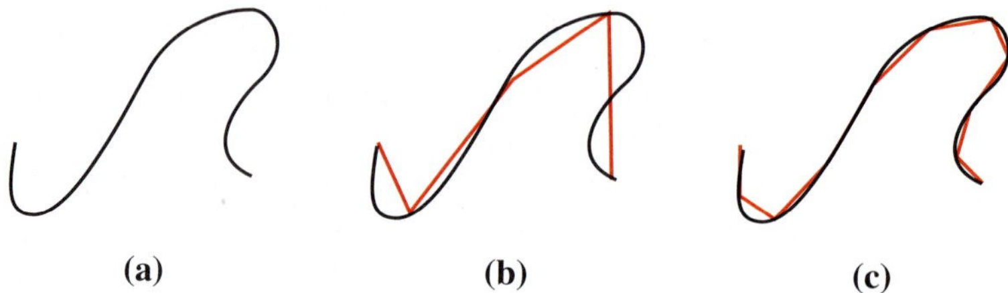

Figure 12.12
The length of the curve in (a) is estimated by measuring the length of a polygonal approximation, as in (b). Increasing the number of vertices gives an improved estimate, as in (c)

The length of a curve can also be measured by first laying a string along the curve and then straightening the string along a ruler. This is the principle that makes the flexible tape measure used for sewing so useful.

EXAMPLE 12.12 DETERMINING THE LENGTH OF A CYCLOID

A circle and the cycloid it generates (see Example 12.8) are shown below. Use a marker pen and a piece of string (or thin strip of paper) to make a tape measure, where the unit of length is the diameter d of the circle.

(a) According to your tape measure, what is the length from point A to point B along the cycloid?
(b) What is the approximate length of the line segment $\overline{AB}$?

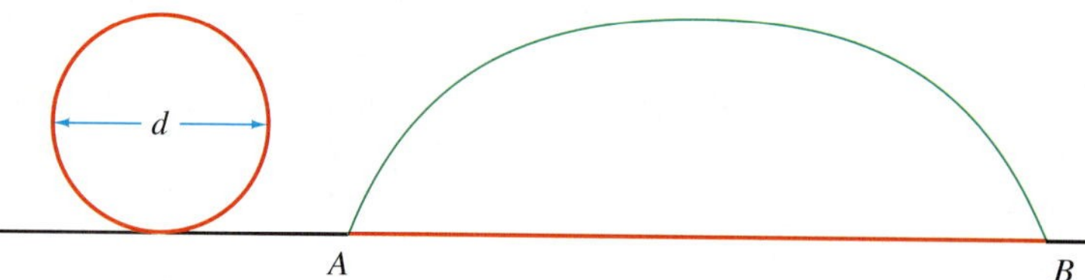

Solution

(a) The tape measure shows that the length of the cycloid is very nearly four diameters of the circle. In 1658, Christopher Wren (1632–1723) proved that the length of a cycloid is *exactly* four diameters. Wren is perhaps best known as the architect of St. Paul's Cathedral in London.
(b) The segment $\overline{AB}$ is a bit over three diameters. Because $\overline{AB}$ is covered by rolling the circle once around, AB is the length around the circle; that is, AB is the circumference, and "a bit over three" is the famous value now written as π.

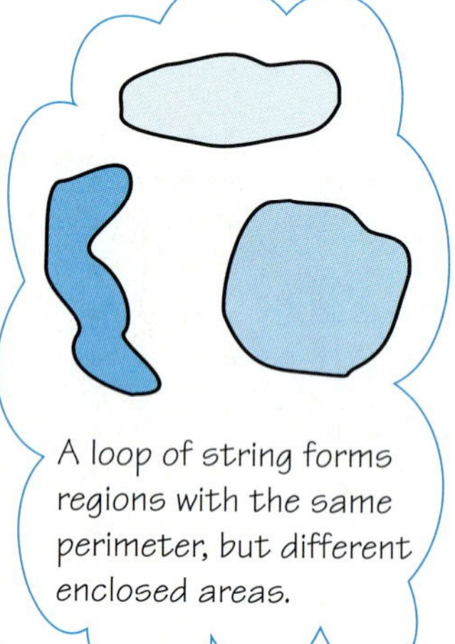

A loop of string forms regions with the same perimeter, but different enclosed areas.

Perimeter

The length of a simple closed plane curve is called the curve's **perimeter.** Therefore, perimeter is a *length* measurement and is given in centimeters, inches, feet, meters, and so on. It is important not to confuse the *area* of the region enclosed by a simple closed curve with the perimeter of the figure. Area is given in cm^2, in^2, ft^2, m^2, and so on. In summary, perimeter is the measure of the distance around a region, and area is the measure of the size of the region within a boundary.

EXAMPLE 12.13 FINDING PERIMETERS

The following figures have been drawn on a square grid, where each square is 1 cm on a side. Give the perimeter and area of each figure.

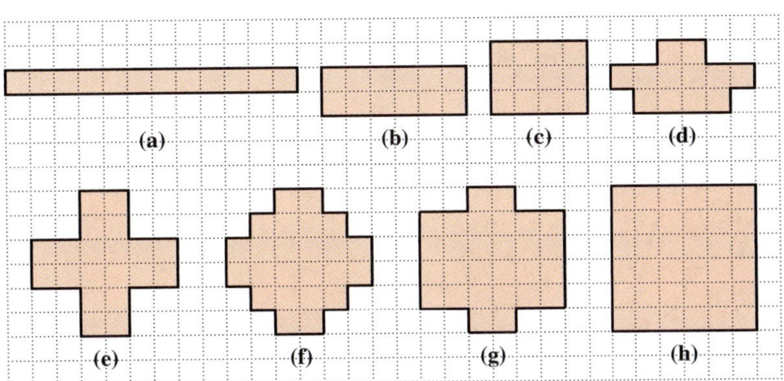

Solution

Figure	(a)	(b)	(c)	(d)	(e)	(f)	(g)	(h)	
Perimeter	26	16	14	18	24	24	24	24	centimeters
Area	12	12	12	12	20	24	28	36	square centimeters

Figures (a), (b), (c), and (d) have the same area, but different perimeters. Figures (e), (f), (g), and (h) have the same perimeter, but different areas.

The Circumference of a Circle

The perimeter of a circle is called the circle's **circumference.** By using a piece of string or a tape measure, or by rolling a disc along a line (as in Example 12.12 (b)), it is easy to rediscover a fact known even in ancient times: The ratio of the circumference of a circle to the circle's diameter is the same for all circles. Two examples are shown in Figure 12.13.

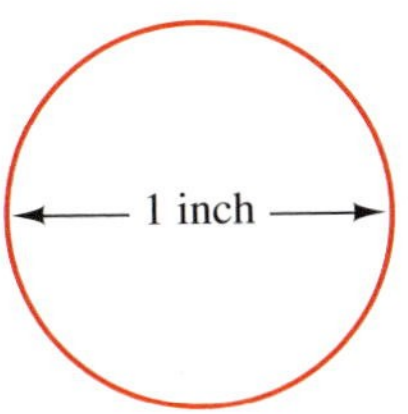

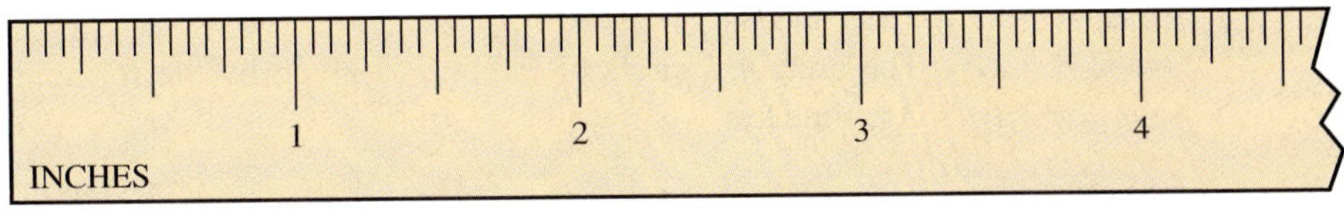

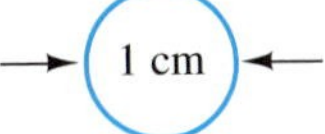

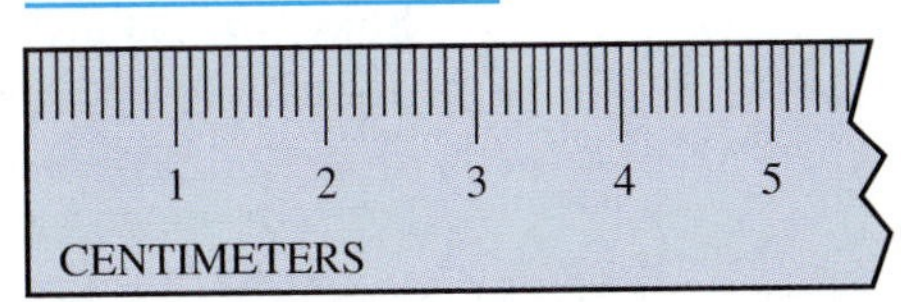

Figure 12.13
The ratio of the circumference C to the diameter d is the same for all circles: $C/d = \pi$, *or* $C = \pi d$

Just for Fun

Square-Inch Mysteries

From a sheet of graph paper cut an 8-by-8 square into the four pieces *A, B, C,* and *D* as shown.

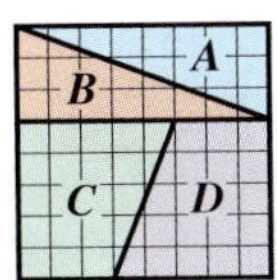

Now rearrange the four pieces into a rectangle as shown here. Does the rectangle have the same area as the square?

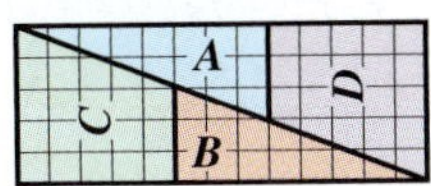

Before you run out to buy 8-by-8 bricks of gold, rearrange the four pieces into a "propeller."

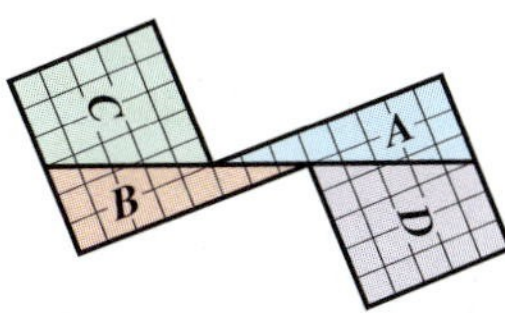

Does it have the same area as the square? Can you account for the changes in area of the three figures?

This ratio, which is somewhat larger than 3, is given by the symbol π, the lowercase Greek letter *pi*.

DEFINITION π
The ratio of the circumference C to the diameter d of a circle is π. Therefore,

$$\frac{C}{d} = \pi \text{ and } C = \pi d.$$

Since the diameter d is twice the radius r of the circle, we also have the formula $C = 2\pi r$.

In 1761 John Lambert proved that π is an irrational number, so it is impossible to express π exactly by a fraction or as a terminating or repeating decimal. The values $3\frac{1}{7}$ and 3.14 are useful approximate values, but precision measurements require the use of more decimal places in the unending decimal expansion $\pi = 3.1415926\ldots$. A circle 100 feet in diameter has an *approximate* circumference of 314 feet, but the exact circumference is 100π feet. It is acceptable to use the symbol π to express results, since this gives exact values. When an approximate numerical value is needed, an appropriate estimate of π such as 3.1416 can be used in the calculations.

EXAMPLE 12.14 CALCULATING THE EQUATORIAL CIRCUMFERENCE OF THE EARTH

The equatorial diameter of the earth is 7926 miles. Calculate the distance around the earth at the equator, using the following approximations for π: **(a)** 3.14 **(b)** 3.1416.

Solution

(a) $(3.14)(7926 \text{ miles}) \doteq 24{,}887.64$ miles
(b) $(3.1416)(7926 \text{ miles}) \doteq 24{,}900.322$ miles

The two different approximations of π account for the difference of about 12.7 miles in the answers.

The Area of a Circle

The area of a circle of radius r is given by the formula πr^2, first proved rigorously by Archimedes.

FORMULA *Area of a Circle*
The area A enclosed by a circle of radius r is $A = \pi r^2$.

Since π is defined as a ratio of lengths, it seems surprising to find that π also occurs in the formula for the area of a circle. A convincing, but informal, derivation of the formula $A = \pi r^2$ is shown in Figure 12.14. The circle of radius r and circumference $C = 2\pi r$ is dissected into congruent sectors that are rearranged to form a "parallelogram"

HIGHLIGHT from HISTORY

A Brief History of π

In the third century B.C., Archimedes showed that π is approximately $3\frac{1}{7}$. To estimate π, Archimedes inscribed a regular polygon in a circle and then calculated the ratio of the polygon's perimeter to the diameter of the circle. An inscribed hexagon shows π is about 3, but by using a 96-gon, Archimedes proved that $3\frac{10}{71} < \pi < 3\frac{10}{70}$. The same idea was used by the Dutch mathematician Van Ceulen (d. 1610), who used a 32,212,254,720-gon to calculate π to 20 decimals. A century later, the English mathematician John Machin took advantage of the invention of calculus to calculate π to 100 decimal places. Machin's method, with some minor variations, was used well into the twentieth century. When this problem was implemented on the ENIAC, the first electronic computer, in 1949, the computer spent 70 hours to calculate π to 2037 decimal places.

Recent records in the calculation of π take advantage of both the extremely high speed of supercomputers and the implementation of highly efficient algorithms of calculation. For example, in 1999, Yasumasa Kanada of the Computer Center at the University of Tokyo computed over 1.2411 trillion decimal digits of π. The calculation took about 400 hours of computer time on a supercomputer capable of performing 2 trillion calculations per second.

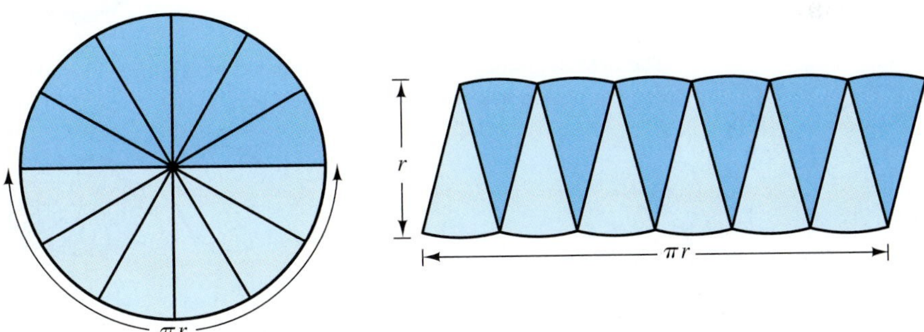

Figure 12.14 The sectors of a circle can be rearranged to approximate a parallelogram of area πr^2

of base $\frac{1}{2}C = \pi r$ and altitude r. By the formula for the area of a parallelogram, the wavy-based "parallelogram" has area $\pi r \times r = \pi r^2$. If the number of sections is made larger and larger, the thin sectors form an increasingly exact approximation to a true parallelogram of area πr^2.

EXAMPLE 12.15 DETERMINING THE SIZE OF A PIZZA π

A 14″ pizza has the same thickness as a 10″ pizza. How many times more ingredients are there on the larger pizza?

Solution

Pizzas are measured by their diameters, so the radii of the two pizzas are 7″ and 5″, respectively. Since the thicknesses are the same, the amount of ingredients used is proportional to the areas of the pizzas. The larger pizza has area $\pi(7 \text{ in})^2 = 49\pi \text{ in}^2$, and the smaller pizza has area $\pi(5 \text{ in})^2 = 25\pi \text{ in}^2$. The ratio of areas is $49\pi \text{ in}^2/25\pi \text{ in}^2 = 1.96$, showing that the 14″ pizza has about twice the ingredients of the 10″ one.

Measurements in Beanland

Materials Needed

Dry beans (small red kidney or white navy beans); enlarged copies of the figures shown in the two activities below.

Directions

In Beanland, the lengths of curves are measured in *beanlengths,* abbreviated "bl." Similarly, the areas of regions are measured in *beanareas,* abbreviated "ba." The diagram below shows that the length of the curve is about 15 bl.

The next diagram shows a region bounded by a simple closed curve. By counting the beans, we see that the region has an area of about 55 ba.

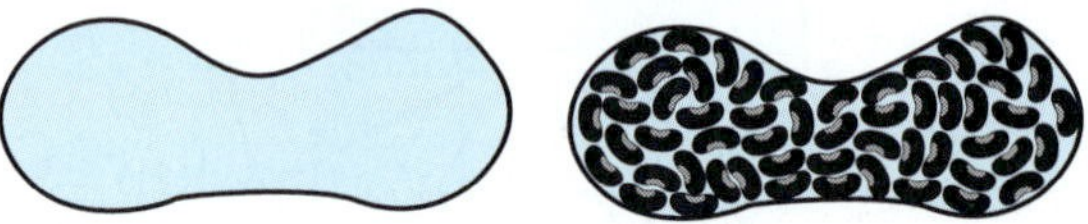

Carry out the following activities in pairs.

Activities

1. Consider this system of squares and circles. (Use an enlarged copy of this figure, with the larger circle about 14 to 16 beanlengths in diameter.)

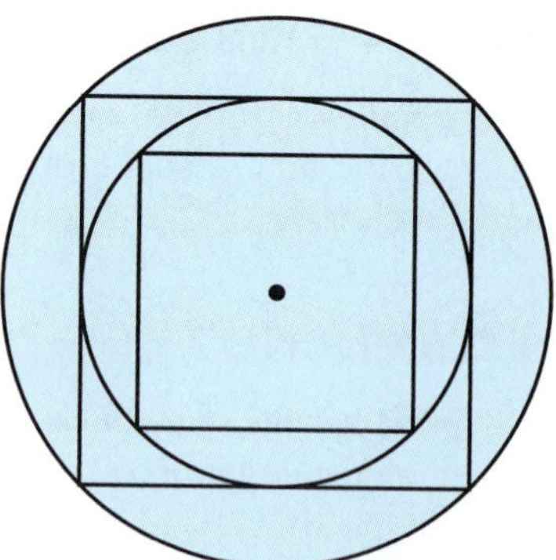

- **(a)** Measure the area of the ring-shaped region between the two circles, using your beans.
- **(b)** Measure the area of the smaller circle, and then compare this area with that of the ring.
- **(c)** Measure the area of the small square and then the area of the region between the two squares. How do these areas compare?
- **(d)** Measure the perimeter of (the distance around) the small square in beanlengths. Next measure the length of the diagonal of the large square. How do these lengths compare?

(e) Measure the circumference and the diameter of the large circle. What is the ratio of the circumference to the diameter? Compare with other groups and determine the average ratio.

2. Consider an equilateral triangle ABC and its circumscribed and inscribed circles. (Again, use an enlargement so the larger circle has a diameter of about 14 to 16 beanlengths.)

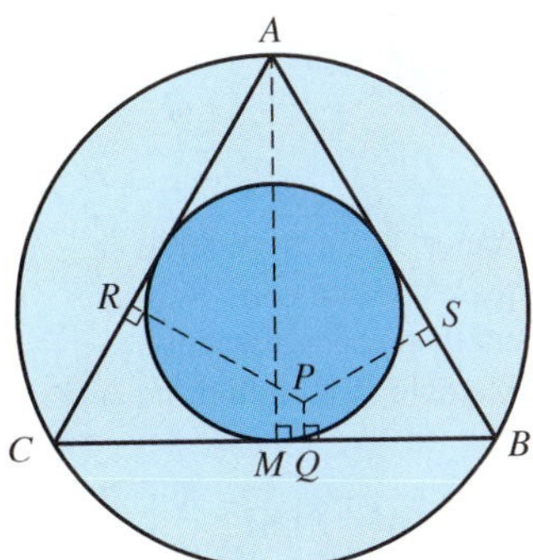

3. Measure the area of the small circle and the area of the ring-shaped region between the two circles. What is the ratio of the area of the ring to that of the small circle?
4. Choose an arbitrary point P inside the triangle and measure the three distances PQ, PR, and PS to the sides of the triangle. How does the sum $PQ + PR + PS$ of these three distances compare with the length AM of the altitude of the triangle?

Problem Set 12.2

Understanding Concepts

1. Botanists often need to measure the rate at which water is lost by transpiration through the leaves of a plant. For this measurement, it is necessary to know the leaf area of the plant. Estimate the area of the leaf shown. It has been overlaid with a grid of squares 1 cm on a side, shown at reduced scale.

2. Cut a convex quadrilateral from card stock, locate the midpoints of its sides, and then cut along the segments joining successive midpoints to give four triangles T_1, T_2, T_3, and T_4 and a parallelogram P.

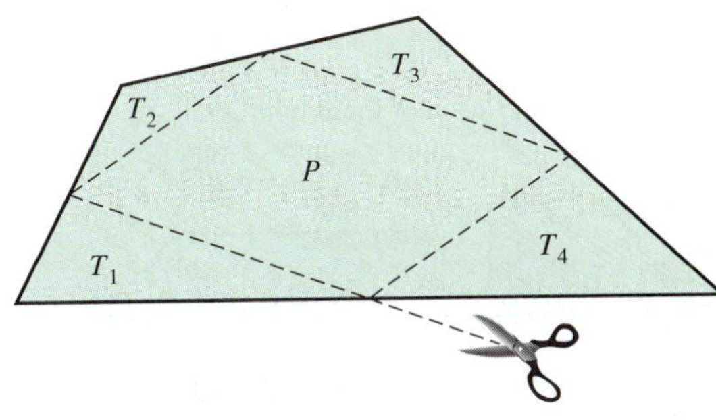

(a) Show that the four triangles can be arranged to cover the parallelogram.

(b) How does the area of the parallelogram compare with the area of the original quadrilateral?

3. The regular dodecagon shown below is dissected into six subregions. Carefully trace the pattern, cut out the subregions, and show how the pieces can be

reassembled into a square. How are the areas of the dodecagon and square related?

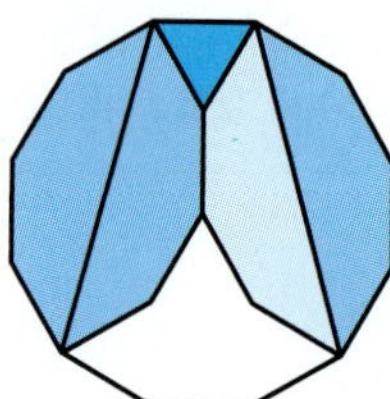

4. This goblet is drawn with circular 90° arcs centered at the black dots in a grid of squares of unit area. Redraw the vase on squared paper, and then use a dissection argument similar to that in Example 12.7 to find the area of the goblet.

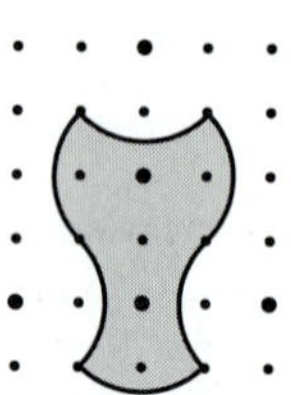

5. Measure the figure F shown below in each of the three nonstandard units of area **(a)**, **(b)**, and **(c)** shown to the right of F. Do so by tracing F and then dissecting the region into subregions with the unit area shape.

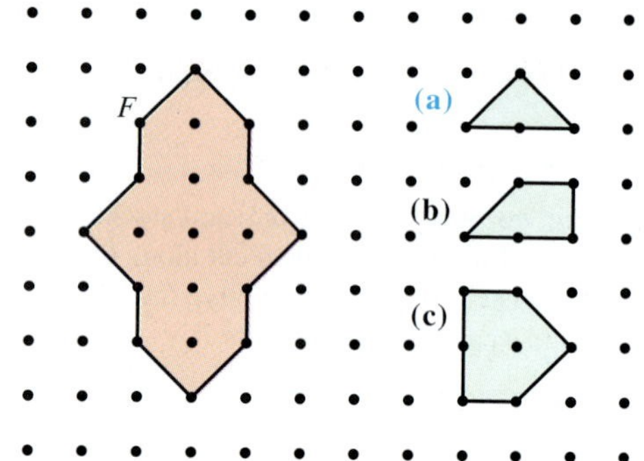

6. Find the area of each of these figures.

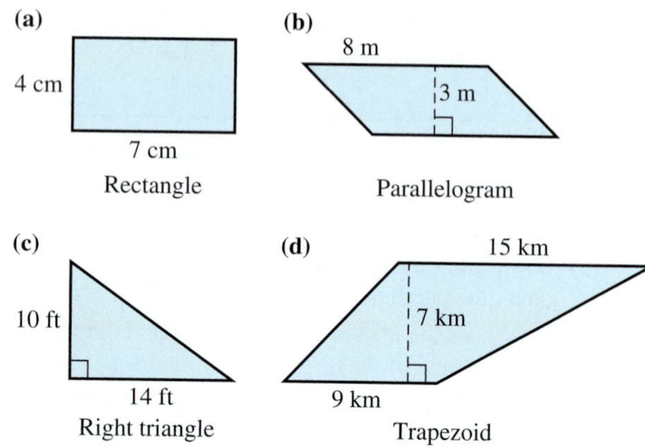

7. Find the area of each of these figures.

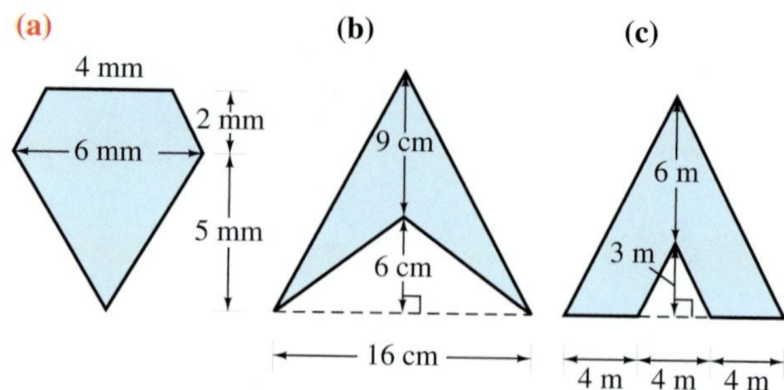

8. Fill in the blanks.
   (a) $3.45 \text{ m}^2 =$ ____ cm^2
   (b) $56{,}000 \text{ mm}^2 =$ ____ cm^2
   (c) $56{,}700 \text{ ft}^2 =$ ____ yd^2
   (d) $0.085 \text{ mi}^2 =$ ____ ft^2
   (e) $47{,}000$ a $=$ ____ ha $=$ ____ m^2
   (f) $5{,}800{,}000 \text{ m}^2 =$ ____ ha $=$ ____ km^2

9. **(a)** A rectangle has area 36 cm² and width 3 cm. What is the length of the rectangle?
   (b) A rectangle has area 60 cm² and perimeter 38 cm. Use the "guess and check" method to find the length and width of the rectangle.

10. Twenty-four 1-cm by 1-cm squares are used to tile a rectangle.
    (a) Find the dimensions of all possible rectangles.
    (b) Which rectangle has the smallest perimeter?
    (c) Which rectangle has the largest perimeter?

11. Find the areas and perimeters of the following parallelograms. Be sure to express your answer in the appropriate units of measurement.

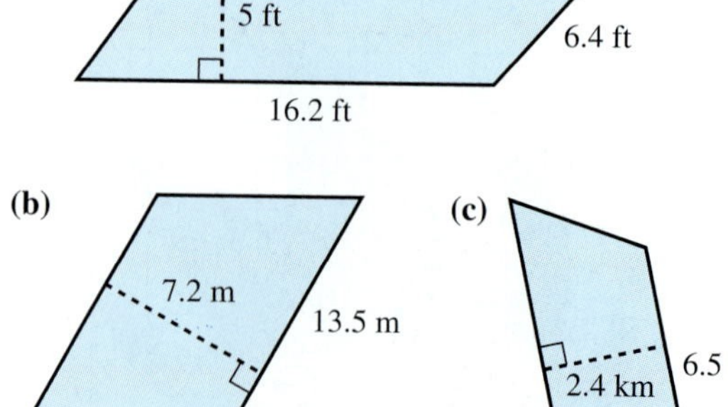

12. Find the areas and perimeters of these triangles.

(a)

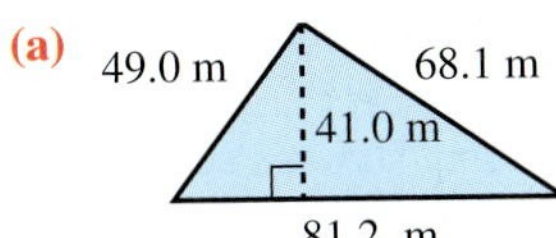

(b)

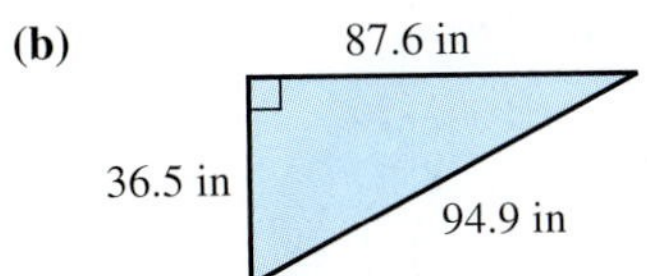

(c)

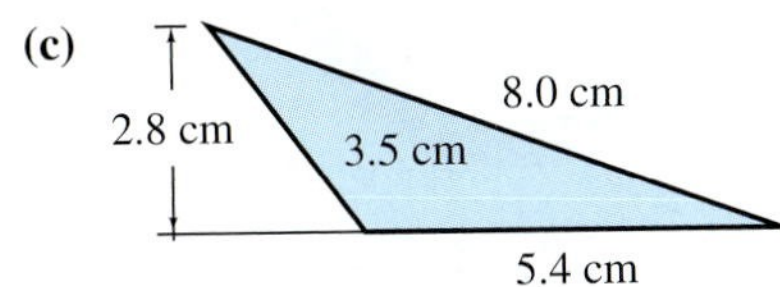

13. Find the areas of these figures. Express the area in square units.

(a)

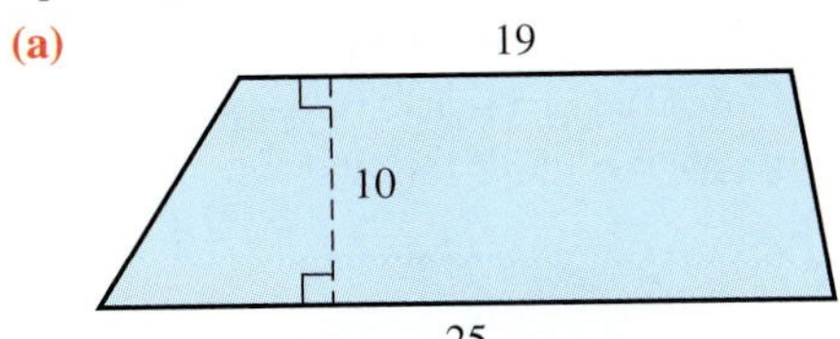

(b)

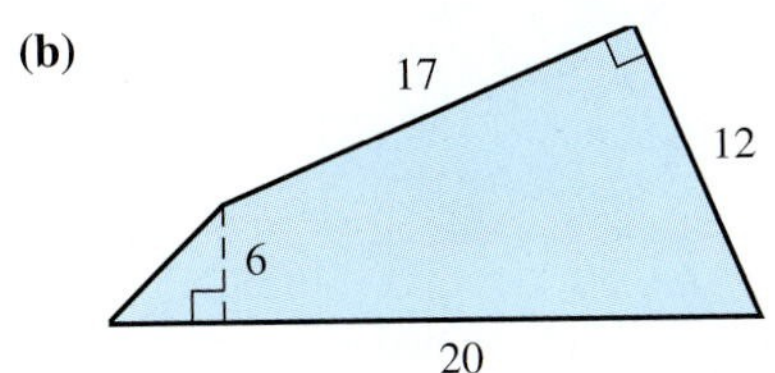

(c)

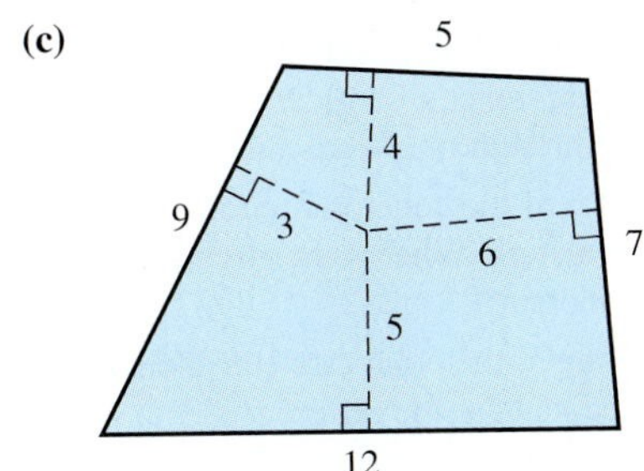

(d)

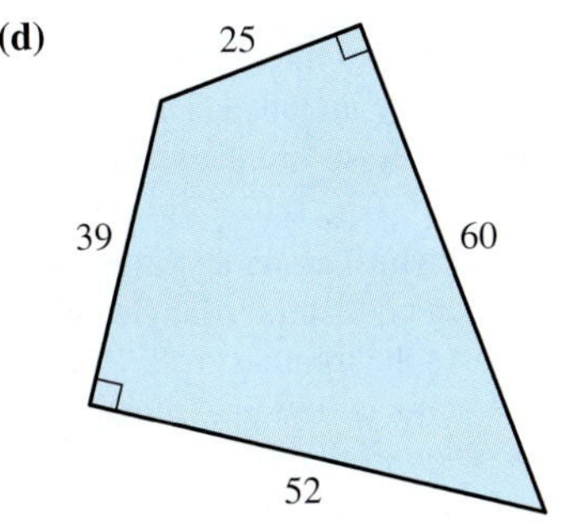

14. Lines k, l, and m are parallel to the line containing the side $\overline{AB}$ of the triangles shown here.

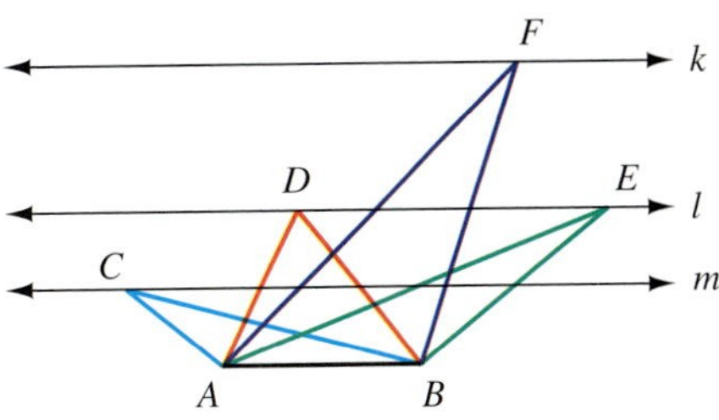

(a) What triangle has the smallest area? Why?

(b) What triangle has the largest area? Why?

(c) Which two triangles have the same area? Why?

15. Lines k, l, and m are equally spaced parallel lines. Let $ABCD$ be a parallelogram of area 12 square units.

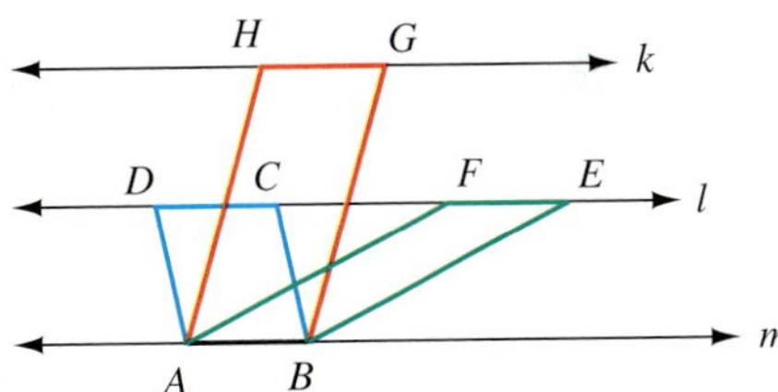

(a) What is the area of the parallelogram $ABEF$?

(b) What is the area of the parallelogram $ABGH$?

(c) If AB = three units of length, what is the distance between the parallel lines?

16. Find the area of each lattice polygon shown below.

(a)

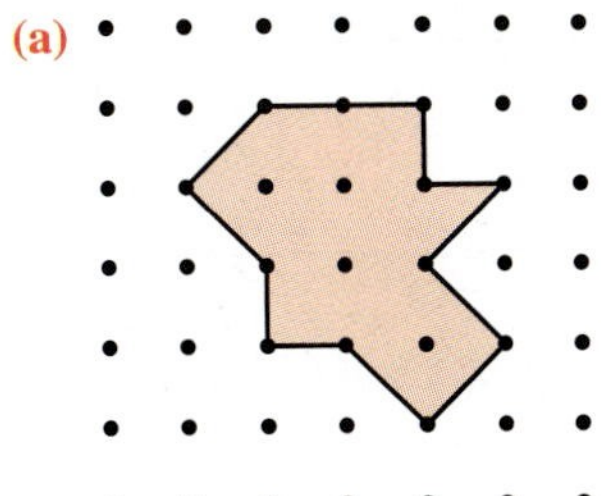

(b)

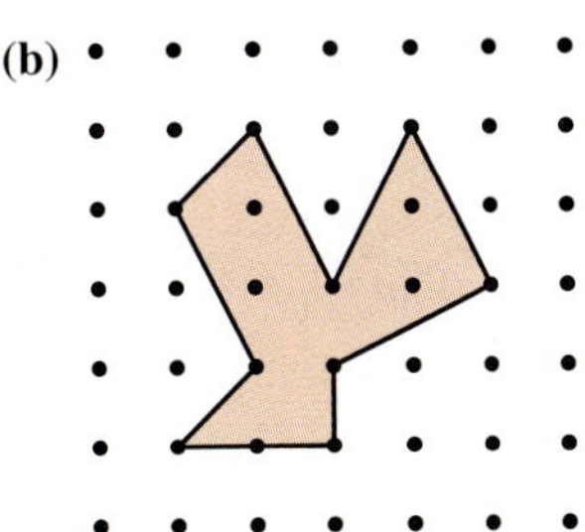

(c)

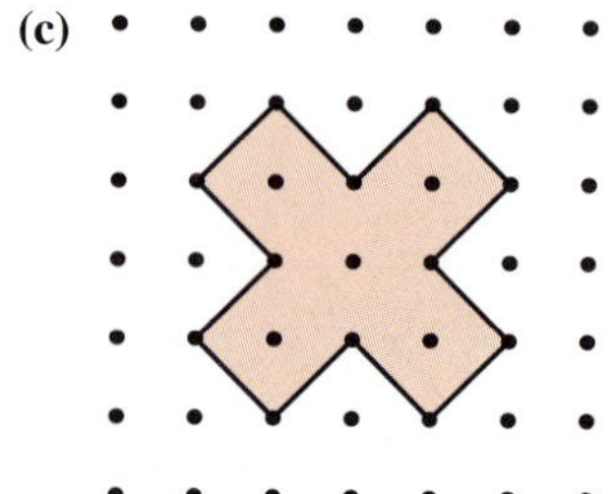

17. An oval track is made by erecting semicircles on each end of a 50-m by 100-m rectangle.

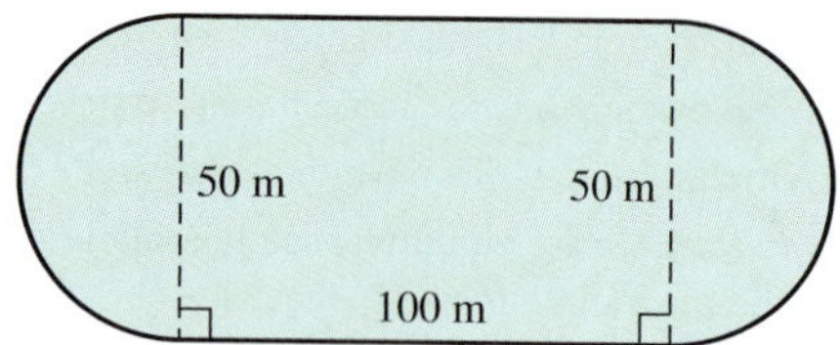

(a) What is the length of the track?

(b) What is the area of the region enclosed by the track?

18. A track has lanes 1 meter wide. The turn radius of the inner lane is 25 meters. To make a fair race, the starting lines in each lane must be staggered so each competitor runs the same distance to the finish line. Find the distance between the starting line in one lane to the starting line in the next lane. Is the same distance used between the first and second lane and between the second and third lane?

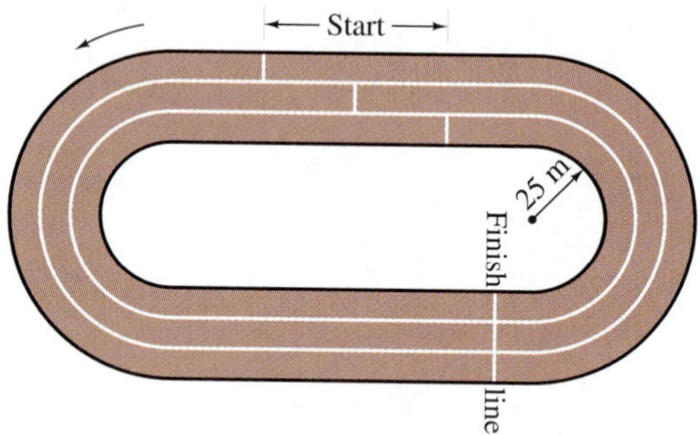

19. An **annulus** is the region bounded by two concentric circles.

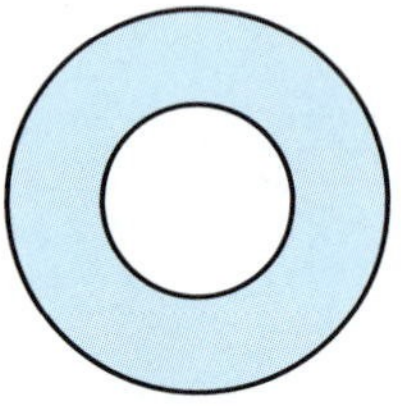

(a) If the radius of the small circle is 1 and the radius of the larger circle is 2, what is the area of the annulus?

(b) A dartboard has four annular rings surrounding a bull's-eye.

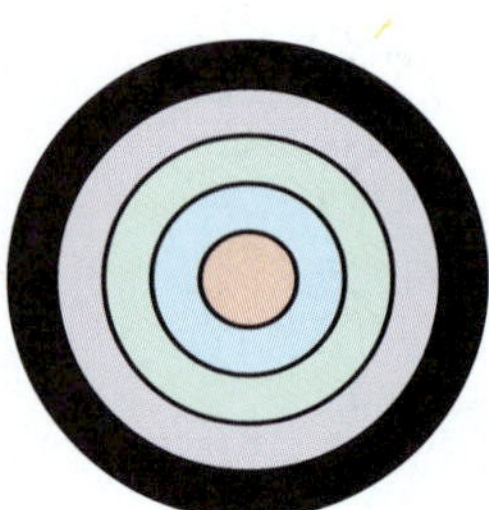

The circles have radii 1, 2, 3, 4, and 5. Suppose a dart is equally likely to hit any point of the board. Is the dart more likely to hit in the outermost ring (shown black) or inside the region consisting of the bull's-eye and the two innermost rings?

20. A circle is inscribed in a square. What percentage of the area of the square is inside the circle?

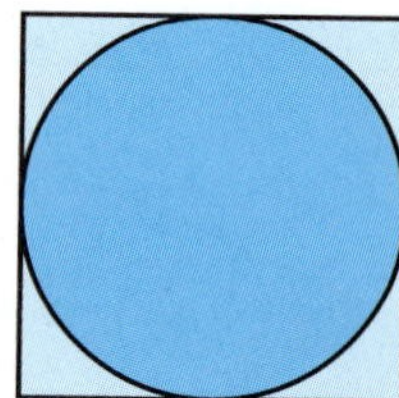

21. The meter was originally defined as one ten-millionth of the distance from the North Pole to the equator.

(a) Assuming the earth is a perfect sphere, what would be the circumference of a great circle on the earth that passes through the North and South Poles?

(b) The diameter of the equatorial circle of the earth is 12,755 kilometers. What is the circumference of the equator in meters?

(c) Which is longer, the polar circle or the equator? Can you account for the difference?

Teaching Concepts

22. In teaching any topic, it is extremely helpful if surprising results of interest can be included. The result of Example 12.8 is such a case, as are the results of the following problems.

(a) Instead of rolling a wheel along a straight line, imagine "rolling" an equilateral triangle as shown on the next page. As the triangle "rolls" from left to right, the point A starts on the line, moves to the top in the middle position of the triangle, and again comes to rest on the line at the right end of the figure. Measure the figure shown *very carefully* with a metric ruler (estimate your measurements to the nearest millimeter), and then determine the area in square centimeters of the equilateral triangle and of the triangle AAA formed by the three locations of the point A. Then calculate the ratio of the area of triangle AAA to the area of the equilateral triangle.

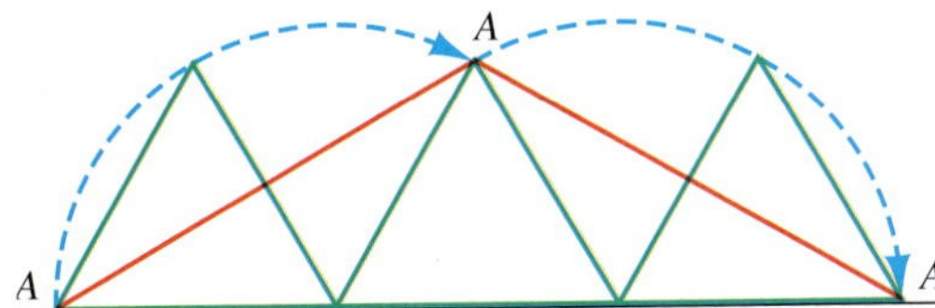

(b) Instead of rolling an equilateral triangle, this time "roll" a square as shown. Again, measure this figure with a metric ruler, and then determine the area of the square and the area under the polygonal arch $AAAA$. As before, also determine the ratio of the area under the polygonal arch to the area of the square. (See Example 12.6, part (a).)

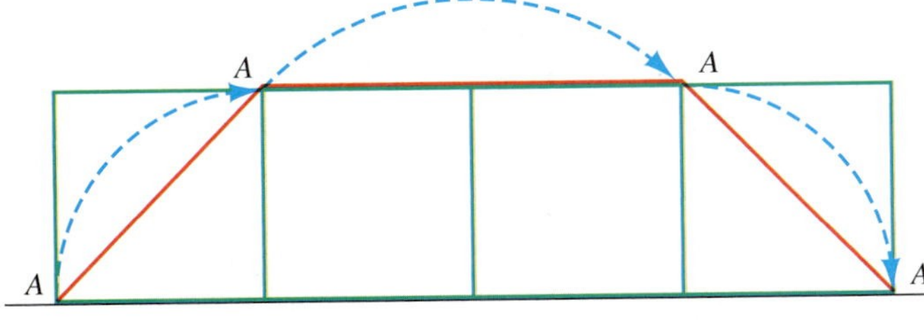

(c) Repeat parts (a) and (b), but this time roll a regular pentagon as shown below.

(d) Repeat parts (a), (b), and (c), but for a regular hexagon. (See Example 12.6. part (b).)

(e) Make a conjecture based on parts (a) through (d).

Thinking Critically

23. Two regions A and B are cut from paper. Suppose the area of region A is 20 cm² larger than that of region B. If the regions are overlapped, by how much does the area of the nonoverlapped part of region A exceed the nonoverlapped part of region B? Explain carefully.

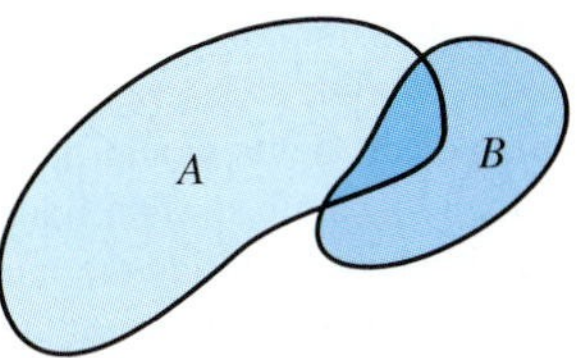

24. Four mutually tangent circles of diameter 10 cm with their centers at the vertices of a square are used to draw a vase as shown below.

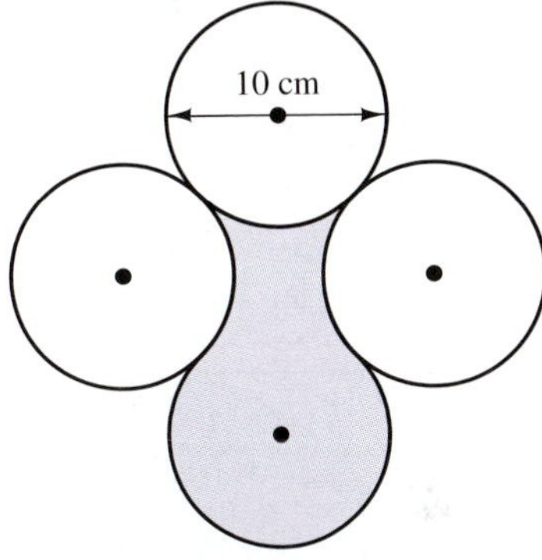

(a) Use dissection and rearrangement to form a square of the same area as the vase. (Show a sequence of steps similar to the solutions in Example 12.7.)

(b) Show that the vase has area 100 cm².

(c) As an extra challenge, see if you can do part (a) by cutting the vase into just three pieces.

25. A square cake measures 8″ by 8″. A wedge-shaped piece is cut by two slices meeting at 90° at the cake's center. What is the area of the top of the piece? Explain your reasoning carefully.

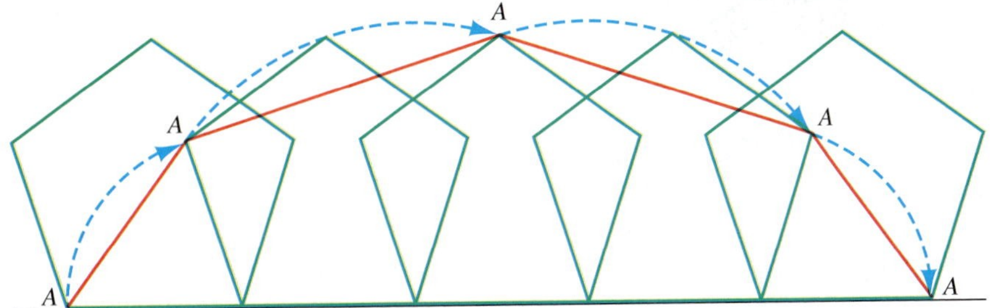

Illustration for Problem 22(c)

26. For reasons lost in history, the two cornfields R and S were divided by two line segments, $\overline{AB}$ and $\overline{BC}$. The friendly owners of the fields would like to divide their adjoining fields by a single straight boundary line. Carefully describe how to divide the quadrilateral into two fields R' and S' with a single segment so the area of each new cornfield is the same as before.

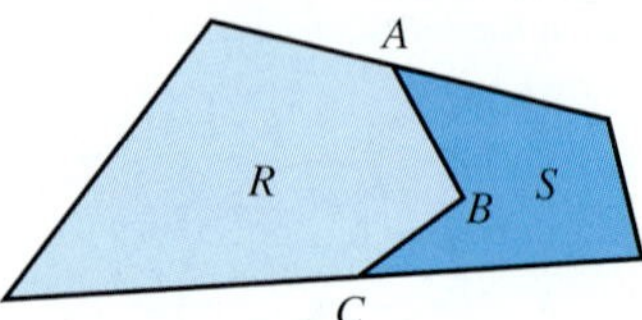

27. **(a)** The colored region shown below is formed by circular arcs drawn from two opposite corners of a 1-by-1 square. What is the area of the region?

(b) Four semicircles are drawn with centers at the midpoints of the sides of a 1-by-1 square. What is the area of the shaded region?

28. A sidewalk 8 feet wide surrounds the polygonally shaped garden of perimeter 300 feet as shown. The sidewalk makes circular sectors of 8-foot radius at the vertices of the polygon. Explain why the area covered by the walk is $2400 + 64\pi$ square feet.

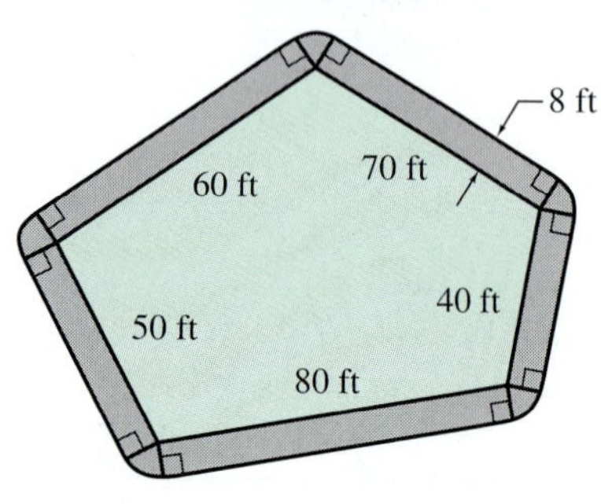

29. Erin walks her dog Nerd with a leash of length L. Nerd is very obedient and always walks directly to Erin's right at the end of his leash. Erin follows several different routes and wants to compare the length of her walk with that of Nerd. For each route below, how much farther does Nerd walk than Erin?

(a) Around a circle of radius R

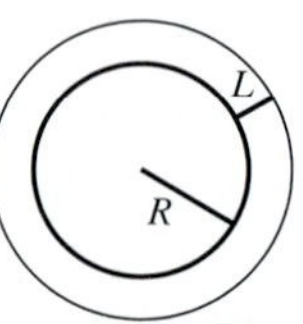

(b) Around a square with sides of length S

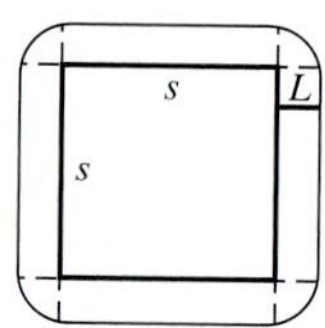

(c) Around a track shaped like a rectangle of length A with half circles of radius R on each end of the rectangle

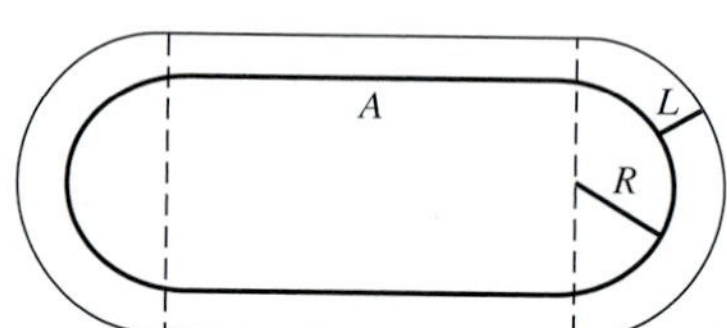

(d) Around a triangle with sides of length A, B, and C

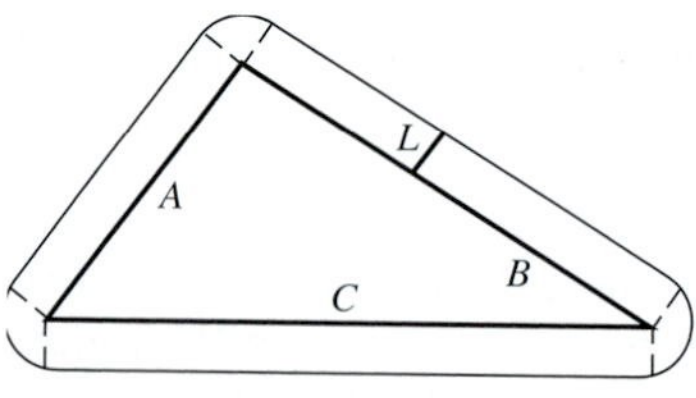

(e) Around any simple closed curve

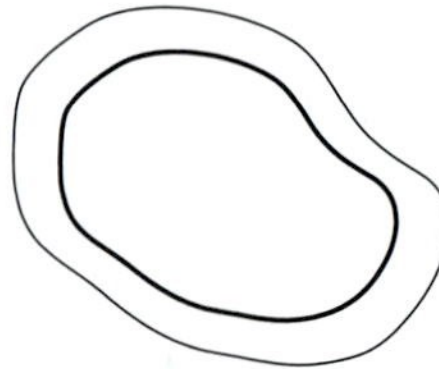

30. Problems 48 and 50 of the Rhind papyrus suggest that the ancient Egyptians approximated π with $(16/9)^2$, which to two decimal places is the quite accurate value 3.16. Use the following sequence of figures to explain the reasoning that may have been used to derive this estimation. Notice that that area of the circle with diameter 9 is approximated by an octagon and that the octagon's area is then approximated by a square of side length 8.

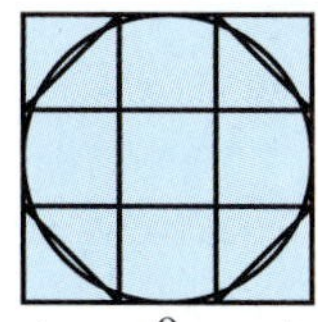

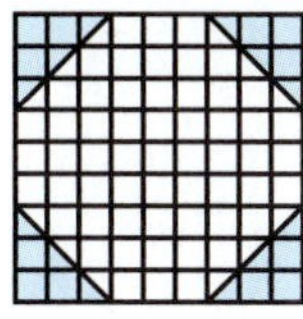
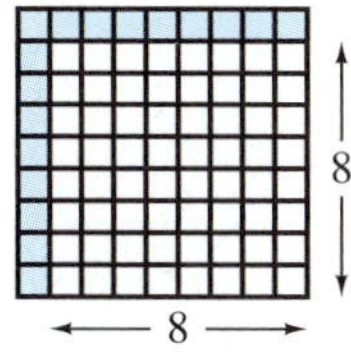

31. The commentaries of the Talmud (Tosfos Pesachim 109a, Tosfos Succah 8a, Marsha Babba Bathra 27a) present a nice approach to the formula $A = \pi r^2$ for the area of a circle. Imagine that the interior of a circle is covered by concentric circles of yarn. The yarn circles are clipped along a vertical radius, and each strand is straightened to cover an isosceles triangle as shown in the accompanying figure. Find the area of the triangle, and then explain how the area formula for a circle follows.

32. According to Jewish history, the Tabernacle of Moses was a 50-cubit by 100-cubit rectangle. The Jerusalem Talmud suggests that a square with sides of length $70\frac{2}{3}$ cubits has nearly the same area, being short by just $\frac{19}{3} - \frac{1}{9}$ square cubits. Show that this is true.

33. Let P be an arbitrary point in an equilateral triangle ABC of altitude h and side s as shown. What is the sum $x + y + z$ of the distances to the sides of the triangle? (*Hint:* The areas of $\triangle ABP$, $\triangle BCP$, and $\triangle ACP$ add up to area($\triangle ABC$).)

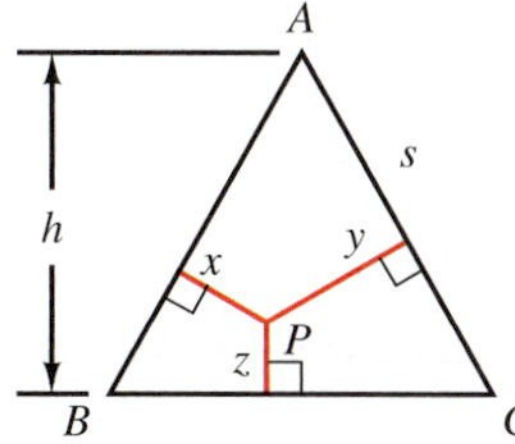

34. Joining each vertex of a triangle shown here to the midpoint of the opposite side divides the triangle into six small triangles.

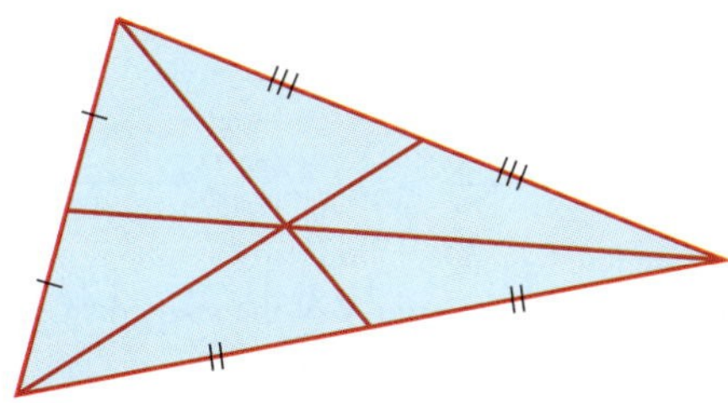

Show that all six triangles have the same area. (*Hint:* Look for pairs of triangles with the same base and height.)

Thinking Cooperatively

For the next four problems and activities, you will need several sheets of paper (including a sheet of ruled notebook paper), scissors, and a ruler.

35. The derivation of the formula for the area of a parallelogram depicted in Figure 12.9 does not apply to a tall, slanted parallelogram, since more than two pieces are required to form a rectangle. Draw a parallelogram something like the one shown, where the base is, say, three vertical ruled lines long. Cut out the parallelogram, and make vertical cuts along every third ruled line. Show that the pieces you obtain can be reassembled into a rectangle.

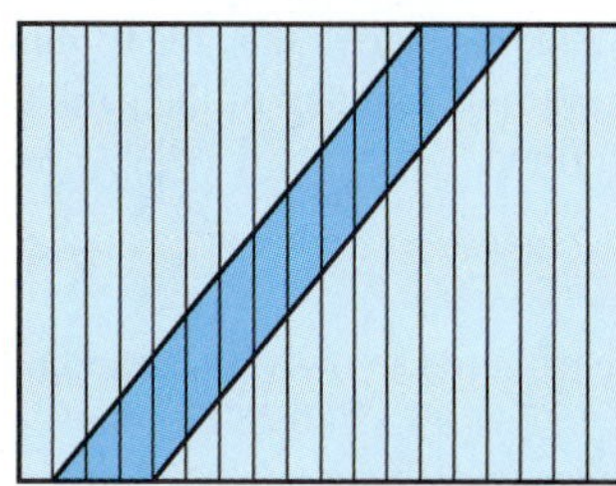

36. Fold a sheet of paper in half and cut out a pair of congruent triangles.

(a) Show that the two triangles can be arranged to form a parallelogram.

(b) Use the construction in part (a) to obtain a new explanation of how the formula for the area of a triangle follows from the area formula for a parallelogram.

37. Cut several triangles as illustrated from paper.

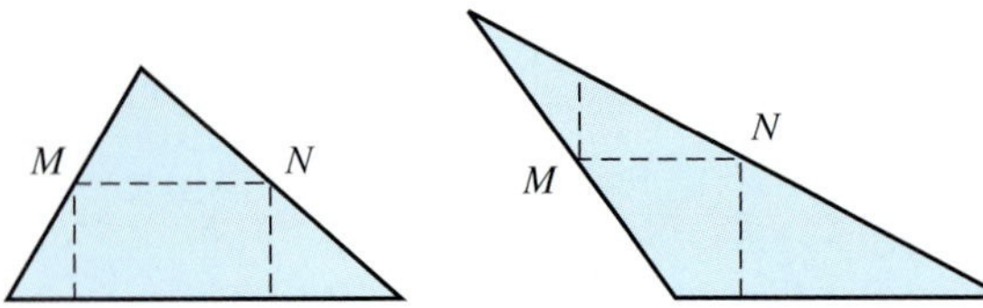

Find the midpoints M and N of the slanted sides (that is, the sides, not the base) by folding.

(a) Fold on the horizontal and vertical lines (shown dashed) to form a doubly covered (two layers of paper) rectangle.

(b) If the triangle has base b and altitude h, what are the lengths of the sides of the rectangle you formed by folding? Obtain the triangle area formula $A = bh/2$.

38. The formula for the area of a trapezoid can be obtained in several ways using paper folding and cutting. Discuss how to obtain the formula by each of these methods.

(a) Fold a sheet of paper in half and cut out, simultaneously, a pair of congruent trapezoids. Show how to arrange the two trapezoids into a parallelogram, and then explain how the area formula for a trapezoid can be derived from the area formula for a parallelogram.

(b) Fold one of the bases of a paper trapezoid onto the other, and crease along the midline between the two bases. Now cut along the crease to create two trapezoids. Show how to arrange them into a parallelogram, and then derive the area formula for the original trapezoid from that of the parallelogram you have formed.

Making Connections

39. Kelly has been hired to mow a large rectangular lawn measuring 75 feet by 125 feet. The lawn mower cuts a path 21 inches wide. Estimate how far (in feet) Kelly must walk to complete the mowing job.

40. Roll ends of carpet are on sale for six dollars per square yard. To finish the rough-cut edges, edging material costing 10 cents per foot is glued in place. Compute the total cost of a roll end measuring 8 feet by 10 feet.

41. A carpet is made by sewing a 1-inch-wide braid around and around until the final shape is an oval with semicircular ends as shown. Estimate the length of braid required. (*Hint:* Estimate the area of the carpet.)

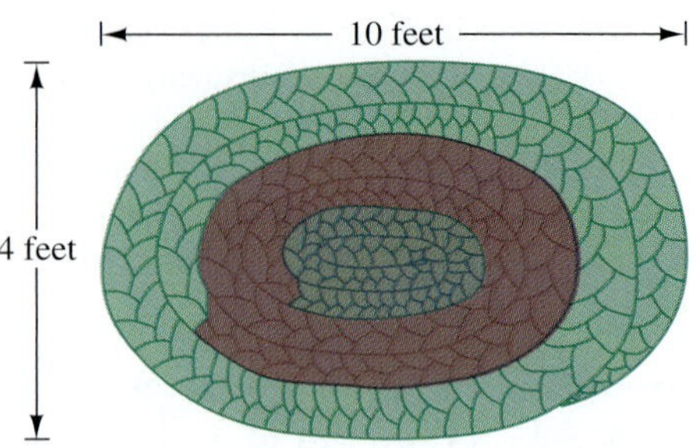

42. An L-shaped house, walkway, garage, and driveway are shown situated on a 70′ by 120′ lot. How many bags of fertilizer are needed for the lawn? Assume the bags are each 20 pounds, and 1 pound of fertilizer will treat 200 square feet of lawn.

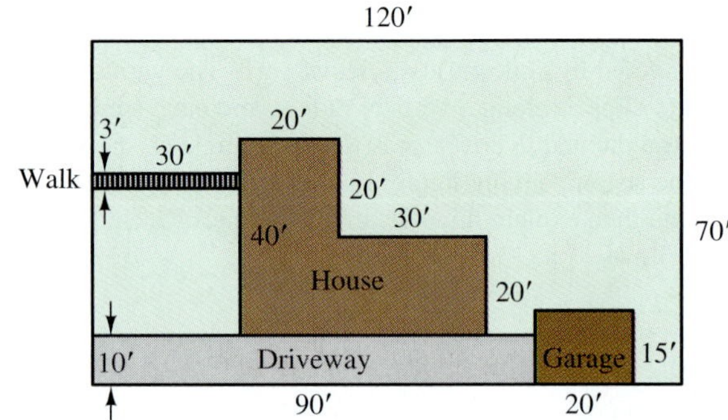

43. A 10′ by 12′ kitchen floor is to be tiled with 8″ square tiles. Estimate the number of tiles this will require.

44. (a) The normal-sized tires on a truck have a 14-inch radius. If oversized tires of 15-inch radius are used, how much farther does the truck travel per revolution of the wheel?

(b) If the speedometer indicates that a truck is traveling at 56 miles per hour, what is the true speed when the truck is running on the oversized tires?

45. Sunaina has 600 feet of fencing. She wishes to build a corral along an existing high, straight wall. She has already decided to make the corral in the shape of an isosceles triangle, with two sides each 300 feet long. What is the measure of $\angle A$ that will give Sunaina the corral of most area? (*Hint:* Consider one of the sides of length 300 feet as a base of the triangle.)

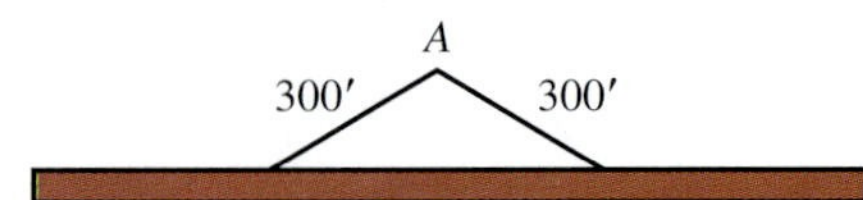

Using a Calculator

46. Let a triangle have sides of lengths a, b, and c. Also, let s denote half the perimeter of the triangle, so $s = \frac{1}{2}(a + b + c)$. The area of the triangle is given by **Hero's** (or **Heron's**) **formula:**

$$A = \sqrt{s(s-a)(s-b)(s-c)}.$$

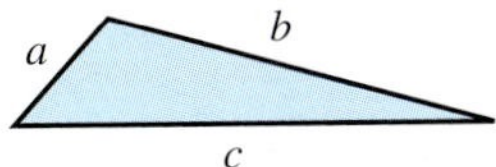

The formula is named for the Greek mathematician Hero of Alexandria, who lived about A.D. 50, but the formula was discovered much earlier by Archimedes. Use Hero's formula to compute the area of these triangles.

(a)

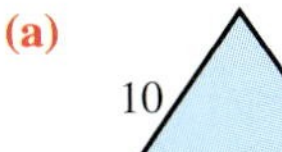

(b)

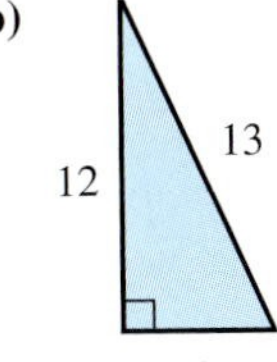

(c)

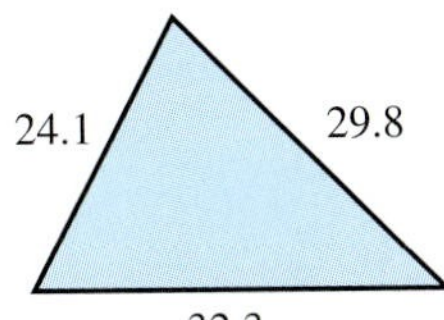

Using a Computer

47. Use geometry software to construct any convex quadrilateral. Next, join the successive midpoints of the sides of the quadrilateral to form a parallelogram. Use the software to compute both the area of the quadrilateral and the area of the parallelogram. What relationship seems to exist between these areas?

48. Draw an equilateral triangle and its inscribed and circumscribed circles. How do the areas of the two circles compare? Investigate with your software.

49. Draw a regular hexagon and its inscribed and circumscribed circles. Use your software to compare the areas of the two circles.

From State Student Assessments

50. (Washington State, Grade 4)
Casey is making a quilt. Quilts are made up of quilt blocks. Each quilt block will look like the one below.

Her quilt will have 25 blocks. Casey knows how much fabric she needs to make the patterned inside squares. Tell the steps she could take to figure out how much fabric she will need to make all of the shaded corner pieces.
Explain your thinking using words, numbers, or pictures.

51. (Washington State, Grade 4)
Which of the following is closest to the distance around the middle of an unsharpened pencil?

A. 25 millimeters

B. 25 centimeters

C. 25 meters

52. (Illinois, Grade 5)
The width of a rectangular rug is 4 feet. If the perimeter is 20 feet, what is the length of the rug?

○ 4 feet
○ 5 feet
○ 6 feet
○ 16 feet
○ 80 feet

For Review

53. Convert these measurements to the unit shown.

(a) 1 m = ____ cm

(b) 352 mm = ____ cm

(c) 1 m^2 = ____ cm^2

(d) 1 m^3 = ____ liters

Discovering Pick's Formula

In 1899, the German mathematician Georg Pick discovered a remarkable formula for the area of a polygon drawn on square dot paper. Polygons of this special type are also known as lattice polygons. Stretching a rubber band onto a geoboard is another easy way to form lattice polygons.

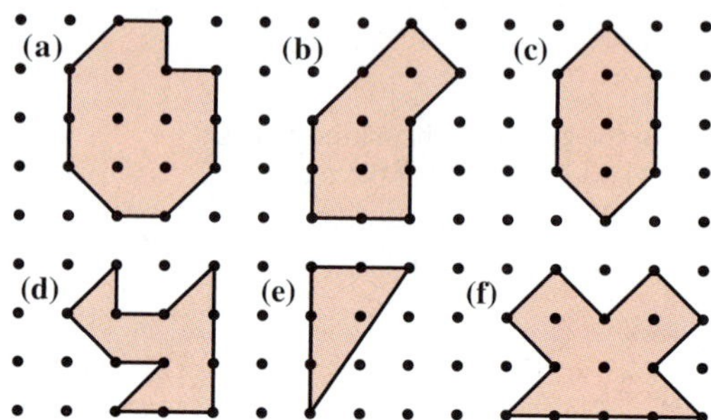

1. Complete the table of values for each polygon, where

b = number of dots on the boundary of the polygon,
i = number of dots in the interior of the polygon, and
A = area of the polygon.

Polygon	b	i	A
(a)	11	5	$9\frac{1}{2}$
(b)			
(c)			
(d)			
(e)			
(f)			

The values of b, i, and A for polygon **(a)** are given as an example.

2. Try to guess a formula for A in terms of b and i. (If you have trouble, add a column of the values of $b/2$. You may also want to obtain more data by drawing other lattice polygons.)

3. Search for an extension of Pick's formula to regions with a hole—that is, a region inside one lattice polygon, but outside a second lattice polygon. Two Examples, (g) and (h) are shown on the next page. Now b is the number of dots on both rubber bands, and i is the number of dots between the two rubber bands.

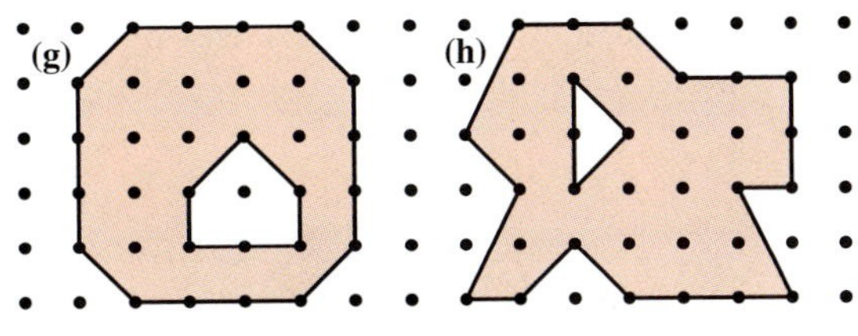

Polygon	b	i	A
(g)	22	9	20
(h)			

12.3 THE PYTHAGOREAN THEOREM

The Pythagorean theorem is the single most remarkable result in geometry. The theorem is aesthetically pleasing and very useful in solving practical problems.

THEOREM *The Pythagorean Theorem*
If a right triangle has legs of length a and b and its hypotenuse has length c, then

$$a^2 + b^2 = c^2.$$

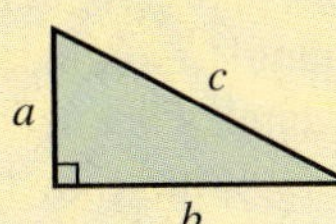

Proving the Pythagorean Theorem

By erecting squares on the sides of a right triangle, the Pythagorean relation $a^2 + b^2 = c^2$ can be interpreted as a result about areas: *The sum of the areas of the squares on the legs of a right triangle is equal to the area of the square on the hypotenuse.* The area interpretation of the Pythagorean theorem is shown in Figure 12.15.

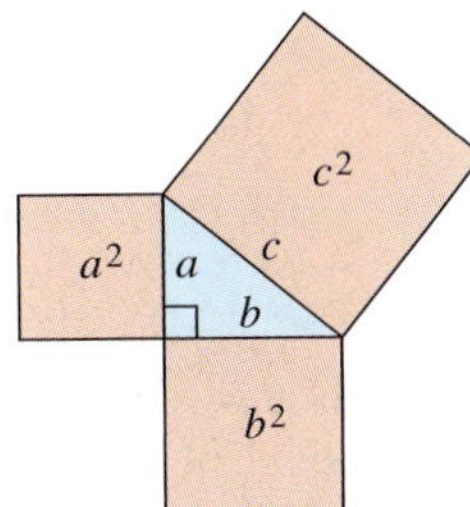

Figure 12.15
The sum of the areas of the squares on the legs of a right triangle equals the area of the square on the hypotenuse

The area interpretation probably led to the discovery of the theorem, at least in special cases. For example, the Babylonian clay tablet in Figure 12.16 shows a large square subdivided by congruent isosceles right triangles. It is apparent that the area of the square on the hypotenuse equals the area of the two squares on the legs.

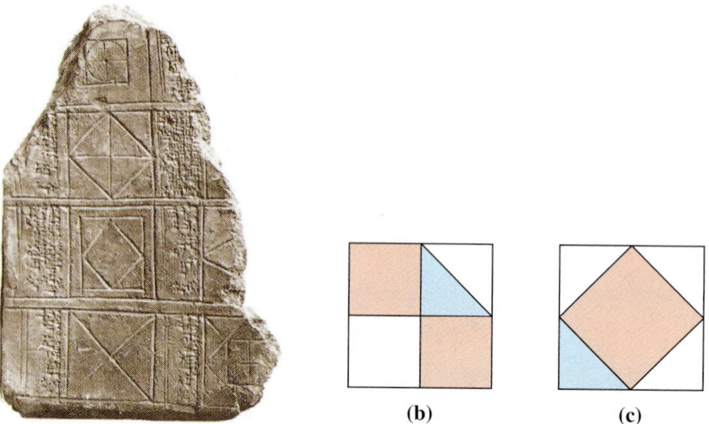

Figure 12.16
A Babylonian tablet suggesting a special case of the Pythagorean theorem

The demonstration depicted in Figure 12.16 is not a general proof of the Pythagorean theorem, because the right triangle is isosceles. However, a similar idea can be followed for arbitrary right triangles. In Figure 12.17(a), we begin with any right triangle, letting a and b denote the respective lengths of the legs and c the length of the hypotenuse. Next consider a square with sides of length $a + b$. Four congruent copies of the right triangle are placed inside the squares in two different ways. In Figure 12.17(b), the four triangles leave two squares uncovered, with respective areas a^2 and b^2. In Figure 12.17(c), the four triangles leave one square of area c^2 uncovered. Since the four triangles must leave the same area uncovered in both arrangements, we conclude that $a^2 + b^2 = c^2$.

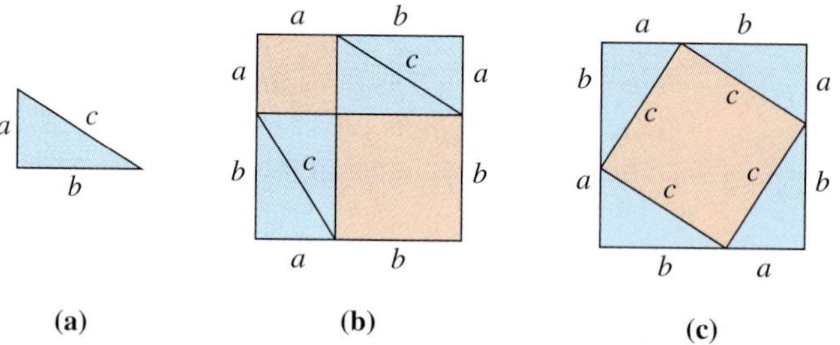

Figure 12.17
A dissection proof of the Pythagorean theorem

No records have survived to indicate what proof, if any, Pythagoras (ca. 572–501 B.C.) may have offered. The dissection proof requires showing that the inner quadrilateral of Figure 12.17(c) is actually a square (why is it?), and Pythagoras's knowledge of angles in a right triangle was sufficient to do this. Since the time of Pythagoras, a tremendous number of proofs have been devised. In the second edition of *The Pythagorean Proposition,* E. S. Loomis catalogs 370 different proofs.

Applications of the Pythagorean Theorem

EXAMPLE 12.16 USING THE PYTHAGOREAN THEOREM

Find the lengths x and y in the following figures.

(a)

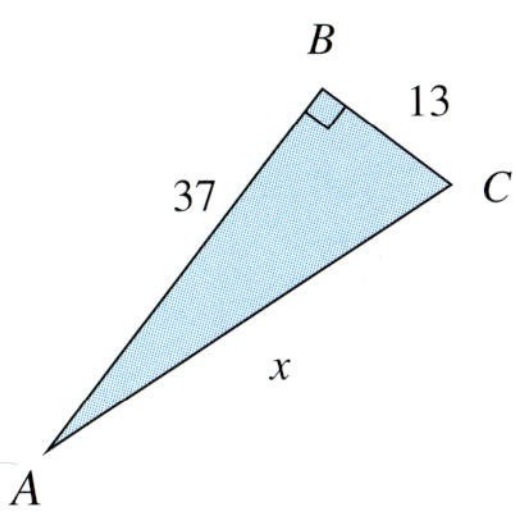

(b)

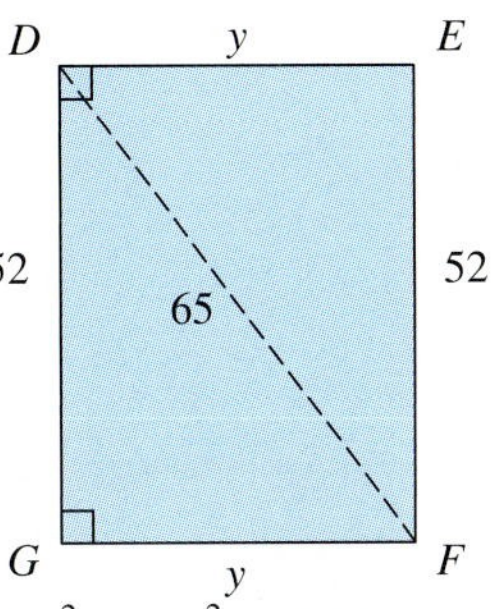

Solution

(a) By the Pythagorean theorem, $x^2 = 13^2 + 37^2 = 169 + 1369 = 1538$. Therefore $x = \sqrt{1538} \doteq 39.2$.

(b) The diagonal $\overline{DF}$ of rectangle $DEFG$ is the hypotenuse of the right triangle DEF. The Pythagorean theorem, applied to $\triangle DEF$, gives $y^2 + 52^2 = 65^2$. Therefore $y^2 = 65^2 - 52^2 = 4225 - 2704 = 1521$, and $y = \sqrt{1521} = 39$.

For many applications it is necessary to write and solve equations based on the Pythagorean relation. Here is an example.

EXAMPLE 12.17 DETERMINING HOW FAR YOU CAN SEE

Imagine yourself on top of a mountain, or perhaps in an airplane, at a known altitude given in feet. Approximately how far away, in miles, is the horizon?

Solution

Understand the Problem

Altitude is a measure of the distance above the surface of the earth. The horizon is the circle of points where our line of sight is tangent to the sphere of the earth's surface. The problem is to derive a formula that expresses, or at least approximates, the distance to the horizon as it depends on the altitude of the observer. Since the altitude is given in feet, while the distance to the horizon is to be given in miles, special care must be taken to handle the units of measure properly.

Devise a Plan

The earth is very nearly a sphere. A line of sight to the horizon forms a leg of a right triangle, as the diagram shows. Since the radius r of the earth is about 4000 miles, and the altitude s is known, the Pythagorean theorem can be used to solve for the distance to the horizon. In the diagram, all distances, including s, are expressed in miles; if h is the altitude in feet, we can use the conversion formula $h = 5280s$ (recall that 1 mile = 5280 feet).

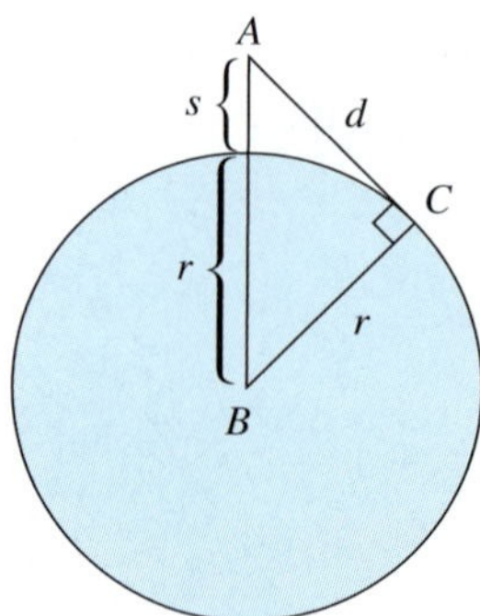

Carry Out the Plan

Applying the Pythagorean theorem to the right triangle ABC gives $d^2 + r^2 = (s + r)^2$. The squared term on the right can be written $s^2 + 2sr + r^2$, so $d^2 + r^2 = s^2 + 2sr + r^2$. Subtracting r^2 from both sides shows that $d^2 = s^2 + 2sr$. Therefore the exact distance d, in miles, is given by the formula

$$d = \sqrt{s^2 + 2rs}.$$

Since $r = 4000$ miles, the formula can also be written

$$d = \sqrt{s^2 + 8000s} \quad \text{or} \quad d = \sqrt{s(s + 8000)}.$$

From the top of a mountain (say, $s = 2$), or from an airplane (say, $s = 7$), or even from the International Space Station ($s = 200$ miles), it is evident that the altitude s is much smaller than 8000. Therefore, little accuracy is lost if the term $(s + 8000)$ in the exact formula is replaced with simply 8000. This gives us the approximate equation

$$d \doteq \sqrt{8000s}.$$

Using the equation $s = \dfrac{h}{5280}$ gives $d \doteq \sqrt{\dfrac{8000}{5280}h}$. Finally, since $\sqrt{\dfrac{8000}{5280}} \doteq 1.2$, we obtain a simple formula for the miles d to the horizon as seen from an altitude of h feet:

$$d \doteq 1.2\sqrt{h}.$$

For example, the distance to the horizon as seen from an airplane flying at 40,000 feet is about $1.2\sqrt{40{,}000} = (1.2)(200) = 240$ miles.

Look Back

This problem involved several steps that are typical of the way the Pythagorean theorem is used:

- draw a figure and label all the distances,
- identify all the right triangles in the drawing,
- write the Pythagorean relationships for all of the right triangles, and
- solve the Pythagorean formulas to determine unknown values needed for the solution of the problem.

The Converse of the Pythagorean Theorem

The numbers 5, 12, and 13 satisfy $5^2 + 12^2 = 13^2$. Is the triangle with sides of length 5, 12, and 13 a right triangle? The answer is yes, since the Pythagorean relation $a^2 + b^2 = c^2$ holds if, *and only if*, a, b, and c are the side lengths of a right triangle. That is, the converse of the Pythagorean theorem is true, as stated without proof in the following theorem.

THEOREM *Converse of the Pythagorean Theorem*
Let a triangle have sides of length a, b, and c. If $a^2 + b^2 = c^2$, then the triangle is a right triangle and the angle opposite the side of length c is its right angle.

EXAMPLE 12.18 CHECKING FOR RIGHT TRIANGLES

Determine if the three lengths given are the lengths of the sides of a right triangle.

(a) 15, 17, 8 **(b)** 10, 5, $5\sqrt{3}$ **(c)** 231, 520, 568

Solution

(a) $8^2 + 15^2 = 64 + 225 = 289 = 17^2$, so 8, 15, and 17 are the lengths of the sides of a right triangle.

(b) $5^2 + (5\sqrt{3})^2 = 25 + 25 \cdot 3 = 25 + 75 = 100 = 10^2$, so 5, $5\sqrt{3}$, and 10 are the lengths of the sides of a right triangle.

(c) $231^2 + 520^2 = 53{,}361 + 270{,}400 = 323{,}761 \neq 322{,}624 = 568^2$, so 231, 520, and 568 are not the lengths of sides of a right triangle. This would be difficult to see by measuring angles with a protractor, since this triangle closely resembles the right triangle with sides of length 231, 520, and 569. Note that $569^2 = 323{,}761$.

Problem Set 12.3

Understanding Concepts

1. Find the distance x in each figure.

(a)
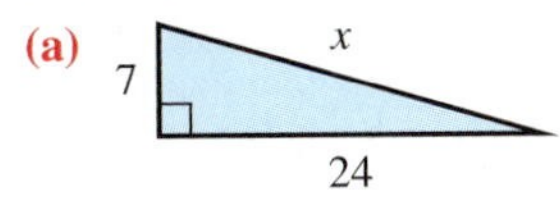

(b)
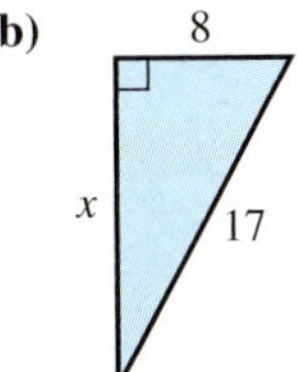

(c)
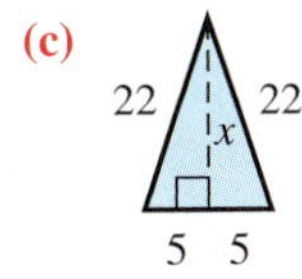

(d)
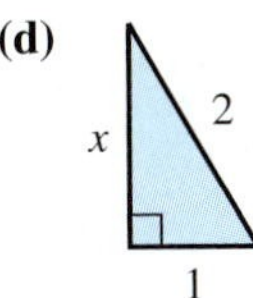

(e)
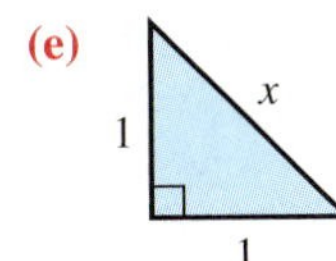

(f)
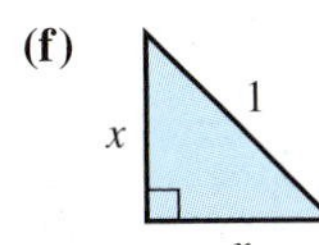

2. Find the distance x in each figure.

(a)

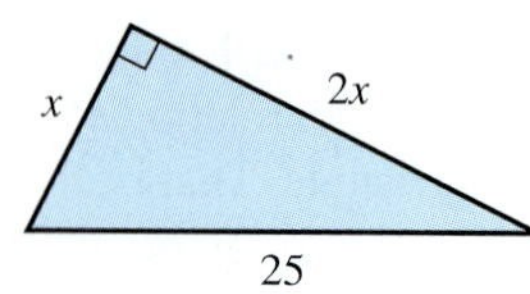

(b)

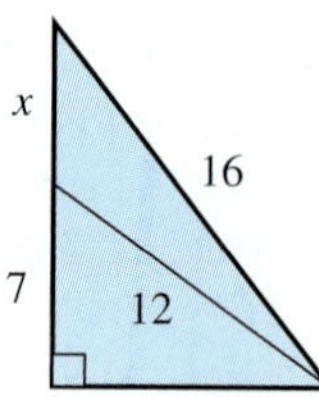

(c)

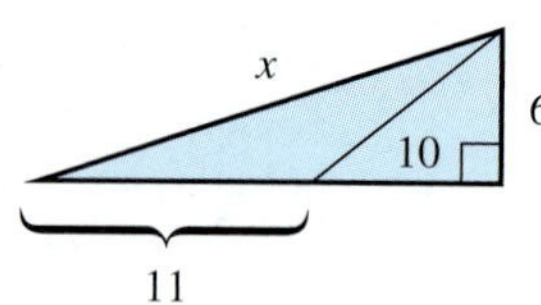

3. Find the distances x and y in the rectangular prism and the cube.

(a)

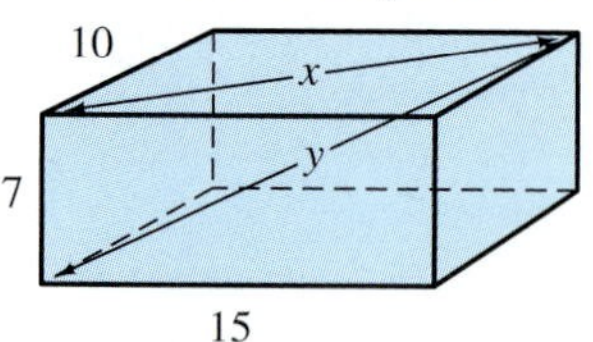

(b)

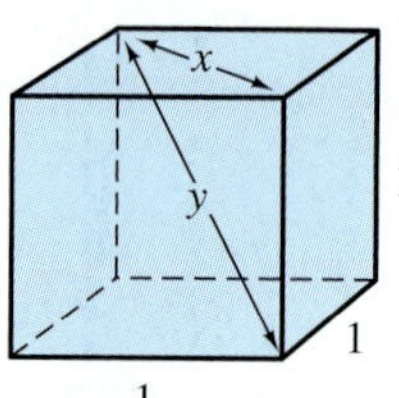

4. Find the distance x in these space figures.

(a)

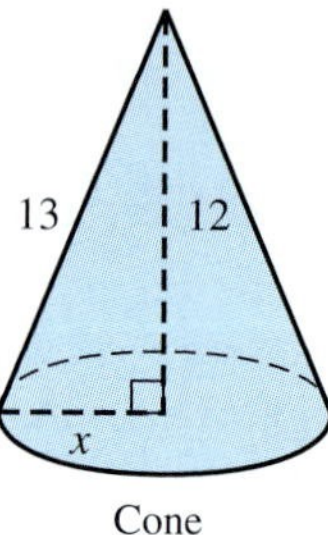

Cone

(b)

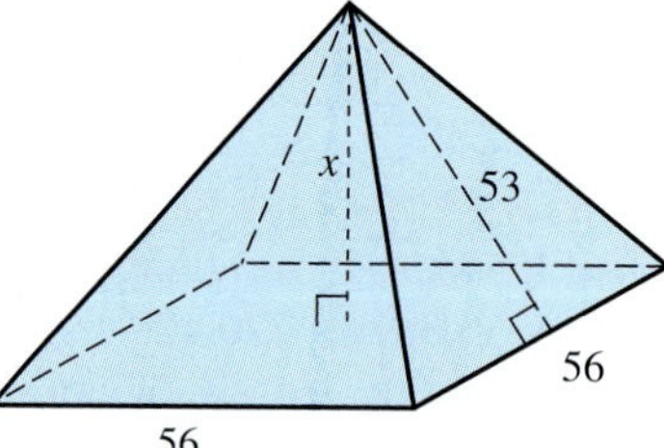

Square-based right regular pyramid, with sides 56 and slant height 53

(c)

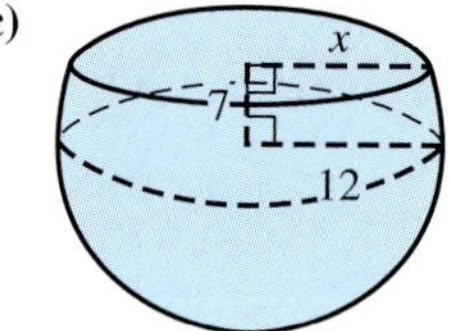

Sphere cut by plane

5. Find the areas of these figures.

(a)

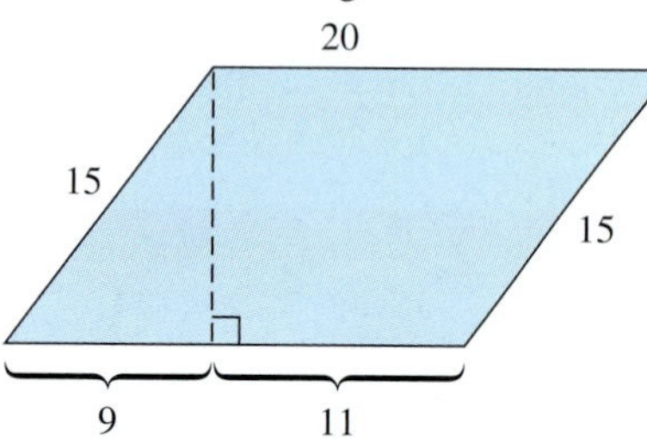

(b)

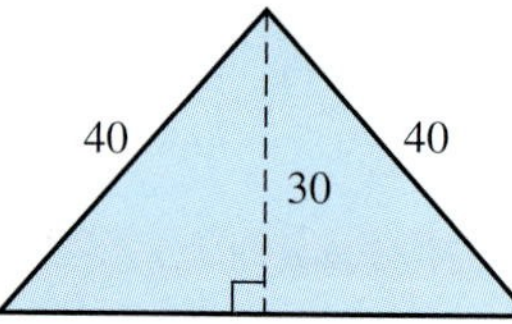

(c)

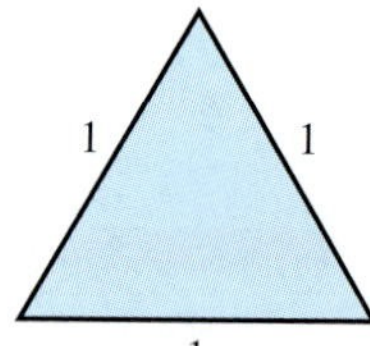

6. Françoise and Maurice cut diagonally across a 50-foot by 100-foot vacant lot on their way to school. How much distance do they save by not staying on the sidewalk?

7. A square with sides of length 2 is inscribed in a circle and circumscribed around another circle. Which is larger, the area of the region between the circles or the area inside the smaller circle?

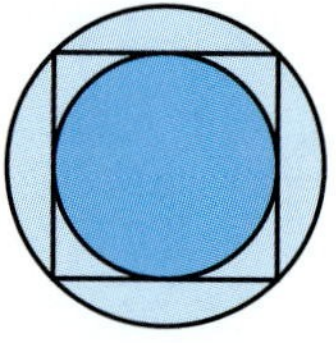

8. At noon, car A left town heading due east at 50 miles per hour. At 1 P.M., car B left the same town heading due north at 40 miles per hour. How far apart were the two cars at

(a) 2 P.M.?

(b) 3:30 P.M.?

9. Find AG in this spiral of right triangles.

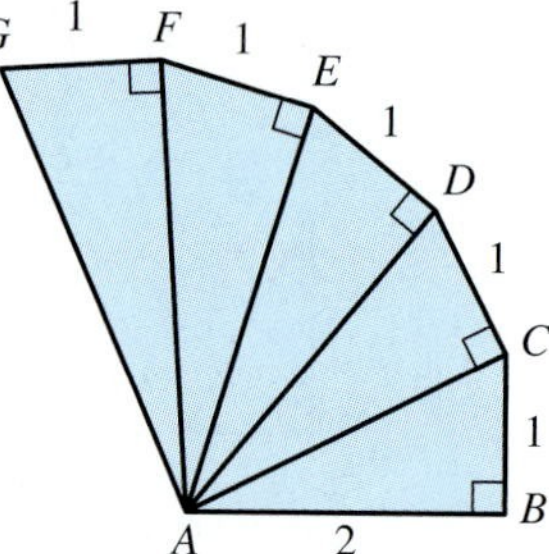

10. **(a)** What is the length of the side of a square inscribed in a circle of radius 1?

(b) What is the length of the side of a cube inscribed in a sphere of radius 1?

11. What is the distance between the centers of these circles?

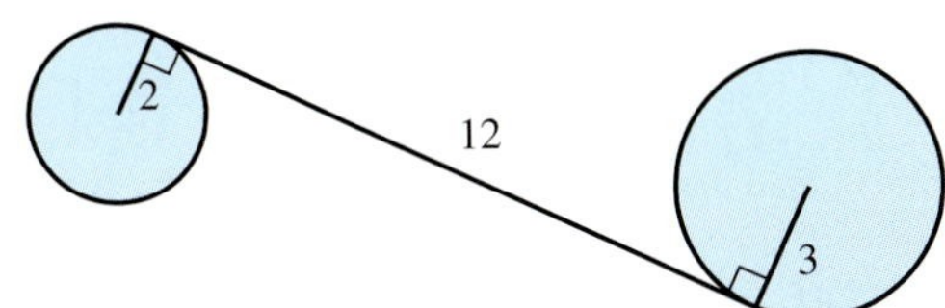

12. What is the radius of the circle shown below?

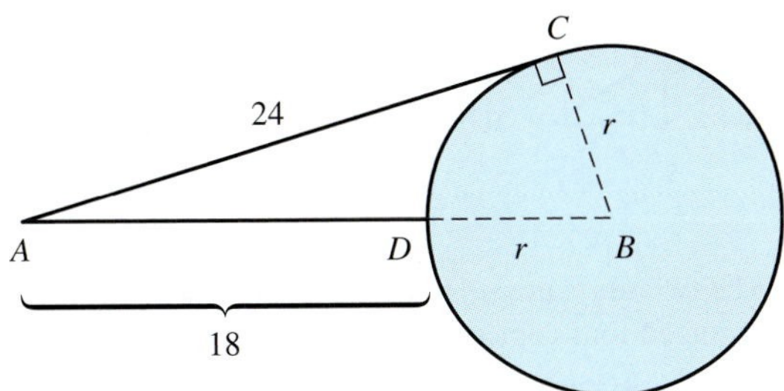

13. Which of the following can be the lengths of the sides of a right triangle?

(a) 21, 28, 35 **(b)** 9, 40, 41

(c) 12, 35, 37 **(d)** 14, 27, $\sqrt{533}$

(e) $7\sqrt{2}, 4\sqrt{7}, 2\sqrt{77}$ **(f)** 9.5, 16.8, 19.3

14. If a, b, and c are the lengths of the sides of a right triangle, explain why $10a$, $10b$, and $10c$ are also the lengths of the sides of a right triangle.

Teaching Concepts

15. During a lesson on the Pythagorean theorem, Sean said, "The sum of the areas of the squares on the sides of a right triangle is equal to the area of the square on the hypotenuse. Would the same thing be true of semicircles instead of squares?"

(a) Is the short answer to Sean's question "yes" or "no"? Justify your answer.

(b) How would you respond to Sean? Would you simply tell him the answer, or would you suggest that he see if he can discover the answer for himself? Or might you perhaps challenge the entire class to try to discover the answer to Sean's question?

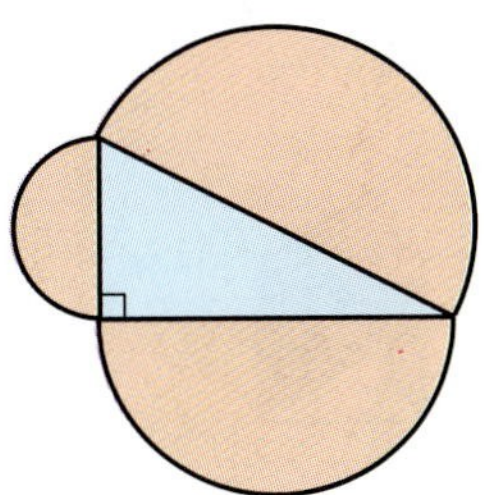

16. Tia says, "Suppose we draw semicircles on the sides of a triangle and compute their areas. If the sum of the areas of two of the semicircles equals the area of the third, is the triangle a right triangle?"

(a) Is the short answer to Tia's question "yes" or "no"? Justify your answer.

(b) How would you respond to Tia?

17. Repeat problem 15, but for equilateral triangles drawn on the sides of the right triangle. (*Suggestion:* Use the assertion that the altitude h of an equilateral triangle of side s is $h = \frac{\sqrt{3}}{2}s$ as in part (a) of problem 19 below.)

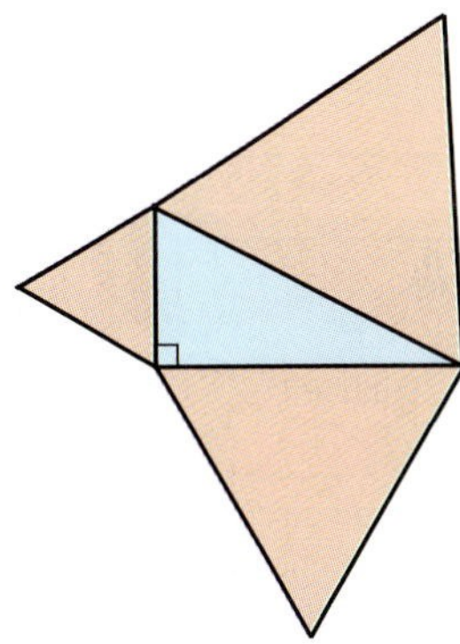

18. Repeat problem 16, but for equilateral triangles drawn on the sides of the original triangle. (*Suggestion:* Again use the assertion of problem 19(a) below.)

Thinking Critically

19. **(a)** Show that the altitude h of an equilateral triangle with sides of length s is given by $h = \frac{\sqrt{3}}{2}s$.

(b) Find a formula for the area of an equilateral triangle of side length s.

(c) Find a formula for the area of a regular hexagon of side length s.

(d) Show that the area of the inscribed circle of a regular hexagon is $\frac{3}{4}$ the area of the circumscribed circle.

20. There are five different lengths between pairs of nails on a 3-nail by 3-nail geoboard:

F
D E
A B C

$AB = 1,\quad AC = 2,\quad AD = \sqrt{2}$
$AE = \sqrt{5},\quad AF = 2\sqrt{2}$

How many different lengths can you find on a 5-nail by 5-nail geoboard?

21. Find the length of the diagonals of this isosceles trapezoid.

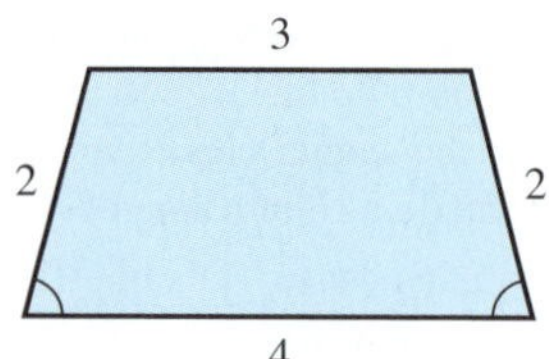

22. An ant is at corner A of a shoebox that is 9 inches long, 5 inches wide, and 3 inches high. What route should the ant follow over the surface of the box to reach the opposite corner C in the shortest distance? (*Suggestion:* It will help to tear along the vertical edges of the box and flatten the top.)

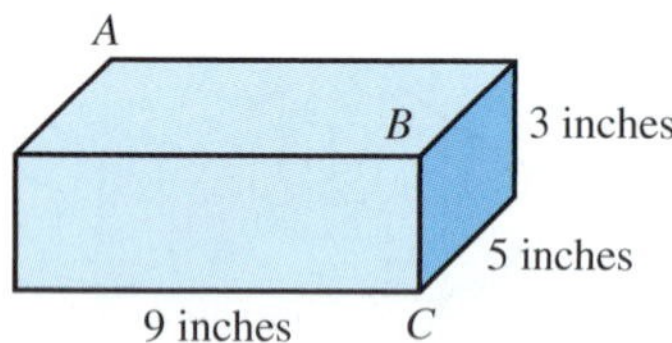

23. A chord of the large circle is tangent to the inner concentric circle. If the chord is 20 cm long, what is the area of the annulus (the region between the two circles)?

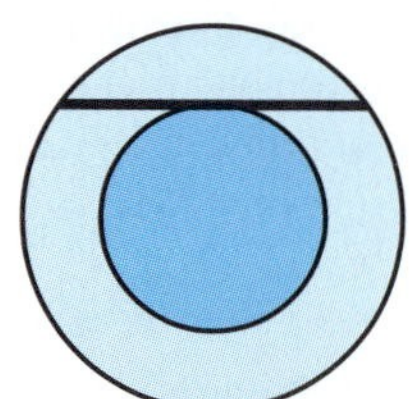

24. Justify why the light-brown regions (which are all parallelograms) in the following sequence of diagrams have the same area. Since the same reasoning shows that the dark-brown regions also have the same area, this provides a striking dynamic proof of the Pythagorean theorem.

Thinking Cooperatively

For problems 25 through 27, you will need several sheets of paper and scissors. Begin by folding a sheet of paper in half twice and then cutting a diagonal to obtain four congruent right triangles. Next, use one of your triangles as a pattern to cut four paper squares whose sides match the sides a, b, $b - a$, and c of the right triangles. Be sure to cut your pieces with care and precision.

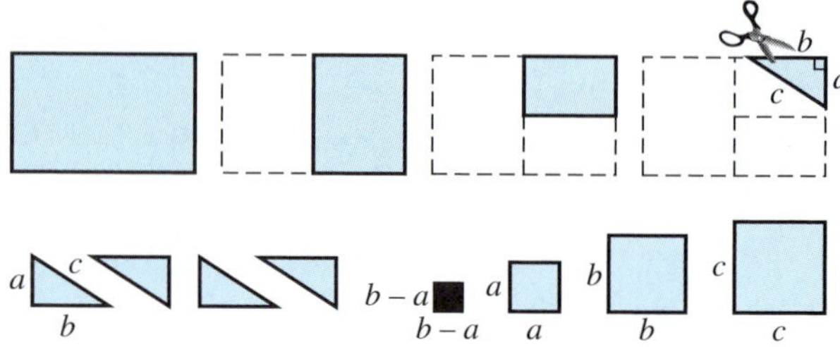

25. The twelfth-century Hindu mathematician Bhaskara arranged four copies of a right triangle of side lengths a, b, and c into a c-by-c square, filling in the center with the $b - a$ sided square.

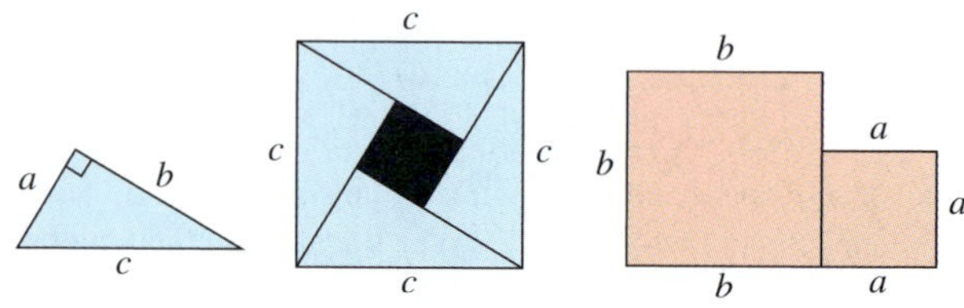

(a) Show how the five pieces in the c-by-c square can be arranged to fill the "double" square region on the right.

(b) Explain how the Pythagorean theorem follows from part (a).

26. Tile the pentagon shown on the next page in two ways:

(a) with two triangles and the squares of sides a and b;

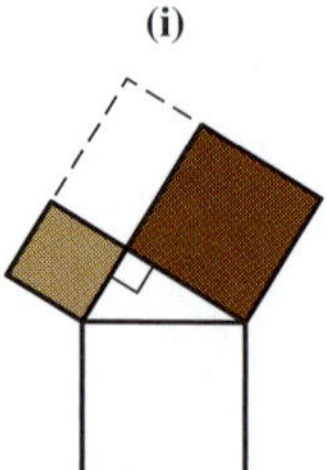

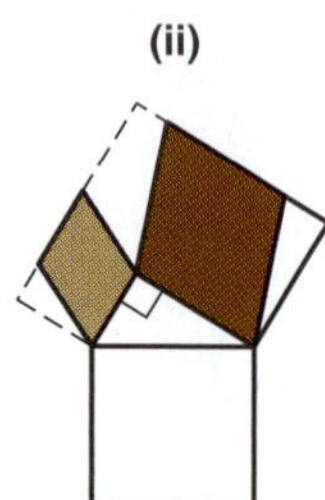

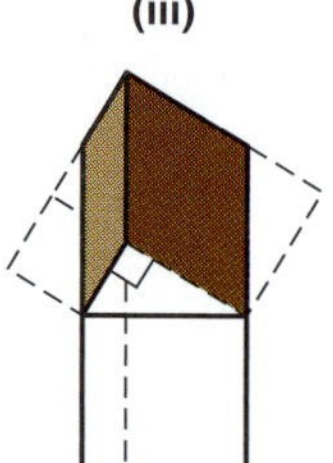

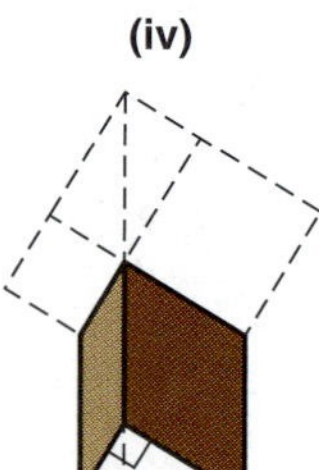

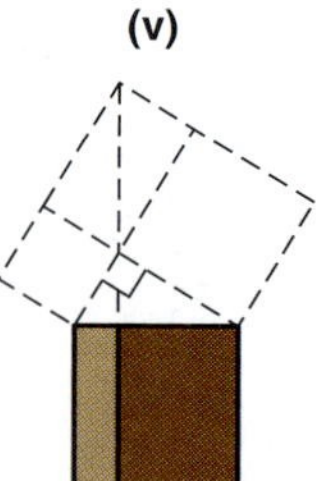

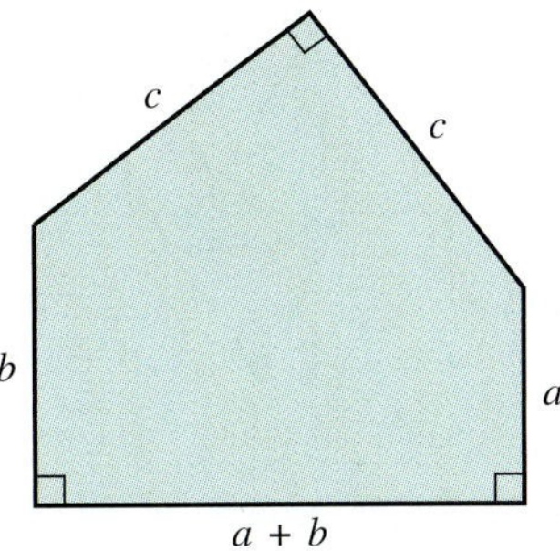

(b) with two triangles and the square of side c.

(c) Explain why the two tilings in parts (a) and (b) prove the Pythagorean theorem.

27. In the nineteenth century, Henry Perigal, a London stockbroker and amateur astronomer, discovered a beautiful scissors-and-paper demonstration of the Pythagorean theorem. Follow these steps to complete your own demonstration: Through the center of the larger square on the leg of the right triangle, draw one line perpendicular to the hypotenuse and a second line parallel to the hypotenuse. Cut along these two lines to divide the square into four congruent pieces, and then show how to arrange these four pieces, together with the square on the shorter leg, to form a square on the hypotenuse of the right triangle.

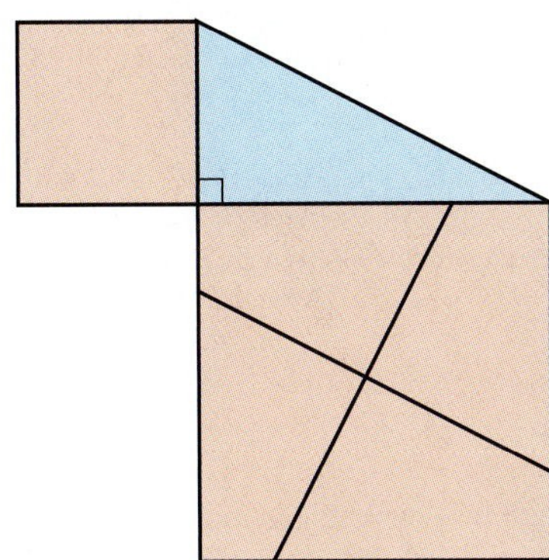

Making Connections

28. A baseball diamond is actually a square 90 feet on a side. What distance must a catcher throw the ball to pick off a runner attempting to steal second base?

29. Approximately what height can be reached from a 24-foot ladder? What assumptions have you made to arrive at your answer?

30. The ancient Egyptians squared off fields with a rope 12 units long, with knots tied to indicate each unit. Explain how such a rope could be used to form a right angle. What theorem justifies their procedure?

31. A water lily is floating in a murky pond rooted on the bottom of the pond by a stem of unknown length.

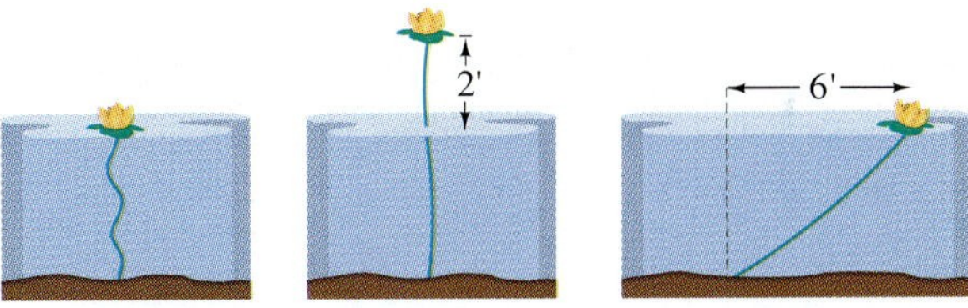

The lily can be lifted 2 feet over the water and moved 6 feet to the side. What is the depth of the pond?

32. A stop sign is to be made by cutting off triangles from the corners of a square sheet of metal 32 inches on a side. What length x will leave a regular octagon? Give your answer to the nearest eighth inch.

33. A 12-foot-wide crosswalk diagonally crosses a street whose curbs are 40 feet apart, intersecting the opposite curb with a 30 foot displacement in the direction of the street. Reflective striping tape will be applied to all four sides of the crosswalk. How many feet of striping should be ordered? (*Suggestion:* First find the length x of the crosswalk. Next, find the area of the crosswalk in order to help you calculate the curb length y.)

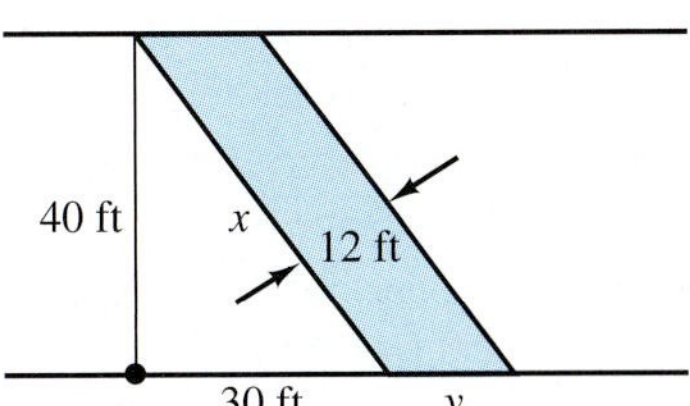

Using a Calculator

34. Any large structure made of steel must be designed to accommodate the expansion of the metal when heated. Thus, pipelines, steel bridge decking, and train track rails have expansion joints. To understand why, consider a mile-long steel rail built without any expansion joint. Suppose on a warm day the rail lengthens by one inch.

If its ends are firmly anchored and the track bows to one side, the amount of deflection can be estimated by considering a right triangle with one leg $\frac{1}{2}$ mile long and a hypotenuse $\frac{1}{2}$ mile + $\frac{1}{2}$ inch long. What is the deflection x? Convert all dimensions to feet to do your calculation.

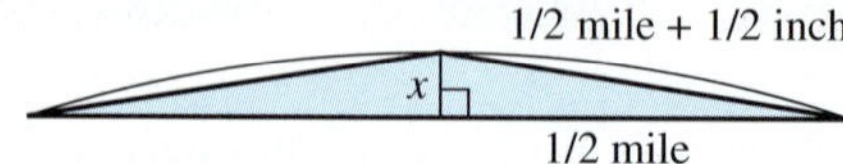

35. A **Pythagorean triple** is a triple of positive integers a, b, and c that satisfy the Pythagorean relationship $a^2 + b^2 = c^2$. For example, 3–4–5 is a Pythagorean triple, since $3^2 + 4^2 = 5^2$.

- **(a)** Verify that 5–12–13, 8–15–17, and 7–24–25 are Pythagorean triples.
- **(b)** If a–b–c is a Pythagorean triple, is $2a$–$2b$–$2c$ also a Pythagorean triple? Explain why or why not.
- **(c)** If a–b–c is a Pythagorean triple, is the triple obtained by adding 2 to each number also a Pythagorean triple? That is, is $(2 + a)$–$(2 + b)$–$(2 + c)$ also a Pythagorean triple? Explain why or why not.
- **(d)** Let u and v be integers, with $1 \leq u < v$, and set $a = v^2 - u^2$, $b = 2vu$, and $c = v^2 + u^2$. Show that a–b–c is a Pythagorean triple.
- **(e)** Use the formulas given in part (d) to compute the Pythagorean triples for $1 \leq u < v \leq 6$. (This can also be done on a spreadsheet.) Use the data to respond to these questions: Which triples a–b–c are primitive, in the sense that a, b, and c have no common divisor other than 1? Is at least one of a, b, or c divisible by 5?

36. Use the approximate formula $d \doteq 1.2\sqrt{h}$ of Example 12.17 to answer these questions.

- **(a)** On a cliff top 100 feet over the ocean, what is the distance to the horizon?
- **(b)** The observation deck of the Sears Tower in Chicago is 1353 feet above ground level. How far can you see across Lake Michigan?
- **(c)** In the Dr. Seuss book *Yertle the Turtle,* Yertle stands on the backs of other turtles and can see 40 miles. How high is Yertle?

Using a Computer

37. Use geometry software to draw any right triangle. On each side, draw outward-pointing equilateral triangles. Use your software to calculate the areas of the triangles.

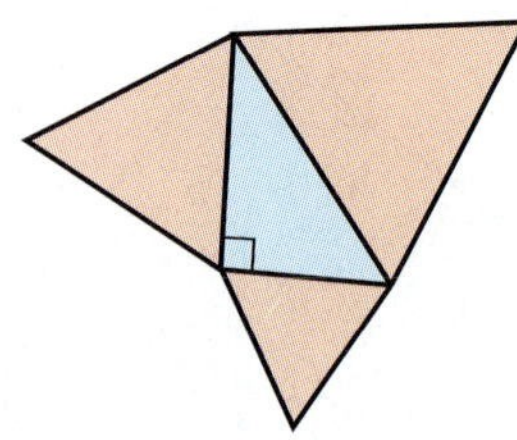

How does the sum of the areas of the triangles on the two legs compare with the area of the equilateral triangle on the hypotenuse?

38. Using geometry software, construct outward squares on each edge of a general triangle ABC, as shown here. Measure the areas of the squares and angle $\angle ABC$, and calculate the sum of the squares on sides $\overline{AB}$ and $\overline{BC}$. Manipulate the triangle until the sum of the areas of the squares closely agrees with the area of the square on side $\overline{AC}$. What is the measure of the angle $\angle ABC$? Write a paragraph describing your exploration.

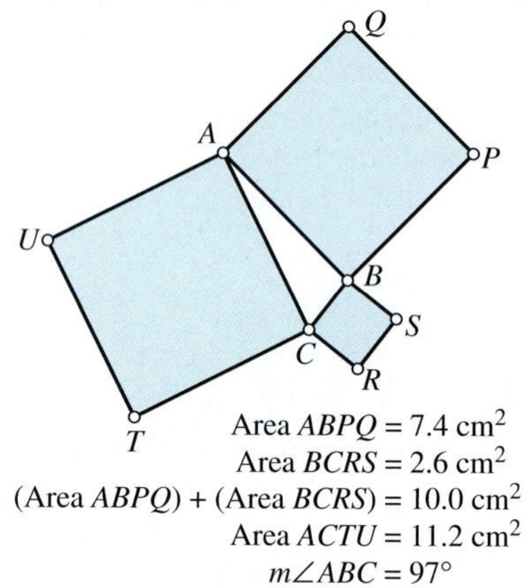

For Review

39. The length of a rectangle is increased by $33\frac{1}{3}$ percent, and the width is decreased by 25 percent. By what percent does the area of the rectangle change?

40. Find the areas of these figures.

(a)

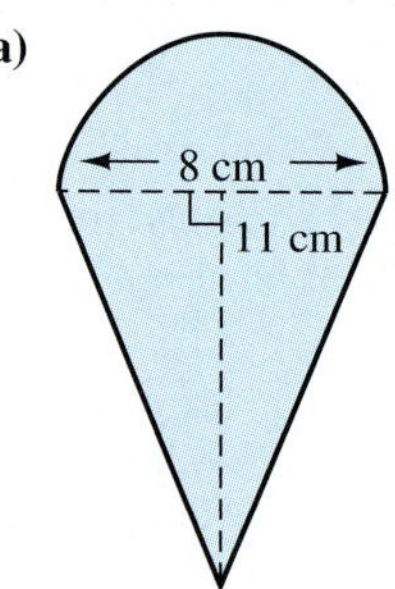

(b)

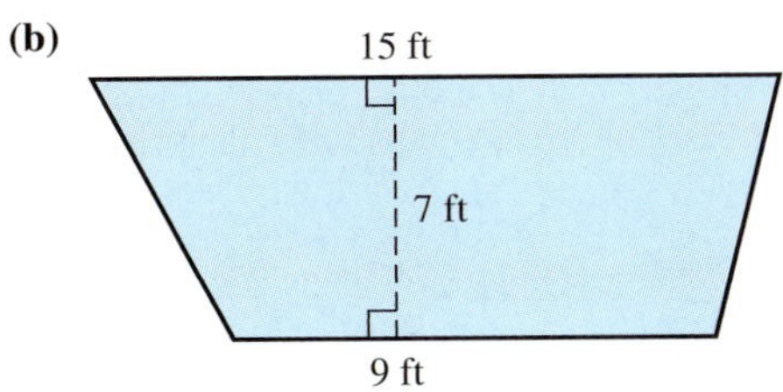

41. Two semicircular arcs, of radius 3 m and 5 m, are centered on the diameter $\overline{AB}$ of a large semicircle as shown. Which route from A to B is shorter: along the large semicircle, or along the two smaller semicircles that touch tangentially at C?

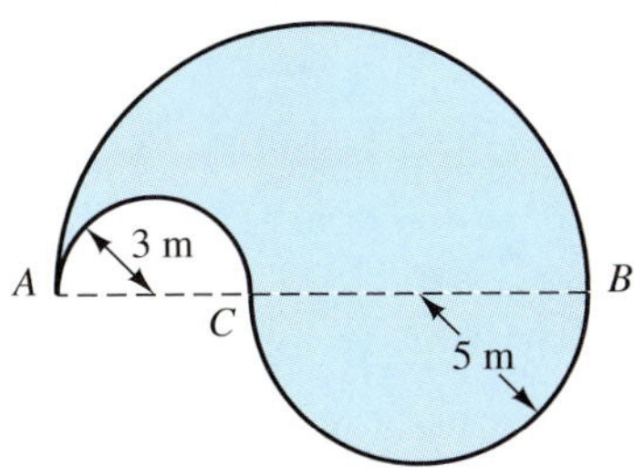

12.4 SURFACE AREA AND VOLUME

The two most important measures of a figure in space are its surface area and volume. These are the counterparts of the perimeter and area, respectively, of a figure in the plane. The **surface area,** often denoted by *SA*, or just *S*, measures the boundary of the space figure. The **volume,** often denoted by *V*, measures the amount of space enclosed within the boundary. It is important to understand how surface area and volume are different. For example, the amount of aluminum in a can of soda is closely related to the can's surface area, whereas the amount of fluid within the can is given by the volume. Even the units are different. Surface area is given in in^2, cm^2, and so on. Volume is given in^3, cm^3, and the like, or possibly in a unit of capacity such as fluid ounces, gallons, milliliters, and so on.

We begin our discussion with surface area. For a polyhedron, the surface area is simply the sum of the areas of each of its faces. It's often useful to imagine cutting a polyhedron apart and placing its faces in a single plane. In many cases, a plane figure or set of figures is formed in this manner with areas that are easy to determine. For space figures with curved boundaries, it is still sometimes possible to imagine cutting and unrolling the surface to create plane figures whose areas can be easily found. For example, to find the surface area of a tin can, we can imagine cutting off both ends and down the seam of the can. This set of steps creates two discs and, when the can is unrolled, a rectangle, all of whose areas are easy to calculate. For a sphere and other more general solids in space, it becomes necessary to approximate the figure with a sequence of a polyhedra whose surface areas approximate that of the space figure. The surface area of the space figure is the limiting value of the surface areas of the approximating polyhedra.

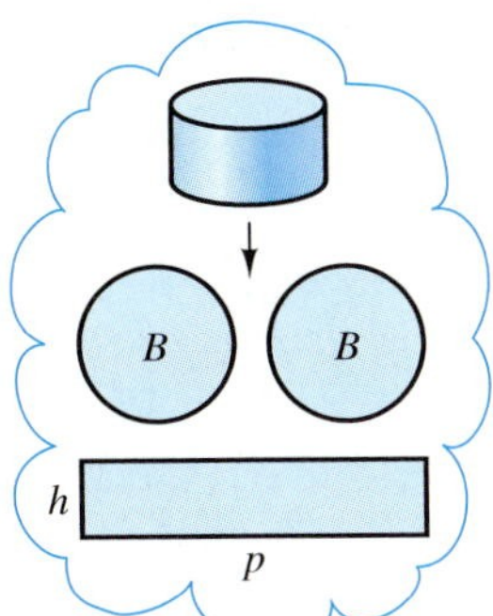

Surface Area of Right Prisms and Cylinders

Figure 12.18 shows how the surface of a right prism is cut into two congruent bases, with the lateral surfaces of the prism unfolded to form a rectangle. If the right prism has height h and the perimeter of the base is p, then the rectangle has area hp.

The same reasoning applies to any right cylinder, where the lateral surface is imagined to be unrolled onto the plane to form a rectangle. The formula obtained below, and indeed all of the formulas in this section, are of far less importance than the ideas used to derive the formulas.

In particular, it is helpful to understand that the total surface area of a three-dimensional solid is often a sum of the **base area** and the **lateral surface area** of the solid (recall that "lateral" means "side").

FORMULA *Surface Area of a Right Prism or Right Cylinder*
Let a right prism or cylinder have height h and bases of area B, and let p be the perimeter of each base. Then the surface area SA is given by

$$SA = 2B + ph.$$

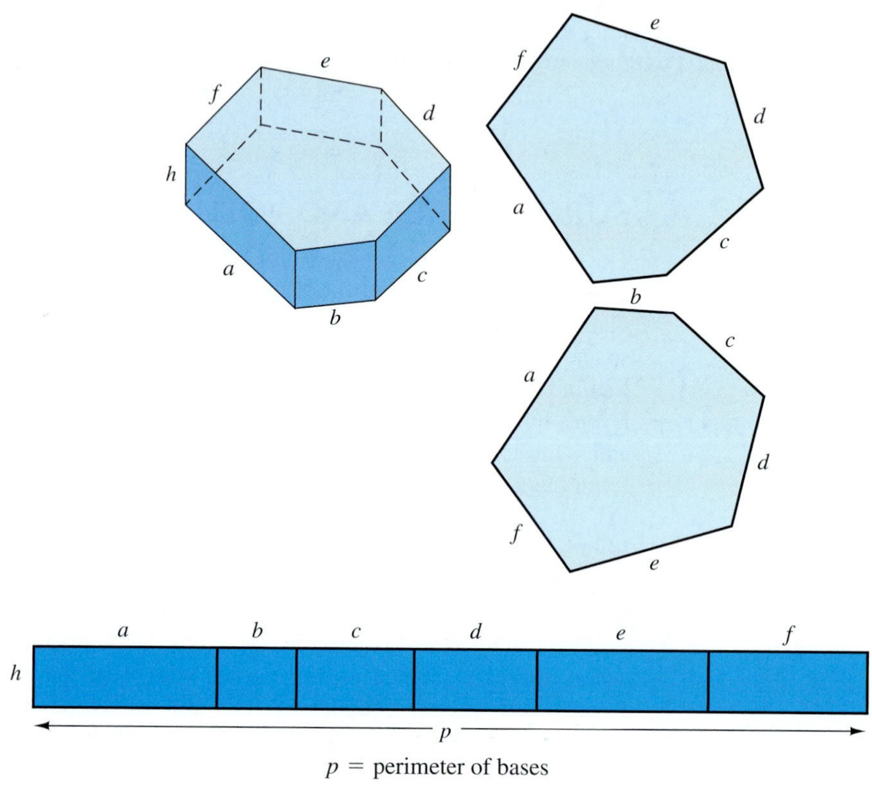

Figure 12.18
The surface of a right prism can be cut and unfolded to form two congruent bases and a rectangle

EXAMPLE 12.19 | FINDING THE SURFACE AREA OF A PRISM-SHAPED GIFT BOX

A gift box has the shape shown below. The height is 10 cm, the longer edges are 20 cm long, and the short edges of the square corner cutouts are each 5 cm long. What is the surface area of the box?

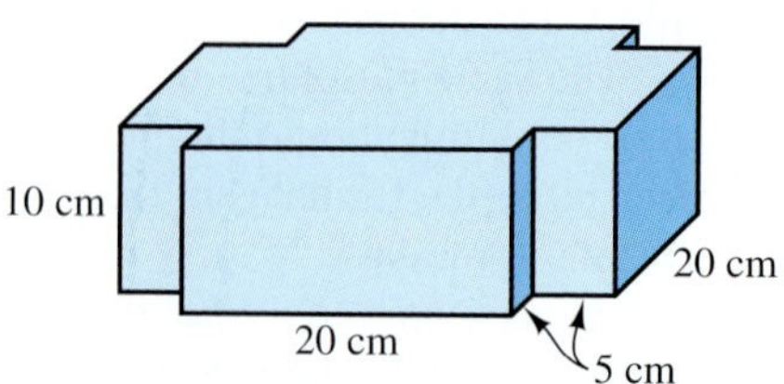

Solution

Each base has the area $B = 800\ \text{cm}^2$. The lateral surface area is that of a rectangle 10 cm high and 120 cm long. That is, the lateral surface area is $1200\ \text{cm}^2$. Altogether, the surface area of the box is $SA = 2 \times 800\ \text{cm}^2 + 1200\ \text{cm}^2 = 2800\ \text{cm}^2$.

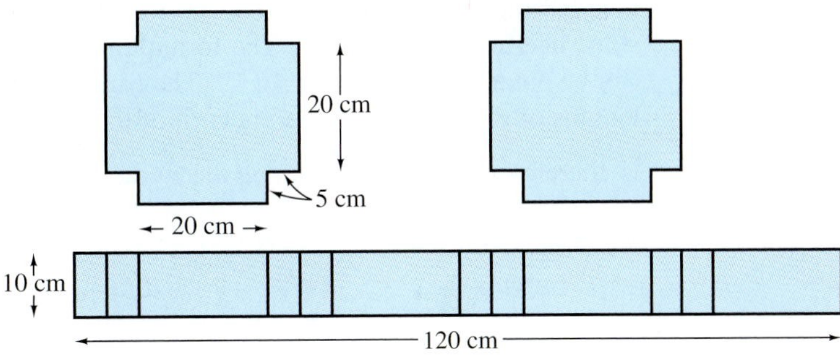

EXAMPLE 12.20 FINDING THE SURFACE AREA OF A CYLINDRICAL JUICE CAN

A small can of frozen orange juice is about 9.5 cm tall and has a diameter of about 5.5 cm. The circular ends are metal, and the rest of the can is cardboard. How much metal and how much cardboard are needed to make a juice can?

Solution

The rectangle below is 9.5 cm wide and 5.5π cm long, so the area of cardboard is $(9.5) \cdot (5.5)\pi\ \text{cm}^2 = 52.25\pi\ \text{cm}^2$, or about $164\ \text{cm}^2$. The circles have a radius of 2.75 cm, so each circle has area $\pi(2.75\ \text{cm})^2$. Twice this is $15.125\pi\ \text{cm}^2$, so the area of the two metal ends is about $47.5\ \text{cm}^2$.

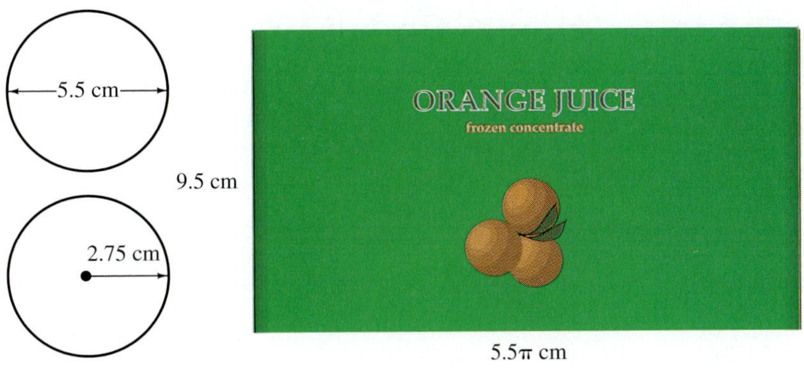

Surface Area of Pyramids

The surface area of a pyramid is computed by adding the area of the base to the sum of the areas of the triangles forming the lateral surface of the pyramid. Of special importance is the **right regular pyramid,** for which the base is a regular polygon and the lateral surface is formed by congruent isosceles triangles. The altitude of the triangles is called the **slant height** of the pyramid. The formula for the surface area of a right regular pyramid can be obtained from Figure 12.19. The triangles each have altitude s, and the sum of the lengths of the bases is the perimeter p of the base polygon. The total area of the triangles is therefore $\frac{1}{2}ps$. If the base of the pyramid has area B, then the total surface area is $SA = B + \frac{1}{2}ps$.

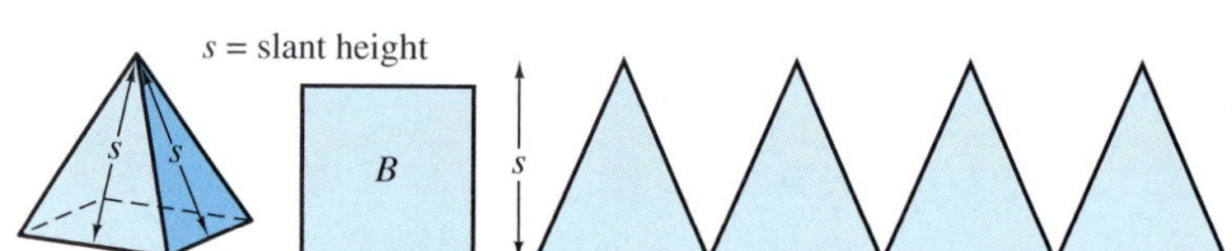

Figure 12.19

A right regular pyramid has total surface area $SA = B + \frac{1}{2}ps$

FORMULA *Surface Area of Right Regular Pyramid*

Let a right regular pyramid have slant height s, and a base of area B and perimeter p. Then the surface area SA of the pyramid is given by the formula

$$SA = B + \frac{1}{2}ps.$$

Once again, the reasoning that leads to the formula is much more important than the formula itself.

EXAMPLE 12.21 FINDING THE SURFACE AREA OF A RIGHT REGULAR PYRAMID

A pyramid has a square base that is 10 cm on a side. The edges that meet at the apex have length 13 cm. Find the slant height of the pyramid, and then calculate the total surface area (including the base) of the pyramid.

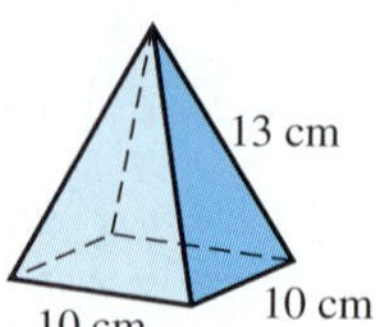

Solution

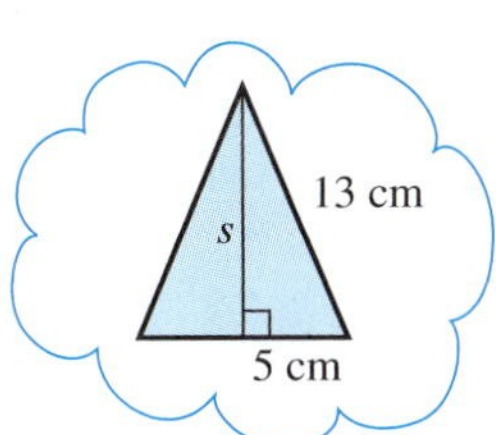

It is clear that the base of the pyramid has area $B = 100 \text{ cm}^2$ and perimeter $p = 40$ cm. The slant height can be calculated with the Pythagorean theorem, which shows that

$$s = \sqrt{13^2 - 5^2} = \sqrt{169 - 25} = \sqrt{144} = 12 \text{ cm}.$$

Thus, the surface area of the pyramid is $SA = 100 \text{ cm}^2 + \frac{1}{2} \cdot 40 \cdot 12 \text{ cm}^2 = 340 \text{ cm}^2$.

Surface Area of Right Circular Cones

Consider a right circular cone of slant height s and with a circular base of radius r. The formula for the surface area of the cone can be derived from the surface area formula for a pyramid. To see how, imagine that the cone is closely approximated by a right regular pyramid. An example is shown in Figure 12.20, where a cone is approximated by a pyramid with a dodecagon (12-gon) as its base.

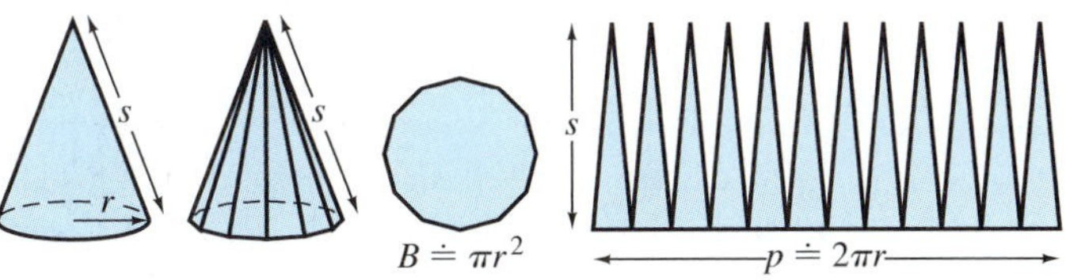

Figure 12.20
A right circular cone has total surface area SA $= \pi r^2 + \pi r s$

The area of the circular base of the pyramid is $B \doteq \pi r^2$, and the perimeter of the base is $p \doteq 2\pi r$. Therefore, the surface area of the pyramid is $SA = B + \frac{1}{2}ps \doteq \pi r^2 + \pi r s$. The pyramid's approximation to the cone becomes increasingly exact as the number of sides in the base polygon is increased, giving us the following formula.

FORMULA *Surface Area of a Right Circular Cone*
Let a right circular cone have slant height s and a base of radius r. Then the surface area SA of the cone is given by the formula

$$SA = \pi r^2 + \pi r s.$$

EXAMPLE 12.22 **FINDING THE LATERAL SURFACE AREA OF AN ICE CREAM CONE**

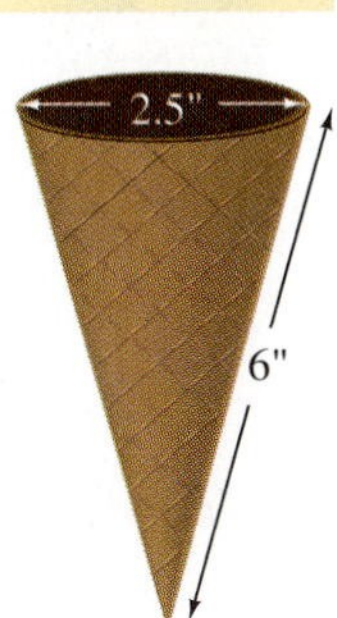

An ice cream cone has a diameter of 2.5 inches and a slant height of 6 inches. What is the lateral surface area of the cone?

Solution

Only the lateral surface area is required, so we need to compute $\pi r s$, where $r = 2.5/2$ inches $= 1.25$ inches and $s = 6$ inches. The lateral surface area is therefore $\pi(1.25 \text{ in}) \cdot (6 \text{ in}) = \pi \cdot (7.5 \text{ in}^2)$.

Using 3.14 to approximate π, we find that the lateral surface area is about 23.6 square inches.

Volumes of Right Prisms and Right Cylinders

The volume of the rectangular box shown in Figure 12.21 is given by lwh, where l, w, and h are, respectively, the length, width, and height of the box. Since lw gives the area B of the base of the box, the volume can also be written in the form $V = Bh$, where B is the area of the base and h is the height.

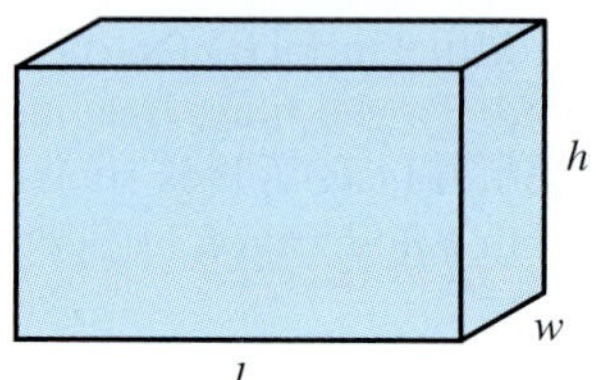

Figure 12.21
A rectangular box has volume $V = lwh$. *Equivalently,* $V = Bh$, *where* B *is the area of the base and* h *is the height.*

Figure 12.22(a) shows a solid composed of many (say, n) small right rectangular prisms, all of the height h. If $B_1, B_2, \ldots, B_n$ are the areas of the bases, the total volume V of the prisms is $B_1h + B_2h + \cdots + B_nh = (B_1 + B_2 + \cdots + B_n)h$. That is, the volume V is given by $V = Bh$, where $B = B_1 + B_2 + \cdots + B_n$ is the total area of the base. The right cylinder depicted in Figure 12.22(b) can be approximated to arbitrary accuracy by prisms of height h as shown.

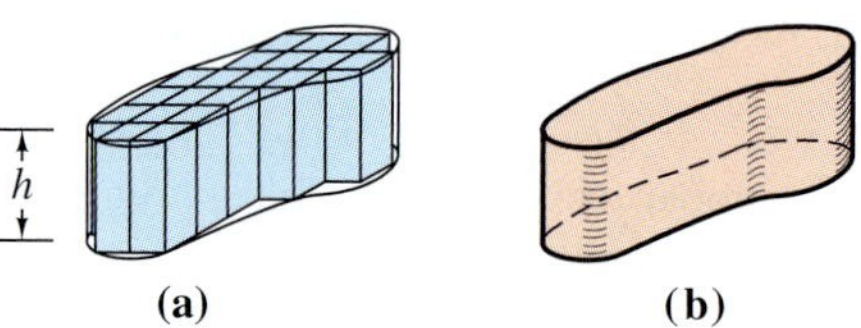

Figure 12.22
Right rectangular prisms can approximate a right prism or a right cylinder

FORMULA *Volume of a Right Prism or a Right Cylinder*
Let a right prism or right cylinder have height h and a base of area B. Then its volume V is given by

$$V = Bh.$$

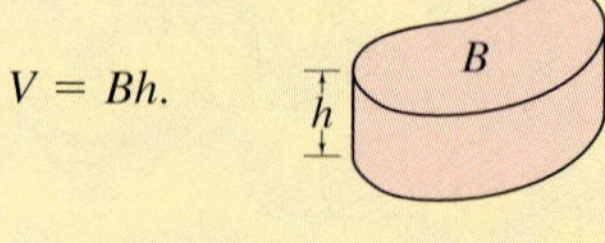

B = area of base

EXAMPLE 12.23 COMPUTING THE VOLUME OF A RIGHT PRISM AND A RIGHT CYLINDER

Find the volume of the gift box and the juice can.

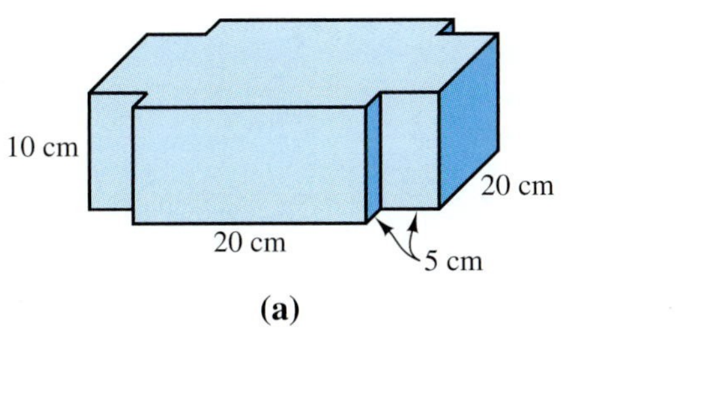

Solution

(a) The base area, $B = 800\text{ cm}^2$, of the gift box was calculated in Example 12.19. The height is $h = 10$ cm, so the volume is $V = Bh = 8000\text{ cm}^3$. This can also be expressed as 8 liters.

(b) The area of the circular base of the juice can is $\pi(2.75\text{ cm})^2 = 7.5625\pi\text{ cm}^2$. The height is $h = 9.5$ cm, so the volume $V = Bh = 71.84375\pi\text{ cm}^3$, or about 226 cm^3.

Volumes of Oblique Prisms and Cylinders

A deck of neatly stacked playing cards forms a right rectangular prism as shown in Figure 12.23. The total volume of the deck is the sum of the volumes of each card. If the cards slide easily on one another, it is easy to tilt the deck to form an oblique prism of the same base area B and the same height h. The solid is still made up of the same cards, so the oblique prism still has volume Bh.

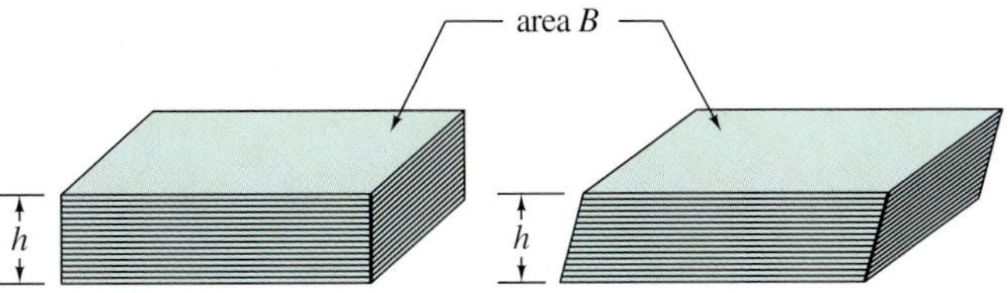

Figure 12.23
An oblique prism of base area B and height h has the same volume V = Bh as the corresponding right prism

Any oblique prism or cylinder can be imagined as a stack of very thin cards, all shaped like the base of the solid. With no change of volume, the oblique stack can be straightened to form a right prism or right cylinder of the same height h and base area B. Both the right and oblique shapes therefore have the same volume, namely, $V = Bh$. This is illustrated in Figure 12.24.

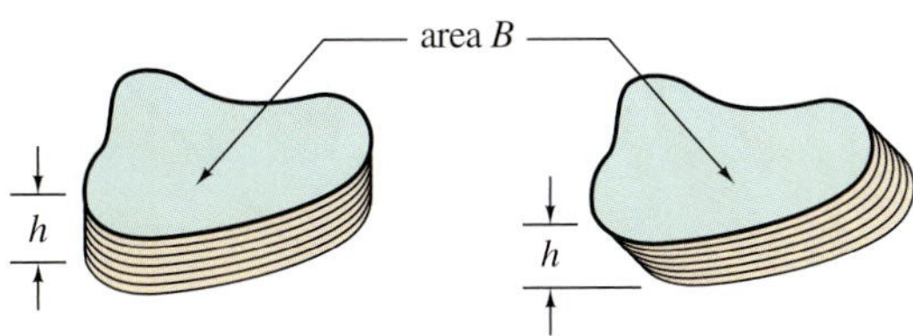

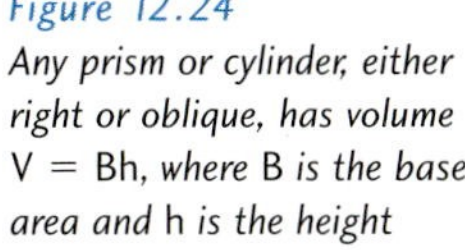

Figure 12.24
Any prism or cylinder, either right or oblique, has volume V = Bh, where B is the base area and h is the height

Into the Classroom

Problem Solving with Measurement

Each pair of students has an orange and a sheet of centimeter-squared graph paper. The following challenge is then given: *Find the area of the peel of the orange.* Students well versed in the basic principles of measurement may solve the problem in a direct, yet appropriate, way: the orange is peeled, and then the peeling is cut or torn into small pieces to tile a region of the graph paper; the region's boundary is traced, and then its area, which equals that of the orange peel, is estimated by counting the number of square centimeters covered.

Problem solving with measurement reinforces both the principles and the processes of measurement. On the other hand, overemphasis on exercises that require only a routine application of a formula reduces measurement to a mechanistic level. Here are two examples illustrating the difference between a routine exercise and a problem.

Exercise: A right triangle has legs of length 6″ and 10″. What is the area of the triangle?

Problem: Two straws, of lengths 6″ and 10″, are joined with paper clips to form a flexible hinge. At what angle should the straws meet to form the sides of a triangle of largest possible area?

Exercise: A rectangular solid has length 6 cm, width 2 cm, and height 2 cm. What is the surface area of the solid?

Problem: The Math Manipulative Supply House sells wooden centimeter cubes in sets of 24 cubes. What shape of a rectangular box that holds one set of cubes requires the least amount of cardboard?

FORMULA *Volume of a General Prism or Cylinder*

A prism or cylinder of height h and base area B has volume $V = Bh$.

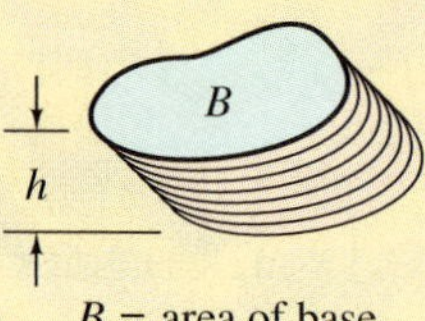

B = area of base

Volumes of Pyramids and Cones

In two-dimensional space (that is, in the plane), a diagonal dissects a square into two congruent right triangles. Thus, the area of each triangle is one-half that of the corresponding square.

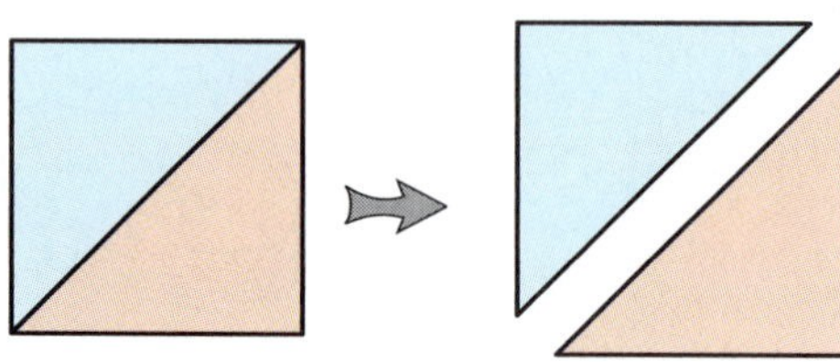

In three-dimensional space, the diagonals from one corner of a cube form the edges of three congruent pyramids that fill the cube. Therefore, each pyramid has one-third the volume of the corresponding cube, as shown in Figure 12.25.

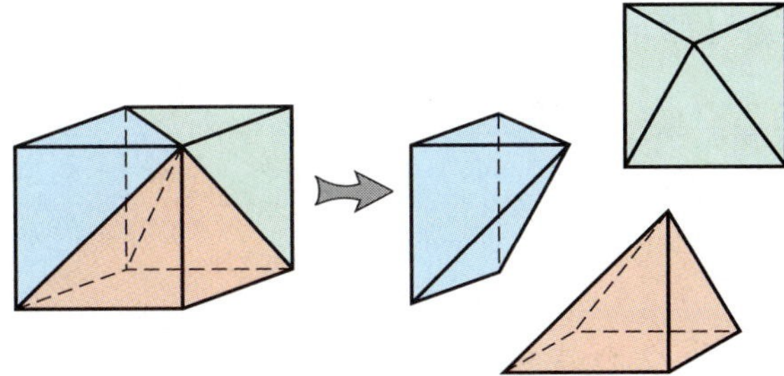

Figure 12.25
A cube can be dissected into three congruent pyramids

Instead of a cube, consider a rectangular solid, and use the diagonals from one corner to decompose the solid into three pyramids. An example is shown in Figure 12.26. In general, the three pyramids are not congruent to one another. However, it can be shown that the volumes of the three pyramids are equal. Therefore, if the prism has base area B and height h, we conclude that each pyramid has volume $\frac{1}{3}Bh$.

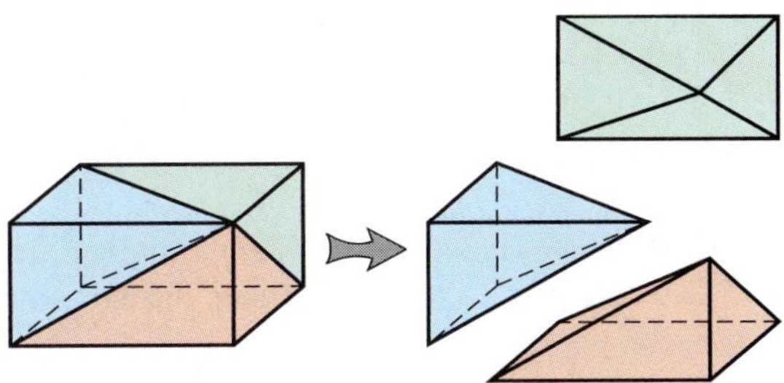

Figure 12.26 A rectangular prism can be dissected into three pyramids of equal volume

Similar reasoning shows that *all* pyramids of base B and height h have volume $\frac{1}{3}Bh$. The base can be any polygon, and the apex can be any point at distance h to the plane of the base, as shown in Figure 12.27.

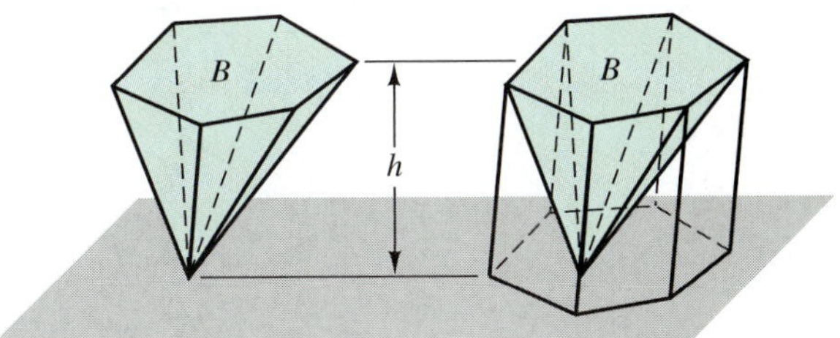

Figure 12.27
A pyramid of height h and base of area B has one-third the volume of a corresponding prism of base area B and height h. Therefore, the pyramid has volume $\frac{1}{3}$Bh

The base of a cone can be approximated to arbitrary accuracy by a polygon with sufficiently many sides, so the volume of a cone of base area B and height h is also given by the formula $\frac{1}{3}Bh$.

FORMULA *Volume of a Pyramid or Cone*
The volume V of a pyramid or cone of height h and base of area B is given by

$$V = \frac{1}{3}Bh.$$

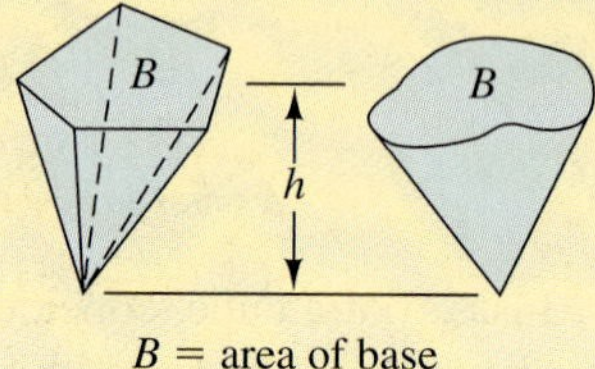

EXAMPLE 12.24 DETERMINING THE VOLUME OF AN EGYPTIAN PYRAMID

The pyramid of Khufu is 147 m high, and its square base is 231 m on each side. What is the volume of the pyramid?

Solution

The area of the base is $(231 \text{ m})^2 = 53{,}361 \text{ m}^2$. Therefore, the volume is

$$\frac{1}{3}(53{,}361 \text{ m}^2) \cdot (147 \text{ m}) = 2{,}614{,}689 \text{ m}^3.$$

If the stones were stacked on a football field, a rectangular prism nearly 2000 feet high would result. For comparison, the 110-story Sears Tower in Chicago reaches 1454 feet.

Volume of a Sphere

Suppose that a solid sphere of radius r is placed in the right circular cylinder of height $2r$ that just contains it. Filling the remaining space in the cylinder with water, we find that removing the sphere leaves the cylinder one-third full, as illustrated in Figure 12.28. This means that the sphere takes up two-thirds of the volume of the cylinder. Since the volume of the cylinder is $Bh = (\pi r^2)(2r) = 2\pi r^3$, the experiment suggests that the volume of a sphere of radius r is given by $\frac{2}{3}(2\pi r^3) = \frac{4}{3}\pi r^3$. The first rigorous proof of this remarkable formula was given by Archimedes.

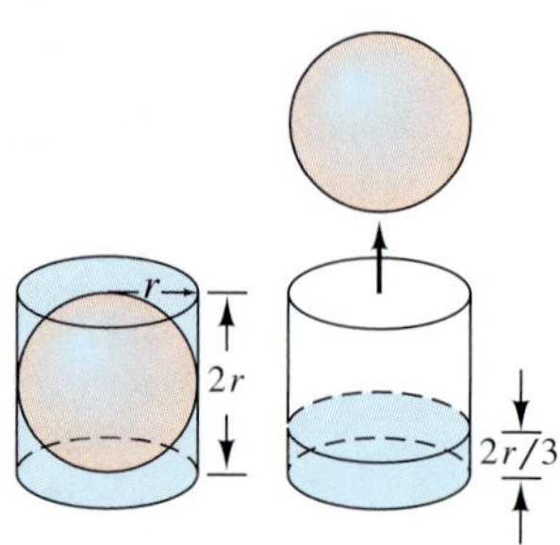

Figure 12.28
A sphere fills two-thirds of the circular cylinder containing the sphere

FORMULA *Volume of a Sphere*
The volume V of a sphere of radius r is given by the formula

$$V = \frac{4}{3}\pi r^3.$$

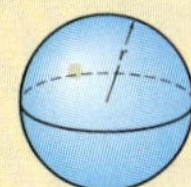

EXAMPLE 12.25 USING THE SPHERE VOLUME FORMULA

An ice cream cone is 5 inches high and has an opening 3 inches in diameter. If filled with ice cream and given a hemispherical top, how much ice cream is there?

Solution The hemisphere has radius 1.5 inches, so its volume is $\frac{2}{3}\pi(1.5 \text{ in})^3 = 2.25\pi \text{ in}^3$. The cone has volume $\frac{1}{3}Bh = \frac{1}{3}\pi(1.5 \text{ in})^2 \cdot (5 \text{ in}) = 3.75\pi \text{ in}^3$. Thus, the total volume is $2.25\pi \text{ in}^3 + 3.75\pi \text{ in}^3 = 6\pi \text{ in}^3$, or about 19 in^3. Since a gallon is 231 in^3, we see that the cone holds very close to a third of a quart of ice cream.

HIGHLIGHT *from* HISTORY

Sophie Germain (1776–1831)

Sophie Germain grew up in a time of social, political, and economic upheaval in France. To shield Sophie from the violence in the streets of Paris during the time of the fall of the Bastille, her wealthy parents confined their 13-year-old daughter to the family's library. There, she chanced upon J. E. Montucla's *History of Mathematics,* which recounts the legend of Archimedes' death. The story tells how a Carthaginian soldier, heedless of orders to spare the renowned mathematician, killed the unsuspecting Archimedes, who remained absorbed in a geometry problem. Sophie wished to explore for herself a subject of such compelling interest.

Sophie's family initially resisted her determination to study mathematics, but eventually they gave her the freedom to follow her intellectual instincts. Since women were not permitted to enroll in the École Polytechnique, which opened in Paris in 1794, Sophie resorted to collecting lecture notes from various professors at the university. Her absence of a formal mathematical education was compensated by her courage to overcome strenuous challenges.

Sophie's early research was in number theory. She corresponded regularly with the great Carl Friedrich Gauss, who gave her work high praise. At the turn of the century, Sophie turned her attention increasingly to the mathematical theory of vibrating elastic surfaces. Her prizewinning paper on vibrating elastic plates in 1816 placed her in the ranks of the most celebrated mathematicians of the time. Gauss recommended that she be awarded an honorary doctorate from the University of Göttingen, but unfortunately Sophie Germain's death came too soon for the awarding of the degree.

The Surface Area of a Sphere

A formula for the surface area of a sphere of radius r can be discovered by dividing the sphere's surface into many (say, n) tiny regions of area B_1, B_2, B_3, . . . , B_n. The sum $B_1 + B_2 + B_3 + \cdots + B_n$ is the surface area S of the sphere. Each region can also be viewed as the base of a pyramidlike solid whose apex is the center of the sphere. Each of the pyramids has height r, so the pyramids have volumes $\frac{1}{3}B_1r$, $\frac{1}{3}B_2r$, $\frac{1}{3}B_3r$, and so on. An example is shown in Figure 12.29.

$S = B_1 + B_2 + B_3 + \cdots + B_n$

area B_1
area B_2
area B_3
r
r

$V = \frac{1}{3}B_1r + \frac{1}{3}B_2r + \frac{1}{3}B_3r + \cdots + \frac{1}{3}B_nr$

Figure 12.29
A solid sphere can be viewed as made up of pyramidlike pieces

We have the following relationships:

$$S = B_1 + B_2 + B_3 + \cdots + B_n \qquad \text{(surface area of the sphere)}$$

and

$$V = \frac{1}{3}B_1r + \frac{1}{3}B_2r + \frac{1}{3}B_3r + \cdots + \frac{1}{3}B_nr. \qquad \text{(volume of the sphere)}$$

Therefore, since $\frac{r}{3}$ is a common factor,

$$V = \frac{r}{3}(B_1 + B_2 + B_3 + \cdots + B_n) = \frac{r}{3}S.$$

Since $V = \frac{4}{3}\pi r^3$, this gives us the equation

$$\frac{4}{3}\pi r^3 = \frac{r}{3}S.$$

Multiplying both sides by 3, and dividing both sides by r, we can solve for the surface area S.

FORMULA *Surface Area of a Sphere*
The surface area S of a sphere of radius r is given by the formula

$$S = 4\pi r^2.$$

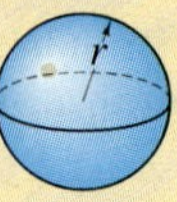

EXAMPLE 12.26 COMPARING EARTH TO JUPITER

The diameter of Jupiter is about 11 times larger than the diameter of our planet Earth. How many times greater is **(a)** the surface area of Jupiter? **(b)** the volume of Jupiter?

Solution

(a) Let r denote the radius of Earth and R the radius of Jupiter. Therefore, $R = 11r$. Using the formula for the surface area of a sphere, we find that the ratio of the surface area of Jupiter to that of Earth is

$$\frac{4\pi R^2}{4\pi r^2} = \frac{R^2}{r^2} = \left(\frac{R}{r}\right)^2 = (11)^2.$$

That is, the surface area of Jupiter is 11^2, or 121, times the surface area of Earth.

(b) Using the sphere volume formula, we find that the ratio of volumes is

$$\frac{\frac{4}{3}\pi R^3}{\frac{4}{3}\pi r^3} = \frac{R^3}{r^3} = \left(\frac{R}{r}\right)^3 = (11)^3.$$

Therefore, the volume of Jupiter is about 11^3, or 1331, times the volume of Earth. Precise measurements of the two not quite spherical planets show that the volume ratio is 1323.3, which agrees closely with 1331.

Comparing Measurements of Similar Figures

Two figures are similar if they have the same shape, but possibly different size. The ratio of all pairs of corresponding lengths in the two figures is a constant value called the scale factor, which we will denote by the letter k. The ratio of *any linear measurement* of the two figures—perimeter, height, diameter, slant height, and so on—is also the scale factor k. For example, since the diameter of Jupiter is 11 times the diameter of Earth, the scale factor is $k = 11$. We then also know that the equator of Jupiter is 11 times as long as Earth's equator.

The ratio of *areas* of similar figures is given by the *square*, k^2, of the scale factor k. The ratio of volumes is given by the *cube*, k^3, of the scale factor. This basic fact is evident for the cubes shown in Figure 12.30.

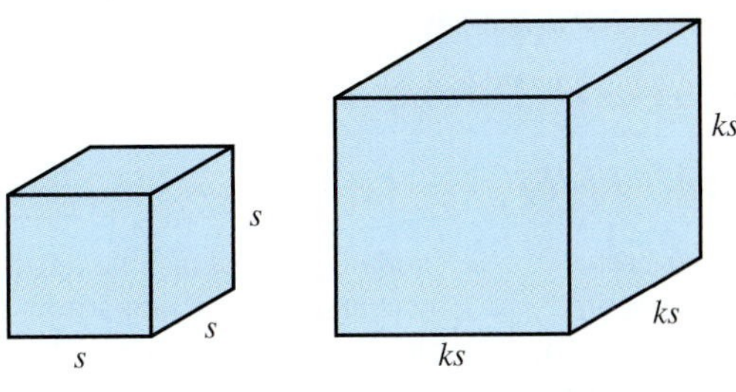

Scale factor = k

	Cube I	Cube II
Length of edge	s	ks
Area of each face	s^2	$k^2 s^2$
Volume	s^3	$k^3 s^3$

Figure 12.30 Area varies by the square, k^2, of the scale factor k, and volume varies by the cube, k^3, of the scale factor

The same comparison of areas and volumes holds for any pair of similar figures, not just the cube shown in Figure 12.30. The following theorem is an important principle for the comparison of the measurements of similar figures.

THEOREM *The Similarity Principle of Measurement*

Let Figures I and II be similar. Suppose some length dimension of Figure II is k times the corresponding dimension of Figure I; that is, k is the scale factor. Then

(i) *any* length measurement—perimeter, diameter, height, slant height, and so on—of Figure II is k times that of the corresponding length measurement of Figure I;

(ii) *any* area measurement—surface area, area of a base, lateral surface area, and so on—of Figure II is k^2 times that of the corresponding area measurement of Figure I;

(iii) *any* volume measurement—total volume, capacity, half-full, and so on—of Figure II is k^3 times the corresponding volume measurement of Figure I.

EXAMPLE 12.27 USING THE SIMILARITY PRINCIPLE

(a) Television sets are measured by the length of the diagonal of the rectangular screen. How many times larger is the screen area of a 40-inch model than that of a 13-inch table model?

(b) A 2″ by 4″ by 8″ rectangular brick of gold weighs about 44 pounds. What are the dimensions of a similarly shaped brick that weighs 10 pounds?

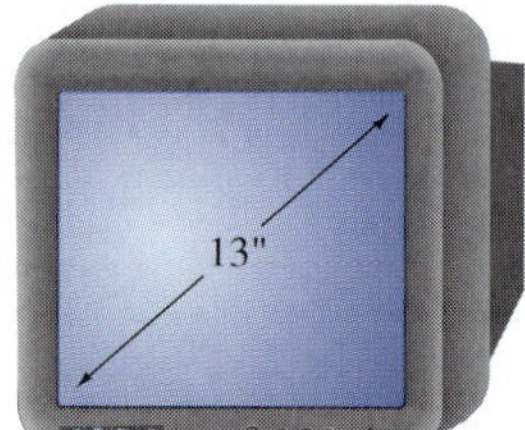

Solution

(a) The scale factor k is $40/13 \doteq 3.08$. Since area varies by $k^2 \doteq (3.08)^2 \doteq 9.5$, the large screen has about 9.5 times the area of the similarly shaped small screen.

2" 4" 8"

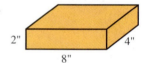

(b) The weight of a gold brick is proportional to its volume, and the volume of similarly shaped bricks varies by the factor k^3, the cube of the scale factor k. Therefore $10 = k^3 44$, so $k^3 = 10/44$, and $k = (10/44)^{1/3} \doteq 0.6$. Multiplying the dimensions of the 44-pound brick by 0.6 gives the approximate dimensions of a similarly shaped 10-pound brick of gold, namely, 1.2″ by 2.4″ by 4.8″.

HIGHLIGHT *from* HISTORY

Jonathan Swift (1667–1745)

Jonathan Swift published his best-known book, *Gulliver's Travels,* in England in 1726. This book has been popular for two-and-a-half centuries because it is both a delightful story about pygmies and giants and a clever satire of many eighteenth-century institutions and scholars. *Gulliver's Travels* stresses the relative contrasts of the pygmies of Lilliput and the giants of Brobdingnag. Swift points out that "nothing is great or little otherwise than by comparison." Comparisons can be fun: if a Lilliputian is 6 inches tall, about how long would his or her shoes be? What is the approximate weight of a Lilliputian? Actually, since strength and weight vary by the very different scale factors of k^2 and k^3, it is impossible for Brobgingnagians and Lilliputians to be similar to Gulliver. The classic essay on this topic is J. B. S. Haldane's essay "On Being the Right Size" (in *Possible Worlds*, New York: Harper, 1928).

Did You Know?

Sizing Up the Universe

We all, children and grownups alike, are inclined to live in our own little world, in our immediate surroundings, or at any rate with our attention concentrated on those things with which we are directly in touch. We tend to forget how vast are the ranges of existing reality which our eyes cannot directly see, and our attitudes may become narrow and provincial. We need to develop a wider outlook, to see ourselves in our relative position in the great and mysterious universe in which we have been born and live.

At school we are introduced to many different spheres of existence, but they are often not connected with each other, so that we are in danger of collecting a large number of images without realizing that they all join together in one great whole. It is therefore important in our education to find the means of developing a wider and more connected view of our world and a truly cosmic view of the universe and our place in it.

This book presents a series of forty pictures composed so that they may help to develop this wider view. They really give a series of views as seen during an imaginary and fantastic journey through space—a journey in one direction, straight upward from the place where it begins. Although these views are as true to reality as they can be made with our present knowledge, they portray a wonderland as full of marvels as that which Alice saw in her dreams.

Kees Boeke, a sixth-grade teacher in Holland, worked with the children in his class to make a picture book that takes the reader on an imaginary journey through the universe. Going from the picture on one page to that on the next page changes the scale of view by a factor of 10, so in 40 jumps the journey moves from the gamma ray to dots representing clumps of distant galaxies. Two similar books that more fully develop Boeke's idea are *Powers of Ten,* by Philip and Phyllis Morrison (San Francisco: W. H. Freeman & Co., 1994), and *Powers of Ten Flipbook,* by Charles and Ray Eames (San Francisco: W. H. Freeman & Co., 1998). Web sites about powers of ten include www.powersof10.com and www.wordwizz.com/pwrsof10.htm.

SOURCE: From Cosmic View: The Universe in 40 Jumps, *by Kees Boeke, with an introduction by Arthur H. Compton. Copyright ® 1957 by Kees Boeke. By permission of Harper Collins.*

Problem Set 12.4

Understanding Concepts

1. Identify which measure—surface area or volume—is most important to know

(a) about a room, to buy the correct amount of paint;

(b) about a chair, to determine the amount of stuffing required to reupholster the chair;

(c) about a swimming pool, to add the correct amount of chlorine;

(d) about a lawn, to apply the correct amount of weed killer.

2. Find the surface area of each of these prisms and cylinders.

(a)

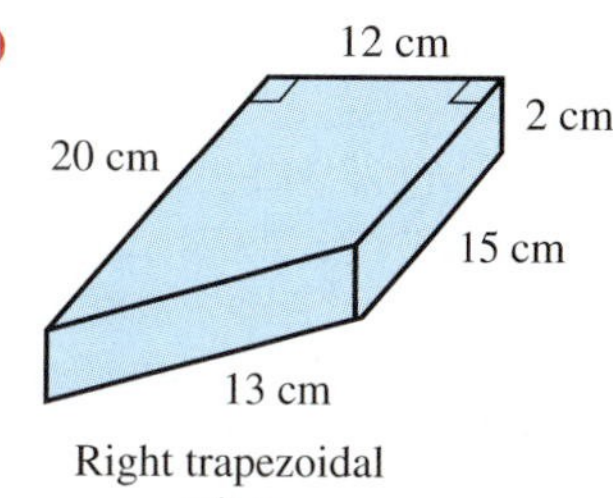

Right trapezoidal prism

(b)

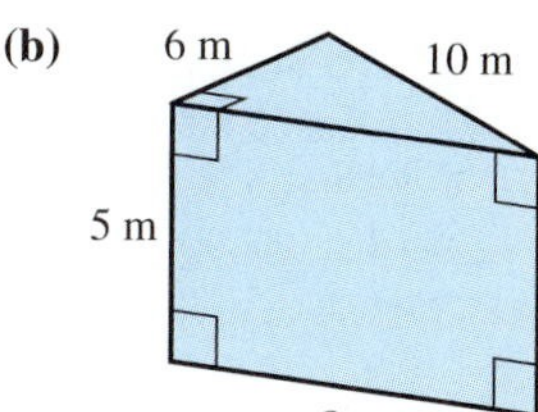

Right triangular prism

(c)

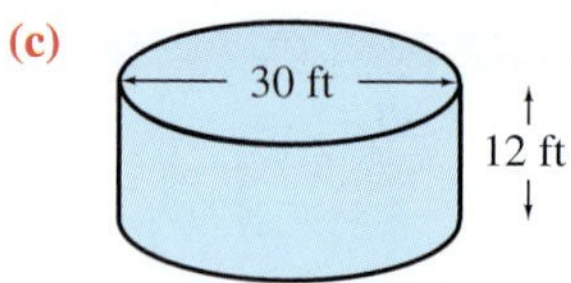

Right circular cylinder

(d)

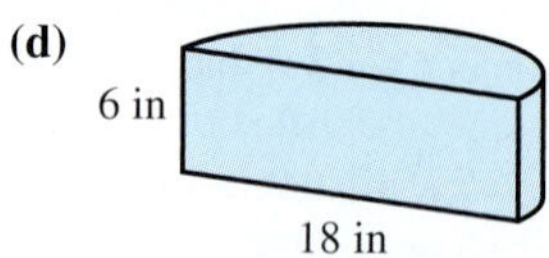

Semicircular right prism

3. Find the surface area of each of these right regular pyramids and right circular cones.

(a)

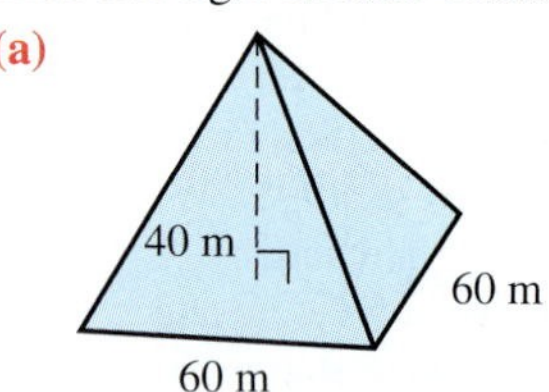

Right square pyramid

(b)

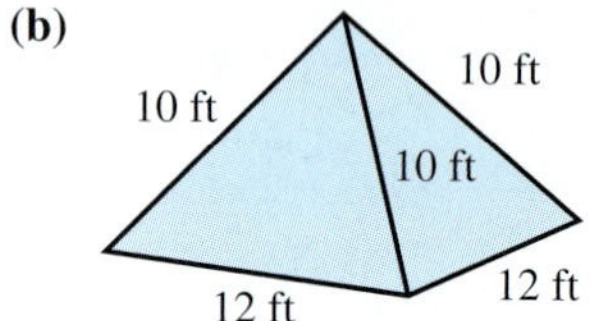

Right square pyramid

(c)

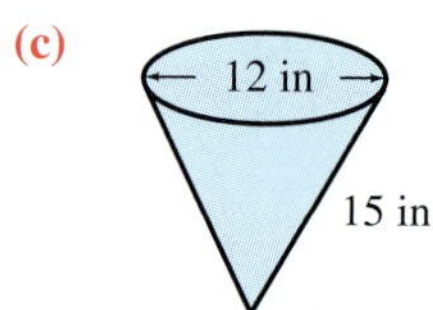

Right circular cone

(d)

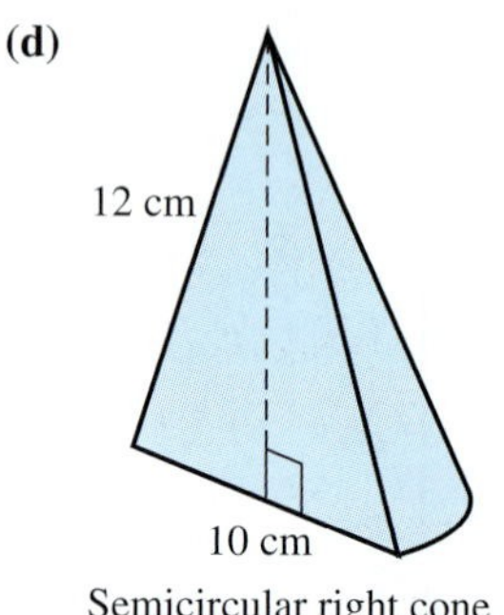

Semicircular right cone

4. Find the volume of each of these prisms and cylinders.

(a)

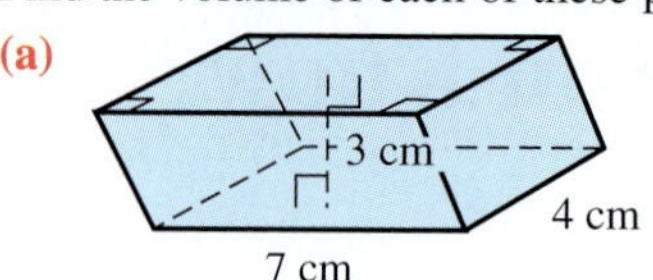

Oblique prism

(b)

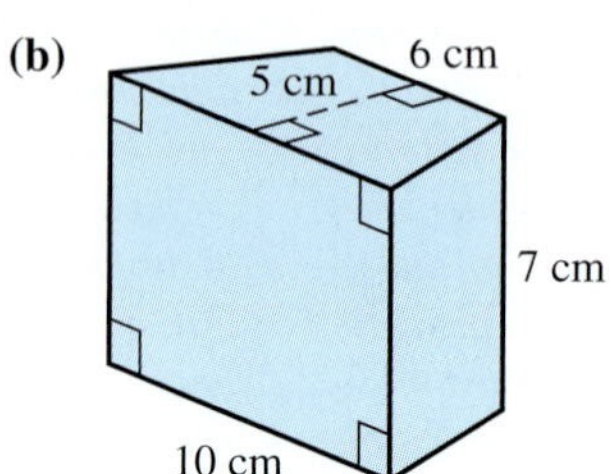

Trapezoidal right prism

(c)

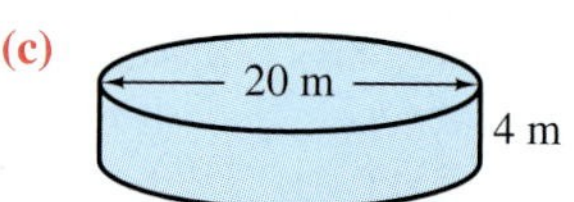

Right circular cylinder

(d)

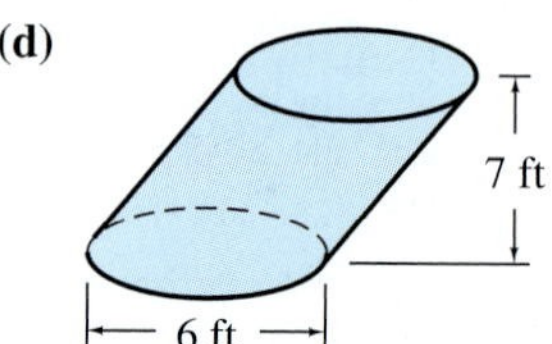

Oblique circular cylinder

5. Find the volume of each of these pyramids and cones.

(a)

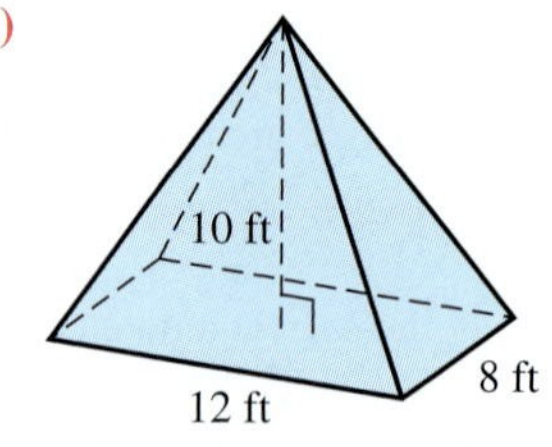

Rectangular pyramid

(b)

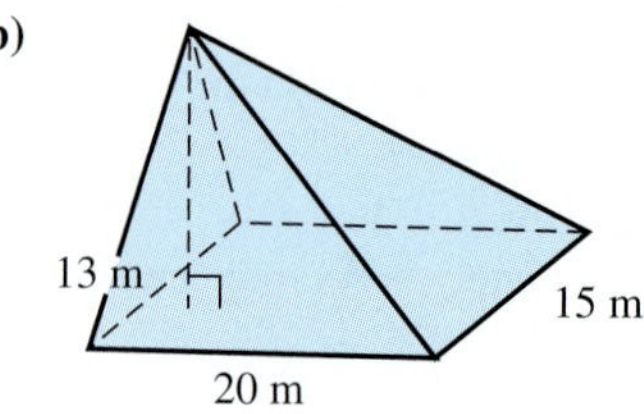

Rectangular pyramid

(c)

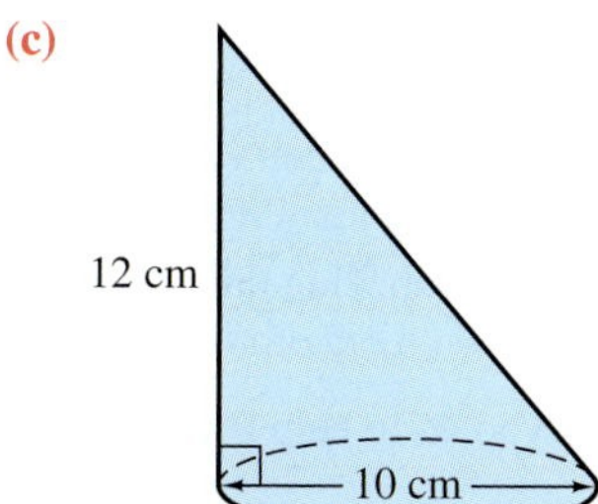

Oblique circular cone

(d)

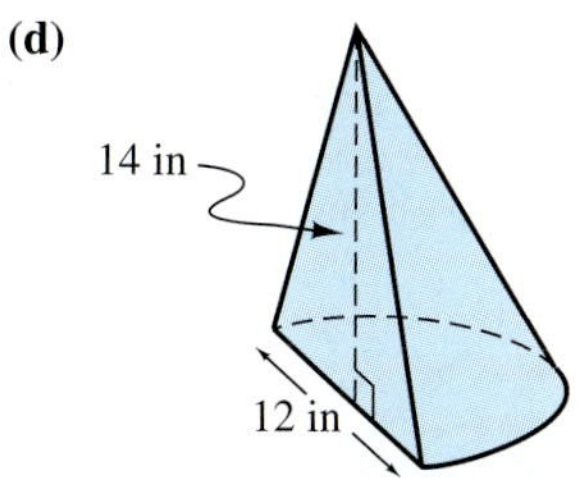

Semicircular cone

6. Find the surface area and volume of each of these solids.

(a)

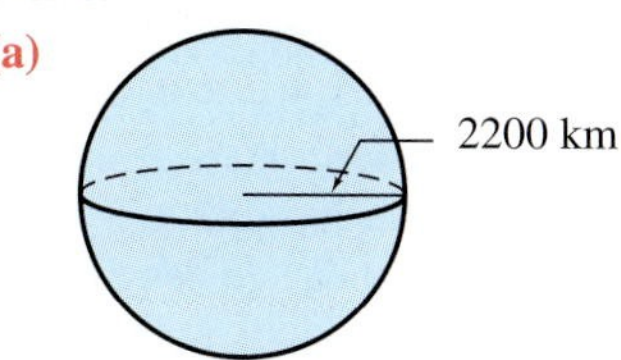

Sphere

(b)

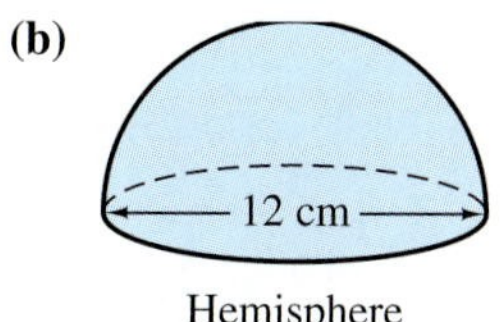

Hemisphere

(c)

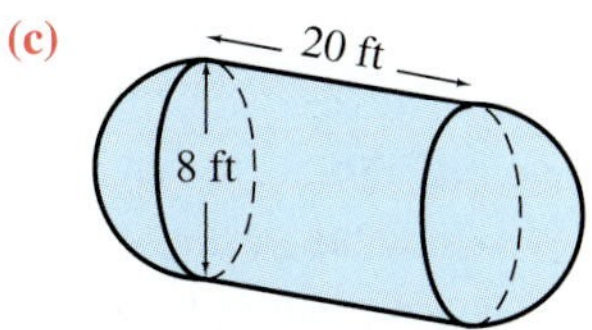

Cylindrical storage tank with hemispherical ends

(d)

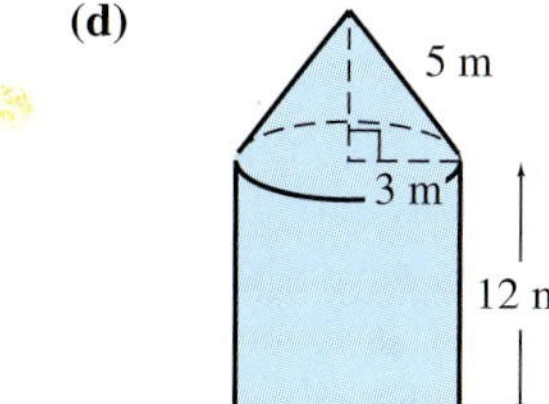

Cylindrical grain silo with conical top

7. An aluminum soda pop can has a diameter of 6.5 cm and a height of about 11 cm. If there are 30 milliliters in a fluid ounce, verify that the capacity of the can is 12 fluid ounces, as printed on the can's label.

8. If it takes a quart of paint to cover the base of a hemisphere, how many quarts does it take to paint the spherical part of the same hemisphere?

9. Archimedes showed that the volume of a sphere is two-thirds the volume of the right circular cylinder just containing the sphere. Show that the area of the sphere is also two-thirds the surface area of the cylinder. (Archimedes was so pleased with these discoveries that he requested that the figure shown be placed on his tombstone.)

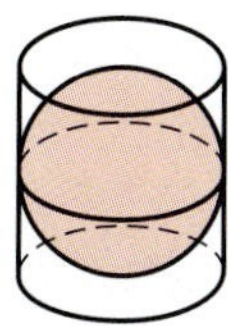

10. A square right regular pyramid is formed by cutting, folding, and gluing the following pattern.

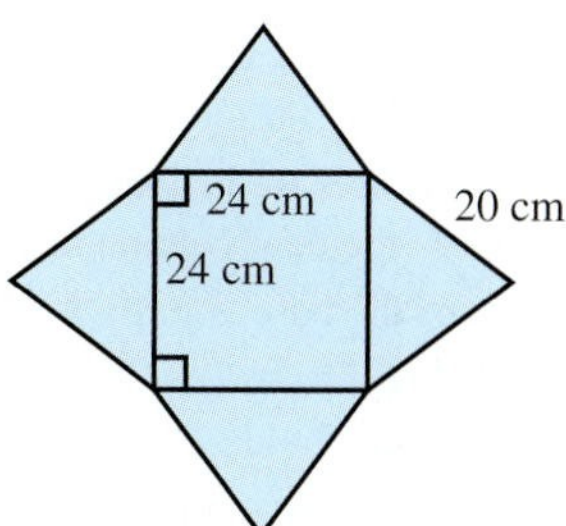

(a) What is the slant height of the pyramid?

(b) What is the lateral surface area of the pyramid?

(c) Use the Pythagorean theorem to find the height of the pyramid.

(d) What is the volume of the pyramid?

Use the similarity principle to answer problems 11 through 16. Explain carefully how the principle is used.

11. (a) An 8″ (diameter) pizza will feed one person. How many people will a 16″ pizza feed?

(b) Is it better to buy one 14″ pizza at \$10 or two 10″ pizzas at \$6 each? (*Hint:* 1.4^2 is about 2.0.)

12. (a) Eight spherical lead fishing sinkers are melted to form a single spherical sinker. If the small sinkers each have diameter of 1/4 inch, what is the diameter of the new large one?

(b) How many small sinkers would it take to make a sinker 1 inch in diameter?

13. What fraction of the area of the large circle is shaded in each figure?

(a)

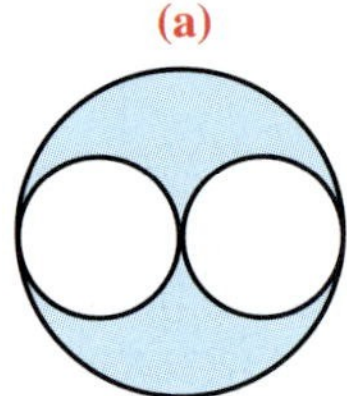

(b)

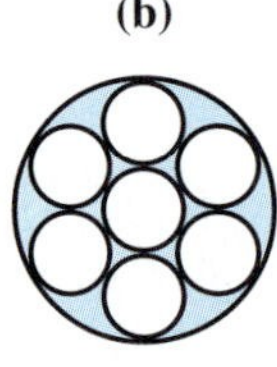

(*Hint:* First compare each unshaded circle with the large circle.)

14. Cones I, II, and III are similar to one another. Fill in the measurements left blank in the following table.

	I	II	III
Height	6	18	cm
Perimeter of base		30	15 cm
Lateral surface area	40		cm^2
Volume			10 cm^3

15. A cylindrical can holds 100 milliliters.

(a) If the radius of the base is doubled and the height halved, what is the new volume of the can?

(b) If the radius of the base is halved and the height is doubled, what is the new volume of the can?

16. A cube 10 cm on a side holds 1 liter.

(a) How many liters does a cube 20 cm on a side hold?

(b) What is the length of each side of a cube that holds 2 liters?

Teaching Concepts

17. Felipe is having trouble understanding why the formula for the volume of a rectangular box is $V = lwh$. What might you do to help Felipe?

18. Eno is having trouble understanding that the volume of a solid nonrectangular shape (say, a pyramid, cone, or sphere) can be given in terms of a unit cube. How would you help Eno understand that this can be so?

19. Ruth Ann is having trouble believing that the volume of *any* pyramid or cone is given by the formula $V = \frac{1}{3}Bh$. What might you do to help Ruth Ann?

Thinking Critically

20. A right circular cone has height r and a circular base of radius $2r$. Compare the volume of the cone to that of a sphere of radius r. Sketch both solids, using the same scale.

21. The right circular cylinder and cone shown in the accompanying figure both have a base of radius r and height $2r$, and the sphere has radius r.

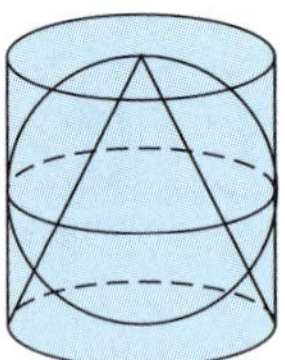

(a) Show that the ratio of the volumes of the cone to the sphere to the cylinder is 1 to 2 to 3.

(b) Show that the ratio of the surface areas of the cone to the sphere to the cylinder is τ to 2 to 3, where $\tau = (1 + \sqrt{5})/2$ is the famous golden ratio.

22. A birthday cake has been baked in a 7″ by 7″ by 2″ pan. Frosting covers the top and sides of the cake.

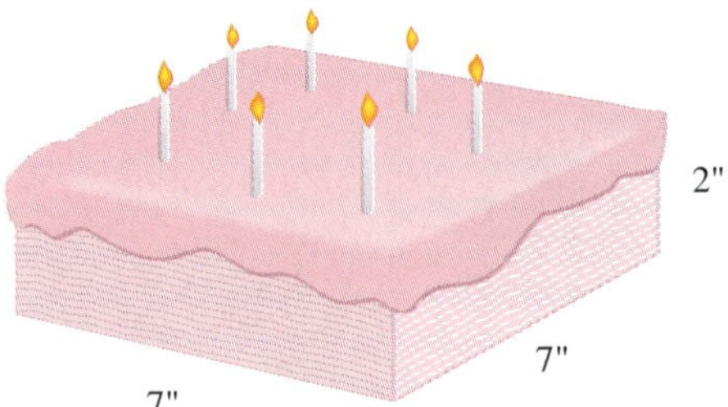

(a) What is the volume of the cake?

(b) What is the area covered by the frosting?

(c) Describe how to cut the cake into eight pieces, so that each piece has the same size (measured by volume) *and* the same amount of frosting (measured by area covered with frosting).

(d) Describe how to cut the cake in seven pieces, each with the same size and amount of frosting. (*Hint:* Consider a slice made by two vertical cuts from the center that intercepts 4 inches of the perimeter.)

23. The cube *ABCDEFGH* with edges of length s contains the tetrahedron *ACEG*.

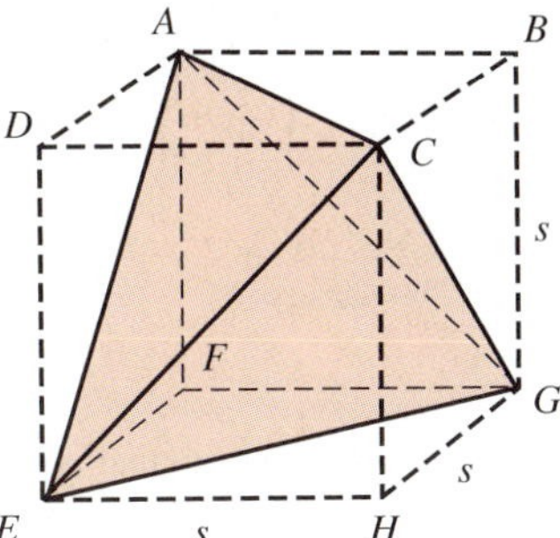

(a) Explain why *ACEG* is a regular tetrahedron with edges of length $\sqrt{2}s$.

(b) Show that the volume of tetrahedron *ACDE* is $\frac{1}{6}s^3$.

(c) Show that the volume of *ACEG* is $\frac{1}{3}s^3$.

(d) Use the similarity principle to explain why a regular tetrahedron with edges of length b has volume $\sqrt{2}b^3/12$.

24. Similar figures are erected on the three sides of a right triangle, as shown in the accompanying figure. What formula relates the areas of the three figures? Explain your reasoning carefully.

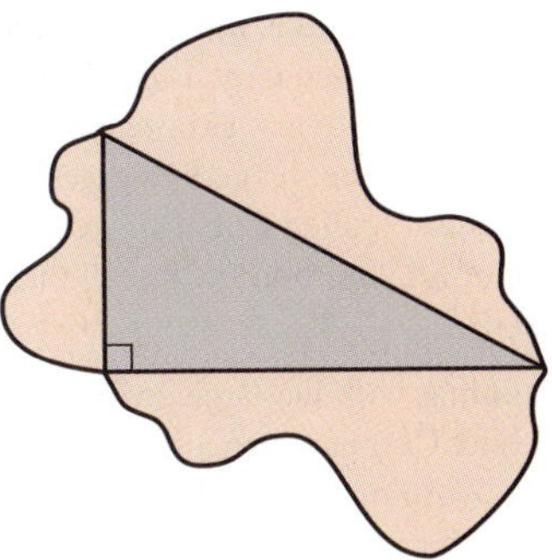

25. An ice cream soda glass is shaped like a cone of height 6 inches and has a capacity of 16 fluid ounces when filled to the rim. Use the similarity principle to answer the following questions, using the fact that the cone of liquid is similar to the cone of the entire region inside the glass.

(a) How high is the soda in the glass when it contains 2 fluid ounces?

(b) How much soda is in the glass when it is filled to a level 1 inch below the rim?

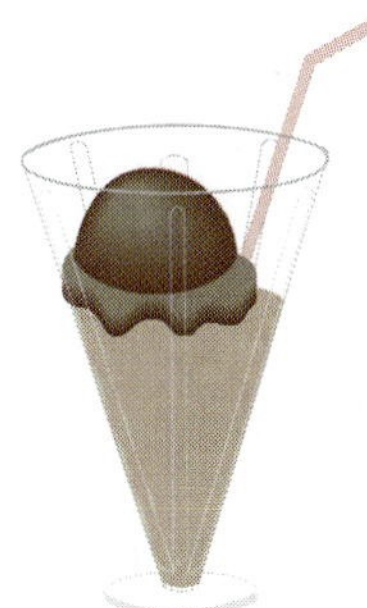

Thinking Cooperatively

In these problems, you will need paper, tape, scissors, and drawing tools (ruler, compass, and pencils). Work in pairs, measuring your constructed models and verifying your observations with calculations.

26. A pyramid is formed by joining a vertex of a cube of side length 8 cm to the four vertices of an opposite face.

(a) Use your drawing tools to accurately make a pattern that, when cut and folded, will form the pyramid. What are the lengths of each edge in your pattern? Use the Pythagorean theorem to find out. Give your answer both exactly, using square roots, and as a decimal approximation.

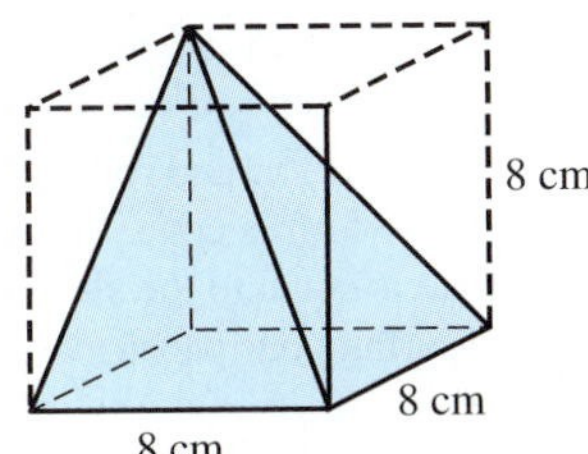

(b) Use the lengths of the edges found in part (a) to obtain the surface area of the pyramid, giving both an exact answer, using square roots, and a decimal approximation.

(c) Trace your pattern onto heavy paper and cut, fold, and tape three paper models of the pyramid. Show that the three congruent copies can be arranged to form a cube. What is the volume of the cube? What is the volume of each pyramid?

27. A sheet of $8\frac{1}{2}''$ by $11''$ notebook paper can be rolled into a cylinder in either of two ways.

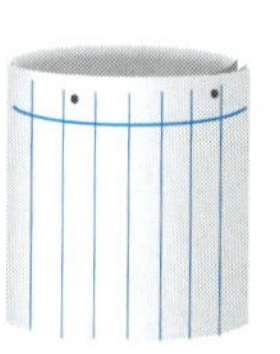

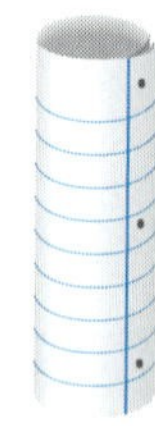

Which way encloses the largest volume? Make a prediction, and then check it.

28. Cut out the circular sector shown in the accompanying figure, and roll it into a right circular cone in which the two 4-inch radial segments are joined.

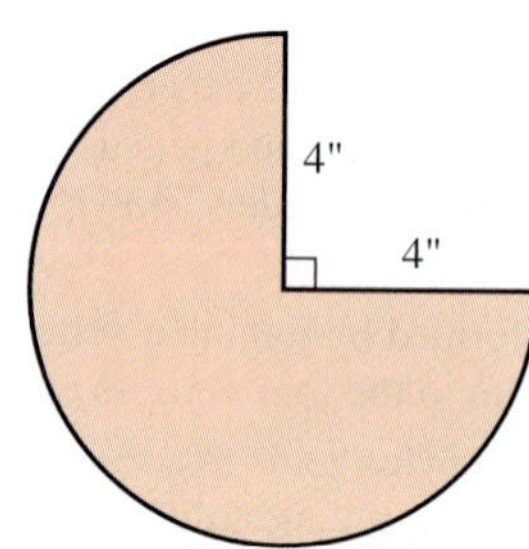

Find the following measurements both by exact computation and by measuring your paper model.

(a) The radius of the base of the cone.

(b) The height of the cone.

(c) Use your answers to parts (a) and (b) to compute the volume of the cone.

29. Cut out a semicircular sector, and roll it up and join the two radial segments to form a cone. Show that the diameter of the cone is equal to the slant height of the cone, both by measuring your paper model and by making a calculation.

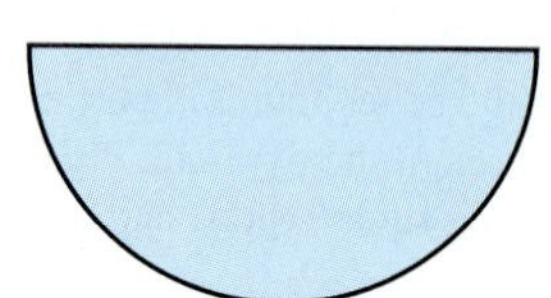

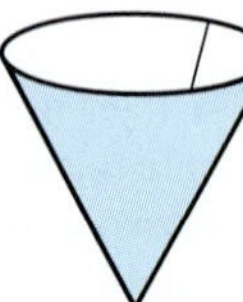

Making Connections

30. A napkin ring is being made of cast silver. It has the shape of a cylinder 1.25 inches high, with a cylindrical hole 1 inch in diameter and a thickness of 1/16 inch. How many ounces of silver are required? It will help to know that silver weighs about 6 ounces per cubic inch.

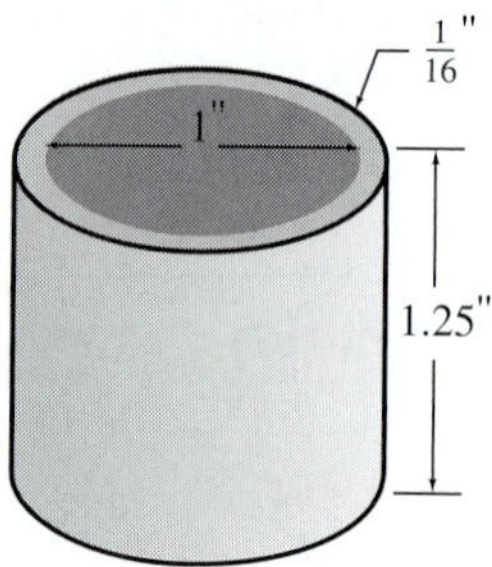

31. Give the dimensions of a rectangular aquarium 40 cm high that holds 48 liters of water.

32. A theater sells $4''$ by $5''$ by $8''$ boxes of popcorn for \$1.75. It also sells cylindrical "tubs" of popcorn for \$3.50, where the tub is $10''$ high and has a diameter of $6''$. Is it better to buy one tub or two boxes of popcorn?

33. Small grapefruit of diameter 3 inches are on sale at five for a dollar. The large 4-inch-diameter grapefruit are three for a dollar. If you are buying \$5 worth of grapefruit, should you choose small ones or large ones in order to get a better deal?

34. **World Records.** *The Guinness Book of World Records,* published annually by Facts on File, New York, contains a fascinating collection of measurements.

(a) The world's largest flawless crystal ball weighs 106.75 pounds and is 13 inches in diameter. What is the weight of a crystal ball 5 inches in diameter?

(b) The largest pyramid is the Quetzalcoatl, 63 miles southeast of Mexico City. It is 177 feet tall and covers an area of 45 acres. Estimate the volume of the pyramid. By comparison, the largest Egyptian pyramid of Khufu (called Cheops by the Greeks) has a volume of 88.2 million ft^3. Recall that an acre is 43,560 ft^2.

(c) The building with the largest volume in the world is the Boeing Company's main assembly plant in Everett, Washington. The building encloses 472 million ft^3 and covers 98.3 acres. What is the size of a cube of equal volume?

35. A water pipe with an inside diameter of 3/4 inches is 50 feet long in its run from the hot-water tank to the faucet. How much hot water is wasted when the water inside the pipe cools down? Give your answer in gallons, where 1 gallon = 231 in^3.

36. In Jonathan Swift's satirical novel *Gulliver's Travels,* Dr. Lemuel Gulliver encounters the tiny Lilliputians. A

Lilliputian is similar to Gulliver, but is 6 inches tall, compared with Gulliver's 6 feet.

(a) Explain why the Lilliputians ordered 1728 rations for Gulliver's dinner.

(b) If the material from Gulliver's shirt was cut up to make shirts for the Lilliputians, how many shirts could be made?

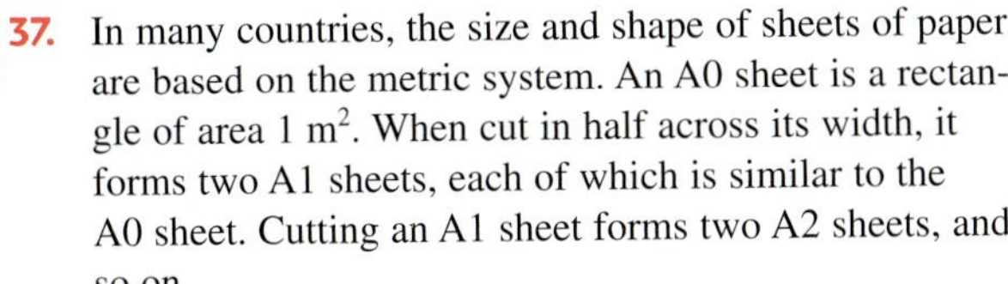

37. In many countries, the size and shape of sheets of paper are based on the metric system. An A0 sheet is a rectangle of area 1 m^2. When cut in half across its width, it forms two A1 sheets, each of which is similar to the A0 sheet. Cutting an A1 sheet forms two A2 sheets, and so on.

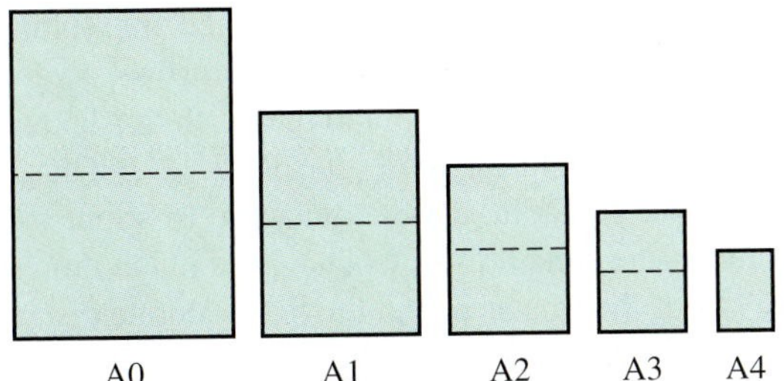

(a) What is the scale factor by which the linear dimensions of an A0 sheet are multiplied to give the corresponding dimensions of an A1 sheet?

(b) Find the width and length, in centimeters, of an A0 sheet.

(c) Find the width and length of an A4 sheet.

(d) What metric-sized paper do you think is used in place of the $8\frac{1}{2}''$ by $11''$ sheets in common use in the United States?

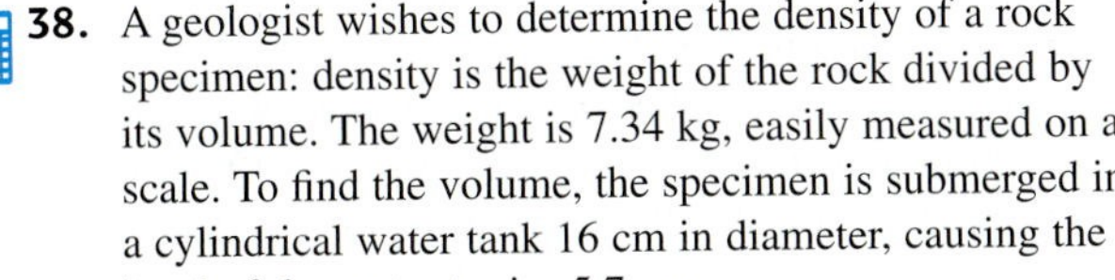

38. A geologist wishes to determine the density of a rock specimen: density is the weight of the rock divided by its volume. The weight is 7.34 kg, easily measured on a scale. To find the volume, the specimen is submerged in a cylindrical water tank 16 cm in diameter, causing the level of the water to rise 5.7 cm.

(a) What is the volume of the specimen?

(b) What is the density of the specimen? Give your answer in grams per cubic centimeter.

39. A rain gauge has a funnel 6 inches in diameter at the top, tapering into a Plexiglas™ collection cylinder whose inside diameter is 2 inches. How far apart should marks be placed on the cylinder to indicate the number of inches of rainfall? (*Hint:* Use the similarity principle.)

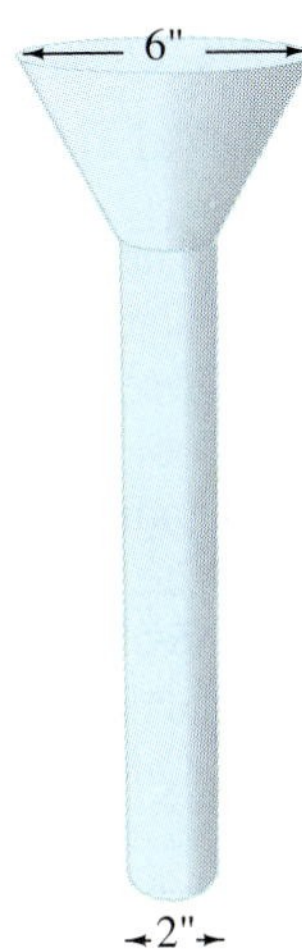

From State Student Assessments

40. (Illinois, Grade 5)
The scale drawing for a new clubhouse is drawn so that 1 cm = 1 meter.
If the scale drawing of the clubhouse is drawn with an area of 100 square cm, what will be the area of the actual clubhouse?

- ○ 1 square meter
- ○ 10 meters
- ○ 10 square meters
- ○ 100 square meters
- ○ 100 meters

For Review

41. Which of the following triples could be the lengths of sides of a right triangle?
(a) 30, 72, 78 (b) 12, 35, 37 (c) 2.0, 2.1, 2.9

42. Find formulas for the perimeter and area of a regular hexagon with sides of length b.

43. Find the perimeter and area of this lattice polygon.

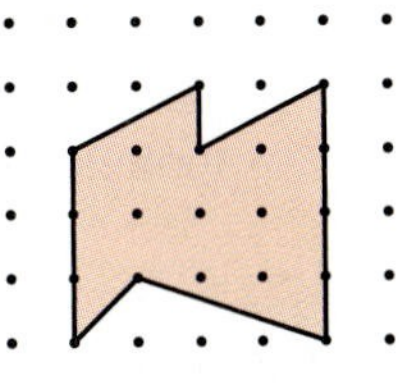

44. What is the speed equivalent to 50 inches per second when expressed in miles per hour?

45. Convert these measurements to the unit shown.

(a) 0.278 m = ____ cm

(b) 2.3 km^2 = ____ m^2

(c) 68,532 cm^3 = ____ L

Epilogue That's about the Size of It!

The size of our everyday world spans about six orders of magnitude; that is, the longest distances of interest are about 10^6, or a million, times larger than the smallest-sized items we deal with. At arm's length—that is, on a scale of about a meter—we find most of the objects (chairs, beach balls, and so on) and life-forms (dogs, cats, horses, and so on) familiar in daily life. A thousandfold increase—to the scale of kilometers—encompasses the distances of ordinary travel, whether across town or cross-country. A thousandfold decrease—to the scale of millimeters—encompasses all that can be seen easily with the unaided eye. The period at the end of this sentence is several tenths of a millimeter across.

To a great extent, the story of science and technology is told by the increasing number of orders of magnitude required to encompass newly discovered objects and phenomena. The first major step to the measurement of the large was Eratosthenes' measurement (ca. 240 B.C.) of the earth's circumference. At noon during the summer solstice, a vertical rod at Syene (now Aswan) cast no shadow, whereas 5000 stadia to the north in Alexandria a vertical rod made an angle of "1/50 of four right angles" (that is, 7°12′). Thus, the earth's circumference is 50×5000, or 250,000 stadia, a value accurate to about 6 percent. The measurement of the solar system—the sun and its planets—was much more difficult: The Copernican model was only 1/7 of the true size, and the first accurate distances to the planets (to within 10 percent) were made in 1672 by the newly created French Academy of Science. The first accurate distance to a star was given in 1837 by Friedrich Wilhelm Bessel, who found that 61 Cygni was 619,000 times as far from Earth as the sun. In 1924 Edwin Hubble proved that William Herschel's "island universes" were separate galaxies far from our own Milky Way galaxy. To measure the universe, where distant quasars are 10 billion light years away, we must measure on the astonishingly large scale of about 10^{25} meters.

Recent breakthroughs in microscopy reveal new images of the very small. At 10^{-6} m we find bacteria, at 10^{-9} m a single sodium atom, and at 10^{-12} m the nucleus of the sodium atom. Current thought suggests that 10^{-16} m—the scale of quarks—may present all there is to see, at least until we reach 10^{-31} m.

This chapter has introduced the basic notions of measurement. Measurement is concerned with how size is determined and communicated. Generally, measurement is an approximation, calling for appropriate judgments to be made about the selection of measurement tools and the level of precision required. For some ideal shapes—triangles, prisms, pyramids, circles, and spheres, to name a few—the measurement process is supplemented by the use of formulas. Measurement, however, is not simply a collection of formulas; often we must return to the basic principles of the measurement process.

Chapter Summary

Key Concepts

Section 12.1 The Measurement Process

- Measurement process. Step 1: Choose the property (length, area, volume, and so on) to be measured. Step 2: Select a unit of measurement. Step 3: Compare the size of the object being measured with the size of the unit by covering, filling, and so on. Step 4:

Express the measurement as the number of units used. A measurement is most often only an approximation, and decisions must be made to choose appropriate measurement tools and units to provide necessary accuracy and precision.

- Standardized system. A measurement system with well-defined units used to communicate size and magnitude. The United States uses the customary system, and the entire world, including the United States, uses the metric system.
- The metric (SI) system. A system with units related by powers of 10 described by a prefix system; for example, $milli = \frac{1}{1000}$, $centi = \frac{1}{100}$, and $kilo = 1000$.
- Metric units. The basic unit of length is the meter, with area given in square meters or hectares (1 hectare $= 10{,}000\text{ m}^2$), and volume in cubic meters. Capacity is also given in liters (1 liter $= 1000\text{ cm}^3$). Mass is given in grams.

Section 12.2. Area and Perimeter

- Area. Area is the amount of the plane covered by a plane region. The unit of area is arbitrary, but usually a square one unit of length on a side is chosen. To compare the area of one region with that of another, the congruence and addition properties are often useful.
- Area formulas for common polygons.

  Rectangle of length l and width w: $A = lw$.

  Parallelogram of base b and height h: $A = bh$.

  Triangle of base b and altitude h: $A = \frac{1}{2}bh$.

  Trapezoid with bases a and b and altitude h: $A = \frac{1}{2}(a + b)h$.
- Length. Length is the distance along a curve.
- Perimeter. Perimeter is the length of a simple closed curve.
- Measurements of a circle. The perimeter of a circle is called the circumference of the circle, given by $2\pi r$, where r is the radius of the circle and π (pi) by definition is the ratio of the circumference to the diameter of a circle. The area of a circle is πr^2.

Section 12.3 The Pythagorean Theorem

- Pythagorean relationship. Three numbers a, b, and c satisfy the Pythagorean relationship if $a^2 + b^2 = c^2$.
- Pythagorean theorem. The lengths a, b, and c of the sides of a right triangle satisfy $a^2 + b^2 = c^2$. Therefore, the sum of the areas of the squares on the legs of a right triangle equals the area of the square on the hypotenuse. The converse of the Pythagorean theorem also holds: if $a^2 + b^2 = c^2$, then a triangle with sides of length a, b, and c is a right triangle.

Section 12.4 Surface Area and Volume

- Surface area of a polyhedron. The surface area of a polyhedron is the sum of the areas of the plane faces. For some polyhedra, such as right prisms and right regular pyramids, it is useful to imagine that the surface is cut and unfolded onto the plane.
- Surface area of a general space figure. This quantity is found as the limiting value of the surface areas of approximating polyhedra.
- Surface area formulas.

  Right prism or right cylinder of height h, bases of area B, and perimeter p: $SA = 2B + ph$.

  Right regular pyramid of slant height s, base of area B, and perimeter p: $SA = B + \frac{1}{2}ps$.

Right circular cone of slant height s, and base radius r: $SA = \pi r^2 + \pi rs$.

- Volume formulas:

  Prism or cylinder of base area B and height h: $V = Bh$.

  Pyramid or cone of base area B and height h: $V = \frac{1}{3}Bh$.
- Volume and surface area of a sphere of radius r: $V = \frac{4}{3}\pi r^3$ and $SA = 4\pi r^2$.
- Similarity principle of measurement: If two space figures are similar with a scale factor k, then all corresponding linear measurements vary by the factor k, all area measurements vary by k^2, and all volume measurements vary by k^3.

Vocabulary and Notation

Section 12.1

Measurement process
Unit of measure
U.S. customary (English) system
  Units of length (inch, foot, yard, mile), area (in^2, ft^2, yd^2, acre, mi^2), and volume and capacity (in^3, ft^3, yd^3, quart, gallon)
Metric (SI) system
  Prefixes: m = *milli* (1/1000), c = *centi* (1/100), k = *kilo* (1000)
  Units of length (meter, centimeter, kilometer), area (m^2, cm^2, km^2, hectare), and volume and capacity (m^3, cm^3, km^3, liter)
Tangram
Weight
Mass
Fahrenheit temperature scale
Celsius temperature scale
Unit (dimensional) analysis

Section 12.2

Area
Unit of area
Dissection of a plane region into subregions
Congruence property of area
Addition property of area
Altitude and base of a parallelogram or triangle
Length of curve
Perimeter
Circumference of circle; π (pi)
Congruent
Lattice polygon

Section 12.3

Pythagorean theorem
Converse of the Pythagorean theorem

Section 12.4

Surface area of a surface in space
Base area
Lateral surface area
Right regular pyramid
Slant height
Volume of a solid
Similarity principle of measurement

Chapter Review Exercises

Section 12.1

1. Select an appropriate metric unit of measurement for each of the following.
   (a) The length of a sheet of notebook paper
   (b) The diameter of a camera lens
   (c) The distance from Los Angeles to Mexico City
   (d) The height of the Washington Monument
   (e) The area of Central Park
   (f) The area of the state of Kentucky
   (g) The volume of a raindrop

(h) The capacity of a punchbowl

2. Give the most likely answer:
   (a) A bottle of cider contains: 30 mL, 4L, 15L
   (b) Cross-country skis have length: 190 cm, 190 km, 190 m
   (c) The living area of a house is: 2000 cm^2, 1.2 ha, 200 m^2

3. An aquarium is a rectangular prism 60 cm long, 40 cm wide, and 35 cm deep. What is the capacity of the aquarium in liters?

4. A sailfish off the coast of Florida took out 300 feet of line in 3 seconds. Estimate the speed of the fish in miles per hour.

Section 12.2

5. Let M be the midpoint of side $\overline{AD}$ of the trapezoid $ABCD$. What is the ratio of the area of triangle MBC to the area of the trapezoid? (*Hint:* Dissect the trapezoid by a horizontal line through M, and rearrange the two pieces to form a parallelogram.)

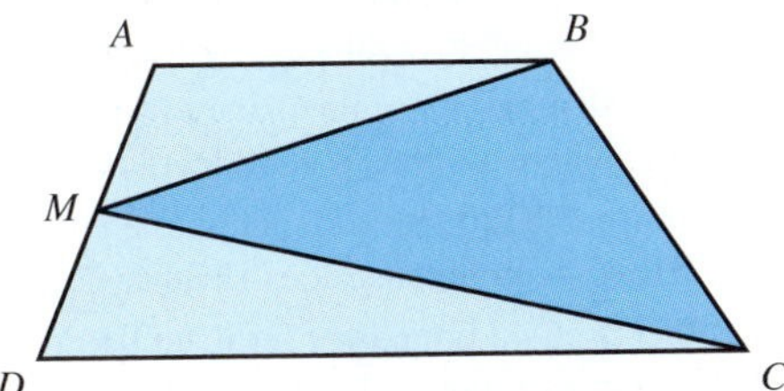

6. A square cake 20 inches on a side is being shared evenly among 5 people. Two cuts have been made to the center as shown.

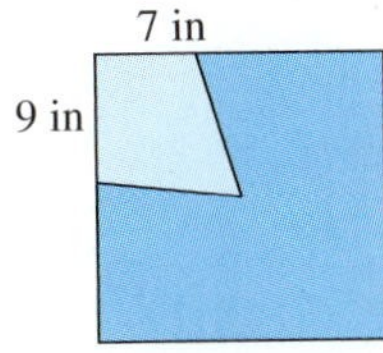

   (a) Is the piece of cake shown a fair piece?
   (b) Draw a figure showing where three more cuts from the edge to the center create pieces of the same size.

7. Find the areas of these figures.
   (a)

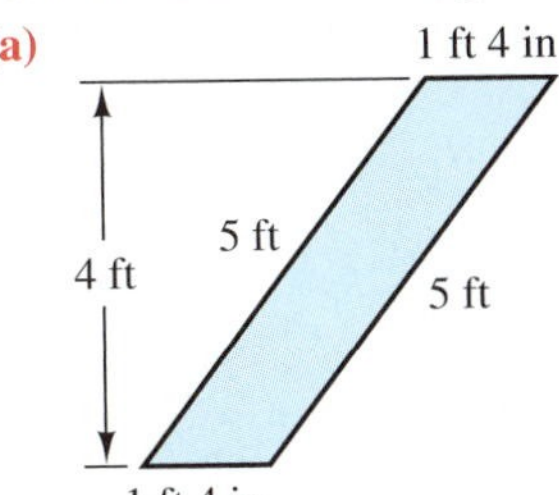

   (b)

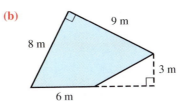

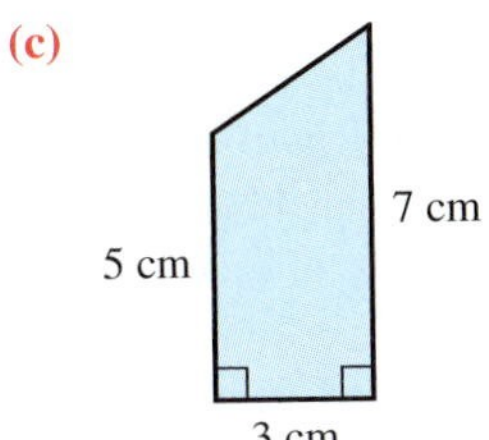

   (c)

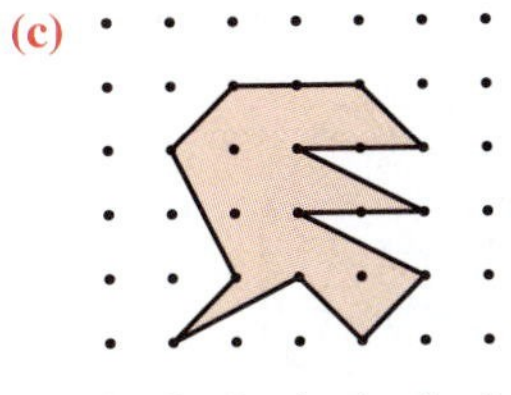

8. Find the area of each lattice polygon.
   (a) (b)

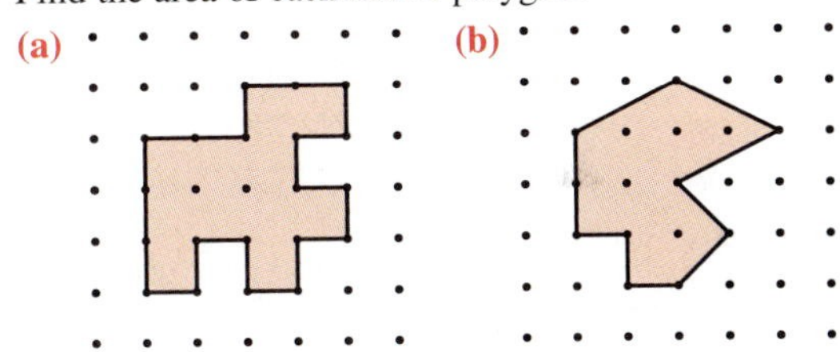

   (c)

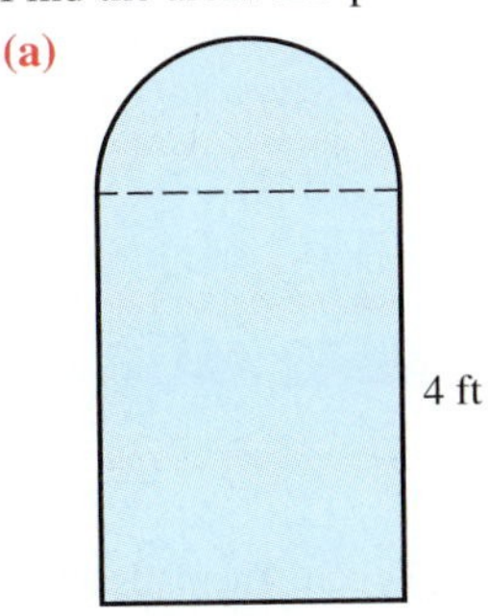

9. Find the areas and perimeters of these figures.
   (a)

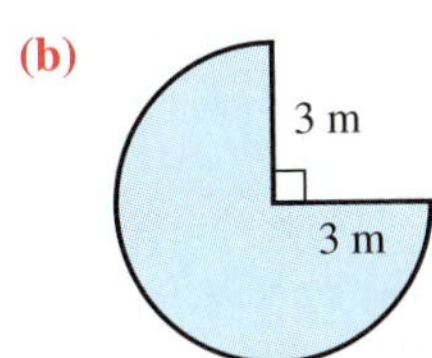

   (b)

Section 12.3

10. Solve for x and y in the figure.

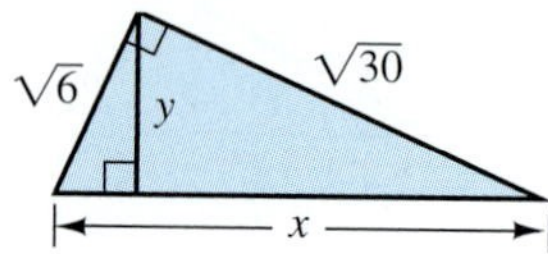

11. A right circular cone has slant height 35 cm and a base of diameter 20 cm. What is the height of the cone?

12. A rectangular box has sides of length 4 inches, 10 inches, and 12 inches. What are the lengths of each of the four diagonals of the box?

13. Find the perimeter of the following lattice polygon.

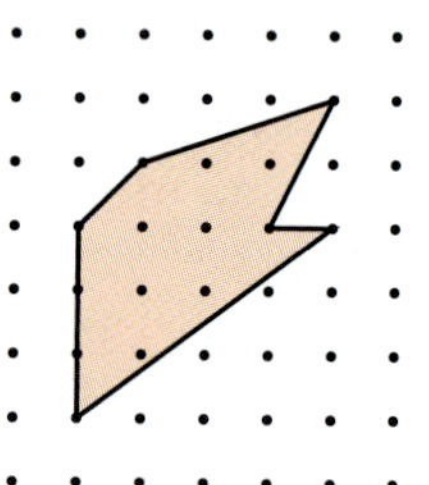

Section 12.4

14. Find the volume and surface area of these figures.

(a)

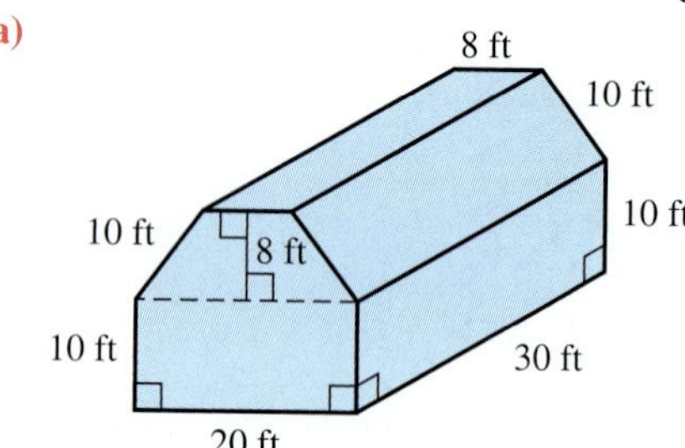

(b)

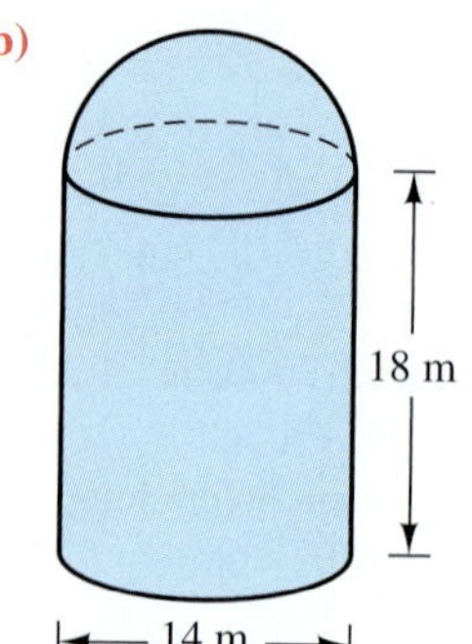

(c)

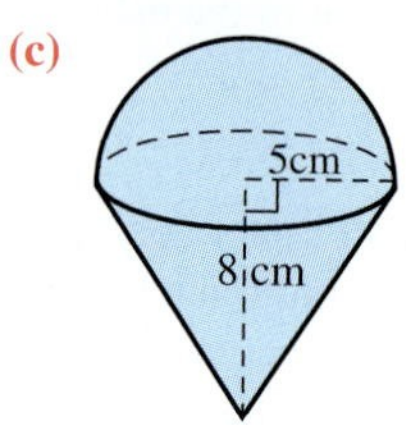

15. Which has the largest volume, a sphere of radius 10 meters or four cubes with sides of length 10 meters?

16. Heather's garden has a similar shape to Johan's, but is 75 feet long, whereas Johan's is 50 feet long.

(a) Johan needed 180 feet of fencing to enclose his garden. How much fencing does Heather need?

(b) Heather used 45 pounds of fertilizer. How much will Johan use, assuming it is applied at the same number of pounds per square foot?

Chapter Test

1. Find the area of each lattice polygon. The nails are 1 cm apart.

(a)

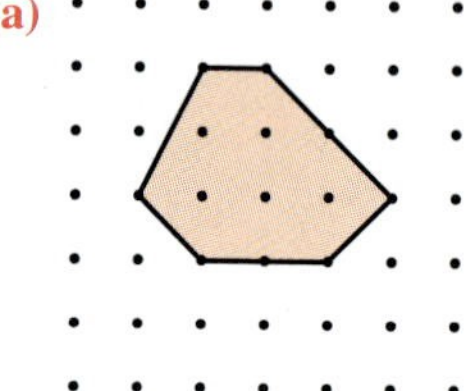

(b)

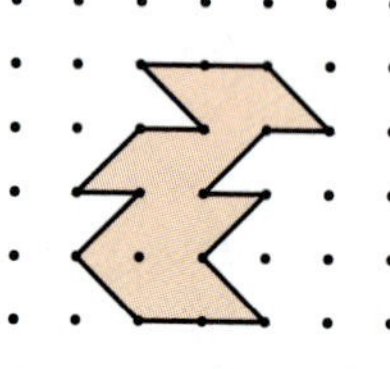

(c)

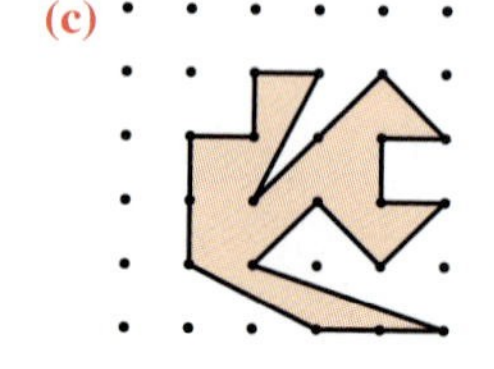

2. (Washington State Student Assessment, Grade 4)
Jim says that the area of shape A is equal to the area of shape B.

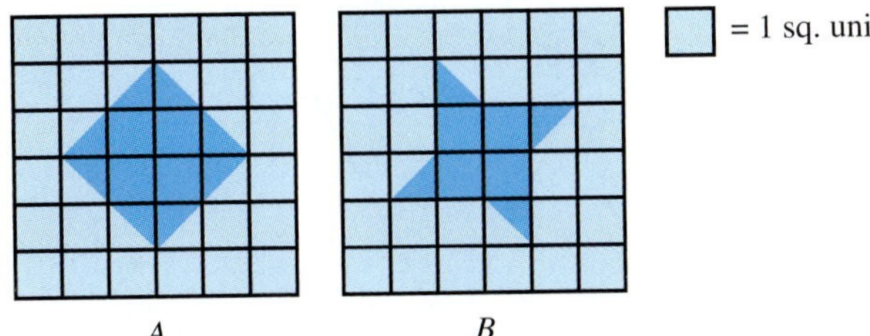

Explain why Jim's thinking is *wrong* using words, numbers, or pictures.

3. Find the surface area and volume of these figures.

(a)
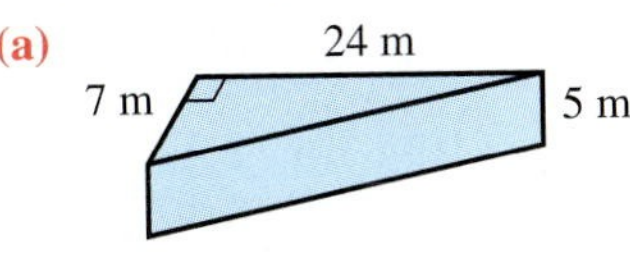

Right triangular prism

(b)
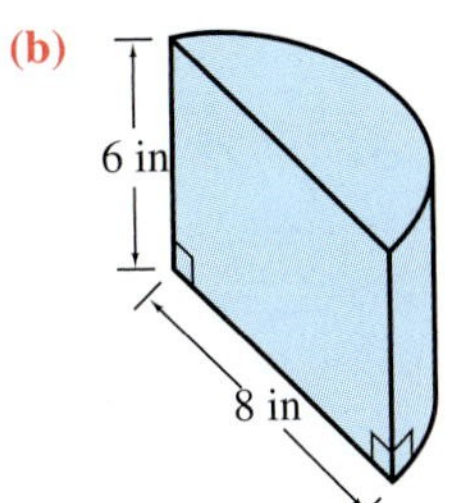

Cylinder with semicircular base

(c)
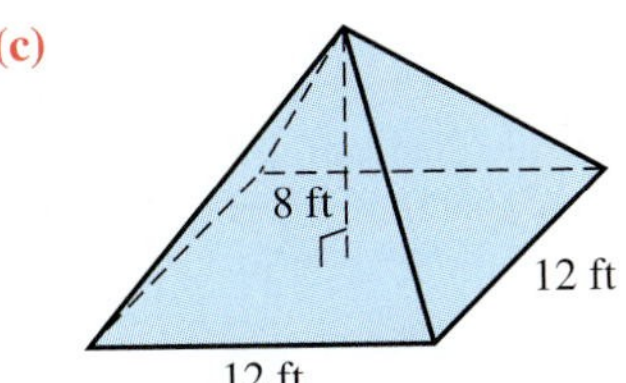

Right regular square pyramid

(d)
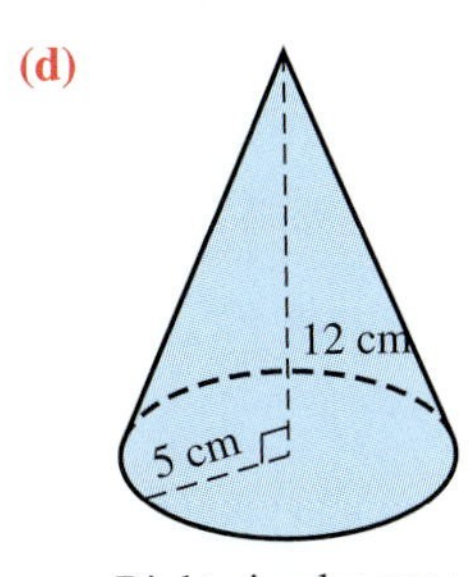

Right circular cone

4. Fill in the blank with the metric unit of measurement that makes the statement reasonable.
   (a) The haze filter on Donna's camera has a diameter of 52 ______ .
   (b) A gray whale has length 18 ______ .
   (c) In the 1968 Olympics, Bob Beaman had a long jump of 8.90 ______ .
   (d) The Mississippi River has a length of 1450 ______ .
   (e) A cup of coffee contains about 250 ______ .
   (f) A fill-up at the gas station took 46 ______ .

5. A lens is made by cutting a section from a sphere. If the lens has diameter 10 mm and height 4 mm, what is the radius of the sphere from which it was cut?

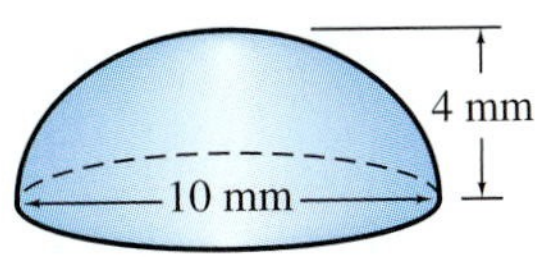

6. (a) Find the perimeter of the lattice triangle PQR.

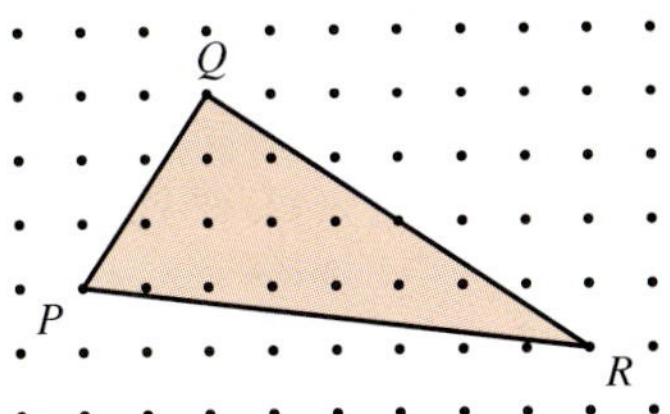

   (b) Is $\triangle PQR$ a right triangle?

7. Fill in the blanks:
   (a) 2161 mm = ______ cm
   (b) 1.682 km = ______ cm
   (c) 0.5 m^2 = ______ cm^2
   (d) 1 ha = ______ m^2
   (e) 4719 mL = ______ L
   (f) 3.2 L = ______ cm^3

8. What is the ratio of the area of the inscribed square to the area of the square circumscribed about the same circle?

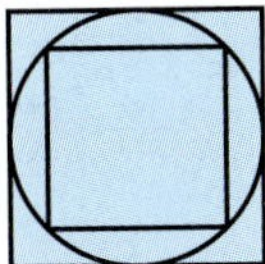

9. A ladder 15 feet long rests against a vertical wall. If the bottom of the ladder is 6 feet from the base of the wall, how high does the ladder reach?

10. Complete the conversions of the measurements in the U.S. customary system, using your calculator when convenient.

(a) 1147 in = ______ yd
(b) 7942 ft = ______ mi
(c) 32.4 yd^2 = ______ ft^2
(d) 9402 acres = ______ mi^2
(e) 7.6 yd^3 = ______ ft^3
(f) 5961 in^3 = ______ ft^3

11. Papa Bear, Mama Bear, and Baby Bear have similar shapes, except Papa Bear is 5 ft tall, Mama Bear is 4 ft tall, and Baby Bear is 2 ft tall. Fill in the values left blank in the following chart.

	Papa Bear	Mama Bear	Baby Bear
Length of suspenders		40 in	
Weight			30 lb
Number of fleas	6000		

12. Find the area and perimeter of the following figure.

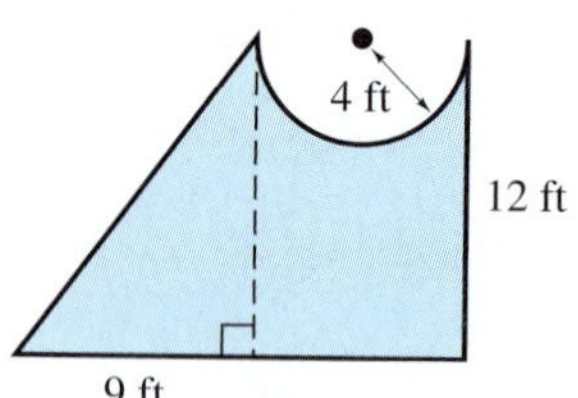

13. Find the area of each figure.

(a)

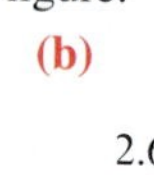

(b)

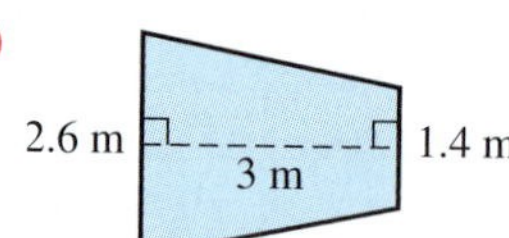

(c)

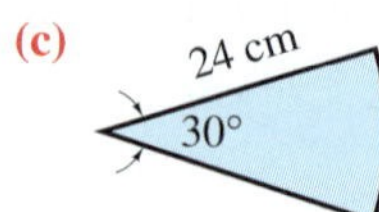

14. Find the area and perimeter of the following kite.

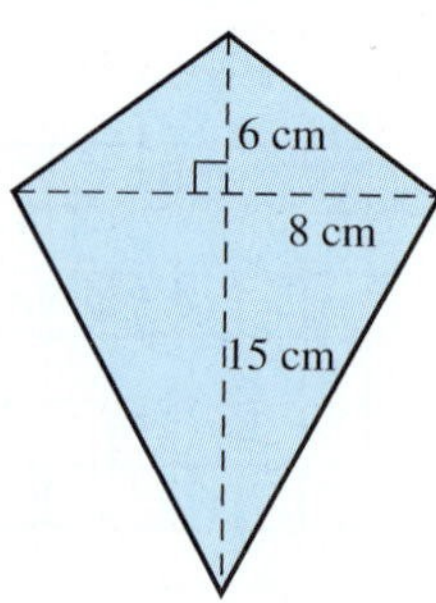

15. Explain how the formula for the area of a triangle can be derived from that of a parallelogram by cutting the triangle along the segment joining the midpoints of the sides opposite the base.

16. Find the surface area and volume of each figure.

(a)

20 m

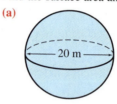

Sphere

(b)

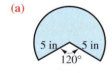

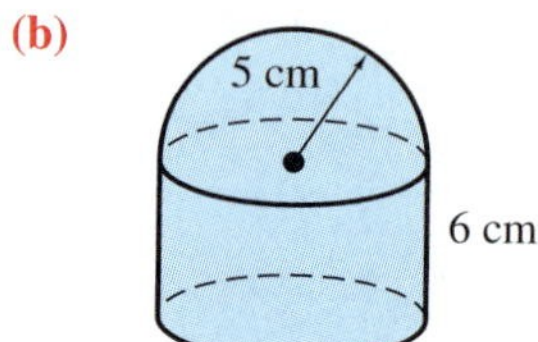

Cylindrical tank with hemispherical top

17. A grapefruit has an outside diameter of 5 inches. When cut open, it is discovered that the peel is 3/4 inches thick. What percentage of the grapefruit's volume is peel?

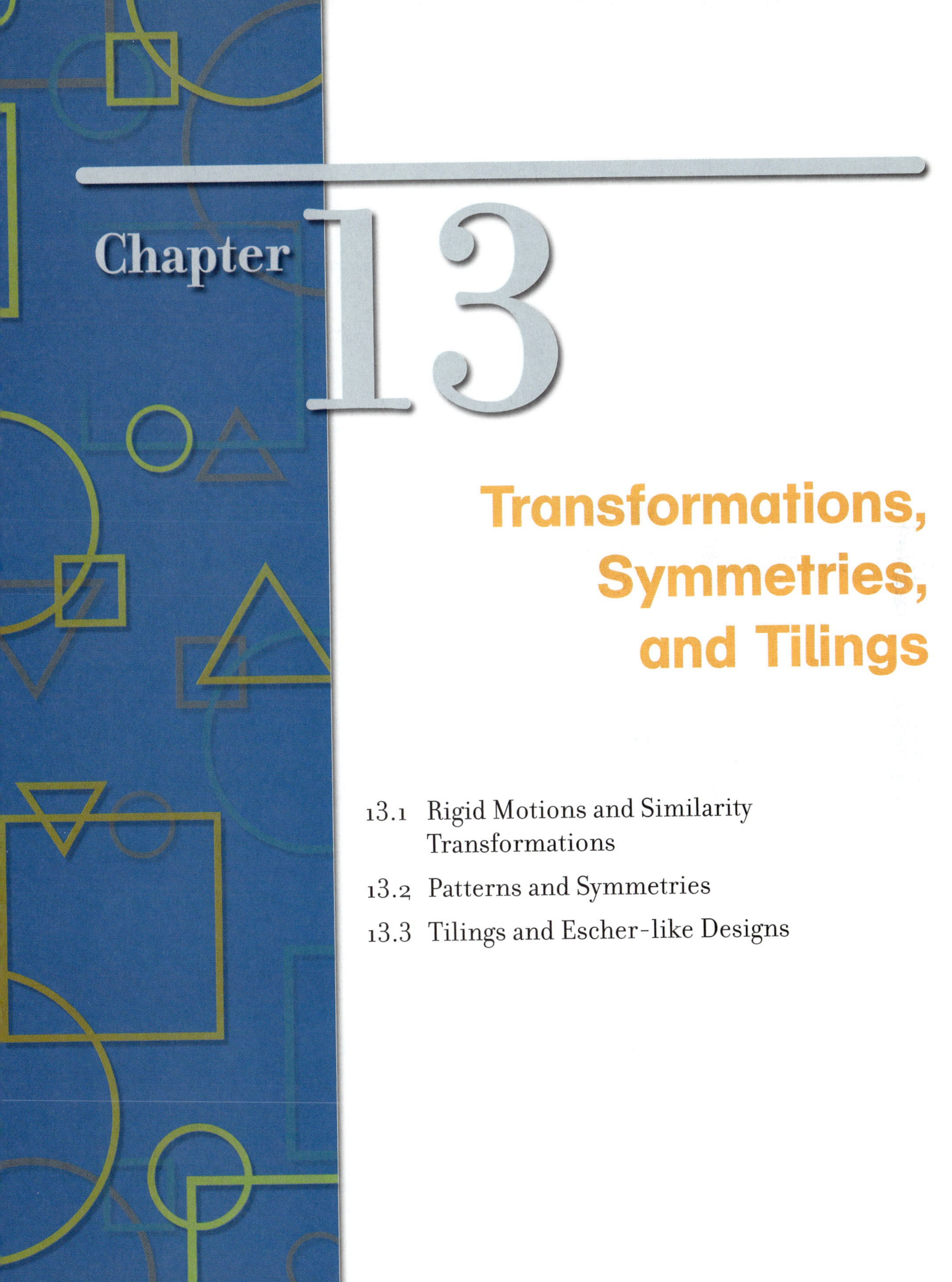

Chapter 13

Transformations, Symmetries, and Tilings

Exploring Reflection and Rotation Symmetry

Materials Needed

Clear acetate sheets, overhead transparency pens, tissues to clean acetate sheets for reuse, and Mira® (if available).

How to Check for Reflection and Rotation Symmetry

A figure has **reflection** (or **line**) **symmetry** if there is a mirror line that reflects the figure onto itself. For example, the parafoil kite below has a vertical line of symmetry. The wheel cover at the right does not have reflection symmetry, but it does have **rotation symmetry,** since the figure turns onto itself when rotated through 72° about the center point.

Reflection symmetry can be verified by the "trace-and-flip" test. The figure and its line of symmetry are traced onto an acetate sheet. The acetate is then flipped across the proposed symmetry line, turning the sheet upside down, to check that the points of the traced figure coincide with the original figure. Alternatively, reflection symmetry can be investigated by placing the drawing line of a Mira over a proposed line of symmetry of a figure.

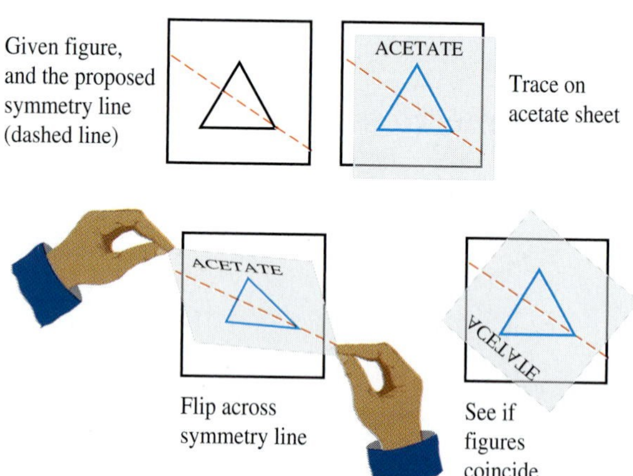

Rotation symmetry can be investigated by the "trace-and-turn" test, illustrated next. The point held fixed is the **center of rotation.** Since the tracing coincides with the original figure after a 120° turn, the trace-and-turn test shows that an equilateral triangle has 120° rotation symmetry.

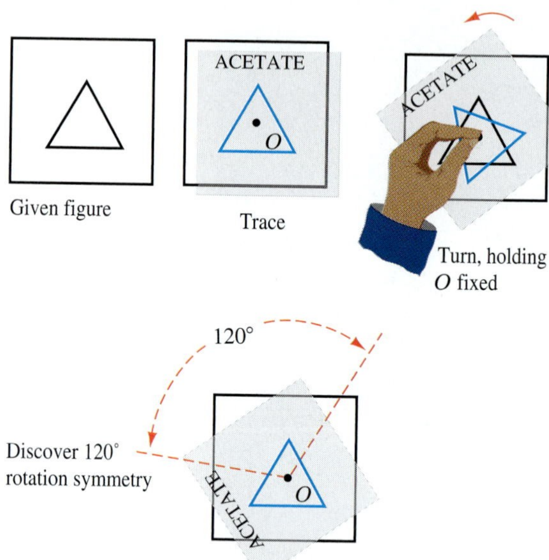

Activities

1. Use either a Mira or the trace-and-flip test to find all lines of symmetry of the following figures. Use dashed lines to draw the symmetry lines.

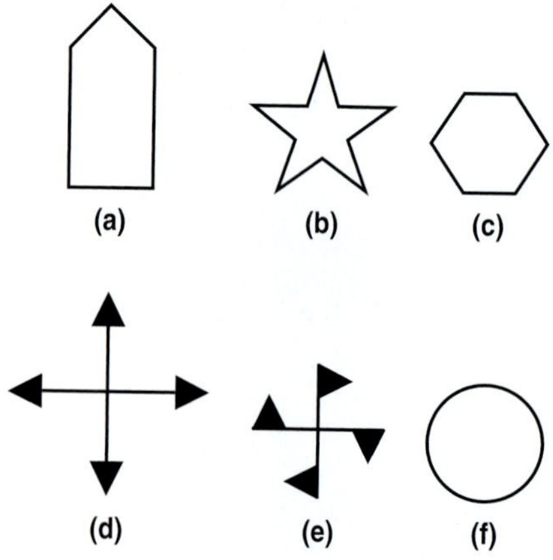

2. Use the trace-and-turn test to describe the rotation symmetries of the figures in Activity 1. Indicate the center of rotation and the angle measure of the rotation.

3. Sketch all lines of symmetry and describe all rotation symmetries for the following polygons.
(a) Triangles: equilateral, isosceles, scalene
(b) Quadrilaterals: square, rhombus, rectangle, parallelogram, trapezoid, isosceles trapezoid, kite

Connections

Transformational Geometry, the Mathematics of Patterns and Symmetries

The Northwest Coast Native Americans occupy a narrow band of land along the western coasts of Washington state, Canada, and southeast Alaska. The tribal groups of this region are renowned for their striking graphic art depicting whales, seals, eagles, bears, and other animal forms. Often the design is highly symmetric, such as the example of Haida art shown in Figure 13.1.

Figure 13.1
Sea lions on a Haida dance tunic

A trip to your local history museum will show that symmetry is a common element found in the decorations and artwork created by all the world's cultures. Moreover, while a design from China is easily distinguished from a design from Central America, it is often evident that the two designs are based on the same underlying pattern of symmetry.

Most people, including schoolchildren, have an intuitive sense of symmetry, but usually the sense is too vague for precise classification and understanding. In this chapter we will see that the concept of a geometric transformation makes it possible to give a precise meaning to symmetry.

Section 13.1 defines and investigates the rigid motions. It is discovered that any rigid motion of the plane is one of four basic types—slides, turns, flips, and glide reflections. Slides, turns, and flips are also called translations, rotations, and reflections, respectively. Dilations are also considered, leading to the concept of a similarity transformation.

Rigid motions are used as a tool to explore symmetry in Section 13.2. Of special interest are the patterns in the plane created by endlessly repeating a single motif.

Section 13.3 investigates the patterns in the plane formed with polygonal tiles. Escher-like designs are then created by modifying polygonal tiles to assume more general shapes.

13.1 RIGID MOTIONS AND SIMILARITY TRANSFORMATIONS

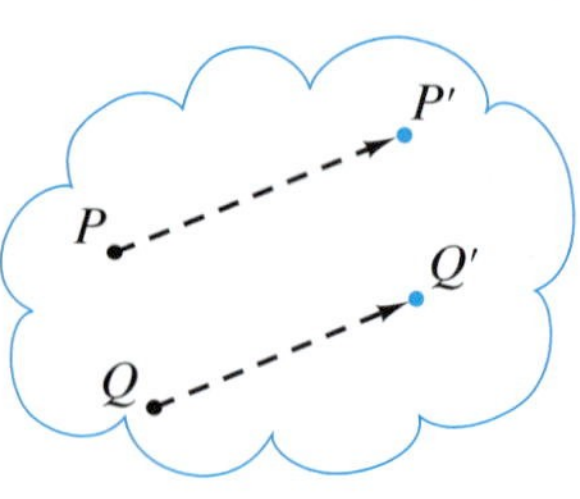

Imagine that each point P of the plane is "moved" to a new position P' in the same plane. Call P' the **image** of P, and call P the **preimage** of P'. If distinct points P and Q have distinct images P' and Q', and every point has a unique preimage point, then the association $P \leftrightarrow P'$ defines a one-to-one correspondence of the plane onto itself. Such a correspondence is called a **transformation of the plane.**

DEFINITION *Transformation of the Plane*
A one-to-one correspondence of the set of points in the plane to itself is a **transformation of the plane.** If point P corresponds to point P', then P' is called the **image** of P under the transformation. Point P is called the **preimage** of P'.

A type of transformation called a **rigid motion** is of special importance. As the name implies, a rigid motion does not allow stretching or shrinking.

DEFINITION *Rigid Motion of the Plane*
A transformation of the plane is a **rigid motion** if, and only if, the distance between any two points P and Q equals the distance between their image points P' and Q'. That is, $PQ = P'Q'$ for all points P and Q.

A rigid motion is also called an **isometry,** meaning "same measure" (*iso* = same, *metry* = measure).

A useful physical model of a rigid motion of the plane can be realized with a sheet of clear acetate and a sheet of paper on which figures with points labeled $A, B, C, \ldots$ are drawn. The figures are traced onto the transparency, and a rigid motion is modeled by moving the transparency to a new position in the plane of the paper. An example is shown in Figure 13.2, where primed letters $A', B', C', \ldots$ indicate the points in the image figure that correspond to the respective points $A, B, C, \ldots$ in the original figure. A rigid motion actually maps *all* of the points of the plane, but usually it is enough to show how a simple figure such as a triangle is moved to describe the motion. It is allowable to turn the transparency upside down before it is returned to the plane of the paper, since the definition of rigid motion is still satisfied.

Original figure

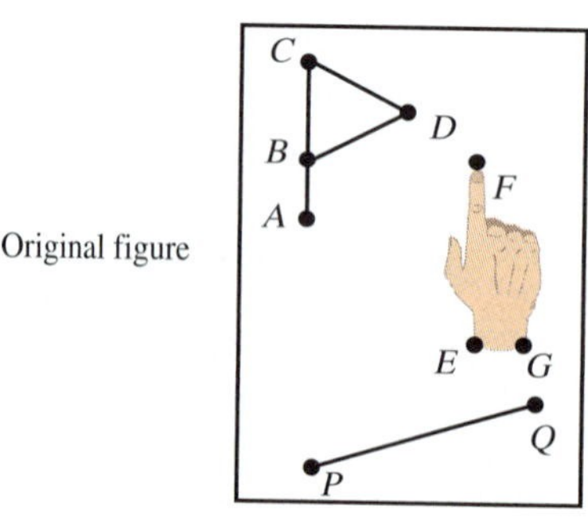

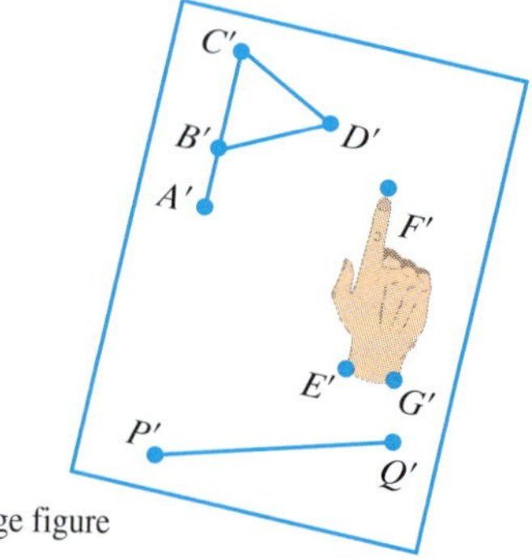

Image figure

Figure 13.2
Illustrating a rigid motion of the plane

Geometric Transformations with Geometry Software

Accurate drawings of rigid motions are not easy to create, especially for youngsters. The following three suggestions may be helpful:

- Use graph paper or square dot paper (or a triangular grid of lines or dots).
- Use a "tile" of a basic figure (say, cut from card stock or made from snap cubes) that can be traced to show its new positions under a motion or sequence of motions.
- Use a reflective drawing tool such as a Mira.

These techniques and tools continue to be valuable in the classroom. However, geometry software (see Appendix D) offers an exciting new way to explore rigid motions and other transformations of the plane. For example, the Transform menu in The Geometer's Sketchpad, shown below, provides the commands to translate, rotate, dilate, or reflect a selected figure. Even quite young children will quickly understand how to use these commands to perform the basic motions.

Transform
Mark Center
Mark Mirror
Mark Angle
Mark Ratio
Mark Vector
Mark Distance
Translate...
Rotate...
Dilate...
Reflect
Iterate...

As an example, suppose that a flag, *FLAG*, has been drawn, as well as points *P* and *Q* used to define a translation. To translate the flag with the slide arrow from *P* to *Q*, select points *P* and *Q* (in that order) and execute the **Mark Vector** command. Now select the flag figure, and execute **Translate . . . By Marked Vector** to obtain the image $F'L'A'G'$ of the flag.

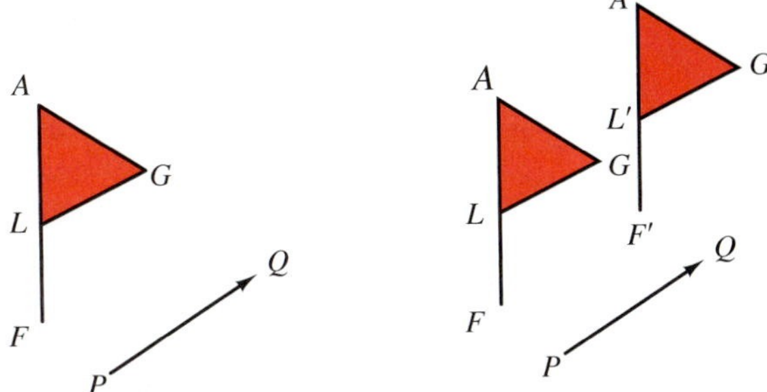

To perform a rotation, first select a point C and mark it as the center of rotation. A selected figure can then be rotated by an angle given either in degrees or as a marked angle. In the same way, a flip is performed by marking a line segment such as $\overline{MN}$ as the mirror line and then reflecting a selected figure across the line.

Once the basic motions are understood, it is fun for students to create more complex figures to demonstrate both their

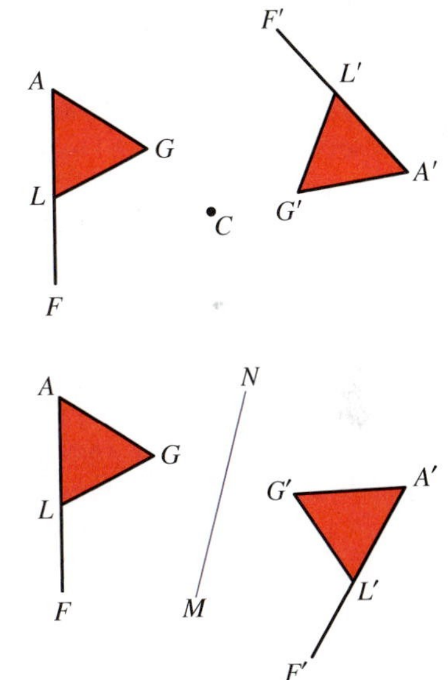

geometric understanding and their skills with the software. For example, a "smiley face" figure below is constructed by drawing first the left side of the face and then reflecting across a vertical centerline.

It needs to be emphasized that only the initial and final positions of a transformation are of interest. The transparency could have been taken for a roller coaster ride before reaching its final position. When the net outcomes of two motions are the same, the transformations are said to be **equivalent.**

The Four Basic Rigid Motions

Four transformations of the plane have special importance. They are the four **basic rigid motions of the plane:** translations, rotations, reflections, and glide reflections.

Translations

A **translation,** also known as a **slide,** is the rigid motion in which all points of the plane are moved in the same direction and the same distance. A translation is illustrated by the "trace-and-slide" model in Figure 13.3. An arrow drawn from a point P to its image point P' completely specifies the two pieces of information required to define a translation: the direction of the slide is the direction of the arrow, and the distance moved is the length of the arrow. The arrow is called the **slide arrow** or **translation vector.**

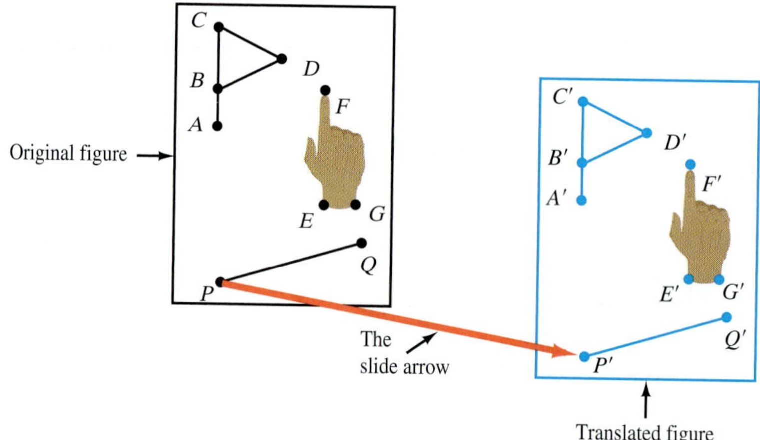

Figure 13.3
A slide, or translation, moves each point of the plane in the same direction and through the same distance

EXAMPLE 13.1 FINDING THE IMAGE UNDER A TRANSLATION

Find the image of the pentagon $ABCDE$ under the slide that takes the point C to C'.

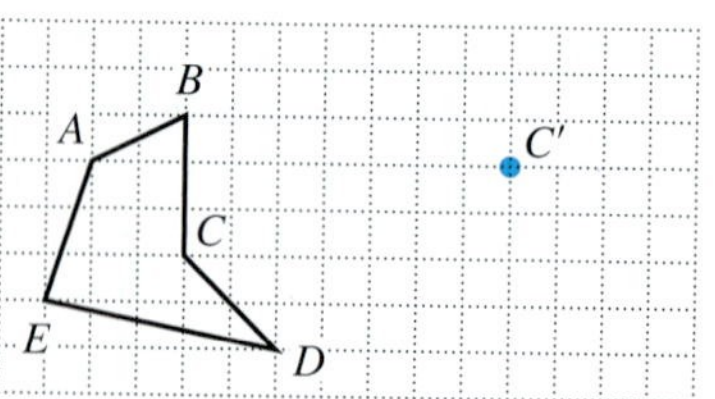

Solution The slide arrow from C to C' is seven units to the right and two units up. Therefore A' is found by moving seven units to the right of A and then two units up. The remaining points are found in the same way, always using the same slide arrow.

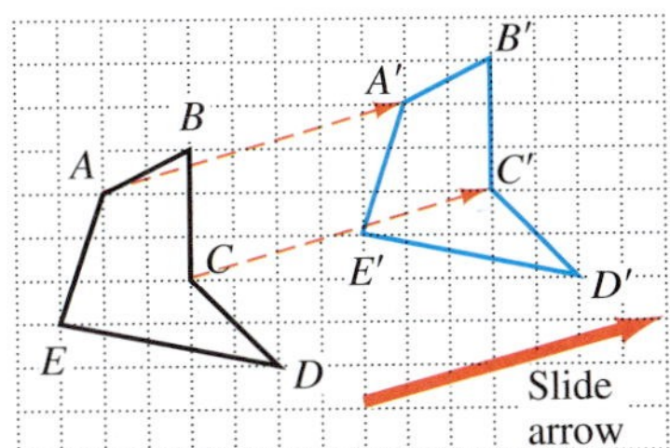

Just for Fun

Inverting the Tetractys

Ten pennies can be tightly packed to form an equilateral triangle. This pattern was called the "tetractys" by the ancient Pythagoreans, and is still familiar today in 10-pin bowling alleys. The problem is to turn the tetractys upside down by sliding one penny at a time to a new position in which it touches two other pennies. It's easy to do moving four pennies, but you should be able to find a three-move solution.

A three-coin triangle can be inverted in one move, a six-coin triangle in two moves, and, as noted above, a 10-coin triangle in three moves. It would be reasonable to guess that a 15-coin triangle can be inverted in four moves, but it actually takes five moves. Try it!

Rotations

The **rotation,** also called a **turn,** is another basic rigid motion. One point of the plane—called the **turn center** or the **center of rotation**—is held fixed, and the remaining points are turned about the center of rotation through the same number of degrees—the **turn angle.** A counterclockwise turn about point O through 120° is shown in Figure 13.4. Note that the right-hand EFG is taken to the right-hand $E'F'G'$.

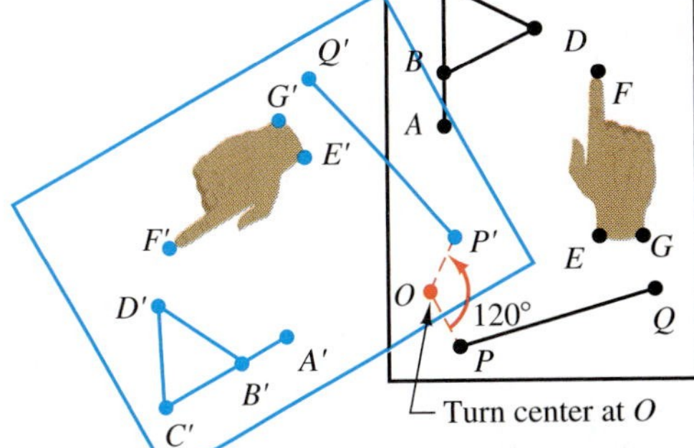

Figure 13.4
A turn, or rotation, rotates each point of the plane about a fixed point O—the turn center—through the same number of degrees and in the same direction of rotation

A rotation is determined by giving the turn center and the directed angle corresponding to the turn angle. This information can be pictured by a **turn arrow,** as shown in Figure 13.5.

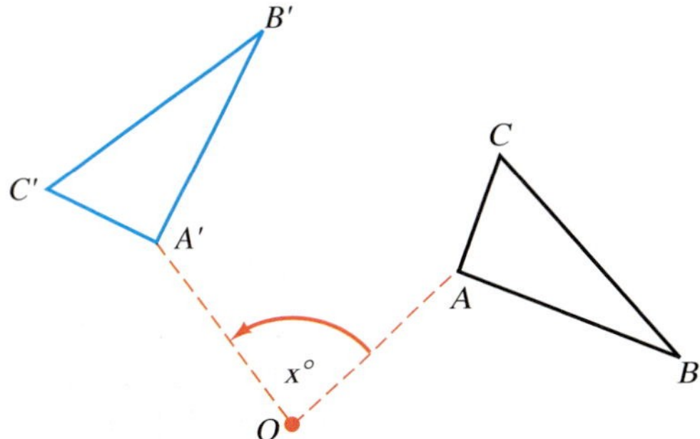

Figure 13.5
The rotation about center O by x° can be indicated by a turn arrow

Usually, counterclockwise turn angles are assigned positive degree measures, whereas negative measures indicate that the rotation is clockwise. In this way a $-120°$ turn is equivalent to a $240°$ turn about the same center. Remember that only the initial and final positions are considered, not the actual physical motion.

EXAMPLE 13.2 FINDING IMAGES UNDER ROTATIONS

Find the image of each figure under the indicated turn.

(a) 90° rotation about *P* **(b)** 180° rotation about *Q* **(c)** $-90°$ rotation about *R*

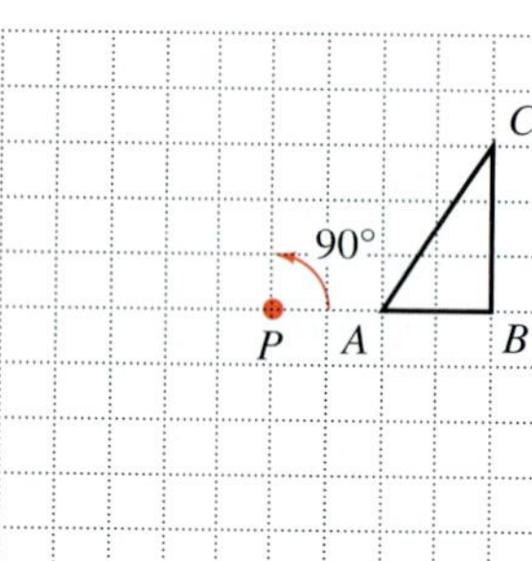

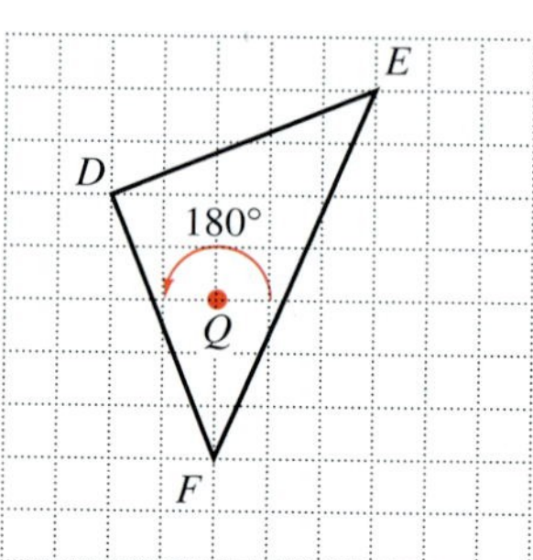

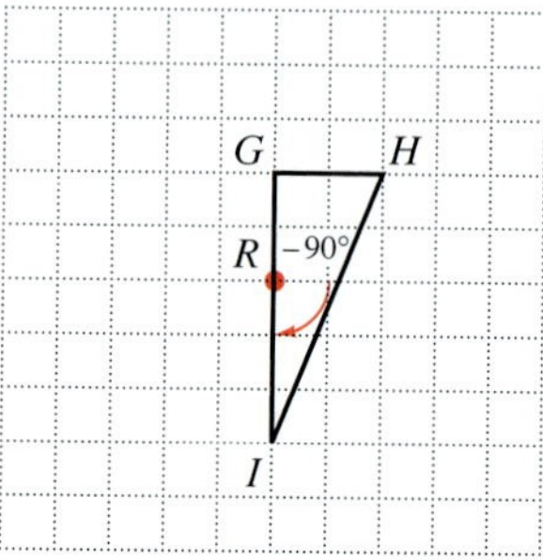

Solution

(a) **(b)** **(c)**

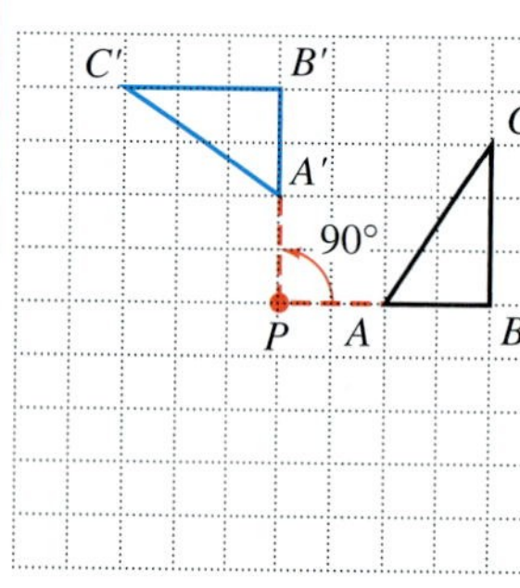

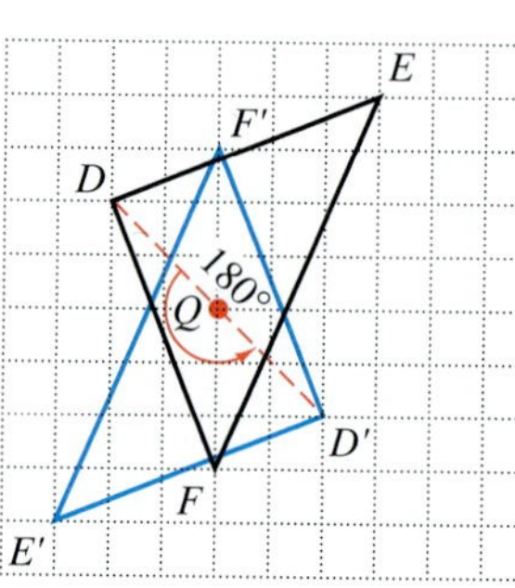

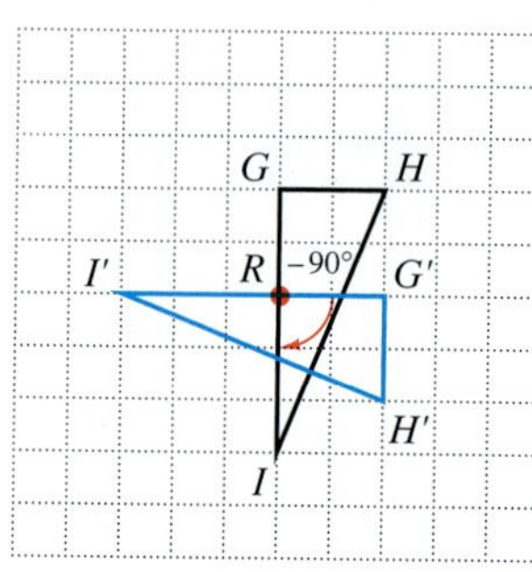

Reflections

The third basic rigid motion is a **reflection,** which is also called a **flip** or a **mirror reflection.** A reflection is determined by a line in the plane called the **line of reflection,** or the **mirror line.** Each point P of the plane is transformed to the point P' on the opposite side of the mirror line m and at the same distance from m, as shown in Figure 13.6. Note that P' is located so that m is the perpendicular bisector of $\overline{PP'}$. Every point Q on m is transformed into itself; that is, $Q' = Q$ if Q is any point on m. Note that a right hand is reflected to a left hand, and vice versa.

Reflections can be performed with a trace-and-flip procedure using an acetate transparency. First, the original figure is traced, including the line of reflection and a reference point (such as Q in Figure 13.6). The transparency is then turned over to perform the flip, and the points along the line of reflection are placed over their original position. The alignment of the reference point ensures that no sliding along the reflection line occurred.

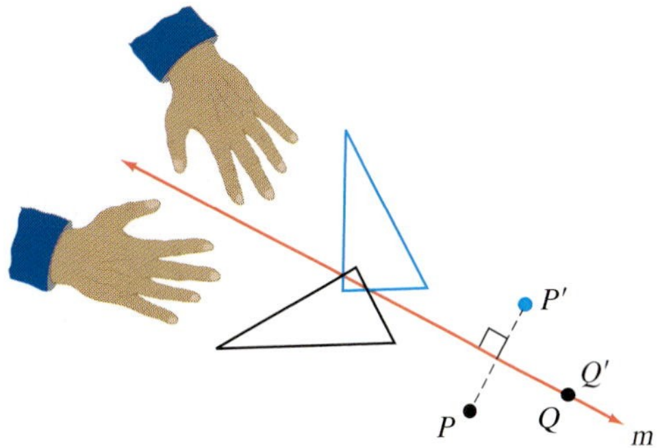

Figure 13.6
A flip, or reflection, about a line m *transforms each point of the plane to its mirror image on the opposite side of* m

The Mira is ideally suited to draw reflections. As shown in Figure 13.7, the plastic surface of a Mira both reflects a figure in front and still allows points behind the surface to be seen. This makes it simple to draw the reflected image of a given figure, with the bottom edge of the Mira acting as the line of reflection.

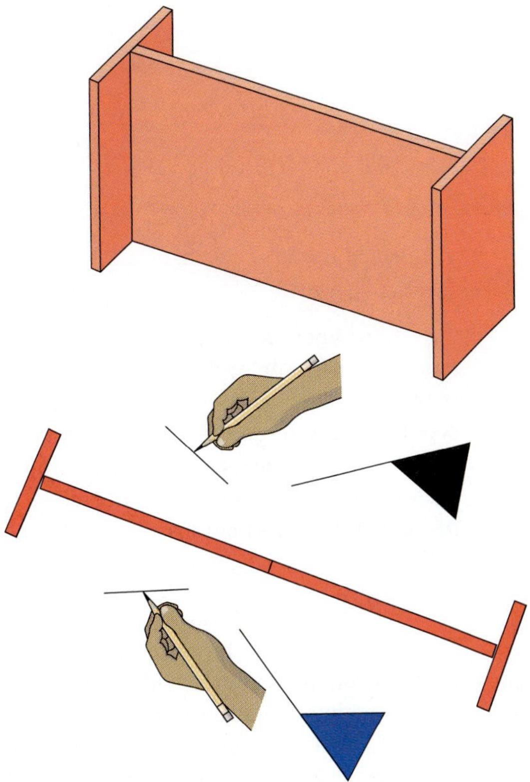

Figure 13.7
The Mira on the top can be used to draw reflections, as shown on the bottom

EXAMPLE 13.3 FINDING IMAGES UNDER REFLECTIONS

Sketch the image of the flag under a flip across line m.

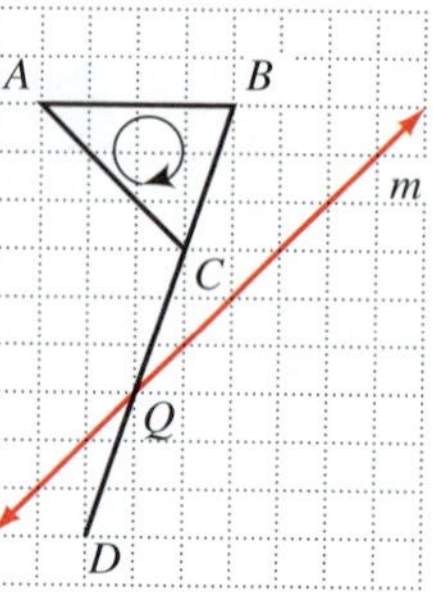

Solution

We can follow the trace-and-flip method if an acetate sheet or tracing paper is available, or we can use a Mira. Alternatively, the point A' that is the mirror point of A across line m can be plotted. Similarly B', C', and so on can be plotted, until the entire image can be sketched accurately.

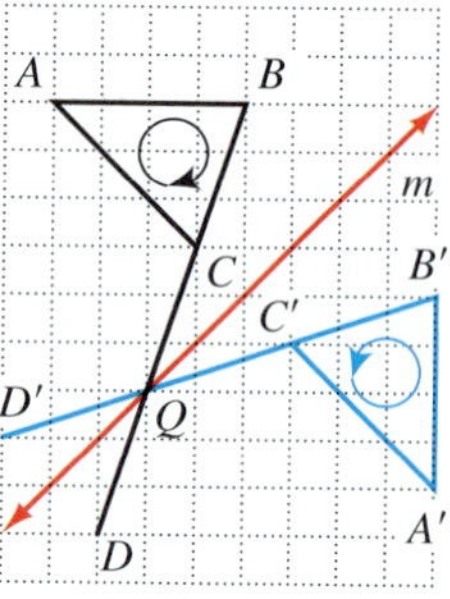

It's important to notice that a reflection reverses left-handed and right-handed orientations. For example, the left-pointing flag in Example 13.3 is transformed into a right-pointing flag, and the clockwise-pointing arrow on the circle becomes a counterclockwise-pointing arrow in the image. A rigid motion that interchanges "handedness" is called **orientation reversing.** Thus, a reflection is orientation reversing. Translations and rotations, since they do not reverse handedness, are **orientation-preserving** transformations.

Glide Reflections

The fourth, and last, basic rigid motion is the **glide reflection.** As the name suggests, a glide reflection combines both a slide and a reflection. The example most easily recalled is the motion that transforms a left footprint into a right footprint, as depicted in Figure 13.8. It is required that the line of reflection be parallel to the direction of the slide. In Figure 13.8 the slide came before the reflection, but if the reflection had preceded the slide the net outcome would have been the same.

A glide reflection changes handedness, so it is an orientation-reversing rigid motion. This is due to the reflection part of the motion.

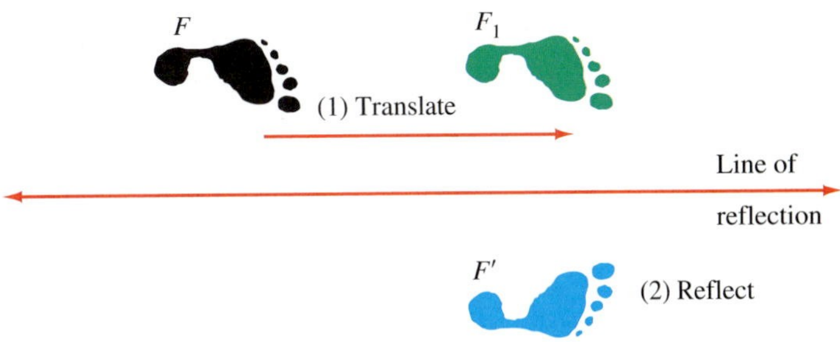

Figure 13.8
A glide reflection combines (1) a slide and (2) a reflection, where the line of reflection is parallel to the direction of the slide

To determine a glide reflection, it is useful to observe from Figure 13.9 that the midpoint M of the segment $\overline{PP'}$ lies on the mirror line of the reflection. This information is the key to solving the problem in the next example.

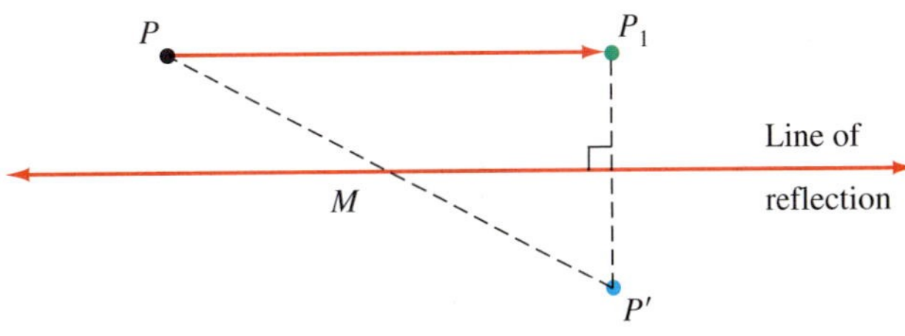

Figure 13.9
If points P and P′ correspond under a glide reflection, then the midpoint M of PP′ lies on the line of reflection

EXAMPLE 13.4 DETERMINING A GLIDE REFLECTION

A glide reflection has taken points A and B of triangle ABC to the points A' and B', as shown below. Find the line of reflection and the slide arrow of the glide reflection, and then sketch the image triangle $A'B'C'$ under the glide.

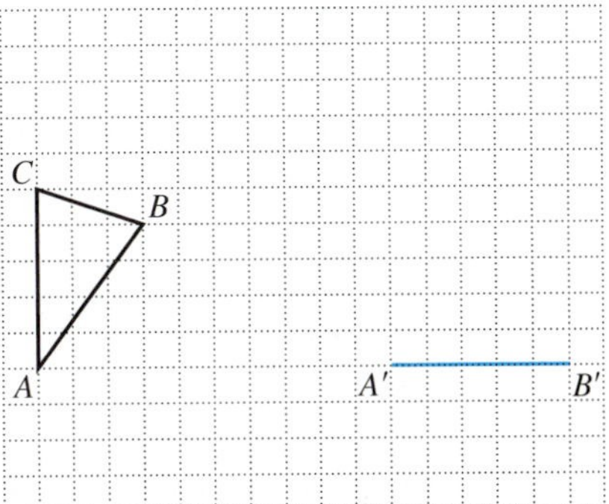

Solution The square grid makes it easy to draw the respective midpoints M and N of the line segments $\overline{AA'}$ and $\overline{BB'}$. Since both M and N lie on the mirror line, $m = \overleftrightarrow{MN}$ is the line of reflection of the glide. Reflecting A' across m determines the point A_1, and the slide arrow of the glide is drawn by connecting A to A_1. The slide arrow is eight units to

School Book Page

The Rigid Motions in Grade Four

PRACTICE AND APPLY

Connect

Two figures that have the same size and shape are **congruent**.

Congruent Not congruent

You can **flip**, **turn**, and **slide** a figure to test if two figures are congruent.

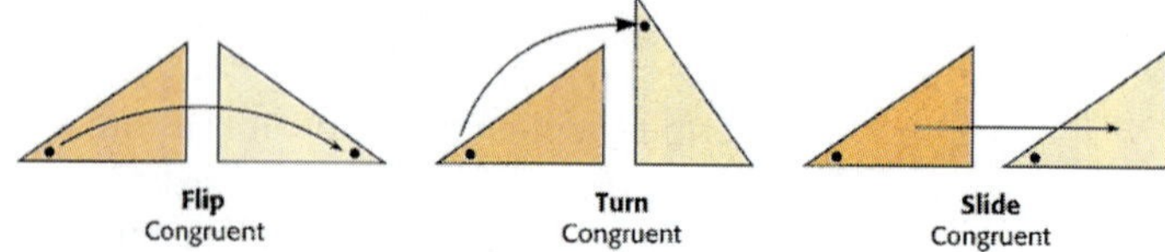

Each pair of triangles fit on top of each other exactly.

Practice

Write whether each picture shows a slide, flip, or turn.

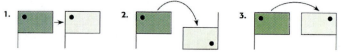

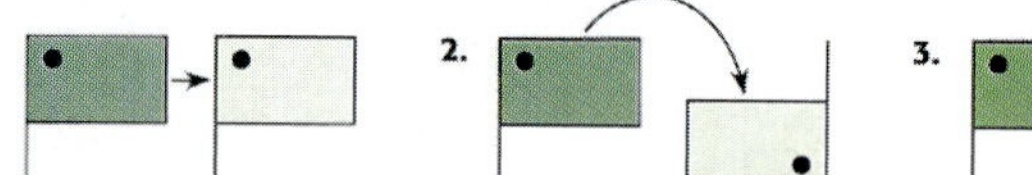

4. **Reasoning** Brian says that a square looks the same whether he slides, flips, or turns it. Is he right? Explain.

5. **Fine Arts** Draw a part of the design in the tile that shows:
   a. a slide.
   b. a flip.
   c. a turn.

Mexican tile makers are known for the beauty of their designs.

6. Do two triangles drawn side by side always show a slide? Explain.

7. **Language Arts** If you turn a lower-case b, you get a q. Find letters that look like other letters after they are flipped or turned. (Don't forget capital letters.)

8. **Journal** How can you use slides, flips, and turns to decide whether two figures are congruent?

Questions for the Teacher

1. Flips, turns, and slides are illustrated with scalene triangles, and flags *with flag poles* are shown in Practice Questions 1, 2, and 3. Why are these shapes preferred to either an equilateral triangle or a rectangle?
2. The Teacher's Edition claims that the answer to Practice Question 4 is "Not always true; a square has equal sides and angles, so it looks the same when you slide it or flip it. It could look like a rhombus if it were turned." Comment on the accuracy of this answer.
3. A student asks you if it is possible to flip the flag with the pole downward so the image pole points upward. What's a good response to this student?
4. Another student asks if it is possible to flip the green flag with the pole pointing downward so the image pole is horizontal. How should you respond?

the right and four units up, which allows us to find C_1. Reflecting C_1 across the line of reflection locates C', and therefore $\triangle A'B'C'$ can be completed.

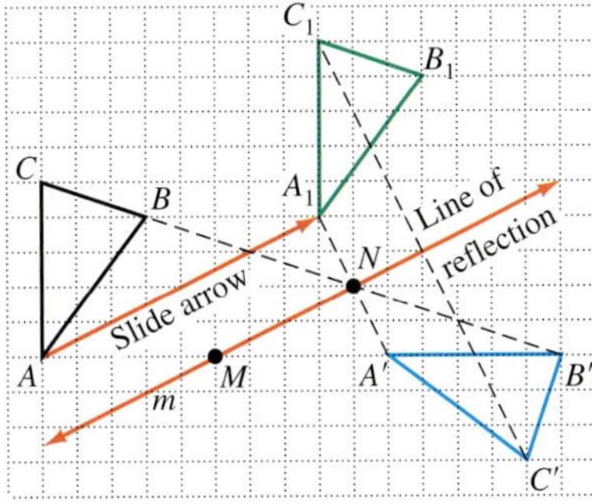

Table 13.1 summarizes useful information about the four basic rigid motions.

TABLE 13.1 The Four Basic Rigid Motions

Name (alternate name) and Sketch	Information Needed	Description	Orientation Property
Translation (slide) 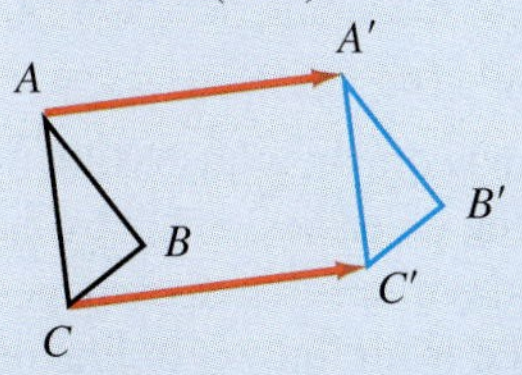	Slide arrow, indicating distance and direction	Every point of the plane is moved the same distance in the same direction.	Orientation is preserved.
Rotation (turn) 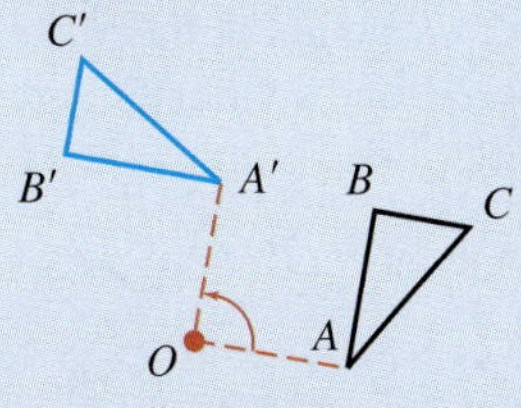	Turn arrow, indicating turn center and turn angle	Every point of the plane is rotated through the same directed angle about the turn center.	Orientation is preserved.
Reflection (flip) 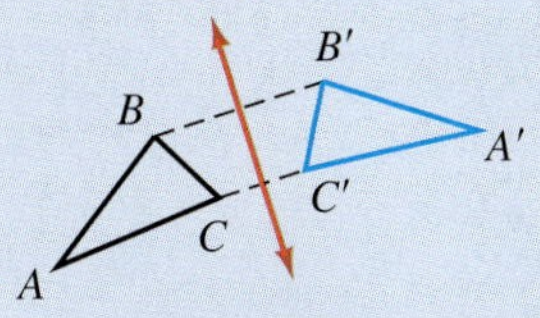	Line of reflection	Every point of the plane is moved to its mirror image on the opposite side of the line of reflection.	Orientation is reversed.
Glide reflection (glide) 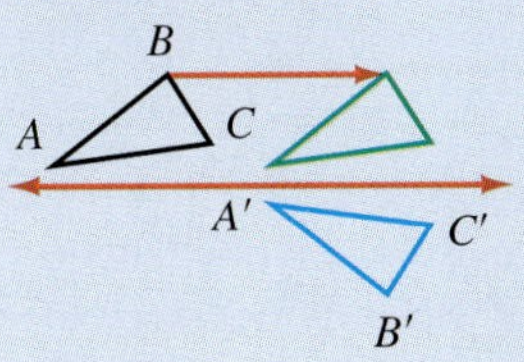	Slide arrow and a line of reflection parallel to the slide direction	Every point of the plane is moved by the same translation and reflected across the same line parallel to the slide direction.	Orientation is reversed.

The Net Outcome of Two Successive Reflections

Recall that any two rigid motions that have the same net outcome are called equivalent. For example, rotations of $-120°$ and $+240°$ about the same center O are equivalent. Similarly the motion consisting of two consecutive 180° rotations about a point O is equivalent to the **identity transformation,** which is the rigid "motion" that leaves all points of the plane fixed.

Two consecutive reflections across the same line of reflection also bring each point back to its original position, so any "double flip" is also equivalent to the identity transformation. Suppose, however, that two flips are taken in succession across two *different* lines of reflection, say, first over m_1 and next over m_2. There are two cases to consider, where m_1 and m_2 are parallel and where m_1 and m_2 intersect.

EXAMPLE 13.5 EXPLORING CONSECUTIVE REFLECTIONS ACROSS PARALLEL LINES

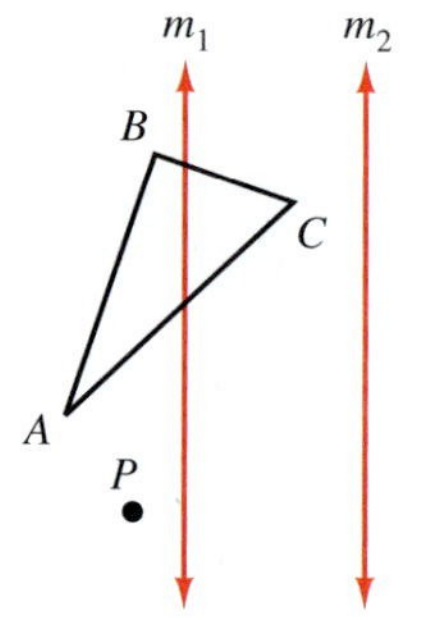

Let m_1 and m_2 be parallel lines of reflection.

(a) Sketch the image of $\triangle ABC$ and point P under the reflection across m_1; let the image be labeled $\triangle A_1B_1C_1$ and P_1.
(b) Sketch the image of $\triangle A_1B_1C_1$ and P_1 under reflection across line m_2; let the image be labeled $\triangle A'B'C'$ and P'.
(c) Describe the net outcome of the rigid motion consisting of the two successive reflections, first across m_1 and next across m_2.

Solution

(a) and **(b)** Each reflection can be drawn with a Mira, by the trace-and-flip method with an acetate sheet, or, easiest of all, by using geometry software. Whatever method is used will result in the images shown below.

(c) If d is the directed distance from line m_1 to line m_2, we see that point P is moved a distance $2d$ in the direction perpendicular to m_1 and m_2 and pointing from m_1 toward m_2. In fact *all* points of the plane are moved in this direction through the same distance $2d$, and so the net outcome of two successive reflections across a pair of parallel lines is equivalent to the translation shown on the right below.

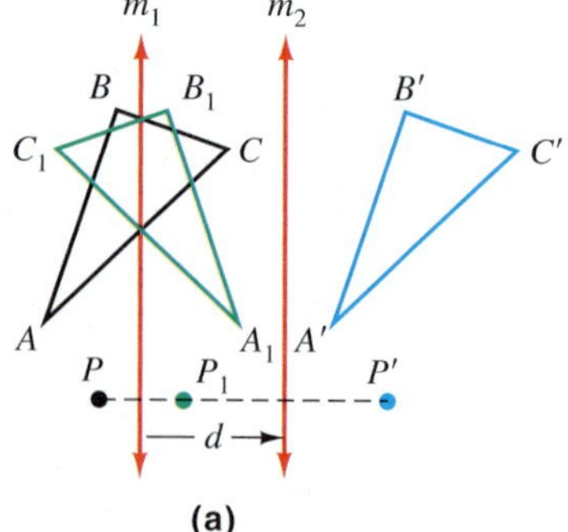

(a)
Two reflections across parallel lines

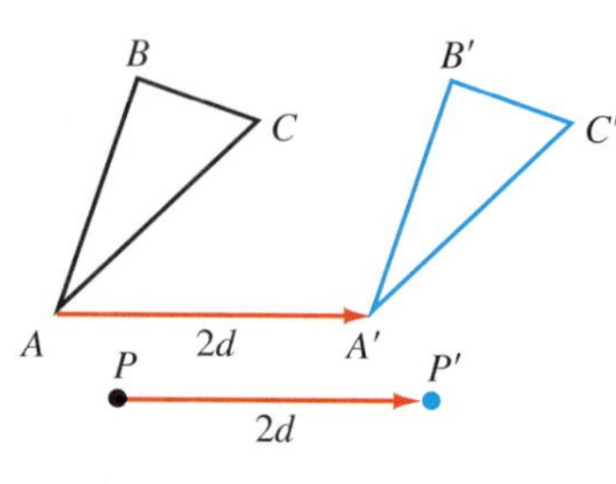

(b)
The equivalent translation

A similar investigation can be carried out for the motion consisting of two successive reflections across lines m_1 and m_2 that intersect at a point O. Most people find the result very surprising: *the net outcome of the two reflections across intersecting lines is equivalent to a rotation about the point O of intersection of m_1 and m_2. The angle of rotation is twice the measure of the directed angle that turns line m_1 onto line m_2.* This is illustrated in Figure 13.10.

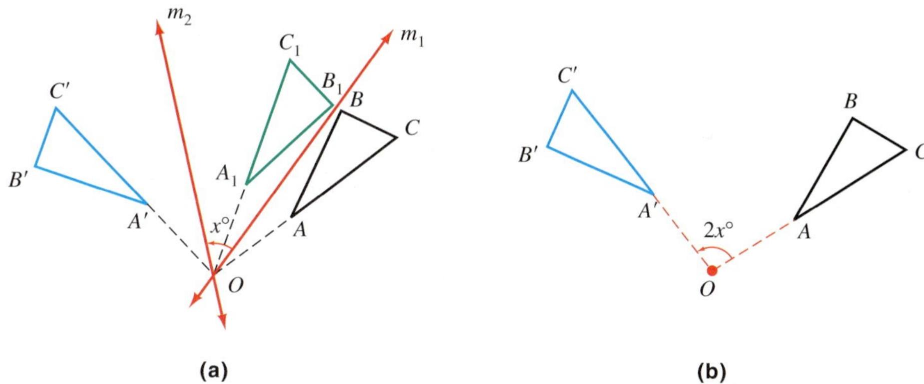

Figure 13.10
Two reflections across intersecting lines are equivalent to a rotation about the point of intersection of the two lines

The following theorem summarizes the two possible net outcomes of a pair of successive reflections.

THEOREM *The Net Outcome of Two Reflections*

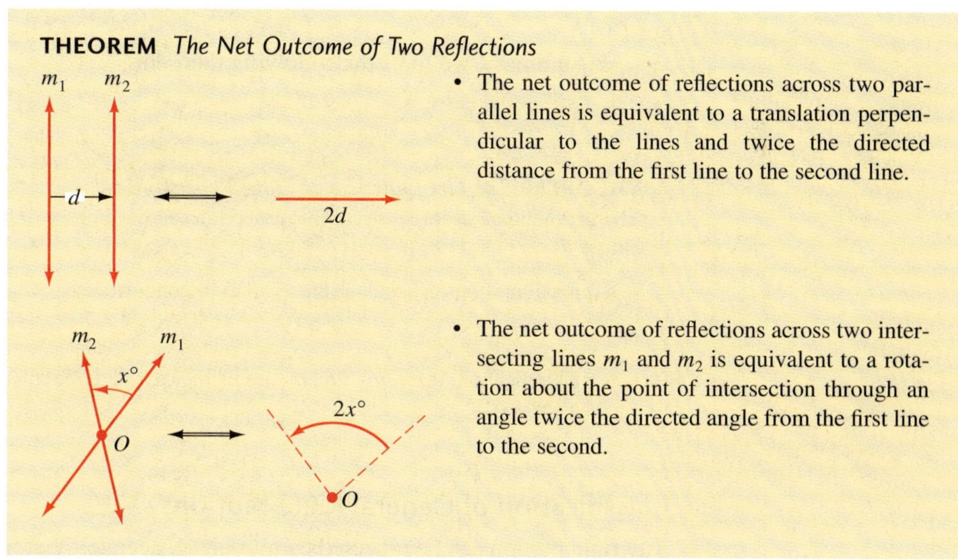

- The net outcome of reflections across two parallel lines is equivalent to a translation perpendicular to the lines and twice the directed distance from the first line to the second line.
- The net outcome of reflections across two intersecting lines m_1 and m_2 is equivalent to a rotation about the point of intersection through an angle twice the directed angle from the first line to the second.

The Net Outcome of Three Successive Reflections

Three lines can be arranged in several ways in the plane. For example, the lines m_1, m_2, and m_3 in Figure 13.11(a) are parallel to one another. The successive images of $\triangle ABC$ across the lines are shown, with $\triangle A'B'C'$ the image at the completion of all three reflections. In

Figure 13.11(b), we see that $\triangle ABC$ can be taken to $\triangle A'B'C'$ by a *single* reflection over the line l. Line l is the image of line m_1 under the translation that takes m_2 to m_3. Thus, three successive reflections across parallel lines are equivalent to a single reflection.

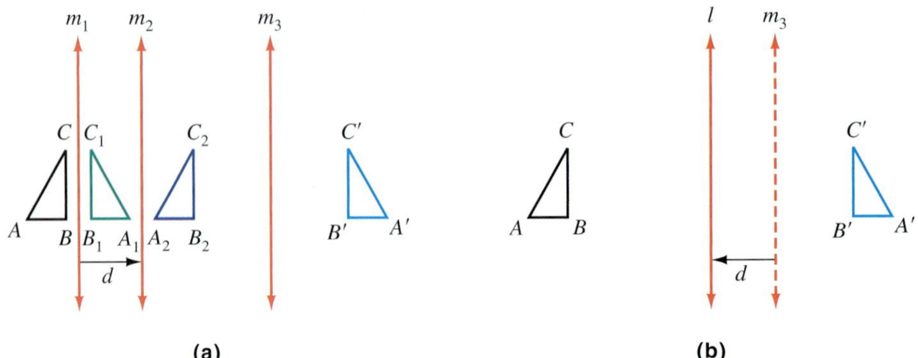

Figure 13.11
Three reflections across parallel lines m_1, m_2, *and* m_3, *are equivalent to a reflection across one line* l

Three reflections across concurrent lines m_1, m_2, and m_3 can also be discovered to be equivalent to a single reflection across a certain line l that passes through the point O of concurrence. (See problem 24 in Problem Set 13.1.) In all other cases, where the three lines are neither parallel nor concurrent, it can be shown that the net outcome of three successive reflections is equivalent to a glide reflection. (See problem 25 in Problem Set 13.1.)

In summary, we have the following theorem.

THEOREM *The Net Outcome of Three Reflections*
The net outcome of three successive reflections across lines m_1, m_2, and m_3 is equivalent to either

- a reflection, if m_1, m_2, and m_3 are parallel or concurrent;

or

- a glide reflection, if m_1, m_2, and m_3 are neither parallel nor concurrent.

Classification of General Rigid Motions

Any rigid motion can be modeled by moving an acetate sheet to a new position in the same plane. If the rigid motion takes $\triangle ABC$ to $\triangle A'B'C'$, the final position of the acetate sheet is uniquely determined by aligning point A with A', point B with B', and point C with C'. Let's now see how the three points can be aligned by a sequence of at most three reflections, as illustrated in Figure 13.12. Beginning with $\triangle ABC$ and its image $\triangle A'B'C'$, the first reflection across the perpendicular bisector of $\overline{AA'}$ maps $\triangle ABC$ to $\triangle A'B_1C_1$. The second reflection across the perpendicular bisector of $\overline{B_1B'}$ maps $\triangle A'B_1C_1$ to $\triangle A'B'C_2$. The third, and last, reflection across the perpendicular bisector $\overline{C_2C'}$ maps $\triangle A'B'C_2$ to $\triangle A'B'C'$.

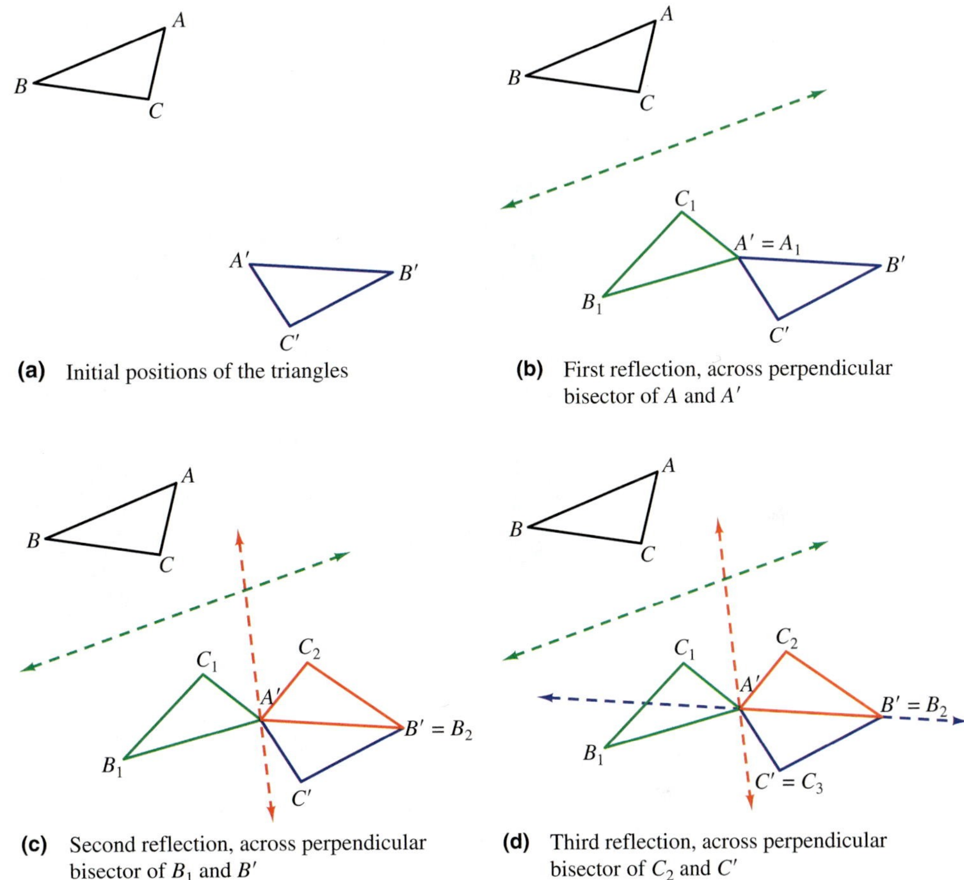

Figure 13.12
Any rigid motion is equivalent to a sequence of at most three reflections

If it happens that $A = A'$, the first reflection is omitted. Similarly, if $B_1 = B'$, the second reflection is omitted, and the third reflection is omitted if $C_2 = C'$.

Since two reflections are equivalent to a translation or rotation, and three reflections are equivalent to a single reflection or a glide reflection, we have proved the following remarkable theorem.

THEOREM *Classification of General Rigid Motions*
Any rigid motion of the plane is equivalent to one of the four basic rigid motions: a translation, a rotation, a reflection, or a glide reflection.

Rigid motions have many applications in geometry. For example, the informal definition of congruence to mean "same size and shape" can now be made precise.

DEFINITION *Congruent Figures*
Two figures are **congruent** if, and only if, one figure is the image of the other under a rigid motion.

The periodic drawings and block prints of M. C. Escher show how the plane can be tiled by congruent figures. Figure 13.13 illustrates a two-motif pattern of fish of two types.

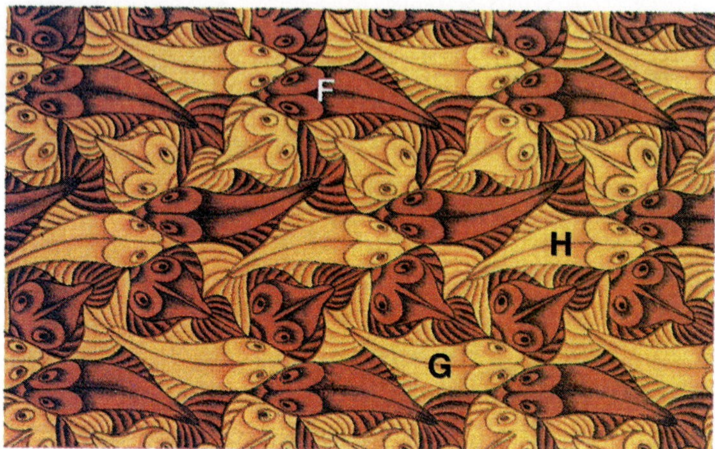

Figure 13.13
A two-motif tiling of the plane by M. C. Escher

EXAMPLE 13.6 CLASSIFYING RIGID MOTIONS

Examine the tiling of M. C. Escher shown in Figure 13.13. Three of the large fish are labeled F, G, and H.

(a) What type of rigid motion takes F onto G?
(b) What type of rigid motion takes F onto H?
(c) What type of rigid motion takes G onto H?

Solution

(a) Since F and G have the same orientation (both bend the tail to the left), they are related by an orientation-preserving transformation. Since the fish face in opposite directions, the motion is not a translation, and so it must be a rotation. (Can you identify the turn center and the size of the angle of rotation?)
(b) F and H have opposite orientation, so either a reflection or a glide reflection takes F onto H. H is not a reflection of F, so it must be a glide reflection. (Can you determine the line of reflection of the glide?)
(c) A glide reflection takes G onto H. The slide is horizontal to the right.

Dilations and Similarity Motions

A rigid motion takes any two points P and Q to the image points P' and Q', respectively, without changing their distance apart. That is, $PQ = P'Q'$. Therefore, any figure mapped by a rigid motion is unchanged in both size and shape. Suppose, however, we wish to find transformations of the plane that preserve shape, but change the size of any figure. Transformations with this property are quite common: consider making an enlargement or reduction on a photocopy machine, or using the Zoom command from the View menu of a computer program.

The simplest transformation to change the size of a figure, but preserve its shape and orientation, is a **dilation** (also called a **size transformation**). A point O is chosen as the center, and the points of the plane are all moved toward $(k < 1)$ or away $(k > 1)$ from the center O by the same proportional factor k. More precisely, we have the following definition.

DEFINITION *Dilation, or Size Transformation*

Let O be a point in the plane and k a positive real number. A **dilation,** or **size transformation,** with **center O** and **scale factor k** is the transformation that takes each point $P \neq O$ of the plane to the point P' on the ray $\overrightarrow{OP}$ for which $OP' = k \cdot OP$, and takes the point O to itself.

Two examples of dilations are shown in Figure 13.14.

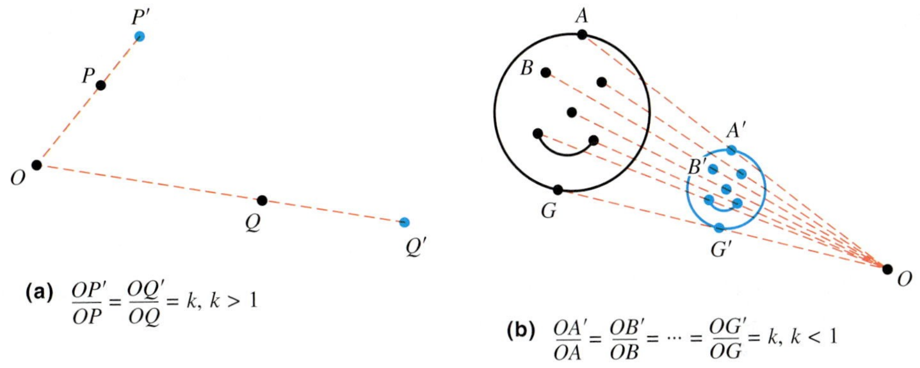

Figure 13.14
Two dilations, or size transformations

When the scale factor k is larger than 1, the image of a figure is larger than the original and the dilation is an **expansion.** If $k < 1$, the dilation is a **contraction.** If $k = 1$, then all points are left unmoved—that is, $P = P'$ for all P—and the dilation is the identity transformation.

The most important fact about dilations is contained in the following theorem. Recall that if a point P is taken onto the image point P', we say that P is the preimage of P'.

THEOREM *Distance Change Under a Dilation*

Under a dilation with scale factor k, the distance between any two image points is k times the distance between their preimages. That is, for all points P and Q, $P'Q' = k \cdot PQ$.

A sequence of dilations and rigid motions performed in succession is called a **similarity transformation.**

DEFINITION *Similarity Transformation*
A transformation is a **similarity transformation** if, and only if, it is a sequence of dilations and rigid motions.

We can now give a precise definition of similarity of figures.

DEFINITION *Similar Figures*
Two figures F and G are **similar,** written $F \sim G$, if, and only if, there is a similarity transformation that takes one figure onto the other figure.

The fish F and G in Figure 13.15 are similar to one another. A dilation centered at point O maps F to F', and a reflection maps F' to G.

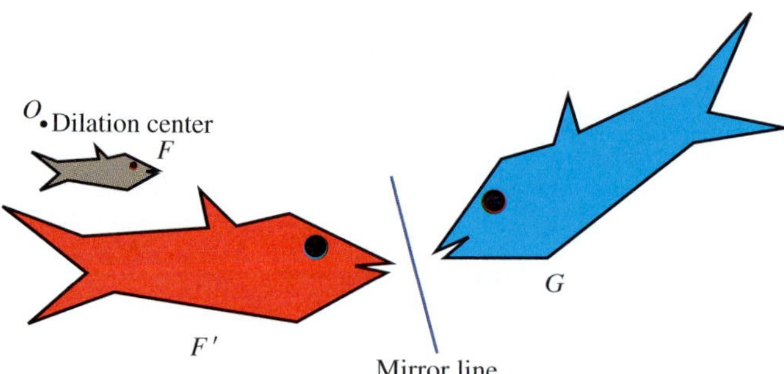

Figure 13.15
A dilation centered at O followed by a reflection define a similarity transformation taking figure F onto figure G. Therefore, figures F and G are similar

Did You Know?

You Can't Listen for Congruence

In 1966 Mark Kac of Rockefeller University asked an apparently simple question: "Can you hear the shape of a drum?" A drum, for Kac, can have any shape. All that's required is that it be a two-dimensional figure having an interior (the drumhead) and a boundary (the rim). The shape of the boundary determines an infinite set of characteristic frequencies at which the interior drumhead will vibrate. Two drums of the same shape generate the same set of frequencies, but Kac wanted to know if the converse is true: If you hear the same set of frequencies from two drums, are the drums necessarily the same shape?

In 1991 Carolyn Gordon and David Webb, formerly at Washington University in St. Louis, and Scott Wolpert at the University of Maryland showed that you cannot hear the shape of a drum. They did so by finding two noncongruent shapes that, if made into drums with drumheads of the same material stretched at the same tension, would vibrate at exactly the same frequencies.

SOURCE: Adapted from Barry Cipra, "You Can't Hear the Shape of a Drum," Science, *Vol. 255: 1642, March 27, 1992.*

EXAMPLE 13.7 VERIFYING SIMILARITY

Show that the small letter F and the larger letter F are similar by describing a similarity transformation that takes the smaller figure onto the larger one.

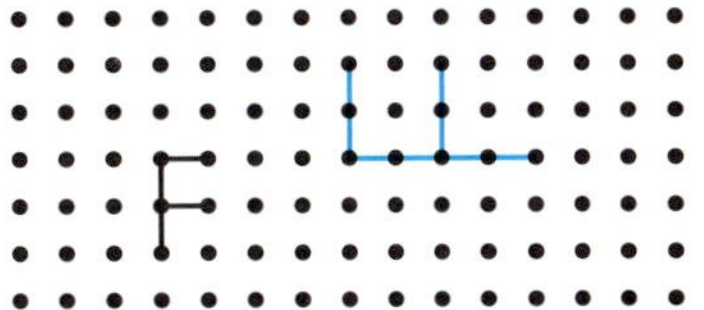

Solution

We need a sequence of dilations and rigid motions that rotate, stretch, and position the smaller letter onto the larger. This can be done in three steps: (a) make a 90° rotation, (b) do a dilation with scale factor $k = 2$, and (c) perform a translation.

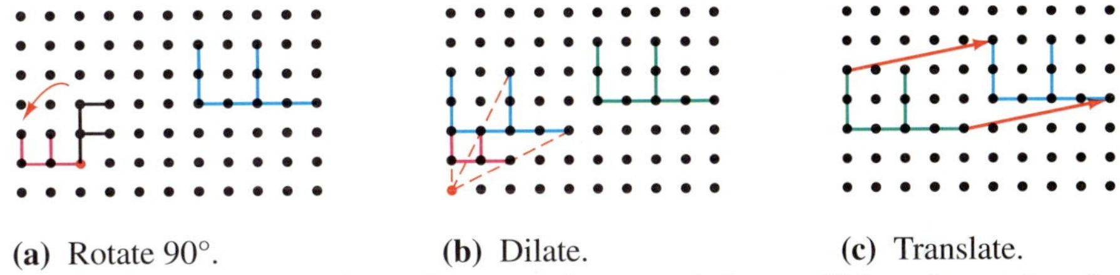

(a) Rotate 90°. **(b)** Dilate. **(c)** Translate.

These three steps are not unique. For example, a translation could have been taken first, then a 90° rotation, and finally a dilation with a properly chosen center. Showing that *some* similarity transformation takes one figure onto the other is all that is required.

Did You Know?

Self-Similarity and Fractals

A line is one dimensional, and a plane is two dimensional. But what if someone asks for a geometric figure whose dimensionality is *between* one and two? The question may seem unanswerable. However, beginning in the late nineteenth century, mathematicians have created numerous objects of uncertain dimension. The Koch curve is an example, named after the Swedish mathematician Helge von Koch, who first described it in 1904. To construct the Koch curve, begin with a line segment. Next replace the middle third by an equilateral triangular bump, resulting in a four-segment polygonal curve. In the third step, add a triangular bump to each side of the curve, which yields a curve with 16 sides.

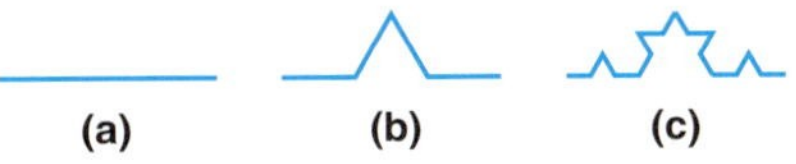

Repeating the process infinitely often gives the Koch curve.

The Koch curve has many interesting properties, but of special interest is its self-similarity: if fragments of the curve are viewed with microscopes of 100 power, 1000 power, or any power whatever, the enlargements all appear identical. This unlimited "roughness" of the Koch curve suggests that its dimension is larger than one. In 1975, Benoit B. Mandelbrot, a fellow at IBM's Thomas J. Watson Research Center, published the first comprehensive study of the geometry of self-similar shapes such as the Koch curve, the Sierpinski gasket, and the Menger sponge. According to Mandelbrot's definition, the Koch curve has dimension D given by the equation $4 = 3^D$. That is, $D = 1.2619\ldots$. Since self-similar objects may have nonintegral dimension, Mandelbrot called such objects fractals, from the same verb *frangere* ("to break") that is the source of our word "fraction."

Fractals are not simply abstract creations by mathematicians. Indeed, intense ongoing scientific research suggests that many, and possibly most, of the physical and life processes are what can be described as chaotic, and the geometric shapes of chaos are fractals. (For a nontechnical introduction to self-similarity, fractals, and chaos, consult James Gleick's *Chaos: Making a New Science*, Penguin Books, New York, 1987.)

Problem Set 13.1

Understanding Concepts

1. Which of the following "transformations" correspond to a rigid motion? Explain the reasoning you have used to give your answer.
   (a) A deck of cards is shuffled.
   (b) A completed jigsaw puzzle is taken apart and then put back together.
   (c) A jigsaw puzzle is taken from the box, assembled, and then placed back in its box.
   (d) A painting is moved to a new position on the same wall.
   (e) Bread dough is allowed to rise.

2. For each figure shown, find its image under the translation that takes P onto P'.

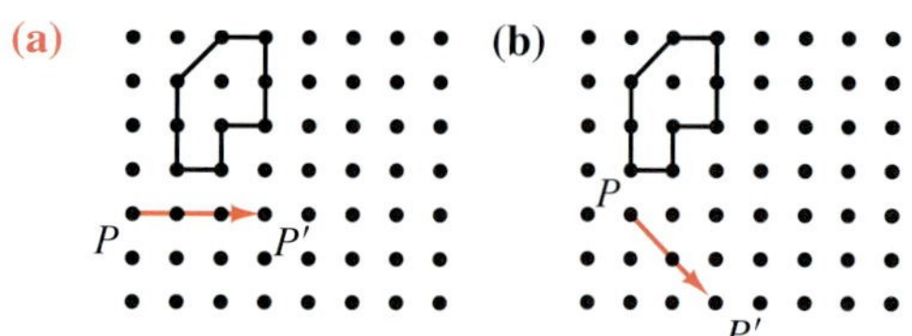

3. The translation that takes point P onto P' has transformed triangle ABC (not shown) into its image $A'B'C'$ (shown in the upcoming figure).
   (a) Draw triangle ABC.
   (b) Describe the rigid motion that transforms $\triangle A'B'C'$ into $\triangle ABC$, and compare it with the translation that takes $\triangle ABC$ onto $\triangle A'B'C'$.

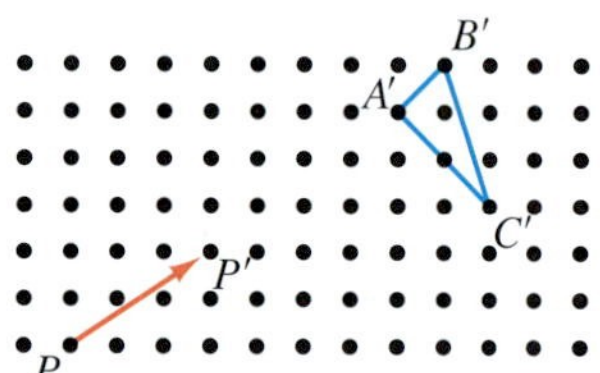

4. In each of the following statements, give an equivalent answer between 0° and 360°.
   (a) A clockwise rotation of 60° is equivalent to a counterclockwise rotation of ___________.
   (b) A clockwise rotation of 433° is equivalent to a clockwise rotation of ___________.
   (c) A clockwise rotation of 3643° is equivalent to a clockwise rotation of ___________.
   (d) A sequence of two consecutive clockwise rotations, first of 280° and next of 120°, is equivalent to a single clockwise rotation of ___________.
   (e) A rotation of $-260°$ is equivalent to a rotation of ___________.

5. Sketch the image of $\triangle ABC$ under the given rotations.

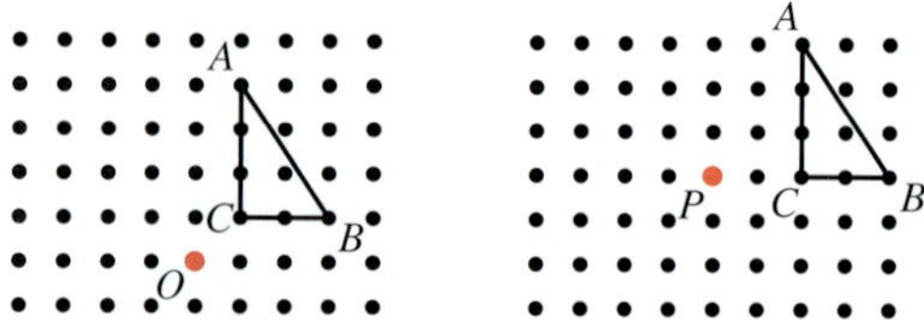

   (a) 90° counterclockwise about O
   (b) 180° about P

6. A rotation has sent A to A' and B to B', as shown below.

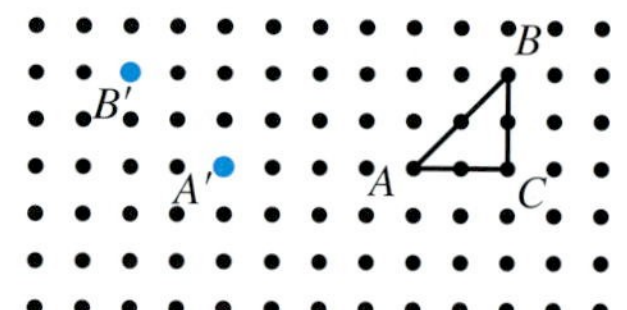

   (a) Find the center of rotation.
   (b) Find the turn angle.
   (c) Sketch the image triangle $A'B'C'$.

7. Trace the following figure, which shows $\triangle ABC$ and its image under a rotation. Use any drawing tools you wish (Mira, compass, or straightedge) to construct the center of rotation. (*Hint:* Why is the center of rotation on the perpendicular bisector of $\overline{AA'}$?)

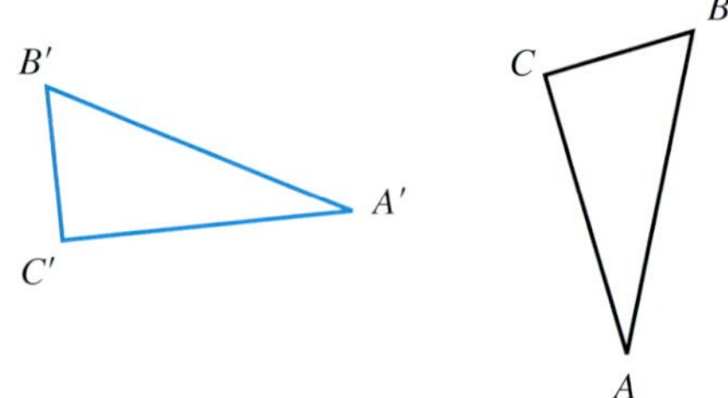

8. Redraw the figure below on graph paper. Then sketch the reflection of $\triangle ABC$ across the mirror line m.

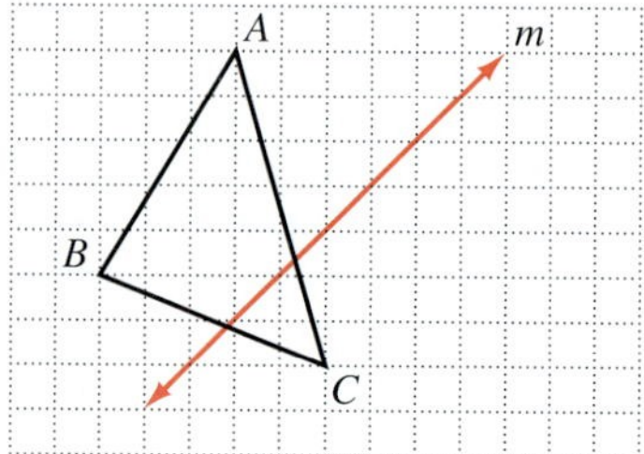

9. In the image below, a reflection has sent P to P'.
   (a) Find the line of reflection.
   (b) Find the image of the polygon $PQRST$ under the reflection that takes P to P'.

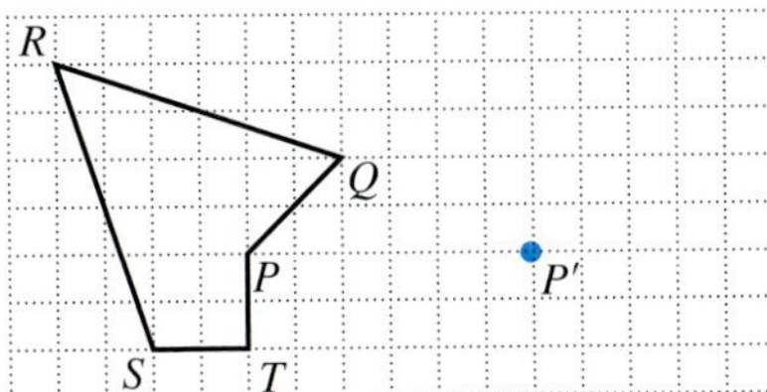

10. (a) A reflection across line m leaves point A fixed, so that $A' = A$. What can be said about A's relationship to m?
    (b) A reflection across line m leaves two points A and B fixed, so that $A' = A$ and $B' = B$. What can you say about A, B, and m?
    (c) A reflection takes point C to point D. Where does the reflection take point D?

11. A glide reflection is defined by the slide arrow and line of reflection m shown. Draw the following images of the polygon $ABCDE$.
    (a) The image $A_1B_1C_1D_1E_1$ under the slide.
    (b) The image $A'B'C'D'E'$ under the glide reflection.

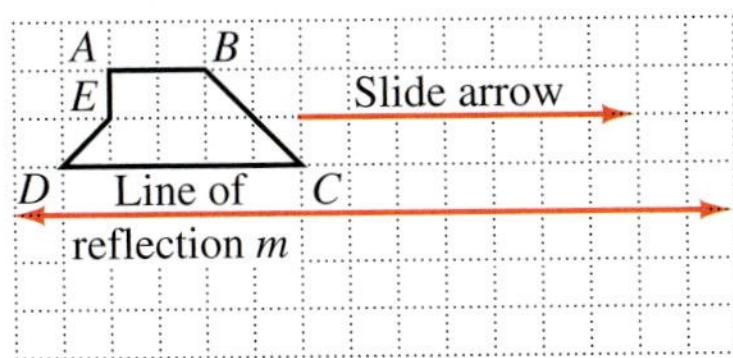

12. A glide reflection has taken B to B' and E to E'. Find
    (a) the line of reflection of the glide.
    (b) the slide arrow.
    (c) the image $A'B'C'D'E'$ of the polygon $ABCDE$.

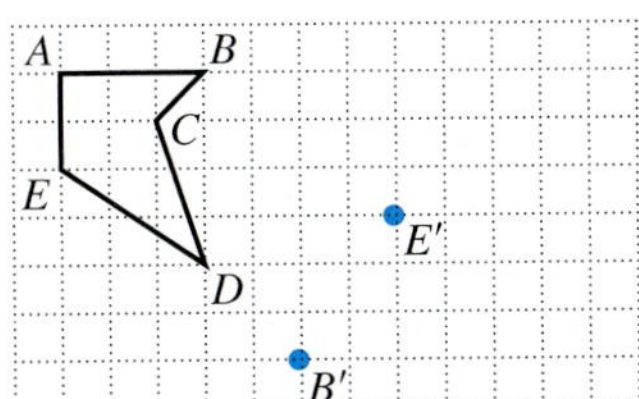

13. In each part below, draw a line m_2 so that the net outcome of successive reflections about m_1 and then m_2 is equivalent to the translation specified by the slide arrow shown.

(a)

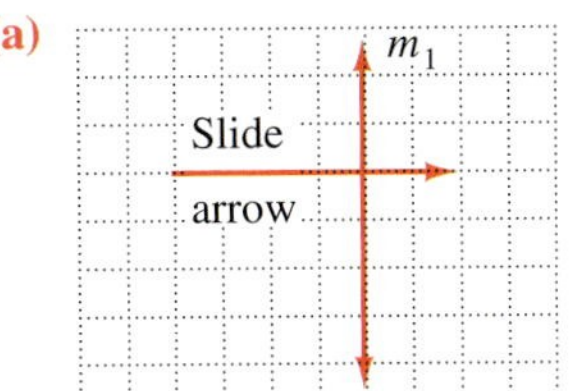

(b)

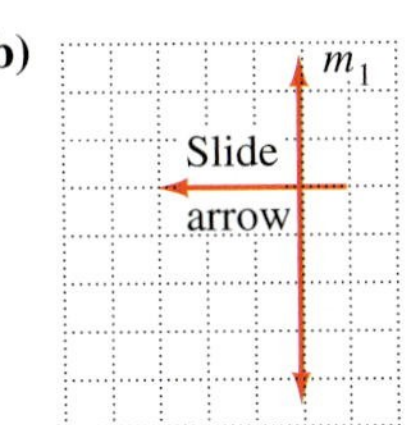

(c)

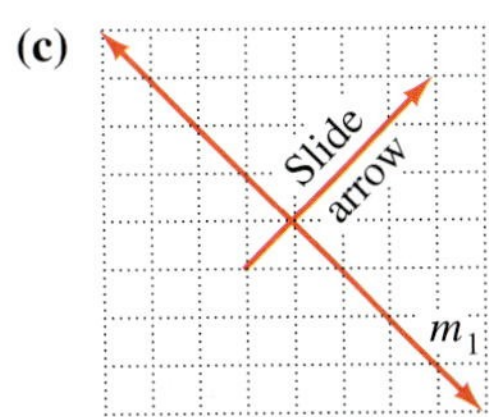

14. A reflection across line j followed by a reflection across line k as shown below is equivalent to the translation by four units to the right, as shown in the table.

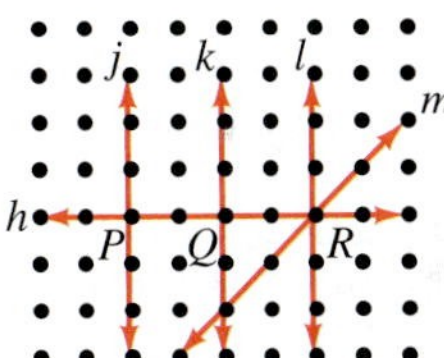

Fill in the missing entries (a) through (f) in the table.

REFLECTION LINES

First	Second	Equivalent Transformation
j	k	Translate right four units
l	m	Rotate clockwise 90° around point R
j	l	(a)
k	(b)	Translate left four units
(c)	k	Translate left four units
h	m	(d)
m	(e)	Rotate 180° about point P
k	(f)	Identity transformation (all points fixed)

15. Use dot or graph paper to copy $\triangle ABC$ and points O and P. Then draw the image of $\triangle ABC$ for

(a) the dilation with center O and scale factor 2.

(b) the dilation with center P and scale factor $1/2$.

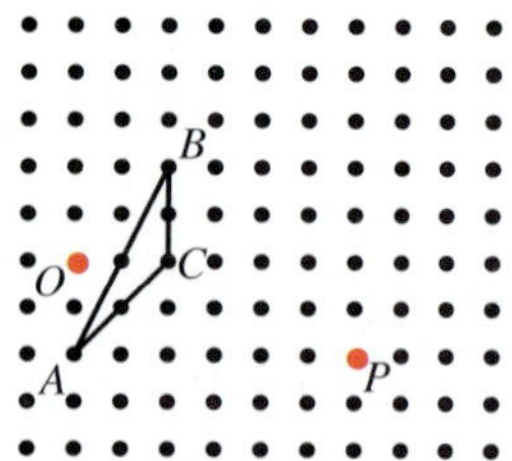

16. In the figure shown below, $\triangle DEF$ is the image of $\triangle ABC$ under the dilation with center O and scale factor 2.

(a) Describe the size transformation that takes $\triangle DEF$ onto $\triangle GHI$. Show the center of the transformation on a sketch, and give the scale factor.

(b) Describe the dilation that takes $\triangle ABC$ onto $\triangle GHI$ by locating the center and giving the scale factor.

(c) The Pythagorean theorem shows that $AB = \sqrt{5}$, so the perimeter of $\triangle ABC$ is $3 + \sqrt{5}$. Explain how to use the scale factors determined in parts (a) and (b) to obtain the perimeters of $\triangle DEF$ and $\triangle GHI$.

(d) A dilation with scale factor 4 takes $\triangle ABC$ onto $\triangle JKL$. What is the perimeter of $\triangle JKL$?

(e) Find the areas of the three triangles. Explain how the areas are related to the scale factors.

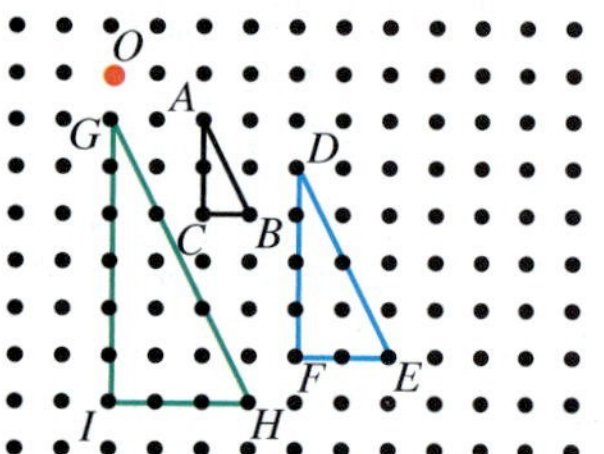

17. Sketch the image of $\triangle JKL$ under the similarity transformation composed of a dilation centered at point P with scale factor $2/3$, followed by a reflection across line m.

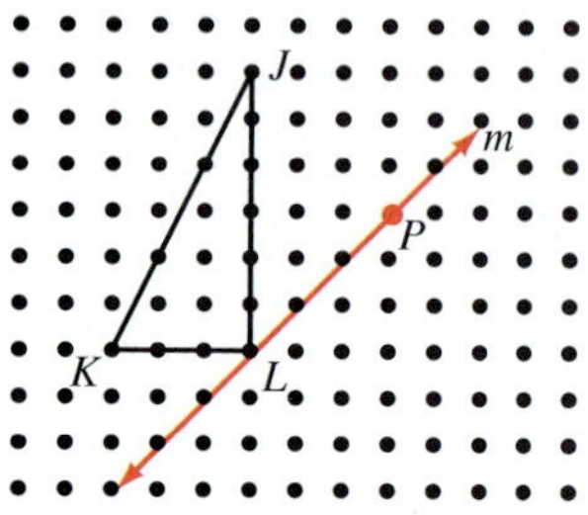

18. Describe a similarity transformation that takes quadrilateral $ABCD$ onto quadrilateral $A'B'C'D'$ as shown. Sketch the intermediate images of the dilation and rigid motions that compose the similarity transformation.

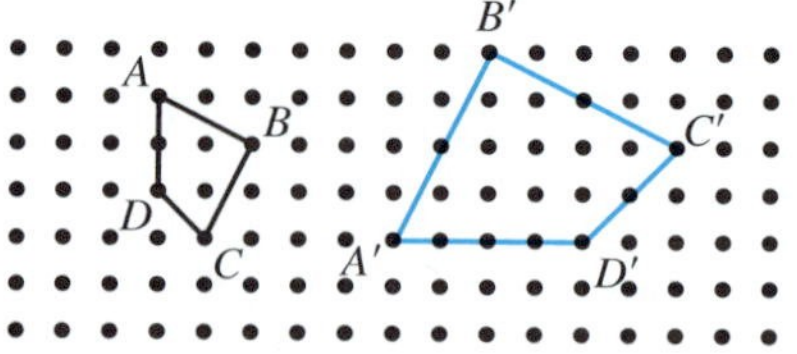

Teaching Concepts

19. Respond to these questions from students working with transformational geometry.

(a) "I reflected point A across the mirror line m, so now it's farther away from point B than before. Isn't a reflection supposed to keep all distances the same?"

(b) "When I rotated my square 90°, it was still a square, but when I rotated it 45°, it turned into a diamond. Isn't the rotated shape supposed to be congruent to the starting shape?"

(c) "Why does the slide arrow have to be parallel to the mirror line in a glide reflection?"

20. A student believes that all rectangles are similar, since all rectangles have four right angles and opposite sides of the same length. How would you respond to this student? In particular, how does the geometrical meaning of "similar" differ from its more general meaning?

Thinking Critically

21. A rigid motion takes points A, B, C, and P onto the respective image points A', B', C', and P', where $AB = 3$ cm, $AC = 4$ cm, $AP = 2$ cm, $BC = 2$ cm, $BP = 4$ cm, and $CP = 4$ cm. In each part, use a compass and ruler to draw the smallest set of points that you know must contain the point P' when you start with

(a) only point A'. (*Hint:* The answer is a circle.)

(b) points A' and B'.

(c) points A', B', and C'.

22. Two successive 90° rotations are taken, first about center O_1 and then about center O_2, where $O_1O_2 = 2$ cm. Describe the basic rigid motion that is equivalent to the successive rotations. Explain carefully, using words and sketches.

23. Suppose that the 90° rotations in problem 22 are each replaced with a 120° rotation. What basic rigid motion is equivalent to the net outcome of the two rotations? Explain with words and sketches.

24. (a) Find the image $\triangle A'B'C'$ of $\triangle ABC$ under the rigid motion consisting of three consecutive reflections across the concurrent lines m_1, m_2, and m_3.

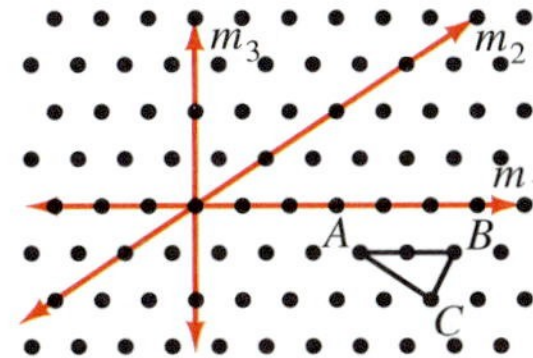

(b) Find a line l so that $\triangle ABC$ is taken onto $\triangle A'B'C'$ by one reflection across l.

25. (a) Find the image $\triangle A'B'C'$ of $\triangle ABC$ under the rigid motion consisting of three consecutive reflections across lines m_1, m_2, and m_3.

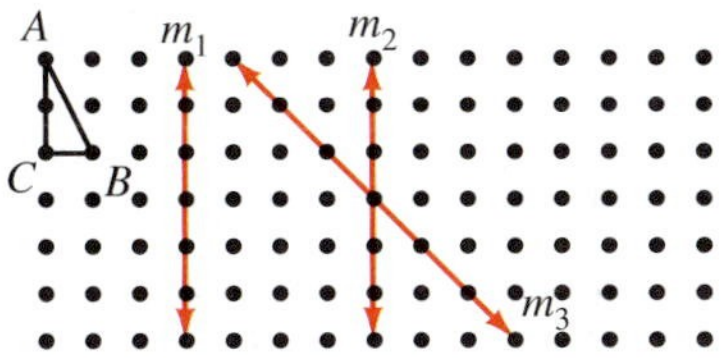

(b) Find the image of $\triangle ABC$ under the glide reflection whose slide arrow extends from P to P' and whose line of reflection is line l.

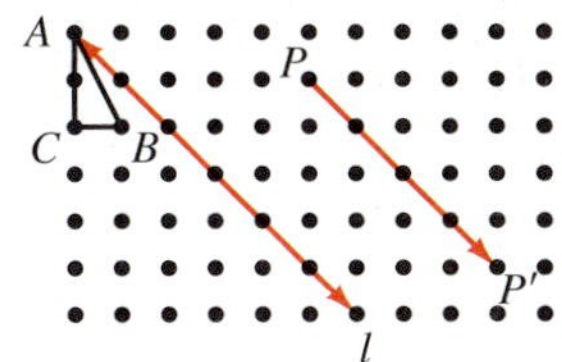

(c) What conclusion can you make about the two rigid motions described in (a) and (b)?

26. Trace the figure below, where $\triangle ABC$ is congruent to $\triangle A'B'C'$.

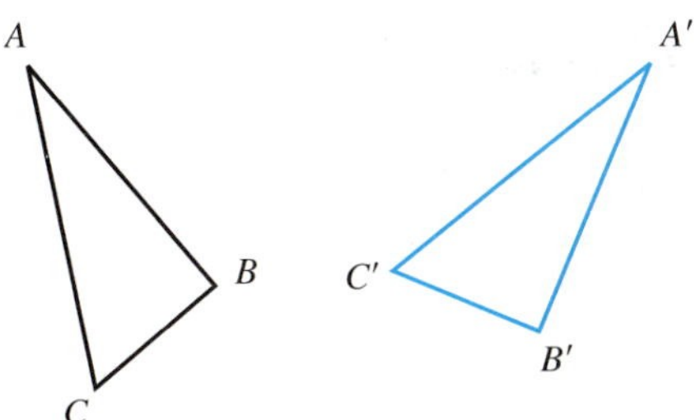

(a) Use a Mira (or other drawing tools) to draw the line m_1 across which A is reflected onto A'. Also, draw the images of B and C and label them B_1 and C_1, respectively.

(b) Draw the line m_2 across which B_1 reflects onto B'. What is the image of C_1 across m_2?

(c) Use the lines m_1 and m_2 to describe the basic rigid motion that takes $\triangle ABC$ to $\triangle A'B'C'$.

27. Trace the figure below, where $\triangle ABC$ is congruent to $\triangle A'B'C'$.

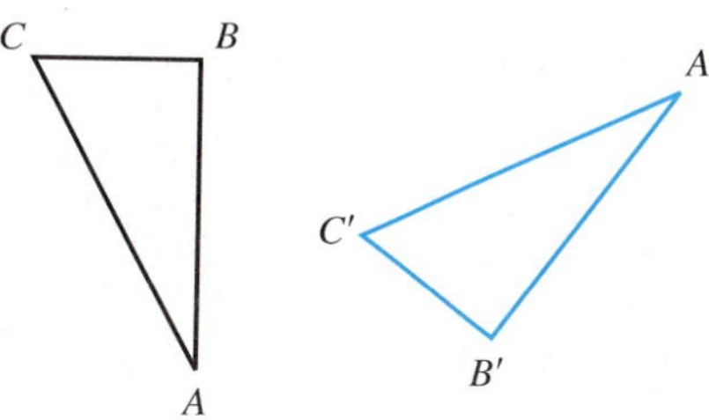

(a) Use a Mira (or other drawing tools) to draw three lines of reflection—m_1, m_2, and m_3—so that

(i) reflection across m_1 takes A onto A' (and B and C are taken to B_1 and C_1, respectively).

(ii) reflection across m_2 takes B_1 onto B' (and C_1 is taken onto C_2).

(iii) reflection across m_3 takes C_2 onto C'.

(b) Describe the type of basic rigid motion that takes $\triangle ABC$ onto $\triangle A'B'C'$.

28. In each of the following parts, a complicated sequence of rigid motions is described. Explain how you know what type of basic rigid motion is equivalent to the net outcome of the motion described.

(a) Reflections are taken across six lines, and no point is taken back onto its original position.

(b) Reflections are taken across 11 lines, and there are points that are taken back onto their original positions.

(c) Two different glide reflections are taken in succession, with the net outcome taking some point back onto its original location.

29. Let C and D be the two circles on the same side of mirror m. Construct a tangent line to circle C that, after reflection, is also tangent to circle D. How many solutions can you find?

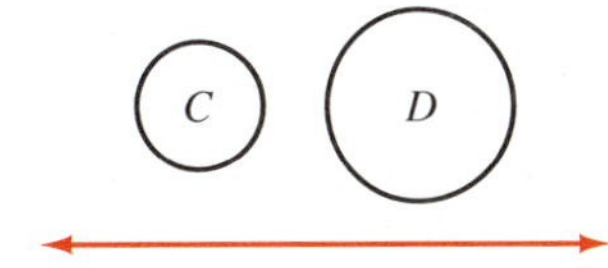

30. A graphic by M. C. Escher is shown below.

(a) What rigid motion takes figure A onto figure B?

(b) What rigid motion takes figure A onto figure C?

(c) What rigid motion takes figure C onto figure D?

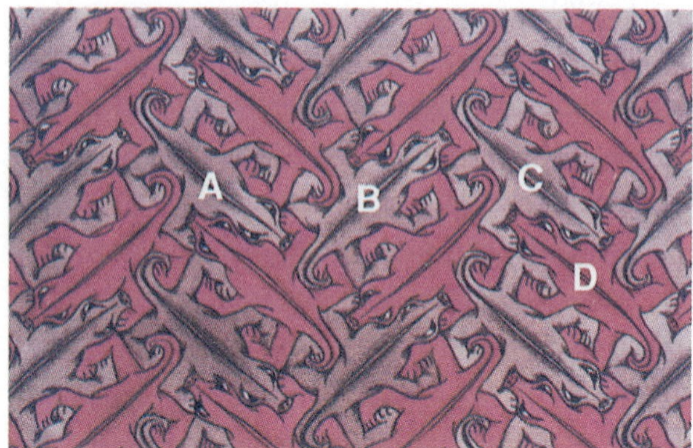

31. A size transformation centered at some point O of line l has taken point P onto P'. Explain how to draw (a) the center O of the transformation, and (b) the image Q' of the point Q, where Q lies on l.

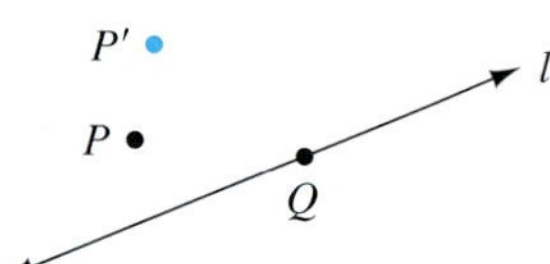

32. A size transformation takes P onto P' and Q onto Q', where the four points P, P', Q, and Q' are collinear as shown here. Explain **(a)** how to draw the image R' of point R, and **(b)** how to locate the center O of the size transformation. (*Hint:* What line through P' must contain R'?)

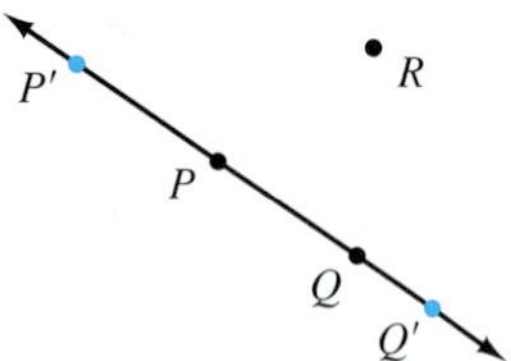

33. A mattress should be turned periodically in order to wear evenly. The mattresses shown below have "flip marks," indicating the axis over which the mattress is flipped. The dashed marks are on the reverse side of the mattress. For example, each mattress below is first flipped over the horizontal axis.

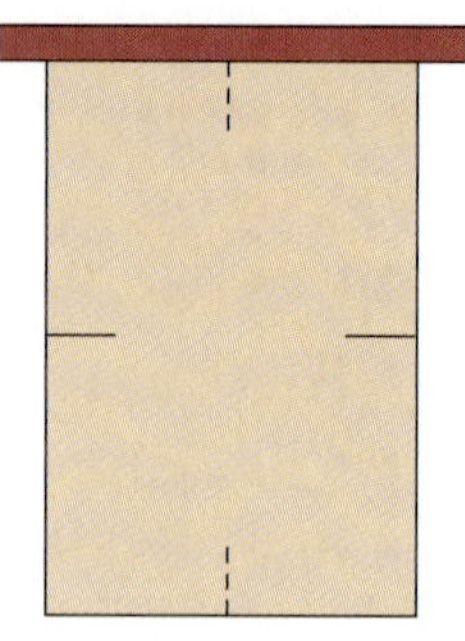

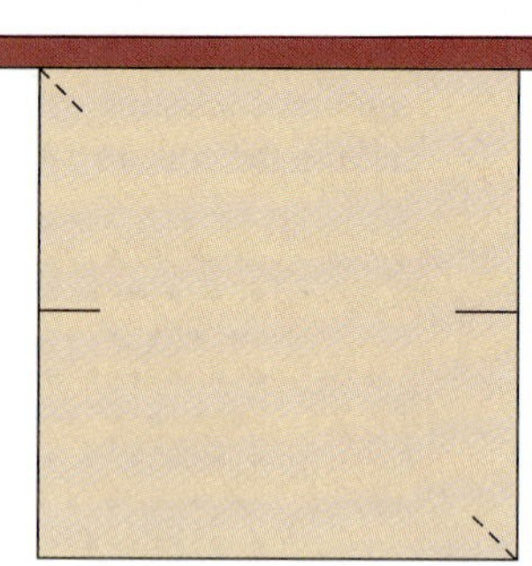

(a) Will the sequence of flips on the left-hand rectangular mattress cycle through all of the positions the mattress can be put on the bed? (*Suggestion:* Model the mattress with an index card, with marks on one side indicating a flip from head to foot, and marks on the opposite side indicating a side-to-side flip.)

(b) Answer part (a) again, but for the square mattress shown.

Making Connections

34. **Fermat's Principle.** Suppose that a ray of light emanating from point P is reflected from a mirror at point R toward point S. It was known even in ancient times that the incident and reflected rays of light make congruent angles to the line m of the mirror. Pierre de Fermat (1601–1665) proposed an important principle to explain why this is so: *light follows the path of shortest distance.* According to Fermat's principle, if Q is some point on the mirror other than R, then the distance $PQ + QS$ must exceed the distance $PR + RS$ traveled by the light.

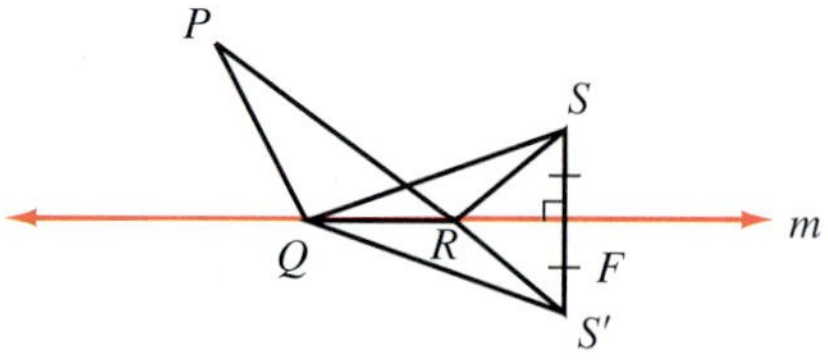

Answer the following questions to verify that $PQ + QS > PR + RS$. Let S' be the image of S under reflection across line m.

(a) Why are $RS' = RS$ and $QS' = QS$? (*Hint:* The reflection across m is a *rigid* motion that preserves distances.)

(b) Why is $PQ + QS' > PS'$?

(c) How does the inequality of part (b) give the desired result that $PQ + QS > PR + RS$?

35. Let P and S be two points on the same side of mirror m. If S' is the point of reflection of point S across m, then the line drawn from P to S' intersects m at the point R of reflection. (See the figure in problem 34.) Suppose P and S are between *two* mirrors m and l as shown below.

(a) Construct a doubly reflected light path $PQRS$ that is reflected off mirror m at Q and then off mirror l at R.

(b) Construct another doubly reflected path $PABS$ that first reflects off mirror l at A and then off mirror m at B.

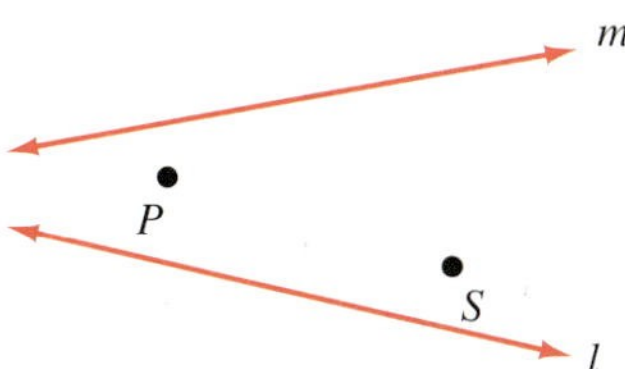

36. A **corner mirror** is formed by placing two mirrors together at a right angle. Explain why looking at yourself in a corner mirror is quite different from seeing yourself in an ordinary mirror. Use sketches to make your ideas clear.

37. A billiard ball is located at a point P along an edge of a rectangular billiard table. Show that there is a billiard shot that strikes the cushions on the other three rails (sides) of the table and then returns to bounce at P.

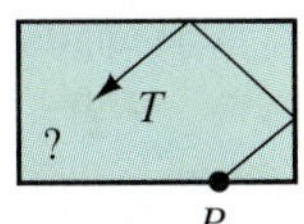

(*Hint:* Suppose the table T is reflected across its sides successively, forming the images T', T'', and T'''.) Explain how a billiard path $PQRSP$ can be found by drawing the line segment $\overline{PP'''}$. Sketch the path $PQRSP$ on the original table T.

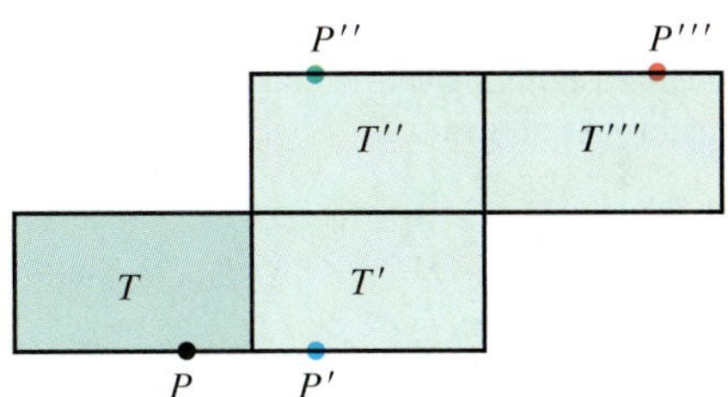

38. A **pantograph** is a mechanical device used to draw enlargements. A simple version can be constructed from cardboard strips that are hinged with brass fasteners. The pivot point at O is held fixed as P is moved over the figure. The pencil point at point P' traces out an enlargement.

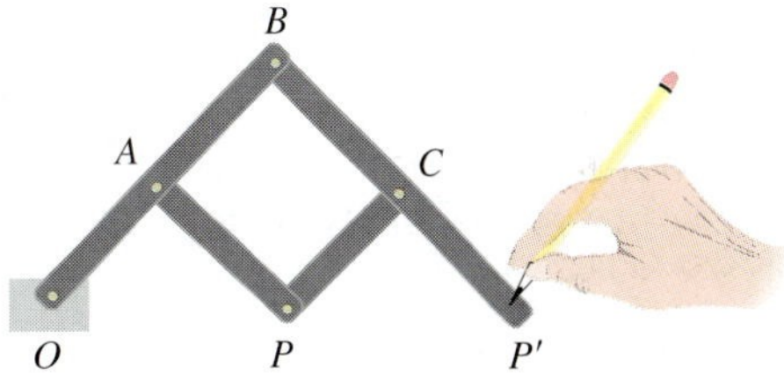

(a) Explain why the pantograph mechanically gives a size transformation.

(b) What is the scale factor of the size transformation? Assume that all of the adjacent points along a strip are the same distance apart.

Using a Computer

39. Draw any convex quadrilateral $ABCD$ with geometry software, and construct the midpoint M of side $\overline{AB}$. Mark point M as the center of rotation, and rotate the quadrilateral 180° about M to form the hexagon $AC'D'BCD$. Show that translations of the hexagon will fit together to tile (cover without overlaps) the plane.

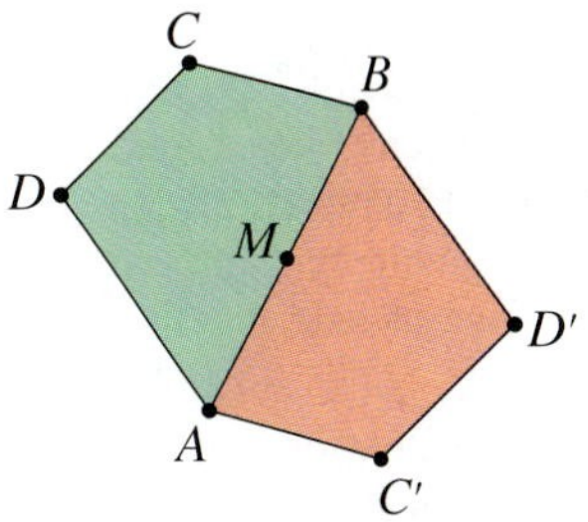

40. Use geometry software to construct a triangle ABC and the outward-pointing equilateral triangles with centers X, Y, and Z as shown.

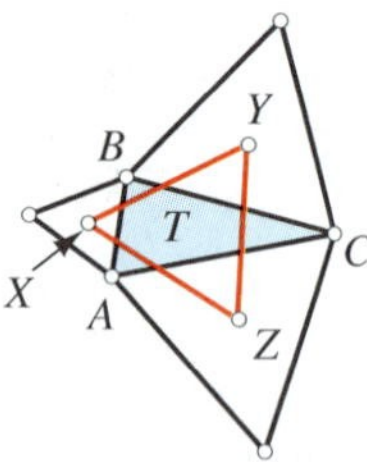

(a) Construct the triangle XYZ. Use your software to investigate the properties of $\triangle XYZ$.

(b) Mark X as a rotation center, and rotate the entire figure shown above by 120°. Similarly, rotate the figure 120° about the centers of the other equilateral triangles. What pattern is formed by the centers X, Y, and Z of all the equilateral triangles and their rotations?

41. Use geometry software to construct a parallelogram $ABCD$ and the outward squares with centers X, Y, Z, and W as shown below.

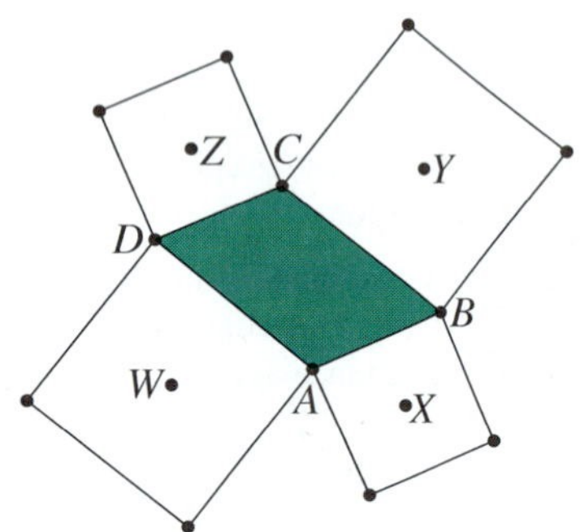

(a) Construct the quadrilateral $XYZW$. Use your software to investigate the properties of $XYZW$.

(b) Mark X as a rotation center, and rotate the entire figure shown by 90°. Similarly, rotate the figure 90° about the centers of the other squares. What pattern is formed by the centers of all the squares and their rotations?

Communicating

42. Explore what happens when translations are taken in succession, writing a report on your conclusions. Give both specific examples and any general principles you find. In particular, include answers to the following questions:

(a) Why is the net outcome of two successive translations equivalent to another translation?

(b) If the first translation takes A onto A_1 and the second translation takes A_1 onto A', why is the arrow from A to A' the slide arrow of the combined motion of the two translations?

(c) What happens to the net outcome if the order in which the two translations are taken is reversed?

From State Student Assessments

43. (Florida, Grade 5)
Carla created the figure below in her art class.

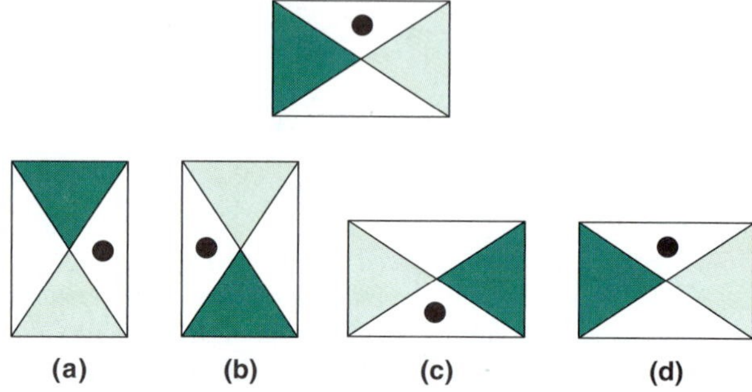

Which of these shows the figure rotated 90° clockwise?

44. (Oregon, Grade 5)
Which two of these figures are congruent?

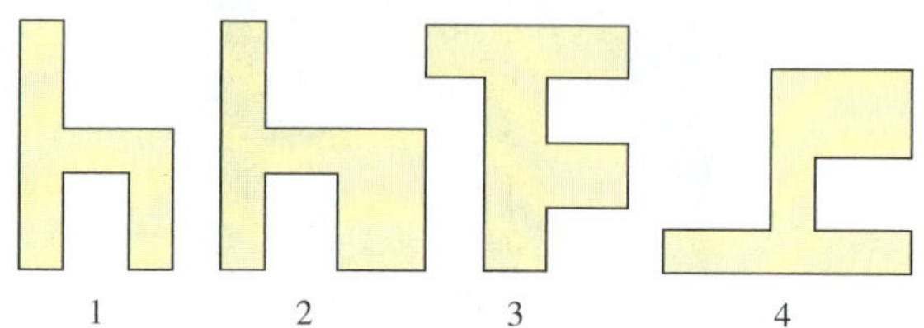

A. 1 and 3
B. 1 and 2
C. 2 and 3
D. 2 and 4

For Review

45. The measures of the three angles in a triangle are $2x°$, $3x°$, and $4x°$. What are the three angle measures?

46. A triangle RST has area 87 cm^2, and $RS = 3$ cm. What is the distance from T to the line containing RS?

47. Give a counterexample which shows that the following conjecture is false: *the midpoints of the sides of every rhombus are the vertices of a square.*

48. If the word "square" in the conjecture stated in problem 47 is replaced by "rectangle," do you think the conjecture is true? Give a proof or provide a counterexample.

13.2 PATTERNS AND SYMMETRIES

Symmetry is a universal principle of organization and form. The circular arc of a rainbow and the hexagonal symmetry of an ice crystal are visible expressions of the symmetry of many, indeed most, of the physical processes of the universe. A seashell and the fanned tail of the peacock are spectacular examples of biological symmetry. Symmetry is the norm of nature and natural law, not the exception.

In the human domain, all cultures of the world, even those in prehistoric times, developed a useful intuitive understanding of the basic concepts of symmetry. Decorations on pottery, walls, tools, weapons, musical instruments, and clothing are often highly symmetric. Buildings, temples, tombs, and other structures are usually designed with an eye to symmetry and balance. Music, poetry, and dance frequently incorporate symmetry in their underlying structure.

While people have long had an informal understanding of symmetry, it is only more recently that mathematics has provided a deeper understanding of what symmetry means and how different kinds of symmetry can be described and classified. In the classroom, youngsters are drawn to this artistic and aesthetic aspect of mathematics.

What Is Symmetry?

The concept of a rigid motion, which we defined and explored in the last section, makes it possible to give a precise definition of a symmetry of a geometric figure in the plane.

DEFINITION *A Symmetry of a Plane Figure*
A **symmetry** of a plane figure is any rigid motion of the plane that moves all the points of the figure back to points of the figure.

Thus, all points P of the figure are taken by the symmetry motion to points P' that also are points of the figure. The identity motion is a symmetry of any figure, but of more interest are figures that have symmetries other than the identity. Under a nonidentity symmetry, some points in the figure move to new positions in the figure, even though the figure as a whole appears unchanged by the motion.

The classification theorem of the preceding section tells us that there are just four basic rigid motions. Therefore, any symmetry of a figure is one of these four basic types, and the symmetry properties of a figure can be fully described by listing all of the symmetries of each type.

from **The NCTM Principles and Standards**

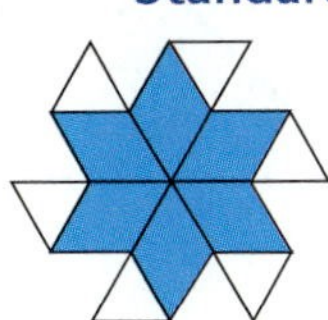

Rotational Symmetry

Students can naturally use their own physical experiences with shapes to learn about transformations such as slides (translations), turns (rotations), and flips (reflections). They use these movements intuitively when they solve puzzles, turning the pieces, flipping them over, and experimenting with new arrangements. Students using interactive computer programs with shapes often have to choose a motion to solve a puzzle. These actions are explorations with transformations and are an important part of spatial learning. They help students become conscious of the motions and encourage them to predict the results of changing a shape's position or orientation but not its size or shape.

Teachers should guide students to recognize, describe, and informally prove the symmetric characteristics of designs through the materials they supply and the questions they

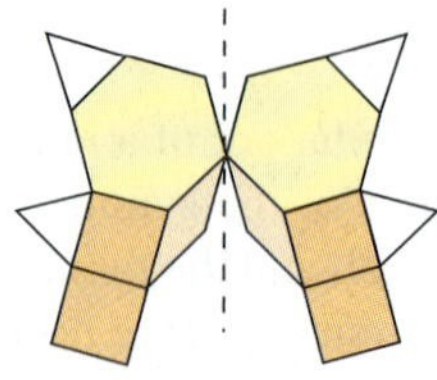

Line Symmetry

ask. Students can use pattern blocks to create designs with line and rotational symmetry (see figure at left) or use paper cutouts, paper folding, and mirrors to investigate lines of symmetry.

SOURCE: *Reprinted with permission from* Principles and Standards for School Mathematics, *pages 99–100.* *Standards are listed with the permission of the National Council of Teachers of Mathematics (NCTM). NCTM does not endorse the content or validity of these alignments.*

Reflection Symmetry

A figure has **reflection symmetry** if a reflection across some line is a symmetry of the figure. The line of reflection is called a **line of symmetry** or a **mirror line** of the figure. Each point P of the figure on one side of the line of symmetry is matched to a point P' of the figure on the opposite side of the line of symmetry. Some figures and their lines of symmetry (shown dashed) are displayed in Figure 13.16.

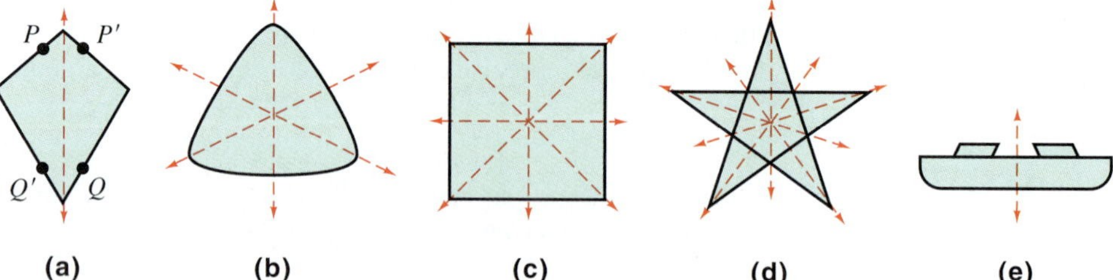

Figure 13.16
Plane figures and their lines of symmetry

Reflection symmetry is also called **line symmetry** or **bilateral symmetry.** Bilateral symmetry is also used to describe figures in space that have a plane of symmetry. For example, ferries (as suggested in Figure 13.16(e)) are often bilaterally symmetric across midships to simplify loading and unloading of their cargo of cars and trucks. Infrequent passengers on such ferries can find it very disorienting when the bow and stern are indistinguishable.

EXAMPLE 13.8 IDENTIFYING LINES OF SYMMETRY

How many lines of symmetry does each letter shown below have?

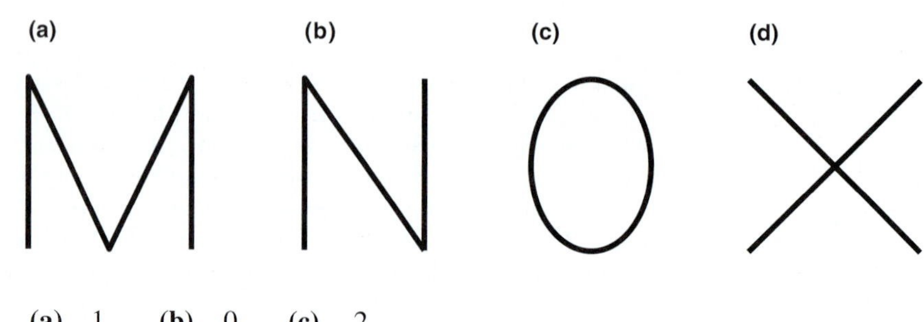

Solution

(a) 1 **(b)** 0 **(c)** 2
(d) 4, since the segments are congruent and intersect at right angles.

HIGHLIGHT from HISTORY

George W. Brainerd and Anna O. Shepherd, Pioneers in Mathematical Anthropology

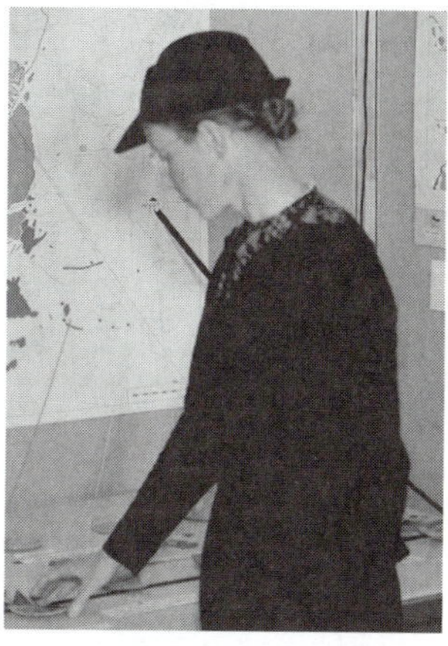

George W. Brainerd, a North American archeologist, was the first person to use the principles of symmetry as a tool for anthropological study. His analysis of the designs on pottery of the prehistoric Anasazi of Monument Valley, Arizona, and of the Maya of the Yucatan Peninsula, led him to formulate a number of principles for the use of symmetry classifications in pattern analysis. Brainerd published his ideas in the article "Symmetry in Primitive Conventional Design," which appeared in 1942 in *American Antiquity,* the leading journal of North American anthropology. Unfortunately, Brainerd's work was almost completely neglected, although it did attract the attention of Anna O. Shepherd, a geologist at the Carnegie Institution in Washington, D.C. In her monograph *The Symmetry of Abstract Design with Special Reference to Ceramic Decoration,* published in 1948, Shepherd discusses how certain symmetries predominate within a specific culture, and how changes within a culture can be identified by symmetry. Shepherd's work, like that of Brainerd, had to wait until the mid-1970s to be fully appreciated. Today, the anthropological significance of symmetry analysis is well established.

Rotation Symmetry

A figure has **rotation symmetry,** or **turn symmetry,** if the figure comes back to itself when it is rotated through a certain angle between 0° and 360°. The center of the turn is called the **center of rotation.** Some examples of figures with rotation symmetry are shown in Figure 13.17.

(a) 90° symmetry

(b) 72° symmetry

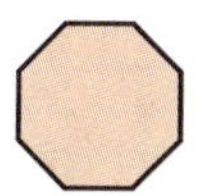

(c) 45° symmetry

(d) 180° symmetry

Figure 13.17
Figures with rotation symmetry

A figure with 90° rotation symmetry automatically has 180° and 270° rotation symmetry. For this reason it is customary to give just the *smallest* positive angle measure that turns the figure back to itself. The only exception is for figures composed of concentric circles, which turn back to themselves after *any* turn about their center. Such figures have **circular symmetry.**

EXAMPLE 13.9 FINDING ANGLES OF ROTATION SYMMETRY

Determine the measures of the angles of rotation symmetry of these figures.

(a) 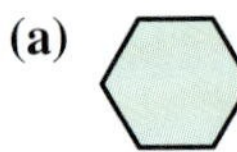 **(b)**  **(c)** 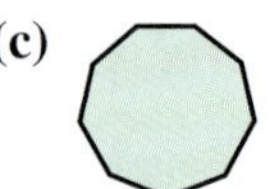 **(d)** 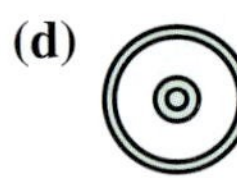

Solution

(a) The smallest positive angle of rotation of a regular hexagon measures 60°, so the turn angles of all of the rotation symmetries are 60°, 120°, 180°, 240°, and 300°.

(b) The only rotation angle measures 180°.

(c) The smallest amount of turn of a regular 9-gon is $360°/9 = 40°$, so the turn angles are 40°, 80°, 120°, 160°, 200°, 240°, 280°, and 320°.

(d) This figure has circular symmetry.

Point Symmetry

A figure has **point symmetry** if it has 180° rotation symmetry about some point O, as shown in Figure 13.18. This means that a half-turn takes the figure back onto itself, and every point P of the figure has a corresponding point P' of the figure that is directly opposite the turn center O, with $OP = OP'$.

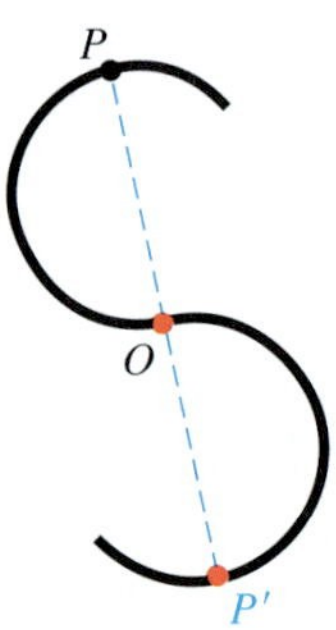

Figure 13.18
The letter S has point symmetry

EXAMPLE 13.10 IDENTIFYING POINT SYMMETRY

What letters, in uppercase block form, can be drawn to have point symmetry?

Solution

H, **I**, **N**, **O**, **S**, **X**, and **Z**.

Patterns: Figures with Translation Symmetries

A **pattern** is a figure with a translation symmetry. To avoid considering the whole plane as a pattern, or even just some set of horizontal lines, it is assumed that there is some minimum positive distance required to translate a pattern back onto itself. Thus, a pattern must be an infinite figure (why?) with motifs repeated infinitely often. Patterns are common on wallpaper, decorative brick walls, printed and woven fabrics, ribbons, and friezes (ceiling

Into the Classroom

Mathematics in Motion

The words *transformation* and *symmetry* often suggest advanced topics best left for gifted middle school students or postponed to the high school curriculum. Quite the opposite is true, however, since there are activities, games, artistic constructions, and problems—all exploring "motion geometry"—that are suitable for students at every grade level. Primary-school children can work with Miras, pattern blocks, and geoboards, if available, to create and investigate symmetric patterns. For example, each student can create a "half" figure with rubber bands on the upper half of a geoboard. Boards are then exchanged, and the students are challenged to complete a mirror-image figure in the lower half of the geoboard. Older children could replace the geoboard with graph paper or dot paper and investigate rotations and point symmetry, as well as reflections and line symmetry. Geometry software also offers exciting possibilities for investigations in transformation geometry.

Here are three more ideas, suggesting how patterns and motions can be approached in the classroom.

- *Follow the leader.* Draw a line with a ruler down a blank sheet of paper. In pairs of students, the "leader" slowly draws a curve, and simultaneously the "follower" draws the reflected curve across the line of symmetry. The students can interchange roles of leader and follower. To explore point symmetry, a prominent dot can be drawn at the center of the sheet. Some students, with a pencil in each hand, might like to attempt a solitaire game.

- *Punchy puzzles.* If the square shown on the far left is folded along the dashed lines, then the pattern of holes can be seen to be created with just one punch. How can a square sheet of paper be folded and punched *one time only* to create the other hole patterns shown?

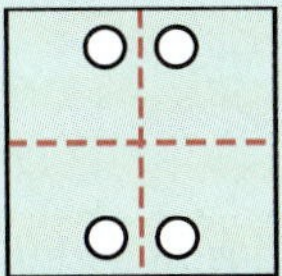 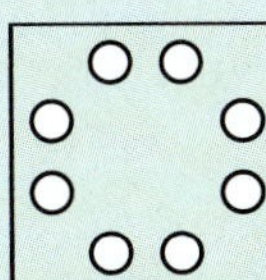 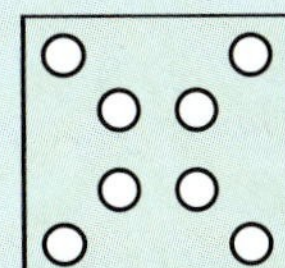 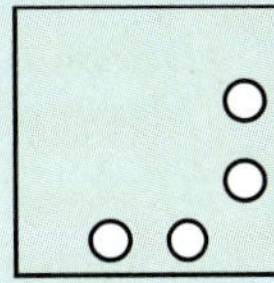 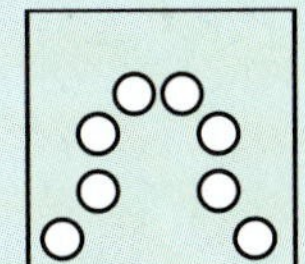

- *Stained-glass window search.* Eight congruent isosceles right triangles, with four of each color, will form a square window. Three windows are shown here, but a reflection and rotation show that the first two windows are really the same. How many different window patterns are there, each with four panes of each of two colors? (You should be able to find 13 distinct patterns, with no two patterns the same, under either a rotation or a reflection.)

border decorations in older buildings). Enough of the pattern must be shown to make it clear how to extend the pattern indefinitely.

There are two types of patterns in the plane, **border patterns** and **wallpaper patterns.** As their names suggest, a border pattern has a repeated motif that has been translated in just one direction to create a strip design, whereas a wallpaper pattern has a motif translated in two nonparallel directions to create a wall design.

Border Patterns and Their Classification

Seven examples of border patterns from a variety of cultures are shown in Figure 13.19.

Figure 13.19
Border patterns from around the world

Some border patterns may have other symmetries in addition to their translation symmetries. However, the possibilities are limited. For example, the only possible rotation symmetry is a half-turn. A careful study has shown that every border pattern has the same symmetries as one of the seven types shown in Figure 13.20.

The two-symbol name assigned by the International Crystallographic Union is shown at the left of each pattern in Figure 13.20. To find the classification symbol of any border, follow these steps.

First Symbol: m, if there is a vertical line of symmetry

1, otherwise

Second Symbol: m, if there is a horizontal line of symmetry

g, if there is a glide reflection (but no horizontal line of symmetry)

2, if there is a half-turn symmetry (but no horizontal line of symmetry or glide symmetry)

1, otherwise

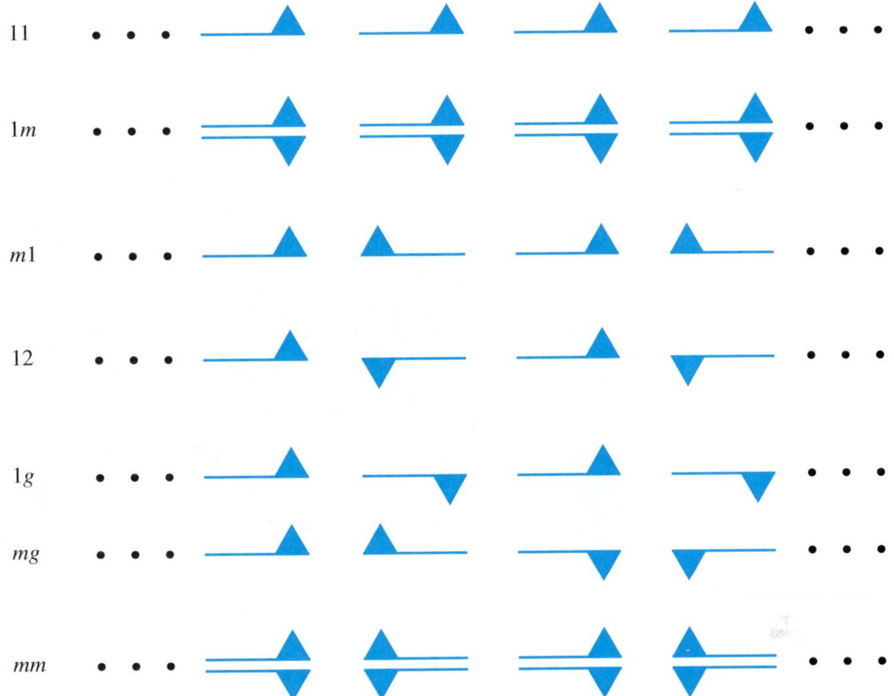

Figure 13.20
The seven symmetry types of border patterns

To check your understanding, cover up the two-symbol name in Figure 13.20 and then follow the two foregoing steps to see if you obtain the correct classification symbols. With practice, you'll be able to determine the two-symbol name for even more complicated border patterns.

EXAMPLE 13.11 CLASSIFYING BORDER PATTERNS

Classify the symmetry type of the following border patterns by assigning the appropriate two-symbol notation.

(a) **(b)** **(c)**

Solution

It is helpful to view the patterns upside down, in a mirror, with a Mira, and so forth, since this will help you discover and verify what symmetry motions are present. A transparency copy of the pattern , if available, is an almost ideal tool to explore and classify patterns of symmetry.

(a) This border has both a vertical and horizontal line of reflection, so the symmetry is of type *mm*.

(b) There is no vertical symmetry line, so the first symbol is 1. There is no horizontal symmetry line, but there is a glide symmetry, so the second symbol is *g*. Altogether, the symmetry is type 1*g*.

(c) There are no lines of reflection, nor is there glide-reflection symmetry. However, there is half-turn symmetry, so the symbol is 12.

Wallpaper Patterns

Two examples of wallpaper patterns are shown in Figure 13.21.

Figure 13.21
Examples of wallpaper patterns

The Arabian pattern on the left has centers of both 60° and 120° rotational symmetry. This pattern also has a 180° rotational symmetry at the center of each of the *Z*-shaped black bars. The Egyptian pattern on the right has centers of 90° rotational symmetry. It can be shown that 60°, 90°, 120°, and 180° are the only possible angle measures of rotational symmetry of any wallpaper pattern, a result called the *crystallographic restriction.* This and other restrictions limit the number of symmetry types to be found in a wallpaper pattern. Indeed, it has been shown that any wallpaper pattern is one of just 17 distinct types.

In the next section, several wallpaper patterns will be created by covering the plane with tiles.

HIGHLIGHT *from* HISTORY

Classifying Symmetry

In the years 1230–1354, the Moors constructed the magnificent Alhambra, a group of palatial buildings in the hills overlooking Granada. The walls and ceilings are decorated with striking patterns formed by regularly repeated motifs. Thirteen of the 17 types of plane symmetry can be found. It is thought that the Islamic ban on human and animal motifs gave rise to the creation of such intricate abstract geometric decoration.

Similarly, artisans from other world cultures have discovered and used repeated-motif designs. Indeed, numerous examples representing all 17 symmetry types have been identified. It took mathematical methods, however, to *prove* that no more than 17 patterns of plane symmetry exist.

The first step toward classification was made by the Russian crystallographer Evgraf Federov in 1891. His work was made more widely available in 1924 through the work of P. Niggli and George Pólya. Fedorov, as a crystallographer, was also interested in patterns of symmetry of space figures. He was able to show that 230 patterns of symmetry exist in three-dimensional space.

Other generalizations have been investigated. For example, colors may be used in a systematic way, such as the red-and-black coloration of a checkerboard. The two-colored patterns were classified by a mathematically knowledgeable textile worker in the 1930s. He discovered that there are 17 two-colored border patterns and 46 two-colored wallpaper patterns.

Did You Know?

Symmetries of Culture

In this book we demonstrate how to use the geometric principles of crystallography to develop a descriptive classification of patterned design. Just as specific chemical assays permit objective analysis and comparison of objects, so too the description of designs by their geometric symmetries makes possible systematic study of their function and meaning within cultural contexts.

This particular type of analysis classifies the underlying structure of decorated forms; that is, the way the parts (elements, motifs, design units) are arranged in the whole design by the geometrical symmetries which repeat them. The classification emphasizes the way the design elements are repeated, not the nature of the elements themselves. The symmetry classes which this method yields, also called motion classes, can be used to describe any design whose parts are repeated in a regular fashion. On most decorated forms such repeated design, properly called pattern, is either planar or can be flattened (e.g., unrolled), so that these repeated designs can be described either as bands or strips (one-dimensional infinite) or as overall patterns (two-dimensional infinite) in a plane.

This excerpt is from the introduction to *Symmetries of Culture: Theory and Practice of Plane Pattern Analysis.* Nearly every page of *Symmetries of Culture* is graced by beautiful photographs and drawings that illustrate the principles of symmetry discovered and utilized by contemporary and historic cultures from around the world.

SOURCE: From Symmetries of Culture: Theory and Practice of Plane Pattern Analysis, *by Dorothy K. Washburn and Donald W. Crow, page ix.*

Problem Set 13.2

Understanding Concepts

1. Carefully trace each figure and draw all of its lines of symmetry.

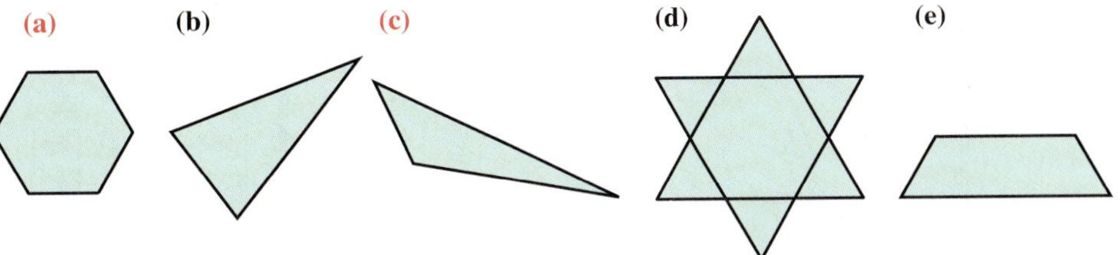

2. Carefully trace each figure and draw all of its lines of symmetry.

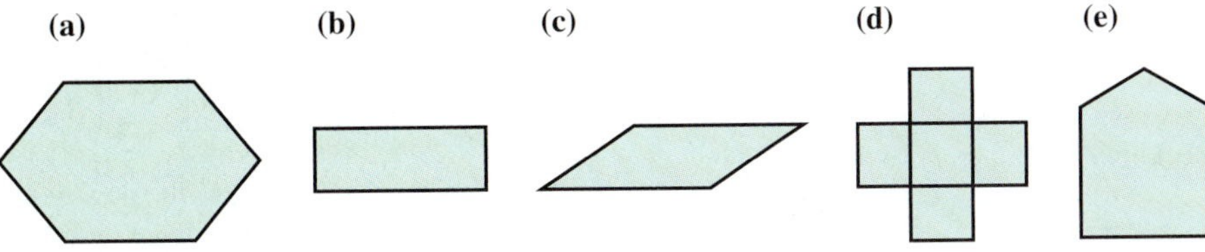

3. Draw polygons with the following symmetries, if possible.
   (a) One line of symmetry, but no rotation symmetry
   (b) Rotation symmetry, but no reflection symmetry
   (c) One line of symmetry and rotation symmetry
4. Describe the most general quadrilateral having the given symmetry property.
   (a) A line of symmetry through a pair of opposite vertices
   (b) A line of symmetry through a pair of midpoints of opposite sides
   (c) Two lines of symmetry, each through a pair of opposite vertices
   (d) Two lines of symmetry, each through a pair of midpoints of opposite sides
   (e) Four lines of symmetry
   (f) A center of 180° rotational symmetry
5. Complete each figure to give it reflection symmetry about line m.

(a)

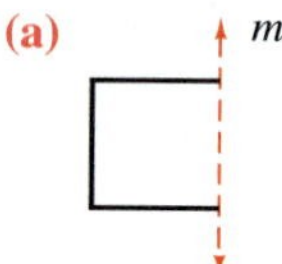

(b)

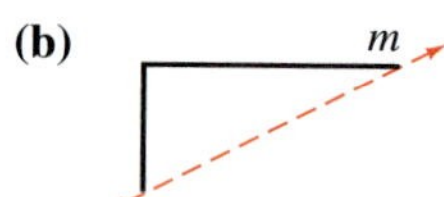

(c)

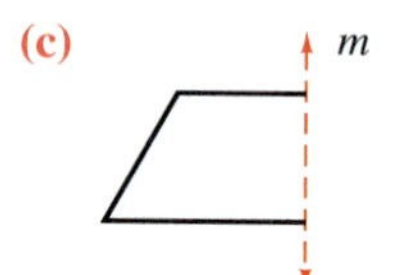

(d)

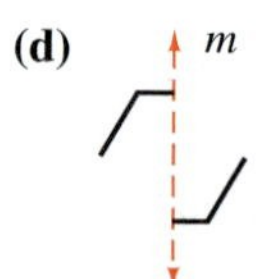

6. Complete each of these figures to give it point symmetry about point O.

(a)

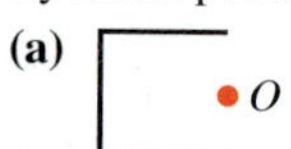

(b)

(c)

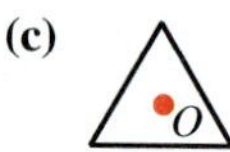

(d)

7. Copy the following figures onto graph paper. Then complete each figure to give it reflection symmetry across the dashed line.

(a)

(b)

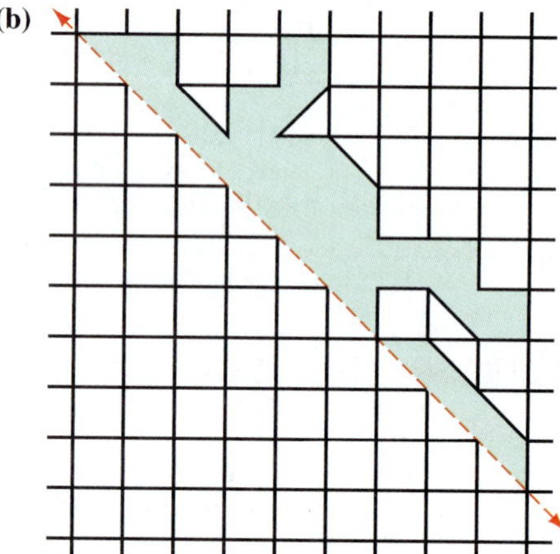

8. A valentine heart is easy to make symmetric: Cut it from a piece of construction paper folded in half once.

   (a) Suppose the paper is folded in half twice and the same cut is made. Sketch the shape you obtain when you unfold the cut pattern.

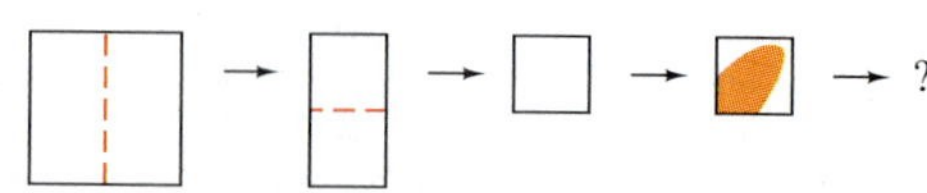

   (b) Describe how to make a six-fold symmetric snowflake by folding and cutting a sheet of paper.

9. Describe all symmetries of each of the following company logos.

(a)

(b)

(c)

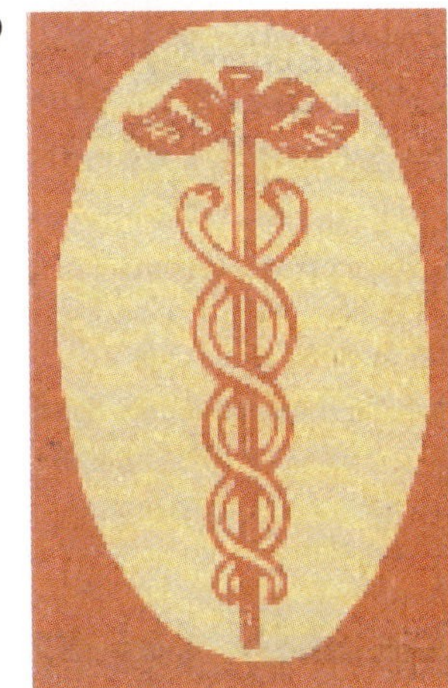

(d)

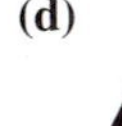

(e)

10. Describe the symmetries of the wheel covers shown.

(a)

(b)

(c)

(d)

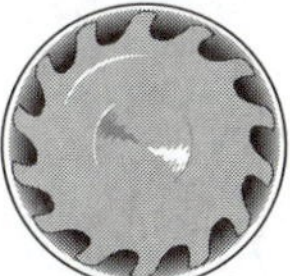

11. (a) Complete the figure shown to give it 90° rotation symmetry about point O.

(b) Repeat part (a), but giving the resulting figure 60° symmetry.

12. Identify the regular n-gons in each part that have the given symmetries.
   (a) There are exactly three lines of symmetry.
   (b) There are exactly four lines of symmetry.
   (c) There are exactly 19 lines of symmetry.
   (d) The polygon has 10° rotation symmetry.
   (e) The polygon has both 6° and 15° rotation symmetry.

13. List all the digits out of 0, 1, 2, 3, 4, 5, 6, 7, 8, and 9 that have
   (a) vertical reflection symmetry.
   (b) horizontal reflection symmetry.
   (c) vertical and horizontal reflection symmetry.
   (d) point symmetry.

   Write the digits in the most symmetric way you can.

14. Repeat problem 13, but for the uppercase capital letters, A, B, . . . , Z, written as symmetrically as possible.

15. Repeat problem 13 for the lowercase letters, a, b, . . . , z, written as symmetrically as possible.

16. Describe all the symmetries of each border pattern, and classify each by the two-symbol notation used in crystallography.
   (a) . . . A A A A A A . . .
   (b) . . . B B B B B B . . .
   (c) . . . N N N N N N . . .

17. Describe all the symmetries of each border pattern, and give its two-symbol classification used by crystallographers.
   (a) . . . H O H O H O . . .
   (b) . . . M W M W M W . . .
   (c) . . . 9 6 9 6 9 6 . . .

Teaching Concepts

18. **Pattern-Block Symmetries.** Children enjoy creating symmetric designs with pattern blocks. Two simple examples are shown on the following page. This intrinsic interest can be utilized effectively to explore many of the fundamental concepts of symmetry.

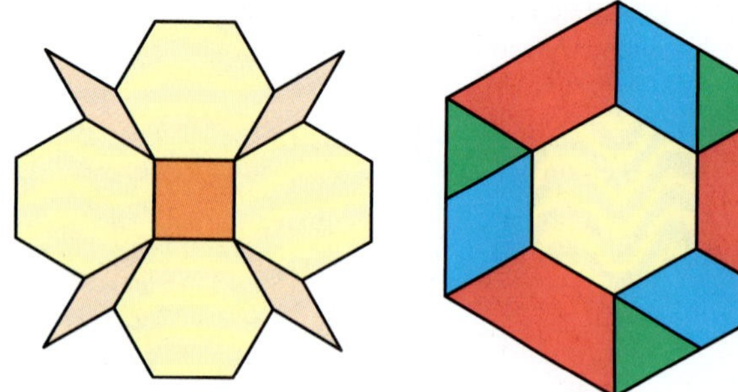

(a) Describe the symmetry of each of the six individual pattern-block shapes (the triangle, square, hexagon, trapezoid, and the large and small rhombuses).

(b) Describe the symmetries of the two foregoing patterns.

(c) Create a pattern that has 60° rotational symmetry and includes squares.

(d) Create a pattern with 30° rotational symmetry that includes squares.

19. Symmetry Search. The following line grid can be made easily from graph paper. Using only segments in the grid, find polygons with the given symmetries. There are many answers, and you (and youngsters) are free to be creative.

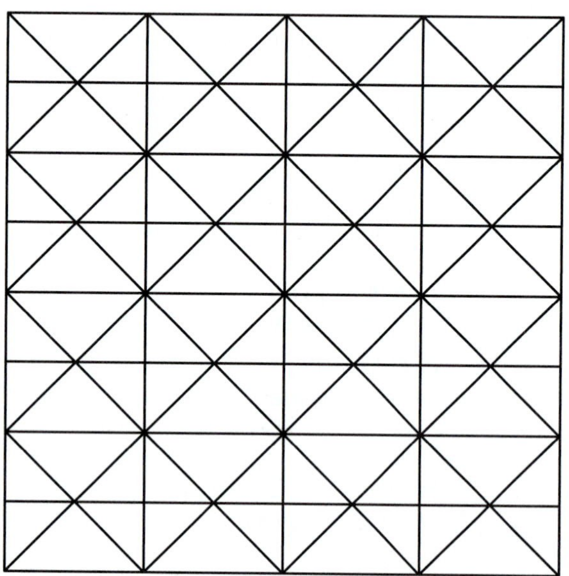

(a) Find an octagon with two lines of symmetry.

(b) Find a quadrilateral with no symmetries.

(c) Find a pentagon with a line of symmetry.

(d) Find a heptagon with a line of symmetry that is neither horizontal nor vertical.

Thinking Critically

20. The Penny Game. Lynn and Kelly are playing a game with just a few, simple rules. Each player takes turns placing a penny on a rectangular tabletop. Each new penny must be flat on the table and cannot touch any of the pennies already on the table. The first player unable to put another penny on the table according to these rules is the loser. Lynn, who gets to make the first move, puts the first penny at the center of the table and is confident of a win. What is Lynn's strategy?

21. A **palindrome** is a word, phrase, sentence, or numeral that is the same read either forward or backward. Examples are WOW, NOON, and TOOT.

(a) If a word written in all capital letters has a vertical line of symmetry, why must it be a palindrome?

(b) NOON does not have a line of symmetry. Find another palindrome with no line of symmetry.

(c) What symmetry do you see in the word "pod"?

22. Describe what symmetries you see in these statements.

(a) "Sums are not set as a test on Erasmus."

(b) "Is it odd how asymmetrical is 'symmetry'? 'Symmetry' is asymmetrical. How odd it is."

(c) "Able was I ere I saw Elba." (attributed to Napoleon)

23. Carefully explain why no border pattern has the symbol "*m*2." (*Hint:* If a border pattern has a vertical line of symmetry and 180° rotation symmetry, what other symmetry must it also have?)

24. Give the two-symbol classification of each of the seven border patterns shown in Figure 13.19.

25. Classify the following Maori rafter patterns. The Maori, the indigenous people of New Zealand, used principles of symmetry to express their belief system. Disregard the color scheme when you classify the symmetry type of the pattern.

(a)

(b)

(c)

(d)

26. Classify the following Inca border patterns.

(a)

(b)

27. In each strip of rectangles shown below, a certain rigid motion applied to the leftmost rectangle takes the figure to the next rectangle. Apply the same motion, but to the second rectangle, to draw the image of the second rectangle in the third rectangle. Continue to use the same motion to fill in the successive rectangles, and then classify the border pattern that is produced.

(a)

p	p	p			

(b)

p	q				

(c)

p	d				

(d)

p	b				

28. The following border is of type 11. For each of the other six types of border symmetry, complete the pattern to give it the corresponding symmetry type by adding as little as possible.

		p				p				p			

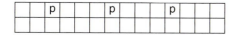

29. These drawings were made by George Pólya for his 1924 paper classifying the 17 wallpaper pattern types.

(a)

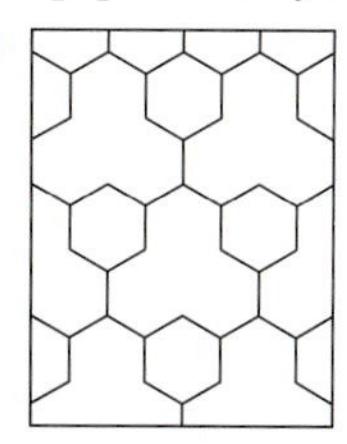

(b)

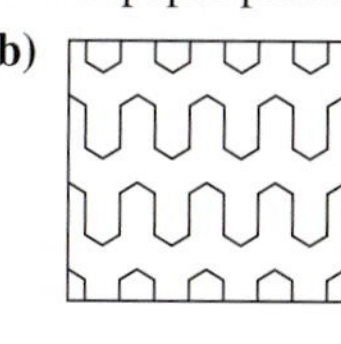

(c)

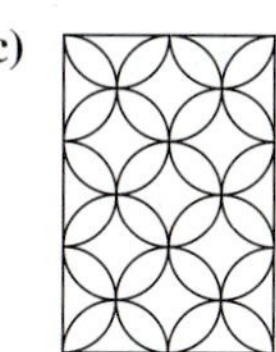

For each pattern, give

(i) the number of directions of reflection symmetry;

(ii) the number of directions of glide symmetry;

(iii) the sizes of angles of rotation symmetry.

30. A three-dimensional figure has **bilateral symmetry** if there is a plane for which every point P of the figure has a mirror image P' of the figure on the opposite side of the plane. For example, a right prism with an equilateral triangular base has four planes of symmetry. Find the number of planes of symmetry of these space figures:

(a) a $1 \times 2 \times 3$ rectangular prism

(b) a $1 \times 2 \times 2$ rectangular prism

(c) a cube

(d) a square-based right regular pyramid (the apex is equidistant from the vertices of the base)

Making Connections

31. Describe what symmetries, or lack of symmetry, you see in the following forms and objects.

(a) a pair of scissors

(b) a T-shirt

(c) a dress shirt

(d) a golf club

(e) a tennis racket

(f) a crossword puzzle

32. Describe the symmetry you find in

(a) an addition table.

(b) a multiplication table.

(c) Pascal's triangle.

33. The figures below result from a famous experiment in physics, the Chladni plate. A square metal plate is supported horizontally at its center, sprinkled with fine dry sand, and then vibrated at different frequencies. The sand migrates to the *nodal lines,* where there is no movement of the plate. In the dark regions between the nodal lines, the plate is in vertical vibrational motion.

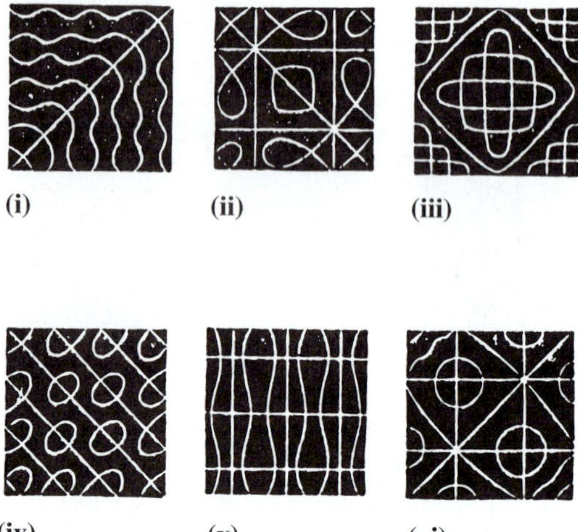

(a) Describe the symmetries of each of the six Chladni plates shown.

(b) The apparatus of the experiment has the symmetries of a square. Do the nodal lines always have the symmetries of a square, or can the vibrational pattern "break" square symmetry? (The Chladni plates shown were published in 1834 in *Of the Connexion of the Physical Sciences,* by Mary Somerville, one of the great female mathematicians of the nineteenth century.)

Thinking Cooperatively

34. ***Mu Torere.*** The Maori people of New Zealand play the two-person game *mu torere* on a board shaped in a regular eight-pointed star. The points of the star are the *kawai,* and the center the *putahi.* Each player has four counters, say, red beans and white beans, arranged initially so the red beans of one player are in the four upper positions and the opponent's white beans are in the lower positions. Players alternate moving one of their beans either into an empty adjacent *kawai* or into (or out of) the *putahi.* A bean can move into the *putahi* only if the bean is adjacent to an opponent's bean. If not, the move is *tapu* (not allowed). The object of the game is to move your beans to a position where the opponent is blocked: any move of the opponent is *tapu.* Work in pairs to make, play, and investigate *mu torere.*

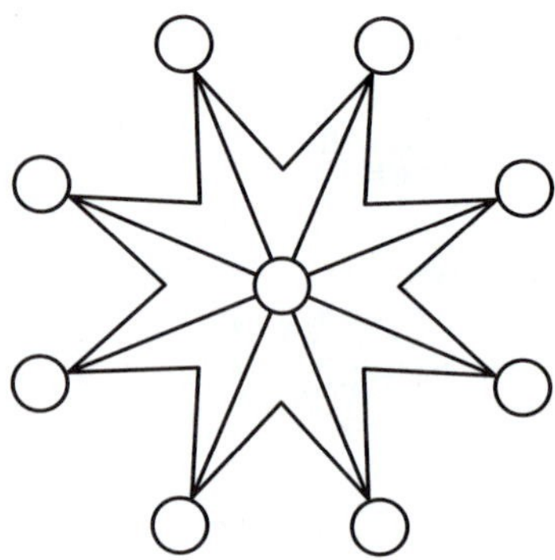

(a) Describe how to make a *mu torere* board by folding and cutting a sheet of paper.

(b) What is the reason a bean cannot be moved to the center unless it is adjacent to an opponent's bean?

(c) Play several games of *mu torere.* Describe the formation that wins the game.

(d) Make a list of games that use a symmetric board. Describe the type of symmetry found in the boards.

35. Cut out an interesting shape from cardboard or heavy paper. A border pattern can be drawn on a strip of paper by repeatedly tracing around the template.

Template

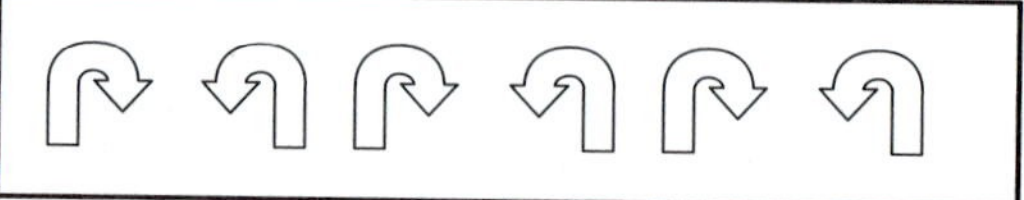
Border Pattern Drawn on Paper Strip

(a) Use the shape to create seven border patterns, drawing one pattern of each symmetry type on a separate strip of paper. Do not write the symmetry symbol on the strip.

(b) Pair up with another student. Match each of your strips to the corresponding strip of your partner having the same symmetry type.

Communicating

36. Write an illustrated short report entitled "Examples of Symmetry in ______," where the blank is filled in with your choice of topic. For example, you might choose "Sports," "Board Games," "Jewelry," "Musical Forms," "Navaho Blankets," "Native American Art," or "Flowers." Use your imagination, and draw on your outside interests and hobbies to be creative. You should include drawings, photocopies, pictures cut from discarded magazines, and so on. Be sure to identify and classify the type of symmetry found in each example.

37. Go on a symmetry hunt across campus, looking for striking examples of symmetry in buildings, decorative brickwork, sculptures, gardens, or wherever you may find it. Provide photos or drawings of three or four examples that you find especially interesting. Describe and classify the types of symmetry found in your examples. Include a border pattern and a wallpaper pattern.

Using a Computer

38. Use geometry software to create a bilaterally symmetric "funny face." An example was shown in the Window on Technology box in Section 13.1.

39. Use geometry software to create wheel-cover patterns (see problem 10) having the following required properties.
    (a) 45° rotational symmetry only
    (b) 60° rotational symmetry *and* bilateral symmetry

From State Student Assessments

40. (Florida, Grade 5)
After studying the Aztec civilization of central Mexico, a student drew one half of an Aztec medallion as shown below. The dotted line represents a line of symmetry. Which of the following tells how you must move the drawing to make a symmetrical medallion?
    A. turn
    B. flip
    C. slide
    D. slide and turn

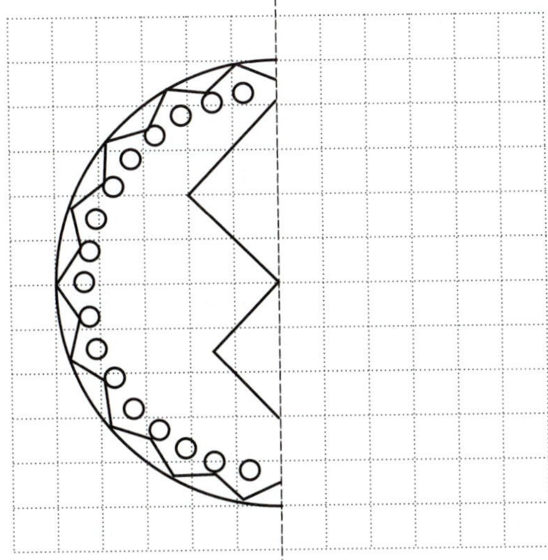

41. (Minnesota, Grade 5)
Ken folded a piece of paper in half and then folded it in half again. He cut out a shape and threw it away. This is how the paper looked when he unfolded it.

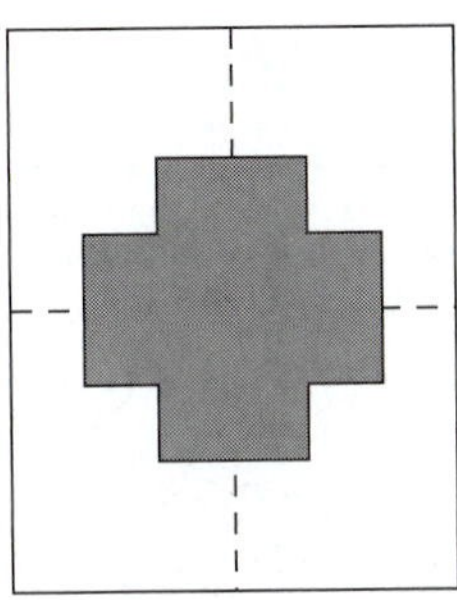

Which drawing below shows how his paper looked **before** he unfolded it?

A. 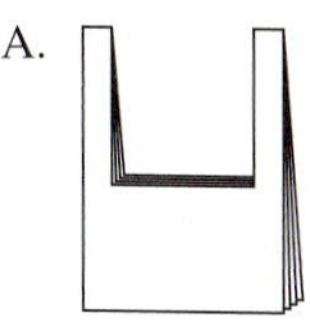

B. 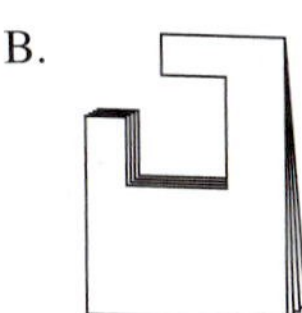

C. 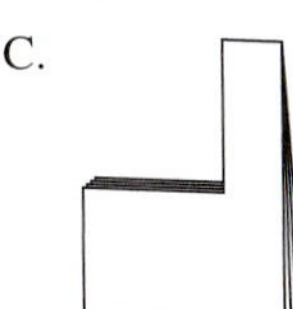

D. 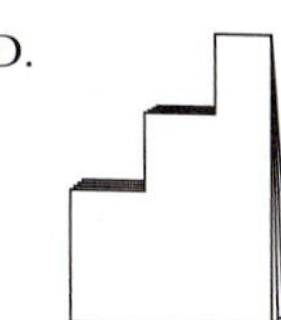

For Review

42. Fill in the blanks in the two statements below.
    (a) A counterclockwise rotation of size 130° about point O followed by a counterclockwise rotation of 220° also about O is equivalent to a single counterclockwise rotation of _____° about O.
    (b) The net outcome of the two rotations described in (a) is equivalent to a clockwise rotation of _____° about O.

43. (a) Triangle ABC is equilateral with sides of length two units. Find the images of points A, B, and C under reflection across the three successive lines m_1, m_2,

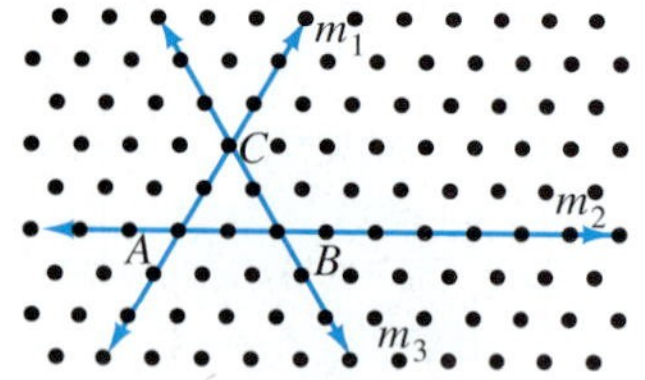

and m_3 shown. Label the respective image points A', B', and C'.

(b) Describe the basic rigid motion that takes $\triangle ABC$ onto $\triangle A'B'C'$.

13.3 TILINGS AND ESCHER-LIKE DESIGNS

This section explores patterns in the plane that are formed by combining repeated shapes. The art of tiling has a history as old as civilization itself. In virtually every ancient culture, the artisan's choices of color and shape were guided as strongly by aesthetic urges as by structural or functional requirements. Imaginative and intricate patterns decorated baskets, pottery, fabrics, wall coverings, and weapons. Some examples of ornamental patterns from different cultures are shown in Figure 13.22, on page 000.

In recent times the interest in tiling patterns has gone beyond their decorative value. For example, metallurgists and crystallographers wish to know how atoms can arrange themselves in a periodic array. Similarly, architects hope to know how simple structural components can be combined to create large building complexes, and computer engineers hope to integrate simple circuit patterns into powerful processors called neural networks. The mathematical analysis of tiling patterns is a response to these contemporary needs. At the same time, the creation and exploration of tilings provides an inherently interesting setting for geometric discovery and problem solving in the elementary and middle school classroom. In particular, children enjoy learning how to create their own periodic drawings in the style of pioneering Dutch artist M. C. Escher.

Tiles and Tilings

The precise meaning of a tile and a tiling is given in the following definition.

DEFINITION *Tiles and Tiling*
A simple closed curve, together with its interior, is a **tile.** A set of tiles forms a **tiling** of a figure if the figure is completely covered by the tiles without overlapping any interior points of the tiles.

Since all points in the figure are covered there can be no gaps between tiles. Tilings are also known as **tessellations,** since the small square tiles in ancient Roman mosaics were called *tessella* in Latin.

Regular Tilings of the Plane

Each tiling shown in Figure 13.23 is a **regular tiling:** the tiles are regular polygons of one shape, and they are joined edge to edge.

Pre-Inca fabric from Peru

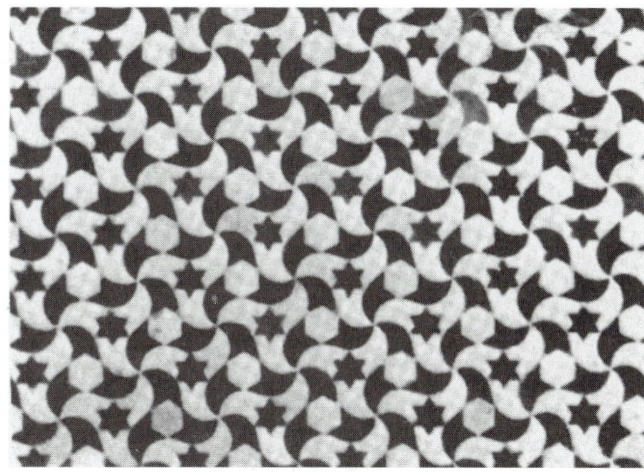

Detail of a tiled wall in the Alhambra

Window from Khirbet-et-Mefdjer, Jericho, Islamic period, eighth century A.D.

Mosaic floor of the fourteenth and fifteenth centuries in the Basilica of Saint Marks Cathedral, Venice

Tilings from Portugal, fifteenth to sixteenth centuries

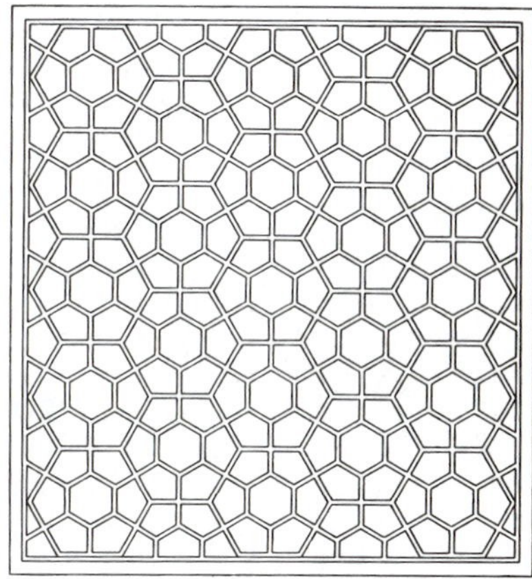

Chinese lattice-work, used to support paper windows

Figure 13.22
Tiling patterns from various cultures

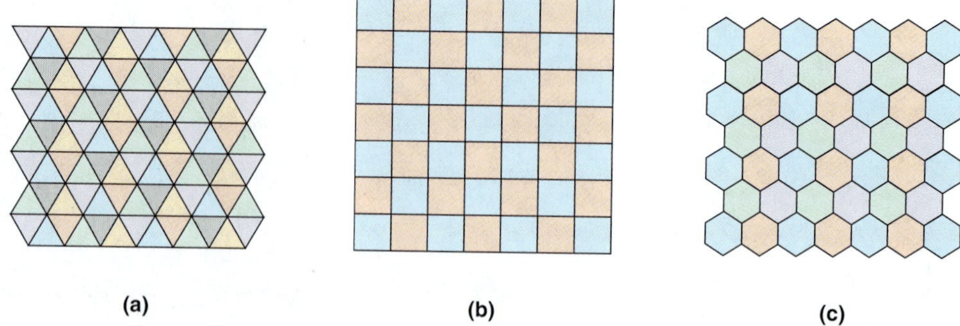
(a)
(b)
(c)

Figure 13.23
The three regular tilings of the plane

Any arrangement of nonoverlapping polygonal tiles surrounding a common vertex is called a **vertex figure.** Thus, four squares form each vertex figure of the regular square tiling, and three regular hexagons form each vertex figure of the hexagonal tiling. The measures of the interior angles meeting at a vertex figure must add to 360°. For example, in the square tiling, $90° + 90° + 90° + 90° = 360°$.

Suppose we attempt to form a vertex figure with regular pentagons as shown in Figure 13.24. The interior angles of a regular pentagon each measure $(5 - 2) \cdot 180°/5 = 108°$. Thus, three regular pentagons fill in $3 \cdot 108° = 324°$ and leave a 36° gap. On the other hand, four regular pentagons create an overlap, since $4 \cdot 108° = 432° > 360°$. Since a vertex figure cannot be formed, no tiling of the plane by regular pentagons is possible.

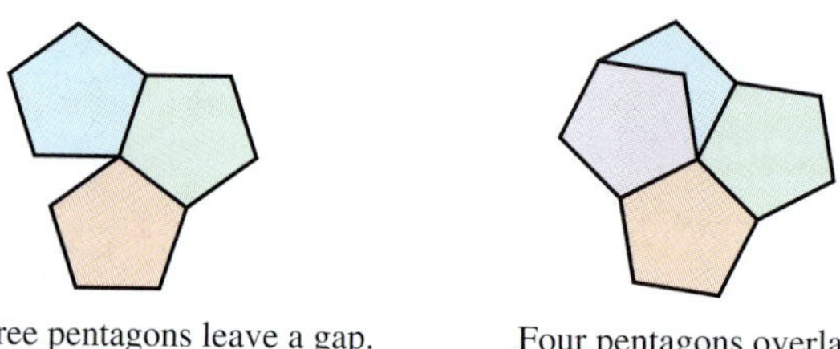
Three pentagons leave a gap.
Four pentagons overlap.

Figure 13.24
Regular pentagons do not tile the plane

Similarly a regular polygon of seven or more sides has an interior angle larger than 120°. Thus, two such polygons leave a gap, but three overlap. We have the following theorem.

> **THEOREM** *The Regular Tilings of the Plane*
> There are exactly three regular tilings of the plane: (a) by equilateral triangles, (b) by squares, and (c) by regular hexagons.

Semiregular Tilings of the Plane

A regular tiling uses congruent polygons of one type to tile the plane. What if regular polygons of several types are allowed? An edge-to-edge tiling of the plane with more than one type of regular polygon *and* with identical vertex figures is called a **semiregular tiling.** It is important to understand the restriction made about the vertex figures—*the same types of polygons must surround each vertex, and they must occur in the same order.* The two vertex figures in Figure 13.25 are not identical, since the two triangles and the two hexagons in the left-hand figure are adjacent, but on the right side the triangles and the hexagons alternate with one another.

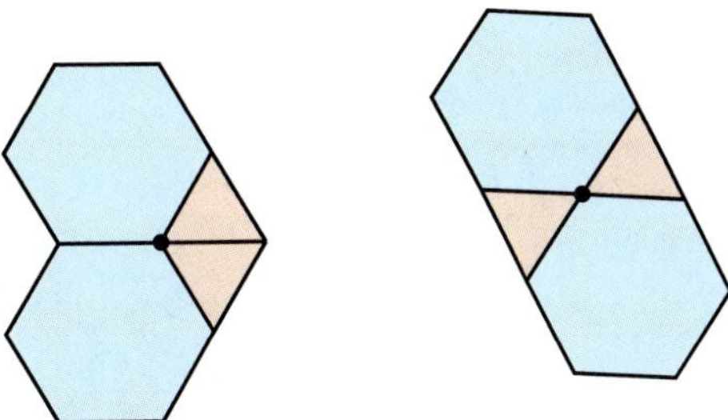

Figure 13.25
There are two distinct types of vertex figures formed by two equilateral triangles and two regular hexagons

To see if the vertex figures in Figure 13.25 can be extended to form a semiregular tiling, we must check to see if the pattern can be completed to make *all* of the vertex figures match the one shown. It is soon discovered that the pattern with adjacent triangles cannot be extended (try it!). On the other hand, the vertex figure of alternating triangles and hexagons extends to a semiregular tiling. You should be able to find it in Figure 13.26.

It can be shown that there are 18 ways to form a vertex figure with regular polygons of two or more types. Of these, eight extend to a semiregular tiling, shown in Figure 13.26.

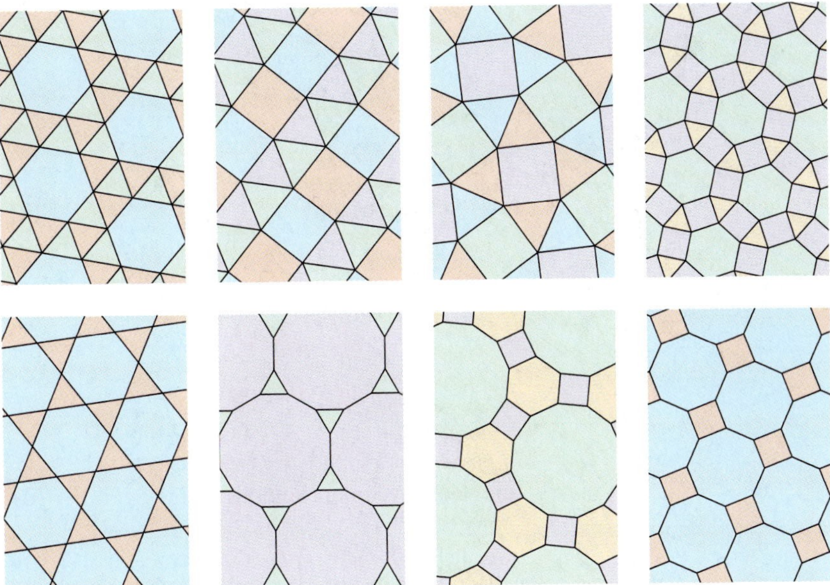

Figure 13.26
The eight semiregular tilings

HIGHLIGHT from HISTORY

Johannes Kepler and Tiling Patterns

The astronomer Johannes Kepler (1571–1630) is celebrated in scientific history for his identification of the elliptical shape of the orbits of the planets about the sun. Less known is Kepler's contribution to the theory of tiling. Here are some drawings from Kepler's book *Harmonice Mundi,* which he published in 1619.

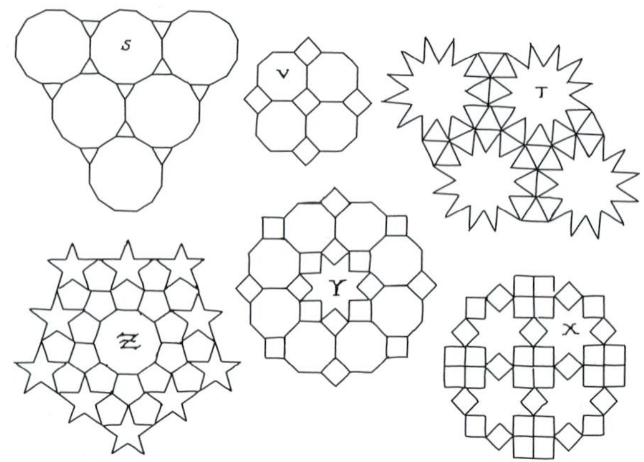

SOURCE: Branko Grünbaum and G. C. Shepard, Tilings and Patterns, *W. H. Freeman and Co., 1987. Reprinted with permission of G. C. Shepard, Emeritus Professor of Mathematics, University of East Anglia, Norwich NR4 7TJ, England.*

Tilings with Irregular Polygons

In the following example it is helpful to cut tile patterns from cardboard or heavy card stock. You can then trace around them to form, if possible, a tiling of the plane. Better yet, create and explore tilings with geometry software.

EXAMPLE 13.12 **EXPLORING TILINGS WITH IRREGULAR POLYGONS**

Which of the polygons below tile the plane?

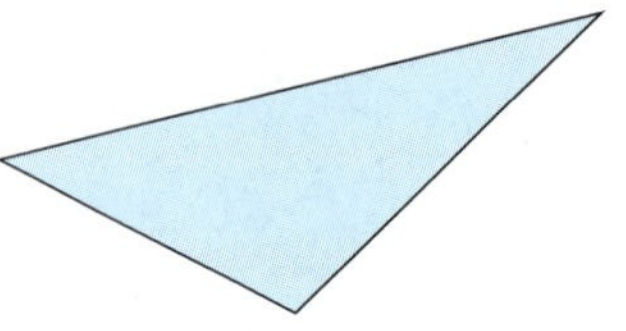

(a) Scalene triangle

(b) Convex quadrilateral

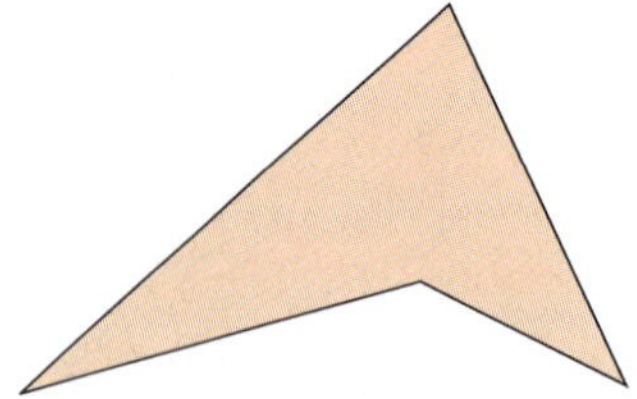

(c) Concave quadrilateral

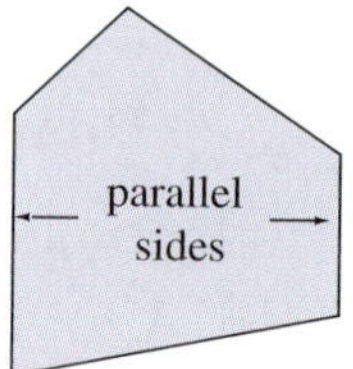

(d) Pentagon with a pair of parallel sides

Solution

(a) Two triangles of identical size and shape can be joined along a corresponding edge to form a parallelogram. Since it is evident that parallelograms tile the plane, it follows that *any triangle will tile the plane.*

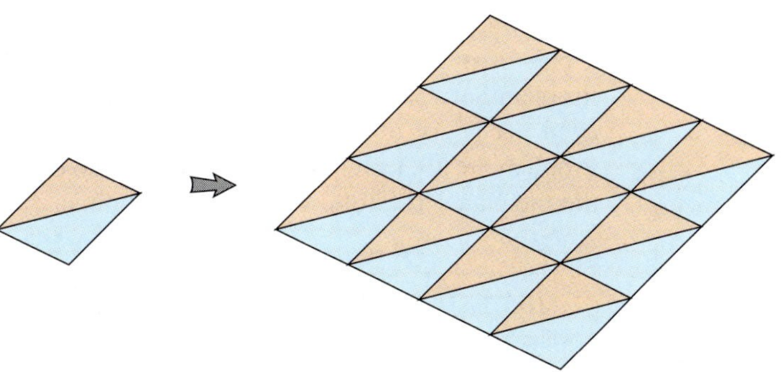

(b) and **(c)** As illustrated on the next page, any quadrilateral will tile the plane. A 180° turn about the midpoint of a side rotates the quadrilateral from one position to an adjacent position. Notice that each vertex of the tiling is surrounded by angles congruent to the four angles of the quadrilateral, whose measures add up to 360°.

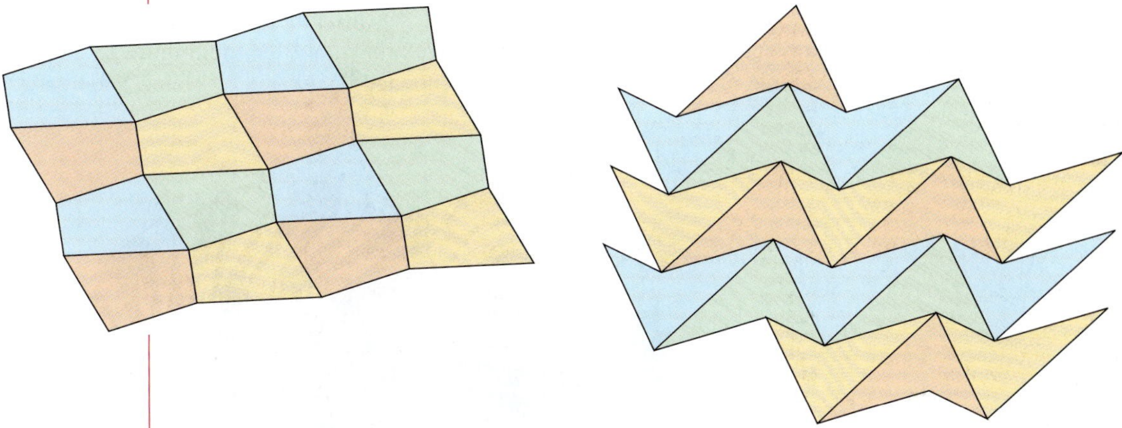

(d) A pentagonal tile with two parallel sides can always tile the plane. If the parallel edges are congruent, the tiling will be edge to edge.

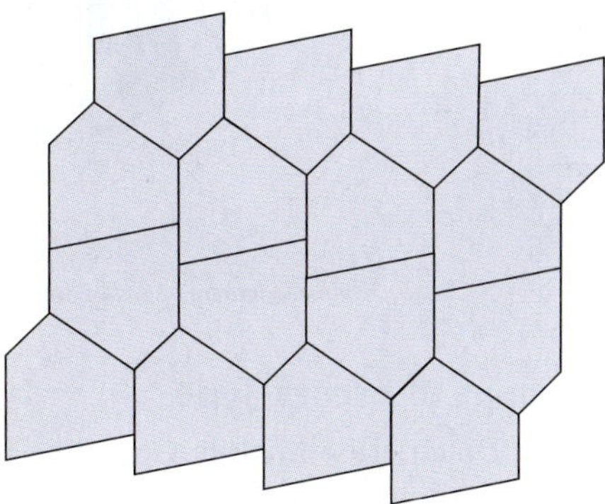

Much like the quadrilateral tiling shown above, a hexagon will tile the plane if it has a pair of opposite sides that are parallel and of the same length. (See problem 11 of Problem Set 13.3.) If all three pairs of opposite sides are congruent and parallel, it is not even necessary to rotate the tile from one position to any other. (See problem 10 of Problem Set 13.3.) It has been shown that no convex polygon with seven or more sides can tile the plane. The following theorem summarizes these discoveries.

THEOREM *Tiling the Plane with Congruent Polygonal Tiles*
The plane can be tiled by

- any triangular tile;
- any quadrangular tile, convex or not;
- certain pentagonal tiles (for example, those with two parallel sides);
- certain hexagonal tiles (for example, those with two opposite parallel sides of the same length).

The plane cannot be tiled by any convex tile with seven or more sides.

Although no convex polygon of seven or more sides can tile the plane, there are many interesting examples of nonconvex polygons that tile. Figure 13.27 shows a striking example of a spiral tiling by 9-gons (nonagons) created by Heinz Voderberg in 1936.

Figure 13.27
Heinz Voderberg's spiral tiling with nonagons

Tilings of Escher Type

The Dutch artist Maurits Cornelius Escher (1898–1972) has created a large number of artistic tilings. His designs have great appeal to artists and the general public, and have also captured the interest of professional geometers. Escher's graphic work is most often based on modifications of known tiling patterns. However, he also discovered new principles of pattern formation that mathematicians had overlooked.

To see how Escher created his print of the birds on the left side of Figure 13.28, we begin by identifying the underlying grid of parallelograms shown on the right.

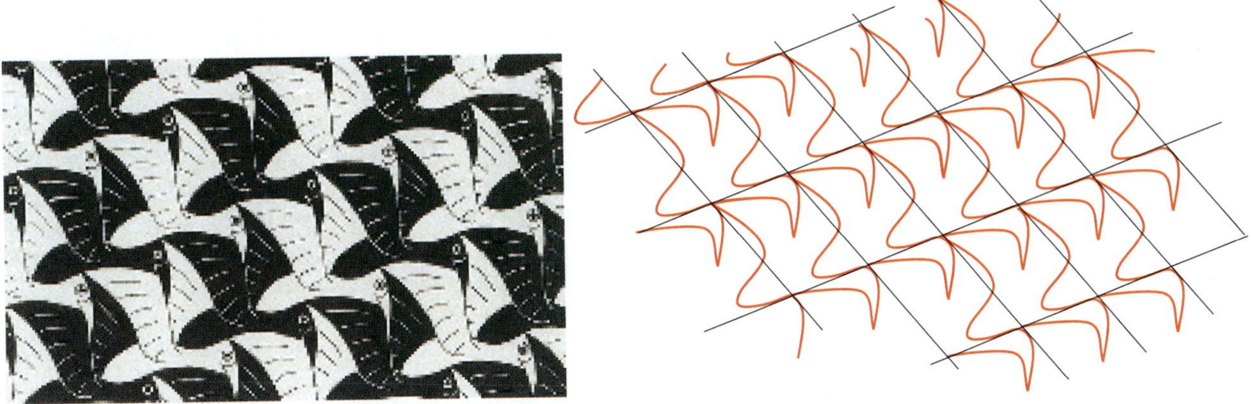

Figure 13.28
M. C. Escher's birds and its grid of parallelograms

The concept of a translation, discussed in Section 13.1, helps us understand how the parallelogram has been modified to become the bird-shaped motif of the tiling. First, imagine replacing the upper edge of the parallelogram with the *V*-shape separating the wings. The *V*-shape is then translated to replace the opposite edge of the parallelogram. Similarly, one of the two remaining straight edges of the parallelogram is modified to form the leading edge of the forward wing, and this curve is then translated to replace the opposite edge of the parallelogram. Finally, the outline is filled in with details such as feathers and an eye to complete the bird motif. The steps modifying the parallelogram into the bird motif are shown in Figure 13.29.

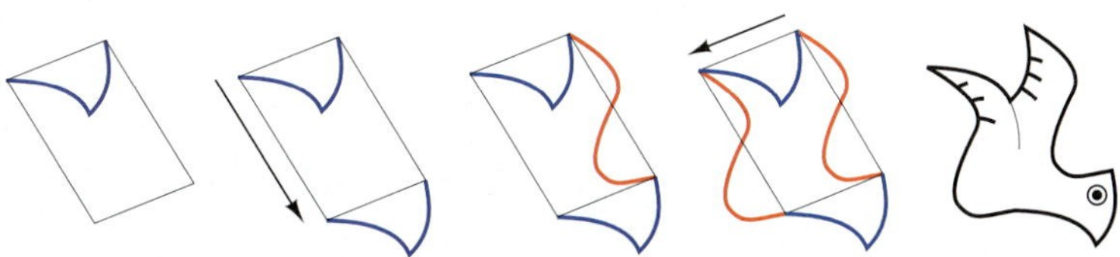

Figure 13.29
Modifying a parallelogram with two translations

Similar procedures will transform any polygonal tiling to an Escher-like tiling. For example, sixth-grade teacher Nancy Putnam used translations to modify each of three pairs of opposite parallel congruent sides of hexagon *ABCDEF*. Her whale tiling is shown in Figure 13.30.

Rotations can also be used to create **Escher-like designs** with interesting symmetries. Figure 13.31, on page 000, shows how a lizard tile can be created by modifying a regular hexagon *ABCDEF*. Side $\overline{AB}$ is first modified and then rotated about vertex *B* to

Did You Know?

Escher's Method of Artistic Tilings

"How did he do it?"

The work of M. C. Escher provokes that irrepressible question. A recurring theme as well as a device in his work from 1937 onward was, in his words, the "regular division of the plane." We see the jigsaw puzzle–like interlocking of birds, fish, lizards, or other creatures in his work; their rigid paving is usually just a fragment, a pause in a transition from two to three dimensions, a springboard from lockstep order to freedom. Escher confesses that the subject is for him a passion.

In his 1958 book *Regelmatige Vlakverdeling (Regular Division of the Plane),* he tells us much about why he uses regular division, explains some of the geometric elements of regular division, addresses the central question of figure and ground, and leads us through the development of a metamorphosis of form. But when we persist in asking "How did he do it?"—that is, how did he make those interlocking creatures—we do not find answers. We do find a few tantalizing hints in the book that Escher did study some technical papers and that he worked out his own theory:

At first I had no idea at all of the possibility of systematically building up my figures. I did not know any "ground rules" and tried, almost without knowing what I was doing, to fit together congruent shapes that I attempted to give the form of animals. Gradually, designing new motifs became easier as a result of my study of literature on the subject, as far as this was possible for someone untrained in mathematics, and especially through the formulation of my own layman's theory, which forced me to think through the possibilities. It remains an extremely absorbing activity, a real mania to which I have become addicted, and from which I sometimes find it hard to tear myself away.

(From the preface to *M. C. Escher: Visions of Symmetry,* by Doris Schattschneider, with Freeman, 1990, Harry N. Abrams, 2004. Reprinted with permission.)

Professor Schattschneider's book is an account of Escher's discovery of the world of geometry and how he used his knowledge to create his intriguing interlocking figures.

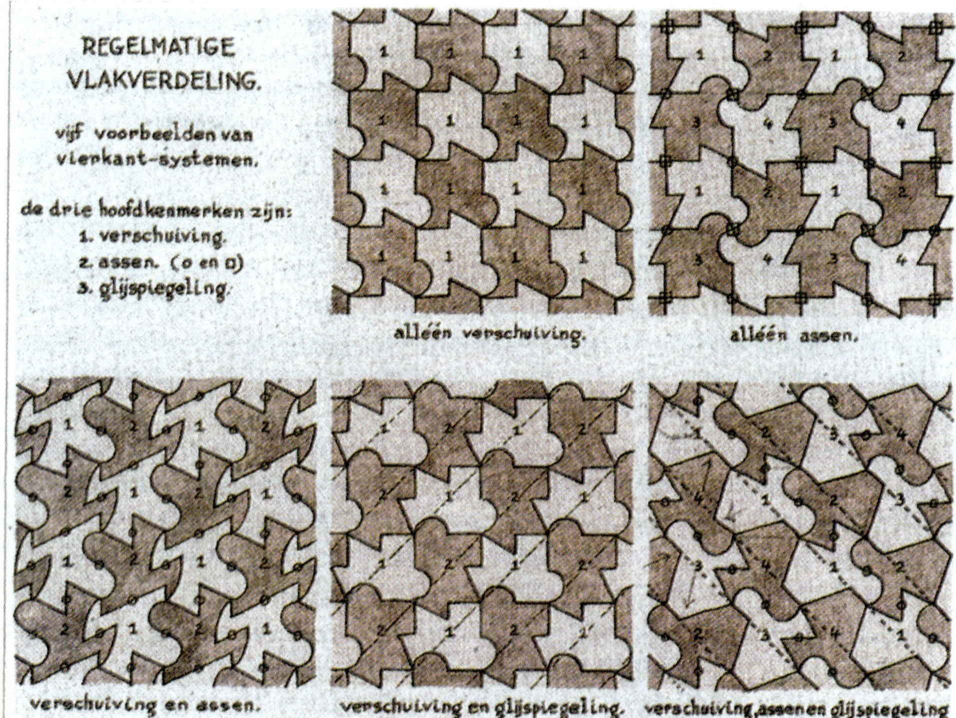

A lecture poster made by Escher to explain the role of isometries in his tilings, which he called regular divisions of the plane (regelmatige vlakverdeling).

SOURCE: From M. C. Escher: Visions of Symmetry, *by Doris Schattschneider. Copyright © 1990 by W. H. Freeman and Company, Harry N. Abrams, 2004. Reprinted by permission.*

Cooperative Investigation

Creating an Escher-like Design

Materials Needed

1. Note cards, 3″ × 5″ (or other card stock)
2. Scissors
3. Pencils and colored markers
4. Blank sheets of paper

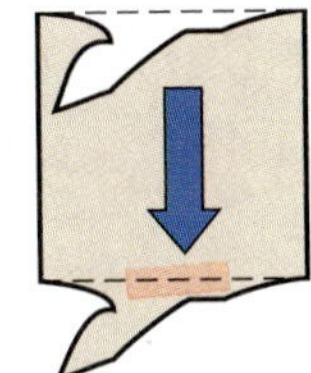

Directions

Step 1. Cut a small (say, $2\frac{1}{2}''$-by-$3''$) rectangle from a note card or card stock.

Step 2. Cut out one side of the rectangle. Translate the cutout piece to the opposite side and tape it in place.

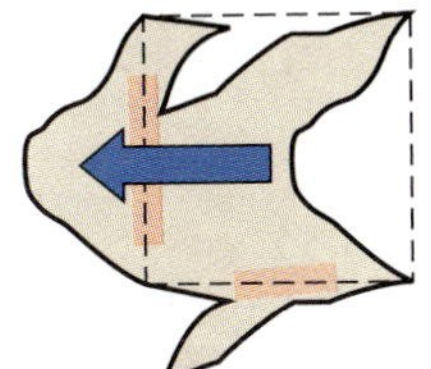

Step 3. Repeat Step 2 for the remaining two parallel sides of the rectangle, as shown.

Step 4. Do an "inkblot" test. Is your shape a frog, a bird, a face, a ___________? Brainstorm with a partner. It may help to rotate your shape or flip it over. Add eyes, mouth, nose, ears, feet, beaks, horns, clothing, scales, fur, and other imaginative details to make your tiling template recognizable and interesting.

Step 5. Trace around the template to create your Escher-like design on a blank sheet of paper. Use colored markers to fill in the details and make your design attractive.

Extensions

Instead of a rectangle, start with any parallelogram and follow the directions given above. It is also possible to adapt this method to create templates based on other tilings of the plane. Several suggestions are described in problems 14 through 17 of Problem Set 13.3.

modify side $\overline{BC}$. The remaining two pairs of sides are modified similarly, resulting in the outline of the lizard tile. The resulting lizard tiling is also shown in Figure 13.31.

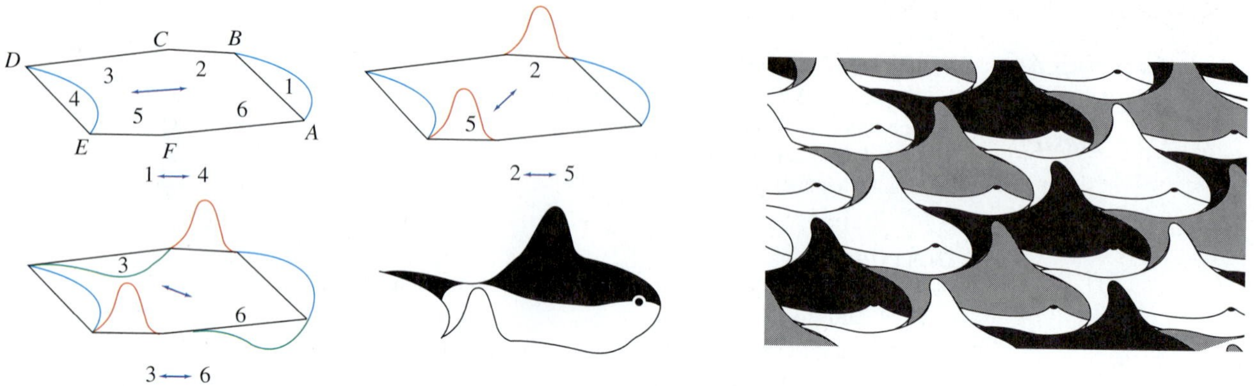

Figure 13.30
Sixth-grade teacher Nancy Putnam modified a hexagon with opposite parallel congruent sides to create an Escher-like tiling

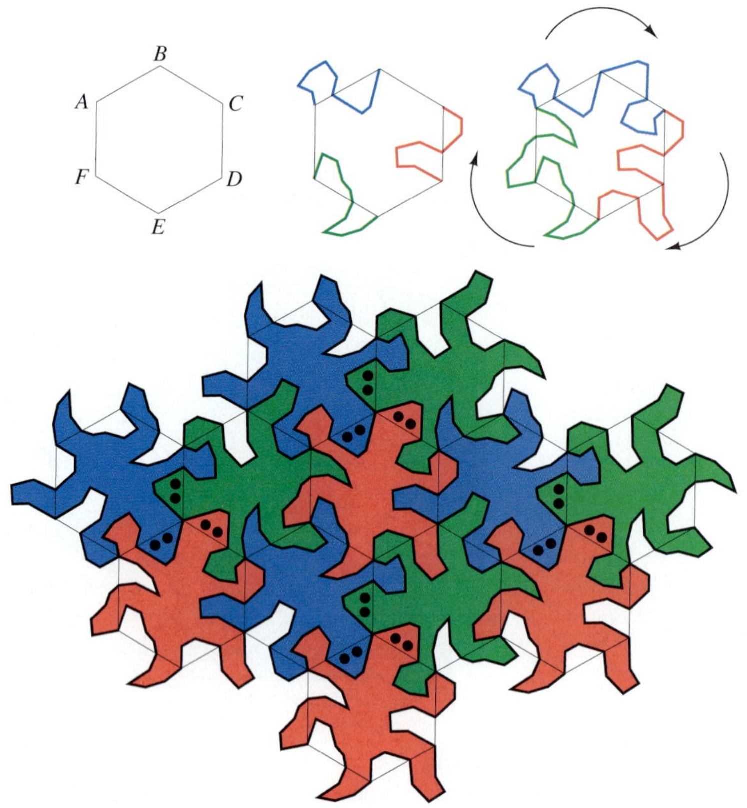

Figure 13.31
Modifying a regular hexagon with rotations to create a lizard tiling

Problem Set 13.3

Understanding Concepts

1. On dot paper arranged in a square grid, show that the given shape will tile the plane.

(a)

(b)

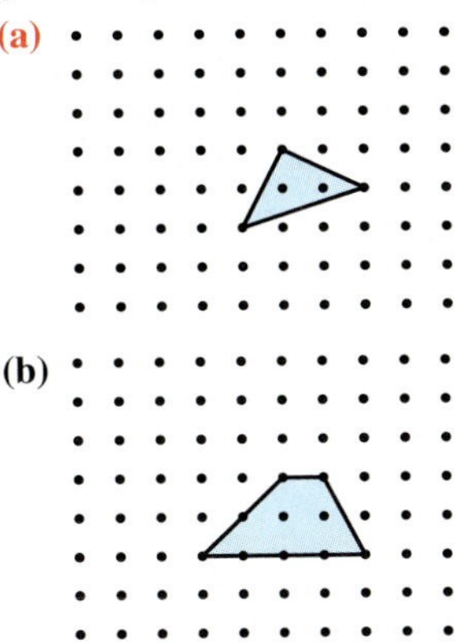

2. On "isometric" dot paper (arranged in a grid of equilateral triangles), show that the given shape will tile the plane.

(a)

(b)

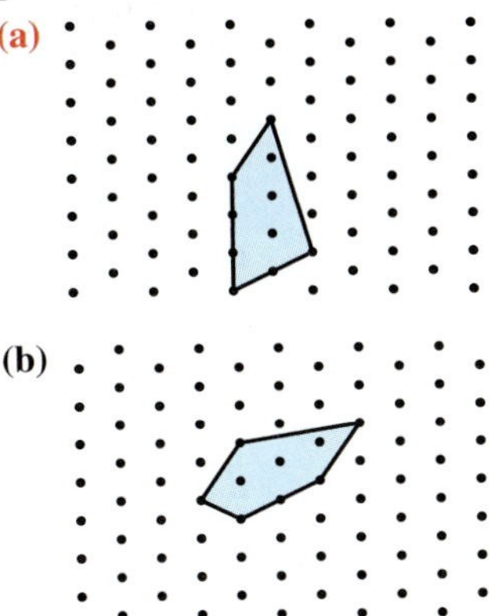

3. Fold a sheet of paper in half, and then use scissors to cut a pair of congruent convex quadrilaterals. Cut one of the quadrilaterals along one of the diagonals, and cut the second quadrilateral along the other diagonal. Show that the four triangles can be arranged to form a parallelogram and therefore tile the plane with repeated copies of the four triangles.

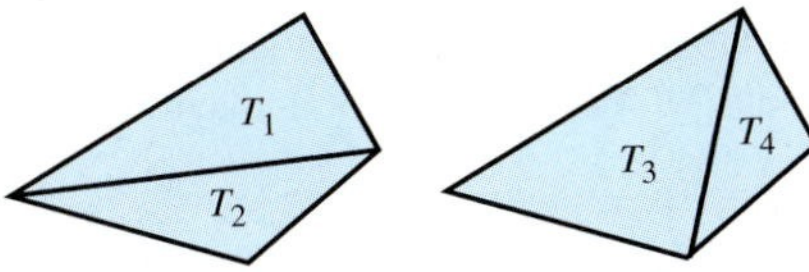

SOURCE: From Quantum Magazine, *September/October, 1992, p. 31. Reprinted with permission from Springer-Verlag.*

4. Branko Grünbaum and G. C. Shephard (*Tilings and Patterns,* W. H. Freeman and Co., 1987) discovered the tiling shown here in the children's coloring book *Altair Design* (E. Holiday, London: Pantheon, 1970).

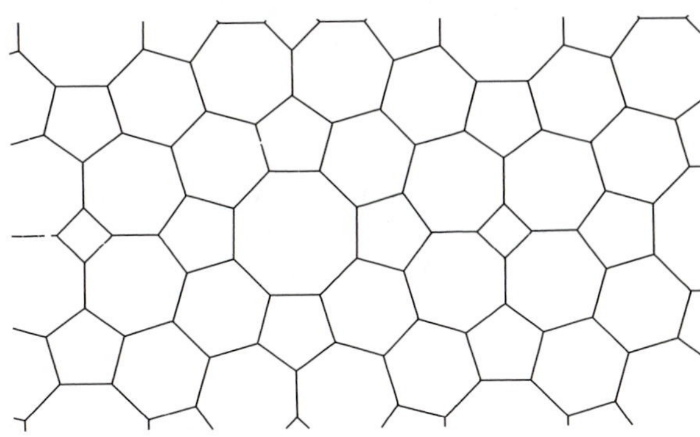

(a) What kinds of polygons appear?

(b) Grünbaum and Shephard claim this is a "fake" tiling by regular polygons. Explain why.

5. A vertex figure of regular polygons is shown.

(a) Find the angle measures of each polygon and directly verify that they add up to 360°.

(b) Explain why the vertex figure does not extend to form a semiregular tiling. (*Suggestion:* Attempt to form the vertex figure at the other two vertices of the triangle.)

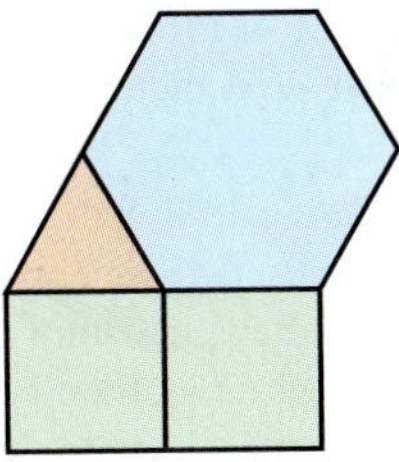

6. Consider the vertex figure formed by a square, a regular pentagon, and a regular 20-gon. Find the measures of the interior angle of each polygon and show that these three measures add up to 360°.

7. Some "letters" of the alphabet will tile the plane. For each letter shown, create an interesting tiling on square dot paper. Look for different patterns that use the same tile.

(a) (b) (c) (d) (e)

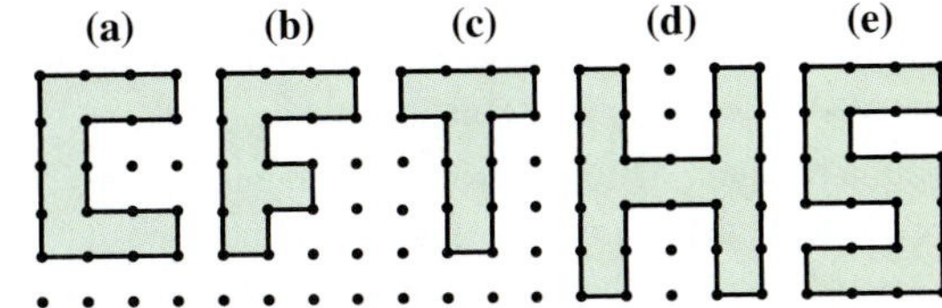

8. A tetromino is a tile formed by joining four congruent squares edge to edge, where adjacent squares must share a common edge. Two tetrominoes and two nontetrominoes are shown below.

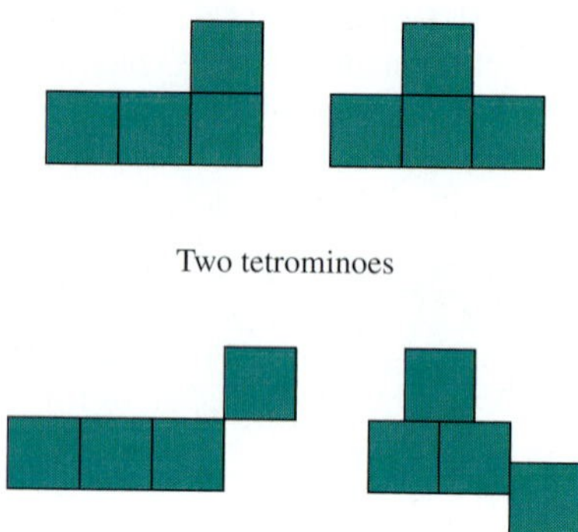

Two tetrominoes

Two nontetrominoes

(a) There are five noncongruent tetrominoes altogether. Find the other three.

(b) Which tetrominoes tile the plane, using unlimited congruent copies of one tetromino?

(c) The five noncongruent tetrominoes have a total area of 20 square units. Can the five shapes tile a 4-by-5 rectangle? (*Suggestion:* Imagine that the rectangle is colored in a red-and-black checkerboard pattern of unit squares. How many red and how many black squares are covered by each tetromino?)

9. Tiles formed by joining five congruent squares edge to edge are called **pentominoes.** The 12 pentominoes are shown in problem 25 of Problem Set 11.3. Use graph paper (or dot paper) to decide which pentominoes tile the plane.

10. Fold a 3″-by-5″ note card in half, and then use scissors to cut (simultaneously) two general quadrilaterals. Rotate one quadrilateral a half-turn and tape the two quadrilaterals together along their corresponding edges to form a hexagonal tile.

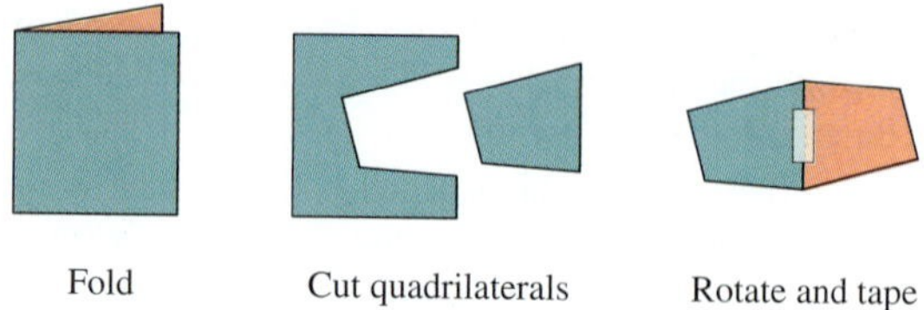

Fold — Cut quadrilaterals — Rotate and tape

Show, using the paper model as a template, that the hexagon tiles the plane. Must the tile be rotated?

11. Cut a hexagon $ABCDEF$ from a rectangular piece of card stock, using a ruler to ensure that the opposite sides $\overline{AB}$ and $\overline{DE}$ have the same length. No restriction is placed on the position of points C or F. Use the paper template to illustrate that a hexagon with a congruent and parallel pair of opposite sides can tile the plane. (*Suggestion:* Half-turns of the template will be required.)

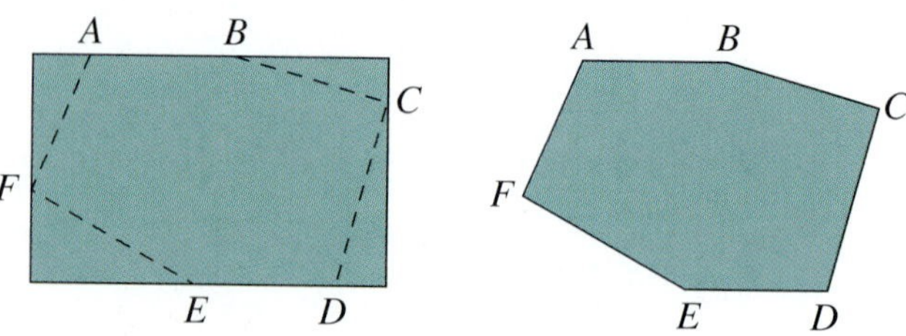

Teaching Concepts

12. **Patterns in World Cultures.** Children's interest in patterns and symmetries can be heightened by incorporating examples from around the world. For example, the pattern below is a pattern on *kente* cloth, a fabric woven by the Ashanti and the Ewe in Togo, West Africa.

Conduct a search in your library or on the Web to discover four or five examples of patterns from a variety of world cultures. Give a careful description of the symmetry properties of each example.

13. **Connections with Art.** The concepts of symmetry and pattern have close ties with art. For example, consider this assignment in an elementary school classroom: Each student is given a square of paper. Each square is to be decorated with a design using three colors and having 180° symmetry. The decorated squares are then used to tile a large poster (or several posters) that is hung on the classroom wall.

(a) Carry out the design project described above in your own class.

(b) Create a similar lesson, but using a different tiling shape with different color and symmetry conditions.

Thinking Critically

14. Construct a paper hexagonal tile with each pair of opposite sides parallel and congruent. The template can be cut from a note card following the method described in problem 10. Make cutouts on three adjacent sides.

Translate each cutout to the opposite side and tape along the corresponding edges to form a template. (See the instructions in the Cooperative Investigation box "Creating an Escher-like Design.")

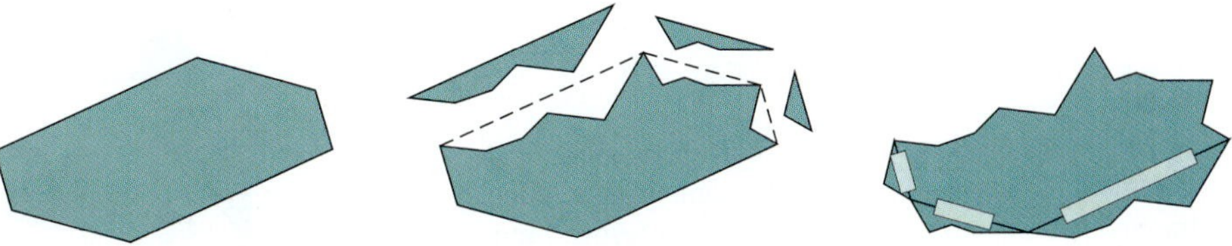

Create a design with your template, adding details such as eyes, mouths, and so forth, to give added interest to your design.

15. Cut an accurate square from a note card. Make cutouts on opposite sides. Rotate each cutout 90° and tape as shown. Use the paper template to create an Escher-like design.

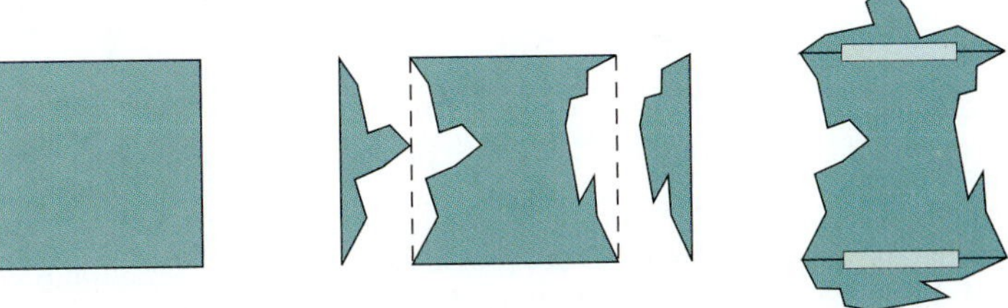

16. Cut an arbitrary triangle from a note card, and lightly fold (do not make a heavy crease) one vertex to another to determine the midpoint of the side between the vertices. The side can be modified by making a cutout on one side of the midpoint, rotating it 180° about the midpoint, and taping it in place. The steps to modify one side of a triangle are shown below.

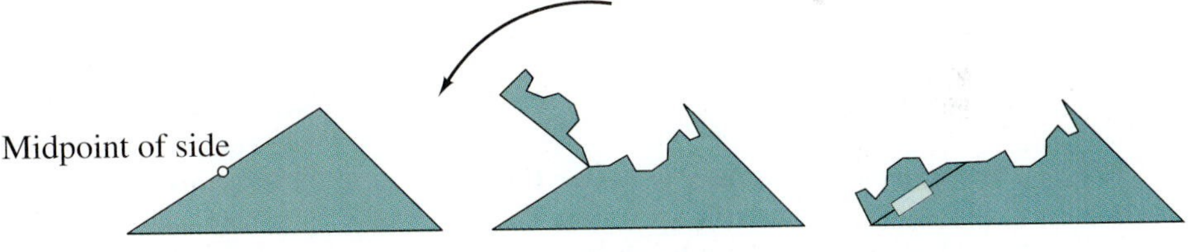

Modify the remaining two sides of the triangle, and use the resulting template to create an Escher-like design.

---

17. Cut a convex quadrilateral from a note card. Make midpoint modifications, as described in problem 16, to each of the four sides. Use the resulting template to create an Escher-like design.

18. For any integer n, $n \geq 3$, show that there is some n-gon that tiles the plane. (*Suggestion:* Consider the midpoint modification described in problems 16 and 17.)

19. Suppose a vertex figure of regular polygons includes a regular octagon. Show that the figure must include another octagon and a square.

20. An equilateral triangle and a parallelogram are each examples of "reptiles," short for "repeating tile." In each case, copies of the tile can be arranged to form its own similar shape.

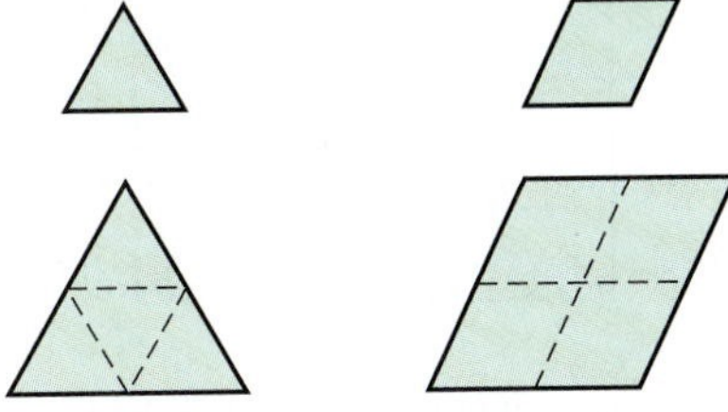

Use square dot paper to show that each of the shapes on the next page is a "reptile."

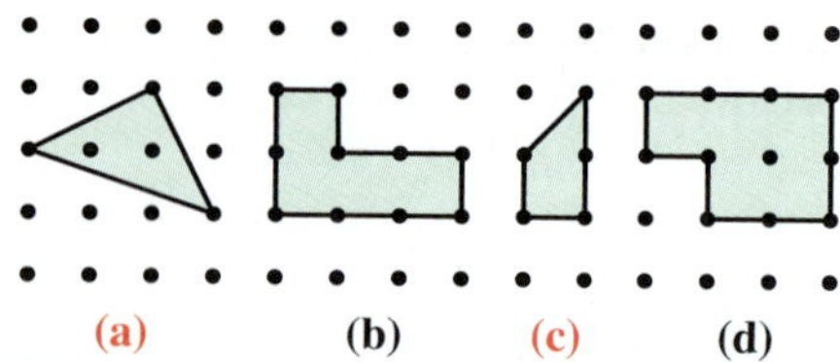

21. A hexiamond is formed from six congruent equilateral triangles. There are 12 different hexiamonds, including the Sphinx, Chevron, and Lobster, shown below. Find the remaining nine hexiamonds and see if you can match their shape to their names: Hexagon, Crook, Crown, Hook, Snake, Yacht, Bar, Signpost, and Butterfly.

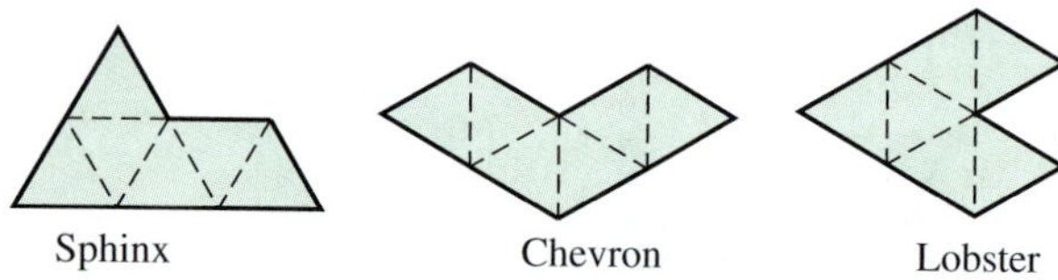

22. **(a)** Show that the Sphinx is a reptile. (See problems 20 and 21 for the definition of a reptile and a diagram of the Sphinx.)

(b) It requires four copies of the Sphinx to form a second-generation Sphinx. How many copies of the original Sphinx are required to form a third-generation Sphinx? Explain your reasoning and provide a sketch.

(c) Explain why any reptile provides a tiling of the plane.

Thinking Cooperatively

23. The seven **tangram** pieces originated in ancient China. As shown below, there are five triangles, a square, and a parallelogram. A serviceable set can be cut from a square of cardboard (see Section 12.1), although plastic and wooden sets are easy to buy or make.

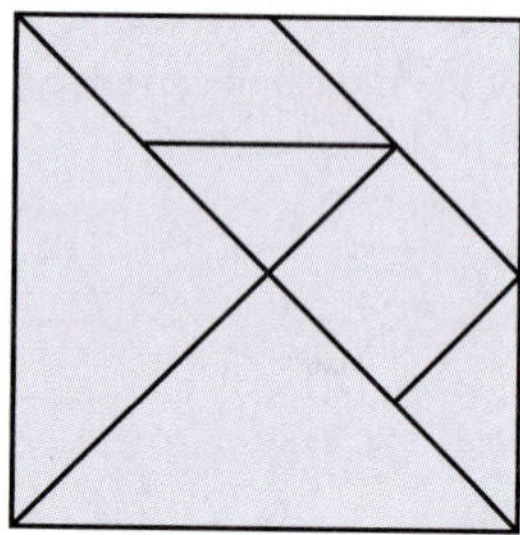

(a) The most common activity is the Chinese ***tangram puzzle.*** A figure is shown in outline, and the challenge is to tile the figure using *all* seven tangram pieces. Form the following animals (taken from the Multicultural Poster Set, National Council of Teachers of Mathematics, 1984), working in pairs.

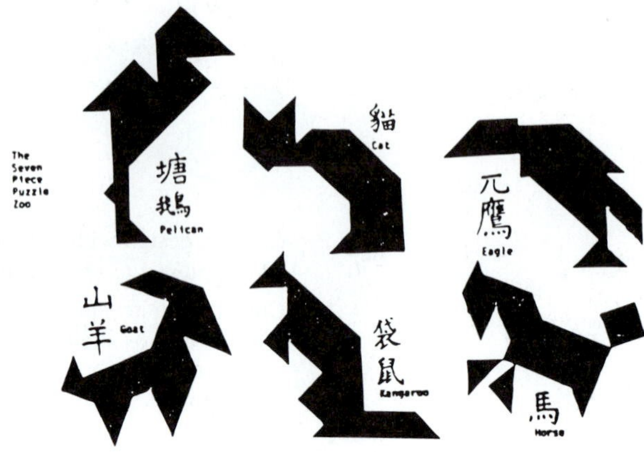

(b) Use the seven tangram pieces to form other recognizable shapes. Use a marker to draw only the outline. Then trade the tangram puzzles between groups and solve the puzzles.

24. Work in groups to find all of the different convex figures that can be tiled by tangrams (see problem 23). Be sure to use all seven tangram pieces in each of the convex figures. There are 13 noncongruent figures in all, and most of the figures can be tiled in several ways.

Using a Computer

25. The Geometer's Sketchpad geometry software includes tools (or scripts) that enable the user to quickly construct a variety of polygons. In particular, the regular polygons of sides 3, 4, 5, 6, and 8 are easily constructed by selecting two points to be adjacent vertices. (Details are found in the *Learning Guide* that accompanies the software; also see Appendix D.)

(a) Use the software to create examples of some of the semiregular tilings of the plane.

(b) Write a brief report that shows your tiling examples and discusses what steps you discovered to make your tilings.

26. Use geometry software to create an Escher-like design based on the tiling of the plane by equilateral triangles, following these steps: Begin by constructing an equilateral triangle. Next, modify one side of the triangle and rotate it 60° about an endpoint vertex to modify a second side of the triangle. Alter the remaining side of the triangle by constructing the midpoint and doing a midpoint modification (see problem 16 and the figure below). Hide unwanted lines and add decorations to complete the tile. To use the tile to make your design, first use rotations to form a six-tile arrangement as shown below. Translations of the six-tile arrangement will complete your design.

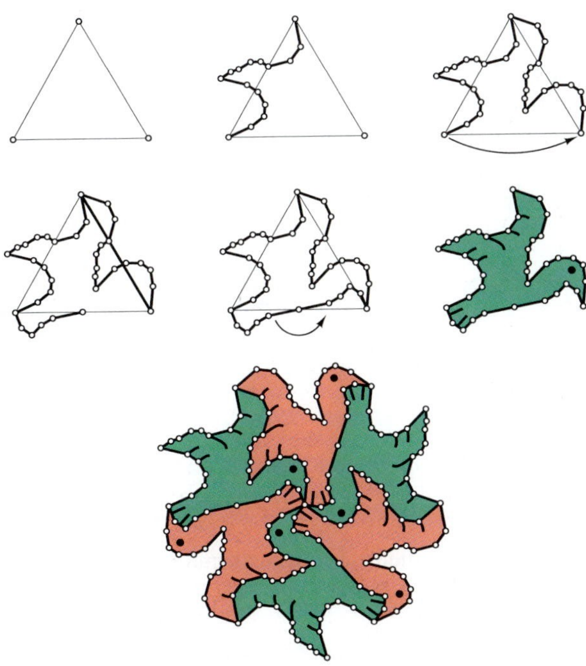

From State Student Assessments

27. (New Jersey, Grade 4)
Which group of shapes was used to form the figure below? Use your colored shapes to help you.

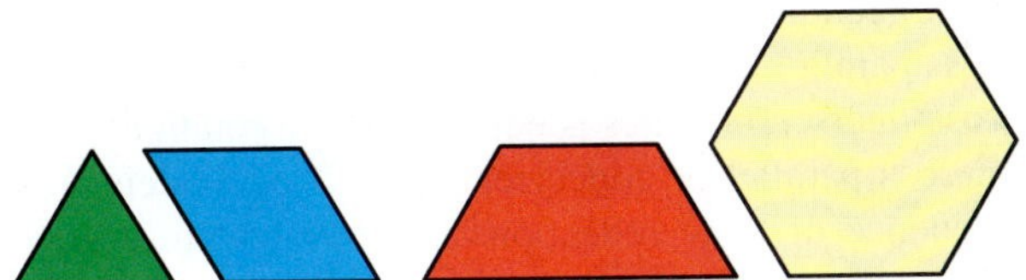

A. 2 yellow hexagons and 1 green triangle
B. 3 red trapezoids
C. 1 yellow hexagon, 1 red trapezoid, and 1 green triangle
D. 1 yellow hexagon and 3 green triangles

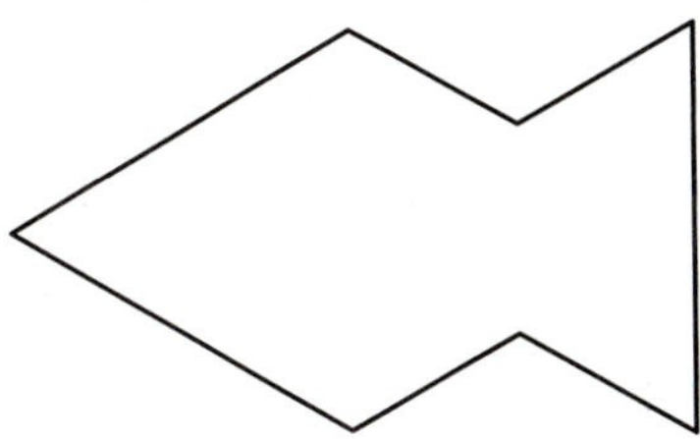

For Review

28. What rule is used to separate the letters of the alphabet in the following arrangement?

ABCDE HI K MNO STUVWXYZ
FG J L PQR

29. What is the common name for these polygons?

(a) a triangle with no line of symmetry
(b) a triangle with exactly one line of symmetry
(c) a triangle with three lines of symmetry
(d) a kite with two lines of symmetry
(e) a regular polygon with six lines of symmetry

30. Identify these polygons.

(a) a triangle with 120° rotational symmetry
(b) a quadrilateral with 180° but not 90° rotational symmetry
(c) a regular polygon with 40° as its smallest angle of rotational symmetry

Epilogue The Dynamical View of Geometry

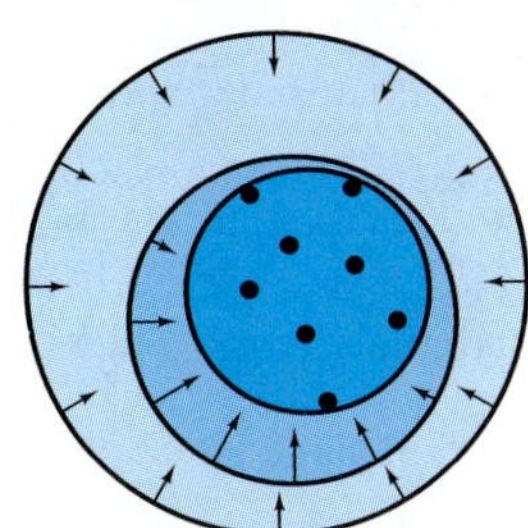

Geometry in Euclid's time presented a static view of shape, as if only still pictures of figures were to be seen in the mind's eye. The concept of a geometric transformation, introduced in the latter part of the nineteenth century, has provided a dynamic view of geometry. Figures are allowed, and even invited, to move and perhaps even change size. The mind's eye sees an animated world of shapes in action.

The dynamic viewpoint is a useful tool for problem solving and discovery. Here's an example: Show that any finite set of points in the plane is contained inside (or on) some circle of smallest radius. To see why, first imagine surrounding the points with a very large circle. Now let the circle shrink around the set of points as tightly as possible, allowing the center of the moving circle to move as well.

As a second example, imagine that a small circle within a triangle is allowed to grow until it just touches all three sides of the triangle. This makes it clear that every triangle has an inscribed circle.

Recent developments in computer graphics, both hardware and software, provide new opportunities to explore which properties change, and which remain invariant, as a figure is altered.

Chapter Summary

Key Concepts

Section 13.1 Rigid Motions and Similarity Transformations

- Transformation. A transformation of the plane is a one-to-one correspondence of the points of the plane. If point P corresponds to point P', P' is the image of P and P is the preimage of P'.
- Rigid motion. A rigid motion, also called an isometry, is a transformation of the plane that preserves distance: $PQ = P'Q'$ for all points P and Q and corresponding respective image points P' and Q'.
- The four basic rigid motions: translation (or slide), rotation (or turn), reflection (or flip), and glide reflection (or glide).
- Identity transformation. The rigid motion for which each point corresponds to itself: $P = P'$ for all points P in the plane.
- Equivalent transformations. Two transformations are equivalent if both transformations take each point P onto the same image point P'.
- Equivalence properties of multiple reflections:
  A sequence of two reflections across parallel lines is equivalent to a translation.
  A sequence of two reflections across intersecting lines is equivalent to a rotation.
  A sequence of three reflections across parallel or concurrent lines is equivalent to a single reflection. Otherwise, a sequence of three reflections is equivalent to a glide reflection.
- Classification theorem for rigid motions. Every rigid motion is equivalent to one of the four basic motions: a translation, a rotation, a reflection, or a glide reflection.
- Congruence. Two figures are congruent if, and only if, one figure is the image of the other under a rigid motion.
- Dilation. A dilation, or size transformation, with center O and scale factor k takes each point P other than O to the point P' on the ray $\overrightarrow{OP}$ so that $OP' = kOP$ and leaves O fixed.
- Distance property of a dilation. A dilation multiplies all distances by the scale factor k, so $P'Q' = k \cdot PQ$ for all P and Q.
- Similarity transformation. A similarity transformation is a sequence of dilations and rigid motions.
- Similar figures. Two figures are similar if, and only if, there is a similarity transformation that takes one figure onto the other.

Section 13.2 Patterns and Symmetries

- Symmetry. A symmetry of a figure is a rigid motion that takes every point of the figure onto an image point that is also a point of the figure.

- Types of symmetry. A figure has reflection symmetry if reflection across some line m called a line of symmetry takes a figure onto itself;
  rotation symmetry if rotation about some center O takes the figure onto itself;
  point symmetry about point O if it has half-turn (180°) symmetry about O.
- Pattern. A pattern is a plane figure with translation symmetries, including some translation of smallest positive distance.
- Border pattern. A pattern with translations in one direction only. There are seven symmetry types of border patterns.
- Wallpaper pattern. A pattern with translation symmetries in more than one direction. There are 17 symmetry types of wallpaper patterns.
- Crystallographic restriction. A wallpaper pattern can have rotation symmetries only of size 60°, 90°, 120°, or 180°.

Section 13.3 Tilings and Escher-like Designs

- Tile. A tile is a simple closed curve in the plane together with its interior.
- Tiling. A covering of a figure with tiles, with neither gaps nor overlaps.
- Tilings of the plane. There are three regular tilings: by equilateral triangles, squares, and regular hexagons. There are eight semiregular tilings that use regular polygon tiles of two or more types, where all vertex figures are identical.
- Tilings of the plane with general polygons. Arbitrary triangles and quadrilaterals, and certain pentagons and hexagons, tile the plane. No convex n-gon with $n \geq 7$ tiles the plane.
- Escher-like designs. Designs resembling the work of M. C. Escher can be created by modifying the straight sides of a polygonal tile to assume more general curves, following rules that preserve the tiling property of the modified shape.

Vocabulary and Notation

Section 13.1

Transformation of the plane
Preimage, image
Rigid motion, or isometry
Basic rigid motion: translation (slide), rotation (turn), reflection (flip), glide reflection (glide)
Slide arrow (or translation vector)
Center of rotation
Turn angle, turn arrow
Line of reflection (mirror line)
Orientation reversing/preserving transformation
Equivalent transformations
Identity transformation
Congruent figures, $F \cong G$
Dilation, or size transformation
  Center, O
  Scale factor k ($0 < k < 1$: contraction; $k > 1$: expansion)
Similarity transformation
Similar figures

Section 13.2

Symmetry of a figure
Reflection symmetry (line symmetry, bilateral symmetry)
  Line of symmetry (mirror line)
  Rotation symmetry (turn symmetry)
Center of rotation
Circular symmetry
Point symmetry
Pattern
  Border pattern
  Wallpaper pattern

Section 13.3

Tile
Tiling, or tessellation
Regular tiling of the plane
Vertex figure
Semiregular tiling of the plane
Escher-like design

Chapter Review Exercises

Section 13.1

1. Draw the image of $ABCDE$ under the translation that takes A onto A'.

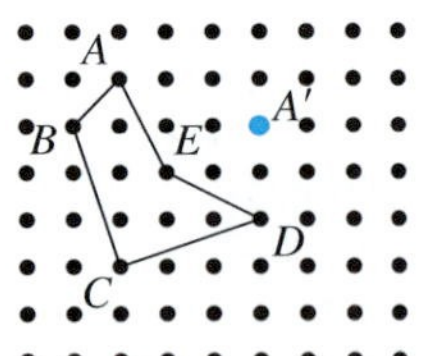

2. Determine the center and turn angle of the rotation that takes A onto A' and B onto B'. Use a protractor, Mira, ruler, or whatever drawing tools you wish.

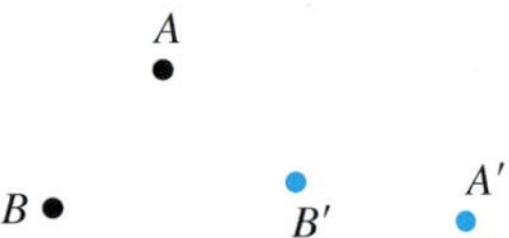

3. Describe the basic rigid motion that takes A, B, and C onto A', B', and C', respectively. Use any drawing tools you wish.

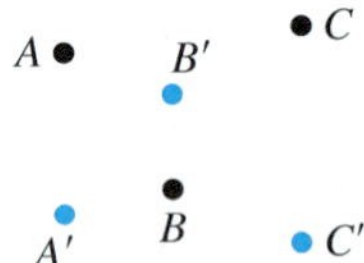

4. A glide reflection has a horizontal line l as its line of reflection and translates 4 inches to the right. Draw three lines of reflection—m_1, m_2, and m_3—so that successive reflections across m_1, m_2, and m_3 result in a motion equivalent to the glide reflection.

5. Sketch the image of the square $ABCD$ under each of these transformations:

   (a) the dilation centered at O with scale factor 2.

   (b) the dilation centered at A with scale factor $1/3$.

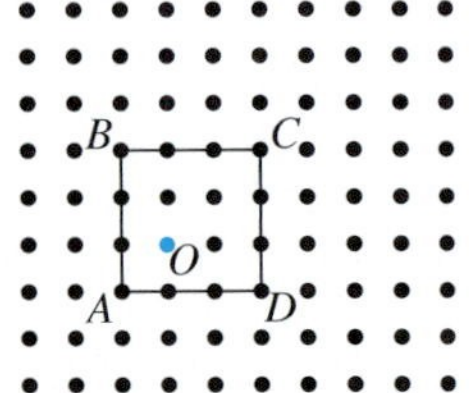

6. Describe the similarity transformation that takes the square $ABCD$ onto the square $JKLM$, where J, K, L, and M are the midpoints of square $ABCD$ as shown.

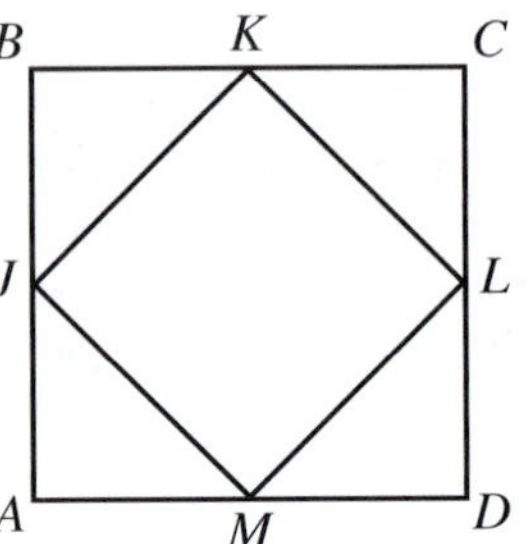

Section 13.2

7. The geometric forms shown are from African art. How many lines of symmetry does each figure have?

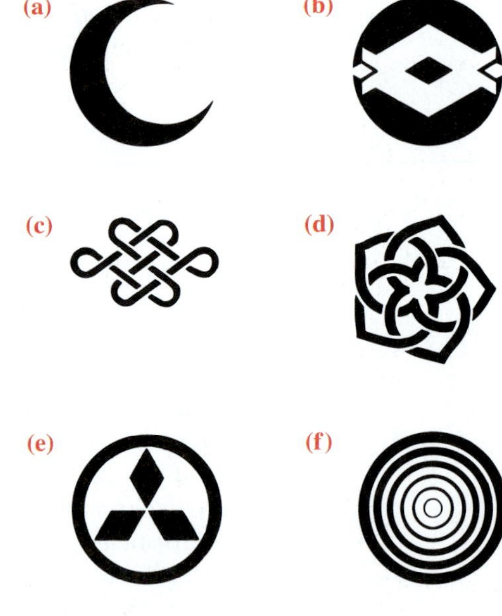

8. For each of the figures shown in problem 7, give all of the angles of rotation symmetry.

9. Describe the symmetries of each of these border patterns.

   (a)

FRENCH RENAISSANCE ORNAMENT FROM CASKET

(b)

STAINED GLASS, CATHEDRAL OF BOURGES

Section 13.3

10. Four regular polygons form a vertex figure in a tiling of the plane. Three of the polygons are a triangle, a square, and a hexagon. What is the fourth polygon?

11. Draw two different vertex figures that each incorporate three equilateral triangles and two squares.

12. Show that the shape shown will tile the plane.

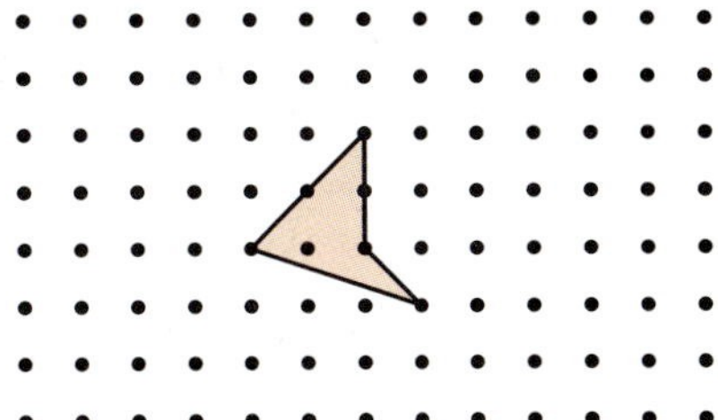

Chapter Test

1. Which of the following shapes will tile the plane?

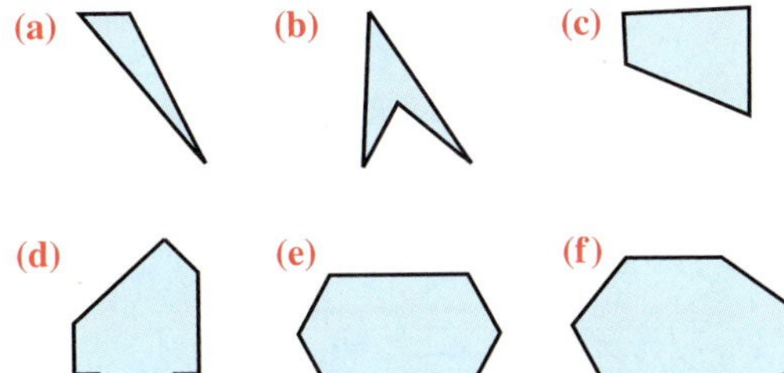

2. Draw two parallel lines m_1 and m_2 so that the sequence of reflections across m_1 and m_2 will map point P to point P'.

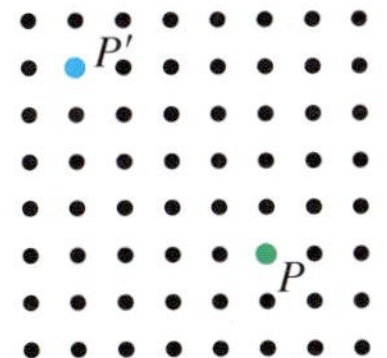

3. A translation takes points A, B, and C onto A', B', and C', respectively. Show the location of C' and B.

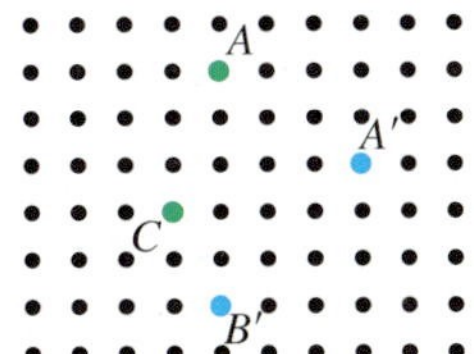

4. In the Escher tiling shown, what type of rigid motion

(a) takes figure A onto figure B?

(b) takes figure B onto figure C?

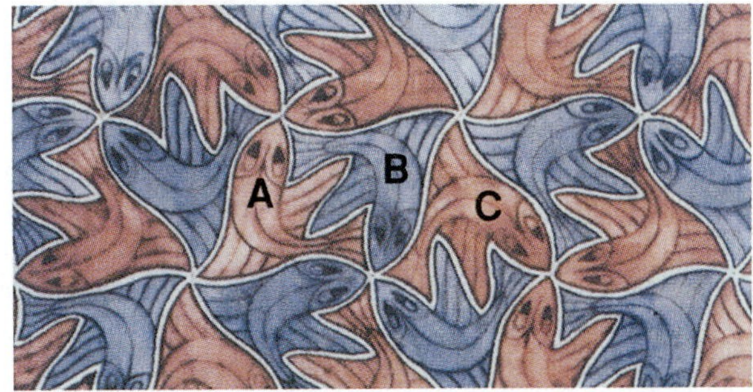

5. Three lines m_1, m_2, and m_3 are shown in each part. What type of rigid motion is equivalent to a sequence of reflections across m_1, m_2, and m_3?

(a)

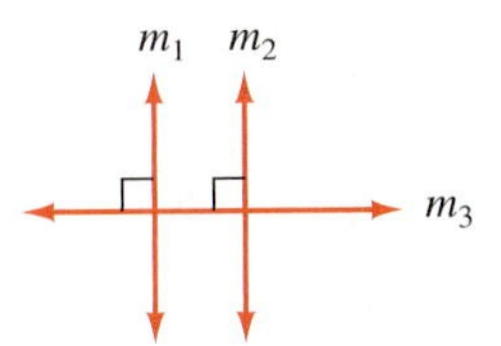

(b) Parallel lines

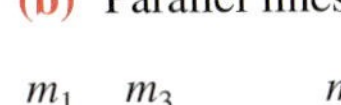

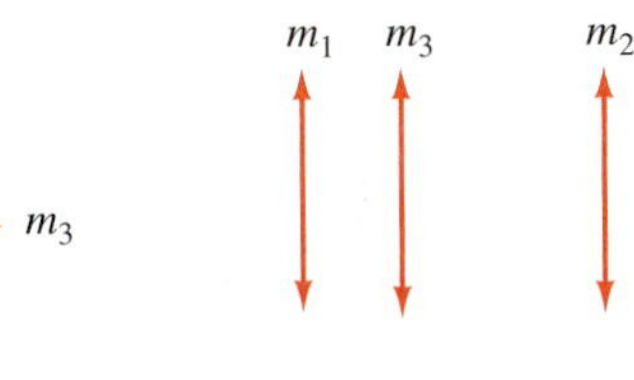

(c)

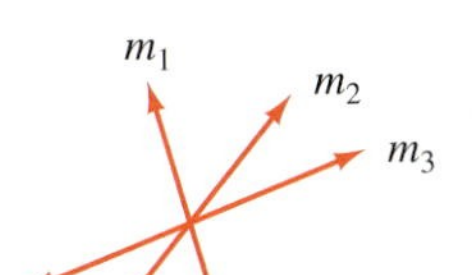

(d)

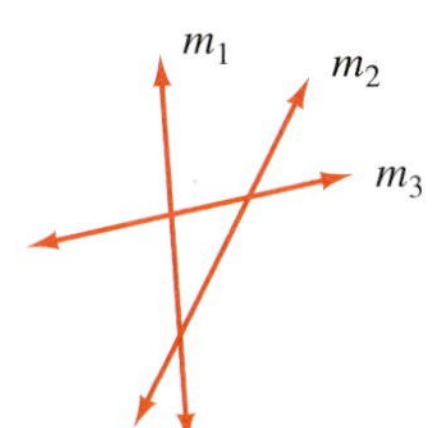

6. List the symmetries found in each of the border patterns shown.

(a)

INDIAN PAINTED LACQUER WORK

(b)

MALTESE LACE

(c)

ANCIENT GREEK SCROLL BORDER

(d)

ITALIAN DAMASK OF THE RENAISSANCE

7. Describe a similarity transformation that takes the square $ABCD$ onto the square $A'B'C'D'$.

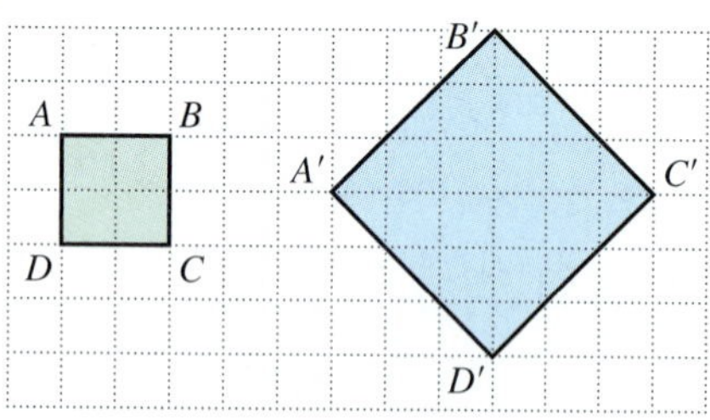

8. Copy the rectangle $ABCD$ and points A' and B', repectively, onto squared paper.

(a) Show that A' and B' are the image of A and B, respectively, under a rotation. Give the center of rotation and the size of the rotation angle.

(b) Draw the image rectangle $A'B'C'D'$.

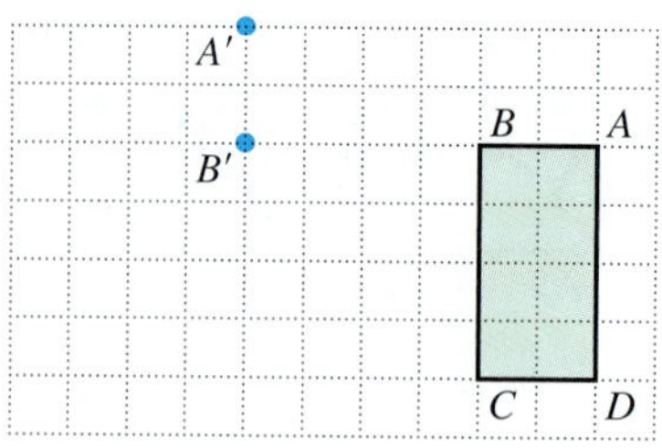

9. Sketch a portion of the semiregular tiling of the plane that uses square and octagonal tiles with a common side length.

10. Trace the drawing of $\angle A$. Suppose A' is the image of A under a reflection. Explain how to draw the image of $\angle A$ under the reflection.

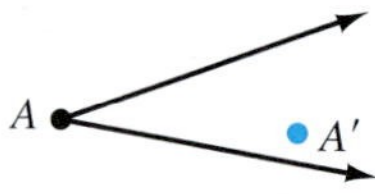

11. A dilation has taken $ABCDEF$ onto $AVWXYZ$ as shown, where $AV = 8$, $VB = 4$, $BC = 20$, and $AZ = 18$.

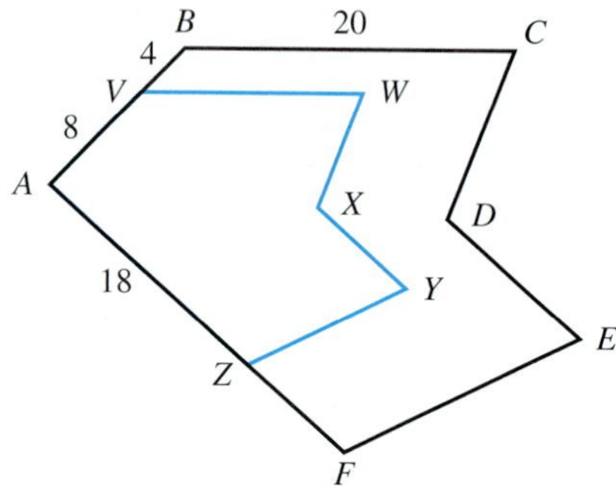

(a) What is the center of the dilation?

(b) What is the scale factor?

(c) What is the distance VW?

(d) What is the distance ZF?

12. Draw two lines l_1 and l_2 so that a sequence of reflections across l_1 and l_2 will rotate point Q to point Q' about the turn center O.

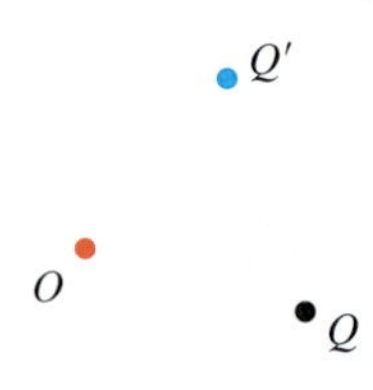

13. Two blank crossword puzzles are shown. What symmetries are found in the grid of black and white squares used by the puzzle maker?

(a)

(b)

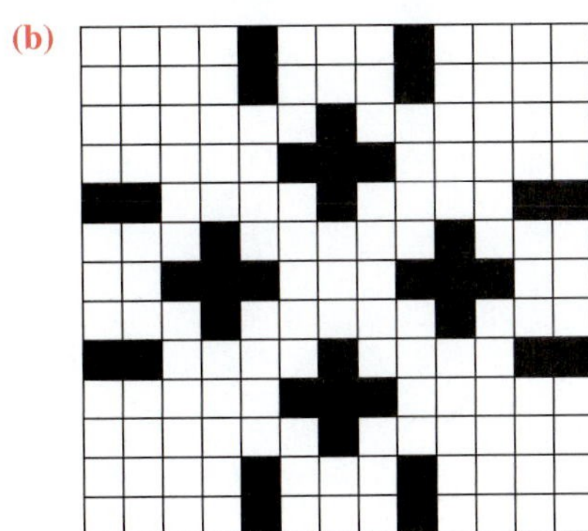

14. Find the line of reflection and slide arrow of the glide reflection that takes rectangle $ABCD$ onto $A'B'C'D'$.

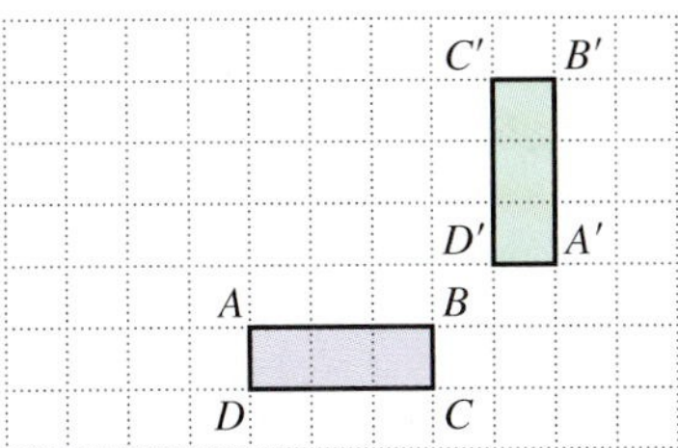

15. Draw all of the lines of symmetry for these figures.

(a) 

(b) 

(c) 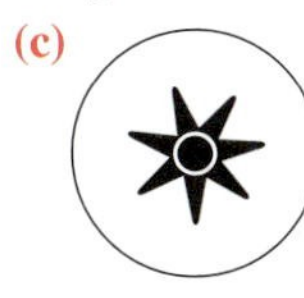

16. What is the size of the smallest positive angle of rotation symmetry in each figure shown in problem 15?

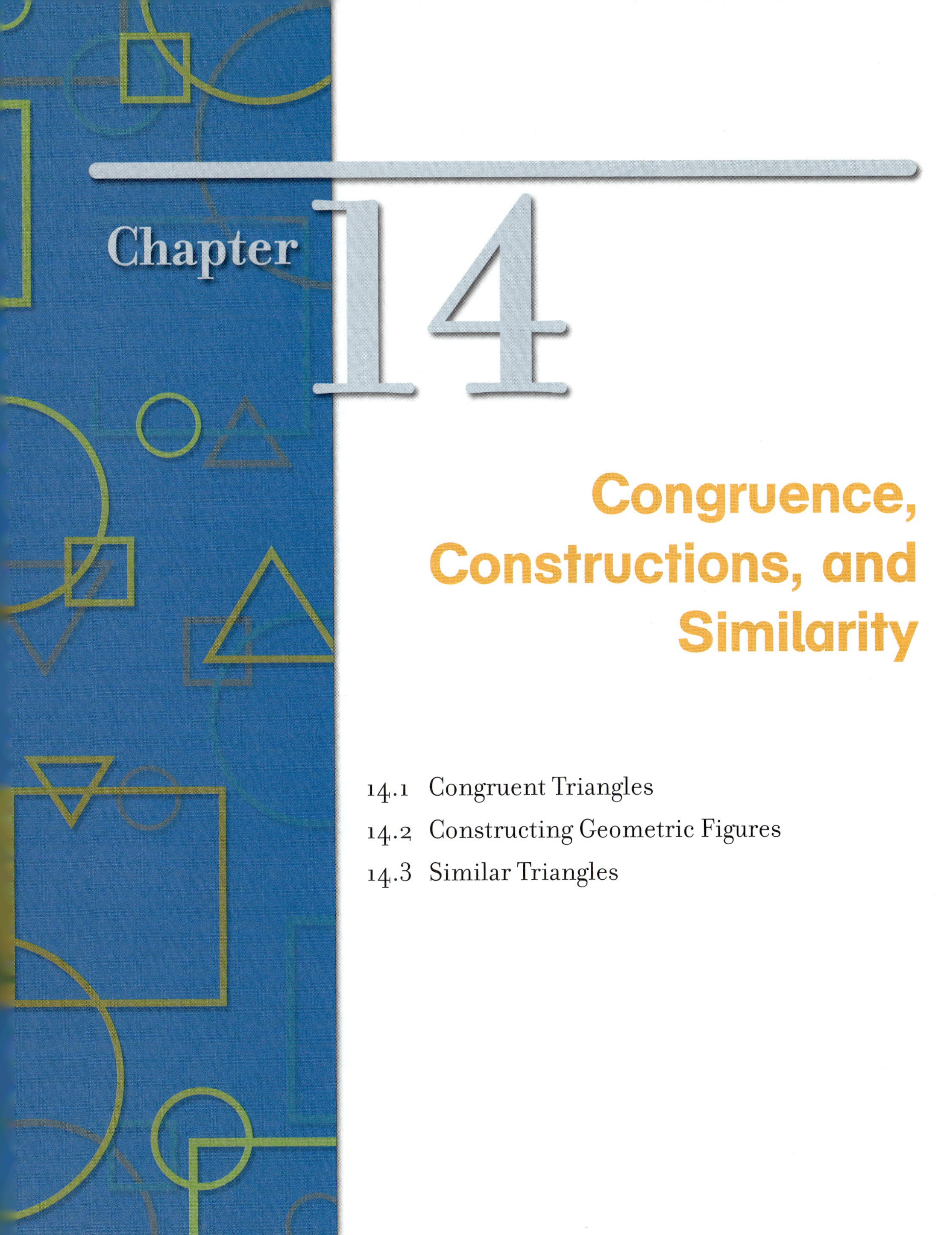

Chapter 14

Congruence, Constructions, and Similarity

14.1 Congruent Triangles

14.2 Constructing Geometric Figures

14.3 Similar Triangles

Hands On Getting Rhombunctious! Folding Paper Polygons

Materials

Rectangular sheets of paper (thin, colorful paper works best, about 4" by 6" or 5" by 7"), rulers, protractors, tape or glue sticks, scissors (optional)

Directions

The goal of this activity is to construct rhombuses and related polygons with paper folding. Some properties of the figures will be investigated by measuring lengths and angles. Unwanted flaps that are not part of the desired final figure can be taped or glued down (or cut off with scissors). It is very important that the folds are made with considerable care. Work in small groups to answer the questions about the properties of the polygons created by your group members.

I. A General Rhombus

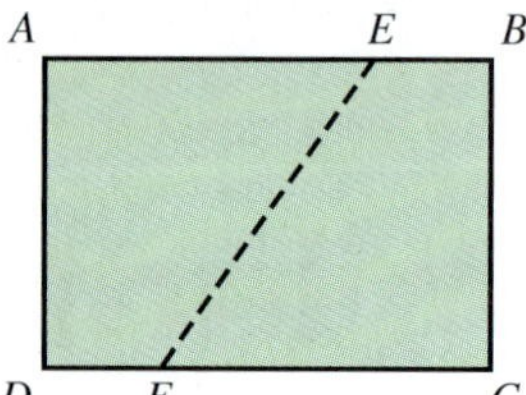

1. Fold C to A to construct segment $\overline{EF}$.

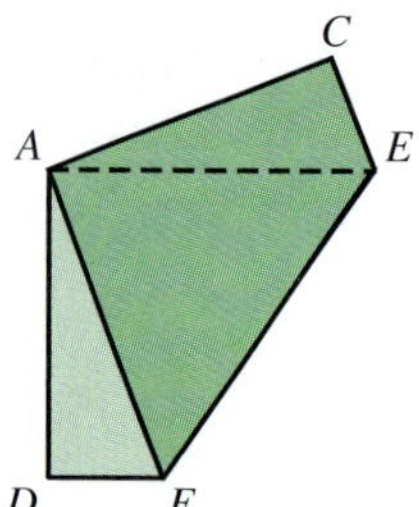

2. Fold up along $\overline{AE}$ and down along $\overline{AF}$.

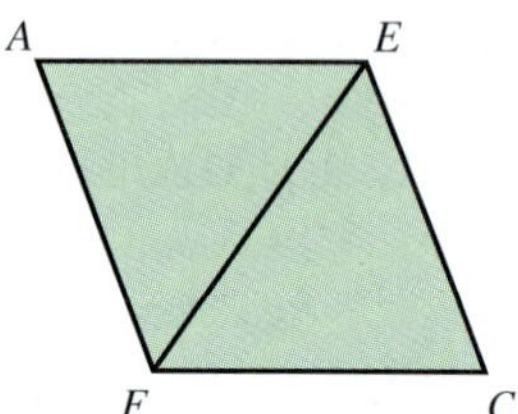

3. Unfold along $\overline{EF}$ to create the rhombus $AECF$ (with the unneeded triangles ACE and ADF taped or glued to the backside).

Questions

1. Use a ruler to measure the sides of $AECF$. Are they equal?
2. Measure the interior angles of $AECF$. How are they related?
3. Measure $\angle EFC$. Is $\overline{EF}$ an angle bisector?
4. Fold F on E and then unfold to create the segment $\overline{AC}$. At what angle do $\overline{AC}$ and $\overline{EF}$ intersect?

II. A Special Rhombus

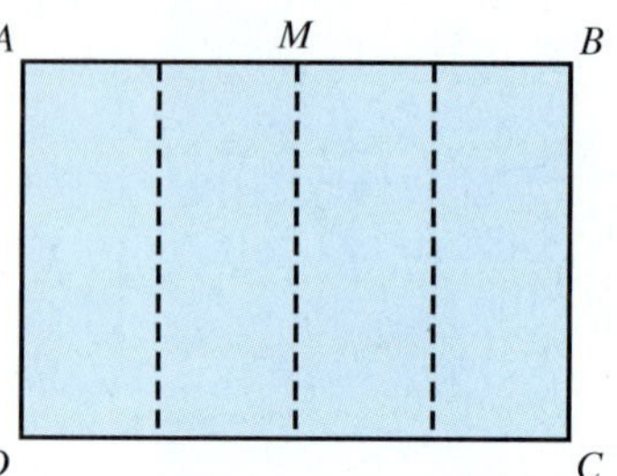

1. Fold a rectangle in half twice, and then unfold.

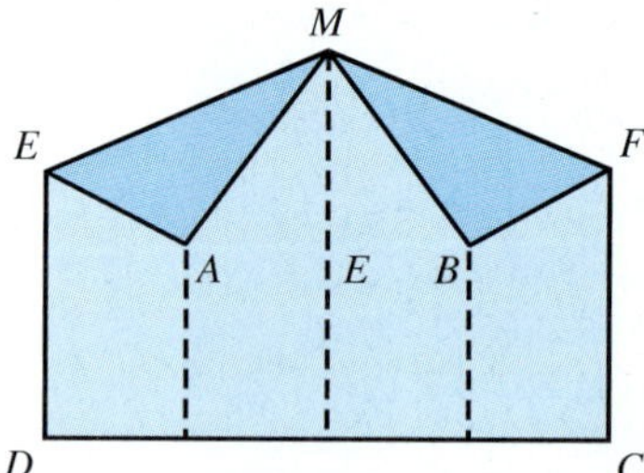

2. Fold A to the 1/4 vertical crease and B to the 3/4 crease, with both folds through the midpoint of side $\overline{AB}$.

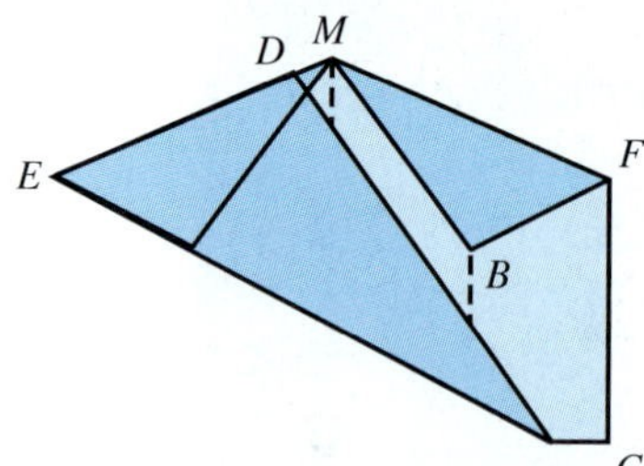

3. Fold D onto $\overline{EM}$ and glue the fold down.

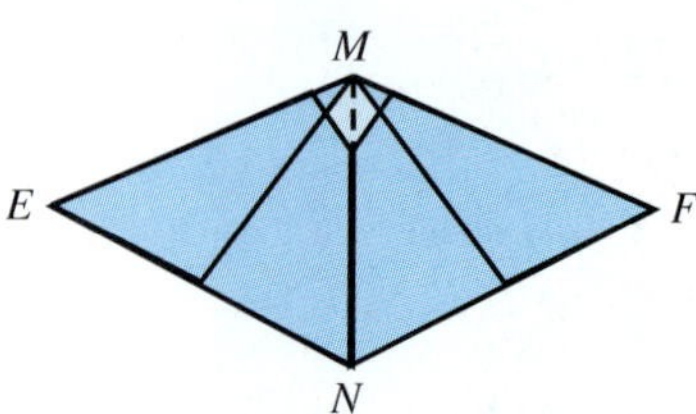

4. Fold C onto $\overline{MF}$ and glue the fold down.

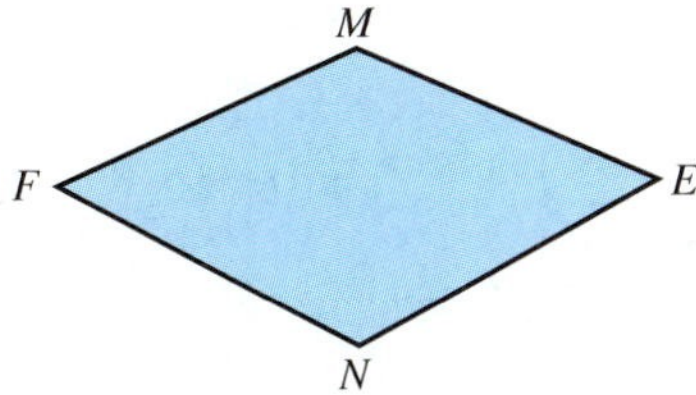

5. Turn the paper over to obtain the rhombus *FMEN*.

Questions

1. What special angles do you find in *FMEN*?
2. If *FMEN* is folded across $\overline{MN}$, what special triangle is formed?
3. If points *F* and *M* are both folded to the midpoint of $\overline{MN}$, what special polygon is constructed?

Connections

Creating and Relating Geometric Figures

This chapter investigates three topics—congruent triangles, geometric constructions, and similar triangles.

Two triangles are **congruent** if they have the same size and shape. In Section 14.1, we investigate what measurements for two triangles are sufficient to guarantee that they are congruent to one another. A constructive approach is taken, using the traditional tools of the compass and straightedge introduced by Euclid. In particular, two basic constructions are introduced—a congruent segment and a congruent angle.

Additional constructions using compass and straightedge are taken up in Section 14.2. This section also introduces several new tools and methods of construction, including paper folding, reflective drawing tools such as the Mira™ and Image Reflector™, and geometry software for the computer.

The concluding Section 14.3 investigates **similar triangles**—triangles with the same shape, but not necessarily the same size.

The topics of this chapter—congruence, constructions, and similarity—have increasingly many connections to our modern technological world. For example, a company strives to manufacture a product that is uniform in quality, performance, and reliability—each item should replicate the same design. Modern industry's ability to produce congruence makes it possible to manufacture affordable automobiles, computers, household appliances, and other sophisticated devices. Congruence and similarity are also a part of our personal lives. Some familiar examples related to congruence are cookie cutters, rubber stamps, and clothing patterns. Examples of items in daily life that are dependent on similarity include maps, scale models, floor plans, the enlargement feature of a photocopy machine, and digital photographs that have been cropped or resized.

14.1 CONGRUENT TRIANGLES

Before considering triangles, it is helpful to investigate line segments. Given two line segments, it is enough to know their lengths to decide if they are congruent: two line segments are congruent if, and only if, they have the same length. We can also take a constructive approach. Construction 1 shows how a compass and straightedge are used to construct a line segment that is congruent to a given segment.

Construction 1

Construct a Congruent Line Segment

Construct a line segment that is congruent to a given line segment $\overline{AB}$.

A B

Procedure

P Z	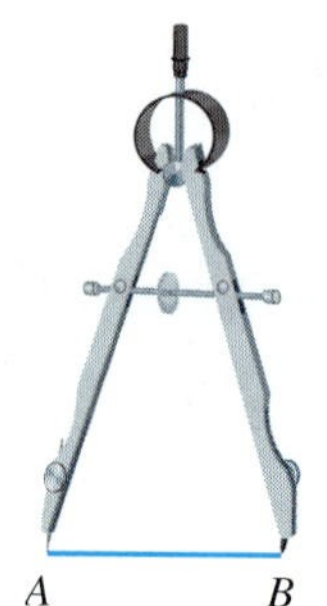	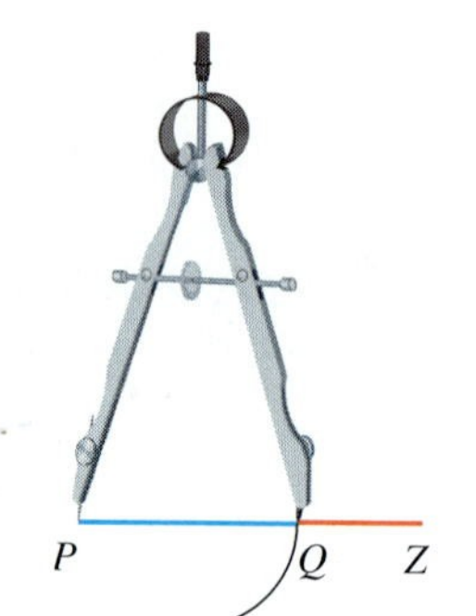
Step 1 Draw segment $\overline{PZ}$, longer than AB.	**Step 2** Put the point of the compass at A and the pencil point on B.	**Step 3** Without changing the opening of the compass, move the point of the compass to P and draw a circular arc. The point of intersection° determines the point Q for which $\overline{PQ} \cong \overline{AB}$.

With this prologue, the investigation of congruent triangles will consider these questions:

- What measurements of a triangle completely describe its size and shape?
- Given two triangles, what subsets of measurements are sufficient to decide if the triangles are congruent to one another?
- Given certain measurements of a triangle ABC, how can a compass and straightedge be used to construct a triangle $\triangle PQR$ that is congruent to $\triangle ABC$?

The size and shape of a triangle are described completely if we specify the **six parts of a triangle,** namely, the three sides $\overline{AB}$, $\overline{BC}$, and $\overline{CA}$ and the three angles $\angle A$, $\angle B$, and $\angle C$. A second triangle PQR is congruent to triangle ABC if there is a matching of vertices $A \leftrightarrow P$, $B \leftrightarrow Q$, and $C \leftrightarrow R$ under which *all six* parts of triangle ABC are congruent to the corresponding six parts of triangle PQR. This is illustrated in Figure 14.1.

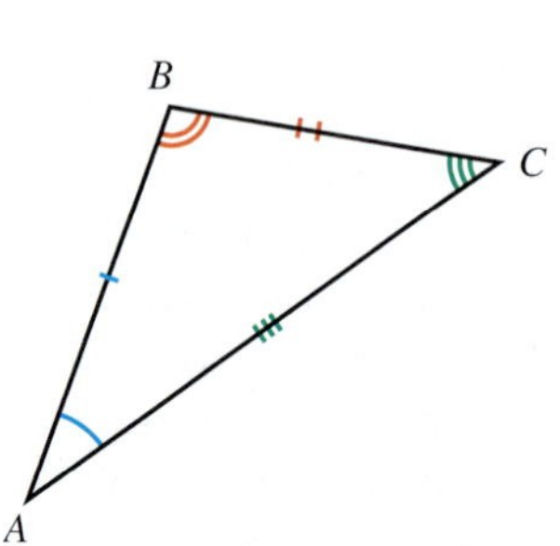

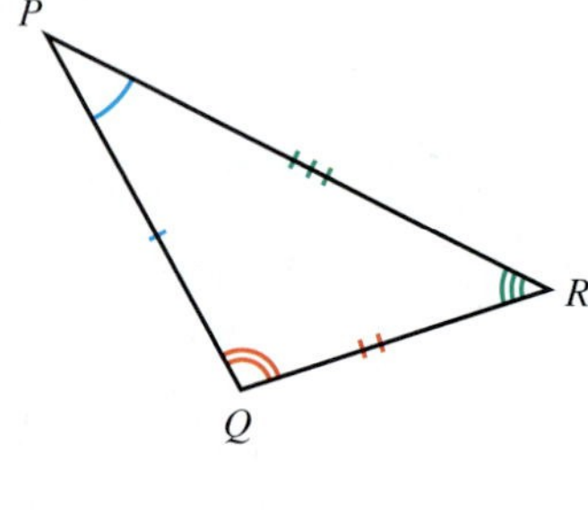

Figure 14.1
Triangles ABC and PQR are congruent under the vertex correspondence $A \leftrightarrow P$, $B \leftrightarrow Q$, and $C \leftrightarrow R$ if, and only if, $\overline{AB} \cong \overline{PQ}$, $\overline{BC} \cong \overline{QR}$, and $\overline{CA} \cong \overline{RP}$, and $\angle A \cong \angle P$, $\angle B \cong \angle Q$, and $\angle C \cong \angle R$

DEFINITION *Congruent Triangles*
Two triangles are **congruent** if, and only if, there is a correspondence of vertices of the triangles such that the corresponding sides and corresponding angles are congruent.

The notation $\triangle ABC \cong \triangle PQR$ is read "Triangle *ABC* is congruent to triangle *PQR*" and conveys the following information:

- The vertex correspondence is $A \leftrightarrow P$, $B \leftrightarrow Q$, and $C \leftrightarrow R$.
- The corresponding sides are congruent: $\overline{AB} \cong \overline{PQ}$, $\overline{BC} \cong \overline{QR}$, and $\overline{CA} \cong \overline{RP}$.
- The corresponding angles are congruent: $\angle A \cong \angle P$, $\angle B \cong \angle Q$, and $\angle C \cong \angle R$.

It is important to notice that the order in which the vertices are listed specifies the vertex correspondence. For the triangles depicted in Figure 14.1, we see that $\triangle ABC \ncong \triangle QRP$ ($\ncong$ is read "is not congruent to"). However, it is correct to say that $\triangle BCA \cong \triangle QRP$.

EXAMPLE 14.1 EXPLORING THE CONGRUENCE RELATION

Use a ruler and protractor to find the two pairs of congruent triangles among the six triangles shown. State the two congruences in the symbolic form $\triangle$____ $\cong$ $\triangle$____.

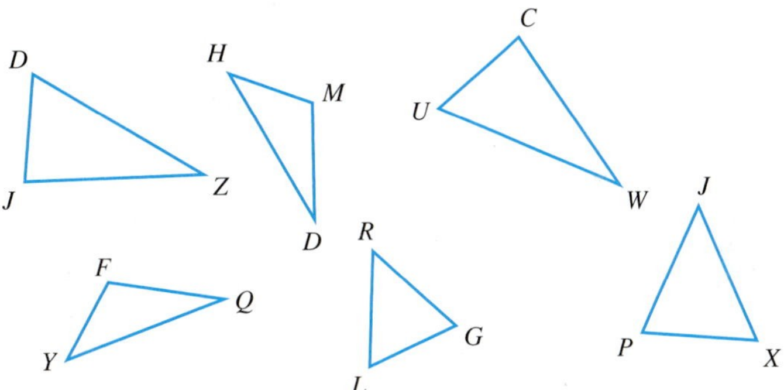

Solution $\triangle DJZ \cong \triangle UCW$ and $\triangle HMD \cong \triangle YFQ$. The order of the vertices can be permuted in each statement. For example, it would also be correct to express the first congruence as $\triangle ZDJ \cong \triangle WUC$.

Suppose that we have only some of the measurements of a triangle *ABC*. For example, suppose that we know the lengths *AB*, *BC*, and *CA* of the three sides, but we are not given any information about the angles. Or suppose we are given a length *AB* and the measurements of two angles $\angle A$ and $\angle B$. Is the information we have sufficient to construct a triangle *PQR* that is necessarily congruent to $\triangle ABC$? These questions will be explored constructively; that is, we will attempt to use a compass and straightedge to construct a triangle *PQR* that is congruent to $\triangle ABC$.

The Side–Side–Side (SSS) Property

EXAMPLE 14.2 EXPLORING THE SIDE–SIDE–SIDE PROPERTY

The three sides of triangle ABC are given as shown. Construct a triangle PQR that is congruent to $\triangle ABC$.

Solution

Step 1 Construct a segment $\overline{PQ}$ of length $x = AB$ using Construction 1.

Step 2 Set the compass to radius $y = BC$ and draw a circle of radius y centered at Q.

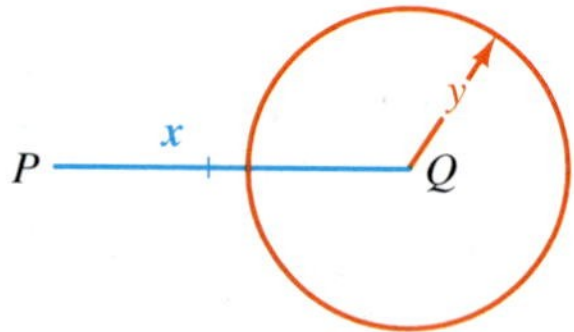

Step 3 Set the compass to radius $z = AC$ and draw a circular arc of radius z centered at P. Let R be either point of intersection with the circle drawn in Step 2.

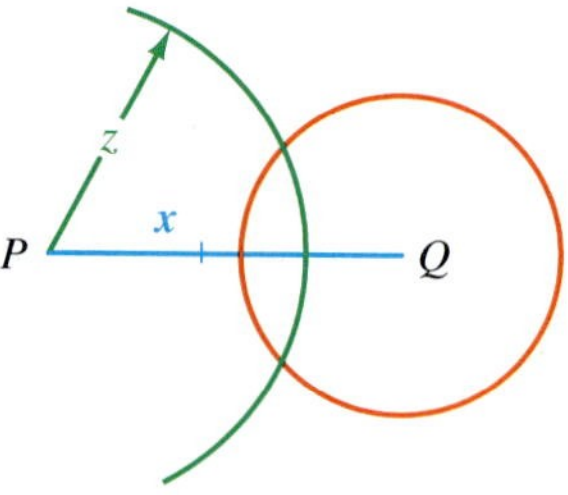

Step 4 Draw the segments $\overline{PR}$ and $\overline{RQ}$. Then $\triangle PQR \cong \triangle ABC$.

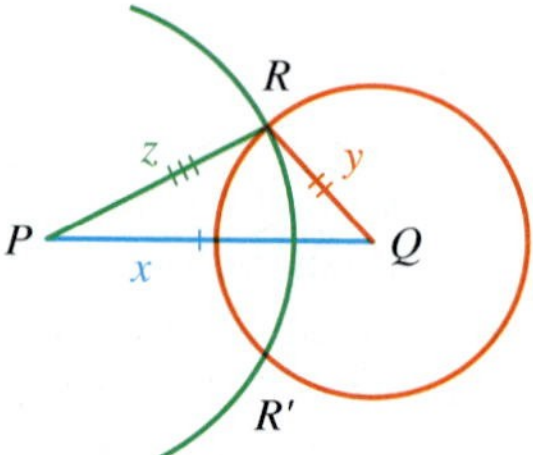

The size and shape of $\triangle PQR$ are uniquely determined. Even if we had chosen the second point of intersection, R', $\triangle PQR'$ would still have been the same size and shape as $\triangle PQR$. Therefore, we are led to the following basic property.

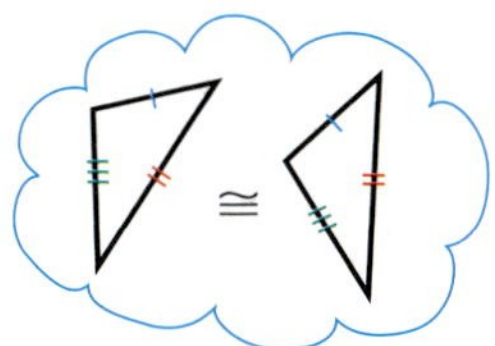

PROPERTY *Side–Side–Side (SSS)*
If the three sides of one triangle are respectively congruent to the three sides of another triangle, then the two triangles are congruent.

In many formal treatments of Euclidean geometry, the SSS property is adopted as a postulate. That is, SSS is true by assumption, not by proof.

EXAMPLE 14.3 USING THE SSS PROPERTY

Let $ABCD$ be a quadrilateral with opposite sides of equal length: $AB = DC$ and $AD = BC$. Show that $ABCD$ is a parallelogram.

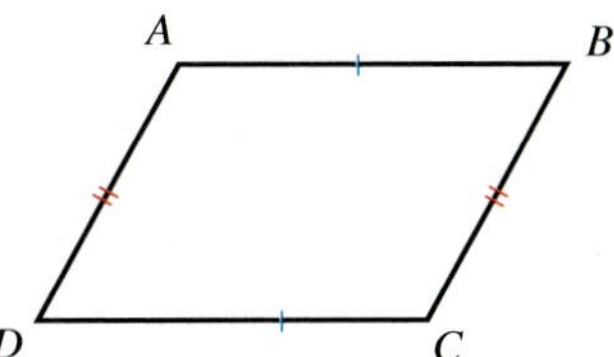

Solution

Many problems in geometry are solved by constructing additional lines or arcs to reveal features of the original figure that would otherwise remain hidden. In this case we construct the diagonal $\overline{AC}$.

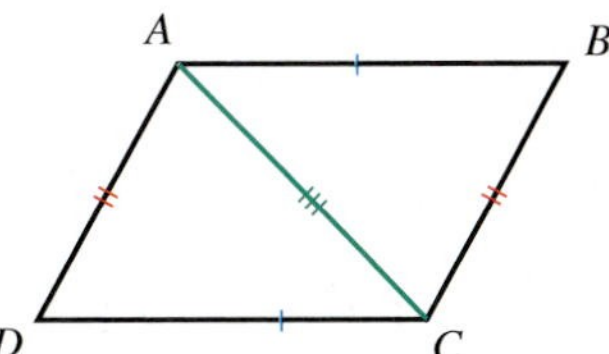

Since $\overline{AC}$ is congruent to itself, we see that $\triangle ABC \cong \triangle CDA$ by the SSS property. Thus, the corresponding angles $\angle BAC$ and $\angle DCA$ are congruent. By the alternate-interior-angles theorem of Chapter 11, we conclude that $\overline{AB} \parallel \overline{DC}$. Similarly, the congruence $\angle BCA \cong \angle DAC$ shows that $\overline{AD} \parallel \overline{BC}$.

An important consequence of the SSS property is the following construction of a congruent angle.

Construction 2

Construct a Congruent Angle

Construct an angle that is congruent to a given angle $\angle D$ as shown.

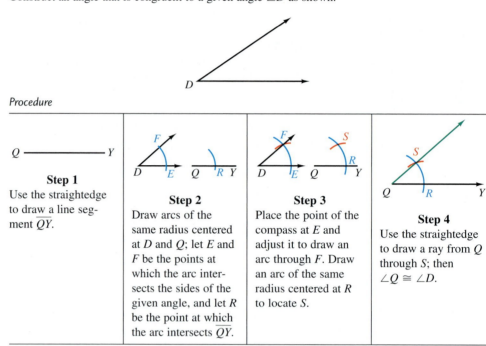

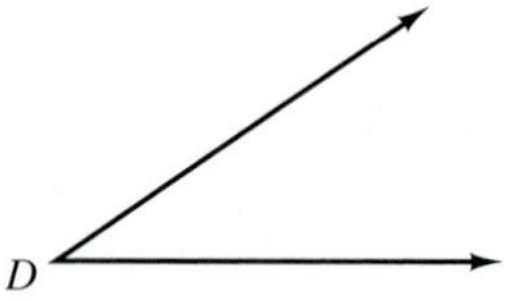

Procedure

Q — Y **Step 1** Use the straightedge to draw a line segment $\overline{QY}$.	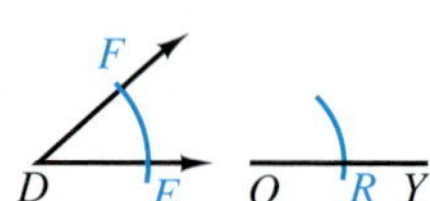 **Step 2** Draw arcs of the same radius centered at D and Q; let E and F be the points at which the arc intersects the sides of the given angle, and let R be the point at which the arc intersects $\overline{QY}$.	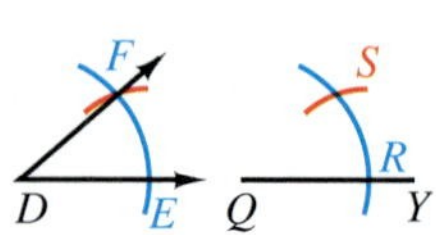 **Step 3** Place the point of the compass at E and adjust it to draw an arc through F. Draw an arc of the same radius centered at R to locate S.	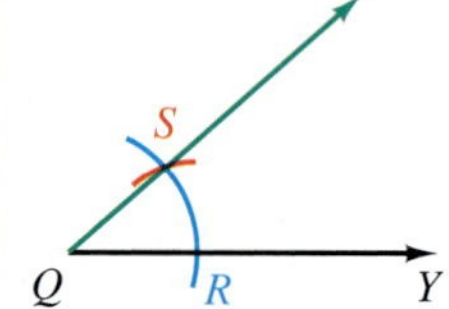 **Step 4** Use the straightedge to draw a ray from Q through S; then $\angle Q \cong \angle D$.

The construction procedure shows that $DE = QR$, $DF = QS$, and $EF = RS$. Therefore, $\triangle DEF \cong \triangle QRS$ by the SSS property, and we see that the corresponding angles $\angle D$ and $\angle Q$ are congruent.

The Triangle Inequality

The SSS property guarantees that two triangles are congruent if they have corresponding sides of the same length. However, not every triple of given lengths corresponds to a triangle, since the length of any side must be less than the sum of the lengths of the other two sides. An example is shown in Figure 14.2.

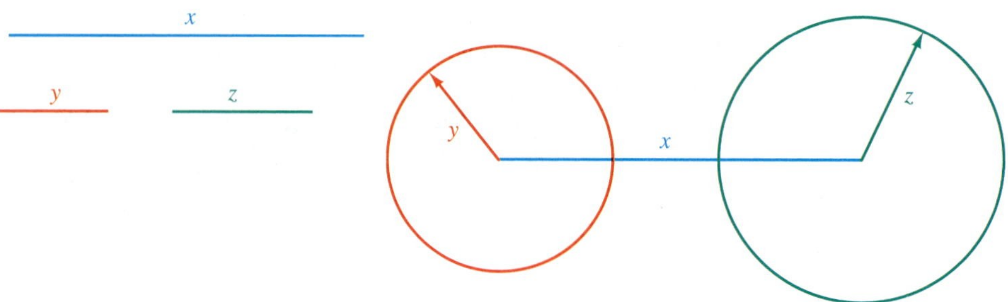

Figure 14.2
If $x \geq y + z$, there is no triangle with sides of length x, y, and z

The lengths of the sides of a triangle must satisfy the following property.

Cooperative Investigation

Exploring Toothpick Triangles

Materials Needed

Toothpicks of equal length (about 20 per person)

Directions

Three toothpicks, placed end to end, form a triangle in one way—an equilateral triangle. Four toothpicks do not form a triangle. Five and six toothpicks each form just one triangle. Two different (that is, noncongruent) triangles can be formed with seven toothpicks.

Triangles					
Number of Toothpicks, n	3	4	5	6	7
Number of Triangles, $T(n)$	1	0	1	1	2

Explore how many noncongruent triangles you can form with 8, 9, 10, 11, or 12 toothpicks. Extend the table above to include your results.

Questions for Consideration

1. How many isosceles toothpick triangles are there for which the two sides of equal length each use four toothpicks?
2. One side of a toothpick triangle uses three toothpicks, and a second side uses five toothpicks. What are the possible numbers of toothpicks in the third side?
3. If two sides of a toothpick triangle together use 11 toothpicks, what is the largest number of toothpicks that can be used in the third side?
4. Suppose toothpicks form a triangle with p, q, and r toothpicks on its three sides. What can you say about the integer r in terms of the numbers p and q?
5. In your table of the number of triangles, suppose $T(n)$ is the number of different toothpick triangles formed from n toothpicks. For odd n, compare $T(n)$ with $T(n + 3)$. For example, compare $T(3)$ with $T(6)$, and compare $T(5)$ with $T(8)$. What pattern do you observe?

PROPERTY *Triangle Inequality*
The sum of the lengths of any two sides of a triangle must be greater than the length of the third side.

A triangle gives rise to three inequalities, as shown in Figure 14.3.

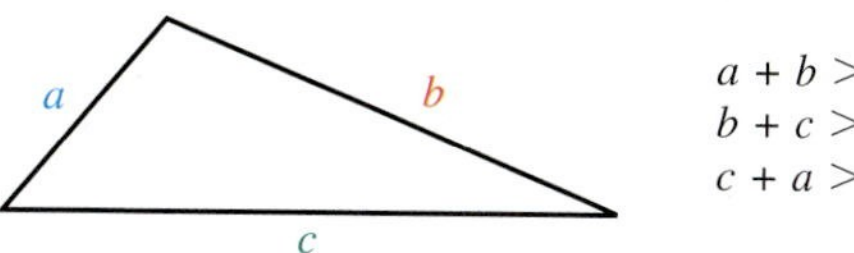

Figure 14.3
The lengths of the sides of any triangle satisfy the triangle inequalities

$a + b > c$
$b + c > a$
$c + a > b$

Cooperative Investigation

Random Spaghetti Triangles

Materials Needed

Uncooked spaghetti, two strands per student
Calculator with random-number generator
Metric rulers
Pencils or marking pens

Directions

Use the calculator to generate pairs of random numbers x and y in the intervals $0 < x$ and $y < L$, where L is the length of the spaghetti in millimeters. Mark the distances x and y, in millimeters, from one end on each strand of spaghetti. Break the strand at these two points, and (if possible) form a triangle from the three lengths of broken spaghetti.

Repeat with the second strand, using another pair of random distances. Gather data from the class in a table.

Number of Students	Number of Triangles	Number of Acute Triangles	Number of Obtuse Triangles

Work in small collaborative groups to answer these questions.

Questions

1. What fraction of the broken spaghetti strands formed a triangle?
2. Assume that x and y are ordered so $0 < x < y < L$. Then the lengths of the three pieces of spaghetti are x, $y - x$, and $L - y$. What inequalities must be satisfied for the pieces to form a triangle with sides x, $y - x$, and $L - y$?
3. Among the triangles that are formed with the broken spaghetti, which type of triangle seems to be most likely—acute or obtuse?

EXAMPLE 14.4 | APPLYING THE TRIANGLE INEQUALITY

The four towns of Abbott, Brownsville, Connell, and Davis are building a new power-generating plant that will serve all four communities. To keep the costs of the power lines at a minimum, the plant is to be located so that the sum of the distances from the plant to the four towns is as small as possible. An engineer recommended locating the plant at point E. A mathematician, seeing that the four towns form a convex quadrilateral as shown, recommended that the plant be built at point M, at which the diagonals of the quadrilateral intersect. Why is location M better than E?

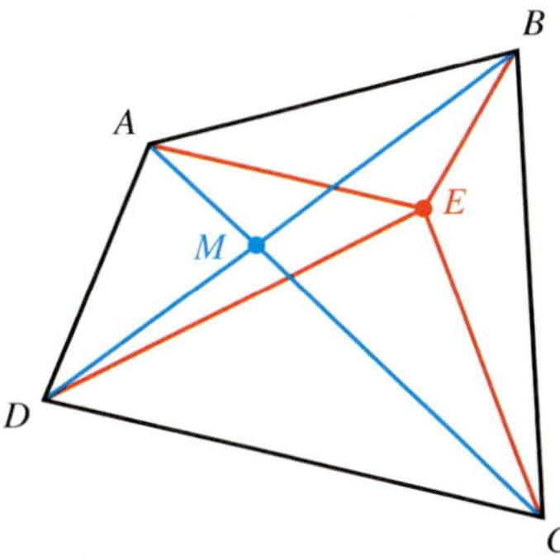

Solution The triangle inequality, applied to $\triangle ACE$, gives $EA + EC > AC$. Since M is on the diagonal $\overline{AC}$, we also know that $AC = MA + MC$, and therefore

$$EA + EC > MA + MC.$$

If we apply the triangle inequality to $\triangle BDE$, the same reasoning gives us the inequality

$$EB + ED > MB + MD.$$

Adding these two inequalities gives us

$$EA + EB + EC + ED > MA + MB + MC + MD.$$

This shows that the sum of the distances to the towns from point E is greater than the sum of the distances to the towns from point M.

The Side–Angle–Side (SAS) Property

The next example explores how to construct a triangle with two given sides and the angle included between the given sides.

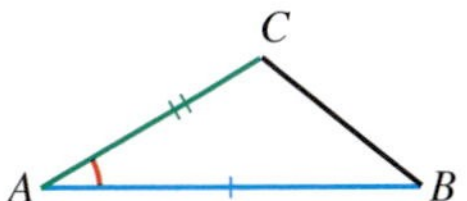

EXAMPLE 14.5 EXPLORING THE SIDE–ANGLE–SIDE CONDITION

Two sides $\overline{AB}$ and $\overline{AC}$ and the angle $\angle A$ included by these sides are given for $\triangle ABC$ as shown. Show that a triangle PQR can be constructed for which $\triangle PQR \cong \triangle ABC$.

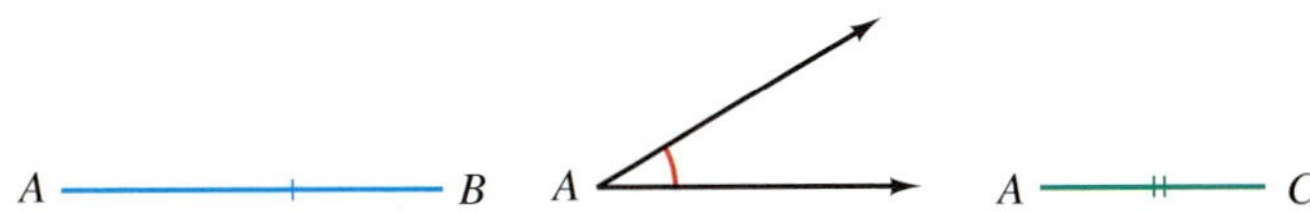

HIGHLIGHT from HISTORY

Christine Ladd-Franklin (1847–1930)

The Metaphysical Club of the Johns Hopkins University met for the first time in October 1879. The club's founder, the great logician C. S. Pierce, read the paper "Non-Euclidean Space." What was unusual was that the paper's author, seated in the audience, was a newly admitted graduate student named Christine Ladd—and Johns Hopkins University did not admit women! Some earlier background explains why an exception was made.

Following her 1869 B.A. from Vassar, Christine had hoped to continue studies in physics, but was denied access to laboratories, largely because of her gender. She therefore turned to mathematics, which she studied on her own as she took a succession of positions teaching science. Shortly after the founding of Johns Hopkins University in 1876, Christine's application to admission to graduate studies was given a favorable report by Fabian Franklin, a young member of the mathematics department who was impressed by several articles and solutions to problems she had published in some English periodicals.

Admitted "on a special status" in 1879, Christine Ladd studied mathematics, philosophy, and psychology and wrote papers in logic, algebra, geometry, and other subjects. By 1882 she had married Fabian Franklin and completed the requirements for a Ph.D., but, as a woman, could not be given the degree. (This was finally rectified in 1924, when she received the degree at age 76.)

Christine Ladd-Franklin continued to write papers in logic throughout her life, but after 1882 turned her attention again to experimental science, especially the psychology of vision. Her lifetime work includes about 20 papers in mathematics, another 20 in logic, and about 50 papers and a book on the theory of vision.

Solution

Step 1 Follow the steps of Construction 2 to construct an angle congruent to $\angle A$; let P denote its vertex.

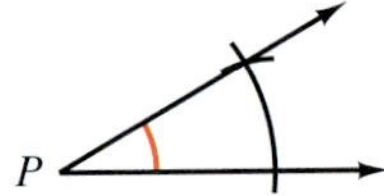

Step 2 Use Construction 1 to construct segments $\overline{PQ}$ and $\overline{PR}$ along the sides of $\angle P$ that are, respectively, of length AB and AC.

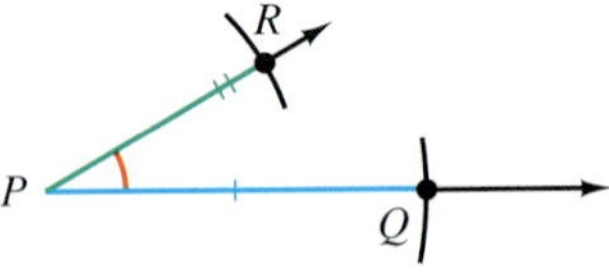

Step 3 Draw segment $\overline{QR}$; then $\triangle PQR \cong \triangle ABC$.

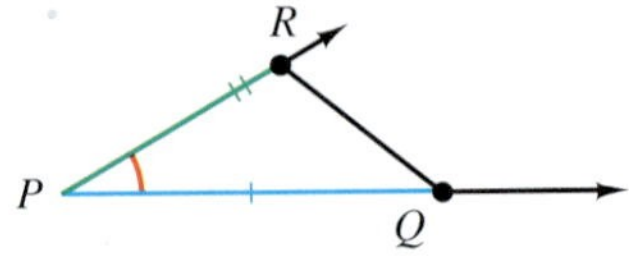

The procedure described in Example 14.5 uniquely determines the size and shape of $\triangle PQR$ when we are given the three parts, side–angle–side, of $\triangle ABC$. The angle has to be the **included angle,** the angle between the given sides. This property is often abbreviated as **SAS (side–angle–side).**

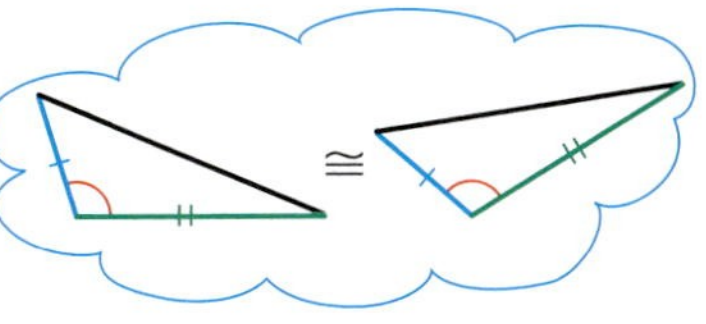

PROPERTY *Side–Angle–Side (SAS)*
If two sides and the included angle of one triangle are congruent to two sides and the included angle of another triangle, then the two triangles are congruent.

EXAMPLE 14.6 USING THE SAS PROPERTY

Two line segments $\overline{AB}$ and $\overline{CD}$ intersect at their common midpoint M. Show that $\overline{AD}$ and $\overline{BC}$ are parallel and have the same length.

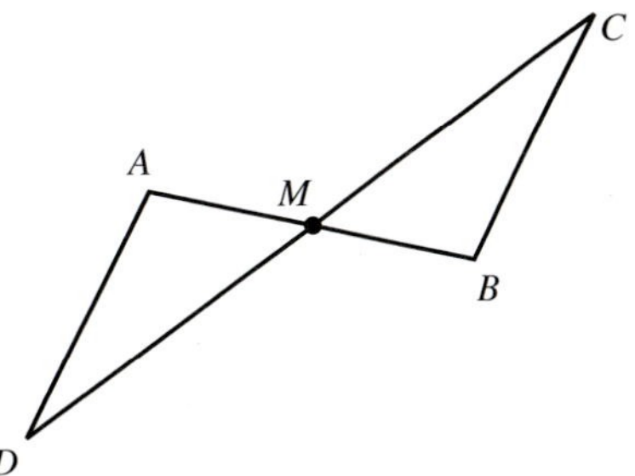

Solution

It is useful to add tick marks and arcs to your drawing to summarize the given information. In this problem, M is the midpoint of $\overline{AB}$, so $AM = BM$. We indicate this on the drawing by putting a single tick mark on each of the segments $\overline{AM}$ and $\overline{MB}$. Similarly $CM = DM$, and we put double tick marks on each of the segments $\overline{CM}$ and $\overline{MD}$. We also use single arcs to indicate the congruence of the vertical angles at M formed by the intersecting segments.

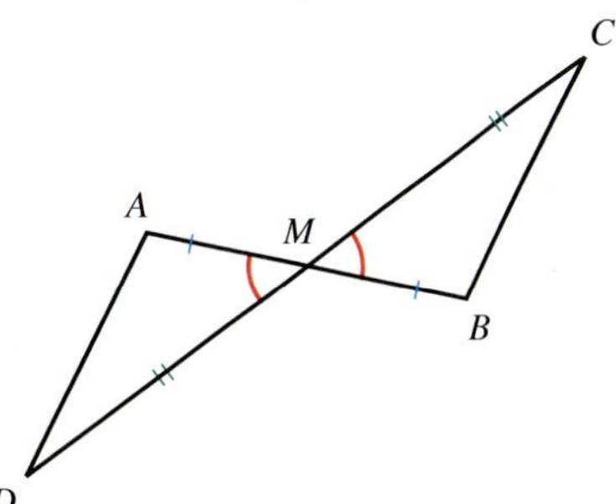

It is now apparent that the SAS property gives us the congruence $\triangle AMD \cong \triangle BMC$. It follows that the corresponding sides $\overline{AD}$ and $\overline{BC}$ are congruent, so that $AD = BC$. We also have that $\angle A \cong \angle B$, so the alternate-interior-angles theorem of Chapter 11 guarantees that $\overline{AD}$ and $\overline{CB}$ are parallel.

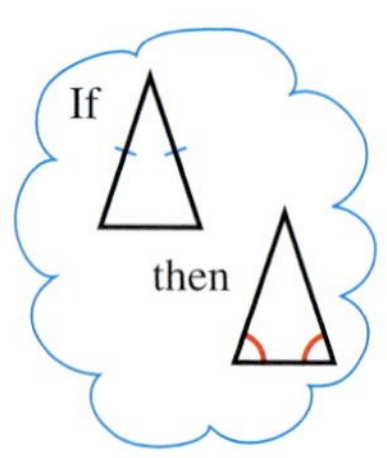

Just for Fun

Twice Around a Triangle

Draw any triangle *ABC* and let *P* be any point on side $\overline{AB}$. Draw an arc centered at *B* to determine the point *Q* on side $\overline{BC}$ for which $BP = BQ$. In the same way, draw an arc at *C* to locate point *R* on $\overline{CA}$, and then draw an arc centered at *A* to determine point P' on $\overline{AB}$. In three more steps, go around the triangle a second time, constructing Q', R', and finally P'' on $\overline{AB}$. What do you find interesting about P''?

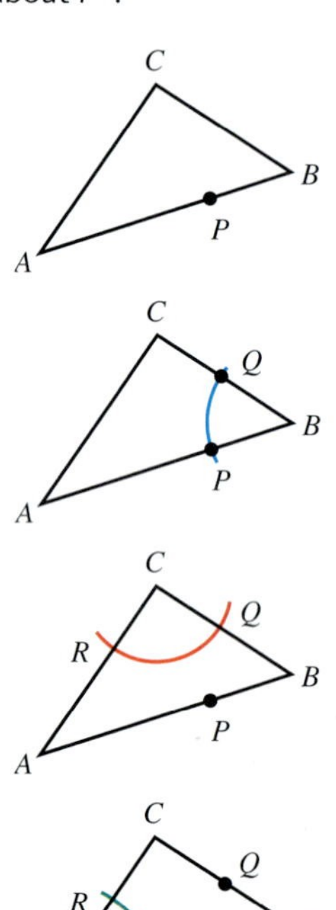

The following theorem about isosceles triangles is an important consequence of the SAS property.

THEOREM *Isosceles Triangle Theorem*
The angles opposite the congruent sides of an isosceles triangle are congruent.

Proof Let $\triangle ABC$ be isosceles, with $\overline{AB}$ and $\overline{AC}$ congruent. Consider the vertex correspondence $A \leftrightarrow A$, $B \leftrightarrow C$, and $C \leftrightarrow B$ (which amounts to looking at the same triangle from the back, so to speak!).

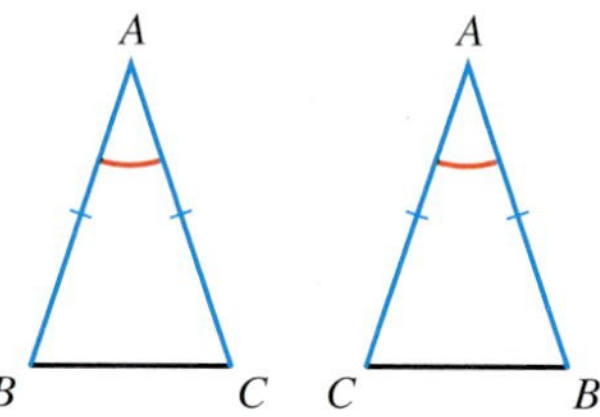

Since $\overline{AB} \cong \overline{AC}$ and $\angle A \cong \angle A$, it follows from the SAS property that $\triangle ABC \cong \triangle ACB$. But then all six corresponding parts of $\triangle ABC$ and $\triangle ACB$ are congruent, including $\angle B \cong \angle C$.

The isosceles triangle theorem has many uses. In particular, it gives a simple way to prove **Thales' theorem.**

THEOREM *Thales' Theorem*
Any triangle *ABC* inscribed in a semicircle with diameter $\overline{AB}$ has a right angle at point *C*.

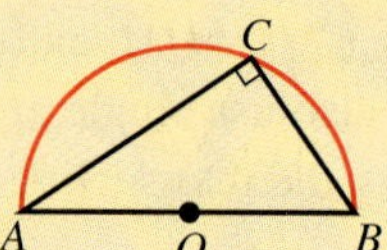

Proof Draw the radius $\overline{OC}$ as shown in the figure below. This divides $\triangle ABC$ into two isosceles triangles, $\triangle AOC$ and $\triangle COB$. The isosceles triangle theorem tells us that the measures of the base angles of $\triangle AOC$ are equal, say, x. Likewise, the base angles of $\triangle COB$ are equal, say, y. Since the sum of the measures of the interior angles of $\triangle ABC$ is $180°$, we have $x + y + (x + y) = 180°$. But this equation tells us that $m(\angle C) = x + y = 90°$.

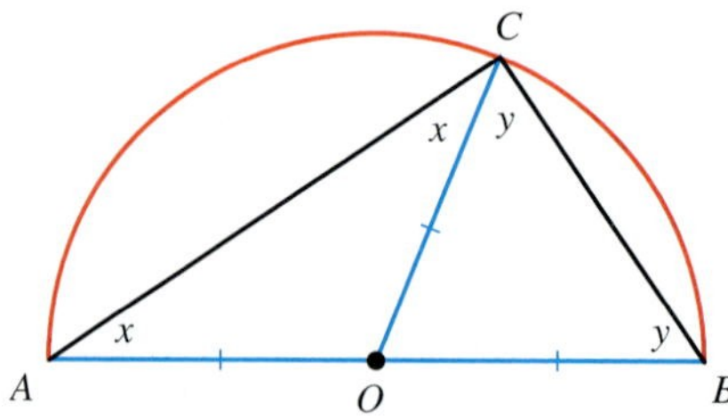

The Angle–Side–Angle (ASA) Property

In the next example, we suppose that two angles of a triangle, and the side included between these angles, are given. Is this information sufficient to be able to construct a congruent triangle?

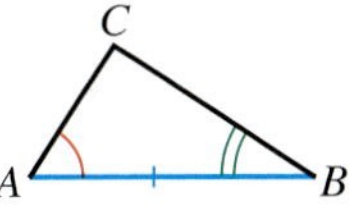

EXAMPLE 14.7 EXPLORING THE ANGLE–SIDE–ANGLE PROPERTY

Two angles and their **included side** are given for $\triangle ABC$, as shown. Construct a triangle PQR that is congruent to triangle ABC.

Solution

Step 1 Use Construction 2 to construct $\angle P \cong \angle A$.

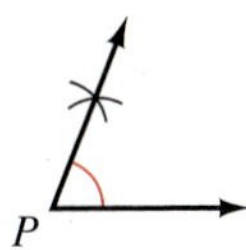

Step 2 Use Construction 1 to construct segment $\overline{PQ}$ on a side of $\angle P$ so that $PQ = AB$.

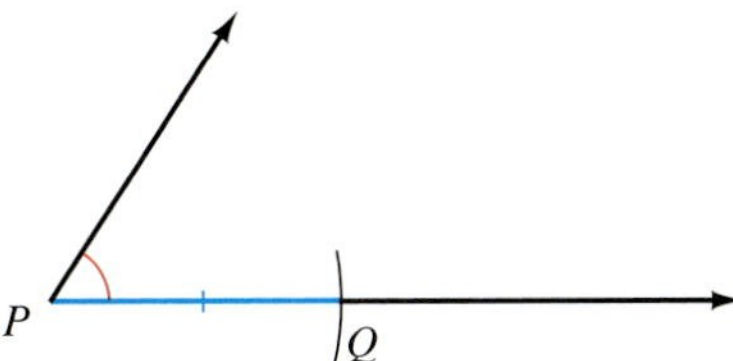

Step 3 Construct an angle congruent to $\angle B$ at vertex Q, with one side containing P and the other side intersecting $\angle P$ to determine point R. Then $\triangle PQR \cong \triangle ABC$.

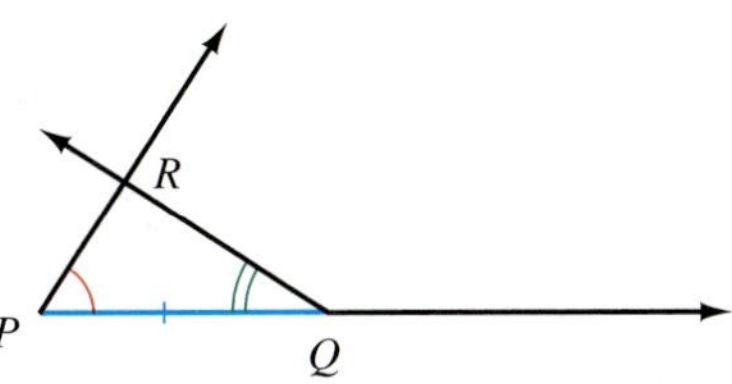

The construction just shown illustrates the **angle–side–angle property,** abbreviated as **ASA.**

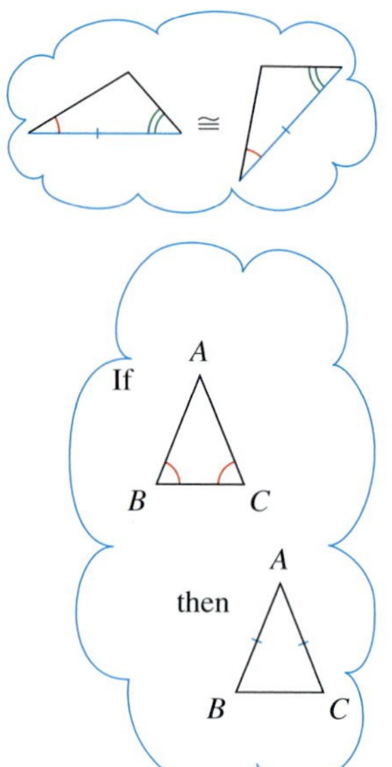

PROPERTY *Angle–Side–Angle (ASA)*
If two angles and the included side of one triangle are congruent to the two angles and the included side of another triangle, then the two triangles are congruent.

The ASA property allows us to prove that any triangle with two angles of the same measure is isosceles.

THEOREM *Converse of the Isosceles Triangle Theorem*
If two angles of a triangle are congruent, then the sides opposite them are congruent; that is, the triangle is isosceles.

Proof Let $\triangle ABC$ have two congruent angles, say, $\angle B \cong \angle C$. We know that $\overline{BC} \cong \overline{CB}$, since a line segment is congruent to itself. By the ASA property, it follows that $\triangle ABC \cong \triangle ACB$. This means that the corresponding sides of $\triangle ABC$ and $\triangle ACB$ are congruent, so $\overline{AB} \cong \overline{AC}$.

The Angle–Angle–Side (AAS) Property

The side in the ASA theorem is the one included by the two angles. However, if *any* two angles of one triangle are congruent to two angles of a second triangle, then all three pairs of corresponding angles are congruent. This follows from the fact that the measures of the three angles of a triangle add up to 180°, so the third angle is uniquely determined by the other two angles. This gives us the **angle–angle–side property,** abbreviated as **AAS.**

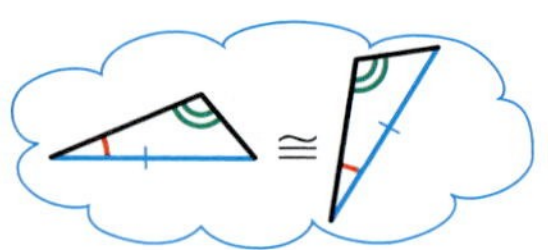

PROPERTY *Angle–Angle–Side (AAS)*
If two angles and a nonincluded side of one triangle are respectively congruent to two angles and the corresponding nonincluded side of a second triangle, then the two triangles are congruent.

Are There SSA and AAA Congruence Properties?

There is no "SSA" congruence property, since it is possible for two noncongruent triangles to have two pairs of congruent sides and a congruent nonincluded angle. An example is shown in Figure 14.4.

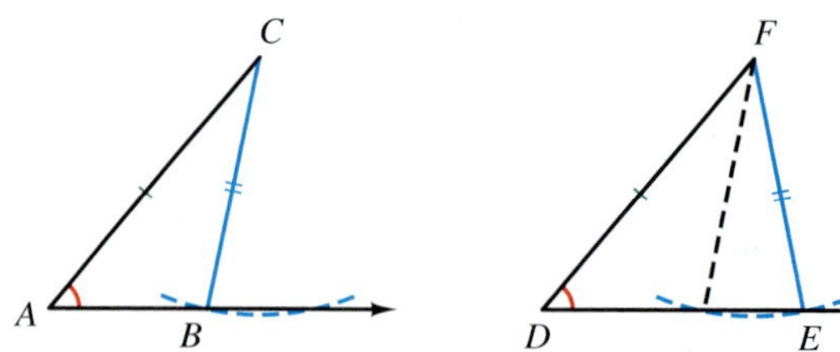

Figure 14.4
Triangles ABC and DEF are not congruent even though AC = DF, BC = EF, and $\angle A \cong \angle D$

Similarly, there is no "AAA" congruence property. For example, Figure 14.5 shows two triangles with three pairs of congruent angles. However, the triangles are not congruent since they are of different size. On the other hand, the shapes of the two triangles are the same, so they are called similar triangles. The properties and applications of similar triangles will be discussed in Section 14.3.

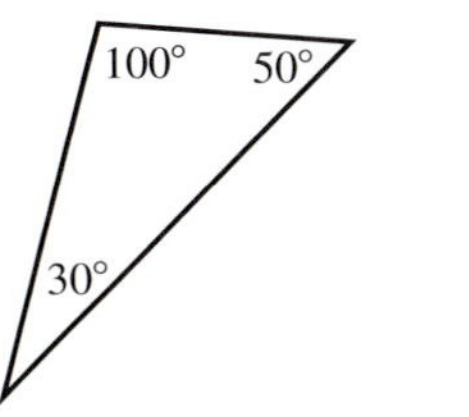

Figure 14.5
The AAA condition guarantees similarity, but not congruence

Problem Set 14.1

Understanding Concepts

1. The two triangles shown below are congruent.

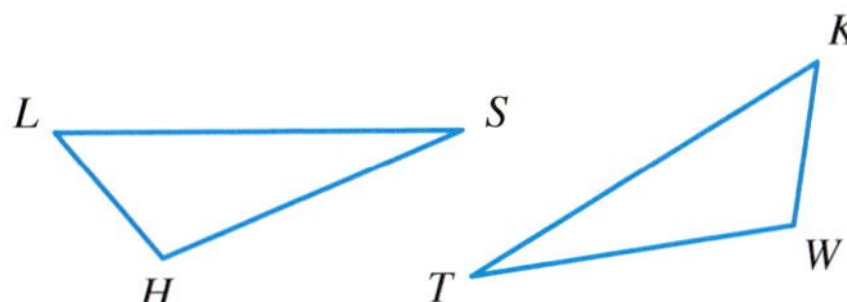

Determine the following:

(a) Corresponding vertices
$L \leftrightarrow$ ____, $H \leftrightarrow$ ____, $S \leftrightarrow$ ____

(b) Corresponding sides
$\overline{LH} \leftrightarrow$ ____, $\overline{HS} \leftrightarrow$ ____, $\overline{SL} \leftrightarrow$ ____

(c) Corresponding angles
$\angle L \leftrightarrow$ ____, $\angle H \leftrightarrow$ ____, $\angle S \leftrightarrow$

(d) $\triangle LHS \cong \triangle$____.

2. Suppose $\triangle JKL \cong \triangle ABC$, where $\triangle ABC$ is shown below.

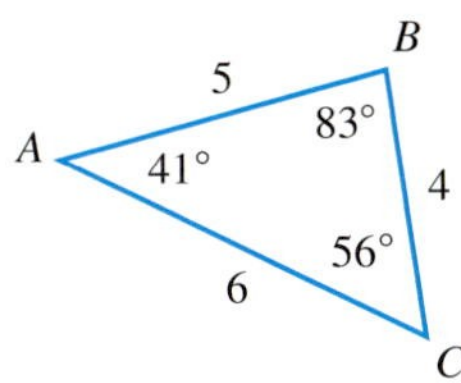

Find the following:

(a) KL **(b)** LJ **(c)** $m(\angle L)$ **(d)** $m(\angle J)$

3. Segments of length x and y are shown below.

Describe procedures, using only a straightedge and a compass, to construct the following:

(a) a line segment $\overline{EF}$ of length $x + y$;

(b) a line segment $\overline{GH}$ of length $x - y$.

4. Trace the angle $\angle D$ shown below. Then use a compass and straightedge to construct an angle $\angle Q$ congruent to $\angle D$. Use a protractor to measure each angle, and report on how closely the measurements of the two angles agree.

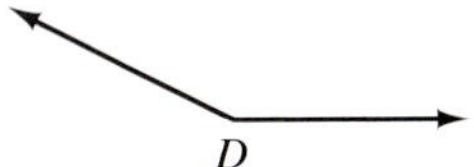

5. Two angles $\angle A$ and $\angle B$ are shown below.

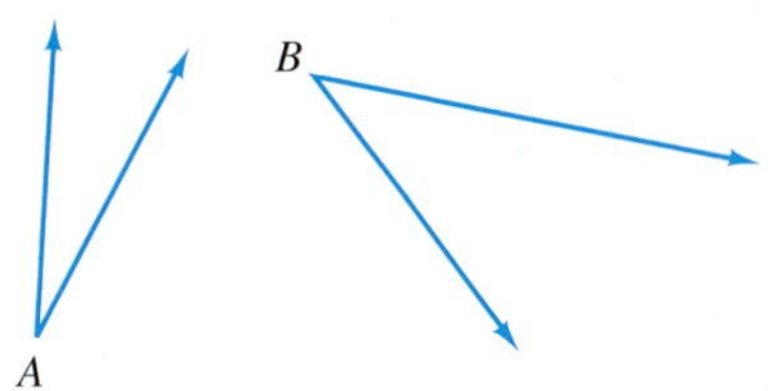

Describe procedures, using only a straightedge and a compass, to construct

(a) $\angle C$ so that $m(\angle C) = m(\angle A) + m(\angle B)$.

(b) $\angle D$ so that $m(\angle D) = m(\angle B) - m(\angle A)$.

(c) $\angle E$ so that $m(\angle E) + m(\angle A) + m(\angle B) = 180°$.

6. Use a ruler, protractor, and compass to construct, when possible, a triangle with the stated properties. If such a triangle cannot be drawn, explain why. Decide if there can be two or more noncongruent triangles with the stated properties:

(a) an isosceles triangle with two sides of length 5 cm and an apex angle of measure 28°

(b) an equilateral triangle with sides of length 6 cm

(c) a triangle with sides of length 8 cm, 2 cm, and 5 cm

(d) a triangle with angles measuring 30° and 110° and a nonincluded side of length 5 cm

(e) a right triangle with legs (the sides including the right angle) of length 6 cm and 4 cm

(f) a triangle with sides of length 10 cm and 6 cm and a nonincluded angle of 45°

(g) a triangle with sides of length 5 cm and 3 cm and an angle of 20°

7. Each part below shows two triangles, with arcs and tick marks identifying congruent parts. If it is possible to conclude that the triangles are congruent, describe what property or theorem you used. If you cannot be sure that the triangles are congruent, state, "No conclusion possible." The first one is done for you.

(a)

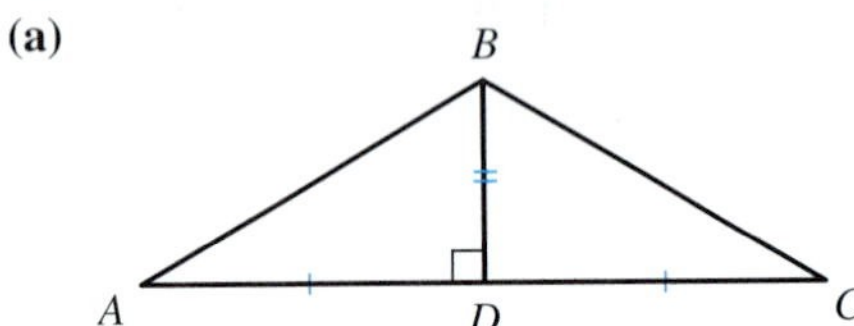

Answer: $\triangle ABD \cong \triangle CBD$ by SAS

(b)

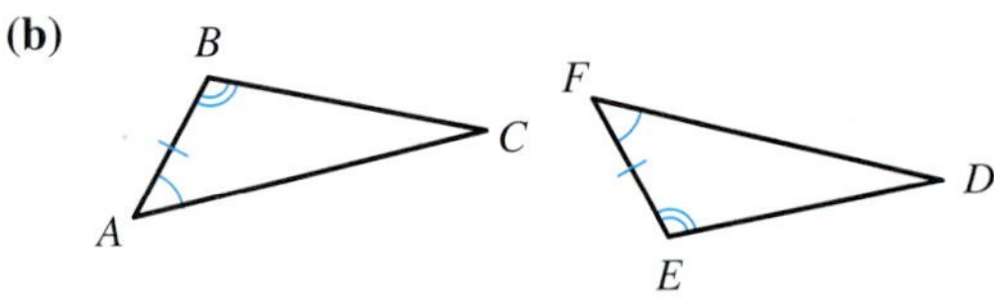

(c)

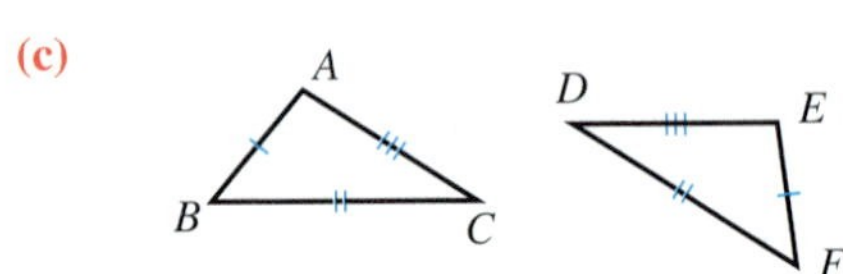

(d)

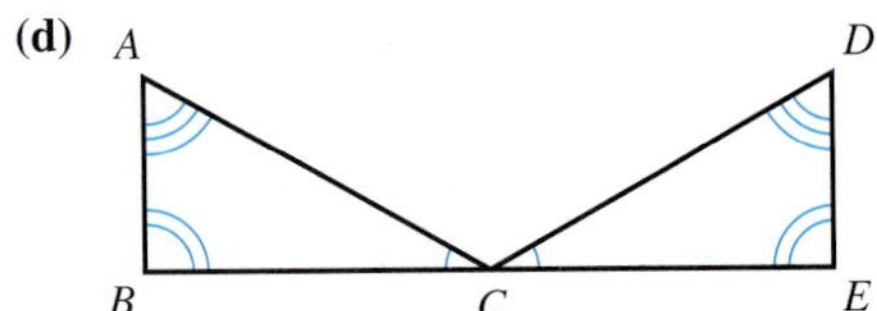

(e)

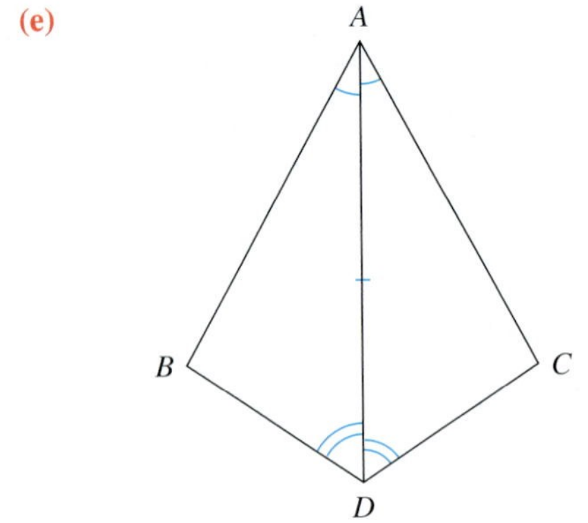

(f)

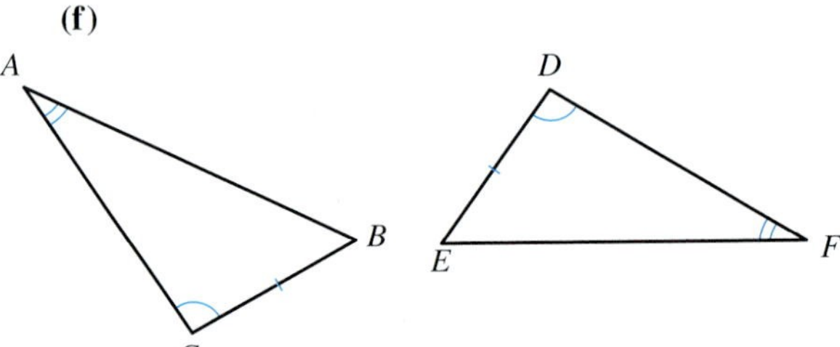

(g)

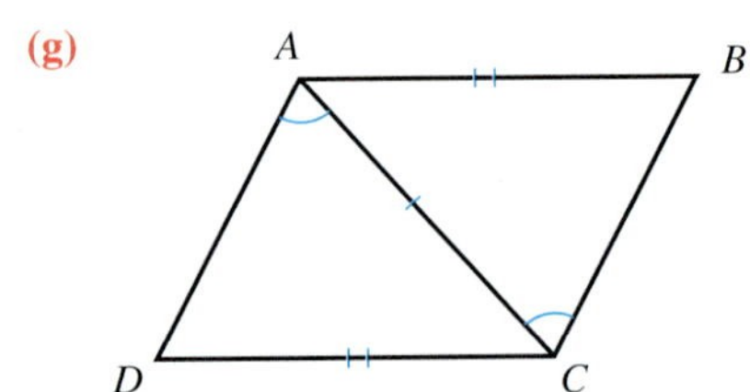

(h)

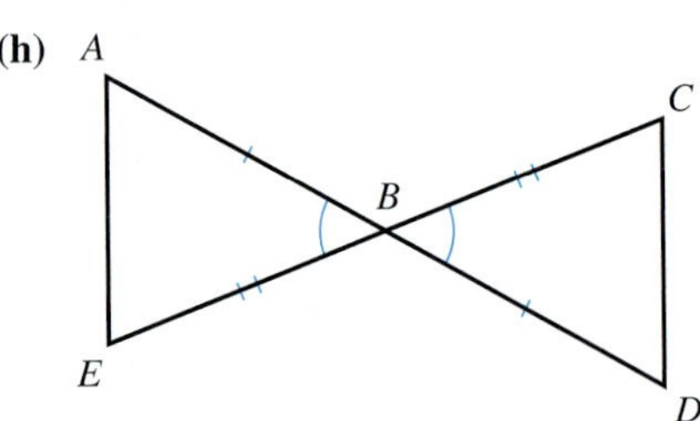

(i)

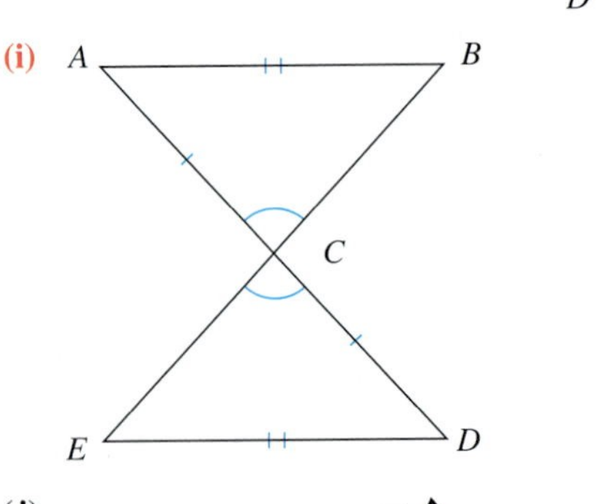

(j)

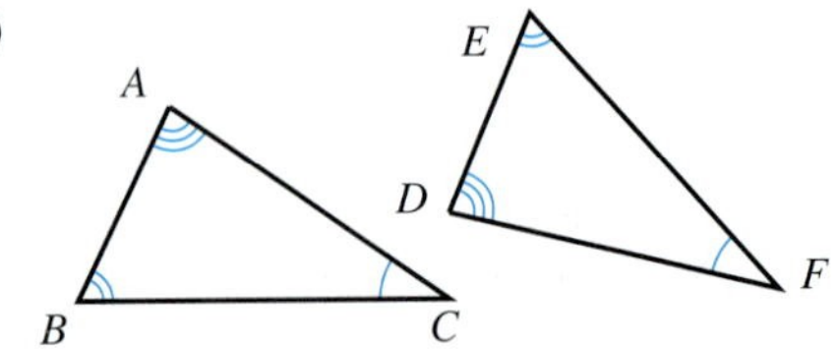

8. Prove that an equilateral triangle is equiangular.
9. Prove that an equiangular triangle is equilateral.
10. Draw an angle $\angle BAC$ and use a protractor to measure the angle. Next construct an arc at A to determine points D and E. Finally, draw arcs of equal radius centered at D and E, denoting the point of intersection of the arcs as F.
    (a) Measure angles 1 and 2. How do they compare with the measure of $\angle BAC$?
    (b) Prove that $\triangle AFD$ is congruent to $\triangle AFE$.
    (c) Explain why angles 1 and 2 are congruent, using the result of part (b).

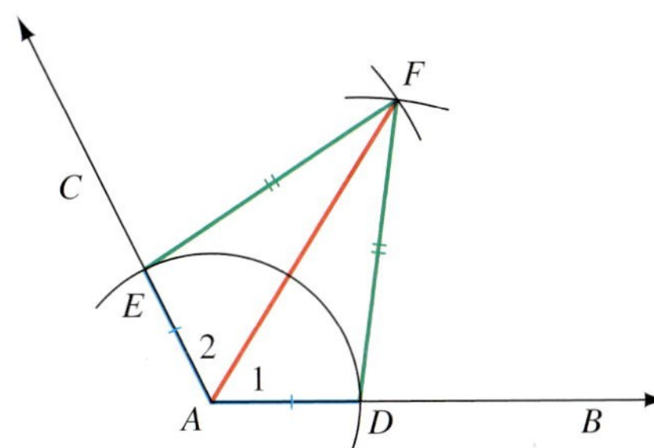

11. Draw a line m and a point P not on the line. Construct an arc centered at P that intersects the line in two points Q and S. Next, draw two arcs of equal radius centered at Q and S, labeling their intersection as point T. Finally, construct the segment from P to T and let V be its intersection with the line m.

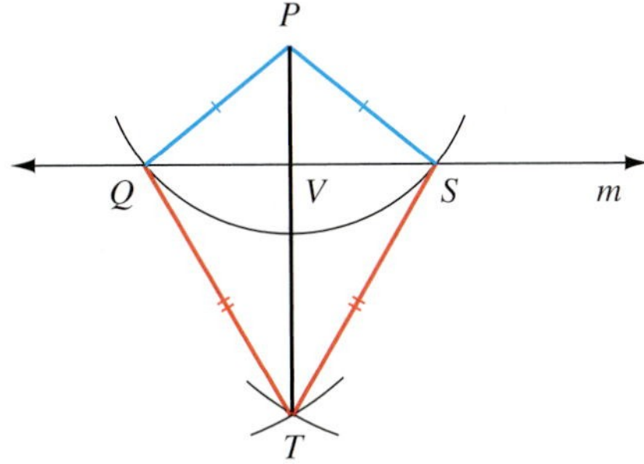

Give reasons that these relationships hold.
(a) $\triangle QPT \cong \triangle SPT$
(b) $\angle QPT \cong \angle SPT$
(c) $\triangle QPV \cong \triangle SPV$
(d) $\angle QVP$ is a right angle. Therefore, the construction gives a line $\overleftrightarrow{PT}$ perpendicular to m that passes through point P.

12. Let $ABCD$ be a parallelogram.

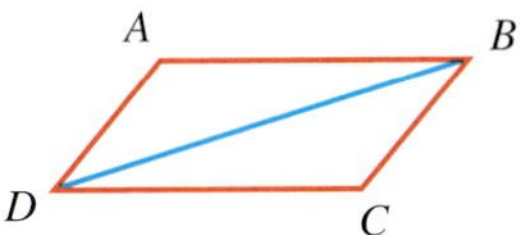

(a) Prove that $\triangle ABD \cong \triangle CDB$. (*Hint:* Use the ASA property.)
(b) Prove that opposite sides of a parallelogram have the same length.
(c) Prove that opposite angles of a parallelogram have the same measure.

13. Let the two diagonals of parallelogram $ABCD$ intersect at point M.

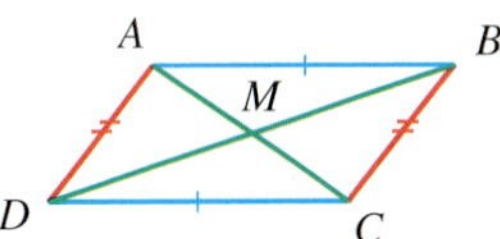

(a) Use the fact that $AB = CD$ (shown in problem 12(b)) to prove that $\triangle ABM \cong \triangle CDM$.
(b) Use part (a) to explain why M is the midpoint of both diagonals of the parallelogram.

14. Let the two diagonals of a rhombus $ABCD$ intersect at M.

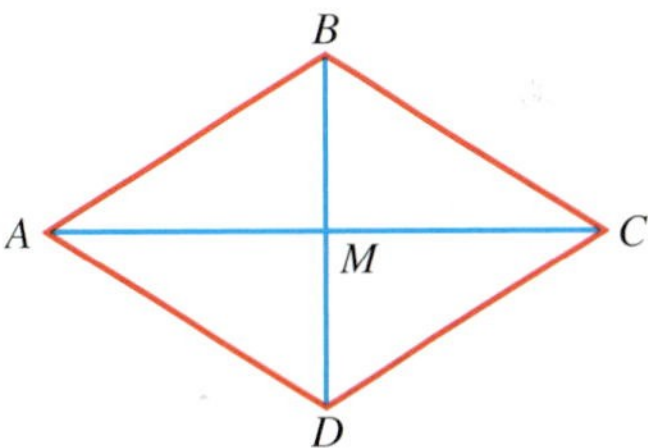

(a) Show that the triangles ABM, CBM, CDM, and ADM are congruent to one another.
(b) Use part (a) to explain why the diagonals of a rhombus bisect the interior angles of the rhombus and intersect at a right angle at M.

15. Let ABC be a right triangle with hypotenuse $\overline{AB}$ and right angle at vertex C. Explain why the circle centered at the midpoint O of the hypotenuse that passes through point C also passes through points A and B. This gives the

    Converse of Thales' Theorem: The hypotenuse of a right triangle inscribed in a circle is a diameter of the circle.

    (*Suggestion:* Imagine that the right triangle ABC is created by constructing the diagonal $\overline{AB}$ of a rectangle $ACBD$.)

16. Suppose you drew a circle by tracing around a bowl. Explain how you can locate the center of the circle with a piece of notebook paper. (*Hint:* Use the converse of Thales' theorem, stated in problem 15.)
17. A triangle has sides of length 4 cm and 9 cm. What can you say about the length of the third side?
18. (a) A quadrilateral has sides of length 2 cm, 7 cm, and 5 cm. What inequality does the length of the fourth side satisfy?
    (b) Let A, B, C, and D be any four points in the plane. Explain why $AD \leq AB + BC + CD$.

Teaching Concepts

19. Copycat Congruence Activity. Write a detailed lesson plan that expands on the following idea:

> The class is divided into small groups. Each group cuts out a large triangle from a sheet of colored construction paper. The triangles are taped to the chalkboard. Each group, using rulers and protractors, chooses two or three triangles from other groups to measure at the board. The group's task is to construct congruent triangles cut from fresh sheets of construction paper, using their measurements. Each group then places each new triangle over the corresponding triangle on the board to test the accuracy of its constructions. Finally, each group describes, in written and oral form, what measurements were taken and how congruence properties were used.

Your lesson plan should clearly state the activity's goals, the standards it addresses, materials required, directions, assessment, and extensions of the activity for further investigation.

Thinking Critically

20. Two angles and a *nonincluded* side of $\triangle ABC$ are drawn below.

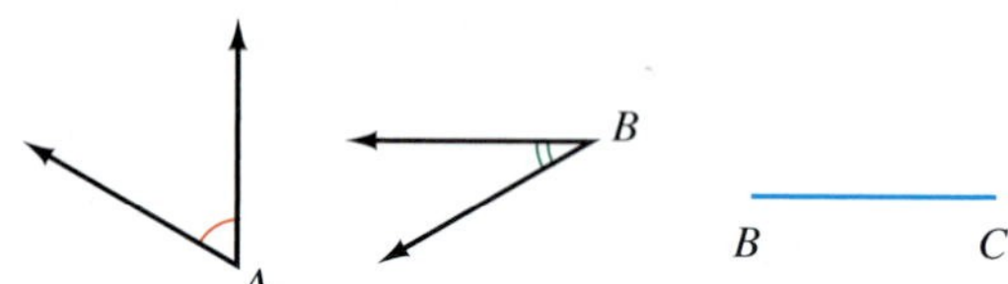

Describe and show the steps of a straightedge-and-compass construction of a triangle PQR that is congruent to $\triangle ABC$.

21. In the figure below, $AB = AE$ and $AC = AD$.

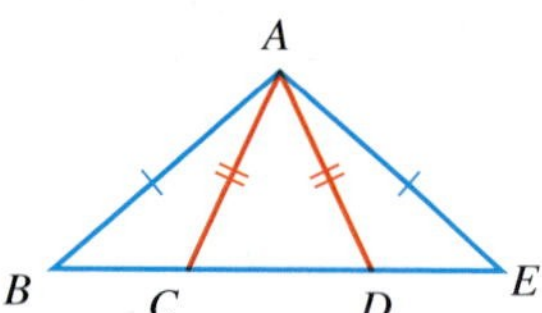

(a) Why is $\angle B \cong \angle E$?

(b) Why is $\angle ACD \cong \angle ADC$?

(c) Use the AAS property to prove that $\triangle ABC \cong \triangle AED$.

(d) Prove that $BC = DE$.

22. Recall that a trapezoid with a pair of congruent angles adjacent to one of its bases is called an isosceles trapezoid.

(a) Prove that the sides joining the bases of an isosceles trapezoid are congruent. The following figure may help you show that $AD = BC$.

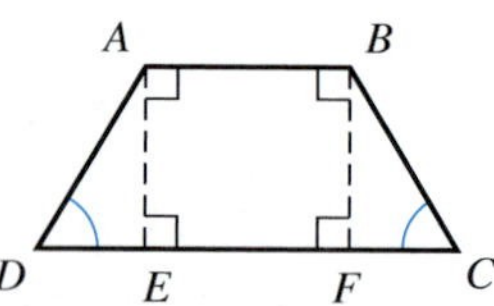

(b) Prove that the diagonals of an isosceles trapezoid are congruent.

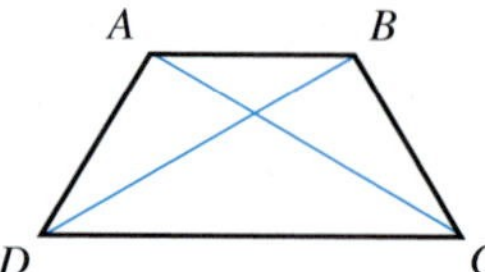

23. For each part, decide if the given conditions are sufficient to conclude that $\triangle ABC \cong \triangle ADE$. If so, give a proof; if not, draw a figure that satisfies the information, but shows that $\triangle ABC$ is not congruent to $\triangle ADE$.

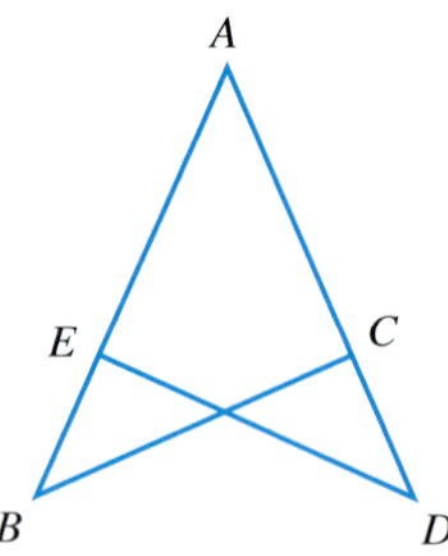

(a) $AB = AD$ and $m(\angle B) = m(\angle D)$

(b) $AB = AD$ and $BC = DE$

(c) $AB = AD$ and $AE = AC$

(d) $EB = CD$ and $BC = DE$

24. Each edge of a tetrahedron is congruent to its opposite edge. For example, $\overline{AB} \cong \overline{CD}$. Prove that the faces of the tetrahedron are congruent to one another.

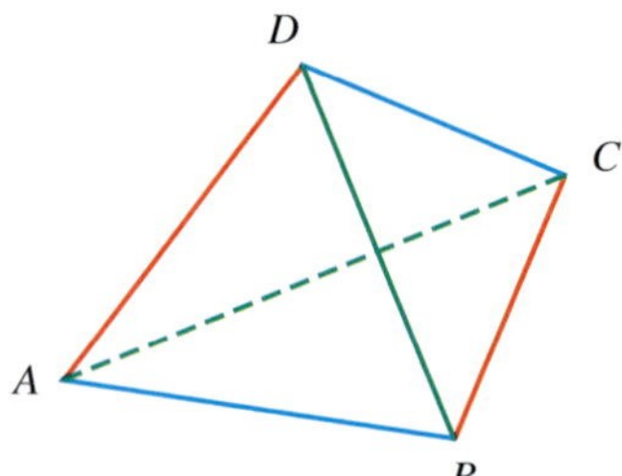

25. (a) Let T be a point on the side $\overline{QR}$ of triangle PQR. Use the triangle inequality to explain why $QP + QR > TP + TR$.

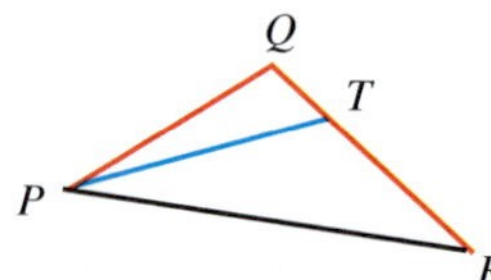

(b) Let S be a point in the interior of triangle PQR. Use part (a) to explain why $QP + QR > SP + SR$. This shows that the sum of distances from a vertex Q to the endpoints of the opposite side $\overline{PR}$ is larger than the sum of distances to the same two points from a point S within the triangle.

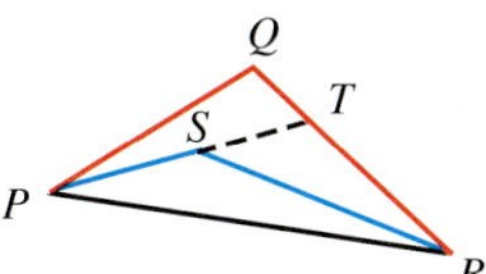

26. In Example 14.4, the four towns form a convex quadrilateral. Suppose instead that Davis is in the interior of the triangle formed by the other three towns, as shown in the figure below. Show that the power station serving the four towns is best located at Davis, instead of an alternative point such as point E, by answering these questions.

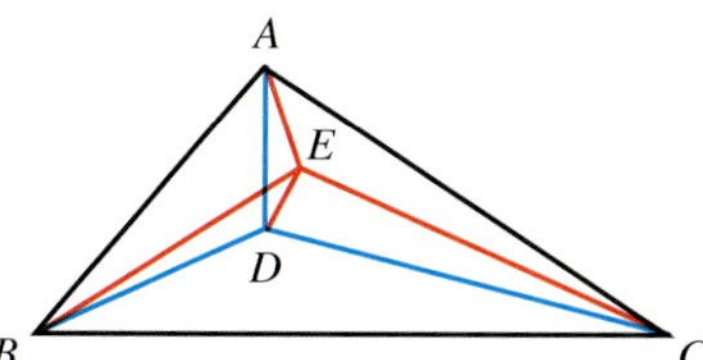

(a) Why is $EA + ED > DA$?

(b) Why is $EB + EC > DB + DC$? (*Suggestion:* Use the result of problem 25(b).)

(c) Why is $EA + ED + EB + EC > DA + DB + DC$?

27. Six towns are located at the vertices A, B, C, D, E, and F of a regular hexagon. A power station located at Q would require $QA + QB + QC + QD + QE + QF$ miles of transmission line to serve the six communities. Describe a better location P for the station, and prove it has the least possible sum of distances to A, B, C, D, E, and F.

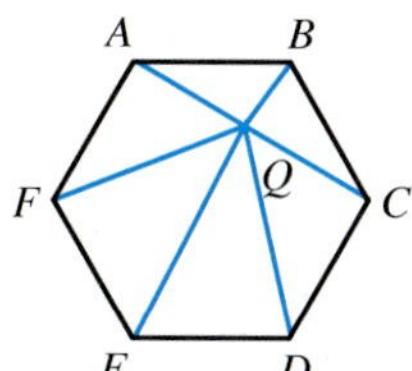

Making Connections

28. A bicycle rack is made from three pieces of metal box tubing. There are bolts at A, B, C, and D that join the tubing and attach the rack to the bumper of a car.

(a) Why is the top of the rack likely to shift sideways?

(b) If a fourth piece of tubing is available, where can it be attached to make the rack rigid? Explain why this works.

(c) Would a rope from A to C make the rack rigid? How about two pieces of rope, from A to C and from B to D?

29. Carpenters construct a wall by nailing studs to a top and bottom plate, as shown below.

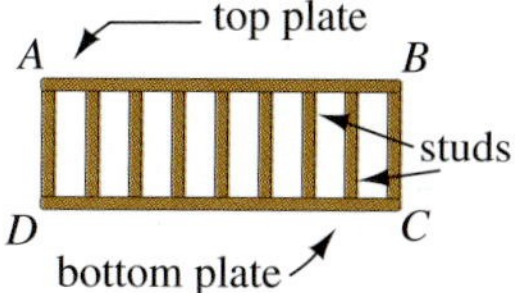

The studs are cut to the same length, and the top and bottom plates are the same length.

(a) If the pieces are properly cut and nailed, is $ABCD$ necessarily rectangular, or are there other shapes the framework can take?

(b) Carpenters frequently "square up" a stud wall by adjusting it so that it has diagonals of equal length. Prove that a parallelogram with congruent diagonals is a rectangle.

(c) Once the wall is "squared up," a diagonal brace is nailed across the frame. Why is this? What geometric principle is involved?

30. The two frameworks shown below are constructed with drinking straws and pins. The triangle is a rigid framework by the SSS property, but the quadrilateral is flexible.

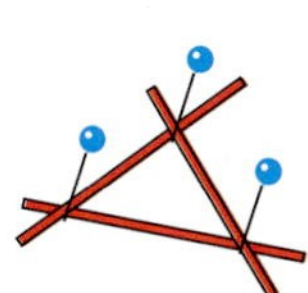

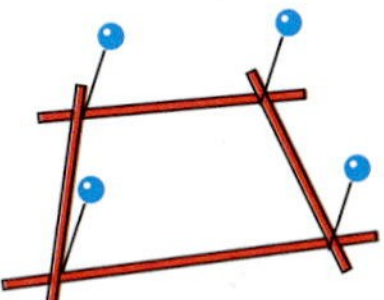

Decide if the straw-and-pin frameworks below are rigid or flexible. It may be helpful to build the frameworks to check your reasoning.

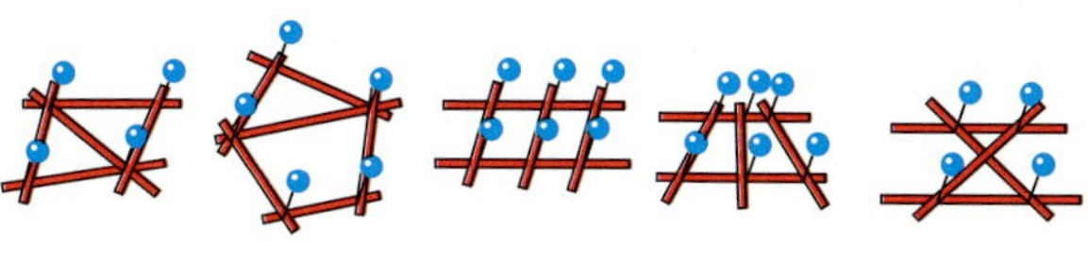

Using a Computer

31. In the power plant location problem of Example 14.4, suppose the town at D (Davis) drops out. This leaves three towns, at A, B, and C, that wish to jointly build a power plant at some location P serving the three communities. To minimize the cost of power lines, P should be situated so the total length $PA + PB + PC$ is as small as possible. Use geometry software to duplicate the figure shown below. Use the **Calculate** . . . command found under the Measure menu to compute $PA + PB + PC$. Drag point P, watching the sum, to determine what seems to be the best position for point P.

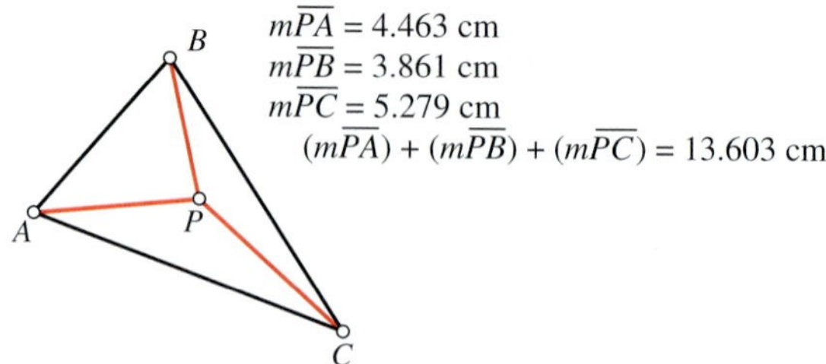

Describe the optimal location of P by considering two types of triangles.

(a) $\triangle ABC$ has an angle at one of its vertices measuring at least 120°.

(b) No angle of $\triangle ABC$ has measure larger than 120°. In this case, describe the location of P by measuring the three angles at P made by the segments joining P to A, B, and C.

(c) (*For more advanced software users*) Construct an outward-pointing equilateral triangle on each of the three sides of $\triangle ABC$ considered in part (b). Next, construct the center of each equilateral triangle, and the circle through the endpoints of the corresponding side of $\triangle ABC$. In the figure below, all three equilateral triangles and two of the circles are constructed. After constructing the third circle, see if you now know how to locate the power plant serving the towns at A, B, and C.

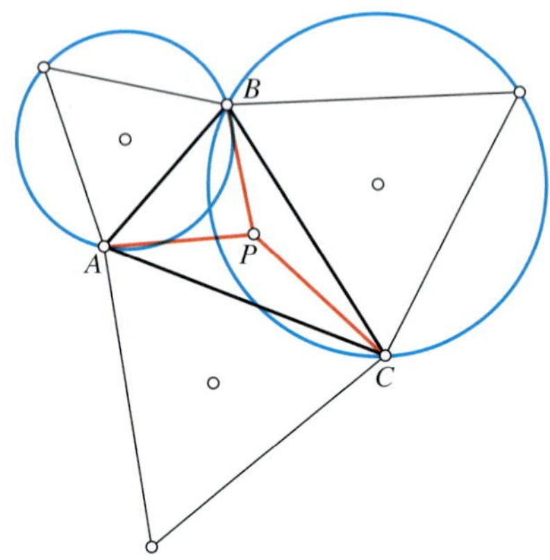

32. Construct a circle, and label its center O. Next construct three points A, B, and P on the circle and draw two angles $\angle AOB$ and $\angle APB$ that both intercept the arc of the circle between A and B.

(a) Measure $\angle AOB$ and $\angle APB$. What relationship do you observe? Move P around the circle and investigate what happens to the measure of $\angle APB$.

(b) Make a conjecture that relates the measures of $\angle AOB$ and $\angle APB$.

(c) Justify your conjecture. (*Hint:* Draw the diameter through P and then mimic the proof of Thales' theorem.)

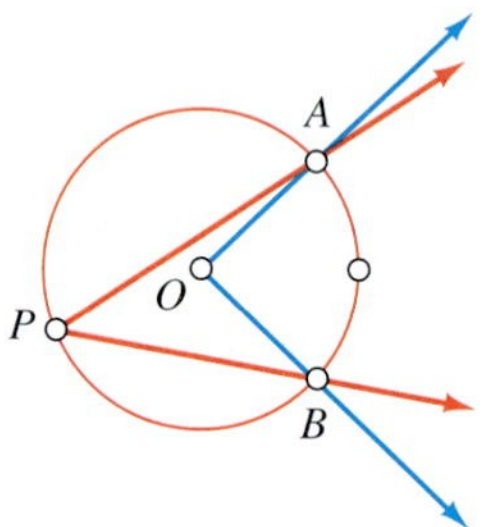

33. Draw a circle. Next construct four points A, B, C, and D on the circle that are joined by line segments to form the inscribed quadrilateral $ABCD$.

(a) Measure $\angle A$ and $\angle C$. What relationship do you observe? Move some of the points of your quadrilateral and see if the relationship between $\angle A$ and $\angle C$ is preserved.

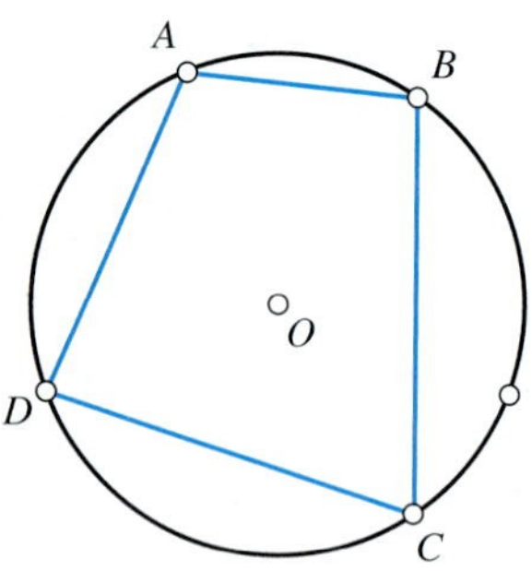

(b) Make a conjecture concerning the relationship of opposite angles of an inscribed quadrilateral.

(c) Justify your conjecture. (*Hint:* Draw the radii $\overline{OA}$, $\overline{OB}$, $\overline{OC}$, and $\overline{OD}$. This creates four isosceles triangles.)

From State Student Assessments

34. (Colorado, Grade 5)

Your teacher told you in class today that the square shown is divided into 8 congruent triangles. After you got home your best friend called and said he did not know what that meant. What would you say to your friend to help him out?

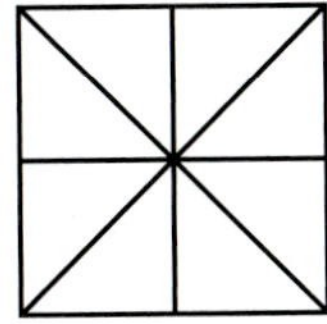

35. (Illinois, Grade 5)
Which tangram pieces are congruent triangles?

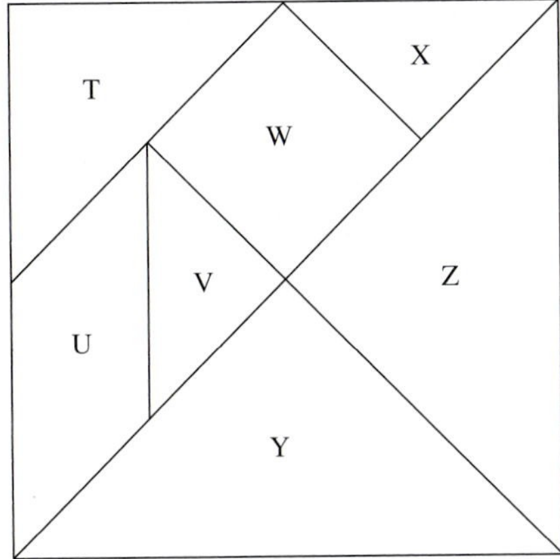

A. T and W
B. T and Z
C. U and X
D. V and X
E. X and Y

For Review

36. A partially completed jigsaw puzzle and the remaining pieces are on a table. Each remaining piece can be imagined to be put into its correct position by one of the four basic rigid motions in the plane: a translation, a reflection, a rotation, or a glide reflection.

(a) Which motions will properly position a right-side-up piece of the puzzle?

(b) Which motions will properly position an upside-down remaining piece?

37. **(a)** Show that translations of the numeral 4, shown below left, can tile the plane.

(b) Show that translations and rotations of the numeral 5, shown below right, can tile the plane.

Sketch each tiling on graph or dot paper.

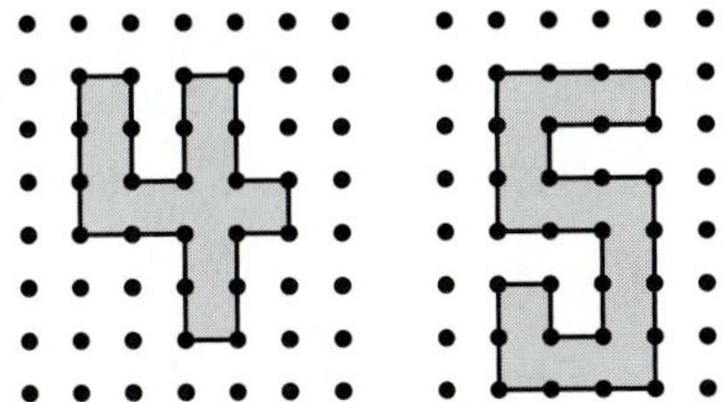

38. For each whole number $n = 0, 1, 2, 3, \ldots$, draw (if possible) a hexagon with exactly n lines of symmetry.

14.2 CONSTRUCTING GEOMETRIC FIGURES

In the last section, two basic constructions were described:

- Construction 1 Construct a congruent line segment.
- Construction 2 Construct a congruent angle.

Since only the straightedge and compass were used, these are examples of Euclidean constructions. In this section we describe a number of other Euclidean constructions and explore some related applications and theorems. We also investigate constructions with the Mira and by paper folding. The constructions shown in the examples and called for in the problems can also be done with geometry software.

To be certain that a construction results in a figure that has a desired property, a proof of the validity of the construction must be given. For example, Construction 2 of a congruent angle is a consequence of the SSS property. Many constructions can be verified by appealing to the properties of a rhombus listed in Figure 14.6.

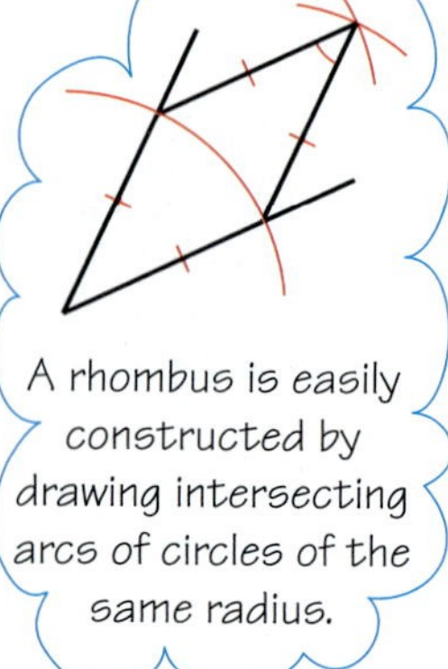

A rhombus is easily constructed by drawing intersecting arcs of circles of the same radius.

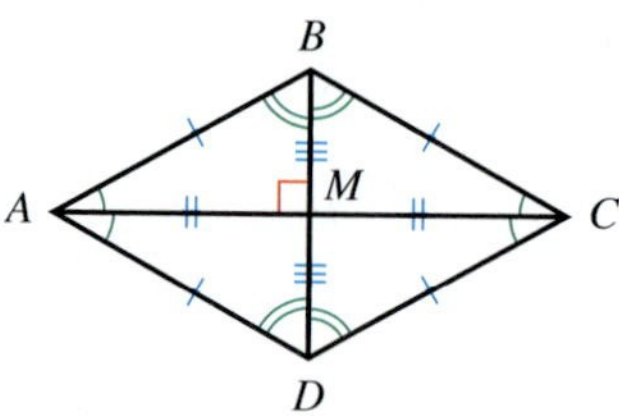

Figure 14.6
The rhombus ABCD has many useful properties:

- *The diagonals are angle bisectors.*
- *The diagonals are perpendicular.*
- *The diagonals intersect at their common midpoint M.*
- *The sides are all congruent to one another.*
- *The opposite sides are parallel.*

Constructing Parallel and Perpendicular Lines

If l is a given line and P is a given point not on l, it is useful to know how to construct the line through P that is either parallel or perpendicular to l. There are alternative procedures that can be devised, and it is interesting to invent some of your own. The following constructions each take advantage of the properties of a rhombus.

Construction 3

Construct a Parallel Line

Given P and l as shown, construct a line through P that is parallel to l.

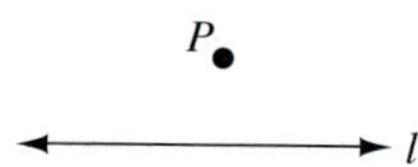

Procedure

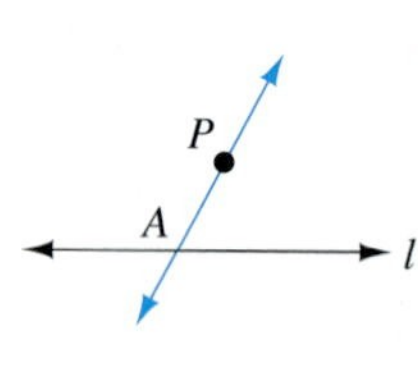	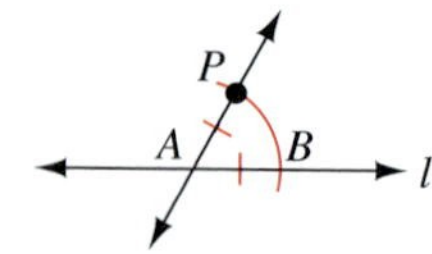	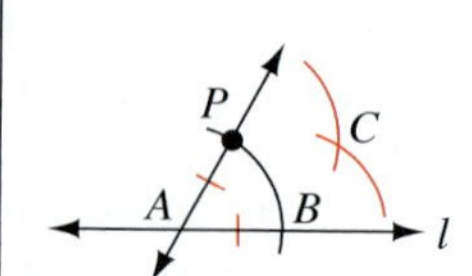	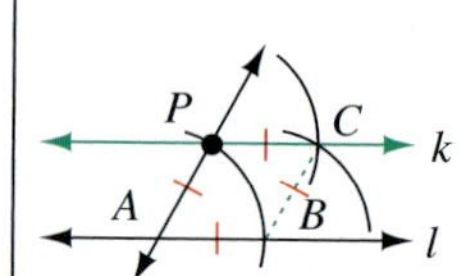
Step 1 Draw a line through P that intersects l at a point labeled A.	**Step 2** Draw an arc centered at A through P, and let B denote its intersection with l.	**Step 3** With the same radius AB, draw arcs with centers at P and B; let C be the intersection of the two arcs.	**Step 4** Draw the line $k = \overleftrightarrow{PC}$; since $ABCP$ is a rhombus, its opposite sides are parallel, and so $k \parallel l$.

Into the Classroom

Constructions in Space

Constructions using a compass and a straightedge lead to pictorial figures, confined to a sheet of paper. Interest and excitement can also be generated by constructing figures in space, using sticks, brass fasteners, paper clips, string, cut paper, multilink cubes, or indeed whatever is available. Such figures can be held and felt, literally giving students a feel for shape. In some cases the shapes can be bent or flexed to give a dynamic liveliness to figures that would remain of lesser interest when only drawn on a sheet of paper.

Books, pamphlets, and journals published by the National Council of Teachers of Mathematics and other publishing companies provide the teacher with a wide variety of ideas and resources of three-dimensional constructions and related activities. Every teacher will want to gather a personal collection of favorite hands-on constructions suitable for his or her classroom. Here are three suggestions for constructions to do in the classroom:

- *Hinged polygons.* Strips of cardstock can be joined with brass fasteners through holes at the ends of each strip. Any triangle is rigid, demonstrating the SSS congruence property. Any quadrilateral, however, is flexible. As the quadrilateral flexes, the sum of the angle measures remains at 360°, as can be checked with a protractor.

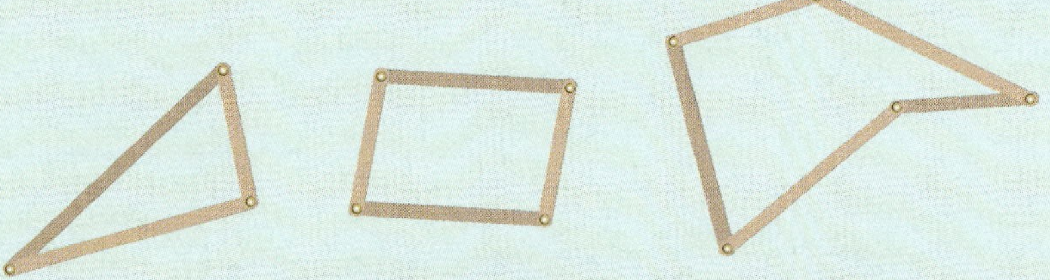

- *Space polygons and polyhedra.* Thin wooden sticks, say, from a "pick-up-sticks" game, can be cut to differing lengths and joined by short pieces of rubber tubing at their endpoints. More than two sticks can meet at a single vertex by inserting a length (or several lengths) of tubing through a hole punched sideways through another section of tubing. It is easy to form quadrilaterals that flex in space, and joining the midpoints of the four sides by elastic bands shows that a parallelogram is always formed. Properties of cubes, tetrahedra, and other polyhedra can also be explored with easily constructed skeletal models.

- *Geodesic domes.* Even primary-grade children will enjoy constructing simple geodesic dome models with toothpicks and mini marshmallows. The pattern for a small model is shown in its flattened position. The framework is then positioned vertically and wrapped around to insert toothpicks b and c into marshmallow A, toothpick a into marshmallow B, and the four upward-pointing toothpicks into marshmallow *C*. The photograph shows first graders with the domes they've made. A full discussion is found in the article "Marshmallows, Toothpicks, and Geodesic Domes," by Stacy Wahl (in *Geometry for Grades K–6: Readings from the Arithmetic Teacher,* National Council of Teachers of Mathematics, 1987).

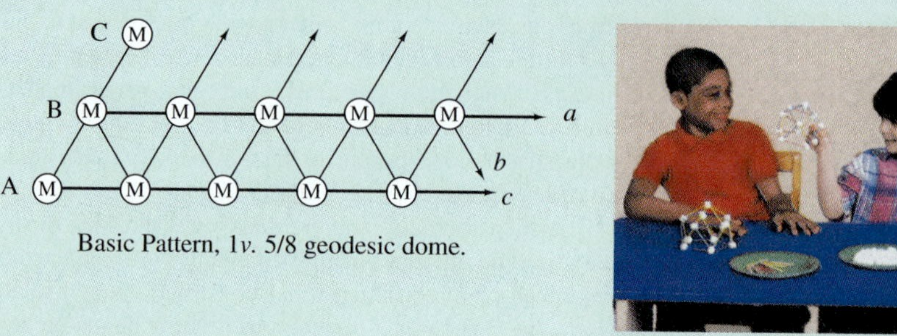

Basic Pattern, 1*v*. 5/8 geodesic dome.

Construction 4

Construct a Perpendicular Line Through a Point Not on the Given Line

Given line *l* and point *P* not on *l* as shown, construct a line through *P* that is perpendicular to *l*.

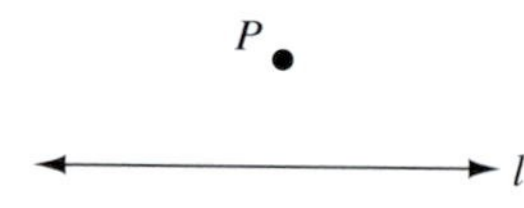

Procedure

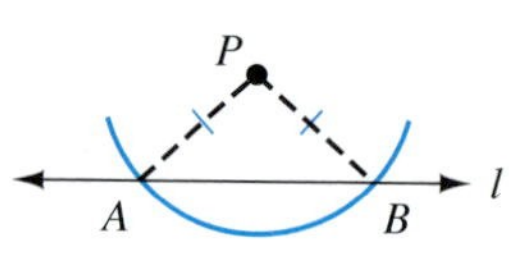 P, A, B, l **Step 1** Draw an arc at *P* that intersects *l* at two points *A* and *B*.	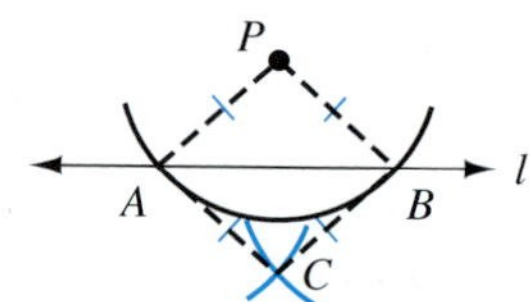 P, A, B, C, l **Step 2** With the compass still at radius *AP*, draw arcs at *A* and *B*, and let *C* be their point of intersection.	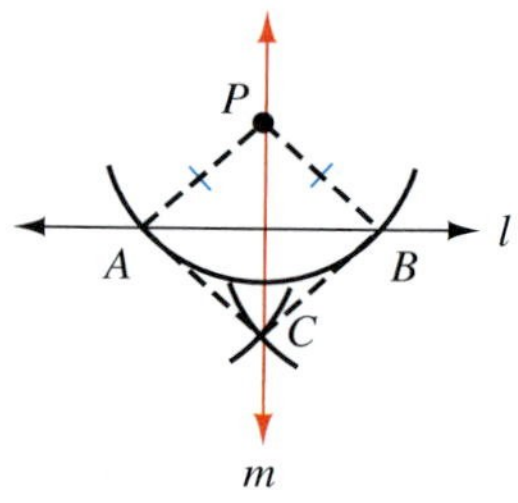 P, A, B, C, l, m **Step 3** Draw the line $\overleftrightarrow{PC}$; since $\overline{PC}$ is a diagonal of rhombus *ABPC*, it is perpendicular to $\overline{AB}$.

The construction of perpendicular lines has several applications:

- **Nearest point *F* on a line *l* from a point *P*.** The line through *P* that is perpendicular to line *l* intersects *l* at the point *F* of *l* that is closest to *P*. *F* is called the **foot** of the perpendicular line from *P*, and *PF* is called the **distance from *P* to *l*.**
- **Point of reflection *P′* of *P* across mirror line *l*.** The point *P′* on the perpendicular to *l* through *P* for which $PF = P'F$, where *F* is the point of intersection of *l* and the perpendicular through *P*, is called the **point of reflection of *P* across *l*.**

- **Altitudes of a triangle.** The perpendicular line segment through a vertex of a triangle to the line containing the opposite side of the triangle is called an **altitude** of the triangle. Its length is an altitude of the triangle when the opposite side is taken as the base of the triangle.

These applications are illustrated in Figure 14.7.

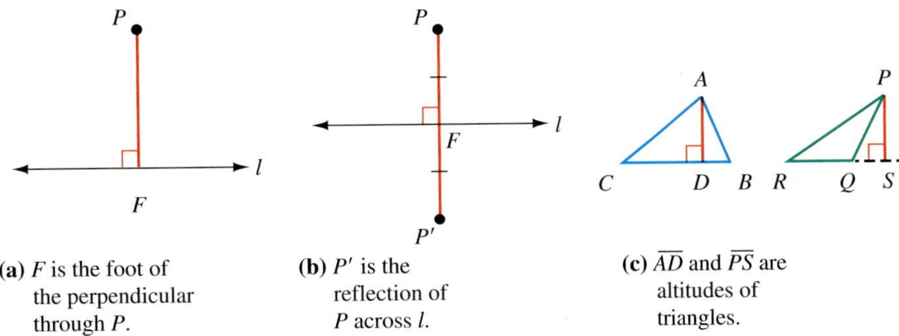

(a) F is the foot of the perpendicular through P.

(b) P' is the reflection of P across l.

(c) $\overline{AD}$ and $\overline{PS}$ are altitudes of triangles.

Figure 14.7
Applications of perpendicular lines

If point P lies on line l, a small modification in the second step of the procedure above is required to construct the line perpendicular to l at P.

Construction 5

Construct the Perpendicular Line Through a Point on a Given Line

Given line l and point P on l as shown, construct the line through P perpendicular to l.

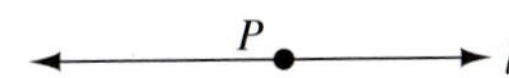

Procedure

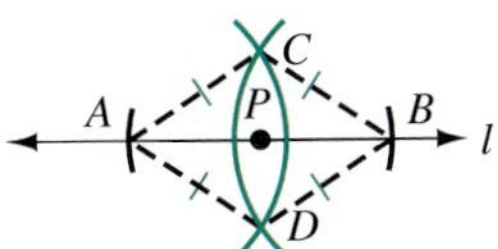

Step 1
Draw two arcs of equal radius centered at P; let A and B be their points of intersection with l.

Step 2
Draw arcs centered at A and B with a radius *greater* than AP; let C and D be their points of intersection.

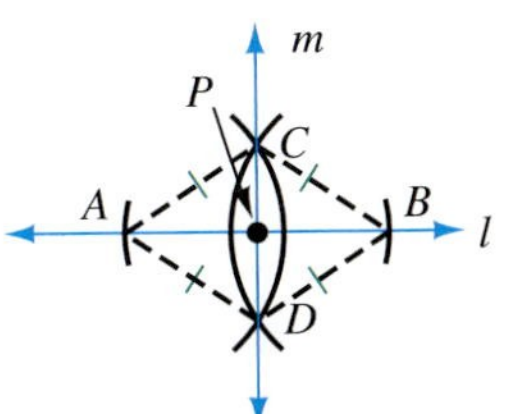

Step 3
Draw line $m = \overleftrightarrow{CD}$; since $\overline{CD}$ is a diagonal of the rhombus $ADBC$ and P is the midpoint of diagonal $\overline{AB}$, m passes through P and is perpendicular to $l = \overleftrightarrow{AB}$; that is, m is perpendicular to l.

Constructing the Midpoint and Perpendicular Bisector of a Line Segment

The line perpendicular to a segment at its midpoint is called the **perpendicular bisector** of the segment. The following construction is also justified by properties of the rhombus.

Construction 6

Construct the Midpoint and Perpendicular Bisector of a Line Segment

Construct the midpoint and perpendicular bisector of the segment $\overline{AB}$ shown.

A B

Procedure

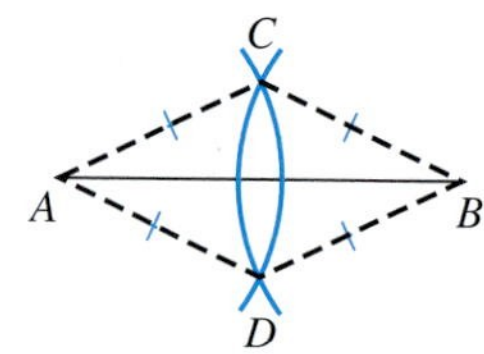

Step 1
Draw arcs of the same radius centered at A and B and intersecting in points C and D.

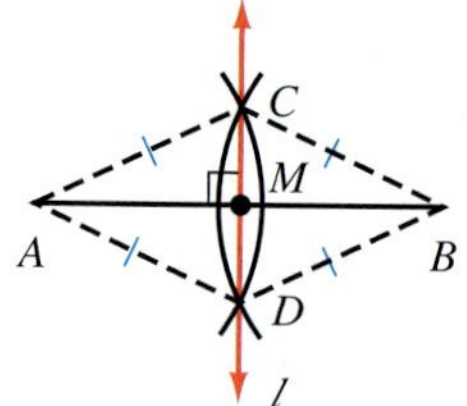

Step 2
Draw the line $\overleftrightarrow{CD}$. Since $ADBC$ is a rhombus, $\overleftrightarrow{CD}$ intersects $\overline{AB}$ at a right angle at the midpoint M of $\overline{AB}$.

The midpoint of a line segment M is the same distance from A and B. Indeed, every point P of the perpendicular bisector of a segment $\overline{AB}$ is equidistant from the endpoints of the segment. That is, $PA = PB$.

THEOREM *Equidistance Property of the Perpendicular Bisector*
A point lies on the perpendicular bisector of a line segment if, and only if, the point is equidistant from the endpoints of the segment.

Proof Let l be the perpendicular bisector of segment $\overline{AB}$. Thus, l intersects $\overline{AB}$ at right angles at the midpoint M, as shown on the left side of the diagram below. Let P be any point of l. By the SAS property, $\triangle PMA \cong \triangle PMB$. This means that the corresponding sides $\overline{PA}$ and $\overline{PB}$ are congruent. Hence, $PA = PB$ as claimed.

The proof that equidistant points from A and B lie on the perpendicular bisector is similar. (See problem 28 of Problem Set 14.2.)

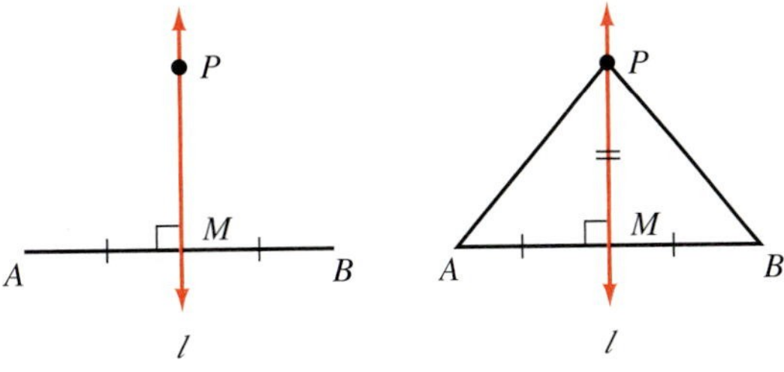

EXAMPLE 14.8 LOCATING AN AIRPORT

The Tri-Cities Airport Authority wishes to locate a new airport to serve its three cities, located at A, B, and C as shown. If possible, it would like a site P that is the same distance from A, B, and C. How can P be located?

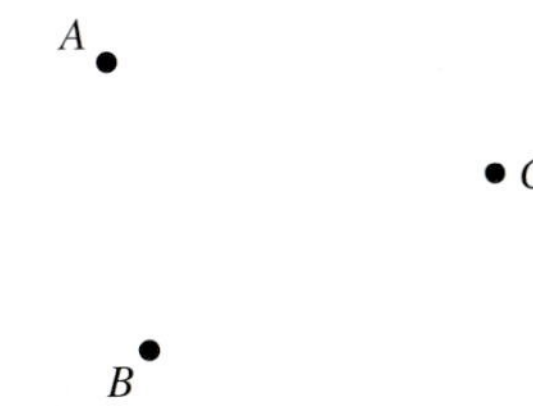

Solution To be equidistant from A and B, P must be on the perpendicular bisector of the segment $\overline{AB}$. Similarly, P must be on the perpendicular bisector of $\overline{BC}$. Since A, B, and C are not collinear, the perpendicular bisectors to $\overline{AB}$ and $\overline{BC}$ are not parallel and we can choose P as their point of intersection. Since $PA = PB$ and $PB = PC$, we have that $PA = PC$. Therefore, P is also equidistant from A and C, so P is also on the perpendicular bisector of $\overline{AC}$. Therefore, P is equidistant from A, B, and C, as desired.

Point P is the center of a unique circle containing A, B, and C as shown in Figure 14.8. The circle is called the **circumscribing circle** of $\triangle ABC$, and P is called the **circumcenter** of $\triangle ABC$. Frequently the circumscribing circle is called the **circumcircle** of the triangle.

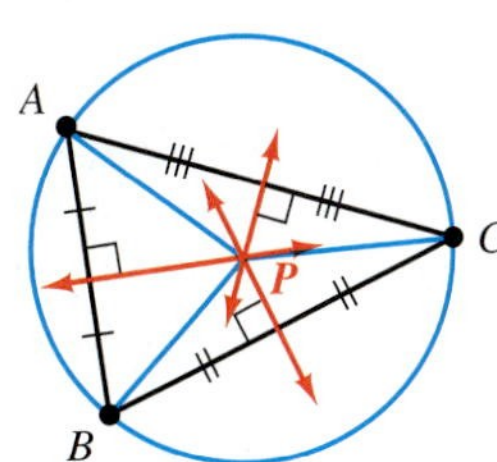

Figure 14.8
The perpendicular bisectors of the sides of $\triangle$ABC are concurrent at a point P equidistant from A, B, and C. Point P is the center of the circumscribing circle of $\triangle$ABC

Constructing the Angle Bisector

Given $\angle ABC$ (see Figure 14.9), we wish to construct the ray $\overrightarrow{BD}$ that forms congruent angles with the sides $\overrightarrow{BA}$ and $\overrightarrow{BC}$. If $\angle ABD \cong \angle CBD$, then $\overrightarrow{BD}$ is the **angle bisector** of $\angle ABC$.

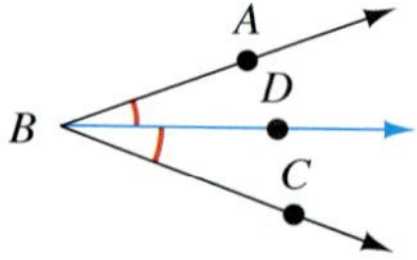

Figure 14.9
$\overrightarrow{BD}$ *is the angle bisector of* $\angle ABC$ *if* $\angle ABD \cong \angle DBC$

Once again, the properties of a rhombus justify the following construction.

Construction 7

Construct the Angle Bisector

Construct the angle bisector of $\angle E$ shown.

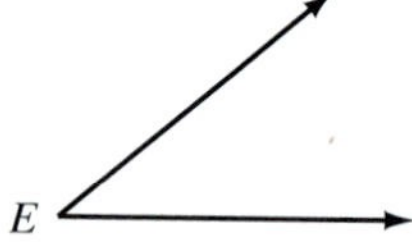

Procedure

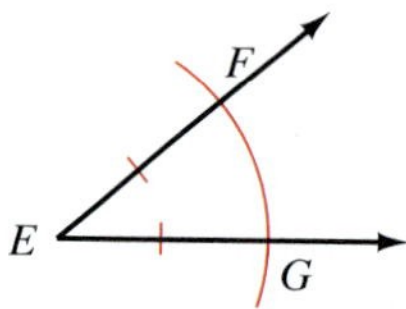	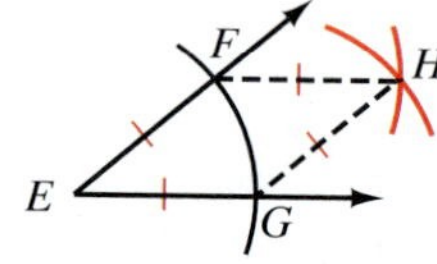	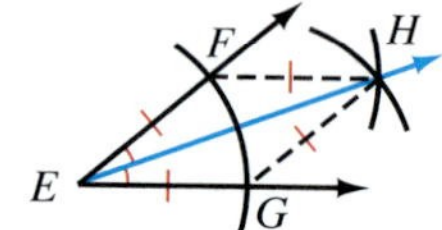
Step 1 Draw an arc centered at E; let F and G denote the points at which the arc intersects the sides of $\angle E$.	**Step 2** Draw arcs, centered at F and G, of radius EF. Let H be the point of intersection of the two arcs.	**Step 3** Draw the ray $\overrightarrow{EH}$. Since the diagonal $\overline{EH}$ forms congruent angles to the sides $\overline{EG}$ and $\overline{EF}$ of the rhombus $EGHF$, $\overrightarrow{EH}$ is the angle bisector of $\angle E$.

The following theorem tells how far the points on the angle bisector are from the sides of an angle.

THEOREM *Equidistance Property of the Angle Bisector*
A point lies on the bisector of an angle if, and only if, the point is equidistant from the sides of the angle.

Proof Let P be any point on the bisector of $\angle A$ as shown. Let F and F' be the feet of the perpendiculars from P to the sides of $\angle A$.

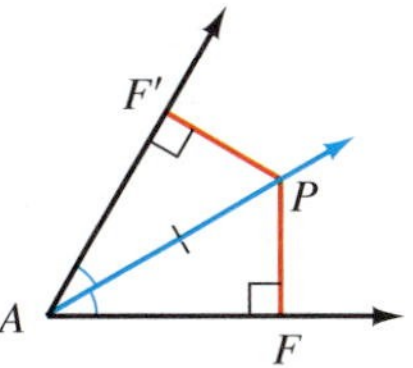

By the AAS congruence property, $\triangle PAF \cong \triangle PAF'$. Therefore, $PF = PF'$. The proof of the converse is similar and left to the reader.

By mimicking the solution of the airport problem in Example 14.8, it can be shown that the bisectors of the interior angles of a triangle are concurrent at the point I that is equidistant to three sides of the triangle (see Figure 14.10). Point I, called the **incenter** of $\triangle ABC$, is the center of the **inscribed circle,** or **incircle,** of $\triangle ABC$.

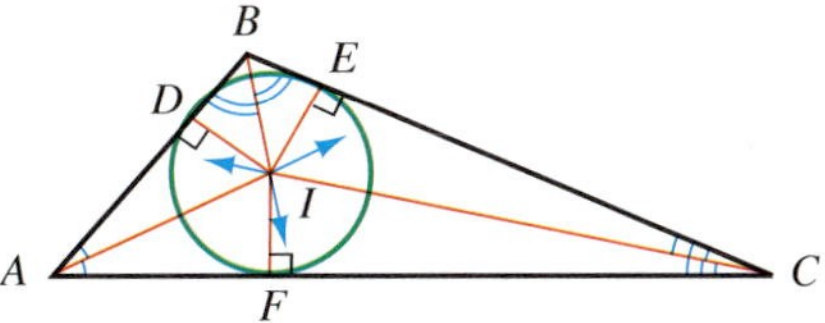

Figure 14.10
The bisectors of the interior angles of a triangle are concurrent at a point I equidistant from the sides of the triangle. I is the center of the inscribed circle of the triangle, which is tangent to the sides of the triangle at points D, E, and F

Constructing Regular Polygons

A square is easily constructed with compass and straightedge. For example, construct two perpendicular lines and a circle centered at the point of intersection, as shown in Figure 14.11(a).

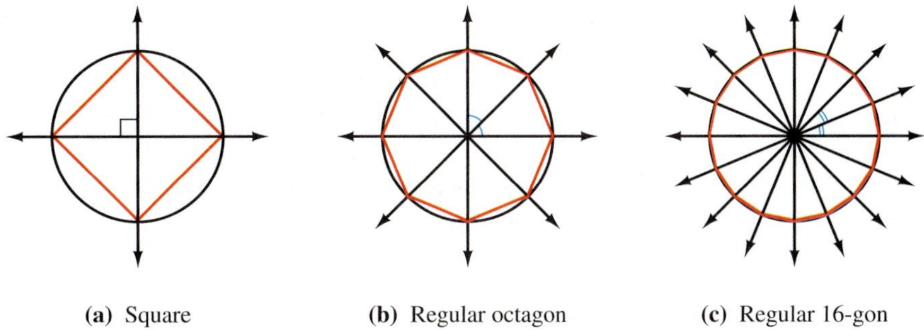

Figure 14.11
Constructing angle bisectors doubles the number of sides of a regular polygon inscribed in a circle

By constructing angle bisectors from the center of a square, the eight vertices of a regular octagon are constructed. Angle bisectors can then be constructed to the octagon of

Figure 14.11(b) to yield the regular 16-gon of part (c) of the figure. When the vertices of a polygon are all points of a given circle, the polygon is called an **inscribed polygon.**

A regular hexagon is particularly easy to inscribe in a given circle with a compass and straightedge: pick any point *A* on the circumference of the circle centered at *O* and then successively strike arcs of radius *OA* around the circle to locate *B*, *C*, *D*, *E*, and *F*. The hexagon *ABCDEF* is regular, since joining the sides to the center *O* forms six congruent equilateral triangles *OAB*, *OBC*, . . . , and *OFA*. As shown in Figure 14.12, connecting every other vertex gives a construction of the inscribed equilateral triangle *ACE*. On the other hand, constructing angle bisectors of the central angles of the hexagon produces an inscribed regular dodecagon.

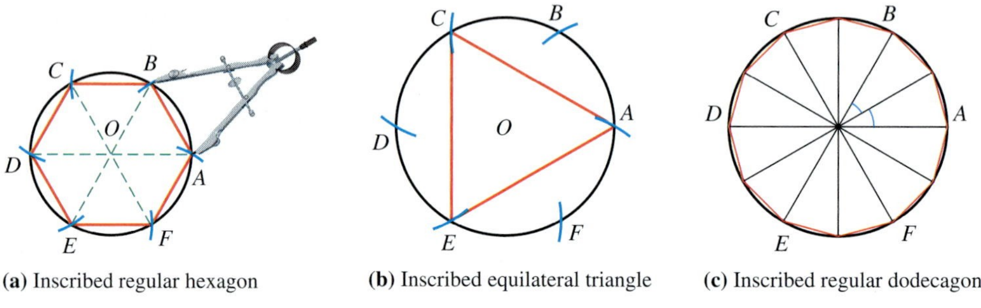

Figure 14.12
The regular 3-, 6-, and 12-gons can be inscribed with compass and straightedge in a given circle

A compass-and-straightedge construction of a regular pentagon requires more ingenuity. (One method is outlined in problem 21 at the end of this section.)

The ancient Greek geometers knew how to construct the regular polygons shown so far. They also knew that the regular 15-gon can be constructed with compass and straightedge. (The construction is outlined in problem 31 at the end of this section.) For over two thousand years, the only regular polygons known to be constructible with compass and straightedge were the ones contained in Book IV of Euclid's *Elements:* the regular 3-, 4-, 5-, and 15-gons and, by angle bisection, the regular polygons obtained by successively doubling the number of sides.

On March 30, 1796, one month before his nineteenth birthday, Carl Friedrich Gauss (1777–1855) entered into a notebook his discovery that a number of other regular polygons were constructible, including the 17-gon, the 257-gon, and the 65,537-gon. The numbers 3, 5, 17, 257, and 65,537 are prime numbers of the special form $F_k = 2^{2^k} + 1$, where k is a nonnegative integer. For example, $F_0 = 2^{2^0} + 1 = 2^1 + 1 = 3$, $F_1 = 2^{2^1} + 1 = 2^2 + 1 = 4 + 1 = 5$, and so on. Numbers of this form had been studied earlier by Pierre de Fermat (1601–1665), and prime numbers of the form $2^{2^k} + 1$ are known as **Fermat primes.**

Here is the remarkable theorem of Gauss. The proof of the "only if" part of the theorem is due to Pierre Wantzel (1814–1848).

THEOREM *The Gauss–Wantzel Constructibility Theorem*

A regular polygon of *n* sides is constructible with compass and straightedge if, and only if, *n* is

1. 4;
2. a Fermat prime;
3. a product of distinct Fermat primes;
4. a power-of-two multiple of a number that is one of the types from 1., 2., or 3. above.

For example, the regular heptagon is not constructible, since 7 is not a prime number of the form $2^{2^k} + 1$. A regular nonagon (9-gon) is also not constructible, since 9 has two factors of 3. On the other hand, a regular polygon of $1020 = 2^2 \cdot 3 \cdot 5 \cdot 17$ sides is constructible, since its odd prime factors are distinct Fermat primes.

Fermat believed that all numbers of the form $F_k = 2^{2^k} + 1$ were prime, but Euler proved this assertion to be false by showing that $F_5 = 2^{2^5} + 1$ is composite; in fact, $F_5 = 4{,}294{,}967{,}297 = (641) \cdot (6{,}700{,}417)$. Likewise, $F_6, F_7, \ldots, F_{32}$ are now known to be composite. It is generally believed that there are only five Fermat primes, namely 3, 5, 17, 257, and 65,537.

Mira™ and Paper-Folding Constructions

Figure 14.13 shows a Mira, which was used in Chapter 13 to draw images under reflection and to investigate lines of symmetry. The Mira can also be used to construct perpendicular lines, midpoints, angle bisectors, and so on. With a little practice, constructions with a Mira are quick and yet very accurate.

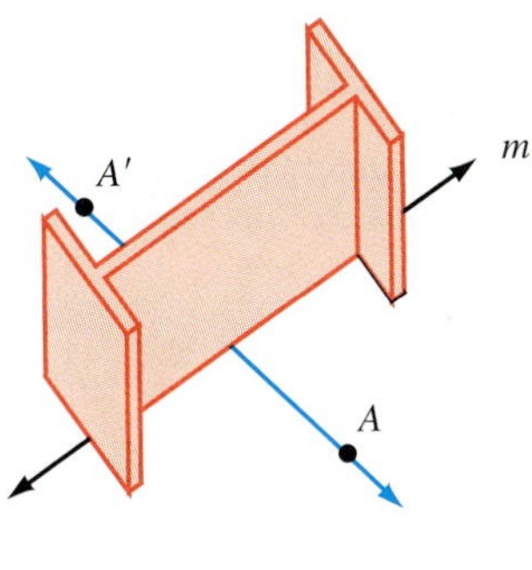

(a) The Mira

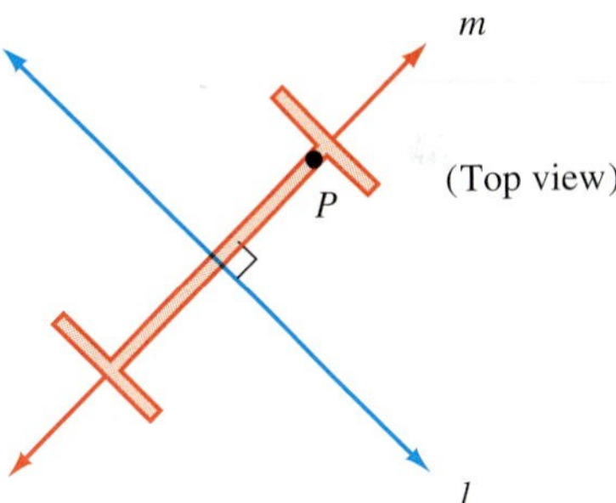

(b) Pivot the Mira about P until line l coincides with its reflection to construct the line m through P that is perpendicular to l.

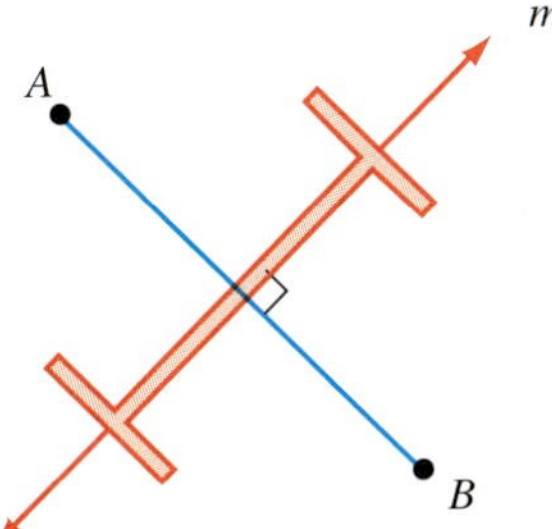

(c) When the reflection of B coincides with A, the drawing edge of the Mira determines the perpendicular bisector of $\overline{AB}$.

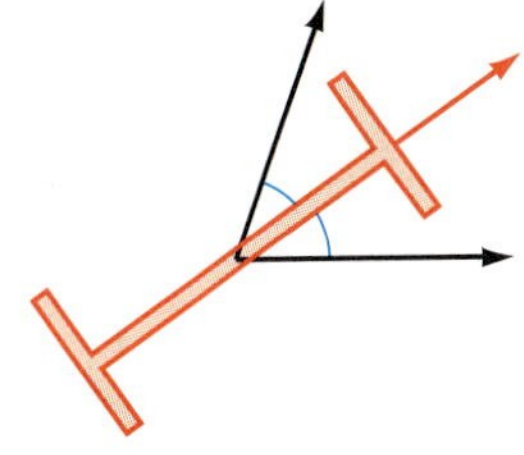

(d) Pivot the Mira about the vertex until the reflection of the near side coincides with the far side to construct the angle bisector.

Figure 14.13
The Mira and its use in three basic constructions

Mira constructions can usually be converted into equivalent paper-folding procedures. The drawing line of the Mira is replaced by the crease line of a fold. It is helpful to

Highlight from History

Three Impossible Construction Problems

The straightedge allows us to draw a line of indefinite length through any two given points, and the compass* allows us to draw a circle with a given point as its center and passing through any given distinct second point. It then becomes a challenge to find procedures to construct a figure using only these simple tools. Many important contributions to geometry were inspired by attempts to solve the following famous problems, each of which arose in antiquity.

1. *The trisection of an angle:* Divide an arbitrary given angle into three congruent angles.

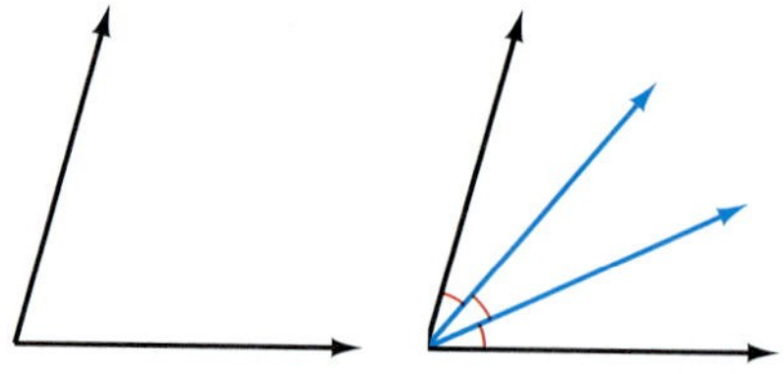

2. *The duplication of the cube:* Given a cube, construct a cube with twice the volume.

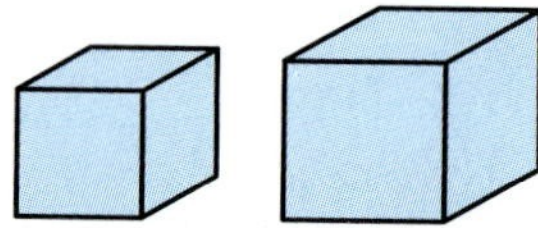

3. *The squaring of the circle:* Given a circle, construct a square of the same area as the circle.

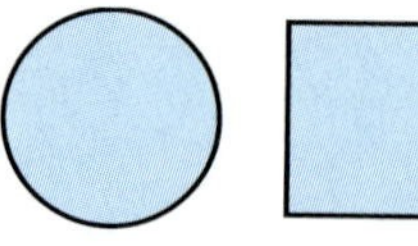

Extensive efforts for over two thousand years failed to solve any of these problems. It was not until the early 1800s that it was shown that these problems were impossible to solve when only the straightedge and compass were allowed. It is interesting to note that methods of algebra were used to prove the impossibility of these geometric construction problems.

*The usage "the compass" is common, but some texts and authors still prefer "compasses" or even "a pair of compasses."

draw lines and points very dark, so they can be seen from the reverse side of the paper. A folding construction of the perpendicular bisector is shown in Figure 14.14.

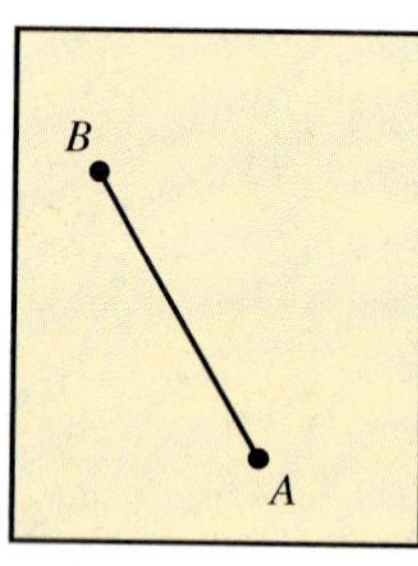

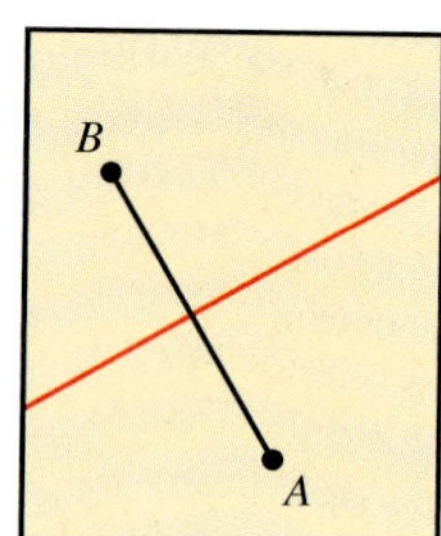

Figure 14.14
Folding point A onto point B forms a crease on the perpendicular bisector of $\overline{AB}$

Constructions with Geometry Software

Constructions made with tangible materials—paper, dowels, rubber bands, and so on—continue to have an important role to play in teaching and learning the principles of geometry. However, increasingly many classrooms are also taking advantage of the computer. Geometry software, as described in Appendix D, allows figures to be constructed on the

Exploring Quadrilaterals and Circles

Every triangle has both an inscribed circle and a circumscribed circle, but most quadrilaterals have neither an inscribed nor a circumscribed circle. Let's investigate how to construct some quadrilaterals that have at least one of these special circles. Work in pairs to discuss and compare your discoveries.

Materials Needed

Spaghetti noodles (uncooked), 3″ × 5″ cards, scissors, drawing and measuring tools (ruler, compass, Mira, and so on)

Exploring Spaghetti Quadrilaterals

Break each of two equally long spaghetti noodles into two pieces. Form a quadrilateral with the four lengths of spaghetti, with opposite sides coming from the same noodle. Put points at *A*, *B*, *C*, and *D*, and then remove the noodles and draw the quadrilateral *ABCD*.

1. Assuming that the quadrilateral *ABCD* has an inscribed circle, discuss how you can construct its center and radius. Carry out your procedure: does it appear that *ABCD* has an inscribed circle?
2. Make another quadrilateral of a different shape, but using the same pieces of spaghetti. Does it have an inscribed circle?

Exploring Note Card Quadrilaterals

Cut a 3″ × 5″ note card diagonally into two pieces. Arrange the pieces as shown and use your ruler to extend the sides to form a quadrilateral *APBQ*.

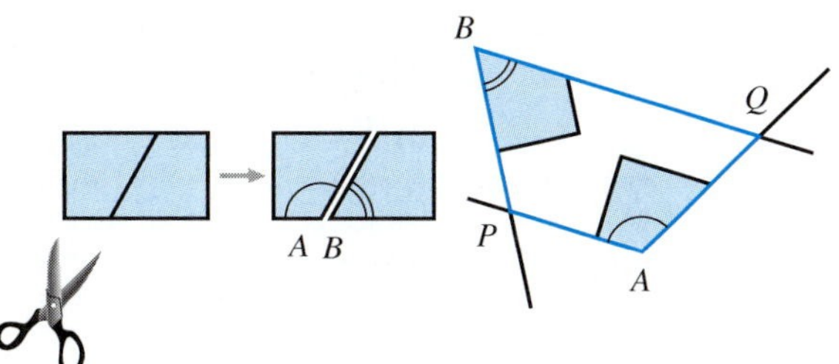

3. Assuming that the quadrilateral *APBQ* has a circumscribed circle, discuss how you can construct its center. Carry out your procedure. Does it appear that there is a circle through all the vertices of *APBQ*?
4. Rearrange the two note card pieces to form another quadrilateral of a different shape. Does it have a circumscribed circle?

A Further Investigation

5. Draw a quadrilateral that has both an inscribed and a circumscribed circle. Describe your method, and draw the two circles.

screen, colored, manipulated, explored with measurement tools, and printed out for display and further investigation. The spirally tiled regular hexagon shown in Figure 14.15 is tedious to construct with compass and straightedge, but enjoyable to create with The Geometer's Sketchpad.

Figure 14.15
A figure constructed with geometry software

from The NCTM Principles and Standards

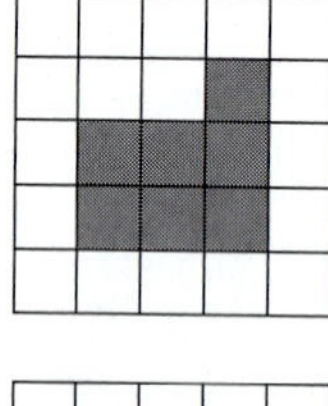

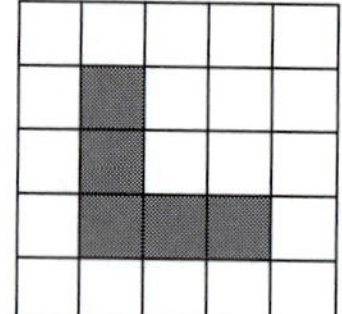

Figure 14.16
Make a building out of 10 cubes by looking at the three pictures of it above.

Students in grades 3–5 should examine the properties of two- and three-dimensional shapes and the relationships among shapes. They should be encouraged to reason about these properties by using spatial relationships. For instance, they might reason about the area of a triangle by visualizing its relationship to a corresponding rectangle or other corresponding parallelogram. In addition to studying physical models of these geometric shapes, they should also develop and use mental images. Students at this age are ready to mentally manipulate shapes, and they can benefit from experiences that challenge them and that can also be verified physically. For example, "Draw a star in the upper right-hand corner of a piece of paper. If you flip the paper horizontally and then turn it 180°, where will the star be?"

Much of the work students do with three-dimensional shapes involves visualization. By representing three-dimensional shapes in two dimensions and constructing three-dimensional shapes from two-dimensional representations, students learn about the characteristics of shapes.

Students should become experienced in using a variety of representations for three-dimensional shapes, for example, making a freehand drawing of a cylinder or cone or constructing a building out of cubes from a set of views (i.e., front, top, and side) like those shown in Figure 14.16 at left.

SOURCE: Reprinted with permission from Principles and Standards for School Mathematics, *pages 168–169.*

Problem Set 14.2

Understanding Concepts

1. The sequence of steps shown below outlines the "corresponding angles" construction to draw a line k that is parallel to a line l and passes through a point P not on l.

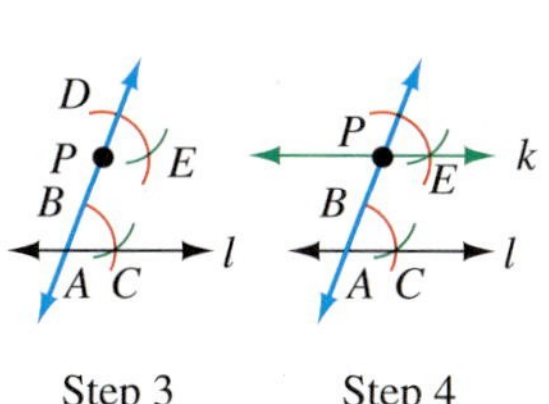

(a) Give a written description of each of the four steps.

(b) Explain why the construction gives the desired line k.

2. (a) Describe, in words and drawings, a Mira construction that gives the line k parallel to a line l and through a point P not on l.

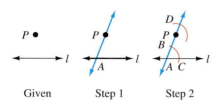

(b) Answer part (a), but use paper folding instead of the Mira.

3. The drafting triangle and straightedge can be used to construct parallel lines, as shown below.

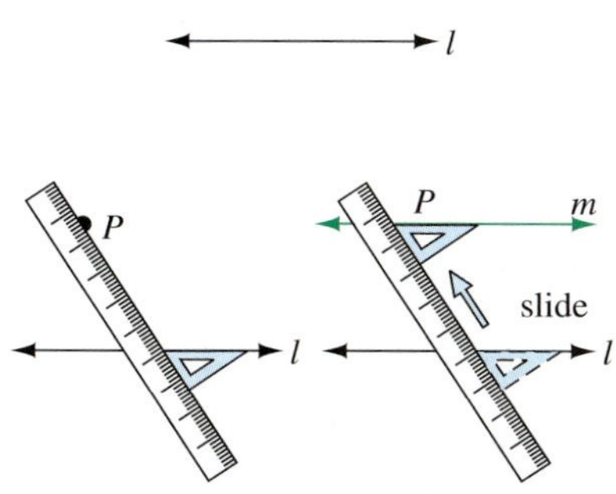

(a) What geometric principle justifies this construction?

(b) Describe, in words and drawings, a procedure using a straightedge and drafting triangle to construct the line m perpendicular to a given line l and passing through a given point P.

4. Use compass and straightedge to construct the following:

(a) Line perpendicular to l through P

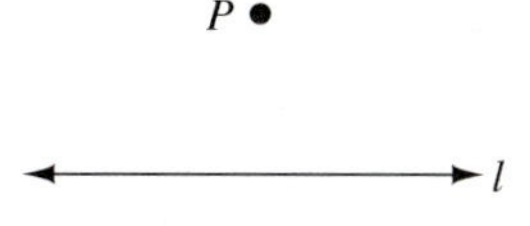

(b) Line perpendicular to l through Q

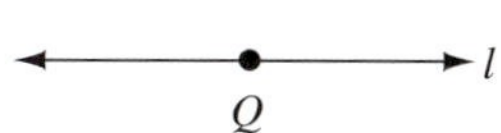

(c) Perpendicular bisector to $\overline{ST}$

(d) Bisector of $\angle A$

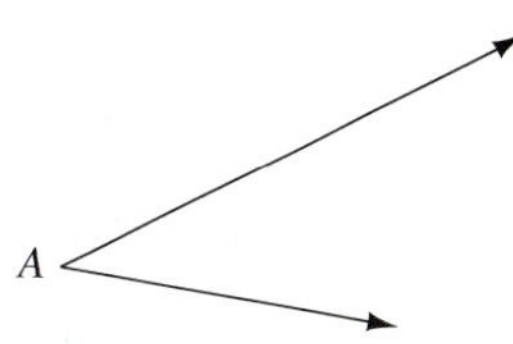

5. Repeat the constructions in problem 4 with a Mira (if available).

6. Repeat the constructions in problem 4 with paper folding.

7. Use a jar lid (or other handy circular object) to draw any three congruent circles through a common point P. Let A, B, and C denote the three points of intersection of pairs of your circles other than P. Now use the jar lid to discover something amazing about the circumscribing circle of $\triangle ABC$.

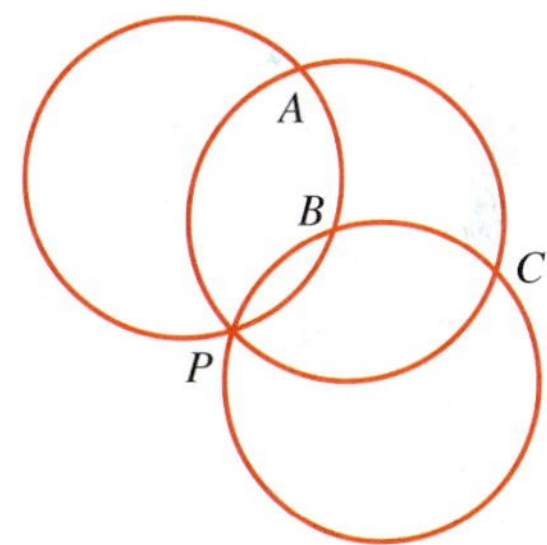

8. Construct an angle ABC and a point T on side $\overrightarrow{BA}$. Show, in words and drawings, how to construct a circle that is tangent to both sides of the angle, with T as one of the points of tangency.

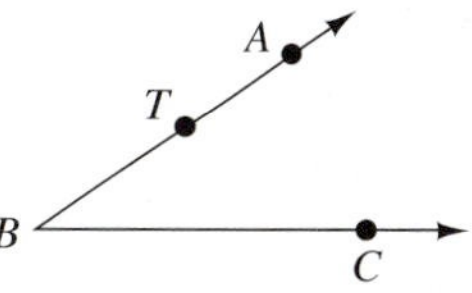

9. Construct the circumcenters and circumscribing circles of triangles of the three types shown. Use any tools you

wish to draw the perpendicular bisectors of the sides of the triangles.

(a) Acute triangle

(b) Right triangle

(c) Obtuse triangle

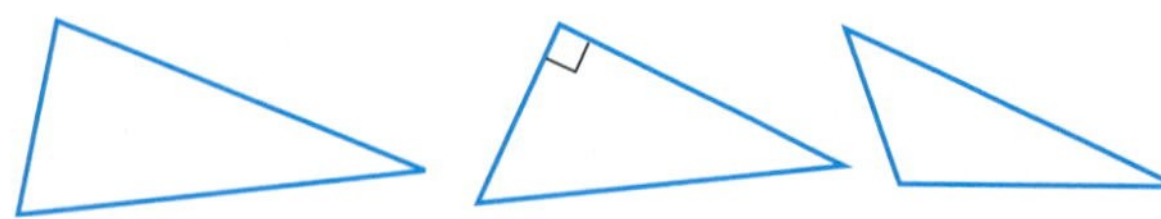

(d) Make a conjecture about where—inside, on, or outside the triangle—the circumcenter of a triangle will be located.

10. Construct the incenters and inscribed circles of the three types of triangles shown in problem 7, using

(a) a compass and straightedge.

(b) a Mira (if available) and compass.

(c) paper folding (copy and cut the triangles from paper with scissors and then fold) and compass.

11. A point P is exterior to a circle centered at Q. Use Thales' theorem to justify why drawing the circle with diameter $\overline{PQ}$ gives a construction of the two lines $\overleftrightarrow{PS}$ and $\overleftrightarrow{PT}$ that are tangent to the circle centered at Q.

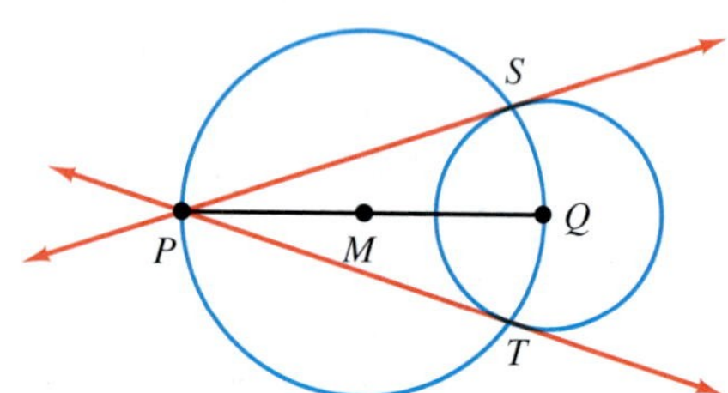

12. Draw any triangle ABC. Using any tools you wish, draw both the inscribed and the circumscribed circles of your triangle. Now choose an arbitrary point X on the larger circle and draw the chords $\overline{XY}$ and $\overline{XZ}$ that are tangent to the smaller circle. Finally, draw chord $\overline{YZ}$ of the larger circle. Start with a new point X' and again draw a triangle $X'Y'Z'$ (colored pencils are helpful). Make a conjecture about all such triangles.

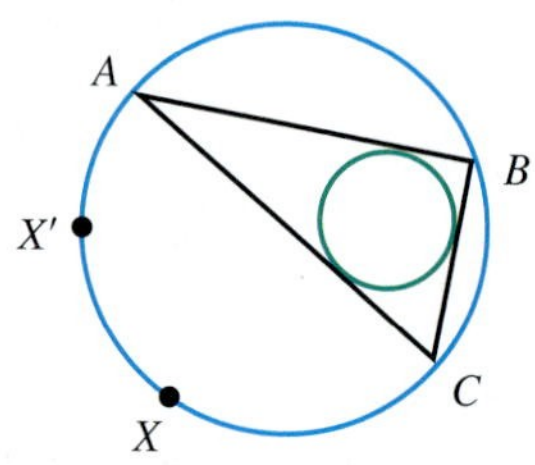

13. (a) Use any drawing tools you wish to construct the altitudes of the three types of triangles in problem 9.

(b) Make a conjecture about the three altitudes of an acute triangle.

(c) Make a conjecture about the three altitudes of a right triangle.

(d) Make a conjecture about the three lines containing the altitudes of an obtuse triangle.

14. Justify the following construction of an equilateral triangle inscribed in a given circle.

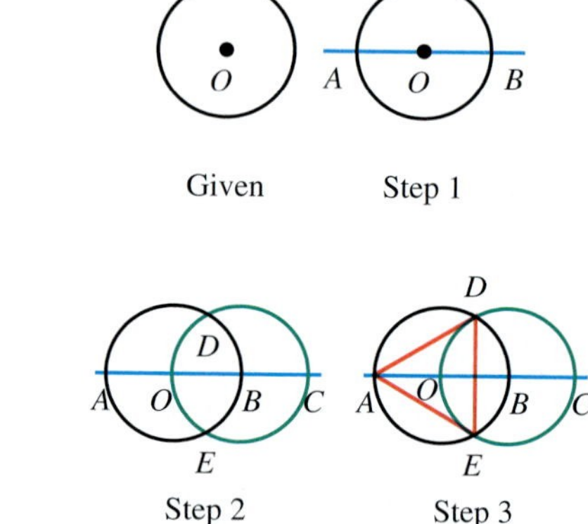

15. A **median** of a triangle is a line segment from a vertex to the midpoint of the opposite side. Using any drawing tools you wish, draw the three medians in each of several triangles. Give a statement that describes what you observe.

16. Let the lines k and l intersect to form two pairs of vertical angles. Prove that the lines m and n that bisect the vertical angles are perpendicular.

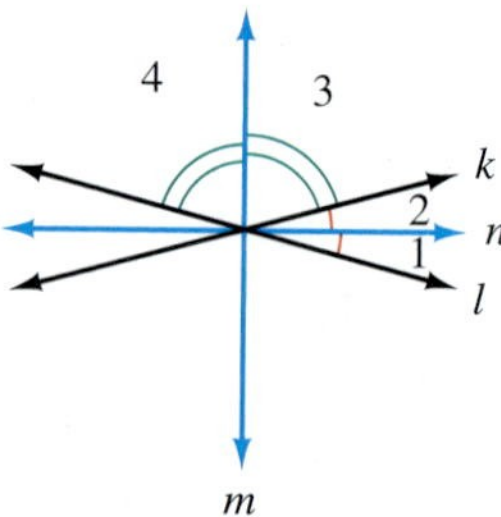

17. (a) Prove that the perpendicular bisector of any chord of a circle contains the center of the circle.

(b) Trace partway around a cup bottom to draw a circular arc. Then explain how to construct the center of the arc. (*Hint:* Use part (a), twice!)

(c) The three congruent circles below are centered at points A, B, and C on the large circular arc, and the circles at centers A and C contain point B. Where do the dashed lines intersect? Explain why.

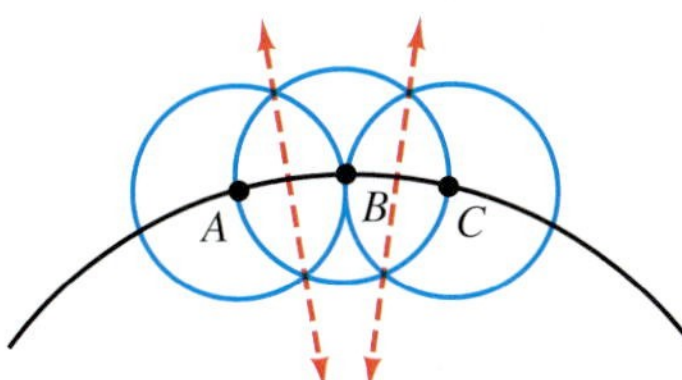

18. Prove that the angle bisectors of a triangle are concurrent, using the equidistance property of the angle bisector. The discussion in Example 14.8 can be used as a model for your proof.

19. The diagram shows how to erect an equilateral triangle ABC on a given line segment $\overline{AB}$, using a compass and straightedge.

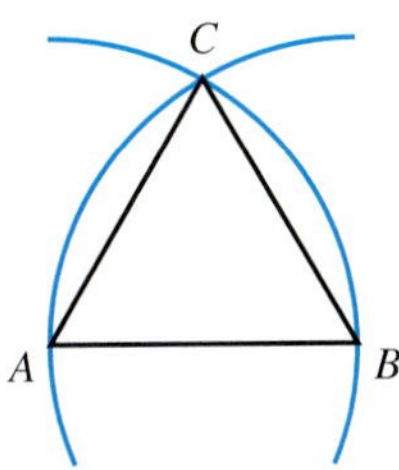

Give careful, step-by-step instructions to construct these polygons erected on a given side $\overline{AB}$.

(a) a square **(b)** a regular hexagon

20. Reflecting point B to fall on the perpendicular bisector of $\overline{AB}$ shows how a Mira can be used to draw an equilateral triangle ABC on a given side $\overline{AB}$.

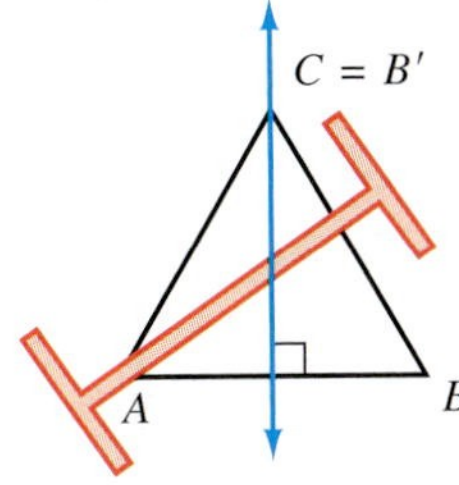

Give careful step-by-step instructions to construct these polygons on a given side with a Mira.

(a) a square **(b)** a regular hexagon

21. **(a)** Construct a regular pentagon inscribed in a circle by following the steps outlined.

(b) Use a ruler and protractor to check that $PENTA$ is regular.

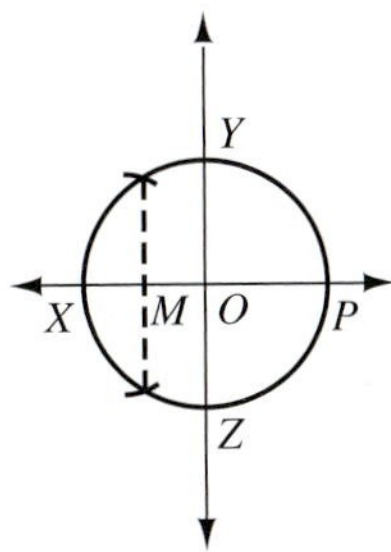

Construct perpendicular diameters, and construct midpoint M of radius $\overline{OX}$.

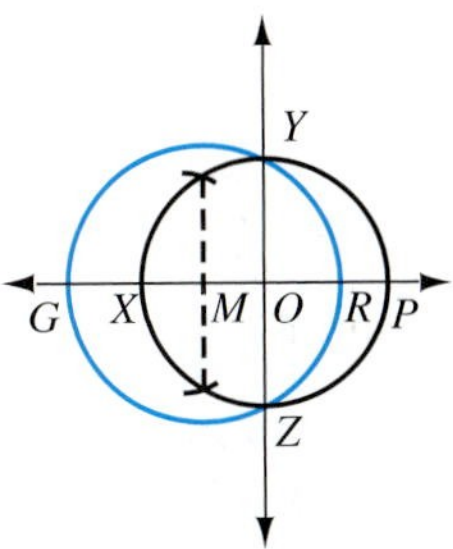

Construct circle at M through Y to locate G and R.

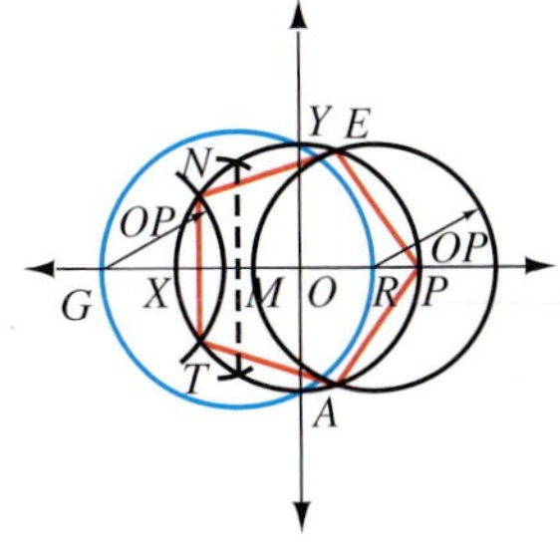

Construct circles at G and R of radius OP to locate the vertices of the regular pentagon $PENTA$.

22. **(a)** Construct a heptagon inscribed in a given circle by following the steps outlined below.

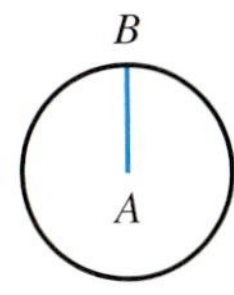

Construct a radius $\overline{AB}$.

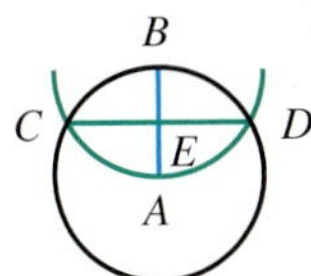

Construct an arc at B through A and th draw chord $\overline{CD}$.

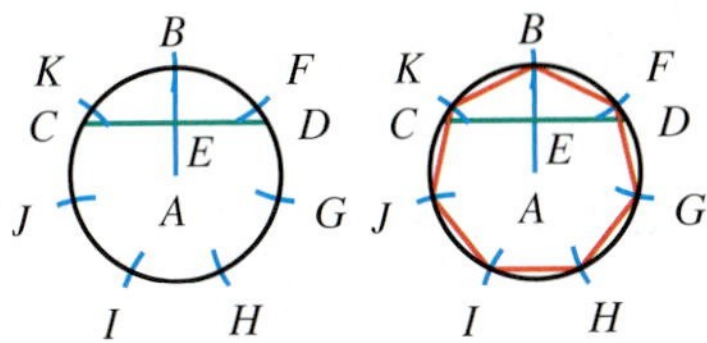

Lay off arcs of radius DE starting at B.

Construct the heptagon $BFGHIJK$.

(b) Is it possible for $BFGHIJK$ to be a regular heptagon, or is it just a close approximation?

23. List the constructible regular n-gons up to $n = 100$, using the Gauss–Wantzel theorem.

Teaching Concepts

24. Kerry has created a new construction for the line through point P perpendicular to line k, where P is not on k. As shown in the given figure, construct two circles through P centered at two distinct points A and B on k. If Q is the second point of intersection of the circles, then $\overleftrightarrow{PQ}$ is the required perpendicular line. Write a response to Kerry. In particular, if the construction is incorrect, explain why. If it is correct, guide Kerry through a proof that justifies the construction.

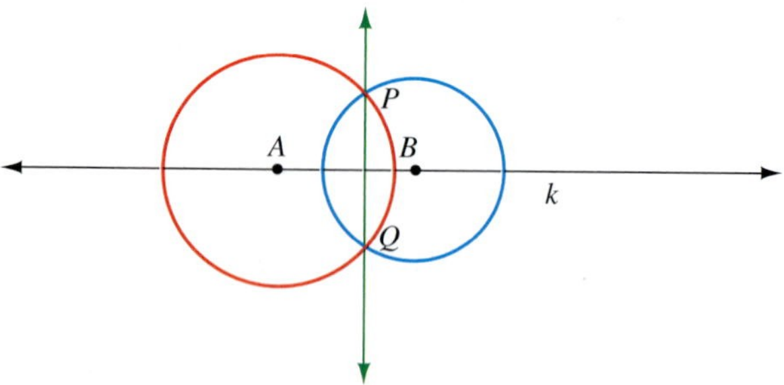

25. Someone claims that trisecting the chord $\overline{BC}$ of an arc centered at A gives a compass-and-straightedge trisection of $\angle A$. How would you respond to this assertion? How well does the method appear to work on an angle of measure 150°? Make a drawing and use a protractor to measure the angles.

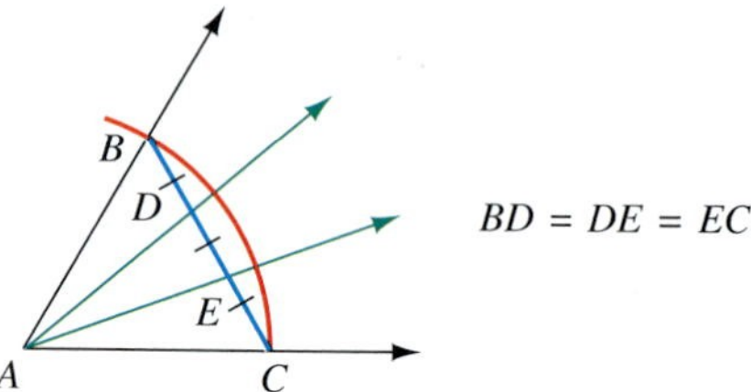

$BD = DE = EC$

26. Leanne believes she has constructed a triangle with two right angles. Her construction is shown below. Draw two circles that intersect at P and Q and two diameters $\overline{QR}$ and $\overline{QS}$. Let $\overline{RS}$ intersect the circles at A and B. Then $\angle QBR$ and $\angle QAS$ are both right angles by Thales' theorem. Thus, $\triangle QAB$ has two right angles. How would you respond to Leanne's assertion?

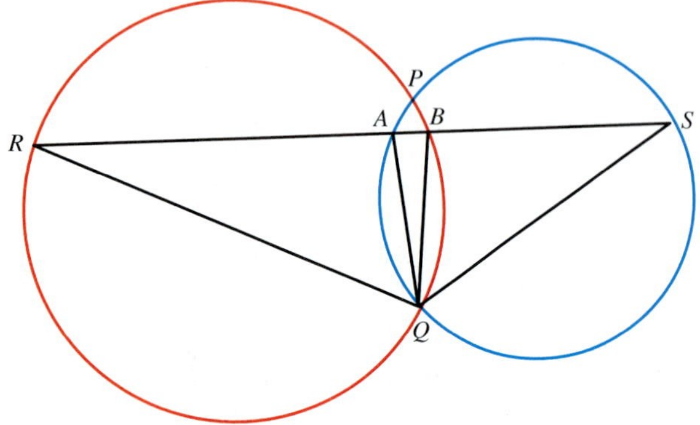

Thinking Critically

27. Trace the segment $\overline{AB}$ and line l shown below.

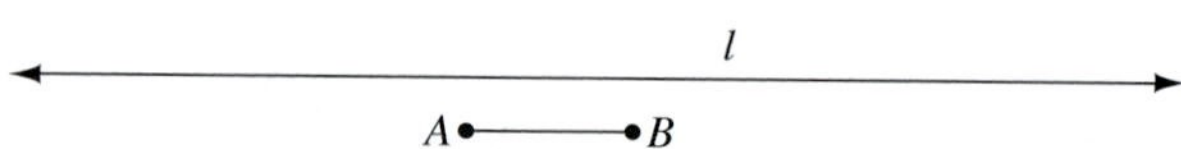

(a) Construct *all* points C on line l for which triangle ABC is isosceles.

(b) Construct *all* points D on line l for which triangle ABD is a right triangle.

28. Complete the proof of the equidistance property of the perpendicular bisector. Do so by showing that if P is equidistant from A and B, then the line $\overleftrightarrow{PM}$ containing P and the midpoint M of $\overline{AB}$ is the perpendicular bisector of $\overline{AB}$.

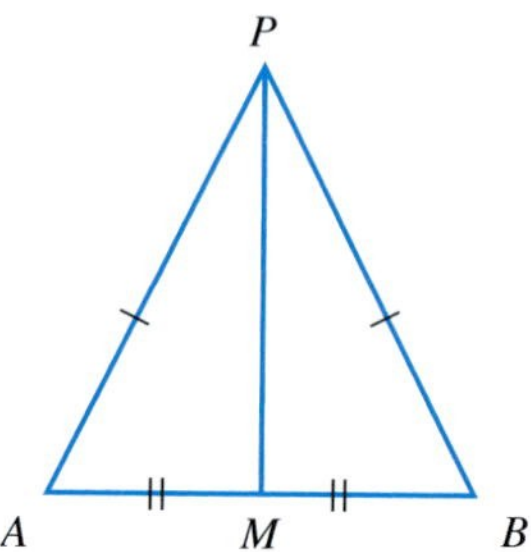

29. An **altitude** of a triangle is a line through a vertex of the triangle that is perpendicular to the line containing the opposite side of the triangle. The altitude through vertex A has been constructed in this figure:

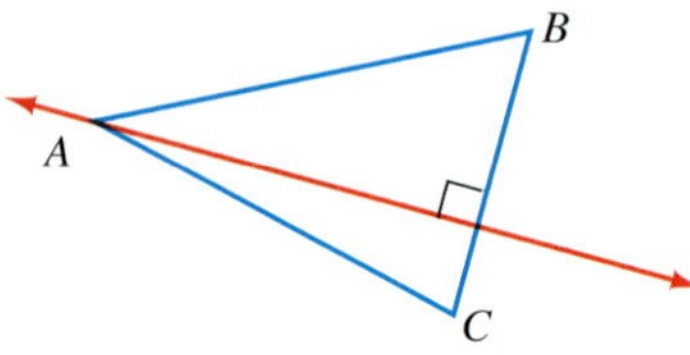

(a) Construct any triangle ABC and all three of its altitudes. What property do you discover about the altitudes of a triangle?

(b) Append three congruent copies of $\triangle ABC$ to form $\triangle PQR$, as shown. It becomes apparent that the altitudes of $\triangle ABC$ are simultaneously the perpendicular bisectors of the sides of $\triangle PQR$. Use this fact to explain why the altitudes of any triangle are concurrent. The common point of intersection of altitudes is called the **orthocenter** of the triangle.

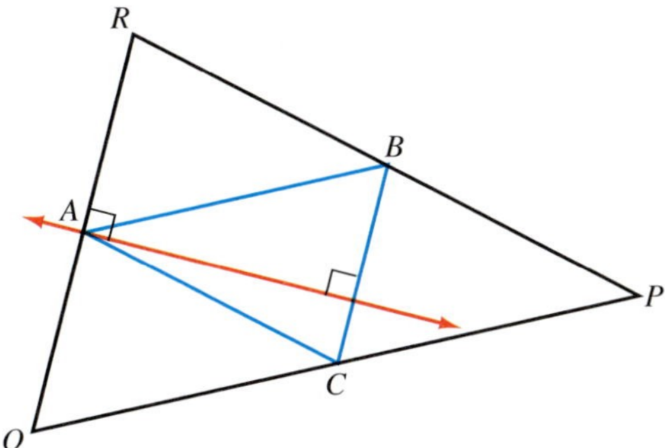

30. Draw any triangle ABC. Using any tools you wish, draw its orthocenter (that is, the point of concurrence of the three altitudes of $\triangle ABC$; see problem 29). Label the orthocenter D. Now trace the four points A, B, C, and D on a clean sheet of paper, and draw the orthocenter of $\triangle BCD$. Are you surprised? Guess, and then check, what the orthocenters of $\triangle ACD$ and $\triangle ABD$ are.

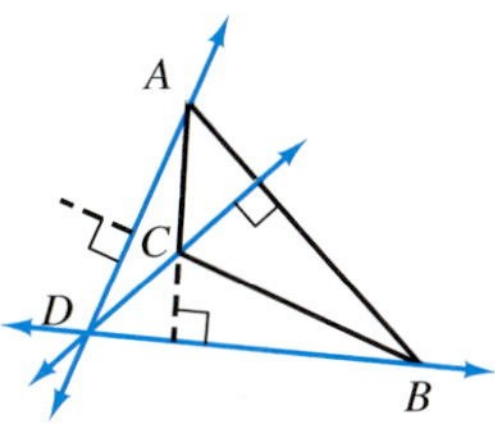

31. A regular pentagon $PENTA$ and an equilateral triangle PQR are both inscribed in the same circle centered at O, with P a common vertex. Calculate $m(\angle NOQ)$ and explain why laying segments off of length QN about the circle constructs a regular inscribed 15-gon.

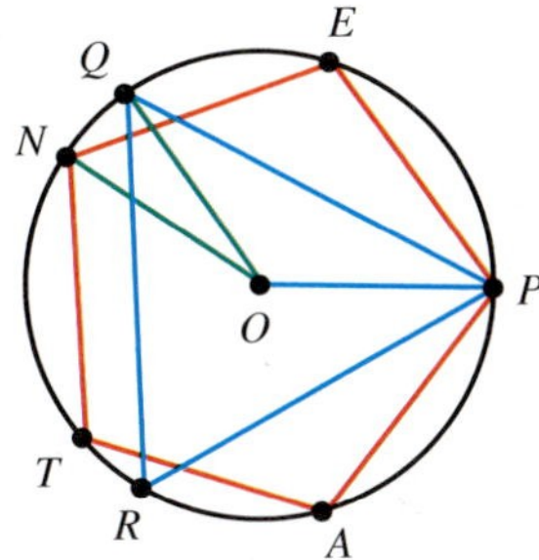

Thinking Cooperatively

The following paper-folding constructions are enjoyable group activities. Each collaborative group needs several sheets of blank paper (origami squares or patty paper work well), scissors, ruler, protractor, and colored pencils.

32. Folding an Angle Trisection. Construct an angle ABC, using the lower edge of a piece of rectangular notebook paper as the side $\overline{AB}$. Fold the lower edge to create two equally spaced folds parallel to $\overline{AB}$, and let D and E denote the fold points along the left edge of the paper. Now fold the lower corner B so that it lies along the lower horizontal fold and, at the same time, point E' lies along the side $\overrightarrow{BC}$ of $\angle ABC$. Mark the points B' and D' as shown. Finally, unfold the paper and construct the two rays from B that pass through B' and D'. Measure $\angle ABC$, $\angle ABB'$, $\angle B'BD'$, and $\angle D'BC$. Your group should discover that the smaller angles trisect $\angle ABC$. Report your measurements and compare the success of your folded angle trisection with other groups.

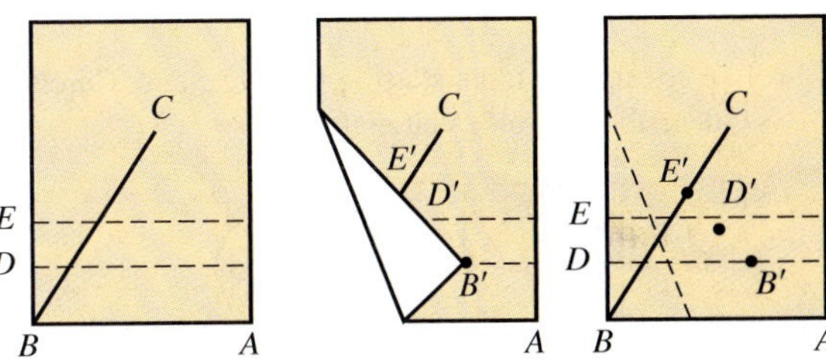

33. Folding Special Points in a Triangle. Cut out several paper triangles with scissors. Discuss procedures to use folding to construct the following points and segments.

(a) The midpoints of the sides of the triangle.

(b) The medians (the segments connecting a vertex to the midpoint of the opposite side). What property do you observe is satisfied by the three medians?

(c) The angle bisectors of a triangle. What special point of the triangle is constructed with the folds?

(d) The perpendicular bisectors of the sides of an acute triangle. What special point is constructed with these folds?

(e) The altitudes (an altitude is a line through a vertex that is perpendicular to the opposite side) of an acute triangle. What special point is constructed with these folds?

34. Folding Regular Polygons. A square can be folded from a rectangular sheet of paper by following the procedures illustrated in this sequence of steps.

Discuss procedures and give demonstrations to construct these regular polygons with paper folding.

(a) An equilateral triangle. Carefully describe the sequence of folds you make.

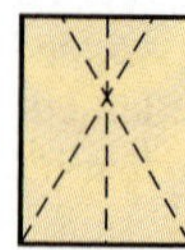
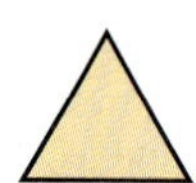

(b) A regular hexagon, starting with the equilateral triangle cut from your construction in part (a). (*Suggestion:* First, use folding to construct the center of the equilateral triangle.)

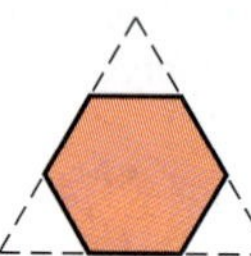

(c) A regular octagon, starting with a square. Carefully describe the folds you make.

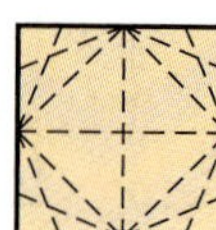
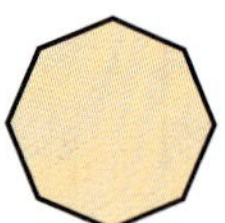

Using a Calculator

35. Verify that $F_3 = 2^{2^3} + 1$ and $F_4 = 2^{2^4} + 1$ are given decimally by 257 and 65,537.

36. **(a)** Verify that the Fermat number $F_5 = 2^{2^5} + 1$ is 4,294,967,297.

(b) Verify that the Fermat number $F_5 = 4{,}294{,}967{,}297$ is composite by computing $(641)(6{,}700{,}417)$.

Using a Computer

37. Draw any triangle ABC. Construct the following three points:
G, the **centroid** (intersection of the medians; see problem 15);
H, the orthocenter (intersection of the altitudes; see problem 29); and
P, the circumcenter (intersection of the perpendicular bisectors of the sides).

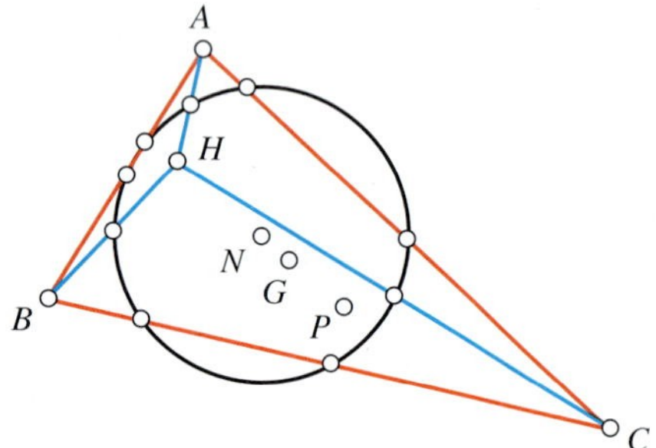

(a) Euler discovered an interesting fact about G, H, and P. What do you suppose the "Euler line" might be?

(b) Measure the distances GH and GP, and then make a conjecture concerning the ratio GH/GP of these distances.

(c) Find the midpoint N of the segment $\overline{PH}$, and draw the circle centered at N that passes through the midpoint of a side of your triangle. Where does it intersect the other sides of the triangle?

(d) Describe how the circle at N intersects the segments $\overline{HA}$, $\overline{HB}$, and $\overline{HC}$.

38. Draw any triangle ABC. On each side, construct outward-pointing equilateral triangles BCR, CAS, and ABT. Also construct the incenters X, Y, and Z of the equilateral triangles.

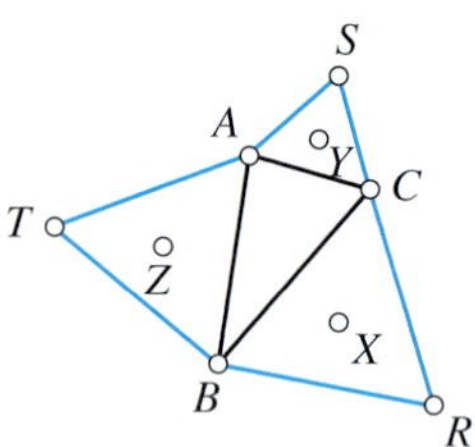

(a) What kind of triangle is XYZ? Measure lengths and angles to support your guess.

(b) Construct the segments $\overline{AR}$, $\overline{BS}$, and $\overline{CT}$, showing they are concurrent at a point F. (F is called the **Fermat point,** and in an acute triangle ABC it is the point with the smallest sum $FA + FB + FC$ of distances to the vertices of $\triangle ABC$).

(c) At what angles do the lines drawn in part (b) intersect at F? Measure to verify your guess.

(d) Draw the segments $\overline{AX}$, $\overline{BY}$, and $\overline{CZ}$, showing they are concurrent at a point N. (N is called the **Napoleon point;** supposedly it was Napoleon who discovered the theorem that you likely discovered in answering part (a).)

(e) Construct the circumcenter P of $\triangle ABC$. What can you conjecture about the three points F, N, and P?

39. Construct three equilateral triangles that share a common vertex A. Let the triangles, labeled counterclockwise around their respective interiors, be $\triangle ABC$, $\triangle AB'C'$, and $\triangle AB''C''$. Draw the midpoint T of $\overline{BC''}$, as shown. Similarly, draw the midpoint R of $\overline{B'C}$ and the midpoint I of $\overline{C'B''}$. Finally, draw the triangle TRI. What kind of a triangle does TRI seem to be? Measure TRI to check your conjecture.

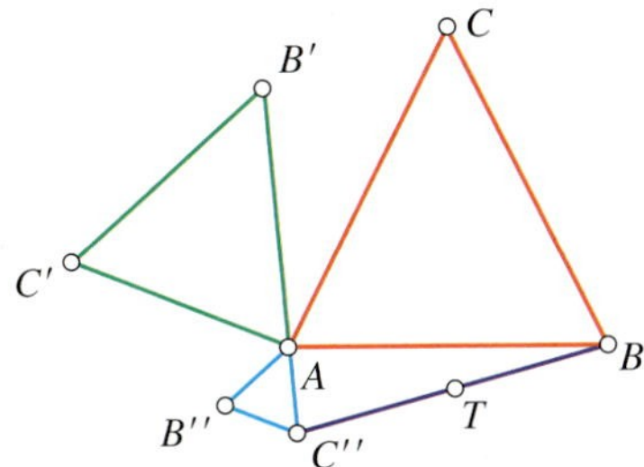

40. Construct a circle and the radii to four points A, B, C, and D on the circle. Then construct tangent lines to the circle at A, B, C, and D. Let Q, R, S, and T denote the points at which successive pairs of tangent lines intersect, to give a quadrilateral $QRST$ that is circumscribed about the circle.

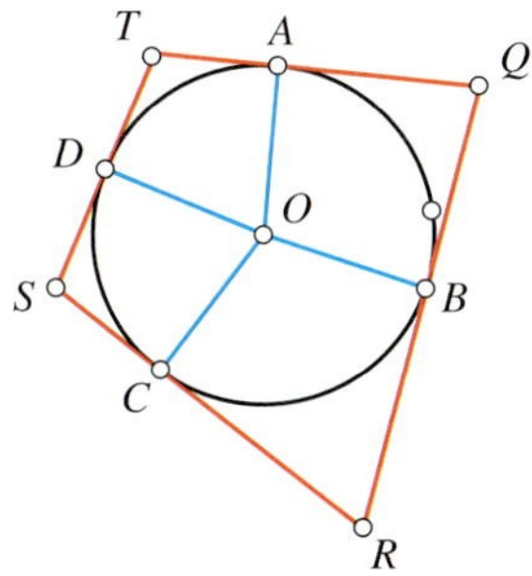

(a) Investigate how the sums of opposite lengths of the sides of $QRST$ compare. Make a conjecture about $QR + ST$ and $RS + TQ$.

(b) Use your geometry software to check that $QA = QB$, $RB = RC$, $SC = SD$, and $TD = TA$. Now prove your conjecture in part (a), using these equations.

From State Student Assessments

41. (Massachusetts, Grade 4)
I have four sides. Two of my sides are parallel. My other two sides are not parallel. Draw me.

For Review

42. Consider the two triangles shown below.

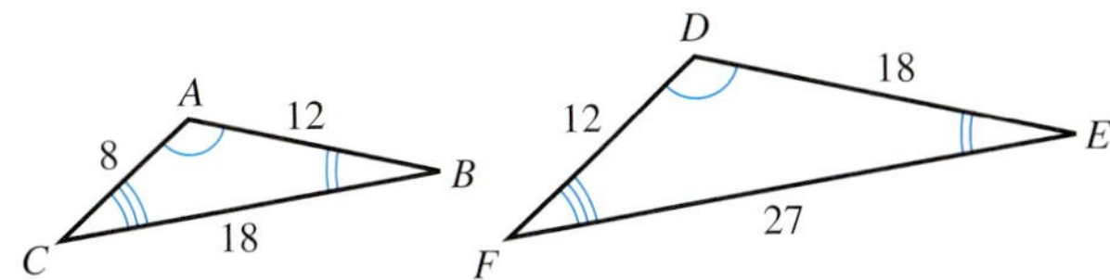

Are the following assertions *true* or *false?* Explain your answer.

(a) Five parts of $\triangle ABC$ are congruent to five parts of $\triangle DEF$.

(b) $\triangle ABC$ is congruent to $\triangle DEF$.

43. A Girl Scout troop needed to determine the width of a river. Alicia followed this plan: first she paced off equally spaced markers at A, B, and C, where A is across the river from a tree at point T. Then Alicia walked 126 feet directly away from the river until she reached point D, at which the tree at T was in line with the stake at B. What is the width of the river, and what geometric property is Alicia relying on?

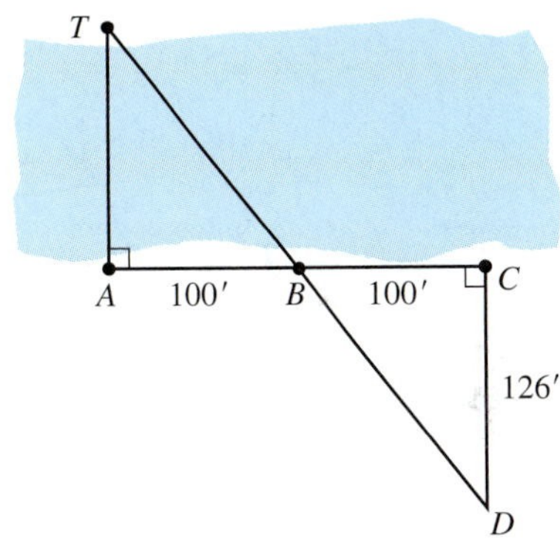

44. Two segments $\overline{AB}$ and $\overline{CD}$ are congruent, parallel, and not on the same line. Prove that the endpoints of the segments are the vertices of a parallelogram.

14.3 SIMILAR TRIANGLES

Two figures are **similar** if they have the same shape, but not necessarily the same size. For example, an overhead projector forms an image on the screen that is similar to the figure on the transparency. The same is true of a photocopy made using the enlarge or reduce setting. Figures in space can also be similar to one another. Design engineers often build small-scale models of buildings, airplanes, or ships and then make tests and measurements on the model to predict whether or not the design objectives will be met in the full-scale structure.

A map or model will indicate how its size compares with the actual size by giving a **scale factor.** For example, a ship model may be scaled at 1 : 100, meaning that two points at a distance x on the model correspond to points on the real ship at a distance $100x$. Conversely, any length on the actual ship divided by 100 gives the corresponding length for the model.

The following definition gives an exact description of similarity for triangles.

DEFINITION *Similar Triangles and the Scale Factor*
Triangle ABC is **similar** to triangle DEF, written $\triangle ABC \sim \triangle DEF$, if, and only if, corresponding angles are congruent and the lengths of corresponding sides have the same ratio. That is, $\triangle ABC \sim \triangle DEF$ if, and only if, $\angle A \cong \angle D$, $\angle B \cong \angle E$, $\angle C \cong \angle F$, and

$$\frac{DE}{AB} = \frac{EF}{BC} = \frac{DF}{AC}.$$

The common ratio of lengths of corresponding sides is called the **scale factor** from $\triangle ABC$ to $\triangle DEF$.

The scale factor from $\triangle DEF$ to $\triangle ABC$ is the reciprocal of the scale factor from $\triangle ABC$ to $\triangle DEF$. For example, if the sides of $\triangle DEF$ are three times the length of the sides of the similar triangle ABC, then the sides of $\triangle ABC$ are one-third the length of the sides of $\triangle DEF$.

Two examples of similar triangles and their scale factors are shown in Figure 14.17.

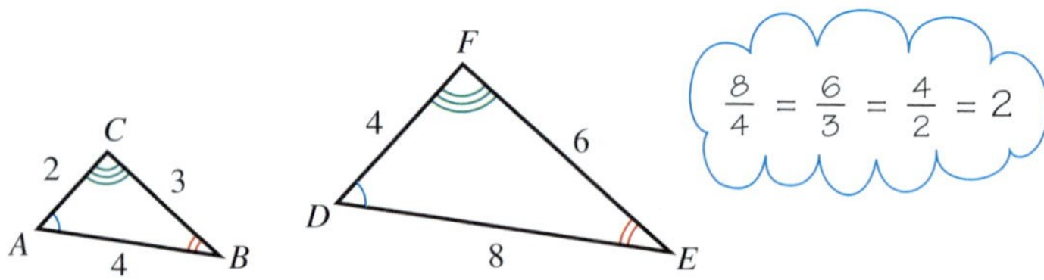

$\triangle ABC \sim \triangle DEF$
Scale factor from $\triangle ABC$ to $\triangle DEF = 2$

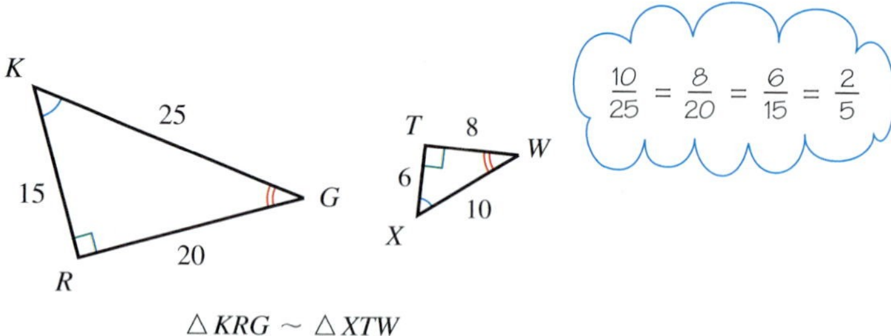

$\triangle KRG \sim \triangle XTW$
Scale factor from $\triangle KRG$ to $\triangle XTW = 2/5$

Figure 14.17
Two pairs of similar triangles

It is possible to conclude that two triangles are similar even when we have incomplete information about the sides and angles of the triangles. The most commonly used criteria for similarity are the angle–angle (AA), side–side–side (SSS), and side–angle–side (SAS) properties.*

*In some books, some, or even all, of these properties are proved on the basis of other assumptions, making the properties theorems. In other texts, the properties are adopted as postulates. In this text, in keeping with an informal treatment of Euclidean geometry, we will refer to the AA, SSS, and SAS criteria for similarity as *properties*.

The Angle–Angle–Angle (AAA) and Angle–Angle (AA) Similarity Properties

In Figure 14.18, $\triangle ABC$ and $\triangle DEF$ have corresponding angles measuring 110°, 40°, and 30°. We see that $\triangle ABC \sim \triangle DEF$. The scale factor can be determined by measuring the lengths of two corresponding sides and forming their ratio. For example, the scale factor is DE/AB.

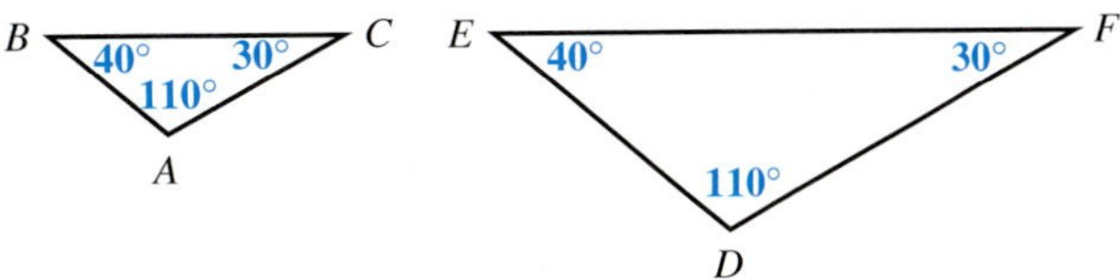

Figure 14.18
Two triangles with three congruent angles are similar by the AAA similarity property

In general, two triangles with three pairs of congruent angles are similar, an observation known as the **angle–angle–angle (AAA) property** of similar triangles. However, all three angles of a triangle are determined once we know two of its angles, since the three measures of the angles add up to 180°. Therefore, the more general property is called the **angle–angle (AA) property of similarity.**

PROPERTY *The AA Similarity Property*
If two angles of one triangle are congruent respectively to two angles of a second triangle, then the triangles are similar.

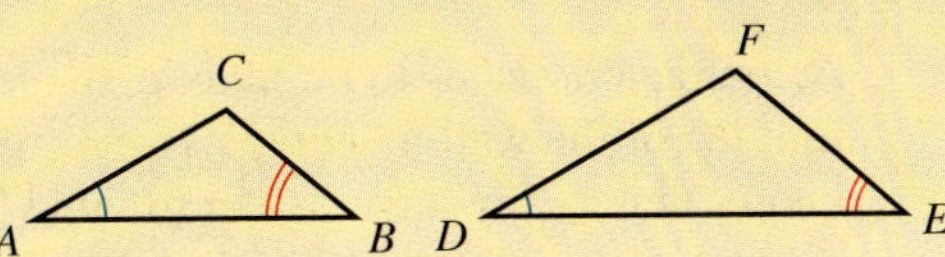

EXAMPLE 14.9 MAKING AN INDIRECT MEASUREMENT WITH SIMILARITY

A tree at point T is in line with a stake at point L when viewed from point N. Use the information in the diagram to measure the distance across the river.

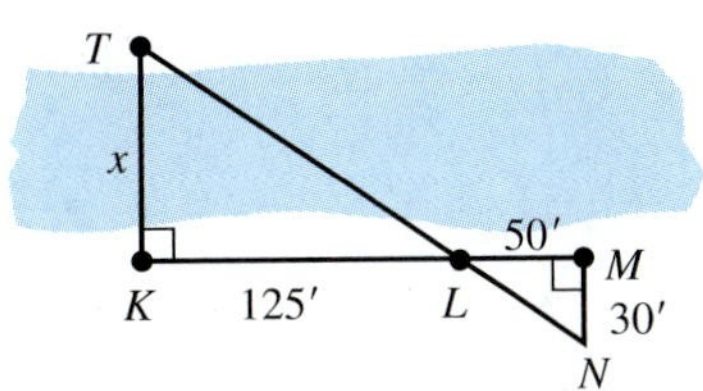

Solution By the vertical-angles theorem, $\angle TLK \cong \angle NLM$. Also, $\angle K \cong \angle M$, since both are right angles. By the AA similarity property, $\triangle TLK \sim \triangle NLM$. Thus, $\frac{TK}{NM} = \frac{KL}{ML}$, since the lengths of corresponding sides have the same ratio. Since $TK = x$, $NM = 30'$, $KL = 125'$, and $ML = 50'$, we have the proportion $\frac{x}{30'} = \frac{125'}{50'}$. Therefore,

$x = \frac{30' \cdot 125'}{50'} = 75'$. We have found the distance across the river by an indirect measurement using similar triangles.

The Side–Side–Side (SSS) Similarity Property

PROPERTY *The SSS Similarity Property*
If the three sides of one triangle are proportional to the three sides of a second triangle, then the triangles are similar. That is, if $\frac{DE}{AB} = \frac{EF}{BC} = \frac{DF}{AC}$, then $\triangle ABC \sim \triangle DEF$.

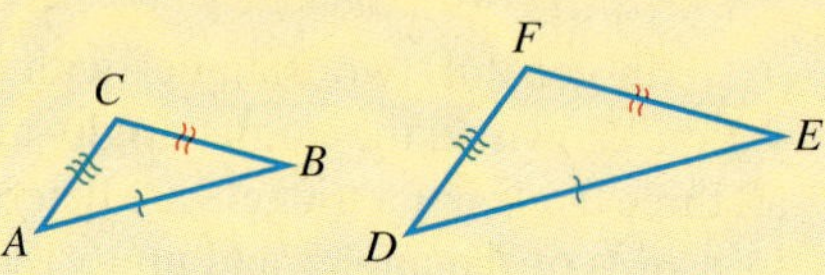

Wavy tick marks are helpful to identify corresponding proportional sides of similar figures, as shown above.

EXAMPLE 14.10 APPLYING THE SSS SIMILARITY PROPERTY

A contractor wishes to build an *L*-shaped concrete footing for a brick wall, with a 12-foot leg of the wall meeting a 10-foot leg of the wall at a right angle. The contractor knows that a 3-by-4-by-5-foot triangle has a right angle opposite the 5-foot side. How can the contractor place stakes at points *X*, *Y*, and *Z* to form a right angle at point *Y*? The wall will be built along string lines stretched from *X* to *Y* and from *Y* to *Z*.

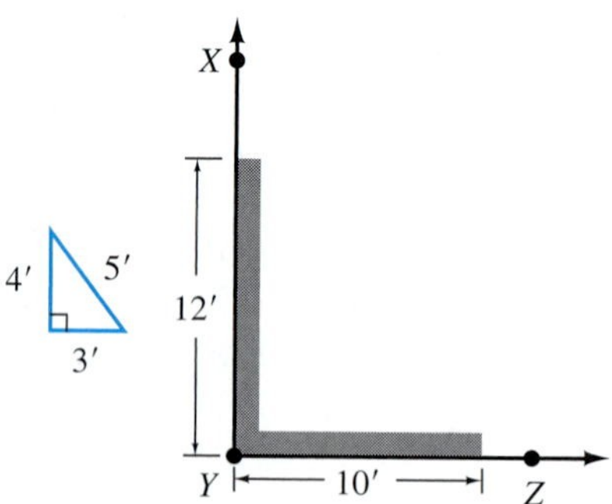

Solution By the SSS similarity property, the 3-4-5-foot right triangle can be magnified by a convenient scale factor to give an accurate right triangle. A scale factor of 4 gives a 12–16–20-foot right triangle. The contractor can place a stake at *X* that is 16 feet from the corner point *Y*. By using two measuring tapes, a stake is placed at the point *Z* where the 20-foot mark on the tape from *X* crosses the 12-foot mark on the tape from *Y*.

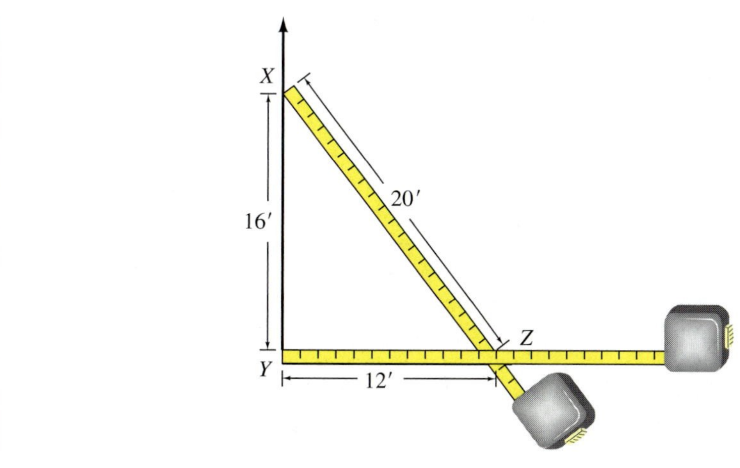

The Side–Angle–Side (SAS) Similarity Property

PROPERTY *The SAS Similarity Property*
If, in two triangles, the ratios of any two pairs of corresponding sides are equal and the included angles are congruent, then the two triangles are similar. That is, if $\frac{DE}{AB} = \frac{DF}{AC}$ and $\angle A \cong \angle D$, then $\triangle ABC \sim \triangle DEF$.

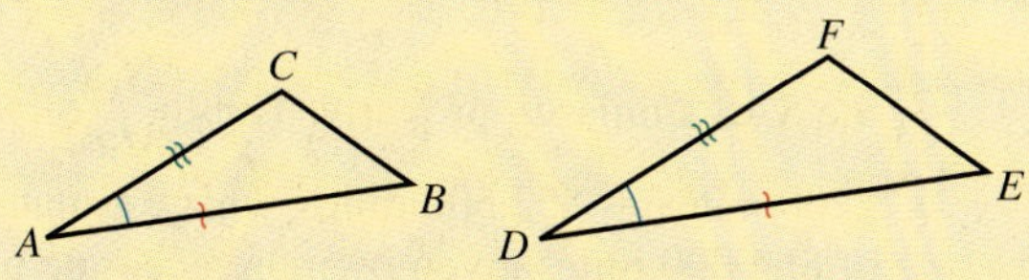

EXAMPLE 14.11 **APPLYING THE SAS SIMILARITY PROPERTY**

The sun is about 93 million miles from the Earth, and the moon is about 240,000 miles distant. If the diameter of the moon is 2200 miles, what is the approximate diameter of the sun? (*Hint:* From the Earth, the sun and moon appear to have the same diameter.)

Solution Because the moon (M) appears to have the same diameter as the sun (S) during an eclipse, they form nearly congruent angles when viewed from the earth (E). This illustration is far from a true scale drawing, but it does show that $\triangle EM_1M_2 \sim \triangle ES_1S_2$ by

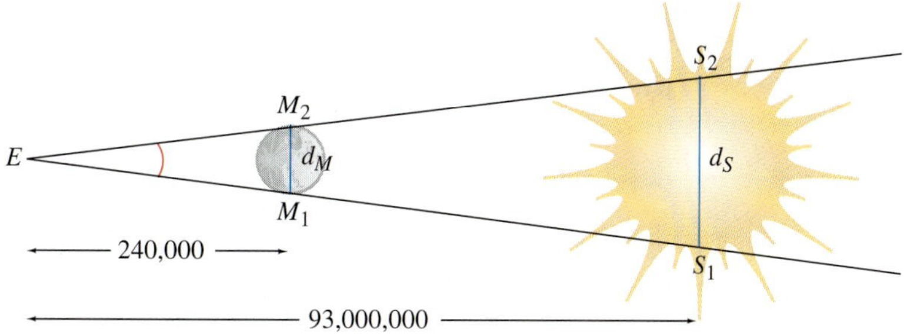

Did You Know?

The Navajo Organization of Space and Time

The Navajo believe in a dynamic universe. Rather than consisting of objects and situations, the universe is made up of processes. Central to our Western mode of thought is the idea that things are separable entities that can be subdivided into smaller discrete units. For us, things that change through time do so by going from one specific state to another specific state. While we believe time to be continuous, we often even break it into discrete units or freeze it and talk about an instant or point in time. Or, we often just ignore time. For example, when we speak of a boundary line dividing a surface into two parts or a line being divided by a point, we are describing a static situation, that is, one in which time plays no role whatever. Among the Navajo, where the focus is on process, change is everpresent; interrelationship and motion are of primary significance. These incorporate and subsume space and time.

. . . To us, the significant aspect of a boundary is that it is a spatial divider; to the Navajo, the significance is the processes of which the boundary is a part and how it affects and is being affected by those processes. The Navajo react quite negatively when fences are placed upon their reservation land. One major reason is their belief that space should not be segmented in an arbitrary and static way.

These excerpts, contrasting Western and Navajo organization of space and time, are from Marcia Ascher's book *Ethnomathematics: A Multicultural View of Mathematical Ideas*. The book gives a fascinating introduction to multicultural mathematical ideas from peoples such as the Inuit, Navajo, and Iroquois of North America; the Incas of South America; the Malekula, Warlpiri, Maori, and Caroline Islanders from Oceania; and the Tshokwe, Bushoong, and Kpelle of Africa.

SOURCE: Marcia Ascher, Ethnomathematics: A Multicultural View of Mathematical Ideas *(Pacific Grove, CA: Brooks/Cole Publishing Company, 1991), pp. 128–129.*

the SAS similarity property. Thus, $\frac{S_1S_2}{M_1M_2} = \frac{ES_1}{EM_1}$, so $S_1S_2 = \frac{ES_1}{EM_1} \cdot M_1M_2 = \frac{93{,}000{,}000}{240{,}000} \cdot$ 2200 miles = 852,500 miles. This estimate compares well with the sun's actual diameter of 864,000 miles given by more accurate methods.

Geometric Problem Solving Using Similar Triangles

The following sequence of examples illustrates how similar triangles can be used to explore the properties of geometric figures. These examples are unified by exploring figures constructed by joining the midpoints of sides of triangles or quadrilaterals.

EXAMPLE 14.12 **EXPLORING THE MEDIAL TRIANGLE**

If X, Y, and Z are the midpoints of the sides of $\triangle ABC$, then $\triangle XYZ$ is called the **medial triangle** of $\triangle ABC$.

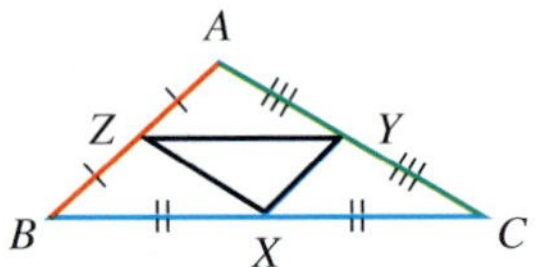

School Book Page

Similarity in Grade Six

Similar Figures

10-7

▶ Lesson Link You've learned about geometric figures. Now you'll explore a class of figures whose dimensions are proportional. ◀

You'll Learn ...

- what similar figures are
- to use proportions to find lengths in a similar figure

... How It's Used

Drivers use similar figures when calculating the actual distance between two locations on a road map.

Vocabulary

similar

Explore Similar Figures

The Same ... But Different!

Materials: Centimeter ruler, Protractor

1. Draw a large triangle, $\triangle A$, with three unequal sides. Measure and label the length of each side and the width of each angle.
2. Draw a smaller triangle whose angles have the same measures as the angles of $\triangle A$. Call the new triangle $\triangle B$. Measure and label the length of each side.
3. Find these ratios:

   a. $\frac{\text{longest side of } \triangle A}{\text{longest side of } \triangle B}$ b. $\frac{\text{mid-length side of } \triangle A}{\text{mid-length side of } \triangle B}$

   c. $\frac{\text{shortest side of } \triangle A}{\text{shortest side of } \triangle B}$
4. Are the longest sides proportional to the shortest sides? Explain.
5. Are the longest sides proportional to the mid-length sides? Explain.
6. Are the mid-length sides proportional to the shortest sides? Explain.

Learn Similar Figures

Recall that figures with the same size and shape are *congruent*. The symbol ≅ means "is congruent to."

Figures that have the same shape but not necessarily the same size are **similar** figures. The symbol ~ means "is similar to."

The objects pictured are similar because their shapes are the same, even though their sizes are different.

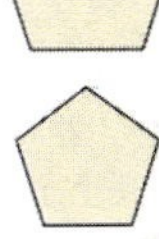

10-7 • Similar Figures **543**

SOURCE: From Scott Foresman–Addison Wesley Middle School Math, *Grade 6, Course 1*, p. 543, by Randall I. Charles et al.

Questions for the Teacher

1. What similarity property is being explored in the sequence of six instructions and questions on the School Book Page?
2. Design a similar sequence of instructions and questions that explores the SAS similarity property.

Show that $\triangle ABC \sim \triangle XYZ$, with a scale factor of $\frac{1}{2}$, and that each side of the medial triangle is parallel to the corresponding side of $\triangle ABC$. For example, $\overline{XY} \parallel \overline{AB}$ and $\frac{XY}{AB} = \frac{1}{2}$.

Solution

From the figure shown above, it is seen that $\frac{CY}{CA} = \frac{CX}{CB} = \frac{1}{2}$. Therefore, by the SAS similarity property, $\triangle ACB \sim \triangle YCX$, with a scale factor of $\frac{1}{2}$. In particular, $\frac{XY}{BA} = \frac{1}{2}$. Moreover, $\angle CAB \cong \angle CYX$, so $\overline{XY} \parallel \overline{AB}$ by the corresponding-angles property. By the same reasoning, the remaining two sides of $\triangle XYZ$ are also parallel and half of the length of the corresponding sides of $\triangle ABC$. Since the sides of $\triangle XYZ$ are half the length of the corresponding sides of $\triangle ABC$, the two triangles are similar by the SSS similarity property.

EXAMPLE 14.13 CLASSIFYING THE MIDPOINT FIGURE OF A QUADRILATERAL

Two quadrilaterals are shown. It appears that joining successive midpoints of the sides of these quadrilaterals forms a parallelogram. Prove that this is indeed the case.

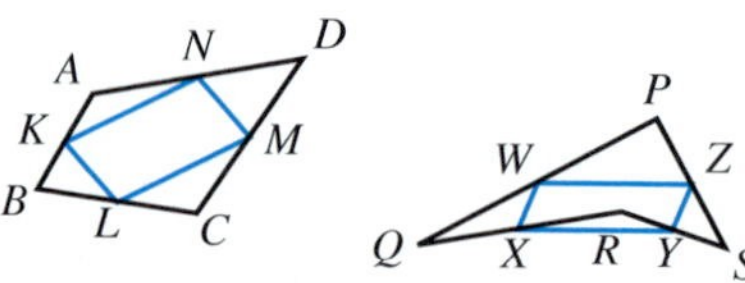

Solution

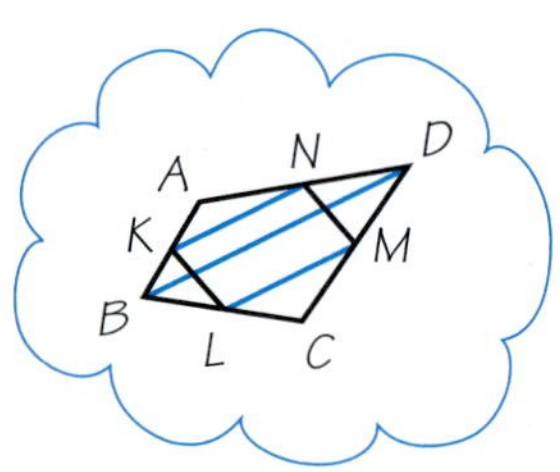

Let $KLMN$ be the midpoint figure of the quadrilateral $ABCD$. Draw the diagonal $\overline{BD}$, and consider $\triangle ABD$ and $\triangle CBD$. By Example 14.12, we know that $\overline{KN}$ and $\overline{LM}$ are both parallel to $\overline{BD}$ and have length $BD/2$. But then $\overline{KN}$ and $\overline{LM}$ are congruent and parallel segments. The same reasoning shows that $\overline{KL}$ and $\overline{NM}$ are congruent and parallel, since both segments are parallel to $\overline{AC}$ and have length $AC/2$. By definition, $KLMN$ is a parallelogram.

The proof just given remains valid for *space* quadrilaterals, for which the four vertices are not necessarily in the same plane. For example, the quadrilateral $PQRS$ shown above may be easily visualized as a nonplanar quadrilateral. However, the midpoint quadrilateral $WXYZ$ is a parallelogram, so it lies in a plane. It's interesting to confirm this result with a quadrilateral made of sticks whose midpoints are joined by elastic bands to form the midpoint parallelogram.

EXAMPLE 14.14 DISCOVERING THE CENTROID OF A TRIANGLE

A **median** of a triangle is a line segment joining a vertex to the midpoint of the opposite side as shown. Prove that the three medians of a triangle are concurrent at a point G that is 2/3 of the distance along each median from the vertex toward the midpoint. G is the **centroid,** or **center of gravity,** of the triangle.

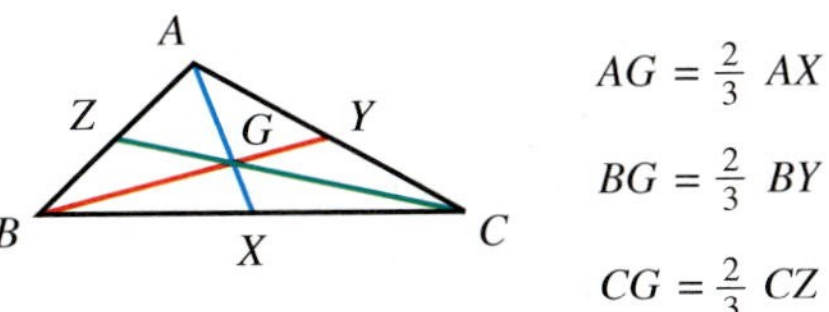

Solution

Understand the Problem

There are really *two* problems to be solved. First, there is a *distance* problem: if G is the point where the two medians $\overline{BY}$ and $\overline{CZ}$ intersect, then we must show that $BG = 2/3\ BY$, or equivalently $BG = 2GY$. Second, there is a *concurrence* problem: if G is the point of intersection of $\overline{BY}$ and $\overline{CZ}$, then we must show that the third median $\overline{AX}$ also passes through G.

Devise a Plan

Since we hope to show that $BG = 2GY$ and $CG = 2GZ$, it may be useful to consider the midpoints M of $\overline{BG}$ and N of $\overline{CG}$. The distance problem for medians $\overline{BY}$ and $\overline{CZ}$ will be solved if it can be shown that M and G trisect $\overline{BY}$, and N and G trisect $\overline{CZ}$. Since Z, M, N, and Y are the successive midpoints of the quadrilateral $ABGC$, we should gain important information by constructing the midpoint figure $ZMNY$, which we know is a parallelogram by Example 14.13.

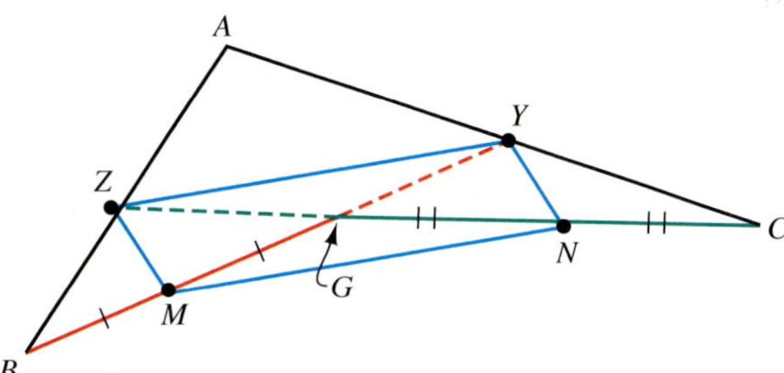

Carry Out the Plan

Because $ZMNY$ is a parallelogram, the point G at which the diagonals intersect is the midpoint of the diagonal $\overline{MY}$ of the parallelogram. Thus, $MG = GY$. But M is the midpoint of $\overline{BG}$, so $BM = MG$. This shows that G is $2/3$ of the distance from B to Y along the median $\overline{BY}$. The same reasoning shows that G and N trisect the median $\overline{CZ}$. The concurrence problem is now solved by symmetry: if G' is the point of intersection of the medians $\overline{BY}$ and $\overline{AX}$, then G' is $2/3$ of the distance from a vertex along either median; therefore, $G = G'$.

Look Back

The problem-solving strategies used in this example are likely to be helpful for other problems:

- *Divide the problem into simpler parts:* we solved a distance problem and a concurrence problem.
- *Consider a simpler problem first:* G was defined as the intersection of *two* medians, and it was to be shown G was $2/3$ of the distances along the two medians from the vertices.
- *Use a related result:* the previous example, showing that the midpoints of sides of any quadrilateral form a parallelogram, was a key idea in the solution.

Cooperative Investigation

Explorations with Centers of Gravity

The **center of gravity,** or **centroid,** of a figure is its point of balance. A plane figure, cut from cardboard, will hang horizontally when it is suspended by a string pinned at the centroid. If pushed slowly over the edge of a table, the figure will begin to teeter just as the centroid reaches the table's edge.

Materials Needed

Heavy cardboard, scissors, ruler, pins, and strings

Directions

1. Cut a large (sides ranging from 6″ to 12″) triangle from cardboard. Hang the triangle from string pinned at some point of the triangle (alternatively, balance the figure on the upright point of a pin). Move the pin to new points to determine the centroid experimentally. Balance the triangle on the edge of a table, and see if the point you found experimentally lies directly over the table's edge.

2. Draw the medians on the triangle, and see if they intersect at the point found experimentally.
3. Cut a large convex quadrilateral from cardboard, and find its centroid experimentally with both the string and table-edge methods.
4. Devise a method to *draw* the centroid of a convex quadrilateral. (*Hint:* Each diagonal of the quadrilateral forms two triangles.)
5. Use the method you've devised to draw the centroids of the cardboard quadrilaterals. Do your points coincide with the experimentally found centroids?

Archimedes made many discoveries about the center of gravity for both plane and solid figures. He was well aware that the medians of a triangle intersect at the centroid.

Problem Set 14.3

Understanding Concepts

1. Which of the following pairs of triangles are similar? If they are similar, explain why, express the similarity with the ~ notation, and give the scale factor.

(a)

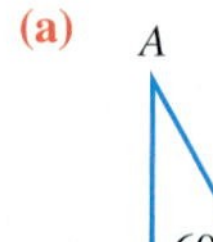

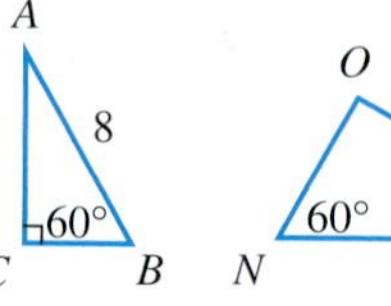

(b)

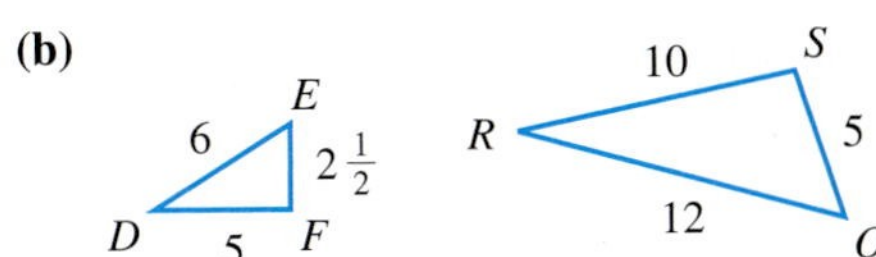

(c)

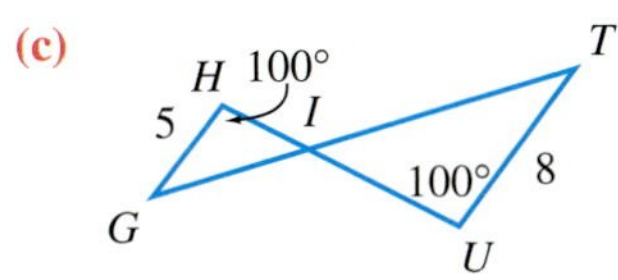

(d)

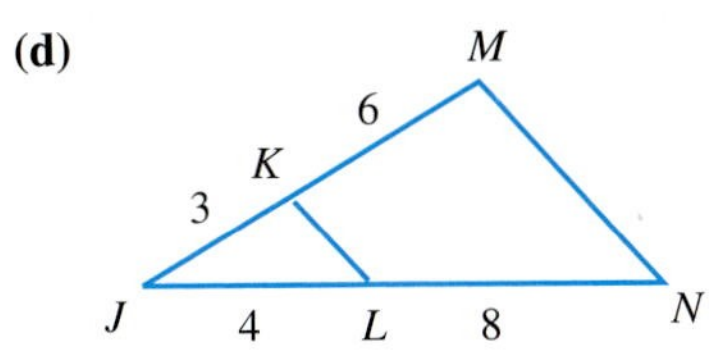

(e)

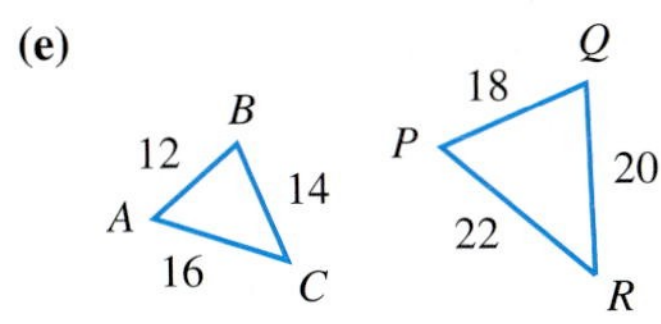

2. Are the following figures necessarily similar? If so, explain why. If not, draw an example to show why not.

- **(a)** Any two equilateral triangles
- **(b)** Any two isosceles triangles
- **(c)** Any two right triangles having an acute angle of measure 36°
- **(d)** Any two isosceles right triangles
- **(e)** Any two congruent triangles
- **(f)** A triangle with sides of lengths 3 and 4 and an angle of 30° and a triangle with sides of lengths 6 and 8 and an angle of 30°

3. A pair of similar triangles is shown in each part. Find the measures of the segments marked with a letter a, b, c, or d.

(a)

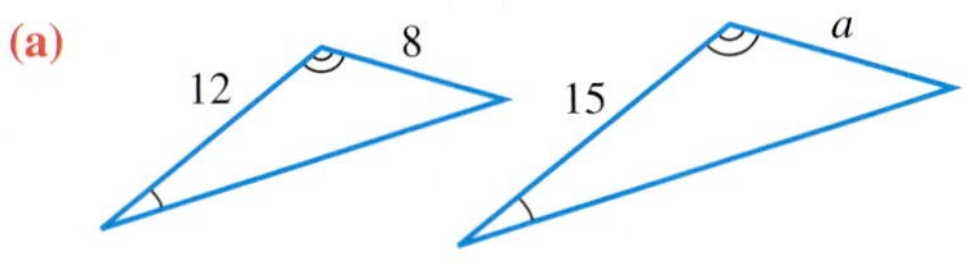

(b)

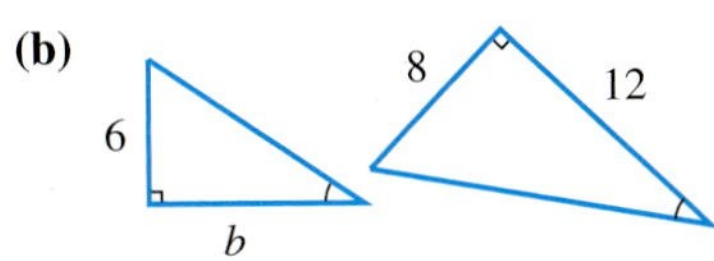

(c)

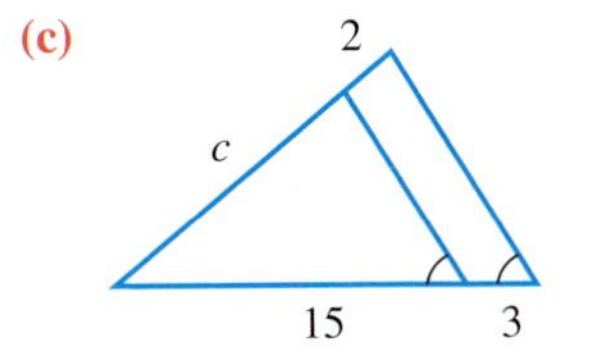

(d)

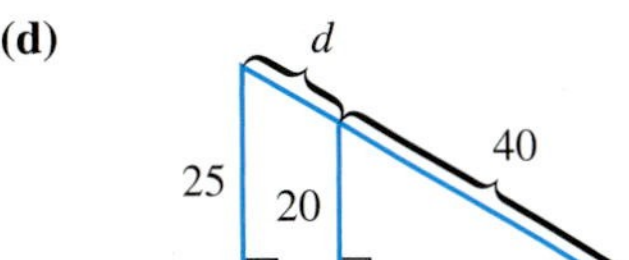

4. **(a)** Two convex quadrilaterals $ABCD$ and $EFGH$ have congruent angles at their corresponding vertices: $\angle A \cong \angle E$, $\angle B \cong \angle F$, $\angle C \cong \angle G$, and $\angle D \cong \angle H$. Can you conclude that the two convex quadrilaterals are similar? Explain.

(b) Two convex quadrilaterals have corresponding sides in the same ratio. Are the quadrilaterals necessarily similar? Explain.

5. Suppose $\triangle ABC \sim \triangle DEF$, where only points D and E are shown. Find all possible locations for F, and draw the corresponding triangles DEF.

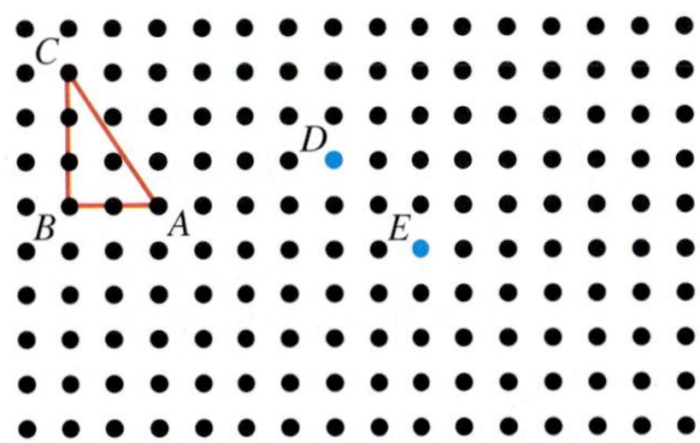

6. The diagonals of the trapezoid $ABCD$ shown below intersect at E, where $\overline{AB} \parallel \overline{CD}$.

(a) Explain why $\triangle ABE \sim \triangle CDE$.

(b) Determine x and y.

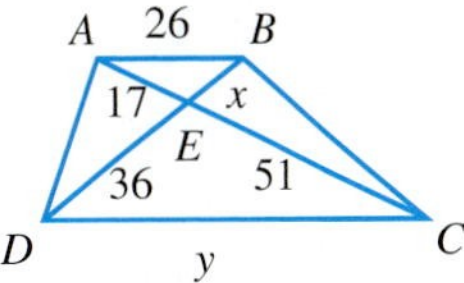

7. Let $\overline{AB}$ and $\overline{CD}$ be parallel, and let $\overline{AD}$ and $\overline{BC}$ intersect at E. Prove that $a \cdot y = x \cdot b$.

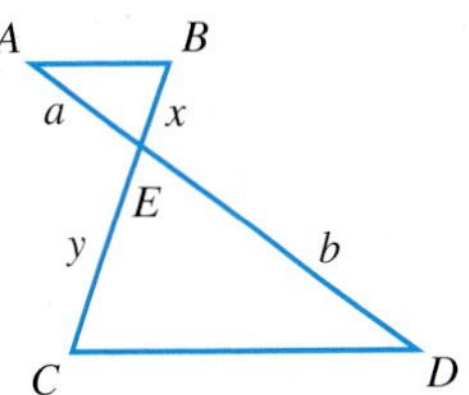

8. $\triangle ABC$ is a right triangle with altitude $\overline{CD}$.

(a) Explain why $\triangle ADC \sim \triangle CDB$.

(b) Find an equation for h and solve it to show that $h = \sqrt{9 \cdot 25} = 3 \cdot 5 = 15$.

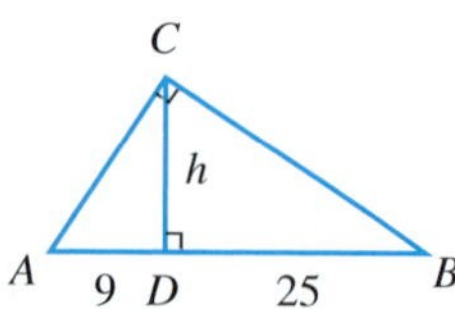

9. An isosceles triangle whose apex angle measures 36° is sometimes called a **golden triangle.**

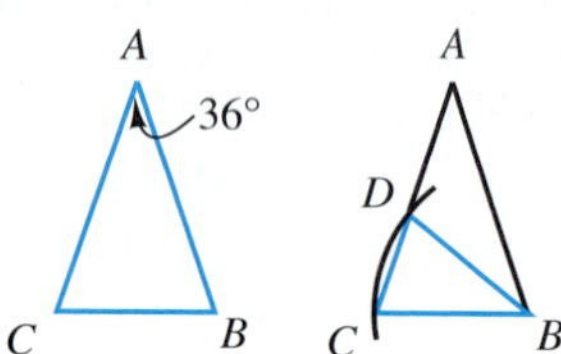

(a) Draw an arc centered at B and passing through C. If D is the point at which the arc intersects $\overline{AC}$, prove that $\triangle BCD$ is also a golden triangle. (*Hint:* What is $m(\angle C)$?)

(b) Construct three more golden triangles CDE, DEF, and EFG, where each contains the next.

10. $\triangle ABC$ is an isosceles triangle with apex C. It has the unusual property that the circular arc centered at A intersects the opposite side at a point D for which $\triangle ABC \sim \triangle BCD$. Use this property to find the measures of the base and apex angles of $\triangle ABC$.

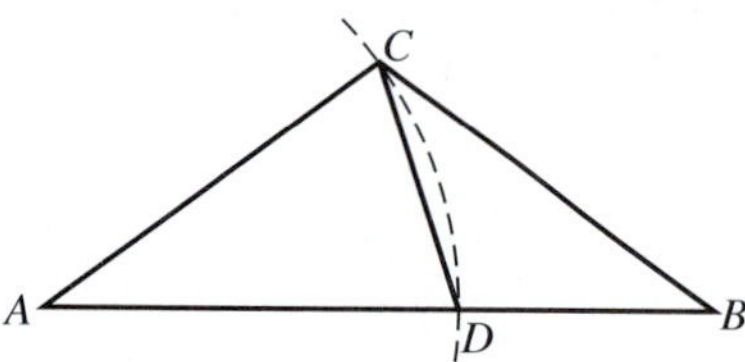

11. Suppose $\triangle ABC \sim \triangle BCA$. What more can you say about $\triangle ABC$?

12. Lined notebook paper provides a convenient way to subdivide a line segment into a specified number of congruent subsegments. The diagram shows how swinging an arc of radius AB subdivides the segment into five congruent parts: $\overline{AX} \cong \overline{XY} \cong \overline{YZ} \cong \overline{ZW} \cong \overline{WB}$.

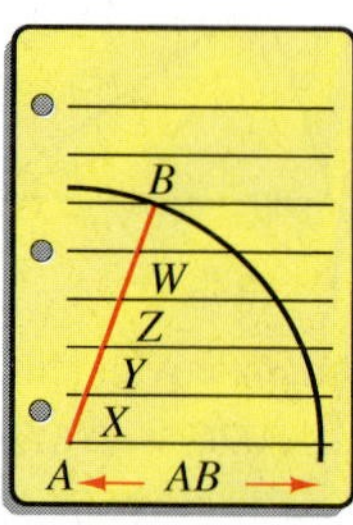

(a) Show how to subdivide a segment $\overline{AB}$ into seven congruent segments.

(b) Carefully explain why the procedure is valid.

Teaching Concepts

13. **Barbie and GI Joe Similarity.** There are many connections of similarity to everyday events. For example, reading maps accurately and drawing maps to scale is a valuable skill. Similarity concepts can often be taught best with frequent applications and references to the students' world. An interesting example is to see how similar, in the mathematical sense, Barbie and GI Joe dolls are to realistic women and men.

(a) Measure the bust, waist, and hips of a Barbie doll. What are the corresponding measurements of a mathematically similar woman who is 5′9″ tall?

(b) Measure the chest, waist, hips, and biceps of a GI Joe figure. What are the corresponding measurements of a mathematically similar man who is 6′2″ tall?

Thinking Critically

14. Let $\overline{CD}$ be the altitude drawn to the hypotenuse of the right triangle ABC shown.

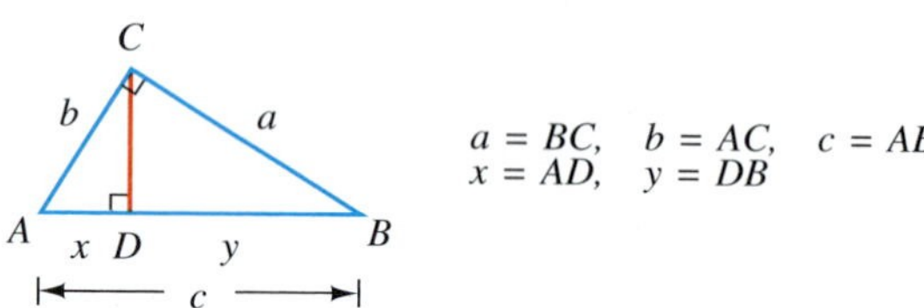

(a) Explain why $\triangle ABC \sim \triangle ACD$ and $\triangle ABC \sim \triangle CBD$.

(b) Explain why $\frac{x}{b} = \frac{b}{c}$ and $\frac{y}{a} = \frac{a}{c}$.

(c) Use part (b) to show that $c^2 = a^2 + b^2$. (This gives a proof of the Pythagorean theorem.)

15. Prove that the square shown inscribed in a right triangle has sides of length $x = \frac{ab}{a + b}$, where a and b are the lengths of the legs of the triangle.

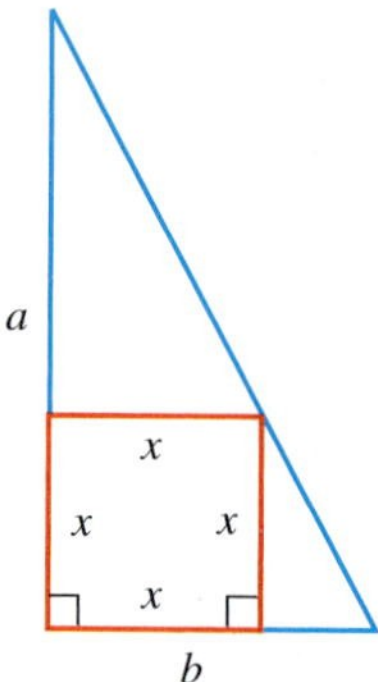

16. The square $ABCD$ has sides of unit length and midpoints at J, K, L, and M. The four segments that join vertices of the square to the midpoints of opposite sides form a smaller inner square $PQRS$.

(a) Use the Pythagorean theorem to show that $DJ = \frac{1}{2}\sqrt{5}$.

(b) Observe that the segment $\overline{ST}$ has length $\frac{1}{2}$ and creates $\triangle STP$ that is similar to $\triangle DJA$. Use this fact to compute the length PS.

(c) What is the area of the inner square?

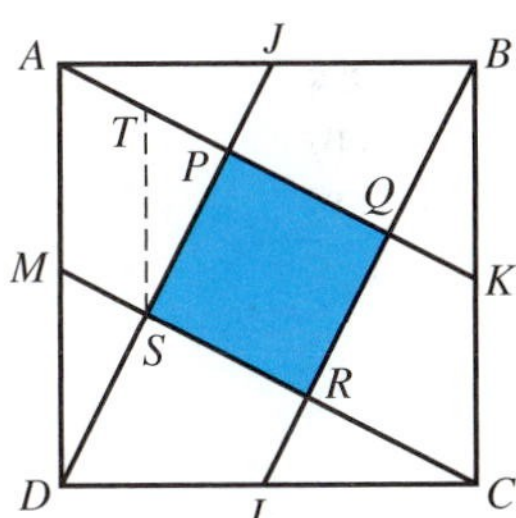

17. The square $ABCD$ has sides of unit length and "one-third" points at J, K, L, and M. Join the vertices of the square to successive "one-third" points to form the smaller inner square $PQRS$. Follow the steps in problem 16 to show that the inner square formed with "one-third" points has 40% of the area of the larger square $ABCD$.

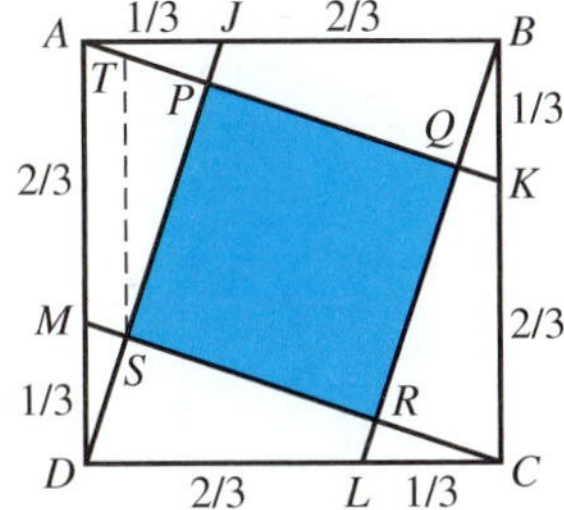

18. Let three arbitrary perpendiculars to the sides of $\triangle ABC$ of the diagram be drawn, meeting in pairs at the points P, Q, and R. Prove that $\triangle PQR \sim \triangle ABC$.

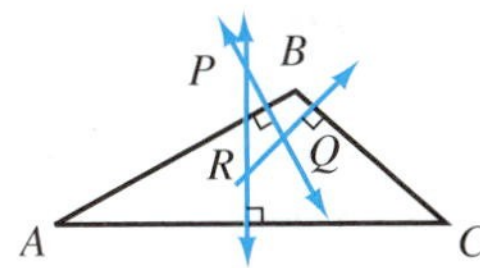

19. Let $PQRS$ be a space quadrilateral, and let W, X, Y, and Z be the midpoints of successive sides. Explain why $\overline{WY}$ and $\overline{XZ}$ intersect at their common midpoint M.

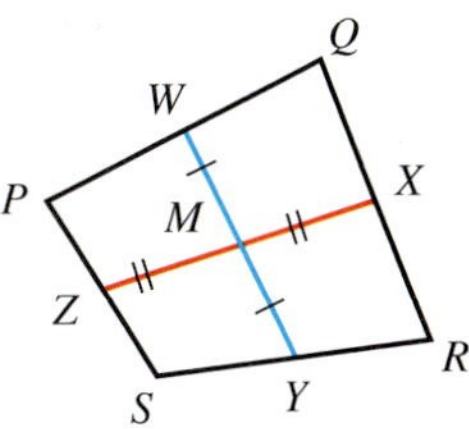

20. Let $ABCD$ be a trapezoid as illustrated, with bases of length $a = AB$ and $b = CD$. Let the diagonals $\overline{AC}$ and $\overline{BD}$ intersect at P, and suppose $\overline{EF}$ is the segment parallel to the bases that passes through P. Show that $EP = FP = \frac{ab}{a + b}$. (Thus, $EF = \frac{2ab}{a + b}$, which is called the **harmonic mean** of a and b.)

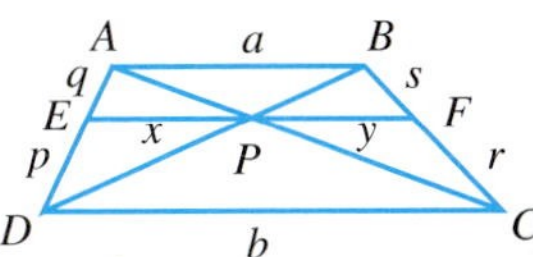

(*Hint:* Explain why $x/a = p/(p + q)$ and $x/b = q/(q + p)$. What happens when these equations are added?)

21. Let M and N be the midpoints of the sides of parallelogram $ABCD$ opposite A as shown. Show that $\overline{AM}$ and $\overline{AN}$ divide the diagonal $\overline{BD}$ into three congruent segments: $BP = PQ = QD$. (*Hint:* Construct $\overline{AC}$, and see Example 14.14.)

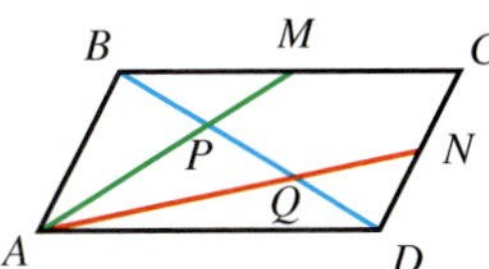

22. Use similarity to find the distances AP, BP, CP, and DP in the figure shown. The smallest squares on the lattice have sides of unit length.

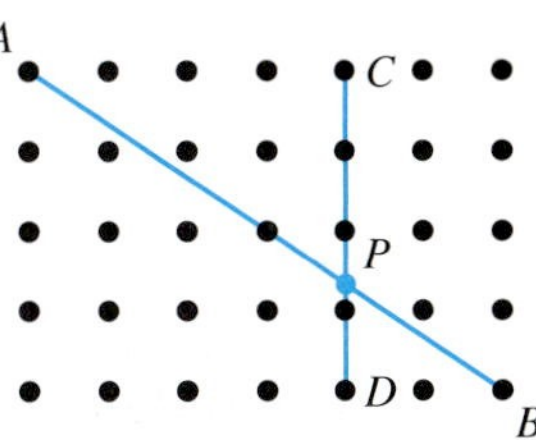

Making Connections

23. Mingxi is standing 75 feet from the base of a tree. The shadow from the top of Mingxi's head coincides with the shadow from the top of the tree. If Mingxi is 5′9″ tall and his shadow is 7′ long, how tall is the tree?

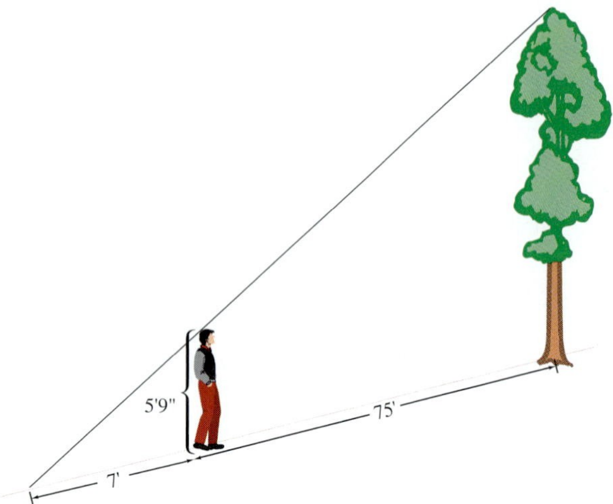

24. Mohini laid a mirror on the level ground 15 feet from the base of a pole as shown. Standing 4 feet from the mirror, she can see the top of the pole reflected in the mirror. If Mohini is 5′5″ tall, how can she estimate the height of the pole? What must she allow for in her calculation?

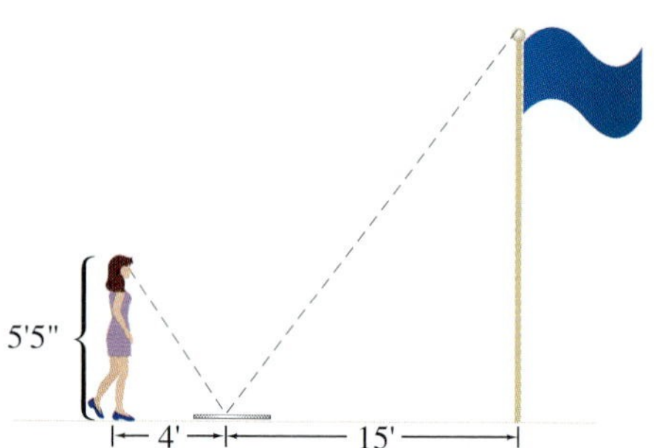

25. By holding a ruler 2 feet in front of her eyes as illustrated in the diagram, Ginny sees that the top and bottom points of a vertical cliff face line up with marks separated by 3.5″ on the ruler. According to the map, Ginny is about a half mile from the cliff. What is the approximate height of the cliff? Recall that a mile is 5280 feet.

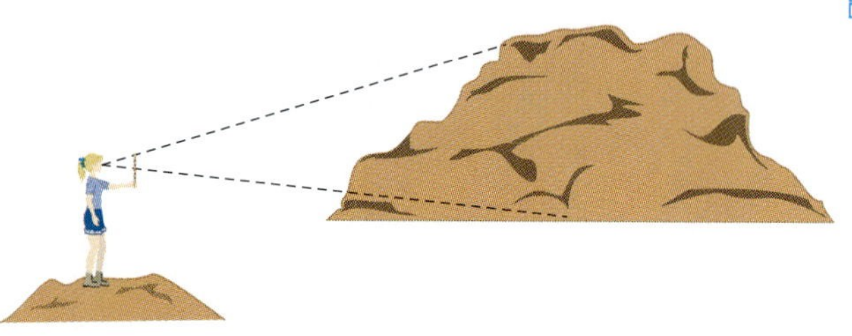

26. A vertical wall 18 feet high casts a shadow 6 feet wide on level ground. If Lisa is 5′3″ tall, how far away from the wall can she stand and still be entirely in the shade?

27. A drive from Prineville to Queenstown currently requires passing through Renton, due to a steep intervening hill. This is shown on the rough map drawn below. The two roads are level and straight and meet at 55° at Renton.

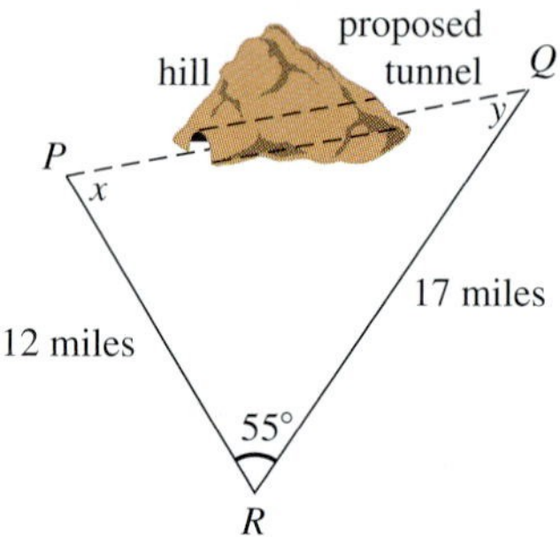

(a) How many miles would be saved by boring a tunnel through the hill? Make an accurate scale drawing, and take measurements off of your drawing.

(b) Two construction crews will dig the tunnel from opposite sides of the hill. What angle measures x and y will ensure that the two crews meet properly at the center of the hill? Take measurements from your scale drawing to give accurate estimates.

28. A sloping ramp is to be built with vertical supports placed at points B, C, and D as shown. The supports at A and E are 8 and 12 feet high, respectively. How high must the supports be at the points B, C, and D?

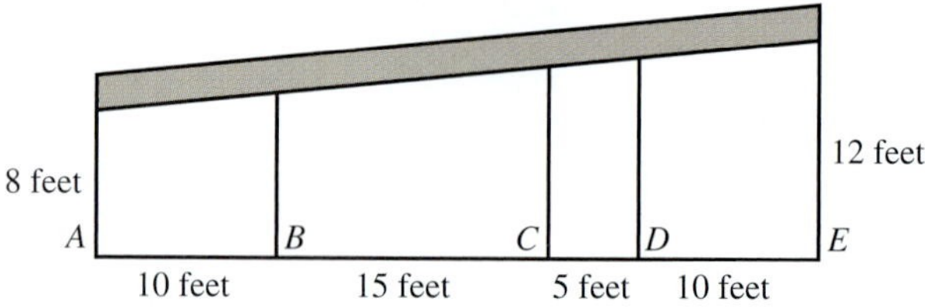

Using a Computer

29. Construct a circle and two chords $\overline{AB}$ and $\overline{CD}$ that intersect at a point P. Measure the angles in the triangles BCP and DAP.

(a) What conclusion about $\triangle BCP$ and $\triangle DAP$ is suggested by your measurements?

(b) Measure the lengths of $\overline{PA}$, $\overline{PB}$, $\overline{PC}$, and $\overline{PD}$, and then compare $PA \cdot PB$ with $PC \cdot PD$. Does your answer to part (a) justify your observation about the two products?

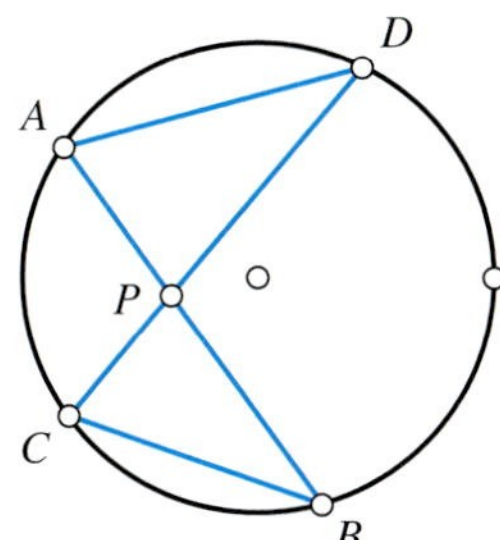

30. Construct a circle and a point P that is outside of the circle as shown. Next, draw two rays from P that intersect the circle at the points A, B, C, and D. Measure the angles in $\triangle BCP$ and $\triangle DAP$.

(a) What conclusion about $\triangle BCP$ and $\triangle DAP$ is suggested by your measurements?

(b) Measure the lengths of $\overline{PA}$, $\overline{PB}$, $\overline{PC}$, and $\overline{PD}$, and then compare $PA \cdot PB$ with $PC \cdot PD$. Does your answer to part (a) justify your observation about the two products?

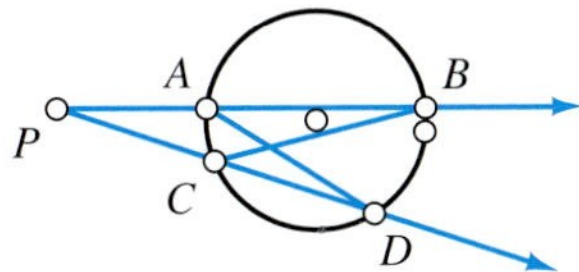

Communicating

31. **(a)** In Example 14.13, it was shown that joining the successive midpoints of a quadrilateral forms a parallelogram. Explore what you find by *reversing* the construction. That is, given a parallelogram $KLMN$, can you construct a quadrilateral $ABCD$ for which $KLMN$ is the midpoint figure? Is $ABCD$ unique?

(b) Write a report discussing your results of part (a).

From State Student Assessments

32. (Illinois, Grade 5)
$\triangle PQR$ is similar to $\triangle STU$. Find the length of TS.

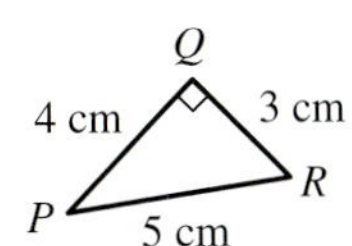

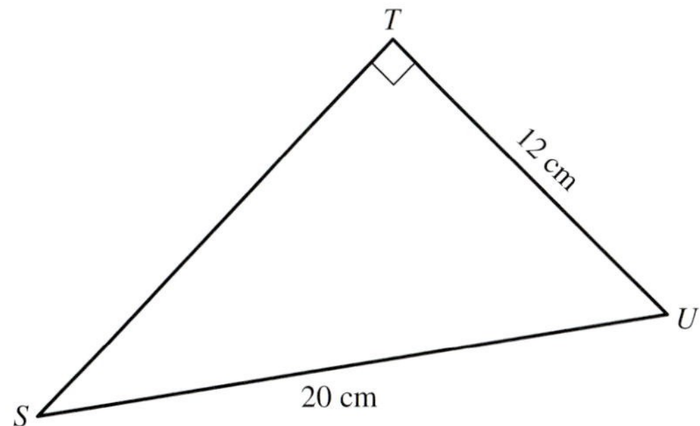

A. 4 cm B. 14 cm
C. 16 cm D. 26 cm
E. 32 cm

For Review

33. Find three pairs of congruent triangles in this figure.

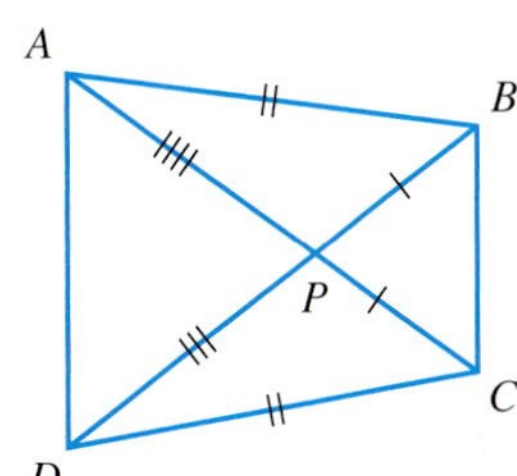

34. Find the measures of the interior angles of $\triangle RST$.

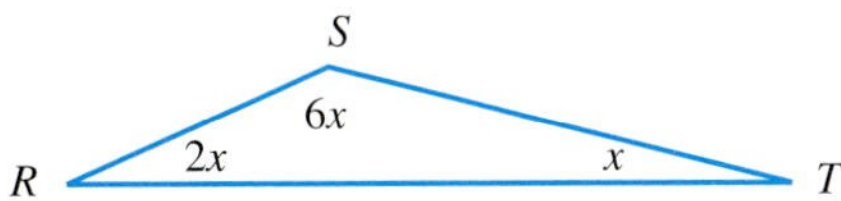

35. Determine the measures of the interior, exterior, and central angles of this regular polygon.

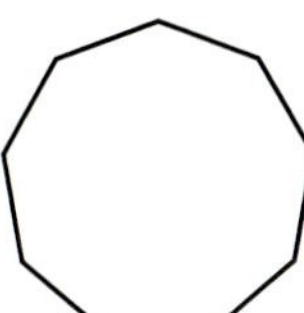

Epilogue The Themes of Geometry

This chapter concludes our study of informal geometry. Several themes have cut across these geometry chapters. These themes, which could properly form the basis for the geometry portion of the elementary school curriculum, are as follows:

- *Congruence*

  The notion of congruence (that is, that one figure is exactly the same size and shape as another) is of great importance.

- *Similarity*

  This embodies the idea that two figures are the same *shape,* but of different size, as if one figure is simply a magnification of the other. Perhaps the most important consequence of similarity is that corresponding lengths in similar figures are proportional.

- *Constructions*

  Traditionally, constructions were restricted to figures and geometric objects that can be created with the Euclidean tools of the straightedge and compass. In recent times, other drawing instruments such as the Mira may be allowed. Figures constructible with one tool may easily differ from the figures constructible with a different tool. For example, an angle trisector is constructible with the Mira, but not with straightedge and compass. Computer geometry programs have also enlarged the set of constructible objects, with fractals as an important example.

- *Invariance*

  Invariance involves the idea that while many properties of geometric figures change from figure to figure, some, often surprisingly, do not. Indeed, much of the interest in, and utility of, geometry derives from this fact. For example, the ratio of the circumference of any circle to its diameter is always $\pi = 3.1415\ldots$; the volume of any cone is always one-third the volume of the corresponding cylinder; the sum of the exterior angles of any convex polygon is always 360°; and so on. Geometry is replete with remarkable and useful invariances.

- *Symmetry*

  There are many kinds of symmetry in geometry. There is the repetitive symmetry that manifests itself in tilings and tessellations; the symmetry of an object as if it were reflected in a mirror (that is, symmetry with respect to a line); symmetry of an object through a point; and rotational symmetry. There is also a sort of symmetry in many of the formulas of geometry. For example, the Pythagorean expression $c^2 = a^2 + b^2$ is unchanged if a and b are interchanged, and similar symmetries are exhibited by the distance and slope formulas in coordinate geometry.

- *Loci*

  A *locus* is a set of points that satisfy certain conditions. For example, a circle is the set of all points in a plane that are equidistant from a fixed point O. As another example, the set of points equidistant from the points A and B is the perpendicular bisector of the segment $\overline{AB}$.

- *Maxima and minima*

  Of all triangles of fixed perimeter, the equilateral triangle has maximum area. Of all rectangles of fixed area, the square has the smallest perimeter. Such questions of maxima and minima often arise in geometry.

- *Limit*

  As an example of the notion of limit, as n gets larger and larger a regular n-gon more and more closely approximates a circle. Indeed, we would say that the limit of a regular n-gon as n tends to infinity *is* a circle. This notion was used in developing the formula for the area of a circle.

- *Measurement*

  This notion needs no comment. The ability to measure and communicate information concerning size and amount is basic to geometric thinking and applications of geometry in the real world.

- *Coordinates*

  This theme stresses the idea that geometric objects can be viewed as sets of points determined by ordered pairs of numbers that satisfy certain conditions. This powerful notion makes it possible to use methods of arithmetic and algebra to obtain geometric results.

- *Logical structure*

  As in the rest of mathematics, geometric ideas do not stand alone. Even in informal geometry, it is important to see how some results follow from others and to realize that guessing or conjecturing alone is not enough.

Chapter Summary

Key Concepts

Section 14.1 Congruent Triangles

- Congruent triangles. Two triangles are congruent if, and only if, there is a correspondence of vertices and sides so that corresponding angles have the same measure and corresponding sides have the same length. That is, $\Delta ABC \cong \Delta DEF$ if, and only if, $m(\angle A) = m(\angle D)$, $m(\angle B) = m(\angle E)$, $m(\angle C) = m(\angle F)$, $AB = DE$, $BC = EF$, and $CA = FD$.
- Congruence properties. Two triangles are congruent if they satisfy one of the following properties: SSS (side–side–side), SAS (side–angle–side), ASA (angle–side–angle), and AAS (angle–angle–side).
- Triangle inequality. The length of the side of a triangle is less than the sum of the lengths of the other two sides.
- Isosceles triangle theorem and converse. A triangle has two sides of the same length if, and only if, their opposite angles are congruent.
- Thales' theorem. A triangle inscribed in a semicircle is a right triangle with the diameter as its hypotenuse.

Section 14.2 Constructing Geometric Figures

- Basic compass-and-straightedge constructions: copy a line segment; copy an angle; construct a line parallel to a line through a point not on the line; construct a perpendicular line from a point to a line; construct a perpendicular to a line through a point on the line; construct a perpendicular bisector; construct an angle bisector.

- Equidistance property of the perpendicular bisector. A point is on the perpendicular bisector of a segment if, and only if, the point is equidistant from the endpoints of the segment.
- Equidistance property of the angle bisector. A point in the interior of an angle is on the angle bisector if, and only if, it is equidistant from the sides of the angle.
- Gauss–Wantzel theorem. A regular n-gon has a compass-and-straightedge construction if, and only if, n is a power of 2 that is 4 or larger or a power-of-two multiple of a product of distinct Fermat primes (a Fermat prime is a number of the form $2^{2^k} + 1$, only five of which—3, 5, 17, 257, and 65,537—are known to be prime).

Section 14.3 Similar Triangles

- Similar triangles. Two triangles are similar if corresponding angles are congruent and corresponding sides have the same ratio of lengths of sides. The common ratio of lengths of corresponding sides is the scale factor.
- Similarity properties. Two triangles are similar if they have one of these properties: AA (angle–angle) similarity (two pairs of congruent angles), SSS (side–side–side) similarity (three proportional sides), or SAS (side–angle–side) similarity (two proportional pairs of sides include congruent angles).

Vocabulary and Notation

Section 14.1

Congruent
Parts of a triangle
Congruent triangles, $\Delta ABC \cong \Delta DEF$
Congruence properties of triangles: SSS, SAS, ASA, AAS
Incenter
Inscribed circle, or incircle
Altitude of a triangle
Perpendicular bisector
Angle bisector
Fermat prime

Section 14.2

Constructions: compass and straightedge, Mira, paper folding, geometry software
Foot of the perpendicular
Circumscribing circle, or circumcircle
Circumcenter

Section 14.3

Similar figures
Similar triangles, $\triangle ABC \sim \Delta DEF$
Scale factor (ratio of similitude)
Similarity properties of triangles: AA, SSS, SAS

Chapter Review Exercises

Section 14.1

1. In each figure, find at least one pair of congruent triangles. Express the congruence using the $\cong$ symbol, and justify why the triangles are congruent.

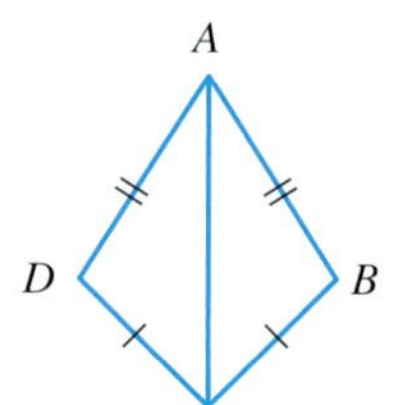

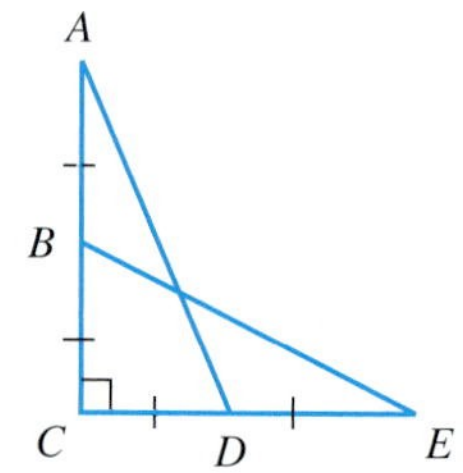

(c)

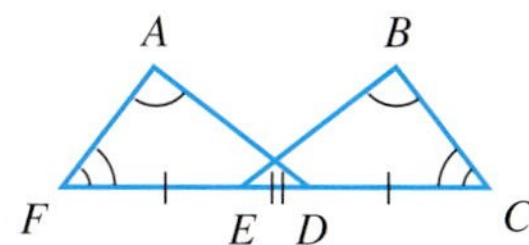

(d)

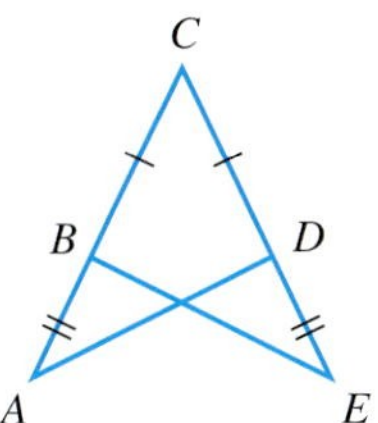

(e)

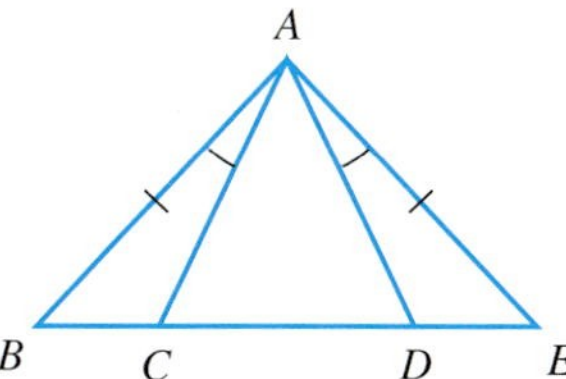

(f)

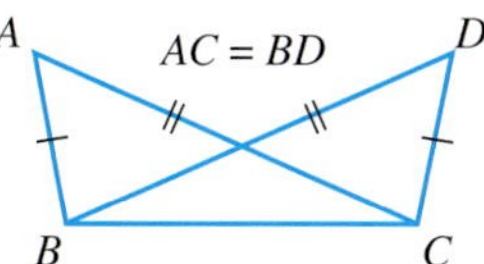

2. Without measuring, fill in the blanks below these figures.

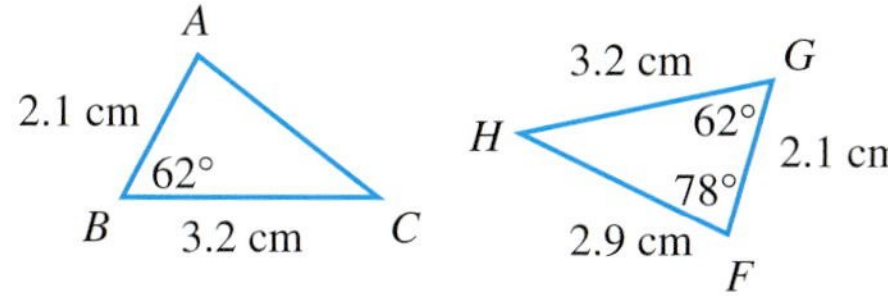

(a) $AC =$ ____
(b) $m(\angle H) =$ ____
(c) $m(\angle A) =$ ____
(d) $m(\angle C) =$ ____

3. Let D be any point of the base $\overline{BC}$ of an isosceles triangle ABC. Locate E on $\overline{AC}$ and F on $\overline{AB}$ so that $EC = BD$ and $BF = DC$. Draw the figure as it is described and then prove that $DE = DF$.

Section 14.2

4. Construct the following with a compass and a straightedge. Show and describe all of your steps.

(a)

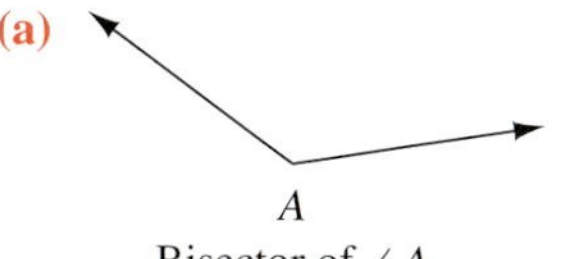

Bisector of $\angle A$

(b)

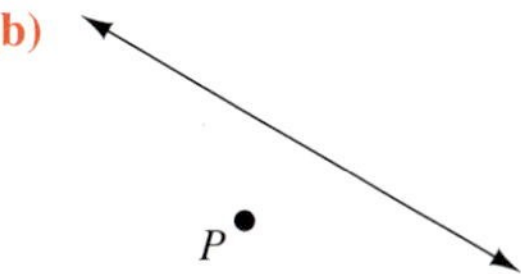

Perpendicular through P

(c)

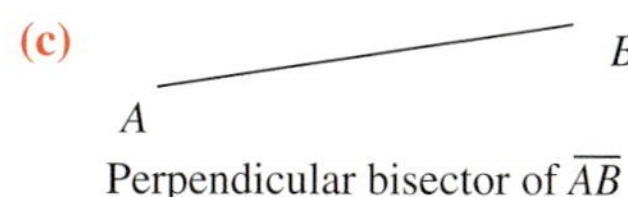

Perpendicular bisector of $\overline{AB}$

(d)

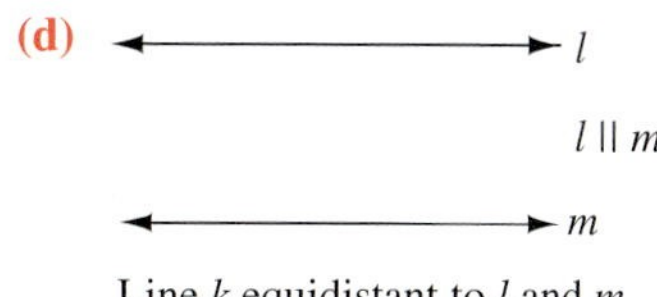

Line k equidistant to l and m

5. Show and describe the position of a Mira that performs each of the constructions of problem 4 in one step.

6. (a) Construct $\triangle ABC$, where $\angle A$, $\overline{AB}$, and $\overline{BC}$ are the parts shown below. Is the shape of $\triangle ABC$ uniquely determined?

(b) Is the shape of $\triangle ABC$ uniquely determined if $\angle C$ is obtuse?

7. Construct a regular hexagon $ABCDEF$, where diagonal $\overline{AD}$ is given below.

Section 14.3

8. (a) If only a ruler is available, is it possible to determine whether two triangles are similar?

(b) If only a protractor is available, is it possible to determine whether two triangles are similar?

9. Triangle ABC is given below. Construct a triangle DEF for which $\triangle ABC \sim \triangle DEF$ and $DE = (3/2)AB$, using a compass and a straightedge.

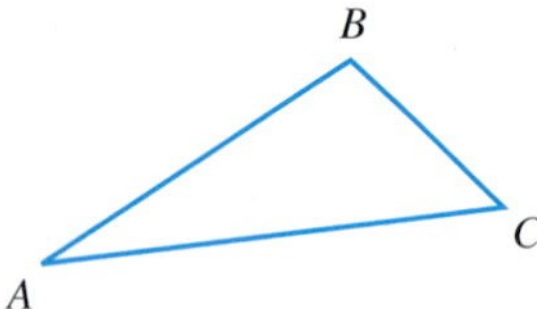

10. Explain why each pair of triangles is similar. State the similarity using the ~ symbol, and give the scale factor.

(a)

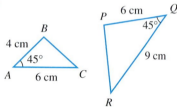

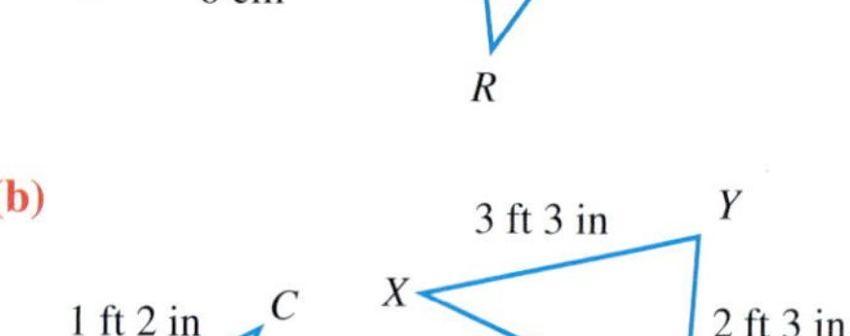

(b)

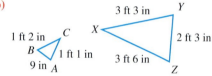

(c)

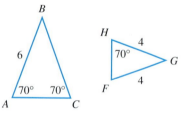

(d)

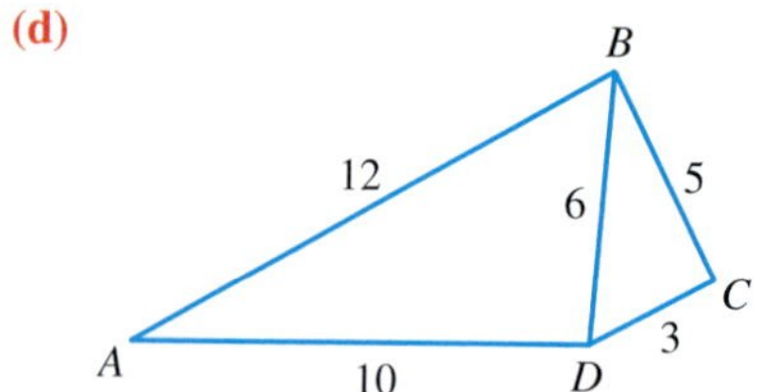

11. Lines k, l, and m are parallel. Find the distances x and y, using similar triangles.

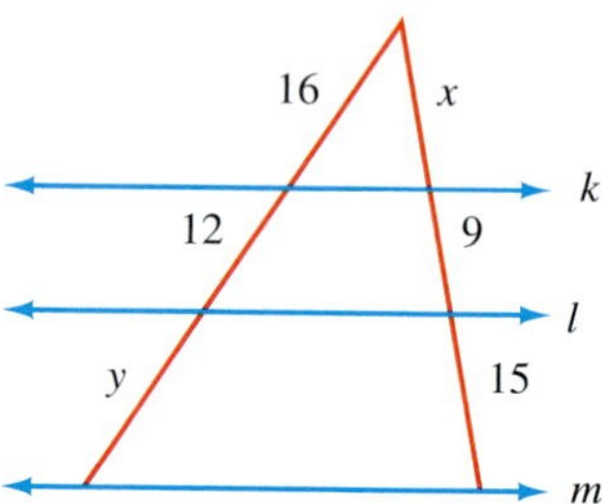

12. In the morning, the shadows cast by the top of a vertical stick 3 feet high and the top of a pyramid were at points S_1 and P_1, respectively, on level ground. That afternoon, because of the motion of the sun in the sky, the points were at S_2 and P_2. If $S_1S_2 = 2$ feet and $P_1P_2 = 270$ feet, what is the height of the pyramid?

Chapter Test

1. A person 6 feet tall casts a shadow 7 feet long, and a tree casts a shadow 56 feet long.

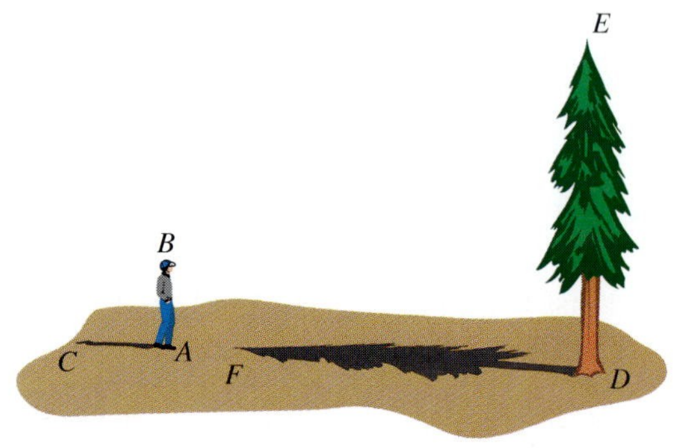

(a) What assumption can you make about the sun's rays?

(b) What other assumption can you make to conclude that $\triangle ABC \sim \triangle DEF$?

(c) How tall is the tree?

2. In $\triangle ABC$, the midpoint M of side $\overline{AB}$ satisfies $MC = MA$. Prove that $\triangle ABC$ is a right triangle.

3. Explain why each pair of the following triangles is similar. Express the similarity with the ~ notation.

(a)

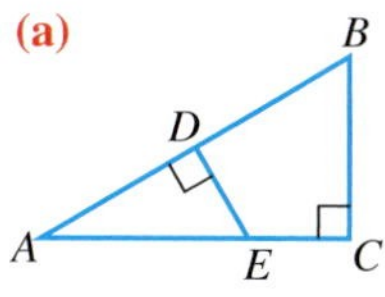

(b)

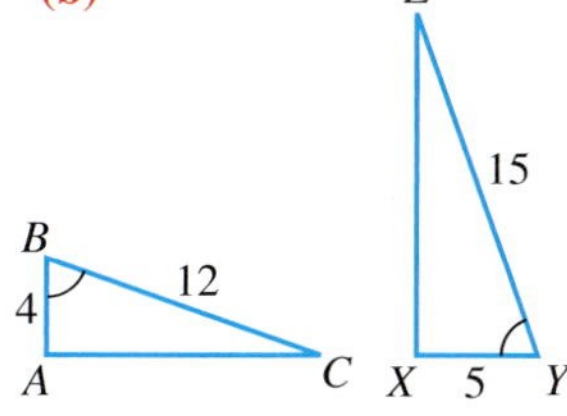

(c)

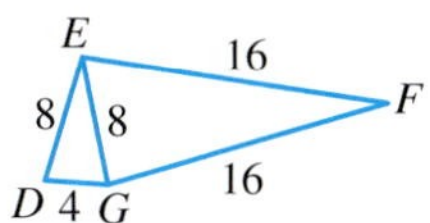

(d)

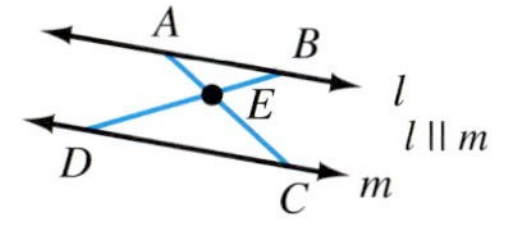

4. In each figure, find a pair of congruent triangles. State what congruence property justifies your conclusion, and write the congruence using the congruence symbol $\cong$.

(a)

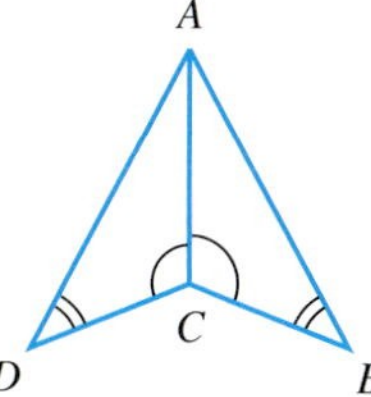

(b)

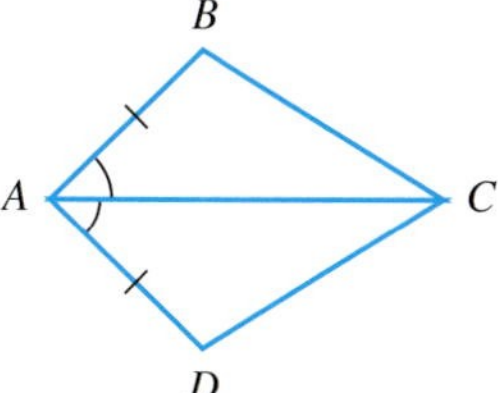

(c)

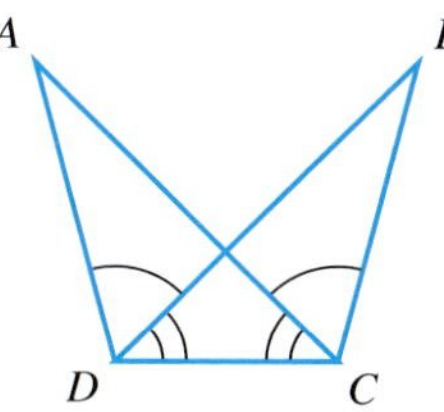

(d)

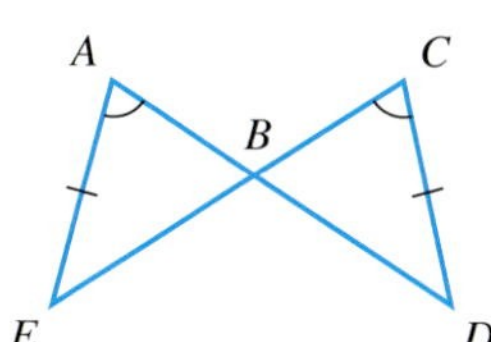

(e)

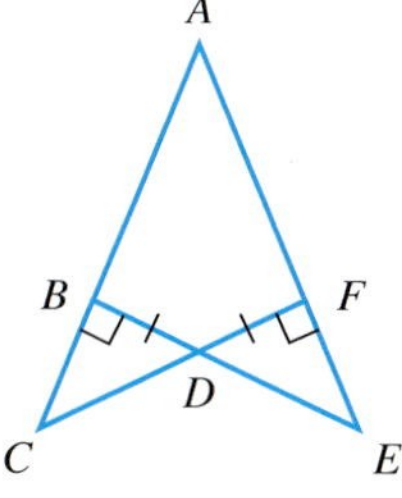

(f)

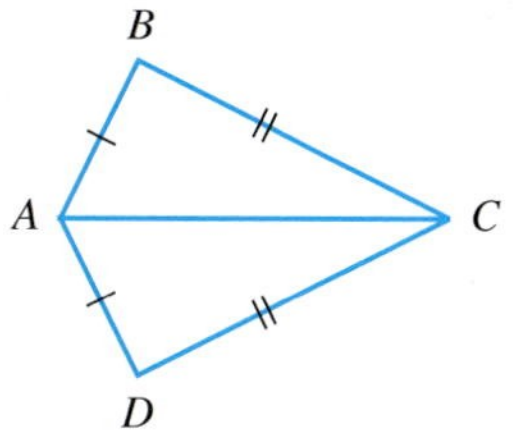

5. If $AB = 12$ and $AD = 2DC$, find AE and EB.

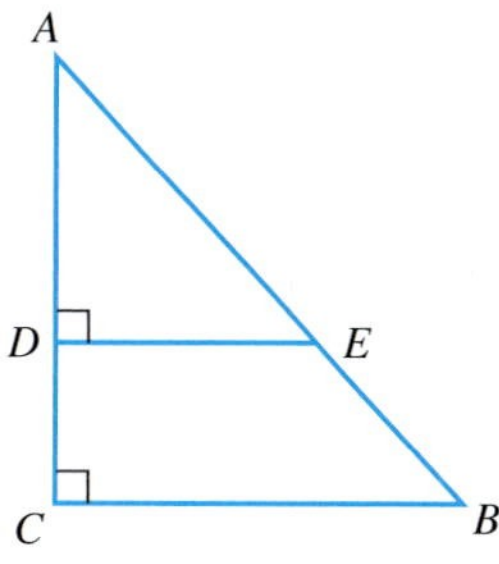

Explain how you found your answer.

6. Fill in the blanks below, where $\triangle KLM \sim \triangle UVW$.

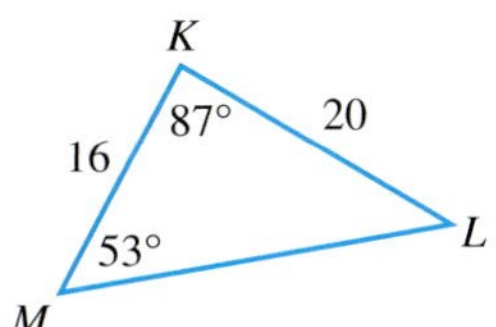

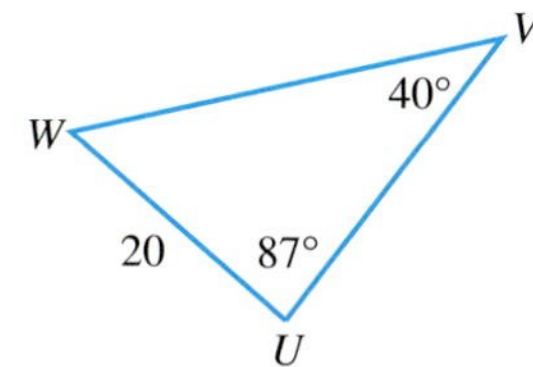

(a) $m(\angle W) =$ ____

(b) Scale factor = ____

(c) $UV =$ ____

7. A triangle has sides of length 10 feet and 16 feet. What is the range of lengths of the third side?

8. Let $ABCDE$ be a regular pentagon. Let $PQRST$ be inscribed so that $AP = BQ = CR = DS = ET$. Prove that $PQRST$ is a regular pentagon.

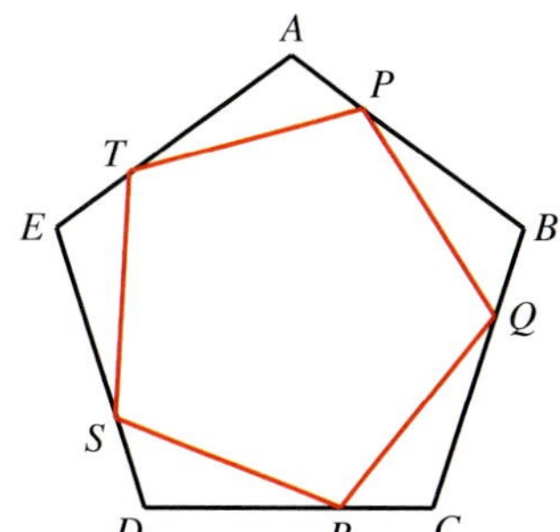

9. Let $\overline{PQ}$ be perpendicular to line l.

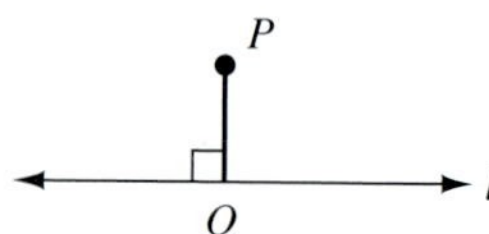

(a) Use a compass and a straightedge to construct two equilateral triangles $\triangle PQR$ and $\triangle PQS$, each with $\overline{PQ}$ as a side.

(b) Use a compass and a straightedge to construct an equilateral triangle PTU for which $\overline{PQ}$ is an altitude. Describe your procedure.

10. Construct a triangle DEF so that $\triangle DEF \cong \triangle ABC$ and $\angle A$, $\overline{AB}$, and $\angle B$ are given as shown below. List the steps you follow.

11. In $\triangle FGH$, $\angle F \cong \angle G$ and $FG = FH$. What can you conclude about the triangle?

12. Decide if the following pairs of triangles are necessarily congruent. If so, state why, and give the vertex correspondence. If not, give a counterexample.

(a)

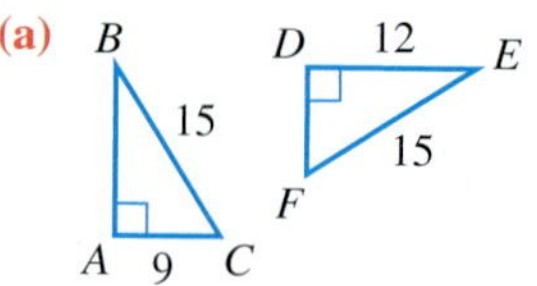

(b)

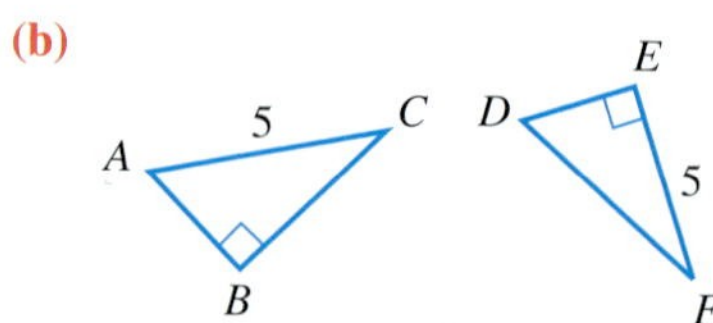

(c)

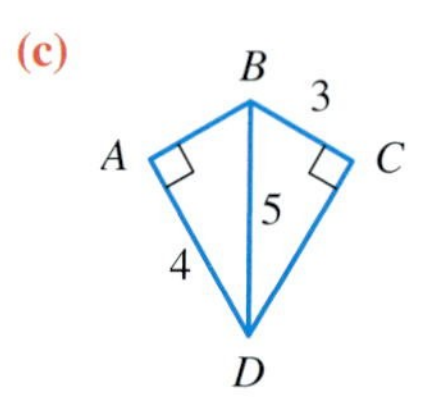

(d)

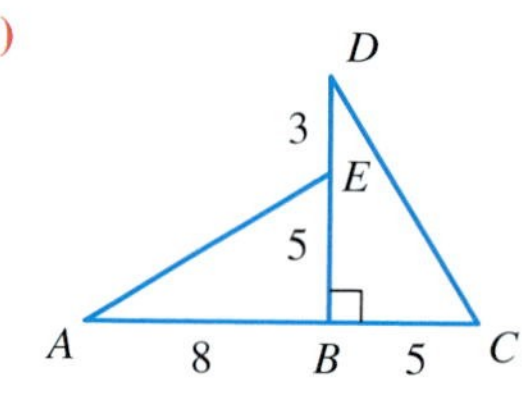

Just for Fun

Thales' Puzzle

Thales is reputed to have calculated the height of the pyramids in Egypt by comparing the shadow cast by the pyramid to the shadow cast by a vertical stick. By similar right triangles $H/h = y/x$, where H is the height of the pyramid, h is the height of the stick, x is the length of the shadow of the stick, and $y = PB$ is the length of the shadow of the pyramid from a point P on the ground to the point B directly beneath apex A. Thus, the height of the pyramid is given by the formula $H = yh/x$. Unfortunately, Thales still had a difficulty: he could easily measure h and x, but y was not simple to measure since point B is somewhere *inside* the pyramid!

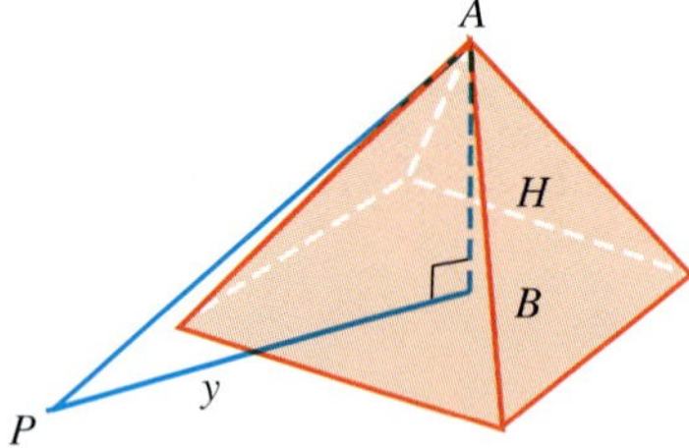

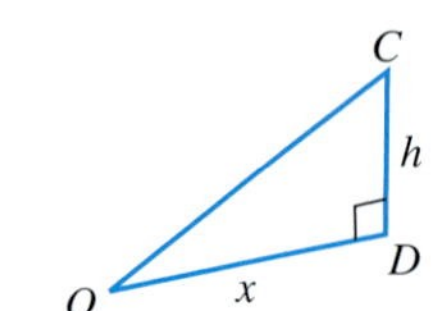

After marking points P and Q one morning, Thales returned a few hours later and realized he could now calculate the height of the pyramid. How did Thales solve his problem?

Appendix A Manipulatives in the Mathematics Classroom

WHAT ARE MANIPULATIVES?

More and more, we are hearing calls for greater use of manipulatives in elementary mathematics teaching. Manipulatives are concrete materials that are used for modeling or representing mathematical operations or concepts. In much the same way that children make models of airplanes or clipper ships so that they can study and learn about them, your students can make and learn from models of two-digit numbers or division. The difference between the two situations is that while the airplane and boat models are smaller versions of actual concrete things, the number models and division models are concrete models of abstract concepts.

When children use bundles of Popsicle™ sticks and single sticks to represent 10s and 1s, or stretch rubber bands around nails on a geoboard to show squares and triangles, they are using manipulatives to model mathematical ideas. Technically, even when young children count on their fingers, they are concretely modeling numbers!

It is important to note that some frequently used "manipulative" materials fail to fit this definition of manipulatives. Flashcards, for instance, even though they are certainly manipulated, are never used to model mathematical ideas. Other objects, like checkers in a checker game, also are manipulated, but are not used to directly represent mathematical ideas.

WHAT KINDS OF THINGS CAN BE USED AS MANIPULATIVES?

Good teachers use a great variety of things as manipulatives. Popsicle™ sticks, dried beans, smooth stones, egg cartons, and poker chips are all inexpensive and effective materials. Some teachers even like to use the students in their classes themselves as models for sets and numbers.

There are also, however, commercial materials that serve more specific modeling purposes. Some of these are base-ten blocks, geoboards, fraction strips or pieces, and algebra tiles. What's important about the materials you select is not their cost, but that they accurately represent the concept or operation that your students are ready to learn about.

WHAT DOES RESEARCH SAY ABOUT USING MANIPULATIVES?

For the last 20 years, research support for using manipulatives in elementary math teaching has been growing. In 1977, Suydam and Higgins were able to identify and review 23 research studies that addressed the question of the effectiveness of manipulative materials for instruction. They concluded that instruction using manipulative materials had a higher probability of producing greater mathematics achievement than nonmanipulative instruction (Suydam & Higgins, 1977).

Twelve years later, Sowell (1989) had 60 studies to work with in a similar review. These studies were conducted at all educational levels, from kindergarten to college, and focused on many different mathematical topics. After using a sophisticated statistical procedure called meta-analysis, Sowell was able to offer a much stronger conclusion than Suydam and Higgins: that manipulative materials do, indeed, have a significant positive effect on achievement, especially when they are used over a long period of time (Sowell, 1989). In addition, Sowell found that the use of concrete materials for instruction was effective in improving students' attitudes toward mathematics.

WHAT DO THE PROFESSIONAL ORGANIZATIONS SAY?

Because of growing research support, professional organizations and leading educators have been urging increased use of concrete materials in teaching. Manipulatives play a very prominent role in the National Council of Teachers of Mathematics *Principles and Standards for School Mathematics.* Even though there is no single standard which says that manipulatives should be used more frequently, their use is supported for many different topics at all levels, from kindergarten through twelfth grade. In the area of numbers and operations, for example, the *Standards* suggest the following:

> *Representing numbers with physical materials should be a major part of mathematics instruction in the elementary grades.* (Standards, *p. 33*)

The *Standards* also make the following assertion:

> *Students' understanding and ability to reason will grow as they represent fractions and decimals with physical materials and on number lines as they learn to generate equivalent representations of fractions and decimals.* (Standards, *p. 33*)

Representations that include concrete materials are discussed in the *Standards* for grades 6–8, which describes their applicability and desirability in the upper grades as well as in the lower elementary school grades:

> *Representations—such as physical objects, drawings, charts, graphs, and symbols—also help students communicate their thinking. Representations are ubiquitous in the middle grades mathematics curriculum proposed here.* (Standards, *p. 280*)

HOW DOES CONCRETE MODELING HELP CHILDREN LEARN MATHEMATICS?

The mechanism by which concrete modeling promotes the learning process is still somewhat of a mystery to us, but the fact that it does is becoming more certain all the time. Most research shows that the best instructional sequence to follow for the presentation of elementary mathematical material is concrete–pictorial–symbolic. Activities with concrete materials should precede those that show pictured relationships, and the latter type of activity should, in turn, precede formal operations with symbols. Ultimately, students need to reach that final level of symbolic proficiency with many of the mathematical skills that they master, but the meanings of those symbols and abstract operations must be firmly rooted in experiences with real objects. Otherwise, their performance of the symbolic operations will simply be rote repetitions of meaningless, memorized procedures.

Concrete and pictorial models are, at best, imperfect representations of abstract mathematical ideas and concepts. Not every characteristic of the concrete model is important mathematically. For example, many teachers use a yellow wooden rod to represent the number 5, but "yellowness" certainly has nothing to do with "fiveness." To prevent the child from abstracting inappropriate characteristics of the model as characteristics of the mathematical idea, multiple models should be used for important concepts. Multiple models of the same idea that are perceptually very different from each other direct the student to abstract from them only what they have in common—the mathematical concept. This abstraction is what leads to meaningful mathematics learning.

WHEN SHOULD MANIPULATIVES BE USED?

Manipulatives and concrete models can be used almost anywhere in the elementary mathematics curriculum and should be used with all students. The most frequent occurrence of modeling in the curriculum will probably be at the point of introduction of new topics. The use of concrete materials while introducing a new topic allows students to gain that necessary foundational experience before trying to demonstrate their understandings symbolically. But intelligent use of manipulatives can also provide the most effective form of remediation. Engaging students in concrete activities related to the mathematics that they are struggling with will frequently help them to identify exactly which part of the process is causing the confusion, and then to work through it.

Use of manipulatives is also not just important for younger children. Older children can benefit just as much from appropriate concrete activity. The skills and concepts that they are asked to deal with are increasingly complex, and concrete introductions can frequently pave the way toward true understanding. Good concrete and pictorial models are available to help students deal with percents, ratios, geometric formulas, integers, and even the solution of algebraic equations.

MANIPULATIVE *DO'S* AND *DON'TS*

Elementary mathematics teachers who have been using manipulatives for many years have learned some rules that make their use in classrooms more effective. Don't use the materials exclusively for demonstrations and teacher explanations. The most effective way to use manipulatives is to have the children work directly with the materials. It is through touching and moving that the learning takes place. Watching the teacher do the manipulation is much less successful.

Discuss appropriate behavior before distributing the materials, and allow plenty of free exploration time after. Children need to have time to do what they want with the materials before they will do what you want with them. This exploration time is also learning time. The more familiar they are with the materials, the more effective their later use of them will be.

Have children work with the materials in small groups and explain their thinking to each other as they do so. Asking children to verbalize their thoughts for others gives them the opportunity to clarify their own thinking. By listening to the discussions, you can make judgments about how well they understand the concepts.

HOW CAN I GET MORE INFORMATION ABOUT USING MANIPULATIVES?

This list of references is a good place to start. The first two citations will be helpful for the practitioner, while the second two are the research reviews mentioned earlier.

National Council of Teachers of Mathematics. *Principles and Standards for School Mathematics.* Reston, VA: NCTM, 2000.

Reys, Robert E., Marilyn N. Suydam, and Mary Montgomery Lindquist. *Helping Children Learn Mathematics,* 2d ed. Englewood Cliffs, NJ: Prentice Hall, 1989.

Sowell, Evelyn J. "Effects of Manipulative Materials in Mathematics Instruction." *Journal for Research in Mathematics Education,* Vol. 20, No. 5, November 1989: pp. 498–505.

Suydam, Marilyn N., and Jon L. Higgins. *Activity-Based Learning in Elementary School Mathematics: Recommendations from Research.* Columbus, OH: ERIC Center for Science, Mathematics, and Environmental Education, 1977.

Van de Wall, John. *Elementary and Middle School Mathematics.* Boston: Allyn & Bacon, 2004.

Appendix B Spreadsheets

SPREADSHEET BASICS

Spreadsheets are computer programs* that allow you to organize and analyze information arranged in a rectangular table. A spreadsheet document, as shown in Figure B.1, is arranged in **columns** that are lettered A, B, C, . . . , from left to right, and **rows** that are numbered downward 1, 2, 3, ... The rectangular box at the intersection of a row and a column is called a **cell.** Each cell is uniquely identified by its **cell address.** For example, the cell in column D and row 3 has the cell address D3. In Figure B.1, the cursor has been clicked on cell D3, turning it into the **active cell.** The active cell is outlined by a heavy border, and its address is displayed at the left of the formula bar.

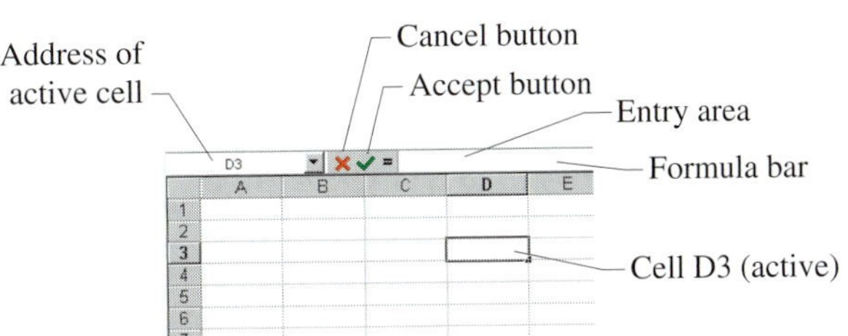

Figure B.1
The spreadsheet screen

Two types of data can be entered into the active cell:

- A **constant value** can be either text or a numerical value such as a fraction, monetary value, time, and so on. A constant value is either typed directly into the cell or entered by positioning the cursor in the entry area of the formula bar and typing. When the typing is completed, click on the Accept button on the formula bar. Alternatively, you can press Enter, Tab, or an arrow key to enter the typed value into the cell.
- A **formula** is an expression that creates a value in the active cell that depends on the current values of one or more other cells in the spreadsheet. Formulas always begin by typing in an equal sign, =, followed by the expression that defines the function. For example, if cell D3 is to be the sum of the values in cells A3, B3, and C3, type in either "= A3 + B3 + C3" or use the SUM function by typing in "= SUM(A3:C3)" to indicate the sum over the range of cells from A3 to C3. (*Note:* Type what is between the quotation marks, not the quotation marks themselves.) After typing the formula, click the Accept button or press Enter, Tab, or an arrow key to accept the formula. Changes or corrections in a value or formula can be made by selecting the cell and typing the changes in the entry area of the formula bar.

*The screens and commands in this appendix refer most directly to Microsoft® Excel; however, only minor changes, and often none at all, are necessary for users of other spreadsheets.

EXAMPLE B.1 SETTING UP A SPREADSHEET GRADE BOOK

Mr. Akmal will be giving 100 possible points for homework, a 25-point quiz, a 50-point project, and a 150-point final exam. He wants to record the scores of his students, total their individual scores, and convert their scores to percents to help him assign grades. Design an electronic grade sheet to record and process the grades.

Solution

The students' names are entered into column A, and the graded items and their total possible points are entered into rows 1 and 2, respectively, of columns B, C, D, and E. To allow the longer names to show, the width of a column can be adjusted by dragging the boundary on the right side of the column heading until the column is the width you want. The scores are entered into the corresponding array of cells, as shown below. The total points available are placed into cell F2 by entering the formula "= SUM(B2:E2)". When the Accept button is clicked, the total point value of 325 points will show in cell F2. The sum of Akiko's scores could be entered by typing "= SUM(B3:E3)" into cell F3, and similar formulas can be entered into the cells F4, F5, and F6. However, it is much easier to use a cut-and-paste procedure. First, select cell F2 and do **Copy.** Next, select *all* of the cells in the range F3 to F6 and do **Paste.** Another method to copy values or formulas is to use the **Fill** command*: first, use the mouse to drag from cell F2 downward to cell F6, thereby selecting a range of cells; then execute the **Edit/Fill ▶ Down** command.

G3 = =F3/F$2

	A	B	C	D	E	F	G
1		Homework	Quiz	Project	Final Exam	Total Points	Percent
2	Points	100	25	50	150	325	100%
3	Akiko	85	22	45	136	288	89%
4	Chad	78	19	46	122	265	82%
5	Maia	93	24	44	150	311	96%
6	Tariq	91	20	41	146	298	92%
7							

The final task is to set up column G to compute the overall percent scores. Since the numerical values within this column are percents, select the column by clicking on the G at the top of the column, and then click the Percent Style button found on the format toolbar. Next, select cell G2 and enter the formula "= F2/F$2". A "100%" now appears in cell G2. The $ symbol was typed in front of the row number of the divisor to fix the row number. Now, select and copy cell G2, and paste (or use **Fill/Down**) to enter the desired formulas into the cell range G3:G6. Each student's total points are divided by the value given in cell F2, and the result is written as a percent. If the $ symbol had been omitted, Akiko's percentage would be given by the incorrect formula "= F3/F3" rather than the correct formula "= F3/F$2", and similar errors would occur for Chad, Maia, and Tariq.

*In Excel, it is even easier to drag the copy handle at the lower right corner of a cell to fill downward or across.

SPREADSHEET FUNCTIONS

Spreadsheets have a variety of built-in functions that can be inserted into formulas. These include the arithmetic operations, using the familiar symbols +, −, *, /, and ^ for addition, subtraction, multiplication, division, and power, respectively. A few frequently used functions are shown in Table B.1.

TABLE B.1 Some Commonly Used Spreadsheet Functions

Function	Outcome
ABS(*number*)	absolute value of the number
SQRT(*number*)	square root of the number
SUM(*number 1, number 2, . . .*)	sum of the numbers
PI()	value of π
FACT(*number*)	factorial of the number
COMBIN(*n, r*)	number of combinations of n objects r at a time
PERMUT(*n, r*)	number of permutations of n objects r at a time
GCD(*number 1, number 2, . . .*)	greatest common divisor of the numbers
LCM(*number 1, number 2, . . .*)	least common multiple of the numbers
MIN(*number 1, number 2, . . .*)	minimum of the numbers
MAX(*number 1, number 2, . . .*)	maximum of the numbers
AVERAGE(*number 1, number 2, . . .*)	average (mean) of the numbers
MEDIAN(*number 1, number 2, . . .*)	median of the numbers
MODE(*number 1, number 2, . . .*)	mode of the numbers
STDEVP(*number 1, number 2, . . .*)	standard deviation of the population
STDEV(*number 1, number 2, . . .*)	standard deviation of the sample
QUARTILE(*array, k*)	kth quartile of the numbers in the array, where $k = 0, 1, 2, 3$, or 4

EXAMPLE B.2 SETTING UP DIFFY ON A SPREADSHEET

In DIFFY, one begins with a row of four numbers, say, 2, 23, 14, and 12. The next row of numbers is created by taking each successive difference of the numbers in the first row, always subtracting the smaller number from the larger. Thus, the first three numbers in the second row are $21 = 23 - 2$, $9 = 23 - 14$, and $2 = 14 - 12$. The last number of the row is obtained by using the first and last numbers of the first row, to give $10 = 12 - 2$. The third row is obtained from the second row in the same way that the second row was obtained from the first row, and additional rows are formed using the same pattern until something interesting happens. Set up a spreadsheet to enable you to explore DIFFY for many different choices of the four numbers in the first row.

Solution The four starting numbers are entered into columns A, B, C, and D and row 1. To ensure that the result of a subtraction is entered as a positive value, the absolute value of the difference is taken. For, example, cell A2 contains the formula "= ABS(A1 − B1)". Copying cell A2 and pasting into B2, C2, and D2 will enter the correct formulas into cells B2 and C2, but cell D2 will incorrectly read "= ABS(D1 − E1)". To correct the error, select cell D2 and edit the formula in the entry area of the formula bar so that it reads "= ABS(D1 − A1)". To enter subsequent rows, select and copy the range A2:D2, and paste into a rectangular range of selected cells with A3 at the upper left corner and a cell in column D (say, D10) at the lower right corner (or use the **Fill/Down** command).

D2 =ABS(D1-A1)

	A	B	C	D
1	2	23	14	12
2	21	9	2	10
3	12	7	8	11
4	5	1	3	1
5				

Once your spreadsheet has been set up, you are ready to explore DIFFY with ease. Try replacing the initial row with other starting values. Don't be afraid to insert some really wild choices, with negative numbers or numbers up in the millions or billions that may require you to widen the columns. If necessary, extend the formulas into new rows to see what eventually happens. Are you surprised at the results? In particular, try the beginning entries 31, 57, 105, and 193.

OTHER SPREADSHEET FEATURES

There are a large number of additional capabilities of a spreadsheet, including many that are useful in the classroom. For example, a spreadsheet can be used as a handy calculator, with the added advantages of being able to display and arrange the calculations and even save and print them.

A spreadsheet can also make some interesting graphs. For example, as shown in Figure B.2, the constant value 1 is entered into both cells A1 and A2. The formula

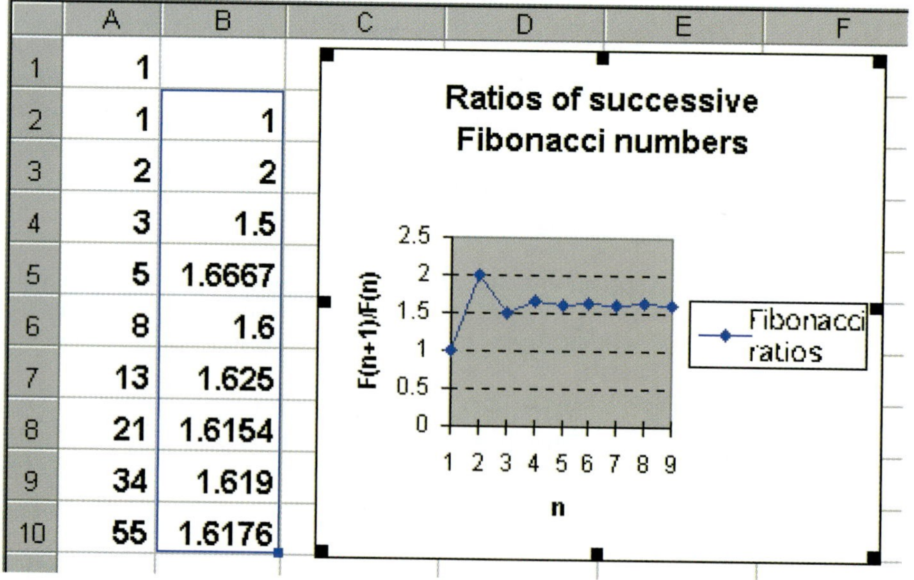

Figure B.2
A line graph created on a spreadsheet

"= A1+ A2" is entered into cell A3, and the formula is then copied down column A. This creates the Fibonacci number sequence 1, 1, 2, 3, 5, 8, 13, 21, 34, 55 in column A. Next, enter the formula "= A2/A1" into cell B2 and copy the formula downward into the rest of column B. It appears that the values in column B are becoming increasingly close to a limit value of about 1.61. This is shown visually by the line graph in the figure, which also reveals that the values in column B alternate between values greater and less than the limit value. To create the line graph, select the cell range B2:B10, click the chart wizard button, and follow the on-screen instructions.

CONCLUSION

Just a few spreadsheet features have been discussed in this appendix. Even so, we have been able to set up a useful electronic grade book and provide a tool to investigate number patterns. With a little instruction and encouragement, students in the upper elementary grades can quickly become proficient in spreadsheet basics. It is not surprising that more and more teachers are incorporating the spreadsheet as a valuable tool for teaching and learning.

Problem Set B

1. **Tracking Expenses.** Arik, Kaia, and Lena went to a teachers' conference for which they were to be reimbursed for their travel expenses by their school district. Arik traveled 252 miles and had meals and lodging expenses of \$57 and \$89, respectively. Kaia's expenses were 382 miles, \$72, and \$66, and Lena's expenses were 124 miles, \$48, and \$56. Mileage was reimbursed at 31.5¢ per mile.
   (a) Set up a spreadsheet for the three teachers' expenses. Compute each teacher's total dollar amount and the total amount of expense to the district.
   (b) Suppose Lena's mileage was mistakenly reported as one way, when in fact she should have reported round-trip mileage of 248 miles. Make this correction, and find Lena's correct total expense and total expense to the district.
2. **Exploring Number Sequences.** Place a 1 into cell A1, enter the formula "= 1 + A1" into cell A2, and copy the formula into the range A3 through A15. This will place the sequence 1, 2, 3, . . . , 15 into column A.
   (a) Enter the formula "= SUM(A\$1:A1)" into cell B1 and copy the formula downward into column B. Describe the number sequence you obtain.
   (b) Enter a 1 into cell C1, enter the formula "= C1 + A2" into cell C2, and copy the formula into column C. How does column C compare with column B?
   (c) Enter a 1 into cell D1, enter the formula "= C1 + C2" into cell D2, and copy the formula into column D below D2. What sequence of numbers is obtained? Describe how the same sequence can be generated in column E by referring to the values of column A.
3. **The Lucas Sequence.** Set up the spreadsheet shown in Figure B.2, giving the Fibonacci numbers and their ratios in columns A and B. To reproduce the line graph, select the range B2:B10 and click the Chart button (or use **Insert . . . /Chart**). Follow the on-screen instructions to finish the plot.
   (a) Replace the 1 in cell A2 with a 3, so that the Lucas sequence 1, 3, 4, 7, 11, . . . appears in column A. What is the tenth Lucas number? What is the ratio of the tenth to the ninth Lucas number? Does the sequence of ratios of successive Lucas numbers seem to have a limit value? How does it compare with the sequence of ratios of successive Fibonacci numbers shown in Figure B.2?
   (b) Replace the values in cells A1 and A2 with any two numbers of your choice. Does this appear to affect the limit value of the ratios?
4. **Fibonacci Fractions.** Select column A in a new spreadsheet document, and under **Format/Cells . . .** select the Custom category under the Number tab. In the box labeled Type, enter the code ??????/?????? to allow fractions with up to six digits in both numerator and denominator. Click the OK button to return to the main screen. Now place a 1 into cell A1, enter the formula "= 1 + 1/A1" into cell A2, and copy the formula down the column. Describe the number sequence that you

observe in the numerators and denominators of the values in column A.

5. **Fibonacci Identities.** Create a spreadsheet that contains the Fibonacci numbers 1, 1, 2, 3, 5, 8, . . . in column A.
   (a) In column B, calculate the sums 1, 1 + 1, 1 + 1 + 2, . . . of successive Fibonacci numbers. (*Suggestion:* Use the formula "= SUM(A$1:A1)" in cell B1 and fill downwards using this formula.) What connection to the Fibonacci numbers in column A do you observe?
   (b) In column C, calculate the squares of the Fibonacci numbers: 1, 1, 4, 9, 25, ... Similarly, in column D, calculate the sequence of sums of the squares—1, 1 + 1, 1 + 1 + 4, 1 + 1 + 4 + 9, . . . —and in column E calculate the sequence of products of successive pairs of Fibonacci numbers—1*1, 1*2, 2*3, 3*5, 5*8, . . . Describe how columns D and E are related, and use your observations to state a Fibonacci number identity.

6. **Statistical Computations.** Consider the data 3, 15, 37, 22, 5, 35, 19, 53, 44, 22, 45, 26, and 37. Use a spreadsheet to find
   (a) the mean and population standard deviation.
   (b) the lower and upper quartiles, and the median of the data.

7. **Problem Solving with a Spreadsheet.** Use a spreadsheet to answer the following problem: *What is the shape of the rectangular field of largest area that can be enclosed with 40 10-foot-long sections of fence?* Take advantage of several spreadsheet features, including graphing (see problem 3), to give a thorough analysis of the problem's solution.

Appendix C *Graphing Calculators*

INTRODUCTION

In this appendix, some of the most useful features of the graphing calculator are discussed and used to solve problems. By following along with a graphing calculator, and perhaps consulting its user's manual occasionally, you will learn some basic graphing-calculator skills. We hope you will be enticed to continue to develop your skills, enabling you to use the graphing calculator effectively both as a tool for solving mathematical problems and as an adjunct to mathematics instruction.

The graphing calculator is easily identified by its large screen, which can display multiple lines of text, graphs of equations, and statistical plots. In this appendix we will refer most directly to the Texas Instruments TI-73 Explorer™, a calculator expressly designed for use in upper elementary and middle school mathematics and science classrooms. Users of other graphing calculators, such as the TI-82 or TI-83, or even calculators from other manufacturers, should not find it too difficult to modify the procedures described here to apply to their particular model. The TI-73 is shown in Figure C.1.

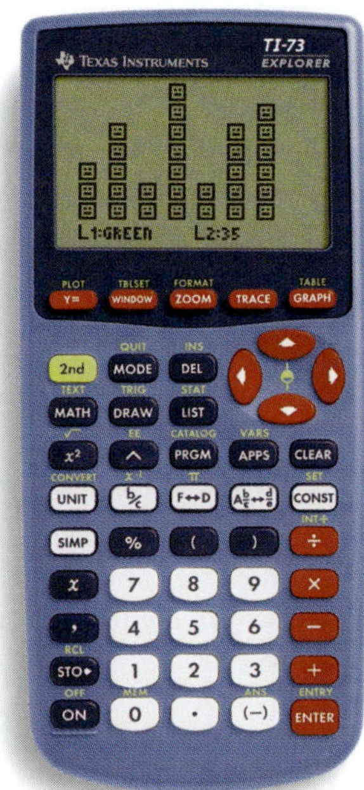

Figure C.1
The TI-73 graphing calculator

ENTERING, EVALUATING, AND EDITING EXPRESSIONS

The primary screen is called the **Home screen.** To return to the Home screen from a menu or application, press 2nd QUIT. Press CLEAR to show a blank screen. From the Home screen, expressions can be entered and evaluated as on any calculator. As a specific example, suppose that we wish to work with the expression $1.2\sqrt{(X)}$, finding its value for various choices of X. This expression is interesting, since it gives the approximate distance to the horizon in miles as observed from a height of X feet. For example, to estimate the distance to a ship on the horizon when observed from a 400-foot-high cliff, we can key in

1.2 [×] [2nd] [√] 400 [)] [ENTER].*

The output shows us that the ship is about 24 miles away. Unlike on a simple calculator, the input expression remains in view on a graphing calculator. Better yet, the expression can be edited and reused. For example, suppose you hike to a higher point at an elevation of 600 feet and want to know how far you can now see to the horizon. Press [2nd] [ENTRY], use the left arrow key to move the cursor over the 4, and then type "6" to replace the 4 with a 6. Pressing [ENTER] now gives the output 29.39387691. This answer contains more decimals than reasonable for an approximate value, so you may find it useful to open the [MODE] menu and use the arrow and Enter key to select, say, 1 decimal place. Return to the Home screen by pressing [2nd] [QUIT], and then redo your calculation by pressing the keys [2nd] [ENTRY] [ENTER] to get the value 29.4. That is, you can see 29.4 miles to the horizon from a height of 600 feet.

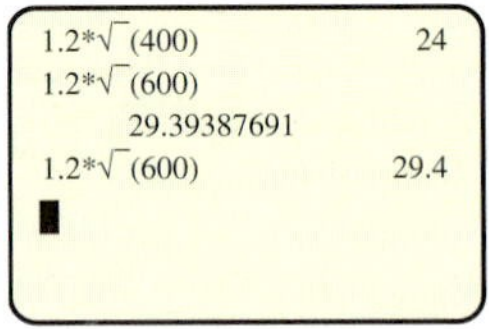

Expressions can be edited by using the arrow keys to position the cursor at the point of the desired change and typing over unwanted symbols. You can use the [DEL] (*delete*) key to remove unwanted symbols, and use [2nd] [INS] (*insert*) sequence to insert new symbols between symbols that you want to remain in place.

ENTERING AND USING LISTS IN THE HOME SCREEN

A **list** is an ordered sequence of numerical values. For example, suppose that we wish to create a list of the altitudes 100, 400, and 900 feet above sea level. The following entry string will create the list from the Home screen:

[{] 100 [,] 400 [,] 900 [}] [DONE]

On the TI-73, press [2nd] [TEXT] to obtain the required curly braces { and } from the text editor. (On the TI-82/83, press [2nd] [{] and [2nd] [}] to enter left and right braces.) The list can be stored to one of the six built-in list names†: L1, L2, L3, L4, L5, and L6. On the TI-73, use the commands [STO→] [2nd] [STAT] 1 [ENTER]. (On the TI-82/83, use [STO→] [2nd] [L1] [ENTER].) The part of the list that has scrolled off the screen to the right may be viewed by pressing the right-arrow key.

*On the TI-73, the square root appears with a left parenthesis automatically in place. A right parenthesis should be keyed in to enclose the expression whose square root is desired.

†On the TI-73, you can also type in a list name of your own choice.

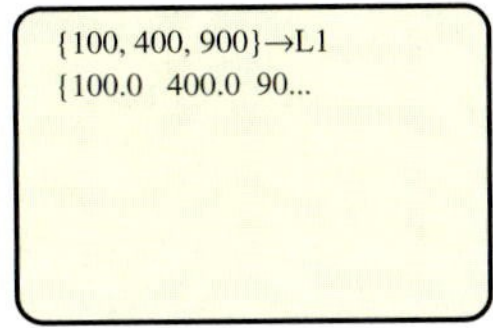

Now we can evaluate the function $1.2\sqrt{(X)}$, with X replaced by the list L1. This is shown on the next screen, where we have entered

1.2 [×] [2nd] [√‾] [2nd] [STAT] 1 [)] [ENTER]

on the TI-73. (On a TI-82/83, use 1.2 [×] [2nd] [√‾] [2nd] [L1] [ENTER].)

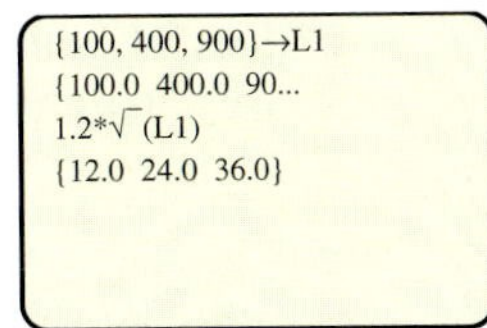

We see that the distances are 12.0, 24.0, and 36.0 miles to the horizon as viewed from the respective heights of 100, 400, and 900 feet.

Lists are a convenient way to generate a sequence of function values with little more effort than generating just a single value of the function. Lists are also used to store and manipulate data for statistical analyses and plots.

ENTERING AND GRAPHING EQUATIONS

As its name suggests, a graphing calculator can graph equations. There are three essential steps to follow:

Step 1 **Enter the function.** Press [Y =] to enter, edit, and select (or deselect) functions in the [Y =] editor. (*Note:* To select or deselect a function in the [Y =] editor, position the cursor over the = sign and press [ENTER]. Selected functions have a dark box containing the = sign and look like ■.)

Step 2 **Set windows values.** Press [WINDOW] to set the range of desired x- and y-values to show in your graph.

Step 3 **Display the graph.** Press [GRAPH] to see the selected function's graph.

EXAMPLE C.1 GRAPHING THE DISTANCE-TO-THE-HORIZON FUNCTION

Graph the distance-to-the-horizon function $y = 1.2\sqrt{X}$, for $0 < X < 1000$.

Solution

Step 1 Press the [Y =] key in the top row of your calculator. This opens the [Y =] editor, showing a list Y1 =, Y2 = ... (On a TI − 82/83, you may need to press and select Func (function) mode). A function of the variable X can then be entered into the space to the right of each of the equal signs. Use the arrow keys to move the cursor around the screen. The function $1.2\sqrt{X}$ has been entered on the first line shown here.

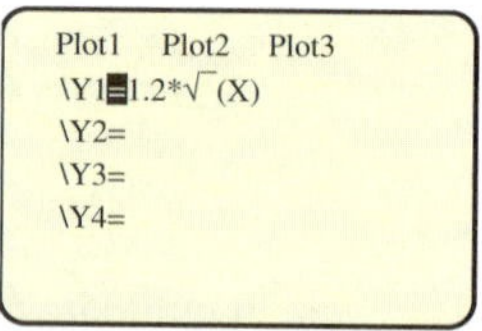

The variable *X* is entered into the formula using the [x] key (see Figure C.1 for the TI-73). (Use the [X,T,θ] key on the TI-82/83.) The dark box containing the equal sign indicates that the Y1 function is selected.

Step 2 Press the [WINDOW] key and enter appropriate minimum and maximum values of both *x* and *y*, as shown on the next screen. The spacing of tick marks along the *x*- and *y*-axes is set by choosing respective values of Xscl (*x-scale*) and Yscl (*y-scale*). (*Note:* ΔX will not appear on the TI-82/83.*)

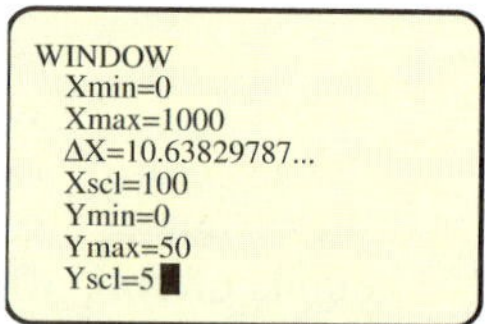

Step 3 Press [GRAPH] to display the graph. If another graph appears on the screen, it is necessary to deselect (turn off) any unwanted active functions in the [Y =] editor or statistical plots in the [2nd] [STAT] editor (or [2nd] [STAT PLOT] editor on the TI-82/83).

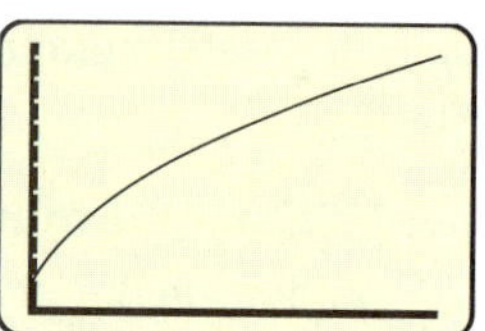

EXPLORING A FUNCTION GRAPH WITH [TRACE]

To move the cursor along points of the graph, press the [TRACE] key and use the left and right arrows to move the blinking cursor along the graph of the function. The coordinates of the cursor are displayed at the bottom of the screen. Continuing with Example C.1, the following screen shows that to see 30 miles to the horizon, you need a height of about 630 feet. Pressing 225 [ENTER] moves the cursor to the point on the graph where $X = 225$, and we see $Y = 10$. That is, the horizon is 10 miles away at a height of 225 feet.

*On the TI-73, setting Xmin and Xmax will automatically set the value of ΔX according to the formula $\Delta X = (\text{Xmax} - \text{Xmin})/94$. The value of ΔX determines the *x*-coordinates of a trace, as described below.

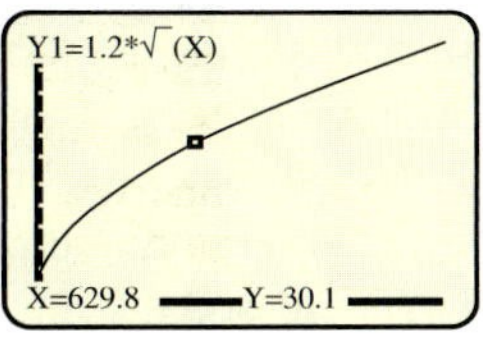

The next example will review and extend your graphing skills.

EXAMPLE C.2 GRAPHING THE ACHILLES-VERSUS-THE-TORTOISE RACE

Achilles and the Tortoise have agreed to compete in a 1000-yard race. At the starting signal, the tortoise lumbers off at a steady 3.8 feet per second (which is actually extremely fast for a tortoise!). Achilles, who knows he runs much faster than the tortoise, grandstands in front of the crowd and finally, after horsing around for 10 minutes, takes off for the finish line at a steady 14.7 feet per second. Does Achilles win the race, or has he been overconfident?

Solution

The tortoise will have traveled $Y1 = 3.8{*}X$ feet at a time of X seconds into the race. Achilles, however, doesn't start the race for 10 minutes (that is, for 600 seconds), so his distance toward the finish line is $Y2 = 14.7{*}(X - 600)$ for $X > 600$.* The distance to the finish line is $Y3 = 3000$ feet. All three functions are entered and selected as shown in the left screen below. The window editor is shown in the center screen, and the three graphs are shown on the right. It's a tight race!

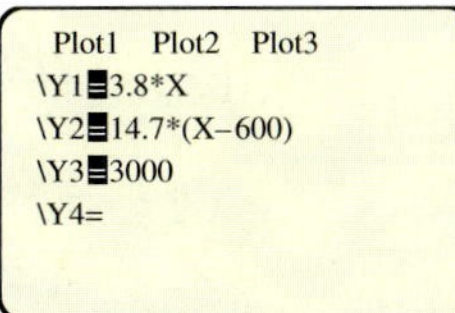

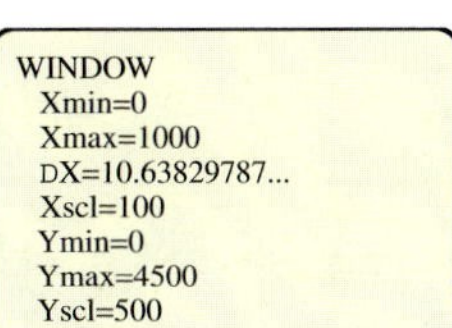

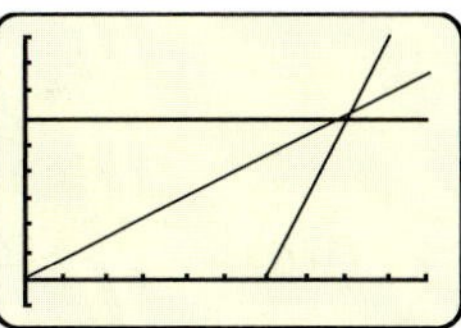

To get a closer look, use ZOOM and TRACE. First, press ZOOM and choose Zbox, as shown on the left below. To draw the small rectangular zoom box shown in the center screen below, use the arrow keys to position the cursor at a corner of the box and press ENTER. Then move the cursor with the arrow keys to the opposite corner of the box and press ENTER once more. This will give you the expanded graph shown in the rightmost screen.

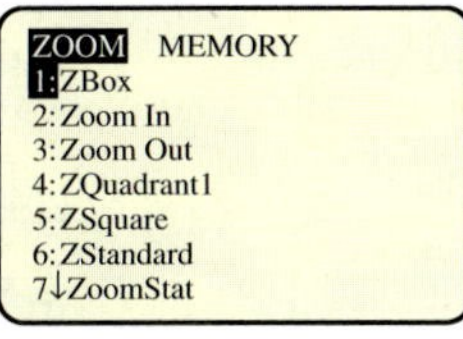

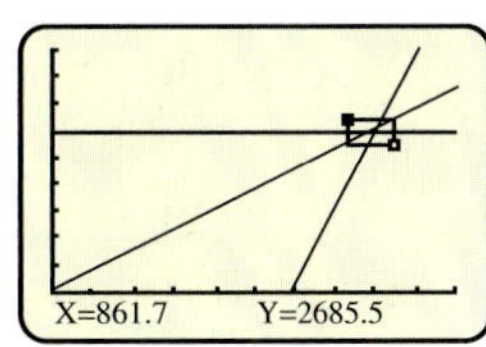

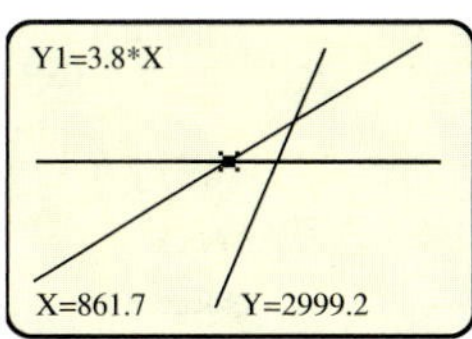

*Better yet, enter $Y2 = 14.7{*}(X - 600)\,(X > 600)$. Including the parenthetical condition $(X > 600)$ restricts the domain of Y2 to $X > 600$.

We now see clearly that Achilles has been overconfident. Pressing TRACE, we can position the cursor along the Y1 graph to show that the tortoise crossed the finish line at about 790 seconds. Pressing the down arrow puts the cursor on the graph of the Y2 function, and moving the cursor to the finish line with the right-arrow key, we see that Achilles crossed the finish line at about 804 seconds.

ENTERING AND EDITING DATA IN THE LIST EDITOR

There are five steps to define a list:

Step 1 Open the list editor by pressing LIST. (Use STAT **Edit . . .** on the TI-82/83.)

Step 2 Use arrow keys to position the cursor at a list, one of L1 through L6. To clear a list, select the list name at the top and press CLEAR ENTER.

Step 3 Enter the list values. Press ENTER or the down arrow to go to the next value.

Step 4 Edit the list as necessary, using 2nd INS, DEL, or CLEAR.

Step 5 Exit the list editor by pressing 2nd QUIT.

EXAMPLE C.3 CREATING LISTS OF STUDENT GRADES

Professor Garza wants to analyze how her students' performances on the 100-point quiz taken early in the semester compare with the overall percents received at the end of the semester. Help her enter the grade-book data of her 13 students into a graphing calculator.

Student	1	2	3	4	5	6	7	8	9	10	11	12	13
Quiz	88	58	92	82	84	60	94	100	74	40	78	96	94
Overall %	85	71	98	80	89	74	82	90	77	74	69	83	98

Solution

To analyze grades, it is enough to round values to the nearest integer, so first set the number of decimals to 0 in the MODE menu. Next, press LIST (or STAT **Edit . . .** on the TI-82/83) to open the list editor, as shown in the upcoming left-hand figure. The highlighted first line of the list also appears at the bottom of the screen, where we see that the first entry of list L1 is denoted by L1(1) and has the value of 100.0. List L1 was defined earlier and is no longer needed, so delete it by first using the up arrow to highlight the list name L1 on the top line and then pressing CLEAR ENTER. This will clear the list L1, as shown in the upcoming right-hand figure.

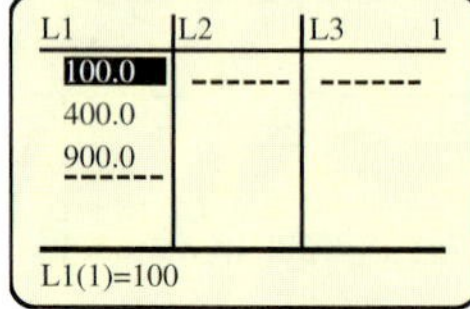

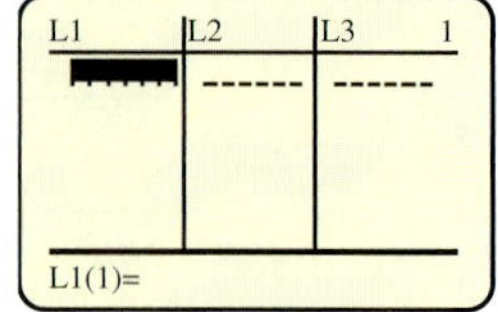

Now enter the quiz scores into list L1. When finished with list L1, use the right arrow to move to position L2(1). Now type in the corresponding overall-percent values. If you make a typing error, it can be corrected using the DEL, 2nd INS, and CLEAR editing keys. Your lists should appear as shown below.

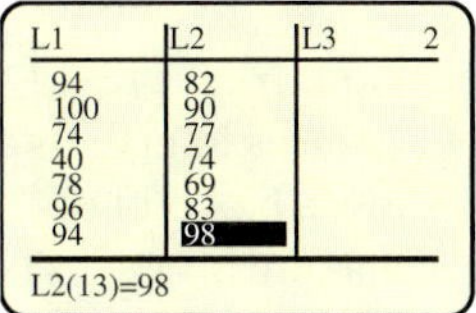

CREATING STATISTICAL PLOTS

A large number of statistical plots are easily created with a graphing calculator. Here we'll demonstrate two types, the scatter plot and the box plot, using the data entered into the lists shown in Example C.3. Pie charts, bar graphs, and histograms can also be plotted with a graphing calculator.

Scatter Plot

Press 2nd PLOT (use STAT PLOT on the TI-82/83) to open the Stat Plot Menu screen shown in the screenshot on the left. Press 1 to open the Stat Plot Editor of Plot 1, as shown in the screenshot on the right.

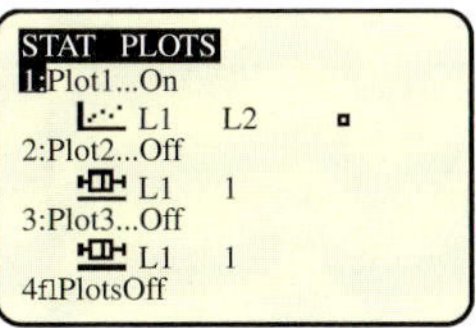

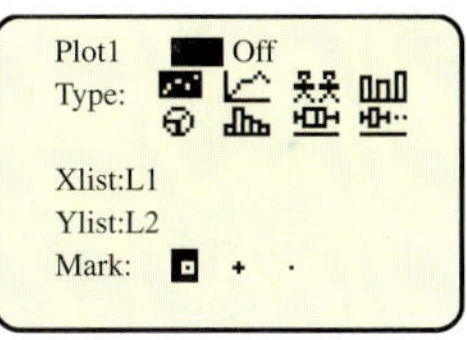

In the Stat Plot Editor, we have selected (turned on) Plot 1, selected a scatter plot from among the types of plots available, used 2nd STAT 1 (2nd L1 on the TI-82/83) to specify L1 as the Xlist, specified L2 as the Ylist, and chosen small open squares to mark the points in the plot.

There are three remaining steps:

- Turn off (deselect) any selected functions in the Y = editor.
- Set the parameters in the WINDOW values screen.
- Press GRAPH to display the scatter plot.

The window values screen is shown on the left, and the scatter plot is shown on the right.

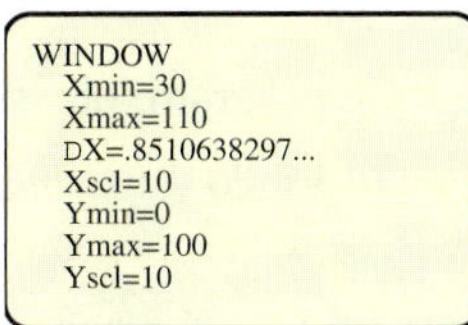

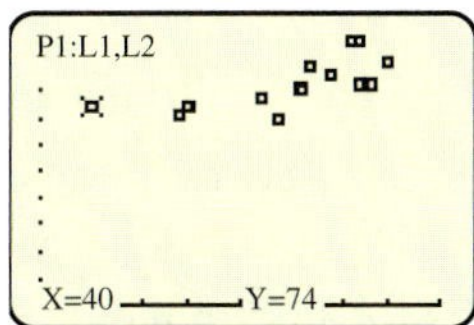

The TRACE key and arrow keys were used to identify the leftmost point in the plot. Its coordinates show that the student starting out with a 40% score on the first quiz was still able to achieve an overall 74% in the course.

Box Plots

Press 2nd PLOT 1 and turn off Plot 1. To look at Plot 1 again at a later time, we can turn Plot 1 back on. Next press 2nd PLOT 2 and turn on Plot 2. The leftmost screen below shows our choices to give us a box plot of the quiz scores contained in list L1. Similarly, set Plot 3 to show a box plot of the list L2. To enter the list name, highlight the list name shown by default and press 2nd STAT 2 to insert the list name L2, as shown on the right below.

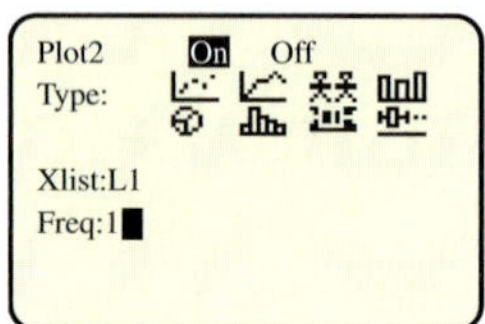

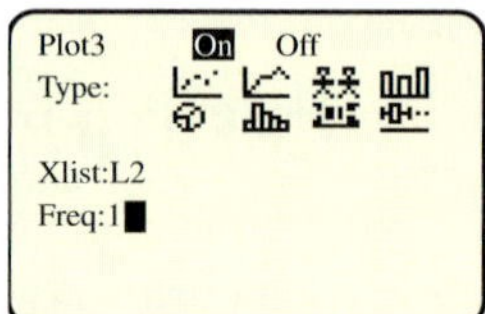

To show the two box plots, press WINDOW and select appropriate Xmin, Xmax, and Xscl values (the y-values are not used). Then press GRAPH. We have also pressed TRACE, so we can read off the maximum, minimum, median, and upper and lower quartile values, using the arrow keys. We see that the overall course median is 82%, just slightly lower than the quiz median of 84%. We also see how the minimum and lower quartiles improved from the first quiz to the final overall percentages earned.

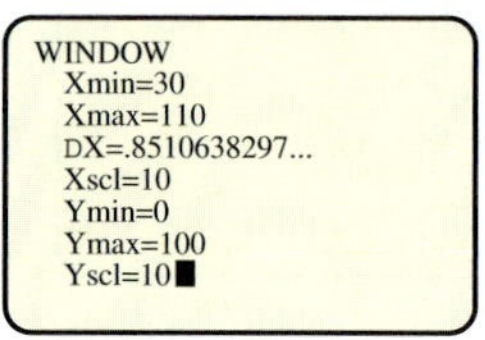

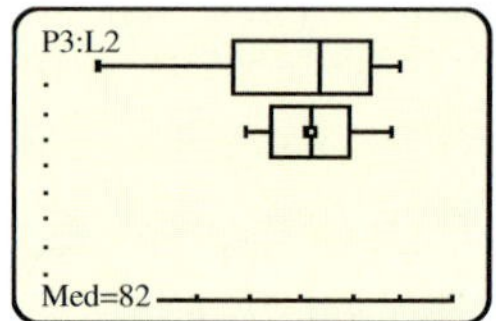

PROGRAMMING THE GRAPHING CALCULATOR

A **program** is a sequence of commands to be executed by the calculator. The procedures to create, edit, and execute programs can be found in the user's guide accompanying your calculator.

EXAMPLE C.4 WRITING A PROGRAM TO CALCULATE THE VOLUME OF A RIGHT CIRCULAR CYLINDER

Write a program named CYLINDER that computes the volume of a right circular cylinder. The program will ask for the radius R and height H of the cylinder to be input, and the program will return the volume of the cylinder, using the formula $\pi R^2 H$.

Solution To name and enter the program, press PRGRM **NEW/Create New.** Use 2nd TEXT to open the text editor, type in the program name CYLINDER, and then press ENTER. The cursor will now be positioned just after the colon at the first line of the program.

```
PROGRAM:CYLINDER
:Input "RADIUS
", R
:Input "HEIGHT
", H
:Disp "VOLUME", π
R²*H
:Pause
```

To enter the first command line, press PRGRM **I/O** (input/output) **Input.** You are automatically returned to the program editor, where "Input" now appears on the first line. Complete the first line by opening the Text editor and typing "RADIUS",R where the quotation marks designate a text entry. The DONE command in the Text editor returns you to the program editor. Then press ENTER to take you to the next line of the program. Enter the second line, **Input** "HEIGHT",H, the same way as the first line. The third line is entered by pressing PRGRM I/O **Disp** (display), then typing "VOLUME", and finally pressing , 2nd π × R 2nd x^2 × H to enter the expression $\pi R^2 H$, which computes the volume of a cylinder. The last line is entered by pressing PRGRM CTL (control) **Pause.** The **Pause** command gives you an opportunity to see the display before the program ends. The program is now complete, so exit the program editor by pressing 2nd QUIT.

To execute your program, press PRGRM EXEC, use the down arrow key to select the program CYLINDER, and then press ENTER. Press ENTER once more to start the program. The program will soon pause, giving you the opportunity to input the value of the radius of the cylinder followed by ENTER. You will next enter the cylinder's height, again followed by ENTER. The program then displays the volume of the cylinder of the radius and height you have entered. To run the program again, press ENTER. To quit a program at any time, press ON QUIT. The result of running CYLINDER for a cylinder with a radius of 5 and a height of 10 is shown below.

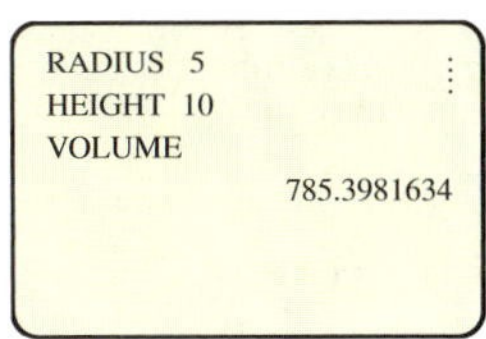

PROGRAMS FOR THE GRAPHING CALCULATOR

The programs for the TI-73 listed below can be typed into your calculator. You may also copy programs from one calculator to another with the unit-to-unit link cable that accompanies your calculator. Users of other TI graphing calculators (or even other brands) will require only minor changes of syntax to adapt these programs to their machine. For example, the **Clear Home** command of the TI-73 must be replaced with the **ClrHome** command on the TI-82/83. When the calculator pauses, say, to let you enter a value or view a displayed value or graph, you must press ENTER to execute the next step of the program. To quit a program press the ON key and then press 1 or ENTER (or, on the TI-82/83, press 2 or ENTER). You can also use the **Goto** command to edit the program you are quitting.

Program	Inputs Required	Output of Program
COINTOSS	Number of coins N and number of tosses T	How often x heads appeared in the T tosses, for $x = 0, 1, 2, \ldots, N$
DICE	Number of dice N and number of tosses T	How often the sum x appeared in the T tosses, for x between N and 6N. The values are displayed in a line graph. Use TRACE to read values on the plot.
DIFFY	Four numbers	See the Cooperative Investigation in Section 2.3.
DIVISORS	Two positive integers A and B	The divisors of A and B are given in side-by-side lists.
DIVVY	Four positive numbers	See problem 32, Problem Set 6.2.
EUCLID	Two positive integers A and B	Steps in Euclidean algorithm, leading to computation of GCD(A, B) and LCM(A, B)
FACTOR	Positive integer N	Prime power factorization of N
KAPREKAR	Initial positive integer	See problem 6, Problem Set 1.5.
MULTIPLS	Two positive integers A and B	The multiples of A and B are given in side-by-side lists.
PALINDRM	Positive integer	See problem 5, Problem Set 1.5.
RANDOM	A, B, and N	N random numbers between A and B

COINTOSS

```
Input "NUMBER OF COINS",N
Input "NUMBER OF TOSSES",T
ClrList L1
N + 1 → dim(L1)
For(X,1,T)
sum(seq(iPart(2*rand),X,1,N,1) → H
1 + L1(H + 1) → L1(H + 1)
End
ClrScreen
For(X,1,N + 1)
Output(2,1,X − 1)
Output(2,5,"HEADS")
Output(3,1,"CAME UP")
Output(4,1,L1(X))
Output(4,5,"TIMES")
Output(7,5,"PRESS ENTER")
Pause
ClrScreen
End
```

DICE

```
Input "NUMBER OF DICE",A
Input "NUMBER OF TOSSES",T
ClrList L1
6*A → dim(L1)
For(K,1,T)
sum(seq(iPart(6*rand + 1),X,1,A))
   → H
L1(H) + 1 → L1(H)
End
ClrScreen
For(X,A,6*A)
Output(2,1,"TOTAL")
Output(2,7,X)
Output(4,1,"OCCURRED")
Output(4,11,L1(X))
Output(5,1,"TIMES")
Pause
End
PlotsOff
seq(X,X,1,6*A) → L2
Plot1(xyLine,L2,L1)
ZoomStat
DispGraph
Pause
```

DIFFY

```
Clear Home
Disp "ENTER FOUR"
Disp "NUMBERS"
Input A
Input B
Input C
Input D
0 → I
Clear Home
Disp A,B,C,D
Repeat A + 0 and B + 0 and C + 0 and
D + 0
Pause
Clear Home
abs(A − B) → W
abs(B − C) → X
abs(C − D) → Y
abs(D − A) → Z
W → A
X → B
Y → C
Z → D
Disp A,B,C,D
I + 1 → I
End
Disp "TOTAL STEPS"
Disp I
Pause
```

DIVISORS

```
Repeat A > 0 and B > 0 and
   fPart(A) = 0 and fPart(B) = 0
Clear Home
Disp "ENTER TWO"
Disp "POSITIVE"
Disp "INTEGERS"
Input A
Input B
End
1 → C
0 → dim(L1)
0 → dim(L2)
For(I,1,A/2,1)
If fPart(A/I) = 0
Then
I → L1(C)
C + 1 → C
End
End
A → L1(C)
1 → C
```

```
For(I,1,B/2,1)
If fPart(B/I) = 0
Then
I → L2(C)
C + 1 → C
End
End
B → L2(C)
Clear Home
Output(1,1,"DIVISORS OF")
Output(2,1,A)
Output(3,1,"AND")
Output(4,1,B)
Pause
Clear Home
If dim(L1) > dim(L2)
Then
dim(L1) → K
Else
dim(L2) → K
End
0 → I
While(I < K)
For(C,1,8,1)
If I ≤ dim(L1) − 1
Then
Output(C,1,L1(I + 1))
End
If I ≤ dim(L2) − 1
Then
Output(C,9,L2(I + 1))
End
I + 1 → I
End
Pause
Clear Home
End
```

DIVVY

```
Float
Clear Home
Disp "ENTER FOUR"
Disp "NUMBERS"
Input A
Input B
Input C
Input D
0 → 1
Clear Home
Disp A,B,C,D
Repeat abs(A − 1) < 10^ −9 and
   abs(B − 1) < 10^ −9 and
   abs(C − 1) < 10^ −9 and
   abs(D − 1) < 10^ −9
Pause
Clear Home
If A ≥ B
Then
(A/B) → W
Else
(B/A) → W
End
If B ≥ C
Then
(B/C) → X
Else
(C/B) → X
End
If C ≥ D
Then
(C/D) → Y
Else
(D/C) → Y
End
If A ≥ D
Then
(A/D) → Z
Else
(D/A) → Z
End
W → A
X → B
Y → C
Z → D
Disp A,B,C,D
I + 1 → I
End
Disp "TOTAL STEPS"
Disp I
Pause
```

EUCLID

```
Repeat A > 0 and B > 0 and
  fPart(A) = 0 and fPart(B) = 0
Clear Home
Disp "INPUT TWO"
Disp "POSITIVE INTS"
Input A
Input B
End
If B > A
Then
A → C
B → A
C → B
End
A → X
B → Y
While 1
iPart(A/B) → Q
round(B*fPart(A/B)) → R
Clear Home
Output(1,1,A)
Output(1,iPart(log(A) + 3), "=")
Output(2,1,Q)
Output(2,iPart(log(Q) + 3), "*")
Output(2,iPart(log(Q) + 5),B)
Output(3,1, "+")
Output(3,3,R)
Pause
B → T
B → A
R → B
If R < .1
Then
Clear Home
Output(1,1,"A =")
Output(1,3,X)
Output(2,1,"B =")
Output(2,3,Y)
Output(3,1,"GCD(A,B) =")
Output(4,3,T)
Output(5,1,"LCM(A,B) =")
Output(6,1,"A*B/GCD(A,B) =")
Output(7,3,(X*Y/T))
Pause
Stop
End
End
```

FACTOR

```
Prompt N
ClrList(L1,L2)
1 → S
2 → P
0 →
√(N) → M
While P ≤ M
While fPart(N/P) = 0
E + 1 → E
N/P → N
End
If E > 0
Then
P → L1(S)
E → L2(S)
0 → E
S + 1 → S
√(N) → M
End
If P + 2
Then
3 → P
Else
P + 2 → P
End
End
If N ≠ 1
Then
N → L1(S)
1 → L2(S)
End
Disp "PRIME FACTORS",L1
Disp "POWERS",L2
Pause
```

KAPREKAR

```
Repeat N > 0 and fPart(N) = 0
Clear Home
Disp "ENTER A"
Disp "POSITIVE"
Disp "INTEGER"
Prompt N
End
While N ≠ 0
0 → A
0 → D
0 → dim(L1)
For(C,1,iPart(log(N)) + 1,1)
round(10*fPart(N/10)) → L1(C)
iPart(N/10) → N
End
SortA(L1)
For(C,1,dim(L1),1)
D + L1(C)*10^ (C − 1) → D
A + L1(C)*10^ (dim(L1) − C) → A
End
Clear Home
Output(1,3,D)
Output(2,1, "−")
Output(2,3,A)
Output(3,3,D − A)
Pause
D − A → N
End
```

MULTIPLS

```
Repeat A > 0 and B > 0 and
  fPart(A) = 0 and fPart(B) = 0
Clear Home
Disp "ENTER TWO"
Disp "POSITIVE"
Disp "INTEGERS"
Input A
Input B
End
0 → dim(L1)
0 → dim(L2)
If B > A
Then
For(I,1,B,1)
A*I → L1(I)
B*I → L2(I)
End
Else
For(I,1,A,1)
A*I → L1(I)
B*I → L2(I)
End
End
Clear Home
Output(1,1,"POSITIVE")
Output(2,1,"MULTIPLES OF")
Output(3,1,A)
Output(4,1,"AND")
Output(5,1,B)
Pause
Clear Home
1 → I
While(I ≤ dim(L1))
For(C,1,8,1)
If I ≤ dim(L1)
Then
Output(C,1,L1(I))
End
If I ≤ dim(L2)
Then
Output(C,9,L2(I))
End
I + 1 → I
End
Pause
Clear Home
End
```

PALINDRM

```
Repeat N > 0 and fPart(N) = 0
Clear Home
Disp "ENTER A"
Disp "POSITIVE"
Disp "INTEGER"
Prompt N
End
0 → R
N → T
While 1
For(C,iPart(log(N)),0, − 1)
R + (round(10*fPart(T/10))*10^
  C) → R
iPart(T/10) → T
End
If N + R
Then
Clear Home
Output(1,1,N)
Output(2,1,"IS A
  PALINDROME")
Pause
Stop
End
Clear Home
Output(3,12 − iPart(log(N)),N)
Output(4,12 −
  (iPart(log(R)) + 2), "+")
Output(4,12 − iPart(log(R)),R)
Output(5,12 −
  iPart(log(N + R)),N + R)
N + R → N
N → T
0 → R
Pause
End
```

RANDOM

```
Clear Home
Disp "RANGE MIN"
Input A
Disp "RANGE MAX"
Input B
Disp "HOW MANY"
Disp "NUMBERS?"
Input N
For(X,1,N,1)
Disp((B − A)*rand + A)
Pause
End
```

Problem Set C

1. **Calculating Values.** According to Newton's inverse-square law of gravity, an object's weight is inversely proportional to the square of the object's distance from the center of the Earth. Assuming the Earth's radius is 4000 miles, this means that a person who weighs 150 pounds on the surface of the Earth will weigh $150\left(\dfrac{4000}{4000 + X}\right)^2$ at an altitude of X miles above the Earth's surface. The function giving the person's weight is therefore Y1 $= 150 * (4000/(4000 + X))^2$, as written in the form entered into a calculator.
   (a) Find the person's weight at an altitude of 500 miles.
   (b) Find the weights of the person at the heights 1000, 2000, and 3000 miles, using a list to replace the single numerical value.

(c) Use guess and check to estimate the altitude at which the person's weight is 75 pounds.

2. **Making a Table.** Enter the function Y1 = 150 * (4000/(4000 + X))^2, introduced in problem 1, using the [Y =] editor. Press [2nd] [TBLSET] (*table setup*) and define TblStart = 0 and ΔTbl = 500. To see the table of Y1 values this creates, press [2nd] [TABLE] Use the arrow keys to find a value in the table that estimates the altitude at which the person's weight is just 15 pounds, 10% of that on the Earth's surface.

3. **Making a Graph.** Consider again the weight function of problem 1.

   (a) Graph the function Y1 = 150 * (4000/(4000 + X))^2.

   (b) Use [TRACE] to estimate the altitude at which the person weighs 25 pounds.

   (c) Use [TRACE] to estimate the person's weight at an altitude of 4000 miles.

4. **Making a Piecewise Graph.** Imagine a deep well, so deep it extends clear to the center of the Earth. The weight of a 150-pound person lowered into the well is given by Y2 = 150 * (X/4000) (X < 4000), where X is the distance from the center of the Earth. For values of X larger than the 4000-mile radius of the Earth, the person's weight varies by the more familiar inverse-square law. That is, the person's weight is given by the function Y3 = 150 * (4000/X)^2 (X > 4000). (*Note:* The conditions (X < 4000) and (X > 4000) listed in parentheses limit the definitions of Y2 and Y3 to their respective domains.)

   (a) Enter the two functions Y2 and Y3 given above and obtain their graphs. (The symbols < and > are found in the text editor, [2nd] [TEXT] on the TI-73 and [2nd] [TEST] on the TI-82/83.)

   (b) Use [TRACE] to find the two distances from the Earth's center, where the person's weight is about 32 pounds.

5. A ball is thrown upward (and slightly outward) from the roof of a 100-foot-high building with an initial upward velocity of 135 feet per second. Its height above the ground after X seconds is given by the function Y1 = 100 + 135 * X − 16 * X^2. Use the graphing calculator to

   (a) graph the time-versus-height function of the ball's motion,

   (b) estimate the maximum height reached by the ball and the number of seconds into the motion when this height is reached, and

   (c) estimate the time into the motion, to the nearest second, when the ball strikes the ground.

6. Graph the following four functions:

   Y1 = X   Y2 = 2 * X   Y3 = (1/2) * X   Y4 = −X

   To see the true shape of the graphs, use [ZOOM] **ZSquare** so the scales of both the x- and y-axes are the same. Consider Y1 as the basic function and Y2, Y3, and Y4 as modifications of the basic function.

   (a) Carefully describe what is similar about the graphs of the modified functions and the graph of the basic function.

   (b) How are the graphs of the modified functions different from that of the basic function?

   (c) Predict the appearance of the graphs of the following functions, and then check the accuracy of your predictions by editing the functions in your calculator and redrawing the graphs.

   Y2 = 5 * X   Y3 = (1/5) * X   Y4 = −5 * X

7. Graph the function Y1 = {1, 5, 1/5, −1} * X^2. Using the list {1, 5, 1/5, −1} as the coefficient of x^2 gives a simple way to simultaneously graph the four functions $y = x^2$, $y = 5x^2$, $y = (1/5)x^2$, and $y = -x^2$. Sketch the four parabolic graphs, and associate each with a corresponding parameter 1, 5, 1/5, or −1. Give reasons for your choices.

8. **Making a Pie Chart on the TI-73.** The students in Mr. Hsu's fifth-grade class took a poll of their favorite ice cream flavors. They discovered that vanilla was the favorite of 9 students, chocolate was favored by 15 students, strawberry by 4 students, and raspberry by 8 students. Enter the categories in list L1, either by number (1, 2, 3, and 4) or (using the text editor) by name, say, "VAN", "CHO", "STRA", and "RAS" (the quotation marks distinguish a text entry from a numerical entry). In list L2, enter the corresponding sequence of numbers: 9, 15, 4, and 8. Now press [2nd] [PLOT] 1 and turn on Plot 1, set the plot type as a pie chart, set CategList to L1, set Data List to L2, and choose either **Number** or **Percent.** Press [GRAPH] to see your pie chart. What percent of the children favor chocolate?

9. **Making a Bar Chart on the TI-73.** Mrs. O'Leary's class has 4, 12, 9, and 10 students whose respective favorite flavors of ice cream are vanilla, chocolate, strawberry, and raspberry. Mr. Hsu's class has the corresponding numbers 9, 15, 4, and 8. (See problem 8.) Enter the values 9, 15, 4, and 8 into list L2 and 4, 12, 9, and 10 into list L3. Use the Stat Plot editor to make a bar chart to compare the two classes. Set CategList to L1, DataList1 to L2, DataList2 to L3, and the number of datalists to 2 so that the choice of DataList3 is not required. Do both classes have the same most common favorite flavor? the same least common favorite flavor?

10. Making a Histogram. The table below shows the grades for Mr. Hsu's fifth-grade class project. Two students received 100, one student had a 97, three students received a 94, and so on, as shown by the frequency list. Scores 90–100 will receive an A, 80–89 a B, 70–79 a C, 60–69 a D, and 0–59 an F. Make a histogram of the grade distribution by entering the scores in list L1 and the corresponding frequencies in list L2. Turn off all plots except Plot 1, and choose list L1 as the Xlist and list L2 as the Freq list in the plot editor. To group the scores by the grade intervals, set the window values to Xmin = 40, Xmax = 100, Xscl = 10, Ymin = −5, Ymax = 15, and Yscl = 1. The number of values within any interval can be displayed by tracing the histogram. Which letter grade is most common on the project? How many A grades are given?

Score	100	97	94	92	91	88	87	84	83	82	80	77	74	71	69	62	53	47
Frequency	2	1	3	2	2	1	4	2	3	1	2	3	4	2	1	1	1	1

Appendix D A Brief Guide to The Geometer's Sketchpad

Exciting new possibilities for exploring geometric concepts on the computer are available with geometry software. This software allows the user to construct geometric figures with both speed and precision. The software will also give the measures of angles, lengths of segments, and areas of regions. Once a figure is constructed, it can be manipulated to a continuum of new shapes that preserve the geometric relationships used to construct the original figure. In this way, the geometric properties of the configuration can be explored in a dynamic environment not possible with traditional paper-and-pencil sketches.

Several geometry programs are currently available, including the following:

Cabri Geometry™ (Texas Instruments)
The Geometry Inventor (Broderbund)
The Geometer's Sketchpad (Key Curriculum Press, Inc.)

This appendix provides a short introduction to *The Geometer's Sketchpad.* More advanced features of Sketchpad are described in the *Learning Guide* and *Reference Manual* that accompany the software. Users of other programs will need to refer to the manual pertinent to their own software. Even so, there is much common ground, and it should not be difficult to modify the procedures in the examples below to accommodate the software being used.

THE SKETCH WINDOW

The Sketch Window (Windows platform) is shown in Figure D.1.

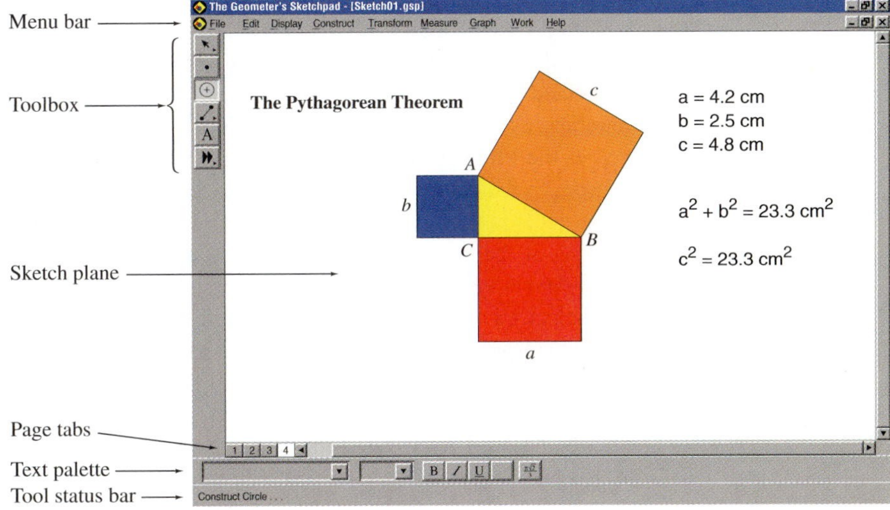

Figure D.1
*The Sketch Window (*Windows *platform) in* The Geometer's Sketchpad, *showing a caption; a labeled sketch; measurements of lengths* a, b, *and* c; *and calculations that illustrate the Pythagorean theorem*

- **Menu bar** Dragging downward accesses a list of commands available to create and investigate figures.
- **Toolbox** Clicking the icon activates the corresponding tool.
- **Sketch plane** The area where drawings appear.
- **Page tabs***
- **Text palette***
- **Tool status bar** Shows which tool is active.

USING THE TOOLBOX

The Toolbox, shown in Figure D.2, is located along the left side of the sketch window. It contains tools for selecting; dragging; creating basic objects such as points, circles, and lines; labeling objects and writing captions; and accessing a menu to use or create custom tools. Clicking on a tool icon activates the corresponding tool.

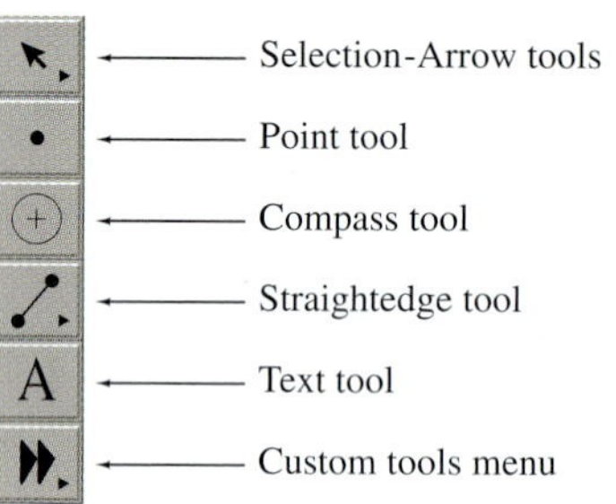

Figure D.2
The Toolbox

When the Selection Arrow is the active tool, clicking on an object in the sketch plane will cause the object to be outlined in pink, indicating that the object is selected. For example, if you select a line segment, you can then open the Display Menu to make the segment thick or dashed or change its color. Clicking on a sequence of objects will create a group of selected objects. For example, if you select two points, you can open the Construct Menu and use the **Circle by Center + Point** command to construct the circle that is centered at the first selected point and passes through the second selected point. Notice that the order in which objects are selected may be quite important. An easy way to select multiple objects is to use the Selection Arrow to click and drag a box, since this action will select all of the objects within the box. If an unwanted object is inadvertently selected, it can be deselected by clicking on it, which will also cause its pink outline to be removed. To deselect all objects, click on any blank space in the sketch plane.

Selection-Arrow tools
Translate Rotate Dilate

There are actually three Selection-arrow tools, accessed by clicking and dragging the arrow icon. All three tools can be used to select an object or group of objects in the figure. The **Translate tool** is used to translate the selected objects to other positions in the sketch window. The **Rotate tool** is used to rotate the selected object around a point previously marked as the rotation center. Similarly, the **Dilate tool** changes the size of the selected objects by performing a size transformation about the marked center. The center

*Available in version 4 of The Geometer's Sketchpad.

point of a rotation or dilation is chosen by selecting the point and executing the **Mark Center** command under the Transform Menu. Alternatively, double-clicking a point will mark it as the center.

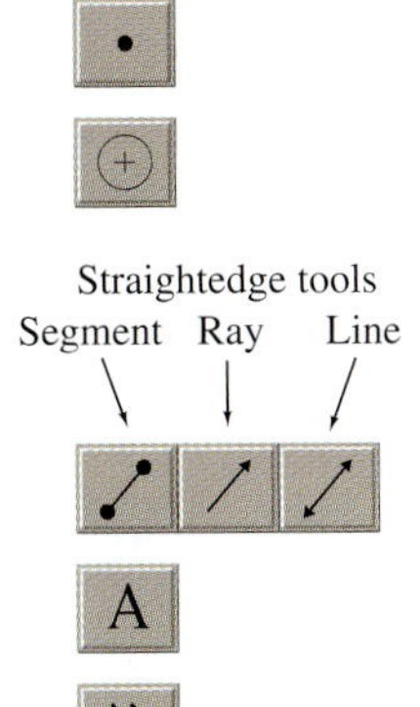

The **Point tool** creates points. Each click constructs a point at the cursor position in the sketch plane.

The **Compass tool** creates a circle by clicking on a point and dragging the cursor to a second point in the sketch plane. The first point is the center of the circle, and the point at which the mouse button is released is a point on the circle.

There are three **Straightedge tools,** activated by clicking and dragging the straightedge tool icon in the toolbox. The **Segment, Ray,** and **Line tools** create, respectively, a segment, ray, or line by clicking on a point and dragging the cursor to a second point in the sketch plane.

The **Text tool** enables you to label objects in the sketch plane. You can also add captions and titles by double-clicking in an empty part of the sketch plane; this opens a text box in which you type information.

The **Custom-Tools menu** contains a list of available custom tools* and the commands for creating new tools.

EXAMPLE D.1 DRAWING A TRIANGLE WITH THE TOOLBOX

Draw a triangle and investigate how it can be labeled and manipulated.

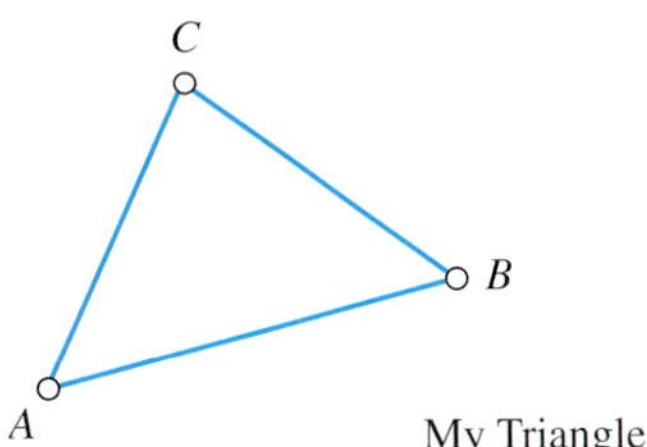

Solution Begin with a clear sketch plane. (*Note:* The **New Sketch** command in the File Menu will create a new sketch window with a blank sketch plane.) Activate the Segment tool, and move the cursor back to the sketch plane. Now click and drag from point to point to draw a triangle. Next, return to the toolbox and activate the Text tool. Clicking the pointer when it turns black provides a label of the vertex or side of your triangle. A second click will hide the label. Double-clicking a label allows you to edit the label. With the Text tool still active, click and drag at an empty place in the sketch plane to create a box in which text can be typed.

The use of the toolbox can be explored in "free play" experimentation. You will quickly discover how to construct circles, rays, line segments, and polygons. However, for more sophisticated constructions, you will want to use commands from the Construct and Transform Menus.

*A *custom tool* carries out a saved subroutine of construction steps. For example, if a figure you wish to construct requires several squares, it is much easier to use a custom tool than to carry out all of the steps to create one square at a time. Information on using and creating custom tools is found in both the help menu and the *Learning Guide* that accompanies the software. Custom tools are new to version 4 of Sketchpad; users of earlier versions of the software can use the similar idea of a script.

Construct
Point On Object Midpoint Intersection
Segment Ray Line Parallel Line Perpendicular Line Angle Bisector
Circle By Center + Point Circle By Center + Radius Arc On Circle Arc Through 3 Points
Interior
Locus

USING THE MENUS

The Construct Menu

An alternative method to create a line segment is to select the desired endpoints of the segment in the sketch plane. The **Segment** command will construct the segment between the selected points. Rays and lines are constructed similarly, where the first point selected is the endpoint of the ray. Each command in the Construct Menu requires certain objects in your sketch to be selected in advance. When the command appears a light gray color in the menu, you do not have the proper objects selected to execute that command.

EXAMPLE D.2 CONSTRUCTING AN EQUILATERAL TRIANGLE, THE MEDIANS, AND THE CENTROID WITH THE CONSTRUCT MENU

Construct any two points A and B. Then draw an equilateral triangle ABC, and construct the midpoints of the sides. Construct each segment from the vertex of the triangle to the midpoint of the opposite side. These three segments, the medians, are concurrent at the centroid G of the triangle.

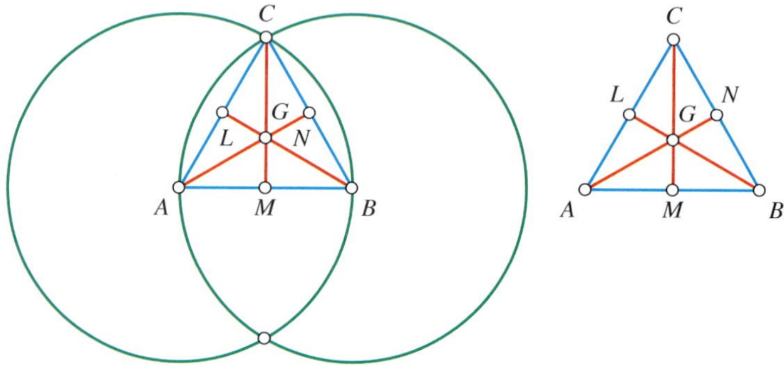

Solution Activate the Point tool from the toolbox, and place two points A and B in the sketch plane. Choose the Select tool from the toolbox and select A and then B. Use the **Circle By Center + Point** command to construct the circle centered at A that passes through B. In the same way, select point B and then point A to construct the circle centered at B that passes through point A. Next, select the two circles and use the **Intersection** command. The two points at which the circles intersect will be constructed. Use the Text tool to label one of these point C. Now select A, B, and C, and execute the **Segment** command from the Construct Menu. This will construct the sides of the desired equilateral triangle ABC. The three sides just drawn are automatically selected, so the **Midpoint** command will construct the midpoints of the sides of the triangle. Label them L, M, and N, with the Text tool. The centroid G of the triangle is now easy to construct: draw $\overline{AN}$ and $\overline{BL}$ and then construct the point, labeled G, at which these two segments intersect. Selecting the circles and executing the **Hide** command from the Display Menu will hide the circles from view.

THE DIFFERENCE BETWEEN A CONSTRUCTION AND A DRAWING

The triangle drawn in Example D.1 has no special property other than being a triangle. By dragging any vertex or side, the triangle can be manipulated to assume any triangular shape (including the degenerate case where the three vertices lie on the same line). It is possible to drag point C so that $\triangle ABC$ appears to be an equilateral triangle. In this way, you will have *drawn* an "equilateral" triangle. However, dragging a vertex quickly changes the triangle to become nonequilateral, so you can easily check that you have not constructed a true equilateral triangle. On the other hand, the triangle created in Example D.2 is equilateral by *construction.* Once points A and B were given, the point C was *constructed* to make $\triangle ABC$ equilateral. You can still drag either point A or B, but point C will always move so that $\triangle ABC$ remains equilateral. It is often best to use commands from the Construct Menu to construct a figure, since figures created with the Toolbox are frequently just drawings.

IMPORTANT TIP: *Construct Your Figure, Don't Just Draw It*
Determine whether you have made a construction or a drawing by dragging several different test points in your figure and checking angle and length measurements.

- **The Drag Test.** If the correct geometric relationships are retained in the figure throughout the manipulation, you likely have constructed the figure.
- **The Measurement Test.** If incorrect geometric relationships become apparent (for example, a "right angle" changes size), you know that you have made a drawing, but not a construction.

Transform
Mark Center Mark Mirror Mark Angle Mark Ratio Mark Vector Mark Distance
Translate . . . Rotate . . . Dilate . . . Reflect
Iterate . . .

THE TRANSFORM MENU

The Transform Menu allows objects to be translated, rotated, dilated, and reflected. For example, to perform a reflection, select the desired line of reflection and execute the **Mark Mirror** command. Next, back in the sketch plane, select all of the objects that you wish to reflect. Now execute the **Reflect** command. This will construct the reflection of the selected objects across the mirror line. Translations, rotations, and dilations are performed in a similar way, and again some experimentation will quickly make it clear how the commands in the Transform Menu are used in constructions and investigations.

EXAMPLE D.3 CREATING A HEXAGONAL TILING WITH THE TRANSFORM MENU

Construct a hexagon $ABCDEF$ with a pair of opposite sides $\overline{AB}$ and $\overline{DE}$ that are parallel and congruent. Construct the midpoint M of $\overline{BC}$, and rotate the hexagon by 180° about the midpoint to create a 10-sided polygonal tile. Then show that translations of the decagon will tile the plane.

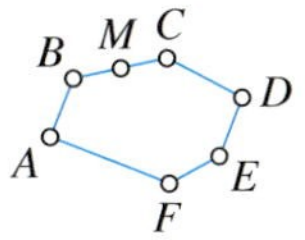

B M C D A E F

Solution

Use the Segment tool to construct three sides $\overline{AB}$, $\overline{BC}$, and $\overline{CD}$. Select points B and D, in that order, and execute the **Mark Vector** command in the Transform Menu. Next, select point A and segment $\overline{AB}$, and then execute the **Translate . . . By Marked Vector** command in the Transform Menu. This will extend $ABCD$ to become $ABCDE$, where $\overline{DE}$ is parallel to and the same length as $\overline{AB}$. Now complete the hexagon by constructing point F and segments $\overline{EF}$ and $\overline{FA}$. Construct the midpoint M by selecting the segment $\overline{BC}$ and executing the **Midpoint** command from the Construct Menu. Now select M and execute the **Mark Center** command to choose M as the center of rotation. Next, select the hexagon (all six sides and all six vertices) and execute the **Rotate . . . By Fixed Angle** (namely, by 180°) command. This creates the desired 10-sided tile. Selecting the tile and repeatedly executing the **Translate . . . By Marked Vector** command will quickly produce a partial tiling of the plane.

Measure

Length
Distance
Perimeter
Circumference
Angle
Area
Arc Angle
Arc Length
Radius
Ratio

Calculate . . .

Coordinates
Abscissa (x)
Ordinate (y)
Coordinate Distance
Slope
Equation

THE MEASURE MENU

Distances, lengths, areas, perimeters, angles, and so forth can be measured by executing commands in the Measure Menu. For example, selecting three points A, B, and C (in that order) and executing the **Angle** command will display the measure of $\angle ABC$ in the sketch plane. The measurement caption can be dragged to any convenient position in the sketch plane.

Algebraic and trigonometric expressions in terms of measurements of the figure under investigation can be formed by selecting the relevant measures and then invoking the **Calculate . . .** command of the Measure Menu. The expression and its value are then displayed in the sketch plane.

EXAMPLE D.4 EXPLORING THE CONVERSE OF THE PYTHAGOREAN THEOREM WITH THE MEASURE MENU

Draw any triangle, and use the Text tool to label the three vertices A, B, and C and the three respective opposite sides a, b, and c. Measure the lengths a, b, and c of the three sides and measure $\angle ACB$. Use the **Calculate . . .** command to display $a^2 + b^2 - c^2$. What kind of a triangle is ABC when $a^2 + b^2 - c^2$ is zero? positive? negative?

$a = 7.2$ cm
$b = 3.4$ cm
$c = 6.5$ cm
$a^2 + b^2 - c^2 = 22.0\text{ cm}^2$
$m\angle ACB = 63.7°$

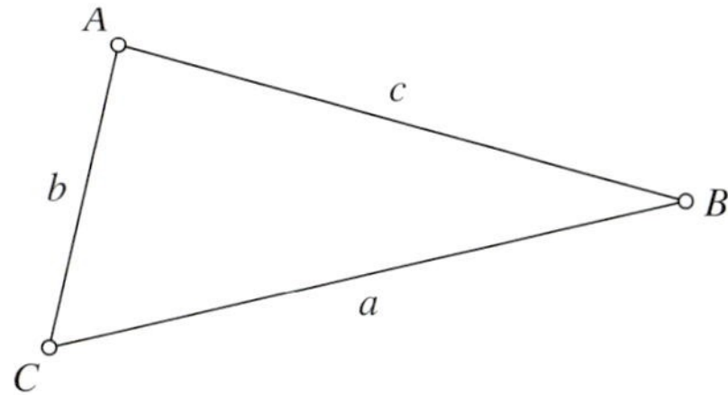

Solution We have already described how a triangle can be drawn. Manipulating the triangle should reveal that $\angle ACB$ is a right angle precisely when the expression $a^2 + b^2 - c^2$ is 0. If $a^2 + b^2 - c^2 > 0$, the angle measure is less than 90°. On the other hand, if $a^2 + b^2 - c^2 < 0$, then the angle measure at vertex C is greater than 90°.

Graph
- Define Coordinate System
- Mark Coordinate System
- Grid Form ▸
- Show/Hide Grid
- Snap Points
- Plot Points . . .
- New Parameter . . .
- New Function . . .
- Plot New Function . . .
- Derivative
- Tabulate
- Add Table Data . . .
- Remove Table Data . . .

THE GRAPH MENU

The **Define Coordinate System** command turns the sketch plane into a coordinate plane, showing the coordinate x- and y-axes and points at the origin, $(0, 0)$, and at $(1, 0)$. Dragging these points allows you to reposition the origin and vary the scale of the coordinate system. The **Grid Form** command permits you to choose among three grid forms. In the square grid form, the x- and y-axes have the same scale. In the rectangular form, the two axes can be scaled independently of one another. In the polar grid form, the grid lines are circles and rays from the origin, and points are plotted by their distance r from the origin and their counterclockwise angle θ from the positive x-axis.

The **Plot Points . . .** command opens a dialog box in which you enter the coordinates of the points you wish to plot in the coordinate plane.

EXAMPLE D.5 PLOTTING POINTS AND FINDING EQUATIONS

Plot the points C, D, E, and F, as $(3, 4)$, $(9, 1)$, $(-1, 1)$, and $(2, 7)$, respectively, and then construct the lines $\overleftrightarrow{CD}$ and $\overleftrightarrow{EF}$ and the point G where the lines intersect. Find the equations and slopes of the two lines, and show that the product of the slopes is -1. Next, plot the point H as $(4, 1)$ and construct the circle centered at H that passes through G. Find the equation of the circle to verify that the radius is 5, and check that the circle also passes through D and E.

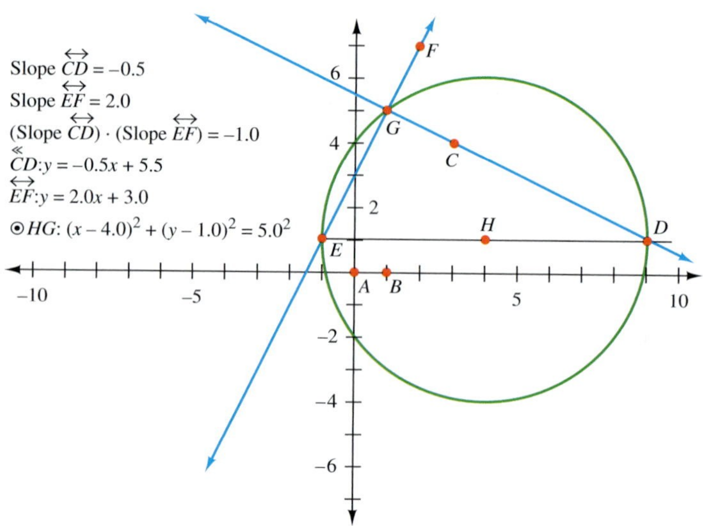

Solution The points can be plotted by opening the **Plot Points . . .** dialog box. Alternatively, the points can be plotted directly by selecting the Point tool and clicking on the appropriate intersections of the grid lines. Use the Text tool from the toolbox to label all the points in your sketch, starting with A as $(0, 0)$ and B as $(1, 0)$. The two lines, $\overleftrightarrow{CD}$ and $\overleftrightarrow{EF}$, can be constructed using the Straightedge tool. With both lines selected, use the **Slope** and **Calculate** commands from the Measure Menu to display the slopes of the lines and compute their product. Clicking on the intersection of the two lines with the Selection-arrow tool will construct their point of intersection, which is then labeled G. Next, use the Compass tool to construct the circle centered at H that passes through G. With the two lines and the circle selected, use the **Equation** command in the Measure Menu to obtain their equations. In particular, the radius of the circle is seen to be 5. Since both D and E are five units from H, we see that $\overline{DE}$ is a diameter of the circle.

Graphs of functions can also be plotted on coordinate axes in the sketch plane. The **New Function . . .** command opens an entry box into which a function expression is typed. To graph the function, select the function expression in the sketch window and execute the **Plot Function** command. To examine the values of the function, select the plot and execute the **Point on Function Plot** command in the Construct Menu. With that point selected, the x- and y-values along the plot can be displayed by using the **Coordinates** command, or the **Abscissa (x)** and **Ordinate (y)** commands, under the Measure Menu. (This is much like using the Trace feature of a graphing calculator.) The function values can be displayed in a table in Sketchpad: select the measures and use the **Tabulate** command in the Graph Menu.

EXAMPLE D.6 FINDING THE RECTANGLE OF LARGEST AREA IN A SEMICIRCLE

Find the rectangle of largest area that can be inscribed in a semicircle, with one side of the rectangle along the diameter.

Solution

We may as well assume that the radius of the semicircle is 1. The following diagram shows an inscribed rectangle whose side along the diameter has length $2x$:

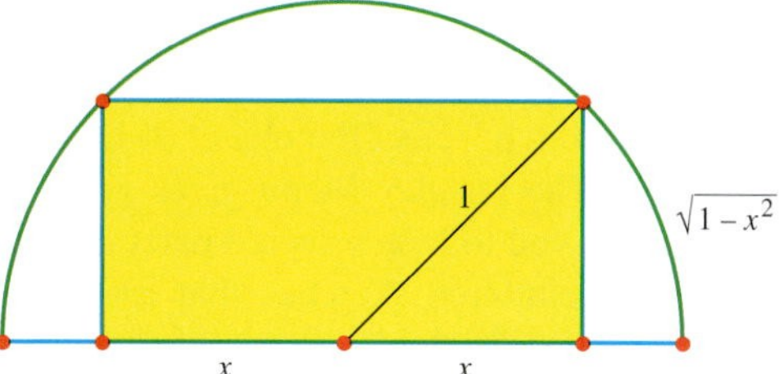

By the Pythagorean theorem, the vertical side has length $\sqrt{1 - x^2}$. The area of the rectangle is therefore given by the function $f(x) = 2x\sqrt{1 - x^2}$, where $0 < x < 1$. The figure below shows the graph of this function as plotted with The Geometer's Sketchpad. The coordinates of point P have been tabulated at several sample locations along the plot. The largest ordinate is $y = 1.0$, occurring when x is about 0.71. A more detailed analysis would show that the exact maximum area is indeed 1, occurring when the rectangle has width $2x = \sqrt{2}$ and height $\sqrt{1 - (1/\sqrt{2})^2} = \sqrt{1 - 1/2} = \sqrt{1/2} = 1/\sqrt{2}$. That is, the rectangle of largest area is the one shaped like a domino (a double-square), with its length twice that of its width.

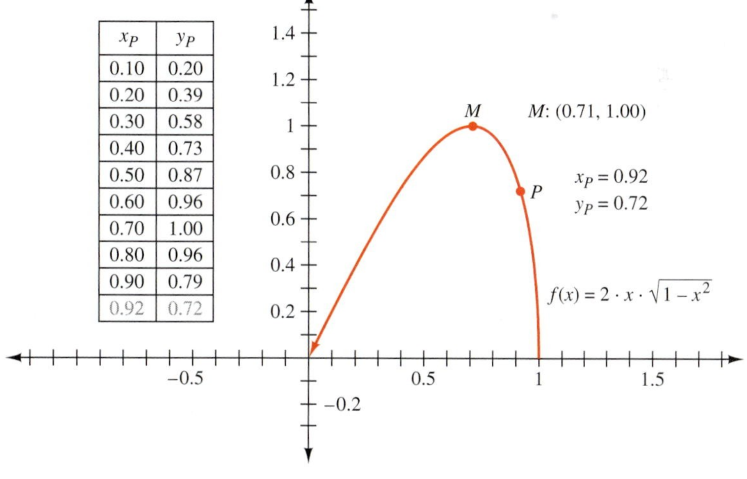

x_P	y_P
0.10	0.20
0.20	0.39
0.30	0.58
0.40	0.73
0.50	0.87
0.60	0.96
0.70	1.00
0.80	0.96
0.90	0.79
0.92	0.72

THE FILE, EDIT, AND DISPLAY MENUS

The File Menu allows you to create a new sketch window or open a saved sketch. Newly created sketches, and modified sketches can be saved for future use. Additional pages can be added to a sketch by using the **Document Options . . . /Add Page** command. This is particularly useful when you wish to create a sequence of related sketches.

The Edit Menu permits you to undo and redo steps in your constructions, as well as perform the usual cuts, copies, and pastes. Deleting or clearing an object will also remove those objects in the sketch that depend on the deleted object, so these functions must be used with care. Sketchpad is a powerful drawing tool, and copied graphics can be inserted into other documents. To find out how a point, line, or other object is related to the rest of your sketch, select the object of interest and use the **Properties . . .** command.

The Display Menu gives you choices of line styles, colors, and text styles. It also allows you to hide selected parts of your sketch from view, which is often useful for making uncluttered sketches that are easy to interpret and understand. The **Hide Objects** command, as the name suggests, retains the objects and simply hides them from view. In this way, it is very different from the **Cut** and **Clear** commands in the Edit Menu. The **Show/Hide Text Palette** command allows you show or hide a menu bar that helps you format captions and labels. The Text palette is shown in Figure D.1.

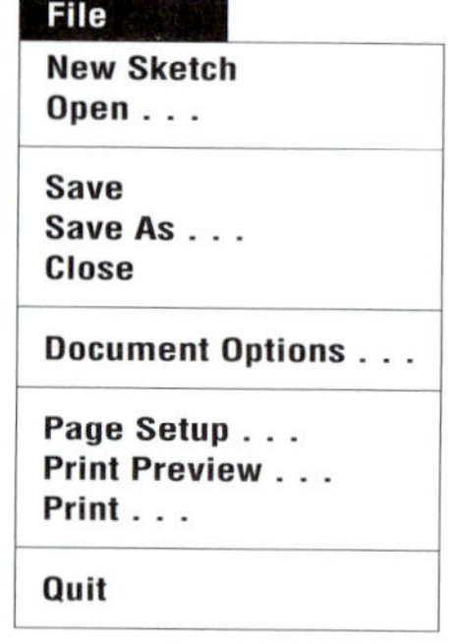

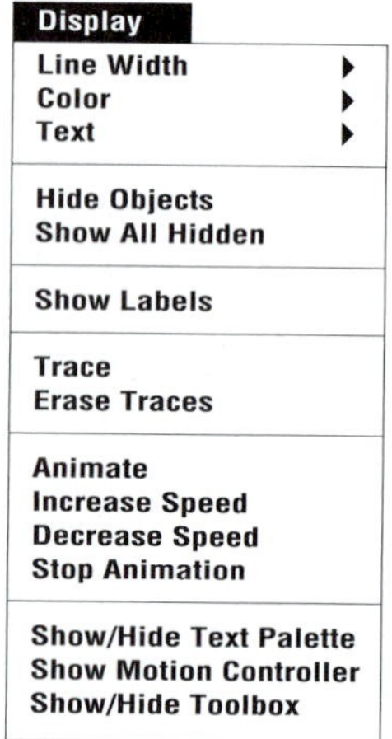

ADVANCED FEATURES

The construction methods described and illustrated in the examples above will enable the user of The Geometer's Sketchpad to create a wide variety of interesting figures to investigate. However, Sketchpad has other features that more experienced users will find valuable. For example, Sketchpad can add an exciting dynamic dimension to a sketch by incorporating animation, movement, hide/show buttons, and so on. The user's manuals for the software describe these features and how they are employed.

Problem Set D

1. **Drawing Figures.** Draw a general quadrilateral, and perform the following manipulations.
   (a) Drag the vertices of the quadrilateral until your quadrilateral appears to be a rhombus. See how close you've come by measuring the lengths of the four sides. Why is the rhombus a drawing and not a construction?
   (b) Drag the vertices of the quadrilateral until your quadrilateral appears to be a rectangle. What measurements can be made to see how close your drawing comes to a true rectangle?
2. **Constructing Figures.** Construct each figure listed below, using commands from the Construct Menu. In words, describe the construction steps that you followed. Hide any midpoints, perpendicular bisectors, parallel lines, circles, and so on that you used to construct your figure. Check that your figure is a construction by dragging various points and observing if the figure retains all of the necessary properties that define the type of figure. Look for alternative constructions of the figures.
   (a) isosceles triangle (given base $\overline{AB}$)
   (b) kite *KITE* (given *K* and *T*)
   (c) isosceles trapezoid *ABCD* (given base $\overline{AB}$)
   (d) parallelogram *ABCD* (given *A*, *B*, and *C*)
   (e) rhombus *ABCD* (given *A* and *C*)
   (f) square *ABCD* (given *A* and *B*)
3. **A Hexagonal Tiling.** Construct any quadrilateral and construct the midpoint of one of its sides. Mark the midpoint as a rotation center. (Alternatively, just double-click the desired rotation center.) Select the entire quadrilateral, and from the Transform Menu, use **Rotate . . .** (by 180° about the marked center) to create a six-sided polygonal tile. Finally, translate the hexagonal tile repeatedly to create a tiling of the plane. Does your quadrilateral still tile the plane if you drag one of its vertices to form a concave quadrilateral?
4. **Measuring and Calculating.** Draw a general quadrilateral *ABCD*. Then construct the midpoints *J*, *K*, *L*, and *M* of the sides and join them with line segments to construct the inscribed quadrilateral *JKLM* as shown below.

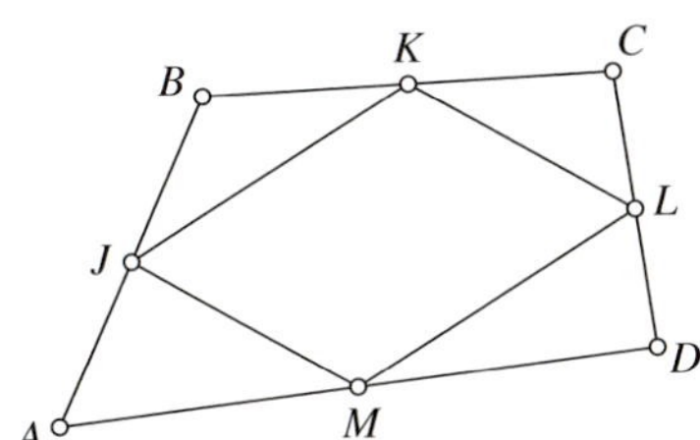

   (a) What is the sum of the measures of the interior angles of *ABCD*?
   (b) What relationships hold for the measures of the interior angles of the quadrilateral *JKLM*?
   (c) What relation exists between the lengths of the sides of *JKLM* and the diagonal distances *AC* and *BD* across the quadrilateral?
   (d) Select (in order) the vertices of *ABCD* and use **Polygon Interior** from the Construct Menu to construct the interior of the quadrilateral. Similarly, construct the interior of *JKLM*. Select the interiors of both *ABCD* and *JKLM* and measure their respective areas with **Measure/Area.** What relationship do you observe between the areas?
5. **Constructing Figures with Coordinates.** To keep the points visible in the plots requested below, select cm (centimeter) as the unit of distance measure under **Preferences . . .** in the Edit Menu. Also select **Show Grid** and **Snap Points** under the Graph Menu.
   (a) Plot the points $(0, 6)$, $(-3, 0)$, and $(9, -6)$, and show that they are the vertices of a right triangle.
   (b) Show that $(6, 3)$, $(-3, 0)$, $(-5, -4)$, and $(4, -1)$ are the vertices of a parallelogram.
   (c) Show that $(13, 1)$, $(-3, 9)$, and $(0, -12)$ are the vertices of a triangle inscribed in a circle centered at $(2, -1)$.
6. **Creating Graphs of Functions.**
   (a) Plot the graph of the function $f(x) = x + \dfrac{2}{\sqrt{x}}$, and estimate the coordinates of the point *M* corresponding to the minimum value of the function.
   (b) On the same axes used in part (a), plot the graph of the function $g(x) = \dfrac{x^2}{2} - 1$, and estimate the coordinates of the point *I* at which the two graphs intersect.

Appendix E Resources

The organizations, materials sources, journals, and Web sites listed below are of value to both preservice and in-service teachers. They represent, however, only a small sample of what is available. We recommend that you create and maintain your own personal list of resources.

ORGANIZATIONS

National Council of Teachers of Mathematics (NCTM)
  1906 Association Drive, Reston, VA 20191
  www.nctm.org
Local affiliates of the NCTM
  (see listing at www.nctm.org/affiliates/)

CONFERENCES

Annual meeting of the NCTM, April
Regional and affiliate conferences
  (see www.nctm.org/meetings/)

NCTM *PRINCIPLES AND STANDARDS FOR SCHOOL MATHEMATICS*

Print form: ISBN 0–87353–480–8
Electronic form: www.nctm.org/standards/

JOURNALS

Journal of Computers in Mathematics and Science Teaching
Journal of Research in Mathematics Education
The Mathematics Teacher
Mathematics Teaching in the Middle School
Principles Practice in Mathematics & Science Education
School Science & Mathematics
Teaching Children Mathematics
Teaching K–8

TECHNOLOGY

Geometer's Sketchpad: Key Curriculum Press (www.keymath.com)
Texas Instruments (http://education.ti.com/)

INTERNET SITES: GENERAL

Explore Math (http://www.exploremath.com/)
Math Archives (http://archives.math.utk.edu/)
Math Central (http://mathcentral.uregina.ca/)

Math Forum (http://mathforum.org/)
MathWorld (http://mathworld.wolfram.com/)
Suzanne's Math Lessons (http://mathforum.org/alejandre/)

INTERNET SITES: BY TOPIC

Border and Wallpaper Patterns
- Jill Britton's Home Page (http://ccins.camosun.bc.ca/~jbritton/Home.htm)
- Kali Java Applet (http://www.geom.umn.edu/java/Kali/program.html)
- Steffen Weber's Page (http://jcrystal.com/steffenweber/)

Fibonacci Numbers
- Fibonacci Association Official Web Site (http://www.mscs.dal.ca/Fibonacci/)
- Ron Knott's Site (http://www.mcs.surrey.ac.uk/Personal/R.Knott/Fibonacci/fib.html)

Fractions
- Visual Fractions (http://www.visualfractions.com/)
- Geoboard (applet) and Pick's Formula (http://www.cut-the-knot.org/ctk/Pick.shtml)

Geometry
- Geometry Games (http://www.gamequarium.com/geometry.html)
- The Geometry Center (www.geom.umn.edu/)

History
- The MacTutor History of Mathematics Archive (http://turnbull.dcs.st-and.ac.uk/~history/)
- Biographies of Women Mathematicians (http://www.agnesscott.edu/lriddle/women/women.htm)

Integer Sequences
- Encyclopedia of Integer Sequences (http://www.research.att.com/~njas/sequences/)

Measurement
- Powers of Ten (http://www.powersof10.com/)

Pattern Blocks, Fraction Bars, etc. (applets)
- Arcytech (http://arcytech.org/java/)

Pi
- The Pi Pages (http://www.cecm.sfu.ca/pi/pi.html)

Prime Numbers
- The Prime Pages (http://www.utm.edu/research/primes/)

Tilings, Tessellations, Escher-like designs, and Geometric Dissections
- Totally Tessellated (http://library.thinkquest.org/16661)
- Tesellations.org (http://www.tessellations.org/)
- Geometric dissections on the web (http://www.cs.purdue.edu/homes/gnf/book/webdiss.html)

Answers to Selected Problems

Chapter 1

Problem Set 1.1 (page 6)

1. (a) 21 bikes, 6 trikes

4. (a)

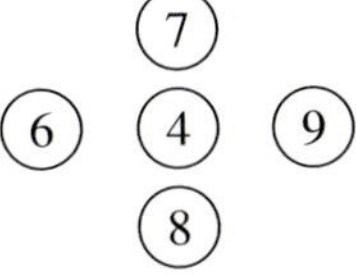

5. 12 (Work backward from 52.)

9. (a)

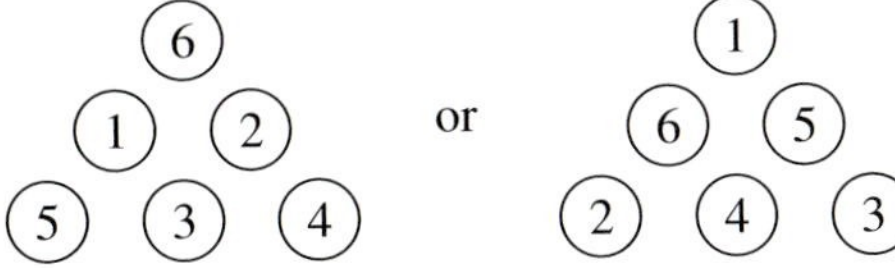

11. (d) There is no solution. **12. (a)** 1, 2, 3, 5, 8, 13, 21 **(c)** 3, 5, 8, 13, 21, 34, 55 **(e)** 2, 1, 3, 4, 7, 11 **14. (b)** Each result is 5.

Problem Set 1.2 (page 18)

1. No. When 10 is multiplied by 5 and 13 is added, the result is 63, not 48.

3. 3

5. (a) Yes. Yes. The rules are really the same.

6. (a) **(c)**

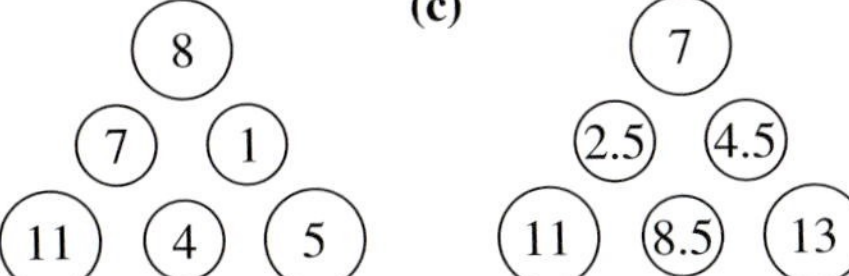

8. 49

10.

1357	3157	5137	7135
1375	3175	5173	7153
1537	3517	5317	7315
1573	3571	5371	7351
1735	3715	5713	7513
1753	3751	5731	7531

12. Assuming that the bags are identical, the possibilities are as shown.

Bag 1	Bag 2	Bag 3
23	1	1
21	3	1
19	5	1
19	3	3
17	7	1
17	3	5
15	9	1
15	7	3
15	5	5
13	11	1
13	9	3
13	7	5
11	11	3
11	9	5
11	7	7
9	9	7

14. 4 and 24, 6 and 16, 8 and 12 **16.** 9 minutes

19. Dawkins, Chalmers, Ertl, Albright, Badgett

21. 46 square meters

Problem Set 1.3 (page 35)

1. (a) 14, 17, 20 **(c)** 10, 15, 15 **(e)** 162, 486, 1458

2. (a)
• • • •   • • • • •   • • • • • •
• • • •   • • • • •   • • • • • •

(d) The nth even number **(f)** 1,443,602

4. (a) 16 **5. (a)** 320 **(c)** 595

7. (a) $1 + 2 + 3 + 4 + 5 + 4 + 3 + 2 + 1 = 25$
$1 + 2 + 3 + 4 + 5 + 6 + 5 + 4 + 3 + 2 + 1 = 36$

10. (a) 86 **(d)** 42

12. (a) $1 - 4 + 9 - 16 + 25 = 15$
$1 - 4 + 9 - 16 + 25 - 36 = -21$
(b) $1 - 4 + 9 - 16 + 25 - 36 + 49 = 28$
$1 - 4 + 9 - 16 + 25 - 36 + 49 - 64 = -36$

(c) For even n, $1 - 4 + 9 - \cdots - n^2 = -\dfrac{n(n+1)}{2}$.

For odd n, $1 - 4 + 9 - \cdots + n^2 = \dfrac{n(n+1)}{2}$.

18. (a) 6 **(c)** 4950

22. (a)

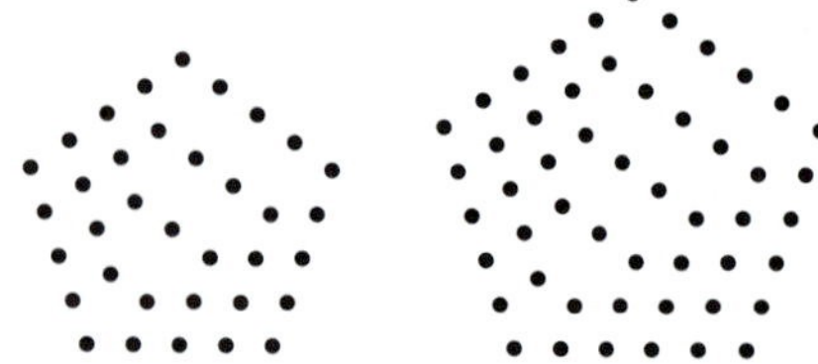

(b) 1, 5, 12, 22, 35, 51
(c) $1 + 4 + 7 + 10 + 13 = 35$
$1 + 4 + 7 + 10 + 13 + 16 = 51$

(d) 28 **(e)** 145 **(f)** $3n - 2$ **(g)** $p_n = \dfrac{n(3n - 1)}{2}$

26. (a) 30; 2100; 29,400

Problem Set 1.4 (page 48)

1. Yes. The second player can add enough tallies to make a multiple of 5 at each step, forcing the first player to be the one to exceed 30.
3. (a) \$28 **6.** Moe was wearing Hiram's coat and Joe's hat. Hiram was wearing Joe's coat and Moe's hat.
7. (a) 25. Not all of the information was needed.
8.

	Choc. Malt	Straw. Shake	Banana Split	Walnut cone
Aaron	X	X	X	O
Boyd	X	X	O	X
Carol	O	X	X	X
Donna	X	O	X	X

Aaron had the walnut cone. Boyd had the banana split. Carol had the chocolate malt. Donna had the strawberry shake.
11. (a) 3 **13.** Since there are only 10 digits (0, 1, 2, . . . , 9) in any collection of 11 natural numbers, there must be two that have the same units digit. The difference of these two numbers must have a 0 as its units digit and is thus divisible by 10.
15. If five points are chosen in a square with diagonal of length $\sqrt{2}$, then, by the pigeonhole principle, at least two of the points must be in or on the boundary of one of the four smaller squares shown. The farthest these two points can be from each other is $\sqrt{2}/2$ units if they are on opposite corners of the small square.

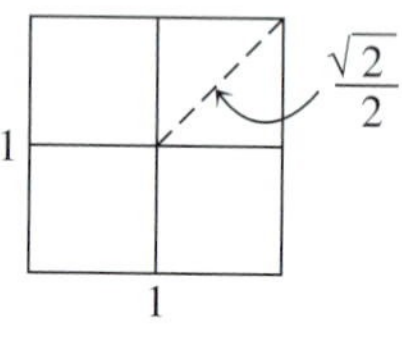

17. If the cups of marbles are arranged as described, each cup will be part of three different groups of three adjacent cups. The sum of all marbles in all groups of three adjacent cups is $3 \cdot (10 \cdot 11/12) = 165$, since each cup of marbles is counted three times. With the marble count of 165 and 10 possible groups of three adjacent cups, by the pigeonhole principle at least one group of three adjacent cups must have 17 or more marbles, since $165 \div 10 = 16.5 > 16$.
19. (i) Since there are 20 people at the party, if each person has at least one friend at the party, each must have either one or two or so on, up to 19, friends at the party. Since $20 > 19$, it follows from the pigeonhole principle that at least two of the 20 people must have the same number of friends at the party. **(ii)** In this case, 19 people have from one to 18 friends at the party. Thus, again by the pigeonhole principle, at least two people must have the same number of friends at the party. **(iii)** In this case, since at least two people have no friends at the party, they have the same number of friends at the party.

Problem Set 1.5 (page 60)

1. (a) 81, 711, 6111 **3. (a)** 9, 98, 987 **7. (a)** Answers will vary. For 2561,

$$6521 - 1256 = 5265$$
$$6552 - 2556 = 3996$$
$$9963 - 3699 = 6264$$
$$6642 - 2466 = 4176$$
$$7641 - 1467 = 6174$$

10. (a) There are $F_5 = 5$ arrangements of five logs, supporting the generalization:

However, there are nine arrangements of six logs, instead of 8 as suggested by the Fibonacci pattern:

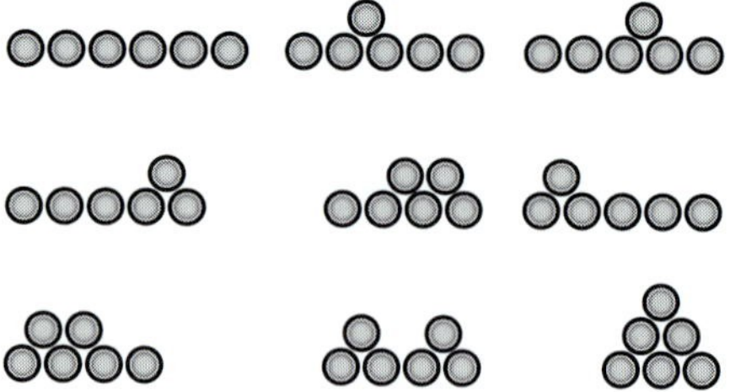

12. (a) Two pennies must be moved.
13. (a) The 1 is represented by the dot in the middle of the bottom row, the 4 by the 4 adjacent dots, the 7 by the next layer of dots, and so on.

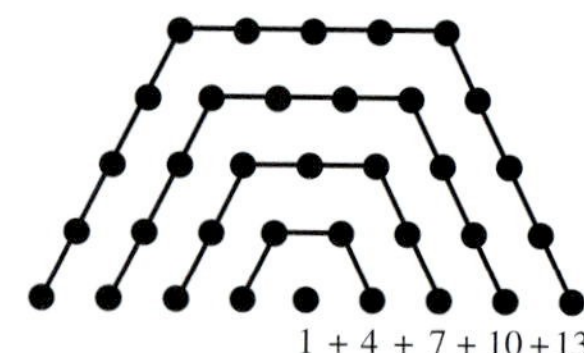

Table for Problem 16

n	1	2	3	4	5	6	7	8	9	10
n^5-n	0	30	240	1020	3120	7770	16800	32760	59040	99990
(n^5-n)/30	0	1	8	34	104	259	560	1092	1968	3333

15. **(a)** The starting position and the 15 moves to interchange the frogs are shown in a table:

○	○	○		●	●	●
○	○	○	●		●	●
○	○		●	○	●	●
○		○	●	○	●	●
○	●	○		○	●	●
○	●	○	●	○		●
○	●	○	●	○	●	
○	●	○	●		●	○
○	●		●	○	●	○
	●	○	●	○	●	○
●		○	●	○	●	○
●	●	○		○	●	○
●	●	○	●	○		○
●	●	○	●		○	○
●	●		●	○	○	○
●	●	●		○	○	○

16. **(a)** Additional examples, such as shown on a spreadsheet (enter the formula "=B1^5-B1" into cell B2 and "=B2/30" into cell B3, and then copy these formulas into the remaining cells in the row), support the conclusion that all such numbers are divisible by 30.

Chapter 1 Review Exercises (page 68)

1. 28

2. **(a)** Answers will vary. One solution is as follows:

$$\begin{array}{r} 179 \\ 368 \\ +452 \\ \hline 999 \end{array}$$

(b) Yes. The digits in any column can be arranged in any order. **(c)** No. The hundreds-column digits must add up to 8 to allow for a carry from the tens column. If the digit 1 is not in the hundreds column, the smallest this sum can be is $2 + 3 + 4 = 9$. Thus, the digit 1 must be in the hundreds column. **3.** 9 **4.** 60 **5.** 88 square feet **6.** 9

7. **(a)** Multiply by 5 and then subtract 2. **(b)** Answers will vary. A good strategy is to give Chanty consecutive integers starting with 0. **8.** **(a)** 3, 6, 12, 24, 48, 96 **(b)** 4, 8, 16, 32, 64, 128 **(c)** 1, 6, 36, 216, 1296, 7776 **(d)** 2, 10, 50, 250, 1250, 6250 **(e)** 7, 7, 7, 7, 7, 7 **9.** Kimberly is the lawyer and the painter; Terry is the engineer and the doctor; Otis is the teacher and the writer.

10. **(a)** $14 + 16 + 18 + 20 = 4^3 + 4$

$22 + 24 + 26 + 28 + 30 = 5^3 + 5$

$32 + 34 + 36 + 38 + 40 + 42 = 6^3 + 6$

(b) $92 + 94 + 96 + 98 + 100 + 102 + 104 + 106 + 108 + 110 = 10^3 + 10$

11. **(a)** 25 **(b)** 1075 **12.** **(a)** 11th **(b)** 6141 **(c)** Duly observed. **(d)** $(6 + 12 + \cdots + 3072 + 6144) - (3 + 6 + \cdots + 3072) = 6144 - 3 = 6141$

13. 442,865 **14.** **(a)** $\dfrac{n(n+1)}{2} + 1$ **(b)** $\dfrac{n(n-1)}{2}$ **(c)** n^2 **15.** The product of the squared entries equals the product of the circled entries. **16.** **(a)** 3 **(b)** 9 **(c)** 27 **(d)** $P_0 + P_1 \cdot 2^1 + P_2 \cdot 2^2 + \cdots + P_n \cdot 2^n = 3^n$, where P_k is the kth element of the nth row of Pascal's triangle. **(e)** 4, 16, 64 **(f)** $P_0 + P_1 \cdot r^1 + P_2 \cdot r^2 + \cdots + P_n r^n = (r+1)^n$

17. **(a)** 5 **(b)** 9 **(c)** 50 **18.** 17

19. **(a)**

$67 \times 677 = 4489$

$667 \times 667 = 444{,}889$

$6667 \times 6667 = 44{,}448{,}889$

(b) 44,444,448,888,889 No, as noted above, patterns can break down.

20. **(a)**

$1 \times 142{,}857 = 142{,}857$

$2 \times 142{,}857 = 285{,}714$

$3 \times 142{,}857 = 428{,}571$

$4 \times 142{,}857 = 571{,}428$

$5 \times 142{,}857 = 714{,}285$

(b) All of the above answers are obtained by starting at an appropriate place in the following circle:

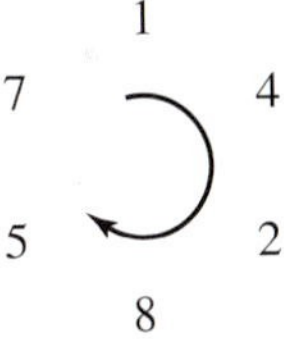

(c) The answer to $7 \times 142{,}857$ is not clear. In fact, $7 \times 142{,}857 = 999{,}999$. **(d)** Apparent patterns may be misleading.

21. One possibility is shown. In fact, P, Q, and R are collinear for every placement of $\triangle ABC$ and $\triangle A'B'C'$.

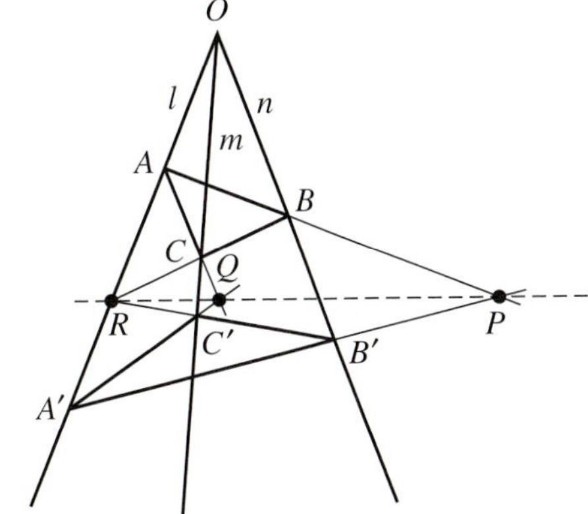

22. (a) Results will vary. It very likely always stops. The process has been checked by computer for the first several million natural numbers, and it has always terminated. However, it has not yet been proved that it will always terminate. **(b)** 19 **23.** Since one of any two consecutive integers must be even, $\frac{s(s+1)}{2}$ must be an integer, say q. Thus, $n^2 = 8q + 1$, as was to be shown. **24.** No, 12 is a multiple of six but the sum of its digits is three, which is not divisible by six.

Chapter 1 Test (page 71)

1. 3 **2.** 21,111,111; 12,111,111; 11,211,111; 11,121,111; 11,112,111; 11,111,211; 11,111,121; 11,111,112
3. (a) $2 \times 100 - 1 = 199$ **(b)** 10,000
4. (a) $2 + 5 + 8 + 11 + 8 + 5 + 2 = 41 = 3^2 + 2 \cdot 4^2$
$2 + 5 + 8 + 11 + 14 + 11 + 8 + 5 + 2 = 66$
$= 4^2 + 2 \cdot 5^2$
(b) $2 + 5 + 8 + \cdots + 26 + 29 + 26 + \cdots + 8 + 5 + 2 = 281 = 9^2 + 2 \cdot 10^2$. **5.** 10 days **6.** The sum of the elements of the nth row of Pascal's triangle is 2^n.
$S = 1 + 2 + 4 + 8 + 16 + 32 + 64 + 128 + 256 + 512 = 1023$. **7.** 36
8. (a) $17 + 18 + 19 + 20 + 21 + 22 + 23 + 24 + 25 = 64 + 125 = 225 - 36$;
$26 + 27 + 28 + 29 + 30 + 31 + 32 + 33 + 34 + 35 + 36 = 125 + 216 = 441 - 100$.
(b) $82 + 83 + 84 + \cdots + 98 + 99 + 100 = 729 + 1000 = 3025 - 1296$
(c) $[(n-1)^2 + 1] + [(n-1)^2 + 2] + \cdots + n^2 =$

$$(n-1)^3 + n^3 = \left(\frac{n(n+1)}{2}\right)^2 - \left(\frac{(n-1)(n-2)}{2}\right)^2$$

9. (a)

57	2	37
12	32	52
27	62	7

(b)

7	2	27
12	32	52
37	62	57

10. It appears that the sum always equals the quotient of the two numbers in the denominator of the last fraction being added on. Thus, we would guess that

$$\begin{aligned} S_n &= \frac{1}{n(n+1)} + \frac{n-1}{n} \\ &= \frac{1}{n(n+1)} + \frac{(n-1)(n+1)}{n(n+1)} \\ &= \frac{1 + n^2 - n + n - 1}{n(n+1)} \\ &= \frac{n^2}{n(n+1)} \\ &= \frac{n}{n+1}, \end{aligned}$$

as expected. Thus,

$$\frac{1}{1 \cdot 2} + \frac{1}{2 \cdot 3} + \cdots + \frac{1}{n(n+1)} = \frac{n}{n+1}.$$

Chapter 2

Problem Set 2.1 (page 83)

1. (a) {Arizona, California, Idaho, Oregon, Utah}
2. (a) {l, i, s, t, h, e, m, n, a, o, y, c}
3. (a) {8, 9, 10, 11, 12} **(c)** {3, 6, 9, 12, 15, 18}
4. Answers will vary. **(a)** $\{x \in U \mid 11 \le x \le 14\}$ or $\{x \in U \mid 10 < x < 15\}$ **(c)** $\{x \in U \mid x = 4n \text{ and } 1 \le n \le 5\}$
5. Answers may vary. **(a)** $\{x \in N \mid x \text{ is even and } x > 12\}$ or $\{x \in N \mid x = 2n \text{ for } n \in N \text{ and } n > 6\}$
6. (a) No, it is not clear which cities with these names are meant. **(b)** Yes, it is clear which cities with these names are meant. **7. (a)** True **(c)** True **(e)** True
8.

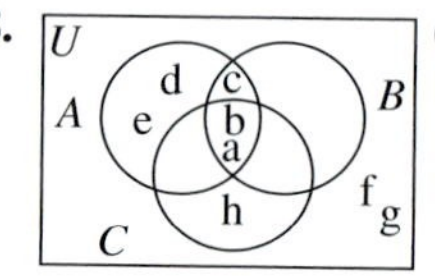

(a) $B \cup C = \{a, b, c, h\}$

(c) $B \cap C = \{a, b\}$ **(e)** $\overline{A} = \{f, g, h\}$ **9. (a)** $M = \{45, 90, 135, 180, 225, 270, 315, \ldots\}$ **(b)** $L \cap M = \{90, 180, 270, \ldots\}$ = the set of simultaneous multiples of 6 and 45 = the set of multiples of 90. **(c)** 90
11. (a)

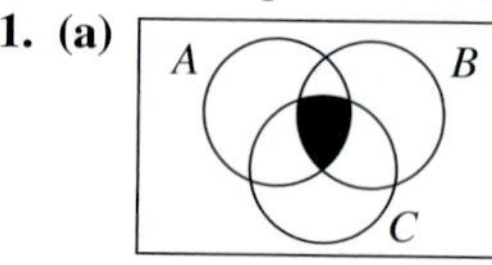

(c)

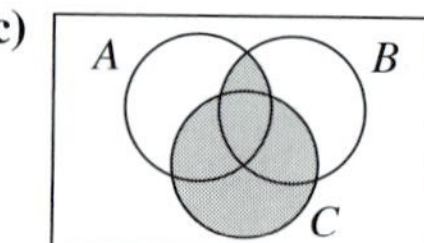

(e)

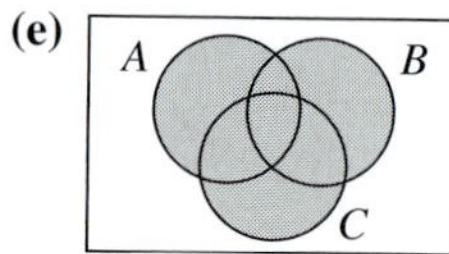

12. (a)

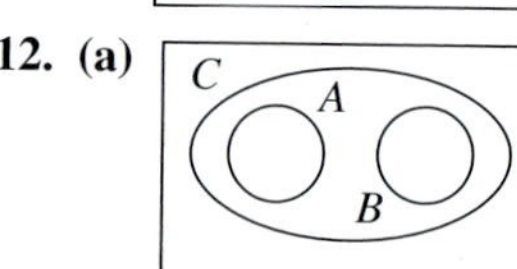

13. No, it is possible that there are elements of A that are also elements of B but not C, or C but not B. For example, let $A = \{1, 2\}$, $B = \{2, 3\}$, and $C = \{3\}$.
14. (a) $\overline{A \cap B} = \{1, 2, 3, 4, 5, 7, 8, 9, 10, 11, 13, 14, 15, 16, 17, 19, 20\}$
$\overline{A} \cup \overline{B} = \{1, 2, 3, 4, 5, 7, 8, 9, 10, 11, 13, 14, 15, 16, 17, 19, 20\}$
$\overline{A \cup B} = \{1, 5, 7, 11, 13, 17, 19\}$
$\overline{A} \cap \overline{B} = \{1, 5, 7, 11, 13, 17, 19\}$
(b) $\overline{A \cap B} = \overline{A} \cup \overline{B}$ and $\overline{A \cup B} = \overline{A} \cap \overline{B}$
15. (a)

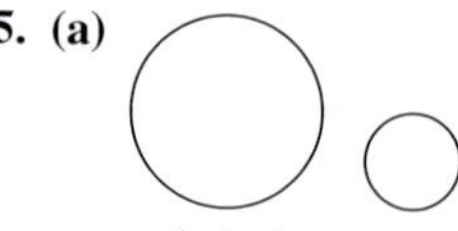

red circles

(c)

triangles and hexagons

(e) 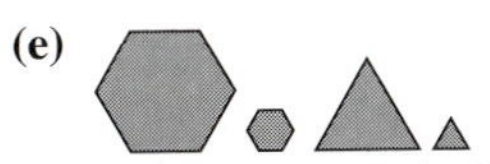

blue figures not circles

16. **(a)** $L \cap T$ **(c)** $S \cup T$ **17.** **(a)** Answers will vary. One possibility is B = set of students taking piano lessons, C = set of students learning a musical instrument.

19. **(a)** 8 regions

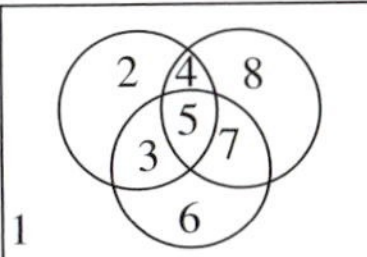

(c) $\overline{A} \cap B \cap C \cap \overline{D}$

20. **(a)**

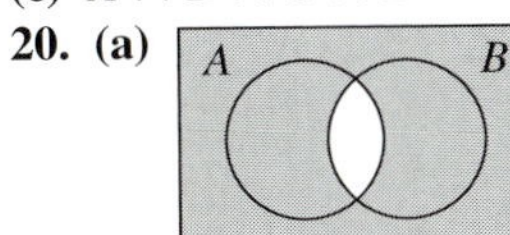

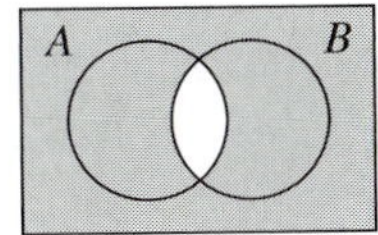

21. **(a)** $6 \times 2 \times 3 = 36$

22. **(a)** Only one way, blue shapes **24.** **(a)** There are eight subsets: $\varnothing$, {P}, {N}, {D}, {P, N}, {P, D}, {N, D}, {P, N, D}.

25.

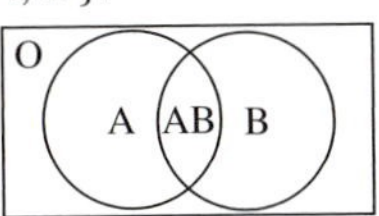

27. Answers will vary. Consult a thesaurus.

31. Answers will vary. One solution is 8, 1, 15, 10, 6, 3, 13, 12, 4, 5, 11, 14, 2, 7, 9.

Problem Set 2.2 (page 97)

1. **(a)** 13: ordinal first: ordinal

2. **(a)** Equivalent, since there are five letters in the set {A, B, M, N, P} **4.** **(a)** 15 **(c)** 2 **6.** **(a)** The correspondence $0 \leftrightarrow 1, 1 \leftrightarrow 2, 2 \leftrightarrow 3, \ldots, w \leftrightarrow w + 1, \ldots$ shows $W \sim N$. **7.** **(a)** Finite **8.** **(a)** Answers will vary. For example, $Q_1 \leftrightarrow Q_2$, $Q_3 \leftrightarrow Q_4$, and so on.

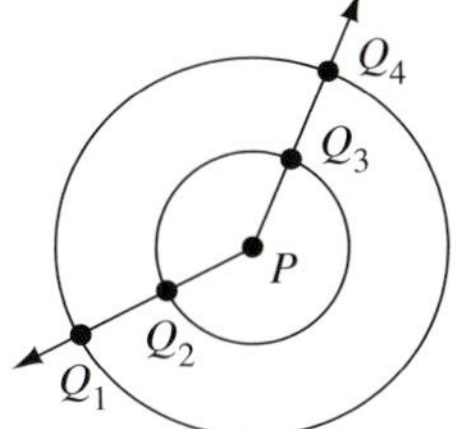

(c) Answers will vary. For example $Q_1 \leftrightarrow Q_2$, and so on.

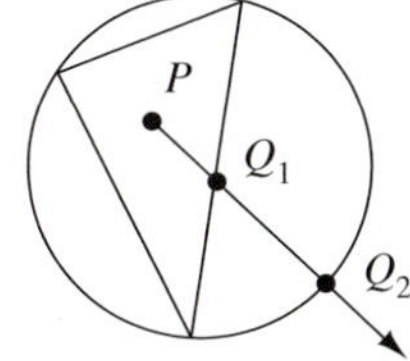

9. **(a)** True **(c)** True **10.** **(a)** $n(A \cap B) \leq n(A)$. The set $A \cap B$ contains only the elements of A that are also elements of B. Thus, $A \cap B$ cannot have more elements than A.

12. **(a)** $1000 \div 6 = 166.666\ldots$, so the largest element of S is $166 \cdot 6 = 996$. Therefore, $n(S) = 166$.

13.

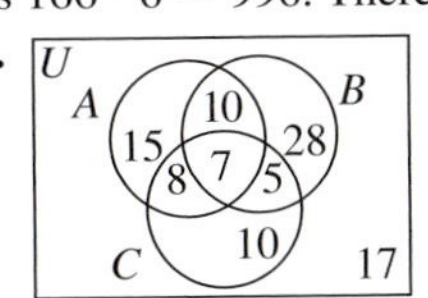

15. **(a)**

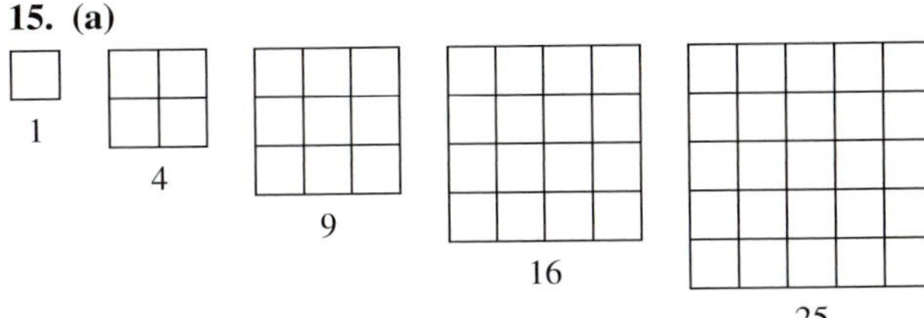

16. **(a)** 0101101 **17.** **(a)** 1110100* (last bit corrected)

21. **(a)** Let each element $a \in A$ be matched to itself: $a \leftrightarrow a$. This gives a one-to-one correspondence from A to itself, so $A \sim A$. **(b)** If $A \sim B$, there is a one-to-one correspondence that matches each element $a \in A$ to an element $b \in B$: $a \leftrightarrow b$. Since each element of B is matched exactly once, the correspondence can be reversed: $b \leftrightarrow a$. This shows that $B \sim A$, and therefore set equivalence is symmetric. **(c)** If $A \sim B$ and $B \sim C$, then there are two one-to-one correspondences: $a \leftrightarrow b$ and $b \leftrightarrow c$. Combining these correspondences gives us a one-to-one correspondence $a \leftrightarrow c$ from A to C. Thus, $A \sim C$, and so set equivalence is transitive. **22.** **(a)** Six ways: yrg, rgy, gyr, ygr, gry, ryg

23. **(a)**

Row 0:	1
Row 1:	1 1
Row 2:	1 2 1
Row 3:	1 3 3 1
Row 4:	1 4 6 4 1

24. Since we are given that $k < l$ and $l < m$, we can choose sets K, L, and M satisfying $K \subset L \subset M$ and $n(K) = k$, $n(L) = l$, and $n(M) = m$. By the transitive property of set inclusion (see Section 2.1, or just look at a Venn diagram), we know that $K \subset M$ and so $k < m$. **28.** Using a Venn diagram and guess and check, we discover that **(a)** four students have visited all three countries. **(b)** 14 students have only been to Canada. **29.** **(a)** All three will meet every $3 \times 4 \times 5$ = 60th day, so there will be 6 days when all three meet. **33.** As the Venn diagram shows, four people have type AB blood.

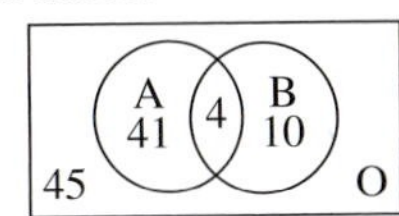

36.

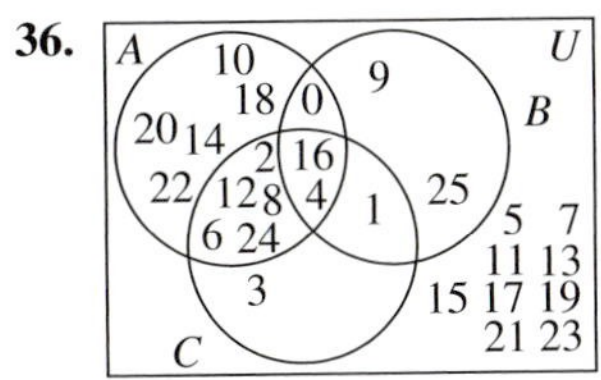

37. **(a)** $\overline{A} = \{\text{b, c, f, g}\}$ **(c)** $A \cup \overline{B} = \{\text{a, b, d, e}\}$
(e) $\overline{A} \cap \overline{B} = \{\text{b}\}$ **(g)** $\overline{A \cap B} = \{\text{a, b, c, d, f, g}\}$

38. **(a)**

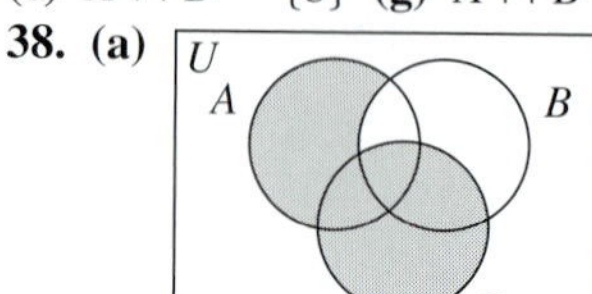

Problem Set 2.3 (page 111)

1. **(a)** **(i)** 5 **(ii)** 5 **(iii)** 4 **(b)** **(ii)** and **(iii)** **3.** **(a)** 4, 5, 6, 7, or 8 **(b)** 4

4. **(a)**

5. **(a)**

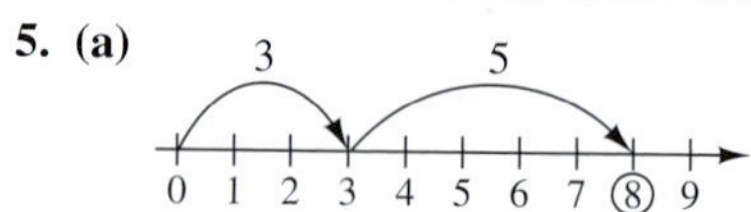

(c)

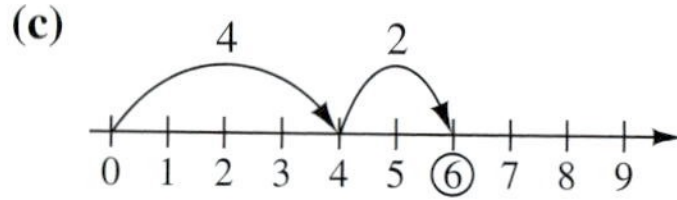

(e)

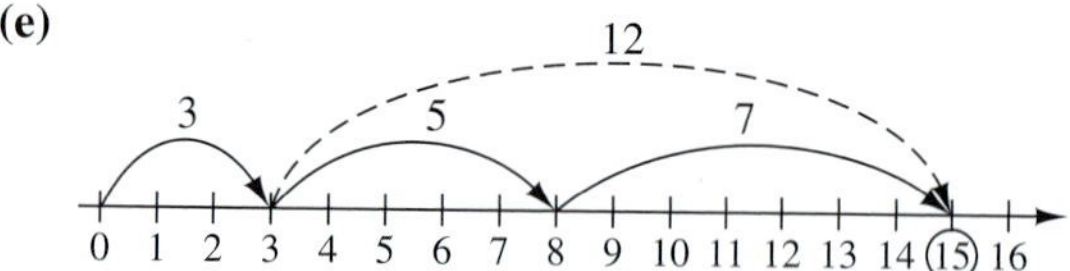

8. **(a)** Closed **(c)** Closed **(f)** Closed **9.** **(a)** Commutative property of addition **(c)** Additive-identity property of zero

11. **(a)**

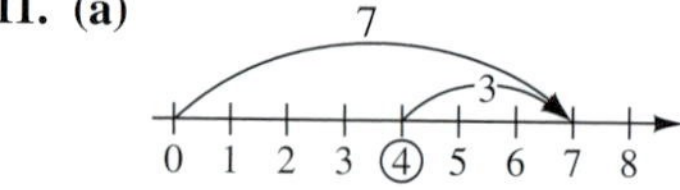

(c)

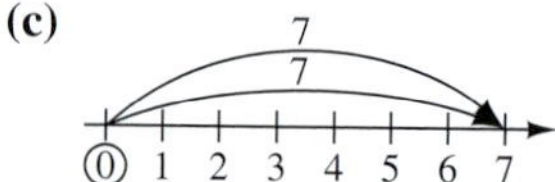

12. **(a)** $5 + 7 = 12$ $\quad 12 - 7 = 5$
$7 + 5 = 12$ $\quad 12 - 5 = 7$
(b) $4 + 8 = 12$ $\quad 12 - 8 = 4$
$8 + 4 = 12$ $\quad 12 - 4 = 8$

13. **(a)**

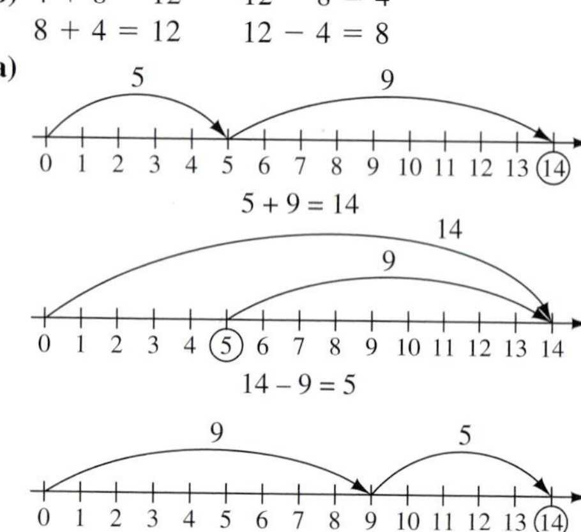

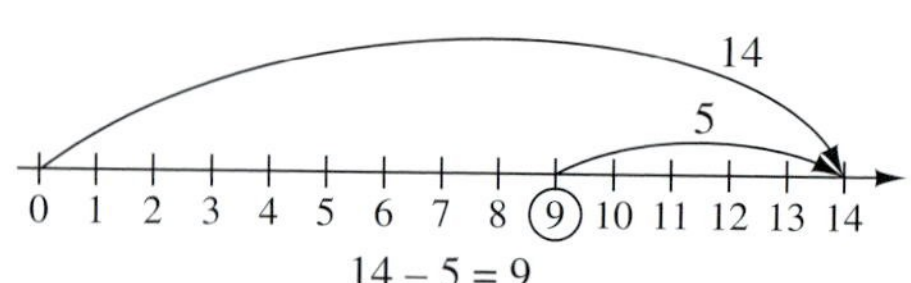

16. $257 - 240 = 17$ pages **18.** **(a)** $(8 - 5) - (2 - 1) = 2$ **(c)** $((8 - 5) - 2) - 1 = 0$

19. **(a)** **20.** **(a)**

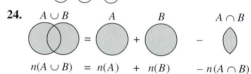

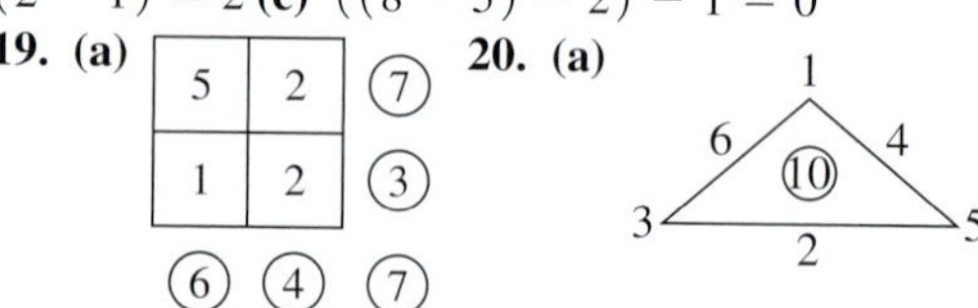

24. $A \cup B$ = A + B − $A \cap B$

$n(A \cup B) = n(A) + n(B) - n(A \cap B)$

29. $\{0\}$

31. **(a)**

n	1	2	3	4	5	6	7	8
t_n	1	3	6	10	15	21	28	36

9	10	11	12	13	14	15
45	55	66	78	91	105	120

(b) $11 = 10 + 1, 12 = 6 + 6, 13 = 10 + 3,$
$14 = 10 + 3 + 1, 15 = 15, 16 = 15 + 1,$
$17 = 15 + 1 + 1, 18 = 15 + 3, 19 = 10 + 6 + 3,$
$20 = 10 + 10, 21 = 21, 22 = 21 + 1, 23 = 10 + 10 + 3,$
$24 = 21 + 3, 25 = 15 + 10$

33.

+	5	4	1	6	9	2	0	8	7	3
3	8	7	4	9	12	5	3	11	10	6
9	14	13	10	15	18	11	9	17	16	12
6	11	10	7	12	15	8	6	14	13	9
4	9	8	5	10	13	6	4	12	11	7
0	5	4	1	6	9	2	0	8	7	3
7	12	11	8	13	16	9	7	15	14	10
5	10	9	6	11	14	7	5	13	12	8
2	7	6	3	8	11	4	2	10	9	5
1	6	5	2	7	10	3	1	9	8	4
8	13	12	9	14	17	10	8	16	15	11

35. **(a)** $0 = 5 - (1 + 4), 1 = 5 - 4, 2 = (1 + 5) - 4,$ $3 = 4 - 1, 4 = 5 - 1, 5 = 1 + 4, 6 = 1 + 5$

38. Statement B

39. **(a)** $n(A \cap B) = 0 + 2 = 2$
(c) $n(\overline{A} \cup C) = 1 + 2 + 3 + 4 + 6 + 7 = 23$
(e) $n(A \cap \overline{C}) = 5 + 0 = 5$

40. 50 students take only geometry and 80 students are not enrolled in any of the three courses.

Problem Set 2.4 (page 130)

1. (a) $3 \times 5 = 15$, repeated addition **(c)** $4 \times 10 = 40$, array model **(e)** $3 \times 6 = 18$, number-line model
2. (a) 55 dominoes, array model **(c)** 50 miles, number-line model **4. (a)** Each of the a lines coming from set A intersects each of the b lines coming from set B.
(b)

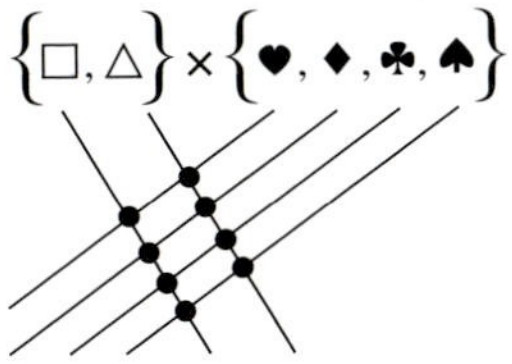

5. (a) Not closed. $2 \times 2 = 4$ and 4 is not in the set.
(c) Not closed. $2 \times 4 = 8$, and 8 is not in the set.
(e) Closed. The product of any two odd whole numbers is always another odd whole number. **(g)** Closed. $2^m \times 2^n = 2^{m+n}$ for any whole numbers m and n. **7. (a)** Commutative property of multiplication **(c)** Multiplication-by-zero property **(e)** Associative property of multiplication
8. (a) Commutative property: $5 \times 3 = 3 \times 5$
(b) Commutative property: $6 \times 2 = 2 \times 6$
9. (a)

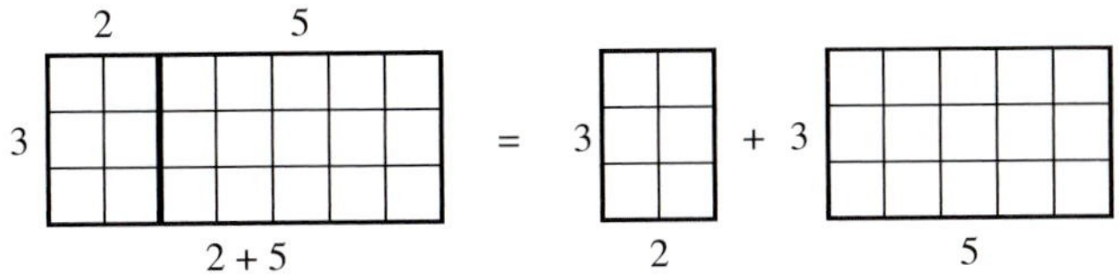

10. The rectangle is $a + b$ by $c + d$, so its area is $(a + b) \times (c + d)$. The rectangle labeled F is a by c, so its area is ac. Similarly, the area of the other three rectangles are given by ad, bc, and bd. So $(a + b) \times (c + d) = ac + ad + bc + bd$.
11.

a	$a \cdot d$	$a \cdot e$	$a \cdot f$
+			
b	$b \cdot d$	$b \cdot e$	$b \cdot f$
+			
c	$c \cdot d$	$c \cdot e$	$c \cdot f$
	d +	e +	f

$(a + b + c) \cdot (d + e + f) = a \cdot d + a \cdot e + a \cdot f + b \cdot d + b \cdot e + b \cdot f + c \cdot d + c \cdot e + c \cdot f$
13. Each nut-and-bolt pair costs 86¢ + 14¢ = \$1.00. Therefore, 18 pairs cost \$18.00. **14. (a)** Distributive property **15. (a)** $18 \div 6 = 3$ **16. (a)** $4 \times 8 = 32$, $8 \times 4 = 32$, $32 \div 8 = 4$, $32 \div 4 = 8$ **17. (a)** Repeated subtraction **18. (a)** 6 **(c)** 6 R 12 **19. (a)** 29
20. (a) 3^{35} **(c)** 3^{10} **(e)** $(yz)^3$ **21. (a)** 2^3 **(c)** 2^{10}
22. (a) 4 **(c)** 10 **24.** Answers will vary. **28. (a)** "How many tickets must still be sold?" **30. (a)** $4 \cdot (5 - 2) = 4 \cdot 3 = 12$, $4 \cdot 5 - 4 \cdot 2 = 20 - 8 = 12$. Therefore, $4 \cdot (5 - 2) = 4 \cdot 5 - 4 \cdot 2$. **(b)** The diagram below makes it clear that $a(b - c) = ab - ac$.

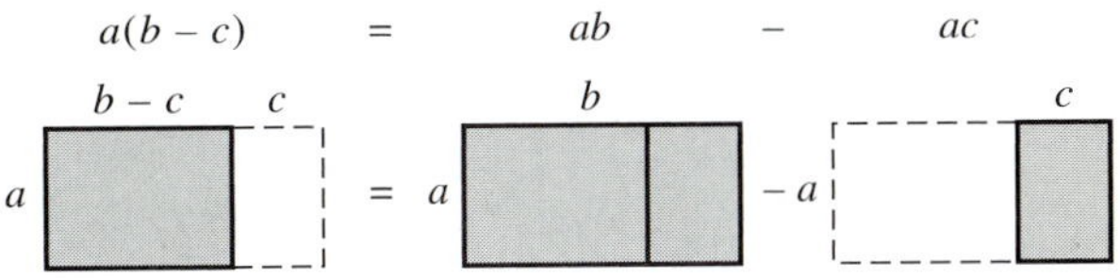

31. $2(1 + 2 + 3 + 4 + 5 + 6 + 7 + 8 + 9 + 10 + 11 + 12) = 2 \cdot \frac{12 \cdot 13}{2} = 156$ times
34. $0 = 2 \times (3 - 3)$, $1 = 2^{(3-3)}$, $2 = 2 + (3 - 3)$, $3 = 3^{(3-2)}$, $4 = 3 + (3 - 2)$, $5 = 2^3 - 3$, $6 = 3^2 - 3$, $7 = 3 \times 3 - 2$, $8 = 3 + 3 + 2$, $9 = ?$, $10 = ?$, $11 = 2^3 + 3$ or $(3 \times 3) + 2$, $12 = 3^2 + 3$, $13 = ?$, $14 = ?$, $15 = 3 \times (3 + 2)$, $16 = ?$, $17 = ?$, $18 = 2 \times 3 \times 3$.
41. $2 + 7 = 9$, $7 + 2 = 9$, $9 - 7 = 2$, $9 - 2 = 7$
43. $2 \times 3 = 6$, $3 \times 2 = 6$

Chapter 2 Review Exercises (page 137)

1. (a) $S = \{4, 9, 16, 25\}$
$P = \{2, 3, 5, 7, 11, 13, 17, 19, 23\}$
$T = \{2, 4, 8, 16\}$
(b) $\overline{P} = \{4, 6, 8, 9, 10, 12, 14, 15, 16, 18, 20, 21, 22, 24, 25\}$
$S \cap T = \{4, 16\}$
$S \cup T = \{2, 4, 8, 9, 16, 25\}$
$S \cap \overline{T} = \{9, 25\}$
2.

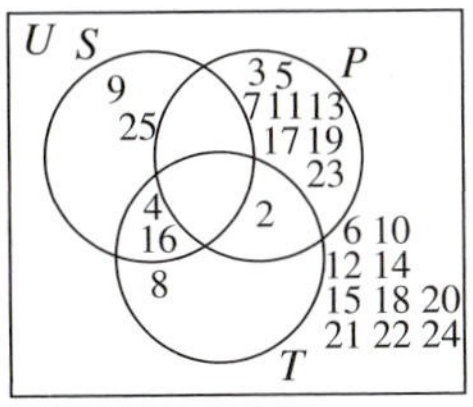

3. (a) $\subseteq$ **(b)** $\subset$ **(c)** $\cap$ **(d)** $\cup$ **4.** $n(S) = 3$, $n(T) = 6$, $n(S \cup T) = 7$, $n(S \cap T) = 2$, $n(S \cap \overline{T}) = 1$, $n(T \cap \overline{S}) = 4$
5.

1	4	9	16	25	36	49	64	81	100
$\updownarrow$	$\updownarrow$	$\updownarrow$	$\updownarrow$	$\updownarrow$	$\updownarrow$	$\updownarrow$	$\updownarrow$	$\updownarrow$	$\updownarrow$
a	b	c	d	e	f	g	h	i	j

6. There is a one-to-one correspondence between the set of cubes and a proper subset. For example,

$$\begin{array}{cccccc} 1 & 8 & 27 & 64 & 125 \ldots & k^3 \ldots \\ \updownarrow & \updownarrow & \updownarrow & \updownarrow & \updownarrow & \updownarrow \\ 1 & 2 & 3 & 4 & 5 \ldots & k \ldots \end{array}$$

7.

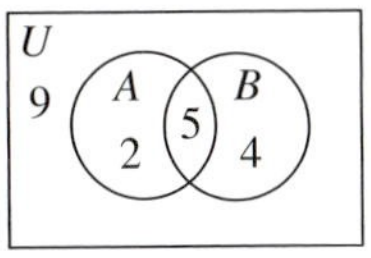

8. (a) Suppose $A = \{a, b, c, d, e\}$ and $B = \{■, ★\}$. Then $n(A) = 5$, $n(B) = 2$, $A \cap B = \emptyset$, and $n(A \cup B) = 5 + 2 = 7$. **(b)**

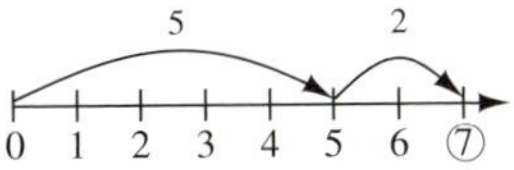

9. **(a)** Commutative property of addition: $7 + 3 = 3 + 7$
(b) Additive-identity property of zero: $7 + 0 = 7$
10. **(a)**

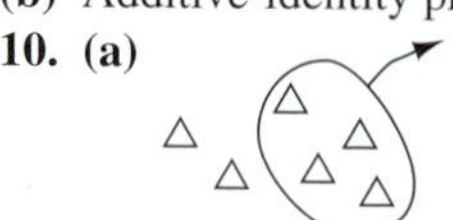

(b)

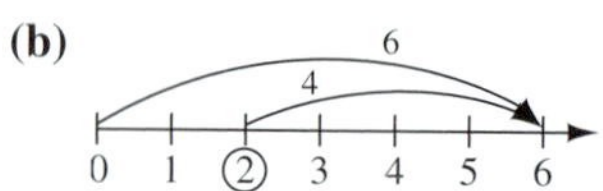

11. **(a)**

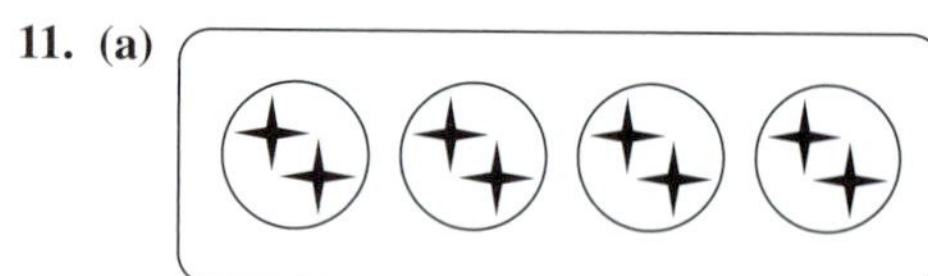

(b) **(c)** **(d)**

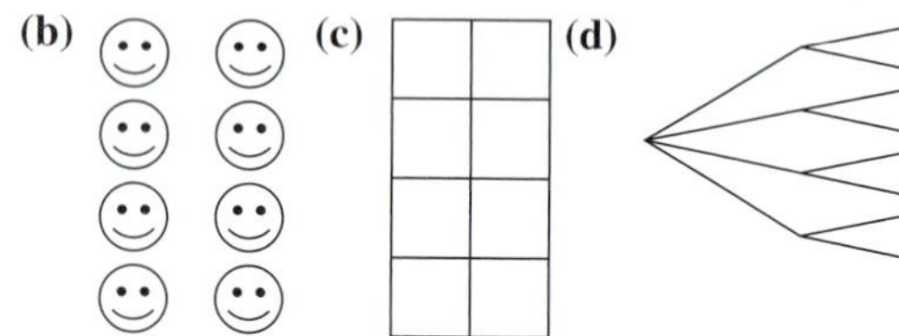

(e)

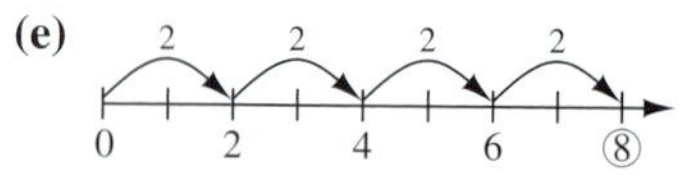

12. **(a)** $A \times B = \{(p, x), (p, y), (q, x), (q, y), (r, x), (r, y), (s, x), (s, y)\}$ **(b)** $4 \times 2 = 8$
13. $6'' \times 6'' \times 8''$
14. Eight rows, with seven full rows and eight soldiers in the back row
15. **(a)**

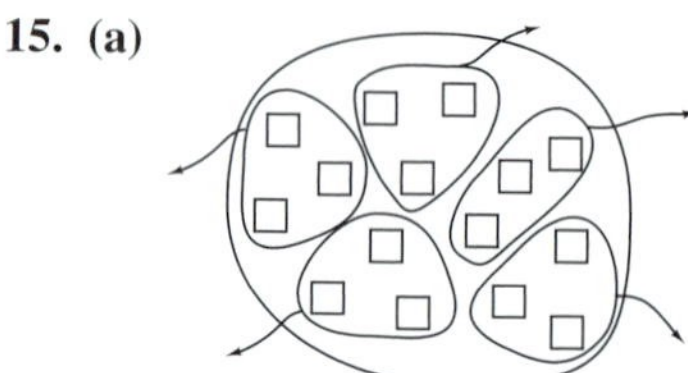

(b) **(c)**

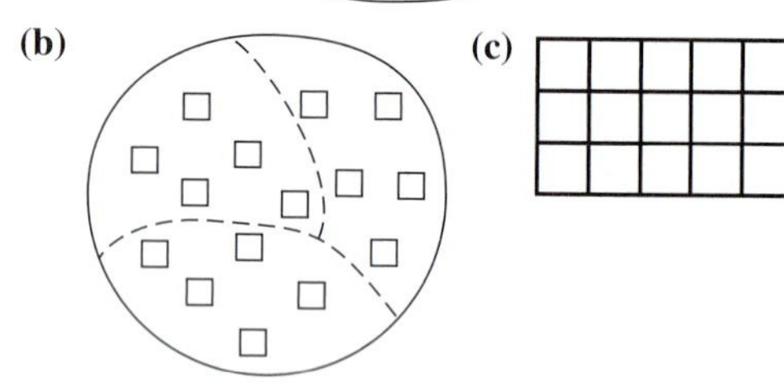

Chapter 2 Test (page 138)

1. **(a)** $4 \times 2 = 8$ **(b)** $12 \div 3 = 4$
(c) $5 \cdot (9 + 2) = 5 \cdot 9 + 5 \cdot 2$ **(d)** $10 - 4 = 6$
2. 15th—ordinal; 1040—nominal; $253—cardinal
3. **(a)** Yes **(b)** No **4.** Answers can vary. For example, let $A = \{a, b, c, d, e\}$ and $B = \{a, b, c, d, e, f, g, h\}$. Then $A \subset B$, so $n(A) < n(B)$. That is, $5 < 8$.
5. **(a)** W is closed under &, since $a + b + ab$ is an element of W for any two whole numbers a and b. **(b)** Using the properties of whole-number arithmetic, $a \& b = a + b + ab = b + a + ba = b \& a$ for all whole numbers a and b, so & is commutative. **(c)** Using the properties of whole-number arithmetic, $a \& (b \& c) = a + (b \& c) + a(b \& c) = a + b + c + bc + a(b + c + bc) = a + b + c + ab + bc + ac + abc$. A similar calculation shows that $(a \& b) \& c$ has the same expanded form, so & is associative. **(d)** $0 \& a = 0 + a + 0(a) = a$ and $a \& 0 = a + 0 + (a)0 = a$ for all whole numbers a, showing that 0 is an identity for &.
6. **(a)** Number line **(b)** Comparison **(c)** Missing addend
7. **(a)** Associative property of addition **(b)** Distributive property of multiplication over addition **(c)** Additive-identity property of zero **(d)** Associative property of multiplication
8. **(a)** **(b)** **(c)**

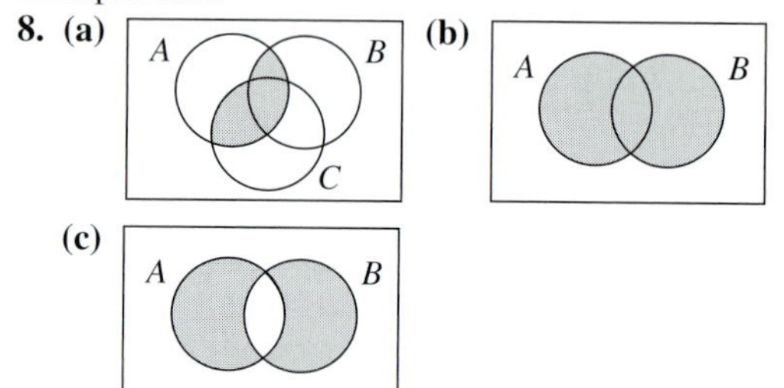

9. **(a)** **(b)**

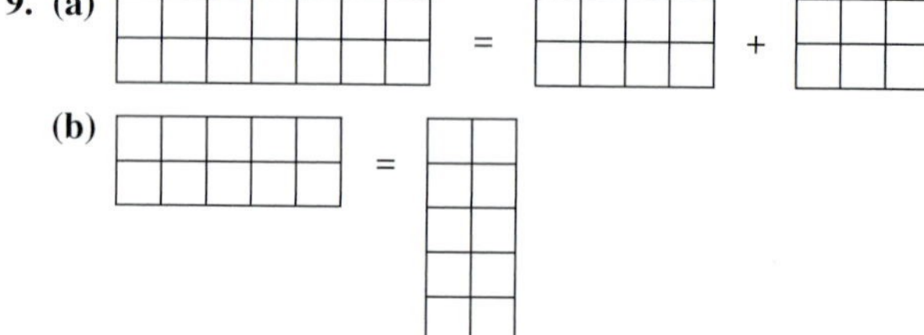

10. 1, 3, 7, or 21 **11.** **(a)** 10 **(b)** 4 **(c)** 2 **(d)** 20
12. $n(A \cap B) = 5, n(\overline{A \cap B}) = 21$ **13.** $A \cap \overline{B} = \varnothing$
14. **(a)** 64 bottles **(b)** Grouping

Chapter 3

Problem Set 3.1 (page 154)

1. **(a)** 2137 **(c)** 120,310 **(e)** 697 **(g)** 60 **(i)** 152
(k) 16,920 **2.** **(a)** ∩ | **3.** **(a)** IX
4. **(a)** ▼▼▼▼‹▼ **5.** **(a)** ••, ═
9. **(a)** MMII, MMIII, MMIV **10.** **(a)** 3795 **(c)** 6048
11. **(a)** ═ ╥ ⊥ ||| **(c)** ╥ ||||
12. **(a)** ⊤ ≣ ||||| $374 + 281 = 655$ **(c)** ≣ |||| ≜ ||
$6224 - 732 = 5492$
13.

17. First trade 10 of your units for a strip to get 3 mats, 25 strips, and 3 units. Then trade 20 strips for 2 mats to get 5 mats, 5 strips, and 3 units. **25.** **(a)** 14 **(c)** 13
27. **(b)** 413

29. **(a)** **(c)** **(e)** **(g)**

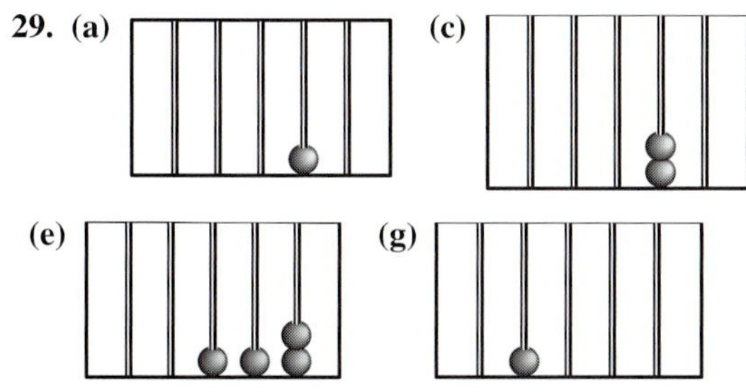

30. **(a)** The "fives" wire **36.** **(a)** $\{1, 2, 3, 4, 5, 7, 9\}$ **(c)** $\{2, 4, 6, 8\}$ **(e)** $\{1, 3\}$ **(g)** $n(A) = 5, n(C) = 4, n(A \cup C) = 9$ **37.** **(a)** Distributive property of multiplication over addition
(c) Commutative property of addition
38. **(a)** $10 - 3 = 7$; $10 - 7 = 3$

Problem Set 3.2 (page 161)

1. 0, 1, 2, 3, 4, 10, 11, 12, 13, 14, 20, 21, 22, 23, 24, 30, 31, 32, 33, 34, 40, 41, 42, 43, 44, 100

5. **(a)** 108 **(c)** 5 **(e)** 125
6. **(a)** 153 **(c)** 6 **(e)** 216
7. **(a)** 591 **(c)** 12 **(e)** 1728
8. **(a)** 2422_{five} **(c)** 10_{five}
9. **(a)** 1330_{six} **(c)** 10_{six}
10. **(a)** 1707_{twelve} **(c)** 100_{twelve}
11. **(a)**

One Thousand Twenty-Fours	Five Hundred Twelves	Two Hundred Fifty-Sixes
1024	512	256
2^{10}	2^9	2^8

One Hundred Twenty-Eights	Sixty-Fours	Thirty-Twos
128	64	32
2^7	2^6	2^5

Sixteens	Eights	Fours	Twos	Units
16	8	4	2	1
2^4	2^3	2^2	2^1	2^0

(c) **(i)** $11{,}000_{two}$ **(ii)** $10{,}010_{two}$ **(iii)** 10_{two} **(iv)** 1000_{two}
16. **(a)** $100{,}000{,}000_{two}$ **17.** **(a)** $7_{ten} = 111_{two}$, $2^3 = 8$ odd entries
18. **(a)** 000, 100, 010, 110, 001, 101, 011, 111. Append a 0 onto the end of the four two-digit sequences; then append a 1 onto the end of the four two-digit sequences. **(c)** 32
19. **(a)** 0, 1, 2, 3, 4, 5, 6, 7 **(c)** The whole numbers from 0 to $2^n - 1$. There are 2^n of these whole numbers, each with a different n-digit base-two representation corresponding to one of the n-digit sequences of 0s and 1s.
20. **(a)** $2 \cdot 2 \cdot 2 = 2^3 = 8$ **(b)** 8 **(c)** $2^4 = 16$

21. **(a)** 000, a purple rod
001, a light-green rod followed by a white rod
010, two red rods
011, a red rod followed by two white rods
100, a white rod followed by a light-green rod
101, a white rod, followed by a red rod, followed by a white rod
110, two white rods followed by a red rod
111, four white rods

As shown here, [¦ ¦ ¦], there are three places where a 4-train may or may not be broken into separate cars. A particular train is formed by deciding to remove a dotted line (choose 0) or make the dotted line solid (choose 1). Since there are two choices for each dotted line, there are $2 \cdot 2 \cdot 2 = 2^3 = 8$ possible different trains. **23.** **(a)** $51 \div 3 = 17, 51 \div 17 = 3$ **24.** Answers will vary. **(a)** $11 \times 31 = 341 \div 31 = 11$ **26.** **(a)** $n + 6$ **(b)** $m - 6$

Problem Set 3.3 (page 177)

1. **(a)**

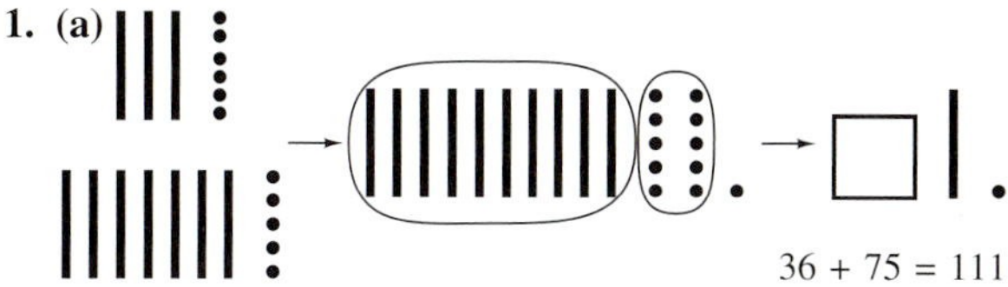

$36 + 75 = 111$

2. **(a)**

$$\begin{array}{r} 23 \\ +\ 44 \\ \hline 7 \\ 60 \\ \hline 67 \end{array}$$

5. **(b)**

100s	10s	1s
○○	○○○○○ ○○	○○○○○

exchange

100s	10s	1s
○○	○○○○○	○○○○○ ○○○○○ ○○○○○

$275 - 136 = 139$

6. **(a)**

$$\begin{array}{r} 78 \\ -\ 35 \\ \hline 43 \end{array}$$

9. In these problems, we must exchange 60 seconds for one minute and 60 minutes for one hour or vice versa.

(a)
3 hours, 24 minutes, 54 seconds
+ 2 hours, 47 minutes, 38 seconds
5 hours, 71 minutes, 92 seconds
= 5 hours, 72 minutes, 32 seconds
= 6 hours, 12 minutes, 32 seconds

(c)
$$\begin{array}{rl} & \text{5 hours, 24 minutes, 54 seconds} \\ - & \text{2 hours, 47 minutes, 38 seconds} \\ \hline & \text{4 hours, 84 minutes, 54 seconds} \\ - & \text{2 hours, 47 minutes, 38 seconds} \\ \hline & \text{2 hours, 37 minutes, 16 seconds} \end{array}$$

11. (a) 0, 1, 2, 3, 10, 11, 12, 13, 20, 21, 22, 23, 30, 31, 32, 33

13. (a) 1012_{four} **(c)** 2120_{four} **(e)** 111_{four} **(g)** 113_{four}

17. (a) $\begin{array}{r} 6763 \\ +\ 5519 \\ \hline 12{,}282 \end{array}$ **(c)** $\begin{array}{r} 881 \\ +\ 362 \\ \hline 1243 \end{array}$ **(e)** $\begin{array}{r} 4002 \\ -\ 1843 \\ \hline 2159 \end{array}$

18. (a) $\begin{array}{r} 2437 \\ 281 \\ +\ 3476 \\ \hline 6194 \end{array}$ **(c)** $\begin{array}{r} 3891 \\ 2493 \\ +\ 5125 \\ \hline 11{,}509 \end{array}$

19. (a) $\begin{array}{r} 835 \\ -\ 241 \\ \hline 594 \end{array}$ **(c)** $\begin{array}{r} 7342 \\ -\ 6534 \\ \hline 808 \end{array}$

20. (a) Five **(c)** Seven or greater **(e)** Seven **(g)** Twelve

24. (a)
$$\begin{array}{ll} 1 \cdot s = 19{,}998 & 6 \cdot s = 119{,}988 \\ 2 \cdot s = 39{,}996 & 7 \cdot s = 139{,}986 \\ 3 \cdot s = 59{,}994 & 8 \cdot s = 159{,}984 \\ 4 \cdot s = 79{,}992 & 9 \cdot s = 179{,}982 \\ 5 \cdot s = 99{,}990 & 10 \cdot s = 199{,}980 \end{array}$$

(d) $14 \cdot s = 279{,}972$, $23 \cdot s = 459{,}954$, $32 \cdot s = 639{,}936$ **(f)** Arithmetic progressions

27. (a)
$$\begin{array}{r} 3'\ 8'' \\ 4'\ 2'' \\ 6'\ 10'' \\ +\ 5'\ 11'' \\ \hline 18'\ 31'' \\ = 20'\ 7'' \end{array}$$

34. (a) $1001 \div 91 = 11$; $1001 \div 11 = 91$; $91 \times 11 = 1001$

35. (a) $143 \times 7 = 1001$; $1001 \div 143 = 7$

37. (a)
$$\begin{array}{ll} A \cup B \cup C &= \{1, 2, 3, 4, 5, 6, 7, 8\} \\ A \cap B &= \{2, 4\} \\ A \cap C &= \{3, 4, 5\} \\ B \cap C &= \{4, 6\} \\ A \cap B \cap C &= \{4\} \end{array}$$

Problem Set 3.4 (page 188)

1. (a)

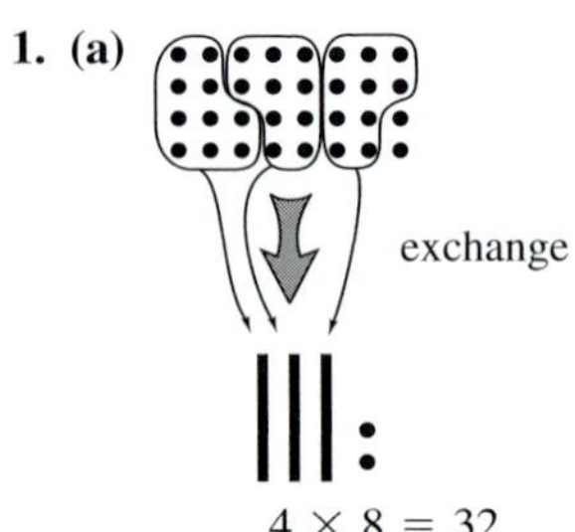

$4 \times 8 = 32$

3. (a) The number of hundreds in $30 \times 70 + 100$, i.e., 200

5. (a) Distributive property of multiplication over addition **(c)** Associative property of addition

9. (a) $27 = 4 \cdot 6 + 3$

10. (a)
$$\begin{array}{r} 21 \\ \underline{1} \\ 20 \\ 351\overline{)7425} \\ \underline{7020} \\ 405 \\ \underline{351} \\ 54 \end{array} \qquad 7425 = 351 \cdot 21 + 54$$

11. (a) $\begin{array}{r} 1\ 7\ 4\ \text{R}\ 3 \\ 5\overline{)8^{3}7^{2}3} \end{array}$

14. (a) $\begin{array}{r} 23_{\text{five}} \\ \times\ \ 3_{\text{five}} \\ \hline 124_{\text{five}} \end{array}$ **15. (a)** $\begin{array}{r} 31\ \text{R}\ 2 \\ 4\overline{)231} \\ \underline{22} \\ 11 \\ \underline{4} \\ 2 \end{array}$

18. (a) Yes **(b)**
$$\begin{array}{r} 285 \\ \times\ \ 362 \\ \hline 855 \\ 1710 \\ 570 \\ \hline 103{,}170 \end{array}$$

20. (a) $34 \cdot 54 = (17 \cdot 2) \cdot 54 = 17 \cdot (2 \cdot 54) = 17 \cdot 108$, since 2 evenly divides 34.

21. (a) 1256_{seven}, $1 \cdot 7^3 + 2 \cdot 7^2 + 5 \cdot 7 + 6 = 482_{\text{ten}}$
(c) $111{,}100{,}010_{\text{two}} = 2^8 + 2^7 + 2^6 + 2^5 + 2^1 = 482_{\text{ten}}$

22. (a) $\begin{array}{r} 7531 \\ \times\ \ \ \ 9 \\ \hline 67{,}779 \end{array}$

29. (a) 912,375 **31. (a)** Without clearing the calculator and reentering the numbers after the equal signs, we obtain the following results:
276,523 [÷] 511 [=] 541.1409 [−] 541 [=] 0.14909002 [×] 511 [=] 71.999997. Therefore, $q = 541$ and $r = 72$. To check, note that $541 \cdot 511 + 72 = 276{,}523$.

32. (a) $141{,}111_{\text{five}}$ **(c)** 3419_{twelve}

36. (a)
$$\begin{array}{rcl} 634 &=& \text{6 hundreds} + \text{3 tens} + \text{4 ones} \\ +\ 163 &=& \text{1 hundred} + \text{6 tens} + \text{3 ones} \\ \hline && \text{7 hundreds} + \text{9 tens} + \text{7 ones} \\ &=& 797 \end{array}$$

(c)
$$\begin{array}{rcl} 363 &=& \text{3 hundreds} + \text{6 tens} + \text{3 ones} \\ +\ 532 &=& \text{5 hundreds} + \text{3 tens} + \text{2 ones} \\ \hline && \text{8 hundreds} + \text{9 tens} + \text{5 ones} \\ &=& 895 \end{array}$$

(e)
$$\begin{array}{rcl} 725 &=& \text{7 hundreds} + \text{2 tens} + \text{5 ones} \\ -413 &=& -(\text{4 hundreds} + \text{1 ten} + \text{3 ones}) \\ \hline && \text{3 hundreds} + \text{1 ten} + \text{2 ones} \\ &=& 312 \end{array}$$

37. (a)
$$\begin{array}{rcl} 374 &=& \text{3 hundreds} + \text{7 tens} + \text{4 ones} \\ +\ 483 &=& \text{4 hundreds} + \text{8 tens} + \text{3 ones} \\ \hline && \text{7 hundreds} + \text{15 tens} + \text{7 ones} \\ &=& \text{8 hundreds} + \text{5 tens} + \text{7 ones} \\ &=& 857 \end{array}$$

(c)
$$\begin{array}{rrl} & 724 = & 7 \text{ hundreds} + 2 \text{ tens} + 4 \text{ ones} \\ + & 532 = & 5 \text{ hundreds} + 3 \text{ tens} + 2 \text{ ones} \\ \hline & & 12 \text{ hundreds} + 5 \text{ tens} + 6 \text{ ones} \\ & & = 1 \text{ thousand} + 2 \text{ hundreds} + 5 \text{ tens} + 6 \text{ ones} \\ & & = 1256 \end{array}$$

(e)
$$\begin{array}{rl} 367 = & 3 \text{ hundreds} + 6 \text{ tens} + 7 \text{ ones} \\ -249 = & -(2 \text{ hundreds} + 4 \text{ tens} + 9 \text{ ones}) \\ \hline 367 = & 3 \text{ hundreds} + 5 \text{ tens} + 17 \text{ ones} \\ -249 = & -(2 \text{ hundreds} + 4 \text{ tens} + 9 \text{ ones}) \\ \hline & 1 \text{ hundred} + 1 \text{ ten} + 8 \text{ ones} \\ = 118 & \end{array}$$

38. (a)
$$\begin{array}{rl} 213 = & 2 \text{ twenty-fives} + 1 \text{ five} + 3 \text{ ones} \\ +\ 131 = & 1 \text{ twenty-five} + 3 \text{ fives} + 1 \text{ ones} \\ \hline & 3 \text{ twenty-fives} + 4 \text{ fives} + 4 \text{ ones} \\ & = 344_{\text{five}} \end{array}$$

(c)
$$\begin{array}{rl} 142 = & 1 \text{ twenty-five} + 4 \text{ fives} + 2 \text{ ones} \\ +\ 123 = & 1 \text{ twenty-five} + 2 \text{ fives} + 3 \text{ ones} \\ \hline & 2 \text{ twenty-fives} + 6 \text{ fives} + 5 \text{ ones} \\ & = 2 \text{ twenty-fives} + 7 \text{ fives} + 0 \text{ ones} \\ & = 3 \text{ twenty-fives} + 2 \text{ fives} + 0 \text{ ones} \\ & = 320_{\text{five}} \end{array}$$

(e)
$$\begin{array}{rl} 344 = & 3 \text{ twenty-fives} + 4 \text{ fives} + 4 \text{ ones} \\ -\ 232 = & -\ (2 \text{ twenty-fives} + 3 \text{ fives} + 2 \text{ ones}) \\ \hline & 1 \text{ twenty-five} + 1 \text{ five} + 2 \text{ ones} \\ = 112_{\text{five}} & \end{array}$$

40. (a) 112 **(c)** 241 **(e)** 233 **(g)** 12 R 13

Problem Set 3.5 (page 205)

1. (c) 92 **(e)** 240 **2. (c)** 138 **(e)** 576 **3. (a)** 787 **(e)** 1026 **4. (a)** 240,000 **5. (a)** 900 **(c)** 27,000,000 **8. (a)** 52,000 **(c)** 49,000 **(e)** 13,000 **9. (a)** 90,000 **10. (a)** 750 **15. (a)** 3 **16. (a)** 27,451 since the last digit should be 1. **18. (a)** $(24)(678) = 16{,}272$ **(c)** $(2467)(8) = 19{,}736$
19. (a) $(88) + 8 + 8 + 8 + 8 + 8 + 8 = 136$
(c) $(888) + 8 + 8 + 8 + 8 + 8 = 928$
20. (a) $(844{,}422) \div 1 = 844{,}422$
(c) $(84 \div (44 \div 22)) \div 1 = 42$

30. (a)
$$\begin{array}{r} 2742 \\ 415 \\ 6943 \\ +\ 2718 \\ \hline 12{,}818 \end{array}$$
31. (a)
$$\begin{array}{r} 2734 \\ -\ 2643 \\ \hline 91 \end{array}$$
32. (a)
$$\begin{array}{r} 347 \\ \times\ 42 \\ \hline 694 \\ 1388 \\ \hline 14{,}574 \end{array}$$

Problem Set 3.6 (page 218)

1. (a) 641 **(c)** 101,388 **(e)** 770 **(g)** 770 **(i)** 237 **2. (a)** 98,915 **(c)** 32 **3. (a)** 49 **(c)** 13 **4. (a)** 841 **(c)** 33
5. (a) [(] [√] 784 [)] [−] 91 [÷] 13 [)] [÷] [(] 8 [×] 49 [−] 11 [×] 35 [)] [=] **7. (a)** $1831 - (17 \times 28) + 34$
13. (a) The [=] key needs to be pressed 14 times. 5 [M+] [+] 3 [=] [M+] [=] [M+] [=] · · · [=] [M+] [MR], 390
16. (a) The following algorithm generates the Lucas numbers. We show at each step the entry, the value of x in the display, and the value M in the memory. The Lucas numbers are printed in bold.

Entry	1	[M+]	3	[+]	[M+]	[MR]
x	**1**	1	3	**3**	3	**4**
M	0	1	1	1	4	4

[+]	[M+]	[MR]	[+]	[M+]	[MR]	[+]	. . .
7	7	**11**	**18**	18	**29**	**47**	. . .
4	11	11	11	29	29	29	. . .

17. (a) $\dfrac{1 + \sqrt{5}}{2}$. The computed number is located in both the display and the memory.
20. (a) To compute $y = \sqrt{5 \cdot 2^2 - 4}$, for example, we enter the string [√] 5 [×] 2 [x^2] [−] 4 [)] [=] to obtain 4. Since this is a whole number, we enter it into the table below the x-value 2. Completing the table, we obtain the results shown.

x	1	2	3	4	5	6	7	8
y	1	4			11			

23. $17 + 18 = 35$; $35 - 17 = 18$; $35 - 18 = 17$
25. $11 \cdot 27 = 297$; $297 \div 27 = 11$; $297 \div 11 = 27$
27.

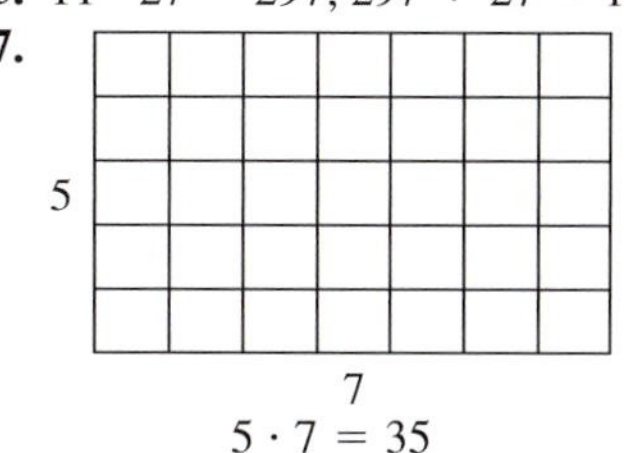

5

7

$5 \cdot 7 = 35$

Chapter 3 Review Exercises (page 223)

1. (a) 2353 **(b)** 58,331 **(c)** 1998 **2.**

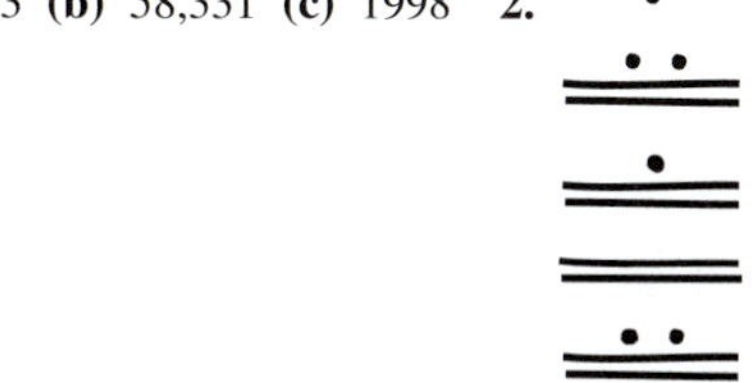

3. Exchange 30 units for 3 strips, and then exchange all 30 strips for 3 mats. The result is 8 mats, 0 strips, and 2 units.
4. (a) 45_{ten} **(b)** 181_{ten} **(c)** 417_{ten} **5. (a)** 2122_{five} **(b)** $100{,}011{,}111_{\text{two}}$ **(c)** 560_{seven}

6.

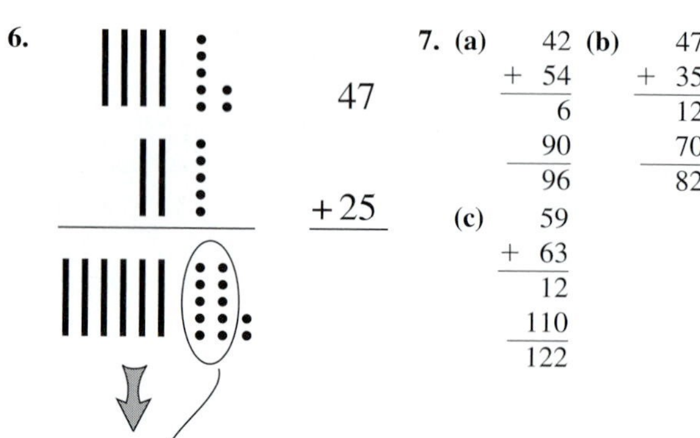

7. (a)
$$\begin{array}{r} 42 \\ +\ 54 \\ \hline 6 \\ 90 \\ \hline 96 \end{array}$$
(b)
$$\begin{array}{r} 47 \\ +\ 35 \\ \hline 12 \\ 70 \\ \hline 82 \end{array}$$
(c)
$$\begin{array}{r} 59 \\ +\ 63 \\ \hline 12 \\ 110 \\ \hline 122 \end{array}$$

8. (a)

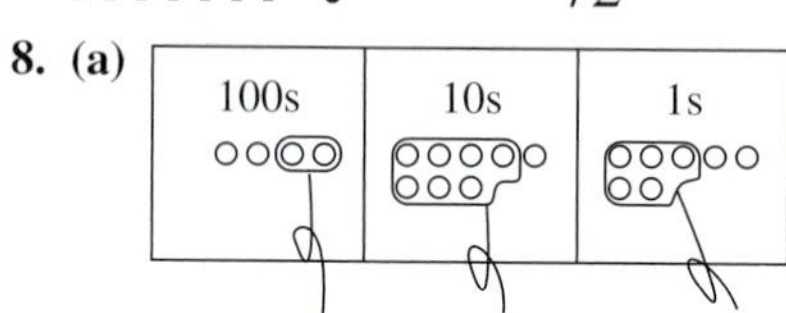

487 − 275 = 212

(b)

100s 10s 1s

100s 10s 1s

547 − 152 = 395

9. (a)
$$\begin{array}{r} 2433 \\ +\ 141 \\ \hline 3124 \end{array}$$
(b)
$$\begin{array}{r} 2433 \\ -\ 141 \\ \hline 2242 \end{array}$$
(c)
$$\begin{array}{r} 243 \\ \times\ \ 42 \\ \hline 1041 \\ 21320 \\ \hline 22411 \end{array}$$

10. (a)
$$\begin{array}{r} 357 \\ \times\ \ 4 \\ \hline 28 \\ 200 \\ 1200 \\ \hline 1428 \end{array}$$
(b)
$$\begin{array}{r} 642 \\ \times\ \ 27 \\ \hline 14 \\ 280 \\ 4200 \\ 40 \\ 800 \\ 12000 \\ \hline 17{,}344 \end{array}$$

11. (a)
$$\begin{array}{r} 127 \\ \hline 7 \\ 20 \\ 100 \\ 7\overline{)895} \\ \underline{700} \\ 195 \\ \underline{140} \\ 55 \\ \underline{49} \\ 6 \end{array}$$
(b)
$$\begin{array}{r} 79 \\ \hline 9 \\ 70 \\ 347\overline{)27483} \\ \underline{24290} \\ 3193 \\ \underline{3123} \\ 70 \end{array}$$

12. (a) $\begin{array}{r} 5487 \text{ R } 1 \\ 5\overline{)27436} \end{array}$ **(b)** $\begin{array}{r} 4948 \text{ R } 0 \\ 8\overline{)39584} \end{array}$

13. (a) 2121_{five} **(b)** $2{,}023{,}221_{\text{five}}$ **14.**

~~42~~	~~35~~
21	70
~~10~~	~~140~~
5	280
~~2~~	~~560~~
1	1120
	1470

15. (a) 300,000 **(b)** 270,000 **(c)** 275,000 **16.** 657 rounds to 700, 439 rounds to 400, 1657 rounds to 2000, and 23 rounds to 20. Thus, **(a)** 657 + 439 is approximately 700 + 400 = 1100. The actual sum is 1096. **(b)** 657 − 439 is approximately 700 − 400 = 300. The actual answer is 218. **(c)** 657 · 439 is approximately 700 · 400 = 280,000. The actual answer is 288,423. **(d)** 1657 ÷ 23 is approximately 2000 ÷ 20 = 100. The actual answer to the nearest hundredth is 72.04. **17.** 2 **18. (a)** 8088 **(b)** 49,149 **19. (a)** 11, 18, 29, 47 **(b)** The integer nearest $[(1 + \sqrt{5})/2] \cdot L_n$ is L_{n+1}. **(c)** No. It does not hold for L_1, L_2, and L_3. **(d)** The nearest integer to $[(1 + \sqrt{5})/2] \cdot L_n$ is L_{n+1} for $n \geq 4$. **20.** Replace the Lucas numbers by Fibonacci numbers throughout. In part (a), replace 7, 11, 18, 29 by 3, 5, 8, 13, respectively. The integer nearest $[(1 + \sqrt{5})/2] \cdot F_n = F_{n+1}$ for all $n \geq 2$.

Chapter 3 Test (page 224)

1.

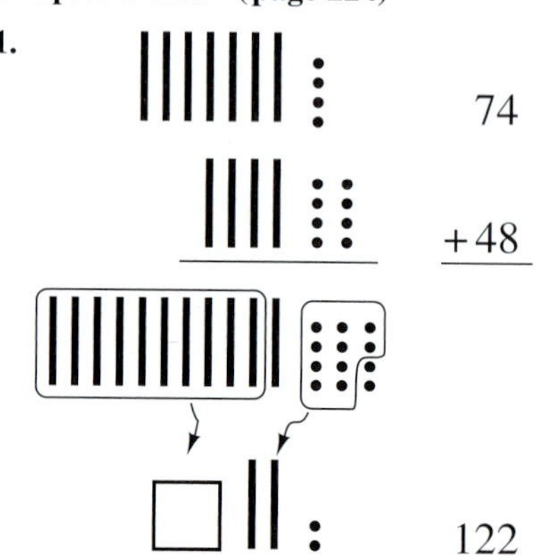

2. (a) 3,000,000 **(b)** 3,400,000 **(c)** 3,380,000 **(d)** 3,377,000 **3.**
$$\begin{array}{r} 751 \\ \times\ \ 93 \\ \hline 69{,}843 \end{array}$$
4. (a) 1, 2, 5, 12 **(b)** 29, 70 **(c)** $f_n = 2f_{n-1} + f_{n-2}$ for $n \geq 3$. **5. (a)** 197_{ten} **(b)** 207_{ten} **(c)** 558_{ten} **6.**
$$\begin{array}{r} 2837 \\ +\ 7224 \\ \hline 10{,}061 \end{array}$$
7.
$$\begin{array}{r} 8236 \\ -\ 3542 \\ \hline 4694 \end{array}$$
8. 4800

9. (a) 1 and 1
9 and 9
36 and 36
100 and 100

(b) 1, 3, 6, 10 **(c)** $\left(\dfrac{n(n+1)}{2}\right)^2$

10.
```
 ————
• • • •
  • •
 ════
 ————
```
11. 1575 **12.**
```
   468
×   20
  9360
```
13. Answers may vary. One possibility is as follows:
2 [M+] 5 [+] [M+] [MR] [+] [M+] [MR] [+] ··· with the repeating sequence [M+] [MR] [+]. The [MR] merely repeats the preceding term, but the [MR] and [+] cause the display to exhibit the next two desired terms in the sequence.
14. **(a)** 340 **(b)** 144 **(c)** 23,111
15. 1,464,843
16. **(a)** 2111_{five} **(b)** $100{,}011{,}001_{\text{two}}$ **(c)** $1E5_{\text{twelve}}$

Chapter 4

Problem Set 4.1 (page 242)

1. **(a)**

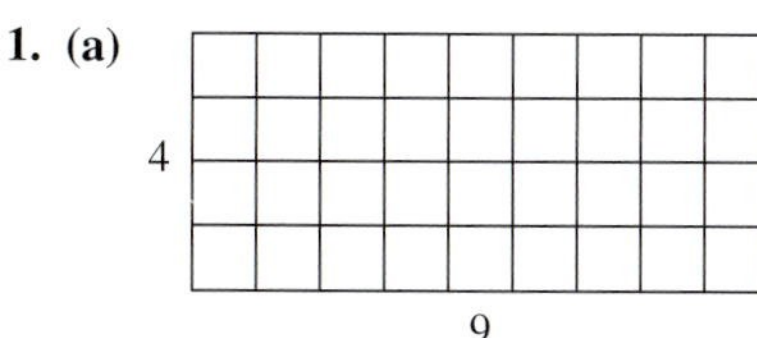

$36 = 4 \cdot 9$
4 divides 36.
3. **(a)** 0, 8, 16, 24, 32, 40, 48, 56, 64, 72
5. **(a)**
```
     72
    /  \
   8    9
  /\   /\
 2  4 3  3
   /\
  2  2
```
(c)
```
    264
    /  \
   8    33
  /\    /\
 2  4  3  11
   /\
  2  2
```
6. **(a)**
```
     5
  5)25
  2)50
 2)100
 7)700
```
$700 = 2 \cdot 2 \cdot 5 \cdot 5 \cdot 7$

(c)
```
     2
   3)6
  3)18
  5)90
 5)450
```
$450 = 5 \cdot 5 \cdot 3 \cdot 3 \cdot 2$

7. **(a)** 1, 2, 3, 4, 6, 8, 12, 16, 24, 48
8. **(a)** $136 = 2^3 \cdot 17^1$, $102 = 2^1 \cdot 3^1 \cdot 17^1$
(b) The divisors of 136 are
$2^0 \cdot 17^0 = 1$, $2^1 \cdot 17^0 = 2$, $2^2 \cdot 17^0 = 4$,
$2^3 \cdot 17^0 = 8$, $2^0 \cdot 17^1 = 17$, $2^1 \cdot 17^1 = 34$, and
$2^2 \cdot 17^1 = 68$, $2^3 \cdot 17^1 = 136$.
9. **(a)** $48 = 2^4 \cdot 3^1$ **(c)** $2250 = 2^1 \cdot 3^2 \cdot 5^3$ **10.** **(a)** Yes, because $28 = 2^2 \cdot 7^1$, so all the prime factors of 28 appear in a and to at least as high a power. **(c)** $2^1 \cdot 7^2$ **13.** **(a)** No. For example, $10 = 2 \cdot 5$ with 2 and 5 both primes. Yet $5 > 3.162\ldots = \sqrt{10}$. **15.** **(a)** True. $n \cdot 0 = 0$ for every natural number n. **(c)** True. $1 \cdot n = n$ for every natural number n. **(e)** False. $0 \div 0 = q$ if, and only if, $0 \cdot q = 0$ for a *unique* integer q. However, this is true for *every* integer q.

17. **(a)** $1, 3, 3^2 = 9$ **18.** **(a)** $496 = 2^4 \cdot 31^1$
(b) $1 + 2 + 4 + 8 + 16 + 31 + 62 + 124 + 248 = 496$
19. **(a)** Deficient **(b)** Abundant **23.** Yes. If $p|bc$, then p must appear in the prime factorization of the product bc and hence in the prime factorization of b or c. But then $p|b$ or $p|c$, as claimed. **25.** **(a)** $N_4 = 11 \cdot 101$, $N_6 = 11 \cdot 10{,}101$, $N_8 = 11 \cdot 1{,}010{,}101$ **27.** **(a)** No **(c)** Yes
31. **(a)** $2^2 \cdot 137^1$ **(c)** $2^1 \cdot 137^1$ **(e)** $(2^1 \cdot 137^1)$ divides $(2^2 \cdot 137^1)$, and $(2^3 \cdot 3^2 \cdot 13^1)$ divides $(2^3 \cdot 3^2 \cdot 7^2 \cdot 13^1)$.
32. **(a)** $2^2 \cdot 3^3 \cdot 7^2 \cdot 13^2$ **(c)** $3^4 \cdot 5^6$ **(e)** The exponents in the prime-power representation are even.
36. **(a)** False. The sum of two odd natural numbers is an even natural number. **(c)** True: $a + b = b + a$ if a and b are any natural numbers. **(e)** True: $a + (b + c) = (a + b) + c$ if a, b, and c are any natural numbers. **(g)** True: $a(b + c) = ab + ac$ if a, b, and c are any natural numbers. **(i)** True: 1 is an element of S. **37.** **(a)** 299
38. **(a)** 19, 23, 27 **(c)** 15, 21, 28 **(e)** 48, 96, 192
39. **(a)** 4082

Problem Set 4.2 (page 252)

1. **(a)** Divisible by 2 and 3 **(c)** Divisible by 5 **2.** **(a)** 1554 **(c)** None **3.** **(a)** Divisible by 7 and 13 **(c)** Divisible by 7 **4.** **(a)** None **(c)** None **9.** **(a)** For any palindrome with an even number of digits, the digits in the odd positions are the same as the digits in the even positions, but with the order reversed. Thus, the difference of the sums of the digits in the even and odd positions is zero, which is divisible by 11.
14. When testing for divisibility by 7, 11, and 13, the digits of the number are broken up into three-digit groups. If a number has the form *abc,abc*, then the difference between the sums of the three-digit numbers in odd positions and even positions will be zero, which is divisible by each of 7, 11, and 13. **16.** **(e)** No **19.** No. He or she could have made other kinds of errors that by chance resulted in the record being out of balance by an amount that is a multiple of 9.
20. **(a)** $2^7 \cdot 3^2 \cdot 7^1$
21. **(a)**
```
        7000
       /    \
     70      100
    /  \    /   \
   10   7  10    10
  /\      /\    /\
 2  5    2  5  2  5
```
$7000 = 2^3 \cdot 5^3 \cdot 7^1$

23. $2^2 \cdot 3^2 \cdot 5^1 \cdot 7^1$

Problem Set 4.3 (page 264)

1. **(a)** 3 **2.** **(a)** 216
3. **(a)** $\text{GCD}(24, 27) \cdot \text{LCM}(24, 27) = 3 \cdot 216 = 648 = 24 \cdot 27$ **4.** **(a)** $\text{GCD}(r, s) = 2^1 \cdot 3^1 \cdot 5^2 = 150$, $\text{LCM}(r, s) = 2^2 \cdot 3^3 \cdot 5^3 = 13{,}500$

12. (a) $\text{GCD}(a, b, c) = 2^0 \cdot 3^1 \cdot 5^1 \cdot 7^0 = 15$, $\text{LCM}(a, b, c) = 2^2 \cdot 3^3 \cdot 5^3 \cdot 7^1 = 94{,}500$

13. (a) $D_{18} = \{1, 2, 3, 6, 9, 18\}$
$D_{24} = \{1, 2, 3, 4, 6, 8, 12, 24\}$
$D_{12} = \{1, 2, 3, 4, 6, 12\}$
$\text{GCD}(18, 24, 12) = 6$
$M_{18} = \{18, 36, 54, 72, 90, \ldots\}$
$M_{24} = \{24, 48, 72, 96, \ldots\}$
$M_{12} = \{12, 24, 36, 48, 60, 72, 84, \ldots\}$
$\text{LCM}(18, 24, 12) = 72$

15. (a) $\text{GCD}(24, 18) = 6$, $\text{GCD}(\text{GCD}(24, 18), 12) = \text{GCD}(6, 12) = 6$
$\text{LCM}(24, 18) = 72$, $\text{LCM}(\text{LCM}(24, 18), 12) = \text{LCM}(72, 12) = 72$

The final results are the same.

16. (a) The 1-, 3-, and 9-rods **(d)** The 1-, 2-, 3-, 6-, 9-, and 18-trains **17. (a)** 12 **(b)** 12 **21. (a)** 18; 76; 1364 **(c)** No: $F_{19} = 4181 = 37 \cdot 113$. **(h)** If $\text{GCD}(F_{16}, F_{20}) = 4$, then the conjecture must be false. **24. (a)** 1224 seconds, $1224 = \text{LCM}(72, 68)$ **(b)** 17 laps **26. (a)** $\text{LCM}(220, 264) = 1320$; $\text{LCM}(220, 275) = 1100$; $\text{LCM}(264, 275 = 6600$ **28. (a)** 3 **29. (a)** 720 **30. (a)** $205{,}800 = 2^3 \cdot 3^1 \cdot 5^2 \cdot 7^3$; $31{,}460 = 2^2 \cdot 5^1 \cdot 11^2 \cdot 13^1$, $25{,}840 = 2^4 \cdot 5^1 \cdot 17^1 \cdot 19^1$

32. (a)

n	$n^2 - 81n + 1681$	Prime?
1	1601	yes
2	1523	yes
3	1447	yes
4	1373	yes
5	1301	yes

(c) No. If a conjecture is true for several cases, it does not mean it is true for *every* case.
(e) The conjecture may seem more probable, but we still cannot say if it is always true.
33. (a) No. By the division algorithm, n leaves a remainder of 1 when divided by 2, 5, or 7. **34. (a)** No. If $p \mid a$ and $p \nmid b$, then $p \nmid (a + b)$.

Problem Set 4.4 (page 274)

1. (a) 1
2. (a)

3. (a) Check the sum of the digits: $1 + 0 + 5 + 7 + 4 + 8 + 6 + 5 + 3 + 7 = 46$. This is not a multiple of 10, so the code is incorrect. **4. (a)** MasterCard **(d)** VISA
5. (a) Not valid **(d)** Not valid **6. (a)** 5
7. (a) Congratulate him. He is calculating the correct check sum. **11.** No, since the fourth code group would have to be I I I I I. **13.** The error would not be caught. **15. (a)** 90 **16. (a) (i)** Not correct **(ii)** Correct **19. (a)** $\frac{1}{1}, \frac{2}{1}, \frac{3}{2}, \frac{5}{3}, \frac{8}{5}$
(b) The compound fraction with n 1s down the left-hand diagonal simplifies to $\frac{F_{n+1}}{F_n}$, where F_n denotes the nth Fibonacci number. **20. (a)** 1.6180 **(c)** 1.6180

Chapter 4 Review Exercises (page 279)

1.

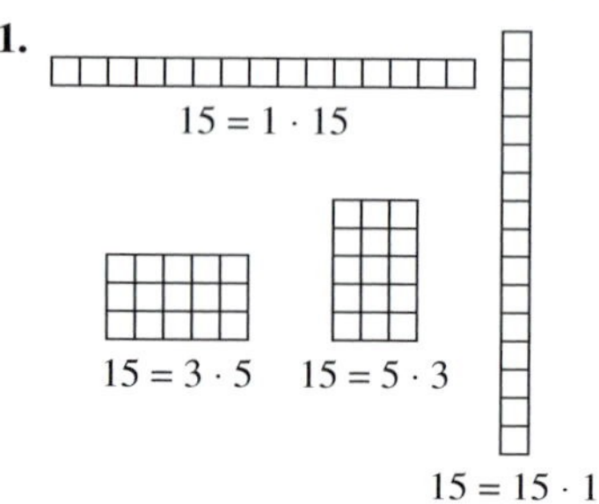

2.

```
     96
    /  \
   8    12
  /\    /\
 2  4  3  4
    /\    /\
   2  2  2  2
```

3. (a) $D_{60} = \{1, 2, 3, 4, 5, 6, 10, 12, 15, 20, 30, 60\}$
(b) $D_{72} = \{1, 2, 3, 4, 6, 8, 9, 12, 18, 24, 36, 72\}$
(c) $D_{60} \cap D_{72} = \{1, 2, 3, 4, 6, 12\}$, so $\text{GCD}(60, 72) = 12$.
4. (a) $1200 = 2^4 \cdot 3^1 \cdot 5^2$ **(b)** $2940 = 2^2 \cdot 3^1 \cdot 5^1 \cdot 7^2$
(c) $\text{GCD}(1200, 2940) = 2^2 \cdot 3^1 \cdot 5^1 \cdot 7^0 = 60$; $\text{LCM}(1200, 2940) = 2^4 \cdot 3^1 \cdot 5^2 \cdot 7^2 = 58{,}800$.
5. Composite; $847 = 11 \times 77$ **6. (a)** Answers will vary. For example, $15 = 3 \cdot 5$; $5 > \sqrt{15}$. **(b)** Yes. $3 \leq \sqrt{15}$.
7. Answers will vary. For example, 8 divides 16 and 4 divides 16, but 32 does not divide 16. **8.** Since $n = 2536$, the prime, 2, divides $3 \cdot 5 \cdot 7 + 11 \cdot 13 \cdot 17$.
9. (a) Divisible by 2 and 5 **(b)** Divisible by 3 and 11 **(c)** Divisible by 5 **(d)** Divisible by 5 and 11
10. (a) Divisible by 11 **(b)** Divisible by 7 and 13 **(c)** Divisible by 11 and 13 **11. (a)** False **(b)** True **(c)** False **(d)** True **12. (a)** $(1 + 4)(1 + 2) = 15$
(b) 1, 3, 7, 9, 21, 27, 49, 63, 81, 147, 189, 441, 567, 1323, 3969 **13.** $d = 5$ **14. (a)** $2^3 \cdot 3^5 \cdot 7^3 \cdot 11^3 \cdot 13^1 = 11{,}537{,}501{,}976$ **(b)** $2^2 \cdot 3^5 \cdot 7^2 \cdot 11^1 = 523{,}908$
15. (a) $D_{63} = \{1, 3, 7, 9, 21, 63\}$, $D_{91} = \{1, 7, 13, 91\}$, and $D_{63} \cap D_{91} = \{1, 7\}$, so $\text{GCD}(91, 63) = 7$. **(b)** $M_{63} = \{63, 126, 189, 252, 315, 378, 441, 504, 567, 630, 693, 756, 819, 882, 945, 1008. \ldots\}$, $M_{91} = \{91, 182, 273, 364, 455, 546, 637, 728, 819, 910, \ldots\}$, and $M_{63} \cap M_{91} = \{819, 1638, \ldots\}$, so $\text{LCM}(63, 91) = 819$. **(c)** $7 \cdot 819 = 5733 = 63 \cdot 91$
16. (a) $2 \cdot 11^2 = 242$ **(b)** $2^3 \cdot 3^2 \cdot 5^2 \cdot 7^1 \cdot 11^3 = 16{,}770{,}600$

17. (a) $12{,}100\overline{)119{,}790}$ = 9 R 10,890

$10{,}890\overline{)12{,}100}$ = 1 R 1210 $1210\overline{)10{,}890}$ = 9 R 0

Thus, GCD(119,790, 12,100) = 1210.
(b) LCM(119,790, 12,100) = 119,790 · 12,100/1210 = 1,197,900. **18.** 2192 **19. (a)** 4 **(b)** 9
20. (a)

(b)

21. (a) 80321-1589 **(b)** 60648-9960 **22. (a)** Not valid **(b)** Valid **23. (a)** Possible solution pairs for *d* and *e*, respectively, are 0 and 7, 1 and 5, 2 and 3, 3 and 1, 5 and 6, 6 and 4, 7 and 2, and 8 and 0. **(b)** The only three possible solution pairs for *d* and *e*, respectively, are 0 and 2, 5 and 1, and 1 and 0. **24.** The check sum should catch each transposition error.

Chapter 4 Test (page 280)

1. (a) Divisible by 2 and 3 **(b)** Divisible by 3 **2. (a)** No. The prime-power representation of *r* contains two 7s, but the prime-power representation of *m* contains only one 7, so *r* does not divide *m*. **(b)** $(3 + 1)(2 + 1)(1 + 1)(4 + 1) = 120$ **(c)** $2^2 \cdot 5^0 \cdot 7^1 \cdot 11^3 = 37{,}268$ **(d)** $2^3 \cdot 5^2 \cdot 7^2 \cdot 11^4 = 143{,}481{,}800$ **3. (a)** F **(b)** T **(c)** T **(d)** F **4.** 7
5. Since $\sqrt{281} \doteq 16.7$, we must check for divisibility by 2, 3, 5, 7, 11, and 13. Standard divisibility tests immediately rule out divisibility by 2, 3, and 5. Also, $281 = 40 \cdot 7 + 1$, $281 = 11 \cdot 25 + 6$, and $281 = 13 \cdot 21 + 8$. Therefore, 281 is a prime. **6. (a)** $0 + 6 + 9 + 9 + 2 + 7 + 5 + 4 + 8 = 50$, so the zip code is correct. **(b)** $8 + 4 + 2 + 3 + 2 + 7 + 6 + 1 + 2 + 0 = 35$, so the zip code is incorrect.
7. Since $533 = 3 \cdot 154 + 91$, $154 = 1 \cdot 91 + 63$, $91 = 1 \cdot 63 + 28$, $63 = 2 \cdot 28 + 7$, and $28 = 4 \cdot 7$, it follows that GCD(154, 553) = 7 and LCM(154, 553) = (154 × 553) ÷ 7 = 12,166.

8. (a) $13{,}534\overline{)997{,}476}$ = 73 R 9494 $9494\overline{)13{,}534}$ = 1 R 4040

$4040\overline{)9494}$ = 2 R 1414 $1414\overline{)4040}$ = 2 R 1212

$1212\overline{)1414}$ = 1 R 202 $202\overline{)1212}$ = 6 R 0

GCD(997,476, 13,534) = 202
(b) LCM(997,476, 13,534) = 997,476 · 13,534/202 = 66,830,892
9. (a) Valid **(b)** Not valid **10. (a)**

```
      8532
     /    \
   12      711
  /  \    /   \
 3    4  9    79
     / \ / \
    2  2 3  3
```

(b) $8532 = 2^2 \cdot 3^3 \cdot 79^1$ **(c)** 4266 **(d)** 17,064

Chapter 5

Problem Set 5.1 (page 291)

1. (a) **(c)**

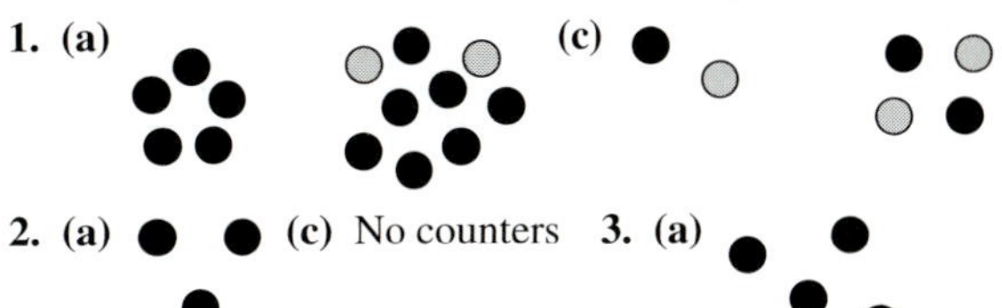

2. (a) **(c)** No counters **3. (a)**

(c) Opp 17 **4. (a)** At mail time, you are delivered a check for \$14. **6. (a)** −15 **7. (a), (c), (e)**

8. (a) 4 **(c)** 6

9. (a)

(c)

10. (a) 34 **(c)** 76 **11. (a)** 13, −13 **15.** Remind them that the absolute value of a number is the number's distance on the number line from zero and that −*n is n* units from 0.
19. (a) Four red counters
20. (a) −12, −10, −8, −6, −4, −2, 0, 2, 4, 6, 8, 10, 12
22. (a) $\{n \mid n \text{ is an integer and } -12 \le n \le 12\}$
24. (a) 2^{20}
25. (a) 100 black, 110 red **30. (a)** $2^5 = 32$ **31. (a)** 90
32. (a) No. The prime-power representation of *c* contains more 3s than the prime-power representation of *a*.
33. (a) $1400 = 2^3 \cdot 5^2 \cdot 7^1$
34. (a) $4554\overline{)5445}$ = 1 R 891 $891\overline{)4554}$ = 5 R 99

$99\overline{)891}$ = 9 R 0 GCD(4554, 5445) = 99

Problem Set 5.2 (page 310)

1. (a) **(c)**

$8 + (-3) = 5$ $-8 - (-3) = -5$

(e) **(g)**

$9 + 4 = 13$ $(-9) + 4 = -5$

2. (a) At mail time, you receive a bill for \$27 and a bill for \$13; $(-27) + (-13) = -40$. **(c)** The mail carrier brings you a check for \$27 and a check for \$13; $27 + 13 = 40$. **(e)** At mail time, you receive a bill for \$41 and a check for \$13; $(-41) + 13 = -28$. **(g)** At mail time, you receive a bill for \$13 and a check for \$41; $(-13) + 41 = 28$.

3. (a)

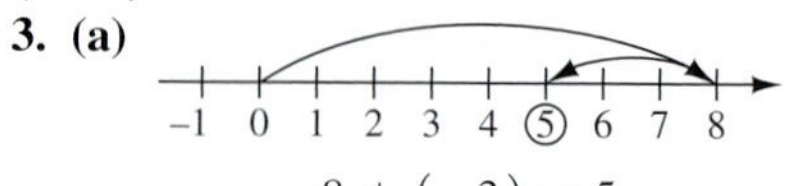

$8 + (-3) = 5$

(c)

$-8 + 3 = -5$

(e)

$4 + (-7) = -3$

(g)

$(-4) + 7 = 3$

4. (a) $13 + (-7)$ **(c)** $(-13) + (-7)$ **(e)** $3 + (-8)$ **(g)** $(-8) + (-13)$ **5. (a)** 40 **(c)** -27 **(e)** -135 **(g)** -135 **9. (a)** More, by \$106 **11. (a)** $-117 < -24$ **13.** Keyshawn has drawn the diagram for $7 + 3$. In a sense, he is correct, since $7 - (-3) = 7 + 3$. **17. (a)** and **(c)** are true. **19.** Not necessarily. If $a \geq b$, then it is possible that $a = b$, and so $a > b$ is false. **21.** $-6, -5, -4, -3, -2, -1, 0, 1, 2, 3, 4, 5, 6$ **23. (a) (i)** 6 **(iii)** 15

(b) (i)

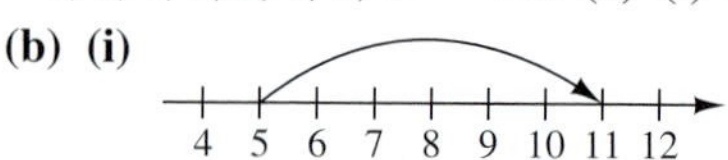

Distance is 6.

(iii)

Distance is 15.

24. (a) (i) $|7 + 2| = 9, |7| + |2| = 9$
(iii) $|7 + (-6)| = 1, |7| + |-6| = 13$

26. (a)

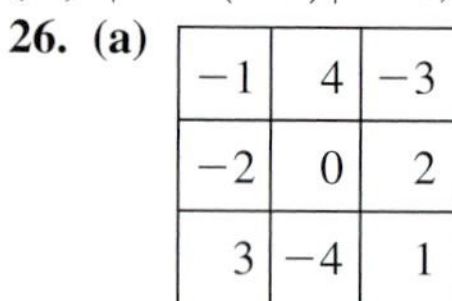

−1	4	−3
−2	0	2
3	−4	1

27. (a)

−1 0 1

−2

2

28. (a) $7 - (-3) = 10, (-3) - 7 = -10$
32. (a) $F_1 + F_3 + F_5 + F_7 = 1 + 2 + 5 + 13 = 21 = F_8$
$F_1 + F_3 + F_5 + F_7 + F_9 = 1 + 2 + 5 + 13 + 34 = 55 = F_{10}$
35. (a) Yes. $F_0 + F_1 = 0 + 1 = 1 = F_2$.

36. (a) $101 + 3 = 104$

39. (a)

t	h
0	0
1	80
2	128
3	144
4	128
5	80
6	0
7	−112

41. (a) 50 **42. (a)** 1575 **(c)** -5909 **(e)** 2053 **45. (a)** Divisible by 3 **46. (a)** Divisible by 7 and 13

Problem Set 5.3 (page 320)

1. (a) 77 **(c)** -77 **(e)** 108 **(g)** -108 **(i)** 0 **2. (a)** 4 **(c)** -4 **(e)** -13 **(g)** 16 **(i)** 36
3. $(-25{,}753) \cdot (-11) = 283{,}283$
$283{,}283 \div (-11) = -25{,}753$
$283{,}283 \div (-25{,}753) = -11$
5. (a) Richer by \$78; $6 \cdot 13 = 78$ **6. (a)** $6 \cdot 3 = 18$
9. (a) Multiplicative property of zero
Distributive property of multiplication over addition
Definition of negative

10. (a)

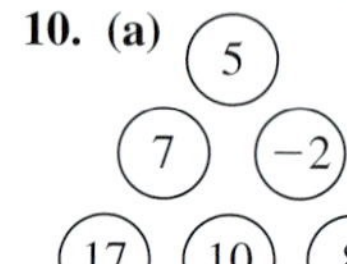

(c)

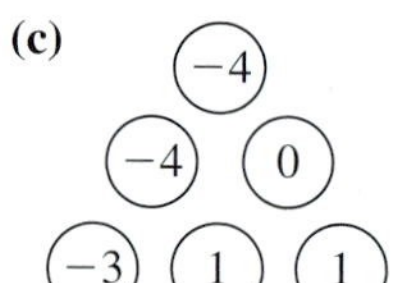

11. (b)

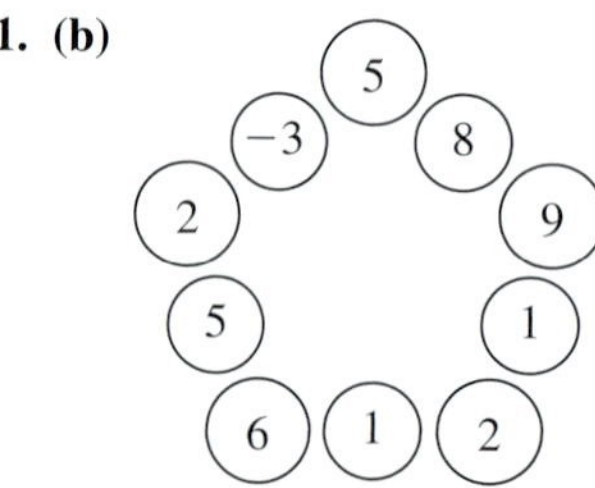

17. (a) \$7 **(b)** $(-105) \div 15 = -7$.
A loss of \$105 shared among 15 people results in each person losing \$7. **19. (b)** 59 **21. (b)** 2,621,440
22. (a) 5, 12, 29, 70, 169, 408

Problem Set 5.4 (page 331)

1. (a) 2 **(c)** 0 **(e)** 8 **2. (a)** 2 **(c)** 9 **(e)** 6 **3. (a)** 2 **(c)** 8 **(e)** 3 **4. (a)** 5 **(c)** 3

5. (a) $9 -_{12} 7 = 9 +_{12} 5 = 2$
(c) $5 -_{12} 9 = 5 +_{12} 3 = 8$
(e) $2 -_{12} 11 = 2 +_{12} 1 = 3$ **6. (a)** 11 **(c)** 0 **(e)** 0
7. (a) 11 **(c)** Undefined **(e)** 9 **8. (a)** 1, 5, 7, 11
10. (a) 2 **(c)** 3 **(e)** 1 **(g)** 4 **(i)** 0 **(k)** 4 **11. (a)** 0; $n +_5 0 = n$ for n in $\{0, 1, 2, 3, 4\}$ **12. (a)** 4, 3, 2, 1, 0
13. (a) 1, 2, 3, 4 **14. (a)** $4 \div_{12} 7 = 4$
(c) $3 \div_5 2 = 4$ **(e)** $2 \div_5 4 = 3$ **(g)** $4 \times_{12} 7^{-1} = 4$
(i) $3 \times_5 2^{-1} = 4$ **(k)** $2 \times_5 4^{-1} = 3$
15. (c)

$$\begin{aligned} (y +_{12} 2) \div_{12} 11 &= 3 \\ y +_{12} 2 &= 3 \times_{12} 11 \\ y +_{12} 2 &= 9 \\ y &= 9 +_{12} 10 \\ y &= 7 \end{aligned}$$

16. (b) (i) 1 **(iii)** 12 **(v)** 11 **(vii)** 9 **20. (a)** 3 **(c)** 3 **(e)** 4 **(g)** 4
21. (a)

n	$2^n - 2$	$2^n -_n 2$
2	2	0
3	6	0
4	14	2
5	30	0
6	62	2
7	126	0
8	254	6
9	510	6
10	1022	2
11	2046	0
12	4094	2
13	8190	0

28. (a) $D_{60} = \{1, 2, 3, 4, 5, 6, 10, 12, 15, 20, 30, 60\}$
$D_{150} = \{1, 2, 3, 5, 6, 10, 15, 25, 30, 50, 75, 150\}$ so $\text{GCD}(60, 150) = 30$.
29. (a)

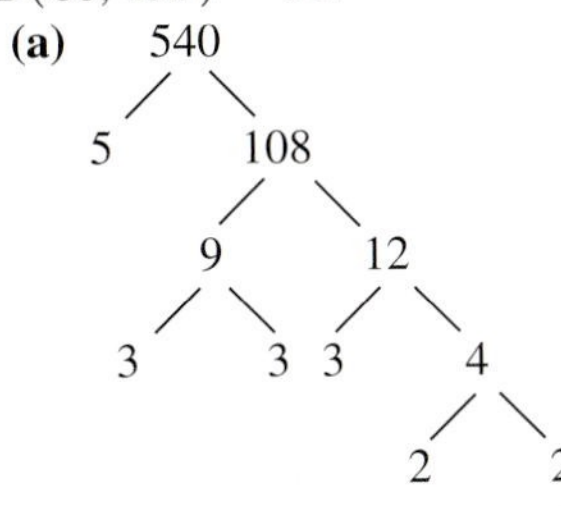

$540 = 2^2 \cdot 3^3 \cdot 5^1$

30. (a) $540 = 2^2 \cdot 3^3 \cdot 5^1$; $600 = 2^3 \cdot 3^1 \cdot 5^2$

Chapter 5 Review Exercises (page 335)

1. (a) -1 **(b)** 5 **(c)** $-15, -13, -11, \ldots, 11, 13, 15$
2. (a) Richer by \$12; 12 **(b)** Poorer by \$37; -37
3. (a) 12 **(b)** -24 **4. (a)** Answers will vary. Any drop that shows five more red counters than black counters represents the integer -5. **(b)** Any drop that shows six more black counters than red counters represents the integer 6.
5. Answers will vary. **(a)** At mail time, you receive a bill for \$114 and a check for \$29. **(b)** The mail carrier brings you a bill for \$19 and a check for \$66. **6. (a)** -44 **(b)** 61
7. $2 + (-4) = -2$ **8.** $(-1) - (-3) = 2$
9. (a) $45 + (-68) = -23$ **(b)** $45 - (-68) = 113$
10. (a) $6 + 3 = 9$ **(b)** $8 + (-4) = 4$ **(c)** $7 - 5 = 2$
(d) $3 - 9 = -6$ **(e)** $(-3) + (-4) = -7$
(f) $(-3) - (-7) = 4$ **(g)** $(-5) - 6 = -11$
11. (a) -2 **(b)** -22 **(c)** -32 **(d)** 12 **(e)** 20 **(f)** -4
12. (a) 27° below zero **(b)** $(-15) - 12 = -27$
13. (a) \$25 **(b)** $(-12) + 37 = 25$
14. (a)

−9 −5 −2 0 2 7

(b) $-9, -5, -2, 0, 2, 7$ **(c)** $-9 + 4 = -5, -5 + 3 = -2, -2 + 2 = 0, 0 + 2 = 2, 2 + 5 = 7$ **15. (a)** $3 \cdot 4 = 12$
(b) $3 \cdot (-4) = -12$ **(c)** $(-3) \cdot 4 = -12$
(d) $(-3) \cdot (-4) = 12$
16. (a)

$$\begin{aligned} 3 \cdot 0 &= 0 && \text{multiplicative property of 0} \\ 3 \cdot [5 + (-5)] &= 0 && \text{definition of negative} \\ 3 \cdot 5 + 3 \cdot (-5) &= 0 && \text{distributive property} \end{aligned}$$

(b) $3 \cdot (-5) = -(3 \cdot 5)$ definition of negative
(c)

$$\begin{aligned} 0 \cdot (-5) &= 0 && \text{multiplicative property of 0} \\ [3 + (-3)] \cdot (-5) &= 0 && \text{definition of negative} \\ 3 \cdot (-5) + (-3) \cdot (-5) &= 0 && \text{distributive property} \\ -(3 \cdot 5) + (-3) \cdot (-5) &= 0 && \text{by part (b)} \end{aligned}$$

(d) $(-3) \cdot (-5) = -(-(3 \cdot 5)) = 3 \cdot 5$ definition of negative and theorem on page 289
17. (a) 56 **(b)** -56 **(c)** -56 **(d)** -7 **(e)** -12 **(f)** 12
18. (a) At mail time, you receive 7 checks, each for \$12.
(b) The mail carrier takes away 7 checks, each for \$13.
(c) The mail carrier takes away 7 bills, each for \$13.
19. If d divides n, there is an integer c such that $dc = n$. But then $d \cdot (-c) = -n$, $(-d) \cdot (-c) = dc = n$, and $(-d) \cdot c = -dc = -n$. Thus, d divides $-n$, $-d$ divides n, and $-d$ divides $-n$. **20. (a)** 3 **(b)** 11 **21.** By the division algorithm, n must be of one of these forms: $6q$, $6q + 1$, $6q + 2$, $6q + 3$, $6q + 4$, or $6q + 5$. If n is not divisible by 2, however, then n cannot be of any of the forms $6q$, $6q + 2$, or $6q + 4$. Likewise, if n is not divisible by 3, n cannot be of the form $6q + 3$ either. Thus, there must be an integer q such that either $n = 6q + 1$ or $n = 6q + 5$.
Case 1

$$\begin{aligned} n &= 6q + 1 \\ n^2 - 1 &= (6q + 1)^2 - 1 \\ &= (36q^2 + 12q + 1) - 1 \\ &= 36q^2 + 12q \\ &= 12q(3q + 1) \end{aligned}$$

If q is even, then 24 divides $12q$ and $n^2 - 1$ is divisible by 24. If q is odd, then $3q + 1$ is even, so again 24 divides $12q(3q + 1)$ and hence $n^2 - 1$ is divisible by 24.

Case 2 $\quad n = 6q + 5$

$$\begin{aligned} n^2 - 1 &= (6q + 5)^2 - 1 \\ &= (36q^2 + 60q + 25) - 1 \\ &= 36q^2 + 60q + 24 \\ &= 12(3q^2 + 5q + 2) \\ &= 12(3q + 2)(q + 1) \end{aligned}$$

If q is even, $3q + 2$ is even. If q is odd, $q + 1$ is even. In either case, it follows that $n^2 - 1$ is divisible by 24.

22. (a) 1 **(b)** 5 **(c)** 0 **(d)** Undefined **(e)** 5 **(f)** 0 **(g)** Undefined **(h)** 3 **(i)** 3 **23. (a)** 4 **(b)** 1 **(c)** 2 **(d)** 4 **24.** 2, 4, 5, 6, 8, 10

Chapter 5 Test (page 337)

1. \$381; $129 + 341 - 13 - 47 - 29 = 381$
2. Richer by \$135; $(-5) \cdot (-27) = 135$
3. (a)

$$\begin{aligned} 1 &= 1 \\ 1 - 4 &= -3 \\ 1 - 4 + 9 &= 6 \\ 1 - 4 + 9 - 16 &= -10 \\ 1 - 4 + 9 - 16 + 25 &= 15 \end{aligned}$$

(b) $1 - 4 + \cdots + n^2 = \dfrac{n(n + 1)}{2} = t_n$ if n is odd.

$1 - 4 + \cdots - n^2 = -\dfrac{n(n + 1)}{2} = -t_n$ if n is even.

4. (a) $-18, -17, -16, \ldots, -3, -2, -1, 0, 1, 2, 3, \ldots,$ 16, 17, 18. **(b)** Yes **5. (a)** -26 **(b)** 12 **(c)** 26 **(d)** -12 **(e)** -361 **(f)** -408 **(g)** -864 **(h)** 105 **(i)** 0 **6.** At mail time the mail carrier delivers a check for \$7 and takes away a bill for \$4. You are \$11 richer. **7. (a)** $-5, -3, -8, -11, -19, -30$ **(b)** $7, -5, 2, -3, -1, -4$ **(c)** $6, -8, -2, -10, -12, -22$ **8.** Poorer by \$9; $(-27) \div 3 = -9$
9. (a) 4 **(b)** 0 **(c)** 5 **(d)** 3 **(e)** 3 **(f)** 7
10. (a) $(-7) + 10 = 3$

(b) $10 - (-7) = 17$

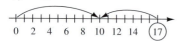

(c) $7 \cdot (-5) = -35$

11. 2160, the same as LCM(240, 54)

Chapter 6

Problem Set 6.1 (page 354)

1. (a) $\frac{1}{6}$ **(c)** $\frac{0}{1}$ **(e)** $\frac{2}{6}$

2. (a) **(c)**

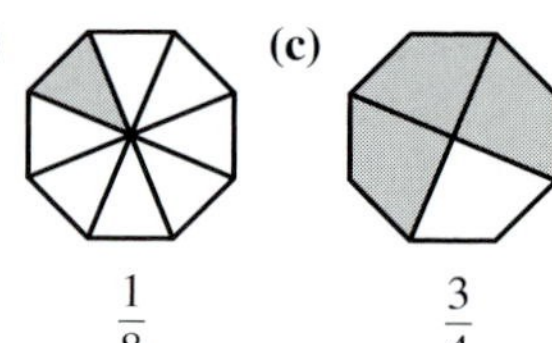

4. (a) $A: \frac{1}{4}, B: \frac{3}{4}, C: \frac{3}{2}$ or $\frac{6}{4}$

(c) $G: \frac{0}{5}, H: \frac{2}{5}, I: \frac{8}{5}$

5. (a)

6. (a) $\frac{1}{3}$ **(c)** $\frac{5}{7}$ **(e)** $\frac{1}{4}$ **(g)** $\frac{2}{3}$

8. (a) $\frac{3}{6} = \frac{1}{2}$

10. (a)

11. (a) 24 **(c)** -140 **12. (a)** Equivalent **(c)** Not equivalent **13. (a)** Equivalent

14. (a) Yes **(c)** Yes **15. (a)** $\frac{7}{12}$ **(c)** $\frac{-31}{43}$

16. (a) $\dfrac{96}{288} = \dfrac{2^5 \cdot 3^1}{2^5 \cdot 3^2} = \dfrac{1}{3}$

17. Answers will vary. Possible answers are shown.

(a) $\frac{15}{55}$ and $\frac{22}{55}$ **(c)** $\frac{32}{24}, \frac{15}{24}$, and $\frac{4}{24}$

18. (a) $\frac{9}{24}$ and $\frac{20}{24}$ **(c)** $\frac{136}{96}$ and $\frac{21}{96}$

19. (a) $\frac{7}{12}, \frac{2}{3}$ **(c)** $\frac{29}{36}, \frac{5}{6}$

21. (a) True. Given two fractions, two equivalent fractions with a common denominator may be found, by finding a common multiple of the two original denominators. Once these fractions, say, $\frac{a}{c}$ and $\frac{b}{c}$, are found, infinitely many more pairs of equivalent fractions can be found, namely, $\frac{a \cdot n}{c \cdot n}$ and $\frac{b \cdot n}{c \cdot n}$ for any integer n other than 0. **(c)** False. Given any positive fraction, a smaller positive fraction may be found by multiplying the denominator by 2. So there cannot be a least positive fraction.

26. (a) $\frac{4}{8}$

(d)

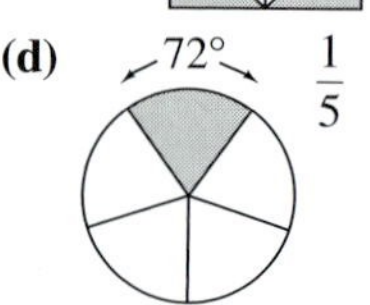

27. The four triangles can be reattached to form a figure with 5 squares, each congruent to the shaded inner square. Therefore, 1/5 of the large square is covered by the inner shaded square.

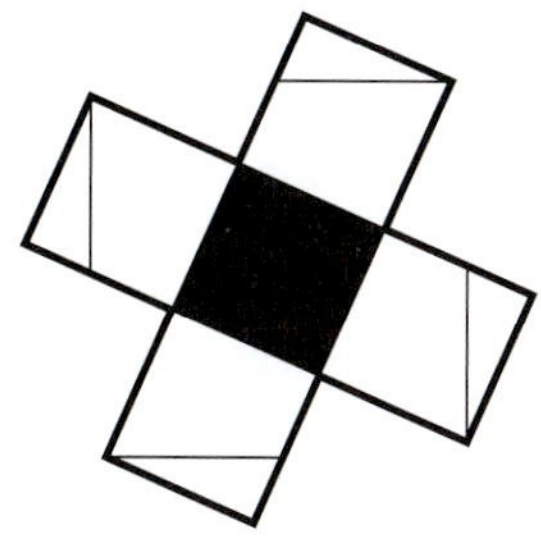

30. (a) Add tick marks to show that the rope was originally 60 feet long.

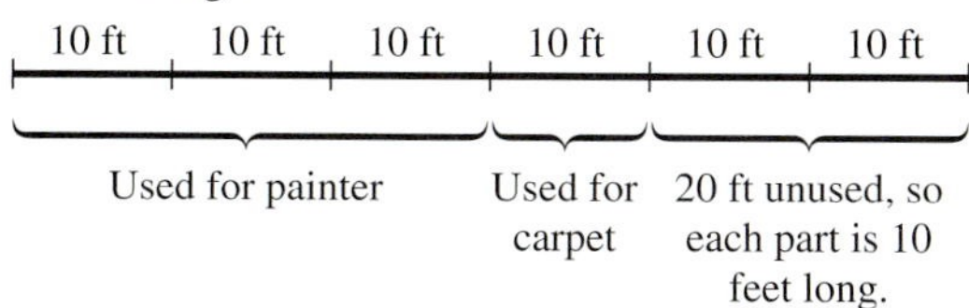

33. (a) $\frac{1}{2}, \frac{1}{6}, \frac{1}{3}$ **36. (a)** 18 gal. **37. (a)** $\frac{3}{5}$ **(c)** $\frac{1}{4}$

38. Yes. Lakeside won $\frac{19}{25}$ of its games, while Shorecrest won $\frac{16}{21}$ of its games, and $\frac{16}{21} > \frac{19}{25}$. **39. (a)** Carol; Carol

44. (a) 5, −8, 13, −21, 34, −55, 89, −144, 233, −377

45. (a) $m = 7, n = 5$ **46.** $d = 5$

Problem Set 6.2 (page 375)

1. (a) $\frac{1}{3} + \frac{1}{2} = \frac{5}{6}$

2. (a)

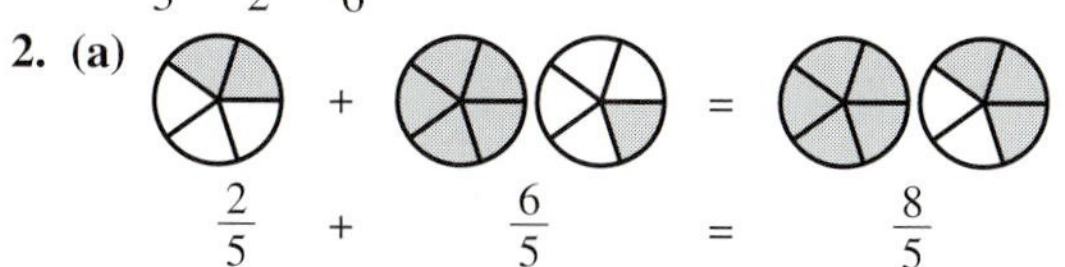

$\frac{2}{5} + \frac{6}{5} = \frac{8}{5}$

4. (a) $\frac{1}{8}$, $\frac{3}{8}$; 0, $\frac{1}{8}$, $\frac{2}{8}$, $\frac{3}{8}$, $\frac{4}{8}$ (circled), $\frac{5}{8}$, $\frac{6}{8}$, $\frac{7}{8}$, $\frac{8}{8}$

5. (a) 0, 1/4 (circled), 2/4, 3/4, 4/4; $\frac{3}{4} + \frac{-2}{4} = \frac{1}{4}$

6. (a) $\frac{5}{7}$ **(c)** $\frac{5}{6}$ **(e)** $\frac{19}{15}$ **(g)** $\frac{73}{100}$

7. (a) $2\frac{1}{4}$ **(c)** $4\frac{19}{23}$

8. (a) $\frac{19}{8}$ **(c)** $\frac{557}{5}$ **9. (a)** $\frac{5}{6} - \frac{1}{4} = \frac{7}{12}$

10. (a) $\frac{3}{4} - \frac{1}{6} = \frac{7}{12}$

11. (a) $\frac{3}{4} - \frac{1}{3} = \frac{9}{12} - \frac{4}{12} = \frac{5}{12}$

12. (a) $\frac{3}{8}$ **(c)** $\frac{4}{3}$ **(e)** $\frac{1}{3}$ **(g)** $\frac{625}{642}$ **13. (a)** $3 \times \frac{5}{2} = \frac{15}{2}$

14. (a) $2 \times \frac{3}{5} = \frac{6}{5}$

16. (a) $\frac{8}{3}$ **(c)** $\frac{4}{9}$ **(e)** $\frac{1}{5}$ **17. (a)** $\frac{8}{15}$ **(c)** $\frac{10}{11}$ **(e)** $\frac{4}{7}$

18. (a) 1 **19. (a)** $2\frac{1}{2} \times 3\frac{3}{4} = \frac{5}{2} \times \frac{15}{4} = \frac{75}{8} = 9\frac{3}{8}$ miles

20. (a) $\frac{3}{4} - \frac{2}{3} = \frac{9}{12} - \frac{8}{12} = \frac{1}{12} > 0$

24. Since the books are neatly in order, Volume 1 is just to the left of Volume 2. Therefore, the bookworm eats only through the front cover of Volume 1 and the back cover of Volume 2, for a total travel distance of $\frac{1}{8} + \frac{1}{8} = \frac{2}{8} = \frac{1}{4}$ inch.

25. (a)

$\frac{1}{2}$	$\frac{1}{12}$	$\frac{5}{12}$
$\frac{1}{4}$	$\frac{1}{3}$	$\frac{5}{12}$
$\frac{1}{4}$	$\frac{7}{12}$	$\frac{1}{6}$

26. (a) $3\frac{1}{2} + 1\frac{2}{5} = \frac{7}{2} + \frac{7}{5} = \frac{35}{10} + \frac{14}{10} = \frac{49}{10}$ and $3\frac{1}{2} \times 1\frac{2}{5} = \frac{7}{2} \times \frac{7}{5} = \frac{49}{10}$

30. (a) **31. (a)** $\frac{1}{5} + \frac{1}{45} = \frac{9}{45} + \frac{1}{45} = \frac{10}{45} = \frac{5 \cdot 2}{5 \cdot 9} = \frac{2}{9}$

33. (a) It terminates.

$\frac{3}{4}$ $\frac{2}{3}$ $\frac{1}{2}$ $\frac{1}{4}$
$\frac{1}{12}$ $\frac{1}{6}$ $\frac{1}{4}$ $\frac{1}{2}$
$\frac{1}{12}$ $\frac{1}{12}$ $\frac{1}{4}$ $\frac{5}{12}$
0 $\frac{1}{6}$ $\frac{1}{6}$ $\frac{1}{3}$
$\frac{1}{6}$ 0 $\frac{1}{6}$ $\frac{1}{3}$
$\frac{1}{6}$ $\frac{1}{6}$ $\frac{1}{6}$ $\frac{1}{6}$
0 0 0 0

34. **(a)**

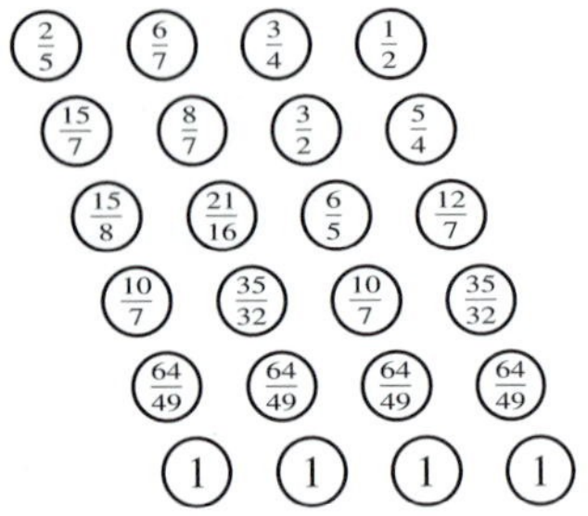

35. 3, 3, 4, 5, 6, 6 **38.** 23 bows **40.** $42

46. **(a)** $\frac{7}{2}$ **48.** 180, $\frac{203}{180}$

Problem Set 6.3 (page 389)

1. Commutative and associative properties of addition: $(3 + 2 + 8) + \left(\frac{1}{5} + \frac{2}{5} + \frac{1}{5}\right)$

3. **(a)** $\frac{-4}{5}$

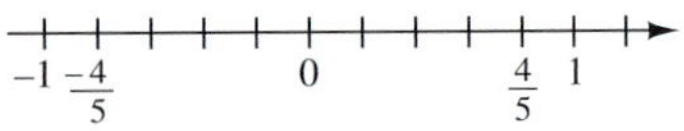

(c) $\frac{8}{3}$

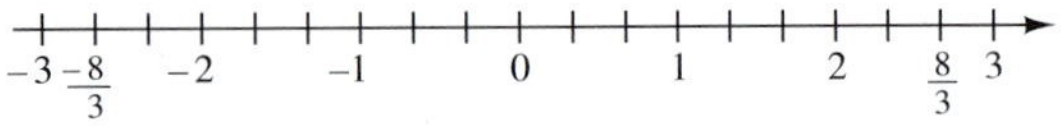

4. **(a)** $\frac{-1}{2}$ **(c)** $\frac{11}{8}$ **5.** **(a)** $\frac{-7}{20}$ **(c)** $\frac{7}{24}$ **(e)** $\frac{-41}{12}$ **6.** **(a)** $\frac{7}{8}$ **(c)** $\frac{1}{2}$ **(e)** 8

7. **(a)** $\frac{2}{3}$

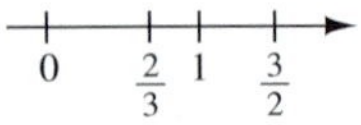

(c) $\frac{-11}{-4}$ or $\frac{11}{4}$

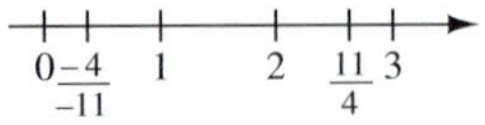

(e) $-\frac{1}{2}$

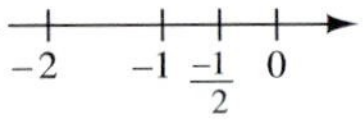

8. **(a)** $\frac{2}{3}$ **(c)** 0 **9.** **(a)** Addition of rational numbers—definition **10.** $\frac{7}{6}$, since $\frac{a}{b} = \frac{2}{3} \div \frac{4}{7} = \frac{2}{3} \cdot \frac{7}{4} = \frac{14}{12} = \frac{7}{6}$.

11. **(a)** $x = \frac{-3}{4}$ **(c)** $x = \frac{-6}{5}$

12. **(a)** Closure property for subtraction and the existence of a multiplicative inverse

13. **(a)** $-\frac{1}{5}, \frac{2}{5}, \frac{4}{5}$

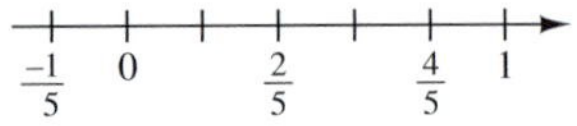

(c) $\frac{3}{8}, \frac{1}{2}, \frac{3}{4}$

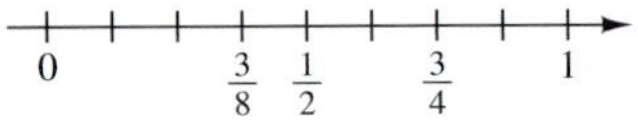

14. **(a)** $-4 \cdot 4 = -16 < -15 = 5 \cdot (-3)$

15. **(a)** $x + \frac{2}{3} > -\frac{1}{3}$

$$x > -\frac{1}{3} - \frac{2}{3} = -1$$

(c) $\frac{3}{4}x < -\frac{1}{2}$

$$x < -\frac{1}{2} \div \frac{3}{4}$$

$$x < -\frac{1}{2} \cdot \frac{4}{3}$$

$$x < -\frac{2}{3}$$

16. **(a)** Answers will vary. One answer is $\frac{1}{2}$, since $\frac{4}{9} < \frac{1}{2} < \frac{6}{11}$. **(c)** Answers will vary. Since $\frac{7}{12} = \frac{14}{24} = \frac{28}{48}$ and $\frac{14}{23} = \frac{28}{46}$, one answer is $\frac{28}{47}$. **19.** **(a)** $\frac{1}{4}$ **(c)** -1

20. **(a)** 9 **21.** **(a)** $\frac{3}{2}$ **(c)** $\frac{3}{5}$ **(e)** 40 **(g)** $-2\frac{1}{8}$

22. **(a)** How many acres does Hal own after his new purchase? **23.** Answers will vary. Possible answers include the following: **(a)** Two pizzas were ordered. $\frac{3}{4}$ of one pizza and $\frac{1}{2}$ of another were eaten. How much pizza was eaten in all? **28.** $\frac{1260}{250} = 5\frac{1}{25}$

29. **(a)** 8 **(c)** $\frac{3}{5}$ **(e)** $\frac{56}{9}$ **34.** **(a)** 4, since $\frac{1}{2} + \frac{2}{8} + \frac{1}{4} = \frac{2}{4} + \frac{1}{4} + \frac{1}{4} = \frac{4}{4}$ **37.** Cut the 8-inch side in half and the 10-inch side in thirds to get six $3\frac{1}{3}$-by-4-inch rectangles. **39.** **(a)** $\frac{-3}{125}$ **40.** **(a)** The replacement rule gives $\frac{7 + 2 \cdot 5}{7 + 5} = \frac{17}{12}$. This is a good approximation of $\sqrt{2}$, since $\left(\frac{17}{12}\right)^2 = \frac{289}{144} = 2 + \frac{1}{144}$. **43.** **(a)** 2 square yards

44. $230 to $345

Chapter 6 Review Exercises (page 397)

1. **(a)** $\frac{2}{4}$ **(b)** $\frac{6}{6}$ **(c)** $\frac{0}{4}$ **(d)** $\frac{5}{3}$

2.

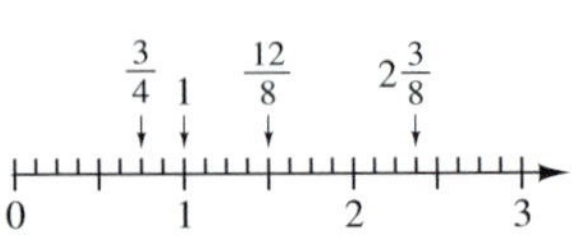

3. **(a)** $\frac{1}{3}$ **(b)** $\frac{4}{33}$ **(c)** $\frac{21}{4}$ **(d)** $\frac{297}{7}$ **4.** $\frac{13}{30}, \frac{13}{27}, \frac{1}{2}, \frac{25}{49}, \frac{26}{49}$

5. **(a)** 36 **(b)** 18

6.

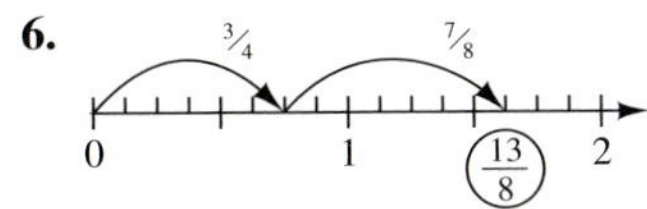

7.

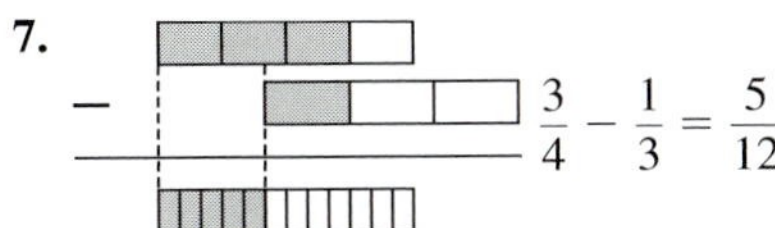

8. **(a)** $\frac{5}{8}$ **(b)** $\frac{-7}{36}$ **(c)** $\frac{2}{15}$ **(d)** $\frac{41}{12}$ or $3\frac{5}{12}$

9. **(a)** **(b)** **(c)**

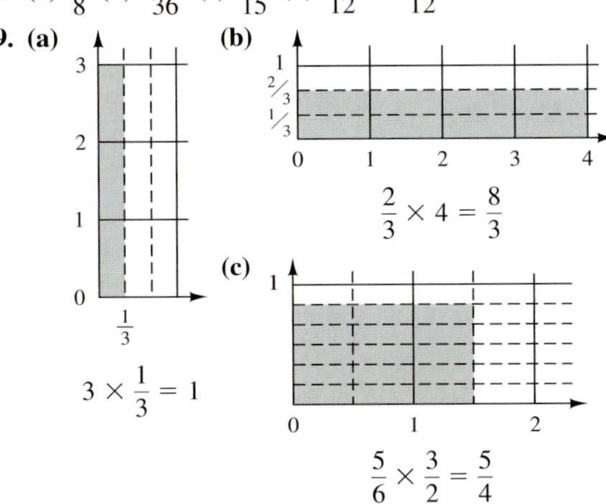

10. $\frac{3}{4}$; this problem corresponds to the sharing (partitive) model of division. **11.** $4\frac{1}{2}$; this problem corresponds to the grouping (measurement) model of division. **12.** $\frac{57}{10}$ miles $= 5\frac{7}{10}$ miles **13.** **(a)** $\frac{-1}{8}$ **(b)** $\frac{3}{2}$ **(c)** $\frac{-1}{2}$ **(d)** $\frac{-7}{10}$ **14.** **(a)** $x = 2$ **(b)** $x = \frac{-1}{6}$ **(c)** $x = \frac{5}{18}$ **(d)** $x = \frac{9}{16}$ **15.** Write $\frac{5}{6} = \frac{55}{66}$ and $\frac{10}{11} = \frac{60}{66}$. Then $\frac{56}{66}$ and $\frac{57}{66}$ are between $\frac{5}{6}$ and $\frac{10}{11}$. Other answers may be given.
16. **(a)** 4 **(b)** 5 **(c)** 20

Chapter 6 Test (page 398)

1. $\frac{a+b}{b} = \frac{c+d}{d}$ if, and only if, $(a+b)d = b(c+d)$— that is, if, and only if, $ad + bd = bc + bd$. But this is so if, and only if, $ad = bc$. And this is so if, and only if, $\frac{a}{b} = \frac{c}{d}$.

2. **(a)** $\frac{1}{8}$ **(b)** $\frac{-7}{9}$ **(c)** 3 **(d)** 1 **3.** **(a)** 0, since $\frac{2}{3} + \frac{-4}{6} = 0$. **(b)** 2, since $\frac{5}{6}\cdot\frac{36}{15} = \frac{5}{15}\cdot\frac{36}{6} = \frac{1}{3}\cdot 6 = 2$. **(c)** 1, since $\frac{9}{5} - \frac{1}{5} = \frac{8}{5}$ is the reciprocal of $\frac{5}{8}$. **(d)** $\frac{1}{3}$, since $\frac{2}{\not 3}\cdot\frac{\not 3}{\not 4}\cdot\frac{\not 4}{\not 5}\cdot\frac{\not 5}{6} = \frac{2}{6} = \frac{1}{3}$. **4.** **(a)** 40 acres **(b)** $\frac{1}{4}$ mile

5. **(a)** If $\frac{a}{b}$ and $\frac{c}{d}$ are two rational numbers with $\frac{a}{b} < \frac{c}{d}$, then there is a rational number $\frac{e}{f}$ such that $\frac{a}{b} < \frac{e}{f} < \frac{c}{d}$. **(b)** Write $\frac{3}{5} = \frac{18}{30}$ and $\frac{2}{3} = \frac{20}{30}$ to see that $\frac{19}{30}$ is between $\frac{3}{5}$ and $\frac{2}{3}$.

6. **(a)** **(b)** **(c)** **(d)**

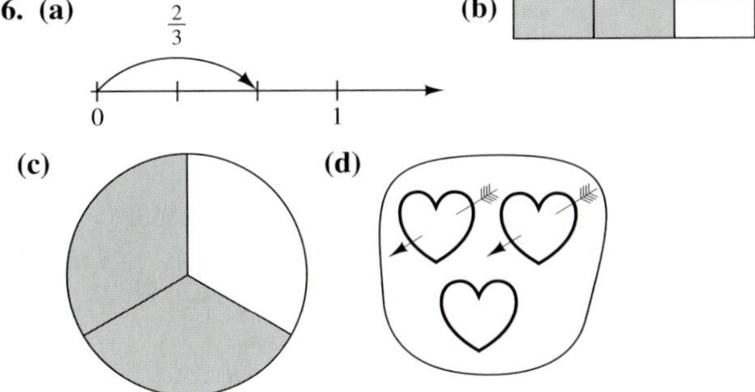

7. **(a)** $11\frac{1}{2}$ **(b)** 48 **(c)** 23 **8.** **(a)** If $\frac{a}{b}$ is a rational number with $a \neq 0$, then its multiplicative inverse is the rational number $\frac{b}{a}$. **(b)** $\frac{2}{3}, -\frac{5}{4}, \frac{-1}{5}$ **9.** Answers will vary. Possibilities include $\frac{-6}{8}, \frac{-9}{12}$, and $\frac{-12}{16}$. **10.** **(a)** $x > \frac{-3}{2}$ **(b)** $x = \frac{-2}{9}$ **(c)** $x > \frac{-4}{15}$ **(d)** $x > \frac{-1}{12}$ **11.** **(a)** If $\frac{a}{b}$ and $\frac{c}{d}$ are rational numbers with $\frac{c}{d} \neq 0$, then $\frac{a}{b} \div \frac{c}{d} = \frac{e}{f}$ if, and only if, $\frac{a}{b} = \frac{c}{d}\cdot\frac{e}{f}$. **(b)** If $\frac{a}{b} \div \frac{c}{d} = \frac{e}{f}$, then $\frac{a}{b} = \frac{c}{d}\cdot\frac{e}{f} = \frac{e}{f}\cdot\frac{c}{d}$. So $\frac{a}{b}\cdot\frac{d}{c} = \frac{e}{f}\cdot\frac{c}{d}\cdot\frac{d}{c} = \frac{e}{f}$. Thus, $\frac{a}{b} \div \frac{c}{d} = \frac{a}{b}\cdot\frac{d}{c}$. **12.** **(a)** Answers will vary. **(b)** Answers will vary. **13.** **(a)** If $\frac{a}{b}$ is a rational number, then its additive inverse is the rational number $\frac{-a}{b}$. **(b)** $\frac{-3}{4}, \frac{7}{4}, \frac{8}{2}$
14. $-3, -1\frac{1}{2}, 0, \frac{5}{8}, \frac{2}{3}, 3, \frac{16}{5}$

Chapter 7

Problem Set 7.1 (page 421)

1. **(a)** $273.412 = 200 + 70 + 3 + \frac{4}{10} + \frac{1}{100} + \frac{2}{1000}$; $273.412 = 2\cdot 10^2 + 7\cdot 10^1 + 3\cdot 10^0 + 4\cdot 10^{-1} + 1\cdot 10^{-2} + 2\cdot 10^{-3}$
2. **(a)** $\frac{81}{250}$; $250 = 2\cdot 5^3$ **3.** **(a)** 0.35 **(c)** 0.04
4. **(a)** $\frac{107}{333}$ **(e)** $\frac{1}{7}$
5. **(a)** $\frac{358}{999} = 0.\overline{358}$ **6.** **(a)** 0.007, 0.017, $0.01\overline{7}$, 0.027
8. Assume $3 - \sqrt{2}$ is rational; then $3 - \sqrt{2} = q$, where q is

rational. This implies that $\sqrt{2} = 3 - q$. But $3 - q$ is rational, since the rational numbers are closed under subtraction. This can't be true, since we know that $\sqrt{2}$ is irrational. Thus, the assumption that $3 - \sqrt{2}$ is rational must be false. So $3 - \sqrt{2}$ is irrational. **14. (a)** $0.1 = \frac{1}{10}$ **15. (a)** $0.\overline{09}$ **(c)** $0.\overline{0009}$ **16. (a)** $\frac{74}{99}$ **17. (a)** $0.\overline{5}$ **(d)** $0.\overline{51}$ **18.** Answers will vary. **(a)** One example is $\sqrt{2} + (3 - \sqrt{2}) = 3$. $\sqrt{2}$ and $(3 - \sqrt{2})$ are both irrational. (See the answer to problem 8.) **(b)** One example is $\sqrt{3} + \sqrt{3} = 2\sqrt{3}$. $\sqrt{3}$ and $2\sqrt{3}$ are both irrational. **21.** Answers will vary. For example, $\frac{\sqrt{2}}{2\sqrt{2}} = \frac{1}{2}$, and $\frac{1}{2}$ is rational. **32. (a)** $\frac{0}{1}, \frac{1}{4}, \frac{1}{3}, \frac{2}{5}, \frac{1}{2}, \frac{3}{5}, \frac{2}{3}, \frac{3}{4}, \frac{1}{1}; \frac{0}{1}, \frac{1}{5}, \frac{1}{4}, \frac{2}{7}, \frac{1}{3}, \frac{3}{8}, \frac{2}{5}, \frac{3}{7}, \frac{1}{2}, \frac{4}{7}, \frac{3}{5}, \frac{5}{8}, \frac{2}{3}, \frac{5}{7}, \frac{3}{4}, \frac{4}{5}, \frac{1}{1}$ **33. (a)** $\frac{7}{6}$ **(c)** $\frac{1}{3}$ **(e)** $\frac{6}{5}$

Problem Set 7.2 (page 433)

1. (a) 403.674 **(c)** 1.137
2. (a) 174.37 **(c)** 26.1 **3. (a)** 35, 35.412 **(c)** 124, 124.7056 **4. (a)** $0.8\overline{3}$ **5. (a)** 2.77×10^{8}
6. (a) 1.05×10^{-10} **(c)** 1.29×10^{-13} **7. (a)** 1.53×10^{10}

11.

0.492	1.107	0.246
0.369	0.615	0.861
0.984	0.123	0.738

15. (c) 2.374 0.041 5.267
2.415 5.308
7.723

16. (a) 3.4, 4.3, 5.2, 6.1, 7.0, 7.9 **(c)** 0.0114, 0.1144, 0.2174, 0.3204, 0.4234, 0.5264 **17. (a)** 2.11, 2.321, 2.5531, 2.8084, 3.0893

21. (a)

4.73
1.32 3.41
8.42 7.10 10.51

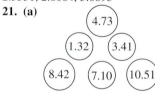

(c)

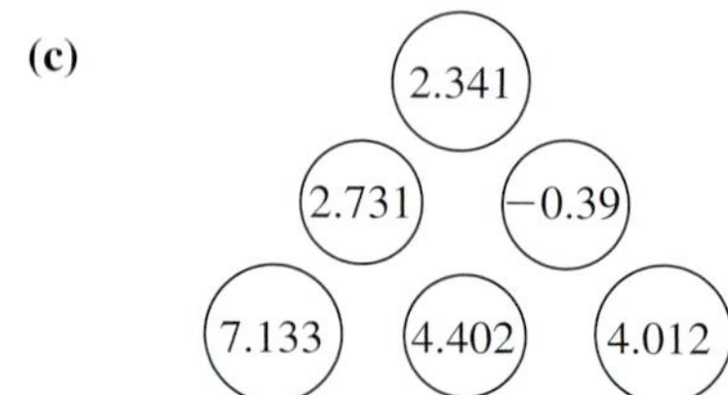

22. (a)

1.33 0.41 1.64
0.92 1.23
1.64 0.72 1.95

(c) Answers will vary. However, all possible solutions have 5.01 in the lower right corner. **23.** \$29.16

26. (a) 210.375 in^2

28. (a)

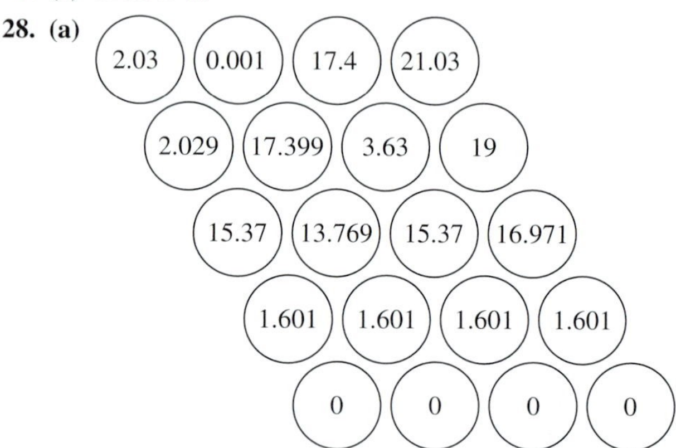

(c) 5, 9, 17, and 31
30. (a) Working to nine decimal places, this yields 1, 1, 1, 1 after four steps.
36. (a)

38. (a)

39. (a) 2303

Problem Set 7.3 (page 445)

1. (a) $\frac{7}{5}$ **(c)** $\frac{7}{12}$ **(e)** $\frac{12}{5}$
2. (a) Yes **(c)** No **(e)** No **(g)** Yes **3. (a)** $r = 9$
4. (a) $\frac{3}{2}$ **(c)** $\frac{2}{3}$
5. (a) \$19.25 **(c)** $\frac{3.5}{19.25} = 0.1\overline{8}, \frac{5}{27.5} = 0.\overline{18}$
8. 19 ft
14. (a)

$$\frac{a}{b} = \frac{c}{d}$$
$$ad = bc$$
$$da = cb$$
$$\frac{d}{c} = \frac{b}{a}$$

(c)
$$\begin{aligned} \frac{a}{b} &= \frac{c}{d} \\ ad &= bc \\ ac + ad &= ac + bc \\ a(c+d) &= (a+b)c \\ \frac{a}{a+b} &= \frac{c}{c+d} \end{aligned}$$

15. (a) $y = 108$ **16. (a)** $y = 4$ **21.** 59 to 58 **23. (a)** 32 ounces for 90¢ **28.** \$13.01 **31.** \$139.93 **37. (a)** -3 **(c)** -12 **(e)** 12 **(g)** -4 **(i)** 7 **(k)** -3 **39.** 72 **41.** $2b = 2^1 \cdot 3^2 \cdot 5 \cdot 11^3$

Problem Set 7.4 (page 457)

1. (a) 18.75% **(c)** 92.5% **(e)** $36.\overline{36}$% **(g)** 22.86% **2. (a)** 19% **(c)** 215% **3. (a)** $\frac{1}{10}$ **(c)** $\frac{5}{8}$ **4. (a)** 196 **(c)** 285.38 **(e)** 5.4962 **5. (a)** 420 **(c)** 6 **(e)** 112 **6. (a)** 50%

8. (a)

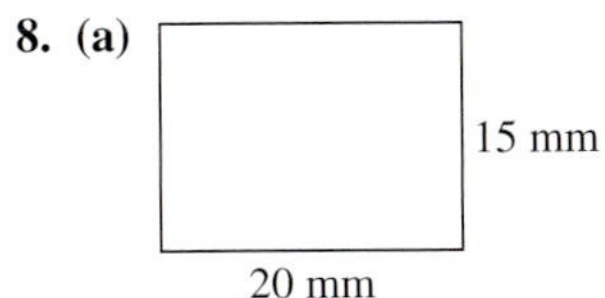

(c)

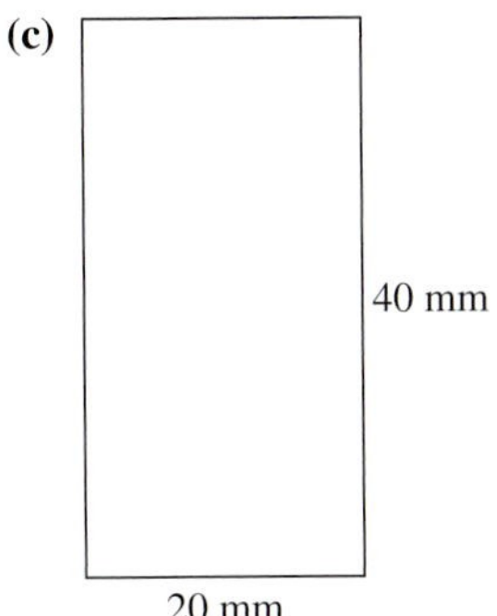

9. (a) $\frac{1}{8} = 0.125 = 12.5\%$ **(c)** $\frac{7}{18} \doteq 0.3888\ldots = 39\%$ **10. (a)** 100 **(c)** 25 **12. (a)** 25% **(c)** 50% **17.** 34.29% **20.** 42% **22. (a)** 60% **27.** \$17,380 **29. (a)** \$62.40 **32.** \$11,956.13 **34. (a)** 15 **(c)** 6 **35. (a)** 5,100,000 **39. (a)** $\frac{17}{23}$ **40. (a)** $\frac{1}{3}$ **(c)** $\frac{7}{45}$ **(e)** $\frac{41}{224}$ **41. (a)** $\frac{1}{6}$ **(c)** $\frac{27}{10}$ **(e)** $\frac{17}{48}$ **42. (a)** $r = 10$

Chapter 7 Review Exercises (page 464)

1. (a) $2 \cdot 10^2 + 7 \cdot 10^1 + 3 \cdot 10^0 + 4 \cdot 10^{-1} + 2 \cdot 10^{-2} + 5 \cdot 10^{-3}$ **(b)** $3 \cdot 10^{-4} + 5 \cdot 10^{-5} + 4 \cdot 10^{-6}$ **2. (a)** 0.056 **(b)** 0.08 **(c)** 0.1375 **3. (a)** $\frac{63}{200}$ **(b)** $\frac{603}{500}$ **(c)** $\frac{2001}{10{,}000}$ **4.** $\frac{2}{66}$, 0.33, $\frac{4}{12}$, 0.3334, $\frac{5}{13}$ **5. (a)** $\frac{3451}{333}$ **(b)** $\frac{707}{330}$ **6.** Irrational. The decimal expansion of a does not have a repeating sequence of digits, and it does not terminate. **7. (a)** $\frac{2}{9}$ **(b)** $\frac{4}{11}$ **8. (a)** 96.1885 **(b)** 20.581 **(c)** 83.898 **(d)** 7.0 **9. (a)** 34.9437 **(b)** 27.999 **(c)** 109.23237 **(d)** 6.0 **10. (a)** About 60, 60.384 **(b)** About 40, 39.813 **(c)** About 500, 620.5815 **(d)** About 3, 3.5975 **11. (a)** 2.473×10^7 **(b)** 1.247×10^{-5} **12. (a)** 8.52×10^9 **(b)** 8.81×10^8 **13.** Suppose $3 - \sqrt{2} = r$, where r is rational. Then $3 - r = \sqrt{2}$. But this implies that $\sqrt{2}$ is rational, since the rationals are closed under subtraction. This is a contradiction, since $\sqrt{2}$ is irrational. Therefore, by contradiction, $3 - \sqrt{2}$ is irrational. **14.** $(3 - \sqrt{2}) + \sqrt{2} = 3$ **15.** The decimal expansion of an irrational number has no repeating sequence of digits and is nonterminating. **16. (a)** 928.125 ft^2 **(b)** 8.4375 qts **17.** Answers will vary. For example, 4123/9999 is such a fraction. **18. (a)** $5/18 = 0.2\overline{7}$; the period starts in the second decimal place. $41/333 = 0.\overline{123}$; the period starts right after the decimal point. $11/36 = 0.30\overline{5}$; the period starts in the third decimal place. $7/45 = 0.1\overline{5}$; the period starts in the second decimal place. $13/80 = 0.1625$; this decimal is terminating. **(b)** Consider the prime-factor representation of the denominator in each of the above. If the highest power of 2 and/or 5 appearing in this prime-factor representation is r, the period begins in the $(r+1)$st decimal place. **19.** 11 to 9 **20. (a)** Yes **(b)** No **(c)** Yes **21.** \$7.88 **22.** 13 gal **23.** $y = \frac{35}{3}$ **24.** 43.2 feet **25. (a)** 62.5% **(b)** 211.5% **(c)** 1.5% **26. (a)** 0.28 **(b)** 0.0105 **(c)** $0.\overline{3}$ **27.** \$3.53 **28.** 8% **29.** 55% **30.** \$3514.98

Chapter 7 Test (page 465)

1. 5 years ago **2.** Answers will vary. A suitable choice is 125/999. **3.** 17 yrs **4. (a)** 0.48 **(b)** $0.\overline{24}$ **(c)** $0.\overline{63}$ **5.** \$1196.41 **6. (a)** $\frac{5}{11}$ **(b)** $\frac{284}{9}$ **(c)** $\frac{7}{20}$ **7.** 5 **8. (a)** 17 to 15 **(b)** 53.125% **9.** 7.30×10^{-13} **10.** 15%

Chapter 8

Problem Set 8.1 (page 479)

1. (a) Constant **(b)** Variable **4. (a)** $p + 5$ **(c)** $\frac{1}{2}(p + 2) - 2$ **5. (a)** 98 **(b)** $(x^2 + 5) \cdot 7$

8. (a) Each additional table increases the number of seats by 4. Thus, the number seated at n tables has the form $a + 4n$, for some a. Since 6 are seated at the first table (when $n = 1$), we see that $a = 2$. Thus, $2 + 4n$ people can be seated at n tables. **(b)** To seat 24 people, n must be the smallest integer for which $2 + 4n$ is at least 24. This occurs at $n = 6$, and leaves two empty places. **9. (a)** 3 **(b)** Since each new

pattern increases the number of hexagons by 3, we seek an expression of the form $a + 3n$. Since there are 4 hexagons in the first case where $n = 1$, we see that $a = 1$. Therefore the nth hexstar has $1 + 3n$ hexagons. **10. (a)** Add 3 to both sides to get $(5x - 3) + 3 = 17 + 3$, or $5x = 20$. Divide both sides by 5 to get $5x/5 = 20/5$, or $x = 4$. **12. (a)** The entries in the second row are $x + 8$ and 9, so $(x + 8) + 9 = 23$, or, equivalently, $x + 17 = 23$. Subtracting 17 from both sides gives $x = 6$. The entries in the second row are then 14 and 9, respectively.
13. Answers may vary, but suppose that x denotes the value in the lower small circle. Then the entries in the other small circles are $17 - x$ and $26 - x$, giving the equation $(17 - x) + (26 - x) = 11$. This simplifies to $43 - 2x = 11$, or $2x = 43 - 11 = 32$. Therefore $x = 16$, and the entries in the other two circles are $17 - 16 = 1$ and $26 - 16 = 10$. Alternatively, one can work clockwise to see that the upper left small circle is $17 - x$ and therefore the remaining small-circle value is $11 - (17 - x) = x - 6$. Then $(x - 6) + x = 26$, or $2x - 6 = 26$. As before, $x = 16$. **14.** In clockwise order from the bottom, the entries are 12, 1, 25, 6, and 11. **16. (a)** Add $3x$ to both sides of the first equation to get $y = 3x + 2$. Subtract 4 from both sides of the second equation to get $y = 6x - 4$.
17. (a) Using distance = rate × time ($d = rt$), we find that the average speed is distance divided by time. Therefore, her average speed was 15 mi ÷ 3 hr = 5 mph. **(b)** Since $t = \frac{d}{r}$, it took 15 mi ÷ 15 mph = 1 hour to get home.
19. Answers can vary. A useful start is to show that the two expressions do not have the same value when, say, $a = 4$ and $b = 3$. For these values of the variables, $(4 + 3)^2 = 7^2 = 49$, but $4^2 + 3^2 = 16 + 9 = 25$. It can then be shown that $(4 + 3)^2 = (4 + 3)(4 + 3) = 4 \cdot 4 + 4 \cdot 3 + 3 \cdot 4 + 3 \cdot 3 = 4 \cdot 4 + 2 \cdot (4 \cdot 3) + 3 \cdot 4$. Eventually, it can be shown that $(a + b)^2 = (a + b)(a + b) = (a + b)a + (a + b)b = a^2 + ba + ab + b^2 = a^2 + 2ab + b^2$, showing the student how the general formula $(a + b)^2 = a^2 + 2ab + b^2$ is obtained. **22.** Let t and u respectively denote the tens and units digits of n. Then $n = 10t + u$. The sum of the digits is $t + u$ and the product of the digits is tu, so we get the condition $10t + u = t + u + tu$. If t and u are subtracted from each side of the equation, we get $9t = tu$. Since t is not 0, the equation can be divided by t to get $9 = u$. We conclude that *every* two-digit number ending in 9 has the property. That is, the solution set is {19, 29, 39, 49, 59, 69, 79, 89, 99}. **24. (a)** There are n vertical toothpicks in each of the $n + 1$ vertical columns, so there are $n(n + 1)$ vertical toothpicks altogether. The same reasoning shows that there are also $n(n + 1)$ horizontal toothpicks. The entire pattern therefore requires $2n(n + 1)$ toothpicks to construct. **(b)** Imagine that each square is constructed individually, so that the n^2 squares are made with $4n^2$ toothpicks. Each of the inner sides of the pattern is now covered with 2 toothpicks, but just 1 toothpick is along the outer sides of the pattern. Since there are $4n$ outer sides, $4n^2 + 4n$ toothpicks cover each side of the pattern with 2 toothpicks. **26.** We assume the commuter travels the same distance d in both directions. The time to work, using the $d = rt$ formula of Table 8.1, is therefore given by the expression d/u. Similarly, the time to return home is given by d/v. The round-trip distance $2d$ is traveled in the total time of $d/u + d/v$, so the average speed of the round-trip is given by $\dfrac{2d}{\frac{d}{u} + \frac{d}{v}}$. These steps can simplify this expression:

$$\frac{2d}{\frac{d}{u} + \frac{d}{v}} = \frac{2d}{\frac{dv + du}{uv}} = \frac{2duv}{dv + du} = \frac{2uvd}{(v + u)d} = \frac{2uv}{v + u}.$$

It wasn't necessary to know the value of the variable d to solve the problem. **28.** Let s be Dana's average isscore on the fourth exam, so that the overall $\dfrac{78 + 86 + 94 + s}{4}$. For a score of at least 90, the inequality must be $\dfrac{78 + 86 + 94 + s}{4} \geq 90$, or $258 + s \geq 360$. That is, $s \geq 360 - 258 = 102$. Unless Dana can get 2 extra credit points, an A grade is out of reach. **30.** Let L and W denote the length and width, respectively, of the rectangle, and S the length of the sides of the square. Since the rectangle is 3 times as long as it is wide, $L = 3W$. Therefore, the perimeter of the rectangle is $2L + 2W = 6W + 2W = 8W$, and its area is $LW = 3W^2$. The perimeter of the square is $4S$, and its area is S^2. We know both the rectangle and the square have the same perimeter, so $8W = 4S$, or $2W = S$. Also, the area of the square is 4 square feet larger than the area of the rectangle, so $S^2 = 3W^2 + 4$. By substitution, $(2W)^2 = 3W^2 + 4$, or $4W^2 = 3W^2 + 4$. Therefore, $W^2 = 4$, and the positive width of the rectangle is $W = 2$. Its length is $L = 3W = 6$. The square has sides of length $S = 2W = 4$.
31. (a) The approximate formula gives that area as $\left(\frac{16}{9}\right)^2 100^2 = 31{,}604.9\ldots$. The area as calculated with the more exact value of π used by a calculator is $\pi\, 100^2 = 31415.9\ldots$. The error is about 189 square feet.
32. (a) \$148, \$219 **34.** Since there are 9 square feet in a square yard, the cost of a remnant is \$3.60/9 = \$0.40 per square foot. An L-by-W remnant will therefore cost (\$0.40)LW. The perimeter of the remnant is $2L + 2W$, so the cost to finish the edges is (\$0.12)($2L + 2W$). This gives a total cost of (\$0.40)LW + (\$0.12)($2L + 2W$). **36.** For t hours on-line, the cost will be \$10 + $\lceil t \rceil$(\$0.20). **40.** If $x = 3.\overline{14}$, then $100x = 314.\overline{14}$. Subtraction gives $99x = 314.\overline{14} - 3.\overline{14} = 314 - 3 = 311$. Therefore $x = \frac{311}{99}$.
41. $5\frac{1}{4} \div 3\frac{1}{2} = \frac{21}{4} \div \frac{7}{2} = \frac{21}{4} \cdot \frac{2}{7} = \frac{21}{7} \cdot$

$\frac{2}{4} = 3 \cdot \frac{1}{2} = \frac{3}{2}$ **42.** Since $24/36 = 2/3$, the second wheel is turning at 2/3 the rate of the first wheel. Since $36/18 = 2$, the third wheel is turning at 2 times the rate of the second wheel. Therefore, the third wheel is turning at $2 \times 2/3 = 4/3$ the rate of the first wheel. Finally, since the first wheel is turning at 300 rpm, the third wheel is turning at $4/3 \times 300 = 400$ rpm.

Problem Set 8.2 (page 489)

1. (a) Not in general, since many students take more than one class **(c)** This is a function. **3. (a)** Not a function, since element c is associated with more than one element of set B and element d is not associated with any elements of set B. **(b)** A function, with range $\{p, q, r\} \subset B$
4. (a) not a function

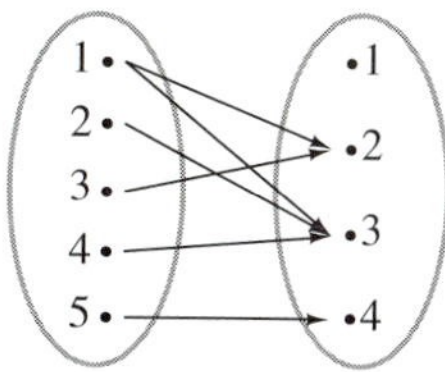

5. (a) Not the graph of a function: for example, there are three y-values associated with the value $x = 3$ **(c)** Not the graph of a function, since the two y-values 1 and 3 correspond to $x = 2$ **8. (a)** $g(0) = 5, g(1) = 4, g(2) = 5, g(3) = 8, g(4) = 13$ **(b)** $\{4, 5, 8, 13\}$ **10. (a)** 12:45 **(b)** 1:00 **11. (a)** G3
12. (a)

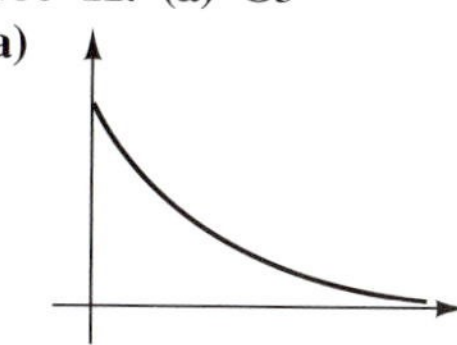

13. (a) $A = 2x^2$ **(b)** $P = 6x$ **(c)** $D = \sqrt{5}x$ **14. (a)** Add 5: $y = x + 5$ **16. (a)** 6, 8, 10, 12 **17. (a)** $F = 10M$
20. (a) $S_5 = 1^2 + 2^2 + 3^2 + 4^2 + 5^2 = 55 = \frac{10 \cdot 11 \cdot 12}{24}$

$S_6 = 1^2 + 2^2 + 3^2 + 4^2 + 5^2 + 6^2 = 91 = \frac{12 \cdot 13 \cdot 14}{24}$

(b) $S_n = \frac{2n(2n + 1)(2n + 2)}{24}$

21. (a)

Size of the Array	Number of Squares of Size 1 × 1	2 × 2	3 × 3	4 × 4	5 × 5	Total Number of Squares
1 × 1	1					1
2 × 2	4	1				5
3 × 3	9	4	1			14
4 × 4	16	9	4	1		30
5 × 5	25	16	9	4	1	55

24. $m = 2, b = 10$ **25.** $m = \frac{9}{5}, b = 32$ **27.** \$13,640
28. $d = \frac{760t}{60 \cdot 60} \doteq \frac{t}{5}$ **29. (a)** \$7.15 **30. (a)** 60¢
34. (a) The calculator shows 5, which corresponds to Friday. **(b)** Sunday **(c)** Tuesday
35. (a)

X	1	2	3	4	5	6	7	8	9	10	11	12
Y_1	1	3	4	6	8	9	11	12	14	16	17	19
Y_2	2	5	7	10	13	15	18	20	23	26	28	31

(b) $Y_2(X) = Y_1(X) + X$ **(c)** The ranges are disjoint, so $R_1 \cap R_2 = \varnothing$. The combined ranges include all of the natural numbers, so $R_1 \cup R_2 = N$. **37.** Let x and $3x$ denote the width and length of the pen, so the perimeter is $2x + 6x$. Since the perimeter is to be 80, this gives the equation $8x = 80$, or $x = 10$. Therefore, the pen is a rectangle 10 feet wide and 30 feet long. **38.** The terms increase by 6, so we look for an expression of the form $6n + a$. Choosing $n = 1$, the first term has the value of 5 if a is chosen so that $6 + a = 5$. That is, $a = -1$, and the expression is $6n-1$. This can be checked to give the subsequent terms in the sequence. **39.** Choosing $n = 0$ gives the equation $c = 1$, so the expression has the form $an^2 + bn + 1$. Next, choosing $n = 1$ we want $a + b + 1 = 0$, and choosing $n = 2$ we also want $4a + 2b + 1 = 1$. That is, subtracting 1 from both sides and dividing by 2, we obtain $2a + b = 0$. Therefore, $b = -2a$. By substitution, $0 = a + b + 1 = a - 2a + 1 = -a + 1$. That is, $a = 1$, and so $b = -2$. The desired expression is then $n^2 - 2n + 1$, which can also be written $(n - 1)^2$. When $n = 4$ and 5, the respective expression values are 9 and 16.

Problem Set 8.3 (page 512)

1. (a), (c), (e), (g), (i)

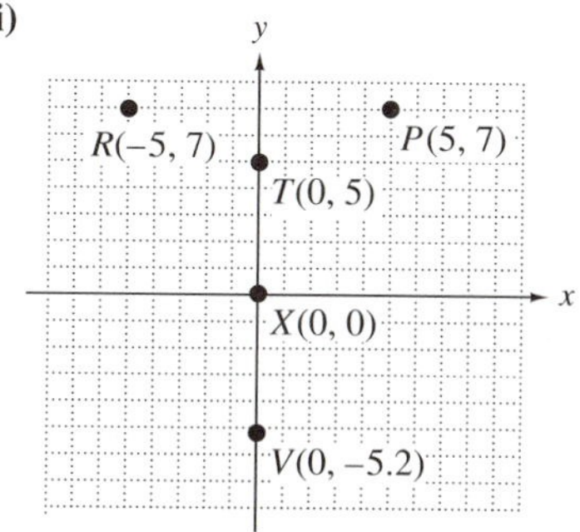

4. (a) $\sqrt{(4 - (-2))^2 + (13 - 5)^2} = \sqrt{36 + 64} = \sqrt{100} = 10$
(c) $\sqrt{(8 - 0)^2 + (-8 - 7)^2} = \sqrt{64 + 225} = \sqrt{289} = 17$
5. (a) $(RS)^2 = (\sqrt{(7 - 1)^2 + (10 - 2)^2})^2 = (\sqrt{36 + 64})^2 = (\sqrt{100})^2 = 100$

$(RT)^2 = (\sqrt{(5-1)^2 + (-1-2)^2})^2 = (\sqrt{16+9})^2 = (\sqrt{25})^2 = 25$
$(ST)^2 = (\sqrt{(5-7)^2 + (-1-10)^2})^2 = (\sqrt{4+121})^2 = (\sqrt{125})^2 = 125$
Since $(RS)^2 + (RT)^2 = 100 + 25 = 125 = (ST)^2$, by the Pythagorean theorem, $\triangle RST$ is a right triangle.

6. (a) 2; upward **(c)** 2; upward **(e)** 1; upward

7. 2 **9.** $a = \frac{9}{2}$

11. (a), (c)

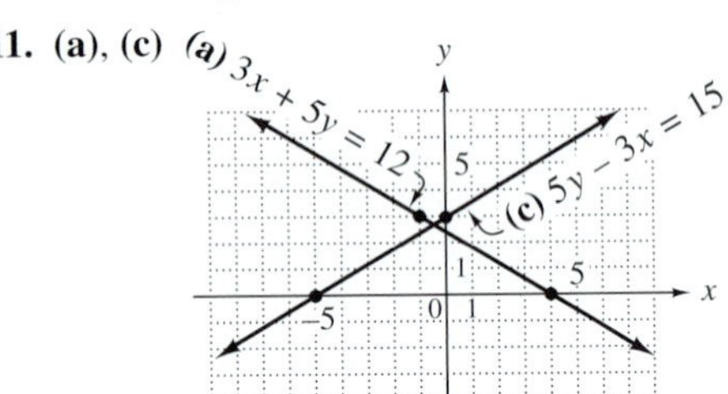

13. (a) $-\frac{35}{3}$ **(c)** $\frac{3}{5}$

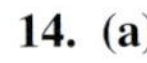

14. (a)

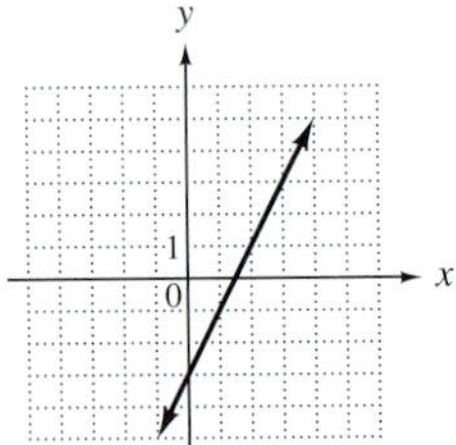

16. (a)

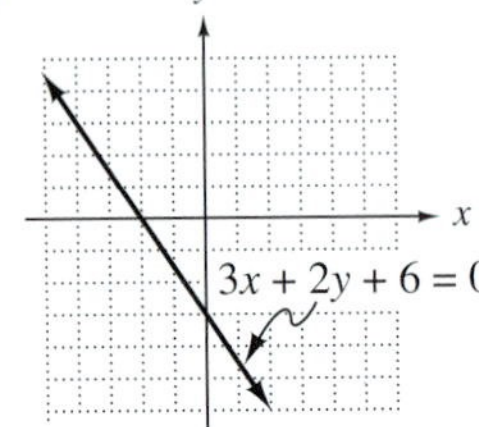

(b) Yes. The function is $y = -\frac{3}{2}x - 3$.

(d) Yes. The function is $y = \frac{5}{3}x - 5$.

18. (a)

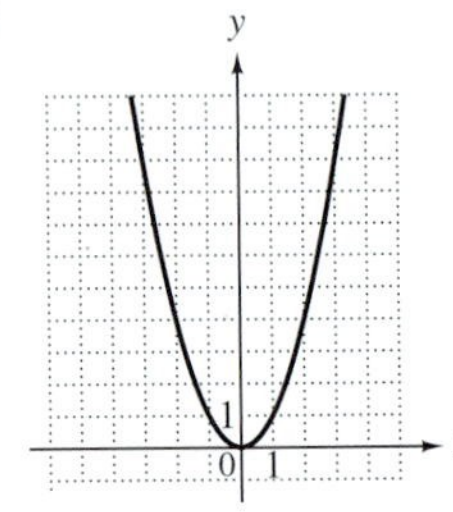

19. (a)

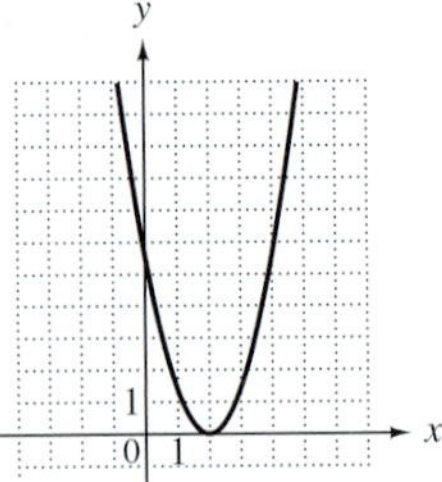

21. (a) The minimum value of $x^2 + 10x$ is -25. It occurs when $x = -5$.

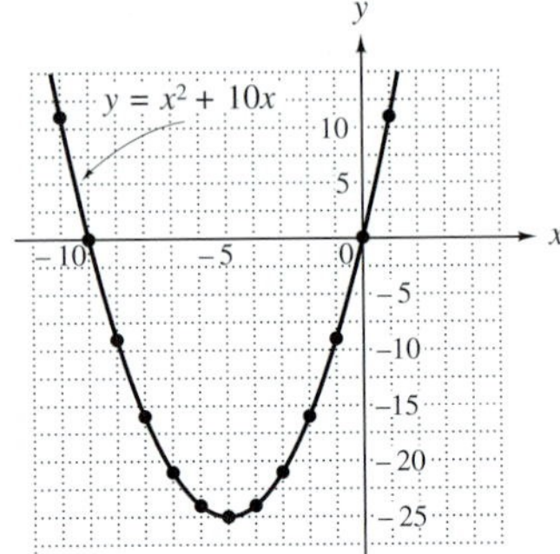

24. (a) $x = 5$ **(c)** $y = \frac{1}{2}x + 1$

25. $r = 10, s = 5$ **27. (a)** $y = \frac{-1}{2}x$

30. (a) $PB^2 = (x-1)^2 + (y-1)^2 = x^2 - 2x + 1 + y^2 - 2y + 1 = x^2 + y^2 - 2x - 2y + 2$
$PC^2 = (x-0)^2 + (y-1)^2 = x^2 + y^2 - 2y + 1 = x^2 + y^2 - 2y + 1$
$PD^2 = (x-0)^2 + (y-0)^2 = x^2 + y^2$
(b) $PA^2 + PC^2 = (x^2 + y^2 - 2x + 1) + (x^2 + y^2 - 2y + 1) = 2x^2 + 2y^2 - 2x - 2y + 2$
$PB^2 + PD^2 = (x^2 + y^2 - 2x - 2y + 2) + (x^2 + y^2) = 2x^2 + 2y^2 - 2x - 2y + 2$
These equations show that $PA^2 + PC^2 = PB^2 + PD^2$ for all x and y. That is, *all* points $P(x, y)$ in the plane satisfy the condition. **34. (a)** The run must be at least 12 times the 30-inch = 2.5-foot rise, so the minimum allowable run is $12 \times 2.5 = 30$ feet. **(b)** The rise is $\frac{1800}{5280}$ miles in a run of 7 miles, so the slope is $\frac{1800}{5280} \div 7 = 0.05$, or about 5%.

36. (a), (b), (c) Answers will vary, but the plot and slope should be reasonable estimates of the regression line and slope given by a graphing calculator (see the solution to part (d)). **(d)** Jennifer is getting about 21.5 miles per gallon.

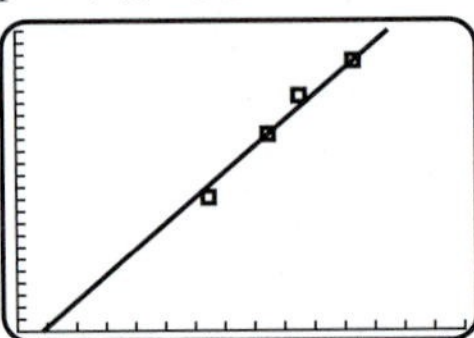

39. **(a)** $f(0.2) = 5$, $f(1) = 1$, $f(5/3) = 3/5 = 0.6$, $f(4) = 1/4 = 0.25$, and $f(5) = 1/5 = 0.2$ **(b)** $r = 0.25$, $s = 2$, $t = 0.5$, $u = 7/4 = 1.75$, and $v = \pi$ **(c)** The range is the interval $0.2 \le y \le 5$. **40.** **(a)** If x denotes the input, the output seems to be given by $-(2x + 1)$, or, equivalently, $-2x - 1$. **(b)** Multiply the input by the value 3 larger than the input; that is, $x(x + 3)$.

Chapter 8 Review Exercises (page 578)

1. **(a)** $a + 5$ **(b)** $b < c$ **(c)** $c - b$ **(d)** $(a + b + c)/3$
2. **(a)** $a + 2 = 11$ **(b)** $b - 3 = \frac{1}{2}(c - 1)$ **(c)** $a = (b + c)/2$ **(d)** $(a + b + c)/3 = 10$
3. **(a)** A one-car train uses 6 toothpicks to form the hexagon. Adding a square + hexagon combination requires an additional 8 toothpicks, so the trains with 1, 3, 5, 7, . . . cars use 6, 6 + 8, 6 + 8 + 8, 6 + 8 + 8 + 8, . . . toothpicks. In general, a train with $2m + 1$ cars will require $6 + 8m$ toothpicks, where $m = 0, 1, 2, 3, \ldots$. A two-car train uses 9 toothpicks, so trains with 2, 4, 6, 8, . . . cars use 9, 9 + 8, 9 + 8 + 8, 9 + 8 + 8 + 8, . . . toothpicks. In general, a train with $2m + 2$ cars uses $9 + 8m$ toothpicks for $m = 0, 1, 2, 3, \ldots$. **(b)** Since $9 + 8m$ is always an odd number, a train with 102 toothpicks has an odd number of cars, say, $2m + 1$. Then $6 + 8m = 102$, or $8m = 96$. Therefore, $m = 12$, and there are $2(12) + 1 = 25$ cars in the train. **4.** **(a)** Not a function, since two y-values are assigned to $x = 3$ **(b)** Function, with domain $\{3, 4, 6, 7, 8\}$ and range $\{1, 2, 3, 4, 5\}$ **(c)** Function, with domain $\{3, 4, 6, 7, 8\}$ and range $\{3, 4\}$
5. **(a)** $y = 3x + 2$ **(b)** $y = x(x + 1)$, or $y = x^2 + x$
6. **(a)** $f(3) = 0$, $f(0.5) = -2.5$, $f(-2) = 20$ **(b)** $x = 0$ or $x = 3$
7. **(a)**

(b) Number of paths with n reflections = number of paths with $n - 1$ reflections + number of paths with $n - 2$ reflections. That is, the number of paths with n reflections is the Fibonacci number F_{n+2}.
8. **(a)**

(b) B is in the first quadrant, C is on the positive y-axis, D is in the second quadrant, E is in the third quadrant, and F is on the negative y-axis.
(c) slope $\overline{BC} = \frac{0}{3} = 0$, slope $\overline{CD} = \frac{1}{2}$, slope $\overline{DE} = \frac{4}{1} = 4$, slope $\overline{EG} = \frac{0}{3} = 0$, slope $\overline{FG} = \frac{2}{4} = \frac{1}{2}$, slope $\overline{GA}$ is undefined, $\overline{BC}$ is parallel to $\overline{EF}$, $\overline{CD}$ is parallel to $\overline{FG}$.

9.

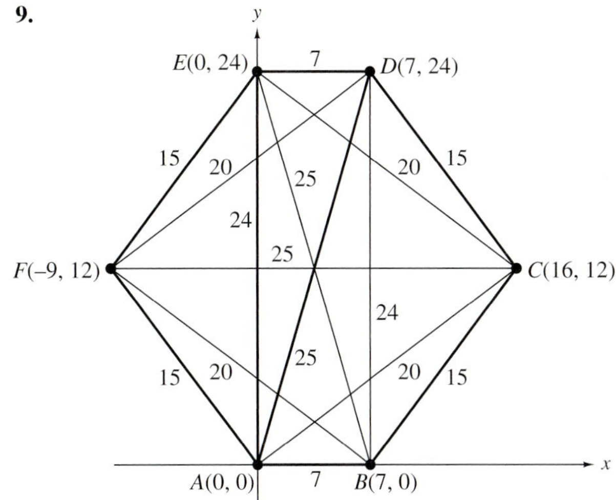

(a) $AB = 7$, $AC = \sqrt{16^2 + 12^2} = \sqrt{400} = 20$, $AD = \sqrt{7^2 + 24^2} = \sqrt{625} = 25$, $AE = 24$, $AF = \sqrt{(-9)^2 + 12^2} = \sqrt{225} = 15$, $CF = 16 + 9 = 25$ **(b)** The lengths of all the sides and diagonals of the hexagon are positive integers. **10.** **(a)** $y = 4 + 2(x - 3)$, using the point–slope form of the equation of a line. In slope–intercept form, the equivalent equation is $y = 4 + 2(x - 3) = 2x - 2$.
(b) $y = -1 + \frac{5 - (-1)}{(-2) - 6}(x - 6) = -1 - \frac{3}{4}(x - 6)$, using the two-point form. In slope–intercept form, the equivalent equation is $y = -\frac{3}{4}x + \frac{7}{2}$. **(c)** $y = 3x - 4$, using the slope–intercept form of the equation of a line.
11. **(a)**

x	−3	−2	−1	0	1	2	3	4	5
$f(x)$	−7	0	5	8	9	8	5	0	−7

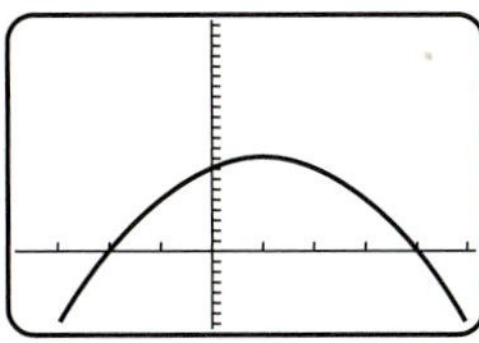

(b)

x	−4	−3	−2	−1	0	1	2	3	4
$g(x)$	1.3	0.9	0.0	−2.0	−4.0	−2.0	0.0	0.9	1.3

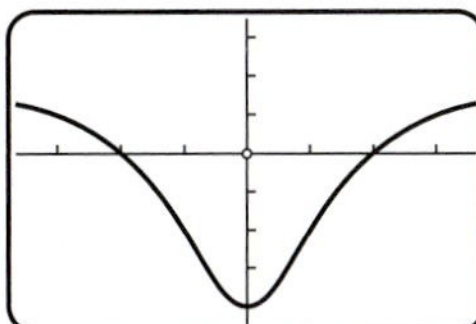

12. **(a)** Define $F(0) = F(2) - F(1) = 1 - 2 = 0$. That is, a Fibonacci number is the *difference* of the two numbers to the right. This rule gives the doubly infinite sequence of Fibonacci numbers $\ldots, 5, -3, 2, -1, 1, 0, 1, 1, 2, 3, 5, \ldots$. In general, the pattern shows that $F(-n) = (-1)^{n+1} F(n)$ for $n = 0, 1, 2, 3, \ldots$. **(b)** Define $L(0) = L(2) - L(1) = 3 - 1 = 2$, and in general $L(n) = L(n+2) - L(n+1)$ for $n = -1, -2, -3, \ldots$. This generates the doubly infinite sequence $\ldots, -11, 7, -4, 3, -1, 2, 1, 3, 4, 7, 11, \ldots$. The pattern shows that $L(-n) = (-1)^n L(n)$ for $n = 1, 2, 3, \ldots$.

Chapter 8 Test (page 520)

1. **(a)** $5x - 7 = x + 9$

$4x - 7 = 9$	[subtract x from both sides]
$4x = 9 + 7 = 16$	[add 7 to both sides]
$x = \frac{16}{4} = 4$	[divide both sides by 4]

(b) $\frac{5 + 7x}{5x - 2} = 2$

$5 + 7x = 2(5x - 2)$	[multiply by $5x - 2$]
$5 + 7x = 10x - 4$	[expand]
$5 = 3x - 4$	[subtract $7x$ from both sides]
$9 = 3x$	[add 4 to both sides]
$3 = x$	[divide both sides by 3]

2. **(a)**

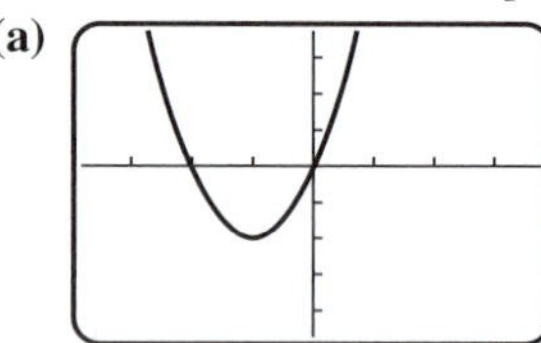

(b) The minimum function value is $y = -2$, the value of the function at $x = -1$. **3.** **(a)** Linear, equivalent to $y = 3x - 4$ **(b)** Nonlinear, since x^2 appears in the equation **(c)** Linear, equivalent to $y = x + 5$ **4.** **(a)** Let t denote the charged time, in total minutes for the month. Then $5 + 0.12t$ is the phone bill (in dollars) for that month. **(b)** Let n = *number* of checks. Then the cost of checking is $8 + 0.15n$. **(c)** Let a and c denote the number of adults and children, respectively. Then the admission fee is $4a + 2.50c$, in dollars. **5.** **(a)** $y = 5$ **(b)** $y = -x - 2$, since the slope is -1 and the line intersects the y-axis at -2 **(c)** The line has slope $\frac{1}{3}$ and passes through $P(2, 2)$, so its equation in point–slope form is $y = 2 + \frac{1}{3}(x - 2)$.

6. **(a)** G3 **(b)** G1 **(c)** G2 **7.** **(a)** 5 **(b)** The bee can enter cell $n + 2$ from either cell $n + 1$ or cell n. If there are $F(n+1)$ ways to get to cell $n + 1$ and $F(n)$ ways to get to cell n, then the number of ways to get to cell $n + 2$ is $F(n+1) + F(n)$. **(c)** 144 **8.** **(a)** $a = 0$ and $b = 5$, since each side either has a horizontal run of 3 and a rise of 4 or a horizontal run of 4 and a fall of 3. **(b)** $\sqrt{4^2 + 3^2} = \sqrt{25} = 5$ **9.** $\{1, 2, 5, 10\}$ **10.** **(a)** $d_1 = 1, d_2 = 5, d_3 = 13, d_4 = 25, d_5 = 41, d_6 = 61$
(b)

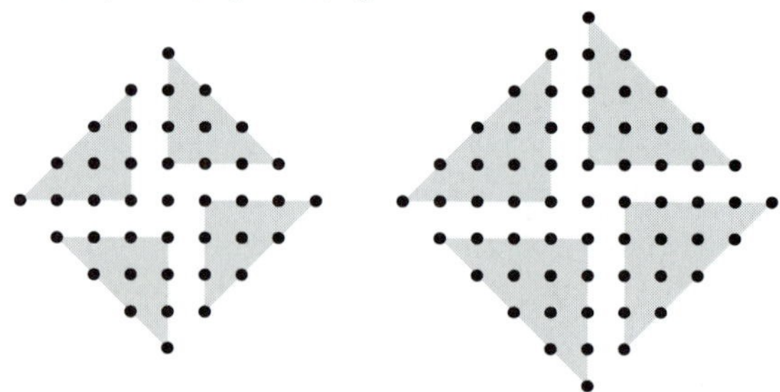

(c) $d_n = 1 + 4t_{n-1} = 1 + 4 \cdot \frac{1}{2}(n-1)n = 1 + 2(n-1)n$

(d)

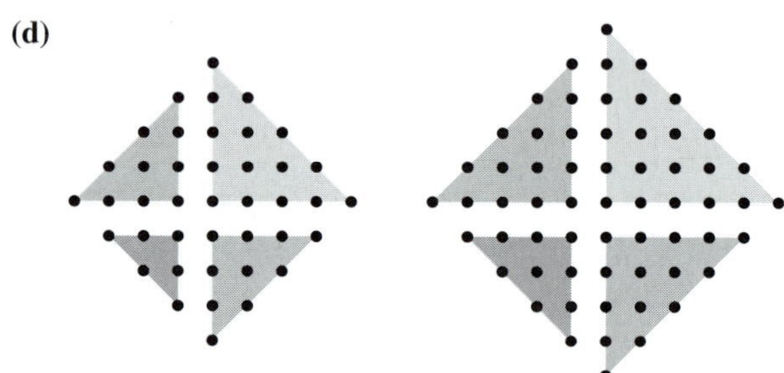

(e) Since $n^2 = t_n + t_{n-1}$ and $d_n = t_n + t_{n-1} + t_{n-1} + t_{n-2}$, $d_n = (t_n + t_{n-1}) + (t_{n-1} + t_{n-2}) = n^2 + (n-1)^2$.
(f) $n^2 + (n-1)^2 = n^2 + (n^2 - 2n + 1) = 2n^2 - 2n + 1 = 1 + 2(n-1)n$

Chapter 9

Problem Set 9.1 (page 938)

3.

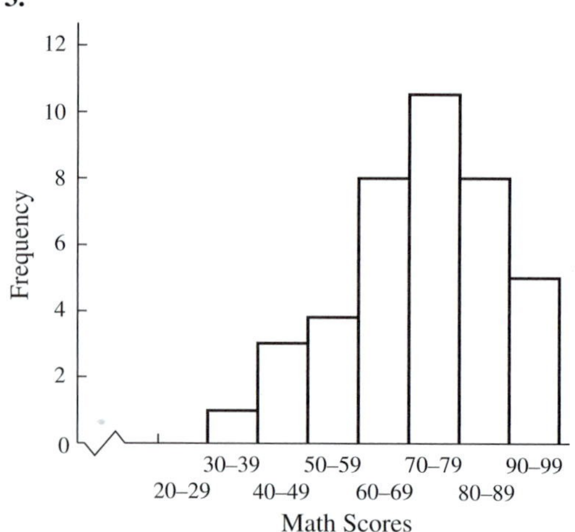

7.

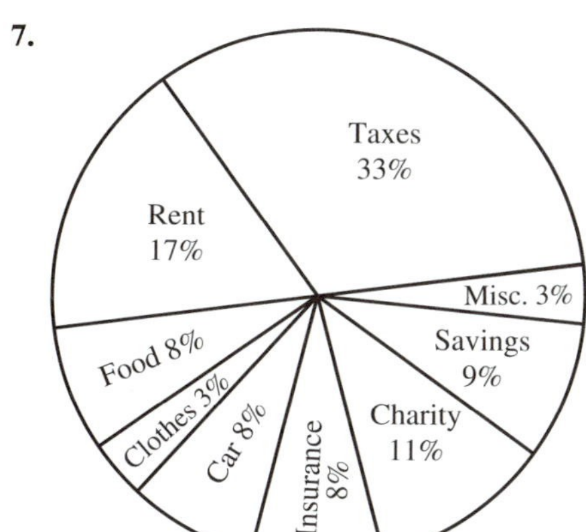

9. (a) 12 **16. (a)** Histogram (A) emphasizes the changes in the daily Dow Jones average by its choice of vertical scale. The changes appear to be large in histogram (A), thus exaggerating the report of stock activity on the evening news. Histogram (B) makes it clear that the changes are minimal. **(b)** $\left(\frac{36}{10086}\right) \cdot 100\% \doteq 0.36\%$, which an investor probably would not worry about. **18. (b)** First can: $V = \pi(4 \text{ cm})^2 (8 \text{ cm}) \doteq 402 \text{ cm}^3$; second can: $V = \pi(3 \text{ cm})^2 (14 \text{ cm}) \doteq 396 \text{ cm}^3$. **21. (a)** 49.6%

26. (a)

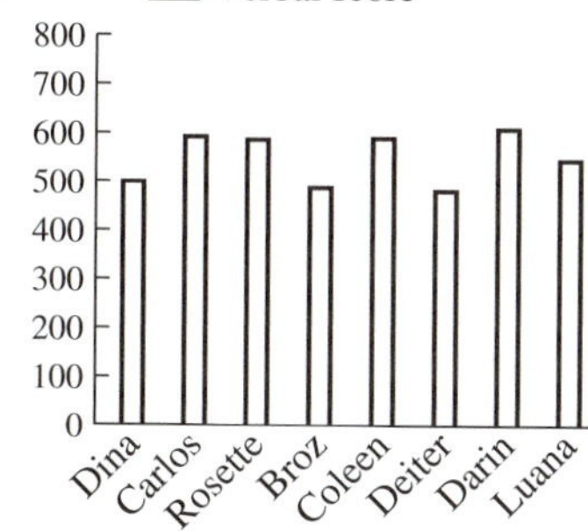

27. (a)

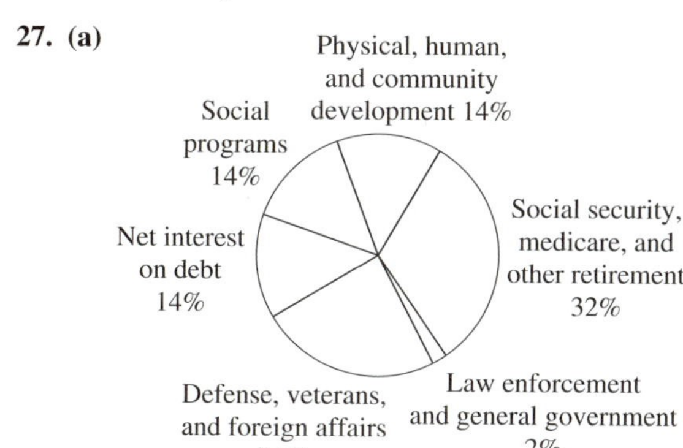

29.

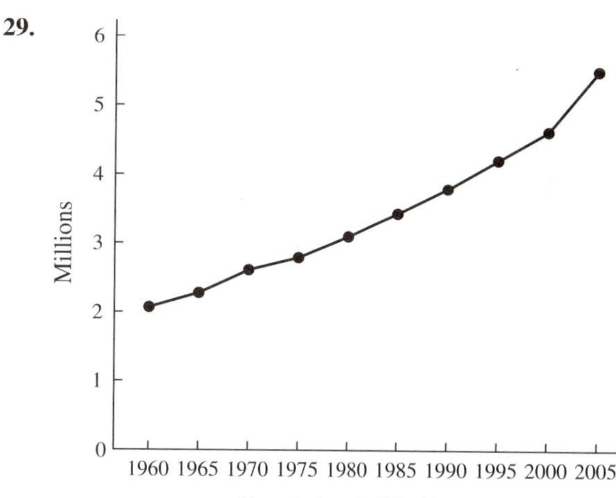

36. (a) 0.125 **(c)** 0.068 **37. (a)** $0.\overline{1}$ **(c)** $0.2\overline{142857}$ **38. (a)** $\frac{3}{8}$ **(c)** $\frac{111}{250}$ **39. (a)** $\frac{337}{90}$ **(c)** $\frac{1}{45}$ **(e)** $\frac{47{,}267}{9990}$

Problem Set 9.2 (page 561)

1. $\bar{x} \doteq 20.6$, $\hat{x} = 19$, mode $= 18$ **3. (a)** $Q_L = 70$, $Q_U = 80$ **(b)** $64 - 70 - 77 - 80 - 86$ **5. (a)** 20.56 **(b)** 4.44 **(c)** 69% **(d)** 94% **(e)** 100% **7. (a)** Except possibly for round-off error, the sum of the deviations equals zero for every data set. **(b)** Compute the sum in part (a) and point out that the negative terms just balance out the positive terms so that the sum is 0. **8. (a)** 4.9 **(b)** 0.41 **(c)** Compliment Leona on a good idea that avoids the canceling out of the effects of the various terms. **10. (a)** $\bar{x} = 27$, $s \doteq 8.4$ **(b)** $\bar{x} = 32$, $s \doteq 8.4$ **13. (a)** $\bar{x} \doteq 36{,}900$, $\hat{x} = 30{,}000$, mode $= 22{,}000$ **14.** Many examples satisfy these conditions. **(a)** 5, 7, 10, 14, 14 **15. (a)** If $s = 0$, then all the data values are equal. **17. (a)** All 3s **18. (a)** 32 **(c)** 32 **21.** The total of data values in A is $30 \cdot 45 = 1350$. The total of data values in B is $40 \cdot 65 = 2600$. For the combined data, $\bar{x} = \frac{1350 + 2600}{30 + 40} \doteq 56.4$. **30. (a)** $\bar{x} \doteq 76.17$ **(b)** $\hat{x} = 78$ **(c)** 78 and 79 **(d)** $s \doteq 11.53$ **(e)** $Q_L = 68$ **(f)** $Q_U = 86$ **(g)**

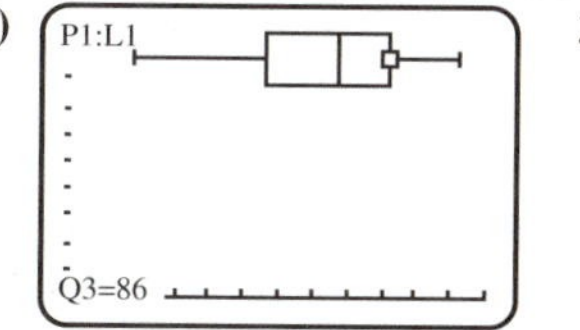

36. There are many possible answers. One pair would be $\frac{11}{16}$ and $\frac{21}{32}$. **38.** Assume that $\sqrt{3} + r = s$ where s is rational. Then $\sqrt{3} = s - r$, and $s - r$ is rational, since the rational numbers are closed under subtraction. But this says $\sqrt{3}$ is rational, which is a contradiction. Therefore, $\sqrt{3} + r$ is irrational.

Problem Set 9.3 (page 576)

1. (a) All freshmen in U.S. colleges and universities in 2004 **(c)** All people in the United States **2.** Yes. Many poorer people cannot afford telephones, many people are irritated by telephone surveys and sales pitches, and so on. These factors could certainly bias a sample. **5. (a)** This is surely a poor sampling procedure. The sample is clearly not random. The selection of the colleges or universities could easily reflect biases of the investigators. The choices of the faculty to be included in the study almost surely also reflect the bias of the administrators of the chosen schools.
(b) Presumably, the population is all college and university faculty. But the opinions of faculty at large research universities are surely vastly different from those of their colleagues at small liberal arts colleges. Indeed, there are almost surely four distinct populations here.
11. (a) Between 21.8 and 27.2 **12. (a)** $\bar{x} \doteq 20.2$, $s \doteq 2$. The z scores corresponding to the data in the order listed are -1.6, 0.9, 0.4, -0.6, 1.4, and -0.6. **13. (a)** -0.1
14. (a) $83\frac{1}{3}$ percentile **15. (a)** $0.04 = 4\%$ to the nearest hundredth **16. (a)** $0.96 - 0.04 = 0.92 = 92\%$ to the nearest hundredth **21.** Yes, since all sides of the die are equally likely to come up, all sequences of 0s and 1s are equally likely to appear. **24. (b)** 4 **25. (a)** Katja's time going upstream is $4 \div 2 = 2$ hours, and her time going downstream is $4 \div 4 = 1$ hour. Thus, the average speed is $8 \div (2 + 1) = 2.\overline{6}$ miles per hour. **(b)** $\dfrac{2}{\frac{1}{2} + \frac{1}{4}} = \dfrac{8}{3} = 2.\overline{6}$
27. (a) Since the z scores indicate the location of data points in a data set, in this case they are likely to be the same as in problem 12(a). **(b)** -1.6, 0.9, 0.4, -0.6, 1.4, -0.6
28. Consider the data set $\{1, 2, 3\}$. Then $\bar{x} = \dfrac{1 + 2 + 3}{3}$, and the z scores are $\dfrac{1 - \bar{x}}{s}$, $\dfrac{2 - \bar{x}}{s}$, and $\dfrac{3 - \bar{x}}{s}$. Therefore, their sum is

$$\begin{aligned} \frac{1 - \bar{x}}{s} + \frac{2 - \bar{x}}{s} + \frac{3 - \bar{x}}{s} &= \frac{1 + 2 + 3 - 3\left(\frac{1 + 2 + 3}{3}\right)}{s} \\ &= \frac{\frac{3(1 + 2 + 3)}{3} - 3\left(\frac{1 + 2 + 3}{3}\right)}{s} \\ &= 0, \end{aligned}$$

and the same computation would be true for $x_1, x_2, \ldots, x_n$.
29. (b) -1.6, 0.9, 0.4, -0.6, 1.4, -0.6 **34. (a)** $\bar{x} = 4.43$ **(b)** $s \doteq 0.68$ **38. (a)** 23 **(b)** 23

Chapter 9 Review Exercises (page 583)

1. (a)

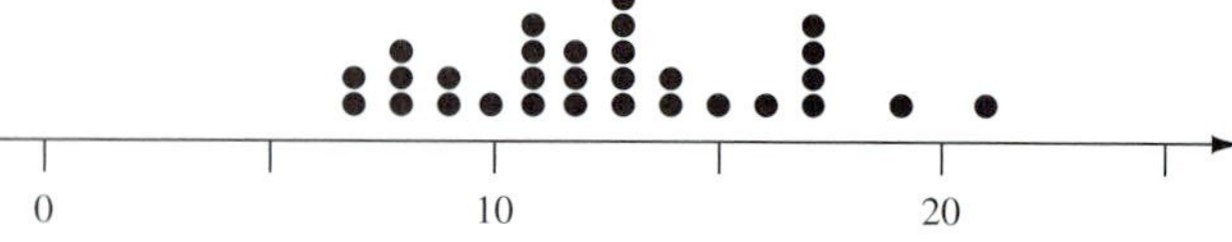

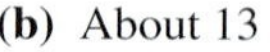
(b) About 13

2.

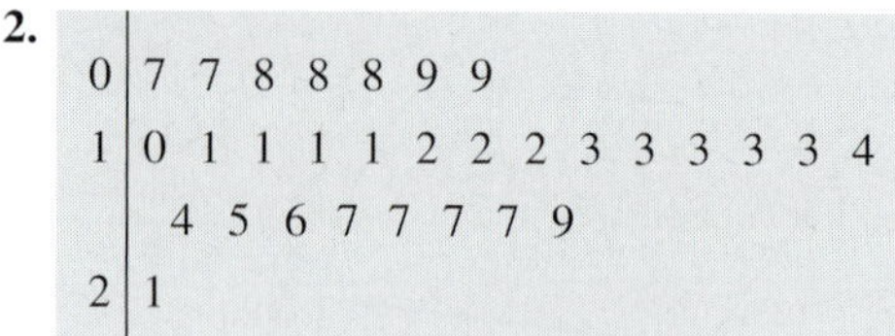

0	7 7 8 8 8 9 9
1	0 1 1 1 1 2 2 2 3 3 3 3 3 4 4 5 6 7 7 7 7 9
2	1

3.

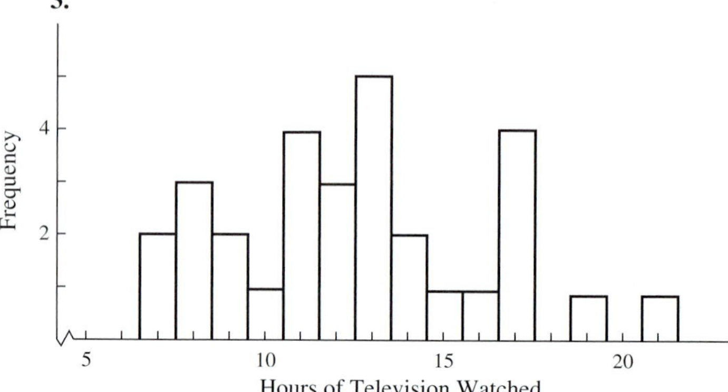

4.

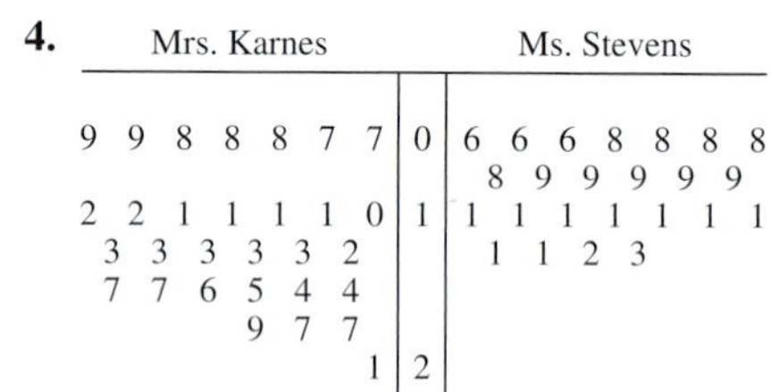

Mrs. Karnes		Ms. Stevens
9 9 8 8 8 7 7	0	6 6 6 8 8 8 8 8 9 9 9 9 9
2 2 1 1 1 1 0 3 3 3 3 3 2 7 7 6 5 4 4 9 7 7	1	1 1 1 1 1 1 1 1 1 2 3
1	2	

5. (a)

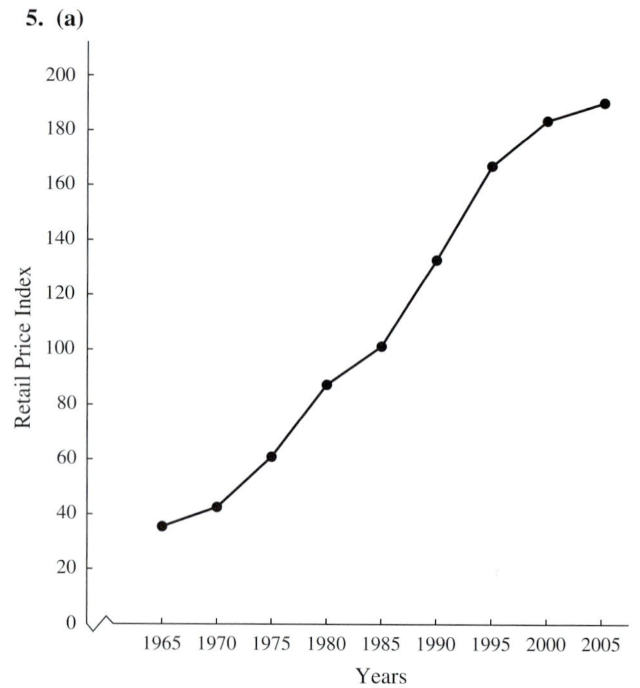

(b) About 48 **(c)** About 200 **6.** Many examples exist. One example is 1, 2, 3, 4, 90, with a mean of 20.

7.

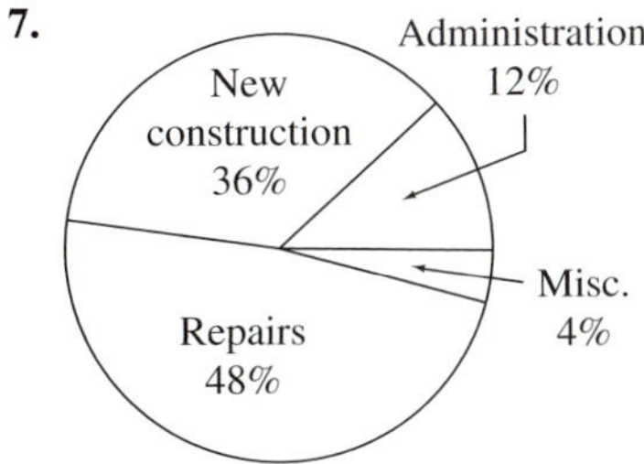

8. (a) The volume of the larger box is about double that of the smaller box. But the length of a side of the larger box is less than double the length of a side of the smaller box, suggesting that the change was less than doubling. **(b)** The volume of the larger is about twice that of the smaller, which defends the pictograph. **9.** How is "medical doctor" defined? Does this include all specialists? osteopaths? naturopaths? chiropractors? acupuncturists? How was the sampling done to determine the stated average?
10. $\bar{x} \doteq 12.6, \hat{x} = 12.5$, mode $= 13, s \doteq 3.6$
11. (a) $Q_L = 10, Q_U = 15$ **(b)** $7 - 10 - 12.5 - 15 - 21$ **(c)** There are no outliers. **(d)** $\hat{x} = 9, Q_L = 8, Q_U = 11$ **(e)** $6 - 8 - 9 - 11 - 13$ **(f)** There are no outliers.
(g)

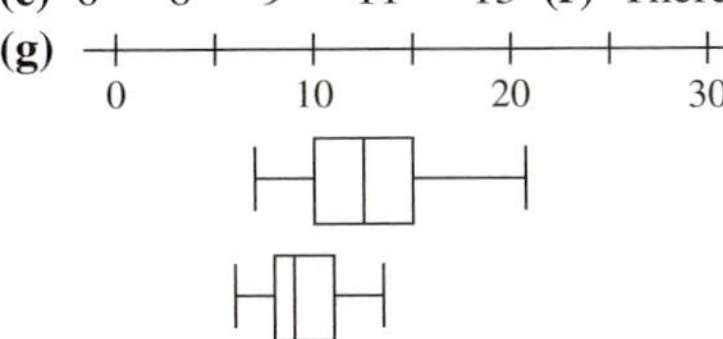

12. (a) $\bar{x} \doteq 27.8, s \doteq 2.0$ **(b)** $\bar{x} = 27.3, s \doteq 1.6$ **(c)** The means are about the same for the two sets of data, but the standard deviation is smaller for the second histogram since the data are less spread out from the mean. **13.** There are $21 \cdot 77 = 1617$ points for the 21 students. So there are 1839 points for all 24 students. Thus, the average is $1839 \div 24 \doteq 76.6$. **14.** There are $27 \cdot 75 = 2025$ points for the second-period students and $30 \cdot 78 = 2340$ points for the fourth-period students. Thus, the average for all students is $(2340 + 2025)/57 \doteq 76.6$. **15.** No. The sample represents only the population of students at State University, not university students nationwide. **16.** You might want to limit your sample to persons 20 years old and older who want to work. Alternatively, you might want to define several populations and determine figures for each: persons 20 years old and older who want to work, teenagers who want full-time employment, teenagers who want part-time employment, adults 20 years old and older who want part-time employment, and so on. **17.** Telephone polls sample only persons sufficiently affluent to own a telephone. They are also biased by the fact that many people do not like telephone polls or telephone commercial solicitations and so refuse to respond or respond inaccurately because of anger. **18.** Voluntary response to mailed questionnaires tends to come primarily from those who feel strongly (either positively or negatively) about an issue or who represent narrow special-interest groups. They are rarely representative of the population as a whole. **19.** One way is to number the students in alphabetical order and select the sample by using a spinner or a random-number generator on a computer. Alternatively, one might print the names of all students on slips of paper, place them into a container, mix them well, and have someone close his or her eyes and select from the container the names of those to be in the sample. **20.** Yes. Just continue taking samples until one finally shows up with 8 out of the 10 in the sample preferring Whito Toothpaste.
21. $\bar{x} \doteq 9.3$ and $s \doteq 3$. The z scores in order are $-0.8, -0.1, -1.1, 0.9, 1.9, -0.8, -0.1$. **22.** Since 12 is the sixth largest number in the data set, and $6/7 = 0.8571$ to the nearest ten thousandth, 12 is the 85.71 percentile to the nearest hundredth. **23.** From Table 9.4, we see that 0.9505 corresponds to 1.65. Therefore, 1.65 is essentially at the 95th percentile. **24.** From Table 9.4, the entry for -0.9 is 0.1841, and the entry for 0.9 is 0.8159. Therefore, the desired percentage is $81.59\% - 18.41\% = 63.18\%$.

Chapter 9 Test (page 587)

1. $(0.40) \cdot (360°) = 144°$ **2.** To the nearest tenth, the z scores for 42, 86, and 80 are -2.8, 0.6, and 0.1 respectively. With a normal distribution, 68% of the data will be within one standard deviation of the mean, 95% of the data will be within two standard deviations of the mean, and 99.7% of the data will be within three standard deviations of the mean.
3.

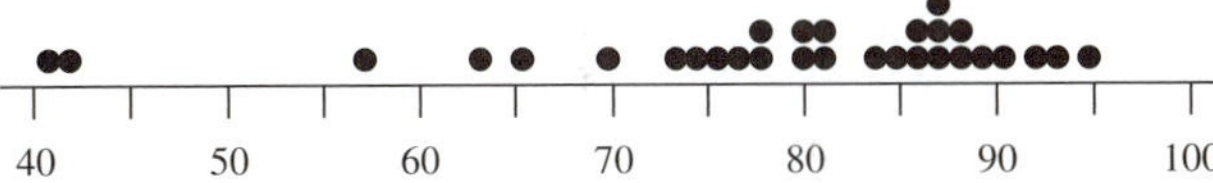

4. (a) $\bar{x} \doteq 78.5$ **(b)** $\hat{x} = 81$ **(c)** mode $= 87$ **(d)** $s \doteq 13.2$
5.

4	1 2
5	7
6	3 6
7	0 4 5 6 7 8 8
8	0 0 1 1 4 5 6 6 7 7 7 8 8 9
9	0 2 3 5

6. (a) $41 - 75 - 81 - 87 - 95$ **(b)** 41 and 42 are outliers, since they are less than $Q_L - 1.5 \cdot \text{IQR} = 57$.

40 50 60 70 80 90 100

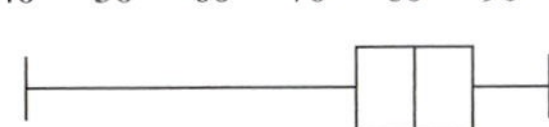

7. If the sample is not chosen at random, it is quite likely to reflect bias—bias of the sampler, bias reflecting the group from which the sample was actually chosen (e.g., views of

teamsters or AARP members), and so on. **8.** A random sample is a sample chosen in such a way that every subset of the population has an equal chance of being included.
9. $\frac{17}{30} = 0.5\overline{6}$ so, to the nearest hundredth, 17 is the 57th percentile. One way would be to number the students and select 200, using random numbers from a random-number generator to determine which students are included in the sample. **10.** To average 80% on all five tests, she must score at least 400. Thus, the total score for her last two tests should be $400 - (77 + 79 + 72) = 172$.

Chapter 10

Problem Set 10.1 (page 601)

1. $P_e = \frac{4}{35} \doteq 0.11$ **4.** **(b)** 1 **7.** **(a)** iv **(c)** ii **(e)** ii
9. **(b)** $2 + 6, 3 + 5, 4 + 4, 5 + 3, 6 + 2$
11. **(a)** Answers will vary. One solution is $P_e(A) = \frac{12}{60} = 0.6$, $P_e(B) = \frac{6}{20} = 0.3$, and $P_e(C) = \frac{2}{20} = 0.1$.
(b) About right, since there are 360° in a complete revolution and $\frac{240}{360} = 0.67$, $\frac{90}{360} = 0.25$, and $\frac{30}{360} = 0.08\overline{3}$
14. Answers will vary. Using the answer to 13(a), we obtain $P_e = (\text{first 6 on fourth roll}) = 0.2$. **20.** Answers will vary. When we did the experiment, we obtained four hearts including one face card, seven diamonds including three face cards, five spades including two face cards, and four clubs with no face cards. This yielded the following results: **(a)** $P_e(R) = 11/20 = 0.55$ **(b)** $P_e(F) = 6/20 = 0.3$ **(c)** $P_e(R \text{ or } F) = 13/20 = 0.65$ **(d)** $P_e(R \text{ and } F) = 4/20 = 0.2$ **(e)** $P_e(R) + P_e(F) - P_e(R \text{ and } F) = 0.55 + 0.3 - 0.2 = 0.65$ **(f)** This suggests that $P_e(R \text{ or } F) = P_e(R) + P_e(F) - P_e(R \text{ and } F)$, unlike the result suggested by problem 19. **21.** Answers will vary. When we did the experiment, H occurred all told 12 times, 5 occurred all told three times, and 5 and H occurred together two times. This yielded the following results: **(a)** $P_e(\text{H}) = 12/20 = 0.6$ **(b)** $P_e(5) = 3/20 = 0.15$ **(c)** $P_e(\text{H and } 5) = 2/20 = 0.1$ **(d)** $P_e(\text{H}) \cdot P_e(5) = (0.6)(0.15) = 0.09$ **(e)** Yes. Since the events H and 5 are independent, the number of simultaneous occurrences of H and 5 should be about $P_e(\text{H}) \cdot (\text{the number of occurrences of } 5)$. But then

$$P_e(\text{H and } 5) \doteq \frac{P_e(\text{H}) \cdot (\text{the number of occurrences of } 5)}{20} = P_e(\text{H}) \cdot P_e(5).$$

28. **(a)** $\frac{73}{162} \doteq 0.45$ **(b)** $\frac{73}{162} \cdot 9 \doteq 4$
29. **(a)** $(0.45) \times (0.45) = 0.2025$, or about 20% of all couples **30.** **(a)** $\frac{9806}{10,000} = 0.9806$
32. **(a)** $n = 10$

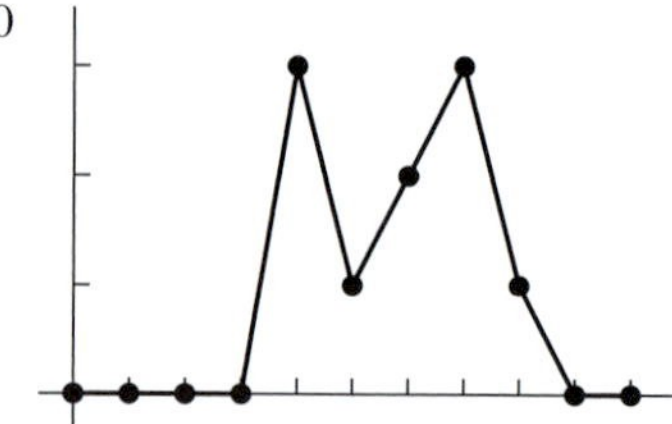

36. $s \doteq 2.3$ **38.** Since this range extends two standard deviations on either side of the mean, $P(\text{an individual falls in the range}) = 0.95$.

Problem Set 10.2 (page 618)

1. $4 = 1 + 3 = 2 + 2 = 3 + 1$; $6 = 1 + 5 = 2 + 4 = 3 + 3 = 4 + 2 = 5 + 1$. There are 8 ways to roll a 4 or 6.
3. **(a)** 3 ways: H and 2, H and 4, H and 6 **5.** The Venn diagram shows that four students speak both English and Japanese.

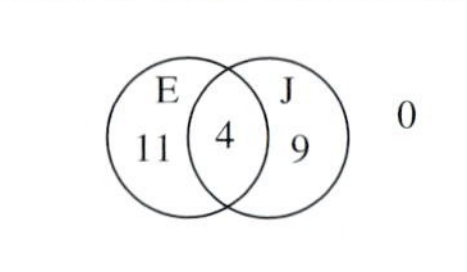

7. Let R be the set of red face cards and let A be the set of black aces. Since $R \cap A = \varnothing$, the addition principle of counting for mutually exclusive events applies, and $n(R \text{ or } A) = n(R) + n(A) = 6 + 2 = 8$.
8. **(a)** $6 \cdot 5 \cdot 4 \cdot 3 = 360$ **(b)** $3 \cdot 5 \cdot 4 \cdot 3 = 180$ **(c)** $3 \cdot 2 \cdot 4 \cdot 3 = 72$
10.

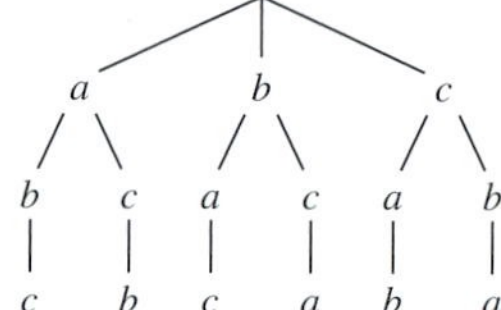

The code words are *abc*, *acb*, *bac*, *bca*, *cab*, and *cba*.
13. **(a)** 5040 **(c)** 72 **(e)** 35,280
14. **(a)** $13 \cdot 12 \cdot 11 \cdot 10 \cdot 9 \cdot 8 \cdot 7 \cdot 6 = 51,891,840$ **(c)** $15 \cdot 14 = 210$ **(e)** $\frac{15!}{15!} = 1$ **16.** **(a)** $5 \cdot 5 \cdot 5 \cdot 5 \cdot 5 = 3125$ **17.** **(a)** $P(6, 6) = 6! = 720$ **19.** **(a)** $\frac{4!}{2!2!} = 6$
25. **(a)** To obtain an even sum, two odd faces or two even faces must come up. This can happen in $3 \cdot 3 + 3 \cdot 3 = 18$ ways. **28.** **(a)** If we think of *ab* as a single entity, then there are four things to put in order. This can be done in $4! = 24$ ways. **(b)** There are 24 with b immediately following a and, by the same reasoning as in part (a), 24 with a immediately following b. Therefore, there are 48 with a and b adjacent. **(c)** By symmetry, in half the possible arrangements a would precede e, and in half e would precede a. Therefore, a precedes e in $5!/2 = 60$ arrangements.

Alternatively, if a is first, a precedes e in $4! = 24$ arrangements. If a is second, a precedes e in $3 \cdot 3! = 18$ arrangements. If a is third, a precedes e in $2 \cdot 3! = 12$ arrangements, and if a is fourth, it precedes e in $3! = 6$ arrangements. Adding, we again obtain 60 as before.

31. The girls can be picked in $C(6,2) = \frac{6 \cdot 5}{2 \cdot 1} = 15$ ways, and the boys can be picked in $C(5,2) = \frac{5 \cdot 4}{2 \cdot 1} = 10$ ways. By the multiplication principle for independent events, the team can be formed in $15 \cdot 10 = 150$ ways.
33. **(a)** $C(8,5)$ **(b)** $C(7,5)$, since there are 7 marbles that are not red. **34.** **(a)** There are 16-trains of length 5. **(b)** There are 1 one-car, 4 two-car, 6 three-car, 4 four-car, and 1 five-car trains of length 5.

36. **(a)** $C(13,3) \cdot C(11,3) = \frac{13 \cdot 12 \cdot 11}{3 \cdot 2 \cdot 1} \cdot \frac{11 \cdot 10 \cdot 9}{3 \cdot 2 \cdot 1} =$ 47,190 **38.** **(a)** 2^4, or 16, words **39.** **(a)** 362,880 **(c)** 2520 **(d)** 20,160 **(e)** 35 **40.** **(a)** $10! = 3{,}628{,}800$ **(c)** $C(13,8) = 1287$ **43.** It is incorrect, since the trials are completely independent. What happens on one trial has no effect on any other trial. She may continue to lose all evening!
45. **(a)**

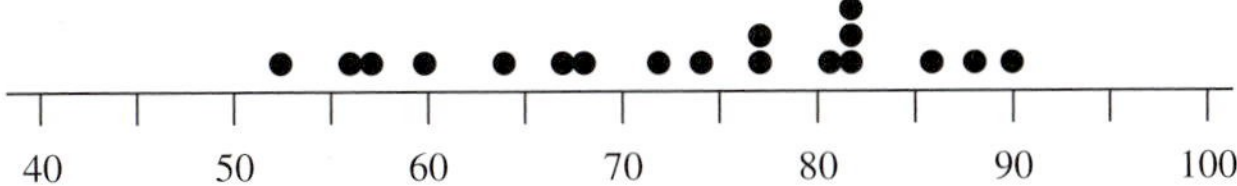

Problem Set 10.3 (page 638)

1. **(a)** Outcomes listed in order of penny, nickel, dime, and quarter are as follows:

HHHH	THHH	HTHT	TTTH	TTTT
	HTHH	HHTT	TTHT	
	HHTH	TTHH	THTT	
	HHHT	THTH	HTTT	
		THHT		
		HTTH		

(c) $P(\text{2 heads and 2 tails}) = \frac{6}{16} = 0.375$

4. **(a)** $\frac{4}{22} \doteq 0.18$ **(c)** $\frac{6}{22} \doteq 0.27$ **5.** **(a)** $\frac{4}{12} \doteq 0.33$

6. **(a)** $\frac{7}{20} = 0.35$ **7.** **(a)** 0.19 **(c)** 0.33 **(e)** 0.25

8. **(a)** $\frac{4}{50} = 0.08$ **(c)** $\frac{0}{50} = 0$

10. $P(3H) = \frac{C(13,3)}{C(52,3)} = \frac{13 \cdot 12 \cdot 11}{52 \cdot 51 \cdot 50} \doteq 0.013$

11. These answers are determined by ratios of angular measures of appropriate regions.

(a) $P(\text{shaded area}) = \frac{3}{10} = 0.3$

(c) $P(\text{region 10 or region 6}) = \frac{2}{10} = 0.2$

(e) $P(8 \mid \text{shaded area}) = \frac{1}{3}$

(g) $P(\text{a vowel or odd-numbered region}) = \frac{7}{10} = 0.7$

12. The probabilities below are ratios of regional areas.
(a) $P(1) = \frac{16}{25} = 0.64$ **(c)** $P(5) = \frac{1}{25} = 0.04$

13. **(a)** $P(\text{b}) = \frac{1}{20}$

(b) $P(\text{a or c or d or e} \mid \text{a or b or c or d or e}) = \frac{4}{5}$

14. **(a)** 5:31 or, equivalently, $\frac{5}{31}$

16. $P(A \text{ or } C) = P(A) + P(C) = \frac{1}{2} + \frac{1}{6} = \frac{4}{6} = \frac{2}{3}$. Thus, the odds in favor of A or C are $\frac{\frac{2}{3}}{1 - \frac{2}{3}} = \frac{2}{1}$, or 2 : 1.

18. $E = \$4 \cdot \frac{1}{36} + \$6 \cdot \frac{2}{36} + \$8 \cdot \frac{3}{36} + \$10 \cdot \frac{4}{36} + \$20 \cdot \frac{5}{36} + \$40 \cdot \frac{6}{36} + \$20 \cdot \frac{5}{36} + \$10 \cdot \frac{4}{36} + \$8 \cdot \frac{3}{36} + \$6 \cdot \frac{2}{36} + \$4 \cdot \frac{1}{36} = \frac{\$600}{36} = \$16.67$ to the nearest penny

22. **(a)** $P(\text{2 white}) = \frac{C(6,2)}{C(14,2)} = \frac{\frac{6 \cdot 5}{2 \cdot 1}}{\frac{14 \cdot 13}{2 \cdot 1}} \doteq 0.16$

23. **(a)** $P(\text{all four red}) = \frac{C(8,4)}{C(19,4)} = \frac{\frac{8 \cdot 7 \cdot 6 \cdot 5}{4 \cdot 3 \cdot 2 \cdot 1}}{\frac{19 \cdot 18 \cdot 17 \cdot 16}{4 \cdot 3 \cdot 2 \cdot 1}} \doteq 0.02$

24. **(a)** $P(\text{a code word begins with } a) = \frac{25 \cdot 24 \cdot 23 \cdot 22}{26 \cdot 25 \cdot 24 \cdot 23 \cdot 22} \doteq 0.04$
27. **(a)** $P(\text{a five-card hand contains exactly two aces}) = \frac{C(4,2) \cdot C(48,3)}{C(52,5)} = \frac{\frac{4 \cdot 3}{2 \cdot 1} \cdot \frac{48 \cdot 47 \cdot 46}{3 \cdot 2 \cdot 1}}{\frac{52 \cdot 51 \cdot 50 \cdot 49 \cdot 48}{5 \cdot 4 \cdot 3 \cdot 2 \cdot 1}} \doteq 0.04$

28. The seven numbers with two alike in a fixed order appear with probability $(1/6)^7$. To compute the number of ways seven such numbers can appear, note that we can choose the number to appear twice in six ways *and* each of the other numbers in only one way *and* we can then order these numbers in $7!/2!$ ways. Thus, the desired probability is

$$6 \cdot \frac{7!}{2!} \cdot \left(\frac{1}{6}\right)^7 \doteq 0.05.$$

29. **(a)** There are five patterns: four 2-loops (that is, the small loops formed by two strings), two 2-loops and a 4-loop, one 2-loop and a 6-loop, two 4-loops, and one 8-loop. **(c)** There are $C(8,2) = 8 \cdot \frac{7}{2} = 28$ ways to choose the pair of 8 strings to be tied. Four of these pairs give a small loop. Thus, the first knot yields a small 2-loop with probability $\frac{4}{28} = \frac{1}{7}$. In Example 10.29, it was shown

that the next three knots each give small loops with probability $\frac{1}{15}$. The probability of four small loops is therefore $\left(\frac{1}{7}\right)\cdot\left(\frac{1}{15}\right) = \frac{1}{105}$. **31. (a)** $P(W_3) = \frac{7\cdot 6\cdot 5}{7\cdot 7\cdot 7} = \frac{30}{49}$, $P(\overline{W}_3) = 1 - P(W_3) = 1 - \frac{30}{49} = \frac{19}{49}$
$P(\overline{W}_3) = 1 - P(W_3) = 1 - \frac{30}{49} = \frac{19}{49}$
32. (a) $P(Y_{23}) = 0.493$; $P(\overline{Y}_{23}) = 1 - P(Y_{23}) = 0.507$. In a group of 23 people, it is more likely than not that two or more share a common birthday.
34. (a) $\frac{1}{C(49, 6)} = \frac{1}{13{,}983{,}816}$, or about 1 in nearly 14 million **35. (a)** $\frac{C(8, 5)\cdot C(72, 15)}{C(80, 20)} \doteq 0.0183$
38. $C(7, 4) + C(8, 4) = \frac{7\cdot 6\cdot 5\cdot 4}{4\cdot 3\cdot 2\cdot 1} + \frac{8\cdot 7\cdot 6\cdot 5}{4\cdot 3\cdot 2\cdot 1} = 105$
40. $5\cdot 21\cdot 4\cdot 20\cdot 3 + 21\cdot 5\cdot 20\cdot 4\cdot 19 = 184{,}800$

Chapter 10 Review Exercises (page 644)

1. Answers will vary. When we did the experiment, three heads and a tail occurred five times, so that $P_e = 5/20 = 0.25$. **2.** Answers will vary. When we did the experiment, we obtained the following:

2	3	4	5	6	7	8	9	10	11	12
I	I	II	II	III	III	IIII	I	II	I	

Using these data, we obtain the following: **(a)** $P_e(3 \text{ or } 4) = 3/20 = 0.15$ **(b)** $P_e(\text{score at least } 5) = 16/20 = 0.80$
3. Answers will vary. Using the above data, we obtain $P_e(5 \text{ or } 6 \text{ or } 7 \mid 5 \text{ or } 6 \text{ or } 7 \text{ or } 8 \text{ or } 9) = 8/13 \doteq 0.6$.
4. Answers will vary. In our study, seven out of 20 chose chocolate, so that $P_e = 7/20 = 0.35$. **5.** Answers will vary. When we did the experiment, we obtained the data shown below.

Point Up	5	4	3	5	2	4	3	2	2	3	2	3
Head Up	0	1	2	0	3	1	2	3	3	2	3	2

Point Up	3	1	2	0	4	4	3	2
Head Up	2	4	3	5	1	1	2	3

Using these data, we obtain the following: **(a)** $P_e(3 \text{ tacks land point up}) = 6/20 = 0.30$. **(b)** $P_e(2 \text{ or } 3 \text{ tacks land point up}) = 12/20 = 0.60$. **6.** Answers will vary. Doing the experiment, we obtained the following data, with the results indicated in (a) and (b).

Number of Trials to Get a 5 or 6	1	2	3	4	5	6	7
			II	IIII	III		I

(a) Average number of rolls required is
$$\frac{3 + 3 + 4 + 4 + 4 + 4 + 5 + 5 + 5 + 7}{10} = 4.4.$$
We guess that it should take four or five rolls to get a 5 or 6. **(b)** $P_e(\text{it takes precisely five rolls to obtain 5 or 6}) = 3/10 = 0.30$. **7.** Answers will vary. Doing the experiment, we obtained the following data, yielding the results in (a), (b), and (c).

	Heart	Nonheart
Ace	I	II
Nonace	𝍸 III	𝍸 IIII

(a) $P_e(\text{ace or heart}) = 11/20 = 0.55$. **(b)** $P_e(\text{ace and heart}) = 1/20 = 0.05$. **(c)** $P_e(\text{ace} \mid \text{heart}) = 1/9 \doteq 0.11$
8. A match will occur with a probability of about 0.63. Thus, you will usually get around 15 or 16 matches in 25 trials.
9. (a) HHH HTT **(b)** 3
HHT THT
HTH TTH
THH TTT
10. (a) $C(15, 9) = \frac{15\cdot 14\cdot 13\cdot 12\cdot 11\cdot 10\cdot 9\cdot 8\cdot 7}{9\cdot 8\cdot 7\cdot 6\cdot 5\cdot 4\cdot 3\cdot 2\cdot 1} = 5005$ **(b)** $C(2, 1)\cdot C(3, 1)\cdot C(13\cdot 7) = \frac{2}{1}\cdot\frac{3}{1}\cdot\frac{13\cdot 12\cdot 11\cdot 10\cdot 9\cdot 8\cdot 7}{7\cdot 6\cdot 5\cdot 4\cdot 3\cdot 2\cdot 1} = 10{,}296$
11. Fill in the Venn diagram from the inside out.

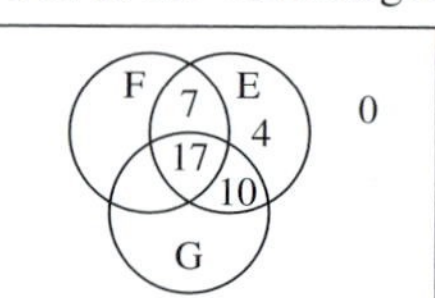

(a) $90 - 4 = 86$ **(b)** 7

12. (a) $C(13, 2) = \frac{13\cdot 12}{2\cdot 1} = 78$ **(b)** $C(12, 2) = \frac{12\cdot 11}{2\cdot 1} = 66$ **(c)** $C(13, 2) + C(12, 2) - C(3, 2) = 78 + 66 - 3 = 141$ **13. (a)** $26\cdot 25\cdot 24\cdot 23\cdot 22 = 7{,}893{,}600$ **(b)** $5\cdot 4\cdot 24\cdot 23\cdot 22 = 242{,}880$ **(c)** $3\cdot 23\cdot 22 = 1518$
14. (a) $\frac{7!}{2!2!2!1!} = 630$ **(b)** $\frac{6!}{2!2!1!1!} = 180$
(c) $\frac{6!}{2!2!1!1!} = 180$
15. (a) HH1, HH2, HH3, HH4, HH5, HH6
HT1, HT2, HT3, HT4, HT5, HT6
TH1, TH2, TH3, TH4, TH5, TH6
TT1, TT2, TT3, TT4, TT5, TT6
(b) $P(\text{TT5}) = 1/24 \doteq 0.04$ **16.** $P(5 \mid TT) = 1/6 \doteq 0.17$
17. $P(\text{sum at most } 11) = 1 - P(\text{sum is } 12) = 1 - \frac{1}{36} \doteq 0.97$ **18. (a)** $[C(5, 2) + C(6, 2) + C(4, 2)]/C(15, 2) = \left(\frac{5\cdot 4}{2\cdot 1} + \frac{6\cdot 5}{2\cdot 1} + \frac{4\cdot 3}{2\cdot 1}\right)\bigg/\frac{15\cdot 14}{2\cdot 1} \doteq 0.30$ **(b)** $P($both

white $= \dfrac{C(5,2)}{C(15,2)} = \dfrac{\frac{5\cdot 4}{2\cdot 1}}{\frac{15\cdot 14}{2\cdot 1}} \doteq 0.10$ **(c)** $P(\text{both white}\mid\text{both the same}) = \dfrac{C(5,2)}{C(5,2)+C(6,2)+C(4,2)} = \dfrac{10}{10+15+6} \doteq 0.32$

19. (a) $P(\text{b and }8) = 0$ **(b)** $P(\text{b or }8) = P(\text{b}) + P(8) = \dfrac{90}{360} + \dfrac{30}{360} = \dfrac{1}{3} \doteq 0.33$ **(c)** $P(\text{b}\mid 8) = 0$
(d) $P(\text{b and }2) = \dfrac{30}{360} \doteq 0.08$
(e) $P(\text{b or }2) = P(\text{b}) + P(2) - P(\text{b and }2) = \dfrac{90}{360} + \dfrac{30}{360} - \dfrac{30}{360} = 0.25$
(f) $P(2\mid\text{b}) = 30/90 \doteq 0.33$ **20. (a)** $3:5$ or $3/5$ **(b)** $1:7$ or $1/7$
21. (a) $P(A)/[1 - P(A)]$; that is, $\dfrac{0.85}{1-0.85} = \dfrac{0.85}{0.15} = \dfrac{17}{3}$, or $17:3$.
(b) $P(A) = 17/25 = 0.68$ **22. (a)** E $= \$5\cdot(.50) + \$10\cdot(.25) + \$20\cdot(.10) = \$7$ **(b)** No. On average, you expect to lose \$3 per game. **23.** Since the sexes of children are independent, $P(\text{other two children are boys}) = \dfrac{1}{2}\cdot\dfrac{1}{2} = 0.25$.

Chapter 10 Test (page 646)

1. (a) 5040 **(b)** 504 **(c)** 40,320 **(d)** 5040 **(e)** 6720 **(f)** 40,320 **(g)** 84 **(h)** 1 **2. (a)** $C(8,5) = \dfrac{8\cdot 7\cdot 6\cdot 5\cdot 4}{5\cdot 4\cdot 3\cdot 2\cdot 1} = 56$ **(b)** $C(5,5)\cdot C(8,5) = \dfrac{5\cdot 4\cdot 3\cdot 2\cdot 1}{5\cdot 4\cdot 3\cdot 2\cdot 1}\cdot\dfrac{8\cdot 7\cdot 6\cdot 5\cdot 4}{5\cdot 4\cdot 3\cdot 2\cdot 1} = 56$ **(c)** $C(5,5) + C(8,5) = \dfrac{5\cdot 4\cdot 3\cdot 2\cdot 1}{5\cdot 4\cdot 3\cdot 2\cdot 1} + \dfrac{8\cdot 7\cdot 6\cdot 5\cdot 4}{5\cdot 4\cdot 3\cdot 2\cdot 1} = 57$
3. $P(2\text{ yellow balls}\mid 3\text{ green balls}) = C(5,2)/C(14,2) \doteq 0.11$ **4.** Prepare a card as shown and ask a number of people to choose a number. Calculate the empirical probability of choosing 3 as the number of times 3 is chosen divided by the number of people questioned. **5.** 7 to 13 **6. (a)** $7^4 = 2401$ **(b)** $7\cdot 6\cdot 5\cdot 4 = 840$
7. (a) $2\cdot 6\cdot 5\cdot 4 = 240$ **(b)** Choose *cd* as a single unit in 1 way, choose two more letters in $C(5,2) = 10$ ways, and arrange these three items in order in $3! = 6$ ways. Proceed similarly for *dc*. Therefore, the desired number is $2\cdot 1\cdot 10\cdot 6 = 120$. **8.** This would be an empirical probability obtained by keeping records for a large number of trials of treating strep throat with penicillin.
9. (a) Fill in the regions in the Venn diagram, starting with the innermost region.

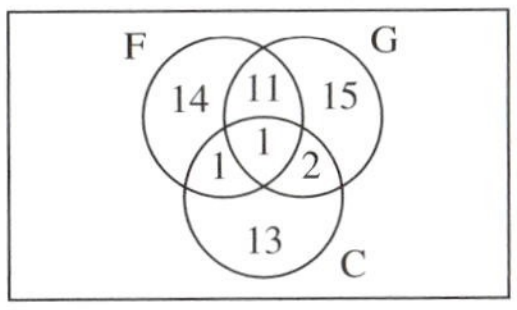

Adding all the counts, we find that there are 57 students in all. **(b)** 13 students **(c)** 11 students **10.** 5 to 12

Chapter 11

Problem Set 11.1 (page 667)

1. (a) $\overleftrightarrow{AB}$ or $\overleftrightarrow{BA}$
2. (a)

3. (a)

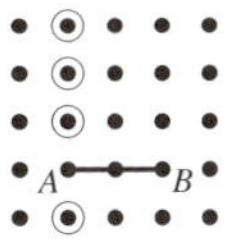

4. (a) 10 angles: $\angle APB$, $\angle APC$, $\angle APD$, $\angle APE$, $\angle BPC$, $\angle BPD$, $\angle BPE$, $\angle CPD$, $\angle CPE$, $\angle DPE$
5. $m(\angle AXB) = m(\angle AXE) - m(\angle BXE) = 180° - 132° = 48°$
$m(\angle CXD) = m(\angle BXD) - m(\angle BXC) = 90° - 35° = 55°$
$m(\angle DXE) = m(\angle BXE) - m(\angle BXD) = 132° - 90° = 42°$
7. The three lines are concurrent. **10. (a)** Opposite angles are supplementary: $m(\angle A) + m(\angle C) = 180°$ and $m(\angle B) + m(\angle D) = 180°$. **11. (a)** Measurements will vary. You should find $m(\angle APB) = m(\angle AQB) = m(\angle ARB)$, and this measure is half of the measure of $\angle AOB$. **12.** Ten times: once between 1 and 2, once between 2 and 3, . . . , and once between 10 and 11.
13. (a) 360° **(d)** 30° **14. (a)** 90° **(c)** The minute hand is on the 6 and the hour hand is halfway between the 4 and the 5. The angle between two consecutive numbers is $\frac{1}{12}$ of a revolution, or 30°. So the angle is $(1.5)(30°) = 45°$.
16. Draw a horizontal ray $\overrightarrow{PQ}$ at P, in the opposite direction of $\overrightarrow{AB}$ and $\overrightarrow{CD}$. The opposite-interior-angles theorem then gives $m(\angle APQ) = 130°$ and $m(\angle CPQ) = 140°$. Thus, $m(\angle P) = 360° - 130° - 140° = 90°$. **17. (a)** 40°, since the measures of the interior angles of a triangle add up to 180°. **(c)** 49°; since the interior angles of a triangle add up to 180°, and a right angle has measure 90°, we have $m(\angle 3) = 180° - (41° + 90°) = 49°$.
18. (a) $x + x + 30° = 180°$, so $x = 75°$. The interior angles measure 75°, 75°, and 30°. **19. (a)** No, because an obtuse angle has measure greater than 90° and adding two such measures would exceed 180°, which is the sum of all three interior angle measures for any triangle. **23. (a)** Zero intersection points of the five lines are parallel to each other.

24. (a) 6 lines **26.** The pencil turns through each interior angle of the triangle. Since the pencil faces the opposite direction when it returns to the starting side, it has turned a total of 180°. This demonstrates that the sum of measures of the interior angles of a triangle is 180°. **28. (a)** By trial and error, 10 taxi segments. Alternatively, this is the number of sequences of 3 E(east) and 2 N(north) segments, and there are 10 ways to form such sequences: NNEEE, NENEE, . . . , EEENN. **31.** The earliest clocks were sundials used in the northern hemisphere. The shadow cast by the gnomon follows an arc in the direction we call clockwise. **34. (a)** A full revolution takes 24 hours. In 1 hour, the earth turns $\frac{1}{24}$ of a revolution, or 15°. **35.** The angle of latitude is equal to the angle of elevation to Polaris. **37. (a)** $58^\circ\ 36'\ 45'' = 58^\circ + \left(\frac{36}{60}\right)^\circ + \left(\frac{45}{3600}\right)^\circ = 58.6125^\circ$ **(c)** $71.32^\circ = 71^\circ + (0.32)(60)' = 71^\circ + 19.2' = 71^\circ + 19' + (0.2)(60)'' = 71^\circ\ 19'\ 12''$ **39. (a)** Q, R, and S are collinear. **40. (a)** The lines $\overleftrightarrow{BB'}$ and $\overleftrightarrow{DD'}$ intersect at a right angle at P.

Problem Set 11.2 (page 688)

1.

	(a)	(b)	(c)	(d)
Simple Curve		✓		
Closed Curve		✓	✓	✓
Polygonal Curve	✓	✓		
Polygon		✓		

2. An example of each figure is given.

(a) **(c)**

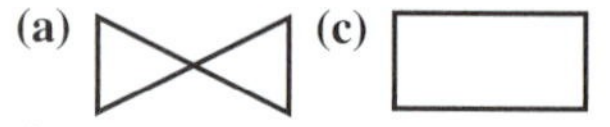

4. (a) Convex **(b)** Concave

5. (a) **(c)**

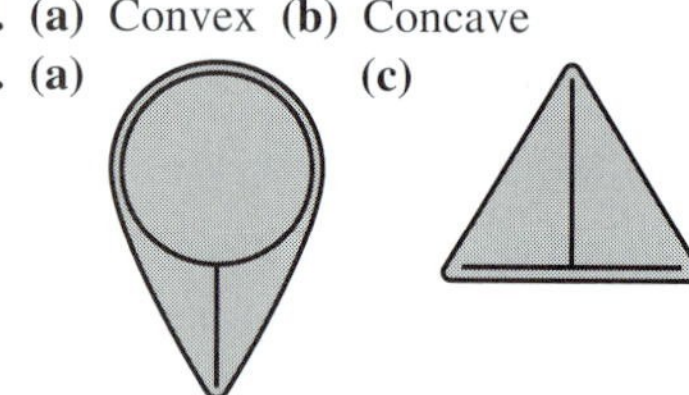

6. (a) 6 **(c)** 8 **7.** $2x + 5x + 5x + 5x + 5x + 2x = (6 - 2)(180^\circ)$, or $24x = 720^\circ$, so $x = 30^\circ$. The angles measure 60°, 150°, 150°, 150°, 150°, and 60°.

9. (a) $(5 - 2)(180)^\circ = 540^\circ$ **(c)** $(6 - 2)(180)^\circ = 720^\circ$

10. Pictures will vary.

(a) Triangle **(c)** Decagon **12. (a)** 360° **(c)** 0°

14. (a) *DAF* and *DBC*

16.

	n	Interior Angle	Exterior Angle	Central Angle
(a)	5	108°	72°	72°
(c)	7	$128\frac{4}{7}^\circ$	$51\frac{3}{7}^\circ$	$51\frac{3}{7}^\circ$

17. (a) $\frac{360^\circ}{n} = 15^\circ$, so $n = 24$. **20. (a)** Any curve from K to D "crosses" the wall an odd number of times, so the dragon is on the opposite side of the wall from the knight and therefore presents no danger. **23. (a)** The boat can drift to any position inside the circle centered at A, where the radius of the circle is the length of the anchor rope.

24. (a) C could be any point (other than A or B) on either of the two lines drawn through A and through B that are perpendicular to $\overline{AB}$.

(b) C could be any point (other than A or B) on the circle with $\overline{AB}$ as its diameter.

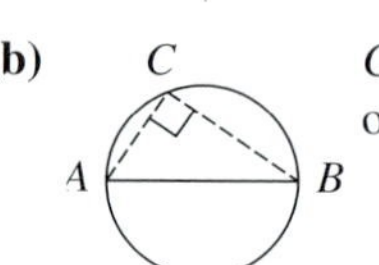

25. At a vertex, the interior angle and the conjugate angle add up to 360°. For an n-gon, the sum of all interior and all conjugate angles is $n \cdot 360^\circ$. All the interior angles add up to $(n - 2) \cdot 180^\circ$, so all the conjugate angles add up to $360^\circ n - (n - 2)\,180^\circ = 360^\circ n - 180^\circ n + 360^\circ = 180^\circ n + 360^\circ = (n + 2) \cdot 180^\circ$. **27.** Such a point S allows for n triangles to be formed, all with the vertex S. The sum of all the interior angles of these n triangles is $n \cdot 180^\circ$, which is equal to the sum of all the interior angles of the n-gon plus 360° for the angles that surround the point S. Thus, the sum of the interior angles of the n-gon is $n \cdot 180^\circ - 360^\circ = (n - 2) \cdot 180^\circ$. **29. (a)** The total turn made when tracing the star is $2 \times 360^\circ$, so the turn made at each point of the star is $7 \times 360^\circ/16 = 157.5^\circ$. Therefore, the angle at each point measures $180^\circ - 157.5^\circ = 22.5^\circ$. **30.** Each new circle creates a new region each time it intersects a previously drawn circle. Since the new circle intersects each of the old circles in two points, this creates the following pattern:

NUMBER OF Circles	NUMBER OF Regions	NUMBER OF Circles	NUMBER OF Regions
1	$2 \qquad = 2$	6	$22 + 2 \cdot 5 = 32$
2	$2 + 2 \cdot 1 = 4$	7	$32 + 2 \cdot 6 = 44$
3	$4 + 2 \cdot 2 = 8$	8	$44 + 2 \cdot 7 = 58$
4	$8 + 2 \cdot 3 = 14$	9	$58 + 2 \cdot 8 = 74$
5	$14 + 2 \cdot 4 = 22$	10	$74 + 2 \cdot 9 = 92$

33. (a) The sum of the interior angles is $360^\circ = m(\angle P) + m(\angle Q) + m(\angle R) + m(\angle S)$. We also know that $m(\angle P) = m(\angle R)$ and $m(\angle Q) = m(\angle S)$, so $360^\circ = m(\angle P) + m(\angle Q) + m(\angle P) + m(\angle Q)$, giving us $180^\circ = m(\angle P) + m(\angle Q)$. **(b)** $m(\angle Q) + m(\angle q) = 180^\circ$ and $m(\angle P) + m(\angle Q) = 180^\circ$, so $m(\angle q) = m(\angle P)$. $\angle q$ and $\angle P$ are corresponding angles, so segments $\overline{PS}$ and $\overline{QR}$ are parallel. $m(\angle q) = m(\angle P)$ and $m(\angle P) = m(\angle R)$, so $m(\angle q) =$

$m(\angle R)$. $\angle q$ and $\angle R$ are alternate interior angles, so their congruence gives $\overline{PQ}$ parallel to $\overline{SR}$. Hence, the figure is a parallelogram.

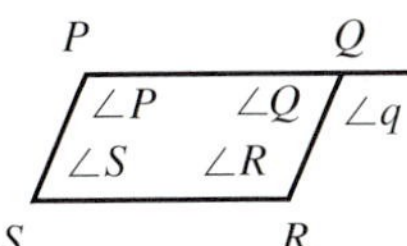

35. (a) Drawings will vary. **(b)** Suppose the black curve has been drawn. It will be seen that it is easy to shade some of the regions in such a way that exactly one of any two regions sharing a common boundary is shaded and the other region is left unshaded. Now imagine adding the red closed curve. Each time the red curve crosses the black curve, it alternates between shaded and unshaded regions. Since the red curve is closed, the red curve closes upon itself in a region of the same type as its starting point. Thus, the number of color alternations must be even. That is, the number in intersection points of the red and black curves is necessarily even. **36.** The number of regions interior to both simple curves is always equal to the number of regions exterior to both curves. The number of regions inside the red curves and outside the black curves is always equal to the number of regions inside the black curves and outside the red curves. There is no connection between the two pairs of equal numbers.

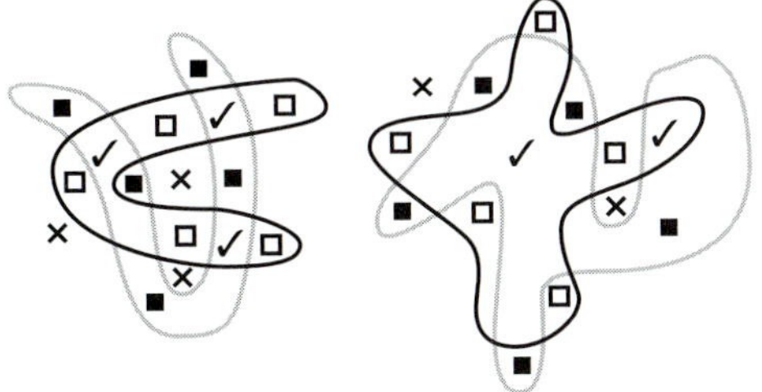

39. (a) $SQRE$ is square. **(b)** No

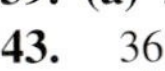
43. 36°

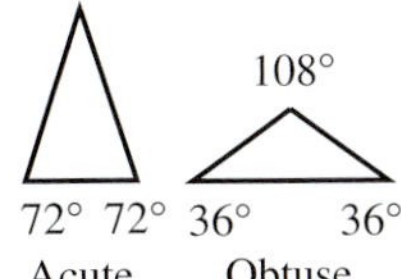

45. One angle is 90°; call the other two angles A and B. The interior angles of a triangle add up to 180°, so $90° + m(\angle A) + m(\angle B) = 180°$, or $m(\angle A) + m(\angle B) = 90°$.

Problem Set 11.3 (page 709)

1. (a) Polyhedron **(c)** Polyhedron **(e)** Not a polyhedron
2. (a) Pentagonal prism **(c)** Oblique circular cone
(e) Right rectangular prism **3. (a)** 4 **(c)** A, B, C, D

6. (a)

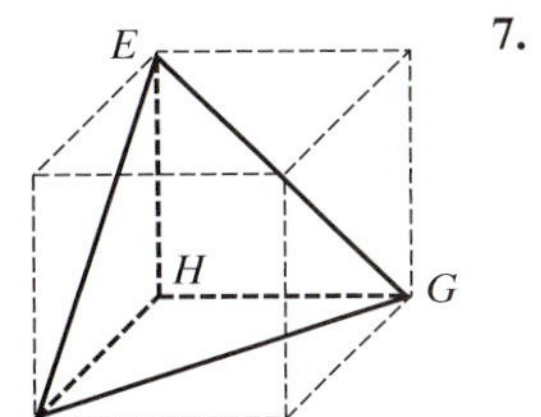

7. (a)

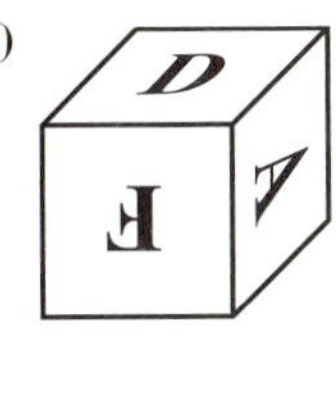

8. (a) 45°, since the dihedral angle between the adjacent sides of the cube is 90° and another pyramid would fit into the gap between the original pyramid and a vertical side of the cube. **(b)** Filling the cube with six such pyramids, one sees that three pyramids surround the edge between the cube's center and any corner. Thus, three copies of the dihedral angle give a full revolution of 360° around this edge, so the dihedral angle measures 120°. **15. (a)** $F = 10$, $V = 7$, and $E = 15$, so $V + F = E + 2$, since $7 + 10 = 15 + 2$.

17. (a)

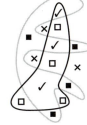

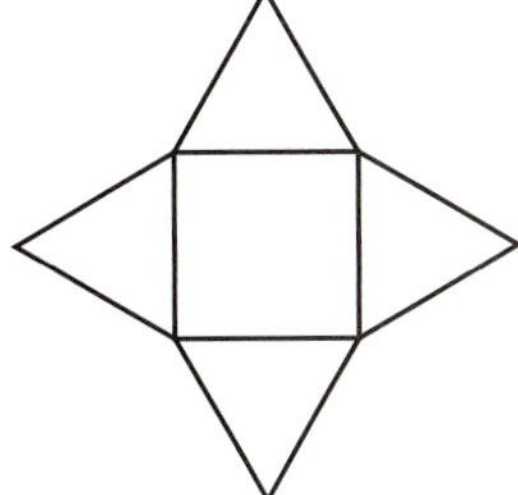

18. (a) $V = 10$, $F = 12$, and $E = 20$.
$V + F = 22 = E + 2$, so Euler's formula holds.

21. (a) Suppose the faces of a polyhedron consist of a p-gon, a q-gon, an r-gon, and so on. Since each edge of the polyhedron borders two faces, the sum $p + q + r + \cdots$ is twice the number of edges. That is, $p + q + r + \cdots = 2E$. Since there are F faces and $p, q, r, \ldots$ are all 3 or greater, we get $2E \geq 3 + 3 + \cdots = 3F$. **(b)** Each of the V vertices of a polyhedron is the endpoint of three or more edges that meet at the vertex. Thus, $3V$ is less than or equal to the total number of ends of the edges. But each of the E edges has two ends, so there are $2E$ ends of edges. We see that $3V \leq 2E$. **(c)** Adding $3V \leq 2E$ and $3F \leq 2E$ shows that $3V + 3F \leq 4E$. But $V + F = E + 2$ (Euler's formula), so $3V + 3F = 3E + 6$. Comparing this result with the inequality, we see that $3E + 6 \leq 4E$. Subtracting $3E$ from both sides shows that $6 \leq E$.
(d) Suppose $E = 7$. Since $3F \leq 2E = 14$, we see that F is no larger than 4. ($F \geq 5$ would give $3F \geq 15$.) Similarly, $3V \leq 2E = 14$ means that $V \leq 4$. Since both $V \leq 4$ and $F \leq 4$, $V + F \leq 8$. But $V + F = E + 2$ (Euler's formula), and $E = 7$, so $V + F = 9$. This contradicts $V + F \leq 8$, so our assumption, $E = 7$, is not possible. **(e)** A pyramid with a base of $3, 4, 5, \ldots, n, \ldots$ sides has $6, 8, 10 \ldots, 2n, \ldots$ edges, respectively. Slicing off a tiny corner at one vertex

somewhere on the base of the pyramid adds three new edges, giving us polyhedra with 9, 11, 13, . . . , $2n + 3$, . . . edges. Altogether, the pyramids and pyramids with a truncated base corner give us polyhedra with 6, 8, 9, 10, 11, . . . edges.

23. **(a)** The base is a pentagon.

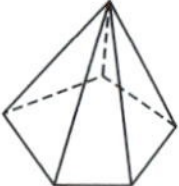

(b) The center polygon is a triangle.

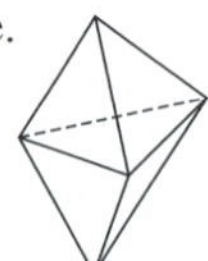

24. **(a)**

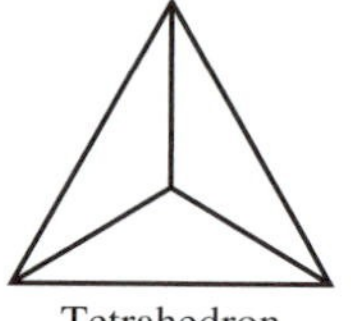

Tetrahedron

(b)

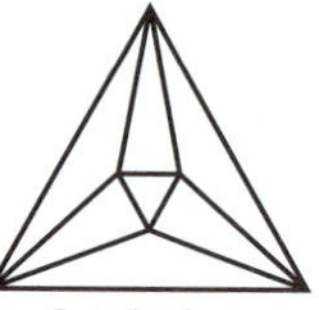

Octahedron

(c)

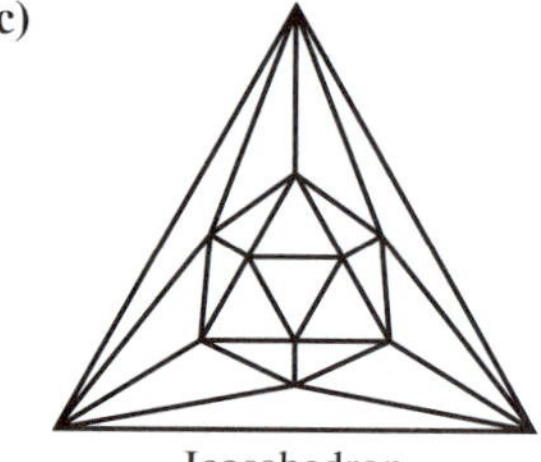

Icosahedron

28. **(a)** When folded, these edges coincide and so must be the same length in the net. The dashed segment $\overline{AP}$ is perpendicular to edge $\overline{BH}$ because folding along $\overline{BH}$ moves point A along a circle that is in a plane perpendicular to the axis $\overline{BH}$ of the fold. **29.** The axis of all the hinges on a door must be along the intersection of the planes of the wall and the plane of the opened door. Since planes meet in a line, the axes of the hinges must be along a single line.

33. A, C (if the cylinder can be triangular), or D (if the cylinder can be hexagonal).

35. $m(\angle x) = 50°$, $m(\angle y) = 50°$, $m(\angle z) = 130°$

36. **(a)**

n	3	4	5	6	7	8
Diagonals	0	2	5	9	14	20

(c) 54 **37.** **(a)** Many polygons are possible. One example is the following:

Problem Set 11.4 (page 722)

1. The networks I and III each represent the information correctly, since there is an edge between vertices when, and only when, the individuals have met. Network II does not represent the information given; for example, it incorrectly indicates that Coralee has met Bianca. **2.** **(a)** Yes. *AHGFEDCBHFDBA* is one Euler path. **(c)** No. **(e)** Yes. *CEAEFBFDBACD* is one Euler path. **4.** **(a)** $D = 24$, $E = 12$ for (a); $D = 22$, $E = 11$ for (b); $D = 24$, $E = 12$ for (c); $D = 30$, $E = 15$ for (d); $D = 22$, $E = 11$ for (e); $D = 30$, $E = 15$ for (f) **5.** **(a)** Yes, since exactly two vertices are odd **(b)** $D = 2 + 2 + 4 + 8 + 3 + 6 + 6 + 1 = 32$, so $E = 16$.

7. **(a)**

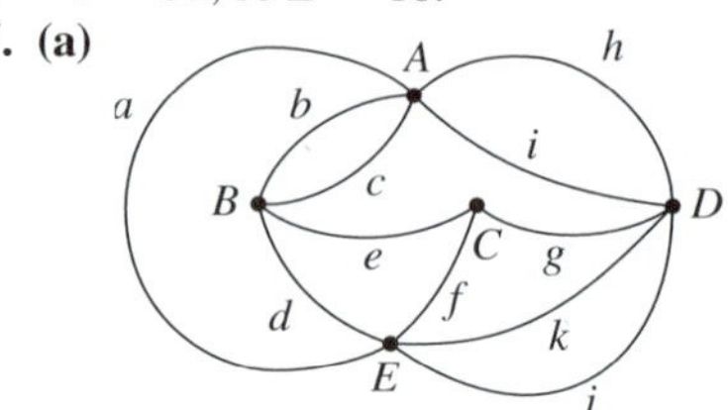

8. **(a)**

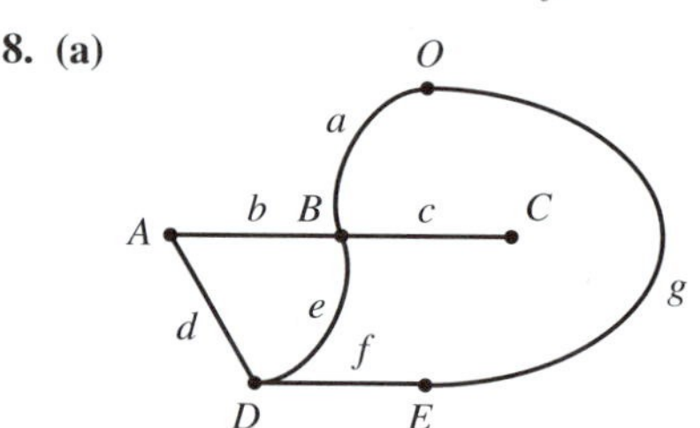

9. **(a)** $V = 6$, $R = 7$, and $E = 11$ **12.** **(a)** The triangle in the network shows that at least three different times must be scheduled, say, science at 2:00, math at 3:00, and basketball at 4:00. Then band can meet at 2:00 and history can meet at 3:00 with no student having a conflict. **13.** **(a)** Combining these two paths gives a closed path with distinct edges. **14.** The traversable trees are those of the form

•, •——•, •——•——•, •——•——•——•, . . .

17. Consider a network with a vertex for each person and an edge between two vertices for each time the corresponding persons have shaken hands. By the result stated in problem 15, there are an even number of people with odd numbers of handshakes. **18.** **(a)** One possibility is *AB*, *JFBGJIDAEI*, *EFGHCKH*. **(b)** Temporarily add $m - 1$ edges between $m - 1$ distinct pairs of the odd vertices. The new edges make these vertices into even vertices. Only two odd vertices remain, so the network is traversable. Following one of the added edges is equivalent to lifting the pencil. The $m - 1$ lifts mean that there are m strokes. **21.** **(a)** 14, as can be counted from a diagram

22. **(a)**

24. One collection of edges is *HA*, *AB*, *BG*, *GF*, *FD*, *DC*, and *DE*. **27.** A square **29.** $x = 30°$, $y = 45°$, and $z = 135°$.

Chapter 11 Review Exercises (page 731)

1. **(a)** $\overleftrightarrow{AC}$ **(b)** $\overline{BD}$ **(c)** AD **(d)** $\angle ABC$ or $\angle CBA$ **(e)** $m(\angle BCD)$, or 90° **(f)** $\overrightarrow{DC}$ **2.** **(a)** $\angle BAD$ **(b)** $\angle BCD$ **(c)** $\angle ABC$, $\angle ADC$ **3.** **(a)** 143° **(b)** 53° **4.** $p = 55°$, $r = 55°$, $s = 125°$, $q = 35°$ **5.** $x = 45°$, $y = 33°$, $z = 147°$ **6.** **(a)** (iv) **(b)** (i) **(c)** (vi) **(d)** (v) **(e)** (ii)

(f) (iii) **7. (a)** No, because obtuse angles have a measure greater than 90°, and the sum of the three interior angles of a triangle is 180°. **(b)** Yes; try angles of 100°, 100°, 100°, and 60°. **(c)** No, because acute angles have a measure less than 90°, and the sum of the interior angles of a quadrilateral must be 360°. **8.** The interior angles add up to $(6 - 2)(180)° = 720°$, so $16x = 720°$ and $x = 45°$. The angles are 135°, 135°, 135°, 45°, 225°, and 45°. **9.** 360° **10. (a)** 6 **(b)** $\overline{CD}, \overline{EF}, \overline{GH}$ **(c)** $\overline{DH}, \overline{GC}, \overline{EH}, \overline{FG}$ **(d)** 45° **11.** Square right prism; triangular pyramid or tetrahedron; oblique circular cylinder; sphere; hexagonal right prism. **12. (a)** **(b)**

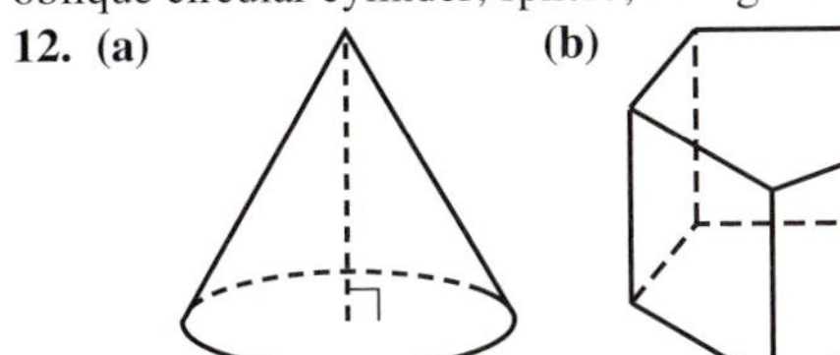

(c)

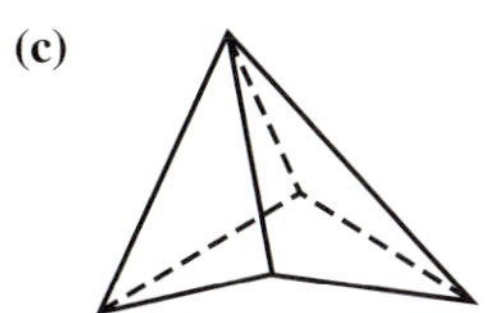

13. (a) See Table 11.4 **(b)** $V = 6$, $F = 8$, and $E = 12$. Thus, $V + F = 14$ and $E + 2 = 14$, so Euler's formula holds. **14.** By Euler's formula, $V + 14 = 24 + 2$, so $V = 12$. **15. (a)** It has four odd vertices. **(b)** An edge between any two of the vertices A, B, C, and E. Many Euler paths are possible. **16.** Construct a network with vertices A, B, C, D, and E and edges corresponding to bridges. Since just two vertices, A and D, have odd degree, there is an Euler path. The Euler path corresponds to a walking path that crosses each bridge exactly once. **17.** $V = 11$, $R = 7$, and $E = 16$. $V + R = 18$ and $E + 2 = 18$, so $V + R = E + 2$ holds.

Chapter 11 Test (page 733)

1. The average interior angle measure for an n-gon is $\frac{(n-2)(180°)}{n}$. We want this value to be 174°, so $\frac{180(n-2)}{n} = 174$. This gives $n = 60$. **2.** Many examples of each are possible. **(a)** **(b)** **(c)** **(d)**

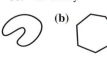

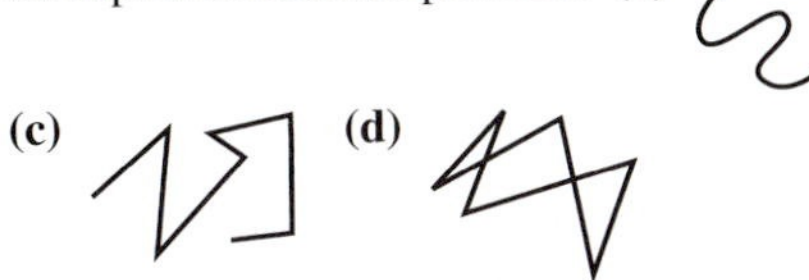

3. (a) 24. Each edge borders a square face and a triangular face. So just count the edges of all the square faces (or all the edges of the triangular faces). **(b)** 12. Either count in the diagram or use Euler's formula.

4.

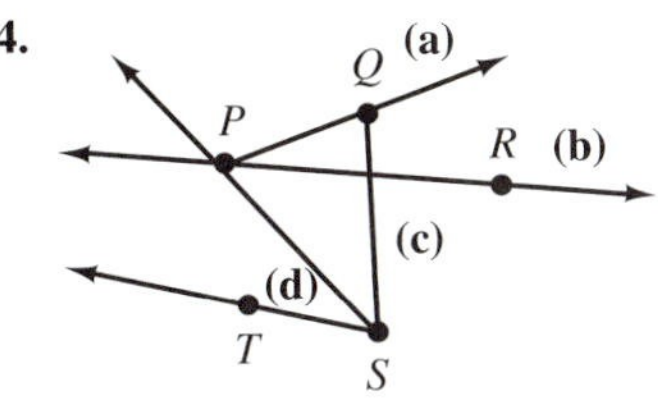

5. (a) By Euler's formula, $V + 7 = 11 + 2$, so $V = 6$. **(b)** Many such networks can be drawn. **6. (a)** C, A **(b)** F **(c)** D **(d)** E **(e)** B **7. (a)** True **(b)** False **(c)** True **(d)** True **8.** The total-turn angle is 1080°, so the turn at each of the seven vertices is $1080°/8 = 135°$. This means that the interior angle at each point is $180° - 135° = 45°$. **9. (a)** Decagon **(b)** Octagon **(c)** Nonagon **10. (a)** Diagrams will vary. **(b)** $V = 16$, $F = 21$, $E = 35$ **(c)** $V + F = 37$ and $E + 2 = 37$, so Euler's formula holds. **11.** The same as an interior angle of a regular pentagon, which is $\frac{3 \cdot 180°}{5} = 108°$ **12.** $t = 40°$, $r = 80°$, $s = 100°$ **13. (a)** (iii) **(b)** (iv) **(c)** (i), (ii) **(d)** (iii); add an edge between any two of the four odd vertices.

Chapter 12

Problem Set 12.1 (page 752)

1. (a) Height, length, thickness, area, diagonal, weight **(c)** Height, length, depth **4. (a)** Answers will vary anywhere from 40 to 54. **(b)** The circular portions of area don't fit well together, but leave gaps between them. **6. (a)** $1 \text{ acre} = \frac{1}{640}\text{ mi}^2 = \left(\frac{1}{640}\text{ mi}^2\right)\left(\frac{5280 \text{ ft}}{1 \text{ mi}}\right)^2 = 43{,}560 \text{ ft}^2$ **8. (a)** Between 165.5 km and 166.5 km **(c)** Between 0.495 mm and 0.505 mm **9. (a)** $33 \text{ cL} = 33 \times 10^{-2} \text{ L} = 330 \times 10^{-3} \text{ L} = 330 \text{ mL}$ **(b)** Not quite; $1 \text{ L} = 1000 \text{ mL}$ and $3 \times 33 \text{ cL} = 990 \text{ mL}$ **10. (a)** 58.728 kg **(c)** 230 g **11. (a)** 3.5 kg **13. (a)** About 28 cm by 22 cm **(c)** About 2 cm **15. (a)** $1 \text{ li} = 15 \text{ yin} \times \frac{10 \text{ zhang}}{1 \text{ yin}} \times \frac{2 \text{ bu}}{\text{zhang}} \times \frac{5 \text{ chi}}{\text{bu}} = 1500 \text{ chi}$ **(c)** $3 \text{ chi} = 3 \text{ chi} \times \frac{1 \text{ zhang}}{10 \text{ chi}} \times \frac{1 \text{ yin}}{10 \text{ zhang}} \times \frac{1 \text{ li}}{15 \text{ yin}} \times \frac{500 \text{ m}}{1 \text{ li}} = 1 \text{ m}$ **16. (a)** one centipede **(b)** one microphone **(c)** one megaphone **(d)** one decacards (deck of cards) **(e)** one kilohertz **(f)** two kilo mockingbirds *(To Kill a Mockingbird)* **20. (a)** 8, 16, 32 **23.** $1 \text{ ha} \doteq (10{,}000 \text{ m}^2)\left(\frac{1 \text{ km}}{1000 \text{ m}}\right)^2\left(\frac{1 \text{ mi}}{1.6 \text{ km}}\right)^2\left(\frac{640 \text{ acres}}{1 \text{ mi}^2}\right) = 2.5 \text{ acres}$ **25.** $\left(\frac{25 \text{ in}}{1 \text{ min}}\right)\left(\frac{60 \text{ min}}{1 \text{ hr}}\right)\left(\frac{24 \text{ hr}}{1 \text{ day}}\right)\left(\frac{14 \text{ day}}{1 \text{ fortnight}}\right)\left(\frac{1 \text{ ft}}{12 \text{ in}}\right)\left(\frac{1 \text{ furlong}}{660 \text{ ft}}\right) \doteq 63.6 \text{ furlong/fortnight}$

27. (a) kilo **(c)** No prefix **(e)** milli
28. $\left(\frac{100 \text{ km}}{9 \text{ L}}\right)\left(\frac{3.7854 \text{ L}}{1 \text{ gal}}\right)\left(\frac{1 \text{ mi}}{1.6 \text{ km}}\right) \doteq 26.3 \text{ mi/gal}$
30. (a) $(5 \text{ gal})\left(\frac{4 \text{ qt}}{1 \text{ gal}}\right)\left(\frac{32 \text{ oz}}{1 \text{ qt}}\right) = 640 \text{ oz}$ Since $\frac{640}{80} = 8$, add 8 liquid ounces of concentrate. **(b)** Add 80×65 mL = 5200 mL = 5.2 L of water. **32.** A league varied from time to time in history, but was usually close to 3 miles. Thus, the *Nautilus* traveled about 60,000 miles. **33. (a)** 4917 liters were added. **38.** Interior angle measures of a triangle add up to 180°, so $8x + 6x + 4x = 180°$ and $x = 10°$. The angles measure 80°, 60°, and 40°. **39.** Since $m(\angle 3) = m(\angle 1) + m(\angle 2)$, we see that $m(\angle 3) > m(\angle 1)$ and $m(\angle 3) > m(\angle 2)$.

Problem Set 12.2 (page 769)

3. The dodecagon and square have the same area, equal to the sum of the areas of the same subregions in their dissections.

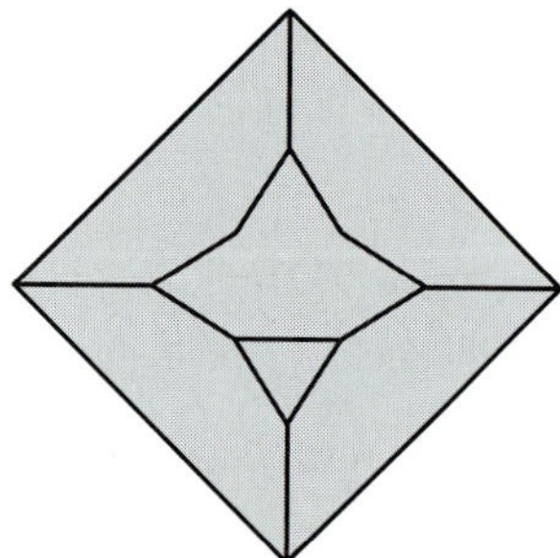

5. (a) 12 units

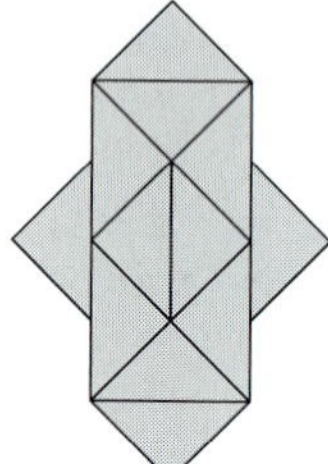

7. (a) $\frac{1}{2} \times (4 \text{ mm} + 6 \text{ mm}) \times 2 \text{ mm} + \frac{1}{2} \times 6 \text{ mm} \times 5 \text{ mm} = 25 \text{ mm}^2$ **9. (a)** 12 cm **10. (a)** 1 cm by 24 cm; 2 cm by 12 cm; 3 cm by 8 cm; 4 cm by 6 cm. The dimensions can also be given in opposite order. **11. (a)** 81 ft², 45.2 ft **12. (a)** 1664.6 m², 198.3 m **13. (a)** 220 square units **14. (a)** $\triangle ABC$; all triangles have the same base and $\triangle ABC$ has the smallest height. **(b)** $\triangle ABF$; it has the largest height. **(c)** $\triangle ABD$ and $\triangle ABE$; they have equal heights and the same base. **16. (a)** 9 square units **17. (a)** $100 \text{ m} + 100 \text{ m} + 2\pi \cdot 25 \text{ m} = (200 + 50\pi)\text{m} \doteq 357 \text{ m}$ **(b)** $(50 \text{ m})(100 \text{ m}) + \pi(25 \text{ m})^2 = (5000 + 625\pi)\text{m}^2 \doteq 6963 \text{ m}^2$ **19. (a)** $\pi(2)^2 - \pi(1)^2 = 3\pi$ square units $\doteq$ 9.4 square units **21. (a)** 40,000,000 m **(b)** $12{,}755\pi$ km $\doteq$ 40,071,000 m **(c)** The equator is larger, since the earth bulges slightly at the equator and is slightly flattened at the poles. **23.** 20 cm². The common overlap reduces the area of both regions by the same amount, so the difference in area is unchanged.

27. (a)

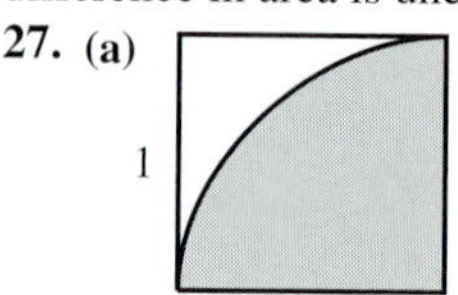

The unshaded region at the left has area

$1^2 - \frac{1}{4}\pi(1^2) = 1 - \pi/4$

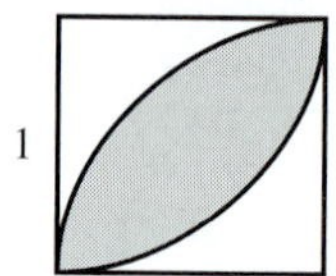

The shaded region at the left has area

$1 - (1 - \pi/4) - (1 - \pi/4) = (\pi/2) - 1$

28. The areas of the rectangular portions of the sidewalk total 2400 ft². The pieces formed with circular areas have total turning of 360°, so when placed together form a circle with radius 8 ft and area $\pi(8 \text{ ft})^2 = 64\pi \text{ ft}^2$. Total area is $(2400 + 64\pi) \text{ ft}^2$. **29. (a)** Erin walks $2\pi R$ and Nerd walks $2\pi(R + L)$, so Nerd walks $2\pi L$ farther than Erin.

33. Draw $\overline{AP}$, $\overline{BP}$, and $\overline{CP}$. Then area $(\triangle ABC) = \frac{1}{2}sh =$ area $(\triangle ABP)$ + area $(\triangle BPC)$ + area $(\triangle CPA) = \frac{1}{2}sx + \frac{1}{2}sz + \frac{1}{2}sy = \frac{1}{2}s(x + y + z)$. Therefore, $\frac{1}{2}sh = \frac{1}{2}s(x + y + z)$, and $h = x + y + z$. *Alternate visual proof.*

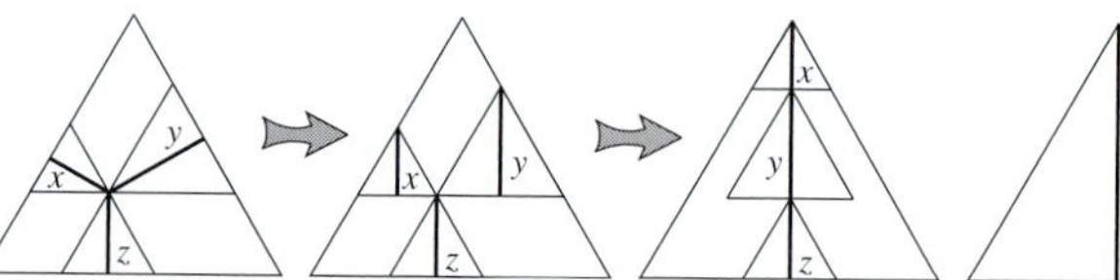

39. The lawn has an area 75 ft × 125 ft = 9375 ft². Since 21 in = $\frac{7}{4}$ ft, the lawn area is equivalent to a rectangle 21 in wide and $9375 \div \frac{7}{4} = 5357.14\ldots$. That is, Kelly will walk about 5357 feet, a bit over a mile (1 mile = 5280 ft). **41.** Consider the carpet as a 6-ft by 4-ft rectangle with semicircular ends of radius 2. Then the carpet's area is $(6 \text{ ft})(4 \text{ ft}) + \pi(2 \text{ ft})^2 = 36.57 \text{ ft}^2 \doteq 5266 \text{ in}^2$. The carpet contains about 5266 inches of braid, or about 439 ft.
43. Since each tile measures $\frac{8}{12} = \frac{2}{3}$ feet on a side, the dimensions of the kitchen in tiles are $10 \div \frac{2}{3} = 15$ and $12 \div \frac{2}{3} = 18$. Therefore, the numbers of tiles needed is $15 \times 18 = 270$. **45.** 90°, since if we view one side of length 300′ as the base, the altitude of the triangle is greatest if the angle is 90° **46. (a)** 48 square units **48.** The circumscribed circle has four times the area of the inscribed circle. **53. (a)** 100 cm **(c)** 10,000 cm²

Problem Set 12.3 (page 783)

1. (a) $x^2 = 7^2 + 24^2 = 49 + 576 = 625$, so $x = \sqrt{625} = 25$. **(c)** $x^2 + 5^2 = 22^2$, so $x = \sqrt{459}$. **(e)** $x^2 = 1^2 + 1^2 = 1 + 1 = 2$, so $x = \sqrt{2}$. **2. (a)** $x^2 + (2x)^2 = (25)^2$, $5x^2 = 625$, $x = \sqrt{125} = 5\sqrt{5}$ **3. (a)** $x^2 = 10^2 + 15^2 = 325$, so $x = \sqrt{325}$; $y^2 = x^2 + 7^2 = 325 + 49 = 374$, so $y = \sqrt{374}$. **4. (a)** $x = \sqrt{13^2 - 12^2} = 5$ **5. (a)** Height $= \sqrt{15^2 - 9^2}$, so area $= (20)(12) = 240$ square units. **7.** The areas are equal. The small circle of radius 1 has area π, and the large circle of radius $\sqrt{2}$ has area $\pi(\sqrt{2})^2 = 2\pi$, so the area between the circles is $2\pi - \pi = \pi$. **9.** $AG = 3$, since $AC = \sqrt{5}$, $AD = \sqrt{6}$, $AE = \sqrt{7}$, $AF = \sqrt{8}$, and $AG = \sqrt{9} = 3$.
11.

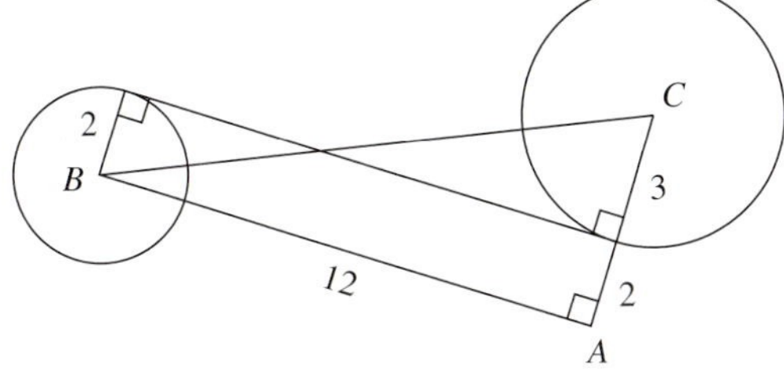

The distance between centers B and C is $\sqrt{12^2 + 5^2} = 13$. **13. (a)** $(21)^2 + (28)^2 = 1225 = (35)^2$; yes. **(c)** $(12)^2 + (35)^2 = 1369 = (37)^2$; yes. **(e)** $(7\sqrt{2})^2 + (4\sqrt{7})^2 = 210 \neq 308 = (2\sqrt{77})^2$; no. **15. (a)** Yes
19. (a)

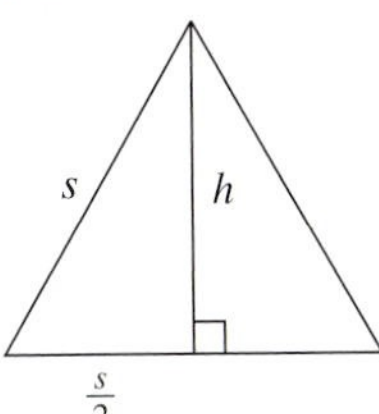

$\left(\frac{s}{2}\right)^2 + (h)^2 = s^2$, by the Pythagorean theorem.

Therefore, $h = \frac{\sqrt{3}}{2}s$. **22.** Flattening the top of the box as suggested gives the diagram shown. By the Pythagorean theorem, $AC'' = \sqrt{12^2 + 5^2} = \sqrt{169} = 13$ and $AC'' = \sqrt{9^2 + 8^2} = \sqrt{145} = 12.04\ldots$. The shortest distance is therefore about 12 inches, crossing over the 9-inch edge.

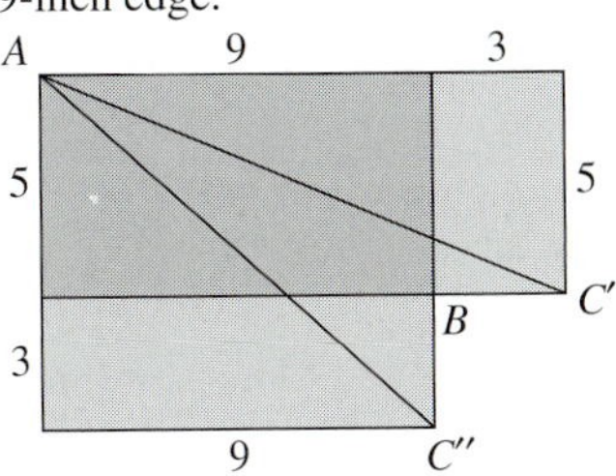

25. (a)

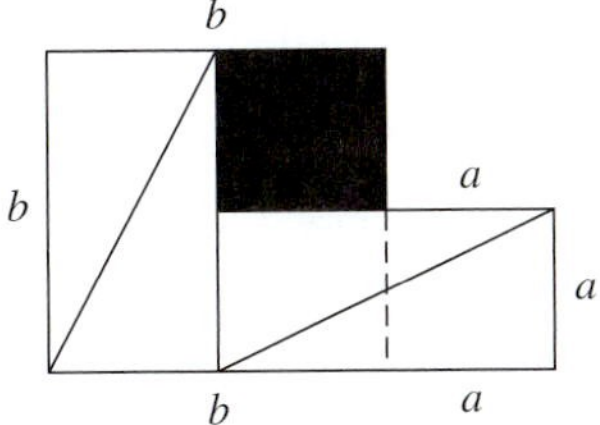

(b) Since the square and double-square are covered by the same five shapes, their areas are equal. The respective areas are c^2 and $a^2 + b^2$, so $c^2 = a^2 + b^2$.
27.

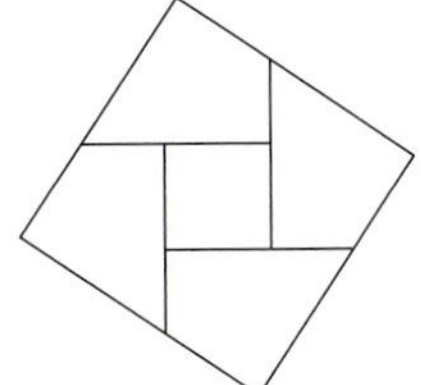

29. Answers will vary, but your ladder cannot be vertical, so the height is less than 24 feet. If the base is 7 feet from the wall, the top of the ladder is still nearly 23 feet off the ground. **31.** Let d be the depth of the pond. The stem length is $d + 2$ (in feet). With the stem held to the side, a right triangle is formed with legs of length 6 and d and hypotenuse of length $(d + 2)$. Then $6^2 + d^2 = (d + 2)^2$, so $d = 8$. **36. (a)** $d \doteq 1.2\sqrt{100} = 1.2(10) = 12$ miles **(b)** $d \doteq 1.2\sqrt{1353} \doteq 1.2(36.78) = 44.1$ miles **(c)** Yertle is approximately 1111 feet high. **37.** The sum of the areas of the equilateral triangles on the legs equals the area of the equilateral triangle on the hypotenuse. **39.** Let the original dimensions be l and w. The changed dimensions are $\frac{4}{3}l$ and $\frac{3}{4}w$. So area $= \left(\frac{4}{3}l\right)\left(\frac{3}{4}w\right) = lw$, the same as the original area. **41.** Along the large semicircle: $\frac{1}{2}(2 \cdot \pi \cdot 8\text{ m}) = 8\pi$ m; along the two smaller semicircles: $\frac{1}{2}(2 \cdot \pi \cdot 3\text{ m}) + \frac{1}{2}(2 \cdot \pi \cdot 5\text{ m}) = 8\pi$ m. The distances are the same.

Problem Set 12.4 (page 803)

2. (a) SA $= 2 \cdot \frac{1}{2}(20\text{ cm} + 15\text{ cm})(12\text{ cm}) + (2\text{ cm})(60\text{ cm}) = 540\text{ cm}^2$ **(c)** SA $= 2 \cdot \pi(15\text{ ft})^2 + 2\pi(15\text{ ft})(12\text{ ft}) = 810\pi\text{ ft}^2 \doteq 2545\text{ ft}^2$
3. (a) Slant height $= \sqrt{(40\text{ m})^2 + (30\text{ m})^2} = 50$ m; SA $= (60\text{ m})^2 + 4 \cdot \frac{1}{2}(60\text{ m})(50\text{ m}) = 9600\text{ m}^2$ **(c)** SA $= \pi(6\text{ in})^2 + \pi(6\text{ in})(15\text{ in}) = 126\pi\text{ in}^2 \doteq 396\text{ in}^2$ **4. (a)** $V = Bh = (7\text{ cm})(4\text{ cm})(3\text{ cm}) = 84\text{ cm}^3$ **(c)** $\pi \times (10\text{ m})^2 \times 4\text{ m} = 400\pi\text{ m}^3 \doteq 1257\text{ m}^3$
5. (a) $\frac{1}{3}(12\text{ ft})(8\text{ ft})(10\text{ ft}) = 320\text{ ft}^3$

(c) $\frac{1}{3} \times \pi(5\text{ cm})^2 \times (12\text{ cm}) = 100\pi\text{ cm}^3 \doteq 314\text{ cm}^3$
6. (a) $S = 4\pi(2200\text{ km})^2 = 19{,}360{,}000\pi\text{ km}^2 \doteq 6.08 \times 10^7\text{ km}^2$; $V = \frac{4}{3}\pi(2200\text{ km})^3 \doteq 4.46 \times 10^{10}\text{ km}^3$.
(c) $SA = 4\pi(4\text{ ft})^2 + (20\text{ ft})(2\pi)(4\text{ ft}) = 224\pi\text{ ft}^2 \doteq 704\text{ ft}^2$; $V = \frac{4}{3}\pi(4\text{ ft})^3 + \pi(4\text{ ft})^2(20\text{ ft}) \doteq 1273\text{ ft}^3$.
9. Area (sphere) $= 4\pi r^2$ square units. Area (cylinder) $= 2 \cdot \pi \cdot r^2 + 2\pi r \cdot (2r) = 6\pi r^2$ square units. Thus, $4\pi r^2/6\pi r^2 = 2/3$. **10. (a)** 16 cm **11. (a)** 4, considering area **(b)** One 14″ pizza is nearly the same amount of pizza, but will save \$2. **13. (a)** $1 - \frac{1}{4} - \frac{1}{4} = \frac{1}{2}$ (each small circle has $\frac{1}{2}$ the diameter, so $\frac{1}{4}$ the area)
15. (a) 200 mL (doubling the radius increases the volume by a factor of 4; halving the height halves the volume)
18. Have Eno carefully fill an oddly shaped container with an open top completely and exactly full of rice. Then have him pour the rice into a suitable rectangular box and measure the volume. **24.** Suppose the area of the similar figure with straight side of length 1 is A. By the similarity principle, the areas of the figures erected on the sides of the triangle are then a^2A, b^2A, and c^2A, where a, b, and c are the respective scale factors. Since $a^2 + b^2 = c^2$ (Pythagorean theorem), we get $a^2A + b^2A = c^2A$, showing that the sum of the areas of the figures erected on the legs is equal to the area of the figure erected on the hypotenuse. **25. (a)** 3″, since $\frac{2}{16} = \frac{1}{8} = \left(\frac{1}{2}\right)^3$ and $\frac{1}{2} \cdot 6'' = 3''$.
26. (a)

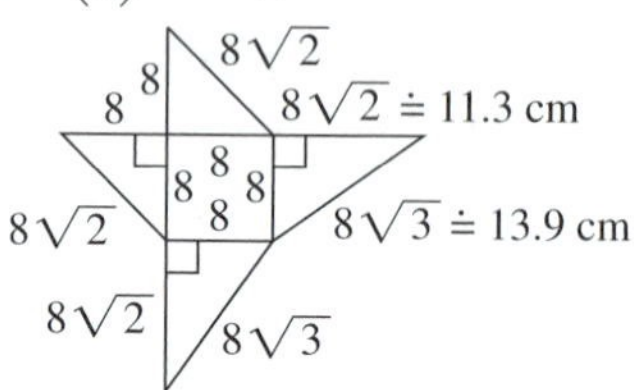

28. (a) Circumference of the cone is $\frac{3}{4} \cdot 2 \cdot \pi(4\text{ in}) = 6\pi$ in, so the radius is 3 in. **29.** Let s be the radius of the semicircle. Then the slant height of the cone is s. Let d be the diameter of the cone. Then $\pi d = \frac{1}{2}2\pi s$, so $d = s$.
30. $V(\text{ring}) = \pi\left(\frac{9}{16}\text{ in}\right)^2 \cdot \left(\frac{5}{4}\text{ in}\right) - \pi\left(\frac{1}{2}\text{ in}\right)^2\left(\frac{5}{4}\text{ in}\right) \doteq 0.26\text{ in}^3$, so about 1.56 ounces. **32.** $V(\text{box}) = 160\text{ in}^3$ and $V(\text{tub}) = \pi(3\text{ in})^2(10\text{ in}) \doteq 283\text{ in}^3$. Two boxes is a better buy. **34. (a)** $\left(\frac{5}{13}\right)^3(106.75) \doteq 6.07$ pounds
36. (a) The scale factor is 12, and volume would be increased by the factor $12^3 = 1728$. **37. (a)** $s = \frac{1}{\sqrt{2}}$, so $s^2 = \frac{1}{2}$, showing that the area of an A1 sheet is half that of an A0 sheet.
(b)

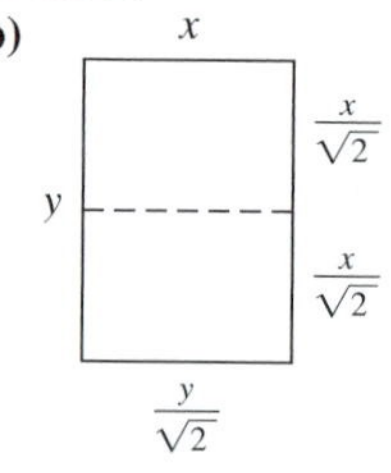

Using part (a), we see that the width x of an A0 sheet is $y/\sqrt{2}$, where y is the length. Then $xy = y^2/\sqrt{2} = 1\text{ m}^2$, so $y = 2^{1/4}\text{ m} \doteq 1.2\text{ m}$ and $x = 1/2^{1/4}\text{ m} \doteq 0.84\text{ m}$. **39.** The circular cross-sections are related by a scale factor of 3, so 9 inches of water in the cylinder corresponds to 1 inch of rainfall. **41. (a)** Yes, $30^2 + 72^2 = 6084 = 78^2$.
43. $P \doteq 8 + 2\sqrt{5} + \sqrt{2} + \sqrt{10} \doteq 17$, $A = 12$ square units **44.** $\frac{50\text{ in}}{1\text{ sec}} \cdot \frac{1\text{ ft}}{12\text{ in}} \cdot \frac{1\text{ mi}}{5280\text{ ft}} \cdot \frac{60\text{ sec}}{1\text{ min}} \cdot \frac{60\text{ min}}{1\text{ hr}} \doteq 2.84\frac{\text{mi}}{\text{hr}}$ **45. (a)** 27.8 cm

Chapter 12 Review Exercises (page 812)

1. (a) Centimeters **(b)** Millimeters **(c)** Kilometers **(d)** Meters **(e)** Hectares **(f)** Square kilometers **(g)** Milliliters **(h)** Liters **2. (a)** 4 L **(b)** 190 cm **(c)** 200 m^2 **3.** 84 L
4. $\frac{300\text{ ft}}{3\text{ sec}} \cdot \frac{1\text{ mile}}{5280\text{ ft}} \cdot \frac{60\text{ sec}}{1\text{ min}} \cdot \frac{60\text{ min}}{1\text{ hr}} \doteq 68$ miles per hour
5.

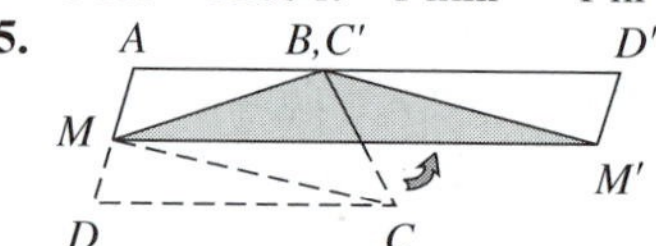

The triangle has half the area of the parallelogram $AD'M'M$, so it also has half the area of the trapezoid $ABCD$.
6. (a) A line from the center to the corner dissects the piece into two triangles of altitude 10, one with base 9 inches and the other with base 7 inches. The area of the piece is

$$\frac{1}{2} \cdot 7 \cdot 10 + \frac{1}{2} \cdot 9 \cdot 10 = \frac{1}{2} \cdot 16 \cdot 10 = 80\text{ in}^2,$$

which is one fifth of the total area, $20^2\text{ in}^2 = 400\text{ in}^2$, of the cake. **(b)** The reasoning shown in part (a) demonstrates that any piece with 16 inches along the perimeter has area 80 in^2. Therefore, the cuts are arranged as shown in the figure.

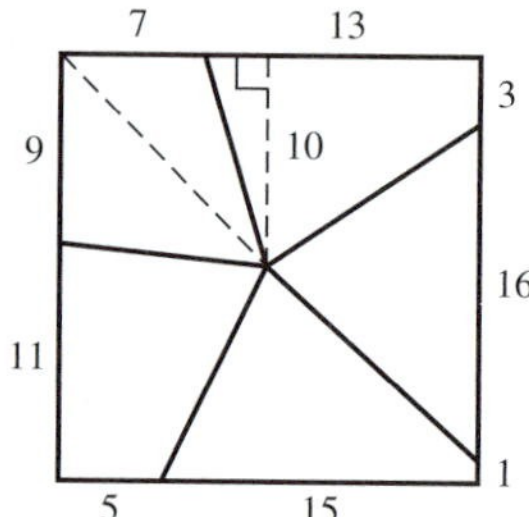

7. (a) $768 \text{ in}^2 = 5\frac{1}{3} \text{ ft}^2$ **(b)** $\frac{1}{2}(8 \text{ m})(9 \text{ m}) + \frac{1}{2}(3 \text{ m}) \cdot (6 \text{ m}) = 45 \text{ m}^2$ **(c)** $\frac{1}{2}(5 \text{ cm} + 7 \text{ cm})(3 \text{ cm}) = 18 \text{ cm}^2$

8. (a) 11 square units **(b)** 9 square units **(c)** $9\frac{1}{2}$ square units **9. (a)** $A = (3 \text{ ft})(4 \text{ ft}) + \frac{1}{2}\pi(1.5 \text{ ft})^2 \doteq 15.5 \text{ ft}^2$, $P = 11 \text{ ft} + \pi(1.5 \text{ ft}) \doteq 15.7 \text{ ft}$ **(b)** $A = \frac{3}{4}\pi(3 \text{ m})^2 \doteq 21.2 \text{ m}^2$, $P = \frac{3}{4} \cdot 2\pi(3 \text{ m}) + 6 \text{ m} \doteq 20.1 \text{ m}$ **10.** $x = 6$, $y = \sqrt{5}$ (*Methods to solve:* Find the area via one of two ways, or use similar triangles.) **11.** $\sqrt{1125} \text{ cm} \doteq 33.5 \text{ cm}$ **12.** $\sqrt{116}$ in, $\sqrt{160}$ in, $\sqrt{244}$ in, $\sqrt{260}$ in **13.** $9 + \sqrt{2} + \sqrt{10} + \sqrt{5} \doteq 15.8$ units **14. (a)** $V = [(10 \text{ ft})(20 \text{ ft}) + \frac{1}{2}(8 \text{ ft} + 20 \text{ ft})(8 \text{ ft})](30 \text{ ft}) = 9360 \text{ ft}^3$; $\text{SA} = 2 \cdot \frac{1}{2}(8 \text{ ft} + 20 \text{ ft})(8 \text{ ft}) + 2 \cdot (10 \text{ ft})(20 \text{ ft}) + 2 \cdot (10 \text{ ft})(30 \text{ ft}) + 2 \cdot (10 \text{ ft})(30 \text{ ft}) + (8 \text{ ft})(30 \text{ ft}) + (20 \text{ ft})(30 \text{ ft}) = 2664 \text{ ft}^2$

(b) $V = \pi(7 \text{ m})^2(18 \text{ m}) + \frac{1}{2} \cdot \frac{4}{3}\pi(7 \text{ m})^3 \doteq 3489 \text{ m}^3$; $\text{SA} = \frac{1}{2} \cdot 4\pi(7 \text{ m})^2 + 2\pi(7 \text{ m})(18 \text{ m}) + \pi(7 \text{ m})^2 \doteq 1253 \text{ m}^2$ **(c)** $V = \frac{1}{3}\pi(5 \text{ cm})^2(8 \text{ cm}) + \frac{1}{2} \cdot \frac{4}{3}\pi(5 \text{ cm})^3 \doteq 471 \text{ cm}^3$; $\text{SA} = \frac{1}{2} \cdot 4\pi(5 \text{ cm})^2 + \pi(5 \text{ cm})(\sqrt{25 + 64} \text{ cm}) = \pi(5 \text{ cm})(\sqrt{25 + 64} \text{ cm}) = (50 + 5\sqrt{89})\pi \text{ cm}^2 \doteq 305 \text{ cm}^2$

15. $V(\text{sphere}) = \frac{4}{3}\pi(10 \text{ m})^3$ and $V(\text{four cubes}) = 4(10 \text{ m})^3$. Since $\pi > 3$, $\frac{4}{3}\pi > 4$, showing that the sphere has the larger volume. **16. (a)** $(180 \text{ ft})(1.5) = 270 \text{ ft}$, since $k = 1.5$ is the scale factor. **(b)** $\frac{45 \text{ pounds}}{(1.5)^2} = 20$ pounds, since Johan's garden area is $(1/1.5)^2$ times that of Heather's.

Chapter 12 Test (page 814)

1. (a) 8 cm^2 **(b)** 8 cm^2 **(c)** 8.5 cm^2 **2.** Each figure contains four full units of area. Figure A contains an additional eight half-units of area, and figure B contains an additional four half-units of area. Jim should conclude that Figure A is four half-units, or two units, of area larger than B.

3. (a) $\text{SA} = 2 \cdot \frac{1}{2}(7 \text{ m})(24 \text{ m}) + (7 \text{ m} + 24 \text{ m} + 25 \text{ m}) \cdot (5 \text{ m}) = 448 \text{ m}^2$, since the diagonal is 25 m; $V = \frac{1}{2}(7 \text{ m})(24 \text{ m})(5 \text{ m}) = 420 \text{ m}^3$. **(b)** $\text{SA} = 2 \cdot \frac{1}{2}\pi(4'')^2 + \frac{1}{2} \cdot 2\pi(4'')(6'') + (6'')(8'') \doteq 173.7 \text{ in}^2$; $V = \frac{1}{2}\pi(4'')^2(6'') \doteq 150.8 \text{ in}^3$. **(c)** Slant height is 10 ft. $\text{SA} = 4 \cdot \frac{1}{2}(12 \text{ ft})(10 \text{ ft}) + (144 \text{ ft})^2 = 384 \text{ ft}^2$; $V = \frac{1}{3}(12 \text{ ft})^2(8 \text{ ft}) = 384 \text{ ft}^3$. **(d)** Slant height is 13 cm. $\text{SA} = \pi(5 \text{ cm})^2 + \pi(5 \text{ cm})(13 \text{ cm}) = 90\pi \text{ cm}^2 \doteq 283 \text{ cm}^2$; $V = \frac{1}{3}\pi(5 \text{ cm})^2(12 \text{ cm}) = 100\pi \text{ cm}^3 \doteq 314 \text{ cm}^3$.

4. (a) mm **(b)** m **(c)** m **(d)** km **(e)** mL **(f)** liters

5. Cross-sectional view:

$(r - 4)^2 + (5)^2 = r^2$, or $r^2 - 8r + 16 + 25 = r^2$. Then $r = (16 + 25)/8 \text{ mm} = 5.125 \text{ mm}$.

6. (a) $P = \sqrt{52} + \sqrt{13} + \sqrt{65} \doteq 18.9$ units **(b)** Yes, $(\sqrt{52})^2 + (\sqrt{13})^2 = 65 = (\sqrt{65})^2$.

7. (a) 216.1 cm **(b)** 168,200 cm **(c)** 5000 cm^2 **(d)** $10{,}000 \text{ m}^2$ **(e)** 4.719 L **(f)** 3200 cm^3

8. Let the radius of the circle be r. Then the circumscribed square has sides of length $2r$ and the inscribed square has sides of length $\sqrt{2}r$.

$$\frac{\text{area (inscribed)}}{\text{area (circumscribed)}} = \frac{(\sqrt{2}r)^2}{(2r)^2} = \frac{2r^2}{4r^2} = \frac{1}{2}$$

Alternative solution: The scale factor of the large to the small square is $1/\sqrt{2}$, so the small square has $(1/\sqrt{2})^2 = 1/2$ the area of the large square.

9. $\sqrt{189} \text{ ft} \doteq 13.7 \text{ ft}$, by the Pythagorean theorem.

10. (a) Approximately 31.86 yd **(b)** Approximately 1.5 mi **(c)** 291.6 ft^2 **(d)** Approximately 14.69 mi^2 **(e)** 205.2 ft^3 **(f)** Approximately 3.45 ft^3

11.

	Papa	Mama	Baby
Length of Suspenders	50	40	20
Weight	468.75	240	30
Number of Fleas	6000	3840	960

Scale Factors	
PB to MB	4/5
MB to BB	1/2

12. $A = \frac{1}{2}(9 \text{ ft})(12 \text{ ft}) + (8 \text{ ft})(12 \text{ ft}) - \frac{1}{2}\pi(4 \text{ ft})^2 \doteq 125 \text{ ft}^2$; $P = 15 \text{ ft} + 9 \text{ ft} + 8 \text{ ft} + 12 \text{ ft} + \frac{1}{2} \cdot 2\pi(4 \text{ ft}) \doteq 56.6 \text{ ft}$

13. (a) $\frac{2}{3}\pi(5 \text{ in})^2 \doteq 52.4 \text{ in}^2$ **(b)** $\frac{1}{2}(2.6 \text{ m} + 1.4 \text{ m})(3 \text{ m}) = 6.0 \text{ m}^2$ **(c)** $\frac{1}{12} \cdot \pi(24 \text{ cm})^2 \doteq 151 \text{ cm}^2$

14. $A = \frac{1}{2} \cdot (6 \text{ cm}) \cdot (16 \text{ cm}) + \frac{1}{2}(15 \text{ cm})(16 \text{ cm}) = 168 \text{ cm}^2$. Sides are 10 cm and 17 cm by the Pythagorean theorem, so $P = 54$ cm.

15.

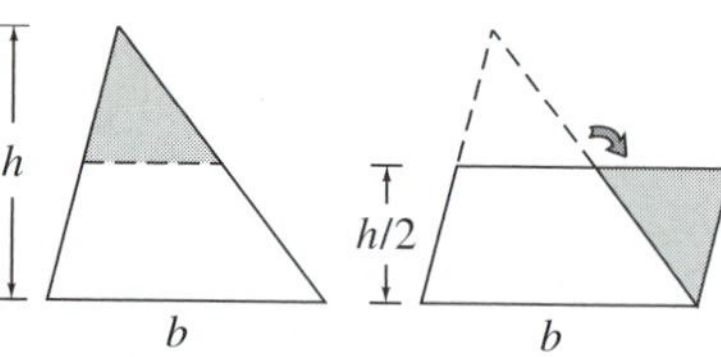

area (triangle) = area (parallelogram) $= (b)(h/2) = \frac{1}{2}bh$

16. (a) $\text{SA} = 4\pi(10 \text{ m})^2 = 400\pi \text{ m}^2 \doteq 1257 \text{ m}^2$; $V = \frac{4}{3}\pi(10 \text{ m})^3 \doteq 4189 \text{ m}^3$. **(b)** $\text{SA} = \pi(5 \text{ cm})^2 + 2\pi(5 \text{ cm})(6 \text{ cm}) + \frac{1}{2} \cdot 4\pi(5 \text{ cm})^2 = 135\pi \text{ cm}^2 \doteq 424 \text{ cm}^2$; $V = \pi(5 \text{ cm})^2(6 \text{ m}) + \frac{1}{2} \cdot \frac{4}{3}\pi(5 \text{ cm})^3 \doteq 733 \text{ cm}^3$.

17. $V(\text{peel}) = \frac{4}{3}\pi(2.5 \text{ in})^3 - \frac{4}{3}\pi(1.75 \text{ in})^3 \doteq 43.0 \text{ in}^3$, and $V(\text{grapefruit}) = \frac{4}{3}\pi(2.5 \text{ in})^3 \doteq 65.4 \text{ in}^3$. About 66 percent is peel. *Alternative solution:* The scale factor is $\left(2\frac{1}{2} - 3/4\right)\Big/2\frac{1}{2} = 0.7$. Since $(0.7)^3 = 0.343$, it follows that 34.3 percent of the grapefruit is not peel, and 65.7 percent is peel.

Chapter 13

Problem Set 13.1 (page 838)

1. (a) Not a rigid motion; distances between particular cards will change. **(c)** No, distances between particular pieces almost certainly will have changed.

2. (a)

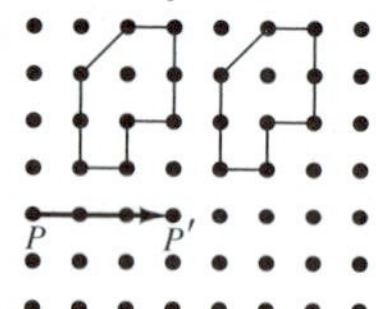

4. (a) 300° **(c)** 43°

5. (a)

6. (a) Center O is the intersection of $\overleftrightarrow{AB}$ and $\overleftrightarrow{A'B'}$

9. (a) Draw a vertical line through the midpoint of $\overline{PP'}$.

11. (a), (b)

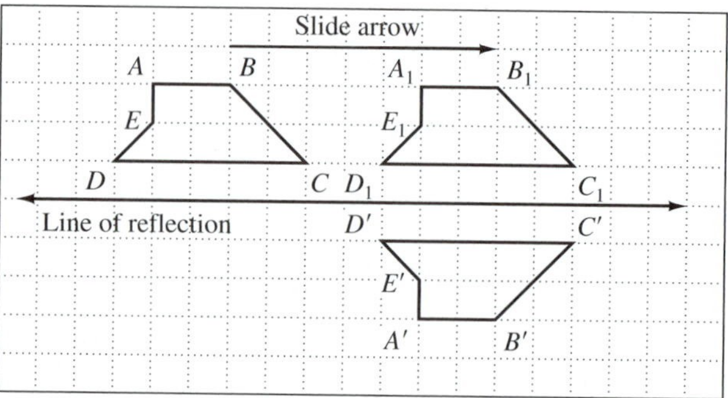

13. (a)

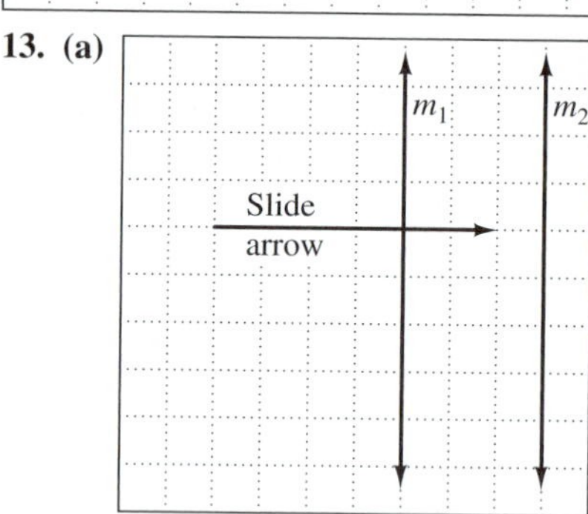

15. (a)

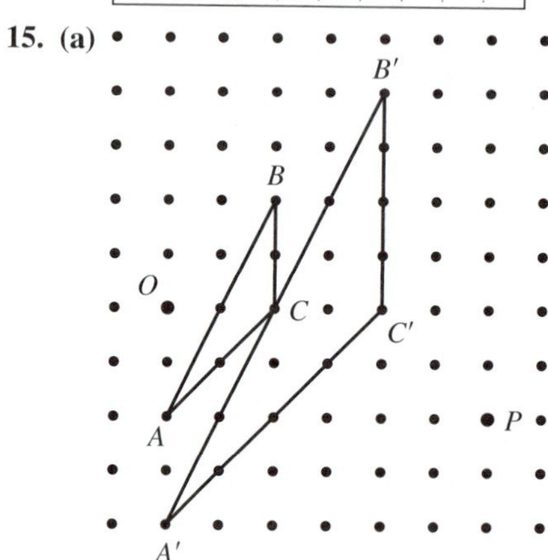

(b)

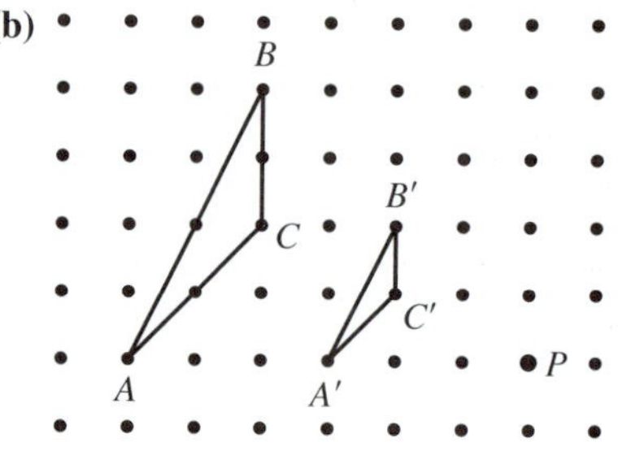

16. (a) Center P, scale factor $\frac{3}{2}$

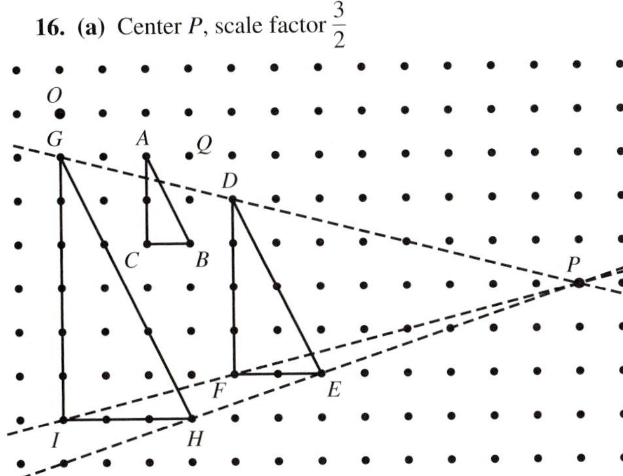

18. Rotate 90° counterclockwise about B. Then perform a size transformation centered at P with scale factor 2. (Other sequences will also work.)

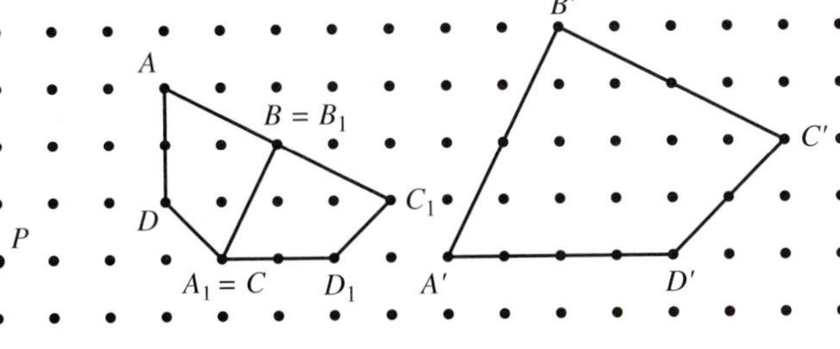

21. (a) P' is some point on the circle of radius 2 cm centered at A'.

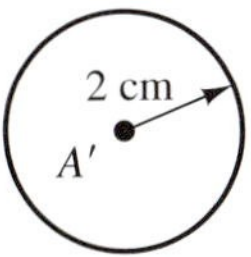

22. On a 1-cm-square grid, the two 90° rotations take O_1 to O'_1 and O_2 to O'_2. This motion is equivalent to a 180° rotation about the point O.

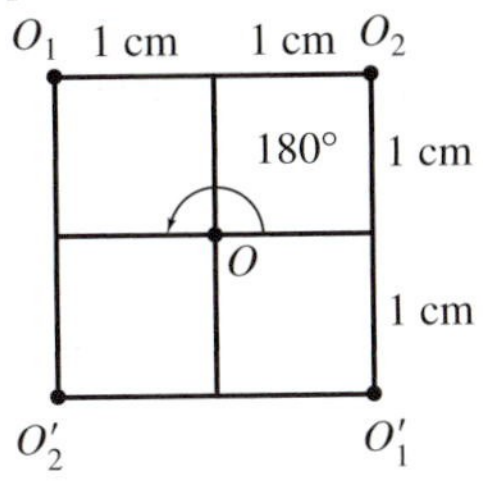

24. (a), (b)

26. (a)

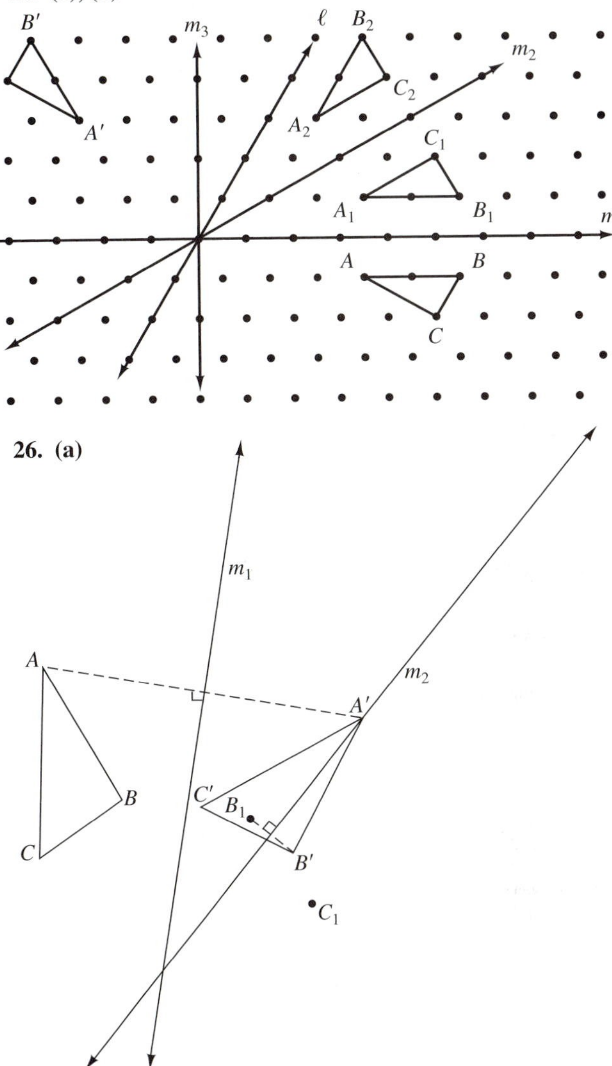

(b) C' **(c)** Reflection first across line m_1 and then across line m_2 is equivalent to a rotation about the point P of intersection of m_1 and m_2, through an angle twice the measure x of the directed angle from line m_1 toward line m_2.

28. (a) A translation: Six reflections give an orientation-preserving rigid motion, so it is either a rotation or a translation. Since a rotation has a fixed point (namely, the rotation center), the motion is a translation.

30. (a) A glide reflection **31.** The line PP' passes through O, so construct-ing this line determines point O. $\overline{PQ}$ and $\overline{P'Q'}$ are parallel, so the line through P' that is parallel to $\overline{PQ}$ will determine Q'.

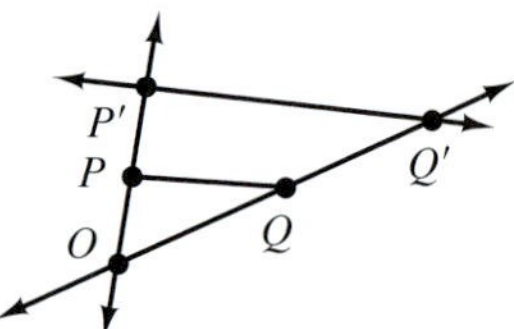

34. (a) $\overline{RS'}$ is the reflection of $\overline{RS}$ across m, so $RS' = RS$, since a reflection preserves all distances. Similarly, $QS' = QS$. **36.** A single mirror reverses orientation, so the double reflection seen in a corner mirror preserves orientation. The corner-mirror reflection of your right hand will appear as a right hand.

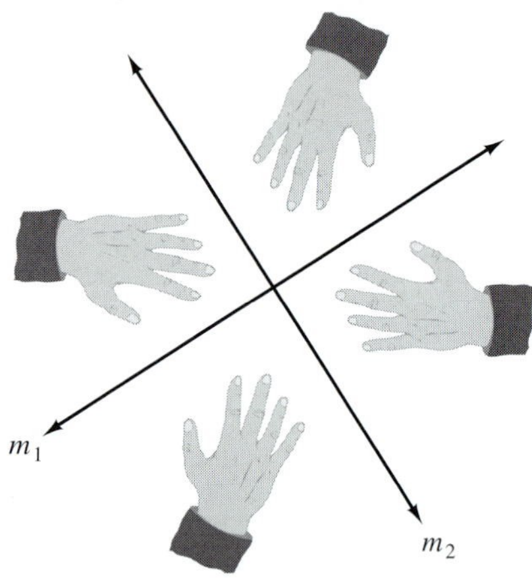

40. (a) Measurements show that triangle XYZ is an equilateral triangle. **(b)** The centers of the rotated equilateral triangles form an equilateral triangular grid. **45.** $2x + 3x + 4x = 180$, so $x = 20$, giving angles of measure 40°, 60°, and 80°. **47.** Midpoints form a 2 by 4 rectangle $KLMN$, for the lattice rhombus $ABCD$.

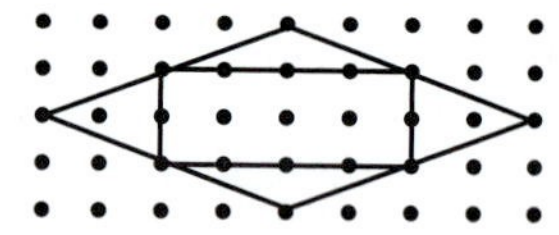

Problem Set 13.2 (page 853)

1. (a) **(c)**

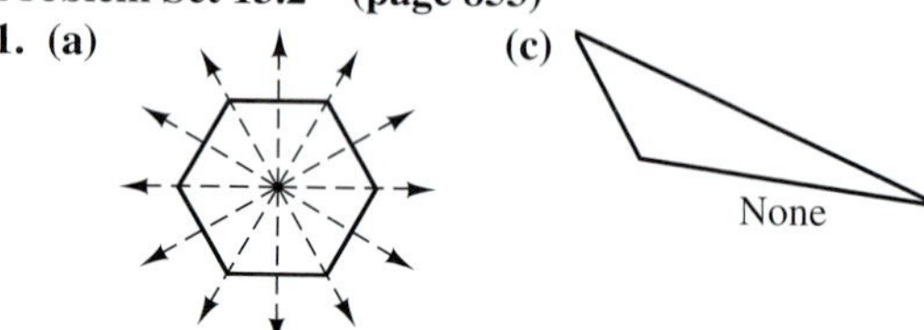

3. (a) Many figures are possible.

4. (a) a kite **(b)** an isosceles trapezoid

5. (a)

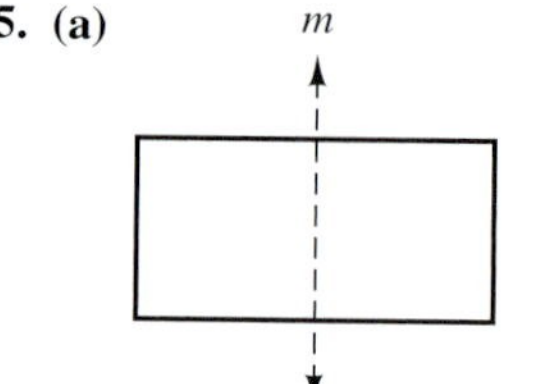

(c)

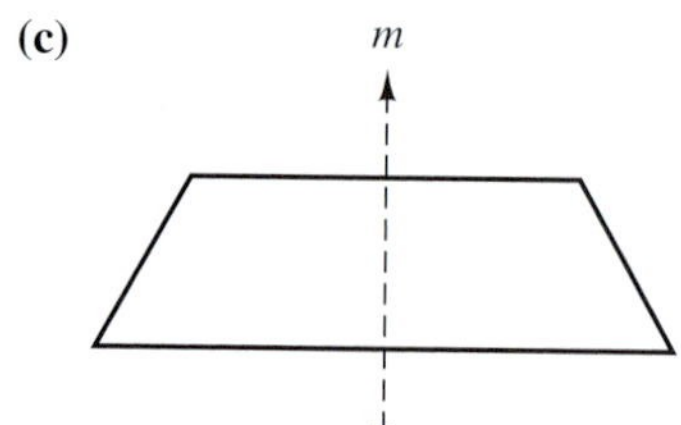

8. (a)

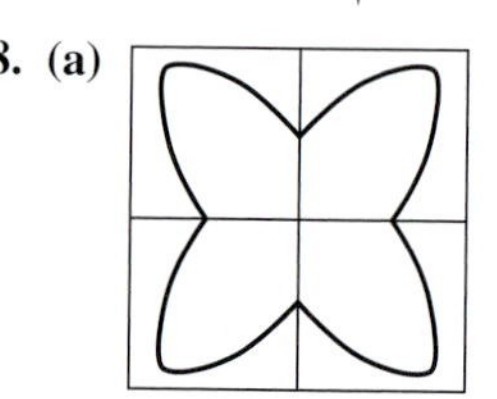

9. (a) One line of symmetry **10. (a)** Five lines of symmetry and 72° rotation symmetry **(b)** 72° rotation symmetry

11. (a)

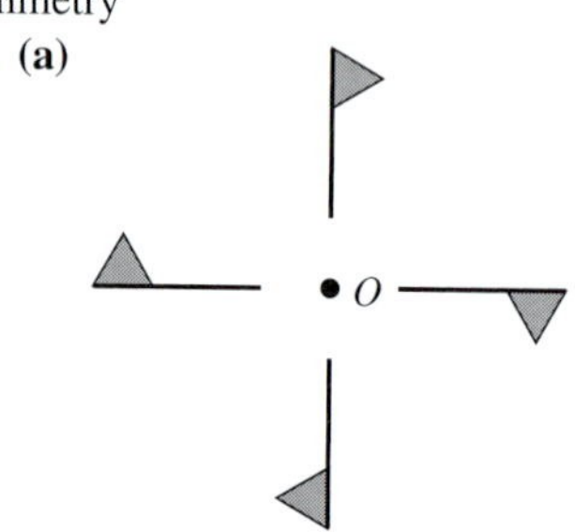

12. (a) Equilateral triangle **(b)** Square **13. (a)** 0, 8 **(b)** 0, 3, 8 **(c)** 0, 8 **(d)** 0, 8 **16. (a)** $m1$ **17. (a)** There are vertical lines of symmetry through the center of each letter, and there is a horizontal line of symmetry. The symbol type is mm. **21. (a)** No letter or digit reflects vertically onto a different letter or digit, so it must reflect onto itself. **23.** The pattern must have a horizontal line of symmetry, so it would be an mm pattern. **25. (a)** $1g$ **(c)** mg

27. (a) 11

p	p	p	p	p	p

(c) 12

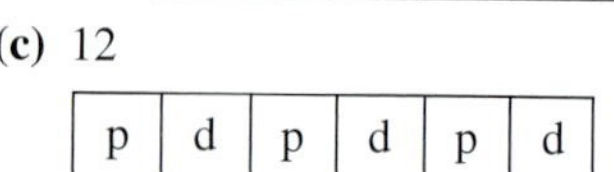

p	d	p	d	p	d

29. (a) Three directions of reflection symmetry; three directions of glide symmetry; 120° rotation symmetry

30. (a) 3 **31. (a)** As left-handed people know well, not all scissors are symmetric. **(c)** A man's dress shirt is not quite symmetric, since it buttons right handed. **(e)** Tennis rackets have two planes of bilateral symmetry. **32. (a)** Across the line of diagonal entries **42. (a)** 350°

43. (a)

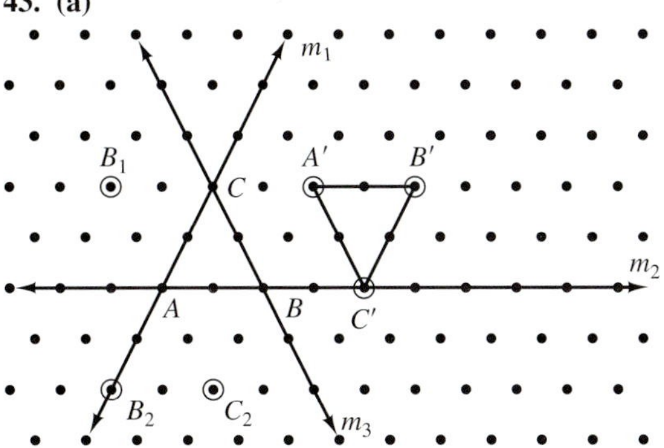

(b) Glide reflection, three units to the right and reflecting across the line l parallel to $\overline{AB}$ and midway between C and $\overline{AB}$.

Problem Set 13.3 (page 871)

1. Many different tilings can be formed. For example:

(a)

2. Yes, tilings are possible. For example:

(a)

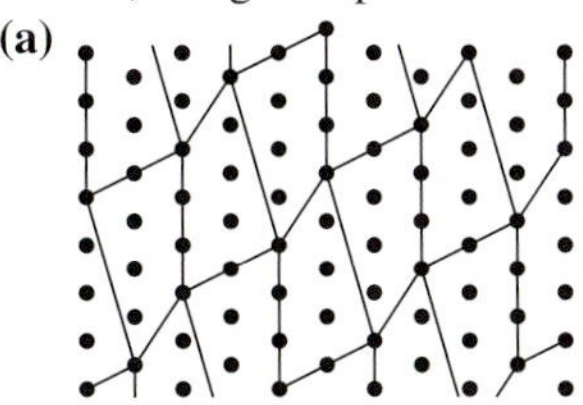

4. (a) Squares, pentagons, hexagons, heptagons, octagons **(b)** Many vertex figures that appear in the tiling cannot correspond to regular polygons. For example, regular 5-, 6- and 8-gons have interior angles of measure 108°, 120°, and 135°. These add up to 108° + 120° + 135° = 363° ≠ 360°.
18. Here's one way to make a tiling 12-gon, by modifying a square tile.

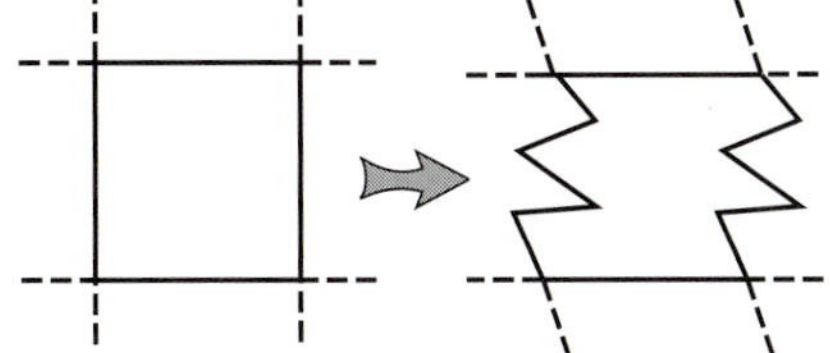

Similarly, modifying an equilateral triangle by a midpoint modification with opposite parallel congruent sides will form an 11-gon that tiles as shown.

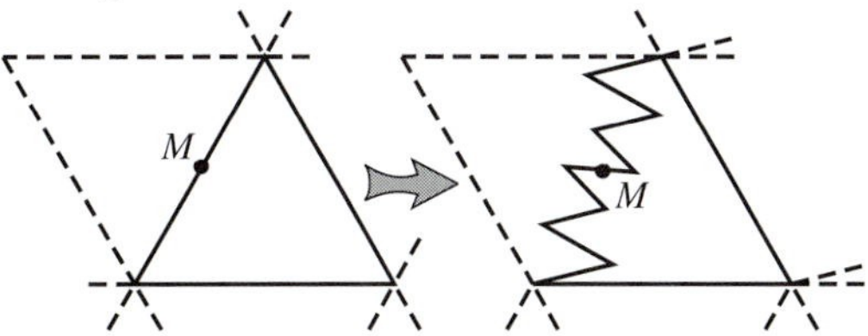

The same idea can be used for any $n \geq 6$, using a modified square for even n and a modified equilateral triangle for odd n.

20. (a) **(c)**

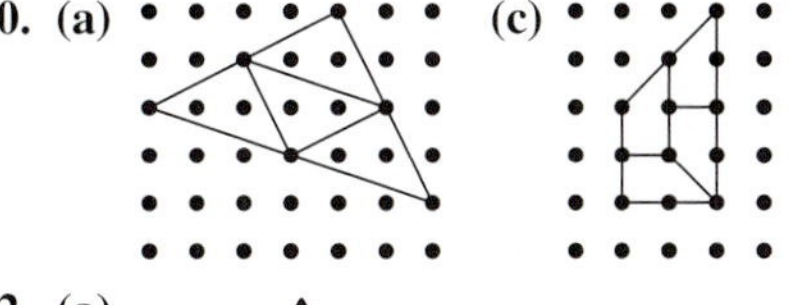

22. (a)

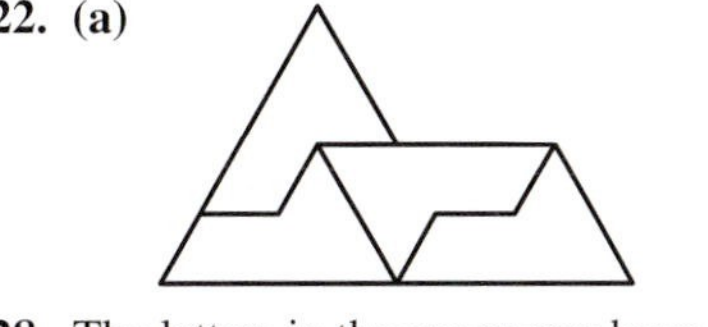

28. The letters in the upper row have either rotational or mirror symmetry, unlike those in the second row.
29. (a) scalene triangle **(b)** isosceles but not equilateral triangle

Chapter 13 Review Exercises (page 878)

1.

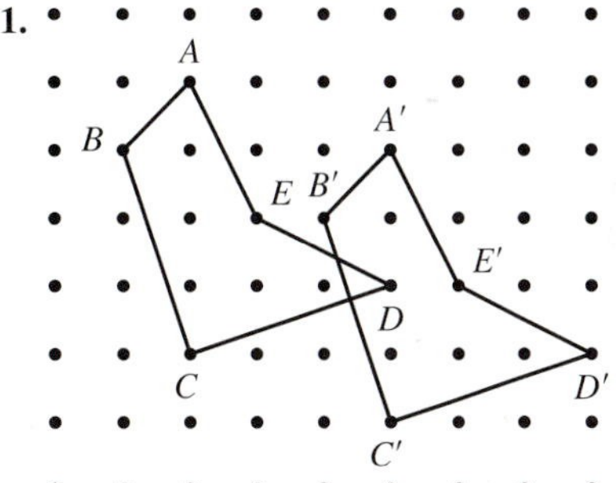

2. Find the perpendicular bisectors of segments $\overline{AA'}$ and $\overline{BB'}$. Their intersection is the turn center O, and the measure of $\angle AOA'$ is the turn angle. **3.** Reflection across line l, the perpendicular bisector common to all three segments AA', BB', and CC'.

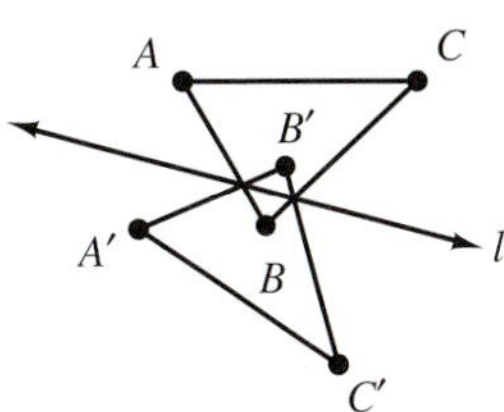

4. Draw any two vertical lines 2 inches apart. There are then three ways to choose the successive lines of reflection.

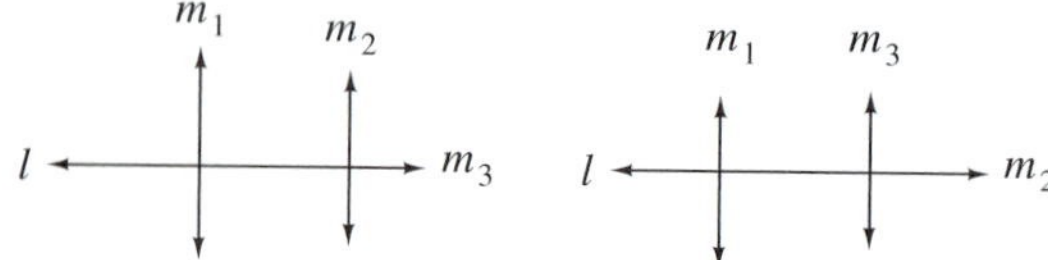

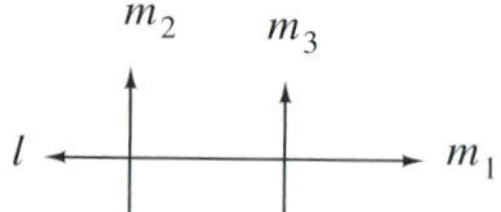

5.

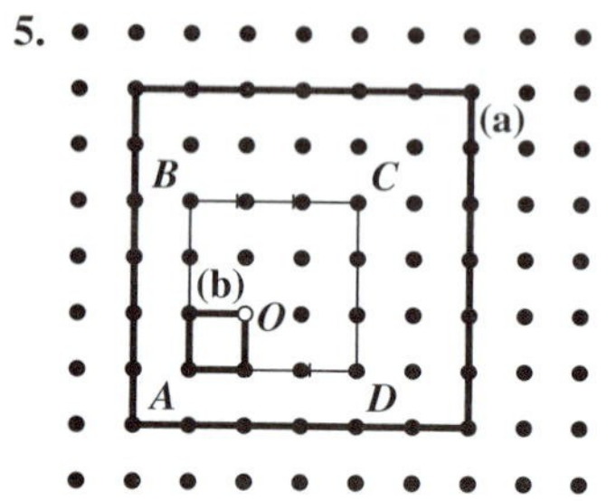

6. Rotate $ABCD$ 45° about the center P of the square. Then do a dilation about P with scale factor $\frac{\sqrt{2}}{2}$. **7. (a)** 1 **(b)** 2 **(c)** 0 **(d)** 0 **(e)** 3 **(f)** All lines through the center point, since the figure has circular symmetry **8. (a)** None **(b)** 180° **(c)** 180° **(d)** 72°, 144°, 216°, 288° **(e)** 120°, 240° **(f)** Any angle **9. (a)** Vertical line of symmetry **(b)** Horizontal line of symmetry **10.** The angles are 60° for the triangle, 90° for the square, and 120° for the hexagon. The angles of the four polygons must add up to 360°, so the fourth angle is 90°, and therefore the fourth polygon is a square.

11.

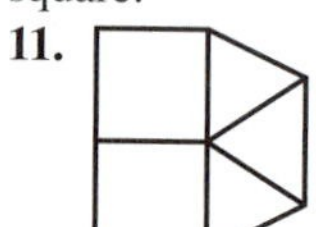

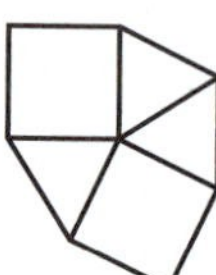

12. Use 180° rotations of the tile about the midpoints of the sides.

Chapter 13 Test (page 879)

1. (a), (b), (c), (d), and (e) tile the plane.
2. Various pairs of lines are possible. The distance between the lines must be half the distance between P and P'.

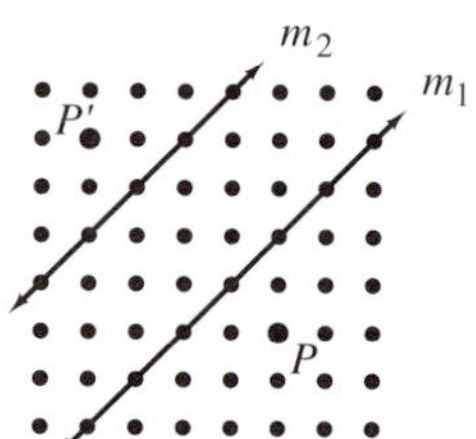

3. C' is two units right of B'; B is two units left of C.
4. (a) Rotation **(b)** Rotation **5. (a)** Glide reflection **(b)** Reflection **(c)** Reflection **(d)** Glide reflection
6. (a) Glide reflection (1*g*) **(b)** Vertical and horizontal lines of symmetry (*mm*), and glide and half-turn symmetries **(c)** Half-turn symmetry (12) **(d)** Half-turn symmetry, vertical line of symmetry, glide reflection (*mg*) **7.** Many transformations are possible. Here is one sequence: translate the square so A is taken to A'; rotate about A' by 45°; perform a dilation about A' with scale factor $3\sqrt{2}/2$ (since $A'B' = 3\sqrt{2}$ and $AB = 2$).
8. (a) Center P and rotation angle 90°
(b)

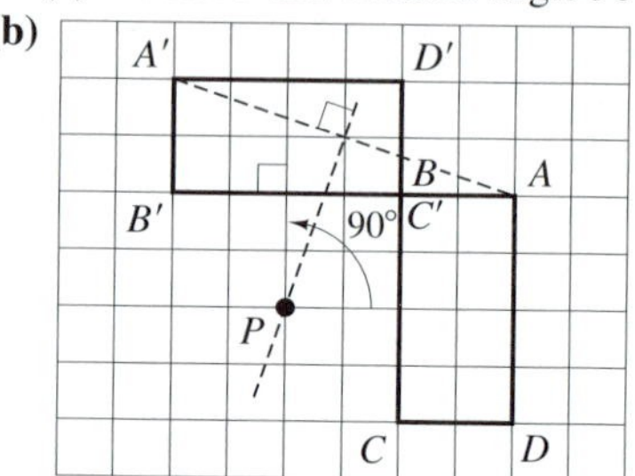

9. A vertex figure uses two octagons and one square.
10. Construct the perpendicular bisector of $\overline{AA'}$, and suppose it intersects the sides of $\angle A$ at points labeled B and C. Then $\angle BA'C$ is the desired angle.

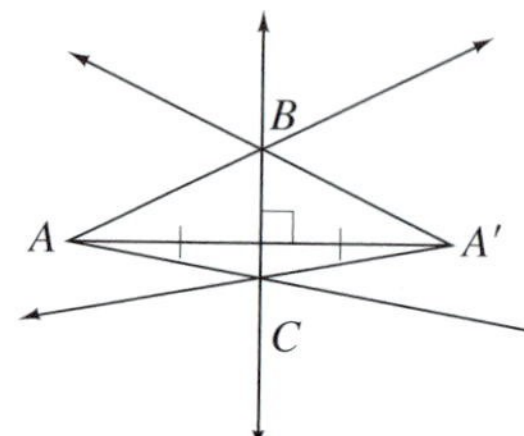

11. (a) A **(b)** $\frac{2}{3}$ **(c)** $\frac{40}{3}$ **(d)** 9
12. Many pairs of lines are possible. The two lines need to intersect at point O, and the directed angle between the lines should be half the measure of $\angle QOQ'$.

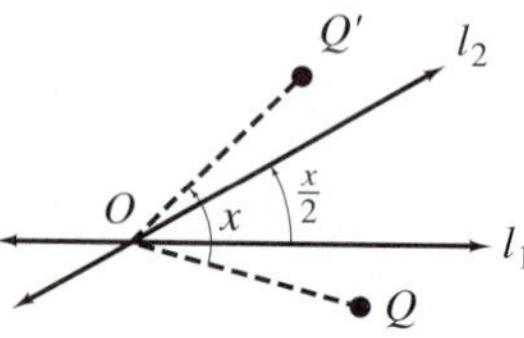

13. (a) Point (180° rotation) symmetry, and two diagonal lines of symmetry **(b)** Four lines of symmetry, and 90° rotation symmetry

14.

15. (a) **(b)**

(c)

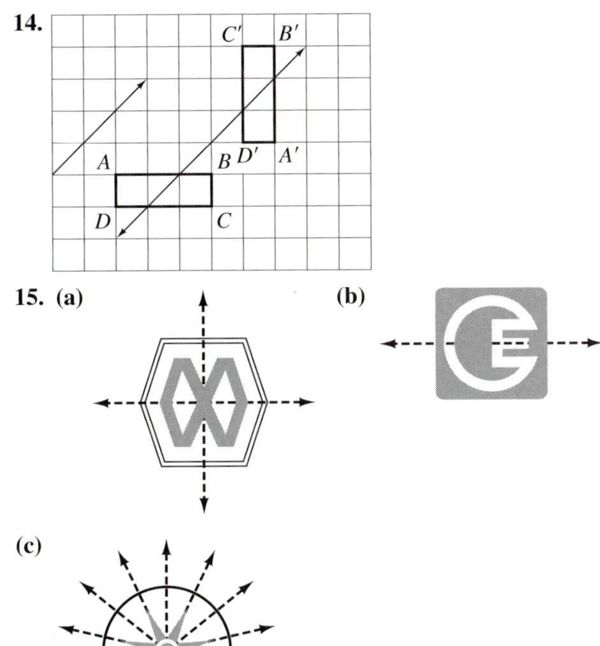

16. (a) 180° **(b)** None **(c)** 360°/7

Chapter 14

Problem Set 14.1 (page 899)

1. (a) $L \leftrightarrow K, H \leftrightarrow W, S \leftrightarrow T$ **(b)** $\overline{KW}, \overline{WT}, \overline{TK}$ **(c)** $\angle K, \angle W, \angle T$ **(d)** $\triangle KWT$ **3. (a)** Draw a line l and mark any point as E. Set your compass to AB, and determine a point G on your line segment with an arc centered at E. Set your compass to distance CD and draw an arc centered at G to determine F on l, away from E. **(b)** Begin as in (a). Set your compass to CD and draw an arc centered at G to determine F on l, back toward E. **6. (a)** One such triangle **(c)** Impossible by the triangle inequality, since $2 + 5 < 8$ **(e)** One such triangle **7. (c)** $\triangle ABC \cong \triangle EFD$ by SSS **(e)** $\triangle ABD \cong \triangle ACD$ by ASA **(g)** No conclusion possible **(i)** No conclusion possible **8.** Let $\triangle ABC$ be equilateral. Since $AB = BC$, it follows from the isosceles triangle theorem that $\angle A \cong \angle C$. In the same way, since $BC = CA$, it follows that $\angle B \cong \angle A$. Thus, all three angles are congruent. **11. (a)** $\triangle QPT \cong \triangle SPT$ by the SSS congruence property **12. (a)** $\angle ABD \cong \angle CDB$, as alternate interior angles between parallel lines. Likewise, $\angle ADB \cong \angle CBD$, and $DB = BD$. By ASA, $\triangle ABD \cong \triangle CDB$. **16.** Place a corner of the rectangular sheet of paper at a point C on the circle, and mark the points A and B where the edges of the paper cross the circle. By the converse of Thales' theorem, $\overline{AB}$ is a diameter of the circle. Repeating the procedure at a second point C' will allow you to construct a second diameter $\overline{A'B'}$. The center of the circle is where the two diameters intersect.

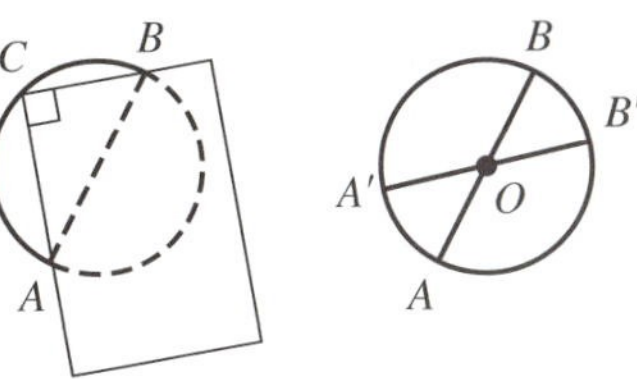

17. It is longer than 5 cm and shorter than 13 cm. (Remember the triangle inequality.) **18. (a)** $0 < s < 14$ cm, where s is the length of the fourth side **21. (a)** $\triangle ABE$ is isosceles, with $AB = BE$, so the base angles are congruent by the isosceles triangle theorem. **(b)** $\triangle ACD$ is isosceles, so its base angles are congruent. **(c)** $\angle ACB$ and $\angle ACD$ are supplementary, as are $\angle ADC$ and $\angle ADE$. Since $\angle ACD \cong \angle ADC$, we get $\angle ACB \cong \angle ADE$. By using $\angle B \cong \angle E$ and $AC = AD$, the AAS property gives $\triangle ABC \cong \triangle AED$. **(d)** From part (c), $\overline{BC} \cong \overline{ED}$. **23. (a)** Yes, using ASA **(c)** Yes, using SAS **24.** If $AB = CD = a$, $BC = AD = b$, and $AC = BD = c$, then each face of the tetrahedron is a triangle with sides of length a, b, and c. By the SSS property, the faces of the tetrahedron are congruent to one another. **25. (a)** By the triangle inequality, $QP + QT > TP$. Therefore, $QP + QT + TR > TP + TR$. But $QT + TR = QR$, so $QP + QR > TP + TR$. **26. (a)** $EA + ED > DA$ by the triangle inequality **28. (a)** The angles at the vertices of a quadrilateral can change even though the lengths of the sides are fixed. (There is no "SSSS congruence property" for a quadrilateral.) **29. (a)** The framework forms a parallelogram, but not necessarily a rectangle. **32. (a), (b)** The measure of $\angle APB$ is always a constant, always satisfying the equation $m(\angle AOB) = 2 \cdot m(\angle APB)$. **(c)** Draw the diameter $\overline{PQ}$. There are two cases to consider.
Case 1:

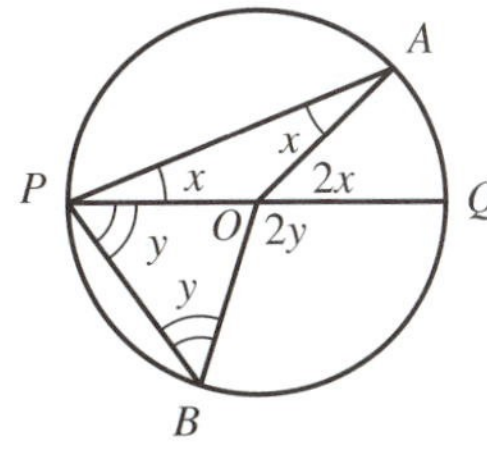

A and B lie on opposite sides of $\overline{PQ}$. We see that $\triangle POA$ is isosceles, so the base angles are congruent. That is, $m(\angle APO) = m(\angle OAP) = x$. Therefore, $m(\angle AOQ) = 2x = 2m(\angle APO)$. By the same reasoning, $m(\angle BOQ) = 2y = 2m(\angle BPO)$. Thus, $m(\angle AOB) = 2x + 2y = 2m(\angle APB)$.

Case 2:

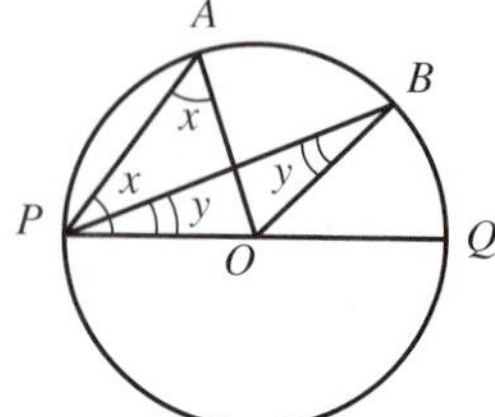

A and B are on the same side of $\overline{PQ}$. Nearly the same analysis holds, but now $m(\angle AOB) = 2x - 2y = 2m(\angle APB)$. **36. (a)** The motion is either a translation or a rotation, since only these motions keep the picture side of the piece facing upward. **(b)** The motion must flip the puzzle piece over, so it is either a reflection or a glide reflection.

Problem Set 14.2 (page 918)

1. (a) Step 1: Draw a line through P that intersects l. Label the intersection point A.

Step 2: Draw arcs of equal radius centered at A and P. Label as B the intersection point of the arc at A with $\overleftrightarrow{AP}$. Label as C the intersection point of the arc at A with l. Label as D the intersection point of the arc at P with $\overleftrightarrow{AP}$.

Step 3: Set the compass to radius BC, and draw the arc centered at D. Label as E the intersection with the arc drawn at P.

Step 4: Construct the line k through P and E.

(b) The construction gives the congruence of the corresponding angles, $\angle PAC \cong \angle DPE$. Therefore, $k \parallel l$ by the corresponding-angles property. **3. (a)** The corresponding-angles property guarantees that m is parallel to l. **(b)** Align the ruler with the line; slide the drafting triangle, with one leg of the right triangle on the ruler, until the second leg of the triangle meets point P. **7.** The jar lid will also give the circle through A, B, and C. When it is drawn, each of the four points A, B, C, and P is the intersection of three of the circles. **8.** Construct the perpendicular to side $\overrightarrow{BA}$ at T and the angle bisector. Let O be their point of intersection. The circle centered at O and passing through T is the desired circle.

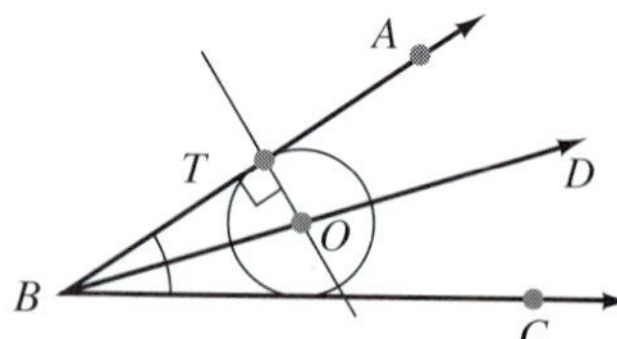

9. (a) The circumcenter will be inside the acute triangle. **(b)** The circumcenter will be at the midpoint of the hypotenuse of a right triangle. **(c)** The circumcenter will be outside an obtuse triangle. **11.** Since $\triangle PQS$ is inscribed in the circle with diameter $\overline{PQ}$, it has a right angle at S by Thales' theorem. Thus, $\overline{PS} \perp \overline{SQ}$. Similarly, $\overline{PT} \perp \overline{TQ}$. **14.** By Thales' theorem, $\angle ADB$ is a right angle. We also see that $\triangle ODB$ is an equilateral triangle, since all sides have the length of the radius. Moreover, $ODBE$ is a rhombus, so the side $\overline{DE}$ is a bisector of the 60° angle $\angle ODB$. Thus, $m(\angle ADE) = m(\angle ADB) - m(\angle EDB) = 90° - 30° = 60°$. Similarly, $m(\angle AED) = 60°$. Therefore, all angles of $\triangle ADE$ have measure 60°, so $\triangle ADE$ is equilateral. **16.** Since $m(\angle 1) + m(\angle 2) + m(\angle 3) + m(\angle 4) = 180°$, $m(\angle 1) = m(\angle 2)$, and $m(\angle 3) = m(\angle 4)$, it follows that $m(\angle 2) + m(\angle 3) = 180°/2 = 90°$. **17. (a)** Suppose the perpendicular bisector of chord $\overline{AB}$ intersects the circle at a point C. Then the circle is the circumscribing circle of $\triangle ABC$. The center of the circumscribing circle is the point of concurrence of the perpendicular bisectors of all three sides of $\triangle ABC$. In particular, the perpendicular bisector of side $\overline{AB}$ contains the center of the circle. **19. (a)** Extend $\overline{AB}$, and construct the line at A that is perpendicular to $\overline{AB}$. Set the compass to radius AB, and mark off this distance on the perpendicular line to determine a point C for which $AC = AB$. Similarly, construct a perpendicular line at B to $\overline{AB}$, and determine a point D (on the same side of $\overline{AB}$ as C) on this perpendicular so that $BD = AB$. $ABDC$ is a square with given side $\overline{AB}$. (Other constructions also work.) **20. (a)** Extend $\overline{AB}$ to a longer segment. Erect perpendicular rays to $\overline{AB}$ at both A and B to the same side of $\overline{AB}$. Bisect the right angle at A, and let its intersection with the ray at B determine point C. Erect the perpendicular at C to $\overline{BC}$, and let D be the intersection with the ray constructed at A. Then $ABCD$ is a square erected on the given side $\overline{AB}$. **(b)** Construct the equilateral triangle $\triangle ABO$ with the given side $\overline{AB}$. The reflection of A across $\overline{OB}$ determines the point C. Repeat reflections to determine the remaining vertices D, E, and F to construct the regular hexagon $ABCDEF$. **23.** Constructible: 3, 4, 5, 6, 8, 10, 12, 15, 16, 17, 20, 24, 30, 32, 34, 40, 48, 51, 60, 64, 68, 80, 85, 96 **27. (a)** There are five points C on line l for which triangle ABC is isosceles. As shown below, they are constructed by the circle at A through B, the circle at B through A, and the perpendicular bisector of $\overline{AB}$.

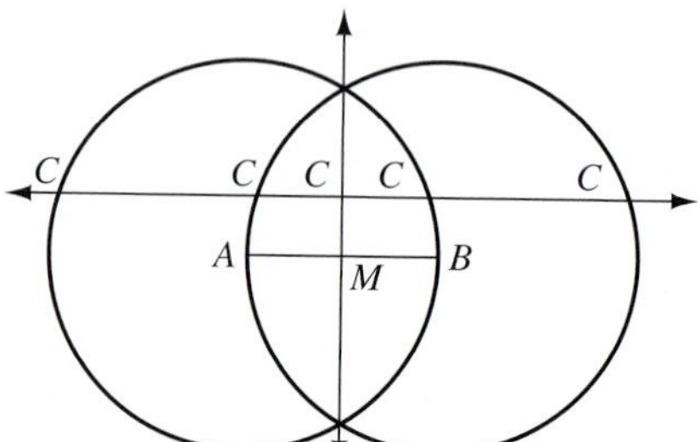

29. (a) The three altitudes of a triangle are concurrent (pass through a single point). **(b)** The perpendicular bisectors of

$\triangle PQR$ are concurrent, since they intersect at the center of the circle that circumscribes $\triangle PQR$. Since the perpendicular bisectors of $\triangle PQR$ are also the altitudes of $\triangle ABC$, the three altitudes are concurrent. **31.** $m(\angle NOQ) = m(\angle PON) - m(\angle POQ) = 144° - 120° = 24°$, since a central angle for a regular pentagon is 72° and a central angle for an equilateral triangle is 120°. A regular 15-gon has central angle $\frac{360°}{15} = 24°$, so laying off segments of length QN would give 15 equally spaced points around the circle. **36. (a)** $F_5 = 2^{2^5} + 1 = 2^{32} + 1 = 4{,}294{,}967{,}296 + 1 = 4{,}294{,}967{,}297$ **37. (a)** G, H, and P are collinear. The Euler line passes through G, H, and P. **(b)** $\frac{GH}{GP} = 2$. Thus, G is one-third of the distance from P to H along the Euler line. **(c)** The circle intersects all sides of $\triangle ABC$ at their midpoints. **(d)** The circle bisects each of the segments $\overline{AH}$, $\overline{BH}$, and $\overline{CH}$. **39.** $\triangle TRI$ is equilateral. **42. (a)** True. Three pairs of angles and two pairs of sides are congruent. **(b)** False. After pairing up congruent angles, the sides with equal lengths are not corresponding sides in the triangles, so the two triangles are not congruent.

Problem Set 14.3 (page 934)

1. (a) First notice that $m(\angle O) = 180° - 60° - 30° = 90°$. Therefore, by the AA similarity property, $\triangle ABC \sim \triangle PNO$. The scale factor from $\triangle ABC$ to $\triangle PNO$ is $\frac{12}{8} = \frac{3}{2}$. **(c)** By the AA similarity property, $\triangle GHI \sim \triangle TUI$, with scale factor $\frac{8}{5}$. **2. (a)** Yes, by AA; all angles are the same, 60°. **(c)** Yes, by the AA similarity property **(e)** Yes, by the SSS similarity property, or AA or SAS **3. (a)** $\frac{12}{15} = \frac{8}{a}$, so $a = 10$. **(c)** $\frac{c}{15} = \frac{c+2}{18}$; $18c = 15c + 30$; $c = 10$ **4. (a)** No. A square and a nonsquare rectangle are convex quadrilaterals with congruent angles, yet are not similar. **6. (a)** $\overline{AB}$ is parallel to $\overline{CD}$ so, by alternate interior angles, $\angle BAE \cong \angle DCE$. Also, $\angle AEB \cong \angle CED$, being vertical angles, so the AA similarity property gives $\triangle ABE \sim \triangle CDE$. **(b)** By similarity, $\frac{x}{36} = \frac{17}{51}$, so $x = 12$. Also, $\frac{26}{y} = \frac{17}{51}$, so $y = 78$. **8. (a)** $\angle CAD \cong \angle BAC$, since they are the same angle and $m(\angle ADC) = m(\angle ACB) = 90°$. By the AA similarity property, $\triangle ADC \sim \triangle ACB$. Likewise, $\triangle CDB \sim \triangle ACB$. Thus, $\triangle ADC \sim \triangle CDB$. **12. (a)** Draw an arc of large enough radius so that point B is on the seventh line above the line with point A. **14. (a)** Use the AA similarity property. **(b)** Using $\triangle ACD \sim \triangle ABC$, we get $\frac{AD}{AC} = \frac{AC}{AB}$, or $\frac{x}{b} = \frac{b}{c}$. Similarly, $\triangle CBD \sim \triangle ABC$ gives $\frac{BD}{BC} = \frac{CB}{AB}$, or $\frac{y}{a} = \frac{a}{c}$. **(c)** $x = \frac{b^2}{c}$ and $y = \frac{a^2}{c}$. Also, $x + y = c$, so $c = \frac{b^2}{c} + \frac{a^2}{c}$, or $c^2 = a^2 + b^2$. **16. (a)** $DJ = \sqrt{1^2 + \left(\frac{1}{2}\right)^2} = \sqrt{\frac{5}{4}} = \frac{1}{2}\sqrt{5}$ **(b)** $PS = AM$; $\frac{TS}{JD} = \frac{1}{\sqrt{5}}$ **(c)** $\text{Area}(PQRS) = \text{Area}(PQRS) = (PS)^2 = \frac{1}{5}$ **19.** The midpoints W, X, Y, and Z are the vertices of a parallelogram, and we know that the diagonals of any parallelogram intersect at their common midpoints. **21.** Let L denote the midpoint of $\overline{AC}$, which is also the midpoint of $\overline{BD}$. By Example 13.14, P is the centroid of $\triangle ABC$ and $BP = \frac{2}{3}BL$. Since $BL = \frac{1}{2}BD$, $BP = \frac{2}{3} \cdot \frac{1}{2}BD = \frac{1}{3}BD$. By the same reasoning, Q is the centroid of $\triangle ADC$ and $QD = \frac{1}{3}BD$. Finally, $PQ = BD - BP - QD = \left(1 - \frac{1}{3} - \frac{1}{3}\right)BD = \frac{1}{3}BD$. **22.** The right triangles $\triangle ACP$ and $\triangle BDP$ have congruent vertical angles at P. Thus, $\triangle ACP \sim \triangle BDP$ by AA similarity. Since $AC/BD = 4/2$, the scale factor is 2. Therefore, $CP = 2DP$. Since $CD = 4$ and $CD = CP + DP$, we see that $CP = \frac{2}{3}(4) = 8/3$ and $DP = \frac{1}{3}(4) = 4/3$. By the Pythagorean theorem, $AP = \sqrt{4^2 + (8/3)^2} = (4/3)\sqrt{13}$, and so $BP = (2/3)\sqrt{13}$. **24.** The right triangles also have a congruent angle at the vertex of the mirror, so the triangles are similar by the AA property. Assuming Mohini's eyes are 5″ beneath the top of her head, this gives the proportion $\frac{h}{5'} = \frac{15'}{4'}$, making the height of the pole $h = (5')\left(\frac{15'}{4'}\right) = 18'9''$. **26.** By similar triangles, $\frac{6' - x}{6'} = \frac{5.25'}{18'}$. Therefore, $x = 6'\left(1 - \frac{5.25'}{18'}\right) = 4.25' = 4'3''$.

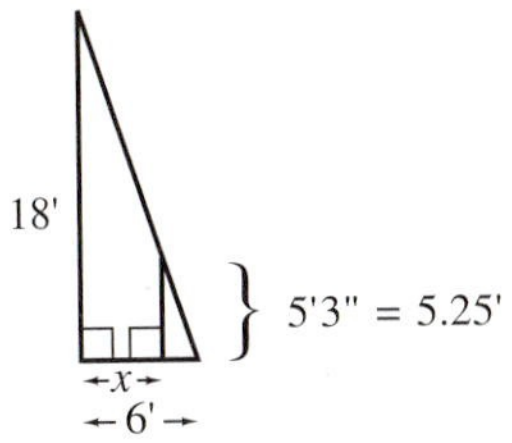

18'
5'3" = 5.25'
x
6'

29. **(a)** $\triangle BCP \sim \triangle DAP$ **(b)** $\dfrac{PA}{PC} = \dfrac{PD}{PB}$, or $PA \cdot PB = PC \cdot PD$ **33.** $\triangle APB \cong \triangle DPC$; $\triangle ABC \cong \triangle DCB$; $\triangle ABD \cong \triangle DCA$

Chapter 14 Review Exercises (page 942)

1. **(a)** $\triangle ACD \cong \triangle ACB$ by SSS **(b)** $\triangle ACD \cong \triangle ECB$ by SAS **(c)** $\triangle ADF \cong \triangle BEC$ by AAS **(d)** $\triangle ACD \cong \triangle ECB$ by SAS **(e)** $\angle B \cong \angle E$, since $\triangle ABE$ is isosceles. Therefore, $\triangle ABC \cong \triangle AED$ by ASA, and $\triangle ABD \cong \triangle AEC$ by ASA. **(f)** $\triangle ABC \cong \triangle DCB$ by SSS **2.** **(a)** 2.9 cm **(b)** 40° **(c)** 78° **(d)** 40° **3.** $\angle B \cong \angle C$, since $\triangle ABC$ is isosceles. By construction, $BF = DC$ and $BD = EC$. Therefore, $\triangle BDF \cong \triangle CED$ by SAS, so $DE = DF$. **4.** **(a), (b), (c)** Standard constructions as in Section 14.2. **(d)** Draw a circle at any point A on line m, and let it intersect line l at B and C. Construct $\overline{AB}$ and $\overline{AC}$. Draw circles of the same radius at B and C to determine the respective midpoints M and N of $\overline{AB}$ and $\overline{AC}$. Then $k = \overleftrightarrow{MN}$ is the desired line. Alternatively, construct a perpendicular line to l at a point on l. This determines a perpendicular segment between l and m. The perpendicular bisector of the segment is the desired line k.

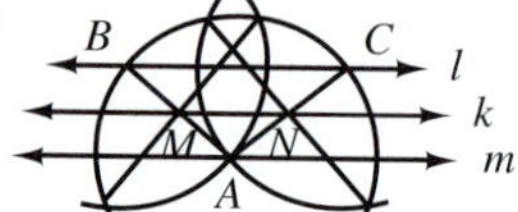

5. **(a)**

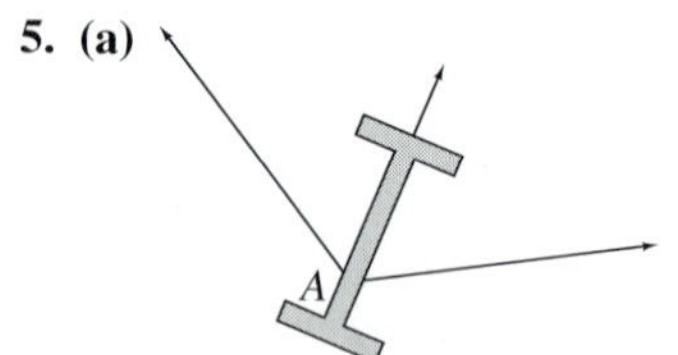

Reflect one side of $\angle A$ onto the other.

(b)

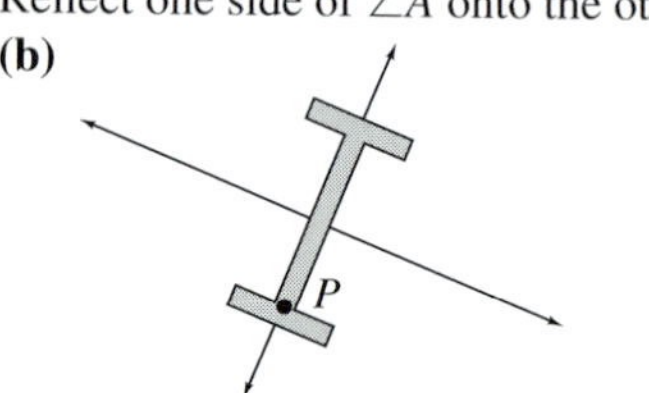

Pivot the Mira about P until the line reflects onto itself.

(c)

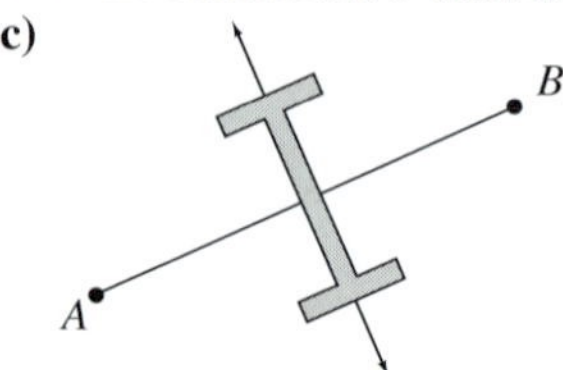

Reflect A onto B.

(d)

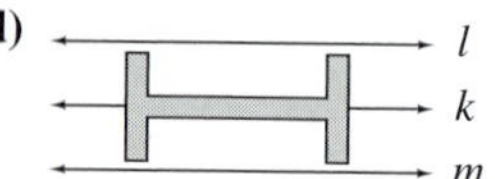

Reflect line m onto line l.

6. **(a)** Construct $\angle A$, lay off length AB, and draw a circle at B of radius BC. The circle intersects the other ray from A at two points C_1 and C_2, giving two triangles $\triangle ABC_1$ and $\triangle ABC_2$.

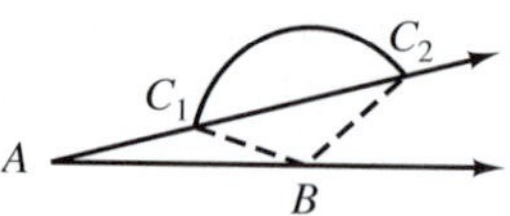

(b) Only $\triangle ABC_1$ has $\angle C = \angle C_1$ obtuse. **7.** Find the midpoint M of $\overline{AD}$. Then draw circles of radius AM centered at A, D, and M.

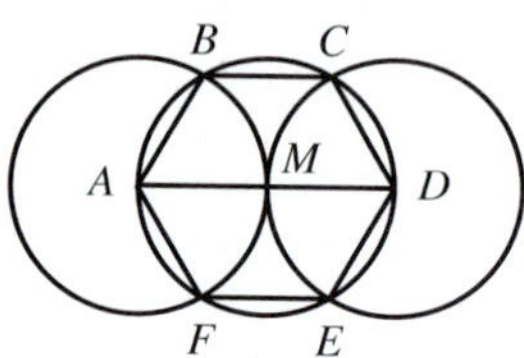

8. **(a)** Yes, using the SSS similarity property **(b)** Yes, using the AA similarity property **9.** Bisect sides $\overline{AB}$ and $\overline{AC}$ to determine their respective midpoints M and N. Extend $\overline{AB}$ beyond B, and draw the circle at B through M. Let E be the intersection with the extension. Similarly, extend $\overline{AC}$ beyond C, draw the circle at C through N, and let F be the intersection of this circle with the extension. By choosing $D = A$, the SAS similarity property guarantees that $\triangle ABC \sim \triangle DEF$. **10.** **(a)** $\triangle BAC \sim \triangle PQR$ by SAS similarity. The scale factor is 6 cm/4 cm = 3/2. **(b)** $\triangle ABC \sim \triangle YZX$ by SSS similarity. The scale factor is 42″/14″ = 3. **(c)** $\triangle ABC \sim \triangle HGF$ by AA similarity. The scale factor is $4/6 = 2/3$. **(d)** $\triangle ADB \sim \triangle BCD$ by SSS similarity. The scale factor is $5/10 = 1/2$. **11.** Draw additional line segments parallel to the given transversals. This creates similar triangles from which it follows that $x/9 = 16/12$, so $x = 9(16/12) = 12$ and $y/12 = 15/9$, and so $y = 12 \cdot (15/9) = 20$.

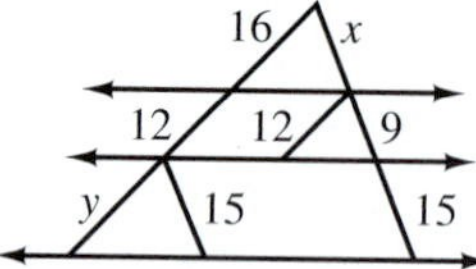

12. Let K be the top of the stick and Y the top of the pyramid. Then $\Delta KS_1S_2 \sim \Delta YP_1P_2$, with scale factor $P_1P_2/S_1S_2 = 270/2 = 135$. Therefore, the height of the pyramid is 135×3 feet = 405 feet.

Chapter 14 Test (page 944)

1. **(a)** That they form the same angle relative to the ground, namely the angle of elevation, since the distant sun's rays are parallel **(b)** That the person and the tree stand at the same angle with the ground; for example, both vertical. Then $\triangle ABC \sim \triangle DEF$ by the AA similarity property.

(c) $\frac{DE}{6'} = \frac{56'}{7'}$, so $DE = 48'$. **2.** Draw the circle with radius MC centered at M. The circle passes through A, B, and C. Thus, by Thales' theorem, $\triangle ACB$ is a right triangle because it is inscribed in a semicircle of diameter $\overline{AB}$.
3. (a) $\triangle ADE \sim \triangle ACB$ by the AA similarity property, since both triangles contain $\angle A$ and a right angle. **(b)** $\triangle ABC \sim \triangle XYZ$ by the SAS similarity property, since $\frac{5}{4} = \frac{15}{12}$. **(c)** $\triangle DEG \sim \triangle EFG$ by the SSS similarity property, since $\frac{16}{8} = \frac{16}{8} = \frac{8}{4}$. **(d)** $\triangle AEB \sim \triangle CED$ by the AA similarity property, with $\angle AEB \cong \angle CED$ being vertical angles and $\angle EBA \cong \angle EDC$ being alternate interior angles between parallel lines. **4. (a)** $\triangle ADC \cong \triangle ABC$ by AAS **(b)** $\triangle ABC \cong \triangle ADC$ by SAS **(c)** $\triangle ADC \cong \triangle BCD$ by ASA **(d)** $\triangle ABE \cong \triangle CBD$ by AAS **(e)** $\triangle BDC \cong \triangle FDE$ by ASA and $\triangle ABE \cong \triangle AFC$ by ASA **(f)** $\triangle ABC \cong \triangle ADC$ by SSS **5.** $\triangle ADE \sim \triangle ACB$ by the AA similarity property, since both triangles contain $\angle A$ and a right angle. Thus, $AE/AB = AD/AC = AD/(AD + DC) = 2DC/(2DC + DC) = 2/3$. Then $AE = \frac{2}{3}(AB) = \frac{2}{3}(12) = 8$ and $EB = AB - AE = 12 - 8 = 4$. **6. (a)** $m(\angle W) = 180° - 87° - 40° = 53°$ **(b)** $\frac{20}{16} = \frac{5}{4}$ **(c)** $UV = \frac{5}{4}KL = \frac{5}{4}(20) = 25$ **7.** The third side is greater than 6 feet and less than 26 feet by the triangle inequality. **8.** The small triangles are congruent to one another by the SAS congruence property, so $PQRST$ is equilateral. Let x and y be the measures of the acute angles in $\triangle APT$. Then each interior angle of $PQRST$ has measure $180° - x - y$, so $PQRST$ is equiangular. Altogether, $PQRST$ is regular.
9. (a) Draw circles of radius PQ, one centered at P and one centered at Q. The circles intersect at points R and S for which $\triangle PQR$ and $\triangle PQS$ are equilateral. **(b)** Construct the angle bisectors of $\angle QPR$ and $\angle QPS$, and denote their intersection with line l as T and U, respectively. $\triangle PTU$ is the desired equilateral triangle. **10.** Construct a segment $\overline{DE}$ that is congruent to $\overline{AB}$. Construct rays at D and E to form angles that are respectively congruent to $\angle A$ and $\angle B$. Let F be a point of intersection of the rays. Then $\triangle DEF \cong \triangle ABC$.
11. $\triangle FGH$ is equilateral. **12. (a)** The Pythagorean theorem shows that $AB = 12$ and $DF = 9$. Therefore, $\triangle ABC \cong \triangle DEF$ by SSS or SAS. **(b)** Not congruent. The hypotenuse of $\triangle ABC$ is 5. Since $EF = 5$, the hypotenuse of $\triangle DEF$ is larger than 5, so the hypotenuses cannot correspond. **(c)** The Pythagorean theorem shows that $AB = 3$ and $CD = 4$. Therefore, $\triangle ABD \cong \triangle CBD$ by SSS or SAS. **(d)** $\triangle ABE \cong \triangle DBC$ by SAS.

Appendices

Problem Set B (page 955)

1. (a) Arik: \$225.38; Kaia: \$258.33; Lena: \$143.06; district: \$626.77 **(b)** Lena corrected: \$182.12; district corrected: \$665.83 **3. (a)** $L(10) = 123$. $L(10)/L(9) = 123/76 = 1.61842\ldots$. The ratio of Lucas numbers appears to approach the same limit value as the ratio of Fibonacci numbers.
5. (a) The value of cell Bn is 1 less than the Fibonacci number in cell A$(n + 2)$. This demonstrates the identity $1 + 1 + 2 + 3 + 5 + \cdots + F(n) = F(n + 2) - 1$.
6. (a) Mean = 27.9; population standard deviation = 14.3 **(b)** Lower quartile = 19; median = 26; upper quartile = 37

Problem Set C (page 968)

1. (a) 119 pounds
3. (a)

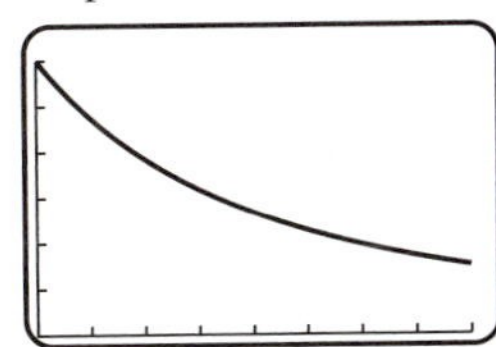

(b) 5745 miles, where Xmin = 0 and Xmax = 6000. The more exact value 5798 is found by setting Xmin = 5700 and Xmax = 5800. **4. (b)** 853 and 8660 miles from the earth's center **6. (a)** All graphs are straight lines through the origin. **(b)** The y-values of Y_2 are twice that of the basic function at the same x-coordinate, giving Y_2 a steeper positive slope than Y_1. The function Y_3 has half the y-values, giving it a more shallow slope. The function Y_4 is the reflection of the basic function across the horizontal x-axis.
9. Chocolate is the most frequent favorite flavor of both classes. Vanilla is the least frequent favorite in Mrs. O'Leary's class, but it is strawberry in Mr. Hsu's class.

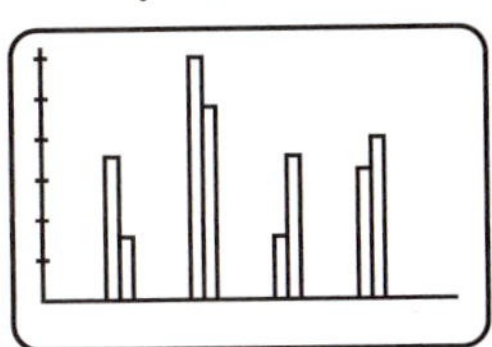

10. The most common grade is a B, earned by 13 students. The trace also shows that 8 students earned between 90 and 99, so together with the 2 students earning 100, there are a total of 10 A grades.

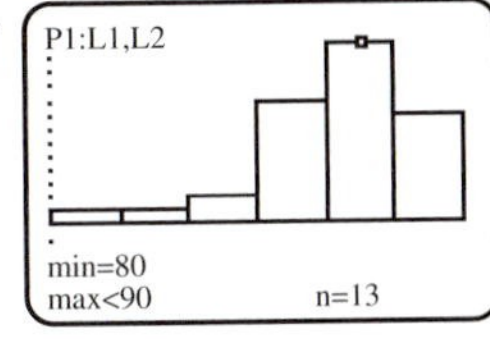

Problem Set D (page 981)

1. (a) The sides are only approximately, not exactly, all the same length. **2. (a)** Construct the midpoint M of the segment $\overline{AB}$. Construct the perpendicular bisector to the segment through M. Construct any point $C \neq M$ on the perpendicular bisector. The triangle ABC is isosceles. **(c)** Construct any point M on the perpendicular bisector of segment $\overline{AB}$. Construct a line parallel to $\overline{AB}$ through point M, and construct any circle centered at M. If the circle intersects the parallel line at points C and D, then $ABCD$ is an isosceles trapezoid. **(e)** Construct the segment $\overline{AC}$ and the perpendicular bisector of the segment. Construct a point B on the perpendicular bisector. Construct a circle centered at A through B. If the circle intersects the perpendicular bisector at D, then $ABCD$ is a rhombus with opposite vertices at A and C. **4. (a)** The angle measures sum to 360°. **5. (a)** Let the respective points be C, D, and E. There are several methods to show that triangle CDE is a right triangle with the right angle at vertex D. *Method One:* Use **Length** to measure the lengths CD, DE, and CE, and use **Calculate** . . . to show that $CD^2 + DE^2 = CE^2$. Triangle CDE is a right triangle by the converse of the Pythagorean theorem. *Method Two:* Measure the slopes of $\overline{CD}$ and $\overline{DE}$ and show that the product of the slopes is -1. This shows that the segments are perpendicular to one another. *Method Three:* Measure $\angle CDE$ and see that it is 90°. **6. (a)** The point M is $(1, 3)$.

Acknowledgments

Chapter 1

Page 10, photo of George Pólya. AP Wide World Photos. Page 27, Carl Friedrich Gauss. Culver Pictures. Page 48, photo of Henri Poincare. Library of Congress. Page 61, Garfield Cartoon. GARFIELD © 1989 Paws, Inc. All Rights Reserved.

Chapter 2

Page 75, photo of Rhind papyrus. Bridgeman Art-New York. Page 75, photo of Incan quipu. Museo Nacional De Arqueologica, Antropologia E Historia Del Peru. Page 77, photo of Georg Cantor. Baveria-Verlag. Page 81, reprinted with permission of ETA/ Cuisenaire, © 2004. All rights reserved. Page 95, photo of Richard Hamming. John Sanders, Naval Postgraduate School. U.S. Navy photo. Used with permission.

Chapter 3

Page 159, photo of Benjamin Banneker stamp. U.S. Postal Service. Page 195, Calvin and Hobbes cartoon. CALVIN AND HOBBES © 1990 Waterson. Reprinted with permission of Universal Press Syndicate. All rights reserved. Page 202, Emily Noether. Professor Gottfried E. Noether.

Chapter 4

Page 232, cartoon of "Hold on there Mr. Webster . . .". © 1999 by Sydney Harris. Used with permission. Page 260, photo of Julia Robinson. Photo courtesy of the American Mathematical Society. All rights reserved. Page 261, photo of Dr. Andrew Wiles. Copyright Denise Applewhite/Corbis.

Chapter 5

Page 288, photo of Edward A. Bouchet. Yale University Library. Page 316, photo of Charlotte Angas Scott. Library of Congress.

Chapter 6

Page 342, Hagar the Horrible cartoon by Dik Browne. © 1995. Reprinted with special permission of King Features Syndicate.

Chapter 7

Page 407, Sacagawea dollar, © Beth Anderson; dime and penny, Comstock. Page 416, "I thought it was a breakthrough, but it was only a misplaced decimal" cartoon. © 1999 by Sydney Harris. Used with permission. Page 419, photo of Pythagoras bust. © Baldwin H. Ward & Kathryn C. Ward/Corbis.

Chapter 8

Page 503, Maria Agnesi. © Bettman/Corbis.

Chapter 9

Page 534, pictograph of "World Population." From TIME, June 1, 1992, p. 54. © 1992, Time, Inc. Reprinted by permission. Page 542, "Percent of all public schools and instructional rooms having internet access: Fall 1994 to Fall 2001." National Digest of Educational Statistics, 2002, p. 10. National Center for Educational Statistics, U.S. Department of Education. Page 542, "Highest level of education attained by persons 25 years and older: March 2001." National Digest of Educational Statistics, 2002, p. 486. National Center for Educational Statistics, U.S. Department of Education. Page 543, "Google's Net Income/Loss (2002/2003)," from Newsweek, May 10, 2004. © 2004 Newsweek, Inc. All rights reserved. Reprinted by permission. Page 543, Two graphs. "More going out than coming in . . ." and ". . . Will deplete Social Security assets by 2032." READER'S DIGEST, December 1998, p. 75. Page 566, cartoon showing bizarre sequence of computer-generated number. © 1999 by Sydney Harris.

Chapter 10

Page 592, Ratio of the number of heads to the number of tosses in Kerrich's coin-tossing experiment. Adapted from Figure 2 in STATISTICS, Second Edition by David Freeman, Robert Pisani, Roger Purves, and Ani Adhikari. Reprinted by permission of W.W. Norton & Company, Inc. Page 593, photo of Jakob Bernoulli. Museum fur Volkskunde von Schweizerriches. Pages 628–629, illustration of Probability and Paradox. Adapted from "Mathematical Games" by Martin Gardner, SCIENTIFIC AMERICAN, March 1976. Copyright © 1976 by Scientific American, Inc. All rights reserved. Page 635, photo of roulette wheel. Arthur Tilley/Stone (Getty Images). Page 637, B.C. cartoon, "State Lottery." © 1989. By permission of Johnny Hart and Creators Syndicate, Inc.

Chapter 11

Page 653, photo of fissures in a gelatinous preparation of the tin foil. Manfred P. Kage/Peter Arnold, Inc. Page 653, photo of butterfly wings. John Bove/Photo Researchers. Page 653, M.C. Escher's "Wall Mosaic in the Alhambra," © 2004 The M.C. Escher Company—Baarn—Holland. All rights reserved. Page 653, photo of a snow crystal. SS/Photo Researchers. Page 653, A fractal, "Tail of the Seahorse," Mandlebrot set from H.O. Pietgen and P.H. Richter, THE BEAUTY OF FRACTALS, Heidelberg, Springer-Verlag, 1986. Page 653, photo of sunflower. Ray E. Ellis/Photo Researchers. Page 656, B.C. cartoon, "Vanishing point." © 1990. By permission of Johnny Hart and Creators Syndicate, Inc. Page 659, photos of Circle Master Compass, Protractor, and Drafting Triangles. Reprinted with permission of ETA/Cuisenaire. © 2004. All rights reserved. Page 659, photos of safety compass and ruler. © Beth Anderson. Page 685, photo of Baha'i House of Worship in Wilmette, Illinois. National Baha'i Headquarters. Page 696, photo of Buckyball. S. Camazine/Photo Researchers. Page 696, Seashell and computer-drawn ideal representation. Dr. Guiseppe Mazza. Page 696, daVinci's drawings of an icosahedron and a dodecahedron. Rare Book Room, New York Public Library. Astor, Lenox, and Tilden Foundations. Page 696, A partial filling of space by truncated octahedra. D'Arcy W. Thompson, ON GROWTH AND FORM, New Edition, Cambridge University Press, Cambridge and New York, 1948. Page 696, skeletons of Microscopic radiolara. ERNST HAECKEL, CHALLENGER, Monograph, 1887. Page 703, A Flexible Polyhedron, Figures 21 and 22 from MATHEMATICS MAGAZINE, Volume 52, Number 5, November 1979, p. 281.

Copyright © 1979 by The Mathematical Association of America. Reprinted by permission. Page 706, photo of Epcot Center. Nik Wheeler/Corbis. Page 706, Polyhedron. D'Arcy W. Thompson, ON GROWTH AND FORM, New Edition, Cambridge University Press, Cambridge and New York, 1948. Page 713, Pentominoes. Figure 10, "The pentominoes" from POLYOMINOES by Solomon W. Golomb, Page 23. (Charles Scribner's Sons 1965; Revised Edition, Princeton University Press, 1994). Copyright © 1965 by Solomon W. Golomb. Reprinted with permission of the author. Page 714, photo of pyrite crystal. H. Chaumeton/Photo Researchers.

Chapter 12

Page 744, Metric Clock Cartoon. © 1999 Sydney Harris. Page 766, photo of the earth. NASA. Page 780, photo of Babylonian Tablet. The British Museum. Page 799, photo of Sophie Germain. Ecole Sophie Germaine, Paris. Page 802, photo of Jonathan Swift. National Portrait Gallery, London.

Chapter 13

Page 819, sea lions on a Haida dance tunic. Thomas Burke Memorial/Washington State Museum. Page 834, M.C. Escher's "Two-motif tiling of the plane," © 2004 The M.C Escher Company—Baarn—Holland. All rights reserved. Page 836, photo of Carolyn Gordon and David Webb. Washington University of Saint Louis. Page 842, Escher tiling. M.C. Escher's "Symmetry Drawing," © 2004 The M.C Escher Company—Baarn—Holland. All rights reserved. Page 847, photo of George W. Brainerd. Southwest Museum, Los Angeles. Page 847, photo of Anna O. Shepherd. Carnegie Institute of Washington, D.C. Page 852, border patterns from around the world. "The 7 Species of One-Dimensional Ornaments" reprinted by permission of the publisher from AESTHETIC MEASURE by George Birkoff, Plate VII, p. 57, Cambridge, Mass.: Harvard University Press, Copyright © 1933 by the President and Fellows of Harvard College. Page 853, Pfembe maternity statue. © Royal Museum for Central Africa; Tervuren, Belgium. Page 855, Mercedes Benz emblem, Courtesy DaimlerChrysler of North America, Inc. Page 855, "The Chevrolet Bow Tie emblem and the Oldsmobile Rocket emblem are trademarks of the Chevrolet Motor Division and Oldsmobile Motor Division, respectively, General Motors Corporation, used with permission. General Motors, however, does not endorse or assume any responsibility for the text, errors, or omissions of Pearson Education, its officers, agents, employees, or other representatives." Page 855, Sterling logo, Sterling Savings Association. Colfax, Washington. Page 858, problem 13.2.33 illustration, 6 Chladni Plates from FEARFUL SYMMETRY © 1992 by Ian Stweart and Martin Golubitsky. Printed by Blackwell Publishers. Page 861, Pre-Inca fabric from Peru. ISUMI. Page 861, Window of a fourteenth-century mosque in Cairo. Israel National Museum, Jerusalem. Page 861, Mosaic floor of the fourteenth and fifteenth century in the Basilica of Saint Marks Cathedral. Saint Marks Cathedral, Venice. Page 861, Tilings from Portugal, fifteenth to sixteenth centuries. SIMOES. Page 861, Chinese lattice work. Daniel Sheets Dye, A GRAMMAR OF CHINESE LATTICE, Figures C9b, S12a, Harvard-Yenching Institute Monograph V, Cambridge, MA. 1937. Page 864, Kepler's tiling patterns. From TILINGS AND PATTERNS by Branko Grunbaum and G.C. Shepard. © 1987 W.H. Freeman and Company. Used by permission. Page 866, Heinz Voderberg's spiral tiling. From Martin Gardner's PENROSE TILES TO TRAPDOOR CIPHERS, W.H. Freeman and Company/Worth Publishers, 1989. Page 867, M.C. Escher's "Birds and its grid of parallelograms," © 2004 The M.C Escher Company—Baarn—Holland. All rights reserved. Page 868, M.C. Escher's "Regular Divisions of the Plane (5 examples)," © 2004 The M.C Escher Company—Baarn—Holland. All rights reserved. Page 870, Escher-like tiling by Nancy Putnam. © Nancy Putnam. Page 871, Fake tiling by regular polygons. From TILINGS AND PATTERNS by Branko Grunbaum and G.C. Shepard. © 1987 W.H. Freeman and Company. Used by permission. Page 874, Chinese tangram puzzle. Reprinted with permission from CURRICULUM AND EVALUATION STANDARDS FOR SCHOOL MATHEMATICS, copyright 1989 by the National Council of Teachers of Mathematics. All rights reserved. Pages 878–879, Chapter review problem #9 a-b. From AESTHETIC MEASURE by George Birkoff, © 1933/Harvard University Press, Cambridge, MA. Page 879, Chapter test problem #4. © 2004 The M.C Escher Company—Baarn—Holland. Private Collection. Page 880, Chapter test problem #6 a-d. From AESTHETIC MEASURE by George Birkoff, Cambridge, Mass.: Harvard University Press, Copyright © 1933 by the President and Fellows of Harvard College.

Chapter 14

Page 894, photo of Christine Ladd Franklin. Archives of the History of American Psychology.

Mathematical Lexicon

Many of the words, prefixes, and suffixes forming the vocabulary of mathematics are derived from words and word roots from Latin, Greek, and other languages. Some of the most common terms are listed below, to serve as an aid to learning and understanding the terminology of mathematics.

acute from Latin *acus* ("needle") by way of *acutus* ("pointed, sharp")

algorithm distortion of Arabic name *al-Khowarazmi* ("the man from Khwarazm"), whose book on the use of Indo-Arabic numeration was translated into Latin as *Liber Algorismi,* meaning "Book of al-Khowarazmi"

angle from Latin *angulus* ("corner, angle")

apex from Latin word meaning "tip, peak"

area Latin *area* ("vacant piece of ground, plot of ground, open court")

associative from Latin *ad* ("to") and *socius* ("partner, companion")

axis from Latin word meaning "axle, pivot"

bi- from Latin prefix derived from *dui-* ("two"); *bi*nary, *bi*nomial, *bi*sect

calculate from Latin *calc* ("chalk, limestone") and diminutive suffix *-ulus* [a *calculus* was a small pebble; *calculare* meant "to use pebbles" = to do arithmetic]

cent- from Latin *centum* ("hundred"); *cent*imeter, per*cent*

circum- from Latin *circum* ("around"); *circum*ference, *circum*scribe

co-, col-, com-, con- from Old Latin *com* ("together with, beside, near"); *com*mutative, *col*linear, *com*plement, *con*gruent

commutative from Latin *co-* ("together with") and *mutare* ("to move")

concurrent from Latin *co-* ("together with") and *currere* ("to run")

conjecture from Latin *co-* ("together with") and *iactus* ("to throw") [conjecture = throw (ideas) together]

cylinder from Greek *kulindros* ("a roller")

de- Latin preposition *de* ("from, down from, away from, out of"); *de*nominator, *de*duction

deca-, deka- from Greek deka- ("ten"); *deca*gon, do*deca*hedron, *deka*meter

deci- from Latin *decimus* ("tenth"); *deci*mal, *deci*meter

diagonal from Greek *dia-* ("through, across") and *gon-* ("angle")

diameter from Greek *dia-* ("through, across") and *metron* ("measure")

digit from Latin *digitus* ("a finger")

distribute from Latin prefix *dis-* ("apart, away") and Latin *tribu* ("a tribe [of Romans]")

empirical from Latin *empiricus* ("a physician whose art is founded solely on practice")

equal from Latin *æquus* ("even, level")

equilateral from Latin *æquus* ("even, level") and *latus* ("side")

equivalent from Latin *æquus* ("even, level") and *valere* ("to have value")

exponent from Latin *ex* ("away") and *ponent-* = present participial stem of *ponere* ("to put")

figure from Latin *figura* ("shape, form, figure")

fraction from Latin *fractus,* past participle of *frangere* ("to break")

geometry from Greek *geo-* ("earth") and *metron* ("measure")

-gon from Greek *gonia* ("angle, corner"); poly*gon,* penta*gon*

-hedron from Greek *hedra* ("base, seat"); poly*hedron,* tetra*hedron*

hept-, sept- Greek *hept,* from prehistoric Greek *sept,* meaning "seven"; *hepta*gon

heuristic from Greek *heuriskein* ("to find, discover")

hex- from Greek *hex,* from prehistoric Greek *sex,* meaning "six"; *hex*agon, *hex*omino

icosahedron from Greek *eikosi* ("twenty") and *hedra* ("bases, seat")

inch from Latin *uncia,* a unit of weight equal to one twelfth of the *libra,* or Roman pound

inscribe from Latin *in* ("in") and *scribere* ("to scratch"); hence, "to write"

integer from Latin *in-* ("not") and Indo-European root *tag-* ("to touch") [an integer is untouched, hence "whole"]

inverse from Latin *in* ("in") and *versus,* past participle of *vertere* ("to turn")

isosceles from Greek *isos* ("equal") and *skelos* ("leg")

kilo- from Greek *khiloi* ("thousand"); *kilo*gram, *kilo*meter

lateral from Latin *latus* ("side")

lb. abbreviation for pound, from *libra,* the Roman unit of weight

line from Latin *linum* ("flax") [The Romans made *linea,* linen thread, from flax.]

median from Latin *medius* ("in the middle")

meter from Greek *metron* ("measure, length")

milli- from Latin *mille* ("one thousand")

multiply from Latin *multi* ("many") and Indo-European *pel* ("to fold")

nonagon from Latin *nonus* ("ninth") and Greek *gon* ("angle")

number from Latin *numerus* ("number")

obtuse from Latin *ob* ("against, near, at") and *tusus* ("to strike, to beat") [*obtusus* = beaten down to the point of being dull]

oct- from Greek *octo* ("eight"); *oct*agon, *oct*ahedron

parallel from Greek *para* ("alongside") and *allenon* ("one another")

pent- Greek *pent* ("five"); *pent*agon, *pent*agram, *pent*omino

percent from Latin *per* ("for") and *centum* ("hundred") [percent = for (each) hundred]

peri- from Greek *peri* ("around"); *peri*meter

plane from Latin *planus* ("flat")

poly- from Greek *polus* ("many"); *poly*gon, *poly*hedra, *poly*omino

prism from Greek *prisma* ("something that has been sawed")

quadr- Latin *quadr-* ("four"); *quadr*ant, *quadr*ilateral

rectangle from Latin *rectangulus* ("right-angled")

-sect from Latin *sectus,* past participle of *secui* ("to cut"); bi*sect,* inter*sect*

surface from Latin *super* ("over") and *facies* ("form, shape")

symmetric from Greek *sun-* ("together with") and *metron* ("measure")

tetra- Greek *tetra-* ("four"); *tetra*hedron, *tetr*omino

trans- Latin *trans* ("across"); *trans*itive, *trans*lation, *trans*versal

tri- from Latin *tri* ("three"); *tri*angle, *tri*sect

vertex from Latin verb *vertere* ("to turn")

zero from Arabic *çifr* ("empty")

Index